DICTIONARY
OF
SCIENTIFIC BIOGRAPHY

PUBLISHED UNDER THE AUSPICES OF
THE AMERICAN COUNCIL OF LEARNED SOCIETIES

The American Council of Learned Societies, organized in 1919 for the purpose of advancing the study of the humanities and of the humanistic aspects of the social sciences, is a nonprofit federation comprising forty-one national scholarly groups. The Council represents the humanities in the United States in the International Union of Academies, provides fellowships and grants-in-aid, supports research-and-planning conferences and symposia, and sponsors special projects and scholarly publications.

MEMBER ORGANIZATIONS

AMERICAN PHILOSOPHICAL SOCIETY, 1743
AMERICAN ACADEMY OF ARTS AND SCIENCES, 1780
AMERICAN ANTIQUARIAN SOCIETY, 1812
AMERICAN ORIENTAL SOCIETY, 1842
AMERICAN NUMISMATIC SOCIETY, 1858
AMERICAN PHILOLOGICAL ASSOCIATION, 1869
ARCHAEOLOGICAL INSTITUTE OF AMERICA, 1879
SOCIETY OF BIBLICAL LITERATURE, 1880
MODERN LANGUAGE ASSOCIATION OF AMERICA, 1883
AMERICAN HISTORICAL ASSOCIATION, 1884
AMERICAN ECONOMIC ASSOCIATION, 1885
AMERICAN FOLKLORE SOCIETY, 1888
AMERICAN DIALECT SOCIETY, 1889
AMERICAN PSYCHOLOGICAL ASSOCIATION, 1892
ASSOCIATION OF AMERICAN LAW SCHOOLS, 1900
AMERICAN PHILOSOPHICAL ASSOCIATION, 1901
AMERICAN ANTHROPOLOGICAL ASSOCIATION, 1902
AMERICAN POLITICAL SCIENCE ASSOCIATION, 1903
BIBLIOGRAPHICAL SOCIETY OF AMERICA, 1904
ASSOCIATION OF AMERICAN GEOGRAPHERS, 1904
HISPANIC SOCIETY OF AMERICA, 1904
AMERICAN SOCIOLOGICAL ASSOCIATION, 1905
AMERICAN SOCIETY OF INTERNATIONAL LAW, 1906
ORGANIZATION OF AMERICAN HISTORIANS, 1907
COLLEGE ART ASSOCIATION OF AMERICA, 1912
HISTORY OF SCIENCE SOCIETY, 1924
LINGUISTIC SOCIETY OF AMERICA, 1924
MEDIAEVAL ACADEMY OF AMERICA, 1925
AMERICAN MUSICOLOGICAL SOCIETY, 1934
SOCIETY OF ARCHITECTURAL HISTORIANS, 1940
ECONOMIC HISTORY ASSOCIATION, 1940
ASSOCIATION FOR ASIAN STUDIES, 1941
AMERICAN SOCIETY FOR AESTHETICS, 1942
METAPHYSICAL SOCIETY OF AMERICA, 1950
AMERICAN STUDIES ASSOCIATION, 1950
RENAISSANCE SOCIETY OF AMERICA, 1954
SOCIETY FOR ETHNOMUSICOLOGY, 1955
AMERICAN SOCIETY FOR LEGAL HISTORY, 1956
AMERICAN SOCIETY FOR THEATRE RESEARCH, 1956
SOCIETY FOR THE HISTORY OF TECHNOLOGY, 1958
AMERICAN COMPARATIVE LITERATURE ASSOCIATION, 1960

DICTIONARY
OF
SCIENTIFIC BIOGRAPHY

CHARLES COULSTON GILLISPIE

Princeton University

EDITOR IN CHIEF

Volume 13

HERMANN STAUDINGER – GIUSEPPE VERONESE

CHARLES SCRIBNER'S SONS · NEW YORK

Copyright © 1970, 1971, 1972, 1973, 1974, 1975, 1976, 1978, 1980
American Council of Learned Societies.
First publication in an eight-volume edition 1981.

Library of Congress Cataloging in Publication Data

Main entry under title:

Dictionary of scientific biography.

"Published under the auspices of the American Council
of Learned Societies."
Includes bibliographies and index.
1. Scientists—Biography. I. Gillispie, Charles
Coulston. II. American Council of Learned Societies
Devoted to Humanistic Studies.
Q141.D5   1981   509′.2′2 [B]   80-27830
ISBN 0-684-16962-2 (set)

ISBN 0-684-16963-0 Vols. 1 & 2   ISBN 0-684-16967-3 Vols. 9 & 10
ISBN 0-684-16964-9 Vols. 3 & 4   ISBN 0-684-16968-1 Vols. 11 & 12
ISBN 0-684-16965-7 Vols. 5 & 6   ISBN 0-684-16969-X Vols. 13 & 14
ISBN 0-684-16966-5 Vols. 7 & 8   ISBN 0-684-16970-3 Vols. 15 & 16

5 7 9 11 13 15 17 19   V/C   20 18 16 14 12 10 8 6 4

Printed in the United States of America

Editorial Board

Editorial Staff

Panel of Consultants

Contributors to Volume 13

The following are the contributors to Volume 13.   Each author's name is followed by the institutional affiliation at the time of publication and the names of the articles written for this volume. The symbol † means that an author is deceased.

GIORGIO ABETTI
University of Florence
TACCHINI; TOSCANELLI DAL POZZO

MICHELE ALDRICH
Aaron Burr Papers
F. B. TAYLOR

ADEL ANBOUBA
Institut Moderne de Liban
AL-ṬŪSĪ

TOBY A. APPEL
Johns Hopkins University
VALENCIENNES

WILBUR APPLEBAUM
Illinois Institute of Technology
STREETE

A. ALBERT BAKER, JR.
California State University, Fullerton
TIEMANN

JOHN R. BAKER
University of Oxford
TREMBLEY

MARGARET E. BARON
TODHUNTER

DONALD G. BATES
McGill University
SYDENHAM

HANS BAUMGÄRTEL
TSCHERMAK

ROBERT P. BECKINSALE
University of Oxford
K. M. VON STERNBERG

LUIGI BELLONI
University of Milan
TRULLI

OTTO THEODOR BENFEY
Guilford College
F. K. J. THIELE

MICHAEL BERNKOPF
Pace College
STIELTJES

RICHARD BIEBL
University of Vienna
TSCHERMAK VON SEYSENEGG

R. P. BOAS, JR.
Northwestern University
TITCHMARSH

WALTER BÖHM
STEFAN

MARY A. B. BRAZIER
University of California, Los Angeles
UKHTOMSKY

GERT H. BRIEGER
University of California, San Francisco
THAYER

T. A. A. BROADBENT †
VENN

W. H. BROCK
University of Leicester
TILLOCH; E. TURNER

SANBORN C. BROWN
Massachusetts Institute of Technology
B. THOMPSON

THEODORE M. BROWN
City College, City University of New York
STUART

VIGGO BRUN
THUE

JED Z. BUCHWALD
University of Toronto
W. THOMSON

K. E. BULLEN
University of Sydney
H. H. TURNER

VERN L. BULLOUGH
California State University, Northridge
VARENIUS

IVOR BULMER-THOMAS
THEAETETUS; THEODORUS OF CYRENE;
THEODOSIUS OF BITHYNIA

WERNER BURAU
University of Hamburg
STAUDT; STUDY; F. O. R. STURM

JOHANN JAKOB BURCKHARDT
University of Zurich
STEINER

JOHN G. BURKE
University of California, Los Angeles
VALMONT DE BOMARE

J. C. BURKILL
University of Cambridge
VALLÉE-POUSSIN

E. ALFRED BURRILL
VAN DE GRAAFF

H. L. L. BUSARD
State University of Leiden
VER EECKE

JAMES H. CASSEDY
National Library of Medicine
STILES

CARLO CASTELLANI
STRUSS; TROJA

ROBERT A. CHIPMAN
University of Toledo
STEINMETZ

STIG CLAESSON
University of Uppsala
SVEDBERG

EDWIN CLARKE
Wellcome Institute for the History of Medicine
TWORT

GEORGE W. CORNER
American Philosophical Society
STREETER

ALBERT B. COSTA
Duquesne University
SWARTS; J. F. THORPE

PIERRE COSTABEL
École Pratique des Hautes Études
VARIGNON

CHARLES COURY †
TESTUT

E. HORNE CRAIGIE
University of Toronto
TÜRCK

J. K. CRELLIN
Wellcome Institute for the History of Medicine
THURNAM

M. P. CROSLAND
University of Kent at Canterbury
THENARD

KARL H. DANNENFELDT
Arizona State University
STEPHANUS OF ALEXANDRIA; SYNESIUS OF CYRENE

STACEY B. DAY
Memorial Sloan-Kettering Cancer Center
STEVENS; TASHIRO

GAVIN DE BEER †
VENETZ

ALLEN G. DEBUS
University of Chicago
VALENTINE

SALLY H. DIEKE
Johns Hopkins University
STRÖMBERG; TRUMPLER

J. G. DORFMAN †
STOLETOV; TAMM; VAVILOV

HAROLD DORN
Stevens Institute of Technology
S. G. THOMAS

K. C. DUNHAM
Institute of Geological Sciences, London
TEALL; TYRRELL

A. HUNTER DUPREE
Brown University
TORREY

JOY B. EASTON
West Virginia University
TUNSTALL; P. TURNER

FRANK N. EGERTON III
University of Wisconsin-Parkside
J. TOWNSEND

GUNNAR ERIKSSON
Umeå University
SWARTZ; C. P. THUNBERG

I. ESTERMANN †
STERN

JOSEPH EWAN
Tulane University
J. TRADESCANT, SR.; J. TRADESCANT, JR.;
TRELEASE

V. A. EYLES
STRACHEY; TOULMIN

SISTER MAUREEN FARRELL,
F.C.J.
University of Manchester
STEIN

I. FEDOSEYEV
Academy of Sciences of the U.S.S.R.
TILLO; VERNADSKY

BERNARD T. FELD
Massachusetts Institute of Technology
SZILARD

JEAN FELDMANN
Pierre and Marie Curie University
THURET

EUGENE S. FERGUSON
University of Delaware
THURSTON

MARTIN FICHMAN
Glendon College, York University
TABOR; ULSTAD

BERNARD FINN
Smithsonian Institution
STURGEON

MENSO FOLKERTS
Technische Universität Berlin
TITIUS

PIETRO FRANCESCHINI
University of Florence
VASSALE

EUGENE FRANKEL
VERDET

H.-CHRIST. FREIESLEBEN
STEINHEIL

HANS FREUDENTHAL
State University of Utrecht
SYLOW

JOHN E. FREY
Northern Michigan University
A. STOCK

JOSEPH S. FRUTON
Yale University
J. B. SUMNER

GERALD L. GEISON
Princeton University
THISELTON-DYER

ARTHUR C. GIESE
Stanford University
C. V. TAYLOR

NEAL W. GILBERT
University of California, Davis
TELESIO

C. STEWART GILLMOR
Wesleyan University
THÉVENOT

THOMAS F. GLICK
Boston University
TEILHARD DE CHARDIN; UNANUE;
VELLOZO

MARIO GLIOZZI
University of Turin
TORRICELLI

MARTHA TEACH GNUDI
University of California, Los Angeles
TORRE

STANLEY GOLDBERG
Hampshire College
TROUTON

D. C. GOODMAN
Open University
TENNANT

JUDITH R. GOODSTEIN
California Institute of Technology
TOLMAN

H. B. GOTTSCHALK
University of Leeds
STRATO OF LAMPSACUS

FRANK GREENAWAY
Science Museum, London
STEAD; W. E. S. TURNER

JOSEPH T. GREGORY
University of California, Berkeley
C. STOCK

A. T. GRIGORIAN
Academy of Sciences of the U.S.S.R.
THĀBIT IBN QURRA; TSIOLKOVSKY;
TUPOLEV; VEKSLER

M. D. GRMEK
*Archives Internationales d'Histoire des
Sciences*
VERANTIUS

HENRY GUERLAC
Cornell University
VAUBAN

FRANCISCO GUERRA
VALVERDE

KARLHEINZ HAAS
TAURINUS

WILLEM D. HACKMAN
Museum of the History of Science, Oxford
SWINDEN

MARIE BOAS HALL
*Imperial College of Science and
Technology*
TACHENIUS

JOHN HALLER
Harvard University
TERMIER

SAMI HAMARNEH
Smithsonian Institution
ABU'L-ḤASAN AḤMAD IBN MOḤAMMAD
AL-ṬABARĪ; ABU'L-ḤASAN 'ALI IBN SAHL
RABBAN AL-ṬABARĪ; IBN AL-TILMĪDH

OWEN HANNAWAY
Johns Hopkins University
TURQUET DE MAYERNE

NIKOLAUS M. HÄRING
*Pontifical Institute of Mediaeval Studies,
Toronto*
THIERRY OF CHARTRES

JOHN L. HEILBRON
University of California, Berkeley
SYMMER; J. J. THOMSON

ERICH HINTZSCHE †
VALENTIN

E. HLAWKA
University of Vienna
TAUBER

M. HOCQUETTE
TURPIN

E. DORRIT HOFFLEIT
Yale University
TROUVELOT

J. E. HOFMANN †
SUTER; TROPFKE; TSCHIRNHAUS

GEORGE F. HOURANI
State University of New York at Buffalo
IBN ṬUFAYL

CONTRIBUTORS TO VOLUME 13

WŁODZIMIERZ HUBICKI
Marie Curie-Skłodowska University
SUCHTEN; THURNEYSSER

KARL HUFBAUER
University of California, Irvine
TROMMSDORFF

G. L. HUXLEY
Queen's University of Belfast
THEON OF SMYRNA; THEUDIUS OF
MAGNESIA; THYMARIDAS

REESE V. JENKINS
Case Western Reserve University
TALBOT

RICHARD I. JOHNSON
Museum of Comparative Zoology
STIMPSON

PHILLIP S. JONES
University of Michigan
B. TAYLOR; VANDERMONDE

P. JOVET
Centre National de Floristique
THOUIN; S. VAILLANT

GEORGE KAHLSON
University of Lund
T. L. THUNBERG

T. N. KARI-NIAZOV
Academy of Sciences of the U.S.S.R.
ULUGH BEG

MARSHALL KAY
Columbia University
STILLE

BRIAN B. KELHAM
STONEY

A. G. KELLER
University of Leicester
STELLUTI

HUBERT C. KENNEDY
Providence College
VAILATI

PEARL KIBRE
Hunter College, City University of New York
THOMAS OF CANTIMPRÉ

F. KLEIN-FRANKE
AL-TĪFASHĪ

FRIEDRICH KLEMM
Deutsches Museum
VALTURIO

AKIRA KOBORI
Université Sangyo de Kyoto
TSU CH'UNG-CHIH

M. KOCH †
TREBRA

SHELDON J. KOPPERL
Grand Valley State Colleges
T. E. THORPE; URBAIN

JAN KREJČÍ
Purkyně University
STERNBERG

VLADISLAV KRUTA
Purkyně University
TEICHMANN; TIEDEMANN; UNZER

P. G. KULIKOVSKY
Academy of Sciences of the U.S.S.R.
P. K. STERNBERG; SUBBOTIN; TIKHOV;
TSERASKY

PAUL KUNITZSCH
University of Munich
AL-SŪFĪ

V. I. KUZNETSOV
Academy of Sciences of the U.S.S.R.
VAGNER

I. M. LAMB
Harvard University
THAXTER

EDWIN LAYTON
University of Minnesota
F. W. TAYLOR

JEAN F. LEROY
Muséum National d'Histoire Naturelle
TOURNEFORT

JACQUES R. LÉVY
Paris Observatory
STEPHAN; THOLLON; TISSERAND

E. B. LEWIS
California Institute of Technology
STURTEVANT

G. A. LINDEBOOM
Free University, Amsterdam
F. D. B. SYLVIUS

STEN LINDROTH
University of Uppsala
STELLER; SWEDENBORG

R. BRUCE LINDSAY
Brown University
J. W. STRUTT; SWANN

JAMES LONGRIGG
University of Newcastle Upon Tyne
THALES

JOHN B. McDIARMID
University of Washington
THEOPHRASTUS

ROBERT M. McKEON
Babson College
VERNIER

DUNCAN McKIE
University of Cambridge
TUTTON

VICTOR A. McKUSICK
Johns Hopkins Hospital
SUTTON

SAUNDERS MAC LANE
University of Chicago
VEBLEN

ROY MacLEOD
University of Sussex
TYNDALL

NORA F. McMILLAN
Merseyside County Museums
SWAINSON

M. MALLET
Centre National de Floristique
THOUIN; S. VAILLANT

NIKOLAUS MANI
University of Bonn
SUDHOFF

ARNALDO MASOTTI
Polytechnic of Milan
TARTAGLIA

KIRTLEY F. MATHER
Harvard University
VAN HISE

OTTO MAYR
Smithsonian Institution
STODOLA

GENEVIEVE MILLER
Case Western Reserve University
G. N. STEWART

ERIC L. MILLS
Dalhousie University
STEBBING

M. G. J. MINNAERT †
STEVIN

GIUSEPPE MONTALENTI
University of Rome
VALLISNIERI

ELLEN J. MOORE
U.S. Geological Survey
G. TROOST

J. B. MORRELL
University of Bradford
T. THOMSON

DALE M. J. MUELLER
Texas A&M University
TOZZI

D. MÜLLER
University of Copenhagen
STEENSTRUP

LETTIE S. MULTHAUF
STRATTON

JOHN E. MURDOCH
Harvard University
SWINESHEAD

J. A. NANNFELDT
Institute for Taxonomic Botany, Uppsala
SVEDELIUS

SEYYED HOSSEIN NASR
University of Teheran
AL-TŪSĪ

CONTRIBUTORS TO VOLUME 13

CLIFFORD M. NELSON
University of California, Berkeley
ULRICH

AXEL V. NIELSEN†
STRÖMGREN; T. N. THIELE

W. A. NIEUWENKAMP
State University of Utrecht
VENING MEINESZ

JOHN D. NORTH
University of Oxford
SYLVESTER; TAIT

A. NOUGARÈDE
Faculté des Sciences, Paris
VAN TIEGHEM

MARY JO NYE
University of Oklahoma
L. J. TROOST

ROBERT OLBY
University of Leeds
STAUDINGER; UNGER

C. D. O'MALLEY †
VAROLIO

JANE OPPENHEIMER
Bryn Mawr College
TENNENT

A. PAPLAUSCAS
Academy of Sciences of the U.S.S.R.
URYSON

JOHN PARASCANDOLA
University of Wisconsin
VAN SLYKE

E. M. PARKINSON
Worcester Polytechnic Institute
STOKES

KAI O. PEDERSEN
University of Uppsala
SVEDBERG; TISELIUS

JEAN PELSENEER
University of Brussels
TILLY; VERHULST

ENRIQUE PEREZ ARBELAEZ †
TRIANA

STUART PIERSON
Memorial University of Newfoundland
TILLET

DAVID PINGREE
Brown University
'UMAR IBN AL-FARRUKHĀN AL-ṬABARĪ;
VARĀHAMIHIRA VATEŚVARA

D. ANTON PINSKER, S.J.
*Archivar des Österreichischen Provinz des
Jesuitenordens, Vienna*
STEPLING

M. PLESSNER †
AL-TĪFASHĪ

LORIS PREMUDA
University of Padua
VALSALVA

RHODA RAPPAPORT
Vassar College
A.-R.-J. TURGOT; É.-F. TURGOT

ABRAHAM ROBINSON †
STOLZ; TOEPLITZ

GLORIA ROBINSON
Yale University
STRASBURGER

FRANCESCO RODOLICO
University of Florence
TARGIONI TOZZETTI

B. VAN ROOTSELAAR
State Agricultural University, Wageningen
TURING

PAUL LAWRENCE ROSE
James Cook University
TACCOLA

EDWARD ROSEN
City University of New York
TARDE; VANINI

B. A. ROSENFELD
Academy of Sciences of the U.S.S.R.
THĀBIT IBN QURRA

G. RUDOLPH
TRAUBE

A. S. SAIDAN
Jordanian University
AL-UMAWĪ; AL-UGLĪDISĪ

BETTINA F. SARGEANT
E. I. du Pont de Nemours and Company
STINE

CARL SCHALÉN
University of Lund
SUNDMAN

GUSTAV SCHERZ †
STENSEN

BRUNO SCHOENEBERG
University of Hamburg
STEINITZ

E. L. SCOTT
Stamford High School, Lincolnshire
URE

E. M. SENCHENKOVA
Academy of Sciences of the U.S.S.R.
TIMIRYAZEV; TSVET

ELIZABETH NOBLE SHOR
Scripps Institution of Oceanography
STEJNEGER; F. B. SUMNER

ROBERT R. SHROCK
Massachusetts Institute of Technology
TWENHOFEL

DANIEL M. SIEGEL
University of Wisconsin
B. STEWART

DIANA M. SIMPKINS
Polytechnic of North London
H. H. THOMAS; VAUCHER

W. A. SMEATON
University College, London
VAUQUELIN; VENEL

P. SMIT
G. R. TREVIRANUS; L. C. TREVIRANUS

CYRIL STANLEY SMITH
Massachusetts Institute of Technology
THEOPHILUS

IAN N. SNEDDON
University of Glasgow
M. STEWART

H. A. M. SNELDERS
State University of Utrecht
TEN RHYNE; TROOSTWIJK; VAN'T HOFF

Z. K. SOKOLOVSKAYA
Academy of Sciences of the U.S.S.R.
STRUVE FAMILY

A. I. SOLOVIEV
Academy of Sciences of the U.S.S.R.
TANFILEV

J. W. T. SPINKS
University of Saskatchewan
STEACIE

PIERRE SPEZIALI
University of Geneva
J. C.-F. STURM; J. TANNERY

NILS SPJELDNAES
Aarhus University
STØRMER; SVERDRUP; C. J. THOMSEN;
TILAS

C. G. G. J. VAN STEENIS
Rijksherbarium, Leiden
TREUB

PER STRØMHOLM
University of Oslo
TACQUET; VALERIO

CHARLES SÜSSKIND
University of California, Berkeley
S. P. THOMPSON; E. THOMSON

KENNETH M. SWEZEY †
TESLA

EDITH DUDLEY SYLLA
North Carolina State University at Raleigh
SWINESHEAD

F. SZABADVÁRY
Technical University, Budapest
SZEBELLÉDY; SZILY; THAN

CHARLES H. TALBOT
*Wellcome Institute for the History of
Medicine*
STEPHEN OF ANTIOCH

CONTRIBUTORS TO VOLUME 13

DICTIONARY
OF
SCIENTIFIC BIOGRAPHY

DICTIONARY OF
SCIENTIFIC BIOGRAPHY

STAUDINGER — VERONESE

STAUDINGER, HERMANN (*b.* Worms, Germany, 23 March 1881; *d.* Freiburg im Breisgau, Germany, 8 September 1965), *organic and macromolecular chemistry.*

Staudinger studied at the Gymnasium in Worms. Then, after a brief period at the University of Halle, he transferred to the technical university at Darmstadt when his father, the neo-Kantian philosopher Franz Staudinger, was appointed to a teaching post in that town. Although Staudinger wished to study botany, his parents were advised to give him first a thorough training in chemistry to prepare him for a career in botany. This excellent advice was followed; and from Darmstadt, Staudinger went on to study in Munich and Halle. His dissertation, on the malonic esters of unsaturated compounds, was written under D. Vorländer and was completed in 1903. But it was in Strasbourg, under Johannes Thiele, that Staudinger made his first and unexpected discovery—the highly reactive ketenes. These formed the subject of his *Habilitation* in 1907, the year in which he was appointed associate professor at the Technische Hochschule in Karlsruhe.

Five years later Staudinger succeeded Willstätter at the great Eidgenössische Technische Hochschule of Zurich, where he remained until his call to Freiburg im Breisgau in 1926. Three years after his retirement from the Freiburg chair he was awarded the Nobel Prize in chemistry. It was fitting, although unintentional, that this recognition of his work on macromolecular chemistry should have come in 1953, at a time when the molecular biology that he had glimpsed more than two decades before was taking shape. Staudinger married the Latvian plant physiologist Magda Woit in 1927.

In Karlsruhe, Staudinger achieved a new and simple synthesis of isoprene, from which polyisoprene (synthetic rubber) had previously been formed; and with C. L. Lautenschläger, he synthesized polyoxymethylenes. These discoveries later served him in his studies of polymer chemistry in Zurich and Freiburg im Breisgau. Staudinger's friends urged him to avoid so difficult a field as the chemistry of polymers, but he was not to be dissuaded. He realized that with polyisoprene he could devise a crucial experiment by which he might be able to confirm either the aggregate or long-chain-molecule theory of the structure of polymers. After synthetic rubber he turned to polyoxymethylene, which he saw as a model for the natural polymer cellulose.

In 1920 Staudinger first expressed his preference for the long-chain-molecule conception of polymers. Six years later he predicted the important role that such macromolecular compounds would be found to play in living organisms, especially in proteins. When he met his future wife, Staudinger's attention was drawn to the role of macromolecules in structural substances like the plant cell-wall constituent, cellulose. Henceforth he sought to introduce the macromolecular concept into biological chemistry.

In the 1920's there existed two conceptions of polymer structure. According to Samuel Pickles and K. Freudenberg, these substances consisted of long-chain molecules held together by "primary" or "Kekulé" bonds; but C. Harries and R. Pummerer believed the real molecules in polymers to be small. These and other authorities held that a polymer is formed by the binding action of the residual forces of unsaturated compounds. These "secondary" valency forces were responsible for the apparent nonstoichiometry and strange physical properties—the nonlinear relation between viscosity and concentration, the tendency to form colloidal solutions, and the failure to yield crystals. Supporters of this aggregate hypothesis argued that the true molecules of a substance like rubber were small and that in the free state they did obey the laws of physical chemistry; in particular, they could be crystallized. Destroy their aggregation and the molecules would be freed. Only thus could

1

the organic chemist be confident that he had a pure compound.

Harries suggested that the secondary forces that held the butadiene molecules together in natural rubber owed their presence to the unsaturated state of these molecules. He stated that hydrogenation of rubber should yield a product with a low boiling point since it would involve saturation of the forces of affinity within the molecules, and this process would destroy the secondary or residual forces between them. In 1922 Staudinger and J. Fritschi produced hydrorubber. Its properties differed little from natural rubber; in particular, it could not be distilled and, like rubber, it gave a colloidal solution. In their paper of that year they used the term "macro-molecular association" for the first time. Two years later Staudinger defined the macromolecule: "For such colloidal particles in which the molecule is identical with the primary particles, in other words, where the single atoms of the colloidal molecule are bound together by normal valency activities, we suggest the term *Makromolekül*."[1] He went on to point out that, since these colloidal particles are the true molecules, no attempts to produce typical, low-molecular solutions with other solvents would succeed.

In the 1920's Staudinger extended these studies to polystyrene and polyoxymethylene, showing that a whole range of products can be produced that, like the members of a homologous series, show a serial order in the viscosity of their solutions. This work was described on three important occasions—in 1924 at the Innsbruck meeting of the Deutsche Naturforscher und Aerzte, in 1925 at a meeting of the Zurich Chemical Society, and in 1926 at the Düsseldorf meeting of the Deutsche Naturforscher und Aerzte. Especially on the occasions in Zurich and Düsseldorf, Staudinger encountered vigorous opposition from the exponents of the aggregate theory. Viscosity measurements, it was argued, did not give direct evidence of molecular weights but, rather, reflected the state of colloidal aggregates in solution. No reliable data on such compounds would be forthcoming until genuine solutions and crystals had been formed. Moreover, the unit cells derived from fiber diagrams of these polymers were far too small to accommodate a macromolecule; and the mineralogist Paul Niggli assured Staudinger that molecules larger than the unit cell did not exist.

It was at this juncture that Theodor Svedberg and Robin Fåhraeus made their first successful measurements of the equilibrium sedimentation of oxy- and carbonylhemoglobin in the ultracentri-

fuge. The result indicated a molecular weight between 3.73 and 4.25 times the minimum value of 16,700 obtained from elementary analysis.[2] This work laid the basis for the recognition of high-molecular compounds in protein chemistry. Meanwhile, Staudinger battled on against the upholders of the aggregate and micellar theories for synthetic polymers, cellulose, and rubber.

So long as there existed no theoretical relationship between molecular weight and viscosity for nonspherical particles (exhibiting non-Newtonian flow), Staudinger's viscosimetric data were thought to be unsatisfactory as evidence for the existence of macromolecules. But in 1929–1930 R. Nodzu and E. Ochiai, working under Staudinger, showed that for low molecular compounds with linear-shaped molecules the viscosity of their solutions is proportional to the number of residues in the chain.

Staudinger hoped to achieve independent evidence for the macromolecular structure of synthetic and natural polymers by installing an ultracentrifuge in Freiburg, but in 1929 the Notgemeinschaft der Deutsche Wissenschaft refused him the necessary funds. In desperation he returned to viscosimetry and succeeded in deriving a relationship known as the Staudinger law, between specific viscosity η_{sp} and molecular weight, where η_{sp} represents the increase in viscosity of a solvent caused by the addition of solute. The solvent constant Km in the equation was evaluated by using solutes of known molecular weight and was then used for polymers of unknown molecular weight. He took the precaution of extrapolating to infinite dilution.

Here was a simple and quick method for obtaining molecular weights, which, unlike the ultracentrifuge, did not require costly and elaborate apparatus. Although further objections were voiced against it, viscosimetry—as based on Staudinger's law—was widely used in industry wherever polymer research was in progress.

To establish the conception of long-chain molecules by independent lines of evidence, Staudinger asked R. Signer to examine the shape of macromolecules in solution. Signer accomplished this task by using the technique of flow birefringence. He devised a simple apparatus for the rapid measurement of the approximate length:breadth ratio of long-chain molecules.

Meanwhile, Staudinger's conception of macromolecules received further support: X-ray crystallographers realized that the symmetry of crystals could be achieved on the basis of rope-like bundles of chains that stretched in the direction of the fiber axis far beyond a single-unit cell, even beyond a crystallite. Furthermore, in America, W. H. Carothers was achieving polymerization by a condensation reaction in which the eliminated water could be measured and the number of residues in the product estimated.

As early as 1926, Staudinger had appreciated the importance of macromolecular compounds in living organisms. He had seen how the traditional methods of isolation and identification of organic compounds inhibited the study of these sensitive and awkward compounds. Consequently, chemists were only "standing on the threshold of the chemistry of organic compounds."[3] Life processes, he argued, were bound up with high polymers in the shape of proteins and enzymes. In a lecture delivered in Munich in 1936 he returned to this theme. "Every gene macromolecule," he declared, "possesses a quite definite structural plan, which determines its function in life."[4] Such giant protein molecules had innumerable possible structures which chemical techniques were then too crude to reveal.

In 1947, in his book *Makromolekulare Chemie und Biologie*, Staudinger again visualized the molecular biology of the future. He reported the first attempts by Linderstrøm-Lang to arrive at amino acid sequences. This was the kind of problem that Staudinger wished to tackle in his later days, but he found no methods suited to the task. In this book Staudinger computed the molecular weight of a bacterium, from which we may conclude that he did allow his enthusiasm for macromolecules to carry him too far. Long before this time C. F. Robinow had demonstrated an organization in bacteria, including a nuclear structure revealed by differential staining.[5] Such an organism could hardly represent a single macromolecule.

Although Staudinger surely comprehended the conception of chemical individuality, it is understandable that he lacked any appreciation of the nature of information transfer from nucleic acids to proteins or, indeed, of the storage of such information in the nucleic acids rather than in the proteins. Nonetheless, his pioneer work in macromolecular chemistry constituted a major foundation for the molecular biology that was to be built upon it. The debates that took place between Staudinger and the champions of the aggregate theory furnished an interesting conflict of paradigms that only the further development of several sciences could resolve. As a result Staudinger achieved recognition for his work belatedly, receiving the Nobel Prize at the age of seventy-three.

Staudinger tried to produce visual evidence of the existence and form of macromolecules. The ultraviolet phase-contrast microscope and the electron microscope were used to this end in Freiburg. Magda Staudinger began such work in 1937. She and G. A. Kausche described spherical molecules of glycogen two years later. In 1942 Staudinger's colleagues E. Husemann and H. Ruska obtained electron micrographs of glycogen particles with a diameter of 10 mμ. From the osmotic pressure of the corresponding glycogen in solution, they concluded that these were the molecules of glycogen with a molecular weight of one and a half million. This work was brought to an abrupt end when the greater part of the chemistry institute was destroyed during the bombing of Freiburg in 1944. By the time normal working conditions were restored, Staudinger's vigorous powers had been spent; but it was due to him that in 1947 a new journal, *Makromolekulare Chemie*, was published by the firm of Karl Alber in Freiburg; the earlier *Journal für makromolekulare Chemie* appeared only from 1943 to 1945. Both journals were edited by Staudinger. On his retirement in 1951, Staudinger's department became the State Research Institute for Macromolecular Chemistry; five years later an associate professorship in macromolecular chemistry was established for the director of this institute.

NOTES

1. H. Staudinger, "Ueber die Konstitution des Kautschuks," in *Berichte der Deutschen chemischen Gesellschaft*, 57 (1924), 1206.
2. T. Svedberg and R. Fåhraeus, "A New Method for the Determination of the Molecular Weight of the Proteins," in *Journal of the American Chemical Society*, 48 (1926), 430–438.
3. H. Staudinger, "Die Chemie der hochmolekularen organischen Stoffe im Sinne der Kekuléschen Strukturlehre," in *Berichte der Deutschen chemischen Gesellschaft*, 59 (1926), 3019–3043.
4. H. Staudinger, "Ueber die makromolekulare Chemie," in *Angewandte Chemie*, 49 (1936), 801.
5. C. F. Robinow, "A Study of the Nuclear Apparatus of Bacteria," in *Proceedings of the Royal Society*, 130 B (1942), 299.

BIBLIOGRAPHY

For a comprehensive bibliography of 644 works of Staudinger, see H. Staudinger, *Arbeitserinnerungen* (Heidelberg, 1961), trans. as *From Organic Chemistry to Macromolecules* (New York, 1970).

Staudinger's work on macromolecular chemistry is discussed in several papers: J. T. Edsal, "Proteins as Macromolecules: An Essay on the Development of the

Macromolecule Concept and Some of Its Vicissitudes," in *Archives of Biochemistry and Biophysics*, supp. 1, pp. 12–20; H. Mark, "Polymers—Past, Present and Future," in an unpublished symposium of the Welch Foundation on polymer science (1965); and R. C. Olby, "The Macromolecule Concept and the Origins of Molecular Biology," in *Journal of Chemical Education*, **47** (1970), 168–174; and *The Path to the Double Helix*, chs. 1, 2 (London, 1974).

ROBERT OLBY

STAUDT, KARL GEORG CHRISTIAN VON (*b.* Rothenburg-ob-der-Tauber, Germany, 24 January 1798; *d.* Erlangen, Germany, 1 June 1867), *mathematics.*

Staudt was the son of Johann Christian von Staudt, a municipal counsel, and Maria Albrecht. Rothenburg, famous for its many antiquities, was then a free imperial German city. The family had settled in Rothenburg as craftsmen as early as 1402. Various members became municipal councilmen in the sixteenth century and received a coat of arms. In 1700 Leopold I ennobled the family. Staudt's maternal ancestors, the Albrechts, also served as councilmen and burgomasters in the seventeenth and eighteenth centuries. Staudt's father was appointed a municipal legal officer by the Bavarian government in 1805, the year Rothenburg became part of the Kingdom of Bavaria.

After carefully supervising his early education, Staudt's parents sent him to the Gymnasium in Ansbach from 1814 to 1817. Then, drawn by the great reputation of Gauss, Staudt attended the University of Göttingen from 1818 to 1822. As a student he was surely well acquainted with Gauss's studies in number theory. His chief concern in these years, however, was theoretical and practical astronomy, to which he was also introduced by Gauss, who was then director of the observatory. As early as 1820 Staudt observed and computed the ephemerides of Mars and Pallas. His most comprehensive work in astronomy was the determination of the orbit of the comet discovered by Joseph Nicollet and Jean-Louis Pons in 1821. His precise calculations were highly praised by Gauss, and later observations led to only minor improvements. Staudt never returned to the field of astronomy, but it was on the basis of this early work that he received the doctorate from the University of Erlangen in 1822.

In the same year Staudt qualified at Munich as a mathematics teacher. His first assignment was at the secondary school in Würzburg. But with

Gauss's intervention he was also able to lecture at the University of Würzburg. His lectures dealt with rather elementary topics. Because of insufficient support from the university, he transferred in 1827 to the secondary school in Nuremberg and taught there and at the Nuremberg polytechnical school until 1835. He finally achieved his primary goal when, on 1 October 1835, he was appointed full professor of mathematics at the University of Erlangen, where he remained until his death. He was unquestionably the leading mathematician at Erlangen, not least because of his outstanding human qualities. The latter, indeed, brought him many honorary posts in the university administration.

As at most German universities during this period, the level of mathematics instruction at Erlangen was not high, nor did the subject attract many students. It was not yet customary for mathematicians to discuss their own research in the classroom—a practice first introduced by Jacobi, at Königsberg. Accordingly, it was not until 1842–1843 that Staudt gave special lectures on his new geometry of position.

In 1832 Staudt married Jeanette Drechsler. They had a son, Eduard, and a daughter, Mathilde, who became the wife of a burgomaster of Erlangen. Staudt's wife died in 1848, and he never remarried. In his last years he suffered greatly from asthma.

Staudt was not a mathematician who astounded his colleagues by a flood of publications in a number of fields. He let his ideas mature for a long period before making them public, and his research was confined exclusively to projective geometry and to the only distantly related Bernoullian numbers. His fame as a great innovator in the history of mathematics stems primarily from his work in projective geometry, which he still called by the old name of "geometry of position," or *Geometrie der Lage*, the title of his principal publication (1847). This work was followed by three supplementary *Beiträge zur Geometrie der Lage* (1856–1860), which together contain more pages than the original book (396 as compared with 216).

After centuries of dominance, Euclidean geometry was challenged by Poncelet and Gergonne, who created projective geometry during the first third of the nineteenth century. These two mathematicians found that, through the use of perspective, circles and squares and other figures could be transformed into arbitrary conic sections and quadrilaterals and that a metric theorem for, say, the circle could be transformed into a metric theorem for conic sections. The most important contributions made by Poncelet (whose main writings appeared between 1813 and 1822) and Gergonne were the polarity theory of the conic sections and the principle of duality. Jakob Steiner, in his fundamental work *Systematische Entwicklung der Abhängigkeit geometrischer Gestalten voneinander* (1832), then introduced the projective production of conic sections and second-degree surfaces that is now named for him.

In their writings, however, all three of these pioneers failed to adhere strictly to the viewpoint of projective geometry, which admits only intersection, union, and incidence of points, straight lines, and planes. Staudt, in his 1847 book, was the first to adopt a fully rigorous approach. Without exception, his predecessors still spoke of distances, perpendiculars, angles, and other entities that play no role in projective geometry. Moreover, as the name of that important relationship indicates, in accounting for the cross ratio of four points on a straight line, they all made use of line segments. In contrast, Staudt stated in the preface to his masterpiece his intention of establishing the "geometry of position" free from all metrical considerations, and in the body of the book he constructed a real projective geometry of two and three dimensions.

Naturally, in Staudt's book these geometries are not founded on a complete axiom system in the modern sense. Rather, he adopted from Euclid's system everything that did not pertain to interval lengths, angles, and perpendicularity. Although it was not necessary, he also retained Euclid's parallel postulate and was therefore obliged to introduce points at infinity. This decision, while burdening his treatment with a constant need to consider the special positions of the geometric elements at infinite distance, altered nothing of the basic structure of geometry without a metric. Using only union and intersection of straight lines in the plane, Staudt constructed the fourth harmonic associated with three points on a straight line. Correspondingly, with three straight lines or planes of a pencil he was able to construct the fourth harmonic element. Although he did not give the theorem that name, he used Desargues's theorem to prove that his construction was precise.

Using the relationship of four points in general position on a plane to four corresponding points or straight lines on another plane or on the same plane, Staudt defined a collineation—or, as the case may be, a correlation—between these planes. Analogously he also pointed out spatial collineations and correlations. In this instance he made use of Möbius' network construction, which enabled him to obtain, from four given points of a plane, denumerably many points by drawing straight lines through point pairs and by making straight lines intersect. He then associated the points derived in this way with correspondingly constructed points and straight lines of the other plane. Felix Klein later noted that a continuity postulate is still required in order to assign to each of the infinitely many points of the first plane its image point (or image lines) on the other plane.

From the time of his first publications, Staudt displayed a grasp of the importance of the principle of duality. For every theorem he stated its converse. (As was customary, he generally gave the theorem on one half of the page and its converse on the other half.) In discussing the autocorrelations of P_2 and P_3, he succeeded in obtaining the polarities and also the null correlations that had previously been discovered by Gaetano Giorgini and Möbius. For example, he described a plane polarity as a particular type of autocorrelation that yields a triangle in which each vertex is associated with the side opposite. On this basis, Staudt formulated the definition of the conic sections and quadrics that bears his name: they are the loci of those points that, through a polarity, are incident with their assigned straight lines or planes. This definition is superior to the one given by Steiner. For instance, in Staudt's definition the conic section appears as a point locus together with the totality of its tangents. Steiner, in contrast, required two different productions: one for the conic section K and another for the totality of its tangents, that is, for the dual figures associated with K. A conic section defined in Staudt's manner can consist of the empty set; that is, it can contain no real points—accordingly, Klein applied the term *nullteilig* to it.

In a coordinate geometry it is easy to extend the domain of the real points to the domain of the complex points with complex coordinates. Employing the concepts of real geometry, Staudt made an essential contribution to synthetic geometry through his elegantly formulated introduction of the complex projective spaces of one, two, and three dimensions. This advance was the principal achievement contained in his *Beiträge zur Geometrie der Lage*. He conceived of the complex points of a straight line P_1 by means of the so-called elliptic involution of the real range p_1 of P_1, which can also be described as those involuted autoprojectivities of p_1 among which pairs of corresponding points intersect each other. It can be shown by calculation that such an elliptic involution has two complex, conjugate fixed points; and Staudt had to furnish the elliptic involutions with two different orientations, so that ultimately he could interpret the oriented elliptic involutions on the real range of P_1 as points of P_1. The degenerate parabolic involutions are to be associated with the real points of P_1. In this way he also extended the real projective planes p_2 and p_3 to complex P_2 and P_3. He then showed—not an easy feat—that P_2 and P_3 satisfy the connection axioms of projective geometry. Among the lines P_1 of P_3 he found three types: those with infinitely many real points, those with only one real point, and those with no real point. He carefully classified the quadrics of P_3 according to the way in which straight lines of these three types lie on them.

Staudt favored the use of the second type of complex line of P_3 as a model of the complex numbers P_1. He applied the term *Wurf* ("throw") to a point quadruple and gave the procedure for finding sums and products in the set of these throws—or, more precisely, the set of the equivalence classes of projectively equivalent throws. Here he approached the projective foundation of the complex number field and the projective metric determination. Staudt termed certain throws neutral: those with real cross ratio, a property that can be determined computationally. Then, for three given points—A, B, C, in P_1—he designated as a chain the set of all those points of P_1 that form a neutral throw with A, B, C. These sets and their generalization to complex P_n are called "Staudt chains." In part three of the *Beiträge*, Staudt also dealt with third- and fourth-order spatial curves in the context of the theory of linear systems of equations.

At the time of their publication, Staudt's books were considered difficult. This assessment arose for several reasons. First, since he sought to present a strictly systematic construction of synthetic geometry, he did not present any formulas; moreover, he refused to employ any diagrams. Second, he cited no other authors. Finally, although his theory of imaginaries was remarkable, it was extremely difficult to manipulate in comparison with algebraic equations. Accordingly, little significant progress could have been expected from its adoption in the study of figures more complicated than conic sections and quadrics.

Staudt is also known today for the Staudt-Clausen theorem in the theory of Bernoulli numbers. These numbers—B_n ($n = 1, 2, \cdots$)—appear in the summation formulas of the nth powers of the first h natural numbers; they also arise in analysis, for example, in the series expansion $x \cot x$. The B_n are rational numbers of alternating sign, and the Staudt-Clausen theorem furnished the first significant indication of the law of their formation. In formulating the theorem, for the natural number n there is a designated uneven prime number, p, called Staudt's prime number, such that $p - 1$ divides the number $2n$. Then, according to the theorem, $(-1)^n B_n$ is a positive rational number, which, aside from its integral component, is a sum of unit fractions, among the denominators of which appear precisely the number 2 and all Staudt prime numbers for n. Staudt published his theorem in 1840; it was also demonstrated, independently, in the same year by Thomas Clausen, who was working in Altona. Staudt published two further, detailed works in Latin on the theory of Bernoulli numbers (Erlangen, 1845); but these writings never became widely known and later authors almost never cited them.

BIBLIOGRAPHY

I. ORIGINAL WORKS. Staudt's major works are "Beweis eines Lehrsatzes, die Bernoullischen Zahlen betreffend," in *Journal für die reine und angewandte Mathematik*, **21** (1840), 372–374; *Geometrie der Lage* (Nuremberg, 1847), with Italian trans. by M. Pieri (see below); and *Beiträge zur Geometrie der Lage*, 3 vols. (Nuremberg, 1856–1860).

II. SECONDARY LITERATURE. See G. Böhmer, *Professor Dr. K. G. Chr. von Staudt, Ein Lebensbild* (Rothenburg-ob-der-Tauber, 1953); M. Noether, "Zur Erinnerung an Karl Georg Christian von Staudt," in *Jahresbericht der Deutschen Mathematikervereinigung*, **32** (1923), 97–119; "Nekrolog auf K. G. Chr. von Staudt," in *Sitzungsberichte der Bayerischen Akademie der Wissenschaften zu München*, **1** (1868), repr. in *Archiv der Mathematik*, **49** (1869), 1–5; and C. Segre, "C. G. C. von Staudt ed i suoi lavori," in *Geometria di posizione de Staudt*, M. Pieri, ed. and trans. (Turin, 1888), 1–17.

WERNER BURAU

STEACIE, EDGAR WILLIAM RICHARD (*b.* Westmount, Quebec, Canada, 25 December 1900; *d.* Ottawa, Ontario, Canada, 28 August 1962), *physical chemistry.*

Steacie was the only child of Captain Richard Steacie and Alice Kate McWood. His father was born in Ballinasloe, Northern Ireland, from which he emigrated at about the age of twenty to settle in Montreal. Edgar was brought up in comfortable circumstances but was only fourteen when his father died. He was a student at the Royal Military

College in Kingston from 1919 to 1920 and then transferred to McGill University, where he obtained his B.Sc. (1923), M.Sc. (1924), and Ph.D. (1926). He pursued postdoctoral study in Frankfurt and Leipzig and at King's College, London (1934–1935). He was made associate professor at McGill in 1937 but in 1939 he moved to the National Research Council of Canada. He was president of the council from 1952 to 1962.

In 1925 Steacie married Dorothy Catalina Day; they had two children, Diane Jeanette (Mrs. W. A. Magill) and John Richard Brian.

Steacie published many articles in scientific journals and wrote three books: *Introduction to Physical Chemistry*, with O. Maass (1926); *Atomic and Free Radical Reactions* (1946); and *Free Radical Mechanisms* (1946).

In the early stages of his career, Steacie established an enviable reputation in the field of photochemistry, especially in the kinetics of gas reactions and their interpretation by free radical mechanisms. He maintained this interest in photochemistry even after taking up administration duties. A steady stream of postdoctoral fellows came to work in his Ottawa laboratory, and a whole generation of young chemists were strongly influenced by him.

As early as 1934 Steacie used reaction mechanisms involving free radicals to interpret the kinetics of thermal decomposition reactions of organic compounds. In 1937 he obtained further evidence on the mechanisms of organic pyrolyses using deuterated compounds. He then extended his interest in reaction mechanisms to photolytic decompositions of a variety of simple organic compounds and to photosensitization by vapors of metals such as mercury, cadmium, and zinc. Using these various techniques he obtained kinetic data on the rates of elementary chemical processes. The extensive studies by Steacie and his many students formed the basis for his authoritative *Atomic and Free Radical Reactions*.

The development of Steacie's administrative career was linked to the development of the National Research Council of Canada. Founded in 1916, the Research Council underwent an enormous expansion during World War II under the direction of C. J. Mackenzie and emerged with an established reputation for good work. During his tenure as president of the council, Steacie helped bring to fruition long-range plans for the growth of Canadian science, partly by the development of the Research Council laboratories themselves and partly by the encouragement of strong scientific centers in Canadian universities. Here his emphasis was on scholarship programs and the support of qualified individuals doing fundamental research. He will be especially remembered for his enthusiastic support of the postdoctoral fellowship program and for the huge increase in university support that took place under him. As chairman of the council, Steacie handled its proceedings with exceptional skill—never insisting on his own viewpoint but often making a masterly summary of the situation and coming forward with an acceptable proposal at the right time.

Outside the council, Steacie defended the interests of its scientists with great skill and vigor. His presidency coincided approximately with the period in which nations and governments became aware not only of the increasing importance of science to national well-being and security but also of the importance of developing an appropriate relationship between science and government.

BIBLIOGRAPHY

I. ORIGINAL WORKS. L. Marion (see below) lists some 230 scientific papers by Steacie. His works include *Laboratory Exercises in General Chemistry* (Montreal, 1929), written with W. H. Hatcher and N. N. Evans; *An Introduction to the Principles of Physical Chemistry* (New York, 1931; 2nd ed., 1939), written with O. Maass; *Atomic and Free Radical Reactions* (New York, 1946; 2nd ed., 1954); and *Free Radical Mechanisms* (New York, 1946). See also *Science in Canada. Selections From the Speeches of E. W. R. Steacie*, J. D. Babbitt, ed. (Toronto, 1965).

II. SECONDARY LITERATURE. On Steacie and his work, see L. Marion, "E. W. R. Steacie. 1900–1962," in *Biographical Memoirs of Fellows of the Royal Society*, **10** (1964), 257–281, with bibliography; and W. A. Noyes, Jr., convocation address at the opening of the E. W. R. Steacie building for chemistry at Carleton Univ., 22 Oct. 1965.

J. W. T. SPINKS

STEAD, JOHN EDWARD (*b.* Howden-on-Tyne, Northumberland, England, 17 October 1851; *d.* Redcar, Yorkshire, England, 31 October 1923), *metallurgy, analytical chemistry.*

Stead was the younger brother of the journalist W. T. Stead (1849–1912).

Educated privately because he was not robust, Stead was apprenticed at the age of sixteen to John Pattinson, the chemical analyst of Newcastle-upon-Tyne. Three years later he worked at Bolckow, Vaughan and Company, an iron manufacturing firm in Garston, from which post he was able to attend evening classes at Owens College (later the University of Manchester). At the age of twenty-five he became a partner to Pattinson and remained an analyst for the remainder of his life. His original work arose from metallurgical problems encountered in the course of his large commercial practice.

Stead studied eutectics in steels and used his findings to interpret such phenomena as the occurrence of blast-furnace "bears," which were large inclusions of metal formed in the hearth. He also studied the crystalline structure of metals, and was one of the first to recognize the significance of Sorby's work in metallography, which he advanced by such techniques as the heat-tinting of specimens. Stead learned a great deal about the effects of phosphorus on steel. Early in his career he explained the afterblow—essential in the basic Bessemer process, for the complete dephosphorization of phosphoric iron—as a result of the removal of phosphorus by iron oxide after the depletion of all the carbon. Another phenomenon connected with phosphorus was that of "ghosts," superficial markings on forgings, which cause the forgings to be suspect. Stead showed that the markings are due to the difference in solubility of carbon in parts of a steel of different phosphorus content. He also showed that these differences, although visible, are not detrimental in the absence of slag inclusions.

Stead's experimental studies were not confined to the laboratory, and many steelworks on Tyneside offered him facilities for study under working conditions. In 1901 Stead was honored by the Bessemer Medal of the Iron and Steel Institute, and two years later he became a fellow of the Royal Society. He also received honorary doctorates from the universities of Manchester, Leeds, and Sheffield.

BIBLIOGRAPHY

Obituary notices on Stead are H. C. H. Carpenter, in *Proceedings of the Royal Society*, **106A** (1924), i–v, with portrait; and an anonymous author, in *Engineering*, **116** (1923), 598–600, also with portrait.

FRANK GREENAWAY

STEBBING, THOMAS ROSCOE REDE (*b.* London, England, 6 February 1835; *d.* Tunbridge Wells, England, 8 July 1926), *zoology*.

Stebbing called himself "a serf to Natural History, principally employed about Crustacea." He was the fourth son of Henry Stebbing, poet, historian, clergyman, and editor of the *Athenaeum*, and Mary Griffin. Several of the thirteen Stebbing children became writers, and Thomas' brother William, a barrister, was a leader-writer and assistant editor of *The Times* (London).

Beginning his education at King's College School and King's College, London (B.A. 1855), Stebbing matriculated at Lincoln College, Oxford, in 1853 and became a scholar of Worcester College the same year. He received the B.A. in 1857 and the M.A. in 1859, remaining as fellow (1860–1868), tutor (1865–1867), vice-provost (1865), and dean (1866) of Worcester College until resigning his fellowship in 1868. Samuel Wilberforce, bishop of Oxford, ordained him priest in 1859.

When Stebbing took a house for tuition at Reigate, Surrey, in 1863 and met the entomologist William Wilson Saunders, his interest in science began. Upon marrying Saunders' daughter Mary Anne (*d.* 1927), also a naturalist, in 1867, he moved to Torquay, Devon, as a tutor. There, under the influence of the naturalist William Pengelly, he began a long series of writings on natural history, Darwinian evolution, and theology.

In 1877 the Stebbings moved to Tunbridge Wells, Kent, where they lived until their deaths. Although a teacher part of this time, Stebbing devoted most of these years to the study of amphipod Crustacea and to writing. Because of his work on the taxonomy of Crustacea, he became a fellow of the Linnean Society in 1895, fellow of the Royal Society in 1896, and was awarded the Gold Medal of the Linnean Society in 1908.

Stebbing's scientific writings date from 1873; most of them are devoted to the taxonomy of amphipod Crustacea, on which he published about 110 papers and two major monographs. He received the Amphipoda collected by H.M.S. *Challenger* on the recommendation of a marine biologist, the Canon A. M. Norman, and published a large monograph on these creatures in 1888. The 600-page annotated bibliography beginning this work is the definitive history of the classification of Amphipoda and quotes the original definition of each known genus. In a short introduction he discussed the ancestry of the Amphipoda, pointing out how small variations could account for its evolutionary radiation. An ancestor of the group had "simplicity" and "completeness" of characters, that is, structures common in many families and also structures that had disappeared in the more specialized ones. According to his discussion of the known families, the classification of Amphipoda was centered on the ancestral family Gammaridae. Also, among Stebbing's writings are many popular works on Crustacea and other arthropods. He wrote with simplicity, grace, wit, and erudition. His knowledge of Crustacea was widely respected and resulted in a worldwide correspondence with specialists.

Stebbing was an early convert to Darwinism, and many of his essays were written in support of it. He was years ahead of other carcinologists in realizing the importance of Darwinian evolution in the Amphipoda, although this is seldom evident in his routine taxonomic papers. He also subscribed to Herbert Spencer's view that natural selection shaped behavior and accounted for human moral progress. The logic of science was applied by Stebbing to theological dogmas of the Church of England. He doubted the literal truth of Genesis, the accuracy of prophecy, miracles, the doctrine of the Trinity, and many of the Thirty-nine Articles. Shorn of dogma and superstition by science, his religion was based on an omniscient and loving God and on the power of unselfishness. He held this view to be perfectly compatible with the evolutionary science given form by James Hutton, William Smith, Charles Lyell, Charles Darwin, and Herbert Spencer, which was the main philosophy of his life.

BIBLIOGRAPHY

I. ORIGINAL WORKS. Stebbing's works include *Essays on Darwinism* (London, 1871); "Report on the Amphipoda Collected by H.M.S. Challenger During the Years 1873–1876," in *Report on the Scientific Results of the Voyage of H.M.S. Challenger During the Years 1873–1876*, Zoology, XXIX (London, 1888), i–xxiv, 1–1737; *A History of Crustacea. Recent Malacostraca* (London, 1893); "Amphipoda I. Gammaridae," in *Tierreich*, 21 (1906), i–xxix, 1–806; "An Autobiographic Sketch," in *Transactions and Proceedings. Torquay Natural History Society*, 4 (1923), 1–5, with portrait; and *Plain Speaking* (London, 1926).

II. SECONDARY LITERATURE. See W. T. Calman, "T. R. R. Stebbing—1835–1926," in *Proceedings of the Royal Society*, 101B (1926), xxx–xxxii, with portrait; "Rev. Thomas Roscoe Rede Stebbing," in *Proceedings of the Linnean Society of London*, session 139 (1926–1927), 101–103; and E. L. Mills, "Amphipods and Equipoise. A Study of T. R. R. Stebbing," in *Transactions of the Connecticut Academy of Arts and Sciences*, 44, 239–256, with portrait.

ERIC L. MILLS

STEENSTRUP, (JOHANNES) JAPETUS SMITH (*b.* Vang, Denmark, 8 March 1813; *d.* Copenhagen, Denmark, 20 June 1897), *zoology.*

Steenstrup was the son of a vicar in northern Jutland; from his youth he hunted, fished, and collected fossils. Although he took no university degree, he taught for six years at the Sorö Academy, where in 1842 he published two works that brought him scientific fame. The first of these, "Geognostisk-geologisk Undersögelse af Skovmoserne Vidnesdam og Lillemose i det nordlige Sjaelland . . .," is a classic work in Scandinavian bog research. In it, Steenstrup compared the results of his own observations on the forest bogs of northern Zealand and on the bogs of the Danish forests, moors, and fens. He noted the postglacial succession and alteration of flora in the bogs and, recognizing that the formation of peat had taken at least five thousand years, was inclined to believe that such changes reflected changes in climate. "We may," he wrote, "consider the bogs as annual reviews in which we can see how the flora and fauna of our country have developed and changed. . . . The further we go back in time, the colder was the climate."

The second major publication of 1842 was *Om Forplantning og Udvikling gjennem vexlende Generationsraekker, en saeregen form for Opfostringen i de lavere Dyreklasser*, Steenstrup's comprehensive presentation of the form of reproduction that he called "alternation of generations," that is, the alternation of asexual and sexual reproduction, or metagenesis. This phenomenon had previously been described by Chamisso, but Steenstrup included a greater number of observations, based on a significantly wider range of subjects, and provided an important chapter on its meaning. Steenstrup's growing reputation, based largely upon these two publications, won him an appointment as professor of zoology at the University of Copenhagen, where he taught from 1846 until 1885.

Steenstrup returned to the study of Scandinavian fossils in 1850 when, with the archaeologist Jens Worsaae, he demonstrated that the shell heaps on the Danish seashores were man-made. He coined the term *Kjökkenmödding* ("kitchen midden") to describe these 4,000–7,000-year-old remains. He also did important taxonomic work on the Cephalopoda, and described many new genera. In a memoir of 1856, "Hektokotyldannelsen hos Octopodslaegterne Argonauta og Tremoctopus . . .," he reported the surprising finding that the arm of the male octopus is modified to fulfill a reproductive function. In addition, he published a number of short papers on a wide variety of zoological subjects.

It is possible that Steenstrup might have accomplished more if he had chosen to concentrate on fewer topics. His significance to science, however, should not be measured by his writings alone; rather, it should be remembered that for a period of fifty years he initiated and guided Danish research in natural history. His work was influenced strongly by the German school of natural science, and, although he corresponded with Darwin (they had both worked with Cirripedias), Steenstrup was never able to accept Darwin's theory of evolution. In a letter to him of 1881, Darwin expressed disappointment that this should have been so.

BIBLIOGRAPHY

I. Original Works. A list of Steenstrup's writings, containing 239 titles, is S. Dahl, "Bibliographia Steenstrupiana," in H. F. E. Jungersen and J. E. B. Warming, *Mindeskrift i Anledning af Hundredaaret for Japetus Steenstrups Fødsel*, I (Copenhagen, 1914). Among the most important are "Geognostisk-geologisk Undersögelse af Skovmoserne Vidnesdam og Lillemose i det nordlige Sjaelland, ledsaget af sammenlignende Bemaerkninger, hentede fra Danmarks Skov-, Kjaer- og Lyngmoser i Almindelighed," in *Kongelige Danske Videnskabernes Selskabs Skrifter*, 4th ser., **9** (1842), 17–120; *Om Forplantning og Udvikling gjennem vexlende Generationsraekker, en saeregen form for Opfostringen i de lavere Dyreklasser* (Copenhagen, 1842); and "Hektokotyldannelsen hos Octopodslaegterne Argonauta og Tremoctopus, oplyst ved lignende Dannelser hos Blaeksprutterne i Almindelighed," in *Kongelige Danske Videnskabernes Selskabs Skrifter*, 5th ser., **4** (1856), 185–216.

II. Secondary Literature. See C. Lütken, "Steenstrup," in *Natural Science*, **11** (1897), 159–169; and R. Spärck, "Japetus Steenstrup," in V. Meisen, ed., *Prominent Danish Scientists Through the Ages* (Copenhagen, 1932), 115–119.

D. Müller

STEFAN, JOSEF (*b.* St. Peter, near Klagenfurt, Austria, 24 March 1835; *d.* Vienna, Austria, 7 January 1893), *physics.*

Stefan's parents were of Slovene origin. An excellent student at the elementary school and later the Gymnasium in Klagenfurt, he enrolled at the University of Vienna in 1853 and became a *realschule* teacher there four years later. He worked with Carl Ludwig, in the latter's laboratory, on the flow of water through tubes. In 1858 he qualified as lecturer at the University of Vienna. He became full professor of higher mathematics and physics in 1863, and three years later was appointed director of the Institute for Experimental Physics, founded by Doppler in 1850. Stefan was a brilliant experimenter and a well-liked teacher. He was dean of the Philosophical Faculty in 1869–1870 and *rector magnificus* in 1876–1877. In 1860 he became a corresponding member, and in 1865 member, of the Imperial Academy of Sciences. He was named secretary of the mathematics-science class of the academy in 1875 and served as vice-president from 1885 until his death. In 1883 he presided over the scientific commission of the International Electricity Exhibition in Vienna. Two years later he held the same position at the International Conference on Musical Pitch in Vienna. He also belonged to several foreign scientific academies, held numerous Austrian and foreign honors, and was both royal and imperial privy councillor.

Stefan's most important work deals with heat radiation (1879). Newton had stated a priori a law of cooling for the temperature loss of incandescent iron in a constant stream of air, and Richmann had restated it in the following form: The speed of cooling is proportional to the difference in temperature between the heated body and the surrounding atmosphere. In equal periods of time, Newton stated, equal quantities of air are heated by quantities of heat proportional to those that they remove from the iron (*Opuscula* [1744], II, 423). G. W. Krafft and Richmann verified this law for temperature differences up to 40° or 50°. Yet as early as 1740 George Martine the younger and others realized its inaccuracy and attempted to replace it with another law according to which the heat losses increased more rapidly.

Nevertheless, physicists still considered Newton's law to be exact. Dalton sought to save it by introducing a new temperature scale. F. Delaroche was aware that the heat losses due to radiation increase much more rapidly than in proportion to the temperature difference, but he did not isolate the radiation from the other heat losses—as Dulong and Petit attempted to do. For radiation in empty space they propounded a rather more complicated law, introducing an absolute temperature scale and extending Newton's law. As was later seen, however, their law also possessed only limited validity and did not agree with measured results even up to 300°C. A. Wüllner remarked in his *Lehrbuch der Experimentalphysik*: . . . "the quantity of heat emitted increases considerably more quickly than does the temperature, especially at higher temperatures." This followed from Tyndall's experiments on the radiation of a platinum wire heated to incandescence by an electric current. From the weak red glow (about 525°C.) up to the full white glow (about 1200°C.), the intensity of the radiation increases almost twelvefold, from 10.4 to 122 (exactly 11.7-fold).

"This observation," Stefan said, "caused me at first to take the heat radiation as proportional to the fourth power of the absolute temperature." (The ratio of the absolute temperatures 273 + 1200 and 273 + 525 yields, in the fourth power, 11.6.) By means of a thorough discussion of the experiments of Dulong and Petit and of other researchers, Stefan showed that this formula agreed with the results of measurements in all temperature ranges. The theoretical deduction of this relationship was first achieved in 1884 by Boltzmann, who also recognized that it is exact only for completely black bodies (Stefan-Boltzmann law of radiation). Moreover, with the aid of his new formula Stefan could calculate, on the basis of Pouillet's and Violle's actinometric observations, a value for the surface temperature of the sun—approximately 6000°C.

Other important work by Stefan concerns heat conduction in gases, a subject on which reliable measurements were lacking because of extreme experimental difficulties. For this purpose Stefan devised a "diathermometer," which was widely used to measure the heat conductivity of clothing materials. His measurements agreed fairly well with those calculated on the basis of the kinetic theory of gases, especially in the cases of air and hydrogen. Stefan explained the variations as resulting from the movements of the atoms against each other within the molecules. He resumed his investigations on heat conduction in 1876 and 1889. In analogous work on the diffusion of gases (1871, 1872), he calculated the theoretical coefficients of diffusion and of friction and their dependence on the absolute temperature, showing that the calculated values were in agreement with experimental results obtained by Maxwell, Graham, and J. Loschmidt. He also demonstrated that the apparent adhesion of two glass plates is a hydrodynamic phenomenon (1874).

Stefan published further experimental and theoretical works on the kinetic theory of heat: on evaporation (1873, 1881); on heat conduction in fluids, on ice formation, on dissolving (1889); and on diffusion of and in fluids (1878, 1879); and especially on the relationship between surface tension and evaporation, which included "Stefan's number" and "Stefan's law" (1886). He also published many works on acoustics.

In the theory of the mutual magnetic effects of two electric circuits, Stefan succeeded in showing, in opposition to Ampère and Grassmann, that clear results can be achieved only by means of the theory of continuous action. Stefan and Helmholtz were then the the only Continental proponents of the Faraday-Maxwell theory of continuous action. Stefan, in fact, made important calculations in the theory of alternating currents, especially regarding the induction coefficients of wire coils. Many of his experimental and theoretical works dealt with difficult and subtle problems in physics—for example, the discovery of the secondary rings in Newton's experiments and other optical problems.

At the International Conference on Musical Pitch, held at Vienna in 1885 (over which Stefan presided), the proposals of the Austrian commission of experts were generally adopted in central and eastern Europe; the standard pitch was established at 435 cycles per second (as had already been done in France in 1859 and in Austria in 1862). For the production of this tone the conference prescribed, according to Stefan's account, the standard tuning fork constructed to replicate the tuning forks of K. R. König.

BIBLIOGRAPHY

The Royal Society *Catalogue of Scientific Papers*, V, 806; VIII, 1003–1004; XI, 480–481; and XVIII, 921; lists 73 works, most of them published in *Sitzungsberichte der k. Akademie der Wissenschaften in Wien*, Math.-wiss. Kl. See also *Bericht über die von der wissenschaftlichen Commission (der Internationalen Elektrizitätsausstellung, Wien, 1883) an Dynamomaschinen und elektrischen Lampen ausgeführten Messungen* (Vienna, 1886)—the commission was mainly under Stefan's direction; and, for his work in establishing the standard pitch, *Beschlüsse und Protokolle der Internationalen Stimmton-Conferenz 1885* (Vienna, 1885).

For Wüllner's remarks on heat emission, see his *Lehrbuch der Experimentalphysik*, 3rd ed., III (Leipzig, 1875), 214–215.

There are obituaries in *Almanach der Akademie der Wissenschaften in Wien*, **43** (1893), 252–257; *Elektrotechnische Zeitschrift*, **14** (1893), 31; and *Leopoldina*, **29** (1893), 53–54. See also C. von Wurzbach, ed., *Biographisches Lexikon des Kaiserthums Oesterreich*, XXXV (Vienna, 1877), 284–286, with bibliography, and Albert von Obermayer, *Zur Erinnerung an Josef Stefan* (Vienna–Leipzig, 1893).

WALTER BÖHM

STEIN, JOHAN WILLEM JAKOB ANTOON, S. J. (*b*. Grave, Netherlands, 27 February 1871; *d*. Rome, Italy, 27 December 1951), *astronomy*.

Stein was the son of Maria Waltéra Boerkamp and Johan Hendrik Stein, a teacher in Grave. He entered the Society of Jesus in 1889 and studied physics and astronomy at the University of Leiden (1894–1901) under Lorentz. His doctoral dissertation was concerned with the Horrebow method for determining latitude. In 1901 he taught mathematics and physics at St. Willebrord's College in Katwijk; two years later he was ordained a priest in Maastricht.

In 1906 Stein was appointed assistant to J. G. Hagen, S.J., director of the Vatican observatory, who was then engaged in compiling an atlas of variable stars. Stein contributed to the studies on these stars and published papers on variable and double stars. He became interested in Hagen's work on the axial rotation of the earth and in 1910 translated into French Kamerlingh Onnes' doctoral dissertation (1879) on that subject.

Stein left Rome in 1910 to teach mathematics and physics at St. Ignatius College, Amsterdam. He remained in close contact with Hagen and in 1924 they published *Die veränderlichen Sterne*, an authoritative work in two volumes: the second volume, by Stein, deals with the mathematical-physical aspect of variable stars. Stein was well known to professional astronomers in the Netherlands and was a member of many astronomical societies. He edited the astronomical section of the journal *Hemel en dampkring* and was known for his friendly help to amateur astronomers.

In 1930 Stein succeeded Hagen as director of the Vatican observatory, where his main task was to supervise its transfer to Castel Gandolfo; the new observatory with its astrophysical laboratory was equipped during the years 1932–1935. Stein continued to direct the work of the observatory until his death in 1951.

Stein kept up a lifelong interest in the history of astronomy and in 1910, on the reappearance of Halley's comet, published new data on the comet of 1066 from a manuscript found in the cathedral at Viterbo. Other subjects of his historical research were the astronomical ideas of Albertus Magnus and the relationship between Galileo and Clavius.

BIBLIOGRAPHY

I. ORIGINAL WORKS. Stein's major printed works are *Die veränderlichen Sterne*, 2 vols., *Mathematisch-physikalischer*, vol. II (Fribourg, 1924); *Stelle doppie nel Catalogo . . . Vaticano* (Rome, 1930); *Catalogo di 982 stelle degli ammassi η e χ Persei* (1928); *Atlas Stellarum Variabilium*, VIII (Rome, 1934), IX (Rome 1941). He contributed to various astronomical journals; and Poggendorff, V, 1203; VI, 2531, gives an extensive list of his scientific papers. A more detailed list covering Stein's publications before 1936 is given in *Notizie e pubblicazioni scientifiche* (Rome, 1936). Stein contributed 106 articles to the Dutch journal *Hemel en dampkring* between 1910 and 1930. Archival material relating to Stein is held by the Vatican observatory at Castel Gandolfo.

II. SECONDARY LITERATURE. Obituary notices include J. de Kort, S.J., in *Specola Vaticana, Ricerche Astronomiche*, **2**, no. 16 (1952), 372–374, with portrait; and A. Pannekoek, in *Jaarboek der Koninklijke Nederlandsche akademie van wetenschappen* (1951–1952), 1–4. An account of Stein, with photograph, is found in *Enciclopedia cattolica*, XI (1953). An account of the history of the Vatican Observatory from its inception until 1951 is found in *La Specola vaticana* (Rome, 1952), also available in German.

SISTER MAUREEN FARRELL, F.C.J.

STEINER, JAKOB (*b*. Utzensdorf, Bern, Switzerland, 18 March 1796; *d*. Bern, Switzerland, 1 April 1863), *mathematics*.

Life. Steiner was the youngest of the eight children of Niklaus Steiner (1752–1826), a small farmer and tradesman, and the former Anna Barbara Weber (1757–1832), who were married on 28 January 1780. The fourth child, Anna Barbara (1786–1870), married David Begert; and their daughter Elisabeth (*b*. 1815) married Friedrich Geiser, a butcher, in 1836. To this marriage was born the mathematician Karl Friedrich Geiser (1843–1934), who was thus a grandnephew of Steiner.[1]

Steiner had a poor education and did not learn to write until he was fourteen. As a child he had to help his parents on the farm and in their business; his skill in calculation was of great assistance. Steiner's desire for learning led him to leave home, against his parents' will, in the spring of 1814 to attend Johann Heinrich Pestalozzi's school at Yverdon, where he was both student and teacher. Pestalozzi found a brilliant interpreter of his revolutionary ideas on education in Steiner, who characterized the new approach in an application to the Prussian Ministry of Education (16 December 1826):

> The method used in Pestalozzi's school, treating the truths of mathematics as objects of independent reflection, led me, as a student there, to seek other

grounds for the theorems presented in the courses than those provided by my teachers. Where possible I looked for deeper bases, and I succeeded so often that my teachers preferred my proofs to their own. As a result, after I had been there for a year and a half, it was thought that I could give instruction in mathematics.[2]

Steiner's posthumous papers include hundreds of pages of manuscripts containing both courses given by his fellow teachers and his own ideas. These papers include the studies "Einige Gesetze über die Teilung der Ebene und des Raumes," which later appeared in the first volume of Crelle's *Journal für die reine und angewandte Mathematik.* Steiner stated that they were inspired by Pestalozzi's views.

In the application of 1826 Steiner also wrote:

> Without my knowing or wishing it, continuous concern with teaching has intensified by striving after scientific unity and coherence. Just as related theorems in a single branch of mathematics grow out of one another in distinct classes, so, I believed, do the branches of mathematics itself. I glimpsed the idea of the organic unity of all the objects of mathematics; and I believed at that time that I could find this unity in some university, if not as an independent subject, at least in the form of specific suggestions.

These two statements provide an excellent characterization of Steiner's basic attitude toward teaching and research. The first advocates independent reflection by the students, a practice that was the foundation of Steiner's great success as a teacher. At first, in Berlin, he was in great demand as a private teacher; among his students was the son of Wilhelm von Humboldt. Steiner often gave his courses as colloquiums, posing questions to the students. This direct contact with the students was often continued outside the classroom. The second statement expresses the idea that guided all his work: to discover the organic unity of all the objects of mathematics, an aim realized especially in his fundamental research on synthetic geometry. Steiner left Yverdon in the autumn of 1818 and went to Heidelberg, where he supported himself by giving private instruction. His most important teacher there was Ferdinand Schweins, whose lectures on combinatorial analysis furnished the basis for two of Steiner's works.[3]

At this time Steiner also studied differential and integral calculus and algebra. In addition, lectures at Heidelberg stimulated the careful work contained in manuscripts on mechanics from 1821, 1824, and 1825, upon which Steiner later drew for investigations on the center of gravity.[4]

Following a friend's advice, Steiner left for Berlin at Easter 1821. Not having passed any academic examinations, he was now obliged to do so in order to obtain a teaching license. He was only partially successful in his examinations and therefore received only a restricted license in mathematics, along with an appointment at the Werder Gymnasium. The initially favorable judgment of his teaching was soon followed by criticism that led to his dismissal in the autumn of 1822. From November 1822 to August 1824 he was enrolled as a student at the University of Berlin, at the same time as C. G. J. Jacobi. He again earned his living by giving private instruction until 1825, when he became assistant master (and in 1829 senior master) at the technical school in Berlin. On 8 October 1834 he was appointed extraordinary professor at the University of Berlin, a post he held until his death.

Steiner never married. He left a fortune of about 90,000 Swiss francs, equivalent to 24,000 thaler.[5] He bequeathed a third of it to the Berlin Academy for establishment of the prize named for him,[6] and 60,000 francs to his relatives. In addition he left 750 francs to the school of his native village, the interest on which is still used to pay for prizes awarded to students adept at mental computations.[7] Steiner, with a yearly income of between 700 and 800 thaler, amassed this fortune by giving lectures on geometry.[8]

Students and contemporaries wrote of the brilliance of Steiner's geometric research and of the fiery temperament he displayed in leading others into the new territory he had discovered. Combined with this were very liberal political views. Moreover, he often behaved crudely and spoke bluntly, thereby alienating a number of people. Thus it is certain that his dismissal from the Werder Gymnasium cannot have been merely a question of his scholarly qualifications. Steiner attributed this action to his refusal to base his course on the textbook written by the school's director, Dr. Zimmermann. The latter, in turn, reproached Steiner for using Pestalozzi's methods, claiming that they were suitable only for elementary instruction and therefore made Steiner's teaching deficient. Steiner also experienced difficulties at the technical school, where he was expected to follow, without question, the orders of the director, K. F. von Klöden. Klöden, however, felt that Steiner did not treat him with proper respect, and made exacting demands of him that were of a magnitude and severity that even a soldier subject to military discipline could hardly be expected to accept.

Steiner's scientific achievements brought him an honorary doctorate from the University of Königsberg (20 April 1833) and membership in the Prussian Academy of Sciences (5 June 1834). He spent the winter of 1854–1855 in Paris and became a corresponding member of the French Academy of Sciences. He had already been made a corresponding member of the Accademia dei Lincei in 1853. A kidney ailment obliged him to take repeated cures in the following years, and he lectured only during the winter terms.

Mathematical Work. Having set himself the task of reforming geometry, Steiner sought to discover simple principles from which many seemingly unrelated theorems in the subject could be deduced in a natural way. He formulated his plan in the preface to *Systematische Entwicklung der Abhängigkeit geometrischer Gestalten voneinander, mit Berücksichtigung der Arbeiten alter und neuer Geometer über Porismen, Projections-Methoden, Geometrie der Lage, Transversalen, Dualität und Reciprocität* (1832), dedicated to Wilhelm von Humboldt:

> The present work is an attempt to discover the organism [*Organismus*] through which the most varied spatial phenomena are linked with one another. There exist a limited number of very simple fundamental relationships that together constitute the schema by means of which the remaining theorems can be developed logically and without difficulty. Through the proper adoption of the few basic relations one becomes master of the entire field. Order replaces chaos; and one sees how all the parts mesh naturally, arrange themselves in the most beautiful order, and form well-defined groups. In this manner one obtains, simultaneously, the elements from which nature starts when, with the greatest possible economy and in the simplest way, it endows the figures with infinitely many properties. Here the main thing is neither the synthetic nor the analytic method, but the discovery of the mutual dependence of the figures and of the way in which their properties are carried over from the simpler to the more complex ones. This connection and transition is the real source of all the remaining individual propositions of geometry. Properties of figures the very existence of which one previously had to be convinced through ingenious demonstrations and which, when found, stood as something marvelous are now revealed as necessary consequences of the most common properties of these newly discovered basic elements, and the former are established a priori by the latter.[9]

Also in the preface Steiner asserted that this work would contain "a systematic development of the problems and theorems concerning the intersection and tangency of the circle in the plane and on spherical surfaces and of spheres." The plan was not carried out, and the manuscript of this part was not published until 1931.[10] But many of the observations, theorems, and problems included in it appeared in "Einige geometrische Betrachtungen" (1826), Steiner's first long publication.[11]

The earliest detailed account of some of the sources of Steiner's concepts and theorems can be found in the posthumously published *Allgemeine Theorie über das Berühren und Schneiden der Kreise und der Kugeln, worunter eine grosse Anzahl neuer Untersuchungen und Sätze, in einem systematischen Entwicklungsgange dargestellt. . . .*[12] The headings of the sections describe its contents: "I. Of Centers, Lines, and Planes of Similitude in Circles and Spheres. II. Of the Power and the Locus of Equal Powers With Respect to Circles and Spheres. III. Of the Common Power in Circles and in Spheres. IV. Of Angles at Which Circles and Spheres Intersect."

In the foreword to *Allgemeine Theorie*, F. Gonseth stated in current terminology the basic principle on which many of Steiner's theorems and constructions are founded: the stereographic projection of the plane onto the sphere.[13] Section 4 of this work contains the following problem (§ 29, X, p. 167): "Draw a circle that intersects at equal angles four arbitrary circles of given size and position." The new methods were applied to the solution of Apollonius' problem (§ 31, II, p. 175): "Find a circle tangent to three arbitrary circles of given size and position." Another problem (§ 31, III, p. 182) reads: "Find a circle that intersects three arbitrary circles of given size and position at the angles α_1, α_2, α_3." Analogous problems for spheres are given in chapter 2, where the theorems and problems are presented systematically according to the number of spheres involved (from two to eight), with size and position again given—for example (p. 306): "Draw a sphere that intersects five arbitrary spheres of given size and position at one and the same angle" and (p. 333) "Find a sphere that is tangent to a sphere M_1 of given size and position and that cuts at one and the same angle three pairs of spheres, of given size and position, M_2 and M_3, M_4 and M_5, M_6 and M_7, each pair taken singly."

At Berlin, Steiner became friendly with Abel,

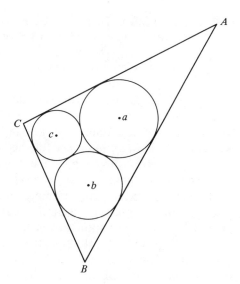

FIGURE 1

Crelle, and Jacobi; and together they introduced a fresh, new current into mathematics. Their efforts were considerably aided by Crelle's founding of the *Journal für die reine und angewandte Mathematik*, to which Steiner contributed sixty-two articles. In the first volume (1826) he published his great work "Einige geometrische Betrachtungen."[14] It contains a selection from the *Allgemeine Theorie* and the first published systematic development of the theory of the power of a point with respect to a circle and of the points of similitude of circles; in his account Steiner mentions Pappus, Viète, and Poncelet. As the first application of these concepts Steiner states, without proof, his solution to Malfatti's problem (§ IV, no. 14). In a given triangle *ABC* draw three circles *a*, *b*, and *c* that are tangent to each other and such that each is

tangent to two sides of the triangle (Figure 1). Steiner then remarks that this is a special case of the next problem (no. 15): "Given three arbitrary circles, M_1, M_2, M_3, of specified size and position, to find three other circles m_1, m_2, m_3 tangent to each other and such that each is tangent to two of the given circles, and that each of the given circles M_i is tangent to two of the circles m_k that are to be found" (Figure 2).

Steiner did not prove his solution. Examination of his posthumous papers shows that he knew of the principle of inversion and that he used it in finding and proving the above and other theorems.[15]

It was likewise by means of an inversion that Steiner found and proved his famous theorem on series of circles (§ IV, no. 22; see Figure 3):

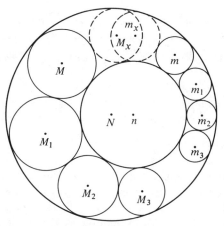

FIGURE 3

Two circles *n*, *N* of assigned size and position, lying one within the other, are given. If, for a definite series of circles M, M_1, $\cdots$, M_x, each of which is tangent to *n* and *N* unequally and that are tangent to each other in order, the interval between *n* and *N* is *commensurable*, that is, if the series consists of $x + 1$ members forming a sequence of *u* circuits such that the last circle M_x is tangent to the first one *M*: then this interval is commensurable for any series of circles m, m_1, $\cdots$, m_x; and the latter series also consists of $x + 1$ members forming *u* circuit, as in the first series.

In this same work (§ VI; see Figure 4), he proves a theorem of Pappus, in the following form:

Given two circles M_1, M_2, of assigned size and position that are tangent to each other in *B*. If one draws two arbitrary circles m_1, m_2 that are tangent to each other externally in *b* and each of which is tangent to the two given circles, and if one drops the perpendiculars $m_1 P_1$, $m_2 P_2$ from the centers m_1, m_2 on the axis $M_1 M_2$ of the given circles and divides these perpendiculars by the radii r_1, r_2 of the circles m_1, m_2:

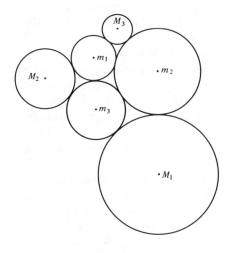

FIGURE 2

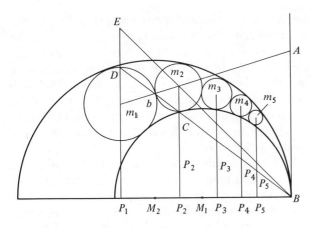

FIGURE 4

lines. Therefore, a plane is divided by n arbitrary straight lines into at most $2+2+3\cdots+(n-1)+$

$$n = 1 + \frac{n(n+1)}{2} = 1 + n + \frac{n(n-1)}{1 \cdot 2} \text{ parts.}$$

He then subdivided space by means of planes and spherical surfaces.

In the following years Crelle's *Journal* and Gergonne's *Annales de mathématiques* published many of Steiner's papers, most of which were either problems to be solved or theorems to be proved.[18] In this way Steiner exerted an exceptionally stimulating influence on geometric research that was strengthened by the publication of his first book, *Systematische Entwicklung* (1832).[19] It was originally supposed to consist of five sections, but only the first appeared. Some of the remaining sections were published in the *Vorlesungen*.[20]

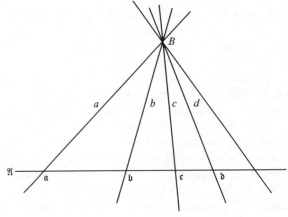

FIGURE 5

then the quotient corresponding to the circle m_2 is greater by 2 than that corresponding to the former; that is, $\frac{m_1P_1}{r_1} + 2 = \frac{m_2P_2}{r_2}$. Or, as Pappus expressed it: the perpendicular m_1P_1 plus the diameter of the corresponding circle m_1 is to that diameter as the perpendicular m_2P_2 is to the diameter of the corresponding circle m_2—that is, $\frac{m_1P_1 + 2r_1}{2r_1} = \frac{m_2P_2}{2r_2}$.

Steiner furnishes a proof of this proposition, which consists, essentially, of the following steps:

1. The straight line $AB \perp BM_1$ is the line of equal powers with respect to the circles M_1 and M_2.

2. AB passes through the exterior center of similitude A of the circles m_1 and m_2.

3. $Am_2 : Am_1 = r_2 : r_1 = BP_2 : BP_1$ and $AB = Ab$.

4. The points B, C, b, D lie on one straight line.

5. The assertion follows from similarity considerations. In a later paper Steiner applied this "ancient" theorem of Pappus to the sphere.[16]

Also in the first volume of Crelle's *Journal*, Steiner published an expanded version of considerations that stemmed from the period that he was in Yverdon: "Einige Gesetze über die Teilung der Ebene und des Raumes."[17] In it he expressly stated that his ideas were inspired by Pestalozzi's ideas. The simplest result in this paper was presented in the following form:

A plane is divided into two parts by a straight line lying within it; by a second straight line that intersects the first, the number of parts of the plane is increased by 2; by a third straight line that intersects the two first lines at two points, the number is increased by 3; and so forth. That is, each successive straight line increases the number of parts by the number of parts into which it was divided by the preceding straight

Steiner believed that the fundamental concepts of plane geometry are the range of points considered as the totality of points $\mathfrak{a}, \mathfrak{b}, \mathfrak{c}, \cdots$ of a straight line $\mathfrak{A}$ and the pencil of lines $a, b, c, \cdots$ through a point B (Figure 5). Since the latter are the intersection points of $a, b, c, \cdots$ with straight line $\mathfrak{A}$, an unambiguous relationship is established between the pencil of lines and the range, a relationship that he called projectivity. In volume II, § 2, of the *Vorlesungen über synthetische Geometrie* (1867), Steiner expressed this property through the statement that the two constructs are of the same *cardinality*, an expression that Georg Cantor adopted and generalized.[21]

In the first chapter of this part of the *Vorlesungen*, Steiner discusses the elements of projectivity, emphasizing the duality between point and straight line. In particular he proves the harmonic properties of the complete quadrangle and of the complete quadrilateral. In the second chapter he treats the simple elements of solid geometry. At the center of the epochal work stands the theory of conic sections in the third book. Here Steiner proved his fundamental theorem: The intersection points of corresponding lines of two projective pencils of lines form a conic section. In its metric formulation this theorem was essentially known to Jan de Witt and Newton.[22] Steiner, however, was the first to recognize that it was a theorem of projective geometry, and he made it the cornerstone of the projective treatment of the theory of conic sections.

Steiner knew of the significance of his discovery.

The above investigation of projective figures, by placing them in oblique positions deliberately avoided closer research into the laws that govern the projective rays for two straight lines A, A_1 [Figure 6]. We shall now proceed to this examination. It leads, as will be seen, to the most interesting and fruitful properties of curves of the second order, the so-called conic sections. From these almost all other properties of the conics can be developed in a single, comprehensive framework and in a surprisingly simple and clear manner. This examination shows the necessary emergence of the conic sections from the elementary geometric figures; indeed, it shows, at the same time, a very remarkable double production of these [sections] by means of projective figures. . . . When one considers with what ingenuity past and present mathematicians have investigated the conic sections, and the almost countless number of properties that remained hidden for so long, one is struck that, as will be seen, almost all the known properties (and many new ones) flow from their projective generation as from a spring; and this generation also reveals the inner nature of the conic sections to us. For even if properties are known that are similar to those named here, the latter have never, in my opinion, been explicitly stated; in no case, however, has anyone until now recognized the importance that they derive from our development of them, where they are raised to the level of fundamental theorems.[23]

In proposition no. 37 of this volume, Steiner stated and proved his fundamental theorem for the circle (Figure 6): "Any two tangents A, A_1 are projective with respect to the corresponding pairs of points in which they are cut by the other tangents; and the point of intersection $\mathfrak{d}$, e_1, of the tangents corresponds to the points $\mathfrak{d}_1$, e, where they touch the circle." He also gave the dual of this theorem (Figure 7): "Any two points $\mathfrak{B}$, $\mathfrak{B}_1$ of a circle are the centers of two projective pencils of lines, the corresponding lines of which intersect in the remaining points of the circle; and the reciprocal tangents d_1, e at the points $\mathfrak{B}$, $\mathfrak{B}_1$ correspond to the

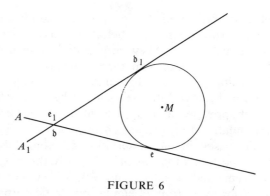

FIGURE 6

lines d, e_1." Applying these theorems to the second-degree cone, Steiner obtained the following result: "Any two tangents of a conic section are projective with respect to the pair of points in which they are intersected by the remaining tangents; and conversely."

Steiner emphasized here that "these new theorems on the second-degree cone and its sections are more important for the investigation of these figures than all the theorems previously known about them, for they are, in the strict sense, the true fundamental theorems."

From these fundamental theorems, Steiner derived consequences ranging from the known theorems on conic sections to the Braikenridge-Maclaurin theorem. Propositions 49–53 deal with the production of projective figures in space. An important group of propositions (54–58) contains previously known "composite theorems and problems" that Steiner was the first to derive in a uniform manner from one basic principle. An example is the following problem taken from Möbius: "Given an arbitrary tetrahedron, draw another the vertices of which lie in the faces of the first and the faces of which pass through the vertices of the

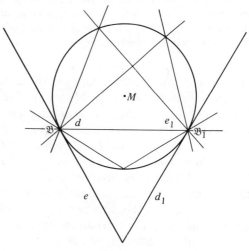

FIGURE 7

first, two vertices of the second tetrahedron being given." Proposition 59, labeled "general observation," contains the "skew projection," a quadratic relationship in space, sometimes called the "Steiner relationship," which had been noted by Poncelet.[24]

The eighty-five "Problems and Theorems" that Steiner appended in a supplement proved especially stimulating to a generation of geometers. They are discussed in the dissertation of Ahmed Karam,[25] who found that, as of 1939, only three problems remained unsolved: no. 70, "What are the properties of a group of similar quadratic surfaces that pass through four or five points of space?"; no. 77, "Does a convex polyhedron always have a topological equivalent that can be either circumscribed about or inscribed in a sphere?"; and no. 76, "If polyhedra are distinguished solely according to their boundary surfaces, there exist only one with four faces, two with five faces, and seven with six faces. How many different bodies are possible with 7, 8, $\cdots$, n faces?"

The last problem was posed by Steiner's teacher at Heidelberg, Ferdinand Schweins.[26] It was partially solved by Otto Hermes in 1903, and further elements of it have been solved by P. J. Frederico.[27]

Of Steiner's work Jacobi stated:

> Starting from a few spatial properties Steiner attempted, by means of a simple schema, to attain a comprehensive view of the multitude of geometric theorems that had been rent asunder. He sought to assign each its special position in relation to the others, to bring order to chaos, to interlock all the parts according to nature, and to assemble them into well-defined groups. In discovering the organism [*Organismus*] through which the most varied phenomena of space are linked, he not only furthered the development of a geometric synthesis; he also provided a model of a complete method and execution for all other branches of mathematics.[28]

Only a year after the appearance of *Systematische Entwicklung*, Steiner published his second book: *Die geometrischen Konstruktionen ausgeführt mittelst der geraden Linie und eines festen Kreises, als Lehrgegenstand auf höheren Unterrichtsanstalten und zur praktischen Benützung* (1833).[29] He took as his point of departure Mascheroni's remark that all constructions made with straightedge and compass can be carried out using compass alone. As a counterpart to this statement, Steiner proved that all such constructions can also be carried out with the straightedge and one fixed

circle. To this end he devoted the first chapter to rectilinear figures and especially to the harmonic properties of the complete quadrilateral. In the third chapter he proved his assertion in a way that enabled him to solve eight fundamental problems, to which all others can be reduced. For example (no. 1): "Draw the parallel to a straight line through a point" and (no. 8) "Find the intersection point of two circles." In the intervening chapter 2, he considered centers of similitude and the radical axis of a pencil of circles. This work, which enjoyed great success, contained an appendix of twenty-one problems that were partly taken from *Systematische Entwicklung*. The first one, for example, was "Given two arbitrary triangles, find a third that is simultaneously circumscribed about the first and inscribed in the second."

At this point we shall present a survey of Steiner's further research, published in volume II of the *Gesammelte Werke* (the page numbers in parentheses refer to that volume). A fuller description of its contents can be found in Louis Kollros' article on Steiner, cited in the bibliography.

Steiner pursued the investigation of conic sections and surfaces in some dozen further publications. Sometimes he merely presented problems and theorems without solutions or proofs. In part the material follows from the general projective approach to geometry; but some of it contains new ideas, as the examination of the extreme-value problem: "Determine an ellipse of greatest surface that is inscribed in a given quadrangle" (333 f.) and "Among all the quadrangles inscribed in an ellipse, that having the greatest perimeter is the one the vertices of which lie in the tangent points of the sides of a rectangle circumscribed about the ellipse. There are infinitely many such quadrangles. . . . All have the same perimeter, which is equal to twice the diagonal of the rectangle. All these quadrangles of greatest perimeter are parallelograms; and they are, simultaneously, circumscribed about another ellipse the axes of which fall on the corresponding axes of the given ellipse and that is confocal with the latter. . . . Among all the quadrangles circumscribed about a given ellipse, the one with least perimeter is that in which the normals at the tangent points of its sides form a rhombus" (411–412).

In a paper on new methods of determining second-order curves, Steiner considered pairs of such curves and demonstrated propositions of the following type: "If two arbitrary conic sections are inscribed in a complete quadrilateral, then the eight points in which they are tangent to the sides lie on another conic section" (477 f.). To the theory of second-degree surfaces, he contributed the geometric proof of Poisson's theorem: The attraction of a homogeneous elliptical sheet falls on a point P in the axis of the cone that has P as vertex and is tangent to the ellipsoid.

Steiner dealt on several occasions with center-of-gravity problems. One of the simplest follows.

If, in a given circle $ADBE$, one takes an arc AB, of which one end point, A, is fixed, and lets it increase steadily from zero, then its center of gravity C will describe a curved line ACM. What properties will this barycentric curve possess? . . . The same question can be phrased generally, if instead of the circle an arbitrary curve is given. . . . Questions like those of the above problem occur if one considers the center of gravity of the segment (instead of the arc) ADB. Other questions of the same sort arise regarding the center of gravity of a variable sector AMB, if M is an arbitrary fixed pole and one arm of the sector is fixed, while the other, MB, turns about the pole M [p. 30; see Figure 8].

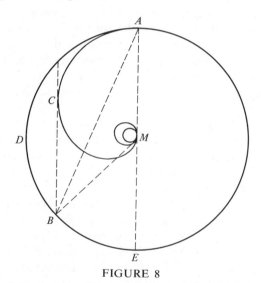

FIGURE 8

Steiner developed a general theory of the center of gravity of mass points in "Von dem Krümmungsschwerpunkt ebener Kurven" (97–159). It led to the pedal curve of a given curve and its area, and was followed by the important memoir "Parallele Flächen" (171–176), which generalized a theorem proved in the preceding paper for plane curves: Let A and B be two parallel polyhedra (surfaces) separated by a distance h. Then it is true for A and B that $B = A + hk + h^2e$; and for the volume I between A and B that $I = hA + h^2k/2 + h^3e/3$, where k is "the sum of the edge curvature" and e "the sum of the vertex curvature."

Steiner's great two-part paper "Ueber Maximum

und Minimum bei den Figuren in der Ebene, auf der Kugelfläche und im Raume überhaupt" (177–308, with 36 figures) was written in Paris during the winter of 1840–1841. It shows the tremendous achievements of which he was still capable—given the necessary time and freedom from distractions. His basic theorem states: "Among all plane figures of the same perimeter, the circle has the greatest area (and conversely)." He gives five ways of demonstrating it, in all of which he assumes the existence of the extremum. All five, moreover, are based on the inequalities of the triangle and of polygons. The first proof proceeds indirectly: Assume that among all figures of equal perimeter the convex figure $EFGH$ possesses the greatest area and that it is not a circle (Figure 9).

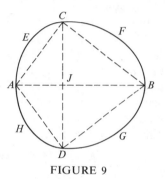

FIGURE 9

Let A and B be two points that bisect the perimeter. Then the surfaces $AEFB$ and $AHGB$ are equal. For if one of them were smaller, then the other could be substituted for it, whereby the perimeter would remain equal and the surface would be increased. These two surfaces should be considered to have the same form; for if they were different, the mirror image of AB could be substituted for one, whereby the perimeter and area would remain the same. According to "Fundamental Theorem II" on triangles, for the extremal figure the angles at C and D must be right angles. Since this consideration holds for every point A of the perimeter, the figure sought is a circle. This is the first occasion on which Steiner employed his principle of symmetrization.

In the fifth proof the basic theorem takes the form that among all plane figures of the same area, the circle has the least perimeter. The principle of the proof is again symmetrization with respect to the axis X.

Steiner next effects the transformation of the pentagon $ABCDE$ (Figure 10) into the pentagon of equal area $abcde$ in such a way that each line segment $B_1B = b_1b$; and bb_1 is bisected by X. As a result, the perimeter decreases. Steiner then turns

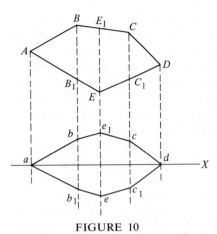

FIGURE 10

to the extremal properties of prisms, pyramids, and the sphere (269–308).

In 1853, while investigating the double tangents of a fourth-degree curve, Steiner encountered a combinatorial problem (435–438): What number N of elements has the property that the elements can be ordered into triplets (t-tuples) in such a way that each two (each t-1) appear in one and only one combination? For the Steiner triple system, N must have the form $6n + 1$ or $6n + 3$, and there

exist $\dfrac{N(N-1)}{2 \cdot 3}$ triples, $\dfrac{N(N-1)(N-3)}{2 \cdot 3 \cdot 4}$ quadruples,

and so on. For example, for $N = 7$ there is only one triple system: 123, 145, 167, 246, 257, 347, 356. For $N = 13$ there are two different triple systems. Steiner was unaware of the work on this topic done by Thomas Kirkman (1847).[30]

In a short paper of fundamental importance, published in 1848 and entitled "Allgemeine Eigenschaften der algebraischen Kurven" (493–500), Steiner first defined and examined the various polar curves of a point with respect to a given curve. He then introduced the "Steiner curves" and discussed tangents at points of inflection, double tangents, cusps, and double points. In particular he indicated the resulting relationships for the twenty-eight double tangents of the fourth-degree curve. Luigi Cremona proved the results and continued Steiner's work in his "Introduzione ad una teoria geometrica delle curve piane."[31]

The desire to find, with the methods of pure geometry, the proofs of the extremely important theorems stated by the celebrated Steiner in his brief memoir "Allgemeine Eigenschaften der algebraischen Kurven" led me to undertake several studies, a sample of which I present here, even though it is incomplete.

In 1851 Steiner wrote "Über solche algebraische Kurven, die einen Mittelpunkt haben . . ." (501–596), a version of which he published in Crelle's *Journal*. In the *Gesammelte Werke* this paper is followed by "Problems and Theorems" (597–601). Steiner's results include the following example: "Through seven given points in the plane there pass, in general, only nine third-degree curves possessing a midpoint." Next follows a discussion of the twenty-eight double tangents of the fourth-degree curve (603–615). In January 1856 Steiner delivered a lecture at the Berlin Academy, "Ueber die Flächen dritten Grades" (649–659), in which he offered four methods of producing these cubic surfaces. The first states: "The nine straight lines in which the surfaces of two arbitrarily given trihedra intersect each other determine, together with one given point, a cubic surface." Aware that Cayley already knew the twenty-seven straight lines of this surface, Steiner introduced the concept of the "nuclear surface" and investigated its properties (656).

Steiner's correspondence with Ludwig Schläfli reveals that the latter discovered his "Doppelsechs" in the course of research on this topic undertaken at Steiner's request.[32] Again, it was Cremona who proved Steiner's theorems in his "Mémoire de géométrie pure sur les surfaces du troisième ordre" (1866).[33] Cremona began his memoir by declaring that "This work . . . contains the demonstration of all the theorems stated by this great geometer [Steiner] in his memoir *Ueber die Flächen dritten Grades*."

A second treatment of Steiner's theorems appeared in Rudolf Sturm's *Synthetische Untersuchungen über Flächen dritter Ordnung* (Leipzig 1867). In the preface Sturm wrote: "Steiner's paper contains a wealth of theorems on cubic surfaces, although, as had become customary with this celebrated geometer, without any proofs and with only few hints of how they might be arrived at." The works of both Cremona and Sturm were submitted in 1866 as entries in the first competition held by the Berlin Academy for the Steiner Prize, which was divided between them.

During his stay at Rome in 1844, Steiner investigated a fourth-order surface of the third class (721–724, 741–742), but it became known only much later through a communication from Karl Weierstrass (1863).[34] The surface, since called the Roman surface or Steiner surface, has the characteristic property that each of its tangent planes cuts it in a pair of conics. On the surfaces there lie three double straight lines that intersect in a triple point of the surface. The surface was the subject of many studies by later mathematicians.

NOTES

1. See F. Bützberger, "Biographie Jakob Steiners."
2. J. Lange, "Jacob Steiners Lebensjahre in Berlin 1821–1863. Nach seinen Personalakten dargestellt"; E. Jahnke, "Schreiben Jacobis . . .," in *Archiv der Mathematik und Physik*, 3rd ser., **4** (1903), 278.
3. *Jacob Steiner's Gesammelte Werke*, K. Weierstrass, ed., I, 175–176, and II, 18.
4. *Ibid.*, II, 97–159.
5. E. Lampe, "Jakob Steiner"; J. H. Graf, "Beiträge zur Biographie Jakob Steiners."
6. K.-R. Biermann, "Jakob Steiner."
7. *Ibid.*
8. J. Lange, *op. cit.*
9. *Gesammelte Werke*, I, 233–234.
10. Steiner, *Allgemeine Theorie über das Berühren und Schneiden der Kreise und der Kugeln*; B. Jegher, "Von Kreisen, die einerlei Kugelfläche liegen. Jakob Steiners Untersuchungen über das Schneiden und Berühren von Kegelkreisen. . . ."
11. *Gesammelte Werke*, I, 17–76.
12. Steiner, *Allgemeine Theorie*.
13. *Ibid.*, xiv–xvi.
14. *Gesammelte Werke*, I, 17–76.
15. See F. Bützburger, "Jakob Steiners Nachlass aus den Jahren 1823–1826," § 11, "Die Erfindung der Inversion"; A. Emch, "The Discovery of Inversion"; and Mautz, *op. cit.*
16. *Gesammelte Werke*, I, 133.
17. *Ibid.*, 77–94.
18. *Ibid.*, 121–228.
19. *Ibid.*, 229–460.
20. The MS material is in Steiner, *Allgemeine Theorie*, and in Jegher, *op. cit.*
21. See G. Cantor, *Gesammelte Abhandlungen* (Berlin, 1932), 151.
22. See W. L. Schaaf, "Mathematicians and Mathematics on Postage Stamps," in *Journal of Recreational Mathematics*, **1** (1968), 208; and I. Newton, *Principia mathematica*, 2nd ed. (Cambridge, 1713), Bk. I, 72; and *Universal Arithmetick* (London, 1728), probs. 57, 95.
23. *Vorlesungen*, II, no. 35.
24. V. Poncelet, *Traité des propriétés projectives des figures* (Paris–Metz, 1822), sec. III, ch. 2.
25. A. Karam, *Sur les 85 problèmes de la "dépendance systématique" de Steiner*. See also L. Kollros, "Jakob Steiner," p. 10.
26. F. Schweins, *Skizze eines Systems der Geometrie* (Heidelberg, 1810), 14–15.
27. See R. Sturm, "Zusammenstellung von Arbeiten, welche sich mit Steinerschen Aufgaben beschäftigen," in *Bibliotheca mathematica*, 3rd ser., **4** (1903), 160–184; P. J. Frederico, "Enumeration of Polyhedra," in *Journal of Combinatorial Theory*, **7** (1969), 155–161.
28. See F. Bützberger, "Biographie Jakob Steiners," 109; and K.-R. Biermann, *op. cit.*, 38.
29. *Gesammelte Werke*, I, 469–522.
30. T. Kirkman, in *Cambridge and Dublin Mathematical Journal*, **2** (1847), 191–204.
31. *Opere matematiche di Luigi Cremona*, I (Milan, 1914), 313–466.
32. See J. H. Graf, *Der Briefwechsel Steiner-Schläfli*.
33. *Opere matematiche di Luigi Cremona*, III (Milan, 1917), 1–121.
34. In *Monatsberichte der Deutschen Akademie der Wissenschaften zu Berlin* (1863), 339, repr. in *Mathematische Werke von Karl Weierstrass*, III (Berlin, 1902), 179–182.

BIBLIOGRAPHY

I. ORIGINAL WORKS. Many of Steiner's writings are in *Jacob Steiner's Gesammelte Werke*, K. Weierstrass, ed., 2 vols. (Berlin, 1881–1882). The major ones include *Jacob Steiners Vorlesungen über synthetische Geometrie*: I, *Die Theorie der Kegelschnitte in elementarer Darstellung*, C. F. Geiser, ed. (Leipzig, 1867; 3rd ed., 1887), and II, *Die Theorie der Kegelschnitte gestützt auf projektive Eigenschaften*, H. Schröter, ed. (Leipzig, 1867; 3rd ed., 1898); *Allgemeine Theorie über das Berühren und Schneiden der Kreise und der Kugeln*, R. Fueter and F. Gonseth, eds. (Zurich–Leipzig, 1931); and Barbara Jegher, "Von Kreisen, die in einerlei Kugelfläche liegen. Jakob Steiners Untersuchungen über das Schneiden und Berühren von Kugelkreisen . . .," in *Mitteilungen der Naturforschenden Gesellschaft in Bern*, n.s. **24** (1967), 1–20.

Two letters from Steiner to Rudolf Wolf, dated 25 July 1841 and 5 Aug. 1848, are in the autograph collection of the Schweizerische Naturforschende Gesellschaft, at the Bern Burgerbibliothek (nos. 110 and 588 under MSS Hist. Helv. XIV, 150).

The following works by Steiner appeared in Ostwald's Klassiker der Exakten Wissenschaften: *Die geometrischen Konstruktionen, ausgeführt mittels der geraden Linie und eines festen Kreises* (1833), no. 60 (Leipzig, 1895), which contains a short biography of Steiner by the editor, A. J. von Oettingen, pp. 81–84; *Systematische Entwicklung der Abhängigkeit geometrischer Gestalten voneinander* (1832), nos. 82–83 (Leipzig, 1896); and *Einige geometrische Betrachtungen* (1826), no. 123, R. Sturm, ed. (Leipzig, 1901).

II. SECONDARY LITERATURE. Undated MS material is F. Bützberger, "Kleine Biographie über Jakob Steiner," Bibliothek der Schweizerischen Naturforschenden Gesellschaft, in the Bern Stadt und Universitätsbibliothek, MSS Hist. Helv. XXIb, 347; "Biographie Jakob Steiners," in the same collection, MSS Hist. Helv. XXIb, 348; and "Jakob Steiners Nachlass aus den Jahren 1823–1826," Bibliothek der Eidgenössischen Technischen Hochschule, Zurich, Hs. 92, pp. 30–223.

On Steiner's youth and years in Yverdon, see especially F. Bützberger, "Zum 100. Geburtstage Jakob Steiners," in *Zeitschrift für mathematischen und naturwissenschaftlichen Unterricht*, **27** (1896), 161 ff.; on the years in Berlin see Felix Eberty, *Jugenderinnerungen eines alten Berliners* (Berlin, 1878; repr. 1925), 238–243; and Julius Lange, "Jacob Steiners Lebensjahre in Berlin 1821–1863. Nach seinen Personalakten dargestellt," in *Wissenschaftliche Beilage zum Jahresbericht der Friedrichs-Werderschen Oberrealschule zu Berlin, Ostern 1899*, Program no. 116 (Berlin, 1899). See also three short obituary notices: C. F. Geiser, "Nekrolog J. Steiner," in *Die Schweiz: Illustrierte Zeitschrift für Literatur und Kunst* (Nov. 1863), 350–355; Otto Hesse, "Jakob Steiner," in *Journal für die reine und angewandte Mathematik*, **62** (1863), 199–200; and Bernhard Wyss, "Nekrolog J. Steiner," in *Bund* (Bern) (9 Apr. 1863).

The first detailed biography of Steiner was written by his grandnephew: Carl Friedrich Geiser, *Zur Erinnerung an Jakob Steiner* (Schaffhausen, 1874). Steiner's correspondence with Schläfli was edited by Schläfli's student J. H. Graf: *Der Briefwechsel Steiner-Schläfli* (Bern, 1896); see also the following three works by Graf: *Der Mathematiker Steiner von Utzensdorf* (Bern, 1897); "Die Exhuminierung Jakob Steiners und die Einweihung des Grabdenkmals Ludwig Schläflis . . . am 18. März 1896," in *Mitteilungen der Naturforschenden Gesellschaft in Bern* (1897), 8–24; and "Beiträge zur Biographie Jakob Steiners," *ibid.* (1905). Another of Schläfli's students, F. Bützberger, examined Steiner's posthumous MSS and reported on them in "Zum 100. Geburtstage Jakob Steiners" (see above) and in his long MS "Jakob Steiners Nachlass aus den Jahren 1823–1826" (see above).

Recent accounts of Steiner's life and work are Louis Kollros, "Jakob Steiner," supp. 2 of *Elemente der Mathematik* (1947), 1–24; and J.-P. Sydler, "Aperçus sur la vie et sur l'oeuvre de Jakob Steiner," in *Enseignement mathématique*, 2nd ser., **11** (1965), 240–257. Valuable corrections of errors in earlier accounts are given by Kurt-R. Biermann, "Jakob Steiner," in *Nova acta Leopoldina*, n.s. **27**, no. 167 (1963), 31–47.

For further information see the following: F. Bützberger, "Jakob Steiner bei Pestalozzi in Yverdon," in *Schweizerische pädagogische Zeitschrift*, **6** (1896), 19–30; and *Bizentrische Polygone, Steinersche Kreis- und Kugelreihen und die Erfindung der Inversion* (Leipzig, 1913); Moritz Cantor, "Jakob Steiner," in *Allgemeine deutsche Biographie*, XXXV (1893), 700–703; A. Emch, "Unpublished Steiner Manuscripts," in *American Mathematical Monthly*, **36** (1929), 273–275; and "The Discovery of Inversion," in *Bulletin of the American Mathematical Society*, **20** (1913–1914), 412–415, and **21** (1914–1915), 206; R. Fueter, *Grosse schweizer Forscher* (Zurich, 1939), 202–203; C. Habicht, "Die Steinerschen Kreisreihen" (Ph.D. diss., Bern, 1904); A. Karam, *Sur les 85 problèmes de la "dépendence systématique" de Steiner* (Ph.D. diss., Eidgenössische Technische Hochschule, Zurich, 1939); E. Kötter, "Die Entwicklung der synthetischen Geometrie. Dritter Teil: Von Steiner bis auf Staudt," in *Jahresbericht der Deutschen Mathematikervereinigung*, **5**, no. 2 (1898), 252 ff.; Emil Lampe, "Jakob Steiner," in *Bibliotheca mathematica*, 3rd ser., **1** (1900), 129–141; Otto Mautz, "Ebene Inversionsgeometrie," in *Wissenschaftliche Beilage zum Bericht über das Gymnasium Schuljahr 1908–1909* (Basel, 1909); R. Sturm, "Zusammenstellung von Arbeiten, welche sich mit Steinerschen Aufgaben beschäftigen," in *Bibliotheca mathematica*, 3rd ser., **4**

(1903), 160–184; and a series of articles by Rudolf Wolf in *Vierteljahrsschrift der Naturforschenden Gesellschaft in Zürich*: **9** (1864), 145 ff.; **13** (1868), 110 ff.; **19** (1874), 325 ff.; **25** (1880), 215 ff.; and **35** (1890), 428 ff.

JOHANN JAKOB BURCKHARDT

STEINHEIL, KARL AUGUST (*b*. Rappoltsweiler [now Ribeauvillé], Alsace, 12 October 1801; *d*. Munich, Germany, 12 September 1870), *physics, astronomy.*

Steinheil was the son of Karl Philipp Steinheil, administrator of the estates of a German prince (later Maximilian I of Bavaria), and the former Christine Maria Franziska von Biarowsky. She bore their son after twenty-four years of matrimony. The family moved to Munich in 1807 but the boy did not attend public school because of his delicate health. He was educated by private tutors; and after two years spent at Nancy and Tours with relatives of his mother's, he passed the examination for the school certificate. In 1821 Steinheil began to study law at Erlangen, but in 1823 he turned to science, especially astronomy, at Göttingen and Königsberg. At the latter he received the Ph.D. on 12 October 1825, under Bessel's supervision, for the dissertation "De specialibus coeli chartis elaborantis."

Steinheil continued his astronomical and optical studies at his estate near Munich. He became known to experts in the field for editing three hours of the *Berliner akademische Sternkarten* and, later, for his prize-winning paper "Elemente der Helligkeitsmessungen am Sternenhimmel" (1837). In 1832 he was named ordinary professor of mathematics and physics at Munich University. From 1849 to 1852 he organized telegraph communications in Austria, then returned to Munich, where he worked in the Ministry of Commerce as a technical consultant. Fulfilling a special desire of Maximilian I, Steinheil founded an optical workshop (1854) that was continued by his son and grandson. In his last years he was conservator of the mathematical and physical collections of the Bavarian Academy at Munich.

Steinheil exerted great influence on the scientific life at Munich. He was a keen-witted discoverer and inventor, especially in optics and telegraphy. In 1838 he theorized that the earth could serve as the second half of a telegraphic circuit; and in 1842 he constructed an ingenious photometer. Steinheil also improved the achromatic telescope (some of this work was done with P. L. von Seidel), designed silvered mirrors for reflectors, and stimulated progress in the establishment of standard weights, electrodeposition, daguerreotypy, and fire prevention.

BIBLIOGRAPHY

Steinheil's writings include "Elemente der Helligkeitsmessungen am Sternenhimmel," in *Abhandlungen der Bayerischen Akademie der Wissenschaften*, Math.-phys. Kl., **2** (1837), 1–140; "Über quantitative Analyse durch physikalische Beobachtungen," *ibid.*, **3** (1842), 689–716; "Über die Bestimmung der Brechungs- und Zerstreuungsverhältnisse verschiedene Medien," *ibid.*, **5**. Abt. 2 (1850), 253–268, written with P. L. von Seidel; "Beschreibung und Vergleichung der galvanischen Telegraphen Deutschlands nach Besichtigung vom April 1849," *ibid.*, Abt. 3, 779–840; "Beiträge zur Photometrie des Himmels," in *Astronomische Nachrichten*, **48** (1858), 369–378; "Über Berichtigung des Äquatoreals," *ibid.*, **52** (1860), 129–140; "Das Chronoskop," in *Abhandlungen der Bayerischen Akademie der Wissenschaften*, Math.-phys. Kl., **10**, Abt. 2 (1868), 357–387; and "Methode, eine Grundlinie durch Fortbewegung eines zylindrischen Massstabes in Gestalt eines Rades zu messen," in *Astronomische Nachrichten*, **72** (1868), 369–378.

A biographical notice is in *Allgemeine deutsche Biographie*, XXXV, 720–724.

H.-CHRIST. FREIESLEBEN

STEINITZ, ERNST (*b.* Laurahütte, Silesia, Germany [now Huta Laura, Poland], 13 June 1871; *d.* Kiel, Germany, 29 September 1928), *mathematics*.

Steinitz began the study of mathematics in 1890 at the University of Breslau (now Wrocław, Poland). A year later he went to Berlin and in 1893 returned to Breslau, where he received the Ph.D. in 1894. Two years later he began his teaching career as *Privatdozent* in mathematics at the Technical College in Berlin-Charlottenburg. In 1910 he was appointed professor at the Technical College in Breslau. He assumed a similar post in 1920 at the University of Kiel, where his friend Otto Toeplitz was teaching, remaining there until his death.

In his most important publication, "Algebraische Theorie der Körper" (1910), Steinitz gave an abstract and general definition of the concept of a "field" (*Körper*) as a system of elements with two operations (addition and multiplication) that satisfy associative and commutative laws (which are joined by the distributive law), the elements of which admit unlimited and unambiguous inversion up to division by zero. Steinitz sought to discuss all possible types of fields and to ascertain their relationships. By means of a systematic development of the consequences of the axioms for commutative fields, he introduced a series of fundamental concepts: prime field, separable elements, perfect fields, and degree of transcendence of an extension. His most important achievement was undoubtedly the proof that for every base field K there exist extension fields L in which all polynomials with coefficients in K decompose into linear factors, and that the smallest possible such field is virtually determined up to isomorphism. Because this smallest field possesses no genuine algebraic extension, Steinitz called it algebraically closed, proving its existence with the aid of the axiom of choice; this is now done by means of Zorn's lemma.

In his basic approach Steinitz was influenced primarily by Heinrich Weber and, in his methods, by Leopold Kronecker; Hensel's discovery in 1899 of the field of *p*-adic numbers provided the direct stimulus for his work. His polished and fully detailed treatment of the subject was the starting point for many far-reaching studies in abstract algebra, including those by E. Artin, H. Hasse, W. Krull, E. Noether, and B. L. van der Waerden. The general concept of the derivative or of differentiation, which Steinitz introduced in special cases, is essential in modern algebraic geometry.

In addition to his epochal paper, Steinitz wrote on the theory of polyhedra, a topic of lifelong interest. He gave two lectures on it at Kiel and prepared a comprehensive treatment during his last years. An almost complete manuscript of a planned book was found among his papers; it was completed and edited by Rademacher in 1934. Dealing chiefly with convex polyhedra and their topological types, the book also includes a detailed historical survey of the development of the theory of polyhedra.

BIBLIOGRAPHY

Steinitz' works are listed in Poggendorff, IV, 1435; V, 1203; and VI, 2534. His most important writings are "Algebraische Theorie der Körper," in *Journal für die reine und angewandte Mathematik*, **137** (1910), 137–309, which was also separately published, R. Baer and H. Hasse, eds. (Berlin–Leipzig, 1930; New York [in German], 1950); and *Vorlesungen über die Theorie der Polyeder, unter Einschluss der Elemente der Topologie*, Hans Rademacher, ed. (Berlin, 1934; repr. Ann Arbor, Mich., 1945).

BRUNO SCHOENEBERG

STEINMETZ, CHARLES PROTEUS (*b.* Breslau, Germany [now Wrocław, Poland], 9 April 1865; *d.* Schenectady, New York, 26 October 1923), *engineering.*

Steinmetz' given name was Karl August Rudolf. He was the only son of Karl Heinrich Steinmetz, a government railway employee, and his first wife, Caroline Neubert. After early education at the local Gymnasium, he entered the University of Breslau, where he remained for five years. During this time, he became an ardent socialist. Placed under police surveillance in 1887, he eventually fled to Zurich, shortly before completing the Ph.D. in mathematics, and there studied mechanical engineering.

In 1889 Steinmetz immigrated into the United States and in 1894 became an American citizen, whereupon he took the name Charles Proteus. He never married but in 1905 legally adopted Joseph Le Roy Hayden as his son. Steinmetz received an honorary M.A. from Harvard (1902), and honorary Ph.D. from Union College, and served as president of both the American Institute of Electrical Engineers (1901) and the Illuminating Engineering Society (1915).

Upon entering the United States, Steinmetz was employed as a draftsman by Rudolf Eickemeyer, a prospering manufacturer of motors and machinery and a versatile inventor and pioneer in electrical research. Four years later the newly founded General Electric Company acquired many of Eickemeyer's electrical patents, along with the services of Steinmetz, who by 1892 had already earned a considerable reputation because of two long papers on the mathematical law of magnetic hysteresis, which was praised by contemporaries as "doing for the magnetic circuit what Ohm did for the electric circuit." Steinmetz' first textbook on electricity, *Theory and Calculation of Alternating Current Phenomena* (1897), written with E. J. Berg, described the complex number technique for analyzing alternating-current circuits that he had first presented to the International Electrical Congress in Chicago in 1893. This work played a decisive role in the turn-of-the-century debate between alternating- and direct-current technologies and the technique is still universally used.

To relieve him of administrative duties, Steinmetz was made a consulting engineer at General Electric after 1895. His principal contributions to the advancement of the company in the ensuing twenty-eight years were covered by 195 patents and included the magnetite arc-lamp electrode; two-phase to three-phase transformation; major improvements in motors, generators, and transformers; developments in mercury-arc lighting; and the analysis and design of high-voltage, alternating-current transmission techniques. Transient phenomena in the latter received his particular attention and were explored in a much-publicized, high-voltage "artificial lightning" testing laboratory. He was one of the earliest advocates of atmospheric pollution control, research on solar energy conversion, nationwide electrical networks, electrification of railways, synthetic production of protein, and electric automobiles. While employed by General Electric, Steinmetz also served on the Union College faculty (1903–1913), where he created the electrical engineering department and started his transmission-line research.

Apart from his technical achievements, Steinmetz' unique contribution to the developing electrical engineering profession was his repeated demonstration of the profitability of applying sophisticated mathematical methods to practical problems. Most of his ten technical books were widely used in colleges and had a tremendous influence on curricular development.

Physically small and crippled from birth, like his father and grandfather, Steinmetz had several unusual hobbies and personal idiosyncrasies, about which many legends accumulated. He retained a lifelong interest in socialism; and after the election of a socialist city government in Schenectady in 1911, he served with distinction in several civic positions.

BIBLIOGRAPHY

I. ORIGINAL WORKS. Steinmetz' works include *Theory and Calculation of Alternating Current Phenomena* (New York, 1897), written with E. J. Berg; *Theoretical Elements of Electrical Engineering* (New York, 1901); *General Lectures on Electrical Engineering* (New York, 1908), J. L. Hayden, ed.; *Theory and Calculation of Transient Electrical Phenomena and Oscillations* (New York, 1909); *Radiation, Light and Illumination* (New York, 1909); *Engineering Mathematics* (New York, 1911); *Elementary Lectures on Electrical Discharges, Waves and Impulses and Other Transients* (New York, 1911); *Theory and Calculation of Electric Circuits* (New York, 1917); and *Theory and Calculation of Electric Apparatus* (New York, 1917). The above were published as a nine-volume *Electrical Engineering Library* (New York, 1921).

II. SECONDARY LITERATURE. See Ernest Caldecott and P. L. Alger, *Steinmetz- The Philosopher* (Schenectady, N.Y., 1965); J. W. Hammond, *Charles Proteus*

Steinmetz (New York, 1924); and J. N. Leonard, *Loki, The Life of Charles Proteus Steinmetz* (New York, 1929).

ROBERT A. CHIPMAN

STEJNEGER, LEONHARD HESS (*b.* Bergen, Norway, 30 October 1851; *d.* Washington, District of Columbia, 28 February 1943), *ornithology, herpetology.*

Having left his native country, his father's mercantile business, and his first wife in 1881, Stejneger moved to the United States to seek employment to study his favorite subject, birds. (In 1880 the once prosperous business of his father, Peter Stamer Steineger, had gone into bankruptcy.) At the urging of his mother, Ingeborg Catharina Hess Steineger, Leonhard had aimed toward a medical career but graduated in law (1875) at the University of Kristiania (Oslo). He had long been interested in birds, and first published a work on them when he was nineteen; he also corresponded with ornithologists throughout the world.

After arriving in the United States, Stejneger went directly to Spencer F. Baird at the Smithsonian Institution and was promptly hired to work under the curator of birds, Robert Ridgway. To fill the vacancy left by H. C. Yarrow, Stejneger was appointed acting curator of reptiles and amphibians at the Smithsonian in 1889 and ten years later became curator. From 1911 until his death, he was curator of the department of biology. (He was exempted from retirement by presidential order.)

Stejneger became a citizen of the United States in 1887, and in 1892 he married Helene Maria Reiners. In addition to becoming an honorary member of many scientific societies, Stejneger was elected to the National Academy of Sciences (1923), received an honorary Ph.D. from the University of Oslo, and was elected honorary president for life of the American Society of Ichthyologists and Herpetologists (1931). He was also made a commander of the Royal Norwegian Order of St. Olav (1939), a member of the International Committee on Zoological Nomenclature (1898), and a member of the Permanent Committee of One Hundred of the International Ornithological Congresses (1905). An accomplished linguist, his skills proved useful at the many international meetings to which he was sent.

Stejneger began his extensive collecting in 1882 with a trip to the Komandorskiya Ostrova Islands to help set up weather stations for the United States Signal Service; there, he also gathered bones of the extinct Pallas's cormorant. He returned to these islands during the period 1895–1897 for the International Fur-Seal Commission. Stejneger collected specimens in the American Southwest, in the South Dakota Badlands, in Puerto Rico, and, on a number of trips to Europe for scientific meetings, he collected specimens and made detailed studies of life zones of the Alps.

Stejneger produced a number of descriptive and classificatory publications on birds; but his curatorial appointment turned his interests to reptiles and amphibians, so that he became an expert on the systematics of those groups. With Thomas Barbour, he completed five editions of the highly useful *Check-list of North American Amphibians and Reptiles* (Cambridge, Mass., 1917).

BIBLIOGRAPHY

I. ORIGINAL WORKS. Stejneger's 400 publications are almost equally divided between ornithology and herpetology, but include other subjects as well, for example, a meticulously researched biography of the first Arctic naturalist, Georg Wilhelm Steller (Cambridge, Mass., 1936), and significant publications on fur seals. In 1885 he published "Results of Ornithological Explorations in the Commander Islands and in Kamtschatka," in *Bulletin. United States National Museum*, **29** (1885), 1–382. He contributed extensively to John S. Kingsley, ed., *Natural History of Birds* (Boston, 1885).

In addition to the fifth ed. of Stejneger's *Check-list*, which was in *Bulletin of the Museum of Comparative Zoology at Harvard College*, **93** (1943), 1–260, "Poisonous Snakes of North America," in *Report of the United States National Museum* for 1893 (1895), 337–487, is a classic in herpetology. Regional studies of birds and of reptiles and amphibians are listed in the full bibliography accompanying Wetmore's biography of Stejneger (see below).

II. SECONDARY LITERATURE. Alexander Wetmore wrote an account of Stejneger's life, background, and scientific contributions in *Biographical Memoirs. National Academy of Sciences*, **24** (1945–1947), 143–195, with complete bibliography. A shorter appreciation, with no bibliography, was written by Waldo L. Schmitt, in *Systematic Zoology*, **13**, no. 4 (1964), 243–249. An account of Stejneger's accomplishments and personality, by A. K. Fisher, appeared in a tribute issue of *Copeia*, (Oct. 1931), 75–83.

ELIZABETH NOBLE SHOR

STEKLOV, VLADIMIR ANDREEVICH (*b.* Nizhni Novgorod [now Gorky], Russia, 9 January 1864;

d. Gaspra, Crimea, U.S.S.R., 30 May 1926), *mathematics, mechanics.*

Steklov's father, Andrey Ivanovich Steklov, a clergyman, taught history and was rector of the Nizhni Novgorod seminary; his mother, Ekaterina Aleksandrovna Dobrolyubov, was a sister of the revolutionary-democratic literary critic Nikolay Dobrolyubov. In 1874–1882 Steklov studied at the Alexander Institute in Nizhni Novgorod; after graduation he entered the department of physics and mathematics at Moscow University, transferring a year later to Kharkov. A. M. Lyapunov, who had been lecturing there since 1885, soon became his scientific supervisor. In 1887 Steklov passed his final examinations, and the following summer it was suggested that he remain at the university to prepare for an academic career. He was appointed university lecturer in mechanics in 1891; two years later he presented his master's thesis, and in 1896 he was named extraordinary professor of mechanics. After defending his doctoral dissertation in 1902, Steklov was elected professor; Lyapunov then moved to St. Petersburg, and Steklov obtained the chair of applied mathematics. An active member of the Kharkov Mathematical Society, he served successively as secretary (1891), deputy chairman (1899), and chairman (1902–1906).

In 1906 Steklov transferred to the chair of mathematics at St. Petersburg University. His profound lectures, open sympathy with the aims of progressive students, and acute criticism of the tsarist order—especially at the universities—added new dimensions to the scientific and educational activity of the department of physics and mathematics and attracted numerous students. In St. Petersburg, Steklov laid the foundations of the school of mathematical physics that achieved considerable distinction, particularly after the October Revolution. Among his pupils were such prominent scientists as A. A. Friedmann, V. I. Smirnov, and Y. D. Tamarkin.

In 1910 V. Steklov was elected a member of the Academy of Sciences (he had been a corresponding member since 1902), and in 1916 he became a member of its board of directors. From then on, especially after becoming vice-president of the Academy in 1919, Steklov devoted most of his time to that organization. During the civil war, military conflicts, economic decline, and the early phases of reconstruction, he proved to be a brilliant scientific administrator. For eight years he worked tirelessly to maintain, and later to enlarge, the activity of the Academy and to reorganize it in

order to bring science and practical requirements closer together. This work embraced all aspects of academic activity, from repairing old buildings and restoring the network of seismic stations to publishing academic proceedings and books, providing foreign periodicals for libraries, and organizing new institutes within the Academy. The Institute of Physics and Mathematics was organized in 1921 on Steklov's suggestion, and he served as its director until his death. In 1934 this institute was divided into the P. N. Lebedev Institute of Physics and the V. A. Steklov Mathematical Institute, both of which became centers of scientific activity.

Along with organizational work, Steklov continued his scholarly pursuits. In his later years he produced a series of articles on the theory of quadratures and on Chebyshev's polynomials, a monograph on mathematical physics, a popular book on the importance of mathematics for mankind, and biographies of Galileo and Newton.

Steklov's early works were devoted mostly to mechanics. In his master's thesis he pointed out the third case in which the integration of equations of a solid body moving in an ideal nonviscous fluid (under certain suppositions) is reduced to quadratures. The two earlier cases were described in 1871 by Rudolf Clebsch, and the fourth (and last) by Lyapunov in 1893. Steklov also treated problems of hydromechanics.

Steklov's principal field of endeavor, however, was mathematical physics and corresponding problems of analysis. Many problems of potential theory, electrostatics, and hydromechanics are reduced to the boundary-value problems of Dirichlet and Neumann when it is necessary to find a solution of Laplace's differential equation satisfying some boundary conditions on a surface S enclosing the region under consideration. Although Neumann, Hermann Schwarz, Poincaré, G. Robin, E. Le Roy, and others had suggested methods of solving such problems, they did not elaborate their rigorous grounding, and their methods were applied to relatively restricted classes of surfaces. The precision of analysis in the general investigation of the problem was first achieved by Lyapunov and Steklov. Steklov presented the first summary of his studies in this field in his doctoral thesis and in the articles "Sur les problèmes fondamentaux de la physique mathématique" and "Théorie générale des fonctions fondamentales." He made a valuable contribution to the theory of fundamental functions (Poincaré's term) or, to use a contemporary expression, the theory of eigen functions depending in a particular way upon the character of the surface S and forming on the surface a normal and orthogonal system; the solution of the boundary-value problems of Dirichlet and Neumann is expressed in terms of these eigen functions. Steklov was the first to demonstrate strictly for a very broad class of surfaces the existence of an infinite sequence of (proper) eigen values and corresponding eigen functions defining them in a way different from Poincaré's. Using a method going back to Fourier, Steklov also solved new problems of the theory of heat conduction subject to some boundary, and initial conditions.

When boundary value problems are considered, an especially difficult problem arises when one wishes to expand an arbitrary function, for example, $f(x)$, subject to certain restrictions, into a convergent series of the form $\sum_{k=0}^{\infty} A_k U_k(x)$ where each A_k is a constant and the eigen functions $U_k(x)$ form a normal and orthogonal system. Particular cases of this kind had occurred since the latter half of the eighteenth century. From 1896, Steklov devoted numerous works to the elaboration of a general method of solving this problem in one, two, and three dimensions; this work resulted in the creation of the general "theory of closedness," the term he introduced in 1910. The condition of closedness established by Steklov is the generalization of Parseval's equality (1805) in the theory of Fourier series. The closed systems are "complete": they cannot be extended by adding new functions without loss of orthogonality; only closed systems may be used for solving the mentioned problem. In the simplest case, when it is necessary to expand a continuous function $f(x)$ on the segment (a, b) into a series of functions of one normal and orthogonal system $\{U_k(x)\}$ with respect to a weight $p(x) \geq 0$, so that $f(x) = \sum_{k=0}^{\infty} A_k U_k$ and the coefficients $A_k = \int_a^b p(x)f(x)U_k(x)dx$, the condition of closedness takes the form $\sum_{k=0}^{\infty} a_k^2 = \int_a^b p(x)f^2(x)dx$. Steklov investigated the closedness of diverse concrete systems and defined certain conditions under which the expansions in question really occur. In 1907 he began to use in the theory of closedness an important "smoothing method," which consisted of replacing the function under study—for example, $f(x)$—with some other mean function—$F_h(x)$ $= \frac{1}{h} \int_x^{x+h} f(t)dt$—that in some sense has more convenient characteristics. For example, it is continu-

ous, whereas $f(x)$ is only integrable. This device is now widely used in mathematical physics. Steklov investigated expansions into series not only with the theory of closedness but also by means of asymptotic methods or by direct evaluation of the remainder term in the series.

The rise of the theory of integral equations at the beginning of the twentieth century, which led to new, general, and effective methods of solving the problems of mathematical physics and expansions of functions on orthogonal systems, inspired Steklov to improve the theory of closedness, although he did not participate in the elaboration of the theory of integral equations itself.

BIBLIOGRAPHY

I. ORIGINAL WORKS. Steklov produced 154 works. Bibliographies are in *Pamyaty V. A. Steklova*, G. I. Ignatius, *Steklov* and (most complete) in V. S. Vladimirov and I. I. Markush, *Akademik V. A. Steklov* (see below). Among his writings are *O dvizhenii tverdogo tela v zhidkosti* ("On the Motion of a Solid Body in a Fluid"; Kharkov, 1893), his master's thesis; *Obshchie metody reshenia osnovnykh zadach matematicheskoy fiziki* ("General Methods of Solving Fundamental Problems of Mathematical Physics"; Kharkov, 1901), his doctoral dissertation; "Sur les problèmes fondamentaux de la physique mathématique," in *Annales de l'École normale supérieure*, 3rd ser., **19** (1902), 191–259, 455–490; "Théorie générale des fonctions fondamentales," in *Annales de la Faculté des sciences de Toulouse*, 2nd ser., **6** (1905), 351–475; "Sur les expressions asymptotiques des certaines fonctions définies par les équations différentielles du second ordre et leurs applications au problème du développement d'une fonction arbitraire en série procédant suivant les dites fonctions," in *Soobshchenia Kharkovskogo matematicheskogo obshchestva*, **10** (1907), 97–201; "Problème du mouvement d'une masse fluide incompressible de la forme ellipsoïdale dont les parties s'attirent suivant la loi de Newton," in *Annales scientifiques de l'Ecole normale supérieure*, 3rd ser., **25** (1908), 469–528, and **26** (1909), 275–336; "Une application nouvelle de ma méthode de développement suivant les fonctions fondamentales," in *Comptes rendus . . . de l'Académie des sciences*, **151** (1910), 974–977; "Sur le mouvement d'un corps solide ayant une cavité ellipsoïdale remplie par un liquide incompressible et sur les variations des latitudes," in *Annales de la Faculté des sciences de Toulouse*, 3rd ser., **1** (1910), 145–256; *Osnovnye zadachi matematicheskoy fiziki* ("Fundamental Problems of Mathematical Physics"), 2 vols. (Petrograd, 1922–1923); and *Matematika i ee znachenie dlya chelovechestva* ("Mathematics and Its Importance for Mankind"; Berlin–Petrograd, 1923).

II. SECONDARY LITERATURE. See G. I. Ignatius, *Vladimir Andreevich Steklov* (Moscow, 1967); V. S.

Vladimirov and I. I. Markush, *Akademik V. A. Steklov* (Moscow, 1973); *Istoria otechestvennoy matematiki* ("History of Native Mathematics"), I. Z. Shtokalo, ed., II and III (Kiev, 1967–1968), see index; *Pamyaty V. A. Steklova* ("Memorial to . . . Steklov"; Leningrad, 1928), with articles by N. M. Gyunter, V. I. Smirnov, B. G. Galiorkin, I. V. Meshchersky, and R. O. Kuzmin—Gyunter's article, "Trudy V. A. Steklova po matematicheskoy fizike" ("Steklov's Works on Mathematical Physics") is reprinted in *Uspekhi matematicheskikh nauk*, **1**, nos. 3–4 (1946), 23–43; Y. V. Uspensky, "Vladimir Andreevich Steklov," in *Izvestiya Akademii nauk SSSR*, 6th ser., **20**, nos. 10–11 (1926), 837–856; and A. Youschkevitch, *Istoria matematiki v Rossii do 1917 goda* ("History of Mathematics in Russia Before 1917"; Moscow, 1968), see index.

A. P. YOUSCHKEVITCH

STELLER, GEORG WILHELM (*b.* Windsheim, Germany, 10 March 1709; *d.* Tyumen, Siberia, Russia, 12 November 1746), *geography, biology.*

Steller (whose real name was Stöller) was born in a small town in Franconia, where his father was an organist. He first studied Lutheran theology at Wittenberg but in 1731 entered the medical faculty of the University of Halle and began to devote himself to botanical research. Apparently he never passed a formal medical examination; but in 1734 he arrived in Danzig via Berlin, was accepted as physician in the Russian army, which was stationed there, and then continued on to St. Petersburg. Steller then worked as an assistant to Johann Amman, a botanist with the Russian Academy of Sciences, and gained a powerful patron in Archbishop Feofan Prokopovich, in whose house he lived.

Steller's desire to join the elaborately planned second Bering expedition as a research member was realized in 1736 when he was nominated a member. In the spring of 1738 he was appointed an assistant at the St. Petersburg Academy and left to join the other members of the expedition. In Yeniseysk, he met the botanist J. G. Gmelin and G. F. Müller. In 1740 they arrived at Okhotsk, and Steller continued with Bering to Kamchatka to study nature and folklore.

In 1741 Steller sailed under Bering's command on the *St. Peter* for America. He was the first natural historian to land on the coast of Alaska, where he collected specimens for several hours before the ship turned west. The return journey met with great difficulties. The crew, decimated by scurvy, was forced to pass the winter under great hard-

ships on Bering Island. Steller survived; and returning to Petropavlovsk in August 1742, he lingered further in Kamchatka, where he pursued his research in natural history and tried to complete his manuscripts. In 1746, en route to St. Petersburg, he was arrested for alleged insubordination as he was approaching the Ural Mountains. He was soon released but died of a fever in the fall of the same year.

Steller was a pioneer in the study of the natural history and geography of Kamchatka and Alaska. His great collections and many manuscripts were sent to the academy in St. Petersburg, and several manuscripts were published posthumously. His *Beschreibung von dem Lande Kamtschatka* (1774) is valuable especially because of his description of the life and habits of the inhabitants and because of his detailed illustrations. His journal of the expedition to Alaska was published by Pallas. On Bering Island, Steller studied the large sea mammals, including the remarkable Steller's sea cow, *Hydrodamalis gigas*, soon to be extinct. Botanical and zoological material from his collections has been published also by Gmelin in his *Flora sibirica* and by Linnaeus and Pallas.

Steller had a difficult and disharmonious character and a violent temper; but as an explorer and field worker, he was rugged, enthusiastic, and indomitable.

BIBLIOGRAPHY

I. ORIGINAL WORKS. Steller's major work is *Beschreibung von dem Lande Kamtschatka* (Frankfurt–Leipzig, 1774; St. Petersburg, 1793). His travel journal with the description of Bering Island and the voyage to Alaska was published by Pallas in *Neue Nordische Beyträge zur physikalischen und geographischen Erd- und Völkerbeschreibung . . .,* **2** (1781), 255–301; **5** (1793), 123–236, also issued separately; and **6** (1793), 1–26. An English trans. is F. A. Golder, *Bering's Voyage . . .,* II (New York, 1925). An abbreviated popular ed. is M. Heydrich, *Von Kamtschatka nach Amerika* (Leipzig, 1926).

Steller's description of the sea animals near Bering Island is "De bestiis marinis," in *Novi Commentarii Academiae Scientiarum Petropolitanae,* **2** (1751), 289–398, with German trans., *Beschreibung von sonderbaren Meerthieren* (Halle, 1753). Letters from Steller to Gmelin are in G. H. T. Plieninger, ed., *Joannis Georgii Gmelini reliquiae quae supersunt* (Stuttgart, 1861).

II. SECONDARY LITERATURE. The major source is Leonhard Stejneger, *Georg Wilhelm Steller, The Pi-*

oneer of Alaskan Natural History (Cambridge, Mass., 1936).

STEN LINDROTH

STELLUTI, FRANCESCO (*b.* Fabriano, Italy, 12 January 1577; *d.* Fabriano, November 1652), *microscopy, scientific organization.*

Stelluti's parents, Bernardino Stelluti and Lucrezia Corradini, belonged to patrician families of Fabriano. They intended him for the law, which subject he went to study at Rome toward the end of the sixteenth century. In Rome he came under the influence of Federico Cesi and Johannes Eck; and with them he helped found the Accademia dei Lincei in August 1603. His academic name was Tardigradus, and his emblem was the planet Saturn, which suggests that he was less quick mentally than his colleagues; he always appears as the loyal helper and companion rather than as the initiator. Stelluti lectured on his specialties, mathematics and astronomy. His classes opened with a general outline of geometry and moved straight on to a mechanized scaling ladder. Stelluti published nothing in mathematics. The Lincei suffered much harassment during their early "secret brotherhood" days, and in 1604 Stelluti was forced to leave Rome for his native Fabriano. He then moved to Parma, attaching himself to the ducal court there.

In October 1605, when Eck returned from his travels, he stayed in Parma with Stelluti, who illustrated the classification of butterflies from the specimens that Eck had brought back with him. This work was Stelluti's introduction to entomology. Once the ground had been cleared for a revival of the Academy, Stelluti and Eck returned to Rome (1608 or 1609).

Stelluti's first reaction to Galileo's *Sidereus nuncius* (1610) was rather disparaging, perhaps out of loyalty to the priority claim made by his colleague Giambattista della Porta. But he was soon won over to wholehearted admiration, and several of his letters to Galileo are preserved. Apart from a few comments on telescopic observations, the letters contain only academic business or personal gossip; Stelluti appreciated the intellectual gap between them and made no attempt to bridge it.

In 1612 Stelluti was elected procurator, or business manager, of the Academy and was entrusted with the negotiations for the purchase of property, where research could be carried on, and the publication of books under the auspices of the Lynxes. He was involved in the publication and distribu-

tion of Galileo's *Istoria e dimostrazioni intorno alle macchie solari* and *Il saggiatore* and wrote introductory verses for both.

In 1625 Stelluti made the first microscopic observations to be published, probably with an instrument that Galileo had sent to Cesi in 1624. Cesi had decided to bring out a short treatise on bees, a fragment from his projected encyclopedia, in compliment to Cardinal Francesco Barberini, of whose support the Academy had high hopes. The frontispiece of this *Apiarium* shows three views of a bee (magnified ten times) with insets of the whole head, eye, antenna and mouthparts (displaying the labia and galea), the rear legs, branched hairs, and the sting. Apparently these observations were checked by his fellow Lynx Fabio Colonna. In 1630 Stelluti published his translation, with commentary, of the satires of Persius. A reference to Arezzo, in which the Barberini family supposedly originated, was pretext enough to insert a *Descrizzione dell'ape*, illustrated by woodcuts based on the *Apiarium*, but magnified only six times, with a short account of the organs (shown separately). Persius' allusion to a grain weevil is illustrated by a microscopic representation (magnified ten times); the tip of the snout with its mandibles (magnified twenty times); and a view of the whole (life-size). Although there are numerous medical and botanical footnotes, they evince only casual observation.

When Cesi died in 1630, Stelluti tried to keep the Academy alive by proposing the election of a new prince, possibly Barberini. Although the cardinal was willing to patronize Stelluti, whose books were dedicated to him, he would not help the Academy to continue. In 1637 Stelluti produced synoptic tables of Porta's *De humana physiognomia*. In the introduction he promised that he would later add a treatise on "the hand of man compared to the feet of some quadrupeds and birds"; but this work never appeared. In the same year he published an account of fossilized wood found in the region of Todi. This work was based closely on Cesi's theories of a class of "metallophytes" intermediate between metals and plants. Stelluti explained how he had abandoned his own *prima facie* assumption that these fossils were simply buried, mineralized tree trunks. For most of the latter part of his life Stelluti served as the faithful adviser to Cesi's widow in her various troubles. He had the satisfaction of seeing the descriptions by the Academy of Mexican flora and fauna through the press in 1651. This publication also contained the synoptic tables of Cesi's system of taxonomy, for which Stelluti wrote an introduction, the melancholy last word of the first Academy of the Lynxes.

BIBLIOGRAPHY

I. ORIGINAL WORKS. Stelluti's works are *Persio tradotto in verso sciolto . . .* (Rome, 1630); *Della fisionomia de tutto il corpo humano . . . in tavole sinottiche ridotta et ordinata* (Rome, 1637); and *Trattato del legno fossile minerale nuovamente scoperto* (Rome, 1637).

II. SECONDARY LITERATURE. See G. Gabrieli, "Il carteggio scientifico ed accademico fra i primi Lincei 1603–30," in *Atti della Reale Accademia Nazionale dei Lincei*. Memorie della classe di scienze morali, storiche e filologiche, ser. 6, **1** (1925), 137–219; "Il carteggio linceo della vecchia accademia di Federico Cesi (1603–30)," *ibid.*, **7** (1938–1939), 1–535; G. Gabrieli, "Francesco Stelluti, Linceo Fabrianese," *ibid.*, ser. 7, **2** (1941), 191–233; B. Odescalchi, *Memorie istorico-critiche dell'Accademia dei Lincei* (Rome, 1806); C. Ramelli, "Discorso intorno a Francesco Stelluti da Fabriano," in *Giornale arcadico di scienze, lettere ed arte*, **87** (1841), 106–135; and C. Singer, "The Earliest Figures of Microscopic Objects," in *Endeavour*, **12** (1953), 197.

A. G. KELLER

STENSEN, NIELS, also known as **Nicolaus Steno** (*b*. Copenhagen, Denmark, 1/11 January 1638; *d*. Schwerin, Germany, 25 November/5 December 1686), *anatomy, geology, mineralogy.*

Stensen was the son of Sten Pedersen, who came from a family of preachers and was a goldsmith. After graduating from the Liebfrauenschule, Stensen entered the University of Copenhagen in 1656. There he studied medicine and came under the special influence of Simon Paulli and Thomas Bartholin. The customary study journey took him at the end of March 1660 to Amsterdam and, on 27 July for his matriculation, to the University of Leiden, from which, after three years of diligent research, he was called home because of the death of his stepfather.

The University of Copenhagen failed to enlist Stensen's services, and so he went to Paris; he is known to have been there as late as November 1664 and to have received on 4 December his M.D. from the University of Leiden in absentia. After a fruitful year in the circle of Thévenot, the king's chamberlain, Stensen went in the autumn of 1665 to Montpellier; and from there he went to Pisa, where he stayed until the beginning of March 1666. He remained in Tuscany until July 1668, mostly at the court of Grand Duke Ferdinand II in Florence.

Although he came from a deeply religious Lutheran family, Stensen became a convert to Catholicism on All Souls' Day 1667 during a period of research in anatomy and geology. This research was interrupted, probably by a summons to Denmark, from August 1668 to June 1670. Upon his return to Florence, Stensen again worked in Tuscany, exploring two alpine grottoes at Lake Garda and Lake Como for the Accademia del Cimento.

Following a call to return to Denmark as royal anatomist, Stensen arrived at Copenhagen on 3 July 1672. Here, mostly for a circle of interested friends, he held a series of anatomical demonstrations. But he left his native city on 14 July 1674 to return to Florence, where he was consecrated a priest, probably in the middle of April 1675, and where he worked for two years as educator and tutor of the crown prince. Upon the invitation of Duke Johann Friedrich of Hannover, he went to Rome, where he was appointed apostolic vicar of northern missions by Pope Innocent XI on 21 August 1677 and was consecrated titular bishop of Titiopolis on 19 September. Until the end of June 1680, he ministered to the scattered remnants of Catholicism in northern Germany, Denmark, and Norway. Then after the death of the duke, he was appointed assistant bishop of Münster in Westphalia, where he was very active. On 1 September 1683 he left the city in protest against the simoniac election of the bishop's successor. After two years of apostolic activity in Hamburg and Schwerin, he died in acute pain from gallstones.

Two educational influences upon Stensen's youth deserve special attention. Since his father's goldsmith shop was near the Round Tower, his scientific interest may have been directed at an early age into technical-mathematical channels to which he wished to return even in the midst of his first period of biological investigations in Leiden. He was interested in minerals and metals; in lenses and light refraction; and in telescopes, microscopes, and thermoscopes. Thomas Walgesten, the inventor of the *laterna magica*, belonged to the circle of his acquaintances. Stensen, of course, knew the medical authors Thomas Bartholin, Pierre Borel, Henricus Regius, Paracelsus, Helmont, and Santorio. Stensen's physical-mathematical interests are indicated by his reading not only the works of Galileo (*Sidereus nuncius* [1610]) and Kepler (*De nive sexangula* [1611]) but also those of Gassendi, Clavius, Gaspar Schott, Snel, and Varenius. Stensen had a precocious desire for methodically founded knowledge, and he was critical of analogies and purely authoritarian statements. He also asked for frequent observations and correct conclusions, and he declared himself in favor of Descartes's method in order to secure the greatest possible certainty. Stensen's writings also contain many passages that show a deeply religious nature and a highly ethical character.

During the second half of his life, in which he attained the highest fame and then—for the sake of God and of human souls—renounced his scientific research, there can be differentiated, both in time and in subject matter, four great periods of research. Each period began with almost accidental individual observations but led to an abundance of important discoveries and basic laws, many of which were recognized only in later centuries.

The first of these periods was devoted to the glandular and lymphatic system. In April 1660, three weeks after his arrival at Amsterdam, where he studied under the direction of Gerhard Bläes (Blasius), Stensen made his first known discovery: while dissecting the head of a sheep, he found the duct of the parotid gland (Stensen's duct), which is a principal source of saliva for the oral cavity.

In Leiden, then the most important university on the Rhine, Stensen sought contact with the two leading medical professors: Johannes van Horne, who independently of Pecquet had discovered the chief thoracic lymph passage, and Franciscus Sylvius, famous as an iatrochemist and for his studies on the brain. A warm and stimulating friendship with Jan Swammerdam also began in Leiden. Soon after Stensen's arrival, van Horne demonstrated on humans the course of the parotid duct and declared it to be Stensen's discovery, although Blasius, in his *Medicina generalis* (1661), not only claimed it for himself but incited his friends to slander Stensen, his former student. There ensued a long quarrel that Stensen settled both objectively and devastatingly in his *Apologiae prodromus* (1663).

The controversy spurred Stensen to the further investigation of the glands. He wrote: "I owe much to the famous man Blasius because he not only gave me cause to assert my property rights, but also to discover other new things."

The glands and lymph vessels were then a new and exciting subject for investigation. In 1622 Aselli had demonstrated the lacteal vessels in the mesentery of a dog; in 1642 Johann Georg Wirsung had shown the excreting duct of the pancreas; and in 1651 Pecquet had demonstrated the *cisterna chyli* and its continuation, the thoracic lymph duct; he also had realized that the latter poured its

contents into the veins. In 1653 Thomas Bartholin demonstrated the thoracic lymph passage and the lymphatic system in humans. He also showed that the lymph vessels connecting the liver to the thoracic duct carried lymph away from the liver, thereby throwing doubt on the Galenic doctrine that blood originated in the liver. When Thomas Wharton published his systematic presentation of the contemporary theory of glands in his *Adenographia* (1656), he announced the discovery of the duct of the submaxillary salivary gland; he also designated the brain and tongue as glands.

In contrast, Stensen very soon advanced from his "little discovery," as he called his first one, to a basic understanding of the whole glandular lymphatic system, which he counted among the most sublime artifices of the Creator. Without changing the names of the conglomerate and conglobate glands, the terms by which Sylvius had already distinguished the anatomical form of the real glands from that of the lymph nodes, Stensen distinguished them according to their function. Arguing against such contemporaries as Bils, Anton Deusing, and Everaertz, on the basis of his observations Stensen stated in his Leiden dissertation (1661):

> I gather from this that the saliva consists of the fluid secreted in the oral glands from the arterial blood which is carried through the lymph ducts with the aid of the *Spiritus animales* [a term then used for the nerves] into the mouth and the closely adjoining muscles, but that the round or conglobate glands in the proximity carry the lymphs received from the outer parts back to the veins so that it becomes mixed with the blood streaming back to the heart.

This discovery led Stensen to consider every fluid in the body as a glandular secretion. He then found a series of glands furnishing fluid to each of the body cavities. He likewise sought the afferent and efferent ducts of secretion. Stensen still used the name "lymph" for all watery glandular secretions, because he was not yet able to differentiate between them and to specify them chemically and physiologically.

In the course of this basic research Stensen presented in his Leiden dissertation new discoveries of glands in the cheeks; beneath the tongue; and in the palate, whose structure of veins, arteries, nerves, and lymph vessels he also described. In his *Observationes anatomicae* (1662), dealing with his new discoveries concerning the glands, he described the lachrymal apparatus in great detail. Stensen determined the purpose of the lachrymal fluid—to facilitate the movement and cleansing of the eyelids on the same principle that applied to the saliva and the mucous membrane of the intestinal canals. He grouped the afferent and efferent lachrymal ducts around the tear gland proper and what was then called *glandula lacrimalis*, in the inner eye corner. The moisture of the nose led him, in this connection, to the discovery of the nasal glands. Stensen considered the possibility that the moisture necessary for the nose could come from the ears through the eustachian tubes, and from the eyes through the nasal duct, but decided that the principal source was the nasal glands. He assumed that this fluid disappeared again through the nostrils and also through an opening to the gullet, the *ductus nasopalatini*, also called *ductus Stenoniani*.

In his survey *De musculis et glandulis* (1664) Stensen enumerated all his new observations and individual discoveries, especially those he had made during the first half of his stay in Leiden: earwax duct, ducts of the cheek glands, the smaller gland ducts under the tongue, the glandular ducts of the palate, the glandular ducts of the epiglottis, the nasal gland, the nasal gland peculiar to sheep, the passage from the eyelids to the nose, the lachrymal ducts, and the gland ducts that lubricate the exterior surface of the ray. In 1673 Stensen found, independently of Peyer, the accumulations of lymph follicles in the small intestine (named for Peyer), but published his findings four years later.

The second period of research began with a challenge to the traditional overestimation of the heart: "One has glorified the heart as the sun, even as the king while upon closer examination, one finds nothing but a muscle," which was directed not only against Aristotle, who saw the heart as the seat of the soul, the source of life, and the central organ of all sensation and motion, but also against Galen, who, following the authority of Plato, assigned the life forces of blood motion and heat distribution to the left ventricle. Even Harvey, who first recognized the purely muscle activity of the heart in maintaining the circulation of the blood, did not abandon the idea of a vital warmth within the heart, or did so only very late.

In a letter of 26 August 1662, Stensen told Thomas Bartholin how fascinated he was by the independent motions of the vena cava, which continued even after the stopping of the heartbeat; this stimulated Stensen to make many investigations of the heart and respiratory organs. On 5 March he had spoken of a careful investigation of the heart musculature, and on 30 April he had stated: "As to the substance of the heart, I think I am able to prove that there exists nothing in the heart that is not found also in a muscle, and that there is nothing missing in the heart which one finds in a muscle."

De musculis et glandis (1664) shows an abundance of new observations and discoveries concerning the anatomy and physiology of individual muscle, and the triangularis, which leads from function of the intercostal muscles, the sacrolumbar muscle, and the triangulararis, which leads from the bony end of the true ribs to the central line of the sternum. He described the role of the diaphragm and several other muscles during respiration; classified the tongue as a muscle; and also described the temporal muscle, and the muscle layer of the esophagus, which has its fibers arranged spirally.

From this research Stensen drew comprehensive conclusions concerning the structure of the muscles: that in each muscle there are arteries, veins, fibers and fibrils, nerves, and membranes; that each muscle fiber ends in a tendon on both sides; that no muscle tissue is a parenchyma (*caro*) but consists instead of closely woven fibers; and that the contractility lies in the muscle substance proper. He then applied all his finding to the heart and proved its muscle structure from both positive and negative evidence. He stated that the heart possesses all the characteristics of a muscle structure and that it is neither the seat of joy nor the source of the blood or of the *spiritus vitales*. The automatic movement, independent of the will, is shared by the heart with other muscles. The findings were new, and even ten years later Bartholin, in a new edition of his *Institutiones anatomicae* (1611), did not accept them. Croone revised his *De ratione motus musculorum* (1664) according to Stensen's findings in the second edition (1670).

The controversy over his views caused Stensen, during his first year in Italy (1666/1667) to publish his *Elementorum myologiae specimen*, which dealt chiefly with the questions: Does the muscle increase in size during contraction? Are hardness and swelling of the muscle signs of an increase in volume? These were acute questions at the time, when even Borelli, one of the leading members of the Accademia del Cimento, still believed that swelling was caused by the influx of nerve fluid. Stensen first provided clear concepts and a clear-cut terminology of the parts of the muscle. Then he characterized the individual muscle fiber and the muscle itself as a parallepiped bordered by six parallelograms. In the second part of the *Elementorum* he dealt with objections against the new knowledge about muscles, and lamented the insufficient knowledge of the muscle fluid. Later, Stensen (before the theory of irritability proclaimed by Haller) discovered that a muscle contraction can also result from direct stimulation of the muscle.

Stensen's muscle research was also a symptom of his philosophical-religious struggle. At the turn of 1662/1663 he had wanted to abandon anatomy for mathematics and physics, a wish probably fostered by the spirit of the times and by his desire for certainty. The need for quantitative knowledge was indicated by Stensen's early enthusiasm for Descartes's method of attaining certainty. During his years in Leiden, Stensen became friendly with Spinoza, whose rationalism may have influenced Stensen so strongly that his Christian belief was endangered. His discovery of the muscle structure of the heart showed him the fallacy of the Cartesian view of the heart as a hearth of fire and made him skeptical of their whole seemingly firmly anchored geometrical philosophy. As he admitted later to Leibniz: "If these gentlemen have been so mistaken with material things which are accessible to the senses, what warranty can they offer that they are not mistaken when they talk about God and the soul?"

The third period of research extended from 1665 to 1667, the last great anatomical-physiological period of his life. The period can be divided into three parts: brain anatomy, embryology, and comparative anatomy.

Stensen's study of brain anatomy was confined almost entirely to his *Discours . . . sur l'anatomie du cerveau* (1669), which he presented in Paris to Thévenot's circle, among whom were many Cartesians. Stensen was stimulated to undertake his brain studies not only by his teacher, the brain anatomist Sylvius, but also by Descartes's *Traité de l'homme*. Vigorously but tactfully, Stensen opposed Descartes's mechanical theory and revealed his anatomical errors, refuting especially his theory concerning the epiphysis. According to Stensen, the epiphysis could not possibly carry out the slightest motion and thereby contribute to one's actions; whereas the Cartesian view was that it inclined itself to one side and then to the other side. In the *Discours* Stensen calls for a sober terminology and proposes new methods of dissection and the preparation of specimens. Stensen demands that the investigation trace the course of the brain fibers and that the investigator strive for a secure knowledge of the anatomical parts before interpreting their functions. In recent times his drawings of the brain have shown that he had a very rich knowledge of its anatomy.

From 1667 to 1669 Stensen contributed two concepts about the ovum and the ovary, oviduct, and uterus. The ovaries were considered to be *testes muliebres*, a type of semen producer. In a recently hatched chick Stensen discovered the oviduct and recognized that it was a canal destined to conduct the yolk directly into the intestine. In the *Elementorum* Stensen says that since he had found the true ova in the female testes, he concluded that they were really ovaries. Johannes Peter Müller credited Stensen with a discovery made by Aristotle but then forgotten: that in the so-called smooth sharks (*Mustelus laevis*) the eggs are not deposited, but the embryos remain connected to the uterus by a placenta, similar to that of mammals. Stensen's embryological observations were not published until 1675, but he had communicated them to De Graaf and Swammerdam. Therefore, the Royal Society of London, in the priority dispute concerning the procreative organs, assigned the credit to Stensen.

Among Stensen's unpublished observations are those on rays and sharks. He established the mucous canal system of these fishes and recognized the significance of the spiral fold in the intestine as a substitute for its greater length in other creatures. He also observed the eyeball stalk, the optical nerve, and the crossing of the optic nerves.

The technical side of Stensen's research was highly developed. He employed simple but effective means, such as the induction of bristles into the gland ducts or the expansion of vessels by inflation. It is not known to what extent he used the microscope, but he knew the optical experts of his time and such microscopists as Swammerdam and Malpighi. He was also a skilled draftsman. His diagram of the blood circulation, which for the first time revealed the heart as two relatively independent hearts or pumps, enabled his pupil Caspar Bartholin, Jr., to develop further the concept of circulation.

The fourth and greatest period of Stensen's research began in Florence at the end of October 1666, when he received the head of a gigantic *Carcharodon rondeletii* that had been caught near Leghorn. He made acute observations of its skin, its canals, the brain and nerves, the Lorenzinian ampullae, and the eyes. The rows of pointed teeth in the mouth, however, led him to a thorough study of their number and substance and also placed immediately before him the question of the relation of these teeth to the so-called *glossopetra* or tongue-stones, which were common on Malta and were considered *lapides sui generis*. Stensen concluded that they were fossil shark's teeth. This led to his paleontological, geological, and mineralogical discoveries.

Scarcely eighteen months later his great work, the *Prodromus*, which outlines the principles of modern geology, was printed. The book presents only the outlines of a discussion, yet almost every sentence or paragraph contains new insights.

After an introductory methodological discussion, Stensen states his purpose: "to find, in the case of a body possessed of a certain shape and produced on the basis of natural laws, the proofs in the body itself which reveal the site and the type of its origin." There follow three important sections: the first concerning the relationship in time between the enclosing body and the body enclosed; the second, the determination of the site and origin of a solid body; and the third, the role of fluids in nature.

In the third section Stensen states such important findings as the fundamental difference between inorganic bodies formed by apposition, and organic bodies formed by internal susception. In the third part the individual enclosing and enclosed bodies are considered. In the central section on geological strata Stensen presents his sediment theory, the time sequence and material of the strata, and data on the site of the stratification. After general observations concerning the effects on the strata of the changes of form through the forces of water and fire, there follows a special section on the origin of mountains. He discusses the sites of ores and minerals and includes an interesting section on crystals. At the end of the section, in drawings and two brief sentences, he states the law of the constancy of crystallic angles. The fourth main part of the work offers an application of the new finding to the geology of Tuscany, which is summarized in six stages of development and illustrated with six drawings. Finally he suggests the possible adaptation of all new findings to the generally prevailing world picture. Although Stensen introduced the concept of chronology and the history of the earth, he had little awareness of the actual duration of geological time.

A glance back to the first main part of the *Prodromus* shows the value of Stensen's methodological directives. He demanded that in the solution of a problem the questions connected with it be considered, that facts be distinguished from assumptions, and that the individual result be examined in connection with the history of science.

In the preface to his last great dissection of a female body (1673) Stensen states the spiritual side of his point of view. He calls the anatomist the index finger of God, addresses science as the servant of life, and declares himself a member of three realms (nature, mind, and faith) and does so in the name of beauty.

BIBLIOGRAPHY

I. ORIGINAL WORKS. Collected works of Stensen's writings are *Nicolai Stenonis opera philosophica*, Vilhelm Maar, ed., 2 vols. (Copenhagen, 1910), which includes the scientific works; *Nicolai Stenonis opera theologica . . .*, Knud Larsen and Gustav Scherz, eds. (Copenhagen, 1941, 1947), the theological and ascetical writings; *Nicolae Stenonis epistolae et epistolae ad eum datae . . .*, Gustav Scherz, ed., 2 vols. (Copenhagen-Fribourg, 1952), which includes Stensen's correspondence with an introduction on the correspondents, notes, and documents; and *Pionier der Wissenschaft, Niels Stensen in seinen Schriften* (Copenhagen, 1963), with a short biography and selected texts with introductions.

Stensen's most important writings include *Observationes anatomicae, . . .* (Leiden, 1662), which consists of four treatises, including Stensen's *Disputatio anatomica de glandulis oris . . .* (1661); *De musculis et glandulis* (Copenhagen, 1664), which includes two letters: "De anatome rajae epistola" and "De vitelli in intestina pulli transitu epistola"; *Elementorum myologiae specimen* (Florence, 1666/1667), also published as *Bibliotheca anatomica* (London, 1709–1714), an abridged English version; *Discours . . . sur l'anatomie du cerveau* (Paris, 1669), also published in G. Scherz, *Nicolaus Steno's Lecture on the Anatomy of the Brain* (Copenhagen, 1965); and *De solido intra solidum naturaliter contento dissertationis prodromus* (Florence, 1669), trans. by H. O[ldenburg], *The Prodromus to a Dissertation Concerning Solids Naturally Contained Within Solids . . .* (London, 1671), and by John G. Winter, *The Prodromus of Nicolas Steno's Dissertation Concerning a Solid Body Enclosed by Process of Nature Within a Solid* (New York, 1916).

See also *Prooemium demonstrationum anatomicarum in Theatro Hafniensi anni 1673*, in *Acata Faniensia* (1673); *Nicolai Stenonis ad novae philosophiae reformatorem de vera philosophia epistola* (Florence, 1675); *Nicolai Stenonis de propria conversione epistola* (Florence, 1677); and *Parochorum hoc age seu evidens demonstratio quod parochus teneatur omnes alias occupationes dimittere et suae attendere perfectioni ut commissas sibi oves ad statum salutis aeternae ipsis a Christo praeparatum perducat* (Florence, 1684).

II. SECONDARY LITERATURE. General biographies about Stensen are M. Bierbaum, *Niels Stensen. Von der Anatomie zur Theologie* (Münster, 1959); R. Cioni, *Niels Stensen. Scientist-Bishop* (New York, 1962); A. D. Jörgensen, *Niels Stensen* (Copenhagen, 1958); and

G. Scherz, *Niels Stensen. Forscher und Denker im Barock* (Stuttgart, 1964). More scholarly works on Stensen include G. Scherz, *Nicolaus Steno and His Indice* (Copenhagen, 1958), with a biography and various studies of Stensen's work; and *Nicolaus Steno and Brain Research in the Seventeenth Century* (London, 1967), from Proceedings of the International Symposium on N. Steno held in Copenhagen, August 1965.

On Stensen's work in anatomy, see P. Franceschini, "Priorita del Borelli e dello Stenone nella conoscenza dell' apparato motore," in *Monitore zoologico italiano* (1948), and in *Rivista di storia delle scienze mediche e naturali* (1951), 1–15; A. Krogh, "Biologen Niels Stensen. Trehundrede År.," in *Nordisk tidsskrift for terapi* (1937), 565–578; V. Maar, "Om Opdagelsen af Ductus Vitello-Intestinalis," in *Kongelige Danske Videnskabernes Selskabs Skrifter* (1908), 233–265; M. T. May, "On the Passage of Yolk Into the Intestines of the Chick," in *Journal of the History of Medicine* (1950), 119–143; H. P. Philipsen, "Ductus parotideus Stenonianus," in *Tendlaegetidende*, **64** (1960), 221–248; C. Schirren, "Niels Stensen entdeckte vor 300 Jahren die später nach Fallot benannte Tetralogie," in *Medizinische Welt* (1965), 278–280; Th. Schlichting, "Das Tagebuch von Niels Stensen," in *Centaurus* (1954), 305–310; C. M. Steenberg, "Niels Stensen som sammenlignende Anatom og Embryolog," in *Naturens Verden* (1938), 202–209; and E. Warburg, "Niels Stensen Beskrivelse af det første publicerede Tilfaelde af Fallots Tetradé," in *Nordisk medicin*, **16** (1942), 3550.

On Stensen's geological work, see E. Becksmann, "N. Steno (1638–1686) und seine Stellung in der Geschichte der Geologie," in *Zeitschrift der Deutschen geologischen Gesellschaft*, **91** (1939), 329–336; A. Garboe, "Niels Stensens (Stenos) geologiske Arbejdes Skaebne," in *Danmarks geologiske undersøgelse*, **4** (1948), 1–34; A. Johnsen, "Die Geschichte der kristall-morphologischen Erkenntnis," in *Sitzungsberichte der Preussischen Akademie der Wissenschaften zu Berlin* (1932), 404; Hj. Oedum, "Niels Stensens geologiske Syn og videnskabelige Tankesaet," in *Naturens Verden* (1938), 49–60; F. Rodolico, "L'evoluzione geologica della Toscana secondo N. Stenone," in *Memorie della Società toscana di scienze naturali*, **60**, ser. A (1953), 3–7; H. Schenk, "Applied Paleontology," in *Bulletin of the American Association of Petroleum Geologists* (1940), 1752; G. Scherz, "Niels Stensens Smaragdreise," in *Centaurus* (1955), 51–57; and H. Seifert, "Nicolaus Steno als Bahnbrecher der modernen Kristallographie," in *Sudhoffs Archiv für Geschichte der Medizin und der Naturwissenschaften* (1954), 29–47.

The more general works on Stensen and his work are A. Faller, "Die philosophischen Voraussetzungen des Anatomen und Biologen Niels Stensen," in *Arzt and Christ* (Salzburg, 1962); K. Larsen, "Stenos Forfold til Filosofi og Religion," in *Kirkehist. Saml.* (Copenhagen, 1938), 511–553; J. Nordström, "Antonio Magliabechi och Nicolaus Steno," in *Lychnos*, **20** (1962), 1–42; R. Rome, "Nicolas Sténon (1638–1686). Anatomiste,

etc.," in *Revue des questions scientifiques* (1956), 517–572; "Nicolas Sténon et la Royal Society," in *Osiris*, **17** (1956), 244–268; and G. Scherz, "Danmarks Stensen-Manuskript," in *Fund og Forskning* (Copenhagen, 1958–1959), 19–33; and "Niels Stensen's First Dissertation," in *Journal of the History of Medicine and Allied Sciences*, **15** (1960), 247–264.

GUSTAV SCHERZ

STEPANOV, VYACHESLAV VASSILIEVICH (*b.* Smolensk, Russia, 4 September 1889; *d.* Moscow, U.S.S.R., 22 July 1950), *mathematics.*

Stepanov was the son of Vassily Ivanovich Stepanov, who taught history and geography at high schools in Smolensk; his mother, Alexandra Yakovlevna, was a teacher at a girls' school. An honor graduate of Smolensk high school in 1908, Stepanov entered the department of physics and mathematics of Moscow University later that year; his scientific supervisor was Egorov. In 1912, when he was about to graduate, it was suggested that he remain at the university to prepare for a professorship. After spending some time at Göttingen, where he attended lectures by Hilbert and E. Landau, Stepanov returned to Moscow and became lecturer at Moscow University in 1915. At that time he published his first scientific work, an article on Paul du Bois-Reymond's theory of the growth of functions.

From the first Soviet years, Stepanov participated in the organization of new types of university work, especially in the training of young scientists at the Research Institute of Mathematics and Mechanics, established at Moscow University in 1921. He was director of the Institute from 1939 until his death. He was also one of the most influential and active leaders of the Moscow Mathematical Society, owing, among other things, to his exceptional erudition and memory. In 1928 Stepanov became a professor, and in 1946 he was elected corresponding member of the Academy of Sciences of the U.S.S.R.

Stepanov's scientific interests were formed first under the influence of Egorov and Luzin, founders of the Moscow school of the theory of functions of a real variable. In works published in 1923 and 1925 Stepanov established the necessary and sufficient conditions under which a function of two variables, defined on a measurable plane set of finite measure greater than zero, possesses a total differential almost everywhere on that set. These works laid the foundations for the studies of I. Y. Verchenko, A. S. Cronrod, and G. P. Tolstov in the theory of functions of n variables. In his most widely known works, Stepanov treated the theory of almost periodic functions, introduced a short time earlier by H. Bohr; he also constructed and investigated new classes of generalized almost periodic functions.

Stepanov's interest in applications of mathematics and his work at the State Astrophysical Institute in 1926–1936 led him to study the qualitative theory of differential equations. In this field his principal works are related to the general theory of dynamic systems that G. D. Birkhoff elaborated, extending the work of Poincaré. Besides writing articles on the study of almost periodic trajectories and on generalization of Birkhoff's ergodic theorem (which found an important application in statistical physics), Stepanov organized a seminar on the qualitative methods of the theory of differential equations (1932) that proved of great importance for the creation of the Soviet scientific school in this field.

BIBLIOGRAPHY

I. ORIGINAL WORKS. Stepanov's writings include "Über totale Differenzierbarkeit," in *Mathematische Annalen*, **90** (1923), 318–320; "Sur les conditions de l'existence de la différentielle totale," in *Matematicheskii sbornik*, **32** (1925), 511–527; "Über einige Verallgemeinerungen der fast periodischen Funktionen," in *Mathematische Annalen*, **95** (1925), 473–498; also in French in *Comptes rendus . . . de l'Académie des sciences*, **181** (1925), 90–94; "Über die Räume der fast periodischen Funktionen," in *Matematicheskii sbornik*, **41** (1934), 166–178, written with A. N. Tikhonov; "Sur une extension du théorème ergodique," in *Compositio mathematica*, **3** (1936), 239–253; *Kachestvennaya teoria differentsialnykh uravneny* (Moscow, 1947; 2nd ed., 1949), written with V. V. Nemytsky, translated into English as *Qualitative Theory of Differential Equations* (2nd ed., Princeton, 1960, 1964); and *Kurs differentsialnykh uravneny* ("Lectures on Differential Equations"; Moscow, 1936; 6th revised ed., 1953), translated into German by J. Auth *et al.* as *Lehrbuch der Differentialgleichungen* (Berlin, 1956).

II. SECONDARY LITERATURE. See P. S. Aleksandrov and V. V. Nemytsky, *Vyacheslav Vassilievich Stepanov* (Moscow, 1956); *Istoria otechestvennoy matematiki* ("History of Native Mathematics"), I. Z. Shtokalo, ed., III–IV (Kiev, 1968–1970), see index; *Matematika v SSSR za sorok let* ("Forty Years of Mathematics in the U.S.S.R."), 2 vols. (Moscow, 1959), see index; and *Matematika v SSSR za tridtsat let* ("Thirty Years of Mathematics in the U.S.S.R."; Moscow–Leningrad, 1948), see index.

A. P. YOUSCHKEVITCH

STEPHAN, ÉDOUARD JEAN MARIE (*b*. Ste.-Pezenne, Deux Sèvres, France, 31 August 1837; *d*. Marseilles, France, 31 December 1923), *astronomy*.

Stephan was admitted first in his class to the École Normale Supérieure in 1859. Upon graduating in 1862 he was invited by Le Verrier to the Paris observatory, where he learned observational techniques. At the same time he worked on his doctoral thesis, on second-order partial differential equations, which he defended in 1865.

Around this time Le Verrier had founded a branch of the Paris observatory at Marseilles. The city was selected as the appropriate site for the eighty-centimeter reflecting telescope, for which Léon Foucault had just constructed the mirror. In 1866 Stephan was assigned to equip and direct the observatory, and when it became independent of the Paris observatory in 1873, he was named director. In 1879 he became professor of astronomy at the University of Marseilles, holding both posts until his retirement in 1907. He became a corresponding member of the Bureau des Longitudes in 1875 and a corresponding member of the Académie des Sciences in 1879.

The scientists at the Marseilles observatory devoted their efforts primarily to the exploration of the sky. Stephan's collaborators, Alphonse Borrelly and J. E. Coggia, discovered a great number of asteroids and comets. Stephan directed his attention mainly to the search for nebulae and to the determination of their positions. He discovered approximately 350 of them, including a compact group known as "Stephan's quintet," which consists of five galaxies, one of which has a radial velocity very different from that of the others. The existence of this group poses two problems that are still unsolved: that of the instability of clusters of galaxies and that of abnormal red shifts. For his work as a whole, Stephan was awarded a prize by the Académie des Sciences in 1884.

The first to study stellar diameters, Stephan used a procedure suggested by Fizeau in 1868, in which the surface of the mirror of a telescope is masked by a screen, except for two separated areas that play the role of the slits in Young's experiment. The image of a point source is formed by the superposition of two groups of diffraction fringes; if the source has a perceptible apparent diameter, the fringes disappear when the distance between the areas attains a certain value, which is inversely proportional to this diameter. Stephan employed the procedure with the eighty-centimeter reflector in 1873 and observed the principal bright stars.

The fringes did not disappear, and he concluded that the stellar diameters were less than 0.16″. The experiment was repeated in 1920 by Michelson, whose determinations of diameters confirmed the upper limit obtained by Stephan.

Among his other accomplishments, Stephan contributed to the geodetic connection of Africa and Europe. In collaboration with Maurice Loewy and François Perrier, he determined the differences of longitude between Algiers, Marseilles, and Paris (1874–1876). In this undertaking the astronomical measurements were associated with telegraphic transmissions of time signals.

BIBLIOGRAPHY

I. ORIGINAL WORKS. For Stephan's observations of asteroids and comets, see "Discovery of a New Planet," in *Monthly Notices of the Royal Astronomical Society*, **27** (1867), 15; and "Comet I 1867, Discovered at Marseilles, 27 January 1867," ibid., 255; as well as thirty-five notes in the *Comptes rendus hebdomadaires des séances de l'Académie des sciences*, **63–128** (1866–1899). On the observations of eclipses, see "Voyage de la Commission française . . . éclipse totale de soleil du 19-8-1868," in *Annales scientifiques de l'École normale supérieure*, **7** (1870), 99–162; and four notes in the *Comptes rendus hebdomadaires des séances de l'Académie des sciences*, **99** (1884), **106** (1888), **130** (1900), and **136** (1903). On the positions and discoveries of nebulae, see "Nebulae Discovered at Marseilles," in *Monthly Notices of the Royal Astronomical Society*, **32–34** and **37** (1872–1877); the last article reports on "Stephan's quintet," which is catalogued under the numbers 19–22, corresponding to the objects NGC 7317, 7318 A and B, 7319, and 7320. See also "Note sur les nébuleuses découvertes à l'Observatoire de Marseille," in *Bulletin astronomique*, **1** (1884), 286–290; "Nébuleuses découvertes et observées à l'Observatoire de Marseille," in *Astronomische Nachrichten*, **105** (1883), 81–90, and **111** (1885), 321–330; and eleven notes in *Comptes rendus hebdomadaires des séances de l'Académie des sciences*, **74–100** (1872–1885). For his work on stellar diameters, see "Sur les franges d'interférence . . .," ibid., **76** (1873), 1008–1010; and "Extrême petitesse du diamètre apparent . . .," ibid., **78** (1874), 1008–1012.

Three other publications should be noted: "Équations aux dérivées partielles du second ordre," in *Annales scientifiques de l'École normale supérieure*, **3** (1866), 7–53, his dissertation; "Détermination de la différence des longitudes entre Paris et Marseille et Alger et Marseille," in *Travaux de l'Observatoire de Marseille*, **1** (1878), 1–214, written with M. Loewy; and "Notice sur l'Observatoire de Marseille," in *Bulletin astronomique*, **1** (1884), 122–132.

II. SECONDARY LITERATURE. See J. Bosler, "Édouard Stephan," in *Bulletin astronomique*, **4** (1924), 5–8, with portrait; G. Bigourdan, "Annonce de la mort de Stephan," in *Comptes rendus hebdomadaires des séances de l'Académie des sciences*, **178** (1924), 21–24; and M. Hamy, "La détermination interférentielle des diamètres des astres," in *Annuaire publié par le Bureau des Longitudes* (1919), B9–B15.

JACQUES R. LÉVY

STEPHANUS, CAROLUS. See **Estienne (Stephanus), Charles.**

STEPHANUS (or STEPHEN) OF ALEXANDRIA (*fl.* first half of seventh century A.D.), *philosophy, mathematics, astronomy, alchemy.*

Stephanus was a public lecturer in Constantinople at the court of Emperor Heraclius (A.D. 610–641). Although primarily a mathematician, he apparently also taught philosophy, astronomy, and music in addition to arithmetic and geometry. Commentaries on Aristotle have come down to us under his name, but Stephanus is also reported to have written on other subjects, including astronomy. He has been identified, probably incorrectly, by some authorities with Stephanus of Athens, a medical writer; and commentaries on Galen and Hippocrates have been attributed to both these authors. The *Opusculum apotelesmaticum*, ascribed to Stephanus of Alexandria but probably dating from the eighth century, deals with Islam in astrological terms.

Considerable attention has been given to a long Greek treatise on alchemy, *De chrysopoeia*, which has been ascribed to Stephanus and which was much praised by later alchemists. Consisting of nine mystical lectures, this uncritical, rhetorical, and theoretical document gives no evidence of experimental work. Indeed, in the first lecture the author writes, "Put away the material theory so that you may be deemed worthy to see the hidden mystery with your intellectual eyes." The work may be dated later than the seventh century, but it is mentioned in an Arabic bibliography of A.D. 987, *Kitāb-al-Fihrist*, where the author is known as Stephanus the Elder, who is said to have "translated for Khālid ibn Yāzid alchemical and other works." This Umayyad prince, much interested in science and especially alchemy, died in A.D. 704.

BIBLIOGRAPHY

I. ORIGINAL WORKS. See *Democritus Abderita, De arte magna, sive de rebus naturalibus. Nec non Synesii, et Pelagii, et Stephani Alexandri, et Michaelis Pselli in eundem commentaria*, Dominic Pizimentus, ed. (Padua, 1573); "De chrysopoeia," Julius L. Ideler, ed., in *Physici et medici graeci minores*, II (Berlin, 1842), 199–253; *Opusculum apotelesmaticum*, Hermann Usener, ed. (Bonn, 1879); "In librum Aristotelis de interpretatione commentarium," Michael Hayduck, ed., in *Commentaria in Aristotelem graeca*, XVIII, 3 (Berlin, 1885); "Anonymi et Stephani in artem rhetoricam commentaria," H. Rabe, ed., *ibid.*, XXI, 2 (Berlin, 1896); and "The Alchemical Works of Stephanos of Alexandria," trans. and commentary by F. Sherwood Taylor, in *Ambix, the Journal of the Society for Study of Alchemy and Early Chemistry*, 1 (1937), 116–139; 2 (1938), 38–49.

II. Secondary Literature. See Marcellin P. E. Berthelot, *Les origines de l'alchemie* (Paris, 1885), 199–201 and *passim*; *Introduction à l'étude de la chimie des anciens et du moyen age* (Paris, 1938), 287–301 and *passim*; Lucien Leclerc, *Histoire de la médecine arabe* (Paris 1876; New York, 1961); Hermann Usener, *De Stephano Alexandrino* (Bonn, 1880); Edmund O. von Lippmann, *Enstehung und Ausbreitung der Alchemie* (Berlin, 1919), 103–105; and George Sarton, *Introduction to the History of Science*, I (Baltimore, 1927), 472–473.

KARL H. DANNENFELDT

STEPHEN OF ANTIOCH (*fl.* first half of the twelfth century), *translation*.

According to a gloss on the *Diete universales* of Isaac Judaeus (Ishāq al-Isrā'īlī),[1] written by a certain Magister Mattheus F.,[2] Stephen of Antioch was a Pisan who went to Syria, learned Arabic, and translated the *Kitāb al-mālikī* of Haly Abbas ('Alī ibn al-'Abbās). This work, known more commonly in the medieval period as the *Liber regius*, was called by Stephen *Regalis dispositio*. Stephen undertook this task because, in his opinion, the text had been incompletely translated and grossly distorted by Constantine the African.[3] In order to prevent confusion between his own translation and that of Constantine, Stephen affixed his name to practically every one of the twenty books contained in the *Regalis dispositio*, adding the date on which each part of the work had been completed and several times naming the place where it had been made. Although the dates he gives are conflicting, the year 1127 is constant and can be reliably accepted. The place of translation, Antioch, has been questioned by some writers,[4] but in view of Stephen's remarks at the end of his *Synonima*, "these are the things we have found in Syria," and the description of him as "the nephew of the Patriarch of Antioch,"[5] there can be little doubt

that the place of translation is correct. The Latin patriarch of Antioch at that time was Bernard, previously bishop of Arethusa in Syria, who died in 1134.[6] Antioch also fits in with what we know of the Pisan contribution to the First Crusade and of the existence of a Pisan quarter in Antioch dating from 1108.[7]

In the prologue to the second part of the *Liber regalis* Stephen promised to assist his readers in their understanding of Arab materia medica by compiling a list of synonyms in three columns: Arabic, Latin, and Greek.[8] This list, which can be found in some manuscripts, does not appear in the printed editions and has been supplanted by an alphabetical list compiled by Michael de Capella.[9]

Stephen's probable connection with Salerno is indicated not only by his remark, made at the end of the *Synonima*, that if readers have difficulty in understanding the latinized Arab words they can consult Sicilian and Salernitan scholars who know both Greek and Arabic, but also by several quotations made from his work by Giovanni Platearius in his *Practica*[10] and by the Salernitan treatise *De aegritudinum curatione*.[11]

Stephen's accusation that Constantine the African was a plagiarist and a distorter of Haly Abbas' text has been uncritically accepted by later writers. A close comparison of the two translations reveals that Stephen's slavishly literal and verbose text closely follows its source, while Constantine's free paraphrase is easier to understand, gives better sense, and does great justice to the original.

Since Stephen says that his translation of the work of Haly Abbas was his first undertaking and that he intended to produce others, it has been suggested that he may be identical with a certain Stephanus Philosophus, who wrote several works on astronomy based on Arabic and Greek sources.[12] A comparison of the literary styles and vocabulary of their works makes the suggestion probable, but further investigation is needed.[13]

NOTES

1. W. M. Schum, *Beschreibendes Verzeichniss der Amplonianischen Handschriften-Sammlung zu Erfurt* (Berlin, 1887), 719; Hs. Amplon. O. 62ª, fols. 49–83v.

2. V. Rose, *Verzeichniss der lateinischen Hss. der königlichen Bibliothek zu Berlin*, II, pt. 3 (Berlin, 1905), 1059–1065, expands the letter *F* to Ferrarius.

3. *Liber totius medicine necessaria continens quem sapientissimus Haly filius abbas discipulus abimeher moysi filii seiar edidit: regique inscripsit. vnde et regalis dispositionis nomen assumpsit. Et a Stephano philosophie discipulo ex arabica lingua in latinam satis ornatam reductas . . . .* (Lyons, 1523), fol. 5r, col. 2.

4. See M. Steinschneider, in *Virchows Archiv für patholo-gische Anatomie und Physiologie und für klinische Medizin*, **39** (1867), 333.

5. B. M. Sloane MS 2426, fol. 8r.

6. P. B. G. Gams, *Series episcoporum* (Regensburg, 1873–1886), 433. For the bishopric of Arethusa in Syria, see M. Lequien, *Oriens christianus, in quattuor patriarchatus digestus; quo exhibentur ecclesiae, patriarchae, caeterique Praesules totius Orientis* (Paris, 1740), II, 915; III, 1190–1191.

7. V. Balaguer, *Historia de Cataluña y de la Corona de Ara-gon*, I (Barcelona, 1863), 620 ff.

8. *Liber regalis* (Lyons, 1523), fol. 136r, col. 1.

9. This list appears at the beginning of the 1523 edition and is not paginated.

10. Printed with *Practica Jo. Serapionis dicta breviarium* (Venice, 1503). The relevant passages can be found on fols. 180r, 180v, cols. 1 and 2, 183v, col. 1.

11. S. de Renzi, *Collectio Salernitana*, II (Naples, 1853), 266, 267, 270, 326–327, with the attribution to M[agister] Pla-tearius.

12. C. H. Haskins, *Studies in Mediaeval Science* (Cambridge, Mass., 1927), 99–103, 135.

13. R. Ganszyniec̆, "Stephanus de modo medendi," in *Archiv für Geschichte der Medizin*, **14** (1923), 110–113, claims him as the author of a text that appeared in a fifteenth-century MS of Cracow and tries, unconvincingly, to make him a pupil of Copho of Salerno.

CHARLES H. TALBOT

STEPLING, JOSEPH (*b.* Regensburg, Germany, 29 June 1716; *d.* Prague, Bohemia [now Czechoslovakia], 11 July 1778), *astronomy, physics, mathematics.*

Stepling's father came from Westphalia and was a secretary to the Imperial Embassy at Ratisbon (Regensburg). His mother's homeland was Bohemia. After his father's early death, the family moved to Prague. There Stepling began his studies at the Gymnasium run by the Jesuits. He soon demonstrated an extraordinary gift for mathematics, and a certain Father Sykora successfully endeavored to bring out his protégé's talent. When Stepling was only seventeen he calculated with great accuracy the lunar eclipse of 28 May 1733.

Despite a frail physical constitution, Stepling was admitted to the Jesuit order in 1733. After a biennial novitiate at Brno, he attended a three-year course of philosophy (1735 to 1738). His pupil, and later biographer, Stanislaus Wydra, states that Stepling, even in his early studies, transposed Aristotelian logic into mathematical formulas, thus becoming an early precursor of modern logic. Having already adopted the atomistic conception of matter (hyle), he radically refused to accept Aristotelian metaphysics and natural philosophy (the hylomorphic system). From 1738 to 1741 Stepling was a teacher at the Gymnasiums of Glatz (now Kłodzko) and Schweidnitz (now Świdnica). During the period 1741–1743, he devoted himself to special studies in mathematics, physics, and astronomy in Prague. From 1743 to 1747 he studied theology there and in 1745 took holy orders. The last year of his training began a tertianship (special studies of the law of the Jesuit order) in Gitschin (now Jičín), after which he declined a professorship of philosophy at the University of Prague in favor of the chair of mathematics. In 1748, at the request of the Berlin Academy, he carried out an exact observation of a solar and lunar eclipse in order to determine the precise location of Prague.

During Stepling's long tenure at Prague, he set up a laboratory for experimental physics and in 1751 built an observatory, the instruments and fittings of which he brought up to the latest scientific standard. In 1753 the Empress Maria Theresia, as part of her reform of higher education, appointed Stepling director of the faculty of philosophy at Prague. In this capacity he modernized the entire philosophical curriculum, which in those days embraced the natural sciences. He was particularly intent on cultivating the exact sciences, including physics and astronomy; and, following the example of the Royal Society in London, he founded a scientific study group. In their monthly sessions, over which he presided until his death, the group carried out research work and investigations in the field of pure mathematics and its application to physics and astronomy. A great number of treatises of this academy were published.

Stepling corresponded with the outstanding contemporary mathematicians and astronomers: Christian Wolf, Leonhard Euler, Christopher Maire, Nicolas-Louis de Lacaille, Maximilian Hell, Joseph Franz, Rudjer Bošković, Heinrich Hiss, and others. Also, Stepling was particularly successful in educating many outstanding scientists, including Johann Wendlingen, Jakob Heinisch, Johannes von Herberstein, Kaspar Sagner, Stephan Schmidt, Johann Körber, and Joseph Bergmann. After his death, Maria Theresia ordered a monument erected in the library of the University of Prague.

BIBLIOGRAPHY

I. ORIGINAL WORKS. Stepling's works include *Eclipsis lunae totalis Pragae octava Augusti 1748 observata* (Prague, 1748); *De actione solis in diversis latitudinibus* (Prague, 1750); *Exercitationes geometrico-analyticae de angulis aliisque frustis cylindrorum, quorum bases sunt sectiones conicae infinitorum generum* (Prague, 1751); *Observationes baroscopicae, thermoscopicae, hyetome-*

tricae (Prague, 1752); *De pluvia lapidea anni 1753 ad Strkow et ejus causis* (Prague, 1754); *Brevicula descriptio speculae astronomicae Pragae instructae* (Wittenberg, 1755); *De terrae motus causa discursus* (Prague, 1756); *Liber II. Euclidis algebraice demonstratus* (Prague, 1757); *Solutio directa problematis de inveniendo centro oscillationis* (Prague, 1759); *Contra insignem superficiei oceani et marium cum eo communicantium inaequalitatem a V. Cl. Henrico Kuehnio assertam* (Prague, 1760); *Beantwortung verschiedener Fragen über die Beschaffenheit der Lichterscheinung Nachts den 28 Hornungstage, und über die Nordlichter* (Prague, 1761); *De aberratione astrorum et luminis; item de mutatione axis terrestris historica relatio* (Prague, 1761); *Adnotationes in celebrem transitum Veneris per discum solis anno labente 6. Jun. futurum* (Prague, 1761); *De terrae motibus . . . adnexa est meditatio de causa mutationis Thermarum Töplicensium . . .* (Prague, 1763); *Vergleichungstafeln der altböhmischen Maasse und deren Preis mit den neu Oestreichischen und deren Preis* (Prague, 1764); *Differentiarum minimarum quantitatum variantium calculus directus, vulgo differentialis* (Prague, 1765); and *Clarissimi ac magnifici viri Iosephi Stepling . . . litterarum commercium eruditi cum primum argumenti* (Prague, 1782).

Stepling published many papers in *Nova acta eruditorum* (1750, 1761), and *Abhandlungen einer Privatgesellschaft in Böhmen zur Aufnahme der Mathematik, der vaterländischen Geschichte, und der Naturgeschichte* (1775–1784).

A bibliography appears in Poggendorff, II, 1004.

II. SECONDARY LITERATURE. On Stepling and his work, see Ludwig Koch, S.J., *Jesuitenlexikon* (Paderborn, 1934), cols. 1692–1693; Franz Martin Pelzel, *Boehmische, Maehrische und Schlesische Gelehrte und Schriftsteller aus dem Orden der Jesuiten* (Prague, 1786), 227–230; Carlos Sommervogel, S.J., *Bibliothèque de la Compagnie de Jésus*, VII (Brussels, 1896), cols. 1564–1568; Stanislaus Wydra, *Laudatio funebris Jos. Stepling coram senatu populoque academico . . . dicta* (Prague, 1778); *Vita Admodum Reverendi ac magnifici viri Iosephi Stepling* (Prague, 1779); and *Oratio ad monumentum a Maria Theresia Aug. Josepho Stepling erectum . . .* (Prague, 1780); and *Abbildungen böhmischer und Mährischer Gelehrter und Künstler nebst kurzen Nachrichten von ihren Leben und Werken*, IV (Prague, 1782), 164–172, which appeared without the name of the author but has the printed signature "Franz Martin Pelzel und die übrigen Verfasser" on the dedication page.

D. ANTON PINSKER, S.J.

STERN, OTTO (*b*. Sohrau, Upper Silesia, Germany [now Zory, Poland], 17 February 1888; *d*. Berkeley, California, 17 August 1969), *physics.*

Stern was the oldest of five children (two sons and three daughters) of Oskar Stern and Eugenie

Rosenthal. Before he reached school age, the family moved to Breslau (now Wrocław, Poland), where Otto received his primary and secondary education at the Johannes Gymnasium. After graduating in 1906, he continued his studies at the universities of Freiburg im Breisgau, Munich, and Breslau, from which he received the Ph.D. in physical chemistry in 1912.

Stern's parents belonged to a prosperous Jewish family of grain merchants and flour millers who were content to let their children satisfy their thirst for knowledge without immediate, professional goals. Even while attending the Gymnasium, which emphasized the classics at the expense of mathematics and the sciences, Stern supplemented his education by perusing various books that his father put at his disposal; and during his university studies, he explored several fields of science before deciding on a career. This approach to learning was in accordance with German academic tradition in the period before World War I, when young men of means could migrate from one university to another and attend lectures on a variety of subjects without regard to curricula or to the time needed for completion of promotion requirements. Thus, Stern attended lectures on theoretical physics by Arnold Sommerfeld, one of the most brilliant lecturers of his generation, and on experimental physics by Otto Lummer and Ernst Pringsheim, both of whom were famous for their elegant research on blackbody radiation.

Stern's real interest, however, was aroused more by his private readings than by his formal studies. The books of Boltzmann on molecular theory and statistical mechanics and of Clausius and Nernst on thermodynamics appear to have greatly influenced his choice of career. Returning to Breslau to complete his university studies, Stern decided to major in physical chemistry because two professors in that department, R. Abegg and O. Sackur, were more closely concerned with his interest in thermodynamics and molecular theory than were the professors of physics. His doctoral dissertation on the osmotic pressure of carbon dioxide in concentrated solutions was both theoretical and experimental and set the style for his future research, which he himself later described as that of an "experimenting theoretician."

Stern's scientific activity can be divided into two distinct periods: the theoretical (1912–1919) and the experimental (1919–1945). During the first period, he was strongly influenced by his contacts with Einstein, whom he joined as a postdoctoral associate in Prague immediately after graduating from Breslau and with whom he moved to Zurich in 1913; by Ehrenfest and Laue, with whom he became acquainted in Zurich; and finally, but to a lesser extent, by Max Born, with whom he began to work after his return to Frankfurt in 1919. During these years, he took advantage of his financial independence, which allowed him to select a place of work without regard to the availability of a remunerative position. He received the *venia legendi* at the Eidgenössische Technische Hochschule in Zurich in 1913 and transferred it to the University of Frankfurt in 1914, thus achieving the status of *Privatdozent*, which carried the right to lecture in the university without salary. From the outbreak of the war in August 1914 until the German defeat in 1918, he served in the German army, first as a private and later as a noncommissioned officer, in various technical assignments. After his demobilization he returned to Frankfurt.

This period can best be described as Stern's *Lehr- und Wanderjahre*. Its most important result was not the production of scientific papers, although Stern's papers published during these years were by no means negligible, but the development of a certain attitude toward the selection of research problems. As Stern told this writer, he was less attracted to Einstein because of his spectacular achievement in formulating the theory of relativity than by his work in molecular theory, particularly the application of the then imperfectly understood quantum concepts to the explanation of the curious temperature behavior of the specific heat of crystalline bodies.

An early paper published with Einstein contributed to one aspect of this problem, namely, the question of the existence of the so-called zero-point energy: Are the atoms of a body at rest at the absolute zero of temperature, or do they oscillate around their equilibrium positions with an energy of $h\nu/2$? But what Stern really learned from Einstein was the evaluation of the importance of current physical problems, which questions to ask, and what experiments should be undertaken at a given time. His association with Einstein developed into a lifelong friendship and planted the seed for the major accomplishments of Stern's later career, culminating in his winning the Nobel Prize in physics for 1943.

Stern's work during the years 1912–1918 was concerned with various problems in statistical thermodynamics. Two papers published during this period are worthy of mention, one because of its scientific merit and the other because of the unusual circumstances of its origin. The first dealt with the absolute entropy of a monoatomic gas. The expression for the entropy of a gas obtained by classical theories contains an arbitrary constant that cannot be computed but that greatly affects such properties as the vapor pressure of a solid or the chemical equilibrium of reacting gases. The importance of this constant had been pointed out by Nernst in the formulation of the third law of thermodynamics (also known as Nernst's theorem).

It was fairly obvious that quantum theory held the key to the solution of this problem, but the methods for the application of quantum concepts to a perfect gas had not yet been discovered. Otto Sackur and Tetrode had already published papers giving a theoretical expression for the entropy constant; and while their result was correct, their derivation was open to justifiable criticism. Stern avoided the need of applying quantum theory to a gas by considering the equilibrium of a solid crystal with its vapor at a high temperature. Under these conditions it was perfectly correct to use classical statistical mechanics for the gas and to apply quantum concepts only to the solid, where the theory provided the necessary guidance. Using Einstein's theory of the specific heat and Nernst's theorem, Stern obtained the same result as Sackur and Tetrode, but this time the derivation was unobjectionable.

The second paper carries the dateline "Lomsha, Russian Poland," probably the only scientific paper that ever originated in this small town. Stern was stationed there during 1916 as a military weather observer, and his duties of recording twice daily the readings of a few instruments left him plenty of free time. To escape boredom he tackled the tedious problem of calculating the energy of a system of coupled mass points—doing all the computations in longhand.

During the last years of the war, many German physicists and physical chemists were assigned to military research at Nernst's laboratory at the University of Berlin. There Stern met James Franck and Max Volmer, both excellent experimenters; and it is very probable that his shift from theoretical to experimental work was due to the influence of these two scientists, who also became his lifelong friends.

After returning to Frankfurt, Stern continued to work on similar theoretical problems; and he published a paper on the surface energy of solids with Max Born, director of the Institute for Theoretical Physics, to which Stern was attached. Shortly thereafter, Stern felt compelled to provide experimental proof for the fundamental concepts used in molecular theory. For this purpose, he began to develop the molecular-beam method. In 1911 Dunoyer had shown that atoms or molecules introduced into a high-vacuum chamber travel along straight trajectories, forming beams of particles that in many respects are similar to light beams. His work had been practically forgotten until Stern realized that it was a very powerful tool for the investigation of properties of free atoms. Thus began the second period of Stern's scientific career, which secured for him a place of honor in the history of physics.

The first application of this method was still concerned with molecular theory, namely, the measurement of molecular velocities in a gas. These quantities had been computed theoretically around 1850; and although the result had been generally accepted, no one had succeeded in providing experimental proof. In 1919 Stern performed an elegant experiment, using beams of silver atoms, and confirmed the theoretical values within the limits of experimental error. Although a nice achievement, it was not exactly world-shaking. Its real importance was that it demonstrated the usefulness of the method and encouraged its further application. (Einstein's teaching of how to recognize the really important problems is clearly visible in this work.)

At that time, Bohr's theory of the atom had undergone rapid development, particularly in the hands of Sommerfeld, who concluded that certain atoms—for example, those of hydrogen, the alkali metals, or silver—should possess magnetic moments of the magnitude of

$$M = \frac{eh}{4\pi mc},$$

where e is the electronic charge, m is the mass of an electron, c is the velocity of light, and h is Planck's constant. Moreover, if such an atom were placed in a magnetic field, it should be able to assume only two distinct orientations, with its axis and magnetic moment either parallel or opposed to the direction of the field. (A third possible orientation—magnetic moment perpendicular to the field—was forbidden by a special selection rule.) While the first conclusion was at least compatible with the classical theory, the second was not, and very few physicists of that time were inclined to take this spatial quantization seriously. Stern recognized that the molecular-beam method was capable of giving a clear yes-or-no answer to this question: If the classical theory were correct, a narrow beam of silver atoms should be broadened

when passing through a nonhomogeneous magnetic field; but if the spatial quantization theory were correct, the beam should be split into two separate beams.

In 1920 this experiment, although simple in concept, was difficult to perform. Not particularly skillful in handling experimental techniques (as opposed to designing experiments), Stern asked Walther Gerlach, a colleague at the Institute for Experimental Physics in Frankfurt, to join in this work. Together they succeeded in proving the reality of space quantization and in measuring the magnetic moment of the silver atom. The five papers reporting this work, which soon became known as the Stern-Gerlach experiment, received wide attention and established Stern's rank among physicists.

In 1921, when these experiments were practically completed, Stern received his first formal academic appointment as associate professor of theoretical physics at the University of Rostock. There he was joined by this writer, who had just completed his doctorate under Stern's friend Max Volmer. At that time, an appointment in Rostock was mainly a stepping-stone to greener pastures, and it was not long before Stern received a call to the University of Hamburg as professor of physical chemistry and as director of the Institute for Physical Chemistry, still to be erected. In the meantime, Stern and this writer, who were later joined by a few additional assistants, postdoctoral guests, and graduate students, were assigned temporary quarters in the Institute for Experimental Physics.

The period 1923–1933 marks the peak of Stern's contributions to physics. Shortly after assuming his post at Hamburg on 1 January 1923, he set out to organize a laboratory specially equipped for molecular-beam research and to devise a program for conducting this research, which was executed, to a large degree, with remarkable success. The first part of the program was concerned with completing and expanding Stern's previous work and with developing new and improved techniques; the second, with demonstrating the wave nature of particles—a revolutionary assumption introduced in 1924 by Louis de Broglie that became the foundation of modern quantum mechanics; and the third, with measuring the magnetic moment of the proton and deuteron. The significance of the work on the wave nature of particles is similar to that of the Stern-Gerlach experiment: each provided unambiguous, direct, and thoroughly convincing proof of revolutionary concepts introduced into the foundations of physics. These experiments were essential for the acceptance of new ideas that had previously been regarded with considerable skepticism.

The last part of Stern's program had a completely different outcome. Dirac had promulgated a theory according to which the ratio of the magnetic moment of the proton to that of the electron should have been the same as the inverse ratio of their masses. This theory was believed so generally that when Stern, O. R. Frisch, and this writer began the very difficult experiments, they were told more than once by eminent theoreticians that they were wasting their time and effort. But Stern's perseverance paid off. Measurements showed a proton magnetic moment two or three times larger than expected. While that result has since been reproduced with greater accuracy, a really satisfactory theoretical explanation is still outstanding. It is this work that was specifically mentioned in Stern's Nobel Prize citation.

With the advent of the Nazi regime in 1933, the work in Hamburg came to an abrupt end. Several of Stern's closest collaborators, who happened to be of Jewish origin, were summarily dismissed, and in protest Stern submitted his resignation before his own foreseeable dismissal. Stern and this writer received invitations to come to the United States, to the Carnegie Institute of Technology, where they began to build a molecular-beam laboratory. Stern was appointed research professor of physics, but the means put at his disposal during the depression years were rather meager. The momentum of the Hamburg laboratory was never regained, although a number of significant papers originated at Carnegie. Stern retired in 1946 to Berkeley, California, where he continued to maintain some contact with local physicists but shunned public appearances. Stricken by a heart attack, he died on 17 August 1969 at the age of eighty-one.

Stern was a member of both the National Academy of Sciences and the American Philosophical Society in 1945 after having received the Nobel Prize. He was also a member of the Royal Danish Academy and received honorary doctorates from the University of California and the Eidgenössiche Technische Hochschule in Zurich.

BIBLIOGRAPHY

I. ORIGINAL WORKS. A complete bibliography of the papers published by Stern and his associates between 1926 and 1933 is listed in Estermann (see below). His early paper on the absolute entropy of a monoatomic gas is "Zur kinetischen Theorie des Dampfdrucks einatomiger fester Stoffe und über die Entropiekonstante einatomiger Gase," in *Physikalische Zeitschrift*, **14** (1913), 629–632; the memoir on the energy of a system of coupled mass points is "Die Entropie fester Lösungen," in *Annalen der Physik*, 4th ser., **49** (1916), 823–841. Subsequent works cited above are "Über die Oberflächenenergie der Kristalle und ihren Einfluss auf die Kristallgestalt," in *Sitzungsberichte der Preussischen Akademie der Wissenschaften zu Berlin* (1919), 901–913, written with Max Born; and "Eine direkte Messung der thermischen Molekulargeschwindigkeit," in *Zeitschrift für Physik*, **2** (1920), 49–56, and **3** (1920), 417–421.

The results of the Stern-Gerlach experiment were reported in "Ein Weg zur experimentellen Prüfung der Richtungsquantelung im Magnetfeld," *ibid.*, **7** (1921), 249–253; "Der experimentelle Nachweis des magnetischen Moments des Silberatoms," *ibid.*, **8** (1921), 110–111, written with W. Gerlach; "Der experimentelle Nachweis der Richtungsquantelung im Magnetfeld," *ibid.*, **9** (1922), 349–352, written with Gerlach; "Das magnetische Moment des Silberatoms," *ibid.*, 353–355, written with Gerlach; and "Über die Richtungsquantelung im Magnetfeld," in *Annalen der Physik*, 4th ser., **74** (1924), 673, written with Gerlach.

Stern's Nobel Prize lecture, "The Method of Molecular Rays," is in *Nobel Lectures 1942–1962 (Physics)* (Amsterdam–London–New York, 1964), 8–16, with biography on 17–18.

II. SECONDARY LITERATURE. See I. Estermann, "Molecular Beam Research in Hamburg, 1922–1933," in his *Recent Research in Molecular Beams* (New York–London, 1959), 1–7, with bibliography.

I. ESTERMANN

STERNBERG, KASPAR MARIA VON (*b.* Prague, Bohemia [now in Czechoslovakia], 6 January 1761; *d.* Březina castle, Radnice, 20 December 1838), *botany, geology, paleontology.*

Sternberg was the scion of an old landed family that took its title from Sternberg castle in the Sázava valley forty-five kilometers southeast of Prague. His eldest brother, Joachim, was fond of mathematics; he also founded an iron factory and took a keen interest in mining and metallurgy, on which he wrote prolifically. Strongly influenced by the activities and philanthropy of his family, Sternberg studied theology at first privately and later at Rome. He then pursued a celibate ecclesiastical career, being appointed successively canon and counselor to the court of the prince-bishop of Regensburg. In 1791 he was nominated counselor to the court at Freisingen.

His duties at Regensburg involved the control of

woods and forests and induced him to study botany and later to found a botanical garden there. In 1805 and 1806 he accompanied the prince-bishop to Paris, where he met many prominent scientists including Faujas de Saint-Fond, who initiated him in the study of fossil plants. Here he received and studied carefully, with the aid of collections in Paris, Ernst von Schlotheim's book (1804) on fossil plant impressions in Coal Measures. In 1808 he inherited the family estate at Radnice with Březina castle and thereafter devoted himself to botanical studies and the promotion of natural science in Bohemia. When, shortly before his death thirty years later, he presided over a large congress of naturalists at Prague, he and Brongniart were recognized as the two leading paleobotanists in the world.

Sternberg studied especially the Carboniferous phytopaleontology but he also published some papers dealing with the trilobites and Pleistocene fauna. He had ready access to fossils and fossil impressions found in the coalfields on his own estate and in the "transitional rocks" near Prague. He tried to interpolate the species of fossil plants into the botanical system by discarding the old names given to the fossil forms and applying existing botanical correlations to them. Thereby he greatly increased the proper botanical significance of fossil floras and paved the way for a scientific treatment of paleobotany. His chief work, the seven-volume *Versuch einer geognostisch-botanischen Darstellung der Flora der Vorwelt* (1820–1833), described two hundred fossil species of plants with the aid of sixty folio plates. The ideas expressed in this work expanded those of Ernst von Schlotheim but were based on a narrower range of material than that contained in the contemporary publications of Brongniart, one of the Paris naturalists who helped to turn Sternberg to botanical pursuits. William Buckland sized up the situation fairly when he wrote: "We owe to the labours of Schlotheim, Sternberg and Ad. Brongniart the foundation of such a systematic arrangement of fossil plants, as enables us to enter, by means of the analogies of recent plants, into the difficult question of the Ancient Vegetation of the Earth, during those periods when the strata were under the process of formation" (*Geology and Mineralogy*, I [London, 1837], 454).

Sternberg's name is commemorated in the technical terminology for fossil organisms, both animal and vegetable, *Sternbergia, Sternbergella,* and *Parasternbergella*. The plants include several small crocuslike species native to Europe, such as *Sternbergia lutea*. The mineral sternbergite is a natural sulfide of silver and iron ($AgFe_2S_3$) crystallized in orthorhombic prisms and first discovered in the mines at Joachimsthal (Jáchymov), Bohemia.

In 1818–1821 Sternberg was one of the chief founders of the Bohemian National Museum, Prague. He acted as its president from 1822 to his death and bequeathed his library and his geological and botanical collections to it. He had close dealings with Goethe and their correspondence has been published. Today his botanical, geological, and paleontological collections remain in the National Museum. His written works are in the Museum of National Literature, Prague, and his diplomas, correspondence, and other literary remains are in the department of the State Archives at Benešov.

BIBLIOGRAPHY

I. ORIGINAL WORKS. Sternberg wrote all his books and articles in German. F. Palacký, *Leben des Grafen K. Sternberg* . . . (see below), contains a bibliography with 74 titles, among which the most important are *Galvanische Versuche in manchen Krankheiten; herausgeben mit einer Einleitung in Bezug auf die Erregungstheorie von J.-U.-G. Schaeffer* (Regensburg, 1803); *Botanische Wanderungen in den Böhmerwald* . . . (Nuremberg, 1806); *Reisen in die rhaetischen Alpen, vorzüglich in botanischer Hinsicht* (Nuremberg–Regensburg, 1806); *Reise durch Tyrol in die Oesterreichischen Provinzen Italiens in Frühjahr 1804* . . . (Regensburg, 1806); *Revisio saxifragarum iconibus illustrata* (Regensburg, 1810, 1822); *Abhandlung über die Pflanzenkunde in Böhmen*, 2 vols. (Prague, 1817–1818); *Versuch einer geognostisch-botanischen Darstellung der Flora der Vorwelt* . . ., 7 vols. (Leipzig–Prague, 1820–1833); and *Umrisse einer Geschichte der böhmischen Bergwerke*, 2 vols. (Prague, 1836–1838). His correspondence with Goethe was published as *Briefwechsel zwischen Goethe und Kaspar Graf von Sternberg (1820–1832)*, F. T. Bratranek, ed. (Vienna, 1866).

II. SECONDARY LITERATURE. The chief biographies are F. Palacký, *Leben des Grafen K. Sternberg* . . . *nebst einem akademischen Vortrag über der Grafen K. und F. Sternberg Leben und Wirken* . . . (Prague, 1868), and V. Zázvorka, "Kašpar Maria hrabě Šternberk, jeho život a význam," in *Zvláštni otisk z časopisu Národniho musea*, **113** (Prague, 1939), 1–22, with portraits. See also W. Whewell, in *Proceedings of the Geological Society* (London), **3** (1838–1842), 72–74; and C. von Wurzbach, *Biographisches Lexikon des kaiserthums Oesterreich*, XXXVIII (Vienna, 1879), 252–266.

ROBERT P. BECKINSALE
JAN KREJČÍ

STERNBERG, PAVEL KARLOVICH (*b.* Orel, Russia, 2 April 1865; *d.* Moscow, U.S.S.R., 31 January 1920), *astronomy*, *gravimetry*.

The son of a petty tradesman, Sternberg became interested in astronomy as a child; at fifteen, having received a spyglass as a gift, he converted it into a telescope and began observations. After graduating from the Orel Gymnasium in 1883, he entered the Faculty of Physics and Mathematics of Moscow University. In his first year he began regular observations at Moscow observatory. Its director, F. A. Bredikhin, attracted Sternberg to his program of cometary research. In his final year Sternberg received the gold medal of the faculty for his paper "O prodolzhitelnosti vrashchenia Krasnogo pyatna Yupitera" ("On the Duration of Rotation of the Red Spot of Jupiter"). After graduating in 1887 with the degree of candidate in mathematical sciences, he remained at the university to prepare for an academic career. In March 1888 he became a supernumerary assistant at the observatory and subsequently participated in Bredikhin's gravimetric expeditions in European Russia.

In 1890 Sternberg was confirmed as astronomer-observer at Moscow observatory. In 1903, after defending his thesis, "Shirota Moskovskoy observatorii v svyazi s dvizheniem polyusov" ("The Latitude of the Moscow Observatory in Relation to the Motion of the Poles"), he was awarded the master's degree. Ten years later, for his dissertation "Nekotorye primenenia fotografii k tochnym izmereniam v astronomii" ("Certain Uses of Photography for Precise Measurements in Astronomy"), Sternberg received the doctorate in astronomy. In July 1916 he succeeded V. K. Tserasky as director of the university observatory.

Sternberg's teaching career had begun in 1887 at the private Kreiman Gymnasium, where he taught physics until 1909. In 1890, as a *Privatdozent* at the University of Moscow, Sternberg gave his first course, on the general theory of planetary perturbations, which was later followed by courses in celestial mechanics, advanced geodesy, spherical astronomy, and descriptive astronomy. From 1901 to 1917 Sternberg was on the staff of the higher courses for women, organized in St. Petersburg, Moscow, and Kiev by groups of progressive professors in the 1870's because tsarist law did not permit women to study at the universities. In 1914 he was elected extraordinary professor of astronomy and geodesy, and the following year he was named honored professor.

From 1905 Sternberg was an active Social Democrat (Bolshevik) and was a member of the clandestine Military-Technical Bureau, which undertook technical preparation for an armed uprising. Only after February 1917 was he openly involved in the political struggle, and in October 1917 he led the revolutionary forces of the Zamoskvoretsky district of Moscow. After the establishment of Soviet power in Moscow, Sternberg was named military commissar. Later he became a member of the Board of the People's Commissariat of Education and was active in preparing for the reform of higher education. In the fall of 1918 he traveled to the Eastern Front as a political commissar of the Second Army, and in 1919 he covered the entire Eastern Front. Having caught a cold during the forced crossing of the Irtysh River near Omsk, he became seriously ill and was evacuated to Moscow, where after two operations he died.

Sternberg was a distinguished researcher and innovator in three fields: gravimetry, the variations of latitude in relation to the motion of the earth's poles, and photographic astrometry.

His numerous gravimetric expeditions, lectures, and practical studies with students aided the formation of a scientific school in Moscow, which after Sternberg's death carried out a wide-ranging gravimetric study of the Soviet Union, development of the theory, and gravimetric prospecting for useful fossils.

During many years of measuring the latitude of the Moscow observatory, Sternberg carefully and precisely investigated the motion of the poles and recommended the organization of a network of stations forming an "international service of polar altitude." This has become the International Latitude Service.

As a pioneer in the use of photography for precise measurements of stellar position, Sternberg used the fifteen-inch double astrograph of the Moscow observatory for an extremely comprehensive study of the possibilities of the new method and the possible sources of systematic errors.

In April 1917, Sternberg was elected president of the All-Russian Congress of Astronomers. In 1931 his name was given to the astronomical institute at Moscow University, which brought together three scientific institutions, including the university observatory.

BIBLIOGRAPHY

I. ORIGINAL WORKS. Sternberg's writings include "Sur la durée de la rotation de la tache rouge de Jupiter," in *Annales de l'observatoire astronomique de Mos-*

cou, 2nd ser., **1**, no. 2 (1888), 91–128; "Observations faites à l'aide du pendule à réversion de Repsold," *ibid.*, **2**, nos. 1–2 (1890), 94–132; "Observations photographiques de l'étoile γ Virginis," *ibid.*, **3**, no. 1 (1893), 120–123; "Dvizhenia zemnykh polyusov" ("The Motion of the Earth's Poles"), in *Nauchnoe slovo*, **3** (1903), 112–115; "Shirota Moskovskoy observatorii v svyazi s dvizheniem polyusov" ("The Latitude of the Moscow Observatory in Relation to the Motion of the Poles"), *Uchenye zapiski Moskovskogo universiteta*, Fiz.-mat. otd. (1904), no. 22; "Application de la photographie aux mesures des étoiles doubles," in *Annales de l'observatoire astronomique de Moscou*, 2nd ser., **5** (1911), 42–71; "Nekotorye primenenia fotografii k tochnym izmereniam v astronomii" ("Some Applications of Photography to Precise Measurements in Astronomy"); *Bulletin de la Société des naturalistes de Moscou* for 1913, nos. 1–3 (1914), his doctoral dissertation; *Kurs sfericheskoy astronomii . . .* ("Course in Spherical Astronomy . . ."; Moscow, 1914), compiled by a student with the initials L. D. from lectures given by Sternberg at Moscow University in the academic year 1913–1914; *Kurs opisatelnoy astronomii* ("Course in Descriptive Astronomy"; Moscow, 1915), compiled by A. A. Sokolov from a course of lectures by Sternberg; "Détermination relative de la pesanteur à Moscou," in *Annales de l'observatoire astronomique à Moscou*, 2nd ser., **8**, no. 1 (1925), 3–17; and "Sila tyazhesti v Moskovskom rayone i ee anomalii" ("The Force of Gravitation in the Moscow Region and Its Anomaly"), *ibid.*, nos. 2–3 (1926), 43–83, observations of Sternberg, edited by I. A. Kazansky.

II. SECONDARY LITERATURE. See S. N. Blazhko, "Pamyati P. K. Sternberga" ("Recollections of Sternberg"), in *Mirovedenie*, **25**, no. 2 (1936), 81–89; O. A. Ivanova, "Moskovskoe voenno-tekhnicheskoe byuro RSDRP (1906–1907)" ("Moscow Military-Technical Bureau . . ."), in *Moskva v trekh revolyutsiakh* ("Moscow in Three Revolutions"; Moscow, 1959); 108–127; Yu. G. Perel, ed., in S. N. Blazhko, *Vydayushchiesya russkie astronomy* ("Outstanding Russian Astronomers"; Moscow, 1951), 141–175; K. A. Timiryazev, "Ucheny-geroy. Pamyati P. K. Sternberga" ("A Scientist-Hero. Recollections of Sternberg"), in his *Sochinenia* ("Works"), IX (Moscow, 1939), 415–424; and P. G. Kulikovsky, *Pavel Karlovich Sternberg. 1865–1920* (Moscow, 1965).

P. G. KULIKOVSKY

STEVENS, EDWARD (*b.* St. Croix, Virgin Islands, *ca.* 1755; *d.* St. Croix, 26 September 1834), *medicine, physiology.*

Stevens' father, Thomas, a prosperous merchant, was reputedly also the father of Alexander Hamilton. Nothing is known of his mother. In his youth Stevens moved with his family to New York. He

graduated A.B. from King's College (now Columbia University) in 1774; and the following year he began studies at the University of Edinburgh, enrolling in the medical school in 1776 and again in 1777. He graduated M.D. on 12 September 1777.

Stevens' inaugural dissertation, "De alimentorum concoctione," presented with ingenuity and insight his experiments and observations on gastric digestion, and clearly confirmed him as the first investigator to isolate human gastric juice. It removed the confusion and contradictions presented in the doctrines of fermentation and trituration, the latter championed by Leeuwenhoek, Borelli, Pitcairn, and Pecquet, and decried by Astruc and Stephen Hales. It also repudiated such views as those of John Pringle and David Macbride. Stevens' work formed a vital bridge linking the experiments of Réaumur before him and Spallanzani and later workers after him. Réaumur had shown, in 1752, that digestion was due to the solvent power of gastric juice. Stevens confirmed this, isolated human gastric juice, and performed experiments both *in vitro* and *in vivo* in man and animals.

Stevens was admitted to the Royal Medical Society (Edinburgh) on 20 January 1776 and served as its president in 1779 and 1780. At Edinburgh he was awarded the Harveian prize for an experimental inquiry on the red color of the blood. He returned to St. Croix about 1783 and practiced medicine there for ten years. In 1793 Stevens moved to Philadelphia, where he received public support from Alexander Hamilton and became embroiled in a controversy with Benjamin Rush over methods of treating yellow fever in the great epidemic of that year. On 18 April 1794 he was admitted to the American Philosophical Society, and the following year he was appointed professor of the practice of medicine in King's College (later Columbia University). It is probable that Stevens' presence and reputation in Philadelphia, as well as his contributions in gastric physiology, contributed to the marked interest in gastric studies that took place round the turn of the century in that city. Of these studies, that of John R. Young is best-known. He undoubtedly was familiar with Stevens' work; indeed, his experiments with bullfrogs and small frogs are reminiscent of Stevens' observations of partially digested small fish inside larger ones.

Stevens was United States consul-general to Santo Domingo from 1799 to 1800. His consular dispatches to Timothy Pickering, Adams, Jefferson, and other leaders, revealing a critical, observant mind, outlined the geopolitical problems facing the United States in the Caribbean at that time. Controversy marred his political life, however, and he returned to the United States in 1801. He made appearances at the American Philosophical Society meetings in 1803 and 1804, probably returning to St. Croix shortly thereafter. Little is known of Stevens' last years. David Hosack wrote to him in St. Croix on yellow fever in 1809, and in 1823 he wrote Hosack a letter introducing his son, who had also graduated at Edinburgh.

Stevens' fundamental and sound gastric studies were confirmed by Spallanzani, who augmented and added to them in masterly fashion, assuring that from then on, gastric physiology would be a well-founded science.

BIBLIOGRAPHY

I. ORIGINAL WORKS. "Dissertatio inauguralis de alimentorum concoctione" (1777) is transcribed, in the original Latin, in *Thesaurus Medicus*, III (Edinburgh, 1785), a selection of medical dissertations from Edinburgh, repr. by William Smellie. An incomplete English trans. of the experimental section is appended to Spallanzani's *Dissertations Relative to the Natural History of Animals and Vegetables* (London, 1784). In 1778 Stevens read a paper to the Royal Medical Society of Edinburgh entitled "What Is the Cause of the Increase of Weight Which Metals Acquire During Their Calcination." A MS copy is in the library of Edinburgh University. No copy appears to have survived of Stevens' Harveian prize thesis, "An Experimental Inquiry Concerning the Red Colour of the Blood."

II. SECONDARY LITERATURE. Stevens has received remarkably little attention. Stacey B. Day appears to be the only one who has endeavored to piece together his life and to correct the fragmentary and incorrect notes that are commonly found. See Stacey B. Day and Roy A. Swanson, "The Important Contribution of Dr. Edward Stevens to the Understanding of Gastric Digestion in Man and Animals," in *Surgery*, **52**, no. 5 (1962), 819–836. The most comprehensive account available is Stacey B. Day, *Edward Stevens, Gastric Physiologist, Physician and American Statesman, With a Complete Translation of His Inaugural Dissertation De Alimentorum Concoctione and Interpretive Notes on Gastric Digestion Along with Certain Other Selected and Diplomatic Papers* (Cincinnati–Montreal, 1969), which presents most of the biographical details of Stevens' life known today. Possibility of his kinship with Alexander Hamilton is examined. The book presents the first complete English trans. of "De alimentorum concoctione"; a trans. of the German précis by Friedrich August Weiz (1782); the exchange of letters between Stevens and Benjamin Rush; the controversy between these two

physicians over the treatment of yellow fever; and a review of Stevens' role as United States consul-general in Santo Domingo. The book provides reference sources: Jefferson MSS, Pickering MSS, Hamilton Papers, Stephen Girard Papers, National Archives records, and source materials in Edinburgh and elsewhere.

STACEY B. DAY

STEVIN, SIMON (*b.* Bruges, Netherlands [now Belgium], 1548; *d.* The Hague, Netherlands, *ca.* March 1620), *mathematics, engineering.*

Stevin was the illegitimate son of Antheunis Stevin and Cathelijne van de Poort, both wealthy citizens of Bruges. There is little reliable information about his early life, although it is known that he worked in the financial administration of Bruges and Antwerp and traveled in Poland, Prussia, and Norway for some time between 1571 and 1577. In 1581 he established himself at Leiden, where he matriculated at the university in 1583. His religious position is not known, nor is it known whether he left the southern Netherlands because of the persecutions fostered by the Spanish occupation. At any rate, in the new republic of the northern Netherlands Stevin found an economic and cultural renaissance in which he at once took an active part. He was first classified as an "engineer," but after 1604 he was quartermaster-general of the army of the States of the Netherlands. At the same time he was mathematics and science tutor to Maurice of Nassau, prince of Orange, for whom he wrote a number of textbooks. He was often consulted on matters of defense and navigation, and he organized a school of engineers at Leiden and served as administrator of Maurice's domains. In 1610 he married Catherine Cray; they had four children, of whom one, Hendrick, was himself a gifted scientist who, after Stevin's death, published a number of his manuscripts.

Stevin's work is part of the general scientific revival that resulted from the commercial and industrial prosperity of the cities of the Netherlands and northern Italy in the sixteenth century. This development was further spurred by the discovery of the principal works of antique science—especially those of Euclid, Apollonius, Diophantus, and Archimedes—which were brought to western Europe from Byzantium, then in a state of decline, or from the Arabic centers of learning in Spain. A man of his time, Stevin wrote on a variety of topics. A number of his works are almost wholly original, while even those that represent surveys of science as it existed around 1600 contain his own interpretations; all are characterized by a remarkably lucid and methodical presentation. Stevin chose to write almost all of his books in the vernacular, in accordance with the spirit of self-confidence of the newly established republic. In the introduction to his *De Beghinselen der Weeghconst* of 1586, he stated his admiration for Dutch as a language of wonderful power in shaping new terms; and a number of the words coined by Stevin and his contemporaries survive in the rich Dutch scientific vocabulary.

Stevin's published works include books on mathematics, mechanics, astronomy, navigation, military science, engineering, music theory, civics, dialectics, bookkeeping, geography, and house building. While many of these works were closely related to his mercantile and administrative interests, a number fall into the realm of pure science. His first book, the *Tafelen van Interest* (1582), derives entirely from his early career in commerce; in it Stevin set out the rules of single and compound interest and gave tables for the rapid computation of discounts and annuities. Such tables had previously been kept secret by big banking houses, since there were few skilled calculators, although after Stevin's publication interest tables became common in the Netherlands.

In *De Thiende*, a twenty-nine-page booklet published in 1585, Stevin introduced decimal fractions for general purposes and showed that operations could be performed as easily with such fractions as with integers. He eliminated all difficulties in handling decimal fractions by interpreting 3.27, for example, as 327 items of the unit 0.01. Decimal fractions had previously found only occasional use in trigonometric tables; although Stevin's notation was somewhat unwieldy, his argument was convincing, and decimal fractions were soon generally adopted. At the end of the tract, Stevin went on to suggest that a decimal system should also be used for weights and measures, coinage, and divisions of the degree of arc.

In *L'arithmétique*, also published in 1585, Stevin gave a general treatment of the arithmetic and algebra of his time, providing geometric counterparts. (An earlier work, the *Problemata geometrica* of 1583 had been entirely devoted to geometry; strongly marked by the influence of Euclid and Archimedes, it contained an especially interesting discussion of the semi-regular bodies that had also been studied by Dürer.) Stevin was of the opinion that all numbers—including squares, square roots, and negative or irrational quantities—were of the same nature, an opinion not shared by contemporary mathematicians but one that was vindicated in

the development of algebra. Stevin introduced a new notation for polynomials and gave simplified and unified solutions for equations of the second, third, and fourth degrees; in an appendix published at a later date he showed how to approximate a real root for an equation of any degree.

De Deursichtighe is a mathematical treatment of perspective, a subject much studied by artists and architects, as well as mathematicians, in the fifteenth and sixteenth centuries. Stevin's book gives an important discussion of the case in which the plane of the drawing is not perpendicular to the plane of the ground and, for special cases, solves the inverse problem of perspective, that is, of finding the position of the eye of the observer, given the object and the perspective drawing of it. A number of other works are also concerned with the application of mathematics to practical problems, and in these the instances in which Stevin had to perform what amounts to an integration are particularly interesting. While mathematicians up to his time had followed the Greek example and given each proof by *reductio ad absurdum*, Stevin introduced methods that, although still cumbersome, paved the way toward the simpler methods of the calculus.

De Beghinselen der Weeghconst is Stevin's chief work in mechanics. Published in 1586, some fifty years before Galileo's discoveries, it is devoted chiefly to statics. From the evidence that it provides, Stevin would seem to be the first Renaissance author to develop and continue the work of Archimedes. The book contains discussions of the theory of the lever, the theorems of the inclined plane, and the determination of the center of gravity; but most particularly it includes what is perhaps the most famous of Stevin's discoveries, the law of the inclined plane, which he demonstrated with the *clootcrans*, or wreath of spheres.

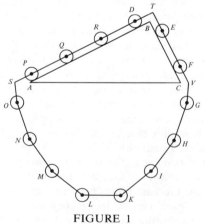

FIGURE 1

The *clootcrans*, as conceived by Stevin, consists of two inclined planes (*AB* and *BC*), of which one is twice the length of the other. A wreath of spheres placed on a string is hung around the triangle *ABC*, all friction being disregarded. The wreath will not begin to rotate by itself, and the lower section *GH · · · MNO*, being symmetrical, may be disregarded. It is thus apparent that the pull toward the left exerted by the four spheres that lie along *AB* must be equal to the pull to the right exerted by the two spheres that lie along *BC* — or, in other words, that the effective component of gravity is inversely proportional to the length of the inclined plane. If one of the inclined planes is then placed vertically, the ratio between the component along the inclined plane and the total force of gravity becomes obvious. This is, in principle, the theory of the parallelogram of forces.

Beneath his diagram of the *clootcrans* Stevin inscribed a cherished maxim, "Wonder en is gheen wonder" — "What appears a wonder is not a wonder" (that is, it is actually understandable), a rallying cry for the new science. He was so delighted with his discovery that he used the diagram of his proof as a seal on his letters, a mark on his instruments, and as a vignette on the title pages of his books; the device also appears as the colophon of this *Dictionary*.

Stevin's next work on mechanics, *De Beghinselen des Waterwichts*, is the first systematic treatise on hydrostatics since Archimedes. In it, Stevin gave a simple and immediately comprehensible explanation for the Archimedean principle of displacement; before a body *C* is immersed, consider a volume of water equal to that of *C*. Since the latter body was at rest, it must have experienced, on displacement, an upward force equal to its weight, while *C* itself will, upon being placed in the water, experience the same degree of buoyancy. Stevin similarly chose to explain the hydrostatic paradox by imagining parts of the water to be solidified, so that neither equilibrium nor pressure was disturbed. He also wrote a number of shorter works in which he applied the principles of mechanics to practical problems of simple machines, balances, the windlass, the hauling of ships, wheels powered by men, the block-and-tackle, and the effect of a bridle upon a horse.

Stevin's chief book on astronomy, *De Hemelloop*, was published in 1608; it is one of the first presentations of the Copernican system, which Stevin unconditionally supported, several years before Galileo and at a time when few other scientists could bring themselves to do likewise. Calling the Copernican hypothesis "the true theory," Stevin demonstrated that the motions of the planets can be inductively derived from observations; since there were no complete direct observations, he used the ephemerides of Johann Stadius in their stead. He first explained the Ptolemaic model (in which the earth is at the center and the sun and planets move in epicycles) by this means, then offered a similar explanation of the Copernican system, in which he improved on the original theory in several minor points.

In a seafaring nation like the Dutch republic matters of navigation were, of course, of great importance. In addition to his astronomical works, Stevin gave a theory of the tides that was — as it must have been, fifty years before Newton — purely empirical. He also, in a short treatise entitled *De Havenvinding*, approached the subject of determining the longitude of a ship, a problem that was not fully solved until the nineteenth century. Several previous authors had suggested that longitude might be determined by measuring the deviation of the magnetic needle from the astronomical meridian, a suggestion based on the assumption that the earthwide distribution of terrestrial magnetism was known. Since the determination of latitude was well known, such a measurement would allow the sailor to chart longitudinal position against the latitudinal circle.

Stevin, in his booklet, gave a clear explanation of this method; he differed from Petrus Plancius and Mercator in that he did not rely upon a priori conceptions of the way in which geomagnetic deviation depends upon geographical position. Although he was willing to offer a conjecture about this dependence, Stevin insisted on the necessity of collecting actual measurements from all possible sources and urged the establishment of an empirical, worldwide survey. His method was sound, although as data began to accumulate it became clear that the magnetic elements were subject to secular variation. The problem of determining longitude was at last solved more simply by the invention of the ship's chronometer.

In *Van de Zeijlstreken*, Stevin set out a method, based on one proposed by Nuñez in 1534, of steering a ship along a loxodrome, always keeping the same course, to describe on the globe a line cutting the meridians at a constant angle. Although the feat was beyond the grasp of the seaman of Stevin's time, his exposition nonetheless contributed to a clear formulation of the principles upon which it was based and helped make the method itself better known both in the Netherlands and abroad.

A considerable body of Stevin's other work developed from his military duties and interests. The Dutch army had been completely reorganized through the efforts of Maurice and counts William Louis and John of Nassau; their innovations, which were widely adopted by other countries, included the establishment of regular drills and maneuvers, the development of fortifications (combined with new methods of attacking a besieged city), and army camps planned after those of the Romans. As quartermaster general, Stevin observed these reforms, as well as actual battles, and wrote in detail, in his usual lucid and systematic style, of sieges, camps, and military equipment.

Stevin's *De Sterctenbouwing* is a treatise on the art of fortification. Although cost prohibited the implementation of the ideas Stevin set out in it, these notions were put to practical effect a century later by Vauban and Coehoorn. *De Legermeting* is a less theoretical work, a description of field encampment during Maurice's campaigns, with the encampment before the Battle of Juliers (in 1610) as a particular example. Stevin gave an account of the layout of the camp, inspired by the writings of Polybius (since later Roman authors were not then known), together with the modifications made by Maurice. He listed all the equipment required in the campaign, and gave detailed instructions concerning the building of huts and the housing of dependents and suppliers. In the last section of the work he made a comparative study of the different methods of deploying soldiers in files and companies, and again recommended distribution by a decimal system. All told, his book gives a vivid impression of the army life of his period.

Of his works on engineering, two books are devoted to the new types of sluices and locks that Stevin himself had helped to devise. He cites their particular usefulness in scouring canals and ditches through the use of tidal action, and cites their application for the waterways of Danzig and other German coastal cities. In these short works he also discusses the formation of sandbanks, peat and quicksand, and the modifications of the course of a river; his explanation of the changes of the surface of the earth, which he attributes to natural forces only, is quite modern.

In *Van de Molens*, Stevin discusses wind-driven drainage mills, crucially important to the flat regions of the Netherlands. Stevin proposed the construction of a new type of mill with more slowly revolving scoop wheels and a smaller number of wider floats, and he further modified the means of transmission of power by making use of conical toothed wheels. A number of mills were built or rebuilt according to his specifications; that they were not completely successful may lie in imperfections in the execution of his design. Stevin also applied the principles of mechanics to windmills, in a series of computations that allowed him to determine, given the size and the number of the cogs, both the minimum wind pressure required on each square foot of the sails to lift the water to the necessary height and how much water is raised by each revolution of the sails. He gave the results of his measurement of fifteen mills.

In another book, *Van de Spiegeling der Singconst*, Stevin turned to the theory of musical tuning, a subject that had enthralled mathematicians from antiquity on. Musicians had also long been concerned with devising a scale in which the intervals of the pure octave (2:1), the pure fifth (3:2), and the pure third (5:4) could be rigorously combined. The chief problem lay in the resolution of the progression by four fifths (96:144:216:324:486) and the interval with the double octave ($96 \times 2 \times 2 = 384$); this ratio, which should be the third, 480:384, is rather the imperfect ratio 486:384. While a number of other mathematicians and musicians had attempted to reach a resolution by minor modifications in the scale, Stevin boldly rejected their methods and declared that all semitones should be equal and that the steps of the scale should each correspond to the successive values of $2^{n/12}$; he dismissed the difference between the third and the fifth as unimportant. Stevin's scale is thus the "equal temperament" now in general use; at the time he proposed it he had been anticipated only by Vincenzio Galilei (1581) and the Chinese prince Chi Tsai-Yü (1584). It is unlikely that Stevin knew the latter's work.

Another of Stevin's many publications was a book on civic life, *Het Burgherlick Leven*. The work is a handbook designed to guide the citizen through periods of civil disorder, a matter of some concern in a nation that had only recently won its freedom through rebellion, and in which religious freedom was still a matter for discussion. Stevin only rarely refers to these circumstances in his book, however; he rather presents his precepts as being completely objective and derived from common sense. The first of his tenets is that the citizen should obey anyone in a position of *de facto* authority, no matter how this authority has been obtained. Since history consists of a succession of princes, Stevin questions how historical rights can be established, then goes on to state that the citizen's duty is to obey the laws, no matter if they appear wrong or unjust. He cites the necessity of religion as a means of instilling virtue in children, but adds that if a man's religion is different from that of his countrymen, the dissenter should either conform or leave. All told, his views are typical of those current in a post-revolutionary period in which consolidation was more important than individual freedoms.

In the last years of his life, Stevin returned to the study of mathematics. He reedited his mathematical works and collected them into the two folio volumes of his *Wisconstighe Ghedachtenissen* (published in 1605–1608). These mathematical memoirs were also published, at almost the same time, in Latin and French translations.

Stevin's writings in general are characterized by his versatility, his ability to combine theory and practice, and the clarity of his argument. They demonstrate a mind confident of the prevalence of reason and common sense and convinced of the comprehensibility of nature. His style, especially the personal way in which he addresses the reader, is particularly charming.

BIBLIOGRAPHY

I. ORIGINAL WORKS. A committee of Dutch scientists has edited *The Principal Works of Simon Stevin*, 5 vols. (Amsterdam, 1955–1968), which contains a bibliography and extensive introductions to each of the works.

Stevin's major works include *Tafelen van Interest* (Antwerp, 1582; Amsterdam, 1590), a French trans. of which appears in *De Thiende* (Leiden, 1585); *Problematum geometricorum-Libri V* (Antwerp, 1583); and *De Thiende* (Leiden, 1585; Gouda, 1626, 1630; Antwerp–The Hague, 1924). Translations of *De Thiende* are H. Gericke and K. Vogel, *De Thiende von Simon Stevin* (Frankfurt am Main, 1965); Robert Norton, *Disme, the Art of Tenths* (London, 1608); J. Tuning, "La disme," in *Mémoires mathématiques* (Leiden, 1608), also reprinted in *Isis*, **23** (1925); Henry Lyte, *The Art of Tenths* (London, 1619), and "The Disme of Simon Stevin," in *Mathematics Teacher*, **14** (1921), 321, also in D. E. Smith, *Source Book of Mathematics* (New York–London, 1929).

Subsequent writings include *L'arithmétique* (Leiden, 1585, 1625), which contains French translations of *Tafelen van Interest* and *De Thiende*; *Vita Politica, Het Burgherlick Leven* (Leiden, 1590; Amsterdam, 1939); *De Stercktenbouwing* (Leiden; 1594; Amsterdam, 1624), also trans. by G. A. von Dantzig (Frankfurt, 1608, 1623), and by Albert Girard, *Les oeuvres mathématiques de Simon Stevin* (Leiden, 1634); *Castrametatio, Dat is legermeting. Nieuwe Maniere van Sterctebou door Spilseuysen* (Rotterdam, 1617), trans. in French (Leiden–Rotterdam, 1618) and in Albert Girard, trans.,

Les oeuvres mathématiques de Simon Stevin (Leiden, 1634), and in German (Frankfurt, 1631); *Van de Spiegeling der Singconst. Van de Molens* (Amsterdam, 1884).

II. SECONDARY LITERATURE. A bibliography of Stevin's works is in *Bibliotheca Belgica. Bibliographie générale des Pays-Bas*, ser. 1, XXIII (Ghent–The Hague, 1880–1890). On Stevin and his work, see R. Depau, *Simon Stevin* (Brussels, 1942), which is in French and contains a bibliography of the articles on Stevin; E. J. Dijksterhuis, *Simon Stevin* (The Hague, 1943), in Dutch, with bibliography of Stevin's works, and which is in an abbreviated English version: R. Hooykaas and M. G. J. Minnaert, eds., *Simon Stevin: Science in the Netherlands Around 1600* (The Hague, 1970); and A. J. J. van de Velde, "Simon Stevin 1548–1948," in *Mededelingen Kongelige Vlaamse Academie*, **10** (1948), 10.

M. G. J. MINNAERT

STEWART, BALFOUR (*b*. Edinburgh, Scotland, 1 November 1828; *d*. Drogheda, Ireland, 19 December 1887), *physics, meteorology, terrestrial magnetism*.

Son of William Stewart, a tea merchant, and named Balfour after his grandmother's family, Balfour Stewart was educated at the universities of St. Andrews and Edinburgh, and then embarked upon a mercantile career. His interest in the physical sciences had, however, been sparked in the natural philosophy class of James D. Forbes at Edinburgh; and after ten years in the business world, Stewart sought a career in science—first, briefly, as an assistant at the Kew observatory and then as an assistant to Forbes at Edinburgh, where he soon made original contributions to the study of radiant heat. In 1859 Stewart returned to Kew as director; and for twelve years he was involved with the continuing missions of that institution, including the study of meteorology, solar physics, and terrestrial magnetism. During this period he was elected a fellow of the Royal Society of London; he married Katharine Stevens, daughter of a London lawyer; and he was awarded the Rumford Medal by the Royal Society for his earlier work on radiant heat. In 1870 Stewart was appointed professor of natural philosophy at Owens College, Manchester. He was subsequently president of the Manchester Literary and Philosophical Society, the Physical Society, and the Society for Psychical Research.

The heart of Stewart's contributions to the study of radiant heat (infrared radiation), and to thermal radiation in general, came in "An Account of Some Experiments on Radiant Heat, Involving an Extension of Prévost's Theory of Exchanges"

(1858). Building in part on suggestions by Forbes and results of Macedonio Melloni and others, he had investigated the abilities of various materials to emit and absorb radiation of various wavelengths. Stewart found, in particular, that a material that radiates heat of a certain wavelength preferentially tends also to absorb heat of that same wavelength preferentially. He dealt with these results by extending a theory that had been developed by Pierre Prevost in the years around 1800. Considering thermal equilibrium to be dynamic, involving continual exchange of the heat substance between the bodies in equilibrium, Prevost had been able to derive relationships between the emissive and reflective powers of surfaces. Extending these arguments from surfaces to volumes, and specializing to individual wavelengths, Stewart derived from the "theory of exchanges" a result in accord with his experimental data on the emissive and absorptive powers of thin plates of various materials: "The absorption of a plate equals its radiation [emission], and that for every description [wavelength] of heat" ("Account," Brace ed., p. 39).

Unfortunately, this original and significant work had little influence on the subsequent development of science. Two years after Stewart's paper had been submitted to the Royal Society of Edinburgh, Gustav Kirchhoff, not knowing Stewart's work and proceeding from experiments on optical spectra, came to similar conclusions. Kirchhoff's results were more solid: the experiments were cleaner, the derivations more rigorous, and the results more clearly and generally stated. Moreover, the particular experimental context of Kirchhoff's work suggested immediate and extremely important applications in chemistry and astronomy. Thus, although Stewart's work was meritorious and prior, as he and his supporters vigorously argued for decades after, it was Kirchhoff's work that had decisive influence throughout the physical sciences.

Upon becoming director of the Kew observatory, Stewart turned his attention to the continuing missions of that institution. (In 1867 Kew was designated the central meteorological observatory of Great Britain; and the annual operating grant from the British Association was then augmented by direct governmental support of the meteorological activities, bringing the yearly budget over £1,000.) Extending investigations of the relationships between terrestrial magnetism and the sunspot cycle, he also studied correlations of these phenomena with various other terrestrial and "cosmical" cycles. Later, in an influential review article on ter-

restrial magnetism (published in the *Encyclopaedia Britannica*), Stewart proposed mechanisms for some of these correlations. In particular, certain variations in the geomagnetic field were to be referred to varying electric currents in the upper atmosphere: Assuming that the rarefied air of the upper atmosphere possessed appreciable electrical conductivity, thermal and tidal air currents, moving the conducting air through the earth's magnetic field, would generate electric currents. By this mechanism the well-known daily variation of the geomagnetic field was explained. Also explained were seasonal variations of the geomagnetic field, analogies between global wind patterns and global magnetic patterns, and lunar correlations. Stewart's hypotheses informed subsequent research in geomagnetism, and strong confirmation ensued.

In later years at Owens College, Stewart lectured on elementary physics and established a teaching laboratory; J. J. Thomson, one of his most illustrious students, traced his initial enthusiasm for research to that laboratory. Textbooks and popularizations by Stewart were widely read; Ernest Rutherford, at age ten, owned one of them. Another widely read book, written in collaboration with Peter Guthrie Tait, was intended to demonstrate the compatibility between science and religion. Entitled *The Unseen Universe*, the book argued that the individual soul was immortal, existing after death in the context of the subtle media of nineteenth-century physics, including the luminiferous ether, the ultramundane (gravitation-causing) particles of Le Sage, and the ubiquitous fluid substratum of William Thomson's vortex atoms; thermodynamics and evolution also supported the argument.

"Have great [physical] vitality, restless," but "no power of great amount of [mental] work," Stewart wrote of himself (Hilts, "Guide"). In the domain of science broadly defined, he participated in many institutions, programs, and discoveries but left no single monument.

BIBLIOGRAPHY

I. ORIGINAL WORKS. The following are representative of Stewart's publications: "An Account of Some Experiments on Radiant Heat, Involving an Extension of Prévost's Theory of Exchanges," in *Transactions of the Royal Society of Edinburgh*, **22** (1857–1861), 1–20, repr. in D. B. Brace, ed., *The Laws of Radiation and Absorption: Memoirs by Prévost, Stewart, Kirchhoff, and Kirchhoff and Bunsen* (New York, 1901), 21–50; *The Unseen Universe: or Physical Speculations on a Future*

State (London, 1875), written with P. G. Tait; and "Terrestrial Magnetism," an appendix to the article on meteorology in *Encyclopaedia Britannica*, 9th ed., XVI, 159–184.

II. SECONDARY LITERATURE. A list of Stewart's publications, as well as a review of his scientific career, is furnished in Arthur Schuster, "Memoir of the Late Professor Balfour Stewart, LL.D., F.R.S.," in *Memoirs and Proceedings of the Manchester Literary and Philosophical Society*, 4th ser., **1** (1888), 253–272. See also Philip J. Hartog, "Balfour Stewart," in *Dictionary of National Biography*.

Aspects of Stewart's activities are treated in the following: Joseph Agassi, "The Kirchhoff-Planck Radiation Law," in *Science*, **156** (1967), 30–37; Sydney Chapman and Julius Bartels, *Geomagnetism*, II (Oxford, 1940), 750–752; P. M. Heimann, "*The Unseen Universe*: Physics and the Philosophy of Nature in Victorian Britain," in *British Journal for the History of Science*, **6** (1972), 73–79; Victor L. Hilts, "A Guide to Francis Galton's *English Men of Science*," in *Transactions of the American Philosophical Society* (in press); Hans Kangro, "Kirchhoff und die spektralanalytische Forschung," editor's *Nachwort* in Gustav Robert Kirchhoff, *Untersuchungen über das Sonnenspectrum und die Spectren der chemischen Elemente und weitere ergänzende Arbeiten aus den Jahren 1859–1862*, Milliaria, no. 17 (Osnabrück, 1972), 17–26; Robert H. Scott, "The History of the Kew Observatory," in *Proceedings of the Royal Society*, **39** (1885), 37–86; Daniel M. Siegel, "Balfour Stewart and Gustav Kirchhoff: Two Independent Approaches to 'Kirchhoff's Radiation Law,'" in *Isis* (in press); and J. J. Thomson, *Recollections and Reflections* (New York, 1937), 18–22.

A large collection of unpublished source materials is held at the Royal Society.

DANIEL M. SIEGEL

STEWART, GEORGE NEIL (*b.* London, Ontario, 18 April 1860; *d.* Cleveland, Ohio, 28 May 1930), *physiology*.

Although Stewart, the son of James Innes and Catherine Sutherland Stewart, was born in Canada, his parents returned to Scotland and he grew up in Lybster, a fishing village in Caithness. His studies at the University of Edinburgh included classics, philosophy, history, and mathematics, the latter leading him to physics and, in 1879, an assistantship with Peter Guthrie Tait. His introduction to physiology came from William Rutherford. Being especially interested in electrophysiology, Stewart spent 1886–1887 studying with du Bois-Reymond at Berlin. After receiving the M.A., B.S., and D.Sc. (1887) degrees at Edinburgh, he became senior demonstrator of physiology at Owens College, Victoria University, Man-

chester (1887–1889), where he learned from William Stirling the value of the illustrative experiment in teaching science.

His decision to study medicine led to the M.B. and M.D. at Edinburgh (1889, 1891) and the D.P.H. at Cambridge (1890) while he was George Henry Lewes student at Downing College (1889–1893). He was also an examiner in physiology at Aberdeen (1890–1894). In 1893, at Henry P. Bowditch's invitation, Stewart went to Harvard as instructor in physiology; the following year he was appointed professor of physiology and histology at Western Reserve University School of Medicine, in Cleveland, where with the exception of four years (1903–1907) as professor of physiology at the University of Chicago, he remained for the rest of his life. In 1907 he became professor of experimental medicine and director of the H. K. Cushing Laboratory.

Stewart's major contribution was in transmitting modern methods of teaching and research in physiology to American medical education. During his first year in Cleveland, using improvised laboratory equipment, he began illustrating lectures with experiments; simultaneously he wrote a 796-page *Manual of Physiology* (1895), in which, for the first time, practical exercises for students were appended to each chapter and experiments on mammals were included. The practice spread to other schools, and the *Manual* became a standard text.

After earlier work on color vision, electrophysiology, Talbot's law, cardiac nerves, otoliths, muscle proteins, and permeability of blood corpuscles, Stewart later experimentally investigated such clinical problems as the effect of total anemia on the brain, resuscitation, the measurement of blood flow by the calorimetric method, and the estimation of pulmonary blood capacity and cardiac output by indicator-dilution techniques. With Julius M. Rogoff he studied the functions of the adrenal medulla and cortex, including the epinephrine output and the usefulness of cortex extracts to treat Addison's disease. They discovered that the adrenal cortex was indispensable to the life of higher animals. Stewart was described as a brilliant teacher, witty, and possessed of prodigious energy and an amazing memory. A perfectionist, he had little pleasure outside his laboratory. The University of Edinburgh awarded him the honorary LL.D. in 1920.

Pernicious anemia and progressive spinal degeneration afflicted Stewart during his later years; but he remained mentally alert until the end, making notes about his own condition.

BIBLIOGRAPHY

Stewart's major work was *A Manual of Physiology With Practical Exercises* (London, 1895; 8th ed., 1918).

Secondary literature includes Torald Sollmann, "George Neil Stewart, Physiologist, April 18, 1860, to May 28, 1930," in *Science*, **72** (1930), 157–162; and Carl J. Wiggers, "The Evolution of Experimental Physiology at the University of Michigan and Western Reserve University," in *Bulletin of the Cleveland Medical Library*, **10** (1963), 14–15. See also the faculty minutes and biography file of the Case Western Reserve University Archives.

GENEVIEVE MILLER

STEWART, MATTHEW (*b*. Rothesay, Isle of Bute, Scotland, January 1717; *d*. Catrine, Ayrshire, Scotland, 23 January 1785), *geometry, astronomy, natural philosophy.*

Stewart's father, Dugald, was minister of the parish of Rothesay; his mother was Janet Bannatyne. Intending to follow his father's career, he entered the University of Glasgow in 1734, soon coming under the influence of Robert Simson, professor of mathematics, and Francis Hutcheson, professor of moral philosophy.

Simson was then attempting to reconstruct Euclid's lost book on porisms; and he communicated his enthusiasm for this project—and for the study of Greek mathematics in general—to Stewart, who soon developed his own approach to the subject. Aware that new horizons were opening in mathematics, Simson conscientiously instructed his students in the newer subjects of calculus and analytical geometry. It was at his suggestion that Stewart went to the University of Edinburgh to work under Colin Maclaurin, himself a pupil of Simson's. Although he was studying calculus, higher plane curves, and cosmogony with Maclaurin, Stewart continued to correspond with Simson and, under his general direction, to carry on his work in pure geometry.

Simson's investigations were slow and laborious, and he was disinclined to publish findings that he regarded as incomplete—after the publication of a paper on porisms in 1723, nothing by him on the subject was published until eight years after his death—but he made his work freely available to Stewart. He actively encouraged Stewart to publish his celebrated series of geometrical propositions, *General Theorems*, in 1746 because the chair of mathematics at Edinburgh was vacant as a result of Maclaurin's service with the government

troops in the Jacobite Rebellion of 1745 and subsequent death from an illness contracted during that campaign. Stewart was then largely unknown in Scottish academic circles; and the chair was offered to James Stirling, who already enjoyed a European reputation as a mathematician of distinction. Stirling declined the invitation; and when the electors to the chair re-assessed the situation at the end of 1746, the reception accorded the publication of Stewart's book had been so favorable that they were encouraged to offer him the chair. Stewart had only recently (May 1745) been ordained minister of the parish of Roseneath, Dunbartonshire, on the nomination of the Duke of Argyll; but he had no hesitation in changing his career and was duly elected to the chair of mathematics at Edinburgh in September 1747.

Stewart's reputation as a mathematician was established overnight by the publication of the *General Theorems*. John Playfair, himself a scientist of distinction, claimed that Stewart's results were "among the most beautiful, as well as most general propositions known in the whole compass of geometry, and are perhaps only equalled by the remarkable *locus* to the circle in the second book of Apollonius, or by the celebrated theorems of Mr. Cotes. . . . The unity which prevails among them is a proof that a single though extensive view guided Mr. Stewart in the discovery of them all" ("Memoir of Matthew Stewart," 59–60). Simson's influence is obvious throughout the work. Several of Stewart's theorems are in fact porisms, although he refrains from calling them by that name, probably through fear of seeming to anticipate Simson. Several of their contemporaries assert in their memoirs that Simson, singularly lacking in personal ambition, was so keen for Stewart to succeed to Maclaurin's chair that he allowed him to incorporate in his book results that were originally Simson's; it is fairly clear that what is usually described as "Stewart's theorem" was demonstrated in lectures by Simson several years before the publication of Stewart's book.

After his election to the chair, Stewart's interests turned to astronomy and natural philosophy; and he displayed great ingenuity in devising purely geometrical proofs of results in these subjects that had previously been established by the use of algebraic and analytical methods. Examples of this kind are to be seen in his *Tracts, Physical and Mathematical* (1761). In a work published in 1763 he extended these methods to provide a basis for the approximate calculation of the distance of the earth from the sun. He derived a value of 29,875 radii of the earth for this distance—a result that was shown shortly afterward (1768) by John Dawson to be greatly in error; Stewart's mistake had been his failure to realize that his geometrical methods did not indicate how small arithmetical errors could grow in the course of his calculation.

Stewart bore the attacks on this work rather badly, and as a result his health began to fail. In 1772 he retired to his country estate at Catrine in Ayrshire, leaving the duties of his chair to his son Dugald, who was elected joint professor with him in 1775.

BIBLIOGRAPHY

I. Original Works. Stewart's books include *General Theorems of Considerable Use in the Higher Parts of Mathematics* (Edinburgh, 1746); *Tracts, Physical and Mathematical . . .* (Edinburgh, 1761); *Distance of the Sun From the Earth . . .* (Edinburgh, 1763); and *Propositiones geometricae more veterum demonstratae ad geometriam antiquam illustrandam et promovendam idoneae* (Edinburgh, 1763).

II. Secondary Literature. See John Playfair, "Memoir of Matthew Stewart," in *Transactions of the Royal Society of Edinburgh*, **1** (1788), 57–76; and Matthew Stewart, *Memoir of Dugald Stewart* (Edinburgh, 1838).

Ian N. Sneddon

STIELTJES, THOMAS JAN (*b.* Zwolle, Netherlands, 29 December 1856; *d.* Toulouse, France, 31 December 1894), *mathematics.*

To the majority of mathematicians, Stieltjes' name is remembered in association with the Stieltjes integral,

$$\int_a^b f(u)\,dg(u),$$

a generalization of the ordinary Riemann integral with wide applications in physics. Yet in his own day, he was renowned as a versatile mathematician whose publications include papers in almost every area of analysis. He is the father of the analytic theory of continued fractions, and his integral was developed as a tool for its study.

Stieltjes, the son of a distinguished Dutch civil engineer, received his principal schooling at the École Polytechnique in Delft. He left the École in 1877 to take up a post at the observatory in Leiden, where he served six years. Evidently he kept up his mathematical studies, since he left Leiden to accept a chair in mathematics at the University of Groningen. Honors came to Stieltjes early. In 1884 the University of Leiden awarded him an honorary doctorate, and in 1885 he was elected to membership in the Royal Academy of Sciences of Amsterdam. Disappointed at Groningen, Stieltjes went to live in Paris, where he received his doctorate of science in 1886. In the same year he was appointed to the faculty at the University of Toulouse, where he remained until his death eight years later. Although not elected to the Academy of Sciences in Paris, Stieltjes was considered for membership in 1892 and won its Ormoy Prize in 1893 for his work on continued fractions.

Stieltjes' published work encompasses almost all of analysis of his time. He made contributions to the theory of ordinary and partial differential equations, studied gamma functions and elliptical functions, and worked in interpolation theory. His thesis was on asymptotic series. A special and increasing interest was the evaluation of particular integrals such as

$$\int_0^\infty \frac{\sin(xu)}{1+u^2}\,du$$

or

$$\int_0^\infty \frac{u\cos(xu)}{1+u^2}\,du$$

and series of the general form $\sum^\infty a_n/x^n$, which arise in a natural way from such integrals. These series also occur in the study of continued fractions, and this may have led Stieltjes to analytic continued fraction theory.

His first paper on continued fractions, published in 1884, proves the convergence of

$$\cfrac{2}{z-\cfrac{1\cdot1}{1\cdot3-\cfrac{3\cdot3}{5\cdot7-\cfrac{4\cdot4}{7\cdot9-\cdots}}}}$$

in the slit complex z-plane excluding the interval $(-1,1)$, with use of the series in decreasing powers of z. This convergence, which is locally uniform, was established by transforming the fraction into a definite integral.

Yet Stieltjes' monument is his last memoir, "Recherches sur les fractions continues," written just before he died, and published in two parts (*Annales de la Faculté des sciences de l'Université de Toulouse pour les sciences mathématiques et physiques*, **1**, ser. 1 [1894], 1–122; **1**, ser. 9 [1895], 5–47), the second posthumously. In it he polished and refined all of his previous work on the subject, and here is the first appearance of his integral. The memoir is a beautiful piece of mathematical writing—clear, self-contained, almost lyric in its style.

In this paper the fraction

$$
\cfrac{1}{a_1 z + \cfrac{1}{a_2 + \cfrac{1}{a_3 z + \cfrac{1}{a_4 + \cfrac{\ddots}{\cfrac{1}{a_{2n-1} z + \cfrac{1}{a_{2n} + \ddots}}}}}}}
\tag{1}
$$

is considered. The a_i's are assumed to be known real positive quantities, and z is a complex variable. Fraction (1) will be said to converge, or otherwise, according to the convergence or not of the sequence of "approximates" $P_n(z)/Q_n(z)$. Each approximate is the rational function formed by considering only the first n terms of (1) and simplifying the resulting compound fraction. Thus, $Q_{2n}(z)$ and $P_{2n+1}(z)$ are polynomials of degree n in z, while P_{2n} is of degree $n-1$, and Q_{2n+1} is of degree $n+1$.

Stieltjes began by studying the roots of the polynomials $P_n(z)Q_n(z)$, which are all real. He proved a whole series of theorems concerning the interlacing of their roots; for example, the roots of $Q_{2n}(z)$ separate the roots of $Q_{2n-2}(z)$. This was then used to prove that the roots of $P_n(z)$ and $Q_n(z)$ ($n=1, 2, \cdots$) are all nonpositive and distinct. Thus, the approximates have the following partial fraction decomposition:

$$
\frac{P_{2n}(z)}{Q_{2n}(z)} = \sum_{k=1}^{n} \frac{M_k}{z + x_k}
\tag{2}
$$

$$
\frac{P_{2n+1}(z)}{Q_{2n+1}(z)} = \frac{N_0}{z} + \sum_{k=1}^{n} \frac{N_k}{z + y_k},
\tag{3}
$$

where in (2) $\{x_1, x_2, \cdots, x_n\}$ are the (positive) roots of $Q_{2n}(-z)$ and in (3) $\{0, y_1, y_2, \cdots, y_n\}$ are the roots of $Q_{2n+1}(-z)$.

Next Stieltjes was able to show that

$$
\frac{P_n(z)}{Q_n(z)} = \sum_{k=1}^{n} (-1)^{k-1} \frac{c_{k-1}}{z^k} + \sum_{k=n+1}^{\infty} \frac{\alpha_{n,k-1}}{z^k},
\tag{4}
$$

where the $\{c_k : k = 1, 2, \cdots, n\}$ depend only upon the original fraction (1) and not upon n. Formula (4) led to the definition of the development of (1) in decreasing powers of z:

$$
\sum_{k=1}^{\infty} (-1)^{k-1} \frac{c_{k-1}}{z^k}
\tag{5}
$$

The c_k are all real and positive, and

$$
\frac{c_{n+1}}{c_n} < \frac{c_{n+2}}{c_{n+1}}.
\tag{6}
$$

Either the sequences of ratios is unbounded, in which case (5) diverges for all z, or (6) is bounded, in which case there is a $\lambda > 0$ with the property that (5) converges for all z satisfying $|z| > \lambda$. Stieltjes then proved that, in the latter case, if the (necessarily positive) roots of $Q_n(-z)$ are ordered according to size, and if the largest is, say, $x_{n,k}$, then $\lim_{n \to \infty} x_{n,k} = \lambda$.

oscillation of (1). Here Stieltjes showed that for all z with positive real part,

$$
\lim_{n \to \infty} \frac{P_{2n+1}(z)}{Q_{2n+1}(z)} = F_1(z)
\tag{7}
$$

and

$$
\lim_{n \to \infty} \frac{P_{2n}(z)}{Q_{2n}(z)} = F(z).
\tag{8}
$$

Furthermore, for z real ($=x$), $F(x)$ and $F_1(x)$ are real, and $F_1(x) \geq F(x)$. Equality in the right half plane, including the positive real axis, was proved to hold if and only if the series

$$
\sum_{k=1}^{\infty} a_k
\tag{9}
$$

formed from the terms of fraction (1) is divergent. (Recall that the a_k are all positive.) Also, since the convergence of (7) and (8) is locally uniform, the functions $F(z)$ and $F_1(z)$ are analytic in the right half plane. Thus, to sum up, Stieltjes had shown that the continued fraction (1) was convergent when and only when the series (9) diverged; otherwise the fraction oscillated. The remaining problem was to extend this result to the z in the left half plane, except for certain points on the negative real axis.

To this end, Stieltjes showed that the limits

$$\lim_{n\to\infty} P_{2n}(z) = p(z) \qquad \lim_{n\to\infty} P_{2n+1}(z) = p_1(z)$$

$$\lim_{n\to\infty} Q_{2n}(z) = q(z) \qquad \lim_{n\to\infty} Q_{2n+1}(z) = q_1(z)$$

all exist, that p, q, p_1, q_1 are all analytic, and that

$$p_1(z)q(z) - p(z)q_1(z) \equiv 1.$$

Then, again he supposed that the roots $\{x_{n,1}, x_{n,2}, \cdots, x_{n,n}\}$ of $Q_{2n}(-z)$ were ordered according to increasing size for each n. Stieltjes then proved that $\lim_{n\to\infty} x_{nk} = \lambda k$, that $\{\lambda_k; k = 1, 2, \cdots\}$ are all distinct real and positive, and that the λ_k are the only zeros of $q(z)$. Similar results hold for $q_1(z)$, $p(z)$, and $p_1(z)$. Next, μ_k is defined by

$$\mu_k = \frac{p(-\lambda_k)}{q'(-\lambda_k)}, \qquad (10)$$

and it was proved that $\mu_k = \lim_{n\to\infty} M_k$ $(k = 1, 2, \cdots)$, where the M_k are from expression (2). Furthermore,

$$S(z) = \sum_{k=1}^{\infty} \frac{\mu_k}{z + x_k} \qquad (11)$$

is meromorphic in the plane, $S(z) = F(z)$ (see [8]), and finally that for each i

$$c_i = \sum_{k=1}^{\infty} \mu_k \lambda_k^i, \qquad (12)$$

where the c_i's are from (5).

In precisely the same way, an infinite set of pairs of positive real numbers similar to $\{(\mu_k, \lambda_k)\}$ above was associated with the sequence $\{p_{2n+1}(z)/Q_{2n+1}(z)\}$. In particular $F_1(z)$ was shown to be meromorphic, and if $\nu_k = \lim_{n\to\infty} N_k$ $((k = 0, 1, 2, \cdots)$; see [3]), and if $\theta_k = \lim_{n\to\infty} y_{n,k}$ $(k = 1, 2, \cdots)$, then

$$c_0 = \sum_{k=0}^{\infty} \nu_k \text{ and } c_i = \sum_{k=1}^{\infty} \nu_k \theta_k^i \ (i = 1, 2, \cdots), \text{ and}$$

$$F(z) = \frac{\nu_0}{z} + \sum_{k=1}^{\infty} \frac{\nu_k}{z + \theta_k}. \qquad (13)$$

In this way he established the analyticity of $F_1(z)$ and $F(z)$ in the slit plane.

Observe that the above systems (12) (or [13]) can be considered as the equations of the moments of all orders of a system of masses μ_k (or ν_k) placed at a distance λ_k (or θ_k) from the origin, and that in either case the ith moment is c_i. Of course, if Σa_k is divergent, $\nu_k = \mu_k$ and $\lambda_k = \theta_k$ for all k, since $F(z)$

and $F_1(z)$ are the same function. But if Σa_k is convergent, the equalities do not hold, even though the c_i are the same in each case.

The further study of the nature of $F_1(z)$ and $F(z)$ in more detail led Stieltjes to the "moment problem": that is, to find a distribution of mass (an infinite set of ordered pairs of positive numbers) whose moments of all orders are known. If this problem can be solved, then $F_1(z)$ and $F(z)$ will be known, since the c_i's can be calculated from the a_k's of fraction (1). However, it is immediately evident that if Σa_k is convergent, there can be no unique solution, as there are at least two. But Stieltjes was able to show (later on) that if Σa_k diverges, there is a unique solution.

It was to solve the moment problem that Stieltjes introduced his integral. First he considered an increasing real-valued function φ defined on the positive real axis, and gave a lengthy discussion of one-sided limits. For example, he showed that φ is continuous at x, if and only if $\varphi^+(x) = \varphi^-(x)$. Only bounded functions with countably many discontinuities on the positive axis were considered. Next he supposed that φ was a step function, with $\varphi(0) = 0$. Then a finite mass condensed at each point of discontinuity can be given by $\varphi^+(x) - \varphi^-(x)$, and $\varphi(b) - \varphi(a)$ is the total mass between a and b; in particular $\varphi(x)$ is the total mass between x and the origin. Also, changing the value of φ at a point of discontinuity does not change the associated mass distribution there.

Stieltjes then defined the integral

$$\int_a^b f(x)\,d\varphi(x) \qquad (14)$$

to be the limit, as max $(x_{i+1} - x_i) \to 0$ of

$$\sum_{i=1}^{n} f(\zeta_i)[\varphi(x_i) - \varphi(x_{i-1})], \qquad (15)$$

where $a = x_0 < x_i < \cdots < x_n = b$ and $x_{i-1} \le \zeta_i \le x_i$. Stieltjes then established the formula for integration by parts

$$\int_a^b f(x)\,d\varphi(x) =$$

$$f(b)\varphi(b) - f(a)\varphi(a)$$

$$- \int_a^b \varphi(x)\,df(x),$$

defined the improper integral $\int_a^\infty f(a)\,d\varphi(x)$ in the usual way, and established many properties of the integral.

Next he considered the function φ_n defined from the even-order approximates by

$$\varphi_n(0) = 0 \quad 0 \le u < x_1$$

$$\varphi_n(u) = \sum_{i=1}^{k} M_i \quad x_k \le u < x_{k+1} (1 \le k \le n-1)$$

$$\varphi_n(u) = \sum_{i=1}^{n} M_i = \frac{1}{a_1} \quad x_n \le u < \infty,$$

where the M_i and the x_k are defined as in (2), and where a_1 is the first term of (1). After a lengthy discussion of "lim inf" and "lim sup" (the ideas were new then), he defined, for each u: $\psi(u) = \lim$ sup $\varphi_n(u)$, $\chi(u) = \lim \inf \varphi_n(u)$, and $\Phi(u) = \frac{1}{2} (\psi(u) + \chi(u))$. Φ was shown to have the property that

$$F(z) = \int_0^\infty \frac{d\Phi(u)}{z+u}$$

and also that the distribution of mass represented by Φ solved the moment problem, since

$$c_k = \int_0^\infty u^k d\Phi(u).$$

A function $\Phi_1(u)$ with similar properties was constructed from the odd-order approximates. He also undertook the study of the inverse problem; that is, given an increasing function $\psi(u)$, with $\psi(0) = 0$, then, by setting

$$c_k = \int_0^\infty u^k d\varphi(x)$$

a fraction like (1) can be determined with the property that

$$F(x) \le \int_0^\infty \frac{d\psi(u)}{x+u} \le F_1(x).$$

Stieltjes' paper, of which only a portion has been summarized here, is a mathematical milestone. The work represents the first general treatment of continued fractions as part of complex analytic function theory; previously, only special cases had been considered. Moreover, it is clearly in the historical line that led to Hilbert spaces and their generalizations. In addition, Stieltjes gave a sort of respectability to discontinuous functions and, together with some earlier work, to divergent series. All together these were astonishing accomplishments for a man who died just two days after his thirty-eighth birthday.

BIBLIOGRAPHY

All Stieltjes' published papers, with some letters, notes, and incomplete works found after his death, are in *Oeuvres complètes de Thomas Jan Stieltjes* (Groningen, 1914–1918). An annotated bibliography of his published works appears with his obituary in *Annales de la Faculté des sciences de l'Université de Toulouse pour les sciences mathématiques et physiques*, **1**, ser. 1 (1895), 1–64.

MICHAEL BERNKOPF

STIFEL (STYFEL, STYFFEL, STIEFFELL, STIFELIUS, or STIEFFEL), MICHAEL (*b*. Esslingen, Germany, *ca.* 1487; *d*. Jena, Germany, 19 April 1567), *mathematics, theology.*

Stifel was the son of Conrad Stifel. Nothing is known about his education except that, on his own testimony, he knew no Greek.[1] He was a monk at the Augustinian monastery of Esslingen, where he was ordained priest in 1511. Reacting to the declining morality of the clergy and the abuses committed in the administration of indulgences, Stifel became an early follower of Luther. While studying the Bible he came upon the numbers in Revelation and in the Book of Daniel.[2]

After 1520 he became increasingly preoccupied with their cabalistic interpretation, for which he used a "word calculus" (*Wortrechnung*). In the malevolent great beast designated in Revelation by the number 666 he saw Pope Leo X.[3] Stifel aroused the suspicion of the bishop of Constance and of his vicar-general by giving absolution without receiving indulgence money and by composing a song in honor of Luther.[4] Realizing that his life was in danger, Stifel escaped in 1522 to Kronberg in the Taunus Mountains, seeking refuge in the castle of a knight named Hartmut, a relative of Franz von Sickingen. He soon had to flee again[5] and went to Wittenberg, where Luther lodged him in his own house. The two became friends, and in 1523 Luther obtained Stifel a post as pastor at the court of the count of Mansfeld. Two years later Stifel became pastor and tutor at Castle Tollet in Upper Austria, in the service of the widow of a nobleman, Wolfgang Jörger.[6] The persecution unleashed by Ferdinand I of Bohemia against the new religious teaching forced Stifel to return to Luther, who procured him a parish at Annaberg. Luther accompanied Stifel there on 25 October 1528 and married him to the widow of the previous incumbent.

At Annaberg, Stifel resumed his dabbling in number mysticism, an activity that, if nothing else, revealed his skill in detecting number-theory relationships. From his reading of the Bible, he thought that he had discovered the date of the end of the world; and in *Ein Rechen-büchlin Vom End Christ* (1532) he prophesied that the event would occur at 8 o'clock on 18 October 1533.[7] On 28 September 1533 Luther implored him not to spread his fantastic notions. Stifel could not be dissuaded, however; and as he vainly warned his assembled congregation of the coming of the end, he was arrested and subsequently dismissed from his post.[8] Through the intervention of Luther—who forgave his "little temptation" (*kleine Anfechtlein*)—and Melanchthon, he finally received another parish, at nearby Holzdorf, in 1535.

Now cured of prophesying, Stifel devoted himself to mathematics. He enrolled at, and received his master's degree from, the University of Wittenberg, where Jacob Milich was lecturer on mathematics.[9] Stifel gave private instruction in mathematics, and among his pupils was Melanchthon's son-in-law Kaspar Peucer. The years at Holzdorf were Stifel's most productive period. At the urging of Milich he wrote *Arithmetica integra* (1544), in which he set forth all that was then known about arithmetic and algebra, supplemented by important original contributions.[10] In his next work, *Deutsche arithmetica* (1545), Stifel sought to make his favorite branch of mathematics, the coss (algebra) or "artful calculation" (*Kunstrechnung*), more accessible to German readers by eliminating foreign words.[11] His last book written at Holzdorf was *Welsche Practick* (1546).[12]

The peaceful years in Holzdorf ended suddenly after the Schmalkaldic War (1547), for the "Hispanier" drove off all the inhabitants.[13] Stifel fled to Prussia, where he finally found a position in 1551 as pastor at Haberstroh, near Königsberg.[14] He lectured on theology and mathematics at the University of Königsberg and brought out a new edition of Christoph Rudolff's *Coss*, which first appeared in 1525 and had since become unavailable. He undertook the republication at the request of a businessman named Christoff Ottendorffer, who paid the printing costs. Stifel reproduced Rudolff's text in its entirety, as well as all 434 problems illustrating the eight rules of the *Coss*. To each chapter of the original text he appended critical notes and additional developments, most of which he drew from his *Arithmetica integra*.[15] Stifel's additions are much longer than the corresponding sections of Rudolff's book.[16]

Stifel returned to playing with numbers, as is evident from his next published book, *Ein sehr wunderbarliche Wortrechnung* (1553).[17] At odds with his colleagues, especially Andreas Osiander, as a result of theological controversies, and urged to return to Holzdorf by his former congregation, he returned to Saxony in 1554. His first post there was as pastor at Brück, near Wittenberg. He then went to Jena, following his friend Matthias Flacius,[18] and lectured on arithmetic and geometry at the university. In 1559 he was mentioned in the register as "senex, artium Magister et minister verbi divini."[19] By this time he apparently had given up his pastorate. Stifel's life in Jena was made difficult by theological disputes until Flacius, from whom Stifel had become alienated, found a successor (Nikolaus Selnecker) for him in 1561.[20] Stifel bequeathed the latter a long work on *Wortrechnung* that was never printed.[21]

In his books Stifel offered more than a methodical exposition of existing knowledge of arithmetic and algebra:[22] he also made original contributions that prepared the way for further progress in these fields. A principal concern was the establishment of generally valid laws. He contended that to improve algebra, it was necessary to formulate rules the validity of which was not limited to special cases, and which therefore could advance the study of the entire subject.[23] He was, in fact, the first to present a general method for solving equations, one that replaced the twenty-four rules traditionally given by the cossists (and the eight that appeared in Rudolff's *Coss*).[24] For example, he pointed out that basically there was nothing different about problems with several unknowns, a type for which Rudolff had introduced a special name.[25] Similarly, he asserted that the symbol "dragma" for the linear member could simply be omitted.[26] Stifel introduced into western mathematics a general method for computing roots that required, however, the use of binomial coefficients. He had discovered these coefficients only with great difficulty, having found no one to teach them to him nor any written accounts of them.[27] Stifel also surpassed his predecessors in the division of general polynomials and extraction of their roots, as well as in computing with irrational numbers.

The second chapter of *Arithmetica integra* is devoted entirely to the numerical treatment of Euclidean irrationals (binomials, residues, and so forth), a topic that Fibonacci had planned to discuss.[28] Stifel's exceptional skill in number theory is evident in his investigation of numerical relationships in number sequences, polygonal numbers, and magic squares.[29] Particularly noteworthy is his contribution to the preliminary stages of logarithmic computation. The starting point for this type of computation was attained by correlating a geometric series with an arithmetic series. This can be seen in Stifel's explanation of the cossists' symbols:

$$0 \cdot 1 \cdot 2 \cdot 3 \cdot 4 \cdot 5 \cdot 6 \cdot 7$$
$$1 \cdot 1x \cdot 1z \cdot 1c \cdot 1zz \cdot 1\beta \cdot 1zc \cdot 1b\beta.$$

On this occasion Stifel introduced the term "exponent" for the numbers of the upper series.[30] In correlating the two number sequences he extended the use of exponents to the domain of the negative numbers in the manner shown below.

−3	−2	−1	0	1	2	3	4	5	6
1/8	1/4	1/2	1	2	4	8	16	32	64

Stifel suspected the importance of his innovation, stating: "A whole book might be written concerning the marvellous things relating to numbers, but I must refrain and leave these things with eyes closed."[31] He did, however, provide a table that made it possible to carry out logarithmic calculations. Stifel approached the concept of the logarithm from another direction as well. He was aware of both inversions of the power, the root, and the "logarithmus." He discussed "division" of a ratio by a number, obtaining, for example, in the case of (27:8) "divided" by 3/4: $\sqrt[3/4]{27:8}$ $= \sqrt[3]{(27:8)^4} = 81:16$.[32] In contrast, Stifel saw the "division" of a ratio by a ratio as a way of finding exponents. Thus, by means of continuous "subtraction," he obtained from the equality $(729:64) = (3:2)^x$ the "quotient" $x = 6$; and in the case of $(2187:128) = (27:8)^x$, the result was $x = 2 \, 1/3$.[33]

Stifel was also a pioneer in the development of algebraic symbolism. To designate the unknowns he used A, B, C, D, and F, as well as the traditional x.[34] For the powers he employed Az, $x^3 = AAA$, $x^4 = FFFF$, and so forth—not just the traditional z (census).[35] He simplified the square root sign from $\sqrt{z}$ to $\sqrt{.}$, and later to $\sqrt{}$ alone.[36] In one instance he closely approached modern symbolism, writing $\sqrt[11]{38}$ as $11\sqrt{38}$.[37] Other, more cumbersome designations for unknowns and the root used in *Deutsche arithmetica* were not adopted by later mathematicians.[38] The only operational signs that Stifel employed were $+$ and $-$; other operations were indicated verbally. Equality was designated either in words or by a point, as in $1x$. $\sqrt{48}$.[39]

Stifel made a thorough study of magic squares and polygonal numbers for a nonmathematical purpose.[40] He correlated the twenty-three letters of the alphabet with the first twenty-three triangular numbers, thereby establishing connections between words and numbers.[41] He called such *Wortrechnung* "the holy arithmetic of numbers."[42] For example, from the number 666 he derived the sentence "Id bestia Leo"; and the equality 2.5 ages = 1,260 days that he found in the book of Daniel yielded the sentence "Vae tibi Papa, vae tibi." Although Stifel's work at Holzdorf is most admired today, he declared that he valued his "word calculus" above all the computations he had ever made.[43]

The development of Stifel's scientific ideas was decisively influenced by Jacob Milich, who recommended that he study Campanus of Novara's translation of Euclid.[44] He also proposed that Stifel write a comprehensive work on arithmetic and algebra (which became *Arithmetica integra*).[45] To prepare for the latter project Stifel worked through Rudolff's *Coss* without assistance.[46] He had already studied proportions in the writings of Boethius, and had long been acquainted with contemporary arithmetic books, such as the *Margarita phylosophica* of Gregor Reisch (1503) and the works of Peter Apian (*Eyn newe unnd wolgegründete Underweysung aller Kauffmanns Rechnung*, 1527) and Adam Ries (*Rechnung auf der Lynihen un Federn in Zalmass und gewicht auff allerley handierung gemacht*, 1522).[47] (He especially admired Ries's book.)[48] Stifel conscientiously named the authors from whom he had taken examples and never neglected to express his appreciation.[49] The enthusiasm with which he followed Milich's advice[50] to collect mathematical writings is obvious from the large number of authors he cited.[51]

Stifel's achievements were respected and adopted by contemporary mathematicians, although, like Clavius, they often did not cite him.[52] The last edition of *Arithmetica integra* appeared in 1586, and the last of the *Coss* in 1615. After that, mathematicians surpassed Stifel's level of knowledge in symbolism and logarithms, and they opened new fields of research. It is for both these reasons that his work fell into neglect. He was, in fact, the greatest German algebrist of the sixteenth century.

NOTES

1. Stifel admitted that he was "ignarus linguae graecae." Since he knew Euclid only from the translations of Campanus

and Zamberti, he turned to other scholars for assistance. See *Arithmetica integra* (cited below as *AI*), fol. 143v.

2. In these numbers Stifel saw "sealed words" (*versiegelte Worte*). See *Wortrechnung* (cited below as *WR*), fol. B2ʳ.

3. The numbers (LDCIMV) that Stifel obtained from the name "Leo DeCIMVs" yielded 1656, 1,000 too much and 10 too little; but he manipulated them to obtain 666 by the addition of *decimus* = 10 and by setting M = *Mysterium*.

4. "Von der Christförmigen rechtgegründten leer Doctoris Martini Luthers . . ." (1522).

5. The castle was besieged by Franz von Sickingen's enemies and was taken on 15 Oct. 1522. A sermon that Stifel gave on 8 Sept. is still preserved; see J. E. Hofmann, "Michael Stifel, 1487?–1567," n. 14.

6. In the Grieskirchen congregation; see Ritter's *Geographisch-statistisches Lexicon*, II (Leipzig, 1906), 1051.

7. On the Biblical passages see J. E. Hofmann, "Michael Stifel," in n. 43. Other dates also were mentioned: 3 Oct. (J. H. Zedler, *Universallexikon*, XL [1744], 22); 16 Oct. (see Treutlein, p. 17, and Poggendorff, II, 1010–1011); and 19 Oct. (Giesing, p. 11). *Der Biograph*, p. 473, mentions the 282nd day and the 42nd week of the year.

8. For an eyewitness report in a letter from Petrus Weller to Ioannes Briessmann, see Strobel, pp. 74–84; the German trans. of the letter is given by Grosse, pp. 19 ff.

9. With Luther and Melanchthon, Milich was also friendly with Stifel at Annaberg and acted as the family's physician. See *AI*, fol. (α4)r, entry of 25 Oct. 1541, *Album der Universität Wittenberg*, I (Leipzig, 1841), 195a: "Gratis inscripti . . . Michael Stifel pastor in Holtzdorff."

10. A detailed description of the contents of the book can be found in Kaestner, *Geschichte der Mathematik*, I, 112 ff.; and in Treutlein; in Cantor; and in Hofmann, *op. cit.*

11. Stifel stated that the *regula falsi* is related to the coss as a point is to a circle. *AI*, 227r.

12. In the *Welsche Practick* (cited below as *WP*) Stifel objected to Apian's problems, which were correct but not comprehensible to everyone. He did not wish to blame Apian but to "diligently expound" his work (see pp. 293, 337). Stifel also drew on the "*praxis italica*" in *AI*, fols. 83v ff.

13. See *WR*. fol. A(1)v.

14. He went to Memel in 1549 and to Eichholz in 1550.

15. The title page bears the date 1553, the preface 1552, and the *explicit* 1554. Stifel made changes in the course of the printing, and thus the table of contents must be corrected. See *Coss*, fol. 179r.

16. Rudolff's 208 pages grew to 494 in Stifel's edition.

17. Osiander wrote on 19 Feb. 1549: ". . . Commentus est novos alphabeti numeros scil. triangulares et delirat multo ineptius quam antea" ("He has devised new numbers for the alphabet, namely the triangular numbers, and his fantasies are more absurd than before"). See B. F. Hummel, *Epistolarum historico-ecclesiasticarum saeculo XVI a celeberrimis viris scriptorum semicenturia altera* (Halle, 1780), 70 ff.

18. Like Stifel and Ottendorffer, Flacius was an opponent of Osiander.

19. Entry in *Die Matrikel der Universität Jena*, I (Jena, 1944), 320.

20. See J. E. Hofmann, "Michael Stifel," p. 59.

21. The MS, "Explicatio apocalypseos," is now in the Karl Marx University Library, Leipzig. The Lutheran congregation of Kronberg possesses a microfilm copy of the work. See W. Meretz, "Aus Stiefels Nachlass," p. 5.

22. The first book of *AI* is devoted to the fundamental operations—including roots, properties of numbers, series, magic squares, proportions, the rule of three, false substitution, and the Welsh practice; the second book treats computation with irrationals, corresponding to the tenth book of Euclid; the third book takes up algebra and equations of higher degree, such as were found in the work of Rudolff

and Cardano and that could be solved by employing certain devices.

The first part of the *Deutsche arithmetica* (cited below as *DA*) is devoted to "household computations" (*Hausrechnung*): carrying out on the abacus fundamental operations and the rule of three using whole numbers. The second part is concerned with computation with fractions, with the German *Coss* or "Kunstrechnung," and with extracting roots on the abacus. The third part, on "church computations" (*Kirchrechnung*), treats the division of the church year.

The division of the *Coss* is the same as in Rudolff's original edition. Stifel also reproduced Rudolff's *Wortrechnung*, but he did not agree with its contents. Among the new elements that he added were remarks on the higher-degree equations that Rudolff had presented; on the rules of the *Cubicoss* formulated by Scipione dal Ferro and Cardano that had been published in the meantime; and a procedure for computing

$$\sqrt[3]{a+\sqrt{b}}$$

Stifel's edition of the *Coss* also contained diagrams for verifying solutions. These had been drawn by Rudolff but did not appear in the original edition. Stifel obtained them from Johann Neudörfer, a brother-in-law of the printer Johannes Petrejus. See *Coss*, fol. 172r.

23. *DA*, fol. 72v.

24. Stifel reduced the three cases of the quadratic equation, $x^2 + a = b$, $x^2 + b = ax$, and $x^2 = ax + b$, to the standard form $x^2 = \pm ax \pm b$. By "extracting roots with cossic numbers" he obtained his rule called AMASIAS: $x = \sqrt{\left(\frac{a}{2}\right)^2 \pm b} \pm \frac{a}{2}$, where the plus and minus signs correspond to those of the standard form. See *AI*, fols. 240r f.; and Treutlein, *op. cit.*, 79. Stifel knew of the double solution only for $x^2 = ax$ (*AI*, fol. 243v). He avoided negative solutions, although he recognized negative numbers as those less than zero. (*AI*, fol. 48r). An equation of which the solution happened to be zero can be found in *AI*, fol. 283r.

25. For the term *quantitas*, see *AI*, fol. 257v; it was also used by Cardano (see *AI*, fol. 252r).

26. *DA*, fol. 17v.

27. *Ibid.*, fol. 72v. The table with binomial coefficients can be found in *AI*, fol. 44v; *DA*, fol. 71v; and *Coss*, fol. 168r.

28. See article on Fibonacci in this Dictionary, IV, 612, n. 7.

29. A detailed account is given by Hofmann in "Michael Stifel 1487?–1567," 13 ff.

30. *AI*, fol. 235.

31. "Posset hic fere novus liber integer scribi de mirabilibus numerorum sed oportet ut me hic subducam et clausis oculis abeam" (*AI*, fol. 249v). Translation from D. E. Smith, *History of Mathematics*, II (New York, 1925), 521.

32. Stifel distinguished between fraction (*Bruch*) and ratio (*Verhältnis*), and wrote the latter as a fraction without a fraction line. Nevertheless, he conceived of the ratio as a fraction; the quotient was its "name." Thus, 4:3 had the name 1 1/3. See *WP*, pp. 36 ff.; and *Coss*, fol. 135v.

33. First Stifel obtained, as the result of two "subtractions," $(2187:128) = (27:8)^2 (3:2)$ and then, because $(3:2) = (27:8)^{1/3}$, $(2187:128) = (27:8)^{2\,1/3}$. The details are in *AI*, fols. 53v ff. On computation with fractional power exponents and fractional radical indices, see *Coss*, fols. 138r f.

34. *AI*, fol. 254r.

35. *DA*, fol. 74v.

36. As Rudolff originally had it in the *Coss*.

37. *DA*, fol. 71r.

38. See *DA*, fol. 61v. Stifel extended the cumbersome symbols for the square root ($\sqrt[2]{\ }$) and cube root ($\sqrt[3]{\ }$) as far as the sixth root (*DA*, fol. 62r). The designations for the unknowns in the *DA* (see fols. 20 ff.) are $x = 1$ Sum:or 1 Sum A:, $x^2 = 1$ Sum:Sum; and so on, up to x^{11} (*DA*, fol. 70v).

39. *Coss*, fol. 351v. He also uses points to indicate inclusion of

several elements in the same operation, as in $\sqrt{z}.\ 6+\sqrt{z2}.=\sqrt{6}+\sqrt{2}.$ *AI*, fol. 112v.

40. The number 666 of the "great beast" appeared in *Ein Rechen Büchlin vom End Christ* (fols., H 4v and 5r) as the sum of all the cells of a magic square.

41. See *WR*, fol. D2r. For Stifel, i = j and u = v = w; Rudolff's alphabet, however, had twenty-four letters, since he included w.

42. See *WR*, fol. A(1)r.

43. See *Coss*, fol. 487v.

44. See *AI*, fol. 226v.

45. *AI*, fols. (α4)v.

46. See *WR*, fol. B(1)r; *Coss*, fol. A2r.

47. *AI*, fols. 55 f., 250r: *Coss*, fol. 23r; *AI*, fol. 102r; *WP*, fol. A2v.

48. *AI*, fol. 226v; *DA*, fol. 31r.

49. The problems come from Peter Apian, Cardano, Johann Neudörfer, Adam Ries, Adamus Gigas, Rudolff, and Widmann.

50. *AI*, fol. (α4)v.

51. Stifel names the following: Apian, Boethius, Campanus, Cardano, Nicholas Cusa, Dürer, Euclid, Gemma Frisius, Faber Stapulensis, Jordanus de Nemore, Neudörfer, Ptolemy, Reisch, Ries, Rudolff, Sacrobosco, Schöner, Theon of Alexandria, Zamberti, and Widmann.

52. See Hofmann, "Michael Stifel 1487?–1567," 31, n. 94.

BIBLIOGRAPHY

I. ORIGINAL WORKS. Stifel's works include *Ein Rechen Büchlin vom End Christ, Apocalypsis in Apocalypsin* (Wittenberg, 1532); *Arithmetica integra*, with preface by Melanchthon (Nuremberg, 1544; 1545; 1546; 1548; 1586); *Deutsche arithmetica, inhaltend Haussrechnung, deutsche Coss, Kirchrechnung* (Nuremberg, 1545); *Rechenbuch von der welschen und deutschen Practick . . .* (Nuremberg, 1546); *Ein sehr wunderbarliche Wortrechnung. Sampt einer mercklichen Erklärung ettlicher Zahlen Danielis unnd der Offenbarung Sanct Iohannis* ([Königsberg], 1553); and *Die Coss Christoffs Rudolffs. Die schönen Exempeln der Coss. Durch Michael Stifel gebessert und sehr gemehrt* (Königsberg, 1552–1553 [colophon dated 1554], 1571; Amsterdam, 1615).

Lists of Stifel's theological writings and songs are in J. E. Hofmann, "Michael Stifel 1487?–1567" (see below), and in the articles by G. Kawerau and W. Meretz cited below. Illustrations of the title pages of Stifel's books are given by Hofmann and Meretz.

II. SECONDARY LITERATURE. See *Allgemeine deutsche biographie*, VI (1893), 208–216; *Der Biograph*, VI (Halle, 1807), 458–488; F. J. Buck, *Lebensbeschreibung der preussischen Mathematiker* (Königsberg, 1764), 34–38; M. Cantor, *Vorlesungen über Geschichte der Mathematik*, 2nd ed., II (Leipzig, 1913), 430–449; C. J. Gerhardt, *Geschichte der Mathematik in Deutschland* (Munich, 1877), 60–74; J. Giesing, *Stifels Arithmetica integra* (Döbeln, 1879); J. Grosse, "Michael Stiefel, der Prophet," in *Westermanns Monatshefte*, no. 85 (Oct. 1863), 1–40; J. E. Hofmann, "Michael Stifel 1487?–1567," *Sudhoffs Archiv . . .*, supp. no. 9 (1968); and "Michael Stifel," in *Jahrbuch für Geschichte der Oberdeutschen Reichsstädte*, Esslinger Studien, **14** (1968), 30–60; G. Kawerau, "Stiefel (Styfel)," in *Realencyclopädie für protestantische Theologie und Kirche*, 3rd ed., X (Leipzig, 1907), 74–88, and XXIV (Leipzig, 1913), 529; A. G. Kaestner, *Geschichte der Mathematik*, I (Göttingen, 1796), 112–128, 163–184; W. Kaunzner, "Deutsche Mathematiker des 15. und 16. Jahrhunderts und ihre Symbolik," *Veröffentlichungen des Forschungsinstitutes des Deutschen Museums für die Geschichte der Naturwissenschaften und der Technik*, ser. A, no. 90 (1971); W. Meretz, "Uber die erste Veröffentlichung von Kronbergs erstem Pfarrer Michael Stiefel," in *Jahresberichte des Kronberger Gymnasiums, die Altkönigsschule* (Jan. 1969), 15–20; and "Aus Stiefels Nachlass," *ibid.* (Feb. 1969), 5–6; J. E. Montucla, *Histoire des mathématiques*, 2nd ed., I (Paris, 1799), 614; G. C. Pisansky, *Historia litteraria Prussiae* (Königsberg, 1765), 228; G. T. Strobel, *Neue Beiträge besonders zur Literatur des 16. Jahrhunderts*, I, pt. 1 (Nuremberg–Altdorf, 1790), 3–90; P. Treutlein, "Das Rechnen im 16. Jahrhundert," in *Abhandlungen zur Geschichte der Mathematik*, 2 (1879), 1–124, esp. 17 f., 33 ff., 42 f., 48 ff., 77 ff.; and F. Unger, *Die Methodik der praktischen Arithmetik* (Leipzig, 1888), see index, 239. Concerning a MS from 1599 with solutions to problems from the *Coss*, see D. E. Smith, *Rara arithmetica* (Boston–London, 1908), 493.

For further bibliographical information, see especially Hofmann, Kawerau, and Meretz.

KURT VOGEL

STILES, CHARLES WARDELL (*b.* Spring Valley, New York, 15 May 1867; *d.* Baltimore, Maryland, 24 January 1941), *zoology, public health.*

Stiles was the son of Samuel Martin Stiles, a Methodist minister, and Elizabeth White Stiles, both of whom belonged to old New England families. After attending high school in Hartford, Connecticut, he attended Wesleyan University for one year before going to Europe in 1886. General studies in Paris and Göttingen, and science studies at the University of Berlin, were followed by a concentration in zoology at the University of Leipzig with the parasitologist Rudolf Leuckart. After receiving the Ph.D. at Leipzig in 1890, Stiles studied in several important European laboratories before returning to the United States in 1891 to work in Washington as principal zoologist at the Bureau of Animal Industry of the Department of Agriculture.

At the Bureau of Animal Industry, Stiles's work included investigation of a wide range of animal parasites. Among these, his studies of trichinosis led to his assignment in 1898 and 1899 as science attaché at the American embassy in Berlin to investigate German allegations that imported American pork was unhealthy. Meanwhile, in 1893 and 1894 Stiles was instrumental in organizing the contingent of American scientists in residence at the Naples zoological station, and he served as secretary of its advisory committee for many years. Also during the 1890's he introduced medical zoology into the curricula of several Eastern medical schools. This and other health-related work led to Stiles's transfer in 1902 from the Bureau of Animal Industry to the Hygienic Laboratory of the United States Public Health Service, where he remained as chief of the division of zoology for the next thirty years.

Stiles's preeminent contributions to health were made in connection with hookworm disease. In 1902 he not only discovered a new variety of hookworm (*Uncinaria americana* [or *Necator americanus*]) but also showed it to be endemic among poor whites of the South. His subsequent efforts to obtain action against the parasite resulted in formation of the Rockefeller Sanitary Commission in 1909. With Stiles as medical director, the commission conducted a five-year campaign that resulted in noticeably improved sanitation and health. In this and in his later public health work, Stiles effectively combined the roles of health educator and epidemiologist with his principal work as a laboratory investigator.

As virtual successor to Joseph Leidy, Stiles contributed to both basic and applied zoology. A prodigious worker and keen observer, he systematically rearranged the principal American helminthological collections, and also identified and reported many new species of parasitic worms. Elected to the International Commission on Zoological Nomenclature in 1895, and its secretary from 1898 to 1936, he exerted great influence on the orderly development of the field. Of even greater importance was his publication, with Albert Hassall, of the monumental *Index-catalogue of Medical and Veterinary Zoology*, which, with its associated key catalogs of insects, parasites, protozoa, crustacea, and arachnids, was a continuing task from the 1890's until the mid-1930's.

BIBLIOGRAPHY

I. ORIGINAL WORKS. Relatively few of Stiles's personal papers have come to light thus far. MSS pertaining to certain aspects of his career are to be found, however, in the archives of the Smithsonian Institution; and there is also some material in the United States National Archives (Department of State files). Much of Stiles's voluminous scientific writing is in the form of reports or other official publications of the Bureau of Animal Industry, the Public Health Service, and the Rockefeller Sanitary Commission. See especially C. W. Stiles and Albert Hassall, *Index-catalogue of Medical and Veterinary Zoology* (Washington, D.C., 1908–   ). A short autobiographical account is "Early History, in Part Esoteric, of the Hookworm (*Uncinariasis*) Campaign in Our Southern United States," in *Journal of Parasitology*, **25** (1939), 283–308.

II. SECONDARY LITERATURE. While there is no book-length study of Stiles, several articles are available. Among the fullest are F. G. Brooks, "Charles Wardell Stiles, Intrepid Scientist," in *Bios* (Mt. Vernon, Iowa), **18** (1947), 139–169; and Mark Sullivan, *Our Times*, III (New York, 1930), 290–332. Shorter accounts are in *National Cyclopaedia of American Biography*, Current vol. D (1934), 62–63; *Who Was Who in America*, *1897–1942*; and *New York Times*, 25 Jan. 1941, 15.

Studies of particular phases of Stiles's career include James H. Cassedy, "The 'Germ of Laziness' in the South 1900–1915: Charles Wardell Stiles and the Progressive Paradox," in *Bulletin of the History of Medicine*, **45** (1971), 159–169; and "Applied Microscopy and American Pork Diplomacy: Charles Wardell Stiles in Germany 1898–1899," in *Isis*, **62** (1971), 4–20; and Benjamin Schwartz, "A Brief Résumé of Dr. Stiles's Contributions to Parasitology," in *Journal of Parasitology*, **19** (1933), 257–261.

JAMES H. CASSEDY

STILLE, WILHELM HANS (*b.* Hannover, Germany, 8 October 1876; *d.* Hannover, 26 December 1966), *tectonic geology.*

Stille was the son of Eduard and Meta Hankes Stille. His father was an army officer and later a manufacturer in Hannover. Hans married Hanna Touraine, of Huguenot ancestry, in 1903; their sons were Wilhelm, a lawyer, and Hans, a banker. Stille was a man of sturdy physique and vigorous health.

He began his studies of stratigraphy and tectonics near his home, continuing fieldwork in the area, with few interruptions, until later in life. Exploration in Colombia when he was a young man introduced him to a continent to which he devoted particular attention. He spent little time in research abroad but directed studies by many students in the western Mediterranean region. His wide and thorough reading helped him to become the leader in synthesizing global tectonics and to recognize some significant relationships.

Stille graduated from the Leibniz School in Hannover in 1896. He entered the Technische Hochschule there as a chemistry student but soon transferred to the University of Göttingen, where, under the influence of Adolf von Koenen, he studied geology until his graduation in 1899. He then worked for the Prussian Geological Survey until 1908, when he was summoned to teach in Hannover. In 1912 he replaced H. Credner as professor of geology and director of the Royal Saxon Geological Survey at Leipzig. The following year he was appointed successor to Koenen in Göttingen, where he established a reputation as an outstanding teacher and philosopher of global tectonics. He was named professor at the University of Berlin in 1932, remaining until he became emeritus in 1950.

Stille was a leader in German geology, an outstanding investigator and collator of the history of global tectonic events, and a highly admired teacher. From the time of his doctorate, he was interested in the chronological sequence of mountain-building events in time; his dissertation concerned late Jurassic orogeny in the Teutoburg Forest region. His concern was to date unconformities as evidence of orogenic events; he considered his "geotectonic classification of geologic history" to be his major contribution. From a review of world literature and his observations, Stille listed some fifty orogenic phases in Paleozoic and later time. He thought each to be essentially synchronous throughout the earth. The phases were smaller pulses in his tectonic eras—Precambrian Assyntic, Caledonic, Variscic, and Neoidic or Alpidic. Each orogenic phase made a part of the crust less mobile, the consolidations progressively enlarging the continents. He referred to progressive stages of European consolidation as Ur-, Pal-, Meso- and Neoeuropa.

Stille's synthesis of global tectonics has been considered a worthy successor to that of Eduard Suess. Stille conceived of the crust as separated into mobile orthogeosynclinal belts with marginal oceanic low cratons and continental high cratons, the latter with subsiding regions that he called parageosynclinal. Thus he had more intense "alpinotype" or orthotectonic deformation contrasting with the "germanotype" or paratectonic, which characterized western Germany. The terms were subsequently misconstrued. Stille later divided his original orthogeosynclinal zones into eugeosynclinal (the pliomagmatic zones or internides) and the relatively amagmatic miogeosynclinal zones (the externides). Although he coined the terms in 1941, he used them very rarely. His magmatic or volcanic geosynclinal belts of the earth (eugeosynclinal belts) gained wide application. During his lifetime it was generally accepted that the low cratons

(ocean basins) were essentially permanent, in contrast with the present view of their more dynamic and transient nature. Stille thought that some ocean basins might be foundered continental "high cratons" but that the Pacific was permanent. In his later years he accepted the idea of large-scale underthrusting of oceanic regions beneath the continents, as on the Pacific margin of South America.

Study of eugeosynclinal belts led to interest in their magmatic history. The progression that Stille described passed from an "initial" basic submarine volcanism (the extrusion of ophiolites) through successive sialic magmatic intrusions during and following deformation, and culminated in the final extrusion of surficial volcanic rocks on consolidated craton.

Pulses of Stille's orogenic time scale are recognized locally, but establishment of the universality of the many phases is beyond the resolution of present stratigraphic methods. Stille astutely recognized the restriction of magmatism to orogenic belts and the magmatic succession through time, although this latter was not wholly a novel concept. He accepted the growth of continents by consolidation of mobile belts, but he was hampered by the conventional dogma of the stability of oceanic cratons. As a pioneer in many aspects of relating the larger tectonic features of the earth, Stille directed attention to the explanation of relationships among large crustal features that are now becoming understood through the advent of geophysical techniques that he did not possess.

Stille received honorary doctorates from the universities of Bucharest, Hannover, Jena, Sofia, and Tübingen, and was elected to honorary membership in many academies of science, geological societies, and other scientific organizations. He was honorary president of the German Geological Society, which awarded him its Leopold von Buch Medal and established the Hans Stille Medal in his honor.

BIBLIOGRAPHY

I. ORIGINAL WORKS. Stille wrote 200 publications. His reputation was most firmly established by *Grundfragen der vergleichenden Tektonik* (Berlin, 1924). A second large work, *Einführung in den Bau Amerikas* (Berlin, 1941), was distributed to only a few close friends; although it is rarely seen, the conclusions were rather widely disseminated in other publications.

Stille's principal field studies were on the region of moderate folding in western Germany; an early work on this region was his doctoral dissertation, *Der*

Gebirgsbau des Teutoburger Waldes zwischen Altenbecken und Detmold (Berlin, 1900). He expressed his first interest in South America in *Geologische Studien im Gebiete des Rio Magdalena* (Stuttgart, 1907). Discussion of larger problems in tectonics include *Tektonische Evolutionen und Revolutionen in der Erdrinde* (Leipzig, 1913); *Die Begriffe Orogenese und Epirogenese* (Berlin, 1919); and "Present Tectonic State of the Earth," in *Bulletin of the American Association of Petroleum Geologists*, **20** (1936), 847–880. "Die Entwicklung des amerikanischen Kordillerensystems in Zeit und Raum," in *Sitzungsberichte der preussischen Akademie der Wissenschaften zu Berlin*, Math.–Phys. Kl. (1936), 134–155, was one of several papers concerned with the Americas. Later histories of specific regional tectonic systems are "Die assyntische Ära und der vormit-, und nachassyntische Magmatismus," in *Zeitschrift der Deutschen geologischen Gesellschaft*, **98** (1948); *Die saxonische Tektonik im Bilde Europas* (Hannover, 1949); and *Der Geotektonische Werdegang der Karpaten* (Hannover, 1953).

II. Secondary Literature. The full bibliography through 1955 was published in *Geotektonisches Symposium zu Ehren von Hans Stille*, Franz Lotze, ed. (Stuttgart, 1956), on his eightieth birthday; S. von Bubnoff prepared an appreciative note in *Geologie*, **5** (1956), 528–529: and A. Pilger, on his ninetieth birthday, in *Geologisches Jahrbuch*, **84** (1967), i–vii. H. J. Margini wrote an obituary in *Geologisches Jahrbuch*, **84** (1967), viii–ix; W. Carle, in *Jahresberichte und Mitteilungen des oberrheinischen geologischen Vereins*, **49** (1967), 17–19; and Roland Brinkmann, in *Proceedings. Geological Society of America* for 1967 (1970), 263–267. Later publications are listed in a short biographical note by Hans Hitlermann in *Berliner Naturhistorische Gesellschaft*, **112** (1968), 5–8. Stille directed more than 100 research students, of whom the earlier are listed in *Festchrift zum 60. Geburtstag von Hans Stille* (Stuttgart, 1936).

MARSHALL KAY

STIMPSON, WILLIAM (*b*. Roxbury, Massachusetts, 14 February 1832; *d*. Ilchester, Maryland, 26 May 1872), *marine zoology.*

Stimpson was the son of Herbert Hawthorne Stimpson and Mary Ann Devereau Brewer. The Stimpsons were an Episcopalian family which had settled in Massachusetts during the seventeenth century. The Brewers were an old Virginia family. During the middle of the nineteenth century the father was a prosperous stove merchant in Boston, having invented the "Stimpson range," which became well-known throughout New England. Stimpson's boyhood was spent near Harvard College in Cambridge, which was then a village with green fields and shaded groves. Here he developed such an interest in natural history that, at the age of fourteen, he presented himself to Augustus A. Gould, the author of the *Invertebrata of Massachusetts* (1841). Gould was so impressed with young Stimpson that he gave him a copy of his book and brought him to the attention of Louis Agassiz, William G. Binney, and other members of the Boston Society of Natural History. He soon began assisting Binney in the study of land snails.

Stimpson graduated from Cambridge High School in 1848, winning the school's highest academic award. Stimpson's father, a practical man with little education, could not envision the study of natural history as a profession. William therefore went to work for a firm of civil engineers, but his employer reported that he was too fond of collecting land snails to make a good surveyor. He then spent one year at the Cambridge Latin School. After a trip to the island of Grand Manan, New Brunswick, to dredge for marine invertebrates, he was reluctantly permitted to become a special student in Agassiz's laboratory at Harvard College in October 1850. On 4 December 1850 he was appointed curator of mollusks at the Boston Society of Natural History. He held this post until 18 May 1853, when, at the age of twenty-one, he was chosen as naturalist for the United States North Pacific Exploring and Surveying Expedition, commanded by Cadwalader Ringgold and, later, by John Rodgers.

The expedition, which lasted until 1856, visited Madeira, South Africa, Australia, the Coral Sea, Hong Kong, Japan, and the Aleutian Islands. Stimpson collected over 5,000 specimens, mostly invertebrates, and made notes and drawings of over 3,000 of them. From the time the expedition returned until 1865, he was in charge of the invertebrate section of the Smithsonian Institution. Stimpson described the Crustacea and other invertebrates collected by the expedition except for the mollusks, which were sent to A. A. Gould. In 1860, Columbia University awarded Stimpson an honorary M.D. in recognition of his knowledge of marine invertebrates. Five years later, he was appointed director of the Chicago Academy of Sciences, which was then moving into a new fireproof building.

Stimpson took with him to Chicago ten thousand jars of Crustacea, at that time the largest collection of its kind in the world: the invertebrates, except mollusks, of the North Pacific Exploring and Surveying Expedition, including the types of J. D. Dana; and his own collection of shells dredged from Maine to Texas; he later received the United States Coast Survey collection of deep-sea Crustacea and mollusks dredged in the Gulf Stream by L. F. de Pourtalès in 1867 and 1868. All these, including Stimpson's notes and drawings and the plates and text of the reports that were to augment the brief descriptions of the northern Pacific Crustacea and Mollusca, which he and Gould had published without figures, were destroyed in the Chicago fire of 1871.

Stimpson's health had never been very good, and he succumbed to tuberculosis in 1872, at the age of forty. He was survived by his wife, Annie Gordon, and a son, Herbert.

Aside from his faunal studies and monographs, Stimpson is remembered as the first naturalist to dredge systematically along the Atlantic coast and for the description of 948 new species of marine invertebrates.

BIBLIOGRAPHY

I. ORIGINAL WORKS. An annotated list of Stimpson's published works is available in Mayer (see below). Among his more important works are "Synopsis of the Marine Invertebrata of Grand Manan; or the Region About the Mouth of the Bay of Fundy, New Brunswick," *Smithsonian Contributions to Knowledge*, **6**, no. 5 (1854); "Researches Upon Hydrobiinae and Allied Forms," Smithsonian Miscellaneous Collection, **7**, no. 4; "Preliminary Report on the Crustacea Dredged in the Straits of Florida, by L. F. de Pourtalès, Pt. 1, Brachyura," in *Bulletin of the Museum of Comparative Zoology at Harvard College*, **2** (1871), 109–160.

II. SECONDARY LITERATURE. See W. H. Dall, "Some American Conchologists," in *Proceedings of the Biological Society of Washington*, **4** (1888), 129–133; W. K. Higley, "Historical Sketch of the Academy," in *Special Publications of the Chicago Academy of Sciences*, no. 1 (1902), 14–26; R. I. Johnson, "The Recent Mollusca of Augustus A. Gould," in *Bulletin. United States National Museum*, no. 239 (1964), 19–28, which contains excerpts from Stimpson's unpublished journal made on the North Pacific Exploring and Surveying Expedition; and A. G. Mayer, "Biographical Memoir of William Stimpson," in *Biographical Memoirs. National Academy of Sciences*, **8** (1918), 419–433.

RICHARD I. JOHNSON

STINE, CHARLES MILTON ALTLAND (*b.* Norwich, Connecticut, 18 October 1882; *d.* Wilmington, Delaware, 28 May 1954), *organic chemistry.*

Stine was the elder son of Milton Henry Stine, a Lutheran clergyman, and Mary Jane Altland Stine. He attended Gettysburg College, from which he received the B.A. in 1901, the B.S. in 1903, the M.A. in 1904, and the M.S. in 1905. He was awarded the Ph.D. by Johns Hopkins University in 1907, the same year in which he joined E. I. du Pont de Nemours and Company. During the early years of his lifelong association with Du Pont, Stine developed a number of products and processes in organic chemistry. He was one of the few American researchers familiar with the German synthetic dye process and his expertise enabled Du Pont to build, during World War I, a dye intermediates plant that was the first major synthetic organic chemical facility in the United States.

Stine became assistant chemical director of Du Pont in 1919; he was made director five years later. He had long been convinced that the company should undertake a program of fundamental research like those sponsored by the German chemical industry. Such programs were at that time virtually unknown in the United States outside the universities, and Stine—having developed research curricula in organic chemistry, chemical engineering, and catalytic processes, and having built a staff of exceptional men—was able to establish, by 1927, the research policy that made Du Pont an industrial pioneer in American science. Within a decade Du Pont's chemical department, under Stine's direction, had produced neoprene, the first general-purpose synthetic rubber, and nylon, the first noncellulosic synthetic fiber. More important, Stine's program gave new impetus and practical application to polymer chemistry and led to the creation of whole new families of fibers, films, plastics, paints, and elastomers and related products.

Stine was named vice-president and Executive Committee adviser on research in 1930. During his tenure he inaugurated and promoted research in agriculture and in animal nutrition and medicine. The animal research facilities that were completed in 1952 near Newark, Delaware, were named in his honor. He retired as vice-president in 1945, because of ill-health, but remained as a director of Du Pont until his death from a heart attack nine years later. He was survived by his wife, Martha E. Molly, whom he had married in 1912, and by their two daughters.

Stine's ability to organize and direct research on a wide variety of subjects brought him international recognition. He was a member of many scientific organizations, and particularly active in the American Association for the Advancement of Science, the American Chemical Society, the American

Institute of Chemical Engineers, and the Franklin Institute. He was a member of Phi Beta Kappa and a trustee of Gettysburg College and the University of Delaware, from both of which he received honorary degrees. The American Section of the Society of the Chemical Industry awarded Stine its Perkin Medal for 1939, "for valuable work in applied chemistry."

BIBLIOGRAPHY

I. ORIGINAL WORKS. Stine's writings include "Effect of One Salt on Hydrating Power of Another Salt Present in the Same Solution," in *American Chemical Journal*, **39** (1908), 313–402, written with Harry C. Jones; "Organic Synthesis and the du Pont Company," in *Chemical and Metallurgical Engineering*, **19** (1918), 569–571, written with C. L. Reese; "Chemical Engineering in Modern Industry," in *Transactions of the American Institute of Chemical Engineers*, **21** (1928), 45–54; "The Chemist's Aid to Agriculture," in *Chemicals*, **30**, no. 25 (1928), 7–9, and no. 26 (1928), 7–8; "Industrial Chemistry," in F. W. Wile, ed., *A Century of Industrial Progress* (Garden City, N.Y., 1928), 335–359; "Recovery of Bromine From Sea Water," in *Industrial and Engineering Chemistry*, **21** (1929), 434–442; "Structure of an Industrial Research Organization," *ibid.*, 657–659; "The Use of Power in Chemical Industries," in *Chemical Age* (London), **21** (1929), 237, also in *National Engineer*, **33** (1929), 273–274; "Chemical Research: A Factor of Prime Importance in American Industry," in *Journal of Chemical Education*, **9** (1932), 2032–2039; "Coordination of Laboratory and Plant Effort," in *Industrial and Engineering Chemistry*, **24** (1932), 191–193; and "Fundamental and Applied Chemical Research," in Malcolm Ross, ed., *Profitable Practice in Industrial Research* (New York–London, 1932), 104–118.

See also "Relation of Chemical to Other Industry," in *Industrial and Engineering Chemistry*, **25** (1933), 487–495; "Approach to Chemical Research Based on a Specific Example," in *Journal of the Franklin Institute*, **218** (1934), 397–410; "The Place of Fundamental Research in an Industrial Research Organization," in *Transactions of the American Institute of Chemical Engineers*, **32** (1936), 127–137; "The Value of Fundamental Research to Industry," in Carnegie Institute of Technology, Coal Research Laboratory, *Proceedings of Technical Meeting . . . December 3, 1936* (Pittsburgh, 1937), 48; "Training Tomorrow's Industrial Leaders," in *Transactions of the American Institute of Chemical Engineers*, **34** (1938), 643–656; and "The Rise of the Organic Chemical Industry in the United States," in *Smithsonian Report* for 1940 (1941), 177–192, reprinted as *Smithsonian Institution Publication* no. 3611 (1941).

II. SECONDARY LITERATURE. See A. D. McFadyen, "Personalities in Chemistry (Charles M. A. Stine)," in *Chemical Industries*, **46**, no. 6 (June 1940), 742; and "Perkin Medal," in *Industrial and Engineering Chemistry*, **32**, no. 2 (Feb. 1940), 137.

BETTINA F. SARGEANT

STIRLING, JAMES (*b.* Garden, Stirlingshire, Scotland, 1692; *d.* Edinburgh, Scotland, 5 December 1770), *mathematics.*

Stirling was the third son of Archibald Stirling and his second wife, Anna Hamilton, and grandson of Lord Garden of Keir. The whole family supported the Jacobite cause, and Archibald Stirling was in prison on a charge of high treason (of which he was later acquitted) while his son attended Glasgow University. James Stirling matriculated at Balliol College, Oxford, in 1711, without taking the oath. He himself was acquitted of the charge of "cursing King George" at the assizes. He seems to have left Oxford in 1716, after refusing to take the oaths needed to continue his scholarship. He did not graduate.

The previous year John Keill had mentioned Stirling's achievements in a letter to Newton, and at about the same time Stirling became acquainted with John Arbuthnot, the well-known mathematician, physician, and satirist. Such connections enabled him to publish (in Oxford) his first book, *Lineae tertii ordinis Neutonianae, sive illustratio tractatus D. Neutoni de enumeratione linearum tertii ordinis* (1717). The eight-page subscription list included Newton himself, besides many Oxford men. The book was dedicated to Nicholas Tron, the Venetian ambassador, who had become a fellow of the Royal Society in 1715, the same year in which Newton's correspondent, the Abbé Conti, was also elected. Stirling may then have held a teaching appointment in Edinburgh,[1] but the fame brought him by his book and the influence of his Venetian friends soon secured him a post in Venice. In 1718 Stirling submitted, through Newton, his first Royal Society paper, "Methodus differentialis Newtoniana illustrata," and in August 1719 he wrote from Venice thanking Newton for his kindness and offering to act as intermediary with Nikolaus I Bernoulli.

Little else is known about Stirling's stay in Venice, although his return to Britain is supposed to have been hastened because he had learned some secrets of the glass industry and may have feared for his life. By mid-1724 he had returned to Scotland, and a few months later he settled in London. In 1726 Newton helped secure Stirling's fellowship in the Royal Society and at about this time

Stirling succeeded Benjamin Worster as one of the partners of the Little Tower Street Academy,[2] conducted by William Watts. This was one of the most successful schools in London; and, although he had to borrow money to pay for the mathematical instruments he needed, Stirling's finances improved. He helped to prepare *A Course of Mechanical and Experimental Philosophy* (to give it the title of a syllabus published in 1727) that included mechanics, hydrostatics, optics, and astronomy, a course very much in the tradition of Keill and Desaguliers, the leading scientific lecturer at that time. Stirling gave up some of his leisure to write his main work, *Methodus differentialis*, which appeared in 1730. A little later, through his friend Arbuthnot, Stirling was brought in as an adviser to Henry St. John, Lord Bolingbroke, since he was considered to be one of the few persons capable of understanding the financial calculations of Sir Robert Walpole. The latter's electoral victory of 1734 led to Bolingbroke's retirement to France.[3]

Given his reputation it was not surprising that Stirling was asked to reorganize the work of the Scottish Mining Company in the lead mines at Leadhills, Lanarkshire, near the border with Dumfries. Stirling was a successful administrator and spent most of his time after 1735 in the remote village. He married Barbara Watson of Thirty-acres, near Stirling; their only child, a daughter, married her cousin Archibald Stirling, who succeeded Stirling as manager at Leadhills.

Although Stirling continued his mathematical correspondence—with John Machin, Alexis-Claude Clairaut, Leonhard Euler, and Martin Folkes, among others—it is clear that most of his energy was spent in mining affairs. His most influential mathematical correspondent, Colin Maclaurin, died in 1746, largely as a result of his efforts in defending Edinburgh against the Jacobite rebellion of the preceding year; Stirling's own political principles prevented him from succeeding to the Edinburgh chair left vacant at Maclaurin's death.[4] In 1748 Stirling was elected to the Berlin Academy of Sciences, even though his directly mathematical activities had ceased; he resigned his fellowship in the Royal Society in 1754. In 1752 he was presented with a silver teakettle for conducting the first survey of the Clyde by the town council of Glasgow, where he also apparently acted as a teacher of bookkeeping, navigation, geography, practical mathematics, and French.[5] In his later years he became too frail to move about easily; he died on a visit to Edinburgh for medical treatment.

Stirling's *tractatus* of 1717 won him a considerable reputation. In it, after a considerable amount of introductory material, Stirling proved Newton's enumeration of seventy-two species of cubic curves and added four more. François Nicole and Nikolaus I Bernoulli then added two more curves, in 1731 and 1733, respectively, the latter in a letter to Stirling.

Stirling next turned from cubics to differences, the other main topic of Newton's *Analysis* (1711). But these studies were interrupted by his moving from Oxford to Venice, from which he wrote to give permission for publication (without an intended supplement) of his 1719 paper "Methodus differentialis." This paper should not be confused with a later book of similar title, but it may be considered a precursor to it, since the book represents the further development and fuller treatment of the same ideas. Some of the same results are given in both; the so-called Newton-Stirling central difference formula,[6] which was also discussed by Cotes, is especially noteworthy.

The *Methodus differentialis: sive tractatus de summatione et interpolatione serierum infinitarum* of 1730 consists of a relatively brief introduction followed by two parts, on summation and interpolation. The work was sufficiently important to be reprinted twice during Stirling's lifetime, in 1753 and 1764, and to be published in an English translation in 1749.[7] The translation was made by Francis Holliday, who was then master of a grammar school near Retford, Nottinghamshire, as well as editor of *Miscellanea curiosa mathematica*, one of a number of relatively short-lived popular mathematical serials[8] published during the mid-eighteenth century. (The translator's preface shows that Holliday had originally intended to publish the translation in his serial and indicates that he planned to follow *The Differential Method* with other translations of Stirling's work as well; perhaps the reception of the book was insufficiently favorable for these other plans to materialize.)

In his preface, Stirling indicated that Newton, too, had considered the problem of speeding the convergence of series by transformations involving differences. De Moivre had made progress with a recurring series, but his methods could be generalized to other series in which "the relation of the terms is varied according to some regular law." The most useful representation of terms was in a series of factorials, positive or negative. Manipulation often required conversion of factorials into powers, and Stirling gave tables of the coefficients for this conversion. He then showed that the columns of the tables gave the coefficients for the inverse expressions, of powers in factorials; those for positive (negative) powers are now called "Stirling Numbers of the first [or second] kind" in his honor.[9] The so-called Stirling series

$$\frac{1}{x-a}=\frac{1}{x}+\frac{a}{x(x+1)}+\frac{a(a+1)}{x(x+1)(x+2)+\cdots}$$

is equivalent to the expansion of $(z^2+nz)^{-1}$ in negative factorials, which is the last example given in his introduction.

Stirling explained that part one of the *Methodus differentialis*, "on the summation of series," was designed to show how to transform series in order to make them converge more rapidly and so to expedite calculation. As an example[10] he gave the series

$$\sum_1^\infty 1/(2n-1)2n,$$

studied by Brouncker in connection with the quadrature of the hyperbola; Stirling concluded that "if anyone would find an accurate value of this series to nine places . . . they would require one thousand million of terms; and this series converges much swifter than many others. . . ." Another example[11] was the calculation—"which Mr. Leibnitz long ago greatly desired"—of

$$\frac{\pi}{4}=1-\frac{1}{3}+\frac{1}{5}-\frac{1}{7}+\cdots.$$

Stirling's sixth proposition was effectively an early example of a test for the convergence of an infinite product; he gave many examples of problems, now solved by the use of gamma functions, that illustrated his aim. The last section of the first part of the book contains an incomplete development of De Moivre's principles used in recurring series; for linear relations with polynomial coefficients connecting a finite number of terms, Stirling reduced the solution to that of a corresponding differential equation.

Stirling continued to show his analytical skill in part two, on the interpolation of series. As an example[12] of interpolation at the beginning of a series, he took the gamma series $T_{n+1}=nT_n$, with $T_1=1$, to find the term $T_{3/2}$ intermediate between the two terms T_1 and T_2 and calculated the result to ten decimal places: his result is now written $\Gamma(1/2)=\sqrt{\pi}$. Stirling's other results are now expressed using gamma functions or hypergeometric series. He also discussed[13] the sum of any number of logarithms of arguments in arithmetical progres-

sion and obtained the logarithmic equivalent of the result, sometimes called Stirling's theorem, that

$$n! \sim n^{n+1/2}e^{-n}.$$

Just before leaving London, Stirling contributed a short article to the *Philosophical Transactions of the Royal Society* entitled "Of the Figure of the Earth, and the Variation of Gravity on the Surface." In it he stated, without proof, that the earth was an oblate spheroid, supporting Newton against the rival Cassinian view. This paper was unknown to Clairaut, who submitted a paper partly duplicating it from Lapland, where he was part of the expedition under Maupertuis that proved Newton's hypothesis.[14] Although Stirling contributed another technical paper ten years later, it is clear that his new post in Scotland did not give him an opportunity to pursue his mathematical activities in any depth and that his significant work was confined to the 1720's and 1730's.

NOTES

1. W. Steven, *History of George Heriot's Hospital*, F. W. Bedford, ed. (Edinburgh, 1859), 307, mentions James Stirling as assistant master, elected 12 August 1717.
2. N. Hans, *New Trends in Education in the Eighteenth Century* (London, 1951), 82–87, gives the best account of the Academy, but his dates for Stirling and Patoun are unreliable.
3. The connection with Bolingbroke is given by Ramsay (see bibliography), 308–309, but is ignored in most accounts.
4. A. Grant, *The Story of the University of Edinburgh*, II (London, 1884), 301.
5. Glasgow City Archives and *Glasgow Courant*, Nov. 1753, Nov. 1754, and Nov. 1755, reported by M. J. M. McDonald and J. A. Cable respectively.
6. D. T. Whiteside, ed., *The Mathematical Papers of Isaac Newton*, IV (Cambridge, 1971), 58, n. 19.
7. The Latin and English versions had 153 and 141 pages, respectively; references in Tweedie to the Latin ed. can be converted to those to the latter, given here, by subtracting about ten.
8. R. C. Archibald, "Notes on some Minor English Mathematical Serials," in *Mathematical Gazette*, **14**, no. 200 (April 1929), 379–400.
9. Stirling, *Differential Method*, 17, 20. A useful, modern textbook, C. Jordan's *Calculus of Finite Differences*, 2nd ed. (New York, 1947), devotes ch. 4 to Stirling's numbers. Jordan and Tweedie give details of the articles by N. Nielsen that stress the significance of Stirling.
10. *Differential Method*, 23–25.
11. *Ibid.*, 27–28.
12. *Ibid.*, 99–103.
13. *Ibid.*, 123–125.
14. I. Todhunter, *A History of the Mathematical Theories of Attraction*, I (London, 1873), ch. 4.

BIBLIOGRAPHY

I. ORIGINAL WORKS. Stirling's works are listed in the text. His major work is *Methodus differentialis: sive tractatus de summatione et interpolatione serierum infinitarum* (London, 1730). The family papers are at the General Register House, Edinburgh; they contain disappointingly few mathematical papers, but more about Stirling's mining activities.

II. SECONDARY LITERATURE. The main authority is the unindexed volume C. Tweedie, *James Stirling: A Sketch of His Life and Works Along With His Scientific Correspondence* (Oxford, 1922); also J. O. Mitchell's *Old Glasgow Essays* (Glasgow, 1905), repr. from "James Stirling Mathematician," *Glasgow Herald* (1886); and J. Ramsay's *Scotland and Scotsmen in the Eighteenth Century*, A. Allardyce, ed., II (Edinburgh, 1888), 306–326. Other works are detailed in the notes and in Tweedie.

P. J. WALLIS

STOCK, ALFRED (*b.* Danzig, West Prussia [now Gdansk, Poland], 16 July 1876; *d.* Aken, Germany, 12 August 1946), *chemistry.*

The son of a bank executive, Stock received a Gymnasium education in Berlin. He developed an interest in science as a schoolboy and matriculated as a chemistry student at the University of Berlin in 1894. Attracted to Emil Fischer's institute, he began his doctoral research in 1895 under Oscar Piloty. Following his graduation, magna cum laude, in 1899, Stock spent a year in Paris as an assistant to Moissan. On 21 August 1906 he married Clara Venzky, who bore him two daughters. Stock pioneered in the development of the chemistry of the boron and silicon hydrides, developed the chemical high-vacuum technique, held numerous high positions in chemical organizations and educational institutes, and made important contributions to chemical education and nomenclature. Some of his work had important technological applications. His intensive investigations into the pathology and detection of mercury were prompted by his desire to spare others the suffering he endured from the effect of chronic mercury poisoning.

Stock devoted the first nine years (1900–1909) of his professional career at the University of Berlin to the preparation and characterization of the modifications of phosphorus, arsenic, and antimony, and their compounds with hydrogen, sulfur, and nitrogen. He identified an unstable yellow form of antimony, and a new polymeric hydride (P_2H_9) and nitride (P_3N_5) of phosphorus. His research on the phosphorus-sulfur system resolved many ambiguities in the literature and established the existence of three of the four well-established phosphorus sulfides. His study of the kinetics of the decomposition of antimony hydride was a classic example of an autocatalytic, heterogeneous decomposition.

In July 1909, Stock was appointed to organize and direct the new Inorganic Chemistry Institute at Breslau, where he began an experimental program inspired by the belief that boron ought to have an extensive and diverse chemistry analogous to that of the hydrocarbons. Previous attempts, by Ramsay and others, to produce hydroboranes by the reaction of magnesium boride with mineral acids had led to incomplete and erroneous conclusions because of the difficulty of isolating and characterizing the small amounts of volatile, unstable, and highly reactive compounds formed. The problem was complicated by the presence of silanes and other gases in the reaction mixture. Stock devised a high-vacuum apparatus that permitted the quantitative manipulation and fractionation of small amounts of gaseous and volatile materials in the absence of air and lubricating greases. He soon established the existence of three boranes—B_2H_6, B_4H_{10}, $B_{10}H_{14}$—and produced evidence for others that later proved to be B_5H_9 and B_6H_{10}.

In 1916 Stock moved to the Kaiser Wilhelm Institute in Berlin. During this time he concerned himself with problems related to the war and to the restoration of German chemistry afterward. His research efforts were devoted primarily to the study of the reactive and volatile silicon hydrides. At the time of Stock's entry into this field, only SiH_4 had been well characterized. His high-vacuum technique enabled him to purify and characterize Si_2H_6 and two new silanes, Si_3H_8 and Si_4H_{10}, and to establish the existence of Si_5H_{12} and Si_6H_{14}. Stock prepared numerous halogen derivatives of these compounds that in turn were used to produce many new and interesting compounds, such as siloxanes, silyl amines and amides, and alkyl silanes. Supported by the experimental knowledge obtained through work on the silanes, he prepared very pure samples of B_2H_6 and B_4H_{10} and isolated B_5H_9 and B_6H_{10}. In 1924 he discovered yet another borane, B_5H_{11}.

Ever since his early postdoctoral years, Stock had suffered from headaches, vertigo, numbness, catarrhs, poor hearing, and ailments of the upper respiratory tract that would not respond to any medical treatment. During 1923 he suffered an almost total loss of memory and hearing. In March 1924, after an unbearable winter, he discovered that his difficulties stemmed from mercury poisoning, caused by prolonged exposure to mercury vapors (several micrograms of mercury per cubic meter) in poorly ventilated laboratories. After a brief rest Stock began a program of research into the analysis and pathology of mercury that continued to the end of his active life. He devised techniques capable of detecting 0.01 microgram of mercury, which he used to determine mercury concentrations in natural materials and common reagents. He examined the modes of ingestion of mercury and traced its path through the body and its accumulation in various organs, often using himself as an experimental subject. He wrote numerous articles warning of the dangers of mercury and suggested precautionary measures.

After his appointment as director of the Chemical Institute at Karlsruhe in October 1926, Stock constructed a model chemical laboratory designed to minimize mercury contamination. Here he determined the genetic relationships between the boranes and developed more effective preparation techniques; studied the reactions of boranes with active metals and prepared compounds that were later recognized as borohydrides, now an important class of reducing agents in organic chemistry; and prepared borazine $B_3N_3H_6$, an important inorganic analogue of benzene.

In addition to his hydride research, Stock investigated a number of compounds of carbon with oxygen, sulfur, selenium, and tellurium; introduced many improvements in apparatus, including the mercury valve, automatic Töpler pump, vapor-tension thermometer, zinc arc, and gas buoyancy balance; and devised numerous accessories for the widely used Stock high-vacuum technique. He drew up a system of chemical nomenclature that has been universally adopted by inorganic chemists. Stock perfected an instructional epidiascope that could project chemical demonstrations on a large screen by means of incident or reflected light. Some of his research led to industrial applications: P_4S_3 was used as a substitute for phosphorus in matches, and P_3N_5 was used to remove traces of oxygen from incandescent lamps. His technique for the electrolytic preparation of beryllium became the basis for the first commercial production of that metal. His work on silicon helped establish the chemical foundations for the technologically important silicone plastics.

Stock not only opened up two important fields of chemistry but also devised apparatus and techniques of great versatility and exactness that made the achievement possible, and established the associated health hazards and precautionary measures necessary for safety. E. Wiberg described him as punctual, neat, exact, and patient, yet witty and sociable. Although his work stimulated a great deal of theoretical speculation, and in some cases led to

practical applications, Stock always remained an experimentalist who was driven by the pure joy of discovery. He placed as little value on speculative opinion as he did on seeking practical applications of his research.

Stock received many honors, including the presidency of both the Verein Deutscher Chemiker (1926–1929) and the Deutsche Chemische Gesellschaft (1936–1938). He was appointed to many major policy committees and represented Germany at several international conferences. His deteriorating health and increasing difficulties with political authorities caused him to retire in October 1936, at the age of sixty. He returned to Berlin, where he continued his research on mercury. By 1940 his movements had become so restricted by the onset of myogelosis (hardening of the muscles) that he was almost completely confined to his home. In May 1943 he gave up his small laboratory, which was needed for war work, and retired to Bad Warmbrunn in Silesia. With Russian cannon rumbling in the distance, he and his wife packed the few belongings they could carry and undertook a grueling train trip to Aken, a small city on the Elbe. Stock died there in obscurity, after a life of great accomplishment and tragic suffering.

BIBLIOGRAPHY

I. ORIGINAL WORKS. A complete bibliography of Stock's 274 publications is in Wiberg (see below). His books include *Praktikum der quantitativen anorganischen Analyse* (Berlin, 1909), written with Arthur Stähler; *Ultra-Strukturchemie* (Berlin, 1920); and *Hydrides of Silicon and Boron* (Ithaca, N.Y., 1933).

II. SECONDARY LITERATURE. The major biography is E. Wiberg, "Alfred Stock," in *Chemische Berichte*, **83** (1950), xix–lxxvi; a short summary of this article is in E. Farber, ed., *Great Chemists* (New York, 1961), 1423–1432.

JOHN E. FREY

STOCK, CHESTER (*b.* San Francisco, California, 28 January 1892; *d.* Pasadena, California, 7 December 1950), *paleontology*.

Stock's parents, John Englebert Stock and the former Maria Henriette Meyer, were natives of Germany; and he attended a Gymnasium as well as public schools until the earthquake and fire of 1906 destroyed the family home and business and obliged him to go to work. In 1910 he entered the University of California, where the odors of the zoology laboratory discouraged him from premedical studies and John C. Merriam's lectures whetted his interest in prehistoric animals. He published a paper on the ground sloths from the tar pits of Rancho La Brea before graduating in 1914; in 1917 he received the Ph.D. and joined the university faculty. When Merriam left Berkeley in 1921, Stock, by then assistant professor, took over his courses. He was called to a professorship at California Institute of Technology in Pasadena in 1926, and in 1947 he became chairman of its Division of Earth Sciences. He also served on the staff of the Los Angeles County Museum, becoming senior curator of earth sciences in 1944 and chief of the Science Division in 1949.

Stock married Clara Margaret Doud in 1921; they had a daughter, Jane Henriette, and a son, John Chester. His wife died in 1934; and in 1935 he married Margaret Wood, by whom he had a second son, James Ellery.

Stock's early studies were on Pleistocene vertebrates, especially ground sloths, on which he published a major monograph in 1925; he also collaborated with John C. Merriam in a monographic study of saber-toothed cats. He continued Merriam's program of exploration for vertebrate fossils in the Great Basin as well as along the Pacific coast, and extended these studies into northern Mexico. Stock's discovery of Eocene land mammals in the Sespe formation of the Ventura basin and near San Diego, and his description of Oligocene mammals from Death Valley, extended the record of land animals on the Pacific coast far earlier than previously known Miocene remains.

Stock's scientific publications consist largely of meticulous descriptions of fossil material, including careful documentation of its geologic occurrence and sound systematic conclusions. He also published several interpretive studies of the Rancho La Brea fauna and of various earlier mammalian assemblages.

Stock attracted many more students than could enter the field of vertebrate paleontology. He was always eager to explain his work to nonprofessional audiences; and his enthusiasm for the life of the past and his warm, cheerful personality won him many friends, within and outside the geological profession, who often supplied him with leads to new fossil occurrences. At the Los Angeles County Museum he was particularly concerned with developing an exhibition of fossil bones in situ at the tar pits in Hancock Park and with plans for the Hall of Evolving Life.

Stock's honors included membership in the National Academy of Sciences and presidencies of the Paleontological Society of America (1945), the Society of Vertebrate Paleontology (1947), and the Geological Society of America (1950), the latter a few weeks before his death.

BIBLIOGRAPHY

I. ORIGINAL WORKS. Stock's writings include "Cenozoic Gravigrade Edentates of Western North America With Special Reference to the Pleistocene Megalonychinae and Mylodontinae of Rancho La Brea," which is *Carnegie Institution of Washington Publication* no. 331 (1925); "Rancho La Brea, a Record of Pleistocene Life in California," which is *Los Angeles County Museum Publication* no. 1 (1930); 6th ed., 1956); and "The Felidae of Rancho La Brea," which is *Carnegie Institution of Washington Publication* no. 422 (1932), written with John C. Merriam. His technical publications on Tertiary and Pleistocene faunas are in *University of California Publications in Geological Sciences* and *Carnegie Institution at Washington Contributions to Paleontology*. Eocene fossils are described in *Proceedings of the National Academy of Sciences of the United States of America*. Full bibliographies are given in Simpson and in Woodring (see below).

II. SECONDARY LITERATURE. See John P. Buwalda, in *Bulletin of the American Association of Petroleum Geologists*, **35** (Mar. 1951), 775–778, with portrait; R. W. Chaney, in *Yearbook. American Philosophical Society* for 1951 (1952), 304–307; Hildegard Howard, in *News Bulletin. Society of Vertebrate Paleontology*, **31** (1951), 32–34, and in *Quarterly. Los Angeles County Museum*, **8**, nos. 3–4 (1951), 15–18, with portrait; George G. Simpson, in *Biographical Memoirs. National Academy of Sciences*, **27** (1952), 335–362, with portrait; and Wendell P. Woodring, in *Proceedings. Geological Society of America* for 1951 (1952), 49–50, 149–156.

JOSEPH T. GREGORY

STODOLA, AUREL BOLESLAV (*b.* Liptovský Mikuláš, Hungary [now Czechoslovakia], 10 May 1859; *d.* Zurich, Switzerland, 25 December 1942), *mechanical engineering.*

Stodola was born in a small Slovakian town at the foot of the High Tatra mountains, the second son of a leather manufacturer, Andreas Stodola, and his wife, Anna Kovač. After attending secondary school, he studied at the Budapest Technical University (1876–1877), the University of Zurich (1877–1878), and the Eidgenössische Technische Hochschule (1878–1880), from which he graduated as a mechanical engineer. He completed his theoretical and practical education informally: he spent two years as a volunteer in the machine factory of the Hungarian State Railways, studied in Berlin (1883) and Paris (1884), and worked in his father's factory and in a machine shop in Brno. In his first permanent position (1886–1892), as chief engineer with Ruston & Co. in Prague, Stodola was responsible for the design of a great variety of steam engines. In 1892 the Eidgenössische Technische Hochschule of Zurich, where his exceptional performance as a student was still remembered, appointed him to its newly founded chair for thermal machinery. Stodola held this position until his retirement in 1929, after which he remained in Zurich. He was married to Darina Palka in 1887; they had two daughters.

During his lifetime Stodola's reputation was very great. In the technical sense his role is best described by saying that during the decades of the main growth of the steam turbine, he was the leading authority on that machine. He also had remarkable personal charm. The importance of his accomplishments, the broad range of his learning, his friendly but aristocratic personality, and his ascetic appearance made him seem the embodiment of the central European ideal of a professor; the loyalty of his friends (among them Einstein) and disciples was extraordinary; and the eulogistic writing devoted to him is not only remarkable in quantity but also uncommonly persuasive.

In Stodola's academic career teaching, industrial consultation, engineering design, and scientific research were intimately blended; perhaps most important was his scientific work, which was characterized by a combination of high mathematical competence with an explicit devotion to practical utility.

Stodola's first publications dealt with the theory of automatic control. He derived the differential equations for the speed-regulating systems of hydroelectric power plants, analyzing them with respect to dynamic stability and checking his results at the power plant itself. In this work he enlisted the help of a colleague, Adolf Hurwitz, who was led through it to the discovery (1895) of the stability criterion named after him.

At the turn of the century Stodola's attention shifted to the steam turbine, which, after its invention in the 1880's, was gradually coming into industrial use and was soon to displace the reciprocating steam engine. It became the subject of his lifework. A lecture given in 1902 before the Society of German Engineers became the book *Die Dampfturbinen und die Aussichten der Wärmekraftmaschinen* (1903), which in later editions, enti-

tled *Dampf- und Gasturbinen*, developed into an authoritative work that holds a unique place in the literature of engineering. Although Stodola also published many articles, most results of his researches were incorporated directly into the various editions of this book.

Perhaps the book's most important innovation concerned the basic thermodynamic treatment of the flow of steam through the turbine. It had become clear that it was impractical to describe the properties of steam, which changed with temperature and pressure, in terms of mathematical formulas; instead these data, which were determined empirically, were presented in graphic charts and printed tables. Recognizing the superiority of entropy charts, which were easy to manipulate and at the same time facilitated the comparison of actual and ideal processes, Stodola was the first to base the analysis of heat engines upon these charts. After having first used the older temperature-entropy charts, he soon turned to the more convenient enthalpy-entropy charts proposed by Richard Mollier (1904). Shortly thereafter this approach was adopted universally.

Stodola also did pioneer work in the flow of steam through Laval nozzles. The possibility of supersonic flow in divergent nozzles was then still a matter of controversy. In careful experiments Stodola studied the conditions under which supersonic flow does occur and obtained results that for the first time could be reconciled with theoretical predictions. Through this line of inquiry he also became one of the first to investigate shock waves.

Although thermodynamics and fluid mechanics and their applications were Stodola's true domain, he also did original work in pure mechanics. He investigated, for example, the strength of disks rotating at high speed and problems of vibration and critical speed.

Stodola's innumerable more technical contributions to steam turbine design are reflected in the fact that the Swiss steam turbine manufacturers, who retained him as a consultant and employed his students, became international leaders in this field. They were also among the pioneers of the gas turbine, a machine in which Stodola had been interested since the beginning of his career, when its prospects had seemed dim indeed.

Stodola had a strong sense of social responsibility. During World War I he worked on the problem, totally unrelated to his basic interests, of improving artificial limbs. Social responsibility is also shown in a later philosophical and reflective work, *Gedanken zu einer Weltanschauung vom Stand-*

punkt des Ingenieurs (1931), which went through several editions.

BIBLIOGRAPHY

I. ORIGINAL WORKS. Stodola's publications are listed in Poggendorff, VI, 2551, and VIIa, pt. 4, 550; in his *Festschrift* (see below), xxi–xxiii; and in *Schweizerische Bauzeitung*, **121** (1943), 77–78. His major book is *Die Dampfturbinen und die Aussichten der Wärmekraftmaschinen* (Berlin, 1903), 5th ed. retitled *Dampf- und Gasturbinen* (Berlin, 1922; 6th ed., 1925); 1st ed. translated by Louis C. Lowenstein as *Steam Turbines, With an Appendix on Gas Turbines, and the Future of Heat Engines* (New York, 1905; 2nd ed., 1906), and the 5th ed. by Lowenstein as *Steam and Gas Turbines* (New York, 1927) and by E. Hahn as *Turbines à vapeur et à gaz*, 2 vols. (Paris, 1925). Other books are *Gedanken zu einer Weltanschauung vom Standpunkt des Ingenieurs* (Berlin, 1931); and *Geheimnisvolle Natur: Weltanschauliche Betrachtungen* (Zurich, 1937).

II. SECONDARY LITERATURE. The most useful biographical treatments of Stodola are the following, listed chronologically: W. G. Noack, "Prof. Dr. Aurel Stodola," in E. Honegger, ed., *Festschrift Prof. Dr. A. Stodola zum 70. Geburtstag* (Zurich–Leipzig, 1929), ix–xx; G. Eichelberg, "Aurel Stodola," in *Schweizerische Bauzeitung*, **121** (1943), 73–74; H. Quiby, "Aurel Stodola, sein wissenschaftliches Werk," *ibid.*, 74–77; E. Sőrensen, "Aurel Stodola," in *Zeitschrift des Vereins deutscher Ingenieure*, **87** (1943), 169–170; Anton Turecký, ed., *Aurel Stodola 1859–1942, pamiatke storočnice narodenia* (Bratislava, 1959); C. Keller, "Zum 100. Geburtstag von Aurel Stodola am 10. Mai 1959," in *Zeitschrift des Vereins deutscher Ingenieure*, **101** (1959), 558–560; and A. Sonntag, "Aurel Stodola 1859/1942," in *Brennstoff-Wärme-Kraft*, **11** (1959), 211–212.

OTTO MAYR

STOKES, GEORGE GABRIEL (*b.* Skreen, County Sligo, Ireland, 13 August 1819; *d.* Cambridge, England, 1 February 1903), *physics, mathematics.*

Stokes was born into an Anglo-Irish family that had found its vocation for a number of generations in the established Church of Ireland. His father, Gabriel Stokes, was the rector of the parish of Skreen in County Sligo. His mother, Elizabeth Haughton, was the daughter of a rector. The youngest of six children, Stokes had three brothers, all of whom took holy orders, and two sisters. He received his earliest education from his father and the parish clerk in Skreen. Stokes then attended school in Dublin before going to Bristol College in Bristol, England, to prepare to enter university. Later in life Stokes recalled that one of his teachers at Bristol, Francis William Newman, a classicist and mathematician, had influenced him profoundly. In 1837 Stokes entered Pembroke College, Cambridge, where during his second year he began to read mathematics with William Hopkins, an outstanding private tutor whose influence on Stokes probably far outweighed that of the official college teaching. When he graduated as senior wrangler and first Smith's prizeman in 1841, Pembroke College immediately elected him to a fellowship.

Stokes became the Lucasian professor at Cambridge in 1849, rescuing the chair from the doldrums into which it had fallen, and restoring it to the eminence it had when held by Newton. Since the Lucasian chair was poorly endowed, Stokes taught at the Government School of Mines in London in the 1850's to augment his income. He held the Lucasian chair until his death in 1903. In 1857 he married Mary Susanna, daughter of the Reverend Thomas Romney Robinson, the astronomer at Armagh Observatory in Ireland. Stokes had to relinquish his fellowship to marry, but under new regulations he held a fellowship again from 1869 to 1902. A very active member of the Cambridge Philosophical Society, he was president from 1859 to 1861. Always willing to perform administrative tasks, Stokes became a secretary for the Royal Society of London in 1854, conscientiously carrying out his duties until 1885 when he became president of the society, a post he held until 1890. The society awarded him the Copley Medal in 1893. From 1887 to 1891 he represented the University of Cambridge in Parliament at Westminster; and from 1886 to 1903 he was president of the Victoria Institute of London, a society founded in 1865 to examine the relationship between Christianity and contemporary thought, especially science. Stokes was universally honored, particularly in later life, with degrees, medals, and membership in foreign societies. He was knighted in 1889. The University of Cambridge lavishly celebrated his jubilee as Lucasian professor in 1899, and three years later Pembroke College bestowed on him its highest honor by electing him master.

As William Thomson commented in his obituary of Stokes, his theoretical and experimental investigations covered the entire realm of natural philosophy. Stokes systematically explored areas of hydrodynamics, the elasticity of solids, and the behavior of waves in elastic solids including the diffraction of light, always concentrating on physically important problems and making his mathematical analyses subservient to physical requirements. His few excursions into pure mathematics were prompted either by a need to develop methods to solve specific physical problems or by a desire to establish the validity of mathematics he was already employing. He also investigated problems in light, gravity, sound, heat, meteorology, solar physics, and chemistry. The field of electricity and magnetism lay almost untouched by him, however; he always regarded that as the domain of his friend Thomson.

After graduating, Stokes followed Hopkins' advice to pursue hydrodynamics, a field in which George Green and James Challis had recently been working at Cambridge. Thus in 1842 Stokes began his investigations by analyzing the steady motion of an incompressible fluid in two dimensions. In one instance, for motion symmetrical about an axis, he was able to solve the problem in three dimensions. In the following year he continued this work. Some of the problems that Stokes tackled had already been solved by Duhamel in his work on the permanent distribution of temperature in solids. Despite this duplication, which Stokes mentioned, he deemed the application of the formulas to fluid flow instead of heat flow sufficiently different to warrant publication. Stokes had not yet analyzed the motion of a fluid with internal friction, later known as viscosity, although references to the effects of friction continually appear in his papers. The problem, however, of the motion of a fluid in a closed box with an interior in the shape of a rectangular parallelepiped, which Stokes solved in 1843, was attacked partly with an eye to possible use in an experiment to test the effects of friction. By 1846 he had performed the experiment, but to Stokes's disappointment the differences between the experimental results and the theoretical calculations that excluded friction were too small to be useful as a test of any theory of internal friction.

Stokes's analysis of the internal friction of fluids appeared in 1845. Navier, Poisson, and Saint-Venant had already derived independently the equations for fluid flow with friction, but in the early 1840's Stokes was not thoroughly familiar with the French literature of mathematical physics, a common situation in Cambridge. Stokes said that he discovered Poisson's paper only after he had derived his own equations. He insisted, however, that his assumptions differed sufficiently from Poisson's and Navier's to justify publishing his own results. One novel feature of Stokes's derivation was that instead of using the Frenchmen's ultimate molecules he assumed that the fluid was infinitely

divisible, for he was careful not to commit himself to the idea that ultimate molecules existed. Another novel feature was his treatment of the relative motion of the parts of the fluid. He was able also to use these equations and the principles behind them to deduce the equations of motion for elastic solids, although he introduced two independent constants for what were later called the moduli of compression and rigidity, instead of one independent constant to describe elasticity as Poisson had. Stokes noted that the equations of motion he obtained for an elastic solid were the same as those that others had derived for the motion of the luminiferous ether in a vacuum. He then justified the applicability of these equations to the ether partly on the basis of the law of continuity, which permitted no sharp distinction between a viscous fluid and a solid, and which he believed held throughout nature.

Stokes became well known in England through a report on recent developments in hydrodynamics, which he presented in 1846 to the British Association for the Advancement of Science. So perceptive and suggestive was his survey that it immediately drew attention to his abilities and further enhanced his reputation as a promising young man. The report shows Stokes's increasing familiarity with the French literature on hydrodynamics and reveals his admiration for the work of George Green.

Stokes then pursued (1847) the topic of oscillatory waves in water, which he had suggested in his report merited further investigation. Poisson and Cauchy had already analyzed the complicated situation in which waves were produced by arbitrary disturbances in the fluid, but Stokes ignored the disturbances to examine the propagation of oscillatory waves the height of which is not negligible compared with their wavelength. Much later, in 1880, Stokes examined the shape of the highest oscillatory waves that could be propagated without changing their form. He showed that the crest of these waves enclosed an angle of 120°, and proposed a method for calculating the shape of the waves.

In one of his most important papers on hydrodynamics, presented in 1850, Stokes applied his theory of the internal friction of fluids to the behavior of pendulums. Poisson, Challis, Green, and Plana had analyzed in the 1830's the behavior of spheres oscillating in fluids, but Stokes took into account the effects of internal friction, including both spherical bobs and cylindrical pendulums. He then compared his theoretical calculations with the results of experiments conducted by others, including Coulomb, Bessel, and Baily. In the same paper he showed that the behavior of water droplets in the atmosphere depended almost completely on the internal friction of air and so explained how clouds could form in the atmosphere of the earth.

On account of his theoretical analysis and experimental observations of pendulums combined with his study of gravity at the surface of the earth, Stokes became the foremost British authority on the principles of geodesy. In his study of 1849 he related the shape of the surface of the earth to the strength of gravity on it without having to adopt any assumptions whatsoever about the interior of the earth. He obtained Clairaut's theorem as a particular result. Stokes assumed merely that the earth has a surface of equilibrium, one perpendicular to the gravity on it, whereas previously assumptions about the distribution of matter in the earth were always introduced to derive Clairaut's theorem. One result of his analysis was an explanation of the well-known observation that gravity is less on a continent than on an island. When the pendulum observations for the Great Trigonometrical Survey of India were conducted from 1865 to 1873, his expertise, together with his position as secretary to the Royal Society, made him an obvious person for the surveyors to turn to for advice, even though numerical calculations based on some of Stokes's own formulas would have been too laborious to carry out.

Occasionally Stokes studied problems in sound, which he considered a branch of hydrodynamics. In 1848 and 1849 he replied to Challis' claim of a contradiction in the commonly accepted theory, and in doing so Stokes introduced surfaces of discontinuity in the velocity and density of the medium. But later, on the basis of the argument by William Thomson and Lord Rayleigh that the proposed motion violated the conservation of energy, he retracted the idea that such motion, later called shock waves, could take place. (Stokes frequently crossed swords with Challis publicly in the *Philosophical Magazine*. They disagreed over the basic equations of fluid flow [1842, 1843, 1851], the theory of aberration [1845, 1846, 1848], and the theory of colors [1856].) In 1857 Stokes explained succinctly the effect of wind on the intensity of sound. Also, using a sphere to represent a bell and an infinite cylinder to represent a string or wire, he analyzed mathematically the production of sound by the transmission of motion from a vibrating body to a surrounding gas (1868). Poisson had already solved the case of the sphere, but Stokes

was quick to point out that Poisson had examined a different problem. Stokes's analysis explained John Leslie's observation that hydrogen or a mixture of hydrogen and air transmitted the sound of a bell feebly, and why sounding boards were necessary for stringed instruments to be heard, the vibrations being communicated to the board and then to the air. In a manner typical of Stokes, he then proceeded to explain how sound was produced by telegraph wires suspended tightly between poles.

The wave theory of light was well established at Cambridge when Stokes entered the university, and he seems to have embraced it right from the beginning of his studies. His earliest investigations in this field centered on the nature of the ether, beginning in 1845 with a proof that the wave theory was consistent with a theory of aberration in which the earth dragged along the ether instead of passing freely through it, as Fresnel had suggested. In 1846 Stokes showed that when the motion of the earth through the ether was not ignored, the laws of reflection and refraction remained unchanged in his own theory as well as in Fresnel's theory, thus offering no way to decide between the two theories of the interaction of the ether with the earth. In 1848 Stokes examined mathematically the properties of the ether, and by analogy with his own theory of the motion of fluids with internal friction he combined in his ether the seemingly contradictory properties of fluidity and solidity. He maintained that to examine the motion of the earth, the ether must be viewed as a very rarefied fluid, but to examine the propagation of light the same ether must be regarded as an elastic solid. To illustrate his view Stokes suggested that the ether is related to air in the same way as thin jelly is to water. Also in 1848 Stokes employed the wave theory of light to calculate the intensity of the central spot in Newton's rings beyond the critical angle of the incident light at which the rings vanish, leaving only the central black spot. He also examined the perfectly black central spot that results when the rings are formed between glasses of the same material. Fresnel had already analyzed this phenomenon, but Stokes's assumptions and derivation differed from his.

In a major paper on the dynamical theory of diffraction (1849), Stokes treated the ether as a sensibly incompressible elastic medium. Poisson had already calculated the disturbance at any point at any time resulting from a given initial disturbance in a finite portion of an elastic solid; but Stokes presented a different derivation, which he deemed simpler and more straightforward than Poisson's. Stokes also determined the disturbance in any direction in secondary waves, upon which the dynamical theory of diffraction depends, not limiting himself, as others had, to secondary waves in the vicinity of the normal to the primary wave. Moreover, by comparing his theory with the results of diffraction experiments that he conducted with a glass grating, Stokes answered the vexing question about the direction of vibrations of plane-polarized light by concluding that they were perpendicular to the plane of polarization.

At this time, both Stokes's theoretical analyses and his experiments covered a broad area of optics. In addition to his experiments on diffraction, he conducted experiments on Talbot's bands (1848), on the recently discovered Haidinger's brushes (1850), on phase differences in streams of plane-polarized light reflected from metallic surfaces (1850), and on the colors of thick plates (1851). Occasionally he invented and constructed his own instruments, as he did to facilitate measurements of astigmatism in the human eye (1849). In 1851 Stokes devised and largely constructed an instrument for analyzing elliptically polarized light. Here we see an excellent example of his theoretical studies complementing his experimental and instrumental work. In 1852 he published a mathematical analysis of the composition and resolution of streams of polarized light originating from different sources; the four parameters by which he characterized polarized light in this study became known as the Stokes parameters.

Stokes's explanation of fluorescence, published in 1852, for which the Royal Society awarded him the Rumford Medal, arose from his investigations begun the previous year into the blue color exhibited at the surface of an otherwise colorless and transparent solution of sulfate of quinine when viewed by transmitted light. Sir John Herschel had described this phenomenon in 1845, and Sir David Brewster had also examined it. Stokes, who had started by repeating some of Herschel's experiments and then had devised his own, rapidly concluded that light of a higher refrangibility, which corresponded to light of a higher frequency, produced light of lower refrangibility in the solution. Thus the invisible ultraviolet rays were absorbed in the solution to produce blue light at the surface. Stokes named this phenomenon fluorescence. Always looking for applications of optics, he quickly devised a method for exhibiting the phenomenon that did not require direct sunlight and so would render a chemist independent of the fickle

British weather in utilizing fluorescence to distinguish between various chemicals. In opening up the entire field of fluorescence to investigation, Stokes showed how it could be used to study the ultraviolet segment of the spectrum. By 1862 Stokes was using the spark from an induction coil to generate the spectra of various metals employed as electrodes. The invisible rays of the spectra were then examined and recorded systematically by means of fluorescence, although Stokes knew that photography was already beginning to replace fluorescence as a tool for mapping out spectra. Through his studies on fluorescence Stokes in 1862 began to collaborate with the Reverend W. Vernon Harcourt, who was one of the few people at that time attempting to vary the chemical composition of glass to produce new glasses with improved optical properties. Hoping to make glasses that would allow them to construct a perfectly achromatic combination, they collaborated until Harcourt's death in 1871.

While studying spectra by means of fluorescence, Stokes speculated on the physical principles of spectra, a topic of growing interest in the 1850's. Although Stokes always disclaimed priority in developing the principles of spectrum analysis, William Thomson insisted vigorously that Stokes taught him the principles in their conversations no later than 1852. They were discussing the topic in their correspondence in 1854 and speculating on the possibility of employing spectra to identify the chemical constituents of the sun. But Stokes did not publish anything on these ideas at that time, so the credit for the development of the principles of spectrum analysis went later to Kirchhoff and Bunsen.

Stokes's use of fluorescence in the 1850's as a tool for investigation typified his increasing emphasis on the exploitation of light to study other aspects of nature than light itself. In the 1860's, for instance, he drew the attention of chemists to the value of optical properties such as absorption and colored reflection as well as fluorescence in discriminating between organic substances. He was also a pioneer in combining spectrum analysis with chemical reactions to study blood.

Stokes's final major mathematical study on light was his classic report of 1862 on the dynamical theory of double refraction, presented to the British Association. He reviewed the theories of Fresnel, Cauchy, Neumann, Green, and MacCullagh, showing his preference for the ideas of Green and pointing out that he thought the true dynamical theory had not yet been discovered. Continuing his study of the dynamical theories, Stokes later showed experimentally that double refraction could not depend on differences of inertia in different directions, an idea W. J. M. Rankine, Lord Rayleigh, and Stokes had all entertained. He concluded that Huygens' construction for the wave fronts should be followed. A very brief summary of his experiments and conclusion was published in 1872, but a detailed account that he promised to present to the Royal Society was never published.

Stokes's papers on pure mathematics were tailored to his requirements for solving physical problems. His paper on periodic series (1847) consisted of an examination of various aspects of the validity of the expansion of an arbitrary function in terms of functions of known form. The expansions are now called Fourier series. In the paper Stokes applied his findings to problems in heat, hydrodynamics, and electricity. In 1850 he calculated the value

of $\int_0^\infty \cos \frac{\pi}{2}\left(x^3 - mx\right)dx$, when m is large and real,

an integral that had arisen in the optical studies of G. B. Airy. The method employed by Stokes for expanding the integral in the form of power series that initially converge rapidly and ultimately diverge rapidly was the one he afterward used in 1850 to determine the motion of a cylindrical pendulum in a fluid with internal friction. In 1857 he solved the

equation $\frac{d^2w}{dz^2} - 9zw = 0$ in the complex z-plane,

which was equivalent to calculating the definite integral above. He also showed that the arbitrary constants forming the coefficients of the linear combination of the two independent asymptotic solutions for large $|z|$ were discontinuous, changing abruptly when the amplitude of z passed through certain values. The discontinuous behavior became known as the Stokes phenomenon, and the lines for which the amplitude of z has a constant value at which the discontinuities occur became known as the Stokes lines. He later examined (1868) a method of determining the arbitrary constants for the asymptotic solutions of the Bessel

equation, $\frac{d^2y}{dx^2} + \frac{1}{x}\frac{dy}{dx} - \frac{n^2}{x^2}y = y$, where n is a real

constant. These studies in mathematics, however, formed only one small area of Stokes's publications.

In the early years of his career, through the Cambridge Philosophical Society, his teaching, and the examinations he composed, Stokes was a pivotal figure in furthering the dissemination of French mathematical physics at Cambridge. Partly because of this, and because of his own researches,

Stokes was a very important formative influence on subsequent generations of Cambridge men, including Maxwell. With Green, who in turn had influenced him, Stokes followed the work of the French, especially Lagrange, Laplace, Fourier, Poisson, and Cauchy. This is seen most clearly in his theoretical studies in optics and hydrodynamics; but it should also be noted that Stokes, even as an undergraduate, experimented incessantly. Yet his interests and investigations extended beyond physics, for his knowledge of chemistry and botany was extensive, and often his work in optics drew him into those fields.

Stokes's output of papers dropped rapidly in the 1850's, while his theoretical studies gradually gave way to experimental investigations. This occurred partly when he became a secretary to the Royal Society in 1854 and partly after he married in 1857. He often took on heavy administrative duties, which prevented him from conducting any research; and so from the 1860's many of his publications related to points arising from his official duty of reading papers submitted to the Royal Society. Stokes's papers eventually became a guide to other people's problems and interests. This is also seen in his correspondence with Thomson, for whom Stokes was a lifelong sounding board.

Throughout his life Stokes invariably took time to reply in detail to private as well as official requests for aid in solving problems, a frequent occurrence. A good example is his paper (1849) on the solution of a differential equation representing the deflection of iron railroad bridges, which Robert Willis, who was on a royal commission looking into the behavior of iron in various structures, had asked him to examine.

Although Stokes never fulfilled the expectations of his contemporaries by publishing a treatise on optics, his Burnett lectures on light, delivered at the University of Aberdeen from 1883 to 1885, were published as a single volume. The Gifford lectures on natural theology, which he delivered at Edinburgh in 1891 and 1893, were also published. A devoutly religious man, Stokes was deeply interested in the relationship of science to religion. This was especially true toward the end of his life, although he did not feel qualified to do justice to his Gifford lectureship.

BIBLIOGRAPHY

I. ORIGINAL WORKS. A comprehensive list of Stokes's papers appears in the Royal Society *Catalogue of Scientific Papers*, V, 838–840; VIII, 1022–1023;

XI, 505–506; XVIII, 977. Almost all of his published papers are included in *Mathematical and Physical Papers*, 5 vols. (Cambridge, 1880–1905); vols. I–III were edited by Stokes, and vols. IV–V posthumously by Sir Joseph Larmor. Vol. V also contains a previously unpublished MS on waves in water, as well as Smith's Prize examination papers and mathematical tripos papers set by Stokes at Cambridge.

A list of lectures and addresses on scientific topics, which were not printed in *Mathematical and Physical Papers*, is included in Larmor's preface to vol. V. A second ed. with a new preface by C. Truesdell appeared as *Mathematical and Physical Papers by the Late Sir George Gabriel Stokes, Bart. . . . Second Edition, Reprinting the Former of 1880–1905, Prepared by the Author (Volumes 1–3) and Sir J. Larmor (Volumes 4–5). With Their Annotations and the Obituary Notices by Lord Kelvin and Lord Rayleigh, and Also Including the Portions of the Original Papers Which Were Omitted From the Former Edition . . .*, 5 vols. (New York–London, 1966). *Memoirs and Scientific Correspondence of the Late Sir George Gabriel Stokes . . . Selected and Arranged by Joseph Larmor . . .*, 2 vols. (Cambridge, 1907; repr., New York–London, 1971), contains selected correspondence of Stokes, memoirs by his daughter Mrs. Laurence Humphrey and some of his colleagues, and miscellaneous material about Stokes's life and work.

Cambridge University Library, England, holds an extensive collection of Stokes's MSS, especially the Stokes Papers, which include his scientific, miscellaneous, family, Royal Society, and religious correspondence; notes for lectures; notes taken in lectures; and material concerning university administration. Add. MS 7618 at Cambridge contains the Stokes-Kelvin correspondence. The Scientific Periodicals Library, Cambridge, holds a number of Stokes's notebooks, some containing records of his experiments.

From the journal's inception in 1857 to 1878, Stokes, with A. Cayley and M. Hermite, assisted editors J. J. Sylvester and N. M. Ferrers of the *Quarterly Journal of Pure and Applied Mathematics* (London). He contributed articles, mostly on physical optics, and revised others on physical topics taken from the *Penny Cyclopaedia*, for *The English Cyclopaedia. A New Dictionary of Universal Knowledge. Conducted by Charles Knight. Arts and Sciences*, 8 vols. (London, 1859–1861). The three series of Burnett lectures were issued separately, and then published together as *Burnett Lectures. On Light. In Three Courses Delivered at Aberdeen in November, 1883, December, 1884, and November, 1885* (London–New York, 1887; 2nd ed., 1892), with a German trans. by O. Dziobek appearing as *Das Licht* (Leipzig, 1888).

Apart from his contributions to the *Journal of the Transactions of the Victoria Institute* (London), Stokes's principal writings on religion and on aspects of its relationship to science are *Natural Theology. The Gifford Lectures Delivered Before the University of Edinburgh in 1891* (London, 1891), *Natural Theology.*

The Gifford Lectures . . . 1893 (London, 1893), and *Conditional Immortality. A Help to Sceptics. A Series of Letters Addressed . . . to James Marchant (With a Prefatory Note by the Latter)* (London, 1897).

The *Transactions of the Cambridge Philosophical Society*, **18** (1900), consists of memoirs presented to the society to celebrate Stokes's jubilee as Lucasian professor.

II. SECONDARY LITERATURE. The two most important obituaries assessing Stokes's scientific work are Lord Kelvin, in *Nature*, **67** (1903), 337–338, also in *Mathematical and Scientific Papers*, 2nd ed., V, xxvii–xxxii, and Lord Rayleigh, *Proceedings of the Royal Society*, **75** (1905), 199–216, repr. in both eds. of *Mathematical and Physical Papers*, V, ix–xxv.

Since these obituaries and Larmor's *Memoir and Scientific Correspondence . . .*, little has been published on Stokes's scientific work. A few recent accounts are Truesdell's preface to the *Mathematical and Physical Papers*, 2nd ed., I, IVA–IVL; I. Grattan-Guinness, *The Development of the Foundations of Mathematical Analysis From Euler to Riemann* (Cambridge, Mass.–London, 1970), 113–120; and David B. Wilson, "George Gabriel Stokes on Stellar Aberration and the Luminiferous Ether," in *British Journal for the History of Science*, **6** (1972–1973), 57–72.

E. M. PARKINSON

STOLETOV, ALEKSANDR GRIGORIEVICH (*b.* Vladimir, Russia, 10 August 1839; *d.* Moscow, Russia, 27 May 1896), *physics.*

Stoletov came from a merchant family that had been exiled to Vladimir for sedition by Ivan the Terrible. His father, Grigory Mikhailovich, owned a small grocery store and a tannery; his mother, Aleksandra Vasilievna, was intelligent and well-read. While still a schoolboy he studied French, English, and German.

In 1856 Stoletov entered the Faculty of Physics and Mathematics at Moscow University as one of the first students from the merchant class to receive a government scholarship. His instructors in physics were M. F. Spassky and N. A. Lyubimov; and he received solid mathematical preparation under N. E. Zernov and N. D. Brashman. After graduating in 1860, he remained in the physics department to prepare for an academic career.

In 1862 Stoletov traveled abroad on a fellowship and spent three and a half years in Germany, where he attended the lectures of Helmholtz, Kirchhoff, and Wilhelm Weber and worked in H. G. Magnus' laboratory. At the end of 1865 he became a physics teacher at Moscow University. Four years later he defended his master's thesis, "Obshchaya zadacha elektrostatiki i privedenie ee k prosteyshemu vidu" ("The General Problem of Electrostatics and Its Reduction to the Simplest Form").

In 1871 Stoletov completed plans for his doctoral dissertation, which was based on his experimental research on the magnetic properties of iron. Since there was no laboratory at Moscow University—although Stoletov tried to have one established—he was again obliged to go abroad. At Heidelberg, Kirchhoff offered him the necessary conditions for working on his dissertation, which he defended in Moscow the following year, "Issledovanie o funktsii namagnichenia myagkogo zheleza."

He became extraordinary professor, and since a physics laboratory had in the meantime been opened at the university, he did his research there. In addition to his lectures, Stoletov also popularized science through his work in the Society of Amateurs of Science, the Society of Amateurs of Natural Sciences, and the Russian Physical-Chemical Society. He attracted a large group of talented young people to the physics laboratory and was the first to organize systematic training of scientific teams in physics. His protracted efforts to establish a physics institute at Moscow University were finally realized in 1887.

In his master's thesis Stoletov examined the theoretical equilibrium of electricity in an arbitrary number of isolated conductors (continuous and field) in an arbitrarily given field created by arbitrarily complex stationary electric poles. He continued to develop the mathematical method of successive approximation of Robert Murphy, Lipschitz, and William Thomson (Lord Kelvin) for the case of an arbitrary number of conductors. Stoletov had already written his thesis when Lipschitz's work on the same subject appeared abroad, but the latter was significantly less complete.

In his doctoral dissertation Stoletov examined the relation of the magnetization of iron to the strength of the external magnetic field. Although it was then known that the course of magnetization is not linear, the form of the function in the area of relatively weak fields remained unexplained. A thorough analysis of the experimental material led Stoletov to conclude that previous investigations, carried out in cylindrical models, could not yield satisfactory results because of the demagnetizing effect of the flat end. His use of Kirchhoff's theory led to the possibility of studying the true relation of magnetization to the field in closed rings magnetized along the perimeter. For this investigation Stoletov developed the method now generally accepted for measuring induction in rings with the aid of a ballistic galvanometer. He was the first to show that magnetic permeability of a ferromagnet increases in proportion to the intensification of the magnetizing field, attains a maximum, and then decreases. Stoletov emphasized the importance of applying these results to electrotechnology.

In his most distinguished work, *Aktinoelektricheskie issledovania* ("Actinoelectric Investigations," 1889), Stoletov experimentally established the basic laws of the external photoelectric effect and certain fundamental regularities of electrical discharge in rarefied gases. The first to develop an experimental method of studying the photoeffect, he showed that during the illumination of metals by ultraviolet light, there is a loss of negative electrical charge. By illuminating the negative plate of a condenser through a latticed-plate positive electrode, he observed a continuous electrical current in the circuit containing the condenser, a cell battery, and a galvanometer. The strength of the current appeared to be proportional to the intensity of the incident light and to the area illuminated. In 1888–1889 Stoletov was the first to show that through the presence of contact difference in potentials between the lattice and the plate, the photoelectric apparatus can, without a supplementary battery, serve as the source of current, converting light energy into electricity. This was the first photoelement, but Stoletov did not patent his invention. In 1890 the German physicists Elster and Geitel independently had the same idea, and received a patent for the photoelement and several of its applications.

Investigating the relation of the photocurrent to the external difference in potential, Stoletov discovered the existence of a saturation current. Further experimental study of this phenomenon at different degrees of rarefaction of air led him to discover an important regularity. If p_m is the pressure of gas at which the current attains the maximum value, l the distance between electrodes, and v the potential difference between them, then,

$$\frac{p_m l}{v} = \text{constant.}$$

The theory of this phenomenon was provided in 1910 by Townsend, who proposed calling the quantity the Stoletov constant.

In his important four-part *O kriticheskom sostoyanii tel* ("On the Critical State of Bodies," 1882–1894), Stoletov thoroughly analyzed the experimental data and the theoretical opinions of various authors, introduced clarity into the discussion of the critical state, and showed the correctness of the ideas of Thomas Andrews and van der Waals.

Stoletov's interest in the history of physics was reflected in an extensive study, *Ocherk razvitia nashikh svedeny o gazakh* ("Sketch of the Development of Our Information on Gases," 1879), and in articles on Newton, da Vinci, and others.

BIBLIOGRAPHY

Stoletov's writings were published as *Sobranie sochineny* ("Collected Works"; Moscow, 1941) and *Izbrannye sochinenia* ("Selected Works"), A. K. Timiryazev, ed. (Moscow, 1950). The Royal Society *Catalogue of Scientific Papers*, VIII, 1024; XI, 507; XII, 707; and XVIII, 978–980, lists 29 of his memoirs and several obituary notices.

On his life and work, see A. I. Kompaneets, *Mirovozrenie A. G. Stoletova* ("The World View of A. G. Stoletov"; Moscow, 1956), with a comprehensive bibliography of his writings and correspondence, pp. 281–286; M. S. Sominsky, *Aleksandr Grigorievich Stoletov* (Leningrad, 1970); and G. M. Teplyakov and P. S. Kudryavtsev, *Aleksandr Grigorievich Stoletov* (Moscow, 1966).

J. G. Dorfman

STOLZ, OTTO (*b*. Hall [now Solbad Hall in Tirol], Austria, 3 July 1842; *d*. Innsbruck, Austria, 25 October 1905), *mathematics*.

Stolz was the son of a physician who later achieved some prominence as a psychiatrist. After graduating from the Gymnasium at Innsbruck, he studied mathematics and natural sciences at the University of Innsbruck and later at Vienna. In 1864 he received the Ph.D. at the University of Vienna, where he was subsequently a *Privatdozent* until 1869, when he obtained a scholarship for further study at Berlin and Göttingen.

From 1869 to 1871 Stolz attended courses given by Weierstrass, Kummer, and Kronecker at Berlin and by Clebsch and Klein at Göttingen. Weierstrass made the greatest impression on him and led him to extend his research from geometry to analysis.

In July 1872 Stolz was appointed associate professor at the University of Innsbruck. He became a full professor in 1876 and married in the same year. He remained in Innsbruck for the rest of his life.

Stolz's earliest papers were concerned with analytic or algebraic geometry, including spherical trigonometry. He later dedicated an increasing part of his research to real analysis, in particular to convergence problems in the theory of series, including double series; to the discussion of the limits of indeterminate ratios; and to integration. Stolz was the first to formulate the counterpart, for double series, of Cauchy's necessary and sufficient condition for convergence. He also generalized Abel's theorem on the behavior of a power series in radial approach to the circle of convergence ("regularity of Abelian summability") to approach in an angular region with a vertex on the circle of convergence.

During his lifetime, and for some time afterward, Stolz was known as the author of several carefully written textbooks, of which *Vorlesungen über allgemeine Arithmetik* (1885–1886) and *Theoretische Arithmetik* (1900–1902) in particular gained wide recognition. The latter work was written with his student J. A. Gmeiner. Stolz is known today for his contributions to many questions of detail rather than for any major single achievement. For example, he is credited by K. Knopp with having been the first to show that every irrational number has a unique representation in decimal notation.

Stolz was greatly interested in the history of mathematics. After the Weierstrass ϵ,δ approach had found general acceptance in the early 1870's, he was the first to point out that Bolzano had suggested essentially the same approach even before Cauchy introduced his own, less rigorous method. Under the influence of P. du Bois-Reymond, Stolz also reexamined the theory of infinitely small and infinitely large quantities that had been used, on shaky foundations, until the advent of Weierstrass' method.

BIBLIOGRAPHY

I. Original Works. Stolz's writings include "Beweis einiger Sätze über Potenzreihen," in *Zeitschrift für Mathematik und Physik*, **20** (1875), 369–376; "B. Bolzano's Bedeutung in der Geschichte der Infinitesimalrechnung," in *Mathematische Annalen*, **18** (1881), 255–279; *Vorlesungen über allgemeine Arithmetik*, 2 vols. (Leipzig, 1885–1886); and *Theoretische Arithmetik*, 2 vols. (Leipzig, 1900–1902), written with J. A. Gmeiner.

II. Secondary Literature. See J. A. Gmeiner, "Otto Stolz," in *Jahresberichte der Deutschen Mathematikervereinigung*, **15** (1906), 309–322; and K. Knopp, *Theorie und Anwendung der unendlichen Reihen*, Grundlehren der mathematischen Wissenschaften in Einzeldarstellungen, no. 2 (Berlin, 1922).

Abraham Robinson

STONEY, GEORGE JOHNSTONE (*b.* Oakley Park, Kingstown [now Dún Laoghaire], County Dublin, Ireland, 15 February 1826; *d.* London, England, 5 July 1911), *mathematical physics.*

Stoney was the eldest son of George Stoney and his wife, Anne, who were Protestant landowners. The family was a talented one: Stoney's younger brother, Bindon Blood Stoney (1828–1909); his son, George Gerald Stoney (1863–1942); and his nephew, George Francis Fitzgerald (1851–1901), made significant contributions to science and technology, and were fellows of the Royal Society.

Stoney graduated from Trinity College, Dublin, in 1848 and became assistant to Lord Rosse in his observatory at Parsonstown (now Birr). After failing to obtain a fellowship at Trinity College, he obtained the chair of natural philosophy at Queen's College, Galway, which he held for five years. In 1857 he returned to Dublin as secretary to Queen's University, in which post he spent the rest of his working life. Stoney was a member of the Royal Irish Academy and the Royal Dublin Society, and was secretary of the latter for over twenty years. In 1893 he moved to London and became involved in the affairs of the Royal Society, of which he had been elected a fellow in 1861; in 1898 he was vice-president and a member of the Council.

Stoney had an interest in all fields of science and, like many of his Irish contemporaries, applied mathematics to the solution of scientific problems. He was particularly interested in spectrum analysis. A paper he wrote in 1868 suggested that spectral lines were due to periodic motions inside the atom rather than to the translational motion of molecules. Stoney continued this line of thought for a number of years and, as a result, put forward important ideas on atomic structure. In 1891 he explained the presence of double and triple lines in spectra by apsidal and precessional motions of orbital electrons.

Since his early work in Rosse's observatory, Stoney maintained an interest in astronomy and wrote many papers on the subject. Using the kinetic theory as a basis of his work, he reached certain conclusions concerning the atmospheres of planets. Stoney's paper of 1897 suggested that if the velocity of molecules exceeded a limit set by the force of gravity, then the molecules would fly off into space. By this means he explained the absence of an atmosphere on the moon.

Stoney probably is best-known for having coined the term "electron." He hoped that by a careful choice of fundamental units, science would be simplified; and at the 1874 meeting of the British Association, he presented the paper "On the Physical Units of Nature." One of the basic units he suggested was the charge carried on a hydrogen ion, which he determined from experimental data. The weight of hydrogen liberated on electrolysis by a given quantity of electricity was known; and by calculating the number of atoms associated with this weight of hydrogen, Stoney found the electric charge associated with each atom. A similar theory of electrical atomicity was advanced by Helmholtz in his Faraday lecture in 1881, and ten years later Stoney introduced the word "electron" for this fundamental unit. The term later came to be used for the "corpuscles" discovered by J. J. Thomson.

BIBLIOGRAPHY

I. Original Works. Stoney's writings include "The Internal Motions of Gases Compared With the Motions of Waves of Light," in *Philosophical Magazine*, 4th ser., **36** (1868), 132–141; "On the Physical Units of Nature," *ibid.*, 5th ser., **11** (1881), 381–389; "On the Cause of Double Lines and of Equidistant Satellites in the Spectra of Gases," in *Scientific Transactions of the Royal Dublin Society*, **4** (1891), 563–608; "Of the 'Electron' or Atom of Electricity," in *Philosophical Magazine*, 5th ser., **38** (1894), 418–420; and "Of Atmospheres Upon Planets and Satellites," in *Scientific Transactions of the Royal Dublin Society*, **6** (1897), 305–328.

II. Secondary Literature. Biographies are by F. T. Trouton in *Nature*, **87** (1911), 50–51; and by an anonymous author in *Proceedings of the Royal Society*, **86** (1912), xx–xxv.

BRIAN B. KELHAM

STØRMER, FREDRIK CARL MÜLERTZ (*b.* Skien, Norway, 3 September 1874; *d.* Oslo, Norway, 13 August 1957), *mathematics, geophysics.*

Størmer's father, Georg Størmer, was a pharmacist; his mother was the former Elisabeth Mülertz. When he was twelve, the family moved to Oslo (then Christiania). As a young boy he was interested in botany, which remained a lifelong hobby. During his high school years, Størmer's interest and ability in mathematics became apparent; and through a friend of the family, who was a professor of mathematics at the University of Oslo, he received instruction in that science. His first publication was published while he was still in high school (1892). Størmer entered the University of Oslo in 1892, received the master's degree in 1898, and was awarded the doctorate in 1903. In the latter year he became professor of pure mathematics, a post he held until his retirement in 1944.

Størmer's first papers were on number theory; but in 1903 he met the physicist Kristian Birkeland, who studied the polar aurora. Birkeland approached the problem experimentally, by bombarding a magnetic sphere in a vacuum with cathode rays. In this way it was possible to observe phenomena resembling the polar aurora. Størmer made the field observations and the theoretical calculations of the charged particles. The observations of the polar aurora were made photographically, by taking parallactic pictures along a base line. Størmer thus accumulated an enormous amount of observational material, not only on the altitude but also on the size, shape, and periodicity of the polar aurora. In the course of this work he also acquired interesting information on noctilucent and mother-of-pearl (nacreous) clouds. He constructed the instruments and worked out the procedures for these observations himself, showing a gift for experimentation that is rare among pure mathematicians.

Størmer's other approach to the study of the polar aurora was a mathematical analysis of the trajectories of charged particles in the earth's dipole magnetic field. It included the numerical integration of series of differential equations—an enormous task before the advent of electronic computers—in which Størmer was assisted by many of his students. His analysis showed that only some trajectories are possible, others being "forbidden." Størmer was also led to postulate a circular electric current in the equatorial plane of the earth and showed that electrons may be trapped into oscillatory trajectories in the earth's dipole field. Although his calculations were made in the course of studying the polar aurora, they became important for other areas of cosmic geophysics. When the latitudinal variation in cosmic radiation was discovered in the 1930's, it could be explained by Størmer's calculations; and the discovery of the Van Allen belts confirmed, to a surprising degree, his theoretical analysis of the trajectories of charged particles from the sun in the dipole field of the earth.

Størmer was an old-fashioned scientist who worked by himself. He mastered the field and often did the manual labor connected with his experiments and calculations. His last book, *The Polar Aurora* (1955), is not only his final summary, but also an up-to-date and authoritative study. Størmer

had no direct followers; but he exerted a profound influence during his forty-one-year teaching career. He also was an excellent popularizer.

BIBLIOGRAPHY

Most of Størmer's scientific publications appeared in the *Norske Videnskabsakademiets Skrifter*; lists can be found in *Årbok. Norske videnskapsakademi i oslo* (1892–1953). His books include *Fra verdensrummets dybder til atomenes indre* (Oslo, 1923), which went through 4 eds. in Norwegian and was translated into 5 foreign languages; and *The Polar Aurora* (Oxford, 1955).

An obituary is L. Harang, "Minnetale over professor Carl Størmer," in *Årbok. Norske videnskapsakademi i Oslo* (1958), 81–85.

Nils Spjeldnaes

STRABO (*b.* Amasia, Asia Minor, 64/63 B.C.; *d.* Amasia, *ca.* A.D. 25), *history, geography.*

Strabo was the son of wealthy parents. He was Greek by language and education; in his youth he studied under the rhetorician Aristodemus at Nysa in Caria, and he may also have known the Stoic polymath Posidonius. In about 44 B.C. he went to Rome to study with the geographer Tyrannion and the philosopher Xenarchus. He became a convert to Stoicism, probably through the offices of the philosopher Athenodorus Cananites, the friend and teacher of the emperor Augustus, although he continued to distrust popular religion (however useful it might be) and to believe in Providence as a first cause. He was again in Rome in 35 B.C. and in 31 B.C. He visited Crete, journeyed through Corinth in 29 B.C., and spent five years, from 25 to 20 B.C., in Alexandria, where he may have studied in the great library. In 25 or 24 B.C., he made a journey from Alexandria up the Nile to Aswan and the Ethiopian frontier in the company of the Roman governor, Marcus Aelius Gallus.[1]

An admirer of the Roman empire, Strabo may have been politically motivated in the writing of his works, although they also contain a great deal of knowledge presented for its own sake. Of these works, only one, the *Geographica*, is extant. It is known that Strabo composed a number of historical works, including a *Hypomnemata historica* in which he recounted incidents in the lives of famous men. He intended a fuller work, incorporating some of the same material, to be a continuation of the work of Polybius, whose history concluded with the years 146/145 B.C. Strabo's work appeared in either forty-three or forty-seven books, and brought Polybius up to date at least as far as the troubles following the assassination of Julius Caesar in 44 B.C.; it may even have extended to about 27-25 B.C.[2]

Strabo apparently published some of his surviving *Geographica* about 7 B.C.; a partially revised version of this appeared in about A.D. 18, and a finished but still incompletely revised work in seventeen books was published later, perhaps after Strabo's death. The place of its publication must have been far from Rome, since the work was not known there; indeed, it was not generally known until the fifth century. It was addressed to men in elevated stations in life; although Strabo stated that it should be of some general interest, he particularly recommended it to statesmen, rulers, and soldiers, as well as to those who wanted an account of known lands (especially those prominent in the history of civilization).

Although by his own statement Strabo traveled from Armenia to Etruria, and from the Black Sea to Ethiopia—and although he knew many parts of Asia Minor from Pontus to Syria—his *Geographica* was based less upon his personal observations than on his reading. He knew little of Italy, except for the areas along the Roman roads in the southern and central parts, and he took little advantage of Roman sources, although he knew Caesar; Calpurnius Piso Frugi,[3] governor of Libya; Aelius Gallus; and an unnamed, but probably Roman, chorographer. He was apparently unfamiliar with Marcus Vipsanius Agrippa's chart of the Roman empire and its adjacent countries.

The bulk of Strabo's material came (although it is not possible to ascertain how directly) from a number of Greek sources that are now lost. Among other writers Strabo drew upon Eratosthenes for mathematical geography and cartography and information about India; upon Eudoxus of Cnidus for astronomy; on Hipparchus, whose astronomical material Strabo used only for mathematical cartography; on Posidonius, especially for information on Spain and Gaul; on Polybius, especially for material concerning Europe; on Artemidorus, for Asia Minor and Egypt; and on Apollodorus of Athens, perhaps for Greece (of which Strabo knew less than he did of Italy). Despite his borrowings from mathematicians and astronomers, and despite his recognition of the importance of the principles of mathematics and physics,[4] Strabo's scientific skills were limited, and he tended to underestimate science, being more sympathetic to "human" interests.

The first two books of the *Geographica* contain a somewhat rambling but still useful survey of earlier geographic theories and serve as an introduction to the rest of the work, wherein Strabo attempted to set out an account of the physical features, products, and national character of each country. His presentation is at once mathematical, chorographical, topographical, physical, political, and historical. He was, of course, dealing with the known world; much of northern Europe and of Africa south of the Mediterranean coastal regions and Egypt was still unexplored, while Asia was known only as far as India and Ceylon. He therefore ignored the great unknown stretches of eastern and northern Asia, and treated Africa as an area smaller than Europe, lying wholly north of the equator; outside Egypt, he noted, it was largely desert.[5]

Within these limitations, and given his reliance on reports of varying degrees of accuracy, Strabo produced an excellent account of parts of western Europe, Asia Minor, and Egypt. He was also good on Gaul, although he relied too little on recent Roman records (including the narratives of Caesar) and made the Pyrenees run from north to south, while the coast (largely because Strabo distrusted the explorer Pytheas, discoverer of Britain) runs northeast from the Pyrenees to the Rhine. His account of the British Isles is understandably weak, as is his treatment of the Baltic region and Scandinavia. His account of Greece is also disappointing.

Strabo followed Eratosthenes in showing the known world as a single ocean-girt landmass (*oikoumene*, that is, inhabited) composed of Europe, Asia, Africa, and their associated islands. The *oikoumene* occupies less than one-half of one quadrilateral on a sphere (about 25,200 geographical miles in circumference, according to Eratosthenes' good calculation) that remains motionless within a revolving spherical universe. Strabo represented the *oikoumene* as being entirely north of the equator of the sphere, occupying one-quarter of the whole; but he surmises that there may be other inhabited land continents, as yet unknown.[6]

Strabo stated that the *oikoumene* should be drawn on one-quarter of a globe not less than ten feet in diameter in order to render it in sufficient detail. He further discussed projecting the *oikoumene* on a plane surface, noting that it made little difference whether the meridians remained parallel to each other, since it was scarcely worthwhile to make them converge even slightly toward the pole. He relied upon established astronomical observations and conclusions to fix the equator, the ecliptic, and the tropics, and was aware that longitudes could be determined accurately only through comparing observations made during a suitable eclipse. He accepted the system of dividing the equator into 360 degrees and of establishing, by astronomical observations, parallels of north latitude, including a main one that intersects, at Rhodes, a main meridian line of longitude.[7] He also adopted the notion of five zones or "belts" in latitude: north frigid (uninhabitable), north temperate (inhabited), torrid (partly uninhabited—Strabo rejected the idea that the unknown southern part of this zone might be habitable), south temperate (habitable), and south frigid (uninhibited).[8]

Strabo represented the *oikoumene* itself as mantle-shaped, tapering toward the east and west. He showed it as extending in length for about 7,000 miles along a parallel drawn from Spain, through Rhodes, and to the Ganges, and in breadth for about 3,000 miles along the main meridian drawn through Rhodes (these distances were an unwise reduction of those set out by Eratosthenes). The encircling ocean intrudes into the landmass in a number of gulfs, especially the Caspian Sea (this error was not originally Strabo's), the Persian Gulf, the Arabian Gulf (the Red Sea), and, largest of all, the "Inside Sea," or "Our Sea" the Mediterranean.[9] It is interesting in this context to note that Strabo considered a long, varied coastline, together with a temperate climate, to be one of the factors important to the rise of civilization; for this reason the Mediterranean lands, especially Greece, developed a culture superior to lands at the outer edges of the *oikoumene*, where the coastline, although extensive, lacks variety of contour.

Strabo devoted much discussion to the forces that had formed the *oikoumene*. A number of the conclusions that he recorded had probably derived from Posidonius; he was also interested in Aristotle's theories about earthquakes and volcanic activity, and in Strato's notion that the Mediterranean had once been a lake that, overfilled by rivers, broke through the Straits of Gibraltar. Strabo suggested that some islands were torn from the mainland by earthquakes, while others (including Sicily) were thrown up by volcanic action. He gave examples of both local and widespread land subsidence and alluded to the uprising of seabeds with consequent flooding; he further described the silting of rivers that form alluvial plains and deltas. His acceptance of the long-held notion that inland regions containing salt marshes, salt beds, sand, and seashells and other marine débris had arisen from the bottom of the sea led him to conclude that Egypt—and the greater part of the *oikoumene*—had once been submerged.[10]

Strabo's whole work is not orderly, perhaps because he was not able to give it a final form. He was fond of historical and mythological digressions, and on some subjects argumentative and obsessive. The *Geographica* is nonetheless highly valuable in its exposition of the development of geography. It marked the first attempt to assemble all available geographical knowledge into a single treatise. A philosophy of geography, it is utterly unlike the mathematical geography of Ptolemy, the geographical parts of Pliny the Elder's *Natural History*, or, indeed, any other surviving work of ancient geography.

NOTES

1. On his birthplace, see Strabo, *Geography*, 12.3.39, and Stephen of Byzantium, under Ἀμάσεια; on his ancestry, Strabo, 10.4.10, 11.2.18, 12.3.33, 12.3.53; on his education and philosophy, 14.1.48, 14.5.4, 12.6.2, 12.3.16, 13.1.54, 16.2.24, 7.3.4, 1.2.34, 2.3.8, 16.4.21, 1.2.8, 17.1.36, 4.1.14—also Athenaeus, *Deipnosophistae*, 14.75.657, and Stephen of Byzantium, *loc. cit.*; on his admiration of Rome, 6.4.2, 1.1.16, 3.2.5; on his travels, 2.5.11; to Rome, 6.2.6, 8.6.23, 5.3.8—compare Dio Cassius, L.10; to Corinth, 10.4.10 and 10.5.3; in Egypt, 2.3.5, 2.5.12, 11.11.5, 17.1.24, 17.1.50, 15.1.45; to Italy, 2.5.11 and 5.2.6.
2. 1.1.22–23, 2.1.9, 11.9.3; Plutarch, "Lucullus," 28; "Sulla," 26; *Suda Lexicon*, "Λούκουλλος"; Flavius Josephus, *Antiquities*, 13.286–287, 319, 347.14.35, 68, 104, 111, 114, 118, 138–139, 15.9–10; and *Contra Apionem*, 2.84.
3. Calpurnius Piso: Strabo, 2.5.33.
4. Strabo, 1.1.16–18, 1.1.22–23, 2.5.1.
5. See especially 2.5.26–33; 17.3.1.
6. 1.1.8; compare 1.4.6, 2.3.6, 2.5.5–7, 2.5.34.
7. 2.5.10, 1.4.1 ff., 1.1.12, 2.1.1, 2.1.10, 2.1.12–13, 2.5.4, 2.5.34–42.
8. 2.3.1–2, 2.5.3, 2.5.5. A zone in latitude must not be confused with a κλίμα "clima" in latitude, which was the "inclination" of a place's horizon to the earth's axis.
9. 2.5.6–9, 2.5.14, 2.5.18.
10. 1.3.4–5, 1.3.8, 1.3.10, 5.4.8, 2.5.18.

BIBLIOGRAPHY

I. Original Works. Strabo's history, now lost, consisted of historical sketches (or memoirs), Ὑπομνήματα Ἱστορικά, of which he cites his "Deeds of Alexander" as an example, and a continuation of Polybius, probably entitled Ἱστορίαι or Ἱστορία, in 43 or 47 bks., of which bk. 2 was identical with bk. 6 of the sketches. The geography, which is extant, was entitled Γεωγραφικά ("Geography" or "Matters Geographical"), and is in 17 bks.

The *Geography* was little read until the late fifth century, and even then copies were rare. We do have in three portions (Codex Vat. gr. 2306, Codex Crypt. Zα xliii, and Codex Vat. 2061A) a palimpsest written at that time and still showing legible remains of the original text of Strabo; we also have quotations by Stephen of Byzantium from another early source. All extant later MSS of Strabo's text and direct medieval quotations are derived from a lost archetype of about the mid-ninth century. For the first nine bks. Codex Parisinus gr. 1397 (late tenth century) is the best MS but lacks bks. 10–17, for which the best MSS are Codex Vat. gr. 1329 (lacking bks. 10, 11, and beginning of 12; late thirteenth or early fourteenth century) and Codex Marc. gr. 640 (A.D. 1321).

The latest treatment of the early tradition is by F. Lasserre in Germaine Aujac and F. Lasserre, *Strabon, Géographie*, I, pt. 1 (Paris, 1969), xlviii ff., which is criticized by D. R. Dicks in *Classical Review,* n.s. **21** (1971), 188 ff.

The 1st ed., the Aldine (Venice, 1516), is based on the corrupt Codex Parisinus gr. 1393. The eds. by Guilielmus Xylander (Basel, 1549, 1571) were revised with commentary by I. Casaubon (Geneva, 1587), who later (Paris, 1620) issued his own ed. with Latin translation by Xylander and notes by F. Morrellius. The pages of his text are often cited (C and a numeral) in references to Strabo's text, rather than those of T. J. van Almaloveen's reissue of Casaubon's ed. (Amsterdam, 1707). Casaubon's was the base for further eds.: L. G. de Bréquigny (Paris, 1763), bks. 1–3 only; J. P. Siebenkees, 7 vols. (Leipzig, 1796–1818); T. Falconer, (Oxford, 1807); and the outstanding one of Adamantios Corais (Coraës, Coray), 3 vols. of text and 1 vol. of notes in modern Greek (Paris, 1815–1819). G. Kramer put the text on a better basis in 3 vols. (Berlin, 1844–1852). Also important are the eds. of A. Meineke, 3 vols. (Leipzig, 1852, 1866–1877) and of C. Müller and F. Dübner, 2 vols. (Paris, 1853–1858).

Several important series are in progress. W. Aly, *Strabonis Geographica in 17 Büchern*, IV, *Strabon von Amaseia. Untersuchungen über Text, Aufbau, und Quellen der Geographica* (Bonn, 1957), is one of several vols. planned to include text, translation, and commentary; *Strabonis Geographica*, I, text of bks. 1 and 2 (Bonn, 1968), was edited after his death by E. Kirsten and F. Lapp. F. Sbordone was responsible for *Strabonis Geographica*, I, bks. 1 and 2 (Rome, 1963). The following have appeared in the Budé series, giving introductions, text, French translation, and short notes; Germaine Aujac and F. Lasserre, I, pt. 1, intro. and bk. 1 (Paris, 1969); Germaine Aujac, I, pt. 2, bk. 2 (Paris, 1969); F. Lasserre, II, bks. 3 and 4 (Paris, 1966); III, bks. 5 and 6 (Paris, 1967); VII, bk. 10 (Paris, 1971).

Still appreciated are the French translations by A. Coray and G. La Porte du Theil (bks. 1–15) and A. Letronne (bks. 16 and 17) (Paris, 1805–1819). The German translation of C. G. Groskurd, 4 vols. (Berlin–Stettin, 1831–1834), is monumental. English translations include those of H. C. Hamilton and W. Falconer, 3 vols. (London, 1892–1893), and of H. L. Jones, with

text, in the Loeb Classical Library series, 8 vols. (London, 1917–1932).

Much other modern work on Strabo, which has been done and is being done, is recorded yearly in J. Marouzeau, *L'année philologique*. Fragments of Strabo's historical work were edited by P. Otto in *Leipziger Studien zur classischen Philologie*, **11** (1889); and are also in C. Müller, *Fragmenta historicorum Graecorum*, III (Paris, 1841), 490 ff., and in F. Jacoby, *Die Fragmente der griechischen Historiker*, IIA (Berlin, 1926), 430–436, and II C (Berlin, 1926), 291–295.

II. SECONDARY LITERATURE. For further study of Strabo, see Germaine Aujac, *Strabon et la science de son temps* (Paris, 1966); H. Berger, *Geschichte der wissenschaftlichen Erdkunde der Griechen* (Leipzig, 2nd ed., 1903), 327–582; E. H. Bunbury, *History of Ancient Geography*, II (London, 1879; 2nd ed., New York, 1959), 209–337; A. Calzon, *Conception de la géographie d'après Strabon* (Fribourg, 1940); M. Dubois, *Examen de la Géographie de Strabon* (Paris, 1891); W. Heidel, *The Frame of Ancient Greek Maps* (New York, 1937), *passim*, but 30–46, 104–122; E. Honigmann, in Pauly-Wissowa, *Real-Encyclopädie der classischen Altertumswissenschaft*, IVa, pt. 1 (1931), 76–155; J. O. Thomson, *History of Ancient Geography* (Cambridge, 1948), 182–186, 286–289, 188–198; H. F. Tozer, *Selections from Strabo* (Oxford, 1893), with intro.; and *A History of Ancient Geography*, 2nd ed., by M. Cary (Cambridge, 1935), 238–260.

E. H. WARMINGTON

STRACHEY, JOHN (*b*. Sutton Court, Chew Magna, Somerset, England, 10 May 1671; *d*. Greenwich, England, 11 June 1743), *geology*.

Strachey was the only son of John and Jane Strachey. He inherited his father's estate when three years old. Little is known about his education and early life. On 26 November 1686, he matriculated at Trinity College, Oxford, but left to study law at Middle Temple, London, in 1687. In 1692 Strachey married Elizabeth Elletson, who bore him eighteen children, and, two years after her death in 1722, Christina Stavely, by whom he had a son. He resided mainly at Sutton Court but spent the latter part of his life in Edinburgh. In 1719 he was elected fellow of the Royal Society.

In the early eighteenth century there were several coal mines near Sutton Court; and it was probably Strachey's interest in them that led him to study geology, for he published two geological papers and a pamphlet, all relative to occurrences of coal in Somerset and elsewhere. In the first two he included diagrams showing a cross section several miles long of the strata in the region of Sutton

Court, based on an intimate knowledge of the coal workings and an examination of rock outcrops in the surrounding countryside. They showed nearly horizontal Jurassic, Triassic, and (in his 1725 paper) Cretaceous strata resting on steeply inclined Coal Measures (Pennsylvanian). This was the first clear demonstration by a British author of an angular unconformity. Within the Coal Measures he recorded seven named seams, one of which was characterized by the fossil shells and plants occurring above it, a very early use of fossils to identify a specific horizon. The diagram also showed the effect of a fault on the continuity of the coal seams.

In his 1727 pamphlet Strachey restated his earlier observations and added details of the strata in other coalfields in England and Scotland, based partly on his own observations. He described the lithological characteristics of the coal seams and associated sediments, but seems to have had little idea how they were formed and hence did not realize the significance of the unconformity he recorded. Strachey recognized the importance of dip and strike, noting that across a broad belt of England, strata in general dip to the east and southeast. To account for it he accepted the suggestion already made by some British authors that all rock formations continue downward to the center of the earth, as a result of the west-to-east rotation of the earth acting on strata that originally were soft and unconsolidated.

Strachey left a number of unpublished manuscripts, one of which, since published, "Of Stones," is of considerable geological interest as a record of early eighteenth-century mining and quarrying operations in England. He also published a map of Somerset showing the sites of coal and metalliferous mines.

Strachey's work is of value as an early attempt to establish a stratigraphical succession, but the importance of the unconformity he recorded was not realized until much later.

BIBLIOGRAPHY

I. ORIGINAL WORKS. Strachey's writings include "A Curious Description of the Strata Observ'd in the Coal-Mines of Mendip in Somersetshire," in *Philosophical Transactions of the Royal Society*, **30** (1719), 968–973; "An Account of the Strata in Coal-Mines, &c.," *ibid.*, **33** (1725), 395–398; *Observations on the Different Strata of Earths, and Minerals. More Particularly of Such as Are Found in the Coal-Mines of Great Britain* (London, 1727); *Somersetshire Survey'd and Protracted* (London,

1736), a map of Somerset, on the scale of about 1/2 inch to the mile; and "Of Stones" (*ca.* 1736), from the MS of John Strachey's proposed "Somersetshire Illustrated," annotated by R. D. Webby, in *Proceedings of the Bristol Naturalists' Society*, **31**, no. 3 (1967), 311–330. Strachey's MSS are preserved in the Somerset Record Office, Taunton, England.

II. SECONDARY LITERATURE. See J. G. C. M. Fuller, "The Industrial Basis of Stratigraphy: John Strachey, 1671–1743, and William Smith, 1769–1839," in *Bulletin of the American Association of Petroleum Geologists*, **53**, no. 11 (1969), 2256–2273; and J. D. Webby, "Some Early Ideas Attributing Easterly Dipping Strata to the Rotation of the Earth," in *Proceedings of the Geologists' Association*, **80**, pt. 1 (1969), 91–97.

V. A. EYLES

STRASBURGER, EDUARD ADOLF (*b.* Warsaw, Poland, 1 February 1844; *d.* Poppelsdorf, Germany, 19 May 1912), *botany*, *plant cytology*.

Strasburger, who clarified the phenomena of cell division and the role of the nucleus and chromosomes in heredity, was born in Warsaw, when the city was under Russian rule. He was the eldest son of Eduard Gottlieb Strasburger, a merchant, and Anna Karoline von Schütz. Both parents were of German descent. He received his early schooling in Warsaw and, upon completing his Gymnasium studies, went to Paris in 1862 and studied for two years at the Sorbonne. He continued in botany under Hermann Schacht at the University of Bonn and gained the technical skill in microscopy that proved invaluable to him. Julius von Sachs lectured at the Agricultural Academy at Poppelsdorf, a suburb of Bonn, providing further stimulus to Strasburger's botanical interests. Strasburger met Nathanael Pringsheim when the latter visited Bonn. Upon the unexpected death of Schacht, Strasburger decided to go to Jena and accept an offer to become an assistant in Pringsheim's laboratory.

Pringsheim in time became his friend as well as his teacher, and Strasburger valued his critical mind; but his imagination was taken by the more speculative approach of the professor of zoology, Ernst Haeckel, especially by Haeckel's enthusiasm for Darwin's theory of evolution. Haeckel's lectures and their long discussions on development determined the direction of Strasburger's work, and he always recalled the influence of Darwin's *Origin of Species*. Thereafter Strasburger was an evolutionist and applied phylogenetic interpretations to the structure and developmental history of plants.

As Pringsheim's student, Strasburger also studied chemistry and zoology. He received the doctorate at Jena in 1866 with the thesis "Asplenium bulbiferum, ein Beitrag zur Entwicklungsgeschichte des Farnblattes mit besonderer Berücksichtigung der Spaltöffnungen und des Chlorophylls." The dissertation was not published, but his results appeared that year in an article in Pringsheim's *Jahrbuch für wissenschaftliche Botanik* and were used when he habilitated in 1867 at the University of Warsaw as *Privatdozent*.

Through the offices of Haeckel, Strasburger was appointed professor extraordinarius at Jena following Pringsheim's retirement and was made director of the botanical gardens; he became full professor two years later, at twenty-seven. In 1870 he married Alexandrine Wertheim, from Warsaw; they had a daughter and a son, Julius, later professor of medicine at Breslau. Strasburger accompanied Haeckel on a scientific trip to Egypt and the Red Sea in 1873, and it was Haeckel who first described to Strasburger the beauties of the Italian Riviera, where he later vacationed. Always interested in the plants of the regions he visited, Strasburger wrote popular articles and a book, *Streifzüge an der Riviera* (Berlin, 1895, Jena, 1904), which was later translated as *Rambles on the Riviera* (London, 1906).

During his twelve years at Jena, Strasburger published his botanical investigations. The development of his cytological observations can be followed in the three editions of his *Zellbildung und Zelltheilung* (1875, 1876, 1880).

Strasburger succeeded Johannes von Hanstein at the University of Bonn as ordinary professor, and in 1881 he transferred to Bonn his morphological and physiological work and investigations of plant reproduction and cytology. He spent the rest of his career there, and his laboratory became the leading center for the study of plant cytology.

Strasburger noted the appearance of plant cells even in his early research, reporting in his doctoral thesis that in the cells of a fern he had seen the nucleus divide during cell division. At that time the nucleus of the dividing cell was generally thought to disappear. The year he came to Jena he published a book on fertilization in conifers, but although his investigations followed upon Wilhelm Hofmeister's work on the alternation of generations, Strasburger wrote, "I was never closely associated with Hofmeister. Unfortunately, during the latter part of his life, Hofmeister became very sensitive and was angry because in 1869 in my work on *Befruchtung bei den Coniferen* I sought to prove that the 'corpuscula' do not correspond to the embryo sacs of angiosperms, but are archegonia" (Chamberlain, *Botanical Gazette*, **54** [1912], 70).

In *Die Coniferen und die Gnetaceen* (1872), Strasburger discussed the morphology of the flower and the origin of tissues. While zoologists independently observed and described the stages of cell division in 1873, using animal cells that were especially suitable because of their transparency, Strasburger, studying the embryogeny of the Coniferae, noticed and followed the formation of the nuclear spindle. He pioneered methods of fixing and hardening tissues, using pure alcohol. His observations on the phenomena of plant cells were included in his *Zellbildung und Zelltheilung*, (1875–1880), and in such papers as "Über Befruchtung und Zelltheilung" (1877) and "Über Polyembryonie" (1878).

The embryo sacs of certain plants permitted numerous nuclei to be seen at the same time in different phases of division. In his illustrations and descriptions, he noted the formation of the equatorial plate, the longitudinal extension of rods that met at the two poles, and the succession of events that paralleled the phenomena being reported by his colleagues in the division of animal cells. This suggested to Strasburger the common descent of vegetable and animal cells, which Haeckel had maintained. As he examined *Spirogyra* and other simple forms, he was again impressed by the difficulties in differentiating the plant from the animal kingdom. He was also carrying on physiological experiments, studying the reactions of swarm spores when he changed the conditions of light and temperature. Strasburger originated a number of terms in the course of his work, among them "phototaxis" and "chloroplast" and later, "cytoplasm," "nucleoplasm," "haploid," and "diploid." He saw the relation between his physiological and morphological investigations. He no doubt had his swarm spore researches in mind when in 1894 he reported that he had unsuccessfully tried to arrest the division of nuclei by lowering the temperature or varying the light. He felt that they might thus be more conveniently observed, especially in multinucleate forms, than when the successive stages of division occurred at chance times.

In 1875 Strasburger thought that free cell formation took place; in the 1880 edition of *Zellbildung und Zelltheilung* he no longer held this view. Meanwhile, in the 1870's there had been a continuing series of observations: Oscar Hertwig in 1875 had seen the spermatozoon as it entered the ovum and inferred that the fusion of the nuclei was the aim of fertilization; and Walther Flemming, aided by stains that distinguished the fibrils within the nucleus, had described in 1879 the nuclear threads with illustrations, and had clearly shown that these chromatin-staining bodies split longitudinally during division. In 1880 Strasburger still maintained that there was a transverse division of the bodies within the nucleus (first to be called chromosomes by Wilhelm Waldeyer in 1888), but as he often did when the evidence required, he changed his interpretation. Wilhelm Pfitzner in 1882 showed in detail the longitudinal segmentation, and the next year Wilhelm Roux pointed out the implications of the process—then called indirect division—for heredity.

In 1884, from his observations of plants, Strasburger independently concluded, as did Hertwig, August Weismann, and Albert von Koelliker at about the same time, that the nucleus was responsible for heredity, and that it contained within the filaments the substance that was divided in halves between the daughter cells and bore the characters of the parents and earlier ancestors.

For Strasburger the fertilization processes of the phanerogams demonstrated the role of the nucleus of the spermatozoon. He considered the male and female germ cells to be equivalent and the bearers of heredity. He now maintained that the longitudinal splitting assured the proportionate distribution of the substances in the nuclear threads, setting forth his views in "Die Controversen der Indirecten Kerntheilung" and *Neue Untersuchungen über den Befruchtungsvorgang bei den Phanerogamen als Grundlage für eine Theorie der Zeugung* (Jena, 1884). Strasburger gathered observations from his own and other researches to show that "fertilization depends only on the cell nuclei." He believed that there was a reciprocal dynamic action between the cell nucleus and the surrounding cytoplasm, but although both exerted influence, the nucleus in a way governed the cytoplasm and controlled metabolism and growth, among other processes.

Strasburger studied the formation of the cell plate in plant material and the growth of the cell wall, which he considered to occur through apposition. Throughout his life he continued his cytological researches; he was interested in the role of the centrosome and the problem of protoplasmic connections between cells; and he proposed (1892) that protoplasm was composed of structurally different substances, a more active kinoplasm in the

fibrils and other structures, and trophoplasm, which was more concerned with nutrition.

Strasburger investigated reproduction in plants from algae and mosses to cryptogams and phanerogams, and from an evolutionary standpoint viewed asexually reproducing organisms as the earlier forms from which sexually differentiated organisms descended (although some forms again lost this differentiation). He presented his ideas in his article "Ueber periodische Reduktion der Chromosomenzahl im Entwicklungsgang der Organismen" (*Biologische Centralblatt*, 14 [1894]). From the life histories of plant forms, and from the simpler to more complex sequences in which sexual or asexual generations developed, Strasburger inferred the courses through which the plants must have evolved. He pointed out the advantages of asexual reproduction under favorable external conditions for the rapid increase of individuals of the species, while sexual reproduction provided for the species to meet unfavorable conditions, and he interpreted the significance of the alternation of generations as it appeared in the higher plants.

Strasburger investigated the anatomy, life history, and aspects of physiology of plants. References to his extensive and careful work can be found throughout the botanical literature of his day and afterward. His textbooks appeared in many editions and translations, and have been revised in recent years. He also carried on investigations of the movement of sap, and in 1891 he studied the movement of sap in trees and the forces propelling it.

Strasburger's famous laboratory at the Botanical Institute of the University of Bonn was housed in the former palace of the electors of Cologne, in Poppelsdorf. The Institute also provided a residence for the professor of botany, with a laboratory and library. Strasburger directed the botanical gardens, which included outdoor collections, arboretums, and greenhouses, and sections containing special displays in which plants were grouped with reference to the solution of certain biological problems.

Under Strasburger, Bonn was foremost in plant cytology. Students came to his laboratory from all over Europe and as far away as Japan; but over half of the foreign students were American, in whom Strasburger took a special interest. They included Douglas H. Campbell, Charles J. Chamberlain, and B. M. Duggan. Bent from long hours spent over the microscope, Strasburger was an impressive teacher who gave each student his careful attention, yet steadily continued his own research. His clear and thoughtful lectures, his comprehensive views both of the broad principles and the fine details of morphology, his active involvement in the problems of the day, and his grasp of plant physiology exerted a lasting influence on the younger botanists. In addition, the laboratory and his textbooks provided methods and techniques essential to accurate experiment and observation. His texts included *Das botanische Practicum*, first published in 1884, and his renowned *Lehrbuch der Botanik für Hochschulen*, written with F. Noll, H. Schenck, and Andreas Schimper (Jena, 1894, and subsequent editions).

Strasburger had many interests beyond the Botanical Institute. He was rector of the University of Bonn in 1891–1892 (his inaugural address was entitled "Das Protoplasma und die Reizbarkeit"). Pringsheim died in 1894, and Strasburger became coeditor with Wilhelm Pfeffer of the *Jahrbucher für wissenschaftliche Botanik*. He belonged to the German Botanical Society, the Prussian Academy of Sciences, and many other scientific societies in Germany. He was a foreign member of the Royal Society of London, the Linnean Society, the American Academy of Arts and Sciences, and learned societies in Italy, France, Belgium, and Ireland.

The government honored Strasburger with the title of *Geheimer Regierungsrat* in 1887, and he received honorary degrees in medicine from the University of Göttingen (1887) and in law from the University of Oxford (1894). The Linnean Society of London awarded Strasburger a medal in 1905 for his contributions to botanical histology and morphology and another medal three years later at the Darwin-Wallace celebration, in recognition of his contribution to the study of evolution. Strasburger was still teaching and engaged in research, with especial interest in questions of sex determination, when he died in 1912. He had been active in research and teaching for nearly fifty years and had linked his extensive botanical and cytological observations to the solution of central problems of heredity and evolution.

BIBLIOGRAPHY

I. ORIGINAL WORKS. For a retrospective view of Strasburger's thought on the cell, see "The Minute Structure of Cells in Relation to Heredity," in *Darwin and Modern Science*, A. C. Seward, ed. (Cambridge, 1909), 102–111. Strasburger's many publications are listed by Beauverie, Tischler, and Küster, and by Cle-

mens Müller (appended to Karsten's obituary in the *Berichte der Deutschen botanischen Gesellschaft*, 80–86), in the articles cited below.

For Strasburger's address upon accepting the Darwin-Wallace medal in 1908, see *The Darwin-Wallace Celebration Held on Thursday, 1st July, 1908, by the Linnean Society of London* (London, 1908), 22–24. A letter to Haeckel, recalling his debt to his teacher upon Haeckel's sixtieth birthday, is found in Uschmann (see below), pp. 67–68. Strasburger's autobiographical letter (Chamberlain, p. 70, see below) recounts his differences with Hofmeister and is a short but interesting personal account.

II. SECONDARY LITERATURE. For discussions of Strasburger and his work, see J. Beauverie, "Édouard Strasburger," in *Revue générale de botanique*, 24 (1912), 417–452, 479–493; Charles Chamberlain, "Eduard Strasburger," in *Botanical Gazette*, 54 (1912), 68–72; Bradley M. Davis, "Eduard Strasburger," in *Genetics*, 36 (1951), 1–3; L. F., "Professor Dr. Eduard Strasburger," in *Lotos*, 60 (1912), 170–171; and J. B. F., "Prof. Eduard Strasburger," in *Nature*, 80 (1912), 379–380.

For a student's description of Strasburger as a teacher and of the laboratory at Bonn, see James Ellis Humphrey, "Eduard Strasburger," in *Botanical Gazette*, 19 (1894), 401–405. Other accounts of Strasburger, his life and work, are B. D. J., "Eduard Strasburger," in *Proceedings of the Linnean Society of London* (1911–1912), 64–66; G. Karsten, "Eduard Strasburger," in *Berichte der Deutschen botanischen Gesellschaft*, 30 (1912), 61–80; and "Eduard Strasburger," in *Biographisches Jahrbuch und Deutscher Nekrolog*, XVII (Berlin, 1915), 25–39; Ernst Küster, "Eduard Strasburger," in *Muenchener medizinische Wochenschrift*, 59 (1912), 1445–1447; and obituary in *Sitzungsberichte der Niederrheinischen Gesellschaft für Natur- und Heilkunde zu Bonn* (1912), 5–18. See also W. J. V. Osterhout, "Eduard Strasburger (1844–1912)," in *Daedalus*, 51 (1916), 927–928; G. Tischler, "Eduard Strasburger," in *Archiv für Zellforschung*, 9 (1913), 1–40; and Georg Uschmann, *Geschichte der Zoologie und der zoologischen Anstalten in Jena 1779–1919* (Jena, 1959), 60–61, 67–69.

A contemporary account of Strasburger's and his colleagues' botanical work is Sydney Howard Vines, *Lectures on the Physiology of Plants* (Cambridge, 1886), 658–660. A valuable overview placing Strasburger's cytological contributions in the context of his time is William Coleman, "Cell, Nucleus, and Inheritance: An Historical Study," in *Proceedings of the American Philosophical Society*, 109 (1965), 124–158, esp. pp. 126, 128–133, 145, 149–151, for Strasburger's work and the development of his thought on the cell. Further background is Arthur Hughes, *A History of Cytology* (London, 1959), 62–63, 65–66, 71–72, 82, 134.

GLORIA ROBINSON

STRATO OF LAMPSACUS (*b*. Lampsacus, Mysia; *d*. Athens, 271/268 B.C.), *natural philosophy*.

Very little is known of Strato's life. His father was Arcesilaus. His birthplace, a small town on the Asian coast of the Hellespont (Dardanelles), had a certain tradition of philosophical studies: Anaxagoras (*d. ca.* 428 B.C.) spent his last years there and may have founded a school, and Epicurus taught there for some years before founding his school at Athens in 306 B.C. Thus the first impulse to scientific work may have come to Strato in his native town. The dominant influence on his thought was that of Aristotle's school at Athens. It is not known when he joined it, but he must have spent a considerable time studying there under Theophrastus; it is just possible, but unlikely, that Aristotle was still at Athens when he entered. The next known event in Strato's life was his appointment as tutor to the future Ptolemy II Philadelphus, who ruled Egypt from 283 to 246 B.C.; this implies a period of residence at Alexandria, where he may have been concerned in the establishment of the museum and perhaps met such important intellectual figures as Diodorus Cronus, the "Megarian" philosopher, and the anatomist and surgeon Herophilus. Strato seems to have maintained a correspondence with Ptolemy and his queen Arsinoë after his return to Athens. On Theophrastus' death *ca.* 287, Strato was elected to succeed him as head of the Lyceum, a position he held until his death eighteen years later.

Since Strato's writings are not extant, knowledge of his work is derived from later sources. These fall into two classes. The first consists of passages in later authors where Strato is quoted or referred to by name; they are mostly quite short and generally report his conclusions without the reasoning that led to them. The most important are in a commentary on Aristotle's *Physics* by Simplicius, a Neoplatonist writing in the sixth century A.D.

The second consists of writings in which Strato is not named but which can be shown to be derived from his work. The importance of these sources is that some of them have preserved the reasoning by which Strato arrived at his theories. The most important texts of this class are the introduction to the *Pneumatica* of Hero of Alexandria and a long extract from a book *On Sounds* (Περὶ Ἀκουστῶν), wrongly attributed to Aristotle, included in Porphyry's commentary on the *Harmonica* of Ptolemy (third century). They are supported by shorter texts from the pseudo-Aristotelian *Problems*, the Greek commentators on Aristotle, and others. The

evidence in favor of attributing their contents to Strato is so strong as to make this virtually certain. Nevertheless, in the following reconstruction of Strato's teaching, the two kinds of evidence will be distinguished as far as possible.

From an incomplete list of his books preserved by Diogenes Laërtius (5.59 ff.), we learn that Strato wrote on logic, ethics, cosmology (including meteorology), psychology, physiology, zoology, and even a book on inventions—in fact, on most of the subjects included in the Aristotelian corpus. But his interest was centered on physics, in the wide sense the ancients gave to that word—the study of the natural world in all its aspects. All his significant work was done in this field, and it is the only part of his teaching about which we have any real information. Strato's preoccupation with this kind of work, in an age when philosophers generally were more concerned with problems of ethics and the theory of knowledge, earned him the sobriquet ὁ φυσικός to distinguish him from others of the same name.

Modern historians generally have represented Strato as an eclectic trying to combine elements from the systems of Aristotle and Democritus. Nothing could be further from the truth. In reality his thought was a one-sided but legitimate, consistent, and often brilliant development of Aristotle's views, along lines that to a large extent had been marked out by Theophrastus. The occasional resemblances between Strato's doctrine and that of the atomists are fortuitous. What Strato did was to strip the Aristotelian world picture of its transcendental elements—or, to put it another way, he refused to acknowledge the reality of anything not subject to the natural laws seen to apply in the sublunary world. In doing this he was carrying to its logical conclusion a process begun by Aristotle himself.

Aristotle had criticized Plato's conception of a world of Forms above and behind the sensible world on the ground that he had failed to show how these two systems were related. Nevertheless, he continued to teach the existence of an Unmoved Mover outside the physical world and above its laws; and within the physical world he distinguished two systems subject to different laws, one extending from the center of the universe to the sphere of the moon's revolution, the other embracing the heavenly bodies. The difficulties entailed by the second division can be seen by studying Aristotle's *Meteorologica*; those involved by the first were stated by Aristotle's pupil Theophrastus in his *Metaphysica*, in which the Unmoved Mover and the teleological view of the world are subjected to a critique that in many ways is a continuation of Aristotle's critique of the Platonic Forms.

It was left to Strato, however, to construct a new cosmology in which the world was explained as the product of immanent forces only. The ancient authorities tried to express this by saying that "everything that exists is the result of natural weights and movements" (fr. 32, from Cicero); that "all divine power is contained in nature, which contains in itself the causes of coming-into-being, growth and decay, but has no consciousness or shape" (fr. 33, from Cicero); that "natural processes are governed by 'what happens spontaneously' (τὸ αὐτόματον)" (fr. 35, from Plutarch); and even that Strato regarded heaven and earth as gods (frs. 37 and 39, both derived from Seneca). The preference for an explanation of the world that does not depend on transcendent causes was shared by Strato with the other chief philosophical schools of his day, the Stoics and the Epicureans; but their teaching was in some ways more naïve and less radical. Strato differed from the Stoics in denying that nature is conscious and provident, and from the atomists in positing a regulating principle to which all the processes that constitute the world are somehow subordinated.

The tendency to bring all reality under the same laws appears again in Strato's description of the physical universe. Like Aristotle, he seems to have regarded it as unique and finite; at least we are told he believed that "void can exist within the universe, but not outside it" (frs. 54–55). But he rejected Aristotle's doctrine of "natural places," and said that all bodies have weight, a natural tendency toward the center of the universe (frs. 50–53); light substances move away from the center because they are "squeezed out" by the heavier. (This was also the view of the atomists and is the only point of any substance on which Strato and they agreed.) The stars are composed of fiery stuff, not, as Aristotle thought, of the mysterious "fifth substance," and are subject to the same laws of gravitation as everything else (fr. 84); a different report, according to which Strato held that the stars derive their light from the sun (fr. 85), probably refers to the moon and planets only.

The immediate cause of physical phenomena is the interplay of natural forces, especially the traditional elementary forces hot, cold, moist, and dry; the hot had a special preeminence. Strato, like Theophrastus, tended to identify these forces with the substances in which they inhere, and treated even heat and light as material emanations (frs. 42–49, 65). Aristotle's distinction between matter and form was blurred in theory by Strato's argument that in any process, the starting point and end point (the formal element), as well as the substrate, are subject to movement (fr. 72; against Aristotle, *Physics*, 224b4), and was ignored in practice. Strato's approach is revealed very clearly in his doctrine of animal reproduction. Aristotle had taught that the male parent contributes the active, formal principle and the female, the material principle. Since the perfection of anything depends on the dominance of form over matter, and males are more perfect than females, the offspring would be male if the (father's) formal principle completely mastered the material contributed by the mother, female if it did not. Strato, however, held (frs. 94–95) that both parents produce "semen" (γονή, the word used by Aristotle to denote the male active-formal principle only), and that the offspring would be male if the father's, female if the mother's "semen" predominated; he also believed that the "power" (δύναμις) of the semen is material, being "of the nature of breath" (πνευματική). Thus the Aristotelian interplay of form and matter was replaced by a tug-of-war between forces, which, although opposed, were fundamentally of the same kind.

Another feature that distinguished Strato's theory from Aristotle's was his belief that bodies are not continuous but consist of tiny particles with void interstices where the particles do not fit together exactly. This scheme was used to account for compression and rarefaction, and for the ability of one substance to penetrate another. Material substances could act on each other only if they first interpenetrated; surface contact was not enough, in Strato's view, to ensure a reaction. Thus the interactions of substances were determined by two factors, their dynamic compatibility and their structure. The same idea is found in some of the writings of Theophrastus; but whereas Theophrastus apparently thought that only solids have this discrete structure, Strato extended it to all bodies, including liquids and gases—even light rays, he says, must consist of particles separated by void, for they are seen to interpenetrate (fr. 65). The existence of void was, however, closely circumscribed. Hero tells us that bodies tend naturally to fill all empty space; no "continuous void" can exist naturally in the world or, as we have seen, outside it. Void interstices or "pores" are found in nature only where the shape of the particles composing bodies does not allow them to fit together exactly. A large void can, however, be produced artificially, as when air is sucked out of a sealed container and, conversely, some bodies can be compressed, that is, their particles can be made to pack more closely into their "pores." But both conditions are unnatural, and the bodies concerned will return to their normal state as soon as the force that produced the artificial conditions is removed. This is the famous law of *horror vacui*, which Hero — or, rather, the engineering school of Ctesibius, of which Hero is a late representative — took from Strato as the theoretical basis for the construction of hydropneumatic machines.

Since Strato's admission of void has been taken as chief evidence for direct atomist influence, it is necessary to emphasize the difference between his doctrine and atomism. This can be expressed by saying that the atoms and void were the starting point of atomist physics, while the particles and interstices were the end point of Strato's. Instead of being the eternal prime constituents of the world, Strato's particles were mere divisions of matter, with no stable individual existence; the qualities hot, cold, moist, and dry, which Strato regarded as primary, were for the atomists epiphenomena of the shapes and sizes of atoms; the void, which for the atomists was infinite in extent and within which the atoms moved freely, existed in Strato's world only as discrete interstices, imprisoned in matter like the bubbles in foam rubber; lastly, the Democritean atoms are in constant movement and tend to repel each other, while Strato's particles hang closely together unless forcibly separated.

To complete the framework of his cosmology, Strato criticized and modified Aristotle's teaching about place, time, and motion. The most fundamental change concerned the nature of place, which Aristotle had defined as the boundary at which a containing body is in contact with a body it contains (*Physics*, 212a6). This definition had been criticized by Theophrastus (frs. 21–22), and Strato went back to an earlier conception that Aristotle had considered and rejected: the place of anything is the hypothetical space between the extremities of that thing. Space is coextensive with the universe; qua space it is empty, but in fact all space is always filled with matter of some kind (frs. 59–60). In this definition Strato seems to have disregarded the existence of void interstices in bodies, presumably because he considered them as part of the substances in which they occur. In regard to motion and time he made only minor changes. He

emphasized the continuity of both and the uniformity of time, and stressed that time is a concomitant of physical things, not an independently existing reality. He also seems to have replaced Aristotle's explanation of the "unnatural" movement of bodies through space with one that closely approximates the modern theory of inertia, and demonstrated experimentally that freely falling bodies accelerate (fr. 73).

Most of our information about the detailed working of Strato's system comes from Hero. Like Aristotle, Strato believed that one element can be "overpowered" and changed into its opposite, a process that can be observed in combustion, evaporation, condensation, and the absorption of water by earth. This occurs on a cosmic scale in the "exhalation" from the earth produced by subterranean heat, which gives rise to wind, rain, and most other meteorological phenomena. Another principle that played a major part in Strato's meteorology was the "antiperistasis," or "circular displacement" of heat and cold, which is explained by Seneca in a passage referring specifically to Strato (fr. 89):

> Hot and cold always go in opposite directions and cannot be together; the cold gathers in places from which the hot has departed, and vice versa. . . . Wells and underground cavities are hot in winter when there is cold on the surface of the earth, because the hot has taken refuge there, yielding the upper regions to the cold to occupy; the hot for its part, when it has entered the lower regions and penetrated there as far as possible, becomes all the more powerful for being concentrated.

If such a concentration of heat is disturbed by an influx of cold, it will break out violently. This can happen in the earth, causing earthquakes, or in clouds, causing lightning, thunder, and typhoons. Probably Strato thought that the "exhalation" from the earth was produced by heat concentrated underground by "antiperistasis" when the surface of the earth cools at night. The direction of winds apparently was governed by *horror vacui*: when the "exhalation" disturbs the overlying air and produces pockets of high and low density, the operation of this law ensures that air flows into the low-pressure areas.

While many elements of Strato's theory, including the "exhalation" and "antiperistasis," had already appeared in the teaching of Aristotle and Theophrastus, Strato's account is distinguished by being based on a limited number of principles, which — unlike some of those invoked by Aristotle — are natural laws in the modern sense. He succeedéd in giving a unified explanation that was flexible enough to account for the phenomena in all their variety.

Strato seems to have devoted a good deal of attention to the nature of light and sound. Light, in his view, is a material emanation, which travels through the "pores" of air and other transparent substances and is reflected by any solid particles in its path, being modified by the colors of the surfaces from which it is reflected and probably also those of the media through which it passes. Transparency is a function of structure: bodies are transparent if they have continuous straight "pores" for light to pass through; provided this condition is satisfied, their density is immaterial. Strato's explanation of sound is more interesting. He discovered that each sound is the result of many separate beats in a stationary medium (usually air), its pitch being governed by the frequency of the beats; thus he anticipated the true account in the most important respects. We do not know how he came to this conclusion; but he could have done so by reinterpreting, in the light of accurate observations of lyres and other sounding objects, Aristotle's doctrine that sound is a "movement of alteration" traveling through a stationary medium from the object to the ear.

Strato's psychology is consistent with his cosmology. He denied the existence of an immaterial or immortal soul—he argued at considerable length against the doctrine of Plato's *Phaedo* (frs. 118, 122–127)—or even of a transcendent element in the soul such as Aristotle's "Active Reason"; he insisted, rather, that mental activity is not essentially different from any other kind of movement (fr. 74). The carrier of all psychic activity is the "breath" or "spirit" ($\pi\nu\epsilon\hat{v}\mu\alpha$), which has its center in the brain behind the eyebrows and spreads to all parts of the body. The soul is a single entity. All mental processes, including sensation, take place at the center; the sense organs, which Strato described as "windows" of the soul, only receive and transmit stimuli; and when the act of perception is complete, the sensation is projected back to the organ where the stimulus originated in the same way as sounds are located at their points of origin (frs. 108–112). Sleep is due to the temporary withdrawal of the "spirit" from the extremities of the body to the psychic center; the complete withdrawal of the "spirit" results in death (frs. 128–129). Dreams are caused by residual movements in the "spirit" arising from past stimuli (frs. 130–131). As usual, most of the constituents of Strato's

doctrine have precedents in the Peripatetic school. The concept of the "Active Reason" had been questioned, but not definitely abandoned, by Theophrastus. The theory of "pneuma" had been originated by Aristotle and further developed by Theophrastus and by Diocles of Carystus, a famous physician and pupil of Aristotle. Only the location of the psychic center in the brain was entirely new; it came from Herophilus, the discoverer of the nervous system.

A noteworthy feature of Strato's work was his extensive use of experiments. While not the first to perform them, he was the first to use experiments systematically to establish a fundamental cosmological doctrine. Strato's experiments are not isolated, but form a progressive series in which each is based on the result of the previous one. Also characteristic of Strato are the care taken to define the conditions in which the experiment takes place and to eliminate all possible alternative explanations of the result, and the practice of pairing controlled experiments with observations of similar phenomena occurring under natural conditions. His purpose presumably was to avoid the charge that his experiments distorted nature.

That Strato's system was in many ways a continuation of those of Aristotle and Theophrastus should not be allowed to obscure his very real originality. This lay in his ability to combine philosophical and scientific reasoning to produce a unified explanation of the world in which theories were shaped by observed facts and the facts themselves interpreted in the light of simple and consistent theories. The need for such a synthesis was particularly acute in the early third century B.C., a time when, largely as a result of Aristotle's teaching and example, factual knowledge was increasing immensely and the natural sciences were asserting their independence. Nevertheless, Strato's influence was limited. The centrifugal tendency among the sciences could not be arrested, and philosophical cosmology came to be dominated increasingly by religious considerations. Signs of a reaction against Strato in his own school appear in the last chapter of the pseudo-Aristotelian *De spiritu*, perhaps written within thirty years of his death. A few Peripatetics of the first century B.C. are reported to have held views similar to some of Strato's; but it is not clear whether they were influenced by his writings, and it is very unlikely that they tried to revive his system as a whole. His name occurs rarely in the extant remains of philosophical writers belonging to other schools. The only pupils of

Strato who made fruitful use of his ideas seem to have been the scientists of Alexandria, Ctesibius the engineer and Erasistratus the physician, and perhaps the astronomer Aristarchus of Samos.

BIBLIOGRAPHY

I. ORIGINAL WORKS. Editions of Strato's works are F. Wehrli, *Die Schule des Aristoteles*, V (Basel, 1950; 2nd ed., 1969), which includes the named fragments and a selection of the others (references in this article are to this ed.); and H. B. Gottschalk, "Strato of Lampsacus: Some Texts," in *Proceedings of the Leeds Philosophical and Literary Society*, Lit.-Hist. Sec., **11**, pt. 6 (1965), 95–182, which deals chiefly with the longer texts in which Strato's name is not mentioned. Both eds. include a commentary and bibliography.

The best text of the *De audibilibus* is in I. Düring, "Porphyrius' Kommentar zur Harmonielehre des Ptolemaios," in *Göteborgs högskolas årsskrift* (1932), pt. 2, 67–77. The authorship of this work has been discussed by H. B. Gottschalk, "The De Audibilibus and Peripatetic Acoustics," in *Hermes*, **96** (1968), 435–460.

II. SECONDARY LITERATURE. See W. Capelle, "Straton 13," in Pauly-Wissowa, *Real-Encyclopädie der classischen altertumswissenschaft*, 2nd ser., IVa (1931), 278–315; and M. Catzemeier, *Die Naturphilosophie des Str. von L.* (Meisenheim, 1970). The older literature has been superseded, but the following article is still worth reading: H. Diels, "Ueber das physikalische System des Straton," in *Sitzungsberichte der Preussischen Akademie der Wissenschaften zu Berlin*, Phil.-hist. Kl. (1893), 101–127, also available in *Kleine Schriften zur Geschichte der antiken Philosophie* (Hildesheim, 1969), 239–265.

H. B. GOTTSCHALK

STRATTON, FREDERICK JOHN MARRION (*b.* Birmingham, England, 16 October 1881; *d.* Cambridge, England, 2 September 1960), *astronomy.*

After attending school in Birmingham, Stratton entered Gonville and Caius College, Cambridge. He was third wrangler in the mathematical tripos of 1904 and in the same year was elected a fellow of his college. His first paper, "On Planetary Inversion" (1906), was essentially a mathematical work concerned with the possibility that tidal forces can produce substantial changes in the obliquities of planetary orbits. Subsequently, however, Stratton joined the Cambridge observatory as an honorary member and worked on the proper motions of faint stars. In 1913 he was appointed assistant director of the Solar Physics Observatory at Cambridge. His astronomical work was interrupted by the First World War, in which he served with great distinction. At the end of the war Stratton relinquished his position at the observatory to become a tutor at Gonville and Caius College. This post gave him the opportunity to direct bright young men into astronomy; and eventually the offices of astronomer royal, astronomer royal for Scotland, and H.M. astronomer at the Cape of Good Hope were filled with Gonville and Caius men.

Stratton went on an expedition to Sumatra with C. R. Davidson to observe the total solar eclipse of 14 January 1926. They obtained excellent spectra of the chromosphere and conducted observations of the intensity of the Balmer lines and of the distribution of intensity of the corona. His report was published in 1929.

In 1928 Stratton succeeded Newall as professor of astrophysics and director of the Solar Physics Observatory. He also was active in the organization of science and was general secretary of the British Association for the Advancement of Science from 1930 to 1935 and general secretary of the International Astronomical Union from 1925 to 1935.

BIBLIOGRAPHY

Stratton's published works include "On Planetary Inversion," in *Monthly Notices of the Royal Astronomical Society*, **66** (1906), 374–402; "Proper Motions of Faint Stars in the Pleiades," in *Memoirs of the Royal Astronomical Society*, **57** (1908), 161–184; "The Constants of the Moon's Physical Libration," *ibid.*, **59** (1909), 257–290; "On Possible Phase-Relations Between the Planets and Sun-Spot Phenomena," in *Monthly Notices of the Royal Astronomical Society*, **72** (1912), 9–26; "Preliminary Note on the Later Spectrum of Nova Geminorum No. 2," *ibid.*, **73** (1913), 72; "On Enhanced (Spark) Lines in the Early Spectra of Nova Geminorum No. 2," *ibid.*, 380–382; "The Spectrum of Nova Geminorum II, 1912," in *Annals of the Solar Physics Observatory* (Cambridge), **4**, pt. 1 (1920), 1–71; *Astronomical Physics* (London–New York, 1925); "Das Sternsystem, zweiter Teil," VI, pt. 2 of *Handbuch der Astrophysik* (Berlin, 1928); "Report on the Total Solar Eclipse of 1926 January 14," in *Memoirs of the Royal Astronomical Society*, **64** (1929), 105–148, written with C. R. Davidson; "The Absorption Spectrum of Nova Herculis, 1934: The First Phase," in *Annals of the Solar Physics Observatory* (Cambridge), **4**, pt. 4 (1936), 131–161; "On Some Spectrograms of Nova Persei 1901," *ibid.*, pt. 2 (1936), 73–84; "The History of the Cambridge Observatories," *ibid.*, **1** (1949), 1–26; "Prof. Megh Nad Saha, F. R. S.," in *Nature*, **177** (1956), 917; "Thomas Royds," in *Monthly Notices of the Royal Astronomical Society*, **116** (1956), 156–158; "John Evershed," in *Biographical Memoirs of Fellows of the Royal Society*, **3** (1957), 41–48; and "Henry Norris Russell," *ibid.*, 173–185.

Stratton was editor of *Scientific Papers by Sir George Howard Darwin*, 5 vols. (Cambridge, 1907–1916); and coeditor of *Observatory*, **36–48** (1913–1925).

Richard v. d. R. Woolley's obituary in *Quarterly Journal of the Royal Astronomical Society*, **2** (1961), 44–49, served as the main basis of this article.

LETTIE S. MULTHAUF

STREETE, THOMAS (*b.* Cork, Ireland [?], 15 March 1622; *d.* Westminster, London, England, 27 August 1689), *astronomy*.

Streete spent most of his life in London, where he was employed as a clerk in the Excise Office under Elias Ashmole. He frequented Gresham College and numbered the leading astronomers both in England and abroad among his friends and acquaintances, often assisting them in observing eclipses, transits, comets, and other unusual astronomical phenomena. For many years Streete published highly regarded ephemerides and worked intensively on the problem of determining longitude at sea. After the great fire of 1666, he was engaged in the resurvey of London.

Streete's most important work was *Astronomia Carolina* (1661). One of the most popular expositions of astronomy in the second half of the seventeenth century, it served as a textbook for Newton, Flamsteed, and Halley. Its tables, constructed from a large number of observations, were generally conceded to be the best of their time. The lunar tables, in particular, were felt to mark an advance over previous ones. The book went through many editions and continued in use well into the eighteenth century.

The *Astronomia Carolina* was an important vehicle for the dissemination of Kepler's astronomical ideas, which were as yet by no means generally accepted. Kepler's first and third laws of planetary motion were clearly stated, and it was from the *Astronomia* that Newton learned of them. In place of Kepler's second law, however, Streete, using as a basis the planetary theories of Seth Ward and Ismael Boulliau, developed an equant construction in which the planets generate equal angles in equal times about the "empty" focus of the elliptical orbit. His explanation of the physical cause of planetary motion employed both Cartesian vortices and

Kepler's concept of quasi-magnetic solar attraction.

BIBLIOGRAPHY

Streete's major works are *Astronomia Carolina: A New Theorie of the Celestial Motions* (London, 1661); *An Appendix to Astronomia Carolina* (London, 1664); and *The Description and Use of the Planetary Systeme* (London, 1674).

There is no adequate and reliable extended account of his life and work.

WILBUR APPLEBAUM

STREETER, GEORGE LINIUS (*b.* Jamestown, New York, 12 January 1873; *d.* Gloversville, New York, 27 July 1948), *embryology*.

Streeter, the son of George Austin Streeter, a glove manufacturer, and Hannah Green Anthony Streeter, was graduated from Union College in 1895 and from the College of Physicians and Surgeons of Columbia University, where he took his M.D. degree, in 1899. After an internship at Roosevelt Hospital in New York City, he became assistant to Henry Hun, a prominent neurologist, in Albany and also taught anatomy at Albany Medical College. In 1902 he studied at Frankfurt with Ludwig Edinger and at Leipzig with Wilhelm His. Under the latter's influence Streeter devoted himself thereafter to embryology and particularly to the development of the human nervous system.

Returning to the United States in 1904, Streeter joined the department of anatomy of the Johns Hopkins Medical School in Baltimore, under the leadership of Franklin P. Mall. There he published his first contribution to embryology, on the development of the cranial and spinal nerves of the human embryo. In this work he showed a talent for three-dimensional analysis of complex microscopic structures, accurate observation, and skilled draftsmanship. His only experimental investigation was a study of the early development of the internal ear in frog embryos (tadpoles) by removal and transplantation of the ear vesicles. His interest in the auditory apparatus led to many years' work on the embryology of the human ear, crowned in 1918 by publication of a distinguished monograph on the development of the labyrinth.

In 1906–1907 Streeter was assistant professor at the Wistar Institute of Anatomy and Biology in Philadelphia and then went to the University of Michigan as professor of anatomy. There he continued his embryological research on the human brain and auditory apparatus with such success that he was recalled to Baltimore to join Mall in the newly organized department of embryology of the Carnegie Institution, located at the Johns Hopkins Medical School. When Mall died in 1917, Streeter succeeded him as director, taking over a well-organized laboratory with the world's largest collection of human embryological material. Continuing Mall's program, he not only carried on his own wide-ranging investigations but also, with generous and enthusiastic leadership, promoted the work of a highly competent staff and numerous guest investigators.

Streeter's paper on the weight and dimensions of human embryos and fetuses at successive stages of development (1920) is the classic account of this subject. His comprehensive chapter on the development of the human brain in Franz Keibel and Mall's *Manual of Human Embryology* (1912) has not been superseded. Mall had been deeply interested in the pathological aspects of human prenatal development. Working before the great twentieth-century advance of genetics, he attributed prenatal retardation, malformation, and intrauterine death to defects or disease of the uterine environment. Streeter, on the other hand, tended to attribute these disasters to genetic factors. It is now known that both environmental and genetic causes are operative in prenatal pathology, but Streeter's skilled exposition of his views was valuable in turning the attention of physicians and biologists to the importance of genetic factors.

Although primarily a morphologist, Streeter always saw the embryo as a living, growing individual, and its organs and tissues as carrying on physiological functions appropriate to its stage of growth. He therefore opposed the kind of embryology represented in its extreme form by Haeckel's law of recapitulation, which regarded every embryonic structure as no more than a record of phylogeny. He was especially critical of the concept of the mammalian branchial arches as merely rudimentary gills rather than as preliminary stages of the external ears and other definitive parts of the head and neck; and he denied the supposed metamerism of the early vertebrate brain. To combat these older ways of thinking, often supported by oversimplified diagrams, he illustrated his own descriptions as far as possible by photomicrographs or realistic drawings based on actual sections.

Streeter's joint studies of the early embryology of the pig, with Chester H. Heuser, and of the rhesus monkey, with Heuser and Carl G. Hartman,

published in the *Contributions to Embryology* of the Carnegie Institution between 1927 and 1941, are among the most accurate descriptions of early mammalian development ever published and stand in the front rank of American scientific achievements. With his great descriptive talent, his long experience in the most exacting kind of morphological study, and the skilled assistance of a technical staff that he himself had largely trained, Streeter was well fitted to begin (about 1935) the great advance in early human embryology made by his staff in cooperation with John Rock and Arthur T. Hertig of Boston. Before this time the human embryo was known only after about the eleventh day following conception. New specimens obtained by Hertig and Rock carried the story back to the beginning and made the earliest stages of human development as well known as the earliest stages of other mammals.

Streeter's last major undertaking, unfinished at his death, was a systematic catalog and analytic description of the human embryo, classified by stages up to the end of the embryonic period—about the forty-eighth day of gestation. The "developmental horizons," as he called them, were published, as far as completed, in the Carnegie Institution *Contributions to Embryology* from 1942 to 1951.

Devoting his entire career to intensive research in a difficult field, Streeter had little time or inclination for popularizing his findings or for general scientific affairs. He published no textbooks nor comprehensive reviews; his talents for exposition were demonstrated largely through numerous papers presented at professional meetings. He was elected to the National Academy of Sciences in 1931 and to the American Philosophical Society in 1943, and was president of the American Association of Anatomists in 1926–1928. He married Julia Allen Smith of Ann Arbor, Michigan, in 1910. Their son and one daughter became physicians, and the other daughter took her doctorate in chemistry.

Streeter retired in 1940 from the directorship of the Carnegie department of embryology but continued his work on human embryology until, with his microscope and drawing board at hand and another section of "Horizons of Human Development" in preparation, he died suddenly of coronary occlusion at the age of seventy-five.

BIBLIOGRAPHY

A complete bibliography of Streeter's more than 130 publications follows the obituary by George W. Corner,

"George Linius Streeter, 1873–1948," in *Biographical Memoirs. National Academy of Sciences*, **28** (1954), 261–287, with portrait.

GEORGE W. CORNER

STRÖMBERG, GUSTAF BENJAMIN (*b.* Göteborg, Sweden, 16 December 1882; *d.* Pasadena, California, 30 January 1962), *astronomy.*

Strömberg spent thirty years, the greater part of his working life, at the Mount Wilson observatory. His main contributions to astronomy were statistical analyses of stellar motions, made at a time when both the size and the manner of rotation of our galaxy were still uncertain quantities.

The son of Bengt Johan Gustaf Lorentz Strömberg and Johanna Elisabeth Noehrman, Strömberg prepared in Göteborg for study at the universities of Kiel and Stockholm. He was an assistant in the Stockholm observatory for eight years, ending in 1914, the year he married Helga Sofia Henning. In 1916, at age thirty-three, he obtained a Ph.D. from the University of Lund; his dissertation was written under the direction of Carl Vilhelm Ludwig Charlier.

In June 1916 Strömberg arrived at Mount Wilson. He served one year as a volunteer assistant in stellar spectroscopy and then was appointed to the staff as astronomer, a position he held until his retirement in 1946. The director in 1916 was Walter S. Adams, who had just developed a technique for estimating a star's distance from its spectrum—referred to as a spectroscopic parallax—thus providing for the first time a way to convert large numbers of stellar brightnesses as observed into intrinsic luminosities, that is, absolute magnitudes. Strömberg's first assignment was to help Adams analyze data on some 1,300 stars, looking for a statistically valid relation between motions through space and absolute magnitudes. They found that faint dwarf stars seemed to move faster than bright giant stars of the same spectral class.

Strömberg continued to analyze the wealth of observational material being obtained at Mount Wilson, and also took his turn at the telescope acquiring it. He was looking for large-scale preferential motions of stars. Having obtained a statistical estimate of how the sun moves through space, he used this to calculate the peculiar motions of stars (across the line of sight), and also the so-called K-term in radial velocities, which for any selected group of stars refers to a net speed either toward or away from us. His discovery of what was

referred to at the time as "Strömberg's asymmetry" confirmed tentative conclusions reached earlier by Boss, by Adams, and by Joy; he found that stars in the plane of the Milky Way had a marked preferential motion, directed around a galactic center approximately coincident with the one proposed by Shapley in 1918. Strömberg thus provided an early confirmation of Shapley's theory, and also supplied the basic data to be used by Lindblad and by Oort in developing the presently accepted picture of galactic rotation.

BIBLIOGRAPHY

Strömberg's publications include "The Relationship of Stellar Motions to Absolute Magnitude," in *Astrophysical Journal*, **45** (1917), 293–305, written with Walter S. Adams; "A Determination of the Solar Motion and the Stream-Motion From Radial Velocities and Absolute Magnitudes of Stars of Late Spectral Types," *ibid.*, **47** (1918), 7–37; "Space Velocities of Long-Period Variable Stars of Classes Me and Se," *ibid.*, **59** (1924), 148–154, written with Paul W. Merrill; "The Asymmetry in Stellar Motions and the Existence of a Velocity-Restriction in Space," *ibid.*, **59** (1924), 228–251; "Analysis of Radial Velocities of Globular Clusters and Non-Galactic Nebulae," *ibid.*, **61** (1925), 353–362; and "The Asymmetry in Stellar Motions as Determined From Radial Velocities," *ibid.*, **61** (1925), 363–388.

Strömberg also wrote two books dealing with the philosophical implications of science: *The Soul of the Universe* (New York, 1940; 2nd ed., 1948) and *The Searchers* (New York, 1948).

Eighty-seven publications by Strömberg are listed in the *Yearbook of the Carnegie Institution of Washington*, **15** (1916)–**46** (1946–1947). In addition Strömberg wrote "Angenäherte allgemeine Störungen des Planeten 471 Papagena und 123 Brunhild," in *Astronomische Nachrichten*, **195** (1913), cols. 129–140, written with Vilhelm Hernlund; Strömberg's dissertation, "On a Method for Studying a Certain Class of Regularities in a Series of Observations, With Application to the Temperature Curve of Uppsala" (Lund, 1915); and the two books mentioned above.

SALLY H. DIEKE

STRÖMGREN, SVANTE ELIS (*b.* Hälsingborg, Sweden, 31 May 1870; *d.* Copenhagen, Denmark, 5 April 1947), *astronomy.*

Strömgren studied astronomy at the University of Lund, where he received his doctorate in 1898. He spent the years 1901–1907 at Kiel, as assistant to the editor of *Astronomische Nachrichten*, and beginning in 1904 he was a lecturer at Kiel University. From 1907 to 1940 he was professor of astronomy at Copenhagen University and director of the observatory. Strömgren's research belongs to the tradition of classical celestial mechanics, even though he used elaborate mathematical apparatus to make numerical computations more than most members of this school did.

In his dissertation Strömgren derived the definitive orbit of Comet 1890 II, which, like most comets, followed an approximately parabolic path relative to the sun. This investigation was the first of a series of papers on the original orbits of such comets—open or closed—and was crucial for work on the cosmogony of comets. In "Ueber den Ursprung der Kometen" (1914), Strömgren concluded that all comets of which the orbits had been determined with sufficient accuracy for a decision about their return to be possible, have followed closed orbits; the hyperbolic motion derived for several comets was a consequence of the perturbations of the large planets during their passage through the internal regions of the solar system. About thirty years later his investigations formed the starting point for J. H. Oort's and A. J. J. van Woerkom's "discovery" of the comet cloud far outside the planetary orbits.

Strömgren's survey of the three-body problem was concerned with two equal masses moving in circular orbits around a common center of gravity and with a body without mass—a *problème restreint*. The movement of the third body was derived by numerical integration with the aim of determining the orbits, which are periodic relative to the coordinate system, in which the main bodies are relatively at rest. With J. Fischer-Petersen and J. P. Møller, Strömgren pointed out a number of groups of simple-periodic orbits. Strömgren's pioneer work has recently been the starting point for further surveys because electronic computers have greatly simplified the routine computational work. In "Connaissance actuelle des orbites dans le problème des trois Corps" (1935), Strömgren gave a general view of the work done at Copenhagen.

Strömgren held a major position in international astronomical collaboration. From 1922 to 1947 he was director of the telegram bureau for astronomical news sponsored by the International Astronomical Union; and as early as 1914, at the outbreak of World War I, Strömgren had established a "neutral bureau" at the Copenhagen observatory, as a branch of the old Kiel bureau. During World War II he maintained the news service at the Lund observatory as a neutral branch bureau. From 1921 to 1930 he was president of the Astronomische Gesellschaft.

Strömgren's textbook of astronomy for students at Copenhagen University was translated into German and enlarged by Bengt Strömgren in 1933. Unique for its time, the work deals with both classical and modern stellar astronomy and astrophysics. From 1920 to 1947 Strömgren was president of the Astronomisk Selskab (Copenhagen) and was editor-in-chief of *Nordisk astronomisk tidsskrift*. He also took a strong interest in the Variable Star Section sponsored by the Astronomisk Selskab. A distinguished Latin scholar, Strömgren had a major role in the translation into modern languages of Tycho Brahe's book on his own instruments and of Peder Horrebow's description of Ole Römer's instruments.

BIBLIOGRAPHY

I. Original Works. A bibliography of Strömgren's papers and books (1896–1940) was edited by K. Lundmark in *Astronomical Papers, Dedicated to Elis Strömgren* (Copenhagen, 1940). His principal works are "Ueber den Ursprung der Kometen," in *Kongelige Danske Videnskabernes Selskabs Skrifter*, 7th ser., Naturv.-mathem. Afd., **11**, p. 4 (1914), 189–251; "Connaissance actuelle des orbites dans le problème des trois corps," in *Bulletin astronomique*, 2nd ser., **9** (1935), 87–130; and *Ole Rømer som Astronom* (Copenhagen, 1944).

With Bengt Strömgren he was coauthor of *Lehrbuch der Astronomie* (Berlin, 1933); and with Bengt Strömgren and Hans Raeder, *Tycho Brahe's Description of His Instruments and Scientific Work . . .* (Copenhagen, 1946).

See also *Publikationer og mindre Meddelelser fra Kjøbenhavns Observatorium* (1910–1947) and "Tre Aartier celest Mekanik paa Kjøbenhavns Observatorium," in *Festskrift, Copenhagen University* (Nov. 1923).

II. Secondary Literature. Obituaries are J. M. V. Hansen, in *Nordisk astronomisk tidsskrift* (1947), 41–43; in *Observatory*, **67** (1947), 142–143; and in *Popular Astronomy*, **55** (1947), 341–343; B. Lindblad, in *Nordisk astronomisk tidsskrift* (1947), 43; and in *Populär astronomisk tidsskrift*, **28** (1947), 59–60; K. Lundmark, in *Monthly Notices of the Royal Astronomical Society*, **108** (1948), 37–41; and N. E. Nørlund, in *Festskrift, Copenhagen University* (Nov. 1947), 149–152; and in *Oversigt over det K. Danske Videnskabernes Selskabs Forhandlinger* (1948), 73–77. See also R. Grammel, "E. Strömgrens Arbeiten zum Dreikörperproblem," in *Vierteljahrsschrift der Astronomischen Gesellschaft*, **64** (1929), 90–100.

Axel V. Nielsen

STRUSS (or **STRUTHIUS**), **JÓZEF** (*b.* Poznan, Poland, 1510; *d.* Poznan, 6 March 1568), *medicine.*

The son of Nicolas Strusiek, a wealthy comb manufacturer, Struss completed his primary education at the parish school of St. Mary Magdalene in Poznan, then went on to Lubranski College, a secondary school of arts. He next moved to Cracow, where in 1531, after seven years of study, he obtained a diploma in the seven liberal arts. In particular Struss had studied Greek and classical philosophy, subjects that then included material of a more strictly scientific nature, and thus was able to increase his knowledge of Aristotle's mathematical and astrological works. At the same time he also studied medicine; but since the University of Cracow did not confer degrees in that subject, he went to Padua, where many Polish and German medical students completed their studies. Vesalius was his fellow pupil and later his teacher there.

Struss's graduation in medicine on 26 October 1535 is recorded in documents preserved in the Old Archive of the University of Padua. He began to teach theoretical medicine at that university on 12 November and remained there until 1537, when he was asked to teach medicine at the University of Cracow, with the particular task of illustrating Galen's *De differentiis morborum* for his students. Struss soon left Cracow; his precise destination is not known but he probably returned to Padua, since it was there, at about this time, that some of his works were published. About 1540 he was again in Poznan, where he had entered the court of Andrei Gorka, then the governor of Greater Poland. The following year he accompanied Gorka to Hungary on a mission of mediation between Ferdinand I of Hungary and Isabella, the widow of John Zápolya, king of Hungary, who was besieged in Buda. On this occasion Struss was called upon to treat both Isabella and Sultan Suleiman I. In 1545 he returned to Poznan, where he established a successful practice and became personal physician to King Sigismund Augustus.

Struss's main work is *Sphygmicae artis*, an accurate clinicophysiological study of the pulse and its alterations; it was perhaps the first work in the history of medicine that suggested the pulse as a reliable source of clinical data and of diagnostic and prognostic information. The physiology on which Struss's study is based is Galen's, as are the plan and framework. But Struss maintained an independent judgment, often contradicting Galen's teaching and following concepts and approaches that were entirely new. *Sphygmicae artis* is a classic

example of the clinical concept of medicine that was then developing in Padua.

According to Manget (*Bibliotheca scriptorum medicorum* . . ., XI [Geneva, 1731], 330) and other writers, the work was first published in 1540, but there is no trace of this edition; and it is more likely, also on the basis of internal evidence, that the first edition was produced at Basel in 1555. In addition to his editions and translations of various classical medical works of Galen and treatises, Struss also published literary works.

BIBLIOGRAPHY

I. Original Works. Struss's writings include *Ad medicum hisce temporibus maximum atque celeberrimum D. Cyprianum de Lowicz de medica arte excellentis carmen elegiacum authore Josepho Struthio Posnaniense* (Cracow, 1529); *Sanctissimi Petris et Domini d. Joannis a Lasco archiepiscopi et primatis regni Epicedion elegiacis versibus confectum* (Cracow, 1531); *Sphygmicae artis iam mille ducentos annos perditae et desideratae libri V* (Padua, 1540 [?]; Basel, 1555); and *Giuseppe Struzio: Dell'arte sfigmica*, C. Castellani and G. Invernizzi, eds. (Turin, 1961), which includes the Latin text, an Italian translation, an introduction, and notes on cardiology and on the history of medicine.

II. Secondary Literature. See H. Barycz, *Historja Uniwersytetu Jagiellońskiego w epoce humanizmu* (Cracow, 1935), 241–242; W. Bugiel, *Un célèbre médecin polonais au XVI siècle: Joseph Struthius* (Paris, 1901); and G. Sterzi, *Josephus Struthius, lettore nello studio di Padova* (Venice, 1910).

CARLO CASTELLANI

STRUTT, JOHN WILLIAM, THIRD BARON RAYLEIGH (*b.* Langford Grove, near Maldon, Essex, England, 12 November 1842; *d.* Terling Place, Witham, Essex, England, 30 June 1919), *experimental and theoretical physics.*

Lord Rayleigh (as he is universally known in scientific circles) was one of the greatest ornaments of British science in the last half of the nineteenth century and the first two decades of the twentieth. A peer by inheritance, he took the unusual course of devoting himself to a scientific career and maintained his research activity continuously from the time of his graduation from Cambridge University in 1865 until almost literally the day of his death. Rayleigh's investigations, reported in 430 scientific papers and his monumental two-volume treatise *The Theory of Sound* (1877–1878), covered every field of what in the twentieth century is commonly referred to as "classical" physics; at the same time he kept abreast of, and made incisive critical comments on, the latest developments of quantum and relativistic physics. Not in any sense a pure mathematician, Rayleigh applied mathematics with great skill and accuracy to a host of problems in theoretical physics. In addition he was an ingenious and resourceful experimentalist, with the uncanny ability to extract the most from the simplest arrangements of apparatus. The discovery and isolation of argon, usually considered by the lay public as his greatest scientific achievement, was a triumph of both careful logical reasoning and patient and painstaking experimentation.

At Cambridge, Strutt became a pupil of the mathematician E. J. Routh and profited greatly from his thorough coaching. This and the inspiration gained from the lectures of Sir George Stokes, at that time Lucasian professor of mathematics, paved the way in part at least for Strutt's emergence as senior wrangler in the mathematical tripos as well as Smith's Prizeman. He became a fellow of Trinity College, Cambridge, in 1866; and from that time on, there was no doubt that he was headed for a distinguished scientific career.

Strutt varied the usual custom of a tour of the Continent after graduation with a visit to the United States, then recovering from the Civil War. On his return to England in 1868 he purchased a set of experimental equipment and proceeded to carry out some investigations at the family seat in Terling Place. This was the genesis of the famous laboratory in which most of his later scientific work was done. Strutt early formed the habit of getting along with very simple scientific apparatus and made much of it himself. It is clear that he was considered somewhat of a freak by members of his family and friends for his determination not to be contented with the life of a country gentleman. It is equally clear that Strutt did not feel he was violating any strongly entrenched custom. He simply wanted to be a scientist; and with typical British stubbornness he pursued this course, feeling that there was nothing unusual or blameworthy in his action.

In 1871 Strutt married Evelyn Balfour, sister of Arthur James Balfour, who became a celebrated scholar, philosopher, and statesman. A serious attack of rheumatic fever occurred shortly after the marriage, and as a recuperative measure Strutt undertook a trip up the Nile. It was on this journey that the *Theory of Sound* had its genesis, although the first volume was not completed and published

until 1877. Shortly after returning to England in 1873, Strutt succeeded to the title and took up residence at Terling. He then began serious experimental work in the laboratory attached to the manor house. He had already developed considerable theoretical interest in radiation phenomena and had published papers on acoustics and optics in the late 1860's and early 1870's. One of these, on the theory of resonance, extended in important fashion the work of Helmholtz and established Rayleigh as a leading authority on sound. Another paper from this early period resolved a long-standing puzzle in optics, the blue color of the sky. In this research, published in 1871, Rayleigh derived the well-known law expressing the scattering of light by small particles as a function of the inverse fourth power of the wavelength of the incident light. It is of interest to note that in this work he used the elastic-solid theory of light and not the recently introduced electromagnetic theory of Maxwell.

In his laboratory at Terling, Rayleigh embarked on a series of experimental studies of optical instruments that apparently originated in his attempts to manufacture cheap diffraction gratings by photographic means. Although not very successful, these early experiments led him to the very important study of the resolving power of gratings, a matter that was then poorly understood by optical experts. It seems clear that Rayleigh was the first to publish formally a clear definition of resolving power of an optical device. He proved that the resolving power of a plane transparent grating is equal to the product of the order of the spectrum and the total number of lines in the grating. This work was continued with a series of fundamental researches on the optical properties of the spectroscope, an instrument that in the late 1870's was becoming increasingly important in the study of the solar spectrum as well as of the spectra of the chemical elements. In his study of optical diffraction and interference, Rayleigh anticipated the French physicist Charles Soret in the invention of the optical zone plate, with its interesting light-focusing property.

During the late 1870's Rayleigh's laboratory in his home at Terling became well established as the seat of his researches, and it appeared likely that he would spend the rest of his career there without serious interruption. The fates decreed otherwise, however, for in 1879 James Clerk Maxwell, the first Cavendish professor of experimental physics at Cambridge, died. Sir William Thomson (later Lord Kelvin), at that time professor of natural philosophy at the University of Glasgow, refused to be considered for the post in succession to Maxwell. Rayleigh, the next obvious choice, accepted the appointment in December 1879 — not without some reluctance, since his natural preference was to continue the Terling routine. The professorial salary was not unwelcome, however, in the face of falling revenues from his estate due to the severe agricultural depression then prevailing in Britain.

Rayleigh remained as professor at Cambridge until 1884. Although admittedly not a brilliant lecturer, he was an effective instructor and, moreover, succeeded in putting laboratory instruction in elementary physics on a firm basis. This was a revolutionary accomplishment in England, and the influence of Rayleigh's pioneer efforts was ultimately felt in higher educational institutions throughout the country. A rather elaborate research program was also set up with the help of his assistants Glazebrook and Shaw, both of whom later became scientists of note. This program involved the redetermination of three electrical standards: the ohm, the ampere, and the volt. Work of this sort had already been started by Maxwell for the British Association for the Advancement of Science. Rayleigh's continuation and development demanded the construction of more precise equipment than Maxwell's, as well as meticulous care and patience in its use. When the investigation was completed in 1884, the results stood the test of time remarkably well. The realization of the importance of standards in physical measurements that this work implied undoubtedly influenced Rayleigh favorably toward the establishment of a government standards laboratory in Britain, which eventually (1900) took the form of the National Physical Laboratory at Teddington, Middlesex.

In 1884 Rayleigh served as president of the British Association for the Advancement of Science, which held its annual meeting that year in Montreal, the first outside the United Kingdom. It provided the occasion for a second trip to the North American continent, and Rayleigh took advantage of it to increase his acquaintance with prominent physicists in the United States and Canada. Immediately after his return to Britain he resigned his professorship at Cambridge and retired to his laboratory at Terling, which remained his scientific headquarters for the rest of his life. Rayleigh did accept a professorship at the Royal Institution of Great Britain in London, and served from 1887 to 1905. This post, however, involved residence in London for only a short time each year and the presentation of a certain number of lectures on topics of his research interest. It did not seriously disturb the continuity of his research program at Terling.

The late 1880's saw the establishment of a more or less definite pattern of research activity. Preferring to have several irons in the fire at the same time, Rayleigh divided his time rather evenly between experimental work in the laboratory and theoretical investigations in his study. An avid reader of the technical literature, he found the origin of many of his researches in questions suggested to him by his reading. He had an uncanny knack of putting his finger on a weak or difficult point in another man's research results and of building an important contribution of his own on it. Rayleigh's grasp of such widely diverse fields as optics and hydrodynamics, acoustics and electromagnetic theory, was phenomenal; and only Maxwell, Kelvin, and Helmholtz came near him in this aspect of his genius.

During the middle and late 1880's Rayleigh's increasing tendency to extend his research net became apparent. His published papers from this period report results of experimental and theoretical work on radiation both optical and acoustical, electromagnetism, general mechanical theorems, vibrations of elastic media, capillarity, and thermodynamics. To this period belongs his pioneer work on the filtration (selective transmission) of waves in periodic structures, as well as his first precise measurements of the density of gases, which led to the discovery of argon. It was also the period in which Rayleigh apparently first became interested in the problem of the complete radiation law, which governs the distribution of energy in the spectrum of blackbody radiation. His work here was tentative, but he fully recognized the physical significance of this puzzling problem to which Planck, Wien, and others were devoting considerable attention. What is now known as the Rayleigh-Jeans law was first enunciated by Rayleigh in 1900.

The discovery and isolation of argon was undoubtedly Rayleigh's most dramatic and famous accomplishment. It emerged as the solution to a scientific puzzle, and Rayleigh was usually at his best when faced by a puzzle. The difficulty was encountered in high-precision measurements of the density of nitrogen, undertaken in the first instance with the aim of obtaining better values of the atomic weight of that element. It was found that the density of nitrogen prepared from ammonia was about one part in two hundred less than the density of nitrogen obtained from air. Repeated reweighings only confirmed the difference and led to Rayleigh's publishing in *Nature* (1892) a short note citing the apparent dilemma and asking for suggestions for its resolution. In a certain sense this was unfortunate, in the light of the priority problem involved in the subsequent discovery. It does, however, illustrate Rayleigh's single-minded devotion to science as a social profession and what may appropriately be called his scientific unselfishness.

The ultimate solution to the peculiar problem of the density of nitrogen was suggested by the reading of a paper published by Henry Cavendish in 1795. He had oxidized the nitrogen in a given volume of air by sparking the air with a primitive static machine. Cavendish found that no matter how long he conducted the sparking, there was always a small residue of gas that apparently could not be further oxidized. He abandoned the research at that point. Had he continued, he presumably would have been the discoverer of argon. Rayleigh decided to push Cavendish's experiment to a conclusion, acting finally on the conviction that there really was another constituent of atmospheric air in addition to the commonly accepted ones.

Rayleigh used an induction coil to provide the electrical discharge for the oxidation of nitrogen, but the process of accumulating enough of the new gas to test its properties was a slow one. In the meantime Sir William Ramsay, having noted Rayleigh's nitrogen-density problem, proceeded to attempt the isolation of the unknown gas by much faster chemical means. Ramsay kept Rayleigh thoroughly informed of his activities, but some confusion and uncertainty still exist over whether Rayleigh actually gave Ramsay his scientific blessing. In the end both shared in the recognition for the discovery of argon and presented their results in a joint paper. There was the usual skepticism over the validity of the result, especially on the part of chemists, who found it hard to believe that a genuinely new element could have remained undetected for so long. The relative chemical inertness of argon was, of course, the explanation. Sooner or later spectroscopic analysis would in any case have revealed its existence. Rayleigh and Ramsay were led to take the hard way in its recognition.

It was largely because of this discovery that Rayleigh was awarded the Nobel Prize in physics in 1904, while Ramsay received the Nobel Prize in chemistry the same year. It is rather ironic that Rayleigh received the prize for work as relevant for chemistry as for physics, when he never felt he had much competence as a chemist. And indeed

there seems little question that his other contributions to physics were vastly more significant than the discovery of argon. The latter caught both the scientific and the popular fancy, however. Although Rayleigh took the discovery very seriously—as he did all his research—and worked very hard at it, it seems clear that once the existence of the new gas and the demonstration of its properties were irrefutably established, Rayleigh was disinclined to go on with this kind of research. Even during the three years of the argon research (1892–1895) he found time to contribute to the scientific literature some twelve papers dealing with the interference and scattering of light, the telephone and its technical problems, and the measurement of the minimum audible intensity of sound.

An illustration of Rayleigh's uncanny ability to forecast developments in physics is provided by his 1899 paper "On the Cooling of Air by Radiation and Conduction and on the Propagation of Sound." In this he faced the problem of the anomalously high sound attenuation observed in air (much greater than that predicted by the transport properties of viscosity and heat conduction). He predicted that the solution to the difficulty might well be found in a relaxation mechanism involving reciprocal transfer of energy between translational and internal energy states of the molecules of the gas through which the sound passes. This suggestion was adopted by various later investigators and has led to the establishment of the vigorous field of molecular acoustics, which by the second half of the twentieth century has thrown new and important light not only on ultrasonic propagation but also on the structure and interaction of molecules.

Any appraisal of Rayleigh's scientific achievements must include mention of his relation to modern physics and, in particular, to the formulation and development of quantum and relativity theories. This poses an interesting but somewhat puzzling problem. In his reading and his association with other scientists, Rayleigh kept fully abreast of all the important activity in physics. He keenly realized the difficulties that classical physics (electromagnetic theory, thermodynamics, and statistical mechanics) was encountering near the end of the nineteenth century in the attempt to explain the experimental phenomena of radiation spectra. But he refused to give up hope that adequate solutions would be forthcoming within the framework of traditional physical theories. Revolutionary ideas evidently were distasteful to him. He could never develop much enthusiasm for Planck's quantum theory and its subsequent development. He never attacked the theory with any vehemence but simply felt it was not to his taste.

His derivation of what came later to be called the Rayleigh-Jeans radiation law (published in 1900, a few months before Planck's famous paper on the distribution law) reflects Rayleigh's general attitude very well. The statistical principle of equipartition of energy among resonators worked very well for long wavelengths. One has the impression that Rayleigh felt a secret longing that with some ingenious maneuvering it might be made to work for the short wavelengths as well. Of course it never has! But he certainly cannot be accused of allowing any nostalgia for traditionalism in physics to keep him from seriously considering the problem and its importance.

Somewhat similar remarks apply to the problem of the unraveling of the intricacies of atomic spectra. Rayleigh fully realized the ultimate significance of this in connection with atomic constitution and tried his hands at numerous calculations of vibratory systems that might possess frequencies in accord, for example, with the Balmer formula for the emission spectrum of hydrogen. He admitted freely that the failures of these attempts indicated the need for new approaches. At the same time Bohr's theory was too radical and revolutionary for his liking.

Rayleigh also was much concerned with the physical problems that ultimately led to the theory of relativity. As far back as 1887 he was interested in astronomical aberration and its bearing on the theory of a luminiferous ether. At that time he indicated a preference for Fresnel's assumption of a stationary ether, despite the presumably null results of Michelson's famous 1881 experiment. Rayleigh was skeptical of the validity of Michelson's early work. Here again it seems clear that he was much disturbed by the possibility that the ether would have to be abandoned as an unworkable hypothesis. His loyalty to the classical wave theory of light was very great. Rayleigh saw the necessity for further experiments, however, and in 1901 undertook to detect possible double refraction in a material medium due to motion through a presumptive stationary ether. The negative results added to the mounting evidence that no physical phenomenon can enable one to distinguish between the motion of two inertial systems so as to say that one is at rest while the other is moving in an absolute sense. Rayleigh contributed nothing to the Einstein theory of relativity as such, although it is evident that he followed its developments with in-

terest. Here again his rather conservative nature asserted itself.

The pace of Rayleigh's research activity did not slacken as he approached his later years. In the last fifteen years of his life he produced ninety papers, of which some reported notable work. For example, to this period belongs a paper on sound waves of finite amplitude, in which the earlier investigations of W. J. M. Rankine and Hugoniot on what came to be called shock waves were much extended. This laid the groundwork for much future development. Other important contributions to acoustics after 1905 were concerned with the binaural effect in human hearing, in which Rayleigh's pioneer investigations paved the way for the relatively enormous amount of interest in this problem in the later twentieth century, and with the filtration and scattering of sound.

The Theory of Sound was kept up-to-date with appropriate revisions and is still a vade mecum in every acoustical research laboratory. The scattering of light from a corrugated surface also provided new insight into a difficult problem.

Along with this intense research activity, Rayleigh devoted considerable attention to professional scientific societies and governmental applied science. The details of the life of a research scientist working at his desk or in his laboratory often seem to offer little of dramatic character. But Rayleigh became an important public figure in his lifetime and devoted much energy to the promotion of science as a whole and physical science in particular. He early became interested in the affairs of the British Association for the Advancement of Science. His first research results were presented at a meeting of the Association at Norwich in 1868, and he served as president of Section A (Mathematics and Physics) at the Southampton meeting in 1882. His presidency of the entire Association for the Montreal meeting in 1884 has already been mentioned.

Elected to the Royal Society in 1873, Rayleigh served as secretary (succeeding to Sir George Stokes) from 1885 to 1896. He took his duties very seriously and made some interesting discoveries in the archives of the Society, including the neglected paper by the Scottish engineer J. J. Waterston, pioneer in the molecular theory of gases. In 1905 Rayleigh was elected president of the Royal Society and served until 1908. Because he never treated any organizational post as a sinecure, he was much in demand when advice and active work on difficult problems were sought.

In 1896 Rayleigh accepted appointment as scientific adviser to Trinity House, a post Michael Faraday had held some sixty years previously. This organization, dating to the time of Henry VIII, has as its function the erection and maintenance of such coastal installations as lighthouses and buoys. Rayleigh served this organization for fifteen years. Much of his later work in optics and acoustics was suggested by problems arising in connection with tests of fog signals and lights. This work for Trinity House is an illustration of his willingness to give freely of time and energy to scientific committees of government and professional organizations in the interests of applied science. A leader in the movement culminating in the establishment of the National Physical Laboratory at Teddington (the British counterpart of the United States National Bureau of Standards), he presided over its executive committee until shortly before his death. Other examples of Rayleigh's public service are his chairmanships of the Explosives Committee of the War Office and his long tour of duty as chief gas examiner of the London gas supply.

Despite the relative shortness of his own career as a university teacher, Rayleigh took a great interest in educational problems and served on the governing boards of several educational institutions. From 1908 to his death in 1919 he served as chancellor of Cambridge University.

The bulk of Rayleigh's experimental notebooks, calculations, and the original MSS of his published papers have been acquired by the United States Air Force Cambridge Research Laboratories in Bedford, Massachusetts, and are now housed there as the Rayleigh Archives. Photostat copies have been distributed to other libraries, particularly the Niels Bohr Library of the American Institute of Physics in New York, and are available for scholarly study.

Public recognition of his scientific achievements came to Rayleigh in full measure. After receiving the Nobel Prize in 1904, he donated its cash award, amounting to about $38,500, to Cambridge University to improve the Cavendish Laboratory and the University Library. Rayleigh was one of the first members of the new Order of Merit when it was established in 1902. He also became a privy councillor in 1905. He was the recipient of thirteen honorary degrees and held honorary memberships in, or received special awards from over fifty learned societies.

Rayleigh may justly be considered the last great polymath of physical science. He outlived his closest rivals Helmholtz, Gibbs, Kelvin, and Poincaré

by a measurable span of years and remained professionally active to the end of his life. At the time of his death he left three completed professional papers unpublished. The amount of work he accomplished in the roughly fifty-five years of his professional career can only be regarded as prodigious. By nature he was not a profoundly or boldly imaginative scientist who would initiate a wholly new idea like the electromagnetic theory of radiation, the quantum theory, or relativity. In this respect he differed from Maxwell, Planck, Bohr, and Einstein. But he did advance enormously the power and scope of applicability of practically every branch of classical physics. He was admired and respected for his sound scientific judgment and his ability to penetrate to the heart of any scientific problem he encountered. Above all, Rayleigh was a modest man. Typical of this was the remark he made in his speech accepting the Order of Merit: "The only merit of which I personally am conscious is that of having pleased myself by my studies, and any results that may have been due to my researches are owing to the fact that it has been a pleasure to me to become a physicist."

BIBLIOGRAPHY

I. ORIGINAL WORKS. Lord Rayleigh's complete bibliography includes one book and 430 articles. All the articles have been published in his *Scientific Papers*, 6 vols. (Cambridge, 1899–1920), repr., 3 vols. (New York, 1964). The scope of Rayleigh's research activity is indicated by the following. His book is *The Theory of Sound*, 2 vols. (London, 1877–1878). His articles include "On Some Electromagnetic Phenomena Considered in Connexion With the Dynamical Theory," in *Philosophical Magazine*, **38** (1869), 1–14; "On the Theory of Resonance," in *Philosophical Transactions of the Royal Society*, **161** (1870), 77–118; "On the Light From the Sky, Its Polarization and Colour Appendix," in *Philosophical Magazine*, **41** (1871), 107–120, 274–279; "On the Scattering of Light by Small Particles," *ibid.*, 447–454; "Investigation of the Disturbance Produced by a Spherical Obstacle on the Waves of Sound," in *Proceedings of the London Mathematical Society*, **4** (1872), 253–283; "On the Application of Photography to Copy Diffraction-Gratings," in *British Association Report* (1872), 39; "On the Diffraction of Object-Glasses," in *Astronomical Society Monthly Notes*, **33** (1872), 59–63; "Some General Theorems Relating to Vibrations," in *Proceedings of the London Mathematical Society*, **4** (1873), 357–368; "On the Manufacture and Theory of Diffraction-Gratings," in *Philosophical Magazine*, **47** (1874), 81–93, 193–205; "General Theorems Relating to Equilibrium and Initial and Steady Motions," *ibid.*, **49** (1875), 218–224; "On the Dissipation of Energy," in *Nature*, **40** (1875), 454–455; "On Waves," in *Philosophical Magazine*, **1** (1876), 257–259; "Our Perception of the Direction of a Source of Sound," in *Nature*, **41** (1876), 32–33; "On the Application of the Principle of Reciprocity to Acoustics," in *Proceedings of the Royal Society*, **25** (1876), 118–122; "Acoustical Observations. I," in *Philosophical Magazine*, n.s. **3** (1877), 456–464; "Absolute Pitch," in *Nature*, **17** (1877), 12–14; "On the Relation Between the Functions of Laplace and Bessel," in *Proceedings of the London Mathematical Society*, **9** (1878), 61–64; "On the Capillary Phenomena of Jets," in *Proceedings of the Royal Society*, **29** (1879), 71–97; and "Acoustical Observations. II," in *Philosophical Magazine*, **7** (1879); 149–162.

Later articles are "On Reflection of Vibrations at the Confines of Two Media Between Which the Transition is Gradual," in *Proceedings of the London Mathematical Society*, **9** (1880), 51–56; "On the Resolving-Power of Telescopes," in *Philosophical Magazine*, **10** (1880), 116–119; "On the Electromagnetic Theory of Light," *ibid.*, **12** (1881), 81–101; "On the Determination of the Ohm [B.A. Unit] in Absolute Measure," in *Proceedings of the Royal Society*, **32** (1881), 104–141, written with Arthur Schuster; "Experiments to Determine the Value of the British Association Unit of Resistance in Absolute Measure," in *Philosophical Transactions of the Royal Society*, **173** (1882), 661–697; "On the Specific Resistance of Mercury," *ibid.*, **174** (1882), 173–185, written with Mrs. H. Sidgwick; "Address to the Mathematical and Physical Science Section of the British Association," in *British Association Report* (1882), 437–441; "On an Instrument Capable of Measuring the Intensity of Aerial Vibrations," in *Philosophical Magazine*, **14** (1882), 186–187; "On the Maintained Vibrations," *ibid.*, **15** (1883), 229–235; "Distribution of Energy in the Spectrum," in *Nature*, **27** (1883), 559–560; "On the Crispations of Fluid Resting Upon a Vibrating Support," in *Philosophical Magazine*, **16** (1883), 50–58; "On Laplace's Theory of Capillarity," *ibid.*, 309–315; "On the Circulation of Air Observed in Kundt's Tubes and on Some Allied Acoustical Problems," in *Philosophical Transactions*, **175** (1883), 1–21; "The Form of Standing Waves on the Surface of Running Water," in *Proceedings of the London Mathematical Society*, **15** (1883), 69–78; "On the Constant of Magnetic Rotation of Light in Bisulphide of Carbon," in *Philosophical Transactions of the Royal Society*, **76** (1884), 343–366; "On Waves Propagated Along the Plane Surface of an Elastic Solid," in *Proceedings of the London Mathematical Society*, **17** (1885), 4–11; "On the Maintenance of Vibrations by Forces of Double Frequency and on the Propagation of Waves Through a Medium Endowed With a Periodic Structure," in *Philosophical Magazine*, **24** (1887), 145–159; "On the Relative Densities of Hydrogen and Oxygen (Preliminary Notice)," in *Proceedings of the Royal Society*, **43** (1887), 356–363; "On the Free Vibrations of an Infinite Plate of Homogeneous Isotropic Elastic Matter," in *Proceedings of the London*

Mathematical Society, **20** (1889), 225–234; "On the Character of the Complete Radiation at a Given Temperature," in *Philosophical Magazine*, **27** (1889), 460–469; "Foam," in *Proceedings of the Royal Institution*, **13** (1890), 85–97; "On the Tension of Water Surfaces, Clean and Contaminated, Investigated by the Method of Ripples," in *Philosophical Magazine*, **30** (1890), 386–400; "On The Theory of Surface Forces," in *Philosophical Magazine*, **30** (1890), 285–298, 456–475; "On the Virial of a System of Hard Colliding Bodies," in *Nature*, **45** (1891), 80–82; "On the Relative Densities of Hydrogen and Oxygen. II," in *Proceedings of the Royal Society*, **50** (1892), 448–463; and "On the Physics of Media That are Composed of Free and Perfectly Elastic Molecules in a State of Motion," in *Philosophical Transactions of the Royal Society*, **183A** (1892), 1–5.

See also "Density of Nitrogen," in *Nature*, **46** (1892), 512–513; "On the Reflection of Sound or Light From a Corrugated Surface," in *British Association Report* (1893), 690–691; "On an Anomaly Encountered in Determinations of the Density of Nitrogen Gas," in *Proceedings of the Royal Society*, **55** (1894), 340–344; "An Attempt at a Quantitative Theory of the Telephone," in *Philosophical Magazine*, **38** (1894), 295–301; "On the Amplitude of Aerial Waves Which Are But Just Audible," *ibid.*, 365–370; "Argon, a New Constituent of the Atmosphere," in *Philosophical Transactions of the Royal Society*, **186A** (1895), 187–241, written with William Ramsay; "Argon," in *Proceedings of the Royal Institution*, **14** (1895), 524–538; "On the Propagation of Waves Upon the Plane Surface Separating Two Portions of Fluid of Different Vorticities," in *Proceedings of the London Mathematical Society*, **27** (1895), 13–18; "On Some Physical Properties of Argon and Helium," in *Proceedings of the Royal Society*, **59** (1896), 198–208; "On the Propagation of Waves Along Connected Systems of Similar Bodies," in *Philosophical Magazine*, **44** (1897), 356–362; "Note on the Pressure of Radiation, Showing an Apparent Failure of the Usual Electromagnetic Equations," *ibid.*, **45** (1898), 522–525; "On the Cooling of Air by Radiation and Conduction and on the Propagation of Sound," *ibid.*, **47** (1899), 308–314; "On the Transmission of Light Through an Atmosphere Containing Small Particles on Suspension, and On the Origin of the Blue of the Sky," *ibid.*, 375–384; "On the Calculation of the Frequency of Vibration of a System in Its Gravest Mode, With an Example from Hydrodynamics," *ibid.*, 566–572; "The Law of Partition of Kinetic Energy," *ibid.*, **49** (1900), 98–118; "Remarks Upon the Law of Complete Radiation," *ibid.*, 539–540; "On the Magnetic Rotation of Light and the Second Law of Thermodynamics," in *Nature*, **64** (1901), 577–578; "On the Pressure of Vibrations," in *Philosophical Magazine*, **3** (1902), 338–346; "Is Rotatory Polarization Influenced by the Earth's Motion?" *ibid.*, **4** (1902), 215–220; "Does Motion Through the Aether Cause Double Refraction?" *ibid.*, 678–683; "On the Bending of Waves Round a Spherical Obstacle," in *Proceedings of the Royal Society*, **72** (1903), 401–441; "On the

Acoustic Shadow of a Sphere," in *Philosophical Transactions of the Royal Society*, **203A** (1904), 87–110; "The Dynamical Theory of Gases and of Radiation," in *Nature*, **71** (1905), 559; **72** (1905), 54–55, 243–244; "On Electrical Vibrations and the Constitution of the Atom," in *Philosophical Magazine*, **11** (1906), 117–123; "On the Experimental Determination of the Ratio of the Electrical Units," *ibid.*, **12** (1906), 97–108; "On Our Perception of Sound Direction," *ibid.*, **13** (1907), 214–232; "Note As to the Application of the Principle of Dynamical Similarity," in *Report of the Advisory Committee for Aeronautics* (1909–1910), 38; "Aerial Plane Waves of Finite Amplitude," in *Proceedings of the Royal Society*, **84A** (1910), 247–284; "On the Propagation of Waves Through a Stratified Medium, with Special Reference to the Question of Reflection," in *Proceedings of the Royal Society*, **86A** (1912), 207–266; "On the Motion of a Viscous Fluid," in *Philosophical Magazine*, **26** (1913), 776–786; "The Pressure of Radiation and Carnot's Principle," in *Nature*, **92** (1914), 527–528; "Some Problems Concerning the Mutual Influence of Resonators Exposed to Primary Plane Waves," in *Philosophical Magazine*, **29** (1915), 209–222; "The Principle of Similitude," in *Nature*, **95** (1915), 66–68, 644; "The Theory of the Helmholtz Resonator," in *Proceedings of the Royal Society*, **92A** (1915), 265–275; "The Le Chatelier-Braun Principle," in *Transactions of the Chemical Society*, **91** (1917), 250–252; "The Theory of Anomalous Dispersion," in *Philosophical Magazine*, **33** (1917), 496–499; "On the Pressure Developed in a Liquid During the Collapse of a Spherical Cavity," *ibid.*, **34** (1917), 94–98; "On the Scattering of Light by a Cloud of Similar Small Particles of Any Shape and Oriented at Random," *ibid.*, **35** (1918), 373–381; "Propagation of Sound and Light in an Irregular Atmosphere," in *Nature*, **101** (1918), 284; "On the Problem of Random Vibrations, and of Random Flights in One, Two, or Three Dimensions," in *Philosophical Magazine*, **37** (1919), 321–347; "Presidential Address," in *Proceedings of the Society for Psychical Research*, **30** (1919), 275–290; and "On Resonant Reflexion of Sound From a Perforated Wall," in *Philosophical Magazine*, **39** (1920), 225–233.

II. SECONDARY LITERATURE. See the obituary notice by Sir Arthur Schuster, in *Proceedings of the Royal Society*, **98A** (1921), 1; Robert John Strutt, *Life of John William Strutt, Third Baron Rayleigh, O.M., F.R.S.* (London, 1924); 2nd augmented ed. with annotations by the author and foreword by John N. Howard (Madison, Wis., 1968); and R. Bruce Lindsay, *Lord Rayleigh, the Man and His Works* (Oxford–London, 1970).

R. B. LINDSAY

STRUTT, ROBERT [ROBIN] JOHN, FOURTH BARON RAYLEIGH (*b.* Terling Place, Witham, Essex, England, 28 August 1875; *d.* Terling, 13 December 1947), *physics*.

Rayleigh is best-known for his work on atmospheric optics. The first child of the renowned physicist J. W. Strutt, third Baron Rayleigh, he attended Eton from 1889 and matriculated at Trinity College, Cambridge, in 1894. After studying experimental physics, he began research at the Cavendish Laboratory in 1899 and published his first scientific paper. Strutt was elected a fellow of the Royal Society in 1905; and from 1908 he was professor of physics at Imperial College, London, until becoming Baron Rayleigh in 1919.

Strutt's early research was on radioactivity. In 1900 he had suggested that alpha radiation might consist of charged particles, a fact verified in 1902 by E. Rutherford. From measurements on the helium content of minerals, Strutt proved that their minimum age was significantly greater than that allowed by Kelvin's geological timetable. Following up the 1884 observations of Emil Warburg on the afterglow in electric discharge and the assertion in 1900 by E. Percival Lewis that the cause of the airglow was nitric oxide, Strutt in 1911 confirmed that discharge airglow was due to a chemically active modification of nitrogen, which he called "active nitrogen." His suggestion that the effect was an atomic phenomenon was not accepted until about 1925. The Lewis-Rayleigh afterglow opened up an important line of optical research.

By comparing absorption spectra, Strutt and Alfred Fowler confirmed in 1916 the presence of ozone in the atmosphere. Strutt investigated its optical effects, and from the nonuniform density distribution he estimated the upper boundary of high-altitude ozone. He examined luminous glows in other gases and metallic vapors and investigated the airglow of the night sky. Strutt determined that this effect was not due to (third Baron) Rayleigh-scattered sunlight and that it had a green line spectrum, later shown to be characteristic of a transition of atomic oxygen.

Rayleigh distinguished the atmospheric airglow generally observable, especially at low latitudes, from the high-latitude night glow associated with magnetic disturbances and auroral phenomena. Because of his pioneering quantitative research, the unit of sky brightness is called the rayleigh; the order of magnitude for the airglow is less than 10^2 rayleighs and that for auroral phenomena lies between 10^2 and 10^4 rayleighs.

Until his death Rayleigh was occupied with matters pertaining both to science and to his many public duties, serving as foreign secretary of the Royal Society, president of the Royal Institution, and chairman of the governing board of Imperial College.

BIBLIOGRAPHY

Rayleigh published over 300 scientific papers. A complete list, in John N. Howard, ed., *Summaries and Abstracts of the Scientific Writings of Robert John Strutt, Fourth Baron Rayleigh, From Notes Written by Him, 1945–1947* (Bedford, Mass., 1969), 63–88, is based on the bibliography included in Alfred C. Egerton, "Lord Rayleigh, 1875–1947," in *Obituary Notices of Fellows of the Royal Society of London*, 6 (1949), 503–538. A selected list is in Charles R. Strutt, "The Optics Papers of Robert John Strutt, Fourth Baron Rayleigh," in *Applied Optics*, 3 (1964), 1116–1119.

Additional biographical material dealing with his family life and background is in Guy Robert Strutt, "Robert John Strutt, Fourth Baron Rayleigh," *ibid.*, 1105–1112. An expanded version of this appeared in John N. Howard, ed., *The Airglow Rayleigh; Robert John Strutt, Fourth Baron Rayleigh, A Memoir by Guy Robert Strutt* (Bedford, Mass., 1969), 1–24. Howard included several contemporary biographical notes relating to Rayleigh that appeared in *Nature*, 125 (1930), 420, and 140 (1937), 456; *The Times* (London) (15 Dec. 1947), 6d; *Proceedings of the Society for Psychical Research*, 48 (1948), 330–331; and *Proceedings of the Royal Institution of Great Britain*, 34 (1948), 156–158. See also A. C. Egerton's notice, in *Dictionary of National Biography* (1959), 850–852.

From personal recollection Sydney Chapman contributed "On the Influence of the Fourth Baron Rayleigh on Air Glow and Auroral Research," in John N. Howard, ed., *The Rayleigh Archives Dedication*, Air Force Cambridge Research Laboratories, Special Report no. 63 (Bedford, Mass., 1967), 46–53. Charles R. Strutt wrote "The Optics Research of Robert John Strutt, Fourth Baron Rayleigh," in *Applied Optics*, 3 (1964), 1113–1115. The results of Strutt and Lewis are contrasted in E. P. Lewis, "The Origin of the Bands in the Spectrum of Active Nitrogen," in *Philosophical Magazine*, 25 (1913), 826–832.

Howard has brought together nearly all the primary and secondary literature pertaining to Strutt in "The Scientific Papers of the Lords Rayleigh," in *Actes du XIe congrès international d'histoire de sciences, 1965*, IV (1968), 315–318. A short obituary notice in *Isis*, 39 (1948), 69, refers to a review by N. R. Campbell, *ibid.*, 8 (1926), 177–181, of Strutt's model scientific biography of his father. This biography, *Life of John William Strutt: Third Baron Rayleigh, O.M., F.R.S.*, was reedited by John N. Howard (Madison, Wis., 1968) and includes 33 pages of annotations by R. J. Strutt keyed to the text.

The early research of R. J. Strutt was considered by C. T. R. Wilson in Cavendish Laboratory, *A History of the Cavendish Laboratory 1871–1910* (London, 1910), 211–215. Part of Strutt's work in its larger scientific context is considered in A. N. Wright and C. A. Winkler, *Acitve Nitrogen* (London, 1968), *passim*. The technical context is considered in F. E. Roach and J. L. Gordon, *The Light of the Night Sky* (Dordrecht, 1973). L. Badash deals with Strutt's helium method of geological dating in "Rutherford, Boltwood and the Age of the Earth: The Origin of Radioactive Dating Techniques," in *Proceedings of the American Philosophical Society*, 112 (1968), 157–169. See also S. I. Levy, *The Rare Earths* (London, 1915), ch. 8. John N. Howard described the 22 notebooks of the fourth Baron Rayleigh including over 4,000 experiments from 1916 to 1944 in "The Rayleigh Notebooks," in *Applied Optics*, 3 (1964), 1132–1133.

Several previously unpublished MSS and the text of Strutt's House of Lords debates and letters to the *Times* (London) are included in John N. Howard, ed., *Robert John Strutt, Fourth Baron Rayleigh, Unpublished Manuscripts and Reviews of His Work* (Bedford, Mass., 1971), which also contains contemporary reviews of Strutt's four books. Additional correspondence of Strutt is in the Lodge collection, Library, University College, London.

THADDEUS J. TRENN

STRUVE, FRIEDRICH GEORG WILHELM (or **Vasily Yakovlevich**) (*b.* Altona, Germany, 15 April 1793; *d.* Pulkovo, Russia, 23 November 1864), *astronomy, geodesy.*

Struve's father, Jacob Struve, came from a peasant family, but by diligence and ability was able to graduate from the University of Göttingen. He held a number of teaching positions in various German towns and then, after ten years, settled in Altona as professor in the Christianeum classical Gymnasium, a school that he had himself attended. He was made principal after three years, and retained that post for about forty years more. Struve's mother, Maria Emerenzia Wise, was the daughter of a preacher. There were six other children.

When Struve reached the age of conscription his parents decided to send him out of Germany. He thus went to Dorpat, Russia, where his older brother Karl had been living for some time. In 1810 he graduated from the University of Dorpat with a degree in philology, but he was not enthusiastic about becoming a teacher of that subject and proposed instead to devote himself to the study of mathematics, astronomy, and geodesy. With the support of the rector of the university and several of its professors, he received permission to work in the university observatory, where he in-

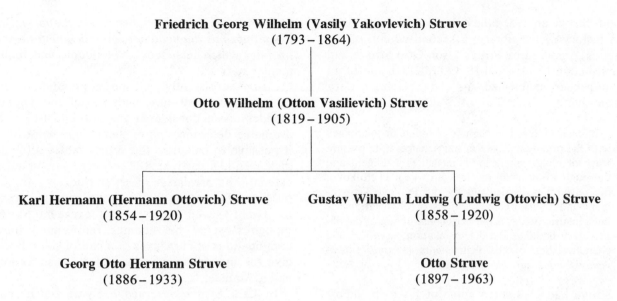

Friedrich Georg Wilhelm (Vasily Yakovlevich) Struve
(1793 – 1864)

Otto Wilhelm (Otton Vasilievich) Struve
(1819 – 1905)

Karl Hermann (Hermann Ottovich) Struve
(1854 – 1920)

Gustav Wilhelm Ludwig (Ludwig Ottovich) Struve
(1858 – 1920)

Georg Otto Hermann Struve
(1886 – 1933)

Otto Struve
(1897 – 1963)

The Struve family, four generations of astronomers who conducted research in Germany, Russia, and the United States for a period of more than 150 years.

stalled a number of previously purchased new instruments and carried out observations by which he was able to make an accurate determination of the latitude and longitude of the building. In 1813 he was awarded the doctorate for this work and was appointed professor extraordinarius of mathematics and astronomy and observing astronomer.

Struve thus began a teaching career that lasted twenty-five years, during which he delivered at Dorpat many courses of lectures on different astronomical and mathematical subjects; he also gave widely popular public lectures and often taught astronomy and geodesy at the general staff headquarters of the Corps of Military Topographers and at the Hydrographic Department. In addition to his teaching, Struve's research was also significant in a number of areas, notably observation of double stars; determination of stellar parallaxes and distribution of stars in space; observation of planets, the moon, comets, and auroras; meridian measurements; statistical techniques; and the design and refinement of astronomical and geodetic instruments.

In January 1814 Struve began to make systematic observations with a transit that he had installed. His results in the next two years were included in the first volume of the *Observationes astronomicas institutas in specula universitatis caesareae Dorpatensis*, published in 1817. (Seven other volumes followed, containing Struve's own observations and those of his assistants and students, which he revised and edited.) He devised new methods of observation that were later widely adopted, and he

improved the Dorpat facility, adding a meridian circle constructed by Georg von Reichenbach in 1822 and a nine-inch refracting telescope made by Fraunhofer (the largest in existence) in 1824.

These efforts brought Struve considerable recognition, and he was invited by the St. Petersburg Academy of Sciences to carry out the organization of a new main observatory, to be constructed at Pulkovo, near St. Petersburg. A preliminary plan for the new observatory had been drawn up by the physicist G. F. Parrot and reviewed by a special commission, of which Struve was a member. The design was approved in April 1834, and the architect A. P. Bryllov was put in charge of construction. Struve himself was appointed director of the new installation, and in June of the same year he went abroad to buy the best instruments to be found. He discussed the new observatory with Bessel, Humboldt, Lindenau, Olbers, Steinheil, H. C. Schumacher, and others, and ordered instruments to be made by Ertel, Repsold, Mertz, Plössl, Pistor, Throughton, and Dent. Among his purchases was a fifteen-inch refractor, the largest made.

The observatory was opened with great ceremony on 19 August 1839, and soon achieved worldwide fame. In 1845 Struve published a detailed description of the installation, including a complete catalog of the instruments, with remarks on their installation, application, and testing procedures. Appended to this description was a catalog of the library of the observatory, enumerating 5,411 astronomical works, of which 3,869 dated from the fifteenth to the seventeenth centuries. In a supplement, Struve analyzed the collection and discussed the history of the library; a second edition of the catalog, prepared by Struve's son Otto Struve and published in 1860, listed 18,890 titles. In a letter to Schumacher, published in 1855, George Airy wrote that

> No astronomer can claim a complete acquaintance with the practical aspect of our science at its present stage of development until he makes a diligent and thorough study of all the treasures kept in Pulkovo; the researches carried out by the Pulkovo astronomers and their precise methods of observation are most instructive, as is the very construction of the observatory building and the installation, selection, and characteristics of the instruments [*Vestnik estestvennykh nauk*, no. 19, 596].

Struve had completed some of his own important research while he was still at Dorpat. He studied the results of Herschel's work on double stars and in 1814 discovered the motions of the satellite of Castor and of η Cassiopeiae by comparing his own recent observations with those made by Herschel between 1780 and 1800. By 1820 he had determined the revolutions of P Ophiuchi and ξ Ursae Majoris, nearly completed since Herschel's observations of 1780. In 1822 he published a catalog of all known double stars, which contained 795 items. He went on to observe some 122,000 stars over two-thirds of the celestial sphere from the north pole to 15° south declination, among which he detected 3,112 double stars, which he classified according to their angular separation. These were incorporated into a new catalog of double stars that he published in 1827.

Struve then undertook to make micrometric measurements of these celestial objects and on 14 November 1834 reported to the St. Petersburg Academy of Sciences that he had so measured 2,736 double stars. He published these results, together with those of a further series of observations, in a paper of 1837 entitled *Stellarum duplicium et multiplicium mensurae micrometricae*. Bessel characterized this paper as "a magnificent work ranking among the greatest performed by astronomical observers in recent times." By 1842 Struve and his assistants had observed the entire northern sky. In addition to the 18,000 stars known, they discovered 518 multiple stars that had not been reported previously, and Struve published data on 514 of them the following year. In 1852, having compared his data with those obtained by James Bradley, Lalande, Piazzi, and Stephen

Groombridge, Struve published his *Stellarum fixarum imprimis duplicium et multiplicium positiones mediae*, which included 2,874 double and fundamental stars.

Struve consistently paid special attention to stellar parallaxes, and may, with Bessel and Thomas Henderson, be considered one of the first to make a reliable determination of them. This work made it possible to calculate the actual values of stellar distances. As early as 1822, Struve attempted to estimate the parallaxes of several dozen stars and published his finding that the parallaxes he derived for twenty-seven stars did not exceed 0.5″. Although most of his findings represented linear combinations of parallaxes of a pair of stars, he did give the individual parallaxes of δ Ursae Majoris and α Aquilae.

In 1836 Struve observed α Lyrae and a small star of 10.5 magnitude, at a distance of 43″. From these observations he derived a series of general conclusions about the parallaxes of fixed stars that he reported to the Academy on 13 January 1837 and included in his *Stellarum duplicium et multiplicium mensurae micrometricae*, presented later that year. "My observations for seventeen dates," he wrote, "show the distance and direction (position angles) of the two stars from which, by calculation using thirty-four equations by the method of least squares, the parallax for the main star proves to be 0.125″, or 1/8 second, with a probable error of 0.055″, or 1/18 second." This result is very close to the modern value of 0.121″, but in the course of verifying his data Struve later reached the less accurate value of $+0.2613″ \pm 0.0254″$.

Struve also conducted investigations on stellar distribution to determine whether there is a statistical dependence between the brightness of stars and their distances. To this end he organized work at Pulkovo on a new star catalog, which he studied closely for changes in the visible star density in the field of the Milky Way. He was thereby led to maintain that the sun was not the center of this system, but lay above its main plane. He was also able to substantiate earlier suggestions by Loys de Chéseaux and Olbers that the interstellar medium is incompletely transparent and to determine the value of the obscuration effect in outer space, results that he published in the preface to Maximilian Weisse's star catalog of 1846 and in his own *Études d'astronomie stellaire* of 1847.

In connection with this study, Struve analyzed the observations that Herschel had made with a twenty-foot telescope and compared the number of stars that Herschel had observed with a theoretical figure to conclude that "the range of Herschel's telescope, determined by observing the sky, scarcely exceeds one-third of the range corresponding to its optical resolution." The reason for this limitation, Struve stated emphatically, is that "the intensity of light decreases in a greater proportion than the ratio of the inverse square of the distance, which means that there is a loss of light, an extinction when it passes through celestial space." This conclusion testified to the presence of rarefied matter in space; in an approximate calculation, Struve also estimated that "The light in passing the same distance of a first-magnitude star is extinguished by almost one-hundredth—in other words, it loses 1/107th of its intensity." The value that he thus obtained is in fair agreement with the currently accepted mean value for absorption near the galactic plane.

Throughout his career Struve was also active in geodetic studies. In 1816 he began three years' work on the triangulation of Livonia, at the request of the Livonian Economic Society, and with K. I. Tenner he sponsored the joint Russian-Scandinavian measurement of 25°20′ of the arc of meridian. He personally supervised the measurement of 9°38′ of the Finnish-Livonian arc between the Dvina and Tornio rivers, and advised Tenner on the measurement of the large and difficult section of the arc (11°10′) between the Dvina and the Danube. The 3°13′ of the arc that lies between the Tornio and Belatsvar, in Sweden, was measured under the supervision of N. H. Selander, while the measurement of 1°46′ of the arc in Sweden, between Atjik and Cape Fuglnes, was carried out under the guidance of Hansteen. The entire project was finished by 1855, and Struve played a large part in organizing and processing the data for publication between 1856 and 1857.

The Russian-Scandinavian measurements were of great value to geodesy and practical astronomy, to which Struve made other important contributions. In 1837 he was involved in determining the difference in level between the Black and Caspian seas, and organized an expedition that investigated some 800 kilometers to yield not only fairly reliable results (26.04 ± 0.25 meters) concerning the difference in level, but also furnished data on terrestrial refraction and produced the first accurate determination of altitude for Mounts Elbrus, Kazbek (eastern and western), Besh-Tau, and Bezymyannaya. Struve's methods of geodetic construction are of considerable interest, and his work on

the altimetry of the Caspian Sea represents the first application of the technique of parallactic polygonometry that was later developed by V. V. Danilov.

Struve further discovered rational methods for the determination of latitude, time, and azimuth that allowed him to eliminate many systematic errors. He participated in a number of chronometric expeditions, among them one conducted by F. F. Shubert for the purpose of correcting the map of the Baltic Sea (1833), in which he was responsible for the astronomical observations. In 1842 he himself organized an expedition that set out from Pulkovo to observe the solar eclipse at Lipetsk—an expedition that, according to Struve, completely eliminated uncertainties about the values for longitudes determined by chronometers that had been transported over long distances by land. In 1843 and 1844 he conducted two other expeditions, between Pulkovo and Altona and between Altona and Greenwich. They established the longitudinal connection between the Pulkovo and Greenwich observatories.

Struve's work brought him international fame, and he was a member of more than forty scientific academies, learned societies, and universities. In particular, he was a corresponding member of the St. Petersburg Academy of Sciences from 1822 and a full member and honorary member for more than forty years. He was also a founding member of the Russian Geographical Society and chairman of its department of mathematical geography.

In 1814 Struve married Emily Wall, the daughter of an Altona merchant; they had twelve children. A year after his first wife died in 1834, Struve married her friend Johanna Bartels, daughter of the mathematician J. M. C. Bartels; they had six children. Twelve of Struve's children survived him, and his son Otto collaborated in his work for a number of years and carried it on after his death.

BIBLIOGRAPHY

I. ORIGINAL WORKS. There is a complete bibliography of Struve's publications (272 titles) in Novokshanova (see below). They include, in addition to works cited in the text, *De geographica positione speculae astronomicae Dorpatensis* (Mitau, 1813), his doctoral thesis: "Catalogus stellarum duplicium a 20° decl. Austr. ad 90° decl. Bor. pro anno 1820," in *Observationes astronomicas institutas in specula universitatis caesareae Dorpatensis*, **3** (1822), 15–24; "Additamentum I. De numero constanti aberrationis et parallaxi annua fixarum ex observationibus stellarum circumpolarium in ascensione oppositarum," *ibid.*, 51–90; "Observationes stellarum duplicium per micrometrum filare Fraunhoferianum tubo mobili Troughtoniano 5 pedum abhibitum," *ibid.*, **4** (1825), 175–195; *Beschreibung des auf der Sternwarte der Kaiserlichen Universität zu Dorpat befindlichen grossen Refractors von Fraunhofer* (Dorpat, 1825); "Micrometer-Beobachtungen des Planeten Saturn mit dem grossen Refractor von Fraunhofer in Dorpat angestellt," in *Astronomische Nachrichten*, **5** (1827), 7–14; **6** (1828), 389–392; "Micrometer-Messungen des Jupiters und seiner Trabanten mit dem grossen Refractor von Fraunhofer angestellt," *ibid.*, **5** (1827), 13–16; *Catalogus novus stellarum duplicium et multiplicium* (Dorpat, 1827); and "Disquisitio de refractione astronomica, stellarum que primariarum declinationibus et ascensionibus rectis, quales sequuntur ex observationibus anni 1822 ad 1826," in *Observationes astronomicas institutas . . . Dorpatensis*, **6** (1830).

Subsequent works are *Anwendung des Durchgangsinstruments für die geographische Ortsbestimmung* (St. Petersburg, 1833); "Expédition organisée par l'Académie impériale des sciences, dans le but de déterminer la différence de niveau de la Mer Noire et de la Mer Caspienne," in *Bulletin scientifique publié par l'Académie impériale des sciences de St.-Pétersbourg*, **1** (1837) 79–80; **2** (1837), 254–270; **3** (1838), 27–31, 127–132, 366–368; "Additamentum in mensuras micrometricas stellarum duplicium editas anno 1837, exhibens mensuras Dorpati annis 1837 et 1838 institutas. Adjecta est disquisitio de parallaxi annua stellae α Lirae," in *Mémoires de l'Académie impériale des sciences de St.-Pétersbourg*, Sciences math. et phys., 6th ser., **2**, no. 4 (1840), 337–358; "Sur les constantes de l'aberration et de la nutation," in *Bulletin scientifique publié par l'Académie impériale des sciences de St.-Pétersbourg*, **8**, no. 13 (1841), 199–206; *Catalogue de 514 étoiles doubles et multiples découvertes sur l'hémisphère céleste boréal . . . et catalogue de 256 étoiles doubles principales où la distance des composantes est de 32″ à 2′ et qui se trouvent sur l'hémisphère boréal* (St. Petersburg, 1843); "Mémoire sur le coefficient constant dans l'aberration des étoiles fixes," in *Bulletin de la classe physique-mathématique de l'Académie impériale des Sciences de St.-Pétersbourg*, **1**, nos. 17–18 (1843), 257–260; and "Table des positions géographiques principales de la Russie," *ibid.*, nos. 19–21 (1843), 289–306.

See also *Expédition chronométrique exécutée entre Poulkova et Altona pour la détermination de la longitude géographique relative de l'observatoire central de Russie* (St. Petersburg, 1844); *Description de l'observatoire astronomique central de Poulkova* (St. Petersburg, 1845), with *Appendice: Catalogus librorum in bibliotheca speculae Pulcovensis contentorum*, 2 vols. (St. Petersburg, 1845; 2nd ed., 1860); "Notice sur la comète à courte période découverte par M. Faye à Paris, d'après les observations faites à l'observatoire de Poulkova," in *Bulletin de l'Académie des sciences de St.-Pétersbourg*,

Cl. phys.-math., **3**, no. 18 (1845), 273–280; "Ueber die im Jahre 1845 auszuführende Chronometer-Expedition ins Innere Russlands," *ibid.*, **4**, no. 3 (1845), written with Otto Struve; *Expédition chronométrique entre Altona et Greenwich pour la détermination de la longitude géographique de l'observatoire central de Russie* (St. Petersburg, 1846); "Obzor geograficheskikh rabot v Rossii" ("Survey of Geographical Work in Russia"), in *Zapiski Russkogo geographicheskogo obshchestva*, **1** (1846) 43–58; *Études d'astronomie stellaire. Sur la Voie Lactée et sur la distance des étoiles fixes* (St. Petersburg, 1847), also in Russian (Moscow, 1953); "Observations de la nouvelle planète (Astrée) faites à l'observatoire central, à l'aide des instruments du méridien, éléments de l'orbite de la planète," in *Bulletin de l'Académie des sciences de St.-Pétersbourg*, Cl. phys.-math., **5**, no. 13 (1847), 193–196; and *Arc du méridien de 25°20′ entre le Danube et la Mer Glaciale, mesuré, depuis 1816 jusqu'en 1855*, 2 vols. (St. Petersburg, 1856–1857; ed. 2, 1861; Moscow, 1957).

II. Secondary Literature. Z. K. Novokshanova, *Vasily Yakovlevich Struve* (Moscow, 1964) is the most complete biography and includes a bibliography of 272 original works (pp. 249–273) and 89 sources. See also A. A. Mikhaylov, ed., *Vasily Yakovlevich Struve. Sbornik statey i materialov k 100-letiyu so dnya smerti* (". . . . Collected Papers and Materials on the Centennial of His Death"; Moscow, 1964). Earlier sources (listed chronologically) are A. N. Savich, *Vospominania o V. Y. Struve. Torzhestvennoe sobranie Akademii nauk 29 dekabrya 1864 g.* ("Memoirs of V. Y. Struve. Grand Meeting of the Academy of Sciences of 29 December 1864"; St. Petersburg, 1865); E. F. Litvinova, *V. Y. Struve, ego zhizn i nauchnaya deyatelnost* ("V. Y. Struve, His Life and Scientific Activity"; St. Petersburg, 1893); A. A. Tillo, "O geograficheskikh zaslugakh V. Y. Struve" ("On the Geographical Merits of V. Y. Struve"), in *Izvestiya Russkogo geograficheskogo obshchestva*, **29**, no. 3 (1893), 151–164; and Otto Struve, *Wilhelm Struve. Zur Erinnerung an den Vater den Geschwistern dargebracht* (Karlsruhe, 1895).

More recent studies include B. A. Vorontsov-Belyaminov, *Zvezdno-statisticheskie raboty V. Struve* ("V. Struve's Works on Stellar Statistics"), in *Trudy soveshchania po istorii estestvoznania 24–26 dekabrya 1946 g.* ("Proceedings of the Conference on the History of Natural Science, 24–26 December 1946"; Moscow–Leningrad, 1948), 132–144; A. N. Deutsch, "Pervoe opredelenie V. Y. Struve parallaksa α Liry" ("V. Y. Struve's First Determination of the Parallax of α Lyrae"), in *Astronomicheskii zhurnal*, **29**, no. 5 (1952), 597–601; B. A. Orlov, "Vasily Yakovlevich Struve," in V. Y. Struve, *Etyudy zvezdnoy astronomii* ("Essays on Stellar Astronomy"; Moscow, 1953), 171–208; T. Rootsmyae, "Akademik V. Y. Struve i ego deyatelnost v Tartuskom universitete" (". . . and His Activity at the University of Tartu"), in *Uchenye zapiski Tartuskogo gosudarstvennogo universiteta*, no. 37 (1955), 30–69;

O. Struve, "The First Determination of Stellar Parallax," in *Sky and Telescope*, **16** (1956), 9–12, 69–72; N. P. Erpylev, "Razvitie zvezdnoy astronomii v Rossii v XIX veke" ("The Development of Stellar Astronomy in Russia in the Nineteenth Century"), in *Istoriko-astronomicheskie issledovaniya*, no. 4 (1958), 43–130; Y. G. Perel, "Vasily Yakovlevich Struve," in *Lyudi russkoy nauki* ("People of Russian Science"; Moscow, 1961), 94–103; and Z. K. Sokolovskaya, "Pervye opredelenia zvezdnykh parallaksov" ("The First Determinations of Stellar Parallax"), in *Vestnik Akademii nauk SSSR*, no. 3 (1972), 132–136.

Works on the Struve family are E. B. Frost, "A Family of Astronomers," in *Popular Astronomy*, **29**, no. 9 (1921), 536–541; A. F. Marshal, "Une dynastie d'astronomes. Les Struve," in *Industrie* (Brussels), **17**, no. 12 (1963), 833–839; and P. van de Kamp, "The Struve Succession," in *Journal of the Royal Astronomical Society of Canada*, **59**, no. 3 (1965), 106–114.

Z. K. SOKOLOVSKAYA

STRUVE, GEORG OTTO HERMANN (*b*. Pulkovo, Russia, 29 December 1886; *d*. Berlin, Germany, 10 June 1933), *astronomy*.

Struve was born at Pulkovo while his father, Karl Hermann Struve, was adjunct astronomer there. He was taken to Königsberg in 1895, when the elder Struve was appointed director of the observatory of that university, and attended the Königsberg Gymnasium, from which he graduated in 1905. He then studied at the universities of Heidelberg and Berlin; in 1910 he was awarded the Ph.D. for a dissertation entitled *Die Darstellung der Pallasbahn durch die Gausssche Theorie für den Zeitraum 1903 bis 1910*, which he had dedicated to the memory of his grandfather Otto Struve. During the next two years he worked as an assistant at the Bonn Observatory, in his father's observatory in Berlin, and at the Bergedorf Observatory, then, in 1913, became astronomer at the Wilhelmshaven naval observatory, where he was placed in charge of chronometers and compasses. In the last post he utilized his spare time and the observatory's meridian circle (made by Repsold) to make observations of Saturn and its satellites, an interest that he shared with his father.

Beginning in 1917 Struve published some ten works on Saturn, including one on the determination of its equator and the orbits of its satellites (1917), an analysis of observations performed by other researchers (1918), an observation of periodic disappearance of some of its rings (1921), the results of observations of its ring system that he

had carried out in South America (1927), and a comparison of the results of both visual and photographic observations of its satellites (1928). This work was summarized in five issues of the sixth volume of the *Veröffentlichungen der Universitätssternwarte zu Berlin-Babelsberg*, published between 1924 and 1933. Struve had begun working as an observer at that observatory in 1919; he became professor there in 1929, and held that post until his death.

In addition to his work on Saturn, Struve also observed the eclipse of Jupiter's satellites and measured the diameter of Venus, applying the theory of contrasts to determine its irradiation. Toward the end of his career he also made observations of Uranus and its satellites, Oberon and Titania, and studied other minor planets, especially Eros, of which an opposition occurred in 1930–1931. He supplemented his data with those that he collected on visits to the Johannesburg Observatory in South Africa, and to the Lick and Yerkes observatories in the United States.

BIBLIOGRAPHY

I. ORIGINAL WORKS. The most complete bibliography of Struve's works (26 titles) is Poggendorff, V, 1224; and VI, 2570–2571. His most important writings are *Die Darstellung der Pallasbahn durch die Gausssche Theorie für den Zeitraum 1903 bis 1910* (Berlin, 1911); "Neue Elemente der inneren Saturnstrabanten," in *Abhandlungen der Preussischen Akademie der Wissenschaften*, Math.-Naturwiss. Kl. (1918), no. 1; and "Neue Untersuchungen im Saturn System," in *Veröffentlichungen der Universitätssternwarte zu Berlin-Babelsberg*, **6**, no. 4 (1930), 82.

II. SECONDARY LITERATURE. There are obituaries by J. Dick, in *Vierteljahrsschrift der Astronomischen Gesellschaft*, **69** (1934), 2–8; and by P. Guthnick, in *Astronomische Nachrichten*, **251**, no. 6003 (1934), 47.

Z. K. SOKOLOVSKAYA

STRUVE, GUSTAV WILHELM LUDWIG (or **Ludwig Ottovich**) (*b*. Pulkovo, Russia, 1 November 1858; *d*. Simferopol, Russia, 4 November 1920), *astronomy, geodesy*.

Struve completed his Gymnasium studies at Vyborg in 1876, then entered the University of Dorpat, where he graduated in 1880 to become a part-time astronomer at Pulkovo Observatory. In 1883 he defended a dissertation entitled "Resultate aus den in Pulkowa angestellten Vergleichungen von Procyon mit benachbarten Sternen," then was sent abroad to work at the observatories of Bonn, Milan, and Leipzig. In 1885 he participated in the general session of the German Astronomical Society held in Geneva and visited the observatories of Paris, Greenwich, Leiden, and Potsdam before returning to Russia in October of the same year. He worked briefly at Pulkovo before assuming the post of observational astronomer at the University of Dorpat. There he received, in 1887, the doctorate in astronomy for his "Bestimmung der Constante der Präcession und der eigenen Bewegung des Sonnensystems."

At Dorpat Struve's chief interest was in the positions and motions of stars. He collaborated with the German Astronomical Society in compiling a catalog of stars between 70° and 75° north declination and later, at Kharkov, in cooperation with N. N. Evdokimov and B. I. Kudrevich, he observed zodiacal stars, reference stars for Eros, and circumpolar stars from 79° to the pole. Many of his publications were concerned with the precession and motion of the solar system.

In 1887 Struve adopted the hypothesis that the rotation of a galaxy is similar to that of a solid body, and derived an angular rotation rate of $-0.41'' \pm 0.42''$ in each hundred years (a value that may be compared to the current one of $-0.028''$ at the distance of the sun from the center of the galaxy). Between 1884 and 1888 he also carried out observations of the occultation of the stars by the moon during total lunar eclipses in the interest of determining the radius of the moon. These results were published in 1893 and earned Struve the first prize of the Russian Astronomical Society. The Society also awarded him its Glazenap Prize for his response (the only one submitted, as it happened) to a competition set, in 1910, on the subject "Treatment of Observations of the Occultation of Stars During the Lunar Eclipses of 1891, 1895, 1898, and 1910." Reviewing Struve's researches, which were published in 1915, F. F. Witram noted that Struve "must be considered the most competent scholar in this field."

In autumn of 1894 Struve moved to Kharkov, where he was first professor extraordinarius and then (from 1897 to 1919) full professor of astronomy and geodesy. He was simultaneously director of the university observatory and, from 1912 until 1919, dean of the faculty of mathematical and physical sciences. He took active part in a number of geodetic projects, including the leveling by which the Kharkov observatory was made part of the Russian state network of altitudes.

In 1919 Struve moved with his family to the Crimea, on the advice of doctors who were treating the illness of his seventeen-year-old son Werner. He became professor of the Tauris University at Simferopol, but his health was weakened by family misfortunes—Werner Struve and a six-year-old daughter died, an elder daughter became ill, and his son Otto had been recalled into the White Army under General Denikin, who was then retreating through the Crimea. Struve died while attending a meeting of the Tauris Learned Association, where he was to have read his paper on the new star in Cygni. His son Otto survived to carry on the family profession.

BIBLIOGRAPHY

I. ORIGINAL WORKS. There is a bibliography of Struve's writings (22 titles) in Slastenov (see below), 171–172. His most important works are "Resultate aus den in Pulkowa angestellten Vergleichungen von Procyon mit benachbarten Sternen," in *Mémoires de l'Académie impériale des sciences de St.-Pétersbourg*, 7th ser., **31**, no. 2 (1883), his master's thesis; "Bestimmung der Constante der Präcession und der eigenen Bewegung des Sonnensystems," *ibid.*, **35**, no. 3 (1887), his doctoral diss.; "Bearbeitung der während der totalen Mondfinsternisse 1884 Oct. 4 und 1888 Jan. 28 beobachteten Sternbedeckungen," in *Beobachtungen der K. Universitäts-Sternwarte zu Jurjew* (Dorpat), **20** (1893), 1–30; *Soedinenie Kharkova s russkoy nivelirnoy setyu tochnoy nivelirovkoy, proizvedennoy professorom L. O. Struve v 1895 i 1899 gg.* ("The Connection of Kharkov With the Russian Vertical Control Network by the Accurate Leveling Made by Professor L. Struve in 1895 and 1899"; St. Petersburg, 1902); and *Obrabotka nablyudeny pokryty zvezd Lunoyu vo vremya polnykh lunnykh zatmeny* ("The Reduction of Observations of the Occultation of Stars by the Moon During Total Lunar Eclipses"; Petrograd, 1915).

II. SECONDARY LITERATURE. On Struve and his work, see A. I. Slastenov, *Astronomia v Kharkovskom universitete za 150 let* ("Astronomy at the University of Kharkov for 150 years"; Kharkov, 1955), 171–172. There are obituaries in *Nauka i ee rabotniki*, no. 3 (1921), 37; by L. Courvoisier, in *Astronomische Nachrichten*, **212** (1921), 351–352; and by N. N. Evdokimov, in *Nauka na Ukraine*, no. 4 (1922), 428–430. On Struve's work in the prize contest of the Russian Astronomical Society, see F. F. Witram, "Otzyv" ("Review"), in *Izvestiya Russkago astronomicheskago obshchestva*, **21**, no. 6 (1915), 144–149.

Z. K. SOKOLOVSKAYA

STRUVE, KARL HERMANN (or **Hermann Ottovich**) (*b.* Pulkovo, Russia, 30 October 1854; *d.*

Neubabelsberg, near Potsdam, Germany, 12 August 1920), *astronomy*.

Hermann Struve received his early education in the Gymnasiums of Karlsruhe (during the winter of 1862–1863) and Vyborg; he took his final examinations in Reval (now Tallin, Estonian S.S.R.) and, in 1872, entered the University of Dorpat. After graduating from the university in 1877 he was admitted to Pulkovo Observatory, where he had passed his childhood. He was then sent to Paris, Strasbourg, Berlin, and Graz to further his education. In Graz, he worked under Boltzmann, who directed his dissertation, "Fresnels Interferenzerscheinungen; theoretisch und experimentell Bearbeitet," which he wrote in the summer of 1881 and submitted to the University of Dorpat soon thereafter. The following year he received the doctorate in astronomy from Dorpat with a study entitled "Über den Einfluss der Diffraktion an Fernröhren auf Lichtscheiben."

In 1874 Struve took part in B. Hasselberg's expedition to eastern Siberia and to the port of Posyet to observe the transit of Venus. In 1877 he became part-time astronomer at Pulkovo; in 1883 he was made adjunct astronomer; and from 1890 until 1895, he served as senior astronomer. Among the works that he published during this tenure, "Bestimmung der Elemente von Japetus und Titan aus der Verbindung dieser Satelliten untereinander" represents an exposition of his method for the micrometric measurement of the satellites of Saturn. He also made thorough investigations of Neptune, Mars, and Jupiter, observed double stars, and wrote a number of theoretical articles on optics.

In 1895, perhaps influenced by his father's move to Germany upon his retirement, Struve accepted a post as director of the observatory at the University of Königsberg. His most important work, "Beobachtungen der Saturnstrabanten am 30-zölligen Pulkowaer Refractor," published in 1898, draws upon the data that he had obtained in his earlier post, and contains, among other things, a list of all the basic constants of Saturn's ring system. Most of Struve's later work on Saturn represents developments of the concepts he presented there.

Struve was awarded the Damoiseau Prize by the Paris Académie des Sciences in 1897 and received the gold medal of the Royal Astronomical Society in 1903. In 1904 he was appointed director of the Berlin-Babelsberg Observatory, and from 1913 until his death he was director of the Neubabelsberg Observatory, which he had helped to found.

Struve was married to Olga Struve, the daughter of his father's cousin. Of their twin sons, one, Georg, also became an astronomer.

BIBLIOGRAPHY

I. Original Works. The most complete bibliography of Struve's works (60 titles) is in Poggendorff, III, 1308; IV, 1457–1458; and V, 1225. They include "Bestimmung der Elemente von Japetus und Titan aus der Verbindung dieser Satelliten untereinander," in *Astronomische Nachrichten*, **111** (1885), 1–10; "Beobachtungen der Saturnstrabanten. Abt. I. Beobachtungen am 15-zölligen Refractor," in *Observations de Poulkova*, supp. 1 (1888); "Beobachtungen des Neptunstrabanten am 30-zölligen Pulkowaer Refractor," in *Mémoires de l'Académie imperiale des sciences de St.-Pétersbourg*, 7th ser., **42**, no. 4 (1894); "Beobachtungen der Saturnstrabanten am 30-zölligen Pulkowaer Refractor," in *Publications de l'Observatoire central (Nicolas) à Poulkova*, 2nd ser., **11** (1898); "Mikrometermessungen von Doppelsternen, ausgefuhrt am 30-zölligen Refractor zu Pulkowa," *ibid.*, **12** (1901); and *Die neue Berliner Sternwarte in Babelsberg* (Berlin, 1919).

II. Secondary Literature. There are obituaries of Struve by F. S. Archenhold. Geh. Reg.-Rat. in *Weltall*, **21**, nos. 5–6 (1920), 35–36; B. Wanach, in *Vierteljahrsschrift der Astronomischen Gesellschaft*, **56** (1921), 4–12; and L. Courvoisier, in *Astronomische Nachrichten*, **212** (1921), 33–38.

Z. K. Sokolovskaya

STRUVE, OTTO (*b.* Kharkov, Russia, 12 August 1897; *d.* Berkeley, California, 6 April 1963), *astronomy*.

Struve was the son of Gustav Wilhelm Ludwig Struve, professor of astronomy and geodesy and director of the observatory at the University of Kharkov. He graduated from the Kharkov Gymnasium with honors in 1914, then entered the university. His studies were interrupted by World War I, and on his father's advice he went to Petrograd in 1915 to enroll in artillery school. He was sent to the Turkish front as a junior officer in the following year. When Russia withdrew from the war in 1918, Struve returned to Kharkov to resume his education, and soon caught up with his former classmates to take a first-class degree in astronomy. He became an instructor at the university, but when civil war broke out in Russia was recalled into the army under General Denikin, who was then retreating before the advancing Red Army.

In 1920 Struve was evacuated from the Crimea, along with the remnants of Denikin's army, and placed on a ship full of starving and dysentery-ridden refugees; no country would admit them, but Turkey allowed them a small territory in which the emigrants lived in tents during the winter of 1920–1921. Struve went to Constantinople in spring of the latter year, but was unable to find housing or employment, since the city was crowded with Russian emigrants.

While in Turkey, Struve, seeking news of his relatives in Russia, wrote to his uncle Georg Hermann Otto Struve, who was then in Berlin. Hermann Struve had died, but his widow wrote to his colleagues of their nephew's plight, and E. B. Frost, director of the Yerkes Observatory in Williams Bay, Wisconsin, invited him to come to the United States as an assistant observer in stellar spectroscopy. After surmounting considerable difficulties—he had no money and no visa—Struve left Turkey and reached Yerkes in November 1921. On Frost's advice, he immediately began spectral investigations of stars and continued his studies and improved his English. In 1923 he received the Ph.D. from the University of Chicago and in the following year became an instructor at the observatory. In 1927 he became an American citizen.

Struve quickly rose to prominence at Yerkes, becoming assistant professor in 1927, associate professor in 1930, assistant director in 1931, and, on Frost's retirement, director in 1932. He held the last post until 1947, when he became chairman and honorary director, a position he held until 1950. From 1932 until 1947 he also served as professor of astrophysics at the University of Chicago, which administered Yerkes.

Struve was also concerned with the founding of the McDonald Observatory in Texas. As he himself recounted, in an article published in *Sky and Telescope* (volume 24 [1962], 316–317),

After becoming a staff member at Yerkes Observatory in 1921, I was most closely associated with George Van Biesbroeck. Together we determined many comet and asteroid positions with the 24-inch reflector. . . . But my main task was observing the spectra of *B* and *A* stars, following a program prepared by Director Edwin B. Frost. . . . It soon became apparent to Van Biesbroeck and me that very faint asteroids and comets could not be observed with the 24-inch. I also realized that the Bruce spectrograph was not suitable for medium or high-dispersion spectra of stars. . . . No wonder Van Biesbroeck and I spent many cloudy night hours trying to think how we could secure a moderately large reflector, preferably in some other location that had more clear nights. . . . Since we were thinking of a small observ-

ing station, equipped with a 60-inch telescope and operated from Yerkes, we wanted it to be not too far from headquarters. We consulted weather data, and noted a good location in the high plains of the Texas Panhandle, near Amarillo.

These plans were not immediately implemented, and it was only in 1932, after Struve had been invited to become assistant director of the Harvard Observatory, that the officials of the University of Chicago decided to offer him what he wanted. Struve became Yerkes' director in the same year, and the university appropriated $15,000 toward a new observatory—far less than the required amount.

The situation was soon resolved, however. A few years earlier, a banker from Paris, Texas, had left $800,000 to the state university at Austin to build and equip a new observatory; his will was contested in the courts for six years, but the money was available to the university by 1932, and Struve, who knew of the bequest, persuaded Robert M. Hutchins, president of the University of Chicago, to meet with the president of the University of Texas to discuss the construction of an observatory in common. "Within ten minutes," according to Struve, "the two university presidents had agreed upon a broad plan of cooperation, whereby Texas would pay for the telescope and retain ownership of it, while Chicago would pay all salaries and most of the operating expenses." Struve himself was to be director of the new observatory, and in the summer of the same year he took part in selecting a site on top of a double hill (named Mt. Locke, after the donor of the land) in west Texas. He immediately began writing to the directors of other large observatories, soliciting their suggestions, and to telescope-makers, asking for bids and proposals.

By 1936 the eighty-two-inch telescope, then the second-largest in the world, was nearly completed; Struve had assembled most of his staff, including C. T. Elvey, W. W. Morgan, G. P. Kuiper, P. Swings, B. G. D. Strömgren, S. Chandrasekhar, and J. L. Greenstein. The observatory was dedicated on 5 May 1939, although a good deal of work was already under way there. Struve remained as its director until 1947, at which time he became honorary director and chairman of the astrophysics department, positions that he held for another two years. In the years just after World War II, he was able to recruit a number of distinguished European scientists, to whom he offered professorships at McDonald or Yerkes.

In 1950 Struve had begun to suffer from overwork and insomnia, and left McDonald and Yerkes to accept a less demanding appointment as head of the department of astronomy at the University of California at Berkeley, and director of its affiliated Leuschner Observatory. He proved to be a gifted teacher, especially of graduate students, and was able to continue his own research, including projects as a guest investigator at Mt. Wilson. He missed the autonomy and the direct access to the president he had had at Yerkes, however, and in 1959 he returned to that institution. In the same year he was offered the post of director of the new National Radio Astronomical Observatory, which was nearly completed at Green Bank, West Virginia, with its giant eighty-five-foot radio telescope. He took the job willingly, in part from a sense of moral obligation—he had himself written of the need to accept "those laborious and often thankless jobs that are needed for the general advancement of science." Within three years he had made the observatory, despite its geographic isolation, a first-rate scientific institution, staffed by a loyal and dedicated group of young scientists.

By 1962, Struve's health forced him to resign from the new observatory, but he was unable to remain idle for long, and soon accepted a joint professorship at the Institute of Advanced Study, Princeton, and the California Institute of Technology. He died soon after, survived by his mother, who had followed him to the United States in 1923, and by his wife, Mary Martha Lanning, a singer whom he had married in 1925. They had no children.

Although much of Struve's career was devoted to organization and administration, he nevertheless found time to conduct his own investigations. T. G. Cowling, in the memorial sketch of Struve published in the *Biographical Memoirs of Fellows of the Royal Society*, cites Swing's account of Struve's demanding schedule:

> While at Yerkes he used to start very early in the morning, finish late and take hardly any time for his meals. At McDonald he could relax somewhat from his administration duties, hence was eager to spend as much time as possible at the telescope. He loved challenges, physical as well as intellectual, and I have often seen him actively measuring at the microscope after only a few hours' sleep following a long winter night at the 82-inch. Occasionally he would drive from Williams Bay to Chicago in the morning, fly from Chicago to Big Spring [Texas], drive then (over 200 miles) to McDonald, and be ready for a whole night's work at the 82-inch.

Struve's chief astronomical and astrophysical interests lay in spectroscopic investigations of binary and variable stars and researches into stellar atmospheres, stellar rotation, the gaseous constituents of cosmic matter, and stellar evolution. The nature of his work was strongly influenced by Henry Norris Russell, and particularly by Russell's "Some Problems of Sidereal Astronomy," published in 1919. As Struve later wrote (in "The General Needs of Astronomy"), "My own work in astrophysics was stimulated and directed by this article, and even today [1955] it forms one of the most inspiring pieces of astronomical literature."

Struve's work in stellar spectroscopy was based upon protracted observations of minute changes in stellar spectra—widened and shifted lines, distortions of their contours, variations in linear intensities, the appearance of new lines and absorption bands and the disappearance of existing ones. From these observations he was able to derive a great number of spectral regularities and to point out the exceptional usefulness of the technique. He himself provided the means for further exploitation of the method by developing a means for obtaining wide-scale images of the spectra; he also, with Elvey, invented a nebular spectrograph that allows the photographic magnification of the contrast lines of weak diffuse nebulae, the spectra of which are not normally visible against the sky.

Struve investigated the gaseous constituents of cosmic matter throughout his career. As early as 1925 he showed that the stationary interstellar lines of calcium, which had been discovered by Henry Plaskett in the preceding year, are created by absorption in the extended gas clouds that are concentrated in the plane of the Milky Way. He demonstrated that these lines become more intense with distance, and, in the 1930's, working in collaboration with B. P. Gerasimovich, found the value for the density of the interstellar gaseous substratum to be 10^{-26} g./cm.^{3}, and its total mass to represent less than one percent of the complete stellar mass per unit of volume. In 1938, using his nebular spectrograph (which had been installed in both the Yerkes and McDonald observatories), Struve discovered the presence of areas of ionized hydrogen in interstellar space, a discovery crucial to modern radio astronomy. He also did research on the influence upon stellar spectra of the physical processes occurring in the stellar atmospheres, and used the division of separate spectral lines to determine that intermolecular electric fields act within the atmosphere of type *A* and *B* hot stars, and that gas turbulence also occurs there.

In 1928, in collaboration with G. A. Shayn, Struve confirmed the axial rotation of single stars that had been suggested as early as the time of Galileo. With Elvey, he investigated the rotation of several thousand stars, and established the relationship between the velocity of stellar rotation and spectral type. He paid particular attention to stars that exhibit an irregular variation in brightness, and attributed their instability to physical processes (for example, flashes, ejection of matter, or pulsation) occurring in either the stars themselves or in their surrounding atmospheres; he further maintained that the rapid rotation (more than 3,000 km./sec.) of certain extremely hot stars produces the effect that a fraction of the stellar atmosphere is thrown by centrifugal force toward the equator of the star. This condition might be expected to produce gaseous rings, and Struve discovered such rings for some stars in 1931. He then, from 1939 until 1949, made a series of meticulous studies of the formation of the gaseous ring around Pleione, in the Pleiades, and was able to detect its pulsation and rapid scattering.

Like his forebears, Struve was interested in double stars; indeed, his first scientific publication, in 1923, was an article "On the Double Star 9 Argus." It contained a considerable amount of data on the masses of components of this visually and spectroscopically binary star; two other articles published by Struve in the following year were devoted to the orbit of the spectroscopically binary star 43 Orionis and to the nature of binary stars of short period. In an article published in 1927 in the Russian journal *Mirovedenie*, Struve noted the time and effort required in observing binary stars, but went on to emphasize the value of such observations. By 1944, he was able to determine the statistical dependence between the periodicity and amplitude of the curve of distribution of line-of-sight velocity of 144 spectroscopically binary stars with periods of longer than 2.4 days. From these data Struve estimated the mean value of the stellar masses to be equivalent to about three solar masses, and he deduced a criterion for distinguishing RR Lyrae variable stars from true binary stars. He also made a study of close pairs and noted that the components of these unstable binary systems are elliptical in shape because of their mutual gravitation.

In his investigations of β Lyrae, Struve found it to be a closely paired, eclipsing binary system composed of a massive, hot, rapidly rotating blue giant and its smaller, cooler yellow satellite, in which the flow of gas from the hotter to the cooler

body surrounds the stars in a kind of circular envelope that extends into space and is partially dispersed there. In a more general investigation conducted with Su-Shu Huang, Struve concluded that the formation of binary stars is accompanied by a partial loss of mass by the parent body and by an exchange of mass and momentum. His observations of close pairs and irregular stars gave him considerable insight into nonstationary star processes and provided him an approach to problems of stellar evolution, an approach that he characterized in the preface to his *Stellar Evolution* of 1950:

The purpose of observational research in astrophysics is to present a unified picture of a series of phenomena and to explain it in terms of a theory or hypothesis. The temptation is always to accumulate more and more factual data and to delay the process of interpretation because we rarely, if ever, feel satisfied that we have enough information to justify a generalization. . . . The Vanuxem Lectures at Princeton in 1949 presented a favorable occasion for taking stock in one field of astrophysics—that of the origin and evolution of single stars and, more especially, of close double stars. This subject has been treated before by other workers, but since their observational basis was different from mine I thought that it would be interesting to present this subject in the light of my own experience at the telescope. . . . I am, however, conscious of the fact that of necessity there is a great deal of speculation in any attempt to discuss the evolution of the stars. . . . The history of previous evolutionary hypotheses teaches us that most of them were wrong. Yet, they have contributed to our understanding of the universe and have, in almost every case, left a permanent imprint upon later hypotheses.

Although Struve's general hypothesis of stellar formation was not widely accepted, a number of aspects of his supporting research were greeted enthusiastically by his fellow astronomers. In particular, his demonstration of the interdependence of the velocity of stellar rotation and other stellar features—especially the positions of stars on the Hertzsprung-Russell diagram—and his detection of a correlation between speed of rotation and the distribution of stars within a cluster, the latter being an indication of the age of these stars, provided astrophysicists with the suggestion that rapid stellar velocities represent the residual motions of the turbulent vortices within the condensing diffuse medium from which the stars had arisen.

Struve was always glad to share his results with fellow workers. He considered international cooperation a necessity, since astronomy is a science of global scope, dependent upon observations made from all over the earth. In a presidential speech before the Dublin meeting of the International Astronomical Union, Struve in 1955 emphasized the problems of the development of an astronomy that would be practical for the launching of man-made satellites and space journeys. He pointed out the growing importance of sophisticated astronomical instrumentation and technology, but maintained that the most important tools of the science are the astronomers themselves.

Struve was also concerned with popularization and with communication among astronomers. From 1932 until 1947 he served as editor-in-chief of the *Astrophysical Journal*, and for more than forty years, from 1923 to 1963, was one of its most prolific authors as well. Between 1949 and 1963 Struve wrote an additional 152 articles (of which fourteen were published in two successive issues) for the more popular *Sky and Telescope*. He there demonstrated his ability to present complicated concepts of astrophysics so as to make them clear to the lay reader—although his pieces were also of interest to professional astronomers. A number of these writings were devoted to the history of science, a bent also apparent in *Stellar Evolution* and in *Astronomy of the Twentieth Century* (1962); and a number concerned the lives and careers of fellow scientists, including E. E. Barnard, the astronomer Charles Darwin, G. A. Shayn, and M. A. Kovalsky. (It may be noted that Struve was careful to acknowledge the work of his colleagues and to assign it its proper value; in particular he considered Barnard, an astronomer who worked at Yerkes until his death in 1923, as "the most capable and productive astronomical observer in the world.")

For several years Struve edited articles and compiled abstracts for *Astronomical Newsletter*. He was especially concerned with Russian works, since he always took a keen interest in developments there, and was vigorous in expounding Russian research.

Struve participated in the activities of many scientific institutions and societies. In addition to the International Astronomical Union, of which he was elected vice-president in 1948 and president in 1952, he held prominent posts in the National Academy of Science, the American Astronomical Society, the American Philosophical Society, and the National Scientific Council. He was a member of the Royal Astronomical Society of London, the Royal Astronomical Society of Canada, the Royal Astronomical Society of New Zealand, and the academies of science of Denmark, Amsterdam, Norway, Sweden, Belgium, and France. He received honorary doctorates from more than ten universities.

In 1944 the Royal Astronomical Society of London gave Struve its highest award, the Gold Medal; he was the fourth member of his family to win it in some 118 years. In presenting the medal, the Society's president, E. A. Milne, noted that the awards council had been "totally uninfluenced by the glamour that surrounds the name Struve. . . . Professor Otto Struve . . . has earned this distinction in his proper right, by the overwhelming significance and value of his brilliant observational and interpretational work in stellar and nebular spectroscopy."

BIBLIOGRAPHY

I. ORIGINAL WORKS. The most complete bibliography of Struve's writings (six books and 446 articles) follows A. Unsöld's obituary in *Mitteilungen der Astronomischen Gesellschaft* (1963), 11–22; it does not, however, include Struve's more than 180 popular scientific articles in *Sky and Telescope*, **6–25** (1946–1963) and *Popular Astronomy*, **32–59** (1924–1951), or his 45 book reviews in *Astrophysical Journal*, **60–110** (1924–1949). His books are *Stellar Evolution. An Exploration From the Observatory* (Princeton, 1950); *The Astronomical Universe* (Eugene, Ore., 1958); *Elementary Astronomy* (New York, 1959), written with B. Lynds and H. Pillans; *The Universe* (Cambridge, Mass., 1961); *Astronomy of the Twentieth Century* (New York–London, 1962), written with V. Zebergs; and *Stellar Spectroscopy*, 2 vols. (Trieste, 1969–1970), written with M. Hack.

II. SECONDARY LITERATURE. There are notices and obituaries on Struve by B. V. Kukarin and P. G. Kulikovsky, in *Astronomicheskii zhurnal*, **40**, no. 6 (1963), 1126–1129; G. J. Odgers, in *Journal of the Royal Astronomical Society of Canada*, **57**, no. 4 (1963), 170–172; C. Payne-Gaposchkin, in *Sky and Telescope*, **25**, no. 6 (1963), 308–310; J. Sahade, in *Ciencia e investigación*, **19** (1963), 195–198; S. Chandrasekhar, in *Astrophysical Journal*, **139**, no. 2 (1964), 423; T. G. Cowling, in *Biographical Memoirs of Fellows of the Royal Society*, **10** (1964), 283–304; L. Goldberg, in *Quarterly Journal of the Royal Astronomical Society*, **5**, no. 3 (1964), 284–290; and P. Swings, in *Bulletin de l'Académie r. de Belgique. Classe des sciences*, 5th ser., **49**, no. 6 (1964), 523–524.

Other sources on Struve's life and work (listed chronologically) are E. A. Milne, "Address. Delivered . . . on the Award of the Gold Medal to Professor Otto Struve," in *Monthly Notices of the Royal Astronomical Society*, **104**, no. 2 (1944), 112–120; A. I. Eremeeva, "Otto Ludwig Struve," in *Vydayushchiesya astronomy mira* ("The World's Leading Astronomers"; Moscow, 1966), 348–357; M. Hack, ed., *Modern Astrophysics*

(Paris – New York, 1967), a memorial volume of collected works dedicated to Struve; P. G. Kulikovsky, *O. Struve i V. Zebergs. Astronomia XX veka* (". . . Twentieth-Century Astronomy"; Moscow, 1968), 5–11; and G. H. Herbig, ed., *Spectroscopic Astrophysics. An Assessment of the Contribution of Otto Struve* (Berkeley, Calif., 1970).

Z. K. SOKOLOVSKAYA

STRUVE, OTTO WILHELM (or Otton Vasilievich) (*b.* Dorpat, Russia [now Tartu, Estonian S.S.R.], 7 May 1819; *d.* Karlsruhe, Germany, 14 April 1905), *astronomy, geodesy.*

The son of F. G. W. Struve and his first wife, Emily Wall, Struve entered the University of Dorpat in 1834 and three years later, while still a student, began to work as an assistant at the university observatory, where he carried out observations under the guidance of his father. In 1839 he graduated with a dissertation entitled "Reduction der am 19/7 März in Dorpat beobachten Plejadenbedeckung" and became adjunct astronomer at the new Pulkovo Observatory, of which his father was director. He himself spent fifty years at Pulkovo, becoming vice-director in 1848 and director in 1862.

Struve shared his father's broad astronomical and geodetic interests and collaborated with him on a number of projects, especially during his early years at Pulkovo. He participated in the systematic survey of all the stars from the north pole to 15° south declination and discovered and described several hundred of the double stars that were included in the *Catalogue revu et corrigé des étoiles doubles et multiples découvertes à l'observatoire central de Poulcova*, published in 1853. He simultaneously began the investigation of the motion of binary and multiple stars that occupied him for almost forty years, during which time he made 6,080 micrometric observations of more than 905 systems. The results of these studies were published in the ninth volume of *Observations de Poulkova* in 1878; in a preface, Struve described his research on systematic errors of micrometric measurements and described the formula he had worked out for their correction by means of a model of double star systems. His work on astronomical constants led him to the determination, in the 1850's, of a value for the constant of precession that was used throughout the world until 1895, when Simon Newcomb derived a more accurate one. This work won Struve the gold medal of the Royal Astronomical Society.

In taking up his father's work on stellar parallaxes, Struve chose to concentrate on those stars—α Lyrae, 61 Cygni, η and μ Cassiopeiae, α Aurigae, and α Aquilae, for example—of which the great motion was assumed to account for their relative closeness to the earth. He was also concerned with the structure of the universe and adhered to William Herschel's theories of the development of stars from nebulae. The great variety of nebula forms, he believed, might represent various stages of star formation, and for this reason he paid particularly close attention to nebulae—such as those in Orion—that are relatively close to stars. Struve maintained that a direct physical link existed among stars which appear to be in the center of nebulae and those which had originated from nebular material. Some of his conclusions are obviously wrong, as for example his belief that only one stellar system could exist in the admittedly infinite universe and that this system—the Milky Way—must be a uniform kinematic entity, having no central body and extending, probably infinitely, into the unknown.

Struve took part in a number of scientific expeditions. In 1842 he went with his father to Lipetsk to observe the solar eclipse and to carry out chronometric observations, while in 1842 and 1843 he participated in the elder Struve's expeditions to Altona and Greenwich, by which the longitudinal relationship of the Pulkovo and Greenwich observatories was established. He himself made further trips between 1846 and 1857 to determine the longitudinal relationships of Pulkovo and Moscow, Warsaw, Kazan, Dorpat, and Arkhangelsk, and observed two other solar eclipses (at Lomge in 1851 and at Pobes, Spain, in 1860). From the latter he was led to conclude that the protuberances visible on the solar surface during eclipses originate on the sun itself.

Much of Struve's geodetic work was conducted in the 1850's, prior to his succession to the directorship of the Pulkovo Observatory. He served as astronomical adviser to the Military Topographic Department of General Staff Headquarters in St. Petersburg from 1847 until 1862 and to the Hydrographic Department of the Marine Ministry from 1854 until 1864. In these posts he dealt with problems of practical astronomy, geodesy, and hydrography, particularly those of latitudinal and longitudinal measurement, accurate altimetric measurement of European Russia, and the improvement of instrumentation. He supported the United States' proposal to introduce a common prime meridian, and discussed this project with American scientists in 1884 when he went to the United States to commission a large refractor for the Pulkovo Observatory from Alvan Clark. His interest in instrumentation extended to spectroscopy and photography, and under his guidance an astrophysical laboratory was established at Pulkovo.

Struve took an active part in the affairs of the Russian Geographical Society, and he served as chairman of its department of mathematical geography from 1860 to 1866; he was also a member of the commission that equipped expeditions to the North Urals and to Central Asia to study the old bed of the Amu-Darya River. He was a member of the St. Petersburg Academy of Sciences and of the French Académie des Sciences; an enthusiastic advocate of international scientific cooperation, he was from 1867 to 1878 chairman of the German Astronomical Society. He was also chairman of the Paris congress on establishing the standard meter (1872) and of the International Commission on Sky Photography (1887).

In 1889, at the end of fifty years of service to the Pulkovo Observatory, Struve resigned as director and went to Karlsruhe, where other members of his family were living. Despite his advanced age, he frequently lectured at the Naturalists' Society of the Technische Hochschule there. He was married twice, first to Emily Dirssen and then, following her death, to Emma Yankovskaya. He had seven children, of whom two sons, Hermann and Ludwig Struve, carried the family profession into a new generation.

BIBLIOGRAPHY

I. ORIGINAL WORKS. Struve's more than 130 published works include "Bestimmung der Constanten der Praecession, mit Berücksichtigung der eigenen Bewegung des Sonnensystems," in *Mémoires de l'Académie impériale des sciences de St.-Pétersbourg*, Ser. sci. math. et phys., 6th ser., **5**, no. 1 (1844), 17–124; "Catalogue revu et corrigé des étoiles doubles découvertes à Poulkova," *ibid.*, 6th ser., **7**, no. 4 (1853), 385–405; "Issledovanie o kompensatsii khronometrov" ("Studies on the Compensation of Chronometers"), in *Morskoi sbornik*, **21**, no. 4 (1856), 52–93; "Nouvelle détermination de la parallaxe annuelle des étoiles α Lyrae et 61 Cygni," in *Mémoires de l'Academie impériale des sciences de St.-Pétersbourg*, 7th ser., **1**, no. 1 (1859), 1–51; "O zvezdnykh sistemakh i tumannykh pyatnakh" ("On Stellar Systems and Nebular Spots"), in *Zapiski Imperatorskoi akademii nauk*, **1**, no. 2 (1862), 145–161; and *Obzor deyatelnosti Nikolaevskoy Glavnoy (Pulkovskoy) observatorii v prodolzhenie pervykh 25 let ee sushchestvovania* ("Survey of the Activities of the Nikolaev Main [Pulkovo] Observatory During Its First Twenty-Five Years"; St. Petersburg, 1865).

Subsequent works include "Résultats de quelques observations supplémentaires faites sur des étoiles doubles artificielles," in *Bulletin de l'Académie impériale des sciences de St.-Pétersbourg*, **12** (1867), 73–95; "O pervom meridiane" ("On the Prime Meridian"), in *Izvestiya Imperatorskogo Russkogo geografícheskogo obshchestva*, **6**, sec. 2 (1870), 1–14; "Ob uslugakh, okazannykh Petrom Velikim matematicheskoy geografii Rossii" ("On the Service of Peter the Great to the Mathematical Geography of Russia"), in *Zapiski Imperatorskoi akademii nauk*, **20**, no. 1 (1872), 1–19; "O resheniakh, prinyatykh na Vashingtonskoy konferentsii otnositelno pervogo meridiana i vselenskogo vremeni" ("On the Resolutions Adopted at the Washington Conference Regarding the Prime Meridian and Universal Greenwich Civil Time," *ibid.*, **50**, supp. 3 (1885), 1–25; "Die Photographie im Dienste der Astronomie," in *Bulletin de l'Académie impériale des sciences de St.-Pétersbourg*, **30** (1886), 484–500; and "Mesures micrométriques des étoiles doubles," in *Observations de Poulkova*, **10**, pt. 2 (1893), 1–226.

II. SECONDARY LITERATURE. On Struve and his work, see *Otton Vasilievich Struve. Materialy dlya biografícheskogo slovarya deystvitelnykh chlenov Imperatorskoy Akademii nauk* (" . . . Materials for the Biographical Dictionary of Full Members of the Imperial Academy of Sciences"), pt. 2 (Petrograd, 1917), 177–182, which includes a bibliography of 131 of his works; D. I. Dubyago, "O nauchnykh zaslugakh akademika Ottona Vasilievicha Struve" ("On the Scientific Merits of Academician Otton Vasilievich Struve"), in *Sobranie protokolov Obshchestva estestvoispytateley pri Kazanskom universitete*, **5** (1887), 141–151.

There are obituaries by M. Nyren, in *Vierteljahrsschrift der Astronomischen Gesellschaft*, **40** (1905), 286–303; and by A. A. Ivanov, in *Izvestiya Russkago astronomicheskago obshchestva*, **11**, nos. 5–6 (1905), 222–224.

Z. K. SOKOLOVSKAYA

STUART, ALEXANDER (*b.* Aberdeen [?], Scotland, 1673; *d.* London [?], England, September 1742), *physiology.*

Stuart, in 1738 the first Croonian lecturer on muscle physiology at the Royal Society, was a Scotsman of obscure origin and uncertain early history. After receiving his M.A. from Marischal College, Aberdeen, in 1691, he enrolled as a medical student at Leiden, on 14 December 1709, at the age of thirty-six, and graduated M.D. two years later. Appointed physician to Westminster Hospital, London, upon its creation in 1719, he was admitted licentiate of the College of Physicians in June 1720. In 1728 he was designated physician to the queen, admitted to the M.D. at Cambridge (Comitiis Regiis), and elevated to a fellowship in the College of Physicians. Stuart achieved some prominence in the College, serving as censor in 1732 and again in 1741. He was also elected fellow of the Royal Society in 1714, was a recipient of its Copley Medal many years later for his work on muscles, and achieved membership in the Académie Royale des Sciences.

Prior to the publication of his Croonian lectures as a special supplement to the *Philosophical Transactions of the Royal Society* in 1739, Stuart had contributed three well-received papers to the same journal. Two of these papers considered the role of bile as a stimulus to the peristaltic motion of the intestines and raised general questions regarding the "animal oeconomy," and the third reported experiments attempting to demonstrate the existence of a fluid in the nerves. Stuart also published in 1738 a substantial essay on the structure and function of muscles: *Dissertatio de structura et motu musculari*. This essay was an expanded version of his inaugural dissertation for the M.D. at Leiden, presented in 1711.

Stuart's principal concern both in his *Dissertatio*, and in the Croonian lectures based closely on it, was to demonstrate that a strict hydraulic iatromechanism was the best theory by which to account for muscular motion. Unlike contemporary British writers who in the 1730's advanced theories of muscular action based on the wavelike movement of animal spirits and the jiggling of elastic nerve fibers, Stuart insisted that the mechanics of sanguinary and nervous fluids, and of their vessels, alone governs the action of the muscles. The forceful flow of blood in the arteries and veins and the trickle of liquid juice through the nerves suffice to cause and control muscular motion.

Stuart saw the muscles as an elaborate network of vessels and open spaces, in basic design not unlike the lungs. The proximate cause of systole is the elastic restitution of the walls of the muscular blood vessels, which had been expanded in a preceding diastole.[1] Alternate systole and diastole accelerates and retards blood flow through the capillary vessels. This effect in turn allows small quantities of nervous juice to exercise a large control over blood flow and, hence, over systole and diastole. Nervous control is concentrated in the neural fiber, which Stuart thought of as "a chain of distensile vesicles whose sides are covered with a net-work of elastic longitudinal and transverse blood-vessels."[2] Since blood flow is particularly difficult when the capillary vessels have been longitudinally stretched and their transverse diameters shortened, the dripping of nerve juice into the vesicles (around which the capillaries are woven) can quickly reverse the hydraulic circumstances. The dimensions of the neural vesicles will subtly alter with the addition of nerve juice, and this modification will cause capillary diameters to alter too. This latter alteration can quickly intensify because the pressure of inflowing blood will multiply the initial effect. Dramatic alterations of blood flow within the muscles can throw them from systole to diastole, or vice versa. Antagonist muscles can thus be seen as if poised in a fine static balance, with the blood mass shared between them via commonly connected vessels ready to switch from one "balance pan" to another nearly instantaneously.[3] The trickle of nerve juice, initiated by the immaterial spirit or mind, can therefore cause the elevation or depression of the muscle balance pans, that is, the systole and diastole of muscle pairs.

Stuart quite possibly derived his theory from analogous ideas on heartbeat and its nervous control introduced by Boerhaave in his *Institutiones medicinae*.[4] Stuart also resorted to several kinds of direct empirical evidence. He explored muscular anatomy through the microscope and with the help of excarnation and injection techniques.[5] He tried to prove the elasticity of the blood vessels by tying arteries, veins, and nerves next to one another onto a wooden board and watching the blood vessels but not the nerves contract.[6] He dissected out a demonstration muscle pair and attendant vessels; common blood vessels and individual nerve supplies were clearly indicated in the special preparation.[7] Finally, Stuart tried to demonstrate his basic theory of muscle action by suspending a decapitated frog by its forelegs in a frame and then pushing down on the exposed spinal cord with a blunt probe. The hind legs of the frog twitched, and Stuart explained this by claiming that the probe forced a small quantity of nerve juice from the spinal cord into the appropriate muscles, the small increment of nervous liquid being sufficient mechanically to trip muscle contraction.[8]

It was for the decapitated "spinal frog" experiment that Stuart was most widely known by his eighteenth-century successors. It did not, however, originate with him. Decapitated spinal frog experiments can easily be traced into the seventeenth century and, in one form, perhaps even as far back as Leonardo da Vinci.[9] Robert Whytt credited the experiment to Stephen Hales, a contemporary of Stuart and a correspondent of Whytt.[10] But it was

Stuart, because of the presumptive importance of the experiment to his iatromechanical theory, who gave it greatly enhanced attention and by doing so fixed the decapitated frog in the imagination of later eighteenth-century physiologists. Albrecht von Haller, at any rate, referred to Stuart's "very useful facts,"[11] and Robert Whytt often mentioned Stuart's work.[12] Both Whytt and Haller focused on the phenomena and largely ignored Stuart's hypotheses. The experiment was borrowed, repeated, and made increasingly more sophisticated. The clarifying articulation of reflex theory in the latter half of the eighteenth century thus owes an unintended but considerable debt to Stuart, for as J. F. Fulton suggested, the decapitated frog became in that period ". . . one of the first martyrs of science."[13]

NOTES

1. Alexander Stuart, *Dissertatio de structura et motu musculari* (London, 1738), p. 120.
2. Alexander Stuart, *Three Lectures on Muscular Motion,* . . . (London, 1739), p. xiii.
3. *Ibid.*, p. vii.
4. *Dr. Boerhaave's Academical Lectures on the Theory of Physic*, II (London, 1743), pp. 75, sect. 190.
5. See, for example, *Dissertatio*, pp. 36 and 47.
6. *Three Lectures*, p. iii.
7. *Ibid.*, p. vi.
8. *Ibid.*, pp. xxxvii–xxxix.
9. See, for example, Georges Canguilhem, *La formation du concept de réflexe* (Paris, 1955), p. 91.
10. See R. K. French, *Robert Whytt, the Soul, and Medicine* (London, 1969), p. 86.
11. Albrecht von Haller, "A Dissertation on the Sensible and Irritable Parts of Animals," Owsei Temkin, ed., in *Bulletin of the History of Medicine*, 4 (1936), 694.
12. French, *op. cit., passim.*
13. Quoted in Canguilhem, *op. cit.*, p. 89.

BIBLIOGRAPHY

I. ORIGINAL WORKS. Stuart's two principal extant works are *Dissertatio de structura et motu musculari* (London, 1738); and *Three Lectures on Muscular Motion, Read Before the Royal Society in the Year 1738* (London, 1739). The *British Museum Catalogue* also lists *Dissertatio medica inauguralis de structura et motu musculari, etc.* (Leiden, 1711), and *New Discoveries and Improvements in Anatomy and Surgery . . . With Cases and Cures* (London, 1738).

Three of Stuart's major papers are "An Essay Upon the Use of the Bile in the Animal Oeconomy, Founded on an Observation of a Wound in the Gallbladder," in *Philosophical Transactions of the Royal Society*, **36** (1729–1730), 341–363; "Experiments to Prove the Existence of a Fluid in the Nerves," *ibid.*, **37** (1731–1732), 327–331; and "Explanation of an Essay on the Use of the Bile in the Animal Oeconomy," *ibid.*, **38** (1733–1734), 5–25.

II. SECONDARY LITERATURE. Stuart has rarely been studied, although in recent years he has been receiving increasing notice. Biographical essentials can be found in Robert W. Innes-Smith, *English-Speaking Students of Medicine at the University of Leyden* (Edinburgh, 1932), 226; and more extensively in William Munk, *The Role of the Royal College of Physicians of London*, II (London, 1878), 109. More than passing mention of Stuart's work is made in R. K. French, *Robert Whytt, the Soul, and Medicine* (London, 1969), 90, 150–151; Karl E. Rothschuh, *History of Physiology*, Guenter B. Risse, trans. (Huntington, N.Y., 1973), 138, 183; and Robert Schofield, *Mechanism and Materialism: British Natural Philosophy in an Age of Reason* (Princeton, 1970), 192–193. Also of considerable utility in setting the eighteenth-century context for Stuart's work are E. Bastholm, *The History of Muscle Physiology* (Copenhagen, 1950); and Georges Canguilhem, *La formation du concept de réflexe* (Paris, 1955).

THEODORE M. BROWN

STUDER, BERNHARD (*b*. Büren, Switzerland, 21 August 1794; *d*. Bern, Switzerland, 2 May 1887), *geology.*

Studer, the son of Samuel Studer, a Protestant pastor, spent most of his youth in Bern, where his father had been appointed professor of practical theology in 1796 at the Bern Academy. To please his family, Studer agreed to study theology, although at the same time he became increasingly interested in mathematics and science. After earning his theology degree, he decided to study these subjects, first at Bern and then, from 1816 to 1818, at Göttingen. He then returned to Bern to teach mathematics at the municipal Gymnasium. He was later given a post at the Bern Academy, where, in addition to mathematics, he taught physics, mathematical geography, and mineralogy. His teaching and publications made him well known, and, shortly after its founding in 1834, the University of Bern offered him the professorship of geology and mineralogy, which he held until his retirement in 1873.

Studer's scientific writings are devoted to the geology of Switzerland, particularly to the Swiss Alps. In his first major work, on the Molasse (1825), he showed himself to be a master of careful, precise observation and clear presentation. His other major publications include *Geologie der westlichen Schweizer Alpen* (1834) and "Die Gebirgsmasse von Davos" (1837). With Arnold Escher von der Linth, his lifelong friend and collaborator,

Studer wrote *Geologische Beschreibung von Mittel-Bündten* (1839). This work, as well as a number of subsequent ones, culminated in 1853 in a further collaboration with Escher: the first geological map of Switzerland. (It was on a scale of 1:380,000; a second edition was published in 1869.) The map was based on painstakingly detailed observations and thorough preliminary studies made throughout the country. Before its appearance, Studer published a text designed to accompany it, entitled *Geologie der Schweiz* (1851–1853); together, map and text constitute the crowning achievement of Studer's scientific work. The *Geologie* contains the first comprehensive description of the structure of the Swiss Alps, which have been of extraordinary importance for understanding the formation of the former and existing mountains of the world, and represents the point of departure for all later synthetic accounts.

In the following years Studer devoted great energy to editing a geological map of Switzerland on a scale of 1:100,000, four times greater than his earlier map. The research on this map, which was based on Dufour's excellent topographical map, was conducted by a geological commission created for this purpose and placed under Studer's direction. (This commission still exists as the Swiss Geological Commission and is responsible for national geological surveys.) The new map ultimately consisted of twenty-one sheets, the first appearing in 1865, the last in 1887, on the day of Studer's death.

Besides his geological works, Studer wrote textbooks on physics and mechanics and on mathematical and physical geography. He was also the author of an extensive history of the physical geography of Switzerland. With Escher, he was one of the founders of modern Alpine geology in Switzerland.

BIBLIOGRAPHY

I. Original Works. Besides many articles and papers, Studer published a number of larger works: *Beyträge zu einer Monographie der Molasse . . .* (Bern, 1825); *Geologie der Westlichen Schweizer Alpen* (Heidelberg, 1834), with 5 plates and a geological map; "Die Gebirgsmasse von Davos," in *Neue Denkschriften der Allgemeinen schweizerischen Gesellschaft für die gesamten Naturwissenschaften*, **1** (1837), with 3 plates; *Geologische Beschreibung von Mittel-Bündten* (Bern, 1839), with 2 plates and 3 maps, written with A. Escher von der Linth; *Geologie der Schweiz*, 2 vols. (Bern–Zurich, 1851–1853); *Einleitung in das Studium der Physik und Elemente der Mechanik* (Bern, 1859); *Ge-*

schichte der physischen Geographie der Schweiz bis 1815 (Bern–Zurich, 1863); and *Index der Petrographie und Stratigraphie der Schweiz und ihrer Umgebungen* (Bern, 1872).

II. Secondary Literature. On Studer and his work, see A. Daubrée, "Notice sur les travaux de M. Studer," in *Comptes rendus hebdomadaires des séances de l'Académie des sciences*, **104** (1887), 1203–1205; C. W. von Gümbel, in *Neue deutsche Biographie*, XXXVI (Leipzig, 1893), 731–734; Albert Heim, *Geologie der Schweiz*, I (Leipzig, 1919), 7–8, with portrait; H. Hölder, *Geologie und Paläontologie in Texten und ihrer Geschichte* (Freiburg im Breisgau–Munich, 1960), *passim*; J. W. Judd, "B. Studer," in *Quarterly Journal of the Geological Society of London*, **44** (1888), 49–50; R. Lauterborn, "Der Rhein. Naturgeschichte eines deutschen Stromes . . .," in *Berichte der Naturforschenden Gesellschaft zu Freiburg im Breisgau*, **33** (1934), 110–113; L. Rütimeyer, "Prof. Bernhardt Studer," in *Verhandlungen der schweizerischen naturforschenden Gesellschaft*, **70** (1887), 177–205; R. Wolf, "Notizen zur schweizerischen Kulturgeschichte," in *Vierteljahrsschrift der Naturforschenden Gesellschaft in Zürich*, **32** (1887), 90–104; and K. A. von Zittel, *History of Geology and Palaeontology*, Maria M. Ogilvie-Gordon, transl. (London, 1901), *passim*.

H. Tobien

STUDY, EDUARD (*b.* Coburg, Germany, 23 March 1862; *d.* Bonn, Germany, 6 January 1930), *mathematics.*

Study, the son of a Gymnasium teacher, studied mathematics and science, beginning in 1880, at the universities of Jena, Strasbourg, Leipzig, and Munich. One of his favorite subjects was biology, and even late in life he investigated entomological questions and assembled an imposing butterfly collection. He received the doctorate from the University of Munich in 1884 and the following year became a *Privatdozent* in mathematics at Leipzig, where he was influenced chiefly by Paul Gordan, an expert in invariant theory.

In 1888 Study left this post to take a similar one at Marburg. From July 1893 to May 1894 he lectured in the United States, mainly at the Johns Hopkins University. He was appointed extraordinary professor at Göttingen in 1894 and full professor at Greifswald in 1897. In 1904 he succeeded Lipschitz at Bonn, where he remained until his retirement in 1927; he died of cancer three years later.

Study was largely self-taught in mathematics, and his writings reflect a highly individual way of thinking. He worked in many areas of geometry but did not accept the geometric axiomatics that Pasch and Hilbert were then developing. (On this point see Study's remarks in his more philosophical writings [1, 2].) Study mastered Grassmann's *Ausdehnungslehre*, Lie's theory of continuous groups, and the calculus of invariant theory; he was highly skilled at employing related algebraic techniques in the solution of geometric questions.

It was then usual for geometers to state their findings with little concern for exactitude in individual aspects of problems, and many theorems were labeled simply "in general," without any indication of the scope of their validity. Questions concerning real numbers, for example, were not carefully distinguished from those concerning complex numbers. Many of Study's papers were addressed to drawing such distinctions. His objections, buttressed by counterexamples, to Schubert's principle of the conservation of number were particularly well known, and the principle was eventually firmly established with suitable restrictions on its range of applicability [3].

In his own work Study demonstrated what he considered to be a thorough treatment of a problem. Moreover, a number of the problems he chose to discuss—for example, Apollonius' tangent problem [4] and Lie's straight-line–sphere transformation [5]—had long been thought resolved. Study was the first to show how the totality of the conic sections of the plane—that is, the conic sections considered as unions of elements—can be mapped into a point set M_5 of P_{27} [6].

With Corrado Segre, Study was one of the leading pioneers in the geometry of complex numbers. He systematically constructed the analytic geometry of the complexly extended Euclidean spaces R_2 and R_3; and, with Fubini, he was the first to introduce metrics for these spaces [7]. His contributions to complex differential geometry include the first systematic studies of isotropic curves and the introduction of isotropic parameters [8].

Adept in the methods of invariant theory—which are almost completely forgotten today—Study, employing the identities of the theory, sought to demonstrate that geometric theorems are independent of coordinates. This undertaking was not a simple one, but he achieved a number of successes. In a long work [9] he derived the formulas of spherical trigonometry from a new point of view, and in the process created many links between trigonometry and other branches of mathematics. He wrote other works on invariant theory, but they provoked little response even at the time of their publication [10].

Study was the first to investigate systematically all algebras possessing up to four generators over R and C [11], including W. R. Hamilton's quaternions, which interested him chiefly because of their applications to geometry and Lie groups. In his long work *Geometrie der Dynamen* [12] Study made a particularly thorough examination of Euclidean kinematics and the related subject of the mechanics of rigid bodies. Unfortunately, because of its awkward style and surfeit of new concepts, this work has never found the public it merits.

BIBLIOGRAPHY

I. ORIGINAL WORKS.

[1] *Die realistische Weltansicht und die Lehre vom Raume* (Brunswick, 1914; 2nd ed. 1923).

[2] *Denken und Darstellung in Mathematik und Naturwissenschaften*, 2nd ed. (Brunswick, 1928).

[3] "Über das sogenannte Prinzip von der Erhaltung der Anzahl," in *Archiv der Mathematik und Physik*, 3rd ser., **8** (1905), 271–278. "Das Prinzip der Erhaltung der Anzahl," in *Berichte über die Verhandlungen der K. Sächsischen Gesellschaft der Wissenschaften zu Leipzig*, Math.-phys. Kl., **68** (1916), 65–92.

[4] "Das Apollonische Problem," in *Mathematische Annalen*, **49** (1897), 497–542.

[5] "Vereinfachte Begründung von Lie's Kugelgeometrie," in *Sitzungsberichte der Preussischen Akademie der Wissenschaften zu Berlin*, **27** (1926), 360–380.

[6] "Über die Geometrie der Kegelschnitte, insbesondere Charakteristikenproblem," in *Mathematische Annalen*, **27** (1886), 58–101.

[7] "Kürzeste Wege im komplexen Gebiet," *ibid.*, **60** (1905), 327–378.

[8] "Zur Differentialgeometrie der analytischen Kurven," in *Transactions of the American Mathematical Society*, **10** (1909), 1–49.

[9] "Sphärische Trigonometrie, orthogonale Substitutionen und elliptische Funktionen," in *Abhandlungen der Sächsischen Akademie der Wissenschaften*, **20** (1893), 83–232.

[10] *Methoden zur Theorie der ternären Formen* (Leipzig, 1889); and *Einleitung in die Theorie der invarianten linearer Transformationen auf Grund der Vektorrechnung* (Brunswick, 1923).

[11] "Theorie der gemeinen und höheren komplexen Grössen," in *Encyklopädie der mathematischen Wissenschaften*, I, pt. 4 (Leipzig).

[12] *Geometrie der Dynamen* . . . (Leipzig, 1903).

II. SECONDARY LITERATURE. See F. Engel, "Eduard Study," in *Jahresberichte der Deutschen Mathematikervereinigung*, **40** (1931), 133–156; and E. A. Weiss, "Eduard Study, ein Nachruf," in *Sitzungsberichte der Berliner mathematischen Gesellschaft*, **10** (1930), 52–77; "Eduard Study's mathematischen Schriften," in

Jahresberichte der Deutschen Mathematikervereinigung, **43** (1934), 108–124, 211–225.

WERNER BURAU

STURGEON, WILLIAM (*b.* Whittington, Lancashire, England, 22 May 1783; *d.* Prestwick, Manchester, England, 4 December 1850), *physics.*

Sturgeon's father, John Sturgeon, is described as an ingenious but idle shoemaker who poached fish and raised gamecocks. His mother, Betsy Adcock, was the daughter of a small shopkeeper.

After a disagreeable apprenticeship to a shoemaker, beginning in 1796, Sturgeon in 1802 joined the militia and two years later enlisted in the Royal Artillery. He was stationed at Woolwich, where he studied natural science at night and performed occasional electrical experiments, which attracted some attention. He left the service in 1820, at the age of thirty-seven, and took up the trade of bootmaker in Woolwich. About 1804 he married a widow named Hutton; they had three children who died in infancy. In 1829 he married Mary Bromley of Shrewsbury; they had one child who died in infancy and one adopted child, Sturgeon's niece.

Sturgeon had developed mechanical skills useful for making scientific apparatus, and he lectured on science to schools and other groups. He was a member of the Woolwich Literary Society and in 1824 was appointed lecturer in science and philosophy at the East India Company Royal Military College of Addiscombe. In 1832 he was on the lecture staff of the short-lived Adelaide Gallery of Practical Science, and in 1840 he went to Manchester to become superintendent of the Royal Victoria Gallery of Practical Sciences. He held this last post for four years, after which he supported his family from his income as an itinerant lecturer. He was a member of the Manchester Literary and Philosophical Society and through the influence of its president received a grant of £200 from Lord John Russell and later, in 1849, an annuity of £50.

Sturgeon's major achievements concerned electromagnetism. In 1825 he received a silver medal and thirty guineas from the Society of Arts in recognition of his electromagnetic apparatus, including his important refinement of the design of the electromagnet. He placed a bar of soft iron in a solenoid and found that the magnetic effect was greatly increased. A coating of shellac on the bar served as insulation between it and the bare wires; Joseph Henry later insulated the wires themselves, thus allowing many more turns and an additional increase in the magnetic force.

Sturgeon's other contributions were mainly designs of apparatus for displaying electromagnetic phenomena. In this respect he exemplified that small but important group of instrumentmakers and lecturers who sought means of exhibiting electrical science in graphic and exciting ways.

In 1836 Sturgeon established a monthly periodical, *Annals of Electricity*, which lasted through ten volumes until 1843, when he founded a successor journal, *Annals of Philosophical Discovery and Monthly Reporter of the Progress of Science and Art*. This journal was terminated at the end of the same year.

BIBLIOGRAPHY

I. ORIGINAL WORKS. Sturgeon's works include *Experimental Researches in Electromagnetism, Galvanism . . .* (London, 1830); *Lectures on Electricity Delivered in The Royal Victorian Gallery, Manchester* (London, 1842); *Twelve Elementary Lectures on Galvanism* (London, 1843); and *Scientific Researches* (Manchester, 1850), which contains all of his important works. In 1843 he edited a reissue of William Barlow's *Magnetical Advertisements* (London, 1616).

Sturgeon's articles, which number about seventy, are listed in the Royal Society *Catalogue of Scientific Papers*, V, 876–878. The description of his magnet appears in "Account of an Electromagnetic Apparatus," in *Annals of Philosophy*, **12** (1826), 357–361. Apparently none of his manuscripts or apparatus has been preserved.

II. SECONDARY LITERATURE. A relatively lengthy entry by William Dee appears in the *Dictionary of National Biography* and a *Biographical Note* by S. P. Thompson was privately printed in 1891. Other, shorter notices are mentioned in Dee's account.

BERNARD S. FINN

STURM, CHARLES-FRANÇOIS (*b.* Geneva, Switzerland, 29 September 1803; *d.* Paris, France, 18 December 1855), *mathematics, physics.*

Sturm's family, originally from Strasbourg, had lived in Geneva since the middle of the eighteenth century. He was the elder son of Jean-Henri Sturm, a teacher of arithmetic, and Jeanne-Louise-Henriette Gremay. Sturm at first studied classics, a field in which he displayed considerable ability. For example, at age sixteen he improvised Greek and Latin verses without the aid of a dictionary. In order to perfect his German, he attended the Lutheran church to hear sermons given in that language. In 1819, the year of his father's death, Sturm abandoned his literary studies and devoted himself to mathematics. At the Geneva Academy he attended the mathematics lectures of Simon L'Huillier and the physics lectures of Marc-Auguste Pictet and Pierre Prevost. L'Huillier, who in 1821 was preparing to retire, soon discovered Sturm's abilities; he encouraged Sturm, offered him advice, and lent him books. His influence, however, was less decisive than that of his successor, Jean-Jacques Schaub.[1] Sturm also attended a course in mathematics given by Baron Jean-Frédéric-Théodore Maurice and one in astronomy taught by Alfrède Gautier. Among Sturm's fellow students were Auguste de La Rive, Jean-Baptiste Dumas, and Daniel Colladon, his best friend.[2]

Having completed his studies at the Academy, Sturm moved early in May 1823 to the château of Coppet, about fifteen kilometers from Geneva, as tutor to the youngest son of Mme de Staël.[3] About ten people lived at the château, including Duke Victor de Broglie, his wife, the former Albertine de Staël, and their three children. Sturm's duties as tutor left him sufficient free time to write his first articles on geometry, which were published immediately in *Annales de mathématiques pures et appliquées*, edited by J. D. Gergonne. Toward the end of the year, he accompanied the duke's family to Paris for a stay of approximately six months. Through de Broglie's assistance he was able to enter the capital's scientific circles.

During this period Sturm wrote to Colladon: "As for M. Arago, I have two or three times been among the group of scientists he invites to his house every Thursday, and there I have seen the leading scientists, MM Laplace, Poisson, Fourier, Gay-Lussac, Ampère, etc. Mr de Humboldt, to whom I was recommended by Mr de Broglie, has shown an interest in me; it is he who brought me to this group. I often attend the meetings of the Institut that take place every Monday."[4]

In May 1824 Sturm returned to Coppet with the de Broglie family, but toward the end of that year he gave up teaching in order to devote himself to scientific research. With Colladon he undertook a study of the compression of liquids, which had just been set by the Paris Academy as the subject of the grand prize in mathematics and physics for the following year. They decided to measure the speed of sound in water—Lake Geneva was nearby—and then to seek the coefficient of compressibility of water, introduce this coefficient into Poisson's

formula for the speed of sound, and compare their results with those predicted by the formula. The project did not, however, yield the desired results. In addition, Colladon seriously injured his hand during the tests.

On 20 December 1825 Sturm and Colladon left for Paris with the intention of attending physics courses and of finding the instruments needed for the experiments that would enable them to complete their memoir. Arago often invited them to his house, and for a time Sturm gave mathematics lessons to his eldest son. In addition, Ampère offered them the use of his physics laboratory.

At the Sorbonne and at the Collège de France, Sturm and Colladon attended the lectures of Ampère and Gay-Lussac in physics and of Cauchy and Lacroix in mathematics. They also were present during the tests on steam engines that Arago and Dulong conducted near the Paris observatory. In addition they visited Fourier, who at this time was engaged in research on heat. Fourier asked Colladon to measure the thermal conductivity of various substances and, recognizing Sturm's inclination and talent for theoretical work, suggested that the latter make a thorough study of a certain procedure in analysis, later called harmonic analysis, that Fourier believed would be of great use in theoretical physics.

Sturm and Colladon finished their paper on the compression of liquids and submitted it to the Academy, which eventually decided that none of the memoirs it had received merited the prize and that the the same subject would be set for the 1827 award. Meanwhile, Sturm and Colladon had been appointed assistants to Ampère, who suggested that they collaborate on a major treatise on experimental and theoretical physics (the project was never undertaken). In November 1826 Colladon returned to Geneva and measured the speed of sound in water between Thonon and Rolle, situated on opposite banks of Lake Geneva. He obtained a value of 1,435 meters per second. The agreement was good with the theoretical speed determined by Poisson's formula, which gave 1,437.8 meters per second. Upon his return to Paris, he and Sturm completed the new version of their memoir. This time it won the grand prize of 3,000 francs, a sum that enabled them to pay the costs of their experiments and to prolong their stay in Paris.

Henceforth their scientific careers diverged. Even in his physical research, however, Sturm continued to obtain interesting results in geometry, notably on the theory of caustic curves of reflection, the poles and polars of conic sections, Desargues's theorem, and involutions.

In 1829, through Ampère's influence, Sturm was appointed chief editor for mathematics of the *Bulletin des sciences et de l'industrie*. On 13 May of that year he presented to the Academy "Mémoire sur la résolution des équations numériques," containing the famous theorem that perhaps did more to assure his reputation than the rest of his writings together. The founder of the *Bulletin*, André Étienne, Baron d'Audebard de Férussac, invited his principal collaborators to assemble at his Paris residence once a week; and it is possible that Sturm met Niels Abel and Évariste Galois there, as well as Cournot, Coriolis, Duhamel, Hachette, and Lacroix.[5]

Sturm and Colladon wished to obtain posts in the state school system; but even though they had the backing of several influential members of the Academy, they were unsuccessful because they were foreigners and Protestants. The revolution of July 1830 proved beneficial to their cause: Arago was able to have Sturm named professor of *mathématiques spéciales* at the Collège Rollin and Colladon, professor of mechanics at the École Centrale des Arts et Manufactures. (Colladon returned to Geneva in 1839.) It is interesting that the minister of public education after the revolution was Duke Victor de Broglie.[6]

Sturm became increasingly interested in the theory of differential equations; and in September 1833, six months after he had acquired French citizenship, he read a memoir on this subject before the Academy. About this time the Geneva Academy considered offering him a post, and in October 1833 he received official notification through La Rive. But Sturm declined it, for his decision to remain in France was irrevocable. He also rejected an offer from the University of Ghent.

Upon the death of Ampère, a seat in the Académie des Sciences became vacant. On 28 November 1836 Sturm was nominated to it by Lacroix; the other candidates were Liouville, Duhamel, Lamé, and Jean-Louis Boucharlat. At the following meeting, it was announced that Liouville and Duhamel had withdrawn their names, considering it right that the seat go to Sturm; he was elected by forty-six of the fifty-two votes cast.

Sturm's career now progressed rapidly; in 1838 he was named *répétiteur* of analysis in Liouville's course at the École Polytechnique, where he became professor of analysis and mechanics in 1840. Also that year he assumed the chair of mechanics formerly held by Poisson at the Faculty of Sciences. In 1837 Sturm became *chevalier* of the Legion of Honor, and in 1840 he won the Copley Medal of the Royal Society and was elected a member of that body. He was already a member of the Berlin Academy (1835) and the Academy of St. Petersburg (1836).

Sturm was obliged to spend much time preparing his courses on differential and integral calculus and on rational mechanics. An excellent lecturer, he was admired for both his personal qualities and his knowledge. Sturm dedicated his remaining time to research. From analysis he turned to optics, particularly to vision, and to mechanics, in which, independently and by a new method, he derived one of Duhamel's theorems on the variation in *vis viva* resulting from a sudden change in the links of a moving system.

Around 1851 Sturm's deteriorating health obliged him to arrange for a substitute at the Sorbonne and at the École Polytechnique. He became obese, had a nervous breakdown, and no longer derived pleasure from intellectual work. His doctors ordered him to walk a great deal and to move to the country. Two years later Sturm resumed some of his teaching duties, but the illness returned—probably with other complications, the nature of which is not known—and it slowly took his life.

On 20 December 1855 a crowd of scientists, friends, and students accompanied Sturm's body to the cemetery of Montparnasse. Moving speeches were given by a Protestant minister and by Liouville, who called Sturm "a second Ampère: candid like him, indifferent to wealth and to the vanities of the world."[7]

Sturm's moral qualities, his innate sense of duty and of honor, and his devotion to the ideals of friendship brought him the esteem and affection of all who knew him. His life, like his writings, was a model of clarity and rigor. Favorable circumstances smoothed the way and permitted him to display his genius; but his long friendship with Colladon and the patronage of such highly placed persons as de Broglie, Arago, and Ampère are also inseparable from his career and should be taken into account in explaining his success.

In the rest of the article we shall not consider Sturm's earliest works nor, in particular, his many articles on plane geometry—in each of which he made a valuable, original contribution. The essential features of his work in this area were incorporated in later works on geometry, often without mention of their origin. We shall, instead, examine rather closely three other important aspects of his work.

Sturm's Theorem. Although the problem of finding the number of real roots of the equation $f(x) = 0$ had already been encountered by Descartes and by Rolle, it was not investigated systematically until the mid-eighteenth century. Gua de Malves made the first significant attempts in 1741, and in 1767 Lagrange approached the problem by forming the transform with the squares of the differences of the zeroes of the polynomial. Later, Fourier considered the sequence formed by the first member of the equation and its successive derivatives. Poisson suggested the problem to Cauchy, who in 1813 sent three notes on the subject to the Academy and in 1815 discussed it at length in his "Mémoire sur la détermination du nombre de racines réelles dans les équations algébriques."[8] By successive eliminations, Cauchy established a system of rational functions of the coefficients of the given polynomial; and from the sign of these functions he deduced the number of zeroes. His was the first complete solution, but the calculations involved are so long and laborious that it was never adopted.

Sturm used Fourier's method, as well as some unpublished results that Fourier had communicated to him. (Sturm credited Fourier for these in the article published in *Bulletin des sciences et de l'industrie*.) But instead of working with the successive derivatives, he was able to develop his method by using only the first derivative. The essential part of the argument is as follows:[9]

Let $V = 0$ be an equation of arbitrary degree with distinct roots, and let V_1 be the derivative of V. One proceeds as in finding the greatest common divisor of V and V_1, the sole difference being that it is necessary to change the signs of all the remainders when they are used as divisors. Let $Q_1, \cdots, Q_{r-1}$ be the quotients and $V_2, \cdots, V_{r-1}$ the remainders, V_r being a constant. One therefore has

$$V = V_1 Q_1 - V_2$$
$$V_1 = V_2 Q_2 - V_3$$
$$\cdots \cdots$$
$$V_{r-2} = V_{r-1} Q_{r-1} - V_r.$$

The statement of the theorem then reads:

Let us substitute two arbitrary numbers a and b, positive or negative, for x in the sequence of functions $V, V_1, V_2, \cdots, V_{r-1}, V_r$. If a is smaller than b, the number of the variations in the sequence of the signs of these functions for $x = b$ will, at most, be equal to the number of the variations in the sequence of the signs of these same functions for $x = a$. And if it is less, the difference will be equal to the number of real roots of the equation $V = 0$ between a and b.

"Variation" in this statement means "change of sign." The demonstration, which includes an examination of two cases, a scholium, and two corollaries, requires several pages. Sturm's discovery elicited great excitement, and he became famous as the mathematician who had filled a lacuna in algebra. It was not long, however, before voices were raised in support of Cauchy—that, for example, of Olry Terquem, editor of the *Nouvelles annales de mathématiques*, who accorded priority to Cauchy while recognizing that Sturm had found a simpler method. Cauchy himself later asserted his priority. As for Sturm, he was satisfied to speak of the "theorem of which I have the honor to bear the name." Charles Hermite made the following assessment: "Sturm's theorem had the good fortune of immediately becoming classic and of finding a place in teaching that it will hold forever. His demonstration, which utilizes only the most elementary considerations, is a rare example of simplicity and elegance."[10]

Cauchy subsequently found a way to determine the number of imaginary roots of an equation; but here, too, Sturm arrived at the same results by a shorter and more elementary method. The proof of this "Cauchy theorem" was published in *Journal de mathématiques pures et appliquées* for 1836 in an article signed by Sturm and Liouville.

The functions $V, V_1, \cdots, V_r$ are called Sturm functions. J. J. Sylvester discussed them in two articles and expressed them by means of the roots of the given equation.[11]

Differential Equations and Infinitesimal Geometry. On 28 September 1833 Sturm presented a memoir on second-order differential equations to the Académie des Sciences, but it was not published until three years later, in *Journal de mathématiques*. In this work Sturm studied equations of the form

$$L\frac{d^2V}{dx^2} + M\frac{dV}{dx} + N \cdot V = 0,$$

where L, M, and N are given functions of x, and V is the unknown function. The integration is, in general, impossible. Sturm's insight was to determine the properties of V without assigning it in advance to any class. Although used today, this method of proceeding was not at all common at that time. Sturm started by writing the given equation as

$$\frac{d}{dx}\left(K\frac{dV}{dx}\right) + G \cdot V = 0,$$

where K and G are new functions of x that can be determined subsequently. This type of differential equation is encountered in several problems of mathematical physics.

Liouville maintained a special interest in this area of Sturm's research, to which he himself made several important additions in two notes to the Academy in 1835 and 1836. Further, in his *Journal* he published a work written with Sturm on the expansion of functions in series; their paper begins with the differential equation

$$\frac{d}{dx}\left(K\frac{dV}{dx}\right) + (gV - l) = 0.$$

Maxime Bôcher, professor at Harvard University, gave a series of lectures at the Sorbonne in the winter of 1913–1914 on the use of Sturm's methods in the theory of differential equations.

In infinitesimal geometry, Sturm examined the problem of finding the surface of revolution that is at the same time a minimal surface. Delaunay had demonstrated that it is generated by the rotation of the curve described by the focus of an ellipse or of a hyperbola that rolls without sliding on a straight line. His method consisted in imposing on the differential equation of minimal surface the condition that it be a surface of revolution. Sturm handled the problem in another way. He began with an arbitrary surface of revolution; calculated its volume; and sought to determine, with the aid of the calculus of variations, in which cases this volume could become minimum. He thus arrived at the differential equation of the meridian and showed that it is indeed that of the curve described by the focus of a conic section. Furthermore, he demonstrated that in the case of the parabola, the meridian is a catenary curve. He then generalized the question and determined the curve that must be rolled on a straight line in order for a certain point of the plane of this curve to describe another curve the differential equation of which is known. Sturm's solution appeared in Liouville's *Journal* of 1841.

Experimental and Mathematical Physics. Sturm and Colladon's prizewinning "Mémoire sur la compression des liquides et la vitesse du son dans l'eau" consists of three parts. The first contains a description of the apparatus used to measure the compression of liquids, an account of the experiments concerning the compressibility of glass, and the tables of the results for mercury, pure water and water saturated with air, alcohol, sulfuric ether, ethyl chloride, acetic ester, nitrous ester, sulfuric acid, nitric acid, acetic acid, essence of turpentine, carbon disulfide, water partially saturated with ammonia gas, and seawater. The second part records the experiments to measure the heat emitted by liquids following the application of strong and sudden pressures, as well as tests made to determine the influence of mechanical compression on the electrical conductivity of several highly conductive liquids. The third part gives the details of Colladon's experiments on the propagation of sound in water and compares the values obtained experimentally with those resulting from the insertion of the measurements of compressibility in Poisson's formula.

Sturm also published many articles on mechanics and analytical mechanics. Three of the most important deal, respectively, with a theorem of Sadi Carnot's on the loss of *vis viva* in a system of which certain parts are inelastic and undergo sudden changes in velocity; with the motion, studied by Poinsot, of a solid about a fixed point; and with a way of shortening the calculations of W. R. Hamilton and Jacobi for integrating the equations of motion. Further, Sturm's *Cours de mécanique*, like his *Cours d'analyse*, was used by many students and remained a classic for half a century.

In addition to the memoir of 1838 on optics, Sturm earlier wrote many articles and notes on caustics and caustic surfaces. His studies on vision culminated in a long work that displayed a profound knowledge of physiology.

Fourier's influence on Sturm is reflected in a memoir of 1836 on a class of partial differential equations. In it Sturm considers the distribution of heat in a bar, either straight or curved, that is composed of a homogeneous or nonhomogeneous substance, and is of constant or variable thickness but of small dimensions. Under these conditions it may be assumed that all the points of a plane section perpendicular to the axis of the bar are at the same temperature at the same instant. In this work, one of his longest and most important, Sturm exhibits such a richness of ideas and skill in handling mathematics as an instrument for solving a problem in theoretical physics that he may unhesitatingly be placed on the same level as his teacher Fourier.

Sturm, who was so adept at combining mathematics with physics in his work, appears today, by virtue of his modes of thinking, as a very modern scientist. Since 1900 there has been growing interest in his mathematical work, especially in the United States. His contribution to physics, on the other hand, has not yet received the examination it merits. There is still no thorough, full-scale study of his life and work based on the unpublished documents.

NOTES

1. Jean-Jacques Schaub (1773–1825) left MSS on the theory of numerical approximations and on the elementary concepts of the calculus of quaternions. His greatest importance for the history of mathematics is that he was the teacher and patron of Sturm, whose family found itself in financial difficulties after the death of the father.
2. Daniel Colladon (1802–1893), who studied law before turning to physics, played an important role in Sturm's life. A skillful experimenter and brilliant inventor, he conceived the idea of illuminated fountains, which were immediate successes in Paris, London, and Chicago. His research on the action of compressed air led him to construct drilling machines for boring tunnels, and he participated in the cutting of the Mont Cenis and St. Gotthard tunnels. He also was an expert in the building of gasworks.
3. See Countess Jean de Pange, Le dernier amour de Madame de Staël (Geneva, 1944). The author, who died at Paris in 1972, at the age of eighty-four, was the sister of Louis de Broglie.
4. This six-page letter of 26 Apr. 1824 is reproduced in D. Colladon, Souvenirs et mémoires (Geneva, 1893). The original is at the Bibliothèque Publique et Universitaire, Geneva, MS 3255, fols. 219–222.
5. See R. Taton, "Les mathématiques dans le Bulletin de Férussac," in Archives internationales d'histoire des sciences, 1 (1947), 100–125.
6. There is a passage concerning Sturm in Souvenirs du duc de Broglie (Paris, 1886), II, 454. published by his son, C. J. V. A. Albert de Broglie.
7. The complete text of Liouville's speech is in E. Prouhet, "Notice sur la vie et les travaux de Ch. Sturm." On the same day Colladon, who hurriedly left Geneva to attend his friend's funeral, sent Auguste de La Rive a long letter containing much information on Sturm; this unpublished letter is MS fr. 3748, fols. 206–207, at the Bibliothèque Publique et Universitaire, Geneva.
8. Journal de l'École polytechnique, 10 (1815), 457–548; see also Oeuvres de Cauchy, 2nd ser., I, 170–257; II, 187–193; XV, 11–16.
9. The statement of the theorem and the notation we follow is Mayer and Charles Choquet, Traité élémentaire d'algèbre (Paris, 1832). Sturm had given them permission to publish the results of his research.
10. A full study, already outdated, of Sturm's theorem is in Charles de Comberousse, Cours de mathématiques, 2nd ed., IV (Paris, 1890), pt. 2, 442–460.
11. "Memoir on Rational Derivation From Equations of Coexistence, That Is to Say, a New and Extended Theory of Elimination," in Philosophical Magazine, 15 (July–Dec. 1839), 428–435; and "On a Theory of the Conjugate Relations of Two Rational Integral Functions, Comprising an Application to the Theory of Sturm's Functions, and That of the Greatest Algebraic Common Measure," in Abstracts of Papers Communicated to the Royal Society of London, 6 (1850–1854), 324–327.

BIBLIOGRAPHY

I. ORIGINAL WORKS. Sturm's books, both published posthumously, are Cours d'analyse de l'École polytechnique, 2 vols. (Paris, 1857–1859), prepared by E. Prouhet, 8th and subsequent eds. prepared by A. de Saint-Germain—the 14th ed. appeared in 1909—translated into German by Theodor Fischer as Lehrbuch der Analysis (Berlin, 1897–1898); and Cours de mécanique de l'École polytechnique, 2 vols. (Paris, 1861), prepared by E. Prouhet, 5th ed. rev. and annotated by A. de Saint-Germain (Paris, 1905).

Sturm's articles, notes, memoirs, and reports are listed below according to the journal in which they appeared.

In Annales de mathématiques pures et appliquées, edited by J. D. Gergonne: "Extension du problème des courbes de poursuite," 13 (1822–1823), 289–303. In Mémoires présentés par divers savants à l'Académie royale de France: "Mémoire sur la compression des liquides," 2nd ser., 5 (1838), 267–347, written with D. Colladon, who republished it thirty-two years after Sturm's death with his own paper of 1841, "Sur la transmission du son dans l'eau," as Mémoire sur la compression des liquides et la vitesse du son dans l'eau (Geneva, 1887); and "Mémoire sur la résolution des équations numériques," 6 (1835), 271–318, the complete text of the work containing the statement and demonstration of Sturm's theorem.

In Nouvelles annales de mathématiques or Journal des candidats aux écoles polytechnique et normale, edited by Orly Terquem and Camille Christophe Gerono: "Sur le mouvement d'un corps solide autour d'un point fixe," 10 (1851), 419–432.

The Bibliothèque Publique et Universitaire, Geneva, has nine original letters (plus one copy) sent by Sturm to Colladon, La Rive, and other Genevans. Colladon's correspondence contains sixteen letters directly concerning Sturm; among recipients are J. Liouville, Baron J.-F.-T. Maurice, Louis-Albert Necker, and Sturm's sister. All these documents are unpublished, except for two letters from Sturm to Colladon.

II. SECONDARY LITERATURE. The first work on Sturm, appearing a year after his death, was E. Prouhet, "Notice sur la vie et les travaux de Ch. Sturm," in Bulletin de bibliographie, d'histoire et de biographie mathématiques, 2 (May–June 1856), 72–89; repr. in Cours d'analyse, 5th ed. (1877), I, xv–xxix. This article leaves much to be desired: the biographical data are incomplete and the analysis of Sturm's work is superficial; and although the list of writings is complete, it contains many errors. A fuller source is the autobiography of Daniel Colladon, Souvenirs et mémoires (Geneva, 1893), which contains long passages on Sturm's life and on their joint work, as well as the complete text of two long letters from Sturm (Coppet, 1823; Paris, 1824).

See also M. B. Porter, "On the Roots of Functions Connected by a Linear Recurrent Relation of the Second Order," in Annals of Mathematics, 2nd ser., 3 (1901), 55–70, in which the author discusses Sturm's first memoir on second-order homogeneous differential equations (which appeared in *Journal de mathématiques pures et appliquées*, **1** [1836], 106–186); J. E. Wright, "Note on the Practical Application of Sturm's Theorem," in *Bulletin of the American Mathematical Society*, **12** (1906), 246–347; and F. H. Safford, "Sturm's Method of Integrating $dx/\sqrt{X} + dy/\sqrt{Y} = 0$," *ibid.*, **17** (1910–1911), 9–15. With respect to the last article, it may be noted that one of the simplest methods for obtaining the addition theorem for the elliptic integrals of the first type is based on a procedure that appears in Sturm's *Cours d'analyse*, 5th ed., II (1877), 340–343.

Maxime Bôcher, "The Published and Unpublished Work of Charles Sturm on Algebraic and Differential Equations," in *Bulletin of the American Mathematical Society*, **18** (1911–1912), 1–18, is the best study on this subject. See also Bôcher's "Charles Sturm et les mathématiques modernes," in *Revue du mois*, **17** (Jan.–June 1914), 88–104; and *Leçons sur les méthodes de Sturm dans la théorie des équations différentielles linéaires et leurs développements modernes*, Gaston Julia, ed. (Paris, 1917). Gaspare Mignosi, "Theorema di Sturm e sue estensioni," in *Rendiconti del Circulo matematico di Palermo*, **49** (1925), 1–164, is the most complete study of Sturm's theorem from both the theoretical and the historical points of view. It includes a long historical and critical introduction on works concerning the theorem and a chronological list of 65 notes and memoirs (pp. 152–158).

Gino Loria, "Charles Sturm et son oeuvre mathématique," in *Enseignement mathématique*, **37** (1938), 249–274, with portrait, is very good and, despite its title, also deals with Sturm's works on mechanics, optics, and the theory of vision. Loria's chronological list of Sturm's works is partly based on that of Prouhet; although superior, it still contains several errors. Giorgio Vivanti, "Sur quelques théorèmes géométriques de Charles Sturm," *ibid.*, 275–291, was inspired by Sturm's article on regular polygons in *Annales mathématiques*, **15** (1825), 250–256. The first of the three theorems treated was developed by L'Huillier.

See also Henri Fehr, "Charles Sturm 1803–1855," in *Pionniers suisses de la science* (Zurich, 1939), 210–211, with portrait; and Pierre Speziali, *Charles-François Sturm (1803–1855). Documents inédits*, Conférences du Palais de la Découverte, ser. D, no. 96 (Paris, 1964). The latter is fully documented, especially with regard to Sturm's biography; it includes a reproduction of a profile of Sturm at age nineteen, based on a pencil drawing by Colladon. The portrait in the articles by Loria and Fehr was based on this drawing. There are no other likenesses of Sturm.

One may also consult the chapter on the Sturm-Liouville theory of differential equations in Garrett Birkhoff, *A Source Book in Classical Analysis* (Cambridge, Mass., 1973), 258–281.

PIERRE SPEZIALI

STURM, FRIEDRICH OTTO RUDOLF (*b*. Breslau, Germany [now Wrocław, Poland], 6 January 1841; *d*. Breslau, 12 April 1919), *mathematics*.

The son of a Breslau businessman, Sturm attended the St. Maria Magdalena Gymnasium. In the winter semester of 1859 he began to study mathematics and physics at the University of Breslau, where in the summer of 1863 he received his doctorate of philosophy. From then until 1872 he worked as a teaching assistant, part-time teacher, and (from 1866) science teacher in Bromberg (now Bydgoszcz, Poland). With the Easter semester of 1872 he became professor of descriptive geometry and graphic statics at the Technical College in Darmstadt. In 1878 he was appointed full professor at Münster, and in 1892 he accepted a similar post at Breslau, where he taught until his death.

Sturm's principal interest was in pure synthetic geometry. Following Poncelet, Steiner, and von Staudt, the practitioners of this field sought to work with very few or no formulas. At Breslau, Sturm had the good fortune to be taught by Heinrich Schroeter, who, as a student of Steiner, strongly encouraged Sturm to take up this type of geometry. Since at Darmstadt, Sturm was required to teach descriptive geometry and graphic statics, he directed his efforts to these subjects and as early as 1874 wrote *Elemente der darstellenden Geometrie* (Leipzig, 1874, 1900), a textbook on descriptive geometry for his students. Except for this book and another such textbook that he published later, *Maxima und Minima in der elementaren Geometrie* (Leipzig–Berlin, 1910), his work was devoted entirely to synthetic geometry. His first studies in this area concerned the theory of third-degree surfaces in their various projective representations. In his dissertation, "De superficiebus tertii ordinis disquisitiones geometricae," Sturm proved a number of properties of these representations that Steiner had stated without proof. In 1864 Sturm shared with Luigi Cremona the Steiner Prize of the Berlin Academy for further investigations of surfaces, all of which are collected in *Synthetische Untersuchungen über Flächen*, his first textbook on the subject.

Sturm was a prolific writer, but there is no need to mention his many journal articles individually, since he later collected almost all of them in two multivolume textbooks (*Die Lehre von den geometrischen Verwandtschaften*, I, II [Leipzig 1908], III, IV [Leipzig, 1909]) on line geometry and geometric transformations. The three-volume work on line geometry is the most extensive ever written on this specialty. Like Plücker, the author of the first systematic treatment of line geometry in algebraic form, Sturm sought to develop subsets of straight lines of P_3. Accordingly, in the first two volumes Sturm treated linear complexes, congruences, and the simplest ruled surfaces up to tetrahedral complexes, all of which can be particularly well handled in a purely geometric fashion. He did not systematically investigate the remaining quadratic complexes until volume three, where the difficulties of his approach—as compared with an algebraic treatment—place many demands on the reader. Sturm rejected as "unintuitive" the interpretation proposed in the nineteenth century by Felix Klein and C. Segre, who held that the line geometry of P_3 could be considered a point geometry of a quadric of P_5.

Sturm's *Lehre von den geometrischen Verwandtschaften*, which appeared in four volumes with more than 1,800 pages, was even larger than *Liniengeometrie*. In Sturm's use of the expression, geometric relationships encompassed, first, all collineations and correlations of projective spaces (extended to both real and complex numbers) of three dimensions at the most. The work, however, is much more than a textbook of projective geometry; it also contains many chapters on algebraic geometry, and among "geometric relationships," Sturm included correspondences, Cremona transformations, and plane projections of the simplest types of rational surfaces. Volume I deals with (1,1) relationships and also with (*a,b*) correspondences on straight lines, spheres, and the constructs generated from them. Volume II contains a description of collineations and correlations between two-step constructs; Volume III provides a similar treatment for three-step constructs; and Volume IV is devoted to Cremona transformations, several plane projections of rational surfaces, and a number of spatial correspondences. Frequently in the work Sturm touches upon questions related to Schubert's enumerative geometry, for example in the treatment of problems of plane and spatial projectivities.

In *Lehre von den geometrischen Verwandtschaften* synthetic geometry in the style of Sturm and his predecessors was developed virtually as far as it could be. During the final years of Sturm's life, mathematicians became markedly less interested in the large number of detailed geometric questions that are discussed in his writings. Consequently, although he trained many doctoral candidates in the course of his career, Sturm had no successor to continue his mathematical work.

BIBLIOGRAPHY

I. ORIGINAL WORKS. A list of Sturm's works is in Poggendorff III, 1312–1313; IV, 1462; V, 1227–1228; VI, 2576.

Sturm's major works are "De superficiebus tertii ordinis disquisitiones geometricae" (Ph.D. diss., Bratislava, 1863); *Synthetische Untersuchungen über Flächen* (Leipzig, 1867); *Elemente der darstellenden Geometrie* (Leipzig, 1874, 1900); *Die Gebilde 1. und 2. Grades der Liniengeometrie in synthetischer Behandlung*, 3 vols. (1892–1896); *Die Lehre von dem geometrischen Verwandtschaften*, I, II (Leipzig, 1908), III, IV (Leipzig, 1909); and *Maxima und Minima in der elementaren Geometrie* (Leipzig–Berlin, 1910).

II. SECONDARY LITERATURE. For works about Sturm, see W. Lorey, "Rudolf Sturm zum Gedenken," in *Zeitschrift für mathematischen und naturwissenschaftlichen Unterricht*, **50** (1919), 289–293; and W. Ludwig, "Rudolf Sturm," in *Jahresbericht der Deutschen Mathematikervereinigung*, **34** (1926), 41–51.

WERNER BURAU

STURTEVANT, ALFRED HENRY (*b*. Jacksonville, Illinois, 21 November 1891; *d*. Pasadena, California, 5 April 1970), *genetics*.

Sturtevant was the youngest of six children of Alfred Henry Sturtevant and Harriet Evelyn Morse. His grandfather Julian M. Sturtevant graduated from Yale Divinity School and was a founder and later president of Illinois College. Sturtevant's father taught mathematics for a time at Illinois College but subsequently turned to farming, first in Illinois and later in southern Alabama, where the family moved when Sturtevant was seven. His early education was in Alabama in a one-room country school, but for the last three years of high school he went to a public school in Mobile.

In the fall of 1908 Sturtevant entered Columbia University. The choice, a crucial one, was made because Sturtevant's oldest brother, Edgar, was then teaching Latin and Greek at Barnard College; Edgar and his wife made it possible for Sturtevant to attend the university by taking him into their home. Sturtevant was greatly influenced by Edgar, from whom he learned the aims and standards of scholarship and research.

As a boy Sturtevant had drawn up the pedigrees of his father's horses and of his own family. He pursued this interest as a hobby while he was at Columbia. Edgar encouraged him to read works on heredity and to learn more about the meaning of pedigrees. As a result Sturtevant read a book on Mendelism by Punnett that greatly stimulated his interest, since he saw how Mendel's principles could be used to explain the pattern of inheritance of certain coat colors in horses. Edgar suggested that Sturtevant work out the genetic relationships, write an account of his findings, and submit it to Thomas Hunt Morgan, who held the chair of experimental zoology at Columbia and from whom Sturtevant had already taken a course. Morgan clearly was impressed, since he not only encouraged Sturtevant to publish the account, which appeared in *Biological Bulletin* in 1910, but also, in the fall of that year, gave Sturtevant a desk in his laboratory, which came to be known as the "fly room." Only a few months before, Morgan had found the first white-eyed mutant in *Drosophila* and had worked out the principles of sex linkage.

After completing his doctoral work with Morgan in 1914, Sturtevant remained at Columbia as a research investigator for the Carnegie Institution of Washington. He was a member of a research team that Morgan had assembled a few years earlier and that consisted principally of two other students of Morgan's, C. B. Bridges and H. J. Muller. The "fly room" in which they conducted all of their experiments was only sixteen by twenty-three feet, and at times as many as eight people had desks in it. According to Sturtevant, the atmosphere was one of high excitement, each new idea being freely put forth and debated. Morgan, Bridges, and Sturtevant remained at Columbia until 1928; Muller left the group in 1921 to take a position at the University of Texas.

In 1922 Sturtevant married Phoebe Curtis Reed; and in the same year they made their first trip to Europe, visiting museums and laboratories in England, Norway, Sweden, and Holland. They had three children.

In 1928 Sturtevant moved to Pasadena to become professor of genetics in the new division of biology that Morgan had established in that year at the California Institute of Technology. Much of the same stimulating atmosphere and unpretentious way of conducting science that Morgan and his students had practiced at Columbia was transferred to the new Kerckhoff Laboratory at Caltech. Sturtevant became the acknowledged and natural leader of the new genetics group established there. He maintained an active research program in which he often collaborated with other members of the genetics staff, including George W. Beadle, Theodosius Dobzhansky, Sterling Emerson, and Jack Schultz. He gave lectures in the general biology course and taught elementary and advanced courses in genetics and, on occasion, a course in entomology. He remained at Caltech until his death except for a year in England and Germany in 1932, as visiting professor of the Carnegie Endowment for International Peace, and shorter periods when he held visiting professorships at a number of American universities. He received many honors, including the National Medal of Science in 1968.

In addition to his principal publications dealing with the genetics and taxonomy of *Drosophila*, Sturtevant contributed papers on the genetics of horses, fowl, mice, moths, snails, iris, and especially the evening primroses (*Oenothera*). Although his chief contributions are in genetics, he was also a leading authority on the taxonomy of several groups of Diptera, especially the genus *Drosophila*, of which he described many new species. He was much interested in the social insects and published several papers on the behavior of ants.

Sturtevant had a prodigious memory and truly encyclopedic interests. He had a natural bent for mathematics but little formal training in it. He especially enjoyed, and was expert at solving, all kinds of puzzles, especially those involving geometrical situations. For him scientific research was an exciting and rewarding activity not unlike puzzle-solving. A common theme of his investigations was an effort to analyze and explain exceptions to established principles.

Sturtevant knew how to design and execute simple, elegant experiments, describing the results in concise, lucid prose. He set high standards for his own research and expected others to do the same.

Sturtevant's discoveries of the principle of gene mapping, of the first reparable gene defect, of the principle underlying fate mapping, of the phenomena of unequal crossing-over, and of position effect were perhaps his greatest scientific achievements. The account of these and some of his other major contributions to science is arranged in approximate chronological order.

Mendel had found that all of the hereditary factors with which he worked assorted independently of one another at the time of gamete formation. Exceptions to this second Mendelian law began to accumulate in 1900–1909. Morgan was the first to provide a satisfactory explanation for such exceptions in terms of a hypothesis, which assumes that genes tending to remain together in passing from one generation to the next must be located in the same chromosome. He further postulated that the extent to which such linked genes recombine at meiosis is a relative measure of their physical distance.

Sturtevant introduced the concept that the frequency of crossing-over between two genes furnishes an index of their distance on a linear genetic map. He proposed that 1 percent of crossing-over be taken as equal to one map unit. He then reasoned that if the distance between two genes, A and B, is equal to x map units and the distance between B and a third gene, C, is equal to y map units, then the distance between A and C will be $x + y$ if B is the middle gene; $x - y$ if C is the middle gene, and $y - x$ if A is the middle gene. The germ of this idea occurred to Sturtevant in conversation with Morgan. In his *History of Genetics*, Sturtevant recorded that he "went home, and spent most of the night (to the neglect of my undergraduate homework) in producing the first chromosome map, including the linked genes, y, w, v, m, and r, in that order, and approximately the relative spacing, as they still appear on the standard maps" (p. 47).

Sturtevant devised a crucial test of the principles of mapping genes by constructing crosses in which all three genes were segregating simultaneously. In the progeny of such "three-factor" crosses, Sturtevant discovered that double crossing-over can occur and that its frequency is equal to, or less than, the product of the two single crossing-over frequencies. Conversely, the frequency of double crossing-over can be used to deduce the order of the three genes. Sturtevant showed that the order obtained from two-factor crosses was fully confirmed and that the three-factor crosses provided a more powerful method of ordering and mapping genes than did two-factor crosses. He published these findings in 1913. His principles and methods of chromosome mapping have enabled geneticists to map the chromosomes of a wide variety of higher organisms, including man.

Sturtevant was as much concerned with the role of genes in development as with the laws governing their transmission from one generation to the next. In 1915 he published an account of the sexual behavior of *Drosophila* that included a study of sexual selection based on the use of specific mutant genes that altered the eye color or body color of the fly. This work was the forerunner of an extensive line of research by others and constituted one of the first examples of the use of specific mutant genes to dissect the behavior of an organism.

One of the more conspicuous roles that genes play in development is their control of the processes of sexual differentiation. In 1919 Sturtevant reported the first case in which intersexuality could be shown to result from the presence of specific recessive genes. Years later he found a similar type of gene that resulted in the virtually complete transformation of females into males. Mutants of still other "sex genes" have been found in *Drosophila* and in many other organisms, including man. As a result, sex has come to be viewed as a complex trait controlled by a number of different genes, mutants of which can be expected to produce various grades of intersexuality.

Sturtevant pioneered in providing experimental approaches to a central problem in biology—how genes produce their effects. An important breakthrough came in 1920, with his discovery of the first reparable gene defect. In studying gynandromorphs of *Drosophila* in which there was somatic mosaicism for the vermilion eye-color mutant, he noticed that the eyes developed the dark red color of the wild type instead of the bright red color of the vermilion mutant, even when the eye could be shown to be genetically vermilion. Evidently, vermilion eye tissue lacked some substance that could be supplied by genetically nonvermilion tissue from another portion of the body. As G. W. Beadle pointed out, much of modern biochemical genetics stems directly from this early work.

Sturtevant had shown in 1913 that for each of the major chromosomes of *Drosophila* there is a corresponding linkage map. He and others had noticed, however, that excessive variation in the amount of crossing-over sometimes occurs. The factors responsible were isolated by Sturtevant and by Muller around 1915 and were shown to act as dominant cross-over suppressors. The first clue to the nature of these factors came in 1921, when Sturtevant compared the chromosome maps of *Drosophila melanogaster* with those of *D. simulans*, a closely related species that he had first described in 1919. These maps closely paralleled one another except for a region of the third chromosome, in which it appeared likely that the two species differed by an inversion in their gene sequences. It was only later that sufficient numbers of mutants were obtained in the various inversion-containing chromosomes of *D. melanogaster* for Sturtevant to establish that the dominant cross-over suppressors were indeed inversions. What had first been a disturbing exception to the generality of Sturtevant's principles of chromosome mapping became, in his hands, another demonstration of their validity. In 1935, after the discovery of the giant salivary gland chromosomes of the Diptera by Emile Heitz and H. Bauer (1933), T. S. Painter, C. B. Bridges, and others demonstrated the existence of inversions and their points of rearrangement by direct microscopic analysis. These cytological studies fully confirmed the standard and inverted sequences that Sturtevant had deduced by purely abstract genetic analysis.

In 1923 Sturtevant provided the first satisfactory explanation of the puzzling pattern of inheritance that others had found for direction of shell-coiling in snails. He showed that it was sufficient to assume a simple Mendelian gene, with dextrality being determined by the dominant allele and with the direction of coiling in the individual being determined not by its own genetic constitution but by that of its mother. He pointed out that such characters are "fundamental," in the sense that they are impressed on the egg by the action of genes in the mother. In 1946 he showed that intersexuality in a species hybrid—that of the *repleta* and *neorepleta* species of *Drosophila*—is an unusually subtle case of maternal inheritance conditioned by an autosomal dominant gene.

In the early 1920's Sturtevant and Morgan had begun a study of the unstable *Bar* mutant of *Drosophila* in order to learn more about the nature of mutations and the mechanisms by which new ones arise. It was known by then that mutations, in the sense of simply inherited changes, could take the form of changes in numbers of chromosomes (such as trisomy or polyploidy), changes involving several genes at a time (deficiencies or duplications), or changes that appeared to be within the gene (point mutations). Efficient methods for the experimental induction and detection of mutations had yet to be worked out, however. Moreover, spontaneous mutants were too rare, for the most part, to permit practical study of specific genes, except in the case of *Bar*. This small-eye mutant had already been found by C. Zeleny to mutate occasionally to either reverted *Bar*, with eyes of normal size, or *Ultra-Bar*, with eyes distinctly smaller than those of *Bar*. In 1925 Sturtevant demonstrated that these derivatives of *Bar* arise at meiosis and are associated, at the time of their origin, with an unusual type of recombination process that he termed "unequal crossing-over." He postulated that the reverted type had lost the *Bar* gene and that *Ultra-Bar* was a tandem duplication for the *Bar* gene.

After the discovery of the giant salivary gland chromosomes, it was shown by H. J. Muller and A. A. Prokofieva, and by C. B. Bridges, that *Bar* itself is a small tandem duplication of a short section of the sex chromosome of *Drosophila*. The exact nature of the unequal crossing-over process then became evident. If the chromosome containing the *Bar* mutant is symbolized as ABCBCDE · · · , where BC is a small segment that has become tandemly repeated, then, in the germ cells of an individual homozygous for such a chromosome, the leftmost BC region of one chromosome may occasionally come to pair with the rightmost BC region of the other chromosome. If a crossing-over then occurs within such unequally paired BC regions, it is evident that two new types of chromosome sequences will be produced: ABCDE · · · and ABCBCBCDE · · · . The former sequence corresponds to reverted *Bar* and the latter to *Ultra-Bar*. Thus, orthodox crossing-over within unequally paired, tandemly duplicated chromosomal segments accounts for the instability of the *Bar* mutant and provides a mechanism for progressively increasing the number of genes in a chromosome.

The process of unequal crossing-over has come to assume increasing prominence in biology as possibly one of the main forces of evolution. To illustrate, this process may have been involved in the evolution of the cluster of closely linked genes controlling the production of the β, γ, and δ polypeptide chains of the human hemoglobin molecule. The extremely close similarity of these chains at the molecular level strongly implies that the genes determining them all arose from a single ancestral gene, presumably by repeated unequal crossing-over. In turn, the resultant duplicated genes evidently diverged gradually from one another by mutation until the gene cluster acquired its present form.

Sturtevant realized that *Bar* and its derivatives provided a unique opportunity to determine whether the position of a gene in the chromosome can affect its function. He devised a critical test that consisted of comparing the sizes of the eyes of two types of flies: homozygotes for *Bar*, the genetic composition of which can be symbolized as BCBC/BCBC; and heterozygotes for reverted *Bar* and *Ultra-Bar*, of composition BC/BCBCBC. He compared the eye sizes by counting the number of facets and showed that the second type of fly had significantly fewer facets than did the first. Since the total content of genic material in the two chromosomes is the same in both cases, the observed differences in eye size constitute a demonstration that the effect of the *Bar* gene (or *Bar* region) does indeed depend upon the position of that gene in the chromosome. Sturtevant's discovery of this phenomenon of "position effect" was the first demonstration that primary genic interactions occur at the site of the genes in the chromosome, as opposed to elsewhere in the nucleus or cytoplasm.

The position effect was shown by H. J. Muller, J. Schultz, and others to take many forms. Moreover, the more primitive the organism, the more prominent (apparently) is the role played by position effect. Thus, in bacteria the chromosome consists of a series of gene clusters, or operons, that are examples par excellence of the position-effect phenomenon, in the sense that the order of the genes in an operon, as François Jacob and Jacques Monod first showed, directly determines the order in which those genes are expressed.

As is often the case in basic science, Sturtevant's discovery of the position effect of *Bar* was a by-product of another study, in this case of the mutations of *Bar*. His accounts of the quite separate phenomena of unequal crossing-over and of position effect were published in 1925 in a paper that bore the modest title "The Effects of Unequal Crossing Over at the Bar Locus in *Drosophila*."

Sturtevant was able to exploit his early use of somatic mosaics to study the developmental effects of genes when he discovered a way of producing them in large numbers. He found that females homozygous for the claret eye-color mutant of *D. simulans* produce a high proportion of gynandromorphs and other mosaics in their offspring. With the aid of this mutant he showed in 1929 that the degree of resemblance in genetic composition between two tissues in a somatic mosaic can serve to measure the degree to which those tissues have a common embryological origin. This principle, which underlies a kind of embryological-genetic mapping process now known as "fate mapping," has been widely exploited and has become a powerful tool of developmental and behavioral genetics.

In 1929 Sturtevant and S. Emerson showed that much of the extraordinarily complex genetics of the evening primrose (*Oenothera*) could be interpreted on a translocation hypothesis that had first been elaborated by John Belling for the jimsonweed, *Datura*. Many of the puzzling and bizarre "mutations" that Hugo de Vries and others had found in this organism remained disturbing thorns in the side of established genetic theory until Sturtevant and Emerson provided a detailed demonstration that they were not genuine mutations but, rather, the expected segregation products from the complex translocations of chromosome arms that are peculiar to, and widespread in, *Oenothera*.

Sturtevant and Dobzhansky collaborated in studying the plethora of inversions that occur in wild strains of many species of *Drosophila*, especially *pseudoobscura*. This work culminated in a paper (1936) that propounded an ingenious method by which inversions could be used as probes to trace phylogenetic relationships. They then successfully applied the method to constructing a detailed phylogeny of various strains and races of *pseudoobscura*.

In 1936 Sturtevant and Beadle published the results of an exhaustive study of the effects of inversions in *Drosophila* on crossing-over and disjunction. In this work they provided the first satisfactory explanation of the frequency and fate of certain aberrant chromosome types that arise as the result of crossing-over in inversion heterozygotes.

Sturtevant always maintained a keen interest in evolution and constantly examined the consequences for evolutionary theory of each new discovery in the rapidly developing science of genetics. He was an excellent naturalist and, as already noted, a taxonomist in his own right. In 1937 he published three "Essays on Evolution" in *Quarterly Review of Biology*. The first dealt with the effect of selection for mutator genes on the mutation rate of a species. The second pointed out some of the special problems of selection that are presented by the existence of sterile workers among the social insects. In the third essay he formulated a general scheme for interpreting one of the great puzzles of evolutionary theory—the origin of the sterility of hybrids.

In 1941 Sturtevant and Edward Novitski brought together the then-known mutational parallels in the genus *Drosophila*. Their results showed that the major chromosome arms of this organism tend to remain intact throughout the speciation process, although the specific order of genes within an arm gradually becomes scrambled, evidently by successive fixations of inversions.

The tiny fourth chromosome of *Drosophila* for many years resisted all efforts to map it genetically, until Sturtevant discovered special conditions that stimulated recombination to occur in that chromosome. His map of that chromosome appeared in 1951.

After 1951 Sturtevant's publications consisted mainly of original contributions to the genetics of iris; general articles on such topics as genetic effects of high-energy irradiation on human populations, the social implications of human genetics, and the theory of genetic recombination; and several taxonomic studies, including a major monograph, written with Marshall R. Wheeler, on the taxonomy of the Ephydridae (Diptera).

In 1954, in his presidential address before the Pacific Division of the American Association for the Advancement of Science, Sturtevant warned of the genetic hazards of fallout from atmospheric testing of atomic bombs. He felt that although there might be a need for bomb testing, the public should be given the best scientific estimate of the biological hazards of fallout.

Sturtevant's last published work on *Drosophila* (1956) was an account of his discovery of a remarkable mutant gene that was without any obvious effect on the fly by itself but that, in combination with another specific mutant gene (determining a prune-colored eye), killed the organism at an early stage of development. In addition to posing a challenging problem in developmental genetics, such highly specific complementary lethal systems provide an opportunity for effecting the self-destruction of certain undesirable classes of flies.

Sturtevant's last major work, *A History of Genetics* (1965), was the outgrowth of a series of lectures given at several universities and of a lifelong interest in the history of science. True to his early love of pedigrees, he presents, in an appendix to that book, detailed intellectual pedigrees of geneticists of his day.

BIBLIOGRAPHY

I. Original Works. A complete bibliography of Sturtevant's publications through 1960 can be found in the appendix to *Genetics and Evolution* (see below). A collection of thirty-three of his more important papers, reprinted in 1961 to honor Sturtevant on his seventieth birthday, includes brief annotations written by him in 1961 for several articles: *Genetics and Evolution, Selected Papers of A. H. Sturtevant*, E. B. Lewis, ed. (San Francisco, 1961).

Sturtevant was the author of *An Analysis of the Effects of Selection*, Carnegie Institution Publication no. 264 (1918); and *The North American Species of Drosophila, ibid.*, no. 301 (1921); and *A History of Genetics* (New York, 1965). His other works include *The Mechanism of Mendelian Heredity* (New York, 1915; New York, 1972), written with T. H. Morgan, H. J. Muller, and C. B. Bridges; *The Genetics of Drosophila* (Amsterdam, 1925), written with T. H. Morgan and C. B. Bridges; and *An Introduction to Genetics* (New York, 1939), written with G. W. Beadle.

Background material can be found in Sturtevant's *A History of Genetics* and his articles "Thomas Hunt Morgan," in *Biographical Memoirs. National Academy of Sciences*, **33** (1959), 283–325; and "The Early Mendelians," in *Proceedings of the American Philosophical Society*, **109** (1965), 199–204.

II. Secondary Literature. There are biographical accounts of Sturtevant by G. W. Beadle, in *Yearbook. American Philosophical Society* (1970), 166–171; and by Sterling Emerson, in *Annual Review of Genetics*, **5** (1971), 1–4. For a discussion of Sturtevant's work in Morgan's laboratory, see G. E. Allen, "Thomas Hunt Morgan," in *DSB*, IX, 515–526.

E. B. Lewis

SUBBOTIN, MIKHAIL FEDOROVICH (*b.* Ostrolenka [now Ostroleka], Lomzhinsk province, Russia [now Poland], 29 June 1893; *d.* Leningrad, U.S.S.R., 26 December 1966), *astronomy, mathematics.*

Subbotin was the son of an army officer. In 1910 he entered the mathematics section of the Faculty of Physics and Mathematics of Warsaw University, where he received the Copernicus stipend, awarded in competition for works on a subject set by the department. In 1912, while still a student, Subbotin worked as a supernumerary calculator at the university astronomical observatory; and after graduating in 1914, he was promoted to junior astronomer. The following year Subbotin was evacuated with the university to Rostov-on-Don; and from there he went to the Polytechnic Institute in Novocherkassk, where he worked until 1922, first as an assistant, then as a docent, and, finally, as professor of mathematics. His first scientific works of this period are mathematical. In 1917 Subbotin passed his master's examination at Rostov-on-Don.

In 1921 Subbotin was invited to work at the Main Russian Astrophysical Observatory, which soon became the State Astrophysical Institute, in Moscow. Ten years later this institute became part of the P. K. Sternberg Astronomical Institute. Subbotin moved to Tashkent in 1922 and became director of the Tashkent division of the State Astrophysical Institute, created on the basis of the old Tashkent observatory. In 1925 the observatory again became independent, with Subbotin as its director; and until 1930 he did much to revitalize and equip it. On his initiative the Kitab international latitude station was created.

From 1930 Subbotin directed the department of astronomy at Leningrad University. From 1935 to 1944 he was chairman of the department of celestial mechanics; from 1931 to 1934, head of the theoretical section of Pulkovo observatory; and from 1934 to 1939, head of the astronomical observatory at Leningrad University. Seriously ill and emaciated from hunger, Subbotin was evacuated in February 1942 from besieged Leningrad to Sverdlovsk, where, after treatment and convalescence, he accepted an invitation to work at the Sternberg Institute, which had been evacuated from Moscow. He traveled several times to Saratov to lecture and consult at Leningrad University, which had been evacuated there. At the end of 1942 Subbotin was named director of the Leningrad Astronomical Institute, which on his recommendation was reorganized in 1943 as the Institute of Theoretical Astronomy of the U.S.S.R. Academy of Sciences and became the main scientific institution in the Soviet Union for problems of celestial mechanics and ephemerides. On his return to Leningrad, Subbotin continued his professorial activity at the university and also taught at the Institute of Theoretical Astronomy.

From 1928 Subbotin was a member of the International Astronomical Union and, from 1933, president of the Commission on Theoretical Astronomy of the Astronomical Council of the U.S.S.R. Academy of Sciences. In 1946 he was elected corresponding member of the Academy of Sciences of the U.S.S.R. In 1963 he was awarded the Order of Lenin.

Subbotin's first scientific work was devoted to the theory of functions and the theory of probability. Several early articles deal with astrometry, particularly the creation of a catalog of faint stars. Later, however, his attention was devoted entirely to celestial mechanics and theoretical astronomy and to related areas of mathematics. He also wrote valuable works in the history of astronomy.

Subbotin began research on celestial mechanics by dealing with the theory of unperturbed motion. His new and original method of computing elliptical orbits from three observations was based on the solution of the Euler-Lambert equation. The solution of the modified equation yielded a semimajor axis, and then the remaining orbital elements were found. A number of Subbotin's works were devoted to the improvement of orbits on the basis of extensive observations. The last of these works included calculations destined to be carried out by electronic computers. In other writings Subbotin not only showed the possibility of improving the convergence of the trigonometric series by which the behavior of perturbing forces is represented, but also gave an expression for determining Laplace coefficients and presented formulas for computing the coefficients of the necessary members of the trigonometric series.

Subbotin also proposed a new, two-parameter form of equation of the Kepler ellipse, the various values of which lead to a number of anomalies, including one that changes with time more uniformly than the true and eccentric anomalies. This greatly simplified the computational integration of the equation of motion, which was particularly important for comets having large orbital eccentricities.

Subbotin's important three-volume course in celestial mechanics embraced all the basic problems of this science: unperturbed movement, the theory of perturbation and lunar theory, and the theory of figures of celestial bodies.

BIBLIOGRAPHY

I. ORIGINAL WORKS. Subbotin's writings include "Ob opredelenii osobykh tochek analiticheskikh funktsy" ("On the Determination of Singular Points of Analytic Functions"), in *Matematicheski sbornik*, **30** (1916), 402–433; "Sur les points singuliers de certaines équations différentielles," in *Bulletin des sciences mathématiques*, 2nd ser., **40**, no. 1 (1916), 339–344, 350–355; "O forme koeffitsientov stepennykh razlozheny algebraicheskikh funktsy" ("On the Form of Coefficients of Exponential Expansion of Algebraic Functions"), in *Izvestiya Donskogo politekhnicheskogo instituta, Novocherkassk*, **7** (1919); "Determination of the Elements of the Orbit of a Planet or Comet by Means of the Variation of Two Geocentric Distances," in *Monthly Notices of the Royal Astronomical Society*, **82** (1922), 383–390; "On the Law of Frequency of Error," in *Matematicheski sbornik*, **31** (1923), 296–300, in English; "On the Solar Rotation Period From Greenwich Sunspot Measures 1886–1909," in *Astronomische Nachrichten*, **218** (1923), 5–12; "Novaya forma uravnenia Eylera-Lamberta i ee primenenie pri vychislenii orbit" ("New Form of the Euler-Lambert Equation and Its Application in the Calculation of Orbits"), in *Russkii astronomicheskii zhurnal*, **1**, no. 1 (1924), 1–28; "A Proposal for a New Method of Improving the Fundamental Starplaces and for Determining the Constant of Aberration," in *Astronomische Nachrichten*, **224** (1925), 163–172; and "Proper Motions of 1186 Stars of the Cluster NGC 7654 (M52) and the Surrounding Region (First Catalogue)," in *Trudy Tashkentskogo gosudarstvennogo universiteta*, 5th ser. (1927), no. 1, 3–32.

Later works are "Sur les propriétés-limites du module des fonctions entières d'ordre fini," in *Mathematische Annalen*, **104** (1931), 377–386; "O chislennom integrirovanii differentsialnykh uravneny" ("On the Numerical

Integration of Differential Equations"), in *Izvestiya Akademii nauk SSSR*, Otd. matem. i estest. nauk, 7th ser. (1933), no. 7, 895–902; *Nebesnaya mekhanika* ("Celestial Mechanics"), 3 vols. (Leningrad–Moscow, 1933–1949; I, repr. 1941), also *Prilozhenie: Vspomogatelnye tablitsy dlya vychisleny orbit i efemerid* ("Appendix: Supplementary Tables for Computation of Orbits and Ephemerides"; 1941); "O novoy anomalii, zaklyuchayushchey kak chastnye sluchai ekstsentricheskuyu, istinnuyu i tangentsialnuyu anomalii" ("On a New Anomaly, Including Particular Cases of Eccentric, True, and Tangential Anomalies"), in *Doklady Akademii nauk SSSR*, n.s. **4**, no. 4 (1936), 167–169, also in *Trudy Astronomicheskoi observatorii Leningradskogo gosudarstvennogo universiteta*, **7** (1937), 9–20; "Astronomicheskie raboty Lagranzha" ("Astronomical Work of Lagrange"), in A. N. Krylov, ed., *Zhozef Lui Lagranzh (1736–1813). K 200-letiyu so dnya rozhdenia* (". . . Lagrange. . . . On the 200th Anniversary of His Birth"; Moscow–Leningrad, 1937), 47–84; and "Nekotorye soobrazhenia po voprosu o postroenii fundamentalnogo kataloga" ("Some Considerations on the Question of the Structure of the Fundamental Catalog"), in *Astronomicheskii zhurnal*, **14**, no. 3 (1937), 228–245.

Other works are *Mnogoznachnye tablitsy logarifmov* ("Multidigit Tables of Logarithms"; Moscow–Leningrad, 1940); "O nekotorykh svoystvakh dvizhenia v zadache *n*-tel" ("On Certain Properties of Motion in the *n*-Body Problem"), in *Doklady Akademii nauk SSSR*, **27**, no. 5 (1940), 441–443; "K voprosu ob orientirovke fundamentalnogo kataloga slabykh svezd" ("Toward the Question of the Orientation of the Fundamental Catalog of Faint Stars"), in *Uchenye zapiski Kazanskogo gosudarstvennogo universiteta*, **100**, no. 4 (1940), 138–141; "Sur le calcul des inégalités séculaires. I. Solution nouvelle du problème de Gauss," in *Journal astronomique de l'URSS*, **18**, no. 1 (1941), 35–50; "Ob odnom sposobe uluchshenia skhodimosti trigonometricheskikh ryadov, imeyushchikh osnovnoe znachenie dlya nebesnoy mekhaniki" ("On One Method of Improving the Convergence of Trigonometric Series of Basic Importance for Celestial Mechanics"), in *Doklady Akademii nauk SSSR*, **40**, no. 8 (1943), 343–347; *Proiskhozhdenie i vozrast Zemli* ("Origin and Age of the Earth"; Moscow, 1945; 1947; 1950); and "Uluchshenie skhodimosti osnovnykh razlozheny teorii vozmushchennogo dvizhenia" ("Improvement in the Convergence of the Basic Expansions of the Theory of Perturbed Motion"), in *Byulleten Instituta teoreticheskoi astronomii*, **4**, no. 1 (1947), 1–16.

Writings from late in Subbotin's life are "Differentsialnoe ispravlenie orbity s ekstsentrisitetom, malo otlichayushchimsya ot edinitsy" ("Differential Correction of the Orbit With an Eccentricity Slightly Different From Unity"), in *Byulleten Instituta teoreticheskoi astronomii*, **7**, no. 6 (1959), 407–415; "O vychislenii parabolicheskikh orbit" ("On the Calculation of Parabolic Orbits"), *ibid.*, 416–419; "Raboty Mukhammeda Nasireddina po teorii dvizhenia solntsa i planet" ("Work of

Muhammed Nasīr al-Dīn on the Theory of Motion of the Sun and Planets"), in *Izvestiya Akademii nauk Azerbaidzhanskoi SSR* (1952), no. 5, 51–58; "Astronomicheskie i geodezicheskie raboty Gaussa" ("Astronomical and Geodesical Works of Gauss"), in *100 let so dnya smerti (1855–1955)* ("100th Anniversary of His Death . . ."; Moscow, 1956), 241–310; "Raboty Anri Paunkare v oblasti nebesnoy mekhaniki" ("Works of Henri Poincaré in the Area of Celestial Mechanics"), in *Voprosy istorii estestvoznaniya i tekhniki* (1956), no. 2, 114–123; and "Astronomicheskie raboty Leonarda Eylera" ("Astronomical Works of Leonhard Euler"), in *Leonard Eyler. K 250-letiyu so dnya rozhdenia* (". . . Euler. On the 250th Anniversary of His Birth"; Moscow, 1958), 268–376.

II. Secondary Literature. See G. A. Merman, "Ocherk matematicheskikh rabot Mikhaila Fedorovicha Subbotina" ("Sketch of the Mathematical Works of . . . Subbotin"), in *Byulleten Instituta teoreticheskoi astronomii*, **7**, no. 3 (1959), 233–255, with a bibliography; and N. S. Yakhontova, "Mikhail Fedorovich Subbotin (k 70-letiyu so dnya rozhdenia)" (". . . Subbotin [on the 70th Anniversary of His Birth]"), *ibid.*, **10**, no. 1 (1965), 2–5.

P. G. Kulikovsky

SUCHTEN (or **Zuchta**), **ALEXANDER** (*b.* Tczew [?], Poland, *ca.* 1520; *d.* Bavaria, 1590 [?]), *chemistry, medicine.*

Suchten's father, George Suchten, was an assessor for the Gdańsk town court; his mother was Eufemia Schultze. The Suchtens were an important family, possessing houses in Gdańsk and an estate near Tczew, where Alexander was probably born. An uncle, Christopher Suchten, was a secretary to King Sigismund I Jagiello; a grandfather and another uncle had been mayors of Gdańsk. In 1521–1522 an "Alexius Zuchta de Gedano alias etiam Suchten, dictus Kaszuba Polonus" lectured at the University of Cracow, where he held the post of *Extraneus de Facultate.* This academic was certainly a cousin of Alexander Suchten, who pursued his own studies in the years 1535–1539 in Elblag.

Suchten's fortunes were greatly influenced by his maternal uncle, Alexander Schultze (or, in Latin, Sculteti), a canon of Frombork and a friend of Copernicus, who resigned his canonry, with consent from Rome, in favor of his nephew. Since (according to a decree of the bishop of Warmia, Joannes Dantiscus) the higher clerical positions could be filled only by those who had studied at a foreign university for three years and had taken a doctorate, Suchten went to Louvain, where he studied medicine, and then to Rome, Ferrara, Bologna, and Padua. In Padua he received the doctorate with a dissertation entitled "Galeni placita." During this time, Suchten's uncle, a follower of Heinrich Bullinger, was accused of heresy and his estate was confiscated. Suchten became involved in his uncle's trial and in 1545 was deprived not only of his canonicate but also of his paternal inheritance. He then went to Königsberg and was poet and physician at the court of Duke Albrecht of Prussia.

In 1549 Suchten went to the Rhineland, where he was both physician and librarian to the elector of the Palatinate, Ottheinrich, a noted bibliophile and collector of alchemical books. While there Suchten became acquainted with the treatises of Paracelsus. He also became a friend of Michael Schütz, known as Toxites, who shared Suchten's interest in poetry, medicine, and alchemy. In 1554 he returned to Poland and was named physician to Sigismund II Augustus. Despite his influential position, he was unable to recover his property.

About 1564 Suchten wrote his two treatises *De tribus facultatibus* and *Decem et octo propositiones*, which created a storm of protest in the medical world. In them, Suchten stated that doctors of medicine who had obtained degrees from the universities of Bologna, Padua, Ferrara, Paris, Louvain, and Wittenberg were nothing but common frauds, and himself endorsed the medical views of Paracelsus. It is not known whether these tracts were manuscripts or whether they were also printed, since they survive only in later editions. At any rate, they were severely criticized by a number of famous physicians, including Konrad Gesner, Thomas Erastus, Crato von Krafftheim, Lukas Steglin, and Achilles Gasser, whose indignation must have come to the attention of Sigismund II, since Suchten was dismissed from his post.

Suchten went for a short time to Königsberg, as physician to Duke Albrecht, then to the court of the German magnate Johann von Seebach in Bavaria. In 1570 he was in Strasbourg, where he published his *Liber unus de secretis antimonii*. During his stay in the Palatinate, Suchten married; he eventually settled somewhere in Bavaria.

Suchten was a distinguished Paracelsian, who dedicated himself to attacking deceit and charlatanism in medicine. He also wrote on the history of chemistry, perhaps the first scholar to do so, and demonstrated, with the aid of scales, that the transmutation of metals into gold is impossible, so that all claims for "successful" transmutations must be fraudulent. He had a considerable reputation as a Latin poet, and his *Dialogus* seems to be autobiographical. A number of his manuscripts were published only after his death.

BIBLIOGRAPHY

I. ORIGINAL WORKS. A complete bibliography of Suchten's published works is in W. Haberling (see below). His works include *Liber unus de secretis antimonii* (Strasbourg, 1570), with English trans. as *Alex. Van Suchten, of the Secrets of Antimony* (London, 1670); *Clavis alchimiae* (Mümpelgardt, 1604); *Dialogus de hydrope* (Mümpelgardt, 1604); *Concordantia chimica* (Mulhouse, 1606); *Colloquia chemica* (Mulhouse, 1606); *De tribus facultatibus explicatio tincturae Theophrasti Paracelsi, de vera medicina in Benedictus Figulus's Pandora* (Strasbourg, 1608) and *Opera omnia*, U. von Dagitza, ed. (Hamburg, 1680). "De lapide philosophorum," a verse in Latin against the possibility of the transmutation of metals into gold, is in Michael Toxites' *Raymundi lullii . . . vade mecum* (Basel, 1572). Suchten's greatest poetical work, on the mythical Polish princess Wanda, is in his *Vandalus* (Königsberg, 1547).

II. SECONDARY LITERATURE. See Wilhelm Haberling, "Alexander von Suchten . . .," in *Zeitschrift des Westpreussischen Geschichtsvereins*, **69** (1926), 177–230, with bibliography; Włodzimierz Hubicki, "Doktor Aleksander Zuchta. Zapomniany polski chemik, lekarz i poeta XVI wieku," in *Studia i materiały z dziejów nauki polskiej*, **1** (1953), 102–120; "Alexander von Suchten," in *Sudhoffs Archiv für Geschichte der Medizin un der Naturwissenschaften*, **44** (1960), 54–63; and "Chemistry and Alchemy in Sixteenth Century Cracow," in *Endeavour*, **17** (1958), 204–207. Earlier literature is cited by John Ferguson in *Bibliotheca chemica*, II (Glasgow, 1906), 417.

WŁODZIMIERZ HUBICKI

SUDHOFF, KARL FRIEDRICH JAKOB (*b*. Frankfurt am Main, Germany, 26 November 1853; *d*. Salzwedel, Germany, 8 October 1938), *history of medicine*.

Sudhoff, the son of a Protestant minister, attended the elementary school and Gymnasium in Frankfurt until his family moved to Zweibrücken and then to Kreuznach, where he finished his secondary education. He studied medicine at the universities of Erlangen, Tübingen, and Berlin, taking the M.D. in 1875, then did postgraduate work in the hospitals of Augsburg and, for a short time, Vienna. In 1878 he established a general practice in Bergen, near Frankfurt; in 1883 he moved to

Hochdahl, near Düsseldorf, where he practiced until 1905.

During the twenty-seven years of his medical practice Sudhoff devoted his spare time to the study of the history of medicine. His own first major contribution to the field was his two-volume *Bibliographia Paracelsica* (1894–1899), an indispensable guide to the Paracelsian printed and manuscript source material. This work brought him some fame, and Sudhoff in 1901 became instrumental in founding the German Society for the History of Medicine and Science. In 1905 he was offered the chair in the history of medicine at the University of Leipzig that had been endowed by the widow of Theodor Puschmann, who had left her entire fortune to that university for the promotion of the study of medical history. Sudhoff left Hochdahl for Leipzig in the same year, and began to develop the first German department for the history of medicine. He further contributed to this field by editing a series of important periodicals, including the *Mitteilungen zur Geschichte der Medizin und der Naturwissenschaften* (from 1902), the *Studien zur Geschichte der Medizin* (from 1907), and *Sudhoffs Klassiker der Medizin* (from 1910). The *Archiv für Geschichte der Medizin und der Naturwissenschaften* (later called *Sudhoffs Archiv*), of which he was founder and to which he contributed heavily, first appeared in 1907.

Sudhoff's own historical researches were concerned chiefly with the fields of ancient, medieval, and Renaissance medicine and with epidemiology. In 1909 he published *Aerztliches aus griechischen Papyrus-Urkunden*, in which he considered ancient materials concerning water supply, housing conditions, clothing, gymnastics, and cosmetics, as well as medical topics; the following year he brought out *Aus dem antiken Badewesen*, a detailed discussion of ancient hygienic practices. He also traveled widely throughout Europe to examine medieval manuscripts, which he photographed and edited; he was particularly concerned with medical and anatomical iconography, and traced the ramifications of the iconographical tradition back to the early Middle Ages. He edited a large number of medieval texts on medicine, surgery, dietetics, and anatomy, and demonstrated the significance of the Salernitan school as the center of medical lore and training in the Latin West. In Renaissance studies, he returned to the works of Paracelsus in 1922 and began the critical edition that he finished eleven years later; he also published facsimile editions of the anatomical tables of Vesalius and Jost de Neg-

ker. In epidemiology, Sudhoff edited important source materials on the early history of syphilis and the plague.

Sudhoff also wrote two textbooks, an innovative history of dentistry and a medical history that continued Pagel's introduction to the subject. But although his publications and editions were vastly consequential, Sudhoff's chief contribution to science lies in his espousal of a strict historical method, based upon an objective and thorough study of original sources, to which he dedicated both himself and his students. His institute at the University of Leipzig was a center of medico-historical research and served as a model for other such departments both in Europe and elsewhere. Upon his retirement from the university in 1925, he was succeeded by Henry Sigerist, but when Sigerist went to the Johns Hopkins University in 1932 Sudhoff returned as acting director, a post in which he remained until 1934.

BIBLIOGRAPHY

I. ORIGINAL WORKS. A complete list of Sudhoff's writings, compiled by G. Herbrand-Hochmuth, is "Systematisches Verzeichnis der Arbeiten Karl Sudhoffs," in *Sudhoffs Archiv für Geschichte der Medizin und der Naturwissenschaften*, **27** (1934), 131–186; with additions, *ibid.*, **31** (1938), 343–344, and **32** (1939), 279–284.

Of his individual works, see especially *Bibliographia Paracelsica*, 2 vols. (Berlin, 1894–1899); *Kurzes Handbuch der Geschichte der Medizin* (Berlin, 1922), which was a new ed. of J. L. Pagel, *Einführung in die Geschichte der Medizin; Iatromathematiker vornehmlich im 15. und 16. Jahrhundert* (Breslau, 1902); *Tradition und Naturbeobachtung in den Illustrationen medizinischer Handschriften und Frühdrucke vornehmlich des 15. Jahrhunderts* (Leipzig, 1907); *Ein Beitrag zur Geschichte der Anatomie im Mittelalter speziell der anatomischen Graphik und Handschriften des 9. bis 15. Jahrhunderts* (Leipzig, 1908); *Ärztliches aus griechischen Papyrus-Urkunden* (Leipzig, 1909); *Aus dem antiken Badewesen* (Berlin, 1910); *Aus der Frügeschichte der Syphilis* (Leipzig, 1912); *Beiträge zur Geschichte der Chirurgie im Mittelalter*, 2 vols. (Leipzig, 1914–1918); *Des Andreas Vesalius sechs anatomische Tafeln vom Jahre 1538* (Leipzig, 1920), written with M. Geisberg; *Geschichte der Zahnheilkunde* (Leipzig, 1921); *Zehn Syphilisdrucke aus den Jahren 1495–1498*, Monumenta medica, 3 (Milan, 1924); *Erstlinge der pädiatrischen Literatur* (Munich, 1925); *Die ersten gedruckten Pestschriften* (Munich, 1926), written with A. Klebs; and *Die anatomischen Tafeln des Jost de Negker*, 1539 (Munich, 1928), written with M. Geisberg. Sudhoff's critical edition of Paracelsus is *Theophrast von Hohen-*

heim . . . Medizinische, naturwissenschaftliche und philosophische Schriften, 14 vols. (Munich-Berlin, 1922–1933).

A selection of Sudhoff's work, compiled by H. E. Sigerist, is "Ausgewählte Abhandlungen zum 75. Geburtstage," in *Sudhoffs Archiv für Geschichte der Medizin und der Naturwissenschaften*, **21** (1929), 1–332; an autobiographical note is "Aus meiner Arbeit. Eine Rückschau," *ibid.*, 333–387.

II. Secondary Literature. Poggendorff, VIIa, 4, 602, gives a comprehensive list of biographical and bibliographical materials concerning Sudhoff and his work. See also W. Artelt, "Karl Sudhoff," in *Janus*, **43** (1939), 84–91; P. Diepgen, "Zur hundertsten Wiederkehr des Geburtstages von Karl Sudhoff am 26. November 1953," in *Archives internationales d'histoire des sciences*, **6** (1953), 260–265; and "Leben und Wirken eines grossen Meisters," in *Wissenschaftliche Zeitschrift der Karl Marx Universität, Leipzig*, **5** (1955–1956), 23–25; F. H. Garrison, "Karl Sudhoff as Editor and Bibliographer," in *Bulletin of the Institute of the History of Medicine*, **2** (1934), 7–9; J. R. Oliver, "Karl Sudhoff as a Classical Philologian," *ibid.*, 10–15; H. E. Sigerist, "Karl Sudhoff, the Man and the Historian," *ibid.*, 3–6; and "Karl Sudhoff, the Mediaevalist," *ibid.*, 22–25; and O. Temkin, "Karl Sudhoff, the Rediscoverer of Paracelsus," *ibid.*, 16–21.

Nikolaus Mani

SUESS, EDUARD (*b.* London, England, 20 August 1831; *d.* Marz, Burgenland, Austria, 26 April 1914), *geology.*

Although the name Suess had been known in Vienna since the fifteenth century, Eduard's family came from Vogtland, a region surrounded by Bohemia, Saxony, and Bavaria. His father was the son of a minister at Bobenneukirchen, near the Bavarian frontier; his mother, Eleonore Zaekauer, was the daughter of a Prague banker. Suess's father established a wool business in London, where Eduard, the couple's second child, was born. The family left England in 1834 and traveled from the coast of the Netherlands to Prague by horse and carriage. As a child Eduard spoke English, and in order to preserve this knowledge he had an English instructor as well as German and French tutors. He therefore was able to enter the Gymnasium at an exceptionally early age.

In 1845 Suess's father took over a leather factory, located near Vienna, which belonged to an ailing brother-in-law. During the revolutionary disturbances of 1848, Suess participated in demonstrations and joined the Légion Académique. He was elected to its committee and represented it on the Committee of Public Safety (*Sicherheits-ausschuss*), of which he was the youngest member. His frequent participation in arbitration enabled him to learn to make decisions, assume responsibility for them, and justify them.

Treatment for an abscess on his foot obliged Suess to go to Prague, where he lived in his grandparents' house. He thus escaped the upheaval in Vienna that put an end to the Metternich era.

While continuing his university studies at Prague, Suess also frequented the museums and made geological excursions to nearby fossil-rich areas. The samples he collected so stimulated his interest that he devoted himself increasingly to geology. When he returned to Vienna, he presented a manuscript on the graptolites to the Society of the Friends of Science, of which the leading member was Wilhelm Haidinger. This scientific work, Suess's first, involved him in a minor dispute with the paleontologist Joachim Barrande; but their difference of opinion ended in friendship. Suess took charge of the society's foreign correspondence and a portion of that of the Museum of Natural History. This work brought him valuable international contact.

A liver ailment compelled Suess to take a cure at Carlsbad (now Karlovy Vary). The region spurred him to analyze its differences from the countrysides around Prague and Vienna. (Comparisons of this sort long held Suess's interest.) He studied the granites and sketched the shapes of columns. As a result he was invited to write the chapter on geology for a tourist guide, his first published work.

Suess was imprisoned on 16 December 1850, probably as a result of a denunciation, but was released at the request of Haidinger, director of the Geological Survey. He then found regular studies at the Technical University to be impossible and devoted himself instead to geology at the Geological Survey and at the Imperial Mineralogical Collection, where he was assigned to classify the brachiopods and was appointed a paid assistant on 10 May 1852. During the summer he was in charge of surveys for the Semmering Tunnel and in the Mur Valley.

In 1853 Franz von Hauer undertook to make a geological profile across the Eastern Alps from Passau to Duino. Suess asked to be made responsible for the highest section, that of the Dachstein. With a guide he reached the summit, where he enjoyed a vast panorama which he compared with the countrysides around Vienna, Prague, and Carlsbad; the differences were puzzling, and he decided to devote himself to elucidation of this problem.

On 12 June 1855 Suess married Hermine Strauss, the niece of the director of the Museum of Natural History and the daughter of a prominent Viennese physician. Fearing that he would not be able to support a family, his father agreed to the marriage on the condition that his son work at the leather factory in the afternoon, while the mornings were devoted to the classification of brachiopods and alpine rocks. In 1856 Suess applied for the post of *Privatdozent* at the University of Vienna but, lacking the doctorate, he failed to qualify. At the recommendation of Haidinger, the minister of education circumvented this difficulty by naming Suess extraordinary professor of paleontology, thereby freeing him from work at the factory. He became full professor of geology five years later.

Looking back on his life, Suess realized that these youthful years had been decisive. The experiences and impressions determined the goals he was to pursue and the means by which he attained them. The outlook of the high bourgeoisie in a multinational empire favored intellectual life and research. A classical and technical education and the consciously exercised gift of presenting and explaining his ideas so as to make them comprehensible had prepared Suess for a manifold career. His knowledge of languages early brought him wide-ranging personal and literary associations. He made geological excursions with researchers from Switzerland (from 1854), France (from 1856), Germany (from 1856), England (from 1862), and Italy (from 1867). Suess thus had the opportunity to visit important regions in the company of those who had explored them, and to discuss his findings and make comparisons. At an advanced age he learned Russian in order to be able to consult the original literature.

Suess took an early interest in education, at first promoting the popular university (from 1855), and then studying conditions at all levels of the educational system. As a result of his work in education, Suess was elected to the Vienna city council (1863–1886), to the provincial diet (1869–1896), and to Parliament (1872–1896). He was even approached to become secretary of state for education, but he preferred to remain a professor of geology. His vast knowledge enabled him to study conditions in other countries, to compare them with those in the empire, and to propose solutions. Investigating the conditions of workers in Great Britain, he offered predictions about the future status of labor in Austria and attempted to prepare the way for change. His recommendations influenced the passage of social legislation and the appointment of labor inspectors (1873).

On 20 December 1857 the dismantling of the bastions and fortifications of the old city of Vienna was begun, to make way for new neighborhoods and official buildings. Using the exposures presented by the excavations, Suess studied the city's subsoil and in 1862 published *Der Boden der Stadt Wien.*

Two difficult projects brought Suess great fame and the gratitude of his fellow citizens: the diversion of drinking water from the mountains to Vienna and the Danube Canal.

A great part of Vienna was supplied with water from wells. Epidemics, particularly of typhoid fever, were frequent and claimed numerous victims. In 1863 Suess was elected to the city council and was named head of a commission to study the water supply. Of fifty-three competing projects, Suess's proposal, which was one of the boldest, was chosen: it called for bringing water from mountain springs by an aqueduct. Three reasons determined its selection: mountain regions received more precipitation than the plains, they were sparsely populated, and the water could flow by gravity. Suitable springs existed in the Alps only seventy miles from Vienna, and their owner was ready to sell them. After convincing the taxpayers and their representatives, it was necessary to plan and direct the construction of the aqueduct, which began operation on 17 October 1873. The number of deaths from typhoid fever subsequently dropped from 34 to 9 per 1,000.

The low-lying sections of Vienna were often flooded. In 1869 Suess was appointed to a commission for the control of the Danube; and although he was an educational inspector busily engaged in reorganizing public instruction, he accepted. In November of that year he was a member of a delegation accompanying the emperor to Egypt to attend the opening of the Suez Canal. Noting the relevance of the canal to his own project, Suess discussed the technical problems of dredging with the head of these operations. He hired this official to come to Vienna in 1870, with his crew and machinery, to excavate the bed of the Danube Canal. Many difficulties arose, some unpredictable, such as the variations in the river's flow. On 15 April 1875 the canal was opened, and there have been no major floods in Vienna since 1876.

After the Universal Exposition of 1873 a chain of financial failures and misfortunes began. Suess attempted to clarify the phenomena by linking political, economic, and geological problems in a new fashion. He also studied the problems associated with gold (1877) and silver (1892). His book on silver, translated into English at the initiative of the United States government, was published in America.

Suess also played an important role in what is now called the politics of science. In 1860 he was named a corresponding member of the Imperial Academy and a full member in 1867; he was elected vice-president in 1893 and president in 1898. He resigned in 1911, having reorganized the Academy into a more effective body for scientific research. Skilled in obtaining funds for important projects, Suess arranged archaeological expeditions to Egypt and oceanographic cruises in the Mediterranean and the Red Sea. He organized the union of German-language academies that later became the Union Mondiale des Académies; and he was able to orient and direct their collaboration toward great common goals, such as the *Thesaurus linguae latinae.*

Suess likewise wielded great influence at the University of Vienna. He was occupied both with the organization of the curriculum and the location and construction of the new university. Although Suess was a Protestant, his election as rector was supported by the votes of the Catholic faculty. Yet the greater part of his political and technical activities gradually yielded to geology and teaching, and Suess declined very attractive offers in order to devote himself to what he considered his principal responsibility. After reaching retirement age, he gave all his time to his fundamental work.

Suess emphasized the joy derived from contact with nature, considering it a relic of the savage state that had passed through the filter of civilization. It is, he believed, only in nature that the geologist can grasp the scale of the mountains and can immerse himself in their disposition, that he learns to read the explanation of their structure and of its development.

Suess knew most of the European geologists of the older generation—that of Élie de Beaumont, Haidinger, and Escher—as well as those of his own and the following one. He often experienced the strength of scientific ties, even across political frontiers. His friendship with the Italian geologist Quintino Sella (who also was the first minister of finance of the kingdom of Italy) was based on their common views on both science and the conduct of affairs of state.

Each summer was dedicated to geological expe-

ditions. In autumn Suess often participated in meetings and geological excursions with foreign colleagues in the Swiss Alps, Italy, and other countries. All his field excursions were carefully planned and formed part of a flexible research program. Suess also emphasized that his work on cataloging the mineralogical collections and in the library of the Museum of Natural History had enabled him to become familiar with groups of fossil forms from different strata and with the literature. Indeed, his knowledge of the geological literature and his gift for locating information in obscure publications was highly impressive.

Suess began his scientific career as a paleontologist. Graptolites, brachiopods, ammonites, and the mammals of the Tertiary especially attracted his attention. He studied their anatomy and compared their modes of life with those of existing species, engaging in what is now called paleobiology. Emphasizing the unity of the living world, in 1875 he created the concept of the biosphere. Earlier, in 1860, Suess had attempted to use the Paleozoic brachiopods to determine the depth of water at the time of sedimentation. He enlarged the goals of Lyell's stratigraphy by comparing series deposited in different regions, on forelands and in mountain chains, and by distinguishing epicontinental and geosynclinal series. Accordingly, his studies gradually shifted from stratigraphy to tectonics, of which he must be considered a creator.

Suess received no formal training in paleontology or geology. An autodidact, he increased his knowledge through contact with his seniors, friends, and students. In 1865 he was commissioned to write a treatise on the geology of the Austrian Empire, for which project he obtained a leave of absence and funds to travel. He undertook long tours on foot. The comparative method, so advantageous in anatomy, appeared of increasing importance to him. After observing asymmetry in the Carpathians, the Sudetes, and the Apennines, he located it again in the Alps. The initial result of these peregrinations was *Die Entstehung der Alpen* (1875), which contained many of his governing ideas in embryonic form. Rather than a geological description of the Alps, it was an overall view of the genesis and structure of mountain chains.

During this period the ideas of Leopold von Buch and Élie de Beaumont still largely determined the image of mountain chains. For Buch, eruptive rocks were the active elements in mountain chains; they raised the central zones and exerted lateral pressures against the two sides, thereby causing folds and other dislocations. Élie de Beaumont determined the folds of various ages and made a first attempt at a global tectonics. Lack of information obliged him to make extrapolations, such as his proposed system of the pentagonal dodecahedron, the directions of which supposedly corresponded to the ages of the folds.

Suess accepted neither the idea of the raising of the central zone by crystalline rocks nor the notion of symmetry with respect to that zone. He distinguished a foreland and a hinterland, between which movements were tangential and unilateral. There were, in his view, no rectilinear directions, but only an ensemble of curved lines. The mountain chains formed garlands and festoons, and their convex side was usually directed toward the exterior. The curves and the disposition of the folds depended on the form of the foreland and on the resistance of the materials. The crystalline rocks of the central zones were passive.

One of the first tasks of the geologist was to trace the contours of the festoons and garlands and to ascertain their internal organization. A hierarchy of features was thus discerned, enabling him to grasp the geometric anatomy and to locate its details. Next he had to reconstruct the movements that had caused these structures, in order to visualize the youth, maturity, and old age of the chain.

This work marked the beginning of the extraordinary development of tectonics and made possible the work of Marcel Bertrand, Pierre Termier, Hans Schardt, Maurice Lugeon, and Emile Argand. The point was stressed by Bertrand in his preface to the French translation of *Das Antlitz der Erde* when he gave an example of this fruitfulness that has become classic. In 1846 Arnold Escher recognized the great extent of the Permian overthrust on the Tertiary of the Glarus Alps. His student and successor Albert Heim extended the study and in 1870 constructed the "Glarus double fold" (*Glarner Doppelfalte*), thrust from both north and south, a phenomenon that aroused a heated polemic throughout Europe and pitted the partisans of overthrusts against those who interpreted the phenomenon as a stratigraphic superposition. Basing his argument on the principle of asymmetry, Bertrand interpreted the ensemble as a single nappe with a root lying to the south and a head plunging toward the north, a view incorporated into Lugeon's synthesis of the Northern Alps (1902). (This was the origin of the modern concept of the nappe or decke.)

Studies conducted in the Alps and in Calabria convinced Suess that earthquakes are manifesta-

tions of mountain chains in motion and are produced along great faults. Volcanoes and intrusions are only accessory phenomena, most often occurring at the interior of arcs.

The concept of the geosyncline, elaborated in the United States by James Hall and James Dana, had been introduced to Europe. Suess stressed not only the difference in thicknesses, but also the diverse evolution of the stratigraphic series of the forelands and of the folded and overthrust zones from the point of view of lithology and of the facies. This was the first rational classification of the features of terrestrial physiognomy and an attempt to interpret and synthesize them that Suess later took up in greater detail.

In 1883 Suess signed a publication contract for his masterpiece, *Das Antlitz der Erde*, scheduled to appear in three volumes. The first fascicle was published the same year, and the entire work was completed in 1909. Certain of the notions, principles, and ways of reasoning presented in the work have entered so profoundly into the thinking of geologists that many are unaware of their origin and consider them archetypes. It is therefore worth mentioning a few examples.

Suess avoided dogmatism in presenting his views. He sought to assimilate and coordinate a century of observations, attempting to extract order from chaos by means of several governing ideas. Field experience since his youth developed in Suess a profound grasp of the relations between major outlines and details, between the structures and the behavior of materials, and between assemblages of facies and their position in the whole. He thus was able to assimilate observations in the literature, most of which were made from different viewpoints. According to his contemporaries, Suess knew the literature better than any other scientist of his time and was familiar with sources that were difficult to obtain even in major libraries. The result was a three-dimensional picture on which he superposed changes through time.

Because *Das Antlitz der Erde* was the first presentation of global views of this type, it was necessary for Suess to create an appropriate language; and his book therefore contains a considerable number of new terms, many of which are still in use.

The principal value of the "great mosaic" method consists of its manner of observing, of representing, and of transmitting—in a language developed especially for the purpose—the ensemble that is the face of the earth, a multidimensional ensemble of forms and contours arranged hierarchically on different scales. The principal features emerged from a host of details, communicating the author's enthusiasm to the reader. Enthusiasm for his discoveries did not, however, render Suess a prisoner of his ideas: over the years he reworked his syntheses and hypotheses. Some of the views presented in the first volumes are contradicted in the last, a circumstance Suess explained by invoking the image of a mountain climber: as an alpinist climbs from rock to rock, the scientist goes from error to error; and even if he does not reach the summit, at each stage his view encompasses more territory.

At the beginning and at the end of this masterpiece the entire earth revolves before the reader's eye as Suess attempts to show its history by deciphering the features of its surface. He distinguished five ancient continents: Laurentia, Fennoscandia (a term created by Wilhelm Ramsay), Angaraland, Gondwanaland, and Antarctica. Between Eurasia and the Indo-African lands there extended a series of recently joined mountain chains that had originated in an ancient Mediterranean, called Tethys by Suess, extending from Central America to the Sunda Islands. These chains were generally concave toward the north in Asia and toward the south in Europe. Another series of mountain chains in festoons and garlands surrounded the Pacific Ocean. Suess distinguished three principal epochs of folding in Europe, evidence of which can be found in the more resistant rock massifs. Their age decreases as one goes from north to south. The most recent chains, bordering on the Mediterranean, are already subsiding. The ancient chains, eroded and covered by more recent deposits, can be reactivated and are therefore posthumous. Suess compared subsidences and giant grabens to tangential dislocations (folds and overthrusts), believing that subsidences characterize an important part of the surface of the earth.

The evolution of the concept of marine oscillations is an interesting example of the way in which Suess conceived a seminal idea, gathered observations to support it, followed its ramifications and weighed its consequences before proposing it, and, finally, set it against the image of nature at the time.

In 1860 and 1861 Suess studied the littoral formation deposited by the Tertiary Mediterranean on the border of the Bohemian massif north of Vienna. Following these ancient shorelines, he speculated that the observed regularity of their elevations was due not to an uplift of the massif but to a lowering of the sea level. From then on, he gathered observations relevant to this subject. After his visit to the region of the Suez Canal in 1869, Suess asserted that the isthmus was the result not of an elevation but of a relatively recent lowering of the waters. He decided at that time to examine the vestiges of the similar shores of northern Norway but was not able to do so until the summer of 1885, when he made two trips to the interior from Tromsö. He described the countryside, including the solifluction phenomena, in minute detail. The stepped terraces on the flanks of the fjords and valleys strongly impressed him. Trusting his own eye and dispensing with leveling, Suess concluded that the terraces are horizontal and do not intersect. They indicate successively lower sea levels without elevation of the shield. Suess not only studied the terraces but also investigated the Scandinavian literature on the subject dating as far back as the eighteenth century.

From his conclusion that the ancient massifs have remained stable ($\epsilon \grave{\upsilon} \sigma \tau \acute{\alpha} \sigma \epsilon \iota \alpha$), Suess deduced that it was the level of the sea that had varied; as a result he proposed the principle of eustatic levels (*Das Antlitz der Erde*, II, 680). Theoretically, these levels should be found on the peripheries of all seas and oceans, with the exception of some interior seas, such as the Baltic. Suess probably would not have been as affirmative regarding the stability of the old shields had he leveled the traces of the ancient sea levels or had he taken into account the measurements of others.

He did, however, reflect for more than fifteen years on the consequences of this concept. In his view the changes in level resulted from deformations of the ocean floor: "The earth yields; the ocean follows" (*ibid.*). A multitude of phenomena, even the form of the continents, tapered toward the south and were connected to the stated principles. Sedimentary deposits in the seas displace the waters and cause them to rise, and major subsidences lower them. Suess traced the history of the great transgressions since the Silurian, concentrating especially on that of the Middle Cretaceous. "The history of the continents," he wrote, "results from that of the seas" (II, 700). The virtual similarity of the great transgressions on all the continents would explain why the limits of the formations (recognized by William Smith and corresponding to interruptions of sedimentation) are found in so many localities.

In his attempt to learn more about the little-known sea floor Suess was forced to extrapolate from structures known on land. He distinguished the Atlantic coastal type from the Pacific, the former interrupting the structural lines and the latter being more or less parallel to them. Although Suess supposed that the great oceanic basins were formed by collapse between stable blocks, he was aware of the contradiction between the hypotheses of collapse and of contraction, and discussed the problem.

Volcanological studies led Suess to posit that oceanic waters result from the degasification of the earth (III, pt. 2, 631). He introduced the distinction between juvenile and vadose waters (III, pt. 2, p. 630). Juvenile gases and, higher in the earth's crust, thermal waters not only displace and concentrate numerous chemical elements; they are also important in the transport of terrestrial heat. Juvenile gases, Suess thought, originate under the sialic crust, a term created to indicate that silicon and aluminum are the principal and characteristic elements of this terrestrial layer (III, pt. 2, 626). Under this first sphere would be that of sima (characterized by silicon and magnesium) and finally the barysphere or nife (nickel and iron), which is primarily metallic. Studies of abyssal rocks sometimes suggested the presence of a "crofesima" or "nicrofesima" (III, pt. 2, 627). Suess thought that juvenile gases also played a role in lunar volcanism (III, pt. 2, 689).

In a final chapter of *Das Antlitz der Erde* Suess discussed life and its distribution on the earth. The old shields provided refuge for the terrestrial faunas, sheltering them during marine transgressions. The history of communications between the ancient lands and of the invasions and exchanges of faunas constitutes an important chapter that completes the study of the formation of mountain chains and of marine transgressions.

The results of many later methods were not at Suess's disposal for this synthesis, but those that did exist were employed to best advantage so that his work remains the structural model for such a project. His accomplishment is still more impressive when it is appreciated that one man was able to create such a panorama. Suess stressed several times that certain problems could be solved only by new techniques—for example, when he regretted the lack of methods for determining absolute ages (II, 703).

Such a critical analysis combined with a synthesis of broad compass strongly impressed his contemporaries and is reflected in the literature of the period. Marcel Bertrand wrote: "It would almost appear to us that he would be the most advanced in our science who has best understood this

book. . . . It is not only a question of making known the origin of ideas that will have an important place in the history of our science; it is also a matter of bringing within the reach of a great number of readers an almost inexhaustible mine of documents, the primary material in every kind of research and new discoveries. At first readers have been struck by an initial flowering of ideas; others remain in embryo on the same pages." Diverse and insular branches of the earth sciences were linked; and regional geologies were fitted into greater ensembles, thus forming new units that often transcended national frontiers.

The newly created field of structural geology was tied by many threads to stratigraphy—once the central concern of geology—and to geomorphology. Many paths of communication were opened toward the understanding of terrestrial substances and of their behavior, composition, origin, and evolution. Friedrich Becke, another Viennese professor, had developed links to petrography by defining the families of Atlantic and Pacific rocks and especially by creating the notion of zones of depth of metamorphism and the principle of diaphthoresis, thereby making it possible to grasp the upward motion of certain metamorphic rocks. Bruno Sander at Innsbruck and Schmidt at Leoben established relations between the deformations and fine structures of rocks and between internal displacement and mineralogical evolution.

In 1912 Alfred Wegener outlined in several lectures his new global kinematics and mobilism. As Bertrand had affirmed, Suess's work was not an end, for it opened many paths toward new knowledge and explanations. The impetus given in different directions to Austrian science and to the international scientific community can still be detected in the greater part of the earth sciences.

BIBLIOGRAPHY

I. ORIGINAL WORKS. Suess's memoirs, *Erinnerungen* (Leipzig, 1916), discuss his family origins as well as his scientific and political activities to the age of sixty and offer valuable insight into the development of his concepts and the reasoning and motivation for his accomplishments.

His separately published works include *Über das Wesen und den Nutzen palaeontologischer Studien* (Vienna, 1857); *Der Boden der Stadt Wien nach seiner Bildungsweise* (Vienna, 1862); *Die Entstehung der Alpen* (Vienna, 1875); *Die Zukunft des Goldes* (Vienna, 1877); and *Die Zukunft des Silbers* (Vienna, 1892), translated into English as *The Future of Silver* (Washington, 1893).

Das Antlitz der Erde, 3 vols. (Prague – Vienna – Leipzig, 1883 – 1909), was translated into French as *La face de la terre*, 4 vols. (Paris, 1897 – 1918), and into English as *The Face of the Earth*, 5 vols. (Oxford, 1904 – 1924).

The Royal Society *Catalogue of Scientific Papers* lists 81 memoirs published before 1900: V, 883 – 884; VIII, 1043 – 1044; XI, 530; and XVIII, 1034. See also Poggendorff, III, 1313 – 1314; IV, 1464; and V, 1230.

II. SECONDARY LITERATURE. Biographical studies of Suess include the notices by A. Geikie, in *Nature*, **72** (1905), 1 – 3; C. F. Parona, in *Atti dell'Accademia delle scienze* (Turin), **49** (1913 – 1914), 959 – 966; E. von Koerber, in *Almanach der Akademie der Wissenschaften in Wien*, **64** (1914), 349 – 362; N. Krebs, in *Mitteilungen der Geographischen Gesellschaft in Wien*, **57** (1914), 296 – 311; R. Michael, in *Zeitschrift der Deutschen geologischen Gesellschaft*, **66** (1914), 260 – 264; and P. Termier, in *Revue générale des sciences pures et appliquées*, **25** (1914), 546 – 552, translated into English in *Smithsonian Annual Report* for 1914 (Washington, 1915), 709 – 718.

On his importance in the history of geology, see G. Sarton, "La synthèse géologique de 1775 à 1918," in *Isis*, **2** (1914 – 1919), 381 – 392.

E. WEGMANN

AL-ṢŪFĪ, ABU'l-ḤUSAYN ʿABD AL-RAḤMĀN IBN ʿUMAR AL-RĀZĪ (*b.* Rayy, Persia, 903; *d.* 986), *astronomy*.

Little is known of al-Ṣūfī's life and career. He seems to have been closely associated with members of the Buwayhid dynasty in Iran and Baghdad, especially ʿAḍud al-Dawla (*d.* 983); besides that of ʿAḍud al-Dawla, names of three other members of the dynasty occur in connection with some of al-Ṣūfī's writings. Occasionally he mentions a "master" (*ustādh*) and "chief" (*raʾīs*) Abu'l-Faḍl Muḥammad ibn al-Ḥusayn, in whose company he visited Dīnawar in 946/947 and Isfahan in 948/949. This person obviously is Ibn al-ʿAmīd (*d.* 970), vizier of ʿAḍud al-Dawla's father. Rukn al-Dawla, who had contributed a foreword to al-Ṣūfī's book on the astrolabe (which was dedicated to one of ʿAḍud al-Dawla's sons, Sharaf al-Dawla).

Al-Ṣūfī is most renowned for his observations and descriptions of the fixed stars. The results of his investigations in this field are presented in his *Kitāb ṣuwar al-kawākib al-thābita* ("Book on the Constellations of the Fixed Stars"), in which he gives a critical revision of Ptolemy's star catalog, adding the differing or additional results of his own observations. The first such revision of Ptolemy's findings, the work became a classic of Islamic astronomy for many centuries; it even found its way into medieval Western science, where al-Ṣūfī became known as Azophi. In the "Book on the Constellations" the series of the forty-eight Ptolemaic constellations are dealt with according to the following scheme: first, there is a general discussion of all the stars of each constellation, in which al-Ṣūfī introduces his own criticism, mostly concerning positions, magnitudes, and colors; second, an identification of Arabic star names with the stars of the Ptolemaic stellar system contained in that constellation; third, two drawings of the constellation, as it is seen in the sky and as it is seen on the celestial globe; and fourth, a table of the stars of the constellation, giving longitude, latitude, and magnitude for each star. The epoch of the star table is the beginning of the year 1276 of Alexander (1 October 964), adding a constant of $12°42'$ to Ptolemy's longitudes (adopting a precession of one degree in sixty-six years, in accordance with the *Zīj al-Mumtaḥan*, the *Tabulae probatae*, which were prepared in 829/830 by order of the caliph al-Maʾmūn in order to improve certain parameters in the classical tradition). The magnitudes represent the results of al-Ṣūfī's own observations.

The scientific significance of this work lies in the valuable records of real star observations – in contrast with those of most other medieval astronomers, who merely repeated the Ptolemaic star catalog, adding constant values to his longitudes. There is also another important aspect: the exact astronomical identification of the several hundred old Arabic star names, which had been registered and transmitted only in philological works that completely omitted the exact astronomical considerations. Al-Ṣūfī did his best to identify them astronomically, although he was not always successful. His identifications were adopted by most later Islamic writers on astronomy and even penetrated modern stellar terminology: from T. Hyde's quotations from al-Ṣūfī and his follower Ulugh Beg, G. Piazzi selected ninety-four star names and introduced them into common use through his *Praecipuarum stellarum inerrantium positiones . . .* (1814).

Al-Ṣūfī also wrote a rather long and detailed *Kitāb al-ʿamal bi'l-asṭurlāb* ("Book on the Use of the Astrolabe"), an "Introduction to the Science of Astrology" (extant only in manuscript), and a "Book on the Use of the Celestial Globe" (also unpublished). He seems to have constructed astronomical instruments as well, for a silver celestial globe of his manufacture is said to have been extant in Egypt around 1043.

A poem on the constellations, in the *radjaz* meter ("Urjūza fī ṣuwar al-kawākib al-thābita"), by Abū ʿAlī ibn Abi'l-Ḥusayn al-Ṣūfī should be mentioned. He usually is referred to as Ibn al-Ṣūfī, the son of al-Ṣūfī. There is, however, sufficient reason to assume that he could not have been a son of the subject of this article, for the person to whom the poem is dedicated was obviously a prince reigning in the middle of the twelfth century.

BIBLIOGRAPHY

I. ORIGINAL WORKS. Al-Ṣūfī's most widely known work is *Kitāb ṣuwar al-kawākib al-thābita* ("Book on the Constellations of the Fixed Stars"). The Arabic text was widely used and was quoted several times by T. Hyde in the commentary to his ed. of Ulugh Beg's star catalog, *Tabulae longitudinis et latitudinis stellarum fixarum ex observatione Ulugh Beighi* (Oxford, 1665), 2nd ed. by G. Sharpe in *Syntagma dissertationum* (Oxford, 1767). L. Ideler drew many quotations from Hyde in his *Untersuchungen über den Ursprung und die Bedeutung der Sternnamen* (Berlin, 1809). Al-Ṣūfī's intro. appears in French trans. by J. J. A. Caussin de Perceval in *Notices et extraits des manuscrits*, XII (Paris, 1831), 236 ff.; the entire book in French, with selected portions in Arabic and the drawings, was edited from two MSS by H. C. F. C. Schjellerup as *Description des étoiles fixes par Abd-al-Rahman Al-Sûfi* (St. Petersburg, 1874); the complete Arabic text was edited from five MSS and was augmented by the "urjūza" poem of Ibn al-Ṣūfī at the Osmania Oriental Publications Bureau (Hyderabad, India, 1954), with intro. in English by H. J. J. Winter.

Naṣīr al-Dīn al-Ṭūsī translated the book into Persian in 1250 but his translation has not been published. By order of Alfonso X of Castile, an adaptation of the work, in Castilian, was completed around the middle of the thirteenth century and was edited by Manuel Rico y Sinobas as *Los libros del saber de astronomia*, I (Madrid, 1863). There is a critical ed. of the star nomenclature in this Castilian version (and in a succeeding Italian trans.) by O. J. Tallgren as "Los nombres árabes de las estrellas y la transcripción alfonsina," in *Homenaje a R. Menéndez Pidal*, II (Madrid, 1925), 633 ff., with "Correcciones y adiciones" in *Revista de filología española*, 12 (1925), 52 ff.

Traces of al-Ṣūfī's star catalog are also found in some Latin MSS, but there is no trans. of the complete text. See P. Kunitzsch, "Ṣūfī Latinus," in *Zeitschrift der Deutschen Morgenländischen Gesellschaft*, 115 (1965), 65 ff. Peter Apian sometimes quotes al-Ṣūfī, but detailed investigations are still required in order to determine whether he quoted from the Arabic or from a translation.

For medieval Arabic criticism of al-Ṣūfī (by al-Bīrūnī and Ibn al-Ṣalāḥ), see P. Kunitzsch, ed., *Ibn aṣ-Ṣalāḥ. Zur Kritik der Koordinatenüberlieferung im Sternkatalog des Almagest* (Göttingen, 1975), 21, 109–111; and the places from p. 38 to 74 given in the name index *s.v.* aṣ-Ṣūfī.

Al-Ṣūfī's other published work is *Kitāb al-ʿamal bi'l-asṭurlāb* ("Book on the Use of the Astrolabe"). The Arabic text, in 386 chapters, was edited from a Paris MS by the Osmania Oriental Publications Bureau (Hyderabad, India, 1962). An intro. in English, by E. S. Kennedy and M. Destombes, was printed separately (Hyderabad, India, 1967). For a geometrical treatise of al-Ṣūfī see F. Sezgin, *Geschichte des arabischen Schrifttums*, V (Leiden, 1974), 309–310.

II. SECONDARY LITERATURE. See C. Brockelmann, *Geschichte der arabischen Litteratur*, 2nd ed., I (Leiden, 1943); 253–254, and *Supplement*, I (Leiden, 1937), 398; and C. A. Storey, *Persian Literature*, II, pt. 1 (London, 1958), 41–42. For Ibn al-Ṣūfī, see Brockelmann, *Supplement*, I, 863, no. 4a. See also A. Hauber, "Die Verbreitung des Astronomen Ṣūfī," in *Islam*, 8 (1918), 48 ff.; M. Shermatov, "Ash-Shirazi's comments on the star catalogue of as-Sufi," in *Uchenye zapiski Dushanbin. gos. ped. in-t.*, 81 (1971), 73–83 (in Russian); S. M. Stern, "ʿAbd al-Raḥmān al-Ṣūfī," in *Encyclopaedia of Islam*, new ed., I (Leiden–London, 1960), 86–87; J. Upton, "A Manuscript of 'The Book of the Fixed Stars' by ʿAbd Ar-Raḥmān Aṣ-Ṣūfī," in *Metropolitan Museum Studies*, 4 (1933), 179–197; E. Wellesz, *An Islamic Book of Constellations* (Oxford, 1965); H. J. J. Winter, "Notes on Al-Kitab Suwar Al-Kawakib," in *Archives internationales d'histoire des sciences*, 8 (1955), 126 ff. For critical remarks on al-Ṣūfī's attitude and method in identifying the ancient Arabic star nomenclature, see P. Kunitzsch, *Untersuchungen zur Sternnomenklatur der Araber* (Wiesbaden, 1961), 10, 14 ff., 31; and also *Arabische Sternnamen in Europa* (Wiesbaden, 1959), 230–231. A sky map, including the Arabic stellar nomenclature according to al-Ṣūfī, was printed as *Supplément* to *Le Mobacher* (Algiers, Sept. 1881).

PAUL KUNITZSCH

SUMNER, FRANCIS BERTODY (*b.* Pomfret, Connecticut, 1 August 1874; *d.* La Jolla, California, 6 September 1945), *biology*.

Sumner spent a lonely childhood in the bare hills near Oakland, California, to which his parents, Arthur Sumner and Mary Augusta Upton, moved a few months after he was born. His father had been a schoolteacher before taking up, unsuccessfully, small-scale farming. He educated his two sons himself until Francis was ten, and he encouraged their collection and observation of the local wildlife.

In 1884 the family moved to Colorado Springs, Colorado, where Francis discovered himself to be a surprisingly good student. He also became conscious of his lack of social accomplishment, which deficiency he later declared, not quite accurately, a lifelong shortcoming. When the family moved to Minneapolis in 1887, Sumner attended high school and in 1894 received the B.S. at the University of Minnesota. He collected fishes in the remote streams and lakes of that state, under the direction of Henry F. Nachtrieb.

At Columbia University, where he went for graduate work in 1895, Sumner was attracted to zoology by Bashford Dean, Edmund B. Wilson, and Henry Fairfield Osborn, although he was always interested in philosophy as well. He received the Ph.D. in 1901 with a thesis on fish embryology that was written under Dean, who had arranged for Sumner's participation in a disastrous Nile expedition (1899) to collect *Polypterus*. For the next seven years Sumner taught natural history at the College of the City of New York and spent most of his summers at Woods Hole, Massachusetts.

In 1903 Sumner married Margaret Elizabeth Clark; in the same year he took up a summer appointment as director of the laboratory of the United States Bureau of Fisheries at Woods Hole. For three years he conducted a detailed biological survey of that area, working with others in collecting and classifying, and himself summarizing the correlations and drawing generalizations. Sumner concluded that the varying distribution of closely related species is a result of environmental differences. In 1911, as naturalist aboard the *Albatross*, he conducted a similar survey of San Francisco Bay for the Bureau of Fisheries.

In 1913 Sumner went to La Jolla, California, to work at the Scripps Institution for Biological Research. The director of the Institution, William E. Ritter, became interested in Sumner's proposed population studies of the deer mouse, *Peromyscus*, and encouraged him to undertake them. Sumner collected many distinctive subspecies of *Peromyscus* extensively in California and to a lesser degree in other states, to breed through a number of generations in a uniform climate, in order to determine whether they would tend to become less variable. He found that through as many as twelve generations the colors and other measurable characters continued to be distinctive. Reluctantly, he concluded that Mendelian inheritance, which he called "a fad," was a more likely cause of speciation by means of minute, cumulative genetic changes than was the environment.

When the Scripps Institution of Oceanography succeeded the former institution for biological research, Sumner's studies turned from mice to the pigments of fishes, a field in which he had done significant research during a six-month visit in 1910 to the Zoological Station in Naples. His original experiments with a variety of fishes proved the direct effect of the albedo upon the deposition of melanin and of guanine within the chromatophores of the skin. Always engrossed singlemindedly in one subject at a time, he did no further work on *Peromyscus*, but his stocks of mice and unpublished material on them were transferred to Lee Raymond Dice at the University of Michigan.

Sumner was a member of the National Academy of Sciences, the Philadelphia Academy of Sciences, the American Philosophical Society, and Phi Beta Kappa.

BIBLIOGRAPHY

I. ORIGINAL WORKS. Sumner was a prolific writer on a variety of biological subjects. Of his major researches, the biological survey at Woods Hole was reported in "A Biological Survey of the Waters of Woods Hole and Vicinity," in *Bulletin of the Bureau of Fisheries* for 1911, **31** (1913), pts. 1–2, written with R. G. Osburn, L. J. Cole, and B. M. Davis. The results of the survey of San Francisco Bay appeared in "A Report Upon the Physical Conditions in San Francisco Bay, Based Upon the Operations of the United States Fisheries Steamer Albatross, During the Years 1912 and 1913," in *University of California Publications in Zoology*, **14** (1914), 1–198, written with G. D. Louderback, W. L. Schmitt, and E. C. Johnston. In addition to many shorter papers on *Peromyscus*, he summarized the results of his fifteen-year study in "Genetic, Distributional and Evolutionary Studies of the Subspecies of Deer-mice (*Peromyscus*)," in *Bibliographia genetica*, **9** (1932), 1–106.

Sumner's early work on pigments of fishes is described in "The Adjustment of Flatfishes to Various Backgrounds," in *Journal of Experimental Zoology*, **10** (1911), 409–479. His later studies were reported in a number of papers, the most significant being "Quantitative Changes in Pigmentation, Resulting From Visual Stimuli in Fishes and Amphibia," in *Biological Reviews*, **15** (1940), 351–375. A candid account of Sumner's life and convictions is given in his self-critical autobiography, *The Life History of an American Naturalist* (Lancaster, Pa., 1945).

II. SECONDARY LITERATURE. A complete bibliography and review of Sumner's researches is in Charles Manning Child, "Biographical Memoir of Francis Bertody Sumner," in *Biographical Memoirs. National Academy of Sciences*, **25** (1948), 147–173. Letters and personal memorabilia are in the library of the Scripps Institution of Oceanography, La Jolla, California.

ELIZABETH NOBLE SHOR

SUMNER, JAMES BATCHELLER (*b.* Canton, Massachusetts, 19 November 1887; *d.* Buffalo, New York, 12 August 1955), *biochemistry.*

The son of Charles and Elizabeth Rand Sumner, James Sumner grew up in a well-to-do New England family engaged in manufacturing and farming. As a boy, he was an enthusiastic hunter, and in consequence of a shooting accident lost the use of his left arm—a misfortune all the more serious because he had been left-handed. With perseverance and ingenuity, he trained himself to use his right hand not only in normal activities but also in sports; he became an expert tennis player and continued to play into his sixties. In 1915 Sumner married Bertha Louise Ricketts; they had five children. After their divorce in 1930, he married Agnes Pauline Lundquist the following year. His second marriage also ended in divorce, and in 1943 Sumner married Mary Morrison Beyer; they had two children.

Upon completing his schooling at the Eliot Grammar School and the Roxbury Latin School, Sumner entered Harvard College in 1906 to study electrical engineering, but shifted to chemistry before graduating in 1910. He then worked briefly in his uncle's textile plant and taught chemistry at Mount Allison University, Sackville, New Brunswick, and at the Worcester Polytechnic Institute in Massachusetts. He returned to Harvard in 1911 to pursue graduate studies. He received the A.M. degree in 1913 and the Ph.D. degree in biological chemistry in 1914; his dissertation, "The Formation of Urea in the Animal Body," was based on research he conducted under the supervision of O. Folin and in association with C. H. Fiske.

In 1914 Sumner became assistant professor of biochemistry at the Ithaca division of the Cornell University Medical College, and in 1929 he was promoted to full professor. After the division was discontinued in 1938, he was successively a member of the department of zoology and of the department of biochemistry and nutrition in the Cornell School of Agriculture. In 1947 a laboratory of enzyme chemistry was established in the latter department, with Sumner as its director.

Sumner decided, in 1917, to isolate an enzyme; his earlier interest in urea metabolism led him to select urease, which catalyzes the cleavage of urea to ammonia and carbon dioxide. He found this enzyme to be present in relatively large amounts in the jack bean (*Canavalia ensiformis*); because of his conviction that enzymes are proteins, he concentrated on the fractionation of the proteins of

this material. After nine years of effort he succeeded in obtaining a crystalline globulin with high urease activity. He published this result in 1926, a time at which Willstätter was advocating the view that enzymes were low-molecular-weight substances readily adsorbed upon such carrier colloids as proteins. Because of Willstätter's eminence in chemistry (he had won the Nobel Prize in 1915 for his work on chlorophyll), Sumner's claim that urease is a protein was not generally accepted by the scientific community. Although he published a series of papers during the years 1926–1930 providing additional data in support of his position, it was not until J. H. Northrop announced, in 1930, the isolation of pepsin in the form of a crystalline protein that the merit of Sumner's work began to be recognized.

By 1937 several other enzymes had been obtained in crystalline form, and convincing data had been offered for the view that the catalytic activity of enzymes is associated with the integrity of individual proteins. Furthermore, W. M. Stanley had isolated, from plants infected with tobacco mosaic, a crystalline protein (later shown by F. C. Bawden and N. W. Pirie to be a nucleoprotein) that carried the infectivity of the virus. The recognition of Sumner's achievement was underlined in 1946 by the decision of the Nobel Committee in chemistry to award him one-half of the prize for that year for "his discovery that enzymes can be crystallized," the other half being shared by Northrop and Stanley.

Because the Willstätter group emphasized the importance of low-molecular-weight substances as bearers of the catalytic activity of enzymes, the iron-porphyrin-containing enzymes peroxidase and catalase were considered around 1930 to represent examples of the adsorption of such small catalytic substances to protein carriers. In 1937 Sumner (with his student A. L. Dounce) reported the crystallization of catalase, and he later provided data to demonstrate its protein nature. During the succeeding years, he published valuable reports on peroxidases, lipoxidase, and other enzymes.

Sumner's scientific career exemplifies the persistence of the investigator who stubbornly treads a path that his more influential contemporaries consider to be a blind alley. As matters turned out, it was they who were going in the wrong direction, and it was Sumner who had chosen the road that led to the great achievements in enzyme chemistry after 1930.

BIBLIOGRAPHY

I. ORIGINAL WORKS. Sumner's books include *Textbook of Biological Chemistry* (New York, 1927); *Chemistry and Methods of Enzymes* (New York, 1943; 2nd ed., 1947; 3rd ed., 1953), written with G. Fred Somers; and *Laboratory Experiments in Biological Chemistry* (New York, 1944; 2nd ed., 1949), also written with Somers. With Karl Myrbäck, Sumner edited the treatise *The Enzymes, Chemistry and Mechanism of Action*, 2 vols. (New York, 1950–1952). Sumner published about 125 research articles; among the most important are "The Isolation and Crystallization of the Enzyme Urease (Preliminary Paper)," in *Journal of Biological Chemistry*, **69** (1926), 435–441; and "Crystalline Catalase," *ibid.*, **121** (1937), 417–424, written with Alexander L. Dounce.

II. SECONDARY LITERATURE. An appreciation of Sumner's life and work is L. A. Maynard's article in *Biographical Memoirs. National Academy of Sciences*, **31** (1958), 376–396, with portrait and bibliography.

JOSEPH S. FRUTON

SUNDMAN, KARL FRITHIOF (*b*. Kaskö, Finland, 28 October 1873; *d*. Helsinki, Finland, 28 September 1949), *astronomy*.

After studying at the University of Helsinki and the Pulkovo Observatory, Sundman was appointed assistant professor at Helsinki in 1902 and later, in 1907, associate professor. In 1918 he was made full professor and director of the Helsinki observatory, which positions he held until his retirement in 1941.

Sundman's scientific work was devoted principally to problems of celestial mechanics, and his name is connected with two important achievements. First, he extended Laplace's propositions concerning the convergence of the series in unperturbed elliptic motion to the perturbation problem. His second achievement concerned the fundamental problem of celestial mechanics, the so-called three-body problem, which he reported in two papers (1907, 1909) and summarized, in 1912, in "Mémoire sur le problème des trois corps." The method he used results in a general solution to the three-body problem because the series developments he derived of the coordinates of the three bodies represent the future motion even if the convergence of the series becomes slower with increasing time.

Sundman's later papers include a description of a machine for the computation of planetary perturbations (1915), studies of the invariability of the axes of planetary orbits (1940), and a numerical calculation of the motion of the sun and the moon at the time of the total solar eclipse on 9 July 1945 (1948).

Although Sundman's main interest lay in theoretical astronomy, he promoted practical work at the Helsinki observatory. He also continued the work on the photographic star catalogue that had been initiated by his predecessor, A. Donner.

BIBLIOGRAPHY

I. ORIGINAL WORKS. Sundman's works include "Utvecklingarna af *e* och *e²* uti kedjebråk med alla partialtäljare lika med ett," in *Öfversikt af Finska Vetenskapssocietetens förhandlingar*, **38** (1896); "Om den personliga eqvationen vid ringmikrometerobservationer," *ibid.*; "Über die Störungen der kleinen Planeten, speciel derjenigen, deren mittlere Bewegung annähernd das Doppelte Jupiters beträgt" (1901), Sundman's dissertation; "Über eine direkte Herleitung der Gyldén'schen A- und B-Koeffizienten als Funktionen von Θ-Transcendenten," in *Förhandlingar vid Nordiska naturforskare-och läkarmötet* (1902); "Recherches sur le problème des trois corps," in *Finska Vetenskapssocietetens Acta*, **34** (1907), 6; "Nouvelles recherches sur le problème des trois corps," *ibid.*, **35** (1909), 9; "Om planeternas banor," in *Öfversikt af Finska Vetenskapssocietetens forhandlingar*, **51C** (1909), 3; and "Sur les singularités réelles dans le problème des trois corps," in *Comptes rendus du Congrès des mathematiciens scandinaves* (1910).

See also "Mémoire sur le problème des trois corps," in *Acta mathematica*, **36** (1912); "Theorie der Planeten," in *Encyklopädie der mathematischen Wissenschaften*, **6** (1915), 2; "Plan d'une machine destinée à donner les perturbations des planètes," in *Festskrift für A.S. Donner* (1915); "Sur les conditions nécessaires et suffisantes pour la convergence du développement de la fonction perturbatrice dans le mouvement plan," in *Öfversikt af Finska Vetenskapssocietetens förhandlingar*, **58A** (1916), 24; "Observations de l'éclipse du soleil à Kumlinge le 21 août 1914," *ibid.*, **59A** (1916), 1; "Étude d'un cliché pris avec le tube polaire de l'observatoire de Helsingfors," *ibid.*, **59A** (1916), 5; "Om de astronomiska rörelseteorierna," *ibid.*, **61C** (1919), 4; "Uber die Richtungslinien für fortgesetzte Untersuchungen in den Planet- und Trabanttheorien," in *Redog. f. 5 Skand. Matem. Kongr. 1922* (1923); "La gravitation universelle et sa vitesse de propagation," in *Festskrift till Ernst Lindelöf* (1929); "Über die Bestimmung geradliniger Bahnen," in *Vierteljahrsschrift der Astronomischen Gesellschaft*, **70** (1935); "Démonstration nouvelle du théorème de Poisson sur l'invariabilité des grands axes," in *Festschrift für Elis Strömgren* (1940), 263; and "The Motions of the Moon and the Sun at the Solar Eclipse of 1945, July 9th," in

L'activité de la Commission géodésique Baltique pendant les années 1944–47 (1948).

C. SCHALÉN

SURINGAR, WILLEM FREDERIK REINIER (*b.* Leeuwarden, Netherlands, 28 December 1832; *d.* Leiden, Netherlands, 12 July 1898), *botany.*

Suringar was the eldest son of Gerard Tjaard Nicolaas Suringar, a book dealer and publisher, and Alida Boudina Koopmans. The family probably belonged to the Dutch Reformed Church. Suringar received his secondary education at Oostbroek, near The Hague, and at the Gymnasium of Leeuwarden. In 1850 he enrolled at the University of Leiden, planning to study medicine, but soon turned to botany. In 1855 he was awarded a gold medal by the university for an essay on the algae of the Netherlands; the work was the basis for his dissertation in natural philosophy, for which he received his doctorate on 13 March 1857. In November of that year he was appointed extraordinary professor of botany at the University of Leiden, substituting for de Vriese, who was on an extended study trip in the Dutch East Indies. De Vriese died shortly after his return, and in May 1862 Suringar succeeded him as professor, occupying the chair until his death in 1898.

In 1857 Suringar was appointed *conservator herbarii* of the Netherlands Botanical Society and coeditor of its journal, *Nederlandsch kruidkundig archief.* In 1868 he became director of the national herbarium. As director of these institutions, Suringar was a leading plant taxonomist of the Netherlands. Although not published under his name, the *Prodromus florae Batavae* could not have been compiled without Suringar's assiduous work at the herbaria. Intended as a preliminary study for an extensive, national flora of the Netherlands to be published by the Netherlands Botanical Society, the work went through two editions (1850–1866, 1901–1916) but was never completed. Suringar's own flora of the Netherlands, first published in 1870 and frequently reprinted, was the leading work until the turn of the century. The planned national flora was not begun until the 1970's, as it seemed no longer needed after the publication of the floras of Oudemans (1859–1862), Suringar, Heimans and Thysse (1899), and Heukels (1900).

Suringar sustained a continuing interest in algae; and his monograph on Japanese algae, written in 1870, was based on herbarium material available in Leiden. His brief concern with bacteriology was reflected in the publication of his study on *Sarcina ventriculi* (1865), for which he was awarded an honorary doctorate by the University of Munich in 1872. He subsequently investigated the cactus family (Cactaceae), especially the genus *Melocactus*, on which he began a monograph that was never completed. According to his student Hugo de Vries, he also contributed studies of the carpels of Cruciferae, polycephaly of Compositae, pitcher formation of *Ulmus* leaves, and peloric flowers of *Digitalis.*

A long-standing opponent of the theory of evolution, Suringar surprised his friends with *Het plantenrijk* (1895), published shortly before his death. In it he discussed the evolution of plants, based on the work of Darwin and his followers, in addition to his own ideas.

Suringar's students included Melchior Treub, who became director of the botanical garden in Buitenzorg, Java [now Bogor, Indonesia], in 1880. Under Treub's leadership, this institution became one of the world's most important centers of botanical research. Many important posts at Buitenzorg were held by Suringar's pupils, for example, J. G. Boerlage, W. Burck, and J. van Breda.

BIBLIOGRAPHY

I. ORIGINAL WORKS. There is apparently no complete bibliography of Suringar's works. His more important books include *Dissertatio botanica inauguralis continens observationes phycologicae in floram Batavam* (Leeuwarden, 1857), his doctoral diss.; *De beteekenis der plantengeographie en de geest van haar onderzoek* (Leeuwarden, 1857), Suringar's inaugural diss. as professor of botany; *De Sarcine (Sarcina ventriculi, Goodsir)* (Leeuwarden, 1865); *De kruidkunde in hare betrekking tot de maatschappij en de hoogeschool* (Leeuwarden, 1868); *Algae japonicae musei botanici Lugduno-Batavorum* (Haarlem, 1870); *Illustration des algues du Japan*, Musée botanique de Leide, I (Leiden, 1872), 63–69; II (Leiden, 1874), 1–15; *Handleiding tot het bepalen van de in Nederland wild groeiende planten* (Leeuwarden, 1870–1873), the flora of the Netherlands, which went through at least 13 eds.; *De kruidkunde in Nederland* (Leiden, 1889); *Het plantenrijk, philogenetische schets* (Leeuwarden, 1895); and *Illustrations du genre Melocactus*, 5 pts. (Leiden, 1897–1905), continued by his son, J. Valckenier Suringar.

Suringar's memoirs are listed in the Royal Society *Catalogue of Scientific Papers*, V, 888; VIII, 1046; XI, 530; XII, 711; and XVIII, 1039–1040. They include a series of papers on *Melocactus*, in *Verslagen en mededeelingen der K. Akademie van Wetenschappen*, Afd. Natuurkunde, 3rd ser., **2** (1886), 183–195; **6** (1889), 408–437, 438–461; **9** (1892), 406–412; and in *Versla-*

gen van de zittingen der wis- en natuurkundige afdeeling der K. Akademie van Wetenschappen, **4** (1896), 251–252; **5** (1897), 1–46; and **6** (1898), 178–192.

II. SECONDARY LITERATURE. On Suringar's life and work, see the notices by K. Schumann, in *Verhandlungen des Botanischen Vereins der Provinz Brandenburg,* **40** (1898), cxvii–cxviii; and *Monatsschrift für Kakteenkunde,* **8** (1898), 134–137; J. Valckenier Suringar, in *Nieuw nederlandsch biografisch woordenboek,* X (Leiden, 1937), 990–995; and H. G. van de Sande Bakhuysen, in *Verslagen van de gewone vergadering der afdeeling natuurkunde, K. Akademie van wetenschappen te Amsterdam,* **7** (1899), 129–130.

Other notices are by Hugo de Vries, in *Eigen Haard,* **23** (1897), 724–727; and *Berichte der Deutschen botanischen Gesellschaft,* **17** (1899), 220–224; L. Vuyck, in *Nederlandsch kruidkundig archief,* 3rd ser., **1** (1899), i–x; and H. Witte, in *Tijdschrift voor tuinbouw,* **4** (1899), 1–5. Two unsigned articles are "Biographische Mitteilungen (Suringar)," in *Leopoldina,* **34** (1898), 144; and "Necrologio (W. F. R. Suringar)," in *Nuova notarisia,* **10** (1899), 45–46.

PETER W. VAN DER PAS

SUTER, HEINRICH (*b.* Hedingen, Zurich canton, Switzerland, 4 January 1848; *d.* Dornach, Switzerland, 17 March 1922), *mathematics, Oriental studies.*

Suter was the son of a farmer and a keeper of posthorses. In 1875 he married Hermine Frauenfelder, sister of a famous philanthropist and preacher of Schaffhausen cathedral, Eduard Frauenfelder; they had three daughters.

Beginning in 1863 Suter attended the Zurich cantonal school, where he learned Latin and Greek. At the University of Zurich and at the Eidgenössische Technische Hochschule he studied mathematics, physics, and astronomy under Christoffel, K. T. Reye, C. F. Geiser, and Rudolf Wolf; he then completed his training under Kronecker, Kummer, and Weierstrass at the University of Berlin, where he also attended lectures on history and philology. Suter received the doctorate from the University of Zurich in 1871 for the dissertation *Geschichte der mathematischen Wissenschaften,* I; *Von den ältesten Zeiten bis Ende des 16. Jahrhunderts,* in which the significance of mathematics for cultural history was emphasized. Although Suter set forth the goal of treating the history of mathematics in terms of the history of ideas, he was prevented from attaining it because of the paucity of available data.

Following a temporary appointment at the Wettingen teachers' training college (Aargau canton)

and as a part-time teacher at the Gymnasiums in Schaffhausen (1874) and St. Gall (1875), Suter taught mathematics and physics at the cantonal schools of Aargau (1876) and Zurich (1886–1918). At the latter he acquired a thorough knowledge of Arabic under the Orientalists Steiner and Hausheer. His chief studies, in addition to numerous minor publications that appeared mainly in *Bibliotheca mathematica* (1889–1912), are "Das Mathematiker-Verzeichnis im Fihrist des . . . an-Nadîm" and "Die Mathematiker und Astronomen der Araber und ihre Werke." The outstanding expert of his time on Muslim mathematics, Suter was awarded an honorary doctorate of philosophy by the University of Zurich shortly before his death.

BIBLIOGRAPHY

I. ORIGINAL WORKS. Suter's writings include *Geschichte der mathematischen Wissenschaften,* I, *Von den ältesten Zeiten bis Ende Des 16. Jahrhunderts* (2nd rev. ed., Zurich, 1873), and II, *Vom Anfange des 17. bis gegen Ende des 18. Jahrhunderts* (Zurich, 1875); "Die Mathematiker auf den Universitäten des Mittelalters," in *Wissenschaftliche Beilage zur Programm der Kantonsschule* (Zürich, 1887); "Das Mathematiker-Verzeichnis im Fihrist des . . . an-Nadîm," in *Abhandlungen zur Geschichte der mathematischen Wissenschaften,* **6** (1892); "Die Mathematiker und Astronomen der Araber und ihre Werke," *ibid.,* **10** (1900); his edition of "Die astronomischen Tafeln des Muhammad ibn Mûsâ al-Khwârizmî," in *Kongelige Danske Videnskabernes Selskabs Skrifter,* 7th ser., **3**, no. 1 (1914); and "Beiträge zur Geschichte der Mathematik bei den Griechen und Arabern," J. Frank, ed., in *Abhandlungen zur Geschichte der Naturwissenschaften und der Medizin* (1922), no. 4, with autobiographical sketch.

II. SECONDARY LITERATURE. See E. Beck, "Heinrich Suter," in *Jahresberichte des Gymnasiums Zürich* for 1921–1922; J. Ruska, "Heinrich Suter," in *Isis,* **5** (1923), 408–417, with portrait and bibliography; and C. Schoy, "Heinrich Suter," in *Neue Zürcher Zeitung* (8 Apr. 1922), also in *Vierteljahrsschrift der Naturforschenden Gesellschaft in Zürich,* **67** (1922), 407–413, with bibliography.

J. E. HOFMANN

SUTHERLAND, WILLIAM (*b.* Dumbarton, Scotland, 4 August 1859; *d.* Melbourne, Australia, 5 October 1911), *theoretical physical chemistry, molecular physics.*

One of five sons of a Scottish woodcarver, Sutherland immigrated to Sydney in 1864 and settled in Melbourne in 1870. Educated locally at Wesley

College, in 1876 he matriculated in the University of Melbourne, graduating B.A. in 1879. In July, Sutherland left for England. He entered University College, London, on a science scholarship and studied under G. Carey Foster, graduating B.Sc. in 1881 with first-class honors in experimental physics. In 1882 Sutherland returned to Melbourne, where he received the M.A. in 1883. Supporting himself by part-time literary activities, he devoted himself to scientific research.

Although Sutherland remained indirectly attached to the University of Melbourne, he held no regular appointment. From 1885 he published about two or three major scientific articles per year, mostly in the *Philosophical Magazine*. He investigated such interrelated topics as the viscosity of gases and liquids, molecular attraction, valency, ionization, ionic velocities, and atomic sizes. His results were internationally recognized and highly valued.

In 1893 Sutherland introduced a quickly accepted dynamical explanation of the hitherto problematic relationship between the viscosity of a gas and its temperature. The Sutherland constant C increases in general with the size of the molecule and is a measure of the strength of the mutual molecular attraction. From 1900 he put forward the theory that the magnetism of the earth results from its rotating electrostatic field. His calculations were later confirmed by L. A. Bauer. Sutherland speculated in 1901 that the spectra of the elements were a function of their rigidity, and in 1902 he published his electric-doublet theory. His heterodox views (1902) regarding the complete dissociation of strong electrolytes at all concentrations were elaborated by S. R. Milner a decade later and by 1923 had become an integral part of the Debye-Hückel theory. Sutherland's later research dealt chiefly with the electronic theory of matter.

Although J. J. Thomson had, in 1899, disputed Sutherland's electron conceived as a disembodied charge, Sutherland's continued efforts to prove that the various properties of matter were essentially electrical in origin were later eulogized by Thomson for their significance.

BIBLIOGRAPHY

Correspondence from Sutherland during the period 1905–1907 is in the Bragg Collection at the Royal Institution, London; additional material is available at the National Library of Australia, Canberra.

A list of Sutherland's sixty-nine major scientific publications is given in W. A. Osborne, *William Sutherland. A Biography* (Melbourne, 1920), reviewed by G. Sarton in *Isis*, 4 (1922), 328–330. Additional articles are F. Johns, in *Australian Biographical Dictionary* (Melbourne, 1934), 346–347, and in *Australian Encyclopedia*, VIII (Sydney, 1965), 372; and P. Serle, in *Dictionary of Australian Biography*, II (Sydney, 1949), 394–395. The Sutherland constant is discussed in A. von Engel, *Ionized Gases* (Oxford, 1965), 31. Sutherland's speculation on complete dissociation in strong electrolytes at all concentrations is considered in its historical and scientific context by N. Feather, *Electricity and Matter* (Edinburgh, 1968), 218–219. His electrical theory of the atom and eccentric symbolism are considered in N. Feather, "A History of Neutrons and Nuclei," in *Contemporary Physics*, 1 (1960), 191–193.

THADDEUS J. TRENN

SUTTON, WALTER STANBOROUGH (*b*. Utica, New York, 5 April 1877; *d*. Kansas City, Kansas, 10 November 1916), *biology, medicine.*

Sutton was the fifth of the seven sons of William Bell Sutton and Agnes Black Sutton. His father, a farmer, moved from New York to Russell County, Kansas, when Sutton was ten years old. He was educated in the local public schools and entered the engineering school of the University of Kansas at Lawrence in 1896. The death of a younger brother from typhoid fever in the summer of 1897 decided Sutton on a career in medicine. He transferred to the school of arts (later the college of liberal arts) the following fall and embarked on biological studies. Under the influence of Clarence Erwin McClung, then an instructor in zoology, he began cytologic work. He received the B.A. degree in 1900 and, working as McClung's first graduate student, earned the M.A. degree in 1901. In the fall of that year he went to Columbia University to work with Edmund Beecher Wilson. Although Sutton never completed a Ph.D. thesis, during the years 1901–1903 he formulated the theory of the chromosomal basis of Mendelism, his most noteworthy contribution to science.

After two years as foreman in the oil fields of Chautauqua County in southeastern Kansas (1903–1905), Sutton returned to the College of Physicians and Surgeons of Columbia University and completed the requirements of the M.D. degree (1907). He spent the following two years in a surgical house officership at Roosevelt Hospital in New York City. From 1909 until his premature death from a ruptured appendix, he practiced surgery privately in Kansas City, Kansas, and in Kansas City, Missouri.

Sutton was of impressive physical appearance, standing six feet tall and weighting 215 pounds — the basis for his nickname "Bill Taft." E. B. Wilson [9] described "his quiet steadfastness and force and a quality of serenity . . . his clear, direct gaze, his self-possessed and tranquil manner." He was elected to Phi Beta Kappa and to Sigma Xi. At the time of his death, he was an associate professor of surgery at the University of Kansas and a fellow of the American College of Surgeons. He never married.

As a farmboy, Sutton displayed great skill in the repair and operation of agricultural equipment. He also built his own camera, thereby prefiguring his later professional use of photography, and he was adept at drawing. His ingenuity was often evident in the laboratory and in his surgical practice.

In his first publication, "The Spermatogonial Divisions of *Brachystola magna*" [1], Sutton used specimens of grasshoppers that he had collected during the summer of 1899 as he rode the "header box" in the wheat fields of his father's farm. With McClung, he discovered the value of this species for cytologic study, since the large size of its cells made it "one of the finest objects thus far discovered for the investigation of the minutest details of cell-structure." The paper was the basis for his subsequent deductions about the role of the chromosomes in heredity.

By the fall of 1902 Sutton had been associated with E. B. Wilson for a year; they had spent the preceding summer collecting and studying marine specimens in North Carolina and Maine. Sutton's intimate familiarity with the meiotic process, together with the expositions of Bateson, the ardent English protagonist of the newly rediscovered Mendelism, crystallized in his mind the relationship between the behavior of the chromosomes at meiosis and Mendelian segregation and assortment. (Bateson himself was slow to accept the chromosomal basis of Mendelism.) Thus, in a second paper on the chromosomes of the grasshopper [2], published in December 1902, Sutton wrote:

> I have endeavored to show that the eleven ordinary chromosomes (autosomes) which enter the nucleus of each spermatid are selected from each of the eleven pairs which make up the double series of the spermatogonia. . . . I may finally call attention to the probability that the association of paternal and maternal chromosomes in pairs and their subsequent separation during the reducing division as indicated above

may constitute the physical basis of the Mendelian law of heredity.

Cytologists had been aware for some time that at one phase of meiosis whole chromosomes separate, or segregate (the "reducing division" of August Weismann). In 1901 Thomas Montgomery had concluded that "in the synapsis stage is effected a union of paternal with maternal chromosomes." It was known, furthermore, that in fertilization chromosomes are contributed in equal numbers by the two gametes (Van Beneden's law).

On 19 December 1902 Wilson published a short note in *Science* proposing a relationship between the phenomena of meiosis and Mendel's laws. He stated the following reason for the note: "Since two investigators, both students in the University, have been led in different ways to recognize this clue or explanation, I have, at their suggestion and with their approval, prepared a brief note in order to place their independent conclusions in proper relation to each other and call attention to the general interest in the subject." Sutton's fellow student William Austin Cannon, later professor of botany at Stanford University, was working with fertile hybrid cotton plants and had found separation of paternal and maternal elements in meiosis.

Wilson [9] later wrote: "I well remember when, in the early Spring of 1902, Sutton first brought his main conclusion to my attention. . . . I also recall that at that time I did not at once fully comprehend his conception or realize its entire weight." Of their work together in the summer of 1902, Wilson wrote, "It was only then in the course of our many discussions, that I first saw the full sweep, and the fundamental significance of his [Sutton's] discovery." A year or so before, the work of several other cytologists had brought them to the verge of the chromosomal theory. "Sutton, however, was the first clearly to perceive and make it known. . . ."

Sutton's "The Chromosomes in Heredity" (1903) is a major landmark in the biologic literature [3]. Like his 1902 paper, it was intended as a preliminary report. Complete, although concise, it displayed model clarity and logic. Sutton built his argument on six components, of which three were corroborations of the findings or suspicions of his predecessors, while three others were uniquely his own, as was the synthesis. The six points are:

(1) That the somatic chromosomes comprise two equivalent groups, one of maternal derivation and one of paternal derivation;

(2) That synapsis consists of pairing of corre-

sponding (homologous) maternal and paternal chromosomes;

(3) That the chromosomes retain their morphologic and functional individuality throughout the life cycle;

(4) That the synaptic mates contain the physical units that correspond to the Mendelian allelomorphs; that is, the chromosomes contain the genes;

(5) That the maternal and paternal chromosomes of different pairs separate independently from each other—"The number of possible combinations in the germ-products of a single individual of any species is represented by the simple formula 2^n in which n represents the number of chromosomes in the reduced series"; and

(6) That "Some chromosomes at least are related to a number of different allelomorphs . . . [but] all the allelomorphs represented by any one chromosome must be inherited together. . . . The same chromosome may contain allelomorphs that may be dominant or recessive independently."

Sutton thus predicted genetic linkage and pointed out that Bateson and Saunders, in their experiments with *Matthiola*, had detected "two cases of correlated qualities which may be explained by association of their physical bases in the same chromosome." (Bateson himself had an alternative and incorrect explanation.) Although Sutton observed and pictured chiasmata, their specific delineation and the suggestion that parts are exchanged between homologous chromosomes are attributed to F. A. Janssens (1909); the relation to genetic recombination was discovered by Thomas Hunt Morgan and his students.

Cytologists had suspected for fifteen or twenty years before Sutton that hereditary factors are carried by the nucleus and even by chromatin, but Sutton's demonstration of the relationship between the behavior of the chromosomes in meiosis and Mendel's two laws was the first strong evidence specifically in support of the theory. Wilson [9] wrote that subsequent to the appearance of Sutton's papers, Theodor Boveri had stated that he himself had already arrived at the same general conclusion. Consequently, the chromosomal theory of inheritance is sometimes called the Sutton-Boveri theory.

None of Sutton's surgical contributions rank with his main contribution to biology. He did, however, introduce colonic administration of ether for surgery of the head and neck and was further responsible for a number of minor technical innovations.

BIBLIOGRAPHY

[1]. Sutton, "The Spermatogonial Divisions of *Brachystola magna*," in *Kansas University Quarterly*, **9** (1900).

[2]. Sutton, "Morphology of the Chromosome Group in *Brachystola magna*," *ibid.*, **4** (1902).

[3]. Sutton, "The Chromosomes in Heredity," *ibid.*, **4** (1903), 231–251. repr. in J. A. Peters, ed., *Classic Papers of Genetics* (Englewood Cliffs, New Jersey, 1959).

[4]. A. Hughes, *A History of Cytology* (London–New York, 1959).

[5]. The anonymous "Walter Stanborough Sutton," in *Journal of the American Medical Association*, **67** (1916).

[6]. *Walter Stanborough Sutton, April 5, 1877–November 10, 1916* (Kansas City, 1917), published by his family and available from the Library of Kansas University School of Medicine.

[7]. V. A. McKusick, "Walter S. Sutton and the Physical Basis of Mendelism," in *Bulletin of the History of Medicine*, **34** (1960), 487–497.

[8]. B. R. Voeller, ed., *The Chromosome Theory of Inheritance. (Classic Papers in Development and Heredity)* (New York, 1968).

[9]. E. B. Wilson, see [6].

Victor A. McKusick

SVEDBERG, THE (THEODOR) (*b*. Fleräng, Valbo, near Gävle, Sweden, 30 August 1884; *d*. Örebro, Sweden, 25 February 1971), *physical chemistry*.

Svedberg was the only child of Elias Svedberg and Augusta Alstermark. His father, a civil engineer, was a very active man with many interests besides his profession. He was strongly attracted to the study of nature and made long excursions with his son, who shared his enthusiasm. He worked as a manager of iron works in Sweden and Norway, but the family suffered economic problems from time to time. As a Gymnasium student Svedberg was especially interested in chemistry, physics, and biology, especially botany, and finally decided to study chemistry, believing that many unsolved problems in biology could be explained as chemical phenomena. In January 1904 he enrolled at the University of Uppsala, with which he remained associated for the rest of his life. He passed the necessary courses and examinations in record time and received the B.Sc. in September 1905; his first scientific paper was published that December. Two years later he defended his dissertation, "Studien zur Lehre von den kolloiden Lösungen," for the doctorate of philosophy and became docent in chemistry. In 1912 he was appointed to the first Swedish chair of physical chemistry.

When, in August 1949, Svedberg reached the mandatory retirement age for a university professor, a special ruling, unique in Swedish academic administration, allowed him to become head of the new Gustaf Werner Institute of Nuclear Chemistry and to retain that post as long as he desired. He resigned in 1967.

Svedberg enjoyed good health throughout his active life, going for long walks almost every day. He loved making excursions in the open country and collecting wild flowers throughout Sweden. This remained his lifelong hobby, and his herbarium finally contained a complete collection of all the phanerogams in Sweden. His botanical excursions extended as far north as Greenland and Svalbard.

Svedberg received the 1926 Nobel Prize in chemistry for his work on disperse systems and was awarded honorary doctorates by the universities of Groningen, Wisconsin, Uppsala, Harvard, Oxford, Delaware, and Paris. He was elected member or honorary member of more than thirty learned societies, including the Royal Society, the National Academy of Sciences (Washington), and the Academy of Sciences of the U.S.S.R.

Svedberg was little interested in university politics and seldom attended faculty meetings. On the other hand, he was very active in promoting research activities both in industry and at the universities. He played an important role in the creation of the first research council in Sweden, the Research Council for Technology (1942), of which he was a member until 1957. He was also a member of the Swedish Atomic Research Council from its foundation in 1945 to 1959 and, from 1947 to 1956, a member of the board of AB Atomenergi, a company partly owned by the Swedish government.

Very early in his chemical studies Svedberg came across books that greatly stimulated his scientific thought: the 1903 edition of Nernst's *Theoretische Chemie*, in which he found the sections on colloids, osmotic pressure, diffusion, and molecular weights the most interesting; Zsigmondy's *Zur Erkenntnis der Kolloide*; and Gregor Bredig's *Anorganische Fermente*. Svedberg was fascinated by the new field of colloid chemistry, which became the main subject of his scientific activity for almost two decades. He began by studying the electric synthesis of metal sols in organic solvents. Bredig had prepared metal sols by letting a direct-current arc burn between metal electrodes under the surface of a liquid. The sols obtained were, however, rather coarse, polydisperse, and contaminated. Svedberg introduced an induction coil with the discharge gap placed in the liquid. The results were striking. He was able to prepare a number of new organosols from more than thirty different metals, and they were more finely dispersed and much less contaminated. In addition, the method was reproducible, making it possible to use the sols for exact quantitative physicochemical studies. With the ultramicroscope Svedberg began studying the Brownian movements of the particles in these sols and determined the influence of solvent, viscosity, temperature, and other factors. In 1906 these studies provided an experimental confirmation of Einstein's and of Smoluchowski's theories on Brownian movement.

Svedberg retained a lifelong interest in radioactive processes. With D. Strömholm he carried out a pioneering investigation on isomorphic coprecipitation of radioactive compounds. Different salts were crystallized in solutions of various radioelements, and it was determined whether or not the radioelement crystallized out with the salts. Their discovery that thorium X, for example, crystallized with lead and barium salts, but not with others, indicated to them the existence of isotopes before that conception was introduced into chemistry by Soddy.

In his Nobel lecture Soddy referred to these experiments, the results of which were published in 1909: "Strömholm and Svedberg were probably first to attempt to fit a part of the disintegration series into the Periodic Table." Referring to their last paper of that year, he added:

> Nevertheless, in their conclusion, is to be found the first published statement that the chemical non-separability found for the radio-elements may apply also to the non-radioactive elements in the Periodic Table. Remarking on the fact that, in the region of the radio-elements, there appear to be three parallel and independent series, they then say "one may suppose that the genetic series proceed down through the Periodic Table, but that always the three elements of the different genetic series, which thus together occupy one place in the Periodic System, are so alike that they always occur together in Nature and also not have been able to be appreciably separated in the laboratory. Perhaps, one can see, as an indication in this direction, the fact that the Mendeleev scheme is only an approximate rule as concerns the atomic weight, but does not possess the exactitude of a natural law; this would not be surprising if the elements of the scheme were mixtures of several homogeneous

elements of similar but not completely identical atomic weight." Thus Strömholm and Svedberg were the first to suggest a general complexity of the chemical elements concealed under their chemical identity.[1]

During these highly productive years prior to 1914 Svedberg published many scientific papers and two monographs: one (1909) on methods for preparing colloidal solutions of inorganic substances, the other (1912) describing his own experimental contribution to the prevailing but almost resolved discussion of whether molecules existed as particles or as a mathematical conception.

From 1913 to 1923 Svedberg continued his studies of the physicochemical properties of colloidal solutions with various co-workers. Toward the end of this period photochemical problems connected with the formation and growth of the latent image in the photographic emulsion also aroused his interest.

For some time it had been evident to Svedberg that, in order to gain further insight into the properties of colloidal solutions, he would need to know not only the mean sizes of their particles but also the frequency distribution of the particle sizes. For this purpose he and H. Rinde developed a method by which the variation in concentration with height in small sedimenting systems could be followed by optical means. The smallest particles that this method allowed them to determine in their metal sols were on the order of 200 mμ in diameter. Svedberg was interested in studying the formation and the growth of the colloidal particles, however, and thus the very small particles were especially important. To study them it would be necessary to increase the rate of sedimentation, which could be done by introducing centrifugal methods. The first attempt to do so was made in 1923, when Svedberg and J. B. Nichols constructed the first optical centrifuge in which the settling of the particles could be followed photographically during the run. No proper determination of the particle sizes could be made, however, because the particles were carried down both by sedimentation in the middle part of the cell and by combined convection and sedimentation along the cell walls in the nonsectorshaped cell. These experiments were carried out while Svedberg was guest professor at the University of Wisconsin at Madison. In recognition of his earlier work he had been invited by Professor J. H. Mathews to spend eight months in Madison giving lectures and organizing research in colloid chemistry. He accepted this invitation with enthusiasm. It gave him the opportunity to work in

a very stimulating atmosphere and proved to be a turning point in his scientific career.

Returning to Uppsala with many new ideas and great enthusiasm, Svedberg started a more extensive research program, including centrifugation, diffusion, and electrophoresis as methods for studying fundamental properties of colloidal systems. The most urgent problem was to improve and reconstruct the optical centrifuge to render it suitable for quantitative measurement of the sedimentation during the run of the centrifuge. With Rinde he introduced a sector-shaped cell for the solution, one of the requirements for obtaining convection-free sedimentation in a revolving centrifuge cell. They derived the important square-dilution law for sedimentation under these conditions. At first they still had great trouble with heat convection currents in the rotating solutions; in the spring of 1924, however, when the rotor was allowed to spin in a hydrogen atmosphere, the problem disappeared and a new tool was introduced into the study of colloidal solutions. This centrifuge made it possible to follow optically the sedimentation of particles too small to be seen even in the ultramicroscope. In analogy with the ultramicroscope and ultrafiltration, Svedberg and Rinde proposed the name "ultracentrifuge" for the new instrument. This first ultracentrifuge could produce a centrifugal field of up to 5,000 times the force of gravity, and the rate of sedimentation of gold particles could be determined for particles as small as about 5 mμ in diameter. The studies of the metal sols by this method were continued by Svedberg's pupils, especially by Rinde.

Svedberg's interest was now focused on determining whether his ultracentrifuge could be used in the study of other colloidal systems, such as proteins. Convinced that they were polydisperse, he sought to determine the frequency distribution of their particle sizes in solutions. The first experiments were disappointing; no sedimentation could be observed in solutions of egg albumin. Later experiments, conducted in the autumn of 1924, with native casein from milk showed a very broad frequency distribution, with coarse particles having diameters on the order of 10 to 70 mμ.

With Robin Fåhraeus, Svedberg tested hemoglobin, which actually sedimented (October 1924). After about two days of centrifugation, the first sedimentation equilibrium of a protein was established. A molecular weight of about 68,000 could be calculated from the variation in the hemoglobin concentration between the meniscus and the bottom of the cell. Combined with the known analytical value for the iron content of hemoglobin, it showed the presence of four iron atoms in the hemoglobin molecule; and within the experimental error of the method, the molecular weight was found to be constant throughout the cell. This finding came as a great surprise to Svedberg. Was it possible that the protein had a well-defined molecular weight? How could he test this hypothesis? Sedimentation equilibrium measurements might give some indication of the uniformity of the particles, but this method made it difficult to obtain more detailed information about the homogeneity of a dissolved protein. If the sedimentation velocity method could be used, an analysis of the shape of the boundary would reveal the presence of inhomogeneous material. This would demand a considerably increased centrifugal field, however; about 70,000 to 100,000 times the gravitational field would be necessary for a reasonable sensitivity. This involved increasing the centrifugal force then available by fifteen to twenty times. An entirely new type of centrifuge had to be constructed; and a number of new problems, concerning technique and safety, had to be discussed and solved.

On 10 January 1926 the new high-speed oil-turbine ultracentrifuge was ready. The test was disappointing; instead of the desired 40,000–42,000 rpm, only 19,000 rpm was reached. During the next three months the main troubles were overcome; and although a number of minor problems remained, Svedberg could start making routine runs with hemoglobin in the centrifuge and could work on the general question of the uniformity of the protein molecules. From these sedimentation velocity experiments he again concluded that hemoglobin in solution gave monodisperse particles. In the following years a number of different types of proteins were studied in the ultracentrifuge, and in many cases they were found to be paucidisperse (two or more distinctly different size classes were present). By fractionation or changes in pH such solutions often yielded monodisperse proteins. Besides casein, only one polydisperse protein was found: gelatin, which often was used as a "model protein" at that time. The most astonishing result was obtained, according to Svedberg, when the hemocyanin from the land snail *Helix pomatia* was centrifuged. From its copper content a minimum molecular weight of 15,000–17,000 had been calculated. It was expected, therefore, that a gradual change in the concentration in the cell should occur during the run, leading to a sedimentation equilibrium. But, on the contrary, the hemocyanin sedimented rapidly with a knife-sharp boundary,

indicating that the particles from this protein were giant molecules and all of the same size. The molecular weight was found to be on the order of five million.

After some years studying various proteins, Svedberg concluded that certain rules existed for the molecular weights of the proteins. In a letter to *Nature* (8 June 1929) he wrote:

> Our work has been rewarded by the discovery of a most unexpected and striking general relationship between the mass of the molecules of different proteins and the mass of the molecules of the same protein at different acidities, as well as of a relationship concerning the size and shape of the protein molecules.
>
> It has been found that all stable native proteins so far studied, can with regard to molecular mass be divided into two large groups: the haemocyanins with molecular weights of the order of millions and all other proteins with molecular weights from about 35,000 to about 210,000. Of the group of the haemocyanins only two representatives, the haemocyanin from the blood of *Helix pomatia* with a spherical molecule of weight 5,000,000 and a radius of 12.0 $\mu\mu$, and the haemocyanin from the blood of *Limulus polyphemus* with a non-spherical molecule of weight 2,000,000 have been studied so far.
>
> The proteins with molecular weights ranging from about 35,000 to 210,000 can, with regard to molecular weight, be divided into four sub-groups. The molecular mass, size, and shape are about the same for all proteins within such a sub-group. The molecular masses characteristic of the three higher sub-groups are—as a first approximation—derived from the molecular mass of the first sub-group by multiplying by the integers *two*, *three* and *six*.[2]

Many new proteins were subsequently studied, and it was found that some had molecular weights lower than 35,000, previously considered the lowest weight unit. An extended study of pH-influenced dissociation-association reactions also was carried out, particularly with the hemocyanins for which molecular weights in the range of millions were found—a startling discovery at that time.

> Moreover the weights of all the well-defined haemocyanin molecules seem to be simple multiples of the lowest among them. In most cases the haemocyanin components of a certain species are interconnected by reversible, pH-influenced dissociation-association reactions. At certain pH values a profound change in the number and percentage of the components take place. The shift in pH necessary to bring about reaction is not more than a few tenths of a unit. Consequently the forces holding dissociable

parts of the molecule together must be very feeble.

Not only the molecular weight of the haemocyanins, but also the mass of most protein molecules, even those belonging to chemically different substances, show a similar relationship. This remarkable regularity points to a common plan for the building up of the protein molecules. Certain aminoacids may be exchanged for others, and this may cause slight deviations from the rule of single multiples, but on the whole only a very limited number of masses seems to be possible. Probably the protein molecule is built up by successive aggregation of definite units, but that only a few aggregates are stable. The higher the molecular weight, the fewer are the possibilities of stable aggregation. The steps between the existing molecules therefore become larger and larger as the weight increases.[3]

With the new results the basic unit was assumed to be about 17,600 instead of 35,000; and the multiples were 2, 4, 8, 16, 24, 48, 96, 192, and 384.

Svedberg's discovery that in most cases the soluble proteins had molecules with a well-defined, uniform size was received with skepticism by many scientists. Traditionally the proteins had been regarded as colloids and as very complicated substances. Svedberg's hypothesis of the multiple law for the molecular weights of proteins elicited even greater skepticism, and some scientists wondered whether Svedberg and his co-workers were measuring artifacts. At the beginning of the 1930's, however, protein studies by other methods began to corroborate the finding that these substances had well-defined, uniform molecules. Later studies have confirmed the homogeneity of the proteins to an even greater extent than Svedberg had anticipated.

In the late 1930's and early 1940's, when studies of proteins other than the respiratory ones became more extended, severe criticism was raised against the multiple law. Many proteins were found, at Uppsala and elsewhere, with molecular weights considerably lower than 17,600. Other proteins that were studied did not fit into the system of multiples. Eventually it became evident that the multiple law did not have the generality that Svedberg had expected. This hypothesis was very important to the development of protein chemistry, however, especially in the 1930's and early 1940's. It initiated greater interest in proteins among chemists and provided an impetus for new work. The introduction of sedimentation velocity ultracentrifugation, and later of the Tiselius electrophoresis technique, made it possible to visualize much more directly to what extent the isolation of an individual protein had been successful.

Svedberg remained intensively engaged in the development of his ultracentrifuge. For the study of the homogeneity of the proteins he needed a centrifuge that could yield a higher centrifugal field than the 100,000 g attained with the earliest (1926) type of high-speed ultracentrifuge. Furthermore, he was anxious to see to what extent the ultracentrifuge method could be developed. Starting in 1930, the ultracentrifuge machinery was completely reconstructed. Until 1939 its rotors were gradually improved, most of the development being concentrated on those of standard size. The following values for the centrifugal field were attained; 200,000 g (1931), 300,000 g (1932), and 400,000 g (1933). Svedberg then sought to determine whether it was possible to make sedimentation studies in still more intense centrifugal fields, at one million g. In order to do so, it would be necessary to increase the speed of the rotor considerably; this could not be done with the standard-size rotors, because they would break long before the necessary speed was reached.

In 1933 and 1934 experiments were made with three smaller rotors. The first exploded during the test runs. With the second rotor a few successful runs were made at 900,000 g before it exploded. The third rotor was used for a few runs at 710,000 g and for many runs at 525,000 g (120,000 rpm). The solution cells gradually became greatly deformed in the high centrifugal fields, and the use of this rotor was discontinued.

Suspending further work on the small rotors, Svedberg now concentrated on the standard-size ones. The cell holes in the early rotors became deformed during the test runs and had to be ground to cylindrical form before being used for routine runs. Even so, a gradual deformation of the cell holes occurred during the use of the rotor and increased with time. A number of different rotor designs were constructed and tested before a satisfactory type was finally achieved in January 1939. Of the twenty-two rotors previously tested, seven standard-size and two small ones had exploded. After such explosions Svedberg was sometimes about to give up; there seemed to be no hope of finding a satisfactory design, and he wondered whether it was really worthwhile to devote further work to improving the ultracentrifuge, or whether it might not be better to concentrate on other problems. His interest in the proteins and his anxiety to prove or disprove his hypothesis of the multiple system for their molecular weights, however, inspired him to continue; and after his retirement he described this period as the happiest of his scientific life.

Toward the end of the 1930's, Svedberg extended his interest in macromolecules of biological origin to include the polysaccharides. With N. Gralén he found that the sap of the bulbs from various species of Liliifloreae contained soluble high-molecular-weight substances. The various species yielded widely different sedimentation diagrams because they contain proteins and polysaccharides of different properties and in different proportions. Two classes of carbohydrates could be distinguished by their sedimentation behavior, and a similarity among the species of the same genus was generally found with regard to the content of high-molecular-weight material that could elucidate problems in systematic botany.

In the 1940's Svedberg and his co-workers extended their investigations to other natural polysaccharides, primarily to determine parameters of molecular size and shape. The gradual shift of his interest to the study of cellulosic materials, particularly wood cellulose and cellulose nitrates, led to close cooperation with the research laboratories of the biggest cellulose manufacturers in Sweden.

The outbreak of World War II forced Svedberg to take up activities connected with the war. Owing to the blockade, no oil-resistant rubber could be obtained from abroad; and Svedberg was charged with developing Swedish production of synthetic rubber (polychloroprene). For several years more than half of the research facilities of his laboratory was used for this development and for pilot plant work. The project was successful and led to a small production plant in northern Sweden. The government and other state agencies demanded much of Svedberg's time. Institutes and industries closely related to war materials sought his help and his advice. The planning necessary for combining these activities with the work at his institute gave him fewer opportunities than before for his own experimental work and for contact with his co-workers and students.

In connection with the work on synthetic rubber, Svedberg took physicochemical studies on other synthetic polymers. He was always looking for new experimental techniques to be used in the study of these high polymers and of cellulose. With I. Jullander he developed an osmotic balance by means of which low osmotic pressures could be determined by weighing. Electron microscopy was

introduced into the study of the structure of native and regenerated celluloses. X-ray techniques were used for cellulosic fibers and electron diffraction for investigation of micelles and crystallites.

Svedberg's interest in radiation chemistry was revived in the late 1930's with investigations using hemocyanins to study the effect of ultraviolet light, α particles, and ultrasonics on solutions of these proteins. In the 1940's, a small neutron generator was built by one of his collaborators, H. Tyrén, to study the action of an uncharged particle on proteins. After being completed, however, it was used mainly for the production of a small amount of radiophosphorus and a few other radioactive isotopes needed at some of the medical institutes at Uppsala. The question soon arose of how to increase the capacity for making radioactive isotopes. The construction of a cyclotron at Uppsala would immensely increase that capacity and would open new fields for research in radiation chemistry there.

One of Svedberg's old friends at the Medical Faculty proposed that he approached Gustaf Werner, a wealthy industrialist in Göteborg, about the possibility of obtaining financial help to build a large cyclotron. The response was very positive, partly because of Werner's interest in the possible medical application of such research. In the spring of 1946, he promised to give one million Swedish crowns for the construction of a cyclotron. Svedberg immediately made plans for the cyclotron and for the building of an adjoining research institute. He even obtained extra funds from the government, and construction of the building was started in February 1947. Svedberg had now decided to devote his time to the creation of this new research institute and to the planning of the work with the cyclotron. In December 1949, some months after his retirement from the chair of physical chemistry, the Gustaf Werner Institute of Nuclear Chemistry was officially inaugurated, and Svedberg received permission to continue his activities as head of that institute. It took another two years, however, before some important model experiments with the magnet and the oscillator were satisfactory and the final installations could be made. In the late fall of 1951, the necessary equipment had been acquired, and the 185 MEV synchro-cyclotron was in full operation.

Although Svedberg had brought a few co-workers from his old institute, he had to assemble a new staff and find students in order to build the research organization for his new institute. Many different problems were studied. One group worked mainly on the biological and medical application of the cyclotron; others investigated the effect of radiation on macromolecules, problems in radiochemistry, and radiation physics. The institute soon became the center in Sweden for research in the area between high-energy physics, chemistry, and biology. Svedberg's own interest in research remained intense throughout his life, and he followed all the work in progress and advised his students and collaborators. He even took active part in some research projects. The last publication bearing his name (1965) deals with recent developments in high-energy proton radiotherapy.

NOTES

1. *Nobel Lectures in Chemistry 1901–1921* (Amsterdam, 1966), 381.
2. *Nature*, **123** (1929), 871.
3. *Ibid.*, **139** (1937), 1061.

BIBLIOGRAPHY

I. ORIGINAL WORKS. Svedberg published 240 papers and books, only a few of which can be mentioned here. A complete bibliography is in the biography by Claesson and Pedersen (1972). His first paper was "Ueber die elektrische Darstellung einiger neuen colloidalen Metalle," in *Berichte der Deutschen chemischen Gesellschaft*, **38** (1905), 3616–3620. His early work on colloids was summarized in his dissertation, "Studien zur Lehre von den kolloiden Lösungen," in *Nova acta Regiae societatis scientiarum upsaliensis*, 4th ser., **2**, no. 1 (1907), 1–160. His contribution to the preparation of colloidal solutions was given in *Die Methoden zur Herstellung kolloider Lösungen anorganischer Stoffe* (Dresden–Leipzig, 1909; 3rd ed., 1922). The work on isomorphic coprecipitation of radioactive compounds is "Untersuchungen über die Chemie der radioaktiven Grundstoffe. I–II," in *Zeitschrift für anorganische Chemie*, **61** (1909), 338–346, and **63** (1909), 197–206, written with D. Strömholm. His experimental contributions to the discussion of whether molecules exist as particles were published in *Die Existenz der Moleküle* (Leipzig, 1912). The first papers dealing with centrifugal methods were "The Determination of the Distribution of Size of Particles in Disperse Systems," in *Journal of the American Chemical Society*, **45** (1923), 943–954, written with H. Rinde; "Determination of Size and Distribution of Size of Particle by Centrifugal Methods," *ibid.*, 2910–2917, written with J. B. Nichols; and "The Ultra-Centrifuge, a New Instrument for the Determination of Size of Particle in Amicroscopic Colloids," *ibid.*, **46** (1924), 2677–2693, written with H. Rinde.

Svedberg's Wisconsin lectures were published in *Colloid Chemistry* (New York, 1924; 2nd ed., rev. and enl. in collaboration with Arne Tiselius, New York, 1928). The first ultracentrifugal determination of the molecular weight of a protein appeared in "A New Method for the Determination of the Molecular Weight of the Proteins," in *Journal of the American Chemical Society*, **48** (1926), 430–438, written with Robin Fåhraeus. Svedberg's Nobel lecture was originally published in Swedish as "Nobelföredrag hållet i Stockholm den 19 maj 1927," in *Les prix Nobel en 1926* (Stockholm, 1927), 1–16; an English version is "The Ultracentrifuge," in *Nobel Lectures in Chemistry 1922–1941* (Amsterdam, 1966), 67–83. The first paper dealing with the hypothesis of the multiple system for the molecular weights of the proteins is "Mass and Size of Protein Molecules," in *Nature*, **123** (1929), 871; two later papers dealing with the same hypothesis are "The Ultracentrifuge and the Study of High-Molecular Compounds," *ibid.*, **139** (1937), 1051–1062; and "A Discussion on the Protein Molecule," in *Proceedings of the Royal Society*, **A170** (1939), 40–56, also *ibid.*, **B127** (1939), 1–17. A comprehensive account of the ultracentrifuge is given in *The Ultracentrifuge* (Oxford, 1940; repr. New York, 1959), written with K. O. Pedersen. The detailed study of the polysaccharides appeared in "Soluble Reserve-Carbohydrates in the Liliifloreae," in *Biochemical Journal*, **34** (1940), 234–238, written with N. Gralén. The osmotic balance was first reported in "The Osmotic Balance," in *Nature*, **153** (1944), 523, written with I. Jullander. The work on cellulose is described in "The Cellulose Molecule. Physical-Chemical Studies," in *Journal of Physical and Colloid Chemistry*, **51** (1947), 1–18.

II. SECONDARY LITERATURE. A complete bibliography of Svedberg's books and papers is in S. Claesson and K. O. Pedersen, "The Svedberg 1884–1971," in *Biographical Memoirs of Fellows of the Royal Society*, **18** (1972), 595–627. Other publications dealing with Svedberg's life are N. Gralén, "The Svedberg 1884–," in S. Lindroth, ed., *Swedish Men of Science 1650–1950* (Stockholm, 1952), 271–279; P.-O. Kinell, "Theodor Svedberg. Kolloidchemiker-Molekülforscher-Atomfachmann," in H. Scherte and W. Spengler, eds., *Forscher und Wissenschaftler im heutigen Europa—Weltall und Erde* (Oldenburg, 1955), 191–198; A. Tiselius and S. Claesson, "The Svedberg and Fifty Years of Physical Chemistry in Sweden," in *Annual Review of Physical Chemistry*, **18** (1967), 1–8; and A. Tiselius and K. O. Pedersen, eds., *The Svedberg 1884–1944* (Uppsala, 1944), published in honor of Svedberg's sixtieth birthday.

STIG CLAESSON
KAI O. PEDERSEN

SVEDELIUS, NILS EBERHARD (*b.* Stockholm, Sweden, 5 August 1873; *d.* Uppsala, Sweden, 2 August 1960), *phycology.*

Svedelius' father, Carl, was a Supreme Court justice; and his father's family included numerous intellectuals, clergymen, governmental officials, army officers, scholars, and teachers. His mother, Ebba Katarina Skytte of Sätra, belonged to the untitled nobility. At a young age he showed keen interest in botany; and in 1891 he entered the University of Uppsala, where F. R. Kjellman was one of his teachers. The marine algal flora of high latitudes was a main interest of Kjellman, who had been the botanist of the *Vega* expedition, which in 1878–1880 accomplished the Northeast Passage. In Sweden the marine algae had been studied mainly at the North Sea coast, whereas the depauperated flora of the Baltic Sea had been neglected. Svedelius therefore selected as the topic for his doctoral dissertation the algal flora of the southern Baltic coast of Sweden, with special emphasis on their morphological and ecological responses to the decreased salinity. In 1901 he received the doctorate and became a docent the following year. Svedelius was appointed professor of botany in 1914 and retired in 1938. Although he retained a lifelong interest in the algae of the Baltic, Svedelius made his contributions of fundamental importance in a totally different branch of phycology—in the elucidation of the various life cycles in the red algae (Florideae) and their evolutionary value.

In 1902–1903 Svedelius visited the tropics, spending most of his time in Ceylon, particularly at Galle, which has a large coral reef where he surveyed the marine algae and their taxonomy, ecology, and distribution. Rather little was ever published from this journey; but it focused his interest on the red algae, and his experiences in Ceylon were in many ways valuable to his later investigations.

The introduction of the microtome and fixing and staining techniques at that time opened a new era in biology. The alternation between a gametophytic and a sporophytic generation had long been known in ferns and mosses, as had the existence of different kinds of reproductive bodies in algae; but no regular pattern could be discerned in the latter group. In 1904 J. L. Williams established the alternation between two externally similar generations in a brown alga, and two years later S. Yamanouchi found it in a red alga, the formation of tetraspores being preceded by meiosis. Realizing that the technical means were now at hand, Svedelius decided to investigate reproduction in the red algae. After establishing one more case of regular alternation, he turned to *Nitophyllum punctatum* (1914), in which tetraspores were known to occur both on special (sporophytic) and on gametophytic individuals. The latter were, as expected, haploid. Their apparent tetraspores were formed without meiosis and thus were not true tetraspores, but a kind of accessory spore. In 1915 Svedelius showed that *Scinaia furcellata*, a species without tetraspores, does not possess independent alternating generations because the first nuclear division after fertilization is meiotic and the carpospores thus become haploid. For the rest of his life (his last paper was written in 1955) he continued along these lines, studying additional species, finding new complications and modifications in the life cycles, and trying to trace their phylogenetic connections. All his publications are exhaustive, detailed, exact, and profusely illustrated by his own instructive and beautiful drawings.

From the beginning of his life-cycle studies Svedelius was interested in the general significance of the alternation of generations in the plant kingdom and was perhaps the first to point out that the acquisition of a sporophyte must be of immense evolutionary value: in organisms in which the zygote is the only diploid cell, one fertilization is followed by only one meiosis, whereas in organisms with a full diploid generation, one fertilization is followed by numerous meioses, each with its own possibilities of genetic segregations.

Svedelius was not exclusively a phycologist. He also published papers on floral biology and on seed anatomy. His inherent interest in organization, administration, and economics engaged him deeply in the affairs of his university and of several academies and learned societies. In 1935–1950 he was president of the Botanical Section of the International Biological Union.

BIBLIOGRAPHY

I. ORIGINAL WORKS. Svedelius' writings include *Studier öfver Österjöns hafsalgflora* ("Studies of the Marine Algae of the Baltic Sea"; Uppsala, 1901), his dissertation; "Ueber den Generationswechsel bei *Delesseria sanguinea*," in *Svensk botanisk tidskrift*, **5** (1911), 260–324; "Ueber die Tetradenteilung in den vielkernigen Tetrasporangiumanlagen bei *Nitophyllum punctatum*," in *Berichte der Deutschen botanischen Gesellschaft*, **32** (1914), 48–57; "Ueber Sporen an Geschlechtspflanzen von *Nitophyllum punctatum*, ein Beitrag zur Frage des Generationswechsels der Florideen," *ibid.*, 106–116; "Zytologisch-entwickelungsgeschichtliche Studien über *Scinaia furcellata*. Ein Beitrag zur Frage der Reduktionsteilung der nicht tetrasporentragenden Florideen," in *Nova acta Regiae societatis scientiarum upsaliensis*,

4th ser., **4** (1915), 1–55; "Generationsväxlingens biologiska betydelse" ("The Biological Significance of the Alternation of Generations"), in *Svensk botanisk tidskrift,* **12** (1918), 487–490; "Alternation of Generations in Relation to Reduction Division," in *Botanical Gazette,* **83** (1927), 362–384; "The Apomictic Tetrad Division in *Lomentaria rosea* in Comparison With the Normal Development in *Lomentaria clavellosa,*" in *Symbolae botanicae upsalienses,* **2** (1937), 1–54; and "Zytologisch-entwickelungsgeschichtliche Studien über *Galaxaura*, eine diplobiontische Nemalionales-Gattung," in *Nova acta Regiae societatis scientiarum upsaliensis,* 4th ser., **13** (1942), 1–154.

II. SECONDARY LITERATURE. More detailed biographies are G. F. Papenfuss, "Nils Eberhard Svedelius. A Chapter in the History of Phycology," in *Phycologia,* **1**, no. 4 (1961), 172–182, with a complete bibliography of his phycological papers; and C. Skottsberg, "Nils Eberhard Svedelius 1873–1960," in *Biographical Memoirs of Fellows of the Royal Society,* **7** (1961), 295–312, with a complete bibliography of his works.

J. A. NANNFELDT

SVERDRUP, HARALD ULRIK (*b.* Sogndal, Norway, 15 November 1888; *d.* Oslo, Norway, 21 August 1957), *geophysics.*

Sverdrup was the son of Johan Edvard Sverdrup, a fundamentalist clergyman and teacher, and Maria Vollar, who was related to the Grieg family. The Sverdrup family itself contained a number of prominent educators, industrialists, artists, and politicians, including Johan Sverdrup, prime minister of Norway and an important figure in the introduction of parliamentarianism and social reform in the 1870's and 1880's. Sverdrup was also distantly related to the arctic explorer Otto Sverdrup, a companion of Nansen, for whom the Sverdrup Islands in the Canadian Arctic were named.

Sverdrup received his early education at home. He was much interested in evolution and the natural sciences, but on entering high school in 1901 honored his parents' wishes by studying classical languages. In 1906 he passed the elementary examinations at the University of Norway, then entered the military academy there, graduating as a reserve officer in 1908. He then began studies in the faculty of sciences. At first interested primarily in astronomy, Sverdrup was an enthusiastic and able student, and was soon discovered by Vilhelm Bjerknes, professor of physics and the leading authority on atmospheric circulation. Sverdrup became Bjerknes' assistant in 1911 and was one of several students who followed him when he transferred to the University of Leipzig two years later. Sverdrup remained in Leipzig for four years, during which he published some twenty papers, either alone or in collaboration. One of these was his doctoral dissertation, "Der nordatlantische Passat," published in 1917. Sverdrup's work of this period was, not surprisingly, strongly influenced by that of Bjerknes, and much of it was concerned with the circulation of the atmosphere. Some of his models and calculations were remarkably exact, and have since been confirmed by more precise measurements.

By 1917 conditions at Leipzig had become intolerable because of World War I, and Bjerknes returned to Norway to become professor of geophysics at the University of Bergen. Sverdrup also returned to Norway, and was soon engaged by the arctic explorer Amundsen to act as chief scientist on an expedition to the North Pole. Sverdrup was eager to go, since the atmospheric conditions that he was studying could be observed more readily in the uniform arctic climate. Since Amundsen's ship, the *Maud*, was small, each crew member had to assume several jobs, and Sverdrup served as navigator and cook, as well as scientist.

The *Maud* expedition left Norway in the summer of 1918. It failed to reach the North Pole, largely because adverse ice and current conditions prevented a regular drift over the polar basin, but Sverdrup was nonetheless able to conduct important research on atmospheric circulation and the magnetic field of the earth. In addition, he became interested in ethnography when the expedition came into contact with some of the tribes, particularly the Chukchi, of northeastern Siberia. This interest, together with an admiration for the culture of primitive peoples, lasted the rest of his life.

In the summer of 1922 the *Maud* was docked for repairs at Seattle and Sverdrup used the occasion to work at the Carnegie Institution of Washington, where he began to interpret the magnetic observations that he had made. He served as sole leader of the expedition when it was resumed in the same year, and continued making oceanographic and meteorological observations until the venture ended three years later. Even though the ship had not actually crossed the pole, Sverdrup was able to unravel the complicated dynamics of the tides in the polar basin. He transcribed part of his results on shipboard, then published them in 1926, the year in which he returned to Norway to succeed Bjerknes as professor of geophysics at the University of Bergen. His complete account of the *Maud* expedition, in which he drew upon observational data to explain the general features of the arctic atmospheric and oceanic circulation and energy distribution, was published in 1933.

In Bergen, Sverdrup also studied the magnetic and oceanographic data obtained by American expeditions in the Pacific and Antarctic oceans. In 1931 he took up an independent research position at the Christian Michelsens Institute and also participated in George Wilkins' adventurous but premature attempt to reach the North Pole by submarine, a voyage that allowed him to make important observations of the deep sea north of Spitsbergen. In 1934 he studied the glaciers of Spitsbergen and, with H. W. Ahlmann, developed the study of the energy balance of glaciers. The following year he went to La Jolla, California, to become director of the Scripps Institute of Oceanography.

At Scripps Sverdrup was primarily concerned with the turbulent processes in the boundary layer between the atmosphere and the sea, although he also found time to work in other areas of geophysics and to write *The Oceans* (1942), a monumental handbook of oceanography that remains an important introduction to the subject. When, during World War II, Scripps became deeply involved in military research, Sverdrup made a number of significant contributions; his precise predictions of tides and the height of waves were particularly valuable in the course of the Pacific war.

In 1948 Sverdrup assumed the directorship of the Norwegian Polar Institute in Oslo. He reorganized the institute, arranged the Norwegian-British-Swedish expedition to Antarctica of 1949–1952, and organized Norwegian participation in the International Geophysical Year 1957–1958. In 1949 he became professor of geophysics at the University of Oslo, then dean of the faculty of science and vice-director of the university. Interested in curriculum reform, he was chairman of a committee for revising the university course of studies somewhat along the lines of those used in the American university system. His demonstrated administrative abilities led to his appointment as chairman of the Norwegian relief program in India. Under his leadership this program was planned to introduce Norwegian technology, particularly as it applied to the fishing industry, to the underdeveloped areas along the Cochin coast without disrupting existing cultural patterns. This plan met with a degree of success that must be attributed to Sverdrup's diplomatic skills and to his profound sympathy for foreign cultures. In the midst of these activities, he died suddenly of a heart attack while attending a meeting.

BIBLIOGRAPHY

I. ORIGINAL WORKS. Sverdrup's most important scientific works are "Der nordatlantische Passat," in *Veröffentlichungen des Geophysikalischen Instituts der Universität Leipzig*, ser. 2, **B 2** (1917); "Dynamic of Tides on the North Siberian Shelf, Results from the Maud Expedition," in *Geofysiske publikasjoner*, **4**, no. 5 (1927); *Scientific Results. The Norwegian North Polar Expedition with the "Maud,"* 3 vols. (Bergen, 1927–1933); *The Oceans*, with M. W. Johnson and R. H. Fleming (New York, 1942).

A complete list of Sverdrup's scientific papers is given in S. Richter, "Bibliografi over H. U. Sverdrups arbeider," in *Det Norske Videnskaps-Akademi i Oslo Årbok 1958* (Oslo, 1959).

II. SECONDARY LITERATURE. The only real biography of Sverdrup is O. Devik, "Minnetale over professor Harald U. Sverdrup," in *Det Norske Videnskaps-Akademi i Oslo Årbok 1958* (Oslo, 1959).

NILS SPJELDNAES

SWAINSON, WILLIAM (*b*. Newington Butts, London, England, 8 October 1789; *d*. Wellington, New Zealand, 7 December 1855), *zoology*.

William Swainson was the eldest surviving son of John Timothy Swainson II, collector of customs at Liverpool and lord of the manor of Hoylake, Cheshire, and his second wife, Frances Stanway.

At the age of fourteen Swainson entered the service of H.M. Customs and Excise but was handicapped by a serious impediment in his speech. His father therefore obtained for him a post in the army commissariat, and in 1807 Swainson was posted first to Malta and then to Sicily. While in Sicily he made extensive collections of botanical and zoological specimens, especially fishes, and became friendly with the eccentric Constantine S. Schmaltz Rafinesque. Swainson visited Greece and Italy, where he was also stationed for a while. In 1815 he retired on half-pay.

Swainson next visited Brazil with Henry Koster, spending part of the years 1817 and 1818 there. Upon his return to England in 1818 Swainson published a brief note on his travels but, disappointed by lack of encouragement, he did not prepare a full account. When his Brazilian material, after long delays, was distributed to specialists for their use, others had forestalled him in describing the new species he had found and his pioneer work was not recognized.

In 1823 Swainson married Mary Parkes of Warwick and thereafter engaged in scientific writing for a living, producing many books and papers during the next seventeen years. He wrote on vertebrates, mollusks, and insects, and he contributed sections on farm and garden pests to Loudon's *Encyclopaedia of Agriculture* and *Encyclopaedia of Gardening*, illustrating all his own work. Unfortunately, Swainson undertook far too much, because of financial pressures, and at times fell seriously behind schedule. Overwork, his wife's death in 1835, and financial losses, as well as a second unsuccessful attempt to obtain a post in the British Museum, caused his decision to emigrate to New Zealand in 1840, and to abandon his scientific writings.

He did, however, publish a few small papers after his emigration, and in 1851–1853 he reported on the timber trees of Victoria, New South Wales, and Tasmania.

William Swainson was a good zoologist, but his unfortunate adherence to the "quinary system" has distorted some of his work. The quinary or circular system, first suggested by William Sharpe Macleay, and eagerly adopted by Swainson, professed that the relationships within any zoological group could be expressed by a series of interlocking circles, and that the "primary circular divisions of any group were three actually, or five apparently." This extraordinary theory was pertinaciously held by Swainson throughout his zoological career and it certainly impaired much of his work. A. Newton and H. Gadow (*A Dictionary of Birds* [1896], p. 35) stated the matter fairly when they wrote that Swainson's indefatigable pursuit of natural history and conscientious labor on its behalf deserve to be remembered as a set-off against the injury he unwittingly caused by his adherence to the absurd quinary system.

Swainson's artistic achievements were of high merit, and he was a pioneer in the use of lithography. His botanical work is unimportant; his claim to remembrance rests upon his zoological work and upon his fine zoological illustrations.

BIBLIOGRAPHY

I. Original Works. Only separate works are listed here since Swainson's scientific papers are listed in the Royal Society *Catalogue of Scientific Papers.* They are *Instructions for Collecting and Preserving Subjects of Natural History and Botany* (Liverpool, 1808); *Zoological Illustrations*, 3 vols. (London, 1820–1823); 2nd. ser. 3 vols. (London, 1829–1833); *Exotic Conchology* (London, 1821–1822), 2nd ed., S. Hanley, ed. (London, 1841), facs. (with additions), R. T. Abbott, ed. (Princeton, 1968); *The Naturalist's Guide,* 2nd. ed.

(London, 1822); *A Catalogue of the Rare and Valuable Shells of the Late Mrs. Bligh. With an Appendix Containing Scientific Descriptions of many New Species and Two Plates* (London, 1822); *Fauna Boreali-Americana; or the Zoology of the northern parts of British America,* pt. 2. *The Birds* (London, 1831), written with John Richardson; and the following volumes of Lardner's *Cabinet Cyclopaedia* (London); *A Preliminary Discourse on the Study of Natural History* (1834); *Treatise on the Geography and Classification of Animals* (1835); *Natural History and Classification of Quadrupeds* (1835); *Animals in Menageries* (1837); *Natural History and Classification of Birds,* 2 vols. (1836–1837); *Natural History of Fishes, Amphibians and Reptiles,* 2 vols. (1838–1839); *Habits and Instincts of Animals* (1839); *Taxidermy, with the Biography of Zoologists* (1840); *A Treatise on Malacology* (1840); *On the History and Natural Arrangement of Insects* (1840), written with W. E. Shuckard. He also wrote *Elements of Modern Conchology* (London, 1835); *Birds of Western Africa,* 2 vols. (Edinburgh, 1837); and *Flycatchers* (Edinburgh, 1838) for the Naturalist's Library.

Other works to which Swainson contributed include J. C. Loudon, *An Encyclopaedia of Gardening,* new ed. (London, 1834); J. C. Loudon, *An Encyclopaedia of Agriculture,* 4th ed. (London, 1839); and Hugh Murray, *An Encyclopaedia of Geography* (London, 1834).

II. Secondary Literature. No full-scale biography of Swainson has yet been published. Biographical notices in various biographical dictionaries are mostly inaccurate. See obituaries in *Proceedings of the Linnean Society of London* (1855–1856), xlix, and in *Gentleman's Magazine* (1856), 532; and Iris M. Winchester, "William Swainson, F.R.S., 1789–1855 and Henry Gabriel Swainson, 1830–1892," in *Turnbull Library Record,* n.s., **1** (1967), 6–19.

Nora F. McMillan

SWAMMERDAM, JAN (*b.* Amsterdam, Netherlands, 12 February 1637; *d.* Amsterdam, 17 February 1680), *biology.*

Swammerdam's father, Jan Jacobszoon Swammerdam, son of a timber merchant, was an apothecary who in 1632 had married Baertje Jans Corvers. The couple's first two children died in early childhood. In 1640 was born Jan's brother, Jacobus, who became an apothecary, and in 1642 a sister, Jannetje. Jan's mother died in 1661. His father had acquired some fame as a collector of curios, including minerals, coins, and animals from all over the world. As a boy, Jan helped care for this collection.

He matriculated in medicine at the University of Leiden on 11 October 1661. Jan's own collection of insects impressed his schoolmates, including Regnier de Graaf, Frederik Ruysch, and Niels Stensen (Steno). Robertus Padtbrugge was a fellow student who later joined the East India Company and sent Swammerdam exotic animals. The eminent professors Franciscus dele Boë Sylvius and Johannes van Horne both refer in their publications to Swammerdam's student researches. He qualified as a candidate in medicine in October 1663, and then spent some time in Saumur, France, staying with Tanaquil Faber, a professor of philology at the Protestant university there.

From about September 1664, Swammerdam lived in Paris as the guest of Melchisédech Thévenot. He was an active member, as was his friend Steno, of Thévenot's scientific academy, an informal club that met to watch experiments and dispute over Cartesian ideas. Returning to Amsterdam about September 1665, Swammerdam joined a group of physicians calling themselves the Private College of Amsterdam, which included Gerhard Blaes (Blasius) and Matthew Slade. The group met irregularly until 1672 and published a description of their dissections. In the winter of 1666–1667, Swammerdam was again in Leiden, where he dissected insects and collaborated with van Horne on the anatomy of the uterus. His medical thesis on respiration, based largely on research carried out in 1663, earned him the M.D. on 22 February 1667. There is no evidence of his ever settling into a medical practice,[1] although in 1670 he was granted the privilege of dissecting bodies in Amsterdam, and he does allude once to being kept from his research by the demands of the seriously ill.[2]

Driven by an inner passion and encouraged by Thévenot and other friends, Swammerdam devoted his life to scientific investigation, but he was interrupted by illness, by his father's insistence that he earn a living, and by periods of depression and religious anxiety. He stayed for a time at The Hague, perhaps as a guest of Maurice, prince of Nassau, who had fishermen bring him specimens, and he occasionally visited Leeuwenhoek in Delft.[3] He lived during the summer of 1670 in the village of Sloten, just outside Amsterdam, where he studied mayflies.

The mystical prophetess Antoinette Bourignon was accompanied in her exile by a friend of Swammerdam's. Jan wrote to her for spiritual comfort on 29 April 1674, and asked her permission before publishing his researches on the mayfly. He visited her in Schleswig-Holstein, between September 1675 and June 1676. On 18 July 1675 Steno sent

Swammerdam's drawings of silkworm anatomy to Malpighi, reporting that the author had destroyed the manuscript text and that they should pray for their friend in his search for God.[4] But Swammerdam's rejection of science was not final, for in January 1680 he provided in his will for the publication of his manuscripts, and these give evidence of having been revised in the last year or two of his life. Not until a half century had passed, however, were his wishes carried out; Boerhaave published the *Biblia naturae* in 1737–1738.

Swammerdam's biological researches fall into two distinct categories, although all were characterized by a preference for mechanistic types of explanation supported by great originality of technique and experiment. His studies of insects have a special quality all their own, while most of his other anatomical and physiological work may be called medical. The medical research was conducted within the fabric of currently fashionable theories, often actually in the company of his colleagues. This category includes his thesis on respiration, his book on the anatomy of the uterus, and other scattered notes. His accomplishments in this area were well known to his contemporaries but occasioned several priority disputes. By contrast, his work on insects was in a sense a private quest and remained largely unpublished during his lifetime. The theme of this work was essentially anti-Aristotelian, for he claimed that insects are no less perfect than higher animals and are not really different in their modes of development.

Swammerdam's medical thesis offers a perfectly Cartesian mechanical explanation of the motion of the lungs and the function of breathing, supplemented by the iatrochemistry of Sylvius. Swammerdam struggled to avoid using any attractive powers, whether of the mouth, of the lungs themselves, or of a partial vacuum, to explain the rushing of air into the lungs. Apparently ignorant of Boyle's idea that air has a springiness, Swammerdam argued that the muscular expansion of the chest outward pushes the ambient air down into the lungs. He dramatized this process with a submerged dog that could breathe through a tube. When the dog inhaled, the level of the water's surface rose, but when the tube was stopped up, the lungs would not follow the chest in its expansion. The normal rising and falling of the lungs was simply the result of air having been pushed into them from outside, not of the motion of the thorax. (John Mayow, using Boyle's ideas, easily destroyed Swammerdam's argument in his tract on respiration in 1668.) Swammerdam described a very curious set of experiments, in which he produced bubbling by drawing air out of sealed containers partly filled with water; the experiments seem to be meant to show that the same effects producible by mouth suction can be duplicated with a syringe. Again, his point was that the action of breathing is mechanical.

Swammerdam seems to have believed that when air in a confined space is pushed, the finer particles of air will run out of the container and the remaining air will then consist of heavier parts and will therefore be dense. This idea of the nature of air pressure allowed him to explain what happens in respiration thus: the expanded lungs press on the air contained in the pleural cavity, thus forcing the subtler parts out of this cavity into the heart. He had already demonstrated this process in January 1663, when his professors and fellow students saw him force air into the lungs of a living dog to produce a visible effervescence in the dog's heart. While agreeing with Harvey that the heart does contract of its own power, Swammerdam integrated this with the Cartesian picture of the blood moving by virtue of its increase in volume as well, first in the right side before being cooled by the lungs, and then again when finally perfected on the left side, under the influence of the fine, subtle particles of air. Like Descartes, Swammerdam suggested that emotions like joy, anger, fear, and happiness depend upon the various degrees of motion of the blood in respiration.

Although so many of Swammerdam's ideas about respiration appear misbegotten in the light of modern theory, or even in comparison to those of his contemporaries at Oxford, the assumptions and experimental technique behind them were essentially the same as those used in his classic proof that muscles do not increase in volume upon contraction. Again he was Cartesian in interpreting the old animal spirit, passing from nerves into muscle, as a very subtle but material fluid. Testing the reaction of various muscles of many different animals, Swammerdam learned that the frog was the best subject for this kind of experiment, and it was with this animal that he gave demonstrations to Steno about 1663 and to the Grand Prince Cosimo III in 1669.

The first part of Swammerdam's demonstration was simply an elaboration of the common observation that a muscle can contract when separated from the body. Using both the heart and a long muscle of the frog separated from the body and the blood vessels, he urged that the fact that they would repeatedly respond to the stimulation of the nerve showed that contraction did not depend on an influx of matter from the brain or of blood, both current theories. Simply placing a muscle in a glass tube made the thickening of its belly and the shortening of its length clearly visible. The enlargement of the belly of a muscle had suggested to a number of men that contraction consists of effervescence or some other increase in volume, but Swammerdam devised an elegant experiment to measure the change in volume, using his favorite instrument. Placing a muscle inside a syringe, the mouth of which was blocked by a drop of colored water, he was able to demonstrate that the contraction of the muscle or heart was not accompanied by an outward motion of the drop; if there was motion at all, it was inward, suggesting a slight decrease in volume. Swammerdam did not suggest an alternate mechanism of contraction, nor did he deny the existence of a matter carried through the nerves, but simply noted that this matter must be of insensible volume. His description and drawings of this investigation were first published in 1738, but they were well known to Steno and through him to others, including Croone and Borelli.[5]

It was at Saumur that Swammerdam perfected a technique for displaying the valves of lymphatic vessels, although the existence of these valves was already known. Ruysch, who at Leiden had also been interested in techniques of anatomical preservation, published a description of a similar preparation in 1665. Swammerdam found that his father had shown his drawings to all who were interested, and although Ruysch was probably innocent, Swammerdam expressed annoyance.[6]

During his second stay at Leiden, in the winter of 1666–1667, Swammerdam and van Horne collaborated on the anatomy of human reproductive organs, both male and female. Swammerdam used wax injection to make the vessels distinct. Although van Horne described this anatomical work in 1668, Swammerdam did not publish his drawings of the preparations until de Graaf made public his own investigations on the same subject. Swammerdam then published *Miraculum naturae*, which he sent, along with the preserved specimens themselves (now lost), to the Royal Society asking that his priority be acknowledged. Besides the technique of wax injection, one of the discoveries in dispute was the very nature of human reproduction. According to van Horne, Steno, Swammerdam, and de Graaf, the organs that had been called female testes were really ovaries, like those of egg-laying animals. Swammerdam claimed to have seen eggs in them. Baer in the nineteenth century is more properly credited with this observation, but the important concept that mammals do have ovaries was accepted in the seventeenth century. Although the Royal Society arbitrated the question of priority in favor of Steno, Harvey's pronouncement that all life originates in eggs made their anatomical search a natural one.

In association with the Private College of Amsterdam, Swammerdam dissected the pancreas of fish and analyzed the pancreatic fluid in the context of Sylvius' theory of digestion, and he also described the formation of hernias. The Private College decided to publish as a body, but Swammerdam later noted that certain of their discoveries were entirely his own. He had shown that the spinal marrow consisted of fibers that terminate in the brain and from which nerves proceed out.

Besides these medical studies, Swammerdam pursued a lifelong inquiry into the nature of lower animals. A visitor of 1662 noted that Swammerdam owned a colored copy of Mouffet's entomology.[7] He was actively collecting, observing, and dissecting insects in Saumur, in and around Paris, in Leiden, and to the end of his days in the countryside around Amsterdam. All he managed to publish was the *Historia insectorum generalis*, Part I, and a monograph on the mayfly, which, in the period of his religious crisis, became the occasion for an extended hymn to the Creator. But he left explicit instructions in his will for the publication of the rest of his entomological studies, and Boerhaave was probably accurately carrying out Swammerdam's intentions when he integrated the text of the *Historia* (slightly revised) with the unpublished manuscripts, using it as a framework that further researches filled in.

Swammerdam's thesis about insects was fundamentally new and significant. For his contemporaries, as for Aristotle, there existed three good arguments that not only placed the insects far from higher animals, but even tended to remove them from the realm of subjects open to scientific study. These arguments were: insects lack internal anatomy; they originate by spontaneous generation; and they develop by metamorphosis. Swammerdam believed that all three arguments were false and devoted a wide variety of investigations to refute these ideas.

The 1669 *Historia* was devoted to the overthrow of the idea of metamorphosis, as its title explains: "General Account of the Bloodless Animals, in Which Will be Clearly Set Forward the True Basis of Their Slow Growth of Limbs, the Vulgar Error of the Transformation, Also Called Metamorphosis, Will be Effectually Washed Away, and Comprehended Concisely in Four Distinct Orders of Changes, or Natural Budding Forth of Limbs." The idea of metamorphosis, which Swammerdam was so determined to refute, was that of a sudden and total change from one kind of creature into another, comparable to the alchemical transmutation of a base metal into gold.

William Harvey, calling the starting point of life an egg, defined two distinct modes of development from an egg. The chick grows in a hen's egg by epigenesis, but the butterfly grows in its "egg" (the chrysalis) by metamorphosis, as does an animal appearing in putrid matter. In epigenesis, the embryo is at first tiny and imperfect; it grows in size while acquiring its parts one after another. In metamorphosis the parts come into existence simultaneously and full-sized. Swammerdam consciously and energetically set out to destroy this supposed difference between the epigenetic development of higher animals and the metamorphic origin of lower animals. He used two kinds of evidence: the dissection of larvae and chrysalides before the final emergence of the adult, and a comparison among various types of insects including some that undergo only partial metamorphosis or none at all. It would seem that Swammerdam caught a clue to the nature of metamorphosis from his observations of the aquatic larvae of mayflies (which he first studied in 1661) and dragonflies (watched at Saumur in 1663 or 1664). The wings, which appear in so impressive a manner after the last molt, can be seen in a late larva, folded up in special protuberances on the back. The gradual growth of the insect can be easily seen in the successive larval stages. There is no difficulty in recognizing this process as the life cycle of one animal changing its form as it grows, just as the chick must change in appearance as well as in size before becoming a hen. Believing that the laws of nature are regular and simple, Swammerdam sought to explain all development according to one model. Those changes that seem metamorphic are really no different from the obviously gradual ones, except that they go on invisibly, under the skin.

Curious to find the growth of a butterfly's wings to be as epigenetic as a dragonfly's, Swammerdam searched for the proper dissecting technique. In his thesis of 1667 he had promised that he would soon explain the transformation of a caterpillar into a chrysalis, and by 1669 he had found that if a mature caterpillar, just preparing to become a chrysalis, is treated first with boiling water, then with wine and vinegar, and if the skin is removed, the rudiments of limbs and wings may be discerned. This demonstration was thought to be significant and exciting, both by Swammerdam and by his contemporaries, but there is no evidence for the dramatic scenes portrayed by Boerhaave, Francis J. Cole, and others. An eyewitness account undermines the picture of Swammerdam as a silent auditor at Thévenot's gatherings.[8] The most dramatic moment for Swammerdam himself may have been when he learned that Malpighi had anticipated him, finding rudiments of wings and legs in a silkworm.[9] Swammerdam claimed to have done his dissection in the presence of Magalotti, which would have been in June or July 1669.[10] Swammerdam regarded this demonstration as a great achievement, for the parts of the butterfly are so soft, tender, and folded that they can be recognized in a late caterpillar on the verge of its change only with difficulty. In a slightly younger caterpillar that is active and feeding they can scarcely be distinguished, because they are even more fluid and confused with the other tissue, he said. Swammerdam did not claim to have detected them in an immature caterpillar. They are at first invisible, in his description, not because of any extreme minuteness but because they are too fluid.

The point that Swammerdam considered most important, since it destroyed the previous ideas of metamorphosis, is that the parts of the butterfly do not come into being suddenly in the chrysalis but are already beginning to grow in the caterpillar. They develop by epigenesis, the process that Harvey described for higher animals, not by metamorphosis:

> [The limbs] which a worm without legs acquires near the chest after its change are not born in the suddenness of changing, or, to speak more exactly, in the quickness of a budding out or rising up of limbs, but these are growing with the worm at their designated places under the skin, one after another by addition, that is, by *epigenesis* . . .[11] and these parts are not born suddenly, but grow on slowly, the one part after the other . . . and they are increased and born in this swelling, budding forth, rising up, budding, and as if stretching of new limbs, gradually by an addition of parts, *epigenesis*, and by no means by a transformation, *metamorphosis*; therein lies the sole foundation of all the changes of bloodless animals.[12]

Swammerdam's principal concern was with development, not origin. After 1669 he studied the process of development in the chrysalis in more detail. He dissected chrysalides after two days, six days, twelve days, and sixteen days, reporting that structures at first so fluid and delicate that they cannot be handled gradually acquire form and firmness.

The sole foundation of all the so-called transformations of insects, according to Swammerdam, is the *Popken* or *Nympha*, that is, a pupa. His definition of this new concept is unclear, but he was fairly consistent in using it for the stage where an insect is preparing to molt by growing a new inner skin. Swammerdam's explanation of the nature of the pupa put great emphasis on the idea that the individual animal always remains the same, that is, we are dealing with the life history of one individual, irrespective of moltings and changes in appearance:

> . . . a nymph, pupa, or chrysalis is nothing other than such a manner of change from a worm and caterpillar, or, to speak more exactly, such a manner of growing, swelling, budding, or protuberance of parts of a worm or caterpillar, as to bear the specific shape of the future animal itself; or, otherwise, that this growth and so forth of parts in a worm or caterpillar is the animal itself in the form of a pupa. So that the matter being properly considered, a worm or caterpillar does not change into a pupa, but becomes a pupa by the growing of parts; so also this pupa, we may add, afterward does not change into a flying beast, but the same worm or caterpillar, having taken on the form of a pupa by shedding its skin, becomes a flying beast. The above changes are nothing else than those of a chick, which is not changed into a hen, but becomes a hen by the growing of its parts: or also, of the young of a frog which does not change into a frog, but becomes a frog by the swelling of its parts.[13]

After describing the mistaken notion of metamorphosis to be found in Thomas Mouffet, Goedaert, and Harvey, and indicating the pupa as the true basis of all insect development, Swammerdam proposed that all the various modes of insect development fall into one of four groups. The first group comprehends those insects that hatch from the egg in their adult form, afterward undergoing no change save growth. Before hatching, the animal lies inside the eggshell motionless, without food, and occupying the entire volume of the egg. Swammerdam regarded the egg itself as really a kind of pupa. "The animal grew within its mother from invisible but nevertheless real beginnings."[14] The animal's first development having taken place within the parent, what looks like an egg in this class is a formed animal lying hidden under a skin, that is, a pupa. In Swammerdam's view, this mode of development is the most obvious and simple to understand.

In the remaining three orders of development, the animal hatches out of its egg before having reached perfection, so it must become a pupa before appearing in its final shape. In the second kind of development, the animal emerges from its egg without wings, but usually with six legs. It acquires its adult structures by a gradual and visible external process of growth. In some members of the group, such as the earwigs, the difference between the young "worm" and the adult are very slight, consisting only in the addition of wings. In others, such as the mayfly, the changes are greater, involving the loss of larval structures as well as the addition of the adult structures. In all cases the adult structures are acquired by visible external growth, and the animal never completely loses its ability to move. The pupa in this second order is simply the last stage before the final molt from which the winged adult emerges.

Swammerdam understood all remaining modes of development to be fundamentally the same, and to differ from the first two groups in the same respect. He defined them in two more groups, while pointing out that the fourth could justifiably have been included within the third. In both these orders the animal leaves its egg in as yet a very imperfect state, lacking many or most of its adult structures. These it acquires in the course of time, gradually, invisibly, under the skin, where the structures lie folded and soft. The adult parts are all existent and recognizable in the pupa, although in some insects they are very obscure. The pupa in both orders does not feed or move.

Swammerdam divided his third order of development into two subgroups, those whose adult parts are obvious in the pupa because of its thin skin and those whose pupa seems to be without parts externally. The nymphs of ants, bees, and beetles are pupae of the first kind, and the chrysalides or aureliae of butterflies and moths are pupae of the second kind. He believed the difference between them to be merely a matter of the thickness of their skin, and even tried to find a simple mechanical explanation for this difference.

Swammerdam's fourth order of insect development is comprised of insects whose pupa, corresponding exactly to the nymph or chrysalis of the third order, is hidden. This is what happens in most flies; their larvae or maggots form a case called a puparium. Swammerdam insisted that the difference between the third and fourth order is one of appearance only, not of essence. The pupa separates from the larval skin, as in the second and third orders, but the animal does not crawl out of this skin, which hardens to form the puparium. The pupa lies within this case just as the bee's pupa lies sealed in its wax cell. The puparium may retain the shape of the larva, that is, be worm-shaped, or it may round up into the shape of an egg. It is of great importance to recognize, wrote Swammerdam, that this worm- or egg-shaped object is not an egg, as Harvey and others had said, nor a kind of pupa, but that it contains a pupa. Yet Swammerdam himself did see all eggs as pupae in one sense, insofar as they contained and concealed an animal.

Swammerdam emphasized the theme and structure of his insect research by appending to the *Historia* a table designed to show that insects develop in essentially the same fashion as do all other living things. All begin in an egg, grow gradually in size and detail until arriving at adulthood, when they are sexually mature. Swammerdam used the frog to illustrate development in higher animals and the carnation to illustrate development in plants. His table presents insects from each of his orders of development, the louse for the first order, the dragonfly for the second, the ant and the moth for the nymph and chrysalis types of the third order, and the dung fly for the fourth order. Five stages of development from egg to adult are numbered, and the numbers correspond to the figures in his tables, which had evidently been carefully planned with this comparison in mind.

Of the three arguments for the imperfection of insects, Swammerdam's *Historia* was concerned with the refutation of only one, metamorphosis as then understood. Working after 1669, Swammerdam attacked the other two arguments as well. In spite of his famous experiments on maggots in meat, Redi decided that insects found inside plants must have appeared there without parents; the Latin edition of Redi's work, appearing in Amsterdam in 1671, stimulated Swammerdam to collect information about insects that cause plant galls. The idea that insects consisted internally of humors was destroyed by Swammerdam's exceptional skills at microdissection; he refined his technique of injecting fluids, including wax, mercury, air, and alcohol; and he often used very fine scissors instead of knives.[15] He used a single-lens microscope made by Jan Hudde of Amsterdam,[16] and another mounted on flexible arms made by Samuel Musschenbroek.[17]

It is not surprising to find such a skilled microscopist as Swammerdam reporting that a frog embryo consists of globules or that there are oval particles in the blood, but historians should note that he attempted no interpretation of these observations. His dissections of various insects, as well as snails, mussels, cuttlefish, a Portuguese man-of-war, and a hermit crab formed a good portion of the study of all invertebrate anatomy before Cuvier.

Swammerdam's four orders of insect development were transformed by John Ray and Martin Lister into four orders of insects.[18] In the subsequent history of classification, types of development played various roles (Linnaeus did not use them), but clearly Swammerdam's own purpose had not been taxonomic.

Swammerdam is commonly called the founder of the theory of preformation. This is ironic because, Swammerdam, like Malpighi, consciously avoided conjecture. The reciprocal intercourse of snails, pictured in his medical thesis, was in a sense the emblem of all his science, for it represented to him the fact that even the safe assumption that animals are either male or female can be destroyed by an observation. He was loath to let his reasoning run beyond his facts. Still, Swammerdam's work did contribute to the development of the idea of preexistence and even emboîtement. His opposition to sudden metamorphosis could easily be read as an opposition to any change at all, for his own concept of the pupa was obscure, and he had no clear notion of the distinction between growth and differentiation.

Swammerdam opposed both spontaneous generation and metamorphosis on the grounds that they led to atheism by allowing chance and accident to rule instead of law and regularity. The basic law of living things was that they came from parents of the same kind by means of eggs. Nor is their growth in the egg subject to chance, as pangenesis, for example, might suggest, but is simply the increase of parts already present. The actual nature of the egg itself was a very difficult question that Swammerdam realized he had no means of solving. But if growth is conceivable while change in substance is not, inevitably the germs of all living things must have been in existence from the time of Creation. Swammerdam alluded to this idea only briefly. It was probable, said Swammerdam in 1669, that there was no generation of the sort that could leave room for chance processes, but only propagation, that is, the growth of something already in existence.[19] Certainly Swammerdam had no full emboîtement theory clearly in mind, for he located the first principles in the egg produced by the female ovary, yet suggested that this concept might explain the biblical statement that Levi was in his father's loins (Hebrews 7:10). He identified the black spot of a frog's egg as itself a frog with all its parts, without claiming to have seen a miniature, but on the contrary expressing wonder that somehow a tiny black spot really is a frog.[20]

Swammerdam said that he communicated his experiments to a learned man, who suggested that they could even explain how Adam's sin affected all his descendants,[21] and that when Eve's eggs were used up the human race would end,[22] but on all such questions Swammerdam forbore to give an opinion himself. It has always been assumed that it was Malebranche to whom Swammerdam referred. Certainly Malebranche was very quick to incorporate Swammerdam's information into his own philosophy.[23] The manuscripts Swammerdam left in 1680 contain no elaboration of the preformationist paragraphs of 1669 and 1672, but did include a careful description of the complete transformations of anatomy to be seen when a beetle larva becomes an adult, the mere watery humor that is visible in bee eggs, and the beginning of a frog in four globules.

Some biographers describe Swammerdam as a mystic, for he was a follower of a woman who purportedly had spoken with God, and he saw in the short flight of the mayfly an image of man's own brief existence. Yet the word *mystical* certainly does not apply to his scientific work, for his was a mechanistic world, instituted by God and operating like clockwork.

NOTES

1. See Engel, "Records," cited in the bibliography.
2. *Book of Nature*, 117.
3. Antoni van Leeuwenhoek, *The Collected Letters*, I (Amsterdam, 1939), 143, letter to Oldenburg of 7 September 1674.
4. Howard B. Adelmann, *Marcello Malpighi and the Evolution of Embryology*, I (Ithaca, 1966), 399.
5. Leonard G. Wilson, "William Croone's Theory of Muscular Contraction," in *Notes and Records. Royal Society of London*, **16** (1961), 158–178.
6. A. M. Luyendijk-Elshout, Introduction to Frederik Ruysch, *Dilucidatio Valvularum in Vasis et Lacteis* (1665), facsimile (1964).
7. Johan Nordström, "Swammerdamiana: Excerpts From the Travel Journal of Olaus Borrichius and Two Letters from Swammerdam to Thévenot," in *Lychnos* (1954–1955), 21–65.
8. *Ibid.*
9. Swammerdam, *Historia*, 131; *Book of Nature*, II, 2. Adelmann, *Malpighi*, II, 844–845.
10. Anxious to establish his priority over Malpighi, whose *De bombyce* appeared in 1669 (see Adelmann, I, 399), Swammerdam would undoubtedly have mentioned any public demonstration of years before. Instead he refers to a visit from Thévenot and Magalotti, which would have been dur-

ing the tour of the Grand Prince (not yet Duke) of Tuscany, Cosimo III. (He does not claim that the Prince himself witnessed this dissection, though he mentions showing him other discoveries). Careful records of the curiosities seen on the tour of 1667–1668 and references to Thévenot on the summer tour of 1669 are the basis for my dating of the visit with Swammerdam. See G. J. Hoogewerff, *De Twee Reizen van Cosimo de' Medici Prins van Toscane door de Nederlanden (1667–1669): Journalen en Documenten* (Amsterdam, 1919), xlix, 45, 319, 392.

11. My literal translation of the Dutch *Historia*, 26.
12. *Ibid.*, 43.
13. *Ibid.*, 8–9. My interpretation differs from those based on the Flloyd translation, *The Book of Nature*.
14. *Ibid.*, p. 100.
15. Boerhaave, *Book of Nature*, xiv–xvi.
16. Swammerdam, *Historia*, 81. See also Balthasar de Monconys, *Journal des Voyages*, II (Lyons, 1665–1666), 161–162.
17. Boerhaave, *loc cit.*
18. John Ray, *Historia Insectorum* (London, 1710).
19. Swammerdam, *Historia*, 51.
20. Swammerdam, *Miraculum*, 21.
21. *Historia*, 52.
22. Swammerdam, *Miraculum*, 22.
23. Paul Schrecker, "Malebranche et le préformisme biologique," in *Revue internationale de philosophie*, **1** (1938), 77–97.

BIBLIOGRAPHY

I. ORIGINAL WORKS. All of Swammerdam's works appeared in various printings and translations. A complete bibliography, including some other authors who cited him, is given in Schierbeek's biography.

Swammerdam's works include *Tractatus physico-anatomico-medicus de respiratione usuque pulmonum* (Leiden, 1667), reprinted, with Dutch trans., in *Opuscula selecta de arte medica neerlandicorum*, VI (Amsterdam, 1927), 46–181; *Historia insectorum generalis, ofte, Algemeene Verhandeling van de Bloedeloose Dierkens . . .* (Utrecht, 1669), French trans., 1682, repr., 1685, and Latin trans., 1685, repr., 1693. Its text is incorporated into the *Biblia Naturae*.

Other works are *Miraculum naturae sive uteri muliebris fabrica* (Leiden, 1672), repr., 1679, 1680, 1717 (and 1729?); *Ephemeri vita of afbeeldingh van 's Menschen Leven, vertoont in de Wonderbaarelijcke en nooyt gehoorde Historie van het vliegent ende een-dagh-levent Haft of Oever-aas* (Amsterdam, 1675), of which the biological portions, without the hymns to the Creator, were published in English by Edward Tyson as *Ephemeri vita* (London, 1681), and in French in Thévenot's *Recueil des voyages; Bybel der Natuure* (Leiden, 1737–1738), with facing pages in Latin, *Biblia naturae, sive historia insectorum*, 3 vols.; German trans. (Leipzig, 1752); in English as *The Book of Nature*, Thomas Flloyd, trans., with footnotes by John Hill (London, 1758).

II. SECONDARY LITERATURE. In 1727 Boerhaave purchased Swammerdam's manuscripts; he also acquired some papers of biographical interest, including forty-one of Swammerdam's letters to Thévenot. These manuscripts and papers are now in the Universitäts-

bibliothek in Göttingen, and a microfilm of them is in Leiden. To this day Boerhaave remains the only scholar to have made use of these papers, so his biography, prefacing the *Biblia naturae* in all its editions, is the chief source for Swammerdam's life. In spite of the fact that parts of it are evidently conjecture, it is on the whole useful and reliable.

Abraham Schierbeek in 1944 combined Boerhaave's biography with later sources and Swammerdam's text when writing his *Jan Swammerdam, zijn Leven en zijn Werken* (N.V. Uitgeversmaatschappij "De Tijdstroom" Lochem [1947]); *Jan Swammerdam (12 February 1637–17 February 1680): His Life and Works* (Amsterdam, 1967).

Sources of information not in Boerhaave, or valuable interpretations, are the following: Francis J. Cole, "The Birthplace of Jan Swammerdam, 1637–1680," in *Isis*, **27** (1937), 452; Hendrik Engel, "Records on Jan Swammerdam in the Amsterdam Archives," in *Centaurus*, **1** (1950), 143–155; and *Observationes anatomicae selectores collegii privat: Amstelodamensi um 1667–1673*, F. J. Cole, ed. (Reading, England, 1938).

On preformation, preexistence, and emboîtement, see Jacques Roger, *Les sciences de la vie dans la pensée française du XVIIIᵉ siècle* (1963), especially pp. 325–384; Howard B. Adelmann, *Marcello Malpighi and the Evolution of Embryology* (Ithaca, N.Y., 1966), 819–886; and Peter J. Bowler, "Preformation and Pre-existence in the Seventeenth Century: A Brief Analysis," in *Journal of the History of Biology*, **4** (1971), 221–244.

MARY P. WINSOR

SWANN, WILLIAM FRANCIS GRAY (*b.* Ironbridge, Shropshire, England, 29 August 1884; *d.* Swarthmore, Pennsylvania, 29 January 1962), *experimental physics, theoretical physics.*

Swann's chief contributions to physics lay in experimental and theoretical studies of cosmic radiation, theoretical research in electromagnetic theory and relativity, and work in the philosophy of physics.

He received his higher education at the Imperial College of Science and Technology, University College, King's College, and the City and Guilds College of London Institute. In 1910 he was awarded the Doctor of Science degree by the University of London. In 1913 Swann went to the United States to become chief of the physics division of the department of terrestrial magnetism at the Carnegie Institution of Washington, where he remained until 1918. For the next nine years he taught physics at the University of Minnesota, the University of Chicago, and at Yale University, where he was director of the Sloane Physics Laboratory from 1924 to 1927. In the latter year Swann became director of the Bartol Research Foundation of the Franklin Institute, the laboratory of which had recently been built on the campus of Swarthmore College. He remained there for the rest of his life, retiring officially in 1959 but staying on as emeritus.

Swann early became interested in geophysics, particularly in regard to the earth's magnetism and atmospheric electricity; and his bibliography shows many papers in the field during the period from 1909 to 1930. The interest in atmospheric electricity naturally led to a growing concern with cosmic radiation, a field he entered vigorously around 1922 and to which he devoted a large part of his research energy. In this area he was equally at home in both theory and experiment, and his frequent summaries of progress were extremely stimulating to all workers in the field. He made the Bartol Research Foundation into one of the world's great centers in cosmic ray studies, a field in which he maintained interest for the rest of his life.

Swann's early fascination with the fundamental problems in physical theory provided by relativity led to many papers on the relation between the latter and electromagnetism and electrodynamics. He was highly critical of most presentations of this subject and was never satisfied until he had probed to the bottom of every difficulty. This attitude induced a concern for the general problems of the philosophy of physics. His lectures in this discipline, always in great demand, led to the preparation of his highly successful book *The Architecture of the Universe* (1934), in which the basic ideas of relativity, thermodynamics, statistical mechanics, and quantum mechanics were set forth with great clarity and charm.

A talented cellist, Swann was a founder and conductor of the Swarthmore Symphony Orchestra. He was a member of the American Philosophical Society and was also prominent in the affairs of the American Physical Society, serving as its president from 1931 to 1933. He received several honorary degrees and was awarded the Elliott Cresson Gold Medal of the Franklin Institute in 1960.

BIBLIOGRAPHY

I. ORIGINAL WORKS. Swann's complete bibliography includes two books and 263 articles. The complete list of papers is available from the Bartol Research Foundation of the Franklin Institute, Swarthmore, Pennsylvania. The following selection is intended to illustrate the breadth of his work.

The books are *The Architecture of the Universe* (New York, 1934) and *Physics* (New York, 1941), written with Ira M. Freeman.

Earlier papers include "The Fitzgerald-Lorentz Contraction, and an Examination of the Method of Determining the Motions of Electrons When Considered Simply as Singularities, Moving so as to Satisfy the Electromagnetic Scheme," in *Philosophical Magazine*, 6th ser., **23** (1912), 86–95; "On the Earth's Magnetic Field," *ibid.*, **24** (1912), 80–100; "The Atmospheric Potential Gradient, and a Theory as to the Cause of Its Connection With Other Phenomena in Atmospheric Electricity, Together With Certain Conclusions as to the Expression for Electric Force Between Two Parallel Charged Plates," in *Terrestrial Magnetism and Atmospheric Electricity*, **18** (1913), 173–184; "On the Origin and Maintenance of the Earth's Charge. Part I," *ibid.*, **20** (1915), 105–126; "The Penetrating Radiation and Its Bearing Upon the Earth's Electric Field," in *Bulletin of the National Research Council. Washington*, no. 17 (1922), 54–77; and "The Fundamentals of Electrodynamics, Part II," in *Bulletin of the National Research Council on Electrodynamics of Moving Media*, no. 24 (1922), 5–74; "The Relation of the Restricted to the General Theory of Relativity and the Significance of the Michelson-Morley Experiment," in *Science*, **62** (1925), 145–148; "The Possibility of Detecting Individual Cosmic Rays," in *Journal of the Franklin Institute*, **206**, no. 6 (Dec. 1928), 771–778; and "Relativity and Electrodynamics," in *Review of Modern Physics*, **2** (July 1930), 243–304.

Later papers are "Electrons as Cosmic Rays," in *Physical Review*, **41**, no. 4 (15 Aug. 1932), 540–542; "On the Nature of the Primary Cosmic Radiation," *ibid.*, **43**, no. 11 (1 June 1933); "The Relation of the Primary Cosmic Radiation to the Phenomena Observed," *ibid.*, **46**, no. 9 (1 Nov. 1934), 828–829; "The Corpuscular Theory of the Primary Cosmic Radiation," *ibid.*, **48** (15 Oct. 1935), 641–648; "Cosmic Ray Observations in the Stratosphere," in *Contributed Technical Papers. Stratosphere Series (National Geographic Society)*, no. 2 (1936), 13–22, written with G. L. Locher, W. E. Danforth, and C. G. and D. D. Montgomery; "The Electrodynamic Force Equation in Its Bearing Upon the Evidence for the Existence of a New Cosmic Ray Particle," in *Physical Review*, **52**, no. 5 (1 Sept. 1937), 387–390; "Showers Produced by Penetrating Rays," in *Physical Review*, **56**, no. 4 (15 Aug. 1939), 378; "The Significance of Scientific Theories," in *Philosophy of Science*, **7**, no. 3 (July 1940), 273–287; "The Relation of Theory to Experiment in Physics," in *Review of Modern Physics*, **13**, no. 3 (July 1941), 190–196; "Mass-Energy Relation in Quantum Theory," in *Physical Review*, **109**, no. 3 (Feb. 1958), 998–1008; and "Certain Matters in Relation to the Restricted Theory of Relativity, With Special Reference to the Clock Paradox and the Paradox of the Identical Twins. I. Fundamentals," in *American Journal of Physics*, **28** no. 1 (Jan. 1960), 55–64.

II. SECONDARY LITERATURE. A biographical sketch

by Martin A. Pomerantz is in *Yearbook. American Philosophical Society* (1962), 178–184. There is also a sketch in *Current Biography, 1960*, 417–419, with photograph.

R. B. LINDSAY

SWARTS, FRÉDÉRIC JEAN EDMOND (*b.* Ixelles, Belgium, 2 September 1866; *d.* Ghent, Belgium, 6 September 1940), *chemistry*.

Frédéric Swarts entered the University of Ghent in 1883 and received doctorates in chemistry (1889) and medicine (1891). His father, Théodore Swarts, had succeeded Kekulé as professor of chemistry at the university in 1871. The younger Swarts spent his entire professional career at Ghent, first as *répétiteur* and then, on his father's retirement in 1903, as professor of chemistry. He was a member of the Académie Royale des Sciences des Lettres et des Beaux-Arts de Belgique, which awarded him its Gold Medal, corresponding member of the Institut de France, president of the Institut International de Chimie Solvay, and charter member and vice-president of the International Union of Pure and Applied Chemistry.

After the discovery of fluorine, few of its compounds had been prepared because of the reactivity and toxicity of the element. Swarts was among the first to study organic fluorine compounds. Unable to use methods of direct fluorination because of the violence of the reactions, he developed a double decomposition process using inorganic fluorides, especially antimony trifluoride and mercurous fluoride, and organic polyhalides, where the halogen atoms are on the same carbon atom (the Swarts reaction, 1892). The first synthesis of an organic fluorine compound was trichlorofluoromethane (1891). Swarts synthesized many aliphatic chlorofluoro and bromofluoro derivatives of hydrocarbons, alcohols, and acids. In 1922 he prepared trifluoroacetic acid, the strongest organic acid known.

The aliphatic chlorofluoro compounds became the first fluorochemicals to be used commercially after Thomas Midgley and A. L. Henne in 1930, using a modified Swarts reaction, prepared the group of fluorinated methanes and ethanes known as the Freons.

Swarts made the first extensive investigations of organic-fluorine compounds. He coupled his syntheses of organic fluorine compounds with physicochemical studies and determined their heats of combustion, molecular refractions, and viscosities, proving that fluorinated organic compounds have weaker intermolecular forces than the corresponding nonfluorinated compounds.

BIBLIOGRAPHY

I. ORIGINAL WORKS. Important papers include "Sur l'acide fluoracétique," in *Bulletin de l'Académie royale de Belgique. Classe des sciences*, **31** (1896), 675–688; "Sur quelques dérivés fluorés du toluol," *ibid.*, **35** (1898), 375–420; "Contribution à l'étude des combinaisons organiques du fluor," in *Mémoires couronnés et mémoires publiés par l'Académie royal des sciences, des lettres et des beaux-arts de Belgique*, **61** no. 4 (1901–1902); "Investigations thermochimiques des combinaisons organiques du fluor," in *Journal de chimie physique et de physico-chimie biologique*, **17** (1919), 3–70; and "Sur l'acide trifluoracétique," in *Bulletin . . .*, **8** (1922), 343–370.

II. SECONDARY LITERATURE. Accounts of the life and work of Swarts are "Frédéric Swarts," in *Bulletin. Société chimique de Belgique*, **49** (1940), 33–35; Marcel Delépine, "Frédéric-Jean Edmond Swarts, 2 Septembre 1866–6 Septembre 1940," in *Comptes rendus hebdomadaires des séances de l'Académie des sciences*, 212 (1941), 1057–1059; Jean Timmermanns, "Frédéric Swarts (1866–1940)," in *Journal of the Chemical Society* (1946), 559–560; and George B. Kauffman, "Frédéric Swarts: Pioneer in Organic Fluorine Chemistry," in *Journal of Chemical Education*, **32** (1955), 301–303.

ALBERT B. COSTA

SWARTZ, OLOF (*b.* Norrköping, Sweden, 21 September 1760; *d.* Stockholm, Sweden, 19 September 1818), *botany*.

Swartz began his studies in the field of medicine at Uppsala University in 1778, the year of Linnaeus' death. He had been interested in botany at an early age, and had already traveled to different parts of Sweden and Finland in order to collect plants and other objects of natural history. His doctoral thesis, written under Linnaeus the younger (who succeeded his father in the chair of botany at Uppsala), and entitled *Methodus muscorum illustrata* (1781), indicated further his scientific devotion. The great adventure in Swartz's life was his journey to the West Indies. He began his trip in 1783, traveling first through eastern North America, stopping at Boston and Philadelphia. Then, during the next two years, he visited Jamaica, Puerto Rico, Haiti, and Cuba. On his way home in 1786 and 1787 he studied the great botanical collections of Linnaeus and Banks in London, comparing his own extensive material with what had already been brought together by botanists of an earlier generation. As a result of studies made during his voyage, Swartz published in 1788 *Nova genera et species plantarum* and other lesser articles and papers. This work was summed up in his magnificent *Flora Indiae Occidentalis I–III* (1797–1806), which included descriptions of all the new genera and species he had found. He described nearly 900 species, most of them new to science.

For several years Swartz lived in Stockholm on a small private income, devoting himself entirely to his botanical research. In 1791 he became Bergian professor and intendant at a newly established school of gardening in Stockholm, owned by the Royal Swedish Academy of Sciences. He received several other appointments in the service of the academy, and finally, in 1811, was elected permanent secretary, the most important position in the academy. He held this office until his death. From 1813 he was professor of botany at the Caroline Institute.

Besides his work on the flora of the West Indies, Swartz is best known for his taxonomic studies of specific plant groups, often in the context of their worldwide distribution. Thus, for example, through his studies of orchids, summarized in *Genera et species Orchidearum* (1805), he was able to improve the systematics of these plants on the basis of the morphological traits of their highly specialized flowers. Swartz's greatest fame, however, rests on his studies of the cryptogams. He broadened greatly the knowledge of Swedish mosses, and he described in his works on the West Indies many new species of lichens and fungi; but, principally, he worked with the ferns of the world. His main works in this field are *Genera et species filicum* (1801) and the monumental *Synopsis filicum* (1806). As in his work on mosses and orchids, Swartz based his fern systematics upon studies of the fructification organs. He tried to deepen the views common in the Linnaean tradition, of which he was a strong adherent, thus opening the way for further study.

Swartz was, along with the much more conservative Thunberg, the most internationally oriented of Swedish botanists. He carried on a huge foreign correspondence and thus had the opportunity to publish his works in Germany, where they found their way to the international scientific community more easily than they would have if published in Sweden. Nevertheless, because of his central position in the Academy of Sciences and because of his universally praised generosity and friendliness, Swartz was the unifying link between the other botanists in his own country.

BIBLIOGRAPHY

I. ORIGINAL WORKS. Swartz's published works in botany are listed in T.O.B.N. Krek, *Bibliotheca botanica suecana* (Uppsala-Stockholm, 1925). Part of his correspondence is in the library of the Royal Academy of Sciences, Stockholm, in the Brinkmanska Arkivet, Trolle-Ljungby, and in the Riksarkivet, Stockholm.

II. SECONDARY LITERATURE. In the posthumous work of Swartz, *Adnotationes botanicae*, J. E. Wikström, ed. (Stockholm, 1829), there are biographies by Wikström, K. Sprengel, and C. A. Agardh. See further S. Lindroth, *Kungl. Svenska Vetenskapsakademiens historia*, II (Stockholm, 1967), 71–75, 229–234, 416–420, and *passim*, and G. Eriksson, *Botanikens historia i Sverige intill år 1800* (Uppsala, 1969), 290–292, 326–328.

GUNNAR ERIKSSON

SWEDENBORG, EMANUEL (*b.* Stockholm, Sweden, 29 January 1688; *d.* London, England, 29 March 1772), *technology, geology, cosmogony, physiology, theology.*

Swedenborg's career is one of the most remarkable in the history of science. In his youth and early manhood he was an enthusiastic scientist and technologist, and published a number of articles in various fields. Almost imperceptibly he turned to religious speculation, which, after a decisive divine revelation, led him to become a visionary and the founder of a religious sect, for which he is best known.

The son of Jesper Swedberg, professor of theology at Uppsala and later bishop of Skara, Swedenborg grew up in Uppsala and studied at its university, specializing in the humanities. He soon turned to the sciences, however, influenced by his brother-in-law, the learned university librarian Erik Benzelius. In the fall of 1710 Swedenborg traveled to England, where he stayed until 1713, mostly in London. During this time he was captivated by what he learned of science. He read Newton. He met Flamsteed, Halley, and John Woodward. He considered the universe to be a problem in mathematics, and, filled with youthful self-confidence, he tried to realize grandiose technical inventions, among them flying machines and submarines. Swedenborg returned to Sweden via Paris and Germany, and in 1716 was appointed extraordinary assessor on the Board of Mines. In this capacity he worked with Christopher Polhem, whom he admired greatly and assisted in far-reaching technical and industrial projects. Many articles in *Daedalus hyperboreus*, Sweden's first, short-lived (1716–1718) scientific journal, which the wealthy Swedenborg published at his own expense, were devoted to Polhem's mechanical inventions. Ennobled in 1719 (until then he signed himself Swedberg), he served for years on the Board of Mines; he was a competent metallurgist and, among other things, experimented with a new process for refining copper ore.

Always manifold in his scientific ambitions Swedenborg during this period wrote many short articles on his observations and theories. They were of varying importance, some indifferent or amateurish in quality, others ingenious and interesting. He was least successful as an astronomer. Swedenborg's attempt to determine longitude at sea by means of the moon (published in 1716 and later several times revised), was submitted in a competition sponsored by the British Parliament. It was rejected by the experts and failed completely. In *Om jordenes och planeternas gång och stånd* ("On the Course and Position of the Earth and the Planets"; Skara, 1718), which was inspired by the Bible, Polhem, and Thomas Burnet's *Telluris theoria sacra*, Swedenborg tried to prove that in earlier times the earth had revolved at a faster rate around the sun. The seasons would have been of similar climate and a paradisiacal spring would have reigned. As the earth slowed down and the length of the year and the seasons increased, the final catastrophe was approaching.

Young Swedenborg was undoubtedly at his best in geology and paleontology. In *Om watnens högd och förra werldens starcka ebb och flod* ("On the Level of the Seas and the Great Tides in Former Times"; Uppsala, 1719) he submitted empirical proof—sedimentary deposits, gravel ridges, fish in landlocked lakes without outlets, and the raising of the land along the Baltic coast—that Scandinavia had once been covered by an ocean from which the land had slowly risen. With the chemist Urban Hiärne, who strongly influenced him, he thus initiated the eighteenth-century debate in Sweden about "water reduction." Swedenborg was very interested in fossils as evidence of a prehistoric flood. He was convinced of their organic origin; and during a journey in 1721–1722, he examined many fossils of plants found near Liège and Aachen. His descriptions of them were published to-

gether with other geological papers in his *Miscellanea observata circa res naturales* (Leipzig, 1722).

Swedenborg's plans were to become increasingly grandiose. In his *Principia rerum naturalium* (Leipzig, 1734), probably conceived as a counterpart to Newton's *Principia*, he sought a comprehensive physical explanation of the world based on mathematical and mechanical principles. While remaining faithful to the general principles of Cartesian natural philosophy, which he had learned while studying at Uppsala, Swedenborg elaborated upon them. According to his cosmogony the physical reality has developed from the mathematical point, which was an entity between infinite and finite. Through a vortical movement implanted on the point, a series of material particles developed (the "first finita," the "second finita," and so on) that eventually led to the cosmos in its present state. In contrast to Descartes, Swedenborg believed that the planets had developed from the chaotic solar mass through expansion of its surrounding shell, which finally joined to form a belt along the equatorial plane of the sun. It then exploded, forming the planets and the satellites. Although the basic construction of Swedenborg's thought heralded the later planetary theories of Buffon, Kant, and Laplace, there is nothing to indicate that it exerted any direct influence on posterity.

In the 1730's Swedenborg pursued his materialistic explanation of the universe to its furthest consequence, concluding that the human soul also derived from the movements of the small particles. But at the same time a disturbing feature emerged in his thought. In speculating on paradise and the nature of angels, Swedenborg became increasingly involved—faithful to the Cartesian way of stating the inquiry—in the body-soul problem; and the soul and the mysteries of organic life soon became his main field of research. He planned enormous works in which physiology step by step was transformed into theology: *Oeconomia regni animalis* (2 vols., London–Amsterdam, 1740–1741) and *Regnum animale* (3 vols., The Hague, 1744–1745).

Swedenborg now sought to explain everything in terms of psyche, considering even the body as a manifestation of divine origin: "Everything lives the life of its soul and the soul lives the life of God's spirit." With the help of Malpighi, Swammerdam, and Vieussens he sought to discover the location of the human soul in the brain and its role as intermediary between mortal and divine. In his *Oeconomia* and in certain manuscripts, especially "De cerebro" (first published 1882–1887), he presented for the first time his theory that the activities of the soul, located in specific centers in the cortex of the brain, were built up from the finest "fibers." In this categorical form it was an original and remarkable hypothesis that remained unnoticed by later physiological researchers.

The religious crisis in Swedenborg's life was now approaching. At the beginning of the 1740's he wrestled with the greatest problem in metaphysics. Wishing to find words for the ineffable, he experimented with a logical-mathematical universal language, a *mathesis universalis* on Leibniz's and Wolff's models, but it turned instead into the theory of correspondence. As worked out in its linguistic and philosophical details, this theory taught that existence was made up of three reciprocal levels—the natural, the psychic, and the divine; each word or concept within a certain level corresponded to a word or concept within another.

A financially independent bachelor, Swedenborg journeyed to Holland and England during this period. Restless and excited, he was plagued by dreams and visions that he described in the peculiar *Drömboken* ("Journal of Dreams"; Stockholm, 1859). At the same time he was working on a great narrative of creation, *De cultu et amore Dei*, but abandoned it when the final vision came upon him at London in the spring of 1745: God revealed himself to Swedenborg and ordered him to interpret the meaning of the Bible; on the same night the world of the spirits, Heaven and Hell, were opened to him.

At the age of fifty-seven Swedenborg abandoned his scientific investigations. For the rest of his life he was purely a visionary and prophet. Many thought him mad. In a stream of Latin works, especially the gigantic commentary on the books of Moses, *Arcana coelestia* (8 vols., London, 1749–1756), he developed his theory of the spiritual world, which was to be the beginning of a new universal religion, represented on earth by the Swedenborgian New Church. But despite its bizarre aspects Swedenborg's theology is by no means a chaos of whims and visions. It is characterized by rigorous logic and obviously is rooted in his previous concern with the physical sciences.

BIBLIOGRAPHY

I. ORIGINAL WORKS. Swedenborg's enormous literary production, only part of which was published during his lifetime, is listed by James Hyde, *A Bibliography of the Works of Emanuel Swedenborg* (London, 1906); and by Alfred H. Stroh and Greta Ekelöf, *An Abridged Chronological List of the Works of Emanuel Swedenborg* (Uppsala, 1910). His most important scientific works are mentioned in the text. Most of Swedenborg's early scientific works, including *Principia* (1734) and letters to Erik Benzelius, among others, are in his *Opera quaedam aut inedita aut obsoleta de rebus naturalibus*, 3 vols. (Stockholm, 1907–1911). Swedenborg's work on longitude was published in Latin as *Methodus nova inveniendi longitudinem locorum . . . ope lunae* (Amsterdam, 1721). He also published *Prodromus principiorum rerum naturalium* (Amsterdam, 1721) and monographs on the metallurgy of iron and copper: *Regnum subterraneum sive minerale de ferro* and *Regnum . . . de cupro et orichalco* (Dresden–Leipzig, 1734). together with *Principia rerum naturalium*, are contained in his *Opera philosophica et mineralia*, 3 vols. (Leipzig, 1734).

Swedenborg's MSS on the physiology of the brain was published by R. L. Tafel as *The Brain Considered Anatomically, Physiologically and Philosophically*, 2 vols. (London, 1882–1887), and as *Three Transactions on the Cerebrum*, Alfred Acton, ed., 2 vols. (Philadelphia, 1938–1940). The unfinished *De cultu et amore Dei* was published at London in 1745. Almost all of Swedenborg's scientific works have been translated into English, most of them in the nineteenth century—for instance, *Principia* (London, 1846). His MSS are in the library of the Royal Swedish Academy of Sciences. Many of them, under the title *Autographa*, have been published by R. L. Tafel, 10 vols. (Stockholm, 1863–1870).

II. SECONDARY LITERATURE. The literature is concerned mainly with his theology and spirit theory. Indispensable, although often unreliable, is R. L. Tafel, *Documents Concerning the Life and Character of Emanuel Swedenborg*, 2 vols. (London, 1875–1877). An excellent introduction, especially to his scientific achievement, is Inge Jonsson, *Emanuel Swedenborg* (New York, 1971). A pioneering work in its time was Martin Lamm, *Swedenborg* (Stockholm, 1915), also in German (Leipzig, 1922) and French (Paris, 1936). Later biographies include Ernst Benz, *Emanuel Swedenborg, Naturforscher und Seher* (Munich, 1948); Cyriel Odhner Sigstedt, *The Swedenborg Epic* (New York, 1952); and Signe Toksvig, *Emanuel Swedenborg, Scientist and Mystic* (New Haven, 1948). Inge Jonsson has also written *Swedenborgs skapelsedrama De cultu et amore Dei* (Stockholm, 1961) and *Swedenborgs korrespondenslära* (Stockholm, 1969).

Various aspects of Swedenborg's scientific thought have been investigated by Svante Arrhenius, "Emanuel Swedenborg as a Cosmologist," in Swedenborg's *Opera quaedam* (see above), II (Stockholm, 1908), xxiii–xxxv; Gustaf Eneström, *Emanuel Swedenborg såsom matematiker* (Stockholm, 1890); Tore Frängsmyr, *Geologi och skapelsetro. Föreställningar om jordens historia från Hiärne till Bergman*, Lychnosbibliotek no. 26 (Stockholm, 1969), on Swedenborg as geologist and cosmologist, with an English summary; N. V. E. Nordenmark, "Swedenborg som astronom," in *Arkiv för matematik, astronomi och fysik*, **23**, ser. A, no. 13 (1933); Gerhard Regnéll, "On the Position of Paleontology and Historical Geology in Sweden Before 1800," in *Arkiv för mineralogi och geologi*, **1** (1949–1954), 1–64; and Hans Schlieper, *Emanuel Swedenborgs System der Naturphilosophie* (Berlin, 1901). Martin Ramström has examined Swedenborg's physiology of the brain in important articles, summarized in "Swedenborg on the Cerebral Cortex as the Seat of Psychical Activity," in *Transactions of the International Swedenborg Congress 1910* (London, 1910), 56–70.

STEN LINDROTH

SWIETEN, GERARD VAN (*b.* Leiden, Netherlands, 7 May 1700; *d.* Schönbrunn Palace, Vienna, Austria, 18 June 1772), *medicine.*

Van Swieten was the son of Thomas Franciscus van Swieten, a notary public, and Elisabeth Loo, who were members of the lesser nobility. During the early years of the Dutch Revolutionary War (1568–1648), one branch of the family became Protestant while the other remained Roman Catholic; van Swieten belonged to the latter.[1]

Like many Dutch Catholics, van Swieten studied at Louvain; matriculating in the autumn of 1714 or the winter of 1714–1715, at the "Falcon" liberal arts college. It is not known how long he stayed there, but it is fairly certain that he left without being awarded a degree.[2] On 26 February 1717, van Swieten enrolled as a medical student at the University of Leiden, attracted by the lectures of Boerhaave. After receiving the M.D. degree on 3 July 1725, he established a medical practice in Leiden, which, although it soon became quite sizable, did not prevent him from continuing to attend every lecture of Boerhaave until the latter's death in 1738. Van Swieten had adapted an existing shorthand system to medical language, so his lecture notes, which still exist, reflect Boerhaave's presentation quite closely. Their mutual respect led Boerhaave to show his most interesting private cases to van Swieten and to express a lively interest in those of van Swieten. At various times Boerhaave stated that van Swieten would be the most suitable person to succeed him as a professor.[3]

A few months after receiving the M.D., van Swieten had started a free *privatissimum.* Although it was not associated with the university in any way, the latter, at the instigation of "the eminent van Royen," forbade van Swieten to continue these lessons in 1734.[4] This action made it abundantly clear that despite Boerhaave's favorable opinion, van Swieten, as a Catholic, could not possibly look forward to a professorate at Leiden. Fortunately, he found a post elsewhere. The empress of Austria, Maria Theresa, invited him to become court physician in 1743, at which time van Swieten declined, and again in 1745, after van Swieten had attended her sister Maria Anna in Brussels. This time he accepted, and was put in charge of all court physicians.

Van Swieten soon made himself useful in many other ways. He reorganized the medical faculty of the University of Vienna, taking Leiden as a model, and added a botanical garden and a chemical laboratory, each headed by a professor. Thus he laid the foundation for the Vienna school of medicine, which became world-famous at the turn of the century. Van Swieten became the president of the medical faculty and taught several courses. In addition to his medical services, he reorganized the censorship of books, reserving books on subjects other than theology, law, and politics for himself. He reorganized the court library and was made chief librarian in 1745.

Despite his many obligations, van Swieten still found time to work on his *Commentaria*, of which he had completed the first two volumes (1742, 1745) while in Leiden. This work, which documents Boerhaave's lectures, is not a straightforward transcription but, rather, a series of commentaries on Boerhaave's *Aphorisms*.[5] Therefore, it is not always clear which parts of the text are based on Boerhaave's lectures and which are van Swieten's own. The *Commentaria*, which greatly contributed to the dissemination of Boerhaave's ideas beyond the circle of his pupils, was reprinted many times and was translated into four languages.

Among van Swieten's contributions to medicine was his modification of the traditional treatment of venereal disease with mercurials; his specific, *liquor Swietenii*, which could be taken orally, made the treatment much less painful.[6] He managed to overcome the aversion of the Vienna physicians to inoculation against smallpox, which was introduced to Austria in 1768 by Jan Ingen-Housz. During the Seven Years War (1756–1763) van Swieten wrote a book on the diseases of the army.

Van Swieten's activities were mainly organizational and political; but his role in establishing the great Vienna school of medicine and his dissemination of Boerhaave's ideas by means of his *Commentaria* entitle him to a place in the history of science.

The name *Swietenia* was given to a genus of the Meliaceae family by Nicolas Jacquin. *Swietenia mahogani* is the mahogany tree.

NOTES

1. An ancestor, Adriaen van Swieten (1532–1584), was one of the signers of the Compromise of Breda ("League of Nobles") in 1566, which heralded the Dutch Revolution, and was responsible for dividing the family into two branches of opposing religions. Catholics were not eligible for positions in the government during the Dutch Republic, and the Roman Catholic religion was forbidden in Holland in 1573. The role of the nobility during the Republic was mainly governmental; therefore, unless a Catholic noble family had landed property, as the van Swietens did not, their noble status became meaningless.
   In some history books, especially on church history, it is stated that van Swieten was a Jansenist. This is, however, by no means proved.
2. According to Baldinger, there were twelve other students in his class, and van Swieten surpassed all of them at the age of sixteen.
3. According to Baldinger.
4. According to Baumann, this was Adriaan van Royen (1704–1779), from 1729 lecturer on botany and from 1732 professor of medicine and chemistry. Brechka states that it probably was David van Royen (1699–1764), who was secretary to the curators of the university from 1725.
5. *Aphorismi de cognoscendis et curandis morbis* (Leiden, 1709). In the 3rd ed. (Leiden, 1715) the number of aphorisms reached 1,495.
6. *Liquor Swietenii* was a solution of about 0.1 percent by weight of mercurous chloride in alcohol.

BIBLIOGRAPHY

I. ORIGINAL WORKS. Van Swieten's stenographic records of Boerhaave's lectures were inherited by Anton von Störck and are now in the Austrian National Library, Vienna. A copy is in the Rijksmuseum voor de Geschiedenis der Natuurwetenschappen, Leiden, which also has transcriptions made by E. C. van Leersum, who broke the code. Van Swieten's extensive library was bought after his death by Empress Maria Theresa for 18,000 florins (Sandifort, **10**, 887), and is now in the Austrian National Library.

Van Swieten's earliest published work was *De arte fabrica et efficacia in corpore humano* (Leiden, 1725), his dissertation. His major book is *Commentaria in Hermanni Boerhaave Aphorismos de cognoscendis et curandis morbis*, 5 vols. (Leiden, 1742–1772). Vol. I went through 4 eds., II, 3 eds., and III, 2 eds. For the numerous other Latin eds., see Lindeboom (1959), nos. 208–235.

The Dutch trans. is *Verklaaring der korte stellingen van Herman Boerhaave over de kennis en de geneezing der ziektens*, 2 vols. (Leiden, 1760–1763), also 10 vols. (1776–1791). The German trans. is *Erläuterungen der Boerhaavischen Lehrsätze von Erkenntniss und Heilung der Krankheiten*, 5 vols. (Vienna–Frankfurt–Leipzig, 1755–1775). The French trans. *Commentaires des Aphorismes d'Hermann Boerhaave sur la connaissance et la cure des maladies*, 7 vols. (Paris, 1765–1768), also 2 vols. (Avignon, 1766). The English trans. is *The Commentaries Upon the Aphorisms of Dr. Herman Boerhaave, Concerning the Knowledge and Cure of Several Diseases, Incident to Human Bodies*, 18 vols.

(London, 1744–1773), also 11 vols. (1754–1759), 14 vols. (1759–1765, and 1771–1773), and 18 vols. (Edinburgh, 1776).

Some parts of the *Commentaria* on specific subjects were published under separate titles. *Aphorismes de chirurgie*, 5 vols. (Paris, 1753), also 7 vols. (1753–1756) and 5 vols. (1768); *Hermann Boerhaave's kurzgefasste Lehrsätze von Erkenntniss und Heilung der sogenannten chirurgischen Krankheiten*, 2 vols. (Danzig, 1753); *Traité du scorbut, devisé en trois parties*, 2 vols. (Paris, 1756, 1783, 1788, 1837); *Traité des maladies des enfants . . .* (Avignon, 1759); *Traité de la péripneumonie* (Paris, 1760); *Traité de la pleurésie* (Paris, 1763); *Traité des fièvres intermittentes* (Paris, 1766); *Maladies des femmes et des enfans, avec un traité des accouchemens*, 2 vols. (Paris, 1769); *Traité de la petite vérole* (Paris, 1776); and *Erläuterung der Boerhaavischen Lehrsätze der Chirurgie*, 2 vols. (Frankfurt, 1778).

The last work published during his lifetime was *Kurze Beschreibung und Heilungsart der Krankheiten, welche am öftesten in dem Feldlager beobachtet werden* (Vienna, 1758), translated into Dutch as *Korte beschrijving en geneeswijze der ziekten die veelzints in de krijgsheirlegers voorkomen* (Amsterdam, 1760, 1764, 1772, 1780, 1790; Bruges, 1765), into Italian as *Breve descrizione delle malattie che regnano piu communemente nella armate e del metodo di trattarle* (Naples, 1761, 1768), and into French as *Description abrégé des maladies qui regnent le plus communément dans les armées, avec la méthode de les traiter* (Bruges, 1765); an English trans. (1762) also is mentioned.

Posthumous works are *Oratio de morte dubia* (Vienna, 1778); *Considerazione intorno alla pretesa magia postuma per servire alla storia di vampiri* (Naples, 1781); and *Constitutiones epidemicae, et morbi potissimum, Lugduni Batavorum observati, ex ejusdem adversariis* (Vienna–Leipzig, 1782; Geneva, 1783).

II. SECONDARY LITERATURE. See the following, listed chronologically: E. G. Baldinger, *Lobreden auf den Freiherrn Gerhard van Swieten . . .* (Jena, 1772), abbrev. Dutch trans. by E. Sandifort in *Natuur en geneeskundige bibliotheek . . .*, **10** (1773), 205–215; I. Wurz, *Trauerrede auf den hochwohlgeborenen Herrn Gerard van Swieten . . .* (Vienna, 1772); A. Fournier, *G. van Swieten als Zensor, nach archivalischen Quellen* (Vienna, 1877); C. von Wurzbach, "Freiherr Gerhard van Swieten," in *Biographisches Lexicon des Kaiserthums Oesterreich*, XLI (1880), 37–50; W. Müller, *Gerard van Swieten, Biographischer Beitrag zur Geschichte der Aufklärung in Oesterreich* (Vienna, 1883); and E. C. van Leersum: "Gerard van Swieten en qualité de censeur," in *Janus*, **11** (1906), 381–398, 446–469, 501–522, 588–606; "A Couple of Letters of Gerard van Swieten on the *liquor Swietenii* and on the Inoculation of Smallpox," *ibid.*, **15** (1910), 345–371; and "Boerhaave's dictaten, inzonderheid zijne klinische lessen. Met een beschrijving van Gerard van Swieten's stenografische nalatenschap," in *Nederlands tijdschrift voor geneeskunde*, **63** (1919), 50–76.

Also see J. J. van der Kley, "G. van Swieten's *Constitutiones epidemicae et morbi potissimum, Lugduni Batavorum observati*," in *Bijdragen tot de geschiedenis der geneeskunde*, **1** (1921), 286–292; V. Kreutzinger, "Zum 150 Todestage Gerhard von Swieten," in *Janus*, **26** (1922), 177–189; H. Pinkhof, "Een advies van Gerard van Swieten," in *Bijdragen tot de geschiedenis der geneeskunde*, **3** (1923), 189–190; G. van Leeuwen, "Gerard van Swieten," in *Nieuw Nederlandsch biografisch woordenboek*, X (1937), 1005–1006; H. T. van Heuveln, *Gerard van Swieten, Leben, Werk und Kampf* (Veendam, 1942); H. Sandra, "De leer der phthisis bij de oude Nederlandsche schrijvers," in *Bijdragen tot de geschiedenis der geneeskunde*, **23** (1943), 40–46; L. Schönbauer, *Das medizinische Wien* (Vienna, 1947). G. A. Lindeboom, "Gerard van Swieten als hervormer der Weensche medische faculteit," in *Bijdragen tot de geschiedenis der geneeskunde*, **30** (1950), 12–20; E. D. Baumann, *Drie eeuwen Nederlandsche Geneeskunde* (Amsterdam, 1951), 222–225; W. Böhm, *Universitas Vindobonensis* (Vienna, 1952); G. A. Lindeboom, *Bibliographia Boerhaaviana* (Leiden, 1959), 47–54; F. T. Brechka, *Gerard van Swieten and His World, 1700–1772* (The Hague, 1970); D. Willemse, "Gerard van Swieten in zijn brieven aan Antonio Nunes Ribeiro Sanches (1739–1754)," in *Scientiarium historia*, **14** (1972), 113–143; and G. A. Lindeboom, "Het consult van Gerard van Swieten voor aartshertogin Marianne van Oostenrijk, de zuster van Maria Theresia," *ibid.*, 97–111; "Gerard van Swieten, Herr und Landstand Von Tirol," in *"Adler," Zeitschrift für Genealogie und Heraldik*, **9** (1972), 187–188; "Acht brieven van Gerard van Swieten uit zijn Hollandsche jaren (1730–1744)," in *Scientiarum historia*, **15** (1973), 73–89; "De Hollandsche tijd van Gerard van Swieten," in *Nederlandsch tijdschrift voor Geneeskunde*, **117** (1973), 1037–1042; and "Gerard van Swieten und seine Zeit," in *Internationales Symposium veranstaltet von der Universität Wien im Institut für Geschichte der Medizin* (Vienna, 1973), 63–79.

PETER W. VAN DER PAS

SWINDEN, JAN HENDRIK VAN (*b.* The Hague, Netherlands, 8 June 1746; *d.* Amsterdam, Netherlands, 9 March 1823), *electricity, magnetism, meteorology, metrology.*

The son of Philippe van Swinden and Anna Maria Tollosan, van Swinden was a prolific scientific writer; many of his works were written in French, the language of his ancestors, who had been driven from France by the revocation of the Edict of Nantes. Most of his experimental work in electricity, magnetism, and meteorology was conducted before 1795, after which year his time was devoted mainly to scientific committee work. In 1798 he was the leading member of the committee

that introduced the metric system into the Netherlands. He married Sara Ribolleau in 1768 and had one son who died before him.

Van Swinden's father, an eminent barrister, wanted his son to study law. Jan, however, from an early age showed more interest in mathematics and mechanics; and although he entered Leiden University to study law in 1763, he soon changed to natural philosophy and mathematics. He was greatly influenced by Newton's *Principia*, as is shown by his doctoral thesis, *De attractione* (1766). He took every opportunity to popularize Newtonian philosophy, and his inaugural lectures at the universities of Franeker and Amsterdam dealt with this topic.

In 1767 van Swinden obtained the chair of philosophy, logic, and metaphysics at Franeker University. He studied all the popular scientific topics of the period, especially magnetism, electricity, meteorology, and chemistry, and corresponded with many leading scientists, including Bonnet, Euler, Deluc, J. C. Wilcke, Bertholon, and Lalande. His *Tentamen de magnete* (1772) dealt with his mathematical theory of magnetism, and in 1777 he and Coulomb shared the gold medal of the Paris Academy of Sciences for a very detailed prize essay on magnetism, *Recherches sur les aiguilles aimantées.*

Van Swinden's best-known work in this field, *Mémoires sur l'analogie de l'électricité et du magnétisme* (1784), included his prize essay on the analogy between magnetism and electricity, in which he compared Mesmer's animal magnetism with electricity. For this work he was awarded the gold medal of the Bavarian Academy of Sciences in 1778. In 1785 van Swinden and van Marum experimented on the influence of electrical discharges on magnets, using the very large electrostatic generator of the Teylers Stichting at Haarlem. Repeating the experiments of Beccaria and others, they noticed that when soft iron or steel bars were placed perpendicular to the line of discharge, the resulting magnetic strength was greater than when the bars were parallel to the direction of the discharge. This puzzling phenomenon could be explained only after Ampère's work of 1820. Incidentally, Ampère at first thought that the field of force moving around the wire through which the electric current passed was also electrical. In 1822, using his large Offerhaus battery, van Marum demonstrated that it was, in fact, magnetic.

Van Swinden was best-known outside the Netherlands for his extraordinarily accurate meteorological observations. Over a ten-year period, he and some of his pupils at Franeker made hourly observations of the terrestrial magnetism and observed the diurnal variation. His thirteen-year record of barometer, thermometer, and hygrometer readings resulted in the publication of some eighteen papers on meteorological topics in various European journals.

In 1785 van Swinden was appointed professor of philosophy, natural philosophy, mathematics, and astronomy at the University of Amsterdam, where he wrote his mathematical textbook *Theoremata geometrica* (1786; enlarged and translated into Dutch, 1790) and the first two volumes of a general textbook on natural philosophy, *Positiones physicae* (1786). The latter ambitious work was never completed. He was appointed chairman of a committee to correct naval charts, which produced a nautical almanac in 1787. He was assisted in this work by his pupil Pieter Nieuwland, who became a lecturer in mathematics, astronomy, and navigation at the University of Amsterdam. The committee's activities also resulted in the publication of a book on the determination of longitude at sea by lunar observations (1787) and work on nautical instruments (1788). In 1798 van Swinden and Henricus Aeneae were sent to Paris by the Dutch government to attend the congress on the introduction of the metric system. He belonged to the commission (other members were J. G. Tralles, Delambre, Legendre) that recalculated the earth's meridian, one forty-millionth part of which was called the meter.

During the French occupation of the Netherlands, van Swinden served on committees dealing with such diverse topics as currency reform and the restructuring of university education. In 1808 he was appointed the first president of the Royal Institution of the Netherlands (now the Royal Netherlands Academy of Sciences and Letters). In 1813, after the overthrow of the French, William I (made king of the Netherlands in 1815) appointed him a councillor of state in recognition of his efforts on behalf of Dutch science.

BIBLIOGRAPHY

I. ORIGINAL WORKS. Van Swinden's main works are cited in the text. His MS material on the history of the microscope was published posthumously in G. Moll, "On the First Invention of Telescopes, Collected From the Notes and Papers of the Late Professor [J. H.] van Swinden," in *Journal of the Royal Institution of Great Britain*, **1** (1831), 319–332, 483–496. He also wrote a detailed study of Eise Eisinga's planetarium at Franeker,

Beschrijving van een kunstuk, verbeeldende een volledig bewegelijk hemelsgestel, uitgedachten en vervaardigd door Eise Eisinga (Franeker, 1780; repr. 1824, 1831; 3rd ed., enl., 1851). His works are listed in D. Bierens de Haan, *Bibliographie néerlandaise historico-scientifique* (Rome, 1883), 273–277, repr. photographically (Nieuwkoop, 1960, 1965). MSS are in various institutions in the Netherlands: Amsterdam University Library has about 50 letters and many MSS on a variety of topics, including education reform and his trip to Paris in 1798; the Koninklijke Bibliotheek, Amsterdam, has a number of notes dealing with scientific topics: weights, measures, and coinage (KA CCC a–g); magnetism and electricity (KA CCCI a–i); meteorology and the northern lights (KA CCCII a–i); mathematics, probability, and population statistics (KA CCCIII a–b); mechanics, astronomy, chronometry, and technology (KA CCCIV a–d); and science and medicine (KA CCCV a–b).

II. SECONDARY LITERATURE. Only short biographies of van Swinden exist to date: A. J. van der Aa, in *Biographisch woordenboek der Nederlanden*, XVII (Haarlem, 1784), 1124–1132, with a good bibliography; and P. C. Molhuysen and P. J. Blok, *Nieuw Nederlandsch biografisch woordenboek*, IV (Leiden, 1918), 1289–1291, which is slightly more detailed. See also G. C. Gerrits, *Grote Nederlanders by de opbouw der natuurwetenschappen* (Leiden, 1948), 236–241. The best biography in English is G. Moll, "A Biographical Account of J. H. van Swinden," in *Edinburgh Journal of Science*, **1** (1824), 197–208.

WILLEM D. HACKMANN

SWINESHEAD (Swyneshed, Suicet, etc.), **RICHARD** (*fl. ca.* 1340–1355), *natural philosophy.*

The name Richard Swineshead is best known to the modern historian of science as that of the author of the *Liber calculationum*, a work composed probably about 1340–1350 and famous later for its extensive use of mathematics within physics. Very little is known about this Richard Swineshead, and furthermore it appears almost certain that the little biographical data that are available about any fourteenth-century Swineshead cannot all be apportioned to one man, but that there were at least two or three men named Swineshead who may have left works in manuscript. In the early twentieth century, Pierre Duhem settled this confusion of Swinesheads to his own satisfaction by asserting that there was a John Swineshead who wrote some extant logical works, *De insolubilibus* and *De obligationibus*, and a Roger Swineshead who wrote a work on physics (found in MS Paris, Bibliothèque Nationale lat. 16621)—variously titled

De motibus naturalibus, Descriptiones motuum,
and *De primo motore* (the latter by Duhem). Duhem concluded that the famous "Calculator," as
the author of the *Liber calculationum* was often
called, was not really named Swineshead at all, but
was rather one Richard of "Ghlymi Eshedi," as he
is called in the explicit of the *Liber calculationum*
in MS Paris, Bibliothèque Nationale lat. 6558, f.
70v. Since Duhem's time, historians have rejected
the supposed name "Ghlymi Eshedi" as a scribal
error and have restored the *Liber calculationum* to
Swineshead. They have not, however, completely
unraveled the problem of the existence of two or
three Swinesheads as authors of several logical and
natural philosophical works.

The most satisfactory theory so far proposed
would seem to be that of James Weisheipl, according to whom there were three fourteenth-century
Swinesheads of note. One, named John Swineshead, was a fellow of Merton College from at least
1343 and pursued a career in law; he died in 1372,
leaving no extant works. A second, named Roger
Swineshead, was also at Oxford, but there is
no record of his having been at Merton College.
This Roger Swineshead wrote the logical works
De insolubilibus and *De obligationibus* and the
physical work *De motibus naturalibus.* He may
have been a Benedictine monk and a master in sa-
cred theology and may have died about 1365. The
third Swineshead, Richard Swineshead, was, like
John, associated with Merton College in the 1340's
and was the author of the famous *Liber calculationum,* and possibly also of two extant *opuscula,*
De motu and *De motu locali,* and of at least a partial *De caelo* commentary.

Given the uncertainty of the biographical data, it
seems proper that all the extant physical works
ascribed to any Swineshead should be included in
this article. This includes, most importantly, the
Liber calculationum, but also the *opuscula* ascribed by Weisheipl to Richard Swineshead and
the *De motibus naturalibus* ascribed by Weisheipl
to Roger Swineshead. All of these works can be
said to fall within the "Oxford calculatory tradition," if not with the works of the so-called Merton
school. The *De motibus naturalibus,* as the earliest
work, will be described first (with folio references
to MS Erfurt, Amplonian F 135), followed by the
Liber calculationum (with folio references to the
1520 Venice edition), to be described in much
greater detail, and finally by the fragmentary *opuscula* (with folio references to MS Cambridge,
Gonville and Caius 499/268), which, although they

most probably were written before the *Liber calculationum,* can more easily be described after that
work. Weisheipl's hypothesis will be followed as to
the correct names of the authors of these works.

De motibus naturalibus

The *De motibus naturalibus* was written at Oxford after the *De proportionibus* of Thomas Bradwardine and at about the same time (ca. 1335) as
the *Regule solvendi sophismata* of William Heytesbury. In the material covered, it is similar to the
latter work, and, in fact, both works treat topics
that were to become standard in treatises *de motu*
in the mid- and late fourteenth century. The *De
motibus naturalibus* has eight parts, called *differ-
entiae*: I. Introduction; II. Definitions of Motion
and Time; III. Generation; IV. Alteration; V.
Augmentation; VI. Local Motion; VII. Causes of
Motion; and VIII. Maxima and Minima. In contrast to Heytesbury's work and to later treatises
de motu, however, the *De motibus naturalibus*
includes large sections of traditional natural philosophy as well as the logicomathematical natural
philosophy typical of the later treatises. It contains
many more facts about the natural world (climates,
burning mirrors, tides, comets, milk, apples, frogs,
worms, etc.) and lacks the strong sophismata character of some of the later works. It represents,
therefore, to some extent, a stage halfway between
thirteenth-century cosmological and fourteenth-
century logicomathematical natural philosophy.

This position, halfway between two traditions, is
represented quite strongly in the organization of
the work: the treatise is fairly clearly divided into
metaphysical-physical discussions and logicomathematical treatments. Thus, for example, the three
parts discussing motion in the categories of quality,
quantity, and place (parts IV, V, VI) each contain
two parts, a first dealing with the physics of the
situation and a second dealing with the quantification of motion in that category. Although the logicomathematical topics that Roger discusses are
generally those discussed by the later authors *de
motu,* the order of topics in his work still reflects
an Aristotelian or medical base. Whereas later authors, especially Parisian-trained authors such as
Albert of Saxony, generally discussed the measures of motion with respect to cause first (*penes
quid attenditur motus tanquam penes causam*) and
then discussed the measures of motion with respect to effect (*tanquam penes effectum*), Roger
begins with the effects of motion (as, indeed, does
the Calculator after him). Furthermore, among

effects, he begins with the effects of alteration rather than with the effects of local motion. In accordance with this order of treatment, Roger's basic notions of the measurement of motion come from the category of quality rather than from causes or from locomotion, as was to be the case in seventeenth-century physics. In line with the earlier medical theory of the temperate, Roger places emphasis on mean degrees, and he considers intension at the same time as remission. When he then goes on to talk about possible mean degrees of local motion, and motions being just as fast as they are slow, and so forth, there seems to be no reasonable explanation except that he has taken "measures" which fit with the then current notions of quality and alteration and has applied them by analogy to local motion, even though the result has no apparent basis in the then current notions of local motion. Finally, to all of this, must be added the fact that Roger's basic theoretical terms for his measurements, namely "latitude," "degree," and the like, are earlier found most prominently in medical theory.

From among all the material in the *De motibus naturalibus* concerned with the "measurement" of motion and therefore most closely related to the work of the Calculator perhaps the two points of greatest interest have to do with two idiosyncratic positions which Roger takes, the first having to do with the function relating forces, resistances, and velocities in motion, the second with the relation of latitudes and degrees.

First, in parts VI and VII, Roger rejects the Aristotelian position that velocity is proportional to force and inversely proportional to resistance. Thus in the first chapter of part VI Roger states five conclusions concerning natural local motion which are all aimed at showing that resistance is not required for natural motion (41va–41vb). In part VII, Roger again repeats this view (43vb–44vb). In fact, he says, the equality or inequality of velocities is caused by the equality or inequality of the proportion of proportions of the mover to the moved, where the moved need not resist. Although Roger then accepts the mathematical preliminaries (for example, definitions of the types of proportionalities) that Bradwardine had set down as requisite for investigating velocities, forces, and resistances, he rejects Bradwardine's function relating these variables (see the article on Bradwardine for a description of his logarithmic-type function). Where there is no resistance, Roger asserts, the proportion of velocities is the same as the proportion of moving powers. Where there are resistances, then the proportion of velocities is the same as the proportion of latitudes of resistance between the degrees of resistance equal to the motive powers and the degrees of the media (this conclusion is equivalent in modern terms to stating that velocity is proportional to the difference between the force and resistance). Concerning cases where one motion is resisted and the other is not, Roger says that the proportion of velocities follows no other proportion, or, in modern terms, that he can find no function relating the velocities to forces and resistances. Although there are some obvious resemblances between Roger's position and the position of Ibn Bājja (Avempace) and the young Galileo, it is not a position at all common in the early fourteenth century.

Second, Roger's combinations of latitudes and degrees for measuring motion are also unique to him, so far as is known. Like the latitudes of earlier authors, Roger's latitudes are ranges within which a given quality, motion, or whatever, may be supposed to vary. Thus in part IV he posits the existence of three latitudes for measuring alteration, each distinguishable by reason into two other latitudes (39ra–39rb). In modern terms the first of these latitudes expresses the range within which the intensities of a quality may vary, the second expresses the range within which velocities of alteration may vary, and the third the range within which accelerations and decelerations of alteration may vary. Similarly, in part VI Roger posits five latitudes for measuring locomotion, all of them distinct from one another only in reason (43ra). In modern terms the first three of these latitudes are the ranges within which velocity or speed may vary and the last two are ranges within which accelerations and decelerations, respectively, may vary. All of these are similar to latitudes posited by the other Oxford calculators, although later there was a tendency to dispense with the latitudes of remissness and tardity that Roger posited (see below).

What is different about Roger's system is his postulation of so-called uniform degrees. Thus Roger defines two types of degrees of heat or any other quality (38rb). One type, the "uniformly difform degree," is a component, divisible part of a latitude of quality and is like the degrees hypothesized by later authors including the Calculator. In calling these degrees "uniformly difform," Roger imagines that each such degree will contain within itself a linearly increasing series of degrees above some minimum and below some maximum, as indeed any segment of a latitude would contain. The other type, the "uniform degree," is not a component part of a latitude, but rather is equally intense throughout, whereas in any part of a latitude the intensity varies. Among the Oxford calculators, only Roger makes such a distinction. Uniform degrees appear again when Roger goes on to discuss the measurement of the velocity of alteration. In the motion of intension of a quality, he says, two velocities of intension can have no ratio to one another if one subject gains a single uniform degree more than the other (39ra). Similarly Roger concludes that some local motions are incomparable to others, and that one latitude of local motion can differ from another by a single uniform degree (43va).

Roger's postulation of uniform degrees having no proportion to latitudes does not seem to be the result of intentional atomism. Rather, the case seems to be that as a pioneer in the effort to find mathematical descriptions and comparisons of concrete distributions of qualities and velocities, he could not devise measures applicable to all cases. Earlier authors had made little attempt to deal with nonuniform distributions of qualities. Roger does try to deal with them, but he has one measure for uniform distributions (the uniform degree), and another for uniformly difform (linearly varying) distributions (the latitude), and none at all for difformly difform (nonlinearly varying) distributions. Rather than stating that he is unable to compare motions or distributions of quality that fall into different categories (which would be, from a modern point of view, the justifiable statement), he says that the motions or distributions themselves have no proportion. It seems very likely that it was exactly the kind of effort to "measure" motion represented by the *De motibus naturalibus* that motivated the Calculator to try to straighten things out in his mathematically much more sophisticated work.

Liber calculationum

The *Liber calculationum* is by far the most famous work associated with the name Swineshead. As it appears in the 1520 Venice edition the *Liber calculationum* contains sixteen parts or *tractatus*. Some of these treatises may have been composed later than others, since they are lacking from some of the extant manuscripts. The emphasis in the *Liber calculationum* is on logicomathematical techniques rather than on physical theory. What it provides are techniques for calculating the values of physical variables and their changes, or for solving problems or sophisms about physical changes. Thus, the order of the treatise is one of increasing complexity in the application of techniques rather than an order determined by categories of subject matter, and the criteria for choosing between competing positions on various topics are often logicomathematical criteria. Thus, it is considered important that theory be complete—that it be able to handle all conceivable cases. Similarly, it is considered important that the mathematical measurements of a given physical variable be continuous, so that, for instance, the mathematical measure of an intensity should not jump suddenly from zero to four degrees (unless there is reason to believe that an instantaneous change occurs physically).

As stated above, as late as the beginning of the fourteenth century natural philosophers dealing with the qualities of subjects (for instance, Walter Burley in his treatises on the intension and remission of forms) assumed tacitly that the individual subjects they dealt with were uniformly qualified. Thus, as in the pharmaceutical tradition, they could talk of a subject hot in the second degree or cold in the third degree (and perhaps about what the result of their combination would be) without questioning whether the individual subjects had qualitative variations within themselves. Roger Swineshead in the *De motibus naturalibus* attempted to deal with variations in distribution, but managed only to establish criteria for uniform and for uniformly varying (*uniformiter difformis*) distributions. Richard Swineshead in his *opuscula De motu* and *De motu locali* declared that difform distributions are too diverse to deal with theoretically (212ra, 213rb). In the *Liber calculationum*, however, he manages to deal with a good number of more complicated (*difformiter difformis*) distributions.

The overall outline of the *Liber calculationum* is as follows. It begins with four treatises dealing with the qualitative degrees of simple and mixed subjects insofar as the degrees of the subjects depend on the degrees in their various parts. Treatise I considers measures of intensity (and, conversely, of remissness, that is, of privations of intensity) per se. Treatise II, on difform qualities and difformly qualified bodies, considers the effects of variations in two dimensions—intensity and extension—on the intensity of a subject taken as a whole. Treatise III again considers two variables in examining how the intensities of two qualities, for example, hotness and dryness, are to be combined in determining the intensity of an elemental subject (this, of course, being related to the Aristotelian theory that each of the four terrestrial elements—earth, air, fire, and water—is qualified in some degree by a combination of two of the four basic elemental qualities—hotness, coldness, wetness, and dryness). Treatise IV then combines the types of variation involved in treatises II and III to consider how both the intensity and extension of two qualities are to be combined in determining the intensity of a compound (mixed) subject. Treatises I–IV, then, steadily increase in mathematical complexity.

In treatises V and VI the Calculator introduces a new dimension, that of density and rarity, and determines how density, rarity, and augmentation are to be measured. Density and rarity are mathematically somewhat more complex than qualitative intensity because, even in the simplest cases, they depend on two variables, amount of matter and quantity, rather than on one. Treatises VII and VIII, then, which consider whether reaction is possible and, in order to answer that question, discuss how powers and resistances are to be measured mathematically, involve all of the variables introduced in the preceding six parts. Treatise IX, on the difficulty of action, and treatise X, on maxima and minima, complete the discussion of the measurement of powers by determining that the difficulty of an action is proportional to the power acting and by considering how the limits of a power are determined with respect to the media it can traverse in a limited or unlimited time. Treatise IX is apparently intended to apply to all types of motion, although the examples discussed nearly all have to do with local motion. In treatise X the preoccupation with local motion becomes complete. This direction of attention to local motion is continued in treatise XI, on the place of the elements, where the contributions of the parts of a body to its natural motion are discussed. Up through treatise XI, then, the three usual categories of motion according to the medieval and Aristotelian view—alteration, augmentation, and local motion—are discussed. It is significant and typical of medieval Aristotelianism that alteration is discussed first as, so to speak, the fundamental type of motion. In treatises XII and XIII the field of attention is extended to include light, treatise XII considering the measure of power of a light source and treatise XIII considering the distribution of illumination in media.

Beginning at about treatise X the tone of the *Liber calculationum* seems to change. Whereas in the first six treatises and again in the ninth several positions are compared, in treatises X through XVI, on the whole (except perhaps for treatise XIII, which is in question form), a single view is expounded. Beginning with treatise XII and continuing at an accelerating pace to the end of the work, the parts consist mostly of long strings of conclusions concerning all the variations on the basic functions of action that can be elicited by Swineshead's mathematical techniques. Treatise XIV consists of conclusions concerning local motion and how its velocity varies depending on the variations of forces and resistances. Treatise XV concerns what will happen if the resistance of the medium varies as the mobile is moving, or if, in a medium with uniformly increasing (*uniformiter difformis*) resistance, an increasing power begins to move. Treatise XVI concerns the various rates at which the maximum degree of a quality will be introduced into a subject depending on its initial state and the varying rate of its alteration, or on the rarefaction of the subject. Why the later treatises of the *Liber calculationum* should differ in tone from the earlier ones is, of course, not explained. It may be simply that the greater complexity involved in the later treatises prevented their being presented in the more usual scholastic question form. But another hypothesis might be that the earlier treatises bear the traces of having been used in university teaching, whereas the later treatises, although in a sense prepared for a similar purpose, never saw actual classroom use. At least the form in which we have them does not seem to reflect that use.

With this sketch of the overall structure of the *Liber calculationum* in hand, a more detailed look at the individual treatises is now in order. Although the *Liber calculationum* is fairly well known to historians of science by title, its contents are to date only very sketchily known, evidently (*a*) because the work is quite difficult and technical and (*b*) because it is not a work known to have influenced Galileo or other figures of the scientific revolution very significantly.

Treatise I: On Intension and Remission. In its structure, treatise I has three basic parts. First, it discusses three positions about the measures of intensity and remissness of qualities; second, it discusses whether and in what way degrees of intensity and remissness of a quality are comparable to each other; and third, it raises and replies to three doubts about rates of variation of quality considered, for instance, as loss of intensity versus increase of remissness or as gain of intensity versus decrease of remissness.

Why should these have been topics of primary interest to Swineshead? Although historians have yet to reveal very much about the connections of the *Liber calculationum* to previous tradition, it is hardly questionable that Richard Swineshead's mathematical inquiries here, like those of Roger Swineshead before him, take place against the background of Aristotelian and medical discussions of qualitative changes, especially changes in hotness, coldness, wetness, and dryness. As in the case of Roger Swineshead, Aristotelian and medical backgrounds may explain why Richard Swineshead starts from the assumption of double measures of quality in terms of intensity and remissness, as related, for instance, to hot and cold, rather than beginning simply from one scale of degrees. And again the medical theory of the temperate, representing health, and of departures from it leading to illness, may similarly explain Richard Swineshead's attention in treatise I to middle or mean degrees of whole latitudes and to degrees which might be said to be "just as intense as they are remiss."

But, furthermore, in the more immediate background of Richard Swineshead's inquiries may have been precisely Roger Swineshead's mathematization of the intensities of qualities, on the one hand, and Richard's own reasoning about the defects of some of Roger's conclusions on the other. Thus, the positions concerning the measures of intensity and remissness that Richard Swineshead considers are (1) that the intensity of any quality depends upon its nearness to the maximum degree of that quality and that remissness depends on distance from that maximum degree; (2) that intensity depends upon the distance from zero degree of a quality and remissness on distance from the maximum degree; and (3) that intensity depends upon the distance from zero degree and remissness upon the nearness to zero degree (2ra–vb). In fact, in the *De motibus naturalibus*, Roger Swineshead had held the second of these positions, and this had led him to various, sometimes peculiar, conclusions comparing the intensity and remissness of degrees (for example, 38va). Thus, Richard Swineshead may well have questioned the wisdom of a position which led to such conclusions and have looked for a better position. Beyond the earlier Aristotelian and medical theories, mathematics might have led him to refer to zero degree and to some small unit as the proper basis for a measurement of intensity. Metaphysics, however, might have led him to refer to the maximum degree of a quality, because any species may be supposed to be defined by its maximum or most perfect exemplar. Mostly on the basis of mathematical consider-

ations, Richard concludes that both intensity and remissness ought to be measured with respect to zero degree (that is, he chooses the third position). (*Tertia positio dicit quod intensio attenditur penes distantiam a non gradu et remissio penes appropinquationem ad non gradum* [2ra].)

A result of Richard's conclusion is that intensity and remissness are no longer symmetrical concepts. Thus, although there can be remission *in infinitum* before zero degree of a quality is reached, there cannot be intension *in infinitum* before the maximum degree of the latitude is reached. This follows as in the case of a finite line (lines often appear in the medieval manuscripts as representations of latitudes), where one can get closer to one extreme *in infinitum* (one can get halfway there, three-fourths of the way there, seven-eighths of the way there, continually halving the distance left), but one cannot get farther and farther from the same end *in infinitum* because one reaches the other end of the line. Consequently, if intensity is measured by distance from zero degree, the maximum degree of a quality must be remiss, which Richard admits (2vb).

In further sections, Swineshead elaborates the concept that remission is a privation with respect to intensity (4rb–4va), and then discusses in more detail the correlations between the latitudes of intensity and remissness and motions of intension and remission. Since remissness is measured by closeness to zero degree, the scale of remissness is an inverse scale (values of remissness are proportional to the inverse of the distance from zero degree) and smaller and smaller distances on the latitude close to zero degree correspond to greater and greater differences in degree of remissness. Swineshead apparently decides in this connection that the easiest solution to the problem of remissness is to label degrees of remissness by the same numbers as the degrees of intensity and merely to say that a degree of two corresponds to twice the remissness of the degree four (5ra). Again, since remission is a privation, if one allowed in imagination intensities beyond the maximum natural degree of the given quality, there would be an infinite latitude between any degree of remission and zero remissness, even though there is only a finite latitude of intensity between any degree of remissness and infinite remissness.

Thus, although it is not stated this way, the net effect of treatise I is to dispense with the need for talking about remissness at all: one can deal with all cases of interest while only considering intensity, and furthermore the intensities one is dealing with will be additive. In this way Richard Swineshead removes what now seems the needless complexity of double measures of quality and at the same time ends up with an additive measure. The treatise by no means provides a complete basis for actual measurements of qualities, but it does help to move in that direction by emphasizing measures of quality that are additive. For the Oxford calculators qualities were not, to use the modern terminology, "intensive magnitudes," but were, even in their intensity, "extensive magnitudes," again to use the modern terminology. In fact, it may be somewhat startling to the modern historian to realize that fourteenth-century authors developed their concept of "dimension" or of additive magnitude in the abstract more often through the discussion of qualitative latitudes than through the discussion of spatial extension, as was to be the case in later science.

It should be noted that treatise I, in addition to determining the proper measures of intensity of qualities, also introduces many of the basic technical terms of the rest of the work. Like the *De motibus naturalibus*, it assumes that any physical variable has a continuous range, called a "latitude," within which it can vary. In the case of qualities, this latitude starts from zero degree (*non gradus*), zero being considered as an exclusive terminus, and goes up to some determinate maximum degree, the exact number of which is usually left vague, but which is commonly assumed to be eight or ten degrees (this number arising out of the previous tradition in which there were, for instance, four degrees of coldness and four degrees of hotness, the two perhaps separated by a mean or temperate mid-degree). Within any latitude there are assumed to be a number of "degrees," these degrees being, so to speak, parts of the latitude rather than indivisibles. Swineshead also makes distinctions between the intensities versus the extensions of qualities. In his use of the terminology of latitudes and degrees, Swineshead was by no means an innovator: he was adopting a familiar set of terms. Among others, Roger Swineshead had talked systematically of the relations of degrees and latitudes before him (although Roger had had his idiosyncratic system of "uniform degrees"). In the thoroughness of his discussion of the pros and cons of various conventions for the measurement of intensity and remissness, Richard Swineshead was, however, outstanding.

Treatises II—IV: On Difformly Qualified Subjects; On the Intension of an Element Possessing Two, Unequally Intense, Qualities; On the Intension and Remission of Mixed Subjects. Directly following the determination of the most appropriate "scale of measure" for intension and remission, in general, in treatise I, treatises II–IV form a single, interrelated whole dealing with the intensity and remissness of simple and mixed subjects insofar as the degrees or "overall measures" of these subjects depend upon the degrees had by their various parts.

Thus, treatise II treats the effects of varying intensities of a single quality as these intensities are distributed over a given subject (and hence considers the two dimensions of intensity and extension) as they bear upon the overall measure of the intensity of the whole, although it only does so for the special cases in which the variation in question is either uniformly difform over the total subject or in which the subject has halves of different, but uniform, intensities. In such cases, Swineshead is in effect asking what measure of intensity is to be assigned the whole. There are, he tells us, two ways (*opiniones* or *positiones*) in which this particular question can be answered: (1) the measure—or as he often calls it, the denomination—of the whole corresponds to the mean degree of the qualified subject (that is, the degree that is equidistant from the initial and final degrees of a uniformly difformly distributed quality or—to take into account the second special case at hand—from the two degrees had by the uniform, but unequally intense, halves of the subject); or (2) the subject should be considered to be just as intense as any of its parts (that is, its overall measure is equivalent to the maximum degree had by the subject [5rb; 6ra]).

Similarly, in treatise III, Swineshead considers how the intensities of two qualities—for example, hotness and dryness—in a given elemental subject (now leaving aside their extension or distribution in this subject) are to be combined in determining the intensity of the whole. We here have to do with three positions: (1) the elemental subject is as intense as the degree equidistant from the degrees of its two qualities, (2) it is as intense as its more remiss quality, or (3) it is as intense as the mean proportional degree between its qualities (9rb).

In treatise IV Swineshead combines the types of variation involved in treatises II and III to consider how both the intensity and the extension of two qualities are to be combined in determining the intensity of a *mixtum* (that is, a compound subject). Four views concerning the measure of such more "complicated" subjects are presented: (1) the intensity of a *mixtum* follows the proportion of the dominant elementary quality to the subdominant elementary quality in it, (2) every *mixtum* is as intense as its dominant elementary quality, (3) every *mixtum* has an intensity in the dominant quality equal to half the difference between its two qualities, (4) the fourth position is presented in two versions: (*a*) The *mixtum* is as intense as the excess of the degree of the dominant quality over the subdominant quality, no account being taken of just what parts of the *mixtum* these qualities are distributed over, or (*b*) it is as intense as the excess between (i) what the dominant quality as extended over such and such a part contributes to the denomination or measure of the whole and (ii) what the subdominant quality as extended over its part of the subject contributes to the denomination of the whole (12va).

However, simply to tabulate the various *positiones* relative to the proper measure of intensity and remissness of the variously qualified simple and mixed subjects that Swineshead presents falls short of representing the substance of treatises II–IV. To begin with, to regard Swineshead's major concern as the unambiguous determination of just which *positio* or theory is the correct one relative to the particular question of measure posed by each treatise is to misrepresent his real interests. At times, Swineshead appears to leave any decision as to the "best theory" an open question. Moreover, even when he does express a preference for a given *positio*, it is seldom without qualifications, and the objections he brings against the opposing, "nonpreferred" *positiones* do not necessarily imply that his primary goal was the "once-and-for-all" rejection of these other *positiones*. His primary concern was rather to show that such and such results follow from this or that *positio*, it being of secondary importance whether these results are, for one reason or another, acceptable or unacceptable (even though they are from time to time so specified); of greater significance was the exhibition of the fact that these results do follow and the explanation of how they follow.

Thus, for example, in treatise III, although Swineshead indicates his preference for the third position (*sustinenda est tertia positio*), it is nonetheless true that the objections or conclusions brought against the second, "less-preferred" position are also relevant to this third position (*sequentur igitur contra istam* [*tertiam*] *positionem inconvenientia sicut contra alias*), with the difference that these same conclusions for the most part are in this instance conceded (11va–12rb). What is more, when we examine Swineshead's procedure

in presenting the objections to the presumably rejected positions, there emerges a more accurate picture of his objectives. Hence, for example, from the view that the overall degree of an element corresponds to the degree midway between the degrees of its two constituent qualities, Swineshead states that there follows the conclusion that there would occur continuously operating infinite velocities of action. This result should be rejected because then the agent in question would suddenly corrupt the patient upon which it acts. Inadmissible as this consequent of the conclusion might be, far more interesting to Swineshead (and hence more deserving of attention) is the fact that this conclusion does indeed follow (*quod tamen ista conclusio sequatur . . .*) from the *positio* under investigation (9va). If all of this is taken into account, one obtains a much better idea of what these treatises of the *Liber calculationum* are all about and is at the same time less puzzled or surprised at Swineshead's lack of emphasis upon the definitive determination of a single, exclusively correct theory or position.

Something more of the general character of this part of the *Liber calculationum* can be derived from a slightly more detailed example drawn from treatise II. The treatise begins by examining the view that the proper measure of the kind of difformly qualified subjects in question corresponds to the mean degree of the subject. Now one of the proofs supporting this view is that, if we take a subject that is, say, either uniformly difformly hot throughout or difformly hot with each half uniformly hot, and remit the more intense half down to the mean degree while equally rapidly intending the more remiss half up to the mean degree, then, since for every part of the subject that is intended there will be a corresponding part remitted equally and no net gain or loss in intensity, it follows that at the beginning the subject contained an intensity equivalent to the mean degree.

However, this proof of the first position or view will not do according to Swineshead, since when combined with the physical assumption that heating rarefies while cooling condenses, the subject will unavoidably be rarefied in one part and condensed in another when the process of equalizing the halves of the subject is carried out; but the moment this equalization commences, the cooler, more remiss, half will, because of the rarefaction caused by heating, become greater than the more intense half, which means that throughout the whole process intension will be occurring over a greater part than is remission; therefore, Swineshead concludes, at the beginning the whole subject must be more remiss than the mean degree. An objection to this procedure is raised, but it is disposed of through a number of replies establishing that the subject must indeed initially have an overall intensity less than the mean degree (5rb – 5vb).

All of this would seem to imply that Swineshead definitely rejected the first "mean degree measure" position, especially if we combine this with the fact that in the following paragraphs he appears to regard the second opposing position (that the subjects in question are just as intense as any of their parts) as acceptable (6ra – 6va). However, such a judgment would be premature. Swineshead has argued not directly against the first position as such, but rather against a proof given of it. Furthermore, in the remaining (and one should note, larger and more impressive) part of treatise II, Swineshead returns to this first "mean degree" position and allows its application to difformly qualified subjects each half of which is uniform and, more generally, to "stair-step qualities" in which the intensities differ, but are uniform, over certain determinate parts of the qualified subjects. This applicability is grounded upon the fact that in a difform subject with uniform halves, a quality extended through a half "denominates the whole only half as much as it denominates the half through which it is extended." Swineshead then generalizes this "new rule" and states that if a quality is "extended in a proportionally smaller part of the whole, it denominates the whole with a correspondingly more remiss degree than it does the part through which it is extended" (6va), thus opening the possibility of considering "stair-step" distributions.

After giving proofs for the special and general cases of his new "rule of denomination," Swineshead raises an objection against it: "If the first proportional part of something be intense in such and such a degree, and the second [proportional part] were twice as intense, the third three times, and so on *in infinitum*, then the whole would be just as intense as the second proportional part. However, this does not appear to be true. For it is apparent that the quality is infinite and thus, if it exists without a contrary, it will infinitely denominate its subject" (6va).

Swineshead shows that this latter inference to infinite denomination does not follow and that it arises because one has ignored the proper denomination criterion he has just set forth (6vb – 7ra). As a preliminary, he devotes considerable space to the important task of establishing that a subject with a quality distribution as specified by the objection is indeed just as intense as its second proportional part, and he presents in detail just how this is so (6va – 6vb). The proportional parts in question are to be taken "according to a double proportion" (that is, the succeeding proportional parts of the subject are its half, fourth, eighth, etc.). Now following the arithmetic increase in intensity over the succeeding proportional parts as stipulated by the objection, it follows that the whole will have the intensity of the second proportional part of the subject. Swineshead proves this by taking two subjects—A and B—and dividing them both according to the required proportional parts. Now take B and "let it be assumed that during the first proportional part of an hour the first [proportional] part of B is intended to its double, and similarly in the second proportional part of the hour the second proportional part of it is intended to its double, and so on *in infinitum* in such a way that at the end [of the hour] B will be uniform in a degree double the degree it now has." Turning then to A, Swineshead asks us to assume that "during the first proportional part of the hour the whole of A except its first proportional part grows more intense by acquiring just as much latitude as the first proportional part of B acquires during that period, while in the second proportional part of the same hour all of A except its first and second proportional parts grows more intense by acquiring just as much latitude as the second proportional part of B then acquires . . . and so on *in infinitum*." Clearly, then, since the whole of A except its first proportional part is equal in extent to its first proportional part, and since the whole of A except its first and second proportional parts is equal to its second proportional part . . . and so on *in infinitum*, it follows that A acquires just as much, and only as much, as B does throughout the hour; therefore, it is overall just as intense as B is at the end of the hour, which is to say that it is doubly intense or has an intensity equivalent to that of its second proportional part [*Q.E.D.*].

In thus determining just how intense A is at the end of its specified intensification, Swineshead has correctly seen that in our terms the infinite geometrical series involved is convergent (if we assume the intensity of the whole of A at the outset to be 1, then $\frac{1}{2} + \frac{2}{4} + \frac{3}{8} + \cdots + \frac{n}{2^n} = 2$). But such an interpretation is misleading. Swineshead gives absolutely no consideration to anything becoming arbitrarily small or tending to zero as we move indefinitely over the specified proportional parts.

Swineshead knows where he is going to end up before he even starts; he has merely redistributed what he already knows to be a given finite increase in the intensity of one subject over another subject, something that is found to be true in most instances of the occurrence of "convergent infinite series" in the late Middle Ages. Yet however Swineshead's accomplishment is interpreted, one should note that his major concern was to show that a subject whose quality was distributed in such a manner *in infinitum* over its parts was in fact consistent with his denomination criterion and did not lead to paradox. It is also notable that, in so increasing the intensity of A, he could have specified that the quality in question was heat, arguing, as he had previously argued against the proof of the first position, that on grounds of the physical assumption that heating causes rarefaction, it followed that A would not be just as intense as its second proportional part. The fact that he did not do so lends further credence to the view that his major interest was in seeing how many "results" could be drawn out of a given position or assumption, the more complicated and surprising the results the better. One such set of results could be derived by applying a physical assumption to the proof of the first position; another, as in the present instance, by ignoring it.

This interest of Swineshead can be even better illustrated if the present example from treatise II is carried yet one step further. Immediately after answering the objection treated above, another objection is put forth claiming that "from this it follows that A is now only finitely intense, yet by means of a merely finite rarefaction will suddenly be made infinitely intense." In Swineshead's reply to the objector's complaint that this is an absurd state of affairs and must be rejected, the important point is again Swineshead's demonstration that this presumably absurd situation can and does obtain (7ra). We can see how this can be so if, Swineshead tells us, we take only every 2nth proportional part of our previously so intensified A and then rarefy the second proportional part of A by any amount howsoever small, while rarefying each of the succeeding proportional parts twice as slowly as the preceding one. Again, in our terms we have to do with a "divergent series," so the conclusion that A is "suddenly made infinitely intense" is a correct one. But to set down the general term of this "series" would be anachronistic and would credit Swineshead with something that was quite outside his thinking. What he should be credited with is ingenious, but much more straightforward. He realized that in selecting only the 2nth proportional parts of A he had chosen parts whose intensities were successively double one another. Therefore, in deliberately specifying that the rarefaction over these parts should be successively "twice as slow," it automatically followed that, considering both the extension and intensity of that amount added to each part by rarefaction, the resulting contribution (no matter how small) to the denomination of the whole would be the same in each instance. And since there were an infinite number of such "added parts," the denomination of the whole immediately became infinite. Once again, in our terms, what Swineshead has done amounts to the adding of a constant amount to each term of an "infinite series." It is more profitable, however, to view Swineshead's concern with the infinite in another, much less modern, way. In the two "objections" that have just been cited, Swineshead has first shown that, astonishing as it might seem, a subject whose quality increases *in infinitum* as distributed over its parts is as a matter of fact only finitely intense overall. One can next take this same finitely intense subject, change it by a finite amount as small as you wish, and it immediately becomes infinitely intense. The switch from infinite to finite and then back to infinite again seems more than incidental. Swineshead was partaking of something that was characteristic of the logical—and by then physical—tradition of solving sophisms. In point of fact, at the end of treatise II he even refers to the conclusions he is dealing with (there are fifteen in all) as *sophismata* (9rb).

The foregoing fairly lengthy discussions of the first four treatises of the *Liber calculationum* should give a good impression of the character of the whole work, not forgetting that the later treatises appear to be slightly more expository in form. The descriptions of some of the special features of the treatises that follow assume the continuation of this same basic character without repeatedly asserting it.

Treatises V–VI: On Rarity and Density; On the Velocity of Augmentation. The fifth and sixth treatises again form a logical unit, this time concerning the quantity or rarity and density of subjects and motions with respect to quantity. The relatively long treatise V (16vb–22rb) has three basic parts. It first addresses directly the question of the proper measures of rarity and density, rejecting the position (1) that rarity depends on the proportion of the quantity of the subject to its matter while density depends on the proportion of matter to quantity, and accepting the position (2) that rarity depends on quantity assuming that the amount of matter remains the same (*raritas attenditur penes quantitatem non simpliciter sed in materia proportionata vel in comparatione ad materiam. Et ponit quod proportionabiliter sicut tota quantitas sit maior manente materia eadem, ita raritas est maior* [17ra]).

It is of interest to the modern historian to realize why Swineshead considered the first position to be significantly different from the second position; it may seem, indeed, to be nothing but an improved and more general version of the second position. The explanation of this point turns out to shed important light on the status of Bradwardine's geometric function relating forces, resistances, and velocities, which had been propounded in his *De proportionibus* in 1328. In fact, for Swineshead, if rarity depended upon the proportion (ratio) of quantity to matter, this would have meant that, for instance, when the proportion of quantity to matter was "doubled" (*dupletur* in his terms, but "squared" in modern terms), then the rarity would be doubled, and the resulting function would have been in modern terms logarithmic or exponential in exactly the same way that Bradwardine's function was logarithmic or exponential. It was their understanding of the meaning of the compounding or "addition" and "subtraction" of proportions (equivalent to multiplying and dividing ratios in the modern sense) that essentially forced fourteenth-century thinkers to this function. Finding it difficult, therefore, to propose the dependence of rarity on the ratio of quantity to matter as this would be understood in the simple modern sense, Swineshead was therefore led to his less elegant second position as a substitute emphasizing quantity and assuming the constancy of matter as a subsidiary consideration in order to avoid the intrusion of a proportion per se.

But having proposed the dependence of rarity on quantity assuming the matter constant, Swineshead enters in the second part of the treatise into a long consideration of how rarity should depend upon quantity. This consideration is subsumed under the question whether both rarity and density are positive entities or whether only one of them is positive, and, if so, which (17ra). Here he can rely in an important way upon his discussion in treatise I of positive and privative entities and their interrelations with regard to intensity and remissness. After an involved discussion, he concludes that density is the positive quality (and rarity privative) and that when a subject is rarefied uniformly for a given period of time it acquires quantity difformly, greater and greater quantities corresponding to equal increments of rarity as the subject becomes more rarefied (*densitas se habet positive et ex uniformi rarefactione alicuius per tempus secundum se totum difformiter acquiritur quantitas et si densius et rarius equalis quantitatis equevelociter rarefierent, rarius maiorem quantitatem acquireret quam densius* [18rb]). Thus, the mathematical characteristics of the measures of rarity become similar to those of the measures of remissness in treatise I, and similar conclusions can be reached. It follows, for instance, that the latitude of rarity between any degree of rarity and zero rarity is infinite (18vb), just as a similar conclusion had followed for remissness.

The third and last section of treatise V raises and replies to doubts, many of which are parallel to earlier considerations concerning quality. Swineshead concludes that a uniformly difformly dense body or a body with unequal degrees of uniform density in its two halves is as dense as its mean degree (18vb–20vb). Similarly, he says that bodies are as rare as their mean degrees provided that it is understood that the latitudes of rarity and density are really the same (19vb), and he disposes of a whole series of doubts about how density and rarity are to be compared by saying that the situation is the same in this case as it is in the case of intensity and remissness (*ad que omnia possunt consimiliter argui et responderi sicut arguebatur ubi tanguntur illa de intensione et remissione, mutatis illis terminis intensio et remissio in istis terminis raritas et densitas* [20vb]). Finally, he replies to a doubt about whether, if there were an infinite quantity with a part which was infinitely dense, the whole would be infinitely dense (21rb–22rb) by saying that just as in similar cases concerning qualities, so here a density extended through only a finite part of an infinite subject would not contribute anything to the denomination of the whole subject (21vb–22ra).

When to the above description of treatise V is coupled the observation that nowhere in treatise V does Swineshead directly inquire into the physical significance to be properly correlated with the concepts density and rarity, it should be clear that Swineshead's real interest here must have been in the mathematical functions involved in the various positions, and in the consequences, whether more or less startling, that could be shown to be consistent with these functions. And again, as in earlier treatises, he concludes by saying that many more sophisms could be developed concerning this material, all of which can easily be solved if the material he has presented is well understood (22rb).

In treatise VI, Swineshead then turns from rarity and density as such to motions of augmentation, where augmentation is considered to be the same as increase of rarity. Like Roger Swineshead before him, he begins by rejecting the position espoused in Heytesbury's *Regule solvendi sophismata* that motion of augmentation is to be measured by the proportion of the new quantity to the old quantity (22va–24va). Second, he turns to the position on augmentation held by Roger Swineshead, namely that augmentation is to be measured by quantity acquired irrespective of the quantity doing the acquiring (24va–vb), but he also rejects this position because it does not adequately handle cases in which a quantity is lost at the same time as one is added. Swineshead then replaces Roger Swineshead's position with an improved version which he accepts, saying that augmentation and diminution should be measured by the net change of quantity of the subject (24vb). In reply to an objection concerning what happens according to this position to the concept of uniform augmentation, Swineshead admits in effect that the concepts of uniform velocities of alteration and motion will not then have an easy parallel in the case of the motion of augmentation, although one could speak of equal parts of a subject gaining equal qualities.

The most striking thing about treatise VI is perhaps the fact that the largest section of the treatise is devoted to the refutation of Heytesbury's view concerning the proper measure of augmentation. And here there are two points to be noted about the arguments provided. First of all, a significantly large proportion of them involve what are in effect augmentations from zero quantity. Since the first position is obviously not applicable to augmentations from zero (since this would put a zero into the denominator of the proportion of quantities that it proposes as the proper measure of augmentation), one might argue that these supposed refutations of the position are misguided. And secondly, Swineshead himself eventually concedes several of the refuting arguments although he did consider the first position to be refuted. Yet these arguments are left to stand as if they were strikes against the first position. Thus, Swineshead says that the inferences which can be drawn from the first position are amazing and contrary to one's idea of what the proper measure of a motion should be (23va)—some of these conclusions being ones that involve the unfair augmentation from zero quantity—but he then also concedes that some of these conclusions are simply true no matter how the velocity of augmentation is measured

(23va). So again one might fairly draw the conclusion that Swineshead's major concern is not really the choice between rival measures, but rather the exhibition of mathematical techniques that one might reasonably use in the discussion of any of the positions.

Treatises VII–VIII: On Reaction; On the Powers of Things. As in the works of the other Oxford calculators where reaction is taken up at a fairly early stage, here too the problem of reaction is really the entire problem of how two qualified bodies act on each other and involves all of the variables discussed in the preceding treatises. For those who, like Richard Swineshead, held the so-called addition of part to part theory of qualitative change, the problem was particularly acute. There were numerous well-known cases (*experimenta*) in which reaction seemed to occur (25va). Furthermore, under the addition theory it seemed that the parts of a quality present should be able to act and react with the other qualitative parts nearby. Yet the previously accepted Aristotelian and medical theory of qualitative change had assumed that the qualities of a given subject could be represented by the single degrees of hot, cold, wet, or dry of the whole, so that if two bodies were brought close together such that one could act on the other, only that body with the higher degree, say of heat, would act as the agent or force causing change, and only that body with the lower degree would act as patient and be changed. Clearly the calculators were in a position in which they had to improve upon previous theory by taking account of distributions of quality, and yet the theoretical situation was so complex and mathematically difficult that they faced an almost impossible task.

After preliminary arguments, Swineshead takes, as the foundation of his solution to the problem, the position that the power of a subject is determined by the multitude of form (*multitudo forme*) in it, where multitude of form is determined not only by the intensity of the form and the extension, but also by what might be called the density of form (26vb). He takes density as the most important factor and asserts that if a foot length of fire were condensed to half a foot, it would still contain the same multitude of form even if the intensity were the same as before. Swineshead next turns to the question of whether the whole patient resists the agent or only the part acted on. After considering various positions, he concludes that although the whole patient does not necessarily resist the agent, the whole part of the patient directly opposite the agent resists, and that all parts do not resist equally, parts further from the agent resisting less than those closer. Unfortunately, arguments can be raised to show that no simple proportionality obtains between distance from the agent and lesser resistance. Swineshead asserts, however, that it will be more clearly understood when he deals with illuminations how resistance decreases with distance (27rb – 28rb).

On the basis of these fundamentals Swineshead concludes that reaction cannot occur between uniform bodies such that the reaction is according to the quality contrary to that of the action (28va – 28vb). (By a uniform body Swineshead means uniform not only according to intensity of quality but also according to the amount of form existing in equal parts of the body.) It is possible for the patient to react according to another quality – so that while the agent heats the patient the patient in turn humidifies the agent. Between difform bodies, on the other hand, there can be action with reaction in the contrary quality in another part. Where such reaction occurs the whole agent and whole patient act and resist according to their power insofar as it is applied in the given situation. Thus, Swineshead appears content in his reply to leave the unspoken and rather improbable implication that in all the observed cases of reaction according to the same quality, the qualities of the two bodies must have been difform.

The rest of treatise VII consists of the solution of three *dubia*. The first, also dealt with by John Dumbleton in his *Summa logicae et philosophiae naturalis*, concerns whether an agent will act more slowly if patients are applied to either side of it than it would if it acted on only one of the patients (29va – 30ra). Swineshead is sure that if two actions concur at the same point the action will be faster, but he is not sure of the solution of the doubt if the two patients are far enough apart so that they do not act on each other. In the latter case, he says, the reader may decide for himself whether the patients will assist each other in resisting, since, although some say they do, it is hard to understand how this could be so (30ra). To a second doubt Swineshead concludes that two difform bodies which are similar in those parts nearest each other can nevertheless still act on each other (30ra – 30va), and to a third doubt he concludes that bodies having maximum degrees of contraries can act on each other (30va – 30vb).

It should be clear that in all of treatise VII one of Swineshead's main questions concerns the additivity or summability of the forces and resistances he is dealing with. Indeed, elsewhere additivity

was one of the major concerns of the Oxford calculators in their efforts at quantification. With respect to single dimensions such as that of intensity, the calculators were adamant that the measure of intensity should be an additive measure. Following the addition theory of the intension and remission of forms, they assumed that an intensity was equivalent to the sum of its parts. In dealing with actions and reactions of bodies, however, the basis for such additivity was not so easily found. Here, and again in treatise XI concerning local motion, Swineshead appears to concede, perhaps to his own disappointment, that difficulties appear to ensue if one attempts to treat subjects as the sums of their parts in any simple fashion. The difficulties of considering not only the forces and resistances of the parts, but also the varying distances of the parts from each other and the possible interactions of the parts on each other, made a detailed part-by-part quantitative treatment practically impossible. We may admire Swineshead's ingenuity in the face of such odds while agreeing that it is unfortunate that the slant of Aristotelian physics towards alteration rather than local motion caused Swineshead to concentrate on such a difficult problem.

Treatise VIII again takes up the question dealt with in treatise VII concerning how the powers of things are to be measured, and it appears probable that parts of treatises VII and VIII represent Swineshead's successive reworkings of the same basic problem. Despite nine arguments against the view that power is to be measured by multitude of form, Swineshead reaffirms his earlier conclusion that it is. The only thing that he adds here is the remark that the amount of form induced in a subject will depend upon the amount of matter present (31rb). He concedes the nine arguments or conclusions *de imaginatione*, their supposed difficulty being based on the view that, for a given form, intensity and extension are inversely related, which, he says, is only accidentally so (31rb – 31va).

If in the work of Oresme, the concept of the "quantity of quality" was to become fundamental, where quantity of quality was the product of intensity times extension, here we see Swineshead's effort to deal with objections based on a similar concept. Whatever objections there may be to Oresme's concept of "quantity of quality" from a modern point of view, students of medieval science have in recent years become so familiar with the concept that there is a tendency to assume the use of a similar concept in other late medieval authors attempting the quantification of qualities or forms. It deserves emphasizing, therefore, that Richard

Swineshead and the other Oxford calculators of 1330–1350 were familiar with concepts like Oresme's quantity of quality, but rejected them in favor of quantifications in terms of intensities alone or in terms of something like Swineshead's multitude of form. This rejection of "quantity of quality" helps explain, among other things, Swineshead's less than total happiness in treatise II with the "mean degree measure" of difform qualities, a measure that he might otherwise have been expected to favor because of its mathematical attractiveness. Swineshead does not always assume, as Oresme's quantity of quality concept implies, that when the extension of a subject is decreased, the form remaining unchanged, the intensity will increase, and he even goes so far as to assert that a form could be condensed to a point without its intensity increasing (31rb).

Treatises IX–X: On the Difficulty of Action; On Maxima and Minima. The ninth and tenth treatises in a sense carry further Swineshead's treatment of action and the forces causing it. Treatise X has a clear precedent in earlier discussions of maxima and minima, in particular in the discussion found in Heytesbury's *Regule solvendi sophismata*, and treatise IX probably is related, although in a nonobvious way, to Bradwardine's *De proportionibus*. The problem covered in treatise IX (and also in Dumbleton's *Summa*, part VI) seems to have arisen because of the Calculator's acceptance of Bradwardine's function for measuring velocities. For the standard Aristotelian position concerning the relation of forces, resistances, and velocities, there was a simple relationship between forces and the velocities produced with a given resistance, such that each equal part of the force could be interpreted as contributing an equal part of the velocity (and here again arose the question of additivity). For the Bradwardinian position, on the other hand, multiples of a force, with the resistance remaining constant, did not produce equal multiples of the velocity. As a result, one needed some other measure of what a force could do. On this subject, Swineshead first rejects two positions: (1) that the action or difficulty produced depends upon the proportion of greater inequality with which the agent acts, so that an agent acting from a greater proportion produces a greater difficulty (31va); and (2) that the action or difficulty produced depends on a proportion of lesser inequality, because an agent closer in power to the strength of the resistance tires more in acting (31va).

The position that Swineshead adopts is the same as that adopted by Dumbleton, namely that the difficulty produced is proportional to the power acting to its ultimate (31vb). In a given uniform resistance there will be an action of maximum difficulty that cannot be produced in that medium, namely the difficulty equal to the power of the resistance (31vb–32ra). When the power of the agent is doubled, the difficulty it can produce is also doubled. The latitude of difficulty or range of all possible difficulties is infinite (32ra).

The remainder of treatise IX consists of the raising of twelve arguments against Swineshead's preferred position and of his replies to them. In these arguments the connection with Bradwardine's function manifests itself. A common assumption behind the objections is that difficulty (or action) and motion (or velocity) ought to be proportional to each other, so that to move something twice as fast is to produce a double action. As stated above, in the commonly assumed Aristotelian "function," a double force, a double velocity, and presumably a double action or difficulty all seem to be correlated with each other. This is not so in the Bradwardinian function, where a doubling of force does not usually correlate with a doubling of velocity. Swineshead simply asserts here, therefore, that difficulty can be correlated with force or power, and that velocity and difficulty produced do not necessarily correspond to each other (32vb–33rb). The main effect of treatise IX, therefore, is to clear away objections that might be raised about Bradwardine's function by those still thinking in an Aristotelian framework.

Treatise X concerns maxima and minima only with respect to the traversal of space in local motion, and the ground it covers is quite standard. Having stated some familiar definitions and suppositions concerning maxima and minima of active and passive powers, Swineshead states two rules: (1) that both debilitatable and nondebilitatable powers have a minimum uniform resistance that they cannot traverse (in familiar scholastic terminology, a *minimum quod non*); and (2) that with respect to media there is a maximum power that cannot traverse a given medium, namely the power equal to the resistance of the medium (a *maximum quod non* [34rb]). Thus, for a power, say, equal to 3, there will be a minimum resistance it cannot traverse, that is, the resistance 3, and, conversely, for the resistance equal to 3, the power equal to 3 will be the maximum power that cannot traverse it.

The rest of treatise X consists of rules for assigning maxima and minima under a variety of possible conditions, that is, when the medium is uniform and when it is difform, when there is a time limit and when there is not, when the power is constant and when it weakens in acting, and when the medium is infinite and when it is finite. Although they may take a while to decipher, these conclusions are mostly the simple results of the assumption that the force must be greater than the resistance for motion to occur. Swineshead himself seems to feel that he is traversing familiar ground, and the treatise is therefore quite short.

Treatise XI: On the Place of an Element. Swineshead's concern in treatise XI is a single problem relating to the motion of a heavy body in the vicinity of its natural place at the center of the universe: whether in free fall, assuming a void or nonresistant medium, the heavy body will ever reach the center of the universe in the sense that the center of the body will eventually coincide with the center of the universe. If we regard the heavy body in question as a thin rod (*simplex columnare*), the variables that Swineshead has to deal with in resolving the problem become evident: as soon as any part of the rod passes the center of the universe, that part may be considered as acting as a resistance against its continued motion. Now one position that can be taken in resolving the problem is that the body acts as the sum of its parts and that therefore the parts of the rod "beyond the center" actually do resist its motion. Assuming this, lengths or segments of the rod will function both as distances traversed as the rod approaches the center and as the forces and resistances involved in determining such a traversal. If we also assume, with Swineshead, Bradwardine's "function" relating velocities with the forces and resistances determining them, then the task to be carried out is to discover a way to apply this "function" to the "distance-determined" forces and resistances acting upon the falling rod in order to calculate the relevant changes in velocity and thus ascertain whether or not the center of the rod ever will reach the center of the universe.

In what is mathematically perhaps the most complicated and sophisticated section of the *Liber calculationum*, Swineshead accomplishes this task and replies that, on the assumption of the rod acting as the sum of its parts, the two centers will never come to coincide. He presents his argument axiomatically, beginning from a number of strictly mathematical *suppositiones* and *regulae* and then moving to their application to the problem at hand. It will be easier to indicate something of the nature of his accomplishment if the order is, at least in part, reversed. Thus, Swineshead clearly realizes and emphasizes the fact that the distance remain-

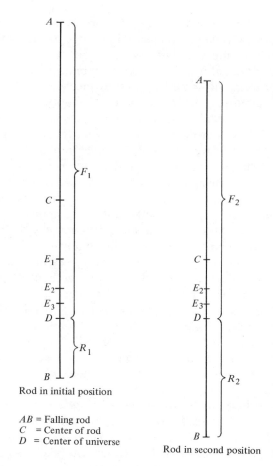

AB = Falling rod
C = Center of rod
D = Center of universe

FIGURE 1

ing between the center of the rod and the center of the universe will always be equal to half the difference between that (greater) part of the rod which is still on this side of the center of the universe and that (lesser) part which is beyond (37ra) (in terms of Figure 1, $CD = (F_1 - R_1)/2$). This obtains no matter what space intervals we consider in the rod's progressive motion toward the center. When this is added to the fact that, with any given motion of the rod, whatever is subtracted from the segment this side of the center of the universe is added to the segment beyond the center, thus determining the "new" forces and resistances obtaining after that motion, Swineshead then has a way to apply Bradwardine's "function" to the whole problem. Divide, for example, the remaining distance (CD) into proportional parts (according to a double proportion); we know the relation of the distance between the centers to the difference between the relevant forces and resistances: $CD = (F_1 - R_1)/2$, but successively following the particular division specified of the remaining distance it is also true that this same half-difference between F_1 and R_1 is equivalent to the excess of F_1 less one-fourth of

$(F_1 - R_1)$ over R_1 plus one-fourth of $(F_1 - R_1)$. But in a strictly mathematical "suppositio" it is stated (and then proved) that the proportion of the thusly decreased F_1 to the thusly increased R_1 will be less than the "sub-double" of the proportion between the original, unaltered F_1 and R_1. (*Si inter aliqua sit proportio maioris inequalitatis, et quarta pars excessus maioris supra minus auferatur a maiori et addatur minori, tunc inter illa in fine erit proportio minor quam subdupla ad proportionem existentem inter ista duo in principio* [35vb].) In modern symbols:

$$\left(F_1 - \frac{F_1 - R_1}{4}\right) : \left(R_1 + \frac{F_1 - R_1}{4}\right) < (F_1 : R_1)^{1/2}.$$

At the same time, the decreased F_1 and the increased R_1 have given us the F_2 and R_2 operative after the rod has moved over the first proportional part of the distance between centers, assuming that its speed is uniform throughout this motion, and, moving over the second proportional part, we can similarly derive F_3 and R_3 and relate them to F_2 and R_2 by means of the same "mathematical supposition," and so on, over succeeding proportional parts and F's and R's. However, the ordering of the force-resistance proportions that is the burden of this mathematical supposition is precisely what is at stake in Bradwardine's "function" claiming that increases and (in the particular problem at hand) decreases in velocity correspond to increases and decreases in the proportion between force and resistance. Bradwardine can therefore be directly applied, yielding resultant velocities over succeeding proportional parts each of which is "more remiss" than half the preceding one ($V_1 > 2V_2 > 4V_3 > \cdots$). Since succeeding proportional parts of the distance decrease by exactly half ($CE_1 = 2E_1E_2 = 4E_2E_3 = \cdots$), it follows that the time intervals for each increment of distance must increase *in infinitum*, which means that the center of the rod will never reach the center of the universe (*maius tempus requireretur ad pertransitionem secunde partis proportionalis quam ad pertransitionem prime . . . et sic in infinitum. Ergo in nullo tempore finito transiret C totam illam distantiam* [37rb]).

Swineshead has reached this result by applying a particular, proportional part division to the distance remaining between the centers, but he provides for the generalization of this division by specifying (and proving) his crucial mathematical supposition in a general form (36va); then we may presumably take any succeeding proportional parts whatsoever in determining the fall of the rod. But his own use of this more general supposition occurs in a second, different proof of the conclusion that the rod will never reach the center of the universe. In our terms, the first proof summarized above assumes a constant velocity—and hence constant force-resistance proportions—over the relevant distance intervals, thus employing a discontinuous, step function in resolving the problem. In a more compact, and more difficult, proof Swineshead comes more directly to grip with his variables as exhibiting a continuous function. Less tractable, and hence more difficult to represent adequately in modern terms, than the first proof, its substance is tied to the proportional comparison of decreases or losses with what we would term rates of decrease or loss (for example, *motus velocius proportionabiliter remittetur quam excessus; ergo excessus tardius et tardius proportionabiliter remittetur* [37rb]).

In both proofs Swineshead has in effect assumed that the rod or heavy body in question is a *grave simplex*, a limitation that he addresses himself to by considering, in reply to several objections, the body as a *mixtum* (37va). Far more important, however, is another objection. It claims, in effect, that the assumption behind Swineshead's whole procedure up to that point—namely, that the rod does act as the sum of its parts, must be false because it implies that there would exist natural inclinations that would be totally without purpose and vain (*appetitus . . . omnino otiosus . . . vanus*), an inadmissible consequent (37rb).

Swineshead therefore sets forth a second, alternative position, one in which the heavy body in question acts as a whole, where its parts contribute to the natural inclination or desire (*appetitus*) of the whole in a manner that is not given precise mathematical determination (37vb–38ra). As might be expected, Swineshead spends far less time treating, and seems much less interested in, this second position, in spite of the fact that it is apparently the true one. This brevity fits well with the whole tenor of the *Liber calculationum* and with what has been noted above of the greater interest in deriving results than in just what the results are. The treatment based on the first, "false" *positio* of the whole body as the sum of its parts also fits well with much of the rest of the *Liber calculationum*, where the mathematical and logical determination of the contribution of parts to wholes is so often a central issue.

Treatises XII–XIII: On Light; On the Action of Light. These two treatises are concerned with light, first with respect to the power of the light source and second with respect to the illumination produced.

The power of a light source, Swineshead states, is measured in the same way as the power of other agents, namely by the multitude of form (38ra). Equal light sources, then, will be those that are not only equal in intensity but also equal in multitude of form (38rb). Thus, if sources with equal multitudes of form are intended by equal latitudes of intensity, they will gain equally in power, but if sources with unequal multitudes of form are intended by equal latitudes of intensity, the one with more form will increase more in power than the other.

On the basis of these presuppositions, Swineshead then draws a number of conclusions or rules treating what happens when either the quantity of light source is varied (by adding or subtracting matter so that the multitude of form is changed) or the intensity of these sources is varied. He concludes, for instance, that if there are two light sources of different intensity, which at the outset are either equal or unequal in quantity, but which then diminish equally in quantity, then proportionally as one is more intense than the other it will diminish in power more rapidly (38rb–38va). Swineshead does not believe that changing the quantity (extension) of a light source without adding or subtracting matter will change its power, but he says that his conclusions can be proved even better by those who hold such a view (38vb).

It should be noted that in this treatise Swineshead does assume that there is a correlation between intensity and multitude of form, something which he felt it necessary to state as an explicit hypothesis when he dealt with similar problems in treatise II (for example, 8va) and something which he, in effect, ignored in treatise VIII. Had his interests been in determining the one correct physical theory, it is hard to believe that he would not have brought these contexts together and somewhere stated what he felt to be the true physical situation with regard to the connections between the intensity, extension, and "density" of a form. With his attention falling as entirely on the quantitative side as it does, he lets apparent inconsistencies slide, covered by the remark that the connection between the intensities and extensions of a given form are only accidental (31rb).

Treatise XIII consists of the solution of two major doubts and a long string of conclusions. In reply to the first doubt, Swineshead concludes that every light source produces its entire latitude, from its maximum degree down to zero degree, in every medium in which it suffices to act, but that a source will cast its light to a greater distance in a rarer medium and to a lesser distance in a denser medium (39vb). Light is remitted (is less intense) at more distant points because of the indisposition caused by the medium between the source and the distant point, so since there is no medium between the source and the point next to it there is no remission at that point.

In reply to the second doubt Swineshead concludes that a light source casts a uniformly difform illumination in a uniform medium (40rb–va). Considering the medium between the source and a distant point as an impediment subtracting from the intensity of illumination, Swineshead comes up with a physically reasonable relation leading to a uniformly difform distribution, avoiding the trap of making the intensity inversely proportional to the distance from the light source. This no doubt is the explanation Swineshead had in mind in treatise VII as being helpful in understanding how distant parts combine their actions and resistances (28rb).

These two replies are then followed by a series of fourteen conclusions (numbered 13–26 in sequence with the conclusions of treatise XII) intended to make clearer what has preceded. They appear quite complex, but we may suppose that he arrived at them by a simple visualization of the situation with few if any mathematical calculations. Thus, Swineshead concludes first that if a light source acts in a uniform medium and if a part of the medium next to the agent is made more dense without changing its quantity (extension), then at every point of the rest of the medium farther from the agent the illumination will be remitted with the same velocity as the illumination at the extreme point of the part made denser is remitted (40vb–41ra). Imagining a graph of the original uniformly

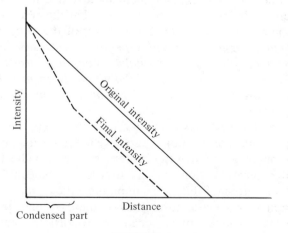

FIGURE 2

difform illumination, we see that this conclusion amounts to saying that when part of the medium is condensed, the slope of decrease of intensity will become steeper in that part, but in the remaining part the slope of decrease will remain the same, being shifted down parallel to itself to connect with the new, more remiss degree at the extreme point of the condensed part (see Figure 2).

Like this first conclusion, Swineshead's other conclusions are easy to understand on the basis of graphs, although he makes no reference to visualizations of the conclusions. In our terms, the variables that he has to work with in the conclusions are that the quantity of the light source determines the rate of decrease of intensity or slope of the distribution of intensities in a given medium, whereas the intensity of the source determines the degree from which that decrease starts. The density of the medium, on the other hand, also determines the slope of the distribution of intensities when the source remains the same. Swineshead can then partially offset changes in the quantity of the light source by changes in the density of the medium, or vice versa, as it suits his purpose (cf. conclusion 19 [41vb]).

It is so natural to us to visualize these conclusions that it is hard to imagine that Swineshead did not do so also. Dumbleton, in the corresponding part of his *Summa*, did make an explicit geometrical analogy, and the tradition of using triangles or cones to represent the dispersal of light in optics would also have prompted mental images of triangles. It is very probable then that the mathematics behind treatise XIII was a simple "visualized geometry."

Treatise XIV: On Local Motion. One of the most exhaustively developed sections of the *Liber calculationum*, treatise XIV begins by stipulating that its contents will be formulated under the assumption "that motion is measured in terms of geometric proportion" (*motum attendi penes proportionem geometricam* [43va]). As immediately becomes obvious, this is Swineshead's elliptical way of informing his readers that he will be accepting Bradwardine's view that variations in velocities correspond directly to variations in the force-resistance proportions determining those velocities (in modern terminology, that arithmetic changes in velocity correspond to geometric changes in the relevant force-resistance ratios). With this as base, what Swineshead accomplishes in setting forth the forty-nine *regulae* that constitute treatise XIV is to give a relatively complete "catalog" of just which *kinds* of changes in velocity correspond to which

kinds of changes in force and resistance and vice versa.

He does this in strict axiomatic fashion, the first three rules presenting what can be regarded as the basic mathematics of proportion which will serve, together with his assumption of Bradwardine's "function," as the key to all that follows. As indicated above, there is no doubt that the medieval tradition of compounding proportions was at the root of Bradwardine's logarithmic-type function, and it is precisely this that Swineshead makes explicit in his initial rules. Thus, the first rule tells us that "whenever a force (*potentia*) increases with respect to a constant resistance (*resistentiae non variatae*), then it will acquire as much proportionally relative to that resistance as it will itself be rendered greater" (43va). That is to say, if some force F_1 acting on a constant resistance R increases to F_2, then the proportional increase in F (the proportion $F_2 : F_1$) is equal to the increase of $F_2 : R$ over $F_1 : R$. As Swineshead makes clear in his proof of this rule, this amounts to a compounding of the proportions involved, that is, when $F_2 : F_1$ is "added to" $F_1 : R$ the result is $F_2 : R$. (Note that following the medieval convention proportions are added to one another [just as are numbers or line segments], when we would say they are multiplied.) In the second and third rules Swineshead establishes corresponding relations for the cases of a decrease in force and an increase or decrease in resistance (the force than being held constant [43va]).

That these rules provide the basis of what Swineshead was attempting to do in treatise XIV can be seen as soon as one introduces the motion or velocities that correspond to or "result from" the force-resistance proportions whose "mathematics of change" he has just established. Thus, to return to the first rule as an example, if we assume that a velocity V_1 corresponds to the proportion $F_1 : R$ and a velocity V_2 to the proportion $F_2 : R$, then, because $F_2 : R$ is greater than $F_1 : R$ by the proportion $F_2 : F_1$ (which is what the compounding of these proportions asserts), it follows that the velocity added is precisely the velocity that results from the proportion $F_2 : F_1$ (and which would result from a proportion $F : R$ equal to $F_2 : F_1$ standing alone). Given this, and the corresponding relations when a resistance is allowed to vary while the force is held constant, Swineshead can determine all that he wishes concerning changes in velocity by paying attention only to the relevant force-force or resistance-resistance proportions representing the changes. When, and only when, these proportions are equal will the corresponding positive or negative increments of velocity be equal.

In deducing his succeeding *regulae* on such a basis, Swineshead does not "calculate" velocity increments from given force-force or resistance-resistance proportions, nor velocities from given force-resistance proportions. In this he was strictly medieval. Rather, he always compares (at least) *pairs* of $F : F$ or $R : R$ proportions. Thus, whenever unequal forces increase or decrease with equal swiftness (*equevelociter*)—which means that $F_2 - F_1 = F_4 - F_3$—then the resultant $F_2 : F_1$, $F_4 : F_3$ proportions will be unequal, whence it follows that the corresponding velocity increments will be unequal. If, on the other hand, the forces increase or decrease proportionally (*eque proportionabiliter*)—which means that $F_2 : F_1 = F_4 : F_3$—then the corresponding velocity increments will be equal. And the same thing holds for increasing or decreasing resistances, the forces being held constant.

Beginning, then, with two rules (4 and 5 [43va]) that apply a change in a *single* force or resistance to, correspondingly, *two* constant resistances or forces (whence the relevant single $F_2 : F_1$ or $R_2 : R_1$ proportions function as pairs since they are applied to pairs of R or F), Swineshead sets forth the implications of his mathematics of force-resistance changes to changes in velocity. In rules 4 and 5, inasmuch as one has a single $F_2 : F_1$ or $R_2 : R_1$ proportion doing double duty, the corresponding velocity increments are naturally the same. (For example, rule 4 reads: "Whenever a force increases or decreases with respect to two equal or unequal, but constant, resistances, it will intend or remit motion with respect to each [of these resistances] with equal swiftness.") Note should be taken of how, throughout treatise XIV, Swineshead handles what we would consider as positive versus negative increments of velocity. Since all determining force and resistance proportions are always of greater inequality (e.g., $F_2 : F_1$ where $F_2 > F_1$ when it is a question of increase in force, $F_1 : F_2$ where $F_1 > F_2$ when decrease is involved), velocity increments are always added, a procedure that follows directly from Swineshead's basic technique of compounding proportions. This means that when it is a question of the remission of motion arising from decreasing forces or increasing resistances, then increments are added to the motion at the end of the change in question, the "sum" of these increments plus the final motion or velocity giving the motion or velocity at the beginning of

the change. When the motion is intended, or the velocity increments are positive, the addition is naturally made to the motion obtaining at the beginning of the change.

In rules 6 through 15 (43va–43vb) Swineshead applies the two kinds of force or resistance change, that is, equally swift (*equevelociter*) or equally proportional (*eque proportionabiliter*) increases or decreases, to *pairs* of changing forces or resistances and infers the corresponding changes in velocity. Thus far, the force and resistance changes can be considered as discrete. However, beginning with rule 19 [43vb] Swineshead faces the case of the uniform and the continuous change of a single force or resistance. Now a single uniformly increasing force acting on a constant resistance, for example, will "generate" pairs of proportions $F_2 : F_1$, $F_3 : F_2$, $F_4 : F_3$, etc., where the succeeding proportions have "common terms" because the increase in force is continuous and where $F_2 - F_1 = F_3 - F_2 = F_4 - F_3 = \cdots$ because the increase is uniform. This latter fact entails that $F_2 : F_1 > F_3 : F_2 > F_4 : F_3 > \cdots$, which in turn implies that the increments of velocity will become successively smaller and smaller. Thus, the first half of rule 19 reads: "If a force increases uniformly with respect to a constant resistance, it will intend motion more and more slowly."

It is important, however, to be able to deal with at least certain kinds of nonuniform or difform changes in force and resistance. Hence, in rules 21 and 22 (43vb–44ra), arguing by a *locus a maiori*, Swineshead shows that if (as he has just established) a uniform gain in force or resistance entails a, respectively, slower and slower intension or remission of motion, then a nonuniform, slower and slower gain in force or resistance will necessarily also entail slower and slower intension and remission. Similarly, if a uniform loss in force or resistance entails a, respectively, faster and faster remission or intension of motion, then a nonuniform faster and faster loss in force or resistance also entails faster and faster remission and intension. The difform changes involved in a faster and faster gain, or a slower and slower loss, of force or resistance are not treated, since in such cases no inferences can be made about the resultant $F : F$ and $R : R$ proportions and, hence, about the resultant velocity increments.

However, the types of difform change in force or resistance that Swineshead can and does treat are precisely those needed in much of the remainder of treatise XIV. Up to this point changes in resistance have been independently given. Beginning with rule 23 (44ra) these changes are ascribed to the medium through which a mobile (represented by the force acting upon it) moves. Furthermore, all increase or decrease of resistance has hitherto been considered merely *relative to time* (whether it be *equevelociter* or *eque proportionabiliter*, no matter). But to ascribe variations in resistance to a medium is to speak of increase or decrease *relative to space*. Consequently, the problem facing Swineshead is to connect increments of resistance with respect to space to increments of resistance with respect to time, which is exactly what rule 23 does. Thus, a uniformly difform medium is one in which equal increments of resistance occur over equal spaces or distances. We also know that any body moving through such a medium in the direction of increasing resistance will move continuously more and more slowly (that is, the spaces S_1, S_2, S_3, $\cdots$ traversed in equal times successively decrease). But these two factors imply that, in equal times, the mobile will encounter smaller and smaller increments of resistance. Hence, equal increments of resistance (Swineshead calls them "latitudes of resistance") over space have been connected to decreasing increments of resistance over time. Accordingly, rule 23 reads: "If some force begins to move from the more remiss extreme of a uniformly difform medium and remains constant in strength, then the resistance with respect to it will increase more and more slowly." However, this slower and slower increase in resistance over time is precisely one of those kinds of difform change Swineshead was able to deal with in rules 21 and 22. This allows him to infer in rule 24 (44rb) that the motion of a mobile under a constant force through a uniformly difform medium in the direction of increasing resistance entails that the motion in question will undergo continuously slower and slower remission.

There are media, however, in which the distribution of resistance over equal spaces is difform, but not uniformly difform. What can be said of them? We know what equal changes of velocity are associated with a resistance that changes uniformly proportionally over time (rules 11 and 27 [43va, 44va–44vb]). Thus, if we imagine a medium of uniform resistance which increases in resistance equiproportionally over time as a constant force mobile moves through it, the mobile will remit its motion uniformly, that is, will undergo equal negative increments of velocity in equal times (rule 28 [44vb–45va]). With this rule in hand, Swineshead then imagines another medium with a resistance constant in time but difform with respect to space

and having at each point the resistance which was at the corresponding point of the first medium when the mobile was at that point. The mobile will then have the same motion in the second medium as it had in the first, that is, a uniformly difform motion. This means that (rule 29 [45va–45vb]) there can be a medium with resistance distributed difformly over space in such a way as to cause a mobile moving in it under a constant force to remit its motion uniformly (even though Swineshead could not describe this distribution).

Nevertheless, what Swineshead has established in rule 29 is of considerable importance for much of the remainder of treatise XIV. Rules 30–43 [45vb–48rb] all have to do with what will, or will not, occur when other constant or changing forces move through a medium in which (again rule 29) a given constant force uniformly remitted its motion. Thus, in rule 30 [45vb–46rb], Swineshead proves that two unequal constant forces cannot both uniformly remit their motion in the same medium. It is worthy of note that to prove this rule, Swineshead has to determine where the mobile that does remit its motion uniformly is at the middle instant of its motion. To do this he uses the ratio of space traversed in the first half of the time to the space traversed in the second half of the time and to find this ratio he uses the famous "Merton mean speed theorem," which he proves for the occasion (45vb–46ra). Apart from the fact that Swineshead gives four different proofs of the theorem, it here appears as a fairly routine lemma. He does not assign it any special importance, and does not even give it the honor of labeling it as a separate rule or conclusion.

Holding in mind the constant force specified in rule 29 as able to cause the uniform remission of motion in a given medium, Swineshead concludes treatise XIV by considering what will transpire when constant forces greater or lesser than that constant force are brought into play and when greater or lesser forces that are undergoing continuous intensification or remission are involved (rules 31–43). The last of these rules points out that a constant force greater than that specified in rule 29, but acting in the same medium, will give rise to a faster and faster remission of motion, that is, to a difformly difform motion. This leads to the final rules (44–49 [48rb–48vb]) of the treatise, which together function as a kind of appendix stipulating various facts and relations concerning difformly difform motions. As a whole, treatise XIV is an extremely impressive exhibition of just which cases of the different kinds of variation in force, resistance, and velocity that can be drawn out of Bradwardine's "function" are amenable to determination and treatment. As in the case of many of the other treatises of the *Liber calculationum*, there is a substantial increase in complexity from the beginning to the end of treatise XIV. But perhaps one of Swineshead's most signal accomplishments is his success in the latter part of the treatise in connecting variations in resistance over time with variations in resistance over space. For to the medieval supporter of Bradwardine's function (or of "Aristotle's function" as well for relating forces, resistances, and velocities), motion in a medium that was nonuniform was exceedingly problematic. As soon as the resistance in a medium was allowed to vary, one had to face the difficulty that the degree of resistance of the medium determined the velocity of the motion, while at the same time the velocity determined where in the medium the mobile would be and hence the resistance it would encounter. One seemed caught in a situation involving a double dependency of the relevant variables on each other. But Swineshead's "translation" of spatial increments of resistance into temporal ones automatically rendered the resistance of the medium time dependent and thus circumvented the troublesome double dependency.

Treatises XV–XVI: On a Nonresisting Medium or on the Increase of Power and Resistance; On the Induction of the Highest Degree. Treatises XV and XVI are continuations of treatise XIV and add ever more complications. In treatise XV Swineshead again considers the local motions of constant or changing powers in extended media, but this time he allows the resistance of the medium to vary while the mobile is moving through it or (we would say) takes the increase of power as an independent variable. In the key rule 29 of treatise XIV Swineshead had considered the motion of a mobile through a uniform medium with resistance changing over time, but this was a tool to allow him to deal with spatially difform resistances, the distributions of which he could not otherwise describe. Treatise XV, however, begins by dealing with temporally changing resistances in their own right.

The first conclusion of treatise XV is an example of Swineshead's mathematical ingeniousness, not in that he does complex mathematics, but in that he sees how to avoid complex mathematics. The conclusion concerns a nonresisting medium (or, in modern terms, a fixed space or vacuum) in which a resistance begins to be generated. The resistance first appears at one end of the medium and

moves progressively across the medium in such a way that the resistance increases uniformly from that end up to the point where the resistance ends. In modern terms, then, we might represent the resistance graphically by a straight line that rotates around the origin, starting in a vertical position, rotating at a decreasing rate (so that any point of the line has a constant horizontal velocity), and increasing in length so that the maximum height of the end point is a constant (see Figure 3). If, Swineshead concludes, a mobile begins to move from the same extreme of the medium at which the resistance begins to be generated, then it will move with a constant velocity (always keeping pace with the progress of a given degree of resistance), provided that the maximum resistance moves away from the mobile faster than the mobile could move with that resistance (48vb). Swineshead proves this conclusion first by showing that there could not have been any initial period of time during which the mobile increased or decreased its velocity, and second by showing that the mobile could not later begin to move faster or more slowly than its given resistance. In the later proof he argues, for instance, that if the mobile were supposed to increase its velocity, then it would immediately begin to encounter greater resistances, implying a decrease rather than an increase in velocity and thus a contradiction; and if the mobile were supposed to decrease its velocity, then conversely it would immediately begin to encounter lesser resistances implying an increase rather than a decrease in velocity and thus another contradiction. So, therefore, it must continue with a constant velocity. As stated above, from a modern point of view, any position that connects resistance with velocity as Bradwardine's function does would seem to be very problematic when applied to difform resistances, given that position (and therefore resistance) would determine the velocity of the

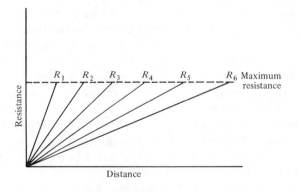

FIGURE 3

mobile, and yet velocity (and initial position) would also determine position, involving a double dependency. We see Swineshead here, however, not only coping with the problems of such a double dependency, but even playing with it and in a sense making sport to come up with ever more intriguing and complicated conclusions. Following the first conclusion, Swineshead proceeds to show that if the motion of the latitude of resistance is accelerated or decelerated the motion of the mobile will accelerate or decelerate also (49ra–50ra), and he goes on from there to prove nine other conclusions concerned with the generation of a latitude of resistance in a nonresisting medium (50vb–51rb).

The second main part of treatise XV concerns the motion of powers augmenting from zero degree in uniformly difform resistances. Again, perhaps the first conclusion of this part may serve as an example of the fourteen conclusions proved. If there is a uniformly difform medium terminated at zero degree in which a power begins to move as it augments uniformly from zero degree (always moving according to the proportion of its power to the point of the medium at which it is), then this power will continually move uniformly (51rb–51va).

The further conclusions of this second part all concern temporally constant uniformly difform resistances, but the powers are allowed to vary in different ways, both uniformly and difformly. These conclusions differ from the conclusions of treatise XIV not only because the resistances involved are uniformly difform rather than difformly difform, but also because the changes of resistance and power are given as the independent variables rather than the velocities, as in treatise XIV.

In this treatise as in the last, the attempt seems to be to give a general description of the possible interrelations of power, resistance, and velocity and their changes. As in the last, too, the main variations of the independent variables considered are uniformly difform variations. This seems to result from the mathematical tractability of such situations rather than from any observational or theoretical context that made such variations likely, as was the case for illumination in treatise XIII, where there was a common belief that the distribution of illumination from a light source was uniformly difform. Within these limitations, Swineshead does, however, manage to build up more and more complex conclusions without doing much explicit mathematics. He rather simply picks those cases, however complex they may appear,

about which something can be said on the basis of the general characteristics of the functions involved.

In the third and last main part of treatise XV Swineshead should then have combined his two variables to allow both resistance and power to vary at the same time and to draw conclusions. The treatise simply ends, however, with the words: "It remains to inquire how both [resistance and power] may simultaneously be acquired." (The 1520 edition adds: "and first are posited rules, etc.") Given the complexity of the conclusions that resulted when only resistance or power was allowed to vary, one could hardly blame Swineshead for failing to push on further.

Treatise XVI is broken down into five chapters, each considering a class of problems concerning the induction of the maximum degree. Chapter 1 considers the alterations of larger and smaller uniformly difform subjects altered either uniformly throughout or by a uniformly difform latitude of alteration. Chapter 2 considers cases where a difformly difform latitude of alteration is extended through the part remaining to be brought to maximum degree in the same way as it was extended through the whole at the start.

Chapters 1 and 2 both consider cases where the alteration is extended at the start through the whole subject. Chapter 3, then, considers cases where the alteration does not extend through the whole subject at the start, but rather begins to be generated at the more intense extreme of the subject. Chapter 4 considers how the induction of the maximum degree is to be measured when the subject is rarefied or condensed during the alteration, deciding that such induction should not be measured by the fixed space outside the subject, but rather, with certain qualifications, by the subject itself. Chapter 5 considers how it may occur, through the successive generation of alteration in a subject, that the subject remains or becomes uniformly difform. Treatise XVI may be the most complex of the *Liber calculationum*, but perhaps enough has been said about the previous treatises so that its character can be imagined without the detailed examination of any of its conclusions.

As in the previous treatises, Swineshead in treatise XVI is preoccupied with uniformly difform alterations and the like, probably because they were well defined, whereas with difformly difform alterations the situation becomes overly complex. It might be remembered that, where alterations are concerned, as in this treatise, there was the common view that all qualitative actions, like light, decrease uniformly difformly (that is, linearly) as one moves away from the agent or source. So here again, as in the treatise on illumination, there might be a physical reason for emphasizing uniformly difform distributions. Nevertheless, if there was such a reason, it is well in the background.

Opuscula

Of the three short *opuscula* that may be assigned to Richard Swineshead, two by explicit ascription to a Swineshead as author, and one by its position between the two others, one is a partial commentary on the *De caelo* and the other two are partially repetitive treatments of motion.

In librum de caelo. Apparently part of a commentary, beginning from text 35 of book I of the *De caelo*, this short fragment is in two main parts, the first dealing with Aristotle's proofs that an infinite body cannot move locally, and the second dealing with the relation of substances to their qualities, in connection with the possibility of action and passion between infinite bodies.

In the first main part, Swineshead considers (1) the proofs that an infinite body cannot rotate, drawing paradoxes concerning the intersections of infinite lines during such rotation; (2) the proofs against translational motion of infinites; and (3) Aristotle's arguments that there cannot be an infinite body so that, *a fortiori*, there cannot be any motion of an infinite body.

In the second main part, Swineshead considers (1) Aristotle's discussion of the possible action and passion of infinites, both with relation to simple subjects and with relation to compounds or mixtures; (2) the question of whether substantial forms can vary within some latitude, apparently with the idea that, if elemental forms can be remitted, then a finite action of an infinite subject might be within the range of possibility; (3) a proof that elements can exist without their qualities; (4) a proof that there cannot be mixed bodies of two or more elements of degrees as remiss as desired (in this part, Swineshead mentions Dumbleton by name and refutes some arguments he makes in part IV of his *Summa*); and (5) arguments concerning the possible perpetuation and duration of compounds. The opinions expressed in this work do not seem to be in conflict with the conclusions of the *Liber calculationum.*

De motu. This second short work (following in Cambridge MS Gonville and Caius 499/268 directly after the *De caelo* fragment) is not explicitly ascribed to Swineshead, but its similarity to the last short work of the manuscript, which is ascribed to Swineshead, is so great that there is every reason to ascribe it also to Swineshead. In the Seville manuscript Colombina 7-7-29 the works (the *De motu* and the following *De motu locali*) appear as one, but this seems hardly plausible since there is a great deal of overlap and repetition between them. It is more likely that they are successive drafts of the same work than that they are both sections of a single larger work.

The *De motu* contains an introduction concerning the material, formal, efficient, and final causes of motion, and two main sections, the first dealing with the measurement of motion with respect to cause and the second with the measurement of motion with respect to effect. Only local motion is considered. The subject matter of both sections is similar to that of treatise XIV of the *Liber calculationum*. Roughly speaking, both the *De motu* and the *De motu locali* seem to occupy an intermediate position between Heytesbury's *Regule* on local motion and the *Calculationes*, a natural supposition being that Swineshead began from Heytesbury's work and went on to develop his ideas from that point.

Concerning the consideration of motion with respect to cause, Swineshead begins by expounding Bradwardine's function relating powers, resistances, and velocities (212ra). This is followed by a number of rules, some of which are the same as, and others similar to, the conclusions of treatise XIV of the *Liber calculationum* (212ra–212va). Swineshead says, for instance, that if a constant power begins to move in the more remiss extreme of a uniformly difform resistance, it will remit its motion more and more slowly (212ra–vb). (Cf. rule 24 of treatise XIV of the *Liber calculationum*.)

In the second section, Swineshead begins with a number of statements concerning the measurement of motion with respect to effect—for instance, that uniform local motion is measured by the line described by the fastest moved point (212vb)—many of which have close analogues in Heytesbury's *Regule*. He states the mean speed rule (213ra) and the rule that in a motion uniformly accelerated from zero or decelerated to zero three times as much is traversed by the more intense half of the motion as by the more remiss half of the motion (213ra), which he derives as a consequence of the mean speed rule.

The second section concludes with five conclusions and a statement concerning the measurement of difform motion. The first three conclusions have to do with the traversal of extended resistances, one of these being the thirtieth rule of treatise XIV of the *Liber calculationum*: if one constant power remits its motion uniformly to zero in a given difform resistance, no other greater or lesser constant power will uniformly intend or remit its motion traversing the same medium (213ra). Concerning velocity in difform motion, Swineshead says that it is not to be measured by the maximum line that is described, but rather by the line that would be described if the velocity were continued for a period of time (213rb).

The *De motu*, then, consists mostly of a series of conclusions along with several stipulations as to how motion is to be measured with respect to cause and with respect to effect. Although the conclusions are so divided, there is little to distinguish them.

De motu locali. This short work receives its title from its explicit, and hence we retain the name despite the fact that this last fragment contains treatments of alteration as well as of local motion. The first two sections of this work correspond to the two main sections of the *De motu*, and many of the rules or conclusions stated are the same.

The work starts with a series of conclusions concerning the effect on velocity of increasing or decreasing the power or resistance in motion (213rb–vb). Many of these conclusions, including eight of the first nine, appear also in the *De motu*, and a similar number, although not always the same ones, appear also in the *Liber calculationum*, treatise XIV. Although these conclusions presuppose Bradwardine's function relating powers, resistances, and velocities, that function is not explicitly stated as it was in the *De motu*.

These rules are followed by a section concerning the measures of resistance, for instance that two resistances of which the most intense degrees are equal must be themselves equal (213vb); and that a motor equal to the maximum degree of a uniformly difform resistance will move in it eternally, never completely traversing it (214ra).

The second section of the *De motu locali* concerns the measures of motion with respect to effect, starting with the stipulation that the velocity of local motion is measured by the line that would be traversed by the fastest moved point (if there is one), provided that it continued its velocity uniformly for a period of time (214rb). Difformly difform motions, Swineshead says, always correspond to some degree within their range of varia-

tion *(ibid.)*, and uniformly difform motions correspond to their middle degrees (214rb–va). On this basis Swineshead mentions four types of sophisms that can arise from the comparison of accelerations to velocities of which some, he says, are possible, that is, not self-contradictory, and some impossible and to be rejected (214va–vb). A mobile can never begin from an infinite part of a magnitude and traverse some part uniformly (215ra), nor can a motion be remitted uniformly from an infinite degree of velocity *(ibid.)*.

If one wants to know how much is traversed by a uniformly difform motion starting and ending at a degree, all one can say in general is that more is traversed than by a mobile moving for the same time with half the maximum degree of the uniformly difform motion (214vb–215ra). All local motions are as fast as any of their parts, and all subjects moved locally are moved as fast as any of their parts, the former being true for all motions but the latter being true only for local motions and not for alterations (215ra).

The third and last section (215ra–215rb) of the *De motu locali* concerns measures of alteration. Like local motion the measure of alteration with respect to cause is the proportion of the power of the altering agent to the power of the altered patient. With respect to effect, the velocity of motion of alteration depends on the maximum latitude of quality that would be acquired by any part of the subject if the velocity were continued for some time period. Irrespective of the measure of velocity of alteration, a subject need not be altered as fast as any of its parts, but often is altered more slowly than a part of it closer to the agent. To determine how fast a subject is altered one has to consider the degree to which it corresponds, calculating the contribution of various degrees to the subject's denomination by the proportion of the subject through which they are extended.

Many sophisms can arise from the comparison of velocities of alteration to the velocities with which subjects are altered. Something may be altered with a faster velocity of alteration and yet be more slowly altered, and hence the two separate measures of alteration must be kept separate. On this basis, it is not contradictory for an agent to alter faster than anything is altered by it. If a uniformly difform alteration is extended throughout a subject, the subject will be altered as the middle degree of that alteration. With brief references to other ways of dealing with alteration the *De motu locali* ends.

Conclusion

As is evident from the discussions of its separate treatises, the *Liber calculationum* places what may seem to be an uncommon emphasis upon the generation of *conclusiones, regulae, objectiones,* and *sophismata*. This unceasing generation of results occurs in other works of the calculatory tradition, but it reaches a high with Richard Swineshead, so much so as to be nearly the defining characteristic of the *Liber calculationum*. Results are drawn to the very limits of manageability. Swineshead's *opuscula, De motu* and *De motu locali*, exhibit the beginnings of this effort to educe results. We, operating with modern mathematics, could generate many more such results, but the subclass that Swineshead himself generates and treats almost completely exhausts the results he could have dealt with, given the techniques at his disposal. And this excogitation of results occurs whether Swineshead is examining two or more *positiones* or *opiniones* relative to a given topic (when, as said above, only slight attention is paid to deciding definitively between *opiniones*) or only one. The major difference between these two cases appears to be that, in the former, Swineshead is more apt to label as *sophismata* the results he is generating. In any event, he makes it quite clear that *conclusiones* can be elicited, objected to, and resolved on both or all sides when a plurality of *opiniones* is at stake (*multe conclusiones possunt elici ex dictis, ad quarum tamen utramque partem probabiles possunt fieri rationes, que per predicta, si bene intelligantur, satis faciliter solvuntur* [34ra]).

Furthermore, there is almost no discussion in the *Liber calculationum* of the contexts in which one might expect the situations represented by these results to occur, nor, indeed, any time spent investigating whether they can occur. Instead, the work proceeds almost entirely *secundum imaginationem*, as Heytesbury's *Regule solvendi sophismata* had before it. In fact, among all other "Mertonian" works, the *Liber calculationum* is most like Heytesbury's *Regule*. Some of the earlier Oxford calculators, like Walter Burley and Roger Swineshead, had fairly frequently considered natural, as well as *de imaginatione*, situations. Richard Swineshead, by contrast, quite consistently imagines situations that will illustrate and draw results out of various theories rather than taking examples from natural occurrence. The emphasis is upon developing and having a set of techniques as complete as one can make it. If we may trust an infer-

ence from the later commentary of Gaetano of Thiene on Heytesbury's *Regule*, one of the reasons for this emphasis was simply the fact that every "calculator" ought to have a system applicable to every conceivable situation (. . . *dicit* [*sc.,* Heytesbury] *quod hoc est tamen impossibile . . . physice loquendo . . . Sed dicit ille magister bene scis hoc, sed quia non implicat contradictionem et est satis imaginabile, ideo calculatores non debent fugere casum* [Venice ed., 1494, 48va]).

The "techniques" that are presented in the *Liber calculationum* may strike the modern reader as basically mathematical. But one must consider such a judgment with care. To begin with, the evidence of the extant manuscripts of the *Liber calculationum* tells us that the medievals themselves did not regard it (or any of the other "calculatory" works, for that matter) as mathematical in the sense of Euclid, Boethius' *Arithmetica*, or Jordanus de Nemore. When the *Liber calculationum* does not take up an entire codex, it or fragments from it invariably appear with other treatises, questions, or notes on natural philosophy or logic. Nevertheless, there is no doubt that, the evidence of medieval codification aside, mathematical functions and considerations pervade Swineshead's major work. They are applied, however, not in order to understand how some phenomenon normally occurs (as was the case in medieval optics, statics, and astronomy), but in a thoroughly *secundum imaginationem* fashion that is totally different from the Greek-based mathematical tradition inherited by the Middle Ages. A good deal of material from the Greek tradition was, of course, utilized by Swineshead, but he utilized it in a most un-Greek way. Mathematical functions are applied in order to determine all conceivable contributions that parts could make to wholes, to distinguish the discontinuous from the continuous, and to encompass situations or results involving infinite intensities, infinite velocities, and other infinite values. It is for these kinds of problems that the *Liber calculationum* contained the required techniques. To the medieval scholar with the patience and ability needed to comprehend this work, the purpose seems to have been that he should learn to operate with the techniques and rules given by Swineshead—just as one was to learn techniques and rules in the tradition of solving sophisms to which the *Liber calculationum* from time to time refers—so expertly as to be able to handle situations and *casus* in every corner of the fourteenth century realm, be it physical, logical, medical, theological, or whatever.

Dissemination and Influence of the Liber calculationum

Although the work of Swineshead's fellow Mertonians and the English "calculatory" tradition in general is generously represented in later fourteenth-century natural philosophy (very notably so, for example, in the work of Oresme), specific evidence of the *Liber calculationum* during this period is not especially plentiful, although shortly after mid-century the English logician Richard Ferebrich appears to employ parts of treatise XIV in his *Calculationes de motu*. Most of the extant manuscripts of Swineshead's major work are of the fifteenth century, and the records we have of its occurrence in library catalogues for the most part date from 1400 and later. It does appear, however, in at least two "student notebooks" of the later fourteenth century: in one merely in terms of fragmentary traces (MS Bibliothèque Nationale, fonds latin 16621), in the other more substantially (MS Worcester Cathedral, F. 35).

If we ask for evidence of the dissemination and influence not of Mertonian ideas in general or of some particular idea like the so-called Merton mean speed theorem, but rather of parts of the *Liber calculationum* itself, the fifteenth and sixteenth centuries are far richer. The first center of interest is Italy, in the middle and toward the end of the fifteenth century, where Swineshead appears as part of the broader preoccupation with English logic and natural philosophy. He appears, moreover, in terms both *pro* and *contra*. He suffers part of the humanist criticism of the "barbari Britanni" that one finds in the likes of Coluccio Salutati; for Leonardo Bruni he is one of those "quorum etiam nomina perhorresco"; and he even gave his name to the "sophisticas quisquilias et *suisetica* inania" complained about loudly by Ermolao Barbaro and others.

Yet it seems fair to claim that one of the major reasons for these humanist complaints was not Swineshead's works themselves, but rather the fact that they and other English "calculationes" had attracted the attention of a fair number of Italian scholars. Thus, his views are found among those of others treated in works *De reactione* written by Gaietano de Thienis, Giovanni Marliani, Angelus de Fossambruno, Vittore Trincavelli, and Pietro Pomponazzi. He also appears to have been very much in the center of the interest in "calculatory" matters in Padua, and especially Pavia, around the mid-fifteenth century, something that jibes extremely well with the number of times that the *Liber calculationum* was published in these cities. We also know that Nicolletto Vernia, earlier a student of both Paul of Pergola and Gaietano de Thienis, went to Pavia to study the *Calculationes Suisset*, information that fits very well indeed with other things we are able to put together about Vernia's interests. Pomponazzi relates that he engaged in a dispute with Francesco di Nardò armed with *argumentis calculatoriis*; but we have more direct evidence of his concern with Swineshead from the fact that he refers to the "Calculator" in his (unedited) *Questio de anima intellectiva* and from the fact that Vernia owned a copy of the *Liber calculationum* (now Biblioteca Vittorio Emanuele, MS 250).

When one turns to the Italian commentaries or questions on the *Liber calculationum*, the first thing to be noted is that most of this literature is preoccupied only with treatise I: *De intensione et remissione formarum*. It was this part of Swineshead that occupied Pomponazzi in his own treatise of the same title written in 1514, and we have similar sixteenth-century works by Tiberio Baccilieri, Cardinal Domenico Grimani, and Hieronymus Picus. Pomponazzi is critical of Swineshead insofar as the "scale of measure" he proposed in treatise I of the *Liber calculationum* maintains the inverse proportionality of intension and remission and ignores their proper ontological status. Intension and remission should be viewed, Pomponazzi felt, as, respectively, perfection and imperfection; this done, one would not, as Swineshead, "measure" remission by nearness to zero degree, but rather, as an imperfection, in terms of its distance from the maximum degree. Pomponazzi refers to a similar disagreement with Swineshead's "non gradum measure" in his *Super libello de substantia orbis*, where he specifically complains that it runs counter to the *via Aristotelis*. Pomponazzi's criticism can be partially explained by the fact that in his (unedited) lectures on Aristotle's *Physics* he felt that Swineshead and other English "calculators" put too much mathematics (*ille truffe spectant ad mathematicum*) and "geometricalia" into natural philosophy, and (as he complained in his *De intensione et remissione formarum*) constructed a *scientia* that was *media inter physicas et mathematicas*.

Italian *expositiones* or *questiones* on other parts of the *Liber calculationum* are rarer than those on treatise I. Bassanus Politus composed a *Tractatus proportionum* specifically claimed to be *introducto-*

rius ad calculationes Suisset; it sets forth in succinct fashion little more than the standard mathematics of proportion and proportionality drawn from such authors as Euclid and Boethius. More to the content of parts of the *Liber calculationum* is Marliani's *Probatio cuiusdam sententie calculatoris de motu locali*. Its concern is with the views of Swineshead on "mean degree measure" in treatise II and his proofs for the "mean speed theorem" in treatise XIV. The two most complete Italian commentaries on the *Liber calculationum* are unedited and unstudied. One is by Christopher de Recaneto, *doctor in artibus* at Padua in 1454, and covers but treatise I and part of treatise II. The other is more extensive, commenting on treatises I–V and VII–VIII, but we know almost nothing of its author, Philippus Aiuta. All we know is that Marliani wrote a *difficultates* sent to Philippus and (as we learn from the incipit of the present commentary) that he was a doctor in arts and medicine. The commentary itself, apparently a compilation of Philippus' view made in 1468 by one Magister Bernardinus Antonius de Spanochiis, relates Swineshead to any number of other mathematicians (especially Euclid) and philosophers, both contemporary, earlier medieval, and ancient. But perhaps its most intriguing aspect is the explicit tendency to render Swineshead more comprehensible by the addition of appropriate figures (*quasdam ymagines in marginibus*).

The second major center of interest in Swineshead was Paris at the beginning of the sixteenth century, where a considerable amount of work was done with the *Liber calculationum*, largely by a group of Spanish and Portuguese scholars. The earliest, and certainly most impressive, among them appears to be Alvaro Thomaz. His *Liber de triplici motu*, published in Paris in 1509, contains an extensive, two-part, preliminary treatise expounding all aspects of the mathematics of proportions, and itself treats all parts of the *Liber calculationum* that deal with *motus*. Alvaro also includes much material drawn from Nicole Oresme, not only explaining, but on occasions expanding what he finds in Swineshead and Oresme. One notable instance of such "expansion" is his treatment of the "infinite series" treated by both of these fourteenth-century authors. Following Alvaro by a few years, both John Dullaert of Ghent (in 1512) and Juan de Celaya (in 1517) include comprehensive expositions of "calculatory" material in their *Questiones* on Aristotle's *Physics*. Celaya treats a good number of *conclusiones* drawn from parts (treatises I, II, IV, VI, XIV, XV) of the *Liber cal-*

culationum and appears to have followed Thomaz in at least the structure of much of what he includes. Another Spanish member of this Paris school was Luis Coronel, who does not, like Celaya, include a lengthy connected exposition of Swineshead material in his *Physice perscrutationes* (published in 1511), but who nevertheless does discuss a fair number of issues and passages from scattered treatises (I, II, V, X, XI) of the *Liber calculationum*. At times he seems to lack a proper understanding of what he is discussing from Swineshead and even complains how *prolixissime et tediose* the reasoning is (in this instance referring to treatise XI). The same kind of complaint and lack of comprehension is probably in part behind the remarks of Diego de Astudillo in his *Questiones* on Aristotle's *Physics* when he excuses his omission of "calculatory disputations" since he would "confound the judgments of beginners . . . ignorant of mathematics." Indeed, if it is a proper appreciation of Swineshead's accomplishments that one has in mind, then none of the "commentators," Italian or Spanish, save Thomaz, qualifies.

Some Renaissance figures continued the fifteenth-century humanist criticism of English "subtilitates" and ridiculed (Luis Vives, for example) Swineshead's work, but others, such as Julius Scaliger and Cardano, praised him as outstandingly acute and ingenious. The most famous later "appreciation" of the Calculator was that of Leibniz. Perhaps initially learning of Swineshead through Scaliger (whom he mentioned in this regard), Leibniz confessed to a certain admiration of Swineshead even before he had had the opportunity to read him. How thoroughly Leibniz finally did read the *Liber calculationum* we do not know; but we do know that he went to the trouble to have the 1520 Venice edition of it transcribed (today Hannover, Niedersächssische Landesbibliothek, MS 615). In Leibniz' eyes, Swineshead's primary accomplishment lay in introducing mathematics into scholastic philosophy, although we should perhaps take "mathematics" to include a certain amount of logic, since at times he coupled Swineshead with Ramon Lull in having accomplished this task. In any event, what Swineshead had done was in his eyes much in harmony with Leibniz' own convictions about the relation of mathematics and "mathematical" logic to philosophy.

Because of the remarks of the likes of Cardano, Scaliger, and Leibniz, Swineshead found his way into eighteenth-century histories of philosophy such as that of Jacob Brucker. After that he was forgotten until Pierre Duhem rediscovered him at the beginning of the twentieth century in his rediscovery of medieval science as a whole.

BIBLIOGRAPHY

I. ORIGINAL WORKS. Manuscripts* of the *Liber calculationum* include Cambridge, Gonville and Caius 499/268, 165r–203v (14c.), tr. I–XI, XV, XII–XIII; *Erfurt, Stadtbibl., Amplon. 0.78, 1r–33r (14c.), contains an abbreviated version of tr. I–II, IV–VIII. Ascribed to one "clymiton" (presumably Killington?) by Schum in his catalogue of the Amplonian MSS, apparently drawing his information from Amplonius Ratinck's fifteenth-century catalogue; *Padua, Bibl. Univ. 924, 51r–70r (15c.), tr. I, VII, VIII, IX; *Paris, Bibl. Nat. lat. 6558, 1r–70v (dated 1375), tr. I–XI, XV, XII–XIII; *Paris, Bibl. Nat. lat. 16621 (14c.), *passim*, fragments (often in altered form); e.g., 52r–v, 212v from tr. XIV; *Pavia. Bibl. Univ., Aldini 314, 1r–83r (15c.), tr. I–XI, XV, XII–XIII, XVI, XIV; Perugia, Bibl. Comm. 1062, 1r–82r (15c.); *Rome, Bibl. Angelica 1963, 1r–106v (15c.), tr. I–IX, XI–XVI; *Rome, Bibl. Vitt. Emanuele 250, 1r–82r (15c.), contains all tr. except III in the following order: I, II, V, VI, XIV (incomplete), XV, XIV (complete), XVI, XII, XIII, VII, VIII, IX, IV, X, XI; belonged to Nicoletto Vernia; *Vatican, Vat. lat. 3064, 1r–120v (15c.), tr. I–X, XVI, XI, XV, XIV, XII–XIII, III (again); *Vatican, Vat. lat. 3095, 1r–119v (15c.), tr. I–X, XVI, XI, XV, XII–XIV; Vatican, Chigi E. IV. 120, 1r–112v (15c.); Venice, Bibl. Naz. San Marco., lat. VI, 226, 1r–98v (15c.). tr. I–II, IV–XII, III (according to L. Thorndike); *Worcester, Cathedral F. 35, 3r, 27r–65v, 70r–75v (14c.), contains, in following order, fragment (Reg. 1–4) of XIV, I–IV, VI–VII, XII–XIII, XVI, V; Cesena, Bibl. Malatest., Plut. IX, sin cod VI.

Editions include Padua *ca.* 1477; Pavia 1498; Venice, 1520. A modern edition of treatise XI has been published in the article of Hoskin and Molland cited below.

The *Opuscula* are *In librum de caelo* (Cambridge, Gonville and Caius 499/268, 204r–211v [15c.]; Worcester, Cathedral F. 35, 65v–69v [14c.], incomplete); *De motu* (Cambridge, Gonville and Caius 499/268, 212r–213r [15c.]; Oxford, Bodl., Digby 154, 42r–44v [14c.]; and Seville, Bibl. Colomb. 7-7-29, 28v–30v [15c.]); and *De motu locali* (Cambridge, Gonville and Caius 499/268, 213r–215r [15c.]; Seville, Bibl. Colomb. 7-7-29, 30v–34r [15c.]).

The *De motibus naturalibus* of Roger Swineshead: Erfurt, Stadtbibl., Amplon. F. 135, 25r–47v (14c.); Paris, Bibl. nat., lat. 16621, 39r, 40v–51v, 54v–62r, 66r–84v (14c.), Fragment of part II, all of parts III–VIII; Venice, Bibl. Naz. San Marco lat. VI, 62, 111r (15c.), definitions of part IV. For MSS of the *De obliga-*

*Many of the MSS do not contain all sixteen *tractatus* of the *Liber calculationum*; the incipits and explicits of each *tractatus* have been checked by the authors for those MSS indicated by an asterisk.

tionibus and *De insolubilibus* ascribed to Roger, see the article by Weisheipl on Roger cited below.

The *Questiones quatuor super physicas magistri Ricardi* (in MSS Vatican, Vat. lat. 2148, 71r–77v; Vat. lat 4429, 64r–70r; Venice, San Marco lat. VI, 72, 81r–112r, 168r–169v) that have been tentatively ascribed to Swineshead by Anneliese Maier are in all probability not his, but rather most likely the work of Richard Killington (or Kilvington).

II. SECONDARY LITERATURE. The two basic biographical-bibliographical sources are A. B. Emden, *A Biographical Register of the University of Oxford to A.D. 1500* (Oxford, 1957–1959), 1836–1837, which includes material on Roger and John; and James A. Weisheipl, "Roger Swynesbed, O.S.B., Logician, Natural Philosopher, and Theologian," in *Oxford Studies Presented to Daniel Callus* (Oxford, 1964), 231–252. These both replace G. C. Brodrick, *Memorials of Merton College* (Oxford, 1885), 212–213. Emden has a long list of variant spellings of Swineshead such as Swyneshed, which he prefers, Suicet, Suincet, etc.

On Swineshead and the calculatory tradition in general, see Pierre Duhem, *Études sur Léonard de Vinci*, III, "Les précurseurs parisiens de Galilée" (Paris, 1913), 405–480; most of this material is reprinted with some additions, omissions, and changes in Duhem's *Le système du monde*, VII (Paris, 1956), 601–653. See also Marshall Clagett, *The Science of Mechanics in the Middle Ages* (Madison, Wis., 1959), chs. 4–7; Anneliese Maier, *Die Vorläufer Galileis im 14. Jahrhundert* (=*Studien zur Naturphilosophie der Spätscholastik*, I), 2nd ed. (Rome, 1966); *Zwei Grundprobleme der scholastischen Naturphilosophie* (=*Studien*, II), 3rd ed. (Rome, 1968); *An der Grenze von Scholastik und Naturwissenschaft* (=*Studien*, III), 2nd ed. (Rome, 1952); John Murdoch, "*Mathesis in philosophiam scholasticam introducta.* The Rise and Development of the Application of Mathematics in Fourteenth Century Philosophy and Theology," in *Arts libéraux et philosophie au moyen âge* (=Acts du quatrième Congrès International de Philosophie Médiévale), (Montreal–Paris, 1969), 215–254; A. G. Molland, "The Geometrical Background to the 'Merton School,'" in *British Journal for the History of Science*, **4** (1968), 108–125; Edith Sylla, *The Oxford Calculators and the Mathematics of Motion, 1320–1350: Physics and Measurement by Latitudes* (unpublished diss., Harvard Univ., 1970); "Medieval Quantifications of Qualities: The 'Merton School,'" in *Archive for History of Exact Sciences*, **8** (1971), 9–39; "Medieval Concepts of the Latitude of Forms: The Oxford Calculators," in *Archives d'histoire doctrinale et littéraire du moyen âge*, **40** (1973), 223–283. More particularly on Swineshead see James A. Weisheipl, "Ockham and Some Mertonians," in *Mediaeval Studies*, **30** (1968), 207–213; Lynn Thorndike, *A History of Magic and Experimental Science*, 8 vols. (New York, 1923–1958), III, 370–385.

See the following on particular treatises of the *Liber calculationum* (although much information is also provided on the individual treatises in some of the more comprehensive literature above): tr. I: Marshall Clagett, "Richard Swineshead and Late Medieval Physics," in *Osiris*, **9** (1950), 131–161; tr. II: John Murdoch, "Philosophy and the Enterprise of Science in the Later Middle Ages," in *The Interaction Between Science and Philosophy*, Y. Elkana, ed. (Atlantic Highlands, N.J., 1974), 67–68; tr. VII: Clagett, *Giovanni Marliani and Late Medieval Physics* (New York, 1941), ch. 2; tr. X: Curtis Wilson, *William Heytesbury: Medieval Logic and the Rise of Mathematical Physics* (Madison, Wis., 1956), ch. 3; tr. XI: M. A. Hoskin and A. G. Molland, "Swineshead on Falling Bodies: An Example of Fourteenth-Century Physics," in *British Journal for the History of Science*, **3** (1966), 150–182; A. G. Molland, "Richard Swineshead on Continuously Varying Quantities," in *Actes du XIIe Congrès International d'Histoire des Sciences*, **4** (Paris, 1968), 127–130; John Murdoch, "Mathesis in philosophiam . . .," 230–231, 250–254; tr. XIV: John Murdoch, *op. cit.*, 228–230; Marshall Clagett, *The Science of Mechanics . . .*, 290–304, for Swineshead's proof of the so-called mean speed theorem.

For functions of *F*, *R*, and *V* similar to Roger Swineshead's, one may consult Ernest Moody, "Galileo and Avempace: The Dynamics of the Leaning Tower Experiment," in *Journal of the History of Ideas*, **12** (1951), 163–193, 375–422. For the "addition" theory of qualitative change mentioned in the discussion of treatise VII, see E. Sylla, "Medieval Concepts of the Latitude of Forms. . . ." This article also contains a more complete discussion of the ideas of Roger Swineshead.

The dissemination and influence of the Liber calculationum. To date, the most adequate treatment of the spread of late-medieval natural philosophy in general is Marshall Clagett's chapter on "English and French Physics, 1350–1600," in *The Science of Mechanics in the Middle Ages* (Madison, Wis. 1959). This chapter contains (630–631) a brief account of a segment of Richard Ferebrich's (Feribrigge) *Calculationes de motu.*

The Italian reception of Swineshead. On the humanist criticism of "Suissetica" and other, equally infamous, "English subtleties," the point of departure is Eugenio Garin, "La cultura fiorentina nella seconda metà del trecento e i 'barbari Britanni,'" as in his *L'età nuova. Ricerche di storia della cultura dal XII al XVI secolo* (Naples, 1969), 139–177; the relevant passages from Coluccio Salutati and Leonardo Bruni Aretino, as well as many other similar sources are cited therein. To this one might add the two anonymous letters of Ermolao Barbaro, edited by V. Branca; *Epistolae* (Florence, 1943), II, 22–23 (on which see the article by Dionisotti below). The various works *De redctione* treating Swineshead's opinion are Angelus de Fossambruno, MS Venice, Bibl. Naz. San Marco, VI, 160, ff. 248–252r; Gaietano de Thienis, *Tractatus perutilis de reactione*, Venice ed., 1491; Giovanni Marliani, *Tractatus de reactione* and *In defensionem tractatus de reactione*, both printed in his *Opera omnia*, II (Pavia, 1482) [on which see Clagett, *Giovanni Marliani and Late Medieval*

Physics, (New York, 1941), ch. 2]; Victorus Trincavellus, *Questio de reactione iuxta doctrinam Aristotelis et Averrois commentatoris*, printed at the end (69r–74r) of the 1520 Venice ed. of the *Liber calculationum*; Pietro Pomponazzi, *Tractatus acutissimi . . . de reactione* (Venice, 1525).

The case for Pavia as the fifteenth-century center of "calculatory," and especially Swineshead, studies is made by Carlo Dionisotti, "Ermolao Barbaro e la Fortuna di Suiseth," in *Medioevo e Rinascimento: Studi in onore di Bruno Nardi* (Florence, 1955), 219–253. Dionisotti briefly treats the evidence for Nicoletto Vernia's studies in Pavia on the Calculator, but see also Eugenio Garin, "Noterelle sulla filosofia del Rinascimento I: A proposito di N. Vernia," in *Rinascimento*, **2** (1951), 57–62. The reference from Vernia's *De anima intellectiva* (MS Venice, Bibl. Naz. San Marco, VI, 105, ff. 156r–160r) was furnished by Edward Mahoney, who is preparing an edition of the text. The relevant bibliography of the Italian commentaries on treatise I, *De intensione et remissione formarum*, of the *Liber calculationum* is Tiberio Baccilieri: Bruno Nardi, *Sigieri di Brabante nel pensiero del Rinascimento italiano* (Rome, 1945), 138–139; Pearl Kibre, "Cardinal Domenico Grimani, 'Questio de intensione et remissione qualitatis': A Commentary on the Tractate of that Title by Richard Suiseth (Calculator)," in *Didascaliae: Studies in Honor of Anselm M. Albareda*, Sesto Prete, ed. (New York, 1961), 149–203; Charles Schmitt, "Hieronymus Picus, Renaissance Platonism and the Calculator," to appear in Anneliese Maier Festschrift; Curtis Wilson, "Pomponazzi's Criticism of Calculator," in *Isis*, **44** (1952), 355–363. Other references to Pomponazzi on Swineshead can be found in Pietro Pomponazzi, *Corsi inediti dell'insegnamento padovano*, I: *Super Libello de substantia orbis, expositio et Quaestiones Quattuor* (1507); *Introduzione e testo a cura di Antonino Poppi* (Padua, 1966); and Bruno Nardi, *Saggi sull' Aristotelismo Padovano dal secolo XIV al XVI* (Florence, 1958).

The relevant sources or literature concerning other parts of the *Liber calculationum* are Bassanus Politus, *Tractatus proportionum introductorius ad calculationes Suiset* (Venice, 1505); Giovanni Marliani, *Probatio cuiusdam sententie calculatoris*, in his *Opera omnia*, II (Pavia, 1482), ff. 19r–25r (on which see Clagett, *Giovanni Marliani and Late Medieval Physics* [New York, 1941], ch. 5); Christopher de Recaneto, *Recolecte super calculationes*, MS Venice, Bibl. Naz. San Marco, Lat. VI, 149, ff. 31r–49v. On Recaneto, see Nardi, *Saggi sull'Aristotelismo Padovano dal secolo XIV al XVI* (Florence, 1958), 117–119, 121–122; Philippus Aiuta, "Pro declaratione Suiset calculatoris," MS Bibl. Vaticana, Chigi E. VI, 197, ff. 132r–149r. Finally, mention should be made of a totally unexamined sixteenth-century work on the same part of the *Liber calculationum* by Raggius of Florence, MS Rome, Bibl. Casan, 1431 (B.VI.7).

The Parisian-Spanish reception of Swineshead. The relevant primary sources are [Alvaro Thomaz], *Liber de triplici motu proportionibus annexis magistri Alvari Thome Ulixbonensis philosophicas Suiseth calculationes ex parte declarans* (Paris, 1509); John Dullaert of Ghent, *Questiones super octo libros phisicorum Aristotelis necnon super libros de celo et mundo* (Lyons, 1512); [Juan de Celaya], *Expositio magistri ioannis de Celaya Valentini in octo libros phisicorum Aristotelis: cum questionibus eiusdem secundum triplicem viam beati Thome, realium et nominalium* (Paris, 1517); Ludovicus Coronel, *Physice perscrutationes* (Paris, 1511); Diego de Astudillo, *Quaestiones super octo libros physicorum et super duos libros de generatione Aristotelis, una cum legitima textus expositione eorundem librorum* (Valladolid, 1532). Of the secondary literature on these—and other related—figures, the two basic articles are by William A. Wallace, "The Concept of Motion in the Sixteenth Century," in *Proceedings of the American Catholic Philosophical Association*, **41** (1967), 184–195; "The 'Calculatores' in Early Sixteenth-Century Physics," in *British Journal for the History of Science*, **4** (1969), 221–232. More detailed bio-bibliographical information on the school of Spanish scholars at Paris in the early sixteenth century can be found in H. Elie, "Quelques maîtres de l'université de Paris vers l'an 1500," in *Archives d'histoire doctrinale et littéraire du moyen âge*, **18** (1950–1951), 193–243; and R. Garcia Villoslada, *La universidad de Paris durante los estudios de Francisco de Vitoria, O.P., 1507–1522, Analecta Gregoriana*, XIV (Rome, 1938). On Alvaro Thomaz' treatment of the "infinite series" in Swineshead (and in Nicole Oresme) see Marshall Clagett, *Nicole Oresme and the Medieval Geometry of Qualities and Motions* (Madison, Wis., 1968), 496–499, 514–516; and H. Wieleitner, "Zur Geschichte der unendlichen Reihen im christlichen Mittelalter," in *Bibliotheca mathematica*, **14** (1913–1914), 150–168. Some biographical details on Alvaro can be found in J. Rey Pastor, *Los matemáticos españoles del siglo XVI* (Toledo, 1926), 82–89.

Later sixteenth- and seventeenth-century appreciation of Swineshead. Appropriate references to Cardano's *De subtilitate* and Julius Scaliger's *Exotericarum exercitationum* are given in Jacob Brucker's *Historia critica philosophiae*, III (Leipzig, 1766), 851. This work contains (849–853) numerous quotations from other authors referring (pro and con) to Swineshead as well as a brief example of the *Liber calculationum* itself. Leibniz' references to Swineshead are too numerous (we know of at least eight of them) to cite completely, but some of the most important of them can be found in L. Couturat's *Opuscules et fragments inédits de Leibniz* (Paris, 1903), 177, 199, 330, 340. Indication of the manuscript copy Leibniz had made of Swineshead can be found in Eduard Bodemann, *Die Handschriften der königlichen offentlichen Bibliothek zu Hannover* (Hannover, 1867), 104–105.

JOHN E. MURDOCH
EDITH DUDLEY SYLLA

SYDENHAM, THOMAS (*b.*, or at least baptized, Wynford Eagle, Dorset, England, 10 September 1624; *d.* London, England, 29 December 1689), *medicine.*

Sydenham was the son of William Sydenham, a Dorset squire, and Mary, daughter of Sir John Geffrey. During the civil war, four or possibly five sons served in the army of Parliament. Two of the sons and the mother lost their lives. After distinguished military service, William, the eldest son, became a close confidant of Cromwell and a prominent figure during the Commonwealth.

Sydenham, himself, saw some military service and eventually attained the rank of captain. It was said that he was left among the dead on the battlefield on one occasion and narrowly escaped death on another. His war service and the prominence of his family in Cromwell's cause gained him political preferences during the Commonwealth period. In what sense their support for Cromwell made Sydenham's family "Puritan" is difficult to say. The major Somerset branches of the family were, by and large, staunchly Anglican and Loyalist; but the Dorset Sydenhams, to whom Thomas belonged, were probably Presbyterian, and his brother William, latterly, Presbyterian-Independent. If a manuscript text entitled "Theologia rationalis" is correctly ascribed to Sydenham, it affords a more precise insight into his particular religious views but nothing on his attitude toward church polity.

The war disrupted Sydenham's earliest education at Oxford soon after his matriculation in Magdalen Hall in 1642. At the end of the first hostilities, he returned to the university, determined to become a physician. Some evidence suggests that he reentered Magdalen Hall and was made master of arts in 1648, possibly so that he could be created bachelor of medicine. In any event, his favor with the parliamentary visitors is clear from his appointment as one of their delegates to Wadham College in 1647 and from his election by them as a fellow of All Souls College in the following year, having been created bachelor of medicine by command of the chancellor of the university a few months before.

Sydenham remained in the academic arena for several years, resigning his fellowship at All Souls in 1655. During this time he may have seen further military service, since Thomas the cornet of the first civil war had become "Captain Sydenham" by 1654. It is known that he received £600 and the promise of employment in view of his financial and military contributions, that he was nominated but not elected to Parliament in 1659, and that on 14

July 1659 he was appointed "comptroller of the pipe."

Within a year or so of his marriage to Mary Gee in 1655, Sydenham began to practice medicine in Westminster, where he remained for the rest of his life. Evidence that he spent some time at Montpellier is no better than, if as good as, evidence that he never left England. It is highly improbable, therefore, that the works of Charles Barbyrac had any influence on him, as has been claimed.

After the Restoration, public political life was likely closed to Sydenham. In any event, he devoted the rest of his career single-mindedly to medicine. He obtained his license to practice from the Royal College of Physicians of London in June 1663. He was admitted member of Pembroke College, Cambridge, on 17 May 1676, and received an M.D. at that time.

In his last years, Sydenham was considerably disabled by gout and renal disease and died at his home in Pall Mall in 1689. He left three sons: William (also a physician), Henry, and James.

Sydenham's depreciation of bookish knowledge and university education for physicians has been linked with the belief that he was himself somewhat untutored and unlettered. But he had ample opportunity for a good education, and a close study of his writings suggests that he was competent in Latin and well versed in contemporary medical thought. Most likely his utilitarian, practical turn of mind owed much to his background, but it would be difficult to say what part was played by native temperament, by religious ethos, by political alignments, or simply by the rural life of a country squire's son. Undoubtedly, this attitude was sustained in later life by the exigencies of an active medical practice. It is much less likely that it can be attributed to a poor education or the diamond-in-the-rough qualities of a cavalry officer, although this image of him has been cultivated. The fragmentary evidence that has survived suggests that his military career during the 1640's and 1650's was a good deal less remarkable than that of his brothers, and his university studies considerably more so. His subsequent career as a noted author, and the associations that he maintained, suggest an intellectual bent, albeit ruggedly individualistic rather than bookishly academic.

Sydenham's associates included Robert Boyle and John Locke. To the former, Sydenham dedicated his first work of 1666, while the latter contributed a commendatory poem to Sydenham in his second edition of 1668. Sydenham's association with Locke was particularly close between 1667 and 1671, when Locke was composing the earliest known drafts of his own *Essay Concerning Human Understanding*. Locke, who eventually qualified himself medically, may have been Sydenham's collaborator, or more likely his student. Certainly Locke served him as an amanuensis, since a number of Sydenham's medical texts are in manuscripts in Locke's handwriting. This fact has raised the question of authorship of the important texts "Anatomia" (1668) and "De arte medica" (1669). Although in Locke's hand, and long thought to be from his head, these texts almost certainly bear the ideas of Sydenham. The two men corresponded through the rest of Sydenham's life, and Locke included him in his *Essay* along with Boyle, Newton, and Huygens as a "master-builder" in the new sciences.

Sydenham, Locke, and Boyle had much in common in their approach to acquiring knowledge of the natural world. They held similar views in epistemology and shared an admiration for Bacon. The question of who might have influenced whom has often been debated, since the results of their respective efforts in medicine, philosophy, and chemistry have been so far-reaching.

Sydenham's philosophy was that of a skeptical physician: skeptical because he thought that human understanding is limited to observing and reasoning about experience and the data of the sensible world; physician because his philosophy was almost exclusively applied to improving medical practice, which made results of treatment the supreme test of the only kind of truth worth having. He was also an optimist, who believed such results to be within human grasp, assured by a nature that is the orderly instrument of a benevolent God.

In his studies of acute diseases, Sydenham began by propounding a method for treating fevers, and he hoped that this method would improve the uncertain and often bad effects of treatment current in his day. The simplicity and naiveté of these early efforts (1666 and 1668) were quickly apparent to him as the Great Plague of London, followed by severe epidemics of smallpox and by puzzling variations in the concurrent continued fevers, demonstrated that the hoped-for reform of therapeutics necessitated closer attention to the differences among diseases confronting him. Therefore, along with notes for a treatise on smallpox, and for a work on medical epistemology, Sydenham kept a notebook (Vaillant Manuscript, Royal College of Physicians, London) of clinical observations from 1669 to 1674. The outcome of this was his *Observationes medicae circa morbor-*

um acutorum historiam et curationem (1676), his magnum opus.

In the preface to this work, Sydenham set forth his premises for more ambitious therapeutic reforms. These premises included his belief in the healing power of nature, nature's orderly production of diseases by species, the need for a more refined delineation of the seasonal and annual variants of these diseases, the unknowability of the insensible causes, the possibility of deriving treatment from the observable phenomena of a disease, and the desirability of making repeated trials of treatment before declaring them effective. In the body of the text, Sydenham exemplified this therapeutic program with respect to acute, epidemic diseases; he paid special attention to one particular factor causing variations in these diseases from epidemic to epidemic. This was the epidemic "constitution," a traditional concept that he had extensively revised.

Sydenham's effort to tie treatment deductively to the observable phenomena, and his concern for epidemic variations, gave an impetus to the more careful bedside observation of disease. Moreover, his insistence that diseases, like plants, have species, was suggestive to eighteenth-century nosographers like Linnaeus, Sauvages, and Cullen. But Sydenham's strong emphasis on the healing power of nature, and methodical treatment designed to modulate that nature, created in his work a systematizing tendency that cast doubt on the need for specific diagnoses. Consequently, almost a century later, he could also be appealed to by such a systematist and vigorous antinosologist as Benjamin Rush, "The Sydenham of America."

For almost two centuries after his death, however, it was particularly as a contributor to therapy that Sydenham acquired his reputation. It was his moderate treatment of smallpox, his use of cinchona, and his invention of liquid laudanum that came to symbolize his contributions to medicine. His renown came chiefly from the fact that he alleviated the suffering of the sick and made ill people well.

In the final decade of his life, Sydenham turned to apply his principles to the chronic diseases and produced a classic description of gout. By this time he had achieved eminence in the world of medical letters, both at home and in continental Europe, having become the "English Hippocrates." Ironically, the only eponymous use of his name that still remains common, "Sydenham's chorea," refers to two paragraphs interjected in one of his treatises, more or less as an aside.

BIBLIOGRAPHY

I. ORIGINAL WORKS. Of the many editions of Sydenham's published works, the best is the *Opera omnia*, William A. Greenhill, ed. (London, 1844; 2nd ed., 1846). It includes a bibliography of the Latin editions of all his writings into the nineteenth century. The standard English trans., *The Works of Thomas Sydenham, M.D.*, 2 vols. (London, 1848), translated from the Latin ed. of Greenhill, with a biography of Sydenham by R. G. Latham, is not always reliable and should be compared with the Latin text, with the nearly contemporary trans. by John Pechey (London, 1696), and with that by John Swan (London, 1742).

A number of Sydenham's MSS and documents relating to his life have been published. The largest collection of these is in Kenneth Dewhurst, *Dr. Thomas Sydenham (1624–1689): His Life and Original Writings* (London, 1966); this includes the controversial "De arte medica" and the "Anatomia," in addition to "Theologia rationalis" and a number of medical fragments. More complete versions of these last works, with other material, are in the unpublished "Medical Observations by Thomas Sydenham," (the so-called Vaillant Manuscript), of various dates, in the library of the Royal College of Physicians, London. This MS is partially in Sydenham's hand, and partially in Locke's. Other literary remains of Sydenham, published and unpublished, are discussed by Dewhurst. See also the appendix to Bates, cited below.

II. SECONDARY LITERATURE. The best known biographies of Sydenham are Joseph F. Payne (London, 1900) and Dewhurst. Payne's biography is not documented, but his article on Sydenham in the *Dictionary of National Biography*, LV (1898), gives extensive references. Payne and Dewhurst differ in a number of details of Sydenham's life, and sometimes both tend to be more assertive than their evidence warrants. L. M. F. Picard, *Thomas Sydenham: Sa vie et ses oeuvres. Thèse pour le doctorat en médecine* (Dijon, 1889), has many details not found in Payne or Dewhurst, and is well documented.

The development and elucidation of Sydenham's concepts of epidemic constitutions, disease species, and methodical treatment can be found in D. G. Bates, "Thomas Sydenham: The Development of His Thought, 1666–1676" (unpublished doctoral diss., The Johns Hopkins University, 1975). A selected bibliography of the secondary literature and an extensive analysis of the Sydenham MSS will also be found there. A full account of the contemporary context for these ideas and Sydenham's influence on eighteenth- and nineteenth-century clinical medicine is still needed.

DONALD G. BATES

SYLOW, PETER LUDVIG MEJDELL (*b.* Christiania [now Oslo], Norway, 12 December 1832; *d.* Christiania, 7 September 1918), *mathematics.*

Sylow was the son of a cavalry captain, Thomas Edvard Sylow, who later became a minister of the government. After graduation from the Christiania Cathedral School in 1850, he studied at the university, where in 1853 he won a mathematics prize contest. He took the high school teacher examination in 1856, and from 1858 to 1898 he taught in the town of Frederikshald (now Halden). Sylow was awarded a scholarship to travel abroad in 1861, and he visited Berlin and Paris. In 1862–1863 he substituted for Ole-Jacob Broch at Christiania University; but until 1898 his only chance for a university chair came in 1869, and he was not appointed. Finally, through Sophus Lie, a special chair was created for him at Christiania University in 1898.

From 1873 to 1881 Sylow and Sophus Lie prepared a new edition of the works of N. H. Abel, and for the first four years Sylow was on leave from his school in order to devote himself to the project. In 1902, with Elling Holst, he published Abel's correspondence. He also published a few papers on elliptic functions, particularly on complex multiplication, and on group theory.

Sylow's name is best-known in connection with certain theorems in group theory and certain subgroups of a given group. In 1845 Cauchy had proved that any finite group G has subgroups of any prime order dividing the order of G. In 1872 Sylow published a 10-page paper containing the first extension of Cauchy's result and perhaps the first profound discovery in abstract group theory after Cauchy. Sylow's main theorem read as follows: First, if p^m is the maximal power of p dividing the order of G, then G has subgroups of order p^i for all i with $0 \leq i \leq m$, and in particular subgroups H of order p^m (called p-Sylow groups); and the index j of the normalizer of H is congruent 1 mod p. Second, the p-Sylow groups of G are conjugate with each other. Sylow's theorems were, and still are, a source of discoveries in group theory and are fundamental to most structural research in finite groups.

BIBLIOGRAPHY

I. ORIGINAL WORKS. Sylow's works are listed in H. B. Kragemo, "Bibliographie der Schriften Ludvig Sylows," in *Norsk matematisk forenings skrifter*, 2nd ser., no. 3 (1933), 25–29. They include "Théorèmes sur les groupes de substitutions," in *Mathematische Annalen*, **5** (1872), 584–594; and "Sur la multiplication complexe des fonctions elliptiques," in *Journal de mathématiques pures et appliquées*, 4th ser., **3** (1887), 109–254.

Sylow's MSS are in the Oslo University Library, U.B. MS, fols. 730–808, and U.B. Brevsamling 7–8.

II. SECONDARY LITERATURE. See T. Skolem, "Ludvig Sylow und seine wissenschaftlichen Arbeiten," in *Norsk matematisk forenings skrifter*, 2nd ser., no. 2 (1933), 14–24; and C. Størmer, "Gedächtnisrede auf Professor Dr. P. L. M. Sylow," *ibid.*, no. 1 (1933), 7–13.

HANS FREUDENTHAL

SYLVESTER, JAMES JOSEPH (*b.* London, England, 3 September 1814; *d.* London, 15 March 1897), *mathematics*.

Although Sylvester is perhaps most widely remembered for his indefatigable work in the theory of invariants, especially that done in conjunction with Arthur Cayley, he wrote extensively on many other topics in the theory of algebraic forms. He left important theorems in connection with Sturm's functions, canonical forms, and determinants; he especially advanced the theory of equations and the theory of partitions.

James Joseph (Joseph then being his surname) was born into a Jewish family originally from Liverpool. The son of Abraham Joseph, who died while the boy was young, James was the sixth and youngest son of nine children, at least four of whom later assumed the name Sylvester for a reason not now apparent.

Until Sylvester was fifteen, he was educated in London, at first in schools for Jewish boys at Highgate and at Islington, and then for five months at the University of London (later University College), where he met Augustus De Morgan. In 1828 he was expelled "for taking a table knife from the refectory with the intention of sticking it into a fellow student who had incurred his displeasure."[1] In 1829 Sylvester went to the school of the Royal Institution, in Liverpool, where he took the first prize in mathematics by an immense margin and won a prize of $500, offered by the Contractors of Lotteries in the United States, for solving a problem in arrangements. At this school he was persecuted for his faith to a point where he ran away to Dublin. There, in the street, he encountered R. Keatinge, a judge and his mother's cousin, who arranged for his return to school.

Sylvester now read mathematics for a short time with Richard Wilson, at one time a fellow of St. John's College, Cambridge, and in October 1831 he himself entered that college, where he stayed until the end of 1833, when he suffered a serious illness that kept him at home until January 1836. After further bouts of illness, Sylvester took the tripos examination in January 1837, placing second. Since he was not prepared to subscribe to the Thirty-Nine Articles of the Church of England, he was not allowed to take the degree or compete for Smith's mathematical prizes—still less secure a fellowship. He went, therefore, to Trinity College, Dublin, where he took the B.A. and M.A. in 1841. (He finally took the equivalent Cambridge degrees in 1872–1873, the enabling legislation having been passed in 1871.)

In 1838 Sylvester went to what is now University College, London, as De Morgan's colleague. He seems to have found the chair in Natural Philosophy uncongenial. In 1839, at the age of twenty-five, he was elected a fellow of the Royal Society on the strength of his earliest papers, written for *Philosophical Magazine* as soon as he had taken his tripos examination. The first four of these concern the analytical development of Fresnel's theory of the optical properties of crystals, and the motion of fluids and rigid bodies. His attention soon turned to more purely mathematical topics, especially the expression of Sturm's functions in terms of the roots of the equation.

From University College, Sylvester moved in 1841 to a post at the University of Virginia. There are many lurid and conflicting reports of the reasons for his having returned to England in the middle of 1843. He apparently differed from his colleagues as to the way an insubordinate student should be treated. He now left the academic world for a time, and in 1844 was appointed Actuary and Secretary to the Equity and Law Life Assurance Company. He apparently gave private tuition in mathematics, for he had Florence Nightingale as a pupil. In 1846, the same year that Cayley entered Lincoln's Inn, Sylvester entered Inner Temple and was finally called to the bar in November 1850. Cayley and Sylvester soon struck up a friendship. At his Oxford inaugural lecture many years later (1885), Sylvester spoke of Cayley, "who though younger than myself, is my spiritual progenitor— who first opened my eyes and purged them of dross so that they could see and accept the higher mysteries of our common mathematical faith." Both men referred on occasion to theorems they had derived separately through the stimulus of their conversations in the intervals between legal business.

In 1854 Sylvester was an unsuccessful candidate both for the chair of mathematics at the Royal Military Academy, Woolwich, and for the professorship in geometry at Gresham College, London. The successful candidate for the former position

soon died, and with the help of Lord Brougham, Sylvester was appointed. He held this post from September 1855 to July 1870. At the same time he became editor, from its first issue in 1855, of the *Quarterly Journal of Pure and Applied Mathematics*, successor to the *Cambridge and Dublin Mathematical Journal*. Assisted as he was by Stokes, Cayley, and Hermite, there was no change in editorship until 1877.

In 1863 Sylvester replaced the geometer Steiner as mathematics correspondent to the French Academy of Sciences. Two years later he delivered a paper on Newton's rule (concerning the number of imaginary roots of an algebraic equation) at King's College, London. A syllabus to the lecture was the first mathematical paper published by De Morgan's newly founded London Mathematical Society, of which Sylvester was president from 1866 to 1868. In 1869 he presided over the Mathematical and Physical Section of the British Association meeting at Exeter. His address was prompted by T. H. Huxley's charge that mathematics was an almost wholly deductive science, knowing nothing of experiment or causation. This led to a controversy carried on in the pages of *Nature*, relating to Kant's doctrine of space and time; Sylvester, however, was not at his best in this kind of discussion. He reprinted an expanded version of his presidential address, together with the correspondence from *Nature*, as an appendix to *The Laws of Verse* (1870). The thoughts of Matthew Arnold, to whom the book was dedicated, are not known. Sylvester had some slight renown throughout his life, especially among his close friends, for his dirigible flights of poetic fancy; and his book was meant to illustrate the quasi-mathematical "principles of phonetic syzygy." Five original verses introduce a long paper on syzygetic relations, and he used his own verse on several other mathematical occasions.

Sylvester translated verse from several languages. For example, under the nom de plume "Syzygeticus," he translated from the German "The Ballad of Sir John de Courcy";[2] and his *Laws of Verse* includes other examples of his work, which is no worse than that of many a nonmathematician. It could be argued, however, that it was worse in a different way. One of his poems had four hundred lines all rhyming with "Rosalind," while another had two hundred rhyming with "Winn." These were products of his later residence in Baltimore. Sylvester had perhaps a better appreciation of music, and took singing lessons from Gounod.

In 1870 Sylvester resigned his post at Woolwich, and after a bitter struggle that involved correspondence in the *Times*, and even a leading article there (17 August 1871), he secured a not unreasonable pension. It was not until 1876, when he was sixty-one, that he again filled any comparable post. When he did so, it was in response to a letter from the American physicist Joseph Henry. The Johns Hopkins University opened in that year, and Sylvester agreed to accept a chair in mathematics in return for his traveling expenses and an annual stipend of $5,000 "paid in gold." "His first pupil, his first class" was G. B. Halsted. A colleague was C. S. Peirce, with whom, indeed, Sylvester became embroiled in controversy on a small point of priority. Peirce nevertheless later said of him that he was "perhaps the mind most exuberant in ideas of pure mathematics of any since Gauss." While at Baltimore, Sylvester founded the *American Journal of Mathematics*, to which he contributed thirty papers. His first was a long and uncharacteristic account of the application of the atomic theory to the graphical representation of the concomitants of binary forms (quadratics). He resigned his position at Johns Hopkins in December 1883, when he was appointed to succeed H. J. S. Smith as Savilian professor of geometry at Oxford.

Sylvester was seventy when he delivered his inaugural "On the Method of Reciprocants" (1 December 1885). By virtue of his chair he became a fellow of New College, where he lived as long as he was in Oxford. He collaborated with James Hammond on the theory of reciprocants (functions of differential coefficients the forms of which are invariant under certain linear transformations of the variables) and also contributed several original papers to mathematical journals before his sight and general health began to fail. In 1892 he was allowed to appoint a deputy, William Esson; and in 1894 he retired, living mainly at London and Tunbridge Wells. For a short period in 1896 and 1897 he wrote more on mathematics (for example, on compound partitions and the Goldbach-Euler conjecture). A little more than a fortnight after a paralytic stroke, he died on 15 March 1897 and is buried in the Jewish cemetery at Ball's Pond, Dalston, London.

Sylvester received many honors in his lifetime, including the Royal Medal (1861) and the Copley Medal (1880). It is of interest that in the receipt of such awards he followed rather than preceded Cayley, who was his junior. Sylvester received honorary degrees from Dublin (1865), Edinburgh (1871), Oxford (1880), and Cambridge (1890).

Sylvester never married. He had been anxious to marry a Miss Marston, whom he met in New York in 1842, on his first visit to America. (She was the godmother of William Matthew Flinders Petrie, from whom the story comes.) It seems that although she had formed a strong attachment for him, she refused him on the ground of religious difference, and neither of them subsequently married.

Sylvester's greatest achievements were in algebra. With Cayley he helped to develop the theory of determinants and their application to nonalgebraic subjects. He was instrumental in helping to turn the attention of algebraists from such studies as the theory of equations—in which he nevertheless did important work—to the theory of forms, invariants, and linear associative algebras generally. His part in this movement is often obscured by his flamboyant style. In 1888 P. G. Tait, in a rather strained correspondence with Cayley over the relations between Tait's solution of a quaternionic equation and Sylvester's solution of a linear matrix equation, wrote with some justice: "I found Sylvester's papers hard to assimilate. A considerable part of each paper seems to be devoted to correction of hasty generalizations in the preceding one!"[3]

A number of Sylvester's early writings concern the reality of the roots of numerical equations, Newton's rule for the number of imaginary roots, and Sturm's theorem. His first published researches into these matters date from 1839, and were followed by a steady stream of special results. In due course he found simple expressions for the Sturmian functions (with the square factors removed) in terms of the roots:

$$f_2(x) = \Sigma (a-b)^2 (x-c) (x-d) \cdots$$
$$f_3(x) = \Sigma (a-b)^2 (a-c)^2 (b-c)^2 (x-d) \cdots .$$

Applying Sturm's process of the greatest algebraic common measure to two independent functions $f(x)$ and $\varphi(x)$, rather than to $f(x)$ and $f'(x)$, he found for the resulting functions expressions involving products of differences between the roots of the equations $f(x) = 0$, $\varphi(x) = 0$. Assuming that the real roots of the two equations are arranged in order of magnitude, the functions are of such a character that the roots of the one equation are intercalated among those of the other.

In connection with Newton's rule, the method of Sturm's proof was applied to a quite different problem. Sylvester supposed x to vary continuously, and investigated the increase and decrease in the changes of sign.[4]

Newton's first statement of his incomplete rule for enumerating imaginary roots dates from 1665–1666.[5] Although valid, the rule was not justified before Sylvester's proofs of the complete rule.

Another problem of great importance investigated in two long memoirs of 1853 and 1864 concerns the nature of the roots of a quintic equation. Sylvester took the functions of the coefficients that serve to decide the reality of the roots, and treated them as the coordinates of a point in n-dimensional space. A point is or is not "facultative" according to whether there corresponds, or fails to correspond, an equation with real coefficients. The character of the roots depends on the bounding surface or surfaces of the facultative regions, and on a single surface depending on the discriminant.[6]

Sylvester showed an early interest in the theory of numbers when he published a beautiful theorem on a product formed from numbers less than and prime to a given number.[7] This he described as "a pendant to the elegant discovery announced by the ever-to-be-lamented and commemorated Horner, with his dying voice"; but unfortunately it was later pointed out to him by Ivory that Gauss had given the theorem in his *Disquisitiones arithmeticae* (1801).[8] It is impossible to do justice in a short space to Sylvester's numerous later contributions to the theory of numbers, especially in the partition of numbers. Sylvester applied Cauchy's theory of residues and originated the concept of a denumerant. He also added several results to Euler's treatment of the "problem of the virgins" (the problem of enumerating positive and integral solutions of indeterminate simultaneous linear equations); but his most novel contributions to the subject are to be found in his use of a graphical method. He represented partitions of numbers by nodes placed in order at the points of a rectangular lattice ("graph"). Thus a partition of 9 $(5 + 3 + 1)$ may be represented by the points of the rows in the lattice. The conjugate partition $(3 + 2 + 2 + 1 + 1)$ is then found by considering the lattice of columns, a fact possibly first appreciated by N. M. Ferrers.[9] This

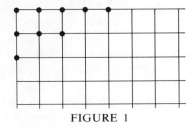

FIGURE 1

representation greatly simplified and showed the way to proofs of many new results in the theory of partitions not only by Sylvester but also by early contributors to his *American Journal of Mathematics*, such as Fabean Franklin.

One of Sylvester's early contributions to the *Journal*, "On Certain Ternary Cubic-Form Equations,"[10] is notable for the geometrical theory of residuation on a cubic curve and the chain rule of rational derivation: From an arbitrary point 1 on the curve it is possible to derive the singly infinite series of points $(1,2,4,5, \cdots 3p \pm 1)$ such that the chord through any two points, m and n, meets the curve again in a point ($m + n$ or $|m - n|$, whichever number is not divisible by 3) of the series. The coordinates of any point m are rational and integral functions of degree m^2 of those of point 1.

Like his friend Cayley, Sylvester was above all an algebraist. As G. Salmon said, the two discussed the algebra of forms for so long that each would often find it hard to say what properly belonged to the other. Sylvester, however, produced the first general theory of contravariants of forms.[11] He was probably the first to recognize that for orthogonal transformations, covariants and contravariants coincide. Moreover, he proved a theorem first given without proof by Cayley, and the truth of which Cayley had begun to doubt. It concerns a certain expression for a number ("Cayley's number") that cannot exceed the number of linearly independent semi-invariants (or invariants) of a certain weight, degree, and extent. Sylvester showed that Cayley's expression for the number of linearly independent ("asyzygetic") semi-invariants of a given type is in fact exact.[12] The result is proved as part of Sylvester's and Cayley's theory of annihilators, which was closely linked to that of generating functions for the tabulation of the partitions of numbers.

Under the influence of Lie's analysis, algebraic invariance was gradually subordinated to a more general theory of invariance under transformation groups. Although Boole had used linear differential operators to generate invariants and covariants, Cayley, Sylvester, and Aronhold were the first to do so systematically. In the calculation of invariants, it may be proved that any invariant I of the binary form (quantic)

$$f = a_0 x^p + p a_1 x^{p-1} y + \cdots + a_p y^p$$

should satisfy the two differential equations

$$\Omega I = 0,$$
$$O I = 0,$$

where Ω and O are linear differential operators:

$$\Omega \equiv a_0 \frac{\partial}{\partial a_1} + 2a_1 \frac{\partial}{\partial a_2} + \cdots + pa_{p-1} \frac{\partial}{\partial a_p},$$

$$O \equiv pa_1 \frac{\partial}{\partial a_0} + (p-1)a_2 \frac{\partial}{\partial a_1} + \cdots + a_p \frac{\partial}{\partial a_{p-1}}.$$

Sylvester called these functions annihilators, built up a rich theory around them, and generalized the method to other forms.[13] With Franklin he exhibited generating functions for all semi-invariants, of any degree, for the forms they studied.[14] Related to these studies is Sylvester's expression, in terms of a linear differential equation, of the condition that a function be an orthogonal covariant or invariant of a binary quantic. Thus the necessary and sufficient condition that F be a covariant for direct orthogonal transformations is that F have as its annihilator

$$y \frac{\partial}{\partial x} - x \frac{\partial}{\partial y} + O - \Omega.$$

Sylvester played an important part in the creation of the theory of canonical forms. What may be his most widely known theorem states that a general binary form of odd order $(2n-1)$ is a sum of n $(2n-1)$-th powers of linear forms. (Thus, for example, a quintic may be reduced to a sum of three fifth powers of linear forms.) Sylvester wrote at length on the canonical reduction of the general $2n$-ic. He showed that even with the ternary quartic, which has fifteen coefficients, the problem was far less simple than it appeared, and that such cannot be simply reduced to a sum of five fourth powers (again with fifteen coefficients). It is here that he introduced the determinant known as the catalecticant, which he showed must vanish if the general $2n$-ic is to be expressed as the sum of n perfect $2n$th powers of linear forms, together with (in general) a term involving the square of the product of these forms.[15]

Early in his study of the effects of linear transformations on real quadratic forms, Sylvester discovered (and named) the law of inertia of quadratic forms.[16] The law was discovered independently by Jacobi.[17] The theorem is that a real quadratic form of rank r may be reduced by means of a real nonsingular linear transformation to the form

$$y_1^2 + \cdots + y_p^2 - y_{p+1}^2 - \cdots - y_r^2,$$

where the index p is uniquely determined. (It follows that two real quadratic forms are equivalent under real and nonsingular transformation if and only if they have the same rank and the same index.)

Another memorable result in the theory of linear transformations and matrices is Sylvester's law of nullity, according to which if r_1 and r_2 are the ranks of two matrices, and if R is the rank of their product,

$$R \leqslant r_1,$$
$$R \leqslant r_2,$$
$$R \geqslant r_1 + r_2 - n,$$

where n is the order of the matrices. For Sylvester the "nullity" of a matrix was the difference between its order and rank, wherefore he wrote his law thus: "The nullity of the product of *two* (and therefore of any number of) matrices cannot be less than the nullity of any factor, nor greater than the sum of the nullities of the *several* factors which make up the product."[18]

Sylvester devised a method (the "dialytic method") for the elimination of one unknown between two equations

$$f(x) \equiv a_0 x^n + a_1 x^{n-1} + \cdots + a_n = 0 (a_0 \neq 0),$$
$$\varphi(x) \equiv b_0 x^m + b_1 x^{m-1} + \cdots + b_m = 0 (b_0 \neq 0).$$

The method is simpler than Euler's well-known method. Sylvester formed n equations from $f(x)$ by separate and successive multiplication by x^{n-1}, $x^{n-2}, \cdots 1$, and m equations from $\varphi(x)$ by successive multiplication by x^{m-1}, $x^{m-2}, \cdots 1$. From the resulting $m+n$ equations he eliminated the $m+n$ power of x, treating each power as an independent variable. The vanishing of the resulting determinant (E) is a necessary condition for f and φ to have a common root, but the method is deficient to the extent that the condition $E = 0$ is not proved sufficient. This type of approach was superseded in Sylvester's lifetime when Kronecker developed a theory of elimination for systems of polynomials in any number of variables, but elementary texts still quote Sylvester's method alongside Euler's and Bezout's.

Sylvester was inordinately proud of his mathematical vocabulary. He once laid claim to the appellation "Mathematical Adam," asserting that he believed he had "given more names (passed into general circulation) to the creatures of the mathematical reason than all the other mathematicians of the age combined."[19] Much of his vocabulary has been forgotten, although some has survived; but it would be a mistake to suppose that Sylvester bestowed names lightly, or that they were a veneer for inferior mathematics. His "combinants," for example, were an important class of invariants of several q-ary p-ics (q and p constant).[20] His "plagiograph" was less obscure under the title "skew pantograph"; but under either name it was an instrument based on an interesting and unexpected geometrical principle that he was the first to perceive.[21] And in like manner one might run through his works, with their "allotrious" factors, their "zetaic" multiplication, and a luxuriant terminology between.

Sylvester thought his verse to be as important as his mathematics; but he was a poor judge, and the two had little in common beyond an exuberant vocabulary. His mathematics spanned, of course, a far greater range than it is possible to review here. One characteristic of this range is that it was covered without much recourse to the writings of contemporaries. As H. F. Baker has pointed out, in projective geometry Sylvester seems to have been ignorant of Poncelet's circular points at infinity, and not to have been attracted by Staudt's methods of dispensing with the ordinary notion of length. Sylvester's papers simply ignore most problems in the foundations of geometry. Remarkable as some of his writings in the theory of numbers, elliptic integrals, and theta functions are, he would have benefited from a closer reading of Gauss, Kummer, Cauchy, Abel, Riemann, and Weierstrass. Neither Lie's work on the theory of continuous groups nor the algebraic solution of the fifth-degree equation elicited any attention from him, and it is perhaps surprising that Cayley did not persuade him of their value. An illustration of Sylvester's self-reliance is found at the end of one of the last lengthy papers he composed, "On Buffon's Problem of the Needle," a new approach to this well-known problem in probabilities.[22] The paper was the outcome of conversations with Morgan Crofton, when Sylvester was his senior at Woolwich in the 1860's; yet an extension of Barbier's theorem, now proved by Sylvester, had been published in 1868 by Crofton himself. Sylvester's strength lay in the fact that he could acknowledge this sort of inadvertent duplication without significantly diminishing the enormous mathematical capital he had amassed.

NOTES

1. H. H. Bellot, *University College, London, 1826–1926* (London, 1929), 38.
2. *Gentleman's Magazine*, n.s., **6** (Feb. 1871), 38–48.
3. C. G. Knott, *The Life and Scientific Work of Peter Guthrie Tait* (Cambridge, 1911), 159.
4. For this and the preceding doctrines, see especially "On a Theory of the Syzygetic Functions . . ." (1853), repr. in his *Collected Mathematical Papers* (henceforth abbreviated as *CMP*), I, no. 57, 429–586; and "Algebraical Researches, Containing a Disquisition on Newton's Rule

. . .," *ibid.*, II, no. 74 (1864), 376–479. On Newton's rule also see *ibid.*, II, no. 81 (1865), 493–494; no. 84 (1865–1866), 498–513; no. 108 (1871), 704–708; and III, no. 42 (1880), 414–425.
5. See D. T. Whiteside, ed., *The Mathematical Papers of Isaac Newton*, I (Cambridge, 1967), 524.
6. See in particular *CMP*, I, no. 57 (1853), 436.
7. *Ibid.*, no. 5 (1838), 39.
8. *Disquisitiones arithmeticae* (1801), 76.
9. *CMP*, I, no. 59 (1853), 597.
10. *Ibid.*, III, no. 39 (1879–1880), 312–391.
11. *Ibid.*, I, no. 33 (1851), 198–202.
12. *Ibid.*, IV, no. 44 (1886), 515–519; see also no. 42 (1886), 458.
13. See, for example, *ibid.*, III, no. 18 (1878), 117–126; no. 27 (1878), 318–340; IV, no. 41 (1886), 278–302, esp. 288; no. 42 (1886), 305–513, esp. 451.
14. See especially *ibid.*, III, no. 67 (1882), 568–622.
15. See the important memoir in *ibid.*, I, no. 42 (1852), 284–327, with its amusing note to 293: "Meicatalecticizant would more completely express the meaning of that which, for the sake of brevity, I denote catalecticant."
16. *Ibid.*, no. 47 (1852), 378–381; no. 57 (1853), 511; IV, no. 49 (1887), 532.
17. *Journal für Mathematik*, **53** (1857), 275–281.
18. *CMP*, IV, no. 15 (1884), 134.
19. *Ibid.*, no. 53 (1888), 588.
20. For further details, see P. Gordan, *Vorlesungen über Invariantentheorie*, II (Leipzig, 1887), 70–78.
21. *CMP*, III, no. 3 (1875), 26–34.
22. *Ibid.*, IV, no. 69 (1890–1891), 663–679.

BIBLIOGRAPHY

I. ORIGINAL WORKS. Sylvester published no lengthy volume of mathematics, although his books on versification and its mathematical principles are numerous, and include *The Law of Continuity as Applied to Versification . . . Illustrated by an English Rendering of "Tyrrhena regnum," Hor. 3, 29 . . .* (London, 1869); *The Laws of Verse, or Principles of Versification Exemplified in Metrical Translations* (London, 1870); *Fliegende Blätter (Rosalind and Other Poems), a Supplement to the Laws of Verse* (London, 1876); *Spring's Debut. A Town Idyll in Two Centuries of Continuous Rhyme* (Baltimore, 1880); *Retrospect. A Verse Composition by the Savilian Professor of Geometry . . . Tr. Into Latin by Undergraduates of New College* (Oxford, 1884); and *Corolla versuum Cantatrici eximiae . . . a professore Saviliano geometriae apud oxonienses* (Oxford, 1895).

Sylvester's mathematical papers are in *The Collected Mathematical Papers of James Joseph Sylvester*, H. F. Baker, ed., 4 vols. (Cambridge, 1904–1912). Thirty of his 87 known letters from American sources are in R. C. Archibald, "Unpublished Letters of James Joseph Sylvester and Other New Information Concerning His Life and Work," in *Osiris*, **1** (1936), 85–154. Archibald gives full bibliographical information on most of the verse writings. Sylvester wrote a sonnet to the Savilian professor of astronomy, Charles Pritchard, on the occasion of his receiving the gold medal of the Royal Astronomical Society, in *Nature*, **33** (1886), 516. One might have imagined that in his literary flamboyance he was imitating Disraeli, had he not also addressed a sonnet to Gladstone (1890).

Thirty-three of Sylvester's lectures were reported by James Hammond in *Lectures Containing an Exposition of the Fundamental Principles of the New Theory of Reciprocants Delivered During . . . 1886 Before the University of Oxford*, repr. from *American Journal of Mathematics* (Oxford–Baltimore, 1888), Lecture 34 is by Hammond. Sylvester contributed well over 300 different mathematical problems to *Educational Times*. These are calendared in *Collected Papers*, IV, 743–747; and several letters concerning the problems are printed in Archibald, *op. cit.*, 124–128.

II. SECONDARY LITERATURE. H. F. Baker included a personal biography of Sylvester in his ed. of the collected papers. R. C. Archibald, *op. cit.*, 91–95, lists 57 publications dealing with Sylvester's life and works. To these may be added R. C. Archibald, "Material Concerning James Joseph Sylvester," in *Studies and Essays Offered to George Sarton* (New York, 1947), 209–217; and R. C. Yates, "Sylvester at the University of Virginia," in *American Mathematical Monthly*, **44** (1937), 194–201.

The most useful discussions of Sylvester's life and work are A. Cayley, "Scientific Worthies XXV: James Joseph Sylvester," in *Nature*, **39** (1889), 217–219, repr. in *The Collected Mathematical Papers of Arthur Cayley*, XIII (Cambridge, 1897), 43–48; F. Franklin, *People and Problems, a Collection of Addresses and Editorials* (New York, 1908), 11–27, first printed in *Bulletin of the American Mathematical Society*, **3** (1897), 299–309; P. A. MacMahon, obituary in *Nature*, **55** (1897), 492–494; and obituary in *Proceedings of the Royal Society*, **63** (1898), ix–xxv; and M. Noether, obituary in *Mathematische Annalen*, **50** (1898), 133–156. Of these, Noether's article is mathematically the most useful. Cayley wrote much invaluable commentary on Sylvester's work, for which see the index to Cayley's collected papers. The *Johns Hopkins University Circulars* are a convenient source of biography, since the editors often reprinted articles on Sylvester that had first been published elsewhere (such as those cited above by Cayley and Franklin, and the first of MacMahon's).

J. D. NORTH

SYLVESTER II, POPE. See **Gerbert**, also known as **Gerbert d'Aurillac.**

SYLVIUS, FRANCISCUS DELE BOË (*b*. Hanau, Germany, 15 March 1614; *d*. Leiden, Netherlands, 15 November 1672), *medicine.*

Sylvius was of southern Flemish extraction. His grandfather, a merchant and descendant of a noble family, emigrated from Kamerijk (Cambria, now in France) to Frankfurt-am-Main. François was the second son of Isaäc dele Boë and Anna de la Vignette, who was from the same area of southern Flanders. For his primary education he was sent to Sedan, where a Calvinist academy had been established and where he received his first medical instruction. He then went to Leiden, where he studied medicine (1633–1635) under Adolph Vorstius and Otto Heurnius. After visiting the universities of Wittenburg and Jena, he received his degree at Basel on 16 March 1637, defending a thesis on "animal movement and its disorders": *De animali motu ejusque laesionibus*. He already signed his name as Sylvius or, in order to avoid confusion with Jacques Sylvius (1478–1555), as dele Boë, Sylvius.

For a year and a half Sylvius practiced medicine in Hanau, but this seems not to have satisfied him. In any case, after a short study trip through France he returned to Leiden, where he hoped to obtain a post with the university. On 17 March 1638, he matriculated again and at his request received permission to give private lectures on anatomy, which met with great approval. He lectured on the *Anatomicae institutiones* of Caspar Bartholin but soon undertook to give anatomical demonstrations in the gallery of the botanic garden that were extended to physiological experiments. He was, in fact, one of the first to defend Harvey's new theory of the circulation of the blood and demonstrate it on dogs. His vigor was so great that Johannes Walaeus (1604–1649), a sharp critic, became a fervent supporter of Harvey, in return making experiments on the new theory. At these demonstrations Sylvius seems to have met Descartes, who as early as 1637 had accepted Harvey's theory of the circulation, although rejecting his concept of the action of the heart.

Because there seemed to be no prospect of an academic career at Leiden, Sylvius decided in the autumn of 1641 to move to Amsterdam, where he soon established a lucrative practice and earned the general esteem of his colleagues. He was appointed physician of the poor-relief board of the Walloon Church, and in 1657 he became a supervisor of the Amsterdam College of Physicians. His medical colleagues included Nicolaas Tulp, Paulus Barbette, and Hendrik van Roonhuyse; and his interest in chemistry brought him the friendship of Otto Sperling and J. R. Glauber. Despite the demands of his practice and professional commitments, he did not neglect scientific research work, performing postmortem examinations and devoting his spare time to chemical experiments.

In 1658, after extended negotiations, Sylvius was persuaded to accept appointment as professor of medicine at Leiden at the high salary of 1,800 guilders—almost twice the usual amount. On 17 September 1658 he delivered his inaugural oration, on the knowledge of man, *De hominis cognitione*. Devoting himself to his new task with great zeal, he proved to be an outstanding faculty member and his eloquence and gift for teaching attracted students from all parts of Europe. In bedside instruction, which he carried out in the old Caecilia Hospital, he showed himself to be an experienced clinician and a devoted teacher, who attracted many students from foreign countries. Although as a rule bedside teaching was limited to two days a week, Sylvius received permission to take his students daily to the hospital, where he performed the autopsies himself. His own ideas on several medical subjects were defended by his students in public disputations, and in 1669–1670 he was vice-chancellor of the university. The first volume of his main work, *Praxeos medicae idea nova*, was published the following year, but he did not live to see the second volume in print.

In 1647 Sylvius married Anna de Ligne, the daughter of a lawyer, who was thirteen years younger than he. One or two children born of this marriage died at a very early age, and in 1657 his wife died. In December 1666 he married a twenty-two-year-old woman, who died three years later; their only daughter died in 1670.

Sylvius' accomplishments were in anatomy and medical chemistry. Although there is some confusion with Jacobus Sylvius of Paris, a skilled anatomist who also worked on the brain, the Leiden Sylvius was responsible for the description of the *fissura Sylvii* and of the *arteria cerebri media Sylvii*, as well as of the fifth ventricle (pseudo ventricle, ventricle of the septum). The *aquaeductus Sylvii* was known to Galen and had been well described by Jacobus Sylvius. Moreover, Sylvius was the first to describe the tubercles in phthisis. In the history of medicine, however, Sylvius was the most brilliant representative of the iatrochemical school, founded by Paracelsus and continued by J. B. van Helmont, which reached its zenith when Sylvius defended the chemiatric conception from the Leiden chair. He was convinced that all physiological and pathological processes could be conceived perfectly in analogy to the processes and experiments observed in the chemical laboratory and could be explained by fermentation, effervescence, and putrefaction. Acid and alkali were considered as fundamental principles in the animal body.

In his therapeutics Sylvius preferred the new chemical medicines to the Galenic ones, using mercury, antimony, and zinc sulfate, among others. His rather speculative and extravagant theories included the belief that the pancreatic juice was acid and effervesced with the alkaline gall in the duodenum, and that ferments went to the heart, where he thought the blood effervesced. Considering the spleen to be the organ in which the blood is purified, he stressed its function to such an extent that he was called *patronus lienis*. In his last years Sylvius encountered public opposition from Anton Deusing, professor at Groningen, who—to Sylvius' great displeasure—was appointed at Leiden University but died before he could assume his duties.

Although Sylvius, with his exaggeration, may have caused much harm in the medical practice of his students, he can nevertheless be considered a promoter of scientific medical research. His enthusiasm inspired several gifted students to valuable anatomical and physiological work: Jan Swammerdam, Nicolaus Steno, and Regnier de Graaf. Sylvius' ideas on the pancreas induced De Graaf to attempt the experiments in which he obtained the pancreatic juice from a dog by means of a fistula.

BIBLIOGRAPHY

I. ORIGINAL WORKS. A more or less complete list of Sylvius' writings is in Baumann's biography (see below). They include *De hominis cognitione*, his inaugural oration (Leiden, 1658; Jena, 1674), reprinted with intro. and Dutch trans. in *Opuscula selecta neerlandicorum de arte medica,* VI (Amsterdam, 1927), 2–45; *Disputationum medicorum decas* (Leiden, 1670, 1674, 1676); and *Praxeos medicae idea nova*, 4 vols. (Leiden, 1671–1674). A collection of his works is *Opera medica* (Amsterdam, 1679).

II. SECONDARY LITERATURE. See L. Schacht, *Oratio funebris in obitum. Nobilissimi, Clarissimi, Expertissimi, D. Francisci de le Boe, Sylvii* (Leiden, 1673); Frank Baker. "The Two Sylviuses. An Historical Study," in *Johns Hopkins Hospital Bulletin*, **20** (1909), 329–339; E. D. Baumann, *François dele Boë Sylvius* (Leiden, 1949), with portrait; Lester S. King, *The Road to Medical Enlightenment 1650–1695* (London–New York, 1970), 93–112; and E. Ashworth Underwood, "Franciscus Sylvius and his Iatrochemical School," in *Endeavour*, **31** (1972), 73–76.

G. A. LINDEBOOM

SYLVIUS, JACOBUS. See **Dubois, Jacques.**

SYMMER, ROBERT (*b*. Galloway, Scotland [?], *ca.* 1707; *d*. London, England, 19 June 1763), *electricity.*

Nothing is known of Symmer's life before 1719, when he matriculated at the University of Edinburgh. Among fellow students who became his friends were Andrew Mitchell, Patrick Murdoch, and the poet James Thomson. All took an interest in contemporary natural philosophy, and all, though born and bred in Scotland, made their careers in England. Mitchell, Murdoch, and Symmer became fellows of the Royal Society of London (in 1735, 1745, and 1752, respectively), while their older contemporary Thomson, the best-known of the group, made his reputation with poems celebrating Newtonian science.

In 1735 Symmer took a belated M.A., perhaps to qualify as traveling companion to Francis Greville, Lord Brooke, later earl of Warwick; it was then a common practice for well-educated impecunious Scots to serve as tutors to influential English families, through whose connections they expected to obtain preferment. Symmer followed or conducted Brooke to London and there rejoined Mitchell, Murdoch (who had left Scotland by the tutorial route in 1729), and Thomson, all of whom had begun to prosper. Symmer found preferment in financial administration, for which he may have prepared under Abraham de Moivre, who gave private lessons in practical arithmetic and who, together with two of his students, signed Symmer's certificate for election to the Royal Society. Symmer eventually became head clerk of the office of the Treasurer of the Chamber, which paid the bills of the King's Household. When he lost this position upon a change of ministries in 1757, he preferred retirement to accepting an offer of the governorship of New Jersey.

Symmer devoted his leisure to wooing a young lady (Eleanora Ross of Balkailley, whom he married in 1760), to keeping Mitchell (then plenipotentiary to the court of Frederick the Great) current with English politics, and to experimenting with electricity. He took up this last study in 1758 on discovering that a pair of silk stockings, one black and one white, worn upon the same leg, would strongly attract one another when removed and separated. Symmer wore this peculiar double hosiery to keep mourning (and warm) during winter. Unfortunately for his reputation, he continued to use his stockings to generate the electricity for his experiments—and this, as he wrote to Mitchell, was "enough to disgust the Delicacy of more than one Philosopher."[1]

From the facts that the stockings attracted strongly when separated and weakly when joined, and that the attraction reappeared on every separation, Symmer concluded that the contrary electricities of Franklin derived from two distinct, opposed, positive principles, perhaps materialized as two essentially different, counterbalancing fluids. His conception therefore differed from Dufay's and Nollet's, in which the two electricities were not contrary, and Franklin's, in which one was a privation of the other. Symmer confirmed this insight by studying the condenser, which, when charged and insulated, bore an evident analogy to his superposed stockings. He found that he could charge a Leyden jar with either a black or a white stocking, and he argued that the circumstances of explosion—the shock or the perforation of paper placed in the circuit[2]—suggested that a real fluid sprang from each surface of a discharging condenser. Another, and more important, experiment was suggested by the analogy: since each stocking substitutes for one surface of a Leyden jar, a charged parallel-plate condenser with glass dielectric should cohere if cut longitudinally through its middle. On the other hand, two complete condensers, electrified in series, should not cohere, since at the interface the independent contrary electricities counterbalance. It was essential to Symmer's argument that they counterbalance but not destroy one another, as appears from the fact that the condensers can be exploded separately.

Symmer hoped that his two powers would not only start a revolution in electricity but also "prove to be the genuine Principle of the Newtonian Philosophy."[3] But the English electricians preferred to follow Franklin or Benjamin Wilson. Symmer therefore decided to try his chances abroad, and sent Mitchell copies of his papers for distribution in the Germanies. He himself took charge of converting Nollet, which proved a consequential move. Although Nollet would not accept Symmer's theory, he endorsed the experiments, which he deemed so many proofs of his own views, and which he improved by replacing the stockings with ribbons resting on a plate of glass. He also sent an account of the modified experiments to Giovanni Francesco Cigna in Turin, in the hope that he would use them to destroy the Franklinist theories of Nollet's rival Beccaria.

Cigna in turn improved the experiments, particularly by substituting an insulated lead plate for

Nollet's glass. He observed that if the ribbon were electrified and removed, and the plate discharged, it could be recharged as often as desired by grounding it when the ribbon was returned. Cigna here hit upon the principle of the electrophore. Volta, to whom it is usually ascribed, also came to it inspired indirectly by Symmer: When Cigna announced that Nollet's experiments did not decide between Franklin and Symmer, Beccaria bent theory and experiment to prove his Franklinist views; and in answering Beccaria, Volta invented the electrophore, a device as important as the Leyden jar for the development of electrical theory.

Symmer's expectations were almost realized. Before the end of the eighteenth century, his dualist theory had captured the Continent; and although his countrymen remained singlist, both sides agreed that no experiment known to them could settle whether electricity came in one power or two.

NOTES

1. Symmer to Mitchell, 7 Apr. 1761. British Museum, Add. MS 6839, fols. 220–221.
2. Franklin helped Symmer strike holes in quires of paper; they hoped that the contours of the punctures would show the direction of the discharge, but the results were ambiguous.
3. Symmer to Mitchell, 19 June 1760. British Museum, Add. MS 6839, fol. 183.

BIBLIOGRAPHY

I. ORIGINAL WORKS. Symmer's only published work is "New Experiments and Observations Concerning Electricity," in *Philosophical Transactions of the Royal Society*, **51**, no. 1 (1759), 340–389, comprising four parts read between Feb. and Dec. 1759. His other accessible work is a volume of MS letters addressed to Mitchell: British Museum, Add. MS 6839.

II. SECONDARY LITERATURE. Biographical information about Symmer has been inferred from the Mitchell letters. See J. L. Heilbron, "Robert Symmer, F.R.S. (*c.* 1707–1763) and the Two Electricities," in *Isis*, which includes an estimate of Symmer's work. See also Joseph Priestley, *The History and Present State of Electricity*, 3rd ed. (London, 1775), I, 308–333, and II, 41, 47; and I. B. Cohen, *Franklin and Newton* (Philadelphia, 1956), 543–546.

J. L. HEILBRON

SYNESIUS OF CYRENE (*b.* Cyrene, *ca.* 370; *d. ca.* 414), *astronomy, physics.*

Of Greek ancestry, Synesius was a man of wide interests. He was a gifted pupil of Hypatia, the beautiful Neoplatonic philosopher, mathematician, and astronomer. After a brief period as a soldier, Synesius settled in his homeland as a landed proprietor. His studies were interrupted in 397 by a three-year mission to Constantinople in order to present a gold crown to Arcadius, the new emperor, in a successful attempt to secure alleviation of taxes for his land. On his return to Cyrene, he married a Christian woman from Alexandria. His literary activity was constantly interrupted by his efforts to organize the military defense of his land against barbarian invaders. Although married and not a Christian, Synesius was, despite his great reluctance, pressured to accept baptism and consecration as bishop of Ptolemais (part of the Pentapolis of Cyrenaica) in 410. The remaining years of his life were busy and made difficult by the death of his sons and the barbarian invasions.

Synesius is well-known for his essays, letters, and hymns, which have been carefully studied; but his scientific attainments are also worthy of respect. He was deeply interested in the theoretical aspects of science, but he also sought to make practical applications of his knowledge. Once, apparently while seriously ill, he requested Hypatia to have Alexandrian metalworkers make him a brass "hydroscope" according to his directions. He needed the instrument in order that he might "ascertain the weight of liquids." The exact nature of the notched cylindrical tube is uncertain. Synesius apparently was not the inventor of this hydrometer or areometer; he merely desired an instrument similar to one he had seen.

Synesius apparently invented a "perfected" astrolabe, an instrument at which, he said, Hipparchus had hinted much earlier. He presented this silver astrolabe (really a planisphere) to a friend in Constantinople. The letter accompanying the gift referred to a full description of the astrolabe and its uses, but that work has been lost. The instrument resulted, he explained, from his having "carefully studied the reduction of the sphere to a plane figure." It must, therefore, have been a representation of the heavenly bodies on a flat surface.

Synesius has been considered by some to be the author of an important Greek alchemic work, an obscure and difficult commentary on Pseudo-Democritus. The tract is in the form of a dialogue with Dioskoros, a priest of Serapis at Alexandria. Other than the similar name, there is nothing to indicate that the bishop of Ptolemais composed this work on alchemy, and there is internal evidence that the work was written before he was born.

BIBLIOGRAPHY

I. ORIGINAL WORKS. The first complete ed. of the writings of Synesius was by Dionysius Petavius (Paris, 1612). Most readily available is the collection in J. P. Migne, *Patrologia Graeca*, LXVI (Paris, 1864), 1021–1756. The *Epistolae* were edited by R. Hercher in his *Epistolographi Graeci* (Paris, 1873). A French trans. of the works of Synesius was published by H. Druon (Paris, 1878). Augustine Fitzgerald published English translations in *The Letters of Synesius of Cyrene* (London, 1926) and *The Essays and Hymns of Synesius of Cyrene* (London, 1930).

II. SECONDARY LITERATURE. Detailed bibliographies of special studies on Synesius are in W. S. Crawford, *Synesius the Hellene* (London, 1901); in José C. Pando, *The Life and Times of Synesius of Cyrene as Revealed in His Works* (Washington, D.C., 1940); and in Richard Volkmann, *Synesius von Cyrene* (Berlin, 1869).

The scientific achievements of Synesius are discussed in B. Kolbe, *Der Bishof Synesius von Cyrene als Physiker und Astronom* (Berlin, 1850); and in Paul Tannery, *Recherches sur l'histoire de l'astronomie ancienne* (Paris, 1893), 50–53.

On the supposed authorship of the alchemic tract, see *Bibliotheca chemica*, John Ferguson, ed., II (Glasgow, 1954), 421–422; and especially Edmund O. Lippmann, *Entstehung und Ausbreitung der Alchemie* (Berlin, 1919), 96–98.

KARL H. DANNENFELDT

SZEBELLÉDY, LÁSZLÓ (*b*. Rétság, Hungary, 20 April 1901; *d*. Budapest, Hungary, 23 January 1944), *analytical chemistry*.

Szebellédy was the son of Ferenc Szebellédy, a pharmacist, and Maria Pohl. He earned a degree in pharmacy at the University of Budapest in 1923, but instead of becoming a pharmacist he turned to scientific research. In 1925 he was named assistant to Lajos Winkler at the Inorganic Chemistry Institute of the University of Budapest. An outstanding analytical chemist, Winkler had become famous for his methods of determining the amount of oxygen dissolved in water (1888) and the iodine-bromine number of fats for his work in precision gravimetry, and for his books. Szebellédy collaborated in Winkler's analytical research. His first independent publications (1929) dealt with the classical methods of analysis. He later spent considerable time away from Budapest working with foreign scientists, notably Wilhelm Böttger at Leipzig and William Treadwell at Zurich. In 1934 Szebellédy qualified as a lecturer in analytical chemistry at the University of Budapest, and in 1939 he was appointed professor of inorganic and analytical chemistry. The extensive program of research that he subsequently undertook was prematurely halted by his death from cancer.

Among the topics that Szebellédy investigated was catalytic ultramicroreactions, introduced into analytical chemistry by I. M. Kolthoff and E. B. Sandell for cases in which it is possible to obtain an accurately measurable endpoint (1937). Most catalytic color reactions, however, proceed continuously. Working with Miklós Ajtai, Szebellédy devised the analytical application for this type of reaction (1939). With the assistance of his young co-worker Zoltán Somogyi (whose death during an air raid preceded his own), Szebellédy invented the coulometric titration method (1938), which is widely used in analytical chemistry.

BIBLIOGRAPHY

Szebellédy's paper on coulometric titration is "Die coulometrische Analyse als Präzisionsmethode," in *Zeitschrift für analytische Chemie*, **112** (1938), 313–336, written with Z. Somogyi; and that on catalytic analysis is "Die quantitative Bestimmung von Vanadin mittels aktivierter Katalyse," in *Mikrochemie*, **26** (1939), 87–94, written with M. Ajtai.

A secondary source is F. Szabadváry, *History of Analytical Chemistry* (Oxford, 1966), 190, 316, 317.

F. SZABADVÁRY

SZILARD, LEO (*b*. Budapest, Hungary, 11 February 1898; *d*. La Jolla, California, 30 May 1964), *physics, biology*.

Szilard, one of the most profoundly original minds of this century, contributed significantly to statistical mechanics, nuclear physics, nuclear engineering, genetics, molecular biology, and political science.

The oldest of three children of a successful Jewish architect-engineer, he was a sickly child and received much of his early education at home, from his mother. His electrical engineering studies were interrupted by World War I; drafted into the Austro-Hungarian army, he was still in officers' school at the end of the war. In 1920 he went to Berlin to continue his studies at the Technische Hochschule. The attraction of physics proved too great, however, and he soon transferred to the University of Berlin, where he received the doctorate in 1922. His dissertation, written under the direction of Max von Laue, showed that the second law of thermodynamics not only covers the mean values of thermodynamic quantities but also determines the form of the law governing the fluctuations around the mean values. The continuation of this work led to his famous paper of 1929, which established the connection between entropy and information, and foreshadowed modern cybernetic theory.

During this period in Berlin, as a research worker at the Kaiser Wilhelm Institute and then as *Privatdozent* at the university, Szilard undertook experimental work in X-ray crystallography with Herman Mark. He also began to patent his long series of pioneering discoveries, including devices anticipating most modern nuclear particle accelerators. With Albert Einstein he patented an electromagnetic pump for liquid refrigerants that now serves as the basis for the circulation of liquid metal coolants in nuclear reactors.

Hitler's assumption of power caused Szilard to leave Germany for England in 1933. There he conceived the idea that it might be possible to achieve a nuclear chain reaction. Szilard's search for an appropriate nuclear reaction (he early realized that the neutron was the key), while a guest at St. Bartholomew's Hospital in 1934 and at the Clarendon Laboratory, Oxford, after 1935, led to the establishment of the Szilard-Chalmers reaction and the discovery of the γ-ray-induced emission of neutrons from beryllium. It was only after he came to the United States, in 1938, that he learned of the discovery of fission in Germany by Hahn and Strassmann.

Szilard instantly recognized—as did nuclear physicists in other countries—that fission would be the key to the release of nuclear energy, and he immediately undertook experiments at Columbia University to demonstrate the release of neutrons in the fission process and to measure their number. With Fermi he organized the research there that eventually led to the first controlled nuclear chain reaction, on 2 December 1942, at Chicago. Probably more than any other individual, Szilard was responsible for the establishment of the Manhattan Project; it was he who arranged for the letter from Einstein to President Roosevelt that brought it about. His contributions to the success of its plutonium production branch, both in physics and in engineering, were manifold, especially in the earliest stages. The basic patent for the nuclear fission reactor was awarded jointly to Fermi and Szilard in 1945, but Szilard never realized any financial profit from it.

The last months of the war found Szilard, with James Frank and other Manhattan Project scientists, engaged in a futile effort to convince President Truman to use the first atomic bomb in a nonlethal demonstration to the Japanese of its destructive power.

After the war Szilard turned to biology. With Aaron Novick he invented and constructed a device for studying growing bacteria and viruses in a stationary state by means of a continuous-flow device, called the chemostat, in which the rate of bacteria growth can be changed by altering the concentration of one of the controlling growth factors. He used it for a number of years in fundamental studies of bacterial mutations and various biochemical mechanisms.

In the late 1950's Szilard became increasingly interested in theoretical problems of biology; his 1959 paper "On the Nature of the Aging Process" still stimulates research and controversy. His last paper, "On Memory and Recall," was published posthumously.

Throughout his life Szilard had a profoundly developed social consciousness. On fleeing Nazi Germany to England, one of his first acts was to inspire the organization of the Academic Assistance Council, to help find positions in other countries for refugee scientists. He was one of the leaders of the successful postwar Congressional lobbying effort by Manhattan Project alumni for a bill establishing civilian control over peaceful development of nuclear energy. Szilard was one of the instigators and active early participants in the international Pugwash Conferences on Science and World Affairs, and he wrote extensively on questions of nuclear arms control and the prevention of war. In 1962 he founded the Council for a Livable World, a Washington lobby on nuclear arms control and foreign policy issues.

Szilard was a fellow of the American Physical Society, the American Academy of Arts and Sciences, and the National Academy of Sciences. He received the Einstein Award in 1958 and the Atoms for Peace Award in 1959.

BIBLIOGRAPHY

Many of Szilard's works are being brought together in *Collected Works of Leo Szilard: Scientific Papers*, Bernard T. Feld and Gertrude W. Szilard, eds. (Cambridge, Mass., 1972–　). His writings include "Über die Ausdehnung der phänomenologischen Thermodynamik auf die Schwankungserscheinungen," in *Zeitschrift für Physik*, **32** (1925), 753–788, his diss.; "Über die Entropieverminderung in einem thermodynamischen System bei Eingriffen intelligenter Wesen," *ibid.*, **53** (1929),

840–856, translated as "On the Decrease of Entropy in a Thermodynamic System by the Intervention of Intelligent Beings," in *Behavioral Science*, **9** (1964), 301–310; "Chemical Separation of the Radioactive Element From Its Bombarded Isotope in the Fermi Effect," in *Nature*, **134** (1934), 462, written with T. A. Chalmers; *The Voice of the Dolphins, and Other Stories* (New York, 1961); a report to the secretary of war (June 1945), written with James Frank, Donald J. Hughes, J. J. Nickson, Eugene Rabinowitch, and Joyce C. Stearns, and a petition to the president of the United States (17 July 1945), in Bulletin of the Atomic Scientists, *The Atomic Age*, M. Grodzins and E. Rabinowitch, eds. (New York, 1963); and reminiscences, in Donald Fleming and Bernard Bailyn, eds., *The Intellectual Migration: Europe and America, 1930–1960* (Cambridge, Mass., 1969).

On his life and work, see the notice by Eugene P. Wigner, in *Biographical Memoirs. National Academy of Sciences*, **40** (1969), 337–341.

BERNARD T. FELD

SZILY, PÁL (*b*. Budapest, Hungary, 16 May 1878; *d*. Mosonmagyaróvár, Hungary, 18 August 1945), *chemistry*.

Szily came from a family of physicians. His father, Adolf Szily, was a doctor and director of a Budapest hospital, and his older brother became a professor of ophthalmology at the University of Budapest. Szily studied medicine at the University of Budapest and after obtaining his medical degree became an assistant at the Institute of Physiology there, where he carried out his fundamental research on the colorimetric determination of hydrogen ion concentration.

Since the time of Robert Boyle various plant juices had been used to determine whether a liquid was acidic or basic. When synthetic substances were introduced as indicators, it was observed that they did not change color at the same level of acidity as the natural juice indicators. On the basis of Arrhenius' theory of ionization (1887), Wilhelm Ostwald introduced the concept of the dissociation constant with a view to ascertaining the strengths of acids and bases as a function of, respectively, hydrogen ion and hydroxyl ion concentration. He also determined the value of the dissociation constant of water. In 1893 Max Le Blanc invented the hydrogen electrode, which made it possible to measure the hydrogen ion concentration electrochemically.

It appears, however, that for a long time chemists were unable to recognize the significance of these developments. They did not comprehend the difference between titrimetrically determinable

amounts of acid and the strengths of acids, for most of them dealt only with the former. Physiologists and biologists, however, were more concerned with the strengths of acids, since small changes in acidity play a great role in various life processes. They were therefore the ones to elucidate the concepts underlying the treatment of these questions and to develop appropriate techniques of measurement. The first to do so was Szily, who in 1903 published "Indikátorak alkalmazásáól állati folyadékok vegyhatásának meghatározására" ("Application of Indicators in the Determination of the Reaction of Animal Fluids"), in *Orvosi Hetilap*, **45** (1903), 509–518. After establishing that the reaction of animal fluids—the hydrogen ion concentration of blood serum, for example—cannot be determined titrimetrically, Szily hit upon the idea of using the indicators for this purpose, since each indicator changes color at a specific hydroxyl ion concentration, regardless of the nature of the base. By using various indicators he was able to establish a scale for estimating acidity. In addition, employing seven different indicators, he devised a scale for making an approximate determination of the acidity of the blood serum. In the course of this research he also determined the resistance of blood serum to the effects of acids and bases (its buffer property).

In 1903 Szily lectured on his results before the Physiology Society of Berlin; and Hans Friedenthal, a lecturer at the University of Berlin, arranged for Szily to continue his research there. Friedenthal began investigations in the same area and perfected Szily's method by using a larger number of indicators and by employing standard (buffer) solutions of precisely known hydrogen ion concentration. In 1904 he reported that he had been unsuccessful in his attempt to produce these solutions by successive dilution of acidic or basic solutions. Szily suggested that he prepare stable solutions of reliable hydrogen ion concentrations by mixing primary and secondary phosphates in different proportions. Szily was, consequently, the inventor of artificial buffer solutions. Research in this area was extended by S. P. L. Sørensen, who introduced the concept of pH in 1909.

In 1905 Szily transferred to the surgery clinic of the University of Budapest, and in 1909 he became director of the serological and bacteriological laboratory of the Budapest Jewish Hospital. Henceforth his research was of a purely medical nature. He investigated the therapeutic effects of Salvarsan and communicated his findings to Paul Ehrlich, who followed with interest the results of the introduction of the drug into Hungary. During World War I, Szily directed an army epidemiological unit and was concerned with typhus therapy and the treatment of equine influenza through vaccination. After the war he published studies on protein therapy, but his scientific activity was steadily eclipsed by the demands of his private practice in Budapest. His practice seems to have disappointed him, for in 1928 he went to the small city of Mosonmagyaróvár as urologist with the state health insurance administration.

After the Nazi assumption of power in Hungary, Szily, a diabetic, was taken in 1944 to the concentration camp at Győr. During the last days of the war in Hungary he escaped deportation to Germany through his professional connections. He returned to his post in Mosonmagyaróvár but, unable to secure sufficient medication, died soon afterward.

BIBLIOGRAPHY

I. ORIGINAL WORKS. Szily was author of thirty chemical and medical publications, including the one mentioned above. His medical publications appeared generally in *Wiener Medizinische Wochenschrift* and *Berliner Klinische Wochenschrift* in 1910–1919.

II. SECONDARY LITERATURE. See F. Szabadváry, "Development of the pH concept. A historical survey," in *Journal of Chemical Education*, **41** (1964), 105–107; *History of Analytical Chemistry* (Oxford, 1966), 363, 376–377; and "Szily Pál (1878–1945)," in *Orvossörténeti közlemények* (*Communicationes de historia artis medicinae*), **38** (1966), 121–130, in Hungarian.

Hans Friedenthal's report is "Die Bestimmung der Reaktion einer Flüssigkeit mit Hilfe von Indikatoren," in *Zeitschrift für Elektrochemie*, **10** (1904), 113–119.

F. SZABADVÁRY

AL-ṬABARĪ, ABUʾL-ḤASAN AḤMAD IBN MU-ḤAMMAD (*b.* Ṭabaristān, Persia, first quarter of the tenth century; *d.* Ṭabaristān, fourth quarter of the tenth century), *philosophy, natural science, medicine.*

Very little is known about al-Ṭabarī's parents or early life. Like his contemporary al-Majūsī (*d.* 994), he studied under the physician Abū Māhir Mūsā ibn Sayyār. After acquiring a good reputation as a physician, al-Ṭabarī became court physician to the Buwayhid king Rukn al-Dawla (reigned 932–976) and his vizier, the literary scholar Abu'l-Faḍl Muḥammad al-Khaṭīb ibn al-ʿAmīd

(*d.* 971). This was a period of great cultural and scientific productivity in Persia and Iraq under the Abbasid caliphate. Several medical authors won wide recognition, not least among them al-Ṭabarī—as is evident from the numerous extant copies of his only known literary contribution, *al-Muʿālajāt al-Buqrāṭiyya,* which consists of ten treatises on Hippocratic medical treatment.

The text of al-Ṭabarī's work sheds much light on his life. It shows that he was a Muslim, deeply influenced by Neoplatonism and Aristotelianism, who respected other religions in the region: Zoroastrianism, Judaism, and Christianity. Unlike many of his contemporaries, al-Ṭabarī excluded much of the religious and Koranic phraseology and jargon used in similar works. He dealt objectively and open-mindedly with such topics as generation and corruption, life and death, marriage and family, vision and thought, pain and pleasure, matter and soul, time and space, temporal and eternal punishment and reward, and Godhood and resurrection. His approach is entirely free from religious bias and theological limitations, and he seems to have been well-acquainted with the writings of Greek philosophers and natural scientists. His treatment of diseases and their medical therapy also bears witness to his appreciation of and indebtedness to the Hippocratic and Galenic tradition.

Al-Ṭabarī nevertheless contributed original ideas and concepts of historical interest. His medical ingenuity led him to become the first practitioner to describe and recommend effective treatment for the itch mite *Sarcoptes scabiei,* the cause of scabies. His theories on health, deontology, medical therapy, and psychotherapy showed his ability to think independently and to make personal observations unhampered by traditional doctrines.

BIBLIOGRAPHY

Al-Ṭabarī's only known work, *al-Muʿālajāt al-Buqrāṭiyya,* which exists in several copies, some incomplete, suggests his encyclopedic approach: treatise 1—definitions and interpretations of natural sciences and phenomena, professional deontology, social behavior, metaphysics, and classification of diseases; treatise 2—on skin diseases of the head and face, and their treatment; treatise 3—on diseases of the head; treatise 4—on the anatomy and physiology of the eye, and its diseases; treatise 5—on diseases of the nose and ear; treatise 6—on the diseases of the mouth, teeth, tongue, uvula, larynx, pharynx, and neck (trachea); treatise 7—on skin diseases of the body; treatise 8—on diseases of the chest, lungs, bronchi, all other members of the respiratory system, diaphragm, and the heart, and their treatment; treatise 9 — on the anatomy and physiology, and diseases of the stomach, and their diagnosis and treatment; treatise 10 — on the anatomy and physiology of the liver, spleen, and intestines, their diseases, and the nutritional values of these same organs such as liver, kidneys, brain, and viscera.

Several catalogs of library MSS and bibliographies of literary contributions of this Islamic period list copies of al-Ṭabarī's *al-Muᶜ-ālajāt*. They include Joseph Aumer, *Die arabischen Handschriften der K. Hof- und Staatsbibliothek* (Munich, 1866), 357; and Carl Brockelmann, *Geschichte der arabischen Literatur*, I (Leiden, 1943), 272, and *Supplement,* I (Leiden, 1937), 422; and "Firdaus'l-Hikmat of Ali b. Rabban al-Ṭabari," in *Zeitschrift Semitisch*, 8 (1932), 270–288.

Ibn Abī Usaybiᶜa in his *ᶜUyūn al-anbāʾ*, 2 vols. (Cairo, 1882), I, 321, was probably the first of the few Muslim biographers to mention al-Ṭabarī. Brief nineteenth-century biographies of al-Ṭabarī based on Ibn Abī Usaybiᶜa are L. Leclerc, *Histoire de la médecine arabe*, I (Paris, 1876), 358; and F. Wüstenfeld, *Geschichte der arabischer Ärzte und Naturforscher* (Göttingen, 1840), 56.

Julius Hirschberg, in *Geschichte der Augenheilkunde bei den Arabern* (Leipzig–Berlin, 1905), 107–108, brought out the importance of al-Ṭabarī's work, especially on eye diseases and treatment. Later, Mohammed Rihab, in "Der arabische Arzt aṭ-Ṭabarī," in *Archiv für Geschichte der Medizin*, 19 (1927), 123–168, and 20 (1928), 27–81, stressed the importance of al-Ṭabarī's contribution to medicine, especially his treatise on skin diseases and his discovery of the itch mite. For the latter see also R. Friedman, "The Story of Scabies; at-Tabari, Discoverer of the *Acarus scabiei*," in *Medical Life*, 45 (1936), 163–176.

For further information consult S. Hamarneh, *Catalogue of Arabic Manuscripts on Medicine and Pharmacy at the British Museum* (Cairo, 1975), no. 70; and G. Sarton, *Introduction to the History of Science*, I, 677, and II, 233.

SAMI HAMARNEH

AL-ṬABARĪ, ABU'L-ḤASAN ᶜALĪ IBN SAHL RABBĀN (*b.* Marw, Persia, *ca.* 808; *d.* Baghdad, *ca.* 861), *medicine, natural science, theology, government.*

Ṭabarī was born into a prominent and religious Syriac Christian family living in Marw in the region of Khurāsān (near present-day Tehran). His father, Sahl, was a highly placed government official who was learned in medicine, philosophy, theology, and astrology. His scholarship and his religious and philanthropic activities won him the prestigious Syriac title of Rabbān, meaning

"teacher." Sahl took a special interest in the upbringing of his son ᶜAlī. Besides giving him a good education, Sahl taught him religion, medicine, and philosophy. The fatherly exhortations and advice led young ᶜAlī to an appreciation and love of learning. When ᶜAlī was ten years old he was taken to Ṭabaristān by his father, who was probably sent there on an assignment for the state. Because of their residence in Ṭabaristān, ᶜAlī became known as al-Ṭabarī. There, he devoted his time to the study of medicine, religion, philosophy, and the natural sciences. Because Ṭabarī excelled in learning and as a counselor to the ruler of Ṭabaristān, he was summoned in 840 to the Abbasid capital to serve in the palace of Caliph al-Muᶜtasim and his successor, Caliph al-Wāthiq (842–847). Under al-Mutawakkil (847–861), Ṭabarī's position was raised to that of companion to the caliph, who "urged and encouraged" him not only to embrace and confess openly his adherence to Islam, but also to defend his new faith against other religions.

In 850 and 855 Ṭabarī wrote his two best-known books, *Firdaws al-ḥikma* and *Al-Dīn wā'l-dawla*, respectively. He dedicated both to his patron and benefactor, al-Mutawakkil.

In *Firdaws* ("Paradise of Wisdom"), Ṭabarī compiled, extracted, and digested information on all aspects of medicine from Greek, Syriac, and Indian medical compendiums. He also added his own observations and interpretations throughout his medical encyclopedia, the first of its kind in Arabic with such scope and comprehensiveness. In addition to medicine, embryology, and surgery, Ṭabarī wrote on toxicology, psychotherapy, cosmogony, and astrology. Ṭabarī also made reference to the outstanding contributions of two of his contemporaries, Yūḥannā ibn Māsawayh (*d.* 857), a pioneer Arabic medical educator and author, and Ḥunayn ibn Isḥāq, the indefatigable translator of medical texts from Syriac and Greek, and one of the foremost Arabic scholars of his time.

Ṭabarī's polemics in *Al-Dīn wā'l-dawla* ("On Religion and Government") shed considerable light on his life, religious beliefs, and philosophy, and reflect on religio-philosophical thought in ninth-century Islam.

It has been erroneously reported that Ṭabarī was a Jew and that he taught the physician Abū Bakr Muḥammad al-Rāzī (865–925). In fact, Ṭabarī must have died in Baghdad before Rāzī was even born. But Rāzī and other medical educators and authors in medieval Islam did benefit from Ṭabarī's works, ingenuity, and ideas, since, through his foresight, industry, and genius, Ṭabarī contributed materially to ninth-century Arabic learning and scholarship.

BIBLIOGRAPHY

I. ORIGINAL WORKS. Muhammad Z. Siddiqi of India, a former student of the late Edward G. Browne, edited Ṭabarī's *Firdaws al-ḥikma* ("Paradise of Wisdom"), with a useful introduction (Berlin, 1928). Since then, other manuscripts—besides those cited in the Berlin and British Museums—have been discovered. Ṭabarī's *Al-Dīn wā'l-dawla* ("On Religion and Government") was edited with an English translation by Alphonse Mingana from a unique manuscript at the John Rylands Library (Cairo–Manchester, 1922–1923).

An Arabic manuscript entitled *Ḥifẓ al-ṣiḥḥa* ("On the Preservation of Health") at the Bodleian Library, Oxford, catalog 1:578, is attributed to Ṭabarī. Other works listed for him by Ibn al-Nadīm in *Al-Fihrist* (completed 987) (Cairo, 1929), p. 426, are *Tuḥfat al-mulūk* and *Kunnāsh al-ḥaḍra* (two medical compendiums dedicated to the Muslim Caliph al-Mutawakkil) and a book on the benefits of solid and liquid diets and drugs.

II. SECONDARY LITERATURE. Several Arabic medieval texts besides *Al-Fihrist* mention Ṭabarī and his works: Ẓahīr al-Dīn 'Alī al-Bayhaqī (*d.* 1170), *Tārīkh ḥukamā' al-Islām*, Muhammad Kurd 'Alī edition (Damascus, 1946), 22–23; Yāqūt al-Ḥamawī (*d.* 1229), *Dictionary of Learned Men*, D. S. Margoliouth, ed., VI (London, 1931), 429, 460. The report by Jamāl al-Dīn al-Qifṭī (*d.* 1248), *Tārīkh al-ḥukamā'*, Julius Lippert, ed. (Leipzig, 1903), 187, errs concerning his religious background and his relation to Rāzī (865–925). Ibn 'Abī Uṣaybi'a (*d.* 1270), *'Uyūn al-anbā'* (Cairo, 1882), 309, quotes both Ibn al-Nadīm and Ibn al-Qifṭī concerning Ṭabarī.

Ṭabarī was mentioned by F. Wüstenfeld, *Geschichte der arabischen Aerzte und Naturforscher* (Göttingen, 1840), 21, and Lucien Leclerc, *Historie de la médecine arabe*, I (Paris, 1876), 292–293. Much attention was paid to Ṭabarī, especially in relation to *Firdaws*, by Edward G. Browne, *Arabian Medicine* (Cambridge, 1921), 37–44; Max Meyerhof, "'Alī at-Ṭabarī's Paradise of Wisdom," in *Isis*, **16** (1931), 6–54; and "Ali-at-Ṭabarī, ein persischen Arzt," in *Zeitschrift der deutschen Morgenländischen Gesellschaft*, **10** (1931), 38–68; J. M. Faddegon, "Notice critique sur le Firdausu'l-hikmat de 'Ali b. Rabban al-Tabari," in *Journal Asiatique*, **218** (1931), 327–352; Alfred Siggel, "Gynäkologie, Embryologie und Frauenhygiene aus dem Paradies der Weisheit über die Ali b. Sahl at-Ṭabari," in *Quellen Studien Geschichte der Naturwissenschafte Medizin*, **8**, pts. 1–2 (1941), 216–272; *Die indischen Bücker aus dem Paradies der Weisheit über die Medizin des Ali b. Sahl Rabban al-Ṭabari* (Wiesbaden, 1950); and *Die propädeutischen Kapitel aus dem Paradies der Weisheit über die Medizin des Ali b. Sahl Rabban al-Ṭabari* (Wiesbaden, 1953).

See also Carl Brockelmann, *Geschichte der arabischen Literatur* (Leiden, 1943), 265; and supp. I (1937), 414–415; Fuat Sezgin, *Geschichte des arabischen Schrifttums*, III (Leiden, 1970), 236–244, with a valuable bibliography; George Sarton, *Introduction to the History of Science*, I (Baltimore, 1927), 546–549, 574; and S. Hamarneh, "Contributions of Ali al-Tabari to Ninth-Century Arabic Culture," in *Folia Orientalia* (Cracow), **12** (1970), 91–101. On particular subjects, see D. V. Subba Reddy, "Indian Medicine in Firdausu'l-hikmat of 'Alī b. Rabban al-Ṭabari," in *Bulletin of the Department of the History of Medicine* (Hyderabad), **1** (1963), 26–49; W. Schmucker, *Die pflanzliche und mineralische Materia Medica in Firdaus al-hikma des Ṭabari* (Bonn, 1969); and S. Hamarneh, *Index of Mss. on Medicine and Pharmacy in the Ẓahiriyyah Library* (Damascus, 1969), 77–82 (Arabic text).

SAMI HAMARNEH

TABOR, JOHN (*b.* Faccombe, Hampshire, England, 1667), *medicine*.

Son of the rector of Faccombe parish, Tabor graduated B.A. at Merton College, Oxford, in 1687 and received his medical degree there on 20 March 1694. An adherent of the English iatro-mathematical school that developed under the aegis of Newtonianism during the early eighteenth century, Tabor attempted to incorporate medical animism into a mathematical framework in his *Exercitationes medicae* (1724). He accepted the rational soul or anima, which presumably induced and controlled the activities of living organisms, as the fundamental cause of physiological processes. Because the anima manifested itself through the movements of bodily parts and fluids according to the established laws of physical motion, however, Tabor held that the primary task of medical theory was the calculation of the size, shape, and movement of organic structures.

Familiar with a wide range of classical and contemporary authors, Tabor owed most to the work of Borelli and employed the computations and mechanical models characteristic of his writings. He devoted considerable attention to detailed, albeit inconclusive, formulations of the shape and elasticity of muscle fibers and offered a comprehensive, if unoriginal, account of the heart's structure and function. While maintaining that the anima was the primary force preserving organic systems against decay, Tabor followed John Freind and James and John Keill in admitting physical attraction as an independent force capable of affecting physiological activity. He construed disease, for example, as the process by which the anima, through muscular spasms, fevers, and similar means, restored bodily equilibrium by counteracting the attractive force of foreign and deleterious particles. Tabor's work had little influence and fell into obscurity with the general demise of iatromathematics in the second half of the eighteenth century. His use of animist hypotheses was intended to emphasize the insufficiency of purely mechanical explanations and reflected Tabor's concern to reconcile reductionist and vitalist traditions in physiological theory.

BIBLIOGRAPHY

I. ORIGINAL WORKS. Tabor's major work was *Exercitationes medicae quae tam morborum quam symptomatum in plerisque morbis rationem illustrant* (London, 1724). An article of antiquarian interest appeared as "An Accurate Account of a Tessellated Pavement, Bath, and Other Roman Antiquities, Lately Discovered Near East Bourne in Sussex," in *Philosophical Transactions of the Royal Society,* **30** (1717–1719), 549–563, 783–802.

II. SECONDARY LITERATURE. The only source for biographical data on Tabor is Joseph Foster, *Alumni Oxonienses,* IV (Oxford, 1892), 1453. A useful account of his medical ideas is Kurt Spengel, *Versuch einer pragmatischen Geschichte der Arzneikunde,* 3rd ed., V (Halle, 1828), 233–234, 349. J. R. Partington, *A History of Chemistry,* II (London, 1961), 623–625, discusses Tabor's opposition to John Mayow's theory of nitro-aerial particles.

MARTIN FICHMAN

TACCHINI, PIETRO (*b.* Modena, Italy, 21 March 1838; *d.* Spilamberto, Modena province, Italy, 24 March 1905), *astronomy, meteorology.*

After obtaining his degree in engineering at the *archiginnasio* of Modena, Tacchini studied astronomy at the observatory of Padua under Giovanni Santini and Virgilio Trettenero. In 1859, following his appointment as deputy director of the observatory of Modena, he established connections with Secchi and Schiaparelli in order to study problems related to the new astrophysics. In 1861 he maintained an active scientific correspondence with Secchi.

Appointed adjunct astronomer at the Palermo observatory in 1863, Tacchini began observations and research on solar physics using the spectroscope. The daily observations of the phenomena on the surface of the sun convinced him of the need for a well-planned national program in order to follow its various phases. He called this effort the study of "solar meteorology," at a time when scientists were just becoming aware that the phenomena of terrestrial meteorology depend upon solar phenomena. The exceptional maximum of the eleven-year sunspot cycle that occurred in 1870, accompanied by the appearance on earth of numerous polar auroras, intensified his observations of the sun at Rome and Palermo.

Secchi and Tacchini decided to found a society to coordinate the work of various Italian observatories and, using standard criteria, to observe solar activity and to promote research on solar physics. Since the spectroscope was virtually the only instrument used for this purpose, the society, founded in 1871, was called the Society of Italian Spectroscopists; its *Memorie,* the first periodical on astrophysics, is still being published by the Italian Astronomical Society, which replaced it in 1920.

A skilled administrator, Tacchini prepared and directed many astronomical expeditions. For the transit of Venus in 1874, Tacchini accompanied his colleagues Alessandro Dorna and Antonio Abetti to Muddapur in Bengal. The principal aim of this expedition was to make extremely accurate observations of the contacts of the disk of Venus with the limb of the sun by using the spectroscope. He traveled to various parts of the world to observe seven total solar eclipses, concentrating his observations especially on the corona and the prominences.

During the eclipse of 1883, which he observed in the Caroline Islands, Tacchini noted white prominences, in contrast with the vivid red ones of hydrogen. Photographic techniques later made it clear that these prominences, which were more extensive than those of hydrogen, are produced by calcium.

In 1879 Tacchini succeeded Secchi as director of the observatory at the Collegio Romano and as director of the Central Meteorological Office, from which he established a vast network of meteorological stations throughout Italy. Fully aware of the advantages of a clear and calm sky, he and Riccò promoted the construction of an observatory on Mt. Etna, at 9,650 feet, for astrophysical and geophysical research. Also with Riccò he enlisted the collaboration of the observatory in Catania for the international project of preparing a chart and a photographic catalog of the sky.

Tacchini was a foreign member of the Royal Astronomical Society and of the Royal Society, and was awarded the Rumford gold medal and the Janssen Prize.

BIBLIOGRAPHY

I. Original Works. Most of Tacchini's numerous publications are in the *Memorie della Società degli spettroscopisti italiani*, in the *Atti dell'Accademia dei Lincei*, and in the publications of the Central Meteorological Office. They include *Il passaggio di Venere sul sole dell' 8 e 9 dicembre 1874 osservato a Muddapur del Bengala* (Palermo, 1875); and "Eclissi totali di sole del dicembre 1870, del maggio 1882, dell'agosto 1886 e 1887," in *Relazioni e note* (Rome, 1888). His correspondence with Secchi is preserved in Italy.

II. Secondary Literature. See G. Abetti, "Celebrazione del primo centenario della nascità di Pietro Tacchini," in *Coelum*, 9 (1939), 81; E. Millosevich, "Necrologia di P. Tacchini," in *Astronomische Nachrichten*, 168, no. 4009 (1905); and A. Riccò, "Necrologia di P. Tacchini," in *Memorie della Società degli spettroscopisti italiani*, 34 (1905), 85.

G. Abetti

TACCOLA, MARIANO DI JACOMO (*b.* Siena, Italy, 4 February 1381; *d.* Siena, 1453/1458), *mechanics.*

The nickname Taccola, meaning "crow" and referring to a talent for woodcarving, was inherited by Mariano from his father, a winegrower. Taccola's first profession was that of sculptor, and he contributed to the carving of the choir of Siena cathedral in 1408. He was active in civic life from at least 1413 and partially qualified as a notary in 1417. From 1424 to 1431 he was chamberlain of the Casa della Sapienza, a residence for scholars at Siena.

By 1427 Taccola seems to have become intensely interested in mechanical technology, a field to which he devoted most of his time for the rest of his life. His earliest dated sketches of machines are from 1427, when he also conducted practical tests of four of his inventions. The trials included a project for erecting a bridge over the Tiber at Rome and one for harborworks at Genoa.

The visit to Siena in 1432–1433 of the future emperor Sigismund brought Taccola a patron for his mechanical inventions. He was appointed one of Sigismund's *nobiles familiares* (1432) and dedicated an elegant book of drawings of machines and exotic animals to him. In this manuscript (Florence copy) Taccola offered to accompany Sigismund to Hungary to fight the Turks; and it seems that he

did so, for in a later manuscript (New York copy) he remarks that he personally fought against the Turks. Certainly by 1435 Taccola had returned to Siena, where he spent the rest of his life working as a sculptor and finishing his "De machinis libri decem" in 1449. He died sometime between 1453 and 1458.

With Brunelleschi and Giovanni Fontana, Taccola was a founder of the Italian school of Renaissance engineers. Although this school was initially influenced by the preceding generation of German engineers (notably Konrad Kyeser), its main inspiration may well have been the Brunelleschian renaissance of architecture. The members accepted the belief of ancient writers that mechanics was a part of architecture because the architect needed a knowledge of the machines necessary to raise building materials and similar devices. Taccola remarks in his notebook (Munich 197) that he discussed engineering matters with Brunelleschi at Siena.

The literacy of the architect-engineers of the Italian Renaissance school has been greatly underrated. Taccola was not simply a craftsman; he had trained as a notary and had been in close contact with scholars during his years at the Casa della Sapienza. Moreover, passages in his writings and in those of other engineers reveal a knowledge of the natural philosophy taught in the universities.

During the Renaissance mechanical technology was of considerable interest for scholars, including Cardinal Bessarion and the university professor Mariano Sozini. Taccola claims to have shown some of his designs to Sozini; and his acclamation as the "Sienese Archimedes" may well have originated among his humanist friends.

Taccola greatly influenced the Italian engineers of the Renaissance. Many of his designs were subsequently incorporated into the works of Francesco di Giorgio Martini and Roberto Valturio, through whom they reached a wider audience. The originality of Taccola's designs is, however, a difficult question. Certainly many devices and processes made their first recorded appearance in his works. At one time or another he has been credited with the invention of the explosive undermining of city walls, the suction pump, underwater breathing apparatus, the box-caisson method for building bridges, water mains and sluice gates, and vertically axled windmills and water mills. Some of these are now known to have been included in earlier treatises; others may have been set down by Taccola after he had seen them in operation. It may be said that Taccola's importance lies in his encyclopedic account of contemporary machine practice rather than in any original invention.

Two ideas of great later significance did, however, make their first known appearance in Taccola's manuscripts: the chain transmission system and the compound crank with connecting rod. By the latter, rotary motion could be converted to reciprocal motion, a technical concept that has been considered crucial for the postmedieval development of Western technology.

BIBLIOGRAPHY

I. ORIGINAL WORKS. Bayerische Staatsbibliothek, Munich, MS Lat. 197, a notebook containing drawings and texts (1427–1441), is to be published at Wiesbaden. Biblioteca Nazionale Centrale, Florence, MS Palat. 766, is bks. 3 (1430–1432) and 4, dedicated to Sigismund, of a treatise entitled "De ingeneis"; the missing books have been reconstructed by Prager and Scaglia (see below). The MS has been edited by James H. Beck, *Liber tertius de ingeneis ac edifitiis non usitatis* (Milan, 1969).

A second MS at Munich, MS Lat. 28800, comprises Taccola's main treatise, "De machinis libri decem" (1449), formerly known as Codex Wilczek I. Codex Wilczek II, a fifteenth-century copy, is now at the New York Public Library, Spencer Collection, MS 136. It has been edited by G. Scaglia as *De machinis. The Engineering Treatise of 1449*, 2 vols. (Wiesbaden-New York, 1971). A splendidly illustrated plagiary by Paolo Santini is at the Bibliothèque Nationale, Paris, MS Lat. 7239. Later copies of excerpts from Taccola's works are at the Biblioteca Marciana, Venice, MS Lat. VIII 40 (2941) and the Biblioteca Nazionale, Florence, MS Palat. 767. Many drawings from the Venice MS are reproduced in G. Canestrini, *Arte militare meccanica* (Milan, [1946?]). At least three autograph MSS seem to have disappeared.

The MSS are described in P. L. Rose, "The Taccola Manuscripts," in *Physis*, **10** (1968), 337–346. A guide to earlier bibliography is Lynn Thorndike, "Marianus Jacobus Taccola," in *Archives internationales d'histoire des sciences*, **8** (1955), 7–26.

II. SECONDARY LITERATURE. For biographical data see James H. Beck, "The Historical 'Taccola' and Emperor Sigismund in Siena," in *Art Bulletin*, **50** (1968), 309–320. See also Frank D. Prager and Giustina Scaglia, *Mariano Taccola and his Book De ingeneis* (Cambridge, Mass., 1972), and Frank D. Prager, "A Manuscript of Taccola, Quoting Brunelleschi, on Problems of Inventors and Builders," in *Proceedings of the American Philosophical Society*, **112** (1968), 131–149, which also reproduces and transcribes the relevant folios from Munich MS Lat. 197. On Taccola's significance, see Bertrand Gille, *The Renaissance Engineers* (Cambridge, Mass.–London, 1966), 81–87; and Lynn

White, Jr., *Medieval Technology and Social Change* (Oxford, 1962), 86, 113.

PAUL LAWRENCE ROSE

TACHENIUS, OTTO (*b.* Herford, Westphalia; *d.* probably Venice, Italy), *medical chemistry, pharmaceutical chemistry.*

The details of Tachenius' life are extremely obscure and are based mainly on statements by his enemies. He is said to have been the son of a miller and to have been apprenticed to an apothecary (which is likely), from whose service he was dismissed for theft. About 1640 he went east to Holstein and Prussia, serving apothecaries in Kiel, Danzig, and Königsberg. In 1644 he went to Italy, acquired an M.D. from Padua in 1652, and settled in Venice, where he sold a "viperine salt" (*sal viperinum*) as a sovereign remedy. While there he wrote a short commentary on J. B. van Helmont's alkahest, in the form of a letter to Duke Frederick of Holstein; he then sent it to Helwig Dieterich, a physician whom he had met in Königsberg, to see through the press. Published as *Epistola de famoso liquore Alkahest*, it is said to have appeared with an appendix criticizing the author's arguments and grammar. (The pamphlet was reportedly published at Venice, but since it now seems to have disappeared, the affair is difficult to disentangle.) Tachenius attacked Dieterich, whom he naturally held responsible, in *Echo . . . de liquore Alcaeist*; it also apparently has disappeared. Dieterich's *Vindiciae adversus Ottonem Techenium* (Hamburg, 1655), however, is extant and gives an account of the whole affair and a scurrilous narrative of Tachenius' youthful career, as well as casting doubt upon the composition of the "viperine salt," which Dieterich claimed was mainly spirit of hartshorn (ammoniacal salt). Tachenius also was attacked by Johann Zwelfer in a new edition of the *Pharmacopoeia Augustana* (1657), on the ground that the viperine salt was no novelty and, in any case, of doubtful efficacy. Tachenius replied in *Hippocrates chemicus*, a defense of the viperine salt that included a discussion of the nature and use of alkalies. In his *Clavis* he further elaborated his theory of alkalies and advanced the theory that acid and alkali are the two principles or elements of all things: acid, hot and dry, provides the masculine principle; alkali, cold and moist, the feminine. According to Tachenius they correspond to the fire and water that Hippocrates found in all things, and hence he claimed to have revived "Hippocratean chemistry"—whatever that may be. His views obviously were derived from the Helmontian theory of acid and alkali as the governing principles of human physiology. Tachenius did not, as historians have claimed, "correctly" define salts as composed of acid and alkali, since to him all matter, animate and inanimate alike, was so composed. *Hippocrates chemicus* is of added interest for its descriptions of industrial methods of the production of soap, sal ammoniac, and corrosive sublimate. The date of his death is unknown; he is variously described as alive in Venice, in either 1669 or 1699, and as having died in 1670.

BIBLIOGRAPHY

I. Original Works. The first two works published by Tachenius are extremely rare and appear not to have been seen by anyone writing on him since the eighteenth century; their titles were then given as *Epistola de famoso liquore Alkahest Helmontii* (Venice, 1652 [or 1655]) and *Echo ad vindicias chyrosophi de liquore Alcaeist* (Venice, 1656). They were followed by a treatise on diseases of the joints (possibly his M.D. thesis), *Exercitatio de recta acceptatione arthritidis et podagrae* (Padua, 1662). Under the anagrammatic pseudonym Marc Antonio Crassellane chinese [sic] there appeared *Lux obnubilata suapte natura refulgens. Vera de lapide philosophico theorica metro italico descripta . . .* (Venice, 1666; Milan, 1968), ascribed to Tachenius by MSS notes in two copies at the Bibliothèque Nationale—French trans. as *La lumière sortant par soy même des ténèbres* (Paris, 1687, 1693); German trans. as *Das aus der Finsterniss von sich selbst hervorbrechende Light* (Langensalza, 1772).

His most popular works, which went through many eds., often together, were *Hippocrates chemicus, per ignem et aquam methodo inaudita novissimi salis viperini antiquissima fundamenta ostendens* (Venice, 1666, 1678, 1697; Brunswick, 1668; Paris, 1669, 1674; Leiden, 1671); and *Antiquissimae Hippocraticae medicinae clavis manuali experientia in naturae fontibus elaborata* (Brunswick, 1668; Venice, 1669, 1697; Frankfurt, 1669, 1673; Leiden, 1671; Paris, 1671, 1672; Lyons, 1671). An English ed. of both is *Otto Tachenius His Hippocrates Chymicus, Which Discovers the Ancient Foundations of the Late Viperine Salt. And His Clavis Thereunto* (London, 1677, 1690). *Tractorum de morborum principe* was published both with *Hippocrates chemicus* (Venice, 1678) and independently (Osnabrück, 1678, 1679).

II. Secondary Literature. Tachenius' acid-alkali theory was first discussed at length in F. Bertrand, *Réflexions nouvelles sur l'acide et sur l'alcalie* (Lyons, 1683). The chief biographical source appears to be J. C. Barchusen, *Historia medicinae* (Amsterdam, 1710), based in turn partly on Dieterich's *Vindiciae*. There are brief biographies in *Allgemeine deutsche Biographie*, XXXVII (Berlin, 1894, 1971), 340; John Ferguson, *Bibliotheca chemica*, II (London, 1906, 1954), 424–425; and Lynn Thorndike, *A History of Magic and Experimental Science*, VIII, 357–361. For a long summary of *Hippocrates chemicus* and *Clavis*, see J. R. Partington, *A History of Chemistry*, II (London, 1961), 291–296.

Marie Boas Hall

TACQUET, ANDREAS (*b.* Antwerp, Belgium, 23 June 1612; *d.* Antwerp, 22 December 1660), *mathematics.*

Tacquet was the son of Pierre Tacquet, a merchant, and Agnes Wandelen of Nuremberg. His father apparently died while the boy was still young but left the family with some means. Tacquet received an excellent education in the Jesuit *collège* of his native town, and a contemporary report describes him as a gifted if somewhat delicate child. In 1629 he entered the Jesuit order as a novice and spent the first two years in Malines and the next four in Louvain, where he studied logic, physics, and mathematics. His mathematics teacher was William Boelmans, a student of and secretary to Gregorius Saint Vincent. After his preliminary training Tacquet taught in various Jesuit *collèges* for five years, notably Greek and poetry at Bruges from 1637 to 1639. From 1640 to 1644 he studied theology in Louvain and in 1644–1645 he taught mathematics there. He took his vows on 1 November 1646 and subsequently taught mathematics in the *collèges* of Louvain (1649–1655) and Antwerp (1645–1649, 1655–1660).

Tacquet's most important mathematical work, *Cylindricorum et annularium*, contained a number of original theorems on cylinders and rings. Its main importance, however, lay in its concern with questions of method. Tacquet rejected all notions that solids are composed of planes, planes of lines, and so on, except as heuristic devices for finding solutions. The approach he adopted was that of Luca Valerio and Gregorius, an essentially Archimedean method. The development of his thought can be seen in the fact that in his *Arithmeticae theoria et praxis* he took the value of ax^n ($x < 1$, $n \to \infty$) to be actually zero. Tacquet's most popular work was *Elementa geometriae*, which went through numerous editions during the seventeenth and eighteenth centuries and was edited and revised by Whiston, Musschenbroek, and Bošković. Although little more than a paraphrase of parts of Euclid and Archimedes, the book was distinguished by its clarity and order. Tacquet's *Opera mathe-*

matica was published posthumously and contained, among other previously printed and unprinted works, his *Astronomia*. In the eighth book of this work he rejected the motion of the earth, first, because there was no proof, physical or philosophical, to prove it; second, because his faith required him to believe in its immobility.

Tacquet's importance was mainly pedagogical and his books taught elementary mathematics to many generations of readers, although his influence on Pascal may have been greater. As a creative mathematician he can hardly be deemed more than minor. He was extremely well-read in mathematics, astronomy, and physics, and seemed to have almost total knowledge of the literature. This makes him appear at times as a typical exponent of the irritatingly erudite eclecticism of seventeenth-century scientific Jesuits. However, most of his works were written as textbooks for the Jesuit *collèges* and had no pretensions to originality. His devotion to his church, his order, and his teaching may explain his relative lack of creativity.

BIBLIOGRAPHY

I. ORIGINAL WORKS. The standard bibliography of Tacquet's works is in C. Sommervogel, *Bibliothèque de la Compagnie de Jésus*, VII (Brussels, 1896), cols. 1806–1811. The most important are *Cylindricorum et annularium* (Antwerp, 1651, 1659), also in the *Opera*; *Elementa geometriae* (Antwerp, 1654, 1665, 1672), which was issued in numerous eds. and revs., including translations into English, Italian and Greek, at least until 1805; and *Arithmeticae theoria et praxis* (Louvain, 1656; Antwerp, 1665, 1682). The *Opera mathematica* (Antwerp, 1669, 1707) contains works on astronomy, spherical trigonometry, practical geometry, and fortification, plus previously published writings on geometry and Aristotle's wheel. Tacquet's correspondence with Huygens is printed in *Oeuvres complètes de Christiaan Huygens, publiées par la Société hollandaise des sciences*, I–III (The Hague, 1888–1890), *passim*.

II. SECONDARY LITERATURE. For biographical information on Tacquet, see H. Bosmans, "Tacquet," in *Biographie nationale*, XXIV (Brussels, 1926–1929), cols. 440–464; and "Le Jésuite mathématicien anversois André Tacquet (1612–1660)," in *Gulden passer*, 3 (1925), 63–87. See also Bosmans' "André Tacquet (S. J.) et son traité d'arithmétique théorique et pratique," in *Isis*, 9 (1927), 66–82. There is no adequate analysis of Tacquet's mathematics and science. For earlier accounts see A. G. Kästner, *Geschichte der Mathematik*, III (Göttingen, 1799), 266–284, 442–449; and J. B. J. Delambre, *Histoire de l'astronomie moderne*, II (Paris, 1821), 531–535. Among modern studies that treat Tac-

quet is C. R. Wallner, "Über die Entstehung des Grenzbegriffes," in *Bibliotheca mathematica*, 3rd ser., **4** (1903), 246–259.

PER STRØMHOLM

IBN ṬAHIR. For a detailed study of his life and work, see **Al-Baghdādī, Abū Manṣūr ᶜAbd al-Qāhir ibn Ṭāhir ibn Muḥammad ibn ᶜAbdallah, al-Tamīnī, al-Shāfiᶜī,** in the Supplement.

TAIT, PETER GUTHRIE (*b.* Dalkeith, Scotland, 28 April 1831; *d.* Edinburgh, Scotland, 4 July 1901), *physics, mathematics.*

Tait was the son of the former Mary Ronaldson and John Tait, who was secretary to the duke of Buccleuch. He was taught first at Dalkeith Grammar School and, after his father's death, at a school in Circus Place and later at the Academy, both in Edinburgh. With his mother and his two sisters Tait lived in Edinburgh with an uncle, John Ronaldson, who introduced the boy to geology, astronomy, and photography. It is interesting to note that the order in the mathematics section of the Edinburgh Academical Club Prize for 1846 was first Tait, then Lewis Campbell, and third J. C. Maxwell. (In the following year Tait was second to Maxwell.) Tait entered Edinburgh University in 1847, and after a session there went in 1848 to Peterhouse, Cambridge, where his tutor was William Hopkins. He graduated as senior wrangler and first Smith's Prizeman in 1852. (Second in the tripos was another student at Peterhouse, W. J. Steele, with whom Tait collaborated on his first book, *Dynamics of a Particle* [1856]. Steele died before completing his portion of the book.)

In 1854 Tait left Cambridge, where he was a fellow of his college, to become professor of mathematics at Queen's College, Belfast. His colleague there was Thomas Andrews, with whom he collaborated in research on the density of ozone and the results of electrical discharge through gases. Other colleagues were Charles Wyville Thomson, who later was scientific leader of the *Challenger* expedition, and James Thomson, brother of William, Lord Kelvin, and discoverer of the effect of pressure on the melting point of ice. Tait's debts to Andrews were undoubtedly great, for the latter introduced him to experimental physics; but he did not, as is occasionally said, introduce Tait to Hamilton's calculus of quaternions, which had occupied Tait while he was at Cambridge.

Tait succeeded J. D. Forbes as professor of natural philosophy at Edinburgh in 1860 and held the chair until shortly before his death. In 1857 he married Margaret Archer Porter, the sister of two Peterhouse friends. One of their four sons, the best amateur golfer of his day, was killed in the Boer War.

At Edinburgh, Tait was confirmed in his recently found liking for experimentation by the duties required of him. In 1862, for example, he wrote a paper jointly with J. A. Wanklyn on electricity developed during evaporation. In 1867, having been greatly taken by Helmholtz's paper on vortex motion, he devised an apparatus for studying vortex smoke rings, thereby giving Kelvin the idea of a vortex atom. His study of vortices was the starting point of a highly important pioneer study of the topology of knots. Tait continued to experiment on thermoelectricity, publishing extensively on the subject and on thermodynamics as a whole. In 1873 he presented a first sketch of his well-known thermoelectric diagram to the Royal Society of Edinburgh. In 1875 he experimented with James Dewar on the behavior of the Crookes radiometer and gave the first satisfactory explanation of it. Between 1876 and 1888, using superb equipment of his own design supplied by the Admiralty, Tait did research on the corrections that it would be necessary to apply to the findings of the *Challenger* expedition regarding deep-sea temperatures. This work led to important experimental studies of compressibility and the behavior of materials under impact. In the same connection Tait wrote a classic paper on the trajectory of a golf ball (1896). The fourth in an important series of papers on the kinetic theory of gases (1886–1892) contained, according to Kelvin, the first proof of the Waterston-Maxwell equipartition theorem.

Tait's life was marked by several controversies, two of which reached a wide public. He felt himself committed to quaternions, having promised Hamilton, only a few days before the latter's death, to publish an elementary treatise on the subject. The work appeared in 1867 and was followed by new editions in 1873 and 1890. Tait disliked intensely the vector methods of J. W. Gibbs and Oliver Heaviside, and in a long exchange of polemics tended to have the worst of the argument. In his controversial *Sketch of the History of Thermodynamics* (1868), a highly prejudiced and pro-British account, the reputations of J. R. Mayer and

Clausius suffer, while Kelvin and Joule are often praised at their expense.

BIBLIOGRAPHY

C. G. Knott, *Life and Scientific Work of Peter Guthrie Tait* (Cambridge, 1911), lists 365 papers and 22 books written wholly or partly by Tait. The last two books listed are collected volumes of Tait's *Scientific Papers* (Cambridge, 1898–1900). His best-known work was vol. I of *Treatise on Natural Philosophy* (Oxford, 1867; Cambridge, 1878, 1883), written jointly with Sir William Thomson and widely known as "T and T'." A promised vol. II failed to appear. Tait and Thomson also collaborated on an elementary version.

Knott's biography, which refers to all the important obituaries, is itself the fundamental biographical source, although very uncritical. See also J. H. Hamilton Dickson, in *Dictionary of National Biography*, 2nd supp., III (1912), 471–474; and A. Macfarlane, "P.G.T.," in *Bibliotheca mathematica*, 3rd ser., 4 (1903), 185–200. For the controversy over the history of thermodynamics, see D. S. L. Cardwell, *From Watt to Clausius* (London, 1971), 282–289.

J. D. NORTH

TALBOT, WILLIAM HENRY FOX (*b.* Melbury House, Dorsetshire, England, 11 February 1800; *d.* Lacock Abbey, Wiltshire, England, 17 September 1877), *photochemistry, mathematics.*

Talbot was the only child of William Davenport Talbot, an officer of dragoons, and Lady Elisabeth Theresa Fox-Strangways, the eldest daughter of the second earl of Ilchester. Four years after his father's death in July 1800, his mother married Rear Admiral Charles Feilding, who established a warm relationship with his stepson. Talbot grew up with two half sisters in an upper-class family that possessed both social position and culture.

While studying at a boarding school in Rottingdean and later at Harrow, Talbot distinguished himself as a scholar. At age seventeen he entered Trinity College, Cambridge, where he studied classical languages and mathematics, receiving in 1820 the Porson Prize for Greek verse, and graduating in 1821 as twelfth wrangler and second chancellor's medalist. Soon thereafter he published a half dozen papers on mathematics but spent much of the next decade traveling on the Continent, living the life of a gentleman scholar. He established himself in the late 1820's at the family estate, historic Lacock Abbey, ran successfully as a Liberal from Chippenham for the first reform Parliament (1833–1834), and on 20 December 1832 married Constance Mundy of Markeaton, Derbyshire. Talbot was an active member of numerous scholarly and scientific societies, including the Royal Society, the Royal Astronomical Society, and the British Association for the Advancement of Science.

Talbot's formative years were spent under the influence of the dominant Romantic atmosphere of England and Western Europe. He frequented Romantic operas and concerts and avidly read the works of Goethe, Byron, and Scott, naming two of his daughters after characters in Scott's works. His love of nature manifested itself in his lifelong interest in flowers and in his penchant for travel. His fondness for the past was stimulated not only by Scott's historical novels and by his own historic estate, but also by Young and Champollion's deciphering of the hieroglyphics on the Rosetta Stone in the early 1820's and Rawlinson and Hincks's deciphering of the Assyrian cuneiform in the middle 1840's. These stimuli merged with his flair for languages to initiate a lifelong series of translations from Assyrian and other ancient languages and of historical and philological studies. Beginning with a book of Greek verse, *Legendary Tales* (London, 1830), these translations and studies included four other books and at least sixty-two articles in scholarly journals.

After completing his study at Cambridge, Talbot continued his work in mathematics, systematically studying elliptic integrals. Building upon the earlier achievements of Fagnano dei Toschi, Euler, Legendre, Jacobi, and Abel, he addressed himself to the problem of summing the integrals of any function. His early mathematical work led to his election as a fellow of the Royal Society, while his work on elliptic integrals brought him the Royal Medal of the Society for the year 1838 and an appointment to the Royal Society council.

During the same period Talbot's interests in chemistry and optics quickened, and he gradually adopted a unified, dynamic view of physical phenomena. The early nineteenth century witnessed the adoption and modification of new theoretical frameworks in chemistry and optics. The discovery of many new substances stimulated increasing concern with chemical composition and structure, while the wave theory of light posed problems with dispersion, absorption, photochemical reaction, and other forms of light-matter interaction. Although Talbot counted Wheatstone, Brewster, and Babbage among his scientific friends, he most closely followed the ideas of his friend John Herschel on light. Adopting the wave theory of light and a kinetic interpretation of light, heat, and matter, he pursued the problem of light-matter interaction through the study of optics, crystallography, and spectra. Intrigued by the similar optical characteristics of light and radiant heat as demonstrated by Melloni and Forbes, he sought to show the unity of the chemical rays with visible rays and heat rays. He also sought to use light and optical properties as analytical tools in order to determine the nature and structure of matter and to develop methods of chemical identification. Utilizing the vibratory theory of molecular behavior in gases, he suggested in 1835 a connection between spectral lines and chemical composition. In an 1836 paper he employed the polarizing microscope as a tool to explore "the internal structure of transparent bodies, even in their minutest visible particles" (*London and Edinburgh Philosophical Magazine* [1836], p. 288). This paper brought him the honor of being named the Bakerian lecturer of the Royal Society for the year 1836.

It was with the development of photography that Talbot's love of nature and landscapes merged with his interests in optics and photochemistry. His efforts to sketch Italian scenery had met with repeated frustration. When he realized that he lacked artistic talent, he turned in 1823 to the use of a camera obscura as a drawing aid, but without satisfaction. Again in October of 1833, while honeymooning on the shores of Lake Como, he met with failure when he used Wollaston's recently developed boon to nature lovers and amateur artists, the camera lucida. At that time it occurred to Talbot to imprint the image on chemically sensitized paper. Returning to England in January 1834, he and his assistant, Nicholaas Henneman, conducted many experiments; by 1835 they were able to obtain "negatives" by employing tiny camera obscuras and paper sensitized with excess silver nitrate and fixed with excess common salt. Between 1835 and 1839, Talbot and Henneman continued their experiments, motivated by a desire for an analytic tool for research on radiant heat and light, as well as by a desire for reproducing images from nature. Following Arago's announcement to the Académie des Sciences 7 January 1839 of the existence of Daguerre's photographic process, Talbot became concerned over the priority of his work; he frantically sought to improve his process prior to the disclosure of Daguerre's. Nevertheless, Daguerre's process proved to be vastly superior to Talbot's in the quality of the image. In September 1840 Talbot discovered that gallic acid would develop a latent image on paper, and he called this new process the calotype. He patented and then disclosed the process in a paper presented to the Royal Society in June of 1841.

Although Talbot's photographic efforts did not meet with major commercial success and, because of his efforts to enforce his patents, did not win him popular acclaim, his paper on the calotype did bring him the honor of the Rumford Medal of the Royal Society (1842) for the most outstanding piece of research on light during the previous two years. In the middle 1840's he published two of the earliest books illustrated with photographs. Although twenty-eight of his fifty-nine scientific papers were published after 1840, most of these were minor papers on photography and mathematics. In 1852 he patented and published a method of photoengraving called photoglyphy. From the mid-1850's, with the increasing public clamor over his patent suits, Talbot's interests shifted increasingly to philological and historical studies. Despite the significant contribution he made in these scholarly pursuits. It was his development of the first negative-positive process in photography—that union of his naturalistic and artistic inclinations with his unitary photochemical interests—that brought him his greatest recognition both during his lifetime and after his death.

BIBLIOGRAPHY

I. ORIGINAL WORKS. Talbot's published scientific work appears exclusively in the fifty-nine articles listed in the *Royal Society Catalogue of Scientific Papers*. His two books illustrated with calotypes are *The Pencil of Nature*, parts I–VI (London, 1844–1846), and *Sun Pictures in Scotland* (London, 1845); his own remarks are contained in Appendix A of G. Tissandier, *A History and Handbook of Photography* (London, 1878).

Considerable data are contained in the legal records of the Court of Chancery, Public Record Office, London: Talbot *v* Colls (1852), and Talbot *v* Henderson (1854).

Manuscripts and artifacts are located at: Lacock Abbey, Wiltshire, England; Science Museum, London; Royal Society, London; Kodak Museum, Harrow, England; George Eastman House, Rochester, New York; Stark Library, University of Texas, Austin, Texas; and Soviet Academy of Sciences, Moscow. Some of the manuscript materials held in the U.S.S.R. have been published in T. P. Kravets, ed., *Dokumenti po istorii izobretenia fotografi* (Moscow, 1949), in English and Russian. See also Wood and Johnston below.

II. SECONDARY LITERATURE. The only biography of Talbot is Arthur H. Booth's *William Henry Fox Talbot . . .* (London, 1965), which is superficial and unreliable.

Even the best sources restrict themselves largely to Talbot's photographic work. These include R. Cull, "Biographical Notice of the Late William Henry Fox Talbot," in *Society of Biblical Archaeology. Transactions,* **6** (1878), 543–549; *Dictionary of National Biography;* H. Gernsheim, "Talbot's and Herschel's Photographic Experiments in 1839," in *Image,* **8** (1959), 132–137; H. and A. Gernsheim, *History of Photography . . .* (New York, 1969); A. Jammes, *William H. Fox Talbot, Inventor of the Negative-Positive Process* (New York, 1973); J. D. Johnston, "William Henry Fox Talbot . . ., Part I," and J. D. Johnston and R. C. Smith, "Part II," in *Photographic Journal,* **87**A (1947), 3–13, and **108**A (1968), 361–371; B. Newhall, "William Henry Fox Talbot," in *Image,* **8** (1959), 60–75; E. Ostroff, "Restoration of Photograph . . .," in *Science,* **154** (7 Oct. 1966), 119–123; M. T. Talbot, "The Life and Personality of Fox Talbot," in *Photographic Journal,* **79** (1939), 546–549; D. B. Thomas, *The First Negatives* (London, 1964); and R. D. Wood, "The Involvement of Sir John Herschel in the Photographic Patent Case, Talbot *v* Henderson, 1854," and "J. B. Reade . . .," in *Annals of Science,* **27** (Sept. 1971), 239–264, and **27** (March 1971), 13–83.

REESE V. JENKINS

TAMM, IGOR EVGENIEVICH (*b.* Vladivostok, Russia, 8 July 1895; *d.* Moscow, U.S.S.R., 12 April 1971), *physics.*

From 1899 Tamm's family lived in the city of Elizavetgrad (now Kirovograd), where his father was a civil engineer. After graduating from the Gymnasium there in 1913, Tamm studied for a year at Edinburgh University; at the beginning of World War I in 1914 he returned to Russia and entered the Faculty of Physics and Mathematics at Moscow University. In 1917 he was a member of the Elizavetgrad City Soviet of Worker and Soldier Deputies and was a delegate to the First Congress of Soviets in Petrograd. The following year Tamm graduated from Moscow University, and in 1921–1922 he worked at the Odessa Polytechnical Institute, where Mandelshtam, who greatly influenced Tamm's later scientific career, was a professor.

From 1922 Tamm worked in Moscow. In 1924 he published his first scientific work, on the electrodynamics of anisotropic media in the special theory of relativity, and became head of the department of theoretical physics at Moscow University, which he directed until 1941. In 1934, after the transfer to Moscow of the Academy of Sciences of the U.S.S.R., Tamm was named director of the theoretical section of its P. N. Lebedev

Physical Institute; and from then on, his activity was concentrated there.

Tamm's first scientific research, begun under the influence of Mandelshtam, was devoted to electrodynamics of anisotropic media, crystal optics in the theory of relativity, quantum theory of paramagnetism, and nonrelativistic quantum mechanics. In a major investigation on the quantum theory of the molecular scattering of light in solid bodies (1930) Tamm conducted the first quantification of elastic (sound) waves in solid bodies and introduced the concept of quanta of sound; later called phonons by Y. I. Frenkel, this concept gained wide acceptance in contemporary physics. In this work he also investigated the Rayleigh scattering of light (the Mandelshtam-Brillouin doublet) and combination scattering (the "Raman effect"), discovered in 1930 by Mandelshtam and G. S. Landsberg in crystals and, at the same time, by C. V. Raman and Krishnan in liquids.

In 1930 Tamm published two works dealing with the phenomenon of relativistic quantum mechanics of the electron, formulated by Dirac: "Über die Wechselwirkung der freien Elektronen . . .," and "Zamechanie dirakovskogo teorii sveta i dispersii" ("A Note on the Dirac Theory of Light and Dispersion"). Dirac's theory received almost general acceptance when it appeared, since it automatically led to the existence of a spin on the electron and allowed a natural interpretation of the fine structure in the spectrum of the hydrogen atom. But the concepts of negative energy states that were involved in the theory appeared unusual until the discovery of the positron and demanded more detailed study of the conclusions from the theory and their experimental verification.

With the aid of a strictly consistent quantum mechanical method, Tamm confirmed the results obtained earlier by Felix Klein and Yoshio Nishina on the basis of the method of correspondence. He also showed that according to Dirac's theory, scattering of even more low-frequency quanta of light from free electrons must occur through intermediate states with the negative electron energies. Thus it was shown that the negative energy states cannot be removed from Dirac's theory. In this work Tamm also proposed a new method of calculation, later developed by H. B. G. Casimir and named for him. Concurrently with Dirac and Oppenheimer, Tamm independently concluded that the fall of the free electron to a negative level was inevitable and determined the probability of annihilation of the electron and "hole."

In 1931–1933 Tamm studied the quantum theory of metals, specifically the external photoeffect in metals and the state levels of the electrons on the surface of the metal. His work with S. P. Shubin was the first to show that the external photoeffect is caused by the presence of a jump in potential on the border of the metal vacuum and is associated with the effect of surface absorption of light, while the optic absorption of light by the metal is associated with the volume effect. In works dating from 1932–1933 Tamm was the first to show, on the basis of quantum mechanics, that, along with the known "zone" electron states inside the crystal, there could also be electron states of a completely different type on the surface of the crystal. Further theoretical and experimental research on semiconductors and dielectrics confirmed that these "Tamm surface levels" of electrons are evident in a great many physical phenomena, particularly in semiconductors, and lead to "barrier" layers.

From 1934 Tamm devoted his research to the atomic nucleus and cosmic rays. Two of the early works were related to the nature of nuclear forces: "Obmennye sily mezhdu neytronami i protonami i teoria Fermi" ("Exchange Forces Between Neutrons and Protons, and Fermi's Theory," 1934) and "β-radioaktivnost i yadernye sili" ("Beta Radioactivity and Nuclear Forces," 1936). Using Fermi's theory of beta radiation, Tamm sought to determine whether the nuclear forces could be caused by an exchange between nucleons, electrons, and neutrinos. In the 1934 work he gave a formula for the potential and evaluated the quantity of force arising in this process, but this kind of exchange interaction proved to be too weak in comparison with the observed nuclear forces and consequently could not serve to explain them. Later it was discovered that the exchange nature of nuclear forces corresponds to the activity; but the exchange is realized not by electrons and neutrinos, as Tamm suggested, but by pi-mesons. All later theories of nuclear forces were constructed according to the theoretical scheme developed in these investigations by Tamm but took into account the role of the pi-meson.

At the same time that he was conducting this research, Tamm and S. A. Altshuler published works on the magnetic moment of the neutron. By analyzing the material he had obtained experimentally, Tamm concluded that although the neutron is a neutral particle, it actually has a magnetic moment. He also correctly estimated the negative sign of this moment. These investigations led Tamm to another important conclusion: that despite current opinion, mesons, which carry an interaction between nucleons, do not have stationary levels in the central Coulomb field. This result led to the work (with L. D. Landau) "O proiskhozhdenii yadernykh sil" ("On the Origin of Nuclear Forces," 1940).

In 1937–1939 Tamm and I. M. Frank developed the theory of radiation of the electron, which moves through a medium with a velocity exceeding the velocity of light in that medium. This theory led to an understanding of the nature of the radiation discovered by S. I. Vavilov and P. A. Cherenkov. For this work Tamm was awarded the Nobel Prize (with I. M. Frank and P. A. Cherenkov) in 1958.

During World War II, Tamm carried out a number of complex practical investigations. In 1945 he returned to the interaction of molecules. In the first of these works he formulated a new method of calculating the interaction not dependent on an expansion in terms of a coupling constant, the large size of which makes perturbation theory inapplicable to quantum meson dynamics. This method, applied by Tamm in 1945 and developed by P. D. Dankov in 1950, is known as the Tamm-Dankov method. It is widely used in the theoretical study of the interaction of mesons with nucleons and of nucleons with each other, particularly in the study of the deuteron.

In 1947 Tamm and V. L. Ginzburg formulated a theory of a molecule that can be found in states with various spins. This work contains the first relativistically invariant wave equations for particles with inner degrees of freedom, described by continuous variables. Tamm's research in the theory of nuclear forces and elementary particles was continued in two directions. One was the construction of a polyphenomenological theory based on the possibility of the existence of isobaric states of nucleons. On this basis Tamm and his colleagues investigated processes of scattering, the photoproduction of pi-mesons by nucleons, and the interaction of nucleons. The other was the development by Tamm, with V. Y. Feynberg and V. P. Silin, of a new form of Tamm's method, proposed by F. J. Dyson and intended particularly for the study of the interaction of pi-mesons with nucleons.

Tamm worked on other questions, including the investigation, according to cascade theory, of cosmic ray showers (with S. Z. Belenky), which were first considered to be ionization losses of particles. Of very great importance was the theory of gas discharge in a powerful magnetic field developed in 1950 by Tamm and A. D. Sakharov. It was the basis of all subsequent Soviet research on guided thermonuclear reactions and led to important results.

An important place among Tamm's investigations was occupied by the work (carried out in 1946 with Mandelshtam) on the meaning of the indeterminacy between time and energy in quantum mechanics. Also worth noting is his work, done in 1948–1949, on several mathematical methods for the theory of particle scattering. During his last years Tamm searched for ways to remove fundamental difficulties in the theory of elementary particles and often presented survey reports at conferences. For example, at the All-Union Conference on Quantum Electrodynamics and the Theory of Elementary Particles (1955), he presented the survey reports "Metod obrezania uravneny po chislu chastits v teorii mezonov" ("Method of Truncating the Equation According to the Number of Particles in the Theory of Mesons"), written with V. P. Silin and V. Y. Feynberg, and "Polufenomenologicheskaya izobarnaya teoria vzaimodeystvia mezonov s nuklonami" ("Semiphenomenological Isobar Theory of Interaction of Mesons With Nucleons"), written with Y. A. Golfand, G. F. Farkov, and others. At the All-Union Conference on the Physics of High-Energy Particles (1956), Tamm, Silin, and Feynberg delivered the survey report "Sravnenie mezonnoy teorii s eksperimentami" ("Comparison of Meson Theory With Experiments").

Tamm's activity was not limited to scientific research. He gave much attention to teaching, to the solution of practical and administrative problems, and to social questions. From 1924 as a docent, and from 1930 as professor and head of the department of theoretical physics at the M. V. Lomonosov Moscow State University, Tamm (in collaboration with Mandelshtam) supervised the orientation and content of all courses in theoretical physics and lectured at the Physics and Mathematics Faculty of the university. During this period he wrote *Osnovy teorii elektrichestva* ("Principles of the Theory of Electricity"), which went through many editions.

The fight for scientific biology, which Tamm led, also is worthy of attention. He maintained the firm conviction that leadership of the natural sciences would pass in the relatively near future from physics to biology. Tamm was active in the Pugwash movement and was awarded the Order of State Prize, First Degree, for his service to science. In 1933 he was elected corresponding member and, in 1953 active member, of the Academy of Sciences of the U.S.S.R.

BIBLIOGRAPHY

I. ORIGINAL WORKS. Tamm's early writings include "Über die Quantentheorie der molekularen Lichtzerstreuung in festen Körpern," in *Zeitschrift für Physik,* **60** (1930); 345–363; "Über die Wechselwirkung der freien Elektronen mit der Strahlung nach der Diracschen Theorie des Elektrons und nach der Quantenelektrodynamik," *ibid.,* **62** (1930), 545–568; "Über eine mögliche Art der Elektronenbindung an Kristalloberflächen," in *Physikalische Zeitschrift der Sowjetunion,* **1** (1932), 733; "Exchange Forces Between Neutrons and Protons, and Fermi's Theory," in *Nature,* **133** (1934), 981; "Nuclear Magnetic Moments and the Properties of the Neutron," *ibid.,* **134** (1934), 380; "Kogerentnoe izluchenie bystrogo elektrona v srede" ("Coherent Radiation of Fast Electrons Passing Through Matter"), in *Doklady Akademii nauk SSSR,* **14**, no. 3 (1937), 107–112; "Svechenie chistykh zhidkostey pod deystviem bystrykh elektronov" ("Luminescence of Pure Liquids Under the Influence of Fast Electrons"); in *Izvestiya Akademii nauk SSSR,* Seria fiz. (1938), nos. 1–2, 29, written with I. M. Frank and P. A. Cherenkov; "The Transmutations of the Cosmic-Ray Electrons and the Nuclear Forces," in *Physical Review,* **53** (1938), 1016–1017; and "Radiation Emitted By Uniformly Moving Electrons," in *Journal of Physics of the U.S.S.R.,* **1**, nos. 5–6 (1939), 439.

Subsequent works are "O proiskhozhdenii yadernykh sil" ("On the Origin of Nuclear Forces"), in *Doklady Akademii nauk SSR,* **29** (1940), 555–556, written with L. D. Landau; "Izluchenie elektrona pri ravnomernom dvizhenii v prelomlyayushchey srede" ("Theory of the Electron in Uniform Motion in a Refracting Medium"), in *Trudy fizicheskago instituta,* **2**, no. 4 (1944), 63, written with I. M. Frank; "The Energy Spectrum of Cascade Electrons," in *Physical Review,* **70** (1946), 660–664, written with S. Z. Belenky; "K teorii spina" ("Toward a Theory of Spin"), in *Zhurnal eksperimentalnogo i teoreticheskogo fizika,* **17**, no. 3 (1947), 227, written with V. L. Ginzburg; and "O nekotorykh matematicheskikh metodakh teorii rasseyania chastits" ("On Certain Mathematical Methods in the Theory of Scattering of Particles"), *ibid.,* **18**, no. 4 (1948), 337–345, and **19**, no. 1 (1949), 74–77.

Later writings include "K relyativistskoy teorii vzaimodeystvia nuklonov" ("Toward a Relativistic Theory of the Mutual Interaction of Nucleons"), *ibid.,* **24**, no. 1 (1954), 3; "Polufenomenologicheskaya teoria vzaimodeystvia π-mezonov s nuklonami. Rasseyanie π-mezonov nuklonami" ("Semiphenomenological Theory of the Mutual Interaction of pi-Mesons With Nucleons. Scattering of pi-Mesons With Nucleons"), *ibid.,* **26**, no. 6 (1954), 649–667, written with Y. A. Golfand and V. Y. Feynberg; "Metod usechennykh uravneny polya i ego primenenie k rasseyaniyu mezonov nuklonami" ("Method of Truncating Field Equations and Its Application to the Scattering of Mesons by Nucleons"), ibid., **29**, no. 1 (1955), 6–19, written with V. P. Silin and V. Y. Feynberg; "O strukture nuklonov" ("On the Structure of Nucleons"), *ibid.,* **32**, no. 1 (1957), 178–180; "Teoria magnitnykh termoyadernykh reaktsy" ("Theory of Magnetic Thermodynamic Reactions"), in *Fizika plazmy i problemy upravlyaemykh termoyadernykh reaktsy* ("Plasma Physics and Problems of Thermodynamic Reactions"), I (Moscow, 1958), 3–19, 31–41; and *Osnovy teorii elektrichestva,* 7th ed. ("Principles of the Theory of Electricity"; Moscow, 1957).

II. SECONDARY LITERATURE. *Igor Evgenievich Tamm, Materialy k biobibliografii uchenykh SSSR.* Seria fizik, no. 9 (Moscow, 1959), with introductory articles by V. L. Ginzburg and E. L. Feynberg and a bibliography; V. L. Ginzburg and E. L. Feynberg, "Igor Evgenievich Tamm," in *Uspekhi fizicheskikh nauk,* **56**, no. 4 (1955), 469–475, with bibliography of Tamm's most important works; V. L. Ginzburg, A. D. Sakharov, and E. L. Feynberg, "Igor Evgenievich Tamm," *ibid.,* **86**, no. 2 (1965), 353–356; I. M. Lifshits and S. I. Pekar, "Tammovskie svyazannye sostoyania elektronov na poverkhnosti kristalla i poverkhnostnye kolebania atomov reshetki" ("Tamm Connected Electron States on the Surface of a Crystal and Surface Vibration of Atoms in a Lattice"), *ibid.,* **56**, no. 4 (1955), 531–568; S. V. Vonsovsky, A. V. Sokolov, and A. Z. Veksler, "Fotoeffekt v metallakh" ("Photoeffect in Metals"), *ibid.,* 477–530; and V. P. Silin and V. Y. Feynberg, "Metod Tamma-Dankova" ("The Tamm-Dankov Method"), *ibid.,* 569–635.

J. G. DORFMAN

TAMMANN, GUSTAV HEINRICH JOHANN APOLLON (*b.* Yamburg [now Kingisepp, R.S.F.S.R.], St. Petersburg gubernia, Russia, 16/28 May 1861; *d.* Göttingen, Germany, 17 December 1938), *physical chemistry.*

Tammann belonged to the generation of scientists who created the discipline of physical chemistry. He was a founder of metallography and of metallurgy, and he pioneered the study of solid-state chemical reactions.

Tammann was born into a German-speaking, Protestant family from Livonia, the members of which had belonged to the untitled Russian nobility since at least the beginning of the nineteenth century. His father, Heinrich Tammann, was municipal physician in Yamburg and in Gorigoretsk, Mogilev gubernia (now Gorki, Belorussian S.S.R.), and professor of practical medicine at the agricultural college (now Belorussian Agricultural Academy) in Gorigoretsk. He was appointed to the faculty of the University of Moscow but died just before assuming the post. Tammann was then only four years old, and the responsibility for his education fell entirely upon his mother, the former Mathilde Schünmann. She returned to her native Dorpat (now Tartu) and skillfully overcame the considerable financial difficulties facing her. After five years of private tutoring, Tammann entered the Dorpat Gymnasium, where, despite poor grades in languages, he was regularly one of the top students in his class. At the age of fourteen, he was offered a promising military career as a protégé of Friedrich Graf von Berg, a Russian field marshal, but his mother rejected the proposal.

Tammann began to study chemistry at Dorpat in 1879. Even before he wrote his *Kandidatenschrift* (on the vapor pressure of solutions), which he submitted in 1883, he succeeded Wilhelm Ostwald as second assistant to Carl Schmidt, a former student of Justus Liebig and of Friedrich Wöhler. He also taught at the local girl's high school and at the county school. A chemist at Dorpat during this period could expect to find work only in problems related to medicine or agriculture. Consequently, at the urging of Johann Lemberg, Tammann decided to specialize in plant physiology, an area to which he was introduced by Gustav von Bunge. In one of a series of papers in this field (1895) he correctly explained the existence of optimal temperatures for enzyme reactions as the result of two independent processes, each of which is temperature-dependent but responds oppositely to temperature changes: the catalytic reaction and the thermal inactivation of the enzyme. Although Tammann retained a strong interest in fermentation processes, membranes, and osmotic pressure, it soon became apparent that he was destined to work in other fields.

Tammann first studied topics on the boundary between chemistry and physics in the autumn of 1883, when, continuing the work of Adolph Wüllner, he began to determine molecular weights from the lowering of vapor pressure. This research provided the basis for his master's essay, which, when he defended it in 1885, incurred strong criticism from Arthur von Oettingen. The high cost of printing prevented Tammann from broadening this research into a doctoral dissertation. Instead, he submitted a thesis on the metamerism of the metaphosphates; and in the same year (1887) he qualified as a university lecturer. In 1889 he worked in Helmholtz's institute at Charlottenburg and with Ostwald at Leipzig. At Leipzig, Tammann met Arrhenius and Nernst, with whom he maintained close friendships throughout his life. His relations with Ostwald, however, always remained distant. Tammann's promotion to first assistant brought him back to Dorpat, where he advanced with un-

usual rapidity. In 1890 he became *Dozent* (refusing, in the same year, an offer of an extraordinary professorship at Giessen). In 1892 he was appointed extraordinary professor and succeeded Schmidt as institute director, and in 1894 he was named full professor.

During his years at Dorpat, Tammann traveled extensively. In 1889–1890 he worked with Nernst at Göttingen on a study of the pressure at which hydrogen is liberated from solutions by the action of metals. He visited Russia several times, partly to learn Russian, on the advice of his friend Mendeleev, who followed his research with great interest. On one of these trips Tammann met Anna Mitscherling, the daughter of a German banker in St. Petersburg. They were married in 1890 and had one son and two daughters. In 1894 Tammann went to the Netherlands, where he made valuable contacts with van't Hoff, Kamerlingh Onnes, Roozeboom, and Jakob van Bemmelen. It was on this occasion that he developed his lifelong interest in heterogeneous equilibria. Also in 1894, during a trip to Nizhni Novgorod, he was stimulated to undertake research on petroleum, during which he discovered several new naphthalenes.[1] In 1897 Tammann traveled to Stockholm, where he saw Berzelius' instruments in such desolate condition that he published an appeal in the *Chemiker-Zeitung*, which prompted the founding of the Berzelius Museum.[2] While visiting the World Exposition at Paris in 1900, he studied the French metallurgical industry and became acquainted with Le Châtelier and his techniques.

At the urging of Nernst, Felix Klein, and the distinguished Prussian minister Friedrich Althoff, a chair of inorganic chemistry (the second in Germany) and a new institute were established at Göttingen in 1903. Tammann was chosen to head the institute, and he remained loyal to Göttingen despite several attractive offers. In 1909 his friend Boris Golitsyn secured his election as full member of the Imperial Academy of St. Petersburg, and, upon his retirement in 1930, he was invited to Riga.

Industry hoped for useful results from the new institute, and Tammann was determined to obtain them by means of a systematic examination of the inorganic materials of daily life. At first he wanted to specialize in silicate chemistry; but a discussion with Arthur Louis Day, who had just received a large grant from the Carnegie Institution for the construction of a geophysical laboratory, convinced him that he should give up this idea and concentrate his attention on metals and glasses.

Tammann continued his program of research when Nernst went to Berlin in 1905, and he succeeded to the latter's professorship at Göttingen in 1907, thus becoming director of the Institute of Physical Chemistry. After his retirement several scientific organizations and metal companies continued to finance Tammann's research, thus enabling him to engage from three to five assistants until 1937.

A giant not only in stature but also in health and capacity for work, Tammann regularly worked in his laboratory for ten hours a day. He was, nevertheless, devoted to his family. Moreover, he possessed an excellent knowledge of Goethe's writings and of Russian history, and he was an ardent swimmer. He restricted his friendships to a few colleagues, with one of whom, Otto Wallach, he would take a walk every Sunday. Tammann was unconventional, avoided all formality, and loved simplicity; yet he inspired respect in all who met him. His relations with co-workers and students, from whom he demanded total commitment, were forthright and often brusque; but basically they were characterized by a deep humanity. Many anecdotes illustrating his unusual sense of humor survive. It served him as an effective weapon when he felt criticism to be in order.

Tammann received many honors, prizes, and awards. He was both a Russian and a Prussian privy councillor. He was awarded four honorary doctorates and was an honorary citizen of the Technische Hochschule of Stuttgart. He belonged to various learned societies, notably the Academies of Berlin, Göttingen, Halle, and Vienna, and was an honorary member of the Russian Academy of Sciences, the Bunsen Society, the German Society for Metallurgy, the Royal Chemical Society, and (with Rutherford and Einstein) the British Institute of Metals.

Tammann's more than five hundred scientific writings extend over a very broad range of subjects and constitute a varied sequence devoted to both fundamental problems and detailed questions. Accordingly, Wilhelm Biltz applied to Tammann a remark of Goethe's: "I have chiseled giants out of marble and cut tiny figurines out of ivory." It is impossible in a few pages to do justice to this rich body of work, which has stimulated research in countless ways.

The theory of dilute solutions proposed by van't Hoff and Arrhenius left an important question unanswered. This deficiency greatly troubled Tammann, who on this question followed the views of Mendeleev. Specifically, in this theory the solutes were considered, by means of a schematic analogy, as ideal gases; however, the reciprocal effects on the solvent were neglected. In order to take these effects into account, Tammann developed the theory of internal pressure in homogeneous systems. According to his theory, the solution behaves like the solvent; the pressure is replaced, in both molecular and ionic solutions, by an attractive force between the solvent and the solute. In experiments begun in 1893 Tammann obtained pressures as high as about 4,000 atmospheres, somewhat greater than those previously reached by Émile Amagat. In 1907 Tammann collected in book form the results of many experiments on various solvents and presented the derivation of his theory. He showed that the temperature-dependence of many properties of solutions (including specific heats, viscosity, surface tension, and optical constants), as well as the pressure-dependence of their conductivity and compressibility, could be understood in the light of his new theory, which has been quite generally confirmed. In a work on compressibility published in 1895, Tammann employed a formula that he attributed to P. G. Tait and that has since been widely known under the latter's name. Only recently has it been pointed out that Tammann's formula is not at all identical with the one given by Tait in 1888.[3]

The field of heterogeneous equilibria—that is, the behavior of matter as a function of pressure, temperature, and chemical composition—was opened by Willard Gibbs's formulation of the phase rule; but it was van der Waals and, above all, Roozeboom who first recognized its outstanding practical significance. When Tammann entered this field in 1895, he treated the problem in its most general form before turning to applications; this allowed him to make substantial contributions to the systematization of inorganic chemistry as well as to the improvement of industrial production methods. The starting point of his research was twofold: the experiments of Thomas Andrews and Louis Cailletet on the equilibrium relationships between the liquid and vapor phases and the theoretical interpretation of the triple point. Many scientists, including Ostwald, believed that a continuous transition exists between the liquid and solid states corresponding to that between the liquid and vapor states. Tammann broke sharply with this conception in 1896. Supporting his argument with experiments and drawing on thermodynamical considerations, he put forward his own views in a monograph published in 1903. He asserted that the melting-point curve cannot end in a critical point; that the pressure-temperature diagram is basically a closed curve; and that, accordingly, all transitions from the crystalline state to other phases must be discontinuous.

Tammann modified his theory in a second monograph (1922), in which he stated that the melting curves must show a maximum (which under certain circumstances might not be a true maximum) and that, as a result, in typical cases the melting temperature ought to decrease again at very high pressures. He confirmed this prediction experimentally in the case of Glauber's salt, but the example was not entirely conclusive. Although several further, unobjectionable examples became known, it was later shown by P. W. Bridgman that up to extremely high pressures, melting-point curve maximums are the exception. This finding meant that Tammann's theory was of little practical significance. Nevertheless, it led him to the insight that the accepted division of matter into three states of aggregation (gas, liquid, and solid) was unsuitable. Anticipating the results of radiography, he postulated instead a twofold division consisting of an isotropic phase (gas, liquid, and amorphous) and an anisotropic phase (crystalline).

Parallel to this research was the series of studies that Tammann began in 1897 on the transition from the isotropic to the anisotropic phase. He showed that crystal growth depends on three independent quantities: the number of nuclei (that is, the number of crystallization centers), the crystallization velocity, and the heat flow. He pursued the study of these fundamental crystallization laws under varied external conditions and on a large number of substances. To a certain extent Tammann's theoretical work in this field culminated in the book *Aggregatzustände* (1922). Starting with notions taken from atomic theory and thermodynamics, he derived a new definition of phase stability; in this undertaking Gibbs's concept of free enthalpy provided the most significant criterion.

Guided by only a few preliminary studies (for example, those of Floris Osmond), Tammann developed a method of determining the chemical composition of a compound from the form of its cooling curve. By 1903 he had perfected the indispensable method of "thermic analysis," which has since been used successfully over a broad range of applications. Tammann employed it to explain systems composed of sulfides and chlorides as well as mixed crystal systems. Most important, however, it provided him with the means of opening up and systematically exploring an important field of inorganic chemistry, the intermetallic compounds.

Until this time it had not been possible to determine whether the fusion of two metals produced compounds of these metals, mixed crystals, or heterogeneous crystal mixtures. When Tammann began his studies of metallic compounds in 1903, little was known about them except what could be found in the works of Le Châtelier, Osmond, Roberts-Austen, C. T. Heycock and F. H. Neville, E. Heyn, F. Wüst, and N. S. Kurnakov, to which he referred. He systematically investigated the field through a combined application of thermic analysis and the microscopic analysis of sections, a technique in use since its development by Schreibers and Widmannstätten. Tammann's goal was to examine the 190 possible series of alloys of twenty common metals taken in mixing proportions that varied in steps of 10 percent by weight. (The total number of alloys was therefore 1,900.) By 1906, with the help of his students the project had progressed to the point that he was able to publish "Ueber die Fähigkeit der Elemente, miteinander Verbindungen zu bilden," which began at the point that Berzelius had reached in his work on the subject. In this article Tammann proved that, in general, the valence relationships and stoichiometric laws of the salts are not valid for metal-compound crystals. He also recognized that the alloys often behave like mixed crystals. World War I was a quiet period for Tammann, who used these years to probe more deeply into the nature of the mixed crystals.

Tammann's research in this area culminated in 1919 in his frequently cited publication on the resistance limits of binary systems as a function of the mixing proportion; in it he set forth the so-called $n/8$ law, which in its most rudimentary form had been used since the Middle Ages in the separation of gold and silver by means of aquafortis (nitric acid). From this law Tammann concluded that the atoms in mixed crystals are not arrayed on a statistical basis but are arranged in accordance with definite mathematical relationships. He realized that his theory of "superlattices" in mixed crystals could be definitively demonstrated only through radiographic inspection. His prediction concerning the superlattices was, in fact, corroborated radiographically for the gold-copper system in 1925 by C. H. Johansson and J. O. Linde. The conjectured relation between resistance limits and superlattices, however, has not proved to be correct; this was first suspected by G. Masing and was later demonstrated, for example, in the gold-silver system.

As Tammann progressed in the interpretation of the chemical properties of alloys, he became increasingly interested in the physical and chemical behavior of metals, in their crystal structure, in their electrical conductivity, and in their mechanical and other properties. Through his study of these topics he opened the field of metal physics. Tammann first addressed himself to two questions that preoccupied him for the rest of his life: What makes it possible for metals in the solid state to be worked? Why do properties of metals change so drastically during the process of cold-working?

Before Tammann began his research in this area, Otto Mügge had shown in the case of salts and transparent minerals that mechanical stresses in crystals produce displacement of their parts along the slip planes. Further, J. A. Ewing and W. Rosenhain had already begun to furnish the first answers concerning the plastic working of metals. Tammann's extended series of works on this subject, which have greatly influenced the techniques of metalworking, began in 1910 with a publication written in collaboration with Otto Faust. Having derived the malleability of metals in the solid state from crystallographic slipping, Tammann, accordingly, saw crystalline rearrangement as the cause of the alterations in the mechanical properties of metals during cold-working, especially the hardening.

It was known that through tempering (heating), the values that certain properties of metals have acquired in the course of being cold-worked return to the levels at which they were before the cold-working began. Tammann explained this phenomenon—which Sorby called recrystallization—as the result of the accumulation of energy during cold-working and of the growth of certain individual crystals at the expense of others. Through repeated recrystallizations Tammann was able to alter the size of crystal grains in metals within broad limits, and under suitable conditions he could even grow single crystals. Such crystals normally do not grow without limit—a fact he attributed to the existence of a *Zwischensubstanz*, a spongelike network consisting of the impurities that always occur along the boundaries of crystal grains. This network, which Tammann isolated in 1921 by elegant etching techniques, could not yet, however, explain other changes that occur during cold-working, such as in the density, color, electrical conductivity, and chemical reactivity of the metals. Tammann conjectured that to account for these changes, it would be necessary to assume the occurrence of alterations within the atoms. Of importance in this connection was his research on the binary-state diagrams of iron and its technically important alloys, on passivity (especially of the iron-chromium alloys), and on iron carbide.

Tammann's illuminating findings about the physical and chemical bases of the metallurgical production processes first emerged from his research on the equilibria between molten metal and slag during the cooling processes in the interior of the earth. Emil Wiechert, on the basis of observations of earthquakes, had deduced the existence of at least three layers in the earth; and in 1924, Tammann postulated the existence of an intermediary sulfide layer between the outer silicate layer and the earth's iron-nickel core. His application of these conceptions to the techniques of steel production was an innovative and valuable contribution. Tammann also studied meteorites and silicates. In a short series of mineralogical and chemical communications issued by the institute of physical chemistry at Göttingen, he published phase diagrams of silicates and discussed the production and thermochemistry of these compounds. He later made a careful study of nontronite and kaolin. In 1925 he recognized that the concept of the molecule must be modified in dealing with silicates and that a chemical constitutional formula is not meaningful for them.

Since the 1890's the examination of the influence of pressure and temperature on matter had directed Tammann's attention to the phenomenon of allotropy or polymorphism. In *Kristallisieren und Schmelzen* (1903) he summarized his experiments on this subject and showed that the phenomenon is much more common than had been expected. His discovery of new modifications of ice (specifically, ice II and ice III) aroused intense interest; and Tammann himself thought that their discovery, along with that of resistance limits, constituted one of his two most noteworthy single scientific contributions. Tammann described the behavior of compressed liquids in 1911 in an approximate but surprisingly simple equation of state similar to one enunciated by O. Tumlirz. In 1915 he solved the much-disputed problem of the flow of glacial ice by showing that the phenomenon was the result of crystalline slipping. He had previously given an elegant theoretical refutation of the explanation proposed by Ostwald, Poynting, and Niggli, who attributed the flow to a pressure-dependent reduction in the melting point of ice.

Tammann's research on the nature of the states of matter had repeatedly led him to consider the glass state—he used the terms "glass" and "amorphous" synonymously—but he does not seem to have started his long series of studies on the glasses until 1925, when the Society of Glass Technology invited him to write a paper on the subject. In 1933 he published a monograph containing the results of his work, most of which was carried out in collaboration with his students. One of the most surprising findings was that the specific volume of a glass depends on the pressure at which it solidifies—dramatic evidence of the extraordinary complexity of the substances. Tammann was particularly intrigued by the "softening interval," the temperature interval within which glass changes from brittle to viscous. According to Tammann's theory of the states of aggregation, no sudden changes in properties should occur in this temperature region. He also showed that although, under certain conditions, many physical properties do change very considerably in this interval, they always change in a continuous manner.

As early as 1911 Tammann's experiments on diffusion in mixed crystals led him into a new field, the study of reactions of solid bodies with other solids and with gaseous substances. His findings led him to break with the old maxim *corpora non agunt, nisi fluida*. He determined the temperature (later named for him) at which mixtures of crystalline powders sinter and thereby laid the foundation of solid-state chemistry, a field later developed by J. A. Hedvall and W. Jander, among others. Tammann's investigation of the tarnish that forms on metallic surfaces was of the greatest importance for the theory of oxidation. In 1919 he stated that the layer of tarnish grows parabolically with time.[4] (This law was later explained theoretically by Carl Wagner.) In 1922, while examining other cases of oxidation, Tammann and Werner Köster found a logarithmic relation between the thickness of the oxidized layer and the time elapsed. The theoretical significance of this relation was not recognized, however, until forty years later.

Although many of Tammann's works had direct technical applications, his personal goal was the development of pure science, and in the final analysis he was concerned only with the search for the laws of nature. He dismissed verbose and flashy scientific writing with the remark, "It records the artist's earthly pilgrimage, which no one needs to know about." He had a remarkable gift for reducing complex problems to simple questions, which he then solved by means of experiments that often were astonishingly simple. This ability was exemplified in his experiments on resistance limits and on the determination of the thickness of oxidation layers. Two other examples are the "Tammann oven," in which a carbon tube serves as both wall and electrical heating element, and his apparatus for measuring outflowing liquids. Tammann's theoretical work did not always attract the interest of his contemporaries; and since later research has often taken paths that diverged from those he followed, many problems that he isolated remain unsolved. Although not an especially talented lecturer, Tammann was an unusually effective teacher. By having trained more than one hundred doctoral candidates and assistants, he helped to determine the conceptions and working methods of an entire generation of chemical physicists and metallurgists.

NOTES

1. The results are set forth in a patent application.
2. See *Chemiker-Zeitung*, **21** (1897), 654.
3. See A. T. J. Hayward, "Compressibility Equations for Liquids—A Comparative Study," in *Journal of Physics*, sec. D, Applied Physics, **18** (1967), 965.
4. A "Commemorative Symposium on the Oxidation of Metals—50 Years of Research" was organized in 1970 in Atlantic City, N.J., by the Electrochemical Society. The introductory lecture, given by C. Wagner, appeared in German in *Werkstoffe und Korrosion*, **21** (1970), 886–894. DECHEMA (Deutsche Gesellschaft für Chemisches Apparatewesen) organized a colloquium on the same subject, held on 23 Oct. 1970; for the lectures presented there, along with other papers on the subject, see *ibid.*, nos. 11–12.

BIBLIOGRAPHY

I. ORIGINAL WORKS. Tammann's posthumous papers and his autobiographical remarks written for his son, Heinrich, are in the possession of the author. Some autobiographical fragments were published in "Jugenderinnerungen eines Dorpater Chemikers," in *Eesti rohuteadlane* (Tartu), nos. 9–10 (1930), 1029–1034; and in "Die Gründung des Instituts für anorganische Chemie," in *Mitteilungen des Universitätsbundes Göttingen*, **16**, no. 1 (1934), 21–25; see also *ibid.*, **17**, no. 2 (1936), 42–45. For an account of his metallurgical works consult "Ueber die im Göttinger Institut für anorganische Chemie ausgeführten metallographischen Arbeiten," in *Zeitschrift für Elektrochemie*, **14** (1908), 789–804. The correspondence between Arrhenius (51 letters) and Tammann (122 letters) is at the Kungliga Vetenskapsakademien, Stockholm.

Tammann's monographs are *Die Dampftensionen der Lösungen*, which is *Mémoires de l'Académie impériale des sciences de St.-Pétersbourg*, 7th ser., **35**, no. 9 (1887); *Kristallisieren und Schmelzen, ein Beitrag zur Lehre der Aenderungen des Aggregatzustandes*

(Leipzig, 1903); *Ueber die Beziehungen zwischen den inneren Kräften und Eigenschaften der Lösungen* (Hamburg–Leipzig, 1907); *Lehrbuch der Metallographie, Chemie und Physik der Metalle und ihrer Legierungen* (Leipzig–Hamburg, 1914; 2nd ed., 1921; 3rd ed., Leipzig, 1923), the 4th ed. of which, *Lehrbuch der Metallkunde* (Leipzig, 1932), was translated from the 3rd ed. into English by R. S. Dean and L. G. Swenson as *A Textbook of Metallography* (New York, 1925) and into Russian (Moscow–Leningrad, 1935); *Die chemischen und galvanischen Eigenschaften von Mischkristallreihen und ihre Atomverteilung, zum Gedächtnis der Entdekkung des Isomorphismus vor 100 Jahren*, a special issue of *Zeitschrift für anorganische und allgemeine Chemie* (Leipzig, 1919); *Aggregatzustände, die Änderung der Materie in Abhängigkeit von Druck und Temperatur* (Leipzig, 1922; 2nd ed., 1923), translated into English by R. F. Mehl as *The States of Aggregation* (Princeton, 1925); *Der Glaszustand* (Leipzig, 1933), translated into Russian (Moscow, 1935); and *Lehrbuch der heterogenen Gleichgewichte* (Brunswick, 1934), translated into Russian (Moscow, 1935).

Tammann prepared a bibliography of his works up to 1901 that was published in G. V. Levitsky, ed., *Biografichesky slovar professorov i prepodavateley Imperatorskago yurievskago, byvshago Dertpskago, universiteta*, I (Yur'ev, 1902), 257–259. An almost complete list of Tammann's journal articles can be found in Poggendorff, IV, 1474; V, 1240–1241; VI, 2610–2612; VIIa, 623–625.

The articles can be grouped into the following main categories: (1) physiology (more than 10 papers); (2) inorganic chemistry (nearly 10 papers, one of which gives for the first time the correct constitutional formula of H_2O_2); (3) solutions, vapor tensions, and osmosis (almost 50 papers); (4) phase rule and state of aggregation (over 70 papers); (5) metallography and metallurgy (a series of 123 [incorrectly numbered 1–121] "Metallographische Mitteilung" were published by him and his co-workers in *Zeitschrift für anorganische Chemie* [1904–1925], plus about 80 additional papers); (6) glasses (about 40 papers); (7) chemical reactions in solids and the oxidation on surfaces (over 20 papers); and (8) geochemistry, silicates, and meteorites (about 15 papers).

From 1904 to 1939 Tammann was coeditor of 200 vols. of *Zeitschrift für anorganische und allgemeine Chemie*.

II. SECONDARY LITERATURE. On the history of Tammann's family see *Deutsches Geschlechterbuch*, CXLII (Marburg, 1967), 373–391. The "Festschrift zum 65. Geburtstag von Gustav Tammann" constitutes all of *Zeitschrift für anorganische und allgemeine Chemie*, **154** (1926). The two best evaluations of Tammann's work as a whole are W. Biltz, "Gustav Tammann zum siebzigsten Geburtstag," *ibid.*, **198** (1931), 1–31; and W. E. Garner, "The Tammann Memorial Lecture," in *Journal of the Chemical Society* (1952), 1961–1973. For discussions of individual aspects

of Tammann's work see A. Portevin, "La méthode d'analyse thermique et les travaux sur les alliages au laboratoire du Professeur Tammann," in *Revue de Métallurgie (Mémoires)*, **4** (1907)–**6** (1909); W. Fraenkel, "Die neuen Forschungen G. Tammanns über Mischkristalle," in *Naturwissenschaften*, **8** (1920), 161–166; F. Körber, "Kristallisieren und Schmelzen," in *Zeitschrift für Metallkunde*, **23** (1931), 134–137; G. Grube, "Die Forschungen G. Tammanns über die Konstitution der Legierungen," *ibid.*, 137–138; G. Masing, "Tammanns Untersuchungen über Kaltreckung, Verfestigung und Rekristallisation," *ibid.*, 139–142; and W. Köster, "Arbeiten von G. Tammann über die chemischen Eigenschaften von Metallen und Legierungen," *ibid.*, 142–146.

The most useful obituaries of Tammann are G. Masing, "Gustav Tammann 1861–1938," in *Berichte der Deutschen chemischen Gesellschaft*, sec. A, **73** (1940), 25–30; and "Gustav Tammann†," in *Zeitschrift für Elektrochemie*, **45** (1939), 121–124, also in *Metall und Erz*, **37** (1940), 189–192; H. O. von Samson-Himmelstjerna, "Gustav Tammann," in *Umschau in Wissenschaft und Technik*, **43** (1939), 88–90; and an unsigned article in *Nachrichten von der Gesellschaft der Wissenschaften zu Göttingen*, *Jahresbericht* for 1938–1939 (1939), 54–66. The satirical remarks on the Third Reich in W. Biltz's obituary, "Gustav Tammann†," in *Zeitschrift für anorganische und allgemeine Chemie*, **240** (Jan. 1939), 114–115, reveal Biltz's political position—which was also Tammann's—and contributed to Biltz's early retirement. Other articles are cited in Poggendorff, VI, 2610, and VIIa, 623.

See also W. Köster, "Zum 100. Geburtstag von Gustav Tammann," in *Metall*, **15** (1961), 704–706, also in *Zeitschrift für Metallkunde*, **52** (1961), 379–381. Several authentic anecdotes about Tammann are reported in J. Hausen, *Was nicht in den Annalen steht*, 2nd ed. (Weinheim, 1958). S. Boström's article "Gustav Tammann," in *Baltische Hefte*, **10** (1964), 139–150, is unreliable and useful, at best, for several anecdotes.

G. A. TAMMANN

TANFILEV, GAVRIIL IVANOVICH (*b.* Tallinn, Russia, 6 March 1857; *d.* Odessa, U.S.S.R., 4 September 1928), *geography, phytogeography, soil science.*

Tanfilev began to study the flora of the chernozem steppes as a student at the Faculty of Physics and Mathematics of St. Petersburg University. After graduating in 1883, he worked in the Department of Agriculture of the Ministry of State Lands (1884–1892), on an expedition to study methods of forest and water management on the Russian steppes (1893–1894), in the St. Petersburg Botanical Garden (1895–1904), and on the Soil

Commission of the Free Economic Society (1888–1905). These activities were accompanied by field investigations of soil and vegetation in European Russia and in western Siberia. Tanfilev participated in the compilation of the soil map of European Russia and was awarded the Great Gold Medal at the Paris World Exhibition in 1900.

In his master's thesis, *Predely lesov na yuge Rossii* ("The Boundaries of the Forest in Southern Russia," 1894), Tanfilev concluded that the absence of forest on the steppe is a result of increased soil alkalinity and bedrock, both due to the dry climate. The lack of forest in the tundra was explained by the marshiness of its soils, the low temperature, and the permafrost, which destroys the roots of trees. In the forest-tundra and forest-steppe border regions, Tanfilev held, there is a constant struggle that results in the dislocation of the zonal boundaries. His conclusions on the battle between forest and steppe and between tundra and forest provoked a heated discussion among geographers, phytogeographers, and soil scientists that has continued unresolved to the present time.

Developing Dokuchaev's ideas on the zonal structure of the Russian landscape, Tanfilev studied the physicogeographical regionalization of European Russia (1897) and five years later published his classic work *Glavneyshie cherty rastitelnosti Rossii* ("Main Features of the Vegetation of Russia"), with a brilliant analysis of the broad zones of plant cover on the plains and the vertical belts in the mountains of the Crimea, Caucasus, and Turkistan. The most important feature of this work is the historical approach to the formation of vegetation zones in post-Tertiary time.

For more than twenty years, Tanfilev studied the cultural geography of plants. His research in this area culminated in *Ocherk geografii i istorii glavneyshikh kulturnykh vastenii* ("Sketch of the Geography and History of the Main Cultivated Plants," 1923).

The breadth of his geographical outlook and his knowledge of natural history based on personal research resulted in a major work on the physical geography of Russia. Four volumes appeared during his lifetime (1916–1924) and the fifth was published posthumously (1931). With its exhaustive bibliography, it was the most detailed and complete collection of information on the natural history of Russia until the early 1930's.

In addition to his research, Tanfilev taught at the universities of St. Petersburg (1895–1903) and Novorossysk, in Odessa (1904–1928).

BIBLIOGRAPHY

I. ORIGINAL WORKS. Tanfilev's *Geograficheskie raboty* ("Geographical Works"; Moscow, 1953) includes a bibliography of his writings.

II. SECONDARY LITERATURE. See S. T. Belozerov, *Gavriil Ivanovich Tanfilev* (Moscow, 1951), which includes a bibliography of his writings and of secondary literature; and "Gavriil Ivanovich Tanfilev," in *Otechestvennye fiziko-geografy i puteshestvenniki* ("Native Physical Geographers and Travelers"; Moscow, 1959); L. S. Berg, "Gavriil Ivanovich Tanfilev," in *Priroda*, **17**, no. 10 (1928); and A. A. Borzov, "Professor Gavriil Ivanovich Tanfilev," in *Zemlevedenie*, **30**, no. 4 (1928).

A. I. SOLOVIEV

TANNERY, JULES (*b.* Mantes-sur-Seine, France, 24 March 1848; *d.* Paris, France, 11 November 1910), *mathematics.*

Tannery was the youngest of the three children of Delphin Tannery, an engineer with the Compagnie des Chemins de Fer de l'Ouest. The eldest child was a daughter and the second was the engineer and historian of science Paul Tannery. The family moved first to Redon, in Ille-et-Vilaine, where his father supervised the construction of a railroad line, and then to Mondeville near Caen.

At the *lycée* in Caen, he was an excellent student, and he won several prizes in the *concours général*. His brother, who was passionately interested in philosophy and Greek antiquity, gave him a taste for these subjects. In 1866 Tannery was admitted with highest standing to the science section of the École Normale Supérieure and, simultaneously, to the École Polytechnique. He decided to enter the École Normale, and in 1869 placed first in the *agrégation*. He was then assigned to teach mathematics at the *lycée* in Rennes, and in 1871 he was named to a post at the *lycée* in Caen, where his former classmate Émile Boutroux was also teaching.

During this period Tannery underwent a religious crisis caused by his profound desire to admire without remorse pagan antiquity, the cult of reason, and the ideas of Lucretius.

Tannery returned to Paris in 1872 as *agrégé-préparateur* of mathematics at the École Normale. Encouraged by Hermite, he began work on a thesis inspired by the works of Fuchs ("Propriétés des intégrales des équations différentielles linéaires à coefficients variables"), which he defended in 1874. Two years later he became editor of the *Bulletin des sciences mathématiques*, on which he collaborated with Darboux, Hoüel, and Picard until his death. He wrote a great number of book reviews for the journal—more than 200 for the years 1905–1910 alone. Characterized by rigorous criticism and an excellent style, the reviews are models of their kind in both form and content.

Tannery taught higher mathematics at the Lycée Saint-Louis and substituted for the professor of physical and experimental mechanics at the Sorbonne. In 1881 he was named *maître de conférences* at the École Normale and, shortly afterward, at the École Normale for women located in Sèvres. From 1884 until his death Tannery served as assistant director of scientific studies at the École Normale; in this post he displayed the full measure of his abilities. At the same time, from 1903, he was professor of differential and integral calculus at the Faculty of Sciences of Paris.

A member of several educational commissions and of the Conseil Supérieur de l'Instruction Publique, Tannery played an important role in the pedagogical reforms in France at the beginning of the twentieth century. Through his lectures and supervisory duties at the École Normale this gifted teacher gave valuable guidance to many students and inspired a number of them to seek careers in science (for example, Paul Painlevé, Jules Drach, and Émile Borel). Tannery was elected *membre libre* of the Académie des Sciences on 11 March 1907, replacing Paul Brouardel.

Tannery possessed considerable gifts as a writer. The pure and elegant style of the poems he composed in his free hours clearly bears the stamp of a classic sensibility. His vast culture, nobility of character, and innate sense of a rationally grounded morality are reflected in each of his *Pensées*, a collection of his thoughts on friendship, the arts, and beauty. Often they exhibit a very refined sense of humor.

Among his scientific publications, the *Introduction à la théorie des fonctions d'une variable* exercised an especially great influence on younger generations of mathematicians. Émile Borel stated that it was a profound, vigorous, and elegant work that taught him how to think. In another book, written with Jules Molk, Tannery presented the results of applying Fuchs's theorems to the linear differential equation that defines the periods of an elliptic function. Tannery also gave a new expansion of the Euler equation. In algebra, following the path opened by Hermite, Tannery studied the similar transformations of the quadratic forms, the invariants of the cubic forms, and the symmetric functions. In geometry, he concentrated his research

on the osculating plane of skewed cubic equations and on a fourth degree surface of which the geodesic lines are algebraic. Poincaré highly esteemed Tannery and commented very favorably on his writings. Tannery's work was known abroad, especially in Germany, where a translation of his book *Notions de mathématiques* was published in 1909.

In 1880 Weierstrass published "Zur Funktionenlehre," in which he dealt with the convergence of a series whose terms are rational functions of one variable. Upon reading it, Tannery sent Weierstrass solutions he obtained in a simpler manner, utilizing elementary theorems of function theory. Weierstrass translated Tannery's letter into German and published it in *Monatsberichte der königlich-preussischen Akademie der Wissenschaften zu Berlin* (1881, 228–230).

Tannery reflected a great deal on the role of number in science, and he sought to show how the entire subject of analysis could be built up on the basis merely of the notion of whole number. In his speculations on the notion of infinity, he arrived at the conclusion that it is equivalent to the simple possibility of indefinite addition. Finally, his interest in the history of science—undoubtedly inspired by his brother—led him to publish Galois's unpublished manuscripts and the correspondence between Liouville and Dirichlet.

Galois had entrusted his manuscripts to his friend Auguste Chevalier, who gave them to Liouville. The latter bequeathed his library to one of his sons-in-law, Célestin de Blignières (1823–1905), a former student at the École Polytechnique and a disciple of Auguste Comte. Mme de Blignières, Liouville's daughter, in turn, gave Galois's papers to Tannery, along with her father's correspondence with Dirichlet.

In his *Éloges et discours académiques* (p. 101), Émile Picard drew the following parallel between Jules and Paul Tannery:

They were extremely close all their lives. Of very different natures, the two brothers complement each other. Paul derived a certain tranquillity from his positivist convictions. A philologist and scholar of extraordinary erudition, he sought to follow, in innumerable notes and articles, the historical evolution of science from Greek antiquity until the end of the seventeenth century. Jules's philosophy, on the other hand, did not free him from intellectual anxiety. His outlook was less universal than his brother's, but also more profound. He had both the subtle mind of the metaphysician and the penetrating insight of the disillusioned moralist.

BIBLIOGRAPHY

I. ORIGINAL WORKS. Tannery's books include *Introduction à la théorie des fonctions d'une variable* (Paris, 1886), 2nd ed., 2 vols. (1904–1910); *Eléments de la théorie des fonctions elliptiques*, 4 vols. (Paris, 1893–1902, with Jules Molk); *Leçons d'arithmétique théorique et pratique* (Paris, 1894; 7th ed., 1917); *Introduction à l'étude de la théorie des nombres et de l'algèbre supérieure*, Émile Borel and Jules Drach, eds. (Paris, 1895), taken from Tannery's lectures at the École Normale Superieure; *Notice sur les travaux scientifiques de M Jules Tannery* (Paris, 1901); *Notions de mathématiques* (Paris, 1903), German trans. (Leipzig, 1909), with historical notes by Paul Tannery; *Leçons d'algèbre et d'analyse à l'usage des classes de mathématiques spéciales*, 2 vols. (Paris, 1906), with Paul Tannery; *Liste des travaux de Paul Tannery* (Bordeaux, 1908), prepared by P. Duhem and preceded by obituaries written by Duhem and J. Tannery; and *Science et philosophie* (Paris, 1912), with a brief article by Émile Borel.

Tannery's articles include "Sur l'équation différentielle linéaire qui relie au module la fonction complète de première espèce," in *Comptes rendus hebdomadaires des séances de l'Académie des sciences*, **86** (1878), 811–812; "Sur quelques propriétés des fonctions complètes de première espèce," *ibid.*, 950–953; "Sur les intégrales eulériennes," *ibid.*, **94** (1882), 1698–1701; and "Sur les fonctions symétriques des différences des racines d'une équation," *ibid.*, **98** (1884), 1420–1422; "Les Mathématiques dans l'Enseignement secondaire," in *La revue de Paris*, **4** (1900), 619–641; "Principes fondamentaux de l'arithmétique," with J. Molk, in *Encyclopédie des sciences mathématiques*, pt. 1, I (Paris, 1904), 1–62; "Sur l'aire du parallélogramme des périodes pour une fonction pu donnée," in *Bulletin des sciences mathématiques*, **28** (1904), 108–117; "Paul Tannery," in *Comptes rendus du IIe Congrès international de philosophie* (Geneva, 1905), 775–797; "Manuscrits et papiers inédits de Galois," in *Bulletin des sciences mathématiques*, **30** (1906), 226–248, 255–263, and **31** (1907), 275–308; "Correspondance entre Liouville et Dirichlet," *ibid.*, **32** (1908), 47–62, 88–95, and **33** (1909), 47–64; "Discours prononcé à Bourg-la-Reine" (at the inauguration of a plaque placed on the house in which Galois was born), *ibid.*, **33** (1909), 158–164; "Pour la science livresque," in *Revue de métaphysique et de morale*, **17** (1909), 161–171; and "Pensées," in *La revue du mois*, **11** (1911), 257–278, 399–435.

II. SECONDARY LITERATURE. The first obituaries of Tannery are the addresses given by P. Painlevé and É. Picard on 13 and 14 November 1910, in *Bulletin des sciences mathématiques*, **34** (1910), 194–197. These were followed by É. Borel, "Jules Tannery, 24 mars 1848–11 novembre 1910," in *La revue du mois*, **11** (1911), 5–16; Émile Hovelaque, "Jules Tannery," in *La revue de Paris*, **1** (1911), 305–322; and A. Châtelet, "Jules Tannery," in *Enseignement mathématique*, **13** (1911), 56–58.

A small book of 140 pages entitled *En souvenir de Jules Tannery MCMXII* was published by subscription by Tannery's friends in 1912; it contains an address by Ernest Lavisse, director of the École Normale, a biographical article by Émile Boutroux, and a selection of Tannery's "Pensées." See also Émile Picard, "La vie et l'oeuvre de Jules Tannery membre de l'Académie," in *Mémoires de l'Académie des sciences de l'Institut de France*, **58** (1926), i–xxxii; the same article, with a few revisions, appeared as "Un géomètre philosophe: Jules Tannery," in *La revue des deux mondes*, **31** (1926), 858–884, and in Picard's *Éloges et discours académiques* (Paris, 1931), 51–104.

See also Poggendorff, III, 1324; IV, 1476; V, 1242; and G. Sarton, "Paul, Jules et Marie Tannery," in *Isis*, **38** (1947–1948), 33–51, which in addition contains a list of Jules Tannery's works on pp. 47–48.

PIERRE SPEZIALI

TANNERY, PAUL (*b.* Mantes-la-Jolie, Yvelines, France, 20 December 1843; *d.* Pantin, Seine–St. Denis, France, 27 November 1904), *history of science, history of mathematics.*

An engineer and administrator by profession, Tannery could devote only his leisure hours to scholarship. Despite this limitation, however, he accomplished a vast amount of penetrating and wide-ranging research and became one of the most influential figures in the rapidly developing study of the history of science at the beginning of the twentieth century. Like his younger brother, Jules, who later became a mathematician, Tannery early received a deeply Christian education from his parents, S. Delphin Tannery, an engineer who worked for railroad companies, and the former E. Opportune Perrier. After proving to be a brilliant pupil at a private school in Mantes, Tannery attended the *lycées* of Le Mans and Caen, where he showed great enthusiasm for the classics, although he had enrolled as a science student. His philosophy teacher, Jules Lachelier, communicated to Tannery a passion for the subject and strengthened his interest in classical antiquity. In 1860 Tannery fulfilled his father's hopes by obtaining one of the highest scores on the competitive entrance examination for the École Polytechnique, where he acquired a solid education in science and technology but devoted much time to other subjects as well. In particular he began to learn Hebrew and developed a strong interest in the teaching of mathematics.

Upon graduating from the École Polytechnique in 1863, Tannery entered the École d'Application des Tabacs as an apprentice engineer. At this time he read Auguste Comte's *Cours de philosophie positive*, an initiation into positivist philosophy that so profoundly influenced him that years later he approached the study of the history of science as a spiritual disciple of Comte.

After working for two years as an assistant engineer at the state tobacco factory in Lille, Tannery was transferred in 1867 to an administrative post at the headquarters of the state tobacco administration in Paris, where he enjoyed a more active intellectual and artistic life. He served in the Franco-Prussian War as an artillery captain and was present during the siege of Paris. An ardent patriot, he was deeply affected by the defeat and never consented to acknowledge the terms of the peace treaty as definitive. Upon demobilization Tannery resumed his former duties. At the same time he eagerly studied philosophy and mathematics, subjects that he discussed with his brother, Jules, who taught at Caen and later at the École Normale Supérieure of Paris, and with such young philosophers as É. Boutroux. In 1872 the tobacco administration sent Tannery to supervise the construction of several buildings in the Périgord region. While there he became seriously ill and was obliged to convalesce for a long period. He used this time to further his knowledge of ancient languages, acquiring a mastery of this field that was evident in his very first publications.

In March 1874 Tannery began to direct an extensive construction project at the state tobacco factory of Bordeaux. This university city had a very active intellectual life, and he soon decided to spend his leisure time investigating various topics in the history of the exact sciences in antiquity, as well as a number of philosophical and philological questions. From 1876 Tannery participated in the work of the Société des Sciences Physiques et Naturelles de Bordeaux and published many studies in its *Mémoires* and in the *Revue philosophique de la France et de l'étranger*, which had recently been founded at Paris. He gradually began to send material to other journals, eventually becoming a fairly regular contributor to about fifteen French and foreign periodicals. He published hundreds of memoirs, articles, notes, and reviews while pursuing a brilliant career in the state tobacco administration. Although many other historians of science have been obliged to conduct their research concurrently with their professional activities, none of them seems to have produced a body

of work comparable to Tannery's in scope and importance.

Although his stay at Bordeaux had proved enriching and fruitful, Tannery soon ended it. In 1877, at his own request, he was appointed engineer at the tobacco factory of Le Havre, a city with intellectual resources far inferior to those of Bordeaux but near the region of Caen, where Tannery's parents lived. He continued, however, to take a lively interest in Greek science; and his survey of mathematics at the time of Plato ("L'éducation platonicienne"), published in the *Revue philosophique de la France et de l'étranger*, was enthusiastically received and was translated into English and German. Meanwhile, his professional obligations, already considerable, become still greater in 1880, when he became acting director of the tobacco factory.

In June 1881 Tannery married Marie-Alexandrine Prisset (1856–1945), daughter of a well-to-do notary in Poitiers. Although she had received only a modest education, his young wife encouraged Tannery to pursue his scholarly research. Several trips abroad during this period enabled Tannery to meet leading scholars, notably J. L. Heiberg, H.-G. Zeuthen, G. Eneström, and M. Cantor, with whom he maintained close and fruitful relationships. Since his situation at Le Havre provided little encouragement for ,his research, however, Tannery soon sought a transfer to Paris.

His request was granted, and in July 1883 he was named appraiser-engineer in a Paris tobacco factory. Once again he was able to devote all his leisure time to scholarship. Although relatively brief, this Paris period was extremely productive. Tannery's principal area of interest was the history of mathematics; he gave a private course on the subject at the Faculty of Sciences in 1884–1885 and published an important series of articles on Greek geometry in the *Bulletin des sciences mathématiques*. He also pursued studies already under way on the origins of Greek science and on various philological questions. In addition he printed previously unpublished Greek texts, as well as original studies on a wide range of topics. Research at the Bibliothèque Nationale and a scholarly visit to Italy enabled him to begin work on two important editorial projects: an edition of the manuscripts of Diophantus, which was entrusted to him in 1883, and one of Fermat's works, for which he received a joint commission with Charles Henry in 1885.

At the end of 1886 Tannery had to leave Paris, in order to direct the tobacco factory at Tonneins

(Lot-et-Garonne). Deprived of the resources of the Bibliothèque Nationale, he was limited to editorial work and to perfecting his manuscripts. He revised and completed a series of articles that had been appearing in the *Revue philosophique de la France et de l'étranger* since 1880 and presented them in book form as *Pour l'histoire de la science hellène. De Thalès à Empédocle*, his first separately printed publication. Tannery also regrouped and completed another series of articles, which had been appearing since 1884 in *Bulletin des sciences mathématiques*, into a second, shorter book, *La géométrie grecque, comment son histoire nous est parvenue et ce que nous en savons*, I. *Histoire générale de la géométrie élémentaire* (the only part to be published). In addition, he continued to prepare the edition of Fermat's works.

Promoted to director of the Bordeaux tobacco factory in January 1888, Tannery spent two years in the city in which he had first become aware of his vocation for history. Renewing contact with intellectual circles there, he became friendly with an amateur scholar, Polydore Hochart, who assisted him in collecting material on the Bordeaux correspondents of Mersenne—the first step in a great project that Tannery was not able to complete. He also worked on a study of Greek astronomy, in which he sought, through a very detailed analysis of the *Almagest*, to gain insight into the different theories outlined by Ptolemy. (The study was published in the *Mémoires de la Société des sciences physiques et naturelles de Bordeaux* in 1893.)

At the beginning of 1890 Tannery returned to the Paris headquarters of the state tobacco authority in order to organize the manufacture of matches and to give instruction in the relevant techniques to the apprentice engineers at the École d'Application des Tabacs. In 1893 he was appointed director of the factory at Pantin, near Paris, a post that he held until his death in 1904. Although this appointment entailed heavy administrative and social responsibilities, Tannery did a remarkable amount of research during this final period of his life. He regularly contributed articles, memoirs, notes, and book reviews to about fifteen journals and completed his editions of the works of Diophantus and of Fermat. Tannery also undertook vast new projects, such as collaborating on the *Histoire générale du IVe siècle à nos jours* of Ernest Lavisse and A. N. Rambaud, teaching at the Collège de France for several years, and preparing a new critical edition of the works and correspondence of Descartes. Through his regular correspondence with French and foreign colleagues and through his activities at several congresses, he laid the foundations for international collaboration in the history of science.

A rapid survey of the various aspects of Tannery's work provides some idea of the scope and importance of his accomplishments during these final years. He wrote some 250 articles, notes, and other communications on the most varied issues in the history of the exact sciences, the history of philosophy, and philology, most of them concerning antiquity, Byzantine civilization, and Western civilization from the Middle Ages to the seventeenth century; they occupy five volumes of his collected *Mémoires scientifiques*.

In the years immediately after 1893, however, Tannery concentrated most of his effort on completing his two major editorial projects. The two volumes of Diophantus' *Opera omnia* appeared in 1893 and 1895. (Tannery also began work on a French translation, but he did not complete it.) The three volumes of the *Oeuvres de Fermat*, which Tannery edited with Charles Henry, were published between 1891 and 1896. The first volume contains mathematical works and "Observations sur Diophante"; the second (1894) contains Fermat's correspondence; and the third consists of French translations of the writings and of Latin fragments by Fermat as well as of several texts by J. de Billy and J. Wallis. Tannery, who played the principal role in editing these volumes, also assembled material for a fourth (*Compléments . . .*), which was completed by Henry and published in 1912. A fifth volume, containing further supplementary material, was published by C. de Waard in 1922.

In 1892 Tannery agreed to substitute for Charles Lévêque in the chair of Greek and Latin philosophy at the Collège de France. Without fundamentally altering the character of the chair, Tannery sought to place greater emphasis on the history of ancient scientific thought and to illustrate its influence on the formation of modern science. Unfortunately, he did not publish any of his courses, and we have a record only of the main subjects he treated. These included Aristotle's *Physics* and *De caelo*, an interpretation of Plato, ancient theories of matter, the commentaries of Simplicius, and atomistic doctrines, as well as various currents of ancient philosophy and even fragments of Orphic poetry. At the end of the academic year 1896–1897, however, a new project began to occupy almost all Tannery's time: an edition of the works of Descartes. Accordingly, he gave up teaching and thereby renounced the possibility of succeeding Lévêque.

Tannery was interested in Descartes during his first stay in Bordeaux, though only in principle, since all his research during that period pertained to ancient thought. But from 1890 on the preparation of his edition of Fermat's work led him to make a thorough study of Descartes's correspondence and to publish a number of items that had remained in manuscript. The rigor of his editorial work and his deep knowledge of seventeenth-century science brought him, in 1894, the co-editorship (with the historian of modern philosophy Charles Adam) of a new critical edition of the works and correspondence of Descartes that was destined to replace the very dated, eleven-volume edition by Victor Cousin (1824–1826). In the last ten years of Tannery's life this undertaking, the scope and importance of which are obvious, absorbed a growing portion of his leisure. His first—and most difficult—task was the preparation of volumes I–V, devoted to Descartes's correspondence (published 1897–1903). He also participated in editing volumes VI (*Discours de la méthode* and *Essais* [1902]), VII (*Meditationes de prima philosophia* [1904]), and IX (*Méditations* and *Principes* [1904]), and left valuable notes for the other volumes published by Henry. This edition, called the "Adam-Tannery" Descartes, is too well-known to require detailed description here. A major contribution to the history of ideas, and especially to the history of science in the seventeenth century, it sparked a renewal of interest in Cartesian philosophy. By its rigor and precision, and the wealth of its documentation, it far surpassed the earlier editions and marked an important step in the elaboration of modern methods of producing critical editions. Only recently has it been necessary to publish a revised and enlarged edition that takes into account the documentary discoveries made since the beginning of the twentieth century.

Although during the final decade of his career Tannery devoted an ever increasing amount of time to this editorial effort, he still managed to publish a large number of studies, chiefly on ancient science and medieval and Byzantine mathematics. Moreover, his vast erudition enabled him to reply to numerous questions posed in the *Intermédiaire des mathématiciens* and to contribute valuable notes to several fascicles of the *Encyclopédie des sciences mathématiques*. Along with his highly specialized works, he wished to produce a more general account of the history of science, the initial outlines of which he had sketched in his chapters

of Lavisse and Rambaud's *Histoire générale*. At the beginning of 1903 it appeared that Tannery would have an especially favorable opportunity to carry out this project. The death of Pierre Laffitte had left vacant the chair of the history of science at the Collège de France, which had been created for him in 1892, and the Assembly of professors at the Collège had voted to maintain the chair. The two consultative bodies, the Assembly and the Académie des Sciences, informed the minister of education that Tannery was their first choice among several candidates. His nomination seemed so certain that he began to write the inaugural lecture of his course. But, for obscure political and philosophical reasons the minister chose the candidate who was second on the list submitted to him: the crystallographer Grégoire Wyrouboff, a positivist philosopher with little competence in the history of science. Strictly speaking the minister was within his rights. Tannery was deeply disappointed by this unjust decision, however, which was vainly opposed by the many French and foreign scholars who considered Tannery one of the leaders in the field. Although the case is not clear, Sarton and Louis have revealed some of Chaumié's motives. First of all it is certain that a militant positivist and freethinker like Wyrouboff fitted more easily into the anticlericalism of Émile Combes's government than a fervid Catholic like Tannery. But it appears also that the minister preferred a course of studies that was oriented toward contemporary science, as Wyrouboff proposed, to the program of general scientific history proposed by Tannery. But there is no doubt that the "scandal of 1903" did great damage to the development of the history of science in France.

Tannery was convinced of the necessity of an international effort to catalog documentary sources and to eliminate the nationalistic interpretations of the history of science that were all too common at that time. The four volumes of his *Mélanges scientifiques* that are devoted to correspondence reveal the extent of his relations with the leading historians of science in France, Germany, Scandinavia, Italy, and elsewhere. Tannery also was active at the international congresses of historical studies (Paris, 1900; Rome, 1903), philosophy (Paris, 1900; Geneva, 1904), and mathematicians (Heidelberg, 1904). Conscious of the interdisciplinary role of the history of science, Tannery wanted the subject to be recognized as a field in its own right by historians and philosophers, as well as by scientists. He also hoped that students of the field would become aware of the distinctive contribution it could make and that close contacts would be established between historians of science in all countries.

This effort was suddenly interrupted a few weeks after Tannery returned from the Geneva congress of 1904. Suffering from cancer of the pancreas, he died at the end of November in that year. A considerable portion of his work was dispersed in various specialized—and sometimes hard-to-find—journals. His widow soon undertook to collect the publications and regroup them according to major subjects: exact sciences in antiquity, in the Middle Ages, and in the Byzantine world; modern science; history of philosophy; philology; and so on. To these she added the book reviews and the correspondence. A number of distinguished historians of science, including Heiberg, Zeuthen, and Loria, assisted her in this project. Through their devotion and hers, the seventeen volumes of Tannery's *Mémoires scientifiques* now include all his works, except for his three books on ancient science and his editions of Diophantus, Fermat, and Descartes. With the aid of C. de Waard, Marie Tannery also began work on the edition of the *Correspondance du P. Marin Mersenne* that her husband had hoped to undertake.

It would be impossible in an article of this length to convey the importance of a body of work as extensive and varied as Tannery's. Perhaps its most notable characteristic is an unwavering concern for rigor and precision. The detailed studies that constituted the bulk of his output were, in Tannery's view, only a necessary stage in the elaboration of much broader syntheses that would ultimately lead to a comprehensive history of science that he himself could only initiate. While some of the results that he published during thirty years of scholarly activity have been brought into question by documentary discoveries or by new interpretations, a large number of his studies retain their value. Even more important, however, is the fruitful influence that Tannery's work has exerted on historians of science in the twentieth century.

BIBLIOGRAPHY

I. Original Works. Lists of Tannery's works were published by Marie Tannery in *Mémoires de la Société des sciences physiques et naturelles de Bordeaux*, 6th ser., **4** (1908), 299–382; and by P. Louis in Tannery's *Mémoires scientifiques*, XVII (Toulouse–Paris, 1950), 61–117. G. Eneström presented "Liste des travaux de Paul Tannery sur les mathématiques et la philosophie des mathématiques," in *Bibliotheca mathematica*, 3rd ser., **6** (1905), 292–304. Shorter bibliographies have

been offered by G. Sarton in *Osiris*, **4** (1938), 703–705; and by R. Taton in *Revue d'histoire des sciences et de leurs applications*, **7** (1954), 369–371. And on the occasion of his candidacy at the Collège de France in Apr. 1903, Tannery drew up "Titres scientifiques de Paul Tannery," reproduced in *Mémoires*, X, 125–136.

Tannery's published work consists of several books, major editions of scientific writings, and a very large number of articles. The three principal books are devoted to ancient science: *Pour l'histoire de la science hellène. De Thalès à Empédocle* (Paris, 1887); 2nd ed. prepared by A. Diès with a pref. by F. Enriques (Paris, 1930); *La géométrie grecque . . .*, I, *Histoire générale de la géométrie élémentaire* (Paris, 1887), the only part to be published; and *Recherches sur l'histoire de l'astronomie ancienne* (Paris, 1893). A fourth, briefer publication concerned the preparation of the ed. of Descartes: *La correspondance de Descartes dans les inédits du fonds Libri* (Paris, 1893).

His major eds. of scientific works are *Oeuvres de Fermat*, 3 vols. (Paris, 1891–1896), edited with C. Henry, plus IV (*Compléments*), published by C. Henry (1912), and V (*Suppléments*) (1922), published by C. de Waard; *Diophanti Alexandrini opera omnia*, 2 vols. (Leipzig, 1893–1895); and *Oeuvres de Descartes*, 12 vols. and supp. (Paris, 1897–1913), with C. Adam—Tannery participated in the editing of vols. I–VII and IX. He also began work on eds. that were continued at the urging of Mme Tannery: *Correspondance du P. Marin Mersenne*, C. de Waard, R. Pintard, and B. Rochot, eds. (Paris, 1932–); and Georgius Pachymeres, *Quadrivium*, E. Stéphanou, ed. (Vatican City, 1940).

Most of Tannery's articles, as well as his correspondence, were collected in the *Mémoires scientifiques*, published by Mme Tannery with the aid of several historians of science. The material is grouped as follows: I–III, *Sciences exactes dans l'antiquité* (Toulouse–Paris, 1912–1915); IV, *Sciences exactes chez les Byzantins* (1920); V, *Sciences exactes au Moyen Âge* (1922); VI, *Sciences modernes* (1926); VII, *Philosophie ancienne* (1925); VIII, *Philosophie moderne* (1927); IX, *Philologie* (1929); X, *Supplément au tome VI. Sciences modernes. Généralités historiques* (1930); XI–XII, *Comptes-rendus et analyses* (1931–1933); XIII–XVI, *Correspondance* (1934–1943); and XVII, *Biographie, bibliographie, compléments et tables* (1950).

The bibliography given by P. Louis in *Mémoires scientifiques*, XVII, 61–117, which indicates the vol. and first pg. of the works reproduced in this ed., also lists articles not in the *Mémoires*, including the 200 articles Tannery wrote for the *Grand encyclopédie* and his notes to certain chapters of the *Encyclopédie des sciences mathématiques*. Vol. XVII of the *Mémoires* also contains a "Table analytique des mémoires scientifiques," 449–494, which considerably facilitates working with this rich and varied collection of studies, as well as an index of Greek words, 495–506.

II. Secondary Literature. See the following, listed chronologically: Charles Adam, "Paul Tannery et l'édition de Descartes," in *Oeuvres de Descartes*, C. Adam and P. Tannery, eds., VIII, v–xviii; H. Bosmans, "Notice sur les travaux de Paul Tannery," in *Revue des questions scientifiques*, 3rd ser., **8** (1905), 544–574; *Discours prononcés aux obsèques de M. Paul Tannery* . . . (Toulouse, 1905); P. Duhem, "Paul Tannery (1843–1904)," in *Revue de philosophie*, **5**, no. 1 (1905), 216–230; F. Picavet, "Paul Tannery, historien de la philosophie," in *Archiv für Geschichte der Philosophie*, 3rd ser., **18** (1905), 293–302; J. Tannery, "Notice sur Paul Tannery," in *Rapports et compte-rendus du IIe Congrès international de philosophie* (Geneva, 1905), 775–797, also in *Mémoires de la Société des sciences physiques et naturelles de Bordeaux*, 6th ser., **4** (1908), 269–293; H.-G. Zeuthen, "L'oeuvre de Paul Tannery comme historien des mathématiques," in *Bibliotheca mathematica*, 3rd ser., **6** (1905), 260–292; G. Milhaud, "Paul Tannery," in *Revue des idées*, **3** (1906), 28–39, also in *Nouvelles études sur l'histoire de la pensée scientifique* (Paris, 1911), 1–20; P. Duhem, "Paul Tannery et la Société des sciences physiques et naturelles de Bordeaux," in *Mémoires de la Société des sciences physiques et naturelles de Bordeaux*, 6th ser., **4** (1908), 295–298; and A. Rivaud, "Paul Tannery, historien de la science antique," in *Revue de métaphysique et de morale*, **11** (1913), 177–210.

Later works are G. Loria, "Paul Tannery et son oeuvre d'historien," in *Archeion* (Rome), **11** (1929), lxxx–xcii; J. Nussbaum, *Paul Tannery et l'histoire des physiologues milésiens* (Lausanne, 1929); F. Enriques, "La signification et l'importance de l'histoire de la science et l'oeuvre de Paul Tannery," in Paul Tannery, *Pour l'histoire de la science hellène*, 2nd ed. (Paris, 1930), xi–xxi; Marie Tannery, P. Boutroux, and G. Sarton, "Paul Tannery," "L'oeuvre de Paul Tannery," and "Bibliographie des travaux de Paul Tannery," in *Osiris*, **4** (1938), 633–705; G. Sarton, "Paul, Jules and Marie Tannery," in *Isis*, **38** (1947), 33–51; and P. Louis, "Biographie de Paul Tannery," in Tannery's *Mémoires scientifiques*, XVII, 1–49. The last two articles contain an account of the Wyrouboff affair. See also a group of articles by H. Berr, S. Delorme, J. Itard, R. Lenoble, P.-H. Michel, G. Sarton, P. Sergescu, J. Tannery, and R. Taton in *Revue d'histoire des sciences et de leurs applications*, **7** (1954), 297–368.

There are accounts of the life and work of Marie Tannery by P. Ducassé, in *Osiris*, **4** (1938), 706–709; P. Louis, in Tannery's *Mémoires scientifiques*, XVII, 51–59; G. Sarton, in *Isis*, **38** (1947), 44–47, 50; and C. de Waard, in *Revue d'histoire des sciences et de leurs applications*, **2** (1948), 90–94.

RENÉ TATON

TARDE, JEAN (*b.* La Roque-Gageac, France, 1561 or 1562; *d.* Sarlat, France, 1636), *astronomy, geography*.

After receiving a doctorate in law from the University of Cahors, Tarde continued his studies at the Sorbonne. Ordained a priest, he was assigned to the parish of Carves, near Belvès, and later rose to be the canon theologian of the cathedral church of Sarlat. In 1594, when the bishop wished to determine the effects of the religious wars in France on the diocese of Sarlat, he designated Tarde vicar-general and commissioned him to make a map of the bishopric.

Tarde charted the neighboring diocese for the bishop of Cahors in 1606. In this topographical survey he used a small quadrant equipped with a compass needle and attached to a sundial. In compliance with the bishop's request that he publish an explanation of this instrument, Tarde wrote and dedicated to the bishop *Les usages du quadrant à l'esguille aymantée* (1621), for which the royal privilege was dated 8 June 1620. On the same day Tarde obtained privileges for the two other works published during his lifetime: *Borbonia sidera* (1620) and his translation of this Latin treatise into French, *Les astres de Borbon* (1622). These works were based on Tarde's conversations with Galileo, whom he had visited in Florence on 12–15 November 1614. Among the numerous subjects they discussed were the recently discovered sunspots. After returning to France, Tarde observed the spots for five years and reached the erroneous conclusion that they were planets, which he proceeded to name in honor of the French royal house, as Galileo had done with the satellites of Jupiter and the Medici family.

Tarde's interpretation of the sunspots was demolished by Gassendi in a letter to Galileo dated 20 July 1625. In that communication he pointed out that despite an assiduous program of observations, Tarde had been unable to identify any sunspot that exhibited the periodic returns characteristic of the true planets, as Tarde himself acknowledged.

BIBLIOGRAPHY

I. ORIGINAL WORKS. Tarde's published works are *Borbonia sidera* (Paris, 1620), translated by Tarde as *Les astres de Borbon* (Paris, 1622, 1623, 1627); *Les usages du quadrant à l'esguille aymantée* (Paris, 1621, 1623, 1627, 1638); and *Les chroniques de Jean Tarde*, Gaston de Gérard and Gabriel Tarde, eds. (Paris, 1887). Unpublished is "Voyage de Jean Tarde dans le midi et en Italie," listed by Philippe Lauer in Bibliothèque Nationale (Paris), *Collections manuscrites sur l'histoire des provinces de France*, inventaire, II (Paris, 1911), p. 60, no. 106, fols. 26–38. For the various eds. of his map of the bishopric of Sarlat, see pp. 395–396 in Dujarric-Descombes, who enumerated Tarde's other geographical works.

II. SECONDARY LITERATURE. See Albert Dujarric-Descombes, "Recherches sur les historiens du Périgord au XVIIe siècle," in *Bulletin de la Société historique et archéologique du Périgord*, 9 (1882), 371–412, 489–497; Antonio Favaro, "Di Giovanni Tarde e di una sua visita a Galileo dal 12 al 15 novembre 1614," in *Bullettino di bibliografia e storia delle scienze matematiche e fisiche*, 20 (1887), 345–371, 374; "Lettre de Claude Aspremont à M. Pichard, chanoine théologal de l'église cathédrale de Périgueux, au sujet du 'docte escrit du télescope de M. Tarde' " (1630), listed in Lauer, *op cit.*, II, p. 51, no. 92, fol. 53; and Gabriel Tarde, "Observations au sujet des astres de Borbon du chanoine Tarde," in *Bulletin de la Société historique et archéologique du Périgord*, 4 (1877), 169–173. Gassendi's letter to Galileo is in Galileo Galilei, *Le opere*, A. Favaro, ed., XIII (Florence, 1903; repr. 1935), no. 1729, lines 56–64; and Gassendi, *Opera omnia* (Stuttgart, 1964; repr. of Lyons, 1658 ed.), VI, 5.

EDWARD ROSEN

TARGIONI TOZZETTI, GIOVANNI (*b.* Florence, Italy, 11 September 1712; *d.* Florence, 7 January 1783), *natural history.*

Targioni's father, Benedetto, was a doctor; his mother, Cecilia, was the daughter of Gerolamo Tozzetti, a jurist. He added her maiden name to his paternal surname, Targioni. In 1734 Targioni Tozzetti received his degree in medicine at Pisa; for the rest of his life he practiced medicine in Florence, where, among other things, he promoted prophylactic inoculation against smallpox. In 1739 he was appointed director of the Magliabechi Library, a position which required his cataloging thousands of books and manuscripts. Targioni Tozzetti was strongly inclined toward natural history. His father, a passionate student of botany, understood and appreciated this interest, and encouraged it by entrusting his son, in 1731, to the botanist Pier Antonio Micheli. For six years Targioni Tozzetti was Micheli's shadow, the latter's pleasure in teaching being matched by the former's in learning. After Micheli's death in 1737 Targioni Tozzetti was deemed worthy to be director of Florence's botanical garden and to teach botany.

A scientific journey from Florence to Cortona, with Micheli in 1732, served to emphasize Targioni Tozzetti's true vocation, that of traveling naturalist. Long journeys undertaken between 1742 and 1745 enabled him to observe natural phenomena and the ancient monuments of considerable sections of Tuscany. The harvest gathered in the field of natural science was truly outstanding, encompassing the three kingdoms of nature.

In his study on the relations between normal hydrography and landform, Targioni Tozzetti made wide-reaching synthetic observations, starting with the most minute analytic observations. Some of the major scientists of his time believed that the erosive action of currents, even though increased by time, was entirely insufficient to account for wide valleys and deep gorges. Assuming a decisive position against the most famous of them all, Buffon, who believed it essential in explaining these phenomena to posit the action of marine currents prior to the emergence of land, Targioni Tozzetti maintained that all the valleys and gorges he had observed were the result of erosion caused by currents. He also noted the enormous quantity of materials altered by the waters themselves in the course of time. On the basis of these observations Targioni Tozzetti was able to outline, for the first time in the history of science, the morphological evolution of certain landscapes, such as that of the hills of the Tuscan ante-Apennines, formed by Pliocene marine sediments. Going back to "very remote times, unknown to us," Targioni Tozzetti said that the hills appeared to have been formed by current-caused erosion on the marine platform that emerged after the sea had receded. He reconstructed the movement of this platform by identifying it with the plane that connects the summits of the existing hills. Considering the future, he stated that these hills would also be "broken up and destroyed" by erosion.

Targioni Tozzetti's interest was also aroused by the disappearance of early Pleistocene lakes, where we find today the characteristic intermountain basins of the Apennines. In the Upper Valdarno he demonstrated the existence of a large lake in antiquity; he then studied the disappearance of the waters as a result of the complete alluvial refilling of the basin. Last, he investigated the regressive erosion that the waters cause on the lacustrine deposits. He also explained with great clarity the origin of the gorges in the wide Arno Valley by the presence of very hard rocky masses which were buried by the lacustrine sediments lacking cohesion. It is in these sediments that one not infrequently finds the bones of elephants and other large mammals. Targioni Tozzetti demonstrated that these fossils are not the remains of elephants that accompanied Hannibal's army during the Second Punic War, as scholars of the period believed; rather, they were a part of the fauna of

Tuscany before the appearance of man in the region.

Throughout his life Targioni Tozzetti hoped to write a full description of Tuscany from both the general synthetic and the regional analytic points of view. The work was never carried out; but his *Prodromo*, published in 1754, is of great interest, for it outlines in minute detail the plan of his work—a plan amazing for the modernity of its conception. The general description would have been developed in much the same manner as would be used today: examination of the relief, hydrography, climate, flora, fauna, and finally human manifestations. Targioni Tozzetti's great understanding of the relations between nature and man extended to such diverse topics as the changes introduced by man in the courses of rivers or at their mouths; the changes in malarial marshlands (a very important economic and social problem in Tuscany in the eighteenth century); the location of population centers in relation to relief or to water; the construction of roads or ports; and the better exploitation by man of all the natural resources. In brief, he saw man as an indefatigable and efficient modifier of terrestrial surfaces: "It would not be useless to consider Tuscany as it indeed was before it was inhabited my man . . . to understand the great changes that have followed successively from industry and later from human negligence." Today Targioni Tozzetti is recognized as one of the precursors of modern human geography.

Targioni Tozzetti was unable to devote much time to scientific research, since he was fully occupied with his duties as a physician and a librarian. He himself felt "condemned to waste his life in studies opposed to his inclination." But the results he achieved, even though limited, justify his being considered, after Lazzaro Spallanzani, as the most active Italian naturalist of the eighteenth century. His scientific passion was transmitted to later generations: his son Ottaviano, his nephew Antonio, and his grandnephew Adolfo were doctors and naturalists of note.

BIBLIOGRAPHY

I. ORIGINAL WORKS. Targioni Tozzetti's writings include *Relazione d'alcuni viaggi fatti in diverse parti della Toscana, per osservare le produzioni naturali, e gli antichi monumenti di essa*, 6 vols. (Florence, 1751–1754; 2nd ed., 12 vols., 1768–1779); *Prodromo della corografia e della topografia fisica della Toscana* (Florence, 1754), and two works valuable for the history of science, which he was one of the first to cultivate, *Notizie sugli aggrandimenti delle scienze fisiche accaduti in Toscana nel corso di anni LX del secolo XVII* (Florence, 1780), and *Notizie della vita e delle opere di Pier' Antonio Micheli* (Florence, 1858). Some of his more interesting scientific material is in the anthology of F. Rodolico, *La Toscana descritta dai naturalisti del settecento* (Florence, 1945), passim.

II. SECONDARY LITERATURE. Biographical and bibliographical information is in *Novelle letterarie*, **14** (1783), col. 97. On Targioni Tozzetti's work, see the following (listed chronologically): O. Marinelli, "Giovanni Targioni Tozzetti e la illustrazione geografica della Toscana," in *Rivista geografica italiana*, **11** (1904), 1–12, 136–145, 226–236; R. Concari, "La geografia umana nei 'Viaggi' di Giovanni Targioni Tozzetti," ibid., **41** (1934), 28–41; and F. Rodolico, "La collezione mineralogica di Giovanni Targioni Tozzetti," in *Catalogo del Museo di storia della scienza* (Florence, 1954), 274–280; and "Lo studio 'fisico' della città di Firenze impostato da Giovanni Targioni Tozzetti nel 1754," in *Rivista geografica italiana*, **64** (1967), 110–113. A detailed description of his journeys, accompanied by itinerary maps, is in F. Rodolico, *L'esplorazione naturalistica dell'Appennino* (Florence, 1963), 133–137, 152–160, 366–369.

FRANCESCO RODOLICO

IBN ṬĀRIQ. See Ya‘qūb ibn Ṭāriq.

TARTAGLIA (also **Tartalea** or **Tartaia**), **NICCOLÒ** (*b*. Brescia, Italy, 1499 or 1500; *d*. Venice, Italy, 13 December 1557), *mathematics, mechanics, topography, military science*.

The surname Tartaglia, which Niccolò always used, was a nickname given to him in his boyhood because of a speech impediment resulting from a wound in the mouth (*tartagliare* means "to stammer"). According to his will, dated 10 December 1557 and now in the Venice State Archives, he had a brother surnamed Fontana, and some historians have attributed that surname to Niccolò as well.

Tartaglia's father, Michele, a postal courier, died about 1506, leaving his widow and children in poverty. Six years later, during the sack of Brescia, Niccolò, while taking shelter in the cathedral, received five serious head wounds. It was only through the loving care of his mother that he recovered. At the age of about fourteen, he went to a Master Francesco to learn to write the alphabet; but by the time he reached "k," he was no longer able to pay the teacher. "From that day," he later wrote in a moving autobiographical sketch, "I never returned to a tutor, but continued to labor by myself over the works of dead men, accompanied only by the daughter of poverty that is called industry" (*Quesiti*, bk. VI, question 8).

Tartaglia began his mathematical studies at an early age and progressed quickly. He moved to Verona, probably sometime between 1516 and 1518, where he was employed as "teacher of the abacus." Certain documents dating from 1529–1533, preserved in the Verona section of the State Archives, testify that he had a family, that he was in reduced financial circumstances, and that he was in charge of a school in the Palazzo Mazzanti. In 1534 he moved to Venice, where he was "professor of mathematics." Tartaglia also gave public lessons in the Church of San Zanipolo (Santi Giovanni e Paolo). Nearly all his works were printed in Venice, where he remained for the rest of his life except for a return to Brescia for about eighteen months in 1548–1549. During this time he taught at Sant'Afra, San Barnaba, San Lorenzo, and at the academy of the nearby village of Rezzato. He died in Venice, poor and alone, in his dwelling in the Calle del Sturion near the Rialto Bridge.

The most important mathematical subject with which Tartaglia's name is linked is the solution of third-degree equations. The rule for solving them had been obtained by Scipione Ferro in the first or second decade of the sixteenth century but was not published at the time. It was rediscovered by Tartaglia in 1535, on the occasion of a mathematical contest with Antonio Maria Fiore, a pupil of Ferro; but Tartaglia did not publish it either. On 25 March 1539, Tartaglia told Girolamo Cardano about it at the latter's house in Milan. Although Cardano had persistently requested the rule and swore not to divulge it, he included it in his *Ars magna* (1545), crediting Ferro and Tartaglia. This breach of promise angered Tartaglia; and in the *Quesiti* (bk. IX), he presented his own research on third-degree equations and his relations with Cardano, whom he discussed in offensive language.

Lodovico Ferrari, who devised the solution of fourth-degree equations, rose to Cardano's defense and sent a notice (*cartello*) of mathematical challenge to Tartaglia. Between 10 February 1547 and 24 July 1548 they exchanged twelve printed brochures (Ferrari's six *Cartelli* and Tartaglia's six *Risposte*, all usually known as *Cartelli*), which are important for their scientific content and are notable for both polemical liveliness and bibliographical rarity. The exchange was followed by a debate' between Tartaglia and Ferrari in the Church of

Santa Maria del Giardino, in Milan, on 10 August 1548. The scientific portion of the dispute consisted of the solution of sixty-two problems that the two contestants had posed to each other. Although centering mainly on arithmetic, algebra, and geometry, the questions also dealt with geography, astronomy, architecture, gnomonics, and optics. They offer a vivid picture of the state of the exact sciences in mid-sixteenth-century Italy.

Tartaglia's other mathematical contributions concern fundamentals of arithmetic, numerical calculations, extraction of roots, rationalization of denominators, combinatorial analysis, and various other problems that are now considered quaint and amusing. "Tartaglia's triangle," the triangular array of binomial coefficients also known as "Pascal's triangle," is found in the *General trattato* (pt. II [1556]) but also appears in earlier works by other authors, although in a different configuration.

$$
\begin{array}{ccccccccc}
 & & & & 1 & & & & \\
 & & & 1 & & 1 & & & \\
 & & 1 & & 2 & & 1 & & \\
 & 1 & & 3 & & 3 & & 1 & \\
1 & & 4 & & 6 & & 4 & & 1 \\
\end{array}
$$
. . . .

The *Cartelli* also contain an extreme-value problem proposed by Ferrari that Tartaglia solved without including the relevant demonstration.

In geometry Tartaglia was a pioneer in calculating the volume of a tetrahedron from the lengths of its sides and in inscribing within a triangle three circles tangent to one another (now called Malfatti's problem). In the *Cartelli* Ferrari and Tartaglia contributed to the theory of division of areas and especially to the geometry of the compass with fixed opening—subjects to which Tartaglia returned in the *General trattato*. Of special importance to geometry, as well as to other fields, was Tartaglia's Italian translation, with commentary, of Euclid's *Elements* (1543), the first printed translation of the work into any modern language.

Tartaglia's contribution to the diffusion of the works of the great classical scientists was not confined to this translation, however. One of the first publishers of Archimedes, he produced an edition (1543) of William of Moerbeke's thirteenth-century Latin version of some of Archimedes' works. Tartaglia returned to Archimedes in 1551, publishing an Italian translation, with commentary, of part of Book I of *De insidentibus aquae* that was included in the *Ragionamento primo* on the *Travagliata inventione*. Material left by Tartaglia provided the basis for Curtius Troianus' publication in

1565 of *De insidentibus aquae* (books I and II) and of Jordanus de Nemore's *Opusculum de ponderositate*. The latter work, entitled *Liber Jordani de ratione ponderis* in various thirteenth-century manuscripts, is important in the history of mechanics because it contains the first correct solution of the problem of the equilibrium of a heavy body on an inclined plane. (Tartaglia had also published such a solution in the *Quesiti*.)

Yet, despite these contributions to the dissemination of knowledge, Tartaglia drew criticism—sharp at times—by apparently presenting William of Moerbeke's translation as his own, by not crediting Jordanus with the solution of the inclined-plane problem, and by proposing in the *Travagliata inventione* a procedure mentioned by others for raising submerged ships. Any unbiased judgment must take into consideration that an extremely easygoing attitude then obtained with regard to literary property.

Tartaglia's contributions to the art of warfare aroused widespread and lasting interest, and the broad range of his competence in nonmathematical areas is also demonstrated in the *Quesiti*. In this work Tartaglia dealt with algebraic and geometric material (including the solution of the cubic equation), and such varied subjects as the firing of artillery, cannonballs, gunpowder, the disposition of infantry, topographical surveying, equilibrium in balances, and statics. His various proposals on fortifications were praised by Carlo Promis. In his attempts at a theoretical study of the motion of a projectile—a study in which he was a pioneer—Tartaglia reached the following notable conclusions: the trajectory is a curved line everywhere; and the maximum range, for any given value of the initial speed of the projectile, is obtained with a firing elevation of 45°. The latter result was obtained through an erroneous argument, but the proposition is correct (in a vacuum) and might well be called Tartaglia's theorem. In ballistics Tartaglia also proposed new ideas, methods, and instruments, important among which are "firing tables."

Problems of gunnery led Tartaglia, in *Nova scientia*, to suggest two instruments for determining inaccessible heights and distances. The historian Pietro Riccardi considered them "the first telemeters" and cited their related theories as "the first attempts at modern tachymetry." In the *Quesiti*, Tartaglia showed how to apply the compass to surveying, and in the *General trattato* he presented the first theory of the surveyor's cross. Hence Riccardi also asserted that he was responsi-

ble for "the major advances in practical geometry of the first half of the sixteenth century."

Tartaglia's attitude toward military matters is shown in his letter dedicating *Nova scientia* to Francesco Maria della Rovere, duke of Urbino; the letter eloquently demonstrates his discreet reticence and effectively reflects his ethical qualities.

The short work *Travagliata inventione* deals not only with raising sunken ships but also with diving suits, weather forecasting, and specific weights. Tartaglia's experiments on the latter are described in Jordanus de Nemore's *De ponderositate*.

Tartaglia's pupils included the English gentleman Richard Wentworth, who was probably the author of an Italian manuscript now at Oxford (Bodleian Library, MS 584), in which Tartaglia is mentioned several times; Giovanni Antonio Rusconi, author of a book on architecture (Venice, 1540); Maffeo Poveiano, author of a work on arithmetic (Bergamo, 1582); and the mathematician and philosopher Giovanni Battista Benedetti, who in his noted work on the geometry of the compass with fixed opening (Venice, 1553) stated that he began the study of Euclid with Tartaglia.

BIBLIOGRAPHY

I. ORIGINAL WORKS. Tartaglia's works are *Nova scientia* (Venice, 1537); *Euclide Megarense* (Venice, 1543); *Opera Archimedis* (Venice, 1543); *Quesiti et inventioni diverse* (Venice, 1546); *Risposte* to Lodovico Ferrari, 6 pts. (1–4, Venice, 1547; 5–6, Brescia, 1548); *Travagliata inventione* (Venice, 1551), with *Ragionamenti* and *Supplimento*; *General trattato di numeri et misure*, 6 pts. (Venice, 1556–1560); *Archimedis De insidentibus aquae* (Venice, 1565); and *Iordani Opusculum de ponderositate* (Venice, 1565). For further information on the various editions see Pietro Riccardi, *Biblioteca matematica italiana* (Modena, 1870–1928; repr. Milan, 1952), I$_2$, 496–507, with supplements in the series of *Aggiunte*.

The original copies of the *Cartelli* are very rare, as is the autographed ed. (212 copies) by Enrico Giordani, *I sei cartelli di matematica disfida . . . di Lodovico Ferrari, coi sei contro-cartelli in risposta di Niccolò Tartaglia* (Milan, 1876). Facs. eds. of the *Quesiti* (Brescia, 1959) and the *Cartelli* (Brescia, 1974) have been published with commentaries by Arnaldo Masotti.

Some of Tartaglia's works on the art of warfare were translated during his lifetime into German (1547) and French (1556). Modern eds. include the following:

1. The new. ed. with English trans. and commentary by E. A. Moody, of Jordanus de Nemore's *De ponderosite*, based on thirteenth–fifteenth-century MSS with Tartaglia's ed. as guide, and prepared with the as-

sistance of R. Clements, A. Ditzel, and J. L. Saunders. It is included in E. A. Moody and Marshall Clagett, *The Medieval Science of Weights* (*Scientia de ponderibus*) (Madison, Wis., 1952; 2nd ed., 1960), 167–227, 330–336, 388–413.

2. The new eds. of Thomas Salusbury's seventeenth-century versions of *Travagliata inventione* with *Ragionamenti* and *Supplimento*, and *Archimedis De insidentibus aquae*, in *Mathematical Collections and Translations. In Two Tomes by Thomas Salusbury. London 1661 and 1665* in facs., with analytical and biobibliographical intro. by Stillman Drake (London–Los Angeles, 1967), II, 331–402, 479–516.

3. The English versions, by Stillman Drake, of long excerpts concerning mechanics, from *Nova scientia* and the *Quesiti*, in *Mechanics in Sixteenth-Century Italy,* selections from Tartaglia *et al.,* translated and annotated by S. Drake and I. E. Drabkin (Madison, Wis., 1969), 61–143.

Tartaglia's correspondence (or extracts from it) are in the *Quesiti* and in the *Terzo ragionamento* on the *Travagliata inventione.* Two letters dealing with fortifications were exchanged in 1549 with the military engineer Jacopo Fusto Castriotto: copies, perhaps from the writers' own time, are at the old city archives, at the University of Urbino. They were published by Vincenzo Tonni-Bazza, "Di una lettera inedita di N. Tartaglia," in *Atti dell'Accademia nazionale dei Lincei. Rendiconti,* 5th ser., **10** (1901), 39–42; and "Frammenti di nuove ricerche intorno a N. Tartaglia," in *Atti del Congresso internazionale di scienze storiche, Roma, 1903,* XII (Rome, 1907), 293–307. Facsimiles of the letters are in Masotti, *Studi su N. Tartaglia* (see below), pls. xxiii, xxiv.

II. Secondary Literature. Works within each section are listed chronologically.

On Tartaglia's life and works, see Baldassarre Boncompagni, "Intorno ad un testamento inedito di N. Tartaglia," in *In memoriam Dominici Chelini–Collectanea mathematica* (Milan, 1881), 363–412, with full-page facs. of his will; Antonio Favaro, "Intorno al testamento inedito di N. Tartaglia pubblicato da Don B. Boncompagni," in *Rivista periodica dei lavori dell' Accademia di Padova,* **32** (1881–1882), 71–108; Vincenzo Tonni-Bazza, "N. Tartaglia nel quarto centenario natalizio," in *Commentari dell'Ateneo di Brescia* (1900), 160–179; Antonio Favaro, "Per la biografia di N. Tartaglia," in *Archivio storico italiano,* **71** (1913), 335–372; and "Di N. Tartaglia e della stampa di alcune sue opere con particolare riguardo alla 'Travagliata inventione,'" in *Isis,* **1** (1913), 329–340; *Ateneo di Brescia–Scoprendosi il monumento a N. Tartaglia* (Brescia, 1918); *Commentari dell'Ateneo di Brescia* (1918), 77–151; Arnaldo Masotti, "Commemorazione di N. Tartaglia," *ibid.* (1957), 25–48; "Sui 'Cartelli di matematica disfida' scambiati fra L. Ferrari e N. Tartaglia," in *Rendiconti dell'Istituto lombardo di scienze e lettere,* Classe di scienze, sec. A, **94** (1960), 31–41; "Su alcuni possibili autografi di N. Tartaglia," *ibid.,* 42–46; and "N. Tar-

taglia," in *Storia di Brescia,* II (Brescia, 1963), 597–617, with 4 full-page plates. Masotti's *Studi su N. Tartaglia* (see below) contains many bibliographical details.

On Tartaglia's works, their translations, and certain MSS by Tartaglia or related to him, see "N. Tartaglia e i suoi 'Quesiti'" and "Rarità tartagliane," in *Atti del Convegno di storia delle matematiche, promosso dall'Ateneo di Brescia nel 1959 in commemorazione del quarto centenario della morte del Tartaglia* (Brescia, 1962), 17–56, 119–160, with 37 full-page plates, which are also in A. Masotti, *Studi su N. Tartaglia* (Brescia, 1962).

Tartaglia's algebra is treated in Pietro Cossali, *Origine, trasporto in Italia, primi progressi in essa dell'algebra,* II (Parma, 1799), 96–158; Silvestro Gherardi, "Di alcuni materiali per la storia della Facoltà matematica nell'antica Università di Bologna," in *Nuovi annali delle scienze naturali* (Bologna), 2nd ser., **5** (1846), 161–187, 241–268, 321–356, 401–436, with additions translated into German by Maximilian Curtze in *Archiv der Mathematik und Physik,* **52** (1870–1871), 65–205; Ettore Bortolotti, "I contributi del Tartaglia, del Cardano, del Ferrari, e della scuola matematica bolognese alla teoria algebrica delle equazioni cubiche," in *Studi e memorie per la storia dell'Università di Bologna,* **10** (1926), 55–108; and *The Great Art or The Rules of Algebra, by Girolamo Cardano,* translated and edited by T. Richard Witmer, with foreword by Oystein Ore (Cambridge, Mass., 1968), 8, 9, 52, 96, 239, as well as the foreword and preface, *passim.*

On his contributions to geometry, see Antonio Favaro, "Notizie storico-critiche sulla divisione delle aree," in *Memorie del R. Istituto veneto di scienze, lettere ed arti,* **22** (1883), 151–152; J. S. Mackay, "Solutions of Euclid's Problems, With a Ruler and One Fixed Aperture of the Compasses, by the Italian Geometers of the Sixteenth Century," in *Proceedings of the Edinburgh Mathematical Society,* **5** (1887), 2–22; W. M. Kutta, "Zur Geschichte der Geometrie mit constanter Zirkelöffnung," in *Nova acta Academiae Caesareae Leopoldino Carolinae germanicae naturae curiosorum,* **71** (1896), 80–91; Giovanni Sansone, "Sulle espressioni del volume del tetraedro," in *Periodico di matematiche,* 4th ser., **3** (1923), 26–27; Harald Geppert, "Sulle costruzioni geometriche che si eseguiscono colla riga ed un compasso di apertura fissa," *ibid.,* **9** (1929), 303–309, 313–317; and Giuseppina Biggiogero, "La geometrica del tetraedro," in *Enciclopedia delle matematiche elementari,* II, pt. 1 (Milan, 1936), 220, 245.

Statics and dynamics are discussed in Raffaello Caverni, *Storia del metodo sperimentale in Italia,* 5 vols. (Florence, 1891–1900), I, 53–54; IV, 190–198; Pierre Duhem, *Les origines de la statique,* I (Paris, 1905), 111–112, 119–120, 199; Alexandre Koyré, "La dynamique de N. Tartaglia," in *La science au seizième siècle–Colloque international de Royaumont 1957* (Paris, 1960), 91–116; and S. Drake, "Introduction" to *Mechanics in Sixteenth-Century Italy* (see above), 16–26, which also includes Tartaglia's links with Archimedes and Euclid as well as with Jordanus de Nemore.

Tartaglia's contributions to the military sciences are treated in Max Jähns, *Geschichte der Kriegswissenschaften*, 3 vols. (Munich–Leipzig, 1889–1891; facs. repr. New York, 1965), xix, 507, 596–605, 626, 707–712, 718, 797–802, 850, 985, 1008.

On fortifications, see Carlo Promis, "Della vita e delle opere degl' italiani scrittori di artiglieria, architettura e meccanica militare da Egidio Colonna a Francesco Marchi 1285–1560," in Francesco di Giorgio Martini, *Trattato di architettura civile e militare*, pt. 2 (Turin, 1841), 69–71, 78; H. Wauvermans, "La fortification de N. Tartaglia," in *Revue belge d'art, de sciences et de technologie militaires*, 1, IV (1876), 1–42; and Antonio Cassi Ramelli, *Dalle caverne ai rifugi blindati—Trenta secoli di architettura militare* (Milan, 1964), 320, 326, 346, 354, 360.

Tartaglia's ballistics is discussed in P. Charbonnier, *Essais sur l'histoire de la balistique* (Paris, 1928), 3, 6, 8–38, 41, 54, 66, 75, 87, 266; A. R. Hall, *Ballistics in the Seventeenth Century* (Cambridge, 1952), 33, 36–43, 45–52, 55, 61, 68–70, 81, 83, 95, 105; and E. G. R. Taylor, *The Mathematical Practitioners of Tudor and Stuart England 1485–1714* (Cambridge, 1954,), which mentions Tartaglia especially in connection with William Bourne and Cyprian Lucar, who translated Tartaglia's writings on ballistics into English—see 17, 30–31, 33, 42, 176, 321, 323, 328, 370.

On Tartaglia's topography, see Giovanni Rossi, *Groma e squadro ovvero storia dell'agrimensura italiana dai tempi antichi al secolo XVII°* (Turin, 1877), 7–8, 115–116, 122–138, 140, 142, 156, 157, 161, 166, 169–171, 213; P. Riccardi, "Cenni sulla storia della geodesia in Italia dalle prime epoche fin oltre la metà del secolo XIX," pt. 1, in *Memorie dell'Accademia delle scienze dell'Istituto di Bologna*, 3rd ser., 10 (1879), 474–478; R. T. Gunther, *Early Science in Oxford*, I (Oxford, 1920; repr. London, 1967), 310, 339, 368; and E. G. R. Taylor, "Cartography, Survey and Navigation," in C. Singer *et al.*, *A History of Technology*, III (Oxford, 1957), 539.

ARNALDO MASOTTI

TASHIRO, SHIRO (in Japanese sources called **Tashiro Shirosuke**) (*b*. Kagoshima prefecture, Japan, 12 February 1883; *d*. Cincinnati, Ohio, 12 June 1963), *biochemistry*.

Tashiro, the son of Shirobe and A. Tashiro, immigrated to the United States in 1901. After graduating B.S. from the University of Chicago in 1909, he served as fellow and assistant in physiological chemistry at Chicago from 1910 until he received the Ph.D. in 1912 at Chicago. In 1913–1914 he was an associate in physiological chemistry, in 1914–1918 instructor, and in 1918 an assistant

professor at Chicago. Tashiro was appointed to the University of Cincinnati College of Medicine in 1919 as associate professor of biochemistry and assistant director of the Biochemistry Service, Cincinnati General Hospital. He returned to Japan for study, and on 26 July 1923 Kyoto University granted him the doctor of medical science degree for a dissertation entitled "Carbon Dioxide Production From Nerve Fibres When Resting and When Stimulated; A Contribution to the Chemical Basis of Irritability; A New Method and Apparatus for the Estimation of Exceedingly Minute Quantities of Carbon Dioxide." Tashiro was appointed full professor at the University of Cincinnati in 1925. He retired in 1952.

Tashiro's career was founded on his invention of the biometer, prior to which no method of analysis was available for minute quantities of carbon dioxide. In conjunction with H. N. McCoy, Tashiro devised an apparatus that could detect carbon dioxide in quantities as small as one ten-millionth of a gram. The fundamental principle of the biometer depended upon the possibility of precipitating exceedingly minute quantities of carbon dioxide as barium carbonate on the surface of a small drop of barium hydroxide solution. When the drop of barium hydroxide is exposed to any sample of a gas free from carbon dioxide, it remains clear; but when more than a definite amount of carbon dioxide is introduced, a precipitate of carbonate appears that is detectable through a lens. Tashiro found that the minimum amount of carbon dioxide that gives a precipitate is 1.0×10^{-7}g. Through the use of the biometer Tashiro was able to conclude that injured living tissue (nerve tissue and dry seeds) had a greater output of carbon dioxide than uninjured tissue, while dead tissue did not emit any carbon dioxide. If injured tissue gave off carbon dioxide, it was still alive.

Tashiro's contributions covered many fields: metabolism in nerves; metabolism gradation in the nerve considered as an organism; metabolism in the growth of tissues; anesthetics; biochemical and physiological factors in the production of a gastric ulcer; dacryohemorrhea; and cholinergics.

Tashiro married Shizuka Kawasaki of Honolulu on 9 November 1915. He contributed to numerous scientific journals, and was a president of the Cincinnati section of Sigma Xi, national honor research society, and the Daniel Drake Society. He received the Crown Prince Memorial Prize from the Imperial Academy of Japan and was a member of many chemical and biological societies in the United States, France, and in the United Kingdom. In 1953 Tashiro became the first Japanese to be admitted to American citizenship at the Cincinnati Immigration and Naturalization office.

In almost all of his undertakings Tashiro's observations were fundamental. His work on the nature of the nerve impulse and its propagation was intimately linked to his invention of the biometer. His studies on bile salts ranged from questioning whether there were bile salts in normal blood to a possible role for these compounds in formation of a gastric ulcer. With N. C. Foot, also of the University of Cincinnati, Tashiro translated an important treatise by Rinya Kawamura on tsutsugamushi disease, scrub typhus (1926).

BIBLIOGRAPHY

I. ORIGINAL WORKS. Tashiro's published writings are *A Chemical Sign of Life* (Chicago, 1917) and numerous papers published in biochemical and physiological journals between 1914 and 1952, including *American Journal of Physiology, Biological Bulletin, Internationale Zeitschrift für physikalisch-chemische Biologie, Journal of Biological Chemistry, Journal of Infectious Diseases, Medical Bulletin of the University of Cincinnati, Proceedings of the American Society of Biological Chemists, Proceedings of the National Academy,* and *Proceedings of the Society for Experimental Biology and Medicine.*

He and N. C. Foot translated *Studies on Tsutsugamushi Fever,* which is *Medical Bulletin, College of Medicine, University of Cincinnati,* **4,** spec. nos. 1 and 2 (1926).

II. SECONDARY LITERATURE. Biographical literature appears to be very scanty. Extremely brief references are made to Tashiro in *Nihon Igakuhakase-Roku,* Munetoshi Konuma, ed. (1954); and *Kyoto Daigaku Gakui-Roku, 1921–1951* ("Directory of Doctorates Granted by Kyoto University"; Kyoto, 1952). He is listed in *American Men of Science,* 10th ed. (1962) and *World Who's Who in Science,* 1949.

Some information on his life and work appeared in newspapers (1927–1963) and are on file at the Public Library of Cincinnati and Hamilton County, Cincinnati, Ohio (Cincinnati *Enquirer,* 1927, 1943, 1953, 1963; New York *Times,* 1932, 1941, 1953; Cincinnati *Post,* 1945, 1953, 1963).

STACEY B. DAY

TAUBER, ALFRED (*b.* Pressburg, Slovakia [now Bratislava, Czechoslovakia], 5 November 1866; *d.* Theresienstadt, Germany [now Terezin, Czechoslovakia], 1942 [?]), *mathematics.*

Tauber entered the University of Vienna in 1884, concentrating on mathematics, physics, philosophy, and political economy. His doctoral dissertation, "Über einige Sätze der Gruppentheorie" (1888), was written under Gustav von Escherich and was intended for publication, although it never appeared in print. In 1891 Tauber qualified as *Privatdozent* with the *Habilitationsschrift* "Über den Zusammenhang des reellen und imaginären Teiles einer Potenzreihe" and subsequently lectured on the theory of series, trigonometric series, and potential theory. From 1895 he also lectured on the mathematics of insurance, a subject of little interest to him. He was subsequently awarded a monthly salary for this work, and from 1899 he also lectured on the subject at the Technical University of Vienna, where he was appointed *Honorardozent* in 1901. Financial responsibilities obliged Tauber to accept the post of head of the mathematics department of the Phönix insurance company in Vienna (1892–1908). After obtaining an assistant professorship at the university in 1908 he remained adviser to the company until 1912. He had an important role in investigations of mortality tables carried out by a group of insurance companies (1903–1907) and was consultant on insurance to the chamber of commerce and legal adviser to the commerce court of Vienna.

Tauber never assumed the duties of a full professor at the University of Vienna, and the title was not formally conferred upon him until 1919. The reasons for his difficulties are not known, but he was apparently not on good terms with some of the professors there. Almost all of his lectures were given at the Technical University. He retired in 1933 but remained as *Privatdozent* at both universities until 1938. Nothing is known about his last days. The central information office of the Vienna police headquarters contains only one entry, dated 28 June 1942: "Departure to Theresienstadt [concentration camp]."

Tauber's scientific work can be divided into three areas. The first comprises papers on function theory and potential theory; those in the latter area, although overshadowed by the work of Lyapunov, are still important. His most important memoir was "Ein Satz aus der Theorie der unendlichen Reihen" (1897). In 1826 Abel had proved a limit theorem on power series (Abel's limit theorem), the converse of which is true, as Tauber demonstrated, only if an additional condition is stipulated; such conditions are now called Tauberian conditions. These theorems are of fundamental importance in analysis, as was shown especially by G. H. Hardy and J. E. Littlewood, who coined the term "Tauberian theorems," and by N. Wiener. Tauber apparently did not follow subsequent developments of this theorem and, remarkably, did not seem to have considered his memoir of particular importance.

The second group includes papers on linear differential equations and the gamma functions. Although of interest, they did not achieve the importance of his other works.

The third group contains papers and reports on the mathematics of insurance. In "Über die Hypothekenversicherung" (1897) and "Gutachten für die sechste internationale Tagung der Versicherungswissenschaften" (1909) he formulated his Risiko equation.

BIBLIOGRAPHY

I. ORIGINAL WORKS. A bibliography of Tauber's works may be found in the article by Pinl and Dick (see below). His outstanding work was "Ein Satz aus der Theorie der unendlichen Reihen," in *Monatshefte für Mathematik und Physik*, **8** (1897), 273–277. See also "Über den Zusammenhang des reellen und imaginären Teiles einer Potenzreihe," *ibid.*, **2** (1891), 79–118, his *Habilitationsschrift*; "Über einige Sätze der Potentialtheorie," *ibid.*, **9** (1898), 74–88; and "Über die Hypothekenversicherung," in *Österreichische Revue*, **22** (1897), 203–205.

II. SECONDARY LITERATURE. On Tauber and his work, see obituaries by E. Bukovics and J. Rybarz in *Festschrift der technischen Hochschule Wien* (Vienna, 1965–1966), I, 344–346; II, 130–132; and Maximilian Pinl and Auguste Dick, "Kollegen in einer dunkeln Zeit: Schluss," in *Jahresbericht der Deutschen Mathematikervereinigung*, **75** (1974), 166–208, especially 202–208, which includes a bibliography.

E. HLAWKA

TAURINUS, FRANZ ADOLPH (*b*. Bad König, Odenwald, Germany, 15 November 1794; *d*. Cologne, Germany, 13 February 1874), *mathematics*.

In F. Engel and P. Stäckel's *Die Theorie der Parallellinien von Euklid bis Gauss* two writings of Taurinus are mentioned as contributions to the subject. Since their book is a collection of documents in the prehistory of non-Euclidean geometry, they reproduce the most important passages of the original works, including extracts from those of Taurinus, which in 1895 were available in only a few copies.

According to the information given by Engel and Stäckel, Taurinus was the son of a court official of the counts of Erbach-Schöneberg; his mother was the former Luise Juliane Schweikart. He studied law at Heidelberg, Giessen, and Göttingen, and from 1822 lived in Cologne as a man of independent means; he thus had the leisure to pursue various scientific interests.

Taurinus presented the results of his mathematical investigations in *Die Theorie der Parallellinien* (1825) and *Geometriae prima elementa* (1826). He received the stimulus for these studies from his uncle F. K. Schweikart (1780–1857), who from 1820 was professor of law at the University of Königsberg, and with whom he corresponded concerning his work. Taurinus also communicated several of his results and demonstrations to Gauss, whose replies are printed in Gauss's *Werke* (VII, 186).

According to Engel and Stäckel, Taurinus' investigations on the theory of parallel lines sought to demonstrate that the sole admissible geometry is Euclidean. As the basis for his argumentation Taurinus used the axiom of the straight line, which postulates that through two points there could be exactly one straight line. In this regard, however, he had no choice but to accept the "internal consistency" of the "third system of geometry," in which the sum of the angles of a triangle amounts to less than two right angles.

His remarks in *Geometriae prima elementa* show that by 1826 Taurinus had clearly recognized the lack of contradiction of this "third system," "logarithmic-spherical geometry," as he called it; had even developed the suitable trigonometry; and had successfully applied trigonometry to a series of elementary problems.

Taurinus' works on the problem of parallel lines, like those of his uncle, Schweikart, represent a middle stage in the historical development of this problem between the efforts of Saccheri and Lambert, on the one hand, and those of Gauss, Lobachevsky, and Bolyai, on the other. Although he sought to preserve the hegemony of Euclidean geometry by reference to the infinite number of non-Euclidean geometries; nonetheless, through an idea that was very close to Lambert's, he moved on to non-Euclidean trigonometry as it was later developed by Bolyai and Lobachevsky.

Moreover, Taurinus presented the idea that elliptical geometry can be "realized" on the sphere. This concept was first taken up again by Bernhard Riemann.

BIBLIOGRAPHY

Taurinus' major works are *Die Theorie der Parallelli-nien* (Cologne, 1825) and *Geometriae prima elementa* (Cologne, 1826).

F. Engel and P. Stäckel, *Die Theorie der Parallelli-nien von Euklid bis Gauss* (Leipzig, 1895), contains selections from Taurinus' works.

KARLHEINZ HAAS

TAYLOR, BROOK (*b*. Edmonton, Middlesex, England, 18 August 1685; *d*. London, England, 29 December 1731), *mathematics*.

Brook Taylor was the son of John Taylor of Bifrons House, Kent, and Olivia, daughter of Sir Nicholas Tempest, Bart. The family was fairly well-to-do, and was connected with the minor nobility. Brook's grandfather, Nathaniel, had supported Oliver Cromwell. John Taylor was a stern parent from whom Brook became estranged in 1721 when he married a woman said to have been of good family but of no fortune. In 1723 Brook returned home after his wife's death in childbirth. He married again in 1725 with his father's approval, but his second wife died in childbirth in 1730. The daughter born at that time survived.

Taylor's home life seems to have influenced his work in several ways. Two of his major scientific contributions deal with the vibrating string and with perspective drawing. His father was interested in music and art, and entertained many musicians in his home. The family archives were said to contain paintings by Brook, and there is an unpublished manuscript entitled *On Musick* among the Taylor materials at St. John's College, Cambridge. This is not the paper said to have been presented to the Royal Society prior to 1713, but a portion of a projected joint work by Taylor, Sir Isaac Newton, and Dr. Pepusch, who apparently was to write on the nonscientific aspects of music.

Taylor was tutored at home before entering St. John's College in 1701, where the chief mathematicians were John Machin and John Keill. Taylor received the LL.B. degree in 1709, was elected to the Royal Society in 1712, and was awarded the LL.D. degree in 1714. He was elected secretary to the Royal Society in January 1714, but he resigned in October 1718 because of ill health and perhaps because of a loss of interest in this rather confining task. He visited France several times both for the sake of his health and for social reasons. Out of these trips grew a scientific correspondence with Pierre Rémond de Montmort dealing with infinite series and Montmort's work in probability. In this Taylor served on some occasions as an intermediary between Montmort and Abraham De Moivre. W. W. Rouse Ball reports that the problem of the knight's tour was first solved by Montmort and De Moivre after it had been suggested by Taylor.[1]

Taylor published his first important paper in the *Philosophical Transactions of the Royal Society* in 1714, but he had actually written it by 1708, according to his correspondence with Keill. The paper dealt with the determination of the center of oscillation of a body, and was typical both of Taylor's work and of the times, in that it dealt with a problem in mechanics, used Newtonian dot notation, and led to a dispute with Johann I Bernoulli.

The period of 1714–1719 was Taylor's most productive, mathematically. The first editions of both his mathematical books, *Methodus incrementorum directa et inversa* and *Linear Perspective*, appeared in 1715. Their second editions appeared in 1717 and 1719 respectively. He also published thirteen articles, some of them letters and reviews, in the *Philosophical Transactions* during the years 1712–1724. These include accounts of experiments with capillarity, magnetism, and the thermometer. In his later years Taylor turned to religious and philosophical writings. His third book, *Comtemplatio philosophica*, was printed posthumously by his grandson in 1793.

Taylor is best known for the theorem or process for expanding functions into infinite series that commonly bears his name. Since it is an important theorem, and since there is disagreement as to the amount of credit that should be given to him for its development, an outline of his derivation of the theorem will be given here. The discussion of Proposition VII, Theorem III of the *Methodus incrementorum* includes the statement:

If z grows to be $z + n\dot{z}$ then x equals

$$x + \frac{n}{1}\dot{x} + \frac{n}{1} \cdot \frac{n-1}{2}\ddot{x} + \frac{n}{1} \cdot \frac{n-1}{2} \cdot \frac{n-2}{3}\dddot{x}, \text{ etc.}$$

Taylor used dots below the variables to represent increments or finite differences, and dots above to represent Newton's fluxions.

The above statement is a notationally improved version of Newton's interpolation formula as given in Lemma 5 of Book III of his *Principia*. This formula had first appeared in a letter from James Gregory to John Collins in 1670.[2] Taylor had derived this formula inductively from a difference table written in terms of x and its successive differences.

Next, Taylor made the substitutions

$$v = nz, \quad \dot{v} = v - z = (n-1)z, \quad \ddot{v} = \dot{v} - z, \text{ etc.},$$

to derive the statement: "as z growing becomes $z + v$, x likewise growing becomes

$$x + x\frac{v}{1 \cdot z} + x\frac{v\dot{v}}{1 \cdot 2z^2} + x\frac{v\dot{v}\ddot{v}}{1 \cdot 2 \cdot 3 \cdot z^3} + \cdots \text{ etc.}"$$

The final step in the derivation and Taylor's original statement of the theorem, which in modern notation is

$$f(x+h) = f(x) + \frac{f'(x)}{1!}h + \frac{f''(x)}{2!}h^2 + \frac{f'''(x)}{3!}h^3$$
$$+ \cdots \frac{f^{(n)}(x)}{n!}h^n + \cdots,$$

is finally derived in Corollary II to Theorem III as follows: "for evanescent increments [write] the fluxions which are proportional to them and make all of $\ddot{v}$, $\dot{v}$, v, $\underset{\cdot}{v}$, $\underset{..}{v}$ equal, then as with time flowing uniformly z becomes $z + v$, so will x become

$$x + \dot{x}\frac{v}{1\dot{z}} + \ddot{x}\frac{v^2}{1 \cdot 2\dot{z}^2} + \dddot{x}\frac{v^3}{1 \cdot 2 \cdot 3\dot{z}^3} + \cdots \text{ etc.}"$$

This becomes the modern form of Taylor's series when we realize that with "time flowing uniformly" $\dot{z}$ is a constant, $\frac{\dot{x}}{\dot{z}} = \frac{dx}{dz}$, and v is the increment in the independent variable.

Taylor's first statement of this theorem had been given in a letter of 26 July 1712 to John Machin, which has been reprinted by H. Bateman. In it Taylor remarked that this discovery grew out of a hint from Machin given in a conversation in Child's Coffeehouse about the use of "Sir Isaac Newton's series" to solve Kepler's problem, and "Dr. Halley's method of extracting roots" of polynomial equations, which had been published in the *Transactions* for 1694.

This shows Taylor's fairness, care, and familiarity with the literature. He used his formula to expand functions in series and to solve differential equations, but he seemed to have no foreshadowing of the fundamental role later assigned to it by Lagrange nor to have any qualms about the lack of rigor in its derivation. Colin Maclaurin noted that the special case of Taylor's series now known as Maclaurin's theorem or series was discussed by Taylor on page 27 of the 1717 edition of the *Methodus*. The term "Taylor's series" was probably first used by L'Huillier in 1786, although Condorcet used both the names of Taylor and d'Alembert in 1784.[3]

Although infinite series were in the air at the time, and Taylor himself noted several sources and motivations for his development, it seems that he developed his formula independently and was the first to state it explicitly and in a general form. Peano based his claim for Johann I Bernoulli's priority on an integration in which Bernoulli used an infinite series in 1694.[4] Pringsheim showed that it is possible to derive Taylor's theorem from Bernoulli's formula by some changes of variable. However, there seems to be no indication that Taylor did this, nor that Bernoulli appreciated the final form or generality of the Taylor theorem. Taylor's Proposition XI, Theorem IV, on the other hand, is directly equivalent to Bernoulli's integration formula. However, Taylor's derivation differs from Bernoulli's in such a way as to entitle him to priority for the process of integration by parts.

Taylor was one of the few English mathematicians who could hold their own in disputes with Continental rivals, although even so he did not always prevail. Bernoulli pointed out that an integration problem issued by Taylor as a challenge to "non-English mathematicians" had already been completed by Leibniz in *Acta eruditorum*. Their debates in the journals occasionally included rather heated phrases and, at one time, a wager of fifty guineas. When Bernoulli suggested in a private letter that they couch their debate in more gentlemanly terms, Taylor replied that he meant to sound sharp and "to show an indignation."

The *Methodus* contained several additional firsts, the importance of which could not have been realized at the time. These include the recognition and determination of a singular solution for a differential equation,[5] a formula involving a change in variables and relating the derivatives of a function to those of its inverse function, the determination of centers of oscillation and percussion, curvature, and the vibrating string problem. The last three problems had been published earlier in the *Philosophical Transactions*, as had been a continued fraction for computing logarithms.

Newton approached curvature by way of the determination of the center of curvature as the limit point of the intersection of two normals. Although this was not published until 1736, Taylor was familiar with Newton's work, since, after applying his own formula, Taylor remarked that the results agreed with those given by Newton for conic sections. Taylor, however, conceived of the radius of curvature as the radius of the limiting circle through three points of a curve, and associated curvature with the problem of the angle of contact dating back to Euclid. He then used curvature and the radius of curvature in giving the first solution for the normal vibrations of the simplest case of the plucked string. In propositions XXII and XXIII he showed that under his conditions each point will vibrate in the manner of a cycloidal pendulum, and he determined the period in terms of the length and weight of the string and a weight supported by the string. There is little doubt that Taylor's work influenced later writers since, for example, Bernoulli cited Taylor in letters to his son Daniel on this topic.

The *Methodus* qualifies Taylor as one of the founders of the calculus of finite differences, and as one of the first to use it in interpolation and in summation of series.

Taylor contributed to the history of the barometer by explaining a derivation of the variation of atmospheric pressure as a logarithmic function of the altitude, and he also contributed to the study of the refraction of light.

Like all of Taylor's writing, his book on linear perspective was so concise that Bernoulli characterized it as "abstruse to all and unintelligible to artists for whom it was more especially written."[6] Even the second edition, which nearly doubled the forty-two pages of the first, showed little improvement in this matter. Its effect, nevertheless, was very substantial, since it passed through four editions, three translations, and twelve authors who prepared twenty-two editions of extended expositions based on Taylor's concepts. He developed his theory of perspective in a formal and rigorous fashion in a sequence of theorems and proofs. The most outstanding and original of his ideas in this field were his definition and use of vanishing points and vanishing lines for all lines and planes, and his development of a theory and practice for the inverse problem of perspective that later served as a basis for work by Lambert and for the development of photogrammetry. Taylor also made free use of the idea of associating infinitely distant points of intersection with parallel lines, and he sought to devise methods for doing geometric constructions directly in perspective.

A study of Brook Taylor's life and work reveals that his contribution to the development of mathematics was substantially greater than the attachment of his name to one theorem would suggest. His work was concise and hard to follow. The surprising number of major concepts that he touched upon, initially developed, but failed to elaborate further leads one to regret that health, family concerns and sadness, or other unassessable factors, including wealth and parental dominance, restricted the mathematically productive portion of his relatively short life.

NOTES

1. W. W. Rouse Ball, *Mathematical Recreations and Essays* (London, 1912), p. 175.
2. H. W. Turnbull, *James Gregory Tercentenary Memorial Volume* (London, 1939), pp. 119–120.
3. Gino Loria, *Storia delle matematiche*, 2nd ed. (Milan, 1950), p. 649.
4. G. Peano, *Formulario mathematico*, 5th ed. (Turin, 1906–1908), pp. 303–304.
5. E. L. Ince, *Ordinary Differential Equations* (New York, 1944), p. 87.
6. *Contemplatio philosophica*, p. 29, quoted from *Acta eruditorum*.

BIBLIOGRAPHY

I. ORIGINAL WORKS. The major source of biographical data as well as the only publication of his philosophical book is *Contemplatio philosophica*: *A Posthumous Work of the late Brook Taylor, L.L.D. F.R.S. Some Time Secretary of the Royal Society to Which Is Prefixed a Life of the Author by his Grandson, Sir William Young, Bart., F.R.S. A.S.S. with an appendix containing Sundry Original Papers, Letters from the Count Raymond de Montmort, Lord Bolingbroke, Mercilly de Villette, Bernoulli, & c.* (London, 1793).

This book and the mathematical letters appended to it are reproduced in Heinrich Auchter, *Brook Taylor der Mathematiker und Philosoph* (Würzburg, 1937). Both of these books have a picture of Taylor as secretary of the Royal Society (1714) as a frontispiece. This picture may be derived from a plaque, since it is signed "R. Earlem, Sculp." It is labeled "From an Original Picture in the Possession of Lady Young." A nearly identical picture labeled "J. Dudley, Sculp." is reproduced in *The Mathematics Teacher*, **27** (January 1927), 4. It is also labeled "London, Published March 26, 1811 by J. Taylor, High Holborn."

Charles Richard Wild, in *A History of the Royal Society* (London, 1848), lists a portrait of Taylor painted by Amiconi among the portraits in possession of the Royal Society, but *The Record of the Royal Society*, 3rd ed. (London, 1912), records in its "List of Portraits in Oil in Possession of the Society" "Brook Taylor L.L.D. F.R.S. (1685–1731). Presented by Sir W. Young, Bart., F.R.S. Painter Unknown."

The two editions of Taylor's *Methodus* cited above were both published in London, as were the editions of his *Linear Perspective*. Complete data on the editions and extensions of this book are contained in P. S. Jones, "Brook Taylor and the Mathematical Theory of Linear Perspective," in *The American Mathematical Monthly*, **58** (Nov. 1951), 597–606.

Additional data on Taylor's correspondence is to be found in H. Bateman, "The Correspondence of Brook Taylor," in *Bibliotheca Mathematica*, 3rd ser., **7** (1906–1907), 367–371; Edward M. Langley, "An Interesting Find," in *The Mathematical Gazette*, **IV** (July 1907), 97–98; Ivo Schneider, "Der Mathematiker Abraham de Moivre," in *Archive for History of Exact Sciences*, **5** (1968/1969), 177–317.

II. SECONDARY LITERATURE. For details of one of Taylor's disputes see Luigi Conte, "Giovanni Bernoulli e le sfida di Brook Taylor," in *Archives de l'histoire des sciences*, **27** (or **1** of new series), 611–622.

The most extensive history of Taylor's theorem is Alfred Pringsheim, "Zur Geschichte des Taylorschen Lehrsatzes," in *Bibliotheca mathematica*, 3rd ser., I (Leipzig, 1900), 433–479.

PHILLIP S. JONES

TAYLOR, CHARLES VINCENT (*b.* near Whitesville, Missouri, 8 February 1885; *d.* Stanford, California, 22 February 1946), *biology.*

Charles Vincent Taylor, the youngest of ten children, was born to Isaac Newton Taylor and Christina Bashor Taylor, on a farm near Whitesville, Missouri. He was a descendant of a family whose members included eminent lawyers, Baptist clergymen, a general, a senator, and a governor. Because his father's farm was not very productive, young Taylor had to work his way through Mount Morris College, Illinois. In college he was prominent in student affairs and, because of his fine singing voice, in musical groups as well. After receiving his B.A. he became principal of a high school in Valley City, North Dakota. His deep interest in biology led him to enroll in 1914 as a graduate student at the University of California, in Berkeley, where he took the M.A. under the supervision of Joseph A. Long, based upon a study of fertilization in the mouse, and the Ph.D. in 1917 under Charles A. Kofoid, based upon a study of the neuromotor apparatus of the ciliate *Euplotes.* He was an instructor of zoology at the university (1917–1918), then Johnston scholar at the Johns Hopkins University (1918–1920). Taylor returned to the University of California, where he taught as assistant professor (1920–1923). In 1921, he married one of his students, Lola Lucille Felder. Spirited and interested in the arts, she brought to the family a measure of gaiety and fun that served as a balance to her husband's more serious nature. Four children resulted from the marriage: Jeanne, Elouise, Lenore, and Isaac Newton.

During the years that he was associated with the University of California, Taylor spent some of his summers at the Marine Biological Laboratory, in Woods Hole, Massachusetts, where he collaborated with Robert Chambers on micromanipulative studies. His microsurgical experiments on the function of motor organelles and of the micronucleus in *Euplotes*, published in 1920 and 1924, are classics. In the summers of 1922 and 1923 he was on the staff at Johns Hopkins Marine Station. Taylor was assistant professor at the University of Michigan (1923–1924), and research associate at the Tortugas laboratory of the Carnegie Institution in the summers of 1924 and 1926. During these years he studied the organization and early development of marine eggs.

In 1925 Taylor was appointed to the staff of the biology department at Stanford University and quickly advanced to full professor. In 1930 he received a particularly attractive offer from the University of Michigan, but decided that biology had a more promising future at Stanford. He took leave that year, however, to teach at the University of Chicago. Upon his return to Stanford, he was made Herzstein professor of biology and chairman of the biology department (both 1931), and then dean of the School of Biological Sciences (1934).

The series of papers that Taylor and his collaborators published on the encystment and excystment of the ciliate *Colpoda duodenaria* has not been superseded or surpassed. From his first experiments published with H. Albert Barker in 1930 to the last papers published shortly before his death in 1946, he demonstrated his progressively more effective control of the physical conditions and state of the experimental organism, a control that necessitated a study of the structure and cytological details of reorganization in order to make certain of the taxonomic position of the experimental animal. Taylor also saw the need for an axenic culture of the organism and made efforts to see this accomplished. As a consequence, one of his collaborators, Laura Garnjobst, finally showed that encystment resulted from the lack of any one of five of the known nutritional factors in the diet. Taylor also suggested a study of stromatogenesis as the basis for resolving the taxonomic position of that most celebrated genus of protozoans, *Tetrahymena*, which was then in a confused state. His earlier interests, in fibrillar systems in protozoans, in solgel transformations in the protoplasm of protozoans, and in the organization of marine eggs and their early development, were essential to his deep interest in reorganization, differentiation, and redifferentiation in protoplasm. The micromanipulator known by his name was an instrument designed to enable him to complete studies on these subjects.

As administrator in the Stanford biology department, Taylor held strong views; he believed that too many scientists emphasized the differences between organisms and their functions, burrowing deeper and deeper into their specialties until they lost sight of what he called the "common denominators" in life. He looked for and emphasized these basic concepts and reorganized the department of biological sciences at Stanford so that such concepts would become the guiding principles. With the help of a sympathetic faculty that he had attracted to the department, he introduced a series of foundation courses to achieve his purpose. Many generations of students benefited from the broad background in basic biology that they received at Stanford. His idealism was misunderstood as a power play by some, however, who resented the engulfment of smaller departments into a larger unit.

In recognition of his many achievements Taylor was elected to the National Academy of Sciences in 1943. Two members of the department, Beadle and Tatum, won the Nobel Prize for work done in the department during Taylor's headship.

Taylor organized a symposium for the celebration of the hundredth anniversary of the cell theory (of Schleiden and Schwann) in 1939. He obtained funds that enabled him to attract an international assembly of eminent scientists. The symposium resulted in the monograph "The Cell and Protoplasm," edited by Forest R. Moulton, and published in 1940.

BIBLIOGRAPHY

I. ORIGINAL WORKS. Taylor's writings include "Demonstration of the Function of the Neuromotor Apparatus in *Euplotes* by the Method of Microdissection," in *University of California Publications in Zoology*, **19** (1920), 403–470; "Fatal Effects of the Removal of the Micronucleus in *Euplotes*," with W. P. Farber, *ibid.*, **26** (1924), 131–144; "Improved Micromanipulation Apparatus," *ibid.*, **26** (1925), 443–454; "An Investigation on Organization of a Sea Urchin Egg," with D. H. Tennant and Douglas M. Whitaker, in *Carnegie Institution of Washington. Publications*, **391** (1929), 1–104; "A Study of the Conditions of Encystment of *Colpoda cucullus*," with H. Albert Barker, in *Physiological Zoology*, **4** (1931), 620–634; "Effects of a Given X-Ray Dose on Cysts of *Colpoda steini* at Successive Stages in their Induced Excystment," with Morden G. Brown and Ar-

thur G. R. Strickland, in *Journal of Cellular and Comparative Physiology*, **9** (1936), 105–116; "Structural Analysis of *Colpoda duodenaria*," with Waldo H. Furgason, in *Archiv für Protistenkunde*, **90** (1938), 320–339; "Growth Studies of *Colpoda duodenaria* in the Absence of Other Living Organisms," with Willem J. Van Wagtendonk, in *Journal of Cellular and Comparative Physiology*, **17** (1941), 349–353.

II. SECONDARY LITERATURE. On Taylor and his work, see Charles H. Danforth, "Biographical Memoir of Charles Vincent Taylor," in *Biographical Memoirs. National Academy of Sciences*, **25** (1948), 205–225; Waldo H. Furgason, "The Significant Cytostomal Pattern of the 'Glaucoma-Colpidium Group' and a Proposed New Genus and Species, *Tetrahymena geleii*," in *Archiv für Protistenkunde*, **94** (1940), 224–266; Laura Garnjobst, "The Effects of Certain Deficient Media on Resting Cyst Formation in *Colpoda duodenaria*," in *Physiological Zoology*, **20** (1947), 5–14; Forest R. Moulton, ed., *The Cell and Protoplasm* (Washington, D.C., 1940); Victor C. Twitty, "Charles Vincent Taylor," in *Anatomical Record*, **98** (1947), 242–243.

ARTHUR C. GIESE

TAYLOR, FRANK BURSLEY (*b.* Fort Wayne, Indiana, 23 November 1860; *d.* Fort Wayne, 12 June 1938), *geology.*

Taylor was the only child of Fanny Wright and Robert Stewart Taylor. His father, a judge and Republican politician, became wealthy from his practice of patent law, specializing in the new electrical and telephone industries. Judge Taylor's business and his interest in flood control on the Mississippi River led him to a considerable mastery of science. Frank Taylor graduated from Fort Wayne High School in 1881 and entered Harvard in 1882. He took courses in geology and astronomy as a special student until ill health forced him to drop out in 1886. Accompanied by F. Savary Pearce, a Philadelphia physician, Taylor traveled through the Great Lakes region for the next several years, studying its postglacial geological history. He married Minnetta Amelia Ketchum of Mackinac Island on 24 April 1899. She accompanied him on field trips, driving the horses and, later, their automobile.

Taylor's father paid his son's field and publishing expenses until May 1900, when Frank obtained his first job, as assistant to Alfred C. Lane at the Michigan Geological Survey. He then became an assistant in the glacial division of the U.S. Geological Survey from June 1900 to 1916, first under the direction of Thomas Chrowder Chamberlin and then of Frank Leverett. His field assignments were in New England and in the Great Lakes area. In 1908, 1909, and 1911 he mapped the moraines of southern Ontario for the Canadian Geological Survey. The American Association for the Advancement of Science awarded Taylor a research grant in 1920 and 1921 to study the New York moraines. Apart from this grant Taylor was again supported by his father's estate from 1916 to 1938. He was an active member of the Geological Society of America, the Michigan Academy of Sciences, and the AAAS.

Before he joined the U.S. Geological Survey in 1900, Taylor had formulated a history for the Great Lakes region from the period of the fullest extension of the Wisconsinian glaciation through the ice sheet's retreat to the present time. Between 1892 and 1896 he argued that old beachlines indicated that a marine submergence of the area had followed the disappearance of the glacier, an idea he dropped when his fieldwork of 1895 on the succession of terminal moraines convinced him that the lakes were formed by ice dams instead. Grove Karl Gilbert had encouraged Taylor to study moraines as the crucial test of his submergence theory. Once he accepted the ice dam theory, his reconstruction of Great Lakes history matched current views on this part of the geological record.

Taylor's studies of beachlines and moraines suggested that retreat of the Wisconsinian glacial field was not steady. Rather, the ice sheet had melted back and then partially readvanced, perhaps dozens of times. Inspired by Chamberlin's reports from the Greenland expeditions of 1896, Taylor decided that Great Lakes topography had controlled the ice sheet, not the other way around. Once local features in the old Midwest were taken into account, the pattern of retreat and partial readvance was a regular series. Taylor's work on glacial history culminated in U.S. Geological Survey monograph 53, *The Pleistocene of Indiana and Michigan and the History of the Great Lakes* (Washington, D.C., 1915), which he wrote with Frank Leverett. This book and Taylor's articles after 1900 show that his federal government experience led him to use the sophisticated geological terminology that marked a professional scientist.

Taylor's scheme included a complicated series of drainage systems to remove glacial meltwater. He paid special attention to the time when the main channel was a strait across what is now Lake Nipissing and its related rivers. During the Nipissing episode, he observed, the flow of Niagara Falls was much less than that over the modern ledge, a hypothesis he supported with studies of the narrower gorge of the past. Gilbert had suggested a lesser flow during some part of the postglacial era; Taylor worked on the details and supplied the reason. One important implication of this diminished flow was that the retreat of the falls could no longer be a reliable geochronological indicator.

Taylor was influenced in his glacial work by the publications of Gerhard de Geer as well as by Gilbert and Chamberlin. He adopted de Geer's concept of an isobase of deformation to describe the uplift of land to the northeast during glacial retreat. Below the isobase no uplift had occurred. Taylor calculated the amount and direction of deformation from the tilting of originally horizontal beachlines of postglacial lakes.

Taylor's courses in astronomy at Harvard inspired some of his later scientific work. He later recalled that on 13 December 1884, it occurred to him that the moon had once been a comet and had been captured by the earth during its last pass through the solar system. In 1898 he elaborated this idea into a theory for the origin of the whole planetary system. Taylor credited Daniel Kirkwood, then an astronomer at Indiana University, with first publishing a similar theory in connection with the origin of the asteroids. According to Taylor, as comets passed close to the sun, they were captured as planets that spun in the orbit of Mercury. Upon each new capture, the planets shifted out one orbit toward Neptune's. His fullest explanation of cosmogony appeared in his privately published book (1903) on the subject, in which he spelled out some of the mathematics and physics of the theory. Taylor felt compelled to challenge Isaac Newton's explanation for the orbit of the moon, because Newton's system did not rigidly determine the moon's course. Taylor believed that the orbits of the moons and planets must be determinate, but the mathematics of the situation seems to have been beyond his grasp. It is doubtful that determinate satellite orbits were required for Taylor's theory.

Taylor's 1898 work on the planetary system, published as a forty-page pamphlet (now extremely rare), also contained his first articulation of a theory of the history of continents. When the moon was captured by the earth, Taylor argued, it created a tidal force on the planet and increased the earth's speed of rotation. These two forces pulled the continents away from the poles toward the equator. He did not return to this point in 1903, but in 1910 published his first detailed arguments for a theory of continental drift in a paper on the origin and arcuate shape of Tertiary mountain ranges. He explicitly acknowledged his debt to Eduard Suess's work on the Asian ranges as a source for much of his theory. Taylor said that as the continents slid toward the equator, they encountered obstructions that created loops of mountain ranges, just as the Wisconsinian glacial sheet had formed lobes as it met local topographical features. Taylor's theory of 1910 also involved movement of crustal material away from the mid-Atlantic ridge.

Taylor published no papers from 1917 through 1920. He spent those years reading and thinking about ways to defend, expand, and clarify his theory of continental drift and mountain creation. His later papers connected geological phenomena, such as earthquakes, to his theory; added studies of other arcuate Tertiary ranges; and argued that the uplift following the Wisconsinian glaciation could be accounted for by crustal creep. These papers also spelled out the differences of Taylor's views from those of Alfred Wegener, whose theory of continental drift was first published in 1912. Taylor rejected isostatic adjustment as a mechanism for crustal movement because he believed its working required a fluid interior for the earth. Wegener's theory used movements toward isostasy on a fluid subcrust to shift continents, while Taylor's slid continents along a narrow shear zone of a fairly rigid earth. Taylor's theory did not fully anticipate the features of plate tectonics, however. The most notable difference is that plate tectonics does not require capture of the moon or other astronomical events to account for earth movements; terrestrial forces are considered adequate.

BIBLIOGRAPHY

I. ORIGINAL WORKS. Leverett (see below) lists Taylor's publications but occasionally omits abstracts of articles published in full elsewhere and, in at least two instances, cites pages incorrectly. Two articles should be added to his compilation: "A Short History of the Great Lakes," in *Inland Educator*, 2 (Mar., Apr., May 1896), 101–103, 138–145, 216–223, respectively; and "Geological History of the Great Lakes," in *Scientific Monthly*, 49 (July 1939), 49–56. According to the National Union Catalog at the Library of Congress, only two copies of Taylor's *An Endogenous Planetary System. A Study in Astronomy* (Fort Wayne, 1898) are in public libraries, one at the John Crerar Library in Chicago and one at the U.S. Geological Survey Library, Reston, Virginia. *The Planetary System; A Study of Its Structure* (Fort Wayne–London, 1903) is a more accessible account of Taylor's astronomical theory; and "Bearing of the Tertiary Mountain Belt on the Origin of the Earth's Plan," in *Bulletin of the Geological Society of America*, 21 (1910), 179–226, is a more available statement on the history of continents than the 1898 pamphlet.

II. SECONDARY LITERATURE. Richard Flint, *Glacial and Quaternary Geology* (New York, 1971), chs. 2, 13, 21, 30, provides helpful background for assessing Taylor's work on recent geological history. For useful critiques of Taylor's theory of continental drift, see Anthony Hallam, *A Revolution in the Earth Sciences: From Continental Drift to Plate Tectonics* (Oxford, 1973), 3–6, and Ursula Marvin, *Continental Drift: The Evolution of a Concept* (Washington, D.C., 1973), 63–64.

Biographical details about Taylor are available in Frank Leverett, "Memorial to Frank Bursley Taylor," in *Proceedings of the Geological Society of America* for 1938 (1939), 191–200. Shorter sketches are by J. H. Bretz, in *Dictionary of American Biography*, supp. 2, 653–654; *American Men of Science*, 5th ed. (1933); *Who's Who in America*, 1936–1937, 2387; by A. C. Lane, in *Proceedings of the American Academy of Arts and Sciences*, 75 (1944), 176–178; by Frank Leverett, in *Science*, n.s. 88 (5 August 1938), 121–122; and Harvard University, Class of 1886, *Secretary's Report*, 7 (1911), 305–307, and 12 (1936), 416–419.

MICHELE L. ALDRICH

TAYLOR, FREDERICK WINSLOW (*b.* Germantown [part of Philadelphia], Pennsylvania, 20 March 1856; *d.* Philadelphia, 21 March 1915), *engineering*.

The scion of an aristocratic Philadelphia family, Taylor seemed destined to follow his father, Franklin, along a well-worn groove to a genteel law practice. But after graduating from Phillips Exeter Academy in 1874 he showed the independence of his strong-minded mother, Emily Annette Winslow, and chose instead to become a mechanical engineer. After serving his apprenticeship as a machinist and pattern maker, Taylor went to work at the Midvale Steel Company in Philadelphia, where he rose from laborer to chief engineer in six years. He obtained the M.E. degree by correspondence from Stevens Institute of Technology in 1883. He married Louise M. Spooner of Philadelphia in 1884.

Taylor was one of a number of engineers who were attempting to convert engineering into a science. He assumed that there were laws (or rational principles) underlying all areas of engineering practice, including management. Beginning at Midvale and continuing through his career as a consultant, Taylor conducted a series of painstaking investigations of metal cutting, tool steel, belting, reinforced concrete, management, and other subjects. These works were highly empirical and owed little to theory, for which Taylor had neither understanding nor sympathy. It was by such a process of cut-and-try experiment that Taylor and J. Maunsel White discovered the process, named for them, for the heat treatment of tool steel (1898). Under the name "high-speed steel" this invention revolutionized machine shop practice by permitting the speed of metal-cutting machinery to be more than doubled.

The achievement for which Taylor is most remembered was his development of "scientific management." He wanted to reduce all aspects of management to "exact science"; and his approach, sometimes termed "task management," was to determine exactly how much each worker should accomplish in a given time. This entailed the discovery of a new measure of human work. At first Taylor attempted to find a correlation between fatigue and foot-pounds of work, but such a direct solution eluded him. Instead, he found what he considered to be the atomistic units of work: "elementary motions." Taylor first broke down a set of operations into these motions and timed them with a stopwatch. He then analyzed the sequence of motions, eliminated the unnecessary ones, and combined the remainder into an optimum series. After adding percentages to cover necessary rest and unavoidable delays, Taylor thought he could calculate the time required for any task.

Time-and-motion study was only the first of a series of managerial innovations. To set times, to assign daily tasks, and to prepare written instructions for each worker required a planning department, which became the nerve center of management under Taylor's system. Precisely determined tasks entailed the complete standardization of tools, operations, and routing. Taylor also devised methods of cost accounting, inventory control, records keeping, and a functional organization of authority that facilitated rational management

Taylor thought that his system of management provided the basis of a scientific ethics. Through it he hoped to end class conflict and establish social justice. Although these larger goals were not achieved, Taylor's system had a profound influence on modern management thought.

BIBLIOGRAPHY

I. ORIGINAL WORKS. The Taylor papers are in the library of Stevens Institute of Technology, Hoboken, N.J.

They include MSS of unpublished addresses as well as correspondence. There is an interesting autobiographical fragment in a letter from Taylor to Morris L. Cooke, 2 Dec. 1910. A printed guide to the papers is available at the repository.

A convenient assemblage of Taylor's most important publications is in Frederick W. Taylor, *Scientific Management: Comprising Shop Management, The Principles of Scientific Management, and Taylor's Testimony Before the Special House Committee* (New York, 1947). Taylor's various papers presented to the American Society of Mechanical Engineers are listed in its *Seventy-Seven Year Index* (New York, 1951). The greatest of these was also published as a book, *On the Art of Cutting Metals* (New York, 1907). In addition he wrote two works with Sanford E. Thompson: *A Treatise on Concrete* (New York, 1905) and *Concrete Costs* (New York, 1912).

II. SECONDARY LITERATURE. The standard biography is Frank B. Copley, *Frederick W. Taylor* (New York, 1923). Memoirs include H. K. Hathaway, ed., "Tributes to Frederick W. Taylor," in *Transactions of the American Society of Mechanical Engineers*, **37** (1915), 1459–1496; "Frederick Winslow Taylor," *ibid.*, 1527–1529; and The Taylor Society, *Frederick Winslow Taylor, A Memorial Volume* (New York [*ca.* 1920]). See also Carl W. Mitman, "Frederick Winslow Taylor," in *Dictionary of American Biography*, XVIII, 323–324.

Modern evaluations of Taylor and his system are in Hugh G. J. Aitken, *Taylorism at the Watertown Arsenal* (Cambridge, Mass., 1960), 13–48; Samuel Haber, *Efficiency and Uplift* (Chicago–London, 1964), 1–30; Edwin Layton, *The Revolt of the Engineers* (Cleveland, 1970), 134–139; Milton J. Nadworny, *Scientific Management and the Unions* (Cambridge, Mass., 1955), 1–33; and Sudhir Kakar, *Frederick Taylor: A Study in Personality and Innovation* (Cambridge, Mass., 1970).

EDWIN LAYTON

TEALL, JETHRO JUSTINIAN HARRIS (*b.* Northleach, England, 5 January 1849; *d.* Dulwich, England, 2 July 1924), *petrography, geology.*

The son of Jethro Teall, a landowner of Sandwich, Kent, and Mary Hathaway of Northleach, Teall was born after the death of his father. He was educated at Northleach Grammar School, at Cheltenham, and at St. John's College, Cambridge, taking the natural sciences tripos in 1872 and receiving the M.A. in 1876. He was inspired to choose geology instead of mathematics at Cambridge by his tutor, T. G. Bonney, and attended Adam Sedgwick's geology course as Woodwardian professor. Teall was the first recipient of the Sedgwick Prize, in 1874, for an essay dealing with the phosphatic deposits near Cambridge. Election to a fellowship of his college followed in 1875, and he retained this post until his marriage to Harriet Cowen of Nottingham, in 1879. The years between 1872 and 1888 were divided between lecturing under the University Extension Scheme in the Midlands, in the north and west of England, and in London, and petrographical research at Cambridge. His lectures attracted large audiences; and his association with the early days of University College, Nottingham, is particularly remembered.

The field of petrography was almost new at this time. The polarizing microscope had been developed by W. H. Fox Talbot in 1834 by making use of William Nicol's single-vision calcite prism (1829). Thin sections were made by H. C. Sorby in the 1850's and J. Clifton Ward had applied the technique to Lake District lavas during the ensuing decade, but a great avenue of research was opening up. Through his work on the pitchstone of Eigg, the whin sill and dikes of northern England, the Cheviot lavas, the Cornish granites, and Lizard gabbros, Teall gave major impetus to the subject. From 1886 to 1888 he was engaged in publishing the definitive work on British petrography with which his name will always be associated. The encouragement given by Bonney and the influence of Rosenbusch, leading petrographical systematist of the time in Europe, had borne good fruit; and Teall had added many new ideas of his own to both igneous and metamorphic petrography.

In 1888 Sir Archibald Geikie invited Teall to join the Geological Survey of Great Britain as a geologist, to take charge of the petrographical work. The primary geological survey of the northwestern Highlands of Scotland had begun in 1885, and its spectacular results demanded close petrographical study. This Teall provided, for the Highlands and for other parts of the United Kingdom, in a series of contributions to official memoirs and a number of important private papers appearing between 1888 and 1903. In 1901 he was chosen to succeed Geikie as director of the Geological Survey. He was at the same time president of the Geological Society of London; and his influential addresses to the Society on the evolution of petrological ideas (1901, 1902) were considered sufficiently important by the Smithsonian Institution to reproduce in its *Report of the Board of Regents.* . . .

During his eleven years as director of the Geological Survey, Teall's output of original work was necessarily curtailed by the demands of administration; but he had the satisfaction of leading the official British geological effort during a productive period. His contributions to science were recognized by his election to the Royal Society in 1890 and by the award of the Bigsby Medal of the Geological Society in 1889 and its highest award, the Wollaston Medal, in 1905. The Paris Academy presented him with its Delesse Prize in 1907; and his official work was recognized with a knighthood in 1916.

BIBLIOGRAPHY

I. Original Works. A full list of Teall's writings is given in the *Geological Magazine* article cited below. They include "On the Chemical and Microscopical Characters of the Whin Sill," in *Quarterly Journal of the Geological Society of London,* **40** (1884), 640–657; "The Metamorphosis of Dolerite into Hornblende-Schist," *ibid.,* **41** (1885), 133–144; *British Petrography, With Special Reference to the Igneous Rocks* (London, 1888); and "The Evolution of Petrological Ideas," in *Quarterly Journal of the Geological Society of London,* **57** (1901), lxii–lxxxvi, and **58** (1902), lxiii–lxxviii.

II. Secondary Literature. Biographies are H. H. Thomas, "Sir Jethro Justinian Harris Teall," in *Dictionary of National Biography* for 1922–1930 (1937), 826–827; and H. Woodward, "Eminent Living Geologists. Jethro Justinian Harris Teall," in *Geological Magazine,* 5th ser., **6** (1909), 1–8. Obituaries are in E. B. Bailey, *Geological Survey of Great Britain* (London, 1952), 144–170; J. S. Flett, *The First Hundred Years of the Geological Survey of Great Britain* (London, 1937), 143–160; *Proceedings of the Royal Society,* B **97** (1925), xv–xvii; *Quarterly Journal of the Geological Society of London,* **81** (1925), lxiii–lxv.

K. C. Dunham

TEICHMANN, LUDWIK KAROL (*b.* Lublin, Poland, 16 September 1823; *d.* Cracow, Poland, 24 November 1895), *anatomy.*

Teichmann's name is associated mainly with his discovery of a simple chemical test for the presence of blood that was widely used. His parents died when Teichmann was six, and his two aunts helped him to complete his secondary education. In 1847 he enrolled at Protestant Theological Faculty at Dorpat, but soon became interested in natural science. In 1850 he was one of the seconds in a duel that ended with the death of a colleague, a circumstance that forced him to flee the country. From Hamburg he went to Heidelberg, where he entered the Faculty of Medicine in November 1850. Working as an anatomical preparator in the department headed by Jacob Henle, he also studied chemistry, physics, and technology. In 1852 he followed Henle to Göttingen, where he graduated M.D. on 18 December 1855 with the dissertation "Zur Lehre der Ganglien." In 1853, while still a student, he published a paper on the crystallization of certain organic compounds of the blood, describing the preparation of the microscopic crystals of hemin (heme chloride). This was the first, and for some ten years the only, test for the presence of blood in suspect stains on clothes, furniture, or other objects. Because of its simplicity and specificity, it became one of the accepted tests in forensic medicine. During the same period he mastered time-consuming anatomical preparations to supply Henle with illustrations for his *Handbuch der systematischen Anatomie*. Henle's failure to acknowledge Teichmann's contributions led to a conflict.

After graduation Teichmann became provisional prosector of anatomy and was granted the Blumenbach fellowship, which allowed him to visit other European anatomical departments and meet prominent anatomists: Koelliker, Johannes Müller, Gegenbaur, A. Retzius, M. P. C. Sappey, Richard Quain, Sharpey. He was particularly well received by Joseph Hyrtl in Vienna, who helped him during an illness and encouraged him to continue his career in anatomy. Hyrtl remained Teichmann's good friend in later years. In 1859 Teichmann obtained the *venia legendi* at Göttingen and in 1861 went to Cracow, where he was offered the chair of pathological anatomy. In 1868 he became professor of descriptive anatomy and retained that post until his retirement in 1894. In 1873 he opened a new anatomical theater and provided many specimens for its museum.

Teichmann's main scientific interest was in the lymphatic vessels and their origin, which he studied by means of an original injection method. He maintained that no direct communication exists between blood capillaries and lymphatics, and discovered the obstruction of lymphatic vessels in elephantiasis.

BIBLIOGRAPHY

I. ORIGINAL WORKS. Teichmann's most important papers are "Über die Krystallisation der organischen Bestandteile des Bluts," in *Zeitschrift für rationelle Medicin*, n.s. **3** (1853), 375–388; and "Über das Haematin," *ibid.*, n.s. **8** (1857), 141–148, in which the method of preparing hemin (heme chloride) crystals is described. *Das Saugadersystem vom anatomischen Standpunkte* (Leipzig, 1861) is his main publication on the lymphatic system. In later years Teichmann is reported to have published 30 papers (in Polish and German) on the lymphatics and others on anatomical techniques. No bibliography of his works seems to have been published.

II. SECONDARY LITERATURE. A biographical sketch based on Teichmann's handwritten autobiography is Leon Wachholz, "Ludwik Teichmann, szkic biograficzno-historyczny," in *Archiwum historji i filozofji medycyny*, **10** (1930), 34–62, summarized in English by A. Laskiewicz, "Professor Ludwik Karol Teichmann, M.D. 1823–1895," in *Bulletin of Polish Medical Science and History*, **8** (1965), 91–92. See also F. Lejars, "Un grand anatomiste polonais Ludwig Teichmann," in *Revue scientifique* (*Revue rose*), 4th ser., **5** (1896), 481–487.

VLADISLAV KRUTA

TEILHARD DE CHARDIN, PIERRE (*b*. near Orcines, Puy-de-Dôme, France, 1 May 1881; *d*. New York, N.Y., 10 April 1955), *paleontology, geology.*

Teilhard was born into an aristocratic family at Sarcenat, the familial home. In 1898 he entered the Jesuit order and, after initial studies at Aix-en-Provence and the Isle of Jersey, was assigned to a Jesuit school in Cairo, where he taught physics and chemistry from 1905 to 1908. In Egypt he acquired his first extensive experience in fieldwork, which awakened his interest in the geology of the Tertiary. Although he published a monograph on Eocene strata in Egypt, Teilhard did not acquire true professional competence until he was transferred to the Jesuit house at Hastings, England, where between 1908 and 1912 he studied not only theology but also vertebrate paleontology. Ordained a priest, he returned to Paris in 1912 to study geology and paleontology with Marcellin Boule at the Museum of Natural History; these researches, interrupted by wartime duty as a stretcher-bearer, led to his Sorbonne thesis (1922) on the mammals of the Lower Eocene in France.

From 1920 to 1923 Teilhard taught geology at the Catholic Institute in Paris. In the latter year he made his first journey to China to participate in a paleontological mission with Émile Licent. By the time he returned to Paris, his habit of interpreting such theological questions as original sin in the light of evolutionary ideas had attracted opposition, with the result that he was forbidden to continue lecturing at the Catholic Institute. In April 1926 he departed for virtual exile in China.

The years 1908–1912 had been ones of critical intellectual formation. At Hastings he became acquainted with the evolutionary philosophy of Henri Bergson, whose book *L'évolution créatrice* proved to be the most important source of Teilhard's emerging world view, although Teilhard's notion of a converging cosmos was antithetical to that of Bergson. By 1916 (in his essay "La vie cosmique") the main lines of Teilhard's idea of a cosmic and directed evolutionary force had already developed. In Paris, before his definitive departure for China, Teilhard further developed his philosophical ideas in conjunction with the Bergsonian philosopher Édouard Le Roy and the Russian geologist V. I. Vernadsky. Le Roy and Teilhard attended Vernadsky's 1922–1923 Sorbonne lectures on geochemistry, in which the Russian explicated his concept of the biosphere. It was this stimulus that led Teilhard, in a series of lectures on evolution in 1925–1926, to develop the concept of the noosphere, or thinking layer of the earth, representing a higher stage of evolutionary development. During his first months in China he set down his ideas in *Le milieu divin*, in which the theme of man as the culmination of the evolutionary impulse emerges in full relief.

The next few years, spent largely in collaboration with Licent, were productive ones for Teilhard. He continued work on Quaternary and Tertiary mammals, participating in 1928 in a joint study with Boule, Licent, and Henri Breuil on the Chinese Paleolithic, his contribution being studies on geology and his specialty, mammalian fauna, which he found to be quite similar to the mammals of the European Pleistocene. In 1929 he was scientific adviser to the Chinese Geological Survey and, in his own words, "heading the geological advance in China." By the end of the decade he had terminated his relationship with Licent and had shifted his scholarly connections from French to Sino-American institutions; it was against this background that the discoveries at Chou-k'ou-tien were made. There, in 1929, a human cranium, that of the celebrated *Sinanthropus*, was unearthed by Pei Wen-chung. Teilhard, as the research team's geologist (as well as coordinator of operations), was able to demonstrate the earliness of the skull, which later proved to be a close relation of the *Pithecanthropus* of Java.

The next phase of Teilhard's career was devoted largely to geological research as he sought to synthesize the continental geology of Asia. In 1929 he began a series of expeditions, including the Roy Chapman Andrews Central Asia expedition (summer 1930) and the Croisière Jaune expedition of 1931–1932 (which Teilhard regarded as a pseudo-scientific venture but one that nevertheless enabled him to complete syntheses of the tectonics

of northern China and of the Pleistocene geology of Central Asia).

By 1934 Teilhard was acting director of the Geological Survey and an active participant in the Cenozoic Research Laboratory in Peking. He next sought to connect the Tertiary and Quaternary geologic structure of northern China with that of the south, following the lines of fissure, a project that led to a trip to India with Helmut de Terra to connect the geology of the subcontinent with that of China. From 1931 to 1938 Teilhard produced a series of essays contributing to his synthesis of Asian geology and paleontology.

After 1938, with the exception of two trips to Africa, Teilhard's fieldwork was over. During the Japanese occupation he wrote several important monographs on human paleontology and fossil mammals in China. It was during this same period that his major philosophical works were conceived, beginning with the first chapter of *Le phénomène humain* in 1938.

The Teilhardian thesis is not a scientific theory but, rather, a philosophical world view based on certain themes drawn from the evolutionary synthesis and expressed in mystical, often poetic, terms. Teilhard discerned in the historical development of the cosmos a law of "complexity-consciousness," a notion reminiscent of Haeckel's views on the psychic unity of the organic world, only extended on a cosmic scale to include the inorganic world. According to Teilhard's law, each successive stage in the evolutionary process is marked, first, by an increasing degree of complexity in organization and, second, by a corresponding increase in degree of consciousness. Evolution thus proceeds in orderly fashion from the inorganic to the organic, from less complex to more highly organized forms of life, through the process of hominization and beyond to "planetization," whereby all the peoples of *Homo sapiens* are to achieve collectively an ultrahuman convergence, seen symbolically as a final "Point Omega."

As a rationale for his view of an anthropocentric universe, Teilhard invoked Heisenberg's uncertainty principle to demonstrate that man is the center of all perspective in the natural world. He then applied the second law of thermodynamics to explain the complexification of the universe over time. Energy, instead of being lost through entropy, is converted into what Teilhard termed "radial energy," a metaphysical construct standing for the evolutionary forces productive of increasing cultural complexity.

In terms of evolutionary process, Teilhard has

been typically interpreted as a neo-Lamarckian exponent of orthogenesis: a process of accretion of changes tending in the same direction, toward a divinely inspired mankind. As T. Dobzhansky has noted, Teilhard's notion of orthogenesis was an eccentric one, the teleological implications of which have been overstressed by religionists to the point of distorting Teilhard's conception of the evolutionary process. That he was a finalist in the philosophical sense cannot be denied; but his finalism was applied to the evolutionary process only in a retrospective way, as a commentary on the cosmic past. His real understanding of evolution was considerably closer to scientific orthodoxy than has generally been supposed, and toward the end of his life it drew closer to the neo-Darwinian synthesis. He saw evolution functioning through a series of purposeful "gropings" (*tâtonnements*) that are random until the "purpose" is achieved. The *hasard dirigé* that governs the process need not be understood in a theologically teleological sense, however, but as a process of response to environmental challenges.

After the war Teilhard returned to Paris, where he labored in vain to publish *Le phénomène humain*. In 1947 he was ordered to refrain from philosophical writing and in 1949 was denied permission by his order to succeed to Breuil's chair of paleontology at the Collège de France. The result of these reverses was a second "exile"—to New York (1951–1955) and a research appointment at the Wenner-Gren Foundation for Anthropological Research. He resumed fieldwork on two trips to South Africa to inform himself of the research on *Australopithecus*. The African experience made it possible for him to complete his scientific synthesis of hominization, in which he characterized anthropogenesis as a bipolar process having an abortive Asian center and another in Africa that led directly to *Homo sapiens*. At the same time Teilhard continued to relate human origins to general geological development. In his final essays he again articulated his view of man as the primary focus of recent evolutionary development.

The diffusion of Teilhardian evolutionism is not a spurious phenomenon, but a further phase in the popularization of the theory of evolution, whereby entire sectors of society previously hostile to Darwinism have been brought into the evolutionary consensus. The primary diffusion began with the publication of *Le phénomène humain* in 1955, peaked around 1967, and, by 1970, had encompassed all the cultures of the Catholic West. In 1957 the Holy Office ordered the works of Teilhard removed from the libraries of Catholic institutions and forbade their sale in Catholic bookstores. This move was a prelude to the *monitum* of 30 June 1962 advising the faithful of errors and ambiguities in Teilhard's philosophical and theological writings. At the same time, however, the Jesuit order relaxed its former stance and now produced the leading ecclesiastical defenders of Teilhard. That the *monitum* proved a dead letter can be seen from the feverish Teilhardian literary activity in France, Spain, and (with stronger opposition) Italy, which has had the effect of depolemicizing evolution in most intellectual and many educational sectors of the Catholic world.

A different phenomenon has been the reception of Teilhardism by European Marxists, especially in the Soviet Union. There, Teilhard's teleological interpretation of evolution has been seen as convergent with dominant concerns of Marxist ideology. The future evolution of the noosphere is to be effected on the basis of the socialization of mankind into ever larger collectivities. Although economic and technological development are indispensable to this movement, both Teilhard and the Marxists believe that "spiritual" (=ideological) factors play a decisive role, particularly a belief in the supreme value of evolution (progress).

BIBLIOGRAPHY

I. ORIGINAL WORKS. The complete scientific works of Teilhard are collected in *Pierre Teilhard de Chardin: L'oeuvre scientifique*, Nicole and Karl Schmitz-Moormann, eds., 10 vols. (Munich, 1971). Of the collected works in the series published in Paris, only vol. II, *L'apparition de l'homme* (1956), translated as *The Appearance of Man* (New York, 1966), contains essays that are, properly speaking, scientific. See also vol. I, *Le phénomène humain* (1955), translated as *The Phenomenon of Man* (New York, 1959); vol. III, *La vision du passé* (1957), translated as *The Vision of the Past* (New York, 1966), and vol. VIII, *Le groupe zoologique humain* (1963), translated as *Man's Place in Nature* (New York, 1966). His fieldwork in China and Africa can be followed in *Lettres de voyages* (Paris, 1956), translated as *Letters From a Traveller* (New York, 1962).

II. SECONDARY LITERATURE. The burgeoning literature on Teilhard and Teilhardism has been cataloged in several bibliographical guides: Joan E. Jarque, *Bibliographie générale des oeuvres et articles sur Pierre Teilhard de Chardin* (Fribourg, 1970); Ladislaus Polgar, *Internationale Teilhard-Bibliographie, 1955–1965* (Munich, 1965); and Daniel Poulin, *Teilhard de Chardin. Essai de bibliographie (1955–1966)* (Quebec, 1966). Selected works on Teilhard are listed yearly in

Archivum historicum Societatis Jesu (Rome). See also Alfred P. Stiernotte, "An Interpretation of Teilhard as Reflected in Recent Literature," in *Zygon*, **3** (1968), 377–425. Useful as critical apparatus are Claude Cuénot, *Nouveau lexique Teilhard de Chardin* (Paris, 1968); and Paul l'Àrchevêque, *Teilhard de Chardin. Index analytique* (Quebec, 1967), a composite index to the first 9 vols. of Teilhard's *Oeuvres*.

Among the periodicals published by societies devoted to the thought of Teilhard, see Société Teilhard de Chardin (Brussels), *Revue* (1960–   ) and *Univers* 1964–   ); Société Pierre Teilhard de Chardin (Paris), *Carnets Teilhard* (1962–   ); Association des Amis de Pierre Teilhard de Chardin (Paris), *Cahiers* (1958–1965) and *Bulletin* (1966–   ); the Teilhard Center for the Future of Man, *Teilhard Review* (1966–   ); and Gesellschaft Teilhard de Chardin (Munich), *Acta Teilhardiana* (1964–   ).

The standard biography of Teilhard is Claude Cuénot, *Pierre Teilhard de Chardin, les grandes étapes de son évolution* (Paris, 1958), also in English (Baltimore, 1965). Works dealing primarily with Teilhard's scientific career are George B. Barbour, *In the Field With Teilhard de Chardin* (New York, 1965); Louis Barjon and Pierre Leroy, *La carrière scientifique de Pierre Teilhard de Chardin* (Monaco, 1964); Henri Breuil, "Les enquêtes du géologue et du préhistorien," in *Table ronde*, no. 90 (June 1955), 19–24; Paul Chauchard, *La pensée scientifique de Teilhard* (Paris, 1965); George Magloire, *Teilhard et le sinanthrope* (Paris, 1964); Henry Fairfield Osborn, "Explorations, Researches and Publications of Pierre Teilhard de Chardin, 1911–1931," *American Museum Novitates*, no. 485 (25 Aug. 1931); Jean Piveteau, *Le Père Teilhard de Chardin savant* (Paris, 1964); and Helmut de Terra, *Mein Weg mit P. Teilhard de Chardin* (Munich, 1962), also in English (London, 1964).

General works on Teilhard's philosophy are Henri de Lubac, *La pensée religieuse du Père Teilhard de Chardin* (Paris, 1962), also in English (New York, 1967); Émile Rideau, *La pensée du Père Teilhard de Chardin* (Paris, 1965), also in English (New York, 1967); and Bernard Towers, *Teilhard de Chardin* (London, 1966). For the "prehistory" of Teilhard's Catholic evolutionism in France, see Henri Begouen, *Quelques souvenirs sur le mouvement des idées transformistes dans les milieux catholiques* (Paris, 1945). Among the explications of Teilhardian evolutionism, see Bernard Delfgaauw, *Teilhard de Chardin* (Baarn, Netherlands, 1961), translated as *Evolution: The Theory of Teilhard de Chardin* (New York, 1969); and Fernando Riaza, *Teilhard de Chardin y la evolución biológica* (Madrid, 1968). Commentary by scientists, from favorable to hostile, includes essays by Theodosius Dobzhansky in *The Biology of Ultimate Concern* (New York, 1969), 213–233, and George Gaylord Simpson in *This View of Life* (New York, 1964), 108–137; and P. B. Medawar's famous critique in *Mind* (Oxford), **70** (1961), 99–106. See also Stephen Toulmin's analysis, "On Teilhard de Chardin," in *Commentary*, **39**, no. 3 (Mar. 1965), 50–55.

Studies of Teilhard's sources are underrepresented in the literature. For his dependence on Bergson, see Madeleine Barthélemy-Madaule, *Bergson et Teilhard de Chardin* (Paris, 1963); on the fleeting, but critical, relationship with Vernadsky, see I. I. Mochalov, *V. I. Vernadsky: Chelovek i myslitel* (Moscow, 1970) 136–138. On Teilhard and Marxist thought, there are numerous articles by Roger Garaudy, including "The Meaning of Life and History in Marx and Teilhard de Chardin: Teilhard's Contribution to the Dialogue Between Christians and Marxists," in *Marxism and Christianity: Studies in the Teilhardian Synthesis* (London, 1967), 58–72; and "Freedom and Creativity: Marxist and Christian," in *Teilhard Review*, **2** (1968–1969), 42–49.

The *monitum* and Catholic opposition to Teilhard are discussed by René d'Ouince, *Un prophète en procès: Teilhard de Chardin dans l'église de son temps*, 2 vols. (Paris, 1970); and by Philippe de la Trinité, *Rome et Teilhard de Chardin* (Paris, 1964) and *Pour et contre Teilhard de Chardin* (Paris, 1970).

The diffusion of Teilhardism has yet to be studied systematically. Representative national studies and bibliographies are (France) J. Hassenforder, *Étude de la diffusion d'un succès de libraire (Le phénomène humain)* (Paris, 1957); (England) Bernard Towers, "The Teilhard Movement in Britain," in *Month*, **36** (1966), 188–196; (Germany) Helmut de Terra, *Bibliographie des deutschsprachigen Schrifttums von und über Pierre Teilhard de Chardin, 1955–1964* (Frankfurt, 1965); (Italy) Elio Gentili, "Pierre Teilhard de Chardin in Italia. Bibliografia," in *Scuola cattolica*, **93** (1965), supp. biblio. 1, 247–334, and **95** (1967), supp. biblio. 2, 138–181; and Marcello Vigli, "Fortuna e funzioni del teilhardismo in Italia," in *Questitalia*, **11** (1968), 352–370; (Soviet Union and eastern Europe) Ladis K. D. Kristof, "Teilhard de Chardin and the Communist Quest for a Space Age World View," in *Russian Review*, **28** (1969), 277–288; and V. Pasika, "Teiiar de Sharden," in *Filosofskaya entsiklopedia*, V (Moscow, 1970), 192–193; (Spain) Miguel Crusafont Pairó, "Teilhard de Chardin en España," in *Acta Teilhardiana*, **5** (1968), 53–63. For a quantitative overview, see José Rubio Carracedo, "Quince años después: Del teilhardismo a Teilhard," in *Arbor* (Madrid), **76** (1970–1971), 43–46.

THOMAS F. GLICK

TELESIO, BERNARDINO (*b.* Cosenza, Italy; 1509; *d.* Cosenza, 1588), *natural philosophy.*

Telesio was one of the group of sixteenth-century Italian speculators known to scholars as "nature philosophers," somewhat to the disparagement of other thinkers who dealt with natural philosophy as then understood but who taught in universities. Telesio's life was relatively uneventful, except for a brief period when he was taken prisoner during the sack of Rome. He was born in Cosenza, near the toe of the Italian boot; and, after schooling elsewhere in Italy, he returned to spend most of his life there. His first training was obtained from his uncle, a capable scholar who taught him Greek as well as Latin. This at once put Telesio into the special category of men who were able to read philosophical or scientific texts from antiquity in the original language and were not forced to depend upon medieval translations or exegesis.

When Telesio arrived at Padua, then, he was able to compare the interpretations of Aristotle or Galen given in his courses with the Greek texts. The Averroist tradition of Aristotelian interpretation did not appeal to him because it was based on Arab sources. Moreover, there is no trace in his writings of any interest in terminist logic or Mertonian physics, as exemplified in the works of "The Calculator" and other Oxford writers, even though these writings had been widely studied in universities of northern Italy during the previous century. Telesio left Padua profoundly dissatisfied with the Aristotelian doctrines presented there. He does not spell out his indictment in detail: in fact, he seldom makes more than passing allusions to contemporary doctrines. When he speaks of "the followers of Aristotle," he is more often referring to ancient Peripatetics than to his contemporaries.

After leaving Padua, Telesio spent some time in a Benedictine monastery developing his own system. He began to write while staying at Naples under the patronage of the Carafa family (Telesio was himself a nobleman, although an impoverished one). He developed his thought in opposition to that of Aristotle and the Greek medical writers, with some use of the Greek commentators. He is considered an arch anti-Aristotelian, yet his own style of thinking is so much like that of Aristotle that he might almost be thought of as an Aristotelian revisionist. Why, otherwise, should Telesio have thought it necessary to make a pilgrimage in 1563 to Brescia in order to explain his views to Vincenzo Maggi, a teacher renowned for his knowledge of Aristotle in the Greek? Maggi listened attentively to his views for several days, then confessed that he could not find anything contradictory about them. This was presumably a tribute to the thoroughness of Telesio's grasp of Aristotle—a grasp so firm that even those most conversant with Aristotle in the Greek were not able to challenge his interpretations or his counterarguments.

Reassured by his interview with Maggi, Telesio published the first version of his major work in 1565, under the title *De rerum natura iuxta propria principia*. The first part of this title, at this late date in the Renaissance, might lead one to expect a presentation of Epicurean physics as found in Lucretius' great poem of the same title, which had been discovered in the previous century. The expectation would be disappointed, however, for Telesio was not an atomist or a corpuscularian at all. This crucial fact must be kept constantly in mind, for it separates Telesio from other so-called "new philosophers" with whom he was later associated (for instance, by Descartes and Leibniz), who were mainly atomists. The subtitle was intended to repudiate principles imposed upon nature and thus expressed Telesio's rejection of what in a later century would have been called a priori theorizing.

In the preface Telesio immediately states that sense is the only valid starting point for speculation on nature. This pronouncement has been taken by most scholars as an empiricist manifesto, but such a view is misleading if one does not recognize that Telesio was not essentially interested in methodology: he never commented on the *Organon* or other logical works. Such epistemology as can be found in his writings occurs in his discussion of the *De anima* and other biological works. Since he was conversant with Galen, Telesio must have encountered that writer's logical empiricism, with its insistence upon the equal roles of sense and reason in science. But there is no stress on reasoning in Telesio's thought: indeed, reasoning is reduced to the detection of similarities among the deliverances of the senses. For this one-sided reliance upon sense alone Telesio was taken to task by another "nature philosopher," Francesco Patrizzi. (Patrizzi's criticisms, with Telesio's rather unconvincing rejoinders, are printed as an appendix in Fiorentino's book, listed in the bibliography).

At any rate Telesio, to judge from his writings, was certainly no more, and probably far less, of a practicing empiricist than his favorite philosophical target, Aristotle. He did not have a theory of experiment such as some scholars claim to find in the works of his contemporary Zabarella, nor did he make any controlled observations like those of Patrizzi. Nor did Telesio have a sophisticated conception of the importance of measurement. Nor, finally, was he a mechanist, in spite of the fact that he rejected action at a distance as being the result of an "occult" quality and in spite of his insistence upon the role of matter as a principle of nature.

Having learned all this, surely the reader must be asking, "Why, then, should Telesio be of any interest whatever to the historian of modern science?" The answer lies in the way in which he deployed certain Aristotelian concepts so as to achieve a new system of physical explanation, rejecting metaphysical entities that had no explanatory role in physics. Telesio's arguments are just about as plausible (or implausible, as the case may be) as Aristotle's, yet they differ drastically in their results. Surely nothing could have been more disturbing to Aristotelians than this!

What, then, are the chief features of Telesio's scheme? His basic explanatory arsenal consists of two opposing factors, heat and cold, to which must be added a third principle termed *materia* but not to be identified with Aristotle's potentiality, at least not officially. "Heat" and "cold" have taken the place of Aristotelian forms. Telesio is quite explicit on this point: He regards "forms" as "snoring"—that is, as otiose and hence dispensable (this may be viewed as one of the earlier attacks on "substantial forms"). Just as "matter" and "form" cannot really be separated in Aristotelian metaphysics, so "heat" and "cold" and "matter" are physically inseparable for Telesio. In general, heat is associated with motion, expansion, light, and life, while cold is associated with immobility, contraction, darkness, and death—although there is considerable wavering as to the precise relationships between these entities.

The following passage (from *De rerum natura* [1586 edition], I, 5) sums up Telesio's scheme: "Three principles [*principia*] of things altogether must be posited: two active natures [*agentes naturae*], heat and cold, and a bodily bulk [*corporea moles*]. . . ." This "bulk" or "matter" is described as being "inert" and "dead," all actions or operations being foreign to it. Aristotle had given a somewhat similar account of heat and cold in *De generatione et corruptione* (at 329b23), where he identifies "the hot" and "the cold," "the wet" and "the dry" as primary (tangible) qualities. In keeping with Aristotelian texts such as this, heat and cold were called *primae activae qualitates* by such contemporary Aristotelians as Zabarella. But for Aristotle the active and the passive qualities make up the four elements (fire being a combination of the hot and the dry, and so on), whereas for Telesio wetness and dryness are not primary but derivative. New explanations are thus required for the traditional "media," water and air.

All of this Telesian doctrine is supposed to rest, as we have seen, on sense—that is, upon observable features of the cosmological landscape. Thus we feel the warmth of the sun and the coldness of the earth. We feel the immobility of the earth (Telesio was no Copernican and, indeed, seems not even to have been acquainted with the heliocentric hypothesis) while we see the sun moving in the sky. As a first crude hypothesis this is perhaps not unpromising—or, any rate, not much less promising than Aristotle's cosmology. But obviously Telesio will have his hands full to explain, on the basis of his two active principles alone, all sorts of changes shown to us by sense. We might note that in making the sun fiery or "igneous," Telesio was, unwittingly, helping to contribute to the breakdown of the barrier that Aristotle had set up between celestial and sublunary physics, the breakdown triumphantly announced by Galileo in his *Dialogue Concerning the Two Chief World Systems.*

Telesio also introduced concepts of space and time that anticipated the absolute space and time of Newtonian physics:

> And thus clearly space can be conceded to be different from the bulk of entities . . . and all entities are located in it. . . . Any given thing is in that portion of space in which it is located, and of which it is the place, and which is completely incorporeal, foreign to all action and all operation, being only a certain aptitude for sustaining bodies and nothing else, ... thus completely different from everything [*De rerum natura*, I, 25].

Thus there can be a void (*vacuum*). Telesio also held that time would continue to flow on even though no motion were observed by man or even existed. He thus broke away from Aristotle's conception of place as the surface of the containing body and of time as the measure of motion.

In keeping with his general approach, Telesio regarded the soul (or *spiritus*) as corporeal—having dispensed with "forms" altogether, he obviously could not accept Aristotle's definition of the soul as "the form of an organic body." This spirit "derived from the seed," for which the body's integument is a container, is distinguished, not very clearly, by Telesio from the soul introduced by God. Sensation, which cannot be the "reception of forms without matter," as defined by Aristotle, is the "perception of the actions of things and impulsions of air and [the soul's] own passions and immutations and motions" (*De rerum natura*, VII, 2). A thoroughly naturalistic ethics is then developed, with virtues being faculties that ensure the conservation and perfection of spirit (*ibid.*, IX, 4). As we have seen, reasoning is subsidiary to sensation, the deficiencies of which it supplies in situations in which the whole of a thing's qualities are not directly observable (*ibid.*, VIII, 3). Telesio's discussions of virtues and vices, as they are displayed by a self-interested creature pursuing its own conservation, anticipate similar treatments in the seventeenth century (Descartes, Hobbes, Spinoza). All in all, perhaps the most judicious verdict on Telesio's thought is that it represented "a robust re-thinking of pre-Socratic naturalism" (Eugenio Garin, *Storia della filosofia italiana*, II [Turin, 1966], 711).

BIBLIOGRAPHY

I. ORIGINAL WORKS. During Telesio's lifetime three eds. of his major work appeared, all under the basic title *De rerum natura iuxta propria principia* but with contents progressively augmented (Rome, 1565; 2nd ed., Naples, 1570; 3rd ed., Naples, 1586). There is a modern ed., by V. Spampanato, 3 vols. (I, Modena, 1910; II, Genoa, 1913; III, Rome, 1923). No one has as yet made a thorough study of Telesio's changes from the earlier to the later eds., even though there are copies of the 1570 ed. with corrections in Telesio's own hand (Naples, Bibl. Naz. XVI E 68), as well as an autograph commentary on the work (Vatican, Cod. Ottob. Lat. 1292). A new and definitive ed. is being prepared by Luigi De Franco (I, Cosenza, 1965). Shorter treatises also exist in MS (for instance, Naples, Bibl. Naz. VIII C 29); some were published by Telesio's disciple Antonio Persio under the title *Varii de naturalibus rebus libelli* (Venice, 1590). A treatise on lightning has been published by Carlo Delcorno, "Il commentario 'De fulmine' di Bernardino Telesio," in *Aevum*, **41** (1967), 474–506.

II. SECONDARY LITERATURE. Still the basic secondary work is Francesco Fiorentino, *Bernardino Telesio*, 2 vols. (Florence, 1872–1874), with an appendix containing previously unpublished material. A bibliography of Telesian scholarship up to 1937 is given in Giovanni Gentile, *Il pensiero italiano del Rinascimento*, 4th ed. (Florence, 1968), 507–522. Of particular interest are the following: Roberto Almagià, "Le dottrine geofisiche di Bernardino Telesio," in his *Scritti geografici (1905–1957)* (Rome, 1961), 151–178; Ernst Cassirer, *Das Erkenntnisproblem* (Berlin, 1906), 212–218; A. Corsano, "La psicologia del Telesio," in *Giornale critico della filosofia italiana*, **21** (1940), 5–12; and Luigi Firpo, "Filosofia italiana e controriforma. IV. La proibizione di Telesio," in *Rivista di filosofia*, **42** (1951), 30–47, which gives the text of a condemnation of Telesio's writings at Padua in 1600.

On Telesio's biology, the most detailed account is Edoardo Zavattari, *La visione della vita nel Rinascimento* (Turin, 1923). On the relation of physics to metaphysics, see Giacomo Soleri, "La metafisica di Bernardino Telesio," in *Rivista di filosofia Neoscolastica*, **34** (1942), 338–356. The most complete work in English is Neil van Deusen, *Telesio, the First of the Moderns* (New York, 1932).

NEAL W. GILBERT

TENNANT, SMITHSON (*b.* Selby, Yorkshire, England, 30 November 1761; *d.* Boulogne, France, 22 February 1815), *chemistry.*

Tennant's father was the Reverend Calvert Tennant, a fellow of St. John's College, Cambridge, and later vicar of Selby; his mother, Mary Daunt Tennant, was the daughter of an apothecary. Both parents had died by the time he was twenty, leaving him an inheritance of land. During 1781 he was a medical student at Edinburgh, where he attended Joseph Black's lectures. In October 1782, he moved to Christ's College, Cambridge, from which he received the M.B. in 1788 and the M.D. in 1796. He was elected fellow of the Royal Society in January 1785 and received the Copley Medal in 1804. In 1799 he was a founding member of the Askesian Society, which soon became the Geological Society. In 1813 he was appointed professor of chemistry at Cambridge. Tennant's travels included a visit to Sweden in 1784, where he met Scheele and Gahn, and a journey to France in 1814–1815 that ended in a fatal riding accident.

Tennant wrote little and consequently was accused of indolence. In 1796 he communicated to the Royal Society his study of the combustion of the diamond. Lavoisier had carried out a series of similar experiments and observed that the gaseous product turned limewater cloudy, as in the burning of charcoal. However, he maintained that this common result merely showed that both charcoal and diamond were in the class of combustibles. Reluctant to stress the analogy further, Lavoisier even wrote that the nature of diamond might never be known. But Tennant insisted that since equal quantities of charcoal and diamond were entirely converted in combustion to equal quantities of fixed air, then both substances must be chemically identical. Certain scientists, notably Humphry Davy, continued to suspect that there were minute chemical differences between these forms of carbon; but Davy soon returned to the interpretation first given by Tennant.

Tennant's most important work, the discovery of two new elements in platinum ore, was described in a paper to the Royal Society in 1804. The extraction of pure, malleable platinum from its crude ore was a problem that taxed eighteenth-century chemists. A notebook preserved at Cambridge on Tennent's travels shows that he had discussed the problem with Gahn and Crell in 1784. At that time the standard procedure was to digest the crude ore in aqua regia; this technique left an insoluble black residue that Proust mistook for graphite. At about the same time Collet-Descotils, Fourcroy, Vauquelin, and Tennent realized that this residue contained something new. Collet-Descotils inferred the existence of a new metal from the red color it gave to platinum precipitates; Fourcroy and Vauquelin called the new metal *ptène* but soon admitted that they had confused two different metals. Tennent alone recognized that the black powder contained two new metals, which he proceeded to isolate and characterize. He called one iridium on account of the variety of colors it produced; the other he named osmium because of the distinctive odor of its volatile compounds.

Tennant had interested William Wollaston in platinum while they were students at Cambridge. By 1800 they had became business partners, selling platinum boilers for the concentration of sulfuric acid and other products made of platinum.

BIBLIOGRAPHY

See Donald McDonald, "Smithson Tennant, F.R.S. (1761–1815)," in *Notes and Records. Royal Society of London*, **17** (1962), 77–94. On Tennent's MSS, see L. F. Gilbert, "W. H. Wollaston MSS. at Cambridge," *ibid.*, **9** (1952), 311–332. His memoirs are listed in the Royal Society *Catalogue of Scientific Papers*.

D. C. GOODMAN

TENNENT, DAVID HILT (*b.* Janesville, Wisconsin, 28 May 1873; *d.* Bryn Mawr, Pennsylvania, 14 January 1941), *biology*.

Tennent was one of four children born to Thomas Tennent, a contractor, and his second wife, Mary Hilt. Thomas also had two children by his first wife. Young David thus grew up in a large family, and lived a rather rigorous and austere life as a child. He became interested in science early and hoped to study medicine, but he was prevented from doing so by the accidental death of his father in 1893. He first became a licensed pharmacist, then in 1895 entered Olivet College in Michigan, where he received the B.S. in 1900. In 1904 he received the Ph.D. from the Johns Hopkins University, where he had studied under W. K. Brooks.

While a graduate student, Tennent spent one year as a substitute instructor at Randolph-Macon College. In 1904 he became instructor in biology at Bryn Mawr College, where he taught for thirty-four years. After his retirement he was a research professor there until his death. In 1909 Tennent married Esther Maddux; their only child, David Maddux, born in 1914, became a research biochemist.

The major part of Tennent's work was in marine biology, and consisted principally of experiments on marine eggs. While a graduate student, he spent several summers at the U.S. Fisheries laboratories in Beaufort, North Carolina; later he spent a number of summers in the Dry Tortugas at the biological laboratory of the Carnegie Institution. He worked also at a number of other marine laboratories in the United States and abroad, and spent two sabbatical leaves in Japan.

Tennent concentrated most of his efforts on the study of echinoderm development. His particular interest was in the role of the nucleus, which he studied both as an embryologist and as a cytologist. His most important investigations dealt with the study of echinoderm hybridization.

Tennent performed cross fertilizations between species, genera, orders, and classes within the Echinodermata. He found that after exposing the eggs of some species (such as *Lytechinus*) to monovalent cations, they could be more easily fertilized by foreign sperm than by sperm of their own species. It was known that chromosomes are often eliminated from the mitotic spindles in hybrids. Tennent, a skilled cytologist, developed an intimate knowledge of the configurations of the chromosome sets and thereby established that it was the paternal chromosomes that were eliminated, presumably as a result of incompatibility between maternal cytoplasm and paternal chromatin. He also showed that, depending on the number of paternal chromosomes eliminated, the embryo develops more or less, sometimes solely, under the influence of the maternal chromosomes. Tennent distinguished for the first time between the autosomes and the sex chromosomes of a number of echinoderms; and by hybridizing two forms that differ in the shape of their sex chromosomes, he established for the first time that the male is the digametic sex in sea urchins.

Like Boveri and Driesch before him, Tennent wished to determine the time at which the influence of the paternal chromosomes first manifests itself in development. Boveri had performed only preliminary experiments, and Driesch only crude ones, to elucidate this point. By crossing species that differed clearly in several features, Tennent confirmed the conclusions of Boveri and Driesch that maternal factors control the earliest phases of development, and that the paternal factors are expressed later in development. Another important contribution to the study of echinoderm hybrids was his demonstration that the direction of development toward the character of the maternal or the paternal species could be altered by changing the hydrogen ion concentration of the seawater in which the hybrids were maintained.

Although he was never a complete innovator in his investigations, the skill, caution, objectivity, and patience with which Tennent worked enabled him to establish on a firm foundation cytological and developmental truths that had merely been intimated by some of his more original predecessors.

BIBLIOGRAPHY

The most detailed biography of Tennent, with a complete bibliography of his articles, is M. S. Gardiner, "Biographical Memoir of David Hilt Tennent 1873–1941," in *Biographical Memoirs. National Academy of Sciences*, **26** (1951), 99–119.

JANE OPPENHEIMER

TEN RHYNE, WILLEM (*b.* Deventer, Netherlands, 1647; *d.* Batavia, Netherlands Indies [now Djakarta, Indonesia], 1 June 1700), *medicine, botany.*

Little is known of the early life of ten Rhyne. After attending the Illustre School in Deventer, he studied medicine from about 1664 to 1666 at the University of Franeker. Then, in 1668, he went to the University of Leiden, where he studied under Sylvius and became an adherent of his iatrochemical school. In 1668 he received his medical degree with the dissertation *De dolore intestinorum a flatu.* In 1669 he published an essay on gout, *Dissertatio de arthritide*, and a treatise on a text of Hippocrates in which he also discussed a number of salts.

In June 1673 ten Rhyne left for Batavia to serve as physician to the Dutch East India Company. During a twenty-six-day stay at the Cape of Good Hope (October–November 1673), he studied the flora and fauna of the area and also the Hottentots.

He published his findings in *Schediasma de promontorio Bonae Speï, ejusque tractus incolis Hottentottis* (1686).

Ten Rhyne arrived in Batavia in January 1674 and in addition to his medical duties he gave anatomy lessons to the local surgeons. He was soon sent by the company to the trading station on Decima in the harbor at Nagasaki. There was a surgeon at Decima, but the emperor had requested that the Dutch East India Company also bring out a qualified physician. Ten Rhyne served in Japan from 1674 until 1676, when he returned to Batavia. In 1677 he was appointed governor of the leper colony. From 1679 until 1681 he was physician on the west coast of Sumatra and from 1681 until his death he was a member of the judicature in Batavia.

During the two years in Japan, ten Rhyne studied tea culture, describing the tea plant, the manufacture and use of tea, and the influence of the stimulant on the body. His findings were published in Jacob Breyn's *Exoticarum plantarum centuria prima* (1678). In this same work of Breyn's, ten Rhyne also gave complete descriptions of the Japanese camphor tree and the plants he had collected at the Cape of Good Hope.

As a result of his work on tropical flora, Hendrik Adriaan van Reede tot Drakestein asked ten Rhyne to collaborate with him on the *Hortus Malabaricus*, which appeared in Amsterdam in twelve volumes between 1679 and 1703.

In 1687 ten Rhyne published his classic work on leprosy: *Verhandelingen van de Asiatise melaatsheid* ("Treatise on Asiatic Leprosy"). He presented an excellent description of the disease and gave an account of its etiology, prophylaxis, and therapy, which is still valid.

Ten Rhyne also published a work (1683) on the practice of acupuncture in Japan.

BIBLIOGRAPHY

I. ORIGINAL WORKS. Ten Rhyne's major works are *Dissertatio de dolore intestinorum a flatu* (Leiden, 1668); *Dissertatio de arthritide* (Leiden, 1669); *Meditationes in Hippocratis textum vigesimum quartum de veteri medicina, cum laciniis de salium figuris* (Leiden, 1669, 2nd ed., 1672); "Excerpta ex observationibus Japonicis de fructice thee, cum fasciculo rariorum plantarum ab ipso inpromontorio Bonae Speï et Sardanha sinu anno 1673 collectarum, atque demum ex India anno 1677 in Europam ad Jacobum Breynium transmissarum," in Jacob Breyn, *Exoticarum plantarum centuria prima* (Danzig, 1678); *Dissertatio de arthritide, mantissa schematica de acupunctura, orationes tres de chymiae et botanicae antiquitate et dignitate, de physiognomia, et de monstris* (London–The Hague, 1683; 2nd ed., Leipzig, 1690); *Schediasma de promontorio Bonae Speï, ejusque tractus incolis Hottentottis* (Schaffhausen, 1686); and *Verhandelingen van de Asiatise melaatsheid* (Amsterdam, 1687), reprinted with English translation in *Opuscula Selecta Neerlandicorum de arte medica*, **14** (1937), 34–113.

II. Secondary Literature. On ten Rhyne and his work, see A. J. van der Aa, *Biographisch Woordenboek der Nederlanden*, X (Haarlem, 1874), 96; H. Kronenberg, *Nieuw Nederlandsch Biographisch Woordenboek*, VI (Leiden, 1924), 1213; L. S. von Römer, *ibid.*, IX (Leiden, 1933), 861–863; and J. M. H. van Dorssen, "Willem ten Rhyne," in *Geneeskundig tijdschrift voor Nederlandsch-Indië*, **51** (1911), 134–228.

H. A. M. Snelders

TERMIER, PIERRE (*b.* Lyons, France, 3 July 1859; *d.* Grenoble, France, 23 October 1930), *metamorphic petrology, structural geology, geotectonics.*

Termier's grandiose synthesis of the geological structure of the Alps (1903) made him the founder of modern tectonics and geodynamics. His works are numerous and stylistically unsurpassed. His vivid essays and biographies of his great teachers are collected in *À la gloire de la terre, La joie de connaître,* and *La vocation de savant;* a fourth volume, *Mélanges,* was published posthumously by his daughter, Jeanne Boussac-Termier.

Termier's interest in literature developed at an early age. His father, Francisque Termier, was often away from the family home in Lyons, and his mother, Jeanne, guided him through his early years. Termier proudly recalled having read poems to his mother at the age of five. A brother, Joseph, was born when Pierre was thirteen.

In 1868 Termier was sent to the Collège des Maristes in Saint-Chamond (Loire), where he developed a passion for mathematics, in addition to his favorite subjects, literature and philosophy. After graduating in 1876 he continued his education in Paris at the École Sainte-Geneviève and, from 1878 to 1880, at the École Polytechnique. On a mountaineering trip to the Belledonne Massif in the western Alps, Termier discovered his vocation for geology. In 1880 he entered the École des Mines in Paris, where he attended the mineralogy course of Ernest Mallard. His thesis was based on a field study of igneous rocks in the Harz Mountains.

In 1883 he married Alice Beylier, and they made their first home in Nice, where Termier was appointed inspector of mines (*ingénieur ordinaire*). Administrative duties took him as far as Corsica. But since he preferred teaching, he applied for the vacant chair at the École des Mines in Saint-Étienne. In 1885 Termier was appointed professor of physics and electricity and later of mineralogy and geology. Under the auspices of the Service de la Carte Géologique de la France he began geologic mapping of the Massif Central (Plateau Central). Although he was attracted by the Tertiary volcanoes of Mont-Mézenc, his work soon focused on the metamorphic Paleozoic basement in the area of Saint-Étienne, especially Mont-Pilat. Termier was greatly stimulated by his colleague Urbain Le Verrier, professor of metallurgy and chemistry, who had also become involved in geologic mapping through his close friendship with August Michel-Lévy, director of the geological survey. Le Verrier concentrated on the problem of granite formation, particularly the process of feldspathization. Termier's field research on the Massif Central was incorporated in five sheets of the *Carte géologique détaillée de la France* (1:80,000).

For thirty years Charles Lory had been the only geologist to carry out field mapping in the French Alps. It was only after Lory's death in 1888 that Marcel-Alexandre Bertrand, who had introduced the nappe theory into Alpine geology, transferred his field research from the Provençal thrust belt to the Alps. For this project, Bertrand assembled a team of geologists. Lory's successor at the University of Grenoble, Wilfrid Kilian, was entrusted with the mapping of the High Calcareous Alps of Savoy. Just at this time the Termier brothers, who were climbing their favorite mountains in the Haute Vanoise, suffered an accident. Receiving word of this and intrigued by Pierre Termier's reputation, Bertrand recognized in him the ideal collaborator for the work in the massifs and the metamorphic zone of Savoy. An extremely fruitful cooperation began, and Termier soon regarded Bertrand as his great mentor and friend. Together they explored the Vanoise in the summer of 1890. Termier went on to complete the mapping of the Haute Vanoise for the 1:80,000 map series. His initial report of 1891 was an epoch-making contribution to Alpine geology, since for the first time the progressive stages of the regional metamorphism were mapped over a considerable distance. The metamorphic grade, transgressive across the stratigraphic sequence, was found to be controlled by the Alpine structure alone. This discovery, in turn, confirmed Lory's early assumption that the axial belt of crystalline schists (*schistes lustrés*) represented a transformed Mesozoic sedimentary sequence. In his report of 1894 on the Grandes Rousses Massif, Termier was able to distinguish also between Hercynian and superimposed Alpine fold structures.

In 1894 Ernest Mallard died suddenly, and Termier was appointed to succeed him as professor of mineralogy and petrology at the École des Mines in Paris. With some regret the Termiers, with their son and five daughters, left the countryside of southern France for Paris, where a second son was born. They maintained a second home at Varces-Allières-et-Risset, near Grenoble. Although the ensuing years brought further triumphs to Termier's successful career, he was spared neither grief nor sorrow. It soon became apparent that his wife suffered from parkinsonism. In 1906 his elder son died in an accident at the age of thirteen, and the following year Bertrand died. Termier found some comfort in a new friendship with the writer Léon Bloy.

After the death of Auguste Michel-Lévy in 1911, Termier was elected director of the Service de la Carte Géologique de la France. The following year his teaching responsibility was increased to include the main course in physical geology at the École des Mines in Paris. Termier attracted many students, for no one could speak of the beauty of the mountains with greater skill and affection. He stressed also his belief in the philosophic and religious values of science, looking upon faith and science as coming from the same source.

In 1914 Termier was elected inspector general of mines. During World War I he served as colonel with the French artillery. In 1916 his wife died in Varces-Allières-et-Risset. Shortly afterward, his son-in-law Jean Boussac died from an injury received near Verdun; Boussac was noted for his works on the Eocene Alpine flysch. Termier suffered yet another bereavement with the death of his younger son from meningitis in 1924.

In 1909 Termier was elected member of the Académie des Sciences. He was made an officer of the Legion of Honor in 1914 and a commander in 1927. He had three terms as president respectively of the Société Minéralogique and the Société Géologique of France, and was vice-president of the Académie des Sciences; at the time of his death, Termier was president elect of the Academy. The Geological Society of London elected him foreign correspondent in 1923 and foreign member in 1929. At the centennial of the Geological Society of France in 1930, he received the degree of doctor *honoris causa* from the University of Innsbruck.

Termier's love of travel enabled him to make the expeditions necessary to unravel the tectonics of Alpine Europe. Research and consulting work took him to Scandinavia, the Urals, Siberia, Mexico, the Colorado Plateau, the Canadian Rockies, and Quebec, as well as to North Africa and Spain. In 1930, on his last trip to Morocco, he fell ill and died shortly after his return to Grenoble.

Termier's field research was apparently directed toward the metamorphic rather than the structural history of the areas he had selected for study. In the Franco-Italian Alps he searched for the causes of regional metamorphism. At first he tried to organize his results so that they would fit into the framework of the widely accepted theory of dynamic metamorphism, in which heat was attributed to the mechanical energy produced during folding and thrusting. By 1903, however, Termier had become a bitter opponent of the theory. Dynamic action deformed but did not transform, and mylonitic rocks were its only product, he argued at the Ninth International Geological Congress in Vienna (1903). He saw the cause of regional metamorphism related, somehow, to the depth of burial in the geosyncline. But geologic depth alone would not be sufficient to cause metamorphism. He assumed the causative factor to be the influx of juvenile liquids and vapors, the *colonnes filtrantes*, which brought with them alkali silicates. Extending the ideas of Le Verrier, he compared the process of feldspathization with the spreading of a grease spot (the *tâche d'huile* mechanism). Termier concluded that the progressive stages of regional metamorphism would ultimately result in massive granitic rocks. The negligible role that he attributed to penetrative movements was subsequently challenged, mainly by the Austrian school, which initiated the study of petrofabrics in support of its case.

Termier's contributions to regional tectonics effected revolutionary changes in the perception of global dynamics, at first only among Alpine geologists. In the dramatic dispute on the structure of the eastern Alps, Termier, as Bertrand's disciple, emerged as the new leader of the French structural geologists, all of whom adhered to the concepts of Eduard Suess as set forth in his *Das Antlitz der Erde*. Suess elucidated for the first time the global grouping of mountain belts in space and time. He argued that the structure of the Alpine-Carpathian mountains could best be explained as the product of a one-sided tectonic drive that thrust the sedimentary fill of the *Tethys* sea onto the old European *Vorland*. Termier first visited Suess late in 1899 for advice concerning French explorative drillings in the Carpathians. The results confirmed Suess's prediction that the Upper Silesian coalfield would extend southward beneath the Carpathian thrust sheets. The French were the first to translate Suess's classic; their elaborate edition (1897–1918) appropriately opens with a preface by Bertrand and concludes with a eulogy by Termier.

Large-scale nappe structures (Bertrand's *nappes de recouvrement*) had been incontestably demonstrated in the Swiss Alps by Hans Schardt (1893) and Maurice Lugeon (1902), and in the Franco-Italian Alps mainly by Termier himself (1899, 1902). It was the metamorphic petrology of the Vanoise that enabled Termier to decipher the tectonics of that region. The key to the structure of the Briançonnais zone in particular was seen in the interpretation of the "thrust slice number four" (*quatrième écaille*); thus the crystalline capping of a few mountaintops west of Briançon became also the cradle of a tectonic principle: that of the "squashing thrust plate" (the *traîneau écraseur*). A rigid mass of schists had been shoved westward over folded flysch and older sediments, which built up the fanlike structure of the Briançonnais. The fold crests of the overridden fan were thrown into tucks and flattened out under the advancing plate. In general, nappe structures were formed either by recumbent folding, gravity sliding, or overthrusting. The *traîneau écraseur*, however, is a thrust plate actively pushed over the top of another block; its width is a measure of the shortening of the crust.

Although the nappe interpretation of the Swiss Alps necessitated corresponding structures in Austria, the constitution of the eastern Alps, some 500 kilometers in length, still remained obscure until, after the International Geological Congress in Vienna (1903), Friedrich Becke led a field trip to Ziller Valley and the Hohe Tauern. On the summit of Amtshorspitze, Termier boldly declared the eastern Alps to be a pile of nappes and the Hohe Tauern in particular to be a window exposing the geosynclinal fill of *schistes lustrés*; he could not help seeing lithological and structural similarities to the western Alps, including the *traîneau écraseur*, which here assumed much larger dimensions. Shortly afterward Termier proved his claim by verifying the window structure of the Hohe Tauern as well as the Lower Engadine in Switzerland. The Hohe Tauern window implied a thrust plate, 120 kilometers in width, similar in size to overthrusts that had already been inferred in Scandinavia. To Termier, the entire structure of the Alps seemed to require a total crustal shortening of 500 kilometers or more. He himself did not fail to consider the geotectonic aspects of this conclusion, and although he did not accept Alfred Wegener's notion of continental drift, he insisted that considerable lateral motion of the crustal blocks had occurred.

Termier's classic perception of the structure of the Alps underwent much modification, but its basic implication of large-scale plate motion was reaffirmed in the later works of Émile Argand, Rudolf Staub, Leopold Kober, and others.

BIBLIOGRAPHY

I. ORIGINAL WORKS. Among Termier's publications are: "Étude sur le massif cristallin du Mont-Pilat, sur la bordure orientale du Plateau Central entre Vienne et Saint-Vallier . . .," in *Bulletin des Services de la carte géologique de France et des topographies souterraines*, no. 1, I (1889), 1–58; "Étude sur la constitution géologique du massif de la Vanoise (Alpes de Savoie)," *ibid.*, no. 20, II (1891), 367–514; "Le massif des Grandes-Rousses (Dauphiné et Savoie)," *ibid.*, no. 40, VI (1894), 169–288; "Les nappes de recouvrement du Briançonnais," in *Bulletin de la Société géologique de France*, 3rd ser., 27 (1899), 47–84; "Quatre coupes à travers les Alpes franco-italiennes," *ibid.*, 4th ser., 2 (1902), 411–433; "Les schistes cristallins des Alpes occidentales," in *International Geological Congress, comptes-rendus 9th, Vienna, 1903* (Vienna, 1904), 571–586, repr. in *La joie de connaître, suite de À la gloire de la terre* (Paris, 1926); "Les montagnes entre Briançon et Vallouise," in *Mémoires pour servir à l'explication de la Carte géologique détaillée de la France* (Paris, 1903); "Sur la structure des Hohe Tauern (Alpes du Tyrol)," in *Comptes rendus hebdomadaires des séances de l'Académie des sciences*, 137 (1903), 875–876; "Les nappes des Alpes orientales et la synthèse des Alpes," in *Bulletin de la Société géologique de France*, 4th ser., 3 (1903), 712–765; "Sur la fenêtre de la Basse-Engadine," in *Comptes rendus hebdomadaires des séances de l'Académie des sciences*, 139 (1904), 648–650; "Roches à lawsonite et à glaucophane et roches à riébeckite de Saint-Véran (Hautes-Alpes)," in *Bulletin de la Société française de minéralogie*, 27 (1904), 265–269; "Les alpes entre le Brenner et la Valteline," in *Bulletin de la Société géologique de France*, 4th ser., 5 (1905), 209–289; *La synthèse géologique des Alpes* (Liège, 1906), repr. in *À la gloire de la terre, souvenirs d'un géologue* (Paris, 1922); "Sur la genèse des terrains cristallophylliens," in *International Geological Congress, comptes rendus 11th, Stockholm, 1910* (Stockholm, 1912), I, 587–595; "Les problèmes de la géologie tectonique dans la Méditerranée occidentale," in *Revue générale des sciences,* **22** (1911), 225–234, repr. in *À la gloire de la terre . . .;* "La dérive des continents," *Bulletin de l'Institut océanographique de Monaco,* no. 443 (15 Apr. 1924), repr. in *La joie de connaître . . .,* translated as "The Drifting of the Continents," in *Smithsonian Report for 1924* (1925), 219–236; *La vocation de savant, suite de À la gloire de la terre et de La joie de connaître* (Paris, 1929); *Mélanges,* pref. by Jeanne Boussac-Termier (Paris, 1932).

II. SECONDARY LITERATURE. The most complete biographies are A. George, *Pierre Termier* (Paris, 1933), and E. Raguin's *éloge* in *Bulletin de la Société géologique de France,* 5th ser., **1** (1931), 429–495, including a complete bibliography. Other notices are by L. Lecornu, in *Comptes rendus hebdomadaires des séances de l'Académie des sciences,* **191** (1930), 685–687; G. Aichino, in *Bollettino. R. Ufficio geologico d'Italia,* **55**, no. 12 (1930), 1–4; E. J. Garwood, in *Proceedings of the Geological Society of London,* **87** (1930–1931), lx–lxii; L. J. Spencer, in *Mineralogical Magazine,* **23** (1933), 359–360; and the pref. by Jeanne Boussac-Termier in *Mélanges* (Paris, 1932).

E. B. Bailey, in *Tectonic Essays, Mainly Alpine* (Oxford, 1935), sets Termier's structural synthesis of the Alps in historical perspective; and H. H. Read evaluates his contribution to metamorphic petrology in *The Granite Controversy* (London, 1957).

JOHN HALLER

TESLA, NIKOLA (*b.* Smiljan, Croatia [now Yugoslavia], 10 July 1856; *d.* New York, N.Y., 7 January 1943), *physics, electrical engineering.*

Tesla was born of Serbian parents in a mountain village that was then part of Austria-Hungary. His father, Milutin Tesla, was a clergyman of the Serbian Orthodox church, while his mother, Djuka Mandić, although illiterate, was a skillful inventor of home and farm implements. Tesla himself was intended for the clergy, but early developed a taste for mathematics and science. When he was seven, the family moved to Gospić, where he finished grammar school and graduated from the Real-Gymnasium. He then attended the Higher Real-Gymnasium in Karlovac and, upon graduation, persuaded his father to let him enter the Joanneum, the polytechnical college of Graz, Austria.

It was while he was a student in Graz that Tesla's attention was first drawn to problems of the induction motor. His observation that a Gramme dynamo that was being run as a motor in a classroom demonstration sparked badly between its commutator and brushes led him to suggest that a motor without a commutator might be devised—an idea that his professor ridiculed. Nothing daunted, Tesla continued to develop the idea. In 1879 he left Graz to enroll at the University of Prague, but left without taking a degree when his father died. He then held a number of jobs; in 1881 he went to Budapest to work for the new telephone company there. During his year there he thought of the principle of the rotating magnetic field, upon which all polyphase induction motors are base . The discovery, by his own account, was instantaneous, complete, and intuitive. Walking in a park with a friend, Antony Szigety, Tesla was moved to recite a passage from Goethe's *Faust* (of which he had the whole by heart) when " . . . the idea came like a lightning flash. In an instant I saw it all, and drew with a stick on the sand the diagrams which were illustrated in my fundamental patents of May, 1888, and which Szigety understood perfectly." It was, however, some time before he was able to exploit his invention commercially.

In 1882 Tesla went to Paris as an engineer with the Continental Edison Company. The following year he was sent to Strasbourg to repair an electric plant, and while there built a crude prototype of his motor. He thus experienced "the supreme satisfaction of seeing for the first time rotation effected by alternating currents without commutator." In 1884 he went to the United States to promote his new alternating-current motor. He arrived in New York with a working knowledge of a dozen languages, a book of poetry, four cents, and an introduction to Thomas Edison. Although Edison was totally committed to direct current, he gave Tesla a job, and for a year Tesla supported himself redesigning direct-current dynamos for the Edison Machine Works. By 1885 he had left Edison and had gone into business developing and promoting an industrial arc lamp. He was forced out of the company when production began, however, and for a time lived precariously, doing odd jobs and day labor. Within two years he was back on his feet, and had formed his own laboratory for the development of his alternating-current motor.

By 1888 Tesla had obtained patents on a whole polyphase system of alternating-current dynamos, transformers, and motors; the rights to these were bought in that year by George Westinghouse, and the "battle of the currents" was begun. Although Edison continued to espouse direct current, Tesla's system triumphed to make possible the first large-scale harnessing of Niagara Falls and to provide the basis for the whole modern electric-power industry. In 1889 Tesla became an American citizen.

During the next few years Tesla worked in his New York laboratories on a wide variety of projects. He was very successful, particularly in his invention of the Tesla coil, an air-core transformer, and in his further research on high-frequency currents. In 1891 he lectured on his high-frequency devices to the American Institute of Electrical Engineers, and this lecture, coupled with a spectacular demonstration of these apparatuses, made him famous. He repeated his performance in Europe, to great acclaim, and enjoyed international celebrity.

In 1893 the Chicago World Columbian Exposition was lighted by means of Tesla's system and work was begun on the installation of power machinery at Niagara Falls. In a lecture-demonstration given in St. Louis in the same year—two years before Marconi's first experiments—Tesla also predicted wireless communication; the apparatus that he employed contained all the elements of spark and continuous wave that were incorporated into radio transmitters before the advent of the vacuum tube. Engrossed as he was with the transmission of substantial amounts of power, however, he almost perversely rejected the notion of transmission by Hertzian waves, which he considered to be wasteful of energy. He thus proposed wireless communication by actual conduction of electricity through natural media, and, working in Colorado Springs, Colorado, in 1899–1900, proved the earth to be a conductor. In a further series of experiments, Tesla produced artificial lightning in flashes of millions of volts that were up to 135 feet long—a feat that has never been equaled. It was at his Colorado laboratory, too, that Tesla, who had become increasingly withdrawn and eccentric ever since the death of his mother in 1892, announced that he had received signals from foreign planets, a statement that was greeted with some skepticism.

Tesla's vision always embraced the widest applications of his discoveries. Of his wireless system, he wrote in 1900: "I have no doubt that it will prove very efficient in enlightening the masses, particularly in still uncivilized countries and less accessible regions, and that it will add materially to general safety, comfort and convenience, and maintenance of peaceful relations." With the financial backing of J. P. Morgan, he began work on a worldwide communications system, and a 200-foot transmission tower was constructed at Shoreham,

on Long Island. By 1905, however, Morgan had withdrawn his support, and the project came to an end. The tower was destroyed by dynamite, under mysterious circumstances, in 1914.

Although he continued to enjoy a measure of fame, Tesla made little money from his inventions, and became increasingly poor during the last decades of his life. His name continued to flourish before the public, however, since he was a reliable source for scientific prophecy, and exploited as such in the popular press. While he gave demonstrations of some of his earlier marvels—his exhibition of a radio-guided teleautomatic boat filled Madison Square Garden in 1898—he became oracular in his later years and, for example, offered no proof of the potent "death-ray" that he announced in 1934, on his seventy-eighth birthday. Nonetheless, Tesla continued to invent devices of commercial and scientific worth, from which, since he seldom bothered to seek a patent, he received little profit.

Tesla was a complete recluse in his last years, living in a series of New York hotel rooms with only pigeons for company. At his death his papers and notes were seized by the Alien Property office; they are now housed in the Nikola Tesla Museum in Belgrade, Yugoslavia, a country in which he is revered as a national hero.

BIBLIOGRAPHY

I. ORIGINAL WORKS. The greatest part of Tesla's notes and correspondence is in the Nikola Tesla Museum, Belgrade, Yugoslavia. That institution has published a selection of source materials, in English, as Leland I. Anderson, ed., *Nikola Tesla, 1856–1943: Lectures, Patents, Articles* (Belgrade, 1956), which includes an autobiographical sketch; another autobiographical segment is "Some Personal Recollections," in *Scientific American* (June, 1915).

II. SECONDARY LITERATURE. A commemorative volume of speeches made on the occasion of the centenary of Tesla's birth is *A Tribute to Nikola Tesla: Presented in Articles, Letters, Documents* (Belgrade, 1961). Full biographies are Inez Hunt and Wanetta W. Draper, *Lightning in His Hand: The Life Story of Nikola Tesla* (Denver, 1964); and John J. O'Neill, *Prodigal Genius: The Life of Nikola Tesla* (New York, 1944). Shorter treatments include Haraden Pratt, "Nikola Tesla, 1856–1943," in *Proceedings of the Institute of Radio Engineers*, **44** (1956), 1106–1108; and Kenneth M. Swezey, "Nikola Tesla, Pathfinder of the Electrical Age," in *Electrical Engineering*, **75** (1956), 786–790; and "Nikola Tesla," in *Science*, **127** (1958), 1147–1159.

KENNETH M. SWEZEY

TESTUT, JEAN LÉO (*b.* Saint-Avit Sénieur, Dordogne, France, 22 March 1849; *d.* Caudéran, near Bordeaux, France, 16 January 1925), *anatomy, anthropology.*

The son of Jean Testut and the former Marie Deynat, Léo Testut began his medical studies at Bordeaux, where he was successively *interne des hôpitaux, préparateur* in physiology, and *chef de travaux* in anatomy (1877). He completed his studies at Paris and in 1877 defended a doctoral dissertation entitled "De la symétrie dans les affections de la peau." As a result of his work in the laboratories of Pierre Paul Broca, Louis Ranvier, Étienne-Jules Marey, Jean-Louis de Quatrefages, and Félix Pouchet, he was able to consolidate his threefold training in anatomy, physiology, and anthropology. In 1880 Testut became *agrégé* in anatomy and physiology in the medical faculty at Bordeaux. He was appointed professor of anatomy at Lille in 1884 and transferred to a similar post at Lyons in 1886. In 1919 he retired and moved to Beaumont. During the Franco-Prussian War Testut won the *médaille militaire*, and he died a commander of the Legion of Honor. He married Jeanne Clissey, whose death preceded his own.

Testut's chief professional activity was the teaching of anatomy through lectures and books. His most important work, the three-volume *Traité d'anatomie humaine* (1889–1892), went through six editions in his lifetime; it was translated into Italian and Spanish and was used throughout the world. Testut also published a *Précis d'anatomie descriptive* (1901) and a two-volume *Traité d'anatomie topographique* (1905–1909; with Octave Jacob). Further, he was the editor of a popular series of condensed medical textbooks. In addition to medicine, Testut wrote on comparative anthropology, paleopathology, archaeology, and local history.

BIBLIOGRAPHY

I. Original Works. *Recherches sur quelques muscles surnuméraires de la région scapulaire antéro-interne* (Paris, 1883); *Les anomalies musculaires chez l'homme expliquées par l'anatomie comparée, leur importance en anthropologie* (Paris, 1884); *Anatomie anthropologique. Qu'est-ce que l'homme pour un anthropologiste?* (opening lesson of the anatomy course, given at the Faculté de Médecine of Lyons, 15 November 1886), (Paris, 1887); *Recherches anthropologiques sur le squelette quaternaire de Chancelade (Dordogne)* (Lyons, 1889); *Traité d'anatomie humaine: anatomie descriptive, histologie, développement,* 3 vols. (Paris, 1889–1892; 9th ed., 1949); *Précis d'anatomie descriptive, aide-mémoire à l'usage des candidats au premier examen de doctorat* (Paris, 1901; 15th ed., 1944); *Traité d'anatomie topographique, avec applications médicochirurgicales,* 2 vols. (Paris, 1905–1909; 4th ed., 1921), written with Octave Jacob.

II. Secondary Literature. A. Latarjet, "Léo Testut," in *Paris médical,* **56** (1925), 199–200; A. Policard, "Léo Testut," in *Presse médicale,* **10** (1925), 157.

See also the editorial "Le professeur Testut, de Lyon," in *Chanteclair,* **107** (1912), 7.

CHARLES COURY

THĀBIT IBN QURRA, AL-ṢĀBIʾ AL-ḤARRĀNĪ (*b.* Ḥarrān, Mesopotamia [now Turkey], 836; *d.* Baghdad, 18 February 901), *mathematics, astronomy, mechanics, medicine, philosophy.*

Life. Thābit ibn Qurra belonged to the Sabian (Mandaean) sect, descended from the Babylonian star worshippers. Because the Sabians' religion was related to the stars they produced many astronomers and mathematicians. During the Hellenistic era they spoke Greek and took Greek names; and after the Arab conquest they spoke Arabic and began to assume Arabic names, although for a long time they remained true to their religion. Thābit, whose native language was Syriac, also knew Greek and Arabic. Most of his scientific works were written in Arabic, but some were in Syriac; he translated many Greek works into Arabic.

In his youth Thābit was a money changer in Ḥarrān. The mathematician Muḥammad ibn Mūsā ibn Shākir, one of three sons of Mūsā ibn Shākir, who was traveling through Ḥarrān, was impressed by his knowledge of languages and invited him to Baghdad; there, under the guidance of the brothers, Thābit became a great scholar in mathematics and astronomy. His mathematical writings, the most studied of his works, played an important role in preparing the way for such important mathematical discoveries as the extension of the concept of number to (positive) real numbers, integral calculus, theorems in spherical trigonometry, analytic geometry, and non-Euclidean geometry. In astronomy Thābit was one of the first reformers of the Ptolemaic system, and in mechanics he was a founder of statics. He was also a distinguished physician and the leader of a Sabian community in Iraq, where he substantially strengthened the sect's influence. During his last years Thābit was in the retinue of the Abbasid Caliph al-Muʿtaḍid (892–902). His son Sinān and his grandsons Ibrāhīm and Thābit were well-known scholars.

Mathematics. Thābit worked in almost all areas of mathematics. He translated many ancient mathematical works from the Greek, particularly all the works of Archimedes that have not been preserved in the original language, including *Lemmata, On Touching Circles,* and *On Triangles,* and Apollonius' *Conics.* He also wrote commentaries on Euclid's *Elements* and Ptolemy's *Almagest.*

Thābit's *Kitāb al-Mafrūḍāt* ("Book of Data") was very popular during the Middle Ages and was included by Naṣīr al-Dīn al-Ṭūsī in his edition of the "Intermediate Books" between Euclid's *Elements* and the *Almagest.* It contains thirty-six propositions in elementary geometry and geometrical algebra, including twelve problems in construction and a geometric problem equivalent to solution of a quadratic equation $(a + x)x = b$. *Maqāla fī istikhrāj al-aʿdād al-mutaḥābba bi-suhūlat al-maslak ilā dhālika* ("Book on the Determination of Amicable Numbers") contains ten propositions in number theory, including ones on the constructions of perfect numbers (equal to the sum of their divisors), coinciding with Euclid's *Elements* IX, 36, on the construction of surplus and "defective" numbers (respectively, those greater and less than the sum of their divisors) and the problem, first solved by Thābit, of the construction of "amicable" numbers (pairs of numbers the sum of the divisors of each of which is equal to the other). Thābit's rule is the following: If $p = 3 \cdot 2^n - 1$, $q = 3 \cdot 2^{n-1} - 1$, and $r = 9 \cdot 2^{2n-1} - 1$, are prime numbers, then $M = 2^n \cdot pq$ and $N = 2^n \cdot r$ are amicable numbers.

Kitāb fī Taʾlīf al-nisab ("Book on the Composition of Ratios") is devoted to "composite ratios" (ratios of geometrical quantities), which are presented in the form of products of ratios. The ancient Greeks, who considered only the natural numbers as numbers, avoided applying arithmetical terminology to geometrical quantities, and thus they named the multiplication of ratios by "composition." Composition of ratios is used in the *Elements* (VI, 23), but is not defined in the original text; instead, only particular cases of composite ratios are defined (*Definitions* V, 9–10). An addition by a later commentator (evidently Theon of Alexandria, in VI, 5) on composite ratios is done in a completely non-Euclidean manner.

Thābit criticizes *Elements* VI, 5, and proposes a definition in the spirit of Euclid: for three quantities *A, B,* and *C,* the ratio A/B is composed of the ratios A/C and C/B, and for six quantities *A, B, C, D, E, F* the ratio A/B is composed of the ratios C/D and E/F, if there are also three quantities *L, M, N,* such that $A/B = L/M$, $C/D = L/N$, $E/F = N/M$. He later defines the "multiplication of quantities by a quantity" and systematically applies arithmetical terminology to geometrical quantities. He also proves a number of theorems on the composition of ratios and solves certain problems concerning them. This treatise was important in preparing the extension of the concept of number to positive real numbers, produced in a clear form in the eleventh century by al-Bīrūnī (*al-Qānūn al-Masʿūdī*) and al-Khayyāmī (*Sharḥ mā ashkhāla min muṣādarāt Kitāb Uqlīdis*).

In *Risāla fī Shakl al-qiṭāʿ* ("Treatise on the Secant Figure") Thābit gives a new and very elegant proof of Menelaus' theorem of the complete spherical quadrilateral, which Ptolemy had used to solve problems in spherical astronomy; to obtain various forms of this theorem Thābit used his own theory of composite ratios. In *Kitāb fī Misāḥat qaṭʿ al-makhrūṭ alladhī yusammā al-mukāfiʾ* ("Book on the Measurement of the Conic Section Called Parabolic") Thābit computed the area of the segment of a parabola. First he proved several theorems on the summation of a numerical sequence from

$$\sum_{k=1}^{n} (2k-1) = n^2 \text{ to } \sum_{k=1}^{n} (2k-1)^2 + \frac{n}{3}$$

$$= \frac{2}{3} \cdot 2n \sum_{k=1}^{n} (2k-1).$$

He then transferred the last result to segments $a_k = (2k-1)a$, $b_k = 2k \cdot b$ and proved the theorem that for any ratio α/β, however small, there can always be found a natural n for which

$$\frac{n}{2n \cdot \sum_{k=1}^{n} (2k-1)} < \alpha/\beta,$$

which is equivalent to the relation $\lim_{n\to\infty} \frac{1}{n^2} = 0$.

Thābit also applied this result to the segments and divides the diameter of the parabola into segments proportional to odd numbers; through the points of division he then takes chords conjugate with the diameter and inscribes in the segment of the parabola a polygon the apexes of which are the ends of these chords. The area of this polygon is valued by upper and lower limits, on the basis of which it is shown that the area of the segment is equal to 2/3 the product of the base by the height. A. P. Youschkevitch has shown that Thābit's computation is equivalent to that of the integral $\int_0^a \sqrt{x}\,dx$ and not $\int_0^b x^2 dx$, as is done in the computation of the area in Archimedes' *Quadrature of the Parabola*. The computation is based essentially on the application of upper and lower integral sums, and the proof is done by the method of exhaustion; there, for the first time, the segment of integration is divided into unequal parts.

In *Maqāla fī Misāḥat al-mujassamāt al-mukāfiya* ("Book on the Measurement of Parabolic Bodies") Thābit introduces a class of bodies obtained by rotating a segment of a parabola around a diameter: "parabolic cupolas" with smooth, projecting, or squeezed vertex and, around the bases, "parabolic spheres," named cupolas and spheres. As in *Kitāb . . . al-mukāfiʾ* he also proved theorems on the summing of a number sequence; a theorem equivalent to $\lim_{n \to \infty} \alpha^n = 0$ for any α, $0 < \alpha < 1$; and a theorem that the volume of the "parabolic cupola" is equal to half the volume of a cylinder, the base of which is the base of the cupola, and the height is the axis of the cupola: the result is equivalent to the computation of the integral $\int_0^a x\,dx$.

Kitāb fī Misāḥat al-ashkāl al-musaṭṭaḥa waʾl-mujassama ("Book on the Measurement of Plane and Solid Figures") contains rules for computing the areas of plane figures and the surfaces and volumes of solids. Besides the rules known earlier there is the rule proved by Thābit in "another book," which has not survived, for computing the volumes of solids with "various bases" (truncated pyramids and cones): if S_1 and S_2 are the areas of the bases and h is the height, then the volume is equal to $V = 1/3h\,(S_1 + \sqrt{S_1 S_2} + S_2)$.

Kitāb fiʾl-taʾattī li-istikhrāj ʿamal al-masāʾil al-handasiyya ("Book on the Method of Solving Geometrical Problems") examines the succession of operations in three forms of geometrical problems: construction, measurement, and proof (in contrast with Euclid, who examined only problems in construction ["problems"] and in proof ["theorems"]. In *Risāla fiʾl-ḥujja al-mansūba ilā Suqrāṭ fiʾl-murabbaʿ wa quṭrihi* ("Treatise on the Proof Attributed to Socrates on the Square and Its Diagonals"), Thābit examines the proof, described by Plato in *Meno*, of Pythagoras' theorem for an isosceles right triangle and gives three new proofs for the general case of this theorem. In the first, from a square constructed on the hypotenuse, two triangles congruent to the given triangle and constructed on two adjacent sides of the square are taken out and are added to the two other sides of the square, and the figure obtained thus consists of squares constructed on the legs of the right triangle. The second proof also is based on the division of squares that are constructed on the legs of a right triangle into parts that form the square constructed on the hypotenuse. The third proof is the generalization of Euclid's *Elements* VI, 31. There is also a generalization of the Pythagorean theorem: If in triangle ABC two straight lines are drawn from the vertex B so as to cut off the similar triangles ABE and BCD, then $AB^2 + BC^2 = AC$ $(AE + CD)$.

In *Kitāb fī ʿamal shakl mujassam dhī arbaʿ ʿashrat qāʿida tuḥīṭu bihi kura maʿlūma* ("Book on the Construction of a Solid Figure . . .") Thābit constructs a fourteen-sided polyhedron inscribed in a given sphere. He next makes two attempts to prove Euclid's fifth postulate: *Maqāla fī burhān al-muṣādara ʾl-mashhūra min Uqlīdis* ("Book of the Proof of the Well-Known Postulate of Euclid") and *Maqāla fī anna ʾl-khaṭṭayn idhā ukhrijā ʿalā zawiyatayn aqal min qāʾimatayn iltaqayā* ("Book on the Fact That Two Lines Drawn [From a Transversal] at Angles Less Than Two Right Angles Will Meet"). The first attempt is based on the unclear assumption that if two straight lines intersected by a third move closer together or farther apart on one side of it, then they must, correspondingly, move farther apart or closer together on the other side. The "proof" consists of five propositions, the most important of which is the third, in which Thābit proves the existence of a parallelogram, by means of which Euclid's fifth postulate is proved in the fifth proposition. The second attempt is based on kinematic considerations. In the introduction to the treatise Thābit criticizes the approach of Euclid, who tries to use motion as little as possible in geometry, asserting the necessity of its use. Further on, he postulates that in "one simple motion" (parallel translation) of a body, all its points describe straight lines. The "proof" consists of seven propositions, in the first of which, from the necessity of using motion, he concludes that equidistant straight lines exist; in the fourth proposition he proves the existence of a rectangle that is used in the seventh proposition to prove Euclid's fifth postulate. These two treatises were an important influence on subsequent attempts to prove the fifth postulate (the latter in particular influenced Ibn al-Haytham's commentaries on Euclid). Similar attempts later led to the creation of non-Euclidean geometry.

Kitāb fī Quṭūʿ al-usṭu wāna wa-basīṭihā ("Book on the Sections of the Cylinder and Its Surface") examines plane sections of an inclined circular cylinder and computes the area of the lateral surfaces of such a cylinder between the two plane sections. The treatise contains thirty-seven propositions. Having shown in the thirteenth that an ellipse is obtained through right-angled compression of the circle, in the next Thābit proves that the area of an ellipse with semiaxes a and b is equal to the area of the circle of radius $\sqrt{ab}$; and in the propositions 15–17 he examines the equiaffine transformation, making the ellipse into a circle equal to it.

Thābit proves that in this case the areas of the segments of the ellipse are equal to the areas of the segments of the circle corresponding to it. In the thirty-seventh proposition he demonstrates that the area of the lateral surface of the cylinder between two plane segments is equal to the product of the length of the periphery of the ellipse that is the least section of the cylinder by the length of the segment of the axis of the cylinder between the sections. This proposition is equivalent to the formula that expresses the elliptical integral of the more general type by means of the simplest type, which gives the length of the periphery of the ellipse.

The algebraic treatise *Qawl fī Taṣḥīḥ masā'il al-jabr bi 'l-barāhī al-handasiyya* ("Discourse on the Establishment of the Correctness of Algebra Problems . . .") establishes the rules for solving the quadratic equations $x^2 + ax = b$, $x^2 + b = ax$, $x^2 = ax + b$, using *Elements* II, 5–6. (In giving the geometrical proofs of these rules earlier, Al-Khwārizmī did not refer to Euclid.) In *Mas'ala fī s'amal al-mutawassiṭayn waqisma zāwiya maʿlū ma bi-thalāth aqsām mutasāwiya* ("Problem of Constructing Two Means and the Division of a Given Angle Into Three Equal Parts"), Thābit solves classical problems of the trisection of an angle and the construction of two mean proportionals that amount to cubic equations. Here these problems are solved by a method equivalent to Archimedes' method of "insertion" which basically involves finding points of intersection of a hyperbola and a circumference. (In his algebraic treatise al-Khayyāmī later used an analogous method to solve all forms of cubic equations that are not equivalent to linear and quadratic ones and that assume positive roots.)

Thābit studied the uneven apparent motion of the sun according to Ptolemy's eccentricity hypothesis in *Kitāb fī Ibṭāʾ al-ḥaraka fī falak al-burūj wa surʿatihā bi-ḥasab al-mawāḍiʿ allatī yakūnu fīhi min al-falak al-khārij al-markaz* ("Book on the Deceleration and Acceleration of the Motion on the Ecliptic . . ."), which contains points of maximum and minimum velocity of apparent motion and points at which the true velocity of apparent motion is equal to the mean velocity of motion. Actually these points contain the instantaneous velocity of the unequal apparent motion of the sun.

A treatise on the sundial, *Kitāb fī ālāt al-sāʿāt allatī tusammā rukhāmāt*, is very interesting for the history of mathematics. In it the definition of height h of the sun and its azimuth A according to its declination δ, the latitude ϕ of the city and the hour angle t leads to the rules $\sin h = \text{dos}(\phi - \delta) - \text{versed}$ $\sin t \cdot \cos \delta \cdot \cos \phi$ and $\sin A = \dfrac{\sin t \cdot \cos \delta}{\cos h}$, which are equivalent to the spherical theorems of cosines and sines for spherical triangles of general forms, the vertexes of which are the sun, the zenith, and the pole of the universe. The rules were formulated by Thābit only for solving concrete problems in spherical astronomy; as a general theorem of spherical trigonometry, the theorem of sines appeared only at the end of the tenth century (Manṣūr ibn ʿIrāq), while the theorem of cosines did not appear until the fifteenth century (Regiomontanus). In the same treatise Thābit examines the transition from the length of the shadow of the gnomon l on the plane of the sundial and the azimuth A of this shadow, which in essence represent the polar coordinates of the point, to "parts of longitude" x and "parts of latitude" y, which represent rectangular coordinates of the same point according to the rule $x = l \sin A$, $y = l \cos A$.

In another treatise on the sundial, *Maqāla fī ṣifat al-ashkāl allatī tahduthu bi-mamarr ṭaraf ẓill al-miqyās fī saṭḥ al-ufuq fī kull yawm wa fī kull balad*, Thābit examines conic sections described by the end of a shadow of the gnomon on the horizontal plane and determines the diameters and centers of these sections for various positions of the sun. In the philosophical treatise *Masā'il su'ila ʿanhā Thābit ibn Qurra al-Ḥarrānī* ("Questions Posed to Thābit . . ."), he emphasizes the abstract character of number (*ʿadad*), as distinct from the concrete "counted thing" (*maʿdūd*), and postulates "the existence of things that are actually infinite in contrast with Aristotle, who recognized only potential infinity. Actual infinity is also used by

Thābit in *Kitāb fi'l qarasṭūn* ("Book on Beam Balance").

Astronomy. Thābit wrote many astronomical works. We have already noted his treatise on the investigation of the apparent motion of the sun; his *Kitāb fī Sanat al-shams* ("Book on the Solar Year") is on the same subject. *Qawl fī īḍāḥ al-wajh alladhī dhakara Baṭlamyūs . . .* concerns the apparent motion of the moon, and *Fī ḥisāb ru'yat al-ahilla*, the visibility of the new moon. In what has been transmitted as *De motu octave spere* and *Risāla ilā Isḥāq ibn Ḥunayn* ("Letter to . . .") Thābit states his kinematic hypothesis, which explains the phenomenon of precession with the aid of the "eighth celestial sphere" (that of the fixed stars); the first seven are those of the sun, moon, and five planets. Thābit explains the "trepidation" of the equinoxes with the help of a ninth sphere. The theory of trepidation first appeared in Islam in connection with Thābit's name.

Mechanics and Physics. Two of Thābit's treatises on weights, *Kitāb fī Ṣifat al-wazn wa-ikhtilāfihi* ("Book on the Properties of Weight and Nonequilibrium") and *Kitāb fi'l-Qarasṭūn* ("Book on Beam Balance"), are devoted to mechanics. In the first he formulates Aristotle's dynamic principle, as well as the conditions of equilibrium of a beam, hung or supported in the middle and weighted on the ends. In the second treatise, starting from the same principle, Thābit proves the principle of equilibrium of levers and demonstrates that two equal loads, balancing a third, can be replaced by their sum at a midpoint without destroying the equilibrium. After further generalizing the latter proposition for the case in which "as many [equal] loads as desired and even infinitely many" are hung at equal distances, Thābit considers the case of equally distributed continuous loads. Here, through the method of exhaustion and examination of upper and lower integral sums, a calculation equivalent

to computation of the integral $\int_a^b x\,n\,dx$. The result

obtained is used to determine the conditions of equilibrium for a heavy beam.

Thābit's work in natural sciences includes *Qawl fi'l-Sabab alladhī ju'ilat lahu miyāh al-baḥr māliḥa* ("Discourse on the Reason Why Seawater Is Salted"), extant in manuscript, and writings on the reason for the formation of mountains and on the striking of fire from stones. He also wrote two treatises on music.

Medicine. Thābit was one of the best-known physicians of the medieval East. Ibn al-Qifṭī, in *Ta'rikh al-ḥukamā*, tells of Thābit's curing a butcher who was given up for dead. Thābit wrote many works on Galen and medicinal treatises, which are almost completely unstudied. Among these treatises are general guides to medicine—*al-Dhakhīra fī 'ilm al-ṭibb* ("A Treasury of Medicine"), *Kitāb al-Rawḍa fi 'l-ṭibb* ("Book of the Garden of Medicine"), *al-Kunnash* ("Collection")—and works on the circulation of the blood, embryology, the cure of various illnesses—*Kitāb fī 'ilm al-'ayn . . .* ("Book on the Science of the Eye . . ."), *Kitāb fi'l-jadarī wa'l-ḥaṣbā* ("Book on Smallpox and Measles"), *Risāla fī tawallud al-ḥaṣāt* ("Treatise on the Origin of Gallstones"), *Risāla fi'l-bayāḍ alladhī yaẓharu fi'l-badan* ("Treatise on Whiteness . . . in the Body")—and on medicines. Thābit also wrote on the anatomy of birds and on veterinary medicine (*Kitāb al-bayṭara*), and commented on *De plantis*, ascribed to Aristotle.

Philosophy and Humanistic Sciences. Thābit's philosophical treatise *Masā'il su'ila 'anhā Thābit ibn Qurra al-Ḥarrānī* comprises his answers to questions posed by his student Abū Mūsā ibn Usayd, a Christian from Iraq. In another extant philosophical treatise, *Maqāla fī talkhīṣ mā aṭā bihi Arisṭūṭālīs fī kitābihi fī Mā ba'd al-ṭabī'a*, Thābit criticizes the views of Plato and Aristotle on the motionlessness of essence, which is undoubtedly related to his opposition to the ancient tradition of not using motion in mathematics. Ibn al-Qifṭī (*op. cit.*, 120) says that Thābit commented on Aristotle's *Categories, De interpretatione*, and *Analytics*. He also wrote on logic, psychology, ethics, the classification of sciences, the grammar of the Syriac language, politics, and the symbolism in Plato's *Republic*. Ibn al-Qifṭī also states that Thābit produced many works in Syriac on religion and the customs of the Sabians.

BIBLIOGRAPHY

I. ORIGINAL WORKS. Thābit's MSS are listed in C. Brockelmann, *Geschichte . . . Literatur*, 2nd ed., I (Leiden, 1943), 241–244, and supp. I (Leiden, 1937), 384–386; Fuat Sezgin, *Geschichte des arabischen Schrifttums*, III (Leiden, 1970), 260–263, and V (Leiden, 1974), 264–272; and H. Suter, *Die Mathematiker und Astronomen der Araber und ihre Werke* (Leipzig, 1900), 34–38, and *Nachträge* (1902), 162–163. Many of his works that are no longer extant are cited by Ibn al-Qifṭī in his *Ta'rīkh al-ḥukamā'*, J. Lippert, ed. (Leipzig, 1903), 115–122.

His published writings include *Kitāb al-Mafrūḍāt*

("Book of Data"), in Naṣīr al-Dīn al-Ṭūsī, *Majmūʿ al-rasāʾil*, II (Hyderabad, 1940), pt. 2; *Maqāla fī istikhrāj al-aʿdād al-mutaḥābba bi-suhūlat al-maslak ilā dhālikâ* ("Book on the Determination of Amicable Numbers by an Easy Method"), Russian trans. by G. P. Matvievskaya in *Materialy k istorii . . .*, 90–116; *Kitāb fī taʾlīf al-nisab* ("Book on the Composition of Ratios"), Russian trans. by B. A. Rosenfeld and L. M. Karpova in the *Fiziko-matematicheskie Nauki v Stranakh Vostoka* ("Physical-Mathematical Sciences in the Countries of the East"; Moscow, 1966), 9–41; *Risāla fī Shakl al-qiṭāʿ* ("Treatise on the Secant Figure"), in Latin trans. by Gerard of Cremona, with notes and German trans.; *Risāla fiʾl-ḥujja al-mansūba ilā Suqrāṭ fiʾl-murabbaʿ wa quṭrih* ("Treatise on the Proof Attributed to Socrates on the Square and Its Diagonals"), Arabic text with Turkish trans. in A. Sayili, "Sābit ibn Kurranin Pitagor teoremini temini," and in English in Sayili's "Thābit ibn Qurra's Generalization of the Pythagorean Theorem"; and *Kitāb fī ʿamal shakl mujassam dhī arbaʿ ʿashrat qāʿida tuḥiṭu bihi kura maʿlūma* ("Book on the Construction of a Solid Figure With Fourteen Sides About Which a Known Sphere Is Described"), ed. with German trans. in E. Bessel-Hagen and O. Spies, "Tābit b. Qurra's Abhandlung über einen halbregelmässigen Vierzehnflächner."

Additional works are *Maqāla fī burhān al-muṣādara ʾl-mashhūra min Uqlīdis* ("Book of the Proof of the Well-Known Postulate of Euclid"), Russian trans. in B. A. Rosenfeld and A. P. Youschkevitch, *Dokazatelstva pyatogo postulata Evklida . . .*, and English trans. in A. I. Sabra, "Thābit ibn Qurra on Euclid's Parallels Postulate"; *Maqāla fī anna ʾl-khaṭṭayn idhā ukhrijā ʿalā zāwiyatayn aqall min qāʾimatayn iltaqayā* ("Book on the Fact That Two Lines Drawn [From a Transversal] at Angles Less Than Two Right Angles Will Meet"), Russian trans. by B. A. Rosenfeld in "Sabit ibn Korra. Kniga o tom, chto dve linii, provedennye pod uglami, menshimi dvukh pryamykh, vstretyatsya," *Istoriko-matematicheskie issledovania*, **15** (1962), 363–380, and English trans. in Sabra, *op. cit.*; and *Qawl fī taṣḥīḥ masāʾil al-jabr biʾl-barahīn al-handasiyya* ("Discourse on the Establishment of the Correctness of Algebra Problems With the Aid of Geometrical Proofs"), ed. and German trans. in P. Luckey, "Tābit b. Qurra über die geometrischen Richtigkeitsnachweis der Auflösung der quadratischen Gleichungen."

Further works are *Qawl fī īḍāḥ al-wajh alladhī dhakara Baṭlamyūs anna bihi istakhraja man taqaddamahu masīrat al-qamar al-dawriyya wa-hiya al-mustawiya* ("Discourse on the Explanation of the Method Noted by Ptolemy That His Predecessors Used for Computation of the Periodic [Mean] Motion of the Moon"), German trans. of the intro. in Hessel-Hagen and Spies, *op. cit.*; *Kitāb fī sanat al-shams* ("Book on the Solar Year"), medieval Latin trans. in F. J. Carmody, *The Astronomical Works of Thabit b. Qurra*, 41–79, and English trans., with commentary, by O. Neugebauer in "Thābit ben Qurra. On the Solar Year and On the Motion of the

Eighth Sphere," in *Proceedings of the American Philosophical Society*, **106** (1962), 267–299; medieval Latin trans. of work on the eighth sphere, "De motu octave spere," in Carmody, *op. cit.*, 84–113, and English trans. in Neugebauer, *op. cit.*, 291–299; *Risāla ilā Isḥāq ibn Ḥunayn* ("A Letter to . . ."), included by Ibn Yūnus in his "Great Ḥakimite *zīj*," Arabic text and French trans. by J. J. Caussin de Parceval, "Le livre de la grande table Hakémite observée par . . . Ebn Younis," 114–118; and *Kitāb fī ālāt al-sāʿāt allatī tusammā rukhāmāt* ("Book on the Timekeeping Instruments Called Sundials"), ed. with German trans. by K. Garbers, ". . . Ein Werk über ebene Sonnenuhren . . .," in *Quellen und Studien zur Geschichte der Mathematik, Astronomie und Physik*, Abt. A, **4** (1936).

Thābit also wrote *Maqāla fī ṣifat al-ashkāl allatī tahduthu bi-mamarr ṭaraf ẓill al-miqyās fī saṭḥ al-ufuq fī kull yawm wa fī kull balad* ("Book on the Description of Figures Obtained by the Passage of the End of a Shadow of a Gnomon in the Horizontal Plane on Any Day and in Any City"), German trans. in E. Wiedemann and J. Frank, "Über die Konstruktion der Schattenlinien von Thābit ibn Qurra"; *Kitāb fī ṣifat al-wazn wa-ikhtilāfihi* ("Book on the Properties of Weight and Nonequilibrium"), included by ʿAbd al-Raḥman al-Khāzinī in his *Kitāb mīzān al-ḥikma* ("Book of the Balance of Wisdom"), 33–38; *Kitāb fiʾl-qarasṭūn* ("Book on Beam Balances"), medieval Latin trans. in F. Buchner, "Die Schrift über der Qarastūn von Thābit b. Qurra," and in E. A. Moody and M. Clagett, *The Medieval Science of Weights*, 77–117 (with English trans.), also German trans. from Arabic MSS in E. Wiedemann, "Die Schrift über den Qarastūn"; and *al-Dhakhīra fī ʿilm al-tibb* ("A Treasury of Medicine"), ed. by G. Ṣubḥī (Cairo, 1928).

Recensions of ancient works are Euclid's *Elements*, ed. with additions by Naṣīr al-Dīn al-Ṭūsī, *Taḥrīr Uqlīdis fī ʿilm al-handasa* (Teheran, 1881); Archimedes' *Lemmata*, Latin trans. with additions by al-Nasawī, in *Archimedis Opera omnia*, J. L. Heiberg, ed., 2nd ed., II (Leipzig, 1912), 510–525; Archimedes' *On Touching Circles* and *Triangles* in *Rasāʾil ibn Qurra* (Hyderabad, 1940); Apollonius' *Conics*, bks. 5–7, Latin trans. in *Apollonii Pergaei Conicorum libri VII* (Florence, 1661), German trans. in L. Nix, *Das fünfte Buch der Conica des Apollonius von Perga in der arabischen Uebersetzung des Thabit ibn Corrah*; *De plantis*, ascribed to Aristotle, ed. in A. J. Arberry, "An Early Arabic Translation From the Greek"; and Galen's medical treatises, in F. Sezgin, *Geschichte des arabischen Schrifttums*, III, 68–140.

II. SECONDARY LITERATURE. See A. J. Arberry, "An Early Arabic Translation From the Greek," in *Bulletin of the Faculty of Arts, Cairo*, **1** (1933), 48–76, 219–257, and **2** (1934), 71–105; E. Bessel-Hagen and O. Spies, "Tābit b. Qurra's Abhandlung über einen halbregelmässigen Vierzehnflächner," in *Quellen und Studien zur Geschichte der Mathematik, Astronomie und Physik*, Abt. B, **2** (1933), 186–198; A. Björnbo, "Thābits Werk über den Transversalensatz . . .," in *Abhandlun-*

gen zur Geschichte der Naturwissenschaften und der Medizin, **7** (1924); F. Buchner, "Die Schrift über der Qarastūn von Thābit b. Qurra," in Sitzungsberichte der Physikalisch-medizinischen Sozietät in Erlangen, **52–53** (1922), 171–188; F. J. Carmody, The Astronomical Works of Thabit b. Qurra (Berkeley–Los Angeles, 1960); J. J. Caussin de Perceval, "Le livre de la grande table Hakémite observée par . . . Ebn Iounis," in Notices et extraits des manuscrits de la Bibliothèque nationale, **7**, pt. 1 (1803–1804), 16–240; D. Chvolson, Die Ssabier und Ssabismus, I (St. Petersburg, 1856), 546–567; and P. Duhem, Les origines de la statique, I (Paris, 1905), 79–92; and Le système du monde, II (Paris, 1914), 117–119, 238–246.

Also see Ibn Abi Uṣaybiʿa, ʿUyūn al-anbāʾfī tabaqāt al-aṭibbāʾ, A. Müller, ed., I (Königsberg, 1884), 115–122; A. G. Kapp, "Arabische Übersetzer und Kommentatoren Euklids . . .," in Isis, **23** (1935), 58–66; L. M. Karpova, "Traktat Sabita ibn Korry o secheniakh tsilindra i ego poverkhnosti" ("Treatise of Thābit ibn Qurra on the Sections of the Cylinder and Its Surface"), in Trudy XIII Mezhdunarodnogo kongressa po istorii nauki (Papers of the XIII International Congress on the History of Science), sec. 3–4 (Moscow, 1974), 103–105; E. S. Kennedy, "The Crescent Visibility Theory of Thābit ibn Qurra," in Proceedings of the Mathematical and Physical Society of the UAR, **24** (1961), 71–74; ʿAbd al-Raḥmān al-Khāzinī, Kitāb mīzān al-ḥikma (Hyderabad, 1940); L. Leclerc, Histoire de la médecine arabe, I (Paris, 1876), 168–172; P. Luckey, "Tābit b. Qurra's Buch über die ebenen Sonnenuhren," in Quellen und Studien zur Geschichte der Mathematik, Astronomie und Physik, Abt. B, **4** (1938), 95–148; and "Tābit b. Qurra über die geometrischen Richtigkeitsnachweis der Auflösung der quadratischen Gleichungen," in Berichte de Sächsischen Akademie der Wissenschaften, Math.-nat. Kl., **13** (1941), 93–114; and G. P. Matvievskaya, Uchenie o chisle na srednevekovom Blizhnem i Srednem Vostoke ("Number Theory in the Medieval Near East and Central Asia"; Tashkent, 1967); and "Materialy k istorii ucheniya o chisle na srednevekovom Blizhnem i Srednem Vostoke" ("Materials for a History of Number Theory in the Medieval Near and Middle East"), in Iz istorii tochnykh nauk na srednevekovom Blizhnem i Srednem Vostoke ("History of the Exact Sciences in the Medieval Near and Middle East"; Tashkent, 1972), 76–169.

Additional works are M. Meyerhof, "The 'Book of Treasure,' an Early Arabic Treatise on Medicine," in Isis, **14** (1930), 55–76; E. A. Moody and M. Clagett, The Medieval Science of Weights (Madison, Wis., 1952); L. Nix, Das fünfte Buch der Conica des Apollonius von Perga in der arabischen Uebersetzung des Thabit ibn Corrah . . . (Leipzig, 1889); S. Pines, "Thabit b. Qurra's Conception of Number and Theory of the Mathematical Infinite," in Actes du XIᵉ Congrès international d'histoire des sciences, III (Wrocław–Warsaw–Cracow), 160–166; B. A. Rosenfeld and L. M.

Karpova, "Traktat Sabita ibn Korry o sostavnykh otnosheniakh" ("Treatise of Thābit ibn Qurra on the Composition of Ratios"), in Fiziko-matematicheskie nauki v stranakh Vostoka ("Physical-Mathematical Sciences in the Countries of the East"), I (Moscow, 1966), 5–8; B. A. Rosenfeld and A. P. Youschkevitch, "Dokazatelstva pyatogo postulata Evklida . . ." ("Proofs of Euclid's Fifth Postulate . . ."), in Istoriko-matematicheskie issledovania, **14** (1961), 587–592; A. I. Sabra, "Thābit ibn Qurra on Euclid's Parallels Postulate," in Journal of the Warburg and Courtauld Institutes, **31** (1968), 12–32; A. Y. Sansur, Matematicheskie trudy Sabita ibn Korry ("Mathematical Works of Thābit ibn Qurra"; Moscow, 1971); G. Sarton, Introduction to the History of Science, I (Baltimore, 1927), 599–600; A. Sayili, "Sābit ibn Kurranin Pitagor teoremini temini," in Türk Tarih Kurumu. Belleten, **22**, no. 88 (1958), 527–549; and "Thabit ibn Qurra's Generalization of the Pythagorean Theorem," in Isis, **51** (1960), 35–37; and O. Schirmer, "Studien zur Astronomie der Araber," in Sitzungsberichte der Physikalisch-medizinischen Sozietät in Erlangen, **58** (1927), 33–88.

See also F. Sezgin, Geschichte des arabischen Schrifttums, III (Leiden, 1970), 260–263; T. D. Stolyarova, "Traktat Sabita ibn Korry 'Kniga o karastune' " ("Thābit ibn Qurra's Treatise 'Book of Qarastūn' "), in Iz istorii tochnykh nauk na srednevekovom Blizhnem i Srednem Vostoke ("History of the Exact Sciences in the Medieval Near East and Central Asia"; Tashkent, 1972), 206–210; and Statika v stranakh Blizhnego i Srednego Vostoka v IX–XI vekakh ("Statics in the . . . Near East and Central Asia in the Ninth-Eleventh Centuries"; Moscow, 1973); H. Suter, "Die Mathematiker und Astronomen der Araber und ihre Werke," in Abhandlungen für Geschichte der mathematischen Wissenschaften, **10** (1900); "Uber die Ausmessung der Parabel von Thābit ben Kurra al-Harrani," in Sitzungsberichte der Physikalisch-medizinischen Societät in Erlangen, **48–49** (1918), 65–86; and "Die Abhandlungen Thābit ben Kurras und Abū Sahl al-Kūhīs über die Ausmessung der Paraboloide," ibid., 186–227; J. Vernet and M. A. Catalá, "Dos tratados de Arquimedes arabe; Tratado de los círculos tangentes y Libro de los triángulos," Publicaciones del Seminario de historia de la ciencia, **2** (1972); E. Wiedemann, "Die Schrift über den Qarastūn," in Bibliotheca mathematica, 3rd ser., **12**, no. 1 (1912), 21–39; and "Über Thābit, sein Leben und Wirken," in Sitzungsberichte der Physikalisch-medizinischen Sozietät in Erlangen, **52** (1922), 189–219; E. Wiedemann and J. Frank, "Über die Konstruktion der Schattenlinien auf horizontalen Sonnenuhren von Thābit ibn Qurra," in Kongelige Danske Videnskabernes Selskabs Skrifter, Math.-fys. meddel., **4** (1922), 7–30; F. Woepcke, "Notice sur une théorie ajoutée par Thābit ben Korrah à l'arithmétique spéculative des grecs," in Journal asiatique, 4th ser., **20** (1852), 420–429; F. Wüstenfeld, Geschichte der arabischen Ärzte (Leipzig, 1840), 34–36; and A. P. Youschkevitch, "Note sur les déterminations infinitésimales chez Thabit ibn Qurra," in *Archives internationales d'histoire des sciences*, no. 66 (1964), 37–45; and (as editor), *Istoria matematiki s drevneyshikh vremen do nachala XIX stoletiya* ("History of Mathematics From Ancient Times to the Beginning of the Nineteenth Century"), I (Moscow, 1970), 221–224, 239–244.

B. A. ROSENFELD
A. T. GRIGORIAN

THADDAEUS FLORENTINUS. See **Alderotti, Taddeo.**

THALES (*b.* Miletus, Ionia, 625 B.C. [?]; *d.* 547 B.C. [?]), *natural philosophy.*

Thales is considered by Aristotle to be the "founder" (ἀρχηγός) of Ionian natural philosophy.[1] He was the son of Examyes and Cleobuline, who were, according to some authorities, of Phoenician origin. But the majority opinion considered him a true Milesian by descent (ἰθαγενής Μιλήσιος), and of a distinguished family. This latter view is probably the correct one since his father's name seems to be Carian rather than Semitic, and the Carians had at this time been almost completely assimilated by the Ionians. According to Diogenes Laërtius, Apollodorus put Thales' birth in Olympiad 35.1 (640 B.C.) and his death at the age of 78 in Olympiad 58 (548–545 B.C.). There is a discrepancy in the figures here: probably 35.1 is a mistake for 39.1 (624), since the confusion of $\bar{\epsilon}$ and $\bar{\vartheta}$ is a very common one. Apollodorus would in that case characteristically have made Thales' death correspond with the date of the fall of Sardis, his *floruit* coincide with the eclipse of the sun dated at 585 B.C.—which he is alleged to have predicted—and assumed his birth to be the conventional forty years before his prime.[2]

Even in antiquity there was considerable doubt concerning Thales' written works. It seems clear that Aristotle did not have access to any book by him, at least none on cosmological matters. Some authorities declare categorically that he left no book behind. Others, however, credit him with the authorship of a work on navigation entitled "The Nautical Star Guide," but in spite of a tradition suggesting that Thales defined the Little Bear and recommended its navigational usefulness to Milesian sailors,[3] it is extremely doubtful that he was the actual author of this work, since Diogenes Laërtius informs us that this book was also attributed to a certain Phokos of Samos. It is most un-

likely that a work of Thales would have been ascribed to someone of comparative obscurity, but not the converse.

Much evidence of practical activities associated with Thales has survived, testifying to his versatility as statesman, tycoon, engineer, mathematician, and astronomer. In the century after his death he became an epitome of practical ingenuity.[4] Herodotus records the stories that Thales advised the Ionians to establish a single deliberative chamber at Teos and that he diverted the river Halys so that Croesus' army might be able to cross. (Herodotus is skeptical about the latter explanation.)[5] Aristotle preserves another anecdote that credits Thales with considerable practical knowledge. According to this account, Thales, when reproached for his impracticality, used his skill in astronomy to forecast a glut in the olive crop, went out and cornered the market in the presses, and thereby made a large profit. Aristotle disbelieves the story and comments that this was a common commercial procedure that men attributed to Thales on account of his wisdom.[6] Plato, on the other hand, whose purpose is to show that philosophy is above mere utilitarian considerations, tells the conflicting anecdote that Thales, while stargazing, fell into a well and was mocked by a pretty Thracian servant girl for trying to find out what was going on in the heavens when he could not even see what was at his feet.[7] It is clear that these stories stem from separate traditions—the one seeking to represent the philosopher as an eminently practical man of affairs and the other as an unworldly dreamer.

Thales achieved his fame as a scientist for having predicted an eclipse of the sun. Herodotus, who is our oldest source for this story, tells us that the eclipse (which must have been total or very nearly so) occurred in the sixth year of the war between the Lydians under Alyattes and the Medes under Cyaxares, and that Thales predicted it to the Ionians, fixing as its term the year in which it actually took place.[8] This eclipse is now generally agreed to have occurred on 28 May 585 B.C. (−584 by astronomical reckoning). It has been widely accepted that Thales was able to perform this striking astronomical feat by using the so-called "Babylonian saros," a cycle of 223 lunar months (18 years, 10 days, 8 hours), after which eclipses both of the sun and moon repeat themselves with very little change. Neugebauer, however, has convincingly demonstrated that the "Babylonian saros" was, in fact, the invention of the English astronomer Edmond Halley in rather a weak moment.[9] The Babylonians did not use cycles to predict solar eclipses but computed them from observations of the latitude of the moon made shortly before the expected syzygy. As Neugebauer says,

> . . . there exists no cycle for solar eclipses visible at a given place; all modern cycles concern the earth as a whole. No Babylonian theory for predicting a solar eclipse existed at 600 B.C., as one can see from the very unsatisfactory situation 400 years later, nor did the Babylonians ever develop any theory which took the influence of geographical latitude into account.[10]

Accordingly, it must be assumed that if Thales did predict the eclipse he made an extremely lucky guess and did not do so upon a scientific basis, since he had no conception of geographical latitude and no means of determining whether a solar eclipse would be visible in a particular locality. He could only have said that an eclipse was possible somewhere at some time in the (chronological) year that ended in 585 B.C. But a more likely explanation seems to be simply that Thales happened to be the *savant* around at the time when this striking astronomical phenomenon occurred and the assumption was made that as a savant he *must* have been able to predict it. There is a situation closely parallel to this one in the next century. In 468–467 B.C. a huge meteorite fell at Aegospotami. This event made a considerable impact, and two sources preserve the absurd report that the fall was predicted by Anaxagoras, who was the Ionian *savant* around at that time.[11]

The Greeks themselves claim to have derived their mathematics from Egypt.[12] Eudemus, the author of the history of mathematics written as part of the systematization of knowledge that went on in the Lyceum, is more explicit. He tells us that it was "Thales who, after a visit to Egypt, first brought this study to Greece" and adds "not only did he make numerous discoveries himself, but he laid the foundations for many other discoveries on the part of his successors, attacking some problems with greater generality and others more empirically." Proclus preserves for us some of the discoveries that Eudemus ascribed to Thales, namely, that the circle is bisected by its diameter,[13] that the base angles of an isosceles triangle are equal,[14] and that vertically opposed angles are equal.[15] In addition he informs us that the theorem that two triangles are equal in every respect if they have two angles and one side respectively equal was referred by Eudemus to Thales with the comment that the latter's measuring the distance of

ships out at sea necessarily involved the use of this theorem.[16]

From the above it can be seen that Eudemus credited Thales with full knowledge of the theory behind his discoveries. He also held that Thales introduced geometry into Greece from Egypt. Our surviving sources of information about the nature of Egyptian mathematics, however, give us no evidence to suggest that Egyptian geometry had advanced beyond certain rule-of-thumb techniques of practical mensuration. Nowhere do we find any attempt to discover why these techniques worked, nor anything resembling a general and theoretical mathematics. It seems most unlikely, then, that the Greeks derived their mathematics from the Egyptians. But could Thales have been the founder of theoretical mathematics in Greece, as Eudemus claimed? Here again the answer must be negative. The first three discoveries attributed to him by the Peripatetic most probably represent "just the neatest abstract solutions of particular problems associated with Thales."[17] Heath points out that the first of these propositions is not even proved in Euclid.[18] As for the last of them, Thales could very easily have made use of a primitive angle-measurer and solved the problem in one of several ways without necessarily formulating an explicit theory about the principles involved.

Van der Waerden, on the other hand, believes that Thales did develop a logical structure for geometry and introduced into this study the idea of proof.[19] He also seeks to derive Greek mathematics from Babylon. This is a very doubtful standpoint. Although Babylonian mathematics, with its sexagesimal place-value system, had certainly developed beyond the primitive level reached by the Egyptians, here too we find nowhere any attempt at proof. Our evidence suggests that the Greeks were influenced by Babylonian mathematics, but that this influence occurred at a date considerably later than the sixth century B.C. If the Greeks had derived their mathematics from Babylonian sources, one would have expected them to have adopted the much more highly developed place-value system. Moreover, the Greeks themselves, who are extremely generous, indeed overgenerous, in acknowledging their scientific debts to other peoples, give no hint of a Babylonian source for their mathematics.

Our knowledge of Thales' cosmology is virtually dependent on two passages in Aristotle. In the *Metaphysics* (A3, 983b6) Aristotle, who patently has no more information beyond what is given here, is of the opinion that Thales considered water to be the material constituent of things, and in the *De caelo* (B13, 294a28), where Aristotle expressly declares his information to be indirect, we are told that Thales held that the earth floats on water. Seneca provides the additional information (*Naturales quaestiones*, III, 14) that Thales used the idea of a floating earth to explain earthquakes. If we can trust this evidence, which seems to stem ultimately from Theophrastus via a Posidonian source, the implication is that Thales displays an attitude of mind strikingly different from anything that had gone before. Homer and Hesiod had explained that earthquakes were due to the activity of the god Poseidon, who frequently bears the epic epithet "Earth Shaker." Thales, by contrast, instead of invoking any such supernatural agency, employs a simple, natural explanation to account for this phenomenon. Cherniss, however, has claimed that Aristotle's knowledge of Thales' belief that the earth floats on water would have been sufficient to induce him to infer that Thales also held water to be his material substrate.[20] But it is impossible to believe that Aristotle could have been so disingenuous as to make this inference and then make explicit conjectures as to why Thales held water to be his ἀρχή. Aristotle's conjectured reasons for the importance attached by Thales to water as the ultimate constituent of things are mainly physiological. He suggests that Thales might have been led to this conception by the observation that nutriment and semen are always moist and that the very warmth of life is a damp warmth. Burnet has rejected these conjectures by Aristotle on the ground that in the sixth century interests were meteorological rather than physiological.[21] But, as Baldry has pointed out, an interest in birth and other phenomena connected with sex is a regular feature even of primitive societies long before other aspects of biology are thought of.[22] However this may be, it is noteworthy that, in view of the parallels to be found between Thales' cosmology and certain Near Eastern mythological cosmogonies,[23] there exists the possibility that Thales' emphasis upon water and his theory that the earth floats on water were derived from some such source, and that he conceived of water as a "remote ancestor" rather than as a persistent substrate. But even if Thales was influenced by mythological precedents[24] and failed to approximate to anything like the Aristotelian material cause, our evidence, sparse and controversial though it is, nevertheless seems sufficient to justify the claim that Thales was the first philosopher. This evidence suggests that Thales' thought shared certain basic characteristics with that of his Ionian successors. These Milesian philosophers, abandoning mythopoeic forms of thought, sought to explain the world about them in terms of its visible constituents. Natural explanations were introduced by them, which took the place of supernatural and mystical ones.[25] Like their mythopoeic predecessors, the Milesians firmly believed that there was an orderliness inherent in the world around them. Again like their predecessors, they attempted to explain the world by showing how it had come to be what it is. But, instead of invoking the agency of supernatural powers, they sought for a unifying hypothesis to account for this order and, to a greater or lesser extent, proceeded to deduce their natural explanations of the various phenomena from it. Two elements, then, characterize early Greek philosophy, the search for natural as opposed to supernatural and mystical explanations, and secondly, the search for a unifying hypothesis. Both of these elements proved influential in paving the way for the development of the sciences, and it is in the light of this innovation that Thales' true importance in the history of science must be assessed.

NOTES

1. *Metaphysics*, A3, 983b17 ff. (DK, 11A12).
2. These datings are now approximately in accordance with the figures given by Demetrius of Phalerum, who placed the canonization of the Seven Sages (of whom Thales was universally regarded as a member) in the archonship of Damasias at Athens (582–581 B.C.).
3. Callimachus, *Iambus*, 1, 52 f. 191 Pfeiffer (DK, 11A3a).
4. See Aristophanes, *Birds* 1009; *Clouds* 180.
5. Herodotus, I, 170; I, 75 (DK, 11A4, 11A6).
6. *Politics*, A11, 1259a6 (DK, 11A10).
7. *Theaetetus*, 174A (DK, 11A9). It is odd that Plato should have applied this story to someone as notoriously practical in his interests as Thales. It makes one think that there may be at least a grain of truth in the story. See my review of Moraux's Budé edition of the *De caelo*, in *Classical Review*, n.s., **20** (1970), 174, and M. Landmann and J. O. Fleckenstein, "Tagesbeobachtung von Sternen in Altertum," in *Vierteljahrsschrift der Naturforschenden Gesellschaft in Zürich*, **88** (1943), 98, notwithstanding Dicks' scornful dismissal of their suggestion. Certainly the motive for this story is clear, but it could have been Thales' practice that determined its form. In general Dicks is far too skeptical in his treatment of the stories told of Thales and relegates them to the status of "the famous story of the First World War about the Russians marching through England with 'snow on their boots.'" But on this latter story see Margo Lawrence, *Shadow of Swords* (London, 1971), in which she reveals that soldiers from Russia, wearing Russian uniform, carrying balalaikas, and singing Slavonic songs, did in fact disembark in 1916 at Newcastle upon Tyne. Admittedly the snow on their boots must be left to folklore.
8. I, 74 (DK, 11A5).
9. O. Neugebauer, *The Exact Sciences in Antiquity*, 141.
10. *Ibid.*, 142.

11. See Diogenes Laërtius, II, 10 (DK, 59A1), and Pliny, *Historia naturalis*, II, 149 (DK, 59A11). See also Cicero, *De divinatione*, I.50.112 (DK, 12A5a), and Pliny, *ibid.*, II, 191, for a sixth-century parallel, where Anaximander is alleged to have predicted an earthquake.

12. See Herodotus, II, 109, who believes that geometry originated from the recurrent need to remeasure land periodically flooded by the Nile; Aristotle, *Metaphysics*, A3, 981b20–25, who believes that mathematics evolved in a highly theoretical way as the invention of a leisured class of Egyptian priests; and Eudemus, who, in spite of being a Peripatetic, sides with Herodotus rather than with Aristotle (see Proclus, *Commentary on Euclid's Elements*, I, 64.16 [Friedlein]).

13. *Commentary on Euclid's Elements*, 157.10 (DK, 11A20).

14. *Ibid.*, 250.20.

15. *Ibid.*, 299.1.

16. *Ibid.*, 352.14.

17. G. S. Kirk, *The Presocratic Philosophers*, 84.

18. T. L. Heath, *Greek Mathematics*, I, 131.

19. B. L. van der Waerden, *Science Awakening*, 89.

20. H. Cherniss, "The Characteristics and Effects of Presocratic Philosophy," in *Journal of the History of Ideas*, 12 (1951), 321.

21. J. Burnet, *Early Greek Philosophy*, 48.

22. H. C. Baldry, "Embryological Analogies in Early Greek Philosophy," in *Classical Quarterly*, 26 (1932), 28.

23. For an excellent account of Egyptian and Mesopotamian cosmogonies, see H. Frankfort, ed., *Before Philosophy* (Penguin Books, London, 1949), pub. orig. as *The Intellectual Adventure of Ancient Man* (Chicago, 1946).

24. Aristotle, it may be noted, cites the parallel in Greek mythology of Oceanus and Tethys, the parents of generation (*Metaphysics*, A3, 983b27ff. [DK,1B10]). But the Greek myth may itself be derived from an oriental source.

25. The gods of whom Thales thought everything was full (see Aristotle, *De anima*, A5, 411a7 [DK, 11A22]) are manifestly different from the personal divinities of traditional mythology.

BIBLIOGRAPHY

For a collection of sources see H. Diels and W. Kranz, *Die Fragmente der Vorsokratiker*, 6th ed., 3 vols. (Berlin, 1951–1952), I, 67–79 (abbreviated as DK above).

See also H. C. Baldry, "Embryological Analogies in Presocratic Cosmogony," in *Classical Quarterly*, 26 (1932), 27–34; J. Burnet, *Greek Philosophy: Part 1, Thales to Plato* (London, 1914); and *Early Greek Philosophy*, 4th ed. (London, 1930); H. Cherniss, "The Characteristics and Effects of Presocratic Philosophy," in *Journal of the History of Ideas*, 12 (1951), 319–345; D. R. Dicks, "Thales," in *Classical Quarterly*, n.s. 9 (1959), 294–309; and "Solstices, Equinoxes and the Presocratics," in *Journal of Hellenic Studies*, 86 (1966), 26–40; J. L. E. Dreyer, *A History of the Planetary Systems from Thales to Kepler* (Cambridge, 1906), repr. as *A History of Astronomy from Thales to Kepler* (New York, 1953); W. K. C. Guthrie, *A History of Greek Philosophy*, I (Cambridge, 1962); T. L. Heath, *Aristarchus of Samos* (Oxford, 1913); and *Greek Mathematics*, I (Oxford, 1921); U. Hölscher, "Anaximander und die Anfänge der Philosophie," in *Hermes*, 81 (1953), 257–277, 385–417; repr. in English in Allen and Furley, *Studies in Presocratic Philosophy*, I (New York, 1970), 281–322; C. H. Kahn, "On Early Greek Astronomy," in *Journal of Hellenic Studies*, 90 (1970), 99–116; G. S. Kirk and J. E. Raven, *The Presocratic Philosophers* (Cambridge, 1957); O. Neugebauer, "The History of Ancient Astronomy, Problems and Methods," in *Journal of Near Eastern Studies*, 4 (1945), 1–38; *The Exact Sciences in Antiquity* (Princeton, 1952; 2nd ed., Providence, R.I., 1957); and "The Survival of Babylonian Methods in the Exact Sciences of Antiquity and Middle Ages," in *Proceedings of the American Philosophical Society*, 107 (1963), 528–535; and B. L. van der Waerden, *Science Awakening*, Arnold Dresden, trans. (Groningen, 1954).

JAMES LONGRIGG

THAN, KÁROLY (*b*. Óbecse, Hungary [now Bečej, Yugoslavia], 20 December 1834; *d*. Budapest, Hungary, 5 July 1908), *chemistry*.

Than was the third son of János Than, an estate manager, and Ottilia Petény; his elder brother, Mór Than, became a well-known painter. He attended several secondary schools, but interrupted his education in 1849, when he was fifteen, to enlist in the Hungarian army during the Hungarian war of independence. At the end of the war, later that same year, he returned home to find his family ruined financially; he therefore worked in a number of pharmacies while completing his secondary studies. A scholarship permitted him to study pharmacy at the University of Vienna, where he took the doctorate in 1858 under Redtenbacher, who engaged him as his personal assistant. Than then received a government subsidy which allowed him to undertake a long educational trip, during which he studied with Bunsen in Heidelberg and with Wurtz in Paris. In 1859 he returned to Vienna to work in Redtenbacher's laboratory.

By 1860 the political situation within the Austrian empire had changed to a degree that Hungarian could be reintroduced as the language of instruction in Hungarian universities. The incumbent professor of chemistry at the University of Budapest, the Austrian Theodor Wertheim, did not have full command of the language, and left to teach at Graz; in 1860 Than was called to replace him. Than also served as director of the chemistry laboratory, and built it into a significant institution. He was active in all the Hungarian scientific organizations, and was vice-president of the Hungarian Academy of Sciences.

Than's own research embraced many fields of chemistry. He introduced the use of potassium bicarbonate (1860) and potassium biiodate (Than's salt, 1890) as standard titrimetric substances in volumetric analysis, and suggested, even before the publication of Arrhenius' ionic theory, that analytical results be reported as carefully determined atom groupings, rather than according to their salts. He discovered carbonyl sulfide (1867), determined the precise vapor density of hydrochloric acid, defined the concept of the molecular volume of a gas (1887), and demonstrated that the anomalous vapor density of ammonium chloride is the result of the thermal dissociation of that compound. His two-volume textbook on general chemistry, *A kisérleti chemia elemei* (1897–1906), was one of the first works to be based on the concepts of physical chemistry.

In 1872 Than married Ervina Kleinschmidt; they had five children. He was created a baron in 1908, shortly before his death.

BIBLIOGRAPHY

I. ORIGINAL WORKS. *A kisérleti chemia elemei* ("Detailed Experimental Chemistry"), 2 vols. (Budapest, 1897–1906). For other publications, see Poggendorff, III, pt. 2, 1333; IV, pt. 2, 1487; V, 1248.

II. SECONDARY LITERATURE. The most complete biography is F. Szabadváry, *Than Károly* (Budapest, 1971), with complete list of Than's writings and a portrait. For discussions of Than and his work, see J. R. Partington, *A Textbook of Inorganic Chemistry*, 6th ed. (London, 1950), 115, 634; F. Szabadváry, *History of Analytical Chemistry* (Oxford, 1966), 252–254; Z. Szökefalvi-Nagy and F. Szabadváry, "Ein Vorschlag zur Darstellung der Analyseergebnisse in 'Ionenform' schon vor der Ausarbeitung der Ionentheorie," in *Talanta*, **13** (1966), 503–506.

F. SZABADVÁRY

THAXTER, ROLAND (*b.* Newtonville [part of Newton], Massachusetts, 28 August 1858; *d.* Cambridge, Massachusetts, 22 April 1932), *cryptogamic botany.*

Thaxter was the son of Levi Lincoln Thaxter and Celia Laighton. On his father's side he was descended from Thomas Thaxter, an Englishman who settled at Hingham, Massachusetts, in 1638. His mother established a considerable reputation in the field of literature by her poems; and his father also was active in literary studies, being an authority on the life and works of Browning. Thaxter attended Boston Latin School and entered Harvard in 1878, graduating with the A.B. in 1882

and the A.M. and Ph.D. in 1888. His chosen field of study was natural history. At the Harvard Graduate School of Arts and Sciences he was able to concentrate on cryptogamic botany under the direction of W. G. Farlow, and from 1886 to 1888 he served as the latter's assistant.

From 1888 to 1891 Thaxter was mycologist at the Connecticut Agricultural Experiment Station; but since his inclination was strongly toward pure, as opposed to applied, research, he was happy to return to Harvard in 1891, at Farlow's invitation, as assistant professor of cryptogamic botany. He became full professor in 1901 and assumed complete responsibility for teaching and research in that field. In 1919, on the death of Farlow, he was nominated professor emeritus and honorary curator of the cryptogamic herbarium, positions which he held until his death.

A Unitarian, Thaxter was married in 1887 to Mabel Gray Freeman of Springfield, Massachusetts, by whom he had four children. Despite poor health, he made several extended study journeys to the West Indies and southernmost South America. He was a member or fellow of numerous learned societies, both American and foreign, and was president of the Botanical Society of America in 1909.

Thaxter's published contributions in mycology were characterized by meticulous accuracy and are classics in their field. His name is associated chiefly with research on the little-known group of entomogenous fungi, which culminated in the publication of a five-volume monograph on the Laboulbeniaceae, a unique and isolated family of fungi that occur as minute parasites on the integuments of various insects; the numerous plates illustrating this treatise, exquisite in their execution, were all by Thaxter. He also published extensively on other groups of fungi, notably the Phycomycetes, and on the hitherto unrecognized assemblage of bacterialike organisms known as the Myxobacteriaceae.

Thaxter's work has had a profound and lasting influence on the development of mycology and of cryptogamic botany generally.

BIBLIOGRAPHY

Thaxter's principal work, *Contribution Towards a Monograph of the Laboulbeniaceae*, was published in 5 pts. by the American Academy of Arts and Sciences (1896–1931).

On his life and work, see the notice by G. P. Clinton, in *Biographical Memoirs. National Academy of Sci-*

ences, **17** (1937), 55–64, with comprehensive bibliography of Thaxter's publications between 1875 and 1931.

I. M. LAMB

THAYER, WILLIAM SYDNEY (*b.* Milton, Massachusetts, 23 June 1864; *d.* Washington, D.C., 10 December 1932), *medicine*.

Thayer, known as "Billy" to his friends and medical colleagues, was the eldest of four children of a prominent New England family. His father, James B. Thayer, was professor of law at Harvard; his mother, Sophia Ripley, was a granddaughter of Gamaliel Bradford and cousin of Ralph Waldo Emerson. His younger brother, Ezra, became dean of the Harvard Law School.

Thayer entered Harvard University at age sixteen, graduated with the B.A. and Phi Beta Kappa honors in 1885, and began medical studies at the Harvard Medical School the same fall. While there, he was particularly stimulated by the professor of pathology, Reginald Heber Fitz, the author of a classic study of appendicitis, who combined clinical and laboratory work, a model Thayer followed in his own career. After receiving his medical degree in 1889, Thayer continued his studies at the principal pathological laboratories of Berlin and Vienna. He returned to Boston in 1890; engaged in private practice very briefly; and later that year accepted a position as resident physician at the Johns Hopkins Hospital, opened the year before. Thayer was recommended for this position by a Harvard classmate, the surgeon J. M. T. Finney, who had preceded him to Baltimore.

At Hopkins, where Thayer spent his entire medical career except for wartime service, he immediately came under the influence and tutelage of William Osler, one of the most impressive and forceful medical teachers of his era. Thayer's solid grounding in the laboratory study of disease fitted well into the Hopkins system, for Osler himself had been a pathologist before turning to clinical medicine. Thayer did most of his important investigative work early in his Baltimore years. Using techniques such as blood staining, which he had learned from Paul Ehrlich in Germany, Thayer investigated a number of diseases and the blood cell's response to them.

The two most prevalent diseases that Thayer encountered in the wards of the Johns Hopkins Hospital when he began his duties were malaria and typhoid fever. Since Civil War days, diagnoses had been confused by the supposed existence of typhomalarial fever. Thayer, fresh from the German laboratories where Virchow, Ehrlich, and others stressed the importance of microscopic study of stained tissues, put these new techniques to good use in an extensive study of malaria in Baltimore. With a colleague from the medical clinic, John Hewetson, Thayer published *The Malarial Fevers of Baltimore* (1895), an analysis of 616 cases of proven malaria seen at the Hopkins in the first five years of the hospital's existence. Besides being an extensive review of the literature and a historical summary of the existing knowledge of malaria, the work clearly differentiated the characteristic fever cycles associated with the distinct species of malaria parasites. Two years later Thayer published *Lectures on the Malarial Fevers*, which quickly became one of the standard works on the subject. He chided his fellow American physicians for being "lamentably backward" in appreciating the advances that had been made since Laveran's discovery of the malaria parasite in 1880.

Under Osler's influence, Thayer slowly changed the emphasis of his work from the science of medicine to its art, from the laboratory to the bedside.

Upon Osler's departure for Oxford in 1905, Thayer was advanced to the position of professor of clinical medicine; but L. F. Barker was chosen to succeed Osler. Both men continued to work together amicably, however; and when Harvard offered Thayer a chair in 1912, he declined. He was a superb diagnostician and clinical teacher, always urging his students to make full use of their senses but not to neglect the laboratory aids to diagnosis. In the first two decades of the twentieth century, he wrote a number of important papers on aspects of cardiology, such as heart blocks, arteriosclerosis, cardiac murmurs, and the third heart sound. He also inspired a number of students and house officers in their work; many later became famous in their own right. For many years Thayer and the professor of pathology, W. G. MacCallum, conducted the clinicopathologic conference attended by most students and house officers.

In 1917, Thayer participated in the Red Cross mission to Russia to study health conditions and to determine what aid was necessary. The task was the more arduous because Thayer had to leave behind his ailing wife, Susan C. Thayer, who died while he was away. They had been married in 1902, and their only child had died in early infancy. Despite his personal anguish, Thayer continued to respond to what he felt was his duty. General Pershing made him director of general medicine for the A.E.F. in 1918. Thayer returned to Hopkins the following year, becoming professor of medicine and physician-in-chief to the Johns Hopkins Hospital. He was named professor emeritus in 1921, but continued his large consulting practice in the Baltimore-Washington area.

Thayer was known as a clinician's clinician, widely sought out for difficult diagnoses, and much respected for his teaching abilities. He was adept in the use of language, his own as well as French, German, and Russian, and urged his students to learn foreign languages. He was a literary man, widely read, and published a volume of poetry. He was known for his dapper dress and the ever-present flower in his lapel. Thayer's greatest influence, like that of Osler before him, was experienced by the many students who learned the art of medicine from him at the bedside. He received numerous foreign medals and citations and honorary degrees. He was also a member of the Board of Overseers of Harvard University.

BIBLIOGRAPHY

I. ORIGINAL WORKS. Thayer's bibliography includes nearly 180 articles on a great variety of clinical subjects. His three books on medical subjects deserve special notice: *The Malarial Fevers of Baltimore* (Baltimore, 1895), written with John Hewetson; *Lectures on Malarial Fevers* (New York, 1897); and *Osler and Other Papers* (Baltimore, 1931).

II. SECONDARY LITERATURE. The only book-length biography is Edith G. Reid, *The Life and Convictions of William Sydney Thayer* (London, 1936), a somewhat gushing portrayal but useful for Thayer the man. For Thayer the physician see especially "The Thayer Memorial Exercises . . .," in *Johns Hopkins Hospital Bulletin*, **55** (1934), 201–219. See also L. F. Barker, "William Sydney Thayer," in *Science*, **76** (1932), 617–619; and Henry M. Thomas, Jr., "Dr. Thayer," in *Johns Hopkins Hospital Bulletin*, **54** (1934), 211–215; and "William Sydney Thayer, Distinguished Physician and Teacher," *ibid.*, **109** (1961), 61–65. Three books not specifically about Thayer are nevertheless important sources: Alan Chesney, *The Johns Hopkins Hospital and the Johns Hopkins University School of Medicine*, I *1867–1893* (Baltimore, 1943), pp. 170–172; Harvey Cushing, *The Life of Sir William Osler*, 2 vols. (London, 1925), *passim*; and J. M. T. Finney, *A Surgeon's Life* (New York, 1940), 309–315.

GERT H. BRIEGER

THEAETETUS (*b.* Athens, *ca.* 417 B.C.; *d.* Athens, 369 B.C.), *mathematics*.

The son of Euphronius of Sunium, Theaetetus studied under Theodorus of Cyrene and at the Academy with Plato. Although no writing of his has survived, Theaetetus had a major influence in the development of Greek mathematics. His contributions to the theory of irrational quantities and the construction of the regular solids are particularly recorded; and he probably devised a general theory of proportion – applicable to incommensurable and to commensurable magnitudes – before the theory developed by Eudoxus and set out in book V of Euclid's *Elements*.

The *Suda* lexicon has two entries[1] under the name Theaetetus:

"Theaetetus, of Athens, astronomer, philosopher, disciple of Socrates, taught at Heraclea. He was the first to write on (or construct) the so-called five solids. He lived after the Peloponnesian war."

"Theaetetus, of Heraclea in Pontus, philosopher, a pupil of Plato."

Some have supposed that these notices refer to the same person, but it is more probable, as G. J. Allman[2] conjectures, that the second Theaetetus was a son or other relative of the first sent by him while teaching at Heraclea to study at the Academy in his native city.

Plato clearly regarded Theaetetus with a respect and admiration second only to that which he felt for Socrates. He made him a principal character in two dialogues, the eponymous *Theaetetus* and the *Sophist*; and it is from the former dialogue that what we know about the life of Theaetetus is chiefly derived.[3] In the dialogue Euclid of Megara gets a servant boy to read to his friend Terpsion a discussion between Socrates, Theodorus, and Theaetetus that Plato recorded soon after it took place on the day that Socrates faced his accusers, that is, in 399 B.C. Since Theaetetus is there referred to as a μειράκιον ("a youth"), it is implied that he was an adolescent, say eighteen years old, that is, he was born about 417 B.C.[4] His father, we are told, left a large fortune, which was squandered by trustees; but this did not prevent Theaetetus from being a liberal giver. Although Theaetetus was given the rare Greek compliment of being καλος τε καὶ ἀγαθός ("a thorough gentleman"), it was the beauty of his mind rather than of his body that impressed his compatriots; for, like Socrates, he had a snub nose and protruding eyes. Among the many young men with whom Theodorus had been acquainted, he had never found one so marvelously gifted; the lad's researches were like a stream of oil flowing without sound. Socrates predicted that Theaetetus would become notable if he came to full years. In the preface to the dialogue Euclid relates how he had just seen Theaetetus being carried in a dying condition from the camp at Corinth to Athens; not only had he been wounded in action, after acquitting himself gallantly, but he had contracted dysentery. This would be in the year 369 B.C., for the only other year in that century in which Athens and Corinth were at war, 394 B.C., would hardly allow time for Theaetetus' manifold accomplishments.[5]

The *Theaetetus* is devoted to the problem of knowledge; and the *Sophist*, apart from a method of definition, to the meaning of nonbeing. Although Theaetetus plays a major part in both discussions, there is no reason to think that he was a philosopher in the usual sense of the word. Plato merely used him as a vehicle for thoughts that he wanted expressed. That the two *Suda* passages use the term "philosopher" proves nothing, since the lexicon regularly calls mathematicians philosophers.[6]

In the summary of the early history of Greek geometry given by Proclus, and probably taken from Eudemus, Theaetetus is mentioned along with Leodamas of Thasos and Archytas of Tarentum as having increased theorems and made an advance toward a more scientific grouping,[7] the zeal for which is well shown in the mathematical passage that Plato introduces into the *Theaetetus*.[8] In this passage Theaetetus first relates how Theodorus demonstrated to him and the younger Socrates (a namesake of the philosopher) in each separate case that $\sqrt{3}, \sqrt{5} \cdots \sqrt{17}$ is a surd. He adds: "Since the number of roots[9] seemed to be infinite, it occurred to us to try to gather them together under one name by which we could call all the roots." Accordingly Theaetetus and the younger Socrates divided all numbers into two classes. A number that could be formed by multiplying equal factors they likened to a square and called "square and equilateral." The other numbers—which could not be formed by multiplying equal factors, but only a greater by a less, or a less by a greater—they likened to an oblong and called "oblong numbers." The lines forming the sides of equilateral numbers they called "lengths," and the lines forming oblong numbers they called "roots." "And similarly," concluded the Theaetetus of the dialogue, "for solids," which can only mean that they attempted a similar classification of cube roots.

The classification may now seem trivial, but the discovery of the irrational was a fairly recent matter[10] and involved a complete recasting of Greek mathematics; and Theaetetus was still only a young man. His more mature work on the subject is recorded in a commentary on the tenth book of Euclid's *Elements*, which has survived only in Arabic and is generally identified with the commentary that Pappus is known to have written. In the introduction to this commentary it is stated:[11]

The aim of Book X of Euclid's treatise on the Elements is to investigate the commensurable and the incommensurable, the rational and irrational continuous quantities. This science had its origin in the school of Pythagoras, but underwent an important development at the hands of the Athenian, Theaetetus, who is justly admired for his natural aptitude in this as in other branches of mathematics. One of the most gifted of men, he patiently pursued the investigation of the truth contained in these branches of science, as Plato bears witness in the book which he called after him, and was in my opinion the chief means of establishing exact distinctions and irrefutable proofs with respect to the above-mentioned quantities. For although later the great Apollonius, whose genius for mathematics was of the highest possible order, added some remarkable species of these after much laborious application, it was nevertheless Theaetetus who distinguished the roots which are commensurable in length from those which are incommensurable, and who divided the more generally known irrational lines according to the different means, assigning the medial line to geometry, the binomial to arithmetic, and the apotome to harmony, as is stated by Eudemus the Peripatetic.

The last sentence gives the key to the achievement of Theaetetus in this field. He laid the foundation of the elaborate classification of irrationals, which is found in Euclid's tenth book; and in particular Theaetetus discovered, and presumably named, the medial, binomial, and apotome. The medial is formed by the product of two magnitudes, the binomial ("of two names") by the sum of two magnitudes, and the apotome (implying that something has been cut off) by the difference of two magnitudes. It is easy to see the correlation between the medial and the geometric mean, for the geometric mean between two irrational magnitudes,[12] a, b, is $\sqrt{ab}$ and is medial. It is also easy to see the correlation between the binomial and the arithmetic mean, for the arithmetic mean between a, b, is $(\frac{1}{2}a + \frac{1}{2}b)$; and this is a binomial. It is not so easy to see the connection between the apotome and the harmonic mean; but a clue is given in the second part of the work, where the commentator returns to the achievement of Theaetetus and observes that if the rectangle contained by two lines is a medial, and one of the sides is a binomial, the other side is an apotome. This in turn recalls Euclid,

Elements X.112, and amounts to saying that the harmonic mean between *a, b*, that is, $2ab/(a + b)$, can be expressed as

$$\frac{2ab}{a^2 - b^2} \cdot (a - b).$$

This leads to the question how much of Euclid's tenth book is due to Theaetetus. After a close examination, B. L. van der Waerden concluded that "The entire book is the work of Theaetetus."[13] There are several reasons, however, for preferring to believe that Theaetetus merely identified the medial, binomial, and apotome lines, correlating them with the three means, as the Arabic commentary says, and that the addition of ten other species of irrationals, making thirteen in all, or twenty-five when the binomials and apotomes are further subdivided, is the work of Euclid himself. A scholium to the fundamental proposition X.9 ("the squares on straight lines commensurable in length have to one another the ratio which a square number has to a square number . . .") runs as follows: "This theorem is the discovery of Theaetetus, and Plato recalls it in the *Theaetetus*, but there it is related to particular cases, here treated generally."[14] This would be a pointless remark if Theaetetus were the author of the whole book.

The careful distinction made in the "Eudemian summary" between Euclid's treatment of Eudoxus and Theaetetus is also relevant. Euclid, says the author, "put together the elements, arranging in order many of Eudoxus' theorems, perfecting many of Theaetetus', and bringing to irrefutable demonstration the things which had been only loosely proved by his predecessors."[15] The implication would seem to be that book V is almost entirely the discovery of Eudoxus save in its arrangement, but book X is partly due to Theaetetus and partly to Euclid himself. The strongest argument for believing that Theaetetus had an almost complete knowledge of the Euclidean theory of irrationals is that the correlation of the apotome with the harmonic mean implies a knowledge of book X.112; but it is relevant that the genuine text of Euclid probably stops at book X.111 with the list of the thirteen irrational straight lines.[16]

A related question is the extent to which the influence of Theaetetus can be seen in the arithmetical books of Euclid's *Elements*, VII–IX. Euclid X.9 depends on VIII.11 ("Between two square numbers there is one mean proportional number . . ."), and VIII.11 depends on VII.17 and VII.18 (in modern notation, $ab : ac = b : c$, and $a : b = ac : bc$). H. G. Zeuthen has argued[17] that

these propositions are an inseparable part of a whole theory established in book VII and in the early part of book VIII, and that this theory must be due to Theaetetus with the object of laying a sound basis for his treatment of irrationals. It is clear, however, as T. L. Heath has pointed out,[18] that before Theaetetus both Hippocrates and Archytas must have known propositions and definitions corresponding to these in books VII and VIII; and there is no reason to abandon the traditional view that the Pythagoreans had a numerical theory of proportion that was taken over by Euclid in his arithmetical books. Theaetetus merely made use of an existing body of knowledge.

Theaetetus' work on irrationals is closely related to the two other main contributions to mathematics attributed to him. The only use made of book X in the subsequent books of Euclid's *Elements* is to express the sides of the regular solids inscribed in a sphere in terms of the diameter. In the case of the pyramid, the octahedron, and the cube, the length of the side is actually determined; in the case of the icosahedron, it is shown to be a minor; and in the case of the dodecahedron, to be an apotome. It is therefore significant that in the passage from the *Suda* lexicon (cited above) Theaetetus is credited as the first to "write upon" or "construct" the so-called five solids (πρῶτος δὲ τὰ πέντε καλούμενα στερεὰ ἔγραψε). It is also significant that at the end of the mathematical passage in the *Theaetetus* he says that he and his companion proceeded to deal with solids in the same way as with squares and oblongs in the plane. Probably on the authority of Theophrastus, Aëtius[19] attributed the discovery of the five regular solids to the Pythagoreans; and Proclus[20] actually attributes to Pythagoras himself the "putting together" (σύστασις) of the "cosmic figures." They are called "cosmic" because of Plato's use of them in the *Timaeus* to build the universe;[21] and no doubt the σύστασις is to be understood as a "putting together" of triangles, squares, and pentagons in order to make solid angles as in that dialogue rather than in the sense of a formal construction. Theaetetus was probably the first to give a theoretical construction for all the five regular solids and to show how to inscribe them in a sphere. A scholium to Euclid, *Elements* XIII, actually attributes to Theaetetus rather than to the Pythagoreans the discovery of the octahedron and icosahedron.[22] On the surface this is puzzling, since the octahedron is a more elementary figure than the dodecahedron, which requires a knowledge of the pentagon; but many objects of dodecahedral form have been found from days much earlier than Pythagoras,[23] and the Pythagorean Hippasus is known to have written on "the construction of the sphere from the twelve pentagons."[24] (It would be in this work, if not earlier, that he would have encountered the irrational, and for his impiety in revealing it, he was drowned at sea.) If the Pythagoreans knew the dodecahedron, almost certainly they knew also the octahedron and probably the icosahedron; and the scholium quoted above may be discounted. The achievement of Theaetetus was to give a complete theoretical construction of all five regular solids such as we find in Euclid, *Elements* XIII; and Theaetetus must be regarded as the main source of the book, although Euclid no doubt arranged the materials in his own impeccable way and put the finishing touches.[25]

The theory of irrationals is also linked with that of proportionals. When the irrational was discovered, it involved a recasting of the Pythagorean theory of proportion, which depended on taking aliquot parts, and which consequently was applicable only to rational numbers, in a more general form applicable also to incommensurable magnitudes. Such a general theory was found by Eudoxus and is embodied in Euclid, *Elements* V. But in 1933[26] Oskar Becker gave a new interpretation of an obscure passage in Aristotle's *Topics*.[27] He suggested that the theory of proportion had already been recast in a highly ingenious form; and if so, the indication is that it was so recast by Eudoxus' older contemporary Theaetetus.

In the passage under discussion Aristotle observes that in mathematics some things are not easily proved for lack of a definition—for example, that a straight line parallel to two of the sides of a parallelogram divides the other two sides and the area in the same ratio; but if the definition is given, it becomes immediately clear, "for the areas have the same ἀνταναίρεσις as the sides, and this is the definition of the same ratio." What does the Greek word mean? The basic meaning is "a taking away," and the older commentators up to Heath

and the Oxford translation supposed that it meant "a taking away of the same fraction." In the figure *EF* is the straight line parallel to the sides *AB*, *DC* of the parallelogram *ABCD*, and *AE*, *BF* are the same parts of *AD*, *BC* respectively as the parallelogram *ABFE* is of the parallelogram *ABCD*. This would be in accordance with the Pythagorean theory of proportion, and the passage would contain nothing significant. But Becker drew attention to the comment by Alexander of Aphrodisias on this passage; he uses the word ἀνθυφαίρεσις and observes that this is what Aristotle means by ἀνταναίρεσις. This might not in itself prove very much—the meaning could still be much the same—if it were not, as Becker also noted, that Euclid, although he does not employ the noun ἀνθυφαίρεσις, does in four places[29] use the verb, ἀνθυφαιρεῖν, and—this is the really significant fact—uses it to describe the process of finding the greatest common measure between two magnitudes. In this process the lesser magnitude is subtracted from the greater as many times as possible until a magnitude smaller than itself is left, and then the difference is subtracted as many times as possible from the lesser until a difference smaller than itself is left, and so on continually (ἀνθυφαιρουμένου δὲ ἀεὶ τοῦ ἐλάσσονος ἀπὸ τοῦ μείζονος). In the case of commensurable magnitudes the process comes to an end after a finite number of steps, but in the case of incommensurable magnitudes the process never comes to an end. A mathematician as acute as Theaetetus would realize that this could be made a test of commensurability (as it is in Euclid, *Elements* X.2) and that by adopting a definition of proportion based on this test he could have a theory of proportion applicable to commensurable no less than incommensurable magnitudes.[30]

It is possible that such a general theory was evolved before Eudoxus by some person other than Theaetetus, but in view of Theaetetus' known competence and his interest in irrationals, he is the most likely author. The attribution becomes even more credible if Zeuthen's explanation of how Theodorus proved the square roots of $\sqrt{3}$, $\sqrt{5}$ . . . $\sqrt{17}$ to be irrational is accepted (see the article on Theodorus of Cyrene); for according to his conjecture Theodorus used this method in each particular case, and Theodorus was the teacher of Theaetetus. Although there is no direct evidence that Theaetetus worked out such a pre-Eudoxan theory of proportion, the presumption in favor is strong; and it has convinced all recent commentators.

It is not known whether Theaetetus made any

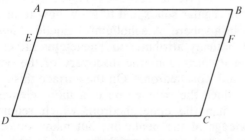

FIGURE 1

discoveries outside these three fields. In the "Eudemian summary" Proclus says:[31] "Hermotimus of Colophon advanced farther the investigations begun by Eudoxus and Theaetetus; he discovered many propositions in the elements and compiled some portion of the theory of loci." While it is clear that Theaetetus studied mathematics under Theodorus, it is uncertain whether he did so at Cyrene or at Athens. It may be accepted that at some time he taught in Heraclea, and he may have been the teacher of Heraclides Ponticus.[32]

NOTES

1. *Suda Lexicon*, Ada Adler, ed., I, pt. 2 (Leipzig, 1931), Θ 93 and 94, p. 689.6–9.

2. G. J. Allman, "Theaetetus," in *Hermathena*, **6** (1887), 269–278, repr. in *Greek Geometry From Thales to Euclid* (London–Dublin, 1889), 206–215.

3. Plato, *Theaetetus, Platonis opera*, J. Burnet, ed., I (Oxford, 1899), 142a–148b; Plato, Loeb Classical Library, H. N. Fowler, ed., VII (London–Cambridge, Mass., 1921; repr. 1967), 6.1–27.24.

4. The birth of Theaetetus has usually been placed in 415 B.C., or even as late as 413, which would make him not more than sixteen years old in 399; but the instances given in H. G. Liddell, R. Scott, and H. Stuart Jones, *A Greek-English Lexicon* (Oxford, 1940), *s.vv.* μειράκιον and ἔφηβος, show clearly that a μειράκιον would not be younger than eighteen and might be nearly as old as twenty-one. A sentence in the *Chronicle* of Eusebius—preserved in the Armenian and in Jerome's Latin version—*Sancti Hieronymi interpretatio chronicae Eusebii Pamphili*, in *Patrologia Latina*, J.-P. Migne, ed., vol. XXVII = S. Hieronymi, vol. VIII (Paris, 1846), cols. 453–454—which would place the central point of Theaetetus' activity in the third year of the 85th Olympiad (438 B.C.)—must be dismissed as an error. Eusebius' statement is repeated by George Syncellus, *Corpus scriptorum historiae Byzantinae*, B. G. Niebuhr, ed., pt. 7.1; *Georgius Syncellus et Nicephorus*, G. Dindorff, ed., I (Bonn, 1829), p. 471.9.

5. It is one of Eva Sachs's principal achievements in her pioneering inaugural dissertation, *De Theaeteto Atheniensi mathematico* (Berlin, 1914), 16–40, to have established this point irrefutably against E. Zeller and others.

6. But Malcolm S. Brown, in "Theaetetus: Knowledge as Continued Learning," in *Journal of the History of Philosophy*, **7** (1969), 359, maintains that Theaetetus "influenced both the course of mathematics and that of philosophy." Brown quotes Sachs, *op. cit.*, p. 69, in support: "Ille re vera philosophus fuit perfectus"; but it is doubtful if Sachs meant the Latin word to imply that Theaetetus was a metaphysician. Brown seizes on the statement of Theaetetus at the beginning of his conversation with Socrates: "When I make a mistake you will correct me" (*Theaetetus*, 146c.). Brown sees in the mathematical work of Theaetetus a process of successive approximations, which can be construed as "containing errors which are being corrected." He holds also that there is an epistemological analogue, "a well-directed discussion of opinions which, even if unsuccessful in arriving at a final answer, would nevertheless permit of an improvement (even an indefinite improvement) of opinion"; and he believes that in this dialogue at least Plato yielded somewhat to the suggestion of Theaetetus that "knowledge is continued learning" (p. 379).

7. Proclus, *In primum Euclidis*, G. Friedlein, ed. (Leipzig, 1873; repr., Hildesheim, 1967), p. 66.14–18; English trans., Glenn R. Morrow, *Proclus: A Commentary on the First Book of Euclid's Elements* (Princeton, 1970), p. 54.11–14.

8. Plato, *Theaetetus, Platonis opera*, J. Burnet, ed., I (Oxford, 1899), 147c–148b; Plato, Loeb Classical Library, H. N. Fowler, ed., VII (London–Cambridge, Mass., 1921; repr., 1967), pp. 24.9–27.24.

9. The Greek word is δυνάμεις, which at a latter date could only mean "squares"; but here its meaning would appear to be "roots," and we can only suppose that at this early stage in Greek mathematics the terminology had not become fixed. It is not necessary with Paul Tannery ("Sur la langue mathématique de Platon," in *Annales de la Faculté des lettres de Bordeaux*, 1 [1884], 96, repr. in *Mémoires scientifiques*, **2** [1912], 92) to alter δύναμις without any MS authority to δυναμένη, the later technical expression for a square root. For a very full discussion of a different interpretation, See Árpád Szabó, *Anfänge der griechischen Mathematik* (Munich–Vienna, 1969), 14–22, 43–57. Szabó holds that δύναμις means *Quadratwert eines Rechtecks* ("square value of a rectangle"), that is, the square equivalent in area to a rectangle. This interpretation has attractions, but the fact that Plato categorically describes δυνάμεις as γραμμαί, "lines," and sets δύναμις in opposition to μῆκος, a rational length, seems fatal to it. But Szabó establishes that δύναμις cannot be power in general.

10. But not so recently as the time of Plato himself. Even if the Athenian stranger in the *Laws* is identified with Plato, it is reading too much into his words αὐτὸς ἀκούσας ὀψέ ποτε τὸ περὶ ταῦτα ἡμῶν πάθος (819d 5–6) to suppose that the irrational was not discovered until the fourth century B.C. Likewise the statement in the "Eudemian summary," in Proclus, *op. cit.*, p. 65.19–21, that Pythagoras "discovered the matter of the irrationals" (τὴν τῶν ἀλόγων πραγματείαν . . . ἀνεῦρεν) must be wrong, and there is almost certainly a textual error—ἀλόγων for ἀναλόγων ("proportionals"). The existence of irrational magnitudes was almost certainly discovered, as Greek tradition asserted, by Hippasus of Metapontum in the middle of the fifth century B.C. The best discussion of the date is Kurt von Fritz, "The Discovery of Incommensurability by Hippasus of Metapontum," in *Studies in Presocratic Philosophy*, David J. Furley and R. E. Allen, eds., I (London–New York, 1970), 382–412. For an attempt to show that the discovery was made in the closing years of the fifth century, see Eric Frank, *Platon und die sogennanten Pythagoreer* (Halle, 1923). Árpád Szabó, *op. cit.*, pp. 60–69, 111–118, 238, seeks to show that the irrational was discovered in the study of mean proportionals as opposed to the prevailing theory that it arose from the study of diagonals of squares after the discovery of "Pythagoras' theorem."

11. The translation is based in the main on that of William Thomson, in William Thomson and Gustav Junge, *The Commentary of Pappus on Book X of Euclid's Elements* (Cambridge, Mass., 1930; repr., New York, 1968), 63; but his "powers" (that is, the squares), although a faithful rendering of the Arabic, has been modified, since "roots" appears to be the meaning. The ambiguity of the Greek δύναμις, before the terminology became fixed, is reflected in the Arabic.

12. It would be going beyond the evidence to attribute to Theaetetus the Euclidean notion (X, *Definition* 3) that a straight line may be rational but commensurable only in square with a rational straight line; that is, that if r is a rational straight line and m, n integers with m/n in its lowest terms not a square, then $\sqrt{m/n} \cdot r$ is rational. T. L. Heath observes, "It would appear that Euclid's terminology here differed as much from that of his predecessors as it does from ours," and he aptly cites the expression of Plato (following the Pythagoreans), in the *Republic* 546c 4–5: ἄρρητος διάμετρος τοῦ πεμπάδος ("the irrational diameter of five") for the diagonal of a square of side five units; that is,

for Plato, and presumably for Theaetetus, as for us, $\sqrt{50}$ is irrational, whereas Euclid would have called it "rational but commensurable in square only." Eva Sachs takes a contrary view, *Die fünf platonischen Körper*, p. 105, but without satisfactory reasons.

13. B. L. van der Waerden, *Science Awakening*, 2nd ed. (Groningen, 1956[?]), p. 172. In full, he writes: "Has the same Theaetetus who studied the medial, the binomial and the apotome, also defined and investigated the ten other irrationalities, or were those introduced later on? It seems to me that all of this is the work of one mathematician. For, the study of the 13 irrationalities is a unit. The same fundamental idea prevails throughout the book, the same methods of proof are applied in all cases. Propositions X.17 and 18 concerning the measurability of the roots of a quadratic equation precede the introduction of binomial and apotome, but these are not used until the higher irrationalities appear on the scene. The theory of the binomial and the apotome is almost inextricably interwoven with that of the 10 higher irrationals. Hence—the entire book is the work of Theaetetus." The conclusion does not follow. The unity may be due to Euclid himself, using some propositions already proved, adding refinements of his own, and welding the whole into one, as Proclus testifies. The division of irrationals into medial, binomial, and apotome can perfectly well be separated from the subdivisions into more complex irrationals. If it were true that X.17 and 18 are not used until after the introduction of the binomial and the apotome, this would prove nothing since they are in their correct logical position; and for that matter, the whole of book X is not used again until book XIII; but, in fact, X.18 is used in X.33, whereas the binomial is not introduced until X.36 and the apotome until X.73.

14. *Euclidis opera omnia*, J. L. Heiberg and H. Menge, eds., V (Leipzig, 1888), Scholium 62 in Elementorum Librum X, p. 450.16–18. There is good reason to believe that the scholiast is Proclus. See H. Knoche, *Untersuchungen über die neu aufgefundenen Scholien des Proklus Diadochus zu Euclids Elementen* (Herford, 1865), p. 24; and J. L. Heiberg, "Paralipomena zu Euclid," in *Hermes*, **38** (1903), p. 341.

15. Proclus, *op. cit.*, p. 68.7–10; Eng. trans. *op. cit.*, p. 56.19–23.

16. J. L. Heiberg gives conclusive reasons for bracketing propositions 112–115, in *Euclidis opera omnia*, J. L. Heiberg and H. Menge, eds., V, p. lxxxv, and concludes: "non dubito, quin hae quoque propositiones 112–115 e doctrina Apollonii promptae sint; nam antiquae sunt et bonae, hoc saltim constare putaverim, eas ab Euclide scriptas non esse."

17. H. G. Zeuthen, "Sur la constitution des livres arithmétiques des Eléments d'Euclide et leur rapport à la question de l'irrationalité," in *Oversigt over det Kongelige Danske Videnskabernes Selskabs Forhandlinger* (1910), 395–435.

18. Thomas Heath, *A History of Greek Mathematics*, I (however, from Oxford, 1921), 211.

19. Aëtius, *Placita*, II, 6, 5, in H. Diels, *Doxographi Graeci* (Berlin, 1879), p. 334; and *Die Fragmente der Vorsokratiker*, H. Diels and W. Kranz, eds., 6th ed., I (Dublin–Zurich, 1951; repr., 1969), p. 403.8–12.

20. Proclus, *op. cit.*, p. 65.20–21; Eng. trans., *op. cit.*, p. 53.5. Morrow translates the Greek word as "structure."

21. Plato, *Timaeus* 53C–55C; *Platonis opera*, J. Burnet, ed., IV (Oxford, 1915); Loeb Classical Library, *Plato, Timaeus etc.*, R. G. Bury, ed. (London–Cambridge, Mass., 1929; repr., 1966), pp. 126.16–134.4.

22. *Euclidis opera omnia*, J. L. Heiberg and H. Menge, eds., V (Leipzig, 1888), Scholium 1 in Elementorum Librum XIII, p. 654.1–10.

23. One, discovered in 1885 at Monte Loffa in the Colli Euganei near Padua, of Etruscan origin, is dated between 1000 and 500 B.C. (F. Lindemann, "Zur Geschichte der Polyeder

und der Zahlzeichen," in *Sitzungsberichte der Bayerischen Akademie der Wissenschaften zu München*, **26** (1897), 725.

24. Iamblichus, *De communi mathematica scientia* 25, N. Festa, ed. (Leipzig, 1891), 77.18–21; *De vita Pythagorica* 18.88, A. Nauck, ed. (Leipzig, 1884; repr., 1965).

25. In the course of a full discussion Eva Sachs, in *Die fünf platonischen Körper* (Berlin, 1917), asserts (p. 105) that the construction of the five solids in Euclid, *Elements*, XIII, 13–17, springs from Theaetetus. She approves H. Vogt, in *Bibliotheca mathematica*, **9**, 3rd ser. (1908–1909), p. 47, for controverting Paul Tannery, *La géométrie Grecque* (Paris, 1887), p. 101, who would ascribe the construction of the five solids to the Pythagoreans while leaving to Theaetetus the calculation of the relation of the sides to the radius of the circumscribing sphere: for how, she and Vogt ask, can the exact construction be accomplished without a prior knowledge of this relation? The question how much of Euclid's book XIII is due to Theaetetus is bound up with the difficult question how much, if any, is due to the Aristaeus who is mentioned by Hypsicles in the so-called *Elements*, Book XIV, J. L. Heiberg and H. Menge, eds., vol. V, p. 6.22–23, as the author of a book entitled *Comparison of the Five Figures*, and whether this Aristaeus is to be identified with Aristaeus the Elder, author of a formative book on solid loci, that is, conics. T. L. Heath, in *The Thirteen Books of Euclid's Elements*, III (Cambridge, 1908; 2nd ed., 1925; repr., New York, 1956), p. 439, following C. A. Bretschneider, *Die Geometrie und die Geometer vor Eukleides* (Leipzig, 1870), p. 171, took the view that "as Aristaeus's work was the newest and latest in which, before Euclid's time, this subject was treated, we have in Euclid XIII at least a partial recapitulation of the contents of the treatise of Aristaeus"; but Eva Sachs, *op. cit.*, p. 107, denies this conclusion.

26. Oskar Becker, "Eudoxos Studien I: Eine voreudoxische Proportionenlehre und ihre Spuren bei Aristoteles und Euklid," in *Quellen und Studien zur Geschichte der Mathematik, Astronomie und Physik*, **2B** (1933), 311–333. To some extent the theory had already been adumbrated independently by H. G. Zeuthen, "Hvorledes Mathematiken i Tiden fra Platon til Euklid," in *Kongelige Danske Videnskabernes Selskabs Skriften*, **5** (1915), 108, and E. J. Dijksterhuis, *De Elementen van Euclides, I* (Groningen, 1929), 71, as Becker himself recognizes in *Das mathematischen Denken der Antike* (Göttingen, 1957), p. 103, *n.* 25. Becker failed to convince T. L. Heath, *Mathematics in Aristotle* (Oxford, 1949; repr., 1970), 80–83, who in the absence of confirmatory evidence could "only regard Becker's article as a highly interesting speculation" (p. 83). It has also been criticized by K. Reidemeister, *Das exakte Denken der Griechen* (Hamburg, 1949), p. 22, and by Árpád Szabó, "Ein Beleg fur die voreudoxische Proportionlehre?" in *Archiv für Begriffsgeschichte*, **9** (1964), 151–171, and in his *Anfänge der griechischen Mathematik* (Munich–Vienna), 134–135, 180–181. The theory received support, however, from a Leiden dissertation by E. B. Plooij, *Euclid's Conception of Ratio as Criticized by Arabian Commentators* (Rotterdam, 1950). Becker rejected the criticisms in *Archiv für Begriffsgeschichte*, **4** (1959), p. 223, and adhered to his theory in his book *Grundlagen der Mathematik in geschichtlicher Entwicklung* (Bonn, 1954; 2nd ed., 1964). His theory has been wholeheartedly endorsed by B. L. van der Waerden, *Science Awakening*, 2nd ed. (Groningen, 1956 [?]), 175–179; by Kurt von Fritz in "The Discovery of Incommensurability by Hippasus of Metapontum," in *Studies in Presocratic Philosophy*, David J. Furley and R. E. Allen, eds., I (London–New York, 1970), 408–410, esp. note 87; but his statement that Heath "still called the definition 'incommensurable'" is unfair, since Heath said it was "'metaphysical' (as Barrow would say)," and in any case this was in *The Thirteen Books of Euclid's Elements*, II (Cambridge, 1908; 2nd ed., 1925; repr., New

York, 1956), p. 121, written before Becker's theory was enunciated; by Malcolm S. Brown, *op. cit.*, pp. 363–364; and by Wilbur Knorr, *The Evolution of the Euclidean Elements* (Dordrecht, 1975).

27. Aristotle, *Topics* VIII.3, 158B 29–159A 1.
28. Alexander of Aphrodisias, *Commentarium in Topica*, Strache and Wallies, eds., in *Commentaria in Aristotelem Graeca*, II (Berlin, 1891), 545.12–17.
29. Euclid, *Elements*, VII.1, VII.2, X.2, and X.3, J. L. Heiberg. ed., II (Leipzig, 1884), 188.13–15, 192.6–7; III (Leipzig, 1886), 12–14, 10.4–5; E. S. Stamatis, ed. (post J. L. Heiberg), II (Leipzig, 1970), 105.8–9, 107.3–4; III (Leipzig, 1972), 3.19–20, 5.8–9.
30. The Arabian commentator al-Māhānī (*fl. ca.* 860), followed by al-Nayrīzī (*fl. ca.* 897), dissatisfied with Euclid's definition, worked out for himself an "anthyphairetic" definition, as was recognized by E. B. Plooij, *op. cit.* For al-Nayrīzī, see *Anaritii in decem libros priores Elementorum Euclidis ex interpretatione Gherardi Cremonensis*, M. Curtze, ed., in *Euclidis opera omnia*, J. L. Heiberg and H. Menge, eds., *Supplementum*, pp. 157–160.
31. Proclus, *op. cit.*, p. 67.20–23; English trans., *op. cit.*, p. 56.9–12.
32. Eva Sachs, *De Theaeteto Atheniensi Mathematico*, p. 64, following Ulrich von Wilamowitz-Moellendorf.

BIBLIOGRAPHY

No original writing by Theaetetus has survived, even in quotation, although his work is undoubtedly embedded in Euclid, *Elements*, X and XIII.

Secondary literature includes G. J. Allman, "Theaetetus," in *Hermathena*, 6 (1887), 269–278, repr. in *Greek Geometry From Thales to Euclid* (London–Dublin, 1889), 206–215; Oskar Becker, "Eudoxos Studien I: Eine voreudoxische Proportionenlehre und ihre Spuren bei Aristoteles und Euklid," in *Quellen und Studien zur Geschichte der Mathematik, Astronomie und Physik*, 2B (1933), 311–333; *ibid.*, 3B (1934), 533–553, repr. in O. Becker, ed., *Zur Geschichte der griechischen Mathematik* (Darmstadt, 1965); in *Archiv für Begriffsgeschichte*, 4 (1959), 223; and in *Grundlagen der Mathematik in geschichtlicher Entwicklung* (Bonn, 1954; 2nd ed., 1964), 78–87; Malcolm S. Brown, "Theaetetus: Knowledge as Continued Learning," in *Journal of the History of Philosophy*, 7 (1969), 359–379; Kurt von Fritz, "The Discovery of Incommensurability by Hippasus of Metapontum," in *Annals of Mathematics*, 46 (1945), 242–264; "Platon, Theatet und die antike Mathematik," in *Philologus*, 87 (1932), 40–62, 136–178; and David J. Furley and R. E. Allen, eds., "The Discovery of Incommensurability by Hippasus of Metapontum," in *Studies in Presocratic Philosophy* I (London–New York, 1970), 382–412.

See also Thomas Heath, *A History of Greek Mathematics*, I (Oxford, 1921), 203–204, 209–212; Pauly-Wissowa, *Real-Encyclopädie der classischen Altertumswissenschaft*, 2nd ser., V, cols. 1351–1372; Eva Sachs, *De Theaeteto Atheniensi mathematico* (Inaugural diss., Berlin, 1914): *Die fünf platonischen Körper* (Berlin, 1917), 88–119; Árpád Szabó, "Ein Beleg für die voreudoxische Proportionenlehre?" in *Archiv für Begriffsgeschichte*, 9 (1964), 151–171; "Die Fruhge-

schichte der Theorie der Irrationalitaten," in *Anfänge der griechischen Mathematik*, pt. 1 (Munich–Vienna, 1969), 38–130; "Die voreuklidische Proportionlehre," *ibid.*, pt. 2, pp. 131–242; Heinrich Vogt, "Die Entdeckungsgeschichte des Irrationalen nach Plato und anderen Quellen des 4. Jahrhunderts," in *Bibliotheca mathematica*, 3 ser., 10 (1909–1910), 97–155; "Zur Entdeckungsgeschichte des Irrationalen," *ibid.*, 14 (1913–1914), 9–29; B. L. van der Waerden, *Ontwakende Wetenschap* (Groningen, 1950), also in English, Arnold Dresden, trans., *Science Awakening* (Groningen, 1954; 2nd ed., [?], 1956), 165–179; A. Wasserstein, "Theaetetus and the History of the Theory of Numbers," in *Classical Quarterly*, n.s. 8 (1958), 165–179; H. G. Zeuthen, "Notes sur l'histoire des mathématiques VIII; Sur la constitution des livres arithmétiques des Eléments d'Euclide et leur rapport à la question de l'irrationalité," in *Oversigt over det Kongelige Danske Videnskabernes Selskabs Forhandlinger* (1910), 395–435; "Sur les connaissances géométriques des Grecs avant la reforme platonicienne de la géométrie," *ibid.* (1913), 431–473; and "Sur l'origine historique de la connaissance des quantités irrationelles," *ibid.* (1915), 333–362; and Wilbur Knorr, *The Evolution of the Euclidean Elements* (Dordrecht, 1975), chs. 7, 8.

See also the Bibliography of the article on Theodorus of Cyrene.

IVOR BULMER-THOMAS

THEGE, MIKLÓS VON KONKOLY. See **Konkoly Thege, Miklós von.**

THEMISTIUS (*b.* Paphlagonia, A.D. 317 [?]; *d.* Constantinople, *ca.* 388), *philosophy, politics.*

Themistius is one of the most interesting representatives of the late Peripatetic school, being at the same time an outstanding Aristotelian scholar, a teacher of philosophy, an eloquent speaker, and an influential politician and diplomat. Some of his ideas are still vital, especially his doctrine of toleration and universal philanthropy. He was born about 320 (presumably in 317) in Paphlagonia, the country of his parents.[1] His father, Eugenius, was a teacher of philosophy, concerned mainly with Aristotle but also with Pythagoras, Plato, Zeno of Citium, and Epicurus. Themistius attended his father's lectures, probably at Constantinople.[2] He himself began teaching philosophy in 345.

As a philosopher Themistius followed in the footsteps of his father, adhering mainly to Aristotle without disregarding Plato. His definition of philosophy as a constant attempt to imitate God, so far as it is possible for man to do so, comes primarily from Plato's *Theaetetus*. His chief philosophical concern was not with logic or metaphysics but with ethics. The teaching of Themistius was very influential,[3] and many students came to Constantinople to attend his lectures. Aiming at the complete education of his students, he provided not only theoretical instruction but also practical preparation for the moral life. He wanted to make Aristotle understandable to everyone, not by an ordinary commentary, but by paraphrasing the texts of the Stagirite and summarizing the philosophical content. In connection with his lecturing, Themistius arranged meetings with the students and discussed particular problems with them. Presumably his paraphrases of Aristotle, as well as his commentaries on Plato, were written between 345 and 355. The paraphrases include the following: that of the *Posterior Analytics*, translated from Arabic into Latin by Gerard of Cremona; of the *Prior Analytics*, not preserved; of the *Physics*; of the *De anima*, translated into Latin by William of Moerbeke (22 November 1267); of the *De caelo* and the twelfth book of the *Metaphysics*, both preserved only in a Hebrew translation; and of the *Categories, Topics, De sensu, De generatione et corruptione*, and possibly of the *Nicomachean Ethics*, none of which has been preserved. The paraphrase of the *Parva naturalia*, attributed to Themistius, was written by Sophonias. None of the commentaries on Plato has been preserved. Many of the paraphrases were translated into Arabic: Themistius was frequently used and quoted by medieval Arabic philosophers. He also wrote some philosophical treatises: Περὶ ἀρετῆς, which has been preserved in a Syrian translation only, and Περὶ ψυχῆς, which is known through some fragments quoted by Stobaeus.[4] The Περὶ γήρως attributed to him is not authentic.

Themistius' political career started in 355, when he was appointed senator on 1 September. The Emperor Valens entrusted him with the education of his son; and Themistius was also appointed the tutor of Arcadius, the son of Theodosius I. He had close relations with the Emperor Julian, who with his help endeavored to revive the ancient Hellenic religion, although he intended to be tolerant toward the Christians. In 383–384 Themistius was *praefectus urbis* and *princeps senatus*. His speeches are closely connected with his political career: thirty-one of them have been completely preserved, two almost completely (*Orationes* XXIII, XXXIII), four are known only through fragments, and the content of three may be reconstructed—the Περὶ ἀρετῆς, his speech on toleration, and his "Epistula ad Julianum." The favorite topics of his speeches are philanthropy, liberty of conscience, the relation between politics and philosophy, the duties of the state, and the ideal of the statesman. The text of *Oratio* XII, entitled "De religionibus" and addressed to the Emperor Valens, is not authentic in the Dindorf edition: it was written by Andreas Dudith (1533–1589) of Breslau. As for *Oratio* XXVI, H. Kesters maintains that it was borrowed entirely from Antisthenes with only minor stylistic modifications.[5] Gregory of Nazianzus called Themistius his friend and praised him as the king of eloquence. Yet Themistius made no effort in his writings to be original, clinging always to classical thought and ancient wisdom, and remaining faithful to the traditional Hellenic religion.

NOTES

1. *Orationes*, Dindorf ed., II.33.28.
2. *Ibid.*, XX.295.3 ff.; XXXIV.460.18; XVII, 261.11–14.
3. See *ibid.*, "Oratio Constantii," 23.31–24.4.
4. Stobaeus, Wachmuth and Hense, eds., III, 468, IV, 530, and V, 1032, 1086–1092.
5. H. Kesters, *Plaidoyer d'un Socratique contre le Phèdre de Platon* (Louvain–Paris, 1959).

BIBLIOGRAPHY

I. ORIGINAL WORKS. Themistius' paraphrases are in *Commentaria in Aristotelem Graeca*, as follows: *In libros Aristotelis De anima paraphrasis*, R. Heinze, ed., V, pt. 3 (Berlin, 1899); *Analyticorum posteriorum paraphrasis*, M. Wallies, ed., V. pt. 1 (Berlin, 1900); *In Aristotelis Physica paraphrasis*, H. Schenkl, ed., V, pt. 2 (Berlin, 1900); *In libros Aristotelis De caelo paraphrasis hebraice et latine*, S. Landauer, ed., V, pt. 4 (Berlin, 1902); and *In Aristotelis Metaphysicorum librum* Λ *paraphrasis hebraice et latine*, S. Landauer, ed., V, pt. 5 (Berlin, 1903).

The *Orationes* have been edited by W. Dindorf (Leipzig, 1832; repr. Hildesheim, 1961) and by H. Schenkl completed by G. Downey (Leipzig, 1965–   ).

Medieval versions in Latin are *Commentaire sur le traité de l'âme d'Aristote, traduction de Guillaume de Moerbeke*, critical ed. by G. Verbeke (Louvain–Paris, 1957; repr., Leiden, 1973); and "Paraphrasis of the *Posterior Analytics* in Gerard of Cremona's Translation," J. R. O'Donnell, ed., in *Mediaeval Studies* (Toronto), **20** (1958), 239–315.

II. SECONDARY LITERATURE. See G. Downey, "Education and Public Problems as Seen by Themistius," in *Transactions and Proceedings of the American Philological Association*, **86** (1955), 291–307; "Education in the Christian Roman Empire. Christian and Pagan Theories Under Constantine and His Successors," in *Speculum*, **32** (1957), 48–61; and "Themistius and the Defence of Hellenism in the Fourth Century," in *Harvard Theological Review*, **50** (1957), 259–274; W. Stegemann, "Themistios," in Pauly-Wissowa, *Real-Ency-*

clopädie der classischen Altertumswissenschaft, 2nd ser., V, pt. 2 (1934), cols. 1642–1680; and G. Verbeke, "Themistius et le *De unitate intellectus* de saint Thomas," in *Revue philosophique de Louvain,* **53** (1955), 141–164.

G. VERBEKE

THENARD, LOUIS JACQUES (*b.* La Louptière [now Louptière-Thenard], Aube, France, 4 May 1777; *d.* Paris, France, 20 or 21 June 1857), *chemistry.*

Thenard was the second son of Étienne Amable Thenard and Cécile Savourat, peasant farmers, who had seven children. He received an elementary education from a local priest, and his obvious intelligence marked him out, so that at the age of eleven he was sent to the *collège* at Sens. He soon had the ambition to become a pharmacist and went to Paris with two friends to take advantage of the educational resources of the capital. Thenard attended the public courses of Vauquelin and Fourcroy and was taken into the Vauquelin household as a bottle washer and scullery boy. Vauquelin eventually allowed him to deputize for him in his lecture course; Thenard's first official appointment came in December 1798, when he was named demonstrator at the École Polytechnique.

When Vauquelin retired from his chair at the Collège de France, Thenard was nominated to succeed him (13 April 1804). Upon the founding in 1808 of the faculties of sciences, Thenard was appointed professor of chemistry at the Paris Faculty. With a secure income and a place in the new scientific-teaching community, he could think of marriage; and in 1814, after four years of negotiating with the family, he married the daughter of Arnould Humblot-Conté. Fourcroy's death in 1809 left a vacancy in the chemistry section of the First Class of the Institute, to which Thenard was elected (29 January 1810). He was a member of the Society of Arcueil and his work in this period (*ca.* 1807–1814) reveals the influence of his colleagues in that group, not least the physical approach of its leader, Berthollet.

Thenard was a lifelong member of the Société d'Encouragement pour l'Industrie Nationale; and when its founder, Chaptal, died in 1832, he was elected president. His interest in applied chemistry also meant that he played a prominent part in judging the national industrial exhibitions held in 1818 and at five-year intervals thereafter. He was also a member of the governing body of the Conservatoire National des Arts et Métiers.

Thenard was appointed dean of the Paris Faculty of Sciences in 1822. In 1830 he was nominated to the Royal Council of Public Instruction, on which he was especially concerned with the teaching of the physical sciences at the university level. From 1845 to 1852 he was chancellor of the University of France—the highest post in the French educational system.

Thenard was made successively a knight (1814), officer (1828), commander (1837), and grand officer (1842) of the Legion of Honor. In 1825 he was given the title of baron. In 1827 and in 1830 Thenard was elected to the Chamber of Deputies, his politics being to the right of center. On 11 October 1832 he was nominated as a peer; and in the upper chamber, as in the lower, his contributions to debates were usually on technical and scientific matters. He was particularly influential in his support of the sugar beet industry, which had begun in Napoleonic times and under the Restoration was threatened by the importation of cane sugar.

Thenard's early scientific work, particularly his interest in plant and animal chemistry, betrays the influence of his first patrons, Vauquelin and Fourcroy. In 1801 he obtained a new acid by distilling tallow. He called it sebacic acid and showed that what Guyton de Morveau had called sebacic acid was only impure acetic acid. Three years later he showed that the acid that Berthollet had named "zoonic acid" (obtained by distilling meat) was, again, really impure acetic acid. In his analysis of bile Thenard obtained a resin and another substance that he named "picromel," which was found to be a good solvent of fats.

His most important organic work was on the esters, then called "ethers." The name "ether" was given to any neutral product formed by the reaction of an acid with alcohol. The only "ether" that had been prepared and studied with any success prior to this time was "sulfuric ether"; but since it was what is now known as diethyl ether and not a true ester, knowledge of this compound tended to hinder rather than help the investigation of other "ethers." It was part of Thenard's achievement to distinguish "sulfuric ether" from the true esters. He made a careful study of the action of nitric, hydrochloric, acetic, benzoic, oxalic, citric, malic, and tartaric acids on ethyl alcohol and prepared the respective esters, many for the first time. His preparation of "nitric ether" (ethyl nitrite) is of value for his concern to obtain a pure product and to determine the yield.

Unknown to Thenard, "muriatic ether" (ethyl chloride) had been prepared slightly earlier by

Gehlen; but Thenard's memoir on this ester is notable for its study of the influence of time on chemical reactions involving organic compounds. He studied the reaction at room temperature between ethyl chloride and a concentrated solution of caustic potash over a period of three months, testing for the decomposition of the ester with silver nitrate solution.

Thenard's quantitative study of "acetic ether" (ethyl acetate) may be regarded as a model for its time. He studied both its preparation (in the presence of concentrated sulfuric acid) and its hydrolysis, always referring quantities of acid to equivalent weights of potash. By distilling the ester with an aqueous solution of caustic potash, he obtained alcohol and potassium acetate. Thenard thus proved conclusively by both analysis and synthesis that the "acetic ether" was a simple compound of acetic acid and alcohol, a valuable datum for organic chemistry. Thenard made the important statement that when alcohol combines with vegetable or mineral acid, the alcohol acts as a "true salifiable base."[1] Thus he drew an extremely useful analogy between the action of acids on bases to form salts in inorganic chemistry and the action of acids on alcohols to form esters in organic chemistry. This analogy was later extended by Chevreul in his study of saponification.

That Thenard was a worthy heir to the analytical skill of Vauquelin is shown by his early study of nickel (1802). He took particular care to obtain nickel free from traces of cobalt, iron, and arsenic. Typically, he announced the discovery of a new, higher oxide of nickel. It was, however, his study of certain cobalt compounds, published in 1804, that brought him greater fame. There was a particular need in France under the Consulate for a new blue pigment; and Thenard was commissioned by the minister of the interior, Chaptal, to obtain one. At one time lapis lazuli had been used, but it had become extremely rare and expensive. Prussian blue was not an effective substitute, and so Thenard experimented with cobalt arsenate, used in the coloration of Sèvres porcelain. He found that alumina heated in certain proportions with the arsenate or phosphate of cobalt produced the most permanent pigment. His final trials on the pigment included exposure to bright light for two months and exposure to acids, alkalies, and hydrogen sulfide. It became known as "Thenard's blue," although a similar color had been obtained earlier by K. F. Wenzel and Gahn. Thenard was helped in his research by the professor of drawing at the École Polytechnique, Léonor Mérimée, who later developed the use of hydrogen peroxide for restoring paintings.

An appreciable part of Thenard's research was concerned with the combining proportions of elements in certain compounds, particularly metal oxides. One of his earliest pieces of work was a report on the existence of six different oxides of antimony, which Proust reduced to two; the correct number (three) was determined later by Berzelius. Thenard announced the existence of four different oxides of cobalt and investigated the oxides and salts of mercury. He did research on the two sulfides of arsenic, realgar and orpiment, and showed that they contain no oxygen. In 1805 he published a memoir on the oxidation of metals in general. Thenard could not agree with Berthollet that oxidation of metals might take place in an indefinite number of stages, yet he believed that there were more different oxides of each metal than most chemists of the time were prepared to admit. He considered the solubility of different oxides in acids, making a particular study of iron and examining the oxidation of freshly precipitated ferrous hydroxide. Thenard established the existence of the unstable white ferrous oxide and thus helped to throw light on the chemistry of iron salts. He investigated phosphates of soda and ammonia, and analyzed phosphorous acid. His analysis of alloys of antimony and tin are a further reminder of his interest in combining proportions (he considered these alloys as compounds).

In 1812 Thenard obtained crystals of ammonium hydrosulfide by mixing ammonia gas and hydrogen sulfide. The proportions of the elements in another sulfide, hydrogen persulfide, had been studied by Berthollet; and Thenard later reexamined this problem. Believing that oxygen and sulfur were analogous elements and having discovered hydrogen peroxide, he considered that hydrogen persulfide was its analogue. He concluded that it varied in composition between extremes of "four atoms of sulfur and one atom of hydrogen sulfide" and "eight atoms of sulfur and one atom of hydrogen sulfide." The apparently variable composition helped to convince Thenard that it was a compound with a variable amount of physically dissolved sulfur. In the later editions of his *Traité* he presented both hydrogen peroxide and hydrogen persulfide as "compounds the elements of which obey forces other than affinity."

Thenard soon acquired a reputation as an analyst and thereby met Biot. In 1803 Biot was nominated by the First Class of the Institute to examine reports of meteorites; samples were brought back to Paris and a chemical analysis was carried out by Thenard. His most important collaboration with Biot was on a comparison of calcite and aragonite, since the two substances presented one of the earliest examples of dimorphism. Thenard's chemical analysis showed no difference between the two minerals despite their different crystalline forms. He and Biot concluded: "The same chemical principles combined in the same proportions can give rise to compounds that differ in their physical properties."[2]

A considerable amount of research was carried out by Thenard in collaboration with Gay-Lussac in 1808–1811, during which period they published about twenty papers. Thenard probably first met Gay-Lussac either when the latter was a student at the École Polytechnique or when they were both on the junior staff of that institution. The earliest record of their collaboration was on the occasion of Gay-Lussac's solo balloon ascent on 16 September 1804. Gay-Lussac took the flask of air he had collected at a high altitude to the laboratory of the École Polytechnique and with Thenard analyzed that air in comparison with ordinary Paris air. It was, however, the news of Davy's isolation of potassium, which reached Paris in the winter of 1807–1808, that prompted them to undertake a sustained collaboration largely in emulation of the English chemist. On 7 March 1808 they announced to the First Class of the Institute that they had prepared potassium by purely chemical means. The method, which involved fusing potash with iron filings in a gun barrel, had the advantage of producing potassium (and similarly sodium) in reasonable quantities, whereas Davy had been able to produce only tiny quantities of the substances. A controversy arose over the nature of potassium and sodium, Davy claiming that they were elements while Gay-Lussac and Thenard gave undue attention to experimental evidence suggesting that they were metal hydrides.

When funds were made available to construct a giant voltaic pile at the École Polytechnique (larger than that used by Davy at the Royal Institution), Gay-Lussac and Thenard were put in charge of the apparatus. Their results, reported in full in their *Recherches physico-chimiques*, are rather disappointing; Davy had effectively creamed the field. The superiority of the French chemists emerges in their investigations of the reactions of potassium metal. By strongly heating it in hydrogen, they prepared potassium hydride; and by heating the hydride in carefully dried ammonia, they obtained the olive-green solid KNH_2. When the solid was heated, it decomposed and ammonia, hydrogen, and nitrogen were released. The action of water on the solid produced potash and ammonia. Thenard and Gay-Lussac went on to use potassium to decompose boric acid and announced the isolation of a new element, boron, in November 1808. They obtained nearly anhydrous hydrofluoric acid by distilling calcium fluoride with concentrated sulfuric acid in a lead retort; and by heating calcium fluoride with boron trioxide, they obtained the gas boron fluoride, which they collected over mercury.

In their work on chlorine Gay-Lussac and Thenard were surprised to find that when the gas was passed over red-hot charcoal, it was not decomposed. This cast doubt on whether the gas then called "oxymuriatic acid gas" really was a compound containing oxygen. The authority of Berthollet persuaded them that this conclusion was not fully justified, and accordingly they mentioned it as only a possibility. It was left to Davy in 1810, after he had read their memoir, to announce that chlorine was in fact an element. Thenard and Gay-Lussac, however, deserve full credit for their pioneering contributions to photochemistry. They investigated the effect of light on mixtures of chlorine and hydrogen and chlorine and ethylene. The extent of the reaction in darkness or in a diffused light was judged by the change in the greenish-yellow color of the chlorine gas. Bright sunlight was found to bring about combination with explosive violence.

Another fruitful collaboration by Gay-Lussac and Thenard was that carried out in 1810 on the combustion analysis of vegetable and animal substances. Lavoisier's published organic analysis had made use of oxygen gas; but the two young chemists greatly extended the generality of this method by using an oxidizing agent, potassium chlorate. On the basis of their analysis they divided vegetable compounds into three classes according to the proportion of hydrogen and oxygen they contained. The class (containing starch and sugar) in which hydrogen and oxygen were in the same proportions as in water corresponds to the carbohydrates. Although in this joint research it is impossible to separate the contributions of Thenard from those of Gay-Lussac, one has the impression that Thenard usually came second to his friend in the quality, originality, and precision of his research.

Inspired by the fundamental work of Lavoisier on alcoholic fermentation, the Institute in 1800 and 1802 offered a prize on the subject. Thenard submitted a memoir and, according to a standard source, it "provided many of the facts upon which

Liebig subsequently based his views."[3] He pointed out that all fermenting liquids deposit a material similar to brewer's yeast and he demonstrated that it contained nitrogen. His study of yeast used to ferment pure sugar showed that it underwent a gradual change and was finally reduced to a white material that contained no nitrogen and produced no reaction with sugar. Thenard had begun by asking, "How is sugary matter changed into alcohol and carbonic acid by means of an intermediate body? What is the nature of this body? How does it act on sugar?"[4] The young chemist was not able to solve these complex problems, which a generation later became a subject of vigorous dispute between biological microscopists and chemists. Berzelius, for example, opposed biological explanations with the theory that fermentation was merely an example of contact catalysis due to a nonliving catalyst—a view that may be traced back to Thenard's work. In 1820, when he was studying the effect of finely divided metals on hydrogen peroxide, Thenard compared this phenomenon to the action of yeast in alcoholic fermentation.[5]

Thenard's greatest single discovery was that of hydrogen peroxide. He read his first paper on the subject to the Académie des Sciences on 27 July 1818, and successive volumes of the *Annales de chimie* contain his researches. The work had its origins in his earlier collaboration with Gay-Lussac, in which they had shown that when potassium or sodium is heated in dry oxygen, a higher oxide is obtained. Heating baryta strongly in oxygen also produced a new higher oxide. In the presence of water all these peroxides decomposed, liberating oxygen. The discovery of hydrogen peroxide seems to have been related to Lavoisier's theory of chemistry, according to which metals combined with acids to form salts only after an initial oxidation reaction; thus it was the metal oxide rather than the metal that dissolved in the acid. The metal should not be too highly oxidized, however, because it would then have little affinity for acids (acids were considered as extreme products of oxidation). Thenard wished to test this idea by seeing whether barium peroxide would dissolve in acids.

Thenard's first paper on hydrogen peroxide announced that he had prepared new oxygenated acids by treatment of barium peroxide with mineral acids. Thus, for instance, barium peroxide dissolved in dilute nitric acid to produce a neutral solution. The barium nitrate was precipitated as barium sulfate, leaving what we recognize as hydrogen peroxide. Unfortunately Thenard used sulfuric acid to remove the barium salt and therefore had an acid product. Using this process, he prepared an "oxygenated acid" containing up to eleven times its own volume of oxygen. Since heating caused decomposition, his method of concentration was to use a vacuum pump at room temperature. By September 1818 Thenard had employed this method in preparing a product containing thirty-two times its own volume of oxygen. He recognized that its decomposition was accelerated by light and found that when the concentrated product came into contact with silver oxide, the oxygen was liberated in a violent reaction. By 23 November he had prepared "oxygenated water" and had begun to doubt whether his "oxygenated acids" were true compounds. Thenard had now, therefore, prepared a second compound of hydrogen and oxygen. It was neutral and could be distilled in a vacuum without decomposition. He found that the decomposition occurring when, for example, manganese dioxide was added, was exothermic. He went on to prepare a very concentrated product containing more than four hundred times its own volume of oxygen and found that it attacked the skin.

A major problem throughout this research had been to discover whether oxygen could combine with acids or water indefinitely, thus supporting the largely discredited ideas of Berthollet. This had been one of Thenard's motives for preparing an increasingly concentrated product. Finally he announced that he had succeeded in reaching the saturation point. This pure "oxygenated water" had a density of 1.455 (modern 1.465) and reacted explosively with various metal oxides. Usually the oxygen evolved consisted of both the "excess" oxygen of the hydrogen peroxide and the oxygen of the metal oxide. In some cases, however, the peroxide acted as an oxidizing agent (for instance, with arsenious oxide). He made the important observation that acids render hydrogen peroxide more stable.

Thenard completed his work on hydrogen peroxide by giving a detailed description of its preparation, starting from pure barium nitrate, which was heated to decompose it; oxygen was passed over the product to convert it into barium peroxide. The latter was then made into a paste in an ice-cooled vessel and just enough sulfuric acid added to precipitate all the barium, which was separated by filtration. Further purification was described to remove alumina, iron, and silica impurities. Thenard's complete work on hydrogen peroxide was summarized in a long article published in 1820 in the *Mémoires* of the Academy. In it he concluded that hydrogen peroxide is a true peroxide (*peroxide d'hydrogène*) and contains twice as much oxygen as water does. He used it to prepare new peroxides and noted its oxidizing action on sulfides. Mérimée, his colleague at the École Polytechnique, suggested applying this reaction to restoration of old paintings.

In Thenard's final and comprehensive paper on hydrogen peroxide (1820), he devoted several pages to the effect of finely divided metals on hydrogen peroxide, distinguishing, for example, between silver in an extreme state of division, finely divided, filings, and massive. Platinum, gold, osmium, palladium, rhodium, and other metals also were listed according to their state of division. Thenard was particularly concerned how these metals could take part in chemical reactions without apparently being affected.

This earlier work will help to explain Thenard's particular interest in catalysis in 1823. Indeed, as early as 1813 he had investigated the effect of the presence of metals in promoting the decomposition of ammonia gas passed through a red-hot glazed porcelain tube.[6] In August 1823 news reached Paris that Döbereiner had shown that spongy platinum at room temperature could bring about the combination of hydrogen and oxygen and that the heat from this reaction was sufficient to make the metal incandescent. Thenard collaborated with Dulong in experiments to confirm this finding. They extended the research by varying the physical state of the metal and by substituting other metals and other gaseous reactions. They demonstrated that the temperatures at which metals showed such effects depended on their state of division. Thenard and Dulong found that palladium, rhodium, and iridium had the same effect as platinum; and they went on to investigate the surface effects of other solids and the conditions under which substances lose their catalytic effect. Thenard, therefore, made significant contributions to knowledge of surface catalysis, although the term "catalyst" was not introduced by Berzelius until 1834.

Thenard was the author of a large and important chemistry textbook that went through six editions and was translated into German, Italian, and Spanish (the section on analysis was translated into English). Through this book he helped restore France to its traditional role as supplier of chemistry textbooks to the rest of the world; his only serious rival was the British chemist Thomas Thomson, who during the first two decades of the nineteenth century produced successive editions of his own textbook. The first edition of Thenard's *Traité de chimie élémentaire* was published in four volumes in 1813–1816. The first two volumes dealt with inorganic chemistry, the third with organic chemistry (divided into vegetable and animal), and the fourth with analytical chemistry. Similar substances were grouped together and discussed in general terms before consideration of their individual properties. In the Lavoisier tradition, oxygen was still considered as a unique element. Besides drawing on previous textbooks—such as those of Lavoisier, Fourcroy, and Thomson—Thenard incorporated the most recent research of his contemporaries. Plates and detailed descriptions of apparatus were provided, and in later editions were published as a separate volume. The detailed index included in each volume makes Thenard's book a particularly useful reference work for the chemistry of its period.

Thenard took great pains to bring successive editions of his textbook up to date. In the sixth and final edition (1834–1836) there is a major rearrangement of the material. Particularly important is the addition of a fourth part that he described as an "Essai de philosophie chimique," in which he dealt with the general principles of chemical combination and classification. He continued, however, to show the same reserve about Dalton's atomic theory that he had expressed in earlier editions.

NOTES

1. *Mémoires de la Société d'Arcueil*, **2** (1809), 24.
2. *Ibid.*, 206.
3. A. Harden, *Alcoholic Fermentation*, 4th ed. (London, 1934), 4.
4. *Annales de chimie*, **46** (1802), 206–207.
5. *Mémoires de l'Académie royale des sciences de l'Institut de France*, 2nd ser., **3**, année 1818 (1820), 487.
6. *Annales de chimie*, **85** (1813), 61.

BIBLIOGRAPHY

I. ORIGINAL WORKS. Thenard's chemistry textbook went through six eds., all published at Paris: *Traité de chimie élémentaire, théorique et pratique*, 4 vols. (1813–1816; 2nd ed., 1817–1818; 3rd ed., 1821; 4th ed., 5 vols., 1824; 5th ed., 1827; 6th ed., 1834–1836). With Gay-Lussac he wrote *Recherches physico-chimiques*, 2 vols. (Paris, 1811).

A selection from Thenard's research papers is presented below. The order follows that of the text. "Sur l'acide sébacique," in *Annales de chimie*, **39** (1801), 193–202; "Observations sur l'acide zoonique," *ibid.*, **43** (1802), 176–184; "Mémoire sur la bile," in *Mémoires de la Société d'Arcueil*, **1** (1807), 23–45; "Mémoire sur les éthers," *ibid.*, 73–114; "Deuxième mémoire sur les éthers . . .," *ibid.*, 115–134; "Troisième mémoire sur les éthers . . .," *ibid.*, 140–160; "De l'action des acides végétaux sur l'alcool . . .," *ibid.*, **2** (1809), 5–22; "Essai sur la combinaison des acides avec les substances végétales et animales," *ibid.*, 23–41.

"Sur le nickel" (1802), in *Annales de chimie*, **50** (1804), 117–133; "Considérations générales sur les couleurs, suivies d'un procédé pour préparer une couleur bleue aussi belle que l'outremer," in *Journal des mines*, **15** (1804), 128–136; "Différents états de l'oxide d'antimoine . . .," in *Annales de chimie*, **32** (1799), 257–269; "Sur l'oxidation des métaux en général et en particulier du fer," *ibid.*, **56** (1805), 59–85; "Observations sur les hydro-sulfures," *ibid.*, **83** (1812), 132–138; "Mémoire sur le soufre hydrogéné ou l'hydrure de soufre," *ibid.*, 2nd ser., **48** (1831), 79–87; "Mémoire sur l'analyse comparée de l'aragonite et du carbonate de chaux rhomboidal," in *Mémoires de la Société d'Arcueil*, **2** (1809), 176–206, written with Biot.

Five memoirs written with Gay-Lussac: "Sur les métaux de la potasse et de la soude," in *Annales de chimie*, **66** (1808), 205–217; "Sur la décomposition et la recomposition de l'acide boracique," *ibid.*, **68** (1808), 169–174; "Sur l'acide fluorique," *ibid.*, **69** (1809), 204–220; "De la nature et des propriétés de l'acide muriatique et de l'acide muriatique oxigéné," in *Mémoires de la Société d'Arcueil*, **2** (1809), 339–358; and "Sur l'analyse végétale et animale," in *Annales de chimie*, **74** (1810), 47–64.

"Sur la fermentation vineuse," in *Annales de chimie*, **46** (1803), 294–320; "Observations sur des nouvelles combinaisons entre l'oxigène et divers acides," in *Annales de chimie et de physique*, 2nd ser., **8** (1818), 306–312; "Nouvelles observations sur les acides et les oxides oxigénés," *ibid.*, **9** (1818), 51–56, 94–98; "Observations sur l'influence de l'eau dans la formation des acides oxigénés," *ibid.*, 314–317; "Nouvelles recherches sur l'eau oxigénée," *ibid.*, 441–443; "Suite des expériences sur l'eau oxigénée," *ibid.*, **10** (1819), 114–115, 335; "Nouvelles observations sur l'eau oxigénée," *ibid.*, **11** (1819), 85–87, 208–216; "Mémoire sur la combinaison de l'oxygène avec l'eau, et sur les propriétés extraordinaires que possède l'eau oxigénée," in *Mémoires de l'Académie royale des sciences de l'Instiut de France, année 1818*, **3** (1820), 385–488.

With Dulong he wrote "Note sur la propriété que possèdent quelques métaux de faciliter la combinaison des fluides élastiques," in *Annales de chimie et de physique*, 2nd ser., **23** (1823), 440–444; and "Nouvelles observations sur la propriété dont jouissent certains corps de favoriser la combinaison des fluides élastiques," *ibid.*, **24** (1823), 380–387.

II. SECONDARY LITERATURE. See M. P. Crosland, *The Society of Arcueil. A View of French Science at the Time of Napoleon*, I (London, 1967), *passim*; F. Du-

bois, *Éloge de M. Thenard, prononcé dans la séance publique annuelle de l'Académie impériale de médecine du 9 Décembre 1862* (Paris, 1863); P. Flourens, "Éloge historique de Louis-Jacques Thenard," in *Éloges historiques*, 3rd ser. (Paris, 1862), 201–248; L. R. Le Canu, *Souvenirs de M. Thenard* (Paris, 1857); J. R. Partington, *A History of Chemistry*, IV (London, 1964), esp. 90–96; and P. Thenard, *Un grand français. Le chimiste Thenard, 1777–1857 par son fils; avec introduction et notes de Georges Bouchard* (Dijon, 1950).

M. P. CROSLAND

THEODORIC BORGOGNONI OF LUCCA. See **Borgognoni of Lucca, Theodoric.**

THEODORIC OF FREIBERG. See **Dietrich von Freiberg.**

THEODORUS OF CYRENE (*b.* Cyrene, North Africa, *ca.* 465 B.C.; *d.* Cyrene [?], after 399 B.C.), *mathematics.*

Theodorus was the mathematical tutor of Plato and Theaetetus and is known for his contribution to the early development of the theory of irrational quantities. Iamblichus includes him in his catalog of Pythagoreans.[1] According to the account of Eudemus as preserved by Proclus,[2] he was a contemporary of Hippocrates of Chios, and they both came after Anaxagoras and Oenopides of Chios. Diogenes Laërtius[3] states that he was the teacher of Plato; and Plato represents him as an old man in the *Theaetetus*, which is set in 399 B.C. Since Anaxagoras was born *ca.* 500 and Plato in 428 or 427, it is reasonable to suppose that Theodorus was born about 465. This would make him sixty-six years old in the fictive year of the *Theaetetus*. According to the dialogue he had been a disciple of Protagoras but had turned at an early age from abstract speculation to geometry.[4] He was in Athens at the time of the death of Socrates.[5] He is also made a character by Plato in the *Sophist* and the *Politicus*. Plato may have sat at his feet in Athens just before the death of Socrates or at Cyrene during his travels after that event. In the dialogue Theaetetus tells Socrates that he learned geometry, astronomy, harmony, and arithmetic from Theodorus.[6] As with Plato, this could have been at Athens or Cyrene.

In the dialogue[7] Theaetetus is made to relate how Theodorus demonstrated to him and to the younger Socrates, a namesake of the philosopher, that the square roots of 3, 5, and so on up to 17 (excluding 9 and 16, it being understood) are incommensurable with the unit; and Theaetetus goes on to say how he and Socrates tried to find a general formula that would comprehend all square roots. Plato clearly purports to be giving a historical account[8] and to be distinguishing the achievement of Theodorus from that of Theaetetus; and it would appear that Theodorus was the first to demonstrate the irrationality of $\sqrt{3}$, $\sqrt{5}$, $\cdots$, $\sqrt{17}$. Two questions immediately arise. Why did he start at $\sqrt{3}$? Why did he stop at $\sqrt{17}$? The answer to the former question must be that the irrationality of $\sqrt{2}$ was already known. It was, indeed, known to the earlier Pythagoreans; and there is a high probability that it was the discovery of the incommensurability of $\sqrt{2}$ with the unit that revealed to the Greeks the existence of the irrational and made necessary a recasting of Greek mathematical theory.[9] After this discovery it would be natural for Theodorus and others to look for further examples of irrationality.

The answer to the second question depends on how Theodorus proved the irrationality of the numbers under examination, and is not so easy. We may rule out at once the suggestion of F. Hultsch that Theodorus tried the method of successively closer approximation, because it would never *prove* irrationality.[10] The answer is dependent also on the meaning given to the words πως ἐνέσχετο. They usually have been translated "for some reason he stopped."[11] A glance at the uses of ἐνέχειν given in the lexicons, however, shows, as R. Hackforth first appreciated, that the Greek must mean "somehow he got into difficulties."[12]

This rules out the possibility that Theodorus stopped at 17 merely because he had to stop somewhere and felt he had proved enough.[13] It also rules out the possibility, despite the contention of A. Wasserstein, that Theodorus merely applied to 3, 5, $\cdots$, 17 the proof of the irrationality of $\sqrt{2}$.[14] This was known to Aristotle and is interpolated in the text of Euclid's *Elements*[15]; it may have been the way in which the irrationality of $\sqrt{2}$ was originally demonstrated. In this proof it is shown that, if the diagonal of a square is commensurable with its side, the same number will be both odd and even.[16] This proof can be generalized for all square roots, and indeed for all roots, in the form "$\sqrt[m]{N}$ is irrational unless N is the m-th power of an integer n."[17] Theodorus would soon have recognized the generality and would have run into no difficulties after 17.

It has been suggested that the Pythagorean de-

votion to the decad may have led Theodorus to stop where he did.[18] For $\sqrt{3}$ can be represented as $\sqrt{2^2 - 1^2}$, and so on for all the odd numbers up to $\sqrt{17} = \sqrt{9^2 - 8^2}$, at which point all the numerals from 1 to 9 would have been exhausted; Theodorus, however, would not have run into any difficulty in proceeding farther by this method, nor does it afford any proof of irrationality.

The above hypothesis is similar to one propounded by an anonymous commentator on the *Theaetetus*.[19] He first says that Plato made Theaetetus start with $\sqrt{3}$ because he had already shown in the *Meno* that the square on the diagonal of a square is double that on the side. He then proceeded to point out that Theaetetus was both a geometer and a student of musical theory. The tone interval has the ratio 9:8. If we double the two numbers we have 18:16; and between these two numbers the arithmetic mean is 17, dividing the extremes into unequal ratios, "as is shown in the commentaries on the *Timaeus*." The comment of Proclus on Plato, *Timaeus* 35B (*Commentarium in Timaeum*, 195A), is relevant, but we need not pursue it because it is clearly a rather farfetched hypothesis to explain why Theodorus stopped at 17.

An ingenious theory has been put forward by J. H. Anderhub.[20] If a right-angled isosceles triangle with unit sides is set out as in Figure 1, its

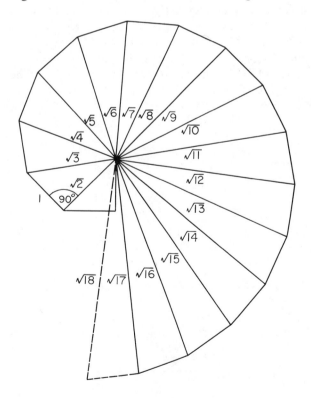

FIGURE 1

hypotenuse is $\sqrt{2}$. If at one extremity of the hypotenuse a perpendicular of unit length is erected, we have a second triangle with hypotenuse $= \sqrt{3}$. The process can be continued with all the hypotenuses $\sqrt{2}, \sqrt{3} \cdots$ radiating from a common point, and the angles at the common point can be shown to be 45°, 35°15', and so on. The total of all the angles up to hypotenuse $= \sqrt{17}$ is approximately 351°10', and the total up to $\sqrt{18}$ is approximately 364°48'—that is, after $\sqrt{17}$ the circle has been completed and the triangles begin to overlap. But although this would have given Theodorus a reason for stopping, he would have had no difficulty in going on; and the method does not prove the irrationality of any hypotenuse.

There is one theory, put forward by H. G. Zeuthen, that satisfies the requirements that there shall be a separate proof for each number $\sqrt{3}, \sqrt{5}, \cdots$ as Plato's text suggests, and that after $\sqrt{17}$ the proof will encounter difficulties.[21] Zeuthen's suggestion is that Theodorus used the process of finding the greatest common measure of two magnitudes as set out in Euclid's *Elements*, X.2, and actually made a test of incommensurability by Euclid: "If when the lesser of two unequal magnitudes is continually subtracted from the greater, the remainder never measures the one before it, the magnitudes will be incommensurable."[22] The method may conveniently be illustrated from $\sqrt{17}$ itself. Let ABC be a right-angled triangle in which $AB = 1$, $BC = 4$, so that $CA = \sqrt{17}$. Let CD be cut off from CA equal to CB so that $AD = \sqrt{17} - 4$, and let DE be drawn at right angles to CA. The triangles CDE, CBE are equal and therefore

$DE = EB$. The triangles ADE, ABC are similar and $DE = 4AD$. We therefore have $DE = 4AD = 4(\sqrt{17} - 4)$. Now from EA let EF be cut off equal to ED and at F let the perpendicular FG be drawn. Then by parity of reasoning

$$AF = AB - BF = AB - 2DE$$
$$= 1 - 8(\sqrt{17} - 4)$$
$$= (\sqrt{17} - 4)(\sqrt{17} + 4)$$
$$- 8(\sqrt{17} - 4)$$
$$= (\sqrt{17} - 4)^2.$$

Obviously, the process can be continued indefinitely, so that ABC, ADE, AFG, $\cdots$ is a diminishing series of triangles such that

$$AB:AD:AF \cdots = 1:(\sqrt{17} - 4):(\sqrt{17} - 4)^2 \cdots$$

and we shall never be left with a magnitude that exactly measures CA, which is accordingly incommensurable.

Theodorus would certainly have used a geometrical proof, but the point can be made as shown below in modern arithmetical notation. The process of finding the greatest common measure of 1 and $\sqrt{17}$ (if any) may be set out as follows:

1) $\sqrt{17}$ (4

$$\frac{4}{\sqrt{17}} - 4$$

$(\sqrt{17} - 4))1(8$

$$\frac{8(\sqrt{17} - 4)}{1 - 8(\sqrt{17} - 4)}$$
$$= (\sqrt{17} - 4)(\sqrt{17} + 4) - 8(\sqrt{17} - 4)$$
$$= (\sqrt{17} - 4)(\sqrt{17} + 4 - 8)$$
$$= (\sqrt{17} - 4)^2.$$

The next stage in the process would be to divide $(\sqrt{17} - 4)^2$ into $(\sqrt{17} - 4)$, but this is the same as dividing $(\sqrt{17} - 4)$ into 1, which was the previous step. The process is therefore periodic and will never end, so that 1 and $\sqrt{17}$ do not have a greatest common measure. It will be recognized as the same process as that for finding a continued fraction equal to $\sqrt{17}$.

It is a powerful argument in favor of this theory that Plato, in the passage of the *Theaetetus* under discussion, for the first time in Greek literature uses the term οὐ σύμμετρος ("incommensurable") for what had previously been described as ἄρρητος ("inexpressible"). This strongly reinforces the conviction that he was doing something new, and that

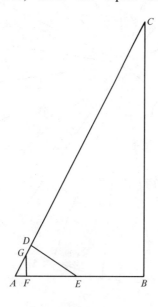

FIGURE 2

the novelty consisted in using the test of incommensurability later found in Euclid.

These proofs, geometrical and arithmetical, are simple; and the former would certainly have been within the grasp of Theodorus. So would the earlier proofs for $\sqrt{3}$, $\sqrt{5}$, and so on.[23] The next case, $\sqrt{18}$, would not call for investigation since $\sqrt{18} = 3\sqrt{2}$; but $\sqrt{19}$ presents difficulties at which even a modern mathematician may quail. Recurrence does not take place until after six stages, which, on the basis of the exposition of B. L. van der Waerden, may be set out as follows.[24] We start by subtracting the appropriate multiple of 1 from $\sqrt{19}$ and get a remainder $\sqrt{19} - 4$. We now divide $\sqrt{19} - 4$ into 1. But

$$\frac{1}{\sqrt{19}-4} = \frac{\sqrt{19}+4}{3}.$$

We treat ($\sqrt{19} + 4$) and 3 in exactly the same way, subtracting $2 \cdot 3$ from $\sqrt{19} + 4$ and getting $\sqrt{19} - 2$. Now.

$$\frac{3}{\sqrt{19}-2} = \frac{\sqrt{19}+2}{5};$$

and we subtract the 5 from $\sqrt{19} + 2$, getting $\sqrt{19} - 3$, and divide this into 5. But

$$\frac{5}{\sqrt{19}-3} = \frac{\sqrt{19}+3}{2};$$

and after subtracting $3 \cdot 2$ from $\sqrt{19} + 3$, we get $\sqrt{19} - 3$ again. But

$$\frac{2}{\sqrt{19}-3} = \frac{\sqrt{19}+3}{5},$$

and subtracting 5 from $\sqrt{19} + 3$ yields $\sqrt{19} - 2$. Now

$$\frac{5}{\sqrt{19}-2} = \frac{\sqrt{19}+2}{3};$$

and by subtracting $2 \cdot 3$ from $\sqrt{19} + 2$ we obtain $\sqrt{19} - 4$. But

$$\frac{3}{\sqrt{19}-4} = \frac{\sqrt{19}+4}{1};$$

subtracting $4 \cdot 1$ from $\sqrt{19} + 4$ leaves us with $\sqrt{19}$, and dividing 1 into $\sqrt{19}$ brings us back where we started. The process is therefore periodic and will never end, so that $\sqrt{19}$ is incommensurable with 1.

This is formidable enough in modern notation, and impossible to set out in a drawing, particularly a drawing in sand. If this is the method that Theodorus used, it is therefore fully understandable why he stopped at $\sqrt{17}$.

Although this is only a hypothesis, there is no other that fits the facts so well; and if his pupil Theaetetus developed a theory of proportion based on the method of finding the greatest common measure, as is argued in the article devoted to him in this *Dictionary*, it becomes virtually certain that this is the method employed by Theodorus.

Proclus, in analyzing curves in the manner of Geminus, criticizes "Theodorus the mathematician" for speaking of "blending" in lines.[25] He is probably to be identified with Theodorus of Cyrene, since in his only other reference Proclus describes the subject of this article. He may also be identified with the Theodorus whom Xenophon held up as a model of a good mathematician.[26]

NOTES

1. Iamblichus, *De vita Pythagorica*, 267; L. Deubner, ed. (Leipzig, 1937), p. 146.8–9.
2. Proclus, *In primum Euclidis*, G. Friedlein, ed. (Leipzig, 1873; repr. Hildesheim, 1967), 65.21–66.7.
3. Diogenes Laërtius, *Vitae philosophorum*, II.103, III.6; H. S. Long, ed., I (Oxford, 1964), 100.9–13, 123.18.
4. Plato, *Theaetetus* 164E–165A, in *Platonis opera*, J. Burnet, ed., I (Oxford, 1899; frequently repr.).
5. It is not obvious why James Gow, *A Short History of Greek Mathematics* (Cambridge, 1884), 164, should flatly contradict the evidence of Plato's dialogue and say, "He does not seem to have visited Athens."
6. Plato, *Theaetetus*, 145C–D.
7. *Ibid.*, 147D–148B.
8. Jean Itard, *Les livres arithmétiques d'Euclide*, Histoire de la Pensée, X (Paris, 1961), is exceptional in regarding as tenable the view that Theodorus and Theaetetus may not be historical persons but "personnages composites nés dans l'esprit même de Platon."
9. The fullest account of this subject is in Kurt von Fritz, "The Discovery of Incommensurability by Hippasus of Metapontium," in *Studies in Presocratic Philosophy*, David J. Furley and R. E. Allen, eds., I (London–New York, 1970), 382–412. But in his earlier paper with the same title in *Annals of Mathematics*, **46** (1945), 242–264, Fritz exposed himself to some strictures, of which he has not taken notice, from A. Wasserstein, "Theaetetus and the Theory of Numbers," 165, n. 3.
10. F. Hultsch, "Die Näherungswerthe irrationaler Quadratwurzeln bei Archimedes," in *Nachrichten von der königlich Gesellschaft der Wissenschaften zu Göttingen*, **22** (1893), 368–428. Hultsch received some support from T. L. Heath in his early work, *The Works of Archimedes* (Cambridge, 1897), lxxix–lxxx, in which he regarded it as "pretty certain" that Theodorus, like Archimedes after him, represented $\sqrt{3}$ geometrically as the perpendicular from an angular point of an equilateral triangle to the opposite side. He also presumed that Theodorus would start from the identity $3 = 48/16 = (49-1)/16$, so that

$$\sqrt{3} < \sqrt{\frac{48+1}{16}} = \frac{7}{4};$$

but in his later work, *A History of Greek Mathematics*, I (Oxford, 1921), 204, he realized that this "would leave Theodorus as far as ever from *proving* that $\sqrt{3}$ is incommensurable." These approximations may, of course, have played a part in Theodorus' researches until he found a demonstrative proof.

11. "There he stopped" (B. Jowett); "here he somehow came to a pause" (B. J. Kennedy); "il s'était, je ne sais pourquoi, arrêté là" (A. Diès); "there, for some reason, he stopped" (F. M. Cornford); "at that he stopped" (H. N. Fowler); "qui, non so come, si fermò" (M. Timpanaro Cardini). But the latest translator, J. McDowell (Oxford, 1973), has the sense right—"at that point he somehow got tied up."

12. *A Greek-English Lexicon*, H. G. Liddell and R. Scott, eds., new ed. by H. Stuart Jones (Oxford, 1940), see ἐνέχω, II, 565. The general meaning of the passive and middle is "to be held, caught, entangled in"; and a particularly relevant example is given in II.2, [κῦρος] ἐνείχετο ἀπορίη̣σι; Herodotus 1.190. *Theatetus* 147D is the only passage quoted for the meaning "come to a standstill" (II.5), and therefore it can hardly determine the meaning of that passage.

R. Hackforth, "Notes on Plato's *Theaetetus*," 128. It is significant that Hackforth's interest is purely literary, and his interpretation is therefore free from any bias in favor of some particular mathematical solution. He is supported by Malcolm S. Brown, "*Theaetetus*: Knowledge as Continued Learning," in *Journal of the History of Philosophy*, 7 (1969), 367.

13. Wasserstein, *op. cit.*, 165, makes this suggestion without necessarily endorsing it. G. H. Hardy and E. M. Wright, *An Introduction to the Theory of Numbers*, 4th ed. (Oxford, 1960), 43, in the light of their views about the difficulty of generalizing the Pythagorean proof for $\sqrt{2}$ (see note 14), regard the suggestion as credible.

14. It is the main burden of Wasserstein's paper (cited above) that this is precisely what Theodorus did. He argues that the difficulties of effecting a valid generalization are such as would have been perceived by Theodorus and that "it was precisely this refusal of the rigorous mathematician to enumerate a general theory based on doubtful foundations that led his pupil Theaetetus to investigate not only the problem of irrationality but also the more fundamental arithmetical questions." Although it may be conceded that Theodorus was an acute mathematician, it is most unlikely that he, or any other ancient mathematician, would have thought about this problem in the manner of G. H. Hardy and E. M. Wright. (See note 16.) Wasserstein's thesis is controverted in detail by Brown, *op. cit.*, 366–367, but in part for an irrelevant reason. Brown accepts van der Waerden's view that bk. VII of Euclid's *Elements* was already in "apple-pie order" before the end of the fifth century, whereas Wasserstein, like Zeuthen, regards it as the work of Theaetetus; but Wasserstein's contention that Theodorus applied the traditional proof for $\sqrt{2}$ can be detached from this belief.

15. Aristotle, *Prior Analytics* 1.23.41a23–30; W. D. Ross, ed. (Oxford, 1949, corr. repr. 1965), English trans. by A. J. Jenkinson as *Analytica priora*, in *The Works of Aristotle*, W. D. Ross, ed., I (Oxford, 1928; repr. 1968). Euclid, *Elements*, X, app. 27; *Euclidis opera omnia*, J. L. Heiberg and H. Menge, eds., III (Leipzig, 1886), 408.1–410.16; E. S. Stamatis, ed., in *Euclidis Elementa* post J. L. Heiberg, III (Leipzig, 1972), 231.10–233.13. An alternative proof is also given. In earlier eds. the proof was printed as Euclid, *Elements*, X.117, but it is now recognized as an interpolation.

16. T. L. Heath, *A History of Greek Mathematics*, I (Oxford, 1921), 205, purports to give a fairly easy generalization; but it is logically defective in that he assumes that "if $m^2 = N \cdot n^2$, therefore m^2 is divisible by N, so that m also is a multiple of N," which is true only if N is not itself the multiple of a square number. Wasserstein, *op. cit.*, 168–169, corrects Heath. Hardy and Wright, *op. cit.*, 40, show that the generalization is not so simple as Heath represented it to be and "requires a good deal more than a 'trivial' variation of the Pythagorean proof." This is true; but Hardy and Wright are working to standards of logical rigor far beyond what

any mathematician of the fifth century B.C. would have demanded, and Theodorus could fairly easily have found a generalization that would have satisfied his own standards.

17. Hardy and Wright, *op cit.*, 41. The authors discuss Theodorus' work helpfully in the light of modern mathematics on 42–45.

18. See Brown, *op cit.*, 367–368, and his n. 26.

19. *Anonymer Kommentar zu Platons Theaetet*, H. Diels and W. Schubart, eds., Berliner Klassikertexte, II (Berlin, 1905). The passage is reproduced and translated, as is Proclus' commentary on the relevant passage in the *Timaeus*, with illuminating notes, in Wasserstein, *op cit.*, 172–179. The commentator does not appear to accept his own suggestion, saying that Theodorus stopped at 17 because that is the first number after 16, and 16 is the only square in which the number denoting the sum of the sides is equal to the number denoting the area $(4 + 4 + 4 + 4 = 4 \times 4)$.

20. Both B. L. van der Waerden, *Science Awakening*, English trans. by Arnold Dresden of *Ontwakende Wetenschap*, 2nd ed., I (Groningen, n.d.), 143; and Árpád Szabó, *Anfänge der griechischen Mathematik* (Munich–Vienna, 1969), 70, attribute this interesting construction to J. H. Anderhub, *Joco-Seria: Aus den Papieren eines reisenden Kaufmannes*, but I have not been able to obtain a copy.

21. H. G. Zeuthen, "Sur la constitution des livres arithmétiques des *Eléments* d'Euclide et leur rapport à la question de l'irrationalité," in *Oversigt over det K. Danske Videnskabernes Selskabs Forhandlinger* for 1910 (1910–1911), 422–426. The theory is taken up again in articles in the same periodical in the volumes for 1913 and 1915. Zeuthen's thesis is supported by O. Toeplitz, 28–29; and by Brown, *op cit.*

22. *Euclidis opera omnia*, J. L. Heiberg and H. Menge, eds., III (Leipzig, 1886), 6.12–8.13; E. C. Stamatis, ed., in *Euclidis Elementa post J. L. Heiberg*, III (Leipzig, 1972), 3.19–4.19.

23. The cases of $\sqrt{5}$ and $\sqrt{10}$ are similar to $\sqrt{17}$. The case of $\sqrt{3}$ is a little more difficult, involving one more step before recurrence takes place. The case of $\sqrt{13}$ is difficult, as may be seen from the process of expressing it as a continued fraction given by G. Chrystal, *Algebra*, II (Edinburgh, 1889), 401–402, where it is shown that recurrence occurs only after five partial quotients:

$$\sqrt{13} = 3 + \cfrac{1}{1+} \cfrac{1}{1+} \cfrac{1}{1+} \cfrac{1}{1+} \cfrac{1}{6+} \cfrac{1}{1+} \cdots,$$

Perhaps it was Theodorus' experience with $\sqrt{13}$ that made him unwilling to embark on $\sqrt{19}$. The same method can be applied to $\sqrt{2}$, although it is probable that $\sqrt{2}$ was originally proved irrational not by this method but by that referred to by Aristotle (see note 15). Zeuthen, *loc. cit.*, and Heath, *loc. cit.*, give the proofs for $\sqrt{5}$ and $\sqrt{3}$; and Heath adds a geometrical proof for $\sqrt{2}$; Hardy and Wright, *op. cit.*, give proofs for $\sqrt{5}$ and $\sqrt{2}$. The method is used by Kurt von Fritz in *Studies in Presocratic Philosophy*, I, 401–406, to prove the incommensurability of the diagonal of a regular pentagon in relation to its side.

24. Van der Waerden, *op. cit.*, 144–146. The method is a simplification of the process of finding the greatest common measure as used in the text for $\sqrt{17}$ by taking new ratios equal to the actual ratios of the process; but van der Waerden's exposition is rather elliptical, and it may more clearly be set out as here.

25. Proclus, *In primum Euclidis*, G. Friedlein, ed., 118.7–9. (The reference in Friedlein's index is incorrect.) In favor of identifying him with Theodorus of Cyrene is the fact that Plato in one place (*Theaetetus* 143B) calls the Cyrenaic "Theodorus the geometer"; and Diogenes Laërtius, *op. cit.*, also calls him "the Cyrenaic geometer" and "the mathematician." Van der Waerden, *op. cit.*, 146, accepts the identification, but it is rejected (by implication) by H. Diels and

W. Kranz, eds., *Die Fragmente der Vorsokratiker*, 6th ed. (Dublin–Zurich, 1954; repr. 1969) and most writers; Glenn R. Morrow, *Proclus: A Commentary on the First Book of Euclid's Elements* (Princeton, 1970), 95, n. 70, thinks the reference is to Theodorus of Soli, who is cited by Plutarch on certain mathematical difficulties in the *Timaeus*. Diogenes Laërtius, *loc. cit.*, refers to twenty persons with the name Theodorus, and Pauly-Wissowa lists no fewer than 203. There was even a second Theodorus of Cyrene, a philosopher of some repute, who flourished at the end of the fourth century B.C. In the passage under discussion Proclus is reproducing Geminus' classification of curves; and in treating mixed curves he says the mixing can come about through "composition," "fusing," or "blending." According to Geminus and Proclus, but not Theodorus, planes can be blended but lines cannot.

26. Xenophon, *Memorabilia*, which is *Commentarii* IV.2, 10; *Xenophontis opera omnia*, E. C. Marchant, ed., II (Oxford, 1901; 2nd ed., 1921; repr. 1942), ll. 25–26.

BIBLIOGRAPHY

The works listed in the bibliography of the article on Theaetetus will serve also for Theodorus. In addition, the following, listed chronologically, may be consulted: T. Bonnesen, in *Periodico di mathematiche*, 4th ser., **1** (1921), 16; H. Hasse and H. Scholz, *Die Grundlagenkrisis der griechisch Mathematik* (Berlin, 1928), 28; K. von Fritz, "Theodorus 31," in Pauly-Wissowa, *Real-Encyclopädie der classischen Altertumswissenschaft*, 2nd ser., V (Stuttgart, 1934), cols. 1811–1825; J. H. Anderhub, *Joco-Seria: Aus den Papieren eines reisenden Kaufmannes* (Wiesbaden, 1941); R. Hackforth, "Notes on Plato's *Theaetetus*," in *Mnemosyne*, 4th ser., **10** (1957), 128; A. Wasserstein, "*Theaetetus* and the History of the Theory of Numbers," in *Classical Quarterly*, n.s. **8** (1958), 165–179; O. Toeplitz, *Kantstudien*, **33**, 28–29; M. Timpanaro Cardini, *Pitagorici, testimonianze e frammenti*, fasc. 2; Bibliotheca di Studi Superiori, **41** (Florence, 1962), 74–81.

IVOR BULMER-THOMAS

THEODOSIUS OF BITHYNIA (*b*. Bithynia, second half of the second century B.C.), *mathematics, astronomy.*

Theodosius was the author of *Sphaerics*, a textbook on the geometry of the sphere, and minor astronomical and astrological works. Strabo, in giving a list of Bithynians worthy of note in various fields, mentions "Hipparchus, Theodosius and his sons, mathematicians."[1] Vitruvius mentions Theodosius as the inventor of a sundial suitable for any region.[2] Strabo's references are usually in chronological order in their respective categories; and since Hipparchus was at the height of his career in 127 B.C., while Strabo and Vitruvius both flourished about the beginning of the Christian era, these statements could refer to the same person

and probably do. They harmonize with the fact that Theodosius is quoted by name as the author of the *Sphaerics* by Menelaus (fl. A.D. 100). To allow sufficient time for his sons to be recognized as mathematicians in their own right before Strabo, Theodosius may best be regarded as a younger contemporary of Hipparchus, born in the second half of the second century B.C. and perhaps surviving into the first century; indeed, it is unlikely that such a work as the *Sphaerics* would have been written long after the development of spherical trigonometry by Hipparchus, for this development makes it look old-fashioned.

Confusion has been created, however, by the notice or notices in the notoriously unreliable *Suda Lexicon*. The passage reads:

> Theodosius, philosopher, wrote *Sphaerics* in three books, a commentary on the chapter of Theudas, two books *On Days and Nights*, a commentary on the *Method* of Archimedes, *Descriptions of Houses* in three books, *Skeptical Chapters*, astrological works, *On Habitations*. Theodosius wrote verses on the spring and other types of works. He was from Tripolis.[3]

It seems probable that the first sentence in this passage confuses the author of the *Sphaerics* with a later skeptical philosopher, for Theudas flourished in the second century of the Christian era;[4] and it also is probable that the second and third sentences should be regarded as a separate notice about a third Theodosius. This would be unimportant if the third sentence had not given rise to the belief that the author of the *Sphaerics* was born at Tripolis in Phoenicia, and in almost all editions until recently he has been described as Theodosius of Tripolis.[5]

Spherics, the geometry of the sphere, was needed for astronomy and was regarded by the ancient Greeks as a branch of astronomy rather than of geometry. Indeed, the Pythagoreans called astronomy "spherics"; and the stereometrical books XII and XIII of Euclid's *Elements*, which lead up to the inscription of the regular solids in a sphere, contain nothing about the geometry of the sphere beyond the proof that the volumes of spheres are in the triplicate ratio of their diameters. Euclid treated this subject in his *Phaenomena*, and just before him Autolycus had dealt with it in his book *On the Moving Sphere*. From a comparison of propositions quoted or assumed by Euclid and Autolycus, it may be inferred that much of Theodosius' *Sphaerics* is derived from some pre-Euclidean textbook, of which some have conjectured that Eudoxus was the author.[6] There is nothing in it that can strictly be called trigonometry, although in III.11 Theodosius proves the equivalent of the formula tan a = sin b tan A for a spherical triangle right-angled at C.

Two of the other works mentioned by the *Suda* have survived. *On Habitations* treats the phenomena caused by the rotation of the earth, particularly what portions of the heavens are visible to the inhabitants of different zones. *On Days and Nights* studies the arc of the ecliptic traversed by the sun each day. Its object is to determine what conditions have to be satisfied in order that the solstice may occur in the meridian at a given place and in order that day and night may really be equal at the equinoxes.

One reason why these three works have survived must be that they were included in the collection that Pappus called "The Little Astronomy"[7] — in contrast with "The Great Astronomy" or *Almagest* of Ptolemy. Pappus also annotated the *Sphaerics* and *On Days and Nights* in some detail.[8] All three works were translated into Arabic toward the end of the ninth century.[9] The translation of the *Sphaerics* up to II.5 is by Qusṭā ibn Lūqā and thereafter by Thābit ibn Qurra. The *Sphaerics* was translated from Arabic into Latin in the twelfth century by Plato of Tivoli and Gerard of Cremona.

There is no reason to doubt that the Theodosius who wrote the *Sphaerics* was also the author of the commentary on the *Method* of Archimedes mentioned by the *Suda*, for the subject matter would be similar. It may be accepted also that he wrote astrological works. It is tempting to think that the Διαγραφαὶ οἰκιῶν, *Descriptions of Houses*, mentioned in the *Suda*, dealt with the "houses of the planets"; but the latter term is always οἶκοι, not οἰκίαι. It must be considered an architectural work, which could, however, be by the author of the *Sphaerics*. The other works mentioned in the *Suda* must be regarded as by another person of the same name. Theodosius' discovery of a sundial suitable for all regions — πρὸς πᾶν κλίμα — may have been recorded in a book, but nothing is known about it.

NOTES

1. Strabo, *Geography* XII.4, 9 c 566, A. Meineke, ed. (Leipzig, 1853), II, 795.13–14.
2. Vitruvius, *De architectura* IX.8,1, F. Krohn, ed. (Leipzig, 1912), p. 218.7.
3. *Suda Lexicon*, under Θεοδόσιος, Ada Adler, ed., II (Leipzig, 1931), Θ 142 and 143, p. 693.
4. Diogenes Laërtius, *Vitae philosophorum* IX.116, H. S.

Long, ed. (Oxford, 1964), II, 493.14. He was, according to Diogenes, the fifth skeptical philosopher in succession to Aenesidemus, who flourished at the time of Cicero.
5. Even the definitive ed. by J. L. Heiberg (1927) is entitled "Theodosius Tripolites Sphaerica" but the first entry in the corrigenda (p. xvi) is "Tripolites deleatur ubique."
6. The following propositions in the *Sphaerics* are certainly pre-Euclidean: bk. I, props. 1, 6, 7, 8, 11, 12, 13, 15, 20; bk. II, props. 1, 2, 3, 5, 8, 9, 10, 13, 15, 17, 18, 19, 20; bk. III, prop. 2.
7. Ὁ μικρὸς ἀστρονομουμενος (sc. τόπος). Pappus, *Collection* VI *titulus*, F. Hultsch, ed., *Pappi Alexandrini Collectionis quae supersunt*, II (Berlin, 1877), 475.
8. *Sphaerics*, ibid., VI.1–33, props. 1–26, F. Hultsch, ed., 475–519; *On Days and Nights*, ibid., VI.48–68, props. 30–36, F. Hultsch, ed., II, 530–555.
9. H. Wenrich, *De auctorum Graecorum versionibus et commentariis Syriacis Arabicis etc.* (Leipzig, 1842), 206; H. Suter, *Die Mathematiker und Astronomen der Araber und ihre Werke* (Leipzig, 1900), 41.

BIBLIOGRAPHY

I. ORIGINAL WORKS. The three surviving works of Theodosius are in many MSS, of which the most important is Codex Vaticanus Graecus 204 (10th cent.). The *Sphaerics* was first printed in a Latin ed. translated from the Arabic (Venice, 1518), which was followed by Voegelin's Latin ed. (Vienna, 1529), also taken from the Arabic. The *editio princeps* of the Greek text (with Latin trans.) is J. Pena, *Theodosii Tripolitae Sphaericorum libri tres* (Paris, 1558). Subsequent eds. are F. Maurolico (Messina, 1558; Latin trans. only); C. Dasypodius (Strasbourg, 1572; enunciations only in Greek and Latin); C. Clavius (Rome, 1586; Latin trans. only, with works of his own); J. Auria (Rome, 1587); M. Mersenne (Paris, 1644); C. Dechales (Lyons, 1674); I. Barrow (London, 1675); J. Hunt (Oxford, 1707); E. Nizze (Berlin, 1852). The definitive ed. is J. L. Heiberg, "Theodosius Tripolites Sphaerica," in *Abhandlungen der Gesellschaft der Wissenschaften zu Göttingen*, phil.-hist. Kl., n.s. **19**, no. 3 (1927), which contains notes on the MSS (i–xv), text with Latin trans. (1–165), and scholia (166–199).

The Greek enunciations of *On Habitations* and *On Days and Nights* were included by Dasypodius in his ed. (Strasbourg, 1572) and Latin translations of the two texts were published by J. Auria (Rome, 1587 and 1591, respectively), but the Greek texts were not printed until the definitive ed. by R. Fecht, "Theodosii *De habitationibus liber De diebus et noctibus libri duo*," in *Abhandlungen der Gesellschaft der Wissenschaften zu Göttingen*, phil.-hist. Kl. n.s. **19**, no. 4 (1927), which contains notes (1–12), text and Latin trans. of *On Habitations* (13–43), scholia on *Habitations* (44–52), text and Latin trans. of *On Days and Nights* (53–155), and scholia on *Days and Nights* (156–176). The scholia were first edited by F. Hultsch, "Scholien zur Sphärik des Theodosios," in *Abhandlungen des philosophisch-historische Classe der K. Sächsischen Gesellschaft der Wissenschaften*, **10**, no. 5 (1887).

There is a German trans. of the *Sphaerics* by E. Nizze, *Die Sphärik des Theodosios* (Stralsund, 1826). There are French translations by D. Henrion (Paris, 1615); J. B. du Hamel (Paris, 1660); and Paul ver Eecke, *Théodose de Tripoli: Les sphériques* (Paris–Bruges, 1927).

II. SECONDARY LITERATURE. The most useful material on Theodosius is Thomas Heath, *A History of Greek Mathematics* (Oxford, 1921), II, 245–252; the Latin intro. to R. Fecht's ed. (see above), 1–12; and K. Ziegler, "Theodosius 5," in Pauly-Wissowa, *Real-Encyclopädie der classischen Altertumswissenschaft*, n.s. V, cols. 1930–1935. Other sources, listed chronologically, are A. Nokk, *Über die Sphärik des Theodosios* (Karlsruhe, 1847); F. Hultsch, "Die *Sphärik* des Theodosios und einige unedierte Texte," in *Berichte der Sächsischen Gesellschaft der Wissenschaften* (1885); R. Carra de Vaux, "Remaniement des *Sphériques* de Théodose par Jahia ibn Muhammed ibn Abī Schukr al-Maghrabī al Andalusī," in *Journal asiatique*, **17** (1891), 287–295; P. Tannery, *Recherches sur l'histoire de l'astronomie ancienne* (Paris, 1893), 36–37; and A. A. Björnbo, "Studien über Menelaos' *Sphärik*: Beiträge zur Geschichte des Sphärik und Trigonometrie der Griechen," in *Abhandlungen zur Geschichte der mathematischen Wissenschaften*, **14** (1902), 64–65; and "Über zwei mathematische Handschriften," in *Bibliotheca mathematica*, n.s. **3** (1902), 63–75.

IVOR BULMER-THOMAS

THEON OF ALEXANDRIA (*fl.* Alexandria, second half of fourth century), *mathematics, astronomy.*

Theon's scholarly activity is firmly dated by his reports of two eclipses that he observed at Alexandria in 364: the solar eclipse of 16 June and the lunar eclipse of 26 November.[1] Other chronological references in his works all point to the 360's and 370's. The solar eclipse of 364 is used as an example of calculation in both the greater and the lesser commentaries on Ptolemy's *Handy Tables*, the lesser commentary providing examples of calculations that correspond to the dates 15 June 360 and 17 November 377.[2] Also, a list of Roman consuls preserved in one manuscript of Theon's edition of the *Handy Tables* stops with the consuls of the year 372.[3] There is only one ancient biographical notice on Theon, and that very brief.[4] It states that Theon lived under the emperor Theodosius I (reigned 379–395), a date consistent with the above evidence. Theon's daughter, Hypatia, who was famous in her own right as a mathematician (she is credited, among other things, with a revision of book 3 of her father's commentary on the *Almagest*) and as a Neoplatonic philosopher, was

torn to pieces by a mob of fanatic Christians at Alexandria in 415. Since there is no mention of Theon in the circumstantial account we have of this event, it is likely that he was already dead. Like his daughter, Theon was certainly a pagan. Whether he, too, favored Neoplatonism cannot be determined.

The ancient biographical notice also informs us that Theon was a member of the "Museum." This was an institution for the support of advanced learning, established at Alexandria about 300 B.C. by Ptolemy I, which had nourished many famous scholars but by Theon's time had declined sadly — if, indeed, it still existed (Theon is the last attested member). Whether he was connected with the Museum or not, Theon was certainly actively engaged in higher education. In the preface to his commentary on the *Almagest*, he says that he has composed the work at the urging of those who attended his lectures on the subject.[5] Indeed, all his extant works are the outcome of his "professorial" activity, being either commentaries on or editions of recognized classics of mathematics and astronomy, intended for the use of students. I will deal with the commentaries in the order in which Theon wrote them, which is established by internal references from one to the other.

Theon's most extensive work is his commentary on Ptolemy's *Almagest*. This was originally in thirteen books, corresponding to the number of books of the *Almagest*; but book 11 is lost, only a fragment of book 5 survives, and there are probably lacunae in other books. The passage in the preface mentioned above suggests that the commentary is a redaction of Theon's lectures, and that is how it reads. It is for the most part a trivial exposition of Ptolemy's text, explaining obvious points at excessive length. Despite Theon's promise to improve over previous commentators on the *Almagest*, "who claim that they will only omit the more obvious points, but in fact prove to have omitted the most difficult,"[6] the commentary is open to precisely this criticism. It is never critical, merely exegetic. To the modern reader it is almost useless for understanding Ptolemy; but it is of value for the occasional information it provides on now-lost mathematical and astronomical works, notably Zenodorus' treatise "On Isoperimetric Figures" in book 1. This passage probably is taken from the earlier commentary on the *Almagest* by Pappus (*fl.* 320), of which only books 5 and 6 survive. Comparison of the two commentaries for book 6 (the only area where they overlap) shows that while Theon borrows much from Pappus, his work is not a mere rewriting of his predecessor's but contains extensive contributions of his own.

Theon also published two commentaries on the *Handy Tables*. The latter, issued by Ptolemy after he had completed the *Almagest*, were meant to provide a convenient means of computing the positions of the heavenly bodies and other astronomical phenomena. In the preface to the larger commentary Theon claims that whereas he had predecessors who commented on the *Almagest*, he is the first to write a commentary on the *Handy Tables*. The earlier of the two commentaries is an extensive one, in five books, addressed to Eulalius and Origenes, whom Theon calls his "companions" (in the Museum ?). In it Theon explains not only how to use the tables but also the reasons for the operations and the basis of the tables' construction, and provides geometrical demonstrations. Thus it frequently covers the same ground as the commentary on the *Almagest*. The second, much smaller, commentary is addressed, like the commentary on the *Almagest*, to Epiphanius, presumably a pupil of Theon's. In the preface Theon refers to the larger commentary as "the more reasoned ($\lambda o\gamma\iota\kappa\omega\tau\acute{\epsilon}\rho\alpha$) introduction to computation with the *Handy Tables*," and explains that he has written this new work for that majority of his pupils in the subject who are unable to follow geometrical proofs.[7] The smaller commentary, then, merely sets out the rules for computation with the tables, adding occasional worked examples but no reasons. Theon's remark indicating the low mathematical caliber of his students is corroborated by what we should surmise from the nature of his works in general.

All other extant works by Theon are editions of previous authors. "Edition" here does not mean an attempt to establish the authentic text but, rather, a reworking of the original in a form considered more suitable for students. The most notable of Theon's editions is that of Euclid's *Elements*, which was so influential that it consigned the original text to near oblivion. Theon himself attests his work in his commentary on the *Almagest*, book 1, chapter 10, where he says: "That sectors of equal circles are in the ratio of the angles [at the centers] we have proved in our edition of the *Elements* at the end of the sixth book."[8] Indeed, nearly all extant manuscripts of the *Elements* have a proposition to that effect attached to book 6, proposition 33; and many of the manuscripts have titles indicating that they are "from the edition of Theon" or even "from the lectures ($\sigma \upsilon\nu o\upsilon\sigma\acute{\iota}\alpha\iota$) of Theon."

It was not until the early nineteenth century, however, when Peyrard discovered that the manuscript Vaticanus Graecus 190, which lacks that proposition and is significantly different from the vulgate in other respects, must be an example of the pre-Theonic text, that it became possible to determine the nature of Theon's alterations of Euclid. They are many but mostly trivial, leaving the essential content of the *Elements* almost unchanged. In many places the wording has been altered or expanded to achieve consistency or perspicuity of expression. Of the occasional changes of mathematical substance, a very few are corrections of real mistakes in Euclid's text. More are due to Theon's misunderstanding the original. In some cases he apparently omits what he considers wrong. He makes frequent additions to fill what he considers gaps in Euclid's reasoning, even interpolating whole propositions, as in the above example. On the whole, his edition can hardly be said to improve on the original, although it may well have fulfilled its purpose of being easier for his students to use.

Other works by Euclid of which Theon produced editions are the *Data* (a treatise on what elements of a geometrical figure must be given to determine it) and the *Optics*, both of which exist in Theonic and pre-Theonic versions. The first of these obviously was intended for more advanced students but shows the same general characteristics as the edition of the *Elements*, except that in it Theon is more inclined to abbreviate Euclid's exposition. The Theonic version of the *Optics*, on the other hand, is so different from the original, not only in its language (which is characteristic of the later *koine*) but also in the form of the proofs, that Heiberg conjectured that the text we have consists of Theon's lectures on the subject as taken down by one of his students. This view is supported by the introduction to the Theonic version, which is an exposition of the principles of optics, mostly in indirect speech, occasionally introduced by "he said" or the like. "He" is not identified in the text; but this part clearly has been taken down from a lecture, and it is a plausible guess that "he" is Theon. There is no direct evidence, however, that Theon was responsible for this version, although he is the most likely candidate.

The same may be said of a treatise on catoptrics (theory of visual reflection) that in the manuscripts is attributed to Euclid but must be judged spurious on stylistic grounds alone. Analysis of the contents shows that it is a late compilation containing a mixture of Euclidean and post-Euclidean optical theory. The style and nature of the treatise would be appropriate for Theon, but that does not prove his authorship. Both the *Optics* and the *Catoptrics* are elementary, and are on a far lower scientific level than Ptolemy's *Optics*, which was, however, neglected in later antiquity and has survived only through the Arabic tradition. If, then, it is correct to associate Theon with these "Euclidean" optical works, we have an example in yet another branch of mathematics of his pedagogical activity, directed toward beginning students.

Theon also produced the version in which Ptolemy's *Handy Tables* have come down to us, according to the superscriptions in the manuscripts. The only evidence we have for the original version is Ptolemy's own introduction giving instructions for their use. From this it appears that the changes introduced by Theon were slight, and confined mostly to the arrangement of the tables and updating the chronological list. No one, however, has yet investigated the problem thoroughly.

Among lost works attributed to Theon by the ancient biographical source is a "Treatise on the Small Astrolabe."[9] Arabic bibliographical works also attribute to him a work entitled "On Operation With the Astrolabe."[10] The term "small astrolabe" evidently is used to distinguish this instrument from the "armillary sphere" (which is always the meaning of $\dot{\alpha}\sigma\tau\rho o\lambda\dot{\alpha}\beta o\nu$ in the *Almagest*). Thus it must refer to the "astrolabe" in the medieval and modern sense, that is, an instrument used to solve problems in spherical astronomy by means of projection of the celestial sphere onto a plane. This interpretation is confirmed by the Arabic sources, which use *asṭurlāb* only in this sense.

No work on the astrolabe predating the sixth century survives, but we do have the treatise of John Philoponus (*fl.* 520) in Greek, and that of Severus Sebokht (written before 660) in Syriac. The latter draws on a previous treatise, the author of which he calls "the philosopher." The historian al-Yaʿqūbī, writing in Arabic about 875, lists the contents of a treatise on the astrolabe that he ascribes to Ptolemy.[11] Neugebauer has shown that Sebokht's treatise corresponds closely to that described by al-Yaʿqūbī. Since Sebokht distinguishes "the philosopher" from Ptolemy (whose tables he quotes by name), and since al-Yaʿqūbī attributes to Ptolemy works (such as "On the Armillary Sphere") that other Arabic bibliographical sources attribute to Theon,[12] Neugebauer concludes, plausibly, that Theon is the author of the astrolabe treatise described by al-Yaʿqūbī and used by Sebokht. It is most unlikely, however, that Theon invented the astrolabe. The essential mathematical theory (of mapping circles of the celestial sphere

onto a plane by stereographic projection) is treated by Ptolemy in his *Planisphaerium*, and the instrument may well predate Ptolemy.

Other lost works attributed to Theon in the Greek biographical source are "On Omens [for weather?] and Examination of Birds and the Cry of Ravens," "On the Rising of the Dog Star," and "On the Rising of the Nile." Nothing is known of these; and some or all should perhaps be attributed to the grammarian Theon of the first century of the Christian era, as should certainly the commentary by "Theon" on Aratus' astronomical poem *Phaenomena*. A slight work on the composition of an astronomical ephemeris found in one manuscript of Theon's commentary on the *Handy Tables* and ascribed to Theon by Delambre[13] certainly belongs to a later period.

Theon was a competent mathematician for his time, but completely unoriginal. He typifies the scholastic of later antiquity who was content to expound recognized classics in his field without ever attempting to go beyond them. The parts of his works that are of most interest for the modern reader, apart from the occasional pieces of historical information, are the worked examples of computations in his commentaries. It is of no small interest to see how the Greeks carried out calculations using their form of the sexagesimal place-value system (Theon provided worked examples of extraction of a square root, as well as multiplication and division). The detailed calculation of the solar eclipse of 364 (which Theon demonstrated both according to the *Almagest* tables and according to the *Handy Tables*) is also most instructive.

For a man of such mediocrity Theon was uncommonly influential. As we have seen, it was his version of Euclid's *Elements* that gained most currency. It was in his edition that the *Handy Tables* passed to Islamic astronomers (among whom it went under his name), and thence (via al-Battānī's work and the Toledan Tables) to Latin Europe in the twelfth century. His commentaries on the *Almagest* and the *Handy Tables* continued to be studied in the Greek-speaking Eastern Empire, and are the basis of at least one Byzantine commentary, that of Stephen of Alexandria on the *Handy Tables*. The work on the astrolabe was probably the main, if not the sole, source of transmission of the theory of that instrument to Islamic astronomy, whence it came to medieval Europe.

One short passage in Theon's shorter commentary on the *Handy Tables* had a remarkable history. He states that "certain ancient astrologers" believed that the tropical and solstitial points had a vibrating back-and-forth motion over eight degrees of the ecliptic. Although not accepting this theory, he explains how to compute the resultant correction to be applied to the positions of the heavenly bodies.[14] According to Theon, the tropical points of "the astrologers" are eight degrees in advance of (to the east of) those of Ptolemy in 158 B.C., and move westward with respect to the latter at a rate of one degree in eighty years. (Thus they would coincide in 483, at which point they would begin to move eastward again.) There is perhaps one other trace of this theory in antiquity, but it was not until Theon's description reached the Islamic astronomers that it bore fruit.

When observational astronomy began to be seriously practiced, under the caliph al-Ma'mūn (early ninth century), it was soon realized that the rate of precession (motion of the tropical points with respect to the fixed stars) as determined by Ptolemy (one degree in one hundred years) was not valid, and that 1.5 degrees in one hundred years was closer to the truth. Rather than impute error to the admired Ptolemy, many preferred to believe that the rate of precession was not a constant, but varied cyclically; the idea undoubtedly came from this passage of Theon's, as is shown by the earliest reference we have, in which the astronomer Ḥabash al-Ḥāsib (ca. 850) is said to have introduced into one set of his astronomical tables "the back-and-forth motion of the ecliptic according to the opinion of Theon."[15] Soon afterward Thābit ibn Qurra (ca. 870) wrote a treatise expounding the theory and proposing a physical model to account for it. This was translated into Latin in the twelfth century under the title "De motu octave spere," and proved enormously influential in western Europe. The theory, usually known as "trepidation," was adopted by the makers of the Alfonsine Tables, and appears in various forms in the works of Peurbach, Johann Werner, and Copernicus. It was still seriously discussed in the late sixteenth century.

NOTES

1. Both in bk. 6 of his commentary on the *Almagest*, Basel ed., 332 and 319, respectively.
2. *Tables manuelles astronomiques*, N. Halma, ed., I, 77–87; examples are on 31 and 74. In the second passage the "90th" year of Diocletian was corrected by H. Usener in his *Kleine Schriften*, III, 22, n. 20, to "94th" on the basis of MS readings. The correction is confirmed by my computations. This passage, however, is probably an interpolation.
3. "Fasti Theonis Alexandrini," H. Usener, ed., 367–368, 381.
4. *Suda Lexicon*, Ada Adler, ed., II, 702.
5. *Commentaires . . . de Théon . . .*, A. Rome, ed., II, 317.
6. *Ibid.*, 318.
7. *Tables manuelles astronomiques*, I, 27.
8. *Commentaires . . . de Théon . . .*, A. Rome, ed., II, 492.
9. *Suda Lexicon*, loc. cit.
10. For instance, *Fihrist*, G. Flügel, ed., I, 268.
11. Translated by M. Klamroth in *Zeitschrift der Deutschen morgenländischen Gesellschaft*, **42** (1888), 23–25.
12. *Ibid.*, 20–23; compare *Fihrist*, loc cit.
13. *Histoire de l'astronomie ancienne*, II, 635.
14. *Tables manuelles astronomiques*, I, 53.
15. Ibn al-Qiftī, J. Lippert, ed., 170.

BIBLIOGRAPHY

I. ORIGINAL WORKS. Bks. 1–4 of the commentary on the *Almagest* were edited by A. Rome as *Commentaires de Pappus et de Théon d'Alexandrie sur l'Almageste*, II and III (Vatican City, 1936–1943), the commentary by Pappus being vol. I (1931)—they are Studi e Testi, nos. 72, 106, and 54, respectively. For the remaining books one must still consult the text in the Greek *Almagest*, *Claudii Ptolemaei magnae constructionis . . . lib. xiii. Theonis Alexandrini in eosdem commentariorum lib. xi* (Basel, 1538). On the relationship of Theon's commentary to that by Pappus, see Rome, *op. cit.*, II, lxxxii–lxxxvi; on Hypatia's supposed revision of bk. 3, *ibid.*, III, cxvi–cxxi.

The longer commentary on the *Handy Tables* has never been printed, except for two short passages published by Usener on pp. 360 and 372–373 of his ed. of Theon's list of consuls, "Fasti Theonis Alexandrini," pp. 359–381 in T. Mommsen, ed., *Chronica minora saeculorum IV. V. VI. VII.* (Berlin, 1898), which is *Monumenta Germaniae historica, auctores antiquissimi*, **13**, pt. 3. I have consulted the longer commentary in MS Nuremberg, Stadtbibliothek Cent. Gr. V, 8, fols. 215r–237v. The shorter commentary is printed in the very bad ed. of the *Handy Tables* by N. Halma, *Tables manuelles astronomiques de Ptolémée et de Théon*, I (Paris, 1822), 27–105. Ptolemy's own intro. to the *Handy Tables* is published by J. L. Heiberg in *Claudii Ptolemaei Opera quae exstant omnia*, II, *Opera astronomica minora* (Leipzig, 1907), 159–185.

On Theon's ed. of Euclid's *Elements*, see J. L. Heiberg, *Litterargeschichtliche Studien uber Euklid* (Leipzig, 1882), 174–180; and esp. Heiberg's prolegomena to his critical ed. of the *Elements*, *Euclidis Opera omnia*, V (Leipzig, 1888), li–lxxvi; for the *Data* see H. Menge, *ibid.*, VI (*Euclidis Data cum commentario Marini*), xxxii–xlix; for the *Optics*, Heiberg, *Litterargeschichtliche Studien*, 138–148; for the *Catoptrics, ibid.*, 148–153, and *Euclidis Opera omnia*, VII (*Optica et Catoptrica*), xlix–l.

II. SECONDARY LITERATURE. A good account is Konrat Ziegler, "Theon 15," in Pauly-Wissowa, *Real-Encyclopädie der classischen Altertumswissenschaft*, 2nd ser., X, cols. 2075–2080. The evidence for Theon's period of activity was collected by H. Usener, "Vergessenes," in his *Kleine Schriften*, III (Leipzig–Berlin,

1914), 21–23. The ancient biographical source is *Suda Lexicon*, Ada Adler, ed., II (Leipzig, 1931), 702, ll. 10–16. On Hypatia, see Socrates Scholasticus, *Historia ecclesiastica*, bk. 7, ch. 15, in J.-P. Migne, ed., *Patrologiae cursus completus*, series Graeca, LXVII, 767–770.

On the sources of the *Catoptrics* see A. Lejeune, "Recherches sur la catoptrique grecque," in *Mémoires de l'Académie royale de Belgique, Classe des lettres*, **52**, no. 2 (1957), 112–151. On Theon and the astrolabe see O. Neugebauer, "The Early History of the Astrolabe," in *Isis*, **40** (1949), 240–256. The treatise of Philoponus was published by H. H. Hase as "Joannis Alexandrini de usu astrolabii . . . libellus," in *Rheinisches Museum für Philologie*, **6** (1839), 127–171; that by Severus Sebokht by F. Nau, "Le traité sur l'astrolabe plan de Sévère Sabokht," in *Journal asiatique*, 9th ser., **13** (1899), 56–101, 238–303.

The passages in al-Ya'qūbī's work referring to the astrolabe and armillary sphere were translated by M. Klamroth as "Ueber die Auszüge aus griechischen Schriftstellern bei al-Ja'qûbî IV," in *Zeitschrift der Deutschen morgenländischen Gesellschaft*, **42** (1888), 1–44. What purports to be the Arabic translation of the work on the armillary sphere is extant in the MS Bombay, Mollā Fīrūz 86. See Fuat Sezgin, *Geschichte des arabischen Schrifttums*, V (Leiden, 1974), 401. Bibliographical works in Arabic mentioning Theon include Ibn al-Nadīm, *Kitāb al-Fihrist*, G. Flügel, ed., I (Leipzig, 1871), 268; and Ibn al-Qiftī, *Ta'rīkh al-ḥukamā*, J. Lippert, ed. (Leipzig, 1903), 108. Ptolemy's *Planisphaerium*, which exists only in the Arabic tradition, is printed (in the medieval Latin trans. from the Arabic) in his *Opera astronomica minora*, 227–259.

The work on the construction of an ephemeris is in Halma, *op. cit.*, III, 38–42, and described by J.-B. J. Delambre in his *Histoire de l'astronomie ancienne*, II (Paris, 1817), 635–638. Examples of division and extraction of a square root in the sexagesimal system, taken from Theon's commentary on the *Almagest*, are reproduced by T. L. Heath in his *A History of Greek Mathematics*, I (Oxford, 1921), 58–62. On Theon's calculation of the solar eclipse of 364 see A. Rome, "The Calculation of an Eclipse of the Sun According to Theon of Alexandria," in *Proceedings of the International Congress of Mathematicians, Cambridge, Mass., 1950*, I (Providence, R.I., 1952), 209–219. On Stephen of Alexandria's commentary on the *Handy Tables*, see H. Usener, "De Stephano Alexandrino," in his *Kleine Schriften*, III, 247–322.

For a possible use in a fourth-century papyrus of the "trepidation" described by Theon, see J. J. Burckhardt, "Zwei griechische Ephemeriden," in *Osiris*, **13** (1958), 79–92. The reference to Ḥabash's use of trepidation is in Ibn al-Qiftī, *op. cit.*, 170. The Latin text of Thābit's *De motu octave spere* has been printed many times; it is best consulted in J. M. Millás Vallicrosa, *Estudios sobre Azarquiel* (Madrid–Granada, 1943–1950), 496–509, trans. and commentary by O. Neugebauer, "Thabit ben Qurra 'On the Solar Year' and 'On the Motion of the

Eighth Sphere,' " in *Proceedings of the American Philosophical Society*, **106** (1962), 264–299. The best discussion of the history of the theory of trepidation in the Latin West is Jerzy Dobrzycki, "Teoria precesji w astronomii średniowiecznej" ("The Theory of Precession in Medieval Astronomy"), with Russian and English summaries, in *Studia i materiały z dziejów nauki polskiej*, ser. C, **11** (1965), 3–47. See also J. L. E. Dreyer, *History of the Planetary Systems From Thales to Kepler* (Cambridge, 1906), index under "Trepidation."

G. J. TOOMER

THEON OF SMYRNA (*fl.* early second century A.D.), *mathematics, astronomy.*

Theon is known chiefly for his handbook, usually called *Expositio rerum mathematicarum ad legendum Platonem utilium*. He may well have been the person called "the old Theon" by Theon of Alexandria in his commentary on the *Almagest*. Ptolemy referred to "Theon the mathematician," who is almost certainly the Theon discussed here, and ascribed to him observations of the planets Venus and Mercury made in 127, 129, 130, and 132 (*Almagest* 9.9, 10.1, 10.2). The latest writers named by Theon were Thrasyllus, who was active under Tiberius, and Adrastus, the Peripatetic and Aristotelian scholar, who flourished not earlier than A.D. 100. A contemporary bust of Theon from Smyrna has an inscription calling him *Platonikos*: he was thus also known as a philosopher, and his philosophical interests are evident in the *Expositio*.

The treatise is valuable for its wide range of citation from earlier sources. There is little evidence of mathematical originality. Despite the title, the book has little to offer the specialist student of Plato's mathematics. It is, rather, a handbook for philosophy students, written to illustrate how arithmetic, geometry, stereometry, music, and astronomy are interrelated. Geometry and stereometry are cursorily treated, however, perhaps because Theon assumed his readers to be adequately acquainted with them. A promise to provide a lengthy treatment of the harmony of the cosmos (p. 17, l. 24, Hiller edition) is not kept in the extant manuscripts; if that part of the treatise was ever written, it may have been lost early.

The arithmetical section treats the types of numbers in the Pythagorean manner; Theon dealt, for example, with primes, geometrical numbers (such as squares), "side" and "diameter" numbers, and progressions.

Music is divided into three kinds: instrumental, musical intervals expressed numerically (theoretical music), and the harmony of the universe. Theon stated clearly that he is not claiming to have discovered any musical principles himself; his aim is to expand the findings of his predecessors. He therefore quoted amply from his authorities—Thrasyllus, Adrastus, Aristoxenus, Hippasus, Eudoxus, and, of course, Plato. In the account of proportions and ratios the discussion concerns the treatment in Eratosthenes' *Platonikos* of the difference between interval and ratio ($\delta\iota\acute{\alpha}\sigma\tau\eta\mu\alpha$ and $\lambda\acute{o}\gamma o\varsigma$). Eratosthenes is also followed in the exposition of the different kinds of means. Some of the musical part descends into mere number mysticism; it is perhaps the least satisfactory feature of the work. A typical remark (p. 106, Hiller ed.) is that "the decad determines number in all respects. It embraces nature entire within itself, even and odd, moving and unmoved, good and bad."

In contrast, the astronomical section, which also depends much on Adrastus, is of great merit. The earth is a sphere; mountains are minute when compared with the earth, which lies at the center of the universe. The several circles of the heavens are explained, as are the assumed deviations in latitude of the sun, moon, and planets. The various views concerning the order of the heavenly bodies are noted; those of the (neo)Pythagoreans are contrasted with the systems of Eratosthenes and "the mathematicians." Some interesting hexameter verses quoted (pp. 138–140, Hiller ed.) on this topic are said to be by Alexander of Aetolia, but are perhaps by Alexander of Ephesus, a contemporary of Cicero; to the other planets, the earth, the sun, the moon, and the sphere of the fixed stars, Alexander gives a tone, so that all are set in an octave by arrangement of the intervals. Eratosthenes did not count the stationary earth, so in his verses he gave a note each to all seven moving bodies and an eighth to the sphere of the fixed stars. This is as close as Theon came to delivering the promised exposition of the harmony of the cosmos.

Theon explained the progressions, stations, and retrogradations of the planets. He described the eccentric and epicyclic hypotheses, and their equivalence. He seemed to consider Hipparchus as the inventor of the epicyclic hypothesis (p. 188, l. 16, Hiller ed.) that "Hipparchos praised as his own"; but there is a misunderstanding, because Apollonius clearly understood the principle of the epicycle before Hipparchus. Apollonius is not among the authorities cited by Theon.

Estimates of the greatest arcs of Mercury and Venus from the sun are given as 20° and 50° (p.

187, ll. 10–13, Hiller ed.). After an extensive account of the systems of rotating spheres worked out by Eudoxus, Callippus and Aristotle (pp. 178 ff., Hiller ed.), Theon turned to conjunctions, transits, occultations, eclipses, and the axis through the poles and the center of the zodiac.

Historically the most valuable part of the concluding pages is the brief fragment from Eudemus on pre-Socratic astronomy, which is full of problems. For example, the extant archetype manuscript here states that according to Anaximander, the earth is "on high" ($\mu\epsilon\tau\acute{\epsilon}\omega\rho o\varsigma$) and "moves" ($\kappa\iota\nu\epsilon\hat{\iota}\tau\alpha\iota$) about the center of the cosmos. Montucla's emendation of $\kappa\iota\nu\epsilon\hat{\iota}\tau\alpha\iota$ to $\kappa\epsilon\hat{\iota}\tau\alpha\iota$ ("rests") (see p. 198, l. 19, Hiller ed.) is attractive but by no means certain, since we do not know what Eudemus wrote, whatever Anaximander's view of the matter may have been. Anaximenes, not Anaxagoras, is here said to have declared that the moon "has her light from the sun."

Other works by Theon are lost. He himself referred to a commentary on Plato's *Republic* (*Expositio* p. 146, l. 4, Hiller ed.). Ibn al-Nadīm's *Fihrist* mentioned a treatise by him on the titles of Plato's writings and the order in which they should be read. He wrote on the ancestry of Plato, but not certainly in a separate treatise; the study may have formed part of the *Republic* commentary (see Hiller ed., p. 146, on Proclus, *On Timaeus*, p. 26A).

BIBLIOGRAPHY

The text of the *Expositio* depends almost entirely on two MSS in Venice: the number theory and the music are in Venet. Marc. 307 (11th–2th cent.), and the astronomy in Venet. Marc. 303 (13th–14th cent.). The first part (pp. 1–119, Hiller ed.) was edited by Ismael Boulliau (Paris, 1644); the other (pp. 120–205, Hiller ed.) by T. H. Martin (Paris, 1849; repr. Groningen, 1971). Both were edited together by E. Hiller in the Teubner version (Leipzig, 1878).

For further discussion, see K. von Fritz, in Pauly-Wissowa, *Real-Encyclopädie der classischen Altertumswissenschaft*, 2nd ser., X (1934), 2067–2075, *s.v.* Theon (14), with bibliography; and T. L. Heath, *A History of Greek Mathematics*, II (Oxford, 1921), 238–244. On the hexameter verses, see E. Hiller, in *Rheinisches Museum für Philologie*, **26** (1871), 586–587; and A. Meineke, *Analecta Alexandrina* (Berlin, 1843; repr. Hildesheim, 1964), 372–374.

G. L. HUXLEY

THEOPHILUS (**Theophilus Presbyter,** also called **Rugerus**) (*fl.* Helmarshausen, Germany [?], early twelfth century), *metallurgy, chemistry.*

The pseudonymous author of *De diversis artibus*, an instructive treatise on practical arts for the adornment of the church, Theophilus wrote in the first quarter of the twelfth century—perhaps, as Lynn White has suggested, in 1122–1123 in answer to Bernard of Clairvaux's animadversions on ecclesiastical luxury. On the basis of a note "qui et Rugerus," written in the seventeenth century on the title page of the Vienna manuscript, Albert Ilg identified Theophilus with the Benedictine monk Roger, who was active as a goldsmith in Helmarshausen around 1100 and slightly after. Although this identification was not accepted by two subsequent editors, Degering and Theobald, the most recent studies support it. Helmarshausen was an important center in northwestern Germany for all of the arts described by Theophilus, and some surviving pieces of ecclesiastical metalwork made by the historic Roger are of a style that almost seems designed to illustrate the metalworking techniques described in the manuscript.

The work is often called *Diversarum artium schedula*, following Lessing, who adopted a phrase from the preface to book I in two manuscripts that lacked titles. The title *De diversis artibus* was found by Dodwell on both the Vienna and the Cambridge manuscripts, and has been used by him and subsequent writers.

There are three parts to the work—book I, on the art of the painter; book II, on the art of the worker in glass; and book III, on the art of the metalworker. The last is most detailed. Art historians have made much use of book I; but the sections on glass and metalwork are of more importance in the history of science and technology, for they constitute the earliest firsthand accounts of many pyrotechnological processes that later bore fruit in chemical science and engineering, as well as in the entire modern materials industry.

With no theoretical speculation whatever, Theophilus recorded intimate practical details of the preparation of pigments, dyes, stained glass, brass, and bronze; the alloying and working of gold and silver; the heat treatment of steel; and the casting of metal objects ranging from small silver chalice handles made by the lost-wax process to huge bronze bells cast in clay molds shaped on a lathe. He described the separation of gold and silver by cementation and by sulfide reaction, as well as the removal of impurities by cupellation. His solder compositions were close to the minimum-melting-point alloys in the series copper-silver and lead-tin as known today. Theophilus gave details for the manufacture on a fairly large scale, for use as pigments, of mercury sulfide, basic copper acetate (verdigris), lead oxides (both PbO and Pb_3O_4), and lead carbonate. Silver sulfide was made and used as niello; linseed oil appeared as a varnish; gold leaf and powder were prepared for manuscript illumination; and vegetable colors sensitive to pH were properly employed. Glass was made by melting a fritted mixture of sand and beechwood ashes in wide-mouthed pots set in a relatively large furnace with a separate fritting hearth. This is the first written reference to the use of wood ashes to produce potassium glass, which two centuries earlier had begun to replace the ancient sodium glass based on natron. The color of Theophilus' glass changed with melting conditions and seems to have depended on the state of oxidation of iron and manganese present as impurities in the raw materials, although coloring through intentional additions of metallic oxides had been well established centuries earlier.

Mechanically, Theophilus made no use of the wind and water power sources that were just being introduced; but he described smaller devices, such as the rotary grindstone, the lathe (in several modifications), an elaborate organ, and a device for crosshatching iron surfaces to receive silver and gold overlay. He described in detail all the metalworker's tools, including the wire-drawing plate. There is little quantitative measurement and no philosophical speculation on the nature of materials; but Theophilus admirably reflects the practical environment and the mental attitude that characterized prescientific technology, and that led to many discoveries of the properties of different kinds of substances, the effects of heat upon them, and the nature of chemical reactions in general. Theophilus was not, of course, the inventor or discoverer of the processes he records, which had existed as practical tradition—in some cases for many centuries—without being reflected in the written record. Virtually everything he says is clear and is confirmed by the evidence of surviving contemporary objects.

Theophilus was an original writer—the first to describe clearly, from personal experience, the practical arts for the purpose of instruction and inspiration. His *De diversis artibus* is an incomparably better source of information on medieval technology than are the compilations of ancient and corrupt recipes, such as the *Mappae clavicula* and the *Compositiones variae* and their derivatives, which were in existence at the same time and were more a product of the acquisitiveness of librarians than of the practical labors of technical men.

BIBLIOGRAPHY

I. ORIGINAL WORKS. The earliest MSS of the *De diversis artibus* are two of the 12th century, one at Wolfenbüttel (Herzog-August Bibliothek 4373) and one at Vienna (Nationalbibliothek 2527). The 13th-century MS in the British Museum (Harley 3915) is the most complete. For a listing of later MSS, see Johnson, Dodwell, and Hawthorne and Smith (below).

The best Latin text, with critical apparatus and English trans., is C. R. Dodwell, *Theophilus, De diversis artibus . . . . Theophilus, The Various Arts . . . .* (London, 1961). The first complete printed ed., based mainly on the Wolfenbüttel MS, was published by G. E. Lessing in his *Zur Geschichte und Literatur aus den Schützen der herzoglichen Bibliothek zu Wolfenbüttel,* VI (Brunswick, 1781), frequently repr. throughout the 19th century in Lessing's *Sämmtliche Schriften.* A critical text and French trans. by Charles de l'Escalopier (Paris, 1843) was soon followed by Robert Hendrie's text (mainly based on the British Museum MS) and English trans., *Theophili, qui et Rugerus, presbyteri et monachi libri III. De diversis artibus: Seu diversarum artium schedula . . .*, with a second title page, *An Essay Upon Various Arts . . . by Theophilus, Called Also Rugerus, Priest and Monk, Forming an Encyclopedia of Christian Art of the Eleventh Century* (London, 1847). A frequently cited, though carelessly edited, text and German trans. is by Albert Ilg: *Theophilus Presbyter Schedula diversarum artium. . . . Revidierter Text, Übersetzung und Appendix,* no. 7 in the series Quellenschriften für Kunstgeschichte und Kunsttechnik des Mittelalters und der Renaissance (Vienna, 1874).

All the above eds. were mainly art-historical or philological in purpose. The first to be edited primarily for its technical content was the text, and German trans., of bks. II and III by Wilhelm Theobald, *Technik des Kunsthandwerks im zehnten* [sic] *Jahrhundert des Theophilus Presbyter Diversarum artium schedula, in auswahl neu herausgegeben, übersetzt und erläutert. . . .* (Berlin, 1933). Then, after over a century of neglect in English, two independent eds. appeared within two years—Dodwell (see above) and J. G. Hawthorne and C. S. Smith, *On Divers Arts. The Treatise of Theophilus. Translated From the Medieval Latin With Introduction and Notes* (Chicago, 1963). There are several other 19th-century eds.

II. SECONDARY LITERATURE. Discussions of Theophilus' treatise and its influence are in the intro. to the printed eds. listed above. The German 1933 and English 1963 translations have extensive technical notes on glass and metalwork. For pigments, see especially H. Rosen-Runge, "Die Buchmalereirezepte des Theophilus," in *Münchner Jahrbuch der bildenden Kunst,* 3rd ser., **3–4** (1952–1953), 159–171; and *Farbgebung und Technik frümittelalterliche Buchmalerei. Studien zu den Traktaten Mappae Clavicula und Heraclius,* 2 vols. (Munich, 1967).

On the MSS and background, see Hermann Degering,

"Theophilus Presbiter qui et Rugerus," in *Westfälische Studien . . . Alois Bömer gewidmet* (Leipzig, 1928), 248–262; R. P. Johnson, "Note on Some Manuscripts of the *Mappae Clavicula,*" in *Speculum,* **10** (1935), 72–81; "The Manuscripts of the *Schedula* of Theophilus Presbyter," *ibid.,* **13** (1938), 86–103; and *Compositiones variae . . . An Introductory Study,* XXIII, no. 3 in Illinois Studies in Language and Literature (Urbana, 1939); D. V. Thompson, "The *Schedula* of Theophilus Presbyter," in *Speculum,* **7** (1932), 199–220; "Theophilus Presbyter . . .," *ibid.,* **42** (1967), 313–339; Lynn White, Jr., "Theophilus Redivivus," in *Technology and Culture,* **5** (1964), 224–233; and C. S. Smith and J. G. Hawthorne, "Mappae Clavicula, a Little Key to the World of Medieval Techniques," in *Transactions of the American Philosophical Society,* **64**, pt. 4 (1974).

CYRIL STANLEY SMITH

THEOPHRASTUS (*b.* Eresus, Lesbos, *ca.* 371 B.C.; *d.* Athens, *ca.* 287 B.C.), *botany, mineralogy, philosophy.*

Theophrastus was associated with Aristotle for more than two decades and succeeded him as head of what came to be known as the Peripatetic school. According to one report (Diogenes Laërtius V, 36), he studied under Plato before joining Aristotle. It is likely that he met Aristotle in Asia Minor (347) or on Lesbos (344–342), went with him to Macedonia (342–335), and then to Athens when Aristotle returned and began to teach in the Lyceum (335). When Aristotle retired to Chalcis shortly before he died (322), Theophrastus became the leader of the scholars and students who had met with Aristotle at the Lyceum.

During his tenure of thirty-five years Theophrastus had two thousand students, among them the physician Erasistratus and the philosopher Arcesilaus. So well was he regarded by the Athenians that an attempt to prosecute him for impiety failed and a restrictive law against him and other philosophers was repealed. On his death he bequeathed the school's property jointly to ten relatives and associates. To one of these, Neleus, he left the library, which would have included not only his own writings and those of Aristotle but also, presumably, the collection of others' writings made by Aristotle and himself. The provisions in his will for repairs to the property, for the use of it by a few friends, and for the disposition of the books suggest that the school had suffered in the tumult through which Athens had passed in the years before his death and that he regarded its future as uncertain. (On the transmission of the Peripatetic texts, see "Aristotle: Tradition and Influence.")

The Hellenistic lists of Theophrastus' writings contain over two hundred titles that cover not only the various branches of science and philosophy but also history, law, literature, music, poetics, and politics. Even if there is duplication in the lists, if some titles belong to parts of longer works, and if some are incorrectly ascribed, it is nevertheless evident that Theophrastus was a man of remarkable learning and industry. From his writings there remain only two longer works on botany, a few short treatises on science, an essay on metaphysics, the *Characters*, and fragmentary excerpts and paraphrases in the works of later writers. In length the two botanical works are about double the remainder.

From such meager and unbalanced evidence a uniform account and just assessment of Theophrastus' scientific accomplishments are impossible. He has generally been considered a botanist whose contributions in other fields were secondary to those of his teacher. His dependence on Aristotle may have been distorted by the transmission of the texts; several of the shorter treatises, notably the *Metaphysics* and *De sensibus*, are found in manuscripts with the longer works of Aristotle on the same subjects, and many of the fragments of lost works are in later Greek commentaries to Aristotle, thus giving the impression that his work was little more than an appendix.

Recent scholarship (Regenbogen and Steinmetz) has done much to correct this impression, but it is doubtful that an exact line can be drawn between the contributions of the two men. Their fundamental agreement is evident from their writing, but it does not follow that their long association was that of teacher and student. That, even when disagreeing, Theophrastus does not mention Aristotle by name is probably a sign not merely of respect but also of his assumption of responsibility for their work as a whole. The books that Neleus inherited from Theophrastus were a Peripatetic corpus. Apparently there was no systematic attempt to divide the works by author until catalogs were made toward the end of the third century B.C.; and even after the editions of Andronicus of Rhodes in the first century B.C., doubt remained. Within the works attributed to Aristotle in antiquity, modern scholars have not only discerned the influence of Theophrastus but also have identified some treatises as partly or wholly his. Questions of authorship are beyond the scope of this article. If in what follows differences are stressed, they should be regarded as due to a continuing process rather than to a radical change in attitude or method.

A further problem in an account of Theophrastus is the state of the preserved evidence. The latest general edition of his work (Wimmer) is based on incomplete knowledge of the manuscripts and fragments, and few of the individual works have received detailed study. It seems best to confine this article to a few works that are accessible in more recent editions with commentary and scholarly translation.

The general trend of Theophrastus' thought is shown by the *Metaphysics*. In the main he does not deviate from Aristotle's assumptions but, rather, points out the difficulties in their application. Thus he agrees that reality is divided into the intelligible and the sensible and that the intelligible cause of the sensible world is an unmoved first mover that causes motion by being the object of desire. But, he asks, if there is only one such mover, why do all the heavenly bodies not have the same motion? Why does desire not presuppose soul and, therefore, a psychical motion better than rotation? Why is rotation limited to the heavenly bodies and shared only incidentally by the sublunar region? If motion is as essential to the heavenly bodies as life is to living things, does it need any explanation (7–11, 27–28)?

With regard to teleology, too, Theophrastus has doubts. He accepts the general principle that nature does nothing in vain, but he questions its applicability. In the heavens, and still more in the terrestrial region, some things seem to be due to coincidence or necessity. What purpose is served by the incursions and refluxes of the sea or by the birth and nutrition of animals—let alone by such superfluous things as the breasts of males or such positively harmful things as the horns of deer? If, he says, nature does desire what is best, most things are recalcitrant; even among animate things, which are a small part of the universe, there are few for which existence is better than nonexistence (15, 28–32).

In the *Metaphysics*, Theophrastus offers no new or conclusive answers to the problems raised—nor is there evidence that he ever did. He does, however, suggest a way of dealing with the phenomena alone. We should perhaps conceive the unity of the universe to be that of a system the various parts of which are fitted together in the greatest possible harmony but differ in the degree to which they possess order, the heavenly bodies possessing more and the sublunar region less (16, 34). Hence there must be different kinds of knowledge, and each kind must use a method appropriate to its object: there must be different methods for the objects of reason; for primary natural objects; and for secondary natural objects, such as animals, plants, and inanimate things (22). When we advance to first principles, we contemplate them with the mind; but the starting point in our search for causes is sense perception, which observes differences among its objects and supplies material for thought (19, 24–25). This emphasis on the differences among objects and methods of knowledge, and on the need to start from observation of the particular, is characteristic of Theophrastus' scientific works as a whole.

On the nature of the primary material substances, Theophrastus accepted Aristotle's theory of four qualitatively distinguished simple bodies. In the introductory sections of his *De igne*, however, he gives a penetrating criticism of the theory. Comparing fire with the other simple bodies, he makes several observations that cast doubt on its status as an elementary substance: the other simple bodies change into one another but cannot generate themselves, but fire both generates and destroys itself; the ways in which fire is generated are for the most part violent; fire is generated in many different ways, but the other simple bodies are generated only by natural change into one another; moreover, these other simple bodies exist by themselves, but fire requires fuel as a substrate and is destroyed when the fuel has been exhausted (1–3).

The last objection is the most serious; if fire cannot exist without fuel, it cannot rightly be called a primary substance or principle, since it is neither simple nor prior to its substrate (4). Theophrastus considers several possibilities: that the first sphere is pure and unmixed heat, that celestial and terrestrial fires are different in kind, and that the sun is a kind of fire. He appears inclined to view heat rather than fire as a principle, for, he says, heat is more widely distributed than fire and is more influential in natural processes. He comes to no conclusion but instead notes further difficulties. If heat is always bound up in a substrate, it appears to be an affection of something else and not a principle. If it is objected that fire cannot exist independently, the same may be said of all the simple bodies, since they are all compounds and are reciprocally involved (8). As he recognizes, his discussion of fire has led to larger questions about the nature of first causes.

Theophrastus declines to attempt answers to the larger questions, and in the main part of *De igne* he concerns himself with terrestrial phenomena of fire. His topics include not only such central matters as the generation, preservation, and extinction of fire but also such farfetched examples as the quenching of fire by salamanders, the melting of coins in the belly, and the jumping of grain on Babylonian threshing floors. In short, he excludes nothing that seems, or is said, to be connected with fire and heat.

In some explanations he relies on two Aristotelian concepts, the interaction of opposites and antiperistasis (the concentration of one thing by another). By the former he explains the generation and destruction of fire: fire is nourished by moist fuel but is destroyed by an excess of moisture or cold, or by a greater fire (10–11, 20, 26–27). By the latter he explains why, in cold weather, fires burn more rapidly, baths are warmer, and our bodies are stronger (12–13, see also 14–18), and why fires are extinguished by excessive compression (11, 58). The difference between the two concepts is that the latter assumes that the qualitative opposites are stable and that the greater does not assimilate the lesser, as it does in combustion. This assumption runs through much of the treatise. He speaks of fire, heat, and flame as if they were composed of discrete particles having different degrees of fineness; and he explains interaction by the symmetry or asymmetry between the particles of one substance and the pores of another (for instance, 42). In so doing he appears to abandon the theory of qualitative elements and return to the pre-Socratic effluence-pore theory. It is clear, however, that the particles are not Democritean atoms, for interaction requires both the appropriate size of particles and pores and the qualitative difference between the substances. Theophrastus has not arrived at a new theory about the essential nature of fire as an element. He has, rather, demonstrated that, as Aristotle had said (*Meteorologica*, 340b21–23), the fire of our experience is different from elemental fire and that the various phenomena associated with it do not have a single explanation. It remained for his successor Strato to formulate a Peripatetic atomism.

Among the extant writings Theophrastus' monograph on petrology, *De lapidibus*, best illustrates his investigation into inanimate compounds of the elements. In this, along with several other works known only from fragments and references, he carries forward the detailed investigation proposed by Aristotle at the end of *Meteorologica* III. His theoretical basis is the classification already made by Plato and adapted by Aristotle: metals are composed of water, and stones and mineral earths are composed of earth (1). On this basis he gives a brief description of the processes by which stones and mineral earths are formed (2–3): their matter is earth that has been purified and made uniform through conflux or filtering or some other kind of separation; the purity and uniformity of the matter determines such qualities as smoothness, density, luster, and transparency; solidification of the matter is due in some cases to heat and in others to cold—although it might seem that all things composed of earth are solidified by heat, since solidification and dissolution are contrary processes.

By thus admitting that heat and cold may have the same effect, Theophrastus recognizes that the classification, as simply stated, is not adequate to account for the diversity of the phenomena. Just as metals are solidified by cold and dissolved by heat, stones and earths ought to be solidified by heat and dissolved by cold; but in fact some stones, such as metal-bearing ores, are dissolved by heat (9). It is noteworthy, too, that although he apparently intends conflux and filtering to account for some distinctive differences between formations (deposits and veins have been suggested by Eichholz), he makes only two inconclusive references to them when he turns to specific instances (50, 61).

The body of the work is a systematic discussion of stones (3–47) and mineral earths (48–69) found around the Mediterranean and in the farther regions traversed by Alexander's army. Substances are distinguished by visual and tactile qualities and by their behavior, particularly in reaction to fire (9–19). Included are the earliest known Greek references to the use of mineral fuel (16), the pearl (36), the touchstone in testing alloys (45–47), and the manufacture of white lead (56). Of particular interest for the history of technology are the accounts of the preparation of pigments (50–60) and the uses of earths dug from pits (61–69).

Theophrastus' purpose is not to give an exhaustive treatment of the subject but to illustrate differences between types and to record unusual cases for further investigation. His descriptions are for the most part brief and restricted to what is readily observable, and he omits many substances that must have been familiar (his list is only one-tenth that of Pliny). How much Theophrastus knew from his own observation is questionable. He probably did not collect specimens systematically or conduct experiments. He frequently makes it clear that he is relying on written documents or hearsay; he has not always observed what he reports as fact, as is shown by his statement that the pearl is transparent. His reliance on the reports of others leads him to treat seriously what might seem too fantastic even for mention, such as stones formed

by the urine of the lynx (28) and pumice formed by sea-foam (19,22). Such instances underscore the factual limitations of Theophrastus' work, but they do not detract from its historical significance. It is the first methodical study of mineralogy and the only one before Agricola's in the sixteenth century that considers mineral substances for themselves rather than for their curative or magical properties.

Theophrastus' works on botany correspond to Aristotle's *Historia animalium* and *De partibus animalium*; in *Historia plantarum* he is concerned with description, classification, and analysis, and in *De causis plantarum* with etiology. In the first book of *Historia plantarum*, taking the tree as his standard, he deals with general matters: permanent and annual parts and their composition; classification into tree, shrub, undershrub, and herb; general and special differences in the plants as wholes and in their parts. From this Theophrastus proceeds to particulars: book II, domesticated trees, their propagation, and their care; book III, wild trees; book VI, undershrubs; books VII and VIII, herbaceous plants. Included are three books on special topics: book IV, trees and plants peculiar to certain regions; book V, woods and their uses; book IX, plant juices and medicinal herbs. The main subjects of *De causis plantarum* are the following: book I, generation and propagation, sprouting and fruiting; book II, effects of natural factors; book III, effects of cultivation; book IV, seeds; book V, alteration, degeneration, and death; book VI, plant juices. (The treatise *De odoribus* and the lost treatise on wine and olive oil may originally have followed book VI.)

Within this framework Theophrastus describes and discusses some 550 species and varieties, extending geographically from the Atlantic through the Mediterranean littoral and as far east as India. Among his literary sources he cites poets, philosophers, and scientists from Homer to Plato (notably Empedocles, Menestor, and Democritus among the pre-Socratics who had written on plants). He makes frequent references to the beliefs and practices of farmers, physicians, root cutters, and other groups, as well as to the inhabitants of various regions (especially Macedonia, Arcadia, and the vicinity of Mt. Ida). His anonymous sources undoubtedly include not only oral reports but also technical writings, such as those by Diocles of Carystus on roots and poisons, and nontechnical writings, such as those by men who accompanied Alexander and noted vegetation of military importance or special interest along the way. (On the last see, for example, *Historia*, bk. IV.)

In his typological procedure Theophrastus makes no fundamental innovations; Aristotle had already used the same procedure in many other subjects, including zoology. Nor does he differ from Aristotle in his physiological theory. He regards plants as living things with a life dependent on the proportion of their innate heat and moisture and on the harmonious relation between them and their environment (*Historia*, I, 1, 1; 2,4 f.; 11, 1; *De causis,* I, 4,6; 10,5; 21,3; 22,2–3). His chief difference is, rather, in perspective. Aristotle regards plants as the lowest members of a system that culminates in man, as sharing with animals the nutritive faculty of the soul, and as illustrating similarities and dissimilarities within the system as a whole. Theophrastus, on the other hand, concentrates on the plants themselves and avoids systematization beyond his immediate subject. He does not speak of the plant's soul; and, although he does use analogy between plants and animals, he emphasizes its limits and says that to strive after comparison where none is possible is a waste of effort and may cause us to abandon the method that is appropriate to the investigation (*Historia*, I,1,4–5).

Theophrastus' insistence on appropriate method follows from his recognition of the differences between plants and animals (*Historia*, I,3,4) and of the manifold nature of plants. Generalization about plants as a whole is difficult because no part is common to them all as the mouth and stomach are common to all animals; they do not all have root, stem, branch, twig, leaf, flower, fruit, bark, core, fibers, and veins, although these and such parts belong to the plant's essential nature (*Historia*, I,1,10 f.). This diversity also makes it difficult to generalize about major classes and even about individual kinds; there is overlapping between classes, some plants seem to depart from their essential nature when they are cultivated (*Historia*, I,3,2), and each kind embraces several different forms (*Historia*, I,14,3). It is now clear why in the *Metaphysics* Theophrastus speaks of a method appropriate to plants as distinct from inanimate substances and even animals. In the study of such diverse material our object is not the universal but the particular, and our instrument is not reason but sense perception (*De causis*, II,4,8); we must pursue the unknown through what is manifest to the senses (*Historia*, I,2,3); and in offering explanations we must use causal principles that are in accord with the particular natures of the plants, for our accounts must agree with our observations (*De causis*, I,1,1).

Consistent with this methodological principle, Theophrastus treats received theory and opinion with respect and skepticism, and seldom commits himself outright on one side or the other. Thus he quotes Aristotle's dictum that nature does nothing in vain, but he does so only in support of what is already evident to perception (*De causis*, I,1,1). He explains the pericarp by anthropocentric teleology as being for man's nourishment, but he goes on to explain it in relation to the seed (*De causis*, I, 16,1; compare I,21,1). The reported infertility of cypress seeds makes him doubt that Aristotle's dictum is true, but he does not renounce it (*De causis*, IV,4,2). So, too, Theophrastus speaks of spontaneous generation and transmutation as if they were simple facts; but again he offers explanations that might have led him to reject these notions, and in the end he leaves the question open (*Historia*, III,1,4–6; VIII,8,3–4). The same noncommittal attitude is evident in his treatment of particular reports. Along with the credible, he includes nonsensical tales, such as that the scorpion is killed by the application of wolfsbane but revived by white hellebore. His comment on this last is significant for his use of all the theories and evidence received from others: "Fabulous tales are not made up without reason" (*Historia*, IX,18,2).

By assembling his data impartially, classifying and discussing them within an elastic system, and withholding judgment when it was not secured by facts, Theophrastus created what he called an appropriate method and laid the groundwork for modern botany. Many of his observations and explanations were necessarily incomplete or erroneous; use of the simplest magnifying lens would have resolved many of his doubts. Among his contributions of lasting interest, his accounts of the following may be mentioned: the "pericarpion," used for the first time as a technical term (*Historia*, I,2,1); parenchymatous and prosenchymatous tissues (*Historia*, I,2,5, f.); petalous and apetalous flowers (*Historia*, I,13,1); hypogynous, perigynous, and epigynous insertions of the corolla (*Historia*, I,13,3); centripetal and centrifugal inflorescences (*Historia*, VII,14,2); angiosperms and gymnosperms (*Historia*, I,11,2); monocotyledons and dicotyledons (*Historia*, VIII, 2,1–4). All except the first of these terms are modern, but there is no doubt that Theophrastus correctly distinguished the features to which they are applied. In the last passage he gives the clearest and most accurate description of germinating seeds before Malpighi in the seventeenth century.

Theophrastus' achievement in botany is all the

more remarkable when we bear in mind that these two works were a small part of his writings. Their preservation does not allow us to suppose that botany was his primary interest; the loss of his works on other subjects may have been due not to their lesser importance but to the chances of manuscript transmission or to the tastes of later antiquity. Nothing in his writing indicates that he thought himself to be—as he has since been called—a professional botanist or that he considered his work comprehensive in detail or in theory. Of the plants that Theophrastus mentions only a third are not attested from other sources, and domesticated and familiar wild varieties are predominant (he says that most of the wild are nameless and little-known; *Historia*, I,14,4); he also omits many plants that he must have known.

Although some of his accounts (such as that of germinating seeds) indicate personal observation, they do not warrant the belief that Theophrastus had an experimental garden or made extensive field trips in Greece, let alone abroad. Nor is there any reason to think that he had collaborators or trained informants either in the Greek part of the Mediterranean or with Alexander's army; if he had, there could hardly be so many gaps and uncertainties in his information. Some of his second-hand information he could not test himself; but, even when he could easily have done so, in some cases he did not (see, for example, *Historia*, VII,1,3–5). It may be asked what Theophrastus' intention was in writing at such length about incomplete and unverified evidence. The answer is probably to be found in his frequent reminders to himself and his readers that there must be further investigation. He was aware that what he wrote was merely the beginning, that more and better data were needed, and that his explanations might need revision. His hopes apparently came to nothing. Later Greek and Roman authors enlarged the stock of useful knowledge and Pliny compiled it, but scientific botany progressed no further until the Renaissance.

Theophrastus was no less influential as a historian and critic of science than as a scientist. Besides several studies of individual pre-Socratics, he wrote a general history in sixteen or eighteen books known as the *Physicorum opiniones*. As Diels has shown, this work was the direct or indirect source of many of the summaries made by the doxographers. The most extensive of these summaries, a handbook known as *Placita philosophorum*, was compiled by Aëtius in the second century from an earlier Stoic summary that in turn was based on Theophrastus' history, with additions of later Stoic and Epicurean material. Through these and related summaries Theophrastus provided not only many of the details of pre-Aristotelian theories but also their selection and schematic arrangement.

It has been supposed that Theophrastus' aim was to write an objective history, but this supposition may be questioned. The extracts on material causes that have been preserved from the first book, although they indicate firsthand knowledge of the pre-Socratic texts, closely follow the summary of causal theories given by Aristotle in *Metaphysics* I; and on other topics there are many similarities between the doxographers and the summary accounts of Aristotle. It would seem that the *Physicorum opiniones* was a compilation of Aristotle's accounts, supplemented by quotations, biographical data, and other information omitted by Aristotle, and arranged under the main topics discussed by Aristotle. Aristotle's purpose in discussing earlier theories is to put them in relation to his own—that is, to show to what extent they anticipated or approximated his and where they were inadequate; hence he reports only what is relevant to his theory and states it in terms of his theory. It may be that Theophrastus intended no more, that his history was a handbook to be used as background for his own exposition of Peripatetic theories. The use to which he might put such a handbook is suggested by his *De sensibus*, in which he reviews and criticizes earlier physiological psychology. Throughout he bases his criticism on Aristotle's doctrine that sense perception involves qualitative change; and, even when he cites the texts of the writers whom he discusses (such as Plato's *Timaeus*), he reformulates the theories so that they may be judged by Peripatetic standards.

BIBLIOGRAPHY

I. Original Works. Editions of Theophrastus' writings are J. G. Schneider, 5 vols. (Leipzig, 1818–1821); F. Wimmer, 3 vols. (Leipzig, 1854–1862), also one vol. with Latin trans. (Paris, 1866); and H. Diels, *De sensibus* and fragments of *Physicorum opiniones*, in *Doxographi Graeci* (Berlin, 1879).

Editions with commentary and translation are *Historia plantarum* (with *De odoribus* and *De signis tempestatum*), A. Hort, ed., 2 vols. (London, 1916); *De causis plantarum*, bk. I, R. E. Dengler, ed. (Ph.D. diss., Univ. of Pa., 1927); *De sensibus*, G. M. Stratton, ed. (New York, 1927); *Metaphysics*, W. D. Ross and F. H. Fobes, eds. (Oxford, 1929); *De lapidibus*, E. R. Caley and J. C. Richards, eds. (Columbus, Ohio, 1956), and

D. E. Eichholz, ed. (Oxford, 1965); and *De igne*, V. Coutant, ed. (Assen, Netherlands, 1971).

II. SECONDARY LITERATURE. See O. Regenbogen, "Theophrastos von Eresos," in Pauly-Wissowa, *Real-Encyclopädie der classischen Altertumswissenschaft*, supp. VII, 1353–1562. Studies that have appeared since Regenbogen's survey include J. B. McDiarmid, "Theophrastus on the Presocratic Causes," in *Harvard Studies in Classical Philology*, **61** (1953), 85–156, and P. Steinmetz, *Die Physik des Theophrast* (Bad Homburg, 1964).

J. B. McDIARMID

THEUDIUS OF MAGNESIA (*fl.* fourth century B.C.), *mathematics.*

Theudius, an early member of the Academy, is known only from a passage in Proclus' commentary on Euclid's *Elements* (*In primum Euclidis Elementorum librum commentarii*, G. Friedlein, ed. [Leipzig, 1883; repr. 1967], I, pp. 67–68). After mentioning Leo (who made an improved collection of the elements of geometry and invented *diorismi*, means of determining when a problem is soluble and when not) and Eudoxus, Proclus says:

> Amyclas [or, better, Amyntas] of Heraclea, one of Plato's friends, Menaechmus, a pupil of Eudoxus who had also studied with Plato, and Dinostratus his brother, made the whole of geometry still more perfect. Theudius the Magnesian had a reputation for excellence in mathematics and in the rest of philosophy; for he ordered . . . the elements carefully and made many of the limiting [or partial] theorems more general.

Proclus then states that another geometer, Athenaeus of Cyzicus, lived at about the same time. "These men associated . . . in the Academy and undertook investigations jointly." Next are mentioned Hermotimus of Colophon, who added to the *Elements*, and Philippus of Medma (or Opus), who is said to have revised and published Plato's *Laws*. All these statements by Proclus may well have originated with an excellent authority, the historian of mathematics Eudemus of Rhodes. Theudius may be placed between Eudoxus and Philippus; that is, he was a contemporary of Aristotle. Indeed, T. L. Heath made the reasonable suggestion that the propositions in elementary geometry that are quoted by Aristotle were taken from Theudius' *Elements*. We have, however, no means of knowing which propositions and theorems were his discoveries; nor are we told which Magnesia was his home.

BIBLIOGRAPHY

See K. von Fritz, in Pauly-Wissowa, *Real-Encyclopädie der classischen Altertumswissenschaft*, XI, pt. 2 (1936), 244–246; T. L. Heath, *A History of Greek Mathematics*, I (Oxford, 1921), 319–321; and *The Thirteen Books of Euclid's Elements* (repr. New York, 1956), 116–117; Glenn R. Morrow, *Proclus: A Commentary on the First Book of Euclid's Elements* (Princeton, 1970), 56, n. 45; and F. Wehrli, *Die Schule des Aristoteles*, VII *Eudemos von Rhodos* (Basel, 1955).

G. L. HUXLEY

THÉVENOT, MELCHISÉDECH (*b.* Paris, France, 1620 or 1621; *d.* Issy [now Issy-les-Moulineaux], France, 29 October 1692), *scientific correspondence and translation, natural philosophy.*

Thévenot was one of the important correspondents linking Paris to the rest of the European scientific world. He influenced the organization and founding of the French Academy of Sciences and was an intimate of Christiaan Huygens, Henry Oldenburg, Adrien Auzout, and numerous other mid-seventeenth-century personages. He was appointed keeper of the royal library in 1684 and admitted as a member of the Academy of Sciences in 1685.

Little is known of Thévenot's personal life. After his formal education he traveled about Europe and made several voyages later in his life, including two diplomatic missions to Italy, but did not leave the Continent ("Thévenot's" voyages to the eastern Mediterranean and to northern Africa were those of his nephew Jean). Thévenot was a bibliophile and man of letters with a personal library of thousands of works, including a rich collection of books and manuscripts in foreign languages. One of the relatively few French scientists who read English, Thévenot also knew Greek, Latin, Hebrew, and several oriental languages, including Arabic and Turkish. His manuscript collection passed to the royal library in 1712.

Groups of savants and amateurs in mid-seventeenth-century Paris met occasionally in private houses to discourse on philosophy and nature. From about 1654, and for a decade thereafter, Henri Louis Habert de Montmor held such meetings with enough regularity and sense of purpose that his group has been called the "Montmor Academy." Thévenot attended Montmor's meetings at least as early as 1658, and regularly for the next several years. Perhaps he began attending as early as 1655, for he met Huygens then, at a time when Huygens is known to have visited often with

Montmor and with Pierre Gassendi, who was then living in Montmor's house. Thévenot's associates at the Montmor meetings included Auzout, Pierre Petit, Ismael Boulliau, Bernard Frenicle de Bessy, Jacques Rohault, and Girard Desargues, as well as most of the important scientific figures who passed through Paris. In these early years the Montmor Academy considered clocks and other mechanical devices, astronomical discoveries, questions of spontaneous generation and other biological matters, recent English and Italian studies on the void, and capillarity. These two latter phenomena especially interested Thévenot as early as 1658, and this interest contributed to his later development of a bubble level. Although there were those at the Montmor meetings, such as Thévenot, who emphasized experiment and observation, other members preferred rhetoric to experiment and the Montmor Academy gained a reputation for disputation and displays of temper.

After about 1662, partly because of these squabbles, Thévenot provided occasions for additional meetings and experimentation in his country house at Issy, several miles south of Paris. There he supported the mathematician Frenicle de Bessy, the Danish anatomist Nils Stensen, and a chemical demonstrator. Also at Issy, Thévenot pursued his studies of the void and made various astronomical and magnetic studies, aided by Petit, Auzout, Frenicle de Bessy, and Huygens. Various problems, including the continuing bickering over the proper emphasis to be placed upon experiment, led to the demise of the Montmor Academy in 1664. Thévenot held his meetings for another year or so, but lack of sufficient funds for apparatus and experimentation hampered his work.

About this time efforts were made to reorganize the Montmor Academy. Samuel-Joseph Sorbière, its permanent secretary, approached the leading minister of Louis XIV, Jean-Baptiste Colbert, and presented him with the draft of a revised constitution and a letter asking for royal protection. At about the same time another plan was proposed for a new academy of scientists. The latter plan, a document of some twenty-six clauses and two thousand words, exists as an unsigned, undated manuscript.[1] Evidence indicates that the plan, entitled "Project de la Compagnie des sciences et des arts," was written about 1663 or 1664 by a group that included Thévenot, Auzout, and Petit. The proposed *compagnie* would perform experiments and make observations to disabuse the world of its "vulgar errors," seek to invent new machines, work especially at discovering the causes of ill health, make lists and tables of practical arts and machines, seek to improve navigation and commerce, publish the memoirs of those who voyaged to foreign lands, and write a universal natural history. It would be composed of

> . . . the most knowledgeable in all the true sciences that one can find, as in geometry, in mechanics, optics, astronomy, geography &c., in physics, medicine, chemistry, anatomy, &c., or in the applied arts, as architecture, fortifications, sculpture, painting, and design, the conduct and elevation of waters [hydraulics], metallurgy, agriculture, navigation &c.[2]

Religion and politics would be excluded. This Baconian program promised to form the *compagnie* for the "perfection of the sciences and the arts and, in general, to search for all that can bring utility or convenience to the human race, and particularly to France."[3] Thévenot's utilitarian project for a *compagnie des sciences et des arts* was quite different from one proposed by Charles Perrault, who envisioned an academy that was less practical and more philosophical and cultural, the pure sciences being allied with belles lettres rather than with engineering.

The Academy of Sciences that emerged in 1666 was more in Perrault's design than in Thévenot's. The amateurs among the Montmor group, including Thévenot, were not members of the new academy. Brown implies that Thévenot, as an amateur, was omitted from consideration.[4] Maury claims that Thévenot, although the "father" of the new academy, was not included among the original members because of "infirmities" that kept him at Issy.[5] This does not accord with the *éloge* in the *Journal des sçavans*, evidently written by a personal friend (possibly Jean Gallois), which states that Thévenot, "being of a robust constitution, . . . enjoyed excellent health" until the month of his death, when, attacked by a fever, he initiated a continuing fast that served further to weaken his health and brought on his death. Recent evidence published by McKeon quotes Thévenot, in a letter to Prince Leopold, as saying he had been invited to become a member.[6] In any case, Thévenot finally became a member of the Academy of Sciences in 1685, filling the place of Pierre de Carcavi, who had died in 1684.

Throughout the 1660's and 1670's Thévenot maintained a wide correspondence with numerous persons. Much of it related to his celebrated translation and publication of voyages of discovery, *Relations de divers voyages* . . .. His only notable direct contribution to science was in instrumentation: a bubble level, later improved by Robert Hooke and by Huygens. Thévenot's work on the level was probably connected with his experiments on capillarity and the siphon, undertaken in 1658–1661. He mentioned his level in 1661 and 1662 in letters to Vincenzo Viviani and to Huygens, and it was publicized in an unsigned pamphlet in 1666.[7] Thévenot's original design probably began to evolve in 1661 and was publicly mentioned at various times between 1666 and 1681, when he described it in his essay "Discours sur l'art de la navigation," published in his *Recueil de voyages* . . .. The original bubble level seemed to lack convexity in the glass tube, a fact remarked upon by Huygens.

Thévenot described and illustrated his level in 1681, still without discussing convexity in the tube, although he did mention the possibility of utilizing Auzout's idea of metal cross hairs with a lens for accurate measurement. The level tube, he said, was about the diameter of one's little finger and its length was seven or eight times its diameter. The level was filled with alcohol and mounted on a stone ruler fitted with a viewing lens. Glass fabrication methods made it difficult to obtain tubes of constant cross section, and Thévenot's design initially was used by others simply to align a plane surface parallel to the horizon. Thévenot's claims that his level would be useful in finding magnetic declination at sea and determining a plane parallel to the horizon at sea seem rather exaggerated. Daumas states that Thévenot's design did not come into common use until the mid-eighteenth century with the development of improved construction techniques.[8]

One other incident made Thévenot a subject of discussion in the scientific salons: his presentation and development of a theory concerning the cause of human and animal respiration. Sometime during 1660 an unnamed person at the Montmor Academy had suggested that a constant atmospheric pulsation caused the "movements of the heart, of the *punctum saliens* of the egg, that of the brain, the diaphragm, or of respiration."[9] Thévenot said that the idea, if not true, at least had some merit; and he proceeded to attempt to demonstrate the atmospheric pulsation experimentally. He claimed that if one put water into a flask with a capillary neck and then tipped the flask so that the water filled part of the neck, one could observe the water pulsating in sympathy with the atmospheric pulsation. After a slight warming, one could observe the "diastolic and systolic [motions] with all the circumstances that one finds in the parts of animals that have this movement," the rate of motion depending on the amount of heat applied.[10]

Thévenot spread this news through Oldenburg and Huygens, and it was widely discussed for a time. This incident is quite typical of the sort of experiment performed and the type of tales told in seventeenth-century science, and Thévenot is not to be imagined a fool. His scientific contributions, however, should be seen as limited to his general encouragement of scientific enterprise through his letters, his personal efforts, and his published translations. In the period after 1685, when he was a member of the Academy of Sciences, he occasionally discussed items at its meetings: for example, a buffalo-like creature in North America, lemon juice as a medicinal cure, the excision and subsequent regeneration of a lizard's tail, and ipecac as useful in treating dysentery.

NOTES

1. See Huygens, *Oeuvres complètes*, IV, 325–329, for the text of this document.
2. *Ibid.*, 328.
3. *Ibid.*, 325.
4. Harcourt Brown, *Scientific Organizations* . . ., 117.
5. L.-F. A. Maury, *L'ancienne Académie des sciences*, 12–13.
6. McKeon, "Une lettre de Melchisédech Thévenot . . .," 2–3.
7. Huygens, *Oeuvres complètes*, XXI, 105–108, presents Huygens' description of Thévenot's level.
8. Maurice Daumas, *Les instruments scientifiques aux XVIIe et XVIIIe siècles*, 77.
9. Thévenot to Huygens, in Huygens, *Oeuvres complètes*, III, 405.
10. *Ibid.*, 406.

BIBLIOGRAPHY

I. ORIGINAL WORKS. Consult the *Catalogue général des livres imprimés de la Bibliothèque nationale*, CLXXXVI, or Michaud's *Biographie universelle*, 2nd ed., XLI, for detailed listing of Thévenot's works. Discussions of the publishing history of Thévenot's *Voyages* appear in Armand Gaston Camus, *Mémoire sur la collection des grands et petits voyages* . . . (Paris, 1802), 279 ff.; and James Lenox, *The Voyages of Thévenot* (New York, 1879).

Thévenot's most famous work was his collection of translations of voyages of discovery, *Relations de divers voyages curieux* . . ., 4 vols. (Paris, 1663–1672); and a small supplement, *Recueil de voyages de Mr Thévenot* (Paris, 1681). The latter contains Thévenot's description of his level in "Discours sur l'art de la navigation." The *Catalogue de la bibliothèque de Thévenot* (Paris, 1694) contains a short autobiographical preface by Thévenot. (I have not examined the latter work.) Drawing on previous books by Everard Digby the elder and by Nicholas Winman, Thévenot wrote a primer on swimming that was published after his death as *L'art de nager* (Paris, 1696) and went through numerous illustrated French and English eds. over the next century.

Coincident with his interests in hydrostatics and mechanics, Thévenot began an ed. of works by Hero and others but died during the preparation. The task was completed by Jean Boivin and Philippe de La Hire: *Veterum mathematicorum, Athenaei, Bitonis, Apollodori, Heronis, Philonis et aliorum opera . . .* (Paris, 1693).

Letters written by and concerning Thévenot are in Christiaan Huygens, *Oeuvres complètes*, 22 vols. (The Hague, 1888–1950), III–VII, IX–X, XVII, XX, XXII; *The Correspondence of Henry Oldenburg*, edited and translated by A. Rupert Hall and Marie Boas Hall, (Madison, Wis., 1965–   ), I–III, VI. Robert M. McKeon published a letter from Thévenot to Prince Leopold in Florence, "Une lettre de Melchisédech Thévenot sur les débuts de l'Académie royale des sciences," in *Revue d'histoire des sciences et de leurs applications*, **18** (1965), 1–6. On Thévenot's influence on the founding of the Academy of Sciences, see also T. McClaughlin, "Une lettre de Melchisédech Thévenot," in *Revue d'histoire des sciences*, **27** (1974), 123–126. Harcourt Brown published an unsigned letter concerning Swammerdam's investigations and properly attributes this to Thévenot in his *Scientific Organizations in Seventeenth Century France (1620–1680)* (Baltimore, 1934), 280–281. Thévenot's comments and reports at the Academy of Sciences for the years 1685–1689 are mentioned in *Histoire de l'Académie royale des sciences depuis 1666 jusqu'à 1699*, I, II (Paris, 1733).

II. SECONDARY LITERATURE. The best general introduction to scientific societies in Thévenot's era, as well as to his career, is Brown, *op cit.* A contemporary *éloge* of Thévenot is in *Journal des sçavans*, **20** (17 Nov. 1692), 646–649 (pub. 1693). A brief biography is in Louis Moréri, *Le grand dictionnaire historique de Moréri*, X (Paris, 1759), 138–139. Subsequent biographical sketches of Thévenot add nothing to these. A brief history of the level and of the acceptance of Thévenot's bubble level is in Maurice Daumas, *Les instruments scientifiques aux XVIIᵉ et XVIIIᵉ siècles* (Paris, 1953), 76–78. For additional discussion of this, see Gilbert Govi, "Recherches historiques sur l'invention du niveau à bulle d'air," in *Bullettino di bibliografia e di storia delle scienze matematiche e fisiche*, **3** (1870), 282–296. McKeon's article (cited above) contains a discussion of events at the time of the founding of the Academy of Sciences in 1666; see also Brown, *op cit.*; Guillaume Bigourdan, "Les premières sociétés savantes de Paris . . .," in *Comptes rendus . . . de l'Académie des sciences*, **163** (1916), 937–943; **164** (1917), 129–134, 159–162, 216–220—also published separately as a pamphlet (Paris, 1919). In addition, consult L.-F. Alfred Maury, *L'ancienne Académie des sciences* (Paris, 1864), 10–13, 31, 37–38; and Roger Hahn, *The Anatomy of a Scientific Institution, the Paris Academy of Sciences,*

1666–1803 (Berkeley, 1971), 6–8. For information concerning Thévenot and his interests in astronomy and scientific instrumentation, see Robert M. McKeon, "Établissement de l'astronomie de précision et oeuvre d'Adrien Auzout" (unpub. diss., University of Paris, 1965).

C. STEWART GILLMOR

THIELE, F. K. JOHANNES (*b.* Ratibor, Upper Silesia, Germany [now Raciborz, Poland], 13 May 1865; *d.* Strasbourg, Germany [now France], 17 April 1918), *chemistry*.

Thiele, a major contributor to our knowledge of nitrogen compounds and to the theory of unsaturated organic molecules, was the second of six children of Friedrich August Thiele, a leading citizen of Ratibor and owner of a publishing house and bookstore. His mother, the former Elfriede Koppe, died when he was six. Thiele studied at the University of Breslau (1883–1884) and then at Halle under Jacob Volhard, obtained his doctorate there in 1890, and taught analytical and organic chemistry. His studies on the nitrogen compounds guanidine and hydrazine and their derivatives, some of them explosive, interested both industry and government.

In 1893 Thiele became associate professor under Adolf von Baeyer at Munich, where he continued his work on nitrogen derivatives and developed his most important research, which dealt with conjugated unsaturated systems—structures involving alternate single and double bonds along a chain of atoms. In 1902 he was appointed professor at Strasbourg, where he expanded and modernized the chemical institute and in 1910 became rector of the university. During World War I, Thiele served for a time as a censor of telegrams and developed a gas mask against carbon monoxide. On Volhard's death in 1910, he became editor of *Justus Liebig's Annalen der Chemie*. In the same year he was elected to the Munich Academy of Sciences. He died, unmarried, of a heart ailment at age fifty-three.

Organic chemistry seemed divided, in the latter nineteenth century, into two parts—the derivatives of benzene, C_6H_6, and the rest, including unsaturated compounds. The latter, containing double or triple bonds, $\diagdown C = C \diagup$ and $-C \equiv C-$, added bromine (and other reagents) readily, forming $\diagdown CBr - CBr \diagup$ and $-CBr_2 - CBr_2-$. Benzene, to which Kekulé in 1865 had assigned a structure with three double bonds

was quite unreactive, however, adding bromine only in the presence of light.

Thiele constructed a bridge between the two realms. Building on Fittig's and A. von Baeyer's discovery that structures $C=C-C=C$ added hydrogen, H_2, not as expected, at a double bond, but at the ends of the chain, thereby yielding $CH-C=C-CH$, with a new double bond in the center, Thiele proposed that "double bond" is a misnomer because the atoms are not bound twice as strongly as when singly bound. Rather, doubly bound carbon atoms retain a "partial valence" that explains their reactivity. In a system of alternating single and double bonds, which Thiele called a "conjugated system," the inner partial valences neutralize each other, leaving the end atoms reactive.

Since benzene has a closed conjugated system

its lack of reactivity is explained.[1] Thiele's theory stimulated extensive research by himself and others, and was a direct precursor of electronic theories of organic reaction mechanisms.[2]

Thiele's extensive research on nitrogen chemistry, particularly on derivatives of hydrazine $[H_2NNH_2]$, and guanidine $[HN=C(NH_2)_2]$, led to the discovery of numerous new compounds and new synthetic processes. He prepared nitramide (NH_2NO_2), which is isomeric with hyponitrous acid $(H_2N_2O_2)$; these were the first examples in inorganic chemistry of compounds having identical molecular formulas but different properties. He also achieved the synthesis of five- and seven-membered ring compounds containing nitrogen.

NOTES

1. J. Thiele, "Zur Kenntnis der ungesättigten Verbindungen," in *Justus Liebig's Annalen der Chemie*, **306** (1899), 87–266; **319** (1901), 129–143.
2. C. K. Ingold, *Structure and Mechanism in Organic Chemistry*, 2nd ed., p. 76.

BIBLIOGRAPHY

I. ORIGINAL WORKS. A full bibliography, listing over 130 articles and the titles of dissertations by Thiele's students, is included in the obituary by F. Straus (see below). A section of Thiele's paper on conjugated systems (*Annalen der Chemie*, **306** [1899], 89–90) is translated in H. M. Leicester and H. S. Klickstein, eds., *A Sourcebook in Chemistry* (New York, 1952), 510–511.

II. SECONDARY LITERATURE. Fritz Straus, "Johannes Thiele," in *Berichte der Deutschen chemischen Gesellschaft*, **60** (1927), 75A–132A, consists of an obituary, extensive discussion of his chemical work, and full bibliography. Poggendorff, IV, 1489–1490, and V, 1249–1250, presents biographical notes and bibliography. J. R. Partington, *A History of Chemistry*, IV (London, 1964), 847–848, summarizes Thiele's chemical contributions and includes important references. Thiele's theory of partial valences is discussed in F. Henrich, *Theorien der organischen Chemie* (Brunswick, 1912), 34–82; in E. Hjelt, *Geschichte der organischen Chemie* (Brunswick, 1916), 459–460, 467–469; and, more recently, in C. K. Ingold, *Structure and Mechanism in Organic Chemistry*, 2nd ed. (Ithaca, N.Y., 1969), 75–77, 184, 957–958.

Thiele's contributions to the chemistry of organic nitrogen compounds are referred to in N. V. Sidgwick, T. W. J. Taylor, and W. Baker, *The Organic Chemistry of Nitrogen* (Oxford, 1937), 274, 286, 287, 297, 348, 361, 378, 384, 446.

OTTO THEODOR BENFEY

THIELE, THORVALD NICOLAI (*b.* Copenhagen, Denmark, 24 December 1838; *d.* Copenhagen, 26 September 1910), *astronomy, mathematics, actuarial mathematics.*

Thiele was the son of Just Mathias Thiele, a well-known Danish folklorist and art historian. While studying at the University of Copenhagen, young Thiele was awarded a gold medal for a paper on the geometry of the apparent course of a solar eclipse. In 1866 he took his doctorate, and from 1875 to 1906 he was professor of astronomy at the university and director of the university observatory.

Thiele's scientific work has been characterized by C. Burrau as "a treatment of numerical values derived from observations." If the word "observation" is taken in its widest sense, and if the word "treatment" is taken to mean a penetrating and original mathematical analysis, then this description must be regarded as apt. The topic of Thiele's dissertation was the determination of the orbit of the visual double star γ Virginis. He developed a new method of orbit determination, now known as the Thiele-Innes method (with some of the formulas later arranged for mechanical computation by Robert Innes). Thiele discussed the systematic errors in the observational material for this star and later for other double stars, using in particular the series of observations published in 1878–1879 by Otto W. Struve.

For many years Thiele continued and intensified his studies of the systematic and accidental errors of observation, thus approaching the field of actuarial mathematics. Indeed, for nearly forty years he was the manager of a life insurance company; in this work he satisfied his interest in the practical use of mathematics and numerical computations. In his scientific work, Thiele tried to discover, by means of numerical calculations, the laws for the distribution of the spectral lines of certain elements; and he was an early pioneer in the numerical search for solutions to the three-body problem, developing the Thiele (or Thiele-Burrau) transformation for this purpose.

The mathematical background of his work is given in his books *Theory of Observations* (1903) and *Interpolationsrechnung* (1909). Because of an eye disease, Thiele was unable to do any practical astronomical observation during much of his career, but he took an early part in the development of photography for astronomical purposes.

BIBLIOGRAPHY

I. ORIGINAL WORKS. Thiele's books include *Undersøgelse af Omløbsbevaegelsen i Dobbeltstjernesystemet Gamma Virginis* (Copenhagen, 1866); *Almindelig Iagttagelseslaere: Sandsynlighedsregning og mindste Kvadraters Methode* (Copenhagen, 1889); *Elementaer Iagttagelseslaere* (Copenhagen, 1897); *Theory of Observations* (London, 1903); and *Interpolationsrechnung* (Leipzig, 1909).

His articles include "Castor, calcul du mouvement relatif et critique des observations de cette étoile double," in *Festskrift, Copenhagen University* (1879); review and discussion of Otto W. Struve's measurements of double stars, in *Vierteljahrsschrift der Astronomischen Gesellschaft*, **15** (1880), 314–348; "Note on the Application of Photography to the Micrometric Measurements of Stars," in *Washington Observations for 1885*, appendix I (1889), 58–67; "Om Nutidens Reform af den iagttagende Astronomie," in *Festskrift, Copenhagen University* (1893); "On the Law of Spectral Series," in *Astrophysical Journal*, **6** (1897), 65–76; "Resolution into Series of the Third Band of the Carbon Band Spectrum," *ibid.*, **8** (1898), 1–27; "Tal og Symboler som Bestemmelser mellem Numeraler," in *Festskrift, Copenhagen University* (1901).

Papers are in *Tidsskrift for Matematik* (Copenhagen, 1859–1903); *Royal Danish Academy*, Oversigter and Skrifter (Copenhagen, 1880–1908); and in *Astronomische Nachrichten*, **48–138** (1858–1895). See also Poggendorff, IV, 1488–1489; V, 1250.

II. SECONDARY LITERATURE. Articles on Thiele are in *Dansk Biografisk Leksikon*, XXIII (1942), 503–506; and C. Burrau, in *Vierteljahrsschrift der Astronomischen Gesellschaft*, **46** (1911), 208–210; J. P. Gram, in *Nyt Tidsskrift for Mathematik*, **21B** (1910), 73–78; N. E. Nørlund, in *Fysisk Tidsskrift*, **9** (1910), 1–7; J. P. Gram, "Professor Thiele som Aktuar," in *Dansk Forsikringsaarbog*, **7** (1910), 26–37.

AXEL V. NIELSEN

THIERRY OF CHARTRES, also known as **Magister Theodoricus Carnotensis** (*b.* Brittany, France, last quarter of the eleventh century; *d. ca.* 1155), *philosophy, theology.*

As early as 1121 Thierry is believed to have taught at the cathedral school of Chartres, together with his brother Bernard, who was acting as chancellor (1119–1126). In 1127 he is recorded as archdeacon of Dreux, near Chartres, and before 1134 he is known to have taught in Paris, where Adalbert, later archbishop of Mainz (1137–1141), studied rhetoric, grammar, and logic under him. Among his students in Paris were also Master Bernard the Breton, later bishop of Quimper (1159–1167), the grammarian Master Peter Helias of Poitiers, Master Ivo of Chartres, Archbishop William of Tyre (1175–1185), John of Salisbury, later bishop of Chartres (1176–1180), and Master Clarembald, archdeacon of Arras.

Although Thierry's fame was based mainly on his courses on the trivium (grammar, logic, rhetoric), he is believed to have taught mathematics with great success. In fact, he is considered to have introduced the use of the *rota* or zero into European mathematics. A *Tractatus de rebus universalibus*, now lost, was dedicated to "Master Thierry." Bernard Silvestre dedicated his *Cosmographia* "to the most famous teacher, Thierry." A translation from Arabic into Latin of Ptolemy's *Planisphaerium*, made in Toulouse in 1144, is dedicated to "Thierry, the Platonist." An epitaph edited by A. Vernet, which celebrates Thierry as "a worthy successor of Aristotle," reveals that he was the first Latin scholar to comment on Aristotle's *Prior Analytics* and *Sophistici Elenchi*.

In the 1130's Thierry began to teach in Paris. He was among the masters who attended the papal consistory at Reims in 1148, where the orthodoxy of Gilbert, bishop of Poitiers, was examined. About the same time, Thierry and Master Gerland of Besançon were guests of Archbishop Albero of Trier. At a later date Thierry seems to have returned to Chartres, for the death roll of Chartres cathedral calls him "chancellor and archdeacon of Notre-Dame [of Chartres]." He bequeathed to the cathedral his Library of the Seven Liberal Arts, called *Eptatheucon* (destroyed by fire in 1944, preserved on film), Justinian's *Institutes, Novellae,* and *Digest*, and forty-five other books. A. Vernet maintains that Thierry retired (*ca.* 1155) to a Cistercian monastery to die and be buried in a monk's habit.

Today Thierry is probably best known for his short commentary on the introductory chapters of Genesis, the *Tractatus de sex dierum operibus*, in the first part of which he explains the unfolding of the universe on the basis of physical laws and provides an analysis of the Biblical text. In the second part he calls upon the quadrivium (arithmetic, music, geometry, astronomy) to lead him to the knowledge of the creator of the universe.

According to Thierry, there are four causes that account for the existence of the universe: (1) God as efficient, (2) His wisdom as formal, (3) His goodness as final, and (4) the four elements (fire, air, water, earth) as material cause created by God "at the beginning" out of nothing. The first three causes reflect the Trinity, for the efficient cause is the Father, the formal cause the Son, and the final cause the Holy Spirit. The four elements were created in a single moment. When Scripture speaks of six days for the creation, we may interpret a natural day as the time in which one whole rotation of the sky, from sunrise to sunrise, is completed or as the period required for the illumination of the air all across the sky.

Thierry held that, once created, the heaven could not stand still because of its extreme lightness. Enveloping all things, it could not move forward from place to place but was bound to rotate in a circular motion. The highest and lightest element, fire, produced both light and heat. The second element, air, conveyed the heat to the third element, water, and by warming it suspended a mass of waters, called firmament, above the air as high as the region of the moon. As a result of this removal of water, the fourth element, earth, appeared in the form of islands. Because of the heat, the earth then conceived the power of producing plants and trees. Acting on the mass of waters suspended in the sky, the heat caused the stellar bodies to be formed, for all stars are made of water and are still nourished on moisture. Because of the greater intensity of heat caused by the stars, the water on earth began to produce such things as water animals and birds. Then earth, too, conceived the power of generating animals—among which was man, made in the image and likeness of God.

All this took place in successive steps during the first six rotations of the heaven. To regulate the orderly succession of time, the various seasons and climates, and the normal process of procreation, the Creator implanted "seminal causes" in the elements. Thierry held that, in addition, a divine power, called the world soul, presides and rules over all matter so as to give it form and order.

BIBLIOGRAPHY

I. ORIGINAL WORKS. Thierry's *Eptatheucon* is still unpublished. A fragment of his commentary on Cicero's *De inventione* has been edited by W. H. D. Suringar in *Historia critica scholiastarum Ratinorum*, I (Leiden, 1834), 213–252. The latest edition of Thierry's *Tractatus de sex dierum operibus* is found in Nikolaus M. Häring, ed., *Commentaries on Boethius by Thierry of Chartres and His School* (Toronto, 1971), 555–575. An analysis is given in N. M. Häring, "The Creation and Creator of the World According to Thierry of Chartres and Clarembald of Arras," in *Archives d'histoire doctrinale et littéraire du moyen âge*, **22** (1955), 137–216, with text edition. Commentaries by Thierry and his school on the *Theological Tractates* of Boethius have been edited by N. M. Häring, *Commentaries on Boethius*, 57–528.

II. SECONDARY LITERATURE. A. Vernet, "Une épitaphe inédite de Thierry de Chartres," in *Recueil de travaux offerts à M. Clovis Brunel*, **2** (1955), 660–670;

E. Jeauneau, "Simples notes sur la cosmologie de Thierry de Chartres," in *Sophia*, **23** (1955), 172–183; and "Mathématiques et trinité chez Thierry de Chartres," in Paul Wilpert, *Miscellanea mediaevalia*, **2** (Berlin, 1963), 289–295; and "Note sur l'école de Chartres," in *Studi medievali*, 3rd ser., V, 2 (1964), 1–45; F. Brunner, "Creatio numerorum, rerum est creatio," in *Mélanges offerts à René Crozet* (Poitiers, 1966), 719–725.

NIKOLAUS M. HÄRING

THIRY, PAUL HENRI. See **Holbach, Paul Henri Thiry, Baron d'.**

THISELTON-DYER, WILLIAM TURNER (*b.* Westminster, England, 28 July 1843; *d.* Witcombe, Gloucestershire, England, 23 December 1928), *botany*.

Although he made no really signal contribution to scientific knowledge, Thiselton-Dyer played a central role in the botanical life of late Victorian England. Through his participation in T. H. Huxley's famous summer course in elementary biology, given at South Kensington for school teachers in the science and art department of the government, he helped extend to botanical circles Huxley's emphasis on evolutionary principles and pioneering efforts in laboratory teaching. As assistant director (1875–1885) and then director (1885–1905) of the Royal Botanic Gardens at Kew, he engineered and oversaw the immense expansion of economic botany throughout the British Empire.

Thiselton-Dyer was the elder son of Catherine Jane Firminger, an accomplished field botanist who introduced him to William Hooker's *British Flora*,[1] and her husband William George Thiselton-Dyer, a physician from Westminster who established a successful practice in Berkeley Street, London. Young Thiselton-Dyer's maternal uncle, T. A. C. Firminger, was a chaplain in Bengal and author of a standard manual of gardening for Bengal and Upper India. His maternal grandfather, Thomas Firminger, served as assistant astronomer royal at Greenwich Observatory from 1799 to 1808. Thiselton-Dyer, who was often addressed as "Dyer" and is sometimes so listed in indexes, derived his name from his paternal grandfather, William Matthew Thiselton, a printer and barrister who took the additional name Dyer by royal license in 1840.

Until 1863, Thiselton-Dyer was educated entirely in London, initially at St. Peter's School in Ea-

ton Square, then at King's College School, and finally at King's College itself, which he entered in 1861 as a medical student. In 1863 he transferred to Christ Church, Oxford, where he read mathematics under Henry J. S. Smith and chemistry under Benjamin Collins Brodie, Jr. After taking his Oxford bachelor of arts with honors in 1865, Thiselton-Dyer turned to natural history under the influence of George Rolleston, Henry N. Moseley, and E. Ray Lankester. In 1867 he won first-class honors in the final Natural Sciences School at Oxford.

By 1870, when he graduated as a bachelor of science from the University of London, Thiselton-Dyer had already served for two years as professor of natural history at the Royal Agricultural College, Cirencester. In 1869 he published (with his friend Henry Trimen) *Flora of Middlesex* and (with Arthur H. Church, a colleague at Cirencester) an English adaptation of Samuel W. Johnson's popular American work, *How Plants Grow*. From 1870 to 1872 he was professor of botany in the Royal College of Science, Dublin. While there he entered into correspondence with Joseph Dalton Hooker, director of the Royal Botanic Gardens at Kew. In 1872, on Hooker's recommendation, he was appointed professor of botany at the Royal Horticultural Society at South Kensington and Chiswick. During the same year he became private secretary and editorial assistant to Hooker.[2]

In the summer of 1873, when T. H. Huxley was abroad because of illness, Thiselton-Dyer assumed general direction of the South Kensington course. With the help of M. A. Lawson of Oxford, he particularly developed the botanical side of the teaching and became the leader of the campaign to bring laboratory training and the new physiological approach into British botany. In the summer of 1874, when Huxley resumed the course, Thiselton-Dyer acted as his botanical demonstrator with assistance from Sydney H. Vines, later professor of botany at Oxford. In the summers of 1875 and 1876 Thiselton-Dyer again offered a series of botanical lectures, with Vines again assisting him in the associated laboratory teaching. His direct participation in the course then effectively ceased, although he did give a final series of botanical lectures in 1880.[3] As the major botanical force behind the course during its early years, he did much to ensure its immense influence and success and to give a new direction to British botany. Toward the same end, he assisted Alfred W. Bennett in an annotated English translation of Sachs's pathbreaking *Textbook of Botany* (1875). He was a natural choice to take charge of the Jodrell Laboratory at Kew upon its completion in 1876. There, under his watchful eye, worked many of the rising stars of the new school of British botany, including Bower, Walter Gardiner, Dukinfield H. Scott, and Marshall Ward.

Overseeing the Jodrell Laboratory was, however, only one aspect of Thiselton-Dyer's new responsibilities at Kew, for he had been named assistant director of the Royal Botanic Gardens in 1875. Hooker had persuaded the government to reactivate this position, vacant since he had succeeded his father, William Jackson Hooker, as director in 1865, chiefly to handle the growing demands on Kew from the colonial office and individual planters throughout the empire. Thiselton-Dyer resigned his professorship at the Royal Horticultural Society and threw himself into his new duties with enormous energy and immediate success. During his first year in office, for example, he sent some Brazilian hevea plants to Ceylon, where they ultimately gave rise to the immense rubber plantations of Ceylon and the Malay Peninsula. Almost equally spectacular results followed his 1880 dispatch of some West Indian cacao plants to his old friend Henry Trimen, who had become director of the botanical gardens at Peradeniya, Ceylon.

In 1877 Thiselton-Dyer married Hooker's eldest daughter, Harriet Ann, who survived him and by whom he had one son and one daughter. In 1885 he succeeded his father-in-law as director of the Royal Botanic Gardens. For the next twenty years his individual contributions were often submerged in and indistinguishable from those of the institution over which he kept so tight a rein. Decisive, direct, methodical, and well-organized, as well as more than a little stubborn, insensitive, and autocratic, he intervened in matters as parochial as the uniforms to be worn by the Kew guards and as expansive as colonial economic development. During his directorship, the Jodrell Laboratory enhanced its reputation as a leading center of botanical research, while the gardens, library, herbarium, and several other buildings at Kew were enlarged and improved. He also paid close attention to the landscaping of the gardens, being particularly concerned with creating a sense of open vistas in the sometimes oppressively luxuriant foliage of Kew.

As director of Kew Gardens, Thiselton-Dyer pursued and expanded the efforts he had made as assistant director to develop economic botany and colonial agriculture. He adapted to the purpose a plan used for more than a century by the East India Company. Officers trained at Kew were placed in charge of botanic stations established throughout the empire. These officers then directed the development of local economic botany with guidance from Kew. To facilitate communication with these officers and other colonial botanists and planters, Thiselton-Dyer founded the *Kew Bulletin* in 1887. In West Africa, by his own account, such efforts increased the value of the rubber export from nothing in 1882 to £500,000 in 1898, while the value of the cocoa export rose from £4 in 1892 to £200,000 in 1904.[4] Partly for the contributions he thus made to the British and colonial economies, Thiselton-Dyer was created Companion of the Order of St. Michael and St. George in 1882, Companion of the Order of the Indian Empire in 1892, and Knight Commander of the Order of St. Michael and St. George in 1899. He served as botanical adviser to the secretary of state for the colonies from 1902 to 1906.

Thiselton-Dyer inherited the onerous task of supervising the great botanical surveys of British colonial territories that had been launched under the two Hookers who preceded him as director of Kew Gardens. In fact, he had been contributing to these surveys since 1872, when he served as editorial assistant to J. D. Hooker, then just beginning the *Flora of British India*. In 1873, in connection with this work, Thiselton-Dyer described the Indian species of six families of flowering plants. When he became director, he and his staff devoted much of their time and energy to the Indian *Flora*, which Hooker continued to edit in retirement and of which the seventh and final volume appeared in 1897. With this project behind them, Thiselton-Dyer and his staff turned to two other major surveys—the flora of South Africa (*Flora capensis*), begun in 1859 and suspended in 1865, and that of tropical Africa (*Flora of Tropical Africa*), begun in 1868 and suspended in 1877. Thiselton-Dyer edited the reports of these surveys throughout the period of his directorship and into retirement, turning the *Flora capensis* over to Sir David Prain in 1913 and completing the *Flora of Tropical Africa* in 1925. From 1895 to 1906 he also took over from Hooker the editorship of the *Icones plantarum* and the *Index Kewensis*, a massive index of the names and bibliographies of all known plants. Eighteen volumes of the *Kew Bulletin* appeared during his directorship. During 1905–1906 he also edited *Curtis's Botanical Magazine*.

Despite the weight of his duties at Kew, Thiselton-Dyer found time to participate in the activities of several botanical and scientific societies. Elected to fellowship in the Linnean Society in 1872, he served on its council from 1874 to 1876 and again from 1884 to 1887, being vice-president from 1885 to 1887. He was also vice-president of the Horticultural Society from 1887 to 1889. At the 1888 meeting of the British Association for the Advancement of Science, he was president of section D (biology), while at the 1895 meeting he presided over the new section K (botany), in the founding of which he played a leading role. From 1886 to 1888 he served on the council and during 1896–1897 as vice-president of the Royal Society, to which he had been elected fellow in 1880.

In 1905 Thiselton-Dyer resigned as director of the Royal Botanic Gardens and retired to Witcombe in Gloucestershire, where he took an active part in county affairs and indulged his passion for ancient botany, his favorite sources being Virgil, Pliny, Theophrastus, Galen, and Dioscorides. He revised the vocabulary of Greek plant names for Liddell and Scott's *Greek-English Lexicon*; contributed the botanical chapters to *A Companion to Greek Studies*, edited by Leonard Whibley (1905), and to *A Companion to Latin Studies*, edited by John E. Sandys (1910); published three articles covering some thirty especially difficult and obscure ancient plant names for the *Journal of Philology*; and assisted Sir Arthur Holt in his two-volume edition of Theophrastus' *Historia plantarum* (1916).

Thiselton-Dyer's heavy editorial and administrative responsibilities sharply restricted his opportunities for original research. His detailed study of the cycads, perhaps his most ambitious work, was incorporated into the *Flora capensis* as a supplement to the fifth volume (1933). His more general scientific papers and addresses reveal him to have been a fervent Darwinian who repeatedly insisted on the utility of specific characters and who defended the principle of natural selection at a time when its sufficiency as an explanation of evolution was under widespread attack.[5] Despite his role in the creation of the new physiological school of British botany, he often deplored the growing disdain for the old natural history method of systematic botany. He particularly emphasized the value of studying geographical distribution, to which he traced Darwin's discovery of the theory of descent, and which was for him still its main support.

NOTES

1. W. T. Thiselton-Dyer, "Plant Biology in the 'Seventies,'" in *Nature*, **115** (1925), 70.
2. Mea Allan, *The Hookers of Kew, 1785–1911* (London, 1967), 224.

3. See Thiselton-Dyer, *op. cit.* More generally on the South Kensington course and Thiselton-Dyer's part in it, see J. Reynolds Green, *A History of Botany in the United Kingdom from the Earliest Times to the End of the 19th Century* (London, 1914), 528–537; S. H. Vines, "The Beginnings of Instruction in General Biology," in *Nature*, **115** (1925), 714–715; and F. O. Bower, *Sixty Years of Botany in Britain (1875–1935): Impressions of an Eyewitness* (London, 1938), *passim*.
4. See *Kew Bulletin*, (1929), 74.
5. See, e.g., W. T. Thiselton-Dyer, "The Utility of Specific Characters," in *Nature*, **54** (1896), 293–294, 435–436, 522; and Ethel Romanes, *The Life and Letters of George John Romanes* (London, 1896), *passim*. On the more general contemporary debate over natural selection, see John E. Lesch, "The Role of Isolation in Evolution: George J. Romanes and John T. Gulick," in *Isis* (in press).

BIBLIOGRAPHY

I. Original Works. No complete bibliography of Thiselton-Dyer's works exists in print. For the period up to 1900, see the Royal Society *Catalogue of Scientific Papers*, VII, 588; IX, 767–768; XII, 728; and XIX, 79–80, which lists 86 papers by him alone and five coauthored with others. Among the more interesting of these are "On Spontaneous Generation and Evolution," in *Quarterly Journal of Microscopical Science*, **10** (1870), 333–354; "The Duke of Argyll and the Neo-Darwinians," in *Nature*, **41** (1890), 247–248; "Acquired Characters and Congenital Variation," *ibid.*, 315–316; "Historical Account of Kew to 1841," in *Kew Bulletin* (1891), 279–327; "Variation and Specific Stability," in *Nature*, **51** (1895), 459–461; and those cited in note 5 above.

For Thiselton-Dyer's more general views on Darwinism, the state of British botany, and the value of studying systematic botany and geographical distribution, see *Report of the British Association for the Advancement of Science*, **58** (1888), 686–701; and *ibid.*, **65** (1895), 836–850. See also the forty-eight-page brochure, Alfred Russel Wallace and Thiselton-Dyer, *The Distribution of Life* (New York, 1885); and Thiselton-Dyer, "The Geographical Distribution of Plants," in *Darwin and Modern Science*, Albert C. Seward, ed. (Cambridge, 1909), 298–318. For access to other works by Thiselton-Dyer, especially several prefaces and introductions, see *British Museum Catalogue of Printed Books*, LVIII, 354–355.

A wealth of manuscript material relating to Thiselton-Dyer is deposited in the herbarium and library of the Royal Botanic Gardens, Kew. Among other items there are a number of his diaries, notebooks, reports, and abstracts of lectures; three bound volumes of letters written by him between 1906 and 1922 (nearly all of them to Sir David Prain); four bound volumes of letters to him; and one bound volume of letters to Lady Thiselton-Dyer in connection with a projected memoir of him, together with her notes on the project and the following draft chapters: F. O. Bower, "The Influence of Sir William Thiselton-Dyer Upon the Teaching of Botany in Britain";

Sir J. Farmer, "The Chelsea Physic Garden"; D. Prain on Thiselton-Dyer's service to Indian botany; H. W. Ridley on his contribution to the Indian rubber industry; A. C. Seward, "Sir William Thiselton-Dyer's Contributions to the Study of Fossil Plants"; S. H. Vines, "Some Account of My Relations With the Late Sir W. Thiselton-Dyer"; and W. Dallimore, "Recollections of Sir William Thiselton-Dyer as Director of the Royal Botanic Gardens, Kew." Seventeen letters between Thiselton-Dyer and T. H. Huxley are preserved in the Huxley Papers at Imperial College, London. For published versions of several letters from G. J. Romanes to Thiselton-Dyer, see Ethel Romanes (note 5 above).

II. SECONDARY LITERATURE. Despite the rich manuscript material described above, and despite Thiselton-Dyer's pertinence to topics as important as the introduction of laboratory teaching into English biology or the relationship between botany and imperialism, no adequate study of him exists. For obituary notices, see *Kew Bulletin* (1929), 67–75; *Nature*, **123** (1929), 212–215; and *Proceedings of the Royal Society*, **106B** (1930), xxiii–xxix. See also D'Arcy Wentworth Thompson, "Sir William Thiselton-Dyer," in *Dictionary of National Biography, 1922–1930*, 830–832; and the sources cited in note 3 above.

GERALD L. GEISON

THÖLDE, JOHANN. See **Valentine, Basil.**

THOLLON, LOUIS (*b*. Ambronay, Ain, France, 2 May 1829; *d*. Nice, France, 8 April 1887), *solar physics.*

Thollon's scientific career began in 1878, when Raphaël Bischoffsheim, the founder of the Nice observatory, requested his assistance in setting up the spectroscopic equipment for the observatory. Thollon designed a high dispersion light spectroscope containing four dispersive prisms, two of which were made with carbon disulfide. Each prism was traversed twice by the ray of light, and the measurement of the position of the lines was effected with the aid of a micrometer. In order to eliminate thermal effects, the apparatus was frequently tested on several solar lines selected as reference points. Work on the spectroscope was completed in 1883.

At the start of his own research, in 1879, Thollon carried out an experiment that has become classic. He demonstrated the Doppler-Fizeau effect in a concrete manner. If one alternately projects on the slit of a spectroscope the eastern and western sides of the sun, the spectral lines exhibit a displacement representing the difference of 4 km per second between the radial velocities of the two sources. The telluric lines that originate in the terrestrial atmosphere remain fixed. This method of using the telluric lines for reference was employed by N. Dunér in 1890 to measure the rotational velocity of the sun.

Thollon established a great chart of the solar spectrum, which was published posthumously, in 1890. The spectrum extends from the visible red to the middle green (from 7,600 Å to 5,100 Å); it contains 3,448 lines, of which 2,336 are solar and 1,112 are telluric, with 246 lines being common. In addition to the position of the lines, the chart gives their intensities at a solar height of 10° and 30° above the horizon, for dry air and saturated air, and also, by extrapolation, for the case of observations made outside the earth's atmosphere. This chart is the last and most important of the documents for which broad data were gathered about the solar spectrum by means of a spectroscope. Such information was subsequently obtained by spectrography.

Thollon was awarded a prize by the Académie des Sciences in 1885.

BIBLIOGRAPHY

I. ORIGINAL WORKS. Thollon published eight notes on spectroscopic technique to the *Comtes rendus hebdomadaires des séances de l'Académie des sciences*, **86–96** (1878–1883); and nine notes on spectroscopic observations of comets, eclipses, and novae, *ibid.*, **92–102** (1881–1886).

Thollon's most important works on solar spectroscopy are "Déplacement de raies spectrales, dû au mouvement de rotation du soleil," *ibid.*, **88** (1879), 169–171; "Constitution et origine du groupe B du spectre solaire," in *Bulletin astronomique*, **1** (1884), 223–230; "Nouveau dessin du spectre solaire," *ibid.*, **3** (1886), 330–343; "Spectroscopie solaire," in *Annales de l'Observatoire de Nice*, **2** (1886), D1–D28; and "Nouveau dessin du spectre solaire," *ibid.*, **3** (1890), A1–A112, with 17 plates in the atlas.

II. SECONDARY LITERATURE. See J. Janssen, "Allocution . . . à l'occasion de la mort de M. Thollon," in *Comptes rendus hebdomadaires des séances de l'Académie des sciences*, **104** (1887), 1047–1048; "Dr. L. Thollon," in *Observatory*, **10** (1887), 207; and E. W. Maunder, "M. Thollon's Atlas of the Solar Spectrum," in *Monthly Notices of the Royal Astronomical Society*, **51** (1891), 260–261. See also A. Cornu, "Sur la méthode Döppler-Fizeau," in *Annuaire publié par le Bureau des longitudes* (1891), D25–D26.

JACQUES R. LÉVY

THOMAS, HUGH HAMSHAW (*b.* Wrexham, Denbighshire, Wales, 29 May 1885; *d.* Cambridge, England, 30 June 1962), *paleobotany*.

Born and brought up in Wales, Thomas spent all of his adult life in Cambridge, apart from service in the two world wars. He was the son of William Hamshaw Thomas and Elizabeth Lloyd. His father, a men's outfitter, was active in the public service of his community. Hugh Hamshaw, the second son, was educated at a private school, Grove Park, until he won a scholarship to Downing College, Cambridge, in 1904. His early interest in fossils, which he collected in the local coalpits, continued at Cambridge, where he read for the natural sciences tripos and took part I but changed to history for part II, although continuing to keep up his interest in botany. He was particularly influenced by A. C. Seward, the professor of botany, and in 1908, the year he graduated, his first scientific paper was published in the *Philosophical Transactions of the Royal Society*.

Intended for the civil service, Thomas took the examination but rejected the post he was offered and remained in Cambridge. He supported himself by elementary tutoring while continuing his research, mainly with E. A. N. Arber of the Sedgwick Museum of Geology. In 1909 he was appointed curator of the Botany School Museum, a post that he held until 1923; during this time he gradually shifted his teaching to botany. In 1911 he worked for a few months with Nathorst in the Stockholm Museum, and he appears to have also visited Russia about the same time. His election as a fellow of Downing College in 1914 gave him financial security.

Thomas spent most of World War I with the Royal Flying Corps in the Near East developing new techniques of mapping from aerial photographs, which were of considerable strategic importance. He continued his work in this field when he returned to Cambridge and kept in touch with research there. His work is still cited today. His interest in aerial photography also brought about his only work on ecology; he wrote on the flora of the Libyan desert and on the aerial survey of vegetation.

In 1923 Thomas was appointed to a university lectureship, and he married Edith Gertrude Torrance from Cape Town, who encouraged him to collect fossils in South Africa. They had one daughter and one son, who studied aircraft engineering and continued some of his father's work.

Thomas spent World War II with the RAF in England, where he was in charge of a specialist unit on photographic interpretation for Intelligence.

Thomas' scientific work was important because of the comprehensiveness of his collecting and fieldwork, his techniques of interpretation, and his detailed work on the systematics of fossil plant groups. Paleontology was mainly carried out in museums at that time, but Thomas collected extensively in England and in many areas abroad. Seward encouraged him to reexamine the Jurassic flora of Yorkshire, which resulted in Thomas' series of papers published between 1911 and 1925.

Work with Nathorst and with Thore Gustaf Halle, whom he also met in Stockholm, showed him the importance of a detailed examination of impressions of the epidermis from fossil plants. He extended this approach to attempts to examine even the compressed internal structures, which was a feature of his most important paper on the Caytoniales in 1925. In this paper he examined the leaves, particularly the petioles, the ovaries, and the pollen organs of hundreds of specimens before concluding that the family, which he named, was closely related to the angiosperms, although not in the direct line of descent. Thomas' paper on Triassic pteridosperms from South Africa, published in 1933, was also important because the pteridosperms were at one time a dominant group in the Southern hemisphere. The publication of this paper was followed by his election in 1934 as a fellow of the Royal Society.

Although Thomas had no formal training in geology, he did some work for the Geological Survey. His later publications on comparative morphology of plants were received with reservations by some of his colleagues. He also wrote on the history and philosophy of science.

BIBLIOGRAPHY

I. ORIGINAL WORKS. The two major papers are "The Caytoniales, a New Group of Angiospermous Plants from the Jurassic Rocks of Yorkshire," in *Philosophical Transactions of the Royal Society*, **213B** (1925), 299–363; and "On Some Pteridospermous Plants from the Mesozoic Rocks of South Africa," *ibid.*, **222** (1933), 193–265.

Thomas appears to have written no books, but there are several monographs listed by Thomas M. Harris, published by the British Museum (Natural History) and the Russian Geological Committee. His work for the Geological Survey is not listed by Harris and comprises

"Refractory materials . . . ," in *Geological Survey Special Reports on the Mineral Resources of Great Britain*, **16** (1920); and "The Geology of the South Wales Coalfield," in *Geological Survey Topographical Memoirs*, **10–12** (1909–1916).

II. SECONDARY LITERATURE. Most comprehensive is the obituary by Thomas M. Harris, in *Biographical Memoirs of Fellows of the Royal Society*, **9** (1963), 287–299, which includes a bibliography of his major works.

DIANA M. SIMPKINS

THOMAS, SIDNEY GILCHRIST (*b*. London, England, 16 April 1850; *d*. Paris, France, 1 February 1885), *metallurgy*.

Thomas invented and commercialized the basic process for dephosphorizing pig iron in the making of wrought iron and steel. As a result of his work, vast deposits of phosphoric iron ores that were previously unsuitable could be converted into steel by either the Bessemer or open-hearth methods. By the early years of the twentieth century most steel was being made by the basic process.

Thomas' father, William Thomas, was a Welshman employed in the civil service. Between the ages of ten and sixteen Thomas was educated at Dulwich College (equivalent to a high school or preparatory school), where he prepared for the matriculation examinations at London University. He intended to study medicine. When his father died in 1867 he abandoned these plans, and after a brief tenure as a teacher of classics at an Essex school he accepted a junior clerkship at a London police court. The following year (1868) he voluntarily transferred to the Thames police court in Stepney, the lower depths of London society, and he remained there for eleven years until the success of his invention enabled him to devote himself completely to metallurgy.

In a full account of his life, his sister Lilian Gilchrist Thompson reported that their mother was a "keen Liberal" and Thomas a "militant Radical" as an adolescent. He was strongly sympathetic to the North in the American Civil War, and he aspired to earn a fortune through applied science and to use it to assist the unfortunate and the neglected. Upon his early death, at the age of thirty-four, he left a sizable estate to his sister with instructions that it be used for "doing good discriminatingly." Lilian Thompson outlived her brother by more than half a century and executed his will scrupulously, largely by sponsoring various efforts to improve the conditions of workers, especially women.

Although Thomas never married, his letters reveal a spirited appreciation of feminine companionship.

During his early years as a police-court clerk, Thomas pursued his interests in metallurgy and applied chemistry by attending courses at the Royal School of Mines and the Birkbeck Literary and Scientific Institution (now Birkbeck College). At the School of Mines he passed all of the examinations, but could not matriculate because he was unable to meet the requirement of regular attendance at the metallurgy lectures. It was in response to a chance remark by a Birkbeck lecturer that Thomas resolved to find a method for producing Bessemer steel from phosphoric ores, a challenge that had been confounding the foremost metallurgists since the introduction of Bessemer's process. (In the older puddling process the phosphorus was readily removed.) By the end of 1875 Thomas had formulated the theory that, in order to remove the phosphorus, it was necessary to provide a strong base so that the phosphoric acid produced in the converter could combine with the base and form a slag. At first he experimented with basic linings for the converter formed chiefly of lime or limestone, but it was found that to preserve the linings from rapid deterioration and to produce a highly basic slag it was necessary to add substantial amounts of basic substances to the molten iron. In the commercial process it was mainly the additives rather than the linings that produced the phosphorus-bearing slag.

To overcome the constraints of both the demands of his clerkship and his lack of adequate experimental facilities, Thomas enlisted the collaboration of his cousin, Percy Carlyle Gilchrist, who was employed as a chemist in a Welsh ironworks. The collaboration proved fruitful and in 1877 the first of many patents was taken out. Although Thomas announced his achievement the following year at a meeting of the Iron and Steel Institute, it was not until 1879 that success was achieved on a commercial scale. At the spring meeting of the Iron and Steel Institute the Thomas-Gilchrist process was fully acknowledged.

So successfully did Thomas manage his patents, and so knowledgeable was he of the intricacies of international patent law, that he was able to defend his rights against formidable challenges. One challenge was, however, irresistible. Unknown to Thomas, the British metallurgist George James Snelus had patented the use of basic linings for the Bessemer converter in 1872. Although Snelus had failed to perfect the process, or even to produce usable linings, Thomas chose to avoid litigation, and a settlement was reached through the arbitration of William Thomson (later Lord Kelvin). Snelus received a share of the profits of the British and American (but not the Continental) rights and in 1883 both men received a Bessemer Gold Medal from the Iron and Steel Institute.

During the final years of his life, Thomas took out several patents on the preparation of the phosphate-rich basic slag for use as fertilizer, a technique that achieved considerable importance after his death.

These later years were also spent in a vain search for health as Thomas traveled extensively, hoping to benefit from climates more congenial than that of Great Britain. He died after a lingering illness diagnosed as emphysema, an ailment disproportionately common among employees of the poorly ventilated London police courts.

BIBLIOGRAPHY

I. ORIGINAL WORKS. Thomas and Gilchrist jointly authored "On the Elimination of Phosphorus," in *Journal of the Iron and Steel Institute*, **14–15** (1879), 120–134; "A Note on Current Dephosphorising Practice," *ibid.*, **19** (1881), 407–412; "The Manufacture of Steel and Ingot-iron from Phosphoric Pig-iron," in *Journal of the Society of Arts*, **30** (1882), 648–660. For a list of Thomas' lesser publications, see Lilian Gilchrist (Thomas) Thompson, *Sidney Gilchrist Thomas: An Invention and Its Consequences* (London, 1940), 56–57.

II. SECONDARY LITERATURE. Two full biographies are R. W. Burnie, *Memoir and Letters of Sidney Gilchrist Thomas* (London, 1891); and Lilian Gilchrist Thompson, *op. cit.* The latter is based heavily on the former but contains additional material on the adoption of the basic process and on his sister's use of the fortune he left. For biographical essays on Thomas and Snelus, see William T. Jeans, *The Creators of the Age of Steel* (New York, 1884).

HAROLD DORN

THOMAS OF CANTIMPRÉ, also known as **Thomas Brabantinus (Brabançon)** (*b.* Leeuw–Saint Pierre, Brabant, Belgium, *ca.* 1186–1210; *d.* Louvain, between 1276 and 1294), *theology, natural history, encyclopedism.*

Thomas of Cantimpré was descended from a noble family named De Monte or Du Mont, residing at Hellenghem (present-day Bellingen) near Leeuw–Saint Pierre, hence his further designation as Thomas van Hellenghem. He entered the abbey of Cantimpré near Cambrai as a novice and re-

mained there for some fifteen years. In 1232 he entered the Dominican Order in Louvain. Thomas was educated in the schools of Liège, Cologne, and Paris (Collège St. Jacques). He presumably was a pupil of Albertus Magnus at Cologne and Paris between 1245 and 1248. The date of his death, as of his birth, is uncertain. He was apparently alive in 1276 but was no longer living in 1294 (Thorndike, II, 374).

The content of Thomas' writings fitted well into the dominant precepts exemplified in the encyclopedic works of his contemporaries, Bartholomaeus Anglicus, Thomas Aquinas, Vincent of Beauvais, and others who adhered to a belief in the essential unity of the natural and the supernatural worlds. Hence, while primarily a theologian who studiously composed biographies of ecclesiastics and saints, Thomas also wrote on natural and pseudo science with the avowed intention of furthering knowledge of theological rather than of scientific truths. His work on bees, *Bonum universale de apibus*, deals with the orders of bees as they compare with the orders or hierarchy of prelates; and his major work on natural phenomena, *De naturis rerum*, on which he toiled for some fourteen or fifteen years and completed between 1228 and 1244, was intended, he informs us, to provide illustrative materials for use in sermons and arguments to bolster the faith. (The title *De naturis rerum* has been utilized, instead of *De natura rerum*, since it has the authority of thirteenth-century manuscripts of the work and most of the later manuscripts. See Thorndike, II, 397–398; and in *Isis*, **54** [1963], 269–277, for lists of MSS with the title *De naturis rerum*.)

Nevertheless, despite his avowed theological bent, Thomas of Cantimpré revealed in the *De naturis rerum* a lively interest in the real and natural world. The subjects treated in the twenty books into which the work is usually divided are largely those of the natural sciences. There are three books on man: one on the parts of the human body, containing a chapter on each part and its ills and their cures (with considerable information on gynecology and obstetrics); another book on the soul; and a third book on strange and monstrous races of men. Then there are seven books on animals: quadrupeds, birds, marine monsters, fish, serpents, and worms; and two books on the vegetable kingdom, including aromatic and medicinal trees and herbs. Finally, there are individual books on fountains and other bodies of water, precious metals, the seven regions of the air, the spheres and planets, meteorology, and the universe and the four elements, with an additional (or twentieth) book on the heavens and eclipses of the sun and moon added in some manuscripts.

Throughout the *De naturis rerum*, Thomas gives evidence of his adherence to the beliefs regarding natural phenomena current in his time, as well as of an undue reliance upon the credibility of his authorities. He also shared with many of his contemporaries a modified belief in astrology. In considering the effects of each of the planets when it was in the ascendant, he noted the moon's influence on the rise and fall of humors and infirmities of the human body; Mercury's purification of the mind of man; the effect of Venus in the generation of all earthly things; the numerous attributes of the sun relating particularly to natural phenomena, such as rain, snow, and the growth and decay of plants; the influence of Mars in causing anger to rise in men, heating the heart and liver, and undermining the health; Jupiter, through its hotness and dryness, causing all living things to spring forth; and Saturn, which causes seeds when planted in the earth to be mortified so that they will bring forth fruit. Throughout his account of the planets Thomas cited supporting passages from Scripture, and he concluded: that "except for human free will and special manifestations of divine will, all nature is placed by God under the rule of the stars. The influence of the Sun and Moon is manifest and 'why should we not with entire reason believe the same of the other planets?'" (Thorndike, II, 393–394).

Thomas also adhered to other beliefs in the occult. He indicated in the book on precious stones his belief in the marvelous powers and occult virtues of stones and gems and in the additional virtues imparted by sculptured gems. Thomas evinced an interest in alchemy; in the book on the seven metals, he alluded to the transmutation of metals (speaking of copper) and cited the *Lumen luminum*, a work on alchemy, which he attributed to Aristotle.

In addition to the instances related to the occult sciences, Thomas also provided interesting items regarding technological improvements. He thus gives, in the book on precious stones (in the discussion of adamant), a description of the mariner's compass, already described by Thomas' earlier contemporary Alexander Neckam (Thorndike, II, 387–388). In the book on the seven metals (in the discussion of tin), Thomas appears to have made one of the first mentions of modern plumbing in his account of the use of molten lead to fuse the pipes of aqueducts. According to Thomas, the pipes "used to be joined with tin, but in 'modern times' human art has thought out a method of uniting them with hot molten lead." Thomas also mentioned steel (Thorndike, II, 392).

As a theologian and encyclopedist, Thomas of Cantimpré did not add to the sum of the scientific knowledge of his time, but he did, in *De naturis rerum*, make clear what the current notions were concerning natural phenomena. He also reported on significant technological developments known to his contemporaries, and thus performed the service of disseminating knowledge that he had culled from his authorities as well as from his own observation.

BIBLIOGRAPHY

I. ORIGINAL WORKS. Information regarding the biographies of churchmen and saints and also the *Bonum universale de apibus* (on prelates) is in W. E. van der Vet, *Het Biënboec van Thomas Van Cantimpré en zijn exempelen* (The Hague, 1902), 29. Book I of the *Bonum universale de apibus* in Middle High German was edited by Nils Otto Heinertz, *Die mittelniederdeutsche Versien des Bienenbuches von Thomas von Chantimpré. Das erste Buch* (Lund, 1906).

The *De naturis rerum* is extant in some 151 MSS. Of these 144 have been enumerated by G. J. J. Walstra, "Thomas de Cantimpré, De naturis rerum. État de la question," in *Vivarium*, **5**, pt. 2 (1967), 146–171; **6**, pt. 2 (1968), 46–61, in preparation for a critical edition of the text. H. Boese, "Zur Textüberlieferung von Thomas Cantimpratensis' Liber de natura rerum," in *Archivum F. F. Praed.*, **39** (1969), 53–68, lists 44 MSS, seven of which, dated in the fourteenth and fifteenth centuries, were not included in Walstra's list. Boese has also published a first volume containing the text: Thomas Cantimpratensis, *Liber de natura rerum. Editio princeps secundum codices manuscriptos*, I, *Text* (Berlin, 1973). Unfortunately this volume contains no critical apparatus, which will apparently follow in a second volume. Earlier both the text and a considerable number of the manuscripts were described by Lynn Thorndike, *History of Magic and Experimental Science*, II (New York, 1923), 396–398; "More Manuscripts of Thomas of Cantimpré, De naturis rerum," in *Isis*, **54**, pt. 2 (1963), 269–277. Both Thorndike and Walstra indicated that the work frequently appears in the manuscripts as anonymous or ascribed to authors other than Thomas of Cantimpré. Also the number of books varies from nineteen to twenty; and there is a difference in their sequence. For example, in some manuscripts, instead of beginning with the book on the parts of the human body (book I), the text begins with the book usually numbered XVI, on the seven regions of the air, and ends with the book on the seven metals (usually book XV). Translations of the work into Flemish are by Jacob van Maerlant, in J. H.

Bormans, *Der naturen bloeme van Jacob van Maerlant*, books I–IV (Brussels, 1857), and E. Verwijs, *Jacob van Maerlant's naturen bloeme* (Groningen, 1878), into Dutch by Broeder Gheraert, in J. Clarisse, *Sterte-en-natuur-kundig* (Leiden, 1847); and into Middle High German by Konrad von Megenberg, in Fr. Pfeiffer, *Dar Buch der Natur* (Stuttgart, 1867).

Portions of the text of Thomas of Cantimpré have also been published separately. Sections from book I have been edited by Christ. Ferckel, *Die Gynäkologie des Thomas von Brabant. Ausgewählte Kapitel aus Buch I de naturis rerum beendet . . . 1240, zum ersten Male herausgegeben* (G. Klein, Alte Meister der Medizin, No. 5) (Munich, 1912), and Alfons Hilka, "Eine altfranzösische moralisierende Bearbeitung des Liber de Monstruosis Hominibus Orientis aus Thomas von Cantimpré, De naturis rerum nach der einzigen Handschrift (Paris, Bibl. Nat. fr. 15106)," in *Abhandlungen der Gesellschaft der Wissenschaften zu Göttingen Philologisch-Historische Klasse*, 3rd ser., no. 7 (1933), 1–73.

II. SECONDARY LITERATURE. Lynn Thorndike, *op. cit.*, has reviewed the principal biographical details, together with previous biographical literature, and has given an analysis of the content of the *De naturis rerum* as contained in the manuscripts. G. J. J. Walstra, *op. cit.*, has similarly analyzed the contents and brought together the listings of some 144 manuscripts so far identified, compiling as well a bibliography of all the previous works treating both the author and *De naturis rerum*. See also a review of the text in H. Boese, "Sur Textüberlieferung . . .," cited above.

PEARL KIBRE

THOMAZ, ALVARO (also known as **Alvaro Tómas** or **Alvarus Thomas**) (*b.* Lisbon, second half of the fifteenth century; place and date of death unknown), *physics, mathematics.*

Biographical data on Thomaz are lacking, save that he was regent of the Collège de Coqueret at Paris on 11 February 1509, as indicated in the colophon of his principal work, and that he is mentioned in the archives of the University of Paris as a master of arts at the same college in 1513. Thomaz is noteworthy for his *Liber de triplici motu proportionibus annexis . . . philosophicas Suiseth calculationes ex parte declarans* ("Book on the Three [Kinds of] Movement, With Ratios Added, Explaining in Part Swineshead's Philosophical [i.e., Physical] Calculations"), printed at Paris in 1509. This work shows Thomaz to be a mathematician and physicist of considerable ability who understood and organized the teachings of fourteenth-century English calculators and Parisian terminists, such as Oresme, making them available to a wide audience of European scholars in the sixteenth century.

Thomaz' work is divided into three parts, the first and second of which are compact expositions of ratios and proportions respectively, while the third is a lengthy application to problems concerning motion. This last part treats, in turn, local motion, augmentation, and alteration. Although designed as a guide to the thought of Richard Swineshead, Thomaz' treatise is not patterned after Swineshead's *Liber calculationum* ("Book of Calculations"), but follows instead an ordering suggested jointly by Thomas Bradwardine's *Tractatus de proportionibus* ("Treatise on Ratios") and by William Heytesbury's *Tractatus de tribus praedicamentis* ("Treatise on the Three Categories"). The first two parts seem to have been inspired by the inferior work of a certain Bassanus Politus, *Tractatus proportionum introductorius ad calculationes Suisset* ("Treatise on Ratios, an Introduction to Swineshead's Calculations"), printed in Venice in 1505, a work that Thomaz effectively castigated as worthless.

Thomaz' citation of authorities was extensive. His mathematics was drawn mainly from Nicomachus, Boethius, Johannes Campanus, and Jordanus de Nemore, while for Euclid he cited "the new translation of Bartholomeus Zambertus." Among the Schoolmen he referred to Thomas Aquinas (and his commentator, Capreolus), Robert Holkot, Duns Scotus, Albert of Saxony, Marsilius of Inghen, Gregory of Rimini, and John Maior. He knew too of the work of Paul of Venice, James of Forli, Cajetan of Thiene, John de Casali, Andrew de Novo Castro, Peter of Mantua, and a writer whom he identified as the Conciliator (Pietro d'Abano). Above all he was conversant with the complex details of Bradwardine's, Heytesbury's, and Swineshead's writings, and also with the little-understood *De proportionibus proportionum* ("On the Ratios of Ratios") of Oresme. Edward Grant states in his edition of the latter work: "Alvarus is the only author known to me who shows an extensive acquaintance with, and understanding of, Oresme's treatise" (p. 71).

Thomaz manifested some originality and considerable independence of judgment, as witness his rejection and reformulation of many of Swineshead's and Oresme's propositions. His treatment of falling bodies was highly imaginative in terms of the types of motive forces and resistive media discussed, but unfortunately was diffuse and inconclusive; contrary to what some scholars have suggested, it contains no explicit adumbration of Galileo's law of uniform acceleration. Thomaz showed facility in the summation of series, generally indicating when they converge and when they do not and, in cases where he cannot determine a precise value, providing limits between which this value must lie.

The influence of Thomaz' work is difficult to assess. Through his colleague at Coqueret, Juan de Celaya, he seems to have assisted in the formation of Celaya's later disciple, Domingo de Soto, who was the first to apply unequivocally the Mertonian "mean-speed theorem" to the case of falling bodies. Thomaz' treatise was cited by many Spaniards, favorably by the Salamancan masters Pedro Margallo and Pedro de Espinosa and by the Dominican Diego de Astudillo, and unfavorably by the Augustinian Alonso de la Veracruz, who blamed Thomaz' "calculatory sophisms" for much wasted time (and midnight oil) on the part of students in arts. At Paris, however, there can be little doubt that Thomaz was the calculator par excellence at the beginning of the sixteenth century, and the principal stimulus for the revival of interest there in the Mertonian approach to mathematical physics.

BIBLIOGRAPHY

I. ORIGINAL WORK. Thomaz' *Liber de triplici motu . . .*, is unavailable in translation; a copy of the original is in the library of the University of Michigan.

II. SECONDARY LITERATURE. For discussions of Thomaz and his work, see Pierre Duhem, *Études sur Léonard de Vinci*, III (Paris, 1913), 531–555, 557, 561; J. Rey Pastor, *Los matemáticos españoles del siglo XVI*, Biblioteca scientia no. 2 (Toledo, 1926), 82–89; Edward Grant, ed. and trans., *Nicole Oresme: De proportionibus proportionum and Ad pauca respicientes* (Madison, 1966), index; and William A. Wallace, "The 'Calculatores' in Early Sixteenth-Century Physics," in *British Journal for the History of Science*, **4** (1969), 221–232.

WILLIAM A. WALLACE, O.P.

THOMPSON, BENJAMIN (COUNT RUMFORD) (*b.* Woburn, Massachusetts, 26 March 1753; *d.* Auteuil, France, 21 August 1814), *physics.*

Thompson's father, Benjamin Thompson, and his mother, Ruth Simonds, were small village farmers in New England. He had little formal schooling, educating himself by self-study with the help of friends and local clergymen. Taking up schoolteaching, he moved to Concord, New Hampshire, where almost immediately (1772) he made an ad-

vantageous marriage with a wealthy widow, Sarah Walker Rolfe, fourteen years his senior. They had one child, Sarah, and permanently separated in 1775. Thompson soon came to the attention of the royal governor of New Hampshire, who commissioned him a major. He became an active Tory in the early part of the American Revolution. He fled to London after the fall of Boston and progressed rapidly in British government circles to become undersecretary of state for the colonies.

After a brief military career in South Carolina and on Long Island in New York, Thompson retired from the British army at the age of thirty-one with the rank of colonel. Knighted by George III, he joined the court of the elector of Bavaria, where he rose to become head of the Bavarian army. Thompson was made a count of the Holy Roman Empire by the Bavarian duke in 1793, taking the name of Count Rumford, after the old name of Concord, New Hampshire.

In 1796 Thompson established the largest prizes that had yet been given for scientific research, singling out the American Academy of Arts and Sciences in Boston and the Royal Society of London as the organizations that would award biennial Rumford premiums for work in heat and light. The prizes are still given. He returned briefly to London around 1800 but settled a few years later in Paris, where he married the widow of Antoine Lavoisier. The marriage was unsuccessful, and he separated from his wife and retired to Auteuil, near Paris, to write and work vigorously on science and technology.

Thompson was an active fellow of the Royal Society of London; he founded the Royal Institution of Great Britain; and in later years he was active in the Institut de France. He was a recipient of many honors and a member of many other scientific societies, including the American Philosophical Society in Philadelphia, the American Academy of Arts and Sciences in Boston, and the Bavarian Academy in Mannheim. He left the residue of his estate to Harvard University to establish a Rumford professorship, which still exists.

Professionally, Thompson was a soldier of fortune. His first serious scientific study was to determine the optimal position of firing vents in cannon and to measure the velocity of the shot as a function of the composition of gunpowder. He used a ballistic pendulum method first introduced by Benjamin Robins. Thompson's long paper on this subject, published in 1781, won him fellowship in the Royal Society of London. Throughout his life he intermittently pursued his studies on the force of gunpowder. In 1797 he published the description of a device for proving gunpowder, which was generally accepted as the standard method by both the British and the Bavarian armies. It was during his investigations of cannon that he was impressed by the large amount of heat generated in cannon barrels by the explosion of gunpowder even when no ball was being fired. He was thus led to accept the vibratory theory of heat, which he championed actively all his life. Thompson's most famous experiment in this area was his demonstration of the process of boring cannon with a dull drill, which he carried out in the arsenal at Munich. Because the heat generated in this process seemed limitless, he reasoned that a fluid caloric did not exist. Thompson carried out many other experiments to demonstrate the reasons for his disbelief in the caloric theory. He unsuccessfully attempted to determine whether heat had weight, which would be an attribute of a fluid; he weighed, at different temperatures, fluids that had markedly different specific heats and heats of fusion. He studied the anomalous expansion of water between 4°C and 0°C to show that the concept that thermal expansion is caused by fluid caloric taking up space was false. He never realized the connection between heat and energy, although he did carry on experiments to demonstrate spontaneous interdiffusion of different density liquids at constant temperature, and he postulated that fluids are in constant random motion.

In his role as military commander for the elector of Bavaria, Thompson demonstrated a genius for technological improvement that was widely recognized at the time, and many of his innovations are still used. In an effort to economize in the military establishment, he made careful studies of the insulating properties of cloth and fur, showing that the principal loss of heat was by convection, and that by inhibiting convection currents greater insulation could be achieved. He also demonstrated that heat passes with great difficulty through vacuums.

Thompson increased the labor force for making military clothing by sweeping Munich clear of beggars and setting them to work in military workhouses. Faced with the problem of feeding this labor force as well as feeding soldiers in the army, he studied the science of nutrition. He developed a theory of the nutritional value of water and introduced soup as a staple diet. He experimented with many cheap foods for mass feeding of the poor, introduced the potato into Central Europe, and published a long essay containing many recipes for the preparation of hearty meals. He searched for a substitute for alcoholic beverages for the common man and wrote extensively on the advantages of drinking coffee, analyzed the brewing of coffee, and designed a large number of drip-type coffee makers. In order to increase the efficiency of cooking devices he studied the insulation properties of solids and invented the concept of enclosing a fire in an insulated box, designing what is now called a kitchen range. His interest in the efficiency of heat transfer led him to design a double boiler and special pots and pans for use on his stoves. He also championed the use of the pressure cooker, invented by Papin. His studies of convection currents led to the design of a roaster for meat and the invention of a calorimeter for measuring the heats of combustion of various fuels to be used in his stoves. To help lower the cost of feeding the Bavarian army, he introduced military gardens where the soldiers could grow their own foods. He provided a demonstration museum for the general Bavarian populace by laying out the large park in the center of Munich known as the English Gardens.

In connection with his military workhouses, Thompson studied the efficiency of illumination, inventing the shadow photometer that bears his name. He also introduced the concept of a standard candle, defining the details of what was the international unit of luminous intensity until the twentieth century. In connection with his studies of illumination, Thompson designed oil lamps, the so-called Rumford lamps. He studied the transmission of light through glass and various translucent substances, including cloth; he also measured the diffusion of light passing through ground glass, silk, and other lamp shade material. As a result of his shadow photometer measurements, he became interested in the phenomena associated with colored shadows and wrote extensively not only about this subject but also about the psychological effects of complementary colors. Thompson's theoretical work on the theory of light centered around its relationship with heat, and its effect on chemical reactions and photosynthesis.

As his international fame for technological innovation grew during his lifetime, Thompson turned his attention to the popularization of technology as a useful endeavor. He established the Royal Institution of Great Britain in London as a museum of technology primarily for the education of artisans and the poor. He hoped to provide them with knowledge and skills in the construction of devices using heat and light. To support this institution financially Rumford initiated scientific lectures and courses of instruction by public subscription for the wealthy London aristocracy, hiring men of the stature of Humphry Davy and Thomas Young as full-time research scientists and lecturers. This effort was so successful that the original purpose of the Institution gradually disappeared and Rumford lost interest although he maintained contact with Davy and his assistant Michael Faraday throughout his life. The Royal Institution has continued as a leading British scientific society.

Thompson was much bothered by the evils of a smoky fireplace. He studied air currents in open fireplaces and introduced the smoke shelf, the throat, and the damper characteristic of the modern chimney. By giving attention to the proper relationship between the size of the fireplace opening and the size of the throat, he increased the efficiency of the fireplace. He also set up equipment for measuring the fundamental properties of heat radiation and demonstrated at the same time as did John Leslie that mat surfaces radiate heat better than shiny ones. His studies led him to design fireplaces with beveled sides and backs to throw more heat into the room. Thompson's studies of the flow of heated fluids allowed him to design steam-heating systems with the proper separation of steam and water. He introduced them not only into private houses but also into the large auditoriums of the Royal Institution in London and the Institut de France in Paris. In addition, Thompson designed commercial distribution systems for steam in the manufacture of soap and in brewing and dyeing.

BIBLIOGRAPHY

I. ORIGINAL WORKS. Thompson published sixty-four papers and essays, and these have been reprinted in five volumes, edited by Sanborn C. Brown and published in Cambridge, Mass.: *The Nature of Heat* (1968); *Practical Applications of Heat* (1969); *Devices and Techniques* (1969); *Light and Armament* (1970); and *Public Institutions* (1970).

II. SECONDARY LITERATURE. The standard biography of Benjamin Thompson is George E. Ellis, *Memoir of Sir Benjamin Thompson, Count Rumford, With Notices of His Daughter* (Boston, 1871), published in connection with *The Complete Works of Count Rumford*, 4 vols. (Boston, 1870–1875).

SANBORN C. BROWN

THOMPSON, D'ARCY WENTWORTH (*b.* Edinburgh, Scotland, 2 May 1860; *d.* St. Andrews, Scotland, 21 June 1948), *natural history, classics, mathematics, oceanography.*

Thompson was the son of D'Arcy Wentworth Thompson, classical master at the Edinburgh Academy and, later, professor of classics at Queen's College, Galway. His mother was Fanny Gamgee, daughter of Joseph Gamgee, a veterinary surgeon. Her brother was Arthur Gamgee, "the first biochemist." Thompson was educated at the Edinburgh Academy and the University of Edinburgh (1877–1880), where he studied anatomy under Turner, chemistry under Crum Brown, and zoology under Wyville Thomson, recently returned from the *Challenger* expedition. He then studied at Trinity College, Cambridge (1880–1883), where he was subsizar and later scholar, and where he read zoology under Francis M. Balfour and physiology under Michael Foster. While there he published his first work, a translation of H. Müller's *Fertilisation of Flowers*, for which Charles Darwin wrote the preface. He gained first-class honors in parts I and II of the natural sciences tripos and for a year taught physiology under Foster. In 1884 Thompson was elected professor of biology at University College, Dundee, and in 1917 was transferred to the chair of natural history at St. Andrews.

Thompson's first marine investigation, that of the fur-seal fisheries, took place in 1896, when he was sent by the British government to the Bering Sea. After representing Great Britain at Washington, D.C., at the Anglo-American commission of inquiry into the Bering Sea seal fishery in 1897, Thompson was made Companion of the Order of the Bath in 1898, in recognition of the success of his mission. A foundation member of the Conseil Permanent International pour l'Exploration de la Mer, he served on the Council from 1902 to 1947, was chairman of the Statistical Committee, and editor of the *Bulletin statistique* (1902–1947).

In 1901 Thompson married Maureen Drury; they had three daughters. In 1916 he was elected fellow of the Royal Society, in 1928 president of the Classical Association (of England and Wales), from 1934–1939 president of the Royal Society of Edinburgh, and in 1936 president of the Scottish Classical Association. He was awarded a knighthood during the coronation honors in 1937.

Thompson did not fit into any particular category; he was equally a scholar, scientist, naturalist, classicist, mathematician, and philosopher. Inheriting a love of the classics from his father and brought up by his scientific grandfather, he straddled two worlds and dominated both.

Thompson's paper "On the Shapes of Eggs and the Causes Which Determine Them," published in

Nature, 1908, shows the direction in which his thought was taking him. In his 1911 presidential address to section D of the British Association, "Magnalia Naturae; or the Greater Problems of Biology," he discussed, for the first time, what he called "the exploration of the borderline of morphology and physics." This was preparatory to the 1917 *On Growth and Form*, his great contribution to scientific literature. In this work Thompson departed from contemporary zoology, which was occupied with orthodox questions of comparative anatomy and evolution, and treated morphological problems by mathematics. The theme was original, unorthodox, and revolutionary. The chapter "On the Comparison of Related Forms" is a demonstration of the orderly deformation of related organic forms mapped out in accordance with Descartes's method of coordinates. The diagrams of transformation have contributed to other work on problems of growth and have influenced research in embryology, taxonomy, paleontology, and ecology.

Thompson initiated no school of research and was followed by no band of disciples. But the indirect influence of *On Growth and Form* is so wide and so important that it is hard to calculate. Scientists, engineers, architects, painters, and poets have acknowledged their indebtedness to this essay that ranges over a wide field of scientific discovery, thought, and history.

BIBLIOGRAPHY

I. ORIGINAL WORKS. For a list of the published writings of D'Arcy Wentworth Thompson, see *Essays on Growth and Form* (cited below), 386–400.

Major works include H. Müller, *The Fertilisation of Flowers*, Thompson, trans. and ed., with a preface by Charles Darwin (London, 1893); *A Bibliography of Protozoa, Sponges, Coelenterata, and Worms for the Years 1861–1883* (Cambridge, 1885); "John Ray," in *Encyclopaedia Britannica*, 9th ed., XX (1886); *A Glossary of Greek Birds* (Oxford, 1895); "On the Shapes of Eggs, and the Causes Which Determine Them," in *Nature*, **78** (1908), 111–113; Aristotle, *Historia animalium*, Thompson, trans. and ed. (Oxford, 1910); "Magnalia Naturae; or the Greater Problems of Biology," in *Nature*, **87** (1911), 325–328; reprinted in *Smithsonian Institution Annual Report* (1911); *On Aristotle as a Biologist*, with an introduction on Herbert Spencer (Oxford, 1911); "Morphology and Mathematics," in *Transactions of the Royal Society of Edinburgh*, **50** (1915), 857–895; *On Growth and Form* (Cambridge, 1917); "Natural Science: Aristotle," in *The Legacy of Greece*, R. W. Livingstone, ed. (Oxford, 1921); *Science and the*

Classics (Oxford, 1940); and *A Glossary of Greek Fishes* (Oxford, 1947).

II. SECONDARY LITERATURE. Important writings on Thompson and his works include Ruth D'Arcy Thompson, *D'Arcy Wentworth Thompson, The Scholar Naturalist, 1860–1948*, with a postscript by P. B. Medawar, "D'Arcy Thompson and Growth and Form" (Oxford, 1958); W. Le Gros Clark and P. B. Medawar, eds., *Essays on Growth and Form* (Oxford, 1945); and Ruth D'Arcy Thompson, *The Remarkable Gamgees* (Edinburgh, 1974).

Brief discussions and obituaries include Clifford Dobell, "D'Arcy Wentworth Thompson," in *Obituary Notices of Fellows of the Royal Society*, **6** (1949), 599–617; W. T. Calman, "Sir D'Arcy Thompson, C.B., LL.D., F.R.S.," in *Royal Society of Edinburgh Year Book* (1946–1948), 44–48; Douglas Young, "Sir D'Arcy Thompson as Classical Scholar," *ibid.*; R. S. Clark, "D'Arcy Wentworth Thompson," in *Journal du Conseil. Conseil permanent international pour l'exploration de la mer*, **16**, no. 1 (1949), 9–13; G. E. Hutchinson, "In Memoriam, D'Arcy Wentworth Thompson," in *American Scientist*, **36** (1948), 577; Lancelot Law Whyte, ed., *Aspects of Form* (London, 1951; 2nd ed., 1968); S. Brody, *Bioenergetics and Growth* (New York, 1945); J. S. Huxley, *Problems of Relative Growth* (London, 1932); and Jack Burnham, *Beyond Modern Sculpture* (Harmondsworth, Middlesex, 1968).

RUTH D'ARCY THOMPSON

THOMPSON, JOHN VAUGHAN (*b.* Berwick-upon-Tweed, England [?], 19 November 1779; *d.* Sydney, New South Wales, Australia, 21 January 1847), *natural history.*

Little is known of Thompson's early life beyond the fact that he spent part of it in the vicinity of Berwick-upon-Tweed, in northern England, where he studied medicine and surgery. In March 1799 he was appointed assistant surgeon to the Prince of Wales Own Fencible Regiment, which had been raised by Sir William Johnston in 1798. Thompson was promoted, after his transfer as assistant surgeon in the 37th Regiment of Foot, on 3 July 1800, to surgeon in June 1803, staff surgeon in December 1812, and then to deputy inspector general of army hospitals in July 1830. In 1835 he was posted as medical officer to the convict settlements in New South Wales, a post he held until February 1844, when he retired on half-pay.

Thompson's early service with the medical department of the British army coincided with the Napoleonic Wars. In December 1799 he went with his regiment to Gibraltar. Early in 1800 he was posted to Guiana and the West Indies, at that time a theater in the war between England, and France and the Netherlands. Thompson's stay gave him the opportunity to familiarize himself with the local fauna and flora. He is said to have returned to England in 1809 (although one of his letters in the archives of the Linnean Society is headed London April 7, 1807, thus suggesting that his service in the West Indies was interrupted by a trip to England).

Three years later Thompson was sent to the Mascarene Islands, where it is claimed one of his duties was to introduce the use of vaccine into Madagascar, although in official correspondence of this period he is styled "Government Agent for Madagascar," which suggests that he was also charged with diplomatic duties. From 1816 to 1835 he was stationed at Cork (often referred to as Cove of Cork) in Ireland, where he undertook much of his research on marine invertebrates. Indeed, this period proved to be the most fruitful of his life. His later post in New South Wales was not distinguished by work in natural history, although he returned to an earlier interest in the introduction of useful plants to the colony.

Thompson's personal life remains largely an enigma. It is not recorded whether he married, and no personal accounts from friends or acquaintances appear to exist. This was perhaps the result of his official duties, which kept him in relatively remote stations, away from the main centers of intellectual and scientific development in the British Isles. He was elected a fellow of the Linnean Society of London on 6 February 1810, and he corresponded—and evidently was on familiar terms—with Alexander MacLeay, a former secretary of the society, who was later appointed colonial secretary to the government of New South Wales. Thompson's work on marine invertebrates was to lead him to a number of revolutionary new concepts in fundamental systematics, which brought him into acrimonious conflict with the zoological establishment in London. His correspondence and published writings of that time suggest a man impatient with the conservatism of his opponents, and eventually embittered by their opposition.

Thompson's biological publications fall into three categories: his earliest writings on botany, a subject to which he returned in later life; his writings of a general nature on zoology; and his absorbing and most valuable contributions in marine zoology. He made little contribution to medical science, although he published in 1832 a pamphlet entitled *The Pestilential Cholera Unmasked . . .*,

a work devoted to diagnosis and treatment of cholera, but exhibiting little understanding of the causative factors involved in the disease. The pamphlet was topical, however, because cholera had been spreading westward from Asia during the previous decade, and the year 1832 saw the first major outbreak in the British Isles.

Thompson's early preoccupation with botany is shown by the publication in 1807 of *A Catalogue of Plants Growing in the Vicinity of Berwick-upon-Tweed*, a competent local flora. That same year two of his botanical communications were read to the Linnean Society in London. The first, "On the Genus *Kaempferia*," was primarily concerned with the systematic arrangement of the genus. The second, "An Account of Some New Species of Piper [pepper], With a Few Cursory Observations on the Genus," was published in 1808. This paper is interesting in that it gives incidentally a list of his travels in the West Indies, and his observations on the genus being made in Trinidad, Saint Vincent, and Grenada. His stay in Trinidad also resulted in an unpublished paper entitled "Description of a New Genus of the Natural Order of Myrti," which was read on 3 March 1812 to the Linnean Society.

During Thompson's stay in Mauritius, he continued his botanical studies and compiled *Catalogue of the Exotic Plants Cultivated in the Mauritius . . .*, which was published anonymously, but according to Thompson's claim after the suppression of the title page bearing his name (presumably someone wanted to deprive Thompson of the credit), in November 1816, soon after he left the island. The catalogue was the first work on the plants of the island to be published locally, and is useful in listing the dates of introduction of many of the exotic forms grown in the Botanic Gardens and elsewhere. It also shows Thompson's interest in the importation of useful plants, an interest he shared with many other colonists of the time. Thompson's last botanical studies, which he wrote after he had settled in New South Wales, dealt with the importation and cultivation of cotton and sugarcane.

Thompson's miscellaneous zoological publications began with his "Description of a New Species of the Genus Mus, Belonging to the Section of Pouched Rats," written in Jamaica; the paper was read in 1812 and published in 1815. In 1829 he published an account of his study of bones of the extinct Mascarene bird fauna made during his stay on the islands; the paper is entitled "Contributions Towards the Natural History of the Dodo . . . a Bird Which Appears to Have Become Extinct

Towards the End of the Seventeenth or Beginning of the Eighteenth Century."

It was apparently while returning from Mauritius in 1816 that Thompson made his first study of marine invertebrates. South of Madagascar he observed a puzzling luminosity in the sea. He trailed a muslin hoop net over the stern of the ship and caught a profusion of small marine animals hitherto invisible in the water. Thompson has been credited with being the first person to use a plankton net, and there is little doubt that his use of it in July or August 1816 was his own idea entirely; but he was anticipated by John Cranch, who used a similar tow net on James Kingston Tuckey's expedition to the River Zaire (or Congo) in April 1816. Thompson, unlike Cranch, lived to use his muslin net to catch marine plankton for many years.

In 1827 Thompson published *Memoir on the Pentacrinus europaeus*, in which he announced the discovery of a shallow-water European species of crinoid echinoderm in Cove harbor. The crinoids, especially the feather stars and sea lilies, were known only from dried specimens found in the West Indies, and their affinities were little known. His note added considerably to what was already known from the work of J. S. Miller. Thompson returned to the subject of *Pentacrinus* in 1836, when he showed that it was the young stage of the European *Comatula*; at the same time he described the then almost unknown polychaete *Myzostomum costatum*, commensal on the feather star, although he referred to it as a "complete zoological puzzle!"

One of Thompson's principal contributions to zoology was the discovery that certain planktonic forms of crustacean, then known by the genus name *Zoea*, undergo metamorphoses until recognizable as the young of the European edible crab (*Cancer pagurus*). He published his findings as the first of his memoirs in the *Zoological Researches* . . ., but he failed to prove the complete metamorphic cycle because his zoea died in the process of change; it was only by comparing them with ova from a berried female crab that he was able to deduce the relationship. Thompson's announcement was accompanied by a second memoir "On the genus *Mysis*," however, in which he showed that the mysidacean crustaceans hatch in a form very similar to that of the adult. Both memoirs appeared only months before Rathke's work on the development of the crayfish (*Potamobius*, formerly *Astacus*), which demonstrated that the young hatch at a late stage of development. It is not sur-

prising then that many established zoologists treated Thompson's claim of metamorphosis in the Crustacea with distinct and often derisory doubt. The issue was a serious one, for the taxonomy in use at the time, as expounded by William Leach and Georges Cuvier, distinguished the Crustacea from the Insecta on the grounds that the development of the former proceeded directly and without metamorphosis. Nevertheless, Thompson was soon able to prove his hypothesis. In his third memoir in the *Zoological Researches* . . . (1829), he described a successful experiment of 1827 in which he kept an ovigerous crab in captivity and examined the newly hatched larvae. In 1830 he claimed to have observed newly hatched larvae of eight genera of brachyuran Crustacea to be zoea; in 1835 he published notes on the natural history of the pea crab *Pinnotheres* (a commensal of the mussel, *Mytilus*), in which he again observed the larvae hatched as zoea; and in 1836 he published a report on his experimental hatching of the eggs of the spider crab *Macropodia rostrata*, as well as other papers on the development of Crustacea. Certain zoologists, notably John O. Westwood, refused to be convinced of the truth of Thompson's observations, even denying altogether the evidence that he presented, and the controversy became heated. Eventually others took up the problem and confirmed Thompson's findings.

Thompson's second important achievement in marine biology was his discovery that cirripeds are Crustacea; in the system proposed by Cuvier, they had been designated as a class of the Mollusca. By using his plankton net, Thompson captured some small translucent crustacean larvae, which he kept alive in captivity. He discovered that these animals metamorphosed and settled as acorn barnacles. His brilliant yet simple demonstration of the systematic position of the cirripeds was published in a fourth memoir in the *Zoological Researches* . . . (1830). Thompson's contribution to the biology of the cirripeds did not end with this study, for in 1836 he contributed a paper on the barnacle *Sacculina*, a parasite of the shore crab *Carcinus maenas*, in which he revealed its true nature and identified and described its larval forms.

Thompson's third major achievement in marine biology was the recognition of the class of animals he named Polyzoa. These animals had been formerly included as part of a heterogeneous collection of enigmatic invertebrates, the so-called zoophytes; but he showed that they were distinct from the colonial hydroids and the ascidians, with which they had been sometimes confused. The term Polyzoa received considerable usage, especially in Great Britain, but it was eventually dropped in favor of Bryozoa, which had been proposed almost contemporaneously.

John Vaughan Thompson was a practical naturalist; his use of the tow net and his observations of the living animals enabled him to make a very real contribution to marine zoology. Moreover, he showed an alert appreciation of the implications of his observations that were quite remarkable in a man untrained in the natural sciences and isolated from the mainstream of zoology. He was largely denied during his lifetime the acknowledgment that he deserved; and it has indeed only been during the second half of the twentieth century that his considerable contributions have been adequately recognized.

BIBLIOGRAPHY

I. ORIGINAL WORKS. The most important of Thompson's works on marine invertebrates are the *Zoological Researches and Illustrations* (Cork, Ireland, 1828–1834), a set of five scarce pamphlets, recently reprinted by the Society for the Bibliography of Natural History (London, 1968); and *The Memoir on the Pentacrinus europaeus: A Recent Species Discovered in the Cove of Cork* (Cork, 1827).

The following papers appeared in serial publications; (letter to the editor) concerning the metamorphosis in brachyuran crustacea *Zoological Journal*, **5** (1831), 383–384; "Memoir on the Metamorphosis and Natural History of the Pinnotheres, or Pea-Crabs," in *Entomological Magazine*, **3** (1835), 85–90; "Of the Double Metamorphosis in Macropodia Phalangium, or Spider-Crab . . .," *ibid.*, **3** (1836), 370–375; "Natural History and Metamorphosis of an Anomalous Crustaceous Parasite of *Carcinus Maenas*, the *Sacculina carcini*," *ibid.*, 452–456; "Memoir on the Star-Fish of the Genus *Comatula*," in *Edinburgh New Philosophical Journal*, **20** (1836), 295–300; and "Memoir on the Metamorphosis in the Macrourae or Long-tailed Crustacea, Exemplified by the Prawn (*Palaemon serratus*)," *ibid.*, **21** (1836), 221–223. The botanical works include *Catalogue of Plants Growing in the Vicinity of Berwick-upon-Tweed* (London, 1807); and *Catalogue of the Exotic Plants Cultivated in the Mauritius . . .* (Mauritius, 1816).

II. SECONDARY LITERATURE. See the introduction by Alwyne Wheeler to the facsimile of *Zoological Researches . . .*, published by the Society for the Bibliography of Natural History (London, 1968), i–vi, for an appreciation of Thompson's work. See T. R. Stebbing, *et al.*, "The Terms Polyzoa and Bryozoa," in *Proceedings of the Linnean Society of London*, session 123 (1910–1911), 61–72 for notes on the proposal of the term Polyzoa; and R. E. Vaughan, "A Forgotten Work by John Vaughan Thompson," in *Proceedings of the Royal Society of Arts and Sciences of Mauritius*, **1** (1953), 241–248, for an account of the catalogue of exotic plants in Mauritius.

ALWYNE WHEELER

THOMPSON, SILVANUS PHILLIPS (*b.* York, England, 19 June 1851; *d.* London, England, 12 June 1916), *applied physics, electrical engineering, history of science and technology.*

Thompson was the second of eight children of Silvanus Thompson, a Quaker schoolteacher, and Bridget Tatham, a member of another distinguished Quaker family. Richard and William Phillips, friends of Faraday, were his great-uncles. After sitting for the external B.A. at the University of London in 1869, Thompson taught at Bootham School in York, where his father was senior master. He continued his scientific studies and earned the B.Sc. (1875) and D.Sc. (1878).

While still a graduate student, he came close to making a discovery of the first order. Edison had noted that trains of sparks associated with electrical apparatus gave rise to effects that could be perceived some distance away, without intervening wires, and had concluded that these effects were a "new force." In a carefully designed experiment, Thompson showed that the mysterious effects were in fact electric; but by not pursuing the matter, he narrowly missed demonstrating experimentally Maxwell's theory of electromagnetic propagation, which Hertz accomplished in 1887–1888.

In 1876 Thompson was appointed lecturer in physics at the newly established University College, Bristol, and in 1878 he became the first professor of physics there. He quickly established himself as a popular and prolific lecturer and author of textbooks on electricity that made his name famous throughout the industrialized world. In 1885 he was appointed principal of Finsbury Technical College in London, one of two polytechnics sponsored by the City and Guilds of London Institute (the other, Central Technical College, later became part of Imperial College of Science and Technology of the University of London). He held this post until his death thirty-one years later.

In London, Thompson participated vigorously in scientific life and became the intimate of Lodge, FitzGerald, Crookes, and other luminaries. The thousands of graduates of Finsbury College, who came along at a time when England had virtually no engineering colleges of university level, represent perhaps his most important contribution. He also took a hand in the development of radiotelegraphy and wrote a privately printed pamphlet in support of Lodge's claims of priority over Marconi in the invention of a crucial feature, resonant tuning. Another popular work (until his death published anonymously), *Calculus Made Easy* (1910), is still in print three generations later.

In addition to his many technical contributions (notably in X rays, luminescence, magnetism, electrical machinery and illumination, and optics), Thompson was a notable historian of science and technology. His biographies of Gilbert, Faraday, and Kelvin are considered excellent of their sort, and he also published a highly polemic account of the life and work of Philipp Reis, a German schoolmaster whom Thompson persisted in regarding as the inventor of the telephone.

Thompson received many British and international honors but missed being named principal of the University of London when it was reorganized in 1901, possibly because of his stand on the Boer War: as an active Quaker, he openly castigated the British government for its inhuman treatment of civilians in the concentration camps of South Africa. In 1881 he married a fellow Quaker, Jane Smeal Henderson. They had four daughters, the second of whom collaborated with her mother in a biography of Thompson.

BIBLIOGRAPHY

I. ORIGINAL WORKS. Thompson was a prolific writer. The list of his books, pamphlets, translations, and privately published works is in the *British Museum General Catalogue of Printed Books* (1964). His most influential textbooks are *Elementary Lessons in Electricity and Magnetism* (London, 1881), *Dynamo-Electric Machinery* (London, 1884), *The Electromagnet and Electromagnetic Mechanism* (London, 1887), *Polyphase Electric Currents and Alternate-Current Motors* (London, 1895), *Light Visible and Invisible* (London, 1896), *Design of Dynamos* (London, 1903), and *The Manufacture of Light* (London, 1906). The ever-popular *Calculus Made Easy* was published first in 1910. He also wrote biographies of Philipp Reis (1883), Faraday (1898), and Kelvin (1910); and in 1901 he translated William Gilbert's *De magnete*. Lodge's priority over Marconi in the use of tuned circuits in radiotelegraphy was lucidly and succinctly argued in Thompson's 38-page pamphlet, *Notes on Sir Oliver Lodge's Patent for Wireless Telegraphy* (1911), which is of considerable historic interest.

II. SECONDARY LITERATURE. The biography by his wife and the second of his four daughters, J. S. and

H. G. Thompson, *Sylvanus Phillips Thompson. His Life and Letters* (London, 1920), contains a complete list of his publications. Short biographies appear in *Dictionary of National Biography* (1912–1921), 528–529; and in *Proceedings of the Royal Society*, **94A** (1917–1918), xvi–xviii, with a portrait.

CHARLES SÜSSKIND

THOMSEN, CHRISTIAN JÜRGENSEN (*b.* Copenhagen, Denmark, 29 December 1788; *d.* Copenhagen, 21 May 1865), *archaeology*.

Thomsen was the son of Christian Thomsen, a merchant and counselor-at-law (*Justitsieråd*), and Hedevig Margaretha Jürgensen. He was privately educated, since his father intended him to take over the family shipping business, and did not attend a university. Thomsen followed his father's wishes and entered a business career, but closed the family firm after the death of his mother in 1840. By that time he was firmly established in a second career as a museum director.

Thomsen had from his youth been interested in collecting art, antiquities, and, especially, coins, and his family's wealth allowed him to indulge these tastes. He was soon known for his numismatic knowledge and for his beautiful and well-organized collection of coins; in 1816, in spite of his youth and his lack of an academic degree, he was made secretary of the Danish Commission for the Preservation of Antiquities and put in charge of their museum. He arranged the collections, which he found in a sorry state, chronologically, and in 1820 divided prehistoric artifacts into representatives of the Stone, Iron, and Bronze Ages, an archaeological sequence that he mentioned in letters as early as 1818. This new chronology was soon widely accepted in English-speaking countries, in which Thomsen's claim to priority was also recognized. It was, however, disputed in Germany, as was Thomsen's assertion of priority, a situation that may have arisen because Thomsen only rarely published articles in scientific journals and, indeed, did not publish his chronology until 1836.

Thomsen was an organizer and modernizer. He was one of the first to use museums as a tool for popular education, and to this end took the almost unprecedented step of opening his collections to the public free of charge. He also lectured widely to lay audiences, hoping thereby both to encourage a general knowledge of antiquities and to create a climate favorable for the working archaeologist. He was successful in these aims, and encouraged a number of enthusiastic amateurs, who made a number of significant archaeological finds. In 1838 he was made inspector of a complex of museums of art, archaeology, and history in Copenhagen, and in 1849 succeeded to their directorship.

In 1841 Thomsen also established the first ethnographical museum, in which he designed exhibits focused upon the tools of ordinary life, showing how peoples really lived, rather than upon rarities. In this museum he also introduced the notion of organizing exhibits according to a "principle of progressive culture," to demonstrate cultural development. His ideas of archaeology, ethnography, and museum direction were all well in advance of his time. That they were widely accepted and emulated is a tribute to Thomsen's persuasive power and to his skill in winning professional agreement to his methods and public support for his interests.

BIBLIOGRAPHY

I. Original Works. Since Thomsen published few works in scientific publications, his ideas are known chiefly through letters, newspaper articles, and museum pamphlets and catalogues of exhibitions. His prehistoric chronology is in *Ledetraad til Nordisk Oldkyndighed* (Copenhagen, 1836), translated into German as *Leitfaden zur nordischen Alterthumskunde*, . . . (Copenhagen, 1837).

II. Secondary Literature. On Thomsen and his work, see V. Hermansen, *C. J. Thomsen and the Founding of the Ethnographical Museum* (Copenhagen, 1941); and "Christian J. Thomsen," in *Dansk Biografisk Lexicon*, XXIII (Copenhagen, 1942), 550–556; B. Hildebrant, "C. J. Thomsen och hans lärda förbindelser i Sverige 1816–1837," in *Kungliga Vitterhets, Historie och Antikvitets Akademiens Handlingar, 1937–1938* (1938); and H. Seger, "Die Anfänge des Dreiperioden-Systems," in *Schumacher-Festschrift zum 70. Geburtstag Karl Schumacher, 14. Oktober 1930* (Mainz, 1930).

Nils Spjeldnaes

THOMSEN, HANS PETER JÖRGEN JULIUS (*b.* Copenhagen, Denmark, 16 February 1826; *d.* Copenhagen, 13 February 1909), *chemistry.*

The son of Thomas Thomsen, a bank auditor, and the former Jensine Friederike Lund, Thomsen left secondary school without graduating. Since he was interested in chemistry, he planned to enter a pharmacy; but before he could do so, the professor of chemistry at the University of Copenhagen gave him the opportunity to work in his laboratory and at the same time to prepare for the entrance examination to the Polytekniske Laereanstalt, practical experience plus the examination being the alternative to the general certificate for admission to that school. Thomsen passed the examination in 1843 and obtained the candidate's degree (M.Sc.) in applied natural sciences in 1846. From 1847 to 1853 he was assistant in the chemical laboratory and, from 1850 to 1856, instructor in agricultural chemistry, at the Polytekniske Laereanstalt. He took time for a study tour to France and Germany in 1853–1854. He was director of weights and measures at Copenhagen from 1856 to 1859 and then, until 1866, taught physics at the Danish Military College. In addition, Thomsen was instructor (1864–1865), lecturer (1865–1866), and professor (1866–1901) of chemistry at the University of Copenhagen and head of the chemical laboratory, at the same time serving as professor of chemistry at the Polytekniske Laereanstalt.

Thomsen's first chemical work was elaboration of a method for the fabrication of soda from cryolite. He obtained a monopoly; and after overcoming some technical difficulties, he opened a factory at Copenhagen (1859), followed by others in Germany, Poland, and the United States. The Solvay method for manufacturing soda soon made cryolite commercially uncompetitive for use in soda production; but it proved to be a valuable raw material for making milk glass, enamel, and especially aluminum. Thus Thomsen's factory could continue separating cryolite from accompanying, less valuable minerals.

This technical achievement was, however, only a minor facet of Thomsen's work; his main activities were in pure science. In 1852 he had submitted "Bidrag til et Thermochemisk System" to the Royal Danish Academy of Sciences and Letters. For this paper he received the Academy's silver medal and 50 rigsdaler (at that time equivalent to $28) "to be used for the purchase of precision equipment." This was the start of a thirty-year program of thermochemical studies, during which Thomsen personally carried out more than 3,500 calorimetric measurements in a room kept at a temperature of 18° C. The results were later collected in the four-volume *Thermochemische Untersuchungen*, of which the three first volumes give the experimental material and the fourth Thomsen's theoretical reflections. His fundamental thought was that the evolution of heat accompanying a chemical reaction (which he called *varmetoning*, equivalent to enthalpy change) is an exact expression of the chemical affinity of the reaction.

Nearly the same theory was advanced a short time later by Berthelot, and a heated discussion between the two scientists took place and continued for several years. Supplementary experiments made by Thomsen (spontaneous reactions accompanied by "negative *varmetoning*," enthalpy loss — that is, endothermic reactions that are nevertheless spontaneous) led him to realize that his theory was only an approximation (it was later found that the theory is valid at absolute zero), and he publicly admitted its inexactitude. Berthelot, however, maintained the theory for many years, despite the facts that told against it.

Thermochemical studies were not Thomsen's only scientific contribution. He determined the heat of neutralization for acids and bases, and used the results to calculate the basicity of polybasic acids; through these and similar measurements he verified Guldberg and Waage's law of mass action. In his chemical experiments performed in galvanic cells, Thomsen found that the electromotive force can be used to calculate the mechanical work necessary for separating a compound into its elementary particles. In many instances, by measuring the electromotive force Thomsen obtained the same value for the affinity as in previous calorimetric experiments, but in other instances a difference was found. It is now known that the electrochemical measurements are theoretically correct, not the calorimetric ones. Thomsen discussed the constitution of benzene, and in a long series of papers he treated the theory that the atoms of an element are composed of smaller elementary particles; these speculations made him suggest a new way of presenting the periodic system of the elements. Just after the discovery of the first noble gas, he predicted the existence of five more members of this group; indicated their place in the periodic system of elements; and predicted their approximate atomic weights. All five were found in the next few years, and Thomsen's predictions of their atomic weights were astonishingly near the values found. It is interesting to note that Thomsen's periodic system was the one used by Niels Bohr when he gave his explanation of the periodic system based upon the number of electrons surrounding the nuclei of the atoms (1922).

The importance of Thomsen's scientific work was rapidly recognized both in Denmark and abroad. In 1860 he was elected a member of the Royal Danish Academy of Sciences and Letters. Nine years later he was nominated as professor of physical chemistry at the University of Leipzig, but he refused the offer. Many foreign scientists asked to work under his guidance, but he was afraid that the comparability of the results obtained would be endangered when more than one person performed the measurements and therefore refused all such requests. Thus no school was formed around him. Thomsen was a foreign member of various academies and honorary member of learned societies, and held honorary doctorates from several universities (but none in France, because of the conflict with Berthelot).

Thomsen was rector of the University of Copenhagen (1886–1887, 1891–1892), principal of the Polytekniske Laereanstalt (1883–1902), and president of the Royal Danish Academy of Sciences and Letters (1888–1909). He was an efficient administrator not only in science and education: He was a member of the Copenhagen city council (1861–1894), and a member of the commission for the reform of the Danish monetary system and system of weights and measures (1863). The conversion of Denmark's money to a decimal system was completed in 1874, and the reform of the system of weights and measures in 1910–1911. Upon his retirement from active service in 1902, Thomsen was appointed titular privy councillor in recognition of his public service.

BIBLIOGRAPHY

I. ORIGINAL WORKS. Thomsen's 227 papers in Danish and foreign scientific journals include "Thermochemische Untersuchungen I–XXXII," in *Poggendorffs Annalen der Physik* (1859–1873); *Journal für praktische Chemie* (1873–1880); and *Berichte der Deutschen chemischen Gesellschaft* (1873–1880); studies on electromotive forces in *Wiedemanns Annalen*, **11** (1880), 246–269; on the constitution of benzene, in *Berichte der Deutschen chemischen Gesellschaft*, **13** (1880), 1808–1811, 2166–2168; "Über die mutmassliche Gruppe inaktiver Elemente," in *Zeitschrift für anorganische . . . Chemie*, **8** (1895), 283–288; and "Systematische Gruppierung der chemischen Elemente," *ibid.*, **9** (1895), 190–193. His books include *Thermochemische Untersuchungen*, 4 vols. (Leipzig, 1882–1886); and *Systematische Durchführung thermochemischer Untersuchungen. Zahlenwerte und theoretische Ergebnisse* (Stuttgart, 1906).

II. SECONDARY LITERATURE. Obituaries are in *Berichte der Deutschen chemischen Gesellschaft*, **42** (1909), 4971–4988; and *Oversigt Dan. Vid. Selsk.* (1909), 27–31. See also *Dansk Biografisk Leksikom*, XXIII (1942), 568–575; and Stig Veibel, *Kemien i Danmark*, I (1939), 202–210; and II (1943), 426–444, with complete bibliography.

STIG VEIBEL

THOMSON, SIR CHARLES WYVILLE (*b*. Bonsyde, Linlithgow, Scotland, 5 March 1830; *d*. Bonsyde, 10 March 1882), *natural history, oceanography*.

Thomson was the son of Andrew Thomson, a surgeon in the East India Company. His earliest education was at Merchiston Castle School. When he was sixteen he matriculated at the University of Edinburgh to study medicine, a field that he was forced to give up after three years because of ill health. Moreover, his primary interest was natural history, especially zoology, botany, and geology. In 1853 he married Jane Ramage Dawson. Their only child, Frank Wyville, became a surgeon captain in the Third Bengal Cavalry.

Thomson held a number of academic positions. In 1851 he was a lecturer in botany at the University of Aberdeen; two years later, in 1853, he was appointed professor of natural history at Queen's College, Cork. In 1854 he became professor of geology at Queen's College, Belfast; and six years later, in 1860, he was named professor of zoology and botany at the same college. In 1868 Thomson accepted the professorship in botany at the Royal College of Science, Dublin, and in 1870 he assumed his last academic position, the regius professorship of natural history at the University of Edinburgh.

While at Belfast, Thomson began to establish himself as a talented marine biologist with his published studies of coelenterates, polyzoans and fossilized cirripeds, trilobites, and crinoids. He also became interested in determining whether life exists at great depths in the sea. Forbes suggested that below 300 fathoms there exists an azoic zone. In 1866, while visiting Michael Sars at Christiania (Oslo), Thomson had the opportunity to examine animals collected at depths below 300 fathoms. Thomson's interest in this question led him to embark upon a series of crucial deep-sea dredging voyages that culminated in the classic *Challenger* expedition of 1872–1876.

In 1868 Thomson and William Benjamin Carpenter, who was at the time vice-president of the Royal Society, persuaded the Society to seek Admiralty support for a deep-sea dredging project in the North Atlantic. Support was granted, and in August of 1868 Thomson and Carpenter began their project on board the paddle steamer H.M.S. *Lightning*. Despite stormy weather they were able to undertake some dredging and to obtain sponges, rhizopods, echinoderms, crustaceans, mollusks, and foraminifers below the 300-fathom mark. Perhaps the most surprising result of this cruise was the discovery of diverse temperatures at similar depths in different regions. The discovery called into question the accepted theory of a relatively constant submarine temperature of 4° C. The success of the *Lightning* cruise led to additional Admiralty support, and in the summer of 1869 the survey ship H.M.S. *Porcupine* was placed at the disposal of the Royal Society. Thomson, Carpenter, and John Gwyn Jeffreys dredged and took serial temperatures off the west coast of Ireland and off the Shetlands. They also began to analyze the composition of seawater from various depths. After dredging in waters over a thousand fathoms below the surface, they obtained on 22 July 1869 samples of mud and marine animals from 2,435 fathoms down. The results of these two cruises clearly cast doubts upon the validity of the azoic theory.

All these findings contributed to a renewed interest in the science of the sea. In 1869 Thomson was made a fellow of the Royal Society for his work. He described the details and accomplishments of the two expeditions in his popular study *The Depths of the Sea* (1873). With the encouragement of Carpenter, the Royal Society again approached the Admiralty for support in extending the scope of the investigations from the North Atlantic to the oceans of the world.

The Admiralty agreed, and on 7 December 1872, H.M.S. *Challenger*, a steam-powered corvette of 2,300 tons, set forth from Sheerness. Thus began a three-and-a-half-year voyage of oceanographic exploration. Since Carpenter did not wish to command the expedition, Thomson was selected as head of the civilian scientists. Once at sea the staff of the *Challenger* began the arduous tasks of sounding, dredging, and taking serial temperatures and water samples. Their dredging confirmed that marine life exists at depths approaching three thousand fathoms. They also discovered nodules of almost pure manganese peroxide on the seafloor. As they dredged and sounded in deeper water, they discovered that a clay bottom is characteristic of great depths. The material of the ocean floor is the residue of a chemical process that removes the carbonate of lime from the calcareous skeletons of foraminifers, mollusks, and other species. In bottom deposits beyond four thousand fathoms in the Pacific Ocean, they discovered a seafloor with new characteristics—radiolarian ooze. John Murray, one of the staff naturalists, uncovered new data on the diurnal migration of plankton and the oceanic distribution of globigerina. The temperature readings at various depths in a number of areas contributed to the growing speculation as to the nature of oceanic circulation. This complex question was not resolved by the scientific staff of the *Challenger*, for there was no physicist aboard to analyze this problem. The *Challenger* expedition was weakest in its examination of the questions of physical oceanography. On 24 May 1876 the ship returned to her berth at Sheerness after a voyage of 68,890 nautical miles and after having made soundings at 362 stations.

Much of the work of the expedition still lay ahead, for the specimens and data collected had to be organized and distributed, and the scientific results published. Publication of this diverse information was an enormous task, one which ultimately cost the British Treasury over £100,000. Queen Victoria conferred a knighthood (1876) upon Thomson for his service to science. While Thomson established the format of the *Challenger Reports*, he did not live to see the completion of the publication of this multivolume work, which chronicled his epic years of oceanographic exploration.

BIBLIOGRAPHY

I. ORIGINAL WORKS. Thomson's scientific papers are listed in the Royal Society's *Catalogue of Scientific Papers*, V, VIII, XI, and XIX. For the cruises of the *Porcupine* and *Lightning*, consult *The Depths of the Sea* (London, 1873). A popular account of the *Challenger*'s activities in the Atlantic may be found in *The Voyage of the Challenger. The Atlantic. A Preliminary Account of the General Results of the Exploring Voyage of H.M.S. Challenger During the Year 1873 and the Early Part of the Year 1876*, 2 vols. (London, 1877). The scientific results were published in *Report on the Scientific Results of the Voyage of H.M.S. Challenger during the Years 1873–1876*, 50 vols. (London, 1880–1895).

II. SECONDARY LITERATURE. For a biographical sketch of Thomson by a former student and assistant, see William Herdman, "Sir C. Wyville Thomson and the 'Challenger' Expedition," in *Founders of Oceanography and Their Work* (London, 1923), 37–67. See also Margaret Deacon, "The Magnificent Generalization" and "The Voyage of H.M.S. Challenger," in *Scientists and the Sea, 1650–1900* (London, 1971), 306–332, 333–365. For an examination of the problems of publishing the reports of the expedition, consult Harold L. Burstyn, "Science and Government in the Nineteenth Century: the Challenger Expedition and its Report," in *Bulletin de l'Institut océanographique*, **2**, spec. no. 2 (1968), 603–611.

PHILLIP DRENNON THOMAS

THOMSON, ELIHU (*b.* Manchester, England, 29 March 1853; *d.* Swampscott, Massachusetts, 13 March 1937), *electrical engineering.*

Thomson was the second of ten children of Scots parents of the skilled artisan class. When he was five the family moved to Philadelphia, where he graduated from the Central High School (then an advanced academy of the Gymnasium type) shortly before his seventeenth birthday. He became the assistant of E. J. Houston and then a teacher at the school. With Houston, Thomson collaborated in experiments that refuted Edison's claim to have discovered a "new force" (wireless transmission) and in other inventions, notably arc lights. They also started the Thomson-Houston Co., a predecessor of the General Electric Co., with which Thomson remained associated throughout his long career.

Thomson's inventions made possible significant improvements in alternating-current motors and transformers, both developments of the first importance in the history of electrical engineering. He invented electric resistance welding and also contributed to improvements in electric control, instrumentation, and radiology. Altogether, Thomson was the recipient of more than seven hundred patents and many awards, not all of which were related to electrical engineering. For instance, he discovered that a mixture of helium and oxygen prevents caisson disease, or bends.

During World War I, Thomson and others attempted to create an engineering school (financed from a bequest by the industrialist Gordon McKay) that would be jointly operated by Harvard University and the Massachusetts Institute of Technology. The project came to nothing when the courts ruled against use of the bequest for that purpose. When the MIT presidency fell vacant in 1919, Thomson was asked to fill the post. He declined but was acting president from 1921 to 1923.

Thomson combined solid achievement with great ingenuity and an uncanny sense for turning ideas and inventions into practical and highly profitable devices. He is also reckoned as one of the pioneers of the electrical manufacturing industry in the United States, in the development of which he took an active part for over half a century.

BIBLIOGRAPHY

I. ORIGINAL WORKS. Thomson wrote no books; his writings are his patents and technical papers, several of which were deemed sufficiently important for inclusion in *Report of the Board of Regents of the Smithsonian Institution*; for example, see (1897), 125–136; (1899), 119–130; (1900), 333–358; (1904), 281–285; and (1913), 243–260. Thomson was also a prolific letter writer. His papers, comprising 13,600 items and 43 vols. of letter books, are in the library of the American Philosophical Society. Some of the letters that he wrote and received have been published in an annotated vol., H. J. Abrahams and M. B. Savin, eds., *Selections From the Scientific Correspondence of Elihu Thomson* (Cambridge, Mass., 1971); but they are limited to his scientific interests and throw only oblique light on his more important activities as a technologist.

II. Secondary Literature. The only full-scale biography is D. O. Woodbury, *Elihu Thomson: Beloved Scientist, 1853–1937* (Cambridge, Mass., 1944, 1960), written in a popular style and lacking all bibliographical apparatus. See also *Dictionary of American Biography*, supp. 2 (New York, 1958), 657–659; *The National Cyclopaedia of American Biography*, XXVII (New York, 1939), 28–30; and Karl T. Compton, in *Biographical Memoirs. National Academy of Sciences*, **21** (1941), 143–179.

Charles Süsskind

THOMSON, JOSEPH JOHN (*b.* Cheetham Hill, near Manchester, England, 18 December 1856; *d.* Cambridge, England, 30 August 1940), *physics.*

Thomson came to physics for want of money to enter engineering. His father, a bookseller, sent him to Owens College to mark time until a leading engineer, to whom he was to be apprenticed, had an opening; but the father died before the vacancy occurred, and the family then could not afford the premium. With the help of small scholarships Thomson continued to an engineering degree at Owens College, which had an excellent scientific faculty including Osborne Reynolds, Henry Roscoe, Balfour Stewart (under whom Thomson did his first experimental work [1]), and Thomas Barker, the professor of mathematics, a former senior wrangler. On Barker's advice Thomson remained at Owens in order to work for an entrance scholarship in mathematics offered by Barker's old college, Trinity (Cambridge). He won a minor scholarship and in 1876 went up to the university where he would spend the rest of his life [58:13–32].

He read for the mathematical tripos, which at that time covered a wide range of pure mathematics as well as applications to many branches of physics. To "wrangle" successfully, that is, to place high on the tripos list, one needed great facility at computation and an ability to cope with the sort of models, or "physical analogies," prized by

the school of Kelvin, Stokes, and Maxwell. But one required neither knowledge nor experience of experimental physics; and so Thomson, who prepared himself by following diligently the advice of his coach, E. J. Routh, did no more than put his foot into the Cavendish Laboratory, and never met Maxwell [58:95, 129], whose work was to inspire his own. He emerged second from the tripos of 1880, after Joseph Larmor, who, like himself, became a Cambridge professor.

Thomson stayed on at Trinity, which awarded him a fellowship in 1881. He followed three lines of mathematical work, apparently diverse in content and style, but forming a coherent group for disciples of Maxwell and continuing research interests for himself. He seldom abandoned an idea he had once developed.

Fellow of Trinity. In Maxwell's practice there is a play, and sometimes a tension, between advancing theory by developing special mechanical models or analogies and deducing basic equations from the most general dynamical relationships. In the first mood, for example, Maxwell reached the equations of electromagnetism via an elaborate picture of a hydrodynamical contrivance supposed responsible for the phenomena; in the second, he obtained them directly from a Lagrangian constructed from known relationships between measurable quantities. The advantage of the second procedure, as Maxwell emphasized, is that one need not know (as one did not know) anything about the "mechanism" at work. The advantage of the first procedure is that, as Thomson would say, it fixes ideas, aids the memory, and, above all, suggests unexpected new directions of experimentation [15:1].

Thomson's earliest effort at designing models was stimulated by the subject set for the Adams Prize of 1882, "a general investigation of the action upon each other of two closed vortices in a perfect incompressible fluid." In his winning *Treatise* [5] Thomson carried the matter further than required, to an application which, for him, gave it "the greater part of the interest it possesse[d]" [5:2], namely Kelvin's theory of the vortex atom (1867). Here the atoms of a gas are represented as reentrant vortices in a frictionless fluid, rather like smoke rings in air; but the vortices, unlike the rings, are eternal, and therefore could reproduce the permanence of the Victorian atom.

The theory appealed to Thomson's romantic strain, to his recurrent wish for a quantitative, mechanistic, and "ultimate" [39:1]—by which he meant not "unique" but parsimonious [5:1]—account of the physical world. In this sense the theory of the vortex atom is perhaps the most fundamental ever started, for it hoped to make do with nothing besides the several perfections of its primitive fluid and pure mathematical analysis. "The difficulties of this method are enormous," Maxwell had written, "but the glory of surmounting them would be unique."[1] Thomson's *Treatise* is perhaps the most glorious episode in this hopeless struggle.

In adapting the pertinent hydrodynamics to the theory of the vortex atom, Thomson was guided, and perhaps even inspired, by the experiments of A. M. Mayer as interpreted by Kelvin [5:107]. Mayer, whose striking results would remain fresh in Thomson's mind [e.g., 26:313–314], had investigated the equilibrium configurations of n vertical magnetized needles floating on water and exposed to the attraction of a large fixed magnet. It appeared that if $n \leq 5$ the magnets would arrange themselves in a single circle, while for large n several concentric rings were required. On the basis of his kinetic representation of magnetism, Kelvin had inferred that to every stable arrangement found by Mayer there must be a counterpart formed by straight columnar vortices. Thomson therefore examined the stability of m vortex rings so coupled that their nearest portions always ran parallel, like threads wrapped symmetrically about a toroid, without crossing one another. Stability required that the vortices be of equal strength, and that $m \leq 6$.

To apply these results to the problem of chemical constitution (to which Thomson, in striking contrast to most physicists, gave continuing consideration [e.g., 39:120–141; 55:28–112]), observe that each of the linked threads in Thomson's arrangement can itself be a combination of n (≤ 6) vortices of equal strength. Let the strengths of all vortex atoms be multiples of that of hydrogen, taken as one. The oxygen atom clearly has strength two. Nitrogen gives trouble, apparently requiring a vortex of strength two in NO and of strength one in NH_3. Carbon likewise has its ambiguities (CO, CH_4); and in general the table of valences with which Thomson concluded his *Treatise* was useless for the chemist. But it was quite characteristic of him—at least in regard to "ultimate" theories— to be satisfied with gross *qualitative* agreement between experiment and the quantitative results he extracted at great labor from simple mechanical representations. Much of his important work on atomic structure, and the theories of chemical action [55:12–26] and of the nature of light [57] which he developed late in life show the same cu-

rious procedure: precise calculation and ingenious analogy applied with great virtuosity to secure only a rough fit with a few data. No doubt this method—or rather mood, for it by no means characterizes all of Thomson's work—helped him to those "happy intuitions" and "inspired generalizations," to that "abundance of ideas" and "endless fertility in invention" which led and impressed his contemporaries.[2] But it was also a method that became sterile in proportion to its success; for its qualitative conquests prepared the way for an exact physics that had no need for it.

A second line in Thomson's early researches descends from Maxwell's phenomenological strain. As part of the dissertation written for his fellowship [58:21], he elaborated an idea that had occurred to him at Owens College, and to which he would return [e.g., 53]—that the potential energy of a given system might be replaced by the kinetic energy of imaginary masses connected to it in an appropriate way. This notion, which anticipated the better-known scheme of Heinrich Hertz, could be made analytic by employing a form of Lagrange's equations worked out by Routh [12:12–15]. From the inspection of a Lagrangian, therefore, one not only cannot determine the underlying mechanism, but in general one cannot tell whether one confronts an ordinary system or one with Thomsonian masses.

Like Maxwell, Thomson was prepared to exploit this result in two ways. First, the fact that one can replace potential energy ("[which] cannot be said, in the strict sense of the term, to explain anything" [12:15]) by kinetic energy supported the hope for a theory based solely on the properties of matter in motion. "When we have done this we have got a complete explanation of any phenomenon and any further explanation must be rather metaphysical than physical" [*ibid.*]. Thomson had in mind a theory like the vortex-ring atom. But, second, the fact that a given Lagrangian is compatible with any number of models is a strong recommendation for avoiding them all, especially since a primary goal of physics—the discovery of new phenomena—can be reached merely by manipulating an appropriate Lagrangian in a prescribed manner.

In a series of papers [7, 9], lectures, and a book, *Applications of Dynamics* [12], Thomson illustrated how to guess at a term in the Lagrangian from a consideration of known phenomena and how, from the term once admitted, to deduce the existence and magnitudes of other effects. He also showed that a time-average of the Lagrangian could play the part of the entropy in certain problems usually handled by the second law of thermodynamics. One of his most important contributions in this line was the development of the notion, perhaps original with him, that electricity flows in much the same way in metals as in electrolytes [e.g., 12: 289–304]. He was to return to this idea in founding the electron theory of metals, first supposing the electrons free [30], as in Arrhenius' picture of a dilute solution, and then, in the hope of accounting for a difficulty in the theory of specific heats, allowing them only intermittent liberty [39:86–102], as in Grotthus' theory of electrolytic conduction [62:419–420, 425].

Three characteristics of these *Applications* deserve notice. For one, Thomson shows himself a master of the literature, not excluding the pertinent papers of German experimentalists. He was to keep fully abreast of the journals (from which he sometimes took ideas whose origin he later forgot) until World War I [59:150–151, 219]. Second, the moderate phenomenology of the *Applications*, a work which eschews the specification of dynamical processes, recurs in much of Thomson's later work. His pioneering theory of the conduction of electricity in gases [24, 28], for example, merely assumes the existence of ions, and describes their behavior not in terms of the electrodynamics of their interactions, but via parameters—especially measures of mobility and recombination—to be fixed by experiment. Only later [e.g., 45] did he sketch a theory of the process of ionization.

Third, Thomson, in common with most of the Cambridge school of mathematical physicists, took it for granted that an appropriate Lagrangian could always be found, or, in other words, that in principle all physical phenomena could be explained mechanically. Further, he thought that one or another of the possible dynamical explanations of a given phenomenon, whose existence is guaranteed by the Lagrangian, ought to be made explicit whenever possible. In this mood, as contrasted to that of the theorist of vortex atoms, Thomson did not require that models of diverse phenomena be consistent among themselves, nor that they avoid action at a distance; but that they admit only those sorts of forces and interactions with which physicists had become familiar since the time of Newton. He never totally abandoned this point of view, which caused him and contemporaries like Lodge and Schuster to deprecate the quantum theory as a screen of ignorance, a cowardly substitute for "a knowledge of the structure of the atom" [50a:27]. Sometimes, as in his theory of the speckled wavefront [34:63–65; 42], designed to explain the

selective ionizing power of X and γ rays, his efforts to cope with the quantum could advance the subject. But after 1910 his schemes for avoiding novelties like Einstein's approach to the photoeffect [49] and Bohr's deduction of the Balmer formula [52], would become increasingly farfetched and fruitless.

The last of Thomson's early research lines was the mathematical development of Maxwell's electrodynamics. His first important results included the discovery of the so-called electromagnetic mass, or extra inertia, possessed by electrified bodies in virtue of their charge [2], and the calculation (in error by a factor of two) of the force—now known as the Lorentz force—exerted by a magnetic field on a moving electrified sphere [*ibid.*]. These results were not only important in themselves: they also marked or sparked the beginning of the rapid harvest of Maxwellian fruits by Fitzgerald, Heaviside, Lamb, Poynting, and Thomson himself [e.g., 6].

One is struck by the literalness with which Thomson at first cultivated Maxwell's theory. Not that he clung slavishly to his model, for his thorough report on electrical theories to the British Association for the Advancement of Science [8] points to obscurities in Maxwell's formulation and discusses competing systems sympathetically. But he tried to remain true to what he considered the peculiar mark of Maxwell's theory: the dielectric "displacement" D, whose divergence represented what other electricians called the electric fluid and whose time rate of change, even in the absence of matter, gave magnetic effects like those of an ordinary current. By obscuring the concept of charge, displacement caused much of the malaise felt by Continental readers of Maxwell, and it could lead even English ones astray. In his important paper of 1881 [2], Thomson reached incorrect results by ascribing the magnetic field of a moving charged sphere solely to D outside it, thereby ignoring the most important factor, the convection of the charge [cf. 62:306–307; 60:24, 55].

In subsequent work Thomson replaced "displacement" (an "unfortunate" term [8:125]) by "polarization," which he represented in terms of electrostatic "tubes of force" supposed to begin and end on "atoms," each tube conferring the electrolytic unit of electricity, or "electron," on its termini [15:1–52]. From this representation, which in its ingenious details shows the hand of the essayist on vortex motion [cf. 58:94], Thomson recovered all the usual formulae of Maxwellian electrodynamics. He also thereby [15:13] stressed the notion of electromagnetic momentum stored in the medium (as a consequence of the translation of the tubes, the cause of magnetism), a notion used by himself [e.g., 39:24] and others to save the equality of action and reaction in electrodynamics [62:366], and to demonstrate the existence of "some invisible [material] universe, which we may call the ether" [36a:235].

It need scarcely be said that, although Thomson believed strongly that students should form some mental picture of the mechanism of the electromagnetic field, he did not urge his own as unique or even as particularly meritorious. "Which particular method the student should adopt is for many purposes of secondary importance, provided that he does adopt one" [15:vii]. "A theory of matter is a policy rather than a creed" [39:1]. He himself used models different in scope and degree of reduction, and, not seldom, conflicting in character; and after the discovery of the electron he freely admitted anti-Maxwellian bugaboos, like electric charge, into his partial pictures of metallic conduction [e.g., 30], atomic structure [e.g., 35], and chemical combination [39:120–139]. Such laxity of course could not be permitted in an "ultimate" theory of electricity [53, 56, 57].

Cavendish Professor. In 1884 Lord Rayleigh, who had succeeded Maxwell, resigned the Cavendish Professorship of Experimental Physics. Thomson had by then completed a few imperfect bits of laboratory work [cf. 44:80; 58:97], including a determination, at Rayleigh's suggestion, of the ratio of the electrostatic to the electromagnetic unit of electricity [4, corrected in 13]. Rayleigh had intended to collaborate in this work which, apart from its imperfection, was typical of the Cavendish during his era; but Thomson, unaware of many of the pitfalls, ran away with the project, published hastily, and gave his colleagues, including the Professor, to doubt that he had any future in experimental physics [59:18–20]. With these credits and his mathematics, he competed for the chair; much to his surprise [58:98], and to the great annoyance of some of his competitors, who included Fitzgerald, Glazebrook, Larmor, Reynolds, and Schuster, he was elected. It says much for the wisdom of the electors, among whom the ancient wranglers Stokes, William Thomson, W. D. Niven, and George Darwin, one of the judges of Thomson's Adams Prize essay, were probably most influential.

Luckily the personnel of the laboratory, including one who had expected to be its chief, remained on; and so the introductory courses set up in Rayleigh's time, and especially those for the many candidates for part I of the Natural Science Tripos, continued to function smoothly while the new Professor found his way. The same staff later (1888) introduced courses for intending physicians, whose fees quickly became an important part of the Cavendish's finances [44:84–89; 61:250–280; 59:19–21].

Thomson chose the phenomena of the gas discharge, whose study Maxwell had recommended, for experimental investigation. The subject had attracted attention in the early 1880's owing largely to the work of Crookes and Goldstein on the cathode rays [62:350–353]. Indeed, the ostensible motivation of Thomson's Maxwellian computations of 1881 was to provide a theoretical guide to the further study of the rays, which he, like Crookes, took to be "particles of matter highly charged with electricity and moving with great velocities" [2:229; 58:91–93]. Two years later Thomson had again turned his attention to the discharge, guided this time by the theory of the vortex atom [3].

The vortical mechanism for chemical bonding, he observed, works only when combining vortex rings have approximately the same size and velocity. Any disturbance in the medium, like the approach of another vortex atom, may alter the critical parameters and prevent linkage or disrupt unions. Now an electric field may be represented by a distribution of velocity in the medium; and the chemical decomposition it stimulates would be the immediate cause of the discharge. In this odd form Thomson introduced an idea of the utmost importance for future work: that the gas discharge proceeds in analogy to electrolysis, by the disruption of chemical bonds. Initially, as was only natural, he regarded the particles into which the molecules separated under the influence of the field as "atoms." Later researches (and, one presumes, a relaxation of his Maxwellian literalness) helped him to see the "atoms" first as "ions," that is, charge carriers of atomic dimensions, and then as mixtures of ions and "bodies much smaller than atoms."

The electrolytic analogy suggested that important clues to the mechanics of gas discharge might come from studies of dielectric breakdown in poorly conducting liquids, or from decomposition of polyatomic gases by sparks. Thomson and his students worked on the one [e.g., 11] and the other [10, the Bakerian lecture of 1887, continued in 16 and 18], and acquired many data without much advance in understanding. By the early 1890's he had concluded that a study of the striated positive light was the most promising avenue to the understanding of the discharge. As for the cathode rays, which had seemed significant in the early 1880's, they now appeared to him but a "local" and "secondary" matter [15:114–115].

Controversy returned the neglected cathode rays to the center of attention. Most English physicists, including Thomson, had taken them to be streams of charged particles, primarily because their paths curved in a magnetic field; while most German physicists, arguing chiefly from their ability to cause glass to fluoresce, had considered them an "aether disturbance" akin to ultraviolet light [62: 351–354]. In the early 1890's the English were put on the defensive by Philipp Lenard, who aggressively pursued the discovery of his master, Heinrich Hertz, that the rays could be passed through thin metal foil impermeable to particles of gas-theoretical dimensions. (Another objection, based on Hertz's inability to deflect the rays in an electrostatic field, was regarded less seriously; what weight it had was largely reduced by Perrin's direct detection of the charge carried by the rays, and vanished altogether when Thomson obtained the deflection in a better vacuum than Hertz had commanded [26:296].) Thomson tried in turn to undermine the position of the etherists by showing that the cathode rays moved at less than the velocity of light [17], but his results—off by two orders of magnitude—did not convince his opponents. At this point Röntgen prepared to enter the fray, and in the process discovered X rays.

Thomson, who had all the apparatus to hand, immediately found that the new rays turned gases they traversed into conductors of electricity [19, 20], and so offered a means much more convenient than disruptive discharge for producing gaseous ions [58:326]. Under his guidance the advanced students at the Cavendish rushed to exploit the new tool, and to make the accurate measurements of ionic parameters on which the Professor built his theory of gas discharge. The first edition of the famous *Conduction of Electricity Through Gases* [33] is a monument to these coordinated researches, in which McClelland [33], Rutherford [24], Townsend, and Zeleny played principal parts [59:74, 125; 60:38–41]. The presence at this time of these "research students"—graduates of other institutions allowed, by a reform introduced in 1895, to work for a research degree without first obtaining a Cambridge B.A.—was a great stroke of luck, as Thomson fully recognized [44:93; 58:325]. It provided talented and highly motivated men who not only developed the Professor's ideas in work

of the highest quality, but also helped raise the enthusiasm of younger recruits to the laboratory [61:269–271].

Thomson also saw in the X rays a possible explanation for the "startling" [37:3] transparency of metal foils to cathode rays: might it not be that in fact no penetration occurs, that cathode rays striking one surface of the metal produce X rays there which in turn create *new* "ions," alias "cathode rays," on the far side? [22] This ingenious subterfuge did not long survive the attacks of Lenard who, at Thomson's invitation [59:55], brought his campaign to the British Association in 1896. Thomson allowed himself to be persuaded of the importance of Lenard's work, and particularly of his discoveries that (*a*) the magnetic "deflectibility" of rays passed outside the tube depends only on the conditions within it, (*b*) that these external "Lenard rays" lost their power of causing fluorescence, that is, were absorbed, in proportion to the density, and independent of the chemical character, of the environment, and (*c*) that the mean free path of the rays outside the tube far exceeded the value to be expected if they consisted of gaseous ions [cf. 25: 430–431]. One suspects that it was in the process of digesting Lenard's results that Thomson first entertained the idea that the cathode rays consisted of bodies smaller than atoms.

To explore the matter further he employed Schuster's old technique of magnetic bending; for from the measured radius of curvature R of a beam of cathode rays deflected by a magnet of strength H one can infer a value for e/m, the ratio of the charge to the mass of the hypothetical cathode-ray particle ($e/m = v/HR$). Since the values for similar ratios, E/M, were known for ions produced in electrolysis, a comparison of e/m to E/M might provide a clue to the nature of the particle.[3]

To obtain e/m by Schuster's method the velocity v of the rays is required. If one takes for v either Thomson's faulty measurement [17] or, as Schuster had done, the mean speed of a gas molecule $(3kT/M)^{1/2}$, $e/m \sim E/M$, that is, one confirms the standard English theory which assimilated the rays to streams of charged particles of atomic dimensions. Lenard's intervention pushed Thomson to devise a more direct way of obtaining v. He found two. In the first [25:432; 26:302–306], the heat $T = nmv^2/2$ delivered by a stream of n "corpuscles" (as Thomson was to call the cathode-ray particle) to a Faraday cup was compared to the total charge, $Q = ne$, simultaneously conveyed, whence $v^2 = (e/m) \cdot (2T/Q)$. In the second [26:307–309], which exploited Thomson's discovery of the electrostatic deflection of the rays, v came from balancing an electric force eF against a magnetic one $(e/c)vH'$ to give no net deviation of the beam, whence $v = cF/H'$. The rough result of these measurements was that e/m exceeded by a factor of 1,000 the E/M for the ion of hydrogen, which has the largest charge-to-mass ratio of the chemical elements. The same anomalous value characterized all the cathode rays Thomson tried, irrespective of the material of the electrodes or of the nature of the gas in the discharge tube in which they were produced [26:306, 309].

At least two other physicists—Emil Wiechert and Walther Kaufmann, both then beginning distinguished careers—had independently obtained the same sort of data about the rays, and had inferred the correct magnitude of e/m by deducing v from the energy which would be acquired by a particle falling through the full potential V of the tube ($mv^2/2 = eV$). The equation could scarcely be justified theoretically, as Thomson liked to observe [e.g., 58:339]; but it gave the right order, and for this reason had been rejected by Schuster before the advent of Lenard. Despite their possessing most or all of the relevant data, neither Wiechert nor Kaufmann discovered the electron. Wiechert came closest: guided by the older Continental ideas about electricity [62:198 ff.], then recently revived by Lorentz, he identified the cathode-ray particle as a disembodied atom of electricity, a fundamental entity distinct from common matter. Kaufmann found nothing at all but an argument against "the hypothesis that assumes the cathode rays to be charged particles shot from the cathode."[4]

When Thomson, following his method, sought a representation of his striking data, he did not forget his old concerns: the vortex atom and Mayer's magnets, the problem of chemical combination and the nature of electricity [cf. 58:94]. It was doubtless these which pushed him to "discover" the electron, that is, to claim far more for the "corpuscle" than the data authorized. For from its large e/m he inferred its small mass (by assuming that its charge was of the order of the electrolytic unit); from its small mass he inferred (what scarcely follows) its small size; from its small size, its penetrability, and an answer to Lenard; from its size again and from the apparent independence of its e/m from the circumstances of its production, that it is a constituent of all chemical atoms. Or, rather, *the* constituent: for, as Thomson pointed out [26:311–314], if the chemical atoms are built up of corpuscles, arranged in rings in the style of Mayer's mag-

nets, one immediately glimpses an electrodynamic explanation of the periodic properties of the elements and saves Prout's vexed hypothesis—that the elements are built up of multiples of a single basic unit, or "protyle"—from the old objections based on deviations of atomic weights from integers. And there is more. Although Thomson called his particle "corpuscle" in order not to prejudge the value of its charge, which he initially believed to be larger than the "electron" [26:312; 60:55], he came quickly to believe that the corpuscle carried the elementary unit of electricity [27:544–545]. Apparently the protyles of matter and electricity were inseparable.

The initial evidence for Thomson's claims consisted primarily of the values of e/m for cathode rays of different provenance and of Lenard's law of absorption, which would follow if atoms contained corpuscles (and nothing but corpuscles capable of slowing cathode rays) in proportion to their weights. (To this might be added the similar law for the absorption of X rays found by McClelland, which Thomson had earlier tied to Prout's hypothesis [21], and which no doubt aided his digestion of Lenard.) Few physicists in 1897 were prepared to believe on this basis that the world was made of corpuscles.

Two years later the claim seemed more than plausible. In the interim, Thomson, following up old experiments of Elster and Geitel [59:108–112], had managed to find other particles—those liberated from metals by ultraviolet light and from carbon filaments by heat—which possessed approximately the same value of e/m as the corpuscle [29]. Moreover, he had succeeded in measuring e alone by exploiting, in a way suggested by Townsend and with an apparatus designed by C. T. R. Wilson, his earlier study of the ability of charged particles to promote condensation of water vapor [14; cf. 58:342–343; 61:195–205; 59:101–105]. The measurement employed X rays or ultraviolet light to create ions in a saturated gas, ions which, as Wilson had painstakingly proved, did indeed serve as condensation nuclei. The gas was then expanded and n droplets formed; from the mass and rate of fall of the fog one can compute n, whence $e = Q/n$, Q being the total charge carried down by the droplets. The result, which agreed to order of magnitude with estimates of the electrolytic "electron" [27:544; 29:562–563], turned out to be 30 percent too high. A second try [31] erred equally by defect. Thomson liked to leave the second decimal place to someone else [60:169]; in this case he left the first as well.

All this evidence, however imposing, left a large logical gap in Thomson's theory; for, strictly speaking, it did not bear on his claim that the *normal* atom consisted of corpuscles. By great good luck this gap was filled even as Thomson prepared his first lecture on the cathode rays [25]. For just then Zeeman established that the particle that, on the theory of Lorentz, gave rise to the spectral lines split by his magnet, possessed an *e/m* 1,000 times that of electrolytic hydrogen. The corpuscle not only belonged in normal atoms, but was responsible for their line spectra. The weight of evidence tipped in Thomson's favor. When he outlined it to the British Association in 1889 [29] it immediately "carried conviction" [58:341]: "The scientific world seemed suddenly to awake to the fact that their fundamental conceptions had been revolutionized."[5]

Atomic Structure. After consolidating his evidence about the electron (as physicists renamed the corpuscle when they came to believe in it) Thomson returned to the problem he had raised when introducing it as Prout's protyle: what causes the electrons of an atom to arrange themselves in the periodic manner implied by the table of the elements? He aimed at a theory that would postulate nothing but a few properties of the universal corpuscle; even the positive electrification apparently needed to retain atomic electrons and to neutralize matter would, he hoped, be reduced to an electronic property [59:140–141]. There were many good reasons, some of which had been identified before 1900 by Fitzgerald, Larmor, and Rayleigh, for believing that such a theory (even admitting two sorts of charges) could not explain important atomic characteristics like the frequencies of spectral lines. Fortunately, Thomson disregarded this counsel of despair, which the failure or barrenness of models proposed by others had made the more compelling when he took up the work in 1903 [34:90–139].

He then represented the positive electrification, which he had not been able to eliminate, in a manner Kelvin had used for a primitive model of a radium atom: a diffuse sphere of constant charge density through which electrons move subject solely to electrostatic forces. Thomson always regarded this model as a *pis aller*, an unsatisfactory incarnation of that "something which causes the space through which the corpuscles [in an atom] are spread to act as if it had a [compensating] charge of positive electricity" [29:565]. But it was easily visualized, and yielded much of what he wanted: for assuming the electrons constrained to circulate in a single plane through the atom's center (another *pis aller*, to ease calculations) Thomson showed that to insure *mechanical* stability the electrons, under the influence of electrostatic forces alone, must distribute themselves into rings in the manner of Mayer's magnets [35].

He drew several important qualitative conclusions. First, since the electrons, unlike the magnets, move in accelerated paths, they must radiate, and consequently no arrangement of them can be permanent. This apparent menace proved a great advantage. The radiation from a ring of *p* electrons *decreases* very quickly as *p* increases, the radius and angular velocity of the particles remaining the same [37]; hence if the rings of an atom are well populated its internal motions might decay very slowly until a critical velocity is reached and the whole explodes. We call such explosions radioactivity. (The obvious inference, that all elements must be radioactive, kept several Cavendish men busy for years [cf. 61:235–237].) The relative stability of matter depends, in this theory, on *n*, the number of electrons in an atom. Since, in 1904, Thomson still took *n* to be on the order of 1,000 times the atomic weight *A*, he did not fear imminent radiation collapse.

Secondly, the electronic distributions calculated by Thomson supported analogies to the behavior of the chemical elements and, in particular, the conclusion that the electronic populations of the atoms of contiguous elements in the periodic table differ by a single unit. Had he made this a principle, it would probably have modified his thinking about the order of *n*; but, as was his practice with such models, he did not take the results of his calculations literally, never assigned the value of *n* for any given element, and, very probably, did not anticipate that exact assignments could soon be made.

In fact the first substantial advances in atomic theory arose from efforts to obtain *n* as a function of *A*. Here Thomson once again led the way by showing how to estimate *n* from measurements of the scattering of light (dispersion), X and β rays [36]. All the data, including some collected at the Cavendish, were interpreted via formulae computed by Thomson under the guidance of his model atom; and the formulae for the scattering of X [33:268] and of β rays [36:773] were the first of their kind. The upshot was that the population of the atom had been grossly overestimated, and that *n* appeared to lie between two-tenths and twice the atomic weight.

A great many experimental studies on the scattering of X, β, and γ rays were then put in train at the Cavendish [61:237]. The multiplication of data prompted Thomson to improve upon his theory of β scattering [43], which, however, rested on an unjustified assumption: that, regardless of the thickness of the scatterer, a β particle acquires a measurable deflection only as a result of encounters with a great many atomic electrons. It was this theory that served Rutherford first as pattern and then as counterfoil during his classical analysis of the scattering of α particles. Although the results of this analysis—the single-scattering theory, the new approximation $n = A/2$, and the nuclear atom—forced the rejection of Thomson's model, they should be viewed not as evidence of the failure, but as proof of the value, of his methods.

Thomson's discovery of the order of n did much more than recommend the cultivation of scattering theory. For one, it undermined the radiative stability of the atom and, by reducing the number of spectral emitters, made what Rayleigh called the "bog of spectroscopy" more mushy than ever. For another, it demonstrated that the chief part of the atomic mass must belong to the positive charge. Thomson tended to ignore the first set of problems, although he once troubled to suggest how a single electron might, during its capture by an ionized atom, emit most or all of the line spectrum [39: 157–162]. It was different with the new-found substantiality of the plus charge, which no longer could be referred persuasively to a "property of the corpuscle."

Thomson's last important experimental work, which extended over many years, was devoted to determining the nature of positive electricity. He concentrated on the "canal" or "positive" rays which can be constructed from the ions in a discharge tube by passing them through a perforated cathode. In earlier studies, by W. Wien in particular, the E/M of the rays had been found by deflecting them in superposed electric and magnetic fields, and catching them on a photographic plate; from the position of the traces on the film it appeared that they consisted of gaseous ions, and especially of hydrogen, which occurred irrespective of the gas filling the tube. Wien inclined to attribute the ubiquitous hydrogen to release of impurities absorbed by the walls of the tube. But this conclusion, as well as the general interpretation of Wien's results, was made problematic by the width of the traces, which Thomson ascribed to the neutralization of ions in the rays by collisions with molecules of the residual gas. His first effort, therefore, was to remove the molecules by realizing the highest vacuum obtainable [cf. 58:350–357]. Wien's bands then broke up, as theory required, into parabolic traces, each deposited by ions possessing different velocities and a common E/M [38].

Many ionic species disclosed themselves, and always H^+, which Thomson accordingly took to be the positive protyle for which he was looking [38: 575; 39:19, 23; 40:12–13]; but after a long exchange with Wien he conceded that the hydrogen was not protyle but impurity [46:248]. During the exchange Thomson introduced many ingenious improvements in experimental technique and in the analysis of the traces. By 1913 his instrument had become sensitive enough to distinguish ionic species of atomic weights 20 and 22 in a neon discharge. At first [48:593] he thought the heavier species a new element, or perhaps a molecular peculiarity, NeH_2^+; but eventually he came around to the new view of Rutherford's school and recognized that he had been the first to isolate isotopes of stable elements [54:88]. In this work Thomson had the help not only of his long-time assistant, E. Everett, but also of Francis Aston, who returned to it after the war and perfected the mass spectroscope, which brought him the Nobel Prize.

Teacher and Administrator. Aston was one of seven Nobel Prizemen, twenty-seven Fellows of the Royal Society, and dozens of professors of physics trained at the Cavendish during Thomson's tenure [58:435–438]. Thomson was an excellent teacher and, when in good form, an unsurpassable lecturer [61:257; 59:42–43], clever, challenging, presuming neither too much nor too little, enthusiastic, and imperturbable. He took pedagogy seriously, on all levels. He interested himself in the improvement of science education in the secondary schools [22, 47] as well as in the universities, for which he and his close friend J. H. Poynting prepared several excellent texts. He kept his own lectures up to date both at the Royal Institution, where he became Professor of Natural Philosophy in 1905 (in addition to his Cambridge post), and at the Cavendish [cf. 61:273–278]. For the benefit of the advanced students in the laboratory he established in 1893 the Cavendish Physical Society, a fortnightly seminar in the German manner in which recent work—including his own—was reviewed and criticized [61:226, 271; 59:41].

Thomson was not himself a good experimentalist, being clumsy with his hands [60:73; 58:118], but he had a genius for designing apparatus and diagnosing its ills [59:175]. This trait, together with his wide and up-to-date interests, his enthusiasm, imagination, and resourcefulness, made him an excellent director of research throughout his tenure of the Cavendish chair. He resigned the professorship in 1919, in favor of Rutherford [59:215–218], before his lack of sympathy for Bohr's new physics could do any damage.

Thomson made every effort to place his best students, and gave generously of his time to keep those who took professorships in the colonies alive professionally. He would see their papers through the press, select demonstrators for them, advise on job openings and laboratory construction, and report recent progress in physics. As administrator of the Cavendish he gave his demonstrators great freedom and interfered as little as possible with laboratory routine [61:226]. He extended the buildings twice, once with accumulated laboratory fees [59:46], and again with Lord Rayleigh's Nobel Prize money, generously given the University for the purpose [59:155–156]. For a time, particularly in the 1890's, the need to save for expansion left little for research [61:270; 59:47–48], and it may be that Thomson could then have done more to improve the finances of the laboratory. He had a good eye for investment himself, and died a moderately wealthy man [59:262].

Thomson received a great many honors, including the Nobel Prize (1906), a knighthood (1908), the Order of Merit (1912), and the Presidency of the Royal Society, which he assumed in 1915. He therefore bore the burden of directing the Society's efforts to assist in the war [cf. 51] and of restraining some of its superpatriots from trying to oust Fellows of German descent like Schuster [59: 181–195]. The tact and energy with which he accomplished these tasks were widely recognized. In 1918 Thomson became Master of his old college, Trinity. He guided its affairs with his wonted geniality and good sense until a few months before his death.

NOTES

1. "Atom," in *Encyclopaedia Britannica* (9th ed., 1875).
2. The first two phrases are A. Righi's, "Sir J. J. Thomson," in *Nature*, **91** (1918), 4–5; the second pair come from N. Bohr, *ibid.*, **118** (1926), 879, and 59:150, respectively.
3. This agrees with the account in 37:3, with the order of ideas in 25:430–432, and with the order of events in 1896–1897. Lenard's role has become less prominent in Thomson's definitive announcement of his discovery [26], and has altogether disappeared from the retrospective account in 44:95, which ascribes the awakening of "doubts" about the ionic interpretation of the rays solely to the results of the bending experiments. Whittaker [62:361], Rayleigh [59:80], and G. P. Thomson [60:44–45] all follow this version, which Thomson enlarged in 58:333–335.

4. W. Kaufmann, "Die magnetische Ablenkbarkeit der Kathodenstrahlen und ihre Abhängigkeit vom Entladungspotential," in *Annalen der Physik*, **61** (1897), 544–552.
5. A. Schuster, *The Progress of Physics 1875–1908* (Cambridge, 1908), 70–71.

BIBLIOGRAPHY

I. Most of Thomson's important papers were published in the *Philosophical Magazine*, which he took to be, and helped to make, the leading English journal for physics. His results would often be reprinted, more or less reworked, in books, two of which became fundamental texts in their fields [33, 50]. No full bibliography of his works exists; the best, that in the obituary notice by the 4th Baron Rayleigh (*Obituary Notices of Fellows of the Royal Society*, **3** [1941], 587–609), contains some 250 items and yet is quite incomplete. It omits letters to *Nature*, at least one of which [19] was important; and misses contributions to cooperative works like the *Encyclopaedia Britannica*, Watt's *Chemical Dictionary*, and the *Recueil des travaux offerts . . . à H. A. Lorentz* (The Hague, 1900). Other important omissions include Thomson's Nobel Prize speech [37], his Rede lecture [21], and his contribution to *James Clerk Maxwell. A Memorial Volume* (New York, 1931). Additional items are supplied by Poggendorff and by Rayleigh [59:292]. A useful but incomplete list of Thomson's publications from 1880 to 1909 may be gathered from 61:285–323.

There follows a list in order of publication of the works of Thomson mentioned in the text (*PM = Philosophical Magazine; PCPS = Proceedings of the Cambridge Philosophical Society; PRI = Proceedings of the Royal Institution; PRS = Proceedings of the Royal Society; PT = Philosophical Transactions of the Royal Society; RBA = Reports of the British Association for the Advancement of Science*).

[1] "Experiments on Contact Electricity Between Non-Conductors," in *PRS*, **25** (1877), 369–372.

[2] "On the Electric and Magnetic Effects Produced by the Motion of Electrified Bodies," in *PM*, **11** (1881), 229–249.

[3] "On a Theory of Electric Discharge in Gases," in *PM*, **15** (1883), 427–434.

[4] "On the Determination of the Number of Electrostatic Units in the Electromagnetic Unit of Electricity," in *PRS*, **35** (1883), 346–347.

[5] *Treatise on the Motion of Vortex Rings* (London, 1883).

[6] "On Electrical Oscillations . . .," in *Proceedings of the London Mathematical Society*, **15** (1884), 197–218.

[7] "On Some Applications of Dynamical Principles to Physical Phenomena," in *PT*, **176** pt. 2 (1885), 307–342.

[8] "Report on Electrical Theories," in *RBA* (1885), 97–155.

[9] "Some Applications of Dynamical Principles to Physical Phenomena," *PT*, **178A** (1887), 471–526.

[10] "On the Dissociation of Some Gases by the Electric Discharge," in *PRS*, **42** (1887), 343–344.

[11] "On the Rate at Which Electricity Leaks Through Liquids Which Are Bad Conductors of Electricity," in *PRS*, **42** (1887), 410–429, written with H. F. Newall.

[12] *Applications of Dynamics to Physics and Chemistry* (London, 1888).

[13] "On Determination of 'v,' the Ratio of the Electromagnetic Unit of Electricity to the Electrostatic Unit," in *PT*, **181** (1889), 583–621, written with G. F. C. Searle.

[14] "The Electrolysis of Steam," in *PRS*, **53** (1893), 90–110.

[15] *Notes on Recent Researches in Electricity and Magnetism* (Oxford, 1893).

[16] "On the Effect of Electrification and Chemical Action on a Steam Jet . . .," in *PM*, **36** (1893), 313–327.

[17] "On the Velocity of the Cathode-Rays," in *PM*, **38** (1894), 358–365.

[18] "On the Electrolysis of Gases," in *PRS*, **58** (1895), 244–257.

[19] "The Röntgen Rays," in *Nature*, **53** (1896), 391–392.

[20] "On the Discharge of Electricity Produced by the Röntgen Rays," in *PRS*, **59** (1896), 274–276.

[21] "The Röntgen Rays," Rede lecture, in *Nature*, **54** (1896), 302–306.

[22] "Presidential Address," section A, in *RBA* (1896), 699–706.

[23] "On the Leakage of Electricity Through Dielectrics Traversed by Röntgen Rays," in *PCPS*, **9** (1896), 126–140, written with J. A. McClelland.

[24] "On the Passage of Electricity Through Gases Exposed to Röntgen Rays," in *PM*, **42** (1896), 392–407, written with E. Rutherford.

[25] "Cathode Rays," in *PRI*, **15** (1897), 419–432.

[26] "Cathode Rays," in *PM*, **44** (1897), 293–316.

[27] "On the Charge of Electricity Carried by the Ions Produced by Röntgen-Rays," in *PM*, **46** (1898), 528–545.

[28] "On the Theory of the Conduction of Electricity Through Gases by Charged Ions," in *PM*, **47** (1899), 253–268.

[29] "On the Masses of the Ions in Gases at Low Pressures," in *PM*, **48** (1899), 547–567.

[30] "Indications relatives à la constitution de la matière," in *Rapports du congrès international de physique* (Paris, 1900), III, 138–151.

[31] "On the Charge of Electricity Carried by Gaseous Ions," in *PM*, **5** (1903), 346–355.

[32] "The Magnetic Properties of Systems of Corpuscles Describing Circular Orbits," in *PM*, **6** (1903), 673–693.

[33] *Conduction of Electricity Through Gases* (Cambridge, 1903).

[34] *Electricity and Matter* (New Haven, 1904).

[35] "On the Structure of the Atom . . .," in *PM*, **7** (1904), 237–265.

[36] "On the Number of Corpuscles in an Atom," in *PM*, **11** (1906), 769–781.

[36a] *On the Light Shown by Recent Investigations of Electricity on the Relation Between Matter and Ether*, Adamson lecture (Manchester, 1907); reprinted in *Annual Report of the Smithsonian Institution* (1908), 233–244.

[37] "Carriers of Negative Electricity," in *Les prix Nobel en 1906* (Stockholm, 1908).

[38] "On Rays of Positive Electricity," in *PM*, **13** (1907), 561–575.

[39] *The Corpuscular Theory of Matter* (London, 1907).

[40] "Presidential Address," in *RBA* (1909), 3–24.

[41] "Positive Electricity," in *PM*, **18** (1909), 821–845.

[42] "On a Theory of the Structure of the Electric Field and Its Application to Röntgen Radiation and to Light," *PM*, **19** (1910), 301–313.

[43] "On the Scattering of Rapidly Moving Electrified Particles," in *PM*, **23** (1912), 449–457.

[44] "Survey of the Last Twenty-five Years," in *A History of the Cavendish Laboratory, 1871–1910* (London, 1910), 75–101.

[45] "Ionization by Moving Electrified Particles," in *PM*, **23** (1912), 449–457.

[46] "Further Experiments on Positive Rays," in *PM*, **24** (1912), 209–253.

[47] "The Functions of Lectures and Textbooks in Science Teaching," in *Nature*, **88** (1912), 399–400.

[48] "Some Further Applications of the Method of Positive Rays," in *PRI*, **20** (1913), 591–600.

[49] "On the Structure of the Atom," in *PM*, **26** (1913), 792–799.

[50] *Rays of Positive Electricity and Their Application to Chemical Analysis* (London, 1913).

[50a] *The Atomic Theory*, Romanes lecture (Oxford, 1914).

[51] "Presidential Address," in *PRS*, **93A** (1916), 90–98; *PRS*, **94A** (1917), 182–90; *PRS*, **95A** (1918), 250–257.

[52] "On the Origin of Spectra and Planck's Law," in *PM*, **37** (1919), 419–446.

[53] "Mass, Energy and Radiation," in *PM*, **39** (1920), 679–689.

[54] "Opening of the Discussion on Isotopes," in *PRS*, **99A** (1921), 87–94.

[55] *The Electron in Chemistry* (Philadelphia, 1923).

[56] "On the Analogy Between the Electromagnetic Field and a Fluid Containing a Large Number of Vortex Filaments," in *PM*, **12** (1931), 1057–1063.

[57] "On Models of the Electric Field and of the Photon," in *PM*, **16** (1933), 809–845.

[58] *Recollections and Reflections* (London, 1936).

II. Thomson's notebooks have been deposited at the Cambridge University Library (Add. 7654/NB), which also has three boxes of his correspondence, primarily incoming (Add. 7654 [ii]) and, in the Rutherford Papers (Add. 7653), some forty letters from him, bits of which were published by A. S. Eve, *Rutherford* (New York, 1939). The Royal Society Library also has a few Thomson autographs, primarily twenty-six letters to Schuster (Sch. 331–356). Indications of other holdings may be found in R. M. MacCleod, *Archives of British Men of*

Science (London, 1972), and in T. S. Kuhn *et al.*, *Sources for History of Quantum Physics* (Philadelphia, 1967).

III. The chief biographies of Thomson are the following:

[59] Lord Rayleigh, *The Life of Sir J. J. Thomson, O.M.* (Cambridge, 1943).

[60] G. P. Thomson, *J. J. Thomson and the Cavendish Laboratory in His Day* (New York, 1965).

For assessments of Thomson's work, see the following:

[61] H. F. Newall, E. Rutherford, C. T. R. Wilson, N. R. Campbell, L. R. Wilberforce *et al.*, *A History of the Cavendish Laboratory* (London, 1910).

[62] E. T. Whittaker, *A History of Theories of Aether and Electricity*. I. *The Classical Theories*, 2nd ed. (New York, 1951).

[63] R. McCormmach, "J. J. Thomson and the Structure of Light," in *British Journal for the History of Science*, **3** (1967), 362–387.

[64] V. M. Dukov, *Elektron: istoria otkritia i izuchenia svoistov* (Moscow, 1966), 108–154.

[65] J. L. Heilbron, "The Scattering of α and β Particles and Rutherford's Atom," in *Archive for History of Exact Science*, **4** (1968), 247–307.

[66] D. Topper, "Commitment to Mechanism: J. J. Thomson, The Early Years," *ibid.*, **7** (1971), 393–410.

J. L. Heilbron

THOMSON, THOMAS (*b.* Crieff, Scotland, 12 April 1773; *d.* Kilmun, Scotland, 2 July 1852), *chemistry*.

Thomson was the seventh child and youngest son of John Thomson, a retired woolman, and Elizabeth Ewan. Having been educated mainly at home by his talented mother and by his brother James, he enjoyed a good classical training at the Burgh School of Stirling from 1786 to 1788. In 1788 he won a bursary to the local University of St. Andrews, where he studied classics, mathematics, and natural philosophy. With a medical career tentatively in mind, in 1791 he began attending a variety of classes at the University of Edinburgh, where in 1795–1796 he was inspired to devote his life to chemistry by Joseph Black's impeccably elegant lectures. Although he did not graduate M.D. until 1799, from 1796 until 1800 he replaced his brother James as assistant editor of the Supplement to the third edition of the *Encyclopaedia Britannica*, to which he contributed extensive articles on chemistry and mineralogy. After 1800 Thomson devoted himself largely to chemistry as teacher, researcher, textbook writer, editor, and historian. Between 1800 and 1811 he made a precarious living as a private lecturer on chemistry in Edin-

burgh. During this period he published, on the basis of his articles in the *Britannica*, his bestselling *A System of Chemistry* (Edinburgh, 1802). In 1805 he acted as a well-paid consultant for the Scottish Excise Board and in 1807, with characteristic enterprise, instituted a private laboratory class in practical chemistry. In 1808, as a pugnacious opponent of Edinburgh's Huttonian geologists, Thomson helped to found the Wernerian Natural History Society. Becoming a fellow of the Royal Society in 1811, Thomson launched himself as a historian with his unofficial *History of the Royal Society* (London, 1812). From 1813 to 1820 he edited, first in London (1813–1817) and then in Glasgow (1817–1820), his own journal *Annals of Philosophy*, which rapidly overtook its proprietary competitors. The last phase of his entrepreneurial career began in 1817, shortly after his marriage to Agnes Colquhoun, when he was unanimously elected lecturer in chemistry at the University of Glasgow. Within seven months his aggressive opportunism had elevated this post into a regius professorship, which he held until his death. Thomson worked hard to reestablish the distinguished tradition of chemistry at the university and to improve its growing medical school. Particularly during the 1830's he was centrally involved in the unsuccessful political attempts made to elevate the status and increase the rights of the regius professors in the university. Tired by his life of unremitting effort, in 1841 Thomson relinquished part of his lecturing and the supervision of the laboratory to his nephew Robert Dundas Thomson, whom he had groomed as his successor. From 1846 until 1852 all his duties were discharged by his nephew, who subsequently failed to acquire the chair.

A self-taught chemist, Thomson first secured his reputation through the publication in 1802 of his *System*, which was so popular that it went through six editions during the next eighteen years; it also received the further accolade of appearing in French, German, and American editions. As the first systematic treatise of a nonelementary kind to break the French monopoly of such works, Thomson's *System* tried patriotically to do justice to the contributions made by British chemists to the new chemistry, which had been established in the late eighteenth century. Unusual for British works of the time, this well-ordered and careful digest was based on a wide range of original and recent papers as well as on the standard works. Quite characteristically, Thomson tabulated numerical data and employed a style that was attractively clear, frequently succinct, and occasionally trenchant. In the third edition of his *System* (Edinburgh, 1807), he began a thirty years' stint as John Dalton's warmest advocate when he extended the latter's chemical atomic theory from gases to include acids, bases, and salts. In January 1808 Thomson was the first to submit an experimental illustration of the law of multiple proportions, doing so at least four months before the publication of Dalton's *New System of Chemical Philosophy* (1808). This paper, "On Oxalic Acid" (*Philosophical Transactions*, **98** [1808], 63–95), also established a useful method of determining empirical formulas. After 1808 Thomson's enthusiasm for chemical atomism and for its mathematical harmonies burgeoned, an enthusiasm palpably displayed in the successive editions of his *System*, and countervailing the cautious skepticism shown by Wollaston and, more strongly, by Davy. His fervor was also apparent in his journal, *Annals of Philosophy*, in which he printed abstracts of currently published research by Continental as well as by British workers; inaugurated annual reports on the progress of science; and offered well-documented scientific biographies. Not unexpectedly, chemistry was the dominant subject of the journal, and its columns a leading vehicle for Dalton's chemical atomic theory. In 1815 Thomson espoused a second and related quantitative chemical cause when he published in his journal William Prout's anonymous paper on the specific gravities of gases. From then on Thomson focused his research upon three connected aims: to put Dalton's theory on a wider and firmer experimental basis; to provide conclusive experimental evidence for Prout's hypothesis that the atomic weights of elements were whole-number multiples of that of hydrogen; and to extend his projected investigation of the composition and formulas of salts to encompass those of all known minerals, particularly those containing aluminum.

During his first five or six years at Glasgow, Thomson and his laboratory students tried to accomplish the first two of these aims; their results were finally revealed in Thomson's *An Attempt to Establish the First Principles of Chemistry by Experiment* (London, 1825). Doubtless Thomson felt that this ambitious work was the culmination not only of his atomic labors but also of a long and busy career. Yet his measurements of the specific gravities of gases, of which he was proud, were shown in 1825 by his friend Harry Rainy to be inaccurate. Worse still, in 1827 the reliability of his gravimetric analyses was severely assailed by Berzelius, at that time the dominant analytical chemist in Europe. His lethal conclusion that many of

Thomson's fundamental experiments were made at the writing table did not enhance the reputation of the dour Scot, although his atomic weights were widely accepted in Great Britain and in the United States between 1825 and about 1835. In 1836 Thomson completed his research program by publishing his *Outlines of Mineralogy, Geology, and Mineral Analysis* (London, 1836), in which he arranged minerals on the basis of their experimentally determined chemical composition and not on the basis of their physical properties. As he discovered no new elements and was working in a research field that had been already intensively developed, the work aroused little controversy or notice.

The outstanding pedagogic feature of Thomson's professorial work was his pioneering emphasis on the laboratory teaching of practical chemistry. By the autumn of 1818 he had extracted from the university a chemical laboratory in which he established the first school of practical chemistry in a British university. He trained his students in the qualitative and quantitative analysis of inorganic substances, particularly minerals. Many of his students seem to have subsequently worked in the Glasgow chemical industry. Some of the more competent pupils were recruited by Thomson to form a small research school. By the mid-1830's his laboratory had become a nursery from which ambitious chemists migrated to other laboratories, including that of Justus Liebig, to complete their training.

Although sometimes arrogant and perpetually sardonic, Thomson became a father figure for Glaswegian scientists during the 1830's. His *History of Chemistry* (London, 1830–1831), unique and authoritative from 1760 onward, was professional propaganda that *inter alia* legitimated chemistry for the educated layman as a noble, rational, and autonomous science. From 1834 until his death Thomson was the president and chief ornament of the Philosophical Society of Glasgow. Indeed, in 1834 he was largely responsible for changing the moribund society dominated by artisans into an intellectually and administratively competent one that under his supervision at last began to publish its *Proceedings* in 1841. Not surprisingly he acted as Glasgow's senior host to distinguished scientific visitors and eagerly supported the activities of the British Association for the Advancement of Science during its first crucial decade. The leadership that Thomson gave to Glaswegian science, like so much else that he did during his career, shows the sort of reward that could be won at that time in British science by an ambitious self-made chemist who was distinguished more for unflagging industry than for brilliant originality.

BIBLIOGRAPHY

I. Original Works. Thomson's important books are listed by J. R. Partington, *A History of Chemistry*, III (London, 1962), 716–721. The *Catalogue of Scientific Papers Compiled by the Royal Society of London* (London, 1871), 5, 970–976, lists 201 items by him.

II. Secondary Literature. The basic biographies remain R. D. Thomson, "Biographical Notice of the Late Thomas Thomson," in *Glasgow Medical Journal*, 5 (1857), 69–80, 121–153, 379–380; and W. Crum, "Sketch of the Life and Labours of Dr. Thomas Thomson," in *Proceedings of the Philosophical Society of Glasgow*, 3 (1855), 250–264. Further secondary sources and a discussion of his professorial career are given by J. B. Morrell, "Thomas Thomson: Professor of Chemistry and University Reformer," in *British Journal for the History of Science*, 4 (1969), 245–265; and "The Chemist-Breeders: the Research Schools of Liebig and Thomas Thomson," in *Ambix*, 19 (1972), 1–46.

J. B. Morrell

THOMSON, SIR WILLIAM (Baron Kelvin of Largs) (*b.* Belfast, Ireland, 26 June 1824; *d.* Netherhall, near Largs, Ayrshire, Scotland, 17 December 1907), *physics.*

Thomson was the son of James Thomson, who, at the time of his son's birth, was professor of engineering at Belfast. In 1832 he became professor of mathematics at Glasgow. He was the author of several noted texts on differential and integral calculus, and he educated William and another son, James, at home. In 1834 both boys matriculated at Glasgow, where the environment was one characteristic of the Scottish universities of the time, which differed greatly from Cambridge. Whereas at Cambridge there was no chair in natural philosophy, nor much interest in the work of the Parisian analysts of the first third of the century, at Glasgow there was a professorship in natural philosophy (held by William Meickleham, who was succeeded by Nichol and then by William Thomson); there was also a chair in chemistry (held by Thomas Thomson).

Meickleham had a great interest in the French approach to physical science and much respect for it. In 1904 Thomson recalled how, "My predecessor in the Natural Philosophy Chair . . . taught his students reverence for the great French mathematicians Legendre, Lagrange, and Laplace. His

immediate successor, Dr. Nichol, added Fresnel and Fourier to this list of scientific nobles."[1] Having been stimulated by Meickleham, Thomson avidly read Fourier's *Théorie analytique de la chaleur* and Laplace's *Mécanique céleste* during a trip to Paris in 1839. Indeed, Thomson's earliest interests centered on questions drawn from both these treatises. His first published paper (except for an early effort concerning the completeness of Fourier series) involved an attempt to find a method for determining the temperature in a heat-conducting solid outside a closed isothermal surface described within the solid. Thomson approached the problem by employing propositions drawn from Laplace's theory of attraction: the theory that treats the forces exerted by shells of attracting matter. In doing so he forged a formal relationship between the theory of the transfer of heat, on the one hand, and on the other, the general class of theories of attraction, in the particular instance of effects exerted by the electrical fluid.[2]

In the course of his analysis Thomson found that the Coulomb force which is exerted by electrical fluid in a state of equilibrium on the surface of a conducting body within which no fluid exists is mathematically parallel to the flow of heat produced by thermal sources distributed, in place of the electrical fluid, over the surface of the conductor. The formal relation assumes that the empty space in and around the conducting body is replaced by a heat-conducting solid, and that the surface of the conductor is itself replaced by a similarly shaped surface over which sources of heat are distributed in equilibrium. (The term "equilibrium" here means that the thermal surface has a constant temperature and does not enclose any sources of heat, all sources being located at the surface.) In stating this formal relation, Thomson was attempting, at this time (1842), only to find a method for solving problems in heat by use of the theory of electricity in equilibrium. The relationship between the theories of the transfer of heat and of electricity was thus purely formal, their connection being mathematical, not physical.

Between 1841 and 1845 Thomson attended Cambridge. His studies did not influence him as deeply as had those during his years at Glasgow, primarily because of the extreme importance attached to finishing in the first rank of the Senate-House examinations. This emphasis required the expenditure of much thought on the particular kinds of mathematical problems asked. Only rarely were those problems related to any physical question that was not contained in Newton's *Principia*.

Thomson did expand his knowledge of the French mathematical techniques and theories during these years, however; and soon after his graduation he journeyed again to Paris, at his father's suggestion, to work in Regnault's laboratory.

On arriving there for the second time, during the summer of 1845, Thomson was warmly received by Liouville and was soon introduced to Cauchy, Sturm, Biot, Dumas, and Regnault. His studies in Paris were crucial for the subsequent development of British physical science. During this period he developed the technique of electrical images, first read Clapeyron's explication of Carnot's theory of the motive power of heat, and formulated a methodology of scientific explanation that strongly influenced Maxwell.

Thomson's extensive contact with Liouville led him to think more deeply about electrical theory. Liouville had heard of Faraday's work in electrostatics, or at least of the aspects in which Faraday claimed to have found that electrical induction occurs in "curved lines." The conception seemed to conflict with the action-at-a-distance approach, and Liouville asked Thomson to write a paper clarifying the differences between Faraday on the one hand and Coulomb and Poisson on the other. This request prompted Thomson to bring together ideas he had been turning over in his mind during the previous three years.

Even in the 1842 paper on isothermal surfaces, Thomson did not treat the electrical fluid as Poisson had done. He knew of Poisson's mode of mathematical development, but he dealt with the fluid more in the manner of Coulomb—that is, without attributing to the imagined fluid the material properties of actual fluids known to experience. Poisson, for his part, had insisted on conceiving of electricity as a fluid that, like other material entities, occupies a finite region of space: the central problem of his theory was the determination of the actual thickness of the electrical layer at any point of a conducting surface.

In 1842, and again early in 1845, Thomson attempted to envision the physical characteristics of the electrical fluid, and reached disquieting conclusions. He found that if electricity is thought of as a fluid the parts of which exert only inverse-square forces upon one another, then the electrical layer at the surface of a conductor can have no physical thickness at all. That result implied that electricity must be a set of point centers of force. At the time he completely rejected that notion. Later, in 1860, he attributed it to Bošković. But this rejection made it increasingly difficult for him to conceive of

Poisson's fluid as a real physical entity at all. Thus, Liouville's request for a discussion of the issue between Faraday's approach and that of action-at-a-distance led Thomson to attempt a restatement of both theories in terms free from physical hypotheses (as he termed hypotheses concerning unobservable entities). In making that effort, Thomson found himself to be constructing an entire methodology for scientific explanation.[3]

He began by distinguishing the "physical" from the "mathematical" content of Poisson's theory, at first implicitly. In the "mechanical theory of electricity," as he termed the physical hypothesis of the electrical fluid,[4] there are fundamental difficulties which create doubt about the adequacy with which the hypothesis represents the nature of all matter. At the same time the "mathematical theory" developed by Poisson did seem to be extremely powerful and to be capable of dealing correctly with a vast range of particular cases. Prompted by the physical difficulty with Poisson's approach and concerned over the apparent conflict with Faraday's ideas, Thomson began to think more carefully about the actual nature of the action-at-a-distance theory. What did it assert that could be accepted without also accepting the electrical fluid as a real entity? For Thomson was caught between his great admiration for the mathematical power of Poisson's approach and his distaste for the electrical fluid. Hoping to resolve this dilemma, he undertook a series of researches that in effect led him to a proof of the equivalence of Faraday's approach with that of the action-at-a-distance school. He achieved that result by excluding all elements that depended on physical hypotheses from both theories.

From Thomson's new point of view, both the French approach to electrical theory and that of Faraday should consist only of sets of mathematical propositions about the "distribution of electricity" on conducting bodies. Of Coulomb, who had never written like Poisson of the "thickness" of the electrical layer, Thomson said that he had "expressed his theory in such a manner that it can only be attacked in the way of proving his experimental results to be inaccurate." He did not, therefore, believe that Coulomb's approach would stand or fall with the fate of the electrical fluid.

Of course, it may be wondered how Thomson could have employed the phrase "distribution of electricity" without believing that some hypothetical entity is implicated. He did not think so, however. Instead, by 1845 he was drawing a distinction between a "physical hypothesis" and an "elementary mathematical law." By a physical hypothesis he meant an assumption concerning the physical existence of an unobservable entity like the electrical fluid or Faraday's contiguous dielectric particles. By an elementary mathematical law he meant a statement that can be directly applied in experiments because its referents are phenomenal entities and mathematical propositions about them. For example, when it is a question of the "distribution of electricity," a phrase that might appear in an "elementary mathematical law," the actual subject concerns the effects produced when a proof-plane is applied to a point of an electrified conductor. The measure of those effects is the twist given to the torsion-bearing thread of an electrometer. Coulomb's laws, therefore, and also those aspects of Poisson's mathematical development of them that do not depend upon the conception of electricity as a physical fluid, were thus actually concise, mathematical laws applicable to the results of such experiments. They were not hypotheses concerning the nature of electricity.

If, however, neither Coulomb's nor Faraday's approaches really did contain physical hypotheses, and if both of them yielded correct laws for the same phenomena, then there should be no conflict. Indeed, thought Thomson, any two theories dealing with the same phenomena, however different they may appear to be, cannot conflict if their most elementary laws can be expressed mathematically with no referents except those that can be interpreted phenomenologically. But having asserted the equivalence of such theories, Thomson had now to provide a method by which equivalence in any particular instance could be demonstrated. He found just such a method in his 1842 connecting of the laws of the transfer of heat with the theory of electricity in equilibrium on conducting surfaces. His reading of George Green's *Essay on the Application of Mathematical Analysis to the Theories of Electricity and Magnetism* (1828) had made the connection even more cogent.

Thomson obtained a copy of Green's work early in 1845. He was especially struck by the proof that a knowledge of the electrical potential at all points suffices to determine both the forces and the distribution of electricity on conducting bodies and permits dispensing with Poisson's postulate of an electrical layer of finite depth. Thomson realized that his own paper of 1842 had actually been founded on a formal equating of temperature with potential, and Green's work convinced him that all of the propositions of the mathematical theory of electricity could be expressed solely in terms of the potential. The relation between the mathematics of heat transfer and of electrical equilibrium thus became even more convincing than he had thought.

This relation made it possible to express the elementary laws of the "mathematical theory" (namely, that containing Poisson's mathematical results but not the conception of electricity as a physical fluid) in the same terms as the laws of the uniform transfer of heat. For Green had actually proved that a knowledge of the potential is sufficient to solve all electrical problems. "We may," Thomson now noted, "employ the elementary principles of one theory, as theorems, relative to the other." That is, the mathematical theory could be expressed in such a fashion that its laws appear as theorems following out of the laws of the transfer of heat when a formal equation is made between temperature and potential.

Construing the mathematical theory in terms of a Fourier heat transfer provided Thomson with a technique for proving the equivalence of Faraday's approach with the mathematical theory. The method was to express the concepts underlying these approaches in propositions drawn from Fourier; if this could be done unambiguously for both, then their equivalence would, he thought, be manifest. If, now, Faraday's principles regarding the induction of electricity on conductors could be connected with the same hypothesis-independent laws to which those of the mathematical theory were linked, then the two methods would come to the same thing. And, indeed, Thomson saw Faraday's system almost as an immediate consequence of the application of Fourier's laws to electrical phenomena (supposing conductors to be separated only by air or a vacuum).

But what of the effects of nonconductors placed between conducting bodies? Thomson had in mind here the properties that Faraday had discovered and had taken for evidence telling in favor of his theory and against that of Poisson. That difficulty would be met, Thomson thought, provided the new law were to be admitted in order to express the effect of a dielectric intermediary, and provided its status were to be that of an elementary mathematical principle, and not of a physical hypothesis. Thomson assumed that the smallest parts of a dielectric under electrical influence possess a "polarity" the laws of which are the same as Poisson's mathematical laws of magnetic polarity. He was thereby able to replace the dielectric by equivalent charged surfaces of electrical equilibrium.[5]

Thomson's conceptions were a powerful technique for analyzing the relationship between theories that seem to conflict. His method for reconciling the systems was to scrutinize them in terms of what each asserted concerning phenomena that could be measured or detected and to try to eliminate whatever was hypothetical. Although there was no guarantee that this critique would always succeed, for Coulomb's and Faraday's theories it did.

Thus it was that Thomson started his policy of eliminating physical hypotheses in order to reconcile Faraday's work with Coulomb's. It was fortified by his aversion to the concept of an imponderable fluid, an idea that was contrary to his beliefs about the nature of matter. For after all his methodology could not be independent of his opinions regarding the actual structure of unobservable entities. Given his particular objections to the electrical fluid, the method that he created between 1842 and 1845 for the purpose of unifying theories had begun to assume a slightly different aspect by 1847. The emphasis now fell upon what the formally equivalent system can tell about the original system as well as upon how it can help eliminate physical hypotheses.

The original purposes, which had been to unify theory and eliminate hypothesis, did not disappear. Instead an element of conceptual elucidation was added that made possible the visualization of the theory in new terms. The elucidatory role of the technique of formal equivalence became very prominent in 1847. In that year Thomson made use of Stokes's 1845 equations for fluids and solids in order to bring out the analogies, first, between the internal linear displacements of an elastic solid and electrostatic force, and, second, between the internal rotational displacements and galvanic and magnetic forces. Thomson's representation was conditioned by Faraday's discovery in 1845 of the rotation of the plane of polarization of light passed through a transparent body subject to magnetic action (the Faraday effect). The rotational nature of the effect led Thomson to characterize magnetic and galvanic forces by internal rotations of elastic media.[6] This new aspect of the process of formal equivalence was elucidatory because Thomson related distinct forces (electric and magnetic) to the internal processes of a single medium. Instead of being entirely distinct phenomena, electrical force and magnetic force were thus linked to a common element.

Yet even here Thomson limited himself to the statement that Faraday's discovery ". . . suggests the idea that there may be a problem connected with the distribution of electricity on conductors, or with the forces of attraction and repulsion exercised by electrified bodies." The purpose of the "mechanical analogy" was different from that of heat representation. The attempt was not to prove the equivalence of theories dealing with the same phenomena, but rather to link theories of different, although related, phenomena (namely, electricity and magnetism) by demonstrating that they could be shown to serve a closely connected set of expressions. The 1845 conception of formal equivalence was not thereby displaced; instead, it was supplemented. Equivalents now made the underlying mathematical connections of phenomena that are quite different more understandable. It was the qualities of clarity and unity that equivalents brought to disparate areas which ultimately had the greatest influence on Maxwell's early work. For it was from Thomson that Maxwell appropriated the equivalence technique as a method for providing unity in place of discord.

After his 1845 sojourn in Paris, Thomson returned to Scotland, where he succeeded to the professorship in natural philosophy at Glasgow, a post that he held for the rest of his life. Neither in Scotland nor in England was there then a university research laboratory or any other in which students could work. Thomson, having had access to Regnault's laboratory, was interested in establishing similar opportunities for students, and he obtained a small sum from the university for that purpose. It made possible the first teaching laboratory in Britain. He was also greatly interested in developing highly accurate measuring instruments, and the facilities of his new laboratory in Glasgow made that possible also.

Thomson's studies in France not only led him to a new approach to electrical theory and to an interest in experimental work and instrumentation; he was also introduced to Sadi Carnot's theory of the motive power of heat, as developed analytically by Clapeyron in 1834. Thomson was deeply impressed by the power of Carnot's theory (although he had read only Clapeyron's explication of it at this date), and especially by the rationalization it afforded for the production of mechanical effect by thermal processes. In an 1848 paper Thomson employed Carnot's theory for the first time in an attempt to establish an "absolute" thermometric scale. The old scale was a merely "arbitrary series of numbered points of reference sufficiently close for the requirements of practical thermometry." By an absolute scale Thomson meant one based on some completely general natural law. That law he took from Carnot—a given amount of heat passing between two given temperatures can produce at most a certain amount of work. In the old scale, based on the air thermometer, the amount of work done by a standard quantity of heat in falling through one degree varied at different points of the scale, as Clapeyron had shown. An absolute scale would be one in which the "value of a degree" would be independent of temperature. Thomson constructed this measure using the results of Regnault, Steele, and others.[7] (The modern Kelvin scale was defined later, following the elucidation of the concept of the conservation of energy.)

Carnot's work was generally unknown, even in Paris; and his theory was new not only to Thomson but to the entire British physical community of the 1840's. A few months after the 1848 paper Thomson presented to the Royal Society of Edinburgh a general account of Carnot's findings. Thomson noted that Carnot's theory was founded on the concept that heat is a substance that, when employed in a complete cycle of operations, enters the body acting as an engine in a given amount. At the end of the cycle it is entirely removed, and the engine remains in its initial state. Carnot had likened the operation of such a heat engine to that of a column-of-water device in which a quantity of water falling through a fixed distance from a given height produces an invariable quantity of motive power, the water being fully transferred from its original height to a reservoir at a lower level. In Carnot's view heat acts in a similar manner: a given quantity of heat abstracted by an engine from a high-temperature reservoir effects changes in the volume of the engine and, at the end of a cycle in which the engine is restored to its original condition, the heat taken from the high-temperature source has been totally transferred to a reservoir at a lower temperature, the amount of power produced being fixed by the temperatures of the reservoirs and the quantity of heat transferred.

Even at this early date, Thomson was ambivalent in his views concerning the concept of heat as substance despite the seeming necessity of accepting that idea if Carnot's theory were to be employed. In his "Account of Carnot's Theory" (1849) he wrote,

> . . . all those assumptions depending on the idea that heat is a *substance*, invariable in quantity, not convertible into any other element, and incapable of being *generated* by any physical agency; in fact the acknowledged principles of latent heat; would require to be tested by a most searching investigation before

they ought to be admitted, as they have usually been, by almost every one who had been engaged on the subject.[8]

Thomson's ambivalence had two distinct sources. The first lay in his skepticism about the propriety of employing imponderable entities. Further, Fourier's theory of the transfer of heat had left open the question of the nature of heat, and Thomson felt this openness to be one of the most important characteristics of the theory. Thus, he would not have accepted the material theory of heat without misgivings, even had he not known the work of Joule, whom he met at the 1847 meeting of the British Association. That work furnished a second set of reasons for Thomson's ambivalence.

Joule believed that heat and mechanical effect are but different aspects of matter, motion, and force. Sensible heat, he supposed, is in reality the "living force of the particles of the bodies in which it is induced"; when in a latent form, heat consists in the "separation of particle from particle, so as to cause them to attract one another through a greater space." Living force and "attraction through space"—by which Joule intended not force alone but a force acting across a given space—are convertible and "equivalent" because "living force may be produced by the action of gravity [or any attractive or repulsive force] through a given distance of space." Conversely, particulate motion can be transformed into particulate arrangement under the action of force. Whenever the living force of macroscopic bodies disappears, an equivalent of either particulate live force or of mechanical rearrangement effected against force is produced.

Thus, according to Joule, heat can be absolutely generated by mechanical action or, indeed, by any physical agency in which mechanical force is the ultimate source of action (as, for example, in an electromagnetic engine turned by hand). Conversely, heat should be absolutely destroyed in those circumstances in which thermal agency effects mechanical action. Joule attempted to support his contentions through a series of experiments involving electromagnetic engines and the internal motions of viscous fluids. In the former case Joule argued that heat is actually generated in a current-bearing conductor by the action of the current directly and not by any transfer of heat from hotter to colder parts of the body. In the latter case, he produced motions in a viscous fluid by means of a paddle wheel run by mechanical force. Although the motions seem to disappear after a time, they are actually converted into the living force particles of the fluid—heat—as is evidenced by a rise in temperature.

These two experiments at best demonstrated, Thomson carefully noted, that heat can be produced by mechanical effect; but they did not prove the converse—that it can be destroyed through conversion into effect. Joule's experimental demonstrations enabled Thomson to keep his distance from the material hypothesis of heat, although he continued to accept what he believed to be Carnot's basic principles. He did not embrace Joule's wider schema of heat as motion and "attraction through space"; indeed, it is highly unlikely that he knew the details of Joule's ideas on these points in 1849 beyond the rather vague assertion that heat is particulate *vis viva*. Thomson probably did not yet fully grasp the conception of latent heat as the arrangement of particles under the action of force. Indeed, his presentation of "Carnot's axioms" would suggest that he did not.

In the 1849 "Account" Thomson rarely referred to the impossibility of perpetual motion, conceived then as the impossibility of obtaining motive power without a corresponding alteration in other conditions. He began directly with his own version of Carnot's two axioms, regarding them as the basis of a theory that makes possible the calculation of the mechanical effect that can be produced *solely* by thermal agency. He gave the following as the principal "questions to be resolved by a complete theory of the subject"; these questions, to which Carnot's theory provides one possible answer, were essentially pragmatic:

1. What is the precise nature of the thermal agency by means of which *mechanical effect* is to be produced, without effects of any other kind?
2. How may the amount of this thermal agency necessary for performing a given quantity of work be estimated?

Thomson answered the first question by a deduction from what he called Carnot's "fundamental axiom," *viz.*, that ". . . at the end of a cycle of operations, when a body is left in precisely its primitive physical conditions, if it has absorbed any heat during one part of the operations, it must have given out exactly the same amount during the remainder of the cycle." On the basis of this axiom, Thomson concluded that "the origin of motive power . . . must be found in the agency of heat entering the body and leaving it" because no other effects are produced during a complete cycle. The

precise mode in which mechanical effect is produced by heat transfer is specified in Carnot's second axiom: "The thermal agency by which mechanical effect may be obtained, is the transference of heat from one body to another at a lower temperature."

Combining these two propositions made it possible to derive a theorem of great importance, which states that a fixed quantity of effect, the maximum which can be obtained, is produced by all reversible engines from the transferral of a given quantity of heat between two specified temperatures: "A perfect thermodynamic engine is such that, whatever amount of mechanical effect it can derive from a certain thermal agency; if an equal amount be spent in working it backwards, an equal reverse thermal effect will be produced." The foregoing has the standing of a theorem, and not a definition, because the criterion of reversibility is implicated in the first axiom through the notion of cyclical processes. The theorem affords the answer to the second question because it is used to calculate the effect produced by the transfer of heat between two fixed reservoirs by a perfect engine.

Both axioms are necessary for the derivation of the perfection theorem, and the first axiom is additionally necessary when analyzing a cycle of operations to ensure that, in a complete cycle, the engine returns to its primitive state. Yet Thomson, troubled by the conception of heat as a substance, felt that he had to justify the fundamental (first) axiom and attempted to do so by asserting that "no operation is known by which heat can be absorbed into a body without either elevating its temperature, or becoming latent, and producing some alteration in its physical condition." But heat engines must return to their initial states at the end of a cycle if—and here is the true motive for Thomson's insistence on the first axiom—calculation of the effects that have purely thermal origins is to be possible. It is again evident that Thomson had not yet grasped Joule's conception of latent heat as converted heat. He had not assimilated the idea that a body does not have to part with heat *as* heat in order to be returned to its primitive state. Instead, it can effect the conversion of heat into "attraction through space," that is, into the configuration of external bodies between which forces act, and thereby return to its original condition. In Thomson's opinion heat had to be transferred as heat in order to appear once again as sensible or latent; it cannot "disappear."

Thus, what Thomson called Carnot's "fundamental axiom" asserted the inconvertibility of heat in the view he then took of it. At this stage in his thought, it was not because convertibility of heat to something else would permit perpetual motion that he objected to it. Rather, he wished to maintain what he took to be Carnot's theory, because abandoning either axiom would ruin any theory of the origins of mechanical effect from thermal agency. He justified the first and fundamental axiom by the lack of any known operation in which heat can be absorbed without the appearance of extraneous effects. It is obvious in retrospect that this position was tenuous because it actually depended upon Thomson's inability to conceive of latent heat as converted heat. It can therefore be seen that in 1849 Thomson did not so much hold to the conception of the materiality of heat as he held to certain propositions that appeared to be founded on that idea, but that are in reality acceptable on empirical and pragmatic grounds. Carnot had himself, toward the end of his short life, come to view the material theory of heat as untenable and had accepted the view that heat is a mode of motion. He had despaired of making his theory of motive power consonant with the kinetic model of heat. Thomson, however, knew nothing of Carnot's later ideas.

Despite Thomson's having accepted the two axioms, in particular the first, he was still not confident of the idea that heat is a substance. He reasoned that Carnot's fundamental axiom, while it "may be considered as still the most probable basis for an investigation of the motive power of heat," might, along with "every other branch of the theory of heat . . . ultimately require to be reconstructed upon another foundation, when our experimental data are more complete." Thomson's ambivalence depended in part upon his aversion to imponderable entities. Joule thought as he did, and wrote

> In our notion of matter two ideas are generally included, namely those of *impenetrability* and *extension*. . . . Impenetrability and extension cannot with much propriety be reckoned among the *properties* of matter, but deserve rather to be called its definitions, because nothing that does not possess the two qualities bears the name of matter. If we conceive of impenetrability and extension we have the idea of matter, and of matter only.

Joule's comments were printed in the Manchester *Courier* of 5 and 12 May 1847, but he may very well have discussed them with Thomson at the British Association meeting of that same year.

Thomson was also impressed by Joule's conception of conservation in general. As Joule had written in the *Courier*:

> . . . the phenomena of nature, whether mechanical, chemical, or vital, consist almost entirely in continual conversion of attraction through space, living force, and heat into one another. Thus it is that order is maintained in the universe—nothing is destroyed, nothing ever lost, but the entire machinery, complicated as it is, works smoothly and harmoniously. And though, as in the awful vision of Ezekiel, "wheel may be in the middle of wheel," and every thing may appear complicated and involved in the apparent confusion and intricacy of an almost endless variety of causes, effects, conversions, and arrangements, yet is the most perfect regularity preserved—the whole being governed by the sovereign will of God.

Thomson was clearly affected by this conception, for in a footnote to his 1849 "Account" he remarked:

> When thermal agency is spent in conducting heat through a solid, what becomes of the mechanical effect it might produce? Nothing can be lost in the operations of nature—no energy can be destroyed. What effect then is produced in place of the mechanical effect which is lost? . . . [A similar problem seems to exist in the question of the] mechanical effect lost in a fluid set in motion in the interior of a rigid closed vessel, and allowed to come to rest by its own internal friction; but in this case the foundation of a solution of the difficulty has been actually found, in Mr. Joule's discovery of the generation of heat by the internal friction of fluids in motion.

It can be seen that in 1849 Thomson was deeply puzzled by the apparently complete disappearance of effect in certain cases. His thoughts were more clearly defined during the following two years as the result of two experiments. One of them clearly lent weight to Carnot's theory, and the other seemed to support that part of Joule's conceptions in which heat is thought to be generated. (Although Thomson had, from 1847 on, been willing to consider the latter possibility, only in 1850 did he accept it as a proven fact.)

The first experiment, performed by his brother, James, confirmed the deduction from Carnot's theory (with both axioms) that the freezing point of water must be lowered when the pressure is increased. One of the most important aspects of that work lay, not so much in its direct test of the two axioms, as in its use of the theorem derived from them regarding the perfection of reversible engines. It was this theorem that James Thomson most frequently employed in his demonstration, thereby giving it a significance greater than that normally associated with a derivative proposition. The work on the lowering of the freezing point of water, therefore, helped to shift attention from the two axioms toward the theorem on the perfection of the reversible engine. This shift was important for William Thomson's subsequent reformulation of Carnot's theory, in that he was to reconstruct the theory by substituting for Carnot's dual axioms a single one from which the perfectibility theorem followed directly.[9]

The second circumstance was more complex than the first and involved the properties of saturated steam under high pressure escaping through an orifice. Rankine had observed that, when saturated steam is permitted to expand, the heat that becomes latent during expansion is greater than the heat that the expanding vapor would normally release as a result of its concomitant drop in temperature. If no part of the vapor is to become liquid, it follows that some external source must supply the extra heat necessary to maintain saturation. More specifically, as a gas expands adiabatically, its temperature drops; the drop in temperature was taken as an indication that a certain amount of the free heat of the gas had become latent. The expansion of the gas was supposed to be the result of this absorption of heat. (The modern explanation of the effect depends upon the negative slope of the temperature-entropy curve in the vapor portion of the liquid-vapor curve. The implication is that the specific heat of saturated steam is negative, meaning that heat is absorbed as its temperature falls. The physical explanation of the effect is that, as the vapor has its temperature lowered, it expands so much to avoid supersaturation that the external work that it performs is greater than the drop in its internal energy. The situation thus requires that heat be absorbed.)

Steam remains saturated during its expansion after its escape at high pressure through an orifice (an expansion rapid enough to be essentially adiabatic). Its condition is evident in that it does not scald, as it would had it become partially liquefied. By mid-1850 Thomson saw this phenomenon as conclusive evidence in favor of Joule's contention that heat can be generated from something else: Thomson thought he knew the source of this extra heat. "There is no possible way," he wrote to Joule, "in which the heat can be acquired except by the friction of the steam as it rushes through the orifice. Hence I think I am justified in saying that your discovery alone can reconcile Mr. Rankine's

discovery with known facts." It was at this time also that Thomson first learned of Clausius' work of the previous April; although he had not read it by October, he commented that Clausius' methods "differ from those of Carnot only in the adoption of your [Joule's] axiom instead of Carnot's. . . ." Within five months Thomson himself had assimilated in full the "dynamical theory of heat" and had so modified Carnot's theory as to be in accord with it—an act accomplished independently, without any detailed knowledge of Clausius' work of May 1850.[10]

Thomson's central concern during these months was the discovery of a principle from which the essential elements of Carnot's theory could be derived while dispensing with the fundamental axiom concerning the conservation of heat *qua* heat. This was not a simple task; without the fundamental axiom, it is extremely difficult to produce a measure of the effect resulting from purely thermal action. The beginning of a solution lay in the new meaning given to the concept of "latent heat" by the dynamical theory, where it is thought of as the work done against the internal, molecular forces of a body. They then store it in the resulting molecular configuration. This notion relieved Thomson of his earlier concern that a body can be returned to its initial state only if the entire amount of heat that has entered it leaves in the same form. He now understood that heat can be converted into a "new" form—"attraction through space" in Joule's terminology—and yet be entirely removed from the transferring engine because the "latent heat" of the engine, its forced molecular configuration, can be directly converted through the performance of work into something of the same kind, namely, the "attraction through space" of external bodies. The problem, therefore, became the discovery of a principle that could be taken as the expression of the essential contents of Carnot's theory once conversion had been admitted. In other words, now that heat could be envisioned as both *vis viva* and molecular configuration, the problem ceased to concern the way in which purely thermal actions could be measured, for thermal actions were reducible to mechanical processes. The problem now was to formulate the theory of heat engines.

As noted above, James Thomson's 1850 paper on the freezing point of water placed great emphasis on the ideal nature of reversible engines. The maximum mechanical effect from the transfer of a quantity of heat between two temperatures will be obtained from engines that yield a mechanical effect when run forward equal to that expended on them when run backward. Although derived from Carnot's two axioms, by 1850 Thomson had come to see this proposition as the central element of the theory. The theorem as stated is insufficient. It is necessary also to stipulate that the engine returns to its initial condition at the end of a cycle. Thomson had earlier believed that this requirement necessitated the acceptance of Carnot's fundamental axiom, but by March 1851 he knew that this was incorrect. Yet the theorem itself could not be derived if either of the two axioms was rejected. Neither was it tenable in its original form, if the dynamical theory was accepted, because it employed the concept of the complete transfer of heat. Thomson was able to reformulate the theorem without referring to transfer of heat. Instead, he referred to heat "quantity": "If an engine be such that, when it is worked backwards, the physical and mechanical agencies in every part of its motion are all reversed, it produces as much mechanical effect as can be produced by any thermodynamic engine, with the same temperatures of source and refrigerator, from a given quantity of heat." (The "quantity of heat" appears to be that which is abstracted from the high-temperature reservoir.) Thomson attributed this formulation to Carnot and Clausius, but the attribution to Carnot is misleading, because Carnot's statement referred to the transfer of heat *in toto* and not simply to heat "quantity" and thermal reservoirs. It was in the proposition on the ideal nature of reversible engines that Thomson located the essential content of Carnot's theory. This proposition, along with Joule's, is one of the two central propositions of the theory of heat engines.

Nonetheless, Thomson was not satisfied with the proposition on the perfection of reversible engines. He felt the need for a more elementary principle because the proposition did not appear to be self-evident, as Carnot's axioms had prior to Joule's work. Working independently of Clausius, Thomson now developed the concept of a new kind of perpetual motion and then deduced the perfection of reversible engines from the postulate of its impossibility. This new perpetual motion, were it possible, would produce useful effects solely by the conversion of heat directly into work—a possibility that is not in conflict with either Joule's proposition or with the impossibility of that kind of perpetual motion in which something for nothing is obtained. Thomson asserted the impossibility of what was later termed perpetual motion of the second kind in the following words: "It is impossible, by means of inanimate material agency, to derive mechanical effect from any portion of matter by cooling it below the temperature of the coldest of the surrounding objects."[11]

After 1844 (at the latest) Thomson felt that Fourier's principles had been overlooked by those "geologists who uncompromisingly oppose all paroxysmal hypotheses, and maintain not only that we have examples now before us, on the earth, of all the different actions by which its crust had been modified in geological history, but that those actions have never, or have not on the whole, been more violent in past time than they are at present." Thomson had early thought the Uniformitarian approach to be untenable. He believed that if both the earth and the sun had once been molten balls cooling through radiation (the earth forming a crust and the sun, because of its peculiarly high temperature and the nature of its substance, remaining an "incandescent liquid mass"), then the dissipation of heat required by Fourier's laws must necessarily have been much more rapid in the past than in the present. If this were so, then clearly such phenomena as the winds, which depend upon thermal gradients, must have been much more vigorous in past times. In 1844 these views were merely beliefs. Because Thomson had not developed in detail the solutions to such problems of heat transfer, he had not investigated the modifications which such circumstances as the solidification of the earth's outer crust might have produced in the rate at which heat is dissipated. Most important, he did not have the full range of data needed for deducing numerical values.

By late 1852 Thomson's beliefs about the inadequacy of Uniformitarian assumptions had been made even more cogent as a result of the second law of thermodynamics. He now thought Fourier's theory of the conduction of heat to be "a beautiful working out of a particular case belonging to the general doctrine of the 'Dissipation of Energy.'" Thomson's earlier beliefs were now reinforced by the energy dissipation conception. He reasoned that geological actions, being ultimately mechanical and due, in cases like volcanism, to internal gradients in the heat of the earth, must have gradually decreased in intensity over time as energy was "dissipated" (or, as he later put it, as "potential energy is exhausted"). Even if the earth as a whole has not cooled appreciably since its formation, it must still follow that volcanic action must have been more intense in the past. For it is a particular case of the conversion of heat into mechanical effect, and the inevitable equalizing of temperature throughout the substance of the earth must ultimately lead to a state of quiescence. By 1852 Thomson also thought that Fourier's laws require both the earth and the sun to have cooled substantially over time (although he had not provided a detailed argument for this latter assertion). And however that may be, he thought that the second law of thermodynamics requires volcanic action to have been much more intense in the past than at present. (If, however, the sun has not cooled appreciably, then atmospheric phenomena on earth—though not necessarily geological—might not have altered radically over time.)

By 1862 Thomson had provided detailed support for the contention that the sun has cooled. It had previously been held that the sun, although an incandescent mass radiating heat, might have remained at about the same temperature, its heat replenished either by the influx of meteors or by the effects of shrinking under the influence of gravity, or by both factors. Thomson discounted these contentions by making numerical estimates. In the first case, he argued, the mass increase attendant on the meteoric influx would have affected planetary motion. In the second case, even if heat is generated by shrinkage—given reasonable estimates of the specific heat of the sun's mass obtained from "Stokes's principles of solar and stellar chemistry" (spectroscopic theory)—the correlation is very poor between the actual amount of heat radiated and the amount that would be provided by gravitational collapse. It is, therefore, most likely, argued Thomson, that the sun has cooled considerably and that it is an incandescent liquid mass receiving no heat from without. On that basis, Thomson further calculated, given his estimate of the solar specific heat and the present rate of radiation, that the sun probably has "not illuminated the earth for 100,000,000, and almost [certainly for not more than] 500,000,000 years."

Thomson's deduction of a maximum limit for the age of the sun was in direct conflict with those geological Uniformitarians who assumed that geological time cannot be given absolute limitations. Thomson soon presented a second paper supporting his earlier belief that the earth also must have been much hotter in the past. He showed that Fourier's laws of the transfer of heat require that, given the present rate of decrease of the heat of the earth with depth, the earth must have solidified from its primordial molten state not less than 20,000,000 and not more than 400,000,000 years ago. These limits were rigorous deductions from Fourier's laws applied to the case of a molten sphere cooling through emission of radiant heat.

They include a probable estimate of the magnitude of the effect due to the formation of the earth's crust; and they hold good provided that the earth has no sources of energy beyond its own central heat.

Thomson's original intention in writing the 1862 papers had been to attack Lyell and the extreme Uniformitarian approach to geology. The Uniformitarians, he felt, looked upon the sun "as Fontenelle's roses looked upon their gardener. 'Our gardener,' say they, 'must be a very old man; within the memory of roses he is the same as he has always been; it is impossible that he can ever be other than he is.' " Lyell had asserted that an absolute geochronology is not useful and most likely not possible. By 1865, however, British geologists were uncertain that Lyell had been correct on this point, and there was disagreement over whether the history of the earth could be absolutely dated. Some believed that it could be grouped in distinct sequences (Pre-Cambrian, Cambrian, etc.) with arbitrary time spans. Thomson unwittingly entered the midst of the geochronology controversy with his insistence that absolute times can, and indeed must, be assigned.

The idea that the central heat of the earth accounts for volcanic action and other geological processes involving thermal variations was assumed by most geologists of the time; and Thomson's criticism implied that, if that idea be accepted, it followed that for a long period the surface of the earth had been the scene of violent and often abrupt changes. This last point was especially damaging to the Darwinians, who had assumed that evolution, being a very slow, gradual process, must occur within the context of uniform geological change. For thirty years the geological and biological community had either to ignore the findings of physics, which few could do comfortably, or else with Huxley attempt to satisfy the demands of Thomson's limitations as best they could.[12]

It was Thomson's acceptance of the dynamical theory of heat and his subsequent reformulation of the axioms of thermodynamics that had led him to several of the conclusions resulting in the geological controversy. The effect of the dynamical theory was not limited to areas directly associated with thermal processes, however. In accepting the conception that sensible heat is particulate *vis viva* and that latent heat is the stored effect of molecular configuration, Thomson, for the first time in his career, deliberately did employ unobservable entities and make use of physical hypotheses. It must be recognized that Thomson's earlier exclusion of unobservables had been dependent upon his inability to attribute truly material properties to such entities. In contrast, the dynamical theory of heat did not require the acceptance of any particular conception of the ultimate nature of material particles. It required only that these particles exist and exhibit the properties of mass, motion, and the power to exert forces (although this latter aspect presents several difficulties). The idea that heat is a mode of particulate motion opened to Thomson a new approach to all physical theory, therefore. He wrote in 1872 of his thoughts before he had accepted the dynamical theory

> . . . [before 1847] I did not . . . know that motion is the very essence of what has hitherto been called matter. At the 1847 meeting of the British Association in Oxford, I learned from Joule the dynamical theory of heat, and was forced to abandon at once many, and gradually from year to year all other, statical preconceptions regarding the ultimate causes of apparently statical phenomena.

In 1855 Thomson wrote John Tyndall a letter in which he referred to the "mechanical qualities" of the medium that pervades all space. He now believed the ultimate explanation of electromagnetic phenomena lay in the structure of that medium, conceived dynamically. On 10 May 1856 Thomson submitted to the Royal Society a paper that employed the dynamical properties of molecular entities to explain the Faraday effect. By then he was willing to employ microstructural entities which possess the requisite material properties, and he argued that, from any galvanic current, there extends a moving spiral that coils about the line of magnetic force passing through the center of the axis of the current. Indeed, he intimated that the current itself consists of the trapping of a segment of this spiral in ponderable matter. Light waves are propagated by transverse vibrations of the particles of the moving spirals, and the plane of polarization of the waves will be rotated in a sense dependent upon the motion of the spiral, which, in turn, depends upon the direction of the magnetic force.[13] (See Fig. 1.)

Thomson believed that this representation afforded the ultimate representation of all electromagnetic effects. Magnetic forces were to be due to the screw motions of the spiraling helices which, when fixed in matter, become currents, and electrostatic forces to their compressions:

> We now look on space as full. We know that light is propagated like sound through pressure and mo-

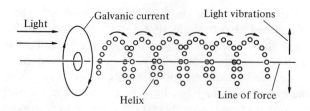

FIGURE 1

tion. . . . If electric force depends on a residual surface action, a resultant of an inner-tension, experienced by the insulating medium, we can conceive that electricity itself is to be understood as not an accident, but an essence of matter. Whatever electricity is, it seems quite certain that electricity in motion IS heat; and that a certain alignment of axes of revolution in this motion is magnetism. . . .[14]

It was this vision of Thomson's that Maxwell seized upon between 1857 and 1862 in his search for a new approach to electromagnetic theory.

Despite his influence upon Maxwell, Thomson was highly skeptical of Maxwell's approach to electromagnetic theory. The reasons for that skepticism lay in Thomson's developing beliefs regarding the nature of the hypotheses to be used in physical explanations. Although the dynamical theory of heat had opened Thomson's mind to the role of unobservable entities in theory, he still had very definite opinions on the nature of the admissible entities. "I have been led," he wrote in 1858 in an unpublished note,[15] ". . . to endeavour to explain some of the known properties of sensible matter by investigating the motion of [a fluid filling the interstices between detached solid particles] on strict dynamical principles." The sole attributes of the space-filling fluid are extension, incompressibility, and inertia; these, together with the laws of mechanics, constitute the "dynamical principles" to be used as the basis of physical theory. Evident in this is a clear reflection of Thomson's early revulsion from imponderable fluids and his strong conviction that it is in extended matter and inertial motion that the ultimate explanation of all physical processes is to be sought.

Yet one phenomenon above all stands in the path of any general theory of the type he desired, and that is elasticity. If nothing is to be admitted beyond matter and motion, then elastic reaction to compression and distortion must be explicable without recourse to "force" in the Newtonian sense. In 1858 Thomson had no theory that could explain elasticity solely on the grounds of extension and inertia. What put him on the track of a solution was Tait's 1867 translation of Helmholtz' 1858 paper on vortex motion in inviscid, incompressible fluids: such fluids are analyzable without recourse to any principles beyond inertia and extension. In his work, Helmholtz, with whom Thomson maintained a close personal friendship, had shown that linear fluid vortices, defined as lines drawn in the fluid along which the angular momentum of differential fluid elements is constant, influence one another's motions through the instantaneously propagated pressures that their existence produces in the fluid as a whole. The effects are instantaneous because the fluid is incompressible. Closed lines of vortex motion exhibit striking patterns: two such figures appear to repel or attract one another in a complex manner dependent upon their mutual orientation and the angular momenta of their constituent elements. These effects are the results solely of pressures produced in the medium.

In 1867 Thomson opened a paper entitled "On Vortex Motion" as follows: "The mathematical work of the present paper has been performed to illustrate the hypothesis that space is continuously occupied by an incompressible frictionless fluid acted on by no force, and that material phenomena of every kind depend solely on motions created in the liquid."[16] In subsequent papers Thomson attempted to demonstrate that closed vortex tubes (tori whose surfaces are formed of closed, linear vortex filaments) behave toward one another much as the material particles that constitute bodies are supposed to act: the tubes "repel" one another, and, under certain conditions, they "attract"—actions that are the results, not of "force," but of material motions alone. Thomson ultimately had to admit that his program could not be expanded to include electromagnetism or the electromagnetic theory of light, gravitation, and chemical phenomena; but throughout the period 1860–1880 he was imbued with the conception of "vortex atoms"— that is, with the idea that all material particles are actually vortex tubes in an all-encompassing medium. Although their initial creation is inexplicable mechanically, he believed that their actions provide the most fruitful basis for theoretical speculation. Helmholtz had shown that vortices, once established by some unknown means in an inviscid medium, have unalterable qualities that can neither be generated nor destroyed by mechanical processes, and so are eternal. Among these qualities, as noted

above, is a pseudoelasticity by which the vortices tend to avoid one another in an apparently elastic manner. Thomson saw in this the possibility of identifying the elastic atoms of kinetic theory with perpetually circulating vortices.

It was the strength of this conviction that accounts for his cold reaction to Maxwell's work. Maxwell, in 1861–1862, and again in 1864, had perforce to employ an unreduced "elasticity," a resistance to distortion of form, in his equations for the dynamical processes of the medium. Further, in their 1864 form the ethereal processes are difficult to visualize—they are dynamical, employing as primordial concepts material substance, momentum, and energy—but the structure of the medium is not necessarily supposed to be continuous, nor is it specified. Thomson regarded elasticity as a property that must be reduced completely to the effects of material motion. By 1867 he was greatly excited by the possibilities of his fluid medium, and he regarded Maxwell's system as at best a way station on the road to a more adequate representation. He could not accept a primordial elasticity. (For that matter, not many of Maxwell's British successors could accept it, and Maxwell himself was troubled by the question.) Also Thomson was not attracted to any dynamical theory that could not be fitted into the framework of a continuous, readily visualized medium. Although these hopes kept him from accepting Maxwell's scheme, and gradually opened a gap between him and the new generation of British "electricians" of the late 1870's and early 1880's, still his conceptions were generally regarded as the best chance for a truly dynamical theory. And Maxwell himself, despite Thomson's disagreement with his theory, was impressed by this vision. He saw in it at least an attempt to achieve ultimate mechanical simplicity; he, too, believed in this goal, having strayed only so far as he felt necessary in his approach to electromagnetism. Maxwell wrote of Thomson's vortices as follows:

> . . . the greatest recommendation of this theory, from a philosophical point of view, is that its success in explaining phenomena does not depend on the ingenuity with which its contrivers "save appearances" by introducing first one hypothetical force and then another. When the vortex atom is once set in motion, all its properties are absolutely fixed and determined by the laws of motion of the primitive fluid, which are fully expressed in the fundamental equations. The disciple of Lucretius may cut and carve his solid atoms in the hope of getting them to combine into worlds; the follower of Boscovich may imagine new

laws of force to meet the requirements of each new phenomenon; but he who dares to plant his feet in the path opened by Helmholtz and Thomson has no such resources. His primitive fluid has no other properties than inertia, invariable density, and perfect mobility, and the method by which the motion of this fluid is to be traced is pure mathematical analysis.[17]

Thomson's concern with the dynamical foundations of physical science, and his insistence that material substance be clearly conceived, are strikingly evident in the *Treatise on Natural Philosophy*, which he wrote with Peter Guthrie Tait in the early 1860's. "Thomson and Tait," as the work is generally known, and Maxwell's later *Treatise on Electricity and Magnetism* were the most influential British physical texts of the last half of the nineteenth century. Originally Thomson had envisioned a multivolume series giving a complete representation of all physical theory, including material processes, heat, light, and electricity and magnetism. He and Tait produced only the first two volumes, however—on kinematics and dynamics. The unfinished character of the work is evident in its continual references to future portions—as, for example, to a planned section on the properties of matter. Thomson and Tait presented in full the kinematics of point particles and the dynamics of motion under force; they placed heavy emphasis upon the dynamics of material media; and they made detailed use both of a new formulation of Lagrangean mechanics and the conservation of energy. The *Treatise on Natural Philosophy* introduced a new generation of British and American physical scientists to the details and concepts of mechanics.

In his attempt to achieve an alternative to Maxwell's theory of light, Thomson could find no clear help in his vortices, and in 1888 he began to look into the reasons for the failure of the elastic-solid theory of the luminiferous ether. He had particularly in view George Green's 1837 medium. Green had been obliged to assume that the ether is incompressible, in order to avoid instability and to remove longitudinal waves. On these grounds, however, he was unable to obtain Fresnel's tangent-law for the reflection and refraction of light for waves polarized normally to the plane of incidence. Encouraged by his vortex theory of the medium, Thomson at first imagined an inviscid fluid permeating the pores of an incompressible, spongelike solid, but he soon found that this structure only augmented the older difficulties. He therefore began to consider whether Green's comments on stability might not be open to doubt. As Thomson considered the problem, it occurred to him that the kind of instability envisaged by Green, that of a spontaneous shrinkage of finite ether volumes, would not occur "provided we either suppose the medium to extend all through boundless space, or give it a fixed containing vessel as its boundary."

By supposing the ether to have no resistance to compression by volume, Thomson was able to show that the hypothesis adopted makes it possible to obtain all of Fresnel's laws, while eliminating the longitudinal wave and avoiding instability. The work had an immediate and widespread impact on the Maxwellian community. Within a month Richard Tetley Glazebrook had written a paper successfully applying the new conception to double refraction, dispersion, and metallic reflection. In the United States, Willard Gibbs compared the new theory to the electrical theory:

> It is evident that the electrical theory of light has a serious rival, in a sense in which, perhaps, one did not exist before the publication of William Thomson's paper in November last. Nevertheless, neither surprise at the results which have been achieved, nor admiration for that happy audacity of genius, which seeking the solution of the problem precisely where no one else would have ventured to look for it, has turned half a century of defeat into victory, should blind us to the actual state of the question.
>
> It may still be said for the electrical theory, that it is not obliged to invent hypotheses, but only to apply the laws furnished by the science of electricity, and that it is difficult to account for the coincidence between the electrical and optical properties of media, unless we regard the motions of light as electrical. . . .[18]

Despite Thomson's deep theoretical concerns, he was always strongly interested in physical instrumentation. He felt that existing instruments were inadequate for precisely determining important physical constants. Instrumentation became even more important as electrical phenomena began to be employed in Britain's increasingly complex industrial economy, and Thomson was involved in the design and implementation of many new devices.

His interest and reputation brought him to the attention of a consortium of British industrialists who, in the mid-1850's, proposed to lay a submarine telegraph cable between Ireland and Newfoundland. Telegraphy by then was a well-developed and extremely profitable business, and the idea of laying such a cable was not new. The undertaking provides perhaps the first instance of a complex interaction between large-scale industrial enterprise and theoretical electricity. Thomson was brought in early in the project as a member of the board of directors, and he played a central role.

The directors had entrusted the technical details of the project to an industrial electrician, E. O. W. Whitehouse; and the many difficulties that plagued it from the outset resulted from Whitehouse's insistence on employing his own system of electrical signaling, despite theoretical objections from Thomson. Thomson had developed a very sensitive apparatus, the mirror-galvanometer, to detect the minuscule currents transmitted through miles of cable, but Whitehouse refused to use it. The Whitehouse-Thomson controversy stemmed primarily from Whitehouse's jealousy of Thomson's reputation. Thomson had asserted that the length of cable would, by a process of statical charging of its insulation, substantially reduce the rate at which signals could be sent unless small voltages were used, so small that only his galvanometer could detect the currents.

The first attempt to lay the cable, in 1857, ended when it snapped and was lost. The second attempt, a year later, was successful, but the large voltages required by the Whitehouse method reduced the ability of the cable to transmit signals rapidly, just as Thomson had predicted. Whitehouse privately recognized the inadequacy of his own instruments and surreptitiously substituted Thomson's galvanometer while claiming success for his own methods. This deception was soon discovered, and the ensuing controversy between Whitehouse, the board of directors, and Thomson combined theoretical science, professional vanity, and financial ignominy. A third cable was laid in 1865, and, with the use of Thomson's instruments, it proved capable of rapid, sustained transmission. Thomson's role as the man who saved a substantial investment made him a hero to the British financial community and to the Victorian public in general; indeed, he was knighted for it. It also was the foundation for a large personal fortune.

As the Atlantic cable affair demonstrates, Thomson was deeply involved in applying instrumentation designed for sensitive measurements to industrial concerns. His involvement in industry did not stem from a great desire to further the application of science to technology, although that motive was certainly there among others. Rather it was a consequence of his interest in instrumentation itself. The two central concerns of his life were the application of the ideas of mechanics to physics and the development of sensitive measuring devices. By the time of his death, Thomson, then Baron Kelvin of

Largs, while perhaps behind the times in his adherence to dynamical modes of thought, was generally looked upon as the founder of British physics. Together with Helmholtz in Germany, he had been the foremost figure in transforming—indeed, in creating—the science of physics as it was known in 1900.

NOTES

1. S. P. Thompson, *The Life of William Thomson*, I.
2. W. Thomson, "On the Uniform Motion of Heat in Homogeneous Solid Bodies, and its Connexion With the Mathematical Theory of Electricity," in *Cambridge Mathematical Journal*, **3** (1843), 71–84.
3. See S. P. Thompson, *op. cit.*, I.
4. W. Thomson, "Demonstration of a Fundamental Proposition in the Mechanical Theory of Electricity," in *Reprint of Papers on Electrostatics and Magnetism*, pp. 100–103.
5. W. Thomson, "On the Mathematical Theory of Electricity in Equilibrium," *op. cit.*, 15–37.
6. W. Thomson, "On a Mechanical Representation of Electric, Magnetic, and Galvanic Forces," in *Mathematical and Physical Papers*, I (Cambridge, 1911), 76–80.
7. W. Thomson, "On an Absolute Thermometric Scale, Founded on Carnot's Theory of the Motive Power of Heat, and Calculated From the Results of Regnault's Experiments on the Pressure and Latent Heat of Steam," *ibid.*, I, 100–106.
8. W. Thomson, "An Account of Carnot's Theory of the Motive Power of Heat, With Numerical Results Deduced From Regnault's Experiments on Steam," in *Transactions of the Royal Society of Edinburgh*, **16** (1849), 541–574.
9. James Thomson, "Theoretical Considerations on the Effect of Pressure in Lowering the Freezing Point of Water," in *Cambridge and Dublin Mathematical Journal*, **5** (1850), 248–255.
10. W. Thomson, "On a Remarkable Property of Steam Connected With the Theory of the Steam-Engine," in *Philosophical Magazine*, **27** (1850), 386–389.
11. W. Thomson, "On the Dynamical Theory of Heat; With Numerical Results Deduced From Mr. Joule's 'Equivalent of a Thermal Unit' and M. Regnault's 'Observations on Steam,'" in *Transactions of the Royal Society of Edinburgh*, **20** (1853), 261–288.
12. On the problems see P. Lawrence, "The Central Heat of the Earth," Ph.D. dissertation, Harvard University, 1973.
13. W. Thomson, "Dynamical Illustrations of the Magnetic and Helicoidal Rotatory Effects of Transparent Bodies on Polarized Light," in *Proceedings of the Royal Society*, **7** (1856), 150–158.
14. W. Thomson, Royal Institution Friday Evening Lecture, in *Reprint of Papers on Electrostatics and Magnetism*, pp. 208–226.
15. Cambridge University Library, MS Add. 7342, box I, notebook IV, pp. I–II. Quoted in Ole Kundsen, "From Lord Kelvin's Notebook: Ether Speculations," in *Centaurus*, **16** (1966), 41–53.
16. W. Thomson, "On Vortex Motion," in *Philosophical Transactions of the Royal Society of London*, **25**, pt. 1 (1868), 217–260.
17. J. C. Maxwell, *Scientific Papers of James Clerk Maxwell* (New York, 1952), 445–484; from "Atomism," in *Encyclopaedia Britannica*, 9th ed., II (1875).
18. J. Willard Gibbs, "A Comparison of the Electric Theory of Light and Sir William Thomson's Theory of a Quasi-Labile Ether," in *American Journal of Science*, **37** (1889), 467–475.

BIBLIOGRAPHY

All of Thomson's papers are collected in two sets, *Reprint of Papers on Electrostatics and Magnetism* (London, 1872); and *Mathematical and Physical Papers*, 6 vols. (Cambridge, 1911). Occasional papers are not reprinted in the former collection, but in their absence a reference to their location in the journals is given. The *Treatise on Natural Philosophy*, 2 vols. (Oxford, 1867), written with P. G. Tait, went through numerous editions.

The only full-scale biography is S. P. Thompson, *The Life of William Thomson, Baron Kelvin of Largs*, 2 vols. (London, 1901). Although uncritical, it contains a vast selection from Thomson's correspondence and includes a complete bibliography of Thomson's papers. Thomson's "green notebooks," kept throughout his life, are unpublished but are available at the University Library, Cambridge. Brief accounts of Thomson's life and work include David Murray, *Lord Kelvin as Professor in the Old College of Glasgow* (Glasgow, 1924); Alexander Russell, *Lord Kelvin, His Life and Work* (London–Edinburgh–New York, 1939); and A. P. Young, *Lord Kelvin, Physicist, Mathematician, Engineer* (London, 1948).

See also Herbert Norton Casson, "Kelvin, His Amazing Life and Worldwide Influence," in *Efficiency Magazine* (1930); John Ferguson, "Lord Kelvin: a Recollection and an Impression," in *Glasgow University Magazine*, **20**, no. 9 (1908); Kelvin Centenary Oration and Addresses Commemorative (London, 1924); Agnes Gardner King, *Kelvin the Man. A Biographical Sketch by His Niece* (London, 1925); Émile Picard, "Notice historique sur la vie et l'oeuvre de Lord Kelvin," in *Annuaire de l'Académie des sciences de Paris* (1920); Robert H. Silliman, "William Thomson: Smoke Rings and Nineteenth-Century Atomism," in *Isis*, **54** (1963), 461–474; E. C. Watson, "College Life at Cambridge in the Days of Stokes, Cayley, Adams, and Kelvin," in *Scripta mathematica*, **6** (1939), 101–106.

JED Z. BUCHWALD

THORPE, JOCELYN FIELD (*b.* London, England, 1 December 1872; *d.* London, 10 June 1940), *chemistry.*

After studying engineering and chemistry at King's College and the Royal College of Science, Thorpe entered the University of Heidelberg in 1892; three years later he received a doctorate in chemistry. He joined the research group of William Henry Perkin, Jr., at the University of Manchester, where he became lecturer in organic chemistry. In 1908 he was elected to the Royal Society and was its Sorby Research Fellow from 1909 to 1913. Subsequently he was appointed to the chair of organic chemistry at the Imperial College of

Science and Technology in London, a post he held until his retirement in 1938. He served on many government and industrial committees and was president of the Chemical Society from 1928 to 1931 and of the Institute of Chemistry from 1933 to 1936. He was knighted in 1939.

Thorpe's first researches in organic chemistry dealt with the synthesis of polybasic acids. He contributed an important method of preparation for dibasic acids, when he found that sodiocyanoacetic ester condenses with the cyanohydrins of aldehydes and ketones to form products that hydrolyze into substituted succinic acids. With Perkin, Thorpe investigated the structure of camphor and related compounds. They synthesized two oxidation products of camphor, camphoronic acid (1897) and camphoric acid (1903), and confirmed the formulas of Bredt for camphor and these acid derivatives. They also confirmed Baeyer's structures for carone and the caronic acids by synthesizing them in 1899.

Thorpe's most valuable contribution to chemistry was his study of the formation and reactions of imino compounds, resulting from his 1904 discovery of the condensation reaction of sodiocyanoacetic ester with the cyano group. He also found imino compounds to be the product of the condensation of nitriles with each other; dinitriles condense intramolecularly to form many new imino compounds.

In 1911 Thorpe established the existence of keto-enol tautomerism between an open chain compound and its cyclic isomer, while in 1919 he and C. K. Ingold proved the presence of "intraannular tautomerism" in bridged rings:

With Ingold, Thorpe collaborated from 1914 to 1928 on a study of valency and chemical bonding; they were the first to propose that the valency angles of the carbon atom may depart from the regular tetrahedral angles in the formation of highly substituted organic compounds.

BIBLIOGRAPHY

I. ORIGINAL WORKS. Thorpe wrote two books on dyes and colors: *The Synthetic Dyestuffs and the Inter-*

mediate Products From Which They Are Derived (London, 1905), written with John C. Cain; and *Synthetic Colouring Matters: Vat Colours* (London, 1923), written with Christopher K. Ingold. With Martha A. Whiteley, Thorpe wrote *A Student's Manual of Organic Chemical Analysis* (London, 1925) and also edited the supplementary vols. (1934–1936) and the 4th ed. (1937) of T. E. Thorpe's *Dictionary of Applied Chemistry.*

Thorpe's important papers include "Synthesis of *i*-Camphoronic Acid," in *Journal of the Chemical Society*, **71** (1897), 1169–1194, written with W. H. Perkin, Jr.; "The Formation and Reactions of Imino Compounds," in three parts, *ibid.*, **85** (1904), 1726–1761, written with H. Baron and F. Remfry; **89** (1906), 1906–1935, written with E. Atkinson; **93** (1908), 165–187, written with C. Moore; "The Formation and Stability of Spiro Compounds," *ibid.*, **115** (1919), 320–383, written with C. K. Ingold; "Ring-Chain Tautomerism," *ibid.*, **121** (1922), 1765–1789, written with C. K. Ingold and E. Perren; "The Chemistry of Polycyclic Structures in Relation to Their Homocyclic Unsaturated Isomerides," *ibid.* (1922), 128–159, written with C. K. Ingold and E. Farmer; and "The Hypothesis of Valency Deflection," *ibid.* (1928), 1318–1321, written with C. K. Ingold.

II. SECONDARY LITERATURE. On Thorpe's life and work, with many references to his papers, see G. A. R. Kon and R. P. Linstead, "Sir Jocelyn Field Thorpe," in *Journal of the Chemical Society* (1941), 444–464, repr. in Alexander Findlay and William Hobson Mills, eds., *British Chemists* (London, 1947), 369–401.

ALBERT B. COSTA

THORPE, THOMAS EDWARD (*b.* Barnes Green, Harpurhey, near Manchester, England, 8 December 1845; *d.* Salcombe, England, 23 February 1925), *chemistry, history of science.*

Thorpe was the son of George Thorpe, a cotton merchant, and Mary Wilde. He received his early education in Manchester and in 1863 entered Owens College, Manchester, as a chemistry student under Roscoe's guidance. Much of his four years at Owens College was spent as Roscoe's private assistant, and he participated in the classical work on vanadium (which resulted in determining its true atomic weight) and on the percentage of carbon dioxide in the atmosphere.

After graduation Thorpe worked with Bunsen at the University of Heidelberg, where he received his doctorate. He then went to Bonn, where he studied ethylbenzoic acid with Kekulé. In 1870 he became professor of chemistry at the Andersonian College, Glasgow. Four years later he was appointed professor of chemistry at the new Yorkshire College of Science at Leeds. He resigned this post in 1885 in order to become Edward Frank-

land's successor at the Royal College of Science in London, which later became the Imperial College of Science and Technology. Between 1894 and 1909 he was director of the government laboratories, responsible for design and equipment. He returned to Imperial College in 1909 but resigned three years later.

Best known for his numerous textbooks and histories of chemistry, Thorpe was also an important figure in inorganic chemical research. His doctoral dissertation on the oxychlorides of chromium and sulfur led to a study of similar phosphorus compounds, resulting in the discovery of thiophosphoryl chloride ($PSCl_3$), phosphoryl fluoride (POF_3), and phosphorus pentafluoride (PF_5). This last discovery was of particular importance, since it necessitated postulating a valence of five for phosphorus. His atomic weight determinations of silicon and gold were the most accurate at the time for those elements. He also determined the weights of titanium, strontium, and radium. He carried out extensive studies of the critical temperatures, viscosities, and molecular volumes of liquids. His investigation of the vapor density of hydrofluoric acid revealed that at lower temperatures it is polymerized.

BIBLIOGRAPHY

I. ORIGINAL WORKS. Thorpe is most remembered for the large number of popular textbooks he wrote. Most of the following went through several editions: *A Dictionary of Applied Chemistry*, 3 vols. (London, 1893); the 3rd ed. (1921–1927), which he was preparing until a few days before his death, contained 7 vols. His other chemistry books were *A Manual of Inorganic Chemistry*, 2 vols. (London, 1898); *Qualitative Chemical Analysis and Laboratory Practice*, 8th ed. (London, 1894), written with M. M. Pattison Muir; *Coal, Its History and Uses* (London, 1879); and *A Series of Chemical Problems With Key* (London, 1907), written with W. Tate.

Thorpe's interest in the history of chemistry led to two books: *Essays in Historical Chemistry* (London, 1894, 1923) and *History of Chemistry* (London, 1909–1910, 1924). In addition he published the following short biographical sketches: *Humphry Davy, Poet and Philosopher* (London, 1901); *Joseph Priestley* (London, 1906); and *The Right Honorable Sir Henry Enfield Roscoe* (London, 1916). After his retirement he enjoyed yachting and wrote two books about his experiences: *The Seine From Havre to Paris* (London, 1913) and *A Yachtsman's Guide to the Dutch Waterways* (London, 1915). His published papers are listed in Poggendorff, III, 1345; IV, 1499; V, 1255.

II. SECONDARY LITERATURE. A detailed obituary notice is P. P. Bedson, "Sir Edward Thorpe," in *Journal of the Chemical Society*, (1926), 1031–1050. See also the *Dictionary of National Biography 1922–1930*, pp. 842–843.

SHELDON J. KOPPERL

THOUIN, ANDRÉ (*b.* Paris, France, 10 February 1747; *d.* Paris, 27 October 1824), *botany.*

Thouin spent his entire life at the Jardin des Plantes in Paris. At age seventeen he succeeded his father as head gardener and became responsible for his family's financial support. He remained unmarried in order to care for his mother, sisters, and brothers, who lived with him in a small apartment in the annex of the old greenhouses. A student of Bernard de Jussieu and Buffon, Thouin assisted the latter in reorganizing the Jardin des Plantes and enriched its greenhouses and the collections of the École de Botanique through many exchanges of plants with foreign botanists. Thouin was a good conversationalist, and his manner was dignified and gentle. His friends included Malesherbes, who filled his pockets with plants and tree branches, and Rousseau, whose misanthropy gave way to openness in Thouin's company. Thouin also maintained contacts with a number of other naturalists, including Bernardin de Saint-Pierre, Bosc, Desfontaines, and Faujas de Saint-Fond.

An unassuming person who detested pomp, Thouin refused to wear the insignia of the Legion of Honor, which he thought should be reserved for soldiers, and donned the costume of the Institut de France only when obliged to do so. Always generous, he made an interest-free loan of 30,000 francs to a friend who was in difficulties. The money came from a sum left to Thouin by someone who was interested in agriculture and had been impressed by his outstanding integrity.

In collaboration with A. P. de Candolle, in 1806 Thouin observed the influence of light on a number of plants. They used six Argand lamps, believing that they thereby attained five-sixths the intensity of sunlight. Thouin made grafts and noted the influence of the stock on certain characteristics of the graft. He was also an early proponent of the teaching of agriculture and horticulture.

In 1802, with Desfontaines, Thouin made an inventory of the convents in the Paris region, as well as of the property of émigrés and of those who had been guillotined. The Convention authorized him to dispose of the fruit trees of the Carthusian convent in Paris; this was the origin of the nursery

of the Muséum d'Histoire Naturelle. Appointed an army commissioner, Thouin confiscated rich collections in the Low Countries in 1794–1795 and others in Italy two years later. Surviving documents unfailingly show that he carried out his assignments with great honesty. He already had a considerable reputation at age thirty-nine, when he became a member of the Academy of Sciences.

In the last year of his life Thouin fell gravely ill. Realizing that he would soon die, he began to write up his observations and correct his manuscripts. His brother Jean (1756–1827) succeeded him as head gardener; and another brother, Gabriel (1747–1829), also devoted himself to horticulture. Thouin's herbarium eventually was left to the Faculty of Sciences of Montpellier.

BIBLIOGRAPHY

Works published during Thouin's lifetime are *Essai sur l'exposition et la division méthodique de l'économie rurale* (Paris, 1805); *Description de l'École d'agriculture pratique de Muséum d'histoire naturelle* (n.p., n.d. [Paris, 1814]); and *Monographie des greffes* ([Paris], 1821 [?]), a technical description of the methods for propagating plants. See also Royal Society *Catalogue of Scientific Papers*, V, 983–984, which lists 32 works.

His posthumously published writings include *Cours de culture et de naturalisation des végétaux* (Paris, 1827), published by his nephew, Oscar Leclerc; *Voyage dans la Belgique, la Hollande et l'Italie* (Paris, 1841), prepared from Thouin's journal by Baron Trouvé; and the supplement to François Rozier's course on agriculture.

Many of Thouin's articles appeared in *Mémoires de la Société d'agriculture; Mémoires de l'Académie des sciences; Dictionnaire d'histoire naturelle*, edited by Déterville; *Annales d'agriculture française;* and the *Annales* and *Mémoires du Muséum national d'histoire naturelle.*

P. JOVET
M. MALLET

THUE, AXEL (*b.* Tönsberg, Norway, 19 February 1863; *d.* Oslo, Norway, 7 March 1922), *mathematics.*

Thue enrolled at Oslo University in 1883 and became a candidate for the doctorate in 1889. From 1891 to 1894 he held a university scholarship in mathematics, and he was a professor of applied mathematics at Oslo from 1903 to 1922.

During 1890–1891 Thue studied at Leipzig under Sophus Lie, but his works do not reveal Lie's influence, probably because of Thue's inability to follow anyone else's line of thought. In 1909 he published his famous article "Über Annäherungswerte algebraischer Zahlen" in Crelle's *Journal*. In 1920 C. L. Siegel found a more precise expression for the approximation of algebraic numbers and K. F. Roth discovered the best possible equation in 1958. Nevertheless, Thue was able to draw a far-reaching conclusion in number theory. He showed that an equation like $y^3 - 2x^3 = 1$ cannot possibly be satisfied by an indefinite number of pairs of numbers x, y, when x and y must be a whole number. Generally formulated, the left side of the equation can be an irreducible homogeneous polynomial in x and y of a degree higher than 2, and the right side can be any whole number. Thue's theorem was characterized by Edmund Landau (1922) as "the most important discovery in elementary number theory that I know."

During 1906–1912 Thue published many articles on series, in one of which he said: "For the development of the logical sciences it will be important to find wide fields for the speculative treatment of difficult problems, without regard to eventual applications." His "Über die gegenseitige Lage gleicher Teile gewisser Zeichenreihen" was characterized as a basic work by G. A. Hedlund (1967).

Thue's most important work in applied mathematics was "De virtuelle hastigheters princip" ("The Principle of Virtual Velocity"), an original statement that has no parallel in the literature. One of the paradoxes Thue liked to state was "The further removed from usefulness or practical application, the more important."

BIBLIOGRAPHY

I. Original Works. A list of 47 articles (1884–1920) is in *Norsk matematisk tidsskrift*, **4** (1922), 46–49. They include "Über Annäherungswerte algebraischer Zahlen," in *Journal für die reine und angewandte Mathematik*, **135** (1909), 284–305; "Eine Eigenschaft der Zahlen der Fermatschen Gleichung," in *Videnskabsselskabets skrifter* (Oslo) (1911), 1–21; and "De virtuelle hastigheters princip," in *Aars Voss'skoles festskrift* (Oslo, 1913), 194–213. *Selected Mathematical Papers of Axel Thue* (Oslo,   ), with an introduction by Carl Ludwig Siegel, is in press.

II. Secondary Literature. There are biographies by C. Størmer, V. Bjerknes, and others in *Norsk matematisk tidsskrift*, **4** (1922), 33–46, with portrait. Viggo Brun and Trygve Nagell published a list of his posthumous works in *Videnskabsselskabets skrifter* (Oslo) (1923), 1–15. Also see G. A. Hedlund, "Remarks on the Work of Axel Thue on Sequences," in *Nordisk matematisk tidsskrift* (1967), 148–150.

Viggo Brun

THUNBERG, CARL PETER (*b*. Jönköping, Sweden, 11 November 1743; *d*. Tunaberg, near Uppsala, Sweden, 8 August 1828), *botany*.

After studying at Jönköping, Thunberg entered Uppsala University in 1761, where he soon came under the influence of Linnaeus. His dissertation for the medical degree, *De ischiade* (1770), was not botanical and Linnaeus did not preside over the disputation; but Thunberg's passion was, nevertheless, natural history and especially botany, and Linnaeus soon considered him a protégé. In order to complete his medical education, Thunberg went to Paris immediately after the disputation, the trip being made possible by a scholarship. Before reaching France he stayed for a while in Holland, where a letter of recommendation from Linnaeus opened the house of Jan and Nikolaus Laurens Burman to him; both were good botanists, and the father was an old friend of Linnaeus.

In Paris, Thunberg received an extraordinary offer: an invitation to follow a Dutch merchant ship to Japan, which was closed to all European nations except Holland. The Burmans had good connections with the rich bourgeoisie of Holland, among whom the enthusiasm for gardening was very great. Thunberg was expected to collect as many Japanese garden plants as possible for his employers. He was, of course, free to make his own purely botanical collections as well, a situation that pleased Linnaeus, with whom Thunberg maintained close correspondence during the voyage. In order to enter Japan, he had to behave in every respect like a good Dutchman. He learned Dutch by stopping off in South Africa, in Cape Colony, where he remained from April 1772 to March 1775, thus fortunately combining language studies with botanical excursions.

There had been very little true botanical investigation of the Cape Colony; during his stay Thunberg made three voyages into the interior, collecting and describing more than three thousand plants, of which about one thousand were new to science. On two trips he was accompanied by the gardener and plant collector Francis Masson, who had been sent to the Cape Colony by Kew Gardens in London. At this time Thunberg began describing species and revising genera from his collections, publishing his papers in the transactions of several Swedish and foreign academies.

In March 1775 Thunberg sailed on a Dutch ship to Batavia. From Java he continued on another Dutch vessel to the island of Deshima in Nagasaki harbor, the only Japanese port open to European trade. For a while Thunberg was able to make short excursions near the town, but bureaucratic difficulties soon curtailed even these few opportunities to collect Japanese plants. A journey to Tokyo with the Dutch ambassador did little to improve the situation. Thunberg nevertheless became the first Western scientist to investigate Japan botanically through the aid of the young Japanese interpreters employed by the traders. Some of them were physicians, eager to learn modern European medicine from Thunberg, who exchanged his knowledge for specimens of Japanese plants. He left the country in December 1776 with a rich collection for further analysis. On his way home he visited Java, Ceylon, and the Cape Colony. Having satisfied his Dutch employers, he went to London, its herbariums and collections being the most important at that time.

Thunberg reached Sweden in 1779 and was appointed botanical demonstrator at Uppsala University. Linnaeus had been succeeded by his son, and it soon became clear that Thunberg and young Linnaeus could not work well together. Although Thunberg was a demonstrator, Linnaeus did not allow him to enter the botanical garden. In 1784 Thunberg succeeded to the professorship of botany (in the Faculty of Medicine), a post he held until his death. He lived quietly on a little estate, Tunaberg, just outside Uppsala, traveling to his daily work in a strange, uncomfortable carriage, well-known among the students as "the rattlesnake." The major event during this long academic period was the transfer of the botanical garden from a low and often flooded region of the town to the much more suitable park of the royal castle, where it is still situated. Most of Thunberg's time as professor was occupied by writing about his extensive collections.

Thunberg's first major work after returning to Sweden was *Flora japonica* (1784), a fundamental account of the floristics and systematics of the vegetation of Japan that describes twenty-one new genera and several hundred new species. His only predecessor of any importance was Engelbert Kaempfer, who traveled in the Far East at the end of the seventeenth century and whose collections Thunberg studied in the British Museum during his visit to London. The vast and important material from the Cape Colony occupied Thunberg for the rest of his life. A preliminary work was *Prodromus plantarum capensium* (1794–1800), a summary of his findings. Much more detailed and important was *Flora capensis* (1807–1823), completed with the help of the German botanist J. A. Schultes. Among his many shorter works are monographs on *Protea*, *Oxalis*, *Ixia*, and *Gladiolus*.

Thunberg was exclusively a descriptive botanist who closely followed Linnaeus, using his methods and his already somewhat outmoded sexual system. He modified the latter slightly, reducing its twenty-four classes to twenty by excluding Gynandria, Monoecia, Dioecia, and Polygamia and distributing their members among the other classes. The aim of his reductions, however, seems to have been more practical than theoretical, the plants of the omitted classes often not being constant in their class characters and fitting more smoothly in other parts of the system. Although Linnaeus had been searching for a truly natural system, Thunberg seemed to have no penchant for speculation. His strong points were his keen eye in the field, his indefatigable spirit and eagerness in collecting, and his concise descriptions. Among his contemporaries he was perhaps the one who described the largest number of new plant genera and species. His aims certainly reached no further.

Thunberg's description of his great voyage, published in four parts in Swedish in 1788–1793 and soon translated into English as *Travels in Europe, Africa and Asia* (1793–1795), as well as in French and German, contains material of great ethnographical interest.

BIBLIOGRAPHY

I. ORIGINAL WORKS. T. O. B. N. Krok, *Bibliotheca botanica suecana* (Stockholm–Uppsala, 1925), 705–716, lists all Thunberg's botanical works. Letters from his many correspondents, both Swedish and foreign, are at the University Library, Uppsala, together with other MSS. His published works include *De ischiade* (Uppsala, 1770); *Flora japonica* (Leipzig, 1784); *Travels in Europe, Africa and Asia*, 4 vols. (London, 1793–1795), first published as 4 vols. (Uppsala, 1788–1793); *Prodromus plantarum capensium* (Uppsala, 1794–1800); and *Flora capensis*, 2 vols. (Uppsala, 1807–1823).

II. SECONDARY LITERATURE. See H. O. Juel, *Plantae Thunbergianae* (Uppsala, 1918), in German; and articles about Thunberg by N. Svedelius in *Isis*, 35 (1944–1945), 128–134; in S. Lindroth, ed., *Swedish Men of Science* (Stockholm, 1952), 151–159; and in *Svenska Linnésällskapets årsskrift*, 27 (1944), 29–64.

Thunberg and Cape Colony botany are treated by M.

Karsten, in *The Old Company's Garden at the Cape and Its Superintendents* (Cape Town, 1951), 132–134, *passim*; "Carl Peter Thunberg. An Early Investigator of Cape Botany," in *Journal of South African Botany*, **5** (1939), 1–27, 87–155; **12** (1946), 127–190. Thunberg and Japan are treated in C. Gaudon, *Le Japon du XVIIIe siècle vu par Ch. P. Thunberg* (Paris, 1966); and in *Forskningsmaterial rörande C. P. Thunberg* (Tokyo, 1953), in Japanese and Swedish.

See also G. Eriksson, *Botanikens historia i Sverige intill år 1800* (Uppsala, 1969), 258–260, 268–270, 321; and S. Lindroth, *Kungliga Svenska Vetenskapsakademiens historia*, II (Stockholm, 1967), 408–412 and *passim*.

GUNNAR ERIKSSON

THUNBERG, THORSTEN LUDVIG (*b.* Torsåker, Sweden, 30 June 1873; *d.* Lund, Sweden, 4 December 1952), *physiology*.

The son of Per Erik Thunberg, a merchant, and of Wendela Maria Elisabeth Hård, Thunberg studied medicine at the University of Uppsala from 1891 to 1900 and obtained the M.D. degree with a dissertation on epidermal sensory perception. He was demonstrator for Olof Hammarsten at the Institute of Physiological Chemistry in 1893–1894 and for Frithiof Holmgren at the Institute of Physiology at Uppsala in 1894–1896. He was reader in physiology at Uppsala in 1897–1904. Immediately after the death of Magnus Blix in 1904, Thunberg went to Lund as temporary occupant of the chair of physiology and embryology; the following year he was appointed to the chair of physiology, from which he retired in 1938.

Thunberg's first work, on sensory physiology, was published in 1893 and his last, in 1953. Thunberg extensively studied the physiology of epidermal sensations, showing, among other things, that a pinprick gives rise to two sensations of pain, the second occurring some seconds after the first. This phenomenon was interpreted much later by Zotterman, who demonstrated the existence of two groups of pain fibers, one of which transmits impulses more slowly than the other. In 1905 Thunberg wrote the chapter "Physiologie der Druck-, Temperatur- und Schmerzempfindungen" for Nagel's *Handbuch der Physiologie des Menschen*. With this work he left a field into which he had been led more by chance than by ability.

In 1903, when Thunberg began to study the elementary processes of metabolism—studies that constitute his major contributions to science—nothing was known of the oxidation processes in the tissue cells. Energy is derived in the cells by combustion (cellular respiration), whereby oxygen is consumed and carbon dioxide produced. Lavoisier had proved that respiration is chemically a combustion process in the 1770's. From then until 1875, the consumption of oxygen and the production of carbon dioxide were thought to be confined to the lungs and the blood. In that year Pflüger proclaimed: "Cells are constantly burning, although we do not see their light with our weak vision." Everyday experience shows that organic material does not burn in oxygen at body temperature, so it was assumed that oxygen was activated in some way. From about 1840 a series of oxygen activation theories appeared; but valid knowledge had to await the work of Thunberg, Otto Warburg, and Heinrich Wieland. Warburg introduced the term *Atmungsferment* and argued that this enzyme that catalyzed oxidation was an iron-pyrrole complex. His idea was based on the orthodox concept that the degradation and combustion of food to carbon dioxide and water took place through the direct attack of oxygen on the carbon atoms of the food. Thunberg and Wieland developed an entirely new conception, rivaling that of Warburg.

In 1908 Thunberg began to study the capacity of cells to burn various organic acids, including acetic, propionic, oxalic, malic, succinic, and citric. These acids were not then recognized as normal constituents of the body that played a role in intracellular metabolism; rather, they were known as products of putrefaction or fermentation. From among all the organic acids Thunberg chose precisely those that proved useful to his aims, thus revealing his unerring sense for the right path to follow.

Wieland turned to this area of investigation just when his cooperation was needed to elucidate the nature of the processes of biological oxidation. In 1912 he found that organic compounds can be oxidized through removal of hydrogen in the presence of a catalyst and that the hydrogen can be taken up by suitable acceptors, such as methylene blue. From 1910 Thunberg integrated Wieland's discoveries with his own, especially those on the biological oxidation of succinate, and initiated the concept of the specific, hydrogen-activating, chain-forming enzyme systems. When Thunberg began this work, all that was known of intracellular enzymes was merely that postmortem autolysis was catalyzed by proteolytic enzymes. Oxidases also had been found.

Thunberg saw that the oxidation of succinate was initiated by an agent in the cells that endowed a hydrogen group in the succinate with a reactivity that it had not previously possessed. The reactivity could easily be demonstrated with methylene blue, which was decolorized by hydrogen uptake. Under the influence of this agent in the cells, the succinate emerged as "hydrogen donator" and released a hydrogen group to the "hydrogen acceptor" methylene blue. The terms "donator" and "acceptor" were introduced by Thunberg and are now in general use. He called the enzymatic agent a dehydrogenase, in this case succinate dehydrogenase. The introduction of the methylene blue method in 1916 opened up a worldwide search for dehydrogenases. The chainlike degradation of the various organic molecules in the organism could now be charted.

Thunberg finally formulated the following generalization concerning the oxidative degradation of food: the degradation is accomplished by a chain of consecutive splittings of hydrogen atoms carried out by a series of dehydrogenases, each with a specific purpose.

With the discovery of a hydrogen-carrying flavoprotein, *das gelbe Atmungsferment*, Warburg in 1932 contributed substantially to Thunberg's conception of hydrogen transport from one system to another as a central mechanism in oxidative metabolism. Thunberg rightfully considered himself responsible for a revolution in the concept of the mechanism of biological oxidation.

In 1905, the year of his appointment to the chair of physiology at Lund, Thunberg recorded that nerve tissue respires, taking up oxygen and giving up carbon dioxide. It had previously been thought that nerve fibers conducted impulses like an electric cable, without measurable energy consumption. Thunberg made this discovery with his microrespirometer, an ingenious device with which he could measure oxygen consumption and carbon dioxide production in small units of tissue. Using this apparatus, he also could demonstrate that traces of certain metals, such as manganese salts, strongly catalyze tissue respiration. This field held his interest throughout his retirement.

Although a scientist first, Thunberg was not unfamiliar with the nonscientific world. As a student at Uppsala he was associated with the radical-liberal group Verdandis, headed by the physiologist Hjalmar Öhrvall. During five decades—through books, popular journals, pamphlets, and the daily press, and as an adviser to the government—he disseminated information on hygiene and medicine that in scope and quality was unparalleled in Sweden.

Thunberg's enormous capacity for work was combined with good health, and he reached the age of seventy before becoming seriously ill, with pneumonia, from which he recovered. He was fully active for more than nine years, until October 1952, when he stumbled over the doorstep of his house and broke his femur. Despite the best care, he died early in December. He had been elected a member of the Royal Swedish Academy of Sciences in 1928, and later of many foreign learned societies.

BIBLIOGRAPHY

Obituaries include Georg Kahlson, "Thorsten Thunberg," in *Acta physiologica scandinavica*, **30** (1953), supp. 111; and F. G. Young, "Prof. T. Thunberg," in *Nature* (12 Dec. 1953).

GEORG KAHLSON

THURET, GUSTAVE ADOLPHE (*b*. Paris, France, 23 May 1817; *d*. Nice, France, 19 May 1875), *botany*.

The third son of Isaac Thuret, consul general of the Netherlands in France, and Jacoba Henrietta van der Paedevoort, Thuret belonged to a French Protestant family that had emigrated to Holland following Louis XIV's revocation of the Edict of Nantes. He received a religious education and remained a strong Protestant. After being tutored in the classics, he studied law at Paris, and obtained the *licence* in 1838. During his youth he made many trips abroad, especially to England; and he spoke English fluently, having been taught it by his mother before he learned French.

Named attaché to the French embassy in Constantinople in 1840, Thuret soon gave up a diplomatic career to devote himself to botanical research. A very wealthy man, he remained an amateur, never seeking a university post. Nevertheless, the importance of his discoveries brought him a corresponding membership in the Académie des Sciences in 1857.

In order to further his research, Thuret hired an assistant—a medical student, Édouard Bornet (1828–1911) who soon became his collaborator and friend. While studying marine algae at Cherbourg, Thuret also drew on the services of the artist A. Riocreux, who skillfully reproduced Thuret's microscopical observations. The latter commenced in 1840 with the discovery of the flagella of the spermatozoids among the Characeae. This encouraged Thuret to look for similar locomotive organs among the other cryptogams, particularly the algae.

Working with Joseph Decaisne, Thuret attempted to study live *Fucus*, obtained from fish markets in Paris. They soon realized, however, that such observations could be made only at the seashore, where much more active live algae were available, along with seawater in which to observe them. Upon their return from a trip to the coast of Normandy, Decaisne and Thuret announced the discovery of the spermatozoids of *Fucus* to the Académie des Sciences on 11 November 1844. They described a red granule (stigma) and two unequal flagella, one pointing forward and the other backward. The subtlety and precision of their observations is remarkable, especially considering the imperfect state of the microscopes at their disposal.

The Académie des Sciences proposed the following subject for the grand prize in the physical sciences for 1847: "L'étude des mouvements des corps reproducteurs ou spores des Algues zoosporées et des corps renfermés dans les anthéridies des Cryptogames." The prize was shared by the two papers submitted, one by Thuret and the other by two naturalists from Marseilles, A. Derbes and A. J. J. Solier. Thuret's paper, however, was incontestably superior. Basing his analysis on the color of the zoospores and of the spermatozoids, as well as on the mode of insertion and orientation of the flagella, Thuret established, among the Zoosporeae classified by Decaisne, a very sharp distinction between the green algae (Chlorosporeae) and the brown algae (Phaeosporeae).

In 1852 Thuret left Paris and moved to Cherbourg, where he completed his initial investigation of *Fucus*. Through skillful fertilization experiments, in the course of which hybrid zygotes were obtained, he demonstrated the role of the spermatozoids in fertilization. Thus in 1854 he clarified the crucial phenomenon of fertilization, which had been shrouded in mystery. He was, however, unable to bring about the fusion of the male and female gametes, observed shortly afterward in another alga (*Oedogonium*) by Nathanael Pringsheim.

Thuret had verified the existence of spermatozoids and of mobile zoospores among green and brown algae, but the reproduction of the blue and red algae remained unknown, since no flagellated cell had been observed in either group. Thuret soon established, however, the absence of sexual reproduction among the blue algae, which multiply only by spores or by fragmentation of the trichomes into hormogonia. The problem of reproduction among the red algae appeared more intractable until 1866, when Thuret and Bornet ascertained that the male gametes (pollinia or spermatia), which never move, cling to and fuse with the hair (trichogyne) surmounting the female cell. They thereby discovered the wholly unexpected fertilization process that precedes the development of the carpospores. The formation of the carpospores, moreover, is accompanied by fusion phenomena that are often complex, involving the mother plant (auxiliary cells). Through the discoveries of Thuret and Bornet, the extremely varied reproductive modes of the different groups of algae, virtually unknown twenty-five years earlier, were definitively elucidated, at least in broad outline.

Thuret confirmed his discoveries by observations of a large number of species. In order to publicize them, he began two books, both of which were completed after his death by Bornet: *Notes algologiques* (1876–1880) and *Études phycologiques* (1878). They were illustrated with remarkable copperplate engravings of Riocreux's drawings that combined scientific exactitude with artistic beauty.

Thuret found it difficult to live in the humid climate of Cherbourg, and in 1856 he moved to the Mediterranean coast at Cap d'Antibes, which was then almost uninhabited. There he bought two fields and built a large villa where he continued his research on Mediterranean algae. His reputation attracted to Antibes a number of foreign botanists who wished to study algae: W. G. Farlow, E. de Janczewski, M. Woronin, and J. Rostafinski, among others. Thuret transformed the land surrounding the villa into a splendid botanical garden into which he introduced many exotic ornamental plants, which are now common in all the gardens along the Côte d'Azur. The "Villa Thuret," which its founder bequeathed to the nation, is now an agronomic research station. Thuret left his algologic collections and library to Bornet, who in turn gave them to the Muséum d'Histoire Naturelle in Paris.

The outstanding characteristic of Thuret's research was his constant concern to observe fully alive algae in their natural environment. This grasp of the necessity for working under conditions that permit the subject of study to remain alive was still unusual in Thuret's time, as was his broad biological conception of the methods suitable for the study of development and of the role of the reproductive organs. Together, these advanced views enabled Thuret to make a wealth of discoveries that revealed the extreme diversity in the reproductive modes of the algae and opened the way for further progress in the subject.

BIBLIOGRAPHY

An obituary is E. Bornet, "Notice biographique sur M. Gustave-Adolphe Thuret," in *Annales des sciences naturelles*, 6th ser., Botanique, **2** (1875), 308–361, and in *Mémoires de la Société impériale des sciences naturelles de Cherbourg*, **20** (1876).

JEAN FELDMANN

THURNAM, JOHN (*b*. Lingcroft, near York, England, 28 December 1810; *d*. Devizes, England, 24 September 1873), *psychiatric medicine, anthropology.*

Thurnam, as the son of a Quaker family, had a characteristically thorough early education. Subsequently, he studied medicine and received his first qualification in 1834. His relevant medical appointments were as medical superintendent at the celebrated asylum the Retreat, at York, between 1839 and 1849, and thereafter at the Wiltshire County Asylum, Devizes.

Thurnam's activities in psychiatry and anthropology mirrored widespread current interest in these subjects, and his studies were a powerful stimulus at the time, even if they are rarely remembered today. In medicine his most important work was *Observations and Essays on the Statistics of Insanity* (London, 1845), which played an immensely important role in the application of statistics to psychiatry. At a period when nonrestraint care was being generally introduced into asylums, when many clinical investigations were being made into psychiatric illnesses, and when numerous new drugs were being tried (at least in the 1860's), a statistical approach to results was essential for making assessments. Thurnam also recognized that such statistical information had to be standardized, and he devised a questionnaire for use at the Retreat.Thurnam contributed to the growing professional organization within psychiatric medicine, using such organization to promote his ideas and his standard questionnaire. He was an original member (1841) of the Medico-Psychological Association and president in 1844 and 1855. The success of Thurnam's work can be seen in innumerable publications, as, for example, W. C. Hood, *Statistics of Insanity* (London, 1855), while in 1882 Daniel Hack Tuke spoke of his work as "a Pharos to guide those who sail on waters where many are

shipwrecked" (*Chapters in the History of the Insane* [London, 1882], 492).

Of his work in anthropology, Thurnam was concerned mostly with the study of skulls, his notable contribution being his co-editorship of *Crania Britannica* (1865).

BIBLIOGRAPHY

I. ORIGINAL WORKS. The full title of Thurnam's celebrated work on statistics is *Observations and Essays on the Statistics of Insanity, and on Establishments for the Insane; to Which Are Added the Statistics of the Retreat Near York* (London, 1845). It superseded an earlier work, *The Statistics of the Retreat; Consisting of a Report and Tables Exhibiting the Experience of That Institution for the Insane From its Establishment in 1796 to 1840* (York, 1841).

Much of Thurnam's significant anthropological work appeared in the 1860's, notably his work with *Crania Britannica. Delineations and Descriptions of the Skulls of the Early Inhabitants of the British Islands* (London, 1865).

II. SECONDARY LITERATURE. There is no modern study of Thurnam, although he is remembered in R. Hunter and I. Macalpine, *Three Hundred Years of Psychiatry, 1535–1860* (London, 1963), 941–945. For useful notices see *Medical Times and Gazette*, **2** (1873), 479; and *Dictionary of National Biography*, **19** (1909), 831–832.

J. K. CRELLIN

THURNEYSSER, LEONHARD (or **Thurnyser, Lienhart**) (*b*. Basel, Switzerland, 5 August 1531; *d*. Cologne, Germany, 8 July 1596), *alchemy.*

Thurneysser was the son of Ursula and Jacob Thurneysser, a goldsmith. He took up his father's profession and also studied with a Dr. Huber, a physician and alchemist resident at Basel; he did not attend a university. He married Margarette Müllerin when he was sixteen but, when he was discovered to be selling gold-covered lead as pure gold, was forced to flee Basel in 1548, leaving his wife behind. He spent some time in Holland, northern Germany, France, and England, then in about 1552 returned to Germany to join the army of Albert, margrave of Brandenburg. He was captured by the Saxon army in the following year and put to work in the mines at Tarenz, in the Inn Valley. Following his release he worked as a goldsmith and smelter in Nuremberg, then returned to the Tyrol, where he was in the service of Archduke Ferdinand from 1560 to 1570. On Ferdinand's instructions he made journeys to England, France, Bohemia, Hungary, Italy, Spain, and North Africa to acquaint himself with metallurgical methods and medicine. His first book, the verse alchemical tract *Archidoxa*, was published in 1569; a similar work, *Quinta essentia*, was issued a year later.

Thurneysser moved in 1571 to Frankfurt an der Oder, where he wrote *Pison*, a kind of textbook of mineral-water analysis. This work came to the notice of Johann Georg, elector of Brandenburg, who summoned Thurneysser to his court. There Thurneysser cured Johann Georg's wife of a serious illness and was made court physician, despite his lack of an academic degree. Enjoying his patron's full confidence, Thurneysser acted as his adviser on metallurgy and mining and took advantage of his position in a masterly way. He established, at the Greyfriars monastery in Berlin, a laboratory— or, indeed, a factory—that employed 300 people in the production of saltpeter, mineral acids, alums, colored glass, drugs, essences, and even amulets. He also founded his own printing house, which published his calendars, prognostications, alchemical and medical tracts, and a wide variety of polemics.

The works that Thurneysser published at this time were impressive examples of the printer's art, illustrated with woodcuts and etchings, and incorporating Greek, Arabic, Syrian, Hebrew, and Chaldean typefaces. Since his books often contained words in languages that he did not know (some of the magic spells and terms that he gave have been identified as common Hungarian swearwords), he was publicly accused of harboring in his inkpot a devil who dictated to him. His chief alchemical works, *Megaln chymia* and *Melisath*, were both first published in Berlin in 1583. The latter is a kind of dictionary directed to clarifying the works and ideas of Paracelsus, whose follower Thurneysser purported himself to be. But although he frequently quoted from Paracelsus, Thurneysser often invented the passages cited himself; and the *Melisath* contains citations of some eighty tracts by Paracelsus that never existed outside Thurneysser's own mind. His works were severely criticized by other physicians, particularly Kaspar Hoffmann, professor of medicine at the University of Frankfurt an der Oder, and Thurneysser would certainly seem to have been a charlatan, whose methods and drugs had at best a dubious power to heal.

Nevertheless, Thurneysser became rich; he owned a large library, collected pictures and other works of art, and established a sort of museum of natural history. He had agents in a number of German and Polish cities, who advertised his wares and sold his drugs, cosmetics, and amulets to the gentry and wealthy burghers of Germany, Poland, and Denmark. He himself courted the favor of royalty, including the deranged Duke Albert Frederick of Prussia, Frederick II of Denmark, and Stephen Báthory, king of Poland. (He dedicated his *Historia sive descriptio plantarum* to the last, but when the king paid him less money than he felt was his due, dedicated the next edition of the work to his patron Johann Georg.)

In addition, Thurneysser conducted a sort of school of alchemy, which numbered the distinguished apothecary Michael Aschenbrenner among its students. He was not without enemies, however, and about 1572 was accused of participating in the murder of the alchemist Sebastian Siebenfreund.

In 1576 there was an outbreak of pestilence in Brandenburg, and Thurneysser, with the court, left Berlin for a period of several months. During this time his second wife, Anna Hüerlin, whom he had married in 1561, died, and his business, which he had left in the hands of his brother Alexander, suffered severely. By 1580 he had decided to move back to Basel, where he purchased an estate called "Zum Thurn," which enabled him to style himself grandly as Leonhard Thurneysser zum Thurn. He had brought a considerable amount of his money to Basel, and he there married his third wife, Marina Herbrodt. In divorce proceedings two years later, the Basel town council made over all of his remaining wealth to his wife, terms against which Thurneysser railed in a number of pamphlets.

Thurneysser returned to Brandenburg and spent the last years of his life attempting to make gold. After he failed to transmute a large quantity of silver, however, he left the service of Johann Georg and traveled to Italy, where he found a patron in Ferdinand de' Medici, grand duke of Tuscany. At this time he became a convert to Roman Catholicism. He returned to Germany shortly before his death in Cologne and in his will asked to be buried there beside Albert the Great, a wish that was never carried out.

BIBLIOGRAPHY

I. ORIGINAL WORKS. Thurneysser's publications include *Archidoxa* (Münster, 1569; Berlin, 1575); *Quinta essentia* (Münster, 1570); *Prokatalepsis* (Frankfurt an der Oder, 1571); *Pison* (Frankfurt an der Oder, 1572);

Chermeneia (Berlin, 1574); *Onomasticon polyglosson* (Berlin, 1574); *Eyporadelosis* (Berlin, 1575); *Bebaiosis agonismoy* (Berlin, 1576); *Historia sive descriptio plantarum* (Berlin, 1578); *Historia und Beschreibung influentischer, elementarische und natürlicher Wirckungen* (Berlin, 1583); *Megaln chymia* (Berlin, 1583); *Melisath* (Berlin, 1583); *Attisholtz oder Attiswalder Badordnung* (Cologne, 1590); *Reise und kriegs Apotecken* (Leipzig, 1602); and *Zehn Bücher von kalten (warmen) mineralischen und metallischen Wassern* (Strasbourg, 1612).

Until 1945 two unedited alchemical manuscripts and a number of letters were to be found in the Berlin Staatsbibliothek.

II. Secondary Literature. On Thurneysser and his work, see *Allgemeine Deutsche Biographie*, XXXVII (1894), 226; Paul Diergart, "Mitteilungen zur Wertung des Paracelsisten. Leonhard Thurnyser," in *Beiträge aus der Geschichte der Chemie* (Leipzig-Vienna, 1909), 306–313; Fritz Ferchl, *Chemisch pharmazeutisches Bio- und Bibliographikon* (Mittenwald, 1937), 536; John Ferguson, *Bibliotheca chemica*, II (Glasgow, 1906), 450–455, which includes excerpts from some of Thurneysser's calendars; Stanisław Kośmiński, *Słownik lekarzów polskich* (Warsaw, 1883), with letter on p. 514 in which Thurneysser described for Stephen Báthory an "alexipharmacum" sovereign against all poisons; Hermann Kopp, *Die Alchemie*, 2 vols. (Heidelberg, 1886), *passim*; J. C. Moehsen, *Beiträge zur Geschichte der Wissenschaften in der Mark Brandenburg* (Berlin-Leipzig, 1783), 55–198; J. R. Partington, *A History of Chemistry*, II (London, 1961), 152–153; Will-Erich Peuckert, *Der Alchemist und sein Weib* (Stuttgart, 1956); Günther Bugge, *Der Alchimist* (Berlin, 1943), a novel based upon Thurneysser's life; B. Reber, "Zwei neue Dokumente über Leonhard Thurneysser zum Thurn," in *Mitteilungen zur Geschichte der Medizin und der Naturwissenschaften*, **5** (1906), 432–439; Karol Christoph Schmieder, *Geschichte der Alchemie* (Munich, 1927), 284–286; Karl Sudhoff, *Bibliographia Paracelsica* (Berlin, 1894), *passim*; and Laszlo Szathmary, *Magyar Alkemistak* (Budapest, 1928), 319–322.

WŁODZIMIERZ HUBICKI

THURSTON, ROBERT HENRY (*b*. Providence, Rhode Island, 25 October 1839; *d*. Ithaca, New York, 25 October 1903), *engineering education, steam engineering, testing of materials.*

Thurston was a mechanical engineer who exerted wide and lasting influence upon the American engineering profession. He was a prolific writer of textbooks, reference works, and technical and popular articles; and he organized and directed two mechanical engineering schools. In addition he organized and was an active member of professional societies and served on industrial and governmental committees.

Thurston was the eldest of the three children of Robert Lawton Thurston, a prominent steam-engine builder in Providence, and Harriet Taylor. Upon completion of high school, Thurston was expected to enter his father's shops; but he and his father were persuaded by one of his teachers that he should continue his education at Brown University, where he graduated in 1859 with a major in science and a minor in civil engineering.

After a short period as draftsman, Thurston represented his father's company in Philadelphia, where in 1861 his first published article appeared in the *Journal of the Franklin Institute*. Throughout the Civil War (1861–1865), he served as an assistant engineer in the navy, and at the end of the war he was assigned to the U.S. Naval Academy in Annapolis as assistant professor of natural and experimental philosophy. The pattern of his subsequent career was evident in his six years of teaching at the academy. Upon the death of the incumbent, he became head of his department. He designed and built a signaling lamp; experimented with lubricants; and published a number of descriptive technical reports and popular articles on naval armament, steam engines, and the manufacture of iron and steel.

Henry Morton, former editor of the *Journal of the Franklin Institute* and president of the newly formed Stevens Institute of Technology in Hoboken, New Jersey, invited Thurston to organize and to direct the school of mechanical engineering. It was during his fourteen years (1871–1885) at the institute that Thurston established his international reputation. In 1874 he persuaded the trustees of Stevens to equip a testing laboratory to serve commercial clients. Subsequently, laboratory work became a part of the engineering curriculum.

While teaching, Thurston invented testing machines and carried out extensive research on strength of materials and lubrication. He was one of the first to demonstrate that the elastic limit of ductile materials can be raised by the application of stress beyond the yield point. His series of public lectures on the history of the steam engine, published several years later, in 1878, was long a standard work on the subject. Major works—which were derived from his classroom lectures on materials of engineering, friction, and lubrication—were published while Thurston was at Stevens. During this period Thurston wrote large books on steam boilers and the steam engine. He served also as secretary of the U.S. Board to Test Iron, Steel, and Other Metals; as U.S. Commissioner to the 1873 Vienna International Exhibition; as an official of the 1876 Philadelphia Centennial Exhibition; and as first president of the American Society of Mechanical Engineers, which in 1880 he helped organize. He also published an extraordinary number of both technical and popular articles.

In 1885 Thurston moved to Cornell University, where he reorganized and directed the Sibley College of Mechanical Engineering. Through his publications and the considerable number of Cornell graduates who became teachers in engineering colleges throughout the country, Thurston influenced greatly the philosophy and direction of engineering education. In an engineering school he expected to treat only professional subjects and to relegate general education to a preparatory school. He considered general education as desirable but not essential; his successors gave only lip service to its desirability. Thurston suggested in 1893 ("Technical Education in the United States," p. 923n) that the ideal technical education was probably to be found in the military academies, where scientific and professional studies and physical training were all given due weight.

Thurston was diligent, enthusiastic, and persistent, and tended toward action rather than reflection. He made massive contributions to the order and the promulgation of the engineering sciences and to the promotion of organizations that were directed toward the increase of material wealth in a progress-oriented social order.

He was a member or honorary member of a score of technical societies in the United States and Europe. He received honorary degrees from Stevens Institute of Technology (1885) and Brown University (1889). Thurston was married in 1865 and, upon the death of his wife, remarried in 1880.

BIBLIOGRAPHY

I. ORIGINAL WORKS. Thurston edited *Reports of the Commissioners of the United States to the International Exhibition Held at Vienna 1873*, 4 vols. (Washington, 1875–1876). His other works include *A History of the Growth of the Steam-Engine* (New York, 1878); *The Materials of Engineering*, 3 vols. (New York, 1883–1884); *A Treatise on Friction and Lost Work in Machinery and Millwork* (New York, 1885); *A Manual of Steam-Boilers, Their Design, Construction, and Operation* (New York, 1888); *Reflections on the Motive Power of Heat* (New York, 1890), a trans. of N. L. S.

Carnot, *Reflexions sur la Puissance Motrice du Feu* (1824); *A Manual of the Steam-Engine: For Engineers and Technical Schools*, 2 vols. (New York, 1891); *Robert Fulton, His Life and Its Results* (New York, 1891); "Technical Education in the United States," in *Transactions of the American Society of Mechanical Engineers*, **14** (1893), 855–1013; and *The Animal as a Machine and a Prime Motor, and the Laws of Energetics* (New York, 1894). A list of Thurston's writings is in Durand's biography (see below), pp. 245–287.

II. SECONDARY LITERATURE. William F. Durand, *Robert Henry Thurston* (New York, 1929), is an uncritical but informative biography by a former associate and friend; a portrait appears as the frontispiece. See also Durand's article on Thurston in *Dictionary of American Biography*, **9** (1936), 518–520. A context is provided for Thurston's work in education and in the American Society of Mechanical Engineers in Monte Calvert. *Mechanical Engineering in America, 1830–1910* (Baltimore, 1967). Thurston's correspondence is in the Cornell University Library, Collection of Regional History and University Archives.

EUGENE S. FERGUSON

THYMARIDAS (*fl.* Paros, first half (?) of fourth century B.C.), *mathematics.*

An early Pythagorean of uncertain date, Thymaridas was a number theorist from the Aegean island of Paros. He defined a unit as a "limiting quantity" (see Iamblichus, *In Nicomachi . . .*, p. 11, 2–3); and he is said to have called a prime number εὐθυγραμμικός ("rectilinear"), because it can only be established one-dimensionally (Iamblichus, *op. cit.*, p. 27, 4–5), since the only measures of a prime number are itself and one. But Thymaridas' chief contribution to number theory was his ἐπάνθημα ("bloom"), which he expressed rather obscurely in generalized form (Iamblichus, *op. cit.*, p. 62, 18 ff.). The rule leads to the solution of a certain set of n simultaneous simple equations connecting n unknowns. The unknown quantity is called "an undetermined number of units" (that is, x); the known quantities are ὡρισμένα ("determined"). The principle of the rule has been explained by Heath as follows:

Let there be n unknown quantities

$$x, x_1, x_2, \cdots x_{n-1}$$

connected by n equations, in such a way that

$$x + x_1 + x_2 + \cdots + x_{n-1} = S$$

$$x + x_1 = a_1,$$

$$x + x_2 = a_2,$$

$$\cdots\cdots\cdots\cdots$$

$$x + x_{n-1} = a_{n-1},$$

the solution is

$$x = \frac{(a_1 + a_2 + \cdots + a_{n-1}) - S}{n - 2}.$$

Iamblichus shows that other equations can be reduced to this form (Iamblichus, *op. cit.*, p. 63, 16 ff.); he gives as an example an indeterminate problem having four unknown quantities in three linear equations. It is not certain that Thymaridas was responsible for the extension of his method.

Of Thymaridas' life we are told only that he fell from prosperity to poverty and that consequently Thestor of Poseidonia sailed to Paros to help him with money specially collected for his benefit (Iamblichus, *De vita pythagorica*, p. 239). "Eumaridas" in a list of Parian Pythagoreans (Iamblichus, *op. cit.*, p. 267) may be a mistake for Thymaridas.

BIBLIOGRAPHY

On Thymaridas and his work, see T. L. Heath, *A History of Greek Mathematics*, I (Oxford, 1921), 69, 72, 94; Iamblichus, in *Nicomachi arithmeticam introductionem*, H. Pistelli, ed. (Leipzig, 1894), 11, 2–3; 27, 4–5; 62, 18 ff.; 63, 16 ff.; and *De vita pythagorica*, L. Deubner, ed. (Leipzig, 1937).

G. HUXLEY

IBN TIBBON, JACOB BEN MACHIR (*b*. Marseilles, France [?], *ca*. 1236; *d*. Montpellier, 1305), *astronomy, science translation.*

In Romance languages Ibn Tibbon is known as Don Profiat or Profeit, and in Latin as Prophatius Judaeus. These names come from translation of the Hebrew *mehir* into the languages of southern France.

His family, commonly designated by the patronymic Ibn Tibbon, came from Granada. His great-grandfather, Judah ben Saul ibn Tibbon (1120–*ca*. 1190), moved in 1150 to the south of France because of the unrest in Granada. He established himself at Lunel, where in 1160 Benjamin of Tudela found him practicing medicine. Having grown up in an Islamic country, Judah ben Saul spoke Arabic and was thus able to translate into Hebrew, for the benefit of his coreligionists, religious and philosophical works written in Arabic by Baḥya ibn Paquda, Solomon ibn Gabirol, Judah ha-Levi, Ibn Janah, and Saadia. His son, Samuel ben Judah (1150–1232), continued this work of translation; but it was his grandson, Moses ben Samuel (*fl*. 1240–1283), and his great-grandson, Jacob ben Machir, who brought the family greatest glory. The translations made by the family were among the most important cultural works of the time, at least from a historical point of view, since they were part of a process by which Arabic learning and, through that, Greek scientific traditions were made available to the scholars of medieval Europe.

Jacob studied medicine at Montpellier, and it appears that from 1266 to 1267 he lived in Gerona, Spain, where he was a follower of Moses ben Nahman. Ideologically he adhered to the Maimonidean philosophy and spent most of his life in Lunel and Montpellier.

Working from texts in Arabic, Ibn Tibbon rendered into Hebrew works by Autolycus of Pitane (*On the Moving Sphere*), Euclid (*Elements, Data*), Menelaus of Alexandria (*Spherics*), Qusṭā ibn Lūqā (*Use of the Celestial Globe*), Ibn al-Haytham (*On the Configuration of the World*), Ibn al-Saffār (*On the Use of the Astrolabe*), al-Zarqālī (*Use of the Ṣafīḥa*), al-Ghazālī (*Balance of Knowledge*), Jābir ibn Aflah (*Correction [iṣlāḥ] of the Almagest*), and Ibn Rushd (compendium of the Organon, commentary on Aristotelian zoology).

His own works, which deal with astronomy, are Prologue to Abraham bar Ḥiyya's *Calculation of the Courses of the Stars*, extracts from the *Almagest*, *Roba' Yisrael*, and *Almanac*.

The *Roba' Yisrael* ("Quadrant of Israel") was written between 1288 and 1293 in Hebrew, and was translated into Latin in 1299 by Armengaud, son of Blaise (*d*. 1314). In 1301 an expansion of this work appeared; the text has been lost but is preserved in the Latin translation (*ca*. 1309) by Peter of St.-Olmer, which was translated back into Hebrew. A new astronomical instrument is described in this work—the so-called *quadrans novus*, as distinct from the *quadrans vetus* of Robert the Englishman (*ca*. 1276) and the tenth-century *vetustissimus*. Examples of the *quadrans novus* have been preserved, and it apparently was much used in its time. It consists of a simplification of the face of the astrolabe by means of two successive rebates that have as their axes the north-south and east-west lines. Whatever connections may exist between this apparatus and similar instruments used by the Arabs have not been clearly established.

The *Almanac* is calculated for Montpellier and dated 1 March 1300 (1301 A.D.). In his prologue Ibn Tibbon notes that he was inspired by the almanac of King Tolomeus, which had been corrected by al-Zarqālī. Those corrections were insufficient, however, and, in addition, introduced new errors. For instance, al-Zarqālī did not take into account that some calculations had been made according to the Coptic calendar. Ibn Tibbon explains that he has followed the method of his predecessors, using the Toledan Tables as a basis; this is not quite correct, however, since his work is actually based on al-Zarqālī's completion of Ammonius' treatise on the almanac. But this distinction between tables and almanac was not usual; and for that reason several writers, among them Andalo di Negro, believed that Ibn Tibbon had used the Alfonsine Tables as a basis, since the prologue was not consistent with the tabular part of the work. The only exception to that model is the collection of lunar tables, inspired by those of al-Khwārizmī, known through al-Zarqālī. In the subsequent development of the work, Ibn Tibbon limits himself to the calculation of ephemerides and modifies the constant to be added at the end of the cycles. Later astronomers—Andalo di Negro, Levi ben Gerson, Abraham Zacuto—found errors that they corrected.

Ibn Tibbon's astronomical work was very highly regarded during the Renaissance and was cited by Copernicus, Reinhold, Clavius, and Kepler.

BIBLIOGRAPHY

I. ORIGINAL WORKS. For the MSS see Moritz Steinschneider, *Die hebräischen Übersetzungen des Mittelalters . . .* (Berlin, 1893; repr. Graz, 1956), index, 1057, and the references given there. For the trans. of the *Ṣafīḥa,* see the ed. and trans. by José María Millás Vallicrosa into Catalan, *Tractat de l'assafea d'Azarquiel* (Barcelona, 1933). One of the texts of the Latin trans. of the work on the *quadrans novus* is G. Bofitto and C. Melzi d'Eril, eds., *Il quadrante d'Israel* (Florence, 1922); the same authors also published *Almanach Dantis Alighieri sive Prophacii Judaei Montispessulani* (Florence, 1908).

II. SECONDARY LITERATURE. See Marion Boutelle, "The Almanac of Azarquiel," in *Centaurus,* **12,** no. 1 (1967), 12–19; P. Duhem, *Le système du monde,* III (Paris, 1915), 298–312; R. T. Gunther, *Early Science in Oxford,* II (Oxford, 1923), 163–169; H. Michel, *Traité de l'astrolabe* (Paris, 1947), 23–24; J. M. Millás Vallicrosa, *Estudios sobre Azarquiel* (Madrid–Granada, 1943–1950), 356–362, 402–404; E. Poulle, "Le quadrant nouveau médiéval," in *Journal des savants* (Apr.–June 1964), 148–167, 182–214; E. Renan and A. Neubauer, *Rabbins français,* repr. from *Histoire littéraire de la France,* XXVII (Paris, 1877); and George Sarton, *Introduction to the History of Science,* II (Baltimore, 1931), 850–853 and index.

J. VERNET

IBN TIBBON, MOSES BEN SAMUEL (*b.* Marseilles, France; *fl.* Montpellier, France, 1240–1283), *medicine, philosophy, translation.*

Moses ben Samuel was related to Jacob ben Machir ibn Tibbon and father of the Judah ben Moses ibn Tibbon who took part in the struggles between Maimonideans and anti-Maimonideans that occurred at Montpellier, aligning himself with the former. He spent most of his life in Montpellier. He wrote a number of works, most of them commentaries on the Bible and the Talmud. Among them is one that deals with the weights and measures mentioned in those works.

Ibn Tibbon is most significant as a translator from Arabic into Hebrew of Ibn Rushd (commentaries on Aristotle: *Physics, De caelo et mundo, De generatione et corruptione, Meteora, De anima, Parva naturalia, Metaphysics, Problems*), Themistius (commentary on Book Λ of the *Metaphysics*), al-Baṭalyūsī (*Kitāb al-ḥadā'iq*), al-Fārābī (*Kitāb mabādi'*), Maimonides (*Kitāb al-sirāj, Kitāb al-farā'iḍ, Maqāla fī ṣina'at al-manṭiq* [dealing with philosophy and religion], and *Regimen Addressed, On Poisons and Antidotes, Commentary on the Aphorisms of Hippocrates* [on medicine]), Euclid (*Elements*), Geminus (*Introduction to Astronomy*), Theodosius of Bithynia (*Spherics*), Jābir ibn Aflaḥ (*Iṣlāḥ al-majisṭī*), Muhammad al-Ḥaṣṣār (treatise on arithmetic and algebra), al-Biṭrūjī (*Kitāb al-hai'a*), Ibn Sīnā (*Canticum, The Small Canon*), Ibn al-Jazzār (*Viaticum peregrinantis*), Ḥunayn ibn Isḥāq (*Isagoge Johannitii ad tegni Galeni*), al-Rāzī (*Antidotary, Division and Distribution of Diseases*).

BIBLIOGRAPHY

For the MSS see Moritz Steinschneider, *Die hebräischen Übersetzungen des Mittelalters . . .* (Berlin, 1893; repr. Graz, 1956), index, 1062, and the references given there.

For general information, see E. Renan and A. Neubauer, *Rabbins français* (Paris, 1877), repr. from *Histoire littéraire de la France;* George Sarton, *Introduction to the History of Science,* II (Baltimore, 1931), 847 and index; and Max Schloessinger, in *Jewish Encyclope-*

dia, VI (New York, 1906), 545–548. Also see references in the article on Jacob ben Machir ibn Tibbon in this Dictionary.

J. VERNET

TIEDEMANN, FRIEDRICH (*b.* Kassel, Germany, 23 August 1781; *d.* Munich, Germany, 22 January 1861), *anatomy, physiology.*

Tiedemann was the son of Dietrich Tiedemann, a professor of philosophy, Greek, and classical literature, who took great interest in his son's early education. He attended the Marburg Gymnasium, then, in 1798, entered the university there to begin medical studies. His chief interest was medical theory, and in 1802 he began to concentrate on theoretical subjects at Bamberg and Würzburg, under the guidance of J. N. Thomann and Kaspar von Siebold; the inability of physicians to prevent his father's death of a contagious disease in 1803 further strengthened his aversion to practical medicine.

Tiedemann graduated from Marburg with the M.D. in 1804, then remained there to attend Franz Joseph Gall's courses on physiology, comparative osteology, and craniology. Gall's exceptionally skillful fine anatomical preparations set an example for Tiedemann's own later work. Tiedemann then returned to Würzburg to study with Schelling. He was, however, able to resist the seductiveness of *Naturphilosophie*—Schelling himself was responsible, Tiedemann later wrote, for his decision to remain faithful to the methods of empirical research and observation—and determined to go to Paris to study with Georges Cuvier. In the course of his journey he met S. T. Soemmerring, who was greatly impressed by Tiedemann's preparation of the nervous system of the pigeon. It was upon Soemmerring's recommendation that Tiedemann was in 1807 called to the Landshut Medical Faculty as professor of anatomy and zoology.

In Paris, under Cuvier's tutelage, Tiedemann became interested in comparative anatomy and morphology, and these studies formed the basis for the first works that he published in Landshut. Among these were three sections of a projected, but never completed, textbook on zoology. Three parts were brought out (in 1808, 1810, and 1814, respectively); of these the first dealt with problems of general zoology, and in particular the classification of mammals, while the second and third were devoted to birds. The published segments of the work were greeted as the most comprehensive compendium of zoological data since Cuvier's own *Leçons d'anatomie comparée.* At the same time, Tiedemann also published a series of monographs summarizing his studies on specific subjects. These included a work on the heart in fish (1809), detailed studies of the anatomy and natural history of the great reptiles (1811 and 1817), and a number of works on the structural anatomy of birds and amphibia, especially the lymphatic vessels and respiratory organs of birds. His 1816 publication on echinoderms (holothurioidea [sea cucumbers], starfish, and sea urchins), a subject set by the Académie des Sciences, won its prize and Tiedemann was elected a corresponding member.

During the same period of his career, Tiedemann did further important research on morphology and embryology. He studied the fetal development of bone tissues and demonstrated beyond doubt that the maternal blood is not transmitted to the fetus, which rather has its own closed circulatory system, separate from (but closely associated with) the maternal one. His most important work in this area lay in his studies of the development of the brain; his *Anatomie und Bildungsgeschichte des Gehirns im Fötus der Menschen, nebst einer vergleichenden Darstellung des Hirnbaues in den Thieren*, dedicated to J. F. Blumenbach, was published in 1816.

In the preface to this work, Tiedemann stated that, since the fully developed fetal brain had been described exhaustively, it was therefore necessary to study its anatomical development in detail; anatomy, he added, would be a mature science only when the different stages of development and mutual relations of various structures had been examined so that the laws that governed them might be understood. Continuity of observation, Tiedemann stated, was essential, and his book represented the results of several years of continuous research. He was thus able to record the exact stage at which a number of structures first arose in the fetal brain, together with the duration of brain growth and the time of its completion. A comparison of the durable forms of the brain of various animals with the transitory embryonic configurations of the human brain allowed him to draw an analogy between the adult animal brain and the human brain at a certain stage of development; he was also able to detect the crossing of the pyramidal tracts in a very young embryo, the changes in the peripheral nerves, and the cranial shift of the sacral part of the spinal cord. His book contained a number of beautiful and accurate illustrations, and was later translated into French and English.

In 1816 Tiedemann went to the University of Heidelberg to take up an appointment as professor of anatomy, comparative anatomy, physiology, and zoology. The new post offered him greater resources (among other things, there was a more generous supply of cadavers for dissection), and brought him into contact with a younger colleague, the chemist Leopold Gmelin, whose work strongly influenced his own. Tiedemann's interests gradually became concentrated on problems of physiology, and with Gmelin he performed a considerable body of research, in which he assumed responsibility for animal experimentation and observation, while Gmelin carried out chemical examinations.

In 1820 Tiedemann and Gmelin published *Versuche über die Wege auf welchen die Substanzen aus dem Magen und Darm ins Blut gelangen, über die Verrichtung der Milz und über die geheimen Harnwege*, an account of their researches on the passage of various substances from the stomach or intestines to the blood, on the function of the spleen, and on the hypothetical hidden urinary ducts. The absorption of nutrients from the alimentary tract had long been of interest to Tiedemann — it had, in fact, motivated his previous studies on the lymphatic vessels — but his work on the spleen was something quite new. The role of the spleen had been hitherto only a subject of speculation; Tiedemann's attempts to prove experimentally, in horses and dogs, that the spleen secreted a fluid that passed through the lymphatic system to participate in the transformation of chyle into blood served as the basis for all further work on the part of the spleen in blood formation. Tiedemann and Gmelin also showed that the hidden urinary ducts posited by other workers could not exist.

In 1822 Friedrich Sigismund Leuckart assumed responsibility for the teaching of zoology and comparative anatomy at Heidelberg and Tiedemann was able to devote all his time to physiology. The following year the Paris Académie des Sciences announced a prize for the best work encompassing the chemical composition of the digestive juices, the digestion of simple and complex foodstuffs in animals (including mammals, birds, reptiles, amphibians, and fish), the contents of the stomach and intestines during periods of fasting and eating, and the passage of digested substances into the blood, lymph, and urine. Under this stimulus Tiedemann and Gmelin decided to expand their earlier work on digestion into a comprehensive study, using chemical methods as well as experimental and microscopic investigations, designed to satisfy the conditions of the award.

Digestion had been thought of as a single process, taking place in a single organ, the stomach. Tiedemann and Gmelin, however, observed the digestion of foodstuffs as they passed through the alimentary tract and investigated the assimilation of nutrients into the blood and lymph and the eventual elimination of waste products to demonstrate that digestion is in fact a complicated series of processes, involving a number of organs. They proved that for some substances digestion represents not a simple dissolution but rather a chemical transformation, as, for example, displayed by the conversion of starch into glucose. They showed experimentally that the digestive juices have discrete properties, as, for instance, the pancreatic juice, formerly thought to be an abdominal saliva, which they obtained through an experimentally induced abdominal fistula and tested chemically to demonstrate its difference from saliva.

Chief among their other findings was their discovery of a number of biliary substances, of which some (such as pigments) were merely excreted, while others played more important physiological roles, as in the absorption of fat, which diminished when bile was not present. They also (shortly after Prout, but independently and by a different method) detected the presence of hydrochloric acid in the stomach, and showed that it increases in quantity with the digestion of certain foods. They further described the color reactions of proteins during pancreatic digestion and discovered glucose, transformed chemically from starch and similar substances, in the blood.

Despite their remarkable achievement, Tiedemann and Gmelin did not win the Académie's prize. That body decided rather to divide the prize between them and two young French scientists, François Leuret and Jean-Louis Lassaigne, giving each entry half the prize and an honorable mention. Resentful, Tiedemann and Gmelin refused to accept any part of the award and announced that they would submit their work to the judgment of the entire scientific world; their *Die Verdauung nach Versuchen* was published in Heidelberg in 1826. The great chemist Berzelius characterized the book as "a long series of investigations about the process of digestion, in which everything that anatomy and chemistry could at present offer to its study was used," and praised it as "uncontestably the most complete physiological examination [of digestion], which has enriched the chemical study of the processes that occur in living animals." And indeed, further advances in the study of digestion became possible only with advances in chemistry.

By 1835 Tiedemann had begun to suffer from the eye complaint that eventually led to the loss of his sight. In that year he was relieved of his courses by his son-in-law, Theodor Bischoff, although he continued to carry out his own research. In 1836 he published, in English, what may be considered one of the earliest basic works of physical anthropology, *On the Brain of the Negro Compared With That of the European and the Orang-Outang*. In this surprisingly modern study, Tiedemann showed that, in contrast to the large difference between the forebrains of apes and men, no substantial differences could be found between the brains of the races of men; although the majority of Negro skulls and cranial cavities that he studied were smaller than those of European specimens, they had, by his measurement, contained brains as large and as heavy. He further stated his finding that there was no area of intellectual activity in which Negroes could not perform as well as European whites, and concluded that there was no natural formation or disposition of the brain in Negroes that would substantiate the notion of their predestined subservient state.

In 1844 Tiedemann took up the task of designing a new anatomical theater. His life was disrupted by the revolution of 1848–1849; not only was he opposed to popular uprisings, for political and philosophical reasons, but also three of his sons were army officers. The eldest, Gustav Nicolaus Tiedemann, was executed under martial law, and the two others had to flee into exile. In 1849 Tiedemann retired from the University of Heidelberg and settled first in Frankfurt am Main, and then, in 1856, in Munich. The fiftieth anniversary of his doctorate was celebrated by his friends and former students in 1854, and a medal was struck for the occasion. In the same year Tiedemann published his last book, a history of tobacco and its use.

BIBLIOGRAPHY

I. ORIGINAL WORKS. A bibliography of Tiedemann's works is found in Theodor Bischoff (see below). Works include *Zoologie*, 3 vols. (Landshut, 1808, 1810, 1814); *Anatomie des Fischherzens* (Landshut, 1809); *Anatomie und Naturgeschichte des Drachen* (Nuremberg, 1811); *Anatomie der kopflosen Missgeburten* (Landshut, 1813); *Anatomie und Bildungsgeschichte des Gehirns im Fötus der Menschen, nebst einer vergleichenden Darstellung des Hirnbaues in den Thieren* (Nuremberg, 1816), French ed. (Paris, 1823), English ed. (Edinburgh, 1826); *Anatomie der Röhrenholothurie, des pomeranzenfarbenen Seesterns und Seeigels* (Landshut, 1816); *Anatomie und Naturgeschichte des Krokodils* (Heidelberg, 1817), with M. Oppel and J. Liboschitz.

Other works include *Abhandlung über das vermeintliche bärenartige Faulthier* (Heidelberg, 1820); *Versuche über die Wege auf welchen die Substanzen aus dem Magen und Darm ins Blut gelangen, über die Verrichtung der Milz und über die geheimen Harnwege* (Heidelberg, 1820), French ed. (Paris, 1821), with Gmelin; *Icones cerebri simiarum et quorundam mammalium variorum* (Heidelberg, 1821); *Tabulae arteriarum corporis humani* (Karlsruhe, 1822); *Tabulae nervorum uteri* (Heidelberg, 1823); *Die Verdauung nach Versuchen* (Heidelberg, 1826, 1827), with Gmelin; *Physiologie des Menschen*: I, *Allgemeine Betrachtungen der organischen Körper* (Darmstadt, 1830); II, not published; III, *Untersuchungen über das Nahrungsbedürfniss, der Nahrungstrieb und die Nahrungsmittel der Menschen* (Darmstadt, 1836); *On the Brain of the Negro Compared With That of the European and the Orang-Outang* (London, 1836), German ed. (Heidelberg, 1837); *Von den Duverneyschen, Bartolinschen oder Cowperschen Drüsen des Weibes und der schiefen Gestaltung und Lage der Gebärmutter* (Heidelberg, 1840); *Von der Verengerung und Verschliessung der Pulsadern in Krankheiten* (Heidelberg, 1843); *Von lebenden Würmern und Insekten in den Geruchs-Organen des Menschen, den Zufällen, welche sie verursachen und den Mitteln sie auszutreiben* (Mannheim, 1844); *Zeitschrift für Physiologie*, ed. with G. R. and L. Ch. Treviranus; and *Geschichte des Tabaks und anderer ähnlichen Genussmittel* (Frankfurt am Main, 1854).

II. SECONDARY LITERATURE. An important study of Tiedemann's life and works is Theodor Bischoff, *Gedächtnissrede auf Friedrich Tiedemann* (Munich, 1861), with bibliography, pp. 36–39. Other biographical studies are P. Flourens, "Tiedemann, Friedrich," in *Biographie universelle ancienne et moderne*, XLI, 526–529, and in *Gazette médicale de Paris*, 29, no. 52 (1861); J. Pagel, in *Allegemeine Deutsche Biographie*, XXXVIII (1894), 577–578; F. Seitz, in *Biographisches Lexikon hervorragender Aerzte aller Zeiten und Völker*, V (1934), 586–587; E. Stübler, in *Geschichte der Medizinischen Fakultät der Universität Heidelberg, 1386–1925* (Heidelberg, 1926), 248–253.

Discussions of Tiedemann's work include (in chronological order) N. Mani, "Das Werk von Friedrich Tiedemann und Leopold Gmelin: 'Die Verdauung nach Versuchen,' und seine Bedeutung für die Entwicklung der Ernährungslehre in der ersten Hälfte des 19. Jahrhunderts," in *Gesnerus*, 13 (1956), 190–214; and H. Hoepke, "Der Streit der Professoren Tiedemann und Henle um den Neubau des Anatomischen Institutes in Heidelberg (1844–1849)," in *Heidelberg Jahrbuch 1961*.

VLADISLAV KRUTA

VAN TIEGHEM, PHILIPPE (*b.* Bailleul, Nord, France, 19 April 1839; *d.* Paris, France, 28 April 1914), *botany.*

Van Tieghem's father, a textile merchant, died in Martinique of yellow fever shortly before the birth of his son; and his mother died immediately afterward. Philippe was their fifth child and was brought up by his uncle and aunt, and later by his sisters. He received his *baccalauréat-ès-sciences* at the *collège* of Bailleul in 1856. As a scholarship student at the *lycée* of Douai, he prepared for the entrance examination to the École Polytechnique but decided instead to take the examination for the École Normale Supérieure, which he entered in 1858. After passing the *agrégation* in physical and natural sciences in 1861, van Tieghem became *agrégé-préparateur* in botany and mineralogy. Under Pasteur's supervision he prepared a dissertation on ammoniacal fermentation. Since his examiners considered that his research pertained essentially to chemistry, he was granted a doctorate in the physical sciences and not, as he had hoped, in the natural sciences. Wishing to become a botanist, he presented a second dissertation on the Araceae, which earned him the desired degree in 1867.

When Payer's chair at the École Normale Supérieure became vacant in 1864, van Tieghem was appointed his successor. Thus, at age twenty-five, he became *maître de conférences* in botany. From his marriage in 1862 to Hélène Sarchi he had four daughters and one son.

Because of the variety and the extent of van Tieghem's 328 recorded writings, a chronological list cannot provide a true picture of his work. Throughout his life he simultaneously studied several subjects; and his research covered five fields of botany: cryptogamy, fermentation, anatomy and biology of phanerogams, the application of anatomy to classification, and plant physiology.

In studying plant evolution and the reproduction of fungi, van Tieghem showed the value of using a pure culture of a single spore kept in a liquid medium. The spore must be protected from contamination by an enclosure yet remain accessible to microscopic observation. For his work on these monosperm cultures, van Tieghem was elected to the Académie des Sciences in 1877 and was appointed professor-administrator at the Muséum d'Histoire Naturelle in May 1879.

Van Tieghem was one of the first to reveal the relationship of blue algae to bacteria. In 1878 he studied "sugar gum" and established that it was a plant that lived on sugar. He described its development and proved that the insoluble substance, an isomer of cellulose, that forms the curd of the gum is excreted by the cells of the organism, which he named *Leuconostos mesenteroides.*

From 1877 to 1879 van Tieghem investigated *Bacillus amylobacter* and butyric fermentation. He demonstrated that only the membranes (whether cutinized, suberized, or lignified) of the submerged sections of aquatic plants can resist this bacterium, the role of which in the decomposition of complex organic substances into simple ones was soon confirmed. In collaboration with Bernard Renault, van Tieghem discovered that organisms closely related to *Bacillus amylobacter* were present in sections of carboniferous plant tissues, where they were preserved by silicification. The two scientists thus established the possibility that coal originated through the same fermentation process occurring in the remote past.

Van Tieghem created an anatomy founded on the homologies of tissues and on their origin from the initial cells. In addition to describing the tissues themselves, he considered the modes of their origin and the pattern of their differentiation. He distinguished three parts of the plant: root, stem, and leaf—a conception that won widespread acceptance largely for its simplicity. The principles of plant symmetry that van Tieghem established are classic. Having defined the organs by their structure, he applied his method to several unanswered questions: the structure of the pistil (his paper on this subject received the Bordin Prize of the Académie des Sciences in 1867), the organization of the ovule, the orientation of the embryo, and the composition of the seed. In 1871 he was awarded a grand prize for his work on the root. Moreover, in his anatomical research on the root and the stem, van Tieghem distinguished between primary and secondary tissues. Starting in 1871, he showed how anatomy could reveal the affinities between plants. He compared organs of the same species in different areas, thus initiating the study of the effect of the environment on plants.

Van Tieghem also did research in plant physiology. His experiments on the effect of cold on seeds and on the resultant "extreme slowing down" of their life processes were crucial. Through his work on the potential of the various parts of the embryo contained in the seed, he proved that each of the embryo's organs is autonomous and can grow without the others, whether or not the seed contains endosperm. He demonstrated that young embryos deprived of endosperm can be nourished by an artificial paste the chemical composition of which is close to that of specific endosperm.

After 1893 van Tieghem turned to work on parasites. He examined many exotic plants on which various degrees of degeneration of the ovule could be detected. His findings concerning this female organ led him to define the structure of the ovule and its integuments in a number of plants. Van Tieghem's work revealed the need for new classification of the phanerogams, based on the ovule and the seed, as well as a complete classification based on the ovule of plants.

Numerous and varied as they are, all van Tieghem's publications are characterized by an important new approach to botany: function is always studied in relation to structure and development.

In addition to conducting scientific investigations, van Tieghem held many official posts. From 1873 to 1886 he was professor of biology at the École Centrale des Arts et Manufactures; from 1885 to 1912, professor at the École Normale Supérieure des Jeunes Filles at Sèvres; and from 1898 to 1914, professor of plant biology at the Institut Agronomique. Students still find his *Traité de botanique* useful.

BIBLIOGRAPHY

I. ORIGINAL WORKS. Many of van Tieghem's writings are listed in *Notice sur les travaux scientifiques de M. Ph. Van Tieghem* (Paris, 1876). See also "Recherches sur la structure du pistil et sur l'anatomie comparée de la fleur, avec un atlas de 16 planches," in *Mémoires des savants étrangers*, **21** (1871), for which van Tieghem was awarded the Bordin Prize in 1867; "Recherches sur la symétrie de structure des plantes vasculaires," in *Annales des sciences naturelles*, Botanique, 5th ser., **12** (1871); *Traité de botanique conforme à l'état présent de la science*, 3rd. ed. (Paris, 1873), trans. into German by J. Sachs; *Traité de botanique* (Paris, 1884); *Éléments de botanique*, I (Paris, 1886), II (Paris, 1888); and "Recherches comparatives sur l'origine des membres endogènes dans les plantes vasculaires," in *Annales des sciences naturelles*, 7th ser., **8** (1889), written with Henri Douliot.

II. SECONDARY LITERATURE. See G. Bonnier, "L'oeuvre de Philippe Van Tieghem," in *Revue générale de botanique*, **26** (1914), 353–441; and J. Costantin, "Philippe Van Tieghem," in *Nature. Revue des sciences et de leurs applications aux arts et à l'industrie*, no. 2137 (1914), 394–396; and "Le rôle de Brongniart, de Renault et de Van Tieghem dans la chaire d'organographie du Muséum," in *Archives du Muséum nationale d'histoire naturelle*, 6th ser., **12** (1935), 319–324.

A. NOUGARÈDE

TIEMANN, JOHANN CARL WILHELM FERDINAND (*b.* Rübeland, Germany, 10 June 1848; *d.* Meran, Austria [now Merano, Italy], 14 November 1899), *chemistry.*

The son of William and Auguste Tiemann, Ferdinand Tiemann is primarily remembered for the synthesis of phenolic aldehydes, known as the Reimer-Tiemann reaction. He studied chemistry and pharmacy in Brunswick at the Collegium Carolinum (now known as the Technische Hochschule); briefly assisted his uncle, Carl Tiemann, a pharmacist; and entered the University of Berlin in 1870 to study chemistry under A. W. von Hofmann. Bertha Tiemann, the younger of his two sisters, married Hofmann in 1877. After serving in the Franco-Prussian War, he spent the rest of his life in the department at Berlin, where he became assistant in 1871, university lecturer in 1878, and professor in 1882. He was editor of the *Berichte der Deutschen chemischen Gesellschaft* from 1882 to 1897.

In 1870 Tiemann published the results of his earliest research on the derivatives of guanidine, trinitrotoluene, and diaminotoluene, and on the synthesis of dinitro- and trinitrobenzoic acids. In 1873 he began the first of many investigations on the purification and analysis of water. Tiemann and Wilhelm Haarmann obtained vanillin from the acid oxidation of conifer glucosides in 1874. During the next year they established the configuration of vanillin, and Tiemann synthesized it from protocatechuic acid. By heating guaiacol with chloroform and aqueous sodium hydroxide, Karl Ludwig Reimer obtained vanillin in 1876; and later that year he and Tiemann extended the reaction as a general method for synthesizing phenolic aldehydes. In 1891 Tiemann discovered the commercial method of preparing vanillin from eugenol (contained in oil of cloves) by the successive processes of rearrangement, acetylation, and oxidation of the side chain followed by hydrolysis. His other work included the synthesis of caffeic acid, the discovery of Ionone (ketones that have a strong odor of violets and are used especially in perfumes), elucidation of the structure of glucosamine, and extensive investigations of terpenes.

Tiemann enjoyed the reputation of being a fine teacher and a popular lecturer. With his students he published many papers on hydroxyaldehydes and related compounds. Failing health forced him to spend the winter of 1899 in the South Tirol, where he died of a heart attack, leaving a widow and three children.

BIBLIOGRAPHY

Some of Tiemann's contributions were first reported in the following papers: "Neue Guanidin-Abkömmlinge," in *Berichte der Deutschen chemischen Gesellschaft*, **3** (1870), 6; "Abkömmlinge des Trinitrotoluols und des Toluylendiamins," *ibid.*, p. 217; "Di- und trinitrirte Benzoesäuren," *ibid.*, p. 223, written with W. Judson; "Ueber das Coniferin und seine Umwandlung in das aromatische Prinzip der Vanille," *ibid.*, **7** (1874), 608, written with W. Haarmann; "Ueber eine synthetische Bildungsweise des Vanillins, über Hydrovanilloin und Vanillylalkohol," *ibid.*, **8** (1875), 1123; "Ueber die Einwirkung von Chloroform auf alkalische Phenolate," *ibid.*, **9** (1876), 824, written with K. L. Reimer; and "Ueber die Einwirkung von Chloroform auf Phenole und besonders aromatische Oxysäuren in alkalischer Lösung," *ibid.*, **9** (1876), 1268, written with K. L. Reimer.

See also "Synthese der Kaffeesäure, Abkömmlinge derselben und der Hydrokaffeesäure," *ibid.*, **11** (1878), 646, written with N. Nagai; "Ueber Vanillin," *ibid.*, **24** (1891), 2870; "Ueber Veilchenaroma," *ibid.*, **26** (1893), 2675, written with P. Krüger; and "Ueber das Glucosamin," *ibid.*, **27** (1894), 138, written with E. Fischer.

For a short biography, a review of his research, and a complete list of his publications, see Otto N. Witt, "Ferdinand Tiemann. Ein Lebensbild," in *Berichte der Deutschen chemischen Gesellschaft*, **34** (1901), 4403–4455.

A. ALBERT BAKER, JR.

TIETZ, J. D. See **Titius, Johann Daniel.**

AL-TĪFĀSHĪ, SHIHĀB AL-DĪN ABU'L-ᶜABBĀS AḤMAD IBN YŪSUF (*b.* Tīfāsh, 1184; *d.* Cairo, 1253/1254), *mineralogy, physiology.*

Al-Tīfāshī began his education in Tīfāsh, three days' journey from Kairouan. At a very early age, he went to Cairo, where one of his teachers was the physician ʿAbd al-Laṭīf al-Baghdādī. After further studies in Damascus, he returned to Tīfāsh and obtained a judgeship. He later settled in Cairo, where he died.

Al-Tīfāshī's book on precious stones bears various names in the manuscripts; the generally accepted title is *Azhār al-afkār fī jawāhir al-aḥjār* ("Blossoms of Thoughts on Precious Stones"). It treats twenty-five stones in as many chapters. No critical edition of the book has been produced, but a text with Italian translation was printed in 1818. The work exists in a longer version, however, in a number of the surviving manuscripts, a fact established by J. J. Clément-Mullet, who used the treatise as a principal source for his *Essai sur la minéralogie arabe*. Although he had the opportunity, Clément-Mullet unfortunately did not prepare a critical edition and omitted from his book much of the important information he had gathered.

Al-Tīfāshī states that he intends to treat stones from five points of view: the generation of the stones at the place of deposit; locations of deposits; types, qualities, and marks of genuineness; magical properties and uses; and prices. Clément-Mullet reproduced only those portions of the manuscripts relating to the first three topics, and even this was done unsystematically. J. Ruska was the first to offer longer selections from the text, based on various recensions in order to show that al-Tīfāshī had drawn on the so-called *Book of Stones* by Aristotle and the *Book of Causes* (or *Secret of the Creation*) attributed to Apollonius of Tyana (Balīnūs).

Al-Tīfāshī also wrote a book on sense perception, of which only the title of the original work is known. An extract is preserved in a manuscript by the fourteenth-century lexicographer Ibn Manzūr, but it has not been carefully studied.

Al-Tīfāshī also produced three books on sexual relations. One of these, which discusses the restoration of potency in old men, was reprinted several times and even appeared in an anonymous English translation. The first part deals with the sex organs, sexual hygiene, and aphrodisiacs, and includes quotations from a number of ancient and Islamic physicians. The second part is a sort of erotic guide for men. A second book, apparently more oriented toward hygiene, has not yet been studied. The third book comprises obscene anecdotes; several stories of this type also appear in the second part of the first of the books mentioned above.

BIBLIOGRAPHY

I. ORIGINAL WORKS. The text of the *Azhār*, with Italian trans., is A. R. Biscia, *Fior di pensieri sulle pietre preziose di Ahmad Teifascite* (Florence, 1818), Italian version also printed separately (Bologna, 1906). J.-J. Clément-Mullet reproduces portions of the MS in his *Essai sur la minéralogie arabe* (Paris, 1868). One of his books on sexual relations appeared as *The Old Man Young Again, Translated From the Arabic by an English Bohemian* (Paris, 1898). J. Ruska includes selections from the *Azhār* in his *Das Steinbuch des Aristoteles* (Heidelberg, 1912), 23–31; and *Tabula Smaragdina* (Heidelberg, 1926).

II. SECONDARY LITERATURE. See C. Brockelmann,

Geschichte der arabischen Literatur, I (Weimar, 1898), 495; and supp., I (Leiden, 1937), 904, and III (1942), 1243; Ibn Farḥūn, *Al-Dībāj al-mudhahhab fī maʿrifat aʿyān ʿulmāʾ al-madhhab* (Cairo, A.H. 1356 [A.D. 1937/1938]); G. Sarton, *Introduction to the History of Science*, II (Baltimore, 1931), 650; M. Steinschneider, "Arabische Lapidarein," in *Zeitschrift der Deutschen morgenländischen Gesellschaft*, **49** (1895), 254 ff.; and M. Ullmann, *Die Medizin im Islam* (Leiden, 1970), 196 f.

M. PLESSNER
F. KLEIN-FRANKE

TIKHOV, GAVRIIL ADRIANOVICH (*b.* Smolevichi, near Minsk, Russia, 1 May 1875; *d.* Alma-Ata, Kazakhstan S.S.R., 25 January 1960), *astrophysics, astrobotany.*

Tikhov was the son of a railway stationmaster. In 1893 he graduated from the Gymnasium in Simferopol with a gold medal and entered the mathematical section of Moscow University. He graduated in 1897 with a first-degree diploma. From 1898 to 1900 he studied in Paris at the Sorbonne and worked as a probationer in Jules Janssen's astrophysics observatory in Meudon. At Janssen's suggestion Tikhov, on 15 November 1899, observed a meteor shower from a balloon. Twice he ascended Mont Blanc to work at Janssen's mountain observatory.

After returning to Russia, Tikhov taught mathematics at a Moscow Gymnasium and then at the Ekaterinoslav Higher Mining School. He also studied meteors and variable stars. In 1906 he was named extra-staff adjunct-astronomer at the Pulkovo observatory. Using the so-called Bredikhin short-focus astrograph, Tikhov began work on photographic astrocolorimetry, using two parts of the spectrum, separated by special color filters. He sensitized films to the necessary portions of the spectrum and prepared gelatin filters colored with various dyes. In 1913 he successfully defended his dissertation for the degree of master of astronomy and geodesy.

From 1914 to 1917 at the Central Aeronautical Navigation Station of the Pilot Observer Military School, Tikhov worked on ways of improving visibility from airplanes and developed light filters and special photographic film for aerial photography. His first monograph on aerial photography (1917) dealt with methods for improving visual and photographic air reconnaissance. It was widely used by pilots of the Allied armies.

After the war Tikhov returned to Pulkovo, and from 1919 to 1931 he also lectured on astrophysics at the University of Leningrad. From 1919 to 1941 he directed the astrophysics laboratory of the P. F. Lesgaft Scientific Institute.

In 1927 Tikhov was elected corresponding member of the Academy of Sciences of the U.S.S.R., and in 1935 he received the degree of doctor of physical and mathematical sciences. He observed the total solar eclipse of 21 September 1941 in Central Asia and subsequently worked at a branch of the U.S.S.R. Academy of Sciences. When the Kazakh S.S.R. Academy of Sciences was organized in 1946, he became an active member. In 1947 Tikhov became head of the section of astrobotany, which he had organized. In 1949 he became professor of astronomy. He was a member of the International Astronomical Union, the American Astronomical Society, and a number of Soviet scientific societies.

By means of colorimetry Tikhov studied the dispersion of light in interstellar space, that is, the rate of diffusion of light at various wavelengths. In this study he chose Algol variable stars, on which he detected a delay in the moment of eclipse in the shorter waves as compared to the longer waves. This phenomenon is now known as the Tikhov-Nordmann effect. (Charles Nordmann discovered this effect some time later.) The Soviet astrophysicist E. R. Mustel subsequently explained it in terms of the distortion of the form of the components and the presence of tidal waves of reduced temperature.

Tikhov later developed the method of the "longitudinal spectrograph," in which the optical imperfections of a poorly achromatized objective lens were used for quickly and conveniently estimating the color of stars; the technique led to the publication of extensive catalogs of star colors.

An extended series of works related to observation of the total solar eclipses of 1927, 1936, and 1941 resulted in the establishment of two structural peculiarities of the solar corona—the "globular," or "dispersed," corona and the "radiant" corona that penetrates the streamers.

Tikhov also made photographic observations of Mars using the thirty-inch refractor with light filters at the Pulkovo observatory. From this research later came the new fields of astrobotany and astrobiology. Tikhov conceived the idea of comparing the spectrophotometric properties of natural formations on the earth (mountains, soils, sands, ice, snow, and plant life) with corresponding properties of various formations on Mars. Specialized investigations of the reflecting abilities of vegetation growing under severe conditions and at high altitudes showed the great degree of adaptability of life and suggested the possible existence of vegetation on Mars. Tikhov himself participated in fifteen expeditions to various mountain regions and the Arctic in support of these ideas.

Tikhov's work on atmospheric optics was closely related to this research. In 1912 he invented an instrument for registering astral scintillation and elaborated a method for measuring the angular diameter of stars on the basis of their scintillation. He later studied the "green flash" that appears at sunset and discovered the anomalous dispersion of light in the earth's atmosphere. He investigated ashen light of the moon, in an attempt to determine how the light of the earth as a planet would be seen from outer space. To study atmospheric optics by daylight, in a clear sky, Tikhov designed an original cyanometer and a one-dimensional colorimeter with a dark blue wedge.

BIBLIOGRAPHY

I. ORIGINAL WORKS. Tikhov's basic works (from his 165 writings) were published as *Osnovnye trudy*, 5 vols. (Alma-Ata, 1954–1960). His autobiography is *Shestdesyat let u teleskopa* ("Sixty Years at the Telescope"; Moscow, 1959). His principal works in astrobotany and astrophysics (1912–1957) have appeared in English trans., 2 vols. (New York, 1960).

II. SECONDARY LITERATURE. On Tikhov's life and work, see the bibliography of his 165 published writings in *Izvestiya Akademii nauk Kazakhskoi SSR*, no. 90, Ser. astrobot., nos. 1–2 (1950), 5–13; N. I. Kucherov, "G. A. Tikhov," in *Izvestiya Glavnoi astronomicheskoi observatorii v Pulkove*, **22**, pt. 2 (1961), 2–5; M. A. Milkhiker and M. A. Daskal, "Gavriil Adrianovich Tikhov," in *Byulleten Vsesoyuznogo astronomo-geodezicheskogo obshchestva*, no. 28 (1960), 56–59; V. V. Sharonov, "Gavriil Adrianovich Tikhov," in *Priroda* (1950), no. 8, 85–88; and N. M. Shtaude, "Osnovnye cherty nauchnogo tvorchestva G. A. Tikhova po lichnym vpechatleniam za 40 let" ("The Basic Features of G. A. Tikhov's Scientific Work From Personal Impressions Over Forty Years"), in *Izvestiya Akademii nauk Kazakhskoi SSR*, no. 90, Ser. astrobot., nos. 1–2 (1950), 19–24.

P. G. KULIKOVSKY

TILAS, DANIEL (*b.* Gammelbo, Västmanland, Sweden, 11 March 1712; *d.* Stockholm, Sweden, 17 October 1772), *geology, mining.*

Tilas' father, Olof Tilas, was an officer and landowner who was knighted in 1719. His mother, Maria Hjärne, was the daughter of the scientist Urban Hjärne. As was usual among the Swedish landed gentry at that time, Tilas received his elementary education at home and was sent to the University of Uppsala in 1723, studying there until 1726 and again from 1728 to 1732. According to his unpublished autobiography, he did not spend much time on studies and gave the university little credit for his education. This is certainly unfair, for he received a good foundation in both Latin and the natural sciences, the latter most probably from his connection with the group around Linnaeus.

Tilas' interest in mining began early, and in 1732 he became assistant (*auskultant*) at the Office of Mines, the state organization that had administrative and legal control of the flourishing Swedish mining industry. He was greatly interested in the inspection of mines and took part in the work at mines and smelting plants. He discovered the need for exact mapping of the mineral veins in the mines, in order to form rational plans, and soon extended this work to mapping the geology of the surrounding areas, to facilitate prospecting for new mines.

Tilas developed an ambitious project to map all of Sweden geologically; and from his correspondence it seems that he also planned a geological map of the world. His inspiration certainly was Linnaeus' plans for similar inventories of the fauna and flora of the world. Tilas produced an unpublished geological map of two provinces of Finland and of a large number of mines. His grand designs were not approved by some of his older colleagues, but his results and ideas made him famous. He was a founding member of the Royal Swedish Academy of Sciences (1739) and served as its president for two terms. From 1741 to 1743 and in 1745 Tilas was a member of the international commission to establish the border between Sweden and Norway. With some of the younger members he walked along the boundary, making a number of interesting geological and other observations in relatively unknown and inaccessible areas.

In 1751 a fire destroyed Tilas' house; and his collections of rocks and minerals, and all his manuscripts and maps, were lost. Loss of the latter was especially disastrous to his great plans, and it was impossible for him to repeat all the fieldwork involved. Some of his results were published in 1765, but much of his scientific work remains in the manuscripts he managed to reconstruct and in the form of unpublished reports. Tilas' great importance lay not only in his plan to make geological maps of large regions but in his activity in other fields of geology. Some of his views on the origin of fossils and erratic boulders seem rather modern; and he introduced a number of terms, including the mineral name feldspar. He was also one of the first geologists to work with oil (1740) and to describe its economic exploitation.

Tilas was married in 1741 to Hedvig Reuterholm (d. 1743) and in 1743 to Anna Catharina Åkerhielm. Through both his marriages he became connected with influential political families and was active at the royal court. Also interested in genealogy and numismatics, Tilas became chamberlain in 1766 and state herald in 1768.

BIBLIOGRAPHY

I. ORIGINAL WORKS. Tilas' works include *En bergmanns rön och forsök i mineralriket* (Stockholm, 1738); "Mineralhistoria øfver Osmundsberget uti Rättviks Socken," in *Kungl. Vetenskaps akademiens Handlingar* (1740); *Stenrikets historia* (Stockholm, 1742); and *Utkast til Sveriges mineralhistoria* (Stockholm, 1765). Complete lists of his published papers are given in the secondary literature. His unpublished diaries and official reports, preserved in Swedish archives, have been used extensively in the study of the politics and culture of his time.

II. SECONDARY LITERATURE. Much of the extensive but rather scattered literature on Tilas is based on his unpublished diaries, MSS, and reports. See E. V. Falk, "Daniel Tilas och Fredrik Gyllenborg," in *Personhistorisk tidsskrift*, **36** (1936), 19–49; G. Regnell, "On the Position of Palaeontology and Historical Geology in Sweden Before 1800," in *Arkiv för Mineralogi och Geologi*, **1**, no. 1 (1949), 1–64; and N. Zenzen, "Geologiska kartor och geologisk kartläggning i Sverige føre upprättande av Sveriges Geologiska Undersökning," in *Geologiska föreningens i Stockholm förhandlingar*, **47** (1925), 311–343; "On the First Use of the Term 'Feldspat' (Feldspar etc.) by Daniel Tilas," *ibid.*, 385–405; and "Daniel Tilas om geologien i svensk-norska gränstrakter," *ibid.*, **53** (1931), 27–46.

NILS SPJELDNAES

TILDEN, SIR WILLIAM AUGUSTUS (*b*. London, England, 15 August 1842; *d*. London, 11 December 1926), *chemistry.*

The eldest son of Augustus Tilden, William had a mixed early education resulting from the numerous moves of his family. He attended East Dereham School briefly before being apprenticed in 1857 to the pharmaceutist Alfred Allchin, who so encouraged Tilden to chemistry that he allowed him to spend the last year of his apprenticeship studying at the Royal College of Chemistry. In 1863 Tilden became a demonstrator at the Pharmaceutical Society, a post he held for nine years while studying at London University for his B.Sc. (1868) and D.Sc. (1871).

Tilden married Charlotte Pither Bush in 1869 (*d*. 1905). He joined Clifton College as science master in 1872 and there began to investigate nitroso derivatives of terpenes, especially α-pinene and limonene. In 1880 he was elected a fellow of the Royal Society and the first professor of chemistry at Mason College. In 1894 he succeeded T. E. Thorpe as professor of chemistry at the Royal College of Science of London. Tilden married his second wife, Julia Mary Ramie, in 1907. He was named emeritus professor of Imperial College, which was formed from the Royal College of Science, and created a knight in 1909. Although retirement ended his bench chemistry, he continued to write and publish books.

Familiarity with aqua regia and nitrosyl chloride led Tilden to study hydrocarbon derivatives of NOCl, first with phenol, then with α-pinene and limonene. During this time the structures of the terpenes including pinene and limonene were subjects of controversy. The ease with which terpenes formed nitrosyl chlorides not only permitted characterization of these compounds but also provided ready intermediates for further reaction. Nitroso terpene derivatives, prepared from the nitrosyl chlorides by base hydrolysis, allowed Tilden to classify terpenes into a limited number of classes bringing some order to the previous chaos. His scheme divided terpenes into three types based on their nitroso derivatives: the turpentines, of which α-pinene is the principal constituent, gave derivatives m.p. 126°C; citrenes, of which limonene is the most commonly occurring, gave derivatives m.p. 76°C; and sylvestrene, which was recognized as different from the other two. Once this division was made, the structure determination of a terpene in a particular class was quite rapid. The first class would be recognized later as containing bicyclic terpenes, and the second and third as monocyclic terpenes, each of a different structure. From his first work in 1877 with the nitroso derivatives Tilden continued throughout his career to study terpenes. One further investigation led him to study the thermal decomposition of terpenes in a red-hot iron tube. He obtained, as one of the many products in the decomposition, isoprene, which is now recognized as the building block not only of the C_{10} terpenes but also the C_{15} sesquiterpenes, the C_{20} diterpenes, and the C_{30} triterpenes. The sticky material he obtained from isoprene exposed to sunlight was a precursor of synthetic rubber, but Tilden never pursued its uses.

The problem of relating specific heat to atomic weight caught Tilden's interest late in his career; he measured specific heat as a function of temperature for a number of metals and found that it was a marked function of temperature. His work demonstrated that specific heat was determined by uncertain physical forces and not the work required to separate the atoms of the substance.

Tilden served the Institute of Chemistry as president from 1891 to 1894, the Chemical Society as treasurer from 1899 to 1903 and as president from 1903 to 1905, and was awarded the Davy Medal in 1905.

BIBLIOGRAPHY

I. ORIGINAL WORKS. No collected volume of Tilden's papers is extant. Most of his terpene papers were published in *Journal of the Chemical Society* (1865–1907) and papers on specific heat in *Philosophical Transactions of the Royal Society.*

Preparation of α-pinene nitrosyl chloride appeared in *Journal of the Chemical Society*, **28** (1875), 514; and preparation of limonene nitrosyl chloride, *ibid.*, **31** (1877), 554.

Tilden wrote several books; the most signal are *A Short History of the Progress of Scientific Chemistry* (London, 1899), *The Elements* (London, 1910), *Chemical Discovery and Invention in the Twentieth Century* (London, 1917), *A Life of Sir William Ramsay* (London, 1918), and *Famous Chemists* (London, 1921).

II. SECONDARY LITERATURE. Two excellent biographical sketches are *Journal of the Chemical Society* (1927), 3190–3202; and *Proceedings of the Royal Society*, **117A**, no. 778 (1928), i–v, with bibliography.

GERALD R. VAN HECKE

TILLET, MATHIEU (*b.* Bordeaux, France, 10 November 1714; *d.* Paris, France, 20 December 1791), *agronomy, chemistry.*

Little is known about Tillet's early life. Like many other French scientists of the period he spent his early manhood in Bordeaux. His interest in science having been stimulated by the local academy, he went to Paris to practice it, and he remained there for the rest of his life. In 1758 he entered the Paris Académie des Sciences as a botanist; but he worked mainly as an assayer and chemist. Tillet enjoyed a string of promotions within the Academy, in the activities of which he took an energetic part. He was director in 1779 and treasurer at the time of his death. He apparently left the accounts (to his successor Lavoisier) in a chaotic state. Before going to Paris, Tillet was director of the mint at Troyes; later he became a *Commissaire du Roi pour les essais et affinage du Royaume.* He was also an enthusiastic member of the Société d'Agriculture.

Tillet's chemical papers, all of which were published by the Academy, sprang directly out of the practical concerns connected with assaying and refining metals for the mint. The papers deal almost exclusively with the metallurgy of gold and silver and are devoted to improving the precision and rationalizing the operations of assay by cupellation. Tillet made no very dramatic innovations in the course of this work, but he contributed in a sober and workmanlike way to the creation of accurate chemical standards, to which the French government scientists directed so much effort in the eighteenth and nineteenth centuries.

Tillet is best remembered for his investigations of wheat smuts and bunts, one genus of which, *Tilletia*, is named for him. In a long series of experiments carried out at Bordeaux and repeated at Paris, Tillet distinguished stinking smut (bunt) from loose smut, disproved the theories that these diseases of wheat were caused by climate or insects, and showed clearly the infectious nature of the dust by planting artificially infected seed among healthy plants and by noting the spread of the disease. He did not know the exact nature of the fungus causing the disease, but he did identify the dust as containing the agent.

BIBLIOGRAPHY

I. ORIGINAL WORKS. See *Dissertation sur la ductibilité des métaux, et les moyens de l'augmenter* (Bordeaux, 1750); *Dissertation sur la cause qui corrumpt et noircit les grains de bled dans les épis; et sur les moyens de prévenir ces accidens. Qui a remporté le Prix au Jugement de l'Académie Royale des Belles-Lettres, Sciences et Arts de Bordeaux* (Bordeaux, 1755), also in German as *Das Hern Tillets, . . . Abhandlung von der Ursache . . .* (Hamburg–Leipzig, 1757), and as *Dissertation on the Cause of the Corruption and Smutting of the Kernels of Wheat in the Head, and on the Means of Preventing These Untoward Circumstances*, H. B. Humphrey, trans. (Ithaca, N.Y., 1937); *Suite des expériences relatives à la meme dissertation* (Bordeaux, 1755); *Précis des expériences qui ont été faites par ordre du Roi à Trianon, sur la cause de la corruption des bleds et sur les moyens de la prévenir; à la suite duquel est une instruction propre à guider les laboureurs dans la manière dont ils doivent préparer le grain avant de le semer* (Paris, 1756); and *Histoire d'un insecte qui dévore les grains de l'Angoumois, avec les moyens que l'on peut employer pour le détruire* (Paris, 1762), written with Duhamel du Monceau.

See also *Essai sur le rapport des poids étrangers avec le marc de France* (Paris, 1766), also in Italian (Florence, 1769); *Observations faites par ordre du Roi sur les côtes de Normandie, au sujet des effets pernicieux qui sont attribués, dans le pays de Caux, à la fumée du Varech, lorsqu'on brûle cette plante pour la réduire en soude* (1772); *Rapport fait à l'Académie royale des sciences, au sujet d'une question relative à l'arpentage* (1772), written with l'Abbé Bossut; *Mémoire sur un moyen nouveau de faire avec exactitude, et tout-à-la-fois, le depart de plusieurs essais d'or, dans un seul et même matras* (Paris, 1779); *Expériences et observations sur le poids du pain au sortir du four et sur le règlement par lequel les boulangers sont assujetis à donner aux pains qu'ils exposent en vente un poids fixe et déterminé* (Paris, 1781); *Projet d'un tarif propre à servir de règle pour établir la valeur du pain proportionnément à celle du blé et des farines, avec des observations sur la mouture économique comme base essentielle de ce tarif, et sur les avantages du commerce des farines, par préférence à celui du blé en nature* (Paris, 1784); and *Rapport fait à l'Académie royale des sciences, relativement à l'avis que le Parlement a demandé à cette Académie sur la contestation qui s'est élevée à Rochefort au sujet de la taxe du pain* (Paris, 1785), written with Le Roy and Demarest.

Other works include *Observations de la Société Royale d'Agriculture sur l'uniformité des poids et mesures* (Paris, 1790), written with Abeille; *Extrait d'une partie du Rapport de l'Académie royale des sciences, fait par MM. Tillet, Leroy, Fourcroy, et Broussonnet, sur le foyer Franklin, perfectionné par M. Desarnad, et sur celui de son invention; Mémoire sur deux machines propres à donner le rapport que les différentes mesures à grains, ou celles des liquides, ont avec le boisseau ou la pinte de Paris* (Paris, n.d.).

To the list of Tillet's papers in Poggendorff, add the following: "Lettre à M. XXX," in *Mercure de France* (1757), 179; "Observations sur la maladie du Mais ou Blé de Turquie," in *Histoire de L'Académie royale des sciences pour l'année 1760* (1766), 254–261; and "Expériences et observations sur la végétation du Blé dans chacune des matières simples dont les terres labourables sont ordinairement composées et dans différens mélanges des ces matières par lesquels on s'est rapproché de ceux qui constituent ces mèmes terres à labour," *ibid.*, 1772 . . . (1775).

II. SECONDARY LITERATURE. See Michaud, *Biographie universelle*, XLI, 545; Poggendorff, II, 1108; and J. R. Partington, *A History of Chemistry*, III, 609–610. For Tillet's work on smuts and bunts, see N. E. Stevens, "Plant Pathology in the Penultimate Century," in *Isis*, 21 (1934), 98–122; and B. Wehnelt, "Mathieu Tillet.

Tilletia," in *Nachrichten über Schädlingsbekämpfung*, 2 (1937), 41–146.

STUART PIERSON

TILLO, ALEKSEY ANDREEVICH (*b.* Kiev, Russia, 25 November 1839; *d.* St. Petersburg, Russia [now Leningrad, U.S.S.R.], 11 January 1900), *geography, cartography, geodesy.*

Tillo's father, an officer in the engineering communications corps, was of French origin. From 1849 to 1859 Tillo studied at the military schools of Kiev and St. Petersburg. In 1862 he graduated from the Mikhaylov Artillery Academy and in 1864 completed the theoretical course of the geodesy section of the General Staff Academy. He subsequently carried out practical geodetic and astronomical work for two years at the Pulkovo observatory under the supervision of Otto Struve. From 1866 to 1871 he was director of the Military Topographical Section of the Orenburg Military District. From 1871 on he held a number of military command positions and received the rank of lieutenant general in 1894. In 1892 he was elected corresponding member of the Académie des Sciences and, in 1894, of the St. Petersburg Academy of Sciences. Tillo's research was associated mainly with the Russian Geographical Society, of which he became a member in 1868. From 1884 he was a member of the council of the society, and from 1889 he was president of its section of mathematical geography.

Tillo's specialty was geodesy, and his chief works were devoted to the study of the hypsometry and orography of Russia. Leveling carried out in the summer of 1874 led Tillo to determine that the level of the Aral Sea had risen seventy-four meters above that of the Caspian Sea. Tillo initiated the major leveling projects carried out in 1875 and 1877 in Siberia; the results of the surveys and the data from his own research provided the basis for his map of the altitudes of European Russia (1884) and his map of the lengths and descents of the rivers of European Russia (1888). Tillo's important hypsometrical map of European Russia (1889) was compiled from data on the altitudes of more than 51,000 points. The map established the presence in European Russia of two broad north-south elevations, which Tillo called the Mid-Russian and the Privolga; the ridges had previously—and incorrectly—been shown on maps as the Ural-Baltic and the Ural-Carpathian.

In an improved version of his map (1896) Tillo included those parts of Germany, Austria-Hungary, and Romania that bordered on Russia. He also studied meteorological phenomena, the hydrology of rivers and lakes, and the magnetism of the earth.

BIBLIOGRAPHY

I. ORIGINAL WORKS. There is a complete bibliography of Tillo's writings in Novokshanova (see below). They include *Opisanie Aralo-Kaspyskoy nivelirovki, proizvedennoy v 1874 godu po porucheniyu Russkogo Geograficheskogo Obshchestva i Orenburgskogo ego otdela* ("A Description of the Aral-Caspian Leveling Carried Out in 1874, Commissioned by the Russian Geographical Society and Its Orenburg Section"; St. Petersburg, 1877); *Opyt svoda nivelirovok Rossyskoy imperii. Materialy dlya gipsometrii Rossii. S Atlasom prodolnykh profiley*, 4 pts. ("An Attempt at a Generalization Comparing the Levels of the Russian Empire. Materials for the Hyposometry of Russia. With an Atlas of Longitudinal Profiles"; St. Petersburg, 1881–1882); *Karta vysot Evropeyskoy Rossii* ("Map of the Altitudes of European Russia"; St. Petersburg, 1884); *Karta dliny i padenia rek Evropeyskoy Rossii* ("Map of the Lengths and Descents of the Rivers of European Russia"; St. Petersburg, 1888); *Gipsometricheskaya karta Evropeyskoy Rossii* ("Hypsometric Map of European Russia"; St. Petersburg, 1889); "Srednyaya vysota sushi i srednyaya glubina morya v severnom i yuzhnom polushariakh i zavisimost sredney vysoty materikov i sredney glubiny morey ot geograficheskoy shiroty" ("The Average Altitude of the Dry Land and the Average Depth of the Sea in the Northern and Southern Hemispheres and the Relation of the Average Altitude of the Continents and the Average Depth of the Seas to Geographical Latitude"), in *Izvestiya Imperatorskogo Russkogo geografichesko obshchestva*, **25**, no. 2 (1889), 113–134; "Raspredelenie atmosfernogo davlenia na prostranstve Rossyskoy imperii i Aziatskogo materika na osnovanii nablyudeny s 1836 po 1885 g." ("Distribution of Atmospheric Pressure in the Area of the Russian Empire and the Asian Continent Based on Observations From 1836 Through 1885"), in *Zapiski Imperatorskogo russkogo geograficheskogo obshchestva*, **21** (1890), 1–308; "Orografia Evropeyskoy Rossii na osnovanii gipsometricheskoy karty" ("The Orography of European Russian Based on a Hypsometric Map"), in *Izvestiya Imperatorskogo Russkogo geografichesko obshchestva*, **26**, pt. 1 (1890), 8–32; *Gipsometricheskaya karta Zapadnoy chasti Evropeyskoy Rossii, v svyazi s prilezhashchimi chastyami Germanii, Avstro-Vengrii i Rumynii* ("Hypsometric Map of the Western Part of European Russia Together with the Adjacent Parts of Germany, Austria-Hungary, and Romania"; St. Petersburg, 1896); *Karta basseynov vnutrennikh vodnykh putey Evropeyskoy Rossii s ukazaniem punktov meteorologicheskikh i vodomernykh nablyudeny* ("Map of the Basins of the Inland Water Routes of European Russia With an Indication of the Points of Meteorological and Water-Measuring Observations"; St. Petersburg, 1897); and *Atlas raspredelenia atmosfernykh osadkov Rossii po mesyatsam i za ves god na osnovanii dvadtsatiletnikh nablyudeny 1871–1890 gg.* ("Atlas of the Distribution of Atmospheric Precipitation in Russia by Months and the Entire Year Based on Twenty-Year Observations 1871–1890"; St. Petersburg, 1897).

II. SECONDARY LITERATURE. See L. S. Berg, "A. A. Tillo," in *Izvestiya Vsesoyuznogo geograficheskogo obshchestva*, **82**, no. 2 (1950), 113–125, published on the fiftieth anniversary of his death; and Z. K. Novokshanova, *Aleksey Andreevich Tillo* (Moscow, 1961).

I. FEDOSEYEV

TILLOCH, ALEXANDER (*b.* Glasgow, Scotland, 28 February 1759; *d.* Islington, London, England, 26 January 1825), *natural philosophy, science journalism.*

The son of John Tulloch, a tobacco merchant and magistrate, Alexander Tilloch attended the University of Glasgow, graduating in 1771. While working in his father's firm he experimented with printing, and in 1781 he rediscovered the method of printing books from plates (stereotyping) instead of from movable type. Stereoprinting had first been developed (unsuccessfully) by the Edinburgh jeweler William Ged in 1725. In 1784 Tilloch patented the process with Andrew Foulis, printer for the University of Glasgow, and together they produced several books. A man of considerable inventiveness, Tilloch filed other patents for mill drives and steam engines, and devised printing methods for the prevention of banknote forgeries.

In 1787, calling himself Tilloch rather than Tulloch, he moved to London, where, with the aid of friends, he bought an evening daily newspaper, *The Star*, which he edited until 1821. From 1809 he used this paper to expound his views on biblical prophecies. He joined the Sandemanian sect (Goswell Street Chapel) but, like Michael Faraday, he reconciled its fundamentalism with a scientific outlook. A gregarious man with pronounced antiquarian tastes, Tilloch belonged to innumerable societies, including William Allen's Askesian group.

In 1797 William Nicholson founded his *Journal of Natural Philosophy*. In June 1798, in direct competition, Tilloch published the monthly *Philosophical Magazine* in order "to diffuse Philosophical Knowledge among every class of Society, and to give the Public as early an Account as possible of every thing new or curious in the Scientific World, both at Home or on the [war-torn] Continent" (preface to the first volume). Launched in the educational spirit of the Scottish Enlightenment, the *Philosophical Magazine* in its early years catered more to artisans than to the educated scientific establishment. (Tilloch founded an explicit mechanics' journal, *Mechanic's Oracle*, in 1824.) From 1810 onward, through skillful journalism, because of the decline and takeover of Nicholson's *Journal* (in 1814), and because of the exciting competition from Thomas Thomson's *Annals of Philosophy* after 1812, Tilloch's *Philosophical Magazine* came to be of major importance for the dissemination of original scientific news.

BIBLIOGRAPHY

I. ORIGINAL WORKS. Eight of Tilloch's post-1800 papers are recorded in the Royal Society *Catalogue of Scientific Papers*, V (London, 1867–1925), 996. His biblical writings are cited in Carlyle (below). Files of the *Star* newspaper and the *Mechanic's Oracle and Artisan's Laboratory and Workshop* (July 1824–1825) are kept at the British Museum. The library of the University of Edinburgh possesses a few letters.

II. SECONDARY LITERATURE. Biographical notices with further references will be found in *Gentleman's Magazine*, 95 (1825), i, 276–281, and by E. Irving Carlyle in *Dictionary of National Biography*, LVI (1898), 391–392. See also W. I. Addison, *Matriculation Albums of the University of Glasgow* (Glasgow, 1913), 96; and J. Harrison, ed., *Printing Patents, 1617–1857* (London, 1859; repr. London, 1969), 93–95.

The foundation of the *Philosophical Magazine* is discussed by Allan and John Ferguson in a bicentenary commemoration number edited by A. Ferguson, *Natural Philosophy Through the Eighteenth Century and Allied Topics* (London, 1948); and S. Lilley, "Nicholson's Journal," in *Annals of Science*, 6 (1948–1950), 78–101.

W. H. BROCK

TILLY, JOSEPH-MARIE DE (*b.* Ypres, Belgium, 16 August 1837; *d.* Schaerbeek, Belgium, 4 August 1906), *geometry.*

Tilly, one of the most profound Belgian mathematicians, attained the rank of lieutenant general by the time of his retirement. As a second lieuten-

ant in the artillery, he was assigned in 1858 to teach a course in mathematics at the regimental school; and it was there that he studied the principles of geometry. In 1860, in *Recherches sur les éléments de géométrie*, Tilly used Anatole Lamarle's methods to criticize Euclid's fifth postulate and achieved results that Lobachevsky had published but of which he was unaware until about 1866. In *Études de mécanique abstraite* (1870), based on the negation of Euclid's postulate, Tilly worked with a Lobachevskian space. He was the first to study non-Euclidean mechanics, a subject he virtually created. His research brought him into contact with Jules Houël, the only French mathematician then interested in the new geometries. Although they never met, their correspondence (1870–1885) was a valuable stimulus to Tilly, who had been working for nine years without guidance.

In *Essai sur les principes fondamentaux de la géométrie et de la mécanique* (1878), Tilly established the Riemannian, Lobachevskian, and Euclidean geometries on the concept of the distance between two points. In his formulation these geometries were based, respectively, on one, two, and three necessary and sufficient, irreducible axioms.

Tilly also wrote on military science and on the history of mathematics in Belgium, including the centenary report on the mathematical activities of the Belgian Royal Academy from 1772 to 1872. These studies were undertaken in the midst of Tilly's demanding professional duties as director of the arsenal at Antwerp and as commandant and director of studies at the École Militaire for ten years. In *Essai de géométrie analytique générale* (1892), the synthesis and crowning achievement of his work, Tilly stressed the fundamental relationship among the ten distances between any two of a group of five points. In brief, he established that geometry is the mathematical physics of distances.

Tilly's last years were marred by his unjust dismissal as commandant of the École Militaire in December 1899 and his forced early retirement in August 1900. The actions of the minister of war were motivated by complaints that Tilly had unduly emphasized the scientific education of future officers. The inspector of studies at the École Militaire, Gérard-Mathieu Leman (later a general), had forbidden Tilly to use the notions of the infinitely small and of the differential.

In 1870 Tilly was elected corresponding member and, in 1878, full member of the science section of the Belgian Royal Academy, of which he was president in 1887.

BIBLIOGRAPHY

See two articles by P. Mansion, in *Annuaire de l'Académie royale de Belgique*, **80** (1914), 203–285, with portrait and bibliography; and in *Biographie nationale . . .*, XXV (Brussels, 1930–1932), 264–269.

J. PELSENEER

IBN AL-TILMĪDH, AMĪN AL-DAWLA ABU'L-HASAN HIBAT ALLĀH IBN ṢAʿĪD (*b.* Baghdad, *ca.* 1073; *d.* Baghdad, 11 February 1165), *medicine, pharmacy, logic, education, literature.*

Ibn al-Tilmīdh's maternal grandfather, Muʿtamad al-Mulk Abu'l-Faraj Yaḥyā ibn al-Tilmīdh, was a physician. He made great efforts to secure a good education for his grandson, who assumed the patronymic Ibn al-Tilmīdh after the grandfather's death. Upon completion of his medical education, Ibn al-Tilmīdh went to Persia, where he practiced for several years in the Khurāsān region. There, according to Ẓahīr al-Dīn al-Bayhaqī (1106–1170), he learned the Persian language. Being a Syriac Christian and vitally interested in his church, its liturgy, and its activities, he also mastered the Syriac (Aramaic) language. Yāqūt al-Ḥamawī (1179–1229) affirmed that Ibn al-Tilmīdh knew Greek as well.

Ibn al-Tilmīdh conducted a lively correspondence with dignitaries, high government officials, colleagues, friends, and members of his family. His letters, collected during his lifetime in a large volume entitled *Tawqīʿāt wa-murāsalāt*, include one of advice and admonition addressed to his son, who does not seem to have been very intelligent. He also wrote numerous short poems on general medicine, the value of learning, dietetics, mental health, friendship, clouds, hospitality, modesty, loneliness, romance, wine, fish, the balance, the astrolabe, armor, and shadows.

After his return to Baghdad, Ibn al-Tilmīdh served under several caliphs, especially al-Muqtafī (1136–1160), who appointed him court physician and chief of the ʿAḍudī hospital, one of the most important institutions of its kind. He also was commissioned by the caliph to conduct licensing examinations for doctors, and he had the largest private medical school in Baghdad in his time. His fame as a medical educator attracted students from far and near.

Ibn al-Tilmīdh enjoyed an excellent reputation not only as an educator but also as a physician. His practice brought him wealth and prosperity, and he was very generous to his students and to the poor. He amassed a large library, most of which was dispersed after his death.

Ibn al-Tilmīdh wrote fourteen books, including pharmaceutical formularies and medical commentaries, some of which were cited by later Arab physicians for more than a century after his death. He was described as a highly respected man — gentle, eloquent, and very friendly — who died at an advanced age without loss of his mental faculties or dignified manners.

BIBLIOGRAPHY

I. ORIGINAL WORKS. Ibn al-Tilmīdh's literary contributions can be classified in four categories:

1. Independent medical works such as *Aqrābādhīn* (a pharmaceutical formulary in 20 chs., compiled from several earlier compendiums); a shorter version, in 13 chs. for use in hospitals only; the (*al-Amīniyya*) *fi'l-faṣd*, on phlebotomy, in 10 chs. (Lucknow, 1890); *Quwa 'l-adwiya al-mufrada* (on the effects of simple drugs used in hospitals), arranged alphabetically, with descriptions, identifications, synonyms (in Syriac, Greek, and Persian), and therapeutic uses of each; and *Mujarrabāt* (on clinical cases that he treated and experimented upon), containing several medical recipes with descriptions of pharmacological effects. Several MSS of these works are extant in many libraries, including the British Museum; Bodleian; Forschungsbibliothek, Gotha; Egyptian National Library, Cairo; and Damascus National Library.

2. Commentaries and selections from Greek medical texts, such as Hippocratic writings and their interpretations by Galen: *Aphorisms, Prognostic,* and *Substitution of Drugs.*

3. Commentaries and abstracts of leading Arabic medical works: as Ḥunayn ibn Isḥāq's *Isagoge* (*al-Masāʾil*); al-Rāzī's *Continens* (*al-Ḥāwī*); Miskawayh's *On Wines and Waters* (*al-Ashriba*); al-Masīḥī's *Hundred Books on Medicine* (*al-Miʾa*); Ibn Sīnā's *Canon* (*al-Qānūn*); and Ibn Jazla's *Minhāj.* Ibn al-Tilmīdh also wrote a commentary entitled *Medicine of the Prophet* (*Ṭibb al-nabī*), thereby becoming the first Christian physician to write on such traditional Muslim books of medical aphorisms. Unfortunately, all these commentaries are lost, except for a few quotations and references preserved by later authors.

4. Collection of his epistles and poems, of which only fragments are still known; see Louis Cheikho, *Al-machriq*, **24** (1921), 251–258, 339–350, and *Catalogue des manuscrits des auteurs chrétiens depuis l'Islam* (Beirut, 1924), 6.

II. SECONDARY LITERATURE. According to Ẓahīr al-Dīn al-Bayhaqī, *Taʾrīkh al-ḥukamāʾ*, M. Kurd ʿAlī, ed. (Damascus, 1946), 144–146, the first to mention Ibn al-Tilmīdh was his contemporary, the historian al-ʿImād al-Isfahānī, in his *Kharīdat al-Qaṣr.* More detailed biographies are given in Yāqūt al-Ḥamawī, *Dictionary of*

Learned Men, D. S. Margoliouth, ed., VII (London, 1931), 243–247; Abu'l-Faraj Ibn al-ʿIbrī (Bar Hebraeus), *Taʾrīkh Mukhtaṣar* (Beirut, 1958), 209–210; Ibn Khallikān, *Wafayāt al-aʿyān*, II (Cairo, 1892), 191–194; Ibn al-Qifṭī, *Taʾrīkh al-ḥukamāʾ*, J. Lippert, ed. (Leipzig, 1903), 340–342; Ibn Abī Uṣaybiʿa, *ʿUyūn al-anbāʾ*, I (Cairo, 1882), 259–295; and Abū Muḥammād ʿAbd Allāh al-Yāfiʿī, *Mirʾāt al-janān*, III (Hyderabad, 1920), 344, based on the above sources.

More modern reference works, listed chronologically, are F. Wüstenfeld, *Geschichte der arabischen Aerzte und Naturforscher* (Göttingen, 1840), 97–98; Lucien Leclerc, *Histoire de la médecine arabe*, II (Paris, 1876), 24–27; George Sarton, *Introduction to the History of Science*, II (Baltimore, 1931), 234; Carl Brockelmann, *Geschichte der arabischen Literatur*, 2nd ed., I (Leiden, 1943), 642, and supp., I (Leiden, 1937), 891; S. Hamarneh, "The Climax of Medieval Arabic Professional Pharmacy," in *Bulletin of the History of Medicine*, **42** (1968), 454–461; *Origins of Pharmacy and Therapy in the Near East* (Tokyo, 1973), 56–64, 87; and *Catalogue of Arabic Manuscripts on Medicine and Pharmacy at the British Library* (Cairo, 1975), nos. Or. 8293–8294.

SAMI HAMARNEH

TIMIRYAZEV, KLIMENT ARKADIEVICH (*b*. St. Petersburg, Russia [now Leningrad, U.S.S.R.], 3 June 1843; *d*. Moscow, U.S.S.R., 28 April 1920), *plant physiology.*

Timiryazev was the youngest of the seven children of Arkady Semenovich, director of the St. Petersburg customshouse, and Adelaida Klementevna, an Englishwoman. The Timiryazevs were fairly well-to-do, but after the father's retirement their financial position changed sharply. A knowledge of foreign languages enabled Timiryazev to set himself up as a reviewer of the English press for the newspaper *Golos* and to help his family by doing literary translations.

Timiryazev was educated at home and in 1860 entered St. Petersburg University in the natural sciences section of the Faculty of Physics and Mathematics. In 1862 he was expelled for participating in student disorders. After a year he renewed his studies at the university, but only as an auditor. He continued his work as a translator and also published an article on Darwin's *Origin of Species* (1864). For work on liverworts (1865) Timiryazev received a gold medal and the degree of Candidate of Sciences. After graduating he worked under Mendeleev in Simbirsk province, studying new methods of agrotechnology. He also carried out experiments in photosynthesis. In 1868, at the

First Congress of Naturalists and Physicians, Timiryazev gave a report on air feeding and the use of artificial light. In the same year, on the recommendation of Beketov he was sent abroad "for preparation for the rank of professor." In 1868–1870 he worked in Germany with Helmholtz, G. Kirchhoff, Bunsen, and Hofmeister; in France, he worked with Berthelot, J. B. Boussingault, and Claude Bernard. On his return to Russia, Timiryazev became a teacher of botany in the Petrov Agricultural and Forestry Academy (now K. A. Timiryazev Moscow Academy of Agriculture), defended his master's dissertation on the spectral analysis of chlorophyll (1871), and became extraordinary professor at the academy. In 1875 he defended his doctoral dissertation on the assimilation of light by plants, and became ordinary professor. In 1877 Timiryazev was elected professor of anatomy and plant physiology at the University of Moscow.

Timiryazev's career was marked by encounters with the government and in 1898 he was discharged as a full-time teacher from Moscow University. He did retain the management of the botanical laboratory. Only in 1917 was he restored to the title of professor at Moscow University, but because of illness he could no longer work in the department. Timiryazev ardently embraced the ideals of the October Revolution.

Timiryazev's basic research was in plant physiology. In the 1860's he began important experimental and theoretical work on photosynthesis. In his master's and doctoral dissertations, and also in a series of later works he first used spectral analysis to study thoroughly the optical properties of chlorophyll and the dependence of photosynthesis on those properties of chlorophyll and on various rays of the solar spectrum. Timiryazev refuted the incorrect opinion that maximum photosynthesis occurs in the yellow-green rays and showed that the process proceeds most intensively in the red part of the spectrum. He substantiated the applicability of the first law of photochemistry and the law of the conservation of energy to the process of photosynthesis.

Timiryazev developed the concept of chlorophyll not only as a physical but also as a chemical sensitizer immediately affected by the oxidation-reduction transformation in the course of photosynthesis. In 1890 he was one of the first to assert the existence of the second absorption band of chlorophyll and, consequently, of the second absorption band of photosynthesis. He studied the relation between photosynthesis and the intensity of illumination (1889) and gave the present generally known graphic expression of the light saturation of photosynthesis, approximately at half of full insolation. Timiryazev stated the basic results of his thirty-five-year research on photosynthesis in the Croonian lecture "The Cosmical Function of the Green Plant," read to the Royal Society of London in 1903. From 1874 to 1903 he prepared a number of works for publication in the collection *Sun, Life, and Chlorophyll* (1923). The theoretical basis he provided for the energetics of photosynthesis is still valid, even though the quantum theory of light has replaced the wave theory.

Timiryazev's success in his research in photosynthesis is explained to a large degree by the fact that he gave much attention to the working out of new methods and the study of physiological processes in plants, for which he himself constructed a number of instruments and devices. He formulated theoretical ideas on the watering and mineral feeding of plants. Timiryazev also felt that it was necessary to apply Darwinian principles, especially natural selection, to explain physiological processes in plants. He himself undertook an experiment to give such an explanation in relation to photosynthesis and the green color of chlorophyll. The fight to introduce the achievements of plant physiology into agriculture took up a large part of Timiryazev's time.

Timiryazev was instrumental in the defense and development of Darwinism, especially in Russia. When he was only a student, he became acquainted with *The Origin of Species* and was able to see in its evolutionary theory the basis of a general theory of the organic world and to understand its materialistic basis. His books "A Short Sketch of the Theory of Darwin" (1865) and "Charles Darwin and His Theory" went through fifteen editions from 1883 to 1941. The ideas of Darwin occupied a central place in the posthumous "Historical Method in Biology" (1922).

BIBLIOGRAPHY

I. Original Works. Timiryazev's writings were published as *Sochinenia*, 10 vols. (Moscow, 1937–1940); see I, 475–495 for a bibliography of his works on plant physiology, Darwinism, and general questions of natural science. His selected works appeared as *Izbrannye sochinenia*, 4 vols. (Moscow, 1948–1949).

His works include "Über die relative Bedeutung von Lichtstrahlen verschiedener Brechbarkeit bei der Koh-

lensäurezersetzung in Pflanzen," in *Botanische Zeitung*, **27** (1869), 169–175; *Spektralny analiz khlorofilla* ("The Spectral Analysis of Chlorophyll"; St. Petersburg, 1871); "Sur l'action de la lumière dans la décomposition de l'acide carbonique par la granule de chlorophylle," in *Atti del Congresso botanico internazionale* (Florence, 1876), 108–114; "Sur le rapport entre l'intensité des radiations solaires et la décomposition de l'acide carbonique par les végétaux," in *Comptes rendus . . . de l'Académie des sciences*, **109** (1889), 379–382; and "The Cosmical Function of the Green Plant," in *Proceedings of the Royal Society*, **72** (1903), 424–461, his Croonian lecture.

II. Secondary Literature. On Timiryazev and his work, see (listed chronologically) V. L. Komarov, N. A. Maksimov, and B. G. Kuznetsov, *Kliment Arkadievich Timiryazev* (Moscow, 1945), with bibliography of secondary literature, 198–211; G. V. Platonov, *Mirovozzrenie K. A. Timiryazeva* ("Timiryazev's World View"; Moscow, 1952), 468–478; A. I. Korchagin, *Kliment Arkadievich Timiryazev. Zhizn i tvorchestvo* (". . . Life and Activity"), 3rd ed. (Moscow, 1957); and E. M. Senchenkova, *K. A. Timiryazev i uchenie o fotosinteze* (". . . and the Theory of Photosynthesis"; Moscow, 1961).

E. M. Senchenkova

TINSEAU D'AMONDANS, CHARLES DE (*b.* Besançon, France, 19 April 1748; *d.* Montpellier, France, 21 March 1822), *mathematics*.

The sixth of the seven children of Marie-Nicolas de Tinseau, *seigneur* of Gennes, and Jeanne Petramand of Velay, Tinseau belonged to the nobility of the Franche-Comté. Admitted to the École Royale du Génie at Mézières in 1769, he graduated as a military engineer at the end of 1771 and until 1791 was an officer in the engineering corps. Gaspard Monge, his professor of mathematics at Mézières, awakened in Tinseau an interest in mathematical research; and in 1772 he presented two memoirs to the Académie des Sciences, one on infinitesimal geometry and the other on astronomy. The following year he was named Bossut's correspondent at the Academy; but after that he seems to have written only one paper, on infinitesimal geometry. Nevertheless, the few items that have survived from his correspondence with Monge before the Revolution attest to a continuing interest in mathematical research.

A participant in the efforts made by the nobility in 1788 to defend the *ancien régime*, Tinseau joined the émigrés gathered at Worms under the leadership of the prince of Condé in 1791.

From then on, he lived in various émigré communities, conducting a very active propaganda campaign against the Revolution and later against the Empire. He attempted to organize uprisings in France, and encouraged and aided the Allied powers in their fight against the French armies. Tinseau also fought in several campaigns and, according to his biographer, provided all the coalitions formed until 1813 with strategic plans of the French army. The intransigence of his anti-Revolution convictions is evident in the dozen political pamphlets that he published between 1792 and 1805 at Worms and London.

Devoted to the Bourbons, whom he considered the sole legitimate dynasty, Tinseau refused an offer of amnesty from Napoleon and rejected offers of naturalization extended by the British government. With the rank of brigadier general in the engineering corps, Tinseau acted as aide de camp to the future Charles X and did not return to France until 1816, at which time he immediately went into retirement.

Tinseau married three times. His first marriage took place in France before the Revolution, the other two in England during his exile. His four children from the first marriage all died without issue. From his third marriage he had a son who died in Africa and a daughter who married the engineer and mathematician François Vallès (1805–1887).

Two of the three memoirs that constitute Tinseau's *oeuvre* deal with topics in the theory of surfaces and curves of double curvature: planes tangent to a surface, contact curves of circumscribed cones or cylinders, various surfaces attached to a space curve, the determination of the osculatory plane at a point of a space curve, problems of quadrature and cubature involving ruled surfaces, the study of the properties of certain special ruled surfaces (particularly conoids), and various results in the analytic geometry of space. In these two papers the equation of the tangent plane at a point of a surface was first worked out in detail (the equation had been known since Parent), methods of descriptive geometry were used in determining the perpendicular common to two straight lines in space, and the Pythagorean theorem was generalized to space (the square of a plane area is equal to the sum of the squares of the projections of this area on mutually perpendicular planes).

Although Tinseau published very little, his papers are of great interest as additions to Monge's earliest works. Indeed, Tinseau appears to have been Monge's first disciple.

BIBLIOGRAPHY

I. ORIGINAL WORKS. Tinseau's two memoirs on infinitesimal geometry were published as "Solution de quelques problèmes relatifs à la théorie des surfaces courbes et des lignes à double courbure," in *Mémoires de mathématique et de physique présentés . . . sçavans*, **9** (1780), 593–624; and "Sur quelques propriétés des solides renfermés par des surfaces composées des lignes droites," *ibid.*, 625–642. An unpublished memoir dated 1772, "Solution de quelques questions d'astronomie," is in the archives of the Académie des Sciences.

Between 1792 and 1805 Tinseau published many violently anti-Revolution and anti-Napoleonic writings, eleven of which are cited in Michaud (see below).

II. SECONDARY LITERATURE. The only somewhat detailed article on Tinseau, by Weiss in Michaud's *Biographie universelle*, XLVI (Paris, 1826), 100–102, deals mainly with his political and military careers. Some observations on his mathematical work are in R. Taton, *L'oeuvre scientifique de Gaspard Monge* (Paris, 1951), see index; and C. B. Boyer, *History of Analytic Geometry* (New York, 1956), 207. Various documents concerning Tinseau are in his dossiers in the archives of the Académie des Sciences and at the Service Historique de l'Armée.

RENÉ TATON

TISELIUS, ARNE WILHELM KAURIN (*b*. Stockholm, Sweden, 10 August 1902; *d*. Stockholm, 29 October 1971), *physical biochemistry*.

Most of Tiselius' ancestors on both sides were scholars and many had shown great interest in science, especially biology. His father, Hans Abraham J:son Tiselius, had taken a degree in mathematics at Uppsala University. His mother, Rosa Kaurin, was the daughter of a Norwegian clergyman. His father died in 1906, and his mother moved with Tiselius and his sister to Göteborg, where her parents-in-law lived and the family had close friends.

Tiselius' profound interest in science was awakened in the grammar school at Göteborg, where he had an inspiring teacher of chemistry and biology. Gradually it became clear to him that he wanted to study at the University of Uppsala with The Svedberg, the leading physical chemist in Sweden. In September 1921 he entered that university, with which he remained associated for the rest of his life. In May 1924 he received the M.A. in chemistry, physics, and mathematics. In November 1930 he presented a doctoral dissertation on electrophoresis and was appointed docent in chemistry. A special chair in biochemistry was created for Tiselius at the Faculty of Science in 1938; he retired thirty years later.

Tiselius married Ingrid Margareta (Greta) Dalén in 1930. He was healthy for most of his life, although during the last few years he was told to reduce his activities in order not to overstrain his heart. It was difficult for him to follow this advice, and after an important meeting at Stockholm he suffered a severe heart attack and died the following morning.

Modest, quiet, and warm-hearted, Tiselius possessed a sense of humor that was both acutely witty and gentle. He was deeply interested in natural history and had a wide knowledge of botany and ornithology. He often made excursions into the countryside to watch and photograph birds.

Tiselius was awarded the 1948 Nobel Prize in chemistry "for his work on electrophoresis and adsorption analysis and especially for his discovery of the complex nature of the proteins occurring in blood serum." He received honorary doctorates from twelve universities and was a member or honorary member of more than thirty learned societies, including the National Academy of Sciences in Washington and the Royal Society.

In the summer of 1944 Tiselius became a member of a governmental committee established to recommend measures for improving conditions for scientific research, especially basic research. Most of the proposals were approved by the Swedish Parliament, and a number of improvements were introduced. When the Swedish Natural Science Research Council was established in 1946, Tiselius was appointed chairman for the first four years.

In 1947 Tiselius became member of the Nobel Committee for Chemistry and vice-president of the Nobel Foundation. At the International Congress of Chemistry held at London in 1947, he was elected vice-president in charge of the section for biological chemistry of the International Union of Pure and Applied Chemistry. Four years later, at the conference in New York, he was elected president of the union. In the 1960's he was active in the creation of the Science Advisory Council to the Swedish government, which, under the chairmanship of the prime minister, deals with Swedish research policy.

In the last decade of his life Tiselius was quite concerned about the problems created by the evolution of science; although eager for society to benefit from the advances, he was aware that scientific developments may present a severe threat to mankind. He gave much thought to the opportunity for the Nobel Foundation to use its unique position and status in a way that would complement its awarding of prizes. He took the initiative by starting Nobel symposia in each of the five prize fields. The participants not only discussed the latest developments but also attempted to assess their social, ethical, and other implications. He firmly believed that the Nobel Foundation could and should play an important role in bringing science to bear on the solution of the most pertinent problems of mankind.

Tiselius entered Svedberg's laboratory as research assistant in July 1925, at the beginning of an outstandingly fruitful intellectual period there. In September 1923 Svedberg had returned to Uppsala with many new ideas after eight months at the University of Wisconsin. He had constructed his first low-speed ultracentrifuge and was then developing the first high-speed device to be used to study the size and shape of protein particles. Svedberg also was interested in determining the electrophoretic properties of proteins, and Tiselius participated in this work. His first paper, published jointly with Svedberg (1926), described a new method for determining the mobility of proteins.

Svedberg was so heavily engaged in the development of his ultracentrifuges that he gave Tiselius free rein to continue the study of electrophoresis. Tiselius began to read biochemistry, which was not included in the chemistry curriculum at that time, and was fascinated by the enormous variability—and especially by the specificity—of biochemical substances.

In his daily work with electrophoresis, Tiselius was often worried by impure or badly defined materials. Even substances that were found to be homogeneous in the ultracentrifuge would often prove inhomogeneous in the electrophoresis experiments. This was particularly true with the serum proteins. He gradually concluded that definition and purification were all-important for the whole of biochemistry. Thus separation became the key problem, and Tiselius was convinced that a solution would require a number of methods besides electrophoresis and ultracentrifugation. He gave some thought to the further development of chromatographic and adsorption methods, and made some preliminary experiments with them. Finally he decided to continue the exploration of electrophoresis and presented his dissertation in November 1930. In the year of his retirement (1968) he described his feelings after the dissertation as follows:

I remember very vividly that I felt disappointed. The method was an improvement, no doubt, but it led me just to the point where I could see indications of very interesting results without being able to prove anything definite. I can still remember this as an almost physical suffering when looking at some of the electrophoresis photographs, especially of serum proteins. I decided to take up an entirely different problem, but a scar was left in my mind which some years later would prove to be significant.

After finishing his dissertation, Tiselius worked on new problems that gave him experience in other fields. He considered this important because he hoped to qualify for a chair in chemistry.

Through his reading Tiselius had learned about the unique capacity of certain zeolite minerals to exchange their water of crystallization for other substances, the crystal structure remaining intact even after the water of crystallization had been removed *in vacuo*. It was known that the optical properties changed when the dried crystals were rehydrated, but until then no quantitative study of the phenomenon had been made. Tiselius saw the possibilities of this accidental observation, found the governing factors, and developed a very elegant and accurate optical method for the quantitative measurement of the diffusion of water vapor and other gases into zeolite crystals. The later portion of the work was carried out at the Frick Chemical Laboratory at Princeton University in 1934–1935, while Tiselius held a Rockefeller Foundation fellowship for study under Hugh S. Taylor.

Even if Tiselius could not concentrate on biochemistry during his stay in the United States, it proved to be a very stimulating year that decisively influenced his career. The atmosphere in the Frick Laboratory was inspiring, but of even greater importance for Tiselius' later work was his frequent contact with research carried on by the Rockefeller Institute in its laboratories in Princeton and New York. This contact led to friendships with J. H. Northrop, W. M. Stanley, and M. L. Anson, as well as the opportunity to meet K. Landsteiner, M. Heidelberger, and L. Michaelis. From discussions with them it became clear to Tiselius that to solve some of their problems they needed some new methods that had been in his mind for years but that he had been unable to realize. Encouraged by the discussions with these friends, he was again convinced that the development of new and more efficient separation processes was a key problem in biochemistry and decided to concentrate on this problem. While still in the United States, he began a total reconstruction of the electrophoresis apparatus.

After his return to Uppsala, Tiselius radically redesigned the experimental technique. The apparatus he developed could be used safely with potential gradients in the U-tube at least ten times those in any earlier electrophoresis apparatus, and made it possible to obtain a much higher resolving power for protein mixtures. The movement of the boundaries could be followed optically by August Toepler's *Schlieren* method. The U-tube could be divided into well-defined sections after the conclusion of the experiments, thus allowing samples to be taken from different parts of the tube for chemical and biological analyses.

The first experiments, carried out with horse serum, immediately demonstrated the advantage of the new instrument. The *Schlieren* pattern showed four protein bands with different mobilities. The fastest-moving band corresponded to the serum albumin boundary; and the next three bands disclosed, for the first time, the presence of at least three electrophoretically different components in serum globulin. Tiselius tentatively named them α, β, and γ globulin. Serum from other animals yielded similar patterns, but with quantitative differences. It also was shown that the antibodies (immunoglobulins) usually were found in the γ globulin or between the β and γ globulin bands.

The method was subsequently tested by Tiselius and his co-workers in all possible ways, new and better refractometric methods were introduced, and various minor changes were made. The new technique also allowed the electrophoretic isolation of the three globulin fractions, and it was thus found that their chemical properties were different. Tiselius soon suspected that each of these three main globulin components consisted of several individual proteins that by chance had similar mobilities. Later methods have verified this suspicion, and many individual proteins have been isolated from the three main globulin fractions.

Tiselius had hoped that his new electrophoresis technique might also be useful in elucidating a problem of great interest to him, the isolation and identification of the large fragments and polypeptides obtained by a mild breakdown of protein molecules. In this respect the method was a disappointment; he felt that electrophoresis was hardly specific enough for separating the multitude of substances occurring in materials of biological origin. He then became interested in adsorption methods, which had been used to some extent in organic and biochemical preparations. The separation had hitherto been studied mainly on the column. Tiselius saw the possibility of developing a new quantitative analytical method in which the separation in the eluate emerging from the column could be observed by refractometric methods similar to those used in electrophoresis. He also gave a theoretical treatment that related the retardation volume of an adsorbed substance to its adsorption coefficient and the mass of adsorbent in the column. He considered the modification of adsorption behavior arising from the presence of a second, more strongly adsorbed solute. Specific retardation volumes were determined for a number of amino acids and peptides; and it was found that the length of the carbon chain had a decisive influence, each additional CH_2 group producing a marked increase. The retardation volumes of neutral amino acids remained unaffected over a wide range of pH, while those of the acidic and basic amino acids showed a strong pH dependence.

A very important technical improvement was made by Tiselius and S. Claesson in 1942 with their introduction of interferometric methods to measure the concentration of the eluate. The object of this development was to overcome the instability arising from the very slight density differences between neighboring layers of eluate by restricting convective mixing to very small volumes. The volume of the interferometric channel was only 0.13 ml. The experimental arrangement was exceptionally well suited for the detailed study of the different types of chromatographic processes and led to important theoretical advances and to their experimental verification.

All the early experiments were carried out by frontal analysis that allowed determination of the concentration of the components in a mixture but did not result in their separation. The latter could be done by using an elution method. The eluted components, however, showed a very marked "tailing." Tiselius showed in 1943 that this could be prevented by adding to the eluting solution a substance with higher adsorption affinity than any of the components in the mixture. The method has since been called displacement analysis.

In the following decade Tiselius and his co-workers made several modifications and improvements in this technique. In most of the work activated charcoal had been used as an adsorbent, and many attempts were made to modify its adsorptive properties by various pretreatments. Tiselius (1954) also tried to use calcium phosphate in the hydroxyl-apatite form as an adsorbent for proteins in conjunction with phosphate buffers as eluting agents, with some degree of success, but the definitive solution to the problem for protein chromatog-

raphy came with the development of the cellulose ion exchangers by E. A. Peterson and H. A. Sober (1956). Tiselius' decisive contribution to chromatography lay in the elucidation of the fundamental processes involved.

Until the mid-1940's Tiselius did a large part of the experimental work by himself, sometimes with the assistance of a technician. After that time great demands were made on him, and he could spend little time in his laboratory. Studies on paper electrophoresis and zone electrophoresis were continued by his co-workers and students under his direction, and much work on other separation problems was delegated to his collaborators. Two important new separation methods originated in Tiselius' laboratory. The dramatic separation of particles and of macromolecules, obtained by P. Å. Albertsson, used partition in aqueous polymer two-phase systems of, for instance, dextran and polyethylene glycol. In the gel-filtration method, devised by J. Porath and Per Flodin, fractionation is obtained according to size and shape of the dissolved molecules.

It was characteristic of Tiselius that he took up well-recognized qualitative experimental phenomena, analyzed them critically, and established their fundamental theoretical basis. As a consequence he was able to introduce essential improvements in experimental technique. His contributions to the development of new methods for analysis and separation of biological systems mark an era in the study of macromolecules and have contributed to the enormous development in biochemistry since the end of the 1930's.

BIBLIOGRAPHY

I. ORIGINAL WORKS. Only a few of Tiselius' 161 published papers can be mentioned here. A complete bibliography is in the biography by Kekwick and Pedersen (see below). They include "A New Method for Determination of the Mobility of Proteins," in *Journal of the American Chemical Society*, **48** (1926), 2272–2278, written with T. Svedberg; "Über die Berechnung thermodynamischer Eigenschaften von kolloiden Lösungen aus Messungen mit der Ultrazentrifuge," in *Zeitschrift für physikalische Chemie*, **124** (1926), 449–463; the revision and enlargement of Svedberg's *Colloid Chemistry* (New York, 1928); "The Moving Boundary Method of Studying the Electrophoresis of Proteins," in *Nova acta Regiae Societatis scientiarum upsaliensis*, 4th ser., **7**, no. 4 (1930), 1–107; "Adsorption and Diffusion in Zeolite Crystals," in *Journal of Physical Chemistry*, **40** (1936), 223–232; "A New Apparatus for Electrophoretic Analysis of Colloidal Mixtures," in *Transactions* *of the Faraday Society*, **33** (1937), 524–531, originally sent for publication to a biochemical journal but refused as being "too physical"; "Electrophoresis of Serumglobulin II. Electrophoretic Analysis of Normal and Immune Sera," in *Biochemical Journal*, **31** (1937), 1464–1477; "A New Method of Adsorption Analysis and Some of Its Applications," in *Advances in Colloid Science*, **1** (1942), 81–98; "Adsorption Analysis by Interferometric Observation," in *Arkiv för kemi, mineralogi och geologi*, **15B**, no. 18 (1942), 1–6, written with S. Claesson; "Displacement Development in Adsorption Analysis," *ibid.*, **16A**, no. 18 (1943), 1–11; "Electrophoresis and Adsorption Analysis as Aids in Investigations of Large Molecular Weight Substances and Their Breakdown Products," in *Les Prix Nobel en 1948*, 102–121, in *Nobel Lectures in Chemistry 1942–1962* (Amsterdam, 1964), 195–215; "Chromatography of Proteins on Calcium-Phosphate Columns," in *Arkiv för kemi*, **7** (1954), 443–449; "Separation and Fractionation of Macromolecules and Particles," in *Science*, **141** (1963), 13–20, written with J. Porath and P. Å. Albertsson; and the autobiographical essay "Reflections From Both Sides of the Counter," in *Annual Review of Biochemistry*, **37** (1968), 1–24.

II. SECONDARY LITERATURE. Tiselius' collaborator S. Hjertén published a biography of his former teacher, "Arne Tiselius 1902–1971," in *Journal of Chromatography*, **65** (1972), 345–348; a bibliography of Tiselius' papers is in R. A. Kekwick and K. O. Pedersen, "Arne Tiselius 1902–1971," in *Biographical Memoirs of Fellows of the Royal Society*, **20** (1974), 401–428. An earlier publication dealing with Tiselius' scientific life is K. O. Pedersen, "Arne Tiselius," in *Acta chemica scandinavica*, **2** (1948), 620–624. The important new approach to the chromatography of proteins is given in E. A. Peterson and H. A. Sober, "Chromatography of Proteins. I. Cellulose Ion-Exchange Adsorbents," in *Journal of the American Chemical Society*, **78** (1956), 751–755. In honor of Tiselius' sixtieth birthday a number of his friends published "Perspectives in the Biochemistry of Large Molecules," which is *Archives of Biochemistry and Biophysics*, supp. 1 (1962).

KAI O. PEDERSEN

TISSERAND, FRANÇOIS FÉLIX (*b.* Nuits-St.-Georges, Côte-d'Or, France, 15 January 1845; *d.* Paris, France, 20 October 1896), *celestial mechanics, astronomy*.

Tisserand was the younger son in a poor Burgundian family. His father, a cooper, died when Tisserand was very young; and his mother brought up the two children. Tisserand was a brilliant student at schools in Beaune and Dijon and later at the École Normale Supérieure, which he entered

at age eighteen. Immediately after his graduation as *agrégé des sciences* in 1866, Le Verrier offered him the post of *astronome-adjoint* at the Paris observatory.

Le Verrier questioned the value of the lunar theory developed by Delaunay, his personal enemy. He entrusted the examination of the theory to Tisserand, who presented Delaunay's results concisely and demonstrated them in a new way, on the basis of Jacobi's principles of analytical mechanics. Tisserand also generalized the results to such an extent that Poincaré wrote: "[Tisserand] has grasped their true significance better, perhaps, than the author" (*Bulletin astronomique*, **13** [1896], 431).

This work constituted Tisserand's doctoral dissertation, which he defended in 1868. He then devoted some time to astronomy, participating in the mission sent to Malacca to observe the solar eclipse of 1868. At the Paris observatory he worked in the Service Méridien, the Service Géodésique, and the Service des Équatoriaux.

In 1873 Tisserand was named director of the Toulouse observatory and professor of astronomy at that city's university. During the next five years, besides reequipping the observatory, he did research on the theoretical and practical determination of the orbits of the asteroids and satellites, and on the perturbations in their orbits. He also contributed to potential theory. Having quickly established a reputation, he was elected a corresponding member of the Académie des Sciences in 1874, although he had not published a major work since his dissertation.

Tisserand was hired by the Paris Faculty of Sciences in 1878 to teach rational mechanics and, from 1883, celestial mechanics. Henceforth he devoted his research primarily to the latter subject, which he examined thoroughly, regularly publishing five or six notes or papers on it every year. In 1885 Tisserand obtained an important result on the three-body problem. Newcomb and, previously, Delaunay (for the case of the moon) had shown that the solutions can be expressed with the aid of purely trigonometric expansions; Tisserand gave a general method that, by means of a contact transformation, allows one actually to compute the expansions. In 1889 he established the relationship known as "Tisserand's criterion," which is applied to the two orbits described by an asteroid or comet before and after it passes close to a planet. The relationship is a function of the orbital elements that is not affected by the perturbation experienced by the body. This criterion is widely used to establish the identity, or lack of identity, of two objects observed at different times and following distinct orbits.

Tisserand's greatest work is his *Traité de mécanique céleste*, the publication of which, begun in 1889, was completed a few months before his death. The four volumes represent an up-to-date version of Laplace's *Mécanique céleste*. In them Tisserand sets forth the general theory of perturbations and the works of Le Verrier on the theory of the planets, and discusses the theories of the moon, the theory of the satellites, the computation of the perturbations of the asteroids, potential theory, and the theory of the shapes of the celestial bodies and of their rotational movements. In addition, he reviews the most significant recent studies on these subjects, up through those of Poincaré. Laplace integrated the works of his predecessors into his text and therefore often was credited with results that were not his own. Tisserand, on the other hand, presents each author's memoir, simplifies its exposition, and integrates into it the fruit of his own research, without always making clear what part of the work is his. Thirty papers are incorporated into the *Traité* in this fashion. Since the authors of modern works on celestial mechanics often derive more of their information from this treatise than from the original papers, Tisserand's contributions are now diffused without being credited to him. A very modest person, Tisserand no doubt would have approved of this situation.

Tisserand was appointed director of the Paris observatory in 1892. In this capacity he pursued the project begun by Admiral Ernest Mouchez of producing the *Catalogue photographique de la carte du ciel*, and in 1896 he presided over the Congrès International de la Carte du Ciel. He succeeded Le Verrier as member of the Académie des Sciences in 1878 and was elected a member of the Bureau des Longitudes in 1879.

Tisserand suffered a fatal stroke in 1896, and was survived by three daughters. He had married while at Toulouse and, having become a widower soon after the birth of his first daughter, remarried in 1885. His collaborators, all of whom were also his friends, described Tisserand as honest, kind, and very solicitous toward young astronomers. His lectures were said to be exceptionally clear. This gift for clarity also is evident in his writings, which are all the easier to read by virtue of the economy and elegance of their demonstrations.

BIBLIOGRAPHY

I. ORIGINAL WORKS. Tisserand's writings include "Extension d'après les principes de Jacobi de la méthode suivie par Delaunay dans sa théorie du mouvement de la lune," in *Journal de mathématiques pures et appliquées*, **13** (1868), 255–303; *Sur la théorie des perturbations planétaires* (Toulouse, 1875), repr. in *Annales scientifiques de l'École normale supérieure*, **7** (1878), 261–274; *Déplacement séculaire du plan de l'orbite du 8ème satellite de Saturne; mouvement des apsides des satellites de Saturne, masse de l'anneau de Saturne* (Toulouse, 1877), repr. in *Annales de l'Observatoire astronomique de Toulouse*, **1** (1880), A1–A71; "Développement de la fonction perturbatrice dans le cas d'une forte inclinaison; perturbation de Pallas par Jupiter," in *Annales de l'Observatoire de Paris*, **15** (1880), C1–C52; "Parallaxe du soleil," *ibid.*, **16** (1882), D1–D48; "Mouvement séculaire des plans des orbites de trois planètes," *ibid.*, E1–E57; "Sur le problème des trois corps," *ibid.*, **18** (1885), G1–G19; *Traité de mécanique céleste*, 4 vols. (Paris, 1889–1896; I and II repr. 1960); and *Leçons sur la détermination des orbites*, J. Perchot, ed. (Paris, 1899).

Tisserand published seventy notes in the *Comptes rendus . . . de l'Académie des sciences* (**70–122** [1870–1896]), most of which were incorporated into the articles and books cited above. In *Bulletin astronomique*, **1–13** (1884–1896), which he founded and edited, he published twenty-eight short articles, the most important of which are "Théorie de la capture des comètes . . .," **6** (1889), 241–257, 289–292, which contains Tisserand's criterion; "État actuel de la théorie de la lune," **8** (1891), 481–503; "Perturbations . . . dans un milieu résistant," **10** (1893), 504–517; "Déplacement séculaire de l'équateur d'une planète et du plan de l'orbite de son satellite," **11** (1894), 337–343; and "Libération des petites planètes," **12** (1895), 488–507.

In the field of teaching and scientific popularization, Tisserand published *Recueil d'exercices sur le calcul infinitésimal* (Paris, 1876; 2nd ed., enl., 1896) and *Leçons de cosmographie* (Paris, 1895), written with H. Andoyer. The latter work contains a chapter on the history of astronomy that consists mainly of the text of three articles that Tisserand had published in *Annuaire du Bureau des longitudes*: "Perturbations et découverte de Neptune" (1885), 805–845; "Mesure des masses en astronomie" (1889), 671–723; and "Accélération séculaire de la lune" (1892), B1–B32. Tisserand wrote four other historical or popular articles for the *Annuaire du Bureau des longitudes*: "Quelques observatoires français du 18ème siècle" (1881), 736–765; "Planètes intramercurielles" (1882), 729–772 (the subject is treated in more detail here than in the *Traité de mécanique céleste*, IV, 524–528); "Petites planètes" (1891), B1–B20; and "Mouvement propre du système solaire" (1897), A1–A32.

II. SECONDARY LITERATURE. See the following, listed chronologically: "Discours prononcés aux obsèques de F. Tisserand," in *Bulletin astronomique*, **13** (1896), 417–439; O. Callandreau, "François Félix Tisserand," in *Monthly Notices of the Royal Astronomical Society*, **57** (1897), 231–233; L. Bassot, H. Poincaré, and M. Loewy, "Discours prononcés à l'inauguration de la statue de F. Tisserand à Nuits-St.-Georges . . .," in *Annuaire du Bureau des longitudes* (1900), E1–E19 (see also *Bulletin astronomique*, **16** [1899], 401); J. Bertrand, "Notice historique sur la vie et les travaux de F. Tisserand," in *Mémoires de l'Académie des sciences de l'Institut de France*, **47** (1904), 269–282; and A. Danjon and A. Léauté, "Inauguration du nouveau monument de F. Tisserand," in *Institut de France, Académie des sciences, notices et discours*, **4** (1962), 559–571.

The *Traité de mécanique céleste* was the subject of many reviews, the most detailed of which is R. Radau, in *Bulletin astronomique*, **6** (1889), 15–26; **7** (1890), 419–424; **11** (1894), 102–110; and **13** (1896), 300–306.

JACQUES R. LÉVY

TITCHMARSH, EDWARD CHARLES (*b.* Newbury, England, 1 June 1899; *d.* Oxford, England, 18 January 1963), *mathematics*.

Titchmarsh was the son of Edward Harper and Caroline Titchmarsh. In 1925 he married Kathleen Blomfield; they had three children. Titchmarsh received his mathematical training at Oxford; and, like most of his contemporaries, he did not take a doctorate. After teaching at University College, London (1923–1929) and the University of Liverpool (1929–1931), he became Savilian professor of geometry at Oxford. He held this position for the rest of his life.

All of Titchmarsh's extensive research was in various branches of analysis; and in spite of his professorial title, he even lectured exclusively on analysis. He made many significant contributions to Fourier series and integrals; to integral equations (in collaboration with G. H. Hardy); to entire functions of a complex variable; to the Riemann zeta-function; and to eigenfunctions of second-order differential equations, a subject to which he devoted the last twenty-five years of his life.

Titchmarsh wrote a Cambridge tract on the zeta-function (1930), and later expanded it into a much larger book (1951) containing practically everything that was known on the subject. His survey of Fourier integrals (1937) is a definitive account of the classical parts of the theory. His work on eigenfunctions appeared in two parts in 1946 and 1958. His text *The Theory of Functions* (1932) was his best-known book; a generation of mathematicians learned the theory of analytic functions and Lebesgue integration from it, and also learned (by observation) how to write mathematics. He also wrote *Mathematics for the General Reader* (1948).

Titchmarsh made many original contributions to analysis, but his influence was at least as great through his systematization of existing knowledge and his improvements of proofs of known results. He saw physics as a source of interesting mathematical problems; but his interest was exclusively in the mathematics, without any regard for its real applicability. The approach, so often sterile, was successful in his case, for it led him into his study of eigenfunctions, in which the importance of his results was less appreciated in Great Britain than in other countries, especially the Soviet Union.

BIBLIOGRAPHY

Titchmarsh's works are *The Zeta-Function of Riemann* (London, 1930); *The Theory of Functions* (Oxford, 1932); *Introduction to the Theory of Fourier Integrals* (Oxford, 1937); *Eigenfunction Expansions Associated With Second-Order Differential Equations*, pt. 1 (Oxford, 1946), pt. 2 (Oxford, 1958); *Mathematics for the General Reader* (London, 1948); and *The Theory of the Riemann Zeta-Function* (Oxford, 1951).

On Titchmarsh and his work, see the obituary by M. L. Cartwright, in *Journal of the London Mathematical Society*, **39** (1964), 544–565.

R. P. BOAS, JR.

TITIUS (TIETZ), JOHANN DANIEL (*b.* Konitz, Germany [now Chojnice, Poland], 2 January 1729; *d.* Wittenberg, Germany, 16 December 1796), *astronomy, physics, biology*.

Titius was the son of Barbara Dorothea Hanow, the daughter of a Lutheran minister, and Jacob Tietz, a draper and Konitz city councillor. His father died when he was young, and Titius was sent to Danzig to be brought up by his maternal uncle, the natural historian Michael Christoph Hanow, who encouraged his interest in natural science. Titius finished his studies at the Danzig grammar school, then, in 1748, entered the University of Leipzig, from which he received the master's degree four years later with a dissertation on Euler's theory of moonlight. In 1755 he became a private lecturer in the Leipzig Faculty of Philosophy; in

April 1756 he accepted an appointment as professor ordinarius for lower mathematics at the University of Wittenberg. In 1762 Titius became professor of physics and Senior of the Faculty of Philosophy at Wittenberg, while in 1768 he was appointed rector of the university. In addition to his courses in mathematics and physics, he also lectured on philosophy, natural theology, and natural law.

Titius was a versatile and industrious man who mastered the natural science of his time without making any significant original contribution to it. Although he only occasionally devoted himself to astronomy, he became famous chiefly for the law—now named for him—governing the distances between the planets and the sun, a law that he may have formulated without making observations. Titius' law was first stated in the 1766 translation of Charles Bonnet's *Contemplation de la nature*, published in Leipzig, and appeared as a note to a number of subsequent editions of this work. It states that the distances between the planets and the sun are laid down in the sequence

$$A = 4 + 2^n \cdot 3 \ (n = -\infty, 0, 1, \cdots, 4).$$

In 1772 this was confirmed by Bode, who placed a hypothetical planet between Mars and Jupiter, in the space that Titius had reserved for a satellite of Mars; it was in this spot that Ceres, the first planetoid, was discovered by Piazzi in 1801. Titius' law is accurate in accounting for the average distances between the planetoids and the sun and is also true for the planet Uranus, discovered by Herschel in 1781; it is, however, absolutely wrong for both Neptune and Pluto.

Titius' chief scientific activity was directed to physics and biology. He published a number of works on physical topics, including a set of conditions and rules for performing experiments. He was particularly concerned with thermometry; in 1765 he presented a survey of thermometry up to that date, with emphasis on the air thermometer, and also wrote a monograph on the metallic thermometer that had been constructed by Hanns Loeser in 1746–1747. In addition, Titius wrote treatises on both theoretical and experimental physics, in which he incorporated the findings of other workers (as, for example, the descriptions of experiments written by Georg Wolfgang Kraft in 1738).

Titius' biological work was influenced by that of Linnaeus. His most extensive publication on the subject, *Lehrbegriff der Naturgeschichte zum ersten Unterrichte* (Leipzig, 1777), is a systematic classification of plants, animals (based in part on the system of Jacob Theodor Klein), and minerals, as well as the elemental substances ether, light, fire, air, and water. In shorter works dealing with the classification of animals (1760) and minerals (1765), Titius attempted to emend Linnaeus' method. He also devoted two other short monographs to specific subjects, the penduline titmouse (1755) and a method for preventing the silting of the split near Danzig by planting acacias, seaweed, and broom (1768).

A number of Titius' other publications are devoted to questions of theology and philosophy as they pertain to science. He also wrote historical works, including a history of West Prussia and Wittenberg, a description of the conquest of West Prussia by Kasimir IV of Poland in 1454–1466, and, on the occasion of the building of a new bridge across the Elbe, a historical survey of earlier bridges at that spot.

Titius was further prominent as the editor of six series of periodicals chiefly concerned with natural science. These were written for the purpose of making new scientific results known to specialists and non-specialists alike and were also designed to entertain; for this reason they achieved considerable popularity. Among them, the *Allgemeines Magazin* was limited to translations of works by foreign authors, while Titius himself contributed articles to the *Neue Gesellschaftliche Erzählungen* and wrote fourteen of the thirty-two articles printed in the *Gemeinnützige Abhandlungen*, which dealt mainly with the natural history of Saxony. In his efforts to make foreign scientific writings available to the greatest number of readers, Titius was also active in promoting cheap reprints of important segments of the *Philosophical Transactions of the Royal Society*.

Although he was offered chairs in other universities, including those of Göttingen, Helmstedt, Danzig, and Kiel, Titius chose to spend forty years at Wittenberg. His son, Salomo Konstantin Titius, also taught at Wittenberg, where he held the third chair of medicine, which embraced anatomy and botany, from 1795 until his own death in 1801.

BIBLIOGRAPHY

I. ORIGINAL WORKS. The most detailed bibliography of Titius' writings is Johann Georg Meusel, *Lexikon der vom Jahre 1750–1800 verstorbenen teutschen Schriftsteller*, XIV (Leipzig, 1815), 74–81. See also Poggendorff and the extensive lists given in *Neues gelehrtes Europa*, pt. 19 (Wolfenbüttel, 1773), 630–642; J. F. Goldbeck, *Litterarische Nachrichten von Preussen*, pt.

1 (Leipzig–Dessau, 1781), 194–200; and F. C. G. Hirsching, *Historisch-litterarisches Handbuch berühmter und denkwürdiger Personen, welche in dem 18. Jahrhundert gelebt haben, fortgesetzt und herausgegeben von J. H. M. Ernesti*, XIV (Leipzig, 1810), 375–376, all of which also give brief biographical notes. An autobiographical note is in *Nachricht von den Gelehrten, welche aus der Stadt Conitz des Polnischen Preussens herstammen* (Leipzig, 1763), 69–74, which also contains Titius' own catalogue of his writings up to that time. His own list of writings up to 1773 is available in the Universitäts- und Landesbibliothek Sachsen-Anhalt, Halle.

Titius' monographs include *Luminis lunaris theoria nova, argumentis Euleri superstructa* (Leipzig, 1752); *Investigatio finium divinorum in rebus naturalibus necessaria, adversus Cartesium Princip. philos. I.28. III.2 defensa* (Leipzig, 1753); *Philosophische Gedanken von dem Wahren Begriffe der Ewigkeit—Eternity is All!* (Leipzig, 1755); *Parus minimus Polonorum remiz Bononiensium pendulinus descriptus* (Leipzig, 1775); *Feyerliches Denkmahl der Ehrfurcht und Treue, dem glorreichen Gedächtnisse Friedrich August's, Königs in Polen, und Kurfürst Friedrich Christian's gewidmet von der Teutschen Gesellschaft in Wittenberg* (Wittenberg, 1763); *Attributorum Dei, apto digestorum ordine, brevis expositio* (Leiden, 1763); *Ortus mundi necessarius a priori assertus* (Wittenberg, 1763); *Thermometri metallici ab inventione illustrissimi atque excellentissimi S.R.I. Comitis Loeseri descriptio* (Leipzig, 1765); *Die gänzliche Ergebung der Lande Preussen an Polen, mittelst des A. 1466, nach der Einnahme von Conitz, zwischen König Casimir dem IV und dem Hohmeister Ludwig von Erlichshausen geschlossenen Friedens, historisch vorgestellet* (Wittenberg, 1766); *Abhandlung über die von der naturforschenden Gesellschaft in Danzig aufgegebene Frage: Welches die dienlichsten und am wenigsten kostbaren Mittel sind, der überhandnehmenden Versandung in der Danziger Nähring vorzubeugen und dem weitern Anwachs der Sanddünen abzuhelfen* (Leipzig, 1768); *Physicae dogmaticae elementa, praelectionum causa evulgata* (Wittenberg, 1773); *Lehrbegriff der Naturgeschichte zum ersten Unterrichte* (Leipzig, 1777); *Grundsätze der theoretischen Haushaltungskunde zum Unterrichte der Anfänger und zur fernern Erklärung entworfen* (Leipzig, 1780); *Physicae experimentalis elementa praelectionum causa in lucem edita* (Leipzig, 1782); and *Nachricht von der vormaligen und der neu erbaueten Elbbrücke bey Wittenberg nebst einigen Beylagen mitgetheilet* (Leipzig, 1788).

The periodicals edited and in part written by Titius are *Allgemeines Magazin der Natur, Kunst und Wissenschaften*, 4 pts. (1753–1754); *Neue Erweiterungen der Erkenntnis und des Vergnügens*, 12 vols. (1753–1762); *Neue Gesellschaftliche Erzählungen für die Liebhaber der Naturlehre, der Haushaltungswissenschaft, der Arztneykunst und der Sitten*, 4 pts. (1758–1762); *Wittenbergisches Wochenblatt zur Aufnahme der Naturkunde,*

und des ökonomischen Gewerbes, 8 vols. (1768–1775); *Gemeinnützige Abhandlungen zur Beförderung der Erkenntniss und des Gebrauches natürlicher Dinge in Absicht auf die Wohlfahrt des Staates und des menschlichen Geschlechts überhaupt*, pt. 1 (1768); and *Nützliche Sammlung von Aufsätzen und Wahrnehmungen über die Witterungen, die Haushaltungskunde, das Gewerbe, die Naturkenntniss, Polizey und andere damit verknüpfte Wissenschaften, als die Fortsetzung des Wittenbergischen Wochenblatts*, 10 vols. (1783–1792).

II. SECONDARY LITERATURE. In addition to the biographical notes in the bibliographies cited above, see the brief notice by R. Knott, in *Allgemeine deutsche Biographie*. A portrait of Titius, engraved by S. Halle, is at the beginning of J. G. Krünitz, *Oeconomische Encyklopädie*, XLV (Berlin, 1789). See also M. M. Nieto, *The Titius-Bode Law of Planetary Distances: Its History and Theory* (Oxford, 1972).

MENSO FOLKERTS

TODHUNTER, ISAAC (*b*. Rye, Sussex, England, 23 November 1820; *d*. Cambridge, England, 1 March 1884), *mathematics*.

Todhunter was the second son of George Todhunter, a Congregational minister in Rye, and Mary Hume. Upon the death of his father, the family moved to Hastings, where his mother opened a school for girls and where Todhunter was educated in private schools. Although he is said to have been extremely backward as a child, Todhunter later made good progress under J. B. Austin, with whom he subsequently obtained employment as a schoolmaster. While teaching at schools in Peckham and in Wimbledon, he enrolled as an evening student at University College, London. In 1842 he was awarded the B.A. (obtaining a mathematical scholarship), and in 1884 he received the M.A. (with gold medal). In the same year—acting on the advice of Augustus De Morgan, professor of mathematics at University College—he entered St John's College, Cambridge, where he graduated B.A. (senior wrangler) in 1848 and was given the Smith Prize. Shortly after graduating he was awarded the Burney Prize for an essay in the field of moral science. The following year he was elected to a fellowship, and he remained at St John's College, where for fifteen years he tutored, lectured, wrote, and examined. According to the rules of the college, he resigned his fellowship upon his marriage in 1864 to Louisa Anna Maria Davies. In 1862 Todhunter was elected fellow of the Royal Society of London, and he served on the council of the society from 1871 to 1873. He was also a founding member of the London Mathematical Society.

Throughout his lifetime Todhunter gave much public service as an examiner for the University of Cambridge in moral sciences and also in the mathematical tripos; he also examined for the University of London and for the Indian Civil Service Commission. Most of his time he devoted to writing, and the formidable series of mathematical textbooks he produced established him as one of the most influential figures in mathematical education of the nineteenth century. The textbooks were full and thorough, and were written with meticulous care. Consequently they were extremely popular with schoolmasters and some titles, in particular the *Algebra* (1858) and the *Euclid* (1862), had fifteen or sixteen editions. Many boys went through school and university studying mathematics entirely from Todhunter's textbooks.

Todhunter had little sympathy for the growing spirit of reform and criticism in mathematical education as evidenced in the formation of the Association for the Improvement of Geometrical Teaching (1871). He resisted all attempts to displace Euclid's *Elements* from its central position in mathematics courses. He also defended vigorously the rigors of the examination system as the only sound basis for obtaining and maintaining high standards in mathematics teaching. In *The Conflict of Studies* . . . (1873) he discussed many matters raised by the new reform movements and defended a point of view that, even at that time, was thought conservative. The attack he made on the teaching of experimental science contains the much-quoted statement, "If he [the boy] does not believe the statements of his tutor—probably a clergyman of mature knowledge, recognized ability and blameless character—his suspicion is irrational and manifests a want of the power of appreciating evidence, a want fatal to his success in that branch of science which he is supposed to be cultivating."

Although Todhunter's textbooks continued in use for many years after his death, his reputation rests on the contribution he made to the history of mathematics. The most important works are *A History of the Progress of the Calculus of Variations During the Nineteenth Century* (1861); *A History of the Mathematical Theory of Probability From the Time of Pascal to That of Laplace* (1865); and *A History of the Mathematical Theories of Attraction and the Figure of the Earth From the Time of Newton to That of Laplace* (1873). A further work, *A History of the Theory of Elasticity*, was published posthumously (1886–

1893). In all of these works, Todhunter gave a close and carefully reasoned account of the difficulties involved and the solutions offered by each investigator. His studies and use of source material were thorough and fully documented.

In 1871 Todhunter won the Adams Prize of the Royal Society, for an essay, *Researches in the Calculus of Variations*. The subject arose out of a controversy that had been carried on in the *Philosophical Magazine* some years before, concerning the nature of discontinuity. Todhunter's thesis illuminated some special cases but was obscured by the lack of any adequate definition of continuity.

Todhunter was not an original mathematician. His textbooks were useful in mathematical education but soon became outdated; the histories are still valuable.

BIBLIOGRAPHY

I. ORIGINAL WORKS. None of Todhunter's biographers have found it worthwhile to compile a full list of his elementary textbooks, which ran into a great many editions in his lifetime and, after his death, were revised by others so that they might continue to be useful in schools. The library of St. John's College, Cambridge, contains most of these books and also a collection of journal articles. There is also a small MS collection, which includes the *Arithmetic* on which Todhunter was working immediately prior to his death.

The more important historical works of Todhunter are *A History of the Progress of the Calculus of Variations During the Nineteenth Century* (Cambridge, 1861); *A History of the Mathematical Theory of Probability From the Time of Pascal to That of Laplace* (Cambridge, 1865); *A History of the Mathematical Theories of Attraction and the Figure of the Earth From the Time of Newton to That of Laplace*, 2 vols. (London, 1873); *A History of the Theory of Elasticity and of the Strength of Materials From Galilei to the Present Time*, K. Pearson, ed., 2 vols. (Cambridge, 1886–1893). Essays on education are contained in *The Conflict of Studies and Other Essays* (London, 1873). The Adams Prize essay was printed as *Researches in the Calculus of Variations* (London, 1871). Todhunter also edited George Boole, *Treatise on Differential Equations* (London, 1865) and *William Whewell. An Account of His Writings, With Selections From His Literary and Scientific Correspondence*, 2 vols. (London, 1876).

II. SECONDARY LITERATURE. On Todhunter and his work, see J. E. B. Mayor, "In Memoriam," in *Cambridge Review*, **5** (1884), 228, 245, 260; E. J. Routh, in *Proceedings of the Royal Society*, **37** (1884), xxvii–

xxxii; and A. Macfarlane, *Lectures on Ten British Mathematicians of the Nineteenth Century* (New York, 1916), 134–146.

MARGARET E. BARON

TOEPLITZ, OTTO (*b.* Breslau, Germany [now Wrocław, Poland], 1 August 1881; *d.* Jerusalem, 19 February 1940), *mathematics.*

Toeplitz' father, Emil Toeplitz, and his grandfather, Julius Toeplitz, were both Gymnasium teachers of mathematics; and they themselves published several mathematical papers. In Breslau, Toeplitz completed the classical Gymnasium and then studied at the university, where he specialized in algebraic geometry and received his Ph.D. in 1905.

The following year Toeplitz moved to Göttingen, where he stayed until he obtained an appointment at the University of Kiel in 1913; he became professor ordinarius in 1920. In 1928 he accepted a chair at the University of Bonn, but soon after Hitler's rise to power in 1933, he was dismissed from the office by the National Socialist regime. For the next few years he was involved in organizational work for the declining Jewish community in Germany. In 1938 he moved to Jerusalem, where he was administrative adviser to the Hebrew University; he also continued to teach in a private seminar, in which he reported the results of his work with G. Köthe.

Toeplitz' chief interest was the theory of infinite linear, bilinear, and quadratic forms, and of the associated infinite matrices, as a framework for concrete problems of analysis. It appears that this interest was sparked by the influence of Hilbert's work on integral equations, which was in the process of publication when Toeplitz arrived in Göttingen; but it was also not unrelated to Toeplitz' earlier work. Thus, following Hilbert, Toeplitz transferred the classical theories on linear, bilinear, and quadratic forms in *n*-dimensional space as far as possible to the infinite-dimensional cases; and he applied the results to the theory of integral equations and to other areas of analysis, such as Fourier series and complex variable theory.

In 1927 Toeplitz published "Integralgleichungen und Gleichungen mit unendlich vielen Unbekannten," written with E. Hellinger, with whom Toeplitz had closely collaborated. Among the important notions and methods that are given in the article, one of the major concepts was that of a normal bilinear form, which is basic in operator theory.

In the 1930's Toeplitz' mathematical research was based on a more general point of view. With G. Köthe, Toeplitz aimed at the development of a general theory of infinite-dimensional coordinate spaces. By this time S. Banach had published his "Théorie des opérations linéaires," but Toeplitz, having himself contributed much to the emergence of a general theory of linear operators, was critical of the work of Banach and his associates, which he considered too abstract. On the other hand, by deemphasizing the importance of the norm in their theory of coordinate spaces, Toeplitz and Köthe helped to develop the even more general theory of locally convex spaces. As an offshoot of his general interest, Toeplitz established, quite early in his career, the "Toeplitz conditions," which are fundamental in the theory of divergent sequences.

Toeplitz was deeply interested in the history of mathematics and held that only a mathematician of stature is qualified to be a historian of mathematics. In particular, he investigated the relation between Greek mathematics and Greek philosophy. He also wrote "Die Entwicklung der Infinitesimalrechnung" (1949), which was intended as an introduction to the calculus on a historical basis; the work is an example of Toeplitz' concern for the teaching of mathematics at the high school and college level. With H. Rademacher, Toeplitz also wrote *Von Zahlen und Figuren* (1930), one of the most successful attempts to bring higher mathematics before the educated public.

Toeplitz was a typical German-Jewish intellectual, who, while retaining an interest in Jewish matters, felt himself to be a part of his country of birth.

BIBLIOGRAPHY

Toeplitz' major works are *Über Systeme von Formen, deren Funktionaldeterminante identisch verschwindet* (Breslau, 1905); "Über allgemeine lineare Mittelbildungen," in *Prace Matematyczno-fizyczne*, **22** (1911), 113–119; "Integralgleichungen und Gleichungen mit unendlich vielen Unbekannten," in *Encyklopädie der mathematischen Wissenschaften*, **2**, pt. 3 (1927), 1395–1597, written with E. Hellinger; *Von Zahlen und Figuren* (1930), written with H. Rademacher; and "Die Entwicklung der Infinitesimalrechnung," in G. Köthe, ed., *Grundlehren*, LXI (1949). See also Poggendorff, V, 1261–1262; VI, 2672; VIIA, 695.

On Toeplitz and his work, see H. Behnke and G. Köthe, "Otto Toeplitz zum Gedächtnis," in *Jahresbericht der Deutschen Mathematikervereinigung*, **66** (1963), 1–16.

ABRAHAM ROBINSON

TOLMAN, RICHARD CHACE (*b.* West Newton, Massachusetts, 4 March 1881; *d.* Pasadena, California, 5 September 1948), *physical chemistry, mathematical physics.*

Tolman came from a prosperous New England family with close ties to the business and academic world. Following in his father's footsteps, Tolman enrolled at the Massachusetts Institute of Technology after attending the public schools in West Newton. He received a bachelor of science degree in chemical engineering in 1903. He spent the following year in Germany, at the Technische Hochschule at Charlottenburg, and later at Crefeld in an industrial chemical laboratory. Upon his return to M.I.T. in 1904 as a graduate student, Tolman joined Arthur Amos Noyes's Research Laboratory of Physical Chemistry and earned his Ph.D. in 1910. Tolman taught briefly at the University of Michigan and the University of Cincinnati before going to the University of California, Berkeley (1912–1916). He became professor of physical chemistry at the University of Illinois in 1916.

In Washington, D.C., in 1918, while serving as chief of the dispersoid section of the Chemical Warfare Service, Tolman crossed paths again with Noyes, then chairman of the Committee on Nitrate Supply. Noyes was already working hard to persuade the government to continue after the war its research program on the nitrogen products used in explosives and fertilizers. His efforts led to the creation of the Fixed Nitrogen Research Laboratory in 1919, and to Tolman's appointment as associate director (1919–1920) and director (1920–1922). The laboratory flourished under Tolman's direction, and became a mecca for bright young physical chemists. In 1922 Tolman joined the faculty of the California Institute of Technology through Noyes's efforts. As professor of physical chemistry and mathematical physics, Tolman served as dean of the graduate school and was a member of the executive council for many years.

The main thrust of Tolman's work in statistical mechanics, relativistic thermodynamics, and cosmology was mathematical and theoretical. His earliest scientific research (1910) involved measuring the electromotive force produced when a centrifugal force is applied to an electrolytic solution. Tolman based the derivation of an expression for the electromotive force on kinetic arguments, in addition to the customary thermodynamic ones, and showed that both yield the same equation. Turning to metallic conductors next, Tolman, working with T. Dale Stewart at Berkeley, demonstrated the production of an electromotive force by measuring the flow of electric current when a coil of wire rotating about its axis is mechanically accelerated and then brought to a sudden halt. In 1916 they made the first laboratory determination of the mass of the electric carrier in metals.

Tolman also published a number of important papers in the field of chemical kinetics in gaseous systems, that is, the problem of accounting for the rate at which chemical reactions take place. His theoretical treatment of monomolecular thermal and photochemical reaction rates underscored the need to clarify the meaning of the loosely defined concept of the energy of activation. This done, Tolman turned to the experimental work of Farrington Daniels and his co-workers on the decomposition of nitrogen pentoxide, the best example of a first-order unimolecular reaction over a range of concentrations and at a series of temperatures, as a check on the proposed mechanisms of chemical reaction then current. In particular, he showed in 1925 that the simple radiation theory of reaction proposed by Jean Baptiste Perrin and W. C. McC. Lewis did not adequately account for known rates of reaction. The papers not only reveal Tolman's precise reasoning and great physical intuition, but also his consuming interest in the application of statistical mechanics to rates of physical-chemical change.

With Gilbert N. Lewis, Tolman published the first American exposition of the special theory of relativity in 1909. Tolman later wrote *The Theory of the Relativity of Motion* (Berkeley, 1917). This early interest in relativity theory was further stimulated by Hubble's discovery in 1929 that red shifts are proportional to distance, and led to a series of studies on the applications of the general theory to the overall structure and evolution of the universe. In his comprehensive treatise on relativistic thermodynamics, Tolman presented his theory of a universe expanding and contracting rhythmically like a beating heart, arguing that gravity has the effect of counteracting the influence of radiation, thus preventing the complete cessation of motion as predicted by the second law of thermodynamics.

During World War II, Tolman served as vice-chairman of the National Defense Research Committee, as scientific adviser to General Leslie R. Groves on the Manhattan Project, and as United States adviser to the wartime Combined Policy Committee. Afterwards, he became scientific adviser to Bernard Baruch on the United Nations Atomic Energy Commission. Honors received during his lifetime included the Medal for Merit and election to the National Academy of Sciences in 1923.

Tolman married Ruth Sherman, a psychologist, in 1924. They had no children. He willed the bulk of his estate to the California Institute of Technology.

BIBLIOGRAPHY

Tolman published four books and over 100 scientific papers, all of which are chronologically listed in the bibliography appended to the biographical introduction prepared by J. G. Kirkwood, O. R. Wulf, and P. S. Epstein, in *Biographical Memoirs. National Academy of Sciences*, **27** (1952), 139–153. In *Principles of Statistical Mechanics* (Oxford, 1938), a monograph that remains a classic in its field, Tolman refashioned statistical mechanics by using quantum rather than classical mechanics as the starting point for the science. Details about his family and childhood can be gleaned from his brother's autobiographical notes, found in B. F. Ritchie, "Edward Chace Tolman," *ibid.*, **37** (1964), 293–324. Bernard Jaffe, *Outposts of Science* (New York, 1935), 506–516, gives a vivid picture of Tolman's work in cosmology at Caltech in the 1930's. His World War II activities are thoroughly covered in Albert B. Christman's *Sailors, Scientists and Rockets*, I (Washington, D.C., 1971).

Manuscript sources include letters in the papers of Gilbert N. Lewis, now in the office of the Chemistry Department, Berkeley, and several boxes of correspondence and unpublished manuscripts in the archives of the California Institute of Technology.

JUDITH R. GOODSTEIN

TORRE, MARCANTONIO DELLA (*b*. Verona, Italy, 1481; *d*. Riva, Italy, 1511), *medicine, anatomy*.

Marcantonio della Torre received the doctorate in philosophy on 22 December 1497 and in medicine on 1 February 1501 at the University of Padua, where his father, Girolamo, was professor of medicine. Immediately appointed public instructor in medicine and later professor of the theory of medicine, Marcantonio della Torre continued to teach in Padua until 1510, when he transferred to the University of Pavia as professor of anatomy. During the following year his promising career was cut short by plague, contracted at Riva on Lake Garda, and his early death at the age of 30 was signaled by numerous humanist obituaries, including a poem by his celebrated compatriot Girolamo Fracastoro.

Little is known of della Torre's medical work, for no manuscripts or published works appear to have survived. His name lives, instead, because of his supposed collaboration with Leonardo da Vinci on a treatise on anatomy. The story of this collaboration, repeated as fact by many later writers, stems primarily from a passage in the second edition of Vasari's *Lives* (1568), added after Vasari had visited Leonardo's heir, Francesco Melzi, and had seen the manuscripts in his possession. That the two men were friends or acquaintances thus rests on a reliable source,[1] but their supposed collaboration, or the influence of Marcantonio upon Leonardo, is open to question on several counts. First of all, their association must have been brief, limited to the time between Marcantonio's move to Pavia in 1510 and his death in 1511. Leonardo was then nearly twice Marcantonio's age, and his interest in anatomy had been aroused as early as 1489, when Marcantonio was seven years old. Leonardo's anatomical dissections in Florence date from 1503, and he was writing about the anatomical text he hoped to publish long before he could have met the younger man. Indeed, about the time their encounter could have occurred, the greater part of Leonardo's anatomical work had already been done. In 1508 he was recording and organizing the results of the dissections done in Florence, and in 1510 wrote that "this winter of the year 1510 I hope to have completed all this anatomy."[2] Moreover, no change in style that might be attributable to Marcantonio's influence is observable in Leonardo's work at this time.[3]

That the two men would have been interested in each other because of their common interest in anatomy is clear. Yet Marcantonio was a classicist who supported traditional Galenism against the newer anatomy of Mondino de' Luzzi, while Leonardo was unfettered by scholasticism, and his anatomical researches sprang largely from his universal curiosity about the workings of nature. Thus the meeting and friendship of Leonardo da Vinci and Marcantonio della Torre appear highly probable, but the chronology of Leonardo's anatomical work and the disparity in their outlook, training, and temperament argue against their having actively collaborated in the preparation of a treatise on anatomy.

NOTES

1. Leonardo referred to a "Marcantonio" in the *Codex Atlanticus*, fol. 20v, 6, datable about 1508–1509; and on Windsor 19102 (C. III. 8), datable about 1510–1512, he wrote "book on water to Messer Marcho Antonio." This was a common name, however, and it is highly uncertain that della Torre was meant.
2. In the phrase "the winter of the year 1510," Leonardo may

have been following the Tuscan usage, meaning 1510–1511, for the Florentine year began March 25.

3. Kenneth Clark, *The Drawings of Leonardo da Vinci in the Collection of Her Majesty the Queen at Windsor Castle. Second Edition Revised with the Assistance of Carlo Pedretti*, I, appendix C, "The Anatomical Studies" (London, 1968–1969), xlvii.

BIBLIOGRAPHY

References to della Torre by sixteenth-century writers include Girolamo Fracastoro, "In obitu M. Antonii Turriani veronensis," in *Opera omnia* (Venice, 1584), ff. 199–200; and Giorgio Vasari, *Delle vite de' più eccellenti pittori scultori et architetti*, 2nd ed., I, pt. 3 (Florence, 1568), 7.

Other sources treating Leonardo's anatomic work and his possible collaboration with della Torre are Gerolamo Calvi, *I manoscritti di Leonardo da Vinci* (Bologna, 1925); Kenneth Clark, *The Drawings of Leonardo da Vinci in the Collection of Her Majesty the Queen at Windsor Castle*, 2nd ed., with Carlo Pedretti (London, 1968–1969); *Leonardo da Vinci. An Account of his Development as an Artist* (New York and Cambridge, 1939), 161; G. B. De Toni, "Frammenti vinciani. I. Intorno a Marco Antonio dalla Torre anatomico veronese del XVI secolo ed all' epoca del suo incontro con Leonardo da Vinci a Pavia," in *Atti del R. Istituto veneto di scienze, lettere ed arti*, 54 (Venice, 1895–1896), 190–203; Kenneth D. Keele, "Leonardo da Vinci's Influence on Renaissance Anatomy," in *Medical History*, 8 (1964), 360–370; C. D. O'Malley and J. B. de C. M. Saunders, eds., *Leonardo da Vinci on the Human Body. The Anatomical, Physiological, and Embryological Drawings. With Translations, Emendations, and a Biographical Introduction* (New York, 1952), 24–25, 31–35; Edmondo Solmi, "Leonardo da Vinci, il Duomo, il Castello, e l'Università di Pavia," in *Scritti vinciani* (Florence, 1924), 67.

Martha Teach Gnudi

TORRES QUEVEDO, LEONARDO (*b*. Santa Cruz de Iguña, Santander, Spain, 28 December 1852; *d*. Madrid, Spain, 18 December 1936), *engineering*.

Born into a family of technicians, Torres Quevedo studied civil engineering (1870–1876) and for a time drew plans for railway lines in southern Spain. Provided with independent means, he traveled throughout Europe. At the request of the mathematician José Echegaray (1832–1916) Torres Quevedo made public his inventions, which won him wide recognition and official support. The Centro de Ensayos de Aeronáutica was created for him by royal decree in 1904, as was the Laboratorio de Mecánica Aplicada (1907, 1911), forerunner of the Centro de Investigaciones Físicas "Leonardo Torres Quevedo" of the Consejo Superior de Investigaciones Científicas.

Torres Quevedo disliked writing – "for me a form of martyrdom," he called it – and thus his scientific contributions must be traced from the few reports he did write and, especially, from the patents he obtained and the machines he built. He was frequently concerned with describing machines, as in his article "Sobre un sistema de notaciones y símbolos destinados a facilitar la descripción de las máquinas" (*Revista de la Real Academia de ciencias . . . de Madrid*, 4 [1906], 429–442).

In algebraic machines, the subject of his inaugural lecture to the Royal Academy of Sciences (1901), Torres Quevedo combined mechanical and electromechanical means to construct a machine that would solve algebraic equations of any degree. The fundamental element is an endless spindle designed to add the construction of one monomial with that of another, automatically carrying out the calculation of Gauss's additive logarithms and working out the formula

$$y = \log(10^x + 1).$$

Telekino, a remote-control system employing Hertzian waves (patented 10 December 1902), with which Torres Quevedo carried out numerous experiments, was completely developed in 1906. Although the military significance of this device did not elude its inventor, he never succeeded in resolving the problem of interference that enemy forces would be able to generate.

Aeronautics was developed in 1902–1909, during which period Torres Quevedo conceived a dirigible system in which three cables, instead of rigid metallic struts, divide the vessel longitudinally into three triangular sections and give the outer covering, after it has been filled with gas under pressure, the physical characteristics required for it to be properly navigable. The *Gaceta* of 31 December 1909 authorized the granting of the patent to the French company ASTRA, and such lighter-than-air craft were satisfactorily employed in the same field as the German zeppelin. (See Espitallier, "Le dirigeable trilobé de l'ingénieur espagnol Torres Quevedo," in *La technique aéronautique*, I [Paris, 1910], 20–28.)

In 1912 Torres built a robot capable of playing the chess endgame of king and rook against king and defeating a human adversary. This device, perfected in 1920, and the Telekino must be recognized as conceptually related to the calculating machine of Charles Babbage, as Torres Quevedo acknowledged in "Ensayos sobre automática. Su definición. Extensión teórica de sus definiciones" (*Revista de la Real Academia de ciencias . . . de Madrid*, **12** [1913], 391–419). His work in this field culminated in an electromechanical calculating machine introduced 26 June 1920, the prototype of which demonstrated that calculations of any kind can be effected by purely mechanical processes. In 1913 Torres Quevedo had established that a machine could proceed by trial and error, in contrast with current belief—"at least when the rules that have to be followed in trial and error are precisely known. . . ."

In 1909 Torres built the funicular railway on Mount Ulía in San Sebastián (280 meters long) and, beginning in 1914, the cable-car line at Niagara Falls, Ontario, inaugurated 10 February 1916 (580 meters long). The method employed was to suspend the car by several cables the tensions of which were made independent of the weight of the car by counterweights borne at the ends of each cable. As a result, the breaking of one cable involved no danger, since there would be no increase in the load carried by the others.

BIBLIOGRAPHY

An interesting biography (a condensed version of a carefully annotated one to be published by the same author) is Leopoldo Rodríguez Alcalde, *Leonardo Torres Quevedo* (Madrid, 1966). The lack of Torres' writings can be overcome, up to 1914, with José A. Sánchez Pérez, "Los inventos de Torres Quevedo," in *Sociedad matemática española* (1914), 24.

J. VERNET

TORREY, JOHN (*b.* New York City, 15 August 1796; *d.* New York City, 10 March 1873), *botany.*

Torrey was the son of William Torrey, a New York merchant of New England ancestry who had fought in the American Revolution. His mother was Margaret Nichols, the daughter of a successful cabinetmaker and owner of real estate. The family lived on the eastern side of the tip of Manhattan. They attended the Presbyterian church, of which John remained a member throughout life. He attended public school in New York and for one year in Boston.

John Torrey first became interested in botany in 1810, when he befriended the scientist Amos Eaton, who was in a prison administered by young Torrey's father, William Torrey. At the time that Torrey's interest in botany first developed, the Linnaean system of classification was still in use in the United States, and the collecting activities of botanists had not as yet exhausted the novelties found even in the environs of New York City. By the time that Torrey's career ended in 1873, a natural system of classification was in use by American botanists, and the range of his own herbarium encompassed the entire North American continent.

Torrey initiated, with the first volume of his *Flora of the Northern and Middle Sections of the United States* (1824), the practice of gathering together in one work all that was known of North American flora. He led American botanists in the adoption of the natural system of classification, developed by Antoine-Laurent de Jussieu and A. P. de Candolle. He edited an American edition of John Lindley's *Introduction to the Natural System of Botany* (1831) and planned a work to be titled the *Flora of North America*, based on the new system. After his protégé Asa Gray joined him as a partner in 1836, he published several fascicles of the *Flora of North America*. He stopped publication in 1843, however, because Gray, who had accepted a position at Harvard College, was no longer in New York, and because both of them were inundated with botanical specimens as a result of western explorations.

In 1843 Torrey published *Flora of the State of New York*, in two volumes, as a part of the New York survey. This work, the most polished and finished to come from Torrey's hand alone, represents him at the height of his powers of taxonomic and nomenclatural discrimination. Between 1843 and his death in 1873 Torrey wrote no fewer than eighteen reports on the dried specimens brought back by explorers, mostly collectors with the topographical engineers, from the western United States. Torrey's herbarium, which went to Columbia College after his death, must be counted as one of his major scientific contributions; it became the foundation for the herbarium of the New York City Botanical Garden. During a period in the 1860's Torrey had in his possession the collections of the Smithsonian Institution, so that his taxonomic work is also embedded in the foundations of the United States National Herbarium.

Torrey wrote few textbooks and did not express himself on the great issues of biology surrounding the publication of Darwin's *Origin of Species*. The limitations preventing a well-rounded career in biology stem from Torrey's incomplete solution of the problem as to what constituted the professional role of a scientist in early- and mid-nineteenth-century America. His research was in botany, but his degree was in medicine (1818), and his teaching was in the fields of chemistry and mineralogy. In the 1820's he taught chemistry at West Point, the College of Physicians and Surgeons in New York, and, for brief periods, at Williams College, New York University, and before general subscription audiences. From 1830 to 1854 he was a professor at Princeton, but he taught there only during the summer term, spending the winters in New York.

In 1851 Torrey began to reorganize his whole pattern of living; he sold his house in Princeton and, a few years later, resigned his teaching posts. In 1853 he became assayer of the United States Mint in New York, which received in those years large shipments of gold from California. In 1856 he became a trustee of Columbia College and in 1860 moved to the campus, to a house that he received in return for his herbarium of 40,000 species and library of 600 volumes. Thus he continued to earn his living teaching chemistry, and to make contributions to science by spending every spare moment on his botanical studies.

Torrey's influence must include his friendships with many of the builders of the American scientific community, beginning with Amos Eaton. Torrey was close to Joseph Henry and played an important role in his appointment to Princeton in 1832. His protection and encouragement gave Asa Gray the status of professional botanist, a status that Torrey himself never achieved. Yet he remained able to work as a peer with his younger colleague to the very end, when Gray edited Torrey's report of the Wilkes expedition from northwestern North America as its author lay dying.

BIBLIOGRAPHY

The two major sources on John Torrey are Andrew Denny Rodgers III, *John Torrey: A Story of North American Botany* (Princeton; 1942); and Christine Chapman Robbins, "John Torrey (1796–1873). His Life and Times," in *Bulletin of the Torrey Botanical Club*, **95** (1968), 515–645. Both have lists of Torrey's works and extensive bibliographies. Robbins has a useful chronology. Extensive MS collections and the Torrey herbarium are at the New York Botanical Garden, Bronx, New York.

A. HUNTER DUPREE

TORRICELLI, EVANGELISTA (*b*. Faenza, Italy, 15 October 1608; *d*. Florence, Italy, 25 October 1647), *mathematics*, *physics*.

Eldest of the three children of Gaspare Torricelli and the former Caterina Angetti, Torricelli soon demonstrated unusual talents. His father, a textile artisan in modest circumstances, sent the boy to his uncle, the Camaldolese monk Jacopo (formerly Alessandro), who supervised his humanistic education. In 1625 and 1626 Torricelli attended the mathematics and philosophy courses of the Jesuit school at Faenza, showing such outstanding aptitude that his uncle was persuaded to send him to Rome for further education at the school run by Benedetto Castelli, a member of his order who was a mathematician and hydraulic engineer, and a former pupil of Galileo's. Castelli took a great liking to the youth, realized his exceptional genius, and engaged him as his secretary.

We have direct evidence on the scope and trend of Torricelli's scientific studies during his stay at Rome in the first letter (11 September 1632) of his surviving correspondence, addressed to Galileo on behalf of Castelli, who was away from Rome. In acknowledging receipt of a letter from Galileo to Castelli, Torricelli seized the opportunity to introduce himself as a mathematician by profession, well versed in the geometry of Apollonius, Archimedes, and Theodosius; he added that he had studied Ptolemy and had seen "nearly everything" by Brahe, Kepler, and Longomontanus. These studies had compelled him to accept the Copernican doctrine and to become "a Galileist by profession and sect"; he had been the first in Rome to make a careful study of Galileo's *Dialogo sopra i due massimi sistemi*, published in February of that year (1632).

After this letter there is a gap in the correspondence until 1640, and it is not known where Torricelli lived or what he did during this period. The most likely hypothesis so far advanced is that from the spring of 1630 to February 1641, he was secretary to Monsignor Giovanni Ciampoli, Galileo's friend and protector, who from 1632 was governor of various cities in the Marches and Umbria (Montalto, Norcia, San Severino, Fabriano). In 1641 Torricelli was again in Rome; he had asked Castelli and other mathematicians for their opinions of a treatise on motion that amplified the doctrine on the motion of projectiles that Galileo had expounded in the third day of the *Discorsi e dimostrazioni matematiche intorno a due nuove scienze* . . . (Leiden, 1638). Castelli considered the work excellent; told Galileo about it; and in April 1641, on his way from Rome to Venice through Pisa and Florence, after appointing Torricelli to give lectures in his absence, submitted the manuscript to Galileo, proposing that the latter should accept Torricelli as assistant in drawing up the two "days" he was thinking of adding to the *Discorsi*. Galileo agreed and invited Torricelli to join him at Arcetri.

But Castelli's delay in returning to Rome and the death of Torricelli's mother, who had moved to Rome with her other children, compelled Torricelli to postpone his arrival at Arcetri until 10 October 1641. He took up residence in Galileo's house, where Vincenzo Viviani was already living, and stayed there in close friendship with Galileo until the latter's death on 8 January 1642. While Torricelli was preparing to return to Rome, Grand Duke Ferdinando II of Tuscany, at Andrea Arrighetti's suggestion, appointed him mathematician and philosopher, the post left vacant by Galileo, with a good salary and lodging in the Medici palace.

Torricelli remained in Florence until his death; these years, the happiest of his life, were filled with the greatest scientific activity. Esteemed for his polished, brilliant, and witty conversation, he soon formed friendships with the outstanding representatives of Florentine culture; the painter Salvatore Rosa, the Hellenist Carlo Dati, and the hydraulic engineer Andrea Arrighetti. In fact, the regular meetings with these friends gave rise to the "Accademia dei Percossi," to whom Torricelli apparently divulged the comedies he was writing, which have not survived but were explicitly mentioned in the memoirs dictated on his deathbed to Lodovico Serenai (*Opere*, IV, 88).

In 1644 Torricelli's only work to be published during his lifetime appeared, the grand duke having assumed all printing costs. The volume, *Opera geometrica*, was divided into three sections: the first dealt with *De sphaera et solidis sphaeralibus libri duo*; the second contained *De motu gravium naturaliter descendentium et proiectorum* (the writing submitted to Galileo for his opinion); and the third section consisted of *De dimensione parabolae*. The work, soon known throughout Italy and Europe, had intrinsic value and, through its clear exposition, diffused the geometry of Cavalieri, whose writings were difficult to read.

The fame that Torricelli acquired as a geometer increased his correspondence with Italian scientists and with a number of French scholars (Carcavi, Mersenne, F. Du Verdus, Roberval), to whom he was introduced by F. Niceron, whom he

met while in Rome. The correspondence was the means of communicating Torricelli's greatest scientific discoveries but also the occasion for fierce arguments on priority, which were common during that century. There were particularly serious polemics with Roberval over the priority of discovery of certain properties of the cycloid, including quadrature, center of gravity, and measurement of the solid generated by its rotation round the base. In order to defend his rights, Torricelli formed the intention of publishing all his correspondence with the French mathematicians, and in 1646 he began drafting *Racconto d'alcuni problemi proposti e passati tra gli matematici di Francia et il Torricelli ne i quattro anni prossimamente passati* (*Opere*, III, 1–32). But while he was engaged in this work he died of a violent illness (probably typhoid fever) lasting only a few days. In accordance with his wish he was buried in the Church of San Lorenzo in Florence, but the location of his tomb is unknown.

Mathematical research occupied Torricelli's entire life. During his youth he had studied the classics of Greek geometry, which dealt with infinitesimal questions by the method of progressive elimination. But since the beginning of the seventeenth century the classical method had often been replaced by more intuitive processes; the first examples were given by Kepler, who in determining areas and volumes abandoned Archimedean methods in favor of more expeditious processes differing from problem to problem and hence difficult to imitate. After many years of meditation, Cavalieri, in his geometry of indivisibles (1635), drew attention to an organic process, toward which Roberval, Fermat, and Descartes had been moving almost in the same year; the coincidence shows that the time was ripe for new geometrical approaches.

The new geometry considered every plane figure as being formed by an infinity of chords intercepted within the figure by a system of parallel straight lines; every chord was then considered as a rectangle of infinitesimal thickness—the indivisible, according to the term introduced by Galileo. From the assumed or verified relations between the indivisibles it was possible to deduce the relations between the totalities through Cavalieri's principle, which may be stated as follows: Given two plane figures comprised between parallel straight lines, if all the straight lines parallel thereto determine in the two figures segments having a constant relation, then the areas of the two figures also have the same relation. The principle is easily extended to solid figures. In essence Cavalieri's geometry, the first step toward infinitesimal calculus, replaced the potential mathematical infinity and infinitesimal of the Greek geometricians with the present infinity and infinitesimal.

After overcoming his initial mistrust of the new method, Torricelli used it as a heuristic instrument for the discovery of new propositions, which he then demonstrated by the classical methods. The promiscuous use of the two methods—that of indivisibles for discovery and the Archimedean process for demonstration—is very frequent in the *Opera geometrica*. The first part of *De sphaera et solidis sphaeralibus*, compiled around 1641, studies figures arising through rotation of a regular polygon inscribed in or circumscribed about a circle around one of its axes of symmetry (already mentioned by Archimedes). Torricelli observes that if the regular polygon has equal sides, one of its axes of symmetry joins two opposite vertices or the midpoints of two opposite sides; if, on the other hand, it does not have equal sides, one of its axes of symmetry joins a vertex with the midpoint of the opposite side. On the basis of this observation he classifies such rotation solids into six kinds, studies their properties, and presents some new propositions and new metrical relations for the round bodies of elementary geometry. The second section of the volume deals with the motion of projectiles, about which more will be said later.

In the third section, apart from giving twenty demonstrations of Archimedes' theorem on squaring the parabola, but without adding anything new of importance, Torricelli shows that the area comprised between the cycloid and its base is equal to three times the area of the generating circle. As an appendix to this part of the work there is a study of the volume generated by a plane area animated by a helicoid motion round an axis of its plane, with the demonstration that it equals the volume generated by the area in a complete rotation round the same axis. Torricelli applies this elegant theorem to various problems and in particular to the surface of a screw with a square thread, which he shows to be equal to a convenient part of a paraboloid with one pitch.

As Torricelli acquired increasing familiarity with the method of indivisibles, he reached the point of surpassing the master—as Cavalieri himself said. In fact he extended the theory by using curved indivisibles, based on the following fundamental concept: In order to allow comparison of two plane figures, the first is cut by a system of curves and the second by a system of parallel straight lines; if each curved indivisible of the first is equal

to the corresponding indivisible of the second, the two figures are equal in area. The simplest example is given by comparison of a circle divided into infinitesimal concentric rings with a triangle (having the rectified circumference as base and the radius as height) divided into infinitesimal strips parallel to the base. From the equality of the rings to the corresponding strips it is concluded that the area of the circle is equal to the area of the triangle.

The principle is also extended to solid figures. Torricelli gave the most brilliant application of it in 1641 by proving a new theorem, a gem of the mathematical literature of the time. The theorem, published in *Opera geometrica*, is as follows (*Opere*, I, 191–213): take any point of an equilateral hyperbola (having the equation $xy = 1$) and take the area comprised by the unlimited section of the hyperbola of asymptote x, asymptote x, and the ordinate of the point selected. Although such area is infinite in size, the solid it generates by rotating round the asymptote, although unlimited in extent, nevertheless has a finite volume, calculated by Torricelli as π/a, where a is the abscissa of the point taken on the hyperbola.

Torricelli's proof, greatly admired by Cavalieri and imitated by Fermat, consists in supposing the solid generated by rotation to be composed of an infinite number of cylindrical surfaces of axis x, all having an equal lateral area, all placed in biunivocal correspondence with the sections of a suitable cylinder, and all equal to the surfaces of that cylinder: the principle of curved indivisibles allows the conclusion that the volume of this cylinder is equal to the volume of the solid generated by rotation of the section of the hyperbola considered. In modern terms Torricelli's process is described by saying that an integral in Cartesian coordinates is replaced by an integral in cylindrical coordinates. Still using curved indivisibles, Torricelli found, among other things, the volume of the solid limited by two plane surfaces and by any lateral surface, in particular the volume of barrels. In 1643 the results were communicated to Fermat, Descartes, and Roberval, who found them very elegant and correct.

The example of the hyperbola induced Torricelli to study more general curves, defined today by equations having the form $x^m y^n = c^n$, with m and n positive whole numbers and $m \neq n$. He discovered that their revolution round an asymptote could generate an infinitely long solid with finite volume and that, under particular conditions, the area comprised between the asymptote and the curve could also be finite. Torricelli intended to coordinate all these results, communicated by letter to various mathematicians in 1646 and 1647, in a single work entitled *De infinitis hyperbolis*, but he died before it could be completed. Only after publication of the *Opere* was it possible to reconstruct the paper from scattered notes.

The geometry of indivisibles was also applied by Torricelli to the determination of the center of gravity of figures. In a letter to Michelangelo Ricci dated 7 April 1646, he communicated the "universal theorem," still considered the most general possible even today, which allows determination of the center of gravity of any figure through the relation between two integrals. Among particular cases mention should be made of the determination of the center of gravity of a circular sector, obtained both by the classic procedure and by the method of indivisibles. Torricelli arrived at the same result, perhaps known to him, that Charles de La Faille had reached in 1632.

Torricelli also directed his attention to rectification of arcs of a curve, which Descartes in his *Géométrie* of 1637 had declared to be impossible, after having learned from Mersenne that Roberval had demonstrated the equality of length of particular arcs of a parabola and of arcs of an Archimedean spiral. Having conceived the logarithmic spiral, which he termed "geometric," he taught a procedure allowing rectification with ruler and compass of the entire section comprised between any point on the curve and the center, to which the curve tends after an infinite number of revolutions. Torricelli further demonstrated that any Archimedean spiral—or "arithmetic spiral," as he called it—can always be made equal to any particular arc of a suitable parabolic curve.

In addition to these contributions to the integral calculus, Torricelli discovered many relationships of differential calculus. Among the applications he made to the concept of derivative, drawn from the doctrine of motion (see below), mention should be made of his research on maxima and minima. He showed that if the sum $x + y$ is constant, the product $x^m y^n$ is maximum if x and y have the same relation as the exponents. He also determined the point still known as Torricelli's point on the plane of a triangle for which the sum of the distances from the vertices is minimum; the problem had been proposed by Fermat.

Torricelli made other important contributions to mathematics during his studies of mechanics. In *De motu gravium* he continued the study of the parabolic motion of projectiles, begun by Galileo, and observed that if the acceleratory force were to cease at any point of the trajectory, the projectile would move in the direction of the tangent to the trajectory. He made use of this observation, earning Galileo's congratulations, to draw the tangent at a point of the Archimedean spiral, or the cycloid, considering the curves as described by a point endowed with two simultaneous motions. In unpublished notes the question is thoroughly studied in a more general treatment. A point is considered that is endowed with two simultaneous motions, one uniform and the other varying, directed along two straight lines perpendicular to each other. After constructing the curve for distance as a function of time, Torricelli shows that the tangent at any point of the curve forms with the time axis an angle the tangent of which measures the speed of the moving object at that point. In substance this recognizes the inverse character of the operations of integration and differentiation, which form the fundamental theorem of the calculus, published in 1670 by Isaac Barrow, who among his predecessors mentioned Galileo, Cavalieri, and Torricelli. But not even Barrow understood the importance of the theorem, which was first demonstrated by Newton.

Full mastery of the new geometrical methods made Torricelli aware of the inherent dangers, so that his manuscripts contain passages against infinites. His unpublished writings, in fact, include a collection of paradoxes to which the doctrine of indivisibles leads when not applied with the necessary precautions.

In *De motu gravium* Torricelli seeks to demonstrate Galileo's principle regarding equal velocities of free fall of weights along inclined planes of equal height. He bases his demonstration on another principle, now called Torricelli's principle but known to Galileo, according to which a rigid system of a number of bodies can move spontaneously on the earth's surface only if its center of gravity descends. After applying the principle to movement through chords of a circle and parabola, Torricelli turns to the motion of projectiles and, generalizing Galileo's doctrine, considers launching at any oblique angle—whereas Galileo had considered horizontal launching only. He demonstrates in general form Galileo's incidental observation that if at any point of the trajectory a projectile is relaunched in the opposite direction at a speed equal to that which it had at such point, the projectile will follow the same trajectory in the reverse direction. The proposition is equivalent to saying that dynamic phenomena are reversible—that the time of Galileo's mechanics is ordered but without direction.

Among the many theorems of external ballistics, Torricelli shows that the parabolas corresponding to a given initial speed and to different inclinations are all tangents to the same parabola (known as the safety parabola or Torricelli's parabola, the first example of an envelope curve of a family of curves).

The treatise concludes with five numerical tables. The first four are trigonometric tables giving the values of sine 2α, sine$^2\alpha$, $\frac{1}{2}$ tan α, and sine α, respectively, for every degree between $0°$ and $90°$; with these tables, when the initial speed and angle of fire are known, all the other elements characteristic of the trajectory can be calculated. The fifth table gives the angle of inclination, when the distance to which the projectile is to be launched and the maximum range of the weapon are known. In the final analysis these are firing tables, the practical value of which is emphasized by the description of their use in Italian, easier than Latin for artillerymen to understand. Italian is also the language used for the concluding description of a new square that made it easier for gunners to calculate elevation of the weapon.

The treatise also refers to the movement of water in a paragraph so important that Ernst Mach proclaimed Torricelli the founder of hydrodynamics. Torricelli's aim was to determine the efflux velocity of a jet of liquid spurting from a small orifice in the bottom of a receptacle. Through experiment he had noted that if the liquid was made to spurt upward, the jet reached a height less than the level of the liquid in the receptacle. He supposed, therefore, that if all the resistances to motion were nil, the jet would reach the level of the liquid. From this hypothesis, equivalent to a conservation principle, he deduced the theorem that bears his name: The velocity of the jet at the point of efflux is equal to that which a single drop of the liquid would have if it could fall freely in a vacuum from the level of the top of the liquid at the orifice of efflux. Torricelli also showed that if the hole is made in a wall of the receptacle, the jet of fluid will be parabolic in form; he then ended the paragraph with interesting observations on the breaking of the fluid stream into drops and on the effects of air resistance. Torricelli's skill in hydraulics was so well known to his contemporaries that he was approached for advice on freeing the Val di Chiana from stagnant waters, and he suggested the method of reclamation by filling.

Torricelli is often credited—although the idea is sometimes attributed to the Grand Duke Ferdinando II—with having converted Galileo's primitive air thermoscope to a liquid thermometer, at first filled with water and later with spirits of wine. On the other hand, there is very good evidence of his technical ability in working telescope lenses, a skill almost certainly acquired during his stay in Florence. By the autumn of 1642 he was already capable of making lenses that were in no way mediocre, although they did not attain the excellence of those made by Francesco Fontana, at that time the most renowned Italian telescope maker. Torricelli had set out to emulate and surpass Fontana. By 1643 he was already able to obtain lenses equal to Fontana's or perhaps even better, but above all he had come to understand that what is really important for the efficiency of a lens is the perfectly spherical machining of the surface, which he carried out with refined techniques. The efficiency of Torricelli's lenses was recognized by the grand duke, who in 1644 presented Torricelli with a gold necklace bearing a medal with the motto "Virtutis praemia."

The fame of Torricelli's excellent lenses quickly became widespread and he received many requests, which he fulfilled at a good profit. He attributed the efficiency of telescopes fitted with his lenses to a machining process that was kept secret at the time but was described in certain papers passed at Torricelli's death to the grand duke, who gave them to Viviani, after which they were lost. An elaborate story has sometimes been woven round this "secret"; but from the surviving documents it seems possible to reconstruct the whole of Torricelli's "secret"—which, apart from the need to enhance the merits of his production in the grand duke's eyes, consisted mainly in very accurate machining of the surfaces, in selecting good-quality glass, and in not fastening the lenses "with pitch, or in any way with fire." But this last precaution—which, according to Torricelli, was known only to God and himself—had been recommended by Hieronymus Sirturi in his *Telescopium* as far back as 1618. In any event, one of Torricelli's telescope lenses, which is now preserved together with other relics at the Museo di Storia della Scienza, Florence, was examined in 1924 by Vasco Ronchi, using the diffraction grating. It was found to be of exquisite workmanship, so much so that one face was seen to have been machined better than the mirror taken as reference surface, and was constructed with the most advanced technique of the period.

The lectures given by Torricelli on various occasions, and collected by Tommaso Bonaventuri in the posthumous volume *Lezioni accademiche*, were by preference on subjects in physics. They include eight lectures to the Accademia della Crusca, of which he was a member (one lecture of thanks for admission to the academy, three on the force of impact, two on lightness, one on wind, and one on fame); one in praise of mathematics, given to the Studio Fiorentino; two on military architecture at the Academy of Drawing, and one of encomium for the "golden century," the fabled epoch of human perfection, delivered to the "Accademia dei Percossi."

From the point of view of physics, the lectures on the force of impact and on wind are of particular interest. In the former he said that he was reporting ideas expressed by Galileo in their informal conversations, and there is no lack of original observations. For example, the assertion that "forces and impetus" (what we call energy) lie in bodies was interpreted by Maxwell in the last paragraph of *A Treatise on Electricity and Magnetism* (1873) as meaning that the propagation of energy is a mediate and not remote action. In the lecture on wind Torricelli refuted the current theory on the formation of wind, which was held to be generated by vaporous exhalations evaporating from the damp earth; on the other hand, he advanced the modern theory that winds are produced by differences of air temperature, and hence of density, between two regions of the earth.

But Torricelli's name is linked above all to the barometric experiment named after him. The argument on vacuum or fullness goes back to the first Greek philosophical schools. In the Middle Ages, Catholic theology replaced Aristotle's doctrine that a vacuum is a contradiction in logic by the concept that nature abhors a vacuum (*horror vacui*). During the Renaissance the argument between supporters of vacuum and those of fullness flared up again. Galileo, joining the rationalist philosophers Telesio and Bruno, opposed Aristotle's arguments against the vacuum and about 1613 experimentally demonstrated the weight of air. But, like the majority of his contemporaries, he believed that an element does not have weight in itself; hence, on the basis of the ascertained weight of air, he was unable to deduce pressures within atmospheric air. To explain the phenomenon that in suction pumps the water does not rise more than eighteen *braccia* (about nine meters), as observed by the Florentine well diggers, Galileo advanced the hypothesis of a force—the "force of vacuum"—that occurred inside the pump and was capable of balancing a column of water eighteen *braccia* high.

In 1630, when Giovanni Battista Baliani asked him why a siphon that was to cross a hill about twenty-one meters high did not work, Galileo replied by reiterating his theory of the force of vacuum. Baliani retorted that in his opinion the failure of the siphon was due to the weight of the air, which by pressing on all sides supported the column of water not under pressure in the top part of the siphon, from which the air had been expelled by the water poured in to fill it. But Galileo did not accept Baliani's ideas, and in the *Discorsi* (1638) he continued to uphold the theory of the force of vacuum. After Galileo's death the discussion continued between his followers in Rome and Florence; and it is probable that the former turned to Torricelli to get his opinion on the working of suction pumps or on a similar experiment that Gasparo Berti is said to have carried out at Rome in 1640 for the purpose of showing that the water in suction pumps rose to more than eighteen *braccia*.

Torricelli, who was perhaps acquainted with Baliani's concept, proceeded to repeat Berti's or Baliani's experiment, using progressively heavier liquids such as seawater, honey, and mercury, which was mined in Tuscany. The use of mercury also allowed him to simplify the filling process by replacing Baliani's or Berti's siphon with a simple glass tube about one meter long. He planned to fill it to the rim with mercury, to close it with one finger and overturn it, and to immerse the open end in mercury in a bowl. To make such a long tube capable of withstanding the weight of mercury was not an easy task at that time (only in 1646 was Mersenne able to obtain a sufficiently strong tube from the French glassworks); Torricelli asked Viviani to make one, and hence the latter was the first to perform the experiment.

In a letter of 11 June 1644 to Michelangelo Ricci, Torricelli described the experiment and, rejecting the theory of the force of vacuum, interpreted it according to Baliani. But even before carrying out the experiment he was aware of the variations in atmospheric pressure, since in the letter he says that he "wished to make an instrument that would show the changes of air, now heavier and denser, now lighter and thinner." According to a fairly well founded hypothesis, he had acquired a knowledge of the variations in atmospheric pressure through skillful observation of the behavior of hydrostatic toys, perhaps invented by him and later called "Cartesian devils." According to Torricelli the force that supports the mercury column is not internal to the tube but external, produced by the atmosphere that weighs on the mercury in the bowl. If, instead of mercury, the tube had contained water, Torricelli predicted that the height of the column would have been greater by the proportion that the weight of mercury exceeds that of water, a result verified by Pascal in 1647. In confirmation of the hypothesis that the cause of support of the mercury is outside and not inside the tube, Torricelli describes other experiments with tubes blown into a sphere at the top, with which equal heights of the mercury column were obtained, so that the force was not due to the volume of vacuum produced and therefore was not a "force of vacuum."

In his reply to Torricelli's letter Ricci put forward three objections showing how difficult it was for contemporaries to understand the transmission of pressure in air: (1) If the bowl is closed with a lid, the air weighs on the lid and not on the mercury, which should therefore fall in the bowl; (2) The weight of the air acts in a vertical direction from top to bottom, so how can it be transmitted from bottom to top inside the tube? (3) Bodies immersed in a fluid are subject to Archimedes' thrust, so the mercury should be pushed upward by a force equivalent to an equal column of air. Torricelli replied in a letter of 28 June 1644, carefully refuting the objections as follows: (1) If the lid does not change the "degree of condensation" of the air locked between the lid itself and the mercury in the bowl, things remain as before—this is shown by the example of a wood cylinder loaded with a weight and cut crosswise by an iron plate, in which the lower part remains compressed as before; (2) Fluids gravitate downward by nature, but "push and spurt in all directions, even upward"; (3) The mercury in the tube is not immersed in air. In substance Torricelli's two letters elaborate the theory of atmospheric pressure, with a hint at what was to be Pascal's principle.

According to the writings of his contemporaries, Torricelli, after succeeding in the experiment, sought to observe the conditions of life of small animals (fish, flies, butterflies) introduced into the vacuum. The results obtained were almost nil, however, because the creatures were crushed by the weight of the mercury before reaching the top part of the tube; and attempts to ascertain whether sound is propagated in a vacuum also appear to have been unsuccessful. In testimony of his great appreciation Grand Duke Ferdinando II issued a decree praising this experiment of Torricelli's very highly.

Copies of Torricelli's two letters were circulated among Italian scientists and were sent to Mersenne, who, traveling to Italy in October 1644, passed through Florence and obtained a repetition of the experiment from Torricelli himself. On his return to France, he informed his friends of Torricelli's experiment, giving rise to flourishing experimental and theoretical activity. Discovery of the barometer, Vincenzo Antinori wrote, changed the appearance of physics just as the telescope changed that of astronomy; the circulation of the blood, that of medicine; and Volta's pile, that of molecular physics.

BIBLIOGRAPHY

I. ORIGINAL WORKS. The writings and scientific correspondence were published in *Opere di Evangelista Torricelli*, Gino Loria and Giuseppe Vassura, eds., 4 vols. in 5 pts. (I–III, Faenza, 1919; IV, 1944).

Individual works are *Opera geometrica. De sphaera et solidis sphaeralibus libri duo . . . De motu gravium naturaliter descendentium et proiectorum libri duo. De dimensione parabolae* (Florence, 1644), the first sec. repr. with its long title, *De sphaera et solidis sphaeralibus libri duo in quibus Archimedis doctrina de sphaera et cylindro denuo componitur, latius promovetur et in omni specie solidorum, quae vel circa, vel intra sphaeram, ex conversione poligonorum regularium gigni possint, universalius propagatur* (Bologna, 1692); *Lezioni accademiche*, Tommaso Bonaventuri, ed. (Florence, 1715; 2nd ed., Milan, 1813); and "Sopra la bonificazione della Valle di Chiana," in *Raccolta d'autori che trattano del moto delle acque*, IV (Florence, 1768). Other short writings were published in historical works, mentioned below.

The majority of Torricelli's MSS, after complicated vicissitudes and some losses, as recounted in the intro. to the *Opere*, are preserved at the Biblioteca Nazionale Centrale, Florence; Angiolo Procissi, in *Evangelista Torricelli nel terzo centenario della morte* (Florence, 1951), 77–109, gives an accurate catalogue raisonné. The autograph works, except for one, and the souvenirs kept at the Torricelli Museum in Faenza were destroyed in 1944.

There are two oil portraits of Torricelli in the Uffizi Gallery in Florence; another portrait, engraved by Pietro Anichini, is reproduced on the frontispiece of the *Lezioni accademiche*.

II. SECONDARY LITERATURE. All histories of mathematics or physics deal more or less fully with Torricelli's life and work. *Opere*, IV, 341–346, contains a bibliography. Some of the most significant works are Timauro Antiate (pseudonym of Carlo Dati), *Lettera ai Filaleti. Della vera storia della cicloide e della famosissima esperienza dell'argento vivo* (Florence, 1663), the first publication of the correspondence with Ricci on the barometric experiment; [Tommaso Bonaventuri], in *Lezioni accademiche*, preface, v–xlix; Angelo Fabroni,

Vitae Italorum doctrina excellentium qui saeculis XVII et XVIII floruerunt, I (Pisa, 1778), 340–399, the appendix of which contains *Racconto di alcuni problemi*; and Giovanni Targioni Tozzetti, *Notizie degli aggrandimenti delle scienze fisiche accaduti in Toscana nel corso di anni LX del secolo XVII*, 4 vols. (Florence, 1780).

See also Vincenzo Antinori, *Notizie istoriche relative all'Accademia del Cimento*, in the series Saggi di Naturali esperienze fatte nell'Accademia del Cimento (Florence, 1841), *passim*, esp. 27; Ernst Mach, *Die Mechanik in ihrer Entwickelung historisch-kritisch dargestellt*, 2nd ed. (Leipzig, 1889), 377 ff.; and Raffaello Caverni, *Storia del metodo sperimentale in Italia*, 6 vols. (Florence, 1891–1900; repr. Bologna, 1970)—vols. I, IV, V have unpublished passages from Torricelli.

After publication of the *Opere*, which contained many unpublished writings, the studies on Torricelli received a new impetus. The following works contain many other bibliographical references: Vasco Ronchi, "Sopra una lente di Evangelista Torricelli," in *l'Universo* (Florence), **5**, no. 2 (1924); Mario Gliozzi, *Origini e sviluppi dell'esperienza torricelliana* (Turin, 1931), repr. with additions in *Opere*, IV, 231–294; C. de Waard, *L'expérience barométrique, ses antécédents et ses explications* (Thouars, 1936); Guido Castelnuovo, *Le origini del calcolo infinitesimale nell'era moderna* (Bologna, 1938; 2nd ed., Milan, 1962), *passim*, esp. 52–53, 58–62; Ettore Bortolotti, "L'opera geometrica di Evangelista Torricelli," in *Monatshefte für Mathematik und Physik*, **48** (1939), repr. in *Opere*, IV, 301–337; Ettore Carruccio, *De infinitis spiralibus*, intro., rearrangement, trans., and notes by Carruccio (Pisa, 1955); Giuseppe Rossini, *Lettere e documenti riguardanti Evangelista Torricelli* (Faenza, 1956); *Convegno di studi torricelliani in occasione del 350° anniversario della nascita di Evangelista Torricelli* (Faenza, 1959); and W. E. Knowles Middleton, *The History of the Barometer* (Baltimore, 1964), ch. 2.

MARIO GLIOZZI

TOSCANELLI DAL POZZO, PAOLO (*b*. Florence, Italy, 1397; *d*. Florence, 1482), *astronomy, geography, medicine*.

Toscanelli's father, Domenico, was a physician. Information on Toscanelli's work is scanty and incomplete, since only a few fragments of his writings are extant. He must have begun his studies in medicine, mathematics, and astronomy at the University of Florence but later transferred to the more famous University of Padua, where he formed a friendship with Nicolas of Cusa. While pursuing his medical studies at Padua, Toscanelli was drawn to astrology but nevertheless achieved important results in astronomy.

On his return to Florence, the *signoria* of the city assigned Toscanelli the treatment of "judicial astrology," then much in vogue. Deemed by Cusa and Regiomontanus as the most learned living mathematician, he was introduced to Brunelleschi, then busy with the construction of the large cupola of the basilica of Santa Maria del Fiore. The great height of the lantern above the cupola gave Toscanelli the idea of placing a gnomon there, the highest ever built. Very little information is available concerning this important astronomical instrument; but the testimony of Egnatio Danti, the cosmographer of Cosimo I de'Medici, states that Toscanelli pierced an opening at the base of the lantern, through which the rays of the sun passed. The purpose was to determine with accuracy the day of the solstice and other astronomical data. The opening is ninety meters above the floor, and at high noon during the summer solstice the sun's rays fall on the marble floor of the basilica.

Stone slabs have been embedded in the floor at various times. The oldest, according to Ximenes, who in 1755 studied and reconstructed this meridian line, was the one that Toscanelli placed there in 1468.

Toscanelli also demonstrated his ability in astronomy through his observations of the comets that appeared in 1433, 1449, 1456, 1457, and 1472. (It was not until 1864 that his manuscripts in the National Library at Florence were discovered.) Although these observations were made without instruments, his methods of cartographic representation were much more accurate than those then in common use. In fact Giovanni Celoria (1842–1920), director of the astronomical observatory at Brera (Milan), was able to calculate the cometary orbits on the basis of Toscanelli's drawings. Thus he ascertained that the comet Toscanelli observed in 1456 was the one now known as Halley's comet.

Cristoforo Landino, professor of rhetoric and poetry at the University of Florence and a friend of Toscanelli, states that the latter held many conversations on geography with travelers and navigators who passed through Florence. It was probably as a result of these conversations that he decided to construct a nautical map of the Atlantic Ocean, even though knowledge of the longitudes of various places was then quite imperfect. Therefore it is not surprising that the positions of Cathay and of the island of Cippangu—that is, of China and Japan—were only vaguely known. They were placed more than one hundred degrees too far to the east, halfway between their correct locations and Lisbon, a displacement toward Europe of about ten thousand kilometers. The purpose of the map was to demonstrate that if one sails west, one can reach the Orient by a shorter route and thus circumnavigate the globe. Documents of the period indicate that the map, which was later reconstructed, was sent by Toscanelli with a letter to Fernando Martins, canon of Lisbon, whom he had met in Italy at the time of Cusa's death. In the letter he demonstrated that it was possible to reach "the most noble and large city of Quinsay" (China) by crossing the Atlantic. At the end of his life, Toscanelli apparently sent a copy of his map to Christopher Columbus, urging him to use it for exploration.

BIBLIOGRAPHY

Gustavo Uzielli, *La vita e i tempi di Paolo dal Pozzo Toscanelli* (Rome, 1893), contains an extensive bibliography and was reprinted in *Pubblicazioni del R. Osservatorio astronomico di Brera*, no. 55 (1921).

See also Carlo Errera, *L'epoca delle grandi scoperte geografiche* (Milan, 1926); G. Fumagalli, *Bibliografia delle opere concernenti Toscanelli e Amerigo Vespucci* (Florence, 1898); Hermann Wagner, "Die Rekonstruktion der Toscanelli Karte von Jahre 1474 und die Pseudo-Facsimilia des Behaim Globus v.j. 1492," in *Nachrichten der K. Gesellschaft der Wissenschaften zu Göttingen*, Phil.-hist. Kl. (1894); and his review of H. Vigaud, "La lettre et la carte de Toscanelli sur la route des Indes par l'ouest . . .," in *Göttingischen gelehrten Anzeigen* (1902), no. 2; and L. Ximenes, *Del vecchio e nuovo gnomone fiorentino* (Florence, 1757).

G. ABETTI

TOULMIN, GEORGE HOGGART (*b.* Southwark, Surrey, England, September 1754; *d.* Wolverhampton, England, July 1817), *geology.*

Toulmin was the eldest son of Robert Toulmin, a prosperous soapmaker whose forebears had lived for many years in Westmorland, in northwest England. Young Toulmin studied medicine at Edinburgh University from 1776 until 1779, when he graduated M.D. with a thesis entitled *De cynanche tonsillari.* Little is known about Toulmin's career after graduation, but it seems likely that he practiced medicine for the rest of his life; at first probably in London, and later in Wolverhampton. He published two unimportant medical works in 1789 and 1810. In the second work he states that he had lectured in London in 1795 on the subjects treated in the book; and a notice of the second book in the *Gentleman's Magazine* (London, 1810) describes him as Dr. G. H. Toulmin of Wolverhampton.

Soon after graduating Toulmin published his only geological work, *The Antiquity and Duration of the World* (London, 1780). This book, reprinted with some changes in title and content in 1783, 1785, and 1789, is chiefly remarkable for having anticipated in a very general way some of the conclusions reached by James Hutton in his *Theory of the Earth* (1788). In his book Toulmin rejected contemptuously earlier attempts to establish a chronology of the earth's history, including, by implication, the Old Testament chronology, and he accepted the Aristotelian belief in the eternity of the world. He claimed that the matter of which the earth is composed, both organic and inorganic, is in a state of constant motion, resulting from decay and erosion; and that new fossiliferous sediments are being deposited in the oceans. He recognized that mountains are destroyed by erosion and supposed that new ones would be formed by elevation. He also claimed that the operations of nature proceed in a slow and uniform manner; and that each part of the universe operates in a manner designed to secure the preservation of both the parts and the whole.

Although reprinted three times, Toulmin's book seems to have been almost completely ignored by contemporary geologists. G. F. Richardson, in his *Geology for Beginners* (London, 1842), stated that "Dr. Toulmin, although doubted and disbelieved in his own day, has expressed opinions which contain the substance of the system of Dr. Hutton, and the principles of Mr. Lyell."

While there are similarities in both the philosophy of the two authors and in the geological conclusions they reached, there is a fundamental difference between the two books. The geological conclusions reached by Hutton are to a large extent based on the extensive studies of rocks in the field which he made before publishing his *Theory*, but Toulmin adduces no evidence at all to suggest that his book was similarly based. On the contrary, he makes much use of previously published literature, notably John Whitehurst's *An Inquiry into the Original State and Formation of the Earth* (London, 1778). This, alone, may have accounted for the neglect of Toulmin's work by geologists; but an additional reason may have been his atheistical tendencies.

Toulmin and his book were forgotten until 1948, when S. I. Tomkeieff commented on the similarities of Toulmin's views to some of the conclusions reached by Hutton. In 1963 D. B. McIntyre drew attention to certain statements in the 1788 version of Hutton's *Theory* that are strikingly similar, textually, to statements to be found in Toulmin's book, and he concluded that Hutton must have read this book before writing his *Theory*. In 1967 G. L. Davies discussed fully the evidence bearing on the question whether Hutton, in compiling his *Theory*, was in any way indebted to Toulmin and concluded that there was no evidence supporting the suggestion; although he surmised that Toulmin, during his stay in Edinburgh, may have read a rough draft of Hutton's *Theory*, which could account for the textual similarities to be found in Toulmin's book. Whatever the truth may be, Toulmin's book can only be regarded as an academic exercise, rather than an original contribution to the development of geological ideas in the eighteenth century.

BIBLIOGRAPHY

I. ORIGINAL WORKS. Toulmin's published works on geology are *The Antiquity and Duration of the World* (London, 1780; repr., 1824); *The Antiquity of the World* (London, 1783); *The Eternity of the World* (London, 1785); and *The Eternity of the Universe* (London, 1789; repr., 1825, 1837). The last three works repeat, with little change, the text of the first.

His published works on medicine are *The Instruments of Medicine, or the Philosophical Digest and Practice of Physic* (London, 1789); *Elements of the Practice of Medicine on a Popular Plan . . . an Elementary Work for Students* (London, 1810).

II. SECONDARY LITERATURE. See S. I. Tomkeieff, "James Hutton and the Philosophy of Geology," in *Transactions of the Edinburgh Geological Society*, **14** (1948), 253–276, and *Proceedings of the Royal Society of Edinburgh*, **5** (1950), 387–400; D. B. McIntyre, "James Hutton and the Philosophy of Geology," in C. C. Albritton, ed., *The Fabric of Geology* (Reading, Mass.– Palo Alto–London, 1963), 1–11; and G. L. Davies, "George Hoggart Toulmin and the Huttonian Theory of the Earth," in *Bulletin of the Geological Society of America*, **78** (1967), 121–124.

V. A. EYLES

TOURNEFORT, JOSEPH PITTON DE (*b*. Aix-en-Provence, France, 3 June 1656; *d*. Paris, France, 28 November 1708), *botany, medicine.*

Tournefort, who had one brother and seven sisters, came from a family of the minor nobility. His father, Pierre Pitton, a lawyer and royal secretary, was *seigneur* of Tournefort; his mother, Aimare de Fagoue, was the daughter of a royal counselor at the chancellery of Provence. Destined at first for the Church, Tournefort received an excellent education in the classical languages and science from the Jesuits. His father's death in 1677 enabled him to prepare for his future vocation, natural history,

especially botany. Until 1683 he divided his time between herborizing and courses in chemistry, medicine, and botany (taught by Magnol) at the University of Montpellier. For several months each year he traveled through the countryside and mountains of the Alps, the Midi, the Pyrenees, and Spain—often accompanied by such botanist friends as Charles Plumier and Pierre Garidel.

In 1683, at the recommendation of Mme de Venelle, a grande dame at court and originally from Aix, Tournefort was chosen as substitute for Guy Fagon, professor at the Jardin du Roi, in Paris. His new post required him not only to teach botany but also to enrich the holdings of live plants in the garden. Each year between 1685 and 1689 he undertook long herborizing expeditions in the Midi and to Holland, England, and especially to the Iberian Peninsula, where he stayed for over a year. In 1691, at the nomination of the Abbé Bignon, he entered the Académie des Sciences; and from 1693 he was Fagon's sole substitute at the Jardin du Roi.

Although Tournefort had published only an edition of his lecture notes entitled *Schola botanica* (1689) and prepared by one of his English students, William Sherard, he was already one of Europe's most noted botanists and had dozens of correspondents, including Magnol at Montpellier and Herman at Leiden.

Between 1694 and his death in 1708 Tournefort's work was marked by two major events. The first was the publication of his *Élémens de botanique* (1694), which appeared in three volumes, two of them consisting of illustrations executed by Claude Aubriet; the work was translated into Latin as *Institutiones rei herbariae* (1700). The second was his voyage to the Levant (1700–1702). Tournefort's account of this voyage, published posthumously in 1717 and translated into several languages, still makes interesting reading. Tournefort's subsequently published writings represent only a small portion of his scientific work, however. At his death, the result of an accident, he left twelve folio volumes on botany, two of which were ready for the press: "Herborisations aux environs de Paris" and "Nomenclature des plantes observées en France, en Espagne, et en Portugal."

The study of living nature, especially of plants, was characterized in the seventeenth century principally by two major currents of research, one dealing with classification and the other with the inner structures of plants and their functions. These movements were not unrelated, and naturalists such as John Ray and Sébastien Vaillant introduced into classification certain fundamental notions from anatomy and biology. Tournefort was less open in this respect: he knew nothing of plant sexuality, refused to employ the microscope, and divided the plant into almost independent parts that he considered as separate entities. These important limitations in his work provoked widespread and often severe criticism. Yet this criticism was unfair to the extent that it failed to take into account Tournefort's overall intentions and the rigor with which he carried them out.

What botany owes to Tournefort is not the invention of the biotaxonomic genus, which had imposed itself empirically on observers since antiquity, but rather the creation of the concept of the genus in the modern sense and its first skillful application. His teacher at Montpellier, Magnol, had refined the concept of family (1689), although no doubt prematurely, for it was not taken up again for more than half a century. The classificatory unit that most attracted Tournefort was the genus—a "cluster of species," as he put it—a natural grouping having a real existence, independent of the observer, and identifiable. His goal was to "reduce each species to its true genus" and to define the genus by character or, as it would be called today, by diagnosis. Once identified, each genus was to receive a name that would evoke only the characteristic expressed in its description, and the name was to be as simple as possible—although tradition should not be completely disregarded. Above all, Tournefort carefully distinguished the act of describing from that of naming.

Tournefort's conception of genus contains a fundamental new contribution that was elaborated in the work of Linnaeus, Bernard de Jussieu, and Adanson. The distinct paths of taxonomy and nomenclature were now acknowledged. Linnaeus defined the species and, in a completely natural way, imposed the binary nomenclature; Jussieu and Adanson made knowledge of natural units at all levels the aim of their research.

In addition to genera, Tournefort suspected the existence of higher units, which he called classes, but in practice he scarcely treated them as such. Rather, like the divisions, he used them most often as quite arbitrary means of identifying species. Absorbed in the huge task of making an inventory of the genera in the material at hand, Tournefort did not have time to consider each species in detail. He was satisfied simply to rely on Gaspard Bauhin's *Pinax*, which, in his opinion, contained a perfectly satisfactory definition of species.

Tournefort's revolution was aided by his exceptionally broad botanical experience, the fame he acquired following his trip to the Levant, and his personality. His deep concern for clarity, simplicity, and rigor in his writings was the source of his attractive style and convincing argumentation. Unfortunately, it also resulted in a certain schematism and, ultimately, led to the superficiality of his method. The *Élémens* and the *Institutiones* are milestones in the history of taxonomy not only for the conceptual advances they reflect but also for the wholly new form in which they are cast. The text of the *Élémens* is in French, accompanied by a technical dictionary, and it is closely related to Aubriet's illustrations. The result is a well-integrated and easily accessible whole that could not fail to produce a sensation. Thus, although Tournefort's work disregarded the major biological discoveries of the seventeenth century, within its self-imposed limits it clearly outlined the avenues of study that led to the modern system of classification.

On the level of principles, Tournefort contributed powerfully and brilliantly to the establishment of objectivity in taxonomy and of research methods suitable for a natural method of classification. (His own system, based upon the flower and the fruit, chiefly the corolla, was, however, highly artificial.) While he was far from discovering the principle of subordination of characteristics, his method displayed a certain subordination related to his recognition of natural units existing at different levels. Further, Tournefort played a decisive role in the emancipation of botany from medicine.

Tournefort's genera—he accepted 725—have largely been retained, through the work of the leading taxonomists of the eighteenth century. Almost one-third of the genera of French flora derive from Tournefort. His herbarium, one of the treasures of the Muséum d'Histoire Naturelle at Paris, contains 6,963 species.

A well-rounded naturalist, Tournefort was also interested in minerals and shells, and his natural history collection included three thousand specimens of shells (G. Brice, cited by G. Ranson in *Tournefort* [Paris, 1957], 106). In addition, Tournefort was a physician with a considerable practice. Through his teaching and his publications he exerted an enormous influence until the end of the eighteenth century.

BIBLIOGRAPHY

I. ORIGINAL WORKS. Tournefort's first publication was a selection from his lectures at the Jardin du Roi, edited by William Sherard under the pseudonym of Simon Warton: *Schola botanica . . .* (Amsterdam, 1689). He subsequently published *Élémens de botanique ou méthode pour connoître les plantes,* 3 vols. (Paris, 1694).

Tournefort's other publications include the four-page pamphlet *Quaestio medica . . . discutienda . . . die Francisco Afforty . . . praeside: An potio e salvia salubris?* (n.p., n.d. [Paris, 1695]); *Histoire des plantes qui naissent aux environs de Paris avec leur usage en médecine* (Paris, 1698), a review of which appeared in *Nouvelles de la république des lettres,* **12** (Mar. 1699); a letter from Tournefort to the author at the beginning of D. Tauvry, *Nouvelle pratique des maladies aiguës et de toutes celles qui dépendent de la fermentation des liqueurs* (Paris, 1698); *Institutiones rei herbariae,* 3 vols. (Paris, 1700); "Observations sur les plantes qui naissent dans le fond de la mer," in *Mémoires de l'Académie royale des sciences,* for 1700 (1703), 27–36; *Corollarium institutionum rei herbariae, in quo plantae 1356 munificentia Ludovici Magni in orientalibus regionibus observatae recensentur* (Paris, 1703), a supp. to the *Institutiones* written by Tournefort after his return from the Levant; "Description du labirinthe de Candie, avec quelques observations sur l'accroissement et sur la génération des pierres," in *Mémoires de l'Académie royale des Sciences* for 1702 (1704), 217–234; "Persicaria orientalis nicotiniae folio, calyce florum purpureo corol. hist. rei herbar," *ibid.,* for 1703 (1705), 302–304; "Description de deux espèces de Chamaerhododendros observés sur les côtes de la Mer Noire," *ibid.,* for 1704 (1706), 345–352; "Description de l'oeillet de la Chine," *ibid.,* for 1705 (1706), 264–266; "Suite de l'établissement de quelques nouveaux genres de plantes," *ibid.,* for 1705 (1706), 83–87, 236–241; "Observation sur les maladies des plantes," *ibid.,* 27–36; "Observations sur la naissance et sur la culture des champignons," *ibid.,* for 1707 (1708), 58–66; and *Materia medica . . .,* (London, 1708), translated from Tournefort's lectures at the Jardin des Plantes and published in English before appearing in French.

Posthumous publications include *Tournefortius contractus sub forma tabularum sistens Institutiones rei herbariae juxta methodum modernarum . . .* (Frankfurt, 1715); *Traité de la matière médicale . . .,* 2 vols. (Paris, 1717); *Relation d'un voyage du Levant fait par ordre du roy . . .,* 2 vols. (Paris, 1717), with maps prepared by Claude Aubriet—this work consists of letters addressed to the count of Pontchartrain during the voyage of 1700–1702 and is preceded by Fontenelle's *éloge* of Tournefort; *Institutiones rei herbariae,* 3rd ed., 3 vols. (London, 1719), with app. by Antoine de Jussieu—this ed. contains the *Isagoge in rem herbarium,* about two-thirds of which (all of the portion on the history of botany) is completely new with respect to the intro., written in French, to the *Élémens de botanique; Histoire des plantes qui naissent aux environs de Paris avec leur usage dans la médecine,* 2nd ed., rev. and enl. by Bernard de Jussieu, 2 vols. (Paris, 1725); *Tournefort's History of Plants Growing About Paris,* John Martyn,

trans. (London, 1732); *Beschryving van eene reize naar de Levant . . . door den Hr Pitton de Tournefort . . .,* 2 vols. in 1 (Amsterdam, 1737); *Abrégé des "Élémens de botanique ou méthode pour connoître les plantes" par M. de Tournefort* (Avignon, 1749); *Matière médicale extraite des meilleurs auteurs et principalement du traité des médicamens de M. de Tournefort et des leçons de M. Ferrein,* 3 vols. (Paris, 1770), which was attributed to C. L. F. Andry, by A. A. Barbier, in *Dictionnaire des ouvrages anonymes . . .,* III (Paris, 1875), col. 82; and *Beschreibung einer auf königlichen. Befechl unternommenen. Reise nach der Levante,* G. W. F. Panzer, trans., 3 vols. plus 1 vol. of plates (Nuremberg, 1776–1777).

Later versions are *Élémens de botanique ou méthode pour connoître les plantes par Pitton de Tournefort . . .,* N. Jolyclerc, ed., 6 vols. (Lyons, 1797); [*Epistola D. D. Volkamero, apud Norinbergenses archiatio*]. *Panzer Georgio Wolfgang Panzero . . . gratulatur simulque quaedam de D. Joanne Georgio Volcamero . . . additis duabus ad illum epistolis Hermann Boerhaave & Jos. Pitt. Tournefort exponit D. Georg. Wolfgang Franciscus Panzer* (Nuremberg, 1802); *Joseph Pitton Tournefort de optima methodo instituenda in re herbaria, ad . . . Gulielmum Sherardum . . . epistola, in qua respondetur dissertationi: D. Raii de variis plantarum methodis* (n.p., n.d.), the letter being dated "Parisiis ex Horto regio, sept. MDCXCVII"; *Réponse de M. Chomel à deux lettres écrites par M. P. C. sur la botanique* (n.p., n.d. [Paris, 1696])—according to J. M. Quérard, the true author is Tournefort (this response is addressed to Philibert Collet, who, in his letters of 1695, spoke of the *Élémens de botanique* as a trans. and abridgement of John Ray's *Historia plantarum*); and "Etablissement de quelques nouveaux genres de plantes par M. Tournefort," in *Mémoires de l'Académie royale des sciences* (1706), 236–241; and *Tableau synoptique de la méthode botanique de Tournefort* (Paris, 1796). See also *Isagoge in rem herbarium (Introduction à la botanique),* translated by G. Becker from the 1719 ed. of the *Institutiones,* in *Tournefort, Muséum d'histoire naturelle* (Paris, 1957), 239–306.

II. SECONDARY LITERATURE. See G. Becker et al., "Les grands naturalistes français," in *Tournefort, Muséum d'histoire naturelle* (Paris, 1957); and H. Daudin, *De Linné à Jussieu* (Paris, 1926).

JEAN F. LEROY

TOWNELEY, RICHARD (*b.* Towneley Hall, near Burnley, Lancashire, England, 1629; *d.* York, England, 22 January 1707), *natural philosophy.*

Towneley was a member of a celebrated Roman Catholic family which, from the reign of Elizabeth I, was burdened by the penal measures of a succession of Protestant rulers, although wealth and ingenuity were sufficient to preserve both their religious integrity and their large estates. These circumstances partly explain Towneley's aversion to publicity and his retiring disposition.

Towneley was the eldest son of Charles Towneley, who was killed at the Battle of Marston Moor (1644). Richard married Margaret Paston, a Norfolk Catholic, and established a large family at Towneley Hall. In spite of the premature accretion of family responsibilities, he was able to devote most of his energies to science; the interest was probably stimulated by his uncle, Christopher Towneley (1604–1674), who had known the northern astronomers Horrocks, Crabtree, and Gascoigne. Following their deaths during the Civil War, Christopher Towneley collected and preserved their manuscripts.

Almost all of Towneley's work took place in his home. He accumulated an outstanding library of scientific works and attracted many local collaborators, mainly Catholic gentry, but also Henry Power and John Flamsteed. With Power, Towneley made early, fruitful investigations concerning air pressure. They repeated and augmented the classical experiments of Torricelli, Pascal, and Pecquet; and in 1660 and 1661 undertook experiments that led to a recognition of the air pressure–volume relationship, subsequently known as Boyle's law. Their discovery was published in 1661, but the law was not generally known until Boyle's *New Experiments Physico-Mechanical* (2nd edition, 1662). Boyle acknowledged Towneley's assistance in arriving at this generalization. Towneley's interest in air pressure continued, with attempts to measure altitudes barometrically and investigations of capillarity and the meteorological use of barometers. The interest widened to include other meteorological records; particularly important were his detailed measurements of rainfall, kept between 1677 and 1704.

Perhaps Towneley's most significant achievement was the improvement of the micrometer. Working from the principle discovered by Gascoigne, Towneley produced a sophisticated micrometer, which he applied to astronomical uses. He introduced this instrument to Flamsteed and the Royal Society. From 1670 he and Flamsteed collaborated on routine astronomical observations.

Towneley's position in English natural philosophy was distinctive, since he was one of the very few thoroughgoing Cartesians.

BIBLIOGRAPHY

I. ORIGINAL WORKS. Towneley's MSS were dispersed at the Victorian sale of the Towneley library; few

of the MSS have been traced. His few publications were communicated by correspondents and friends. *Mercurial Experiments Made at Towneley Hall in the Years 1660 and 1661* was known in September 1661, but no copy has survived; it was reprinted in Henry Power, *Experimental Philosophy* (London, 1664). Towneley's micrometer was described in *Philosophical Transactions of the Royal Society*, **2**, nos. 25, 29 (1667), 457–458, 541–544. For Flamsteed's communication of Towneley's account of the eclipse of 1676, see *Philosophical Transactions of the Royal Society*, **11** (1676), 602–604. For Towneley's rainfall records, see *ibid.*, **18** (1699), 51; **21** (1702), 47; and **25** (1705), 1877–1881.

Towneley's most substantial surviving MSS are "Considerations uppon Mr. Hooke's attempt for ye Explication of ye Expt. of ye waters ascent into small Glasse canes" (1665) and "A preliminarie discourse wherein . . . the existence or qualitie and motion of a subtle matter is proved" (1667), in the Fulton Library, Yale University. For his correspondence, see Webster (1966), below. Particularly important is the Flamsteed-Towneley correspondence, Royal Society MS 243.

II. SECONDARY LITERATURE. Towneley is not mentioned in the standard biographical dictionaries. His scientific work is summarized in A. Wolf, *History of Science and Technology in the Sixteenth and Seventeenth Centuries*, 2nd ed. (London, 1962). For more detailed discussion, see the following articles by C. Webster: "Richard Towneley and Boyle's Law," in *Nature*, **197** (1963), 226–228; "The Discovery of Boyle's Law and the Concept of the Elasticity of Air in the Seventeenth Century," in *Archive for History of Exact Sciences*, **2** (1965), 441–502; "Richard Towneley, the Towneley Group and Seventeenth-century Science," in *Transactions of the Historic Society of Lancashire and Cheshire*, **118** (1966), 51–76; and "Henry Power's Experimental Philosophy," in *Ambix*, **14** (1967), 150–178.

CHARLES WEBSTER

TOWNSEND, JOHN SEALY EDWARD (*b.* Galway, Ireland, 7 June 1868; *d.* Oxford, England, 16 February 1957), *physics.*

Townsend is best known for his research concerning the kinetics of ions and electrons in gases. The son of a college professor, he was educated at Trinity College, Dublin. He studied mathematics and physics, receiving his degree in 1890. After five years of teaching mathematics, in 1895 Townsend became one of the first outside research students to enter the Cavendish Laboratory under J. J. Thomson.

In 1897 Townsend made a direct determination of the absolute unit of charge using an original method, which "included practically all the ideas which were later used in accurate measurements of the charge."[1] Using Stokes's law, Townsend measured the rate of fall of a cloud that had condensed on an electrified gas, which had been liberated in electrolysis and then bubbled through water. By February 1898 he published the unit of charge as 5×10^{-10} esu.[2] In 1898 Townsend proved that the fundamental constant of electrolysis was equivalent to the charge carried by a gaseous ion whatever its mode of production. In that same year he also developed a method for determining the rate of ion diffusion indirectly using the ion mobility. By August 1900, the year of his election as Wykeham professor of physics at Oxford, Townsend had published a preliminary statement of his unique collision theory of ionization. Considering

FIGURE 1: Based upon J. S. Townsend, *The Theory of Ionization of Gases by Collision* (London, 1910), p. 1, and extended on the basis of R. Papoular, *Electrical Phenomena in Gases* (London, 1965), p. 123, and A. von Engel, *Ionized Gases*, 2nd ed. (Oxford, 1965), p. 223.

Schematic current-voltage characteristics for gaseous discharge at low pressure (*ca.* 1 mm Hg).

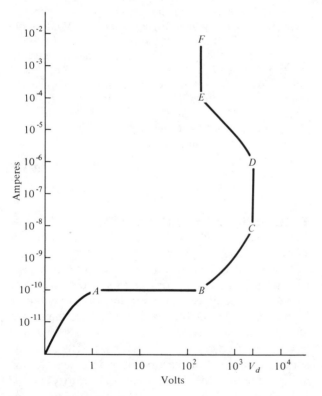

AB = saturation plateau.
BC = collision multiplication.
CD = Townsend discharge.
DE = transition to field-sustained discharge.
EF = glow discharge.
V_d = sparking voltage.

the ionization potential to be less than 15 volts instead of more than 150 volts, as was then commonly held, Townsend established that the motion of ions under the influence of an electric field was sufficient to form secondary ions in the gas. The ionization by collision was caused mainly by the "negative ions," which Townsend considered "the same as the negatively charged particles which are given off when ultra-violet light falls on a zinc plate. It has been shown by Professor [J. J.] Thomson that the mass of these particles is 1/500 of the mass of a molecule of hydrogen."[3] The rate of secondary ionization by electrons was a function of the pressure and the applied voltage. His theory could adequately account for the anomalous increase in conductivity under ultraviolet light observed already in 1890 by Stoletow. By 1903, the year of his election as fellow of the Royal Society, Townsend had included the role of the positive ions in his collision theory. He developed an expression,[4] ultimately containing his two well-known ionization coefficients α and γ, by which he could describe the Townsend discharge and also the breakdown or spark discharge (see Figure 1).

The collision principle was the basis of the 1908 particle detector of E. Rutherford and H. Geiger. Townsend had written to Rutherford that

> . . . the dodge of multiplying small conductivity by collisions works very well. . . . The case of a wire inside a cylinder has been worked out by Kirkby.[5] . . . You are certain to get unsteady effects if you try to multiply by too big a factor as the small variations produced in [the coefficients] α and β by variations of EMF or pressure have a large effect on the multiplier when it rises to the value of 500 or 1000.[6]

Kirkby and Townsend also quantitatively investigated electrochemical effects of the ionization of gases. Townsend studied the motion of electron swarms, noting that individual electrons may have random velocities much greater than the mean drift velocity of the swarm. He also noted that the mean free path of electrons in gases is energy-dependent. By the end of World War I Townsend was fifty, but during the subsequent two decades he averaged over two scientific papers a year and published five books. During the early 1920's Townsend, independently of C. Ramsauer, discovered a new physical effect. He reported in 1924 the fact that the monatomic gases, especially argon and helium, seemed particularly transparent to low energy electrons, since these could traverse such gaseous media without "feeling" its presence. Ramsauer had observed the diminished dispersion for such slow-speed electrons compared with swifter ones. Although Townsend did not himself become involved with quantum theory, this Ramsauer-Townsend effect[7] was the analogue in gases of the results with solid-state targets obtained a few years later by C. J. Davisson and independently by G. P. Thomson, and it thus became important in the understanding of the wave nature of the electron. In 1941 Townsend was knighted in recognition of his many scientific contributions, and he died in his eighty-eighth year.

NOTES

1. E. Rutherford, "The Development of the Theory of Atomic Structure," in J. Needham and W. Pagel, eds., *Background to Modern Science* (Cambridge, 1938), 64–65.
2. This result compared favorably with that of R. A. Millikan, who, like Townsend, had avoided the expansion principle and who had closely approximated by 1911 the accepted value of 4.8×10^{-10} esu. Using an expansion chamber to form a cloud, J. J. Thomson published a result of 6.5×10^{-10} esu in December 1898.
3. J. Townsend, "The Conductivity Produced in Gases by the Motion of Negatively Charged Ions," in *Philosophical Magazine*, **1** (1901), 198–227. Considerable confusion could arise concerning Townsend's term "negative ion," for it could be taken as indicating a massive particle that had acquired a negative charge. However, whether he described the negatively charged particle as having passed from an ionic state into an electronic state (*Electricity in Gases* [1915], p. 119) or as simply electrons (*Electrons in Gases* [1947], p. 92), his negative ions were very small with respect to the molecule of hydrogen.
4. Townsend's first expression was contained in "Some Effects Produced by Positive Ions," in *Electrician*, **50** (1903), 971.

$$n = n_0 \frac{(\alpha - \beta) \exp (\alpha - \beta) a}{\alpha - \beta \exp (\alpha - \beta) a}.$$

The α and β were his first set of ionization coefficients, where α was the number of collisions or ion pairs produced per centimeter of path by an electron, and where β was the number of ion pairs produced per centimeter by a positive ion. The number of electrons produced by the external radiation was "n_0," "n" represented the total number of electrons arriving at the anode, and "a" was the distance between the electrodes. The dominant secondary effect of the positive ions, however, was not in gaseous collisions but was the release of secondary electrons at the cathode. In addition to α (the ionization coefficient representing the electrons released by other electrons colliding with the gas on their way to the anode), Townsend introduced his secondary emission coefficient γ, which represented the ionizing electrons released at the cathode. The general expression accordingly became:

$$n = n_0 \frac{\exp \alpha a}{1 - \gamma (\exp \alpha a - 1)}.$$

The sparking potential was a maximum value, and Townsend emphasized that the potential required for sustained discharge was normally significantly less. The conditions for breakdown were obtained on either expression by letting "n" approach infinity. For a given applied voltage and pressure the critical distance "d" for the spark discharge

(the end of the region of Townsend discharge and the beginning of the region of the field-sustained discharge) was thus described on the earlier formulation by $\alpha = \beta \exp (\alpha - \beta) d$, and on the later formulation by $1 = \gamma (\exp \alpha d - 1)$.

5. P. J. Kirkby, a research student of Townsend, "On the Electrical Conductivities Produced in Air by the Motion of Negative Ions," in *Philosophical Magazine*, **3** (1902), 212–225.
6. Townsend letter to E. Rutherford, 10 March 1908, in Cambridge University Library, Add. MSS 7653/T76.
7. A. von Engel, *Ionized Gases*, 2nd ed. (Oxford, 1965), 31. J. Townsend, *Motion of Electrons in Gases* (Oxford, 1925), 26–29, an address to the Franklin Institute, Philadelphia, Sept. 1924.

BIBLIOGRAPHY

Townsend published over one hundred papers and several books as listed by his biographer A. von Engel, in "John Sealy Edward Townsend, 1868–1957," in *Biographical Memoirs of Fellows of the Royal Society*, **3** (1957), 257–272; and in his notice for the *Dictionary of National Biography: 1951–1960*, pp. 983–985. A small collection of his correspondence exists at the Cambridge University Library Add. MSS 7653/T71–T89; but T. S. Kuhn has noted, in *Sources for History of Quantum Mechanics* (Philadelphia, 1967), p. 92b, that no extensive collection is likely to exist. See also A. von Engel, in *Nature*, **179** (1957), 757–758, and Maurice de Broglie, "Notice nécrologique sur Sir John Townsend," in *Comptes rendus hebdomadaires des séances de l'Académie des sciences*, **244** (1957), 3105–3106. In *A History of the Cavendish Laboratory* (London, 1910), E. Rutherford in ch. 6 and C. T. R. Wilson in ch. 7 discussed the early work of Townsend. After over 40 years of continuous service at Oxford, at his retirement a notice of his work appeared in *Nature*, **157** (1946), 293. A valuable article is by C. A. Russell, in T. I. Williams, ed., *A Biographical Dictionary of Scientists* (London, 1969), 517–518. The work of Townsend concerning diffusion of ions in gases and also regarding the fundamental unit of charge is considered in detail in N. Feather, *Electricity and Matter* (Edinburgh, 1968), 306–313; and in R. A. Millikan, *The Electron* (Chicago, 1963), facsimile of 1917, 34–38, 43–47, 51–52, 123–125, esp. in appendices A and B.

The development of the expression for the Townsend discharge and the breakdown equation is discussed in J. A. Crowther, *Ions, Electrons, and Ionizing Radiations* (London, 1961), 54–56; and is treated technically in A. von Engel, *Ionized Gases*, 2nd ed. (Oxford, 1965), 171–182.

THADDEUS J. TRENN

TOWNSEND, JOSEPH (*b.* London, England, 4 April 1739; *d.* Pewsey, Wiltshire, England, 9 November 1816), *medicine, geology, economics.*

Townsend was the fourth son of Bridget Phipps

Townsend and Chauncy Townsend, a linen merchant, mine inspector, and member of Parliament. He attended Clare Hall, Cambridge University, and received his B.A. in 1762 and his M.A. in 1765. In 1762–1763 Townsend studied medicine in Edinburgh, attending the classes of William Cullen in anatomy, Robert Whytt in physiology, and John Hope in botany.

While a student Townsend came under the influence of Calvinistic Methodism and was ordained a minister in 1765. He was an unusually tall man and a powerful speaker, and for a number of years he traveled through England as an evangelical minister. The experience was unpleasant, and his enthusiasm gradually waned. Richard Graves satirized Townsend's ministry in a novel, *The Spiritual Quixote* (1772). In 1773 he married Joyce Nankivell, by whom he had two daughters and four sons. She died in 1785 and in 1790 he married Lydia Hammond Clerke, widow of Sir John Clerke. She died in 1814.

Townsend traveled to France, Holland, and Flanders in 1770 as chaplain to the duke of Atholl. In 1786–1787 he traveled through Spain and wrote an important account of the Spanish economy, similar to those travel accounts Arthur Young was writing on Britain and France. Townsend's *Journey Through Spain* (3 vols., 1791; 3rd ed. 1814) was popular enough to be translated into German (1792), Dutch (1792–1793), and French (1800).

In 1786 Townsend published an attack on British charity for being too indulgent to the poor, and in his *Journey Through Spain* he extended this attack to Spanish institutions. He argued that when the poor depended upon charity, the increase of their population would deplete the wealth of the country. Realizing that the population of Spain had declined, he analyzed the causes. (The population of Spain had declined sharply in the seventeenth century, but at the time he was writing this, it was actually increasing rapidly.) His discussion contributed to the awareness of the importance of population as an economic factor. Malthus claimed not to have known Townsend's writings on population until after publishing the first edition of his *Essay on the Principle of Population* (1798), but he expressed an appreciation for Townsend's writings in the second and subsequent editions of his *Essay*.

Townsend wrote two very popular manuals for the practice of medicine: *The Physician's Vade Mecum* (1781; 10th ed., 1807) and *Elements of Therapeutics, or a Guide to Health* (1795; 3rd ed., 1801). The latter work contains the first English description of Antonio de Gimbernat's operation for strangulated femoral hernia, and Townsend also published the first English description of pellagra (under the name of "mal de la rosa") in his *Journey* (II, 10).

Townsend developed an early interest in geology and paleontology, perhaps because of his father's influence. The *Journey* contains numerous geologic descriptions and speculations. He became friends with William Smith, who in 1799 explained to him the method of correlating strata by the kinds of fossils in them. Townsend published one of the first and clearest accounts of Smith's discovery in the first volume of *The Character of Moses Established for Veracity as an Historian, Recording Events From the Creation to the Deluge* (2 vols., 1813–1815). Townsend used Smith's discovery to attack James Hutton's assertion that there is no evidence concerning the origin of the earth. In spite of his doctrinaire stand, there is considerable merit to Townsend's geological discussion. At the founding of the Geological Society of London in 1807, Townsend was elected an honorary member.

BIBLIOGRAPHY

I. ORIGINAL WORKS. For a list of Townsend's writings, see the *General Catalogue of Printed Books* of the British Museum. This list does not include his now rare treatise *On the Agency of Vital Air in the Cure of Various Diseases, With Cases*, 2nd ed. (London, 1824), later retitled *Townsend on Vital Air, Being Numerous Cases Showing the Effects of Vital Air and Other Factitious Airs: As Judiciously Practiced by Dr. Thornton*, 9th ed. (London, 1827); and as *Townsend on Pneumatic Medicine*, 10th ed. (London, 1830). The only known copies of the 2nd and 9th editions have been in private hands and are described briefly by A. D. Morris (see below).

II. SECONDARY LITERATURE. The best account of Townsend's life and career and of his medical contributions is by A. D. Morris, "The Reverend Joseph Townsend MA, MGS (1739–1816) Physician and Geologist—'Colossus of Roads,'" in *Proceedings of the Royal Society of Medicine*, 62 (1969), 471–477.

For other biographical information and a discussion of Townsend's geology, see A. G. Davis, "The Triumvirate: A Chapter in the Heroic Age of Geology," in *Proceedings of the Croydon Natural History and Scientific Society*, 11 (1943), 122–146. See also Charles Coulston Gillispie, *Genesis and Geology: A Study in the Relations of Scientific Thought, Natural Theology, and Social Opinion in Great Britain, 1790–1850* (Cambridge, Mass., 1951; New York, 1959), for a discussion of Townsend's *The Character of Moses*. . . .

Townsend's demographic ideas have been discussed by Kenneth Smith, *The Malthusian Controversy*

(London, 1951) and by Ashley Montagu and Mark Neuman in an edition of Townsend's *A Dissertation on the Poor Laws by a Well-Wisher to Mankind* (Berkeley, 1971). On Spain's population at the time of Townsend's visit, see Earl J. Hamilton, "The Decline of Spain," in *Economic History Review*, **8** (1938), 168–179; Massimo Livi-Bacci, "Fertility and Population Growth in Spain in the Eighteenth and Nineteenth Centuries," in *Daedalus*, **97** (1968), 523–535; and Jorge Nadal, *La Poblacion Española, Siglos XVI a XX* (Barcelona, 1966). For a discussion of the Spanish economy, see Earl J. Hamilton, *War and Prices in Spain, 1651–1800* (Cambridge, Mass., 1947; New York, 1969).

FRANK N. EGERTON III

TOZZI, DON BRUNO (*b.* Florence, Italy, 27 November 1656; *d.* Vallombrosa, Italy, 24 January 1743), *botany*.

Tozzi was the son of Francesco Simone Tozzi. Although his family was of modest means, Don Bruno was nevertheless able to pursue a formal study of philosophy and theology before his investiture, at age twenty, as a monk in the order of Vallombrosa. He was successful in the performance of his religious duties but repeatedly refused promotions. He eventually did become procurator general and abbot of the order. At the same time he managed to pursue his interest in botany. His ecclesiastical duties required frequent journeys, and he availed himself of the opportunity to study and collect plants as he made his way from place to place on foot.

An endowment allowed Tozzi to obtain a choice collection of scientific books. He was a teacher and friend of Pier Antonio Micheli, and their lifelong friendship was enhanced by the many excursions they made together collecting plants. Micheli named the rare genus *Tozzia* for his mentor.

Abbot Tozzi not only had a keen eye for finding and collecting plants, but he also became adept at watercolor illustration of phanerogams as well as cryptogamic species. A number of his works were devoted exclusively to fungi, lichens, algae, and bryophytes. He generously shared his collections and drawings with prominent botanical figures of the day, including William Sherard, who sent him books. He also kept an active correspondence and exchange of materials with Hermann Boerhaave, James Petiver, and Hans Sloane. Tozzi became well-known as an able teacher and authority on the Italian flora. As many as 200 plants are illustrated in his *Catalogus plantarum etruriae et insularum adjacentium*. Along with his friend Micheli, he was

a founder of the Società Botanica Fiorentina, and he was elected to the Royal Society of London. He declined offers to teach in London because of advancing age and duties to his order. After retirement he began some folios of birds and insects, and devoted himself largely to his botanical interests until his death.

BIBLIOGRAPHY

I. ORIGINAL WORKS. Tozzi's contributions to botany are too numerous to list individually. Manuscripts by Tozzi not sent to his contemporaries are preserved at the Biblioteca Nazionale Centrale in Florence. Some watercolor drawings sent to Sherard are at the University of Oxford. The Sloane collections in the British Museum (Natural History) include some of Tozzi's color illustrations as well as some letters written to Petiver and Sloane. A nearly complete list of Tozzi's works are cited in P. A. Saccardo and F. Cavara, "Funghi di Vallombrosa," in *Nuovo giornale botanico italiano e Bolletino della Società botanica italiana*, n.s. **7** (1900), 272–310.

II. SECONDARY LITERATURE. The number of works containing information on Tozzi is scarcely indicative of the importance of this man to early European botany. Holdings of Tozzi's at the University of Oxford are listed by H. N. Clokie, *An Account of the Herbaria of the Department of Botany in the University of Oxford*, VIII (Oxford, 1964).

A tribute to Tozzi was made on the occasion of the 200th anniversary of Micheli's death in a small paper presented at the Italian Botanical Society by G. Negri, "Don Bruno Tozzi (1656–1743)," in *Nuovo giornale botanico italiano e Bolletino della Società botanica italiana*, n.s. **45** (1939), cix–cxiv.

A few insights concerning Tozzi and his works can be found in J. Proskauer, "Bruno Tozzi's Little Mystery, or a Quarter Millennium of Confusion," in *Webbia*, **20** (1965), 227–239.

DALE M. J. MUELLER

TRADESCANT, JOHN (*b.* Suffolk [?], England, 1570/1575; *d.* South Lambeth, Surrey, England, 15/16 April 1638), *natural history*.

In 1773 Ducarel wrote that Tradescant may be "justly considered the earliest collector (in this kingdom) of every thing that was curious in Natural History, viz., minerals, birds, fishes, insects, &c . . . coins and medals of all sorts, besides a great variety of uncommon rarities." Tradescant's father, Thomas Tradescant (a descendant of Willelmus Treluskant of Suffolk), in 1578 left Suffolk, where John was probably born, for London; his mother's name is unknown. The first verified record is John's marriage to Elizabeth Day at Meopham, Kent, on 18 June 1607. Their only son, John, was born the following year. Beginning in 1604 Tradescant was gardener for the properties of Robert Cecil, earl of Salisbury. There is some evidence that before 1600 he had been gardener to William Brookes at Cobham Hall.

Tradescant made his first (?) collecting trip to Flanders for Cecil in 1610, and it was probably then that he introduced the large-leaved Brussels strawberry into England. In 1611 he visited the Lamont nursery at Rouen and the apothecary garden of Jean Robin on the Île Notre-Dame. He also introduced species to the gardens at Hatfield, until its sale on the earl's death in 1612.

Six years later Tradescant made the first botanical visit to Russia when he accompanied Sir Dudley Digges, who was sent by James I under the Muscovy Company to negotiate a loan to the emperor of Russia. They reached Archangel on 16 July 1618, and Tradescant noted larch, white hellebore, and other plants (Allan, pp. 84–89), especially on the islands about the mouth of the Dvina River. On 5 August 1618 they set sail for England. Tradescant wrote an account of the expedition (Konovalov, pp. 130–141).

In 1620 Tradescant joined Sir Samuel Argall's expedition against the Barbary corsairs, taking the opportunity to collect garden seeds and fruits (Allan, pp. 101–103) and *naturalia*, including the first specimens of gutta-percha and mazer wood to be seen in England. After a short employment with George Villiers, duke of Buckingham, from 1625 to 1628, during which service he joined a military expedition to La Rochelle, Tradescant used the nucleus of his own collections to set up a garden and a museum in South Lambeth, a pioneer enterprise. His museum was enriched by gifts from virtuosi (sea captains supplied "such toyes as they could bring from other parts") and visited by Charles I and nobility. Doubtless the most famous exhibit was the stuffed dodo, seen by Willughby and Ray.

During this time Tradescant became gardener to Charles I and Queen Henrietta Maria. In 1631 René Morin began to send novelties. The following year Tradescant testified that an alleged unicorn's horn was that of a narwhal "yet very precious against poison."

His *Catalogus* (1634), the only known copy of which is in Magdalen College, lists 750 garden plants grown at South Lambeth, arranged alphabetically, and concludes with a catalog of fruits.

Forty Virginia species have been recognized in this *Catalogus* upon which the *Musaeum Tradescantianum* (1656) was based.

The generic name *Tradescantia* dates from H. B. Ruppius, *Flora Jenensis* (1718), and was accepted by Linnaeus. Tradescant "liv'd till [he] had travelled art and nature thro'" and was buried on 17 April 1638, to the southeast of Lambeth Church.

BIBLIOGRAPHY

I. ORIGINAL WORKS. John Goodyer's copy of Tradescant's *Plantarum in Horto Iohannem* [sic] *Tradescanti nascentium catalogus* (1634) survives as the only known copy in Magdalen College, Oxford; and as Gunther suggested, it is almost certainly a copy made up from printer's proof, judging from the error on the title page. Many Tradescant curiosities, for example, chief Powhatan's feather cape, although often without precise provenance, are preserved in the (new) Ashmolean Museum. An album of Tradescant's "choicest Flowers and Plants exquisitely limned on vellum by Mr. Alex: Marshall" is in the library of Windsor Castle (see A. P. Oppé, *English Drawings, Stuart and Georgian Periods, in the Collection of His Majesty the King at Windsor Castle* (London, 1950), 74–75.

II. SECONDARY LITERATURE. Classic accounts of Tradescant and his work are William Watson, "Some Account of the Remains of John Tradescant's Garden at Lambeth," in *Philosophical Transactions of the Royal Society*, 46 (1752), 160–161; and Andrew Coltee Ducarel's letter to Watson, *ibid.*, 63 (1773), 79–88, plate 4, also published as a separately paginated reprint (London, 1773). Both accounts were used by Richard Pulteney, in his *Historical and Biographical Sketches of the Progress of Botany in England*, I (London, 1790), 175–179.

See also Edward F. Rimbault, "Family of the Tradescants," in *Notes and Queries*, 3 (1851), 353–355; and esp. G. S. B[oulger], in the *Dictionary of National Biography*, 57 (1899), 143–147. The latest biography, emphasizing the genealogy and botany, is Mea Allan, *The Tradescants* (London, 1964). Tradescant's 1618 Russian journal (Ashmole MS 824, xvi) has been published verbatim by S. Konovalov, "Two Documents Concerning Anglo-Russian Relations in the Early Seventeenth Century," in *Oxford Slavonic Papers*, 2 (1951), 130–141. The manuscript's identity was established by Joseph von Hamel, "Tradescant der Aeltere 1618 in Russland . . .," in *Recueil des actes de la séance publique l'Académie impériale des sciences de Saint-Petersbourg* (1847), 85–348, with portrait and map; also issued as a separately paginated reprint, and abridged in book form as *England and Russia: Comprising the Voyages of John Tradescant the Elder* [etc.], translated by John Studdy Leigh (London, 1854), reissued with only title page altered, as *Early English Voyages to Northern Russia* (London, 1857). Marjorie F. Warner, "The Morins," in *National Horticultural Magazine*, **33** (1954), 168–176, summarizes René Morin's contacts with Tradescant.

For documented commentary on relationships with Ashmole, see C. H. Josten, *Elias Ashmole (1617–1692)*, 5 vols. (Oxford, 1966), 2048–2049.

JOSEPH EWAN

TRADESCANT, JOHN (*b*. Meopham, Kent, England, 4 August 1608; *d*. South Lambeth, Surrey, England, 22 April 1662), *natural history*.

Traveler, collector, and gardener, Tradescant carried on the activities of his father of the same name. At the age of eleven he was a pupil at King's School, Canterbury, and at twenty-six was sworn into the Company of Master Gardeners of London. In 1637 he was in Virginia "gathering all varieties of flowers, plants, shells, &c" for his father's museum at South Lambeth. After his father's death he succeeded him as gardener to Queen Henrietta Maria. Tradescant married twice: first Jane Hurte, on 29 February 1627, by whom he had a daughter and son; and, after her death, Hester Pooks, on 1 October 1638. He may have made a second trip to Virginia in 1642.

The first use of "ark" for the museum appeared about 1645 in a poem mentioning "Tradeskin and his ark of novelties." His first meeting with Elias Ashmole was evidently in 1650. A draft of a catalog of the Tradescant collections was prepared by Ashmole and Thomas Wharton (Josten, I, 94), four years before the publication of *Musaeum Tradescantianum* (1656) under Tradescant's name. The *Musaeum* contains references to Aldrovandi, Belon, L'Écluse, Rondelet, Moffett, and Markgraf; and it constitutes an inventory of worldwide curiosities. Of the some two hundred Barbadian and Virginian plants identified in the two lists of 1634 and 1656, ten plants date from before 1600; but the remainder are notable introductions into horticulture.

The circumstances surrounding Tradescant's bequest of the museum to Ashmole are confused (see Josten, II, 768; and Whitehead, 54). In 1659 during a drinking party, Ashmole seems to have obtained a signed bequest from Tradescant, but in 1661 Tradescant willed his "closet of rarities" to his wife Hester, stipulating that on her death they go to Oxford or Cambridge "to which of them shee shall think fitt." In 1674 Ashmole removed the Tradescant collections to his own house and three

years later delivered his collections and Tradescant's "rarities" to Oxford, where they were displayed next to the Sheldonian in a building reputedly designed by Christopher Wren. It was not long before Hester was found drowned in her pond. The "name of Tradescant was unjustly sunk in that of Ashmole" (Pulteney, I, p. 179).

BIBLIOGRAPHY

I. ORIGINAL WORKS. The G. Wharton copy of *Musaeum Tradescantianum: or a Collection of Rarities Preserved at South-Lambeth Near London* (London, 1656) is in Morton Arboretum Library; it is reproduced as Old Ashmolean Reprint I (Oxford, 1925), with omission of pp. 74–178, and in toto in Mea Allan, *The Tradescants* (London, 1964), 247–312. The only authenticated specimen of Tradescant's shell collection is the holotype of *Strombus listeri* (S. Peter Dance, *Shell Collecting* [1966], 37). His miscellaneous numismatic, archaeological, and zoological collections originally in the "Old Ashmolean" were dispersed in 1831 to various Oxford museums. His *hortus siccus* is in the Bodleian (Ashmole MS 1465).

II. SECONDARY LITERATURE. See John Tradescant, *ante*, since most accounts include father and son. See also Richard Pulteney, *Historical and Biographical Sketches of the Progress of Botany in England*, I (London, 1790), 179; and S. W. Singer, "Tradescants and Elias Ashmole," in *Notes and Queries*, 5 (1852), 367–368, 385–387. Résumés of his respective fields are Dorothy Gardiner, "The Tradescants and Their Times," in *Journal of Royal Horticultural Society*, **53** (1928), 308–317: and P. J. P. Whitehead, "Museums in the History of Zoology," in *Museums Journal*, **70** (1970), 50–57. Commentary on Tradescant's "dodar" is given in Masauji Hachisuka, *The Dodo and Kindred Birds* (London, 1953); and references to Tradescant portraits are in David Piper, *Catalogue of the Seventeenth Century Portraits in the National Portrait Gallery, 1625–1714* (Cambridge, 1963), 350–351. For his relationship with Ashmole see C. H. Josten, *Elias Ashmole (1617–1692)*, 5 vols. (Oxford, 1966).

JOSEPH EWAN

TRAGUS, HIERONYMUS. See Bock, Jerome.

TRAUBE, MORITZ (*b*. Ratibor, Silesia [now Racibórz, Poland], 12 February 1826; *d*. Berlin, Germany, 28 June 1894), *physiological chemistry*.

Traube was the son of a Jewish wine merchant. His elder brother Ludwig (1818–1876), a specialist in internal medicine at Berlin, became famous through his work in experimental pathology. His father and brother quickly recognized the scientific abilities of young Moritz, who at the age of sixteen completed his Gymnasium studies with outstanding grades in the humanities.

Traube began his scientific education in Berlin but soon transferred to Giessen, where he studied under Liebig. At twenty-one he earned the doctorate with a work on chromium compounds. He founded a physical society with his brother Ludwig, who also encouraged him to study medicine. In 1849, however, Traube had to abandon his medical studies because of family obligations: his younger brother died unexpectedly, and his father asked him to take over the management of the large family business.

Traube's decision to comply with his father's wishes undoubtedly was a difficult one, and he subsequently led a double existence as a businessman and a researcher. With modest means he fitted out a private laboratory, where he spent his free hours. The first was in an attic room in Ratibor; later he built another in Breslau, where he transferred the wine business in order to facilitate his research; and the third was in Berlin. In Breslau, Traube worked in the laboratory of the physiologist Rudolf Heidenhain but later used his own facilities and engaged two assistants. Despite the great care with which he conducted his research and the originality of his ideas, Traube encountered considerable prejudice because he held no established academic post. Eventually, however, he achieved recognition, and in 1867 the University of Halle granted him an honorary doctorate.

Traube had two sons and three daughters. One of his sons, Hermann, became professor of mineralogy at the University of Berlin. In the last years of his life, which he spent in Berlin, Traube devoted himself entirely to scientific work until he succumbed to diabetes in 1894. He had begun to investigate the biochemical process of diabetes mellitus in man in 1852 and had distinguished two forms of the disease.

Traube's experiments extended over much of physiological chemistry but proved of great importance for general chemistry as well. In his *Handbuch zur Geschichte der Naturwissenschaft und der Technik* (1908), Darmstaedter draws attention to three of Traube's discoveries: oxygen-carrying ferments, or as they are now called, enzymes (1858); semipermeable membranes (1867); and artificial models of the cell (1875). Traube's most interesting studies were on the source of the energy for muscle contraction, on the nature of the ac-

tion of ferments, and on the formation of precipitation membranes.

Traube maintained that the energy source of muscle activity was the combustion of nitrogen-free substances. In 1861 he proposed that the organized portion of the muscle is not destroyed during work, urea is not a measurement of the production of muscle force, and albuminous substances (now called proteins) are not decomposed through muscle activity. This hypothesis, which contradicted Liebig's views, was sharply criticized by Georg Meissner in *Bericht über die Fortschritte der Anatomie und Physiologie* (1861–1863). Traube's position was not vindicated until Pettenkofer and Voit published results showing that carbohydrates could be interpreted as a source of energy. In this area Traube appears even to have anticipated the Bohr effect.

In his investigation of the action of ferments, Traube distinguished two types of "dialysis," one involving reduction and the other oxidation. In the reduction process the ferment removes oxygen from a substance and releases it either to another substance or to the air; whereas in the oxidation, the ferment takes oxygen from the air and releases it to an oxidizable substance. In Traube's view, the effectiveness of the ferment derives solely from the transfer of oxygen (*Platzwechsel*, or "transposition"). He found a simple model for such a process in finely divided platinum (platinum sponge), which decomposes hydrogen peroxide without limit (reduction) and recombines unlimited amounts of hydrogen and oxygen into water (oxidation). According to Traube, biological oxidation takes place in two steps. First, a substance A, together with oxygen and water, forms hydrogen peroxide in the presence of an oxidase; then a substance B is oxidized with the assistance of the hydrogen peroxide thus formed. This scheme can be represented by means of the following equation:

$$A + O_2 + 2H_2O \longrightarrow A(OH)_2 + H_2O_2$$
$$B + H_2O_2 \longrightarrow BO + H_2O$$

If this oxidation of B is viewed as the dehydrogenation of a substance RH_2, then the re-formation of A can be understood by means of an extension of this scheme:

$$RH_2 \longrightarrow R + 2H$$
$$A(OH)_2 + 2H \longrightarrow A + 2H_2O$$

(see W. Bladergroen, *Physikalische Chemie in Medizin und Biologie* [Basel, 1949], 623–624). Traube believed that ferments were specific substances, similar to proteins, that could not be obtained in pure form because of the ease with which they decompose.

Traube's discoveries of the formation of precipitation membranes (1864, 1867) were especially significant. He produced homogeneous membranes in the forms of closed cells by placing gelatin in tannic acid solution. The sphere became covered with a membrane and gradually increased considerably in volume through absorption of water; the contents remained clear, transparent glue, however, and no tannic acid entered. Open glass tubes were closed off by metalliferous membranes formed from cupric ferrocyanide and Prussian blue. According to Traube, the endosmotic process occurring in the case of such homogeneous membranes is the result of the attraction of the solute for the solvent (*endosmotische Kraft*). The most diffusible substances cannot penetrate certain membranes if the molecular interstices of the latter are smaller than the entering molecules. These findings permitted the development of a technique for measuring the magnitude of the attraction of various substances for water. Further, Traube's discovery of the semipermeable membrane provided the basis for Pfeffer's measurement of osmotic pressure and for van't Hoff's far-reaching theoretical interpretation of osmosis.

In the last years of his life Traube worked on problems pertaining to uric acid synthesis and to autoxidation, as well as on developing a process for producing sterile water.

BIBLIOGRAPHY

I. ORIGINAL WORKS. Traube's papers on general chemistry appeared mainly in *Poggendorff's Annalen der Chemie* and in *Berichte der Deutschen chemischen Gesellschaft*, and those on physiology in *Virchows Archiv* and in *Archiv für Anatomie und Physiologie*, as well as in *Zentralblatt für die medicinischen Wissenschaften*. Detailed reports on Traube's earlier works are in the *Bericht* that Henle and Meissner prepared for *Fortschritte der Anatomie und Physiologie*. Some of Traube's studies on ferment chemistry were reviewed in Leo Maly's *Jahresbericht für Thierchemie* by Nathan Zuntz, among others.

Traube published a monograph entitled *Theorie der Fermentwirkungen* (Berlin, 1858). His scattered papers were posthumously edited by H. and W. Traube as *Gesammelte Abhandlungen von Moritz Traube* (Berlin, 1899).

Traube's papers include "Ueber die Beziehung der Respiration zur Muskelthätigkeit und die Bedeutung der Respiration überhaupt," in *Virchows Archiv*, **21** (1861), 386–414; "Experimente zur Theorie der Zellenbildung," in *Medizinische-Zentralblatt*, no. 39 (1864); "Experimente zur Theorie der Zellenbildung und Endosmose," in *Archiv für Anatomie und Physiologie* (1867), 87–165; "Ueber das Verhalten der Alkoholhefe in sauerstoffgasfreien Medien," in *Berichte der Deutschen chemischen Gesellschaft*, **10** (1877), 510–513; "Zur Lehre von der Autoxydation," *ibid.*, **22** (1889), 1496–1514; and "Einfaches Verfahren, Wasser in grossen Mengen keimfrei zu machen," in *Zeitschrift für Hygiene*, **16** (1894), 149–150.

II. SECONDARY LITERATURE. See G. Bodländer, "Nekrolog auf M. Traube," in *Berichte der Deutschen chemischen Gesellschaft*, **28** (1894), 1085; E. von Meyer, "Moritz Traube," in *Handwörterbuch der Naturwissenschaften*, X (Jena, 1915), 43–44; and K. Müller, "Moritz Traube (1826–1894) und seine Theorie der Fermente" (M.D. diss., Zurich, 1970).

Further information on Traube's life and works is in W. Bladergroen, *Physikalische Chemie in Medizin und Biologie* (Basel, 1949), 623–624; L. Darmstaedter, *Handbuch zur Geschichte der Naturwissenschaften und der Technik* (Berlin, 1908), 585, 664, 734; F. Lieben, *Geschichte der physiologischen Chemie* (Leipzig–Vienna, 1935; repr. Hildesheim–New York, 1970), 117, 237–239; and Poggendorff, II, 1126; III, 1363; and IV, 1519.

G. RUDOLPH

TRAVERS, MORRIS WILLIAM (*b.* Kensington, London, England, 24 January 1872; *d.* Stroud, Gloucestershire, England, 25 August 1961), *physical chemistry, cryogenics, industrial chemistry.*

Travers actively collaborated with William Ramsay in the discovery and specification of the inert gases. The second of the four sons of a London physician, Travers was schooled from 1879 to 1882 at Ramsgate and from 1882 to 1885 at Woking. In the spring of 1885 he transferred to Blundell's School (Tiverton), which was equipped with a good chemistry laboratory. From 1889 to 1893 he studied at University College London taking his B.Sc. in 1893. Intending to specialize in organic chemistry, he worked with Alban Haller at the University of Nancy during early 1894. Returning to work with Ramsay in late 1894, just at the time of the discovery of argon, Travers took a position as demonstrator at University College London and continued research in organic chemistry under Collie. In 1898 he took his D.Sc. and became assistant professor.

After the discovery of helium in 1895, Travers

assisted Ramsay in the determination of the properties of argon and helium. They also heated minerals and meteorites for new gases, but inevitably only helium was evolved. For some time it had been suspected that yet another gas might accompany argon, and accordingly they prepared a large quantity of this gas by removing oxygen and nitrogen from air and by forcing the residue into a bulb. In order to apply the technique of fractional distillation, it was first essential to liquefy this argon residue. With the assistance of Hampson, Travers and Ramsay obtained in May 1898 a large quantity of liquid air for this purpose. Practicing with air to prepare themselves for an investigation of the argon, Travers evaporated a small quantity of the liquid air and collected the least volatile fraction. A spectrum analysis of the residue yielded the new lines of krypton. (Travers collected the gas from the last ten cubic centimeters of 750 cc of evaporating liquid air. After removing the oxygen and nitrogen, leaving a gaseous residue of about 26 cc, he observed on 31 May 1898 with Ramsay the new spectral lines. They telegraphed the news to Berthelot in Paris and sent their communication to the Royal Society on 3 June. It was announced on 6 June in the Paris Academy and on 9 June in London.) If the argon residue contained a constituent of lower boiling point (namely, greater volatility), it would distill over first. Accordingly they liquefied the argon and upon evaporation collected the most volatile fraction. A spectrum analysis showed the new lines of neon, the suspected companion of argon. (Travers prepared the most volatile fraction on 11 June. With Ramsay he observed the new spectral lines on 12 June. The paper was read on 16 June to the Royal Society, London, and announced in Paris on 20 June 1898.) A further unexpected companion, "metargon," proved to be a mistake, like the sensational "etherion" of Brush. Metargon was merely argon contaminated by carbon monoxide. C. F. Brush on 23 August announced the existence of an alleged gaseous constituent having a density 10^{-4} that of hydrogen (*Science*, **8** [1898], 485–494).

Travers and Ramsay next turned their attention toward the task of obtaining quantities of each inert gas sufficient to determine the properties. In spite of its low volatility, krypton proved difficult to obtain in isolated quantities. It was found to have an even less volatile companion, xenon, identified spectroscopically about 12 July 1898 and announced to the British Association on 8 September 1898. Since the neon fraction from the argon residue also contained helium, liquid air proved insufficient; it was necessary to employ liquid hydrogen for the condensation.

Although Dewar had successfully liquefied hydrogen by May 1898, Travers independently constructed the required apparatus. With liquid hydrogen they were able to condense the neon portion, while the helium remained gaseous. Evaporating this liquid, they obtained sufficient neon by July 1900 to complete their study of the inert gases. Continuing his cryogenic research, Travers made probably the first accurate temperature measurements of liquefied gases, and set up several experimental liquid air plants in Europe. In November 1903 Travers replaced S. Young as professor of chemistry at University College, Bristol, and contributed to that institution receiving a university charter. He was elected a fellow of the Royal Society in 1904, and two years later went to Bangalore as director of the Indian Institute of Science, which opened to students in 1911. In 1909 Travers married the sister of R. Whytlaw-Gray. During this period Travers conducted an interesting but controversial study on boron.

Returning to England in July 1914, Travers directed glass manufacture at Walthamstow (Duroglass, Ltd.) throughout World War I, and later became president of the Society of Glass Technology. From 1920 he was concerned with high-temperature furnaces and fuel technology, including the gasification of coal, and in 1925 took part in founding the Institute of Fuel. Travers was made honorary professor of chemistry at Bristol in 1927 and established a research group that worked on thermal decomposition of organic vapors and also gaseous and heterogeneous reactions. He also awakened the youthful scientific interest of A. B. Pippard during this period. He was president of the Faraday Society from 1936 to 1938 and retired in 1937. Then Travers began his historical research on Ramsay, organizing the Ramsay papers and compiling and working on twenty-four volumes of documentation. During World War II, Travers served in an advisory capacity and was a consultant to the explosives section of the Ministry of Supply from 1940 to 1945. Traveling extensively, he was a general trouble-shooter concerned with a variety of technical problems. From 1953 Travers resumed his biography of Ramsay and completed it in 1955.

BIBLIOGRAPHY

I. ORIGINAL WORKS. A nearly complete list of Travers' publications is in C. E. H. Bawn, "Morris William

Travers 1872–1961," in *Biographical Memoirs of the Fellows of the Royal Society,* **9** (1963), 301–313. Travers also wrote *William Ramsay and University College London,* a small booklet privately issued by University College London for the Ramsay centenary, and "Sir William Ramsay," in *Endeavour,* **11** (1952), 126–131. The best source on the discovery of the inert gases is *The Discovery of the Rare Gases* (London, 1928). For a brief account see Travers' "The Rare Gases of the Atmosphere," in E. F. Armstrong, ed., *Chemistry in the Twentieth Century* (London, 1924), 82–87. An extensive MS collection of Travers is incorporated with the Ramsay Papers (esp. items 82–105) in the library at University College London. R. M. MacLeod and J. R. Friday, *Archives of British Men of Science* (London, 1972), notes items in the Royal Society.

II. SECONDARY LITERATURE. On Travers and his work, see D. H. Everett, in *Nature,* **192** (1961), 1127–1128; W. L. Hardin, *The Rise and Development of the Liquefaction of Gases* (New York, 1899), which contains a relevant discussion of the then current state of cyrogenics and also of the discovery of krypton, neon, metargon, and etherion; D. McKie, "Morris William Travers 1872–1961," in *Proceedings of the Chemical Society* (1964), 377–378, which mentions an autobiography, the draft of which is in the library of University College London; P. Walden, "Lothar Meyer, Mendelejeff, Ramsay und das periodische System der Elemente," in Günther Bugge. ed., *Das Buch der Grossen Chemiker,* 2nd ed., II (Weinheim, 1955), 229–287; Weeks, *Discovery of the Elements,* 7th ed. (Easton, 1968), 750–773, 868–896; and T. I. Williams in his *A Biographical Dictionary of Scientists* (London, 1969), 518–519.

THADDEUS J. TRENN

TREBRA, FRIEDRICH WILHELM HEINRICH VON

(*b.* Allstedt, Weimar, Germany, 5 April 1740; *d.* Freiberg, Germany, 16 July 1819), *mining.*

Trebra was the eldest son of Christoph Heinrich von Trebra, a courtier at Weimar, and his wife, born Albertina Amalia Karolina von Werder. After attending the monastery school in Rossleben, he studied law, philosophy, mathematics, and natural science for seven semesters at the University of Jena. Pursuing his inclination for science, he enrolled in 1766 in the newly founded mining academy at Freiberg, Saxony.

Only one year later Trebra was serving as assessor for the Bureau of Mines (Oberbergamt) at Freiberg. He rapidly acquired considerable technical knowledge and carried out his assignments to the complete satisfaction of his superiors. As a result he was appointed inspector of mines at Marienberg, Saxony, on 1 December 1767. He was resolute in enforcing the regulations and strict and merciless in punishing infractions. Since he also introduced many technical improvements into silver mining, he brought the industry in this region to a new level of prosperity.

In recognition of his services Trebra was named Commissioner of Mines (*kommissionsrat*) in 1770 and, in 1773, assistant supervisor of mines for the Bureau of Mines at Freiberg. In 1776 the duke of Weimar commissioned him to conduct a study of the underlying copper-bearing schist near Ilmenau. There he met Goethe, who had charge of the Ilmenau mines. The two remained good friends for nearly forty years.

In 1779 Trebra became an inspector of mines for the government of Hannover, working for more than a decade at Zellerfeld, in the Harz Mountains, and furthering the development of the metal-mining industry in the Upper Harz. His tireless efforts to increase the productivity of the mines under his jurisdiction finally led to his appointment, in 1791, as royal supervisor of mines for Great Britain and Electoral Brunswick-Lüneburg in the mining city of Clausthal.

For reasons that have not been established, Trebra resigned his post as inspector of mines in 1795 and retired to his estate, Bretleben, on the Unstrut River. There he devoted himself to breeding sheep, growing hemp, operating a distillery, and producing saltpeter and starch. This retirement was regretted by many, not least by Goethe.

A few years later Trebra accepted an offer from the government of Saxony of a post as chief mining inspector and director of the *Oberbergamt* of Freiberg. Upon assuming this office in 1801 he became head of the Saxon mining industry, retaining the post until his death in 1819. He was highly effective in this capacity; and in recognition of his outstanding accomplishments, he was appointed a commander of the Royal Order of the Saxon Civil Service.

BIBLIOGRAPHY

I. ORIGINAL WORKS. Trebra's books include *Erklärung der Bergwerks-Charte von dem wichtigsten Theile der Gebürge im Bergamtsrefier Marienberg* (Annaberg, 1770), also translated into Dutch; *Erfahrungen vom Innern der Gebirge nach Beobachtungen gesammelt* (Dessau–Leipzig, 1785), also translated into French; *Bergbaukunde,* edited with I. von Born, 2 vols. (Leipzig, 1789–1790); *Mineralien-Kabinet . . .* (Clausthal, 1795); *Das Silberausbringen des Königlich Sächsischen Erzgebirges auf die nächst verflossenen 40 Jahre von 1762 bis 1801* (Freiberg, 1802); *Merkwürdigkeiten der tie-*

fen Hauptstöllen des Bergamtsreviers Freyberg (Freiberg, 1804); and *Bergmeisterleben und Wirken in Marienberg* (Freiberg, 1818).

His articles include "Kalkhaltiges phosphorescirendes Steinmark," in *Chemische Annalen für die Freunde der Naturlehre*, pt. 1 (1784), 387–389; "Über die vom Himmel gefallenen Steine," in *Monatliche Correspondenz zur Beförderung der Erd- und Himmelskunde*, **9** (1804), 137; and "Über die innere Temperatur der Erde," *ibid.*, 349–350.

II. SECONDARY LITERATURE. See von Gümbel's notice on Trebra, in *Allgemeine deutsche Biographie*, XXXVIII (1894), 550–551; "Friedrich . . . von Trebra," in C. Schiffner, *Aus dem Leben alter Freiberger Bergstudenten*, I (Freiberg, 1935), 115–191; "Friedrich . . . von Trebra," in Walter Serlo, *Männer des Bergbaus* (Berlin, 1937), 145–146; Walter Herrmann, "Goethe und Trebra. Freundschaft und Austausch zwischen Weimar und Freiberg," in *Freiberger Forschungshefte*, ser. D, no. 9 (1955); and A. F. Wappler, *Oberberghauptmann von Trebra und die drei ersten sächsischen Kunstmeister Mende, Baldauf und Brendel* (Freiberg, 1906).

M. KOCH

TRELEASE, WILLIAM (*b.* Mount Vernon, New York, 22 February 1857; *d.* Urbana, Illinois, 1 January 1945), *botany*.

Trelease bridged the era of Asa Gray, whom he knew, and the twentieth century to become an internationally known botanist, teacher, and administrator. He was the son of Mary Gandall and Samuel Ritter Trelease, of Dutch and Cornish ancestry. After attending schools in Branford, Connecticut, and Brooklyn, New York, he entered Cornell University, where he studied under the botanist A. N. Prentiss and the entomologist John H. Comstock. While still an undergraduate he published four papers on pollination in *American Naturalist* and the *Bulletin of the Torrey Botanical Club*, and made field studies for the United States government on cotton insects in Alabama. Following his graduation (B.S., 1880), he entered Harvard to study fungi under William G. Farlow. During his Harvard year he came under the influence of Gray, Sereno Watson, George L. Goodale, and Samuel H. Scudder.

Beginning in 1881 Trelease taught systematic botany, horticulture, forestry, and economic entomology at the University of Wisconsin. To these courses he added bacteriology, the first time it was offered in the United States. In 1884 he was awarded the science doctorate at Harvard; his thesis concerned the zoogloeae. George Engelmann

and Gray urged Henry Shaw to appoint Trelease the head of the newly established Missouri Botanical Garden in St. Louis, popularly known as Shaw's Gardens, to be patterned after the Royal Botanic Gardens at Kew. In twenty years Trelease brought together a garden of 12,000 species under cultivation, a notable library of some 70,000 books (including the E. L. Sturtevant pre-Linnaean collection), and a herbarium of 700,000 specimens.

Following his resignation in 1912 and a year in Europe, Trelease assumed the headship of the department of botany at the University of Illinois, where he remained until his retirement in 1926. Among his 300 published papers and books were monographs on the genus *Agave*, mistletoes, and oaks, and an unfinished work on Piperaceae begun when he was seventy-five years old. Trelease described more than 2,500 species and varieties of plants. Addicted to bibliographic thoroughness, he was able to provide in his work a base for modern research involving supplemental techniques. Nonetheless, Trelease found time to serve as president of the leading organizations of his profession.

BIBLIOGRAPHY

I. ORIGINAL WORKS. For a bibliography of Trelease's writings see Kunkel (below). Trelease's noteworthy writings include *Botanical Works of the Late George Engelmann* (Cambridge, 1887), with Asa Gray; articles for L. H. Bailey's *Cyclopedia of American Horticulture* (New York, 1900–1902); *Agave in the West Indies* (Washington, D.C., 1913); *The Genus Phoradendron* (Urbana, Illinois, 1916); and the *American Oaks* (Washington, D.C., 1924).

The Piperaceae of Northern South America (Urbana, 1950) was completed by T. G. Yuncker. Trelease's papers on Mexican botany are listed with annotations in Ida K. Langman, *A Selected Guide to the Literature on the Flowering Plants of Mexico* (Philadelphia, 1964).

Four scrapbooks of letters, clippings, and memorabilia are preserved at the Missouri Botanical Garden.

II. SECONDARY LITERATURE. The fullest sketch is L. O. Kunkel, in *Biographical Memoirs. National Academy of Sciences*, **35** (1961), 307–332, with a portrait and bibliography of his writings by T. G. Yuncker for the years 1879–1950. See also L. H. Bailey in *Yearbook. American Philosophical Society* (1945), 420–425; and J. M. Greenman, in *Missouri Botanical Garden Bulletin*, **33** (1945), 71–72. Personal tributes are J. Christian Bay, *William Trelease, 1857–1945, Personal Reminiscences* (Chicago, 1945); and L. H. Pammel, *Prominent Men I Have Met*, III (Ames, Iowa, 1927), both privately printed.

JOSEPH EWAN

TREMBLEY, ABRAHAM (*b.* Geneva [now Switzerland], 3 September 1710; *d.* Petit Sacconex, near Geneva, 12 May 1784), *zoology.*

Abraham Trembley's father, Jean, was an officer in the Genevan army and rose to be its commander, the syndic of the guard. Abraham was educated at the Academy of Geneva, and in 1733 found employment as a tutor in Holland. His career was much influenced by his residence at Leiden, where he was in close touch with several distinguished scientists in the university. It was probably here that he met Count Bentinck, a curator of the university; and he became tutor to the latter's two sons at the mansion of Sorgvliet, near The Hague. Here he carried out the researches on the hydra that gained him a fellowship in the Royal Society in 1743 and made him famous. The duke of Richmond witnessed some of his experiments at Sorgvliet, and young Trembley made a deep impression on him. On his deathbed in 1750, the duke consigned to Trembley the care and education of his son and heir, then aged fifteen. Trembley conducted the youth on an extensive tour of the Continent. When they parted in 1756, he was so generously rewarded by the young duke that it was never again necessary for him to work for a living. In 1757 he returned to Geneva, married, and bought the country house at Petit Sacconex in which he lived for the rest of his life. Trembley devoted himself mainly to the instruction of his children and to writing books on education, politics, religion, and moral philosophy. His methods of instruction were novel and anticipated many later developments in educational theory. He was deeply religious, a Christian without strictly adhering to any particular sect.

It was in June 1740 that Trembley had his first opportunity to study the hydra. He had taken into his room at Sorgvliet some ditch water containing aquatic plants. He noticed a green hydra (*Chlorohydra viridissima*) attached to one of the plants, and at first took it to be itself a plant. He noticed, however, that the tentacles moved, and soon found that organisms of this species could change their positions; but he still hesitated to call them animals. He decided to determine the matter by cutting one of them in two, on the supposition that if both parts survived, the organism would clearly be classifiable as a plant. The cut was made in such a way that one of the parts possessed all the tentacles. He watched the regeneration of both fragments during the following days until regeneration was complete, when both parts were tentaculate; they were indistinguishable from one another.

Nevertheless, the apparent spontaneity of the hydra's movements made it difficult for him to accept the conclusion that these were in fact plants. To obtain help in reaching a decision, he sent some specimens to Réaumur and wrote him a very long letter about his observations. The letter was read in full at two sessions of the Académie Royale des Sciences in Paris in March 1741. Living specimens were also exhibited at the Academy.

Trembley accepted Réaumur's assurance that the organisms were unquestionably animals. A related species, probably the brown *Hydra vulgaris*, had in fact been discovered by Leeuwenhoek in 1702, and a note on it had been published. Professor Émile Guyénot has shrewdly remarked that if Trembley had been aware that animals of this sort were already known he would probably not have studied them so attentively; and if his first specimen had been *H. vulgaris* or *Pelmatohydra oligactis* (also brownish), he would not have been likely to think it was a plant. It would not then have occurred to him to make the experiment that resulted in the discovery of multiplication by artificial division. Trembley eventually studied all three species, but their modern names were not introduced in his time.

Réaumur checked for himself the truth of Trembley's account, and both Bentinck and Richmond wrote to the Royal Society to confirm it. It is not easy today to recapture the sense of utter amazement caused by the realization that an animal could be multiplied by cutting it in pieces. Voltaire refused to believe it. Long afterwards he still persisted in denying the possibility and dismissed the subject contemptuously, describing the animal in print as "a kind of small rush."

Trembley pursued his investigations on the hydra. He described the characteristics of the hydra's living substance (afterwards named protoplasm); investigated its reactions to light and to changes in temperature; watched it capturing its prey, eating, and digesting; and he found that *vulgaris* and *oligactis* could be colored red, black, or a feeble green by supplying suitable foods. His feeding experiments led to the introduction of a particularly interesting technique. Drawing no sharp distinction between the granules of the endoderm on one hand and the nematocysts of the ectoderm on the other, he thought that the animal might perhaps be able to nourish itself if turned inside out. He succeeded in this operation, despite the almost incredible difficulty presented by the minute size of the animals. He proved that a reversed *H. vulgaris* can, indeed, survive and feed without re-reversal, that is, without returning to its original form. Long afterwards there were still biologists who could not believe this, but the experiment was successfully repeated by M. Nussbaum in 1887. The ectoderm cells of reversed hydras are known nowadays to be able to migrate singly or in groups to the exterior, without any re-reversal of the organism as a whole.

Trembley devoted much time to a detailed study of the budding process, regeneration, and the artificial production of monsters. He then set to work to find out whether multiplication by division could be achieved in reverse, that is, whether two hydras could be transformed into one. This operation was perhaps the most dramatic of all his experiments. He placed one hydra within another, pushing it in tail first through the mouth of the other specimen until its tail projected from near the tail of the other, while its head had still not entered the mouth. The posterior end of the internal animal was permanently grafted, and a single individual with two tails resulted, one derived from each of the two original animals. This, the first permanent graft of animal tissues, was done in October 1742. Both constituent parts of the composite organism multiplied by budding. Several modifications of the grafting process were carried out successfully.

It is noteworthy that Trembley tried on hydra nearly every possible experiment in regeneration and grafting that was likely to give interesting results. He was exploring territory that was almost entirely unknown; yet the planning of his work was so complete, and its execution so thorough, that he might almost have been performing routine experiments on a system established by the long experience of others. The only important exception is that he did not try to find out whether the polarity of a body part could be reversed so that its anterior-posterior axis could be changed.

Trembley was persuaded by Réaumur to bring together the results of his work on the hydra in a book. The result was a beautifully produced volume entitled *Mémoires, pour servir à l'histoire d'un genre de polypes d'eau douce, à bras en forme de cornes* (1744), with very fine illustrations by Pieter Lyonet.

Although Trembley's study of the hydra is the best known of his works, it is arguable that some of his other researches are of equal importance. He was the first to describe multiplication and colony formation in Protozoa and multiplication by budding in Oligochaeta and Polyzoa, and he was also the first to give a full account of the anatomy of the individual in any member of the latter group. Until he discovered reproduction by budding in Oligo-

chaeta and Polyzoa, it had never been seen in any animal other than the hydra. Some twenty years later, while living at Petit Sacconex, he was the first to witness cell division in the strict sense of the term (division of a uninucleate cell). Still, it was not possible in Trembley's time for him or anyone else to recognize the full significance of some of his discoveries. When he discovered that Protozoa multiplied by division, for example, no one had yet suggested the separation of these animals into a distinct phylum; and when he watched cell division in *Synedra*, it was not known that this diatom was a single cell.

Trembley was above all a student of processes, and in this respect he led the way in zoology. It is often said that he was the father of experimental zoology, but it is doubtful whether he would have approved of this description. He used both observation and experiment freely, according to which method was most likely to serve his ends in a particular case. His only purely morphological study, however, was on the anatomy of the *Lophopus* (Polyzoa). G. J. Allman, in his *Monograph of the Fresh-Water Polyzoa* (1856), said of Trembley's work on this animal, "The description is wonderfully accurate, and the anatomical details have been in few points surpassed by subsequent observers." Trembley's accuracy of observation was one of his most striking characteristics. He had a passion for demonstrable facts and perhaps an exaggerated dislike of hypothesis.

BIBLIOGRAPHY

I. ORIGINAL WORKS. Important works by Trembley are *Mémoires, pour servir ǎ l'histoire d'un genre de polypes d'eau douce, à bras en forme de cornes* (Leiden, 1744); and *Instructions d'un père à ses enfans, sur la nature et sur la religion* (Geneva, 1775).

II. SECONDARY LITERATURE. Works on Trembley are J. R. Baker, *Abraham Trembley of Geneva: Scientist and Philosopher, 1710–1784* (London, 1952); [Jean Trembley] *Mémoire historique sur la vie et les écrits de Monsieur Abraham Trembley* (Neuchâtel, 1787); and Maurice Trembley, *Correspondence inédite entre Réaumur et Abraham Trembley comprenant 113 lettres recueillies et annotées par Maurice Trembley*, with an introduction by É. Guyénot (Geneva, 1943).

JOHN R. BAKER

TREUB, MELCHIOR (*b.* Voorschoten, Holland, 26 December 1851; *d.* St. Raphael, France, 3 October 1910), *botany*.

Treub's father was burgomaster of Voorschoten, a small village; both his parents were of Swiss extraction, and spoke French at home. Treub himself was also fluent in English, German, and Dutch, but preferred French for his scientific publications. He attended secondary school at Leiden, then entered the university there, where he won a gold medal for a report on the true nature of lichens, written in response to a prize question. He expanded this work into the thesis for which he was granted the doctorate, *summa cum laude*, in 1873. Treub remained at Leiden for the next seven years as docent and assistant in the botanical institute under his former tutor, Professor W. F. R. Suringar. During this period he wrote twenty-nine papers on a variety of subjects—especially plant cytology, histology, nucleus division, and embryology—that demonstrated his acute powers of observation, manual dexterity in cutting sections, and skill as a draftsman. He was nominated to the Netherlands Academy of Sciences at the unusually early age of twenty-eight.

In 1880 Treub was called to the directorship of the botanical gardens at Buitenzorg (now Bogor), West Java, where he succeeded R. H. C. C. Scheffer. The complex consisted of the gardens themselves; a building housing administrative offices, a museum, an herbarium, and a library; the Cultuurtuin, a garden devoted to the cultivation of plants of economic importance, attached to a school of agriculture; and a mountain garden at Tjibodas (at 1,400 meters). The institution also published two periodicals, the *Annales du jardin botanique de Buitenzorg* and the *Jaarverslag*, its annual report, but these, like the entire establishment, had fallen into disarray. Treub set out to revitalize the institution, and in so doing developed formidable administrative skills and powers of persuasion.

Since a scientific center was, in a colonial commercial society, almost a luxury, Treub sought to publicize its merits. Many decisions about the Buitenzorg garden came from The Hague, which also set the budget, so Treub therefore wrote a number of articles for a literary journal in the Netherlands, by which he made the public aware of the economic potential of the colony. He published the annual reports of the garden punctually, and used these documents as a means of informing both the government authorities and the general public about the work being done at the garden and about its goals. Through the *Annales* and through a series of basic studies (on the embryology of Cycadaceae, Loranthaceae, and Burmanniaceae, and on the latex of Euphorbiaceae, classes of climbing plants, pitcher plants, and the structure of plants infested by ants) Treub also won the attention of the international botanical community.

By 1881 Treub had redesigned the curriculum of the agricultural school and had hired an assistant director, W. Burck, who began to develop the museum along economic botanical lines. Burck also did research on caoutchouc (rubber) trees and on commercial products of the Dipterocarpaceae and cultivated caoutchouc trees in a special garden at Tjipetir. The results of his work were too voluminous to be published in the annual report, so Treub got permission to start a new periodical, the *Mededeelingen*, which first appeared in 1883. As foreign scientists began to visit the gardens in increasing numbers, Treub equipped a nearby abandoned military hospital as a laboratory; this opened in January 1885, and became known as the Treub Laboratory. The results of the work done there were published in the *Annales*; Treub had thus assured himself of the unpaid services of a whole staff of eminent botanists.

Treub's campaign was successful and both the government and the community of planters became convinced of the value of applied scientific research. He secured additional staff, again at no cost to the institution, by suggesting an arrangement whereby the planters assumed the salaries of his assistants, while the government agreed to let their work proceed under Treub's supervision and their results be published in the *Mededeelingen*. Under this system, researches were made concerning diseases of tobacco, tea, coffee, indigo, and cacao plants. A chain of experimental stations was established since much of the necessary work could not be performed in Buitenzorg's climate and soil conditions.

As the facilities and staff for applied botanical research were expanded, Treub began to plan new divisions for basic research. In 1887, while on sick leave in Europe, he approached both the Netherlands and foreign governments and obtained "Buitenzorg funds," annual or biennial grants for prominent botanists to work in Java; during the same leave, he also sought private funds to finance explorations of Indonesia. (The latter effort resulted in the foundation, in 1890, of the Society for the Promotion of the Natural Sciences in the Netherlands Indies, a body that still exists and is commonly called the "Treub Society.")

Following his return to Java, Treub supervised the further expansion of the botanic garden. A phytochemical-pharmacological laboratory was established in 1887 and an agricultural chemistry laboratory in 1890; by 1894 Treub was able to found a zoological museum and to hire an agricultural zoologist, J. C. Koningsberger, whose work eventually led to the foundation of an institute of plant diseases. He further acquired a forest preserve adjacent to the mountain garden at Tjibodas, and in 1891 erected a field laboratory there.

During this time Treub also conducted a major study on rice; his own research on food crops, coupled with his new laboratories, was the nucleus from which the Java General Agriculture Experiment Station grew. In addition, Treub encouraged the publication of the journal *Teysmannia* (begun in 1894), which was devoted largely to practical short notes from the botanic garden; in 1900 he also laid out a series of demonstration fields that were intended to benefit farmers as well as administrators. A new library was built in 1896, and by 1905 the complex that Treub had built up was elevated to a full civil department, of which Treub was named director. He continued to serve at the same time as director of the gardens themselves.

A number of significant botanical studies were published during Treub's tenure as director, chief among them a *Flore de Buitenzorg* and the *Icones Bogoriensis*. In 1894 he started yet another periodical, the *Bulletin du jardin botanique de Buitenzorg*, which was designed to reach the international scientific community. Despite the pressures of his administrative duties, Treub also did research of his own and published his results. These studies include a work on the embryology and biohistory of the club ferns, a study of a previously unknown form of fertilization in Casuarina (chalazogamy), writings on the embryology of Ficus and Elatostema, a theory of the origin of proteins based on the finding of prussic acid in Pangium, and works on the sociology of the rain forest and the new flora of Krakatoa. He also, in 1886, wrote a history of the Buitenzorg botanic garden.

By the end of 1909 Treub's health had begun to fail, and he retired from his posts to the south of France, where he planned to start a garden of exotic plants and to pursue the microscopic study of tropical plants. He was able to publish only one paper, on the embryology of Garcinia, before his death a year later.

BIBLIOGRAPHY

I. ORIGINAL WORKS. The notice by F. A. F. C. Went, cited below, lists some 103 books and papers by Treub; of these the scientific works are in French, while a number of others on a variety of subjects are in Dutch. See especially Treub's commentaries in the *Jaarverslag* for the years 1880 to 1910 and, of the individual works, *Onderzoekingen over de natuur der Lichenen*, his thesis (Leiden, 1873); *Onderzoekingen over serehziek suikerriet gedaan in 's-Lands Plantentuin te Buitenzorg* (Djakarta, 1885); *Geschiedenis van 's-Lands Plantentuin te Buitenzorg* (Djakarta, 1889); *Korte geschiedenis van 's-Lands Plantentuin* (Djakarta, 1892); *De beteekenis van tropische botanische tuinen* (Djakarta, 1892); *Over de taak en werkkring van 's-Lands Plantentuin te Buitenzorg* (Buitenzorg, 1899); and the posthumously published *Landbouw, Januari 1905-October 1909. Beredeneerd overzicht der verrichtingen en bemoeiingen met het oog op de praktijk van land-, tuin-, en boschbouw, veeteelt, visscherij en aanverwante aangelegenheden* (Amsterdam, 1910).

II. SECONDARY LITERATURE. See F. A. F. C. Went, "In Memoriam," in *Annales du jardin botanique de Buitenzorg*, 2nd ser., **9** (1911), i–xxxii.

C. G. G. J. VAN STEENIS

TREVIRANUS, GOTTFRIED REINHOLD (*b.* Bremen, Germany, 4 February 1776; *d.* Bremen, 16 February 1837), *zootomy, physiology.*

Treviranus was the eldest son of Joachim Johann Jacob Treviranus, merchant and later a notary, and brother of the botanist Ludolph Christian Treviranus. He introduced the notion of biology as a distinct discipline into Germany and was one of the first to express the idea that the cell is the structural unit of living matter.

Treviranus was a keen observer and an able theorizer. As an adherent of *Naturphilosophie*, he supposed that the universal laws of nature were represented by general ideas and that these ideas were the instruments to throw the light of knowledge over the darkness of the world of facts and phenomena. All his studies centered on the theme "What laws underlie living nature?" In order to answer this question, he studied life in all its aspects: generation, food and feeding, psychical processes, geographical distribution, the interaction between mind and body, and the relationship between the organism and its environment. Even before Lamarck published his *Philosophie zoologique*, Treviranus stated that changes in the physical environment could induce corresponding changes in organic structures—or, as he wrote in his *Biologie* (1802), any living creature has the ability to adapt its organization to changing external conditions. As a consequence Treviranus was an opponent of Cuvier's theory of catastrophism. Both Ernst Haeckel and August Weismann considered Treviranus as one of their predecessors. He was so, however, only within the limits of *Natur-philosophie*; nowhere did Treviranus explain how changes in organic structures were induced nor how they could become hereditary.

From 1793 to 1796 Treviranus studied medicine and mathematics at the University of Göttingen. In the latter year, after having written his doctoral dissertation, he returned to Bremen, where he established a practice as a consulting physician. In his spare time he made observations on the structure and function of living beings, especially of the invertebrates. He was an accomplished microscopist. He prepared most of his beautiful illustrations himself.

In 1797 Treviranus was appointed professor of mathematics and medicine at the Bremen lyceum, and in the same year he published the first volume of *Physiologische Fragmente*. In 1800 he published the results of a series of experiments concerning the influence of certain chemical substances—particularly of drugs such as opium and belladonna—on plants and animals.

Soon after he had settled in Bremen, Treviranus wrote his magnum opus, *Biologie* (1802–1822), in which he sought to summarize all basic knowledge of his time about the structure and function of living matter. This six-volume work, intended to be a modernization of Haller's *Elementa physiologiae*, greatly influenced his contemporaries. Volume I treats the interpretation of living nature, the basic laws of biology, the empirical basis of biology, the definition of life, and the principles of classification; volume II, the organization of living nature, the distribution of plants and animals, and the influence of external conditions; volume III, growth and development; volume IV, nutrition and the digestive systems in plants and animals, respiration, and circulation; volume V, animal health, animal electricity, phosphorescence, automatic movements in animals and plants, the organization of the nervous system, and instinct; and volume VI, relation between the physical and the psychical worlds, the mind-body problem, and the objective and subjective worlds of the senses.

Treviranus published his ideas in a more condensed form in *Erscheinungen und Gesetze* (1831–1832), a classic in theoretical biology. In this work he also incorporated an account of the advances made in biology during the preceding thirty years. He paid greater attention to mind-body relations, the process of generation, periodicity, constitution, health and disease, and temperament. Some of these subjects were considered in greater detail in *Beiträge zur Aufklärung* . . .

(1835–1837). In the first and third volumes, Treviranus tried to prove mathematically that the structure of the crystalline lens is responsible for man's ability to see things in perspective; the second volume contains microscopical investigations of animal tissues.

These theoretical studies induced Treviranus to undertake research in many fields; the results have been recorded in some of his collections, such as *Vermischte Schriften* (1816–1821); *Beobachtungen aus der Zootomie*, published by his brother (1839); and *Zeitschrift für Physiologie*, which he founded in 1824 with his brother Ludolph Christian and Friedrich Tiedemann.

Of particular interest are Treviranus' anatomical studies on invertebrates, such as those he made on the reproductive organs of worms, mollusks, insects, and arachnids; and on the respiratory and circulatory organs of crustaceans and other invertebrates. Also noteworthy are his monographic studies on the anatomy of the louse, of wingless insects, of snails, and of arachnids. In addition Treviranus performed many microscopic-anatomical studies on vertebrates: the reproductive organs of fishes, amphibians, the tortoise, the mole, the hedgehog, and the guinea pig; the ears of birds and the eye of the narwhal; and the nervous system of birds.

Treviranus paid special attention to the anatomy and physiology of the sensory organs. In *Beiträge zur Anatomie* he formulated mathematical laws of diffraction in order to discover the physical basis of vision. More particularly he was interested in determining which mechanism in the eye is responsible for our seeing things in their relative positions and what the function may be of such structures as the cornea, lens, and retina. This work contains many comparative studies of the visual processes in the various classes of animals.

BIBLIOGRAPHY

I. ORIGINAL WORKS. Treviranus' earlier writings include *De emendanda physiologia* (Göttingen, 1796), his inaugural dissertation; *Physiologische Fragmente*, 2 vols. (Hannover, 1797–1799); "Über den Einfluss des galvanischen Agens . . .," in *Nordisches Archiv für Natur- und Arzneiwissenschaft*, **1**, no. 2 (1800), 240–305; *Biologie, oder Philosophie der lebenden Natur für Naturforscher und Aerzte*, 6 vols. (Göttingen, 1802–1822); "Ueber den innern Bau der Arachniden," in *Abhandlungen der Physikalisch-medizinischen Societät in Erlangen* (1812), no. 1, 1–48; *Vermischte Schriften anatomischen und physiologischen Inhalts*, 4 vols.

(Göttingen–Bremen, 1816–1821), written with L. C. Treviranus; "Ueber die Zeugungstheile und die Fortpflanzung der Mollusken," in *Zeitschrift für Physiologie*, **1** (1824), 1–55; "Ueber den innern Bau der Schnecke des Ohrs der Vögel," *ibid.*, 188–196; "Ueber die Harnwerkzeuge und die männlichen Zeugungstheile der Schildkröten überhaupt und besonders der *Emys serrata*," *ibid.*, **2** (1827), 282–288; *Beiträge zur Anatomie und Physiologie der Sinneswerkzeuge des Menschen und der Thiere* (Bremen, 1828); "Ueber das Gehirn und die Sinneswerkzeuge des virginischen Beutelthieres," in *Zeitschrift für Physiologie*, **3** (1829), 45–61; and "Ueber die Entstehung der geschlechtslosen Individuen bei den Hymenopteren, besonders der Bienen," *ibid.*, 220–234.

Later works are "Ueber die hinteren Hemisphären des Gehirns der Vögel, Amphibien und Fische," in *Zeitschrift für Physiologie*, **4** (1831), 39–67; "Ueber das Nervensystem des Scorpions und der Spinne," *ibid.*, 89–96; "Ueber die Zeugung der Egel," *ibid.*, 159–167, and **5** (1833), 133–136; *Die Erscheinungen und Gesetze des organischen Lebens*, 2 vols. (Bremen, 1831–1832); "Ueber die Verbreitung des Antlitznerven im Labyrinth des Ohrs der Vögel," in *Zeitschrift für Physiologie*, **5** (1833), 94–96; "Ueber die Zeugung des Erdregenwurms," *ibid.*, 154–156; *Beiträge zur Aufklärung der Erscheinungen und Gesetze des organischen Lebens*, 3 vols. (Bremen, 1835–1837); and *Beobachtungen aus der Zootomie und Physiologie*, L. C. Treviranus, ed. (Bremen, 1839).

II. Secondary Literature. Biographies are by G. Barkhausen, G. H. Schumacher, and G. Hartlaub, in *Biographische Skizzen verstorbener Bremischer Aerzte und Naturforscher* (Bremen, 1844), 433–590; and K. F. P. von Martius, *Akademische Denkreden* (Leipzig, 1866), 55–69. There is a critical evaluation of Treviranus' study of mollusk reproduction by Lorenz Oken, in *Isis* (1827), esp. 752–754.

P. Smit

TREVIRANUS, LUDOLPH CHRISTIAN (*b*. Bremen, Germany, 18 September 1779; *d*. Bonn, Germany, 6 May 1864), *plant anatomy, plant physiology.*

Treviranus was particularly interested in the basic structural unit of living beings—the cell—and in the forces causing structural differentiation, and he therefore studied the various aspects of the relation between structure and function during the ontogeny of plants.

Ludolph Christian was three years younger than his brother, Gottfried Reinhold. Their father, Joachim Johann Jacob Treviranus (*d*. 1806), was a merchant who, after a business failure, became a notary.

Treviranus studied medicine at the University of Jena, where F. W. Schelling was one of his teachers. The greatest philosophical influence on him, however, was that of his brother Gottfried. As an adherent of *Naturphilosophie*, Ludolph Christian sought the causes of the vital processes in life itself; and, like his brother, he denied any difference between organic and inorganic forces, and rejected any physical explanation of the phenomena of life.

After completing his doctoral dissertation (1801), Treviranus established himself as a consulting physician in Bremen. On the basis of his work on the inner structure of plants (1806), which contains interesting observations on the structure and function of the reproductive organs, the University of Rostock offered him the chair of botany. In 1816 he went to Breslau; and in 1830 he and Nees von Esenbeck of Bonn exchanged positions. Treviranus showed a lively interest in the organization and scientific function of botanical gardens; his writings in this field (1843, 1848), however, caused him much trouble, for his ideas did not win the support he needed from colleagues in either Breslau or Bonn.

Treviranus' major work is *Physiologie der Gewächse* (1835–1838), a study devoted to the relation between structure and function in plants. It deals with differences between plant and animal life, the elementary structures of plants (cells, sap and spiral vessels), tissues and organs, movements of sap, excretion of water and absorption of light by the leaves, growth and generation, various forms of reproduction, environmental influences, rest and movements of plants, and length of life. The book is still of interest because of the authoritative historical introductions to the subjects, an approach that was characteristic of all Treviranus' publications. Its practical contents were of less value. Treviranus preferred to use simple magnifying glasses, which prevented him from discovering such fundamental phenomena as the process of fertilization and the formation of the seeds; thus many of his observations were superseded during his lifetime. According to his point of view, influenced by *Naturphilosophie*, fertilization—in plants as well as in animals—should be mediated by a "palpable matter" (1822, 1831).

Of more importance from a purely physiological point of view were Treviranus' observations on the influence of chemical substances on plants, a series of experiments that were continued by Heinrich Göppert, his pupil at Breslau; his research on the movement of particles in the cell; and his studies on the movement of sap in trees (1811, 1817).

With his brother Gottfried, Treviranus published a series of essays on anatomy and physiology (1816–1821). The fourth volume (1821) contains contributions by Ludolph Christian on the formation and structure of the epidermis, the structure of stomata, honeydew as an excretion of plant lice and as a symptom of plant disease, sexuality and germination of plants, propagation by means of bulbs, and reproduction in cryptogamic plants.

With Friedrich Tiedemann the brothers founded *Zeitschrift für Physiologie* (1824), which includes essays by Ludolph Christian on the anatomy and function of sap vessels, the structure of the reproductive organs, the evaporation of water in plants, luminescence and the production of heat in plants, and the impossibility of visualizing the act of fertilization (1831).

Treviranus' interest in history and the arts culminated in a booklet on woodcuts (1855), in which he considered their use as an aid to botanical learning in the Renaissance (Otto Brunfels, Jerome Bock, Leonhard Fuchs), their improvement (by Conrad Gesner, Matthioli, Dodoens, L'Obel, L'Écluse [Clusius], and Rudolf Camerarius), their decline in the seventeenth century, and their revival in the eighteenth century (Thomas Bewick). In the last years of his life, Treviranus wrote a series of monographs on plant genera, including *Delphinium*, *Aquilegia*, *Durieua*, *Astilbe*, and *Lindernia*.

BIBLIOGRAPHY

Among Treviranus' earlier works are *Quaedam ad magnetismum sic dictum animalem spectantia* (Jena, 1801), his doctoral dissertation; "Vom Bau der cryptogamischen Wassergewächse," in F. Weber and M. Mohr, eds., *Beiträge zur Naturkunde*, I (1805); *Vom inwendigen Bau der Gewächse und von der Saftbewegung in denselben* (Göttingen, 1806); *Beyträge zur Pflanzenphysiologie* (Göttingen, 1811), which includes the German trans. of a series of essays by Thomas Knight on the movement of sap in trees; "Ueber die Ausdünstung der Gewächse und deren Organe," in G. R. and L. C. Treviranus, *Vermischte Schriften . . .*, I (Göttingen, 1816), 171–188; "Fernere Beobachtungen über die Bewegung der grünen Materie im Pflanzenreiche," *ibid.*, II (Bremen, 1817), 71–92; "Abhandlungen phytologischen Inhalts," *ibid.*, IV (Bremen, 1821), 1–222; *Die Lehre vom Geschlechte der Pflanzen in Bezug auf die neuesten Angriffe erwogen* (Bremen, 1822); "Ueber den eigenen Saft der Gewächse . . .," in *Zeitschrift für Physiologie*, 1 (1824), 147–180; "Bemerkungen über den Bau der Befruchtungstheile und das Befruchtungsgeschäft der Gewächse," *ibid.*, 2 (1827), 185–187; "Etwas über die wässerigen Absonderungen blättriger Pflanzentheile," *ibid.*, 3 (1829), 72–78; and "Entwickelt sich Licht und Wärme beim Leben der Gewächse?" *ibid.*, 257–268.

Later works are *Caroli Clusii Atrebatis et Conradi Gesneri Tigurini, epistolae ineditae* (Leipzig, 1830); "Gelangt die Befruchtungsmaterie der Gewächse zu deren Samen-Anlagen auf eine sichtbare Weise?" in *Zeitschrift für Physiologie*, 4 (1831), 125–145; *Physiologie der Gewächse*, 2 vols. (Bonn, 1835–1838); "Ueber die Gattung *Lindernia* . . .," in *Linnaea*, 16 (1842), 113–126; *Theorie der Gartenkunde . . .* (Erlangen, 1843), German trans. of John Lindley, *The Theory of Horticulture* (London, 1840); *Bemerkungen über die Führung von botanischen Gärten . . .* (Bonn, 1848); *Caricis specierum in imperio Rutheno huiusque lectarum enumeratio* (Stuttgart, 1852); "Ueber die umbelliferen-Gattung *Durieua*," in *Botanische Zeitung*, 11 (1853), 193–195; "Ueber die Gattung *Astilbe*," *ibid.*, 13 (1855), 817–820; *Die Anwendung des Holzschnittes zur bildlichen Darstellung von Pflanzen nach Entstehung, Blüthe, Verfall und Restauration* (Leipzig, 1855; repr. Utrecht, 1949); "Ueber einige Stellen in des älteren Plinius Naturgeschichte der Gewächse," in *Botanische Zeitung*, 17 (1859), 321–325; and "Lebens-Abriss von Ludolph Christian Treviranus," *ibid.*, 24 (1866), supp., with an incomplete bibliography.

A biography is in K. F. P. von Martius, *Akademische Denkreden* (Leipzig, 1866), 523–538.

P. SMIT

TREVISANUS. See Bernard of Trevisan.

TRIANA, JOSÉ GERÓNIMO (or **JERÓNIMO**) (*b.* Zipaquirá, Colombia, 1826; *d.* Paris, France, October 1890), *botany.*

The botanical sciences were not firmly established in Colombia until the 1930's, when the Colombian National Herbarium and the Botanical Institute of the National University were founded at Bogotá. Until then the investigation of the rich and interesting Colombian flora had been carried on by a series of more or less self-taught individuals without institutional affiliation. Working with very little bibliographical information, they had to deal with an extremely mountainous area in which each climatic zone and mountain range posed different problems. Thus it is all the more extraordinary that from the arrival of José Celestino Mutis at Cartagena in 1760 until the 1930's botanical study was uninterrupted in Colombia.

Triana began the study of botany at Bogotá as the private pupil of Francisco Javier Matis, the last survivor of the Mutis botanical expedition. In 1850 the government appointed him associate botanist on the commission (headed by Agustín Codazzi) charged with preparing and publishing a geographical map of Colombia that would establish its borders. Triana traveled with the commission throughout Andean Colombia collecting plants in order to verify his catalog of the flora, and recording the place and date of collection, barometric pressure, common name, and popular use of each species. Triana's herbarium, amassed through tedious observations, was presented to the government in 1855 in thirty-eight volumes, each containing about 100 species. He also gathered data on the ethnobotanic legacy of Colombia: botanic medical traditions of the native medicine men, the raw materials of primitive industries, and the home remedies of the mestizo societies. Together with his herbarium, Triana's data on economic and medicinal botany established his scientific reputation both in Colombia and in Europe.

While still in Colombia, Triana had begun correspondence with foreign botanical explorers of Colombia, including Luis Schlim, J. Linden, Julian Warscewiez, J. J. Jewies, H. Holton, and Herrmann Karsten. Interest in the Colombian flora increased in the botanical centers and gardens of France, England, Belgium, and Germany; and in 1850–1857 Triana benefited from the company of explorers sent from Europe to be trained as botanists.

In 1856 Triana was commissioned by the government to go to Europe for two years in order to publicize Colombian plants of economic value. When he reached Paris, Triana met the botanist Descaine. He collaborated on the publication of the *Flora de la Nueva Granada* with Jules Planchon, with whom he worked at Montpellier in 1858 and 1859; at the end of this time he left ready for the press *Mémoire sur la famille des guttifères*, which appeared in 1860. At the end of his career he won the esteem of Filippo Parlatore, professor of botany at Florence.

In 1865, at the Horticultural Exposition of Amsterdam, Triana presented his *Monografía de las melastomáceas*, which was published at London in 1871 and was awarded the Candolle Prize.

Triana's greatest ambition was to publish a flora of his country; but the government, impoverished by civil wars, withdrew financial support. He therefore chose to prepare a systematic work in collaboration with Planchon and others, *Prodromus florae Novogranatensis, ou énumération des plantes de la Nouvelle Grenade*. Triana used the illustrations and descriptions prepared by Mutis and Francisco José Caldas for their *Historia de los*

árboles de quina, which were in Madrid, when he published his *Nouvelles études sur les quinquinas* (1870).

Triana's study of the bibliographical sources started in London (1865–1867). He traveled to Madrid twice in order to study Mutis' material on quinine and prevailed upon Queen Isabella II to instruct the administrators of the Madrid Botanical Garden to give him access to the icons. Triana immediately made a new and extensive systematic recension of the *Cinchona* species. He also obtained permission to present Mutis' *Quinología* at the Universal Exposition of Paris in 1867. This exhibit brought him many honors and prizes.

Two of Triana's most valuable botanical works are still unpublished, "Catálogo metódico de los dibujos de la Flora de Nueva Granada hechos bajo la dirección de don José Celestino Mutis" and "Catálogo de los ejemplares que componen el herbario formado por José J. Triana. . . ."

In 1889 Triana was struck by a carriage, an accident that apparently aggravated some of his preexisting ailments. He underwent an operation in 1890 and died that October. In 1857, fifteen days before his departure from Bogotá for Europe, he married Mercedes Umaña, who bore him fifteen children.

BIBLIOGRAPHY

Triana's writings include *Nuevos jeneros i especies de plantas para la flora Neo-Granadina* (Bogotá, 1854); *Prodromus florae Novogranatensis*, 2 vols. (Paris, 1862–1867), written with J. E. Planchon, which also appeared as a series of memoirs in *Annales des sciences naturelles* (Botany), 3rd ser., **17**–4th ser., **17** (1862–1873); *Nouvelles études sur les quinquinas, d'après les matériaux présentés en 1867 à l'Exposition universelle de Paris* (Paris, 1870); and "Les mélastomacées," in *Transactions of the Linnean Society of London*, **28** (1871), 1–188, also published separately (London, 1871).

On his life and work, see *Abhandlungen herausgegeben vom Naturwissenschaftlichen Verein zu Bremen*, **3** (1873), 393–403; and **5** (1878), 29–33; and *Journal of Botany, British and Foreign*, **29** (1891), 46–47.

ENRIQUE PÉREZ ARBELÁEZ

TRILIA, BERNARD OF. See **Bernard of Le Treille.**

TRISMEGISTUS. See **Hermes Trismegistus.**

TROJA, MICHELE (*b.* Andria, Apulia, Italy, 23 June 1747; *d.* Naples, Italy, 12 April 1827), *medicine.*

Troja studied medicine at Naples and then won a competition for the post of assistant surgeon at that city's Hospital of San Giacomo degli Spagnoli. In 1774 he obtained a scholarship for postgraduate study that enabled him to go to Paris, where he began the research on the formation of bone callus and bone regeneration that made him famous. For this work, which continued that of Henri-Louis Duhamel du Monceau, he was nominated corresponding member of the Paris Academy and was invited by Diderot to write a number of articles for the supplement of the *Encyclopédie*.

When he returned home in 1779, Troja was given the chair of ophthalmology at the University of Naples, a post established especially for him, since the subject had not been taught there until then. In the following year he became surgeon of the king's chamber for having successfully cured the crown prince of a disease. In 1802 Troja campaigned successfully for the creation of a commission to spread the knowledge and use of Jenner's smallpox vaccination.

Troja is known especially for his studies on the nutrition and regeneration of bones, a subject that was the focus of academic interest at that time. Even Spallanzani became interested in it; and in his letters he referred to the work of Troja, although apparently without great enthusiasm. Such an unfavorable judgment does not appear justified today, for Troja considerably improved the techniques of study. For instance, he used immersion in nitric acid to bring out bone structure and applied zinc sulfate to obtain coloration of the cells and improve the visibility of microscopic structures.

Troja's research on the eye is preserved in a volume of lectures on eye diseases. In urology he studied the diseases of the bladder and of the urinary system in general, and invented the flexible catheter. This invention derived from his active interest in and knowledge of India rubber, which had just been introduced into Europe.

In addition to his own work in anatomy and microscopy, Troja collaborated with G. S. Poli for many years in the latter's investigations on the anatomy of mollusks.

BIBLIOGRAPHY

I. ORIGINAL WORKS. Troja's writings include *De novorum ossium in integris aut maxime ob morbis deperditionibus regeneratione experimenta* (Paris, 1775); *Lezioni intorno alle malattie degli occhi* (Naples, 1780); *Lezioni intorno ai mali della vescica orinaria e delle sue appartenenze . . . Colla giunta di una memoria sulla costruzione dei cateteri flessibili* (Naples, 1785–1793); and *Osservazioni ed esperimenti sulle ossa; in supplemento ad un'opera sulla rigenerazione delle ossa, impressa nel 1775* (Naples, 1814).

II. SECONDARY LITERATURE. See M. del Gaizo, "Della vita e delle opere di Michele Troja," in *Atti della R. Accademia medico-chirurgica* (Naples), **52** (1899), 191; and **53** (1899), 351; Gianni Randelli, "Ripetizione degli esperimenti di Michele Troja sulla rigenerazione delle ossa," in *Physis*, **6** (1964), 45–64; and A. von Schoenberg, *Biographie des Dr. und Professor M. Troja* (Erlangen, 1828).

CARLO CASTELLANI

TROMMSDORFF, JOHANN BARTHOLOMÄUS (*b*. Erfurt, Germany, 8 May 1770; *d*. Erfurt, 8 March 1837), *chemistry, pharmacy*.

Chemistry and pharmacy loomed large in Trommsdorff's youthful environment. Not only was his father, Dr. Wilhelm Bernhard Trommsdorff, the owner of an apothecary shop but he was also a successful chemistry teacher in the University of Erfurt and the chief local representative of chemistry in Erfurt's revitalized Academy of Useful Sciences. However, young Trommsdorff never had the advantage of his father's instruction, for, two days before his twelfth birthday, his father died. Financial difficulties soon arose, compelling Trommsdorff to abandon plans for a higher education and enter pharmacy. He went to Weimar, where he served as an apprentice under his father's old friend the chemist Wilhelm Heinrich Bucholz and Bucholz's employee the chemist Johann Friedrich Göttling. Thanks to their instruction and encouragement, he published his first note in Lorenz von Crell's *Chemische Annalen* in 1787. That same year he also completed his apprenticeship. After two and a half years as a journeyman in Erfurt, Stettin, and Stargard, he took over the family apothecary shop in Erfurt.

Trommsdorff continued publishing articles and small monographs on various chemical and pharmaceutical topics. By early 1793 he felt confident enough to venture into the debate over the composition of mercuric oxide that was raging between the antiphlogistonist Sigismund Friedrich Hermbstaedt in Berlin and the phlogistonist Friedrich Albrecht Carl Gren in Halle. He sided with his boyhood friend Gren, who maintained that completely fresh mercury calx per se did not yield dephlogisticated air when reduced. To his chagrin, he and his allies were soon discredited by Hermbstaedt's party in this crucial controversy in the German antiphlogistic revolution. Trommsdorff's first reaction was to renounce all theorizing. However, influenced by the works of Kant and Georg Christoph Lichtenberg, he was soon calling for the unification of physics, chemistry, and natural history. For the next few years, he showed considerable interest in attaining this elusive goal of the *Naturphilosophen*. In the early 1800's, however, the empiricist in him triumphed. From then on his research was limited to turning out hundreds of useful, but essentially routine, chemical studies.

Although Trommsdorff's work as a chemist was valued highly enough for him to hold a chair of chemistry at Erfurt University and to be offered several other chairs, including that of Martin Heinrich Klaproth in Berlin, his work as a scholarly pharmacist was of greater significance. His numerous texts were exceptionally popular, most of them going through many editions in Germany and abroad. His *Journal der Pharmacie* (1794–1834) was the leading periodical for pharmacy and pharmaceutical chemistry until Justus von Liebig began publishing his *Annalen der Pharmacie* in 1832. And his "Chemical-physical-pharmaceutical Boarding School," attended by over 300 students between 1795 and 1828, played a major role in the training of the founding generation of the German drug industry.

BIBLIOGRAPHY

For a complete bibliography of Trommsdorff's more than four hundred publications, see Adolph Peter Callisen, *Dem Andenken des verdienten Chemiker Dr. Joh. Barthol. Trommsdorff* (Copenhagen, 2nd ed., 1837).

For his life and influence, see Hermann Trommsdorff, "Johann Bartholomä Trommsdorff und seine Zeitgenossen," in *Jahrbuch der Akademie gemeinnütziger Wissenschaften in Erfurt*, new ser., **53** (1937), 5–55; and **55** (1941), 131–234.

Other studies of Trommsdorff include Christa Caumitz, "Johann Bartholomäus Trommsdorff (1770–1837): Ein Begründer der wissenschaftlichen deutschen Pharmazie," a manuscript in the Archives of the Deutsche Akademie der Naturforscher, Leopoldina (Halle, German Democratic Republic); Hermann Gittner, ed., *Die Harzreisen des Johann Bartholomä Trommsdorff 1798 und 1805* (Oberhausen, 1957); and "Die Rheinreise des Johann Bartholomai Trommsdorff zur 13. Naturforscherversammlung in Bonn anno 1835," in *Deutsche Apothekerzeitung*, **99** (1959), 31–36; Horst Rudolf Abe *et al.*, "Johann Bartholomäus Trommsdorff und die Begründung der modernen Pharmazie," in *Beiträge zur Geschichte der Universität Erfurt*, **16** (1971–1972); and

Wilhelm Vershofen, *Die Anfänge der chemisch-pharmazeutischen Industrie*, I (Berlin-Stuttgart, 1949), and II (Aulendorf, 1952).

KARL HUFBAUER

TROOST, GERARD (*b.* 's Hertogenbosch, Netherlands, 15 March 1776; *d.* Nashville, Tennessee, 14 August 1850), *geology, mineralogy, paleontology, natural history.*

Troost was the son of Everhard Joseph Troost and Anna Cornelia van Haeck. He attended the University of Leiden and the Athenaeum in Amsterdam, where he specialized in chemistry, geology, and natural history. He received his master of pharmacy from the Athenaeum and his doctorate of medicine at Leiden. He never practiced medicine, and although he was briefly a pharmacist in Amsterdam and The Hague and later in the United States, his interest in geology gradually became dominant.

In 1807 Louis Bonaparte, the appointed king of Holland, sent Troost to Paris, where, as a colleague of Haüy, he became skilled in mineralogy and crystallography. For two years Troost collected for the king specimens of minerals from various parts of Europe. He also studied with A. G. Werner, whom he accompanied on geologic field trips. He translated Humboldt's *Ansichten der Natur* into Dutch, and as a result the two men became lifelong friends. In 1809 Troost accompanied a Dutch scientific expedition that sailed for Java by way of the Cape of Good Hope. Although the ship was captured by French pirates, Troost eventually made his way back to Europe and soon tried to reach Java again, by way of the United States. But in 1810, while he was in Philadelphia, Louis Bonaparte, who in too many matters put the interest of his subjects ahead of those of his brother Napoleon, was forced to abdicate; and Holland was incorporated for a while into the French empire. Therefore, Troost decided to stay in the United States. He became an American citizen and established a pharmaceutical and chemical laboratory in Philadelphia. In 1811 Troost married Margaret Tage of Philadelphia, by whom he had two children, Caroline and Lewis. She died in 1819, and he then married a Mrs. O'Reilly of Philadelphia.

Troost was short and portly and had a kindly disposition. He was a polished man of the world and a profound scholar, and his manner was always unassuming. He was proficient in several languages, but Americans noted that he always spoke English with a Dutch accent. He won the respect and friendship of all classes of people.

Troost was one of seven men who in 1812 founded the Academy of Natural Sciences of Philadelphia, and he was its first president. In 1826 a geologic map of Philadelphia and environs, which he had prepared, was published by the Philadelphia Society for Promoting Agriculture.

In 1825 Troost joined Robert Dale Owen, who with Maclure, Say, and Lesueur, made up what they described as a "boatload of knowledge." Their boat took them down the Ohio River to New Harmony, Indiana, where they planned to establish a utopian society. Interest waned at New Harmony, however, and in 1827 Troost moved to Nashville, Tennessee, where he lived for the rest of his life.

In 1828 Troost was appointed professor of geology and mineralogy at the University of Nashville, where he also taught chemistry. He held this post until his death and was esteemed by his students. He was state geologist of Tennessee from 1831 to 1839, and he prepared the first geologic map of the state. He was one of the early workers in stratigraphy in the United States and contributed to the knowledge of the mineral resources of Tennessee. He housed his extensive collections of minerals, rocks, fossils, shells, Indian relics, and mounted birds in a private museum at Nashville that was open to the public and considered one of the finest museums west of the Appalachians.

In mid-July 1850, only four weeks before his death, Troost finished the manuscript of his study of the fossil crinoids of Tennessee. He had written it with much difficulty; in his introduction to the manuscript he expressed a fear that since his memory and sight were both sadly impaired, the work might contain some inaccuracies. The manuscript was received 18 July 1850 by the Smithsonian Institution, which undertook to publish it, subject to the editorial approval of Louis Agassiz. After Agassiz had kept the paper for five years without expressing his opinion of its worth, it was sent for review to James Hall, Jr., in Albany, New York. But the manuscript, with Troost's collection of Tennessee crinoids, was still in Hall's possession at the time of his death more than forty years later. The fossils and the manuscript were then returned to the Smithsonian Institution.

Hall and his colleagues may have attributed their unconscionable delay to Troost's misgivings about possible inaccuracies, although Hall had introduced under his own authorship four of Troost's

genera, quoting Troost's descriptions for three of them. Troost's paper was finally published in 1909 with supplementary descriptions and observations by Elvira Wood. But by that time most of the species and genera originally recognized by Troost had been described by others, and his memory was deprived of the credit that would have been accorded it if his important monograph had been printed when it was received more than half a century earlier.

BIBLIOGRAPHY

I. ORIGINAL WORKS. A bibliography of Troost's papers is in L. C. Glenn (see below). Papers on mineralogy, published from 1821 to 1848, are primarily in *American Journal of Science* and *Journal of the Philadelphia Academy of Sciences*. Papers on paleontology, published from 1834 to 1850, are primarily in *Transactions of the Geological Society of Pennsylvania*. Papers on the geology of Tennessee, published from 1835 to 1849, are in *American Journal of Science*, *Journal of the Tennessee Senate* (Nashville), and Tennessee Senate and House documents (Knoxville).

II. SECONDARY LITERATURE. See J. M. Clarke, "Prof. James Hall and the Troost Manuscript," in *American Geologist*, **35**, no. 4 (1905), 256–257, and *James Hall of Albany, Geologist and Paleontologist, 1811–1898* (Albany, N.Y., 1923), 233; L. C. Glenn, "Gerard Troost," in *American Geologist*, **35**, no. 2 (1905), 72–94; Phillip Lindsley, "The Life and Character of Professor Gerard Troost, M.D.," in *The Works of Phillip Lindsley*, I (Philadelphia, 1859), 541–588; Dumas Malone, ed., "Gerard Troost," in *Dictionary of American Biography*, XVIII (1938), 647–648; G. P. Merrill, *The First One Hundred Years of American Geology* (New York, 1964), 111, 138, 215–216; H. G. Rooker, "A Sketch of the Life and Work of Dr. Gerard Troost," in *Tennessee Historical Magazine*, 2nd ser., **3**, no. 1 (1932), 3–19, with portrait opposite p. 3; and Elvira Wood, "A Critical Summary of Troost's Unpublished Manuscript on the Crinoids of Tennessee," in *Bulletin. United States National Museum*, **64** (1909), with portrait.

ELLEN J. MOORE

TROOST, LOUIS JOSEPH (*b.* Paris, France, 17 October 1825; *d.* Paris, 30 September 1911), *chemistry.*

After receiving his *agrégé* at the École Normale Supérieure in Paris in 1851, Troost taught briefly at the Lycée d'Angoulême and at the Lycée Bonaparte in Paris. He received his doctorate in 1857 and continued to work in Deville's thermochemical laboratory alongside researchers who included

Henri Debray, F. Isambert, Paul Hautefeuille, and Alfred Ditte. In 1874 he was appointed to the chair of chemistry at the Sorbonne and in 1884 was elected to the Académie des Sciences.

Troost's first studies concerned preparations of salts of lithium, but the greater part of his research focused on isomerism, allotropy, and dissociation. Together with Deville he studied the variations with rising temperature of vapor densities of a number of substances, including iodine, phosphorus, arsenic, and zirconium. They confirmed Dumas's observation that the vapor density of sulfur near its boiling point corresponds to a hexatomic molecule and found that sulfur vapor becomes diatomic above 800°C. Their determinations indicated analogous vapor-density variations for selenium and tellurium and also aided in the clarification of the chemistry of niobium and tantalum. All their observations were couched in terms of equivalents and volumetric considerations rather than atoms.

In 1868 Troost commenced a long period of collaboration with Hautefeuille. Their researches included the conditions of transformation of cyanogen into paracyanogen and of cyanic acid into cyanuric acid; the allotropic conversions of white phosphorus into red phosphorus and of oxygen into ozone; and the preparations of new compounds of boron and silicon, particularly their chlorides. With Hautefeuille, Troost also analyzed the absorption of hydrogen by sodium, potassium, and palladium; and they studied the roles of manganese and silicon in iron metallurgy. Their investigations of the introduction of nitric acid into hydrocarbons supported Berthelot's conclusion that there is much greater mechanical work available in the nitric ethers (such as nitroglycerine) than in nitrobenzene and similar products.

BIBLIOGRAPHY

I. Original Works. Troost wrote two principal textbooks: *Précis de chimie*, 2nd ed. (Paris, 1867; 45th rev. ed., with E. Pechard, 1932); and *Traité eléméntaire de chimie comprenant les principales applications à l'hygiène, aux arts et à l'industrie*, 2 vols. (Paris, 1865; 24th rev. ed., with E. Pechard, 1948), which became a classic text in secondary education. The most complete list of Troost's publications is in his own *Notice sur les travaux scientifiques de M. Louis Troost* (Paris, 1888). Several of his papers are reprinted in Henri Le Châtelier, ed., *Les classiques de la science*: vol. III, *Eau oxygénée et ozone* (Paris, 1913), and vol. VI, *La fusion du platine et dissociation* (Paris, 1914).

II. Secondary Literature. On Troost and his work, see Armand Gautier, "Séance du lundi 2 octobre 1911," in *Comptes rendus hebdomadaires des séances de l'Académie des sciences*, **153** (1911), 611–615, esp. 613–615; and an unsigned obituary, "Louis Joseph Troost," in *Nature*, **87** (1911), 491–492. There are also discussions of the researches of Troost and Hautefeuille, in Georges Lemoine, "Les travaux et la vie de Paul Hautefeuille," in *Revue des questions scientifiques*, **55** (1904), 5–25; and Alfred Lacroix, "Gabriel Hautefeuille (1836–1902)," in Lacroix's *Figures de savants*, 2 vols. (Paris, 1932), vol. I, 81–89.

Mary Jo Nye

TROOSTWIJK, ADRIAAN PAETS VAN (*b.* Utrecht, Netherlands, 4 March 1752; *d.* Nieuwersluis, Netherlands, 3 April 1837), *chemistry.*

Nothing is known of van Troostwijk's education. From the age of eighteen until his retirement in 1816 he worked in Amsterdam as a merchant. He was also an important Dutch chemist and published thirty-five works between 1778 and 1818. From 1806 to 1816 he was a member of the Royal Institute of Sciences, Literature, and Fine Arts (the present Royal Netherlands Academy of Sciences). In 1816 he moved to Nieuwersluis, where he remained until his death.

Van Troostwijk was greatly influenced by his friend the physician Jan Rudolph Deiman (1743–1808) and by Martinus van Marum, director of Teyler's Museum in Haarlem. From 1778 until 1792 van Troostwijk published the results of his own research and of work done in collaboration with Deiman, van Marum, and Krayenhoff. With Deiman he investigated the improvement of air by growing plants (1778), the nature and properties of carbon dioxide (1781), and the influence of galvanism on both the sick and the healthy. In 1783 he published a work written with van Marum on the electrophorus and, four years later, a joint work on the noxious fumes in swamps, drains, mines, and factories. Using van Marum's electrical machine, they investigated the chemical action of electricity on various gases and on metal oxides. After sparking dephlogisticated air (oxygen), they remarked that it "had acquired a very strong smell, which to us very much resembled the strong smell of electric matter, only much more so than we had ever smelled before" (observation of ozone). When they sparked fixed air (carbon dioxide) its volume was slightly increased. They put equal amounts of electrified and unelectrified fixed air over water and found that after two days only one-tenth of the latter but two-fifths of the former remained. They did not, however, realize that the electrified fixed air was partially decomposed into carbon monoxide and oxygen. During their first experiments on the decomposition of water by electric sparks, Deiman and van Troostwijk also subjected carbon dioxide to violent electric discharges. They found that it produced a much larger volume of inflammable air than water. They did not recognize that the inflammable air obtained in this way was carbon monoxide, and not hydrogen, and they came to the wrong conclusion that fixed air contains water.

In 1785 van Troostwijk published a treatise on the recently discovered types of air, and he was co-author with Krayenhoff of a work on the use of electricity in physics and medicine (1788). A phlogistonist until 1788, van Troostwijk renounced his former belief chiefly through the influence of the publications of van Marum (1787) and Alexander Petrus Nahuys (1788). In 1789 van Troostwijk and Deiman published the results of their experiments on the decomposition of water by static electricity and its synthesis by combustion.

Around 1791 van Troostwijk, Deiman, Pieter Nieuwland, and Nicolaas Bondt founded the Batavian Club, better known as the Society of Dutch Chemists, a circle of friends who met to study chemistry. The apothecary Anthonie Lauwerenburgh and the physician Gerard Vrolik later became members of the group. The society was instrumental in securing recognition in the Netherlands for Lavoisier's discoveries and published many articles in support of the new oxidation theory. Van Troostwijk was, in all probability, the leader of the society, which remained active until 1808. Articles published by the Dutch chemists were collected in the journal *Recherches physico-chimiques* (1792–1794) and later appeared in an enlarged Dutch translation as *Natuur-scheikundige verhandelingen* (Amsterdam, 1799–1801).

Van Troostwijk also investigated the preparation of olefiant gas (ethylene) from the action of concentrated sulfuric acid on ethyl alcohol and the preparation of ethylene chloride from the action of chlorine on ethylene (1794). The latter compound is still known as Dutch oil or Dutch liquid. Also that year he reported on the glow resulting from the mixture of sulfur and various metals heated in the absence of oxygen.

BIBLIOGRAPHY

I. ORIGINAL WORKS. There is a bibliography of van Troostwijk's writings in H. P. M. van der Horn van den Bos, "Bibliographie des chimistes hollandais dans la période de Lavoisier," in *Archives du Musée Teyler*, 2nd ser., 6 (1900), 375–420. His works include *Verhandeling over het nut van den groeij der boomen en planten, tot zuivering der lucht* (Amsterdam, 1780), written with J. R. Deiman; *Verhandeling over de vaste lucht* (Rotterdam, 1781), written with Deiman; *Verhandeling over de vorderingen in de luchtkennis* (Amsterdam, 1785); *De l'application de l'électricité à la physique et à la médecine* (Amsterdam, 1788), written with C. R. T. Krayenhoff; and "Lettre à M. de la Métherie, sur une manière de décomposer l'eau en air inflammable et en air vital," in *Journal de physique, de chimie et de l'histoire naturelle*, 35 (1789), 369–378, written with Deiman.

Later writings are *Beschrijving van een electriseermachine en van proefnemingen met dezelve in het werk gesteld* (Amsterdam, 1790), written with Deiman; "Expériences sur l'inflammation du mélange de soufre et de métaux, sans la présence de l'oxigène," in *Recherches physico-chimiques*, 3 (1794), 71–96, written with Deiman *et al.*; "Recherches sur les divers espèces des gaz, qu'on obtient en mêlant l'acide sulfurique concentré avec l'alcool," in *Journal de physique . . .*, 45 (1794), 178–191, written with Deiman, Bondt, and Lauwerenburgh.

II. SECONDARY LITERATURE. On van Troostwijk and his work, see H. P. M. van der Horn van den Bos: *De Nederlandsche scheikundigen van het laatst der vorige eeuw* (Utrecht, 1881); *Het aandeel dat de Scheikundigen in Frankrijk, Engeland, Duitschland en Noord- en Zuid-Nederland hebben gehad in het tot algemeene erkenning brengen van het Systeem van Lavoisier* (Amsterdam, 1895); "Matériaux pour l'histoire de la chimie dans les Pays-Bas. A. Paets van Troostwijk, un chimiste d'Amsterdam de la fin du 18e siècle, 1752–1837," in *Archives du Musée Teyler*, 2nd ser., 9, pt. 2 (1904), 155–199; and "Adriaan Paets van Troostwijk," in *Chemisch weekblad*, 6 (1909), 1–35. See also E. Cohen, *Das Lachgas. Eine chemisch-kulturhistorische Studie* (Leipzig, 1907), 14–21; and H. A. M. Snelders, "Uiteenzettingen van het stelsel van Lavoisier door Nederlanders in het laatste kwart van de achttiende eeuw," in *Scientiarum historia*, 8 (1966), 89–100.

H. A. M. SNELDERS

TROPFKE, JOHANNES (*b.* Berlin, Germany, 14 October 1866; *d.* Berlin, 10 November 1939), *history of mathematics.*

Tropfke came from a wealthy family. A bright student, he was encouraged to study a number of different subjects at the Friedrichs Gymnasium, and his wide-ranging interests at the University of Berlin are reflected in the list of subjects Tropfke prepared for the state examination in 1889: mathematics, physics, philosophy, botany, zoology, Latin, and Greek. His dissertation dealt with a topic in the theory of functions. An enthusiastic teacher, he was director of the Kirschner Oberrealschule from 1912 to 1932.

Tropfke's program for changing the secondary school mathematics curriculum, presented in 1899, was published in expanded form as *Geschichte der Elementarmathematik in systematischer Darstellung* (1902–1903). Even in this form, however, it was still based mainly on a study of secondary literature that was largely unchecked by an examination of original sources. Tropfke subsequently produced a second, seven-volume edition (1921–1924) that benefited from the constructive and sympathetic criticism of Heinrich Wieleitner and Gustaf Eneström. This work offered what was, at the time, an excellent overall account of the subject, enriched by a wealth of extremely valuable citations and references. Moreover, it exerted a decisive influence on the reorganization of mathematical education, encouraging teachers to devote greater attention to historical development.

Advances in the study of the history of mathematics led Tropfke to undertake a third, revised and enlarged edition of his work; but it was not completed. He died shortly after the beginning of the war; and the remaining volumes, already in manuscript, were destroyed.

BIBLIOGRAPHY

Tropfke's writings include *Zur Darstellung des elliptischen Integrals erster Gattung* (Halle, 1889), his dissertation; *Erstmaliges Auftreten der einzelnen Bestandteile unserer Schulmathematik* (Berlin, 1899); *Geschichte der Elementarmathematik in systematischer Darstellung*, 2 vols. (Leipzig, 1902–1903; 2nd ed., 7 vols., Leipzig–Berlin, 1921–1924; 3rd ed., 4 vols., Leipzig–Berlin, 1930–1940); "Archimedes und die Trigonometrie," in *Archiv für Geschichte der Mathematik, der Naturwissenschaften und der Technik*, n.s. **10** (1928), 432–461; "Zur Geschichte der quadratischen Gleichungen über dreieinhalb Jahrtausende," in *Jahresberichte der Deutschen Mathematiker-vereinigung*, **43** (1933), 98–107; and **44** (1934), 26–47, 95–119; and "Die Siebenecksabhandlung des Archimedes," in *Osiris*, **1** (1936), 636–651.

An obituary by J. E. Hofmann, with portrait and bibliography, is in *Deutsche Mathematik*, **6** (1941), 114–118.

J. E. HOFMANN

TROUGHTON, EDWARD (*b.* Corney, Cumberland, England, October 1753; *d.* London, England, June 1836), *mathematics, optics, physics.*

Troughton was one of the most competent mathematical instrument makers of the late eighteenth and early nineteenth centuries. In many ways his career was parallel to that of Jesse Ramsden, his earlier counterpart, whom he was to replace as the foremost instrument maker of England.

In 1770 Troughton was apprenticed to his elder brother, John, who specialized in dividing and engraving instruments for other makers. His shop was on Surrey Street, in the Strand. In 1779 John and Edward Troughton became partners, and in 1782 they bought the business of Benjamin Cole at "The Sign of the Orrery," at 136 Fleet Street. This was a well-established enterprise, having been founded by John Worgan about 1680 and continued, in turn, by John Rowley, Thomas Wright, and the two Benjamin Coles, father and son.

John Troughton died in 1784; and Edward conducted the business alone until 1826, when he joined with William Simms, a skilled instrument maker. The firm was renamed Troughton and Simms and, after Troughton's retirement in 1831, continued under that name until 1922 when, through a merger, it became Cooke, Troughton and Simms Ltd.

Troughton's reputation rested on the accuracy and beautiful proportions of his instruments. In 1822 he wrote, "The beauty of the instrument lies not in the flourishes of the engraver, chaser and carver but in the uniformity of figure and just proportion alone."

Troughton made many contributions to the development of instrument making: in 1788 an improvement of Hadley's quadrant; in 1790 a mercurial pendulum; and in 1796 a refined version of the Borda, or reflecting circle. He was responsible for substituting spider web filaments for hair or wire in his optical micrometers.

Troughton's most notable achievement was the improvement of the method of dividing a circle. His paper on this in 1809 won him the Copley Medal from the Royal Society of London, which elected him a fellow the following year. In 1822 he was elected a fellow of the Royal Society of Edinburgh. He was a founding member of the Royal Astronomical Society.

Examples of his instruments are to be found in the Kensington Science Museum, London; the museums of the history of science in Oxford and Florence; the Whipple Museum, Cambridge; the National Maritime Museum, Greenwich; the Peabody Museum, Salem, Massachusetts; the Conservatoire des Arts et Métiers, Paris; and the Smithsonian Institution, Washington, D.C.

BIBLIOGRAPHY

I. ORIGINAL WORKS. Troughton's works are "An Account of a Method of Dividing Astronomical and Other Instruments by Ocular Inspection, in Which the Usual Tools for Graduating Are Not Employed, etc.," in *Philosophical Transactions of the Royal Society*, **99**, pt. 1 (1809), 105–145; *On the Repeating and Altitude-Azimuth Circle* (London, 1812); and "An Account of the Repeating Circle and of the Altitude and Azimuth Instrument, Describing Their Different Constructions, Etc.," in *Memoirs of the Astronomical Society*, **33** (1821), and in *Philosophical Magazine* (1822).

II. SECONDARY LITERATURE. On Troughton's life and work, see the *Dictionary of National Biography*, XIX (London, 1917), 1186–1187. Other works include Maria Luisa Bonelli, *Catalogo degli Strumenti del Museo di Storia della Scienza* (Florence, 1954), pp. 67, 204, 206, 225; Maurice Daumas, *Les instruments scientifiques au XVII et XVIII siècles* (Paris, 1953), 320–321; Nicholas Goodison, *English Barometers, 1680–1860* (New York, 1968), 240; Henry C. King, *The History of the Telescope* (London, 1955), 230–236; J. A. Repsold, *Zur Geschichte der astronomischen Messwerkzeuge, 1450–1830* (Leipzig, 1908), 118–122; E. G. R. Taylor, *The Mathematical Practitioners of Hanoverian England* (London, 1966), 298–299; and E. Wilfred Taylor and J. Simms Wilson, *At the Sign of the Orrery*, pp. 24–30.

RODERICK S. WEBSTER

TROUTON, FREDERICK THOMAS (*b.* Dublin, Ireland, 24 November 1863; *d.* Downe, Kent, England, 21 September 1922), *physics.*

Trouton came from a wealthy and prominent Dublin family. He performed brilliantly in his undergraduate work at Trinity College, Dublin, taking degrees in both engineering and physical science. In recognition of this work, he was awarded the Large Gold Medal, an honor rarely bestowed for work in science. Trouton remained at Trinity College as FitzGerald's assistant. Trouton and FitzGerald remained the closest of colleagues and confidants until FitzGerald's death in 1901. In 1902 Trouton was appointed Quain professor of physics, University College London. At London he pursued his interests in both engineering and physics until 1912, when he was struck by a severe illness that led to permanent paralysis in both legs. Trouton's active scientific career was at an end, but his spirit was not broken, and the wit and charm for which he was noted were not dampened even after the loss of his two sons in World War I. He continued to advise students and colleagues from his sickbed until his death at the age of fifty-eight.

Throughout his career, Trouton occupied himself with problems in both engineering and physical science. As an undergraduate, he discovered a relationship—known as Trouton's Law—between the latent heat and the molecular weight of a substance. According to the law, the ratio of the product of the molecular weight and the latent heat to the absolute temperature is a constant. The relationship is not precise and Trouton himself held it to be of little significance. It was also during his undergraduate days that Trouton took an active role in surveying for a railway.

Trouton's dual interests in applied research and physics continued throughout his career. He devoted considerable energy to investigations of the viscosity of pitch and molten glass at a variety of temperatures; the dynamics of the condensation of water vapor on glass, glass wool, and related substances; the effects of surface moisture on the conductivity of glass; and the relationship between the concentration and the adsorption of dyestuffs on sand. The practical implications of these studies were an explicit motivation for carrying them out.

The influence of FitzGerald seems to have been decisive in Trouton's more abstract research. Thus, shortly after Hertz published his startling discovery of the propagation of electromagnetic fields, Trouton and FitzGerald undertook a series of replicate investigations. But the investigations for which Trouton is remembered were those in which he attempted to determine the relative velocity of the earth and ether.

Independent of Hendrik A. Lorentz, FitzGerald had suggested that the null result of the Michelson-Morley experiment could be accounted for if one assumed that material objects contracted in the direction of motion as a result of interaction with ether. In 1903 Trouton and H. R. Noble undertook an experiment to measure the torsional force on a suspended charged plate condenser as a result of the interaction of the charges in the plates with the ether wind. The widely publicized results were, of course, null.

In 1908, in association with Alexander O. Rankine, Trouton undertook yet another ether drift experiment. They attempted to measure the change in resistance of a copper wire when the wire is rotated parallel and transverse to the direction in which the earth moves around the sun. As was the case with the Trouton-Noble experiment, the experiment was an extremely delicate one, calling for a considerable degree of virtuosity and cleverness. But again the results were null.

Such research placed Trouton in the great tradition of nineteenth-century British physics. He was, perhaps, the last of the well-trained British ether-mechanists.

BIBLIOGRAPHY

I. ORIGINAL WORKS. Trouton's published articles include "On Molecular Latent Heat," in *Philosophical Magazine*, **18** (1884), 54–57; "Repetition of Hertz's Experiments and Determination of the Direction of the Vibration of Light," in *Nature*, **39** (1889), 391–393; "Electrolysis Away From Electrodes," in *Electrician*, **43** (1899), 294; "Flow of Liquid Through Partitions," *ibid.*, 596–597; "Effect on Charged Condenser of Motion through the Ether," in *Transactions of the Royal Dublin Society*, **7** (1902), 379–384; "Forces Acting on a Charged Condenser Moving Through Space," with H. R. Noble, in *Proceedings of the Royal Society*, **72** (1903), 132–133; "Forces Acting on a Charged Condenser Moving Through Space," with H. R. Noble, in *Philosophical Transactions of the Royal Society*, **202A** (1904), 165–181; "Viscosity of Pitch-like Substances," with E. S. Andrews, in *Philosophy Magazine*, **7** (1904), 347–355; and in *Proceedings of the Physical Society*, **19** (1904), 47–56; "Coefficient of Viscous Traction and Its Relation to That of Viscosity," in *Proceedings of the Royal Society*, **77A** (1906), 426–440; "Vapour Pressure in Equilibrium With Substances Holding Varying Amounts of Moisture," *ibid.*, 292–314; "Leakage Currents on Glass Surfaces," with C. Searle, in *Philosophy Magazine*, **12** (1906), 336–347; "Condensation of Water Vapour on Glass Surfaces," in *Proceedings of the Royal Society*, **79A** (1907), 383–390; "Condensation of Moisture of Solid Surfaces," in *Chemical News*, **96** (1907), 92–93; "Rate of Recovery of Residual Charge in Electric Condensers," with S. Russ, in *Philosophical Magazine*, **13** (1907), 578–588; "Electrical Resistance in Moving Matter," in *Proceedings of the Royal Society*, **80A** (1908), 420–435; "Mechanism of the Semipermeable Membrane and a New Method of Determining Osmotic Pressure," *ibid.*, **86A** (1912), 149–154.

II. SECONDARY LITERATURE. An obituary may be found in *Nature*, **110** (1922), 490–491. For biographical sketches, see *Proceedings of the Royal Society*, **110A** (1926), iv; and E. Scott Barr, "Anniversaries in 1963," in *American Journal of Physics*, **31** (1963), 85–86.

STANLEY GOLDBERG

TROUVELOT, ÉTIENNE LÉOPOLD (*b.* Guyencourt, Aisne, France, 26 December 1827; *d.* Meudon, France, 22 April 1895), *natural history, astronomy.*

A keen observer and skillful artist, Trouvelot spent several years (1872–1874) working with the fifteen-inch refractor at Harvard Observatory. The drawings he made there and elsewhere are still widely known. Except for Rutherfurd's wet-plate photographs of the sun and moon, made in 1865, Trouvelot's drawings were considered the most accurate pictures of celestial objects available until the perfection of dry-plate photography.

Little is known about Trouvelot's life before he came to America in 1857; there is no indication that he was especially interested in astronomy during his early years. Interested in silkworm culture, he thought the European gypsy moth, *Porthetria dispar*, might serve the same purpose. In 1869 he imported to Medford, Massachusetts, some live egg clusters for experimentation. Unfortunately, a few of the moths escaped, and after a decade began to proliferate alarmingly. This was the origin of the defoliation of trees in the northeastern United States.

A member of the Boston Natural History Society, Trouvelot presented papers on a variety of topics—mostly concerned with zoology—at society meetings. His earliest major contribution to astronomy, published in 1875 in *Silliman's Journal* (now the *American Journal of Science*), was a paper on sunspots. Some one thousand of his sunspot drawings were deposited at the Harvard College Observatory. In 1881 Trouvelot was elected to the American Academy of Arts and Sciences.

In 1882 Trouvelot returned to France to join the staff of the new observatory at Meudon. While at Meudon he was particularly successful in his observations of solar prominences. In 1883 he went to the Caroline Islands to observe the total eclipse of the sun; he searched without avail for a supposed intra-Mercurial planet to account for the anomaly in the motion of Mercury.

Trouvelot published some fifty astronomical papers covering a wide range of topics, the most important being on the sun and Venus. He presented his last paper, concerning the transit of Mercury, on 12 November 1894 at the French Academy.

BIBLIOGRAPHY

I. ORIGINAL WORKS. For reproductions of Trouvelot's astronomical drawings, see *The Trouvelot Astronomical Drawings Manual* (New York, 1882); also see *Annals of Harvard College Observatory*, VIII, pt. 2 (1876), containing engravings prepared under the direction of Joseph Winlock. For references to Trouvelot's published astronomical papers, see Poggendorff, III, 1368; IV, 1526.

II. SECONDARY LITERATURE. Obituaries are contained in *Nature*, **52** (1895), 11; and *Observatory*, **18** (1895), 245–246. For mentions of Trouvelot's work, see Edward H. Forbush and Charles H. Fernald, "The Gypsy Moth," in *Report. Massachusetts Department of Agriculture* (1896); and *Annals of Harvard College Observatory*, **8**, pt. 1 (1876), 53, 55, 64.

E. Dorrit Hoffleit

TROWBRIDGE, JOHN (*b*. Boston, Massachusetts, 5 August 1843; *d*. Cambridge, Massachusetts, 18 February 1923), *physics*.

Trowbridge was a pioneer in the movement that established serious scientific research in America in the latter decades of the 19th century. The son of a prominent New England family, John Trowbridge studied at the Boston Latin School and graduated in 1865 with highest honors from the Lawrence Scientific School of Harvard University. After five years of teaching, he became professor of physics at Harvard (1870), serving in this capacity for forty years until his retirement in 1910.

Trowbridge was a strong advocate of laboratory practice as an integral part of scientific education. To this end he formally urged, as early as 1877–1878, the endowment of a "building devoted to physical investigation," with a staff dedicated to scientific research. He pointed out that while "there are no precedents for the endowment of a Physical Laboratory in connection with an American University, there is the greater honor in becoming the leader." He then designed and carried out the construction at Harvard of the Jefferson Physical Laboratory. He served as titular head upon its completion in 1884 and as director from 1888 until his retirement.

Trowbridge's main line of research was concerned with electrical phenomena. He devised the cosine galvanometer (1871) to measure strong electrical currents. With W. C. Sabine in 1890 he investigated high-frequency electrical oscillations using a revolving mirror. As early as 1890 Trowbridge supported the view that the carriers of electricity in wires were something other than atoms. By 1894 he investigated the magnetic effect of high-frequency oscillations and was one of the many pioneers of remote signaling. In 1895 he investigated with Duane the velocity of propagation of electrical waves. He undertook studies of the discharge of electricity through gases and, from 1896, examined both the production and effect of Röntgen radiation. In 1897 he studied with Theodore W. Richards the spectrum of argon and other gases. Aware of the need for a constant source of high voltage, he developed, also in 1897, a storage battery of 10,000 cells. In addition to his scientific work he was the associate editor of the *American Journal of Science* from 1880 to 1920.

BIBLIOGRAPHY

I. Original Works. Trowbridge published over one hundred original papers and several books dealing with scientific subjects. His most important book is *What is Electricity?* (London, 1897). A partial bibliography is included in Edwin H. Hall, "John Trowbridge: 1843–1923," in *Biographical Memoirs. National Academy of Sciences*, **14** (1930), 185–204, which also includes a discussion of some of the papers. This list is supplemented by the *Royal Society Catalogue of Scientific Papers*, XIX, 214–215.

Trowbridge wrote "Recent Advance in Physical Science," in *International Monthly*, **1** (1900), 123–132; and "Progress of Electricity From 1800 to 1900," in *The Nineteenth Century: A Review of Progress* (New York–London), 417–427. His "The Endowment of the Physical Laboratory at Harvard College" (Cambridge, Mass., *ca.* 1877), in the Harvard University Archives, HUF 693.77.24, includes a bibliography of the published research under the direction of Trowbridge between 1871 and 1877 in support of his proposal. He also described the endowment and construction of the laboratory in "The Jefferson Physical Laboratory," in *Science*, **5** (1885), 229–231.

Trowbridge's correspondence is in the Harvard College Library and in the Harvard University Archives.

II. Secondary Literature. Edwin H. Hall contributed the article on Trowbridge for the *Dictionary of American Biography*. Theodore Lyman wrote the biographical article "John Trowbridge 1843–1923," in *Proceedings of the American Academy of Arts and Sciences*, **60** (1925), 651–654. Further biographical literature is cited in Max Armin, ed., *Internationale Personalbibliographie*, II, 687. The work of Trowbridge on remote signaling is considered by E. Hawks in *Pioneers of Wireless* (London, 1927), 121–128.

Thaddeus J. Trenn

TRULLI, GIOVANNI (*b*. Veroli, Frosinone province, Italy, 1598; *d*. Rome, Italy, 27 December 1661), *medicine*.

The years of Trulli's youth are rather obscure. We only know that he went to France for training in surgery, particularly lithotomy. In 1636 he settled in residence in Rome as surgeon to Cardinal Francesco Barberini, nephew of Pope Urban VIII, and at the University of Rome as special professor of surgery. He was relieved of this post after the death of Urban VIII, whose corpse he dissected on 29 July 1644, finding cardiac ossification (left ventricle), as well as gallstones and kidney stones. He was also surgeon at the Santo Spirito Hospital in Rome.

Trulli probably played a role in the contacts between Francesco Barberini and William Harvey, who was in Rome during September and October 1636. In the winter of 1636–1637, he formed a friendship with the German physician P. M. Schlegel, who gave public anatomical demonstrations in Rome to illustrate the doctrine of circulation of the blood. A firm supporter of this doctrine, Trulli was enthusiastic in his efforts to disseminate it. His opinion is known to us rather by indirect evidence and by his correspondence with M. A. Severino than by writings specifically devoted to the problem of circulation.

Toward the end of 1637 Galileo had become completely blind. On 23 January 1638 P. B. Borghi sent him advice from Rome, praising the surgeon who had given it to him. Galileo then sent a report to Rome on his vision (a document now lost), on the basis of which Trulli was able to provide a full consultation. Forwarded to Galileo by Borghi on 20 February 1638, this consultation constitutes the most important medical document on Galileo's blindness.

BIBLIOGRAPHY

I. ORIGINAL WORKS. Trulli's consultation on Galileo's blindness is in *Le Opere di Galileo Galilei*, Edizione Nazionale, XIX, 552–554. His letter "De serie venarum" is in M. A. Severino, *Seilophlebotome castigata, sive de venae salvatellae usu et abusu* (Hanau, 1654), 150–153.

II. SECONDARY LITERATURE. See Luigi Belloni, "La dottrina della circolazione del sangue e la scuola Galileiana 1636–61," in *Gesnerus*, **28** (1971), 7–34; and Felice Grondona, "In tema di etiogenesi della cecità di Galileo," in *Atti del Symposium internazionale di storia, metodologia, logica e filosofia della scienza "Galileo nella storia e nella filosofia della scienza" (Firenze–Pisa, 14–16 settembre 1964)* (Florence, 1967), 141–154.

LUIGI BELLONI

TRUMPLER, ROBERT JULIUS (*b.* Zurich, Switzerland, 2 October 1886; *d.* Oakland, California, 10 September 1956), *astronomy.*

The third of ten children born to a Swiss industrialist, Wilhelm Ernst Trümpler, and his wife,

Luise Hurter, Trumpler lived and studied in Zurich until age twenty-one. This period included two years at the University of Zurich, which he left in 1908 for the University of Göttingen; two years later he received his Ph.D. with a dissertation written under Leopold Ambronn.

Following a postdoctoral year at Göttingen, Trumpler served four years in the Swiss Geodetic Survey. He determined latitudes and longitudes, but he also became interested in the way stars in the Pleiades cluster move together across the sky. Such annual proper motions also interested Frank Schlesinger, then director of the Allegheny Observatory near Pittsburgh. He met Trumpler in 1913 at a meeting of the German Astronomical Society and invited him to come to Allegheny. Trumpler arrived in 1915 and began making comparative studies of galactic star clusters—so named because they are located in the disk of our galaxy. In August 1916 Trumpler married Augusta De La Harpe; three daughters and two sons were born to them.

At the invitation of W. W. Campbell, Trumpler went to the Lick Observatory in 1919. He joined the staff in 1920 and rose by 1929 to the post of astronomer. In 1938 he was named professor of astronomy on the Berkeley campus of the parent organization, the University of California, a position he retained until his retirement in 1951.

At Lick, Campbell chose Trumpler to assist him, in 1922, in a test of the general theory of relativity. The test involved an expedition to Wallal, on the northwest coast of Australia, to photograph stars in the sky near the totally eclipsed sun, for comparison with their positions as photographed at night four months earlier in Tahiti. The data, after suitable statistical treatment, showed an outward deflection at the edge of the sun of $1.75 \pm 0.09''$ (compared to Einstein's prediction of $1.745''$), a result considerably more accurate than that obtained by Eddington three years earlier.

Trumpler used the thirty-six-inch Lick refractor to study Mars during the favorable opposition of 1924. To his surprise, he concluded that the long dark markings, named canals by Schiaparelli, were real—but neither as straight nor as sharp and narrow as Lowell believed them to be. Trumpler's feeling that a "canal" such as Coprates might be a volcanic fault received some support in 1972 from close-up photographs taken by Mariner 9.

Trumpler's work on galactic star clusters proved to be his most significant contribution. In 1925 he published evidence that the mix of stars in galactic clusters differs markedly; some clusters contain massive blue stars but no yellow or red giants, while in others the opposite is true. In the hands of later workers, such as Baade and Sandage, these findings were developed into the currently accepted picture of how individual stars evolve with time. And in 1930, in a paper including data on 334 galactic clusters, Trumpler showed that distances to galactic clusters were being overestimated because interstellar material, previously thought to be nonexistent, was dimming the starlight by an average of 0.67 magnitudes for every kiloparsec of distance. This discovery brought distances measured in the galactic disk into agreement with those found by Shapley above and below the disk, and showed why estimates of galactic size, as made for instance by Kapteyn, were too small.

Trumpler was elected to the National Academy of Sciences in 1932. The Astronomical Society of the Pacific elected him president in 1932 and again in 1939 and has established an award in his memory, given annually to a promising postdoctoral astronomer.

BIBLIOGRAPHY

I. ORIGINAL WORKS. Trumpler's dissertation was "Eine Methode zur photographischen Bestimmung von Meridian-durchgängen." His work confirming the general theory of relativity appeared in two papers (both written with W. W. Campbell): "Observations or the Deflection of Light in Passing Through the Sun's Gravitational Field, Made During the Total Solar Eclipse of September 21, 1922," in *Lick Observatory Bulletin*, **11** (1923), 41–54; and "Observations Made With a Pair of Five-foot Cameras on the Light Deflections in the Sun's Gravitational Field at the Total Solar Eclipse of September 21, 1922," *ibid.*, **13** (1926), 130–160.

Trumpler's study of Mars, including both photographs and drawings, can be found in "Observations of Mars at the Opposition of 1924," *ibid.*, **13** (1926), 19–45.

Trumpler's first paper on galactic star clusters, "Die relativen Eigenbewegungen der Plejadensterne," is in *Astronomische Nachrichten*, **200** (1915), cols. 217–230. Another early contribution, "Comparison and Classification of Star Clusters," is in *Publications of the Allegheny Observatory, University of Pittsburgh*, **6** (1922), 45–74. His system for classifying galactic clusters on the basis of the spectra of their constituent stars appears in "Spectral Types in Open Clusters," in *Publications of the Astronomical Society of the Pacific*, **37** (1925), 307–318. His discovery of the interstellar absorption of light, including the selective way it operates, is described in "Preliminary Results on the Distances, Dimensions, and Space Distribution of Open Clusters," in *Lick Observatory Bulletin*, **14** (1930), 154–188. A summary of what galactic clusters can—and cannot—reveal about intraga-lactic distances and galactic structure is contained in a paper Trumpler read at the dedication of the McDonald Observatory in 1939, published as "Galactic Star Clusters," in *Astrophysical Journal*, **91** (1940), 186–201.

Trumpler's book, *Statistical Astronomy*, written with Harold F. Weaver (Berkeley, 1953; repr., 1962), is the distillation of a graduate course he taught for over fifteen years. No complete list of Trumpler's publications has been published. Besides the above book, he wrote nine summarizing articles and approximately sixty-five research reports.

II. SECONDARY LITERATURE. An obituary notice on Trumpler, written by Harold and Paul Weaver, appeared in *Publications of the Astronomical Society of the Pacific*, **69** (1957), 304–307, with portrait and facs. signature.

SALLY H. DIEKE

TSCHERMAK, GUSTAV (*b.* Littau [now Litove], near Olomouc, Czechoslovakia, 19 April 1836; *d.* Vienna, Austria, 4 May 1927), *petrography, mineralogy.*

Tschermak grew up in the small Moravian town in which his grandfather had been a teacher and his father, Ignaz Czermak, was a tax collector. Despite the spelling of their name, the family considered themselves German. Tschermak began his secondary studies with a private tutor, then in 1850 entered Olomouc Gymnasium. He found the German-language instruction there to be inadequate, and this, together with his reaction to the rising tide of Czech national consciousness that followed the revolution of 1848 in Bohemia and Moravia, intensified his own feeling of Germanness and led him to found an anti-Slavic German student union. At the same time, he Germanized the spelling of his name. He distinguished himself in science during these years, and also founded a student natural history club. He was encouraged in these interests by one of his teachers, a Dr. Schwippel, and by the astronomer Julius Schmidt, who was then working in the private observatory in Olomouc.

In 1856 Tschermak enrolled in the Faculty of Philosophy of the University of Vienna. He began to study chemistry with Joseph Redtenbacher and learned the techniques of morphological and optical crystallography from Wilhelm Joseph Grailich. Although he attended no mineralogy lectures, he frequently visited the excellent imperial mineral collection. His first petrological work, "Das Trachytgebirge bei Banow," was published in 1858, while he was still a student. In 1860 Tschermak passed his teacher's examinations and was appointed assistant by the mineralogy professor Franz Xaver Zippe; he received the doctorate from the University of Tübingen later in the same year, and in 1861 qualified as *Privatdozent* in chemistry and mineralogy at the University of Vienna. In 1862 he became second assistant curator of the imperial mineralogical collection, and a few years later, first assistant curator.

During the 1860's Tschermak began the series of petrographical researches that, with his later work on meteorites, were to bring him an international reputation. He investigated the paragenesis of minerals in several granites, the quartz content in plagioclase, and the role of olivine in various rocks. He also contributed to the newly emerging methodology of microscopic investigation of rocks by means of using pleochroism to distinguish minerals of the augite, amphibole, and biotite groups. Dissatisfied with current knowledge about the most important rock-forming minerals, he set out to ascertain their crystal forms, their physical properties, and their compositional variations. He was thereby able to present the relationships between these mineral groups and to establish the prerequisites for an exact systematics, to which his chemical analysis of minerals, conducted with his friend Ernst Ludwig, was of fundamental importance.

Tschermak's most significant contribution in this direction lay in his work on feldspar. He realized that the many varieties of this mineral that had previously been distinguished could be derived from three compounds (which also occur in nature in almost pure form)—potash feldspar, albite, and anorthite. From this he was able to demonstrate that the various calcium-sodium feldspars form a homogeneous isomorphous series from pure calcium to pure sodium feldspar, of which the physical properties are a function of the proportion of the end members, that is, albite and anorthite.

Tschermak published his feldspar theory in *Die Feldspatgruppe* (1864), a work by which he, after a long dispute, firmly established his point of view. He made this theory the basis for his subsequent research, investigating almost all the important rock-forming silicates to confirm his idea that the great variety of chemical composition demonstrated in this group may be explained by the isomorphic mixture of simple compounds, from which changes in physical properties emerge naturally and in obedience to a law. A further petrographical book, *Die Porphyrgesteine Österreichs aus der mittleren geologischen Epoche*, embodying his previous research and the results of extensive investigatory travels, won a prize from the Vienna

Academy of Sciences in 1867 and was published as a book two years later.

In spring 1868 Tschermak was named associate professor of petrography at the University of Vienna, and in the autumn of the same year he was made director of the imperial mineral collection. He carried on an active research program, which attracted a number of young scientists, and, in order to make their results more widely known, founded the *Mineralogische Mitteilungen*, of which the first volume appeared in 1871. This periodical soon attracted foreign contributors; its first numbers were published as supplements to the *Jahrbuch der K. K. geologischen Reichsanstalt*, but after 1878 it was issued independently as *Tschermaks mineralogische und petrographische Mitteilungen*.

Tschermak began his work on meteorites in 1870, investigating the mineral content and inner structure of specimens from the imperial collection. He presented a theory of the origin of meteorites whereby these objects were cast off from small celestial bodies by volcanic activity—or, more precisely, by explosions of gases. In 1883 he published *Die mikroskopische Beschaffenheit der Meteoriten*, which became a standard work on the subject.

During this period Tschermak's academic career also advanced. In 1873 he was made full professor of mineralogy and petrography at the university (while retaining his curatorial post); in 1876 he received an offer, which he declined, from the University of Göttingen; and in 1877 he gave up his directorship of the imperial collection to devote all of his time to his work at the university where, the following year, a mineralogy and petrology institute was put at his disposal. He was elected dean of the Faculty of Philosophy in 1883 and rector of the university ten years later. He also continued to do research on silicates and meteorites, supplementing it with other petrological and mineralogical works, until his retirement in 1906. His *Lehrbuch der Mineralogie* was published in 1883, and went through a number of editions, while in 1907, after he had left the university, he was able to demonstrate that certain periodic meteor showers are characterized by petrographical conformity.

Tschermak's works brought him many honors. He became a member of the Vienna Academy of Sciences in 1875, was awarded the title "Hofrat" in 1886, and, on his retirement in 1906, was ennobled (subsequently styling himself "Tschermak von Seysenegg"). He was one of the founders of the Austrian Mineralogical Society and was elected its honorary president in 1910, on the occasion of the fiftieth anniversary of his doctorate. He was either a member or an honorary member of almost every important scientific society and natural history association. Of the four children of his two marriages, one daughter and two sons became distinguished scientists.

BIBLIOGRAPHY

I. ORIGINAL WORKS. Tschermak's writings include "Das Trachytgebirge bei Banow," in *Jahrbuch der Kaiserlichen Königlichen geologischen Reichsanstalt*, **9** (1858), 63–79; *Ein Beitrag zur Bildungsgeschichte der Mandelsteine* (Vienna, 1863); *Chemisch-mineralogische Studien. I, Die Feldspatgruppe* (Vienna, 1864); *Beobachtungen über die Verbreitung des Olivins in den Felsarten* (Vienna, 1867); *Quarzführende Plagioklasgesteine* (Vienna, 1867); *Die Porphyrgesteine Österreichs aus der mittleren geologischen Epoche* (Vienna, 1869); *Beitrag zur Kenntnis der Salzlager* (Vienna, 1871); *Ein Meteoreisen aus der Wüste Atacama* (Vienna, 1871); *Die Bildung der Meteoriten und der Vulkanismus* (Vienna, 1875) ; *Die Glimmergruppe* (Vienna, 1877); *Die mikroskopische Beschaffenheit der Meteoriten* (Vienna, 1883); *Die Skapolithreihe* (Vienna, 1883); and *Lehrbuch der Mineralogie* (Vienna, 1884).

In addition to memoirs in *Mineralogische und petrographische Mitteilungen*, Tschermak published articles in *Justus Liebigs Annalen der Chemie, Neues Jahrbuch für Mineralogie, Geologie und Paläontologie*, Poggendorff's *Annalen der Physik und Chemie, Almanach der Akademie der Wissenschaften in Wien, Denkschriften der Akademie der Wissenschaften* (Vienna), and *Sitzungsberichte der Akademie der Wissenschaften in Wien*.

II. SECONDARY LITERATURE. The most extensive obituary is by Tschermak's student and successor as professor, Friedrich Becke, in *Mineralogische und petrographische Mitteilungen*, **39** (1928), i–x. Other obituaries are by E. S. Dana, in *American Journal of Science*, 5th ser., **14** (1927), also in *American Mineralogist*, **12** (1927); and by J. W. Evans, in *Nature*, **120** (1927), 195–196. *Mineralogische und petrographische Mitteilungen*, **25** (1906), dedicated to Tschermak on his retirement, contains a signed photographic portrait.

HANS BAUMGÄRTEL

TSCHERMAK VON SEYSENEGG, ERICH (*b.* Vienna, Austria, 12 November 1871; *d.* Vienna, 11 October 1962), *botany, genetics.*

Tschermak came from a family of scholars. His father, Gustav Tschermak, was director of the Imperial Mineralogical Museum and, from 1873, professor of mineralogy and petrography at the University of Vienna. He was created a member of the hereditary nobility with the title "von Seysenegg." His mother, Hermine Fenzl, was a daughter of the botanist Eduard Fenzl, director of the Botanical Institute and Garden of the University of Vienna. Tschermak's older brother Armin, who married a daughter of the geologist Albrecht Penck, became professor of physiology at Ferdinand University in Prague and, after 1945, at the University in Regensburg for a few years. His sister Silvia was a mineralogist.

With his brother, Tschermak attended the humanistic Gymnasium at the Kremsmünster monastery (in Upper Austria). In 1891 he enrolled simultaneously at the University of Vienna and at the Hochschule für Bodenkultur at Vienna. After two semesters he volunteered for one year's work on a nobleman's estate near Freiberg, Saxony, in order to learn the basics of agricultural practice. He subsequently continued his studies at the University of Halle, where he took the agricultural examination for the agricultural diploma in 1895 and the following year obtained the Ph.D. with a dissertation in botany.

Tschermak, a Roman Catholic, married twice but had no children. In his autobiography he writes that he was a weak and sickly child; nevertheless, he lived to be nearly ninety-one. Although he suffered greatly in his last years from arthritis in his hands, he never lost his zest for work, his good humor, his kindness, or his enjoyment of social life.

Tschermak held honorary doctorates from the universities of Vienna, Giessen, and Ghent as well as from the agricultural universities of Vienna, Berlin, and Brno. He was a member of the Institut de France, of the Leopoldina Carolina in Halle and of the Max Planck Society, as well as numerous scientific societies and associations. He was awarded the Cothenius Medal and the Goethe Medal.

After completing his studies at Halle, Tschermak was employed at the agricultural stations at Stendal and Quedlinburg, Germany. In 1898, in Ghent, he studied breeding of vegetables and flowers. Stimulated by Darwin's *The Effects of Cross and Self Fertilization in the Vegetable Kingdom* (1876), Tschermak began hybridization experiments with various types of peas at the botanical garden in Ghent. The results of this research led to his rediscovery of Mendel's laws of heredity, which had been disregarded until then and had been rediscovered independently and simultaneously by Hugo de Vries in Amsterdam and by

Carl Correns in Leipzig. The publications of these three investigators appeared nearly simultaneously in 1900. Tschermak's papers appeared in *Zeitschrift für das landwirtschaftliche Versuchswesen in Österreich* and in summary form in *Berichte der Deutschen botanischen Gesellschaft.*

Tschermak's greatest service was his exclusive, immediate recognition of the importance of Mendel's laws of heredity and his application of these laws in his own breeding experiments. In 1900 Tschermak became an academic lecturer at the Hochschule für Bodenkultur, where he taught theory of plant production, commercial cultivation of plants, and production of vegetables in fields. In 1902 he became assistant to the professor of the theory of plant breeding; in 1906 he was appointed extraordinary professor; and in 1909 he was made full professor. In addition to teaching, Tschermak was for several years director of the Royal Institute for Plant Breeding of the Prince of Liechtenstein (later the Mendel Institute) in Eisgrub, Moravia.

Tschermak traveled extensively to acquire new experience in his field. He made four visits to Sweden, where he served as adviser on the expansion of the Swedish Seed Association at Svalöf. He had a warm relationship with the director of this institute, Nilsson-Ehle, and with his successor. In 1909 Tschermak traveled to the United States with his friend Kurt von Rümker, primarily to meet Luther Burbank and to study his methods. He revisited the United States the following year.

While he was at the Hochschule für Bodenkultur, Tschermak made many new crosses of cultivated plants that increased their diversity and value. Since he had only limited experimental facilities available, he frequently turned over the testing and propagation of his new plants to capable agricultural businesses and to interested farmers, many of whom had established test plots at his suggestion. Of special practical importance were Tschermak's new types of rye, wheat, barley, oats, and legumes. His breeding of short-shooted edible types of pumpkins with soft-shelled, oil-rich seeds (later known as Tschermak's pumpkin) made possible a broader use. Also of great value were his new varieties of flowers, such as the gillyflower (*Matthiola*), and many new types of primroses.

Tschermak's more than 100 publications are impressive because of the diversity and originality of his investigations, observations, and theories. His most important achievement, the rediscovery of Mendel's laws, was closely related to his observations of the xenia phenomenon of various plants: in certain instances the seeds in the female plant show effects that are transmitted through the pollen of the male plant. Tschermak investigated the xenia phenomenon in a group of legumes as well as in various types of corn and in the gillyflower. The well-known xenia of the corncob, which manifests itself in various colors and types of surfaces of the kernels, has been extensively investigated by Correns.

Tschermak also was interested in the stimulation effects of alien pollen. He interpreted his findings on fertilization by irritation to mean that nonfertilizing pollen and even dehydrating agents can cause a physiological stimulation of the mature egg cell, and initiate parthenogenetic development.

Tschermak's discovery and elucidation of a cryptomeric heredity among the gillyflowers is a classic work in theoretical genetics. Here, white-flowered hybrids from white-flowered parents can produce "nova" plants bearing colored flowers in the F_2 generation, if each of the parents possesses a dominant and a recessive gene for white flowers.

Tschermak's investigations of intergeneric hybridization led to basic discoveries about hybridization. He suspected that various of his fertile intergeneric hybrids, such as that of wheat and rye—the fertile hybrid of *Aegilops ovata* ($2n = 28$) and *Triticum dicoccoides* ($2n = 28$)—had been produced by the fertilization of unreduced F_1 gametes. For this hybrid, *Aegilotriticum* ($2n = 56$), the additive hybrid number of chromosomes could be demonstrated by cytological examination. This was the first artificially produced and cytologically demonstrated additive intergeneric hybrid of the Gramineae. Today the polyploidization of sterile F_1 hybrids is one of the techniques for the experimental investigation of relationships, as well as for the breeding of artificial, synthetic types of cultivated plants.

Tschermak demonstrated his diverse abilities in the breeding of agricultural and garden plants, for which he used his highly developed technique of hybridization. He was concerned primarily with such difficult problems as combining early ripening with high yield, which he achieved.

The Tschermak-grafted Marchfeld rye, originally bred by selection of the grain and of the offspring without interbreeding with foreign stock, is the only grain variety that has not yet been surpassed in yield (Marchfeld, near Vienna) by any other type, either foreign or domestic. For the self-pollinating types of grain such as wheat, barley, and oats, on the other hand, the new varieties obtained by continuous hybridization yielded 25–50 percent more than the older ones.

Today Tschermak's fundamental idea, the systematic combination of genes, and the method derived from Mendel's "laws of segregation" for the investigation of the offspring of single individuals isolated from hybrid populations, have become a matter of routine in the breeding of cultivated plants.

BIBLIOGRAPHY

A detailed account of Tschermak's life and a complete bibliography of his works are in his autobiography, *Erich von Tschermak-Seysenegg, Leben und Wirken* (Berlin–Hamburg, 1958).

His works include "Über künstliche Kreuzung von *Pisum sativum*," in *Zeitschrift für das landwirtschaftliche Versuchswesen in Österreich*, **3** (1900), 465–555, summarized in *Berichte der Deutschen botanischen Gesellschaft*, **18** (1900), 232–239; "Über die Züchtung neuer Getreiderassen mittels künstlicher Kreuzung," in *Zeitschrift für das landwirtschaftliche Versuchswesen in Österreich*, **4** (1901), 1029–1060, and **9** (1906), 699–743; "Über den Einfluss der Bestäubung auf die Ausbildung der Fruchthülle," in *Berichte der Deutschen botanischen Gesellschaft*, **20** (1902), 7–16; "Über Züchtung landwirtschaftlich und gärtnerisch wichtiger Hülsenfrüchte," *Arbeiten der Deutschen Landwirtschaftsgesellschaft für Österreich*, no. 4 (1920); "Über fruchtbare Aegilops-Weizenbastarde," in *Berichte der Deutschen botanischen Gesellschaft*, **44** (1926), 110–132, written with H. Bleier; "Zur zytologischen Auffassung meiner Aegilotricum-Bastarde und der Artbastarde überhaupt. Theorie der Chromosomenaddition oder Kernchimärie," *ibid.*, **47** (1929), 253–261; "Bemerkungen über echte und falsche Grössen-Xenien," in *Zeitschrift für Pflanzenzüchtung*, **17** (1932), 447–450; "Über einige Blütenanomalien bei Primeln und ihre Vererbungsweise," in *Biologia generalis*, **8** (1932), 337–350; "Der schalenlose Kürbis als Ölfrucht," *Deutsche Landwirtschaftliche Presse*, **61**, no. 3 (1934); and "Reizfruchtung (Samenbildung ohne Befruchtung)," in *Biologia generalis*, **19** (1949), 3–50.

RICHARD BIEBL

TSCHIRNHAUS, EHRENFRIED WALTHER (*b.* Kieslingswalde, near Görlitz, Germany, 10 April 1651; *d.* Dresden, Germany, 11 October 1708), *mathematics, physics, philosophy*.

Tschirnhaus was the youngest son of Christoph von Tschirnhaus, a landowner, and Elisabeth Eleonore Freiin Achyll von Stirling, who belonged to a collateral branch of the mathematically gifted Stirling family. His mother died when he was six, but he was brought up by a loving stepmother. After receiving an excellent education from private tutors, Tschirnhaus entered the senior class of the Görlitz Gymnasium in 1666. In the autumn of 1668 he enrolled at the University of Leiden to study philosophy, mathematics, and medicine. He was deeply impressed by the tolerant atmosphere there, as well as by the fiery philosopher Arnold Geulincx, an occasionalist, and the distinguished physician F. de la Boë (Sylvius), who taught Harvey's theory of the circulation of the blood. The most profound influence on him in these years, however, was that of Descartes's philosophy and mathematics, to which he was introduced in private instruction by Pieter van Schooten.

At the beginning of the war between Holland and France in 1672, Tschirnhaus joined the student volunteer corps but did not see action. Following a short visit to Kieslingswalde in 1674, he returned to Leiden and was introduced by his school friend Pieter van Gent to Spinoza, whose teachings he immediately adopted. With a letter of recommendation from Spinoza, he went to London in May 1675 to see Henry Oldenburg. Tschirnhaus had become an excellent algebraist and was able to make a persuasive presentation of his methods for solving equations. He visited John Wallis at Oxford and held discussions with John Collins, to whom he showed examples of his methods. On closer examination, however, they proved to be special cases of a previously known solution.

Bearing recommendations from Oldenburg addressed to Huygens and Leibniz, Tschirnhaus moved to Paris in the fall of 1675. He did not then know French, and when engaged to teach mathematics to one of Colbert's sons, did so in Latin. In an animated exchange with Leibniz, Tschirnhaus reported in general terms on his own methods but only half listened to what Leibniz told him concerning his recent creation of a symbolism for infinitesimal processes. In fact, Tschirnhaus never did grasp the significance of Leibniz' disclosure, and throughout his life he considered the infinitesimal symbolism to be of limited applicability.

Leibniz introduced Tschirnhaus to Clerselier, who had custody of Descartes's papers and allowed them to look through unpublished manuscripts. The two also had an opportunity to examine the posthumous papers of Pascal and Roberval. Tschirnhaus reported on the progress of his studies at Paris in a number of interesting letters to Pieter van Gent, Spinoza, and the latter's friend G. H. Schuller. In the summer of 1676 he corresponded with Oldenburg concerning Descartes's mathematical methods. Tschirnhaus considered them unsurpassable, but Collins had expressed considerable skepticism about them. Consequently, the reports in which Oldenburg and Newton communicated the results obtained by expansions of series were addressed jointly to Leibniz and Tschirnhaus. In his reply of 1 September, Tschirnhaus judged these results somewhat disparagingly; Collins responded with a strong rebuttal, as did Newton in a second letter to Leibniz and Tschirnhaus (3 November 1676).

Also in 1676 Tschirnhaus accompanied Count Nimpsch of Silesia on a trip to southern France and Italy. Everywhere he went, Tschirnhaus sought contact with leading scientists, collected observations, and reported interesting discoveries to Leibniz. Among the matters he communicated to Leibniz was an algorithmic method of reduction that he wrongly believed could be applied to equations of any higher degree. (This method was published in *Acta eruditorum* in 1683.) He also reported on a supposedly new method of quadrature that was in fact merely the result of recasting a procedure devised by Gregory of Saint-Vincent in 1647 in a form better suited for computation. (The improvement had been effected by the use of indivisibles of zero width.) During his return trip in 1679, Tschirnhaus stopped at Paris, at The Hague (where he saw Huygens), and at Hannover (where he visited Leibniz).

While continuing his mathematical research Tschirnhaus constructed effective circular and parabolic mirrors, with which he obtained high temperatures by focusing sunlight. He also made burning glasses, though not without flaws. During a trip to Paris in the summer of 1682, he became a member of the Académie des Sciences. He did not, however, receive the hoped-for royal pension that would have enabled him to pursue his scientific work free from financial concern. After returning from Paris, Tschirnhaus married Elisabeth Eleonore von Lest, who took over most of the details of managing the estate his father had left him, thus permitting him to devote his time entirely to study. Among his achievements was the rediscovery of the process for making hard-paste porcelain. J. F. Böttger, who is usually given the credit, was a skilled craftsman; but all his work was done under Tschirnhaus' supervision.

Tschirnhaus exhausted his mathematical talents in searching for algorithms. Lacking insight into the more profound relations among mathematical propositions, he was all too ready to assert the existence of general relationships on the basis of particular results that he obtained. Further, he was unwilling to accept suggestions directly from other mathematicians, although he would later adopt them as his own inventions and publish them as such. This tactic led to bitter controversies with Leibniz, Huygens, La Hire, and Jakob I and Johann I Bernoulli; and it ultimately cost him his scientific reputation. Without going into details, we may mention two of these disputes. The first, with Leibniz, concerned the possibility of algebraic quadratures of algebraic curves (1682–1684). The second, with Fatio de Duillier (1687–1689), was provoked by Tschirnhaus' publication of an incorrect method of finding tangents to curves generated by the motion of a drawing pencil within a system of taut threads. The method appeared in a major work of considerable philosophical importance, *Medicina corporis et mentis* (1686–1687), which was influential in the early stages of the Enlightenment. Another work by Tschirnhaus, *Gründliche Anleitung zu den nützlichen Wissenschaften* (1700), was highly praised by Leibniz in 1701. Both books deeply impressed Leibniz' disciple Christian Wolff.

Tschirnhaus was essentially an autodidact. During his university years he lacked the guidance of a kind, experienced, yet strict teacher, who could have restrained his exuberant temperament, moderated his excessive enthusiasm for Descartes's ideas, and instilled in him a greater measure of self-criticism. Even so, Tschirnhaus' achievements—often accomplished with insufficient means—were far more significant than the average contribution made by university teachers of science during his lifetime. Indeed, even his errors proved to be important and fruitful stimuli for other scientists.

BIBLIOGRAPHY

I. ORIGINAL WORKS. The major portion of Tschirnhaus' unpublished papers is in the MS division of the library of the University of Wrocław and among the Leibniz MSS of the Niedersächsische Landesbibliothek, Hannover.

His books are *Medicina corporis et mentis*, 3 pts. in 2 vols. (Amsterdam, 1686–1687; 2nd ed., Leipzig, 1695), also translated into German (Frankfurt, 1688; 2nd ed., Lüneburg, 1705–1708) and recently retranslated into German by J. Haussleiter, in *Acta historica Leopoldina* (Leipzig, 1963), with biography, portrait, and detailed bibliography prepared by R. Zaunick; and *Gründliche Anleitung zu den nützlichen Wissenschaften* (n.p., 1700; 2nd ed., 1708; 3rd ed., 1712).

Tschirnhaus' mathematical papers and reviews appeared in *Journal des sçavans* (Amsterdam), no. 15 (8 June 1682), 210–213; in the following issues of *Acta eruditorum*: (1682), 364–365, 391–393; (1683), 122–124, 204–207, 433–437; (1686), 169–176; (1687), 524–527; (1690), 68–73, 169–172, 481–487, 561–

565; (1695), 322–323, 489–493; (1696), 519–524; (1697), 113, 220–223, 409–410; (1698), 259–261; and in *Mémoires de physique et de mathématique de l'Académie royale des sciences* for 1701 (1704), 291–293, and for 1702 (1704), 1–3.

His papers on the burning glass were published in the following issues of *Acta eruditorum*: (1687), 52–54; (1688), 206; (1691), 517–520; (1696), 345–347, 554; (1697), 414–419; (1699), 445–448; and in *Histoire de l'Académie royale des sciences* for 1699 (1702), 90–94, and for 1700 (1703), 131–134.

Some of his correspondence has been published in Huygens, *Oeuvres complètes*, VIII and IX (The Hague, 1899–1900); Spinoza, *Opera*, 3rd ed., III (The Hague, 1914); and in Leibniz, *Briefwechsel mit Mathematikern*, C. I. Gerhardt, ed. (Berlin, 1899; repr. Hildesheim, 1962). The entire Leibniz-Tschirnhaus correspondence will eventually be published in Leibniz, *Sämtliche Schriften und Briefe*, 3rd ser.

II. SECONDARY LITERATURE. See H. Weissenborn, *Lebensbeschreibung des E. W. v. Tschirnhaus . . . und Würdigung seiner Verdienste* (Eisenach, 1866); and the following, more recent, works of J. E. Hofmann: "Das *Opus geometricum* des Gregorius a S. Vincentio und seine Einwirkung auf Leibniz," in *Abhandlungen der Preussischen Akademie der Wissenschaften*, Math.-naturwiss. Kl. (1941), no. 13, 55–69; *Die Entwicklungsgeschichte der Leibnizschen Mathematik während des Aufenthaltes in Paris (1672–1676)* (Munich, 1949), enl. English ed., *Leibniz in Paris, 1672–1676* (London–New York, 1974); *Über Jakob Bernoullis Beiträge zur Infinitesimalmathematik* (Geneva, 1956); "Aus der Frühzeit der Infinitesimalmethoden: Auseinandersetzung um die algebraische Quadratur algebraischer Kurven in der zweiten Hälfte des 17. Jahrhunderts," in *Archive for History of Exact Sciences*, 2, no. 4 (1965), 270–343; and "Drei Sätze von E. W. v. Tschirnhaus über Kreissehnen," in *Studia Leibnitiana*, 3 (1971), 99–115. See also E. Winter, *E. W. von Tschirnhaus und die Frühaufklärung in Mittel- und Osteuropa* (Berlin, 1960).

J. E. HOFMANN

TSERASKY (or CERASKY), VITOLD KARLO-VICH (*b.* Slutsk, Minsk guberniya, Russia, 9 May 1849; *d.* Meshcherskoe, near Podolsk, Moscow oblast, U.S.S.R., 29 May 1925), *astronomy.*

Tserasky's father was a secondary school geography teacher. The discovery of Donati's comet in 1858 awakened his interest in astronomy. In 1867 he entered Moscow University and while a second-year student worked at the observatory as part-time calculator. Tserasky received a gold medal for a student work on the motion of Mars. After graduating he became a supernumerary assistant at the observatory. In 1874 he took part in an arduous expedition to Kyakhta, on the Chinese border, to observe the transit of Venus.

After returning to Moscow, Tserasky began systematic photographic study of the sun, but in 1877 he became deeply involved in instrumental astrophotometry. Simultaneously with the Potsdam and Harvard observatories, he conducted pioneering work in precise stellar photometry. In 1878 he was designated astronomer-observer at the Moscow observatory. In 1883 he defended his master's dissertation on the determination of the brightness of white stars. The astrophotometer of Zöllner led him to invent an instrument (the Zöllner-Tserasky photometer) that became the subject of his doctoral dissertation (1887).

At the end of the 1870's Tserasky began teaching at the Higher School for Women, which were begun by a number of progressive professors—advocates of higher education for women, who did not have the right to study in universities during the tsarist period. From 1882 he was lecturer at the University of Moscow; in 1884 he became *Privatdozent*; and in 1889 he was elected professor of astronomy. His public lectures also were very successful.

In 1884 Tserasky married Lidia Petrovna Shelekhova, later known for her discoveries of more than 200 variable stars on the negatives of the Moscow Observatory. In 1890 Tserasky succeeded Bredikhin as director of the Moscow observatory, and from 1895–1903 he supervised the reconstruction of the observatory and the renovation of its equipment. In 1914 he was elected corresponding member of the St. Petersburg Academy of Sciences. After retiring in 1916 because of his health, he settled in Feodosiya, in the Crimea. In 1922 he moved to the home of his son, a physician, not far from Moscow, where he died three years later.

In addition to his numerous works in photometry, Tserasky made in 1895 the first experimental determination of the lower limit of the temperature of the sun (3,500°K.), by melting a number of refractory minerals at the focus of a concave mirror having a diameter and focal length both of one meter. (On the basis of these data Scheiner computed the temperature of the surface of the sun at 6,600°K., which approximates the correct value.) Following his discovery, at the end of the last century, of a few variable stars, Tserasky organized systematic astrophotographic studies using a wide-angled, short-focus astrograph (the "equatorial camera") built according to his design. From the resulting negatives L. P. Tseraskaya discovered more than 200 variable stars. In 1885 Tserasky discovered luminous (silver) clouds and with Belopolsky determined their altitude at 80 kilometers (actual value ~ 82 km.). Tserasky studied meteors and invented an original instrument for determining their angular velocity. In 1911 he published the results of his photometric comparison of the stellar magnitude of the sun with that of Venus and a number of bright stars; his value of the stellar magnitude of the sun (−26.50) has become an important weight in world summaries. To determine the degree of flattening of the sun's sphere Tserasky devised an original heliometric objective and, for detailed study of sunspots, a specially constructed eyepiece. In 1908 he constructed a device that used solar energy to set off an electric bell on his desk.

An unpublished photometric catalog of 466 circumpolar stars, found among Tserasky's papers, was edited by G. A. Manova and was included in a collection of his selected works (1953).

BIBLIOGRAPHY

I. Original Works. Tserasky's 92 published writings include *O prokhozhdenii Venery po disku Solntsa v 1874 g.* ("On the Transit of Venus Across the Disk of the Sun in 1874"; Moscow, 1875); "Ob opredelenii yarkosti belykh zvezd" ("On the Determination of the Brightness of White Stars"), in *Uchenye zapiski Moskovskogo universiteta,* Otd. fiz.-matem., no. 4 (1882), 105–176, his master's diss.; "Astronomichesky fotometr i ego prilozhenia" ("The Astronomical Photometer and Its Uses"), in *Matematichesky sbornik,* **13** (1886), 551–632, his doctoral thesis; "Sur les nuages lumineux," in *Annales de l'Observatoire astronomique de Moscou,* 2nd ser., **2,** bks. 1–2 (1890), 177–180; "Études photométriques sur l'amas stellaire χ Persei," *ibid.,* **3,** bk. 2 (1896), 1–24; "Sur la température du soleil," *ibid.,* 121–122, also in J. Scheiner, *Strahlung und Temperatur der Sonne* (Leipzig, 1899); "Ob opredelenii formy solnechnogo diska" ("On the Determination of the Form of the Solar Disk"), in *Izvestia Akademii nauk,* **11,** no. 2 (Sept. 1899), 59–60; "Études photométriques sur l'amas stellaire Coma Berenices," in *Annales de l'Observatoire astronomique de Moscou,* 2nd ser., **4** (1902), 87–120, and **6** (1917), 33–44; "Détermination photométrique de la grandeur stellaire du soleil," in *Astronomische Nachrichten,* **170** (1905), 135–138; "Sur une équation personnelle dans les observations photométriques," *ibid.,* **171** (1906), 135–136; *Sfericheskaya astronomia. Lektsii, chitannye v 1909–1910 gg.* ("Spherical Astronomy. Lectures Given in 1909–1910"; Moscow, 1910); "Détermination photométrique de la grandeur stellaire du soleil," in *Annales de l'Observatoire astronomique de Moscou,* 2nd ser., **5** (1911),

1–30; "Un oculaire pour l'étude détaillée des taches solaires," *ibid.,* 31–33; "Un objectif héliométrique pour la détermination de la forme du disque solaire," *ibid.,* 34–35; "Détermination des erreurs constantes des observations photométriques," *ibid.,* **6** (1917), 45–61; and *Izbrannye raboty po astronomii* ("Selected Works on Astronomy"; Moscow, 1953).

II. Secondary Literature. On Tserasky and his work, see S. Blazhko, "W. Ceraski," in *Astronomische Nachrichten,* **225** (1925), 111–112; "Nauchnye raboty professora V. K. Tseraskogo" ("The Scientific Works . . ."), in *Russky astronomichesky kalendar* for 1925, 128–134; and "Vitold Karlovich Tserasky. Zhizneopisanie" (" . . . A Description of His Life"), in Tserasky's *Izbrannye raboty po astronomii* ("Selected Works on Astronomy," V. V. Podobed, ed., Moscow, 1953), 11–29; I. A. Kazansky's bibliography of Tserasky's works, *ibid.,* 46–52; Y. G. Perel, "K voprosu o mirovozzrenii V. K. Tseraskogo" ("Toward the Question of Tserasky's World Views"), in *Istoriko-astronomicheskie issledovaniya* (1955), no. 1, 323–334; and "Vitold Karlovich Tserasky," in *Vydayushchiesya russkie astronomy* ("Outstanding Russian Astronomers"; Moscow, 1951), 63–84; K. D. Pokrovsky, "V. K. Tserasky," in *Russky astronomichesky kalendar* for 1925, 115–127, published for his seventy-fifth birthday; B. A. Vorontsov-Velyaminov, "Nauchnaya deyatelnost V. K. Tseraskogo" ("The Scientific Activity of V. K. Tserasky"), in Tserasky's *Izbrannye raboty po astronomii,* 30–45; and F. Y. Zotov, "Vospominania o V. K. Tseraskom" ("Recollections. . ."), in *Istoriko-astronomicheskie issledovaniya* (1955), no. 1, 335–342.

P. G. Kulikovsky

TSIOLKOVSKY, KONSTANTIN EDUARDOVICH (*b.* Izhevsk, Ryazan guberniya, Russia, 17 September 1857; *d.* Kaluga, U.S.S.R., 19 September 1935), *mechanics, aeronautics, astronautics.*

Tsiolkovsky was the son of a forester. At the age of nine he became almost completely deaf following a serious illness. Unable to continue in school, he was obliged to study on his own, using his father's library. From the age of thirteen Tsiolkovsky began systematically to study the natural sciences. It was at this time that his inclination for invention became apparent.

In 1873, his father sent him to Moscow to continue his self-education. Studying on his own, Tsiolkovsky completed the entire secondary-school course and a considerable part of the university course.

After having passed the teaching examination in 1879—without attending lectures—Tsiolkovsky was appointed in 1880 to the Borovsk district school, sixty miles southwest of Moscow, where he taught arithmetic and geometry. He devoted most of his free time to scientific investigations.

In the mid-1880's, Tsiolkovsky began research in aerostatics. He worked out a project for an all-metal dirigible with a corrugated metal shell, the volume of which could be varied in flight. Tsiolkovsky further developed his dirigible theory in "Aerostat metallichesky upravlyaemy" ("A Controlled Metal Dirigible," 1892).

In "K voprosu o letanii posredstvom krylev" ("On the Question of Winged Flight"), completed in 1890, Tsiolkovsky investigated the magnitude of forces acting upon a moving disk. In this work he made the first attempt to evaluate quantitatively the influence of the length of the disk on the magnitude of the aerodynamic forces. In 1891, part of the work was printed by the Society of Friends of Natural Science under the title "Davlenie zhidkosti na ravnomerno dvizhushchuyusya v ney ploskost" ("The Pressure of a Fluid on a Plane Moving Uniformly Through It"). It was Tsiolkovsky's first published work.

In 1892 Tsiolkovsky moved to Kaluga, where he continued to teach without interrupting his scientific research. In 1894 his work "Aeroplan ili ptitsepodobnaya (aviatsionnaya) letatelnaya mashina" ("The Aeroplane, or A Flying Birdlike Machine [Aircraft]") was published in *Nauka i zhizn*. In this work he proposed a plan for a plane having a metal frame (similar to contemporary aircraft)—a monoplane with a streamlined fuselage, freely supported wings, a thick profile with a rounded forward edge, a wheeled undercarriage, and an internal combustion engine. He also suggested using twin screw propellers rotating in opposite directions. He expounded the idea of using the gyroscope in aircraft as a simple automatic pilot.

While working on these projects Tsiolkovsky came to grips with the unavoidable necessity of obtaining precise data on the resistance of a medium. A series of experiments under natural conditions led him to test his models in an artificial airstream. In 1897 he constructed a wind tunnel, the first in Russia to be used in aviation.

Of greater significance are Tsiolkovsky's works on outer space. He became interested in questions of interplanetary travel at the age of sixteen. In "Svobodnoe prostranstvo" ("Free Space," 1883), he examined phenomena that occur in a medium in which the forces of gravity—for all practical purposes—are not active. This paper included the first formulation of the possibility of applying the principle of reactive motion for flight in a vacuum, which led to a simple plan for a spaceship. Tsiolkovsky also considered several questions concerning the necessary conditions of life for plants and animals in space.

In 1896 Tsiolkovsky began to explore the possibility of interplanetary travel by means of rockets. In 1897 he formulated his now widely known formula establishing the analytical dependence between the velocity of a rocket at a given moment, the velocity of the expulsion of gas particles from the nozzle of the engine, the mass of the rocket, and the mass of the expended explosive material.

In "Issledovanie mirovykh prostranstv reaktivnymi priborami" ("A Study of Atmospheric Space Using Reactive Devices," 1903), Tsiolkovsky set forth his theory of the motion of rockets, established the possibility of space travel by means of rockets, and adduced the fundamental flight formulas.

Tsiolkovsky contributed to the recently established mechanics of bodies of changing mass. He evolved a theory of rocket flight taking into account the change of mass while in motion; he suggested the concept of gas-driven rudders for guiding a rocket in a vacuum; and he determined the coefficient of a rocket's practical operation.

From 1903 to 1917 Tsiolkovsky offered several plans for constructing rocket ships. He considered such questions as guiding a rocket in a vacuum, the use of a fuel component to cool the combustion chamber walls, and the application of refractory elements.

Tsiolkovsky's advanced ideas did not find acceptance. He was met with indifference and disbelief, and many considered this autodidact to be a rootless dreamer. Having received neither material nor moral support, Tsiolkovsky was left to his own resources. "It has been difficult for me," he wrote with bitterness, "to work alone for many years under unfavorable conditions and not even to see the possibility for hope or assistance."

The conditions of life and work for Tsiolkovsky changed radically after the October Revolution. In 1918 he became a member of the Academy, and in November 1921 he was allotted a personal pension. It became possible for him to devote himself completely to his scientific work.

During the 1920's Tsiolkovsky continued his investigations in aeronautics, and he elaborated his theory of multistage rockets. He also began working out his theory of the flight of jet airplanes, devoting a number of papers to this question.

In 1921 Tsiolkovsky conceived the idea of building a transport vehicle that would be carried on a cushion of air. This idea was further developed in "Soprotivlenie vozdukha i skory poezd" ("Air Resistance and the Express Train," 1927) and "Obshchie uslovia transporta" ("General Conditions for Transport," 1934).

In the mid-1920's, Tsiolkovsky's works on rocket engineering and space flight began to win international recognition. Hermann Oberth, the German rocket technologist, wrote to Tsiolkovsky in 1929: "You have ignited the flame, and we shall not permit it to be extinguished; we shall make every effort so that the greatest dream of mankind might be fulfilled."

Despite old age, Tsiolkovsky continued his scientific work. In "Dostizhenie stratosfery" ("Reaching the Stratosphere," 1932), he formulated the requirements of explosive fuel for use in jet engines. In 1934 and 1935 he proposed using clusters of rockets in order to reach great speeds.

"The main motive of my life," Tsiolkovsky wrote in evaluating his activity, "has been to . . . move humanity forward if only slightly. This is exactly why I have been interested in those things that never yielded either bread or strength. But I hope that my labors perhaps . . . may give to society mountains of bread and infinite power."

BIBLIOGRAPHY

Tsiolkovsky's collected works were published as *Sobranie sochineny*, 5 vols. (Moscow, 1951–1967). On his life and work, see *Konstantin Eduardovich Tsiolkovsky (1857–1932)* (Moscow–Leningrad, 1932), a jubilee collection published to commemorate his seventy-fifth birthday; and M. S. Arlazorov, *K. E. Tsiolkovsky, ego zhizn i deyatelnost* (". . . His Life and Work"; Moscow, 1962).

A. T. GRIGORIAN

TSU CH'UNG-CHIH (*b.* Fan-yang prefecture [modern Hopeh province], China, *ca.* A.D. 429; *d.* China, *ca.* A.D. 500), *mathematics*.

Tsu Ch'ung-chih was in the service of the emperor Hsiao-wu (*r.* 454–464) of the Liu Sung dynasty, first as an officer subordinate to the prefect of Nan-hsü (in modern Kiangsu province), then as an officer on the military staff in the capital city of Chien-k'ang (modern Nanking). During this time he also carried out work in mathematics and astronomy; upon the death of the emperor in 464, he

left the imperial service to devote himself entirely to science. His son, Tsu Keng, was also an accomplished mathematician.

Tsu Ch'ung-chih would have known the standard works of Chinese mathematics, the *Chou-pi suan-ching* ("Mathematical Book on the Measurement With the Pole"), the *Hai-tao suan-ching* ("Sea-island Manual"), and especially, the *Chui-chang suan-shu* ("Mathematical Manual in Nine Chapters"), of which Liu Hui had published a new edition, with commentary, in 263. Like his predecessors, Tsu Ch'ung-chih was particularly interested in determining the value of π. This value was given as 3 in the *Chou-pi suan-ching*; as 3.1547 by Liu Hsin (*d.* 23); as $\sqrt{10}$, or $\frac{92}{29}$, by Chang Heng (78–139); and as $\frac{142}{35}$, that is, 3.155, by Wan Fan (219–257). Since the original works of these mathematicians have been lost, it is impossible to determine how these values were obtained, and the earliest extant account of the process is that given by Liu Hui, who reached an approximate value of 3.14. Late in the fourth century, Ho Ch'ēng-tien arrived at an approximate value of $\frac{22}{7}$, or 3.1428.

Tsu Ch'ung-chih's work toward obtaining a more accurate value for π is chronicled in the calendrical chapters (*Lu-li chih*) of the *Sui-shu*, an official history of the Sui dynasty that was compiled in the seventh century by Wei Cheng and others. According to this work,

> Tsu Ch'ung-chih further devised a precise method. Taking a circle of diameter 100,000,000, which he considered to be equal to one *chang* [ten *ch'ih*, or Chinese feet, usually slightly greater than English feet], he found the circumference of this circle to be less than 31,415,927 *chang*, but greater than 31,415,926 *chang*. [He deduced from these results] that the accurate value of the circumference must lie between these two values. Therefore the precise value of the ratio of the circumference of a circle to its diameter is as 355 to 113, and the approximate value is as 22 to 7.

The *Sui-shu* historians then mention that Tsu Ch'ung-chih's work was lost, probably because his methods were so advanced as to be beyond the reach of other mathematicians, and for this reason were not studied or preserved. In his *Chun-suan-shih Lung-ts'ung* ("Collected Essays on the History of Chinese Mathematics" [1933]), Li Yen attempted to establish the method by which Tsu

Ch'ung-chih determined that the accurate value of π lay between 3.1415926 and 3.1415927, or $\frac{355}{113}$.

It was his conjecture that

"As $\frac{22}{7} > \pi > 3$, Tsu Ch'ung-chih must have set forth that, by the equality

$$\pi = \frac{22x + 3y}{7x + y} = 3.14159265,$$

one can deduce that

$$x = 15.996y, \text{ that is, that } x = 16y.$$

Therefore

$$\pi = \frac{22 \times 16y + 3y}{7 \times 16y + y} = \frac{22 \times 16 + 3}{7 \times 16 + 1} = \frac{355}{113},"$$

For the derivation of

$$\pi = \frac{22x + 3y}{7x + y},$$

when a, b, c, and d are positive integers, it is easy to confirm that the inequalities

$$\frac{a}{b} \geqq \frac{a+c}{b+d} \geqq \frac{c}{d}$$

hold. If these inequalities are taken into consideration, the inequalities

$$\frac{22}{7} \geqq \frac{22x + 3y}{7x + y} \geqq \frac{3}{1}$$

may be derived.

Ch'ien Pao-tsung, in *Chung-kuo shu-hsüeh-shih* ("History of Chinese Mathematics" [1964]), assumed that Tsu Ch'ung-chih used the inequality

$$S_{2n} < S < S_{2n} + (S_{2n} - S_n),$$

where S_{2n} is the perimeter of a regular polygon of $2n$ sides inscribed within a circle of circumference S, while S_n is the perimeter of a regular polygon of n sides inscribed within the same circle. Ch'ien Pao-tsung thus found that

$$S_{12288} = 3.14159251$$

and

$$S_{24576} = 3.14159261,$$

resulting in the inequality

$$3.1415926 < \pi < 3.1415927.$$

Of Tsu Ch'ung-chih's astronomical work, the most important was his attempt to reform the calendar. The Chinese calendar had been based upon a cycle of 235 lunations in nineteen years, but in 462 Tsu Ch'ung-chih suggested a new system, the Ta-ming calendar, based upon a cycle of 4,836 lunations in 391 years. His new calendar also incorporated a value of forty-five years and eleven months a *tu* (365¼ *tu* representing 360°) for the precession of the equinoxes. Although Tsu Ch'ung-chih's powerful opponent Tai Fa-hsing strongly denounced the new system, the emperor Hsiao-Wu intended to adopt it in the year 464, but he died before his order was put into effect. Since his successor was strongly influenced by Tai Fa-hsing, the Ta-ming calendar was never put into official use.

BIBLIOGRAPHY

On Tsu Ch'ung-chih and his works see Li Yen, *Chung-suan-shih lun-ts'ung* ("Collected Essays on the History of Chinese Mathematics"), I–III (Shanghai, 1933–1934), IV (Shanghai, 1947), I–V (Peking, 1954–1955); *Chung-kuo shu-hsüeh ta-kang* ("Outline of Chinese Mathematics"; Shanghai 1931, repr. Peking 1958), 45–50; *Chun-kuo suan-hsüeh-shi* ("History of Chinese Mathematics"; Shanghai, 1937, repr. Peking, 1955); "Tsu Ch'ung-chih, Great Mathematician of Ancient China," in *People's China*, **24** (1956), 24; and *Chun-kuo ku-tai shu-hsüeh shi-hua* ("Historical Description of the Ancient Mathematics of China"; Peking, 1961), written with Tu Shih-jan.

See also Ch'ien Pao-tsung, *Chung-kuo shu-hsüeh-shih* ("History of Chinese Mathematics"; Peking, 1964), 83–90; Chou Ch'ing-shu, "Wo-kuo Ku-tai wei-ta ti k'o-hsüeh-chia: Tsu Ch'ung-chih" ("A Great Scientist of Ancient China: Tsu Ch'ung-chih"), in Li Kuang-pi and Ch'ien Chün-hua, *Chung-kuo k'o-hsüeh chi-shu fa-ming ho k'o-hsüeh chi-shu jēn-wu lun-chi* ("Essays on Chinese Discoveries and Inventions in Science and Technology and the Men Who Made Them"; Peking, 1955), 270–282; Li Ti, *Ta k'o-hsüeh-chia Tsu Ch'ung-chih* ("Tsu Ch'ung-chih the Great Scientist"; Shanghai, 1959); Ulrich Libbrecht, *Chinese Mathematics in the Thirteenth Century* (Cambridge, Mass., 1973), 275–276; Mao I-shēng, "Chung-kuo yüan-chou-lü lüeh-shih" ("Outline History of π in China"), in *K'o-hsüeh*, **3** (1917), 411; Mikami Yashio, *Development of Mathematics in China and Japan* (Leipzig, 1912), 51; Joseph Needham, *Science and Civilization in China*, III (Cambridge, 1959), 102; A. P. Youschkevitch, *Geschichte der Mathematik im Mittelalter* (Leipzig, 1964), 59; and Yen Tun-chieh, "Tsu Keng pieh chuan" ("Special Biography of Tsu Keng") in *K'o-hsüeh*, **25** (1941), 460.

AKIRA KOBORI

TSVET (or TSWETT), MIKHAIL SEMENOVICH (*b.* Asti, Italy, 14 May 1872; *d.* Voronezh, Russia, 26 June 1919), *plant physiology, plant biochemistry.*

Tsvet was the son of Semen Nikolaevich Tsvet, a Russian civil servant, and Maria de Dorozza, an Italian who had been raised in Russia. His parents had stopped in Asti en route to Switzerland for a cure when he was born. His mother died soon after Tsvet's birth, and his father had to leave the infant in Lausanne with a nurse while he returned to Russia.

Tsvet's childhood and youth were spent in Lausanne and Geneva. In 1891 he entered the department of mathematics and physics of the University of Geneva, where his special interests were chemistry, physics, and botany. While still a student, he did his first scientific work on plant anatomy, which received the Davy Prize and was published in 1894. After receiving the baccalaureate in physical and natural sciences in 1893, Tsvet continued work in the general botanical laboratory on his doctoral dissertation, "Études de physiologie cellulaire," which he defended in 1896.

In the summer of 1896 Tsvet moved to Russia, and from the beginning of 1897 he continued his investigations at the laboratory of plant anatomy and physiology of the Academy of Sciences and especially at the St. Petersburg Biological Laboratory, where, however, he did not have an academic post. Not until the autumn of 1897 did he become a botany teacher in the women's courses at the laboratory.

Since foreign scientific degrees were not legally recognized in Russia, in 1901 Tsvet passed the examination for the master's degree in botany at Kazan University and defended his thesis, "Fiziko-khimicheskoe stroenie khlorofilnogo zerna" ("The Physicochemical Structure of the Chlorophyll Grain"). In January 1902 he became supernumerary laboratory assistant in the department of plant anatomy and physiology at Warsaw University, and in 1903 he was appointed *Privatdozent*. Tsvet took on the added duty of teaching botany and microbiology at the Warsaw Veterinary Institute in 1907; a year later he began teaching these subjects at Warsaw Technical University, at which time he resigned from the university. He received the doctorate in botany in 1910 after defending the dissertation on chromophils in the plant and animal kingdoms.

World War I interrupted Tsvet's scientific work. In the summer of 1915 the Warsaw Polytechnical Institute was evacuated to Moscow and, in 1916,

to Nizhni Novgorod (now Gorky). Tsvet devoted much time and effort to organizing the work of the botanical laboratory in both cities and, at Nizhni Novgorod, participated in the organization of the Society of Natural Scientists and of the advanced agricultural courses.

In March 1917 Tsvet became professor of botany and director of the botanical garden at Yuryev (now Tartu) University, and that autumn he began teaching. He worked in Yuryev for only a short time, however. On 23 February 1918 Austrian and German soldiers entered the town, and Yuryev University soon ceased to function as a Russian institution. After the signing of the Treaty of Brest-Litovsk, the university was transferred to Voronezh in August 1918. Tsvet was able to work here for less than a year. His health, uncertain since birth, was finally ruined by excessive work, the displacements, and the hardships of the war years. He died of a chronic heart ailment at the age of forty-seven.

Tsvet's scientific legacy consists of sixty-nine publications, produced in the relatively short period from 1894 to 1916. He began research at a time when the data and methods of chemistry and physics were becoming more widely used for the discovery of the nature of the life processes. This aided the establishment of plant physiology as an independent science in the mid-nineteenth century and, toward the end of the century, contributed to the formation within plant physiology of such areas of research as cytophysiology. It was in the latter field that Tsvet saw the great possibilities for applying the results of chemical research and methods to achieve a better understanding of the nature of the plant organism. Even in his earliest works, on cytology and plant anatomy, he tried not only to describe the structures but also to discover their significance and functions, thereby creating new techniques and methods that were clearly expressed in his doctoral dissertation.

This tendency appeared more broadly and fruitfully in Tsvet's research at St. Petersburg. In his Geneva dissertation the central topic was the structure of chloroplasts, while in his later research the main subject was chlorophyll. In "O khloroglobine" ("On Chloroglobin") and "O prirode khloroglobina" ("On the Nature of Chloroglobin"; both 1900), Tsvet showed that the green pigment is found in the chloroplasts in the form of the chlorophyll-albumin complex, which, in analogy with hemoglobin, he called "chloroglobin." This term is now generally accepted; but at that time it met with sharp criticism from Tsvet's contemporaries, who doubted the precision of his research methods.

In "Khlorofilliny i metakhlorofilliny" ("Chlorophyllins and Metachlorophyllins"; 1900) Tsvet contested the widespread view that only two pigments were present in the leaf: green chlorophyll and yellow xanthophyll. Using the five existing methods of physical analysis (fractional solution, differential solution, fractional precipitation, "wet sublimation," and diffusion) to separate the pigments with the least possible alteration, he established that in leaves there are two green pigments— chlorophyll α and β (now known as chlorophyll a and b), differing in color, fluorescence, and spectral absorption. Tsvet obtained a pure sample of the α form of chlorophyll but not of the β. This led him to attempt to develop a method that would consider the properties of the relationships of the chloroglobin pigments by means of adsorption. Tsvet decided to make the principle of adsorption the basis of this new method that, like filter paper, would allow the extraction from a solution of pigments in unchanged form. He stated his preliminary ideas on this question in his master's thesis, which contains, as he later recognized, the embryonic form of the method of chromatographic adsorption analysis that he soon developed.

On 30 December 1901, Tsvet presented a report to the Eleventh Congress of Russian Natural Scientists and Physicians at St. Petersburg, "Metody i zadachi fiziologicheskogo issledovania khlorofilla" ("Methods and Problems of Physiological Research on Chlorophyll"), in which he revealed his adsorption method and demonstrated its effects. He made a special, detailed report on 8 March 1903 to the biological section of the Warsaw Society of Natural Scientists, "O novoy kategorii adsorbtsionnykh yavleny i o primenenii ikh k biokhimicheskomu analizu" ("On a New Category of Adsorption Phenomena and on Its Application to Biochemical Analysis"). In it Tsvet described how he set himself the problem of creating a physical method that, in distinction from the chemical method, would, by using adsorption, allow the isolation of plant pigments and the separation of a mixture of such pigments in unchanged form. He experimentally substantiated that many substances could be adsorbed and explained the nature of this phenomenon. He also stated the theoretical bases and practical uses of the method.

In 1906, in *Berichte der Deutschen botanischen Gesellschaft*, Tsvet published two articles on his method and the data he had obtained with it on the pigment composition of plant leaves. In it he made the first suggestion to call the new method "chromatography" and formulated the law of adsorption replacement, giving a full description of the entire chromatographic setup, including sketches. He also provided a detailed account of the techniques of chromatographic experiments in general and in the study of chlorophyll in particular. His work on the creation of chromatography and its theoretical basis was summed up in his Russian doctoral dissertation, in which he gave a full demonstration of the difference of the chromatographic method from Friedrich Goppelsröder's capillary analysis.

Having tried 126 different powdered adsorbents, Tsvet found that those most effective for isolating plant pigments were calcium carbonate, sugarcane, and inulin. Grinding fresh leaves in a mixture of petroleum ether and a small amount of alcohol, he obtained an extract that he shook with distilled water to remove the alcohol and then filtered through a tube filled with powdered adsorbent. According to the strength of the various adsorptive capacities, the pigments were distributed into six differently colored layers in the tube. Taking the adsorbent from the tube, he obtained a column of powder that could be cut with a knife, after which each pigment could be washed from it separately. Thus Tsvet obtained in a pure form both chlorophyll a and chlorophyll b, and could separate from brown and diatomic algae the previously unknown chlorophyll c (in Tsvet's terminology, chlorophyll γ) and a number of previously unknown forms of xanthophyll: α, α', α'', and β (xanthophyll α later became known as lutein and xanthophyll β as taraxanthin). In 1911 Tsvet discovered in the leaves of *Thuja*, and isolated in pure form, a red-yellow pigment that he called rodoxanthin. Having discovered many forms of xanthophyll and their chemical relationship with the yellow pigment carotene, Tsvet suggested in 1911 that they be considered as one general group and called "carotenoids"—a term that has now won general acceptance.

Although Tsvet's chromotographic method was known to many of his contemporaries and was even used successfully in several laboratories to obtain pure forms of chlorophylls and carotenoids, its acceptance was very limited. The wide use of chromatography began in the 1930's, when Richard Kuhn, L. Zechmeister, and Paul Karrer simultaneously used it to study the chemistry of carotene and vitamin A. Dozens of other previously unknown forms of carotenoids and their products also were obtained, and colorless substances were isolated and purified: vitamins, hormones, enzymes. On the basis of Tsvet's method of chromatographic adsorption, a number of new forms of chromatography have been developed: ion-replacement, gas, distributive on paper, thin-layer, sedimentary. They have been widely used in biochemistry, analytical chemistry, biology, medicine, agriculture, and in a number of industries—chemical, pharmaceutical, food processing—where it is necessary to obtain absolutely pure substances, to separate complex mixtures, or to identify unknown compounds.

BIBLIOGRAPHY

I. ORIGINAL WORKS. Tsvet's writings include "Études de physiologie cellulaire," in *Bulletin du Laboratoire de botanique générale de l'Université de Genève*, **1**, no. 1 (1896), 123–206, his Geneva doctoral diss.; "Fiziko-khimicheskoe stroenie khlorofilnogo zerna" ("The Physicochemical Structure of the Chlorophyll Grain"), in *Trudy Kazanskogo obshchestva estestvoispytatelei*, **35**, no. 3 (1901), 1–268, his master's thesis at Kazan; "O novoy kategorii adsorbtsionnykh yavleny i o primenenii ikh k biokhimicheskomu analizu" ("On a New Category of Adsorption Phenomena and on Its Application to Biochemical Analysis"), in *Trudy Varshavskago obshchestva estestvoispytatelei*, Otd. biol., **14** (1903), 20–39; "Physikalisch-chemische Studien über das Chlorophyll. Die Adsorptionen," in *Berichte der Deutschen botanischen gesellschaft*, **24** (1906), 316–323; "Adsorptionsanalyse und chromatographische Methode. Anwendung auf die Chemie des Chlorophylls," *ibid.*, 384–393; *Khromofilly v rastitelnom i zhivotnom mire* ("Chromophils in the Plant and Animal Kingdoms"; Warsaw, 1910), for which he received the doctorate in botany; and "Über das makro- und mikrochemischen Nachweis des Carotins," in *Berichte der Deutschen botanischen Gesellschaft*, **29** (1911), 630–636.

II. SECONDARY LITERATURE. See C. Dhéré, "Michel Tswett," in *Candollea* (Geneva), **10** (1943), 23–63; T. Robinson, "Michael Tswett," in *Chimia Annual Studies in the History of Chemistry*, **6** (1960), 146–161; E. M. Senchenkova, *Mikhail Semenovich Tsvet* (Moscow, 1973), with a bibliography of works about Tsvet to 1973; "Otkrytie khromatografii i Akademia nauk" ("The Discovery of Chromatography and the Academy of Sciences," in *Priroda* (1974), no. 5, 92–101; and "Michail Semenovic Tsvet und die Chromatographie," in *Schriftenreihe für Geschichte der Naturwissenschaften, Technik und Medizin*, **12** (1975), 111–126; and R. L. M. Singe, "Tsvet, Willstätter, and the Use of Adsorption of Proteins," in *Archives of Biochemistry and Biophysics*, supp. 1 (1962), 1–6.

E. M. SENCHENKOVA

IBN ṬUFAYL, ABŪ BAKR MUHAMMAD (Latin, **Abubacer**) (*b.* Guadix, Spain, before 1110; *d.* Marrakesh, Morocco, 1185), *medicine, philosophy.*

Ibn Ṭufayl was a Spanish Muslim who received a broad education in the religion of Islam and the Arabic secular sciences. His professional career was that of a physician, first at Granada, Ceuta, and Tangier, and later (1163–1182) as court physician to the Almohad sultan of Morocco and Andalusia. He introduced Ibn Rushd to the sultan (*ca.* 1169) and commissioned him to write his commentaries on the works of Aristotle.

Ibn Ṭufayl is best known for his philosophical book *Ḥayy ibn Yaqẓān* ("The Living, Son of the Wakeful"). After a valuable introduction surveying the rise of philosophy in western Islam, the author presents Neoplatonic philosophy in the form of a myth. Ḥayy is a boy born on a desert island and reared by a doe. As he grows up he teaches himself, entirely by his own observation and reasoning, some practical arts and the rudiments of the empirical sciences. In his adult life he proceeds by reasoning and intuition to an understanding of metaphysics and theology and to an ascetic practice, all of which culminate in mystical visions by his intellect of God, the Necessary Being and Cause of the world. In later experiences he converses with a devout Muslim, and they agree that there is no difference in doctrine between Ḥayy's philosophy and the revealed religion of Islam, but that it is useless to teach philosophy to most people, for whom only the simplest practice of Islam is helpful.

The main aims of the myth appear to be to show (1) that the Neoplatonic philosophy is that which a rational man, undistracted by social interests or prejudices, will naturally arrive at, and (2) that the practice implied in this philosophy leads to the supreme happiness for man, which is the mystical state of the soul. In details Ibn Ṭufayl generally follows Ibn Sīnā, but there are some differences. For example, Ibn Ṭufayl thinks it unproved that the world is eternal rather than created in time, and holds that intelligibles are abstracted by the human intellect, not presented to it by an external Active Intellect.

Ḥayy ibn Yaqẓān has always been widely read in Arabic, and appeared in several European translations from 1671 onward; it probably influenced Defoe's *Robinson Crusoe* (1717).

Another work by Ibn Ṭufayl, *Rajaz ṭawīl fī ʿilm al-ṭibb* ("Long Poem in *Rajaz* Meter on Medical Science"), was discovered recently in manuscript at Rabat. He is also known to have influenced his pupil al-Biṭrūjī to abandon the Ptolemaic astronomy of eccentrics and epicycles in favor of a more Aristotelian system, but no astronomical writings by the master have survived.

BIBLIOGRAPHY

Texts of Ibn Ṭufayl's *Ḥayy ibn Yaqẓān* include the French ed. and trans. of L. Gauthier (2nd ed., Beirut, 1936); that of Aḥmad Amīn (Cairo, 1952); a partial English trans. by G. N. Atiyeh, in R. Lerner and M. Mahdi, *Medieval Political Philosophy* (Chicago, 1963), 134–162. and an English trans., intro., and notes by L. Goodman, *Ibn Ṭufayl's Ḥayy ibn Yaqẓān* (New York, 1971).

For studies of Ibn Ṭufayl and his work, see M. Cruz Hernandez, *Historia de la filosofia Hispano-musulmana*, I (Madrid, 1957), ch. 11; L. Gauthier, *Ibn Thofaïl, sa vie, ses oeuvres* (Paris, 1909); A.-M. Goichon, "Ḥayy b. Yakẓān," and B. Carra de Vaux, "Ibn Ṭufayl," both in *Encyclopaedia of Islam*, new ed. (Leiden, 1960–   ), III; G. F. Hourani, "The Principal Subject of Ibn Ṭufayl's *Ḥayy b. Yaqẓān*," in *Journal of Near Eastern Studies*, **15** (1956), 40–46; A. Pastor, *The Idea of Robinson Crusoe* (Watford, 1930); T. Sarnelli, "Primauté de Cordoue dans la médecine arabe d'Occident," in *Actas del Primer Congreso de Estudios Arabes y Islamicos* (Madrid, 1964), 441–451, describing the medical poem; and S. S. Hawi, *Islamic Naturalism and Mysticism, a Philosophic Study of Ibn Ṭufayl's Ḥayy ibn Yaqẓān* (Leiden, 1974).

George F. Hourani

TULASNE, LOUIS-RENÉ (*b.* Azay-le-Rideau, France, 12 September 1815; *d.* Hyères, France, 22 December 1885), *mycology.*

Tulasne studied law in Poitiers with the intention of becoming a solicitor. In 1839, however, he inherited a considerable sum of money from his father and went to Paris to join his brother Charles, who was conducting medical studies there. The two brothers decided to give up their previous careers and to devote themselves to a life of botany, Christian religion, and charitable activities. They cooperated in these efforts until 1884, when Charles Tulasne died. Of the fifty-seven botanical works that Louis René Tulasne eventually published, his brother, who was also a talented and dexterous draftsman and illustrator, assisted him in the composition of fifteen.

In Paris, Tulasne attended the lectures of Brongniart, A. de Jussieu, and J. H. Léveillé and became a collaborator of Auguste de Saint-Hilaire at the Muséum National d'Histoire Naturelle. His first memoir, on the structure of the *Elaphomyces* fungi, was published in 1841. He then went on to study the results obtained by the British scientist Miles Berkeley, who had demonstrated that there are two fruiting forms in four genera of Gasteromycetes. He confirmed Berkeley's findings and generalized them in a series of ten memoirs, of which the last, *Fungi hypogaei* (Paris, 1851), remains one of the foundations of the modern study of this group.

At this time the Uredinales and Ustilaginales—the rusts and smuts that cause serious plant diseases—were little known, and some taxonomists thought them to be related to the Gasteromycetes. Tulasne began to study these parasites and was able to determine by experimentaton that in them the germination of teleutospores gives an intermediate promycelial stage, homologous to a basidium, which produces basidospores. Cell continuity, he established, was in these instances of prime importance. He then demonstrated the various fruiting forms borne in the growth of a single thallus in many species. He published, in addition to an important memoir on Ustilaginales, a precise explanation of the life-cycle of rye ergot (*Claviceps purpurea*), descriptions of the genus *Hypoxylon*, and various notes about *Erysiphaceae* (the powdery mildews).

Tulasne also did research on the reproduction and physiology of lichens, demonstrating that they display a filamentous habit rather similar to the mycelium of a fungus in their first stages of growth. Working with almost every European species of lichen, he determined the presence of spermatogonia and pycnidia in them and simultaneously succeeded in cultivating some species through sowing spores. In recognition of this work, as well as his important contributions to mycology, Tulasne was in 1854 elected to the Académie des Sciences.

Although his scientific reputation was based largely upon his work on cryptogams, Tulasne conducted other significant research on flowering plants. In 1849, having made a thorough study of the important herbaria of Paris and London, he brought out a paper on the Podostemeae, a group of plants similar to the mosses and ferns in appearance but belonging to a higher order because of their dicotyledonous embryos. He also published, between 1853 and 1855, two memoirs on American Leguminosae, four books on the flora of Colombia, a series of notes on the flora of Madagascar, four papers on the Monimiaceae, a work on American Gnetaceae, and a memoir, with descriptions of new species, on the two American genera *Antidesma* and *Stilaginella*.

Tulasne's major publication, written with his brother, was *Selecta fungorum carpologia*, published in three volumes between 1857 and 1865. At its completion, when he was fifty years old, Tulasne believed that his health was failing. He therefore presented his rich herbarium to the Muséum National d'Histoire Naturelle, gave his library to the Paris Université Catholique, and retired with his brother to the south of France, where they lived quietly for the rest of their lives.

BIBLIOGRAPHY

On Tulasne and his work, see E. Bornet, "Notice sur M. L. R. Tulasne," in *Comptes rendus hebdomadaires des séances de l'Académie des Sciences*, **103** (1886), 957–966; M. Chadefaud, "Mycologie," in A. Davy de Virville, *Histoire de la botanique en France* (Paris, 1954), 219–234; and P. Duchartre, "Notice sur M. L. R. Tulasne et sur son oeuvre botanique," in *Comptes rendus hebdomadaires des séances de l'Académie des Sciences*, **101** (1885), 1438–1444.

G. Viennot-Bourgin

TULP, NICOLAAS (*b.* Amsterdam, Netherlands, 11 October 1593; *d.* The Hague, Netherlands, 12 September 1674), *medicine, anatomy.*

Nicolaas Tulp was the son of Pieter Dirkz, a wealthy merchant of Amsterdam, and Grietje Dirks Poelenburgh. Dirkz, meaning son of Dirk (Henry), was not a family name. According to Busken Huet, Tulp's real name was Claes (Nicolas) Pieterz. He took the name Tulp ("tulip") from the sculpture on the gable of his house on the Keizersgracht in Amsterdam.

On 19 February 1611, Tulp matriculated at the University of Leiden under the name of Nicolaas Petraeus. He obtained his medical degree on 30 September 1614, after defending his dissertation, *De cholera humida*. Tulp then settled in Amsterdam, where he soon had a lucrative practice. It is said that he was the first physician in town to visit his patients in a carriage.

Tulp was appointed *praelector* of anatomy in 1628, succeeding Joannes Fonteyn. He had the personal title *professor anatomiae* and was charged with teaching the surgeons of the city and illustrating the lectures, whenever possible, with public dissections.

Tulp is perhaps best known from the painting by Rembrandt commissioned by the Surgeons Guild in 1632. He is portrayed giving a lecture demonstration with an opened body. The painting, the famous "Anatomy Lesson of Dr. Tulp," is in the Mauritshuis in The Hague.

One of Amsterdam's leading citizens, Tulp served as a member of the city government, including four terms as mayor. He was also a curator of the Athenaeum, the city's school of higher learning.

In 1652, because of the pressures of his civic and professional duties, Tulp resigned his position as *praelector*. He was succeeded by J. Deyman.

Tulp's main work, the *Observationum medicarum libri tres* (1641), contains descriptions of 228 cases. Most of them cannot stand the test of modern criticism, but there are some valuable observations. For example, he described the ileocecal valve, sometimes called Tulp's valve, which is located at the junction of the large and small intestines, and he gave a correct description of its function. He also described the chyle vessels of the small intestine, which actually was a rediscovery. According to Baumann, both Herophilus and Erasistratus knew of these vessels but the knowledge was lost. Tulp also was the first to describe beriberi.

Tulp is also sometimes credited with the first description of the orangutan, about a half century before Tyson (1699), but incorrectly so. What Tulp described was a chimpanzee, which does appear to have been the first in Europe.

A great admirer of Hippocrates, Tulp opposed the new ideas of the iatrochemists, of whom J. B. van Helmont was the most notable. He was especially against the use of antimony, which was beginning to be prescribed.

The first pharmacopoeia of the Netherlands was compiled at Tulp's suggestion, He proposed the idea to a group of six physicians during a dinner at his house on 18 April 1635. A committee was formed, and on 5 May 1636 (date of the preface) the book was published. Since the time was so short, it is generally believed that Tulp had most of the manuscript ready when he made the proposal and that the book was therefore largely from his own hand. The municipality of Amsterdam ruled that this pharmacopoeia was henceforth to be the only one used in the city. Thus was formed the Collegium Medicum Amstelaedamense, set up to enforce the decree. The Collegium soon developed into a municipal committee for health care, the earliest example of governmental concern with public health in the Netherlands.

BIBLIOGRAPHY

I. ORIGINAL WORKS. The *Pharmacopoea Amstelae-damensis, senatus auctoritate munita* (Amsterdam, 1636) went into many editions, the most recent being a facsimile edition of 1961. *Observationum medicarum libri tres* (Amsterdam, 1641) also had several eds., some with slightly changed titles (Amsterdam, 1652, 1672, 1685; and Leiden, 1716, 1739). The fifth ed. has a biography and the sixth ed. has the funeral oration by L. Wolzogen, professor of church history at the Athenaeum of Amsterdam.

Dutch translations are *De drie boecken der medicijnsche aanmerkingen* (Amsterdam, 1650); *Geneeskundige waarnemingen, Naar den zesden druk uit het latijn vertaald, Hier is bijgevoegd de lijkrede van L. Wolzogen* (Leiden, 1740); and *Hippocrates, Aphorismen, of kortbondige spreuken. Beneffens desselfs wet en vermaningen. Alsmede d'aanmaningen van N. Tulp. Vertaald door Steph. Blankaart* (Amsterdam, 1680 [?]; 2nd ed., 1714).

II. SECONDARY LITERATURE. See H. F. Thijssen, "Voorlezing over Nicolaas Tulp," in *Magazijn voor wetenschappen, Kunsten en Letteren van N. G. van Kampen*, 3 (1824); L. H. van Bochove, *Dissertatio historica-medica inauguralis de Nicolao Tulpio, anatomes practicae strenuo cultore* (Leiden, 1846); H. C. Rogge, "Nicolaas Tulp," in *De Gids*, 3rd ser., 18 (1880), 77–125; E. H. M. Thijssen, *Nicolaas Tulp als geneeskundige geschetst. Eene bijdrage tot de geschiedenis der geneeskunde in de achttiende eeuw* (Amsterdam, 1881); and C. Busken Huet, *Het land van Rembrandt*, 2nd ed. (Haarlem, 1886), II B, 57–61.

Other works are E. D. Baumann, "Nicolaas Tulp" in *Nieuw Nederlandsch Biografisch Woordenboek*, III (1914), 1250–1251; J. S. Theissen, "Nicolaas (Claes Pieterz) Tulp," in *Gedenkboek van het Athenaeum en de Universiteit van Amsterdam* (Amsterdam, 1932), 695–696; A. Bredius, *Rembrandt, schilderijen* (Utrecht, 1935); P. van der Wielen, "De eerste Nederlandsche Pharmacopee," in *Bijdragen tot de Geschiedenis der Geneeskunde*, 16 (1936), 57–63; E. D. Baumann, *Uit drie eeuwen Nederlandsche Geneeskunde* (Amsterdam, 1951), esp. 64–71; and A. Querido, "Nicolaas Tulp en zijn manuscript," in *Spiegel Historiael*, 5 (1970), 305–311, on Tulp's own Dutch translation of the *Observationum*.

PETER W. VAN DER PAS

TUNSTALL, CUTHBERT (*b.* Hackforth, Yorkshire, England, 1474; *d.* London, England, 18 November 1559), *theology, diplomacy, mathematics.*

Tunstall was the natural son of Thomas Tunstall and a daughter of Sir John Conyers, and he was later legitimated (in canon law) by their marriage. He attended Oxford (*ca.* 1491) and Cambridge (*ca.* 1496) but removed to Padua in 1499, where he remained for about six years and became doctor of both canon and civil (Roman) laws. He was appointed bishop of London (1522) and later bishop of Durham (1530, deprived 1552, restored 1553, deprived 1559). Although of strong religious convictions, he was humane and moderate, and was respected even by his opponents in matters of religion. While remaining faithful to Roman Catholic dogma, he was aware that reform was needed. He would protest decisions of Henry VIII (who often kept him away from London when unpopular decisions were to be made), but once they had been made, he would submit. Under Mary he refrained from persecuting Protestants. An outstanding classical scholar, Tunstall was a close friend of Sir Thomas More, to whom his arithmetic was dedicated, and of Erasmus, whom he assisted in the preparation of the second edition of his Greek New Testament.

Tunstall's Latin arithmetic, *De arte supputandi* (1522), was published as a farewell to secular writings just before he was consecrated bishop of London. The work made no claim to originality of material but had been compiled over the years from all available works in Latin or other languages that Tunstall understood. As master of the rolls (1516–1522), and on diplomatic missions to the Continent, he had felt the need to refresh his memory of arithmetic to protect himself in monetary transactions. From the material he had collected he determined to write such a clear treatise that no one who knew Latin would lack an instructor in the art of reckoning. The work seems not to have been popular in England. It has never been translated into English, and all editions but the first were printed on the Continent, where it was greatly admired. For example, Simon Grynaeus dedicated the first Greek text of Euclid's *Elements* (Basel, 1533) to Tunstall, since he had explained the calculating of numbers in so excellent a manner. England had lagged behind the rest of Europe in mathematics. Only a chapter on "Arsemetrike and Whereof It Proceedeth," in Caxton's *The Mirrour of the World* (1481), had preceded Tunstall's *De arte supputandi*; and it was not until 1537 that an arithmetic appeared in English.

BIBLIOGRAPHY

In addition to the London eds. of *De arte supputandi* (1522), there were Paris eds. (1529, 1535, 1538) and

Strasbourg eds. (1543, 1544, 1548, 1551). For Tunstall's ecclesiastical writings, see Charles Sturge, *Cuthbert Tunstal* (New York, 1938), which also contains a chapter on the arithmetic. For Erasmian humanism and religious developments in England during Tunstall's lifetime, see L. B. Smith, *Tudor Prelates and Politics, 1536–1558* (Princeton, 1953) and J. K. McConica, *English Humanists and Reformation Politics* (Oxford, 1965).

Joy B. Easton

TUPOLEV, ANDREY NIKOLAEVICH (*b.* Pustomazovo, Tver [now Kalinin] guberniya, Russia, 10 November 1888; *d.* Moscow, U.S.S.R., 23 December 1972), *mechanics, aeronautical engineering.*

The son of a notary, Tupolev studied at the Tver provincial Gymnasium from 1900 to 1908 and in 1909 entered Moscow Technical School (now the N. E. Bauman Moscow Higher Technical School). Under the influence of N. E. Zhukovsky he became interested in aviation and joined a club of aeronautics enthusiasts that fostered the activities of many scientists and aeronautical engineers who later became well known. He designed, built, and flew training gliders; planned and constructed wind tunnels; and participated in the creation of an aerodynamic laboratory at the Technical School. In 1916 Tupolev and other members of the group under Zhukovsky's guidance took part in the creation of a bureau for experimental testing of aeronautical designs, one of the first scientific research institutions of its kind.

After graduating in 1918, Tupolev and the other members of Zhukovsky's collective organized the Central Aerohydrodynamics Institute in Moscow. Tupolev devoted much time and energy to this institute, serving as assistant director from 1918 to 1935. The office of design, headed by Tupolev and established at the institute in 1922, established a project for the first in a series of airplanes built by Tupolev. An independent design office was organized with Tupolev as director in 1936. Tupolev was the first designer in the Soviet Union to use all-metal construction in both civil and military aviation.

In 1933 Tupolev became a corresponding and, in 1953, active member of the Academy of Sciences of the U.S.S.R. His services were also recognized in Italy, England, and the United States. In 1970 an Italian national center for the development of methods of air transport in Italy awarded Tupolev the Leonardo da Vinci Prize "for planning the world's first supersonic passenger airplane, the Tu-144." Also that year the Royal Aircraft Establishment of England elected him honorary member and awarded him a special diploma, and in 1971 he became an honorary member of the American Institute of Aeronautics and Astronautics.

Under Tupolev's leadership more than 100 types of aircraft were designed, from light fighter planes to huge long-range passenger aircraft.

Tupolev's airplanes played a major role in the study of the Arctic. A series of record flights were made in his airplanes, including the 1937 nonstop Moscow–Vancouver flight over the North Pole of V. P. Chkalov in an airplane of the ANT-25 type.

In 1955 Tupolev built the first Russian jet passenger airplane, the Tu-104. In December 1968 he completed the first test flight of a supersonic passenger airplane, the Tu-144. For his military airplanes, which played an important role in World War II, Tupolev was awarded the title of lieutenant general in the engineering-technical service.

Continuing Zhukovsky's work, Tupolev further developed the principles of aerodynamics and the calculation of stability. Although skilled in computation, he could extract from mathematical formulas physical implications and technical ideas and could evaluate them with profound scientific and technical insight. This gift for observing physical phenomena behind a mathematical framework allowed Tupolev to solve the most complex problems of such varied disciplines as gas dynamics, automation, the static and dynamic strength of structures, and radiotechnology—fundamental aspects of modern aeronautical science.

BIBLIOGRAPHY

Tupolev's writings include "Aerodinamichesky raschet aeroplanov" ("An Aerodynamic Account of Aeroplanes"), in *Trudy Aviatsionnogo raschetno-ispytatelnogo byuro,* no. 1 (1917); and "Pervy sovetsky metallichesky samolet ANT-2" ("The First Soviet All-Metal Airplane ANT-2"), in *Samolet,* no. 8 (1924), 12–18. For biographical details, see the article in *Bolshaya sovetskaya entsiklopedia* ("Great Soviet Encyclopedia"), 2nd ed., XLIII (Moscow, 1956), 415.

A. T. Grigorian

TÜRCK, LUDWIG (*b.* Vienna, Austria, 22 July 1810; *d.* Vienna, 25 February 1868), *medicine, laryngology, neurology.*

Türck's father, the jeweler to the Austrian imperial family and nobility, provided for his son generously, so that his economic and social positions were secure. The family was highly cultured and, apparently, devoted to music. Ludwig himself was reputedly a virtuoso cellist. At his death, his estate included two extremely valuable instruments. His brother, Joseph, owned a large and valuable collection of violins, said to be one of the finest in its time.

After study in the Gymnasium and the medical school in Vienna, Ludwig qualified as a physician there in 1836. His economic independence enabled him to devote himself to research, and by 1840 he was deeply involved in intensive studies of the anatomy and pathology of the nervous system. In 1844 he went to Paris to extend his studies under the great French physicians, who led the world in this field at the time. In Austria the outstanding leader in medical education was Baron Türkheim, and Türck's aggressive talents and brilliant intellectual endowments brought him to the baron's attention. Türkheim was director of the General Hospital in Vienna and took Türck under his patronage after his return from Paris. He arranged for a special division for nervous diseases to be established in the hospital with young Türck in charge. Here Türck remained for thirteen years and built up a solid scientific reputation by his intensive investigations in the neurological clinic. He published the results of these investigations both in periodicals and as monographs. Besides his early monograph on spinal irritation (1843), there were numerous contributions on the tracts in the spinal cord and their origins, on the roots of the trigeminal nerve, and on the results of tests of cutaneous sensibility. Türck's name is preserved in the nomenclature of the mammalian temporo-pontine tract, which is termed the bundle of Türck.

In 1857 the largest hospital in Vienna was established and Türck was appointed physician in chief. In the same year the direction of his principal researches shifted to laryngoscopy, to the nearly complete exclusion of other research. Manuel García, the renowned singer and vocal teacher in London, had sought with the aid of a mirror to observe the production of the voice and the visible alterations of the vocal organs that accompany its modulations. Even earlier, Senn, in Geneva, had suggested making visible the interior of the larynx by means of a small mirror inserted in the throat, and in the following decade prominent physicians in France and England, particularly Trousseau and Liston, had employed such instruments, but without useful results. When García published an account of his observations in 1855, Türck realized the possibilities for valuable clinical applications and, although ignorant of the details of the earlier procedures, he constructed an apparatus forthwith and went on to use it for diagnostic and operative purposes. Persistent experiments and abundant observations in the daily routine of hospital practice enabled him to develop and improve his instruments and soon the idea was crowned with brilliant success.

The earlier reports in no way detract from the originality of Türck's discovery. Not only was he unacquainted with the manner of García's investigations, but the aims and methods of the two men were totally different and unrelated. In 1857 Türck displayed to Ludwig the interior of the larynx of a patient in his ward and thereby found a practical solution for a problem that had long been troubling the physiologists and clinicians.

In March 1858 appeared an article by Czermak in the *Wiener medizinischen Wochenschrift* urgently recommending to physicians the practical application of the laryngeal mirror—giving rise to a bitter dispute over priority. This lasted for years until, after Türck's death, a professional declaration affirmed "that the history of medicine must forever link the name of Türck with laryngoscopy. To him alone is due the practical application of the laryngoscope for diagnostic and operative purposes."

BIBLIOGRAPHY

See *Abhandlung über Spinalirritation nach eigenen, grösstentheils im Wiener allgemeinen Krankenhause angestellten Beobachtungen* (Vienna, 1843); *Ph. Ricord's Lehre von der Syphilis. Nach dessen klinischen Vorträgen dargestellt von Ludwig Türck* (Vienna, 1846); "Fortsetzungen zum Gehirns," in *Sitzungsberichte der K. Akademie der Wissenschaften zu Wien*, Math.-nat. Cl., VI, 228; *Praktische Anleitung zur Laryngoskopie* (Vienna, 1860); *Recherches cliniques sur diverses Maladies du Larynx, de la trachée et du pharynx etudiées a l'aide du laryngoscope* (Paris, 1862); *Klinik der Krankheiten des Kehlkopfes und der Luftröhre. Nebst einer Anleitung zum Gebrauche des Kehlkopfrachenspiegels und zur Localbehandlung der Kehlkopfkrankheiten* (Vienna, 1866); *Atlas dazu. In 27 chromolithogr. Tafeln von A. Elfinger und C. Heitzmann* (Vienna, 1866); *Ueber Hautsensibilitätsbezirke der einzelnen Rückenmarknervenpaare. Aus dessen literarischen Nachlasse zusammengestellt von Professor Dr. C. Wedl* (Vienna, 1869); *Gesammelte neurologische Schriften* (Leipzig, 1910). Türck's articles in *Allgemeinen Wiener medizinischen Zeitung* provide a true record of the progress of his research.

E. HORNE CRAIGIE

TURGOT, ANNE-ROBERT-JACQUES (*b.* Paris, France, 10 May 1727; *d.* Paris, 18 March 1781), *economics, philosophy.*

The most famous member of a distinguished family, Turgot was the third son of Michel-Étienne Turgot, *prévôt des marchands de Paris*, city planner, and sponsor of the survey map of Paris known as the Turgot map. An older brother, Étienne-François, served briefly as governor of French Guiana and was a competent botanist and agronomist.

Turgot was originally intended for the priesthood and studied at the Séminaire St.-Sulpice and the Sorbonne. In 1751, recognizing that he had no religious vocation, he decided to follow the family tradition of public service. He was appointed intendant of Limoges in 1761, and his reforms there, over a period of thirteen years, made him a figure of national prominence. Perhaps best known was his abolition of the *corvée* (forced labor on the roads); by using professional rather than peasant labor, he achieved results of such quality that in 1787 the roads of Limousin were still being described as the best in France.[1] Offered transfers to more prosperous regions, Turgot preferred to remain in Limousin; he left the area only in 1774, when he was appointed minister of the navy in the first government formed by Louis XVI. One month later he was named controller general of finance, a post he was to hold for less than two years (August 1774–May 1776). In that capacity he attempted bold reforms on a national scale, thus antagonizing many special interest groups and earning an enduring reputation as the most courageous and enlightened official of the old regime.

Turgot's interests and talents were encyclopedic, extending far beyond his modern image as an economist. He knew five foreign languages well, studied two more, and published poetry and prose translations from English, German, and Latin. He displayed similar versatility in the sciences; his writings dealt with aspects of physics, chemistry, and geology. In addition, he reportedly knew enough astronomy, geography, and navigational theory and practice to be an unusually well-qualified minister of the navy.

Turgot's formal education in the sciences included the study of Newtonian physics with the Abbé Sigorgne and chemistry with Guillaume-François Rouelle. In 1748, while still a student, Turgot drafted an essay on "the causes of progress and decline in the arts and sciences," which he intended to submit for a prize then being offered by the Academy of Soissons. That year he also composed a brief and interesting critique of Buffon's cosmology and geology, the principles of which had just been published in a prospectus for Buffon's *Histoire naturelle*. Turgot's only other foray into geology was a series of field notes based on travels during the year 1760.

The one scientific work by Turgot published during his lifetime was the article "Expansibilité," in the Diderot *Encyclopédie*.[2] Here he was concerned with distinguishing between vaporization and evaporation, the latter defined as a loss of volume from the surface of a liquid or solid exposed to the atmosphere; he defined vaporization as the result of forces of repulsion—he elaborated at length upon this Newtonian theme—which act at the particulate level and produce a change of state. His article apparently had some influence on Lavoisier during the 1760's, and Condorcet later declared that Turgot had "opened new views in natural philosophy."[3] During the last years of his life Turgot was able to return to some of the topics raised in his article, and especially to the study of distillations *in vacuo* and under changing temperature conditions.

Turgot's talents as a chemist are revealed in the letters he exchanged with Condorcet in 1771–1772. Anticipating some of the ideas of Lavoisier, he argued persuasively that the gain in weight of a metallic calx (oxide) should be explained as a combination of the metal with "air." As he admitted, he had not tested his conclusions in the laboratory, and his argument was a logical, inductive one.

More significant than Turgot's own scientific activities was his role as a patron and a public official who regularly sought the advice of scientific experts. His recourse to scientists was both frequent and imaginative, suggesting not a series of ad hoc decisions but rather a general philosophy put into practice. Influenced by both Vincent de Gournay and the physiocrats, Turgot believed it desirable to free agriculture, industry, and commerce from excessive governmental regulation; the inescapable corollary to this doctrine was that laissez-faire would ensure progress and prosperity. But to this idea he added a conviction that was to him of equal weight: the results of scientific research, applied to technology and taught to peasants, craftsmen, manufacturers, and others, could provide a firm foundation for progress. These beliefs are apparent in many of his actions, such as his appointment of geologist Nicolas Desmarest as inspector of manufactures in Limousin[4] and his pioneering efforts to gather accurate statistical data about Limousin and, eventually, all of France.

Turgot's views were not new and, in fact, were shared by contemporaries ranging from the Trudaines to the encyclopedists. If the differences between Turgot and others are matters of degree rather than of kind, the consistency of his policies elevates him to a position of distinction among royal ministers, while his political power separates him from other philosophers. Furthermore, Turgot's own knowledge and ability in the sciences made him an intelligent judge of the ability of others, and he purposely increased his acquaintance with such subjects as agronomy in order to understand more fully the advice given him by the scientists he consulted. His attitude in this respect was succinctly summarized by Condorcet:

> He was not afraid to consult men of science, because he was not afraid of truth. . . . But he knew at the same time that the learned, accustomed to system and demonstration, carried sometimes to excess the spirit of scepticism and uncertainty; and that in consulting them, it is necessary both to seek to understand, and to be capable of understanding them. . . . [He] regarded the *encouragement of the arts and sciences* as an indispensable duty of his office.[5]

Translated into policy, his attitude led Turgot to devise some projects of his own, solicit suggestions from scientists, sponsor translations of scientific treatises, and encourage the work of inventors.

While an intendant, Turgot's major concern was the improvement of agriculture in a region as poor as Brittany and subject to endemic and epidemic famine. Aware of the studies of agronomists, he became convinced that potatoes were hardy enough to thrive where wheat could not, and that the potato was a food nourishing to both human beings and livestock. His efforts to persuade his peasants to make the experiment eventually met with some success, years before the potato became common in other French provinces. Turgot was also the patron of the Society of Agriculture of Limoges, founded by his predecessor, Pajot de Marcheval, in 1759. He doubted that experiments done in common could be done properly, but he attended some meetings, donated equipment, and used his own funds to establish prizes for essays on agricultural subjects. In 1766 he founded a short-lived school of veterinary medicine in Limoges, patterned upon the school in Lyons to which Turgot had sent students since its opening in 1762. When the Collège de Limoges was undergoing reform after 1762, Turgot used his influence to see that the sciences were given some prominence in the new curriculum.

As controller general, Turgot's concern with agriculture was expressed in two especially significant ways. A serious outbreak of murrain in 1774 led him to send Félix Vicq d'Azyr to the afflicted southwestern provinces to study the disease and to recommend remedies. The problems of diagnosis and control suggested to Turgot that the study of epidemics be undertaken systematically, and he and Vicq became founders of a society formed for that purpose and for the improvement of medical education and research. The society, which first met in 1776, later became the Société Royale de Médecine with Vicq d'Azyr its first secretary. At the same time, Turgot continued his efforts to introduce new crops. As an intendant, he had brought to Limousin crops already being raised elsewhere in France, but as controller general he extended his interests to the genuinely exotic. In 1775 he arranged the mission of naturalist Joseph Dombey to South America, instructing Dombey to bring back to France plants of botanical interest and of potential agricultural value. Turgot was intimately involved in planning this voyage, for which he sought the advice of Condorcet and the botanists Antoine-Laurent de Jussieu and André Thouin.

The voyages of Dombey and others were not wholly agricultural in purpose. One of Turgot's broader aims, which he shared with the physiocrats, was to bring about a more rationally organized world trade, with each country growing and producing those goods for which its soil, climate, and technological skills were best suited. But in addition Turgot dreamed of being able to send scientists all over the world as a kind of "traveling academy" for the collection of scientific information.[6] His view, not uncommon at the time, that science belonged to all humanity rather than to particular nations was made explicit in a memorandum written after 1776 and addressed to Louis XVI; Turgot urged the king (who accepted his advice) that, with France on the verge of war with England and with Captain Cook engaged in the third of his voyages, all French ships should be ordered not to molest Cook in any way.

The scientist consulted most regularly by Turgot was Condorcet, who was a personal friend and seems to have served informally as Turgot's liaison with the Académie Royale des Sciences. Among the many projects in which Condorcet played a part was Turgot's plan to introduce a uniform system of weights and measures, a reform long considered desirable. Like the later designers of the metric system, Turgot thought it best to select a natural constant—the length of a seconds pendulum at a given latitude—for the unit of length; a standard of weight should be determined in some comparable way, and the coinage should then be issued in units corresponding to divisions of weight. It was on Condorcet's advice that Turgot asked the astronomer Charles Messier to carry out the preliminary measurements, and in 1775 Turgot and Condorcet drew up instructions for Messier. Although the work was begun, it was discontinued when Turgot fell from power.

Turgot's interest in the system of canals and natural waterways of France originated from his desire to improve internal commerce. In 1775, on the advice of Condorcet, he created a committee of three eminent scientists—Condorcet, Bossut, and d'Alembert—to examine proposals for new canals and to inspect those already under construction. The scientists proposed that they work without salary, but Condorcet was rewarded with an appointment as director of the mint, while Bossut was named to a newly created post of professor of hydrodynamics. The committee was abolished upon Turgot's dismissal from office.

To reform the manufacture of gunpowder, Turgot created the Régie des Poudres in 1775, appointing Lavoisier one of the *régisseurs*. This government agency not only administered nationally the manufacture of saltpeter, but it also encouraged research to improve the quality and increase the quantity of saltpeter used for gunpowder. Research continued—much of it by Lavoisier—long after Turgot's dismissal, with results soon visible during the American Revolution and later during the French Revolution.

Turgot's activities during his twenty months as controller general—and for eight of the twenty months he was confined to bed with severe attacks of gout—were remarkable for their number, variety, and intelligence. His efforts can be considered the logical culmination of his earliest ideas and of his work as an intendant, and they bear witness to the consistency with which he tried to put into practice his ideals as an enlightened reformer. Although few of his reforms outlasted his term in office—several were reintroduced in later decades—his philosophy was to leave its mark upon subsequent generations.

NOTES

1. Arthur Young, *Travels in France During the Years 1787, 1788, 1789* (many editions), entry dated 6 June 1787.
2. Denis Diderot and Jean Le Rond d'Alembert, *Encyclopédie, ou Dictionnaire raisonné des sciences, des arts et des*

métiers, VI (1756), 274–285, and the important errata, VII (1757), 1028–1029. Reprinted in Schelle, *Oeuvres*, I, 538–576.
3. Condorcet, *Life*, 30–31.
4. Letter from Turgot to Daniel Trudaine, 10 September 1762, in Pierre Bonnassieux and Eugène Lelong, *Conseil de commerce et Bureau du commerce 1700–1791. Inventaire analytique des Procès-Verbaux* (Paris, 1900), xlv, col. 2, n. 2.
5. Condorcet, *op. cit.*, 136–137, 144.
6. Dupont de Nemours, *Mémoires*, I, 122. Cf. Schelle, *Oeuvres*, II, 523–533.

BIBLIOGRAPHY

I. ORIGINAL WORKS. 1. *Publications*: The best of several editions of Turgot's works is *Oeuvres de Turgot et documents le concernant*, Gustave Schelle, ed., 5 vols. (Paris, 1913–1923), although the editor's transcriptions are sometimes faulty, and he omitted much of value, including letters addressed to Turgot. Still indispensable, therefore, is the *Correspondance inédite de Condorcet et de Turgot, 1770–1779*, Charles Henry, ed. (Paris, 1883).

The authorship of published geological notes, attributed to Turgot, remains in some doubt, although it is certain that Turgot did make geological observations during his travels. Relevant texts and discussion are in *Oeuvres*, II, 604, and in *Oeuvres de M. Turgot, ministre d'état*, P.-S. Dupont de Nemours, ed., 9 vols. (Paris, 1808–1811), I, 52–53; III, 376–447.

2. *Manuscripts*: There are three major repositories, one of them private; the others are the Archives Nationales, Paris, and the Eleutherian Mills Historical Library, Wilmington, Delaware. Cf. John B. Riggs, *A Guide to the Manuscripts in the Eleutherian Mills Historical Library: Accessions Through the Year 1965* (Greenville, Del., 1970). Most documents were published by Schelle, but the footnotes in his edition provide a guide to some of the manuscripts he decided to omit.

II. SECONDARY LITERATURE. P. F. C. Foncin, *Essai sur le ministère de Turgot* (Paris, 1877); Douglas Dakin, *Turgot and the Ancien Régime in France* (London, [1939]), with valuable bibliography and notes; Henry Guerlac, *Lavoisier—The Crucial Year: The Background and Origin of His First Experiments on Combustion in 1772* (Ithaca, New York, [1961]), ch. 5; Roger Hahn, "The Chair of Hydrodynamics in Paris, 1775–1791: A Creation of Turgot," in *Actes du Xe Congrès international d'histoire des sciences*, II (Paris, 1964), 751–754; Rhoda Rappaport, "Government Patronage of Science in Eighteenth-Century France," in *History of Science*, VIII (1969 [publ. 1970]), 119–136; Jerry Gough, "Nouvelle contribution à l'étude de l'évolution des idées de Lavoisier sur la nature de l'air et sur la calcination des métaux," in *Archives internationales d'histoire des sciences*, **22** (1969), 267–275.

See also Denis I. Duveen and Herbert S. Klickstein, *A Bibliography of the Works of Antoine Laurent Lavoisier 1743–1794* (London, 1954), esp. 219–222; a subject index to this volume is in Duveen, *Supplement to a Bibliography of the Works of Antoine Laurent Lavoisier 1743–1794* (London, 1965).

The many general studies of Turgot tend to minimize or ignore scientific questions, but still valuable are the earliest biographies: P.-S. Dupont de Nemours, *Mémoires sur la vie et les ouvrages de M. Turgot, ministre d'Etat*, 2 vols.-in-1 (Philadelphia, 1782); and Marquis de Condorcet, *Vie de Monsieur Turgot* (London, 1786), translated anonymously as *The Life of M. Turgot* (London, 1787); also published in *Oeuvres de Condorcet*, A. C. O'Connor and F. Arago, eds., 12 vols. (Paris, 1847–1849), V, 1–233.

For discussions of Turgot's relations with the scientific community, see Keith M. Baker, *Condorcet, From Natural Philosophy to Social Mathematics* (Chicago, 1975); and C. C. Gillispie, "Probability and Politics: Laplace, Condorcet, and Turgot," in *Proceedings of the American Philosophical Society*, **116** (1972), 1–20.

RHODA RAPPAPORT

TURGOT, ÉTIENNE-FRANÇOIS (*b.* Paris, France, 16 June 1721; *d.* Château of Bons, Calvados, France, 25 December 1788), *botany, agronomy.*

An older brother of the famous reformer, the Chevalier Turgot, knight of the Order of Malta and Marquis de Soumont, served briefly as the governor of Guiana (1764–1765). After his retirement from public life in 1765, he devoted himself to agricultural experiments and study, interests which had already led him to introduce the cultivation of exotic crops in Malta and Guiana. Like his friends Malesherbes and Duhamel du Monceau, with whom he corresponded on agricultural subjects, Turgot used a large part of his land in Normandy for the naturalization of foreign trees and the cultivation of botanical rarities. He also maintained a residence in Paris, where he was able to introduce one of his protégés, Hector Saint-John de Crèvecoeur, to other naturalists and agronomists. He was a member of the Académie Royale des Sciences (1765), and a founding member of the Société d'Agriculture de Paris (1761).

Turgot is usually said to have died on 21 October 1789. The date 25 December 1788 adopted here is taken from the announcement in the *Journal de Paris* (1 January 1789, p. 4).

BIBLIOGRAPHY

I. ORIGINAL WORKS. Works by Turgot include *Mémoire instructif sur la manière de rassembler, de pré-* parer, de conserver, et d'envoyer les diverses curiosités d'histoire naturelle (Paris–Lyons, 1758); and *Essai sur les arbres d'ornement, les arbrisseaux, et arbustes de pleine terre* (Amsterdam–Paris, 1778). The second of these is a partial translation, with added material, of Philip Miller, *The Gardener's Dictionary*, 7th ed. (London, 1759; repr. Dublin, 1764).

Turgot wrote several articles in *Mémoires de l'Académie royale des sciences*, and in *Mémoires d'agriculture* of the Société Royale d'Agriculture (Paris). Most extant manuscripts are in private collections.

II. SECONDARY LITERATURE. References to Turgot and his work are found in the eulogies by P.-A.-M. Broussonet, in *Mémoires d'agriculture*, trimestre d'automne (1789), and the Marquis de Condorcet, in *Histoire de l'Académie des sciences, 1789* (1793), 31–38; and in Alfred Lacroix, *Notice historique sur les membres et correspondants de l'Académie des sciences ayant travaillé dans les colonies françaises de la Guyane et des Antilles de la fin du XVIIe siècle au début du XIXe* (Paris, 1932); and *Figures de savants*, III (Paris, 1932–1938), 61–69.

RHODA RAPPAPORT

TURING, ALAN MATHISON (*b.* London, England, 23 June 1912; *d.* Wilmslow, England, 7 June 1954), *mathematics, mathematical logic, computer technology.*

Turing was the son of Julius Mathias Turing and Ethel Sara Stoney. After attending Sherborne School he entered King's College, Cambridge, in 1931. He was elected a fellow of the college in 1935 for his dissertation "On the Gaussian Error Function," which won a Smith's prize in the following year. From 1936 until 1938 Turing worked at Princeton University with Alonzo Church.

While at Princeton Turing published one of his most important contributions to mathematical logic, his 1937 paper "On Computable Numbers, With an Application to the *Entscheidungsproblem*," which immediately attracted general attention. In it he analyzed the processes that can be carried out in computing a number to arrive at a concept of a theoretical "universal" computing machine (the "Turing machine"), capable of operating upon any "computable" sequence – that is, any sequence of zeros and ones. The paper included Turing's proof that Hilbert's *Entscheidungsproblem* is not solvable by these means. Church had, somewhat earlier, solved Hilbert's problem by employing a λ-definable function as a precise form of the intuitive notion of effectively calculable function, while in 1936, S. C. Kleene had proved the equivalence of λ-definability and the Herbrand-Gödel theory of general recursiveness. In his "Computability and λ-Definability" of 1937, Turing demonstrated that his and Church's ideas were equivalent.

In 1939 Turing published "Systems of Logic Based on Ordinals," in which he examined the question of constructing to any ordinal number α a logic $L\alpha$, such that any problem could be solved within some $L\alpha$. This paper had a far-reaching influence; in 1942 E. L. Post drew upon it for one of his theories for classifying unsolvable problems, while in 1958 G. Kreisel suggested the use of ordinal logics in characterizing informal methods of proof. In the latter year S. Feferman also adapted Turing's ideas to use ordinal logics in predicative mathematics.

In 1939 Turing returned to King's College, where his fellowship was renewed. His research was interrupted by World War II, however, and from the latter part of 1939 until 1948 he was employed in the communications department of the Foreign Office; he was awarded the O.B.E. for his work there. After the war, he declined the offer of a lectureship at Cambridge and, in autumn of 1945, joined the staff of the National Physical Laboratory to work on the design of an automatic computing engine (ACE).

In 1948 Turing became a reader in the University of Manchester and assistant director of the Manchester automatic digital machine (MADAM). He also continued to work in mathematical theory, and improved E. L. Post's demonstration of the existence of a semigroup with unsolvable word problem by exhibiting a semigroup with cancellation for which the word problem is (recursively) unsolvable. He made further contributions to group theory and performed calculations on the Riemann zeta-function in which he incorporated his practical work on computing machines.

In 1950 Turing took up the question of the ability of a machine to think, a subject that had gained general interest with the increasing application of mechanical computing devices to more and more complex tasks. His "Computing Machinery and Intelligence" was addressed to a broad audience and marked by a lively style. *The Programmer's Handbook for the Manchester Electronic Computer*, produced under his direction, was published in the same year.

Throughout his life Turing was also interested in applying mathematical and mechanical theory to the biological problem of life forms. He made a promising approach to this question in his 1952 publication "The Chemical Basis of Morphogene-sis." In this work he exploited the mathematical demonstration that small variations in the initial conditions of first-order systems of differential equations may result in appreciable deviations in the asymptotic behavior of their solutions to posit that unknown functions might function biologically as form-producers; he was thus able to account for asymmetry in both mathematical and biological form. He was at work on a general theory when he died of perhaps accidental poisoning.

BIBLIOGRAPHY

I. ORIGINAL WORKS. An edition of Turing's collected works is in preparation by Professor Dr. R. O. Gandy. See especially Turing's "On Computable Numbers, With an Application to the *Entscheidungsproblem*," in *Proceedings of the London Mathematical Society*, **42** (1937), 230–265; "On Computable Numbers, With an Application to the *Entscheidungsproblem*. A Correction," *ibid.*, **43** (1937), 544–547; "Computability and λ-Definability," in *Journal of Symbolic Logic*, **2** (1937), 153–163; "Systems of Logic Based on Ordinals," in *Proceedings of the London Mathematical Society*, **45** (1939), 161–228; "The Word Problem in Semigroups With Cancellation," in *Annals of Mathematics*, **52** (1950), 491–505; "Computing Machinery and Intelligence," in *Mind*, **59** (1950), 433–460, repr. as "Can a Machine Think?" in J. R. Newman, *The World of Mathematics*, IV (New York, 1956), 2099–2133; and "The Chemical Basis of Morphogenesis," in *Philosophical Transactions of the Royal Society*, **237** (1952), 37–72.

II. SECONDARY LITERATURE. S. Turing, *Alan M. Turing* (Cambridge 1959), includes a bibliography of works by and about Turing. See also M. Davis, *Computability and Unsolvability* (New York, 1958); S. Feferman, "Ordinal Logics Re-examined," in *Journal of Symbolic Logic*, **23** (1958), 105; "On the Strength of Ordinal Logics," *ibid.*, 105–106; "Transfinite Recursive Progressions of Axiomatic Theories," *ibid.*, **27** (1962), 259–316; and "Autonomous Transfinite Progressions and the Extent of Predicative Mathematics," in B. van Rootselaar and J. F. Staal, eds., *Logic, Methodology and Philosophy of Science*, III (Amsterdam, 1968), 121–135; S. C. Kleene, *Introduction to Metamathematics* (Amsterdam–Groningen, 1952); and *Mathematical Logic* (New York, 1967); G. Kreisel, "Ordinal Logics and the Characterization of Informal Concepts of Proof," in *Proceedings of the International Congress of Mathematicians, 1958* (Cambridge, 1960), 289–299; and M. H. A. Newman, "Alan Mathison Turing," in *Biographical Memoirs of Fellows of the Royal Society 1955*, I (London, 1955), 253–263, which also has a bibliography.

B. VAN ROOTSELAAR

TURNER, EDWARD (*b*. Kingston, Jamaica, July 1796; *d*. Hampstead, London, England, 12 February 1837), *analytical chemistry.*

Turner was the second son of Dutton Smith Turner, a prosperous planter, and Mary Gale Redwar, a Creole of English ancestry. Her relatives raised and educated him in England. After attending Bath Grammar School he was apprenticed to a local country doctor from 1811 to 1814. He then spent two years walking the wards of the London Hospital before studying medicine at the University of Edinburgh from 1816 to 1819. At Edinburgh, where he was president of the student medical society, he formed a deep friendship with the energetic Robert Christison, a fellow medical student.

In August 1820, following an unsatisfactory attempt to practice medicine at Bath, Turner and Christison went to Paris for further study. Here Turner became attracted toward chemistry and physics by the lectures and experimental activities of Gay-Lussac, Pelletier, and Robiquet. From the spring of 1821 until the summer of 1823 he studied mineral analysis and chemistry at Göttingen with Friedrich von Stromeyer, who had a small teaching laboratory. Encouraged by Christison, he returned to Edinburgh in 1823 and became an important private, or extramural, lecturer in chemistry, exploiting Thomas Charles Hope's failure to provide practical chemistry teaching within the university by mounting laboratory classes. He also acted as chemical editor for David Brewster's *Edinburgh Journal of Science.* In 1827, through the influence of Leonard Horner, and supported by Brewster, Christison, Hope, Jameson, and Thomas Thomson, he was appointed professor of chemistry and lecturer in geology at the new University of London (University College, London). His teaching here ensured the popularity of chemistry among London medical students and his own financial success. From the opening of classes in 1829 he held a laboratory demonstration course (assisted by Robert Warrington). The college also provided him with a small research laboratory.

An excellent lecturer, Turner enchanted his listeners, who found him gentle and easy to approach. He was always a devout Christian, but after his health broke in 1834 he underwent evangelical conversion and hoped to impart moral and religious, as well as scientific, instruction to his students. He is distinguished as the author of one of the best nineteenth-century textbooks on chemistry, for his determination of atomic weights, and

for his attitude toward Prout's hypothesis that atomic weights were integral multiples of the atomic weight of hydrogen. His many mineralogical analyses were unexceptional.

He began his career as a naïve disciple of Thomas Thomson, whose *First Principles of Chemistry* (1825) he admired as a work of "profound sagacity." He was, therefore, initially convinced of the soundness of the Prout-Thomson multiple weights hypothesis, although like other chemists he stressed the hypothetical character of Dalton's atomism compared with the laws of chemical combination. His essay on this distinction (1825) was the foundation for his popular up-to-date textbook *Elements of Chemistry* (1827).

Through his mineralogical interests Turner was led to investigate the atomic weight of manganese, which Thomson had determined by the precipitation reaction between manganese sulfate and barium chloride, but which Berzelius had shown to be inaccurate. In 1828 Turner decided to place himself in the delicate position of "umpire between two of the greatest of living chemists" by investigating the discrepancies between Thomson's and Berzelius' atomic weights. He found immediately that Berzelius' criticism of Thomson's careless use of the reagent barium chloride was justified. From 1829 to 1833 he gradually showed that the remarkable edifice raised by Thomson's *Principles* was a house of cards. At the meeting of the British Association for the Advancement of Science held at Oxford in 1832, and at the Royal Society in 1833, Turner demonstrated by careful analyses that Thomson's atomic weights for chlorine, nitrogen, sulfur, lead, and mercury were in serious error, that his own values confirmed those of Berzelius, and, consequently, that although integral atomic weights might be used as convenient approximations by "medical men, students, and manufacturers," the true values were inconsistent with Prout's original hypothesis.

It is clear, however, that Turner believed (perhaps as a result of his friendship with Prout, or because of his own latent enthusiasm for mathematics) that chemists might one day find a simple relationship between atomic weights. He has sometimes been criticized for this wavering conclusion; but it was consistent with his empiricism and his positivistic view of physical theory. Turner believed that analytical chemistry was open to further improvements; until such time, however, chemists had no right to make unqualified guesses either way about mathematical relationships between the elements. Turner's emphasis upon the analyst as critical arbiter of theory, as well as his exacting standards, stimulated the later researches of F. Penny and J. S. Stas.

BIBLIOGRAPHY

I. ORIGINAL WORKS. Forty papers by Turner are listed in the Royal Society *Catalogue of Scientific Papers*, VI (London, 1867–1925). His books are *De Causis Febris Epidemicae nunc Edinburgi grassantis* (Edinburgh, 1819), his M.D. thesis; *Introduction to the Study of the Laws of Chemical Combination and the Atomic Theory* (London, 1825), with German trans. (Tübingen, 1828); *Elements of Chemistry* (Edinburgh, 1827; 2nd ed., London, 1828, and German trans., Leipzig, 1829; 3rd ed., London, 1831; 4th ed., London, 1833, which introduced formulas; 5th ed., London, 1834). Each of these eds. received American printings. There were three posthumous eds. by J. Liebig; W. G. Turner (1811–1855), Turner's brother, an industrial chemist; W. Gregory (London, 1842); and by Liebig and Gregory (London, 1842, 1847). Turner also published with Anthony Todd Thomson, *Two Letters to the Proprietors of the University of London, In Reply to Some Remarks in Mr. [G.S.] Pattison's Statement* (London, 1831), 16 pp.

For letters to Berzelius, see H. G. Söderbaum, ed., *Jac. Berzelius Bref* (Uppsala, 1912–1935), III, vii, 273–285. There is a small collection of Turner's college correspondence at University College, London.

II. SECONDARY LITERATURE. The most informative accounts of Turner are R. Christison, *Biographical Sketch of the late Edward Turner, M.D.*, two eds. (Edinburgh, 1837), 36 pp.; anon., "Edward Turner," in *Gentleman's Magazine*, 1 (1837), 434–435; W. Whewell, "Address to the Geological Society," in *Proceedings of the Geological Society*, 2 (1833–1838), 626–627; and the curious funeral sermon by Rev. Thomas Dale, *The Philosopher Entering . . . the Kingdom of Heaven* (London, 1837). Biographical accounts which assess his analytical work are Henry Terrey, "Edward Turner, M.D., F.R.S. (1796–1837)," in *Annals of Science*, 2 (1937), 137–152, with portrait; and J. S. Rowe, "Chemical Studies at University College, London," unpub. Ph.D. thesis, 1955.

W. H. BROCK

TURNER, HERBERT HALL (*b*. Leeds, England, 13 August 1861; *d*. Stockholm, Sweden, 20 August 1930), *astronomy, seismology*.

Turner, the son of John Turner, an artist, gained high honors in mathematics at Cambridge and was second wrangler in 1882. He was chief assistant at the Royal Observatory, Greenwich, from 1884 to 1893 and Savilian professor of astronomy at Oxford from 1893 until his death in 1930. He became a fellow of the Royal Society of London in 1897 and a correspondent of the Paris Academy of Sciences in 1908.

Turner's fame rests considerably on his ability and energy as an organizer of international scientific projects. He was a principal coordinator of work on the astrographic chart that began in 1887 and made extensive use of the new science of photography. He was active in establishing and contributed to eclipse expeditions to the West Indies in 1886, Japan in 1896, India in 1898, Algiers in 1900, and Egypt in 1905.

Turner's interests turned to seismology following the death in 1913 of his friend John Milne, who had been publishing regular analyses of instrumental records of great earthquakes. Turner took over this task, at first on a temporary basis; but the work soon became one of his major interests and culminated in the publication of the *International Seismological Summary*. This quarterly publication was a compendium of information derived from instrumental records on all well-recorded earthquakes over an uninterrupted period from 1918 to 1963. It included the estimated origin times, epicenters, and focal depths, as well as much auxiliary numerical detail, and has supplied the principal source material over several decades for some of the most important research on earthquakes and the internal structure of the earth.

In computing the origin times of earthquakes, Turner adapted Zöppritz' tables, which gave values for the travel times of earthquake waves through the earth in terms of the distances covered. These tables, known as the Zöppritz-Turner tables, were widely used into the 1930's.

Turner held several high international offices in both astronomy and seismology. He died while presiding at a meeting of the Seismology Section of the International Union of Geodesy and Geophysics.

BIBLIOGRAPHY

Turner's major publication was the quarterly *International Seismological Summary* for the years 1918–1927, published at Oxford under his editorship. He also wrote some 180 scientific papers, most of them published in *Monthly Notices of the Royal Astronomical Society*, and four semipopular books, published in London: *Modern Astronomy* (1901); *Astronomical Discovery* (1904); *The Great Star Map* (1912); and *A Voyage in Space* (1915).

There is a short notice by R. A. Sampson in *Dictionary of National Biography, 1922–1930*.

K. E. BULLEN

TURNER, PETER (*b.* London, England, 1586; *d.* London, January 1652), *mathematics*.

Turner was the son of Dr. Peter Turner and Pascha Parr, and the grandson of William Turner, the physician and naturalist and dean of Wells. He received his B.A. in 1605 and M.A. in 1612 at Oxford, and became a fellow of Merton College in 1607. Turner was the second Gresham professor of geometry (1620–1630) and second Savilian professor of geometry (1630–1648), in both cases succeeding Henry Briggs. He retained his Merton fellowship, going to London for his Gresham lectures in term time. In 1629 he was appointed to a commission charged with the revision of the Oxford statutes. The final draft was largely the work of Brian Twynne, but it was polished for the press by Turner, who was noted for his Latin style.

He was one of the first scholars to enlist for King Charles in 1641. He was captured at the battle of Edgehill and imprisoned for a time. He was ejected both from his fellowship and his professorship by the Parliamentary Visitors in 1648, and retired to live in straitened circumstances with his widowed sister in Southwark. Both Turner and the Savilian professor of astronomy, John Greaves, were replaced by Cambridge men—Turner by John Wallis, and Greaves by Seth Ward.

It is impossible to judge Turner's abilities in mathematics. He left no mathematical writings, and, indeed, seems to have been noted rather as a Latinist and a linguist, being skilled in Greek, Hebrew, and Arabic. According to Wood he destroyed many of his writings, being of too critical a mind. Further, his effects at Oxford were seized during the Civil War. Some translations from Greek to Latin of the church fathers in the possession of his colleague Mr. Henry Jacobs; Latin poems to the memory of Sir Thomas Bodley (1613); and the preface to the revised Oxford Statutes (1634) are all that are known to have survived.

BIBLIOGRAPHY

Turner's life can be found in John Ward, *Lives of the Professors of Gresham College* (1740; Johnson repr., Sources of Science, no. 71). See also C. E. Mallet, *A History of the University of Oxford*, II (Oxford, 1924– 1928; repr., New York, 1968); and the *Calendar of State Papers, Domestic*, during the reign of Charles I.

JOY B. EASTON

TURNER, WILLIAM (*b.* Morpeth, Northumberland, England, 1508; *d.* London, England, 7 July 1568), *natural history, medicine*.

Very little of what is recorded of Turner's family and early life in the northern counties is other than conjectural. His first appearance in the official records is as a student of Pembroke Hall, Cambridge, in 1526. He graduated B.A. in 1529/1530, and M.A. in 1533; he was a fellow of Pembroke Hall between 1530 and 1537, when he married Jane Auder of Cambridge. Under the influence of Hugh Latimer, Turner became an ardent religious reformer. His views were first made apparent in his translation of the *Comparison Betwene the Olde Learnynge and the Newe* (1537) by the German theologian Urbanus Regius.

Turner produced numerous religious tracts of the same fierce and uncompromising tone, concluding with *A Newe Booke of Spirituall Physic* (1555). His career was considerably affected by his extreme religious position. Although inclined from an early age to the tranquil study of natural history, he was forced by the threat of religious persecution to spend long periods in exile. During the first of these periods (1540–1546), Turner studied medicine in Italy, obtaining an M.D. at either Ferrara or Bologna, and traveled extensively in Germany and Switzerland, forming friendships with Konrad Gesner and other European naturalists. Having returned to England, Turner found that recognition came slowly. He was made dean of Wells Cathedral in 1551 and held this position until 1553. He incorporated for an M.D. at Oxford; Turner's work on natural history shows a consistently medical bias, and he seems to have combined medical and clerical duties throughout his career. A second period of exile (1553–1558), following the death of Edward VI, was chiefly spent as a medical practitioner in Weissenburg. Although Turner's foreign journeys were undertaken reluctantly out of necessity, they proved of inestimable value to his scientific work. He became fully acquainted with the Continental literature and with the latest trends in the research being conducted by the flourishing school of humanist naturalists. He was also able to extend his knowledge of the flora and fauna of Europe.

Some time after his second return to England,

Turner was restored to the deanery of Wells. His final years were spent in the undisturbed study of botany, in collaboration with a wide circle of friends who ranged from apothecaries to gentleman patrons.

Turner's vocation as a field naturalist emerged early. His publications include reference to material collected during his youth at Morpeth and as a student in East Anglia. At Cambridge he became dissatisfied with the derivative herbals and natural histories then in use, which gave little real impression of the local flora and fauna. He set out to produce reliable lists of English animals and plants using approved nomenclature based on classical sources and humanist usage. But Turner did not allow the scholarly aspect of his work to overshadow the results of his firsthand observations relating to morphology, distribution, behavior, and pharmacology. His main difficulty in carrying forward his botanical studies, in the troubled period of the Reformation, was in securing a subsistence and adequate toleration of his religious views. He made a modest beginning to his program with *Libellus de re herbaria* (1538), a list of 144 plants, the names of which were given in alphabetical order in Latin, with English and Greek synonyms. At this stage Turner had a good knowledge of the classical languages but not of the most recent work of humanist naturalists. This defect was supplied, and his next works written, during his periods in exile abroad. *Avium praecipuarum* (1544) follows the pattern of the *Libellus* and is a tentative list of birds mentioned in Pliny and Aristotle, with identifications of northern European species. Turner's information provides valuable evidence about the distribution of species during the sixteenth century. A further excursion into ornithology was his edition of the *Dialogus de avibus et earum nominibus*, written by his recently deceased friend Gisbert Longolius. Turner's work on birds progressed no further, his attention turning instead to fishes, on which he composed a preliminary essay addressed to Konrad Gesner. This study was primarily a list of English fishes with notes on their distribution. It shows little acquaintance with the recent writings of Belon and Rondelet in ichthyology.

Turner was more successful in completing his botanical studies. No doubt inspired by the Continental herbalists, he composed a Latin herbal; but he abandoned this project after his first return to England, when he became convinced of the necessity of first studying the British flora in its entirety. He also decided to publish in the vernacular, recognizing the need to diffuse botanical and medical knowledge as widely as possible among his countrymen, even at the cost of limiting the European influence of his work. The *Libellus* was accordingly expanded into *The Names of Herbes* (1549). For this larger collection of species, Turner drew on his Continental experience to include German and French names, with details of distribution. This publication served as an advertisement for the first section of his *New Herball* (1551), in which plants were again listed alphabetically under their Latin names, for the letters from *A* to *F*. Synonyms were given in English and other languages. Turner's descriptions were unorthodox, since he expressed these in vivid vernacular and usually included evidence drawn from firsthand observations. He showed little inclination to follow authorities and was scornful of much long-cherished herbal lore. The medical bias of the book was particularly marked, Turner intending it for use by apothecaries or laymen with medical interests. One unsatisfactory aspect of the *Herball* was the quality of its illustrations, which were poor copies of those in Leonhard Fuchs's *De historia stirpium* (1542). Turner was also arbitrarily selective, including some quite rare herbaceous plants but omitting common trees, grasses, and sedges. His second exile delayed the latter part of the *Herball* until 1562. This section, from *F* to *P*, was improved by his more extensive knowledge of German plants and by his access to Pietro Mattioli's translation of Dioscorides. He further reinforced the medical aspect of the work by including a treatise on baths, the first of numerous English works on this subject. Turner was mainly concerned with Bath, but he also made reference to baths in Italy and Germany. The first complete edition of the *Herball* appeared in 1568. Part I had been revised and expanded, Part II was unaltered, and a new Part III had been added to complete the alphabet. The quality of the descriptions was maintained: Turner wisely resisted the temptation to multiply species in the quest for encyclopedic coverage.

W. A. Cooke has estimated that Turner's pioneering flora provided the first descriptions of a total of 238 species of native plants. The vernacular names coined by Turner for indistinctly recognized species have passed into general use. In spite of its originality, Turner's work was not particularly well known to later botanists; Jean Bauhin and John Ray were exceptional in making active use of his *Herball*. It is evident that Turner's great competence in botany carried over to other spheres of natural history. Only the turmoil of his life prevented him from expanding his informative essays on birds and fishes. The preface to *The Names of Herbes* and his letter appealing for patronage to Sir William Cecil (1550) indicate that he planned books on fishes, stones, and minerals, and also a corrected translation of the New Testament. Turner's liberal interests as a naturalist provided a blueprint of the design for a system of nature that was made manifest by John Ray in the next century.

BIBLIOGRAPHY

I. ORIGINAL WORKS. A complete list of Turner's works, including those not published, or lost, is given by Charles H. Cooper, *Athenae Cantabrigenses*, I (Cambridge, 1858), 256–259. Cooper lists thirty-four titles.

Turner's botanical works are *Libellus de re herbaria* (London, 1538); reprinted with a biographical introduction by B. D. Jackson (London, 1877), who also gives a list of Turner's works; *The Names of Herbes in Greke, Latin, Englishe, Duch and Frenche* (London, 1548/1549); reprinted, J. Britten, ed., English Dialect Society (London, 1882). The editions of Jackson and Britten have been reprinted, with a new introduction by W. T. Stearn, by the Ray Society (London, 1965). Turner's major botanical work is *A New Herball Wherein are Conteyned the Names of Herbes* [first part] (London, 1551); *The Seconde Parte* (Cologne, 1562), including "A Book of the Bath of Baeth"; *The First and Seconde Partes of the Herbal . . . With the Third Parte* (Cologne, 1568).

Turner's zoological writings are *Avium praecipuarum quarum apud Plinium et Aristotelem mentio* (Cologne, 1544); "Epistola Conrarrdo Gesnero," in Konrad Gesner, *Historia animalium*, IV (Zurich, 1558), 1294–1297.

II. SECONDARY LITERATURE. See A. Arber, *Herbals, Their Origin and Evolution*, 2nd ed. (Cambridge, 1938); W. A. Cooke, *First Records of British Flowering Plants*, 2nd ed. (London, 1900); T. P. Harrison, "William Turner, Naturalist and Priest," in *University of Texas Studies in English*, XXXIII (1954), 1–12; R. Pulteney, *Historical and Biographical Sketches of the Progress of Botany*, I (London, 1790), 40–70; C. E. Raven, *English Naturalists from Neckam to Ray* (Cambridge, 1947), by far the most extensive account of Turner as a naturalist.

CHARLES WEBSTER

TURNER, WILLIAM (*b.* Lancaster, England, 7 January 1832; *d.* Edinburgh, Scotland, 15 February 1916), *anatomy, academic administration.*

William Turner was the son of a cabinetmaker. The father died in 1837, and the boy was brought up in poor circumstances by his mother (*née* Margaret Aldren), a woman of strong character and simple faith. He was educated at a private school and apprenticed at the age of fifteen to Christopher Johnson, a local general medical practitioner.

At sixteen he proceeded to St. Bartholomew's Hospital, London, and qualified with the membership of the Royal College of Surgeons of England in the summer of 1853. His subsequent career was distinguished by parallel activities in both academic and administrative spheres.

In 1861 Turner became a fellow of the Royal Society of Edinburgh and followed Kelvin as its president in 1908. He was made fellow of the Royal Society of London in 1877, president of the Royal College of Surgeons of Edinburgh in 1882 (although he never practiced medicine or surgery), president of the Anatomical Society of Great Britain in 1892, and president of the British Association at its Bradford meeting in 1900. He was knighted in 1886 and was president of the General Medical Council of Great Britain from 1898 to 1904.

Turner's portrait, painted by Sir George Reid in 1895, shows a genial shrewd gentleman in characteristic pose. He was also painted in full academic dress by Sir James Guthrie in 1913; the portrait is now in the Court Room of the Old College, Edinburgh University.

In 1862 Turner married Anne Logan; they had three sons, two of whom practiced medicine, and two daughters.

One year after qualifying, on the recommendation of Sir James Paget, Turner was made senior demonstrator to John Goodsir, professor of anatomy at the University of Edinburgh. This school of anatomy was then the most distinguished in Britain. Turner succeeded to the chair in 1867, resigning only in 1903 when he took up the position of principal of the university, which he held until his death. He had then been a member of its Senate for forty-nine years.

In each of the bodies that he led, Turner's intimate knowledge of medical legislation, his scientific attainments, and his force of character were used to carry through important programs and reforms. Thus in Edinburgh he was instrumental in raising funds needed for the building of the new university; and when reform of medical education became imperative in the 1880's, Turner's minority report to the Royal Commission of 1881 (the Medical Acts Commission) was ultimately used as the basis of the Medical Act of 1886, an enactment that still largely governs medical education in Britain. Curiously, for one with so enlightened an outlook, Turner was consistently against the education of women alongside men in medicine, although he would have agreed to the separate establishment of medical colleges for women.

At the same time as these multifarious duties, he conducted full scientific and teaching activities. Arthur Keith wrote "on the thread of his life [are] strung all the beads of British anatomy for half a century and more." From his school at Edinburgh came graduates occupying no fewer than thirty-six chairs of anatomy, from Glasgow to Calcutta.

Turner inaugurated the *Journal of Anatomy and Physiology* in 1867 and was its editor for many years. In 1887 he and his friend and mentor George Murray Humphry, of Cambridge, founded the Anatomical Society of Great Britain and Ireland. These were important events for the teaching and influence of British anatomy.

From 1854 onward Turner published many papers, taking anthropological and comparative anatomy as his main interests. His publications list contains 276 titles. Edinburgh anatomists had long taken a wide view of their subject, based upon comparative studies. Turner broadened this basis and British anthropology attained a new eminence in world science.

Turner used his knowledge of craniology to support Huxley's defense of evolutionary theory, and his correspondence with Darwin (on rudimentary vestigial organs) is quoted by his son Logan Turner. Many of his observations were included in *The Descent of Man*. He was critical of Dubois's work on *Pithecanthropus erectus*, which he did not see as a new species intermediate between man and ape, but, in the modern light, as an early evolutionary form.

This active, conscientious, able, and amiable man was another example of anatomist-administrator. He placed British anthropology on a new footing, and his teaching in the Edinburgh anatomy department greatly influenced other schools. But his most lasting contribution probably lay in the field of medical education, in which he carried his colleagues and the government with him in opposing a "one portal" system of entry into medicine. He insisted on the right of universities to grant their own medical degrees, while supporting the idea of joint qualifying boards outside the universities; for example, the Conjoint Board of the two royal colleges and the triple diploma of the Glasgow colleges. His concept still forms the basis of British qualification in medicine.

BIBLIOGRAPHY

I. ORIGINAL WORKS. Turner's 276 papers are in his *List of Published Writings 1854–1910* (1915), of which copies are in the possession of his grandson and in the library of Edinburgh University. The subjects in this list include anatomy and physiology, comparative anatomy and zoology, pathological anatomy and anthropology, together with his many presidential addresses.

Among the titles may be noted Joseph Lister and William Turner, "Observations on the Structures of Nerve Fibres," in *Quarterly Journal of Microscopical Science*, **8** (1860), 29–34, also published separately (London, 1859); it includes a neat drawing by Lister himself. His long series of classical papers on the comparative anatomy of the placenta is exemplified by "Placentation in the Cetacea," in *Transactions of the Royal Society of Edinburgh*, **26** (1871), 467–504.

The 1875 edition of the *Encyclopaedia Britannica* contains Turner's article on "Anatomy," the historical section of which is described by Garrison as the best monograph on the subject in English.

Anthropology is represented by *The Comparative Osteology of the Races of Man*, constituting parts XXIX and XLVII of the zoological series of the *Scientific Results of the Voyage of H.M.S. Challenger* (Edinburgh, 1884–1886). His many contributions to craniology include those on the "People of Scotland" and the "People of India," in *Transactions of the Royal Society of Edinburgh, passim*.

Turner's collection of Cetacea made over fifty years is described in *The Marine Mammals in the Anatomical Museum of the University of Edinburgh* (Edinburgh, 1913).

A paper on "M. Dubois' Description. . . . of Pithecanthropus Erectus" is in *Journal of Anatomy and Physiology*, **29** (1895), 424–445.

II. SECONDARY LITERATURE. The main source is the biography by his son A. Logan Turner, *Sir William Turner, A Chapter in Medical History* (London, 1919). Many journals published obituary notices, the fullest account being in *British Medical Journal* (1916), **1**, 326–331. Others may be found in *Lancet* (1916), **1**, 484–486; *Nature*, **96** (1916), 79; and *The Student. Turner Memorial Number* (11 July 1916), published by the Students' Council, University of Edinburgh.

K. BRYN THOMAS

TURNER, WILLIAM ERNEST STEPHEN (*b.* Wednesbury, Staffordshire, England, 22 September 1881; *d.* Sheffield, England, 27 October 1963), *glass technology.*

Turner was the eldest son and the second of seven children of William George Turner and Emma Blanche Turner. His working-class parents sacrificed so that he could gain an education. Turner progressed from a Smethwick board school to King Edward VI Grammar School, Birmingham, and then in 1898 to Mason University College (which became Birmingham University in 1900). He graduated with a bachelor of science in 1902 and earned a master of science in 1904. Turner's first post was at Sheffield University under W. P. Wynne, where he early showed his characteristic capacity for organization. He lectured on physical chemistry for metallurgical students. His earlier experimental research was in conventional physical chemistry (for example, solubility and molecular weights in solution).

During this part of his career Turner continually urged the employment of scientists in industry and the establishment of a closer liaison between universities and industry. On the outbreak of war in 1914, he successfully advocated the formation of a Sheffield University technical advisory committee to consider problems raised by the cutting off of supplies from Germany. Although the initial problems lay in the field of metallurgy, glass soon became an issue. Turner drew up a report on the glass industry of Yorkshire, in which he dealt with the poor practical methods then in use, the paucity of technical literature on glass, and the need for teaching and research. He recommended that Sheffield become a center for instruction in glass manufacture. It was a triumph for Turner that, owing to his foresight, a Department of Glass Technology was created in the middle of the war.

At this time the government was promoting research associations (supported by both industry and government), but Turner resisted this type of organization and retained, through an appointed committee known as the Glass Delegacy, a high degree of independence for his department from the university, government, and industry while also securing the support of each. A separate organization, the Glass Research Association, foundered in 1925, and Turner's Department of Glass Technology held unchallenged world leadership in glass research for a generation. The research was directed mainly toward industrial problems, his view being that it was "wiser to tackle the immediate problems first and then let the need for the long dated, fundamental problems grow out of the imperative need for more information or for sounder basic principles" (Mellor Memorial Lecture, 1957). Typical problems had to do with the composition of raw materials for lead crystal, the resistance of chemical glass to reagents, and the design of furnaces. Fundamental studies emerged, for example, on the variation of the physical properties of glass over a range of compositions, and the effect of small quantities of minor additives on these properties. Demands of other industries also raised problems, including that of the design of glass-to-metal seals in electrical construction.

World War II limited the freedom that Turner and his department had hitherto enjoyed. But the department continued to give valuable technical service, and even to increase its student enrollment. The Department of Glass Technology later reverted to a more orthodox form of administration comparable to that found in other university departments.

In 1916 Turner established the Society of Glass Technology, and a *Journal* began appearing in 1917. He was the editor until it ceased publication in 1959. By that time the form and content of the journal seemed no longer to meet the needs of a science and industry that he himself had done so much to change. Turner promoted international exchanges and was president of an International Commission for Glass Technology from 1933 to 1953.

During his retirement Turner encouraged the growing application of science to the archaeology of glass. He was an inveterate traveler, and although physically handicapped by the results of childhood poliomyelitis, he was a vigorous walker, even in the Alps.

Turner was married twice, first to Mary Isobel Marshall (died 1939), who bore him four children, and then in 1943 to Annie Helen Nairn, an artist in glass, who designed the presidential badge of the Society of Glass Technology. He was an officer of the Order of the British Empire (1918), fellow of the Royal Society (1938), and the holder of many foreign honors.

BIBLIOGRAPHY

For a list of Turner's writings, see *Biographical Memoirs of Fellows of the Royal Society*, X (1964), 325–356. This work lists twenty papers from the period 1905–1914, mainly on physical chemistry; some 240 research papers from 1914–1954, on glass technology; some eighty lectures and addresses on general industrial problems; and some thirty papers on glass archaeology.

An obituary notice is in *Glass Technology*, 4 (Dec. 1963), 165–169, with two portraits.

Personal reminiscences are in W. E. S. Turner, "The Department of Glass Technology and Its Work Since 1915," in *Journal of the Society of Glass Technology*,

21 (1937), transactions 5–43; R. W. Douglas, "W. E. S. Turner–Applied Scientist," in *Glass Technology*, **8** (February 1969), 19–28.

FRANK GREENAWAY

TURPIN, PIERRE JEAN FRANÇOIS (*b.* Vire, France, 11 March 1775; *d.* Paris, France, 1 May 1840), *botany.*

Turpin was the son of an impoverished artisan. He studied drawing at the École des Beaux-Arts in Vire, then, in 1780, enlisted as a soldier in the Calvados battalion. In 1794 he was sent to Haiti, where he met Alexandre Poiteau, a gardener at the Paris Muséum d'Histoire Naturelle, who taught him botany. Turpin and Poiteau collaborated in a study of Haitian flora; they collected an herbarium of some 1,200 plants, of which Turpin made drawings of a large number, and of which they together described about 800 species. They took this material to France, but Turpin soon returned to make a further exploration of Hispaniola and of the island of Tortuga, which lies off its northwest coast. In 1800 he made a trip to the United States, where he met Humboldt, but returned again to Haiti to serve as an army pharmacist in the campaign against Toussaint-L'Ouverture that was being conducted by General Leclerc. In 1802 Turpin settled in France to devote himself to botany and botanical illustration.

As a botanical artist Turpin achieved a fame equal to that of Redouté. He collaborated on a number of the most important botanical publications of the early nineteenth century, including Humboldt's *Plantae aequinoctiales . . . in ordem digessit Amatus Bonpland*, Benjamin Delessert's *Icones selectae plantarum pr. part.*, and J. L. M. Poiret's *Leçons de flore*, to which he contributed fifty-seven plates. He also made a number of drawings for the less distinguished *Flore du dictionnaire des sciences médicales* of F. P. Chaumeton, Chambéret, and Poiret. He himself composed a flora of Paris, then collaborated on another with Poiteau, who had also published a previous work on the same subject. Only the first eight parts of their joint work was printed, as *Flore parisienne contenant la description des plantes qui croissent naturellement aux environs de Paris* (1808–1813). Poiteau and Turpin also collaborated on a new, admirably illustrated edition of Duhamel du Monceau's *Traité des arbres fruitiers* of 1768; this work, which was important for distinguishing botanical species from the races or varieties known to gardeners, had lacked good drawings. In the new recension, published in 1808–1835, the work became one of the most beautiful books on fruit trees ever published.

Turpin's own botanical research reflected the broad scientific concerns of his time. A systematist by temperament, he believed in the great chain of being and in the continuity of forms and organs. He thus sought an archetypal model to explain the constitution of plants, and was particularly impressed by Goethe's notion of the leaf as the archetypal organ of the plant. (Charles Gaudichaud-Beaupré had proposed that this fundamental organ be called the "phyton," because it was formed by both the leaf–the phyllome–and the base that forms part of the stem, the phyllopodium.)

Turpin defended the idea of organ types in a number of works, including "Mémoire sur l'inflorescence des Graminées et des Cypéracées" (1819), "Organographie végétale" (1827), *Mémoire sur l'organisation intérieure et extérieure des tubercules du Solanum tuberosum* (1828), "Mémoire de nosologie végétale" (1833), and "Observations générales sur l'organogénie et la physiologie des végétaux" (1835). His *Examen d'une chloranthie ou monstruosité observée sur l'inflorescence du saule marceau* of 1833 is a related teratological study. In 1837 Turpin presented to the Académie des Sciences a drawing, executed in 1804, of the plant type that he had first conceived in Haiti. He intended this to illustrate the unity of organic composition and the original identity of all the foliaceous and lateral appendicular organs of the plant; the engraving was one of several published in C.-F. Martins's edition of the *Oeuvres d'histoire naturelle de Goethe* (1837). Goethe himself had, shortly before his death, asked Turpin to illustrate the theory that he had presented in his *Versuch die Metamorphose der Pflanzen zur erklaeren* of 1790.

Turpin also participated in the elaboration of the cell theory. His writings on the subject, published in 1820, were largely influenced by Sprengel's idea that the "utricle" (cell) contained vesicles and granules, including the chlorophyll granules that played an active part in cellular development. From Sprengel, too, Turpin derived the notion of the plant as an aggregate of independent, completely individualized cells. Turpin did further research on the lower plants, and made a number of contributions to the systematics of freshwater algae. In addition, he was one of the first to confirm the conclusion of C. Cagniard de la Tour and T. Schwann that yeast is a living organism that reproduces by budding.

BIBLIOGRAPHY

Turpin's writings include *Flore parisienne contenant la description des plantes qui croissent naturellement aux environs de Paris*, 8 pts. (Paris, 1808–1813), written with A. Poiteau; "Mémoire sur l'inflorescence des Graminées et des Cyperacées," in *Memoires du Muséum d'histoire naturelle*, **5** (1819), 426–492; "Organographie végétale," *ibid.*, **14** (1827); *Mémoire sur l'organisation intérieure et extérieure des tubercules du Solanum tuberosum* (Paris, 1828); *Examen d'une chloranthie ou monstruosité observée sur l'inflorescence du saule marceau* (Paris, 1833); "Mémoire de nosologie végétale," in *Mémoires présentés par divers savants*, **6** (1835), 217–240; and "Observations générales sur l'organogénie et la physiologie des végétaux," in *Mémoires de l'Académie des sciences*, 2nd ser., **14** (1835). He also collaborated with Poiteau on a new ed. of Duhamel du Monceau's *Traité des arbres fruitiers*, 6 vols. (Paris, 1808–1835).

In addition, Turpin contributed to Humboldt's *Plantae aequinoctiales . . . in ordinem digessit Amatus Bonpland*, 2 vols. (Paris, 1805–1818); F. P. Chaumeton, Chambéret, and J. L. M. Poiret's *Flore . . . médicale*, 8 vols. (Paris, 1814–1820); Poiret's *Leçons de flore*, 3 vols. (Paris, 1819–1820); and Benjamin Delessert's *Icones selectae plantarum pr. part.* (Paris, 1820–1823).

Twenty-five artistic items by Turpin are now in the Lindley Library.

M. HOCQUETTE

TURQUET DE MAYERNE, THEODORE (*b.* Mayerne, near Geneva, Switzerland, 28 September 1573; *d.* London, England, 15/16 March 1655), *medicine, chemistry.*

Turquet de Mayerne was the son of the noted Huguenot historian and political theorist Louis Turquet de Mayerne. Following early schooling at Geneva, he went to the University of Heidelberg and thence to Montpellier, where he graduated M.D. in 1597. In his subsequent career, that of an eminently successful court physician in both France and England, he displayed a remarkable ability to survive professional and political upheaval. Although not a prominent scientific figure in his own right, he was influential in the introduction and support of chemical therapeutics in medicine.

After graduating at Montpellier, Turquet went to Paris, where he became the protégé of Jean Ribit, first physician to Henry IV and a fellow Calvinist. He became a royal physician, and as Ribit's disciple he built up a successful practice that included many notables, particularly but not exclusively among the Huguenots. On Ribit's death in 1605, Turquet inherited their joint clientele. Ribit and

Turquet both endorsed the use of chemical remedies in their practice, and fostered the training of apothecaries in the preparation of these new medicaments. They probably were instrumental in establishing Jean Beguin's chemistry courses in Paris.

This advocacy of chemical therapeutics aroused the hostility of the Paris Medical Faculty, and Turquet became personally embroiled in the bitter polemics that ensued between the Faculty and proponents of Paracelsian therapy. In 1603 yet another Calvinist royal physician, Joseph Duchesne (Quercetanus), wrote a treatise defending Paracelsian-Hermetic medicine, which promptly elicited an anonymous and vituperative reply from the Faculty (most probably written by the elder Jean Riolan). In response to this Faculty-sponsored attack, Turquet published in the same year (1603) a moderate defense of chemical therapeutics arguing that the new remedies did not contravene the principles of medicine as set down by Hippocrates and Galen. This publication, however, was sufficient to bring him the official censure of the Faculty on 5 December 1603. As a privileged royal physician, Turquet was able to continue in practice, despite the fulminations of the Faculty.

Following the assassination of Henry IV, Turquet moved permanently to England in 1611; he had visited there in 1606, when he was incorporated M.D. at the University of Oxford. He came as first physician to James I and later served Charles I and his queen in a similar capacity. His professional career in England was a dazzling success and made him very rich. Turquet was elected a fellow of the Royal College of Physicians in 1616; he bought the *seigneurie* of Aubonne (near Lausanne) in 1621; and he was knighted by James in 1624. Following the outbreak of the English Civil War, he retired to Chelsea, where he died in 1655; he was buried in St. Martin-in-the-Fields. Only one daughter of his seven children by two marriages survived him.

In England, Turquet continued his interest in chemical therapeutics and the training of apothecaries. In association with Henry Atkins, several times president of the Royal College of Physicians, he helped establish the Worshipful Society of Apothecaries, which gave English pharmacists distinct corporate status and distinguished them from the grocers. He also served on the Royal College's committee that produced the first edition of the *Pharmacopoeia Londinensis* (1618), intended to provide the first standardized English formulary. It generally has been assumed that he was influential in the inclusion of chemical remedies in this text. It should be noted, however, that when the first specific proposals for the *Pharmacopoeia* were drawn up in 1589, such remedies were included. Also, Turquet did not join the College's committee on the *Pharmacopoeia* until shortly before its publication. He wrote the dedicatory epistle to James I, which should not be confused with the preface to the reader in which chemical remedies are defended. Turquet developed some chemical prescriptions of his own, including the popular *lotio nigra*, the main ingredient of which was mercuric oxide. He also experimented with pigments and included among his friends the artists Rubens, Van Dyke, Peter Lely, and Jean Petitot. It was also through his efforts that the manuscript of Thomas Moffett's *Theatrum insectorum* was published in 1634.

BIBLIOGRAPHY

I. ORIGINAL WORKS. Turquet de Mayerne published little in his lifetime. An early travel book attributed to him is entitled *Sommaire description de la France, Allemagne, Italie et Espagne* (Geneva, 1591; 1653). His defense of chemical remedies is *Apologia in qua videre est, inviolatis Hippocratis et Galeni legibus, remedia chymice preparata tuto usurpari posse, ad cujusdam anonymi calumnias responsio* (La Rochelle, 1603). Turquet did, however, keep extensive personal and clinical records throughout his life, the bulk of which became the property of Sir Hans Sloane and are now in the Sloane collection of the British Museum (for description see Edward J. L. Scott, *Index to the Sloane Manuscripts in the British Museum* [London, 1904], 349–350).

Turquet's MSS formed the basis of several posthumously published collections of his clinical case histories. The distinction of his patients gives these added historical interest. The first such collection was published, along with a treatise on gout, as *Tractatus de arthritide. Accesserunt ejusdem consilia aliquot medicinalia* (Geneva, 1674), translated into English by Thomas Sherley, 2 vols. (London, 1676–1677). A more comprehensive and systematic collection was published with a preface by Walter Charlton as *Praxeos Mayernianae in morbis internis* (London, 1690, 1695; Augsburg, 1691; Geneva, 1692 [as *Praxis medica*]). The last and most complete collection was edited by Joseph Browne as *Theodori Turquet Mayernii . . . opera medica, complectentia consilia, epistolas, et observationes, pharmacopeam, variasque medicamentorum formulas* (London, 1700, 1701, 1703).

II. SECONDARY LITERATURE. Accounts of Turquet's life and writings include the following, listed chronologically: Norman Moore, "Mayerne, Sir Theodore Turquet de," in *Dictionary of National Biography*; Thomas Gib-

son, "A Sketch of the Career of Theodore Turquet de Mayerne," in *Annals of Medical History*, n.s. 5 (1933), 315–326; "An Account of Dr. Theodore Turquet de Mayerne's 'Praxis Medica.' Augsburg 1691," *ibid.*, 438–443; and "Letters of Dr. Theodore Turquet de Mayerne to the Syndics and Executive Council of the Republic of Geneva," *ibid.*, n.s. 9 (1937), 401–421; and William B. Ober, "Sir Theodore Turquet de Mayerne, M.D., F.R.C.P. (1573–1655); Stuart Physician and Observer," in *New York State Journal of Medicine*, 70 (1970), 449–458. Most of these sources contain factual inaccuracies and must be used with caution.

For Turquet's association with Jean Ribit, see Hugh Trevor-Roper, "The Sieur de la Rivière, Paracelsian Physician of Henri IV," in *Science, Medicine and Society in the Renaissance, Essays to Honor Walter Pagel*, Allen G. Debus, ed., II (New York, 1972), 227–250. The Paris dispute over chemical therapy is discussed in W. P. D. Wightman, *Science and the Renaissance*, I (Edinburgh–London–New York, 1962), 256–263.

For Turquet's relationship with the English apothecaries, see C. Wall, H. Charles Cameron, and E. Ashworth Underwood, *A History of the Worshipful Society of Apothecaries of London*, I (London, 1963), *passim*. His role in the publication of the *London Pharmacopoeia* is discussed by George Urdang in the historical intro. to the facs. repro. of the 1618 ed. of *Pharmacopoeia Londinensis* (Madison, Wis., 1944); also see Urdang's "How Chemicals Entered the Official Pharmacopoeias," in *Archives internationales d'histoire des sciences*, n.s. 7 (1954), 303–314; and Allen G. Debus, *The English Paracelsians* (New York, 1966), 150–156. These should be assessed, however, in the light of Sir George Clark, *A History of the Royal College of Physicians of London*, I (Oxford, 1964), 227–230.

OWEN HANNAWAY

AL-ṬŪSĪ, MUḤAMMAD IBN MUḤAMMAD IBN AL-ḤASAN, usually known as **NAṢĪR AL-DĪN** (*b.* Ṭūs, Persia, 18 February 1201; *d.* Kadhimain, near Baghdad, 26 June 1274), *astronomy, mathematics, mineralogy, logic, philosophy, ethics, theology.*

Life. Naṣīr al-Dīn, known to his compatriots as Muḥaqqiq-i Ṭūsī, Khwāja-yi Ṭūsī, or Khwāja Naṣīr, is one of the best-known and most influential figures in Islamic intellectual history. He studied the religious sciences and elements of the "intellectual sciences" with his father, a jurisprudent of the Twelve Imām school of Shīʿism at Ṭūs. He also very likely studied logic, natural philosophy, and metaphysics with his maternal uncle in the same city. During this period he also received instruction in algebra and geometry. Afterward he set out for Nīshāpūr, then still a major center of learning, to complete his formal advanced education; and it was in this city that he gained a reputation as an outstanding scholar. His most famous teachers were Farīd al-Dīn al-Dāmād, who through four intermediaries was linked to Ibn Sīnā and his school and with whom Ṭūsī studied philosophy; Quṭb al-Dīn al-Maṣrī, who was himself the best-known student of Fakhr al-Dīn al-Rāzī (1148– 1209), with whom al-Ṭūsī studied medicine, concentrating mostly on the text of Ibn Sīnā's *Canon*; and Kamāl al-Dīn ibn Yūnus (1156–1242), with whom he studied mostly mathematics.

This period was one of the most tumultuous in Islamic history: Mongols were advancing toward Khurasan from Central Asia. Therefore, although already a famous scholar, al-Ṭūsī could not find a suitable position and the tranquillity necessary for a scholarly life. The only islands of peace at this time in Khurasan were the Ismāʿīlī forts and mountain strongholds, and he was invited to avail himself of their security by the Ismāʿīlī ruler, Naṣīr al-Dīn Muḥtashim. Al-Ṭūsī accepted the invitation and went to Quhistan, where he was received with great honor and was held in high esteem at the Ismāʿīlī court, although most likely he was not free to leave had he wanted to. The date of his entrance into the service of the Ismāʿīlī rulers is not known exactly but was certainly sometime before 1232, for it was during that year that he wrote his famous *Akhlāq-i nāṣirī* for the Ismāʿīlī ruler. During his stay at the various Ismāʿīlī strongholds, including Alamut, al-Ṭūsī wrote a number of his important ethical, logical, philosophical, and mathematical works, including *Asās al-iqtibās* (on logic) and *Risāla-yi muʿīniyya* (on astronomy). His fame as a scholar reached as far as China.

Hūlāgū ended the rule of the Ismāʿīlīs in northern Persia in 1256. His interest in astrology, and therefore his respect for astronomers, combined with al-Ṭūsī's fame in this field, made Hūlāgū especially respectful toward him after he had captured Alamut and "freed" al-Ṭūsī from the fort. Henceforth al-Ṭūsī remained in the service of Hūlāgū as his scientific adviser and was given charge of religious endowments (*awqāf*) and religious affairs. He accompanied Hūlāgū on the expedition that led to the conquest of Baghdad in 1258 and later visited the Shīʿite centers of Iraq, such as Ḥilla.

Having gained the full confidence of Hūlāgū, and benefiting from his interest in astrology, al-Ṭūsī was able to gain his approval to construct a major observatory at Marāgha. Construction began in 1259, and the Īlkhānī astronomical tables were completed in 1272 under Abāqā, after the death of

Hūlāgū. In 1274, while at Baghdad, al-Ṭūsī fell ill and died a month later. He was buried near the mausoleum of the seventh Shīʿite imām, Mūsā al-Kāẓim, a few miles from Baghdad.

Works. Nearly 150 treatises and letters by Naṣīr al-Dīn al-Ṭūsī are known, of which twenty-five are in Persian and the rest in Arabic. There is even a treatise on geomancy that al-Ṭūsī wrote in Arabic, Persian, and Turkish, demonstrating his mastery of all three languages. It is said that he also knew Greek. His writings concern nearly every branch of the Islamic sciences, from astronomy to philosophy and from the occult sciences to theology. Of the two, Ibn Sīnā was the better physician and al-Ṭūsī the greater mathematician and more competent writer in Persian. But otherwise their breadth of knowledge and influence can be compared very favorably. Moreover, the writings of al-Ṭūsī are distinguished by the fact that so many became authoritative works in the Islamic world.

Al-Ṭūsī composed five works in logic, of which *Asās al-iqtibās* ("Foundations of Inference"), written in Persian, is the most important. In fact, it is one of the most extensive of its kind ever written, surpassed only by the section on logic of Ibn Sīnā's *al-Shifāʾ*. In mathematics al-Ṭūsī composed a series of recensions (*taḥrīr*) upon the works of Autolycus, Aristarchus, Euclid, Apollonius, Archimedes, Hypsicles, Theodosius, Menelaus, and Ptolemy. The texts studied by students of mathematics between Euclid's *Elements* and Ptolemy's *Almagest* were known as the "intermediate works" (*mutawassiṭāt*); and the collection of al-Ṭūsī's works concerning this "intermediate" body of texts became standard in the teaching of mathematics, along with his recensions of Euclid and Ptolemy. He also wrote many original treatises on arithmetic, geometry, and trigonometry, of which the most important are *Jawāmiʿ al-ḥisāb biʾl-takht waʾl turāb* ("The Comprehensive Work on Computation with Board and Dust"), *al-Risāla al-shāfiya* ("The Satisfying Treatise"), and *Kashf al-qināʿ fī asrār shakl al-qiṭāʿ*, known as the *Book of the Principle of Transversal*, which was translated into Latin and influenced Regiomontanus. The best-known of al-Ṭūsī's numerous astronomical works is *Zīj-i īlkhānī* ("The Īlkhānī Tables"), written in Persian and later translated into Arabic and also partially into Latin, by John Greaves, as *Astronomia quaedam ex traditione Shah Cholgii Persae una cum hypothesibus planetarum* (London, 1650). Other major astronomical works are *Tadhkirah* ("Treasury of Astronomy") and his treatises on particular astronomical subjects, such as that on the astrolabe. He also translated the *Ṣuwar al-kawākib* ("Figures of the Fixed Stars" of ʿAbd al-Raḥmān al-Ṣūfī from Arabic into Persian. In the other sciences al-Ṭūsī produced many works, of which *Tanksūkh-nāma* ("The Book of Precious Materials") is particularly noteworthy. He also wrote on astrology.

In philosophy, ethics, and theology al-Ṭūsī composed a commentary on *al-Ishārāt waʾl-tanbīhāt* ("The Book of Directives and Remarks") of Ibn Sīnā; the *Akhlāq-i nāṣirī* (*Nasirean Ethics*), the best-known ethical work in the Persian language, and the *Tajrīd* ("Catharsis"), the main source book of Shiʿite theology, upon which over 400 commentaries and glosses have been composed. Al-Ṭūsī wrote outstanding expositions of Ismāʿīlī doctrine, chief among them the *Taṣawwurāt* ("Notions"), and composed mystical treatises, such as *Awṣāf al-ashrāf* ("Qualifications of the Noble").

Al-Ṭūsī also composed lucid and delicate poetry, mostly in Persian.

Scientific Achievements. In logic al-Ṭūsī followed the teachings of Ibn Sīnā but took a new step in studying the relation between logic and mathematics. He also elucidated the conditional conjunctive (*iqtirānī*) syllogism better than his predecessor. He converted logical terms into mathematical signs and clarified the mathematical signs employed by Abuʾl-Barakāt in his *Kitāb al-muʿtabar* ("The Esteemed Book"). Al-Ṭūsī also distinguished between the meaning of "substance" in the philosophical sense and its use as a scientific term, and clarified the relation of the categories with respect to metaphysics and logic.

In mathematics al-Ṭūsī's contributions were mainly in arithmetic, geometry, and trigonometry. He continued the work of al-Khayyāmī in extending the meaning of number to include irrationals. In his *Shakl al-qiṭāʿ* he showed the commutative property of multiplication between pairs of ratios (which are real numbers) and stated that every ratio is a number. *Jawāmiʿ al-ḥisāb*, which marks an important stage in the development of the Indian numerals, contains a reference to Pascal's triangle and the earliest extant method of extracting fourth and higher roots of numbers. In collaboration with his colleagues at Marāgha, al-Ṭūsī also began to develop computational mathematics, which was pursued later by al-Kāshī and other mathematicians of the Tīmūrid period.

In geometry al-Ṭūsī also followed the work of al-Khayyāmī and in his *al-Risāla al-shāfiya* he examined Euclid's fifth postulate. His attempt to prove it through Euclidean geometry was unsuc-

cessful. He demonstrated that in the quadrilateral *ABCD*, in which *AB* and *DC* are equal and both perpendicular to *BC*, and the angles *A* and *D* are equal, if angles *A* and *D* are acute, the sum of the angles of a triangle will be less than 180°.[1] This is characteristic of the geometry of Lobachevski and shows that al-Ṭūsī, like al-Khayyāmī, had demonstrated some of the properties of the then unknown non-Euclidean geometry. The quadrilateral associated with Saccheri was employed centuries before him by Thābit ibn Qurra, al-Ṭūsī, and al-Khayyāmī.

Probably al-Ṭūsī's most outstanding contribution to mathematics was in trigonometry. In *Shakl al-qiṭāʿ*, which follows the earlier work of Abuʾl-Wafāʾ, Manṣūr ibn ʿIrāq, and al-Bīrūnī, al-Ṭūsī for the first time, as far as modern research has been able to show, developed trigonometry without using Menelaus' theorem or astronomy. This work is really the first in history on trigonometry as an independent branch of pure mathematics and the first in which all six cases for a right-angled spherical triangle are set forth. If c = the hypotenuse of a spherical triangle, then:

$$\cos c = \cos a \cos b \qquad \cot A = \tan b \cot c$$
$$\cos c = \cot A \cot B \qquad \sin b = \sin c \sin B$$
$$\cos A = \cos a \sin B \qquad \sin b = \tan a \cot A.$$

He also presents the theorem of sines:

$$\frac{a}{\sin A} = \frac{b}{\sin B} = \frac{c}{\sin C}.$$

It is described clearly for the first time in this book, a landmark in the history of mathematics.

Al-Ṭūsī is best-known as an astronomer. With Hūlāgū's support he gained the necessary financial assistance and supervised the construction of the first observatory in the modern sense. Its financial support, based upon endowment funds; its lifespan, which exceeded that of its founder; its use as a center of instruction in science and philosophy; and the collaboration of many scientists in its activities mark this observatory as a major scientific institution in the history of science. The observatory was staffed by Quṭb al-Dīn al-Shīrāzī, Muḥyī ʾl-Dīn al-Maghribī, Fakhr al-Dīn al-Marāghī, Muʾayyad al-Dīn al-ʿUrḍī, ʿAlī ibn ʿUmar al-Qazwīnī, Najm al-Dīn Dabīrān al-Kātibī al-Qazwīnī, Athīr al-Dīn al-Abharī, al-Ṭūsī's sons Aṣīl al-Dīn and Ṣadr al-Dīn, the Chinese scholar Fao Mun-ji, and the librarian Kamāl al-Dīn al-Aykī. It had excellent instruments made by Muʾayyad al-Dīn al-ʿUrḍī in 1261–1262, including a giant mural quadrant, an armillary sphere with five rings and an alidade, a solstitial armill, an azimuth ring with two quadrants, and a parallactic ruler. It was also equipped with a fine library with books on all the sciences. Twelve years of observation and calculation led to the completion of the *Zīj-i īlkhānī* in 1271, to which Muḥyī 'l-Dīn al-Maghribī later wrote a supplement. The work of the observatory was not confined to astronomy, however; it played a major role in the revival of all the sciences and philosophy.

Al-Ṭūsī's contributions to astronomy, besides the *Zīj* and the recension of the *Almagest*, consist of a criticism of Ptolemaic astronomy in his *Tadhkira*, which is perhaps the most thorough exposition of the shortcomings of Ptolemaic astronomy in medieval times, and the proposal of a new theory of planetary motion. The only new mathematical model to appear in medieval astronomy, this theory influenced not only Quṭb al-Dīn al-Shīrāzī and Ibn al-Shāṭir but also most likely Copernicus, who followed closely the planetary models of Naṣīr al-Dīn's students. In chapter 13 of the second treatise of the *Tadhkira*, al-Ṭūsī proves that "if one circle rolls inside the periphery of a stationary circle, the radius of the first being half the second, then any point on the first describes a straight line, a diameter of the second."[2] E. S. Kennedy, who first discovered this late medieval planetary theory issuing from Marāgha, interprets it as "a linkage of two equal length vectors, the second rotating with constant velocity twice that of the first and in a direction opposite the first."[3] He has called this the "Ṭūsī-couple" and has demonstrated (see Figures 1 and 2) its application by al-Ṭūsī, Quṭb al-Dīn, and Ibn al-Shāṭir to planetary motion and its comparison with the Ptolemaic model.[4]

This innovation, which originated with al-Ṭūsī, is without doubt the most important departure from Ptolemaic astronomy before modern times. Except for the heliocentric thesis, the "novelty" of Copernicus' astronomy is already found in the works of al-Ṭūsī and his followers, which probably reached Copernicus through Byzantine intermediaries.

The most important mineralogical work by al-Ṭūsī is *Tanksūkh-nāma*, written in Persian and based on many of the earlier Muslim sources, such as the works of Jābir ibn Ḥayyān, al-Kindī, Muḥammad ibn Zakariyyā', al-Rāzī, 'Uṭārid ibn Muḥammad, and especially al-Bīrūnī, whose *Kitāb al-jamāhir fī maʿrifat al-jawāhir* ("The Book of Multitudes Concerning the Knowledge of Precious Stones") is the main source of al-Ṭūsī's work. In fact the *Tanksūkh-nāma*, which derives its name

from the Turco-Mongolian word meaning "something precious," probably is second in importance in the annals of Muslim mineralogy only to al-Bīrūnī's masterpiece.

Al-Ṭūsī's work comprises four chapters. In the first he discusses the nature of compounds; the four elements, their mixture, and the coming into being of a "fifth quality" called temperament (*mizāj*), which can accept the forms of different species; and the role of vapors and the rays of the sun in their formation, in all of this following closely the theories of Ibn Sīnā's *De mineralibus*. An interesting section is devoted to colors, which al-Ṭūsī believes result from the mixture of white and black. In jewels, colors are due to the mixture of earthy and watery elements contained in the substance of the jewel.

The second chapter is devoted exclusively to jewels, their qualities, and their properties. Special attention is paid to rubies, the medical and occult properties of which are discussed extensively. In the third chapter al-Ṭūsī turns to metals and gives an alchemical theory of metallic formation, calling sulfur the father and mercury the mother of metals. He also enumerates the seven traditional metals, including *khārṣīnī*. Like so many Muslim philosopher-scientists, al-Ṭūsī accepts the cosmological and mineralogical theories of alchemy concerning the formation of metals without belonging to the alchemical tradition or even discussing the transmutation of base metal into gold. A section on perfumes ends the book, which is one of the major sources of Muslim mineralogy and is valuable as a source of Persian scientific vocabulary in this field.

Of all the major fields of science, al-Ṭūsī was least interested in medicine, which he nevertheless studied, generally following the teachings of Ibn Sīnā. He also composed a few works on medicine including *Qawānīn al-ṭibb* ("Principles of Medicine") and a commentary on Ibn Sīnā's *Canon*, and exchanged letters with various medical authorities on such subjects as breathing and temperament. He expressed certain differences of opinion with Ibn Sīnā concerning the temperament of each organ of the body but otherwise followed his teachings. Al-Ṭūsī's view of medicine was mainly philosophical; and perhaps his greatest contribution was in psychosomatic medicine, which he discusses, among other places, in his ethical writings, especially *Akhlāq-i nāṣirī* (*Nasirean Ethics*).

Al-Ṭūsī was one of the foremost philosophers of Islam, reviving the Peripatetic (*mashshā'ī*) teachings of Ibn Sīnā after they had been eclipsed for

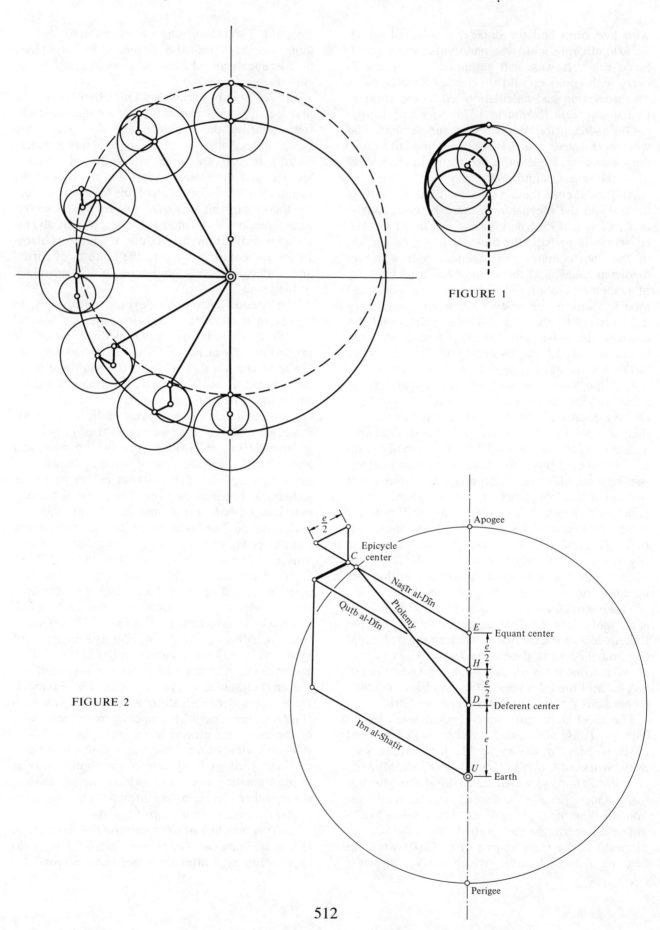

FIGURE 1

FIGURE 2

$\frac{e}{2}$

Epicycle
center

Naṣīr al-Dīn

Qutb al-Dīn

Ptolemy

Ibn al-Shāṭir

Apogee

E — Equant center

$\frac{e}{2}$

H

$\frac{e}{2}$

D — Deferent center

e

U — Earth

Perigee

nearly two centuries by *Kalām*. He wrote a masterful commentary on the *Ishārāt waʾl-tanbīhāt* of Ibn Sīnā, which Fakhr al-Dīn al-Rāzī had attacked severely during the previous century. In this work, which is unusual among Muslim philosophical works for its almost mathematical precision, al-Ṭūsī succeeded in rekindling the light of philosophy in Islam. But while claiming in this work to be a mere follower of Ibn Sīnā, in several places questions of God's knowledge of particulars, the nature of space, and the createdness of the physical world clearly shows his debt to Shihāb al-Dīn al-Suhrawardī and some of the Muslim theologians. Al-Ṭūsī in fact marks the first stage in the gradual synthesis of the Peripatetic and Illuminationist (*ishrāqī*) schools, a tendency that became clearer in the writings of his foremost student, Quṭb al-Dīn al-Shīrāzī. He also wrote many philosophical treatises in Persian, so that his prose in this field must be considered, along with the writings of Nāṣir-i Khusraw, Suhrawardī, and Afḍal al-Dīn al-Kāshānī, as the most important in the Persian language.

In ethics al-Ṭūsī composed two major works, both in Persian: the *Akhlāq-i muḥtashimī* ("The Muḥtashimī Ethics") and the much better-known *Nasirean Ethics*, his most famous opus. Based upon the *Tahdhīb al-akhlāq* ("The Refinement of Character") of Muskūya (Miskawayh), the *Nasirean Ethics* expounds a philosophical system combining Islamic teachings with the ethical theories of the Aristotelian and, to a certain extent, the Platonic traditions. The work also contains an elaborate discussion of psychology and psychic healing. For centuries it has been the most popular ethical work among the Muslims of India and Persia.

In Twelve Imām Shiʿism, al-Ṭūsī is considered as much a theologian as a scientist and philosopher because of his *Tajrīd*, which is still central to Shiʿite theological education. A work of great intellectual rigor, the *Tajrīd* represents the first systematic treatment of Shiʿite *Kalām* and is therefore the foundation of systematic theology for the Twelve Imām Shiʿites. In the history of Islam, which is known for its multitalented figures of genius, it is not possible to find another person who was at once an outstanding astronomer and mathematician and the most authoritative theologian of a major branch of Islam.

Influence. Al-Ṭūsī's influence, especially in eastern Islam, was immense. Probably, if we take all fields into account, he was more responsible for the revival of the Islamic sciences than any other individual. His bringing together so many competent scholars and scientists at Marāgha resulted not only in the revival of mathematics and astronomy but also in the renewal of Islamic philosophy and even theology. Al-Ṭūsī's works were for centuries authoritative in many fields of Islamic learning; and his students, such as Quṭb al-Dīn and ʿAllāma Ḥillī, became outstanding scholars and scientists. His astronomical activities influenced the observatories at Samarkand and Istanbul and in the West to a much greater extent than was thought to be the case until recently; and his mathematical studies affected all later Islamic mathematics. In fact, the work of al-Ṭūsī and his collaborators at Marāgha moved eastward to influence Chinese science, which, as a result of the Mongol invasion, had a much closer relationship with Islam. The school of al-Ṭūsī also influenced later Indian science as cultivated under the Moguls and even as late as the eighteenth century, as can be seen in the observatory constructed by Jai Singh II, which indirectly reflects the observatory of Marāgha.

In the West al-Ṭūsī is known almost entirely as an astronomer and mathematician whose significance, at least in these fields, is becoming increasingly evident. In the Muslim East he has always been considered as a foremost example of the "wise man" (*ḥakīm*), one who, while possessing an acute analytical mind, which he devoted to mathematical, astronomical, and logical studies, extended the horizon of his thought to embrace philosophy and theology and even journeyed beyond the limited horizon of all mental activity to seek ultimate knowledge in the ecstasy provided by gnosis (*ʿirfān*) and Sufism.

NOTES

1. E. S. Kennedy, "The Exact Sciences in Iran Under the Seljuqs and Mongols," 664.
2. E. S. Kennedy, "Late Medieval Planetary Theory," 369.
3. *Ibid.*
4. *Ibid.*, 369, 367.

BIBLIOGRAPHY

Al-Ṭūsī's major published work is *The Nasirean Ethics*, translated by G. M. Wickens (London, 1964).

Secondary literature includes A. Carathéodory Pasha, *Traité de quadrilatère* (Constantinople, 1891); B. Carra de Vaux, "Les sphères célestes selon Nasīr-Eddīn Attūsī," in P. Tannery, ed., *Recherches sur l'histoire de l'astronomie ancienne* (Paris, 1893), app. 4, 337–361; A. P. Youschkevitch, and B. A. Rosenfeld, *Die Mathematik der Lander des Ostens in Mittelalter* (Berlin, 1960), 277–288, 304–308; E. S. Kennedy, "Late Medieval Planetary Theory," in *Isis*, **57** (1966), 365–378; and "The Exact Sciences in Iran Under the Seljuqs and Mongols," in *Cambridge History of Iran*, V (Cambridge, 1968), 659–679; M. Mudarris Raḍawī, *Aḥwal wa āthār-i ustād bashar . . . Khwāja Naṣīr al-Dīn* (Teheran, A.H. 1334, 1955 A.D.); S. H. Nasr, *Three Muslim Sages* (Cambridge, Mass., 1964); and *Science and Civilization in Islam* (Cambridge, Mass., 1968; New York, 1970); G. Sarton, *Introduction to the History of Science*, II, pt. 2 (Baltimore, 1931), 1001–1013; A. Sayili, *The Observatory in Islam* (Ankara, 1960); B. H. Siddiqui, "Naṣīr al-Dīn Ṭūsī," in M. M. Sharif, ed., *A History of Muslim Philosophy*, I (Wiesbaden, 1963), 564–580; A. S. Saidan, "The Comprehensive Work on Computation With Board and Dust by Naṣīr al-Dīn al-Ṭūsī," in *Al-abḥāth*, **20**, no. 2 (June 1967), 91–163, and no. 3 (Sept. 1967), 213–293, in Arabic; and *Yādnāmāyi Khwāja Naṣīr al-Dīn Ṭūsī*, I (Teheran, A.H. 1336, 1957 A.D.), in Persian.

SEYYED HOSSEIN NASR

AL-ṬŪSĪ, SHARAF AL-DĪN AL-MUẒAFFAR IBN MUḤAMMAD IBN AL-MUẒAFFAR (*b.* Ṭūs [?], Iran; *d.* Iran, *ca.* 1213/1214), *astronomy, mathematics.*

The name of Sharaf al-Dīn's birthplace, Ṭūs, refers both to a city and to its surrounding region, which with Mashhad and Nīshāpur formed a very prosperous area in the twelfth century.[1] A century earlier, Ṭūs had given Islam one of its most profound thinkers, al-Ghazālī (*d.* 1111); and it was soon to produce a great astronomer and theologian, Naṣīr al-Dīn (*d.* 1274). Nothing is known about the first years of al-Ṭūsī's life; but it is reported that, faithful to the tradition of medieval scholars, he went on a long journey to some of the major cities of the time. His itinerary can be reconstructed from undated information preserved in biographies of his contemporaries.

Al-Ṭūsī taught at Damascus, probably about 1165.[2] His most distinguished student there was Abu'l-Faḍl (*b. ca.* 1135), an excellent carpenter who helped make the wood paneling of the Bīmāristān al-Nūrī (1154–1159) before discovering the joys of Euclid and Ptolemy.[3] Al-Ṭūsī most probably then stayed at Aleppo, where one of his pupils was a respected member of the city's Jewish community, Abu'l-Faḍl Binyāmīn (*d.* 1207/1208), whom he instructed in the science of numbers, the use of astronomical tables, and astrology, and, at a less advanced level, in the other rational sciences.[4] From the nature of these courses, it is reasonable to suppose that they lasted about three years.

Al-Ṭūsī's most outstanding pupil, however, was Kamāl al-Dīn Ibn Yūnus (*d.* 1243) of Mosul, through whom al-Ṭūsī's teachings passed to Naṣīr al-Dīn and Athīr al-Dīn al-Abharī (*d.* 1263/1265).[5] Al-Ṭūsī was apparently in Mosul in the years preceding 1175,[6] for around this date two physicians from Damascus went there to study with him, but he had already left.[7] One of them then went to the neighboring city of Irbil, where he became a pupil of Ibn al-Dahhān.[8] About this time, however, the latter left Irbil to join Saladin, who had just seized Damascus (1174).[9] Al-Ṭūsī returned to Iran, where he died around 1213, at an advanced age.

Al-Ṭūsī is known for his linear astrolabe (al-Ṭūsī's staff), a simple wooden rod with graduated markings but without sights. It was furnished with a plumb line and a double cord for making angular measurements and bore a perforated pointer. This staff reproduced, in concrete form, the meridian line of the plane astrolabe—that is, the line upon which the engraved markings of that instrument are projected. (These markings are of stars, circles of declination, and heights.) Supplementary scales indicate the right ascensions of the sun at its entry into the signs of the zodiac as well as the hourly shadows. Al-Ṭūsī described the construction and use of the linear astrolabe in several treatises, praising its simplicity and claiming that an amateur could build it in about an hour. His staff made it possible to carry out the observations used to determine the height of the stars, the time, the direction of the Kaʿba, and the ascendants. The instrument, although inexpensive to construct, was less accurate than the ordinary astrolabe. It also was less decorative, and perhaps for this reason it was of little interest to collectors. In any case, not a single linear astrolabe has survived.[10]

Al-Ṭūsī's greatest achievement is recorded in a work that has not yet been analyzed by historians, the manuscript Loth III, 767, in the collection of the India Office, London. This manuscript is actually a reworking of the original by an unknown author who proudly states that he has eliminated the mathematical tables and shortened some of the long passages. He makes no further claims; and even if he had wished to make more substantial changes, the great difficulty of the work would have discouraged him. The entire contents of the work may, therefore, confidently be attributed to al-Ṭūsī. The treatise, which may have been mentioned by al-Sinjārī,[11] is not the first of its kind by an Arab author. A cross check of citations from Jamshīd al-Kāshī and Tāsh Kopru Zādeh reveals that al-Masʿūdī, a disciple of al-Khayyāmī, wrote on the numerical solution of third-degree equations.[12] The existence of an earlier author is not explicitly indicated, but, about 1350, Yaḥyā al-Kāshī noted several similar writings, without specifying dates or names.[13] In the following paragraphs we shall present the most remarkable results in al-Ṭūsī's treatise, but we cannot state the degree of originality for each.

The treatise divides the twenty-five equations of degree $n \leq 3$ into three groups. The first includes twelve equations: those of degree $n \leq 2$ or that reduce to that degree, plus the equation $x^3 = a$. The second contains the eight equations of the third degree that always admit one (positive) solution.[14] The third group is composed of the five equations that can give rise to impossible solutions:[15]

$$x^3 + c = ax^2$$

$$x^3 + c = b^2 x$$

$$x^3 + 3ax^2 + c = 3b^2 x$$

$$x^3 + b^2 x + c = 3ax^2$$

$$x^3 + c = 3ax^2 + 3b^2 x$$

We shall not give details of the geometric solutions, since they do not differ from those presented by al-Khayyāmī. (The care that al-Ṭūsī bestows on the study of the problem of the relative position of two conics is, however, worth noting.) On the other hand, the outstanding discussion of the existence of the roots of the group of equations that can give rise to impossible solutions merits the closest examination. Accordingly, we shall outline, by way of example, al-Ṭūsī's treatment of the fourth equation of this group, which, like the others, is based on the calculation of a maximum. Given that $x^3 < 3ax^2$; therefore $x < 3a$. Then $b^2 x < x^2(3a - x)$, so that $b^2 < x(3a - x)$. The maximum of $x(3a - x)$ is $(3a/2)^2$.[16] Therefore $b < 3a/2$. We consider $x^2 + b^2/3 = 2ax$ and take its root $x_1 = a + \sqrt{a^2 - b^2/3}$. A discussion of its existence does not arise, since $b < 3a/2$. We form $f(x_1) = x_1^2(3a - x_1) - b^2 x_1$. If $f(x_1) = c$, the equation $x^3 + b^3 x + c = 3ax^2$ has a

solution $x = x_1$. If $f(x_1) < c$, there is no solution. If $f(x_1) > c$, the equation has two roots separated by x_1. Turning to an evaluation of al-Ṭūsī's treatment in the light of the differential calculus, we set $f(x) = 3ax^2 - x^3 - b^2 x$; then $f'(x) = 6ax - 3x^2 - b^2$. Thus $f'(x)$ reduces to zero when $x^2 - 2ax + b^2/3 = 0$. Accordingly, the roots x_0 and x_1 are equal to $a \pm \sqrt{a^2 - b^2/3}$. Finally, $f(x_1) > 0$ implies $b < 3a/2$.

x	0		x_0		x_1		$3a$
$f'(x)$		$-$	0	$+$	0	$-$	
$f(x)$	0	$\searrow$	$f(x_0)$	$\nearrow$	$f(x_1)$	$\searrow$	$-3ab^2$

The text does not say what led al-Ṭūsī to such profound and beautiful results. The idea of determining the maximum of $x^2(a - x)$, $x(b^2 - x^2)$, $\cdots$ might have been suggested by the solution of $x(a - x) = b^2$. The value of the maximum of $x^2(a - x)$ might have been borrowed from Archimedes, who, unlike al-Ṭūsī, established it geometrically.[17] Yet, even if al-Ṭūsī started from this point, he still had far to go. Pursuing his solution of the equation $x^3 + bx^2 + c = 3ax^2$, he shows that the two solutions are, respectively, $x_1 + X$, where X is the root of $X^3 + 3(x_1 - a)X = f(x_1) - c$, and $x_1 - X$, where X is the root of $X^3 + f(x_1) - c = 3(x_1 - a)X$. This method contains the genesis of a genuine change of variables, and one must admire the author's intention of interrelating the various equations—an approach quite different from traditional Arab thinking on this topic, which emphasized independent solutions of problems (as in the classic solution of the second-degree equations).

We shall conclude with a very schematic presentation of al-Ṭūsī's solution of the equation $x^3 + 3ax = N$, using the example $x^3 + 36x = 91,750,087$.[18] Let x_1 be the number in the hundreds' place of the root; then x_1^3 will represent millions and $3ax_1$ will represent hundreds. Therefore, we place x_1 in the millions' box (the upper line in Table I) and $a = 12$ in the hundreds' box (on the lower line; actually, since a is greater than nine, it is carried over

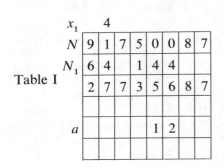

Table I

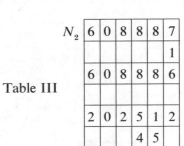

Table II

Table III

into that of the thousands). We then calculate the greatest x_1 such that $x_1^3 \leq 91$; this yields $x_1 = 4$. We remove $x_1^3 + 36x_1$ from N, obtaining $N = 27{,}735{,}687$. We next place $x_1^2 = 16$ under x_1 in the line containing a and decrease the lower line by one rank and x_1 by two. The result is Table II.

We now calculate the figure in the tens' place. It will be the greatest x_2 such that $3x_2$ multiplied by 16 can be subtracted from 277. Accordingly, $x_2 = 5$, and we place it to the right of 4 in the upper line. In the lower line we put $x_1 x_2$ in the position under $x_1 = 4$. We then subtract from N_1 the total of x_2^3 and the product of $3x_2$ times the lower line—that is, $3x_2(x_1^2 + x_1 x_2 + a) = 15(180{,}012)$. This yields N_2. We add $x_1 x_2$ (that is, 20) to the lower line in the position under $x_1 = 4$ and $x_2^2 = 25$ in the position under $x_2 = 5$. The line becomes 202,512. We decrease it by one rank and decrease the upper line by two. The result is Table III.

Finally, we calculate x_3 such that the product of $3x_3$ times $20 \leq 60$. Thus $x_3 = 1$. We place it to the right of 5. To the lower line we add 45 and subtract from N_2 the total of x_3^3 and the product of $3x_3$ times the lower line (202,962). The remainder is 0. The root of the equation is therefore 451. The method is independent of the system of numeration and permits as close an approximation of the root as desired; it suffices to add a row of three to the last remainder and to continue operating in the same manner. The treatise also gives analogous methods of numerical resolution for the other equations, even for those of the second degree.

NOTES

1. Guy Le Strange, *The Lands of the Eastern Caliphate* (Cambridge, 1909). See the chapter on Khurāsān (with references to the Arab geographers).
2. See Ibn Abī Uṣaybiʿa. *ʿUyūn al-anbāʾ*, II, 190–191.
3. This was a hospital built by Sultan Nūr al-Dīn ibn Zenki, famous for his wars against the Crusaders. See Ibn al-Athīr, *al-Tārīkh al-Bāhir fiʾl dawl ʾl-atābikiyya*, A. A. Ṭulaymāt, ed. (Cairo, 1963), 170; and Shawkat al-Shaṭṭī, *Mūjaz tārīkh al-ṭibb ʿind al-ʿArab* (Damascus, 1959), 22. See also Ibn Abī Uṣaybiʿa, *loc. cit.*
4. Ibn al-Qifṭī. *Tārikh* (Cairo, 1948), 278.
5. See Ibn Khallikān, *Wafayāt al-aʿyan*, IV, no. 718; and G. Sarton, *Introduction to the History of Science*, II, 600, and II, pt. 2, 1001–1013.
6. In 1193 Ibn Yūnus went to Baghdad to continue his religious studies; see Ibn Khallikān, *loc. cit.* See also Tāsh Kopru Zādeh, *Miftāḥ al-saʿāda*, II, 214–215.
7. They were Ibn al-Ḥājib and Muwaffaq al-Dīn. See Ibn Abī Uṣaybiʿa, II, 181–182, 191–192.
8. *Ibid.*
9. See Ibn Khallikān, IV, no. 655.
10. See Henri Michel, *Traité de l'astrolabe*, 22. The same point is also made in L. A. Mayer, *Islamic Astrolabists and Their Works* (Geneva, 1956).

11. Al-Sinjārī, *Irshād al-qāṣid* (Beirut, 1904), 124. Although probably valid, the citation raises some doubt. In fact, the title, *Kitāb al-Muẓaffar al-Ṭūsī*, becomes, in certain editions of Ṭāsh Kopru Zādeh's *Miftāḥ al-saʿāda* (for instance, I, 327) which, however, derive from al-Sinjārī: *Kitāb al-Ẓafar of al-Ṭūsī* (Naṣir al-Dīn).
12. Jamshīd al-Kāshī, *Miftāḥ al-ḥisāb*, MS Paris Ar. 5020, fol. 98; and Ṭāsh Kopru Zādeh, *Miftāḥ al-saʿāda*, I, 327. Sharaf al-Dīn Muḥammad ibn Masʿūd ibn Muḥammad al-Masʿūdī is cited in the article on Muḥammad ibn Aḥmad al-Shurwāni in Ṣafadī, *al-Wāfī*, Ritter, ed. (Istanbul), II, 497, as having taught the *Ishārāt* of Ibn Sīnā to Fakhr al-Dīn al-Rāzi (1164–1238) after having studied under al-Khayyām. He is the author of *al-Kifāya fiʾl-hidāya*; see Ḥājjī Khalīfa, *Kashf al-Ẓunūn*, II, col. 1500. Khalīfa also cites his algebra (I, col. 857).
13. Yaḥya al-Kāshī, *al-Lubāb fiʾl-Ḥisāb*, Aya Sofya MS 2757. See fol. 65r, l. 21; fol. 65v, l. 3; and fol. 67r, l. 25. The MS, written in 1373, bears notes in the author's hand. See the article on al-Kāshī in Sarton, *Introduction to the History of Science*, III, pt. 1, 698.
14. Only $x^3 + ax = bx^2 + c$ can admit up to three positive solutions.
15. *Kitāb fiʾl-jabr waʾl-muqābala*, India Office (London), Loth 767. The equations are found on pp. 101r–112r; 112r–121r; 121r–130r; 130r–142v; and 142v–179r.
16. This is an immediate consequence of Euclid's *Elements*, II, 5.
17. T. L. Heath, *The Works of Archimedes* (New York, 1953), 67–72.
18. In the treatise (fols. 54v–55v) the equation actually solved is $x^3 + 36x = 33{,}087{,}717$, the root of which is 321.

BIBLIOGRAPHY

I. Original Works. Al-Ṭūsī's works include the following:

1. *Kitāb fiʾl jabr waʾl muqābala*, India Office (London), Loth 767.

2. *Risāla fiʾl-asṭurlāb al-khaṭṭī*, British Museum, Or. 5479.

3. *Maʿrifat al-asṭurlāb al-musaṭṭaḥ waʾl-ʿamal bihi*, Leiden 1082. The MS does not bear this title, which was erroneously given to it by some bibliographers, and discusses the linear astrolabe, not the plane astrolabe. The third part, containing demonstrations, is missing from the MS.

4. *Kitāb fi maʿrifat al asṭurlāb al-musaṭṭaḥ waʾl ʿamal bihi*, Seray 3505, 2nd. If Max Krause's identification of this MS with Leiden 1082 is correct, it would be necessary to conclude that we do not have al-Ṭūsī's treatise on the plane astrolabe.

5. *Risāla fiʾl-asṭurlāb al-Khaṭṭī*, Seray 3342, 7.

6. *Risāla fiʾl-asṭurlāb al-Khaṭṭī*, Seray 3464, 1.

7. *Jawāb ʿalā suʾāl liʾamir al-umarāʾ Shams al-Dīn*, Leiden 1027; Columbia University, Smith, Or. 45, 2. This work concerns the division of a square into three trapezoids and a rectangle, with the relationships preassigned.

8. *Fiʾl-Khaṭṭayn alladhayn yaqrubān wa la yaltaqiyān*, Aya Sofya 2646, 2, 71r–v, deals with the existence of an asymptote to the (equilateral) hyperbola and contains the same demonstration as in *Kitāb fiʾl-jabr waʾl-muqābala*, (1), fols. 38r–40r.

II. SECONDARY LITERATURE. See the following:

9. Ibn Khallikān, *Wafayāt al-a'yān* (Cairo, 1948).

10. Ibn Abī Uṣaybi'a, *'Uyūn al-anbā'* (Cairo, 1882).

11. Ṭāsh Kopru Zādeh, *Miftāḥ al-sa'āda* (Hyderabad, 1910–1911).

12. Ḥājjī Khalīfa, *Kashf al-ẓunūn* (Istanbul, 1941–1943).

13. H. Suter, *Die Mathematiker und Astronomen der Araber* (Leipzig, 1900), 134 (no. 333).

14. Max Krause, "Stambuler Handschriften islamischer Mathematiker," in *Quellen und Studien zur Geschichte der Mathematik, Astronomie und Physik*, Abt. B, Studien, **3** (1936), 437–532, see 490.

15. C. Brockelmann, *Geschichte der arabischen Literatur*, I, 2nd ed. (Leiden, 1943), 472, and supp. I (Leiden, 1937), 858.

16. G. Sarton, *Introduction to the History of Science*, II, pt. 2 (Baltimore, 1950), 622–623.

17. Carlo Nallino, article on the astrolabe (*asṭurlāb*) in *Encyclopaedia of Islam*, 1st ed., I (1913); and by Willy Hartner, *ibid.*, 2nd ed., I, 722–728.

18. Henri Michel, *Traité de l'astrolabe* (Paris, 1947), 115–122; and "L'astrolabe linéaire d'al-Ṭūsī," in *Ciel et terre* (1943), nos. 3–4. A description, sketch, and note on the use of al-Ṭūsī's linear astrolabe can be found on p. 21.

19. R. Carra de Vaux, "L'astrolabe linéaire ou bâton d'al-Tousi," in *Journal asiatique*, 11th ser., **5** (1895), 464–516. This article reproduces the text of al-Ḥasan al Marrākushī with a French translation.

ADEL ANBOUBA

TUTTON, ALFRED EDWIN HOWARD (*b*. Stockport, Cheshire, England, 22 August 1864; *d*. Dallington, Sussex, England, 14 July 1938), *crystallography*.

The only child of James Tutton, a venetian blind manufacturer, Alfred Tutton left school at fourteen. He subsequently won a scholarship to the Royal College of Science in London, where in 1886 he graduated with the principal prizes for geology, physics, and chemistry. In 1889 he became lecturer in chemical analysis. In 1895 he was appointed inspector of technical schools and served successively in the Oxford, London, and Plymouth districts. Tutton retired to Cambridge in 1924 and occasionally lectured for the university. From 1895 until 1931, when he moved to Dallington, Sussex, he maintained a crystallographic laboratory in his various houses. On his final retirement in 1931 his instruments were purchased by Manchester University, which had awarded him an honorary D.Sc. in 1926.

Tutton married Margaret Loat of Cumnor Place, Oxford, on 18 June 1902. They had two sons and four daughters. Throughout his life he enjoyed excellent health, although he suffered a serious climbing accident in the Alps in 1926, and in his last years was afflicted with failing eyesight.

In 1899 Tutton was elected a fellow of the Royal Society and served as president of the Mineralogical Society of Great Britain from 1912 to 1915. His principal interests besides crystallography were climbing and glaciology; all three were united in his book *The Natural History of Ice and Snow* (1927).

Tutton's earliest researches with Thomas E. Thorpe (1890–1891) were purely chemical and concerned the lower oxides of phosphorus. During this period also he assisted Thorpe and Arthur W. Rücker with the magnetic survey of Scotland and England.

Early in his career Tutton began the program of research that was to occupy him for the next forty years: the precise goniometric and optical study of isomorphous salts. He began with the series R_2XO_4 where $R = K$, Rb, Cs, NH_4, and Tl; and $X = S$, Se. Next came the series $R_2M(XO_4)_2 \cdot 6H_2O$, with R and X the same and $M = Mg$, Zn, Fe, Ni, Co, Mn, Cu, Cd. For each series he demonstrated, using goniometric and optical techniques at the highest level of refinement, that physical properties vary regularly with the atomic properties of the substituent elements. The same general conclusion was verified later for the alkali perchlorates and for the double chromates of the alkalis. Tutton's results were published in about fifty papers between 1890 and 1929. The essentials were republished and his apparatus described in detail in *Crystalline Structure and Chemical Constitution* (1910), *Crystallography and Practical Crystal Measurement* (1911; 2nd ed., 2 vols., 1922), and *Crystalline Form and Chemical Constitution* (1926). He also wrote a general survey of crystallography in *Crystals* (1911).

Out of the highly precise crystallographic measurements in which Tutton was so expert came the interferential comparator that he applied to a comparison of the substandards of the yard with the Imperial standard yard and, following Michelson, the evaluation of the standard yard in terms of the wavelength of cadmium red.

Tutton's reputation rests on the outstanding precision of his goniometric and optical work, notable in that it was achieved while he was busily occupied with official duties. The rapid development of X-ray crystallography diverted his type of work from the mainstream of crystallographic research, but nevertheless his data, although not fundamental, added substantially to the understanding of isomorphism; and his ingeniously designed apparatus formed the basis of later instruments useful in a wide range of fields.

BIBLIOGRAPHY

I. ORIGINAL WORKS. Important works by Tutton are *Crystalline Structure and Chemical Constitution* (London, 1910); *Crystals* (London, 1911); *Crystallography and Practical Crystal Measurement* (London, 1911; 2nd ed., London, 1922); *The Natural History of Crystals* (London, 1924); *Crystalline Form and Chemical Constitution* (London, 1926); and *The Natural History of Ice and Snow, Illustrated From the Alps* (London, 1927). There are also many papers in *Proceedings of the Royal Society, Journal of the Chemical Society*, and other journals.

II. SECONDARY LITERATURE. For discussions of Tutton and his work, see J. R. Partington, *Nature*, **142** (1938), 321–322; and L. J. Spencer, *Mineralogical Magazine and Journal of the Mineralogical Society*, **25** (1939), 301–303. See also *Obituary Notices of Fellows of the Royal Society of London*, **2** (1939), 621–626.

DUNCAN MCKIE

TWENHOFEL, WILLIAM HENRY (*b.* Covington, Kentucky, 16 April 1875; *d.* Atlanta, Georgia, 4 January 1957), *geology.*

Twenhofel's parents, Ernst A. H. J. Twenhofel and Helena Steuwer Twenhofel, of German ancestry, obtained a scant livelihood from cultivating a small farm near Covington. Twenhofel became acquainted at an early age with the rigors of farm life, for he had to help earn his own way. As a result of his work on the farm, he developed the physique and self-reliance that were to serve him well in later life when he carried on his geologic fieldwork in remote parts of North America. He also developed a keen interest in the outdoors, and especially in the reaction of plants and animals to their environment. His interest in nature continued throughout his life, and some of his most important scientific work concerned the formation, erosion, and preservation of soils, the action of plants and animals in producing and modifying sediments, and the ultimate deposition of the sediments and their organic constituents to form sedimentary rocks.

In 1899 Twenhofel married his childhood sweetheart, Virgie Mae Stephens; they had three children, Lilian Helena, Helen Vivian, and William Stephens Twenhofel. Observant and ambitious, and keenly interested in understanding nature, the youthful Twenhofel taught in the local schools for six years until he could earn enough to enter National Normal University at nearby Lebanon, Kentucky, in 1902. Awarded his baccalaureate in 1904, he next taught science and mathematics in East Texas Normal College from 1904 to 1907. He entered Yale University at age thirty-two. While earning three degrees in geology, the B.A. in 1908, M.A. in 1910, and Ph.D. in 1912, he came under the influence of Charles Schuchert and Joseph Barrell. Twenhofel became a devoted follower of both men, finding in their teachings the stimulation to extend and diversify his interest in nature.

Using the college campus as a base of operations, and teaching as a means of livelihood, he began his geological career while still a graduate student at Yale by accepting an assistant professorship at the University of Kansas in 1910. In 1916 he moved on to the University of Wisconsin, where he joined Charles R. Van Hise, Charles K. Leith, Alexander N. Winchell, Warren J. Mead, and Armin K. Lobeck. These men, Twenhofel's colleagues for the next thirty years, had a great effect on his scientific career.

When Twenhofel entered Yale, Schuchert was involved in a controversy concerning the Ordovician-Silurian boundary, and had decided that the solution lay in rocks of the Maritime Provinces of Canada. Twenhofel joined him on a field reconnaissance expedition in 1908; he later chose as the subject of his doctoral thesis the critical section on Anticosti Island, Quebec. Later in his career Twenhofel published important papers and reports on the Silurian section at Arisaig, Nova Scotia (1908), the Ordovician-Silurian strata of Anticosti Island (1927), the geology of the Mingan Islands of Quebec (1938), the Mid-Paleozoic rocks of Newfoundland (1937, 1954) and the Baltic Provinces of Europe (1916), and the Silurian of Maine (1941). These important contributions, which brought him international recognition as an authority on Ordovician and Silurian stratigraphy and paleontology, demonstrated the transitional nature of the Ordovician-Silurian boundary in northeastern North America.

While carrying on his stratigraphic studies and teaching his classes at the University of Wisconsin, Twenhofel also continued his interest in sedimentation. While at the University of Kansas Twenhofel studied the remarkable Wreford and Foraker cherts and the Comanchean deposits of central Kansas. Soon after going to Wisconsin he became involved in a far-reaching controversy with Edward O. Ulrich concerning the stratigraphy of some Upper Cambrian formations of the Upper Mississippi Valley. Ulrich argued for the traditional layer-cake concept, placing the Mazomanie glauconitic beds above the Franconia formation. Twenhofel considered the two sequences essentially contemporaneous and laterally transitional, which was a novel and revolutionary interpretation for the time, but one that he and his graduate students ultimately established by their detailed fieldwork. First as member and later as chairman of the Committee on Sedimentation of the National Research Council during the three decades from 1919 to 1949, he greatly stimulated research in sedimentation, along with T. Wayland Vaughan, Edward M. Kindle, and Arthur C. Trowbridge. His work as author of the monumental *Treatise on Sedimentation* (1926; 2nd ed., 1932) was internationally recognized.

When Raymond C. Moore launched the *Journal of Sedimentary Petrology* in 1931, Twenhofel became one of the associate editors. He served as editor from 1933 to 1946, guiding the struggling publication through difficult times to an established position among geologic journals. Twenhofel's discussions of black shales stimulated renewed interest in those controversial rocks; his papers on ancient coral reefs and related subjects contributed to increased emphasis on paleoecology; and his early concerns over soil erosion and conservation were ahead of their time. Studies that he and his students made of the sediments of Wisconsin lakes greatly stimulated limnological research, and his reports on the black sands of the Oregon beaches called attention to those deposits of winnowed sands as possible sources of certain valuable minerals, as magnetite and ilmenite.

More than anyone else in his time, Twenhofel led and promoted the study of sedimentation as a branch of geology, and the present great strength and importance of the subject are largely the result of his leadership as investigator, teacher, author, and editor.

BIBLIOGRAPHY

I. ORIGINAL WORKS. Twenhofel's scientific writings cover a broad range of geologic subjects and reveal an unusual ability to organize and interpret large amounts of descriptive matter. He wrote more than seventy-five important articles and reports, and five widely used textbooks, all on some aspect of sedimentation, stratigraphy, and paleontology. His writings are listed in the Bibliographies of North American Geology, published as *Bulletins of the United States Geological Survey*. His five major works are *Treatise on Sedimentation* (Baltimore, 1926; 2nd ed., 1932); *Invertebrate Paleontology*, with R. R. Shrock (New York, 1935); *Principles of Sedimentation* (New York, 1939); *Methods of Study of Sediments*, with S. A. Tyler (New York, 1941); and R. R. Shrock and W. H. Twenhofel, *Principles of Invertebrate Paleontology* (New York, 1953).

Major reports dealing with both stratigraphy and paleontology include "The Geology and Invertebrate Paleontology of the Comanchean and 'Dakota' Formations of Kansas," *Kansas State Geological Survey*, bulletin 9 (1924); *et al.*, "Geology of Anticosti Island," *Canadian Geological Survey*, memoir 54 (1927); "The Building of Kentucky," *Kentucky Geological Survey*, ser. 6, **37** (1931); *et al.*, "Geology and Paleontology of Mingan Islands, Quebec . . . ," *Geological Society of America*, special paper 11 (1938); "Soil, the Most Valuable Mineral Resource; Its Origin, Destruction, and Preservation," *Oregon Department of Geology and Mineral Industries Bulletin*, 26 (1944).

II. SECONDARY LITERATURE. Additional biographical and bibliographical data are included in R. R. Shrock, "William Henry Twenhofel—Honorary Member," in *Bulletin of the American Association of Petroleum Geologists*, **31** (1947), 835–840; "Memorial to William Henry Twenhofel (1875–1957)," in *Journal of Sedimentary Petrology*, **27** (1957), 203; "William Henry Twenhofel (1875–1957)," in *Bulletin of the American Association of Petroleum Geologists*, **41** (1957), 978–980; C. O. Dunbar, "Memorial to William Henry Twenhofel (1875–1957)," in *Proceedings. Geological Society of America*; annual report for 1960 (1962), 151–156.

ROBERT R. SHROCK

TWORT, FREDERICK WILLIAM (*b*. Camberley, London, England, 22 October 1877; *d*. Camberley, 20 March 1950), *microbiology*.

Twort was the eldest of the eleven children of William Henry Twort, a medical practitioner in Camberley. He attended St. Thomas' Hospital Medical School in London, where he qualified in 1900 as member of the Royal College of Surgeons of England and licentiate of the Royal College of Physicians of London. He carried out no clinical practice and in 1901 was appointed assistant to Louis Jenner, the superintendent of the Clinical Laboratory of St. Thomas' Hospital. In the following year (1902), Twort entered the field of microbiology, in which discipline he remained for the rest of his professional career, as assistant to William Bulloch, bacteriologist to the London Hospital. He stayed there for seven years, during which time he gained wide experience in hospital bacteriology while beginning his own investigations. By 1907 the latter had become of primary importance to him, and he obtained the position of superintendent of the Brown Animal Sanatory Institution, a veterinary dispensary in London where Charles Sherrington and Victor Horsley had worked. Here Twort isolated himself from his colleagues in bacteriology and engaged only in research. Being by nature a recluse and an outstanding example of the independent research worker, he remained in this post for thirty-five years, except for a period of military service during World War I. He was appointed professor of bacteriology in the University of London in 1919 and was elected a fellow of the Royal Society in 1929.

Twort possessed great industry and powers of concentration as well as unusual ability; his remarkable technical skills were of the utmost advantage in his work. His aloofness and his inability to commit himself in his scientific writings militated against dissemination of his undoubtedly outstanding and original contributions to bacteriology. He was a kind and gentle person with a simple, uncomplicated nature; but he was also shy, naïve, slightly paranoid, and without much humor. Nevertheless, he could be aggressive and obstinate in defense of a principle, as for example the financial support of research. Biographers vaguely mention controversies and a "combative personality" (*Lancet*, 1950), but do not elaborate. He has also been styled "an erratic genius" (*British Medical Journal*, 1950).

Twort married Dorothy Nony Banister, daughter of F. J. Banister, an architect, and she helped him in his work. They had three daughters and a son who also entered the medical profession.

Twort's work in bacteriology was based on the premise that pathogenic bacteria must have evolved from wild, free-living organisms, an original theme at a time when bacteriologists were studying pathogens *per se* almost exclusively. Among several of his original contributions the most important was the discovery of the lytic phenomena now known to be caused by bacteriophage. Twort was working on the purification of vaccinia virus in an attempt to discover nonpathogenic ultramicroscopic viruses when he encountered a substance that could dissolve the bacterium *Staphylococcus aureus* in culture. He isolated it by ultrafiltration, and was able to transmit it indefinitely to further generations of micrococci. It is now known that this substance is a virus that attacks bacteria. Although Twort made this suggestion, he also advanced other suggestions, and, as usual, did not commit himself to any one. His classic paper on ultramicroscopic viruses, published in

Lancet in 1915, was in the form of a preliminary communication, but the expected sequel appeared during World War I and excited little interest. Moreover, in 1915 Twort joined the army, and on his return to his investigations in 1919 he did not continue what eventually became a most fruitful field of research. In 1917 Félix d'Hérelle discovered the same phenomenon and named it bacteriophage. The attempts he made to establish his own priority only served to reaffirm that of Twort.

In his other researches Twort also made original contributions that led eventually to fundamental advances in bacteriology. His first important paper, published in 1907, established the essentials of the adaptation and mutation of bacteria, showing, for example, that a nonpathogenic microorganism could become pathogenic. Once again his work gained little recognition at the time. It was the same with his paper published in 1909 on the factors that affect the growth of bacteria; the significance of these factors was not recognized until twenty years later. Twort's studies of Johne's bacillus, a mycobacterium responsible for a chronic intestinal disease of cattle, carried out with G. L. Y. Ingram in 1912, revealed an essential growth factor that, when present, allowed the organism to be cultured and a protective vaccine to be prepared. It opened up the whole field of the nutritional needs of bacteria, a field that could only be developed as biochemistry advanced. During World War I Twort investigated dysentery in the Middle East and discovered special forms of the causative bacillus; but his speculations on them were vague (1920) and the only result was his demonstration that these bizarre forms could be induced.

After World War I, Twort's researches became increasingly insignificant as he struggled to prove his thesis that pathogens derive from wild ancestors, that bacteria evolve from viruses, and that viruses come from even more primitive forms of life. The outcome of all these studies seems of little importance; they were terminated abruptly in 1944 by the destruction of his laboratory and equipment by enemy action.

BIBLIOGRAPHY

I. ORIGINAL WORKS. Twort's classic paper, on what is now termed bacteriophage, is "An Investigation of the Nature of Ultra-Microscopic Viruses," in *Lancet* (1915), 2, 1241–1243. It is reprinted in full in N. Hayon, ed., *Selected Papers on Virology* (Englewood Cliffs, N.J., 1964), 97–102, and extracts appear in H. A. Lechevalier and M. Solotrorovsky, *Three Centuries of Microbiology* (New York, 1955), 303–306.

A list of Twort's papers, containing twenty-two items, is in P. Fildes' obituary (see below), 517. Those of importance are "The Fermentation of Glucosides by Bacteria of the Typhoid-coli Group and the Acquisition of New Fermenting Powers by *Bacillus dysenteriae* and other Micro-organisms. Preliminary Communication," in *Proceedings of the Royal Society*, series B, **79** (1907), 329–336; "The Influence of Glucosides on the Growth of Acid-fast Bacteria," *ibid.*, **81** (1909), 248; "A Method for Isolating and Cultivating the *Mycobacterium enteritidis chronicae pseudo-tuberculosae bovis*, Johne, . . .," with G. L. Y. Ingram, *ibid.*, **84** (1912), 517–542; and "Researches on Dysentery," in *British Journal of Experimental Pathology*, **1** (1920), 237–243.

II. SECONDARY LITERATURE. The only extended discussion of Twort and his work is by P. Fildes, *Obituary Notices of Fellows of the Royal Society*, **7** (1951), 505–517. Obituaries are in *British Medical Journal* (1950), **1**, 788–789, with a portrait; *Lancet* (1950), **1**, 648–649; and *Nature*, **165** (1950), 874. There is a portrait of Twort when working with W. Bulloch in C. E. Dolman, "Paul Ehrlich and William Bulloch: a Correspondence and Friendship (1896–1914)," in *Clio Medica*, **3** (1968), 65–84, fig. 9.

EDWIN CLARKE

TYNDALL, JOHN (*b.* Leighlinbridge, County Carlow, Ireland, 2 August 1820; *d.* Hindhead, Surrey, England, 4 December 1893), *natural philosophy, microbiology, popularization of science.*

Tyndall was the son of John Tyndall, an ardent Orangeman who was at different times a small landowner, shoemaker, leather dealer, and member of the Irish constabulary. Educated until he was nineteen at the national school in Carlow, Tyndall gained a vision of science, self-instruction, and moral duty shaped by his private reading of Carlyle, Emerson, and Fichte. Tyndall began work as a draftsman and civil engineer in the Irish Ordnance Survey but in 1842 was transferred to the English survey at Preston, Lancashire. In England for the first time, Tyndall witnessed economic depression and civil strife. Carlyle's *Past and Present* moved him deeply and, in 1843, focused his opposition to the oppressive policies and incompetent management of the ordnance survey. His protests rejected, Tyndall was dismissed and returned briefly to Ireland. He then found work in Lancashire and Yorkshire as a surveyor and engineer during the railway mania of 1844–1845. In 1847, once again unemployed, he was befriended by George Edmondson, a Quaker from Preston. Edmondson had recently begun at Queenwood College, Hampshire,

one of the first schools in England to have a laboratory for the teaching of science. At Queenwood, where Tyndall taught mathematics and drawing, he was joined by his friend the geometer Thomas Archer Hirst and the chemist Edward Frankland. Under Frankland's influence, Tyndall was introduced to German science. In 1848, Tyndall left Queenwood with Frankland to study at the University of Marburg. There, living on his savings, Tyndall completed a mathematical dissertation for the doctorate under Friedrich Stegmann, and then entered the laboratory of Karl Herrmann Knoblauch, recently arrived from Berlin, who was at that time extending the work of Faraday and Plücker on diamagnetism. Tyndall's first scientific research was undertaken in collaboration with Knoblauch; his first article, on the behavior of crystalline bodies between the poles of a magnet, was published in the *Philosophical Magazine* in 1851.

From 1851 Tyndall's life in both science and public affairs followed clear lines of development — as researcher, educator, popularizer, and controversial public figure. Like other men of science of his generation Tyndall had great difficulty in obtaining paid work in science. Jobless in England for two years, rejected from two posts in Ireland and (with his close friend T. H. Huxley) by the universities of Sydney and Toronto, Tyndall was obliged, like Huxley, to write, lecture, and examine. By the mid-1850's, Tyndall's prospects improved. In 1851 Huxley was elected a fellow of the Royal Society; Tyndall, aided by the patronage of Faraday, followed in 1852. In the following year, with a growing reputation, Tyndall became professor of natural philosophy at the Royal Institution, where, under Faraday's guidance, Tyndall developed his natural talents for lecturing and research. In 1867 he succeeded Faraday as superintendent of the Royal Institution and as adviser to Trinity House and the Board of Trade. From 1867 to 1885 his position at the Royal Institution gave him a central vantage point in British science.

Tyndall's research can be arbitrarily described in two phases: the first, between 1853 and 1874, witnessed a steady progression within physics, while the second, between 1874 and the early 1880's, saw the amplification of his work in other domains. In the first phase, his work on diamagnetism, involving the effects of compression on hundreds of crystalline substances (1851–1856), led to the study of Penrhyn slate and the problem of "slaty cleavage" (1854–1856). Generalizing from the effects of pressure on slate led him to the study of glacial movement (1856–1859); in turn, glaciers fostered a passion for mountaineering and a fascination for what was to become his major work—the effects of solar and, later, heat radiation on atmospheric gases (1860–1870). He then considered the scattering of light particles in the atmosphere (the "Tyndall effect") and explained the blue color of the sky ("Rayleigh scattering"). The scattering of sunlight by dust particles (much evident in the dust-laden air of Albemarle Street) led him to consider means of destroying airborne organic matter by heat; this in turn kindled his interest in the case against spontaneous generation (1870–1876) and brought him to the defense of Pasteur. This formidable capacity to move from electromagnetism through thermodynamics and into bacteriology was the hallmark of Tyndall's genius. No less formidable were his talents in describing, with charm and lucidity, the phenomena of physics to large audiences.

The enormous range of Tyndall's inquiries reflected many different intellectual influences. An explicit discussion of these influences in their context has yet to be undertaken. We lack any comprehensive review of Tyndall's work on optical and crystalline structure; on magnetism, radiation, and mountaineering; and on the relationship between his several research programs. Tyndall is remembered chiefly for his efforts to verify the high absorptive and radiative power of aqueous vapor; to measure the absorption and transmission of heat by many different gases and liquids; to explain the selective influence of the atmosphere on different sounds; and to establish the principle of "discontinuous heating" ("Tyndallization") as a sterilizing technique. Practical applications of his work in meteorology, fog signaling, and bacteriology were seen within his lifetime. In other ways his work anticipated important later developments. His explanations of diamagnetic phenomena and mechanical pressure—and their relation to molecular forces—could not be confirmed in the absence of a comprehensive theory of atomic structure. Yet his early research, extending the association of diamagnetisim with induced polarity, still relates to problems and techniques in high-pressure research in solid state and applied physics. In other fields his work has worn less well. His explanations of glacial movement by fracture and regelation were not conclusive; the conjectures of his adversary James D. Forbes on the viscosity and plasticity of glacial behavior, for example, were subsequently supported by applications of thermodynamic principles to continuous deformation under stress, and as a result have since widely prevailed.

Public demands on Tyndall's time were enormous, with the Royal Institution absorbing most of his energies. In his thirty-three years there he delivered over fifty Friday discourses, over 300 afternoon lectures, and twelve Christmas courses for young people. In addition he served as examiner for the Royal Military College (1855–1857), professor of physics at the Royal School of Mines (1859–1868), and lecturer at Eton (1856) and the London Institution (1856–1859). Much of his additional lecturing and examining were undertaken to supplement his salary at the Royal Institution, and many of his textbook commissions were accepted for the same reason.

Tyndall's influence upon what *Nature* called the scientific movement was direct and profound. He occupied a unique place in the popular exposition of science. In 1859 he joined with Huxley in writing a regular column for the *Saturday Review*. In 1863–1867 he acted as scientific adviser to *The Reader*, and in 1869 he helped inaugurate the journal *Nature*. In 1871, with Spencer and Huxley, he advised Edward Livingston Youmans on the *International Scientific Series* (to which he contributed Volume I [1872]). Tyndall's prolonged debates with publishers and his evidence to the Royal Commission on Copyright (1878) reveal the difficulties of earning a living as a scientific author. Tyndall contended with this partly by republishing his popular essays and lectures. For example, the American edition of *Fragments of Science*, which appeared in 1871, was sold out on the day of publication; and his *Forms of Water* (1872) went through twelve English editions by 1897.

Among his fellow members of the famous "X Club" (founded 1864) and his scientific contemporaries in the Metaphysical Club (founded 1869), Tyndall became an evangelist for scientific naturalism and the public support of research. He contributed "Science Lectures for the People" (begun in 1862) and gave evidence to the Select Committee on Scientific Instruction (1867–1868). In 1866–1867 he contributed to the British Association committee on the teaching of science; and in 1868 he became president of Section A, and 1874, president of the British Association at Belfast. By 1872, *Vanity Fair* spoke glowingly of his energy, imagination, and rhetoric, "at all times to be envied, and at nearly all times to be admired." This was the spirit in which his American tour (1872–1873) was conducted. The conclusion to his *Lectures on Light* and his bequest for fellowships at Harvard, Columbia, and Pennsylvania resonate with his hopes for the encouragement of scientific research.

With his flair for public debate, Tyndall earned the sobriquet "Xccentric" from the X Club. What Oliver Lodge called Tyndall's "wholesome rightness" came to the defense of many whom he believed ill used. Thus he advanced Monseigneur Rendu's claims against those of J. D. Forbes on the movement of glaciers; and he defended J. D. Hooker against intervention by A. S. Ayrton's Office of Works. His sincerity had the defects of its virtues; conviction bred defensiveness, even obstinacy, in celebrated debates with Forbes (1857–1867), C. A. Akin (1862–1863), P. G. Tait and William Thomson (1873–1874), Henry C. Bastian (1870–1873), and, notoriously, with John Ruskin (1874).

But these intellectual skirmishes were overshadowed by the battle that followed the "Belfast Address," Tyndall's presidential address to the British Association in 1874. Tyndall's quixotic conflicts with religious authority—notably about prayer and miracles—were intense and sustained. His devotion to experiment and verification, and his determination to find the truth, to reject metaphysics, and to reveal the ultimate mechanism of natural phenomena, had impelled his search for "agents of explanation" which would unify the physical relations of heat, magnetism, electricity and sound, and even the "ultra-scientific region" of the mind. This program was hardly new in 1874. But the explicit confrontation between materialism and revealed religion, provoked by the archdemocrat of science before the "parliament of science," left deep scars. Caustically satirized in William Hurrell Mallock's *New Republic* in 1877, Tyndall became to many more villain than hero.

If the year 1874 was a climacteric in Tyndall's public reputation, 1876 was a watershed in his private life. In that year, when he was fifty-six, he married Louisa Charlotte Hamilton, then aged thirty-one, the eldest daughter of Lord Claud Hamilton. The late 1870's and 1880's, however, were years of persistent illness, requiring frequent recuperative trips to his favorite retreat at Bel Alp, above the Rhone Valley. During the early 1880's, he continued to serve as "scientific adviser" to government and undertook fresh responsibilities with the Royal Commission on Accidents in Coal Mines (1879–1886). But by 1884 his relations with government were strained by a violent dispute with Joseph Chamberlain over lighthouse policy, which led to his resignation as scientific adviser to Trinity House and the Board of Trade. Politically, Tyndall always considered himself "in some sense Liberal, in some sense, Radical." As an Orange-

man, he admired Parnell and denounced the "Romish hierarchy of the National League." In 1885, rejecting Gladstone's policies for home rule in Ireland, and outraged by the government's failure in the Sudan, he broke finally with Liberalism. He was even moved to consider standing as a Unionist candidate for a Glasgow constituency in the election of May 1885.

In 1886 Tyndall fell seriously ill, and the following year he retired from the Royal Institution and withdrew to his house at Hindhead, near Haslemere, Surrey. Bedridden by insomnia and indigestion, in 1893 he died from an accidental overdose of chloral, tragically administered by his devoted wife, who survived him by forty-seven years.

Although he received five honorary doctorates and was an honorary member of thirty-five scientific societies, Tyndall was never offered national honors.

Tyndall's contemporaries did not view his work uncritically. Perhaps the least complimentary review of his work appeared in the tenth edition of the *Encyclopaedia Britannica* (1902), in an article by Oliver Lodge. Lodge claimed Tyndall's knowledge was "picturesque and vivid" rather than "thorough and exact"; that Tyndall never popularized anything especially recondite, yet "never hesitated to elaborate the simple"; that his research lacked originality and definition, so that his superficial understanding of physical issues promoted unnecessary disputes. There are difficulties with Lodge's interpretation, especially in relation to Tyndall's work on radiant heat and spontaneous generation. After protests from Tyndall's colleagues, Lodge removed the more inflammatory passages in the eleventh edition of the *Britannica*. Subsequently, little has appeared to qualify the received impression of a sensitive observer, a skillful experimentalist, a dedicated "field physicist" and Alpinist, an inspired communicator, and a "shining beacon to struggling self-taught youth," who, in *Nature*'s words, brought "democracy into touch with scientific research."

BIBLIOGRAPHY

I. ORIGINAL WORKS. Tyndall published more than 180 experimental papers, of which the Royal Society Catalogue lists more than 140, and more than sixty scientific lectures, addresses, and reviews, in addition to a considerable number of popular essays on literature, religion, mountaineering, and travel, many of which appeared in series and embodied material that he repeated in different forms and in different languages.

The most important of Tyndall's essays and lectures

are reproduced in several books, all published during his lifetime. These and other major books are *The Glaciers of the Alps* (London, 1860); *Heat Considered as a Mode of Motion* (London, 1863); *On Sound* (London, 1867; 4th ed., 1883), a later edition of which is *The Science of Sound* (New York, 1964); *Researches on Diamagnetism and Magne-Crystallic Action* (London, 1870); *Hours of Exercise in the Alps* (London, 1871); *Fragments of Science for Unscientific People* (London, 1871); *Contributions to Molecular Physics in the Domain of Radiation Heat* (London, 1872); *The Forms of Water in Clouds, Rivers, Ice, and Glaciers* (London, 1872; 12th ed., 1897); *Six Lectures on Light, Delivered in America, 1872–1873* (London, 1873; 5th ed., 1895); *The Floating Matter of the Air in Relation to Putrefaction and Infection* (London, 1881); *New Fragments* (London, 1892).

An outline of Tyndall's writings would be incomplete without a notice of the many contributions he made to the *Liverpool Mercury* under the name "Spectator," and to the *Carlow Sentinel* and *Preston Chronicle* under the name Wat Ripon between 1843 and 1849.

By far the greatest archive of Tyndall's papers exists at the Royal Institution of Great Britain. This collection, together with his published writings and much of his correspondence, has been catalogued, and the catalogue is available in microfiche with an accompanying printed introduction by J. Friday, R. MacLeod, and P. Shepherd, *John Tyndall, Natural Philosopher, 1820–1893* (London, 1974).

II. SECONDARY LITERATURE. The most accessible sources of biographical material on Tyndall are in *Nature*, **49** (1894), 128; *Dictionary of National Biography*, XIX, 1358–1363; and Oliver Joseph Lodge, in *Encyclopaedia Britannica*, 10th ed., XXXIII (1902), 517–521; and 11th ed., XXVII (1910–1911), 499–500, in which passages offending Mrs. Tyndall were removed.

Tributes to Tyndall are by Herbert Spencer, "The Late Professor Tyndall," in *Fortnightly Review*, **55** (1894), 141–148; T. H. Huxley, "Professor Tyndall," in *Nineteenth Century*, **35** (1894), 1–11; and Edward Frankland, "John Tyndall, 1820–1893," in *Proceedings of the Royal Society*, **55** (1894), xviii–xxxiv.

Mrs. Tyndall hoped, but failed, to complete a life of her husband. Owing to the circumstances of Tyndall's death, this biography was repeatedly delayed. Following Louisa Tyndall's death, a biography was completed by A. S. Eve and C. H. Creasey, *Life and Work of John Tyndall* (London, 1945).

There are several vignettes of Tyndall including D. Thompson, "John Tyndall (1820–1893), A Study in Vocational Enterprise," in *Vocational Aspects of Secondary and Further Education*, **9** (1957), 38–48; and J. G. Crowther, "John Tyndall," in *Scientific Types* (London, 1968).

Many informal details of Tyndall's social and scientific life are revealed in the diaries of Thomas Archer Hirst, perhaps Tyndall's closest friend, which repose in the Royal Institution, and which have been edited by W. H. Brock and R. MacLeod.

There have been few sustained assessments of Tyndall, the exceptions being Lord Rayleigh, "The Scientific Work of Tyndall," in *Royal Institution Library of Science*, **4** (1894), 273–281; and W. Bragg, "Tyndall's Experiments on Magne-Crystallic Action," *ibid.*, **9** (1927), 131–154. Aspects of his work have also been treated by James Bryant Conant, "Pasteur's and Tyndall's Study of Spontaneous Generation," in *Harvard Case Histories in Experimental Science*, case 7 (Cambridge, Mass., 1953), 487–539; E. J. Wiseman, "John Tyndall: His Contributions to the Defeat of the Theory of Spontaneous Generation of Life," in *School Science Review*, **159** (1965), 362–367; J. K. Crellin, "Airborne Particles and the Germ Theory, 1860–1880," in *Annals of Science*, **22** (1966), 49–60; and "The Problem of Heat Resistance of Micro-Organisms in the British Spontaneous Generation Controversy of 1860–1880," in *Medical History*, **10** (1966), 50–59; Glenn Vandervliet, *Microbiology and the Spontaneous Generation Debate During the 1870s* (Lawrence, Kansas, 1971); J. Friday, "A Microscopic Incident in a Monumental Struggle: Huxley and Antibiosis in 1875," in *British Journal for the History of Science*, VII (1974), 61–71; and some more general references that are in William Bulloch, *The History of Bacteriology* (London, 1938).

There has recently been growing interest in the substance of Tyndall's scientific and political controversies. Cf. Bernard Semmel, *The Governor Eyre Controversy* (London, 1962), 123–128; R. MacLeod, "Science and Government in Victorian England: Lighthouse Illumination and the Board of Trade, 1868–1886," in *Isis*, **60** (1969), 5–38; Frank M. Turner, "Rainfall, Plagues and the Prince of Wales: A Chapter in the Conflict of Religion and Science," in *Journal of British Studies*, **13** (May 1974), 46–65. The controversy with Forbes on glacier motion has been described by J. S. Rowlinson, "The Theory of Glaciers," in *Notes and Records of the Royal Society*, **26** (1971), 189–204. The Mayer-Joule controversy is dealt with in T. S. Kuhn, "Energy Conservation as an Example of Simultaneous Discovery," in M. Claggett, ed., *Critical Problems in the History of Science* (Madison, Wis., 1959), and in J. T. Lloyd, "Background to the Joule-Mayer Controversy," in *Notes and Records of the Royal Society*, **25** (1970), 211–235.

A comprehensive review of Tyndall's work on optical and crystalline structure, on magnetism, radiation, and mountaineering, as well as the relationships between his several "research programmes" is lacking. The historical record has stressed his activities as popularizer, lecturer, and man of letters.

ROY MACLEOD

TYRRELL, JOSEPH BURR (*b.* Weston, Ontario, Canada, 1 November 1858; *d.* Toronto, Canada, 26 August 1957), *geology, exploration, mining.*

Tyrrell was the son of William Tyrrell, a building contractor (later reeve of York Township and of Weston) and Elizabeth Burr, the daughter of Rowland Burr, a mill architect. He was educated at the Old Grammar School, Weston, and at Upper Canada College, Toronto; it is significant that he became an expert shot with the pistol at an early age.

At the University of Toronto he had a distinguished undergraduate career, gaining first-class honors in chemistry, biology, mineralogy, and geology, and winning the only natural sciences scholarship awarded in his year. After graduation he began studies leading to a career in the legal profession, but a threat of tuberculosis caused him to change to the outdoor life of a geologist.

His first appointment was with the Geological Survey of Canada, already a well-established organization, directed by A. C. Selwyn. Tyrrell was assigned as field assistant to the assistant director, Dr. George Dawson, who had already made a considerable reputation for his exploration and geological reconnaissance work in the more remote parts of Canada. Together they made traverses through the Crows Nest, Kootenay, and Kicking Horse passes in the Rocky Mountains. Tyrrell also made geological surveys of the Cretaceous coal measures of the foothills between Calgary and Edmonton, discovering in the course of this work the first remains of giant carnivorous dinosaurs found in Canada. In old age he was to describe this as "one of the most pleasurable thrills" of his life.

In the 1890's Tyrrell's attention turned to the Pre-Cambrian shield area of northern and arctic Canada, and he made a series of arduous journeys there by wagon and canoe. These trips were important for the geological surveys achieved, for apart from the sparse indigenous population of Indians and Eskimos, the only previous visitors had been white trappers. The territory included what are now parts of northern Manitoba and Saskatchewan. The two most noteworthy expeditions (1893 and 1894) were his crossings of the Barren Lands, the treeless arctic wastes lying to the west of Hudson Bay. In each case the starting point was the Lake Athabasca Post of the Hudson's Bay Company. The goal was the company station at Fort Churchill. The first expedition reached Hudson Bay at Chesterfield Inlet at latitude 63° 58'. The journey had proved dangerous. Had they not come upon herds of caribou, they would have starved. Winter overtook them early during the 500-mile journey along Hudson Bay to Fort Churchill. Tyrrell noted a great river discharging into Hudson Bay and resolved to return to investigate it the fol-

lowing year. He did so and named the river after R. Munro Ferguson, a member of the group. Maps were prepared by Tyrrell during these expeditions, and the scientific results were embodied in very full reports, parts of which remain the only authoritative accounts of this area.

Tyrrell was also a keen naturalist and his writings include a catalogue of the mammals of Canada, descriptions of the winter home of the caribou, and an account of the distribution of conifers. Scientifically, however, he is remembered for his contributions to glacial geology at this stage, notably the recognition that three major Pleistocene ice sheets—which he called the Labradorean, the Patrician, and the Keewatin—had covered northern and eastern Canada.

Still more significant were his later contributions to economic geology. He had already noted seepages of oil in the foothills zone north of Edmonton before 1893 and had urged investigation by drilling. This petroleum field was not adequately explored until the 1950's, however. He had also discovered amber deposits at Cedar Lake.

In 1898, after writing up his northern expeditions, he was assigned to the Yukon, where the Klondike gold rush, the most spectacular in Canadian history, was at its height. Tyrrell reported on the geological situation for the government, but upon his return to Ottawa at the end of the year he had come to the conclusion that he wished to take a more active part in the mining industry. Accordingly he left the Geological Survey to return in a private capacity to the Yukon. He became a geological and mining consultant and quickly won the confidence of the gold miners. He then spent seven years in the Yukon.

In 1894 Tyrrell married Mary Edith Carey, daughter of the Reverend Dr. G. M. W. Carey, of New Brunswick. After he became established in Dawson City his wife and small daughter Mary joined him.

Tyrrell's career in the Yukon spanned the changeover from streaming to large-scale hydraulic mining, and he learned and contributed much to the geology of alluvial gold. In 1895 he returned to the east, however, and after a short period as mining adviser to Sir William Mackenzie, the railroad developer, he set up as a consultant in Toronto. He was retained by the Anglo-French Mining Company of London, to which he began to make regular visits.

Tyrrell's influence on the development of metalliferous mining in Canada was considerable; he took a great interest in the famous silver mining district of Cobalt, Ontario, but his greatest achievement was the discovery of the Kirkland Lake gold deposit, which he predicted from structural reasoning. The sinking of a 600-meter shaft proved the orebody and led to the founding of the highly successful Kirkland Lake Gold Mining Company. Tyrrell was president of the company until a few years before he died.

BIBLIOGRAPHY

I. ORIGINAL WORKS. Tyrrell's works include a series of accounts of his expeditions in *Report. Geological Survey of Canada.* See especially Assiniboine and Saskatchewan, **2** (1887), E1–152; northwestern Manitoba, **5** (1892), E1–235; country between Athabasca Lake and Churchill, **6** (1893), D1–120; Doobaunt, Kazam, and Ferguson rivers, **9** (1896), F1–218.

Among his numerous published papers the following deserve mention: "The Glaciation of North Central Canada," in *Journal of Geology*, **6** (1898), 147–160; "Natural Resources of the Barren Lands of Canada," in *Scots. Geogr. Mag.*, **15** (1900), 126–138; "The Gold-Bearing Alluvial Deposits of the Klondike District," in *Transactions of the Institution of Mining and Metallurgy*, **8** (1900), 217–229; "Concentration of Gold of the Klondike," in *Economic Geology*, **2** (1907), 393–399; "The Law of the Pay Streak in Placer Deposits," in *Transactions of the Institution of Mining and Metallurgy*, **21** (1912), 593–605. He was responsible for the descriptions of the geology and mineral resources of the Yukon and the Northwest Territories in A. Shortt and A. G. Doughty, eds., *Canada and Its Provinces*, XXII (Ottawa, 1914), 583–660.

II. SECONDARY LITERATURE. For a biography up to year 1930, see W. J. Loudon, *A Canadian Geologist* (Toronto, 1930), with photograph. See also "Award of the Murchison Medal," in *Proceedings of the Geological Society*, **54** (1918), xlii–xliv; "Award of the Wollaston Medal," *ibid.*, **103** (1947), xxxvi–xxxviii; and obituary by D. R. D(erry), *ibid.*, no. 1563 (1958), 130–133.

K. C. DUNHAM

TYSON, EDWARD (*b.* Bristol, England, 20 January 1650/1651; *d.* London, England, 1 August 1708), *comparative anatomy, medicine.*

Tyson was born into a good Church of England family of some means. After attending private schools in Bristol, Tyson entered Magdalen Hall, Oxford, in 1667 where he was strongly influenced by Plot. From Oxford he received the B.A. in 1670 and M.A. in 1673. During this six-year period Tyson performed many dissections on diverse animals and worked in botany, being influenced by Grew's *The Anatomy of Vegetables Begun* (1672).

In 1673 Tyson began medical studies, receiving a bachelor of medicine degree from Oxford in 1677. In the same year his first publication appeared in Plot's *Natural History of Oxfordshire.*

Ready to begin practicing medicine, Tyson moved to London in 1677 and took up residence with his brother-in-law Richard Morton, in whose house he carried on various experiments, particularly in anatomy. He soon became affiliated with several members of the Royal Society and began publishing in the *Philosophical Transactions* in February 1678. Almost immediately Tyson developed a close relationship to Hooke, who made numerous references to Tyson in his *Diary.* During his first several years in London, Tyson published a handful of papers on morbid anatomy and pathological subjects in the *Philosophical Transactions* and Bartholin's *Acta medica et philosophica hafniensa.*

Tyson's first major contribution in comparative anatomy, published in 1680 under the title *Phocaena, or the Anatomy of a Porpess . . .*, was a description of a dolphin. A full quarter of the forty-eight pages is devoted to "A Preliminary Discourse Concerning Anatomy and a Natural History of Animals," in which Tyson presents his ideas on the importance of comparative anatomy and gives an outline for a proposed natural history of animals. Tyson criticizes the earlier encyclopedic style of natural history, which placed more emphasis on other authors than on the natural objects. He argues for beginning with the simplest animals and ascending through each of the tribes of animals. Here and in other of his anatomical works Tyson repeats his belief in the Great Chain of Being as seen in a gradation between all animals and the existence of intermediate types between each of the major groups. Tyson thinks his "Porpess" is the transitional link between the fishes and the land quadrupeds. This "Preliminary Discourse" contains a clear expression of the principles and methodology of comparative anatomy. In these ideas and the role it played the "Discourse" is very similar to the admirable, anonymous, introductory essay by Claude Perrault in *Mémoires pour servir à l'histoire naturelle des animaux* (1671–1676). These two essays did much to set the style and direction for the significant quantity of comparative-anatomical work in the late seventeenth century in Paris and London. Whether Tyson was acquainted with the work of Perrault and "the Parisians" in 1680 is indeterminable, although likely, since he had reviewed for the *Philosophical Collections* the *Mémoires pour servir à l'histoire des plantes* (1679), also produced under the auspices of the Paris Academy of Sciences.

Like "the Parisians," Tyson well recognized that before a general natural history of animals could be written there had to be an accumulation of observations and descriptions of many different kinds of animals. The starting point for such a project should be very good descriptions of representative animals, which can serve as reference points and points of comparison for many similar animals. He thought his description of the "Porpess" might serve as the representative of the cetaceans. Through arrangements made by Hooke—who attended the dissection and did the drawings that were later engraved and published in Tyson's *Phocaena*—Tyson was able to dissect the dolphin in November 1679. Tyson hoped the description he presented of his "Porpess" would serve as a model and provide inspiration for others. It was a good description on which others could build and which could help serve as the basis for a general natural history. The following year, in his preface to the English edition, which he arranged to have translated, of Swammerdam's *Ephemeri vita* (London, 1681), Tyson repeated the idea that studies of individual species would be the basis of any general natural history.

By 1681 Tyson was well established in London as a comparative anatomist and as a physician. He was elected a fellow of the Royal Society (1679), received a doctorate of physics from Cambridge and was admitted a candidate of the Royal College of Physicians in 1680, and was elected to the first of many terms on the council of the Royal Society (1681). In 1683 Tyson was appointed one of the two curators of the Royal Society who were responsible for providing demonstrations at each meeting. Tyson was also elected a fellow of the Royal College of Physicians; later he served as a censor of the college. The following year Tyson was appointed to two positions which had just become vacant—the Ventera readership in anatomy at the Surgeons Hall, from which he retired in 1699, and physician to Bethlehem and Bridewell Hospitals (now the Royal Bethlehem Hospital), where he also later served as a governor. In 1686 Tyson was elected a member of the Philosophical Society of Oxford.

After 1680 Tyson published just over two dozen works, mostly in the *Philosophical Transactions,* about half of which dealt with natural history. The balance of his output described several pathological cases, instances of monstrous development, and abnormal births. In 1683 Tyson described the anatomy of a rattlesnake, which had been brought to England from Virginia. This very thorough description may then have been the most complete study of any reptile. Later in the same year he described both the broad tapeworm and the roundworm of man as well as a specimen of the Mexican warthog. Because decomposition had begun on the latter, Tyson concentrated on the animal's dorsal scent gland. Glands, and particularly scent glands, were a subject of recurring interest to Tyson. His first publication (1677) was on scent glands. Over many years he gathered information on human and nonhuman glands, which he brought together in "Adenologia," the manuscript of which was never published and has been lost.

In 1685 Tyson contributed two descriptions, of a shark embryo and of the lumpfish, to John Ray's edition of Francis Willughby's *History of Fishes* (1686). Also in 1685 Tyson supplied a considerable amount of comparative-anatomical material to Samuel Collins for the latter's *A Systeme of Anatomy.* In his own *Myotomia Reformata* (1694) William Cowper published Tyson's discovery of the preputial and coronal glands in the glans penis of man. This discovery is perhaps the only thing for which Tyson may be remembered in medicine.

A live opossum was taken to London from Virginia in 1697. Its death in April 1698 provided Tyson with the material for one of his best anatomical descriptions. Fortunately for Tyson the specimen was a female, and he focused on the peculiarities of the reproductive system. He described the marsupium and the marsupial bones. In this paper, also published in 1698 as a separate work, he reiterates the principles of comparative anatomy, which he set forth in his *Phocaena.* In 1704 Tyson published an anatomy of a male opossum, on which Cowper collaborated with him.

In 1699 Tyson published his best-known work, *Orang-Outang, Sive Homo Sylvestris: or, the Anatomy of a Pygmie Compared With That of a Monkey, an Ape, and a Man.* The animal described was actually a young (two to three years old) chimpanzee. The term "Orang-Outang" is a native Malaysian term for "man of the woods" and was long used as a generic term for the larger nonhuman primates. Similarly, "ape" was applied to many of the Old World, tailless monkeys, as with the macaque still known as the "Barbary ape." Tyson was ably assisted on the *Orang-Outang* by Cowper, who did the section on myology, did the drawings for all of the plates, and mounted the skeleton, which is now at the British Museum (Natural History).

Generally the anatomical description is quite thorough and competent, particularly considering how little comparative, nonhuman primate material was then known in Europe. For comparative material Tyson relied quite heavily on the description of monkeys in Alexander Pitfeild's English translation of the Parisians' *Mémoires*. Tyson concluded with a table of the ways his "Pygmie" more resembled a man than an ape, and vice versa, concluding that it belonged between man and the apes. He was surprised that the orangutan's brain was so similar to that of man, because there is so much difference between a man's soul and the soul of the brutes that one would expect a greater difference in their respective organs of the soul. For Tyson, comparative anatomy was a means of understanding and of determining the order of the animals on the Great Chain of Being. Tyson repeatedly emphasizes how close the structure of his Pygmie is to that of man and that the Pygmie is the closest approach of the animal kingdom to the rational qualities of man. While the notion of the Great Chain has a history that long antedates Tyson, he did, in this first anatomical description of one of the great apes, clearly identify an occupant for the rung immediately below man. The identification of such gradational links was one of his objectives for comparative anatomy.

The last section of Tyson's *Orang-Outang* is devoted to "A Philological Essay Concerning the Pygmies of the Ancients." In this major, early contribution to the study of the folklore of the primates Tyson tried to demonstrate that the many references to assorted, but similar, creatures in ancient literature really referred to a nonhuman primate such as his orangutan. Two years later Tyson did a similar philological study regarding the mantichora (a mandrill in the genus *Papio*), which remained unpublished until Montague's biography of Tyson. Apparently Tyson intended this and other material on primates to be an appendix to the *Orang-Outang*. After the publication of the *Orang-Outang*, Tyson published several more papers in the *Philosophical Transactions*, the most interesting of which was on the male opossum and a fish, the yellow gurnard (sea robin).

By no means all of Tyson's scientific work appeared in print. There are many references in the records of the Royal Society to the research that Tyson was doing. These are sometimes referring to specimens turned over to Tyson for dissection (in one case the Society bought an ostrich for him to dissect), and sometimes are Tyson's reports. In the "Tyson Folio" at the Royal College of Surgeons are Tyson's research notes from numerous dissections as well as original drawings for his papers, both published and unpublished.

Tyson did not publish a large number of either anatomical or medical works, of which the most important was the *Orang-Outang*. Tyson carried on an active medical practice from 1677 until his sudden death in 1708. During much of this period he lectured on anatomy at Surgeons Hall and served as physician to Bethlehem and Bridewell Hospitals, where a wing is named in Tyson's memory. Tyson was a quiet man of orderly habits who occurs seldom in contemporary references and correspondence. Apparently his chief delight came from his studies.

BIBLIOGRAPHY

I. ORIGINAL WORKS. Tyson's two major writings were *Phocaena, or the Anatomy of a Porpess, Dissected at Gresham College: With a Preliminary Discourse Concerning Anatomy, and a Natural History of Animals* (London, 1680); and *Orang-Outang, Sive Homo Sylvestris: or, the Anatomy of a Pygmie Compared with That of a Monkey, an Ape, and a Man. To Which Is Added, A Philological Essay Concerning the Pygmies, the Cynocephali, the Satyrs, and Sphinges of the Ancients. Wherein It Will Appear That They Are All Either Apes or Monkeys, and Not Men, as Formerly Pretended* (London, 1699). A facs. ed. of *Orang-Outang* (London, 1966) has an introduction by M. F. A. Montague; a 2nd ed., entitled *The Anatomy of a Pygmy* (London, 1751), also included several of Tyson's shorter writings. Most of Tyson's other writings appeared in *Philosophical Transactions of the Royal Society*. A complete bibliography of Tyson's writings appears in Montague (see below).

At the Royal College of Surgeons, in folio vol. no. 324, are preserved a number of MSS by Tyson and many drawings of his dissections. The British Museum (MS Sloane 2770) has Tyson's MS of the myology portion of his anatomy lectures. At least six partial syllabi of his anatomy lectures exist at the British Museum and the Bodleian Library.

II. SECONDARY LITERATURE. The standard source for Tyson's life is M. F. Ashley Montague, *Edward Tyson, M.D. F.R.S. 1650–1708 and the Rise of Human and Comparative Anatomy in England* (Philadeophia, 1943). Montague exhaustively searched the contemporary literature and records for references to Tyson, and has brought his results together in this volume. A bibliography of Tyson's writings is included.

WESLEY C. WILLIAMS

UBALDO, GUIDO. See **Monte, Guidobaldo, Marchese del.**

UKHTOMSKY, ALEXEI ALEXEIVICH (*b*. Rhurik, Russia, 20 September 1875; *d*. Leningrad, U.S.S.R., 31 August 1942), *physiology.*

Ukhtomsky was descended from an ancient princely line whose name sprang from the Ukhtoma river that flowed through their ancestral domain in Rhurik in the province of Yaroslav. After receiving his secondary education at the military college in Nizhny Novgorod (now Gorki), he enlisted in the army as a cadet. Moved by strong religious beliefs, Ukhtomsky resigned from the army to enter the Moscow Theological Academy at Zagorsk. There was little inkling then of a future career in science; theology, church history, psychology, philosophy, and rhetoric absorbed his attention. In this religious climate he developed an intense interest in the faith of the Old Believers, or Roskolniks, a dissenting group in the Russian Orthodox Church. The group held tenaciously to the ritualistic practices that had been abandoned during Patriarch Nikon's reforms in the seventeenth century.

During a pilgrimage to visit their monasteries, he visited Tyumen, Siberia, where he met Rasputin. Ukhtomsky later invited him to St. Petersburg and introduced him into the religious salons of the aristocracy.

In 1902 Ukhtomsky, influenced by Wedensky, entered the University of St. Petersburg. Making a complete change, he entered the physico-mathematical faculty and specialized in animal physiology under Wedensky. He first studied the effects of anemia, oxygen lack, and fatigue on neuromuscular preparations.

After graduating from the university, he took a postgraduate course and soon became assistant in the physiological department, demonstrating experiments for the lecturers. During this time he carried out research with Wedensky on "The Reflexes of Antagonistic Muscles to Electrical Stimulation of a Sensory Nerve."

It was while working as a demonstrator that he stumbled on the phenomenon that led to his theory of a dominant focus of cortical excitation operating to exclude and inhibit other concurrent functions. He had to prepare for the class an experiment on electrical stimulation of the motor cortex of the dog. To his dismay, stimulation produced no movement, even when he increased the strength of the current. Suddenly the dog defecated, and immediately following this, cortical stimulation once again produced a motor response.

Ukhtomsky was so struck by this observation that he put it to experimental investigation and made it the subject of his thesis "On the Dependence of Cortical Motor Effects on Secondary Central Influences," which he defended for a master's degree in 1912. This was his last laboratory work and for the remaining thirty years of his life he made no further experiments, but in this thesis was incorporated the main theme of all Ukhtomsky's future concepts, namely the principle of "dominanta." This term appeared in 1923 in the title of the first of his many publications on the subject.

On gaining his master's degree he was appointed a docent in physiology, responsible for lecturing on such subjects as the physiology of the sense organs and of the central nervous system. In all his scientific teaching he upheld the tenets of the Wedensky school. His broad education in history and philosophy and his wide knowledge of the physiological literature made him a fine lecturer, but he did not refrain, even in the laboratory, from trying, often successfully, to convert his pupils to his religious faith. He was kind and generous to impecunious students, often gaining help for them through his position as churchwarden at St. Nicholas, the Old Believers' church in St. Petersburg.

Ukhtomsky relinquished none of his religious observances. In 1912 he attended the All-Russian Old Believer Congress, at which he reported on "The Splendor of Church Singing." He observed all the religious rites both at home and in the laboratory; during Lent he came to the lecture room carrying a rosary of leather beads and wearing a black wooden cross around his neck.

Simple in tastes, he lived in a two-room flat surrounded by books. Many were physiological texts (some were library books that he could not bring himself to part with), and others were rare, old, handwritten religious books. Even when, in 1922, he was appointed to the chair in physiology and could have afforded more comfortable living quarters, he preferred to remain in his small flat. He never married but lived alone, reading and drawing icons of the Old Believer pattern. He chose to dress as a rich peasant—wearing hunting boots, a coarse jacket, and underclothes made from linen woven by the peasants. He wore his hair long in moujik fashion and, by religious observance, never shaved his beard.

During the October Revolution he was arrested and imprisoned. His many friends at the University of Petrograd appealed on his behalf and he was freed. He was reinstated, and on the death of Wedensky in 1922 succeeded to the chair in physiology. As head of the department he began to develop his own ideas of central nervous system function. He spoke vaguely of processes that could not be observed experimentally, such as "the learning of a rhythm" by nerve cells, nerve endings, and muscles; of "constellations and excitations"; of "active rest"; and so on. His followers, recognizing his brilliance at the same time that they recognized his failings as an experimental scientist, strove to advance his philosophical views by presenting them as close to those of dialectical materialism, and indeed his report to the All-Russian Physiological Congress was published, not in a physiological journal, but in a philosophical one, the *Pod Znamenem Marxisma.*

Ukhtomsky remained in favor and became a member of the Academy of Sciences of the U.S.S.R. and was honored by having the Ukhtomsky Institute of Physiology in Leningrad University named for him.

Known principally for his concept of dominanta, Ukhtomsky's ideas evolved substantially from his academic pedigree. Sechenov (1829–1905) had discovered that inhibition could be an active process in the nervous system (not merely a suppression or occlusion of excitation). His pupil Wedensky developed further the ideas of excitation and inhibition, and his pupil's pupil Ukhtomsky derived his unified theory of dominanta. Sechenov held that all forms of activity of the central nervous system were derivatives of the simplest act—the reflex. He envisioned three main links in the reflex: (1) stimulation of sensory receptors, (2) action of a reflex center, and (3) excitation of the effector systems (muscles and glands). It was the second of these that Ukhtomsky developed into the concept of dominanta.

The rhythm of the respiratory movements and the reciprocal innervation of limb muscles were two of the outstanding physiological systems that suggested a central control mechanism, one which Ukhtomsky envisaged as ensembles of excitation and of inhibition responsible for these rhythmic activations. Both he and Wedensky attempted to apply this principle to the nerves themselves, calling this "the rhythm of nerves."

An important feature of the concept of dominance was the lability by which the controlling nerve center could establish new functional connections between various centers, ensuring a specific reaction of the central nervous system to the particular stimulation it received (a concept one meets again in Pavlovian theories of conditioned reflexes). In other words, the balance of excitation and inhibition in a system was labile and not an automatism. This broke away from the automatic alternation previously conceived as operating in reciprocal innervation as described by Sherrington.

The concept that in the West was to develop into the recognition of homeostasis was called by Ukhtomsky "rest" or "balance." He envisaged it as the goal of a biological system following any stimulation or disturbance. His ideas along these lines extended into metabolism and the phenomena of fatigue states.

During the siege of Leningrad in World War II, Ukhtomsky's institute was evacuated to Saratov on the Volga, but he remained in the city. He studied shock in the wounded and continued his activities at the university. Suffering already from incipient cancer of the throat, chronic hypertension, and gangrene in his legs, his death was hastened by starvation and, on 31 August 1942, he died.

BIBLIOGRAPHY

The complete works of Ukhtomsky have been published in *Sobranie Sochineny* ("Collected Works"), 5 vols. (Leningrad, 1945–1954). An English trans. by D. B. Lindsley is being prepared by the American Psychological Association.

Mary A. B. Brazier

ULLOA Y DE LA TORRE GIRAL, ANTONIO DE (*b.* Seville, Spain, 12 January 1716; *d.* Isla de Léon, Cádiz, Spain, 5 July 1795), *natural history.*

A mariner by profession, Ulloa was commissioned by the government, along with Jorge Juan y Santacilla, to accompany the expedition sent to America by the Paris Academy of Sciences to measure an arc of meridian (1736–1745). While returning to Spain he was captured by the English but took advantage of his stay in Britain to further his education. Upon reaching Madrid, he published, with Jorge Juan, *Relación histórica del viaje a la América meridional* (1748). The sections devoted to geology and other technical matters, written by Ulloa, include the first scientific description of the platinum found in the sands of the Río Pinto (Magdalena) in Colombia in 1736: "a stone of such resistance that it is very difficult to break or shatter it by striking it on a steel anvil." It was studied by William Watson (1750) and by François Chavanneau in the Vergara laboratory (1786). Ferdinand VI sent Ulloa on a mission throughout Europe to learn about the most recent scientific discoveries.

Ulloa participated in the creation of the royal natural history collection (1752) and of the naval observatory at Cádiz (1754). In 1758 he was appointed general manager of the mines of Huancavélica, Peru, and later assumed high posts in the Spanish navy. He also served as first Spanish governor of Louisiana (1766–1768).

Ulloa's observations embraced many fields. In the catalog of nature that he entitled *Noticias americanas: Entretenimiento físico-histórico sobre América meridional y septentrional-oriental* (1772) he discussed "climates and the products of the three kingdoms—vegetable, animal, and mineral." The work called *Observación en el mar de un eclipse de sol* (1778) presents a certain interest. His scientific avocations so absorbed Ulloa that he began to neglect his duties as a high-ranking naval officer and was court-martialed, although he was cleared of the charges. *Conversaciones . . . con sus tres hijos* may be considered a didactic work.

Noticias secretas de América (1826), published after Ulloa's death and that of Juan y Santacilla, is a confidential report on the situation in America that had been sent to the marquis of Ensenada.

BIBLIOGRAPHY

I. ORIGINAL WORKS. Ulloa's writings include *Relación histórica del viaje a la América meridional . . .*, 4 vols. (Madrid, 1748), written with Jorge Juan y Santacilla; *Noticias americanas: Entretenimiento físico-histórico sobre América meridional y septentrional-oriental* (Madrid, 1772); *Observación en el mar de un eclipse de sol* (Madrid, 1779); *Conversaciones de Ulloa con sus tres hijos* (Madrid, 1795); and *Noticias secretas de América* (London, 1826), written with Juan y Santacilla.

II. SECONDARY LITERATURE. Juan Sempere y Guarinos, *Escritores del reinado de Carlos III*, VI (Madrid, 1785), 158–176, is important. See also Julio Fernández Guillén y Tato, *Los tenientes de navío Jorge Juan y Santacilla y Antonio de Ulloa y de la Torre-Giral y la medición del meridiano* (Madrid, 1935). On his description of platinum, see Mary Elvira Weeks, *Discovery of the Elements*, 6th ed. (Easton, Pa., 1956), ch. 16, esp. 409–412.

J. VERNET

ULRICH, EDWARD OSCAR (*b.* Cincinnati, Ohio, 1 February 1857; *d.* Washington, D.C., 22 February 1944), *stratigraphy, paleontology.*

Ulrich's career was typical of those nineteenth-century American paleontologists who, beginning as self-taught amateur collectors, reached professional status through initial independent publication, then commissions for state or territorial geologic surveys, and finally, a permanent position with the federal survey, a large museum, or a major university.[1]

Ulrich was one of eight children of an immigrant Alsation carpenter,[2] later a contractor in the Cincinnati area. Ulrich's formal education was limited to intermittent terms at two Ohio colleges during the 1870's.[3] He resumed carpentry with his father and continued collecting from the extensive and highly fossiliferous Upper Ordovician rocks of Cincinnati. In 1877 the Cincinnati Society of Natural History recognized his abilities with a small-salaried curatorship. He resigned the post in 1879 and spent a lively two years superintending silver mines near Boulder, Colorado.

Relocated at Newport, Kentucky, by 1884 Ulrich had published in the *Journal of the Cincinnati Society of Natural History* six brief paleontological papers and one major systematic investigation of American Paleozoic bryozoans, to which he expanded the new techniques of thin sections.[4]

With contracts from the Illinois and Minnesota geologic surveys for further studies of Paleozoic bryozoans, sponges, and mollusks, he combined efforts in 1885 with Charles Schuchert.[5] Both men became able lithographic illustrators.

Ulrich made fundamental taxonomic contributions to a dozen major marine invertebrate taxa. Preparation of bryozoan collections for the British Museum (Natural History) and the museum of the University of Munich led to his contributions on the bryozoans and ostracod crustaceans in the first English edition of Karl A. von Zittel's *Text-book of Palaeontology*, Charles R. Eastman, trans. (London, 1896). Between 1885 and 1897 Ulrich also described Paleozoic faunas and conducted stratigraphic studies for the Ohio and Kentucky surveys, served five years as curator of geology for the Cincinnati Society, and for nine years was associate editor of the *American Geologist*.

In 1897 Ulrich's reputation brought him a temporary appointment to the U.S. Geological Survey. This post, which became permanent in 1901, marked a significant shift in research emphasis from paleontology to stratigraphy. Relocation at Washington, D.C., renewed Ulrich's association with Schuchert until 1904; they wrote "Paleozoic Seas and Barriers in Eastern North America" during this interval. When Raymond Smith Bassler, who had been Ulrich's assistant during the previous decade at Newport, joined the U.S. National Museum, they continued close professional ties in major studies of bryozoans (1904).[6] Paleozoic ostracods (1923), and the enigmatic conodonts (1926).

In later years Ulrich considered himself first and foremost a stratigrapher and believed stratigraphic syntheses to be his most lasting contributions. Geologic mapping experience in the Upper Mississippi Valley[7] and Appalachians culminated in the article "Revision of the Paleozoic Systems" (1911). Among modifications and proposed additions to the standard geologic time scale were two new lower Paleozoic time-stratigraphic units of systemic rank, the Ozarkian and Canadian, inserted between the restricted Cambrian and Ordovician systems.

In his publications on stratigraphic methods and philosophy, Ulrich wholeheartedly adopted Thomas Chrowder Chamberlin's concept that records of diastrophic events were the natural and ultimate bases for defining and correlating stratigraphic units. In Ulrich's view, the differential isostatic adjustments yielded oscillatory littoral displacements, which produced abrupt discontinuities in geologic and faunal sequences deposited in multiple isolated epicontinental basins. These basins were alternately flooded from opposite directions by oceans flanking the North American continent. Rock units and their faunal assemblages bounded by these disconformities were discontinuous both spatially and temporally. Mutually exclusive, they were terminated by nondeposition rather than facies—lateral gradation to coeval units, formed in different depositional environments, an alternate concept popularized in the United States by Amadeus William Grabau[8] and adopted by Schuchert.

Ulrich's stratigraphic models were initially of value in interpreting the midcontinent region but failed to explain the more complex Appalachian geology. Ulrich used index or guide-fossil techniques at the most discrete taxonomic levels to make the initial identifications of the datum planes, not to define or correlate tectonic event-based broader stratigraphic units. He considered recognition of the introduction of new generic types a complementary and less than precisely coincident verification of physical boundaries, in contrast to Schuchert's insistence that the nonrepetitive aspect of organic evolution made fossils fundamental determinants of time and event.

Albertina Zuest, whom Ulrich married in Cincinnati in 1886, died in 1932 after a decade of disabling illness; their union was childless. The following year he married Lydia Sennhauser, his first wife's nurse during the early stages of her illness. Ulrich was appointed associate in paleontology at the U.S. National Museum in 1914, and occupied that position after his formal retirement from the U.S. Geological Survey in 1932. He continued his studies, accumulating paleontologic evidence for his Ozarkian and Canadian systems, and supplying stratigraphic data for systematic investigations contributed mostly by younger museum associates. These studies failed to provide an acceptable paleontological base[9] for Ulrich's systemic units; the Ozarkian is not part of present usage, which considers the Canadian the basal series of the Ordovician system.

Ulrich was an eminent authority on American lower Paleozoic stratigraphy during the first three decades of the twentieth century. His vast experience and encyclopedic knowledge enabled him to dominate the discipline. Perhaps his most important influence during his years on the U.S. Geological Survey was the controversial aspect of his research and his disputative nature, which caused contemporaries to reexamine critically their own investigations.

NOTES

1. Raymond Smith Bassler, "Development of Invertebrate Paleontology in America," in *Bulletin of the Geological Society of America*, **44** (1933), 268.
2. Edward was the eldest of the five children born to Charles Ulrich and his first wife, Julia Schnell.
3. German Wallace College, now Baldwin-Wallace College, Berea, Ohio, which Ulrich attended in 1874–1875, awarded him the honorary degrees of Master of Arts (1886) and Doctor of Science (1892).
4. Henry Alleyne Nicholson's work in the previous decade had shown the need for continuing investigations of bryozoans by this method, one that disclosed the usefulness of internal structures in recognizing widespread external homeomorphy.
5. Schuchert later recalled their association in Newport (1881–1888), especially the four years during which he worked as Ulrich's assistant. See unpublished MS autobiography, ch. 7, p. 30; box XLIII, Charles Schuchert Papers, Manuscripts and Archives Division, Sterling Library, Yale University.
6. Bassler, who prepared the bryozoan section for the 2nd ed. (1913) of the work by Karl A. von Zittel, culminated his systematic work within the Ulrichian framework of ideas in *Treatise on Invertebrate Paleontology, Part G, Bryozoa* (Lawrence, Kansas, 1953); its excellent organization and bibliography reflect his abilities as a compiler. In Ulrich and Bassler's investigations, thin sections were studied primarily with the 10-power hand lens. Their contemporaries, Edgar Roscoe Cumings and Jesse James Galloway at the University of Indiana, made initial microscopic studies of bryozoan modes of growth and functional morphology, techniques now being applied to investigations of systematics and phylogeny for the revision volume of the *Treatise*.
7. The only major unpublished work from this period is the joint geologic mapping and stratigraphic study with Nevin Melancthon Fenneman in 1904–1914 of the Cincinnati East 15-minute quadrangle. The map and accompanying MS formerly in the survey's Washington open files is no longer ex-

tant. The Department of Geology of the University of Cincinnati presently holds a photostatic copy. Ulrich's adamant positioning of the Richmondian stage as basal Silurian disagreed with the usage of the Geological Names Committee of the U.S. Geological Survey, and most contemporary usage, which held it to be uppermost Ordovician. See William Henry Shideler, "The Ordovician-Silurian Boundary," in *Ohio Journal of Science*, **16**, no. 8 (1916), 329–335.

8. A comparison and critique of Ulrichian and Grabauan stratigraphic methods and philosophies is Carl Owen Dunbar and John Rodgers, *Principles of Stratigraphy* (New York, 1957), 136, 284–288.

9. Charles Schuchert, "Ozarkian and Canadian Brachiopods," in *American Journal of Science*, **237**, no. 2 (1939), 135–138.

BIBLIOGRAPHY

I. Original Works. A large collection of papers, field notebooks, locality registers, monthly reports, a partly published MS, and memorabilia is in the Paleontology and Stratigraphy Branch, U.S. Geological Survey, National Museum of Natural History, Washington, D.C.

A second unpublished MS is held by the Denver Library of the survey, which also preserves the 1912–1917 correspondence between Ulrich and George Willis Stose. Folder 8, Geological Division General File (Chief Geologist), 1901–1916, Records Group 57 (Records of the Geological Survey) of the National Archives, Washington, D.C., contains field correspondence sent by Ulrich during 1902–1915. Letters sent in 1888–1893 by Josua Lindahl, then Illinois state geologist, to Ulrich are on microfilm in the Augustana College Library, Rock Island, Illinois. The department of geology, University of Iowa, Iowa City, holds copies of 1934 correspondence between Ulrich and A. C. Trowbridge.

Ulrich's 141 published titles are listed in the several bibliographies of North American geology, *Bulletin of the United States Geological Survey*, nos. 745 (1923), 1031–1033; 823 (1931), 624–625; 937 (1944), 961–963; and 1049 (1957), 927.

Among his paleontological writings are "American Paleozoic Bryozoa," in *Journal of the Cincinnati Museum of Natural History*, **5** (1882), 121–175, 232–257; **6** (1883), 82–92, 148–168, 245–279; and **7** (1884), 24–51; "American Paleozoic Sponges," in *Illinois Geological Survey*, **8**, pt. 2, *Palaeontology* (1890), 209–241, 243–251, 253–282 (the last section with Oliver Everett); "Paleozoic Bryozoa," *ibid.*, 283–688, pls. XXIX–LXXVIII; "New and Little Known American Paleozoic Ostracoda," in *Journal of the Cincinnati Society of Natural History*, **13** (1890–1891), 104–137, 173–211, pls. 7–18; "On Lower Silurian Bryozoa of Minnesota," in *Geological and Natural History Survey of Minnesota, Final Report*, **3**, *Paleontology*, pt. 1 (1895), 96–332, pls. 1–28 [folio]; "New Lower Silurian Lamellibranchiata of Minnesota," *ibid.*, pt. 2 (1897), 475–628, pls. XXXV–XLII; "The Lower Silurian Ostracoda of Minnesota," *ibid.*, 629–693, pls. XLIII–XLVI; "The Lower Silurian Gastropoda of Minnesota," *ibid.*, 813–1081, pls. LXI–LXXXII, with Wilbur H. Scofield; "A

Revision of the Paleozoic Bryozoa," pt. 1, "On Genera and Species of Ctenostoma," in *Smithsonian Miscellaneous Collections*, **45**, no. 1452 (1904), 256–294, pls. LXV–LXVIII, pt. 2, "On Genera and Species of Trepostoma," *ibid.*, **47**, no. 1470 (1904), 15–55, pls. VI–XIV, with Raymond Smith Bassler; "Paleozoic Ostracoda: Their Morphology, Classification, and Occurrence," in Maryland Geological Survey, *Silurian* (Baltimore, 1923), 271–391, pls. XXXVI–LXV, with Bassler; "A Classification of the Toothlike Fossils, Conodonts, With Descriptions of American Devonian and Mississippian Species," in *Proceedings of the United States National Museum*, **68**, no. 2613 (1926), 1–63, pls. 1–11, with Bassler; "The Cambrian of the Upper Mississippi Valley," pt. 1, "Trilobita; Dikelocephalinae and Osceolinae," in *Bulletin of the Public Museum*, **12**, no. 1 (1930), 1–122, pls. 1–23; pt. 2, "Trilobita; Saukiinae," *ibid.*, **12**, no. 2 (1933), 123–306, pls. 24–45, with Charles Elmer Resser; "Cambrian Bivalved Crustacea of the Order Conchostraca," in *Proceedings of the United States National Museum*, **78**, no. 2847 (1931), 1–130, pls. 1–10, with Bassler; "Ozarkian and Canadian Brachiopoda," in *Special Papers of the Geological Society of America*, no. 13 (1938), viii, 1–323, pls. 1–58, with Gustav Arthur Cooper; "Ozarkian and Canadian Cephalopods," pt. 1, "Nautilicones," *ibid.*, no. 37 (1942), x, 1–157, pls. 1–57, with August Frederick Foerste, William Madison Furnish, and Arthur K. Miller; pt. 2, "Brevicones," *ibid.*, no. 49 (1943), x, 1–240, pls. 1–70, with Foerste and Miller; pt. 3, "Longicones and Summary," *ibid.*, no. 58 (1944), x, 1–226, pls. 1–68, with Foerste, Miller, and Athel Glyde Unklesbay.

Ulrich's stratigraphic publications include "The Lower Silurian Deposits of the Upper Mississippi Province: a Correlation of the Strata With Those in the Cincinnati, Tennessee, New York, and Canadian Provinces, and the Stratigraphic and Geographic Distribution of the Fossils," in *Geological and Natural History Survey of Minnesota, Final Report*, **3**, *Paleontology*, pt. 2 (1897), lxxxiii–cxxviii [folio], with Newton Horace Winchell; "Paleozoic Seas and Barriers in Eastern North America," in *Bulletin of the New York State Museum*, no. 52, Paleontology no. 6 (1902), 633–663, with C. Schuchert; "Columbia Folio, Tennessee," in *U.S. Geological Survey, Geological Atlas*, fol. no. 95 (1903), 1–6, 8 pls., with Charles Willard Hayes; "The Lead, Zinc, and Fluorspar Deposits of Western Kentucky, Part I, Geology and General Relations," in *Professional Papers. U.S. Geological Survey*, no. 36 (1905), 15–105, pls. I–VII; "Revision of the Paleozoic Systems," in *Bulletin of the Geological Society of America*, **22** (1911), 281–680, pls. 25–29, "Index . . .," *ibid.*, **24** (1913), 625–668; "The Chattanoogan Series, With Special Reference to the Ohio Shale Problem," in *American Journal of Science*, 4th ser., **34**, no. 200 (1912), 157–183; "The Ordovician-Silurian Boundary," in *Congrès Géologique International, Compte-Rendu de la XIIIe Session, Canada, 1913* (Ottawa, 1914), 593–669; "Correlation by Displacements of the Strand-line and the Function and Proper Use of Fossils in Correlation," in *Bulletin of the Geological Society of America*, **27** (1916), 451–490; "Major Causes of Land and Sea Oscillations," in *Journal of the Washington Academy of Sciences*, **10**, no. 3 (1920), 57–78, reprinted in *Smithsonian Institution, Annual Report for 1920* (1922), 321–337; "Some New Facts Bearing on Correlations of Chester Formations," in *Bulletin of the Geological Society of America*, **33** (1922), 805–852; "American Silurian Formations," in Maryland Geological Survey, *Silurian* (Baltimore, 1923), 233–270, with Bassler; "Notes on New Names in Table of Formations and on Physical Evidence of Breaks Between Paleozoic Systems in Wisconsin," in *Transactions of the Wisconsin Academy of Sciences, Arts and Letters*, **21** (1924), 71–107; "Relative Values of Criteria Used in Drawing the Ordovician-Silurian Boundary," in *Bulletin of the Geological Society of America*, **37** (1926), 279–348; "Ordovician Trilobites of the Family Telephidae and Concerned Stratigraphic Correlations," in *Proceedings of the United States National Museum*, **76**, no. 2818 (1930), 1–101, pls. 1–8.

II. SECONDARY LITERATURE. On Ulrich and his work, see R. S. Bassler, "Edward Oscar Ulrich (1857–1944)," in *Bulletin of the American Association of Petroleum Geologists*, **28**, no. 5 (1944), 687–689; and "Memorial to Edward Oscar Ulrich," in *Proceedings. Geological Society of America* (1945), 331–351, pl. 23; G. Arthur Cooper, "Edward Oscar Ulrich," in *Dictionary of American Biography*, supp. 3, pp. 782–783; Percy Edward Raymond, "Edward Oscar Ulrich," in *Science*, **99**, no. 2570 (1944), 256; Rudolf Ruedemann, "Biographical Memoir of Edward Oscar Ulrich, 1857–1944," in *Biographical Memoirs. National Academy of Sciences*, **24**, no. 7 (1947), 259–280, with portrait; and *The National Cyclopaedia of American Biography*, XXXIII (1947), 63–64.

CLIFFORD M. NELSON

ULRICH OF STRASBOURG (or **ULRICUS DE ARGENTINA** or **ULRICH ENGELBERTI**) (*b.* early thirteenth century; *d.* Paris, *ca.* 1278), *natural philosophy.*

A student of Albertus Magnus at Cologne (1248–1254), Ulrich became his devoted disciple. He lectured at Strasbourg for many years before serving as provincial of the German Dominicans from 1272 to 1277, when he was sent to Paris to lecture on Peter Lombard's *Sentences* and to obtain the degree of master; but he died shortly after arriving there.

Among Ulrich's writings are a lost treatise on meteors, a commentary on the *Sentences*, and his chief work, the *Summa de summo bono* ("A Summary Concerning the Supreme Good"), projected for eight books but extant only to the fifth treatise

of the sixth book. This is a Neoplatonic work that is metaphysical in tone and is heavily indebted to Arab thought. Duhem translated and analyzed portions concerned with medieval astronomy and found that they are largely derived from Albertus Magnus. For example, Ulrich cited Ptolemy and a spurious work of Alpetragius (al-Bitrūjī); Denys the Carthusian (1402–1471) attempted to correct the latter reference, noting that it should be attributed to al-Fārābī (see Duhem, p. 360). Ulrich probably used only Albertus' report of their teachings. Ulrich held that there are ten celestial spheres, and was mainly concerned with linking them to intelligences as movers, and with explaining the characteristics of the various planets in terms of the primary qualities associated with their spheres. He mentioned the precession of the equinoxes incidentally as being one degree each hundred years, but otherwise adduced no astronomical data; he seems not to have been himself an observer of the heavens. Ulrich's importance for medieval science would seem to reside in the witness he provides to the all-pervading influence of Albertus Magnus as an authority, on the Continent at least, in matters scientific.

BIBLIOGRAPHY

Étienne Gilson, *History of Christian Philosophy in the Middle Ages* (New York–London, 1955), 431–433, 751–753, references the edited portions of Ulrich's *Summa*, summarizes its teaching, and provides a guide to bibliography. See also Pierre Duhem, *Le système du monde*, III (Paris, 1915; repr. 1958), 358–363; and Caroline Putnam, "Ulrich of Strasbourg and the Aristotelian Causes," in *Studies in Philosophy and the History of Philosophy*, **1** (1961), 139–159.

WILLIAM A. WALLACE, O.P.

ULSTAD, PHILIPP (*fl.* early sixteenth century), *medicine, alchemy.*

Little is known of Philipp Ulstad's life other than that he was a Nuremberg patrician who taught medicine at the Academy in Fribourg, Switzerland, during the first half of the sixteenth century. Aside from a small treatise on the plague, he published one book of significance for the history of science: *Coelum philosophorum . . .* (1525). This work was extremely popular, going through more than twenty editions and serving as a standard authority on the preparation and use of distillates for nearly a century. Ulstad emphasized the medical efficacy of chemical distillates, thus departing somewhat from conventional medieval pharmacology and preparing the way, in part, for the more intimate connection between chemistry and medicine effected by his contemporary Paracelsus and the latter's disciples in the second half of the sixteenth century.

Ulstad's work was based mainly on the writings attributed to Ramon Lull, Albertus Magnus, Arnald of Villanova, and John of Rupescissa. He was most clearly indebted to John of Rupescissa's doctrine of the fifth essence, namely, the substance that can be extracted from all mundane bodies by ordinary chemical methods and that is the chemically active principle of each body. Ulstad was primarily concerned with the curative and preservative properties of rectified alcohol—which he denoted variously as the fifth essence, *aqua vite*, *aurum potabile*, and *coelum philosophorum*. He maintained that the fifth essence, although not incorruptible, was less corruptible than the four elements and owed its medical value to its ability to regulate the bodily humors and thereby preserve the human body from decay.

Despite his use of alchemical terminology, Ulstad clearly dissociated himself from the enigmatic aspects of the alchemical tradition in offering his concise and rational account of the preparation of distilled remedies. Concerned with culling from the medieval alchemical corpus those techniques and ideas of practical utility, he ensured that they were made available to as large an audience as possible, including all apothecaries, surgeons, and medical doctors.

The lucidity of his technical directions was a major reason for the influence exerted by Ulstad. His discussion of apparatus and manipulative procedures afforded the sixteenth-century investigator an accurate summary of the best distilling theory then available. Of particular importance is Ulstad's description and woodcut of a distilling column with vertical water-cooled coils that, although not original with him, contributed to the decline of the less efficient and uncontrollable air-cooling methods commonly employed. He also clearly presented a rudimentary dephlegmation technique based on the introduction of oil-soaked sponges into the still head to retain the phlegm and obtain better fractionation. Ulstad's recipes for the extraction of the fifth essence dealt with a wide variety of sources, including gold, spices, herbs, fruits, flowers, precious stones, and metals, and he specified the particular ailments most responsive to each essence.

A codifier rather than an innovator, Ulstad contributed to the rise of iatrochemistry by demonstrating that drugs and other medicinals depend for their efficacy upon pure spirits or essences that can be extracted by the methods of chemistry. His ideas reappear in the writings of many prominent scientific figures, including Konrad Gesner and Andreas Libavius.

BIBLIOGRAPHY

I. ORIGINAL WORKS. Ulstad's major book is *Coelum philosophorum seu de secretis naturae liber* (Fribourg, 1525; 1st French trans., Paris, 1546; 1st German trans., Strasbourg, 1527). The later editions and translations have more elaborate titles. I have used the following edition: *Le Ciel des Philosophes, ou sont contenus les secrets de nature et comme l'homme se peult tenir en santé, et longuement vivre . . . extraict des livres de Arnould de Ville neuve, du grand Albert, Raymont Lulle, Jehan de la Roche Tranchée* (Paris, 1550). Ulstad's other work is *De epidemia tractatus* (Basel, 1526).

II. SECONDARY LITERATURE. There is no full-scale biography of Ulstad. Details concerning his work are found in Lynn Thorndike, *A History of Magic and Experimental Science*, V (New York, 1941), 541–542, 602, 621; James R. Partington, *A History of Chemistry*, II (London, 1961), 84–86; J. Ferguson, *Bibliotheca chemica*, II (Glasgow, 1906), 482–483; and Edward R. Atkinson and Arthur H. Hughes, "The 'Coelum Philosophorum' of Philipp Ulstad," in *Journal of Chemical Education*, **16** (1939), 103–107. R. J. Forbes, *Short History of the Art of Distillation* (Leiden, 1948), 127–130, states that Ulstad was probably an associate of Hieronymus Brunschwig, although he gives no proof for the claim.

MARTIN FICHMAN

ULUGH BEG (*b.* Sulṭāniyya, Central Asia, 22 March 1394; *d.* near Samarkand, Central Asia [now Uzbek S.S.R.], 27 October 1449), *astronomy*.

Ulugh Beg, which means "great prince," was a title that replaced his original name, Muḥammad Taragay. He was raised at the court of his grandfather, Tamerlane, and from 1409 was the ruler of Maverannakhr, the chief city of which was Samarkand.

In contrast with his grandfather, Ulugh Beg was not interested in conquest but gained fame as a scientist. At Samarkand in 1420 he founded a *madrasa*, or institution of higher learning, in which astronomy was the most important subject. Ulugh Beg himself selected the scientists who taught there, first interviewing them to determine their qualifications. His *madrasa* differed from others of that time both in the content and in the level of the subjects taught there. Besides Ulugh Beg, the lecturers included Ṣalāḥ al-Dīn Mūsā ibn Maḥmūd (Qāḍī Zāda), and Ghiyāth al-Dīn Jamshīd al-Kāshī.

Four years after founding the *madrasa*, Ulugh Beg erected a three-story observatory. At the instigation of the jurists, however, the building was reduced to ruins by the beginning of the sixteenth century, and in time apparently disappeared. Its precise location remained unknown until 1908, when the archaeologist V. L. Vyatkin found its remains.

The main instrument of the observatory proved to be—not a quadrant, as Vyatkin thought—but a "Fakhrī sextant." A trench about two meters wide was dug in a hill, along the line of the meridian, and in it was placed a segment of the arc of the instrument. The part that is preserved, which was in the trench, consists of two parallel walls faced with marble, fifty-one centimeters apart.

The main use of the Fakhrī sextant was in determining the basic constants of astronomy: the inclination of the ecliptic to the equator, the point of the vernal equinox, the length of the tropical year, and other constants arising from observation of the sun. Thus it was built chiefly for solar observations in general and for observations of the moon and the planets in particular (an arc of 60° is sufficient). Other instruments used were an armillary sphere, a triquetrum, an astrolabe, and a *shāmila* (an instrument serving as astrolabe and quadrant).

With the aid of the Fakhrī sextant one could determine at noon every day the meridional height of the sun, its distance from the zenith, and its declination; and from this information one could deduce the latitude and the inclination of the ecliptic, such that between the latitude φ, the distance from the zenith z, and the declination δ there is the known relationship $\varphi = z + \delta$. For example, letting ϵ be the inclination of the ecliptic, the distance from the zenith at noon on the day of the summer solstice is $z_1 = \varphi - \epsilon$ and on the day of the winter solstice $z_2 = \varphi + \epsilon$, equations that lead to $\epsilon = 1/2 (z_2 - z_1)$. The value obtained by Ulugh Beg for the inclination of the ecliptic, $\epsilon = 23° 30' 17''$, differs by only 32'' from the true value (for his time).[1] According to him, the latitude of Samarkand was 39° 37' 33''.[2]

The radius of the Fakhrī sextant in Ulugh Beg's observatory was 40.04 meters, which made it the largest astronomical instrument in the world of that type. On the arc of the sextant are divisions in which 70.2 centimeters corresponds to one degree; 11.7 millimeters (or, if rounded, 12 millimeters)

represents one minute; 1 millimeter is five seconds; and 0.4 millimeter is two seconds. It has been experimentally established that with unrestricted time for observation and sufficient training of the observer, the value of the threshold of angular discrimination can be considered as two to five seconds. Thus the choice of the scale of the main instrument, and its smallest divisions, was made with consideration for the limits of angular discrimination.

An important result of the scientific work of Ulugh Beg and his school was the astronomical tables called the *Zīj* of Ulugh Beg or the *Zīj-i Gurgāni* (Guragon, the title of Genghis Khan's son-in-law, was also used by Ulugh Beg). This work, originally written in the Tadzhik language, consists of a theoretical section and the results of the observations made at the Samarkand observatory; the latter include actual tables of calendar calculations, of trigonometry, and of the planets, as well as a star catalog.

The basis of Ulugh Beg's trigonometric tables was the determination of sin 1° with great accuracy. One of the methods of solving this problem was Ulugh Beg's, and another was that of Al-Kāshī.[3] Both lead to the solution of the third-degree algebraic equation with the form

$$x^3 + ax + b = 0,$$

where $x = \sin 1°$. Solving this equation by an original method of subsequent approximations, one obtains

$$x = \sin 1° = 0.017452406437283571.$$

In his trigonometric tables Ulugh Beg gives the values of sines and tangents for every minute to 45°, and for every five minutes from 45° to 90°; the values of cotangents are given for every degree. Comparing the values of the sines of any angles—for example, 20°, 23°, and 26°—with the corresponding true values, we obtain the following:

According to Ulugh Beg	True Value
20° 0.342020142	20° 0.342020143
23° 0.390731129	23° 0.390731128
26° 0.438371147	26° 0.438371147

Also strikingly accurate is the study of the yearly movements of the five bright planets known in the time of Ulugh Beg, as is evident below:[4]

According to Ulugh Beg	True Value	
Saturn	12°13'39''	12°13'36'' (d'Alembert)
Jupiter	30°20'34''	30°20'31'' (d'Alembert)
Mars	191°17'15''	191°17'10'' (Lalande)

Venus   224°17′32″   224°17′30″ (Lalande)
Mercury 53°43′13″   53°43′3″ (Lalande)

Thus the difference between Ulugh Beg's data and that of modern times relating to the first four planets falls within the limits of two to five seconds.

In the case of Mercury the difference is somewhat larger—ten seconds at most—because, of the planets mentioned, Mercury has the greatest orbital velocity. In addition, the eccentricity of its orbit is 0.206—that is, it is considerable in comparison with the eccentricity of the four other planets—while the greatest visible angular distance of Mercury from the sun's disk is only about 28°. These peculiarities of the planet make observation of it with the naked eye fairly difficult, and consequently have an adverse effect on the accuracy of the results of observation. The yearly precession was determined by Ulugh Beg[5] to be 51.4″, while the true value is 50.2″.

The situation is somewhat different with Ulugh Beg's values for the positions of the stars. After that of Hipparchus, the star catalog of Ulugh Beg was the second in seventeen centuries. It contains 1,018 stars, the positions of some of which were determined mainly from observations made at the Samarkand observatory, and others from observations made before the beginning of 1437 (A.H. 841).[6] The latter were taken from the star catalog of al-Sūfī, who apparently borrowed them from Ptolemy. Thus the star catalog of Ulugh Beg has great value, since it is basically original, but nevertheless was influenced by Ptolemy, at least in respect to its coordinates.

In 1941 an expedition under the leadership of T. N. Kari-Niazov discovered the tomb of Ulugh Beg in the mausoleum of Tamerlane in Samarkand. In contrast with the Islamic custom of burying the dead only in a shroud, Ulugh Beg lay fully clothed in a sarcophagus, in agreement with the prescription of the *shariat*: a man who died as a *shakhid* (martyr) had to be buried in his clothes. On the skeleton, traces of his violent death are clear: the third cervical vertebra was severed by a sharp instrument in such a way that the main portion of the body and an arc of that vertebra were cut off cleanly; the blow, struck from the left, also cut through the right corner of the lower jaw and its lower edge.

NOTES

1. Ulugh Beg, *Zīj Guragoni*, Biruni Institute of Oriental Studies, Uzbek S.S.R. Academy of Sciences, MS 2214, 1.11a.

2. *Ibid.*, 1. 102b.
3. Birjantsi, *Sharḥ, Zīj Guragoni*, Biruni Institute of Oriental Studies, Uzbek S.S.R. Academy of Sciences, MS 704, 1. 49a.
4. J. B. J. Delambre, *Histoire de l'astronomie indienne orientale* (Paris, 1787), 155.
5. Ulugh Beg., *op. cit.*, ll. 117b, 118a.
6. *Ibid.*

BIBLIOGRAPHY

See Abū Ṭahir Hoja, *Samaria*, from Tadzhik into Russian, V. L. Vyatkin, trans. (Samarkand, 1899); Z. Babur, *Babur-name*, M. A. Sale, trans. (Tashkent, 1948); V. V. Bartold, *Istoria kulturnoy zhizni Turkestana* ("History of the Cultural Life of Turkestan"; Leningrad, 1927); F. Baily, "The Catalogues of Ptolomey, Ulug-Beigh, Tycho Brahe, Halley and Hevelius, Deduced From the Best Authorities, With Various Notes and Corrections," in *Memoirs of the Royal Astronomical Society* (London, 1843); G. Bigourdan, *L'astronomie* (Paris, 1925); F. Boquet, *Histoire de l'astronomie* (Paris, 1925); C. Brockelmann, *Geschichte der arabischen Literatur* (Weimar, 1898); J. B. J. Delambre, *Histoire de l'astronomie du moyen âge* (Paris, 1819); I. Greave, *Binae tabulae geographicae, una Nassir-Eddini Persae, altera Ulug-Beigi Tartari* (London, 1652); and T. Hayde, *Tabulae longitudinis et latitudinis stellarum fixarum ex observatione Ulug-beighi* (Oxford, 1665).

Also see T. N. Kari-Niazov, *Observatoria Ulugbeka v svete novykh dannykh* ("The Observatory of Ulugh Beg in the Light of New Information"; Tashkent, 1947); and *Astronomicheskaya shkola Ulugbeka* ("The Astronomical School of Ulugh Beg"; Moscow, 1950; 2nd ed., enl., Tashkent, 1967); P. S. Laplace, *Précis de l'histoire de l'astronomie* (Paris, 1865); E. B. Knobel, *Ulughbeg's Catalogue of Stars* (Washington, D.C., 1917); Salih Zaki, *Asar-i Bakiya* ("Eternal Monument"; Constantinople, 1911); G. Sarton, *Introduction to the History of Science*, II (Baltimore, 1931); L. Sédillot, *Prolégomènes des tables astronomiques d'Oloug-Beg* (Paris, 1853); G. Sharpe, *Tabulae longitudinis et latitudinis stellarum fixarum ex observatione Ulugbeighi* (Oxford, 1767); V. P. Shcheglov, "K voprosu o geograficheskikh koordinatakh i azimute sekstanta observatorii Ulugbeka a g. Samarkande" ("Toward the Question of the Geographical Coordinates and the Azimuth of the Sextant at the Observatory of Ulugh Beg and of the City of Samarkand"), in *Astronomicheskiy zhurnal*, **30**, no. 2 (1953); H. Suter, *Die mathematiker und Astronomen der Araber und ihre Werke* (Leipzig, 1900); and V. L. Vyatkin, "Ochet o raskopkakh observatorii Mirza Ulugbeka v 1908 i 1909 godakh" ("An Account of the Excavations of the Observatory of Mirz Ulugh Beg in 1908 and 1909"), in *Izvestiya Russkago komiteta dlya izucheniya srednei i vostochnoi azii*, 2nd ser. (1912), no. 11.

T. N. KARI-NIAZOV

ULYANOV, VLADIMIR ILYICH. See **Lenin (Ulyanov), Vladimir Ilyich.**

ᶜUMAR AL-KHAYYĀMĪ. See **ᶜAl-Khayyāmī (or Khayyām), Ghiyāth al-Dīn Abu'l-Fatḥ ᶜUmar ibn Ibrāhīm al-Nīsābūrī (or al-Naysābūrī),** also known as **Omar Khayyam.**

ᶜUMAR IBN AL-FARRUKHĀN AL-ṬABARĪ (*fl.* Baghdad, Iraq, 762–812), *astrology, astronomy.*

Abū Ḥafṣ ʿUmar was the son of a native of Ṭabaristān, the Iranian province just south of the Caspian Sea, who bore the ancient Persian name Farrukhān; he was thus one of those Persian scholars who made the early Abbasid court a center for the translation of Pahlavi scientific texts into Arabic. He first appears on the scene as one of the group of astrologers, including Nawbakht, Māshāʾallāh, and al-Fazārī, whom al-Manṣūr asked to select an auspicious time for the foundation of Baghdad; they chose 30 July 762[1]. The latest date that we have for him is Shawwāl of A.H. 196—that is, 15 June–13 July of A.D. 812—when he finished his version of Ptolemy's *Kitāb al-arbaᶜa* (*Tetrabiblos*). These dates make it evident that Abū Maᶜshar was wrong in stating, as reported by his pupil Shādhān in his *Mudhākarāt*[2] and repeated by Ṣāᶜid al-Andalusī[3] and Ibn al-Qifṭī,[4] that ʿUmar was called to Baghdad by the wazīr, al-Faḍl ibn Sahl (*d.* 818), and introduced to al-Maʾmūn. Abū Maᶜshar's other statement[5] that he was devoted to Yaḥyā ibn Khālid ibn Barmak (*d.* 807) may well be true.

Of ʿUmar's personal life nothing else is known save that he had a son, Abū Bakr Muḥammad, who also wrote extensively on astrology and astronomy. Unfortunately, Ibn al-Nadīm[6] has often confused the father and son in his lists of their works. The following titles of ʿUmar's works, therefore, belong primarily to those texts known to us from more reliable sources.

1. A *tafsīr* or paraphrase of Ptolemy's *Tetrabiblos* finished between 15 June and 13 July of 812. This is preserved in Uppsala, Universitetsbibliotheket MS Arab. 203. According to the introduction, ʿUmar himself translated the text, presumably from a Pahlavi version; Ibn al-Nadīm states that he used the translation of Abū Yaḥyā al-Baṭrīq, presumably from the original Greek. The truth may be that he wrote his paraphrase, based on the Pahlavi, at the request of al-Baṭrīq.

2. A *tafsīr* of the astrological work of Dorotheus of Sidon, based on a Pahlavi recension of the early fifth century. This is preserved in two manuscripts: Yeni Cami 784 and Berlin or. oct. 2603. The present author is preparing an edition of this text.

3. *Mukhtaṣar masāʾil al-Qayṣarānī* ("Abridgment of the Caesarean (?) Interrogations") in 138 chapters. This work has been preserved in many manuscripts; I have examined Berlin Ar. 5878 and 5879, Escorial Ar. 938, and Beirut, Univ. St. Joseph Ar. 215. Though the name Qayṣarānī remains obscure, it certainly has nothing to do with the *Jāmiᶜ al-kitāb* of Abū Yūsuf Yaᶜqub ibn ᶜAlī al-Qaṣrānī who flourished at the courts of Jurjān and Astarābād in the late ninth century. The *Mukhtaṣar* may be identical with the *Kitāb al-ikhtiyārāt* ("Book of Elections) at Alexandria, MS Ḥurūf 12.

4. *Kitāb fi'l-mawālīd* ("Book About Nativities"), a short treatise on genethlialogy preserved in Arabic in only one manuscript (Nuru Osmaniye 2951, ff. 162v–172). This is probably identical with the Latin *De nativitatibus secundum Omar* in three books, translated by Iohannes Hispalensis (and a second time by Salomon with the help of the son of Abaumet the Jew in 1217?); see F. J. Carmody, *Arabic Astronomical and Astrological Sciences in Latin Translation* (Berkeley–Los Angeles, 1956), 38–39 (Carmody's *De iudiciis astrorum* is obviously al-Farghānī's, and *Laurentius Beham de ascensione termini Haomar* does not necessarily have any connection with ʿUmar ibn al-Farrukhān). I have consulted the edition by Nicolaus Prückner, *Iulii Firmici Materni . . . Libri VIII* (Basel, 1551), pt. 2, pp. 118–141. ʿUmar's sources are Ptolemy, Dorotheus, and Māshāʾallāh, as might have been expected.

5. *Kitāb al-ᶜilal.* This work is known to us only through a citation by al-Bīrūnī in his treatise on the solar equation (*Rasāʾil al-Bīrūnī* [Hyderabad, 1948], pt. I, p. 132), in which he gives approximate methods by which the sine of the solar equation corresponding to α is made to vary with $\sin \lambda (\alpha)$ and by which the solar equation corresponding to α is made to vary with the declination of $\lambda (\alpha)$. These methods are described by E. S. Kennedy and A. Muruwwa, "Bīrūnī on the Solar Equation," in *Journal of Near Eastern Studies,* **17** (1958), 112–121, esp. 118–119. Al-Bīrūnī seems to have devoted a treatise to exposing the ineptitude of ʿUmar's astronomy, as he lists in his bibliography (D. J. Boilot, "L'oeuvre d'al-Beruni: essai bibliographique," in *Mélanges de l'Institut Dominicain d'Études Orientales,* **2** [1955], 161–256) as no. 62 a *Fi'l-faḥs ᶜan nawādir abī Ḥafṣ ᶜUmar ibn al-*

Farrukhān ("On Inquiring About the Rarities of Abū Ḥafṣ ʿUmar ibn al-Farrukhān"), which, he claims, covers 240 folios.

NOTES

1. D. Pingree, "The Fragments of the Works of al-Fazārī," in *Journal of Near Eastern Studies*, **29** (1970), 103–123, esp. 104.
2. I have not succeeded in locating this story in the imperfect manuscripts of the *Mudhākarāt* available to me.
3. *Kitāb ṭabaqāt al-umam*, R. Blachère, trans. (Paris, 1935), 111. Ṣāʿid also elsewhere (p. 117) reports that ʿUmar wrote for al-Maʾmūn.
4. *Taʾrīkh al-ḥukamāʾ*, J. Lippert, ed. (Leipzig, 1903), 242.
5. *Ibid.*, the Greek translation is published by F. Cumont in *Catalogus codicum astrologorum graecorum*, V, pt. 1 (Brussels, 1904), 150–151.
6. *Fihrist*, G. Flügel, ed., 2 vols. (Leipzig, 1871–1872), I, p. 273; copied by Ibn al-Qifṭī, p. 242.

BIBLIOGRAPHY

There are short articles on ʿUmar in H. Suter, *Die Mathematiker und Astronomen der Araber und ihre Werke* (Leipzig, 1900), 7–8; and in C. Brockelmann, *Geschichte der arabischen Literatur*, I (Leiden, 1943), 249, and supp. I (Leiden, 1937), 392, where several additional treatises alleged to exist in manuscript are listed.

DAVID PINGREE

AL-UMAWĪ, ABŪ ʿABDALLĀH YAʿĪSH IBN IBRĀHĪM IBN YŪSUF IBN SIMĀK AL-ANDA-LUSĪ (*fl*. Damascus, fourteenth century), *arithmetic*.

Al-Umawī was a Spanish Arab who lived in Damascus, where he taught arithmetic. On the single authority of Ḥājjī Khalīfa, the year of his death is usually given as A.H. 895 (A.D. 1489/1490). But a marginal note on the ninth folio of his arithmetic (MS 1509, 1°, Carullah), written by him to give license to a copyist to teach his work, is dated 17 Dhuʾl-Ḥijja 774 (9 June 1373). The copyist is ʿAbd al-Qādir ibn Muḥammad ibnʿAbd al-Qādir, al-Ḥanbalī, al-Maqdisī. He states that he finished copying the text at Mount Qāsyūn in Damascus on 8 Dhuʾl-Ḥijja 774.

The text referred to is *Marāsim al-intisāb fī ʿilm al-ḥisāb*. A small work in eighteen folios, it is significant in being written by a western Muslim for Easterners, a circumstance that should not discredit the common belief that arithmetic flourished more in eastern than in western Islam. The work represents a trend of Arabic arithmetic in which, as early as the tenth century, the Indian "dust board" calculations had begun to be modified to suit paper and ink; and arithmetic was enriched by concepts from the traditional finger reckoning and the Pythagorean theory of numbers. The trend seems to have started in Damascus; the earliest extant text that shows it is al-Uqlīdisī's *al-Fuṣūl fiʾl-ḥisāb al-hindī*, written in A.H. 341 (A.D. 952/3). But there are reasons to believe that the trend had greater influence in the West than in the East.

The forms of the numerals used in the West differed from those in the East, but al-Umawī avoids using numerals except in a table of sequences, in which the western forms appear. The attempts to modify the Indian schemes resulted in several methods, especially of multiplication. Al-Umawī, however, says little about these methods and describes the principal operations briefly, as if his aim is to show what in western arithmetic is unknown, or not widely known, in the East. Thus he insists that the common fraction should be written as $\frac{a}{b}$, whereas the easterners continued to write it as $\overset{a}{\underset{b}{}}$, like the Indians, or as $\overset{0}{\underset{b}{a}}$.

He also insists that the numbers operated upon, say, in multiplication, must be separated from the steps of the operation by placing a straight line under them. Such lines appear in the works of Ibn al-Bannāʾ of Morocco (*d*. 1321) but not in the East until late in the Middle Ages.

Like the classical Indian authors, in treating addition al-Umawī dispenses with the operation in a few words and moves on to the summation of sequences. Those he discusses are the following:

1. The arithmetical progression in general and the sum of natural numbers, natural odd numbers, and natural even numbers in particular

2. The geometrical progression in general and 2^r and $\sum_{r=0}^{n} 2^r$ in particular

3. The sequences and series of polygonal numbers, namely $\{1 + (r - 1)d\}$ and $Sn = \sum_{r=1}^{n} \{1 + (r - 1)d\}$

4. The sequences and series of pyramidal numbers, namely $\{S_r\}$ and $\sum_{r=1}^{n} \{S_r\}$

5. Summations of r^3, $(2r + 1)^3$, $(2r)^3$ from $r = 1$ to $r = n$

6. Summations of $r(r + 1)$, $(2r + 1)(2r + 3)$, $2r(2r + 2)$ from $r = 1$ to $r = n$.

The sequences of polygonal and pyramidal numbers were transmitted to the Arabs in Thābit ibn Qurra's translation of Nicomachus' *Introduction to Arithmetic*. Also, al-Karajī had given geometrical proofs of $\Sigma\ r^3$, $(2r + 1)^3$, $(2r)^3$ in *al-Fakhrī* (see T. Heath, *Manual of Greek Mathematics* [Oxford, 1931], 68).

Without symbolism, al-Umawī often takes the sum of ten terms as an example, a practice started by the Babylonians and adopted by Diophantus and Arabic authors.

In subtraction al-Umawī considers casting out sevens, eights, nines, and elevens. All Hindu-Arabic arithmetic books consider casting out nines; and some add casting out other numbers. Some also treat casting out elevens in the way used today for testing divisibility by 11, which is attributed to Pierre Forcadel (1556). Al-Umawī adds casting out eights and sevens, in a way that leads directly to the following general rule:

Take any integer N in the decimal scale. Clearly $N = a_0 + a_1 \cdot 10 + a_2 \cdot 10^2 + \cdots = \Sigma\ a_s \cdot 10^s$. It is required to find the remainder after casting out p's from N, where p is any other integer. Let r_s be the remainder of 10^s, that is $10^s \equiv r_s \pmod{p}$; it follows that if $\Sigma\ a_s \cdot r_s$ is divisible by p so is N. This is a theorem that is attributed to Blaise Pascal (1664); see L. E. Dickson, *Theory of Numbers*, I (New York, 1952), p. 337.

In the text al-Umawī states that the sequence r_s, in the cases he considers, is finite and recurring. Thus for $p = 7$, $r_s = (1, 3, 2, 6, 4, 5)$.

In dealing with square and cube roots, al-Umawī states rules of approximation that are not as well developed as those of the arithmeticians of the East, who had already developed the following rules of approximation.

$$\sqrt{n} = a + \frac{n - a^2}{2a + 1},$$

where a^2 is the greatest integral square in n,

and

$$\sqrt[3]{n} = a + \frac{n - a^3}{3a^2 + 3a + 1},$$

where a^3 is the greatest integral cube in n. These rules do not appear in al-Umawī's text. Instead, we find

$$\sqrt{n} = a + \frac{n - a^2}{2a} \text{ or } (a + 1) - \frac{(a + 1)^2 - n}{2(a + 1)}$$

$$\sqrt[3]{n} = a + \frac{n - a^3}{3a^2} \text{ or } (a + 1) - \frac{(a + 1)^3 - n}{3(a + 1)^2}.$$

Again, al-Umawī does not consider the method of extracting roots of higher order, which had been known in the East since the eleventh century.

For finding perfect squares and cubes, however, he gives the following rules, most of which have not been found in other texts.

If n is a perfect square:

1. It must end with an even number of zeros, or have 1, 4, 5, 6, or 9 in the units' place.

2. If the units' place is 6, the tens' place must be odd; in all other cases it is even.

3. If the units' place is 1, the hundreds' place and half the tens' place must be both even or both odd.

4. If the units' place is 5, the tens' place is 2.

5. $n \equiv 0, 1, 2, 4 \pmod{7}$
   $\equiv 0, 1, 4 \pmod{8}$
   $\equiv 0, 1, 4, 7 \pmod{9}$

If n is a perfect cube:

1. If it ends with 0, 1, 4, 5, 6, or 9, its cube root ends with 000, 1, 4, 5, 6, or 9, respectively. If it ends with 3, 7, 2, or 8, the root ends with 7, 3, 8, or 2, respectively.

2. $n \equiv 0, 1, 6 \pmod{7}$
   $\equiv 0, 1, 3, 5, 7 \pmod{8}$
   $\equiv 0, 1, 8 \pmod{9}$

Evidently al-Umawī's *Marāsim al-intisāb fī ʾilm al-ḥisāb* is worthy of scholarly interest, especially in connection with the early history of number theory.

Another work by the same author is preserved in MS 5174 ḥ in Alexandria under the name of *Rafʿ al-ishkāl fī misāḥat al-ashkāl* (removal of doubts concerning the mensuration of figures); it is a small treatise of seventeen folios in which we find nothing on mensuration that the arithmeticians of the East did not know.

BIBLIOGRAPHY

On al-Umawī and his work, see C. Brockelmann, *Geschichte der arabischen Literatur*, supp. 2 (Leiden, 1938), p. 379, and II (Leiden, 1949), p. 344; L. E. Dickson, *History of the Theory of Numbers*, 3 vols. (New York, 1952); Ḥājjī Khalīfa, *Kashf alẓunūn . . .*, 2 vols. (Constantinople, 1941); T. L. Heath, *A History of Greek Mathematics*, 2 vols. (Oxford, 1921); Ibn al-Nadīm, *Al-Fihrist* (Cairo); Nicomachus, *Al-Madkhal ilā ʿilm al-ʿadad*, Thābit ibn Qurra, trans., W. Kutch, ed. (Beirut, 1958); and H. Suter, *Die Mathematiker und Astronomen der Araber und ihre Werke* (Leipzig, 1950), no. 453, p. 187.

A. S. SAIDAN

UNANUE, JOSÉ HIPÓLITO (*b.* Arica, Peru [now Chile], 13 August 1755; *d.* Lima, Peru, 15 July 1833), *natural history.*

Unanue, the outstanding figure of the Peruvian enlightenment, began ecclesiastical studies at Arequipa and Lima, but abandoned a church career for one in medicine. He received his medical degree under the direction of Gabriel Moreno around 1784 and by 1789 was professor of anatomy at the University of San Marcos.

From the beginning of his medical career, Unanue devoted himself to the cause of reform in medical and scientific education. In 1792 he succeeded in instituting an anatomical amphitheater in Lima, which the government had authorized nearly thirty-five years earlier. Lacking in equipment and funds, the amphitheater was nevertheless able to provide instruction in dissection for medical students, and Unanue lectured on anatomy there in 1793–1794. The following year Unanue inaugurated a series of clinical lectures in medicine, in which specialists instructed the students on specific diseases. During the same period (1791–1794) he was editor of the *Mercurio Peruano*, a prime conduit for the diffusion of modern scientific ideas in Peru and to which Unanue contributed a number of articles on medical and scientific subjects.

Between 1799 and 1805 Unanue collected data for his major work, *Observaciones sobre el clima de Lima*, a treatise in the Hippocratic tradition, purporting to explicate the climatic causes of disease in the city of Lima. To substantiate his thesis, Unanue correlated meteorological data with clinical observations, combined with traditional and modern medical concepts. In spite of his frequent citations of Newton, Boerhaave, and other modern scientists, Unanue's book has an archaic cast. He denied, for example, the relevance of chemistry to medical practice. The book comprises five sections. The first section describes the climate and physical setting of Lima. The second is an ecological discussion of climate and its influence on vegetation, animals, and human beings. The third expounds the influences of climate on disease. The fourth discusses what dietary and other curative recourses could be had to cure climate-induced diseases. The fifth, a medical profile of the year 1799 in Lima, is perhaps the most interesting part of the book, inasmuch as it provides a meteorological and epidemiological chronicle for that year. The book had tremendous influence, especially in Lima itself, where it appears to have been canonized by the local medical intelligentsia to the point that it inhibited the reception of new medical and scientific ideas.

In 1807 Viceroy José Abascal asked Unanue to submit plans for a new medical school and to serve as its first director. Unanue recommended, in a memorial of the same year, that a medical college be created in one of the hospitals of the city, where students could be instructed in anatomy, physiology, surgery, medicine, and pharmacy. The College of San Fernando opened in 1811 with ten professorships, many held by former students of Unanue. The curriculum, which included mathematics, psychology, and experimental physics, was decidedly modern, the texts having been selected by Unanue personally. At the same time (from 1807), Unanue also served Peru as medical inspector (*protomédico general*).

Like many Latin American men of science of the early nineteenth century, Unanue played a prominent role in his country's struggle for independence, serving as negotiator and cabinet minister.

BIBLIOGRAPHY

I. ORIGINAL WORKS. *Observaciones sobre el clima de Lima, y su influencia sobre los seres organizados, en especial el hombre* (Lima, 1806; 2nd ed., Madrid, 1814). A more recent edition is found in *Obras científicas y literarias del doctor don J. Hipólito Unanue*, Eugenio Larrabure y Unanue, ed., 3 vols. (Barcelona, 1914), I. The standard modern edition is by Carlos Enrique Paz Soldán (Lima, 1940).

Articles of scientific interest originally published in the *Mercurio Peruano*, 12 vols. (1791–1794; facs. ed., Lima, 1964–1966), are reprinted in *Obras científicas*, II.

II. SECONDARY LITERATURE. Various aspects of Unanue's scientific career are discussed in Luís Alayza y Paz Soldán, *Unanue, geógrafo, médico y estadista* (Lima, 1954); Juan B. Lastres, *Hipólito Unanue* (Lima, 1955); and "Hipólito Unanue y *El clima de Lima*," in *Boletín de la Sociedad geográfica de Lima*, **54** (1937), 75–87; Hugo Neira Samanez, *Hipólito Unanue y el nacimiento de la patria* (Lima, 1967); Carlos Enrique Paz Soldán, *Hímnos a Hipólito Unanue* (Lima, 1955); Hermilio Valdizán, "El doctor don Hipólito Unanue (Apuntes bio-bibliográficos)," in *Unanue. Revista Trimestral de Historia de la Medicina Peruana*, II (Lima), nos. 1–2 (Mar.–June 1926), 3–57; and John E. Woodham, "The influence of Hipólito Unanue on Peruvian Medical Science: A Reappraisal," in *Hispanic American Historical Review*, **50** (1970), 693–714.

THOMAS F. GLICK

UNGER, FRANZ (*b.* Der Gute Amthof, near Leutschach, Austria, 30 November 1800; *d.* Graz, Austria, 13 February 1870), *botany.*

The son of a Styrian jurist, Unger received a Gymnasium education at Graz. His father intended that he become a lawyer; but, under the influence of his friend Anton Sauter, Unger turned to medicine. He studied at the universities of Vienna and Prague, qualifying in 1827. After two years at Stockerau as a general practitioner, he became physician to the assize court in Kitzbühel. In 1835 Unger moved to Graz as professor of botany and zoology and as director of the botanical garden at the Johanneum. He received an offer from the University of Vienna in 1849, when a new chair of plant anatomy and physiology was established there. Unger held that post until 1866, when he retired to Graz.

In his own time Unger's fame rested chiefly on the *Grundzüge der Botanik* (1843), which he wrote with Stephen Endlicher. The work contains a description of cell multiplication by division that is based on Unger's St. Petersburg prize essay (1840) on the structure of the flowering plant stem and his *Aphorismen* (1838). The *Grundzüge* made him Schleiden's first opponent on the question of the origin of cells, Schleiden advocating free cell formation, while Unger confined this mechanism to the early stages in the formation of organs and put forward as its successor cell multiplication by division. Unger maintained his position in this area of botany with *Der Anatomie und Physiologie der Pflanzen* (1855).

Unger was an evolutionist before 1859. His conception of the succession of species clearly was influenced by *Naturphilosophie*, but it also was supported by experimental studies of the impact of changes in the conditions of life upon plant variability. These studies, carried out at Kitzbühel, Graz, and Vienna in the 1830's and 1840's, led Unger to oppose the popular link between external influences, especially soil conditions, and variability. His views on evolution were presented in a popular form in a series of articles in the local press in Vienna, and subsequently collected in book form as *Botanische Briefe* (1852). This piece of scientific journalism provoked a violent personal attack upon Unger by the Catholic press. His resignation was prevented by a strong student protest.

Unger's evolutionary views also found expression in his attempt to reconstruct the botanical features of the landscape in earlier geological eras, in *Die Urwelt in ihren verschiedenen Uebergangs-*

perioden (1851). This and other works reflect his deep interest in paleobotany. He was also a pioneer in the study of the early history of cultivated plants as revealed through cultural relics.

Unger also contributed to the understanding of fertilization in the lower plants. Although at first holding the mistaken belief that antherozoids were infusorians, in 1837 he argued effectively for the identification of these bodies in the mosses as the male gametes. No longer did he regard the production of antherozoids as signaling the conversion of plant into animal.

Among Unger's students at Vienna was Gregor Mendel. Unger's involvement in the working out of the cell theory and its application to the fertilization process may well have played a crucial role in equipping Mendel for the cytological interpretation of his breeding experiments.

BIBLIOGRAPHY

I. ORIGINAL WORKS. A bibliography of Unger's publications is in A. Reyer's biography cited below. Unger's works include *Über den Einflusse des Bodens auf die Vertheilung der Gewächse* (Vienna, 1836); a series of books on anatomy and physiology: *Aphorismen zur Anatomie und Physiologie der Pflanzen* (Vienna, 1838), *Grundzüge der Anatomie und Physiologie der Pflanzen* (Vienna, 1846), and *Der Anatomie und Physiologie der Pflanzen* (Vienna, 1855); and several books on paleobotany: *Synopsis plantarum fossilium* (Leipzig, 1845); *Genera et species plantarum fossilium* (Vienna, 1850); *Die Urwelt in ihren verschiedenen Bildungsperioden* (Vienna, 1851), translated as *Ideal Views of the Primitive World in Its Geological and Palaeontological Phases* (London, 1863); and *Versuch der Geschichte der Pflanzenwelt* (Vienna, 1852). Unger's important prize essay is *Über den Bau des Dikotyledonenstammes* (St. Petersburg, 1840).

Other works include *Grundzüge der Botanik* (Vienna, 1843), written with S. Endlicher; "Beiträge zur Lehre von der Bodenstätigkeit gewisser Pflanzen," in *Denkschriften der Akademie der Wissenschaften*, **1** (1850), 83–89, written with F. Hruschauer; and *Gratz. Ein naturhistorisch-statistisch-topographisches Gemählde dieser Stadt und ihrer Umgebungen* (Gratz, 1843), written with A. von Muchar, C. Weiglein, and G. F. Schreiner. Unger's best-known popular work is *Botanische Briefe* (Vienna, 1852), translated as *Botanical Letters to a Friend* (London, 1863).

Unger's papers on botany in relation to cultural history appeared in the *Sitzungsberichte der Akademie der Wissenschaften in Wien*, and all the papers are listed in Reyer's biography and in the Royal Society *Catalogue of Scientific Papers*, **6**, p. 87; **8**, p. 1137.

Unger's correspondence with Endlicher is in G. Haberlandt, ed., *Briefwechsel zwischen Franz Unger und Stephen Endlicher* (Berlin, 1899).

II. SECONDARY LITERATURE. The standard biography of Unger is A. Reyer, *Leben und Werke des Naturhistorikers Dr. Franz Unger* (Graz, 1871). Unger's cytological studies are discussed in Julius Sachs, *History of Botany (1530–1860)* (Oxford, 1906), 325–329, 336–340; and in J. Lorch, "The Elusive Cambium," in *Archives internationales d'histoire des sciences*, **20** (1967), 253–283. The relationship between Unger and Mendel is discussed by R. C. Olby, in "Franz Unger and the Wiener Kirchenzeitung: An Attack on One of Mendel's Teachers by the Editor of a Catholic Newspaper," in *Folia Mendeliana*, no. 2 (1967), 29–37; and in "The Influence of Physiology on Hereditary Theories in the Nineteenth Century," *ibid.*, no. 6 (1971), 99–103. For an evaluation of some of Unger's scientific work, see Johanna Enslein, *Die wissenschaftgeschichtliche Untersuchung und Wertung der anatomischen, physiologischen und ökologischen Arbeiten von Franz Unger* (Vienna, 1956).

ROBERT OLBY

UNZER, JOHANN AUGUST (*b.* Halle, Germany, 29 April 1727; *d.* Altona, Germany, 2 April 1799), *physiology*, *medicine*.

Unzer began the study of medicine when he was only twelve; his early teacher was Johann Juncker, an ardent disciple of Stahl. He received the M.D. in 1748, but even before then, in 1746, he had begun to publish metaphysical and philosophical works on such physiological problems as life, the emotions, and sleep, in which he defended Stahl's animistic doctrines. Nonetheless, he also seems to have been influenced by the mechanistic and eclectic medicine of Halle's other leading man, Stahl's opponent Friedrich Hoffmann, who was still teaching at the university when Unzer began his studies.

In 1750 Unzer left Halle to take up a busy medical practice, first in Hamburg and then in Altona. (Although some sources suggest that he was for a time professor at the small University of Rinteln, this cannot now be confirmed.) At the same time, he conducted research and, from 1759 to 1764, edited a popular medical weekly, *Der Arzt*, and a series of collections of medical writings that were translated into Dutch, Danish, and Swedish. His writings of this period indicate his gradual abandonment of the Stahlian system and his attempts to work out his own physiology, or "physiological metaphysics," as it was later characterized. While Unzer remained interested in the basic phenomena of life, especially the higher animal functions, he turned from his earlier animism toward a more anatomical and physiological approach, directed principally to the role of the nervous system in animal functions. His *Erste Gründe einer Physiologie der eigentlichen thierischen Natur thierischer Körper* (1771) is the product of twenty-five years of research and reflection.

The *Erste Gründe* (later translated into English as *The Principles of Physiology of the Proper Animal Nature of the Animal Organism*) marked an attempt to establish the fundamental bases of zoology, considered as a natural science embracing all the animal kingdom, according to the "forces" of each species. In it, Unzer made use of a broad comparative method, which he applied particularly to nerve functions and to motion. He distinguished three types of motion—those that are dependent on the will, those that (although conscious) are independent of the will, and those that are wholly unconscious and involitional. He proceeded from the notion of the animal as a machine to state that some animals (*beseelte Tiere*) have a soul that produces their movements, while others (*unbeseelte Tiere*) have neither soul nor brain, and are instead moved by animal "forces."

Unzer was led to draw this distinction by a series of observations, particularly of decapitated higher animals, which he compared to lower animals that have no brain. He concluded that the brain is the seat of the soul, although animal machines are capable of organic work without the stimulus of a brain or soul, and many animal movements occur through neural stimulus only. He recognized that external stimuli tend to be referred toward the brain but noted that they could be reflected or deviated and localized either in the brain or at a lower level in nerve crossings. He thus distinguished the afferent (*aufleitend*) and efferent (*ableitend*) nerves. He made a thorough study of the nervous reactions, and noted that external stimuli are transmitted in the nervous system by reflection. He also advanced the notion that motor phenomena may be caused by external stimuli that are not consciously perceived and emphasized the difference between voluntary and involuntary movements. Unzer's effort toward defining a rational concept of reflex action was elaborated by Georgius Prochaska in 1784.

Although Unzer's original contribution to science was slight, he nonetheless provided a valuable step in the development of physiology. His careful and essentially correct presentation of the mechanical and material aspects of nerve functions bridged the gap between conflicting views and became the basis of a considerable body of work on the nervous system in the nineteenth century. Unzer's wife, Johanna Charlotte Ziegler, was also a writer on natural history.

BIBLIOGRAPHY

I. ORIGINAL WORKS. Unzer's physiological writings include *Gedanken vom Einflusse der Seele in ihrem Körper* (Halle, 1746); *Gedanken vom Schlaf* (Halle, 1746); *Neue Lehre von den Gemüthsbewegungen* (Halle, 1746); *Dissertatio inauguralis medica de sternutatione* (Halle–Magdeburg, 1748); *Philosophische Betrachtung des menschlichen Körpers überhaupt* (Halle, 1750); and *Grundriss eines Lehrgebäudes von der Sinnlichkeit der thierischen Körper* (Rinteln, 1768). The last was a preliminary study for his most important book, *Erste Gründe einer Physiologie der eigentlichen thierischen Natur thierischer Körper* (Leipzig, 1771), which was trans. into English (London, 1851) and is quoted by Fearing, below, and by Edward E. Clarke and C. D. O'Malley in *The Human Brain and Spinal Cord* (Berkeley–Los Angeles, 1968), 342–345.

His medical books include *Sammlung kleiner physikalischen Schriften*: vols. I and III, *Physikalische Schriften*; vol. II, *Zur speculativen Philosophie* (Leipzig, 1768–1769); *Medizinisches Handbuch*, 3 vols. (Leipzig, 1770; 5th ed. 1794); *Über die Ansteckung, besonders der Pocken, in einer Beurtheilung der neuen Hoffmann'schen Pockentheorie* (Leipzig, 1778); *Einleitung zur allgemeinen Pathologie der ansteckenden Krankheiten* (Leipzig, 1782); and *Verteidigung seiner Einwürfe gegen die Pockentheorie des Hrn. Geh. Rath Hoffmann* (Leipzig, 1783).

II. SECONDARY LITERATURE. There are few biographical works about Unzer, but see T. Kirchhoff, in *Deutsche Irrenärzte*, I (Berlin, 1921), 13–15. On his work, particularly in the development of animal physiology and on reflex action, see Georges Canguilhem, *La formation du concept de réflexe aux XVIIe et XVIIIe siècles* (Paris, 1955), 108–114; Franklin Fearing, *Reflex Action. A Study in the History of Physiological Psychology* (London, 1930), 90–93; C. F. Hodge, "A Sketch of the History of Reflex Action," in *American Journal of Psychology*, **3** (1890), 149–167, 343–363; Thomas Laycock, intro. to *The Principles of Physiology of the Proper Animal Nature of the Animal Organism* (his trans. of Unzer's *Erste Gründe*) (London, 1851); and Max Neuburger, *Die historische Entwicklung der experimentellen Gehirn- und Rückenmarksphysiologie vor Flourens* (Stuttgart, 1897).

VLADISLAV KRUTA

AL-UQLĪDISĪ, ABU'L-ḤASAN AḤMAD IBN IBRĀHĪM (*fl.* Damascus, 952–953), *arithmetic.*

No source book mentions al-Uqlīdisī. He is known only from a unique copy of his work entitled *Kitāb al-fuṣūl fī'l-ḥisāb al-hindī* (MS 802, Yeni Cami, Istanbul), the front page of which bears the author's name and the statement that the text was written at Damascus in 952–953. The manuscript was copied in A.D. 1157. In the introduction the author states that he has traveled extensively, read all books on Indian arithmetic that he has found, and learned from every noted arithmetician he has met. The epithet al-Uqlīdisī generally was attached to the names of persons who made copies of Euclid's *Elements* for sale, so it is possible that he earned his living in that way. Internal evidence shows that he had experience in teaching Indian arithmetic, for he knows what beginners ask and how to answer their questions.

The book is in four parts. In the first, Hindu numerals are introduced; the place-value concept is explained; and the arithmetical operations, including extraction of square roots, are described, with many examples applied to integers and common fractions in both the decimal and the sexagesimal systems.

In the second part the subject matter is treated at a higher level and includes the method of casting out nines and several variations of the schemes of operations explained in the first part. In the introduction the author states that in this part he has collected the methods used by noted manipulators, expressed in the Indian way. This section contains almost all the schemes of multiplication that appear in later Latin works.

In the third part, justifications of the several concepts and steps suggested in the first two parts are given, generally in answer to questions beginning "Why" or "How is it."

A few words may be necessary for an appreciation of the fourth part. The first few lines of the text state that Indian arithmetic, as transmitted to the Arabs, required the use of the dust abacus. Later it is said that the operations depended upon shifting the figures and erasing them. For instance, in the example 329×456, the numbers are written as shown below:

$$3\ 2\ 9$$

$$4\ 5\ 6$$

Then 3 is multiplied by 4 and the product is inserted in the top line as 12; 3 is multiplied by 5, which requires putting 5 above, erasing 2, and putting 3 in its place; 3 is multiplied by 6, making it necessary to remove 3 from the top line and write 8 in its place, to erase the 5 before it, and to put 6 in its place. In preparation for the next step, the lower line is shifted one place to the right. The array is now as shown below:

$$1\ 3\ 6\ 8\ 2\ 9$$
$$4\ 5\ 6$$

456 is to be multiplied by 2, which is above the units place of 456; the position of the units digit of the multiplicand in the lower line indicates the multiplier. The remaining steps can now be followed with ease.

Obviously paper and ink cannot be easily used with such schemes. In the fourth part of the text, modifications of the Indian schemes are suggested whereby the abacus can be dispensed with, and ink and paper used instead. We can now judge that al-Uqlīdisī's modification presents a first step in a long chain of attempts that resulted in discarding the abacus completely, first in western Islam and, many centuries later, in the eastern part.

After suggesting a modification of each operation, al-Uqlīdisī proposed that:

1. Greek letters might replace the nine Indian numerals.

2. The Indian numerals with superimposed dots might form a new Arabic alphabet.

3. There might be calculating dice, with one or two numerals on each face, to use instead of the abacus.

4. There might be a calculating board to be used by the blind.

The second idea is cited in other texts, and the third is reminiscent of Boëthius' apexes. It is as likely as not that here al-Uqlīdisī is describing methods used elsewhere rather than making original suggestions. The book ends with a lengthy discussion of $\Sigma 2^r$ and the method of extracting the cube root.

Al-Uqlīdisī was proud of the following accomplishments in his work:

1. In part 1 he presented the contents of all earlier texts on Indian arithmetic and applied it in the sexagesimal system. We do not have these texts to enable us to judge how far he was correct in this claim. The Latin *Algorismus corpus*, however, indicates that Indian arithmetic as presented by al-Khwārizmī (ninth century) differed basically from that which spread later in the Muslim world. Application of the Indian schemes to the sexagesimal system is found in all later Arabic arithmetic books.

2. In part 2 he gave methods known only to noted arithmeticians, and extended the method of casting out nines to fractions and square roots. On the evidence of later texts, one is inclined to accept this claim of al-Uqlīdisī.

3. In part 4, he showed that Indian arithmetic no longer needed the abacus. This modification was more agreeable to the West than to the East. In support we may note that Ibn al-Bannā' (*d.* 1321) of Morocco included as a curiosity in one of his arithmetical works the statement that the ancients had used dust for calculation, whereas Naṣīr al-Dīn al-Ṭūsī (*d.* 1274) of Persia found the dust abacus still important enough to write a book on it.

4. In discussing $\Sigma 2^r$ he distinguished between the *n*th term and the sum of *n* terms, which he claimed that some manipulators had confused.

5. He claimed to be the first to have written satisfactorily on the cube root.

There are no documents to decide the last two claims, but we have other reasons to consider al-Uqlīdisī's *Kitāb al-fuṣūl fi'l-ḥisāb al-hindī* the most important of some one hundred extant Arabic arithmetic texts.

First, it is the earliest known text that contains a direct treatment of decimal fractions. The author suggests a decimal sign, a stroke over the units' place, and insists that it must always be used. In a process of successive division by 2 he obtains the sequence 13, 6.5, 3.25, 1.625, 0.8125. He knows how to regain 13 by successive multiplication by 2 and by ignoring the zeros to the right. In a process of repeatedly increasing 135 by one-tenth, he obtains the array

135	148.5	163.35
13.5	14.85	16.335 and so on.
148.5	163.35	179.685

Again, in finding the approximate roots of numbers, he uses the rules

$$\sqrt{a} = \sqrt{ak^2}/k, \qquad \sqrt[3]{a} = \sqrt[3]{ak^3}/k$$

and takes *k* equal to a multiple of ten.

Although many other arithmeticians used the same rules, all of them rather mechanically transformed the decimal fraction obtained into the sexagesimal system, without showing any sign of comprehension of the decimal idea. Only al-Uqlīdisī gives the root in the decimal scale in several cases. In all operations where powers of ten are involved in the numerator or the denominator, he is well at home.

Second, al-Uqlīdisī's is the first text to tell us clearly that Indian arithmetic depended on the dust abacus. In his introduction, the author compared the Indian system with the then current finger-reckoning and made a correct evaluation of the merits and drawbacks of each. It is now known that Abu'l-Wafā' (940–997/998) and Ibn al-Bannā' made passing statements about the dust abacus in Indian arithmetic, but these references were too terse to catch the attention of the scholars who first studied their works.

BIBLIOGRAPHY

See A. S. Saidan, "The Earliest Extant Arabic Arithmetic," in *Isis*, **57** (1966), 475–490.

A. S. SAIDAN

URBAN, GEORGES (*b*. Paris, France, 12 April 1872; *d*. Paris, 5 November 1938), *chemistry, mineralogy.*

The son of a professor of chemistry and assistant to Edmond Frémy, Urbain entered the École de Physique et de Chimie in Paris at his father's request and graduated first in his class in 1894. While serving as assistant in the mineral laboratory, he came under the influence of Pierre Curie, who introduced him to scientific research. From 1895 until 1898 he was the private assistant of Charles Friedel and was awarded his doctorate in 1899 from the University of Paris for his thesis on the rare earths. In 1906 Urbain was named assistant professor of analytical chemistry at the Sorbonne and was rapidly promoted to professor of mineral chemistry in 1908. In 1928 he was appointed professor of general chemistry, director of the Institut de Chimie de Paris, and codirector of the Institut de Biologie Physico-chimique. During his term as professor at the Institut he created a mecca for good chemistry students. His lectures were well-delivered and extremely popular.

Urbain's name is linked with his important studies of the rare earths, which occupied him mainly from 1895 until 1912. More than 200,000 fractional crystallizations enabled him to separate rigorously samarium, europium, gadolinium, terbium, dysprosium, and holmium. He found that the ethyl sulfates of the rare earths were the easiest derivatives to separate. The most important single study was his separation of ytterbium (considered an element by Jean Marignac) into ytterbium and the previously unknown lutetium, named after Lutetia, the ancient name of Paris. Urbain's determinations of the atomic weights of these elements were accepted by the authorities and led to his la-ter election as president of the International Committee on Atomic Weights.

In 1911 Urbain observed another element, not of the rare earth family, which he named celtium. Henry Moseley felt that it was the missing element 72, although X-ray evidence was inconclusive. After World War I, Urbain continued his investigations and, in 1922, confirmed the presence of element 72 in his samples. But Hevesy and Coster isolated larger concentrations of the element that same year and are credited with the discovery of hafnium. For a short period of time the committee accepted both the symbols Ct and Hf for the element.

After 1912 Urbain's interests turned increasingly toward theoretical complex chemistry. He critically evaluated and extended Alfred Werner's coordination theory and proposed his own theory of homeomerism (equal properties). This concept extended the idea of isomorphism beyond the limits of the crystalline state by dropping the condition of equality of interfacial angles. Two substances, then, in order to be homeomeric must have equal molecular coefficients of energy—properties that are generally easy to measure. Thus he defined isotopes as elements that are nearly rigorously homeomeric.

Among Urbain's other deep interests were music and the history and philosophy of science. A fine piano player, he read most of the didactic works on music and composed several brilliant pieces as well as a book on the subject. He keenly analyzed the development of chemistry in a work that has been compared to those of Jean-Baptiste-André Dumas. He firmly believed in the essential unity of inorganic and organic chemistry and felt that by modifying Werner's theory he could achieve this unification for theoretical chemistry.

BIBLIOGRAPHY

I. ORIGINAL WORKS. A complete listing of Urbain's papers is found in Paul Job, "Notice sur la vie et les travaux de Georges Urbain," in *Bulletin de la Société chimique de France*, **6** (1939), 745–766.

Urbain wrote seven books on chemistry. They are, in chronological order, *Introduction à l'étude de la spectrochemie* (Paris, 1911), which is not an advanced treatise on spectrochemistry but is rather an introduction for the beginner on methods of special analysis, including a lengthy historical survey of the subject; *Introduction à la chimie des complexes minéraux* (Paris, 1914), with A. Sénéchal, which is based on Urbain's course at the Sorbonne and, despite its title, is a general text on physical and inorganic chemistry; *Les disciplines d'une science. La chimie* (Paris, 1921), which discusses his attempt to unify inorganic and organic chemistry; *L'énergétique des réactions chimiques* (Paris, 1925), which is a theoretical introduction to chemical thermodynamics based on a course given at the Sorbonne; *Les notions fondamentales d'éléments chimiques et d'atomes* (Paris, 1925); and *La coordination des atomes dans la molecule et la symbolique chimique*, 2 vols. (Paris, 1933), which are two purely theoretical treatises; and *Traité de chimie générale, notions et principes fondamentaux* (Paris, 1939), with P. Job, G. Allard, and G. Champetier, but mostly the work of Urbain.

Urbain places his rare earth studies in their historical perspective in his review article, "Research on Yttrium Earths," in *Chemical Reviews*, **1** (1924), 143–185.

Worthy of note is his book *Le Tombeau d'Aristoxène. Essai sur la musique* (Paris, 1924), in which he proposed that music is more intellectual than sensuous and, consequently, can be the object of a methodical (if not truly scientific) study.

II. SECONDARY LITERATURE. The most detailed biographical sketch of Urbain is Paul Job's obituary notice cited above. An interesting English article written by two of Urbain's former students is Georges Champetier and Charlotte H. Boatner, "Georges Urbain," in *Journal of Chemical Education*, **17** (1940), 103–109.

SHELDON J. KOPPERL

URE, ANDREW (*b*. Glasgow, Scotland, 18 May 1778; *d*. London, England, 2 January 1857), *chemistry*.

A pioneer in the teaching of science to artisans, Ure was also one of the first scientists to earn his living as a consultant. Apart from his teaching and some improvements in the techniques of chemical analysis, he made no significant contributions to the advancement of science. Yet he was an indefatigable writer and encyclopedist, and his major work, the *Philosophy of Manufactures . . .* (1835), while containing ridiculous passages, nevertheless embodies the first clear recognition that what came to be called the industrial revolution was a novel and irreversible alteration in the human condition. Unfortunately, his intemperate scorn for his contemporaries and the self-aggrandizement that characterizes much of his writing obscured his positive qualities and made him many enemies.

Ure was the son of a cheesemonger, Alexander Ure, and his wife Anne Adam. He graduated M.D. at the University of Glasgow in 1801, and after a short period as an army surgeon succeeded George Birkbeck in 1804 as professor of natural philosophy at the Andersonian Institution (now the University of Strathclyde), Glasgow, then in its formative years. Ure's evening lectures in chemistry and mechanics for artisans were extremely popular, although his undoubted success as a teacher was marred by the air of conscious superiority that he adopted.

In 1814 Ure began to lecture at the Royal Belfast Academical Institution during the summer vacations. While in Ireland in 1816 he tried unsuccessfully to interest the Linen Board in a new method of determining the strength of alkaline solutions. His innovation, which he did not publish until much later (*Pharmaceutical Journal*, **3** [1844], 430–450), was to employ solutions of acids and alkalies in such concentrations that these could be expressed in terms of their chemical equivalents. Thus, he was led to originate the concept of normality in volumetric analysis (see W. V. Farrar, "The Origin of Normality," in *Education in Chemistry*, **4** [1967], 277–279).

In 1818 Ure created a sensation by conducting in public a gruesome experiment on the activation by electricity of the muscles of an executed murderer. Of Ure's many papers, one describing improvements in the use of copper oxide in organic analysis was of contemporary importance. His most enduring papers were those relating the composition and densities of aqueous solutions of the mineral acids—the results were quoted in chemist's pocket books until nearly the end of the century. Ure became a fellow of the Royal Society in 1822. The great interest he took in the application of science to the arts and industry bore fruit in his *Dictionary of Chemistry* (1821). This work was originally undertaken at the request of the publishers as a revision of William Nicholson's *Dictionary of Chemistry*, but Ure said that so much of the latter was obsolete that the work had largely to be rewritten. He eventually widened its scope and retitled the work the *Dictionary of Arts, Manufactures and Mines* (1839).

In 1829 Ure published his *System of Geology*, an outdated attempt to reconcile contemporary geological discoveries with the Mosaic account of the Creation. He thought that geologists, by questioning the latter, were undermining the bases of religion and morality. The book was the subject of a devastating attack by Adam Sedgwick at a meeting of the Geological Society of London in 1830 (*Proceedings of the Geological Society of London*, **1** [1834], 208–210). Quite apart from defending a lost cause, Ure had not kept himself up to date and many old errors had been repeated. Moreover, he included considerable extracts from other works without any indication of quotation. Sedgwick concluded that Ure had "shown neither the information nor the industry which might justify him in becoming an interpreter of the labours of others, or the framer of a system of his own."

As Ure's extramural activities increased his teaching seriously deteriorated, leading to many disputes with the managers of the Institution. He resigned in 1830, moved to London, and became probably the first consulting chemist in Great Britain, ready to offer advice on any problem connected with science. His investigations and matters of public concern provided materials for an unceasing flow of writings, including a series of tendentious pamphlets, in which his fellow scientists were frequently castigated. His *Philosophy of Manufactures*, intended as an introductory volume to a series dealing with various industries, was based on a tour of the manufacturing districts of Lancashire, Derbyshire, and Cheshire.

Ure married Catharine Monteath in 1807, but he divorced her in 1819 because of her infidelity. They had two sons and a daughter; Alexander, the elder son, became a London surgeon and wrote several papers. He has sometimes been confused with his father.

BIBLIOGRAPHY

I. ORIGINAL WORKS. Ure's books are *A New Systematic Table of the Materia Medica, With a Preliminary Dissertation, Historical, Critical, and Explanatory, on the Operation of Medicines* (Glasgow, 1813); *A Dictionary of Chemistry on the Basis of Mr. Nicholson* (London, 1821; 4th ed., 1835); *A Dictionary of Arts, Manufactures and Mines* (London, 1839; 4th ed., 1853; 7th ed., 1875); *A New System of Geology* (London, 1829), which was an attempt to reconcile contemporary geological discoveries with the Mosaic account of the Creation; *The Philosophy of Manufactures, or an Exposition of the Scientific, Moral and Commercial Economy of the Factory System of Great Britain* (London, 1835); *The Cotton Manufacture of Great Britain* (London, 1836), which was the first and only one published of an intended series—posthumous editions of this and the preliminary volume were published in 1861, with additions by P. L. Simmonds; and a translation of Claude Louis Berthollet and A. B. Berthollet, *Elements of the Art of Dyeing; With a Description of the Art of Bleaching by Oxymuriatic Acid*, 2 vols. (London, 1824).

Ure also published several pamphlets and a large number of papers, most of which were short and trivial; an incomplete list (though omitting little of any importance) is in *The Royal Society Catalogue of Scientific Papers*, VI (London, 1872), 89–90.

II. SECONDARY LITERATURE. The fullest and most accurate account of Ure is W. V. Farrar, "Andrew Ure and the Philosophy of Manufactures," in *Notes and Records of the Royal Society of London*, **27** (1972–1973), 299–324, with portrait of Ure. See also W. S. C. Copeman, "Andrew Ure, M.D., F.R.S. (1778–1857)," in *Proceedings of the Royal Society of Medicine*, **44** (1951), 655–662. The second article, although based on family papers (the author was a descendant of Ure), is unreliable, particularly with regard to Ure's work. *Dr. Andrew Ure: A Slight Sketch Reprinted From the Times and Other Periodicals, of January, 1857* (anonymous, privately printed, 1875) is eulogistic and misleading.

E. L. SCOTT

URYSON, PAVEL SAMUILOVICH (*b.* Odessa, Russia, 3 February 1898; *d.* Batz, France, 17 August 1924), *mathematics.*

The son of a distinguished Odessa financier, Uryson attended a private secondary school in Moscow in 1915 and entered the University of Moscow, intending to study physics. The same year he published his first scientific work, on Coolidge tube radiation, prepared under the guidance of P. P. Lazarev. Fascinated by the lectures of D. F. Egorov and N. N. Luzin, Uryson began specializing in mathematics and in 1919, after graduating from the university, he remained there to prepare for a teaching career. Uryson's works were at first concerned with integral equations and with other problems of analysis; but in the summer of 1921, being engaged in solving two problems presented to him by Egorov, he turned to topology. In June of that year Uryson was appointed assistant professor at the University of Moscow, where, in particular, he lectured on topology and, later, in 1923–1924, on the mathematical theory of relativity. He was also professor at the Second Moscow University (now the Lenin Moscow Pedagogical Institute).

Uryson's publications on topology first appeared in 1922 in the *Comptes rendus* of the Académie des Sciences, as well as in Soviet and Polish journals. His ideas were also presented in lectures, memoirs, and discourses. The reports he delivered at the Mathematical Society of Göttingen in 1923 attracted the attention of Hilbert; and in the summer of 1924, while touring Germany, Holland, and France, he met L. E. J. Brouwer and Felix Hausdorff, who praised his works highly. Uryson drowned off the coast of Brittany at the age of twenty-six while on vacation.

Although his scientific activity lasted for only

about five years, he greatly influenced the subsequent development of topology and laid the foundations of the Soviet school of topology, which was then led by his friend P. S. Aleksandrov, with whom he carried out several investigations.

The two cardinal aspects of Uryson's works on topology are topological space (abstract topology) and the theory of dimensionality.

In abstract topology his main results are the introduction and investigation of a class of the so-called normal spaces, metrization theorems, including a theorem on the existence of a topological mapping of any normed space with a countable base into Hilbert space.

The principal tool used in all the most recent investigations of normed spaces is the classical "Uryson's lemma," which proves the existence, for any two disjoint closed sets of a normed space, of a continuous function $f(x)$ which is defined over the given space. It satisfies the inequality $0 \leq f(x) \leq 1$ within that space and assumes on one of the two given sets the value zero, whereas on the other set it assumes the value of unity. Based on this lemma is Uryson's theorem on the metrizability of normed spaces having a countable base, and the theorem on the possibility of extending any continuous function defined on a closed set of a normed space R, to a function continuous over the entire space R. Both theorems are fundamental in general topology.

The theory of dimensionality created by Uryson in 1921–1922 was presented in his memoirs on Cantorian varieties, published posthumously in 1925–1926. In this work Uryson first presented an inductive definition of dimensionality that proved highly fruitful and became classical. Uryson then established that dimensionality, in the sense of the new definition of the n-dimensional Euclidean space R^n, actually equals n. In the process Uryson obtained a number of important results.

For $n = 4$ the equality dim $R^n = n$ was proved only by going beyond the limits of the inductive definition of dimensionality. Uryson's proof of the "theorem of equivalence" appeared to be a turning point in the development of the theory of dimensionality and of a considerable part of topology in general. The second part of this work is devoted to the creation of the theory of one-dimensional continua, in particular, their indexes of branching and continua of condensation.

Concurrently with and independently of Uryson, the Austrian mathematician Karl Menger was engaged in the same field; and the theory of dimensionality is often referred to as the Uryson-Menger theory.

BIBLIOGRAPHY

I. ORIGINAL WORKS. Uryson's collected works on topology and other branches of mathematics were published as *Trudy po topologii i drugim oblastiam matematiki*, 2 vols. (Moscow–Leningrad, 1951). Separately published writings include "Sur l'unicité de la solution des équations linéaires de M. Volterra," in *Bulletin de l'Académie polonaise des sciences*, ser. A (1922), 57–62; "Sur une fonction analytique partout continue," in *Fundamenta mathematica*, **4** (1923), 144–150; "Ein Beitrag zur Theorie der ebenen Gebiete unendlich hohen Zusammenhanges," in *Mathematische Zeitschrift*, **21** (1924), 133–150; "Zur Theorie der topologischen Räume," in *Mathematische Annalen*, **92** (1924), 258–266, written with P. S. Aleksandrov; "Über die Metrisation der kompakten topologischen Räume," ibid., 275–293; and "Mémoire sur les multiplicitées cantoriennes," in *Fundamenta mathematica*, **7** (1925), 30–137; and **8** (1926), 225–356.

II. SECONDARY LITERATURE. For a detailed bibliography of Uryson's works, see *Matematika v SSSR za 40 let* ("Mathematics in the U.S.S.R. for Forty Years"), II (Moscow, 1959), 696–697. On Uryson's life and work, see P. S. Aleksandrov, in *Uspekhi astronomicheskikh nauk*, **5** (1950), 196–202.

A. PAPLAUSCAS

VAGNER (or **WAGNER**), **EGOR EGOROVICH** (*b.* Kazan, Russia, 30 November 1849; *d.* Warsaw, Poland, 27 November 1903), *chemistry*.

Vagner was a student of A. M. Zaytsev and of A. M. Butlerov and was one of the founders of the chemistry of terpenes. He graduated from the University of Kazan in 1874. In his student works, published with Zaytsev, he proposed a new and effective method for the zinc-organic synthesis of secondary alcohols. In 1882 he became professor at the Novo-Aleksandr Agricultural Institute and, in 1886, at the University of Warsaw.

Vagner's scientific research dealt mainly with the natural mono- and bicyclic terpenes and with the synthesis and study of the properties of alicyclic compounds. From 1882 to 1888 he developed a universal method (Vagner oxidation) for determining the number and location of ethylene bonds in the organic molecule by means of hydroxylation with weak aqueous solutions of potassium permanganate to form glycol. It was this simple and fruitful technique that made it possible in the nineteenth century to study the structure of complex unsaturated organic compounds. Despite the later appearance of new methods of establishing molecular structure, especially Harries' ozonization method, the Vagner oxidization has remained important. Having determined the structure of the molecules of limonene and its acid derivatives, Vagner then clarified the more difficult question of the structure of pinene. He specified the structural formulas of pinol, terpene, terpineol, camphene, borneol, isoborneol, carvone, carone, carvestrene, sobrerol, sobreritrite β-camphors and other terpenes. Prior to his work, erroneous formulas had frequently been assigned to these substances.

Having discovered the genetic connection between the hydrocarbons of the terpene series and their acid and halogen derivatives, Vagner established the mechanism of many reactions of terpenes. His discovery of the camphene regrouping of the first class was important for the theoretical development of organic chemistry. This regrouping was later studied in detail by Hans Meerwein and is now called the Vagner-Meerwein rearrangement:

Vagner's proposal for the rational classification of terpenes on the basis of their structural-chemical peculiarities is now accepted by contemporary chemistry.

BIBLIOGRAPHY

I. ORIGINAL WORKS. Vagner's writings include *K reaktsii okislenia nepredelnykh uglerodistykh soedineny* ("On the Oxidization Reaction of Unsaturated Carbon Compounds"; Warsaw, 1888); "Zur Constitution der Pinens," in *Berichte der Deutschen chemischen Gesellschaft*, **24** (1891), 2187–2190; and "K stroeniyu terpenov i im rodstvennykh soedineny" ("On the Structure of Terpenes and Related Compounds"), in *Zhurnal Russkago fiziko-khimicheskago obshchestva*, **26** (1894), 327–362; and **28** (1896), 56–108, 206, 398, 484–501.

II. SECONDARY LITERATURE. On Vagner and his work, see A. E. Arbuzov, "Egor Egorovich Vagner," in *Trudy Instituta istorii estestvoznaniya i tekhniki. Akademiya nauk SSSR*, **4** (1952), 46–61, and V. Lavrov, "Egor Egorovich Vagner, zhizn i deitelnost" ("Egor Egorovich, Life and Work"), in *Zhurnal Russkago fiziko-khimicheskago obshchestva pri Imperatorskago St. Peterburgskago universitete*, **36** (1904), 1337–1339.

V. I. KUZNETSOV

VAILATI, GIOVANNI (*b.* Crema, Italy, 24 April 1863; *d.* Rome, Italy, 14 May 1909), *logic, philosophy of science, history of science.*

Vailati's parents were Vincenzo Vailati and Teresa Albergoni. After attending boarding schools in Monza and Lodi, he enrolled in the University of Turin in 1880, graduating in engineering in 1884 and in mathematics in 1888. Then followed a period of independent study in which he especially studied languages (his writings show a proficiency in Greek, Latin, English, French, German, and Spanish); this was interrupted by the offer of an assistantship at the University of Turin by his former teacher Giuseppe Peano, professor of infinitesimal calculus. Vailati was Peano's assistant from 1892 to 1895, when he became an assistant in projective geometry and later honorary assistant to Volterra. In 1899 he requested a secondary school appointment and was at first sent to Syracuse, transferring to Bari in 1900, to Como in 1901, and to Florence in 1904.

Vailati came of a Catholic family but lost his faith during his early university years. Throughout his life he had affectionate and devoted friends; he never married. His premature death was attributed to heart trouble, complicated by pulmonitis.

Vailati's first ten publications, dealing principally with mathematical logic, were published in the *Rivista di matematica*, founded by Peano in 1891. He also collaborated, especially with historical notes, in the *Formulario* project announced by Peano in 1892. Vailati gained international recognition with the publication of three essays in the history and methodology of science, originally given as introductory lectures to his course in the history of mechanics at the University of Turin (1896–1898).

Vailati was always concerned with tracing ideas back to their origins, and his intimate knowledge of Greek and Latin was invaluable. (In the analytical index to the *Scritti*, "Aristotle" has twice the space of any other entry.) His work in this area will perhaps be his most lasting contribution.

Vailati received most attention during his lifetime as the leading Italian exponent of pragmatism. After his transfer to Florence in 1904 he collaborated, along with his friend and disciple Mario Calderoni, in the publication of the journal *Leonardo*, founded the year before by G. Papini and G. Prezzolini. His philosophical position was closer to that of Charles Sanders Peirce than to the more popular William James, but it remained distinct, individual, and original.

Vailati's wide range of interest included, at various periods, psychic research, economics, and political science (in which he took socialism seriously, but opposed Marx's theory of value.) In all of these areas his acute critical sense allowed him, as was often said, "to succeed in saying in a few words what others had succeeded in *not* saying in many volumes." When the occasion seemed to call for it, he did not hesitate to criticize sharply the opinions of even eminent scientists (for example, he criticized Poincaré's views on mathematical logic).

Finally, Vailati's pedagogical activities must be noted, in recognition of which he was appointed a member of the commission for the reform of the secondary schools. For the work of the commission he established his residence in Rome in 1906, dividing his time between there and Florence, but in 1908 he voluntarily returned to teaching in Florence.

After his death, Vailati's reputation quickly suffered an eclipse; this was partly the result of the form in which his writings appeared. He never published a book-length monograph. Indeed, many of his original ideas appeared in critical reviews, which occupy, by page count, approximately 43 percent of the *Scritti*. After 1950 there was a revival of interest in his work, centering mainly on his philosophical views, but hindered by the general unavailability of his writings. Vailati also carried on a wide correspondence, which is mostly unpublished. Projects announced in 1958 for the publication of his correspondence and a new edition of his writings were not carried out.

BIBLIOGRAPHY

I. ORIGINAL WORKS. With very minor exceptions the published writings of Vailati were collected in the *Scritti di G. Vailati (1863–1909)* (Leipzig–Florence, 1911). The article "Sulla teoria delle proporzioni," appeared posthumously in *Questioni riguardanti le matematiche elementari*, F. Enriques, ed., I (Bologna, 1924), 143–191. There have been three short anthologies, *Gli strumenti della conoscenza*, M. Calderoni, ed. (Lanciano, 1911); *Il pragmatismo*, Giovanni Papini, ed. (Lanciano, 1911); and *Il metodo della filosofia*, Ferruccio Rossi-Landi, ed. (Bari, 1957; repr. 1967).

II. SECONDARY LITERATURE. The *Scritti* contains a biography by Orazio Premoli. Calderoni's preface to *Gli strumenti . . .* (1909) is also valuable. Essential for any study of Vailati is F. Rossi-Landi, "Materiale per lo studio di Vailati," in *Rivista critica di storia della filosofia*, **12** (1957), 468–485; and **13** (1958), 82–108.

An entire number of the *Rivista critica . . .*, **18** (1963), 275–523, contains papers presented by twenty authors for the centenary of Vailati's birth.

HUBERT C. KENNEDY

VAILLANT, LÉON-LOUIS (*b.* Paris, France, 11 November 1834; *d.* Paris, 27 November 1914), *ichthyology, herpetology.*

Vaillant studied medicine and science at Montpellier and obtained both the M.D. and Ph.D. in Paris with theses on human hair (*Essai sur le système pileux dans l'espèce humaine*, 1861) and on the mollusk Tridacna (*Remarques sur l'anatomie de la Tridacna elongata*, 1865). After several years as instructor at the universities of Paris and Montpellier, he became assistant and then full professor at the Muséum National d'Histoire Naturelle in Paris. In 1875 he obtained the chair of herpetology and ichthyology at the Muséum, from which he retired in 1910; he died at work in 1914.

Vaillant's earliest duties at the museum included the installation of new galleries to replace those destroyed by the Prussians in 1871. He identified, labeled, arranged, and exhibited his collections with care and delight. As head keeper of the reptiles, batrachians, and fish in the menagerie (which contained 616 animals when he retired), he had ample opportunity for observation. He taught ichthyology and herpetology for thirty-five years and became a world-renowned specialist.

Vaillant published almost two hundred papers, each but a few pages long, and each containing a new detail that was significant in determining the precise classification of some little-known species. While he was still at Montpellier, his articles appeared mainly in *Mémoires de la Société de biologie* and in *Annales des sciences naturelles*. After the Franco-Prussian War, he published in *Bulletin de la Société philomathique de Paris*. Only after 1895 did he change to a public journal and publish forty-four more papers in the new *Bulletin du Muséum d'histoire naturelle*. During his entire professional life Vaillant kept up a steady flow of one or two communications a year to the Academy of Sciences; yet the Academy rejected his candidacy twice, in 1886 and 1887. He never tried again.

Vaillant spent his life at the Muséum. He concentrated on the classification of new specimens and, in eight instances, helped publish the results of expeditions that he had either participated in or had watched closely.

In 1863–1864 Vaillant traveled to the Gulf of Suez with a government mission (doubtless related to the construction of the canal). His study of a mollusk (*Recherches sur la famille des Tridacnidés*, 1865) won the first Savigny Prize from the Académie des Sciences. In 1880–1883 he participated, as member of a deep-sea dredging commission, in the voyages of the frigates *Travailleur* and *Talisman*, and published his part of the resulting discoveries in a beautifully illustrated volume (*Poissons* [Paris, 1888]). After that, his travels were vicarious.

Vaillant wrote in 1882 a general scientific account of the French mission to Somaliland, established in 1891 the official list of fish brought back from the Algerian Sahara and, in 1893, a list of fish brought back from the Arctic by the frigate *La Manche*. He wrote in 1905 the section on "Fish" for the report of a French expedition to Antarctica, the section on "Turtles" in 1911 for the volume compiled by a military mission to South America, and he left unfinished a study containing the results of a French expedition to Central America.

All Vaillant's writings are technical throughout, with no comments; yet the excitement of the scientist stalking his prey and nailing a precise fact is always present. He explained his methodology as being based on anatomy illuminated by physiology and ethology and aided by microscopic techniques and histology. Sometimes he dealt with extinct species and studied naturalists' descriptions as far back as antiquity, which enabled him, he said, to settle many a dispute. Finally, he was interested in the distribution of species geographically, historically, and at various depths in the ocean.

Vaillant presented several reports of general significance, based on ichthyologic and herpetologic research, to the Académie des Sciences. The fauna of the ocean floor was still quite unknown, and bathymetric studies were just beginning to permit differentiation of "littoral," "coastal," and "abyssal" regions. He also served as a consultant to a national committee on deep-sea fishing and on juries for the World's Fair of 1878, 1889, and 1900. His only official distinction was the *rosette* of the Legion of Honor.

BIBLIOGRAPHY

I. ORIGINAL WORKS. Works by Vaillant include *Essai sur le système pileux dans l'espèce humaine* (Paris, 1861), the thesis for the M.D.; "Mémoire pour servir à l'histoire anatomique de la Sirène lacertine," in *Annales des sciences naturelles*, 4th ser., **19** (1863), 295–346; "Recherches sur la famille des Tridacnidés," in *Annales des sciences naturelles (Zoologie)*, **4** (1865), 65–172; "Remarques sur l'anatomie de la Tridacna elongata," in *Comptes rendus hebdomadaires des séances de l'Académie des sciences*, **61** (1865), 601–603; "Recherches sur les poissons des eaux douces de l'Amérique Septentrionale, désignés par M. L. Agassiz sous le nom d'Etheostomatidae," in *Nouvelles archives du Muséum d'histoire naturelle*, **9** (1873), 5–154; "Sur la disposition des vertèbres cervicales chez les Chéloniens," in *Annales des sciences naturelles*, 6th ser., **10** (1880), no. 7; *Mission G. Révoil aux pays Somalis. Faune et flore. Reptiles et batraciens* (Paris, 1882); "Catalogue raisonné des reptiles et batraciens d'Assinie, donnés par M. Chaper au Muséum d'histoire naturelle," in *Bulletin de la Société zoologique de France*, **9** (1884), 343–354; "Considérations sur les poissons des grandes profondeurs, en particulier sur ceux qui appartiennent au sous-ordre des abdominales," in *Comptes rendus hebdomadaires des séances de l'Académie des sciences*, **103** (1886), 1237–1239; **104** (1887), 123–126; "Matériaux pour servir à l'histoire ichthyologique des Archipels de la Société et des Pomotous," in *Bulletin de la Société philomathique de Paris*, **11** (1887), 49–62; *Poissons* (Paris, 1888), vol. II of Expéditions scientifiques du *Travailleur* et du *Talisman* pendant les années 1880, 1881, 1882, 1883, sous la direction de A. Milne-Edwards; "Rapport présenté au Comité consultatif des pêches sur les morues, la reproduction de la sardine et les causes probables de sa disparition," in *Revue maritime et coloniale*, **97** (1888), 544–554, with L. F. Henneguy; *Rapport adressé au Ministre de la marine au nom du Comité consultatif des pêches maritimes sur la pêche de la montée d'anguilles* (Paris, 1889); *Sur la présence du saumon dans les eaux marines de la Norvège*, in A. Berthoude, *Le Saumon et la loi sur la pêche* (Versailles, 1889); "Les collections d'herpétologie et d'ichthyologie au Muséum d'histoire naturelle," in *Revue scientifique*, **45** (1890), 513–522.

Other works include *Lombriciniens, hirudiniens, bdellomorphes, térétulariens et planariens*, 3 vols. (Paris, 1889–1890), pt. III of *Histoire naturelle des annelés marins d'eau douce*; *Poissons*, sect. IV, pt. II of *L'extrême sud algérien. Contribution à l'histoire naturelle de cette région. Catalogue raisonné et étude des échantillons recueillis dans le Sahara algérien. Nouvelles archives des missions scientifiques*, I (1891), 360; "Sur la délimitation des zones littorales," in *Comptes rendus hebdomadaires des séances de l'Académie des sciences*, **112** (1891), 1038–1040; *Les Poissons d'aquarium. Conférence faite à la Société nationale d'acclimatation* (Paris, 1893); *Les tortues éteintes de l'île Rodriguez, d'après les pièces conservées dans les galeries du Muséum. Centenaire du Muséum d'histoire naturelle de Paris* (Paris, 1893), 254–288; "Essai monographique sur les Silures du genre Synodontis," in *Nouvelles archives du Muséum d'histoire naturelle*, **7** (1895), 233–284; **8** (1896), 87–178; *La Tortue de Perrault (Testudo indica, Schneider)*, étude historique (Paris, [1896]); *Guide à la* *ménagerie des reptiles.* Muséum d'histoire naturelle (Paris, 1897); "Contribution à l'étude des Emydosauriens. Catalogue raisonné des Jacaretinga et Alligator de la collection du Muséum," in *Nouvelles archives du Muséum d'histoire naturelle,* **10** (1898), 143–212; "Contribution à l'étude de la faune ichthyologique de la Guyane," in *Leyden Museum Notes,* **20** (1898–1899), 1–20; *Notice sur les travaux scientifiques de Monsieur L.-L. Vaillant* (Paris, 1900); *Poissons* (Paris, 1905), in Expédition antarctique française, 1903–1905, commandée par le Dr. Jean Charcot. Sciences naturelles: documents scientifiques; *Crocodiles et tortues* (Paris, 1910), pt. I of *Histoire naturelle des reptiles,* XVII of *Histoire physique, naturelle et politique de Madagascar,* with G. Grandidier; "Chéloniens et batracien modèle recueillis par le Dr. Rivet," in *Zoologie* (Paris, 1911), IX of *Mission du service géographique de l'armée pour la mesure d'un arc de méridien équatorial en Amérique du Sud, sous le contrôle scientifique de l'Académie des sciences; Études sur les poissons,* pt. IV of *Recherches zoologiques publiées sous la direction de M. H. Milne-Edwards* (Paris, 1883–1915), in Mission scientifique au Mexique et dans l'Amérique Centrale.

See also the Royal Society *Catalogue of Scientific Papers,* for a detailed list of Vaillant's published works.

II. SECONDARY LITERATURE. See Louis Roule, "Allocution prononcée aux obsèques de M. Léon Vaillant," in *Bulletin du Muséum national d'histoire naturelle,* **20** (1914), 374–375.

DORA B. WEINER

VAILLANT, SÉBASTIEN (*b.* Vigny, Val d'Oise, France, 26 May 1669; *d.* Paris, France, 20 May 1722), *botany.*

Vaillant, who came from a family of farmers, was interested in plants from the time of his youth. It is reported that he cured himself of an intermittent fever at the age of eight. An organist at Pontoise when he was eleven, he studied medicine at the hospital there and in 1688 began to practice surgery at Évreux. In 1690 he joined the army as a surgeon and was present at the battle of Fleurus on 1 July. The following year Vaillant moved to Paris, and in 1692 he established himself as a surgeon at Neuilly, working also at the Hôtel-Dieu in Paris. Upon learning that Tournefort was giving courses at the Jardin Royal des Herbes Médicales, usually called Jardin du Roi and now the Jardin des Plantes, he became an assiduous auditor, arriving on foot from Neuilly at five in the morning. Tournefort quickly noticed his gifted and ardent disciple, who brought to class plants collected on his professional rounds and during excursions directed by Tournefort himself.

Vaillant soon left Neuilly to become secretary to Père de Valois, at whose home he met Guy Fagon, first physician of the king, demonstrator of plants, and later superintendent of the Jardin du Roi. When Fagon arrived, Vaillant was classifying mosses; and Fagon engaged him as secretary. This post gave Vaillant valuable opportunities for collecting plants. The herbarium that he established and steadily expanded was preserved in the Cabinet du Roi until the beginning of the nineteenth century, when it was dispersed in the general herbarium of the Muséum d'Histoire Naturelle, as the whole institution became in the Revolution. Vaillant was put in charge of the garden itself, but Fagon, who highly esteemed his honesty, discretion, and broad botanical knowledge, had him appointed—without his having to request it—assistant demonstrator of plants and then demonstrator of plants. (The latter post had several times been denied to Tournefort.) Vaillant not only had his students collect plants but also lent them his research notes to aid in identification and even allowed them to read his manuscripts.

Louis XIV ordered Vaillant to create a Cabinet de Drogues. Vaillant wished to cultivate exotic plants, however, and Fagon obtained permission from the king to build France's first greenhouse (1714). When it proved too small, a second greenhouse, twice as large, was built in 1717. In the latter year Vaillant substituted for the titular professor at the Jardin du Roi, and his opening lecture (at six in the morning) drew a large audience. He was so well liked by the students that the professor allowed him to continue to give the course. Many scientists accompanied Vaillant on botanical excursions over a fourteen-year period, notably along the coasts of Normandy and Brittany.

Vaillant's unstinting dedication to his work undoubtedly aggravated his asthmatic condition, and his premature death from an unidentified pulmonary disease prevented the publication of some of his manuscripts, notably his inaugural lecture at the Jardin du Roi. In it Vaillant established, on the basis of irrefutable evidence—and for the first time in France—the existence of plant sexuality. (A pistachio tree used in his demonstrations is still alive in the Alpine garden of the Muséum d'Histoire Naturelle.) Also notable are the posthumously published *Catalogue des plantes des environs de Paris,* often called the "petit botanicon," and the *Botanicon parisiense,* properly so called, which was published by Boerhaave from Vaillant's notes.

The genus *Vaillantia* (Tournefort) and the species *Galium vaillantii* and *Bulliardia vaillantii* are named for Vaillant, who was also interested in mosses, lichens, and fungi. The most favorable and best-founded judgment of Vaillant was made by Linnaeus. Responding to criticisms by a number of botanists, including Dillenius and Jussieu, he declared: "He was a great observer, and every day I become more convinced that no one has been more skillful in establishing genera."

BIBLIOGRAPHY

Botanicon parisiense, ou denombrement . . . des plantes qui se trouvent aux environs de Paris, Hermann Boerhaave, ed. (Leiden–Amsterdam, 1927), contains beautiful plates executed by Claude Aubriet and is preceded by a detailed biography of Vaillant and a full list of his writings.

The most recent article on Vaillant is Jacques Rousseau, "Sébastien Vaillant, an Outstanding 18th-Century Botanist," in *Regnum vegetabile*, **71** (1970), 195–228. See also Jean-François Leroy, "La botanique au Jardin des Plantes (1626–1970)," in *Adansonia*, 2nd ser., **11**, pp. 225–250.

P. JOVET
J. MALLET

VALDEN, PAVEL IVANOVICH. See **Walden, Paul.**

VALENCIENNES, ACHILLE (*b*. Paris, France, 9 August 1794; *d*. Paris, 13 April 1865), *zoology*.

Valenciennes, whose father had been an aide to Daubenton since 1784, was born in the Muséum d'Histoire Naturelle in Paris and spent all his life associated with that institution. He attended the Collège de Rouen, but the premature death of his father cut short his further education and brought to an end his hopes to attend the École Polytechnique. Instead, in order to help support his family, Valenciennes became a *préparateur* at the Muséum in 1812. He aided successively Geoffroy Saint-Hilaire, Lamarck, and Cuvier with their zoological collections, and he eventually became an *aide-naturaliste* associated with the chair of reptiles and fish held in turn by Lacépède and Constant Duméril.

Early in Valenciennes's career Cuvier obtained for him the task of classifying the animals described by Humboldt on his journey to Latin America from 1799 to 1803. This was the beginning of a lifelong friendship between Humboldt and Valenciennes. Humboldt acted as a patron for

Valenciennes, assisting his entry into the Academy of Sciences in 1844.

Valenciennes's major scientific achievement was his collaboration with Cuvier on the classic work *Histoire naturelle des poissons*. After Cuvier chose him, Valenciennes visited the cabinets of Holland, England, and Germany to obtain additional materials for the work. Cuvier and Valenciennes published eight volumes jointly from 1828 to Cuvier's death in 1832. Valenciennes then continued the series until 1849 with fourteen more volumes. Even so, the work remained incomplete, for the classification of cartilaginous fish had not been treated. Cuvier was responsible for the general scheme of classification, a modification of his classification of fish in *Le règne animal*. Valenciennes excelled at the descriptive aspect of the work. Because of his recognized expertise in the classification and description of fish, he was asked to publish the ichthyology of several scientific voyages of navigation.

In 1832, after a bitterly contested election in which he was opposed by Jean René Constant Quoy, Valenciennes succeeded Henri Ducrotay de Blainville in the chair of annelids, mollusks, and zoophytes at the Muséum. Although he greatly increased the collections associated with his chair, Valenciennes did not specialize in the subject area. He continued to write in all areas of zoology, including zoological paleontology. Although best known as an ichthyologist, his memoirs on mollusks and zoophytes include studies of the simplicity of the gills of the corbicula and lucina as compared to other lamellibranch mollusks, a monograph on the panopea, researches with Frémy on the chemical composition of eggshells in the animal series, and his classification of the gorgoniae based on the composition of the axis rather than on the external form.

Although he planned a large work on the sponges, Valenciennes never published a general work on the areas connected with his chair at the Muséum. A protégé of Cuvier, he tended to exaggerate Cuvier's insistence on facts as opposed to hypotheses. Valenciennes was not an original thinker, and most of his work consists of monographs on genera or descriptions and classifications of new species brought to Paris by travelers. He had a wide knowledge of zoology, but he never synthesized it.

In addition to his position at the Muséum, Valenciennes became master of conferences in zoology at the École Normale Supérieure in 1831 and professor of zoology at the École de Pharmacie in

1856. He married Alphonsine Gottis in 1831. Personally a rather brusque man, he was prone to forming enemies. He has been described by detractors as a "bon gros," corpulent, lazy, and hedonistic.

BIBLIOGRAPHY

I. ORIGINAL WORKS. Valenciennes's only major work is the *Histoire naturelle des poissons*, 22 vols. (Paris, 1828–1849). A list of his memoirs can be found in the Royal Society *Catalogue of Scientific Papers*, VI, 95–97, and VIII, 1141. A more complete listing of his works on fish and reptiles, including the ichthyologies he wrote for several voyages of navigation, can be found in Maurice Blanc, "Travaux ichthyologiques et herpétologiques publiés par Achille Valenciennes," in *Mémoires de l'Institut français d'Afrique noire*, **68** (1963), 71–75. Manuscripts left by Valenciennes are catalogued by Théodore Monad, "Achille Valenciennes et l'Histoire Naturelle des Poissons," *ibid.*, 9–45, also see 34–43. Jean Théodoridés has published a series of more than seventy letters written by Alexander von Humboldt to Valenciennes from 1818 to 1858 in "Une amitié de savants au siècle dernier: Alexandre von Humboldt et Achille Valenciennes (correspondance inédite)," in *Biologie médicale*, **54** (1965), i–cxxix.

II. SECONDARY LITERATURE. The entire *Mémoires de l'Institut français d'Afrique noire*, **68** (1963), is devoted to a collection of historical and ichthyological papers dedicated to Valenciennes (*Mélanges ichthyologiques dédiés à la mémoire d'Achille Valenciennes* [1794–1865]). The article by Théodore Monad (see above) contains an account of Valenciennes's entry into the Academy, a useful bibliography (pp. 43–45), an iconography, a list of manuscript sources, and detailed information on the publication of the *Histoire naturelle des poissons*.

Contemporary biographical studies include L. Hallez, "Muséum d'Histoire Naturelle. Cours de M. Lacaze-Duthiers. Valenciennes," in *Revue des cours scientifique*, **3** (1866), 377–384; and Alphonse Milne-Edwards, "Éloge de M. Valenciennes," in *Journal de pharmacie et de chimie*, 4th ser., **5** (1867), 5–17.

TOBY A. APPEL

VALENTIN, GABRIEL GUSTAV (*b*. Breslau, Prussia [now Wrocław, Poland], 8 July 1810; *d*. Bern, Switzerland, 24 May 1883), *embryology, general and comparative anatomy, physiology*.

Valentin is usually known as Purkyně's most important student, but he early cut his ties to his teacher and worked on problems of his own choice. The only child of Abraham Valentin, a silverware merchant and assistant rabbi in Breslau, and Caroline Bloch, Valentin was equally interested in languages and science as a student at the Maria Magdalena Gymnasium. A knowledge of Hebrew enabled him also to study the Talmud. This, together with the religious traditions observed in his parents' home, instilled in Valentin a firm conviction in the beliefs of his forefathers.

At the age of eighteen Valentin began to study medicine at the University of Breslau, where his most influential teachers were the botanist Nees von Esenbeck and the physiologist Purkyně. After four years he received his medical degree with a dissertation on the formation of muscle tissue, and he passed the state medical examination at Berlin in 1833. His father's death obliged Valentin to begin practicing medicine immediately, in order to earn a living. He found greater satisfaction, however, in the time that he was able to devote to microscopic studies.

Perseverance, a gift for observation, and an outstanding memory were the foundations of Valentin's wide-ranging scientific knowledge. A notable additional asset was his mathematical ability, which was of particular service to him in handling physiological problems. Valentin wrote more than two hundred papers and articles, as well as a number of books, some of which are quite long. Although his initial research centered on the formation of plant and animal tissue, he also was interested in the processes of intracellular movement in plants; and in his study of animals he was particularly concerned with embryology. He experimentally produced double malformations in chick embryos, on which he reported to the Versammlung Deutscher Naturforscher und Aerzte meeting at Breslau in 1833. The following year, Valentin undertook a study of the structure of nerve fibers in the brain and spinal cord, in particular measuring their thickness. This research, however, was hampered by a lack of sufficiently developed techniques.

In the spring of 1834, while conducting research designed to detect eggs in vertebrates, Valentin discovered the ciliated epithelium in the oviduct of rabbits; and with Purkyně he investigated its distribution in various classes of vertebrates. They also demonstrated the influence of chemical substances on the ciliary movement and ascertained that the movement is independent of the nervous system. The importance of this research was recognized in Valentin's election to membership in the Leopoldinisch-Karolinische Deutsche Akademie der Naturforscher. Concurrently he wrote *Handbuch der Entwickelungsgeschichte des Menschen* . . . (1835) and worked on a prize question posed by the French Academy of Sciences in 1833: To determine whether the way in which animal tissues develop can be compared with that of plant tissues. In February 1835 Valentin submitted his answer to the Academy under the title "Histiogenia comparata." This Latin manuscript runs to more than 1,000 quarto pages and includes many illustrations by the author.

In the summer of 1835 tensions developed between Valentin and Purkyně over their use of the same microscope in Purkyně's house. Valentin thereupon sought to obtain an independent post. He received an offer from Dorpat; but there, and in Prussia, his Jewish faith proved to be a handicap. His situation soon improved, however; for in December 1835 the jury of the French Academy of Sciences awarded him the Grand Prix des Sciences Physiques for his histology manuscript. The prize, worth 3,000 gold francs, enabled Valentin to buy a large microscope and to travel to Berlin to see Johannes Müller. More important, however, he was recognized as an outstanding microscopist. He accepted a professorship of physiology and zootomy from the University of Bern, after making sure that he would not be required to abandon his religion. At the age of twenty-six Valentin became the first Jewish professor at a German-language university.

Valentin's prize manuscript was not suitable for publication in the form in which he had submitted it, and he was asked to prepare a shortened version. He did not complete it, however, until the beginning of 1838, because of the time taken up by moving. The rapid advances in cytology during this period obliged him to take into account many new findings of other researchers, so that the shorter version no longer answered the original question; and for this reason it was not published. In fact, the essential contents of Valentin's original manuscript were not published until more than a century later.

In his manuscript of 1834 Valentin designated the fundamental structural units of animal tissue as *granula* and *globuli*. Less frequently he called them *corpora* or *corpuscula*, but it is possible that these names refer only to the nuclei. Terms such as the *granulosa* of the ovarian follicle, the "grain" (*Korn*) layers of the cerebellum and of the retina, and the blood "corpuscles" recall the early years of the cell theory. Valentin provided good examples to illustrate cellular structure: the rudimentary form of the hoof (the blastema of the hoof of domestic animals) and fat cells, which he called *cystae et contentum oleosum*. The *cysta* is therefore the cytoplasm surrounding the fat globules. Valen-

tin mentioned that in cartilage and bones he had seen traces of granules, by which he sometimes meant cells and sometimes their nuclei. (Volf mistakenly alleged that Valentin used the term "cell" as early as 1835, but the passage he cited as proof was actually from the shortened version of 1838.) The manuscript of 1835 does contain clear references to certain similarities between animal and plant tissues, but Valentin did not attribute special significance to them; they simply were not important to him. He even concluded from his study that the development of plant tissue is not comparable to that of animal tissue.

At Bern, Valentin had the use of a small but adequately equipped laboratory in the new anatomical institute. He continued to publish a periodical that he had founded at Breslau, *Repertorium für Anatomie und Physiologie*, which appeared from 1836 to 1843. The sole contributor, he reported the results of his own studies and surveyed the latest physiological literature. For example, from 1836 he used the term "cell" in describing many types of epithelium. Among the structural elements of the conjunctiva, he found nuclei and nucleoli; but he did not apply the term "nucleolus" to the latter, calling them, rather, "a kind of second nucleus within the nucleus."

Valentin pursued his microscopic examination of the structure of nerve tissue with great enthusiasm. Since he clung to his notion of terminal loops of nerves and refused to recognize the occurrence of gray (marrowless) nerve fibers, he became involved in controversies with Johannes Müller and Robert Remak, as well as with Friedrich Heinrich Bidder and Alfred Wilhelm Volkmann.

Valentin also made comparative studies of the sea urchin and of the structure of the electric eel. For his research he devised a double-bladed knife for preparing thin sections, and he was the first to use microincineration in the study of animal tissue (bear spermatozoa in 1839). He also made a number of good observations of the structure of the eye, including the cornea and the ganglionic cells in the nerve fiber layer of the retina. His "Grundzüge der Entwicklung der tierischen Gewebe" is still worth reading.

In relation to his teaching duties, Valentin began to do more research in physiology. He designated the glossopharyngeal nerve as the chief nerve of the sense of taste, and he correctly assessed the effect of the electrically stimulated vagus nerve and of the sympathetic nervous system on stomach contraction. He also developed a highly respected method of determining blood volume, but it proved to be inexact. In extensive experiments on himself, Valentin studied the phenomenon of *perspiratio insensibilis* (1843). He also ventured into the unfamiliar fields of biochemistry and biophysics, but in the latter discipline his results were challenged by Emil du Bois-Reymond (1848).

While traveling to Bern, Valentin had made two lifelong friends during a stop at Frankfurt: Gabriel Riesser, a pioneer in the struggle for Jewish emancipation, and the Göttingen mathematician Moritz Abraham Stern. He made other acquaintances in 1837 while in Paris, where, through the recommendation of Humboldt, he met Pierre Flourens, François Magendie, and Gilbert Breschet. In 1839 he traveled to Nice, where his most important contacts were with Rudolf Wagner. Valentin always welcomed visitors, among whom were Jacob Henle of Zurich and Adolph Hannover of Copenhagen.

Valentin was painstaking and conscientious, but he was more critical of the work of others than of his own. The letters that survive give conflicting images of his personality. He joined the Freemasons at an early date, and he was generous and — sometimes, at least — sociable. Occasionally he was boisterous or ironic; yet he himself was very sensitive, easily offended, sometimes mistrustful, and often dissatisfied. He was self-assertive in dealing with his colleagues and co-workers, and thus his relations with them were very strained at times. Valentin served several times as dean of the medical faculty; but despite more than forty years of teaching at Bern, he was never elected rector of the university.

In 1841 Valentin married his cousin Henriette Samosch; they had three children. Over the years the couple became increasingly estranged, and the children had to be sent out to board because of domestic difficulties.

In 1844 Valentin published the two-volume *Lehrbuch der Physiologie des Menschen*, an undertaking for which his previous study of the specialized literature for the *Repertorium* had prepared him well. A novel aspect of the *Lehrbuch* was the frequent attempt to treat problems mathematically. For example, Valentin mentions the then unknown diastasic property of pancreatic juice and reports his observations on the magnitude of the respiratory pressure (see Rothschuh [1952], pp. 13, 51). A second edition of the textbook soon proved necessary (1847–1850), and in it Valentin demonstrated the existence of the threshold of taste (Rothschuh [1952], p. 105). Valentin's *Grundriss der Physiologie des Menschen* (1846), designed for both independent study and use with courses, went through four editions by 1855 and, like the *Lehrbuch*, appeared in several translations. These textbooks were replaced after about a decade of popularity by Carl Ludwig's *Physiologie des Menschen* (1852–1856). Another important source for Valentin's positions on contemporary problems was his reviews of the physiological literature in Canstatt's *Jahresberichte über die Fortschritte der gesammten Medicin in allen Ländern* (1844–1865).

The respect that Valentin enjoyed in Bern is evident in his becoming the first Jew to be granted citizenship by that city. His scientific standing is revealed not only in the visits by Henle and others but also in Alfonso Corti's six-month stay at Bern in 1849 in order to learn microscopy from him. Corti, moreover, continued to consult with him until 1854.

As director of the Bern Anatomical Institute from 1853 to 1863, Valentin sought to make permanent the provisional arrangement under which it functioned, even though the trend elsewhere was toward a separation of anatomy and physiology. Younger colleagues, working under his supervision, took over part of the teaching in anatomy; but they were more held back than encouraged. Valentin had undoubtedly assumed too heavy a load of responsibilities, and he attempted to lighten his teaching duties in the laboratory courses by composing a pamphlet for students entitled *Die kunstgerechte Entfernung der Eingeweide des menschlichen Körpers* (1857). Of greater scientific significance were the studies he began at this time on the hibernation of the marmot ("Beiträge zur Kenntnis des Winterschlafes der Murmeltiere"), which merit an examination with the aid of the more exact techniques now available. Valentin's *Untersuchung der Pflanzen- und der Tiergewebe in polarisiertem Licht* (1861) was subjected by W. J. Schmidt to a thorough analysis that will be of interest to specialists in the field. Schmidt also points out that Valentin made a number of innovations that contributed to the development of the microscope.

In January 1863 Valentin's wife died after a long illness. At about the same time, the medical faculty sought to put an end to the union of the chairs of anatomy and physiology that he had imposed. Valentin, who was dean at the time, sought to maintain the status quo by appealing to higher authority, and he did not shrink from threats. The government, however, confined his responsibilities to physiology and named Christoph Theodor Aeby professor of anatomy. When the medical faculty declared in December that Valentin still had its support as dean, he allowed himself to be mollified and remained in office.

Until the fall of 1881, when a heart attack rendered him incapable of working, Valentin continued his scientific research, devoting these years primarily to polarization and spectroscopic studies. The results are recorded in "Histologische und physiologische Studien" (1862–1882). The series, which includes forty-five publications, was completed by four "Beiträge zur Mikroskopie" (1870–1875). Parallel with this research, Valentin made seven studies on the effects of curare and other arrow poisons, especially on muscles and nerves ("Untersuchungen über Pfeilgifte" [1868–1873]). The last fruits of his long research career are presented in twelve "Eudiometrisch-toxikologische Untersuchungen" (1876–1881).

Such extensive activity naturally brought Valentin many honors. In addition to being a member of the Leopoldinisch-Karolinische Deutsche Akademie der Naturforscher, he was a foreign corresponding member of the Académie Royale de Médecine de Belgique, of which he became honorary member in 1862. He was also a corresponding member of the Académie de Médecine of Paris; associate member of the Académie Royale des Sciences, des Lettres et des Beaux-Arts de Belgique; and honorary member of the medical societies of Stockholm, Erlangen, Hamburg, Budapest, Turin, Heidelberg, and Copenhagen, and of several scientific societies. The philosophy faculty of Bern awarded him an honorary doctorate, and he was presented with *Festschriften* on his jubilee dates.

BIBLIOGRAPHY

I. ORIGINAL WORKS. Valentin's earlier writings include *Historiae evolutionis systematis muscularis prolulsio* (Breslau, 1832), his dissertation; "Entdeckung continuierlicher, durch Wimperhaare erzeugter Flimmerbewegungen, als eines allgemeinen Phänomens in den Klassen der Amphibien, Vögel und Säugethiere," in Johannes Müller's *Archiv für Anatomie, Physiologie, und wissenschaftliche Medicin*, **1** (1834), 391–400, written with Purkyně; *Handbuch der Entwickelungsgeschichte des Menschen mit vergleichender Rücksicht der Entwickelung der Säugethiere und Vögel* (Berlin, 1835); "De motu vibratorio animalium vertebratorum," in *Verhandlungen der Leopoldinisch-Carolinischen Akademie der Naturforscher*, **17**, Abt. 2 (1835), 841–854, written with Purkyně; "Über den Verlauf und die letzten Enden der Nerven," *ibid.*, **18**, Abt. 1 (1836), 51–240; *De functionibus nervorum cerebralium et nervi sympathici libri quatuor* (Bern–St. Gallen, 1839); and "Über die Spermatozoen des Bären," in *Verhandlungen*

der Leopoldinisch-Carolinischen Akademie der Naturforscher, **19**, Abt. 1 (1839), 237–244.

Subsequent works are Valentin's edition of "Hirn-und Nervenlehre," in S. T. von Sömmering, *Vom Baue des menschlichen Körpers*, IV (Leipzig, 1841); "Beiträge zur Anatomie des Zitteraales (*Gymnotus electricus*)," in *Neue Denkschriften der allgemeinen Schweizerischen Gesellschaft für die gesammten Naturwissenschaften*, **6** (1842), 1–74; "Grundzüge der Entwicklung der tierischen Gewebe," in Rudolf Wagner, ed., *Lehrbuch der speziellen Physiologie* (Leipzig, 1842); *Lehrbuch der Physiologie des Menschen*, 2 vols. (Brunswick, 1844; 2nd ed., 1847–1850); *Grundriss der Physiologie des Menschen* (Brunswick, 1846; 4th ed., 1855), translated into English as *A Textbook of Physiology* (London, 1853); *Die kunstgerechte Entfernung der Eingeweide des menschlichen Körpers* (Frankfurt, 1857); and *Die Untersuchung der Pflanzen- und der Tiergewebe in polarisiertem Licht* (Leipzig, 1861).

Later works include "Beiträge zur Kenntnis des Winterschlafes der Murmelliere," in *Untersuchungen zur Naturlehre des Menschen und der Tierre*, **1–13** (1857–1888); "Histologische und physiologische Studien," in *Zeitschrift für rationelle Medicin*, 3rd ser., **14–36** (1862–1869) and in *Zeitschrift für Biologie*, **6–18** (1870–1882); "Untersuchungen über Pfeilgifte," in Pflüger's *Archiv für die gesammte Physiologie*, **1–7** (1868–1873); "Beiträge zur Mikroskopie," in *Archiv für mikroskopische Anatomie*, **6** (1870), 581–597, **7** (1871), 140–156, 220–238, and **11** (1875), 661–687; and "Eudiometrisch-toxikologische Untersuchungen," in *Archiv für experimentelle Pathologie*, **5–13** (1876–1881).

The MS of "Histiogenia comparata" is in the archives of the Paris Academy of Sciences. It has been published by M. B. Volf, in *Věstník Československé zoologické společnosti*, **6–7** (1938–1939), 476–512; and by E. Hintzsche in *Berner Beiträge zur Geschichte der Medizin und der Naturwissenschaften* (Bern), no. 20 (1963).

II. SECONDARY LITERATURE. See Erich Hintzsche, "Gustav Gabriel Valentin (1810–1883)," in *Berner Beiträge zur Geschichte der Medizin und der Naturwissenschaften* (Bern), no. 12 (1953), with complete bibliography of Valentin's writings; Bruno Kisch, "Gabriel Gustav Valentin," in "Forgotten Leaders in Modern Medicine," in *Transactions of the American Philosophical Society*, n.s. **44** (1954), 142–192; W. J. Schmidt, "Gabriel Gustav Valentin," in Hugo Freund and Alexander Berg, eds., *Geschichte der Mikroskopie*, II (Frankfurt, 1964), 413–422; and Karl E. Rothschuh, *Entwicklungsgeschichte physiologischer Probleme in Tabellenform* (Munich–Berlin, 1952).

ERICH HINTZSCHE

VALENTINE, BASIL, or **Basilius Valentinus,** *chemistry, alchemy, iatrochemistry.*
Supposedly a German Benedictine monk born at

Mainz in 1394, Basil Valentine is said to have been a member of St. Peter's in Erfurt in 1413 and to have been elected prior of the same monastery the following year. Other accounts mention that he traveled widely in Europe and that he made a journey to Egypt late in life.

There is no contemporary evidence for any of the facts relating to Basil Valentine and, indeed, the works attributed to him refer to events that occurred after his death (for example, the discovery of America). Although a number of different individuals have been suggested, the authorship of these texts is most commonly attributed to Johann Thölde, a councillor and salt boiler of Frankenhausen in Thuringia, whose *Haligraphia* (1603 and 1612) closely resembles the *Letztes Testament* (1626), one of the principal works ascribed to Basil Valentine.

The actual author was clearly familiar both with laboratory procedures and mining techniques, and he refers to mines located in Central Europe and elsewhere frequently in the *Letztes Testament*. There are long lists of chemical recipes to be found throughout the Basilian corpus, and it is evident that the author was well aware of methods for the preparation of the three mineral acids. He discussed all of the then known metals as well as preparations that might be made from them, and he gave special attention to the precious metals that he believed could be produced through the transmutation of the less perfect metals. He described the precipitation of copper from solution by iron as an example of natural transmutation. Should it be thought that Basil Valentine need be read primarily for his succinct laboratory directions and observations in nature, one need only refer to his *Zwölff Schlüssel* (1602), a traditional book of alchemical symbolism that was to become one of the most frequently reprinted chemical-alchemical treatises of the seventeenth and eighteenth centuries.

Another work attributed to Basil Valentine, the *Triumph Wagen Antimonii* (1604), has a special significance for several reasons. This work contains a wealth of information on antimony, its ores, and other related metals and minerals, as well as on laboratory procedures in general. In addition, the work is important as the primary source for the many controversial antimonial compounds employed as medicines by seventeenth-century chemical physicians.

Indeed, although the chemical compounds described by Basil Valentine have been the subject of considerable research, their medical influence has not been adequately assessed to date. The *Triumph Wagen Antimonii* and other works by this author are clearly related to the Paracelsian treatises of the period. Here may be found a rather typical call for a new investigation of nature so that we might uncover and better understand the secrets of God's creation. The origin of the metals is described in terms of the three principles (mercury, sulfur, and salt), and the origin of these, in turn, is discussed through reference to a macrocosmic distillation. The emphasis is clearly on an understanding of nature as a whole through chemical operations in the laboratory or through chemical analogies. The macrocosm-microcosm analogy is employed throughout and the significance of medicine is sought on both these levels. The true physician is told to seek out the vital spirits in all things with his knowledge of chemistry. Above all, the different preparations of metals are to be found and their effects are to be determined. It is the chemist who can eliminate the poisonous nature of antimony and make it into fit medicines for human ailments. Again in a fashion reminiscent of Paracelsus, Basil Valentine insisted that the physician may utilize these potent chemical medicines only in union with his knowledge of weights; that is, he must pay attention to proper dosage.

Thus, although Basil Valentine may be properly discussed as one of the more significant chemists of the early seventeenth century, he may also be judged in the context of the contemporary iatrochemical literature. In the latter case he emerges as a rather typical Paracelsian who emphasized the macrocosm-microcosm universe with all of its implications. Nevertheless, this corpus of writings, along with those supposedly written by Isaac and John Isaac Hollandus (also then ascribed to the fifteenth century), was employed by chemists of the seventeenth century who sought to destroy the reputation of Paracelsus. Accepting a fifteenth-century date for Basil Valentine, van Helmont and others were able to accuse Paracelsus of having plagiarized the views of his predecessors. It was partially for this reason, partially for their alchemical appeal, and partially for their genuine chemical value that the works attributed to Basil Valentine were frequently published and translated throughout the seventeenth and the eighteenth centuries.

BIBLIOGRAPHY

I. ORIGINAL WORKS. The most thorough bibliography of the various editions of the works of Basil Valentine is found in James R. Partington, *A History of Chemistry*, II (New York–London, 1961), 190–195. The standard collected edition is the *Chymische Schriften*, which appeared first in two volumes at Hamburg in 1677. Corrected editions appeared in 1694, 1700, and 1717, while the fifth edition of 1740 added a 3rd vol. containing additional tracts. The final editions of 1769 and 1775 were essentially reprints of the fifth edition. In addition to the above, the Latin *Basilii Valentini scripta chymica* appeared in Hamburg in 1700.

Of the specific works mentioned, the *Letztes Testament* appeared first at Jena in two parts in 1626. As a five-part work it appeared first at Strasbourg in 1651. An English version by an anonymous translator appeared at London in 1657 and a new translation by J(ohn) W(ebster) was printed in London in 1671. The *Zwölff Schlüssel* was first printed by Johann Thölde in 1599, and it appeared frequently throughout the seventeenth and the eighteenth centuries. It was translated into Latin in Michael Maier's *Tripus Aureus* in 1618, and it was one of the four texts by Basil Valentine to be included in Jean-Jacques Manget, *Bibliotheca Chemica Curiosa*, II (Geneva, 1702), 409–423. This work appeared in French as *Les douze clefs de philosophie* (Paris, 1624), and in a new translation by Eugène Canseliet (Paris, 1956). *The Twelve Keys* appeared in English as the third part of *The Last Will and Testament of Basil Valentine* (London, 1671), and also in A. E. Waite's translation of *The Hermetic Museum*, I (London, 1893; repr. London, 1953), 315–357.

Johann Thölde's edition of the *Triumph Wagen Antimonii* appeared first at Leipzig in 1604 with a forward by Joachim Tanckius. Many separate German editions appeared throughout the seventeenth and the eighteenth centuries until 1770. The first Latin edition was that made by John Fabre (Toulouse, 1646), but the most important is that with the commentary by Theodor Kerckring (Amsterdam, 1671; 1685). The *Triumphant Chariot of Antimony* was translated first into English by I. H. Oxon. (London, 1660), and again by Richard Russell (London, 1678). A third English translation was made by A. E. Waite (London, 1893; repr. London, 1963).

Additional information on these and other works in the Basilian corpus will be found in the Partington bibliography.

II. SECONDARY LITERATURE. By far the most important survey of the chemistry of Basil Valentine will be found in J. R. Partington, *op. cit.*, 183–203, with a survey of the secondary literature, 183, note 1. Those interested in a traditional alchemical life of Basil Valentine may turn to the final leaves (fols. Ccc i and Ccc ii of the 1700 edition) of the *Chymische Schriften* or, in far more detail (and with a bibliography), in Karl Christoph Schmieder's *Geschichte der Alchemie* (Halle, 1832), repr. with an introduction by Franz Strunz (1927), 197–210. The arguments against the fifteenth-century date of the texts were marshaled by Hermann Kopp in his various publications, but see especially his *Beiträge zur Geschichte der Chemie*, 3 pts. (Brunswick, 1869–1875).

Both the references in Partington and in the recently published *ISIS Cumulative Bibliography. A Bibliography of the History of Science formed from ISIS Critical Bibliographies 1–90. 1913–1965*, Magda Whitrow, ed., 2 vols. (London, 1971), testify to the very limited attention given by scholars to the works attributed to Basil Valentine in this century.

ALLEN G. DEBUS

VALERIANUS, MAGNUS. See **Magni, Valeriano.**

VALERIO (or **VALERI**), **LUCA** (*b.* Naples, Italy, 1552; *d.* Rome, Italy, 17 January 1618), *mathematics*.

Valerio was the son of Giovanni Valeri, of Ferrara, and Giovanna Rodomano, of Greek extraction. He was brought up on Corfu and was educated in Rome at the Collegio Romano, where Clavius was one of his teachers. He studied philosophy and theology, although his main interest was mathematics. Most of his life was spent at Rome as a teacher, both private and public. Valerio taught rhetoric and Greek at the Collegio Greco and, from 1600 until his death, mathematics at the Sapienza in Rome. Among his private pupils were the future Pope Clement VIII and the poet Margherita Sarrocchi, with whom he apparently had a love affair. For a time he was also corrector of Greek in the Vatican Library. On 7 June 1612 he was elected a member of the Accademia dei Lincei and was active in its affairs until 1616. Apparently Galileo and Valerio had met about 1590 in Pisa; and around 1610 they were conducting a brisk and friendly correspondence, replete with expressions of mutual admiration. On 24 March 1616, however, Valerio was expelled from the Lincei for reasons that are now obscure. We do know that he objected to its wholehearted support for Galileo's Copernicanism in the controversy of 1616, but all the facts of the case are not available. Valerio spent the last two years of his life in obscurity and disgrace. It is a supreme irony that a few years later Galileo came to a similar end, but for the opposite reason. Galileo, however, rose above their common fate, for in his *Discorsi* of 1638 he called Valerio "greatest geometer, new Archimèdes of our time."

Valerio's *De centro gravitatis* consists of the application of Archimedean methods to the determination of the volumes and centers of gravity of the various solids of rotation and their segments. One of the most interesting lemmas of the book says in effect that if $\lim x = a$ and $\lim y = b$, and if

$\dfrac{x}{y} = c$ = constant, then $\dfrac{a}{b} = \dfrac{\lim x}{\lim y} = \lim \dfrac{x}{y} = c$, which is basically the same as lemma IV of book I of Newton's *Principia* and as Cavalieri's principle. In *Quadratura parabolae* Valerio used the known center of gravity of a hemisphere to find that of a segment of a parabola. He then used this result to determine the area of the segment. Valerio's method was that of Archimedes, although he introduced general lemmas to dispense with the cumbersome *reductio ad absurdum* process. Some of his theorems may be said to make implicit use of a limit approach, but outwardly he was strictly finitist.

Valerio was strongly influenced by Commandino and apparently, in his method for finding centers of gravity, also by Maurolico. Among the mathematicians who studied him and spoke highly of him were Cavalieri, Torricelli, and J. C. de la Faille. He also had a direct influence on Guldin, Gregorius Saint Vincent, and Tacquet.

BIBLIOGRAPHY

I. ORIGINAL WORKS. Valerio published three books: *Subtilium indagationum seu quadratura circuli et aliorum curvilineorum* (Rome, 1582), of which there is an apparently unique copy in the Alexandrine Library in Rome; *De centro gravitatis solidorum* (Rome, 1604; Bologna, 1661 [with *Quadratura*]); and *Quadratura parabolae* (Rome, 1606; Bologna, 1661). The *De piramidis et conis* mentioned in some accounts is probably a bibliographical ghost. Valerio's letters to Galileo, Cesi, and Baldi are printed in the Edizione Nazionale of Galileo's works.

II. SECONDARY LITERATURE. The standard source of information on Valerio is G. Gabrieli, "Luca Valerio Linceo e un episodio memorabile della vecchia Accademia," in *Atti dell'Accademia nazionale dei Lincei. Rendiconti*, Cl. di scienze morali, storiche e filologiche, 6th ser., **9** (1933), 691–728, which has a good bibliography. The first modern historian to draw attention to Valerio as a mathematician was C. R. Wallner, "Über die Entstehung des Grenzbegriffes," in *Bibliotheca mathematica*, 3rd ser., **4** (1903), 246–259. Later accounts include H. Bosmans, "Les démonstrations par l'analyse infinitésimale," in *Annales de la Société scientifique de Bruxelles*, **37** (1913), 211–228; and A. Tosi, "De centro gravitatis solidorum di Luca Valerio," in *Periodico di matematiche*, **35** (1957), 189–201. See also H. Wieleitner, "Das Fortleben der Archimedischen Infinitesimalmethoden bis zum Beginn des 17. Jahrh., insbesondere über Schwerpunktbestimmungen," in *Quellen und Studien zur Geschichte der Mathematik, Astronomie und Physik*, Abt. B, Studien, **1** (1931), 201–220.

PER STRØMHOLM

VALLÉE-POUSSIN, CHARLES-JEAN-GUSTAVE-NICOLAS DE LA (*b.* Louvain, Belgium, 14 August 1866; *d.* Louvain, 2 March 1962), *mathematics*.

Vallée-Poussin's father was for nearly forty years professor of mineralogy and geology at Louvain. Young Vallée-Poussin entered the Jesuit College at Mons, but he found the teaching in some subjects, notably philosophy, unacceptable. He turned to engineering, although, after obtaining his diploma, he devoted himself to pure mathematics. Since boyhood he had been encouraged in mathematics by Louis-Philippe Gilbert and in 1891 he became Gilbert's assistant at the University of Louvain. Gilbert died in 1892 and, at the age of twenty-six, Vallée-Poussin was elected to his chair. He remained all his life at Louvain.

Vallée-Poussin made a very happy marriage with the gifted daughter of a Belgian family whom he met on holiday in Norway in 1900.

As the outstanding Belgian mathematician of his generation, Vallée-Poussin received many tributes. In accordance with custom, he was honored by celebrations at Louvain in 1928 after thirty-five years in his chair, and again in 1943 after fifty years. On the former occasion, the king of the Belgians conferred on Vallée-Poussin the rank of baron. The Belgian Royal Academy elected him a member in 1909, and he became an associate member of the Paris Académie des Sciences in 1945. He was also a commander of the Legion of Honor and honorary president of the International Mathematical Union.

Vallée-Poussin's earliest investigations were concerned with topics of analysis suggested by his own teaching. He proved in an elegant and general form theorems in the differential and integral calculus. In 1892 his memoir on differential equations was awarded a *couronne* by the Belgian Royal Academy. He quickly showed his analytical power in a spectacular way by his researches into the distribution of primes. After nearly a century of conjectures and proofs of partial results, the prime number theorem—that $\pi(x)$, the number of primes $p \leq x$, is asymptotically $x/\log x$—was proved independently by Hadamard and by Vallée-Poussin in 1896. The two proofs look very different, but each is achieved by difficult arguments of complex function theory applied to the zeta function of Riemann. Vallée-Poussin extended his researches to cover the distribution of primes in arithmetical progressions and primes represented by binary quadratic forms. He also made an advance of the first importance in the original prime number theorem by assigning an upper estimate to the difference between $\pi(x)$ and the logarithmic integral li$x$, which remained for twenty years the closest known. Apart from his two later papers on the zeta function in 1916, Vallée-Poussin left to others the development of the ideas that he had introduced into the theory of numbers.

Although the proof of the prime number theorem was Vallée-Poussin's highest achievement, his main impact on mathematical thought was his *Cours d'analyse*, a model of style, economy, and lucidity. The sweeping changes that Vallée-Poussin made in successive editions of his work reflected his current interests. The first edition expounded the traditional calculus, differential equations, and differential geometry; it was just too early for the Lebesgue integral. Two sizes of type were used, the larger for a basic course and the smaller for supplementary matter suited to mathematical specialists. In the second edition the part in small type was greatly expanded to take in set theory, measure and the Lebesgue integral, bounded variation, the Jordan curve theorem, and trigonometric series up to the theorems of Parseval des Chênes and Fejér. The third edition of volume I (1914) introduced the Stolz-Fréchet definition of the differentiability of $f(x,y)$. The third edition of volume II was burned when the German army overran Louvain. It would have pursued the discussion of the Lebesgue integral.

Vallée-Poussin, invited to Harvard and to Paris in 1915 and 1916, expanded this work into the Borel tract, *Intégrales de Lebesgue* . . ., which bears the marks of successive refinements of treatment. The second edition of the tract included analytic sets (Lusin, Souslin) and the Stieltjes integral. Vallée-Poussin's *Cours d'analyse* itself reverted after 1919 to a basic course without the small print.

In the decade after 1908 Vallée-Poussin made fundamental advances in the theory of approximation to functions by algebraic and trigonometric polynomials. The fact that any continuous function $f(x)$ can be thus approximated uniformly in a closed interval had been proved in 1885 by Weierstrass by integrating the product of $f(u)$ and a peak function $K(u,x)$, which rises steeply to its maximum at $u = x$. Vallée-Poussin (and Landau independently) applied this singular integral method with $K(u,x)$ of the form

$$\{1-(u-x)^2\}^n \text{ or } \{\cos \tfrac{1}{2}(u-x)\}^{2n}$$

to obtain results about the closeness of approximation to $f(x)$ by polynomials of assigned degree under hypotheses about f and its derivatives.

The Lebesgue integral gave new life to the theory of trigonometric series, and Vallée-Poussin proved a number of results that have become classic, notably his uniqueness theorem, his test for convergence, and a method of summation that is stronger than all the Cesàro methods.

During the first quarter of the twentieth century, Vallée-Poussin's interests were dominated by the Borel-Lebesgue revolution and were centered on the real variable. (His is the one *Cours d'analyse* that contains no complex function theory.) After 1925 he turned again to the complex variable, in particular to potential theory and conformal representation. He collected his contributions in a book *Le potentiel logarithmique* (1949), the publication of which was held up by the war. By the time the book appeared some of his ideas had been superseded by those of a younger school of French analysts.

BIBLIOGRAPHY

I. ORIGINAL WORKS. A list of Vallée-Poussin's important papers is in *Journal of the London Mathematical Society*, **39** (1964), 174–175.

Books by Vallée-Poussin are *Cours d'analyse infinitésimale*, 2 vols. (Louvain–Paris, 1903–1906; 2nd ed., 1909–1912; 3rd ed., vol. I only, 1914; 4th ed., 1921–1922)—with changes in each of these editions and fewer changes in succeeding editions (vol. 2 of the 7th ed., 1938; vol. I of the 8th ed., 1938); *Intégrales de Lebesgue fonctions d'ensemble, classes de Baire* (Paris, 1916); *Leçons sur l'approximation des fonctions d'une variable réelle* (Paris, 1919); *Leçons de mécanique analytique* (Paris, 1924); *Les nouvelles méthodes de la théorie du potentiel et le problème généralisé de Dirichlet*. Actualités scientifiques et industrielles (Paris, 1937); *Le potentiel logarithmique, balayage et représentation conforme* (Louvain–Paris, 1949).

II. SECONDARY LITERATURE. Obituary notices are by P. Montel, in *Comptes rendus . . . de l'Académie des sciences*, 2 April 1962; and J. C. Burkill, in *Journal of the London Mathematical Society*, **39** (1964), 165–175.

J. C. BURKILL

VALLISNIERI (or **VALLISNERI**), **ANTONIO** (*b*. Trassilico, Garfagnana district, Lucca province, Italy, 3 May 1661; *d*. Padua, Italy, 18 January 1730), *biology, medicine*.

Vallisnieri's father, Lorenzo, was governor of the territory of Camporgiano, Garfagnana, and married Maria Lucrezia de' Davini. He later moved to Trassilico, where Antonio spent his first years and was educated by his father. After attending school in Modena he was sent to a Roman Catholic college in Reggio nell'Emilia, where he studied grammar and rhetoric and received a bachelor's degree in Aristotelian philosophy (1682). On the advice of one of his teachers that Vallisnieri learn other philosophic systems more congenial to his interest in nature, he became a pupil of Malpighi, who was then professor at the medical faculty of the University of Bologna. In 1684 Vallisnieri took his doctorate in medicine and philosophy at Reggio, then returned to Bologna for another year of study with Malpighi. He spent 1687 and 1688 at Venice and Parma, where he completed his medical training. He then settled in Reggio, practicing medicine but spending considerable time collecting and dissecting animals and observing natural phenomena. The results of his experiments on and observations of the generation of insects were published as "Sopra la curiosa origine di molti insetti."

Federico Marcello, a magistrate of the Republic of Venice and one of the "Riformatori dello Studio di Padova," read Vallisnieri's work and persuaded the government of Venice to appoint him to a chair at the University of Padua. On 26 August 1700 Vallisnieri was appointed to the chair of modern experimental philosophy, which was soon changed to that of practical medicine. From 1710 until his death he occupied the first chair of theoretical medicine at Padua. After Lancisi's death in 1720, he declined the posts of first physician to the pope and the first chair of medicine at Turin. Most of Vallisnieri's colleagues at Padua still favored the Scholastic philosophy and were dubious about the experimental method that Vallisnieri used in his scientific investigations. He was clever enough, however, to praise the Scholastic tradition in his inaugural lecture (14 December 1700). The text of the lecture has been lost, but the title is significant: "Studia recentiora non evertunt antiquam medicinam, sed confirmant" ("Recent Studies Do Not Subvert the Old Medicine, They Confirm It"). He was active at Padua for the rest of his life as both teacher and physician, and especially as naturalist; and he assembled a rich library and a large collection of mineralogical, geological, zoological, anatomical, and archaeological objects, which after his death was given by his son to the university.

Known throughout Europe, Vallisnieri was a fellow of many learned academies, including the Royal Society of London (1705). In 1718 Duke Rinaldo I of Modena made him a member of the hereditary nobility (knight), and the city of Reggio included his name on the list of nobility. Vallisni-

eri's exceedingly varied cultural interests were reflected in verses and in his extensive correspondence (both mostly unpublished) with private citizens as well as university professors.

Vallisnieri married Laura Mattacodi; eleven of their children survived childhood but only three daughters and one son survived their father. The son, named for his father, succeeded him in the chair of theoretical medicine and edited his father's collected works (1733).

Vallisnieri's first important scientific contribution was a complement to Redi's demonstration of the fallacy of the hypothesis of spontaneous generation. In 1668 Redi had shown through precise experiments that flies produced in putrefying flesh do not originate by spontaneous generation but derive from eggs previously laid by other flies of the same species. As for the insects forming galls, he was unable to solve the mystery of their origin and left the possibility that they are generated by the vegetative force of the plants that bear the galls. Malpighi, in *Anatomes plantarum* (1675–1679), had already denied the vegetal origin of insects, observing that females lay the eggs in the plant buds by means of a long ovipositor and considering the galls as tumors produced by the presence of the insect eggs. The fact was confirmed by Vallisnieri, who observed that all parasite insects of plants, whether or not they produce galls, derive from eggs. The same is true of entomophagous insects. Thus Vallisnieri enforced his own opinion on the existence of "a perpetual law of nature that like always generates like." His thought was less clear regarding the generation of the parasites of man and domestic animals. The latter belong to two categories: dipteran insect larvae (flies, oestrids) and true worms (ascarids and tapeworms). The former, originating from eggs introduced from outside, eventually metamorphose into flies. While denying the alleged origin of intestinal worms from earthworms or fruit worms Vallisnieri believed that they are transmitted from mother to child "through the passages that carry chyle for the nourishment of the fetus" or through the milk.

His research on the reproductive systems of man and animals led Vallisnieri to observe tubal motility and the movement of the ends of the fallopian tubes to the ovaries. But his general conclusions were mostly mistaken: he denied the function of spermatozoa in fertilization (and was followed by Spallanzani and many others), and he failed to recognize the significance of the Graafian follicle, believing that the mammalian egg is formed in the corpus luteum. In considering the origin of the embryo, he adhered to the ovistic branch of preformism.

Vallisnieri developed to a considerable extent the theory of the "chain of beings"—the progression and connection of all created things, with man at the apex. He was thus a precursor of the "ladder" established by Charles Bonnet (1779). The chain was interpreted not in what modern biologists would call an evolutionary sense, but merely as a realization of a design preexisting in the mind of God. Such ideas were, nevertheless, forerunners of evolutionary concepts. It is remarkable that Vallisnieri considered man to be related to animals. In one of his letters he speculated on the hypothesis that souls progress in such a way that the beast's soul and man's immortal soul are of equal nature. The letter, which remained unpublished by Vallisnieri and by his son because it would have put them into a very difficult position with the Church, was published by G. Brognolico (1895) and then in part by L. Camerano (1905).

Carlo Francesco Cogrossi, a physician from Crema, published *Nuova idea del male contagioso de' buoi* (1714), dedicated to Vallisnieri, in which he proposed the hypothesis of *contagium vivum*—that a contagious disease such as cattle plague is due to microscopic parasites. Vallisnieri replied in a long letter of the same title, published in *Nuove osservazioni fisiche e mediche* (1715), in which he supported Cogrossi's hypothesis with many personal observations and considerations. These works were a very important step toward understanding the etiology of infectious diseases.

Vallisnieri's curiosity about all natural phenomena led him to make excursions on which he did much geological and geophysical work, including a description of salt waters in Emilia-Romagna, studies of a freshwater spring in the Gulf of Spezia, and investigations of earth movements and the origin of alluvial valleys. In *Dei corpi marini* he accepted Fracastoro's concept that fossil shells found on mountains were there because the land had once been under the sea and had not been carried there by the Flood. In *Lezione accademica*, Vallisnieri rejected the theory that spring water originates from seawater that evaporates and recondenses after being percolated through the earth. He demonstrated with sound arguments that it comes from atmospheric precipitation.

In comparison with contemporary scientific thought, Vallisnieri's work has a very modern character. Following—as Redi had already done—the way opened by Galileo (whom he quotes only twice) and Francis Bacon (whom he also quotes) he rejected Scholastic knowledge and trusted solely in direct observation and in experiments. His evaluation of the statements of the ancient naturalists, and especially of Aristotle, was critical and objective. A very important characteristic of Vallisnieri's work was the constant search for general laws and the denial of every miraculous or occultist interpretation of natural phenomena. Even monsters, he believed, must have their laws; even errors have their fixed terms. Vallisnieri was thus one of the first modern naturalists to have a clear awareness of the character of scientific phenomena. Because the static design that he discovered in natural objects reflected the mind of God, Vallisnieri's outlook was essentially still Aristotelian and compatible with Christian belief, while containing the germs of future scientific development.

BIBLIOGRAPHY

I. ORIGINAL WORKS. All of Vallisnieri's works, including many letters on scientific subjects, were reprinted in *Opere fisico-mediche stampate e manoscritte del . . . Antonio Vallisneri raccolte da Antonio suo figliuolo*, 3 vols. (Venice, 1733).

His writings include "Sopra la curiosa origine di molti insetti," in G. Albrizzi, *Galleria di Minerva* (Venice, 1700); *Considerazioni ed esperienze intorno alla generazione dei vermi ordinari del corpo umano* (Padua, 1710, 1726); *Considerazioni intorno al creduto cervello di bue impietrito* (Padua, 1710); *Esperienze ed osservazioni intorno alla origine, sviluppo e costumi dei vari insetti* (Padua, 1713); *Istoria del camaleonte affricano* (Venice, 1713); *Nuove osservazioni ed esperienze intorno alla ovaia scoperta ne' vermi tondi dell'uomo e de' vitelli, con varie lettere* (Padua, 1713); *Varie lettere spettanti alla storia medica e naturale* (Padua, 1713); *Lezione accademica intorno all'origine delle fontane* (Venice, 1715); *Nuove osservazioni fisiche e mediche fatte nella costituzione verminosa ed epidemica seguita nelle cavalle, cavalli e puledri del Mantovano e del Dominio di Venezia* (Venice, 1715); *Dei corpi marini che sui monti si trovano* (Venice, 1721); *Istoria della generazione dell'uomo e degli animali, se sia de'vermicelli spermatici, o dalle uova, . . .* (Venice, 1721); and *Dell'uso e dell'abuso delle bevande e bagnature calde e fredde* (Modena, 1725; Naples, 1727).

A chapter of *Esperienze ed osservazioni intorno all'origine . . .*, "Ragionamento dell'estro dei poeti e dell'estro degli armenti," has been republished separately (Rome, 1885).

The location of the MSS and correspondence of Vallisnieri is given by B. Brunelli (see below).

II. SECONDARY LITERATURE. The best source for Vallisnieri's biography is Giannartico di Porcia, "Notizie della vita e degli studi del Kavalier Antonio Vallisnieri tratte dalle memorie di lui vivente," in *Opere fisico-*

mediche . . ., xli–lxxx. Further information is in Nicolò Papadopoli, *Historia gymnasii Patavini,* I (Venice, 1726), 169 ff.; Angelo Fabroni, *Vitae italorum doctrina excellentium, qui saeculi XVII et XVIII floruerunt* (Pisa, 1778–1805), VII, 9–90; Girolamo Tiraboschi, *Biblioteca modenese* (Modena, 1781–1786), V, 322–338; G. B. Venturi, *Storia di Scandiano* (Modena, 1822), 143 ff.; Camillo Ugoni, biography of Vallisnieri in E. de Tipaldo, ed., *Biografia degli italiani illustri nelle scienze, lettere ed arti,* III (Venice, 1836), 460–466; and Bruno Brunelli, *Figurine e costumi nella corrispondenza di un medico del Settecento (Antonio Vallisnieri)* (Milan, 1938).

On his scientific and medical work, see Luigi Configliachi, *Intorno agli scritti del cav. Antonio Vallisnieri* (Padua, 1836); Ercole Ferrario, *Su la vita e gli scritti di Antonio Vallisneri* (Milan, 1854); Bernardino Panizza, *Di un autografo inedito del Vallisnieri sopra la peste bovina* (Padua, 1864); Lorenzo Camerano, "Antonio Vallisnieri e i moderni concetti intorno ai viventi," in *Atti dell'Accademia delle scienze,* 2nd ser., 55 (1905), 69–112; Joseph Franchini, "Antonio Vallisnieri on the Second Centenary of His Death," in *Annals of Medical History,* n.s. 3 (1931), 58 ff.; and *Il metodo sperimentale in biologia da Vallisneri ad oggi,* symposium held at the University of Padua on the third centenary of Vallisnieri's birth (Padua, 1962), supp. to *Atti e memorie dell'Accademia patavina di scienze, lettere ed arti,* 73.

On the spelling of the name, see D. Carbone and L. Castaldi, "Vallisnieri o Vallisneri?" in *Rivista di storia delle scienze mediche e naturali,* 19 (1937), 306; B. Brunelli and L. Castaldi, "Ancora su Vallisnieri o Vallisneri," *ibid.,* 20 (1938), 37–38; and P. Capparoni, "Di nuovo su Vallisnieri o Vallisneri," *ibid.,* 85.

Giuseppe Montalenti

VALMONT DE BOMARE, JACQUES-CHRISTOPHE (*b.* Rouen, France, 17 September 1731; *d.* Paris, France, 24 August 1807), *mineralogy, natural history.*

The extensive writings and public lectures of Valmont de Bomare made him one of the most influential popularizers of natural history studies in France during the later years of the Enlightenment.

Valmont de Bomare's father was an *avocat* of the *parlement* of Rouen and had planned a legal career for his son, who was a brilliant student, particularly of the classics. The works of Aristotle and Pliny influenced Valmont de Bomare to turn to the sciences, however, and he studied pharmacy and chemistry at Rouen before going to Paris in 1751. There, he formulated a plan for developing a comprehensive course of lectures in natural history.

Aware of the fact that he needed to acquire a broader and deeper knowledge of this vast area, he obtained a commission as traveling naturalist for the government from Voyer d'Argenson, then minister of war. This quasi-diplomatic status permitted him to make extended visits to most of Europe during the next twelve years. On these journeys Valmont de Bomare studied the geology and mineralogy of the countries he visited, and he inspected mines and chemical and metallurgical works in addition to meeting foreign scientists. In July 1756 he introduced at the Jardin des Plantes his projected public course in natural history, which was highly successful, and which he continued to offer yearly until 1788. As a result of his activities he met Buffon, Daubenton, Nollet, Guillaume-Francois Rouelle, d'Holbach, d'Alembert, and Diderot.

In 1762 Valmont de Bomare published a two-volume work entitled *Minéralogie, ou nouvelle exposition du règne minéral.* In it, he described minerals and arranged them into nine classes on the basis of their external characteristics and resistance to the action of fire and water, depending to a great extent on the prior classification of Johan Wallerius. His most important work was his *Dictionnaire raisonné universel d'histoire naturelle,* first published in five volumes in 1764. Four enlarged editions of this work subsequently appeared, the last in fifteen volumes in 1800. Valmont de Bomare's *Dictionnaire* was highly successful in encouraging the popular study of natural history, and it served as a model for all similar works.

In 1769 Valmont de Bomare accepted the position as head of the cabinet of physics and natural history of the Prince de Condé at Chantilly. Unfortunately, he merged his own collections with those of the prince, and during the Revolution they were all confiscated. At that time, afraid of being compromised, he destroyed all of the diaries of his various journeys and his correspondence with such luminaries as Linnaeus and Rousseau.

In 1796 Valmont de Bomare was appointed professor of natural history at the École centrale in the Rue Saint-Antoine, and he remained in that position until 1806 when he became assistant headmaster of the Lycée Charlemagne. When the Institut de France was established in 1795 he was named associate member of the mineralogy section. But he failed in his bid to become a member of the first class after the death of Jean d'Arcet in 1801.

Valmont de Bomare did not produce any original

scientific work. In the 1760's he read three papers to the Académie des Sciences, two of which were subsequently published in the *Mémoires de mathématique et de physique*. The first was a description of certain pyrite and marcasite deposits in the Palatinate; the second concerned a process of refining camphor employed in Holland; and the third treated a Dutch method of refining borax. His scientific reputation was earned, instead, through his lectures and scientific writings.

BIBLIOGRAPHY

I. ORIGINAL WORKS. Valmont de Bomare described the mineral and fossil collection that he had accumulated during his travels in his *Catalogue du cabinet d'histoire naturelle de M. Bomare de Valmont* [sic] (Paris, 1758). His *Minéralogie, ou nouvelle exposition du règne minéral*, 2 vols. (Paris, 1762; 2nd ed., 1774), was translated into German and published (Dresden, 1769). His *Dictionnaire raisonné universel d'histoire naturelle* appeared in five editions, 5 vols. (Paris, 1764), 6 vols. (Verdun, 1768–1770), 9 vols. (Paris, 1775), 15 vols. (Lyons, 1791), and 15 vols. (Lyons, 1800). His two published articles were "Mémoire sur les pyrites et sur les vitriols," in *Mémoires de mathématique et de physique, présentés à l'Académie Royale des Sciences*, **5** (1768), 617–630; and "Mémoire sur le raffinage du camphre," *ibid.*, **9** (1780), 470–480.

II. SECONDARY LITERATURE. There is an obituary of Valmont de Bomare in *Le moniteur universel* (23 September 1807). Other biographical articles are *Biographie générale*, XLV (1870), 894–895; *Biographie universelle*, XLII (1854), 513–514; and *Biographie universelle et portative des contemporains*, IV (1836), 1469.

JOHN G. BURKE

VALSALVA, ANTON MARIA (*b.* Imola, Italy, 17 June 1666; *d.* Bologna, Italy, 2 February 1723), *anatomy.*

Valsalva came from a distinguished and well-to-do family. The son of Pompeo Valsalva, a goldsmith, and Caterina Tosi, he was the third of eight children. Valsalva was educated by the Jesuits in the humanities, mathematics, and natural sciences, the latter arousing his interest in animal morphology and entomology. He subsequently moved to Bologna, where he studied philosophy with Lelio Trionfetti, mathematics with Pietro Mengoli, and geometry with Rodelli.

Valsalva may be considered a Galilean through Borelli and thence through Malpighi, founder of microscopic anatomy and Valsalva's teacher at Bologna University. Malpighi deeply respected Valsalva, who was his favorite pupil and who greatly admired Malpighi.

On 10 June 1687, Valsalva became a doctor of medicine and philosophy, defending the dissertation "Sulla superiorità delle dottrine sperimentali." His name was then entered on the roll of Bolognese doctors and, with Santi Giorgio, Guglielmini, Giacomo Beccari, and Albertini, he began to attend scientific meetings at Eustachio Manfredi's house that led to the founding of the Accademia degli Inquieti. In 1697 he became public engraver of anatomy, and in 1704 he published *De aure humana tractatus*. A year later Valsalva was named lecturer and demonstrator in anatomy, a post he held for the rest of his life.

Valsalva was devoted to teaching and scientific research, as well as to the practice of medicine. He spent much time in the anatomical amphitheater, the unhealthy air of which affected his health. Seized by such a *furor studendi* that he even made an organoleptic evaluation of exudates, Valsalva observed that the serum produced by gangrene was so acrid that, after tasting it, its extreme sourness irritated the papillae of his tongue for an entire day.

Valsalva's scientific integrity was noteworthy. When he was elected, with Vittorio Stancari, by the Bologna Academy as censor of the first volume of Morgagni's *Adversaria anatomica*, he asked for time in order to be able to give a considered and precise opinion. When the objection was raised that this would delay publication of the book, Valsalva replied, "That's how I am . . . . I love Morgagni, but I love the truth more."

On 22 April 1709, at the age of forty-three, Valsalva married Elena Lisi, the seventeen-year-old daughter of a noble Bolognese senatorial family; they had six children, three of whom died young. In 1721, during a consultation with Morgagni in Venice, he suffered a temporary dyslalia, a symptom of the fatal apoplexy that struck him two years later.

Valsalva's famous *De aure humana tractatus*, with the anatomical letters that constitute an extensive commentary, is provided with plates that clearly illustrate the parts of the ear. The preparation of these drawings was probably influenced by Eustachi's plates, which Valsalva greatly admired. His strongest incentive for devoting attention to the human ear probably came from Galileo's new methodological approach and from his own interest in revising acoustics.

The treatise is in two parts, each divided into three chapters; the first part is mainly anatomical and the second physiological, in the Galenic sense of the usefulness of the individual parts. Valsalva divides the ear into the outer, middle, and inner parts. He provides the first detailed and precise description of the outer auricular muscles, based on a wax cast of the external auditory duct, and reproduces its course and diameters. In the middle ear he clearly illustrates the hammer and the tube, which he called the Eustachian tube, in which he recognizes cartilaginous, membranous, and bony components, as well as the muscles of the bones in the middle ear. He describes the morphology of the pharyngeal musculature and the muscle fasciae controlling the Eustachian tube, thus anticipating the concept of the unity of otorhinopharyngeal pathology from the morphological standpoint. (The importance of nasopharyngeal conditions in relation to diseases of the ear is fully recognized today.) He next gives careful measurements of the diameter of the eardrum in relation to important formations, such as the windows of the labyrinth wall and the semicircular canals. He was the first to use the term "labyrinth" for the whole of the inner ear, although his idea of the membranous labyrinth was still confused (he did, however, recognize the existence of the labyrinthine liquid).

Valsalva's treatment of the physiology of hearing contains some aspects of particular interest. He emphasizes the usefulness of the structure of the pinna, the auditory tube, and the ceruminous glands. Valsalva does not attribute special importance to the eardrum; in fact, hearing continues even if the membrane is perforated. In the transmission of sound greater importance is given to the bones of the middle ear, considered as a series of levers transmitting the sound to the labyrinth. The function of the labyrinth, similar to that described by Duverney, is distinguished in Valsalva's conception by his failure to consider the lamina cochleae or the semicircular canals as organs that perceive sound; rather, he attributes this function to the *zonae sonorae*, interpreted as the ultimate branches of the auditory nerve. The semicircular canals, as well as the cochlea, are viewed as purely acoustic organs, in accordance with a theory valid until the early nineteenth century. For Valsalva, as for Socrates, the sense of hearing can receive all sounds without becoming overloaded.

Valsalva was an extremely skilled anatomist and pathologist, a fine physician, and an excellent surgeon for a quarter-century in the Bolognese hospitals, especially Sant' Orsola. He was responsible for establishing the hospital institutes and for regulating the courses of study. As a surgeon he antici-

pated the importance of nephrectomy and splenectomy, and did work in ophthalmology, rhinology, and vasal and tumor surgery.

The procedure described in *De aure* and revived by Morgagni in *De sedibus*, which consists of making the patient exhale violently with mouth and nose closed, is still known as Valsalva's test and has acquired importance in modern cardiovascular symptomatology. It was originally used to remove foreign bodies from the ear and to improve hypacusis.

Valsalva has a place in the history of psychiatry for having been among the first to call for, and in part to implement, humanitarian treatment of the insane, preceding Vincenzo Chiarugi and Philippe Pinel. He considered madness to be analogous to organic disease.

BIBLIOGRAPHY

In addition to *De aure humana tractatus* (Bologna, 1704), a complete edition of Valsalva's writings, edited by Morgagni, was published posthumously in 2 vols. (Venice, 1740).

The first and fundamental biography also was written by Morgagni, *De vita et scriptis Antonii Mariae Valsalvae commentariolum* (Venice, 1740). See also the collection *Terzo centenario della nascita di Antonio Maria Valsalva* (Imola, 1966); G. Bilancioni, "La figura e l'opera di Valsalva," in his *Sulle rive del Lete* (Rome, 1930), 77–100; P. Capparoni, "Antonio Maria Valsalva," in *Profili bio-bibliografici di medici e naturalisti celebri italiani dal secolo XV al secolo XVIII* (Rome, 1932), 92–94; and P. Ravanelli, *A. M. Valsalva (1666–1723) anatomico-medico-chirurgo-primo psichiatra* (Imola, 1966).

LORIS PREMUDA

VALTURIO, ROBERTO (*b.* Rimini, Italy, February 1405; *d.* Rimini, August 1475), *military technology, diffusion of knowledge.*

Little is known about Valturio's life. The son of Cicco di Jacopo de' Valturi, he received a good education at Rimini and quickly mastered Greek and Latin. For a long time he served as apostolic secretary to Pope Eugene IV, a post once held by his father. In 1446 or 1447 Valturio entered the service of the ruler of Rimini, Sigismondo Pandolfo Malatesta. As a private secretary with some influence at court, he was an intermediary between Sigismondo and the artists and scholars attracted to his court. Valturio was not, as has sometimes been supposed, a military engineer or an architect.

Nor did he participate in planning the citadel of Rimini, Rocca Malatestiana.

Nevertheless, at Sigismondo's request, Valturio wrote a treatise on the art of war, *Elenchus et index rerum militarium*. Most likely completed between 1455 and 1460, it is known by the briefer title *De re militari*. Valturio undertook the task more as a man of letters and as a humanist scholar than as an expert in the subject. The work consists of twelve books that treat the art of war both generally and from a historical point of view. The most fully illustrated book is the tenth, on offensive and defensive weapons. The work appeared during the transition from the old military technology to the new one based on gunpowder. Valturio treats mainly Roman and medieval military techniques, more recent ones receiving only cursory coverage in the text, although they are somewhat more adequately presented in the illustrations. Book X contains accounts of siege towers, war chariots, screws for breaking iron gratings, catapults, and battering rams; and book XI covers ships, pontoons, and life belts. Firearms are discussed but are relegated to a subordinate role.

Valturio also presents unusual objects of the kind often found in fourteenth- and early fifteenth-century manuscripts on the art of war: an elbow-shaped weapon in which the bolt and chamber are arranged perpendicular to each other; a storming wagon moved by windwheels; a monstrous war machine with a dragon's head, similar to one depicted several decades later in a relief done at Urbino by Francesco di Giorgio Martini; and a completely sealed submarine propelled by paddle wheels, which certainly was never built. Valturio's sources were primarily ancient authors, but he also drew on a few contemporary and—for the fantastic devices—late medieval writers.

The first printed edition of Valturio's work (1472) was a masterpiece of typography and woodcut. The woodcuts (or at least the drawings) were formerly attributed to Matteo de' Pasti; but they may have been done, as E. Rodakiewicz has proposed, by Fra Giovanni Giocondo Veronese. Military leaders of the period held the book in high esteem, and Leonardo da Vinci copied passages of the text and commented on them. Some of the manuscripts, such as those at Dresden and Munich, which contain very fine drawings, may have been produced after the first printed edition and in fact were based upon it.

After Sigismondo's death in 1468, Valturio remained at the court of Rimini under his son and successor, Roberto. According to the investigation made by A. F. Massèra, Valturio died at Rimini in August 1475. In 1484, during the reign of Roberto's successor, Pandolfo IV Malatesta, Valturio's remains were placed in the Church of San Francesco, which in 1446 had been renovated by Leone Battista Alberti into the Tempio Malatestiano.

BIBLIOGRAPHY

I. ORIGINAL WORKS. *De re militari* exists in 22 MSS held at Cesena (Bibl. Malatestiana), Dresden (Landes-Bibl.), Florence (Bibl. Riccardiana and Bibl. Laurenziana), Milan (Bibl. Ambrosiana), Modena (Bibl. Estense), Munich (Bayerische Staatsbibl.), London (British Museum), and seven other libraries (see Rodakiewicz). It has appeared in various eds. and was first printed as *Elenchus et index rerum militarium* (*De re militari*) (Verona, 1472), with 82 woodcuts (Hain-Copinger 15847 = Klebs 1014/1). The 2nd ed. (Verona, 1483) contains 96 woodcuts, copied from those in the 1472 ed. (Hain-Copinger 15848 + Klebs 1014/2). The work also appeared in an Italian trans. (Verona, 1483; Hain-Copinger 15849 = Klebs 1015/1); further Latin eds. (Paris, 1532, 1533, 1534, 1535, 1555); and a French trans. by Loys Meigret, *Les douze livres de Robert Valturin touchant la discipline militaire* (Paris, 1555).

II. SECONDARY LITERATURE. See the following, listed chronologically: C. Yriarte, *Un condottiere au XVe siècle* (Paris, 1882), 128–132, 263–267; M. Jähns, *Geschichte der Kriegswissenschaften*, pt. 1 (Munich, 1889), 358–362; L. Olschki, *Geschichte der neusprachlichen wissenschaftlichen Literatur*, I (Leipzig, 1919), 131–132; A. F. Massèra, "Quando morì Roberto Valturio?" in *Giornale storico della letteratura italiana*, 75 (1920), 118–119; H. T. Horwitz, "Mariano und Valturio," in *Geschichtsblätter für Technik und Industrie*, 9 (1922), 38–40; L. Hain, *Repertorium bibliographicum*, II, pt. 2 (Berlin, 1925), no. 15847 and supp. edited by W. A. Copinger, pt. 1 (Berlin, 1926); A. F. Massèra, *Roberto Valturio* (Pesaro, 1927); A. C. Klebs, "Incunabula scientifica et medica," in *Osiris*, 4 (1938), no. 1014/1; the entry for Valturio in *Enciclopedia biografica e bibliografica italiana*, ser. 50 (Milan, 1939), 314–315; E. Rodakiewicz, "The *editio princeps* of Valturio's *De re militari* in Relation to the Dresden and Munich mss.," in *Maso Finiguerra*, 5 (1940), 14–82; F. Babinger, *Mehmed der Eroberer und seine Zeit* (Munich, 1953), 210, 214–215, 552; and B. Gille, *Les ingénieurs de la Renaissance* (Paris, 1964), 78–80, 235–236.

FRIEDRICH KLEMM

VALVERDE, JUAN DE (*b.* Amusco, Palencia, Spain, *ca.* 1520; *d.* Rome [?], Italy, *ca.* 1588), *medicine, anatomy.*

Few reliable biographical sources are extant, although Valverde's works contain some information. It is believed that he studied humanities and philosophy at Valladolid University but, like most Spanish scholars of his time, went to Italy after graduation. He received anatomical training at Padua for several years under Vesalius and Colombo until 1543 and became assistant to the latter when he went to Pisa in 1544. It may be assumed that Valverde also accompanied Colombo to Rome in 1548 and that he settled there, for he was one of the prosectors at the autopsy of Cardinal Cibò in 1550. By 1551 Valverde had finished *De animi et corporis sanitate tuenda libellus*, dedicated to Cardinal Verallo. Shortly afterward he became physician to Cardinal Álvarez de Toledo, general inquisitor of Rome; and it was while holding this office that he wrote the anatomical treatise *Historia de la composición del cuerpo humano*. While teaching medicine at the Santo Spirito Hospital in 1555, Valverde was among those considered for the post of papal physician, which was given to another Spaniard, Juan de Aguilera.

It has been stated that in 1558 Valverde visited Amusco, carrying special papal indulgences for the town's church. The credit given by Valverde to Antonio Tabo de Albenga in the first Italian version of his *Historia* (1559) could be interpreted to mean that Valverde had married into the Tabo family. He was alive in 1586, when his engraved portrait first appeared in the edition of the *Historia* published in that year; but he probably was dead by 1589, when Michele Colombo, son of Realdo, published the Latin version of the work. The records for 1602 of the St. Sebastian Brotherhood who cared for the sick poor at Amusco contain a grateful mention of "the late" Doctor Juan de Valverde.

Valverde's *De . . . sanitate tuenda* contains sound doctrine on personal hygiene and shows good knowledge of classical sources; but it was the *Historia*, published thirteen years after Vesalius' *Fabrica*, that brought him fame. Valverde based his illustrations on Vesalius', although he offered fifteen new ones and improved Vesalius' with copperplates engraved by Gaspar Becerra; he also made more than sixty corrections and additions to Vesalius' work, including the description of the stapes of the ear, the short palmar muscle, the human uterus, and in particular the true nature of the cardiac septum. On the basis of experiments performed with Realdo Colombo, Valverde corrected Galen's and Vesalius' idea that blood passed through the septum from the right ventricle to the left, and he gave an accurate and correct description of the pulmonary circuit of the blood. His text ran to thirteen editions and was printed in preference to Vesalius'. Arturo Castiglioni stated that Valverde's *Historia* was the most widely read and studied book of the Renaissance.

BIBLIOGRAPHY

I. ORIGINAL WORKS. Valverde's published writings are *De animi et corporis sanitate tuenda libellus* (Paris, 1552; Venice, 1553); *Historia de la composición del cuerpo humano* (Rome, 1556), Italian trans. by Valverde (Rome, 1559, 1560; Venice, 1586, 1606, 1608, 1682), Latin trans. by Michele Colombo (Venice, 1589, 1607); and *Vivae imagines* (Antwerp, 1566, 1572 [colophon dated 1579]), also in Dutch (Antwerp, 1568, 1647).

II. SECONDARY LITERATURE. See Luis Alberti López, *La anatomía y los anatomistas españoles del Renacimiento* (Madrid, 1948); Victor Escribano García, *La anatomía y los anatomistas españoles del siglo XVI* (Granada, 1902); César Fernández-Ruiz, "Estudio biográfico sobre el Dr. D. Juan Valverde, gran anatómico del siglo XVI, y su obra," in *Clínica y laboratorio*, **66** (1958), 207–240; and Francisco Guerra, "Juan de Valverde de Amusco," in *Clio medica*, **2** (1967), 339–362.

FRANCISCO GUERRA

VAN DE GRAAFF, ROBERT JEMISON (*b.* Tuscaloosa, Alabama, 20 December 1901; *d.* Boston, Massachusetts, 16 January 1967), *physics.*

Van de Graaff was born and raised in the cotton country near Tuscaloosa. He studied engineering at the University of Alabama, where he earned the B.S. in 1922 and M.S. in 1923, and physics at the Sorbonne and at Oxford, where he earned the Ph.D. in 1928, and where he conceived the invention of his belt-charged electrostatic high-voltage generator. He was at the threshold of his scientific career when, as a National Research fellow at Princeton, he constructed the first working model of the generator, operating at 80,000 volts, in 1929.

Under the encouragement of Karl T. Compton, then president of Massachusetts Institute of Technology, Van de Graaff came to MIT as a research associate in 1931 to start a series of developments of his invention, for the precisely controllable acceleration of charged nuclear particles and electrons to high velocities for nuclear-physics research. He became associate professor of physics in 1934, continuing in that position until he resigned from MIT in 1960.

With John G. Trump, later to become professor of electrical engineering, Van de Graaff adapted the principles of his high-voltage generator to the production of intense, penetrating X rays for the precise treatment of deep-seated tumors. The first clinical installation was a huge 1-MeV X-ray generator at the Huntington Memorial Hospital in Boston, in 1937. Van de Graaff's association with Trump developed into a close one, enduring until his death.

During World War II, Van de Graaff was the director of the MIT High Voltage Radiographic Project, sponsored by the Office of Scientific Research and Development. In association with one of his protégés, William W. Buechner, he led in the development of the electrostatic generator for the U.S. Navy to use in the radiographic examination of heavy ordnance. Five 2-MeV X-ray generators were constructed for the navy during the war. The experience gained in this project became a basis for the eventual commercial manufacture of Van de Graaff particle accelerators.

Trump and Van de Graaff founded High Voltage Engineering Corporation in late 1946, again with the capable guidance of Compton. Denis M. Robinson, formerly head of the electrical engineering department at the University of Birmingham, England, was appointed president of the company. Van de Graaff served as director and chief physicist (later chief scientist) of the organization, and he devoted his full time to the company after his resignation from MIT.

High Voltage Engineering Corporation was the first company organized with the express purpose of manufacturing particle accelerators. With the counsel of both Van de Graaff and Trump, and under Robinson's leadership, it made a succession of advances in accelerator technology for nuclear physics, radiation therapy, and the industrial applications of electrons and X rays. Van de Graaff urged the company to undertake the important development of the tandem principle of particle acceleration (originally invented by Willard Bennett in 1937 and rediscovered by Luis W. Alvarez in 1951).

In the late 1950's Van de Graaff invented the insulating-core transformer, which can generate powerful direct currents at higher voltages than possible with the conventional transformer-rectifier systems, for application in industrial processing with high-energy electrons. Modifications of this principle are used also for power-factor correction in the transmission of high-voltage power.

Toward the end of his life, Van de Graaff concentrated on the development of a means for accelerating heavy ions, utilizing the tandem principle, with the objective of providing physicists with a complete freedom in choosing target and projectile nuclei. One ambition, frustrated by his death, was to accelerate uranium nuclei to sufficiently high velocities so that they would coalesce with stationary uranium nuclei, thus possibly opening up the field of synthesizing transplutonium isotopes.

BIBLIOGRAPHY

I. ORIGINAL WORKS. Published works by Van de Graaff are "A 1,500,000 Volt Electrostatic Generator," in *Physical Review*, **38** (1931), 1919–1920; "Experiments on the Elastic Single Scattering of Electrons by Nuclei," *ibid*., **69** (1946), 452–459, with W. W. Buechner and H. Feshbach; "Calorimetric Experiment on the Radiation Losses of 2-MeV Electrons," *ibid*., **70** (1946), 174–177, with W. W. Buechner; "Further Experiments on the Elastic Single Scattering of Electrons by Nuclei," *ibid*., **72** (1947), 678–679, with W. W. Buechner, E. A. Burrill, H. Feshbach, and A. Sperduto; "An Investigation of Radiography in the Range From 0.5 to 2.5 Million Volts," ASTM *Bulletin*, no. 155 (1948), 54–64, with W. W. Buechner, E. A. Burrill, H. Feshbach, L. R. McIntosh, and A. Sperduto; "Electrostatic Generators for the Acceleration of Charged Particles," in *Progress in Physics*, **11** (1948), 1–18, with W. W. Buechner and J. G. Trump; "Irradiation of Biological Materials by High-Energy Roentgen Rays and Cathode Rays," in *Journal of Applied Physics*, **19** (1948), 599–604, with J. G. Trump; "Thick-Target X-Ray Production in the Range From 1250 to 2350 Kilovolts," in *Physical Review*, **74** (1948), 1348–1352, with W. W. Buechner, E. A. Burrill, and A. Sperduto; "Secondary Emission of Electrons by High-Energy Electrons," *ibid*., **75** (1949), 44–45, with J. G. Trump; "Secondary Electron Emission From Metals Under Positive Ion Bombardment in High Extractive Fields," in *Journal of Applied Physics*, **23** (1952), 264–266, with J. G. Trump and E. W. Webster; "Tandem Electrostatic Accelerators," in *Nuclear Instruments and Methods*, **8** (1960), 195–202; "High-Voltage Acceleration Tube Utilizing Inclined-Field Principles," in *Nature*, **195** (1962), 1292–1293, with P. H. Rose and A. B. Wittkower; and "Electrostatic Acceleration of Very Heavy Ions, With Resulting Possibilities for Nuclear Research," in *Bulletin. American Physical Society* (Mexico City meeting, August 29, 1966).

II. SECONDARY LITERATURE. Biographies of Van de Graaff are E. A. Burrill, "Van de Graaff, the Man and His Accelerators," in *Physics Today*, **20** (1967), 49–52; N. Felici, "R. J. Van de Graaff: 1901–1967," in

Bulletin commissariat à l'energie atomique (March-April 1967), p. 20; and P. H. Rose, "In Memoriam: Robert Jemison Van de Graaff," in *Nuclear Instruments and Methods*, **60** (1968), 1–3.

E. ALFRED BURRILL

VANDERMONDE, ALEXANDRE-THÉOPHILE, also known as **Alexis, Abnit,** and **Charles-Auguste Vandermonde** (*b*. Paris, France, 28 February 1735; *d*. Paris, 1 January 1796), *mathematics*.

Vandermonde's father, a physician, directed his sickly son toward a musical career. An acquaintanceship with Fontaine, however, so stimulated Vandermonde that in 1771 he was elected to the Académie des Sciences, to which he presented four mathematical papers (his total mathematical production) in 1771–1772. Later Vandermonde wrote several papers on harmony, and it was said at that time that musicians considered Vandermonde to be a mathematician and that mathematicians viewed him as a musician. This latter view was unfair in that his mathematical work—although small, not generally well known, and a little delayed in publication—was both significant and influential.

Vandermonde's membership in the Academy led to a paper on experiments with cold, made with Bezout and Lavoisier in 1776, and a paper on the manufacture of steel with Berthollet and Monge in 1786. Vandermonde became an ardent and active revolutionary, being such a close friend of Monge that he was termed "femme de Monge." He was a member of the Commune of Paris and the club of the Jacobins. In 1782 he was director of the Conservatoire des Arts et Métiers and in 1792, chief of the Bureau de l'Habillement des Armées. He joined in the design of a course in political economy for the École Normale and in 1795 was named a member of the Institut National.

Vandermonde is best known for the determinant that is named after him:

$$\begin{vmatrix} 1 & a_1 & a_1^2 & \cdots & a_1^{n-1} \\ 1 & a_2 & a_2^2 & \cdots & a_2^{n-1} \\ \hline \\ 1 & a_n & a_n^2 & \cdots & a_n^{n-1} \end{vmatrix} = \prod_{i>j} (a_i - a_j).$$

The determinant does not seem to occur in Vandermonde's work, although his third paper dealt with factorials and he did work with products elsewhere. Lebesgue believed that the attribution of this determinant to Vandermonde was due to a misreading of his notation. Muir (see Bibliography) did not mention this particular determinant, which some also attributed to Cauchy, but Muir asserted that Vandermonde's fourth paper was the first to give a connected exposition of determinants because he (1) defined a contemporary symbolism that was more complete, simple, and appropriate than that of Leibniz; (2) defined determinants as functions apart from the solution of linear equations presented by Cramer but also treated by Vandermonde; and (3) gave a number of properties of these functions, such as the number and signs of the terms and the effect of interchanging two consecutive indices (rows or columns), which he used to show that a determinant is zero if two rows or columns are identical. On this basis, Muir said that Vandermonde was "The only one fit to be viewed as the founder of the theory of determinants." Lebesgue, however, felt that this was neither very original, since there had been earlier workers, nor very important, since others were building equivalent theories, but that Vandermonde's real and unrecognized claim to fame was lodged in his first paper, in which he approached the general problem of the solvability of algebraic equations through a study of functions invariant under permutations of the roots of the equations.

Cauchy assigned priority in this to Lagrange and Vandermonde. Vandermonde read his paper in November 1770, but he did not become a member of the Academy until 1771; and the paper was not published until 1774. During this interval Lagrange published two *mémoires* on the topic. Although Vandermonde's methods were close to those later developed by Abel and Galois for testing the solvability of equations, and although his treatment of the binomial equation $x^m - 1 = 0$ could easily have led to the anticipation of Gauss's results on constructible polygons, Vandermonde himself did not rigorously or completely establish his results nor did he see the implications for geometry. Nevertheless, Kronecker dated the modern movement in algebra to Vandermonde's 1770 paper.

According to Maxwell, Vandermonde's second paper was cited in one of Gauss's notebooks, along with some work of Euler, as being one of two attempts to extend the ideas of Leibniz on the geometry of situation or analysis situs. The paper dealt with the knight's tour and involved the number of interweavings of curves, which Gauss then represented by a double integral and associated with the study of electrical potential.

Unfortunately Vandermonde's spurt of enthusiasm and creativity, which in two years produced four insightful mathematical papers, at least two of which were of substantial importance, was quickly diverted by the exciting politics of the time and, perhaps, by poor health.

BIBLIOGRAPHY

I. ORIGINAL WORKS. Vandermonde's mathematical papers appeared in *Histoire de l'Académie royale des sciences* . . . as follows: "Mémoire sur la résolution des équations" (1771), 365–415; "Remarques sur des problèmes de situation" (1771), 566–574; "Mémoire sur des irrationnelles de différents ordres avec une application au cercle," pt. 1 (1772), 489–498; and "Mémoire sur élimination," pt. 2 (1772), 516–532.

The three algebraic papers were reprinted in C. Itzigsohn, *Abhandlungen aus der reinen Mathematik. In deutscher Sprache herausgegeben* (Berlin, 1888).

II. SECONDARY LITERATURE. The most comprehensive account of Vandermonde's work is Henri Lebesgue, "L'oeuvre mathématique de Vandermonde," in *Thales, recueil des travaux de l'Institut d'histoire des sciences*, IV (1937–1939), 28–42, and in *Enseignement mathématique*, 2nd ser., **1** (1955), 203–223. Also useful are Niels Nielsen, "Vandermonde," in *Géomètres Français sous la revolution* (Copenhagen, 1929), 229–237; Thomas Muir, *The Theory of Determinants in the Historical Order of Their Development*, 2nd ed., I (London, 1906), repr. (New York, 1960), 17–24; and H. Simon, "Vandermondes Vornamen," in *Zeitschrift für Mathematik und Physik*, **41** (1896), 83–85.

PHILLIP S. JONES

VAN DER WAALS, JOHANNES DIDERIK. See Waals, Johannes Diderik van der.

VAN HISE, CHARLES RICHARD (*b.* Fulton, Wisconsin, 29 May 1857; *d.* Milwaukee, Wisconsin, 19 November 1918), *geology*.

A pioneer in the use of the petrographic microscope as a tool for analyzing crystalline rocks and in the application of quantitative methods to the study of geologic phenomena, Van Hise established general principles—still valid a half century later—for deciphering the complexities of Precambrian rocks and understanding the processes of metamorphism. He began his field studies in the Lake Superior region as a geologist of the Wisconsin Geological Survey before 1879, and continued them as a member of the faculty of the department of geology at the University of Wisconsin, where he was an instructor (1879–1883), assistant professor (1883–1886), and professor (1886–1903). His studies were soon extended throughout much of North America under the auspices of the U. S. Geological Survey, by which he was also employed as an assistant geologist (1883–1888), geologist in charge of Lake Superior Division (1888–1900), geologist in charge of Division of Pre-Cambrian and Metamorphic Geology (1900–1908), and consulting geologist (1909–1918).

Van Hise continued his research for several years after he became president of the University of Wisconsin in 1903, but he found it necessary before long to devote his time almost exclusively to administrative responsibilities, which he carried forward with great distinction, and to public affairs.

Early in his career, Van Hise was concerned primarily with mapping ore-bearing formations and determining their structure in order to facilitate mining operations. This soon led him to the enunciation of basic theories concerning ore-deposition and Precambrian history, most of which have been supported by the work of later geologists. His "Principles of North American Pre-Cambrian Geology" (1896) and "Treatise on Metamorphism" (1904) are classics of geologic literature, still useful today. Notable also is the leadership he displayed in the development of a valid rationale for the conservation of natural resources and the wise use of metalliferous ores for human welfare.

BIBLIOGRAPHY

I. ORIGINAL WORKS. Published works by Van Hise include "Crystalline Rocks of the Wisconsin Valley," in *Geology of Wisconsin*, **4** (1882), 627–714, written with R. D. Irving; "The Pre-Cambrian Rocks of the Black Hills," in *Bulletin of the Geological Society of America*, **1** (1890), 203–244; "The Penokee Iron-Bearing Series of Michigan and Wisconsin," in *Monographs of the U.S. Geological Survey*, no. 19 (1892), 1–534, written with R. D. Irving; "Principles of North American Pre-Cambrian Geology," in *Report of the United States Geological Survey*, no. 16 (1896), pt. 1, 573–874; "The Marquette Iron-Bearing District of Michigan," *Monographs of the U.S. Geological Survey*, no. 28 (1897), 1–608, written with W. S. Bayley and H. L. Smyth; "Metamorphism of Rocks and Rock Flowage," in *Bulletin of the Geological Society of America*, **9** (1898), 269–328; "The Iron-Ore Deposits of the Lake Superior Region," in *Report of the United States Geological Survey*, no. 21 (1899–1900), pt. 3 (1901), 305–434; "A Treatise on Metamorphism," in *Monographs of the U.S. Geological Survey*, no. 47 (1904), 1–1286; "Pre-Cambrian Geology of North America," in *Bulletin of* the United States Geological Survey, no. 360 (1909), 1–939, written with C. K. Leith; *The Conservation of Natural Resources in the United States* (New York, 1910); and "The Influence of Applied Geology and the Mining Industry Upon the Economic Development of the World," in *Compte-Rendu, International Geological Congress,* XI (1912), 259–261.

II. Secondary Literature. Biographies of Van Hise are T. C. Chamberlin, "Biographical Memoir of Charles Richard Van Hise, 1857–1918," in *Memoirs of the National Academy of Sciences,* **17** (1924), 143–151, including a bibliography of eighty-four titles; and C. K. Leith, "Memorial of Charles Richard Van Hise," in *Bulletin of the Geological Society of America,* **31** (1920), 100–110, including a bibliography of fifty-nine titles.

Kirtley F. Mather

VANINI, GIULIO CESARE (*b.* Taurisano, Lecce, Italy, *ca.* 1585; *d.* Toulouse, France, 9 February 1619), *philosophy.*

The son of a local official and a Spanish noblewoman, Vanini became a Carmelite friar about 1603. The records of the University of Naples show that a doctorate in both canon and civil law was awarded Giulio Cesare Vanini on 6 June 1606. After selling a house and some personal belongings in Naples (16 May 1608), he enrolled in the faculty of theology at Padua University and preached in various places, including Venice.

Under threat of being banished to Naples by the general of his order, Vanini appealed to the English ambassador to Venice, who recommended him to the archbishop of Canterbury (7 February 1612). Five months later in the Italian church in London Vanini publicly renounced Catholicism in the presence of Francis Bacon. But his adherence to Anglicanism was short-lived. On 10 July 1613 his appeal to Rome for permission to reenter the Catholic church as a secular priest, without being required to rejoin the Carmelite order, was granted by the Holy Office and on 22 August by the pope himself.

After escaping from an English prison in the spring of 1614, Vanini made his way to Paris, where (27 August 1614) he was denied permission to publish his work on the Council of Trent, a work that has not survived. In Lyons, in the summer of 1615, he published his *Amphitheatrum aeternae providentiae . . .* with the approval of the local ecclesiastical censor and also with a royal privilege. Likewise with a royal privilege and with the approbation of the Sorbonne, a Parisian publisher, professing that he had surreptitiously procured a copy of Vanini's sixty dialogues dealing with the secrets of nature, issued (1 September 1616) the philosopher's only other surviving work. A month later the Sorbonne censors claimed that the printed version differed from the manuscript they had approved.

Having studied medicine in Paris, Vanini practiced in Toulouse, where he also taught philosophy privately under an assumed name. In that stronghold of the Catholic counterreformation he was arrested (2 August 1618) and kept in jail more than half a year. On 9 February 1619, exactly nineteen years after Giordano Bruno had been condemned to martyrdom at the stake, Vanini's tongue was pulled out by pincers and cut off; he was strangled and then burned; and his ashes were scattered to the winds. To justify the savage sentence, Vanini's biography was falsified and his character maligned by a host of unscrupulous writers.

Centuries later, after the historical record had been corrected, the authorities in the capital of Vanini's native province of Lecce unveiled (24 September 1868) a bust of the philosopher.

BIBLIOGRAPHY

I. Original Works. Vanini's two published works are *Amphitheatrum aeternae providentiae divino-magicum christiano-physicum nec non astrologo-catholicum adversus veteres philosophos, atheos, epicureos, peripateticos, et stoicos* (Lyons, 1615); and *De admirandis naturae reginae deaeque mortalium arcanis* (Paris, 1616).

The *Amphitheatrum* in its entirety and extracts from the dialogues were translated into French by Xavier Rousselot, *Oeuvres philosophiques de Vanini* (Paris, 1842). Both works were translated into Italian by Guido Porzio, *Le opere di Giulio Cesare Vanini,* 2 vols. (Lecce, 1913).

II. Secondary Literature. Works on Vanini and his philosophy are Don Cameron Allen, *Doubt's Boundless Sea: Skepticism and Faith in the Renaissance* (Baltimore, 1964), 58–74; J.-Roger Charbonnel, *La pensée italienne au XVIe siècle et le courant libertin* (Paris, 1919), 302–383; Luigi Corvaglia, *Le opere di Giulio Cesare Vanini e le loro fonti,* 2 vols. (Milan, 1933–1934); and *Vanini, edizioni e plagi* (Casarano, 1934); Victor Cousin, "Vanini, ses écrits, sa vie et sa mort," in *Revue des deux mondes,* **4** (1843), 673–728; Francesco Fiorentino, *Studi e ritratti della rinascenza* (Bari, 1911), 423–471; William L. Hine, "Mersenne and Vanini," in *Renaissance Quarterly* (forthcoming); Emile Namer, *Documents sur la vie de Jules-César Vanini de Taurisano* (Bari, 1965); and "Vanini et la préparation de l'esprit scientifique à l'aube du XVIIe siècle," in *Revue d'histoire des sciences,* **25** (1972), 207–220; Andrzej Nowicki, *Giulio Cesare Vanini* (Wrocław, 1968);

John Owen, *The Skeptics of the Italian Renaissance*, 2nd ed. (London–New York, 1893); 3rd ed. (London, 1908); reprint (Port Washington, New York, 1970), 343–419; Raffaele Palumbo, *Giulio Cesare Vanini e i suoi tempi* (Naples, 1878); Guido Porzio, *Antologia Vaniniana* (Lecce, 1908); Victor Ivanovich Rutenburg, *Velikii italianskii ateist Vanini* (Moscow, 1959); Giorgio Spini, *Ricerca dei libertini* (Rome, 1950), 117–135; and "Vaniniana," in *Rinascimento*, **1** (1950), 71–90.

EDWARD ROSEN

VAN SLYKE, DONALD DEXTER (*b*. Pike, New York, 29 March 1883; *d*. Garden City, New York, 4 May 1971), *biochemistry*.

Van Slyke was the son of Lucius Van Slyke, a noted chemist who spent most of his career at the agricultural experiment station in Geneva, New York, and Lucy Dexter. After attending high school in Geneva, he studied for a year at Hobart College before entering the University of Michigan, where he received the B.S. degree in chemistry in 1905. He continued his studies at Michigan and in 1907 received the Ph.D. in organic chemistry under Moses Gomberg. From 1907 to 1914 Van Slyke was a research chemist in the biochemical laboratory of Phoebus A. Levene at the Rockefeller Institute for Medical Research. In 1914 he became chief chemist of the hospital of the Rockefeller Institute, a position he held until 1949, when he moved to Brookhaven National Laboratory. He continued his chemical research at Brookhaven almost until his death. Van Slyke also edited the *Journal of Biological Chemistry* from 1914 to 1925. Among the honors that he received were the Willard Gibbs Medal (1939), the George M. Kober Medal (1942), and membership in the National Academy of Sciences (1921).

Van Slyke's remarkable ability to develop analytical apparatus and methods, especially gasometric methods, proved extremely useful in biochemistry and clinical medicine. For example, in 1911 he developed his famous nitrous acid method for determining the number of free amino groups in a peptide or protein. The method is based on the fact that amino groups react with nitrous acid quantitatively to release gaseous nitrogen, which can be measured. Van Slyke's manometric apparatus for the analysis of gases in blood and other solutions was adapted to the quantitative determination of numerous constituents of body fluids and was widely used in biochemical laboratories. His collaborative effort with John Peters, *Quantitative Clinical Chemistry* (1931–1932), was a classic in its field. Many of the analytical methods presented in the work were developed in Van Slyke's laboratory, and most of the other procedures described were tested there before inclusion in the volume on methods. To honor his contributions in this field, the American Society of Clinical Chemistry created the Donald D. Van Slyke Award and appropriately selected Van Slyke as the first recipient in 1957.

Most of Van Slyke's research concerned acid-base, gas, fluid, and electrolyte equilibriums in body fluids and the relation of these equilibriums to disease states. His post at the Rockefeller Institute hospital probably was largely responsible for focusing his attention on problems of clinical biochemistry. His first effort in this area involved the study of acidosis, a condition that aroused his interest because it often develops in diabetes. In 1914 Lawrence J. Henderson and Walter W. Palmer had defined acidosis as a decrease in the bicarbonate concentration of the blood. In 1917 Van Slyke and Glenn Cullen introduced the term "alkaline reserve" to describe the bicarbonate concentration of blood and developed a quick and accurate method, which became the standard procedure, for determining the level of plasma bicarbonate. The definition of acidosis as a decrease in the body's alkaline reserve actually applies only to metabolic acidosis, however, and not to the condition known as respiratory acidosis. Van Slyke also established the normal and abnormal variations that may be encountered in the acid-base balance of the blood, and developed an exact mathematical definition of buffer value.

In 1919 Henderson and his associates began an investigation of the physicochemical equilibriums of the constituents of blood. Van Slyke's collaboration was solicited for this project; and it was agreed that the equilibriums of the gases, fluids, and electrolytes of blood would be studied in his laboratory. Van Slyke and his co-workers established that the distribution of electrolyte ions between plasma and corpuscles occurs in accordance with the Gibbs-Donnan law. They determined experimentally the buffer values of oxyhemoglobin and reduced hemoglobin, determined that hemoglobin is responsible for much of the total buffer value of blood, and showed how the Gibbs-Donnan law and the acid-base properties of hemoglobin explain the unequal distribution of diffusible ions between red cells and plasma (the chloride-bicarbonate shift).

Van Slyke's other important contributions include his studies on nephritis, his discovery and identification of the amino acid hydroxylysine, and his establishment of the fact that urinary ammonia is derived largely from glutamine rather than from urea.

Van Slyke and his colleagues introduced the concept of "blood urea clearance"—the cubic centimeters of blood per minute cleared of urea by renal excretion—as a measure of the functional ability of the kidney. The urea clearance test that they developed proved to be exceedingly useful in clinical work and in laboratory investigations.

BIBLIOGRAPHY

I. ORIGINAL WORKS. There apparently is no published bibliography of Van Slyke's works. The five books that he wrote alone or in collaboration are *Cyanosis* (Baltimore, 1923), written with Christen Lundsgaard; *Factors Affecting the Distribution of Electrolytes, Water, and Gases in the Animal Body* (Philadelphia, 1926); *Observations on the Courses of Different Types of Bright's Disease and on the Resultant Changes in Renal Anatomy* (Baltimore, 1930), written with nine others; *Quantitative Clinical Chemistry*, 2 vols. (Baltimore, 1931–1932; 2nd ed., 1946), written with John Peters; and *Micromanometric Analyses* (Baltimore, 1961), written with John Plazin. Almost all of his important papers were published in the *Journal of Biological Chemistry*. Particularly noteworthy publications in this journal include the series "Studies of Acidosis," **30**–106 (1917–1934); and "Studies of Gas and Electrolyte Equilibria in Blood," **54**–105 (1922–1934).

II. SECONDARY LITERATURE. The most substantial biographical article is A. Baird Hastings, "Donald Dexter Van Slyke, 1883–1971," in *Journal of Biological Chemistry*, **247** (1972), 1635–1640. See also the biographical sketch in *Current Biography* (1943), 50–51; and Van Slyke's autobiographical article in *Modern Men of Science*, I (New York, 1966), 495–496. George Corner, *A History of the Rockefeller Institute, 1901–1953, Origins and Growth* (New York, 1964), discusses the work of Van Slyke and his associates at the Rockefeller Institute and hospital—see esp. 274–280, 483–488. A. Baird Hastings, "A Biochemist's Anabasis," in *Annual Review of Biochemistry*, **39** (1970), 3–7, describes some of the work carried out in Van Slyke's laboratory during the 1920's. Van Slyke was interviewed in detail under the Oral History Program of the National Library of Medicine, Bethesda, Md., and the transcript and tape of this memoir are on file at the library.

JOHN PARASCANDOLA

VAN'T HOFF, JACOBUS HENRICUS (*b.* Rotterdam, Netherlands, 30 August 1852; *d.* Steglitz [now Berlin], Germany, 1 March 1911), *physical chemistry.*

Van't Hoff was the third of seven children born to Jacobus Henricus van't Hoff, a physician, and Alida Jacoba Kolff. In 1867, at the age of fifteen, he completed his elementary schooling and entered the fourth class of the five-year secondary school in Rotterdam. In 1869 he passed the final examination and told his parents that he wished to study chemistry. It was agreed that he would study technology at Delft before going to a university. He completed the usual three-year program at the Polytechnic School at Delft in two years; his teachers there included the chemist A. C. Oudemans and the physicist H. G. van de Sande Bakhuyzen.

At Delft, van't Hoff also studied calculus and became interested in philosophy. He immersed himself in Comte's *Cours de philosophie positive,* Whewell's *History of the Inductive Sciences,* and Hippolyte Taine's *De l'intelligence.* He also read Byron's poetic works with great fervor.

In 1871 van't Hoff entered the University of Leiden, where he studied mainly mathematics. From the autumn of 1872 to the following spring he worked with Kekulé at Bonn. In 1873 he passed the doctoral examination in chemistry at the University of Utrecht and early the following year went to Paris for further study under Wurtz. Here he met Le Bel, who later independently published a theory to explain optical isomerism based on stereochemical considerations.

In the summer of 1874 van't Hoff returned to the Netherlands and in September of that year published his theory of the asymmetric carbon atom, a work that inspired the development of stereochemistry. On 22 December 1874 he obtained the Ph.D. at Utrecht under Eduard Mulder's guidance for an undistinguished dissertation on cyanoacetic and malonic acids. In 1876 he was appointed lecturer in physics at the State Veterinary School in Utrecht and began writing his first book, *Ansichten über die organische Chemie* (1878–1881).

In 1877 van't Hoff was appointed lecturer in theoretical and physical chemistry at the University of Amsterdam, where, from 1878 until 1896, he was successively professor of chemistry, mineralogy, and geology and head of the department of chemistry. His appointment was undoubtedly due to J. W. Gunning, professor of chemistry and pharmacy at Amsterdam, who became his lifelong friend. In his inaugural lecture on 11 October 1878 van't Hoff defended the view that in studying the natural sciences both observation and imagination are necessary. Having studied the lives of many scientists, he concluded that the most prominent among them had been gifted with a highly developed imagination.

After 1877 van't Hoff began his studies in chemical thermodynamics and affinity, and in 1884 he stated his principle of mobile equilibrium. From 1885 to 1890 he published the results of his studies on osmotic pressure and explored the analogy between dilute solutions and gases. In 1887 van't Hoff was named professor at the University of Leipzig. Although this invitation catalyzed the authorities at Amsterdam to provide funds for a new chemical laboratory, which was completed in 1891, van't Hoff moved to Berlin in 1896, having been elected to the Royal Prussian Academy of Sciences and appointed professor at the university. Because he lectured only once a week, he was now able to devote himself completely to research. His lectures appeared in *Vorlesungen über theoretische und physikalische Chemie* (1898–1900), which was translated into many languages, and in *Die chemischen Grundlehren nach Menge, Mass und Zeit* (1912). With Wilhelm Ostwald, he was a cofounder of the *Zeitschrift für physikalische Chemie*, the first issue of which appeared in February 1887.

In 1885 van't Hoff was elected a member of the Royal Netherlands Academy of Sciences. He received honorary doctorates from Harvard and Yale (1901), Victoria University of Manchester (1903), and the University of Heidelberg (1908); and was awarded the Davy Medal of the Royal Society of London (1893) and the Helmholtz Medal of the Prussian Academy of Sciences (1911). In 1901 he became the first Nobel laureate in chemistry for his work on osmotic pressure in solutions and on the laws of chemical dynamics. He was also appointed Chevalier de la Légion d'Honneur (1894), senator of the Kaiser-Wilhelm Gesellschaft (1911), and was a member of the Royal Academy of Sciences of Göttingen (1892), the Chemical Society of London (1898), the American Chemical Society (1898), and the Académie des Sciences (1905). In 1911 he died of pulmonary tuberculosis. His body was cremated and the ashes placed in the cemetery at Berlin-Dahlem. He was survived by his wife, Johanna Francisca Mees, whom he had married in 1878, two daughters, and two sons.

Stereochemistry. In 1873 the German chemist Wislicenus published an article on lactic acids, in which he reiterated the view that the only difference between the two optically active forms of the acid must be in the spatial arrangements of the atoms. After van't Hoff had studied this theory, he published a twelve-page pamphlet, *Voorstel tot uitbreiding der tegenwoordig in de scheikunde gebruikte structuur-formules in de ruimte*, which included a page of diagrams. Van't Hoff's name appeared only at the end of the paper, which was dated 5 September 1874.

At the suggestion of Buys Ballot, professor of physics at Utrecht, the paper was soon translated into French, and the following year van't Hoff published his views in extended form as *La chimie dans l'espace*. His revolutionary ideas on the theory of the asymmetric carbon atom did not attract the attention of chemists, however, until Wislicenus asked van't Hoff's permission for a German translation by one of his pupils, Felix Herrmann. The translation was published in 1877 as *Die Lagerung der Atome im Raume*. An English translation by J. E. Marsh appeared in 1891 as *Chemistry in Space*. Then, in 1887, van't Hoff published *Dix années dans l'histoire d'une théorie*, in which he pointed out that Le Bel had independently arrived at the same idea, although in a more abstract form.

Both van't Hoff and Le Bel showed that arrangements of four different univalent groups at the corners of a regular tetrahedron (which van't Hoff defined as an asymmetric carbon atom) will produce two structures, one of which is the mirror-image of the other. The latter is a condition for the existence of optical isomers, already realized in 1860 by Pasteur, who found that optical rotation arises from asymmetry in the molecules themselves. Van't Hoff stated that when the four affinities of one carbon atom are represented by four mutually perpendicular directions lying in the same plane, then we may expect two isomeric forms from derivatives of methane of the type $CH_2(R_1)_2$. Because such isomeric types do not occur in nature, van't Hoff supposed that the affinities of the carbon atom are directed to the corners of a tetrahedron and that the carbon atom is at the center. In such a tetrahedron a compound of the type $CH_2(R_1)_2$ cannot exist in two isomeric forms, but for compounds of the type $CR_1R_2R_3R_4$ it is possible to construct two spatial models that are nonsuperimposable images of one another. In this case there is no center or plane of symmetry for the tetrahedron.

In the first part of the *Voorstel*, van't Hoff discussed the relationship between the asymmetric carbon atom and optical activity. Drawing on several examples he showed that all the compounds of carbon that, in solution, rotate the plane of polarized light possess an asymmetric carbon atom (for example, tartaric acid, maleic acid, sugars, and camphor). Then van't Hoff showed that while the derivatives of optically active compounds lose their rotatory power when the asymmetry of all of the carbon atoms disappears, in the contrary case they usually do not lose this power. Finally he showed that if one makes a list of compounds that contain an asymmetric carbon atom it appears that in many cases the reverse of his first statement is not true, that is, not every compound with such an atom has an influence upon polarized light.

Van't Hoff's concepts of the asymmetric carbon atom explained the occurrence of many cases of isomerism not explicable in terms of the structural formulas then current. Moreover, it pointed out the existence of a link between optical activity and the presence of an asymmetric carbon atom. Van't Hoff also discussed the relationship between the asymmetric carbon atom and the number of isomers. In *La chimie dans l'espace* he showed that the number of possible isomers of a compound with n inequivalent asymmetric carbon atoms is 2^n, and he indicated how the number of isomers decreased if one or more of the asymmetric carbon atoms is equivalent.

Having introduced the concept of the tetrahedral carbon atom to explain the optical isomerism of a number of organic compounds, van't Hoff turned in the second and third part of *Voorstel* to another type of isomerism, which also appeared to be a consequence of the tetrahedral atom, namely, compounds containing doubly and triply linked carbon atoms. A carbon-carbon double bond of the type $R_1R_2C=CR_1R_2$ is represented by two tetrahedrons with one edge in common, as in the case of maleic and fumaric acids, bromomaleic and bromoisomaleic acids, citraconic and mesaconic acids, crotonic and isocrotonic acids, and chlorocrotonic and chloroisocrotonic acids. Van't Hoff pointed out that when two tetrahedrons are joined on one edge and R_1, R_2, R_3, and R_4 represent the univalent groups that saturate the remaining free affinities of the carbon atoms, possibilities for isomerism occur when R_1 differs from R_2 and when R_3 differs from R_4. This form of isomerism is now called geometric or cis-trans isomerism. In cases when optical activity was found but the formula was symmetrical, van't Hoff postulated (usually correctly) either an error in the formula or the presence of an optically active impurity. In 1894 he ventured the opinion, later confirmed, that the occurrence of optically active substances in nature

might be the consequence of the action of circularly polarized light in the atmosphere on optically inactive substances.

Although van't Hoff and Le Bel shared certain views concerning the carbon atom, van't Hoff was more imaginative and broader in his conceptions and thus incurred harsher criticism, especially from Kolbe, who saw in van't Hoff's work a regression of German chemical research to the speculative aspects of *Naturphilosophie*:

> A Dr. J. H. van't Hoff, of the veterinary school at Utrecht, has as it seems, no taste for exact chemical investigation. He has thought it more convenient to mount Pegasus (obviously loaned by the veterinary school) and to proclaim in his *La chimie dans l'espace* how during his bold flight to the top of the chemical Parnassus, the atoms appeared to him to have grouped themselves throughout universal space ["Zeichen der Zeit," in *Journal für praktische Chemie*, **15** (1877), 473].

He was also criticized by Fittig, Adolf Claus, Wilhelm Lossen, and Friedrich Hinrichsen on the basis that his theories were incompatible with physical laws. Although Wurtz, Spring, and Louis Henry wrote warm acknowledgments, they made no attempt to discuss or criticize his theory. The first to give serious attention to van't Hoff's theory was Buys Ballot, who in the journal *Maandblad voor natuurwetenschappen* (1875) published an open letter to van't Hoff. His reply, in the same journal, discusses a number of interesting points raised in the letter and includes diagrams of the configurations of the ten isomeric saccharic acids.

In volume I of *Ansichten über die organische Chemie* van't Hoff systematically examined the physical and chemical properties of organic substances regarded and classified as derivatives of methane. In volume II he discussed the general relation between the constitution and fundamental properties of organic substances. Especially interested in their physical properties, he attempted to relate stability and reactivity to thermodynamic data, reaction velocities, and chemical equilibriums. Remarkably, van't Hoff made little use here of his stereochemical ideas.

Physical chemistry. In 1884 van't Hoff published *Études de dynamique chimique*, which dealt not only with reaction rates but also with the theory of equilibrium and the theory of affinity based on free energy. In the first section of the book he classified reaction velocities as unimolecular, bimolecular, and multimolecular. He started from the observation (accidentally discovered during his stereochemical researches) that dibromosuccinic acid decomposes at 100°C., a process that he classified as a unimolecular (first-order) reaction. As an example of a bimolecular (second-order) reaction he used the saponification of the sodium salt of monochloric acid, which he had studied in 1883 with his pupil L. C. Schwab: $CH_2ClCOOH + NaOH \rightarrow NaCl + CH_2OHCOONa$.

Van't Hoff recognized the positive-salt effect of the sodium chloride and explained deviations in more concentrated solutions as the variations in volume of the molecules. He also determined the order of chemical reaction for many compounds, for example, the first-order decomposition of arsenic hydride. When arsine is heated, one would expect the chemical equation of its decomposition to indicate a quadrimolecular reaction. But after having determined the velocity of decomposition, van't Hoff found that the reaction is of the first order. Thus he discovered that the order may differ from the molecularity, that is, the number of molecules shown in the ordinary chemical reaction equation. Moreover, van't Hoff found that his researches were complicated by activity factors, reaction milieus, and the movements of the molecules.

Van't Hoff's experiments on the influence of temperature on reaction velocity culminated in his famous thermodynamic relationship between the absolute temperature T and the velocity constant K:

$$\frac{d \ln K}{dT} = \frac{A}{T^2} + B,$$

where A and B are factors dependent on the temperature, and A is now called the activation energy. To make the relation plausible, van't Hoff adopted the notion (first used by Leopold von Pfaundler) of chemical equilibrium as the result of two opposite reactions; but van't Hoff was the first to introduce the double-arrow symbol (still universally used) to indicate the dynamic nature of chemical equilibrium.

After investigating the inflammation temperature at which the reaction takes place, van't Hoff derived the law of mass action on the basis of reaction velocities—the velocities of the forward and reverse reactions being equal at equilibrium. He also established the general equation for the effect of the absolute temperature T on the equilibrium constant K:

$$\frac{d \ln K}{dT} = \frac{q}{2T^2},$$

in which q is the heat of reaction at constant volume. The derivation of this equation is not given in the *Études*. In 1886 van't Hoff showed that the Clausius-Clapeyron equation (in the form given by Horstmann), which related the temperature coefficient of the vapor pressure to the heat of reaction and volume change, can be generalized in terms of the equilibrium constant, as given above. Since $K = k_1/k_2$, where k_1 and k_2 are the reaction velocities of the forward and reverse directions,

$$\frac{d \ln k_1}{dT} - \frac{d \ln k_2}{dT} = \frac{q}{RT^2},$$

so that

$$\frac{d \ln K}{dT} = \frac{A}{T^2} + B.$$

From this so-called van't Hoff isochor it follows that the increase or decrease of the equilibrium constant with the absolute temperature depends upon the sign of the reaction heat q at constant volume. Van't Hoff applied his relation to both homogeneous and heterogeneous equilibriums, to condensed systems (in which no component has a variable concentration), and to physical equilibriums, that is, changes of state.

Van't Hoff formulated his principle of mobile equilibrium in the limited sense that at constant volume the equilibrium will tend to shift in such a direction as to oppose the temperature change that is imposed upon the system: "Every equilibrium at constant volume between two systems is displaced by fall of temperature in the direction of that system in the production of which heat is developed." In 1884 Le Châtelier cast the principle in a general form and extended it to include compensation, by change of volume, for imposed pressure changes. This principle is known as the van't Hoff–Le Châtelier principle.

In the fifth section of his *Études*, which dealt with affinity, van't Hoff defined the work of chemical affinity A as the heat q produced in the transformation, divided by the absolute temperature P of the transition point and multiplied by the difference between P and the given temperature T:

$$A = q\frac{P - T}{P}.$$

The quantity A is now called the maximum external work of the system. By differentiating the equation in respect to T, we find the Gibbs-Helmholtz relation for the dependence of the absolute temperature T on the electromotive force at a constant volume:

$$\frac{dA}{dT} = \frac{q-A}{T}.$$

Van't Hoff also established a simple thermodynamic relationship between the osmotic pressure D of the solution and the vapor pressures of pure water S_e and of the solution S_z: $D = 10.5\ T \log S_e/S_z$.

At first the *Études* received little attention. It was neither a textbook nor a purely scientific treatise; it included many new formulas that were presented and applied without derivation. Although the same subjects were discussed in his *Vorlesungen über theoretische und physikalische Chemie*, the latter work was better arranged and included the results of subsequent research — and thus became a valuable textbook. The proper derivations of the equations in the *Études* appeared in a number of publications. In "L'équilibre chimique dans les systèmes gazeux, ou dissous à l'état dilué" (1886) van't Hoff showed from quantitative experiments on osmosis that dilute solutions of cane sugar obey the laws of Boyle, Gay-Lussac, and particularly Avogadro.

In his study of solutions, van't Hoff also investigated their properties in the presence of semipermeable barriers. He extended the quantitative investigation of the botanist Wilhelm Pfeffer (1877), who had contained solutions of cane sugar, and of other substances, within membranes of hexacyanocopper II ferrate, which he formed in the pores of earthenware pots by soaking them first in a solution of copper sulfate and then of potassium ferrocyanide. Van't Hoff showed that the osmotic pressure P of a solution inside such a vessel immersed in the pure solvent is in apparently direct proportion to the concentration of the solute and in inverse proportion to the volume V of the solution at a given temperature. At a given concentration, P is proportional to the absolute temperature T. The relation serves the general gas law $pV = kT$.

Van't Hoff then applied this law thermodynamically to various solutions. He found that the laws of Gay-Lussac, Boyle, and Avogadro are valid only for ideal solutions, that is, those solutions that are diluted to such an extent that they behave like "ideal" gases and in which both the reciprocal actions of the dissolved molecules and the space occupied by these molecules compared with the volume of the solution itself can be neglected.

To the analogy that exists between gases and solutions van't Hoff gave the general expression $pV = iRT$, in which the coefficient i expresses the ratio of the actual osmotic effect produced by an electrolyte to the effect that would be produced if it behaved like a nonelectrolyte. He also arrived at the important generalization that the osmotic pressure that the dissolved substance would exercise in the gaseous state if it occupied a volume is equal to the volume of the solution. Thus he applied Avogadro's law to dilute solutions. Van't Hoff determined that the coefficient i has a value of nearly one for dilute solutions and exactly one for gases. He reached this value by various methods, including the vapor pressure and Raoult's results on the lowering of the freezing point. For dilute solutions of binary electrolytes, such as sodium chloride and potassium nitrate, he found values ranging from 1.7 to 1.9. Hugo de Vries's experiments with plant cells and Donders' and Hartog Jacob Hamburger's experiments with red blood corpuscles produced isotonic coefficients that agreed with van't Hoff's.

Thus van't Hoff was able to prove that the laws of thermodynamics are valid not only for gases but also for dilute solutions. His pressure law gave general validity to the electrolytic theory of Arrhenius, who recognized in the values of i the magnitude that he had deduced, from experiments on electrical conductance, as the number of ions in which electrolytes are divided in solution. Consequently, van't Hoff became an adherent of the theory of electrolytic dissociation.

In "Lois de l'équilibre chimique dans l'état dilué, gazeux ou dissous" (1886) van't Hoff showed that for many substances the value of i was one, thus validating the relation $pV = RT$ for osmotic pressure. It then became possible to calculate the osmotic pressure of a dissolved substance from its chemical formula and, conversely, the molecular weight of a substance from the osmotic pressure. In "Conditions électriques de l'équilibre chimique" (1886), van't Hoff gave a fundamental relation between the chemical equilibrium constant and the electromotive force (the free energy) of a chemical process:

$$\ln K = \frac{-E}{2\,T},$$

in which K is the chemical equilibrium constant, E is the electromotive force of a reversible galvanic cell, and T is the absolute temperature.

While at Amsterdam, van't Hoff worked on physicochemical problems with a number of his pupils (Johan Eykman, Pieter Frowein, Arnold Holleman, Cohen, and Willem Jorissen) and with foreign chemists who came to Amsterdam to study under him (Arrhenius and Wilhelm Meyerhoffer). Besides his fundamental contributions to thermodynamics of chemical reactions, van't Hoff also studied solid solutions and double salts. In an important paper on solid solutions, "Ueber feste Lösungen und Molekulargewichtsbestimmung an festen Körpern" (1890), he determined, with the aid of his laws, the molecular weights of the dissolved substance—a solution of carbon in iron or a solution of hydrogen in palladium.

In *Vorlesungen über Bildung und Spaltung von Doppelsalzen* (1897) van't Hoff outlined the theoretical and practical treatment of the formation, separation, and conversion of many double salts, especially the tartrates of sodium, ammonium, and potassium. The book also gave a survey of the work in this field by van't Hoff and by a number of his pupils in the laboratory at Amsterdam.

At Berlin, van't Hoff studied the origin of oceanic deposits and the conditions of the formation of oceanic salt deposits, particularly those at Stassfurt, from the point of view of Gibbs's phase rule. He investigated phase equilibriums that form when various quantities of individual salts from the Stassfurt minerals are placed in water that is evaporated at a constant temperature. He also studied the form, order, and quantities of these equilibriums and the effect on them of time, temperature, and pressure. This important theoretical study was of special benefit to the German potash industry. Van't Hoff's method generally consisted in determining the fundamental nonvariant equilibriums (consisting of vapor, solution, and three solid phases) that characterize a four-component system at each particular temperature. In this study he was assisted chiefly by Meyerhoffer. Their results were published in the *Sitzungsberichte* of the Prussian Academy of Sciences and were summarized in van't Hoff's two-volume *Zur Bildung der ozeanischen Salzablagerungen*.

Chemistry is indebted to van't Hoff for his fundamental contributions to the unification of chemical kinetics, thermodynamics, and physical measurements. He was instrumental in founding physical chemistry as an independent discipline.

BIBLIOGRAPHY

I. ORIGINAL WORKS. Van't Hoff's doctoral thesis, *Bijdrage tot de kennis van het cyanazijnzuur en malonzuur* (Utrecht, 1874), was preceded by the publication, a few months earlier, of his important *Voorstel tot uitbreiding der tegenwoordig in de scheikunde gebruikte structuur-formules in de ruimte; benevens een daarmeê samenhangende opmerking omtrent het verband tusschen optisch actief vermogen en chemische constitutie van organische verbindingen* ("Proposal for the Extension of the Formulas Now in Use in Chemistry Into Space; Together With a Related Remark on the Relation Between the Optical Rotating Power and the Chemical Constitution of Organic Compounds"; Utrecht, 1874). It was translated into French as "Sur les formules de structure dans l'espace," in *Archives néerlandaises des sciences exactes et naturelles*, 9 (1874), 445–454; and an English version, "Structural Formulas in Space," appeared in G. M. Richardson, ed., *The Foundations of Stereo Chemistry. Memoirs by Pasteur, van't Hoff, Le bel and Wislicenus* (New York, 1901), 37–46.

Van't Hoff's views were published in extended form as *La chimie dans l'espace* (Rotterdam, 1875), trans. into German by F. Herrmann as *Die Lagerung der Atome im Raume* (Brunswick, 1877, 1894, 1908); and into English by J. E. Marsh as *Chemistry in Space* (Oxford, 1891) and by A. Eiloart as *The Arrangement of Atoms in Space* (London–Bombay–New York, 1898).

An enl. ed. of *La chimie dans l'espace* appeared as *Dix années dans l'histoire d'une théorie* (Rotterdam, 1887); new ed., *Stéréochimie* (Paris, 1892). Van't Hoff's reply to Buys Ballot is "Isomerie en atoomligging," in *Maandblad voor natuurwetenschappen*, 6 (1875), 37–45.

Subsequent writings are *Ansichten über die organische Chemie*, 2 vols. (Brunswick, 1878–1881); *Études de dynamique chimique* (Amsterdam, 1884); and "L'équilibre chimique dans les systèmes gazeux, ou dissous à l'état dilué," in *Archives néerlandaises des sciences exactes et naturelles*, 20 (1886), 239–302; "Lois de l'équilibre chimique dans l'état dilué, gazeux ou dissous," in *Kungliga Svenska vetenskapsakademiens handlingar*, 21, no. 17 (1886), 3–41; "Une propriété générale de la matière diluée," *ibid.*, 42–49; and "Conditions électriques de l'équilibre chimique," *ibid.*, 50–58; were translated into English in *The Foundations of the Theory of Dilute Solutions*, Alembic Club Reprints no. 19 (Edinburgh, 1929), 5–42.

Later works are "Die Rolle des osmotischen Druckes in der Analogie zwischen Lösungen und Gasen," in *Zeitschrift für physikalische Chemie*, 1 (1887), 481–508; *Vorlesungen über Bildung und Spaltung von Doppelsalzen* (Leipzig, 1897); *Vorlesungen über theoretische und physikalische Chemie*, 3 vols. (Brunswick, 1898–1900; 2nd ed., 1901–1903), with English trans. by R. A. Lehfeldt as *Lectures on Theoretical and Physical Chemistry*, 3 vols. (London, 1899–1900); "Ueber feste Lösungen und Molekulargewichtsbestimmung an festen Körpern," in *Zeitschrift für physikalische Chemie*, 5 (1890), 322–339; *Acht Vorträge über physikalische Chemie, gehalten auf Einladung der Universität Chicago, 20 bis 24 Juni 1901* (Brunswick, 1902), with English trans. by A. Smith as *Physical Chemistry in the Service of the Sciences* (Chicago, 1903); *Zur Bildung der ozeanischen Salzablagerungen*, 2 vols. (Brunswick, 1905–1909); and *Die chemischen Grundlehren nach Menge, Mass und Zeit* (Brunswick, 1912).

Van't Hoff contributed to E. Cohen and H. Precht, eds., *Untersuchungen über die Bildungsverhältnisse der ozeanischen Salzablagerungen, insbesondere des Stassfurter Salzlagers* (Leipzig, 1912), published after his death. His 1901 Nobel Prize lecture, "Osmotic Pressure and Chemical Equilibrium," is in *Nobel Lectures. Chemistry, 1901–1921* (Amsterdam–London–New York, 1966), 5–10.

II. SECONDARY LITERATURE. The most comprehensive study of van't Hoff's life and work is E. Cohen, *Jacobus Henricus van't Hoff. Sein Leben und Werken* (Leipzig, 1912), with complete bibliography, 598–622. His professorship at Amsterdam is extensively described in W. P. Jorissen and L. T. Reicher, *J. H. van't Hoffs Amsterdamer Periode 1877–1895* (Den Helder, 1912). Achille Le Bel's 1874 article is "Sur les relations qui existent entre les formules atomiques des corps organiques et le pouvoir rotatoire de leurs dissolutions," in *Bulletin de la Société chimique de Paris*, **22** (1874), 337–347; C. H. D. Buys Ballot's open letter is "Openbare brief aan Dr. J. H. van't Hoff," in *Maandblad voor natuurwetenschappen*, **6** (1875), 21–28.

There are obituary notices by H. C. Jones, in *Proceedings of the American Philosophical Society*, **50** (1911), iii–xii; W. Ostwald, in *Berichte der Deutschen chemischen Gesellschaft*, **44** (1911), 2219–2252; F. G. Donnan, in *Proceedings of the Royal Society of London*, **86A** (1912), xxxix–xliii; and J. Walker, in *Journal of the Chemical Society*, **103** (1913), 1127–1143. See also H. A. M. Snelders, "The Birth of Stereochemistry. An Analysis of the 1874 papers of J. H. van't Hoff and J. A. le Bel," in *Janus*, **60** (1973), 261–278; and "The Reception of J. H. van't Hoff's Theory of the Asymmetric Carbon Atom," in *Journal of Chemical Education*, **51** (1974), 2–7.

H. A. M. SNELDERS

VANUXEM, LARDNER (*b.* Philadelphia, Pennsylvania, 23 July 1792; Bristol, Pennsylvania, 25 January 1848), *geology.*

Vanuxem was the son of James Vanuxem, a shipping merchant of Philadelphia, formerly of Dunkirk, France. Young Vanuxem left his father's business at the age of twenty-four and studied for the next three years in Paris at the École des Mines with the mineralogists Alexandre Brongniart, René-Just Haüy, and others. Upon his return to the United States in 1819, Vanuxem became professor of chemistry and mineralogy at South Carolina College, a post that he held until 1826, when he retired to practice geology as a profession.

Among Vanuxem's activities during the next few years was a visit to Mexico to examine gold-mining properties; he also made geologic investigations in New Jersey, New York, Ohio, Kentucky, Tennessee, and Virginia. In his work Vanuxem

substituted paleontological criteria for classification based on lithology and attitude, correcting Maclure's and Eaton's erroneous assignment of the "western country, and the back and upper parts of New York, with secondary rocks" (1829).

Vanuxem was one of the geologists appointed by Governor William L. Marcy of New York in 1836 to carry out the geologic survey of the state. Of the principal geologists of that famous survey, William Williams Mather, Ebenezer Emmons, Timothy A. Conrad, and James Hall—all of whom proved highly competent—Vanuxem was the oldest, the most experienced in the field, and the only one with any formal training in geology. From this survey came his major scientific contribution: his report on the geology of the third geologic district of New York (1842). The slenderest of the four reports, Vanuxem's was a model of organization, presentation, and economy of words without loss of significant detail. Even more than a century later his report is still the starting point for any geologic work in central New York State. Much of the stratigraphic classification adopted by the survey for the New York rocks, long the standard for the eastern United States, may be attributed to Vanuxem's sensible influence. It was in this report that he introduced into stratigraphy the concept of type locating.

The most important result of Vanuxem's studies of the Atlantic coastal plain was his demonstration in 1828, based upon fossils, of the presence of strata distinct from the Tertiary and equivalent to the Cretaceous of Europe—the first recognition of this system in North America and one of the first secure intercontinental correlations.

BIBLIOGRAPHY

I. ORIGINAL WORKS. *Geology of New York. Part III, Comprising the Survey of the Third Geological District* (Albany, 1842).

II. SECONDARY LITERATURE. "Sketch of Professor Lardner Vanuxem," in *Popular Science Monthly*, **46** (1895), 833–840, with a portrait.

JOHN W. WELLS

VARĀHAMIHIRA (*fl.* near Ujjain, India, sixth century), *astronomy, astrology.*

The best-known and most respected astrologer of India, Varāhamihira was the son and pupil of Ādityadāsa, a Maga Brāhmaṇa and descendant of Iranian Zoroastrians who immigrated to northern India in the centuries about the beginning of the

Christian era and who, while retaining some traces of the solar worship of their forebears, were absorbed into Hinduism. Varāhamihira himself stated that he was a native of Avantī or Western Mālwā (the region about Ujjain) and that he resided in the village Kāpitthaka, which is probably to be identified with the ruins at Kayatha about twelve miles from Ujjain. His date is established by his own adaptation in the *Pañcasiddhāntikā* of Lāta's epoch, 505, and by the references to him as an authority in the *Brāhmasphuṭasiddhānta* composed by Brahmagupta in 628. It has further been suggested that he was connected with the Aulikara court at Daśapura (Mandasor), and in particular with Yaśodharman, who was reigning in 532.

His numerous writings covered all of the traditional fields of astrology and astronomy in India, generally in pairs. It is evident from internal cross-references that he composed the *Pañcasiddhāntikā* and *Bṛhatsaṃhitā* simultaneously toward the beginning of his career, although some additions were made to the latter after his other major works were completed. The *Bṛhajjātaka* was probably composed toward the end of his life, and the other treatises fall somewhere in between.

Varāhamihira was not original in his writings. In genethlialogy he depended primarily on Sphujidhvaja's and Satya's expositions of an Indianized Greek system, in divination on the Indian adaptations by Garga and others of Mesopotamian omenseries, and in astronomy on representatives of three traditions: the Mesopotamian-influenced *vedāṅga*-astronomy as represented in the first century *Paitāmahasiddhānta*, the Indian versions of Greco-Babylonian solar, lunar, and planetary theory in the *Vasiṣṭhasiddhānta* and *Pauliśasiddhānta*, and the essentially Hellenistic astronomy of the *Romakasiddhānta* and Lāta's *Sūryasiddhānta*. Since we have very few other sources for studying these traditions in India in the period before 500, Varāhamihira's work is extremely valuable; and as we know little else about the Greek traditions that the sources of the *Pañcasiddhāntikā* depend on, it affords us a most useful if somewhat problematic insight into pre-Ptolemaic Greek astronomy.

Varāhamihira's works are as follows:

1. The *Pañcasiddhāntikā*, edited with translation and commentary by O. Neugebauer and D. Pingree, 2 pts. (Copenhagen, 1970–1971). This difficult text deals with solar, lunar, and planetary theory; problems of time and terrestrial latitude; eclipses; astronomical instruments; and cosmology. Something has been said of its sources and its importance above.

2. The *Bṛhatsaṃhitā* on divination, edited with the commentary of Utpala (966) by Sudhākara Dvivedin, 2 vols. (Benares, 1895–1897; repr., Benares, 1968); there are several English translations, of which the best is H. Kern, "The Bṛhat-Sañhitā," in *Journal of the Royal Asiatic Society* (1870), 430–479; (1871), 45–90, 231–288; (1873), 36–91, 279–338; and (1875), 81–134; this was reprinted in H. Kern, *Vespreide Geschriften*, 16 vols. (The Hague, 1913–1929), I, 169–319, and II, 1–154. This extensive treatise, besides being one of the most complete extant Sanskrit treatises on divination, is very valuable for the information it contains about Indian geography and society; see, for instance, J. F. Fleet, "The Topographical List of the Brihat-Sanhita," in *Indian Antiquary*, **22** (1893), 169–195; and A. M. Shastri, *India as Seen in the Bṛhatsaṃhitā of Varāhamihira* (Delhi-Patna-Varanasi, 1969).

3. The *Samāsasaṃhitā*, Varāhamihira's shorter work on divination. This is now lost, but many of the quotations from it can be found in A. M. Shastri, "Contribution Towards the Reconstruction of the Samāsa-Saṃhitā of Varāhamihira," in *Bhāratīya Vidyā*, **23** (1963), 22–39.

4. The *Vaṭakaṇikā*, a third work on divination, is also lost save for some quoted verses; see P. V. Kane, "The Vaṭakaṇikā of Varāhamihira," in *Vishveshvaranand Indological Journal*, **1** (1963), 63–65.

5. The *Bṛhajjātaka*, Varāhamihira's major work on genethlialogy; it has often been commented on and often translated. The most useful commentary is that of Utpala (966), published, for example, at Bombay in 1864. This is still the standard work on natal horoscopy in India. For its relation to Greek astrology, see *The Yavanajātaka of Sphujidhvaja*, D. Pingree, ed., which is to appear in the Harvard Oriental Series.

6. The *Laghujātaka* is the shorter treatise on genethlialogy. It also was commented on by Utpala, and it was translated into Arabic by al-Bīrūnī, who inserted it into his *India*. There are many editions: for example, with the Hindī *orṭikā* of Kāśīrāma (Bombay, 1936). Unfortunately, there exists no critical edition of either of these popular textbooks on genethlialogy.

7. The *Bṛhadyātrā* is a major treatise on military astrology. An edition of it with the surviving fragment of Utpala's commentary, prepared by D. Pingree, is in *Bulletin of the Government Oriental Manuscripts Library, Madras*, **20** (1972), 1, app., 1–92; 2, app., i–xiv; and 93–130; repr. (Madras, 1972).

8. The *Yogayātrā* is a shorter text on military astrology. The first nine chapters were published by H. Kern, "Die Yogayātrā des Varāhamihira," in *Indische Studien*, **10** (1868), 161–212; **14** (1876), 312–358; and **15** (1878), 167–184; and the whole, in imperfect fashion, by J. Lal (Lahore, 1944). A critical edition of the text with the commentary of Utpala has been prepared by D. Pingree.

9. The *Ṭikaṇikāyātrā* is a third treatise on military astrology. It was edited by V. R. Pandit, "Ṭikanikayātrā of Varāhamihira," in *Journal of the University of Bombay*, **20** (*Arts*, No. 26) (1951), 40–63.

10. The *Vivāhapaṭala*, a text on astrology as related to marriage, is preserved in a unique manuscript at Baroda. An edition has been prepared by V. R. Pandit.

BIBLIOGRAPHY

Additional bibliographical references to those given above will be found in O. Neugebauer and D. Pingree, *Pañcasiddhāntikā*, II, pp. 152–154; A. M. Shastri, *India as Seen in the Bṛhatsaṃhitā*, pp. 504–515; and D. Pingree, *Census of the Exact Sciences in Sanskrit*, series A, V (forthcoming).

David Pingree

VARENIUS, BERNHARDUS (Bernhard Varen) (*b.* Hitzacker, in the district of Hannover, Germany, 1622; *d.* Amsterdam, Holland, 1650), *physical geography.*

Son of the court preacher to the duke of Brunswick, Varenius spent his early years at Uelzen, the home of the duke. From 1640 to 1642 he studied in the Gymnasium of Hamburg, and from there he went to the universities at Königsberg (1643–1645) and Leiden (1645–1649), devoting himself to mathematics, medicine, and natural history. He took his medical degree at Leiden in 1649, and settled in Amsterdam with the intention of practicing medicine.

From some remarks in his writing it is evident that Varenius felt he had little future in Hannover, which had been devastated by the Thirty Years' War. In Amsterdam the recent discoveries of Abel Tasman, Willem Schouten, and other Dutch navigators, and his friendship with Willem Blaeu as well as other geographers led him to concentrate on geography rather than medicine, much to his own economic detriment.

Varenius was a prolific writer, although two of his early works, an academic treatise on motion according to Aristotle, which he wrote at Hamburg in 1642, and a table of universal history written at Amsterdam in 1649, were probably never published and are not now extant. His first significant works were the *Descriptio regni Japoniae* and the *Tractatus de religione Japoniae.* These works of special geography are usually regarded as a single book in part because they were published under the title of *Descriptio regni Japoniae et Siam, cum brevi informatione de diversis omnium religionibus* (Amsterdam, 1649). In actuality, the volume consisted of five separate works—a description of Japan by Varenius, a Latin translation of Jodocus Schouten's account of Siam, the discourse on religion in Japan by Varenius, excerpts from Leo Africanus on religion in Africa, and a short *Dissertatio de Rebuspublica in genere.*

The best known of Varenius' works is his *Geographia generalis* (Amsterdam, 1650), which established a framework for physical geography capable of including new facts of discovery as they arose. The work became the standard geographic text for more than a century. Varenius believed that there were three ways by which the truth of geographical propositions could be established, first by geometrical, arithmetical, and trigonometrical means; second by astronomical precepts and theorems; and third by experience. In the case of special geography, only celestial properties can be proven, and the remainder must rest on the experience either of the writer or of other observers. His *Geographia generalis* was divided into three sections. In the first he examined the mathematical facts relating to the earth, including its figure, dimensions, motions, and measurements. In the second part he examined the effect of the sun, stars, climates, and seasons on the earth and the differences of apparent time at various places. In the third part he treated briefly the actual divisions of the surface of the earth and laid down the principles of what is now known as regional geography. Each area was to be classified according to terrestrial information, including longitude, the nature of the terrain, and fertility; by celestial information, including the distance of the place from the pole, the climate, and the length of day and night; and finally by human information, including the status of the inhabitants, and their art, trade, virtues, vices, ceremonies, speech, and religion. Varenius' work long held the position as the best treatise on scientific and comparative geography. Humboldt and others were much impressed by it, and Newton revised parts of it for an edition published in England.

BIBLIOGRAPHY

I. ORIGINAL WORKS. Varenius' *Descriptio regni Japoniae et Siam, cum brevi informatione de diversis omnium religionibus* went through four editions, the last published in 1673. There was also a German summary of the description of Japan published at Jena in 1670. The brief treatise on religions of the world was included as an appendix in Alexander Ross, *Pansebeia* (London, 1653).

The *Geographia generalis*, according to the research of Gottfried Lange, went through fifteen complete and four partial editions in five European languages, plus ten in a summarized French version. There was also a summarized Japanese version published in 1932. Four editions were published by Elzevir in Amsterdam in 1650, 1664, 1671, and 1672. Newton's revision was published at Cambridge in 1672 and 1681, and at Jena in 1693. Another edition edited by James Jurin was published at Cambridge in 1712 and at Naples in 1715. The first English translation was by Richard Blome (London, 1682), and this was reprinted twice before 1693. A second English translation based upon Jurin's edition, translated by Dugdale and revised by Shaw, went through four editions between 1733 and 1765. A Dutch translation appeared in 1755 and a French translation by Philippe-Florent de Puisieux in that same year. A Russian version was published in Moscow in 1718 and a second one at St. Petersburg in 1790. The ten French summaries by Guillaume Sanson appeared between 1681 and 1743. In addition, Varenius wrote several brief works on human ecology published in Amsterdam in 1650.

II. SECONDARY LITERATURE. For discussions of Varenius and his work see J. N. L. Baker, "The Geography of Bernhard Varenius," in *Transactions of the Institute of British Geographers*, no. 21 (1955), 51–60; and Gottfried Lange, "Das Werk der Varenius: Eine kritische Gesamtbibliographie," in *Erdkunde*, **15** (1961), 10–18. See also Gottfried Lange, "Varenius über die Grundfragen der Geographie: Ein Beitrag zur Problemgeschichte der geographischen Wissenschaft," in *Petermanns geographische Mitteilungen*, **105** (1961), 274–283; and Hans Offe, "Bernard Varenius (1622–1650)," in *Geographisches Taschenbuch und Jahrweiser zur Landeskunde* (1960–1961), 435–438.

The most detailed study of Varenius' life was done by Siegmund Günther, *Varenius* (Leipzig, 1905), but his study has been much supplemented by works cited above. The brief biography by F. Ratzel, "Bernard Varenius," in *Allgemeine Deutsche Biographie*, XXXIX (Leipzig, 1895), must now be regarded as somewhat inaccurate.

VERN L. BULLOUGH

VARIGNON, PIERRE (*b.* Caen, France, 1654; *d.* Paris, France, 23 December 1722), *mathematics, mechanics.*

It is due to Lagrange that Varignon's name gained recognition in the teaching of mechanics in France in the nineteenth century, and until rather recently his name was linked with a theorem on the composition of forces that is now identified with the properties of the vector product. The passage of time diminishes this kind of fame; but historians are discovering in Varignon's work—which, admittedly, is of second rank with regard to substantive results—an importance for the philosophy of science. Expressive of the attempt to reduce the number of basic principles in mechanics in order to improve the organization of the subject, Varignon's accomplishments illustrate the relationship between this effort and progress made in notation and in operational procedures in pure mathematics.

The son and brother of contracting masons, Varignon stated that his entire patrimony consisted of his family's technical knowledge; it proved, however, to be of considerable importance for his later career. He probably studied at the Jesuit *collège* in Caen, which he would have entered at a relatively late age. The only certain information about this period of his life, however, is that relating to his entrance into the religious life: he submitted to the tonsure on 19 December 1676, earned his Master of Arts degree on 15 September 1682, and became a priest in the St.-Ouen parish of Caen on 10 March 1683. An ecclesiastical career enabled him to study at the University of Caen, where he was certainly one of the oldest students.

One of Varignon's fellow students was Charles Castel, Abbé de Saint-Pierre (1658–1743), who later achieved fame for his philanthropy. Saint-Pierre soon offered to share his lodgings and income with Varignon. The two left Caen for Paris in 1686. When Varignon reached Paris, he had already done considerable scientific research; and the contacts he made through Saint-Pierre accomplished the rest. As early as 1687 he had access to Pierre Bayle's periodical, *Nouvelles de la république des lettres*, for the publication of his memoir on tackle blocks for pulleys, and his first published book, *Projet d'une nouvelle méchanique*, was dedicated to the Académie des Sciences.

Although not to be compared with Newton's *Principia*, which appeared in the same year as Varignon's *Projet*, their simultaneous publication perhaps brought the latter work a greater success among French scientists than it would otherwise have had. In any case, the success of the *Projet* brought Varignon nomination as geometer in the Académie des Sciences in 1688, as well as the first appointment to the newly created professorship of mathematics at the Collège Mazarin. Within two years, therefore, Varignon was set in his career. He taught—and resided—at the Collège Mazarin until his death. In 1704 the former secretary of the Academy, Jean-Baptiste du Hamel, resigned in Varignon's favor from the chair of Greek and Latin philosophy at the Collège Royal (now the Collège de France). The title of the chair in no way restricted the scientific topics that could be taught by its holder, who had sole discretion in this regard.

Fully occupied by his teaching duties and his responsibilities as an academician, Varignon had no leisure to prepare works for publication. After a short second work, *Nouvelles conjectures sur la pesanteur* (1690), his literary production consisted of articles for learned journals and a large number of memoirs submitted to the Academy. His correspondence, however, particularly with Leibniz and Johann I Bernoulli, bears witness to his role in the scientific life of his age. From the papers he left at his death, most of which are now lost, his disciples assembled several posthumous works: *Nouvelle mécanique* (announced in the *Projet* of 1687) and *Éclaircissemens sur l'analyse des infiniment petits*, both published in 1725, and *Élémens de mathématiques* (1731), which was based on his courses at the Collège Mazarin.

Varignon's intense pedagogical activity, extending over more than thirty years, constituted his chief contribution to the progress of science and was the source of his fame. By inaugurating a chair devoted specifically to mathematics at the Collège Mazarin, he joined the handful of men who were then teaching advanced mathematics; and it is in this context that his work was of great importance.

Bossut and Montucla, writing the history of mathematics half a century later, were unable to ignore Varignon; but, lacking the necessary historical distance, they were unjustly severe. Bossut, for example, wrote: "Endowed with an excellent memory, Varignon read a great deal, closely examined the writings of the pioneers [*inventeurs*], generalized their methods, and appropriated their ideas; and some students took disguised or enlarged reformulations to be discoveries." But since the essential precondition of a teacher's effectiveness is that he constantly broaden his knowledge and keep it current, Bossut should have praised Varignon for having done just that, instead of condemning him for not sufficiently citing his sources. The latter judgment is, of course, possible; but Varignon's writings offer no incontrovertible support for it. Montucla's evaluation was more penetrating; he criticized Varignon primarily for what may be called a mania for "generalization." Certainly, Varignon had neither a precise nor an acceptable notion of that process and often confused it with the mere use of algebraic language. Viewed in its historical context, however, this failing is not at all astonishing.

The pejorative assessments of Bossut and Montucla were echoed by Pierre Duhem, who in his *Origines de la statique* (1905–1906) wrote ironically of Varignon's naïve belief in his own originality in mechanics. Yet, like earlier criticisms, Duhem's is not wholly justified. The audacity that average intellects must needs muster in order to fight for progressive ideas always presupposes a certain naïveté on their part. Indeed, the more it becomes evident that Varignon was not a genius, the less the defects of his thought ought to be allowed to weigh against estimates of his real accomplishments.

From this point of view, Lagrange underscored the essential point. In the posthumous edition of the *Nouvelle mécanique* he found the text of a letter from Johann Bernoulli to Varignon (26 January 1717) marking the emergence of the principle of virtual velocities, and he realized that in this matter Varignon deserves credit on two counts: for preparing the way for and eliciting Bernoulli's statement, and for attempting to provide the broadest justification of the principle. Thus the period between the *Projet* of 1687 and the *Nouvelle mécanique* witnessed the development of what appeared a century later to be the very foundation of classical mechanics.

Lagrange was not mistaken, either, about Varignon's active role in the initial development of the principle of the composition of forces. The technique of composing forces by the rule of the parallelogram had undergone more than a century of development when it was published, simultaneously, in 1687 in Newton's *Principia*, Varignon's *Projet*, and the second edition of Bernard Lamy's *Traitez de méchanique*. The enunciation of the principle, which appeared as a consequence of the composition of infinitely small movements and not of finite ones, eliminated a troublesome confusion that had hampered progress in the subject.

The simultaneous publication of the principle makes difficult any judgment regarding priority. Nevertheless, it was Varignon alone who grasped two important points. The first is that the law of the lever does not hold a privileged position in statics, and that the unification of "mechanics" (the science of simple machines) was to be carried out on the basis of the composition of forces. The second concerns the inclined plane: that the real reason for the equilibrium observed is that the resultant of the applied forces is orthogonal to the possible displacement. These two points provide a good indication of Varignon's contribution to the development of the principle of virtual velocities.

It must, of course, be added that his contribution was limited to general statics; but this was the point of departure for d'Alembert's subsequent extension of the principle to dynamics. In the latter field Varignon did not solve any of the important problems of his time—as Bossut correctly observed. Nevertheless, in his memoirs to the Academy, he showed how to apply infinitesimal analysis to the science of motion and how, in specific cases, to use the relationship between force and acceleration. The laborious nature of this work does not detract from its historical importance.

In working with the model of falling bodies, Varignon encountered difficulties in obtaining acceleration as a second derivative. This problem had the advantage, however, of obliging him to reassess the importance of the new differential and integral calculus. His acceptance of the new procedures occurred between 1692 and 1695, and he was among those who gave the most favorable reception to the publication of L'Hospital's *Analyse des infiniment petits* in 1696. The *Éclaircissemens* is composed of critical notes that Varignon, as a professor, considered necessary in presenting L'Hospital's pioneering work to young mathematicians—further evidence of his constructive role in the movement to transform the operations used in mathematics. But Varignon accomplished even more: in 1700–1701 he refuted Rolle's arguments against the new calculus, challenged the cabal that had formed within the Academy, and obliged Leibniz to furnish a more precise account of his ideas. Leibniz, to be sure, did not give him all the aid desired. Nevertheless, he encouraged Varignon to cease debating principles and to start developing mechanical applications of the new mathematics. The questions that Varignon subsequently treated show how faithfully he followed Leibniz' advice.

In his course at the Collège Royal for 1722–1723, Varignon planned to discuss the foundations of infinitesimal calculus but was able to do no more than outline his ideas. Although he died before he could present what was undoubtedly the core of a lifetime's experience, that experience had already borne fruit.

BIBLIOGRAPHY

I. ORIGINAL WORKS. Varignon's books include *Projet d'une nouvelle méchanique* (Paris, 1687); *Nouvelles conjectures sur la pesanteur* (Paris, 1690); *Éclaircissemens sur l'analyse des infiniment petits* (Paris, 1725); *Nouvelle mécanique ou statique . . .*, 2 vols. (Paris, 1725); *Traité du mouvement et de la mesure des eaux coulantes et jaillissantes . . .* (Paris, 1725); and *Élémens de mathématiques . . .* (Paris, 1731).

His major published articles are "Démonstration générale de l'usage des poulies à moufle," in *Nouvelles de la république des lettres* (May 1687), 487–498; a sequel to the preceding article containing "une nouvelle démonstration du paradoxe de M. Mariotte," in *Histoire des ouvrages des sçavans*, **1** (Oct. 1687), 172–176; "Règles du mouvement en général," in *Mémoires de mathématiques et de physique tirés des registres de l'Académie . . .* (1692), 190–195; "Des cycloïdes ou roulettes à l'infini," *ibid.* (1693), 43–47; "Règles des mouvemens accélérés suivant toutes les proportions imaginables d'accélérations ordonnées," *ibid.*, 93–96; "Méthode pour trouver des courbes le long desquelles un corps tombant s'approche ou s'eloigne de l'horizon en telle raison des temps qu'on voudra . . .," in *Mémoires de l'Académie royale des sciences* (1699), 1–13.

Later works are "Du mouvement en général par toutes sortes de courbes et des forces centrales tant centrifuges que centripètes, nécessaires aux corps qui les décrivent," *ibid.* (1700), 83–101; "Application au mouvement des planètes," *ibid.*, 218–237; "De la figure ou curvité des fusées des horloges à ressort," *ibid.* (1702), 192–202; "Du mouvement des eaux . . .," *ibid.* (1703), 238–261; "Du mouvement des planètes sur leurs orbes en y comprenant le mouvement de l'apogée on de l'aphélie," *ibid.* (1705), 347–361; "Différentes manières infiniment générales de trouver les rayons osculateurs de toutes sortes de courbes . . .," *ibid.* (1706), 490–507; "Des mouvements faits dans des milieux qui leur résistent en raison quelconque," *ibid.* (1707), 382–476; "Des forces centrales inverses," *ibid.* (1710), 533–544; "Réflexions sur l'usage que la mécanique peut avoir en géométrie," *ibid.* (1714), 77–121; and "Précaution à prendre dans l'usage des suites ou séries infinies . . .," *ibid.* (1715), 203–225.

II. SECONDARY LITERATURE. See Pierre Costabel, "Contribution à l'histoire de la loi de la chute des graves," in *Revue d'histoire des sciences et de leurs applications*, **1** (1947), 193–205; "Le paradoxe de Mariotte," in *Archives internationales d'histoire des sciences*, **2** (1948), 864–886; "Pierre Varignon et la diffusion en France du calcul différentiel et intégral," *Conférences du Palais de la découverte*, ser. D, no. 108 (1965); and "Varignon, Lamy et le parallélogramme des forces," in *Archives internationales d'histoire des sciences*, **19**, nos. 74–75 (1966), 103–124; J. O. Fleckenstein, "Pierre Varignon und die mathematischen Wissenschaften in Zeitalter des Cartesianismus," *ibid.*, **2** (1948), 76–138; and Bernard de Fontenelle, "Éloge de M. Varignon," in *Histoire et mémoires de l'Académie des sciences* for 1722, 189–204.

PIERRE COSTABEL

VAROLIO, COSTANZO (*b.* Bologna, Italy, 1543; *d.* Rome, Italy, 1575), *medicine.*

Varolio, the son of Sebastiano Varolio, a Bolognese citizen, studied medicine at the University of Bologna. He displayed interest and aptitude in particular for anatomy, which he pursued under the direction of Giulio Cesare Aranzio. Varolio received his medical degree in 1567, and in 1569 was given the extraordinary chair of surgery, which carried with it the responsibility of teaching anatomy as well. He held this position until 1572, when he went to Rome. There is little positive biographical information about Varolio and, consequently, some difference of opinion as to whether or not he went to Rome upon invitation to join the medical faculty of the Sapienza, the papal university. Contrary to former positive assertions, Varolio's name appears not to have been listed in the *rotuli* of that institution. Moreover, there is also question as to whether he had been invited to become the physician or surgeon of Pope Gregory XIII or, indeed, had any appointment at all in the papal medical service. In any case, Varolio was esteemed by the pope and enjoyed his patronage during the three years that he was in Rome. According to a contemporary commemorative inscription—which refers to Varolio's anatomical lectures and demonstrations "*in gymnasio Romano*"—he is declared to have died "from an unknown ailment."

During his short life Varolio wrote two books, *De nervis opticis nonnullisque aliis praeter communem opinionem in humano capite observatis* (Padua, 1573), illustrated with three views of the brain drawn by the author, and the posthumously published *Anatomiae sive de resolutione corporis humani libri IIII* (Frankfurt, 1591), which has been described as a teleologic physiology of man, but which contains a reprint of *De nervis opticis*. It has furthermore been asserted that Varolio was the author of a work entitled *De cerebro*, published at Frankfurt in 1591, and of a large anatomical treatise in four books with many illustrations that was being printed at the time of his death. Neither attribution appears to be correct; in both instances there was apparently confusion with the existing

posthumous *Anatomiae libri IIII*, which contains no illustrations except for the three figures of the brain that appeared originally in the earlier work of 1573.

De nervis optics, the source of Varolio's anatomical reputation, consists of a letter to Girolamo Mercuriale dated 1 April 1572, the latter's reply, and a response by Varolio. The book was published without Varolio's consent or knowledge through the efforts of Paolo Aicardi, a disciple of Mercuriale.

From Galenic times onward the brain had been dissected *in situ* by means of a series of horizontal slices begun at the uppermost part of the cerebral hemispheres. Varolio, "considering most organs of the brain to be near the base of the head, and the brain by its weight, especially in the dead body, to compress them between itself and the skull," judged the usual method of dissection "to be hindered by many obstacles." In consequence he used a new method by which he first removed the brain from the skull, turned it over, and dissected it from below, beginning at its base. "If one proceeds in this way," he wrote, "each of [the brain's] organs may be observed as completely as desirable." Although he referred to his new method of dissection as "unusual" and "very difficult," it did permit a better observation of the structures of the brain, notably the cranial nerves, and was widely followed. It was in consequence of his new technique that Varolio was able for the first time to observe and describe the pons, still known as the *pons Varolii*, so called because the spinal marrow was "carried under this transverse spinal process as a flowing stream is carried under a bridge." Although Varolio considered the pons to be part of the cerebellum, it has more recently been attributed to the brain stem.

As a result of this new method of dissecting the brain, Varolio was able to make some contributions to the knowledge of the course and terminations of the cranial nerves. In the instance of the optic nerve, which provided the title of the book, he traced its course approximately to the true termination. He also suggested that the spinal cord had four tracts, two anterior serving sensation and two posterior for cerebellar functions.

BIBLIOGRAPHY

Varolio's two books, mentioned above, are very rare, *De nervis opticis* especially so. This work has been reprinted in facsimile (Brussels, 1969).

Works giving biographical information about Varolio are scanty and many data must be used with caution. They are, in chronological order, Gaetano Marini, *Degli archiatri pontifici*, I (Rome, 1784), xxxviii, 429; Giovanni Fantuzzi, *Notizie degli scrittori bolognese*, VIII (Bologna, 1790), 158–160; Michele Medici, *Compendio storico della Scuola Anatomica di Bologna* (Bologna, 1857), 84–90; Umberto Dallari, *I rotuli dei lettori legisti e artisti dello Studio Bolognese dal 1384 al 1799*, II (Bologna, 1889), 176, 179, 182; and Ludwig Choulant, *History and Bibliography of Anatomic Illustration*, Mortimer Frank, trans. (Chicago, 1920), 214–215. There is also some information on Varolio as an anatomist in Antoine Portal, *Histoire de l'anatomie et de la chirurgie*, II (Paris, 1770), 28–38.

C. D. O'MALLEY

VARRO, MARCUS TERENTIUS (*b.* Reate, Italy, 116 B.C.; *d.* Rome, 27 B.C.), *encyclopedism, polymathy, biology.*

Varro, whom Quintilian dubbed "the most learned of the Romans," came from an obscure equestrian family at Reate in the Sabine country. He studied under the Stoic grammarian L. Aelius Stilo at Rome and with the Academic philosopher Antiochus of Ascalon at Athens. Varro devoted most of his life to public service. From 86 B.C. to 43 B.C., he proceeded through the ranks of the *cursus honorum* until he reached the office of praetor. During the Civil War he led Pompeian forces in Spain; after the death of Pompey he was pardoned by Caesar. In 47 B.C. Caesar appointed him director of the proposed public library at Rome. Proscribed by Antony in 43 B.C., Varro's villa and library were seized. Having escaped the death sentence, he devoted the remainder of his life to scholarship.

Varro was the most prolific of all Roman authors, writing more than 620 books under seventy-four different titles. Of these works, only the three books of *De re rustica* are substantially complete. Books V to X are extant from the twenty-five-book study *De lingua Latina*. For Varro's other writings, only fragments are extant. His lost works include the 150 books of the popular *Saturae Menippeae*, the treatises of the *Logistoricon libri LXXVI*, the 700 illustrated biographies of famous Greeks and Romans in the *Imagines*, the *Antiquitatum rerum humanarum et divinarum libri XLI*, and the highly influential *Disciplinarum libri IX*.

Varro began his agricultural treatise *De re rustica* when he was eighty years old. Written in the form of a dialogue, this handbook on husbandry discusses general agricultural practices, domestic cattle, and small livestock—poultry, game birds, bees, and fish. The three books of this study were based on Greek and Latin sources and on Varro's own agricultural experiences. It presents a comprehensive view of Roman farming techniques.

De lingua Latina is a philological study of the Latin language. In its entirety it contained an examination of etymology, inflections, and syntax. While its etymology is often fanciful, it does contain perceptive comments on early Latin words. The extant portions of this treatise include approximately half of the books on etymology and inflections.

Varro's most lasting scientific legacy was the orientation he gave to later scholarship. In his lost *Disciplinarum libri IX*, Varro introduced the Greek encyclopedic tradition into Roman thought. The *Disciplinae* was an encyclopedia of the liberal arts based on Greek sources. Containing chapters on the traditional Greek *trivium* (grammar, rhetoric, and logic) and *quadrivium* (arithmetic, geometry, astronomy, and music), along with sections on medicine and architecture, it was designed to provide a survey of the knowledge needed by a free man. Later scholars deleted medicine and architecture, which became professional studies, and retained the subjects of Varro's *trivium* and *quadrivium* as the basis for medieval education.

Varro's work popularized both the encyclopedic tradition and the liberal arts. Ultimately, handbooks and encyclopedias based on excerpts from the writings of earlier authorities became a fundamental part of Roman scientific thought.

BIBLIOGRAPHY

I. ORIGINAL WORKS. From antiquity to the present, Varro's *De re rustica* has usually appeared with Cato's agricultural treatise. The *editio princeps* of Varro's study was published by Nicolas Jenson at Venice in 1472. Other editions soon followed at Bologna in 1494 and Venice in 1514. Modern editions include those published at Leipzig in 1884 and 1929, and at Cambridge in 1935.

De lingua Latina has also appeared in a number of editions. The date and place of publication of the *editio princeps* by Pomponius Laetus is unknown. There were a number of fifteenth- and sixteenth-century editions: Rome, 1474; Venice, 1475, 1483, 1492, 1498, 1513; Milan, 1510; and Paris, 1529. There have also been a number of modern editions: Berlin, 1826; Leipzig, 1910; and Cambridge, 1938.

II. SECONDARY LITERATURE. Accounts of Varro's life and works may be found in Conrad Cichorius, *His-*

torische Studien zu Varro (Bonn, 1922), 189–241; Gaston Boissier, *Étude sur la vie et ouvrages de M. T. Varron* (Paris, 1861); Hellfried Dahlmann, "M. Terentius Varro" in Pauly-Wissowa, *Real-Encyclopädie der classischen Altertumswissenschaft,* supp. vol. VI (1935), 1172–1277; Jens Erik Skydsgaard, *Varro the Scholar* (Copenhagen, 1968); and William H. Stahl, *Roman Science* (Madison, Wis., 1962).

Numerous bibliographical references may be found in *L'année philologique.*

PHILLIP DRENNON THOMAS

VASSALE, GIULIO (*b.* Lerici, Italy, 22 June 1862; *d.* Modena, Italy, 3 January 1913), *endocrinology.*

One of the outstanding endocrinologists of his generation, Vassale contributed to the early development of that science in Italy. He began to study medicine at the University of Modena but transferred to Turin, from which he graduated in 1887 with a thesis on regeneration of the gastric mucous membrane. In the same year he was appointed assistant to the professor of general pathology and pathological anatomy at Modena, and in 1889 he was employed at the asylum of Reggio nell'Emilia as a dissector. In 1894 he was named substitute teacher of general pathology at Modena, and in 1898 full professor. He died, a bachelor, of cancer at the age of fifty.

A pupil of Bizzozero, Vassale began his scientific activity as a histologist at Turin, with microscopic observations of the gastric mucosa, *Sulla riproduzione della mucosa gastrica* (1888). This work, started in 1887 as his doctorate thesis, demonstrated that the canine gastric epithelium and gastric glands were regenerated, after aseptic excision of the mucous coat, by proliferation from elements of the glands contiguous to the edges of the excision. By 1889, however, Vassale had become interested in neurology. He spent twenty years (1890–1910) in experimental research on internal secretions, principally those of the thyroid, parathyroid, and adrenal glands.

In 1890 Vassale studied the changes in the canine pancreas following the ligature of Wirsung's duct, and the following year he demonstrated the independence of the islands of Langerhans from the pancreatic digestive gland. In contrast with the pancreatic digestive gland (which underwent atrophy), the islands of Langerhans were not damaged by ligature of Wirsung's duct; and there was no glycosuria. This observation marked the beginning of the insular theory of diabetes.

In November 1890, Vassale demonstrated that an aqueous extract of the thyroid gland quickly suppresses the serious results of thyroid failure. His conclusions were confirmed in animals in 1891 by M. E. Gley and in man by George R. Murray in the same year. Vassale had been anticipated, however, by Gustavo Pisenti of Perugia, who had administered intravenous injections of aqueous extracts of thyroid early in 1890.

In 1892 Vassale demonstrated the cachexia following experimental destruction of the hypophysis; and his experiments on animals were contemporary with those of Gley and Gheorgi Marinescu. From 1896 to 1906 he studied the effects of a total parathyroidectomy and demonstrated, in dogs, that acute tetany was not due to insufficiency of the thyroid but resulted from the inadvertent ablation of the parathyroid when the thyroid was removed.

From 1898 to 1908 Vassale investigated the physiopathology of the adrenal glands and helped to differentiate their medullary and cortical components. In 1902 he demonstrated that acute adrenal insufficiency is due to insufficiency of the medulla alone. Three years later he showed that adrenaline, the substance active in the elevation of blood pressure, was produced only by the medullar cells of the adrenal glands.

BIBLIOGRAPHY

I. ORIGINAL WORKS. Vassale's writings include *Sulla riproduzione della mucosa gastrica* (Modena, 1888); "Intorno agli effetti della iniezione di succo di tiroide nei cani operati di estirpazione della tiroide," in *Rivista sperimentale di freniatria e medicina legale delle alienazioni mentali,* **16** (1890), 439–455; *Ricerche microscopiche e sperimentali sulle alterazioni del pancreas consecutive alla legatura del dotto di Wirsung* (Reggio nell'Emilia, 1891); "Sulla distruzione della ghiandola pituitaria," *ibid.,* **18** (1892), 525–561; "Sugli effetti dell'estirpazione delle ghiandole paratiroidee," in *Rivista di patologia nervosa e mentale,* **1** (1896), 95–99; "Sugli effetti dello svuotamento delle capsule soprarenali," in *Bollettino della Società medico-chirurgica di Modena* (meeting of 12 Feb. 1898); and *Fisiopatologia delle ghiandole a secrezione interna* (Modena, 1914), a reprint of all Vassale's works on endocrinology, issued by the Medical School of Modena.

II. SECONDARY LITERATURE. See R. Abderhalden, "Le secrezioni interne," in *Ciba Review,* **5** (1951), 916; L. Castaldi, "Il precursore della preparazione della insulina," in *Riforma medica,* **15** (1929); E. Centanni, "Giulio Vassale," in *Biochimica e terapia sperimentale,* **4** (1913), 193–198; J. Derrien, *Étude historique et critique sur le traitement du myxoedème par les injections de liquide thyroïdien* (Paris, 1893), a doctoral dissertation; V. Diamare, "Documenti per la storia della teoria insulare del diabete, e sui precedenti dell'insulina," in *Archivio di fisiologia*, **22** (1924), 141–157; U. Lombroso, "Sugli elementi che compiono la funzione interna del pancreas," in *Archivio di farmacologia sperimentale e scienze affini*, **7** (1908), 170–218, a critical and precise review with a rich bibliography; G. Pisenti, "Le basi dell'opoterapia tiroidea," in *Terapia* (Milan), **20** (1930), 193–195; and M. Segale, "Giulio Vassale," in *Pathologica*, **5** (1913), 137–145, with bibliography and portrait.

PIETRO FRANCESCHINI

VASTARINI-CRESI, GIOVANNI (*b.* Taranto, Italy, 1870; *d.* Naples, Italy, 14 April 1924), *anatomy, histology.*

Vastarini-Cresi was lecturer, assistant professor, and then from 1905 in charge of microscopic anatomy in the Naples Anatomical Institute. In 1919 Vastarini-Cresi was appointed head of the institute and was succeeded by Giovanni Antonelli, Giunio Salvi, and Riccardo Versari. Vastarini-Cresi led an austere life, entirely devoted to scientific activity and to teaching numerous pupils, especially in the histological field. A scrupulous and intelligent researcher, and the author of appreciated works on morphology, he studied arteriovenous anastomosis and the hypopharynx in man. He made excellent contributions to histological technique, especially with the method of glycogen-staining in tissues that bears his name. Lambertini reported that in studying the taste organ Vastarini-Cresi (1) asserted that vallate papillae can also arise from the rear part of the tongue, contrary to His's doctrine, which restricts the territory of origin of the taste buds to the *tuberculum impar* only; (2) demonstrated the presence of a double vallate papilla rising from the declivity of Morgagni's foramen cecum and observed the presence of retrocecal vallate papillae; and (3) described that in the lingual innervation, in addition to branches going to the taste buds on its own sides, the glossopharynx also sends fibers passing to the opposite side and reaching the central circumvallate papilla of the foramen cecum. On Vastarini-Cresi and his work, see Gastone Lambertini, *Dizionario anatomico* (Naples, 1949), 216, 279, 564; and "Necrologio di Giovanni Vastarini-Cresi," in *Monitore zoologico italiano*, **35** (1924), 104.

BRUNO ZANOBIO

VAṬEŚVARA (*b.* 880 at Ānandapura [modern Wadnagar], Gujarat, India), *astronomy.*

The son of Mahadatta Bhaṭṭa, Vaṭeśvara wrote a *Vaṭeśvarasiddhānta* at the age of twenty-four (that

is, in 904), in which he frequently follows, but at times severely criticizes, the *Brāhmasphuṭasiddhānta* of Brahmagupta (born 598); basically Vaṭeśvara's text belongs to the Āryapakṣa (see essay in Supplement). The manuscripts of this work are exceedingly rare so that its contents beyond the first three chapters, which have been published, remain obscure; but what is known makes it clear that this is an extremely important work for understanding the developments that took place in Indian astronomy between Brahmagupta and Bhāskara II (born 1115), particularly because of Vaṭeśvara's explicit criticisms of his predecessors modeled on Brahmagupta's eleventh chapter, the *Tantraparīkṣā*.

The only other work of Vaṭeśvara that we know of is an astronomical handbook, *Karaṇasāra*, which is lost in Sanskrit, but which was often cited by al-Bīrūnī (born 973) in his *India*, his *Transits*, and his *Al-Qānūn al-Masʿūdī*. From the former (chap. 49 and 53) we learn that the epoch of the *Karaṇasāra* was A.D. 899; this is confirmed by the *Karaṇasāra*'s computation of the assumed motion of the Saptarṣi (Ursa Major) in the *India* (chap. 45) and in the *Qānūn* (IX, 1). Al-Bīrūnī, however, calls the author of the *Karaṇasāra* "Batīshfar ibn Mahadatta from the city of Nāgarapūr," which Sachau erroneously translates "Vitteśvara, the son of Bhadatta" (*India*, chap. 14); the name Nāgarapura refers to the fact that Anandapura was the center of the Nāgara Brāhmaṇas. There is, then, no doubt about the identity of the authors of the *Vaṭeśvarasiddhānta* and of the *Karaṇasāra*.

BIBLIOGRAPHY

A first volume of an edition of the *Vaṭeśvarasiddhānta* with Sanskrit and Hindī commentaries was published by Ram Swarup Sharma and Mukund Mishra (New Delhi, 1962); many errors in the commentary are corrected by T. S. Kuppanna Shastri, "The System of the *Vaṭeśvara Siddhānta*," in *Indian Journal of History of Science*, **4** (1969), 135–143. See also R. N. Rai, "Sine Values of the *Vaṭeśvarasiddhānta*," *ibid.*, **7** (1972), 1–15; and "Calculation of *Ahargaṇa*, in the *Vaṭeśvarasiddhānta*," *ibid.*, 27–37; and K. S. Shukla, "Hindu Astronomer Vaṭeśvara and His Works," in *Gaṇita*, **23** (1972), 2, 65–74.

DAVID PINGREE

VAUBAN, SÉBASTIEN LE PRESTRE DE (*b.* St.-Léger-de-Fougeret [now St.-Léger-Vauban], near Avallon, Burgundy, France, 15 May 1633; *d.* Paris, France, 30 March 1707), *military engineering.*

France's greatest military engineer—the famous "taker of cities"—and a dedicated public servant of Louis XIV, Vauban scarcely deserves to be called a scientist. Although made an honorary member of the Royal Academy of Sciences in his old age, it was less for any scientific attainments than for his long, devoted, and manifold services to France. He never distinguished himself in mathematics or physics like Lazare Carnot or Coulomb, both trained as military engineers. Vauban was a practical man of little culture and sparse scientific training who was skilled in the application of simple arithmetic and geometry and the elementary principles of surveying and civil engineering to fortification and siegecraft. Above all he had range and flexibility of mind, common sense, and the tact and insight that comes from long experience.

Vauban's family was of modest origin—notaries and small merchants of Bazoches in the climatically rugged Morvan. In the sixteenth century one member of Vauban's family entered the lesser nobility through the purchase of a small fief. Vauban's father boasted the title of "Écuyer, Seigneur de Champignolle et de Vauban." Saint-Simon from his lofty social eminence called Vauban a "petit gentilhomme de Bourgogne, tout au plus."

Of Vauban's youth little is known. At first he was taught by the village curate of St.-Léger-de-Fougeret, where he was born during the reign of Louis XIII and the administration of Cardinal Richelieu. At the age of ten Vauban was sent to the Carmelite collège of Semur-en-Auxois; here he acquired the rudiments of mathematics, a smattering of history, and showed some talent for draftsmanship. In 1651, when he was seventeen, he entered upon his military career as a cadet in the forces of Louis II de Bourbon, prince de Condé, then in rebellion against the king. Under Condé he served his apprenticeship by working on the fortifications of Clermont-en-Argonne. Captured by the king's forces in 1653, he shared in the pardon of Condé and his troops and entered the royal army. Cardinal Mazarin, learning of Vauban's knack for fortification, placed him under the chevalier de Clerville, a man of mediocre gifts who was then regarded as the foremost military engineer of France. Two years later Vauban earned the rank of ingénieur ordinaire du roi.

Between the end of the war with Spain in 1659 and Louis XIV's first war of conquest in 1667, Vauban worked under Clerville repairing and improving the fortifications of the kingdom. During the brief War of Devolution, he so distinguished himself—notably by his talent for siegecraft—that Louvois, his direct superior, promoted him over the head of Clerville to the rank of commissaire

général, in effect making him the virtual director of all the engineering work in Louvois's department.[1] The conquests of the War of Devolution in Hainaut and Flanders launched Vauban on a building program involving such conquered towns as Bergues, Furnes, Tournai, and Lille, the outposts of the future expansion.

Nine years of peace had succeeded the war with Spain. Vauban returned briefly to his native Morvan, where he married Jeanne d'Osnay, daughter of the baron d'Epiry. Yet soon after the wedding he returned to garrison duty at Nancy, leaving his wife behind. Until late in life, his visits home were rare, although he fathered two daughters and a son who died in infancy. In 1675 he bought the nearby Château de Bazoches, but until his last years his stays were brief in this, his favorite residence. He was constantly in the field. As Daniel Halévy put it, Vauban's true family was the army.[2]

This, then, was the tireless rhythm of Vauban's life: constant repairs and construction of fortress towns in time of peace; in war, sieges and new conquests; then more feverish construction during the ensuing intervals of peace, always with an eye towards the strategic goal of giving France the maximum security.[3] Until the year of his death, Vauban was constantly on the move, traveling from one end of France to the other on horseback, or, in his later years, carried in a famous sedan chair borne by horses. Although he sedulously avoided the court and was rarely seen in Paris or Versailles for any length of time, he kept in steady communication with his superiors, deluging Louvois, for example, with innumerable letters and reports written in pungent and undoctored prose. This "vie errante" of over forty years gave him an unrivaled knowledge of the state of France, and led to many of the proposals he set forth in his letters or in his ironically mistitled Oisivetés, a collection of papers written during intervals of repose at Bazoches or while traveling.

In his long and active life Vauban directed some fifty sieges, and Sir Reginald Blomfield lists nearly a hundred towns and strong points fortified or radically strengthened by Vauban. In the seventeenth and eighteenth centuries, sieges were usually the focal operations of a campaign and the inevitable preliminary to invading foreign territory; indeed they were more frequent than infantry combat in open country, and were begun as readily as pitched battles were avoided or broken off. In siegecraft, the art of reducing fortified places, Vauban made his reputation as early as 1658–1659 at Montmédy, Ypres, Gravelines, and Oudenaarde in the

Low Countries. His most famous success was the siege of Namur in 1692, defended by the great Dutch engineer, Cohorn, and immortalized by Uncle Toby in Laurence Sterne's *Tristram Shandy*. Here the French troops were commanded by the king in person, and Vauban's conduct of the siege was recorded by the official historian, the playwright Racine, in three letters to the poet Boileau. But it was at the earlier siege of Maastricht that Vauban, always concerned with the welfare of his troops, and hoping to reduce casualties, introduced the system of parallel trenches by means of which the assailants could approach under cover within close range of the ramparts of a fortress. [4] Less concerned about the fate of the defenders, at the siege of Philippsburg he first used the ricochet fire of mortars, where the propelling charge was so greatly reduced that the projectile was gently lobbed into the enemy's stronghold, where it would rebound this way and that, a peril to man and machine.

In the design of fortified places, Vauban owed a considerable debt to his predecessors. He was the heir of nearly a century and a half of progress, during which the profile was lowered and the bastioned trace (a polygonal outline marked by projecting bastions) reached a high degree of perfection at the hands of men like Jean Erard of Bar-le-Duc, the founder of the French school of fortification,[5] and Blaise Pagan (1604–1665), a theorist rather than a practical engineer and an astronomer of some repute. Pagan's *Traité des fortifications* (1645) strongly influenced Vauban; indeed his earliest forts were based on Pagan's designs, but with minor improvements and adaptations to differences in terrain, a characteristic feature of Vauban's work.[6]

It has been claimed that Vauban followed three different "systems" in the course of his career. This claim has rightly been challenged. He never published anything about his methods, or wrote anything except to stress the importance of accumulated experience, flexibility of mind, and a distrust of bookish formulas. Many later writers have agreed with Lazare Carnot who admired Vauban, yet who gave him little credit for striking originality. "The fortification of Vauban," Carnot wrote, "reveals to the eye only a succession of works known before his time, whereas to the mind of a good observer it offers sublime results, brilliant combinations, and masterpieces of industry."[7] Later studies, by Lieutenant Colonel Lazard and by Sir Reginald Blomfield have altered our perspective. Blomfield, while unable to praise Vauban's architectural taste (he severely criticizes the design

of gateways into some of Vauban's fortified towns), fully agrees with Lazard that, strictly speaking, Vauban did not follow sharply defined systems; instead, as his experience grew and deepened, we detect distinguishable periods in which he favored modified designs, all variations of the polygonal trace with bastions. His major works were the citadel of Lille (his earliest honor was to be made its governor); Maubeuge, considered one of the best examples of a fortified town; and his masterpiece and last major work, Neuf-Brisach, designed as a wholly new fortified town,[8] octagonal in shape with elaborate outworks. Vauban's tendency was increasingly to rely "on the outworks of his forts more than on the traditional rampart, bastion, moat, covered way and glacis."[9] The main body of his fort was protected by demilunes (detached triangular works erected in the moat), detached bastions, or hornworks (*ouvrages à cornes*),[10] which pushed the defense out into the surrounding country. Such detached works first appear timidly in forts of what was once called his "second system," for example, at Belfort and Besançon; but the culmination of this development is to be seen in the outline of Neuf-Brisach, which deserves also an important place in the history of city planning.

During Vauban's lifetime there was not yet, in any real sense, an organized corps of army engineers. The engineers were men of diverse training and background: some were civilians (architects or mathematicians) of known ability; others held commissions in the infantry or cavalry and, like Vauban, had served a novitiate under an established practitioner. To see created a true corps of engineers, as a regularly constituted arm of the service with its own specially trained officers, with troops (including sappers and miners) under their command, and with a distinctive uniform, was an objective Vauban strove to attain throughout his career, although with little success. His recommendations did not bear fruit until later in the eighteenth century.

Vauban's concerns were not limited to his own branch of the military profession. He was one of the most indefatigable military reformers of his age, and his proposals left few areas untouched. He was keenly interested in improving artillery; disliking bronze cannon, he urged the army to emulate the navy in the use of iron. He was a tireless advocate of the flintlock musket, and wished to abolish the pike and substitute for it the bayonet fitted with a sleeve or socket, in order to align the blade at the side of the barrel, thereby allowing the piece to be fired with bayonet fixed.

In a long paper of the *Oisivetés* on the reorganization of the army, he gives his views on war in general and the French army in particular.[11] War, he wrote, has "interest" (in the sense of our phrase "vested interest") as father; its mother is ambition and for its close relatives it has all the passions that lead us to evil. The French, nevertheless, like war if it can be carried out with honor; they are courageous and intelligent; yet service in the army is hated: desertions are constant, even though they are cruelly punished on the gallows. Recruitment by force and the press gang is a source of constant complaint; and Vauban finds it a grave social evil, taking husbands away from their families and driving some wives to beggary. The mode of recruitment was to be reformed, and the special categories of men who are exempted should be sharply reduced. The pay (*solde*) must be increased and the conditions of service drastically improved. To Vauban's influence is at least partly due the restriction, if not elimination, of the practice of quartering troops on the civilian population. After the treaty of Aix-la-Chapelle, this practice was supplemented by barracks (*casernes*), many of which were built by Vauban, chiefly used in frontier regions and recently conquered territory.

Vauban's reformist impulses were far from limited to military matters. Deeply sensitive to the woes of others, shocked by the inefficiency and distress he encountered in town and country on his travels, his official correspondence and his *Oisivetés* contain blunt criticism on all manner of topics. His proposals were sometimes farfetched but at other times bold and farsighted. He wrote about the need to preserve the forests of France and urged that lumbering be regulated by the state. He proposed that the *caprerie*—those freelance, piratical raids on enemy shipping—should be systematized and supported by the government. Surely his most courageous proposal was to urge the recall of the Huguenots. He pointed out not only that the Revocation of the Edict of Nantes was an injustice to the Protestants, but that it brought grave injury upon France, which thereby lost excellent craftsmen, veteran soldiers, and eight or nine thousand of *the* finest sailors.[12]

Best known, of course, are Vauban's economic views. Like his contemporaries, Pierre Boisguilbert and Henri Boulainvilliers, he had antimercantilist sentiments and was fully as well informed as they about the impoverishment of the kingdom and the need, above all, to reform the system of taxation, the unfairness of which was matched only by the arbitrariness and corruption of the tax collec-

tors (*maltôtiers*). About 1680 Vauban drafted a memoir on the salt tax (*gabelle*), and in 1695 in his *Projet de capitation* he outlined a whole new scheme of direct taxation. His last words on the subject, and his chief published work, was his clandestine *Dîme royale*, published anonymously and without a license in 1707.[13] It reviews the poverty and misery of the French people, the causes of which, he wrote, "are sufficiently well known," and which he documents with statistics. Instead of the current forms of taxation he proposes a single tax to fall on all the king's subjects, taking the form of one-tenth of the produce of landed wealth, or one-tenth of other forms of income. Basing his proposal on historical precedent, including the Bible, but modeling his proposal specifically on the *Dîme ecclésiastique*, by which the church contributed to the national treasury, he argued that this would be the simplest to collect, and a form of levy less likely to arouse the ire of the people. It would, indeed, fall on all persons regardless of privilege or rank with the sole exception of the clergy.

When Fontenelle, who delivered Vauban's eulogy before the Academy, spoke of Vauban as chosen by the Academy of Sciences as a mathematician, who more than anyone else had "drawn mathematics down from the skies," this can be thought of in terms of Vauban's keen interest in, and constant recourse to, statistics. Here, perhaps, he deserves to be mentioned along with his older contemporaries in Britain, William Petty and John Graunt.

To his superiors, and to many who have written about him, he was admired less perhaps for his recognized genius than as a model of the ideal public servant. Indeed the word *citoyen*, in the sense of a good citizen, a man devoted to the welfare of his country, was first used in reference to him. Voltaire said of Vauban: "He proved, by his conduct, that there can be citizens under an absolute government."[14]

Yet despite his achievements, the honors he most desired were slow to come. He was sixty-five when he was made an *académicien honoraire* of the Academy of Sciences at the time of its reorganization (and installation in the Louvre) in 1699.[15] In 1703, while inspecting the fortifications of Namur, he learned that he had been made a marshall of France. This high honor, for which he had been several times passed over, had never, according to Saint-Simon, been attained before by anybody "of this sort," by which the great nobleman must have meant a man of middling origin or a mere engineer. Finally, in 1705, he received the highest distinction his sovereign could bestow, membership in the select Ordre du Saint Esprit.

Despite these belated honors, Vauban's last years were a disappointment. The king refused to assign him to conduct a major siege (that of Kehl), probably because of his age, but officially because the task was unworthy of his new illustrious rank. Although he led a successful attack on another fortress in 1703, he was in effect put out to pasture. His health began to fail and he suffered increasingly from chronic bronchitis and other ailments. The last months of his life were spent in his Paris house near the Palais-Royal. In early 1707 he fell dangerously ill with pneumonia and on Wednesday, 30 March, not long after the publication of his influential *Dîme royale*, he died in the arms of his son-in-law. After a modest funeral service his body was brought to Bazoches and buried, with his heart in a separate lead casket, in the family vault in the chapel that Vauban had himself built.[16]

NOTES

1. The administration of military engineering, which had been centralized in the time of Sully, was divided in 1661. One department was under the minister of war (first Michel Le Tellier, then his son, the Marquis de Louvois). This department supervised work in the provinces of Flanders, Hainaut, Artois, and along the eastern frontier. The other department was headed by Colbert, first as controller general of finance and later (1669) as secretary of the navy. Until his death in 1683, Colbert was responsible for fortification in Picardie, Champagne, the Three Bishoprics, Provence, and Languedoc.

2. Halévy, *Vauban*, p. 104.

3. For the strategic consideration of Vauban's fortress building, and the concept of the *pré carré* (a fleshed-out defensable frontier), see H. Guerlac, in E. M. Earle, ed., *Makers of Modern Strategy: Military Thought From Machiavelli to Hitler*, 44–47.

4. Voltaire ascribed the invention, or the first use of, the system of parallels to Italian engineers serving the Turks in the siege of Candia (now Iráklion) in Crete; see *Oeuvres complètes de Voltaire*, XIV (Paris, 1878), 263–264.

5. Erard's famous work with his *La fortification démontrée et réduite en art* (Paris, 1594; repr., 1604).

6. On Pagan, see Blomfield, ch. 4.

7. See his *Éloge de Vauban* (1784). For this appraisal the reader may consult Charles Coulston Gillispie, *Lazare Carnot Savant* (Princeton, 1971). This passage has been frequently cited.

8. For a plan of the town of Neuf-Brisach, see Bélidor, *La science des ingénieurs*, bk. 4, plate 25, p. 60. Bélidor takes Vauban's scheme for defending the town as his model for military architecture, and illustrates its defenses in bk. 6. plate 52, p. 40. A model of Neuf-Brisach is in the Musée des Plans Reliefs in the Invalides in Paris. A photograph is reproduced by Hebbert and Rothrock, "Marshal Vauban," in *History Today*, **24**, pt. 1, p. 156. A colored aerial view is published in Michel Parent and Jacques Verroust, *Vauban*, p. 15.

9. Blomfield, pp. 53–54.

10. A hornwork is so named because of the acute angles, facing the point of attack, of the two detached half bastions.

11. Rochas d'Aiglun, ed., *Vauban–Oisivetés et Correspondance*, I, p. 267.

12. *Ibid.*, p. 466.

13. The first edition of the *Dîme Royal* was published without indication of place. According to Rochas d'Aiglun, it was probably printed at Rouen in 1706, under the supervision of the Abbé Roget de Beaumont, a longtime collaborator of Vauban. There have been a number of later editions, and an English trans. has appeared.

14. "Catalogue de la plupart des écrivains français qui ont paru dans le siècle de Louis XIV," in *Oeuvres complètes de Voltaire*, XIV (Paris, 1878), 141. Fontenelle had called Vauban "Un Romain, qu'il semblait que notre siècle eut dérobé aux plus heureux temps de la république," in *Éloges de Fontenelle*, Francisque Bouillier, ed. (Paris, n.d.), p. 31. According to James Kip Finch, *Engineering and Western Civilization* (New York, 1951), p. 36, n. 3, it was Saint-Simon who is "said to have coined the word citizen and applied it to Vauban." I have not found that Saint-Simon in fact uses the term, at least in referring to Vauban.

15. The other honorary members besides Vauban were the Abbé de Louvois, the Chevalier Renau, Father Thomas Gouye, Jean Truchet, otherwise known as Father Sébastien, and the aged Malebranche. Most had reputations as mathematicians including, of course, Malebranche. See Ernest Maindron, *L'Académie des sciences* (Paris, 1888), p. 27.

16. For the ceremonial reburial of Vauban's heart, in the Invalides during the reign of Napoleon, see Hebbert and Rothrock, "Marshal Vauban," in *History Today*, **24**, pt. 2 (1974), 264.

BIBLIOGRAPHY

I. ORIGINAL WORKS. Vauban wrote voluminously and published hardly anything in his own lifetime. Many works attributed to him (like the small manual entitled *Directeur général des fortifications* [The Hague, 1785]) are probably spurious. It is debated whether he ever wrote a treatise on fortification. Unquestionably his, however, is the *Traité de l'attaque de la défense des places* (The Hague, 1737; repr., 2 vols., 1742), written for the Duc de Bourgogne. This work was republished by Latour-Foissac in 1795, and by Lieutenant Colonel Augoyat in 1829; a German trans., 2 vols. in 1, appeared in Berlin, 1751. A similar work, written earlier (perhaps as early as 1667) is his *Mémoire pour servir d'instruction dans la conduite des sièges et dans la défense des places* (Leiden, 1740), trans. by George A. Rothrock as *A Manual of Siegecraft and Fortification* (Ann Arbor, Mich., 1968). Of the mass of Vauban MSS that survive, a significant portion have been reprinted, notably in the nineteenth century. The *Abrége des services du maréchal de Vauban, fait par lui en 1703* was printed under the editorship of Augoyat, who also published the first 4 vols. of *Oisivetés* in 1842–1845. Vauban's *De l'importance dont Paris est à la France, et le soin que l'on doit prendre de sa conservation* was first printed in Paris in 1821.

Vauban's most famous book, *Projet d'une dixme royal*, was published first in 4° in 1707 with neither place nor date of publication, then in Brussels in 1708. It was reprinted by Eugène Daire, in *Economistes français du dix-huitième siècle* (Paris, 1843). Its interesting preface was included by Rochas d'Aiglun in his own *Vauban—*

Sa Famille et ses Ecrits—Ses Oisivetés et sa Correspondance, 2 vols. (Paris–Grenoble, 1910), which is a
major source. D'Aiglun's work begins with a genealogy
of Vauban, and in the first volume includes the most
important memoirs in the 12 vols. of MSS to which
Vauban gave the title of Oisivetés. The second volume
of this work is devoted to Vauban's correspondence
with Louvois and others.

II. SECONDARY LITERATURE. For a laudatory view of
Vauban by a contemporary, see Charles Perrault, Les
hommes illustres qui ont paru en France pendant ce siècle, 2 vols. (Paris, 1696). Still valuable is the early
sketch of Vauban given by Fontenelle in his éloge; see
Oeuvres Complètes de Fontenelle, I (Paris, 1818),
95–103, and Éloges de Fontenelle, avec introduction et
notes, Francisque Bouillier, ed. (Paris, n.d.), 22–31. As
indicated in our text there are interesting references to
Vauban by Saint-Simon in his Mémoires and by Voltaire
in his Siècle de Louis Quatorze. Lazare Carnot's appraisal, Éloge de Vauban (Paris, 1784), aroused much
contemporary debate among the military, and was vi-
ciously attacked by Cholerlos de Laclos. A fine study in
English, by an architect and architectural historian, is Sir
Reginald Blomfield, Sébastien le Prestre de Vauban,
1633–1707 (London, 1938), making good use of Colonel
de Génie Pierre Elizir Lazard, Vauban (Paris, 1934),
which covers the many aspects of Vauban's career. Both
works are well illustrated—Blomfield's with his own
drawings of Vauban forts in their modern state of preservation, and Lazard with sketches by Vauban and with
photographs of models from the Musée des Plans Reliefs of the French army's geographical service. Two recent general studies deserve especially to be cited: Marcel Parent and Jacques Verroust, Vauban (Paris, 1971),
sumptuously illustrated and splendidly printed; and F. J.
Hebbert and G. A. Rothrock, "Marshal Vauban," in
History Today, 24 1974), 149–157, 258–264. Among
nineteenth-century works, George Michel, Histoire de
Vauban (Paris, 1879), is a useful connected narrative, but
not much of an improvement on J. J. Roy, Histoire de
Vauban (1844).

More specialized aspects are Jacques Guttin, Vauban
et le corps des ingénieurs militaires (Paris, 1957); Humbert Ricolfi, Vauban et le génie militaire dans les alpes-
maritimes (Nice, 1935); Ferdinand Dreyfus, Vauban
économiste (Paris, 1892); Walter Brauer, Frankreichs
wirtshaftliche und soziale Lage von 1700; dargestellt
unter besonderer Berüchsichtigung der Werke von Vauban und Boisguillebert (Marburg, 1968); and Félix Cadet, Histoire de l'économie politique. Les précurseurs.
Boisguilbert, Vauban, Quesnay, Turgot (New York,
1970), repr. of the early 1869 ed. Henry Guerlac,
"Vauban: The Impact of Science on War," in Edward
Meade Earle, Makers of Modern Strategy (Princeton,
1943), 26–48, with bibliography on pp. 522–523, draws
upon—for its discussion of Vauban strategic aims—
H. Chotard, "Louis XIV, Louvois, Vauban et les fortifications du Nord de la France, d'après les lettres inédites
de Louvois addressées à M. de Chazerat, Gentilhomme

d'Auvergne," in Annales du Comité Flamand de France,
18 (1889–1890), and on Gaston Zeller, L'organisation
défensive des frontières du Nord et de l'Est au XVIIe
siècle (Paris, 1928).

HENRY GUERLAC

VAUCHER, JEAN PIERRE ÉTIENNE (b. Geneva,
Switzerland, 27 April 1763; d. Geneva, 5 January
1841), botany.

Vaucher was the son of a carpenter, and he expected to follow that trade; but by the age of
twelve he knew that he wanted an academic career. He entered college, where he found a general
education, including theology and physical sciences, but not botany, which he began to study as
a leisure pursuit following interest aroused by his
exploration of the mountains around Geneva. He
was ordained in 1787 and worked as a parish priest
from 1797 to 1822, carrying out his duties conscientiously, but always giving much time and energy
to botany, which increasingly fascinated him.

A founder member of the Société de Physique et
d'Histoire Naturelle de Genève, he was asked in
1794 to give a course of lectures on botany for the
Society. These were attended by A. P. de Candolle
who became a close friend. By 1798 Vaucher was
sufficiently recognized as a botanist and an exceptionally good teacher to be appointed honorary
professor of botany in the university, and he held
this post until 1807 when he transferred to the
chair of ecclesiastical history, which he held until
1839, undertaking additional responsibilities as
rector from 1818 to 1821. He married and had
children.

Vaucher's most important work was his observation and interpretation of conjugation and spore
formation in algae, particularly in Ectosperma, later renamed Vaucheria by de Candolle. Although
the cell theory had not yet been developed, he drew
and described the cells showing that they were
bounded by separate walls and had a certain degree of independence. He showed how conjugation
can occur between cells of threads lying side by
side, or cells of a single thread folded over on itself, by means of the communication channel
through which the contents of one cell pass to fuse
with the contents of the other cell. He also showed
male organs, or anthers, protruding to meet other
cells and was sure that this was a sexual act, comparable to that found in higher plants and to the
copulation of animals. His optical equipment did
not allow him to see the actual fertilization, but he

correctly inferred it. The newly formed grains, or spores, dropped to the bottom of the ditches in which the algae were found, and in order to verify that they would germinate to give new filaments of *Ectosperma* in the spring he had to culture them in the laboratory, an uncommon practice at that time. He showed similar conjugation in other primitive algae, and the formation of new nets in old cells of *Hydrodictyon*.

Other work published by Vaucher was wide-ranging. Perhaps the most comprehensive was the *Histoire physiologique des plantes d'Europe* (1804–1841), mainly a taxonomic compilation arranged by the system of Candolle. He wrote monographs on *Equisetum* (1822) and broomrape (1827) and a paper on leaf fall (1826) that was the first to examine in detail the structure of the base of the petiole.

As a member of the agriculture section of the Société des Arts de Genève from 1796, he carried out some useful investigations on diseases of vines and of wheat, the effects of temperature on plant growth, culture of the potato, and management of woodlands.

BIBLIOGRAPHY

I. ORIGINAL WORKS. The descriptions of conjugation and germination in filamentous algae were published as "Sur les graines des conferves," in *Journal de physique*, **52** (1801), 344–358, and more fully in *Histoire des conferves d'eau douce . . . des Tremelles et des Ulves* (Geneva, 1803). *Histoire physiologique des plantes d'Europe*, t.l. (Geneva, 1804), was reissued in 1830 and the complete work was published in 4 vols. in Paris a few days before Vaucher's death in 1841. *Monographie des Prêles* (Geneva–Paris, 1822) was followed by *Monographie des Orobranches* (Geneva–Paris, 1827). The "Mémoire sur la chute des feuilles" was published in *Mémoires de la Société de physique et d'histoire naturelle de Genève*, **1** (1821), 120–136, and a report of the work, followed by discussion, appeared in *Edinburgh Journal of Science*, **5** (1826), 330–338.

Vaucher's sermons were printed as *Souvenir d'un pasteur genevois* (Geneva, 1842).

II. SECONDARY LITERATURE. The most comprehensive biography of Vaucher is an anonymous obituary in *Verhandlungen der Schweizerischen Naturforschenden Gesellschaft bei ihrer Versammlung zu Zürich*, **26** (1841), 308–313. There is another anonymous obituary in *Mémoires de la Société de physique et d'histoire naturelle de Genève*, **10** (1843), xxiv–xxvi, and one by A. de Candolle in *Annals and Magazine of Natural History*, **10** (1842), 161–168, 241–248, translated from the *Bibliographie universelle de Genève* (1841). The *Souvenir* has a 22-page biographical preface by Vaucher's son.

Two modern evaluations of his work are by C. Baehni, "Il y a 150 ans, le Genevois, J. P. Vaucher découvrait la fécondation chez les algues" in *Les Musées de Genève*, **10** no. 10 (1953), and by G. de Morsier, "Contribution à l'histoire de la genétique," in *Physis*, **7** (1965), 497–500.

His relations with the university may be found in C. Borgeaud, *Histoire de l'Université de Genève*, II (Geneva, 1909); and Candolle's tributes are in his *Histoire de la botanique Genevois* (Geneva, 1830), which includes a bibliography of 14 items, and in his *Mémoires et souvenirs de Genève* (Geneva, 1862).

Vaucher's publications can be traced through the *Royal Society Catalogue*, which lists 15 items, and G. A. Pritzel, *Thesaurus literaturae botanicae* (Leipzig, 1872), entries 9704–9710.

DIANA M. SIMPKINS

VAUQUELIN, NICOLAS LOUIS (*b*. St. André d'Hébertot, Normandy, France, 16 May 1763; *d*. St. André d'Hébertot, 14 November 1829), *chemistry*.

The son of Nicolas Vauquelin, an estate manager, and Catherine Le Chartier, Vauquelin became assistant to a pharmacist in Rouen when he was about fourteen, but left after he was reprimanded for taking notes of the scientific lectures given by his master. He went to Paris and eventually worked for a pharmacist named Chéradame, a cousin of the chemist Fourcroy. About 1784 Vauquelin became Fourcroy's laboratory and lecture assistant and at his invitation began to lecture at the lycée. But his voice was weak and he lacked confidence; and although he later occupied several chairs he never achieved fame as a lecturer. It soon became clear that he was a first-class experimental chemist, however, and his relationship with Fourcroy developed into an association of equals. Their first joint research was published in 1790, but political events interrupted their collaboration.

Vauquelin left Fourcroy's laboratory by 1792 and became the manager of a pharmacy. In 1793 he spent several months as a hospital pharmacist at Meaux, near Paris. In September 1793 he was sent to the region around Tours by the government in order to organize the production of saltpeter, urgently needed for gunpowder. Vauquelin returned to Paris and resumed his association with Fourcroy late in 1794, when he was appointed assistant professor of chemistry at the new École Centrale des Travaux Publics (later, École Polytechnique). In 1795 he received the title of master in pharmacy and was elected to the Institut de France. He had apparently been elected to the old Académie des

Sciences on 31 July 1793, a few days before its suppression, but his appointment had not been confirmed by the government.

Vauquelin left the École Polytechnique when its staff was reduced in 1797, but he continued as inspector of mines (a post he had held since 1794), and he retained his position as professor of assaying at the École des Mines, which he had entered after its reorganization in 1795. He also became official assayer of precious metals for Paris, and in 1799 he published a useful *manuel de l'essayeur*. He left the École des Mines in 1801 to succeed Jean d'Arcet as professor of chemistry at the Collège de France; in addition, he became director of the École de Pharmacie on its foundation in 1803.

In 1804 he moved again, from the Collège de France to the Muséum d'Histoire Naturelle, where he followed Antoine-Louis Brongniart as professor of applied chemistry. At the Muséum he was once more a colleague of Fourcroy, whose death in 1809 left a vacant chair in chemistry at the Faculté de Médecine. Much of Fourcroy's research had been done in collaboration with Vauquelin, who was his obvious successor. However, the professor had to be medically qualified, so Vauquelin obtained his doctorate with a thesis on the chemical analysis of the human brain; he received his appointment in 1811. Along with several other professors he was dismissed in 1822 during a politically inspired reform of the Faculté, but he retained his posts at the Muséum and the École de Pharmacie. In 1828 he was elected to parliament as a deputy for Calvados, his native district.

In the course of his numerous analyses of minerals, Vauquelin discovered two new elements in 1798. By boiling the rare Siberian mineral crocoite (lead chromate) with potassium carbonate he obtained the yellow salt of an unknown acid. On reduction with carbon the acid yielded a metal that he named chromium on account of its colorful compounds (Greek, *chroma*-color). In beryl (beryllium aluminum silicate) he found an earth (oxide) that superficially resembled alumina (aluminum oxide) but was insoluble in alkali and did not form alum. At the suggestion of the editors of *Annales de chimie*, he originally called it glucina, from the sweetness of its sulfate, but later it was renamed beryllia. Metallic beryllium was not obtained until 1828, when Friedrich Wöhler and, independently, Antoine Bussy first isolated it.

The most important of Vauquelin's many analyses of vegetable and animal substances were done with Fourcroy. In 1804 the two friends set up a small factory in Paris for the manufacture of high-quality chemicals. Vauquelin, the more active partner, was personally involved in its management for several years and retained a financial interest until 1822.

Vauquelin never married, and Fourcroy's sisters, Madame Le Bailly and Madame Guédon, kept house for him from about 1790 until they died in 1819 and 1824 respectively.

BIBLIOGRAPHY

I. ORIGINAL WORKS. Vauquelin's 305 contributions to periodicals that he wrote himself and the seventy-one that he wrote jointly with Fourcroy or others are listed in *The Royal Society Catalogue of Scientific Papers*, VI (London, 1872), 114–128; a less complete list is given in Poggendorff, II, cols. 1182–1190. His only book is *Manuel de l'essayeur* (Paris, 1799). This was reprinted in 1812 and a new ed., revised by A. D. Vergnaud, appeared in 1836; there is a German trans. by F. Wolff, with notes by M. H. Klaproth (Königsberg, 1800) and a Spanish trans. (Paris, 1826).

While working on the production of saltpeter he wrote, with Trusson as coauthor, a 32-page pamphlet, *Instruction sur la combustion des végétaux, la fabrication du salin, de la cendre gravelée, et sur la manière de saturer les eaux salpêtrées* (Tours, 1794); trans. into Portuguese as *Instrucçao sobre a combustaõ dos vegetaes . . .* (Lisbon, 1798). Vauquelin wrote articles on apparatus in Fourcroy's *Encyclopédie méthodique chimie*, II (Paris, 1792), and he was co-author, with Fourcroy, of VI (Paris, 1815). He also supervised the preparation of the two vols. of plates (1813, 1814). Reports made by Vauquelin to various institutions are listed in the author catalogue of the Bibliothèque Nationale, Paris, vol. 204, cols. 206–207.

II. SECONDARY LITERATURE. The earliest accounts of Vauquelin's life are A. Chevallier and Robinet, *Notice historique sur N. L. Vauquelin* (Paris, 1830); and G. Cuvier, "Éloge historique de Louis Nicolas (*sic*) Vauquelin," in *Mémoires de l'Académie Royale des Sciences de l'Institut*, **12** (1833), xxxix–lvi. Additional information is given by A. Chevallier, "Notice biographique sur M. Vauquelin," in *Journal de chimie médicale*, ser. 3, **6** (1850), 542–549; and E. Pariset, *Histoire des membres de l'Académie Royale de Médecine*, I (Paris, 1850), 317–350.

Vauquelin's baptismal certificate, now in the town hall of St. André d'Hébertot, shows that he was named Nicolas Louis, and not Louis Nicolas, as is often stated. This certificate has been published, with some other manuscripts, by M. Bouvet, "Documents encore ignorés sur Vauquelin," in *Revue d'histoire de la pharmacie*, **16** (1963), 17–20. Vauquelin's election to the Académie des Sciences in 1793 is discussed by M. Bouvet, "Vauquelin fut-il membre de l'Académie des Sciences?" *ibid.*,

12 (1955), 66–70. Information about the chemical factory is given in three papers by G. Kersaint, "L'Usine de Vauquelin et Fourcroy," *ibid.*, **14** (1959), 25–30; "Sur la fabrique de produits chimiques établie par Fourcroy et Vauquelin," in *Comptes rendus hebdomadaires des séances de l'Académie des sciences*, 247 (1958), 461–464; and "Sur une correspondence inédite de Nicolas Louis Vauquelin," in *Bulletin. Société chimique de France* (1958), 1603.

For Vauquelin's work on saltpeter manufacture in 1793–1794, see C. Richard, *Le comité de salut public et les fabrications de guerre sous la terreur* (Paris, 1922), 429. For accounts of his research on chromium and beryllium, see M. E. Weeks and H. M. Leicester, *The Discovery of the Elements* (Easton, Pa., 1968), 271–281, 535–540. See also "Fourcroy," in *DSB, V*, 92–93.

W. A. SMEATON

VAVILOV, SERGEY IVANOVICH (*b*. Moscow, Russia, 24 March 1891; *d*. Moscow, 25 January 1951), *physics*.

Vavilov, the son of a manufacturer and trader, and the youngest brother of the botanist Nikolay Ivanovich Vavilov, received his secondary education at the Moscow Commercial School, concentrating on physics and chemistry. In 1909 he entered the Faculty of Physics and Mathematics of Moscow University, and by the end of the first year he had chosen a subject on which to work in Lebedev's laboratory, under the supervision of P. P. Lazarev. Vavilov did not finish this work, however, for in 1911 Lebedev and Lazarev left Moscow University to protest the violation of the university's autonomy by the minister of education, L. A. Kasso. Vavilov's scientific work was transferred to Shanyansky City University, a small private institution in Moscow, where Lebedev's laboratory was set up.

In 1913 Vavilov published his first scientific work, on the photometry of polychromatic light sources; at the same time he carried out research on a subject proposed by Lazarev, who was then studying the discoloration by light of collodion films colored with cyanine dyes. Assigned to investigate the discoloration of these dyes through heating, Vavilov constructed an ingenious experimental device and discovered the essential differences between the discolorations caused by light and by heat. The work was published in 1914, and Vavilov received a gold medal for it (1915). At the beginning of World War I, he was drafted into the army and sent to the front. Throughout the conflict he served mainly in the technical units; and in his mobile radio laboratory he was able to conduct research in radiotechnology, the results of which were published in 1919.

Discharged from the army in 1918, Vavilov began independent scientific work on photoluminescence and physical optics in general at the Institute of Physics and Biophysics, which was directed by Lazarev. At the same time he was a *Privatdozent* at the Moscow Higher Technical School, where in 1920 he became professor of physics. In 1929 Vavilov was named professor and head of the department of general physics at Moscow University, and from 1930 all of his work was carried out there. He was elected corresponding member of the Academy of Sciences of the U.S.S.R. in 1931, and the following year was named scientific director of the State Optical Institute now named for him. At this time Vavilov moved to Leningrad but retained his affiliation with Moscow University. At the State Optical Institute he conducted wide-ranging and fruitful work in a new laboratory for luminescence.

In 1932 Vavilov was elected full member of the Academy of Sciences of the U.S.S.R. and became director of the Physics Institute of the Academy of Sciences in Leningrad. Two years later, when the Academy of Sciences of the U.S.S.R. was transferred to Moscow, this institute came with it and, through Vavilov's efforts, became the important P. N. Lebedev Physical Institute.

During World War II, Vavilov worked on problems connected with national defense, although the State Optical Institute was evacuated to Yoshkar-Ola and the Physics Institute to Kazan. These institutes provided much of the high-quality optical equipment needed by the Soviet armed forces. In 1943 Vavilov was named commissioner of the State Committee for Defense of the U.S.S.R. and was awarded the Order of Lenin and the State Prize. He was elected president of the Academy of Sciences of the U.S.S.R. while retaining all of his other positions, and in 1945 was awarded a second Order of Lenin. As president of the Academy of Sciences, Vavilov was particularly concerned with the relation of science to the national economy and aided the development of scientific institutions in the national republics and in the major industrial regions.

Vavilov conducted important research aimed at determining what proportion of absorbed light is converted into fluorescent light. Although it had been assumed that this portion is extremely small, Vavilov showed that the energy output of fluorescence reaches 70–80 percent and in some cases approaches 100 percent. In his research on the polarization of fluorescence, he was the first to show that the degree of polarization depends on the wavelength of the light that stimulates the fluorescence. This finding led to the creation of the study of the spectra of fluorescence. Under Vavilov's leadership a detailed method of luminescent analysis was elaborated and luminescent lamps were developed. His research was described in *Mikrostruktura sveta* ("Microstructure of Light"), an original proof of the quantum nature of light.

A major discovery made under Vavilov's supervision was the Vavilov-Cherenkov effect, a special kind of luminescence that occurs when charged molecules in a medium move at a velocity exceeding the velocity of light in that medium. For this work P. A. Cherenkov, I. Y. Tamm, and I. M. Frank were awarded the Nobel Prize in 1958, seven years after Vavilov's death.

Vavilov's works in the history of science include a Russian translation and commentary on Newton's *Optiks* and his *Lectures on Optics*, a work on Galileo's place in the history of optics, and a popular biography of Newton.

BIBLIOGRAPHY

I. ORIGINAL WORKS. Vavilov's collected works were published as *Sobranie sochineny*, 5 vols. (Moscow, 1952–1956), with biographical sketch by V. L. Levshin. See also *Mikrostruktura sveta* ("The Microstructure of Light"; 2nd ed., rev. and enl., Moscow–Leningrad, 1945); and *Glaz i solntse* ("The Eye and the Sun"; Moscow–Leningrad, 1950).

II. SECONDARY LITERATURE. See N. I. Artobolevsky, *Vydayushchysya sovetsky ucheny i obshchestvenny deyatel S. I. Vavilov* ("The Distinguished Soviet Scientist and Social Activist S. I. Vavilov"; Moscow, 1961), which includes a bibliography of his basic scientific works; and E. V. Shpolsky, *Vydayushchysya sovetsky ucheny S. I. Vavilov* (Moscow, 1956). A collection of articles commemorating his seventieth birthday was published in *Uspekhi fizicheskikh nauk*, **75**, no. 2 (1961).

J. DORFMAN

VEBLEN, OSWALD (*b.* Decorah, Iowa, 24 June 1880; *d.* Brooklin, Maine, 10 August 1960), *mathematics.*

Veblen's parents were both children of immigrants from Norway. His mother (1851–1908) was born Kirsti Hougen; his father, Andrew Anderson Veblen (1848–1932), was professor of physics at the University of Iowa; and his uncle, Thorstein Veblen (1857–1929), was famous for his book *The Theory of the Leisure Class.* Veblen himself had

two B.A. degrees (Iowa, 1898; Harvard, 1900) but was most influenced by his graduate study (Ph.D., 1903) at the University of Chicago under E. H. Moore. He had a happy marriage (1908) to Elizabeth Mary Dixon Richardson. His influential teaching career was at Princeton, both at the University (1905–1932) and at the Institute for Advanced Study (1932–1950).

The axiomatic method, so characteristic of twentieth-century mathematics, had a brilliant start in Hilbert's *Grundlagen der Geometrie* (1899). In this book, precise and subtle analysis corrected the logical inadequacies in Euclid's *Elements.* Veblen's work, starting at this point, was devoted to precise analysis of this and many other branches of geometry, notably topology and differential geometry; his ideas have been extensively developed by many younger American geometers.

The initial step was Veblen's thesis (1903, published 1904), which gave a careful axiomatization for Euclidean geometry, different from that of Hilbert because based on just two primitive notions, "point" and "order" (of points on a line), as initially suggested by Pasch and Peano. With this systematic start on the axiomatic method, Veblen's interests expanded to include the foundations of analysis (where he emphasized the role of the Heine-Borel theorem, that is, of compactness) and finite projective geometries. His work in projective geometry culminated in a magnificent two-volume work, *Projective Geometry* (vol. I, 1910; vol. II, 1918), in collaboration with J. W. Young. This book gives a lucid and leisurely presentation of the whole sweep of these geometries over arbitrary fields and over the real number field, with the properties of conics and projectivities and with the classification of geometries by the Klein-Erlanger program. It includes a masterful exposition of the axiomatic method (independence and categoricity of axioms), which had extensive influence on other workers in algebra and geometry.

Veblen's greatest contribution probably lies in his development of analysis situs. This branch of geometry deals with numerical and algebraic measures of the "connectivity" of geometric figures. It was initiated by Poincaré in a famous but difficult series of memoirs (1895–1904). It came naturally to Veblen's attention through his earlier work on the Jordan curve theorem (that is, how a closed curve separates the plane) and on order and orientation to Euclidean and projective geometry. Veblen's 1916 Colloquium Lectures to the American Mathematical Society led to his 1922 *Analysis Situs*, which for nearly a decade was the only systematic treatment in book form of the pioneering ideas of Poincaré. This book was carefully studied by several generations of mathematicians, who went on to transform and rename the subject (first "combinatorial topology," then "algebraic topology," then "homological algebra") and to found a large American school of topology with wide international influence.

Veblen also did extensive work on differential geometry, especially on the geometry of paths (today treated by affine connections) and on projective relativity (four-component spinors). His more important work in this field would seem to be his part in the transition from purely local differential geometry to global considerations. His expository monograph (1927) on the invariants of quadratic differential forms gave a clear statement of the usual formal local theory; and it led naturally to his later monograph with J. H. C. Whitehead, *The Foundations of Differential Geometry* (1932). This monograph contained the first adequate definition of a global differentiable manifold. Their definition was complicated; for example, they did not assume that the underlying topological space is Hausdorff. Soon afterward H. Whitney, starting from this Veblen-Whitehead definition, developed the simpler definition of a differentiable manifold, which has now become standard (and extended to other cases such as complex analytic manifolds). In this case the Veblen-Whitehead book had influence not through many readers, but essentially through one reformulation, that of Whitney.

Veblen had an extensive mathematical effect upon others; in earlier years through many notable co-workers and students (R. L. Moore, J. W. Alexander, J. H. M. Wedderburn, T. Y. Thomas, Alonzo Church, J. H. C. Whitehead, and many others) and in later years by his activities as a mathematical statesman and as a leader in the development of the School of Mathematics at the Institute for Advanced Study.

BIBLIOGRAPHY

I. ORIGINAL WORKS. Veblen's thesis was published as "A System of Axioms for Geometry," in *Transactions of the American Mathematical Society*, **5** (1904), 343–384. His other works include *Projective Geometry*, I (New York, 1910), written with J. W. Young; II (Boston, 1918); *Analysis Situs*, in *Colloquium Publications. American Mathematical Society*, **5**, pt. 2 (1922); 2nd ed. (1931); "Invariants of Quadratic Differential Forms," in *Cambridge Tracts in Mathematics and Mathematical Physics*, **24** (1927); "The Foundations of Differential Geometry," *ibid.*, **29** (1932), written with J. H. C. Whitehead; and "Projective Relativitätstheorie," in *Ergenbnisse der Mathematik und ihrer Grenzgebiete*, **2**, no. 1 (1933).

II. SECONDARY LITERATURE. On Veblen and his work, see *American Mathematical Society Semicentennial Publications*, **1** (1938), 206–211, with complete list of Veblen's doctoral students; Saunders MacLane, in *Biographical Memoirs. National Academy of Sciences*, **37** (1964), 325–341, with bibliography; Deane Montgomery, in *Bulletin of the American Society*, **69** (1963), 26–36; and *Yearbook. American Philosophical Society* (1962), 187–193.

SAUNDERS MAC LANE

VEJDOVSKÝ, FRANTIŠEK (*b.* Kouřim, Bohemia [now Czechoslovakia], 24 October 1849; *d.* Prague, Czechoslovakia, 4 December 1939), *zoology.*

For a detailed study of his life and work, see Supplement.

VEKSLER, VLADIMIR IOSIFOVICH (*b.* Zhitomir, Russia, 4 March 1907; *d.* Moscow, U.S.S.R., 22 September 1966), *physics, engineering.*

Veksler was the son of an engineer, Iosif Lvovich Veksler. After working as an apparatus assembler in a factory, Veksler graduated from the Moscow Energetics Institute, and held a post at the All-Union Electrotechnical Institute and, later, at the Lebedev Physics Institute of the Academy of Sciences of the U.S.S.R.

In 1954 Veksler became director of the electrophysics laboratory of the Academy of Sciences, and from 1956 he headed the high-energy laboratory of the Joint Institute for Nuclear Research. He was elected an associate member of the Academy of Sciences of the U.S.S.R. in 1946 and a full member in 1958.

Veksler began his career with studies of cosmic rays and discovered a new type of interaction between high-energy particles and atomic nuclei. He is best known, however, for his work on the theory of the accelerator, an apparatus for artificially obtaining the charged particles of great energy that usually are necessary for investigations of the atomic nucleus.

In 1944, simultaneously with E. M. MacMillan, Veksler established the principle of phase stability of accelerated particles. This discovery, which is applied to all modern accelerators, proved to be a turning point in nuclear physics and in the physics of elementary particles. The largest Soviet accelerators were planned and constructed under Veksler's direction. For his fundamental investigations regarding accelerators he received the Lenin and State prizes of the U.S.S.R.

Veksler was a creator of the large groups of scientists and engineers at the Joint Institute for Nuclear Research, the Lebedev Physics Institute, and at Moscow University. He also was active in international scientific collaboration, serving for many years as chairman of the Commission on High-Energy Physics of the International Union of Pure and Applied Physics. His services were noted abroad by the award of the U.S. Atoms for Peace Prize in 1963.

BIBLIOGRAPHY

Veksler's writings include *Eksperimentalnye metody yadernoy fiziki* ("Experimental Methods in Nuclear Physics"; Moscow–Leningrad, 1940); "Novy metod uskorenia relyativistskikh chastits" ("A New Method of Accelerating Relativistic Particles"), in *Doklady Akademii nauk SSSR*, **43**, no. 8 (1944), 346; "On the Stability of Electron Motion in the Induction Accelerator of the Betatron Type," in *Fizicheskii zhurnal*, **9**, no. 3 (1945), 153; "Elektronno-yadernye livni kosmicheskikh luchey i yaderno-kaskadny protsess" ("Electronuclear Showers of Cosmic Rays and the Nuclear-Cascade Process"), in *Zhurnal eksperimentalnoi i teoreticheskoi fiziki*, **19**, no. 9 (1949), 135; *Ionizatsionnye metody issledovania izlucheny* ("Ionization Methods in the Study of Radiation"; Moscow, 1950); and "Kogerentny metod uskorenia zaryazhennykh chastits" ("The Coherent Method of Accelerating Charged Particles"), in *Atomnaya energiya*, **11**, no. 5 (1957), 427.

A. T. GRIGORIAN

VELLOZO, JOSÉ MARIANO DA CONCEIÇÃO (*b.* San José, Minas Gerais, Brazil, 1742; *d.* Rio de Janeiro, Brazil, 13 June 1811), *botany.*

Vellozo, the father of Brazilian botany, entered the Franciscan order in 1761 and, although largely self-educated in the sciences, became instructor in geometry at São Paulo ten years later. In 1782 he was commissioned by the viceroy, Luis de Vasconcelos, to prepare a study of the flora of Rio de Janeiro province. He roamed the countryside for eight years, turning his monastic cell into a herbarium where he analyzed his specimens. His companion in the field and fellow Franciscan, Francisco Solano, prepared drawings of many of the plants for the completed work, *Flora fluminensis.*

In 1790, his work on the *Flora* completed, Vellozo accompanied his patron Vasconcelos to Lisbon, where he was named director of the Tipogra-

fia Calcográfica e Tipoplástica do Arco do Cego, a printshop later absorbed into the National Press of Portugal. There he directed the publication of numerous scientific works, many of which he had written or translated. Although he was summoned to organize the herbarium of the Royal Museum in 1797–1798, Vellozo spent most of the 1790's producing works of scientific and economic popularization, the majority of which were characterized by the utilitarian motive so typical of the Enlightenment. Particularly noteworthy was his essay on alkaloids (1798), the first part of which was a treatise on popular chemistry and the second a description of Brazilian plants from which alkaloids could be derived. Between 1798 and 1806 Vellozo published a multivolume work on economic botany, *O fazendeiro do Brazil*, in which he discussed plants that could contribute to the economic development of Brazil: sugarcane, dyes, coffee and cocoa, spices, and textile fibers.

Vellozo also was interested in zoology. In 1800 he published a descriptive treatise on the birds of Brazil, *Aviário brasílico*; in the first chapter, a history of ornithology from the sixteenth through the eighteenth centuries, Vellozo discussed the contributions of European naturalists from Konrad Gesner to George Edwards.

In the two decades following his fieldwork, Vellozo discovered that collecting the data for his *Flora* was a good deal easier than getting the work into print—a fate he shared with the members of the contemporaneous Spanish botanical expeditions. In 1808 the imminent publication of the *Flora* was aborted when Geoffroy Saint-Hilaire went to the National Press and, under orders from General Andoche Junot, took the plates to Paris, where he and Candolle later used them. This bizarre incident delayed publication until 1825, when Emperor Pedro I ordered the printing of the text. Two years later the engravings of 1,640 plants were published at Paris. In the introduction to his masterpiece, Vellozo included a terse description of his achievement: "I observed, had drawn, and reduced to Linnean nomenclature according to the sexual system, fully seventeen hundred species of plants." The *Flora fluminensis* is regarded as the greatest creation of Enlightenment science in Brazil.

BIBLIOGRAPHY

I. ORIGINAL WORKS. *Flora fluminensis, seu descriptionum plantarum praefectura fluminensi* appeared in two parts: text (Rio de Janeiro, 1825), 2nd ed. as *Archuivos do Museu nacional* (Rio de Janeiro), **5** (1881), and plates *Florae fluminensis icones*, Antonio da Arrabida, ed., 12 vols. (Paris, 1827). The ornithological treatise is *Aviário brasílico ou galeria ornitológica das aves indígenas do Brasil* (Lisbon, 1800). Among the more important works on economic botany are *Alografia dos álcales fixos* (Lisbon, 1798); *Memória e extratos sôbre a pipereira negra* (Lisbon, 1798); *O fazendeiro do Brazil*, 11 vols. in 5 (Lisbon, 1798–1806); *Quinografia portuguêsa* (Lisbon, 1799); and the didactic *O naturalista instruído nos diversos métodos, antigos e modernos, de ajuntar, preparar e conservar as producões dos três reinos de natureza* (Lisbon, 1800).

II. SECONDARY LITERATURE. Biographical data are in Augusto Vitorino Sacramento Blake, *Dicionario bibliográfico brasileiro*, 7 vols. (Rio de Janeiro, 1883–1902), V, 64–70; José Saldanha da Gama, "Biographia do botánico brasileiro José Mariano da Conceição Velloso," in *Revista do Instituto histórico e geográfico brasileiro*, **37** (1868), 137–305; Carlos Stellfeld, *Os dois Vellozo* (Rio de Janeiro, 1952); and "A biografia de Frei Velozo," in *Tribuna farmacéutica* (Curitiba), **21** (1953), 119–124. A valuable collection of documents on the *Flora fluminensis* is *Flora fluminensis de Frei José Mariano da Conceição Vellozo. Documentos*, Thomaz Borgmeier, ed. (Rio de Janeiro, 1961). On Vellozo's role in the Brazilian Enlightenment, see José Ferreira Carrato, *Ingreja, iluminismo e escolas mineiras coloniais* (São Paulo, 1968).

THOMAS F. GLICK

VENEL, GABRIEL FRANÇOIS (*b.* Tourbes, near Pézenas, France, 23 August 1723; *d.* Montpellier, France, 29 October 1775), *chemistry, medicine.*

The son of Étienne Venel, a physician, and Anne Hiché, Venel qualified as a doctor of medicine at the University of Montpellier in 1742. About 1746 he went to Paris to gain further experience in hospitals, and there he attended Guillaume-François Rouelle's chemistry lectures and soon decided to devote himself to chemistry. He secured the patronage of Louis, duc d'Orléans, who put him in charge of his laboratory at the Palais Royal, and he was appointed a royal censor for books on natural history, medicine, and chemistry.

Venel was interested in the composition of vegetable matter; in 1752 he compared fire analysis with the recently introduced method of solvent extraction, praising the latter and saying that it showed some principles—extract and resin, for example—to be common to the entire vegetable kingdom, whereas others, including essential oils and gums, were constituents of only certain plants. He never achieved his aim of examining each prin-

ciple separately, for his attention was diverted to the analysis of mineral waters.

In 1750 Venel described his analysis of the effervescent mineral water of Selz, in Germany. Evaporation yielded only common salt and a little lime, and he was more interested in the effervescence, which was, he thought, caused by the escape of common air. All water contained a small amount of dissolved air, but Selzer and other effervescent waters contained superabundant air, as Venel called it. He made artificial Selzer water by adding the correct amounts of marine (hydrochloric) acid and soda to pure water, and he called the product aerated water, a term that is still in use. Stephen Hales had thought that effervescent mineral waters contained "sulphurous spirit"; Venel's experiments proved the absence of the gas now called sulfur dioxide, but he failed to notice that the "superabundant air" differed in any way from common air. It was, of course, carbon dioxide, characterized in 1754 by Joseph Black, who called it fixed air. Fourcroy later commented that no one had ever been so close to making a discovery without actually making it as was Venel.[1]

The government wanted a report on all French mineral waters, and in 1753 Venel and Pierre Bayen were appointed to prepare it. They traveled widely, but when the Seven Years' War started in 1756, Bayen became an army pharmacist[2] and there were insufficient funds for Venel to continue alone.

While living in Paris, Venel met Diderot, one of the editors of the *Encyclopédie*. Most chemical articles in the first two volumes were by Paul-Jacques Malouin, but Venel contributed a few to the second volume (1751), and then became the principal author on chemistry and materia medica. Venel, who was unknown outside a small Parisian circle, was a strange choice as successor to so distinguished an academician as Malouin, but he rose to the occasion and wrote more than 700 articles in volumes three to seventeen (1753–1765). Many were short accounts of preparations, properties, and medicinal uses of substances, but in some articles, including "Menstrue," "Mixte," and "Principes," he expounded his ideas on the nature of matter—ideas that were based on an extensive and critical reading of recent chemical literature. He followed Hales in regarding air as a common constituent of solids, and he generally approved of Stahl's phlogiston theory, but, curiously, he wrote no article on "Phlogistique" (although cross-references were given to such an article), perhaps because of his uncertainty about the properties of phlogiston.

In his longest article, "Chymie," Venel criticized the Newtonians who saw chemistry merely as a branch of physics. The chemist was concerned with relationships between corpuscles and principles, and these were not necessarily subject to the same forces as the large masses studied by physicists. Even though they lacked mathematical precision, chemical theories could be as valid as physical laws, but Venel admitted that chemistry had not yet progressed as far as physics and that it awaited a "new Paracelsus" who would bring about a revolution in his subject.

Upon returning to the Languedoc, his native province, in 1758, Venel was elected to the Société Royale des Sciences de Montpellier; in 1759 he became a professor at the University of Montpellier, where he gave an annual course on materia medica. He recognized two classes of medicament—external and internal—with a further division into those that acted mechanically (including external ligatures or metallic mercury taken internally) and those that affected the body chemically. He admitted that there was not yet a satisfactory theory of the chemical action of medicaments, so the physician had to rely largely on observation.

Each year Venel gave a chemistry course, but this was outside the university, in the laboratory of Jacques Montet, who provided the accompanying demonstrations.[3] In sixty-four lectures Venel gave a comprehensive account of the vegetable, animal, and mineral kingdoms, treating them in the same order as Rouelle. He did not discuss chemical theories in depth, but he expressed an opinion about phlogiston, which, he believed, did not gravitate toward the center of the earth but, on the contrary, levitated and thus diminished the weight of anything containing it.[4] Its release from a metal during calcination, therefore, caused the observed increase in weight. This theory was compatible with Venel's belief that principles did not necessarily behave like large masses, but he never published it and, according to Bayen, did not hold it strongly.[5] By the end of his life Venel had doubts about the whole doctrine of phlogiston, writing in 1775 that it was "beginning to age a little."[6]

Diderot said that Venel "vegetated" in Languedoc.[7] Certainly he lived a comfortable bachelor life and cultivated his tastes for good food, wine, and conversation, but he was not idle between lecture courses, for he carried out agricultural experiments on his estate near Pézenas and he became interested in other practical matters.

Wood was scarce in Languedoc and at the request of the provincial government Venel wrote in 1775 a long treatise on the advantages of coal, which was abundant there. He recommended it for domestic use and also for local industries, including the manufacture of silk and olive oil (both of which required much boiling water) and the distillation of alcohol. He also resumed his work on mineral waters in 1773, and by 1775 had visited all the remaining French provinces, but he died before completing his book on the subject. This book, like other work done in his later years, remained unpublished. Fourcroy wrote a fitting epitaph: "Venel was better known by what he promised the sciences than by what he really did for them."[8]

NOTES

1. Antoine François de Fourcroy, *Encyclopédie méthodique, chimie . . .*, III (Paris, 1796), 364.
2. Bayen was attached to the French expedition to Minorca in March 1756, and became chief pharmacist to the army in Germany in October 1756. See A. Balland, *Les pharmaciens militaires français* (Paris, 1913), 35–36, 39.
3. Jacques Montet (1722–1782), a pharmacist who had attended Rouelle's chemistry lectures, was a leading member of the Société Royale des Sciences de Montpellier. See J. Poitevin, "Éloge de M. Montet," in R. N. D. Desgenettes, ed., *Éloges des académiciens de Montpellier* (Paris, 1811), 242–249.
4. "Cours de chymie fait, chez Monsieur Montet apoticaire; par Monsieur Venel," in the library of the Wellcome Institute for the History of Medicine, London, MS 1516, p. 164.
5. P. Bayen, *Observations sur la physique . . .*, 3 (1774), 282.
6. G. F. Venel, *Instructions sur l'usage de la houille* (Avignon-Lyons, 1775), 495–496.
7. D. Diderot, "Voyage à Bourbonne et à Langres," in *Oeuvres complètes*, R. Lewinter, ed., VIII (Paris, 1971), 600–627 (quotation from p. 606).
8. Fourcroy, *op. cit.*, 262.

BIBLIOGRAPHY

I. ORIGINAL WORKS. Before graduating from Montpellier, Venel defended a thesis, *Dissertatio de humorum crassitudine* (Montpellier, 1741). When applying for the vacant chair in 1759 he wrote *Quaestiones chemicae duodecim* (Montpellier, 1759), twelve essays on medico-chemical topics that were judged better than similar collections submitted by the other candidates.

Venel's account of seltzer water was published in two parts, as "Mémoire sur l'analyse des eaux de Selters ou de Selz, Première (Seconde) Partie," in *Mémoires de mathématique et de physique, présentés à l'Académie Royale des Sciences, par divers Savans*, 2 (1755), 53–

79, 80–112; his first (and only) paper on vegetable analysis is "Essai sur l'analyse des végétaux. Premier mémoire," *ibid.*, 319–332. An abridged version of the work on seltzer water appeared in one of the rare early issues of *Observations sur la physique* . . . (August 1772), 60–71; it was reprinted in *Introduction aux observations sur la physique* . . ., 2 (1777), 331–334. Venel's only other publication on mineral waters is a pamphlet, written with P. Bayen, *Examen chimique d'une eau minérale nouvellement découverte à Passy dans la maison de Monsieur et de Madame de Calsabigi* (1755); repr. with other descriptions of the same water in the anonymously edited *Analyses chimiques des nouvelles eaux minérales, vitrioliques, ferrugineuses, découvertes à Passy dans la maison de Madame de Calsabigi* (1757), 19–52.

More than 700 articles by Venel appeared in Diderot's *Encyclopédie*, from II (1751) to XVII (1765), and he contributed two articles to the *Supplément*, I (1776). These are listed in R. N. Schwab, W. E. Rex, and J. Lough, "Inventory of Diderot's *Encyclopédie*," in *Studies on Voltaire and the Eighteenth Century*, LXXX (1971), LXXXIII (1971), LXXXV (1972), XCI–XCII (1972), with an author index in XCIII (1972). The important article "Chymie" in *Encyclopédie*, III (Paris, 1753), 408–437, was reprinted by A. F. Fourcroy, *Encyclopédie méthodique, chimie*, III (Paris, 1796), 262–303; a substantial extract (omitting Venel's account of the history of chemistry) is in J. Proust, *L'encyclopédisme dans le Bas-Languedoc au XVIIIᵉ siècle* (Montpellier, 1968), 106–140.

The last work by Venel to appear in his lifetime was *Instructions sur l'usage de la houille, plus connue sous le nom impropre de charbon de terre* . . . (Avignon-Lyons, 1775). His course on materia medica was published posthumously as *Précis de matière médicale*, 2 vols. (Paris, 1787), with many notes and additions by Joseph Barthélemy Carrère. Students' MS notes of Venel's courses on materia medica and chemistry are in the libraries of the University of Montpellier, MSS 563–564; and the Wellcome Institute for the History of Medicine, London, MSS 1516–1517 (both courses, dated 1761) and MS 347 (materia medica, dated 1767).

II. SECONDARY LITERATURE. Two valuable biographical accounts by men who knew Venel well are J. J. Menuret de Chambaud, *Éloge historique de M. Venel . . . par M. J. J. M.* (Grenoble-Paris, 1777); and E. H. de Ratte, "Éloge de M. Venel," in *Observations sur la physique* . . ., 10 (1777), 3–14, repr. in J. B. Carrère's edition of G. F. Venel, *Précis de matière médicale*, I (Paris, 1787), vii–xxxviii, and also in R. N. D. Desgenettes, ed., *Éloges des académiciens de Montpellier* (Paris, 1811), 194–203.

Additional information is given by J. Castelnau, *Mémoire historique et biographique sur l'ancienne Société Royale des Sciences de Montpellier* (Montpellier, 1858), 74–76, 154.

Venel's association with the *Encyclopédie* is discussed by J. Proust, *L'encyclopédisme dans le Bas-*

Languedoc au XVIIIᵉ siècle (Montpellier, 1968), 23–27, 33–35. Venel's theory of the levity of phlogiston is put in its historical context by J. R. Partington and D. McKie, "Historical studies in the phlogiston theory. I. The levity of phlogiston," in *Annals of Science*, 2 (1937), 361–404. Venel's criticism of Newtonian chemistry is discussed by Arnold Thackray, *Atoms and Powers* (Cambridge, Mass., 1970), 193–197.

W. A. SMEATON

VENETZ, IGNATZ (*b.* Visperterminen, Valais, Switzerland, 21 March 1788; *d.* Saxon-les-Bains, Valais, Switzerland, 20 April 1859), *civil engineering, glaciology.*

Venetz was the son of Peter Ignatz Venetz, who, although said to be descended from Venetian nobility, was a poor carpenter all his life. His mother was Anna Maria Stoffel. Although his parents wished him to become a priest, Venetz was attracted to science and mathematics and studied these subjects at the Collège de Brig. During the French occupation of the Valais he entered the Service des Ponts et Chaussées, in which he rose to become chief engineer for the district. His engineering career was, however, marred by a disaster in 1818, when the Gétroz glacier in the Val de Bagnes had grown, after three years of heavy snow, to dam up a stream which quickly became a lake. It was feared that warm weather would melt the glacier and cause a flood. Venetz tried to obviate the menace by making a channel through the ice through which the water might drain gradually. Unfortunately, the channel so weakened the ice dam that it gave way. Five hundred million cubic feet of water poured into the valley below, causing great loss of life and property.

Venetz' work in the Val de Bagnes led to a happier result through his conversations with a local chamois hunter, Jean-Pierre Perraudin, who had noticed that the striations left by glaciers extended as far as Martigny, a dozen miles away. Although skeptical, Venetz made his own observations and found that the hunter was correct. He presented his findings in a paper, read in 1821, which included evidence of earlier advances and retreats of glaciers and which won the prize offered by the Swiss Natural Science Society for an explanation of climatic deterioration. The work was nonetheless generally ignored until it was finally published in 1833 by Jean Charpentier, who had been convinced by another work of Venetz (written in 1829 but not published until 1861) that glaciers had extended more widely in earlier ages. Venetz thus provided the link between the observations of a peasant and the great discovery of a scientist—the Ice Age.

BIBLIOGRAPHY

Venetz' works include "Mémoire sur les variations de la température dans les Alpes de la Suisse," in *Denkschriften der allgemeinen Schweizerischen Gesellschaft für die gesammten Naturwissenschaften*, **1**, sec. 2 (1833), 1; and "Mémoire sur l'extension des anciens glaciers, renfermant quelques explications sur leurs effets remarquables," in *Nouvelles mémoires de la Société helvétique des sciences naturelles*, **18** (1861), 1.

On his life and work, see Ignace Mariétan, "La vie et l'oeuvre de l'ingénieur Ignace Venetz (1799–1859)," in *Bulletin murithienne de la Société valaisanne des sciences naturelles*, **76** (1959), 1.

GAVIN DE BEER

VENING MEINESZ, FELIX ANDRIES (*b.* Scheveningen, Netherlands, 30 July 1887; *d.* Amersfoort, Netherlands, 10 August 1966), *geodesy, geophysics.*

The son of Sjoerd Vening Meinesz, burgomaster of Rotterdam and later of Amsterdam, and Cornelia den Tex, Vening Meinesz studied civil engineering at the Technical University of Delft, obtaining a degree in 1910. Shortly afterward J. J. A. Muller of the Rijkscommissie voor Graadmeting en Waterpassing asked him to join that government bureau and participate in a gravimetric survey of the Netherlands. Vening Meinesz agreed immediately; the survey would take a few years at most, and after that there would still be time to go into practice as a civil engineer. Gravity, however, held him in its grip for life.

It is not surprising that Vening Meinesz, who never married, became enamored of geodesy. Imbued as a child with the importance of solemn governmental affairs, he felt at home among the geodesists' formal courtesy in a tradition of international cooperation that was over a century old. In 1911 he participated in the measurement of a base near Lyons by the Service Géographique de l'Armée. He was later elected president of the Association Géodésique Internationale (1933–1955) and of the International Union of Geodesy and Geophysics (1948–1951), in each case for his contributions to these sciences and for his continued efforts to bring order to their entangled relations. That both geodesy and solid-earth geophysics were needed to interpret the results of Vening

Meinesz' gravity measurements, and that geology had its role as well, became abundantly clear after his first cruises. His strenuous efforts to achieve scientific interplay, however, elicited only a meager response.

One is struck by a strictly logical line of development in Vening Meinesz' scientific achievements; everything seems to follow a careful plan, although of course no one could have foreseen that grave obstacles would yield to ingenuity and perseverance, or that salient features of the field of attraction of the earth exist—the Vening Meinesz belts of negative anomalies.

Vening Meinesz' first measurement of gravity in the Netherlands confronted him with what appeared to be an insurmountable obstacle. At many stations it proved impossible to find a stable support for the pendulums; the continuous vibration of the peaty subsoil would lead to unacceptable errors in the results. Vening Meinesz' efforts to overcome this difficulty were typical of the manner in which he confronted a problem. Far from trusting to luck in trying modifications of the experiments, he started with the thorough theoretical investigation embodied in his doctoral dissertation *Bÿdragen tot de theorie der slingerwearnemingen* (1915). In the introduction Vening Meinesz disparaged the originality of his own work; most of the disturbances had already been investigated when various authors had been confronted with them. His work, however, presents a systematic treatment starting from the fundamental equation

$$\frac{g}{l}\vartheta + \ddot{\vartheta} + S = 0,$$

where S represents the sum of all disturbances (*storingen*) the squares and products (second-order corrections) of which generally are negligible, so that, for instance, it is not necessary to introduce the temperature-dependence of the correction for finite amplitude.

For practical observations the most important conclusion of this theoretical work is that the mean of the periods of two isochronous pendulums swinging in the same plane with equal amplitude in opposite phases is not affected by the troublesome S-term, $\ddot{y}$, the horizontal acceleration in that plane. Vening Meinesz had to use an apparatus in which two pairs of pendulums swung in two mutually perpendicular planes. For each pair the above conditions could be fulfilled with sufficient precision; and, using this method, by 1921 he had measured gravity at fifty-one stations covering the territory of the Netherlands at intervals of about forty kilometers. It was claimed that the mean-square error of the difference of the values for g from that obtained at the central station at De Bilt was somewhat less than 2 milligals. Gravity at De Bilt was compared, both before and after the survey, with the absolutely determined value at Potsdam, where Vening Meinesz had been instructed in the practice of pendulum observations by the director of the Geodetic Institute, Ludwig Haasemann. During the survey the chronometers were compared with the clock at the Leiden astronomical observatory by telephone and after 1919 the radio time signals from Paris were used.

While working to solve the problems of unstable support, Vening Meinesz was tempted to direct his efforts more boldly to the apparently overambitious plan of measuring gravity at sea. In his *Theory and Practice of Pendulum Observations at Sea*, he again began with the first equations of motion for two pendulums affected by the same horizontal accelerations:

$$\ddot{\vartheta}_1 + \frac{g}{l_1}\vartheta_1 + \frac{\ddot{y}}{l_1} = 0, \ddot{\vartheta}_2 + \frac{g}{l_2}\vartheta_2 + \frac{\ddot{y}}{l_2} = 0,$$

then jubilantly exclaimed:

It is clear that $\ddot{y}$ can be eliminated from these two equations, and this is the fundamental principle of the method. If the pendulums are isochronous, so that $l_1 = l_2$, the result of this elimination is very simple: the difference of the equations gives

$$(\ddot{\vartheta}_1 - \ddot{\vartheta}_2) + \frac{g}{l}(\vartheta_1 - \vartheta_2) = 0.$$

which has the same shape as the equation of motion of an undisturbed pendulum of the same mathematical length l and an angle of elongation $(\vartheta_1 - \vartheta_2)$. We reach in this way the important conclusion that the difference of the angles of elongation may be considered as the angle of elongation of a fictitious pendulum, which is not disturbed by the horizontal accelerations of the apparatus, and which is isochronous with the original pendulums.

A shipboard attempt proved a complete failure, however, for even tiny surface waves striking the hull cause fairly high accelerations. It was necessary to wait for so exceptional a calm that the method was totally impractical. Vening Meinesz gratefully acknowledged the suggestion of F. K. T. van Iterson, chief engineer of the state coal mines, who, while on a submarine training dive conducted by the Dutch navy, had been struck by the profound tranquillity during submersion. Waves are in fact damped exponentially with depth, so that only 2 percent of the amplitude remains at a depth equal to the wavelength. The brisk movements of shorter waves were therefore imperceptible at the moderate depth then attainable by submarines, while the disturbance caused by the longer waves was handled by the fictitious pendulum. (See Figure 1.)

The first voyage to measure gravity at sea, from the Netherlands to Java in 1923, began under adverse circumstances. The small submarines moved slowly when submerged and could cover only limited distances. As a rule dives were restricted to the demands of the gravity observations. A heavy six-day storm wreaked the usual hardships on a

TABLE I
VENING MEINESZ' GRAVITY EXPEDITIONS AT SEA

Year	Submarine H.M.S.	Route	Number of Observations
1923	K II	Holland – Suez – Java	32
1925	K XI	Holland – Alexandria	10 (experiments with new pendulum apparatus)
1926	K XIII	Holland – Panama – Java	128
1927	K XIII	Java Deep	26
1929 – 30	K XIII	Indonesia	237
1932	O 13	Atlantic	60
1934 – 35	K 18	Holland – Buenos Aires – Cape Town – Freemantle – Java	237
1937	O 16	Holland – Washington – Lisbon	93
1937	O 12	Curaçao – Holland	20 (experiments on wave motion and Browne terms)
1938	O 13	Channel	
1939	O 19	North Sea	
		Total	843

boat of 630 tons, aggravated by the limited room, by water coming through the hatch, and by clothes refusing to dry. But, Vening Meinesz reported, "The worst thing of all was that the pendulum apparatus could not be used." (The rolling at a depth of thirty meters was still heavy enough to endanger the pendulums.) Finally, off the coast of Portugal with a smooth sea, three dives afforded an opportunity for observations. It now became clear that the apparatus, if it was to operate under normal waves, would have to be fitted in a kind of cradle to counteract the roll of the boat.

The harbor of Gibraltar was entered, and Vening Meinesz quickly developed the pendulum records; they showed good results, suggesting that no difficulties would arise if the tilt of the instrument could be kept small enough. The British commander gave permission for the naval yard to construct a suspension about an axis parallel to the keel. On the voyage to Java this tenacity was rewarded with thirty successful observations.

The Sterneck apparatus dealt separately with each pendulum, and the horizontal accelerations gave rise to irregular records that required days to interpret. Vening Meinesz constructed an entirely new instrument that was admirable for its ingenuity as well as for exquisite craftsmanship. The fundamental principle, elimination of $\ddot{y}$ by subtraction,

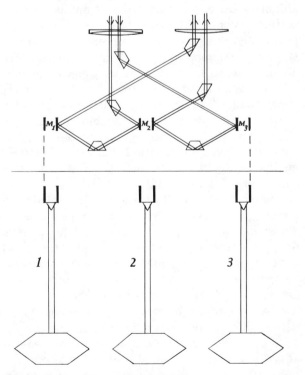

FIGURE 1. Top section represents a plan above pendulum mirrors M_i, prisms, lenses, and light paths. Lower section represents elevation of the three pendulums.

is applied in a beam of light reflected from one pendulum mirror onto the opposite mirror of a second pendulum, thus recording the desired fictitious pendulum (Figure 1). This device proved highly successful; from an engagingly regular pattern the value of gravity can be derived in a few hours. Furthermore, the strictly regular records have the advantage of showing at a glance that some conceivable disturbances affecting amplitude do not exist. Actually there are three pendulums swinging in the same plane; the outer two, each combined with the middle one, yield the records of two fictitious pendulums. The amplitude of the middle pendulum is kept as small as possible, and therefore the periods of the outer pendulums enter preponderantly into the result. Hence their stability can be relied upon as long as the slight difference between the two fictitious periods remains the same. The small amplitude of the middle pendulum is recorded separately against a short auxiliary damped pendulum, and this record is used to evaluate the correction for finite amplitude. A second damped pendulum records the tilt of the swinging plane. (In the Sterneck instrument the latter expedient was unnecessary; each pair of pendulums indicated the tilt of the swinging plane of the other pair.)

A cruise from Holland to Egypt to test the new apparatus (1925) gave satisfactory results. On the first cruise to Java the recording apparatus was separately mounted at the distance of one meter required for an easily measurable deflection of a light spot on the records. Thus a fixed direction was necessary for the registering beam of light, which therefore was directed through the axis of the cradle, made hollow for this purpose. This axis, being constantly perpendicular to the swinging plane, had to be kept horizontal with a tolerance of thirty minutes. This requirement demanded extremely careful trimming of the boat, and placed a great strain on the helmsman and the crew, who were not allowed to move about.

Suspension in gimbals seemed unavoidable, which necessitated rearrangement of the recording piece so that it could be joined firmly to the pendulum box. To retain its full length, the light path was "folded" by introducing more prisms. A good example of Vening Meinesz' attention to details is seen in the length of this path's being 1,162 millimeters, whereas the focal distance of the exit lens is 1,110 millimeters. This excess length compensates for the loss of convergence by refraction of the rays as they enter a prism. (For the collimator these figures are 653 and 616 millimeters.)

His equipment now being shipshape, Vening Meinesz strongly desired to take it to sea. Fortunately he combined technical skill with a rare gift for persuasion. His tenacious politeness succeeded in convincing government officials and the admiralty of the urgent need for gravity values on all oceans. He next inspired the Dutch submarine service to unheard-of achievements. At that time it was still thought impossible for submarines to travel long distances without escort ships on the surface. K XIII, however, crossed both the Atlantic and Pacific oceans unescorted. Several longer and shorter cruises followed over the southern Atlantic and the Indian Ocean and the seas of Indonesia, yielding hundreds of observations (see Table I).

While Vening Meinesz was on his 1937 cruise, B. C. Browne, a young geodesist at Cambridge, sent him a letter that might be summarized as "Dear Sir, you have made a mistake." On studying Browne's arguments, Vening Meinesz had to admit that the numerous regular corrections failed to cover some of the disturbances due to the motions of the boat. He made this admission gracefully, without vexation or excuses. It is even possible that he enjoyed the opportunity to work again with the mechanics of the pendulum and to produce in corroboration three papers on "Browne terms" and the second volume of *Theory and Practice*.

With the primitive cradle, the single axis of which was fixed in the direction of the keel, an acceleration $\ddot{z}$, parallel to the axis and hence to the knife-edges, clearly could not affect the motion of the pendulums. The effect of the other component of horizontal acceleration, $\ddot{y}$, had been dealt with by the fictitious pendulum device. On the other hand, it is equally clear that suspension in a complete gimbal system requires consideration of the resultant of gravity and acceleration given by vector addition: $g + \ddot{x} + \ddot{y} + \ddot{z}$.

A piece of good luck—as Vening Meinesz put it—made it possible for the vertical acceleration to be recovered from the old records. Assuming a circular wave movement, as was concurrently deduced from several wave theories, the horizontal acceleration should be equal—$|\ddot{x}| = |\ddot{y} + \ddot{z}|$—and thus Browne's correction could be introduced. For the future, however, a more direct determination seemed worth trying. If the angle between apparent gravity and the true vertical could be found, the horizontal accelerations would be known. This would be no easy task, for the whole apparatus in its gimbals follows apparent gravity.

Vening Meinesz may well have been equally pleased with the opportunity to show that his skill

in tackling such problems had not deteriorated during twelve years of making routine measurements. In close cooperation with Browne the "slow pendulums" were designed and used as a highly simplified but effective artificial horizon. A slow pendulum consists of a horizontal brass beam balanced on a knife-edge. The center of gravity being only a few tenths of a millimeter beneath this edge, the pendulum's period amounts to about half a minute. Therefore, waves going round in less than ten seconds can set it in motion to only an insignificant degree. Keeping a steady position in space, two such pendulums thus afford a means for registering the tilt of the apparatus in the gimbals (or, rather, of the damped pendulums, which, because of their short periods, follow exactly the changing direction of apparent gravity). For recording the slow pendulums, three lenses and thirteen prisms had to be added to the four mirrors, four lenses, and twenty-three prisms already in use. The housing of the new pendulums could be fitted into the eighty millimeters left between the main box and the recording box, and the whole instrument now appeared to have been made all of a piece.

The instrument was used for only a short time, however, because the outbreak of World War II suspended the peaceful use of submarines. After the war the Royal Dutch Navy was again helpful, although it had a greatly reduced number of submarines. Vening Meinesz, however, had to delegate the work to younger observers. For over thirty years his apparatus (of which some five copies are in existence) provided the only means for measuring gravity at sea, but in the late 1950's it was superseded. Then spring gravimeters mounted on stabilized platforms on surface ships began to record gravity values along continuous profiles. American investigators used the Vening Meinesz method until 1959, the number of stations then amounting to nearly 3,000.

Vening Meinesz' Observations in Relation to Geodesy. Geodesy seeks to ascertain the shape and dimensions of the earth. Altitudes and depths must refer to a certain curved surface called the geoid, which is the gravity-equipotential surface coinciding with the mean surface of the sea and its continuation on land. The geoid is approximately an ellipsoid of revolution, minor deviations being due to irregularities in the distribution of masses. These deviations are therefore closely linked to the values of gravity as measured by pendulum observations. If gravity is determined everywhere on the earth, the shape of the geoid with reference to the ellipsoid can be calculated according to a formula deduced by Stokes. This calculation greatly occupied Vening Meinesz. For instance, *Gravity Expeditions at Sea*, II, 13–17, concludes with "In the future it will become possible to apply the Theorem of Stokes in its full accuracy and to solve in this way the central geodetic problem: the determination of the Figure of the Earth."

Values of gravity averaged over regions extending some tens of kilometers are appropriate for this purpose. Observations on land have the disadvantage that a nearby irregular mass may cause a severe deviation from the mean. At sea this problem does not arise, since no irregularity can be nearer than the sea floor. Through Vening Meinesz' and English and American observations, approximation was already possible for about half the globe. Recently the central problem was solved by observing the trajectories of satellites. Their acceleration at any moment is that of gravity averaged over a region the radius of which is proportional to the distance of the satellite from the earth (about 100 kilometers).

Since the eighteenth century geodesists have asked whether the geoid, instead of being represented by an ellipsoid of revolution, should rather be seen as a triaxial ellipsoid—that is, whether the equator was a circle or an ellipse, the latter shape involving a systematic variation in the values of gravity. Having closed a ring of observations around the globe, Vening Meinesz could rule out the possibility of regular deviations and therefore establish that the equator must be represented by a circle.

Isostasy. The principle of isostasy is usually elucidated by analogy with blocks of wood or a mass of ice floating on stagnant water. The experiments by which we try to understand the origin of the geographic and geologic structure of the earth are perforce on a small scale. Small-scale models are very deceptive because of the strength of materials factor. In order to imitate real rocks on a continental scale the tensile strength has to be practically zero. Hans Cloos and P. H. Kuenen were among the first to experiment with wet clay or soft wax, but in general it is impossible to get all the mechanical parameters to scale and difficulties multiply. These problems do not affect Vening Meinesz' numerous calculations in applied mechanics. For large nonplastic deformations and for disruptions, calculations are hardly possible. For small strains seismic data are very useful and allow fairly accurate estimates of the specific weight of materials at depth.

In the latter half of the nineteenth century the earth came to be seen as originally a molten sphere, which at an early stage of cooling came to be enclosed in a strong crust of consolidation. Vening Meinesz constantly had in mind an earth model with an elastic crust on a viscous substratum. This led to his preference for interpretation in terms of regional isostasy, reducing observed gravity by applying the so-called regional isostatic compensation. For such a calculation a number of assumptions are tried regarding the thickness of the crust (twenty or thirty kilometers) and its strength, resulting in a bend area with a radius of 232, 174, 116, 58, or 29 kilometers (also 0 kilometers: local compensation). An assumption resulting in a complete reduction of the anomaly may be true—at least it cannot be said to be wrong. The great drawback of gravity measurements, however, is that although a value can be calculated from a given repartition of masses, the reverse calculation is impossible. A given set of gravity values can result from a wide range of depth configurations that depart in many ways from a simplified assumption. In addition, in some instances none of the regional isostatic assumptions made by Vening Meinesz can be said to give satisfactory results.

The Vening Meinesz Belts. The most striking feature shown by the earth's gravity field was discovered by Vening Meinesz during his earlier cruises. The "Vening Meinesz negative gravity anomaly belts of island arcs," as they were called by H. H. Hess (*Gedenkboek*, 183), with their bold, steeply descending lows of gravity, are not reduced to zero by any of the common suppositions. Winding for thousands of miles in gentle curves, along deep-sea trenches and rows of mostly volcanic islands, the belts often follow the outcrop of a roughly stratiform cluster of shallow and deep earthquake foci dipping at about forty-five degrees landward.

The regular concurrence of these conspicuous phenomena challenged every inventive mind to discover their significance. Vening Meinesz directed his attention chiefly to the gravity anomalies; his explanation of them by means of the buckling hypothesis brought strong reactions from geologists. Throughout he used the classic tool of compression in a rigid crust. Strong enough to overcome the resistance of the crust, the compression would cause a thickening along a line of weakness that apparently coincides with the frontier between a continent or shallow sea and the ocean. This thickened part of the crust, representing increased mass, naturally sinks to restore isostasy, thus causing the compressive forces to deviate downward. On further compression the crust will be folded and the folded parts pressed downward, thus causing light crustal rocks to take the place of heavy mantle materials. This process explains the observed negative gravity anomalies. The Dutch geologist Kuenen, who often did model experiments to explain geological phenomena, illustrated this buckling by compressing a floating layer of soft wax, which behaved exactly as Vening Meinesz had predicted. The downward buckling was soon interpreted as a geosyncline receiving sediments because of the compensating rise of the borders, which led to emergence of land at the continental edge and its subsequent attack by erosion. The resulting sediments, filling in the depression, would then be folded by continued pressure. The deep-sea trenches often are situated somewhat farther from the continent than the most strongly negative anomaly; this situation is explained by the sediments' having filled part of the trench.

A strong departure from isostasy cannot be of unlimited duration. In due course the base will be buoyed, thus producing an elevated strip of folded sediments. In Vening Meinesz' view, folded mountain chains, such as the Alps, represent buckles of an older period that now reach great altitudes because of isostatic compensation by the remnants of a base of crustal rocks. The view that the Indonesian archipelago was comparable with an initial stage of Alpine mountains had long been considered by geologists. This view, joined with Kuenen's experiment, secured a favorable reception for the buckling hypothesis.

Areas of strong anomalies are necessarily of limited size, at least in breadth. A large, round domain could not depart from isostasy to a degree comparable with the negative anomaly belts, unless one accepted an exorbitant value for the strength of the crust. An example of an extensive area about 1,000 kilometers across having moderate negative anomalies is in Scandinavia and is readily interpreted in connection with the thick ice cap formerly present there. That ice cap would have been brought into isostatic equilibrium during the long glacial epoch by weighing down the crust of the earth. It melted rather suddenly, a short geological time (6,000 years) ago. In correspondence with the remaining negative anomaly, the land is now rising. In combination these two facts, numerically assessed, enabled Vening Meinesz to calculate the resistance of the substratum, which, expressed in terms of viscosity, yields a value of 10^{22} poises. He used this value for calculations concerning the velocity of the movement that other departures from isostatic equilibrium may cause.

Convection Currents in the Mantle. In the 1920's Arthur Holmes, among others, had used convection currents to explain continental drift. By general agreement, however, the geoscientists of the northern hemisphere had also dismissed the sound arguments offered by Wegener, bluntly denying the possibility of wandering continents. For Vening Meinesz the crust was so rigid that he had no choice but to state that the continents were firmly fixed, while the interior of the earth became the site of great activity. He often thought in terms of a cooling earth and accepted an increase in temperature with depth greater than the adiabatic gradient. The resulting departure from equilibrium led to convection currents, about the velocity of which the Scandinavian viscosity could give information. The drag of the currents on the crust, in Vening Meinesz' view, led to no more than a buckle here and there, notably along the Vening Meinesz belts—which, therefore, had to be located near a descending branch of the currents.

Vening Meinesz was strongly convinced that he had derived an unshakable argument for the reality of convection currents from a regularity apparent in the development of the earth's relief in spherical harmonics. He showed this regularity to be present, at first, in the development calculated in 1922 by Adelbert Prey up to the sixteenth order, which, at his request, was extended to the thirty-first order by the Mathematical Center of Amsterdam. Its numerical result reflects mainly the contrasting heights of oceans and continents, whereas it is their distribution on the globe that must be closely linked with the system of convection currents: around sinks in this system patches of light continental rocks had floated together in the formation of the land on the surface of the globe. This all had taken place in a primordial stage, when the earth was still largely molten. Through cooling, the crust soon became so rigid as to resist further rearrangement; throughout the whole of geological time the continents retained their shapes and relative positions despite continued convection underneath. (When, in the early 1960's, continental drift was held to be established by paleomagnetism, the currents were still there, serving as a motor.)

Convection had been challenged; and seismic data led to the view that a lower part of the mantle, because of higher specific weight, could not be involved in those currents. Vening Meinesz argued that the high density at depth was probably due to the presence of olivine in the spinel modification. The reversible transition between the olivine and the spinel phases could not be an obstacle, yet the boundary between them ought to be sharp — and, according to the seismologists, it was not. This objection was met by the plausible assumption of a fayalite component that caused phase equilibrium to persist through a certain range of pressures. These ideas were worked out with the physico-chemists J. L. Meijering and C. J. M. Rooymans, and were likewise taken up by A. E. Ringwood.

Problems like these, clearly needing information from many different fields, strengthened Vening Meinesz' conviction that interdisciplinary cooperation is necessary when fundamental principles are involved. He felt this to be especially the case in geophysics and geology, and he had some success. Certain chapters in volume II of *Gravity Expeditions at Sea* written by the prominent Dutch geologists J. H. Umbgrove and Kuenen resulted in a genetic interpretation of the Indonesian archipelago. Twenty years later B. J. Collette submitted a doctoral dissertation concerning the effect of gravity on the geology of the Sunda Islands. In 1955–1957 Collette measured gravity on the North Sea, the shallower parts with a static gravimeter lowered to the bottom and the deeper northern part, up to 61° north latitude, with the Vening Meinesz instrument.

Vening Meinesz maintained a keen interest in the ocean floor. Later investigators did not forget the grand old pioneer: his honors steadily increased, culminating in the Vetlesen Prize in 1962. In 1963 the new institute for geophysics and geochemistry at Utrecht University was named for him.

BIBLIOGRAPHY

I. ORIGINAL WORKS. A list of Vening Meinesz' publications to 1957 is included in the jubilee volume, *Gedenkboek F. A. Vening Meinesz*, which is *Verhandelingen van het Nederlandsch geologisch-mijnbouwkundig genootschap*, Geol. ser., **18** (1957), with portrait. His most important works are *Theory and Practice of Pendulum Observations at Sea*, 2 vols. (Delft, 1929–1941), with illustrations of apparatus and records; *The Gravity Measuring Cruise of the U.S. Submarine S-21* (Washington, 1930), written with F. E. Wright, which includes illustrations of apparatus and records as well as a full explanation of the instruments and method; and *Gravity Expeditions at Sea*, 5 vols. (Delft, 1932–1960; II, repr. 1964), vol. II written with J. H. Umbgrove and P. H. Kuenen.

II. SECONDARY LITERATURE. See B. J. Collette, "In Memoriam Dr. Ir. Felix Andries Vening Meinesz," in *Geologie en mijnbouw*, **45** (Sept. 1966), 285–290, in English, with portrait and complementary bibliography to

that in the jubilee volume (see above); J. L. Meijering and C. J. M. Rooymans, "On the Olivine-Spinel Transition in the Earth's Mantle," in *Proceedings of the K. nederlandsche akademie van wetenschappen*, Ser. B, **61** (1958), 333–344; and J. Lamar Worzel, "Pendulum Gravity Measurements at Sea, 1936–1959," *Contributions. Lamont Geological Observatory*, no. 807 (1963), which contains the most important continuation of Vening Meinesz' observations and interpretations.

W. NIEUWENKAMP

VENN, JOHN (*b*. Hull, England, 4 August 1834; *d*. Cambridge, England, 4 April 1923), *probability, logic*.

Venn's family was one of a group belonging to the evangelical wing of the Church of England that was noted for its philanthropic work. This group, which included the Macaulays, Thorntons, and Wilberforces, was centered in the London suburb of Clapham and was nicknamed the "Clapham Sect." A pivotal figure was Venn's grandfather, Rev. John Venn, rector of Clapham.

After attending two London schools, at Highgate and Islington, Venn entered Gonville and Caius College, Cambridge, in 1853; took his degree in mathematics in 1857; and was elected a fellow of his college, holding the fellowship until his death. He took holy orders in 1859, but after a short interval of parochial work he returned to Cambridge as college lecturer in moral sciences and played a considerable part in the development of the newly established moral sciences tripos examination. In 1883 Venn resigned his clerical orders, being out of sympathy with orthodox Anglican dogma but remained a devout lay member of the church. He received the Cambridge Sc.D. in 1883 and in that year was elected a fellow of the Royal Society.

Besides his scientific works, Venn conducted much research into historical records and wrote books on the history of his college and on his family. He also undertook the preparation of *Alumni Cantabrigienses*, a tremendous task in which he was assisted by his son, J. A. Venn; two volumes appeared in his lifetime.

Venn's volumes on probability and logic were highly esteemed textbooks in the late nineteenth and early twentieth centuries. The historian H. T. Buckle had discussed the validity of statistical studies of human activities, and De Morgan and Boole had written on the foundations of probability theory; this work stimulated Venn to write his *Logic of Chance*. In this book he disclaimed any attempt to make extensive use of mathematical techniques; he believed that there was a need for a thorough and logical discussion of principles, and his work was an essay in that direction. British predecessors were critically discussed. Venn thought that De Morgan's *Formal Logic* provided a good investigation, but he was not prepared to accept his principles; he was dubious about Boole's *Laws of Thought*, for he was not entirely happy with certain aspects of Boole's algebraic analysis of logic.

Venn attempted to deal with the notorious Petersburg paradox by insisting on the concept of "average gain," which is connected with his revision of the basic definition of probability. The classical definition, given in the early eighteenth century by De Moivre, considers the situation in which there are s successes in m trials and defines the probability of a success as s/m. One weakness of that definition is that if a possible ambiguity is to be evaded by specifying that the m possibilities are all equally likely, we may be led into a circular argument. Venn offered the following definition: If in a large number m of trials there are s successes (and $m - s$ failures), the probability of a success is the limit of s/m as m tends to infinity. This definition avoids the difficulties that arise when the classical definition is applied to, say, a die with bias, but is itself not free from defects. The existence of the limit cannot be proved from the definition; Venn implicitly assumes that a unique limit must exist. In the twentieth century R. von Mises improved Venn's work by adding explicit postulates on the existence of the Venn limit that effectively restrict the nature of the possible trials. The Venn definition also has been criticized on the practical grounds that it alone cannot provide a specific numerical value for the probability and that further hypotheses must be added in order to arrive at such values.

Venn's books on logic were based on a thorough study of earlier works, of which he had a very large collection that is now in the Cambridge University Library. His writings can still be consulted with profit but are chiefly remembered for the use of logical diagrams, although *Symbolic Logic* is largely an attempt to interpret and correct Boole's work. The use of geometric diagrams to represent syllogistic logic has a long history, but Leibniz was the first to use them systematically rather than as casual illustrations. "All A is B" is represented by a circle marked A placed wholly inside another circle marked B; "Some A is B" is represented by two overlapping circles; and the standard syllo-

gisms are depicted by means of three circles. The procedure was further developed by Euler, and in the early nineteenth century many writers offered varieties of diagrammatic representation. In preparing his book Venn had made a careful survey of such writings, and his chapter discussing them is severely critical. Boole's *Laws of Thought* (1854) was the first efficient development of an algebra of logic, but he did not use diagrams.

Venn was strongly influenced by Boole's work, and in his books he clarifies some inconsistencies and ambiguities in Boole's ideas and notations; but his chief contribution to logic was his systematic explanation and development of the method of geometrical representation. He pointed out that diagrams that merely represent the relations between two classes, or two propositions, are not sufficiently general; and he proposed a series of simple closed curves (circles or more elaborate forms) dividing the plane into compartments, such that each successive curve should intersect all the compartments already obtained. For one term, compartments x, $\bar{x}$ (the negation or complement of x) are needed; for two terms, four compartments are needed; for three terms, eight; and so on. By the time five classes are under consideration, the diagram is becoming complicated to the point of uselessness. To illustrate the principles of symbolic logic, Venn deliberately provided a variety of concrete instances, often of the type now found in collections of mathematical puzzles, for he remarked that the subject is sufficiently abstract to present difficulty to the average student, who must be helped by a supply of realizable examples.

Since the null class was not then accepted as a class, Venn also had to discuss whether the diagrams represented compartments or classes—that is, whether compartments could be regarded as unoccupied. A compartment known to be unoccupied could be shown by shading it, and a universal proposition could be represented by a suitable unoccupied compartment: Thus "No A is B" could be represented by shading the area common to the two intersecting circles (or closed curves) representing A and B. Venn's treatment of the "universe of discourse" was somewhat indefinite and was criticized by C. L. Dodgson in his *Symbolic Logic* (published under the pseudonym by which he was better known, Lewis Carroll); Dodgson insisted, in Carollian style, on the use of a closed compartment enclosing the whole diagram to delimit the universe of discourse. Venn diagrams, as they are now generally called, have recently been much used in elementary mathematics to encourage logical thinking at a fairly early stage of a child's education.

BIBLIOGRAPHY

I. ORIGINAL WORKS. Venn's books include *The Logic of Chance* (London–Cambridge, 1866); *Symbolic Logic* (London–Cambridge, 1881); and *The Principles of Empirical Logic* (London–Cambridge, 1889).

II. SECONDARY LITERATURE. A detailed obituary notice of Venn by his son, J. A. Venn, is in *Dictionary of National Biography* for 1922–1930, 869–870. For a critical discussion of definitions of probability, see Harold Jeffreys, *Theory of Probability*, 2nd ed. (Oxford, 1948). A succinct but valuable account of Venn diagrams is M. E. Baron, "A Note on the Historical Development of Logic Diagrams: Leibniz, Euler and Venn," in *Mathematical Gazette*, **53** (May 1969), 113–125.

T. A. A. BROADBENT

VERANTIUS, FAUSTUS (also known as **FAUSTO VRANČIĆ** or **VERANZIO**) (*b.* Šibenik, Dalmatia, 1551; *d.* Venice, Italy, 20 January 1617), *engineering.*

Son of Michael Vrančić, a diplomat and poet, and of Catherine Dobroević, Verantius came from a noble Croatian family; its members, aristocracy of the city of Šibenik, were related to several Church dignitaries and to a viceroy of Croatia. His uncle, Anthony Verantius (1504–1573), was archbishop of Esztergom, primate of Hungary, cardinal, and an influential statesman. He took charge of Verantius' education, sending him to study philosophy and law at Padua (1568–1570) and initiating him into the political intrigues of the day. Although Anthony Verantius was principally a man of letters, he was greatly interested in the art of fortification and supervised the construction of the fortress at Eger. It is possible that his uncle's enthusiasm for technical problems influenced Verantius.

In 1579, Verantius became commander of the citadel at Veszprim. Two years later he resigned this post to accept an offer from Emperor Rudolf II to become secretary of the royal chancellory of Hungary. Thus from 1581 to 1594 Verantius was a diplomat, working at times for the emperor at Prague as well as for Archduke Ernest at Vienna. In his leisure time Verantius studied mechanics and mathematics.

In 1594 Verantius resigned his position at the Hapsburg court. From then until 1598 he lived in Dalmatia and Italy, mainly Venice, where, in 1595, he published a dictionary in five languages (Latin, Italian, German, Croatian, and Hungarian). Verantius had two children, and following the death of his wife, he took religious vows. In 1598 Rudolf II granted him the title of bishop of Csanad, an honorary office since the bishopric was then occupied by the Turks. Nevertheless, Verantius interrupted his literary and scientific work in order to accept an important political assignment, as imperial counselor for Hungarian and Transylvanian affairs. Although he was a skillful courtier and an able administrator, his career was hampered by his impetuous nature. Disappointed in his political ambitions, he left the court at Prague in 1605 and became a member of the Congregation of St. Paul, in Rome.

Verantius became friendly at Rome with Giovanni Ambrogio Mazenta, a Barnabite like himself and, from 1611, general of the Congregation of St. Paul. Very possibly it was Mazenta who interested Verantius in the construction of machines and in architectural problems. Verantius undoubtedly had an opportunity to see many of Leonardo da Vinci's technical drawings, of which Mazenta had prepared a list about 1587. During his stay at Rome, Verantius had drawn and engraved a series of "new machines." At his request on 9 June 1614 Louis XIII granted him a privilege for printing a "book of machines." According to its terms, for fifteen years no one would be permitted to publish another edition; and for thirty years no French subject would be allowed, without Verantius' permission, to "put into use . . . the said machines of his invention [which have] never been seen before." Cosimo II de' Medici, grand duke of Tuscany, granted Verantius an analogous privilege (June 1615) for the book "that the latter wishes to publish."

During this period Verantius fell gravely ill; and his doctors advised him to leave Rome. He drew up his will on 12 June 1615 and decided to return to Šibenik to await his death. His efforts to publish the book on machines in France, Rome, or Florence were unsuccessful. He was so intent on carrying out his project, however, that on his way to Dalmatia he stopped at Venice, where in 1616 he published a treatise on logic (which he called "ars discendi et docendi scientias") and, most important, a splendid folio volume entitled *Machinae novae*. Too ill to continue his trip, Verantius died in Venice; but in accordance with the provisions of his will, his body was taken to Šibenik and placed in the family burial vault on the isle of Prvić.

Machinae novae poses some bibliographic problems, since it is undated and the surviving copies are not identical. The work consists of a title page, forty-nine plates, and five sets of explanations of the plates. Each set has a new pagination and is written in a different language (Latin, Italian, Spanish, French, and German). Although there are at least two different title pages and some copies do not contain explanations in five languages, it can be assumed that the work was published only once and, in particular, that all the plates are from a single printing. The publication dates of 1595, 1605, and 1617 found in the literature and in certain library catalogs must be considered erroneous. The printing of the *Machinae novae* was completed during the first half of 1616 (or perhaps the last two months of 1615), for in July 1616 several of Verantius' friends thanked him for sending them the book.

Some of Verantius' inventions are applicable to the solution of hydrological problems, for example, the project for preventing the Tiber from overflowing its banks at Rome and that of providing Venice with fresh water. Others concern the construction of clepsydras, sundials, mills, presses, and bridges and boats destined for widely different uses. Unlike Leonardo da Vinci, Verantius had no interest in machines of war; rather, he devoted himself especially to perfecting agricultural implements. He foresaw the advantages of the assembly line, and in his many designs for mills he was especially concerned with the rational use of various sources of energy. In this connection his idea of utilizing the motive force of the tides is particularly important. His devices demonstrate an intuitive grasp of the principle of the mechanical moment and of the triangle of forces. His designs for a wind turbine, a funicular railway, and a bridge suspended by iron chains represent a definite advance over contemporary techniques. Although some of Verantius' "machines" are not wholly original or independent inventions, many of them are explained for the first time in print in *Machinae novae*. One example is *homovolens*, the first published mention of a parachute.

BIBLIOGRAPHY

I. ORIGINAL WORKS. The original ed. of *Machinae novae* is rare, but the work is now available in two fasc. eds., edited by F. Klemm and A. Wissner (Munich, 1965) and by U. Forti (Milan, 1968). *Dictionarium quinque nobilissimarum Europae linguarum* (Venice, 1595) is also available in modern editions (Bratislava, 1834; Zagreb, 1971). The treatise on logic is *Logica nova suis ipsius instrumentis formata et cognita* (Venice, 1616).

II. SECONDARY LITERATURE. There is still no critical study of Verantius' life, ideas, and inventions. The best sources for biographical details are the obituary by J. T. Marnavich, *Oratio habita in funere ill. ac rev. viri Fausti Verantij* (Venice, 1617); and the study by G. Gyurikovits, "Biographia Verantii," in Verantius' *Dictionarium pentaglottum* (Bratislava, 1834), ix–xx. On his scientific and technical work, see G. Boffito, *Scrittori barnabiti*, IV (Florence, 1937), 148–152; H. T. Horwitz, "Ueber Fausto Veranzio und sein Werk *Machinae novae*," in *Archeion*, **8** (1927), 169–175; V. Muljević, "Faust Vrančić kao fizičar i konstruktor," in *Hrvatsko sveučilište* (Zagreb), **1**, no. 6 (1971), 13–15; and F. Savorgnan di Brazza, "Un inventore dalmata del 500: Fausto Veranzio da Sebenico," in *Archivio storico per la Dalmazia*, **13** (1932), 55–73.

M. D. GRMEK

VERDET, MARCEL ÉMILE (*b.* Nîmes, France, 13 March 1824; *d.* Nîmes, 3 June 1866), *physics*.

Verdet was one of the outstanding physics teachers of mid-nineteenth-century France, holding professorships at the École normale supérieure, the École polytechnique, and the Faculté des sciences in Paris. He introduced into the French scientific world the thermodynamics of Joule, Clausius, Helmholtz, and William Thomson, and conducted important experiments on the effects of a magnetic field on plane-polarized light.

Little is known of Verdet's background except that he came from a leading Protestant family in the south of France. An early preference for teaching led him to attend the École normale supérieure, rather than the École polytechnique, to which he was also accepted. At the close of his studies he scored so high on the *concours d'agrégation* that he was spared the customary tour of duty in the provinces and appointed directly to the Lycée Henri IV in Paris. In 1848 he received his *doctorat* from the Sorbonne and was appointed lecturer in physics at the École normale, a position he held for the rest of his life. Four years later he became entrance examiner at the École polytechnique; then examiner in physics (1853) and professor of physics (1862). Later his fame as a teacher gained him the chair of mathematical physics at the Paris Faculté des sciences.

Verdet educated his colleagues as well as his students. French physicists of his time were ignorant of much of the research going on outside their country, so Verdet undertook to publish abstracts of the most important articles appearing in foreign journals. From 1852 to 1864 every volume of the *Annales de chimie* contained ten or more of his synopses. Since much of the work being done in England and Germany in this era centered on the development of the mechanical theory of heat, Verdet soon became the French expert in this subject. In 1864–1865 he taught the new thermodynamics at the Sorbonne, and the notes from his course were compiled by two students and published as *La théorie mécanique de la chaleur*, a textbook which has become a classic.

Verdet's original scientific papers, although limited in quantity by his extensive academic duties, his short life, and poor vision, reveal a talent not as an innovator but, rather, as a painstaking investigator who filled in the details of others' discoveries. His early research included a series of experiments on electromagnetic induction and a theoretical treatise on the image-forming power of lenses. In his major effort Verdet investigated the phenomenon now known as the "Faraday effect": the rotation of the plane of polarization of a ray of light by a transparent solid or liquid in a magnetic field (this effect was discovered by Michael Faraday in 1845). Verdet studied the dependence of the Faraday effect on the strength of the magnet causing the rotation, the medium in which the light is traveling, and the color of the light. He found that the magnetic power of rotation was directly proportional to the strength of the magnet, inversely proportional to the square of the wavelength of the light, and related to the index of refraction of the material. In recognition of the importance of this work a measure of the power of magnetic rotation was named "Verdet's constant."

BIBLIOGRAPHY

I. ORIGINAL WORKS. Verdet's entire scientific output is collected in *Oeuvres de Verdet*, 8 vols. (Paris, 1868–1872), which includes all his scientific papers and the lecture notes from his courses. His writings include "Mémoires sur la physique publiés à l'étranger, extraits par M. E. Verdet," in *Annales de chimie*, 3rd ser., **34–69**, and 4th ser., **1–4** (1852–1864); *La théorie mécanique de la chaleur* (Paris, 1868); "Recherches sur les phénomènes d'induction produits par les décharges électriques," in *Annales de chimie*, 3rd ser., **24** (1848), 377–405; "Note sur les courants induits d'ordres supérieurs," *ibid.*, **29** (1850), 501–506; "Recherches sur les phénomènes d'induction produits par le mouvement des métaux . . .," *ibid.*, **31** (1851), 187–217; "Sur l'intensité des images lumineuses formées au foyer des lentilles et des miroirs," *ibid.*, 489–503; and "Recherches sur les propriétés optiques développées dans les corps transparents par l'action du magnétisme," *ibid.*, **41** (1854), 370–412; **43** (1855), 37–44; **52** (1858), 129–168; and **69** (1863), 415–491.

II. SECONDARY LITERATURE. The two best biographies of Verdet are by M. A. Levistal, in *Annales scientifiques de l'École normale supérieure*, **3** (1866), 343–351; and A. de La Rive, in *Mémorial de l'Association des anciens élèves de l'École normale 1846–1876* (Paris, 1877).

EUGENE FRANKEL

VER EECKE, PAUL (*b*. Menin, Belgium, 13 February 1867; *d*. Berchem, Belgium, 14 October 1959), *mathematics.*

Ver Eecke attended the *collège* in Menin until he was fifteen and completed his secondary education at Bruges. After graduating as a mining engineer from the University of Liège in 1891 and following a short period in private industry, he entered the Administration du Travail in 1894, where he served until his retirement in 1932. His many honors included membership in the Société Mathématique de Belgique, the Académie Internationale d'Histoire des Sciences, and the Comité Belge d'Histoire des Sciences.

While quite young, Ver Eecke became interested in ancient Greek mathematics, especially in the works of Archimedes, and his first publication (1921) was a French translation of the complete works of Archimedes. Nearly all his scholarship concerned the translation of Greek mathematical works into French, the only exceptions being his translations from the Latin into French of the *Liber quadratorum* of Leonardo Fibonacci (written in 1225) and a treatise by Vito Caravelli. He carried out this work not as a philologist but as a scientist, adhering closely to the Greek text. His translations were preceded by surveys of the periods in which the mathematicians lived; and in the footnotes Ver Eecke gave the proofs in modern notation. Ver Eecke thus provided historians of science with a fairly accurate reflection of thought in antiquity and the scientific significance of the works.

In addition to translations Ver Eecke also wrote articles. In "Le théorème dit de Guldin considéré au point de vue historique" he defended the assumption that the law bearing his name was original with Guldin. Ver Eecke's arguments were, however, rejected by R. C. Archibald (*Scripta mathematica*, **1** [1932], 267).

BIBLIOGRAPHY

Ver Eecke's works include *Les oeuvres complètes d'Archimède* (Brussels, 1921); *Les Coniques d'Apollonius de Perge* (Bruges, 1923; repr. Paris, 1963); *Diophante d'Alexandrie. Les six livres arithmétiques et le livre des nombres polygones* (Bruges, 1926); *Les Sphériques de Théodose de Tripoli* (Bruges, 1927); *Serenus d'Antinoë. Le livre "De la section du cylindre" et le livre "De la section du cône"* (Paris–Bruges, 1929); "Note sur le procédé de la démonstration indirecte chez les géomètres d l'antiquité grecque," in *Mathesis*, **44** (1930), 382–384; "Note sur la théorie du plan incliné chez les mathématiciens grecs," *ibid.*, **45** (1931), 352–355; "Le théorème dit de Guldin considéré au point de vue historique," *ibid.*, **46** (1932), 395–397; "La mécanique des Grecs d'après Pappus d'Alexandrie," in *Scientia* (Milan), **54** (1933), 114–122; *Pappus d'Alexandrie. La Collection mathématique* (Paris–Bruges, 1933); "Le traité des hosoèdres de Vito Caravelli (1724–1800)," in *Mathesis*, **49** (1935), 59–82; "Le traité du métrage des divers bois de Didyme d'Alexandrie," in *Annales de la Société scientifique de Bruxelles*, ser. A, **56** (1936), 6–16; "Note sur une démonstration antique d'un théorème de lieu géométrique," in *Mathesis*, **51** (1937), 11–14; *Euclide. L'Optique et la Catoptrique* (Paris–Bruges, 1938); "Note sur une interprétation erronée d'une définition pythagoricienne de la ligne géométrique," in *Antiquité classique*, **7** (1938), 271–273; *Les opuscules mathématiques de Didyme, Diophane et Anthemius, suivis du fragment mathématique de Bobbio* (Paris–Bruges, 1940); *Proclus de Lycie. Les commentaires sur le premier livre des Éléments d'Euclide* (Bruges, 1948); and *Léonard de Pise. Le livre des nombres carrés* (Bruges, 1952).

H. L. L. Busard

VERHULST, PIERRE-FRANÇOIS (*b*. Brussels, Belgium, 28 October 1804; *d*. Brussels, 15 February, 1849), *statistics, sociology, probability theory, mathematics*.

According to Adolphe Quetelet, Verhulst, while on a trip to Rome, "conceived the idea of carrying out a reform in the Papal States and of persuading the Holy Father to give a constitution to his people." The project was, in fact, considered; and Verhulst, ordered to leave Rome, was almost besieged in his apartment.

Verhulst first thought of publishing the complete works of Euler but abandoned this idea in order to study with Quetelet, with whom he eventually collaborated on social statistics. The two did not, however, always share the same views in this field, in which the theoretical foundations were uncertain and observations far from abundant. It was generally assumed, following Malthus, that the

tendency of a population to increase follows a geometric progression. Quetelet, however, believed he had grounds for asserting that the sum of the obstacles opposed to the indefinite growth of population increases in proportion to the square of the rate at which the population tends to grow. Verhulst showed in 1846 that these obstacles increase in proportion to the ratio of the excess population to the total population. He was thus led to give the figure of 9,400,000 as the upper limit for the population of Belgium (which, in fact, has grown to 9,581,000 by 1967). Verhulst's research on the law of population growth makes him a precursor of modern students of the subject.

Verhulst was a professor at the Université Libre of Brussels and later at the École Royale Militaire. He was elected to the Académie Royale de Belgique in 1841 and became its president in 1848.

BIBLIOGRAPHY

There are articles on Verhulst by J. Pelseneer, in *Biographie nationale publiée par l'Académie royale de Belgique*, XXVI (Brussels, 1936–1938), cols. 658–663, with bibliography; and A. Quetelet, in *Annuaire de l'Académie r. des sciences, des lettres et des beaux-arts de Belgique*, **16** (1850), 97–124, with a bibliography of Verhulst's works and a portrait.

J. Pelseneer

VERNADSKY, VLADÍMIR IVANOVICH (*b*. St. Petersburg, Russia, 12 March 1863; *d*. Moscow, U.S.S.R., 6 January 1945), *mineralogy, geochemistry, biogeochemistry*.

Vernadsky's father, Ivan Vasilievich Vernadsky, a member of the gentry, was a professor at Kiev and Moscow universities and later at the Main Pedagogical Institute and the Aleksandrov Lycée in St. Petersburg. From 1857 to 1864 he edited liberal economic journals, in which he spoke against serfdom. His mother, Anna Petrovna Konstantinovich, came from a Ukrainian landowning family and taught singing.

While still at the classical gymnasium, Vernadsky became interested in chemistry but was also attracted by history, philosophy, and Slavic languages. From 1881 to 1885 he was a student in the Natural Sciences Section of the Physics and Mathematics Faculty at St. Petersburg University. According to Vernadsky, while he was there, Mendeleev's brilliant lectures awakened a strong desire for knowledge and its application. Dokuchaev, who lectured on mineralogy and crystallography, supervised Vernadsky's first scientific research. After graduation Vernadsky remained at the university to prepare for a teaching career, and from 1886 he was curator of the mineralogical collection at the university. In the latter year he married Natalia Egorovna Staritskaya; they had two children. From 1888 to 1890 Vernadsky traveled abroad. Interested in the structure of crystal substances, he worked in Groth's crystallographic laboratory at Munich and in Le Châtelier's and Fouqué's laboratories at the Collège de France. He also made geological excursions in Italy, Germany, Switzerland, Austria, France, and England. Vernadsky subsequently traveled widely in Europe and North America in order to become acquainted with areas of geological and mineralogical interest, to participate in international scientific congresses, and to study the organization of scientific research and higher education.

In the fall of 1890 Vernadsky became a *Privatdozent* in mineralogy and crystallography at Moscow University. In 1891 he defended his master's thesis, "O gruppe sillimanita i roli glinozema v silikatakh" ("On the Sillimanite Group and the Role of Alumina in Silicates") and, in 1897, his doctoral dissertation, "O yavleniakh skolzhenia kristallicheskogo veshchestva" ("On the Phenomena of Gliding in Crystal Substances"). Vernadsky was appointed professor at Moscow University in 1898. The St. Petersburg Academy of Sciences elected him associate member in 1909 and academician in 1912.

Vernadsky was among those professors and teachers who, in response to the reactionary treatment of the university by the Ministry of National Education, left Moscow in 1911 and moved to St. Petersburg. In 1914 he became director of the Geological and Mineralogical Museum of the Academy of Sciences and a year later joined the Commisssion to Study the Natural Productive Forces of Russia, serving as its president in 1915–1917 and in 1926–1930. From June 1917 to the beginning of 1921 Vernadsky lived in the Ukraine. He was responsible for the creation of the Ukrainian S.S.R. Academy of Sciences in 1919 and was its first president. Following his return to Leningrad from Paris and Prague in 1926, he continued his scientific and organizational activity at the Academy of Sciences. In 1926 he founded and headed the Commission on the History of Knowledge of the Academy of Sciences of the U.S.S.R. and, the following year, the Section on Living Substances (now the V. I. Vernadsky Institute of Geochemistry and Analytical Chemistry). At Ver-

nadsky's suggestion, at the Seventeenth International Geological Congress, held at Moscow in 1937, an international commission was created to determine the absolute age of geological rock by radioactive methods; he was elected vice-president of this commission.

In the first twenty-five years of his scientific career, Vernadsky was concerned with crystallography and, primarily, mineralogy, which he also studied later, when he turned to geochemistry, radiogeology, and biogeochemistry. Two important works devoted to the structure of crystalline substances were Vernadsky's doctoral dissertation (1897) and *Osnovy kristallografii* ("Fundamentals of Crystallography"; 1904), in which he developed the relation of crystal form to physicochemical structure and emphasized the importance of energetics in studying crystals.

Vernadsky opened a new, evolutionary direction in mineralogy. His research was presented in *Opyt opisatelnoy mineralogii* ("Experiment in Descriptive Mineralogy"; 1908–1922) and *Istoria mineralov zemnoy kory* ("History of Minerals of the Earth's Crust; 1923–1936). Defining mineralogy as the chemistry and history of the minerals in the earth's crust, he believed that "mineralogy, like chemistry, must study not only the products of chemical reactions but also the very processes of reaction" (*Izbrannye sochinenia*, II, 9)—and in the concept of mineral he included gases and water. In his works on mineralogy Vernadsky started from the premise that the purposes of mineralogy as a science are to establish the chemical composition of minerals; to explain the conditions necessary for chemical reactions involved in the genesis and paragenesis of minerals, and to study the conditions under which minerals change in the various zones of the earth.

Vernadsky's works stated the laws of paragenesis of minerals, the concept of which had been introduced into mineralogy in 1849 by J. F. A. Breithaupt. In *Paragenezis khimicheskikh elementov v zemnoy kore* ("Paragenesis of Chemical Elements in the Earth's Crust"; 1910), Vernadsky, starting from the capacity of isomorphic compounds to produce isomorphic mixtures—solid solutions—introduced the concept of natural isomorphic series of chemical elements. His table of elements in the earth's crust consisted of eighteen series, each of them related to definite thermodynamic zones. Vernadsky distinguished three such zones: the weathering crust, the area of low temperature and low pressure; the area of metamorphism, the area of high pressure and moderate temperature; and the deep layers of the lithosphere—the area of magmatization—the area of high temperature and high pressure. He emphasized that "isomorphic series are transformed and change under the influence of changes of temperature and pressure" (*Izbrannye sochinenia*, I, 404). Vernadsky's position was the basis for the development of the theory of paragenesis of elements and minerals, which was of great importance in the search for useful mineral deposits.

Vernadsky's great contribution to mineralogy was his research on silicates and aluminosilicate minerals, which constitute a major part of the earth's crust. The aluminosilicates had previously been considered to be salts of silicic acid and their acid properties to be attributable only to alumina. Vernadsky refuted this view in his master's dissertation and showed experimentally a different structure of aluminosilicates, according to which aluminum in the most important rock-forming minerals—feldspars and micas—is chemically analogous to silicon. He proposed the theory of the kaolin nucleus, composed of two atoms of aluminum, two of silicon, and seven of oxygen, and constituting the basis of many minerals. This theory has played an important role in explaining the structure, genesis, and classification of minerals. It was later confirmed by X-ray structural analysis, and it is now considered an established fact that silicon and aluminum in aluminosilicates are joined by atoms of oxygen placed at the points of tetrahedrons, which represent the framework of the aluminosilicates, the cavities of which are filled with large cations.

Vernadsky's basic works in geochemistry are of great importance. Geochemistry is a science of the twentieth century, although the term was introduced into the literature in 1838 by C. F. Schonbein. Defining geochemistry as a science, Vernadsky wrote:

> Geochemistry scientifically studies . . . the atoms of the earth's crust and, as much as possible, of all the planets. It studies their history, distribution, and motion in space-time, their genetic relationships on our planet. It is sharply distinguished from mineralogy, which studies the history of the earth in the same space and the same time only as the history of compounds of atoms—molecules and crystals [*Izbrannye sochinemia*, I, 14].

Vernadsky's geochemical research, which he conducted intensively from 1908 to 1910, was generalized in *Ocherki geokhimii* ("Sketches in Geochemistry"; 1927, 1934), which first appeared in French as *La géochimie* (Paris, 1924). Directly connected with this work is *Biosfera* ("Biosphere"; 1926). Examining the early period of geochemistry, Vernadsky showed the importance of the contributions of many scientists in different countries, including Robert Boyle. In his geochemical works he gave remarkable descriptions of many elements of the earth's crust. He prepared precise data on the chemical composition of the earth's crust to a depth of twenty kilometers and tabulated the chemical elements of this layer in weight percentages.

The history of the chemical elements in the earth's crust, Vernadsky showed, can be reduced to their migrations: the motion of atoms in the formation of compounds; their transformation into mobile liquids, gases, and solid bodies; and their assimilation into the respiration, nourishment, and metabolism of organisms.

These migrations in the crust of the earth create large systems of chemical equilibria or modes of occurrence of chemical elements. Vernadsky distinguished four such modes: molecules and compounds in minerals, rocks, liquids, and gases; elements in living organisms; elements in magmas, occurring under conditions of high pressure and temperature; and dispersed elements. He paid much attention to the geospheres (the earth's layers), between which the migration of chemical elements also occurs. Material on geospheres occupies a large portion of *Ocherki geokhimii* ("Sketches in Geochemistry").

Vernadsky considered inaccurate the widely accepted view that the earth's crust is the remnant of the first crust of a once-liquid molten mass. In studying it, he found it convenient to distinguish, in a simplified form, the following layers: the biosphere, the layer occupied by all living things; the stratisphere, the layer of sedimentary rock; the metamorphic layer; the granite layer; and the basalt layer. The layers have a close genetic relation to each other, and cyclical transfers of chemical elements occur among them. Vernadsky believed that these processes have formed "burial layers," including the granite, that have gone through the biosphere stage and thus should be called ex-biospheres.

Vernadsky classified the elements of Mendeleev's periodic system into six groups according to their role in the geochemical history of the earth's crust: noble gases, noble metals, cyclical elements, "scattered" elements, strongly radioactive elements, and rare earths. The cyclical elements, which constitute about 99.7 percent of the weight of the earth's crust, play the main role in the processes originating there.

Each element in a given geosphere enters into the compounds appropriate to it under given thermodynamic conditions. These compounds are broken down in the transition to another geosphere, where other compounds are formed. In this process, "after more or less prolonged and complex changes the element returns to the first compound and begins a new cycle, which is completed for the element by a new return to its primary condition" (*Izbrannye sochinenia*, I, 40). These cycles are partly reversible, however, and some of the atoms constantly leave the cycle. Vernadsky examined the sources of energy under the influence of which tectonic motions and transfers of substances in the earth's crust take place. On the basis of work by Joly and Strutt (Lord Rayleigh), as well as his own research, he considered the energy released by the radioactive decomposition of elements to be one of the main sources of these processes: "The heat effect of radioactive decomposition has been so substantial that it allows us to discard the hypothesis of the once molten planet and provides a basis for a new scientific consideration of atomic heating of the substance of the planet, sharply distinct in its various localities" (*Izbrannye sochinenia*, I, 225).

Vernadsky was one of the first to recognize radioactivity as a powerful source of energy. When even physicists did not clearly recognize its practical significance, he pointed out the responsibility of scientists for the consequences of the use of their discovery. He wrote in 1922:

> We are approaching a great revolution in the life of humanity, with which nothing . . . earlier . . . can be compared. The time is not far away when man will take atomic energy into his hands. . . . This can occur in the near future; it may happen after a century. But it is clear that it will inevitably happen. Does man know how to use this power, to direct it to good and not to self-destruction? Has he . . . the ability to use this force, which science will inevitably give him? Scientists must not close their eyes to the possible consequences of their . . . work, of . . . progress. They must consider themselves responsible for the consequences of their discoveries. They must relate their work to the best organization of all humanity [*Ocherki i rechi*, II, foreword].

In 1910 Vernadsky and his colleagues began to seek radioactive minerals and to study them in the laboratory. Greatly concerned with radioactivity, Vernadsky later laid the foundations of a new science, radiogeology.

In the last twenty years of his scientific career Vernadsky and his colleagues, especially A. P. Vinogradov, concentrated on the chemical composition of plants and animals. At the same time he clarified the role of living organisms in reactions and transformations of chemical elements in the biosphere. He introduced into geochemistry the concept of living matter as the totality of living organisms expressed in terms of weight, chemical composition, and energy. Such an approach to the study of living matter allowed Vernadsky to express in mathematical form certain regularities of the multiplication of organisms. He showed the primary importance of living organisms as accumulators, transformers, and carriers of solar radiation—the second powerful source of energy in geochemical cycles.

The analysis of biogeochemical processes led Vernadsky to conclude that the main gases of the earth's atmosphere—oxygen, nitrogen, and carbon dioxide—are created by living things. He indicated the immediate role of living matter in the concentration of many chemical elements in the earth's crust, especially carbon, silicon, calcium, nitrogen, iron, and manganese. Living matter, he asserted, influences the entire chemistry of the earth's crust and determines the history of almost every element in it. He also considered the possibility that the quantity of living matter on the earth apparently has been constant throughout geological time. There is another possibility, however: that the mass of living matter grows over geological time. Vernadsky is thus considered the founder of the theory of the biosphere and of a new area of geochemistry: biogeochemistry.

Vernadsky gave great attention to the hydrosphere. His research in this field is presented in *Istoria prirodnykh vod* ("History of the Waters of Nature"; 1931), in which he treats the mineralogy of water and explains the relation between water and the solid crust of the earth. In the first volume of *Opyt opisatelnoy mineralogii* ("Experiment in Descriptive Mineralogy"; 1908–1914), and in later works Vernadsky showed the substantial effect of human activity on the history of the earth. As a result of this activity the face of the planet is increasingly changing as are the chemical properties of its surface.

BIBLIOGRAPHY

I. Original Works. Vernadsky's writings include "O gruppe silliamanita i roli glinozema v silikatakh" ("On the Sillimanite Group and the Role of Alumina in

Silicates"), in *Byulleten Moskovskogo obshchestva ispytatelei prirody*, **5** (1891), 1–100, 165–169; *Ocherki i rechi* ("Sketches and Speeches"), 2 vols. (Petrograd, 1922); *Biosfera* ("The Biosphere"; Leningrad, 1926; Moscow, 1967); *Biogeokhimicheskie ocherki. 1922–1932 gg.* ("Biogeochemical Sketches . . ."; Moscow–Leningrad, 1940); *Izbrannye sochinenia* ("Selected Works"), 6 vols. (Moscow, 1954–1960); and *Khimicheskoe stroenie biosfery Zemli i ee okryzhenia* ("The Chemical Structure of the Earth and Its Environs"; Moscow, 1965). A work still in MS is "Nauchnaya mysl kak planetnoe yavelnie" ("Scientific Thought as a Planetary Phenomenon"), Moscow, Archives of the Academy of Sciences, fond 518.

II. Secondary Literature. See A. E. Fersman, *Zhiznenny put akademika Vladimira Ivanovicha Vernadskogo* ("The Career of Academician . . . Vernadsky"; Moscow, 1946); B. L. Lichkov, *Vladimir Ivanovich Vernadsky. 1863–1945* (Moscow, 1948), with bibliography on 83–102; *Materialy k biobibliografii uchenykh SSSR* ("Materials for a Biobibliography of Scientists of the U.S.S.R."), Chem. ser., no. 6 (Moscow–Leningrad, 1947), devoted to Vernadsky; K. V. Vlasov, "Vladimir Ivanovich Vernadsky," in *Lyudi russkoy nauki* ("People of Russian Science"; Moscow, 1962), 135–157; and A. P. Vinogradov, *Vladimir Ivanovich Vernadsky* (Moscow–Leningrad, 1947).

I. A. Fedoseyev

VERNEUIL, PHILIPPE ÉDOUARD POULLETIER DE (*b.* Paris, France, 13 February 1805; *d.* Paris, 29 May 1873), *geology, paleontology.*

Édouard de Verneuil, the son of Antoine César Poulletier de Verneuil and Genevieve Pauline Flore Laurens de l'Ormeon, prepared for the law, received a license, and in 1833 entered the employment of the French Ministry of Justice. He studied the geological lessons of Élie de Beaumont and then in 1835, being of independent means, devoted himself to science.

Verneuil's scientific career was concerned with describing the geological formations and fossil groups of the Paleozoic era. His extensive knowledge of these fossils enabled him to make his major theoretical contribution to stratigraphy—the hypothesis that the Paleozoic deposits of the United States and Spain are parallel to those of Europe.

Verneuil made his principal researches in five areas in Europe and North America. From 1835 to 1838 he traveled in Europe from England to the Crimea; in 1840 and 1841 he toured European Russia with the English geologist Roderick Murchison; in 1846 he briefly visited the United States; from 1849 to 1862 he traveled almost annually in Spain; and in the years after 1864, he studied the activity of Mount Vesuvius. He made

his first tour in 1835 to Wales. His objective was to study the formations that had recently led Murchison and his colleague Adam Sedgwick to break down Abraham Werner's Transitional Formations (specifically, the graywacke) into the Silurian and Cambrian systems. Verneuil's observations led to his support for an extension of the classification. In the summer of 1836, he traveled along the Danube to Turkey and the Crimea, neither of which had received extensive geological examination. He was primarily interested in the prominent Tertiary terrains, but he described the Silurian system around Constantinople.

Verneuil's adoption of Murchison's theories, his defense of them before the Geological Society of France, and his knowledge of the fossils of the older strata inevitably brought him into personal contact with the English scientist. In 1839 Murchison and Sedgwick were in northern Germany, Belgium, and northern France to gather evidence for a new system—the Devonian—that they had described in England. Verneuil, who had earlier followed the Silurian system in northern France, was invited to join their French tour. Verneuil charmed Murchison with his sophistication and his piano playing, and their professional acquaintance became a lifelong friendship.

The geological tours and perhaps personal contact with Murchison stimulated the beginning of Verneuil's theoretical contribution to geology. In 1840 Verneuil suggested that the Silurian system, which had then been observed in England, Europe, and Turkey, was universal in extent. This suggestion was the germ of his generalization of the uniformity of Paleozoic laws. Proving the generalization provided the research for his mature scientific career.

Verneuil's first opportunity to verify the universality of the new Paleozoic systems came later in 1840. Murchison desired to travel in Russia, where he expected he could easily observe the Silurian, Devonian, and Carboniferous layers; and he asked Verneuil to accompany him. Tours were made in the summer of 1840 and 1841. Not only did the two geologists—who were intermittently joined by the Russian naturalist Alexandr Keyserling—verify Murchison's systems for European Russia, but on the second tour Murchison discovered, and Verneuil made the fossil analysis of, a new formation in Perm, hence called the Permian system. These strata, which paralleled the Zechstein of Germany and Magnesian Limestone of England, were proved by Verneuil to constitute the upper limit of the recently defined Paleozoic era. Verneuil was quickly able to demonstrate the same fossil group in western Europe and thus the existence of the Permian system there as well.

Upon the publication of the results of the Russian expeditions in the two-volume *Geology of Russia in Europe and the Ural Mountains*, with Murchison and Keyserling (London–Paris, 1845), in which he was chiefly responsible for the volume on paleontology, Verneuil could turn his attention to verifying the Paleozoic order elsewhere. The outstanding problem, as he had recognized in 1840, was to bring the geological structure of the Americas into correspondence with that of Europe. In 1846 he journeyed to the United States primarily to study formations in New York and Ohio, where American geologists had already described the strata in detail, although in local perspectives. He sought to answer two questions: Did the fossil species in America and Europe present themselves in the same order? And was it possible to trace between the American Paleozoic stages the divisions established in Europe under the names of the Silurian, Devonian, and Carboniferous systems? Verneuil's research immediately provided affirmative answers. He announced in his important essay "Note sur le parallélisme des roches des dépôts paléozoïques de l'Amerique septentrionale avec ceux de l'Europe" (in *Bulletin de la Société géologique de France*, 2 [1846–1847], 646–709) that the Paleozoic deposits of the two continents are parallel and that the distribution of fossils in each had been made according to a common law.

In his remaining career, Verneuil was largely concerned with the geology of Spain. Spanish stratigraphy had been brought to his attention as early as January 1844 by a colleague who sent fossils to him for identification. Beginning in 1849, Verneuil traveled often to the Iberian Peninsula and was able to prove that its Paleozoic order corresponded to that east of the Pyrenees. In 1858, and frequently after 1864, Verneuil studied Mount Vesuvius.

Verneuil was elected president of the Geological Society of France three times, in 1840, 1853, and 1867. In 1854 he was elected to the Académie des Sciences, and he was a foreign associate of the Royal Society of London and the academies of science of St. Petersburg and Berlin.

BIBLIOGRAPHY

I. ORIGINAL WORKS. A list of the principal publications is given in Daubrée, "Notice nécrologie sur Édouard de Verneuil," in *Bulletin de la Société géologique de France*, **3**, ser. 3 (1875–1876), 317–328. A complete list of articles is in the Royal Society *Catalogue of Scientific Papers (1800–1900)*, VI, VIII. Scattered pieces of correspondence are listed in Ministère de l'éducation nationale, *Catalogue générale des manuscrits des bibliothèques publiques en France* (Paris, 1885–), VIII, XXXIX, XLIX, LV; and the *Catalogue of Manuscripts in the American Philosophical Society Library* (Westport, Conn., n.d.), IX.

Verneuil's fossil collection, left to the École des Mines, Paris, is described by J. Barrande, "Collection paléontologique de M. Edouard de Verneuil," in *Annales des mines, mémoires*, **7**, ser. 4 (1873), 327–338.

II. SECONDARY LITERATURE. The best biographical memoir, from which most other such memoirs are derived, is Daubrée, "Discours sur M. Édouard de Verneuil," in *Annales des mines, mémoires*, **7**, ser. 4 (1873), 318–326. Other memoirs are listed in the Royal Society *Catalogue of Scientific Papers (1800–1900)*, VIII, XI. Verneuil's relationship with Murchison is discussed in Archibald Geikie, *Life of Sir Roderick I. Murchison*, 2 vols. (London, 1875).

RONALD C. TOBEY

VERNIER, PIERRE (*b.* Ornans, France, 19 August 1584; *d.* Ornans, 14 September 1638), *military engineering, scientific instrumentation.*

From his father, Jean, who was castellan of the château of Ornans, a lawyer by training, and probably an engineer, Vernier inherited an interest in all sorts of measuring instruments. At an early age he studied the writings of contemporary scientists and learned how measuring instruments work. After becoming adept at using them, Vernier worked as a military engineer for the Spanish Hapsburgs, then rulers of Franche-Comté. His reading of the works of Nuñez Salaciense (Nonius), Clavius, and Tycho Brahe, combined with his experience in helping his father survey and prepare an excellent map of Franche-Comté, led him to seek a new way to read off angles on surveying instruments.

Around 1540 Nuñez, a Portugese mathematician who was trying to improve the accuracy of the astrolabe, a sixteenth-century preoccupation, hit upon the idea of engraving on the face of the astrolabe a series of scales laid out along concentric circles. The scale on any circle was determined by dividing the circumference of the circle into an equal number of parts, one less than that dividing the next circle out and one greater than that dividing the next circle in. Thus in shooting a star, the line of sight would inevitably fall very close to a whole division on one of the scales. By an elementary calculation or by use of a table, one could easily determine the number of degrees, minutes, and seconds of an angle being measured. In theory Nuñez' method could bring great accuracy, but it was extremely difficult to engrave with precision a different scale on each concentric circle. Brahe remarked that in practice Nuñez' method failed to live up to its promise (*Astronomiae instauratae mechanica* [Nuremberg, 1611], p. 2).

About fifty years later Clavius, who had studied under Nuñez, found a way to facilitate the engraving of the various scales on the concentric circular segments. His associate Jacobus Curtius further simplified Nuñez' method by placing the same scale on the concentric segments in such a way that the zero of the scale on any segment started one degree away from that of the scale on the preceding segment. It thus became possible to read off degrees and minutes directly if the outer scale were divided into degrees and if there were sixty concentric circular segments. These methods were described by Clavius in his *Geometria practica* (Rome, 1604), which Vernier surely read and meditated on.

Vernier replaced the fifty-nine inner concentric circular segments with a mobile concentric segment, thereby giving a mobile scale rather than a series of static ones. Thus he solved the difficulty of engraving many different concentric scales. By this time (1630) he had acquired a reputation as an excellent engineer and was *conseiller et général des monnaies* for the count of Burgundy. He made a special trip to Brussels to present his invention to Isabelle-Claire-Eugénie, the infanta of Spain, the ruler of Franche-Comté, who had him publish a description of it. Throughout the seventeenth century this work remained on the whole unknown to the European scientific community. Certainly the declining importance of the Spanish Hapsburgs within Western Europe did not facilitate diffusion of Vernier's treatise nor did the development of technology favor the vernier because the extra precision it brought could not make seventeenth-century instruments more accurate than was allowed by open sights, which were commonly used, and by the imprecise methods of marking scales. Indeed, at the start of the eighteenth century, as soon as the technological situation became propitious, the vernier began to be used; but Vernier's name did not become associated with his invention until around the middle of the century.

After publishing his treatise in 1631, Vernier returned to Dôle, where he designed and directed the construction of fortifications. His other engineering projects included the design of a building for the harquebusiers of Dôle. In 1636 illness forced him to discontinue the practice of engineering; and he returned to Ornans, where he died a few years later.

BIBLIOGRAPHY

I. ORIGINAL WORKS. Vernier's writings are *La construction, l'usage, et les propriétez du quadrant nouveau de mathématique: Comme aussi la construction de la table des sinus . . .* (Brussels, 1631); "Traité d'artillerie" (MS at the Bibliothèque Nationale, Paris, according to Michel, p. 349); a map prepared by Vernier's father with the aid of Vernier is described in Claude François Rolland, "Études sur la cartographie ancienne de la Franche-Comté . . .," in *Mémoires de la Société d'émulation du Doubs*, 8th ser., **7** (1912), 187–299; and **8** (1913), 375–429 (see 404–421); the plan of a mill that Feuvrier attributes to Vernier in his study on him (see below) figures in J. Feuvrier, "Les derniers moulins à bras et à chevaux en Franche-Comté," in *Procès-verbaux et mémoires de l'Académie des sciences, belles-lettres et arts de Besançon* (1909), 125–131; that of the building for the harquebusiers that he also attributes to Vernier is in J. Feuvrier, "Les chevaliers du noble et hardy jeu de l'arquebuse de la ville de Dôle," in *Mémoires de la Société d'émulation du Jura*, 6th ser., **2** (1897), 1–70, plan on plate between 24–25, discussion on 27–28.

II. SECONDARY LITERATURE. The Archives Départementales du Doubs confirmed the correctness of Vernier's dates of birth and death. Henri Michel, "Le 'vernier' et son inventeur Pierre Vernier d'Ornans," in *Mémoires de la Société d'émulation du Doubs*, 8th ser., **8** (1913), 310–373, gives a comprehensive history of the vernier and the salient features of Vernier's life; Julien Feuvrier, "L'ingénieur Pierre Vernier à Dôle," in *Procès-verbaux et mémoires de l'Académie des sciences, belles-lettres et arts de Besançon* (1912), 293–302, which contains a facsimile of a handwritten report by Vernier, adds to Michel details concerning Vernier's family and his activity as an engineer in Dôle. Wilhelm Lührs, "Ein Beitrag zur Geschichte der Transversalteilungen und des 'Nonius,'" in *Zeitschrift für Vermessungswesen*, **39** (1910), 177–191, 209–223, 241–254, gives details of the background to the vernier; A. Breusing, "Nonius oder Vernier?" in *Astronomische Nachrichten*, **96** (1879), 131–134, is superseded by the above studies; Maurice Daumas, *Les instruments scientifiques aux XVIIe et XVIIIe siècles* (Paris, 1953), 250–255, briefly discusses the history of the vernier; and J. B. J. Delambre, *Histoire de l'astronomie moderne*, II (Paris, 1821), 119–125, gives extensive extracts from Vernier's treatise on the vernier.

ROBERT M. McKEON

VERONESE, GIUSEPPE (*b*. Chioggia, Italy, 7 May 1854; *d*. Padua, Italy, 17 July 1917), *mathematics*.

Giuseppe Veronese, professor of geometry at the University of Padua from 1881 until his death, was one of the foremost Italian mathematicians of his time. He took part also in political life, first as a member of Parliament for Chioggia (1897–1900), then as a member of the City Council of Padua, and finally as a senator (1904–1917).

Veronese's father was a house painter in Chioggia, then a small fishing town not far from Venice; his mother, Ottavia Duse, was a cousin of the celebrated actress Eleonora Duse. In 1885 Veronese married the Baroness Beatrice Bartolini; they had five children. Veronese was a handsome man, tall and commanding, but in his last years his health was undermined by influenza, which he had contracted in 1912 and which left him with grave cardiovascular disorders.

Because of his parents' poverty, Veronese had to interrupt his studies when he was eighteen and take a minor job in Vienna; but through the generosity of Count Nicolò Papadopoli he was able to resume his studies a year later, first at the Zurich Polytechnic, where he studied engineering and mathematics under Wilhelm Fiedler, and later, following a correspondence with Luigi Cremona, at the University of Rome, from which he graduated in 1877. In the previous year he had become assistant in analytical geometry, an unheard-of distinction for an undergraduate, after demonstrating his exceptional abilities in a paper on Pascal's hexagram, a work he had begun at Zurich. In 1880–1881 Veronese did postgraduate study at Leipzig. Immediately afterward, he won the competition for the professorship of complementary algebra and algebraic geometry at the University of Padua, where he succeeded Giusto Bellavitis. The latter had shown personal liking for Veronese but was fiercely opposed to the new approaches to geometry, which Veronese supported.

Veronese published only about thirty papers; but some of them, also available in German, were extremely important in the history of geometry. In particular he may be considered the main founder of the projective geometry of hyperspaces with n dimensions, which had previously been linear algebra presented geometrically, rather than geometry. Hyperspaces began to assume a more truly geometrical aspect when Veronese used an original recursion method to produce them: a plane can be obtained by projecting the points of a straight line from a point outside it, and a three-dimensional space by projecting the points of a plane from a point outside it, and so on. He is also remembered for "Veronese's surface," a two-dimensional surface of a five-dimensional space, which in its simplest expression can be represented by the parametric equations $x_1 = u^2$, $x_2 = uv$, $x_3 = v^2$, $x_4 = u$, $x_5 = v$, where $x_1, \ldots, x_5$ are the nonhomogeneous coordinates of the space and u and v are two independent parameters. The study of this surface is equivalent, from the point of view of projection, to the study of all the conics of a plane; and one of its projections in ordinary space is Steiner's Roman surface.

Veronese was also one of the first to study non-Archimedean geometry, at first arousing strong opposition, and he demonstrated the independence of Archimedes' postulate—which states that among the multiples of a given magnitude there is always one greater than every fixed magnitude—from the other postulates of geometry. Veronese also wrote useful books for the secondary schools.

Veronese was a member of the Accademia Nazionale dei Lincei and of other Italian academies; and his pupils included Guido Castelnuovo and Tullio Levi-Civita.

When a member of Parliament, Veronese campaigned strenuously for the conservation of the Lagoon of Venice.

BIBLIOGRAPHY

An obituary is C. Segre, "Commemorazione del socio nazionale Giuseppe Veronese," in *Atti dell' Accademia nazionale dei Lincei. Rendiconti*, 5th ser., **26**, pt. 2 (1917), 249–258, with a bibliography of Veronese's publications.

F. G. Tricomi

DICTIONARY
OF
SCIENTIFIC BIOGRAPHY

PUBLISHED UNDER THE AUSPICES OF
THE AMERICAN COUNCIL OF LEARNED SOCIETIES

The American Council of Learned Societies, organized in 1919 for the purpose of advancing the study of the humanities and of the humanistic aspects of the social sciences, is a nonprofit federation comprising forty-two national scholarly groups. The Council represents the humanities in the United States in the International Union of Academies, provides fellowships and grants-in-aid, supports research-and-planning conferences and symposia, and sponsors special projects and scholarly publications.

MEMBER ORGANIZATIONS

AMERICAN PHILOSOPHICAL SOCIETY, 1743
AMERICAN ACADEMY OF ARTS AND SCIENCES, 1780
AMERICAN ANTIQUARIAN SOCIETY, 1812
AMERICAN ORIENTAL SOCIETY, 1842
AMERICAN NUMISMATIC SOCIETY, 1858
AMERICAN PHILOLOGICAL ASSOCIATION, 1869
ARCHAEOLOGICAL INSTITUTE OF AMERICA, 1879
SOCIETY OF BIBLICAL LITERATURE, 1880
MODERN LANGUAGE ASSOCIATION OF AMERICA, 1883
AMERICAN HISTORICAL ASSOCIATION, 1884
AMERICAN ECONOMIC ASSOCIATION, 1885
AMERICAN FOLKLORE SOCIETY, 1888
AMERICAN DIALECT SOCIETY, 1889
AMERICAN PSYCHOLOGICAL ASSOCIATION, 1892
ASSOCIATION OF AMERICAN LAW SCHOOLS, 1900
AMERICAN PHILOSOPHICAL ASSOCIATION, 1901
AMERICAN ANTHROPOLOGICAL ASSOCIATION, 1902
AMERICAN POLITICAL SCIENCE ASSOCIATION, 1903
BIBLIOGRAPHICAL SOCIETY OF AMERICA, 1904
ASSOCIATION OF AMERICAN GEOGRAPHERS, 1904
HISPANIC SOCIETY OF AMERICA, 1904
AMERICAN SOCIOLOGICAL ASSOCIATION, 1905
AMERICAN SOCIETY OF INTERNATIONAL LAW, 1906
ORGANIZATION OF AMERICAN HISTORIANS, 1907
COLLEGE ART ASSOCIATION OF AMERICA, 1912
HISTORY OF SCIENCE SOCIETY, 1924
LINGUISTIC SOCIETY OF AMERICA, 1924
MEDIAEVAL ACADEMY OF AMERICA, 1925
AMERICAN MUSICOLOGICAL SOCIETY, 1934
SOCIETY OF ARCHITECTURAL HISTORIANS, 1940
ECONOMIC HISTORY ASSOCIATION, 1940
ASSOCIATION FOR ASIAN STUDIES, 1941
AMERICAN SOCIETY FOR AESTHETICS, 1942
METAPHYSICAL SOCIETY OF AMERICA, 1950
AMERICAN STUDIES ASSOCIATION, 1950
RENAISSANCE SOCIETY OF AMERICA, 1954
SOCIETY FOR ETHNOMUSICOLOGY, 1955
AMERICAN SOCIETY FOR LEGAL HISTORY, 1956
AMERICAN SOCIETY FOR THEATRE RESEARCH, 1956
SOCIETY FOR THE HISTORY OF TECHNOLOGY, 1958
AMERICAN COMPARATIVE LITERATURE ASSOCIATION, 1960
AMERICAN SOCIETY FOR EIGHTEENTH-CENTURY STUDIES, 1969

DICTIONARY

OF

SCIENTIFIC BIOGRAPHY

CHARLES COULSTON GILLISPIE

Princeton University

EDITOR IN CHIEF

Volume 14

ADDISON EMERY VERRILL – JOHANN ZWELFER

CHARLES SCRIBNER'S SONS · NEW YORK

Panel of Consultants

Contributors to Volume 14

The following are the contributors to Volume 14.   Each author's name is followed by the institutional affiliation at the time of publication and the names of the articles written for this volume. The symbol † means that an author is deceased.

HANS AARSLEFF
Princeton University
V. WEIGEL; WILKINS

S. MAQBUL AHMAD
Aligarh Muslim University
YĀQŪT AL-ḤAMAWĪ

LUIS DE ALBUQUERQUE
University of Coimbra
ZACUTO

GARLAND E. ALLEN
Washington University
E. B. WILSON

G. C. AMSTUTZ
University of Heidelberg
ZIRKEL

HENRY N. ANDREWS
University of Connecticut
W. C. WILLIAMSON

WILBUR APPLEBAUM
Illinois Institute of Technology
WING

WILLIAM A. BARKER
University of Santa Clara
A. N. WHITEHEAD

I. G. BASHMAKOVA
Academy of Sciences of the U.S.S.R.
VORONOY; ZOLOTAREV

DONALD G. BATES
McGill University
J. R. YOUNG

HANS BAUMGÄRTEL
J. C. W. VOIGT

JOHN J. BEER
University of Delaware
WEIZMANN

WHITFIELD J. BELL, JR.
American Philosophical Society
WISTAR

OTTO THEODOR BENFEY
Guilford College
WASHBURN

RICHARD BERENDZEN
The American University, Washington, D.C.
C. A. YOUNG

ALEX BERMAN
University of Cincinnati
VIREY

RICHARD BIEBL†
WIESNER

KURT R. BIERMANN
Akademie der Wissenschaften der DDR
WEIERSTRASS

ASIT K. BISWAS
Department of Environment, Ottawa
WILD

MARGARET R. BISWAS
Department of Environment, Ottawa
WILD

L. J. BLACHER
Academy of Sciences of the U.S.S.R.
ZAVADOVSKY; ZAVARZIN

MAX BLACK
Cornell University
WITTGENSTEIN

UNO BOKLUND
Royal Pharmaceutical Institute, Stockholm
WALLERIUS

GERT H. BRIEGER
University of California, San Francisco
WELCH

T. A. A. BROADBENT†
W. WALLACE; J. WILSON

W. H. BROCK
University of Leicester
WANKLYN; A. W. WILLIAMSON

B. A. BRODY
Rice University
WHATELY

JOHN HEDLEY BROOKE
University of Lancaster
WURTZ

STEPHEN G. BRUSH
University of Maryland
WATERSTON

GERD BUCHDAHL
Whipple Science Museum
C. WOLFF

JED Z. BUCHWALD
University of Toronto
VILLARI

K. E. BULLEN
University of Sydney
WEGENER; WIECHERT

IVOR BULMER-THOMAS
ZENODORUS

WERNER BURAU
University of Hamburg
WANGERIN; WEINGARTEN

J. J. BURCKHARDT
University of Zurich
J. R. WOLF

DEAN BURK
National Cancer Institute
O. H. WARBURG

J. C. BURKILL
University of Cambridge
W. H. YOUNG

JOHN C. BURNHAM
Ohio State University
YERKES

H. L. L. BUSARD
State University of Leiden
VIÈTE

ROBERT E. BUTTS
University of Western Ontario
WHEWELL

G. V. BYKOV
Academy of Sciences of the U.S.S.R.
ZININ

JEROME J. BYLEBYL
University of Chicago
WILLDENOW

WILLIAM F. BYNUM
University College London
WERNICKE; WITHERING; ZINSSER

W. A. CAMPBELL
University of Newcastle Upon Tyne
C. WINKLER

LUIGI CAMPEDELLI
University of Florence
ZUCCHI

ROBERT CANTWELL
A. WILSON

GUIDO CAROBBI
University of Florence
ZAMBONINI

CARLO CASTELLANI
University of Parma
ZAMBECCARI

JOHN CHALLINOR
University College of Wales
T. WEBSTER; WHITEHURST

ALLAN CHAPMAN
University of Oxford
G. WHARTON

H. CHIARI†
WEICHSELBAUM

CONTRIBUTORS TO VOLUME 14

EDWIN CLARKE
Wellcome Institute for the History of Medicine
C. WHITE

ALBERT B. COSTA
Duquesne University
WALTER; WISLICENUS

RUTH SCHWARTZ COWAN
State University of New York at Stony Brook
WELDON

GLYN DANIEL
University of Cambridge
WOOLLEY; WORM; WORSAAE

KAREL L. DE BOUVÈRE, S.C.J.
University of Santa Clara
A. N. WHITEHEAD

ALLEN G. DEBUS
University of Chicago
J. WEBSTER

SOLOMON DIAMOND
California State University, Los Angeles
WUNDT

J. DIEUDONNÉ
VON NEUMANN; WEYL

SALLY H. DIEKE
Johns Hopkins University
W. H. WRIGHT

WILLIAM DOCK
State University of New York, Brooklyn
W. C. WELLS

HAROLD DORN
Stevens Institute of Technology
WATT; WEDGWOOD

H. DÖRRIE
University of Münster/Westphalia
XENOCRATES OF CHALCEDON

SIGALIA DOSTROVSKY
Barnard College
VILLARD; VIOLLE; WHEATSTONE

A. HUNTER DUPREE
Brown University
WYMAN

JOY B. EASTON
West Virginia University
WITT

FRANK N. EGERTON III
University of Wisconsin-Parkside
H. C. WATSON

CHURCHILL EISENHART
U. S. Department of Commerce, National Bureau of Standards
WILKS; YOUDEN

VĚRA EISNEROVÁ
ZALUŽANSKÝ

GUNNAR ERIKSSON
Umeå University
WAHLENBERG

V. A. EYLES
J. WOODWARD

VERA N. FEDCHINA
Academy of Sciences of the U.S.S.R.
VYSOTSKY

A. S. FEDOROV
Academy of Sciences of the U.S.S.R.
VIZE

I. A. FEDOSEEV
Academy of Sciences of the U.S.S.R.
VOEYKOV

LUCIENNE FÉLIX
University of Paris
VESSIOT

JAMES W. FELT, S.J.
University of Santa Clara
A. N. WHITEHEAD

KONRADIN FERRARI D'OCCHIEPPO
University of Vienna
E. WEISS

MENSO FOLKERTS
Technische Universität Berlin
J. WERNER

DEAN R. FOWLER
Center for Process Studies, Claremont, Calif.
A. N. WHITEHEAD

ROBERT G. FRANK, JR.
University of California, Los Angeles
T. WILLIS

V. A. FRANK-KAMENETSKY
Leningrad State University
WULFF

FRITZ FRAUNBERGER
O. WIENER

ARTHUR H. FRAZIER
U.S. Geological Survey
WOLTMAN

H.-CHRIST. FREIESLEBEN
M. F. J. C. WOLF

HANS FREUDENTHAL
State University of Utrecht
N. WIENER

B. VON FREYBERG
University of Erlangen-Nuremburg
WALCH

KURT VON FRITZ
University of Munich
ZENO OF ELEA; ZENO OF SIDON

JOSEPH S. FRUTON
Yale University
WILLSTÄTTER

DAVID J. FURLEY
Princeton University
ZENO OF CITIUM

JEAN-CLAUDE GALL
Université Louis Pasteur, Strasbourg
VOLTZ

CHARLES C. GILLISPIE
Princeton University
VOLTAIRE

PAUL GLEES
University of Göttingen
WALDEYER-HARTZ; WIEDERSHEIM

EDWARD D. GOLDBERG
Scripps Institution of Oceanography
WASHINGTON

STANLEY GOLDBERG
Hampshire College
W. VOIGT

D. C. GOODMAN
Open University
W. H. WOLLASTON

I. DE GRAAF BIERBRAUWER-WÜRTZ
L. W. WINKLER

EDWARD GRANT
University of Indiana
VIVES

JOHN C. GREENE
University of Connecticut
VOLNEY

NORMAN T. GRIDGEMAN
National Research Council of Canada
E. B. WILSON

A. T. GRIGORIAN
Academy of Sciences of the U.S.S.R.
VYSHNEGRADSKY; ZHUKOVSKY; V. P. ZUBOV

N. A. GRIGORIAN
Academy of Sciences of the U.S.S.R.
VVEDENSKY

M. D. GRMEK
Archives Internationales d'Histoire des Sciences
VIEUSSENS

ERIC W. GROVES
British Museum (Natural History)
G. WHITE

V. GUTINA
Academy of Sciences of the U.S.S.R.
VINOGRADSKY

KARLHEINZ HAAS
ZEUTHEN

H. R. HAHNLOSER†
VILLARD DE HONNECOURT

A. RUPERT HALL
Imperial College, London
VIGANI

SAMI K. HAMARNEH
Smithsonian Institution
IBN WAḤSHIYYA; AL-ZAHRĀWĪ; IBN ZUHR

RICHARD HART
National Academy of Sciences
C. A. YOUNG

J. L. HEILBRON
University of California, Berkeley
VOLTA; W. WATSON; WILCKE

DIETER B. HERRMANN
Archenhold Observatory
VOGEL; WILHELM IV; ZOLLNER

ERICH HINTZSCHE†
VESLING

HELMUT HÖLDER
University of Münster/Westphalia
ZITTEL

FREDERIC L. HOLMES
University of Western Ontario
VOIT

WILLIAM T. HOLSER
University of Oregon
C. S. WEISS

HO PENG YOKE
Griffith University
YANG HUI

MICHAEL A. HOSKIN
University of Cambridge
T. WRIGHT

PIERRE HUARD
René Descartes University
VICQ D'AZYR

KARL HUFBAUER
University of California, Irvine
WEDEL; C. E. WEIGEL

AARON J. IHDE
University of Wisconsin-Madison
WILEY; R. R. WILLIAMS

M. J. IMBAULT-HUART
René Descartes University
VICQ D'AZYR

DANIEL JONES
Oregon State University
WIELAND

PHILLIP S. JONES
University of Michigan
WESSEL

HANS KANGRO
University of Hamburg
WEHNELT; WIEN

GEORGE B. KAUFFMAN
California State University, Fresno
WAAGE; A. WERNER

ROBIN KEEN
Gillingham Technical High School
F. WÖHLER

MILTON KERKER
Clarkson College of Technology
ZSIGMONDY

GUNTHER KERSTEIN
WIEGLEB

C. G. KING
Columbia University
WU

DAVID A. KING
*Smithsonian Institution Project in Medieval
Islamic Astronomy, Cairo*
IBN YŪNUS

LAWRENCE J. KING
State University of New York at Geneseo
A. WALKER

GEORGE KISH
University of Michigan
WALDSEEMÜLLER

MARC KLEIN†
WICKERSHEIMER

BRONISLAW KNASTER
ZARANKIEWICZ

ELAINE KOPPELMAN
Goucher College
WOODHOUSE

SHELDON J. KOPPERL
Grand Valley State Colleges
WILHELMY

HANS-GÜNTHER KÖRBER
*Zentralbibliothek des Meteorologischen
Dienstes der DDR, Potsdam*
WIEDEMANN; L. C. WIENER; WRÓBLEWSKI

VLADISLAV KRUTA
Purkyně University
WAGNER; E. H. WEBER

WILLIAM LEFANU
Royal College of Surgeons
T. WHARTON

HENRY M. LEICESTER
University of the Pacific
VIRTANEN; WALLACH; WINDAUS

ERNA LESKY
University of Vienna
WAGNER VON JAUREGG; WERTHEIM

JACQUES R. LÈVY
Paris Observatory
C. J. E. WOLF

DAVID C. LINDBERG
University of Wisconsin-Madison
WITELO

STEN LINDROTH
University of Uppsala
WARGENTIN

R. B. LINDSAY
Brown University
R. W. WOOD

ALBERT G. LONG
*Hancock Museum, University of Newcastle
Upon Tyne*
WITHAM

ESMOND R. LONG
University of Pennsylvania
H. G. WELLS

MARVIN W. MCFARLAND
Library of Congress
WRIGHT BROTHERS

LUDOLF VON MACKENSON
Astronomisch-Physikalisches Kabinett
A. WÖHLER

H. LEWIS MCKINNEY
University of Kansas
A. R. WALLACE

SAUNDERS MAC LANE
University of Chicago
E. B. WILSON

DANIEL MARTIN
University of Glasgow
WHITTAKER

ERNST MAYR
Harvard University
WHITMAN

OTTO MAYR
Smithsonian Institution
ZEUNER

DANIEL MERRIMAN
Yale University
WALLICH

WILLIAM J. MORISON
University of Louisville
G. F. WRIGHT

EDGAR W. MORSE
California State College, Sonoma
T. YOUNG

D. MÜLLER
University of Copenhagen
WARMING

LETTIE S. MULTHAUF
ZACH

R. P. MULTHAUF
Smithsonian Institution
ZWELFER

GERALD D. NASH
University of New Mexico
WHITNEY

HENRY NATHAN
WEDDERBURN

A. NATUCCI†
VIVIANI

CLIFFORD M. NELSON
University of California, Berkeley
WHITFIELD

CONTRIBUTORS TO VOLUME 14

J. D. NORTH
University of Oxford
J. H. C. WHITEHEAD; F. WOLLASTON;
YULE

CHRISTOFFER OFTEDAHL
Technical University of Norway
J. H. L. VOGT; T. VOGT

C. D. O'MALLEY†
VESALIUS

ALEXANDER M. OSPOVAT
Oklahoma State University
A. G. WERNER

FRANKLIN PARKER
West Virginia University
A. E. WRIGHT

JEAN PELSENEER
University of Brussels
WENDELIN

FRANCIS PERRIN
Académie des Sciences, Paris
P. WEISS

P. E. PILET
University of Lausanne
C. VOGT; WEPFER; YERSIN

DAVID PINGREE
Brown University
VIJAYANANDA; YA'QUB IBN ṬĀRIQ;
YĀTIVṚṢABHA; YAVANEŚVARA

A. F. PLAKHOTNIK
Academy of Sciences of the U.S.S.R.
N. N. ZUBOV

A. F. PLATÉ
Academy of Sciences of the U.S.S.R.
ZELINSKY

M. PLESSNER†
ZOSIMUS OF PANOPOLIS

HOWARD PLOTKIN
University of Western Ontario
A. WILSON

EMMANUEL POULLE
École Nationale des Chartes
WILLIAM OF SAINT-CLOUD; WILLIAM THE
ENGLISHMAN

LORIS PREMUDA
University of Padua
M. WIELAND

PAUL H. PRICE
*West Virginia Geological and Economic
Survey*
I. C. WHITE

J. A. PRINS
WAALS; ZERNIKE

HANS QUERNER
University of Heidelberg
M. W. C. WEBER

SAMUEL X. RADBILL
College of Physicians of New Jersey
WHYTT

HANS RAMSER
VOLKMANN; E. G. WARBURG

R. A. RANKIN
University of Glasgow
G. N. WATSON

NATHAN REINGOLD
Smithsonian Institution
R. S. WOODWARD

L. M. DE RIJK
Filosofisch Instituut, Leiden
WILLIAM OF SHERWOOD

GUENTER B. RISSE
University of Wisconsin-Madison
VIRCHOW; WILBRAND

ANDRE RIVIER†
XENOPHANES

GLORIA ROBINSON
Yale University
WEISMAN

JACQUES ROGER
University of Paris
WHISTON

GRETE RONGE
WACKENRODER

B. VAN ROOTSELAAR
State Agricultural University, Wageningen
ZERMELO

BERNARD ROTH
Stanford University
R. WILLIS

K. E. ROTHSCHUH
University of Münster/Westphalia
VERWORN

HUNTER ROUSE
University of Iowa
WEISBACH

G. RUDOLPH
University of Kiel
WEIGERT

MICHAEL T. RYAN
University of Chicago
VILLALPANDO

F. SCHMEIDLER
University of Munich Observatory
WILSING

CHARLES B. SCHMITT
Warburg Institute
ZABARELLA

BRUNO SCHOENEBERG
University of Hamburg
H. WEBER

E. L. SCOTT
Stamford High School, Lincolnshire
R. WATSON; WOULFE

HAROLD W. SCOTT
Michigan State University
J. WALKER

J. F. SCOTT†
WARING; WREN

CHRISTOPH J. SCRIBA
University of Hamburg
WALLIS; WIELEITNER

A. SEIDENBERG
University of California, Berkeley
WILCZYNSKI

E. M. SENCHENKOVA
Academy of Sciences of the U.S.S.R.
VORONIN

ELIZABETH NOBLE SHOR
Scripps Institution of Oceanography
VERRILL; S. WATSON; WHEELER; C. D.
WHITE; WILLISTON

DIANA M. SIMPKINS
Polytechnic of North London
VILMORIN

NATHAN SIVIN
Massachusetts Institute of Technology
WANG HSI-SHAN

P. SMIT
Catholic University, Nijmegen
WIGAND

E. SNORRASON
Rigshospitalet, Copenhagen
WINSLØW

GLENN SONNEDECKER
University of Wisconsin-Madison
H. C. WOOD

JAMES BROOKES SPENCER
Oregon State University
ZEEMAN

CURT STERN
University of California, Berkeley
WEINBERG

SHLOMO STERNBERG
Harvard University
WINTNER

FRANS STOCKMANS
University of Brussels
ZEILLER

JANIS STRADINŠ
Academy of Sciences of the U.S.S.R.
WALDEN

D. J. STRUIK
Massachusetts Institute of Technology
VLACQ

JUDITH P. SWAZEY
Boston University School of Music
WALLER

FERENC SZABADVÁRY
Technical University, Budapest
ZEMPLÉN

GIORGIO TABARRONI
Universities of Modena and Bologna
ZANOTTI

xii

CONTRIBUTORS TO VOLUME 14

C. H. TALBOT
Wellcome Institute for the History of Medicine
VIGO

M. TEICH
University of Cambridge
F. WALD

K. BRYN THOMAS
Royal Berkshire Hospital
WATERTON

PHILLIP DRENNON THOMAS
Wichita State University
WALTER OF ODINGTON; T. WHITE; WIED

VICTOR E. THOREN
Indiana University
WARD

THADDEUS J. TRENN
University of Regensburg
WHYTLAW-GRAY; S. YOUNG

F. G. TRICOMI
Academy of Sciences of Turin
VITALI

HENRY S. TROPP
Humboldt State University
J. W. YOUNG

G. L'E. TURNER
University of Oxford
B. WILSON; C. T. R. WILSON; J. WINTHROP

P. W. VAN DER PAS
South Pasadena, Calif.
VRIES

STIG VEIBEL
Technical University of Denmark
ZEISE

J. VERNET
University of Barcelona
IBN WAFĪD; YAḤYĀ IBN ABĪ MANṢŪR; AL-ZARQĀLI

KURT VOGEL
University of Munich
WIDMAN; WITTICH; WOEPCKE

E. VOLTERRA
University of Texas
VOLTERRA

WILLIAM A. WALLACE, O.P.
Catholic University of America
VINCENT OF BEAUVAIS; WILLIAM OF AUVERGNE

P. J. WALLIS
University of Newcastle Upon Tyne
E. WRIGHT

DEBORAH JEAN WARNER
Smithsonian Institution
WINLOCK

AARON C. WATERS
University of California, Santa Cruz
B. WILLIS

CHARLES WEBSTER
University of Oxford
WALTON

MARY A. WELCH
University of Nottingham
WILLUGHBY

JOHN W. WELLS
Cornell University
H. S. WILLIAMS

F. W. WENT
University of Nevada Desert Research Institute
WENT

JOYCE WEVERS
State University of Utrecht
WIDMANNSTÄTTEN

ALWYNE WHEELER
British Museum (Natural History)
WOTTON

RONALD S. WILKINSON
Library of Congress
J. WINTHROP

J. WOLFOWITZ
University of Illinois at Urbana-Champaign
A. WALD

A. E. WOODRUFF
Yeshiva University
W. E. WEBER

ELLIS L. YOCHELSON
U.S. Geological Survey
WALCOTT

H. S. YODER, JR.
Carnegie Institution of Washington
WINCHELL FAMILY; F. E. WRIGHT

A. P. YOUSCHKEVITCH
Academy of Sciences of the U.S.S.R.
V. P. ZUBOV

DICTIONARY
OF
SCIENTIFIC BIOGRAPHY

DICTIONARY OF SCIENTIFIC BIOGRAPHY

VERRILL – ZWELFER

VERRILL, ADDISON EMERY (*b.* Greenwood, Maine, 9 February 1839; *d.* Santa Barbara, California, 10 December 1926), *zoology.*

Verrill's ability to identify and remember a tremendous variety of animals, plants, and minerals appeared at a very early age. He began with rocks and minerals near his home and later collected plants, shells, and animals at Norway, Maine, where the family moved in 1853. His parents, George Washington Verrill and Lucy Hillborn, were both descended from early New England families. In preparation for his boyhood ambition to study and work under Louis Agassiz, Verrill attended Norway Liberal Institute and in 1859 entered Harvard College. Agassiz put him to work studying birds, urged him to take up zoology instead of geology, and arranged summer trips to the Bay of Fundy, Anticosti Island, and the coast of Labrador for him, in the company of Alpheus Hyatt and N. S. Shaler. Verrill's early interest in geology lasted throughout his life as a secondary profession.

Even before his graduation from Harvard in 1862, Verrill worked as assistant to Agassiz at the Museum of Comparative Zoology and continued there until 1864. A request from Yale College for recommendations to the newly established chair of zoology led Agassiz to suggest Verrill in 1864. He held the post until his retirement in 1907. He began the zoological collections of the Peabody Museum and was curator from 1865 to 1910. He also taught large classes in geology at Yale from 1870 to 1894. Simultaneously with his Yale duties, Verrill was in charge of the scientific investigations of the U.S. Commission of Fish and Fisheries, curator of the Boston Society of Natural History for ten years, and professor of comparative anatomy and entomology for one course each spring at the University of Wisconsin from 1868 to 1870. He served as associate editor of the *American Journal of Science* from 1869 to 1920.

Verrill received an honorary M.A. from Yale in 1867. He was an early member of the National Academy of Sciences, president of the Connecticut Academy of Arts and Sciences for some years, a corresponding member of the Société Zoologique de France, and a member of many American scientific societies.

Trained under Agassiz, Verrill became an outstanding taxonomist, for he considered taxonomy the foundation of biology. Until 1870 he collected marine invertebrates by dredging each summer along the Maine coast and in the Bay of Fundy, and completed the identifications of the collections during the school year. Aided by his wife's brother, Sidney I. Smith, Verrill made a detailed ecologic study of the fauna of Vineyard Sound in 1873, the first of this magnitude in the United States and the model for later ones. He was assigned to handle the invertebrates of Spencer F. Baird's Commission of Fish and Fisheries survey of the New England coast, and elected to describe and classify all the groups himself, intending to write monographs covering each major unit.

Because of his remarkable facility for distinguishing significant morphological features, he described, with few errors in judgment, at least one thousand marine invertebrates, in every phylum except Protozoa. His outstanding contribution was in the classification and natural history of corals; but his work on echinoderms, especially starfishes and the very confusing brittle stars, placed that group on a secure taxonomic foundation. He was among the first to make a phylogenetic separation of the echinoderms from the coelenterates, both of which had previously been combined into the Radiata. Verrill also completed three monographs on crustaceans, having taken over that group when the original worker, Sidney Smith, became blind. Aided by Katharine J. Bush, he wrote extensively on the mollusks, especially the cephalopods, and directed the making of a life-size model of a giant squid for the Peabody Museum. In addition to his taxonomic work, he wrote considerably on the habits and natural history of each group he studied.

While participating in the dredging trips of the Commission of Fish and Fisheries, Verrill devised useful marine collecting equipment; and he enlisted students on his own weekend dredging trips by sailboat near New Haven, reaching depths of 4,000 meters. From three trips to Bermuda and extensive study he presented a painstaking report on the zoology, botany, geology, and even history and the effect of civilization on those islands. Interested in many fields of zoology, Verrill introduced new names into entomology and parasitology as well as into marine invertebrate groups; he also alerted the public to the hazards of tapeworms and similar parasites. His type specimens are at the Peabody Museum and the U.S. National Museum. He contributed most of the zoological definitions to the 1890 edition of Webster's *International Dictionary.*

BIBLIOGRAPHY

I. ORIGINAL WORKS. The memoir by Coe, cited below, contains a detailed bibliography, by subject, of over 300 papers by Verrill and includes monographs on specific animal groups. His major report on Bermuda is "The Bermuda Islands: Their Scenery, Climate, Productions, Physiography, Natural History and Geology, With Sketches of Their Early History and the Changes due to Man," in *Transactions of the Connecticut Academy of Arts and Sciences,* **11** (1902), 413–911—further reports on the islands appeared in 1906 and 1907. The meticulous Vineyard Sound study is "Report Upon the Invertebrate Animals of Vineyard Sound and the Adjacent Waters, With an Account of the Physical Characters of the Region," in *Report of the United States Commissioner of Fisheries,* **1** (1873), 295–747.

II. SECONDARY LITERATURE. Wesley R. Coe, "Biographical Memoir of Addison Emery Verrill," in *Biographical Memoirs. National Academy of Sciences,* **14** (1932), 19–66, provides information on Verrill's childhood, personality, and scientific accomplishments. Shorter versions, also by Coe, appeared in *Science,* **66,** no. 1697 (1927), 28–29; and in *American Journal of Science,* 5th ser., **13,** no. 77 (1927), 377–387.

ELIZABETH NOBLE SHOR

VERULAM, BARON. See **Bacon, Francis.**

VERWORN, MAX (*b.* Berlin, Germany, 4 November 1863; *d.* Bonn, Germany, 23 November 1921), *physiology.*

The son of a Prussian civil servant, Verworn completed his secondary education at the Friedrich Wilhelm Gymnasium, occupying his free time with biological experiments. In 1884 he enrolled as a student of zoology and medicine at the University of Berlin, where he attended lectures by F. E. Schultze, Emil du Bois-Reymond, and Rudolf Virchow. He received the Ph.D. in zoology in 1887 and immediately began to study the medical sciences at Jena, where he met Wilhelm Preyer, Wilhelm Biedermann, and Ernst Haeckel, the latter exerting the greatest influence. In 1889, Verworn received the M.D. from Jena, where he remained until 1901, first as an assistant, then as a university lecturer (1891), and finally as associate professor at the physiological institute (1895). His marriage that year to Josephine Huse, whom he had met in Naples, was childless. In 1901 Verworn succeeded Georg Meissner at Göttingen, and in 1910 he replaced Eduard Pflüger at Bonn. In 1911, he gave the Silliman lectures at Yale University. At both Göttingen and Bonn he attracted many young research workers, including Hans Winterstein, August Pütter, and F. W. Fröhlich. He greatly influenced many Japanese students, including H. Nagai and Y. Ishikawa, who spread his teachings in Japan.

During his special experimental investigations Verworn was concerned primarily with the general basic problems of life. This preoccupation can be noticed especially in his biological-physiological studies, where he investigated the basic phenomena of life such as irritation, paralysis, narcosis, biotonus (biological base tension), hypnosis, fatigue, and recovery. In this endeavor he worked with unicellular organisms, such as Protista, or the cells of higher organisms. Verworn was a major advocate of cellular physiology; and his experimental investigations were concerned mainly with the elementary processes of muscle fibers, nerve fibers, and sensory organs. For him each analysis of function always ended with the function of the cells: for secretion, with the glandular cells; for the action of the heart, with the heart muscle cells; for psychical conduction, with the ganglion cells. In general, comparative, and microscopic physiology, Verworn followed the model of Johannes Müller, to whom he dedicated his *Allgemeine Physiologie* (1895). The book was widely read outside the field of physiology, especially since Verworn included clear statements of the controversies of the time, such as vitalism versus mechanism, psychophysical problems, and monism. In 1902 he founded *Zeitschrift für allgemeine Physiologie;* many of his articles had previously appeared in *Pflügers Archiv für die gesamte Physiologie.*

Verworn conducted his first Protista experiments during the years 1887–1891, while on research trips to the Mediterranean and Red seas and working at the Zoological Station in Naples under Anton Dohrn. His studies of one-cell animals encompassed the manifestations of regeneration (1888), the relations between cell nucleus and psyche (1890), the phenomena of stimulation and response by means of galvanic current, and, most important, the polar effects of direct current (1894, 1897). His work on irritation, irritability, and paralysis was published in 1914. Verworn differentiated spontaneous manifestations of life from the "response of stimulation" by means of various external living conditions. He did not limit his concern only to the problems of cellular physiology or theoretical biology but, rather, felt strong need to clarify the fundamental issues and therefore also investigated the relation of body and mind. In his *Psycho-physiologische Protisten-Studien* (1889), Verworn discarded the dualistic point of view in favor of the monistic—not in terms of materialism but of psychomonism. To him the final elements of the being are psychic. "The physical world exists not next to the psychic, on the contrary, within the psychic" (*Naturwissenschaft und Weltanschauung* [1904], 29).

Like Mach and Richard Avenarius, Verworn advocated the principle of conditional research rather than the usual search for causes. This multiconditionalism manifests itself especially in the basic phenomena of life, which he made the main subject of his studies (1918). He also did research on primitive cultures and anthropology, collecting artifacts from primitive cultures, specimens of ethnological importance, prehistoric subjects, and coins. Many of these were used in the writing of *Die Anfänge der Kunst* (Jena, 1909).

In the history of ideas Verworn belongs with such materialists and positivists as Ludwig Büchner, Moleschott, Haeckel, and Mach. Since he did not hesitate to establish new hypotheses—*Zellseele, Atomseele, Biotonus*—and to expound them in his works, he had many opponents in addition to supporters throughout his life.

BIBLIOGRAPHY

I. ORIGINAL WORKS. Verworn's books include *Psycho-physiologische Protisten-Studien. Experimentelle Untersuchungen* (Jena, 1889); *Die Bewegung der lebendigen Substanz* (Jena, 1892); *Allgemeine Physiologie. Ein Grundriss der Lehre vom Leben (Jena, 1895); Beiträge zur Physiologie des Centralnervensystems* (Jena, 1898); *Die Mechanik des Geisteslebens* (Leipzig, 1907); *Physiologisches Praktikum* (Jena, 1907); *Die Entwicklung des menschlichen Geistes* (Jena, 1910); *Narkose* (Jena, 1912); *Irritability* (New Haven, 1913), the Silliman lectures; and *Erregung und Lähmung. Eine allgemeine Physiologie der Reizwirkungen* (Jena, 1914).

Among his articles are "Biologische Protistenstudien," in *Zeitschrift für wissenschaftliche Zoologie*, **46** (1888), 455–470, and **50** (1890), 443–468; "Die polare Erregung . . .," in *Pflügers Archiv für die gesamte Physiologie*, **45** (1889), 1–36; **46** (1890), 267–303; **62** (1896), 415–450; and **65** (1897), 47–62; "Die physiologische Bedeutung des Zellkerns," *ibid.*, **51** (1892), 1–118; "Der körnige Zerfall. Ein Beitrag zur Physiologie des Todes," *ibid.*, **63** (1896), 253–272; "Einleitung" (as editor), in *Zeitschrift für allgemeine Physiologie*, **1** (1902), 1–18; "Die cellularphysiologische Grundlage des Gedächtnisses," *ibid.*, **6** (1907), 119–139; and "Die cellularphysiologische Grundlage des Abstraktionsprozesses," *ibid.*, **14** (1912), 277–296. He also published articles on the effect of strychnine and on fatigue, exhaustion, and recovery in the nerve center of the spinal column, in *Archiv für Anatomie und Physiologie*, Physiol. Abt. (1900), 385–414; and supp. (1900), 152–176.

II. SECONDARY LITERATURE. See Silvestro Baglioni, "Max Verworn," in *Rivista di biologia*, **4** (1922), 126–133, with bibliography; I. Fischer, in *Biographisches Lexikon der hervorragenden Ärzte der letzten 50 Jahre*, II (Berlin–Vienna, 1933), 1616–1617; Friedrich W. Fröhlich, "Max Verworn," in *Zeitschrift für allgemeine Physiologie*, **20** (1923), 185–192; R. Matthaei, "Max Verworn," in *Deutsche medizinische Wochenschrift*, **48** (1922), 102–103, with portrait; A. Pütter, "Max Verworn," in *Münchener medizinische Wochenschrift*, **68** (1921), 1655–1656; W. Thörner, "Max Verworn," in *Medizinische Klinik*, **18** (1922), 130–131; and R. Wüllenweber, *Der Physiologe Max Verworn* (Bonn, 1968), inaugural M.D. diss.

K. E. ROTHSCHUH

VESALIUS, ANDREAS (*b*. Brussels, Belgium, 31 December 1514; *d*. Zákinthos, Greece, 15 October 1564), *medicine*.

The date of Vesalius' birth is derived from a horoscope cast by the Milanese physician Girolamo Cardano, from which it may also be determined that he was born at a quarter to six in the morning. His father, also named Andreas, was an apothecary of the Emperor Charles V and the illegitimate son of Everard van Wesele or Vesalius and, as such, was a humble member of a family already distinguished for several generations in medical circles. The maiden name of Vesalius' mother was Isabel Crabbe, and this resemblance to the name of the English poet has given rise to the legend that she was an Englishwoman. The name Crabbe is in fact common in Brabant.

The young Vesalius received his elementary education in Brussels and matriculated at the University of Louvain in February 1530 to pursue the arts course, the necessary prerequisite for entrance into a professional school. It is not known when he decided to study medicine, but such a decision could have been related to the emperor's legitimization of the young man's father in 1531, which may have encouraged him to carry on his family's traditional profession.

Since at this time the medical school of Louvain had little repute, Vesalius chose to carry on his medical studies at the more illustrious faculty of the University of Paris, matriculating there probably in September 1533, where he studied with Guinter of Andernach, Jacobus Sylvius (Jacques Dubois), and Jean Ferne. Guinter, who in his *Institutiones anatomicae* (1536) spoke very favorably of his student, and Sylvius, an arch-Galenist and later an enemy of Vesalius, each in his own way directed the young man toward anatomical research. Since they were both supporters of the Galenic tradition, it was natural that their student, although he acquired skill in the technique of dissection, remained under the influence of Galenic concepts of anatomy.

The war between France and the Holy Roman Empire compelled Vesalius to leave Paris in 1536. He returned to Louvain, where, with the friendly support of the burgomaster, he was able to reintroduce anatomical dissection, which had not been part of the medical curriculum for many years, and in 1537 he received the degree of bachelor of medicine. While completing his studies he produced his *Paraphrasis in nonum librum Rhazae ad Regem Almansorem* (Louvain, 1537), in which he compared Muslim and Galenic therapy—to the disadvantage of the former—but sought to preserve the reputation of Rhazes and to reconcile him with the Greeks. A youthful work, of no significance except as an indication of Vesalius' continued allegiance to Galen, it was nevertheless important enough to its author for him to reprint it in Basel later in the year, on his way to Italy.

In the autumn of 1537 Vesalius enrolled in the medical school of the University of Padua, then the most famous in Europe, where, after two days of examinations, he received the degree of doctor of medicine *magna cum laude* on 5 December 1537 and on the following day accepted appointment as *explicator chirurgiae* with the responsibili-

ty of lecturing on surgery and anatomy. Immediately thereafter he gave the required annual anatomical lectures and demonstrations, which although Galenic in character were unusual because, contrary to custom, Vesalius himself performed the dissections rather than consigning that task to a surgeon. In addition, he produced four large anatomical charts representing the portal, caval, arterial, and nervous systems, based chiefly on this dissection and intended as a reference work and memory aid for his students when the cadaver was no longer available. These figures were distinctly novel both in their size, which permitted deceptively naturalistic although primarily Galenic portrayals of even the smaller structures, and in their detailed identification of the parts through an elaborately indexed anatomical terminology in Greek, Latin, Arabic, and Hebrew. The theft and subsequent publication of the drawing of the nervous system and the danger of plagiarism of the others led Vesalius to publish the three remaining drawings, together with three views of the skeleton by the Dutch artist Jan Stephen of Calcar, a student in Titian's studio. Although without a general title, they are usually referred to as *Tabulae anatomicae sex* (Venice, 1538).

In the same year Vesalius produced a dissection manual for his students, *Institutiones anatomicae secundum Galeni sententiam . . . per Ioannem Guinterium Andernachum . . . ab Andrea Vesalio . . . auctiores et emendatiores redditae* (Venice, 1538). As the title indicates, this was a revised and augmented edition of the Galenically oriented dissection manual of his former teacher in Paris, Guinter of Andernach. The revisions display a concern with the minutiae of dissection technique, and the augmentations offer several independent anatomical judgments, such as the briefly expressed but clearly anti-Galenic observation that the cardiac systole is synchronous with the arterial pulse.

Also in 1538 Vesalius visited Matteo Corti, professor of medicine in Bologna, and discussed the problems of therapy by venesection. Differences of opinion between the two men seem to have been the impulse behind Vesalius' next book, *Epistola docens venam axillarem dextri cubiti in dolore laterali secundam* (Basel, 1539), written in support of the revived classical procedure first advocated in a posthumous publication (1525) of the Parisian physician Pierre Brissot. In this procedure blood was drawn from a site near the location of the ailment, in contrast to the Muslim and medieval practice of drawing blood from a distant part of the body. As the title of his book indicates, Vesalius sought to locate the precise site for venesection in pleurisy within the framework of the classical method. The real significance of the book lay in Vesalius' attempt to support his arguments by the location and continuity of the venous system rather than by an appeal to earlier authority. Despite his own still faulty knowledge, his method may be called scientific in relation to that of others; certainly it was nontraditional and required that his opponents resort to the same method if they wished to reply effectively. With this novel approach to the problem of venesection Vesalius posed the then striking hypothesis that anatomical dissection might be used to test speculation. Here too he declared clearly, on the basis of vivisection, that cardiac systole was synchronous with arterial expansion and for the first time mentioned his initial efforts in the preparation of the anatomical monograph that was ultimately to take shape as *De humani corporis fabrica*.

These activities and the novelty of Vesalius' teaching were greatly appreciated by both the teachers and students of the university. Indeed, the official document by which the young anatomist was reappointed in 1539 to the medical faculty, with a considerable increase in salary, declared that "he has aroused very great admiration in all the students." Although the opportunities for dissection were limited by the small number of cadavers available, by 1538, from his lecturing on Galen and his own dissecting, Vesalius began to realize that there were contradictions between Galen's texts and his own observations in the human body.

In 1539 his supply of dissection material became much greater when Marcantonio Contarini, judge of the Paduan criminal court, became interested in Vesalius' investigations and made the bodies of executed criminals available to him—occasionally delaying executions to suit the convenience of the young anatomist. For the first time Vesalius had sufficient human material to make and to repeat detailed and comparative dissections. As a result, he became increasingly convinced that Galen's description of human anatomy was basically an account of the anatomy of animals in general and was often erroneous insofar as the human body was concerned. During the winter of 1539 he was sufficiently sure of his position to challenge the validity of Galenic anatomy in Padua and shortly thereafter to repeat the challenge in Bologna.

Vesalius went to Bologna in January 1540, at the invitation of the medical students of that city,

to present a series of anatomical demonstrations, in the course of which he boldly declared that human anatomy could be learned only from the dissection and observation of the human body. As proof of the nonhuman source of Galen's anatomy he articulated ape and human skeletons and demonstrated that Galen's description of the bones agreed only with the former.

On his return to Padua, Vesalius began the composition of the *Fabrica* in its final form. For the next two years, until the summer of 1542, he concentrated his efforts on this huge work, sparing no expenditure of energy or money. He hired the best draftsmen he could find to make the illustrations and the finest Venetian block cutters to reproduce them. To print the book he chose Joannes Oporinus of Basel. In the summer of 1542 Vesalius left Padua for Basel to oversee the printing of his book, and the following May he dissected the body of an executed malefactor and articulated the skeleton, which is still preserved in the University of Basel's anatomical institute.

With the publication of *De humani corporis fabrica* (Basel, 1543)—in August rather than in June as given in the colophon—and of its *Epitome* (Basel, 1543; German translation by Albanus Torinus, Basel, 1543), Vesalius, with youthful impetuosity, decided to relinquish his anatomical studies for the practice of medicine. Since there was a long tradition of imperial service in his family, he applied to the Emperor Charles V and received an appointment as physician to the imperial household. It was an unfortunate decision since much of his time was henceforth devoted to the complaints of the gluttonous emperor and, as Vesalius wrote, "to the Gallic disease, gastrointestinal disorders, and chronic ailments, which are the usual complaints of my patients." The imperial service once entered could not be abandoned; Vesalius remained the emperor's physician until the latter's abdication, thirteen years later.

Despite his renunciation of anatomical studies, it was inevitable that Vesalius would soon return to his first interest. In January 1544 he traveled to Pisa to give a series of demonstrations at the invitation of Cosimo I, grand duke of Tuscany, who sought unsuccessfully to retain his services. Thereafter, while acting as a military surgeon in the course of the emperor's wars, Vesalius never failed to visit any nearby medical school, to participate in postmortem examinations, or to take advantage of any opportunities for anatomical research. In 1546, during an extended visit to Regensburg, he wrote a long letter partly concerned with the discovery and therapeutic use of the chinaroot (*Chinae radix*) in the treatment of syphilis and partly to justify his anatomically heretical activities against the attack of the Galenic anatomists of Paris, most notably those of his former teacher Sylvius. It was published under the title *Epistola rationem modumque propinandi radicis chynae decocti pertractans* (Basel, 1546).

During his service with the imperial army Vesalius was able to apply his unrivaled anatomical knowledge to surgery. He learned the emollient treatment of gunshot wounds from the Italian surgeon Bartolomeo Maggi; and although his surgery seems to have been burdened at first by an academic quality not required or even desirable on the battlefield, he quickly learned existing surgical techniques and went on to develop others. His most notable contribution was the introduction, as early as 1547, of surgically induced drainage of empyema, and he became so proficient in this procedure that he was sufficiently confident of the outcome to recommend it to other surgeons. His account of this operation, written as a letter (1562) to Giovanni Filippo Ingrassia of Sicily, was an outstanding contribution to the surgical literature. His reputation as a surgeon became so great that in 1559, when Henry II of France received what was to be a fatal head wound in a tournament, Vesalius was summoned from Brussels and placed in charge of the patient, despite the presence of the distinguished French surgeon Ambroise Paré. Vesalius wrote the report of the case after its termination.

The qualities of mind that had been responsible for the *Fabrica* brought Vesalius the reputation of being one of the great physicians of his age; his opinion was widely sought in grave medical problems. There are a number of contemporary references to him as "that noble physician" and "the best physician in the world." An instance of what he considered the proper relation of anatomy to medicine was his remarkable diagnosis and correct prognosis in 1555 of an internal aneurysm in a living patient.

As his experience became greater and as he realized the need for correcting errors of fact and faults of composition in the *Fabrica*, Vesalius gave more thought to a new edition. It is not known when an agreement was reached with the publisher Oporinus for the costly enterprise, but it was at some time after 1547; and it seems most likely that Vesalius wrote the revised text during an extended sojourn with the emperor in Augsburg between August 1550 and October 1551. However, it was only after a long delay that the revised edition was published in Basel in August 1555.

With the abdication of Charles V in 1555, Vesalius for unknown reasons took service with his son Philip II of Spain as physician to the Netherlanders at the Spanish court and, from time to time, to the king himself. He remained in Spain from 1559 until the year of his death.

At the close of 1561 Vesalius completed a long reply to the *Observationes anatomicae* (1561) of Gabriele Falloppio, a respectful criticism of certain aspects of the *Fabrica*, which had been sent to him by the author during the preceding summer. Vesalius' reply, later published under the title of *Anatomicarum Gabrielis Falloppii observationum examen* (Venice, 1564), is partly a defense against Falloppio's criticisms and partly an acceptance of them. In addition, it stated Vesalius' desire to return to his former chair of anatomy at Padua. During the spring of 1562, on the command of Philip II, Vesalius joined the physicians involved in the care and treatment of Don Carlos, the king's son and heir, who, as the result of a fall, had received a severe injury to his head and was for long in grave danger.

In 1564 Vesalius left Spain for a trip to the Holy Land. Contrary to various legends, the journey appears to have been undertaken with the friendly approbation of the king, although it is not entirely clear whether Vesalius intended to return to Spain. After a visit to Venice—where he apparently was invited to accept his former chair at Padua in succession to Falloppio, who had died—he set sail in March for the Holy Land by way of Cyprus. It is not known precisely when the return voyage was begun, but in any event his ship was delayed by a violent storm. After much hardship it finally reached the island of Zákinthos in October, where Vesalius died and was buried in an unidentified site.

Vesalius produced only one book of great importance, *De humani corporis fabrica* (1543), to which may be added several complementary works, the *Epitome* (1543), *Epistola rationem modumque propinandi radicis chynae* (1546), the revised edition of the *Fabrica* (1555), and the *Examen* (1564).

Several motives underlay the composition and publication of the *Fabrica*. According to Vesalius medicine was properly composed of three parts: drugs, diet, and "the use of the hands," by which last he referred to surgical practice and especially to its necessary preliminary, a knowledge of human anatomy that could be acquired only by dissecting human bodies with one's own hands. Through disdain of anatomy, the most fundamental aspect of medicine, or, as Vesalius phrased it, by refusal to lay their hands on the patient's body, physicians betray their profession and are physicians only in part.

Vesalius hoped that by his example in Padua and especially by his verbal and pictorial presentation in the *Fabrica* he might persuade the medical world to appreciate anatomy as fundamental to all other aspects of medicine and that, through the application of his principles of investigation, a genuine knowledge of human anatomy would be achieved by others, in contrast to the more restricted traditional outlook and the uncritical acceptance of Galenic anatomy. The very word "fabrica" could be interpreted as referring not only to the structure of the body but to the basic structure or foundation of the medical art as well. Thus, Vesalius directed his work toward the established physician, whom he hoped to attract to the study of anatomy as a major but neglected aspect of a true medicine and, no less important, toward those members of the medical profession who were concerned with the teaching of anatomy and might be induced to forsake their long-accepted traditional methods for those proposed by Vesalius. As anatomy was then taught, he wrote, "there is very little offered to the [students] that could not better be taught by a butcher in his shop."

The *Fabrica* was also written to demonstrate the fallacious character of Galenic anatomy and all that it implied. Since Galen's anatomy was based upon the dissection and observation of animals, it was worthless as an explanation of the human structure; and since previous anatomical texts were essentially Galenic, they likewise were worthless and ought to be disregarded. Human anatomy was to be learned only by dissection and investigation of the human body, the true source of such knowledge. Nevertheless it was desirable that human dissection be accompanied by a parallel dissection of the bodies of other animals in order to show the differences in structure and hence the source of Galen's errors. "Physicians ought to make use not only of the bones of man but, for the sake of Galen, of those of the ape and dog." It was because of Vesalius that Padua became the first great center of comparative as well as of human anatomical studies, a dual interest that continued to develop under his successors Falloppio, Fabrici, and Casserio.

According to Vesalius, the student or physician ought to carry on these activities himself and should personally dissect the human body. The professor or teacher must also descend from his *cathedra*, dismiss the surgeon who had formerly performed the actual anatomy, and undertake his own dissecting. Moreover, it was not sufficient to base judgments upon a single dissection: the same dissection should be repeated upon several bodies until the dissector could be certain that his observations did not represent structural anomaly. Even the reader of the *Fabrica* must not be content to accept Vesalius' descriptions without question but ought to test them by his own dissections and observations. For this purpose the descriptive chapters of the *Fabrica* are frequently followed by directions for making one's own dissection of the part described so as to arrive at an independent conclusion.

Vesalius regarded the *Fabrica* as the gospel of a new approach to human anatomical studies and a new method of anatomical investigation. In Padua both the gospel and its explications were presented directly by the author. For those elsewhere it was presented through the *Fabrica* with its long and complete descriptions, its illustrative and diagrammatic guides to aid recognition of details and to supplement the reader's possible shortage of dissection specimens, and even its indirect encouragement of body snatching if necessary. The work reflects fully Vesalius' method of instruction from about the end of 1539 through 1542 and represents some of it pictorially on the title page.

The presentation of a new anatomy and anatomical method raised several problems, of which the first was that of terminology. As in the *Tabulae anatomicae* (1538), Vesalius continued to use terms from several languages but stressed the Greek form wherever possible. If this was not enough for clarity, an extensive description was given to localize the part with reference to other parts, and illustrations of the particular organ or structure were provided. Additionally, as a mnemonic device and for increased comprehension, anatomical structures were related to common objects, the radius, for example, being compared to the weaver's shuttle and the trapezius muscle to the cowl of the Benedictine monks. Some of Vesalius' terms are still in use, so that this aspect of his pedagogy plays the same role today as it did in the sixteenth century. Thus the names of two of the auditory ossicles, the incus and malleus, are derived from Vesalius' description of them as "that one somewhat resembling the shape of an anvil [*incus*]" and "that one resembling a hammer [*malleus*]." The valve of the left atrioventricular orifice, the mitral valve, "you may aptly compare to a bishop's miter."

Vesalius' greatest contribution to the elucidation of anatomy is to be found in the illustrations to the *Fabrica*. With the exception of those few diagrammatic illustrations that are known to have been drawn by him there is no positive identification of individual draftsmen. The soundest theory is that they were students from Titian's studio in Venice. Possibly among them was Jan Stephen of Calcar, who drew the three figures of the skeleton for the published version of Vesalius' anatomical plates of 1538; but the three skeletons of the *Fabrica* are so greatly superior to those of the earlier work that it seems unlikely that Calcar was responsible for them.

The anatomical detail of the illustrations and their numbered and lettered explanatory legends make it clear that the drawings were made under the supervision of Vesalius for the specific purpose of clarification of a particular portion of the text. Not only is the quality of draftsmanship and precision of detail immensely superior to that of earlier books but the marginal references to the illustrations, which in some instances relate a textual description to several illustrations located in different parts of the work, are also entirely without precedent. For the first time the pedagogic purpose of illustrations was achieved—so well that unfortunately attention has more recently been centered upon the illustrations to the exclusion of the text, thereby nullifying Vesalius' purpose and even damaging his reputation. He has, for example, been criticized for the exaggerated upward extension of the rectus abdominis muscle as it appears in the fifth "muscle man," although the legend accompanying the illustration explains this as having been done deliberately to represent an error of Galenic anatomy. Several such seeming errors are in fact deliberate distortions serving pedagogical purposes; they are not appreciated, however, unless text and illustrations are studied together.

In addition to the title page the most noteworthy illustrations in the *Fabrica* are the three celebrated skeletal figures and the series of "muscle men" which through their postures were given a dynamic quality that was intentional and specifically referred to by Vesalius. The "muscle men," shown from the front, side, and back, and displaying in sequence from the surface downward the underlying layers of muscle, were a novelty, although crudely foreshadowed by the series of figures in Berengario da Carpi's *Commentaria* (1521); the latter, however, were wholly lacking the elegance and detail to be found in the Vesalian figures.

Owing to the larger amount of dissection material available to him, Vesalius was not compelled to follow the traditional pattern of dissection and description originally established by Mondino (1316). Consequently, book I of the *Fabrica* opens with a description of the bones. This arrangement was desirable since according to Vesalius the bones are the foundation of the body, the structure to which everything else must be related; and in his anatomical demonstrations he was accustomed to sketch the position of the bones on the surface of the body with charcoal in order to orient the students. The fundamental significance of the bones was further indicated by his reference to the femur, for example, as either the bone itself or the entire leg of which the bone was the basic structure. Moreover, the bones are not only supports for the body; since by their structure and formation they assist and control movement, it is necessary to recognize in them a dynamic quality that Vesalius sought to emphasize by the suggestion of movement in the poses of the skeletons.

The teleological argument that pervades the *Fabrica*, an inheritance from Galen, is very pronounced in the description of osteology. "By not first explaining the bones anatomists . . . deter [the student] from a worthy examination of the works of God." Vesalius did not allow this doctrine of final causes to control his investigations, however, since unlike his medieval predecessors he sought to discover first structure and related function, and only then the ultimate purpose.

In his description of human osteology, the subject of book I, Vesalius made some of his strongest assaults upon Galenic anatomy. He called attention to Galen's false assertion that the human mandible is formed of two bones and demonstrated the significance of this error as reflecting a dependence upon animal sources. Likewise he pointed to the fact that the Galenic description of the sternum as formed of seven segments is true of the ape but not of the adult human sternum, which has only three. Similarly the "humerus, according to Galen, is with the exception only of the femur, the largest bone of the body. Nevertheless the fibula and tibia are distinctly of greater length than the humerus." In addition to such criticisms, there is extensive description of osteological detail, which, because much of it was wholly novel, required detailed illustrations, elaborately related by letter and number to the text. Despite some errors of description and occasional references to animal anatomy in the Galenic tradition, this first book represents Vesalian anatomy on the highest level. It concludes with a remarkable chapter on the procedure for preparation of the bones and articulation of the skeleton, since it was essential that a skeleton always be available at the dissection. Such a skeleton is a central figure of the title page.

As he had done with the bones, so Vesalius endeavored in book II to identify and give the fullest possible description of every muscle and its function; and an examination of the "muscle men" indicates the thoroughness with which that task was performed. Unfortunately, his system of identifying muscles numerically according to the part they served was cumbersome in comparison with the method of identification by origin and insertion introduced by Sylvius in 1555 and later revised and improved by the Swiss anatomist Gaspard Bauhin. The first two books represent the major Vesalian achievement in terms of accuracy of description and present the most telling blows against Galenic anatomy. In book II Vesalius also most frequently provided chapters dealing with the dissection procedure used to arrive at his conclusions. The description of the vascular system in book III is less satisfactory because of Vesalius' failure to master the complexities of distribution of the vessels and because of the close relationship of the vascular system to Galenic physiology. Vesalius was compelled to subscribe to this for lack of any other theories. The errors in the Vesalian description of the distribution of the vessels are due to his reliance on Galen, as the only other writer to have attempted such a description in detail, and to the difficulty of discovering anew the entire vascular arrangement in rapidly putrefying human material. Although Vesalius was partly successful, as, for example, in his account of the interior mesenteric and the hemorrhoidal veins, there are many indications that he was compelled to rely for much of his account on the anatomy of animals. This is clearly apparent in the illustration of the "arterial man," where the arrangement of the branchings of the aortic arch actually illustrate simian anatomy.

Book IV provides an account of the nervous system. It is introduced by an attempt to clarify and limit the meaning of the word "nerve" to the vehicle transmitting sensation and motion, because "leading anatomists declare that there are three kinds of nerve": ligament, tendon, and aponeurosis. "From dissection of the body it is clear that no nerve arises from the heart as it seemed to Aristo-

tle in particular and to no few others." Although Vesalius was obliged to accept the Galenic explanation of nervous action as induced by animal spirit distributed through the nerves from the brain, his examination of the optic nerve led him to the conclusion that the nerves were not hollow, as Galen had asserted. "I inspected the nerves carefully, treating them with warm water, but I was unable to discover a passage of that sort in the whole course of the nerve."

Vesalius accepted Galen's classification of the cranial nerves into seven pairs even though he recognized more than that number and described a portion of the trochlear nerve. To avoid confusion he declared that he would "not depart from the enumeration of the cranial nerves that was established by the ancients." Although he was not wholly successful in his efforts to trace the cranial nerves to their origins, and despite some confusion about their peripheral distribution, the level of knowledge in the text and illustrations was well above that of contemporary works and was not to be surpassed for about a generation. Vesalius was more successful in tracing the spinal nerves, but on the whole the account of the nerves must be described as being of lesser quality than some of the other books.

The description of the abdominal organs in book V is detailed and reasonably accurate. Since he knew of no alternative Vesalius accepted that aspect of Galen's physiology which placed the manufacture of the blood in the liver. Nevertheless he denied not only that the vena cava takes its origin from the liver but also that the liver is composed of concreted blood. Here his strongest blow against Galen and medieval Galenic tradition was his denial, based on human and comparative anatomy, of the current belief in the liver's multiple (usually five) lobes. According to Vesalius the number of lobes increased with the descent in the chain of animal life. In man the liver had a single mass, while the livers of monkeys, dogs, sheep, and other animals had multiple lobes that became more numerous and more clearly apparent. This difference once again proved the error of dependence upon nonhuman materials.

Vesalius also denied the erroneous Galenic belief that there was a bile duct opening into the stomach as well as one into the duodenum. In regard to the position of the kidneys, he had begun to move away from the erroneous view expressed in the *Tabulae anatomicae* that the right kidney was placed higher than the left. Although this error is illustrated in the *Fabrica*, the text declares that the reverse could also be true. Despite this partial error of traditionalism, Vesalius denied a second traditional opinion that the urine passed through the kidneys by means of a filter device. The filter theory had also been denied by Berengario da Carpi; but Vesalius went a step further by asserting that the "serous blood" was deliberately selected or drawn into the kidney's membranous body and its "branchings" to be freed of its "serous humor" in the same way that the vena cava was able to select and acquire blood from the portal vein, and that the excrement was then carried by the ureters to the bladder.

The book ends with a discussion of human generation and the organs of reproduction. Although Vesalius denied the medieval doctrine of the seven-celled uterus and declared the traditional representation of the horned uterus to result from the use of animal specimens, his description of the fetus and fetal apparatus was of less significance, reflecting, as he admitted, the lack of sufficient pregnant human specimens.

Book VI describes the organs of the thorax. It is chiefly important for the description of the heart, which Vesalius described as approaching the nature of muscle in appearance, although it could not be true muscle since muscle supplied voluntary motion and the motion of the heart was involuntary. In this instance Vesalian principle bowed to Galenic theory, and recognition of the muscular substance of the heart had to await William Harvey's investigations in the next century.

Like all his contemporaries Vesalius regarded the heart as formed of two chambers or ventricles. The right atrium was not considered to be a chamber but rather a continuation of the inferior and superior venae cavae, considered as a single, extended vessel; and the left atrium was thought to be part of the pulmonary vein. According to Galen the ventricles were divided by a midwall containing minute openings or pores through which the blood passed or seeped from the right ventricle into the left, an opinion that Vesalius strongly questioned even though by implication he was casting doubt on Galen's cardiovascular physiology. "The septum of the ventricles having been formed, as I said, of the very thick substance of the heart . . . none of its pits—at least insofar as can be ascertained by the senses—penetrates from the right ventricle into the left. Thus we are compelled to astonishment at the industry of the Creator who causes the blood to sweat through from the right ventricle into the left through passages which escape our sight." Finally Vesalius gave strong expression to his opinion of ecclesiastical censorship over the question of the heart as the site of the soul. After referring to the opinions of the major ancient philosophers on the location of the soul, he continued:

> Lest I come into collision here with some scandalmonger or censor of heresy, I shall wholly abstain from consideration of the divisions of the soul and their locations, since today . . . you will find a great many censors of our very holy and true religion. If they hear someone murmur something about the opinions of Plato, Aristotle or his interpreters, or of Galen regarding the soul, even in anatomy where these matters especially ought to be examined, they immediately judge him to be suspect in his faith and somewhat doubtful about the soul's immortality. They do not understand that this is a necessity for physicians if they desire to engage properly in their art. . . .

The seventh and final book provides a description of the anatomy of the brain, accompanied by a series of detailed illustrations revealing the successive steps in its dissection. Until the time of Vesalius, illustrations of the brain and any accompanying text usually stressed the localization of intellectual activities in the ventricles, with perception in the anterior ventricles, judgment in the middle, and memory in the posterior. Sensation and motion were considered the work of animal spirit produced in a fine network of arteries at the base of the brain, the *rete mirabile*. The existence of the *rete mirabile* in the human brain had been questioned by Berengario da Carpi. It was now firmly denied by Vesalius, who showed the belief in this organ to have been the result of dissection of animals, since such an arterial network does in fact exist in ungulates. Vesalius was also the first to state that the ventricles had no function except the collection of fluid. Moreover, he denied that the mind could be split up into the separate mental faculties hitherto attributed to it. As a corollary he intimated that although animal spirit affected sensation and motion, it had nothing at all to do with mental activity—in short he suggested a divorce between the physical and mental animal. The discussion of the brain is concluded by a chapter on the procedure to be followed for its dissection and by a final, separate section on experiments in vivisection, derived and developed mostly from experiments described by Galen. The separate treatment of this latter material indicated a recognition of physiology as a discipline distinct from anatomy.

In the *Fabrica* Vesalius made many contributions to the body of anatomical knowledge, by description of structures hitherto unknown, by detailed descriptions of structures known only in the most elementary terms, and by the correction of erroneous descriptions. Despite his many errors his contribution was far greater than that of any previous author, and for a considerable time all anatomists, even those unsympathetic to him, were compelled to refer to the *Fabrica*. Its success and influence can be measured by the shrillness of Galenic apologists, by the plentiful but unacknowledged borrowings of many, and by the avowed indebtedness of the generous few, such as Falloppio. Although Colombo, Falloppio, and Eustachi corrected a number of Vesalius' errors and in some respects advanced beyond him in their anatomical knowledge, Colombo published his anatomical studies sixteen years after the appearance of the *Fabrica*, Falloppio eighteen, and Eustachi twenty. Furthermore, they relied heavily upon Vesalius' work, the detailed nature of which made it relatively easy for others to correct or to make further contributions. Although their accomplishments deserve recognition they were built upon Vesalian foundations.

More important than the anatomical information contained in the *Fabrica* was the scientific principle enunciated therein. This was beyond criticism, fundamental to anatomical research, and has remained so. It was not difficult to demonstrate Galen's errors of anatomy, but such a demonstration was only a means to an end. Its significance lay in the reason for those errors: Galen's attempt to project the anatomy of animals upon the human body. From time to time others had pointed to Galenic errors, but no one had proposed a consistent policy of doubting the authority of Galen or of any other recognized authority until the only true source of anatomical knowledge—dissection and observation of the human structure—had been tested. With the publication of the *Fabrica* all major investigators of anatomy were compelled to recognize the new principle, even though at first some paid no more than lip service to it.

For medical students and those with limited or no anatomical knowledge Vesalius composed a briefer work, the misnamed *Epitome* (1543) of the *Fabrica*. In the *Epitome* Vesalius returned to the tradition of the *Tabulae anatomicae* insofar as the illustrations in this work seem to have been considered more important than the text. The text was arranged somewhat differently from that of the *Fabrica*, since, although the first two chapters deal with bones and muscles, respectively, they are followed by chapters on the digestive system, cardio-

vascular system, nervous system (here including the brain), and finally the reproductive system. This is the simplified arrangement that Vesalius advocated "for one wholly unskilled in dissection." Although Vesalius called the work an epitome and declared it a pathway to the *Fabrica*, such is not the case; the vast text of the greater work could not be compressed into such slight dimensions, and it is certainly not, as he also wrote, a summary. At best it is a condensation of selections from the *Fabrica* and, hence, not a major scientific work.

The second edition of the *Fabrica* was considerably altered both in style and to some degree in the arrangement of the contents. The actual alterations of the contents, found chiefly in books V and VI, include the addition of accounts of autopsies performed by Vesalius from 1543 onward, revision and correction of the description of the fetal membranes, and a clear statement that the cardiac septum is impermeable.

The impact of the new Vesalian illustrations was reflected as early as 1538 in such plagiarisms of the *Tabulae anatomicae* as those published in Paris (1538), Augsburg (1539), Cologne (1539), and Strasbourg (1541). The much more remarkable illustrations of the *Fabrica* were subject to even greater plagiarism, the first instance being the excellent copperplates of the Flemish engraver Thomas Geminus (1544), published with a slightly altered text of the *Epitome* under the title *Compendiosa totius anatomie delineatio aere exarata* (London, 1545). If we except the uncompleted work of Canano, Geminus' book has the further distinction of being the first anatomical treatise to contain copper-engraved illustrations. These were republished in London in 1553 and 1559, and by Jacques Grévin in Paris in 1564, 1565, and 1569; the original plates were copied and recopied thereafter for many subsequent editions. One example of the many plagiarisms of the illustrations of the *Fabrica* is the much reduced, crude woodcut copies that are to be found in Bernardino Montaña de Monserrate's *Libro de anathomia del hombre* (1551), where they have, in fact, no relationship to the Galenic text of this first Spanish anatomical treatise in the vernacular. Somewhat better, larger, and more significant copies are to be found in Ambroise Paré's *Anatomie universelle* (1561).

More fundamentally influential were the Vesalian principle underlying anatomical investigation, the method, and the contributions to knowledge of human anatomical structures. These were somewhat slower in diffusion and occasionally met opposition, as in Jacobus Sylvius' violent attack against Vesalius' anti-Galenism, *Vaesani cujusdam calumniarum in Hippocratis Galenique rem anatomicam depulsio* (1551) and the later attack of Francesco dal Pozzo, *Apologia in anatome pro Galeno contra Andream Vessalium Bruxellensem* (1562). Advanced by the successive occupants of the anatomical chair at Padua (Realdo Colombo, Gabriele Falloppio, and Fabrici), the Vesalian principles were thence diffused through Italy and later throughout western Europe. By the beginning of the seventeenth century, with the exception of a few conservative centers such as Paris and some parts of the Empire, Vesalian anatomy had gained both academic and general support.

BIBLIOGRAPHY

The various editions of Vesalius' writings and most of the literature about him are to be found in Harvey Cushing, *A Bio-Bibliography of Andreas Vesalius*, 2nd ed. (Hamden, Conn., 1962); more recent papers and studies of importance, some of them the result of the Vesalian celebrations of 1964, are listed in C. D. O'Malley, "A Review of Vesalian Literature," in *History of Science*, IV (Cambridge, 1965), 1–14. M. H. Spielmann, *The Iconography of Andreas Vesalius* (London, 1925), deals with the various likenesses of Vesalius produced since the sixteenth century, although the only genuine portrait known is that in the *Fabrica*. The standard biography is C. D. O'Malley, *Andreas Vesalius of Brussels, 1514–1564* (Berkeley–Los Angeles, 1964); and particular points of importance have been dealt with by Charles Singer, *A Prelude to Modern Science* (Cambridge, 1946), and Ruben Ericksson, *Andreas Vesalius' first Public Anatomy at Bologna 1540* (Uppsala, 1959). See also Moritz Roth, *Andreas Vesalius Bruxellensis* (Berlin, 1892).

C. D. O'MALLEY

VESLING, JOHANN (*b.* Minden, Germany, 1598; *d.* Padua, Italy, 30 August 1649), *anatomy, botany.*

Vesling's reputation rests on his excellent powers of observation. Nothing certain is known about his parents, although it appears that the Catholic family fled to Vienna to escape religious persecution. All the biographers agree that Vesling attended secondary school and studied medicine in Venice, but his name is not found among the registration records of the university. Since Vesling stated that Everhardius Vorstius of Leiden was his teacher, an examination of the records of the university reveals that Johannes Wesling of Minden enrolled as a student at Leiden at age twenty on 15 November 1619. Vorstius had studied at German and Italian universities, including Bologna, and was especially interested in botany as well as in medicine. Vesling, too, had a predilection for botany, and he also went to Bologna, presumably on Vorstius' advice. Vesling named Fabrizio Bartoletti as his teacher in Italy. Bartoletti, who moved from Bologna to Mantua after 1675, instilled in Vesling an enthusiasm for anatomy and surgery.

The next documented fact that we have concerning Vesling's career dates from the winter of 1627–1628. He performed, in the presence of Venetian physicians, an anatomical demonstration that earned him the right to practice in the areas controlled by Venice. His teaching was so highly esteemed that even Paduan students came to hear him. The Venetian government unfairly refused to reimburse Vesling for the expenses he incurred in conducting his demonstrations. This was probably done in an attempt to drive him from the city in order to protect the much older Paduan professor Caimo from the competition of his younger colleague. For the same reason, Vesling was directed to serve as physician to the patrician Alvise Cornaro during the latter's term as Venetian representative in Cairo. The two men left for Egypt at the beginning of August 1628. Vesling studied the flora of the country with great interest. In many cases his observations were more accurate than those previously made by Prospero Alpini, and his book on the subject is also better illustrated than the latter's. Of particular interest are his comments on the coffee plant.

All of Vesling's biographers assign an earlier date to his trip to Egypt, despite the fact that Haller long ago published the correct information on the basis of letters Vesling sent to Wilhem Fabry (Hildanus Fabricius). Only Adelmann cites this reference and mentions that a date contained in Vesling's working notes makes it certain that he was still in Egypt on 7 May 1632. On that day Vesling repeated his investigations of the development of the chick embryo in artificially hatched eggs. The results of these embryological studies are very fully discussed by Adelmann, and the reader should consult his account for details. The place names that Vesling mentions in his writings show that he was familiar with only a small part of Egypt, that between Rashid (Rosetta) and Memphis. He traveled to Palestine only once, on which occasion he became a knight of the Order of the Holy Sepulcher in Jerusalem.

Because of his stay in Egypt, Vesling escaped the epidemic of plague that ravaged northern Italy in 1629–1631. During the epidemic students avoided Padua, and the chair of anatomy was vacant for a year. On 30 December 1632 Vesling was appointed professor of anatomy and surgery, and at the beginning of 1633 he returned from Egypt. Vesling proved to be a very able teacher and enlivened his lectures with drawings that he himself had prepared and that were later used in his *Syntagma anatomicum*. This textbook, characterized by a concise style, went through many editions and was translated into several languages. Of particular scientific value are his descriptions and illustrations of the chyle vessels (lacteals) and his assertion that four is the normal number of pulmonary veins emptying into the left auricle of the heart. Further, he was the first to see the *ductus thoracicus*, but he did not mention the discovery until 1649, in a letter to Thomas Bartholin.

In 1638 Vesling ceased lecturing on surgery and turned instead to botany. Under his direction the botanical garden in Padua was renovated, as several plant catalogues of the period show. In 1648 Vesling was given a leave of absence that allowed him to undertake a second botanical expedition. He went to Crete but returned ill, and he died soon after. In accordance with his wishes he was buried in the cloister of the church of St. Anthony in Padua. Vesling's posthumous papers contain much remarkable material that he would undoubtedly have formulated in more precise terms. Among the more notable things to be found in these papers (which were published by Thomas Bartholin) is a correction of his initial findings concerning the sexual organs of the viper and the scent glands of the snakes.

BIBLIOGRAPHY

I. Original Works. Vesling's most important anatomical work is *Syntagma anatomicum, publicis dissectionibus in auditorum usum diligenter aptatum* (Padua, 1641). Along with many Latin editions, this work appeared in Dutch, German, and English translations—the last under the title *The Anatomy of the Body of Man*, N. Culpeper, trans. (London, 1653).

Vesling's most important work in botany is *De plantis aegyptiis observationibus et notae ad Prosperum Alpinum, cum additamenta aliarum eiusdem regionis* (Padua, 1638). Vesling's embryological and comparative anatomical investigations were published by Thomas Bartholin as *De pullitione Aegyptiorum et aliae observationes anatomicae et epistolae medicae posthumae* (Copenhagen, 1664).

II. Secondary Literature. Two biographical accounts are Arturo Castiglioni, in *Enciclopedia Italiana*, XXXV (1937), p. 218; and A. Francesco La Cava, in

Castalia (Milan, 1948). For an assessment of Vesling's contribution to the history of embryology see Howard B. Adelmann, *Marcello Malpighi and the Evolution of Embryology* (Ithaca, N.Y., 1966). Vesling's comparative anatomical studies are discussed in F. C. Cole, *A History of Comparative Anatomy* (London, 1944).

Erich Hintzsche

VESSIOT, ERNEST (*b.* Marseilles, France, 8 March 1865; *d.* La Bauche, Savoie, France, 17 October 1952), *mathematics*.

Vessiot's ancestors were farmers near Langres, in the Haute-Marne. The family rose slowly in the social hierarchy, becoming teachers and, later, school principals. Vessiot's father was a *lycée* teacher and subsequently inspector general of primary schools. Vessiot, the third of six children, became a university professor and member of the Académie des Sciences.

A good record at the *lycée* in Marseilles enabled Vessiot to attend the École Normale Supérieure, which he entered second in his class, after Jacques Hadamard. In 1887 he obtained a teaching post at the *lycée* in Lyons. After receiving the doctorate in 1892, he taught at the universities of Lille, Toulouse, Lyons, and, finally, Paris (1910). Vessiot's first assignment at Paris was to prepare students for the *licence*. Later he taught courses in the theory of functions, in analytical mechanics, and in celestial mechanics. He became director of the École Normale Supérieure (serving in this post until his retirement in 1935) and was elected to the mechanics section of the Académie des Sciences in 1943.

Vessiot's research dealt with the application of the notion of continuous groups, finite or infinite, to the study of differential equations. Extending results obtained by Émile Picard, Vessiot demonstrated in his dissertation (1892) the existence of a group of linear substitutions with constant coefficients operating on a system of n independent solutions of a differential equation. The rigor and depth of his work on groups of linear rational transformations allowed Vessiot to put into more precise form and to develop research begun by Jules Drach (1902) and to extend the results of Élie Cartan on the integration of differential systems (1907). He also completed Volterra's study of Fredholm integrals. The extension of these integrals to partial differential equations led Vessiot to obtain original results concerning perturbations in celestial mechanics, the propagation of waves of discontinuity, and general relativity. During World War I, Vessiot was assigned to work on problems in ballistics, and he corrected certain empirical formulas then in use.

A dedicated teacher, Vessiot wrote useful and well-received textbooks. As director of the École Normale Supérieure he supervised the construction of new laboratories in collaboration with his physicist colleagues Henri Abraham, Léon Bloch, and Georges Bruhat, all of whom fell victim to the Nazis during German occupation.

BIBLIOGRAPHY

Vessiot's writings include "Sur l'interprétation mécanique des transformations de contact infinitésimales," in *Bulletin de la Société mathématique de France*, **34** (1906), 230–269; "Essai sur la propagation des ondes," in *Annales scientifiques de l'École normale supérieure*, 3rd ser., **26** (1909), 404–448; "Sur la réductibilité et l'intégration des systèmes complets," *ibid.*, **29** (1912), 209–278; "Sur la théorie des multiplicités et le calcul des variations," in *Bulletin de la Société mathématique de France*, **40** (1912), 68–139; *Leçons de géométrie supérieure* (Paris, 1919); "Sur une théorie nouvelle des problèmes d'intégration," in *Bulletin de la Société mathématique de France*, **52** (1924), 336–395; "Sur la réductibilité des équations algébriques ou différentielles," in *Annales scientifiques de l'École normale supérieure*, 3rd ser., **57** (1940), 1–60; **58** (1941), 1–36; and **63** (1946), 1–23, also in *Bulletin de la Société mathématique de France*, **75** (1947), 9–26; and *Cours de mathématiques générales*, 3 vols. (Paris, 1921–1952), written with Paul Montel.

For a discussion of Vessiot's work see Élie Cartan, "L'oeuvre scientifique de M. Ernest Vessiot," in *Bulletin de la Société mathématique de France*, **75** (1947), 1–8.

LUCIENNE FÉLIX

VICQ D'AZYR, FÉLIX (*b*. Valognes, Manche, France, 28 April 1748; *d*. Paris, France, 20 June 1794), *anatomy, epidemiology, medical education.*

Vicq d'Azyr was a member of the Académie Française and the Académie des Sciences, as well as permanent secretary of the Société Royale de Médecine. He also was substitute professor at the Jardin du Roi and personal physician to Marie Antoinette.

After studying at Caen, Vicq d'Azyr went to Paris in 1765 and developed a marked interest in anatomy and physiology. In 1773 he gave a private course in those subjects, which attracted a large audience. He became the student of Antoine Petit (1722–1794) and of Daubenton, professor at the Jardin du Roi, whose niece he married. In 1774

Vicq d'Azyr earned his medical degree and on 16 March was elected to the Académie des Sciences as *adjoint anatomiste*, replacing Portal.[1] He soon demonstrated his considerable talents and in 1775 was sent by Turgot to stem a serious epizootic disease ravaging the Midi. The mission was successful, and in 1778 Vicq d'Azyr was named permanent secretary of the Société Royale de Médecine. On 12 December 1784 he was elected *associé anatomiste* of the Académie des Sciences, replacing Petit and became *associé* in anatomy on 23 April 1785. He received a double honor in 1788, with his election to membership of the Académie Française and his appointment as physician to the queen, who called him "mon philosophe."

During the Revolution, Vicq d'Azyr's position was ambiguous; although he continued to serve the queen, he remained a guiding spirit in the movement for medical reform, of which he had been a pioneer. Among his friends, Bailly and Lavoisier were guillotined; whereas others, such as Fourcroy, if not responsible for the executions, at least condoned them. Like his contemporary Pierre Desault, Vicq d'Azyr was obsessed by the fear that he was under suspicion and would be taken before the Revolutionary Tribunal. After attending the festival of the Supreme Being he came down with a violent fever and died in a delirium—an external manifestation (it was said) of the inner tensions and obsessions that had made rest impossible for him.

Vicq d'Azyr was greatly interested in comparative anatomy. Before the changes that he and J. F. Blumenbach made in the study of the subject, it was customary to dissect animals of different species and juxtapose their various organs. The species were poorly delimited, and Vicq d'Azyr showed that, in any case, the object of the science was different. According to him, what was most important was to decide the significant characteristics (structures of limbs, of extremities, and of pectoral and pelvic girdles; mode of remaining at rest and of moving; dentition; and so forth), to study their variations systematically throughout the animal kingdom, and to develop a nomenclature based on that used by the chemists. Toward this end he began work on *Système anatomique des vertébrés*, continued by Hippolyte Cloquet (1787–1840).

Vicq d'Azyr devoted particular attention to neuroanatomy, studying the cervical plexus and especially the vertebrate brain. In his research he introduced quantitative data, comparing the weight of the brain with body weight. In morphology he followed the majority of contemporary anatomists in neglecting the ventricles, which in the older theories had been considered reservoirs of the "animal spirits."

Like Steno before him, Vicq d'Azyr attached great importance to the structure of the fibers in the white matter of the brain. He wrote four memoirs on them, sometimes showing them in transverse and frontal sections and sometimes "scraping them without damaging the surface." (The latter technique was later perfected by Gall and Spurzheim.) He also described the mammillothalamic bundle and Reil's ribbon.

Rejecting the views of Malpighi and Vieussens, who attributed no functional importance to the cerebral cortex, Vicq d'Azyr attempted to systematize its complex morphology. In particular he isolated the convolution of the corpus callosum, the cuneus, and the sulcus separating the frontal lobe from the parietal, later described by Rolando (1829).[2]

Despite his enthusiasm for dissection, Vicq d'Azyr was not satisfied with the results obtained by this technique: "Seeing and describing are two things that everyone believes he is capable of doing; yet few people really can do them. The first requires great powers of concentration and insights in dealing with the type of object observed; the second requires method and knowledge of the terms necessary for conveying an exact idea of what one has seen" (*Oeuvres*, IV, 208). This statement probably alludes to the absence of a nomenclature for the cerebral convolutions, which at the time were no better described than those of the intestine. It also reflects a contradictory aspect of the author's thought that has been stressed by G. Lanteri-Laura.[3] Vicq d'Azyr was among those who continued to believe in a very simple neuroanatomical schema in which physiological considerations were predominant. This conception left little room for embryology, comparative anatomy, experimentation, or clinical findings, and led to some very controversial cerebral localizations. The scientists who held this view seem to have been unaware that it was rapidly being undermined by their own discoveries and methods. Paradoxically, while their research was making the complexity of brain structures increasingly evident, their theories helped to delay a thorough study of these structures.

Vicq d'Azyr was an eminent veterinarian. In the eighteenth century, animal medicine was economically important for French farmers and aristocrats. It was no less important ideologically for the two major "philosophical" groups in France: the Physi-

ocrats, who were concerned primarily with agronomic and social questions, and the Encyclopedists, who did not neglect nature but were less interested than the former group in the practical consequences of their teaching. Fortunately, Turgot and Exupère Bertin convinced the government that responsibility for dealing with epizootic diseases could not be entrusted to self-taught farriers. As a result the government founded specialized veterinary schools at Lyons (1762) and at Alfort (1765). In these schools, the term *maréchalerie* was replaced by that of *art vétérinaire* starting in 1767.

Veterinary teaching was then influenced by two contending opinions. The Physiocrats held that veterinary medicine should be allied with agriculture, while the Encyclopedists thought it more properly belonged with human medicine. The latter view, which was shared by Buffon and a number of physicians who dealt with both humans and animals (notably Nicolas Chambon de Montaux, Daubenton, and Vicq d'Azyr at Paris, and Vitet and Jacques Petetin at Lyons), temporarily triumphed following the death of Claude Bourgelat (1779) and the retirement of Bertin (1780). At this time Vicq d'Azyr taught at Alfort and wrote a number of works on comparative anatomy, since the material necessary for studying the subject was readily available. He also conceived a project for reorganizing medical studies, one aspect of which involved combining in a single overall curriculum the teaching of animal medicine (represented by the school at Alfort) and of human medicine (represented by the Faculty of Medicine, the schools of surgery, the Collège Royal, the Jardin du Roi, and the Jardin des Apothicaires).

Vicq d'Azyr was no mere theorist of veterinary medicine. He showed his ability in practice and played a particularly important part in the fight against epizootic diseases started by Turgot in 1775–1776. His *Instruction sur la manière de désinfecter une paroisse* (1775) is particularly interesting in this regard. Taking a realistic view of what was possible, the work delegates more responsibility to the government and the army than to the veterinarians. It recommends an operation encompassing diagnosis of the disease, isolation of the suspected area by a cordon of troops, quarantine, treatment, and in some cases destruction of the animals, followed by repopulation. The basic weapon in this struggle was the disinfection of hides and stables with chemicals (sulfuric acid, sulfur, gunpowder), which were substituted for the traditional perfumes with aromatic plants. This technique proved so effective that the French decrees of 1881 and 1898, which are still in effect, essentially reproduced the recommendations that Vicq d'Azyr made in 1776.

Vicq d'Azyr's publications on human medicine are of little importance. A number of them appeared under "Médecine" in the *Encyclopédie méthodique*, of which he was an editor. On the other hand, he was a pioneer in public health and medical education. As permanent secretary of the Société Royale de Médecine, Vicq d'Azyr maintained a network for exchanging information and conducting studies on a national scale for fifteen years. In this effort he was assisted by practicing physicians, medical schools, and provincial academies. The inquiries he initiated into *topographie médicale et salubrité* encouraged many provincial doctors to take a more modern and nationally oriented view of health problems and helped foster the idea that public health is one of the responsibilities of the medical profession.[4] Furthermore, while acting as medical expert for the Constituent Assembly, Vicq d'Azyr created and guided the activities of its Comité de Salubrité. He called upon the physicians who had been elected deputies to support adoption of a project for the reform of French medicine that he had completed in 1790. The project, approved by the Constituent Assembly, was adopted without major changes by the National Convention in 1794, following Fourcroy's report on it.

NOTES

1. The archives of the Académie des Sciences contain a letter from the duke of La Vrillière, dated 13 Mar. 1774, on the election of Vicq d'Azyr and of Toussaint Bordenave to the Academy.
2. See P. Broca, "Note sur la topographie cérébrale et sur quelques points de l'histoire des circonvolutions."
3. See G. Lanteri-Laura, *L'homme et son cerveau selon Gall. Histoire et signification de la phrénologie.*
4. See D. Weiner, "Le droit de l'homme à la santé."

BIBLIOGRAPHY

I. ORIGINAL WORKS. There is no comprehensive bibliography of the works of Vicq d'Azyr. Many of his papers are in boxes of material from the Société Royale de Médecine, now in the library of the Académie Nationale de Médecine; the catalog of these cartons has not yet been published. They contain letters and autograph memoirs of Vicq d'Azyr from the Société Royale de Médecine (fols. 159, 160; MS 33 [33]). The library of the Muséum National d'Histoire Naturelle has various notes and annotated extracts from the minute books of the Société Royale de Médecine for 1779 and 1780 (MSS 1452–1459), in 8 vols. At the Archives Nationales (ser. AF I^{23}) there are the original versions of the *procès verbaux* of the meetings of the Comité de Salubrité of the Constituent Assembly; cartons F^{17} 1236–1239, F^{17} 1245–1246, and F^{17} 1094 contain reports by and correspondence from Vicq d'Azyr on various matters. The dossier on Vicq d'Azyr at the Académie des Sciences contains a dictated letter of 12 Oct. 1777 to a colleague; an autograph of June 1780, "Examens d'enfants atteints d'anomalies osseuses et de déformations"; an undated letter to a M. Perrier concerning the examination of the water of the Seine by means of a special pump (only the postscript is autograph); a letter concerning the anatomical observation of a mandrill, a callithrix, and a macaque, which was published in the *Mémoires* of the Académie des Sciences for 1780; another undated autograph; an autograph memoir on the anastomoses of Delafosse (undated); and an autograph memoir on the anatomical observation of a thirty-six-year-old woman suffering from pains in the uterus, followed by death.

Vicq d'Azyr published many articles in *Journal des sciences et des beaux-arts, Mercure de France, Clef du cabinet des souverains,* and *Journal des savants.* His most important articles were published in the *Mémoires* of the Académie des Sciences (from 1772 to 1785) and in the *Mémoires* of the Société Royale de Médecine de Paris (from 1775 to 1788). A fairly complete listing of these works is in J. D. Reuss, ed., *Repertorium commentationum,* 16 vols. (Göttingen, 1801–1821; repr. New York, 1962)—see the indexes to vols. II, III, VI, VIII, X, XV, and XVI. The Royal Society *Catalogue of Scientific Papers,* V, 152, lists two papers published in *Bulletin des sciences de la Société philomathique de Paris,* to which should be added "Observations sur un bruit singulier dans la région du coeur d'un particulier" (July 1791), 22.

A collected ed. of Vicq d'Azyr's works, with annotations and a discussion of his life and writings by J. L. Moreau, was published as *Oeuvres de Vicq-d'Azyr,* 6 vols. (Paris, 1805).

Space does not permit the listing of the many separately published works that appeared during Vicq d'Azyr's lifetime; they include translations, dissertations for which he served as chairman, and *éloges,* which are included in his *Oeuvres* (see above). Many of these works are listed in Bibliothèque Nationale, *Catalogue général des livres imprimés,* CCVIII (Paris, 1970), cols. 311–322. The following bibliography (listed in alphabetical order), although incomplete, is representative of Vicq d'Azyr's scientific publications: *Avis aux habitants des campagnes où règne la contagion* (Condom, 1774); *Avis important relativement aux bestiaux atteints de la maladie épizootique* (Condom, 1775); *Consultation sur le traitement qui convient aux bestiaux attaqués de l'épizootie* (Bordeaux, 1775); *De l'influence des marais* sur la santé . . . (Paris, 1790); *Dictionnaire de médecine de l'Encyclopédie méthodique*, 14 vols. (Paris, 1787–1830)–Vicq d'Azyr contributed many articles to the first six vols.; *Discours sur l'anatomie comparée* (Paris, n.d.); *Exposé des moyens curatifs et préservatifs qui peuvent être employés contre les maladies pestilentielles des bêtes à cornes* (Paris, 1776); *Instruction relative à l'épizootie (pour les soldats)* (Rouen, 1775); *Instruction relative à l'épizootie (pour les syndics)* (Rouen, 1775); *Instruction sur la manière de désinfecter les cuirs des bestiaux morts de l'épizootie* . . . (Paris, 1775); *Instruction sur la manière de désinfecter les étables des bestiaux attaqués de l'épizootie* (Paris, 1776); *Instruction sur la manière de désinfecter les villages* (Paris, 1775); *Instruction sur la manière de désinfecter une paroisse* (Paris, 1775); *Instruction sur la manière d'inventorier et de conserver. . . tous les objets qui peuvent servir aux arts* . . . (Paris, 1794), attributed to Vicq d'Azyr; *La médecine des bêtes à cornes*, 2 vols. (Paris, 1781); *Nouveau plan de conduite pour détruire entièrement la maladie épizootique* (Lille, n.d.); *Observations sur les moyens que l'on peut employer pour préserver les animaux sains de la contagion* . . . (Bordeaux, 1774); *Précaution pour la purification des étables* (n.p., n.d.); *Recueil d'observations . . . sur différentes méthodes . . . pour guérir la maladie épidémique qui attaque les bêtes à cornes* . . . (Paris, 1775); *Système anatomique. Quadrupèdes* (Paris, 1792); *Traité d'anatomie et de physiologie* (Paris, 1786); *Traité de l'anatomie du cerveau* (Paris, 1813); and *Table pour servir à l'histoire anatomique et naturelle des corps vivants* (n.p., n.d. [Paris, 1774]).

II. SECONDARY LITERATURE. Articles on Vicq d'Azyr are in Michaud, *Biographie universelle*, XLVIII, 374–378; *Encyclopédie méthodique*, XIII (Paris, 1830), 446–455; Dezeimeris's *Dictionnaire historique de la médecine ancienne et moderne*, IV (Paris, 1839), 330–334; Bayle and Thillaye's *Biographie médicale*, II (Paris, 1855), 718–720; and *Biographisches Lexicon*, V (Berlin–Vienna, 1934), 747–749.

Other works are L. Barbillion, "Vicq d'Azyr," in *Paris médical*, **62** (1926), 309–311 (appendix); F. G. Boisseau and C. Cavenne, "Vicq d'Azyr," in *Dictionnaire des sciences médicales. Biographie médicale*, VII (Paris, 1825), 429–432; Paul Broca, "Note sur la topographie cérébrale et sur quelques points de l'histoire des circonvolutions," in *Bulletin de l'Académie de médecine*, 2nd ser., **5** (1876), 824–834; P. J. G. Cabanis, "Éloge de Vicq d'Azyr," in his *Oeuvres complètes*, V (Paris, 1825), 177–216; G. Cuvier, *Histoire des sciences naturelles depuis leur origine jusqu'a nos jours*, 5 vols. (Paris, 1841–1845), IV, 297–305; V, 45, 379; J. Dobson, *Anatomical Eponyms* (Edinburgh–London, 1962); Michel Dronne, *Bertin et l'élevage français au XVIII siècle* (Alfort, 1965), 145–212; F. Dubois, "Recherches historiques sur les dernières années de Louis et de Vicq d'Azyr, secrétaires perpétuels de la Société royale de médecine et de la Société royale de chirurgie.

L'histoire de la guillotine," in *Journal des connaissances médicales pratiques et de pharmacologie*, **34** (1867), 17–19, 33–37, 49–51; A. J. L. M. Dufresne, *Notes sur la vie et les oeuvres de Vicq d'Azyr (1748–1794). Histoire de la fondation de l'Académie de médecine* (Bordeaux, 1906), diss. for the M.D. (no. 65); and the anonymous "Bibliographical Sketch on Vicq d'Azyr," in *Edinburgh Medical and Surgical Journal*, **3** (1807), 180–185.

See also R. L. M. Faull, D. M. Taylor, and J. B. Carman, "Soemmering and the Substantia Nigra," in *Medical History*, **12** (1968), 297–299; M. Genty, "Vicq d'Azyr commissaire pour l'extraction du salpêtre," in *Progrès médical*, supp. ill. no. 2 (1936), 9–16; C. A. de Gerville, "Vicq d'Azyr," in his *Études géographiques et historiques sur le département de la Manche*, XL (Cherbourg, 1854), 284; Lucien Hahn, "Vicq d'Azyr," in *Dictionnaire encyclopédique des sciences médicales*, 5th ser., III (Paris, 1889), 452–453; H. Hours, *La lutte contre les épizooties et l'école vétérinaire de Lyon au XVIIIe siècle* (Paris, 1957), 90–92; P. Huard, "L'enseignement médico-chirurgical," in R. Taton, ed., *Enseignement et diffusion des sciences en France au XVIIIème siècle* (Paris, 1964), 170–236; Kaime, "Une anecdote de la vie de Vicq d'Azyr," in *Revue de thérapeutique médico-chirurgicale* (1859), 249; Claude Lafisse, *Éloge de Vicq d'Azyr* . . . (Paris, 1797), also in *Recueil périodique de la Société de médecine de Paris*, III (Paris, 1798), 201–226; G. Lanteri-Laura, *L'homme et son cerveau selon Gall. Histoire et signification de la phrénologie* (Paris, 1970); and P. E. Lemontey, *Éloge historique de Vicq d'Azyr, prononcé . . . 23 août 1825* (Paris, n.d.).

Further works are L. F. A. Maury, *Les académies d'autrefois. L'ancienne Académie des sciences* (Paris, 1864); L. Merle, "La vie et l'oeuvre du Dr. Jean Gabriel Gallot (1744–1794)," in *Mémoires de la Société des antiquaires de l'Ouest* (Poitiers), 4th ser., **5** (1962); J. L. Moreau, *Éloge de Vicq d'Azyr suivi d'un précis des travaux anatomiques et physiologiques de ce célèbre médecin* (Paris, 1798); J. Noir, "Un savant, un innovateur et un réalisateur. Félix Vicq d'Azyr," in *Concours médical* (6 Apr. 1927), 927–929; Félix Pascalis Ouvrière, *An Exposition of the Dangers of Interment in Cities* (New York, 1824); J. Roger, *Les médecins normands*, II (Paris, 1895), 169–181; C. A. Sainte Beuve, "Notice sur Vicq d'Azyr," in *Union médicale*, **8** (1854), 355–356, 359–361, 371–372; A. C. Saucerotte, "Vicq d'Azyr," in *Nouvelle biographie générale*, XLV (Paris, 1866), 89–91; J. A. Sharp, "Alex Monro Secundus and the Interventricular Foramen," in *Medical History*, **5** (Jan. 1961), 83–89; W. A. Smeaton, *Fourcroy, Chemist and Revolutionary* (Cambridge, 1962), *passim*; and Dora Weiner, "Le droit de l'homme à la santé. Une belle idée devant l'Assemblée constituante, 1790–91," in *Clio medica*, **5** (1970), 209–223.

P. HUARD
M. J. IMBAULT-HUART

VIDUS VIDIUS. See Guidi, Guido.

VIÈTE, FRANÇOIS (*b.* Fontenay-le-Comte, Poitou [now Vendée], France, 1540; *d.* Paris, France, 23 February 1603), *mathematics.*

Viète's father, Étienne, was an attorney in Fontenay and notary at Le Busseau. His mother was Marguerite Dupont, daughter of Françoise Brisson and thus a first cousin of Barnabé Brisson. Viète was married twice: to Barbe Cothereau and, after her death, to Juliette Leclerc. After an education in Fontenay, Viète entered the University of Poitiers to study law. He received a bachelor's degree in law in 1560 but four years later abandoned the profession to enter the service of Antoinette d'Aubeterre, mother of Catherine of Parthenay, supervising the latter's education and remaining her loyal friend and adviser throughout his life. After Antoinette d'Aubeterre was widowed in 1566, Viète followed her to La Rochelle. From 1570 to 1573 he was at Paris, and on 24 October of that year Charles IX appointed him counselor to the *parlement* of Brittany at Rennes. He remained at Rennes for six years, and on 25 March 1580 he became *maître de requêtes* at Paris (an office attached to the *parlement*) and royal privy counselor. From the end of 1584 until April 1589 Viète was banished from the royal court by political enemies and spent some time at Beauvoir-sur-Mer. He was recalled to court by Henry III when the latter was obliged to leave Paris and to move the government to Tours, where Viète became counselor of the *parlement*. During the war against Spain, Viète served Henry IV by decoding intercepted letters written in cipher. A letter from the liaison officer Juan de Moreo to Philip II of Spain, dated 28 October 1589, fell into Henry's hands. The message, in a new cipher that Philip had given Moreo when he departed for France, consisted of the usual alphabet with homophonous substitutions, plus a code list of 413 terms represented by groups of two or three letters or of two numbers, either underlined or dotted. A line above a two-digit group indicated that it could be ignored. It was not until 15 March 1590 that Viète was able to send Henry the completed solution, although he had previously submitted parts of it. He returned to Paris in 1594 and to Fontenay in 1597. He was in Paris in 1599 but was dismissed by Henry IV on 14 December 1602.

Viète had only two periods of leisure (1564–1568 and 1584–1589). His first scientific works were his lectures to Catherine of Parthenay, only one of which has survived in a French translation: *Principes de cosmographie, tirés d'un manuscrit de Viette, et traduits en françois* (Paris, 1637). This tract, containing essays on the sphere, on the elements of geography, and on the elements of astronomy, has little in common with his "Harmonicon coeleste," which was never published but is available in manuscript (an autograph in Florence, Biblioteca Nazionale Centrale, MSS della Biblioteca Magliabechiana, cl. XI, cod. XXXVI, and a copy in cod. XXXVII; a copy by G. Borelli in Rome, Biblioteca Nazionale Centrale Vittorio Emanuele II, fondo San Pantaleone; and the Libri-Carucci copy in Paris, Bibliothèque Nationale, fonds lat. 7274. Part of the treatise is in Paris, Bibliothèque Nationale fonds Nouv. acqu. lat. 1644, fols. 67^r–79^v; and a French index of the part of Bibliothèque Nationale, fonds lat. 7274, is in Bibliothèque Nationale, Nouv. acqu. franç. 3282, fols. 119^r–123^r. The "Harmonicon coeleste," in five books, is Ptolemaic because Viète did not believe that Copernicus' hypothesis was geometrically valid).

All of Viète's mathematical investigations are closely connected with his cosmological and astronomical work. The *Canon mathematicus, seu ad triangula cum appendicibus*, publication of which began in 1571, was intended to form the preparatory, trigonometric part of the "Harmonicon coeleste." The *Canon* is composed of four parts, only the first two of which were published in 1579: "Canon mathematicus," which contains a table of trigonometric lines with some additional tables, and "Universalium inspectionum ad Canonem mathematicum liber singularis," which gives the computational methods used in the construction of the canon and explains the computation of plane and spherical triangles with the aid of the general trigonometric relations existing among the determinant components of such triangles. These relations were brought together in tables that allow the relevant proportion obtaining among three known and one unknown component of the triangle to be read off directly. The two other parts, devoted to astronomy, were not published. Viète certainly knew the work of Rheticus, for he adopted the triangles of three series that the latter had developed.

The *Canon* has six tables, the first of which gives, minute by minute, the values of the six trigonometric lines. For the construction of this table Viète applied the method given by Ptolemy in his *Almagest*, which was improved by the Arabs and introduced into the West through the translation of al-Zarqālī's *Canones sive regulae super tabulas astronomiae* by Gerard of Cremona; John de Lignères's *Canones tabularum primi mobilis* in the fourteenth century; and John of Gmunden's *Tractatus de sinibus, chordis et arcubus*, which inspired Peurbach and Regiomontanus. All these took as their point of departure an arc of 15° called a *kardaga*. The second table, "Canon triangulorum laterum rationalium," was based on the following proposition: "If there is a right-angled triangle having h for the hypotenuse, b for the base, and p for the perpendicular, and the semi-difference $(h - p)/2 = 1$; then $h = (b^2/4) + 1$ and $p = (b^2/4) - 1$. If b is given successive values of an arithmetical progression, the difference will be constant in the table of values of h and p thus formed. The third table, "Ad logisticem per $E\xi\epsilon\chi o\nu\tau\alpha\delta\alpha\varsigma$ tabella," is a multiplication table in the form of a right triangle that immediately gives, in degrees and minutes, the product $n \cdot n'/60$ for all the numbers n and n' included between 0 and 60. "Fractionum apud mathematicos usitarum, alterius in alterum reductionibus tabella adcommodata," the fourth table, gives the quotients obtaining by dividing the Egyptian year, the day, and the hour, and their principal subdivisions by each other and also by the most commonly used integers. The fifth table, "Mathematici canonis epitome," gives the values of the trigonometric lines from degree to degree and the length of the arc expressed in parts of the radius. The sixth table, "Canon triangulorum ad singulas partes quadranti circuli secundum $E\xi\epsilon\chi o\nu\tau\alpha\delta\omega\nu$ logisticem," gives the value of the six trigonometric lines from degree to degree, the radius 1 being divided into sixty parts, each part into sixty primes, and each prime into sixty seconds.

After the canon of triangles with rational sides, in the second part of the *Canon*, Viète gave as functions of the radius the values of the sides of inscribed polygons with three, four, six, ten, and fifteen sides and the relations that exist among these trigonometric lines, which permit easy calculation of the tables. In his solution of oblique triangles, Viète solved all the cases (except where three sides are given) by proportionality of sides to the sines of the angles opposite the sides; for the case of three sides, he follows the ancients in subdividing the triangle into right triangles. For spherical triangles he employed the same notation as for plane triangles and established that a spherical right triangle is determined by the total sine and two other elements. In spherical oblique triangles, Viète followed the ancients and Regiomontanus in subdividing the triangle into two right triangles by

an arc of a great circle perpendicular to one of the sides and passing through the vertex of the angle opposite. Also in the second part of the *Canon*, Viète wrote decimal fractions with the fractional part printed in smaller type than the integral and separated from the latter by a vertical line.

The most important of Viète's many works on algebra was *In artem analyticem isagoge*, the earliest work on symbolic algebra (Tours, 1591). It also introduced the use of letters both for known quantities, which were denoted by the consonants B, C, D, and so on, and for unknown quantities, which were denoted by the vowels. Furthermore, in using A to denote the unknown quantity x, Viète sometimes employed A quadratus, A cubus . . . to represent x^2, x^3. . . . This innovation, considered one of the most significant advances in the history of mathematics, prepared the way for the development of algebra.

The two main Greek sources on which Viète drew appear in the opening chapter: book VII of Pappus' *Collection* and Diophantus' *Arithmetica*. The point of departure for Viète's "renovation" was his joining of facts, presented by Pappus only in reference to geometric theorems and problems, to the procedure of Diophantus' *Arithmetica*. On the basis of Pappus' exposition, Viète called this procedure *ars analytice*. In chapter 1 he undertook a new organization of the "analytic" art. To the two kinds of analysis mentioned by Pappus, the "theoretical" and the "problematical" (which he called "zetetic," or "seeking [the truth]," and "poristic," i.e.; "productive [of the proposed theorem]"), he added a third, which he called "rhetic" ("telling" with respect to the numbers), or "exegetic" ("exhibiting" in respect to the geometric magnitudes). He defined the new kind of analysis as the procedure through which the magnitude sought is produced from the equation or proportion set up in canonical form.

In chapter 2 Viète amalgamates some of the "common notions" enumerated in book I of Euclid's *Elements* with some definitions and theorems of book V, of the geometric books II and VI, and of the "arithmetical" books VII and VIII to form his stipulations for equations and proportions. In chapter 3 he gives the fundamental "law of homogeneity," according to which only magnitudes of "like genus" can be compared with each other, and in the fourth chapter he lays down "the canonical rules of species calculation." These correspond to the rules for addition, subtraction, multiplication, and division used for instruction in ordinary calculation. In this chapter he presents a mode of calculation carried out completely in terms of "species" of numbers and calls it *logistice speciosa*—in contrast with calculation using determinate numbers, which is *logistice numerosa*. Of significance for formation of the concepts of modern mathematics, Viète devotes the *logistice speciosa* to pure algebra, understood as the most comprehensive possible analytic art, applicable indifferently to numbers and to geometric magnitudes. By this process the concept of *eidos*, or species, undergoes a universalizing extension while preserving its link to the realm of numbers. In this general procedure the species represent simply general magnitudes. Viète's *logistice speciosa*, on the other hand, is understood as the procedure analogous to geometric analysis and is directly related to Diophantus' *Arithmetica*.

In chapter 5 Viète presents the *leges zeteticae*, which refer to elementary operations with equations: to antithesis (proposition I), the transfer of one of the parts of one side of the equation to the other; to hypobibasm (proposition II), the reduction of the degree of an equation by the division of all members by the species common to all of them; and to parabolism (proposition III), the removal of the coefficient of the *potestas* (conversion of the equation into the form of a proportion). The sixth chapter, "De theorematum per poristicen examinatione," deals more with synthesis and its relation to analysis than with poristics. It states that the poristic way is to be taken when a problem does not fit immediately into the systematic context.

In chapter 7, on the function of the rhetic art, Viète treats the third kind of analysis (rhetic or exegetic), which is applied to numbers if the search is for a magnitude expressible in a number, as well as to lengths, planes, or solids if the thing itself must be shown, starting from canonically ordered equations.

In chapter 8, the final one, Viète gives some definitions—such as of "equation": An equation is a comparison of an unknown magnitude with a determinate one—some rules, and some outlines of his works *De numerosa potestatum purarum, atque adfectarum ad exegesin resolutione tractatus, Effectionum geometricarum canonica recensio,* and *Supplementum geometriae.* In 1630 the work was translated into French by A. Vasset (very probably Claude Hardy) as *L'algèbre nouvelle de M. Viette* and by J. L. de Vaulezard as *Introduction en l'art analytique, ou nouvelle algèbre de François Viète.* Both also contain a translation of Viète's *Zeteticorum libri quinque.* A

modern French translation of the work was published by F. Ritter in *Bullettino di bibliografia. . .,* 1 (1868), 223–244. An English version by J. W. Smith appeared as an appendix to Jacob Klein's *Greek Mathematical Thought and the Origin of Algebra* (Cambridge, Mass., 1968).

In 1593 Viète published *Zeteticorum libri quinque,* which he very probably had completed in 1591. In it he offered a sample of *logistice speciosa* and contrasted it directly with Diophantus' *Arithmetica,* which, in his opinion, remained too much within the limits of the *logistice numerosa.* In order to stress the parallelism of the two works, Viète ended the fifth book of his *Zetetics* with the same problem that concludes the fifth book of Diophantus' *Arithmetica.* In other parts of the book he also takes series of problems from the Diophantus work. References by Peletier and Peter Ramus, as well as Guilielmus Xylander's translation (1575), must certainly have introduced Viète to the *Arithmetica,* which he undoubtedly also came to know in the original.

Moreover, as K. Reich has proved in her paper "Diophant, Cardano, Bombelli, Viète, ein Vergleich ihrer Aufgaben," he was acquainted with Cardano's *De numerorum proprietatibus, Ars magna,* and *Ars magna arithmeticae* and mentioned his name in problems II,21 and II,22. According to Reich, however, it is not known whether Viète, in preparing his *Zetetics,* considered Bombelli's *Algebra.* The *Zetetics* is composed of five books, the first of which contains ten problems that seek to determine quantities of which the sum, difference, or ratio is known. The problems of the second book give the sum or difference of the squares or cubes of the unknown quantities, their product, and the ratio of this product to the sum or the difference of their squares. In the third book the unknown quantities are proportional, and one is required to find them if the sum or the difference of the extremes or means is given. This book contains the application of these problems to right triangles. The fourth book gives the solutions of second- and third-degree indeterminate problems, such as IV, 2,3, to divide a number, which is the sum of two squares, into two other squares. The fifth book contains problems of the same kind, but generally concerning three numbers: for instance (V,9), to find a right triangle in such a way that the area augmented with a given number, which is the sum of two squares, is a square.

Viète's notation in his early publications is somewhat different from that in his collected works, edited by F. van Schooten in 1646. For example, the modern (3 $BD^2 - 3 BA^2$)/4 is printed in the *Zetetics* as (*B* in *D* quadratum 3 − *B* in *A* quadratum 3)/4, while in 1646 it is reprinted in the form (*B* in *Dq* 3 − *B* in *Aq* 3)/4. Moreover, the radical sign found in the 1646 edition is a modification introduced by van Schooten. Viète rejected the radical, using instead the letter *l* in the *Zetetics* — for example, $l \cdot 121$ for $\sqrt{121}$. The same holds for Viète's *Effectionum geometricarum canonica recensio*, the outline of which he had given in his *Isagoge*: "With a view to exegetic in geometry, the analytical art selects and enumerates more regular procedures by which equations of 'sides' and 'squares' may be completely interpreted" — that is, it concerns a convenient method for solving geometrical problems by using the coefficients of the equation in question, without solving the corresponding equation. All the solutions he gives in this tract have been carried out by geometric construction with the ruler and compass: for instance, the proof of proposition X, which leads to the equation $x^2 - px = q^2$, and that of proposition XVII, which leads to the equation $x^4 + p^2x^2 = p^2q^2$.

In 1593 at Tours, Jamet Mettayer edited *Francisci Vietae Supplementum geometriae, ex opere restitutae mathematicae analyseos seu algebra nova*. The following statement from proposition XXV — "Enimvero ostensum est in tractatu de aequationum recognitione, aequationes quadratoquadratorum ad aequationes cuborum reduci" — is important because it shows that by 1593 his tract *De aequationum recognitione* had already been completed, long before its publication by Alexander Anderson (1615). The tract begins with the following postulate: A straight line can be drawn from any point across any two lines (or a circle and a straight line) in such a way that the intercept between these two lines (or the line and the circle) will be equal to a given distance, any possible intercept having been predefined. The twenty-five propositions that follow can be divided into four groups:

1. Propositions 1–7 contain the solution of the problem of the mesographicum — to find two mean proportionals between two given straight line segments — and its solution immediately yields the solution of the problem of doubling the cube.

2. Propositions 8–18 contain the solution of the problem of the trisection of an angle and the corresponding cubic equation. The trigonometric solution of the cubic equation occurs twice: in propositions 16 and 17.

3. Propositions 19–24 contain the solution of the problem of finding the side of the regular heptagon that is to be inscribed in a given circle.

4. Proposition 25 explains the importance of the applied method: the construction of two mean proportionals, the trisection of an angle, and all problems that cannot be solved only by means of the ruler and compass but that lead to cubic and biquadratic equations, can be solved with the aid of the ancient *neusis* procedure.

In 1592 Viète began a lively dispute with J. J. Scaliger when the latter published a purported solution of the quadrature of the circle, the trisection of an angle, and the construction of two mean proportionals between two given line segments by means of the ruler and compass only. In that year Viète gave public lectures at Tours and proved that Scaliger's assertions were incorrect, without mentioning the name of the author. For this reason he decided in 1593 to publish book VIII of his *Variorum de rebus mathematicis responsorum Liber VIII, cuius praecipua capita sunt: De duplicatione cubi et quadratione circuli, quae claudit πρόχειρον seu ad usum mathematici canonis methodica*. In chapters 1, 2, and 5 Viète treats the traditional problem of the doubling of the cube, that is, of the construction of two mean proportionals. In the first chapter, on the basis of Plutarch's *Life of Marcellus* (ch. 14), he calls this an irrational problem. In the fifth chapter he treats it synthetically, referring to the "ex Poristicis methodus" that he had presented in the *Supplementum geometriae*. In chapter 3 he is concerned with the trisection of an angle and, in chapter 7, with the construction of the regular heptagon to be inscribed in a given circle, proposed by François de Foix, count of Candale, the most important contemporary editor and reviser of Euclid. Chapters 6 and 14 are related to Archimedes' *On Spirals*, already known in the Latin West through the Moerbeke translation of 1269.

In chapter 8 Viète discusses the quadratrix and, in chapter 11, the lunes that can be squared. He investigates the problem of the corniculate angle in chapter 13 and sides with Peletier, maintaining that the angle of contact is no angle. Viète's proof is new: the circle may be regarded as a plane figure with an infinite number of sides and angles; but a straight line touching a straight line, however short it may be, will coincide with that straight line and will not form an angle. Never before had the meaning of "contact" been stated so plainly. In chapter 16 Viète gives a very interesting construction of the tangent to the Archimedean spiral and, in chapter 18, the earliest explicit expression for π by an infinite number of operations. Considering regular polygons of 4, 8, 16, . . . sides, inscribed in a circle of unit radius, he found that the area of the circle is

$$2 \cdot \cfrac{1}{\sqrt{\tfrac{1}{2}} \cdot \sqrt{\tfrac{1}{2}+\tfrac{1}{2}\sqrt{\tfrac{1}{2}}} \cdot \sqrt{\tfrac{1}{2}+\tfrac{1}{2}\sqrt{\tfrac{1}{2}+\tfrac{1}{2}\sqrt{\tfrac{1}{2}}\cdots}}},$$

from which he obtained

$$\frac{\pi}{2} = \cfrac{1}{\sqrt{\tfrac{1}{2}} \cdot \sqrt{\tfrac{1}{2}+\tfrac{1}{2}\sqrt{\tfrac{1}{2}}\cdots}}.$$

The trigonometric portion of this treatise begins with chapter 19 and concerns right and oblique plane and spherical triangles. In regard to the polar triangle and Viète's use of it, Braunmühl in his *Vorlesungen* assures the reader that Viète's reciprocal figure is the same as the polar triangle. He arrives at this conclusion because Viète's theorems are arranged in such a manner that each theorem is the dual of the one immediately preceding it.

Since Scaliger could not defend himself against Viète's criticism, he left France for the Netherlands, where soon after his arrival in 1594 he published his *Cyclometrica elementa*, followed some months later by his *Mesolabium*. Viète responded with *Munimen adversus cyclometrica nova* (1594) and *Pseudomesolabium* (1595). In the first, through a nice consideration based on the use of the Archimedean spiral, he gives two interesting approximations of a segment of a circle. In the second he seeks those chords cutting the diameter in such a way that the four parts increase in geometric series. In the appendix Viète refutes Scaliger's assertion that in the inscribed quadrilateral the diameter and both diagonals are in arithmetical proportion.

Viète's mathematical reputation was already considerable when the ambassador from the Netherlands remarked to Henry IV that France did not possess any geometricians capable of solving a problem propounded in 1593 by Adrian Romanus to all mathematicians and that required the solution of a forty-fifth-degree equation. The king thereupon summoned Viète and informed him of the challenge. Viète saw that the equation was satisfied by the chord of a circle (of unit radius) that subtends an angle $2\pi/45$ at the center. In a few minutes he gave the king one solution of the problem written in pencil and, the next day, twenty-two more. He did not find forty-five solutions because the remaining ones involve negative sines, which were unintelligible to him.

Viète published his answer, *Ad problema, quod omnibus mathematicis totius orbis construendum proposuit Adrianus Romanus, responsum*, in 1595. In the introduction he says: "I, who do not profess to be a mathematician, but who, whenever there is leisure, delight in mathematical studies. . . ." Regarding Romanus' equation, Viète had seen at once that since $45 = 3 \cdot 3 \cdot 5$, it was necessary only to divide an angle once into five equal parts, and then twice into three, a division that could be effected by corresponding fifth- and third-degree equations. In the above problem he solved the equation $3x - x^3 = a$; using the roots x he determined y by $3y - y^3 = x$, and by the equation $5z - 5z^3 + z^5 = y$ he found the required roots z.

At the end of his work Viète proposed to Romanus, referring to Apollonius' *Tangencies*, the problem to draw a circle that touches three given circles. Romanus was acquainted with Regiomontanus' statement that he doubted the possibility of a solution by means of the ruler and compass only. He therefore solved the problem by determining the center of the required circle by means of the intersection of two hyperbolas; this solution did not, however, possess the rigor of the ancient geometry. In 1600 Viète presented a solution that had all the rigor desirable in his *Apollonius Gallus, seu exsuscitata Apollonii Pergaei Περὶ ἐπαφῶν geometria ad V. C. A. Romanum*, in which he gave a Euclidean solution using the center of similitude of two circles. Romanus was so impressed that he traveled to Fontenay to meet Viète, beginning an acquaintanceship that soon became warm friendship. Viète himself did not publish the book; very probably it was done by Marino Ghetaldi. A Greek letter dedicated to Viète precedes the text in the original edition. In appendix I, confronted with certain problems that Regiomontanus could solve algebraically but not geometrically, Viète provides their geometric construction and notes, by way of introduction, that these geometric constructions are important. In appendix II he vehemently attacks Copernicus, and there is also a reference to a work intended to correct the errors in the work of Copernicus and the defects in that of Ptolemy. It was to have been entitled *Francelinis* and to have contained a composition, "Epilogistice motuum coelestium Pruteniana," based on hypotheses termed Apollonian, such as the hypothesis of the movable eccentric.

In the 1591 edition of the *Isagoge*, Viète had already given the outline of the *De numerosa potestatum purarum, atque adfectarum ad exegesin resolutione tractatus*. The "numerical resolution of powers" referred to in the title means solving equations that have numerical solutions, such as $x^2 = 2916$ or $x^2 + 7x = 60750$. The work was published in 1600 at Paris, edited by Marino Ghetaldi, with Viète's consent. (All information concerning the edition is taken from a letter written by Ghetaldi to Michel Coignet, dated 15 February 1600, which is printed at the end of the work.) Viète gave some of his manuscripts to Ghetaldi when the latter was in Paris. Ghetaldi took them to Rome and allowed his friends there to make a copy. After Viète's death his heirs gave other manuscripts to his friend Pierre Alleaume, who left them to his son Jacques, a pupil of Viète's. Jacques entrusted Anderson with the treatises *De aequationum recognitione, Notae ad logisticem posteriores*, and *Analytica angularium sectionum*.

In *De numerosa potestatum*, Viète gives a method of approximation to the roots of numerical equations that resembles the one for ordinary root extraction. Taking $f(x) = k$, where k is positive, Viète separates the required root from the rest, then substitutes an approximate value for it and shows that another digit of the root can be obtained by division. A repetition of this process gives the next digit, and so on. Thus, in $x^5 - 5x^3 + 500x = 7{,}905{,}504$, he takes $r = 20$, then computes $7{,}905{,}504 - r^5 + 5r^3 - 500r$ and divides the result by a value that in modern notation would be $|(f[r + s_1] - f[r])| - s_1^n$, where n is the degree of the equation and s_1 is a unit of the denomination of the next digit to be found. Thus, if the required root is 243 and r has been taken to be 200, then s_1 is 10; but if r is taken as 240, then s_1 is 1. In the example above, where $r = 20$, the divisor is 878,295 and the quotient yields the next digit of the root, 4. One obtains $x = 20 + 4 = 24$, the required root.

Viète also had a role in the improvements of the Julian calendar. The yearly determination of the movable feasts had long resulted in great confusion. The rapid progress of astronomy led to the consideration of this subject, and many new calendars were proposed. Pope Gregory XIII convoked a large number of mathematicians, astronomers, and prelates, who decided upon the adoption of the calendar proposed by Clavius. To rectify the errors of the Julian calendar, it was agreed to write 15 October into the new calendar immediately after 4 October 1582. The Gregorian calendar met with great opposition among scientists, including Viète and Tobias Müller. Viète valued the studies involved in a reform of the calendar; and toward the end of his life he allowed himself to be carried away by them and to engage in unjustified polem-

ics against Clavius, the result of which was the publication with Mettayer of *Libellorum supplicum in regia magistri relatio kalendarii vere Gregoriani ad ecclesiasticos doctores exhibita pontifici maximo Clementi VIII anno Christo 1600 iubilaeo* (1600). He gave the work to Cardinal Cinzio Aldobrandini, who transmitted it to Clavius. Since Clavius rejected the proposed corrections, Viète and Pierre Mettayer, the son of Jean, published a libel against Clavius that was as vehement as it was unjust: *Francisci Vietae adversus Christophorum Clavium expostulatio* (1602).

Francisci Vietae fontenaensis de aequationum recognitione et emendatione tractatus duo was published in 1615, under the editorship of Viète's Scottish friend Alexander Anderson. The treatise "De emendatione" contains the subject matter of the work as announced in the *Isagoge* under the title "Ad logisticen speciosam notae posteriores" and sets forth a series of formulas (*notae*) concerning transformations of equations. In particular it presents general methods for solving third- and fourth-degree equations. This work reveals Viète's partial knowledge of the relations between the coefficients and the roots of an equation. Viète demonstrates that if the coefficient of the second term in a second-degree equation is minus the sum of two numbers the product of which is the third term, then the two numbers are roots of the equation. Viète rejected all but positive roots, however, so it was impossible for him to perceive fully the relations in question.

Viète's solution of a cubic equation is as follows: Given $x^3 + 3B^2x = 2Z^3$. To solve this let $y^2 + yx = B^2$. Since from the constitution of such an equation B^2 is understood to be a rectangle of which the lesser of the two sides is y, and the difference between it and the larger side is x, $(B^2 - y^2)/y = x$. Therefore $(B^6 - 3B^4y^2 + 3B^2y^4 - y^6)/y^3 + (3B^4 - 3B^2y^2)/y = 2Z^3$. When all terms have been multiplied by y^3 and properly ordered, one obtains $y^6 + 2Z^3y^3 = B^6$. Since this equation is quadratic with a positive root it also has a cube root. Thus the required reduction is effected. Conclusion: If, therefore, $x^3 + 3B^2x = 2Z^3$, and $\sqrt{B^6 + Z^6} - Z^3 = D^3$, then $(B^2 - D^2)/D$ is x, as required.

In the solution of biquadratics, Viète remains true to his principle of reduction. He first removes the term involving x^3 to obtain the form $x^4 + a^2x^2 + b^3x = c^4$. He then moves the terms involving x^2 and x to the right-hand side of the equation and adds $x^2y^2 + y^4/4$ to each side, so that the equation becomes $(x^2 + y^2/2)^2 = x^2(y^2 - a^2) - b^3x + y^4/4 + c^4$. He then chooses y so that the right-hand side of this equation is a perfect square. Substituting this value of y, he can take the square root of both sides and thus obtain two quadratic equations for x, each of which can be solved.

In theorem 3 of chapter VI, Viète gives a trigonometrical solution of Cardano's irreducible case in cubics. He applies the equation $(2 \cos \alpha)^3 - 3 (2 \cos \alpha) = 2 \cos 3\alpha$ to the solution of $x^3 - 3a^2x = a^2b$, when $a>b/2$, by setting $x = 2a \cos \alpha$ and determining 3α from $b = 2a \cos 3\alpha$. In the last chapter Viète resolves into linear factors $x - x_k$ the first member of an algebraic equation $\phi(x) = 0$ from the second up to the fifth degree. Anderson's edition is the only one besides the *Opera* of 1646. There is still a manuscript that contains the text (Paris, Bibliothèque Nationale, Nouv. acqu. lat. 1644, fols. 1^r–31^v, "De recognitione aequationum tractatus," and fols. 32^r–60^v, "De aequationum emendatione tractatus secundus").

In 1615 Anderson published Viète's treatise on angle sections, *Ad angularium sectionum analyticem theoremata καθολικώτερα a Francisco Vieta fontenaensis primum excogitata at absque ulla demonstratione ad nos transmissa, jam tandem demonstrationibus confirmata.* This treatise deals, in part, with general formulas of chords, sines, cosines, and tangents of multiple arcs in terms of the trigonometric lines of the simple arcs. Viète first applies algebraic transformation to trigonometry, particularly to the multisection of angles, but without proofs and calculations, which were added by Anderson. In theorem 6 Viète considers the equations for multiple angles: letting $2 \cos a = x$, he expresses $\cos na$ as a function of x for all integers $n < 11$; and at the end he presents a table for determining the coefficients. In theorem 7 he expresses $2x^{n-2} \sin na$ in terms of x and y using $2 \sin a = x$ and $2 \sin 2a = y$. After theorem 10 Viète states: "Thus the analysis of angular sections involves geometric and arithmetic secrets which hitherto have been penetrated by no one." To the treatises of the *Isagoge* belong "Ad logisticen speciosam notae priores" and "Ad logisticen speciosam notae posteriores," the latter now lost. The first was not published during his life, because Viète believed that the manuscript was not yet suitable for publication. (It was published by Jean de Beaugrand in 1631.) It represents a collection of elementary general algebraic formulas that correspond to the arithmetical propositions of the second and ninth books of Euclid's *Elements*, as well as some interesting propositions that combine algebra with geometry. In propositions 48–51 Viète derives the

formulas for $\sin 2x$; $\cos 2x$; $\sin 3x$; $\cos 3x$; $\sin 4x$; $\cos 4x$; $\sin 5x$ and $\cos 5x$ expressed in $\sin x$ and $\cos x$ by applying proposition 46, "From two right-angled triangles construct a third right-angled triangle," to two congruent right triangles; to right triangles with simple and double angles; with simple and triple angles, and with simple and quadruple angles respectively. He remarks, that the coefficients are equal to those in the expansion $(B + D)^n$ (B being the perpendicular and D the base of the original right triangle), that the various terms must be "homogeneous" and that the signs are alternately + and −. (A French translation of this work was published by F. Ritter in *Bullettino di bibliografia* . . ., **1** [1868], 245–276.) Besides Viète's published works there are manuscripts containing works of him or attributed to him. In addition to Nouv. acqu. lat. 1644, Bibliothéque Nationale, fonds lat., nouv. acqu. 1643, contains few new elements. The author was very well acquainted with Viète's work, particularly with his *De numerosa potestatum . . . ad exegesin resolutione . . .*; he betrays the influence of Simon Stevin's *Arithmétique* because his manner of denoting the powers of the unknown depends on the method used by Stevin and he uses the signs for equality and square root. London, British Museum, Sloane 652, fols. 1–9, contains the *Isagoge*, and fols. 10–40 the *Zetetics*.

BIBLIOGRAPHY

I. ORIGINAL WORKS. Note references in text. Additional editions of Viète's works are *Quinque orationes philosophicae* (Paris, 1555); and *Deschiffrement d'une lettre escripte par le Commandeur Moreo au roy d'Espagne son maître* (Tours, 1590). MSS include "Mémoires de la vie de Jean de Parthenay Larchevêque," Bibliothèque Nationale, coll. Dupuy, vol. 743, fols. 189–219; "Généalogie de la maison de Parthenay," Bibliothèque de la Société d'Histoire du Protestantisme, no. 417: "Discours des choses advenues à Lyon, durant que M. de Soubise y commandait," Bibliothèque Nationale, fonds français 20783; and "Manuscrit sur la ligue," Bibliothèque Nationale, fonds français 15499. Viète's collected works were issued as *Opera mathematica* by Frans van Schooten (Leiden, 1646; repr. Hildesheim, 1970).

II. SECONDARY LITERATURE. The best survey of Viète's life and works is in F. Ritter, "François Viète, inventeur de l'algèbre moderne, 1540–1603. Essai sur sa vie et son oeuvre," in *Revue occidentale philosophique, sociale et politique*, 2nd ser., **10** (1895), 234–274, 354–415.

See also the following, listed chronologically: Florian Cajori, *A History of Mathematics* (New York, 1894;

repr. New York, 1961), 137–139, 143–144; A. von Braunmühl, *Vorlesungen über Geschichte der Trigonometrie*, I (Leipzig, 1900), 157–183; M. Cantor, *Vorlesungen über Geschichte der Mathematik*, II (Leipzig, 1900; repr. Stuttgart, 1965), 582–591, 629–641; H. G. Zeuthen, *Geschichte der Mathematik im 16. und 17. Jahrhundert* (Leipzig, 1903; repr. Stuttgart, 1966), 95–109; 115–126; M. C. Zeller, *The Development of Trigonometry From Regiomontanus to Pitiscus* (Ann Arbor, Mich., 1944), 73–85; H. Lebesgue, *Commentaires sur l'oeuvre de F. Viète*, Monographies de l'Enseignement Mathématique, No. 4 (Geneva, 1958), 10–17; P. Dedron and J. Itard, *Mathématiques et mathématiciens* (Paris, 1959), 173–185; J. E. Hofmann, "Über Viètes Beiträge zur Geometrie der Einschiebungen," in *Mathematische-physikalische Semesterberichte*, VIII (Göttingen, 1962), 191–214; H. L. L. Busard, "Über einige Papiere aus Viètes Nachlass in der Pariser Bibliothèque Nationale," in *Centaurus*, 10 (1964), 65–126; D. Kahn, *The Codebreakers, the Story of Secret Writing* (New York, 1968), 116–118; J. Klein, *Greek Mathematical Thought and the Origin of Algebra* (Cambridge, 1968), 150–185, 253–285, 315–353; K. Reich, "Diophant, Cardano, Bombelli, Viète, ein Vergleich ihrer Aufgaben," in *Rechenpfennige* (Munich, 1968), 131–150; K. Reich, "Quelques remarques sur Marinus Ghetaldus et François Viète," in *Actes du symposium international "La géométrie et l'algèbre au début du XVIIe siècle"* (Zagreb, 1969), 171–174; J. Grisard, "François Viète mathématicien de la fin du seizième siècle," *Thèse de 3e cycle. École pratique des hautes études*, (Paris, 1968); and K. Reich and H. Gericke, "François Viète, Einführung in die Neue Algebra," in *Historiae scientiarum elementa*, V (Munich, 1973).

H. L. L. BUSARD

VIEUSSENS, RAYMOND (*b.* Vigan, Lot, France, *ca.* 1635; *d.* Montpellier, France, 16 August 1715), *anatomy, medicine.*

Vieussens' father, François Vieussens, is known to have been a *bourgeois* of Vigan despite his descendants' claims to be of the nobility and their assertion that Raymond Vieussens was the son of a lieutenant-colonel, Alexandre-Gaspard, *seigneur* of Vieussens, who died at the siege of Barcelona. Vieussens never signed his name with the particle, but it is joined to his name in the posthumous editions of his writings.

Vieussens studied medicine at Montpellier, where he was awarded the doctorate on 9 October 1670, when he was about thirty-five, a surprisingly late age; the absence of information concerning his early years, however, precludes our ascertaining why he did not receive it earlier. In any case, at the time of his graduation he was already well-

known, for almost immediately he was named physician at the Hôtel Dieu St.-Éloi, then the leading hospital of Montpellier. Vieussens subsequently became chief physician there and apparently retained that post for the rest of his life. Between 1679 and 1697 twelve children were born to him and his wife, Elisabeth Peyret.

On several occasions Vieussens left Montpellier for long periods to treat important people in Paris. The publication of his books on the nervous system (1684) and on fermentation (1688) made Vieussens famous. To reward him for his writings, the king granted him the title of royal physician and an annual pension of 1,000 *livres*. Although he never treated the king, he was personal physician to the duchess of Montpensier, the Grande Mademoiselle, from 1690 until her death in 1693. Vieussens was elected to the Académie des Sciences in 1699 as correspondent of P. S. Régis, and on 15 February 1708 he was promoted to associate anatomist. In 1707 the king named him councillor of state. Although prominent in scientific medicine at Montpellier, Vieussens spent his entire career—except for his studies—outside the city's university and sometimes even in opposition to its professors. He founded a virtual dynasty of physicians: two of his sons became royal physicians, and two of his daughters married physicians. His grandson Daniel prepared the posthumous edition of his *Histoire des maladies internes.*

From the time he entered St.-Éloi, Vieussens divided his time and interest between medical practice and anatomical research. The regulations then in effect allowed him, as chief physician of the hospital, to perform a large number of autopsies. Like most of the anatomists of the time, he was as much concerned with what was called normal anatomy as with pathological changes. The study of pathological morphology, however, still lacked a satisfactory unifying theory. In this respect it is significant that while Vieussens sought quick publication of the results of his anatomical research, a large portion of his pathological observations, some of them very original, was not made public until long after his death.

Vieussens' research on the nervous system is of great importance. In *Nevrographia universalis* (1684) he sought to continue the work of Thomas Willis, which he greatly admired. The first to make good use of Steno's suggestion that the white substance in the brain should be studied by tracing the path of its fibers, Vieussens described the olivary nucleus and the centrum semiovale; the latter still bears his name. Moreover, his description of the fine structure of the cerebellum, including the discovery of the dentate nuclei, surpassed all previous publications on the subject. The most original part of the work concerns the paths of the peripheral nerves. Vieussens also studied the structure of the ear and angiology. The weak point of his work on the nervous system is his tendency to conjoin his correct morphological observations with quite fantastic physiological explanations. In his speculations on physiology, Vieussens drew inspiration from both the mechanistic philosophy of Descartes and the iatrochemical ideas of F. de la Boë (Sylvius). He believed that he had demonstrated the existence of the nervous fluid.

One of Vieussens' major areas of study was fermentation. He investigated the chemical composition of the blood with great fervor but an equal lack of success. The discovery of an acidic salt in the blood was the source of a long and painful public polemic with the Montpellier professor Pierre Chirac, who claimed to be the first to have extracted this substance from the blood. The priority dispute was particularly unfruitful in that the discovery was erroneous. Nevertheless, until the end of his life Vieussens considered his chemical research on the blood to be his most important work.

Vieussens greatly underestimated the significance of his cardiological observations, which, in the judgment of posterity, were a truly pioneer effort. Most of his studies on the physiology and pathology of the heart and of the circulation were undertaken during the last decade of his life. While the experimental portion of the work was not published until 1755, in the posthumous *Expériences et réflexions . . .*, the clinical and anatomicopathological observations were published during his lifetime in two cardiological treatises, the more important of which is *Traité nouveau de la structure et des causes du mouvement naturel du coeur* (1715). By injecting mercury into various vessels and internal organs of living animals and fresh human cadavers, Vieussens was able to trace the exact course of the blood's flow in different parts of the body. He confirmed the hypothesis that there is a continuous vascular pathway between the arterial and venous vessels. In cardiac pathology, he was the first to describe mitral stenosis and aortic insufficiency on the basis of both clinical and anatomicopathological observation. Vieussens had already noted that a disease of the aorta manifests itself by a characteristic pulse, which was rediscovered a century later by D. J. Corrigan, whose name it now bears.

BIBLIOGRAPHY

I. ORIGINAL WORKS. Vieussens' principal works are *Nevrographia universalis* (Lyons, 1684), also in French (Toulouse, 1774); *Tractatus duo. Primus: De remotis et proximis mixti principiis in ordine ad corpus humanum spectatis. Secundus: De natura, differentiis, subjectis, conditionibus et causis fermentationum* (Lyons, 1688); *Epistola de sanguinis humani cum sale fixo* (Leipzig, 1698); *Novum vasorum corporis humani systema* (Amsterdam, 1705), also in French (Toulouse, 1774); *Nouvelles découvertes sur le coeur* (Toulouse, 1706); *Traité nouveau de la structure de l'oreille* (Toulouse, 1714); *Traité nouveau de la structure et des causes du mouvement naturel du coeur* (Toulouse, 1715); *Traité nouveau des liqueurs du corps humain* (Toulouse, 1715); *Expériences et réflexions sur la structure et l'usage des viscères, suivies d'une explication physico-méchanique de la plupart des maladies* (Paris, 1755); and *Histoire des maladies internes*, 3 vols. (Toulouse, 1774–1775).

II. SECONDARY LITERATURE. The best biography is L. Dulieu, "Raymond Vieussens," in *Monspeliensis Hippocrates*, **10**, no. 35 (1967), 9–26. A general account can be found in C. E. Kellet, "Life and Work of Raymond Vieussens," in *Annals of Medical History*, 3rd ser., **4** (1942), 31–53. Vieussens's neurological work is discussed in B. Sachs, "Raymond de Vieussens, Noted Neuro-Anatomist and Physician of the XVIIth Century," in *Proceedings of the Charaka Club*, **3** (1910), 99–105; and in E. Clarke and C. D. O'Malley, *The Human Brain and Spinal Cord* (Berkeley–Los Angeles, 1968), 584–591, 636–641. For his cardiological discoveries, see J. J. Philipp, "Raymond Vieussens und J. M. Lancisi's Verdienste um die Lehre von den Krankheiten des Herzens," in *Janus* (Breslau), **2** (1847), 580–598; E. Schroer, *Die Förderung der Kenntnisse der Herzkrankheiten durch Vieussens und Sénac* (Düsseldorf, 1937); and C. E. Kellet, "Raymond Vieussens on Mitral Stenosis," in *British Heart Journal*, **21** (1959), 440–444.

M. D. GRMEK

VIGANI, JOHN FRANCIS (*b.* Verona, Italy, *ca.* 1650[?]; *d.* Newark-on-Trent, England, February 1713 [o.s.]), *chemistry, pharmacy.*

Almost every point concerning Vigani's life is doubtful, except that on 10 February 1702 [o.s.] the senate of the University of Cambridge conferred upon him the title of professor of chemistry "because he has with much praise practised the art of chemistry among us for twenty years (not without great profit to the studious)." He was the first to hold this title. Vigani therefore came to Cambridge about 1682, although he always lived in Newark with his family and had no formal association with any college. He taught at several, including Queens' (where his cabinet of materia medica is still preserved) and Trinity (where a laboratory was built for him by the master, Richard Bentley, in 1707). He was on friendly terms with John Covel and other leading men, including Isaac Newton—who took "much delight and pleasure" in his company until Vigani told him an off-color joke. Vigani described himself as *Veronensis* and mentioned a visit to Parma in 1671; otherwise nothing is known of his origins. He had no recorded degree or medical license. The years before his settlement in England are said to have been devoted to travel, study, and collecting. He ceased to teach at Cambridge in 1708 and was buried 26 February 1713 [o.s.].

The few contemporary references to Vigani's teaching on chemistry, the construction of furnaces, and materia medica are favorable; three anonymous sets of lecture notes survive (University Library, *ca.* 1700; Queens' College, 1707; and a set formerly owned by Sir C. S. Sherrington, 1705). Vigani's chemistry may be further studied in his one published work, *Medulla chemiae* (later *chymiae*; Danzig, 1682; London, 1683, later editions at London, Leiden, Nuremberg and Basel). It is not known why the first publication was at Danzig; an earlier edition (London, 1658) has been recorded and is not wholly implausible but is not now traceable. The first edition consisted of nineteen pages and the second of seventy-one and included plates of furnaces and apparatus. Vigani was above all a practical working chemist and pharmacist with no interest in theory, referring readers who desired more theoretical discussion to the writings of Robert Boyle. He attended carefully to the purity of materials and was generally cautious or even skeptical; he denied, for example, that antimony is not chemically dissolved into "antimony wine." Nevertheless, he was confused about such matters as the distillation of vinegar and the identity of gases, which signify "nothing else but a blast or Vapour," he wrote. Vigani seems to have been completely free of any alchemical tinge. His main object was to teach in a plain and reliable way the methods of preparing useful chemical compounds and pharmaceutical recipes. His cabinet contains an eclectic mixture of vegetable, mineral, animal, and chemical medicaments.

BIBLIOGRAPHY

See L. J. M. Coleby, "John Francis Vigani," in *Annals of Science*, **8** (1952), 46–60; and E. S. Peck, "Vigani and His Cabinet," in *Cambridge Antiquarian Society Communications*, **34** (1934), 34–49.

A. RUPERT HALL

VIGO, GIOVANNI DA (*b.* Rapallo, Italy, 1450; *d.* 1525), *medicine.*

Vigo studied under Battista di Rapallo, surgeon to the marquis of Saluzzo. He is said to have served as surgeon at the siege of Saluzzo (1485–1486), but he makes no reference to it. Vigo practiced at Genoa, where he was befriended by Bendinelli de Saulis, later cardinal of Santa Sabina. About 1495 he went to Savona and found favor with Cardinal Giuliano della Rovere, then captain of papal armies in Umbria. When Giuliano became Pope Julius II in 1503, Vigo was summoned to the papal court as surgeon, at an annual salary of ninety-six ducats. In 1506 he served Julius II in the war against the Borgias, and at Bologna cured the pope of a hard node between two fingers. Later, in the campaign against Ferrara, he cured Julius of an ear infection. Among his patients were the duke of Urbino and many cardinals. After Julius' death (21 February 1513), Vigo became surgeon to his nephew, Sisto della Rovere, who, like his uncle, suffered from gout; his salary was then three hundred gold *scudi* a year.

After Sisto's death in 1517, Vigo retired from public life. He had at least two sons: Ambrogio, provost of Santa Maria Maddelena in Genoa and protonotary apostolic, and Luigi, a surgeon for whom he wrote *Practica in arte chirurgica copiosa* (Rome, 1514). The book, begun in 1503 and completed in 1513, treats anatomy, abscesses, wounds, ulcers, syphilis, fractures, simple medicines, and antidotes; there is also a supplement on spells, aphrodisiacs, cosmetics, cures for obesity and thinness, and a method for extracting a dead fetus. Various parts were written for different people: a *consilium* on cancer of the breast for a noblewoman, on the stone for Tommaso Regis, on syphilis for Giovanni Antracino, on gout for Cardinal Sisto della Rovere, and on antidotes for Vigo's son Luigi. Consequently the work is not well organized: diseases of the ears appear in the chapter on breasts, diseases of the teeth in the chapter on venereal disease, and diseases of the eyes in the treatise on ulcers.

Vigo's wide reading led him to copy much from others. Often overruled by physicians at the papal court, he became a timid surgeon: he left the operation for hernia and extraction of stones and cataracts to itinerant surgeons and, except for trephination and amputation, performed few operations. He relied mainly on cauterization, plasters, and ointments. For gunshot wounds, to which he attributed bruising, burning, and poisoning, his basic treatment was the application of "Egyptian" ointment, cauterization with boiling oil, and oil of roses with egg yolk.

Vigo was among the first to advocate the use of mercury ointment in treating syphilis, although Leoniceno and Cumano dismissed it as inefficacious and Juan Almenar considered it to be the cause of epilepsy and paralysis. Vigo distinguished between the primary and secondary stages of the disease, anticipating the views of Antonio Musa Brassavola of Ferrara.

In 1517, possibly stimulated by the *Compendium in chirurgia* (1514) of his pupil Mariano Santo da Barletta, Vigo published the five-book *Practica in arte chirurgica compendiosa*, in which he amplified and made more precise his teaching on certain topics, particularly on trephination. He was the first during the Middle Ages to describe the crown saw for removing a bone disk from the skull, an instrument known to Hippocrates but long fallen into oblivion. The instrument was illustrated by Andrea della Croce in *Chirurgiae universalis opus absolutum* (Venice, 1573).

Assessments of Vigo's contribution to surgery vary: the Italians consider him an innovator who anticipated many later developments; the Germans and the French are inclined to dismiss him as a mere compiler and propagandist of Arab doctrines.

BIBLIOGRAPHY

I. ORIGINAL WORKS. Vigo's two published works were *Practica in arte chirurgica copiosa* (Rome, 1514; Pavia, 1514), repr. at Lyons in 11 eds. between 1516 and 1582; and *Practica in arte chirurgica compendiosa* (Rome[?], 1517; Pavia, 1518; Venice, 1520; Florence, 1522, 1525); also translated into French by Nicholas Godin (1525), into Italian by Lorenzo Chrysaorio (1540), into English by Bartholomew Traheron (1543), into Spanish by Miguel Juan Pascual (1557), and into German and Portuguese in the seventeenth century. There were many eds. of each trans. at late as the eighteenth century.

II. SECONDARY LITERATURE See the following, listed chronologically: G. L. Marini, *Degli archiatri pontificii*, I (Rome, 1784), 300–303; V. Malacarne, *Delle opere de' medici, e de' cerusici che nacquero o fiorivono prima del secolo XVI negli stati della real casa di Savoia*, I (Turin, 1786), 187; G. G. Bonino, *Biografia medica Piemontese*, I (Turin, 1824), 108–121; B. Mojon, *Ritratti ed elogi di Liguri illustri* (Genoa, 1820); J.-F. Malgaigne, *Oeuvres complètes d'Ambroise Paré*, I (Paris, 1840), clxxv–clxxxii; G. B. Pescetto, *Biografia medica Ligure*, I (Genoa, 1846), 69–87; E. Gurlt, *Geschichte der Chirurgie*, I (Berlin, 1898), 919–942; V. Nicaise, "À propos de Jean de Vigo," in *Bulletin de la Société française d'histoire de la médecine*, 2 (1903), 313–347; G. Davide, "Giovanni da Vigo (1450–1525)," in *Rivista di storia delle scienze mediche e naturali*, 17 (1926), 21–35; H. Frölich, *Biographisches Lexikon der hervorragenden Aerzte*, V (Berlin, 1934), 758; and E. Razzoli, *Giovanni da Vigo, archiatro di Giulio II* (Milan, 1939).

C. H. TALBOT

VIJAYANANDA (or **VIJAYANANDIN**) (*fl.* Benares, India, 966), *astronomy.*

Vijayananda, the son of Jayananda, a Brāhmaṇa of Benares who followed the Saurapakṣa (see essay in the Supplement), wrote a *Karaṇatilaka* the epoch of which is 23/24 March 966; he is, then, obviously not the Vijayananda whose method of computing the longitudes of Jupiter and Saturn is referred to by Varāhamihira (*fl. ca.* 550) in his *Pañcasiddhāntikā* (XVII, 62). The *Karaṇatilaka* is known to us only in the Arabic translation with examples (some for 1025) by al-Bīrūnī, entitled *Ghurrat al-zījāt*, which survives in a unique manuscript at Ahmadabad that can be supplemented by many quotations in al-Bīrūnī's *India, Canon, Transits*, and *Shadows*.

The *Karaṇatilaka* consists of fourteen chapters:

1. On the *ahargaṇa* (lapsed time since the epoch).

2. On the mean and true longitudes of the two luminaries.

3. On the *pañcāṅga* (length of daylight; *naksatras; tithis; yogas;* and *karaṇas*).

4. On the mean longitudes of the five planets.

5. On the true longitudes of the five planets.

6. On the three problems relating to diurnal motion.

7. On lunar eclipses.

8. On solar eclipses.

9. On the projection of eclipses.

10. On the first visibilities of the planets.

11. On conjunctions of the planets.

12. On conjunctions of the planets with the fixed stars.

13. On the lunar crescent.

14. On the *pātas* of the sun and moon.

BIBLIOGRAPHY

An ed. of the *Ghurrat al-zījāt* was prepared by M. F. Quraishi of Lahore but has never been published. Sayyid Samad Husain Rizvi has edited a portion of the work (into ch. 6) with a somewhat cumbersome translation and commentary as "A Unique and Unknown Book of al-Beruni," in *Islamic Culture*, **37** (1963), 112–130, 167–187, 223–245; **38** (1964), 47–74, 195–212; **39** (1965), 1–26, 137–180.

DAVID PINGREE

VILLALPANDO, JUAN BAUTISTA (*b.* Córdoba, Spain, 1552; *d.* Rome, Italy, 1608), *architecture, mathematics, mechanics.*

Little is known about Villalpando's life. After entering the Jesuit order in 1575, he studied under Father Jerome Prado, who was writing a commentary on the book of Ezekiel. Evidently Villalpando's immense erudition was already apparent, and he soon joined Prado in his exegesis. In 1592 the pair moved to Rome to complete their work. Originally commissioned to provide a commentary only on chapters 40, 41, and 42 of Ezekiel, which deal with the architectural description of Solomon's temple, Villalpando suddenly found himself heir to a larger task when Prado died in 1595, having completed only the first twenty-six chapters. Although Villalpando himself died before completing the commentary, he managed to publish three volumes: *Hieronymi Pradi et Ioannis Baptistae Villalpandi e Societate Iesu in Ezechielem explanationes et apparatus urbis ac templi Hierosolymitani* (Rome, 1596–1604).

Like most Renaissance biblical commentaries, Villalpando's *Ezechiel* is the work of a polymath, containing copious information on subjects ranging from astrology, music, mathematical theories of proportion, and ornate reconstructions of Hebrew, Greek, and Roman systems of weights, measures, and currency to more orthodox etymological and scriptural preoccupations. The widely disparate topics that Villalpando considers in his attempt to re-create the temple would seem to express a Vitruvian vision of the architect. The influence of Vitruvius on Villalpando is crucial, and in this sense his work may be seen as a part of the general Renaissance revival of Vitruvius. Villalpando's great achievement was to have demonstrated in systematic fashion how Solomon's temple, as revealed by God to Ezekiel, was constructed according to Vitruvian principles of harmony and proportion, thus endowing classical architecture with divine approbation. Here, too, Villalpando was continuing an older humanist trend. In showing the celestial locus of classical architecture, he provided further evidence for the preestablished harmony between classical pagan culture and Christian civilization. It is significant that both Philip II of Spain and Pope Clement VIII expressed their approval of Villalpando's *Ezechiel* during the peak of the Counter-Reformation.

The third volume of the exegesis contains the bulk of Villalpando's mathematical and mechanical speculations. While his work on proportion and harmony (II, bk. 1, chs. 1–5) follows earlier Renaissance architectural utilizations of Euclid, his twenty-one propositions on "the center of gravity and the line of direction" (ch. 6) were deemed original enough to be reproduced by Mersenne in his *Synopsis mathematica* (1626). Duhem, who "rediscovered" Villalpando, conjectured that the Jesuit pilfered his propositions and their deductive proofs from a no longer extant manuscript by Leonardo dealing with local motion. Although Taylor has suggested that Villalpando may have had access to Leonardo's manuscripts through his mentor Juan de Herrera, other sources seem more plausible. Given Villalpando's lifelong interest in mathematics, it is highly probable that before his departure for Rome, he may have attended the Academia de Mathemáticas in Madrid, where he may have been introduced to the works of Archimedes. More interesting is the possibility that he knew Christoph Clavius, a fellow Jesuit and friend of Galileo, who was teaching in Rome at the same time. Villalpando relied heavily on Clavius' *Elements of Euclid*, speaking of it in terms of endearment, and it is possible that Clavius introduced him to the work of Commandino and Guido Ubaldo del Monte on the center of gravity. Villalpando, then, can be seen as participating in the sixteenth-century revival of Archimedes and Pappus as reconstructed by Commandino and Guido Ubaldo.

Villalpando's influence has been strongest in the history of architecture. The idea of the Escorial in Spain may have been derived from his earlier designs of the temple, and Inigo Jones certainly utilized his conceptions in introducing Palladian architecture into England. But Villalpando does touch the history of seventeenth-century science in a particularly sensitive area. No less a scientist than Isaac Newton used Villalpando's work in his own attempt to construct Solomon's temple and to determine the dimensions of the biblical cubit.

BIBLIOGRAPHY

I. ORIGINAL WORKS. Aside from the commentary on Ezekiel, Villalpando edited and annotated a medieval exegesis of St. Paul: *S. Remigii Rhemensis episcopi explanationes epistolarum B. Pauli Apostoli* (Rome, 1598). There is also in the Biblioteca Nacional, Madrid, a MS entitled "Relacion de la antigua Jerusalén remitida á Felippe II por el Padre J. B. Villalpando," which establishes Villalpando's connections with the royal court.

II. SECONDARY LITERATURE. The most thorough study of Villalpando's life and architectural accomplishments is René C. Taylor, "El Padre Villalpando (1552–1608) y sus ideas estéticas," in *Academia. Anales y boletín de la Real Academia de San Fernando* (1952), no. 2, 3–65. Taylor revised some of his conclusions, stressing the role of the occult in Villalpando's work, in "Architecture and Magic," in *Essays in the History of Architecture Presented to Rudolph Wittkower* (London, 1967), 81–110, and in "Hermetism and Mystical Architecture in the Society of Jesus," in R. Wittkower and I. B. Jaffe, eds., *Baroque Art: The Jesuit Contribution* (New York, 1972), 63–97, and esp. the documents printed in App. B. The mathematical and aesthetic background of Villalpando's discussion of proportion is discussed in Rudolph Wittkower, *Architectural Principles in the Age of Humanism*, 3rd ed. (London, 1962), 121 ff.

Pierre Duhem's observations on Villalpando's mechanics were first published in *Les origines de la statique*, II (Paris, 1906), 115–126; and were substantially repeated in "Léonard de Vinci et Villalpand," in his *Études sur Léonard de Vinci*, I (Paris, 1906), 53–85.

MICHAEL T. RYAN

VILLANOVA. See **Arnald of Villanova.**

VILLARD DE HONNECOURT (*b.* Honnecourt, Picardy, France, *ca.* 1190), *architecture.*

Villard de Honnecourt (who signed himself Wilars de Honecourt) wrote the most important known medieval source by an artist, the *Bauhüttenbuch* (Bibliothèque Nationale, Paris, MS fr. 19093) between about 1225 and 1235. Only thirty-three of the more than fifty parchment folios that he carried with him for years have been preserved. In 207 pen-and-ink drawings Villard brought together models for every type of worker enrolled in a builders' guild: architectural motifs, elements of the applied arts, machines, figures for sculpture and painting, proportion diagrams, and basic construction aids. He later added a detailed title, chapter headings, and long commentaries in the manner of illustrated treatises prepared by builders' guilds.

Hence it was not a mere "album" (Quicherat, 1849) or "sketchbook" (Willis, 1859) but, rather, a lodgebook.

Villard's technical expressions are generally the oldest in Old French and the Picard dialect; moreover, they are the only ones that are illustrated and thus can be determined precisely. The master speaks directly to his students, often expressing the most personal value judgments of the Middle Ages. Thus, he sketched the classical window at Rheims (20b)[1] "because I loved it above all else"; at Laon, "a tower such as I have never seen anywhere else," although "I was in many lands" (18), and a lectern "of the best kind I know" (13). Two successors in his guild (known as Master 2 and Master 3) completed the book with expert additions and extracts from other technical treatises.

Villard's sketches of the most important structures of the time allow us to follow his wanderings. He must have been born around 1190 at Honnecourt, for he drew the ground plan of the neighboring Cistercian abbey of Vaucelles (33a)—where he was undoubtedly a student—and later, with his neighbor Pierre de Corbie, developed a plan for a similar church (29a). At Chartres he sketched the west rose window (30c); at Laon, the west tower (18–19) of the cathedral; at Cambrai, the floor plans (28c); and at Rheims, the completed and planned structural members (20, 30, 60–64). On his way to Hungary, where he had been summoned by the Cistercians and "where I long remained," he sketched an ideal plan for a church of the order (28b) and the south rose window of the Lausanne cathedral (31a). In the Cistercian cloister at Pilis, Laszlo Gerevich discovered floorboards like those Villard sketched in Hungary;[2] and thus it is probable that he was engaged in building churches there and in constructing the tomb of Queen Agnes, who was murdered in 1213.

Important individual Christian and allegorical figures drawn by Villard have been preserved, as have scenes of the Passion and of martyrdom, purely secular scenes, and complicated studies of movements. The *Muldenfaltenstil* (style of deeply molded drapery folds) that Villard employs, which has its origins in antiquity, belongs to the classical transitional phase between high Romanesque and early Gothic that predominated from 1210 to 1235, especially at Rheims. Villard, in fact, borrowed a great number of examples from antiquity, including four partially draped nudes, a Roman tomb, lion fights, and lion-taming. On the other hand, his nature studies, including birds and a lion *peint al vif*, are unique.

Villard's automatons derive in part from ancient sources, such as the spherical handwarmer (17d) and the magic fountain, and in part from Indian and Arabic sources, such as the *perpetuum mobile*. Of fundamental importance is his chapter on portraiture, in which he develops Gothic figures from abstract directrixes or geometric diagrams. His work on masonry contains the basic construction aids used by masons, such as *tierspoint* and *quintpoint*, which are *estraites de iométrie*, and—drawn from ancient sources—the fundamental procedures of bisecting the square and the circle (Plato), the Archimedean spiral, and altimetry of the Roman land surveyors. Consideration of Villard's sketches of the lectern, clock tower, and *perpetuum mobile* makes it possible to determine the laws of construction through which the medieval builders concretized their conceptual images.

NOTES

1. The numbers in parentheses refer to the plates in both of Hahnloser's eds. of Villard's MS.
2. Laszlo Gerevich, "Villard de Honnecourt in Ungarn," in *Müvészettörténeti értesítö*, **20** (1971), 81–104, with German abstract, 104–105.

BIBLIOGRAPHY

I. ORIGINAL WORKS. Eds. of Villard's *Bauhüttenbuch* are by J. B. A. Lassus and A. Darcel, *Album de Villard de Honnecourt, architecte du XIIIᵉ siècle* (Paris, 1858), the 1st illustrated ed., with engraved plates; R. Willis, *Facsimile of the Sketch-Book of Wilars de Honecort* (London, 1859), translated from the Lassus ed., with many additional notes and articles and the same plates—a much improved ed.; H. Omont, *Album de Villard de Honnecourt, architecte du XIIIe siècle* (Paris, n.d. [1906]; 2nd ed., 1927), earliest eds. with photographic repros. but no comments; H. R. Hahnloser, *Villard de Honnecourt, kritische Gesamtausgabe des Bauhüttenbuches MS fr. 19093 der Pariser Nationalbibliothek* (Vienna, 1935), the 1st complete ed., with German trans., glossary, and complete bibliography; the 2nd ed. (Graz–Vienna, 1972), an offprint of the 1st ed. contains many new notes and plates; and T. R. Bowie, *The Sketchbook of Villard de Honnecourt* (Bloomington, Ind., 1959), with a few comments—a 2nd ed. (1962) gives the plates in iconographical order.

II. SECONDARY LITERATURE. See J. Quicherat, "Notice sur l'album de Villard de Honnecourt, architecte du XIIIe siècle," in *Revue archéologique*, **6** (1849), 65–80, 164–188, 211–216, and pls. 116–118, which contains important comments; and N. X. Willemin and A. Pottier, *Monuments français inédits pour servir à l'histoire des arts*, I (Paris, 1825), 62 and pl. 106, only a few preliminary notes.

H. R. HAHNLOSER

VILLARD, PAUL (*b*. Lyons, France, 28 September 1860; *d*. Bayonne, France, 13 January 1934), *physics*.

Villard's main work involved the experimental study of cathode rays, X rays, and radioactivity from 1897 to 1907 or thereabouts.

Villard entered the École Normale Supérieure in 1881 and received the *agrégé* in 1884. After teaching in provincial secondary schools and in Paris, he received permission to work in the chemistry laboratory of the École Normale, where he conducted all his research. Villard received the Wilde and the La Caze prizes of the Paris Academy of Sciences and was elected a member of its general physics section in 1908. During his later years he received a *pension d'honneur* from the Caisse Nationale des Sciences.

Villard's background was in physical chemistry. During his first ten years at the chemistry laboratory of the École Normale, he worked on the hydrates of argon, methane, methylene, and acetylene, and on topics associated with change of phase. Villard retained a chemical point of view after having become involved in radiation physics, considering, for example, the "chemical action" of X rays and the "reducing action" of cathode rays.

Since Villard made all his own apparatus he was quite familiar with all of its technical details and, in addition to his research, he designed instruments and techniques that were useful to practical radiologists.

In his early work on cathode-ray tubes, Villard was interested in the nature of the material that moves in the tube. In this context he also considered the nature of canal rays, a puzzle since Goldstein had observed them in 1886. In 1898 Villard observed that positively charged material moves toward the cathode and that when there are holes in the cathode, a visible stream, causing heating, passes through it. He suggested that the positive material forms the canal rays after it passes through the cathode. He was not, however, able to deflect the stream moving through the cathode by means of electric or magnetic fields; and he assumed that it had lost its charge at the cathode. On the basis of the reducing action of various rays in the cathode tube, Villard concluded that the moving material is hydrogen. (In the same year Wien managed to deflect canal rays with electric and magnetic fields, and he found that their e/m was on the order of that of the hydrogen ion.)[1]

Villard was the first to observe a penetrating radiation, which he named γ radiation, following the pattern of Rutherford's names for α and β rays. In 1900, while studying the secondary emission produced by both cathode rays and radium radiation on passing through a metal sheet, he observed that a component of the radium radiation was sufficiently energetic to pass directly through; and he associated it with the "nondeviable" (uncharged) component. As a further test Villard used a more active radium source given by the Curies and sent the radiation consecutively through two photographic plates. He concluded that the "X rays" emitted by the radium are much more penetrating than the charged rays. (At this time Villard appears already to have associated the penetrating radiation with X rays; the electromagnetic nature of the γ rays was not proved until 1914.)

In 1906 Villard began studying the aurora experimentally by simulating the phenomenon with cathode rays in a magnetic field. (Olaf Birkeland, a Norwegian, had produced the first laboratory model of the aurora in 1896 by sending cathode rays toward a magnetized sphere.) Villard, who had been studying the properties of cathode rays produced when the cathode is in a strong magnetic field, "made some very fine experiments," according to C. Störmer, and "succeeded in producing threadlike currents of cathode rays which made it possible to follow the trajectories in detail."[2] Villard was interested in the theoretical implications of his work and, differing from others, believed the source of the aurora to be terrestrial.

NOTES

1. W. Wien, "Untersuchungen über die electrische Entladung in verdünnten Gasen," in Wiedemann's *Annalen der Physik und Chemie,* n.s. **65** (1898), 440–452; "Die electrostatische und magnetische Ablenkung der Canalstrahlen," in *Verhandlungen der Physikalischen Gesellschaft zu Berlin* (1898), 10–12.
2. C. Störmer, *The Polar Aurora* (Oxford, 1955), 290.

BIBLIOGRAPHY

I. ORIGINAL WORKS. Villard's papers include "Sur les rayons cathodiques," in *Comptes rendus de l'Académie des sciences,* **126** (1898), 1339–1341, 1564–1566; **127** (1898), 173–175; **130** (1900), 1614–1616; "Sur la réflexion et la réfraction des rayons cathodiques et des rayons déviables du radium," *ibid.,* **130**

(1900), 1010–1012; "Sur le rayonnement du radium," *ibid.,* 1178–1179; and "Sur l'aurore boréale," *ibid.,* **142** (1906), 1330–1333; **143** (1906), 143–145.

Villard's papers published before 1900 are listed in the Royal Society *Catalogue of Scientific Papers,* XIX, 352. Those published after 1900 are listed in *Science Abstracts.*

II. SECONDARY LITERATURE. The discussion of Villard's work presented in connection with the award of the La Caze Prize is in *Comptes rendus . . . de l'Académie des sciences,* **145** (1907), 1002–1005. E. Borel's obituary address on Villard is *ibid.,* **198** (1934), 213–215.

SIGALIA DOSTROVSKY

VILLARI, EMILIO (*b.* Naples, Italy, 25 September 1836; *d.* Naples, 20 August 1904), *physics.*

After receiving his secondary education in Naples, Villari studied at the University of Pisa, where he became professor of mathematics and medicine in 1860. Between 1860 and 1871 he lived, successively, in Florence, Berlin, and Florence again, then became professor of experimental physics at the University of Bologna. In 1900 he returned to Naples, where he taught until his death.

Villari's major interest centered on the effects of electromagnetic forces on material media. His work was not deeply mathematical; he was primarily an experimentalist who used theories qualitatively. His physical outlook was eclectic, combining elements of the Weberean action-at-a-distance school with Faraday's empirical discoveries regarding electromagnetic induction. To Villari the central element of electromagnetism was the electrical current; and his most original work consisted of an attempt to explain, from an action-at-a-distance standpoint, the peculiar effects of alternating currents on their conductors.

By 1873 Villari and others had noted that metals emit much more heat when carrying alternating currents than when bearing direct currents. Kelvin and Maxwell explained this increased heat by means of the concept of "self-induction": the continuous change in the magnitude and sign of the alternating current is supposed to produce a changing magnetic field and thus to induce an electromotive force in the conductor, a force that acts to oppose the changes in the current. In order to overcome this opposing force, more energy must be expended than if the force were absent; and this excess energy appears as heat.

To Villari, however, the problem was not one of

the action of the current, mediated by the field, upon itself. Rather he believed that the current affects the body of the conductor, which in turn reacts upon the current. He thought that all metals are composed of innately magnetic molecules that are acted on by an electric current because currents exert magnetic forces. When a circuit is first closed, the initial effect of the current as it begins to flow is to produce a realignment of the conductor's magnetic molecules. As the molecules begin to move under the action of the increasing current, they produce a changing magnetic force in the vicinity of the current. According to Faraday's law of electromagnetic induction, a changing magnetic force induces an electromotive force; as the molecules align, they therefore engender an electromotive force that opposes the force producing the current. With alternating currents the magnetic molecules will be moving constantly, and energy will be required to overcome the opposing forces induced by their motion; this energy appears as heat.

Villari's theory did not have wide influence. His experimental results, however, were well-known and extensively utilized. Nonetheless, even his theoretical opinions are important because they illustrate that, at least on the Continent, certain researchers preferred to work outside a Maxwellian context well into the 1880's. Whereas the self-induction explanation of Kelvin and Maxwell relied on the mediation of a field to produce the effects on the currents, Villari's explanation assumed a direct action between the currents and the magnetic molecules. Villari's magnetic molecules are closely related to the electrically polarizable molecules of Mossotti's dielectrics, and both were action-at-a-distance theories in that they postulated forces acting directly between various kinds of elemental electrical and magnetic fluids.

BIBLIOGRAPHY

Villari's more important works include "Intorno ad alcuni fatti singolari di elettro-magnetismo, ed alla ipotesi di Weber sulle elettro-calamite," in *Nuovo cimento*, **21–22** (1865–1866), 415–427; "Influenza della magnetizzazione sulla conducibilità elettrica del ferro," in *Rendiconti dell'Istituto lombardo di scienze e lettere*, 2nd ser., **1** (1868), 853–862; "Sulle correnti indotte tra il ferro ed altri metalli," in *Nuovo cimento*, 2nd ser., **1** (1869), 218–242; "Ricerche sulle correnti interrote ed invertite, studiate nei loro effetti termici ed elettro-dinamici," in *Memorie della R. Accademia delle scienze dell'Istituto di Bologna*, 3rd ser., **4** (1873), 157–195; and "Sulla diversa tensione delle correnti elettriche indotte fra circuiti totalmente di rame od in parte di ferro," *ibid.*, 449–467. Villari's work on currents is referred to in L. Lorenz, "Ueber die Fortpflanzung der Electricität," in *Annalen der Physik und Chemie*, n.s., 7, no. 6 (1879), 141–192.

JED Z. BUCHWALD

VILLEFRANCHE. See **La Roche, Estienne de**.

VILMORIN, PIERRE LOUIS FRANÇOIS LEVEQUE DE (*b.* Paris, France, 18 April 1816; *d.* Paris, 21 March 1860), *botany*.

Vilmorin's father, Philippe André Leveque de Vilmorin, was president of the distinguished Paris seed firm Vilmorin-Andrieux et Cie; and he brought up Louis, his eldest son, to succeed him when he retired (1843). Louis was physically handicapped and appears to have been educated privately; and although he learned the business thoroughly and was capable of heading the company, he always took more interest in research. His wife, Elisa Bailly, helped him in the business and also conducted research on strawberries. Henri, the eldest of their three sons, succeeded to the business.

The firm of Vilmorin-Andrieux had already established a reputation for breeding improved stock; since 1771 it had regularly published catalogs listing available varieties with instructions for their cultivation. Under Louis de Vilmorin the firm extended and organized the work on breeding and set up an experimental farm at Verrières-le-Buisson.

Vilmorin's first important work was on the breeding of wheat. The Société d'Agriculture had asked his father to investigate the classification of cultivated wheats; but most of the work was done by Louis, who in 1850 published a classified catalog of seven species and fifty-three distinct varieties of wheat, indicating not only their characters but also their relationships. In it he showed awareness that the horticulturist's classification was not the same as the botanist's in dealing with distinct species.

From 1850 until his death Vilmorin published a series of papers on the breeding of cereal grains, potatoes, sugarbeets, and flowers. He was not isolated in the world of commerce, but was a friend of J. B. Boussingault and collaborated with Édouard Duchesne. He also edited the periodical *Bon jardinier* from 1844 until his death.

Vilmorin's work of greatest economic importance was breeding a new variety of sugarbeet with a straight taproot and a sugar content of around 20 percent, nearly double the previous maximum. In this work, published in 1856, he already showed his appreciation of the importance of finding a reliable method of assaying for the sugar content and of conducting the breeding under controlled conditions. The main part of his paper in 1856 was a record of his assay method, but he outlines his breeding techniques, and discusses his results with the frank admission that he does not understand the transmission of the "qualité sucrée." He selected plants with a high percentage of sugar and gathered and sowed the seed separately by a method originated by his father, but he does not record methods of preventing cross-fertilization. He found in the first generation three distinct groups: one of plants that were consistent high yielders, one with variable yield but including those of exceptionally high yield, and another of consistent low yielders. He then showed that the consistently high-yielding group bred true, forming a race that was high in sugar in the second generation and even more in the third. These variations were independent of culture and he believed that his rich race was permanently fixed, though of course yield would then be affected by culture.

Vilmorin's most significant work in breeding was reported by his son Henri in 1877. From 1856 to 1860 he bred plants of *Lupinus hirsutus* with pink or blue flowers, counted the progeny of each color, and tabulated the results. This was the first experimental work since Sageret to show numerical relations of the segregation of characters and was contemporary with Mendel's experiments. The lupines were generally self-fertilized, commonly blue but sometimes pink, and had no intermediate colors. From twenty-seven blue-flowered plants he obtained twenty-five blue that bred true and two pink that in later generations produced plants in ratios approximating three pink to one blue. Since Vilmorin's numbers were small, however, and since it seems likely that some cross-fertilization occurred, he concluded that there was no mathematical relationship and the segregation was due to conflict of vital forces. His theory of centripetal forces (the hereditary influence of parents) and centrifugal ones (the totality of ancestral influence), propounded in 1851, was refined to include the force of individual variation, and he considered that the proportion of pink and blue flowers measured the strength of these forces. The greatest tendency was to resemble parents, and the forces of atavism weakened with distance.

BIBLIOGRAPHY

I. ORIGINAL WORKS. Vilmorin's works include *Essai d'un catalogue méthodique et synonymique des fromens qui composent la collection de L. Vilmorin* (Paris, 1850); "Note sur la création d'une nouvelle race de betteraves à sucre. Considérations sur l'hérédité dans les végétaux," in *Comptes rendus . . . de l'Académie des sciences,* **43** (1856), 871–874; *Notices sur l'amélioration des plantes par le semis et considérations sur l'hérédité dans les végétaux* (Paris, 1869); "Note sur une expérience relative à l'étude de l'hérédité dans les végétaux," in *Mémoires de la Société nationale d'agriculture de France* for 1877 (1879), 223–231, written by his son Henri; and "Tableau des effets de la rouille sur une série de variétés de froments," *ibid.,* **126** (1881), 219–226.

II. SECONDARY LITERATURE. There is a short unsigned obituary in *Gardener's Chronicle* (1860), 366; and an appreciative notice by J. A. Barral in *Revue horticole* (1860), 172–174; there is also a memorial in *Genetics,* **19** (1934), comprising a frontispiece protrait, 3 unnumbered pp. of explanatory text by John H. Parker, and text on the back cover; the note refers to, but does not locate, a bibliography of the publications of the Vilmorins by J. H. Parker, in typescript. Most information and comment on Vilmorin's work is found in surveys of the family. The most critical scientific assessment is in H. F. Roberts, *Plant Hybridization Before Mendel* (New York–London, (1965), 143–151. Also see Gustave Heuzé, "Les Vilmorin," in *Revue horticole,* **71** (1899), 453–459; and "La maison Vilmorin-Andrieux et Cie," in *Revue de l'horticulture belge et étrangère,* **36** (1910), 249–257.

There is a general history of sugarbeet genetics by J. L. de Vilmorin, *L'hérédité chez la betterave cultivée* (Paris, 1923), pp. 69–73 of which relate to the work of Louis de Vilmorin.

DIANA M. SIMPKINS

VINCENT OF BEAUVAIS (*b.* Beauvais, Oise, France, *ca.* 1190; *d.* Beauvais, *ca.* 1264), *natural science, transmission of knowledge.*

Vincent, known in Latin as Bellovacensis, seems to have studied at the University of Paris and to have entered the Dominican order there about 1220. Transferred to the priory in Beauvais around 1233, he became a close friend of Ralph, first abbot of the Cistercian monastery at nearby Royaumont; through Ralph he formed a lifelong friendship with Louis IX. Earlier Vincent had begun his *Speculum maius,* or "great mirror," which was to make available to one and all the hitherto inaccessible wisdom of classical and ecclesiastical authors. The king, hearing of this, desired a copy for himself and supplied the funds necessary for the work's completion. The date of composition of the *Speculum* is difficult to determine, since it went through a series of redactions and has several parts. The first version probably appeared in 1244, followed by a second some three years later and by a third in the 1250's.

Three of its components are of unquestioned authenticity: the "Speculum naturale," or "mirror of nature," an encyclopedia of nature as created by God; the "Speculum historiale," or "mirror of history," giving the history of mankind from the Creation to 1254 (in the final version); and the "Speculum doctrinale," or "mirror of teaching," summarizing all of the learned arts—liberal, mechanical, and medical, among others. To these an anonymous author, writing sometime between 1310 and 1325, added a fourth part, the "Speculum morale," or "mirror of morals," drawn mainly from the writings of Thomas Aquinas. Written in Latin, the work was translated into various vernaculars and even appeared in verse. It went through seven printings; but none of these is reliable, each containing editorial interpolations and rearrangements. Besides the *Speculum*, Vincent wrote several ascetical and theological treatises; he is especially noteworthy for his writings and influence in education (see Gabriel).

The "Speculum naturale" reflects a theological orientation in its plan, which follows the biblical account of the six days of Creation, if not in its content, which reveals it to be "a great storehouse of medieval lore" (Thorndike, 475). Its superiority to other medieval encyclopedias of nature derives from the author's access to larger and better libraries, and from his having the use of secretaries; its weakness is traceable to its being essentially a compilation of excerpts—although made with care and usually assigned to the proper authority—that shows little or no investigative originality, critical sense, or organic unity. Vincent draws heavily from Pliny, Isidore of Seville, Adelard of Bath, and Thomas of Cantimpré; and his work also is interspersed with references to Albertus Magnus and Thomas Aquinas, possibly added in subsequent revisions.

The "Speculum naturale" is composed of thirty-two books, most containing over one hundred chapters. As a preliminary, book 1 treats God, angels, and the original work of Creation; book 2 launches into the work of the first day, considering the material universe and digressing on the text "Let there be light" (Gen. 1:3) to provide thirty-four chapters on optics that deal with the nature of light, the origin of colors, and the properties of mirrors. Books 3 and 4, devoted to the work of the second day, use the formation of the firmament and the heavens to provide treatises on astronomy and meteorology, respectively. The separation of the dry land from the waters on the third day opens a treatise on geology and mineralogy that comprises 95 chapters; similarly, the creation of plants leads to 156 chapters on botany. The works of subsequent days occasion the presentation of all available information on the birds of the air, the fishes of the sea, and the animals that inhabit the dry land, man included.

The "Speculum doctrinale," composed of seventeen books divided into 2,374 chapters, duplicates some of the earlier material but is concerned more with practical matters. It deals with grammar, logic, husbandry, political affairs, trades, medicine, physics, mathematics, astrology, music, weights and measures, and surveying, and even includes a dictionary of some 3,200 entries. For the mathematical arts Vincent relies on Nicomachus of Gerasa, Boethius, and al-Fārābī; his treatment is generally brief, perfunctory, and otherwise uninteresting.

The absence of a reliable text makes it difficult to characterize and evaluate Vincent's science. His views on geography and astronomy seem to derive largely from Ibn Sīnā and are akin to those of Albertus Magnus; thus his text, at least as it has come down to us, gives basically Albertus' account of al-Bitrūjī's theory and concludes in favor of the Ptolemaic conception of the universe. He takes his chemistry from al-Rāzī's *De aluminibus et salibus*, in the translation by Gerard of Cremona, and generally follows Ibn Sīnā's presentation of alchemical doctrines. Vincent's treatment of plants seems to be based on Alfred of Sareschel's *De plantis*. His treatise on falconry he acknowledges as being abridged from "a letter written by Aquila, Symachus, and Theodotion to Ptolemy, king of Egypt, in which they treated of noble birds and medicines for them" (*Speculum naturale*, Douai ed., I, col. 1197); Sarton notes that the original text of this letter has been lost and that apart from Vincent's excerpts it survives only in an early Catalan version, to which he gives a reference (*Introduction . . .*, p. 931). According to Cuvier, his descriptions of fishes are superior to those of Albertus Magnus. Generally, however, Vincent's excerpts are taken from classical authors and do not represent the best material available to the experts of his day; he makes little attempt to be up-to-date or to integrate the new with the old. He is somewhat credulous and occasionally intermingles superstition with verified knowledge. Yet, withal, Vincent's work is truly monumental and is the best encyclopedia to come out of the Middle Ages.

BIBLIOGRAPHY

I. ORIGINAL WORKS. Vincent's works are the *Speculum maius* (Strasbourg, 7 vols., 1473–1476; Basel, 1481 ["Speculum naturale" and "Speculum morale" only]; Nuremberg, 2 vols., 1473–1486; Venice, 1484, 1494, 1591; Douai, 4 vols., 1624, repr. Graz, 1964) and *De eruditione filiorum nobilium*, A. Steiner, ed. (Cambridge, Mass., 1938).

II. SECONDARY LITERATURE. For general accounts and bibliography see W. A. Hinnebusch, *The History of the Dominican Order*, II. *Intellectual and Cultural Life to 1500* (New York, 1973), 421–428; Michel Lemoine, "L'oeuvre encyclopédique de Vincent de Beauvais," *Cahiers d'histoire mondiale*, IX (1966), 483–518, 571–579, repr. *La pensée encyclopédique au moyen âge* (Neuchatel, 1966); G. Sarton, *Introduction to the History of Science*, II, pt. 2 (Washington, 1931), 929–932, and *passim* (see index); and L. Thorndike, *A History of Magic and Experimental Science*, II (New York, 1923), 457–476. More specialized studies are A. Gabriel, *The Educational Ideas of Vincent of Beauvais*, 2nd ed. (Notre Dame, 1962); G. Göller, "Vinzenz von Beauvais und sein Musiktraktat in Speculum doctrinale," in *Kölner Beiträge zur Musikforschung*, **15** (1959), 29–34; and P. Duhem, *Études sur Léonard de Vinci*, II (Paris, 1909), 318–319, and *Le système du monde*, III (Paris, 1915; repr. 1958), 346–348.

WILLIAM A. WALLACE, O.P.

VINCI, LEONARDO DA. See **Leonardo da Vinci**.

VINOGRADSKY, SERGEY NIKOLAEVICH (*b.* Kiev, Russia, 13 September 1856; *d.* Brie-Comte-Robert, France, 24 February 1953), *microbiology*.

Vinogradsky's father, Nikolay Konstantinovich Vinogradsky, was a member of the State Council; his mother, Natalia Viktorovna Skoropadskaya, came from a noble family. In 1866 he entered the Kiev Gymnasium and in 1873 graduated with a gold medal. The same year Vinogradsky enrolled at the Law Faculty of Kiev University. After a month, however, he transferred to the natural sciences division of the Physics and Mathematics Faculty.

An interest in music led Vinogradsky to transfer to the St. Petersburg Conservatory. In 1877, however, he entered the natural sciences department of St. Petersburg University. He graduated in 1881 with the candidate of sciences degree and re-mained at the university to prepare for an academic career.

Vinogradsky was greatly influenced by Pasteur's ideas and experimental research and, attracted by the great opportunities, decided to devote himself to the new science of microbiology. From 1881 to 1884 Vinogradsky did his first experimental work, a study of the influence of external conditions on the development of the fungus *Mycoderma vini*.

After receiving the master's degree in 1884, Vinogradsky went in 1885 to Strasbourg, where he worked under Anton de Bary. In 1890 he moved to Zurich, where he did postgraduate study in chemistry for two years. From 1884 to 1889 Vinogradsky conducted research on the physiology and morphology of sulfur and iron bacteria, and then on nitrifying bacteria. This research brought him a wide reputation.

In 1890, Pasteur invited Vinogradsky to participate in the organization of a bacteriological laboratory at the institute he had established in Paris. In the same year the Institute of Experimental Medicine at St. Petersburg was completed, and in 1891 Vinogradsky became director of its section of general microbiology.

In 1902 he was named director of the entire institute. (He had to resign three years later, however, because of acute nephritis.) In 1903, by a decision of the scientific council of Kharkov University, Vinogradsky received a doctorate in botany without defending a dissertation. In 1912 Vinogradsky moved to the Ukraine, where for ten years he scarcely worked at experimental microbiology, concentrating on the organization of research in land use and soil science.

In 1922 Vinogradsky accepted an invitation from E. Roux to become director of the division of agricultural microbiology of the Pasteur Institute, which built a laboratory for him near Paris.

Vinogradsky's scientific and organizational work received international recognition. In 1894 he became a member of the Russian Academy of Sciences; in 1902 the Académie des Sciences of France elected him a corresponding member; and the National Society of Horticulture of France elected him an active member. Vinogradsky founded the Society of Microbiology in 1903 and was its president for the first two years. In 1923 he was elected an honorary member of the Academy of Sciences of the U.S.S.R.

Vinogradsky's interest centered on complex questions of the physiology and morphology of microorganisms and the development of culture methods for saprophytic and pathogenic microbes.

His most important studies concerned the morphological variability of microbes, the discovery of microbes' capacity for chemosynthesis, and the creation and development of the bases for ecological and soil microbiology.

During the 1870's and 1880's there was much discussion concerning the variability of microorganisms; and two trends, monomorphism and polymorphism, were apparent. Sulfur and iron bacteria were the central objects of research. Vinogradsky was the first researcher to study the morphology of microbes not by investigating fixed preparations but by observing living, normally developed cells in a microculture developed in a drop suspended under a protective glass cover. Studying the culture of sulfur bacteria, which the polymorphist Friedrich Zopf had used to confirm his views, Vinogradsky discovered a mixture of microorganisms. On the basis of these observations he showed that sulfur and iron bacteria are characterized by a strict cycle of development and do not display chaotic variability, as the advocates of polymorphism had asserted. This significantly strengthened the position of monomorphism, which, for the end of the nineteenth century, was progressive. Vinogradsky never advocated the constancy of species, however, and repeatedly criticized the monomorphists.

The study of the morphology of sulfur bacteria and iron bacteria led Vinogradsky to investigate their physiology. He determined that the sulfur appears in the cells of sulfur bacteria through oxidation of hydrogen sulfide. Relating this fact to the energy metabolism of these bacteria, Vinogradsky presented the idea that the oxidation of hydrogen sulfide is analogous to the process of respiration, which provides the cell with its necessary energy. He called this phenomenon "mineral respiration," or, in the terminology of the time (1922), anorgoxidation.

The theory of chemoautotrophic metabolism of substances received convincing proof in Vinogradsky's research on the physiology of nitrifying bacteria. His new method of studying microorganisms—selective cultures—and the concept of chemoautotrophic feeding led to the solution of the problem that had been studied by many of Vinogradsky's predecessors. Using a mineral medium without any organic substances, Vinogradsky obtained pure cultures of two autonomous stimulants of nitrification, *Nitrosomonas* and *Nitrobacter*.

In his works on nitrification, Vinogradsky presented the theory of chemosynthesis as concrete and experimentally based. The chemosynthetic activity of nitrifying bacteria was shown by means of precise quantitative determinations of the relationship of oxidized nitrogen to the assimilated carbon. The discovery of chemosynthesis, an important event in nineteenth-century biological science, is still significant. New data have introduced only corrections and additions but have not changed its scientific basis.

The ecological approach to the study of microorganisms living freely in nature helped Vinogradsky to obtain important data on the metabolism of nitrifying and cellulose-decomposing bacteria. Having created an amosphere of pure nitrogen, he became the first microbiologist to separate pure cultures of a new stimulant of anaerobic nitrogen fixation—*Clostridium pastorianum*. The study of the energy metabolism of this microbe helped to establish that it is the stimulant of soil-oxidizing fermentation.

The study of *Azotobacter* was a continuation of research on biological nitrogen fixation. Vinogradsky studied it in its natural habitats: soil or on plates of silicic acid gel.

In his work on symbiotic nitrogen fixation, Vinogradsky developed extremely accurate methods of chemical analysis that permitted him to collect nitrogen in quantities on the order of several micrograms. The use of this method led to the discovery of the formation of ammonia during the process of symbiotic nitrogen-fixation. Equally fruitful was the application of ecological research to cellulose-decomposing bacteria. Using filter paper placed in a silicic acid gel and small particles of soil as the culture medium, Vinogradsky separated three types of cellulose-composing bacteria—*Cytophaga, Cellvibrio*, and *Cellfalcicula*.

Vinogradsky's works provided a firm scientific base for the ecological approach to the study of soil microflora. His development of special methods for soil microbiology—direct study of cells in the soil, spontaneous culture in dense media, and microbe cultures in soils—were of fundamental importance in the development of that science.

BIBLIOGRAPHY

I. ORIGINAL WORKS. Vinogradsky's writings include *Krugovorot azota v prirode* ("The Circulation of Nitrogen in Nature"; Moscow, 1894); *O roli mikrobov v obshchem krugovorote zhizni* ("On the Role of Microbes in the General Circulation of Life"; St. Petersburg, 1897); "K morfologii organizmov protsessa obrazovania selitry v pochve" ("On the Morphology of the Process of Organisms for Producing Niter in Soil"), in *Arkhiv biologicheskikh nauk*, **1** (1922); "Sur la décomposition de la cellulose dans le sol," in *Comptes rendus . . . de l'Académie des sciences*, **183** (1926), 691–694; "Sur la morphologie et l'oecologie des azotobacter," in *Annales de l'Institut Pasteur*, **60** (1938), 351–400; and *Mikrobiologia pochvy* ("Microbiology of the Soil"; Moscow, 1952), an anthology. These and other works can be found in *Izbrannye trudy*, 2 vols. ("Selected Works"; Moscow, 1953).

II. SECONDARY LITERATURE. On Vinogradsky and his work, see A. A. Imshenstsky, "Pamyati S. N. Vinogradskogo" ("Memories of Vinogradsky"), in *Mikrobiologia*, **22**, no. 5 (1953); and "Vinogradsky. K 100-letiyu so dnya rozhdenia" (". . . on the Centenary of His Birth"), *ibid.*, **26**, no. 1 (1957); M. M. Kononova, "S. N. Vinogradsky," in *Pochvovedenie* (1953), no. 10; S. I. Kuznetsov, "Trudy vydayushchegosya russkogo mikrobiologa" ("Works of the Outstanding Russian Microbiologist"), in *Priroda* (May 1953), 119–120; D. M. Novogrudsky, "S. N. Vinogradsky. Pervy period deyatelnosti" (". . . The First Period of His Activity"), in *Mikrobiologia*, **26** (1956); and V. L. Omelyansky, "S. N. Vinogradsky (po povodu 70-letia)" (". . . on His Seventieth Birthday"), and "Zapiska ob uchenykh trudakh S. N. Vinogradskogo" ("Note on the Scientific Works of Vinogradsky"), in *Izbrannye trudy*.

V. GUTINA

VIOLLE, JULES LOUIS GABRIEL (*b.* Langres, France, 16 November 1841; *d.* Fixin, France, 12 September 1923), *physics.*

Most of Violle's research consisted of the experimental study of topics associated with heat radiation and with high temperatures, and developed out of Violle's interest in the temperature of the sun.

Violle entered the École Normale Supérieure in 1861. He became *agrégé* in 1868, and received the degree of *docteur-ès-sciences* in 1870. (Violle was already interested in heat; for his thesis he determined the mechanical equivalent of heat by rotating a copper disk between the poles of an electromagnet and measuring calorimetrically the heat produced by the induction currents.[1]) Violle taught at the universities of Grenoble and Lyons, and in 1884 he became *maître de conférences* at the École Normale. In 1892 he became professor of physics at the Conservatoire des Arts et Métiers, and in 1897 he was elected to the French Academy of Sciences. Violle was interested in American science and, after visiting the United States and the World's Columbian Exposition in Chicago in 1893, he published some discussions of it.[2]

Early in 1874, in Grenoble, Violle began work on his first major project, the design and use of an actinometer to measure the solar constant and, indirectly, to determine the temperature of the sun. The solar constant had been defined and first measured in 1837 by Pouillet.[3] In Violle's instrument a thermometer is kept in a container at constant temperature, and sunlight is allowed to fall onto the bulb through a small hole. The rise in temperature to an equilibrium value when the hole is opened and the subsequent fall in temperature when the hole is closed again are measured. From this information Violle calculated the initial change in the temperature of the bulb and determined the rate at which heat reaches it.

In order to obtain the solar constant, it is necessary to correct the amount of heat that the instrument actually receives by the amount of heat absorbed by the atmosphere. Before Violle, the atmospheric absorption was found by comparing measurements made at different times of day. Violle compared observations made simultaneously at different altitudes. On 16 and 17 August 1875, he took measurements at the top of Mont Blanc, while others were being taken more than 3.5 kilometers below. Violle found an empirical formula for the atmospheric absorption, and concluded that the solar constant was 2.54 cal./cm.²/ minute.

Without a knowledge of the relation between temperature and energy radiation, it was difficult to deduce the solar temperature from the solar constant. (It was not until 1879 that Stefan showed that the energy varies as T^4, and the proportionality constant was not known until much later.[4]) Violle extrapolated the empirical law of Dulong and Petit. To check on the validity of this extrapolation, Violle used his actinometer to determine the known temperature of molten steel (about 1,000° C). Violle concluded that the effective temperature of the sun is about 1,500° C.

At the time of Violle's work there was much interest in the temperature of the sun. The Paris Academy had proposed the determination of it as a problem for the Bordin prize in 1874 and in 1876. Because of the problems involved in determining atmospheric absorption and extrapolating the law of Dulong and Petit to high temperatures, Violle's work received only a recompense and not the actual prize (which was not awarded at all).[5] Some of the difficulties were solved a few years later by Samuel Pierpont Langley, who showed that the atmospheric absorption varies with frequency, and who designed a bolometer to measure the heat received across the spectrum. When Langley made his first measurements by this method (in 1881, on an expedition to Mt. Whitney), he obtained some of his data with an actinometer sent especially for the purpose by Violle.[6]

As a consequence of the work on solar temperature, Violle became involved in various questions related to the determination of high temperatures. For example, he found the specific heats of platinum, palladium, and iridium up to the highest temperatures that can be measured with a gas thermometer, and then, extrapolating the relationship between specific heats and temperature, he determined their melting points. To learn something about the relation between temperature and radiation, Violle used his actinometer to determine the heat radiation emitted by platinum at various temperatures. He found the rise in energy to be slower than the extrapolation from the law of Dulong and Petit would imply, and also that at high temperatures more of the energy is in the shorter wavelengths. Violle suggested that the light emitted by liquid platinum be used as a photometric standard, and it was adopted by the International Conference on Electrical Units and Standards in 1884.

From about 1885, and continuing for about twenty years, Violle did experiments with Théodore Vautier on the propagation of sound. Violle and Vautier were interested in obtaining an accurate determination of the velocity of sound and in studying various nonlinearities and dispersions in the propagation. They analyzed the propagation of sound along an underground cylindrical pipe built for the Grenoble water system, which provided a path length of more than 12 kilometers. They looked for effects of frequency and amplitude on the velocity and for changes in the form of the disturbance during its propagation.

NOTES

1. "Sur l'équivalent mécanique de la chaleur," in *Annales de chimie*, **21** (1870), 64–97.
2. "L'exposition de Chicago et la science américaine," in *Revue des deux mondes*, **123** (1894), 579–611; "Court aperçu de l'état de l'astronomie aux États-Unis," in *Ciel et terre*, **15** (1894–1895), 223–232; "Le mouvement scientifique aux États-Unis," in *Annales du conservatoire des arts et métiers*, **6** (1894), 253–313.
3. C. Pouillet, "Mémoire sur la chaleur solaire, sur les pouvoirs rayonnants et absorbants de l'air atmosphérique, et sur la température de l'espace," in *Comptes rendus hebdomadaires des séances de l'Académie des sciences*, **7** (1838), 24–65.
4. Max Jammer, *The Conceptual Development of Quantum Mechanics* (New York, 1966), 6–8.
5. *Comptes rendus hebdomadaires des séances de l'Académie des sciences*, **84** (1877), 813–817. On the difficulty of determining the solar constant, S. P. Langley wrote the following:

"We are as though at the bottom of a turbid and agitated sea, and trying thence to obtain an idea of what goes on in an upper region of light and calm." *Researches on Solar Heat and Its Absorption by the Earth's Atmosphere* (Washington, D.C., 1884), 45.
6. S. P. Langley, *ibid.*, 70.

BIBLIOGRAPHY

I. ORIGINAL WORKS. Violle's papers include "Sur la température du soleil," in *Comptes rendus hebdomadaires des séances de l'Académie des sciences*, **78** (1874), 1425–27, 1816–1820; **79** (1874), 746–749; "Mesures actinométriques au sommet du Mont Blanc," *ibid.*, **82** (1876), 662–665; "Résultats des mesures actinométriques au sommet du Mont Blanc," *ibid.*, 729–731; "Conclusions des mesures actinométriques faites au sommet du Mont Blanc," *ibid.*, 896–898; "Chaleur spécifique et chaleur de fusion du platine," *ibid.*, **85** (1877), 543–546; "Sur la loi de rayonnement," *ibid.*, **92** (1881), 1204–1206; "Sur la propagation du son à l'intérieur d'un tuyau cylindrique," in *Annales de chimie et de physique*, ser. 6, **19** (1890), 306–345, with Théodore Vautier.

The following works list publications by Violle: "Notice sur les trauvaux scientifiques de M. Jules Violle" (Paris, 1889); Royal Society *Catalogue of Scientific Papers*, VIII, 1158; XII, 757; XIX, 368–369; J. C. Poggendorff, *Biographisch-Literarisches Handwörterbuch*, III, 1393; IV, 1570–1571.

Violle summarized his work in the autobiographical essay *Notice sur les travaux scientifiques de M. Jules Violle* (Paris, 1889).

II. SECONDARY LITERATURE. P. Villard's obituary address is in *Comptes rendus hebdomadaires des séances de l'Académie des sciences*, **177** (1923), 513–515. Samuel P. Langley discussed Violle's work on the solar constant (and other earlier work) thoroughly in his *Researches on Solar Heat and Its Absorption by the Earth's Atmosphere* (Washington, D.C., 1884).

SIGALIA DOSTROVSKY

VIRCHOW, RUDOLF CARL (*b.* Schivelbein, Pomerania, Germany, 13 October 1821; *d.* Berlin, Germany, 5 September 1902), *pathology, social medicine, public health, anthropology.*

A strong and versatile personality equally interested in the scientific and social aspects of medicine, Virchow was the most prominent German physician of the nineteenth century. His long and successful career reflects the ascendancy of German medicine after 1840, a process that gradually provided the basic underpinnings to a discipline that was still largely clinical. Armed with great self-confidence, aggressiveness, and a deep sense of social justice, Virchow became a medical activist who engaged vigorously in political polemics and participated in social reforms. His elevation of science to the level of quasi-religious dogma and his utopian view of medicine as *the* science of man should be interpreted within the framework of his times, a period that witnessed the effective adoption of scientific method in medicine.

Virchow was born in a small town in backward and rural eastern Pomerania; he was the only son of a modest merchant. He expressed an early interest in the natural sciences and received private lessons in the classical languages. Such a background enabled Virchow to become educationally competitive and in 1835 he successfully transferred to the Gymnasium in Köslin, where he received a broad humanistic training and subsequently demonstrated high scholarly abilities.

Because of his promising aptitudes, Virchow received in 1839 a military fellowship to study medicine at the Friedrich-Wilhelms Institut in Berlin. The institution, popularly known as the "Pépinière," provided educational opportunities for those unable to afford the costs in return for subsequent army medical service.

Although contemporary German medicine was only slowly shifting away from purely theoretical concerns, Virchow had the opportunity to study under Johannes Müller and Johann L. Schönlein, thereby being exposed to experimental laboratory and physical diagnostic methods, as well as epidemiological studies.

In 1843 Virchow received his medical degree from the University of Berlin with a doctoral dissertation on the corneal manifestations of rheumatic disease. Shortly thereafter he received an appointment as "company surgeon" or medical house officer at the Charité Hospital in Berlin, where he rotated through the various services. In addition, with the hospital's prosector, Robert Froriep (1804–1861), Virchow carried on microscopic studies on vascular inflammation and the problems of thrombosis and embolism.

In 1845 two forceful speeches delivered by invitation before large and influential audiences at the Friedrich-Wilhelms Institut revealed young Virchow as one of the most articulate spokesmen for the new generation of German physicians. Rejecting transcendental concerns, Virchow envisaged medical progress from three main sources: clinical observations, including the examination of the patient with the aid of physicochemical methods; animal experimentation to test specific etiologies and study certain drug effects; and pathological

anatomy, especially at the microscopic level. Life, he insisted, was merely the sum of physical and chemical actions and essentially the expression of cell activity.

Virchow's rather provocative ideas generated considerable hostility among his older peers, but he passed his licensure examination in 1846 without difficulties and began teaching pathological anatomy. Under the auspices of Prussia's high military and civilian authorities, he traveled to Prague and Vienna in order to evaluate their programs in pathology. One of the consequences of his trip was Virchow's strong attack on Rokitansky and the Viennese Medical School, whom he indicted for their dogmatism and support of an outdated humoralism.

After completing his *Habilitationsschrift* in 1847, Virchow was officially appointed an instructor under the deanship of Johannes Müller at the University of Berlin; he also succeeded Froriep as prosector at the Charité Hospital. In the same year Virchow launched—ostensibly in order to publish his speeches of 1845—a new scientific journal with Benno Reinhardt, a colleague in pathology. The publication, named *Archiv für pathologische Anatomie und Physiologie, und für die klinische Medizin*, became one of the most prominent medical periodicals of the time; and Virchow remained its editor until his death.

The typhus epidemic that ravaged the Prussian province of Upper Silesia in early 1848 prompted the government to send a team of physicians to the area to survey the disaster. With the pediatrician and bureaucrat Stephan F. Barez, Virchow visited the afflicted region for almost three weeks and came face to face with the backward and destitute Polish minority, who were struggling precariously to survive. According to his own testimony, the impact of that encounter left an indelible mark on his already liberal social and political beliefs. Instead of merely returning with a new set of medical guidelines for the Prussian government, Virchow recommended political freedom, and sweeping educational and economic reforms for the people of Upper Silesia.

Virchow's gradual alienation from the status quo and his political radicalization led him to participate actively in the uprisings of 1848 in Berlin, where he fought alongside his friends on the barricades. As a result, Virchow was thrown into a full schedule of political activities and became a member of the Berlin Democratic Congress and editor of a weekly entitled *Die medizinische Reform*. Virchow's triumph, however, was shortlived. In early 1849 he was suspended from his academic position as prosector at the Charité Hospital because of his revolutionary activities. Although he was partially reinstated as a result of protests from medical circles and students, the defeat of liberalism imposed restrictions on Virchow and created an unfavorable climate for his activities.

Thus, in November 1849 Virchow finally left Berlin and went on to the University of Würzburg in order to assume the recently created chair in pathological anatomy, the first of its kind in Germany. He was temporarily separated from political concerns, and the ensuing years marked Virchow's highest level of scientific achievement and the establishment of the concept of "cellular pathology." He was also deeply engaged in teaching, and among his most famous Würzburg students were Edwin Klebs (1834–1913), Ernst H. P. A. Haeckel (1834–1919), and Adolf Kussmaul (1822–1902). Virchow initiated the publication of the six-volume *Handbuch der speziellen Pathologie und Therapie*, a monumental textbook of pathology and therapeutics; he also edited the famous *Jahresbericht*, a German yearbook depicting medical advances.

In 1856 Virchow accepted an invitation to return to Berlin as professor of pathological anatomy and director of the newly created Pathological Institute. Under Virchow the institution became a famous training ground for a large number of German and foreign medical scientists, including Hoppe-Seyler, Recklinghausen, and Cohnheim. In addition, for almost two decades Virchow remained in charge of a clinical section at the Charité Hospital, thereby carrying out the program of medical progress enunciated in 1845.

Two aspects of Virchow the pathologist should be distinguished: his scientific methodology and his activities in the field of cellular pathology. Without being original, Virchow stressed the importance of observation and experiment, strongly condemning his speculative predecessors. He himself, however, fell prey to the still lingering desire for an overall synthesis of medical knowledge and the establishment of first principles. Imaginative and intuitive, Virchow performed numerous inductive leaps, leaving to others, whose work he often did not acknowledge, the painstaking task of fact collecting.

Virchow expressed an early interest in "pathological physiology" and promulgated a dynamic view of disease processes, which viewed the static structural changes (pathological anatomy) sequentially. Although the idea was by no means novel, with the aid of improved microscopic and bio-chemical techniques, Virchow applied the concept successfully. For Virchow the microscope became the central tool for reducing pathological processes to alterations occurring at the cellular level. Hence, the cell became the fundamental living unit in both health and disease—a biological rather than a mechanical entity. Virchow's notion of cellular pathology implied that all the manifestations of disease could be reduced to disturbances of living cells. Moreover, according to Virchow's famous principle, "omnis cellula e cellula," all cells originated from other cells. Cellular function, in turn, depended on intracellular physicochemical changes, which were reflected in the varying morphology. Finally, all pathological forms were to be viewed as deviations from the normal structures. Virchow's cellular pathology demolished the vestiges of humoral and neural physiopathology, and placed the field on its modern basis.

During the early years of Virchow's second period in Berlin, his interests began to shift gradually from pathology to anthropology, while he was engaged at the same time in a fair amount of political activities. Nevertheless, Virchow published in 1858 his most famous book, *Die Cellularpathologie . . .*, and in 1863 his work on tumors, *Die krankhaften Geschwülste*.

At the suggestion of an old friend, Virchow was appointed in 1859 to the Berlin City Council, where he concentrated his efforts on matters related to public health. Aided by the mayor of Berlin, Karl T. Seydel, who was his brother-in-law, Virchow was instrumental in achieving improvements in both the sewage system and water supply of the rapidly growing metropolis. In 1861 he was elected a member of the Prussian lower house and represented the new liberal Deutsche Fortschrittspartei (German Progressive Party), which he had founded with some friends. As an early leader of the opposition to Bismarck's policy of rearmament and forced unification, Virchow brought down on himself the wrath of his opponent and was challenged by Bismarck to a duel, which he was wise enough to avoid. During the ensuing Franco-Prussian War of 1870, Virchow was active in organizing military hospital facilities and establishing ambulance and train services for the wounded.

Following his experiences in Upper Silesia, Virchow stressed a sociological theory of disease, claiming that political and socioeconomic factors acted as significant predisposing factors in many ailments. He even went so far as to declare that certain epidemics arose specifically in response to some social upheavals. Virchow considered a number of diseases as "artificial" or primarily caused by conditions within society and thus liable to cure or elimination through social change. As early as 1848 Virchow insisted on the constitutional right of every individual to be healthy. Society had the responsibility to provide the necessary sanitary conditions for the unhampered development of its members. Here again, through his work in the Reichstag and the city council of Berlin, Virchow not only espoused lofty ideals but fought hard to achieve the necessary reforms in school hygiene, sewage treatment, pure water control, and hospital construction.

In proclaiming that medicine was the highest form of human insight and the mother of all the sciences, Virchow was following in the footsteps of French social thought and also expressing a postulate of the German philosophers of nature. Although his utopian hopes for medicine as the unified science of man did not materialize, Virchow's efforts were helpful in associating the rapidly developing natural sciences with medical concerns. His attempts to derive an ethical framework from the biological sciences laid the foundations of bioethics.

In his later years Virchow's skeptical attitude toward bacteriology was based, to a large extent, on his belief that there was no single cause of disease. He did not consider any germ to be the sole etiologic agent in an infectious illness. The bacterial agents were, in Virchow's view, only one factor in the causation of disease among a variety of environmental and sociological factors clearly discernible during the typhus and cholera epidemics of 1847–1849. Such broad considerations did not originate with Virchow, but they gained greater attention and significance with his prestigious support.

From 1870 onward, Virchow rather assiduously cultivated another science: anthropology. Cofounder of the German Anthropological Society a year earlier and author of several studies dealing with skull deformities, he studied the physical characteristics of the Germans, especially the Frisians. After performing a nationwide racial survey of schoolchildren, Virchow concluded that there was no pure German race but only a mixture of differing morphological types. On another matter, he questioned Darwinism as an established fact, viewing it rather as a tentative hypothesis in search of adequate proof.

The Darwinian stimulus to archaeological research also affected Virchow, and in 1870 he began his own excavations in Pomerania. His later friendship with Schliemann lent some legitimacy to this enthusiastic dilettante and eventually helped to attract the treasures to Berlin. In 1879 Virchow himself traveled with Schliemann to Hissarlik, where Homer's Troy was being excavated, and in 1888 he participated in another archaeological dig in Egypt.

In 1886 Virchow was instrumental in the erection of the Berlin Ethnological Museum, followed by the Museum of German Folklore in 1888. Throughout the 1880's he continued to play a key role in the budgetary matters of the Reichstag, and he remained chairman of the finance committee until his death.

Virchow's eightieth birthday in 1901 became the occasion for an unprecedented worldwide celebration. A torchlight parade in Berlin and numerous receptions in the leading scientific centers, even as far away as Japan and Russia, gave testimony to his unparalleled international reputation. Never seriously ill throughout his long life, Virchow suffered a broken hip in early 1902 after falling from a streetcar in Berlin. Although seemingly on the mend, the long period of inactivity seriously undermined his health, and he died several months later of cardiac insufficiency.

Virchow's great fame made him a widely respected authority in his numerous fields of endeavor. His penchant, however, for polemics and acrimonious exchanges with colleagues exerted unfavorable influences for the development of certain medical ideas and methods. An example was his opposition to the prophylactic hand washings of Semmelweis for the prevention of puerperal fever. In his later years Virchow displayed a stifling dogmatism and a certain pedantry, which in some measure detracted from his earlier popularity. In spite of these traits he was overwhelmingly self-confident and untiringly persuasive in popularizing his views. Few great men have been privileged to perceive more clearly the fruits of their labors in the autumn of their lives than Virchow. In less than half a century Germany had progressed from speculative and philosophical healing to become the world center of modern scientific medicine, and Virchow had played a decisive role in this crucial transformation.

BIBLIOGRAPHY

I. ORIGINAL WORKS. Most of Virchow's important medical and anthropological writings are enumerated chronologically in a small "Festschrift" edited by J. Schwalbe on the occasion of the physician's 80th birthday: *Virchow-Bibliographie 1843–1901* (Berlin, 1901), which covers close to 2,000 titles, and contains a valuable subject index. Pertinent archival material can be found in the "Nachlass Rudolf Virchow" of the Literatur-Archiv, Institut für deutsche Sprache und Literatur, Deutsche Akademie der Wissenschaften, East Berlin. Thor Jager (Wichita, Kansas) has a large collection of Virchow's original MSS and letters, with many pamphlets and books.

Prominent among Virchow's publications were *Die Cellularpathologie in ihrer Begründung auf physiologische und pathologische Gewebelehre* (Berlin, 1858), representing 20 lectures that he delivered at the Pathological Institute in Berlin between February and April 1858. The 2nd ed. of the work was translated into English by F. Chance, *Cellular Pathology as Based Upon Physiological and Pathological Histology* (London, 1860). Two important collections of Virchow's writings are *Gesammelte Abhandlungen zur wissenschaftlichen Medizin* (Frankfurt, 1856), which contains, among others, articles on white cells and leukemia, thrombosis and embolism, gynecological subjects, and the pathology of the newborn; and *Gesammelte Abhandlungen aus dem Gebiet der oeffentlichen Medizin und der Seuchenlehre*, 2 vols. (Berlin, 1879), dealing with medical reform and public health; epidemics and mortality statistics; hospitals, military and urban sanitation; and legal medicine.

A collection of 30 lectures on tumors given at the University of Berlin during the winter semester 1862–1863 is in *Die krankhaften Geschwülste*, 3 vols. (Berlin, 1863–1867). For other lectures on general pathology, see *Die Vorlesungen Rudolf Virchows über allgemeine pathologische Anatomie aus dem Wintersemester 1855–56 in Würzburg*, E. Kugler, ed. (Jena, 1930).

Some of Virchow's more philosophical essays and sociopolitical speeches have received wider diffusion and have been translated into English. *Die Freiheit der Wissenschaft im modernen Staat* (Berlin, 1877) appeared as *The Freedom of Science in the Modern State* (London, 1878). *Morgagni und der anatomische Gedanke* (Berlin, 1894) was translated by R. E. Schlueter and J. Auer, "Morgagni and the Anatomical Concept," in *Bulletin of the History of Medicine*, 7 (1939), 975–989. A series of talks are in *Disease, Life and Man, Selected Essays by Rudolf Virchow*, trans. and with an introduction by L. J. Rather (Stanford, 1958), including two articles on "Standpoints in Scientific Medicine" (1847, 1877), which also appeared in *Bulletin of the History of Medicine*, 30 (1956), 436–449, 537–543.

For Virchow's critique of Rokitansky's pathology, L. J. Rather, trans., "Virchow's Review of Rokitansky's 'Handbuch' in the Preussische Medizinal Zeitung, Dec. 1846," in *Clio medica*, 4 (1969), 127–140. Virchow's Croonian lecture delivered at the Royal Society of London in 1893 was translated into English and published as "The Place of Pathology Among the Biological Sciences," in *Proceedings of the Royal Society*, 53 (1893), 114–129. Virchow also gave the second Huxley lecture at the opening of Charing Cross Hospital Medical School, London, in 1898, which appeared first in English as "Recent Advances in Science and Their Bearing on Medicine and Surgery," in *British Medical Journal* (1898), 2, 1021–1028.

Virchow's early studies and relationships with his parents are contained in an extensive correspondence, *Rudolf Virchow, Briefe an seine Eltern, 1839 bis 1864*, M. Rabl, ed. (Leipzig, 1906).

Virchow's most notable anthropological writings are *Beiträge zur physischen Antropologie der Deutschen mit besonderer Berücksichtigung der Friesen* (Berlin, 1877) and *Crania Ethnica Americana, Sammlung auserlesener amerikanischer Schädeltypen* (Berlin, 1892). See also *Menschen und Affenschädel* (Berlin, 1870), a lecture that was translated as *The Cranial Affinities of Man and the Ape* (Berlin, 1871). For an English version of Virchow's review, see C. A. Bleismer, "Anthropology in the Last Twenty Years," in *Report of the Board of Regents of the Smithsonian Institution* (1890), 550–570. "Rassenbildung und Erblichkeit" (1896) was translated as "Heredity and the Formation of Race," in *This Is Race*, E. W. Count, ed. (New York, 1950), 176–193.

II. Secondary Literature. The most important work on Virchow is Erwin H. Ackerknecht, *Rudolf Virchow, Doctor, Statesman, Anthropologist* (Madison, Wis., 1953), which is not primarily a biography but rather an analysis of Virchow's ideas, works, and accomplishments. Among some of the more recent biographical works in German are Ludwig Aschoff, *Rudolf Virchow* (Hamburg, 1948); Hellmuth Unger, *Virchow, ein Leben für die Forschung* (Hamburg, 1953); Curt Froboese, *Rudolf Virchow* (Stuttgart, 1953); Kurt Winter, *Rudolf Virchow* (Leipzig, 1956); and Ernst Meyer, *Rudolf Virchow* (Wiesbaden, 1956).

In 1921 a large number of speeches and articles appeared in Germany to commemorate Virchow's 100th birthday. For a list of these works, see *Virchows Archiv für pathologische Anatomie und Physiologie und für klinische Medizin*, 235 (1921), and the *Deutsche medizinische Wochenschrift*, 47, no. 40 (1921), 1185–1195. Several letters written by the young Virchow appear in G. B. Gruber, "Aus der Jungarztzeit von Rudolf Virchow," in *Virchows Archiv . . .*, 321 (1952), 462–481. Brief biographical sketches of Virchow in English are the obituary by F. Semon in *British Medical Journal*, (1902), 2, 795–802; O. Israel, "Rudolph Virchow, 1821–1902," in *Report of the Board of Regents of the Smithsonian Institution* (1902), 641–659; James J. Walsh, *Makers of Modern Medicine* (New York, 1915), 357–430; and Henry E. Sigerist, *Grosse Ärzte, eine Geschichte der Heilkunde in Lebensbildern* (Munich, 1932), English trans. by E. and C. Paul, *The Great Doctors* (Garden City, N.Y., 1958), 319–330.

Recent German works dealing with Virchow's achievements are Felix Boenheim, *Virchow, Werk und Wirkung* (Berlin, 1957); Wolfgang Jacob, *Medizinische Anthropologie im 19. Jahrhundert*; and Gerhard Hiltner, *Rudolf Virchow, ein weltgeschichtlicher Brennpunkt im Werdegang von Naturwissenschaft und Medizin*

(Stuttgart, 1970). Also noteworthy is K. Panne, "Die Wissenschaftstheorie von Rudolf Virchow" (unpublished doctoral dissertation, Univ. of Düsseldorf, 1967). Numerous articles dealing with aspects of Virchow's work are W. Pagel, "Virchow und die Grundlagen der Medizin des XIX. Jahrhunderts," in *Jenaer medizin-historische Beiträge*, **14** (1931), 1–44; P. Diepgen, "Virchow und die Romantik," in *Deutsche medizinische Wochenschrift*, **58** (1932), 1256–1258; L. J. Rather, "Virchow und die Entwicklung der Entzündungsfrage im 19. Jahrhundert," in *Verhandlungen des XX. Internationalen Kongresses für die Geschichte der Medizin* (Hildesheim, 1968), 161–177; and H. M. Koelbing, "Rudolf Virchow und die moderne Pathologie," in *Münchener medizinische Wochenschrift*, **110** (1968), 349–354.

Other valuable references to Virchow are L. S. King, "Cell Theory, Key to Modern Medicine," in *The Growth of Medical Thought* (Chicago, 1963), 207–219; and W. H. McMenemey, "Cellular Pathology, With Special Teachings on Medical Thought and Practice," in *Medicine and Science in the 1860's*, F. N. L. Poynter, ed. (London, 1968), 13–43.

Important journal articles are W. Pagel, "The Speculative Basis of Modern Pathology. Jahn, Virchow, and the Philosophy of Pathology," in *Bulletin of the History of Medicine*, **38** (1945), 1–43; J. W. Wilson, "Virchow's Contribution to the Cell Theory," in *Journal of the History of Medicine*, **2** (1947), 163–178; P. Klemperer, "The Pathology of Morgagni and Virchow," in *Bulletin of the History of Medicine*, **27** (1953), 24–38; D. Pridan, "Rudolf Virchow and Social Medicine in Historical Perspective," in *Medical History*, **8** (1964), 274–284; and L. J. Rather, "Rudolf Virchow's Views on Pathology, Pathological Anatomy and Cellular Pathology," in *Archives of Pathology*, **82** (1966), 197–204.

GUENTER B. RISSE

VIREY, JULIEN-JOSEPH (*b.* Hortes, Haute-Marne, France, 22 December 1775; *d.* Paris, France, 9 March 1846), *natural history, philosophy of nature, pharmacy, anthropology, hygiene, psychology, physiology.*

After serving an apprenticeship in pharmacy with an uncle in Langres, Virey entered military service in 1794 as pharmacist third class. Except for a few brief tours of duty, notably with the Army of the Rhine, his military career was spent at the Val-de-Grâce hospital in Paris, where from 1804 until his retirement in 1813, he was acting chief pharmacist. In 1814 Virey obtained an M.D. degree from the Faculty of Medicine in Paris, and in 1823 he was elected to the Academy of Medicine. He lectured on natural history at the Athénée de Paris in 1814–1815; and for some time after 1830 he represented the Haute-Marne in the Chamber of Deputies.

A remarkably prolific author, Virey produced works encompassing a wide range of interests. His *Traité de pharmacie* (1811) enjoyed considerable authority during the first half of the nineteenth century, appearing in several editions. His *Histoire naturelle des médicamens, des alimens et des poisons* (1820), as well as his numerous descriptive articles on natural products and natural history, demonstrated a high level of practical expertise. Virey also produced a large body of philosophical writings dealing with natural philosophy, anthropology, social hygiene, psychology, and physiology.

Vitalism and teleology are basic components of Virey's natural philosophy and are elaborated at length in his *De la puissance vitale* (1823) and *Philosophie de l'histoire naturelle* (1835). An "intelligence formatrice" directs the organization of life forms, and variations in fixed species are oscillations around primordial types; modification of species can come about only through cosmic change. In his *L'art de perfectionner l'homme* (1808), a work on mental health, Virey attempted to refute the sensationalism of Condillac and the materialism of Cabanis and his fellow Idéologues. Virey's lectures at the Athénée de Paris, published in 1822 as *Histoire des moeurs et de l'instinct des animaux*, sought to demonstrate how instinct and intelligence are related to the structures and functions of nervous systems. Considerable effort and erudition went into his *Histoire naturelle du genre humain* (1801; 2nd ed., 1824), a work typical of early nineteenth-century anthropology in its generalizations on types of man, customs, religion, psychology, language, infancy, women, and social organization. The status of women in society received extended treatment in *De la femme, sous ses rapports physiologique, moral et littéraire* (1823). Virey's last major work, *De la physiologie dans ses rapports avec la philosophie* (1844), attempted to construct a metaphysical foundation for physiological psychology.

But perhaps the two most significant philosophical studies undertaken by Virey concerned circadian rhythms and social hygiene. In his M.D. thesis, *Ephémérides de la vie humaine* (1814), Virey likened the daily-recurring physiological cycles in man to "une sorte d'horloge vivante" and speculated on how diurnal states of health are affected by periodic exogenous phenomena. Although some of the ideas developed in this work had already been discussed by Erasmus Darwin and others, his treatment of this subject nevertheless remains fresh and innovative. In *Hygiène philosophique* (1828), Virey explored in an original, if speculative, manner the influence of social, as well as political, institutions and events on the health of individuals and nations.

BIBLIOGRAPHY

I. ORIGINAL WORKS. There is no complete bibliography of Virey's publications. The most comprehensive is A. C. P. Callisen, *Medicinisches Schriftsteller-Lexicon*, XX (Copenhagen, 1834), 158–177, and XXXIII (Altona, 1845), 159–162. Some other listings are *Catalogue général des livres imprimés de la Bibliothèque nationale*, CCXI (Paris, 1972), 1036–1044; *Exposé des travaux de J.-J. Virey, dans les sciences philosophiques* (Paris, 1842); *Index-Catalogue of the Library of the Surgeon-General's Office*, XV (Washington, D.C., 1894), 768, and 2nd ser., XX (Washington, D.C., 1915), 265; J. M. Quérard, *La France littéraire*, X (Paris, 1839), 232–235; and Royal Society, *Catalogue of Scientific Papers*, VI, 166–172.

Virey wrote hundreds of articles, the bulk of them appearing in *Journal de pharmacie*, on the editorial board of which he served, and in the *Dictionnaire des sciences médicales*, 60 vols. (Paris, 1812–1824). Among the publications to which he contributed prominently were the first two eds. of *Nouveau dictionnaire d'histoire naturelle . . .*, 24 vols. (Paris, 1803–1804) and 36 vols. (Paris, 1816–1819); and to the *Dictionnaire de la conversation et de la lecture*, 52 vols. (Paris, 1832–1839).

II. SECONDARY LITERATURE. See an unsigned obituary in *Archives générales de médecine*, 4th ser., **11** (1846), 116–119; Alex Berman, "Romantic Hygeia: J. J. Virey (1775–1846), Pharmacist and Philosopher of Nature," in *Bulletin of the History of Medicine*, **39** (Mar.-Apr. 1965), 134–142; *Biographie universelle et portative des contemporains, ou dictionnaire historique des hommes morts depuis 1788 jusqu'à nos jours . . .*, V (Paris, 1834), 875–876; Maurice Bouvet, "Les origines de l'hôpital du Val-de-Grâce et ses premiers pharmaciens (de 1793 à 1815)," in *Revue d'histoire de la pharmacie*, **7** (1939), 136–145; J. H. Réveillé-Parise, "Galerie médicale (no. xxvii). Virey (Julien-Joseph)," in *Gazette médicale de Paris*, 3rd ser., **1** (1846), 847–851; [Claude Lachaise], under pseudonym C. Sachaile, *Les médecins de Paris jugés par leurs oeuvres . . .* (Paris, 1845), 628–629; and E. Soubeiran, "Discours prononcé par M. Soubeiran, aux funérailles de M. Virey," in *Journal de pharmacie et de chimie*, 3rd ser., **9** (1846), 277–282.

Virey's vitalism, as propounded in his *De la puissance vitale* (1823), is discussed in J. P. Damiron, *Essai sur l'histoire de la philosophie en France au XIXe siècle*, 3rd ed., II (Paris, 1834), 25–39.

ALEX BERMAN

VIRTANEN, ARTTURI ILMARI (*b*. Helsinki, Finland, 15 January 1895; *d*. Helsinki, 11 November 1973), *biochemistry*.

Virtanen, the son of Kaarlo and Serafina Isotalo Virtanen, received his elementary education at the classical lyceum in Viipuri (now Vyborg, R.S.F.S.R.), after which he entered the University of Helsinki. He received the Master of Science degree in 1916 and for a year served as first assistant in the Central Industrial Laboratory in Helsinki, then returned to the university for the doctorate, which he obtained in 1919. Virtanen did work in physical chemistry at Zurich in 1920, in bacteriology at Stockholm in 1921, and in enzymology at Stockholm with Euler-Chelpin in 1923 and 1924. Between 1918 and 1920 he was associated with laboratories for the control of butter and cheese manufacture, and from 1921 to 1931 he was director of the laboratories of the Finnish Cooperative Dairies Association. In 1931 Virtanen became director of the Biochemical Research Institute at Helsinki, a position he held for life. After 1924 he also held academic posts: *Dozent* at the University of Helsinki and professor of biochemistry at the Technical University in Helsinki, remaining at the latter until 1939. From 1939 to 1948 he was professor of biochemistry at the University of Helsinki. In 1920 he married Lilja Moisio. They had two sons.

Virtanen's broad scientific background led to his interest in theoretical biochemistry, while his experience in the dairy industry acquainted him with agricultural problems. Throughout his life he combined these interests in work that contributed greatly both to academic biochemistry and to agricultural chemistry.

Virtanen's first biochemical studies concerned bacterial fermentations. In 1924 he showed the necessity for the presence of cozymase in lactic and propionic fermentations. Convinced that most of the proteins in plant cells were enzymes, he undertook a comparison of protein content and enzyme activity of the cells. His attention was thus drawn to the nitrogenous substances of plants, and in 1925 he began to investigate their production in the root nodules of leguminous plants. Virtanen recognized that during storage much of the nitrogenous material was lost. This fact was of great practical importance in agriculture, since when fodder was kept for a long period, its value as a cattle food decreased.

These considerations led Virtanen to study methods for preserving the quality of fresh fodder. He soon learned that deterioration was slowed in an acid medium. Careful studies of various methods for producing a nutritionally safe degree of acidity that would preserve quality led him to the discovery of the AIV method of fodder storage (the name being taken from his initials). It consisted in treating the fodder with a specific mixture of hydrochloric and sulfuric acids so that silage would rapidly reach a determined degree of acidity. Fodder treated in this way retained nearly its full content of proteins, carotene, and vitamin C for prolonged periods. Cattle fed on it produced milk rich in protein and vitamin A. The method was introduced on Finnish farms in 1929, and its use gradually spread to other countries. For this discovery Virtanen was awarded the Nobel Prize for chemistry in 1945.

While this work was continuing, Virtanen was also pursuing his purely biochemical studies. He found that the synthesis of nitrogenous compounds in leguminous plant roots by bacteria required the presence of a red pigment resembling hemoglobin. He investigated the methods by which plants synthesize vitamins, and in later years he studied the chemical composition of higher plants, isolating a number of new compounds, some of considerable nutritional importance.

In addition to the Nobel Prize, Virtanen received many honorary degrees and medals, and served on the editorial boards of numerous biochemical journals. He was the Finnish representative on the United Nations Commission on Nutrition, and from 1948 to 1963 was president of the Academy of Finland.

BIBLIOGRAPHY

Virtanen's work on nitrogen fixation was summed up in his book *Cattle Fodder and Human Nutrition With Special Reference to Biological Nitrogen Fixation* (Cambridge, 1938). His account of his studies on the AIV system, as well as his biography, are in *Nobel Lectures in Chemistry 1942–1962* (Amsterdam–London–New York, 1964), 71–105.

HENRY M. LEICESTER

VITALI, GIUSEPPE (*b.* Ravenna, Italy, 26 August 1875; *d.* Bologna, Italy, 29 February 1932), *mathematics.*

Vitali was unusual, in that for most of his life he worked in relative isolation, although he lived in Genoa and thus was not cut off from intellectual life. Nevertheless, he achieved such valuable results in the theory of functions of a real variable that he is considered one of the greatest predecessors of Lebesgue.

Vitali graduated from the Scuola Normale Superiore at Pisa in 1899 and immediately became assistant to Ulisse Dini, then one of the most authoritative Italian mathematicians, whose recommendation and approval could assure a promising career to a young mathematician. Vitali left this coveted post after two years, however, possibly because of financial need, and taught at various secondary schools, ending at the Liceo C. Colombo in Genoa (1904–1923). He also became involved in politics there, as a Socialist town councillor and municipal magistrate. In 1922, after the rise to power of fascism and the dissolution of the Socialist party, Vitali returned to his studies and made such progress that at the end of 1923 he won the competition for the professorship of infinitesimal analysis at the University of Modena. The following year he moved to Padua and, in 1930, to Bologna.

In 1926 Vitali was struck by a serious circulatory disorder. Weakened in body but not in mind, he returned to research and teaching; about half his published works (of which there are not many) were composed after this illness, even though he could not write.

Vitali was essentially self-taught and accustomed to working alone. This isolation sometimes led him inadvertently to duplicate someone else's discoveries, but he also avoided well-trodden paths. He holds undisputed priority in a number of discoveries: a theorem on set-covering, the notion of an absolutely continuous function, a theorem on the analyticity of the limit of certain successions of equilimited analytical functions, and criteria for closure of systems of orthogonal functions.

In his last years Vitali confined himself to problems of less general interest, such as his new absolute differential calculus and, in collaboration with his friend and colleague A. Tònolo (1885–1962), his "geometry" of Hilbert spaces—neither of which has aroused particular interest.

After Vitali's death Giovanni Sansone published, as coauthors, Vitali's useful *Moderna teoria delle funzioni di variabile reale* (Bologna, 1935; 3rd ed., 1952), the first part of which was written mainly by Vitali.

Vitali was a corresponding fellow of the Academy of Sciences of Turin (1928), of the Accademia dei Lincei (1930), and of the Academy of Bologna (1931).

BIBLIOGRAPHY

See the biographies by S. Pincherle, in *Bollettino dell'Unione matematica italiana*, **11** (1932), 125–126, A. Tònolo, in *Rendiconti del Seminario matematico dell' Università di Padova*, **3** (1932), 67–81, which has a bibliography; and F. G. Tricomi, in *Memorie dell' Accademia delle scienze di Torino*, 4th ser., **4** (1962), 115–116.

F. G. Tricomi

VITELO. See **Witelo**.

VITRUVIUS (*fl.* Rome, first century B.C.), *architecture*.

For a detailed account of his life and work, see Supplement.

VIVES, JUAN LUIS (*b.* Valencia, Spain, 6 March 1492; *d.* Bruges, Netherlands [now Belgium], 6 May 1540), *education, philosophy, psychology*.

Probably born to Jewish parents who adopted Catholicism in the oppressive religious atmosphere of fifteenth-century Spain,[1] Vives became one of the greatest Catholic humanists of sixteenth-century Europe. After early schooling in liberal Valencia, he left Spain in 1510 (never to return) and entered the University of Paris, where Spanish masters and students flourished. There, under Gaspar Lax and Jean Dullaert of Ghent, Vives received a Scholastic education that emphasized Aristotelian terminist logic, dialectic, and disputation, a program against which his developing humanist inclinations soon rebelled.

In 1512 Vives was attracted to the Low Countries, especially Bruges, where in 1514 he took up permanent residence (he married Margaret Valdaura of Bruges in 1524), and Louvain, where he attended lectures at the university in 1514 and qualified as lecturer in 1520.

Over the years Vives left Bruges intermittently. Especially significant is the period between 1523 and 1528, when he lectured at Oxford University (Corpus Christi College) and met, or continued earlier friendships with, Thomas More, John Fisher, and Thomas Linacre, and was highly regarded by Henry VIII and his queen, Catherine of Aragon. When Henry sought to divorce Catherine and relations between Henry and Spain soured, Vives fell under a cloud. His lectureship at Oxford was terminated in 1527 and he was banished from England in 1528. Frequently ill and plagued with debt, Vives produced many of his most important works during the last decade of his life.

On intimate terms with the greatest humanists of his day, including Erasmus and Budé, Vives was not only a master of classical Latin literature (he apparently cared much less for the Greek classics) but also wrote on religion, education, rhetoric, philosophy, methodology, science, and politics. Science and philosophy were not of interest for their own sakes, but only insofar as they could prove of practical use in subduing human passions and improving morality. Vives believed that original sin had weakened human reason to the extent that it could not determine nature's primary, necessary principles and was, therefore, incapable of arriving at scientific demonstration in the strict Aristotelian sense. Human knowledge was dependent on experience derived from the five fallible senses. Since the true essences of things transcended experience, knowledge of them lay beyond human reason. Man's knowledge of things was therefore based upon probability, conjecture, and approximation, which were, however, adequate because, despite original sin, God had generously allowed man sufficient reason to master nature, as evidenced by human control over the sublunar region.[2] By assuming that God guaranteed the reliability of human knowledge to whatever extent was necessary, Vives avoided falling into total skepticism. The basic empiricism described here formed the foundation of his theories of education, which emphasized observation, simple experiments, and direct experience.

Vives has been justly hailed as a major figure in the history of psychology. He held that the essence of the soul—mind—was indescribable.[3] It could be known only by its actions, as observed by the internal and external senses. Before Descartes and Francis Bacon, Vives developed an empirical psychology in which he advocated the study of mental activity introspectively and in others. He formulated a theory of association of ideas from an elaborate analysis of memory. If two ideas are implanted in the mind simultaneously, or within a short interval of time, the occurrence of one would cause the recall of the other.[4]

In commemorating the fourth centenary of

Vives' death, the Bibliothèque Nationale exhibited over five hundred editions of his works.[5] They bear witness to his great influence on his own and subsequent centuries.

NOTES

1. Carlos G. Noreña, "Juan Luis Vives," 18–22.
2. *De prima philosophia*, bk. I, in *Opera omnia*, III, 188.
3. *De anima et vita* (Bruges, 1538), in *Opera omnia*, III, 332.
4. *Ibid.*, 349–350.
5. Noreña, *op. cit.*, 1. For the catalog of the exhibition, see J. Estelrich, *Vivès, exposition organisée à la Bibliothèque nationale, Paris, janvier–mars, 1941* (Paris, 1942).

BIBLIOGRAPHY

I. ORIGINAL WORKS. Vives' *Opera omnia* was first published at Basel in 1555. Relying heavily on the Basel ed., Gregorio Mayans y Síscar published the only other ed. of the collected works: *Joannis Ludovici Vivis valentini Opera omnia*, 8 vols. (Valencia, 1782–1790; repr. London, 1964). Although incomplete (as in the earlier Basel ed., it lacks the *Commentaries on Saint Augustine* and perhaps a few other minor works; see Noreña, "Juan Luis Vives," 4), it does include the works relevant to science and philosophy, which appear in vols. III and VI. In addition to a number of brief treatises, vol. III contains *De Aristotelis operibus censura, De instrumento probabilitatis liber unus, De syllogismo, De prima philosophia, sive De intimo naturae opificio* (in three books), and *De anima et vita* (a lengthy treatise in three books, which treats many of the traditional topics in Aristotle's *De anima*; a photocopy repr. of the Basel ed. [1538] of this work was issued by Mario Sancipriano [Turin, 1959]); vol. VI contains the *De disciplinis*, composed of two parts, *De causis corruptarum artium* in seven bks., depicting the low state of the arts in Vives' day (especially relevant are book 3, which treats logic, and book 5, which denounces natural philosophy, medicine, and mathematics), and *De tradendis disciplinis*, in five bks., devoted to the reformation and revitalization of the fallen arts.

For a chronological list of Vives' works, see Carlos G. Noreña, *Juan Luis Vives*, which is vol. 34 in International Archives of the History of Ideas (The Hague, 1970), app. 2, 307–308; app. 1, 300–306, is "Editions of Vives' Main Works From 1520–1650" (also see Sancipriano's bibliography of eds., pp. x–xiv of his repr. ed. of *De anima et vita*, cited above). For the translations into Spanish and English, see Noreña, *op. cit.*, 310–311; and, despite the title, for English translations of Vives' Latin works, see Remigio Ugo Pane, *English Translations From the Spanish 1484–1943: A Bibliography* (New Brunswick, N.J., 1944), 201–202.

II. SECONDARY LITERATURE. Extensive bibliographies of secondary literature appear in Noreña (see above), 311–321; and Sancipriano's ed. of *De anima et vita* (see above), xiv–xviii. Noreña also includes a useful survey of the history of research on Vives in ch. 1: "The Vicissitudes of Vives' Fame," 1–14.

The standard biography and evaluation of Vives' work is Adolfo Bonilla y San Martín, *Luis Vives y la filosofía del renacimiento*, 3 vols. (Madrid, 1903). A briefer, but still substantial account of Vives' life is Lorenzo Riber's intro. to his Spanish trans. of Vives' *Opera omnia*, in *Juan Luis Vives Obras completas*, 2 vols. (Madrid, 1947–1948), 13–255. Critical of previous biographical accounts, especially on the question of Vives' Jewish parentage, is Noreña (see above), pt. 1, "The Life of Juan Luis Vives," 1–6, 1–120; a briefer biographical sketch appears in *Vives' "Introduction to Wisdom," a Renaissance Textbook*, edited, with an introduction, by Marian Leona Tobriner, S.N.J.M., which is no. 35 in the series Classics in Education (New York, 1968), 9–36.

Vives' attitudes toward Scholastic philosophy and science and his own views of science appear to have received little attention. Pierre Duhem, *Études sur Léonard de Vinci*, 3 vols. (Paris, 1906–1913), describes Vives' scornful and vivid denunciation of Scholastic education in medicine, logic, and natural philosophy at the University of Paris (III, 168–172, 180–181, 488, 490). Of substantive scientific ideas, Duhem mentions (III, 144–146) only Vives' acceptance of the much-debated Scholastic "moment of rest" (*quies media*) alleged to occur between the upward violent motion of a projectile and its subsequent downward motion. A sense of Vives' attitude to Scholastic philosophy and science can be gleaned from Noreña (see above), pt. 2, "Vives' Thought," 131–299. For Vives' role as an educational reformer, see William Harrison Woodward, "Juan Luis Vives, 1492–1540," in *Studies in Education During the Age of the Renaissance 1400–1600* (New York, 1965; original publication, 1906), 180–210, and Foster Watson, "Vives On Education," in *Vives: On Education, A Translation of the De tradendis disciplinis of Juan Luis Vives*, with an introduction by Foster Watson and a foreword by Francesco Cordasco (Totowa, N.J., 1971; original publication, 1913), ci–clvii. Contributions by Vives to education and psychology are briefly summarized by Walter A. Daly, *The Educational Psychology of Juan Luis Vives* (Ph.D. diss., Catholic University of America, 1924); and Foster Watson, "The Father of Modern Psychology," in *Psychological Review*, **22**, no. 5 (Sept. 1915), 333–353.

EDWARD GRANT

VIVIANI, VINCENZO (*b.* Florence, Italy, 5 April 1622; *d.* Florence, 22 September 1703), *mathematics.*

Viviani was the son of Jacopo di Michelangelo Viviani, a member of the noble Franchi family, and Maria Alamanno del Nente. He studied the humanities with the Jesuits and mathematics with

Settimi, a friend of Galileo's. His intelligence and ability led to his presentation in 1638 to Ferdinand II de' Medici, grand duke of Tuscany. Ferdinand introduced him to Galileo, who was so impressed by his talent that he took him into his house at Arcetri as a collaborator in 1639. After Galileo's death, Viviani wrote a historical account of his life and hoped to publish a complete edition of his works. The plan, however, could not be carried out because of opposition by the Church—a serious blow not only to Viviani's reputation but even more to the progress of science in Italy. Since he was unable to pursue the evolution of mathematical ideas that were developing during that period, Viviani turned his talent and inventiveness solely to the study and imitation of the ancients.

Although the Medici court gave him much work, Viviani studied the geometry of the ancients. His accomplishments brought him membership in the Accademia del Cimento, and in 1696 he became a member of the Royal Society of London. In 1699 he was elected one of the eight foreign members of the Académie des Sciences in Paris. He declined offers of high scientific positions from King John II Casimir of Poland and from Louis XIV.

Viviani's first project was an attempted restoration of a work by Aristaeus the Elder, *De locis solidis secunda divinatio geometrica*, which Viviani undertook when he was twenty-four. Aristaeus' work is believed to have been the first methodical exposition of the curves discovered by Menaechmus; but since it has been entirely lost, it is difficult to estimate how close Viviani came to the original work.

Viviani also undertook to reconstruct the fifth book of Apollonius' *Conics*, the first four books of which had been discovered and published. While examining the oriental codices in the grand duke's library in Florence, Borelli discovered a set of papers on which was written "Eight Books of Apollonius' *Conics*." (Actually, the manuscript contained only the first seven books.) Since the manuscript was in Arabic, Borelli obtained the grand duke's permission to take it to Rome, where he turned it over to Abraham Ecchellensis, who was competent to translate it into Latin. The contents of the work were kept secret, however, in order to give Viviani time to complete the publication of his *De maximis et minimis*, which finally appeared at Florence in 1659. Two years later the translation of Apollonius' work was published under Borelli's editorship, and it then became possible to ascertain the substantial similarity between the two works.

Another important work was *Quinto libro degli Elementi di Euclide* (1674). With the rigor and prolixity of the ancients, Viviani devoted an appendix to geometric problems, among which was one on the trisection of an angle, solved by the use of the cylindrical spiral or of a cycloid; another was the problem of duplicating the cube, solved by means of conics or of the cubic $xy^2 = k$.

Viviani also produced the Italian version of Euclid's *Elements* (1690) that was reprinted in 1867 by Betti and Brioschi, in order to raise the level of the teaching of geometry in Italy. Following the example of other learned men of the period, Viviani proposed a problem—known as the "Florentine enigma"—that received wide recognition as soon as the foremost mathematicians began to work on it.[1] The problem was to perforate a hemispheric arch, having four equal windows, in such a way that the residual surface could be squared. Viviani solved the problem by a method that became well known.[2] It is accomplished by the intersection of four right cylinders, the bases of which are tangent to the base of the hemisphere.

There is an Italian translation by Viviani of a work by Archimedes on the rectification of a circumference and the squaring of a circle. He also collected and arranged works by Torricelli after the latter's death.[3]

The search for a point in the plane of a triangle such that the sum of the distances from the vertices shall be the minimum was proposed by Fermat to Torricelli, and by Torricelli to Viviani, who solved the problem (appendix to *De maximis et minimis*, p. 144). This problem was also solved by Torricelli and Cavalieri for triangles with angles less than 120°.[4] It led to a correspondence among Torricelli, Fermat, and Roberval to which Viviani refers (*ibid.*, p. 147).

NOTES

1. During a visit to Italy in 1689, Leibniz met Viviani and solved his problem. It was the first example of the calculation of the area of a curved surface by means of integral calculus (*Acta eruditorum* [1692], 275–279). Jakob I Bernoulli solved the problem (*ibid.*), and that work led to his study of the area of quadrics of revolution (*Acta eruditorum* [Oct. 1696]).

2. Guido Grandi demonstrated the correctness of Viviani's solution by applying the method of indivisibles. There is a reference to this solution in a letter written by Huygens to L'Hospital (*Oeuvres de C. Huygens*, X [The Hague, 1905], 829). In an appendix to this work the publishers inserted a previously unpublished passage by Huygens, in which he demonstrates that the solutions proposed by Leibniz and Viviani are identical.

3. See *Opere di Evangelista Torricelli*, I (Faenza, 1919), pt. 1, 329–407, and pt. 2, 3–43, 49–55.
4. B. Cavalieri, *Exercitationes geometricae sex* (Bologna, 1647), 504–510.

BIBLIOGRAPHY

I. ORIGINAL WORKS. Viviani's writings are *De maximis et minimis geometrica divinatio in quintum Conicorum Apollonii Pergaei, adhuc desideratum* (Florence, 1659); *Quinto libro di Euclide, ovvero scienza universale delle proporzioni, spiegate colla dottrina del Galileo* (Florence, 1674); *Diporto geometrico* (Florence, 1676); *Enodatio problematum universis geometricis praepositorum a D. Claudio Comiers* (Florence, 1677); *Discorso intorno al difendersi dai riempimenti e dalla corrosione dei fiumi* (Florence, 1688); *Elementi piani e solidi di Euclide agl'illustrissimi Sig. dell'Accademia de' Nobili* (Florence, 1690); *Formazione e misura di tutti i cieli* (Florence, 1692); and *De locis solidis secunda divinatio geometrica in quinque libros iniura temporum amissos, Aristaei senioris geometrae* (Florence, 1702).

II. SECONDARY LITERATURE. See L. Conte, "Vincenzo Viviani e l'invenzione di due medie proporzionali," in *Periodico di matematiche*, **25**, no. 4 (1952), 185; A. Fabroni, *Vitae italorum doctrina excellentium*, I (Pisa, 1777), 307–344; and Gino Loria, *Curve piane speciali algebriche e trascendenti*, I (Milan, 1930), 373; *Curve sghembe speciali algebriche e trascendenti*, I (Bologna, 1925), 201–233, and II, 63–65; and *Storia delle matematiche*, 2nd ed. (Milan, 1950), see index.

There are biographical articles in L. Berzolari *et al.*, eds., *Enciclopedia delle matematiche elementari*, II, pt. I (Milan, 1937), 61, 172, 193, 523, 524; and II, pt. 2 (Milan, 1938), 48; and Treccani's *Enciclopedia italiana*, XXXV, 529.

A. NATUCCI

VIZE, VLADIMIR YULEVICH (*b.* Tsarskoe Selo [now Pushkin], Russia, 5 March 1886; *d.* Leningrad, U.S.S.R., 19 February 1954), *oceanography, meteorology, glaciology.*

After graduating from the Gymnasium in St. Petersburg Vize studied at the universities of Göttingen and Halle, majoring in chemistry. He soon began to read books on polar expeditions and decided to become an Arctic explorer. In 1910 he returned to Russia, where he studied in the department of physics and mathematics of St. Petersburg University. While still a student (1910–1911) he traveled through the Lovozero and Khibiny tundras in northwestern Russia, studied the life of the Saamian (Lopari or Lapp) tribes, and discovered and mapped a number of lakes. In 1912–1914 he

was a member of Sedov's expedition to the North Pole. From the early 1920's to the mid-1930's he participated in the major Soviet Arctic expeditions, directing most of them.

Vize was responsible for the establishment of many polar stations in the Soviet Arctic and for the development and application of new methods of research, particularly in ice forecasting. In his works he developed the ideas of Carl Wilhelm Brennecke, Paul Gerhard Schott, Wilhelm Meinardus, and Fridtjof Nansen on the influence of atmospheric processes and hydrological conditions on the formation of ice in the Arctic seas. Continuing their work, Vize also investigated the influence of ice formation in Arctic seas on the circulation of the atmosphere. He applied the results of this research to forecast ice drift, ice conditions, hydrodynamics, temperature conditions, and other factors of substantial significance for scientific weather forecasting.

Vize strongly favored the exploitation of a sea route along the northern shore of the Soviet Union. In 1932, under his leadership, the first complete west-east through passage was made by the northern sea route. Two years later the same voyage was completed from east to west. In 1936–1937 Vize led research expeditions to high latitudes that gathered extensive material on hydrology, meteorology, glaciology, hydrochemistry, and other areas of oceanology.

Vize was a corresponding member of the Academy of Sciences of the U.S.S.R. from 1933, president of the scientific council of the Arctic Institute (later the Arctic and Antarctic Scientific Research Institute), Leningrad (1929–1950), and professor at Leningrad University from 1945. He was a member of the American and English geographical societies and the International Meteorological Committee.

BIBLIOGRAPHY

I. ORIGINAL WORKS. Vize's writings include *Osnovy dolgosrochnykh ledovykh prognozov dlya arkticheskikh morey* ("The Bases of Long-Term Ice Predictions for the Arctic Seas"; Moscow, 1944); *Na "Sibiryakove" i "Litke" cherez Ledovitye morya* ("On the 'Sibiryak' and 'Litka' Across the Arctic Seas"; Moscow, 1946); and *Morya Sovetskoy arktiki. Ocherki po istorii issledovania* ("The Seas of the Soviet Arctic. Sketches in the History of Research"; Moscow, 1948).

II. SECONDARY LITERATURE. See V. K. Buynitsky, *Vladimir Yulievich Vize* (Leningrad, 1969); and A. F.

Laktionov, "Vladimir Yulievich Vize," in *Otechestven-nye fiziko-geografy i puteshestvenniki* ("Native Physical Geographers and Travelers"; Moscow, 1959), 759–765.

A. S. FEDOROV

VLACQ (VLACK, VLACCUS), ADRIAAN (*b.* Gouda, Netherlands, 1600; *d.* The Hague, Netherlands, late 1666 or early 1667), *mathematics, publishing.*

A member of a well-to-do family, Vlacq received a good education. Interested in mathematics, he became acquainted with a local surveyor and teacher, Ezechiel De Decker (*ca.* 1595–*ca.* 1657), for whom he translated into Dutch several recent books written in Latin by British authors on the new art of reckoning, notably some by Napier and that by Briggs on logarithms. They decided to publish these and related works in Dutch. *Het eerste deel van de Nieuwe telkonst* appeared in 1626 under the name of De Decker, who in the preface praised Vlacq for his help. It contained Napier's *Rabdologia* in Dutch translation, a paper on business arithmetic by De Decker, and Stevin's *Thiende.* Also that year De Decker published the *Nieuwe telkonst,* a small table of logarithms to base 10 for the numbers from 1 to 10,000, based on Briggs's *Arithmetica logarithmica* (1624). The work promised a full table of logarithms, an accomplishment realized in *Het tweede deel van de Nieuwe telkonst* (1627), again under the name of De Decker with credit to Vlacq. It contained not only the Briggsian logarithms from 1 to 10,000 and from 90,000 to 100,000, already published by Briggs, but also those of all numbers from 1 to 100,000 (to ten decimal places). The latter, the result of Vlacq's computations, did what Briggs had planned to do.

Vlacq took out the privileges on these books and had them published by the Gouda firm of Pieter Rammaseyn, in which he seems to have had a financial interest. Having paid for the publication of tables he himself had computed, Vlacq saw no objection to republishing them under his own name in the *Arithmetica logarithmica* (1628). Although De Decker was not mentioned, there is no indication that he later resented this. Vlacq's fame rests on these tables, which were well received and contain relatively few errors. The *Tweede deel* of 1627, actually the first complete table of decimal logarithms, was long forgotten until a copy was rediscovered in 1920.

To the *Arithmetica logarithmica,* Vlacq added *Canon triangulorum sive tabula artificialium sinuum,* with the decimal logarithms of the trigonometric lines computed from Pitiscus' *Thesaurus mathematicus* (1613). In a letter to John Pell of 25 October 1628, Briggs states that the 1,000 printed copies of this book, with Latin, Dutch, and French prefaces, were almost all sold. The probable reason is that they were used by George Miller for his *Logarithmicall Arithmeticke* (London, 1631), identical with Vlacq's book except for the English preface.

From about 1632 to 1642, Vlacq had a book business in London, which he moved to Paris. After 1648 he was in The Hague, publishing many books and repeatedly involved in business or political quarrels. The books he published include Briggs and Gellibrand's *Trigonometria britannica,* containing the logarithms of the trigonometric lines with angles divided into tenths (Stevin's idea), and his own *Trigonometria artificialis,* using the traditional sexagesimal division of angles. They have log sine, log cosine, log tangent, and log secant for angles increasing by ten seconds. Both books were published by the firm of Rammaseyn (Gouda, 1633).

Since all these tables were large, Vlacq, with his keen business instincts, published the small *Tabulae sinuum, tangentium et secantium et logarithmi sin. tang. et numerorum ab unitate ad 10000* (Gouda, 1636). These tables, carried to seven decimal places, were a great success and were often reprinted and reedited, and were translated into French and German (there is a Leipzig edition of 1821).

From 1652 to 1655 Vlacq waged a pamphlet war, in which he took the English royalist side, thereby provoking an attack by John Milton. In 1654 he is mentioned as successor to Johannes Rammaseyn. Between 1651 and 1662 he was regularly listed as a visitor to the Frankfurt book fair.

BIBLIOGRAPHY

On Vlacq's life and work, see D. Bierens de Haan, "Adriaan Vlack en Ezechiel De Decker," in *Verslagen en mededeelingen der Koninklyke Akademie van weten-schappen,* Afd. Natuurkunde, 2nd ser., **8** (1874), 57–99; and "Adriaan Vlack en zyne logarithmentafels," *ibid.,* 163–199; C. de Waard, "Vlacq (Adriaan)," in *Nieuw nederlandsch biographisch woordenboek,* II (1912), 1503–1506; J. W. L. Glaisher, "Notice Respecting Some New Facts in the Early History of Logarithms," in *Philosophical Magazine,* 4th ser., **44** (1872), 291–

303, and **45** (1873), 376–382; and D. Bierens de Haan, "On Certain Early Logarithmic Tables," *ibid.*, 371–376. The rediscovery of the *Tweede deel* by M. van Haaften is reported in his "Ce n'est pas Vlacq, en 1628, mais De Decker, en 1627, qui a publié une table de logarithmes étendue et complète," in *Nieuw archief voor wiskunde*, **15** (1928), 49–54; he first reported it in *Verzekeringsbode*, **39** (4 Sept. 1920), 383–386. In "Quelques nouvelles données concernant l'histoire des anciennes tables néerlandaises de logarithmes," in *Nieuw archief voor wiskunde,* **21** (1942), 59–64, and in *Nieuw tydschrift voor wiskunde*, **31** (1943–1944), 137–144, van Haaften supplements his account by reference to three documents on the business relationship between De Decker and Vlacq found by P. J. T. Endenburg in the Gouda archives, reported in "De oudste nederlandsche logarithmentafels en hun makers," in *Het Boek*, **25** (1938–1939), 311–320.

D. J. STRUIK

VOEYKOV, ALEKSANDR IVANOVICH (*b*. Moscow, Russia, 20 May 1842; *d*. Petrograd, Russia [now Leningrad, U.S.S.R.], 9 February 1916), *geography, climatology.*

Voeykov's parents died when he was five; and he spent his childhood on an uncle's estate, where he received an excellent education that included English, French, and German. In 1860 he entered the Physics and Mathematics Faculty of St. Petersburg University; but since the university was soon closed by the government because of student disorders, he continued his education at the universities of Heidelberg, Berlin, and Göttingen. At Göttingen and Berlin he studied meteorology, and in 1865 he defended a doctoral dissertation at Göttingen, "Ueber die directe Insolation und Strahlung an verschiedenen Orten der Erdoberfläche."

On his return to Russia in 1866, Voeykov was elected to the Russian Geographical Society, in which he was active for fifty years. In 1870, while helping to outfit an expedition to northern Russia, he emphasized the necessity of studying the meteorology of that area, pointing out that atmospheric processes in the high latitudes must exert a strong influence on the climate and meteorological conditions of the middle latitudes.

A year earlier, at the request of the Geographical Society, Voeykov had visited the network of meteorological observatories in western Europe; this led to extensive ties between the Russian Geographical Society and other European scientific institutions. In the spring of 1872 he studied the chernozem in Galicia, Bukovina, Walachia,

Transylvania, and Austria; that fall and winter he traveled through western Europe, and most of 1873 was spent in the United States and Canada. Voeykov spent the last three months of 1873 in Washington, at the invitation of the secretary of the Smithsonian Institution, Joseph Henry, completing a manuscript by James Coffin entitled "Winds of the Globe," which he supplemented with data on the winds of Russia.

For almost all of 1874 and the beginning of 1875 Voeykov traveled through South America, learning Spanish and Portuguese. In June 1875 he returned to Russia, and that October he began a journey to India, Java, southern China, and Japan. Because he was a scientist, Voeykov obtained the right to travel throughout Japan, accompanied by a young Japanese who knew Russian. On the basis of information supplied by the inhabitants and from his own observations of the vegetation Voeykov compiled a general map of the climate of Japan. In January 1877 he returned to St. Petersburg and published his observations.

In 1870, Voeykov became secretary of a meteorological commission organized within the Russian Geographical Society. He undertook the organization of meteorological observations, especially on precipitation, and sought to attract a large number of volunteer observers. In 1885 Voeykov obtained a government subsidy to organize twelve meteorological stations for gathering information of special value to agriculture. From its founding in 1891 until 1916, Voeykov was editor-in-chief of *Meteorologicheskii vestnik.*

In 1884 Voeykov was elected docent, in 1885 extraordinary professor, and in 1887 ordinary professor at St. Petersburg University. He became director of the Higher Geographical Courses—the first higher educational institution for geography in Russia—in 1915. He was a member or honorary member of many Russian and foreign scientific societies but was not elected a corresponding member of the Academy of Sciences until 1910.

Voeykov's basic work was *Klimaty zemnogo shara, v osobennosti Rossii* ("Climates of the Earth, Particularly Russia"; 1884), in which he generalized achievements in meteorology, climatology, and hydrology, and his own scientific experiments. Although J. von Hann had published *Handbuch der Klimatologie* in 1883, Voeykov's work was published in German, in revised form, in 1887. This circumstance is explained by the novelty of his treatment of the subject: along with descriptions of climates, he demonstrated the reasons for their differences, described the essential meteorological phenomena and climatic processes, and examined their development and interaction with other natural factors.

Before 1884 Voeykov had published several works on the circulation of the atmosphere on the earth's surface and had shown the close relation of climate to the general circulation of the atmosphere. Emphasizing the particular importance of solar radiation as the moving force in that circulation, he believed that one of the most important problems for the physical sciences was the "introduction of an input-output table of solar heat received by the earth, with its spheres of air and water" (*Izbrannye sochinenia*, 167).

Voeykov was the first to establish the role of monsoons in the subtropics. He discovered the crest of the high barometric maximum (the Voeykov axis), formed over Asia in the cold months and extending from the Siberian anticyclone to western Europe.

In *Klimaty zemnogo shara*, Voeykov examined the most important climatic factors, indicating the primary importance of atmospheric moisture, solar radiation, and the circulation of air. He investigated all the stages of the hydrologic cycle and devoted separate chapters to atmospheric moisture, evaporation, cloud formation, precipitation, rivers, and lakes. He treated precipitation as the opposite of evaporation in the earth's hydrologic cycle, maintaining that the relation between these processes directly determines the density of the river network and the pattern of occurrence of rivers and lakes: "Under stable, even conditions, the more abundant the precipitation and the less the evaporation from the surface of the soil and water, as well as from plants, the richer in running water the country will be. Thus rivers may be regarded as the product of climate" (*Izbrannye sochinenia*, I, 243).

Voeykov demonstrated the averaging influence of lower reaches of large rivers on the climatic conditions of the entire basin. Examining certain hydraulic and hydrological aspects of river currents (particularly the influence on the drainage of the permeability of the soil, the level of the river, and the moderating influence of lakes on the level of rivers flowing through them), he classified streams in relation to the character of their sources and, as he said, established nine main types in relation to climate. He also showed that if rivers reflect the climate at a given time, then lakes reflect climatic changes.

Voeykov was one of the first to calculate the annual flow of all the rivers of the earth, although his estimate was substantially less than the actual figure. He made the first scientifically based calculation of the balance between inflow and evaporation in the Caspian Sea.

Many of Voeykov's works were devoted to the snow cover, and he was the first climatologist to detail its important influence on climate. Having established that the snow cover reflects solar radiation into space, he showed that it melts chiefly as a result of the action upon it of warm air masses. His general conclusion regarding the influence of snow on the heat balance of the earth was that for the entire planet, the warming influence of snow greatly exceeds the cooling effect; without the snow cover, the earth would be much cooler than it is. He also compared climates of present glacial regions with climate during the Ice Age.

Voeykov studied periodic and nonperiodic changes in climate; in particular, he disproved the erroneous belief that Central Asia was becoming drier.

BIBLIOGRAPHY

I. ORIGINAL WORKS. Voeykov published over 1,700 works, including "Ueber das Klima von ost-Asien," in *Zeitschrift der Österreichischen Gesellschaft für die Meteorologie*, **5**, no. 1 (1870), 39–42; "Dię atmosphärische Circulation," Supp. 38 (1874) to *Petermanns geographischen Mitteilungen*, also in Russian in his *Izbrannye sochinenia*, II, 159–221; "Raspredelenie osadkov v Rossii" ("The Distribution of Precipitation in Russia"), in *Zapiski Imperatorskago russkago geograficheskago obshchestva po obshchei geografii*, **6**, no. 1 (1875), 1–72; "Puteshestvie po Yaponii, iyuloktyabr 1876 g." ("Journey Through Japan, July–Oct. 1876"), in *Izvestiya Gosudarstvennogo russkogo geographicheskogo obshchestva*, **13**, no. 4, sec. 2 (1877), 195–240; "Klimat oblasti mussonov Vostochnoy Azii: Amurskogo kraya, Zabaykalya, Manchzhurii, Vostochnoy Mongolii, Kitaya, Yaponii . . ." ("Climate of the Monsoon Regions of East Asia: The Amur Region, Transbaikalia, Manchuria, Eastern Mongolia, China, Japan . . ."), *ibid.*, **15**, no. 5, sec. 2 (1879), 321–410; and "Klimaticheskie uslovia lednikovykh yavleny nastoyashchikh i proshedshikh" ("Climatic Conditions of Glacial Phenomena of the Present and Past"), in *Zapiski Imperatorskago Mineralogicheskago obshchestva*, 2nd ser., pt. 16 (1881), 21–90.

See also *Klimaty zemnogo shara, v osobennosti Rossii* ("Climates of the Earth, Particularly Russia"; St. Petersburg, 1884), also in his *Izbrannye sochinenia*, I, 161–750; *Snezhny pokrov, ego vliyanie na klimat i pogodu i sposoby issledovania* ("Snow Cover, Its Influence on Climate and Weather and Methods of Research"; St. Petersburg, 1885); "O klimate Tsentralnoy

Azii na osnovanii nablyudeny chetyrekh ekspeditsy N. M. Przhevalskogo" ("The Climate of Central Asia on the Basis of Observations of N. M. Przhevalsky's Four Expeditions"), in *Nauchnye rezultaty puteshestvy Przhevalskogo po Tsentralnoy Azii* ("Scientific Results of Przhevalsky's Travels Through Central Asia"; St. Petersburg, 1895), 239–281; "Klimat polesya" ("Climate of Woodlands"), in *Prilozhenia k "Ocherku rabot Zapadnoy ekspeditsii po osusheniyu bolot za 1873–1898 gg."* ("Appendix to 'Sketch of the Work of the Western Expedition for Draining Swamps in 1873–1898'"; St. Petersburg, 1899), 1–132; "Klimat Indysko-go okeana i Indii" ("Climate of the Indian Ocean and India"), in *Zapiski po gidrografii* (1908), no. 29, 178–263; and "Oroshenie Zakaspyskoy oblasti s tochki zren-ia geografii i klimatologii" ("Irrigation of the Transcaspi-an Region From the Point of View of Geography and Climatology"), in *Izvestia Gosudarstvennogo russkogo geograficheskogo obshchestva*, **44**, no. 3 (1908), 131–160; *Le Turkestan russe* (Paris, 1914); "Klimaty rus-skikh i zagranichnykh lechebnykh mestnostey" ("Climates of Russian and Foreign Therapeutic Locali-ties"), in *Prakticheskaya meditsina* (1915), no. 6, 87–176, and no. 10, 177–180; *Izbrannye sochinenia* ("Selected Works"), 4 vols. (Moscow, 1948–1957); and *Vozdeystvie cheloveka na prirodu. Izbrannye stati* ("The Influence of Man on Nature. Selected Articles"; Mos-cow, 1949).

II. SECONDARY LITERATURE. See A. A. Grigoriev, "Rukovodyashchie klimatologicheskie idei A. I. Voey-kova" ("Voeykov's Guiding Climatological Ideas"), in Voeykov's *Izbrannye sochinenia*, I, 10–34; K. K. Mar-kov, "A. I. Voeykov kak istorik klimatov Zemli" ("Voeykov as Historian of the Climates of the Earth"), in *Izvestiya Akademii nauk SSSR*, Geog. ser. (1951), no. 3, 46–54; V. V. Pokshishevsky, "A. I. Voeykov i vo-prosy geografii naselenia" ("Voeykov and Questions of the Geography of Population"), in *Voprosy geografii* (1947), no. 5, 33–40; and "A. I. Voeykov kak ekonomi-ko-geograf" ("Voeykov as Economic Geographer"), in *Otechestvennye ekonomiko-geografy* ("Native Eco-nomic Geographers"; Moscow, 1957), 275–283; and G. D. Rikhter, "Zhizen i deyatelnost A. I. Voeykova" ("Voeykov's Life and Work"), in Voeykov's *Izbrannye sochinenia*, I, 35–82.

I. A. FEDOSEEV

VOGEL, HERMANN CARL (*b.* Leipzig, Germany, 3 April 1841; *d.* Potsdam, Germany, 13 August 1907), *astrophysics.*

Vogel was the sixth child of Johann Carl Chris-toph Vogel, principal of a Leipzig Gymnasium, whose friends included Alexander von Humboldt, Robert Bunsen, and Carl Ritter. Vogel's older brother Eduard, who later became an astronomer and African explorer, was a friend of Heinrich Louis d'Arrest, director of the Leipzig astronomical observatory. Through this friendship Vogel came into contact with astronomy while still young.

After graduating from his father's school, Vogel entered the Dresden Polytechnical School in 1860. Before he completed his training there, however, his parents died, leaving him with serious financial problems. He managed to support himself by doing odd jobs, supplemented by aid from his oldest brother.

Vogel returned to Leipzig in 1863 and began to study natural science at the university. He immedi-ately became second assistant at the university observatory, directed by Karl Bruhns. Vogel's remarkable manual dexterity was very helpful in manipulating the instruments used to observe nebulae and star clusters. The Leipzig observatory was participating in the Astronomische Gesell-schaft's "zone project," a great scanning operation of the northern skies, the goal of which was to ascertain the coordinates of all stars down to the ninth magnitude. At Bruhns's suggestion Vogel agreed to make his nebulae observations in the zone +9°30′ to 15°30′, the area assigned to the Leipzig observatory. This work formed the basis of his inaugural dissertation, which contained a report of these observations and a detailed historical sur-vey of the observation of nebulae (Jena, 1870).

While Vogel was a student at Leipzig, J. K. F. Zöllner obtained a professorship there (1866). Scarcely seven years older than Vogel, Zöllner exerted a lasting influence on his career, especially by insistent advocacy of astrophysics (the exami-nation of celestial objects with the then new methods of photometry, spectroscopy, and photog-raphy). The period of Zöllner's most important work in stellar photometry coincided with the years in which Vogel was working on his doctor-ate. At this time, for example, Zöllner proposed his ingenious design for a reversible spectroscope, with which he sought to demonstrate the existence of Doppler shifts in stellar spectra.

At the recommendation of Zöllner and Bruhns, Vogel was named director of the observatory of F. G. von Bülow at Bothkamp, near Kiel, in 1870. Bülow, an ardent amateur astronomer, wished to finance the construction of an observatory suitable for serious scientific research, even though he him-self seldom did such work. Thus Vogel had com-plete scientific freedom and, most important, he had sole discretion in determining the program of research and in procuring the necessary technical equipment to carry it out. The observatory had a considerable number of instruments, including a relatively large refracting telescope with an aperture of 293.5 mm. (11 1/2 inches) and fitted with an automatic guiding mechanism. Vogel soon acquired additional devices for it, including a spectroscope and a camera obscura. Among the other instruments at the observatory were a comet-seeker (aperture, 136 mm.; focal distance, 1670 mm.), a Fraunhofer refracting telescope (aperture, 75 mm.; focal distance, 1160 mm.), a Zöllner photometer, a ten-inch prismatic circle, two good pendulum clocks, and meteorological measuring devices. Vogel's detailed description of the observatory's equipment was published in *Astronomische Nachrichten*, **77** (1871), cols. 289–298.

Bülow's generosity was of great significance for Vogel's scientific development, and Vogel later stated that it was at Bothkamp that he really learned astrophysics. In particular, while there he worked intensively on the spectroscopic analysis of the stars. With the eleven-inch equatorial telescope he investigated the spectra of Mercury, Venus, Mars, Jupiter, and Uranus, as well as those of various nebulae, of Comet III 1871, of the northern lights, and of the sun. In addition, with a reversible spectroscope placed at his disposal by Zöllner, he tried to ascertain the rotation of the sun. Following the attempts made by Huggins in England (1868), Vogel sought to determine spectroscopically the radial velocity of the fixed stars, a project on which he obviously was in close contact with Zöllner. The results, however, were uncertain; and Vogel, dissatisfied, temporarily abandoned this research. In the next two years he published the results he had achieved with his co-worker O. Lohse in *Beobachtungen, angestellt auf der Sternwarte des Kammerhern v. Bülow zu Bothkamp* (1872–1875). Another work on the spectra of the planets won the prize of the Royal Danish Academy of Sciences and Letters at Copenhagen. Through these publications both the Bothkamp Observatory and its director became well-known in the scientific world.

In 1871, Wilhelm Foerster, clearly grasping the importance of astrophysical research, sent a memoir to the German crown prince and the minister of education that urged the construction of an astrophysical observatory at Potsdam, near Berlin. Vogel's name naturally arose in the ensuing discussions, and in 1874 he was asked to become an observer at the future observatory. In his new post Vogel collaborated in planning the equipment for the new institution, a task for which he was well prepared by his experience at Bothkamp. In order to broaden his knowledge of recent developments in astronomy, he traveled to England, Scotland, and Ireland in 1875. On the trip he met and held scientific discussions with the leading astronomers of those countries, especially Huggins and Airy.

Even before the opening of the new observatory, Vogel pursued the astrophysical research he had begun at Bothkamp, although the time he could devote to it was limited by his responsibilities at Potsdam. It was hoped that Kirchhoff, whose research played a major role in the creation of astrophysics, would become director of the new observatory, but he refused, on the ground that the post would not allow him sufficient time for his theoretical studies because its incumbent would have sole responsibility for running the observatory. He agreed, however, to work at the observatory if he could be a codirector. His terms were accepted, and in July 1876 Kirchhoff, Wilhelm Foerster, and Arthur von Auwers (who served as business manager) were appointed codirectors of the observatory.

Vogel turned to research in spectrophotometry, a field that was to become of considerable importance. In 1876, through a study of Nova Cygni, Vogel obtained the first firm evidence of the changes that occur in a nova spectrum during the fading phase. He also began an extensive examination of the solar spectrum, intending to replace the solar spectrum tables of Kirchhoff (1861–1862) and Ångstrom (1868) with more precise ones. Vogel's painstaking measurements, the reliability of which was further increased by the new absolute wavelength measurements of Gustav Müller and Kempf, constituted an outstanding achievement for the period with regard to both exactitude and abundance of lines. Unfortunately, they were soon superseded by Rowland's tables, which were produced with diffraction gratings and therefore contained more precise measurements. In 1879, the year he completed the solar spectrum measurements, Vogel was named full professor; and on 15 March 1882 he was appointed director of the Potsdam Astrophysical Observatory, which had been officially put into service in 1879.

Vogel chose the spectroscopy of the fixed stars as his area of specialization. In response to Secchi's proposal, he sought to classify the spectra of the fixed stars; believing that these spectra would reflect the stages of development through which the stars had passed. He decided to test his classification by a spectroscopic examination of the skies, which he executed during the following years. Vogel was disappointed by the extremely uniform distribution of the individual spectral classes of his rough schema—a distribution based on his own observations (of only 4,000 objects)—for he had hoped to obtain interesting results. Such results were later obtained through a study of the Harvard classes, which were established on the basis of a greater number of objects, some of which were less bright than those Vogel had examined. Dissatisfied with his findings, Vogel abandoned work in this area.

Meanwhile, Vogel returned to a problem that had intrigued him at Bothkamp: the determination of the radial components of stellar velocities from the Doppler shifts detectable in stellar spectra. His crucial fundamental idea was to employ photography, a technique that had recently been highly developed and that he had already used to record stellar spectra. In April 1887 Vogel attended the International Congress for Astrophotography in Paris, where, presumably, he became more convinced of the correctness of his ideas. Scarcely a year later he presented to the Royal Academy of Sciences at Berlin the first results of his research, in "Über die Bestimmung der Bewegung von Sternen in Visions radius durch spektrographische Beobachtung." In it he showed that the spectrographic method yields more exact results than visual measurement of stellar spectra in the eyepiece. In a table derived from work done in collaboration with Julius Scheiner, Vogel reported the Doppler shifts of the hydrogen γ lines in the spectra of Sirius (α Canis Majoris), Procyon (α Canis Minoris), Rigel (β Orionis), and Arcturus (α Bootis). Although this initial research showed traces of haste and was conducted with instruments that were not fully adequate to the task, the results evoked great interest; for this work ended the protracted controversy over the value of Doppler's theory for the investigation of phenomena of motion in the universe and thereby gave astrophysics a new tool of immense value. Vogel worked with his collaborators to improve the observatory's apparatus, and in 1892 he published "Untersuchung über die Eigenbewegung der Sterne im Visionsradius auf spektrographischem Wege" (*Publikationen des Astrophysikalischen Observatoriums zu Potsdam*, 7, no. 25).

Vogel's use of spectrography led to a sensational success: the discovery of the spectroscopic double stars. On the basis of periodic displacements in the spectral lines of Algol (β Persei) and Spica (α Virginis), Vogel proved that these objects are actually eclipsing binary stars, the components of which could not be detected as separate entities by means of optical devices. The establishment of periodic line displacements in the Algol spectrum and their well-defined relationship with the variation of exposure provided the first exact confirmation of the supposition that Algol is a component of a double star system. From the spectrographs of this system, Vogel and Scheiner derived the orbital velocity of the brighter component. Employing data on the variation of exposure and several plausible hypotheses, they also determined the dimensions of the system, the diameter of both components, the total mass of the system, and the distances of the components from each other (1889). This was the first time that important new information was logically deduced by comparing measurements of Doppler shifts with data concerning the variation of exposure of a spectrographic double star system. Vogel's introduction of these new techniques soon led to further discoveries, the first of which were Edward Pickering's findings regarding Menkalinan (β Aurigae) and Mizar (ζ Ursae Majoris). The importance of such research is immediately evident from the relatively high number of spectrographic double stars accessible to the astronomer. (For example, probably half of the stars of spectrographic type B are of this kind.) In 1904 the spectrographic double stars played a decisive role in Hartmann's discovery of the interstellar calcium absorption lines.

As director of the Potsdam Astrophysical Observatory, Vogel had an increasing number of organizational duties, which gradually obliged him to restrict his own scientific work. On the other hand, his role in the expansion of the observatory was of considerable importance for the progress of astrophysical research. His influence and tireless efforts ensured that the observatory equipment was adequate for its ambitious research programs—for instance, a very costly photographic double refractor that was put in service in 1899 and was used to determine the radial velocities of weaker stars. The lens, achromatized for photographic work, had an aperture of 800 mm. and was the largest photographic lens ever made. Shortly after this instrument went into service, Vogel became seriously ill and had to cease working for a while. Although he never completely recovered, he was able to do a further series of studies, including one on the operational possibilities of short-focal-length reflecting telescopes for research on nebulae.

Vogel's scientific achievement brought him many international honors and assured him a place in the history of modern astronomy. The great scientific importance of the observatory he headed was due in large part to his willingness and ability to attract distinguished co-workers. Furthermore, to the extent that it was possible, given the research programs he had selected, Vogel usually assigned his co-workers to projects that corresponded to their scientific interests.

Vogel left a large sum of money to the Potsdam observatory, the interest on it to pay for study abroad and for the support of gifted children.

BIBLIOGRAPHY

I. ORIGINAL WORKS. The writings that Vogel published while at Potsdam are listed in W. Hassenstein, "Das Astrophysikalische Observatorium Potsdam in den Jahren 1875–1939," in *Mitteilungen des Astrophysikalischen Observatoriums* (*Potsdam*), **1** (1941). This article also contains a list of all the works by Vogel's co-workers (pp. 13–55), both those that appeared in the observatory's own publications and also those that appeared elsewhere (pp. 13–53). Vogel's earlier publications and his articles in *Astronomische Nachrichten* are listed in *Generalregister der Bände 41 bis 80 der Astronomischen Nachrichten Nr. 961 bis 1920* (Kiel, 1938), col. 108; and in *Generalregister der Bände 81 bis 120 der Astronomischen Nachrichten Nr. 1921 bis 2880* (Kiel, 1891), cols. 121–122.

II. SECONDARY LITERATURE. See W. Brunner, *Pioniere der Weltallforschung* (Stuttgart, n.d. [1954?]); A. F., "Hermann Carl Vogel," in *Monthly Notices of the Royal Astronomical Society*, **68**, no. 4 (Feb. 1908), 254–257; W. Foerster, *Lebenserinnerungen und Lebenshoffnungen* (Berlin, 1911), esp. 139–140; O. Lohse, "Hermann Carl Vogel (Todes-Anzeige)," in *Astronomische Nachrichten*, **175** (1907), cols. 373–378; and G. Müller, "Hermann Carl Vogel," in *Vierteljahrsschrift der Astronomischen Gesellschaft*, **42** (1907), 323–339. See also D. B. Hermann, "Für Vorgeschichte der Astrophysikalischen Observatoriums Potsdam," in *Astronomische Nachrichten*, **296** (1975), 245–259.

DIETER B. HERRMANN

VOGT, CARL (*b*. Giessen, Germany, 5 July 1817; *d*. Geneva, Switzerland, 5 May 1895), *medicine, natural science.*

Vogt was the son of Philipp Friedrich Vogt, a physician, who from 1835 taught pathology and pharmacodynamics at the University of Berne. He began his studies at Giessen, where he was one of Liebig's best students and was encouraged to become a chemist. In 1835 Vogt went to Berne, where he enrolled in Valentin's courses in anatomy and physiology, and decided to study medicine. He received the medical diploma in 1839, but the natural sciences had a greater appeal for him. He enthusiastically agreed to collaborate on a major treatise on Central European freshwater fish, a project headed by Louis Agassiz; his name appears as author of the volume on Salmonidae (1842). Vogt subsequently went to Paris, where he began to write his *Lettres philosophiques et physiologiques*, intended for friends in Germany. They were published in their entirety as *Physiologische Briefe für Gebildete aller Stände* (Tubingen, 1845–1847).

After returning to Giessen, where he taught zoology and became *Reichregent* (1842), Vogt also was active in the Revolution of 1848. His involvement forced him to flee to Geneva, where he taught geology because the chair of zoology was held by F. J. Pictet de la Rive. Having become a naturalized citizen of Geneva, Vogt entered politics, becoming *conseiller aux états* and later *conseiller national*. He also was influential in transforming the Académie de Genève into a university (1872).

A work from this period, *Köhlerglaube und Wissenschaft* (Giessen, 1853), caused a great stir and went through several editions. Vogt became the staunch supporter of scientific materialism, later made famous by Ernst Haeckel. His gift for polemic and oratory enabled Vogt to exert considerable influence through both his speeches and his numerous publications. In 1872 he was appointed to the chair of zoology and became director of the Institute of Zoology, a post he held until his death.

Above all, Vogt was a distinguished zoologist whose writings did much to further the development of this science. His *Lehrbuch der praktischen vergleichenden Anatomie* (Brunswick, 1855) was long a classic. In 1840, at the age of twenty-three, Vogt published an important work, *Beiträge zur Neurologie des Reptilien*. After having begun to study the embryology of certain freshwater fish under the guidance of Agassiz, he continued this work throughout his life.

The chair of geology, assigned to him by the Geneva government, gave Vogt the opportunity to write *Lehrbuch der Geologie und Petrefaktenkunde*, the four editions of which (1846–1879) demonstrate its wide interest. It was, however, to marine biological research that Vogt devoted his energy, as is shown by his remarkable publications on hectocotyli, Cephalopoda, and Siphonophora. In 1863 Vogt published the two-volume *Vorlesungen über den Menschen, seine Stellung in der Schöpfung und in der Geschichte der Erde*, which assured his reputation as a scholar and materialist philosopher. "Mémoires sur les microcéphales ou hommes-singes" (1866) brought Vogt recognition as one of the first anthropologists. He espoused

Darwin's ideas and became a strong partisan of natural selection.

BIBLIOGRAPHY

Vogt's autobiography is *Aus meinem Leben, Erinnerungen und Rückblicke* (Stuttgart, 1896).

See also H. Buess, *Recherches, découvertes et inventions de médecins suisses*, translated by R. Kaech (Basel, 1946), 65–66; E. Hirschman, *K. Vogt als Politiker* (Ph.D. diss., Berlin, 1924); E. Krause, "Carl Vogt," in *Allgemeine deutsche Biographie*, XL (Leipzig, 1896), 181–189; H. Misteli, *Carl Vogt. Seine Entwicklung vom angehenden naturwissenschaftlichen Materialisten zum idealen Politiker des Paulskirche* (Ph.D. diss., Zurich, 1938); and W. Vogt, *La vie d'un homme* (Paris, 1896).

P. E. PILET

VOGT, JOHAN HERMANN LIE (*b.* Tvedestrand, Norway, 14 October 1858; *d.* Trondheim, Norway, 3 January 1932), *geology.*

Vogt was the son of a physician and, on his mother's side, the nephew of Sophus Lie. After studying for a year at the Dresden Polytechnikum, he transferred to the University of Christiania (Oslo), where he graduated as a mining engineer–geologist in 1880. He did graduate work at Stockholm in 1882, studying geology under W. C. Brøgger, metallurgy under Richard Åkermann, and chemistry under W. Eggerts. Vogt was appointed professor of metallurgy at the University of Christiania in 1886 and in 1912 moved to the Technical University of Norway at Trondheim, where he was professor of geology, ore deposits, and metallurgy (except iron). He retained this post until his retirement in 1929.

Vogt was a pioneer in ore geology and in the physical chemistry of silicates as a basis for igneous rock petrology. His work in the latter field began with studies of slag minerals. In a series of papers, the first of which appeared in 1883, he provided the first descriptions of a number of slag minerals: enstatite, wollastonite (pseudowollastonite), fayalite, monticellite varieties, åkermannite, oldhamite, manganblende (alabandite), troilite, and sphalerite. Vogt soon began using the crystallization of slags as a model for silicate crystallization in igneous rocks, as shown by the title of his papers of 1888–1890: "Beiträge zur Kenntnis der Gesetze der Mineralbildung in Schmelzmassen und in den neovulkanischen Ergussgesteinen." Inspired by his mineral-

ogical studies of slags, he generalized his studies of ores and silicate rocks in "Die Silikatschmelzlösungen" (2 pts., 1902), a pioneer paper in which the crystallization relations of the different minerals and their dependence upon eutectic relations are considered. The lowering of melting points in melts with several components and the importance of eutectic compositions in binary series are discussed. The general relationships are applied to natural silicate melts and specifically to the eutectic relation between quartz and feldspars in igneous rocks. Bearing still more directly on natural relationships is "Physikalisch-chemische Gesetze der Kristallizationsfolge in Eruptiv-Gesteinen" (1905), in which the crystallization within the ternary feldspar system orthoclase-albite-anorthite and the granite system quartz-orthoclase-albite is discussed.

In another classic paper, "Über anchi-monomineralische und anchi-eutektische Eruptiv-Gesteine" (1908), Vogt discussed magmatic differentiation, on the basis of the theory proposed by Brøgger and several other leading petrographers at the turn of the century, treating the parallelism between crystallization and differentiation of silicate melts. In this work he stressed the importance of eutectic crystallization as a major factor in magmatic differentiation. Since he emphasized the physicochemical laws governing the crystallization and development of igneous rocks, Vogt was sharply critical of the static petrographic system developed by Cross, Iddings, Pirsson, and Washington (CIPW). During World War I, Vogt studied mixed systems of silicates and sulfides, stressing the importance of the low mutual solubility of such melts. This result was fundamental to his treatise on magmatic sulfide ore formation, especially nickel ore. Publications from Vogt's last ten years were marked by his continued analysis of the physicochemical laws governing magmatic differentiation.

Vogt's first paper in English appeared in *Journal of Geology*: "The Physical Chemistry of the Crystallization and Magmatic Differentiation of Igneous Rocks" (1921). Nearly the same title was used for the three-volume work published at Oslo (1924–1931). The most important part of the first volume is the discussion of the concentration of mixed crystals with high melting points in early-formed rocks and that of mixed crystals with low melting points in the magmatic end stage. The second volume deals with the relations of the ternary feldspar system orthoclase-albite-anorthite, based on many analyses; and the third, which concentrates on the rocks of the magmatic end stage, also contains a large number of chemical analyses. Not all of Vogt's main conclusions have proved correct, but he applied the principles of physical chemistry to natural silicate systems more intensely than anyone else of his generation and therefore is often called the father of modern physicochemical petrology. Among the many terms Vogt introduced is "cotectic" curves, applied to what were formerly called eutectic lines or reaction lines. His last paper was "What We May Learn from Brøgger's Essexitic Hurum Volcano Concerning Magmatic Differentiation" (*Festschrift Brøgger, Norsk geologisk tidsskrift* [1932]).

Vogt's other main field of interest was ore geology, in which he was considered a leader at that time, although his work was less original. His most important general contribution is the concept of a group of magmatic ore deposits—ilmenite, chromite, and nickeliferous pentlandite—formed early in the crystallization sequence. Vogt's early contributions (1884–1889) concerned important ore deposits in Norway. In these papers he supported the proponents of a sedimentary origin for pyritic sulfide ores. Soon afterward, Vogt changed his views on ore genesis from the syngenetic to the epigenetic theory, which in the 1870's was supported by Theodor Kjerulf and had originated with J. Durocher and Duchanoy around 1850. A magmatic, epigenetic view of sulfide ore deposits was maintained in Vogt's first well-known paper published in *Zeitschrift für praktische Geologie*: "Über die Kieslagerstätten von Typus Röros, Vigsnäs, Sulitjelma in Norwegen und Rammelsberg in Deutschland." Vogt's views were based on sharp field observations, a very good memory, and extensive travel. In the 1880's he visited European universities and observed many important ore deposits. He therefore was well qualified to write *Die Lagerstätten der nutzbaren Mineralien und Gesteine* (1910–1921) with F. Beyschlag and P. Krusch. Important at the time, this two-volume treatise remains a major handbook of ore geology.

Vogt was active in politics for several years and was always eager to apply his theoretical knowledge of ore deposits to practical purposes, seeking new ore fields and helping to develop existing ones. This interest is reflected in a work published in 1895, "Kobberets historie i fortid og nutid og om udsigterne for fremtiden, med saerligt hensyn til den norske bergverksdrift på kobber" ("The History of Copper in the Past and Present and the Prospects for the Future, with Special Regard to the Norwegian Mining of Copper").

Vogt received an honorary doctorate from the University of Aachen in 1911, the Penrose Medal of the Society of Economic Geologists (United States), and the Wollaston Medal of the Geological Society of London.

BIBLIOGRAPHY

There is a complete bibliography of Vogt's more than 200 published works in *Norsk geologisk tidsskrift*, **11** (1932), 454–466. The titles of his most important publications are mentioned in the text.

The only detailed biographies were published in Norwegian, some soon after his death and one commemorating the centenary of his birth—J. A. W. Bugge, in *Kongelige Norske videnskabers selskabs forhandlinger*, **31** (1958).

Christoffer Oftedahl

VOGT, THOROLF (*b.* Vang, Hedmark, Norway, 7 June 1888; *d.* Trondheim, Norway, 8 December 1958), *geology.*

The son of J. H. L. Vogt, Thorolf published his first mineralogical paper at the age of twenty. He became assistant geologist for the Geological Survey of Norway in 1909 and geologist in 1914. From 1915 to 1923 he was research associate at the University of Oslo. Vogt succeeded his father as professor of mineralogy and geology at the Technical University of Norway, at Trondheim, in 1929. In his earlier years he studied in Sweden, Denmark, Great Britain, and the United States; later he led a series of expeditions to Arctic areas, to Spitsbergen (1925, 1928), and to southeastern Greenland (1931). Of his more than 100 papers, many are short and of little importance; and his contribution to geology consists of a few major volumes. A description of the stratigraphy of the arcosic rocks (sparagmites) in central Norway and their relations to marine Lower Cambrian (1923) remains the standard treatise of Eocambrian (now called latest Precambrian) stratigraphy in Scandinavia. Vogt's main work was a monograph on a small area in northern Norway within the Caledonian zone, a petrographic-geologic description of the sulfide-rich Sulitjelma area (1927). Vogt's most important contribution was perhaps the new discussion of metamorphism on the basis of the mineral facies of Pentti Eskola. The monograph was intended as an introduction to a projected work on the description of the sulfide deposits, but the latter never appeared.

During World War II, Vogt drew upon his extensive botanical knowledge in studying plants near the sulfide deposits of Røros, in southeastern Norway. Twelve brief papers resulting from this work, collectively entitled "Geokjemisk og geobotanisk mamleting," are now considered to be a classic in geochemical prospecting. A paper of 1945 describes the stratigraphy and petrography of a small area south of Trondheim in great detail. This area is of fundamental importance for the eugeosynclinal sequence of the Scandinavian Caledonides, and the paper remained the standard reference for twenty years. At the time of his death Vogt was working several fields, including a general study of metamorphic amphiboles. The completed manuscript of the first part, "Constitution and Classification," was printed posthumously.

Vogt's importance in Norwegian geology derives from his publications and his wide-ranging activity. He taught geology and ore geology to mining students for thirty years and participated in many ore-prospecting projects. He also was active in introducing and developing geophysical methods into the practice of ore prospecting in Norway.

BIBLIOGRAPHY

Vogt published some 100 works from 1908 to 1958; three more papers appeared posthumously (1964–1967). The most important are "Sulitelmafeltets geologi og petrografi," which is Norges geologiske undersøkelse, no. 121 (1927); "Geokjemisk og geobotanisk mamleting," in Kongelige Norske videnskabers selskabs forhandlinger, 12–20 (1939–1947); "The Geology of Part of the Hølonda-Horg District, a Type Area in the Trondheim Region," in Norsk geologisk tidsskrift, 25 (1945), 449–528; and "The Amphibole Group, Constitution and Classification," in Kongelige Norske videnskabernes selskabs skrifter, no. 7 (1966), 1–55.

There is a short biography by Ivar Oftedal (in Norwegian) and a complete bibliography in Norsk geologisk tidsskrift, 39 (1959), 1–11.

CHRISTOFFER OFTEDAHL

VOIGT, JOHANN CARL WILHELM (*b.* Allstedt, Germany, 20 February 1752; *d.* Ilmenau, Germany, 1 January 1821), *geology, mining.*

Voigt came from a family of Thuringian civil servants; his father was magistrate of the district of Allstedt, which was then part of the duchy of Weimar. After studying law at the University of Jena (1773–1775), he transferred to the Freiberg mining academy (1776–1779), where he studied mainly with Abraham Gottlob Werner.

When Voigt returned to Weimar, he became acquainted with Goethe, who had already been working for a long time with his elder brother in the Weimar administration and with whom he formed a lifelong friendship. Through Voigt, Goethe learned the systematics and classification of minerals and rocks. Their friendship was of great importance to both men: through Voigt, Goethe learned the fundamentals of mineralogy and became an admirer of Werner; and through Goethe, Voigt began a career as a researcher and mining official.

Voigt was commissioned to tour the duchy of Weimar and describe it mineralogically. The result of this work, done in 1780, was the two-volume *Mineralogische Reisen durch das Herzogthum Weimar und Eisenach*, which Goethe published (1781–1785). At Goethe's behest Voigt traveled through the Fulda region in 1781 and the Harz Mountains in 1782, preparing mineralogical maps of these areas. In 1783 he became mining secretary and later mining director. Much of his time was subsequently devoted to running the silver mine at Ilmenau, which was reopened on Goethe's initiative in 1784 but had to be abandoned by 1800. Voigt's works on the petrography and geology of Thuringia and his contributions to the science of mineral deposits brought him considerable renown. He became known more widely, however, through his geological works, especially his contribution to the debate concerning the origin of basalt.

In his *Mineralogische Reisen* Voigt had presented the Thuringian basalts as being of volcanic origin. This approach provoked a protest from Werner, who had become the leader of the "neptunist" party, which considered basalt to be a marine sediment. The "neptunist controversy" lasted in its public, and sometimes quite sharply conducted, form until about 1795, at which time the neptunists appeared to have won. In the course of this dispute Voigt became the leader of the volcanists in Germany. His interpretation of the basalt mountains as the remnants of the erosion of the eruptions of great volcanoes was almost correct; but he then allowed himself to be influenced by the incorrect views of William Hamilton and August Veltheim, who considered the basalt mountains to be groups of volcanoes exposed by erosion. Voigt did, however, make a number of enduring contributions to scientific volcanism and was the first to draw attention to the phenomena of contact metamorphism.

BIBLIOGRAPHY

I. ORIGINAL WORKS. Voigt's publications include *Mineralogische Reisen durch das Herzogthum Weimar und Eisenach*, 2 vols. (Weimar, 1781–1785); *Petrographische Landkarte des Hochstifts Fulda* (Frankfurt, 1782); *Mineralogische Reise von Weimar über den Thüringer Wald und Meiningen bis Hanau* (Leipzig, 1787); *Mineralogische und bergmännische Abhandlungen*, 3 vols. (Leipzig, 1789–1791); *Praktische Gebirgskunde* (Weimar, 1792); *Kleine mineralogische Schriften* (Weimar, 1799); *Mineralogische Reise nach den Braunkohlenwerken und Basalten in Hessen wie auch nach den Schieferkohlenwerken des Unterharzes* (Weimar, 1802); and *Geschichte des Ilmenauischen Bergbaues* (Sondershausen–Nordhausen, 1821).

II. SECONDARY LITERATURE. There is no full-length biography, but Voigt is often mentioned in biographies of Werner and Goethe, and in works on the controversy over Neptunism—for instance, Walther Herrmann, *Goethe und Trebra*, which is Freiberger Forschungshefte, ser. D., no. 9 (Berlin, 1955), 36, 48–52, 62–64; and Otfried Wagenbreth, *Abraham Gottlob Werner und der Höhepunkt des Neptunistenstreits um 1790*, which is Freiberger Forschungshefte, ser. D, no. 11 (Berlin, 1955), see 183–241. A short biography is Carl Schiffner, *Aus dem Leben alter Freiberger Bergstudenten*, I (Freiberg, 1935), 16–17.

HANS BAUMGÄRTEL

VOIGT, WOLDEMAR (*b*. Leipzig, Germany, 2 September 1850; *d*. Göttingen, Germany, 13 December 1919), *physics*.

Voigt graduated from the Nikolaischule at Leipzig in 1868. He then entered the University of Leipzig, but in 1870 his studies were interrupted by service in the Franco-Prussian War. He resumed his studies in 1871, this time at Königsberg. At first Voigt was undecided between a career in physics and a career in music, for the latter had always played a large role in his life: Felix Mendelssohn and Robert Schumann had been frequent visitors to his parents' house. His musical ear was highly trained; and while in the army he would often pass the time while marching by reciting, note for note, the complete orchestration of entire symphonic pieces. He finally decided on a career in physics, on the ground that, unlike music, in physics there is a reasonable mean, not simply highs and lows.

While at Königsberg, Voigt came under the influence of Franz Neumann, his deep respect and love for whom largely determined his career, in terms of subject matter, the style of his research,

and the manner in which he presented his work to the physics community. His dissertation on the elastic constants of rock salt was completed in 1874. He then returned to Leipzig, where he taught at the the Nikolaischule, but in 1875 was called back to Königsberg as extraordinary professor of physics. In 1883 Voigt was appointed ordinary professor of theoretical physics at Göttingen, with the promise that he and Eduard Reike were to have a new physical institute (which was not ready until 1905). His chief research interests centered on the understanding of crystals, but near the turn of the century he became more and more concerned with the Zeeman effect and the electron theory.

Voigt's interest in crystals was closely related to Neumann's work. At Königsberg, Neumann had worked in both the physics department and the department of mineralogy, so it was quite natural that he should do extensive work on the optical properties of crystals. Neumann had developed a mechanical theory of light propagation that assumed that light oscillations had a mechanical-elastic nature. The oscillations were transmitted through an ether conceived of as an elastic solid. He had not restricted his activities in physics to theoretical work, however, and had initiated a great number of experimental studies; his students spent many hours in his laboratories studying the properties of crystals.

Voigt brought this tradition of theoretical and experimental work to Göttingen. Although for many years he was hampered by lack of adequate facilities, he not only pursued theoretical studies of the properties of crystals but also undertook a host of very delicate experimental investigations in which the physical properties of many crystalline substances were measured.

According to the theories of Poisson and Cauchy, which were based on special molecular assumptions, certain relationships must exist between the constants of a crystal regardless of its classification. Voigt determined the elastic constants for a wide variety of crystals and showed that the predicted relationships were not at all satisfied. While some felt that this work vindicated those who objected to forming special hypotheses about the nature of crystals, Voigt did not accept this point of view and in many of his publications indicated the direction that must be taken in amending the molecular hypothesis.

In 1887, in a paper on the Doppler effect in which he analyzed the differential equations for oscillations in an incompressible elastic medium, Voigt established a set of transformation equations that later became known as the Lorentz transformations.

Voigt's extensive theoretical and physical researches on the nature of crystals were summarized in *Magneto- und Elektro-Optik* (1908) and *Kristallphysik* (1910). These treatises reveal the elegance of his mathematical treatments and the great orderliness that his research had brought to the understanding of crystals. The elastic, thermal, electric, and magnetic properties of crystals were ordered in magnitudes of three types: scalar, vector, and tensor. In fact, it was Voigt who in 1898 had introduced the term "tensor" into the vocabulary of mathematical physics.

Even though Voigt devoted considerable time to his research and his students, and even though he acquired more administrative responsibility at Göttingen, he never gave up an active interest in music and musicology. He was recognized as an expert on Bach's vocal works and in 1911 published a book on Bach's church cantatas. Voigt often referred to the study of physics in musical terms. To him the region of science that represented the highest degree of orchestration and that possessed the utmost in rhythm and melody was crystal physics. It was altogether fitting that on 15 December 1919 his funeral bier was carried from his house to its final resting place to the strains of a Bach chorale.

BIBLIOGRAPHY

I. ORIGINAL WORKS. There is no comprehensive catalog of Voigt's more than 200 publications. Among his most significant works are "Allgemeine Formeln für die Bestimmung der Elasticitäts Constanten von Krystallen durch die Beobachtung der Biegung und Drillung von Prismen," in *Annalen der Physik*, **16** (1882), 273–310, 398–415; "Volumen und Winkeländerung krystallinischen Körper bei all-oder einseitigen Druck," *ibid.*, 416–426; "Theorie des Lichtes für vollkommen durchsichtige Media," *ibid.*, **19** (1883), 873–908; "Zur Theorie des Lichtes," *ibid.*, **20** (1883), 444–452; "Theorie der absorbirenden isotropen Medien insbesonder Theorie der optischen Eigenschaften der Metalle," *ibid.*, **23** (1884), 104–147; "Theorie der electromagnetischen Drehung der Polarisationsebene," *ibid.*, 493–511; "Zur Theorie des Lichtes für absorbirende isotrope Medien," *ibid.*, **31** (1887), 233–242; and "Zur Erklärung der elliptischen Polarisation bei Reflexion an durchsichtigen Medien," *ibid.*, **32** (1887), 526–528.

Also see "Ueber das Doppler'sche Princip," in *Nachrichten von der Königlichen Gesellschaft der Wissenschaften zu Göttingen* (1887), 44–51; "Theorie des Lichtes für bewegte Medien," in *Annalen der Physik*, **35** (1888), 370–396; 524–551; "Zur Theorie des Lichtes," *ibid.*, **43** (1891), 410–437; "Ueber einen einfachen Apparat zur Bestimmung der thermischen Dilation fester Körper, speciel der Krystalle," *ibid.*, 831–834; "Bestimmung der Elasticitätsconstanten einiger quasi-isotroper Metalle durch langsame Schwingungen von Stäben," *ibid.*, **48** (1893), 674–707; "Bestimmung der Constanten der thermische-Dilation und des thermische-Druckes für einige quasi-isotrope Metalle," *ibid.*, **49** (1893), 697–708; "Die specifischen Wärmen c_p und c_v einiger quasi-isotroper Metalle," *ibid.*, 709–718; "Beiträge zur molecularen Theorie der Piëzoelectricität," *ibid.*, **51** (1894), 638–660; "Ueber Medien ohne innere Kräfte und über eine durch sie gelieferte mechanische Deutung der Maxwell-Hertz'schen Gleichungen," *ibid.*, **52** (1894), 665–672; and "Beiträge zur geometrischen Darstellung der physikalischen Eigenschaften der Krystalle," *ibid.*, **63** (1897), 376–385.

Additional works are "Zur kinetischen Theorie idealer Flüssigkeiten," in *Nachrichten von der Königlichen Gesellschaft der Wissenschaften zu Göttingen* (1897), 19–47, 261–272; "Lässt sich die Pyroelectricität der Krystalle vollständig auf piëzoelectrische Wirkungen zuruckführen?" in *Annalen der Physik*, **66** (1898), 1030–1060; "Doppelbrechung von im Magnetfelde befindlichen Natriumdampf in der Richtung normal zu den Kraftlinien," in *Nachrichten von der Königlichen Gesellschaft der Wissenschaften zu Göttingen* (1898), 355–359; "Ueber das bei der sogenannten totalen Reflexion in des zweite Medium eindringende Licht," in *Annalen der Physik*, **67** (1899), 185–200; "Zur Theorie der magneto-optischen Erscheinungen," *ibid.*, 345–365; "Ueber die Proportionalität von Emissions- und Absorptionsvermögen," *ibid.*, 366–387; "Weiteres zur Theorie der Zeemaneffectes," *ibid.*, **68** (1899), 352–364; "Neuere Untersuchungen über die optischen Wirkungen eines Magnetfeldes," in *Physikalische Zeitschrift*, **1** (1899), 116–120, 128–131, 138–143; "Zur Festigkeitlehre," in *Annalen der Physik*, **4** (1901), 567–591; "Beiträge zur Elektronentheorie des Lichtes," *ibid.*, **6** (1901), 459–505; "Elektronenhypothese und Theorie des Magnetismus," in *Nachrichten von der Königlichen Gesellschaft der Wissenschaften zu Göttingen* (1901), 169–200; and "Ueber einige neuere Beobachtungen von magneto-optischen Wirkungen," in *Annalen der Physik*, **8** (1902), 872–889.

Further, see "Ueber das optische Verhalten von Kristallen der hemiëdrischen Gruppe des monokinen Systemes," in *Physikalische Zeitschrift*, **7** (1906), 267–269; "Betrachtungen über die komplizierteren Formen des Zeemaneffektes," in *Annalen der Physik*, **24** (1907), 193–224; "Beobachtungen über natürliche und magnetische Drehung der Polarisationsebene in Krystallen von K. Honda," in *Physikalische Zeitschrift*, **9** (1908), 585–590; *Magneto- und Elektro-Optik* (Leipzig, 1908); *Kristallphysik* (Leipzig–Berlin, 1910; 2nd ed., 1926); "Zur Theorie der komplizierteren Zeemaneffecte," in

Annalen der Physik, **36** (1911), 873–906; "Allgemeines über Emission und Absorption in zusammenhang mit der Frage der Intensitätsmessungen beim Zeeman-Effect," in *Nachrichten von der Königlichen Gesellschaft der Wissenschaften zu Göttingen* (1911), 71–97; "Ueber Emission und Absorption schichtenweise stetig inhomogener Körper," in *Annalen der Physik,* **39** (1912), 1381–1407; "Weiteres zur Polarisation des Rowland Gittern gebeugten Lichtes," in *Nachrichten von der Königlichen Gesellschaft der Wissenschaften zu Göttingen* (1912), 385–417, written with P. Collet; "Ueber die anormalen Zeeman-effekte der Wasserstofflinien," in *Annalen der Physik,* **40** (1913), 368–380; "Weiteres zum Ausbau der Koppelungstheorie der Zeeman-effekte," *ibid.,* **41** (1913), 403–440; "Ueber die Zeeman-effekte bei mehrfachen Serienlinien besonders auch bei dem O-Triplet λ =3947," *ibid.,* **43** (1914), 1137–1164; and "Ueber sekundäre Wirkungen bei piëzoelektrischen Vorgängen, insbesondere im Falle der Drillung und Biegung eines Krieszylinders," *ibid.,* **48** (1915), 433–448.

II. Secondary Literature. See E. T. Whittaker, *A History of the Theories of Aether and Electricity,* 2 vols. (New York, 1960), I, 333, 415; II, 33, 160, 238–239. Obituary notices are by H. L[amb?], in *Proceedings of the Royal Society,* **99A** (1921), xxix–xxx; and by C. Runge, in *Physikalische Zeitschrift,* **21** (1920), 81–82; and in *Nachrichten von der Königlichen Gesellschaft zu Göttingen: Geschäftliche Mitteilungen aus dem Jahre 1920* (Göttingen, 1920), 47–52.

Stanley Goldberg

VOIT, CARL VON (*b.* Amberg, Bavaria, 31 October 1831; *d.* Munich, Germany, 31 January 1908), *physiology.*

Voit was the son of August Voit, a well-known architect. He entered medical school in Munich, in 1848, and completed his training there in 1854, after spending the year 1851 at Würzburg. In 1855 he studied chemistry in Göttingen with Wöhler, and the following year became an assistant to Theodor Bischoff at the Physiological Institute in Munich. In 1859 he became a lecturer at the University of Munich, and in 1863 he was named professor of physiology, a position he held for the rest of his career. During the next three decades Voit became the leader of the dominant school investigating metabolism. He acquired an authoritative position through the technical mastery with which he refined previously developed procedures, and by means of which he was able to resolve fundamental problems into which his predecessors had fallen.

When Voit returned to Munich in 1852, he attended Liebig's chemistry course, and was inspired by Liebig's writings on "animal chemistry" to investigate the "laws" of animal nutrition. For many years Voit was guided by Liebig's theory that the organized parts of the body are formed exclusively by nitrogenous "albuminoid" nutrient substances (plastic aliments), and that non-nitrogenous nutrients (respiratory aliments) are oxidized in the blood to produce animal heat. He also adhered to Liebig's belief that all mechanical work is produced by the "metamorphosis" of the nitrogenous tissue constituents. Liebig's contention that one could measure the amount of tissue metamorphosis by the formation of urea provided the research program to which Voit devoted much of his career. At the time Voit entered this field, Jean-Baptiste Boussingault, Friedrich Bidder and Carl Schmidt, and Bischoff had already carried out extensive comparative measurements of the intake and output of the elements (carbon, hydrogen, oxygen, and nitrogen) constituting the bulk of the food, excretions, and respiratory gases of animals. Their results had created several theoretical and practical dilemmas. They realized that one should ideally measure the composition of the food and excrements simultaneously with the gaseous exchanges, but the differing experimental conditions appropriate respectively to the collection of the excrements and to that of the respiratory gases had prevented this. In 1852 Bischoff encountered another serious setback. Using a simple and reliable new method developed by Liebig for measuring urea, Bischoff found that large portions of the dietary nitrogen were unaccounted for in the urine; this unexplained nitrogen "deficit" seemed to preclude direct measurements of the turnover of nitrogenous tissue constituents.

When he became Bischoff's assistant, Voit continued Bischoff's feeding experiments on dogs. Taking care to assure that the nitrogen content of the meat that was fed to the dogs was uniform, and to collect the feces and urine without losses, Voit found that the nitrogen absorbed in the meat was always nearly equal to that in the urea, or could be accounted for by the weight changes in the animal. This outcome reassured Voit that under rigorously controlled conditions one could rely on the quantity of urea excreted as a measure of nitrogenous metabolism.

Encouraged by this success, Voit carried out extensive further investigations with Bischoff, determining the quantities of urea formed by a dog under various dietary conditions. Between 1857 and 1860 they tried pure meat diets, in which they systematically increased and decreased the daily quantities, and combinations of meat with varying quantities of sugar, starch, fat, and gelatin. The changes in the urea production under these conditions led them to conclude that the rate of decomposition of nitrogenous matter in an animal does not depend directly upon the quantity of nitrogenous nutrient but upon the nutritive condition of the animal. As the nitrogenous mass of the body increases or decreases, so does the rate of metamorphosis of these constituents. The same diet might at one time supply enough nitrogen and at another time not, because the requirement itself varies with the changing condition of the animal. Additions of sugar, starch, or fat decrease the output of urea by a relatively small amount. In 1860 they published their results in a lengthy treatise entitled *The Laws of the Nutrition of Carnivorous Animals*. Practically, they believed they had established methods that could be used for determining the most economical quantitative combination of nitrogenous and non-nitrogenous foods, which would maintain a given animal. Theoretically, they thought they had confirmed Liebig's distinction between plastic and respiratory nutrients.

Then Voit took up on his own the question of whether other factors can influence the rate of decomposition of organic substances. He began by examining the effects of coffee but concluded that it does not significantly affect the nitrogenous metabolism. During this series of experiments, Voit realized that he could distinguish effects of an added factor most clearly if the animal were in a condition of equilibrium between the intake and output of nitrogenous substances. He learned that if a dog is fed a steady nitrogenous daily diet, the nitrogen consumption gradually rises or falls until it balances the intake, and the nutritional condition of the animal thereafter remains constant.

From Liebig's assertion that the metamorphosis of nitrogenous tissue substance is the sole source of motion and his claim that the urea formed is a measure of that metamorphosis, it followed that muscle activity ought to increase the amount of urea excreted. To test that inference, Voit trained a dog to run rapidly, for ten minutes at a time, on a large treadmill. He compared the urea output over three days on which the dog had run six times daily, with the output over three days of rest. He found that the performance of a great amount of mechanical work produced only very small increments of urea excretion. These unexpected results seemed to him at first completely incompatible with his previous conceptions. In order to reconcile the results he assumed that the energy released by nitrogenous decomposition during rest is converted into an "electromotive force," which can be transformed in turn into muscle motion. There is therefore a store of energy available, limited by the amount of nitrogenous matter that decomposes each day. This theory explained why an animal can do only a certain amount of work in a day. Thereafter Voit considered the urea production as a measure not of the muscle activity at any particular time, but of the capacity for such activity over a longer time period.

The rigor and comprehensiveness of Bischoff and Voit's investigations quickly won them a leading position in the field of nutrition, and after 1860 their experimental results were featured in influential physiological textbooks. Their interpretations of their results, however, were not generally accepted. Carl Vogt published a sweeping polemic review in which he tried to show that their analytical foundations were not adequate to support their theories concerning internal processes. In a more penetrating critique, Moritz Traube argued in 1861 that Voit's treadmill experiment repudiated the doctrine that organized muscle is decomposed by its work, and that urea gives a measure of the muscle force expended. Of all the objections that Voit's views encountered, that which caused him the most trouble was the repeated denial of his claim that all of the nitrogen of the substances decomposed in the body is excreted in the urine and feces. From 1860 to 1870 Voit devoted much of his effort to the defense of his methods and conclusions from such criticisms.

In their joint experiments Voit and Bischoff calculated indirectly the quantities of carbon, hydrogen, and oxygen exhaled, by subtracting the amounts of these elements excreted from the amounts ingested. From these differences and a rough estimate of the daily heat production, they judged whether the substances gained or lost, in addition to the nitrogenous constituents, were likely to be fat, water, or both. In order to obviate the uncertainties arising from such estimates, Voit afterward sought means to measure directly the respiratory products. His colleague and former teacher Pettenkofer constructed a respiration chamber large enough to accommodate either a man or an animal for a day. In 1861 they began combined feeding-respiration experiments. In the first group of experiments they measured the carbonic acid exhaled, but not the water, so that they could not calculate the oxygen consumption. The quantity of carbonic acid exhaled daily varied widely and, like urea production, seemed to depend on changes in the nutritional conditions. With a large pure meat diet, less carbon was exhaled than ingested, a result that led Pettenkofer and Voit to think that after the decomposition of the "flesh" in the body, a portion of the non-nitrogenous residue may be deposited as fat. This and similar subsequent results led Voit to question the prevailing view that the fat formed in animals is derived from carbohydrate.

Beginning in 1862 Pettenkofer and Voit measured the water vapor as well as the carbonic acid exhaled. They were then able to calculate the absorbed oxygen from the difference between the initial weight of the dog—together with the food and water it ingested—and its final weight plus the excretions and the expired carbonic acid and water. Their first experiments produced anomalous ratios between the oxygen absorbed and that contained in the exhaled carbonic acid. In February 1863, after making various refinements, they succeeded in attaining a complete balance of the incoming and outgoing elements for their dog at equilibrium on a pure meat diet. The difference between the total measured daily input of carbon, hydrogen, nitrogen, and oxygen, and the totals contained in the excretions and exhaled gases, was less than one percent of the total mass of material exchanged. As Pettenkofer and Voit pointed out, this was the first measurement of the intake and output of an animal for which every single value was ascertained by experiment. They had reached a goal that numerous investigators had pursued for nearly twenty-five years. To do so they had had to combine two types of experiment—nutritional and respirational measurements—which had previously been carried out separately because the conditions required for accurate measurements of one type of experiment seemed incompatible with those required for the other. It had been difficult to extend respirational experiments beyond a few hours, whereas nutritional experiments had appeared reliable only if they lasted several days. Pettenkofer's large, accurate respiration apparatus, in addition to the precise control Voit had attained over the diets and analysis of excretions, enabled the two collaborators to make both sets of measurements on one animal over the same time period.

Throughout the 1860's Voit kept up the same kind of measurements of the nitrogenous *Stoffwechsel* of dogs that he had been making since 1857. In the later experiments he elucidated in closer detail the variables affecting the consumption of nitrogenous substances in particular nutritional states, and he further refined his control over the analytical factors. By 1866 he could show that

the excretion of urea by a dog on a uniform diet is so regular that the daily quantities deviate from the mean by less than three grams. Voit increasingly stressed the primacy of method over theories concerning the internal nature of nutritional phenomena. Gradually he lost his commitment to the central theoretical ideas for the support of which he had originally devised his methods. Yet he never gave up the goal of understanding the intermediate steps in the metamorphosis of nutrient materials within animals.

In 1866 Voit showed that the daily urea production of a dog in periods of hunger declined in diminishing decrements until after the sixth day, when the rate remained nearly steady. The initial decrease was largest when the animal had been best nourished just before the period of abstinence. When he subtracted the nearly constant amount secreted after the fifth day from the amount excreted during the first day of hunger, the remainder was proportional to the quantity of "flesh" stored up in the animal during its preceding period of nourishment. These patterns led Voit to propose that there are two types of protein in the body—a large proportion derived directly from the nourishment, of which about 70 percent undergoes decomposition daily, and a much smaller amount, of which only about 1 percent can decompose during one day. He called the two types storage protein and organ protein, and in his later investigations he sought to determine the effects of various kinds of nourishment on their proportions.

Using the respiration chamber, Voit also continued his collaborative work with Pettenkofer. In August 1865 they applied their methods to a diabetic patient in order to see how the rates of the nutritional decomposition processes are changed by this disease. In 1866 they carried out similar investigations on a leukemia patient. They then realized that they needed bases of comparison with the rates of decomposition in normal humans. Therefore, over the next twelve months they carried out investigations on two healthy men. The papers that resulted from these and their subsequent experiments provided the foundation for the many studies on metabolism in health and disease that were carried out over the rest of the century.

During the 1860's a number of events, including especially the experiments of Fick and Wislicenus, undermined Liebig's theory that the decomposition of nitrogenous substances is the sole source of muscle work. With Pettenkofer, Voit confirmed his own earlier treadmill experiments showing that exercise produces no increase in the formation of urea. With the respiration chamber they also observed the large augmentation in the expiration of CO_2 and water, which they and others linked with the increased consumption of fat or carbohydrate. Voit continued to resist the conclusion that the combustion of non-nitrogenous substances provided mechanical work, for that view ignored the obvious influence of the protein content of the nourishment and of the body on the sustained capacity of the organism for work. Nevertheless, his own belief that there are two sources of urea—organ and storage protein—had rendered nearly meaningless Liebig's definition of nitrogenous nutrients as "plastic" aliments. By 1867 Voit acknowledged that there was no evidence for Liebig's idea that proteins are decomposed only during the activity of the organs, or that nitrogenous nutrients must become part of organized tissue before they are decomposed. The old concept of the *Stoffwechsel* which represented this view had, he said, lost its meaning. At the same time Voit was developing further support for his theory that animal fat was produced from the decomposition products of nitrogenous substances rather than from carbohydrate. Liebig strongly disagreed with Voit's view. By 1869, when Voit presented all of the accumulated evidence for his own position, he had dissociated himself from most of Liebig's theories concerning nutrition. In that year Liebig wrote a defense of his concepts of the source of muscle motion and of the conversion of carbohydrate to fat. In his article he referred to Voit's demonstration of the formation of fat from protein as worthless. Deeply offended, Voit wrote in 1870 a long reply in which he relentlessly exposed the inadequacies of Liebig's theories on nutrition.

Voit's own nutritional theories remained controversial. Pflüger and others opposed his distinction between organ and storage, or "circulating," protein and his arguments for the conversion of protein into fat. Voit's laboratory nevertheless became increasingly the center of activity in the field. Because the methods he had developed were crucial to his success, those hoping to enter the field found it important to work under his direction. A large proportion of the leaders of the era in which metabolic balance investigations reached their high point came out of Voit's laboratory. They included Max Rubner, Joseph Bauer, Friedrich von Müller, Alexander Ellinger, Edward Cathcart, Max Cremer, Graham Lusk, and Voit's successor, Otto Frank.

The work in the Munich laboratory continued along the lines Voit had established during the 1860's. In later investigations purified protein preparations replaced the trimmed whole meat, which Voit had earlier used. Attention focused on defining an adequate nourishment and on determining whether substances such as asparagine and peptones have nutritive value—that is, whether their addition to a non-nitrogenous diet can substitute for protein, and whether their addition to a diet including protein can "spare" protein. Beginning in the 1880's some of the experiments were done on white rats, the small size and omnivorous habits of which made them particularly convenient for such investigations.

In later years Voit spent much of his time on official university duties, and served as secretary for the mathematics and physical sciences section of the Bavarian Academy of Sciences. Voit continued to deliver memorable course lectures up until the last year of his life, when ill health finally forced him to forgo his "greatest pleasure."

BIBLIOGRAPHY

I. ORIGINAL WORKS. Major articles or monographs by Voit include *Physiologisch-chemische Untersuchungen*, I (Augsburg, 1857); *Untersuchungen über den Einfluss des Kochsalzes, des Kaffee's und der Muskelbewegungen auf den Stoffwechsel* (Munich, 1860); "Physiologie des Allgemeinen Stoffwechsels und der Ernährung," in *Handbuch der Physiologie*, L. Hermann, ed., VI, pt. 1 (Leipzig, 1881), and *Die Gesetze der Ernährung des Fleischfressers durch neue Untersuchungen festgestellt* (Leipzig, 1860), written with T. L. W. Bischoff.

The germinal articles by Pettenkofer and Voit using the respiratory chamber are "Untersuchungen über die Respiration," in *Annalen der Chemie und Pharmacie*, supp. 2 (1863), 52–70; and "Ueber die Producte der Respiration des Hundes bei Fleischnährung und über die Gleichung der Einnahmen und Ausgaben des Körpers dabei," *ibid.*, 361–377. For a detailed description of a later, smaller version of the Pettenkofer respiration apparatus, including detailed drawings, see "Beschreibung eines Apparates zur Untersuchung der gasförmigen Ausscheidungen des Thierkörpers," in *Abhandlungen der Bayerischen Akademie der Wissenschaften*, **12** (1876), 219–271.

Beginning in 1865, articles by Voit and his students reporting the research carried out in the Munich Institute of Physiology appeared regularly in the *Zeitschrift für Biologie*, of which he was a founding editor. Preliminary communications were often published in the *Sitzungsberichte der Bayerischen Akademie der Wissenschaften zu München*.

II. SECONDARY LITERATURE. Otto Frank, *Carl von Voit, Gedächtnisrede* (Munich, 1910), a eulogy by

Voit's successor, contains a long, but not exhaustive bibliography of Voit's publications. Graham Lusk, *Nutrition* (New York, 1969), contains a lengthy summary of Voit's contributions, and "Carl von Voit, Master and Friend," in *Annals of Medical History*, **3** (1931), 583–594, is a very informal reminiscence, with four photographs of Voit, two photographs of the Physiological Institute, and transcriptions of letters from Voit to Lusk.

FREDERIC L. HOLMES

VOLKMANN, PAUL OSKAR EDUARD (*b.* Bladiau, near Heiligenheil, Germany, 12 January 1856; *d.* Königsberg, Germany [now Kaliningrad, R.S.F.S.R.], 20 April 1938), *physics, epistemology, history of science.*

The son of a minister, Volkmann attended the Friedrichkollegium in Königsberg from 1864 to 1875. He began to study mathematics and physics in 1875 at the University of Königsberg, where his most important teachers were Heinrich Weber and Woldemar Voigt. He assisted Voigt while still a student; after receiving the doctorate in 1880, he became Voigt's regular assistant.

Volkmann remained at the University of Königsberg throughout his career. He qualified as a lecturer in 1882 and in 1886 succeeded Voigt as assistant professor of theoretical physics; he became full professor in 1894. Volkmann was assigned not only to teach theoretical physics but also to direct the laboratory of thermodynamics and optics that was part of the institute of theoretical physics. Besides providing an introduction to theoretical physics, Volkmann offered seminars in theoretical mathematical physics, practical laboratory periods in mathematical physics, and occasional sessions to perfect the manual skills required in the laboratory.

Volkmann's early publications were devoted exclusively to theoretical and experimental physics; later they dealt increasingly with epistemology, the history of science, and pedagogy. Most of his publications on physics concerned the determination of the surface tension of water and aqueous solutions on the basis of their height in capillary tubes and between flat plates. His careful investigations on this topic found recognition in two papers by Niels Bohr (1909, 1910), who pointed out the agreement between his own results—reached by a different method—and Volkmann's earlier findings. Volkmann also studied the theory of physical systems of measurement, Green's expression for the potential of the luminiferous ether, Mac-

Cullagh's theory of the total reflection of light, the measurement of soil temperatures, and Ohm's law.

Volkmann took a position on the atomic theory in the third thesis of his dissertation: "The acceptance of the absolute indivisibility of the atom is philosophically quite conceivable and leads to no contradiction." In 1897, stimulated by Boltzmann's "Ueber die Unentbehrlichkeit der Atomistik in der Naturwissenschaft," he set forth his own moderate views in "Ueber notwendige und nicht-notwendige Verwertung der Atomistik in der Naturwissenschaft," which was favorably received.

Volkmann considered axiomatics in "Hat die Physik Axiome?" (1894), in which he rejected the idea of axiomatizing physics but clearly grasped the essential aspects of the subject and recognized its importance five years before the publication of David Hilbert's *Die Grundlagen der Geometrie*, through which this method first entered mathematics.

Volkmann presented his epistemological views in *Erkenntnistheoretische Grundzüge der Naturwissenschaft . . .* and *Einführung in das Studium der theoretischen Physik . . ..* His theory states that because of man's limited intellect and understanding, he necessarily has a subjective comprehension of experience. This comprehension is flawed by errors, which must be detected and eliminated; the goal is an objective knowledge of experience. The means to this end is the introduction of postulates, hypotheses, and natural laws, which permit the construction of a system of knowledge that transcends sense perception and enables man to use mathematics to solve physical problems. Once these foundations are laid, an "oscillation" begins between subjective perception and objective reality. There will always be a difference between the object and the subjective conception of it, but man seeks to narrow the gap by constantly reformulating and adapting his ideas.

In his studies on the history of science, Volkmann dealt most fully with Newton and Franz Neumann. *Franz Neumann, Beiträge zur Geschichte der deutschen Wissenschaft* (1896) contains abundant material on nineteenth-century physics and physicists. The essays "Kant und die theoretische Physik der Gegenwart" and "Studien über Ernst Mach vom Standpunkt eines theoretischen Physikers der Gegenwart," both of which appeared in *Annalen der Philosophie* (1924), were Volkmann's last publications.

In 1887, with Ferdinand Lindemann, Volkmann published *Ratschläge für die Studierenden der reinen und angewandten Mathematik*, a work that explicitly presumed four or five years of study by its readers. He also considered his *Einführung in das Studium der theoretischen Physik* as a contribution to the teaching of physics. In 1912 he instituted a refresher course for Gymnasium teachers, which led to the publication *Fragen des physikalischen Schulunterrichtes*, in which Volkmann advocated the principle of teaching by example.

Despite his interest in education, Volkmann exerted only a limited influence as a teacher. His reticent and careful manner offered little excitement to young physicists hoping to hear him discuss new theories and concepts. He found satisfaction in immersing himself in the fund of existing knowledge, so he had little to communicate to students like Hilbert and Sommerfeld. The limited impact of Volkmann's teaching may well account for the fact that, following his death, none of his former students published an obituary of him or a tribute to his work.

BIBLIOGRAPHY

I. ORIGINAL WORKS. Volkmann's writings are listed in Poggendorff, III, 1400; IV, 1578; V, 1317–1318; and VI, 2772. They include *Ueber den Einfluss der Krümmung der Wand auf die Constanten der Capillarität bei benetzenden Flüssigkeiten* (Leipzig, 1880), his doctoral dissertation; *Vorlesungen über die Theorie des Lichtes (unter Rücksicht auf die elastische und die electromagnetische Anschauung)* (Leipzig–Berlin, 1891); *Erkenntnistheoretische Grundzüge der Naturwissenschaften und ihre Beziehungen zum Geistesleben der Gegenwart* (Leipzig–Berlin, 1896; 2nd ed., enl., 1910); *Franz Neumann, Beiträge zur Geschichte der deutschen Wissenschaft* (Leipzig–Berlin, 1896); *Einführung in das Studium der theoretischen Physik, insbesondere das der analytischen Mechanik, mit einer Einleitung in die Theorie der physikalischen Erkenntnis* (Leipzig–Berlin, 1900; 2nd ed., 1913); and *Fragen des physikalischen Schulunterrichtes, Vier Vorträge* (Leipzig–Berlin, 1913).

II. SECONDARY LITERATURE. There is a short biography in *Deutsche Senioren der Physik* (Leipzig, 1936), Karte 19, with portrait. See also B. Bavink, "Formalistisches und realistisches Definitionsverfahren in der Physik," in *Zeitschrift für den physikalischen und chemischen Unterricht*, **31** (1918), 161–172; Niels Bohr, "Determination of the Surface-Tension of Water by the Method of Jet Vibration," in *Philosophical Transactions of the Royal Society*, **209** (1909), 282–317, esp. 315–316; and "On the Determination of the Tension of a Recently Formed Water-Surface," in *Proceedings of the Royal Society*, **A84** (1910), 395–403, esp. 402–403; L. Boltzmann, "Ueber die Unentbehrlichkeit der Atomistik in der Naturwissenschaft," in *Annalen der Physik*, 3rd ser., **60** (1897), 231–247; and "Nochmals über die Atomistik," *ibid.*, **61** (1897), 790–793; A. Höfler, "Zur physikalischen Didaktik und zur physikalischen Philosophie," in *Zeitschrift für den physikalischen und chemischen Unterricht*, **31** (1918), 1–9, 37–46; W. Lorey, *Das Studium der Mathematik an den deutschen Universitäten seit Anfang des 19. Jahrhunderts* (Leipzig–Berlin, 1916), esp. 262, 282, 299; and F. Poske, "Galilei und der Kausalbegriff," in *Archiv für die Geschichte der Naturwissenschaften und der Technik*, **6** (1913), 288–293; "Das Ohmsche Gesetz im Unterricht," in *Aus der Natur* (Leipzig), **14** (1917–1918), 49–59; and "Studien zur Didaktik des physikalischen Unterrichts," in *Zeitschrift für den physikalischen und chemischen Unterricht*, **31** (1918), 191–193.

HANS RAMSER

VOLNEY, CONSTANTIN-FRANÇOIS CHASSE-BOEUF, COMTE DE (*b*. Craon, France, 3 February 1757; *d*. Paris, France, 20 April 1820), *geography, linguistics, sociology*.

As a student in Paris, Volney learned from Holbach, Mme Helvétius, and other French Idéologues the principles and outlook that were to dominate his thought and action. He then embarked on a voyage to the Levant (1783–1785) to gather data for a systematic account relating the political and social state of the Near East to the physical environment. His *Voyage en Égypte et en Syrie . . .* (1787) was a pioneer work in physical and human geography, distinguished from earlier travel accounts by its systematic method and high standards of accuracy. Volney later used similar methods in producing his *Tableau du climat et du sol des États-Unis d'Amérique* (1803), based on his travels in America (1795–1798). This work contained the first colored geological map of the United States and the first general account of the geology of the trans-Allegheny region.

During the French Revolution, Volney won literary fame for his deistic work *Les ruines, ou méditations sur les révolutions des empires* (1791) and served as a delegate to the Constituent Assembly. Imprisoned during the Reign of Terror, he was later appointed professor of history at the École Normale Supérieure, where he urged the study of history as a social science. Elected to the Institut de France in 1797 and made senator and count by Napoleon, Volney became increasingly disaffected with the Napoleonic regime. Gradually he withdrew from public life to devote himself to linguistic and historical studies.

In linguistics Volney pursued the idea of developing a universal alphabet, an idea embodied in his *Simplification des langues orientales . . .*

(1795), *L'alphabet européen appliqué aux langues asiatiques* (1819), and *L'hébreu simplifié* . . . (1820), and in his bequest of 24,000 francs to establish a prize for work in this field. His studies of Greek, Jewish, and Egyptian chronology, collected in *Recherches nouvelles sur l'histoire ancienne* (1813–1814), were erudite but overambitious.

The scholar, the sociologist, the scientific traveler, and the Idéologue were united in Volney. A pioneer in several fields of inquiry, he was master of none. In all he endeavored to liberate the human mind and to rationalize human institutions by means of an *enquête des faits*.

BIBLIOGRAPHY

Incomplete collections of Volney's works are *Oeuvres complètes de Volney . . . mise en ordre et précédées de la vie de l'auteur* [by A. Bossange], 8 vols. (Paris, 1820–1822); and *Oeuvres complètes, avec notice de Bossange et buste de Volney par David d'Angers*, 4 pts. (Paris, 1837), both of which underwent subsequent eds.

The most comprehensive study of Volney, containing an extensive account of the primary sources, a chronological list of some of the eds. of his various works, and a selection of the most useful secondary literature, is Jean Gaulmier, *L'Idéologue Volney (1757–1820). Contribution à l'histoire de l'orientalisme en France* (Beirut, 1951). Gaulmier has also published Volney's *Voyage en Égypte et en Syrie* in a modern ed. with intro. and notes (Paris–The Hague, 1959). See also Gilbert Chinard, *Volney et l'Amérique d'après des documents inédits et sa correspondance avec Jefferson*, Johns Hopkins Studies in Romance Literatures and Languages, I (Baltimore, 1923). George W. White's intro. to the Hafner ed. of Charles Brockden Brown's trans. of Volney's *Tableau du climat et du sol des États-Unis d'Amérique — A View of the Soil and Climate of the United States of America by C. F. Volney Translated With Occasional Remarks by C. B. Brown* . . . (New York–London, 1968)—gives a critical evaluation of Volney's contributions to early American geology.

JOHN C. GREENE

VOLTA, ALESSANDRO GIUSEPPE ANTONIO ANASTASIO (*b.* Como, duchy of Milan, Italy, 18 February 1745; *d.* Como, 5 March 1827), *physics*.

Volta came from a Lombard family ennobled by the municipality of Como and almost extinguished, in his time, through its service to the church. One of his three paternal uncles was a Dominican, another a canon, and the third an archdeacon; his father, Filippo (1692–*ca.* 1752), after eleven years as a Jesuit, withdrew to propagate the line. Filippo

Volta's marriage in 1733 with Maddelena de' conti Inzaghi (*d.* 1782) produced seven children who survived childhood; three girls, two of whom became nuns; three boys who followed precisely the careers of their paternal uncles; and Alessandro, the youngest, who narrowly escaped recruitment by his first teachers, the Jesuits.

The doctrines, social life, and observances of the church of Rome consequently made up a large part of Volta's culture. He chose clerics as his chief friends, remained close to his brothers the canon and archdeacon, and actively practiced the Catholic religion. Examples of his religiosity include a flirtation with Jansenism in the 1790's; a confession of faith in 1815 to help defend religion against scientism (*Epistolario*, V, 290–292); and an appeal in 1794 to his brothers and to the professor of theology at the University of Pavia for advice about marriage. Not that Volta was prudish or ascetic. He was a large, vigorous man, who, in the words of his friend Lichtenberg, "understood a lot about the electricity of women" (*Epistolario*, II, 269). For many years he enjoyed the favors of a singer, Marianna Paris, whom he might have married but for the weight of theological, and family, opinion.

Volta was about seven when his father died. His uncle the canon took charge of his education, which began in 1757 at the local Jesuit college, where his quickness soon attracted the attention of his teachers. In 1761 the philosophy professor, Girolamo Bonensi, tried to recruit him; his suit, sweetened by gifts of chocolates and bonbons, alarmed Volta's uncle, who took him from school. Bonensi continued his campaign in letters (*Epistolario*, I, 6–33) carried secretly by Volta's eccentric friend, the future canon Giulio Cesare Gattoni (1741–1809), until Volta's uncle the Dominican, who shared his order's opinion of Jesuits, put an end to the affair.

Volta continued his education at the Seminario Benzi, where Lucretius' *De rerum natura* made a powerful impression upon him, and at the so-called Gattoni tower, a disused redoubt rented by his richer and older friend as a laboratory and museum. This cabinet, begun about 1765, won a reputation for its collections in natural history. It also sheltered some physics: a joint study by Gattoni and Volta of the electricity brought down by its lightning rod, said to be the first erected in Como, and experiments of Volta's made possible by books, instruments, and encouragement generously supplied by Gattoni. The first fruit of these mixed studies was a Latin poem of some 500 hexameters in which Volta celebrated the discoveries of

Priestley, Nollet, Symmer, and Musschenbroek (*Aggiunte*, 123–135). Several other poems by Volta in French and Italian survive (*Aggiunte*, 136–158); according to Gattoni, he was always "an excellent judge of all kinds of literature,"[1] which, however, did not save his own style from prolixity.

Volta's uncles wished to make him an attorney, a profession well represented on his mother's side of the family. Volta preferred to obey what he called his genius, which directed him, at the age of eighteen, to the study of electricity.

Electrostatics. The chief authorities on electricity in the early 1760's were Nollet and Beccaria, to whom Volta would write whenever questions or suggestions occurred to him. His first letter to Beccaria, inspired in part by Bošković's ideas, announced that electrical phenomena arose entirely from an attractive force operating between the electrical fluid and common matter (*Epistolario*, I, 4). Beccaria, a testy man who held to the original Franklinist theory of a self-repulsive electrical fluid, took a year to reply, and did so only after Volta had apologized for his "very frivolous chatter." As a cure for frivolity Beccaria recommended reading Beccaria and doing experiments (*Epistolario*, I, 33–36; *Opere*, III, 23). Volta followed the advice, without access to the usual apparatus; forced to invent cheap substitutes, he began to develop that genius for inexpensive, effective instrumentation that determined his career.

His earliest results, communicated to Beccaria in April 1765, derived from the discovery, which Volta fancied new, that silk rubbed by hand became plus, and silk rubbed by glass, minus. He designed a machine to capitalize on the electrical properties of silk and drew up a schematic triboelectric series, doubtless independent of Wilcke's. The correspondence lasted until 1769, when Volta published a Latin dissertation, *De vi attractiva*, which boldly reinterpreted Franklin's theory and Beccaria's latest experiments in terms of the unique attractive principle (*Epistolario*, I, 36–43, 64–65; *Opere*, III, 6–7, 10–11, 19–20, 23–24).

Volta observed that Franklinist electrical matter cannot itself be the cause of electrical motions because it courses unidirectionally, from excess to defect, while in the most common of experiments, as Nollet had emphasized, the same electrified body simultaneously imposes both attractions and repulsions. Nor can the effluvia operate indirectly, by impelling the air, for electrical attraction takes place between bodies immersed in oil (an experiment Volta lifted without acknowledgment from Cigna). We must therefore admit short-range at-tractive forces. To the usual objection that multiplying such forces clutters matter with special nonmechanical powers, Volta countered that, since only "mixed bodies" are electric, one need imagine no special virtue of electricity, but merely a net macroscopic force compounded from the different microscopic forces possessed by the particles of pure substances, or from the universal, elemental, multipurpose force of Bošković. Nor should one falter at the great range of electrical attraction: we have, on the one hand, the patent example of magnetism and, on the other, the existence of electrical atmospheres. These, according to Volta's even-handed compromise, consist of surplus electrical fluid, the attraction of which extends a little way beyond their physical limits. "However that might be, for present purposes it need only be granted that attractive forces really exist in bodies" (*Opere*, III, 25–29, 85).

Volta's fundamental concept is that there exists for each body a state of saturation in which the integrated attractions of its particles for electric fluid are precisely satisfied. This integrated attraction may be altered by any process, mechanical or chemical, that displaces the particles relative to one another; friction, pressure, and, perhaps, evaporation electrify bodies by destroying the existing pattern of saturated forces and redistributing the electrical fluid (*Opere*, III, 30–34). In this proposition one sees the seeds of the experiment of Volta, Lavoisier, and Laplace on electrification by evaporation, and, perhaps, of Volta's consequential concept of contact charge. As for the notion of saturation, it vaguely foreshadowed the concept of tension, Volta's qualitative equivalent of the modern potential: the condition of electrical equilibrium between two bodies being not equality of quantity of electric fluid, but of degree of departure from saturation.

For the rest, *De vi attractiva* is an exercise in reducing the standard phenomena—attraction, "repulsion" (really attraction away from the "repelling" body), the Leyden experiment, and the effects of Beccaria's vindicating electricity—to the single attractive force. Again one can see fruitful tendencies, particularly in Volta's analysis of induction in an insulated conductor *B* under the influence of a positively charged body *A*: *A*'s atmosphere supersaturates *B* without altering *B*'s integrated positive force; *B* therefore sheds fluid, which surrounds its far side in an atmosphere. Touch *B*: it loses its surplus, but shows no electrical signs because its residual fluid and *A*'s atmosphere exactly saturate it (bring it to zero potential). Now remove A: B is no longer saturated, and shows itself negative (*Opere*, III, 36–50). Here one sees seeds of the electrophore and the condensator. Although Volta soon acknowledged that the single attractive force could not account for many simple phenomena—for example, the difference between insulators and conductors, and the charging of a Leyden jar—he continued to be guided by it and to ascribe most electrical effects to it, until 1784 or even later (*Opere*, III, 56–71, 85; IV, 410–413).

The reluctance to change or discard a once-useful theory was characteristic of Volta. As he said when describing his slowness to accept Lavoisier's chemistry, he wished to be neither too open nor too resistant to novelties. He remained faithful to the Franklinist hypothesis of a single electrical fluid, "la nostra cara dottrina" (*Opere*, IV, 359, 380), while most important physicists of the Continent preferred the dualistic system of Symmer. Volta eventually was brought to agree that all known electrical effects could be explained on either system; but he preferred the singlist, partly (as he said) because of a reluctance to multiply entities unnecessarily, and mainly because of his scientific conservatism (*Opere*, IV, 269; *Epistolario*, II, 278).

To concoct the electrophore, the most intriguing electrical device since the Leyden jar, Volta had only to combine the insight that resin retained its electricity longer than glass with the fact, emphasized by Cigna and Beccaria, that a metal plate and a charged insulator properly maneuvered can produce many flashes without enervating the electric. Beccaria inspired the combination. In 1772 he published a lengthy, difficult, updated version of *Elettricismo artificiale*, which emphasized more strongly than before his odd view that the contrary electricities destroy one another in the union of a charged insulator with a momentarily grounded conductor, only to reappear, "revindicated," in subsequent separations. Beccaria also criticized the hypothesis of the unique attractive force, without deigning to mention Volta, who in return conceived that, if he could greatly increase the duration of the effects ascribed to vindicating electricity, the implausible theory of alternate destructions and incomplete recuperations would fall to the ground. After many trials Volta found that an insulator made of three parts turpentine, two parts resin, and one part wax answered perfectly; and in June 1775 he informed Priestley of the invention of an *elettroforo perpetuo*, which "electrified but once, briefly and moderately, never loses its electricity,

and although repeatedly touched, obstinately preserves the strength of its signs" (*Opere*, III, 96).

The device consisted of a metal dish containing a dielectric cake, and a light wooden shield covered with tin foil rounded to remove all corners and joined to an insulating handle. The cake is first charged, say negatively, by rubbing. The shield is then set upon it, and momentarily grounded, thereby charging positively by induction. The shield may then be removed and its charge given to, say, the hook of a Leyden jar; then replaced, touched, and again brought to the hook; and so on until the condenser is moderately charged. Any number of jars and electrophores may be electrified without regenerating the original; and if it should decline, it can be reinvigorated by lightly rubbing its cake with the coating of a Leyden jar that the shield had charged through the hook. Volta set great store by this last property, which did seem to vouchsafe eternal life to the electrophore and to justify the term *elettricità vindice indeficiente*, with which he proposed to celebrate his victory over Beccaria (*Opere*, III, 98–105).

The triumph was clouded. Beccaria thundered that the "perpetuity" of the charge of the electrophore proved nothing and that he and Cigna had already described the necessary manipulations. Other claimants came or were thrust forward: Stephen Gray, Aepinus, Wilcke, and the Jesuits of Peking. With his customary good sense (*Opere*, III, 120, 137–143), Volta acknowledged the role of Cigna, but insisted, quite rightly, that he alone had made a usable instrument, and had developed the cake, the armatures, and the play with the bottle. Even Wilcke, who had fully grasped the theory, had not embodied it in the sort of apparatus—sturdy, useful, powerful, intriguing—characteristic of Volta's designs.

The electrophore killed off not only vindicating electricity but also the last vestiges of the old doctrine of literal atmospheres (*Opere*, III, 140n; *Epistolario*, I, 275–280). The only successful theories of the device, for example, those of Ingen-Housz and Wilcke, employed actions at a distance between electrical fluids confined by the surfaces of conductors. Accordingly, as contemporaries recognized,[2] the electrophore caused electricians to take seriously the neglected approach of Aepinus. Volta himself first met with a copy of Aepinus' "incomparably profound book" (*Opere*, III, 210n, 236) in the 1770's, too late to guide his invention but in time to assist his own revision of the concept of atmospheres.

The mid-1770's marked the beginning of Volta's

career. In October 1774 he took his first academic job, principal or regent of the state Gymnasium in Como (*Epistolario*, I, 66–68), then recently taken over from the Jesuits. Next came the electrophore and, at Volta's request, the professorship of experimental physics at the Gymnasium, which he garnered in 1775 without the usual examination (*Epistolario*, I, 99, 100). A sally into pneumatics brought the discovery of methane (1776) and a greater reputation, which helped him in 1777 to obtain state support for a trip to the chief centers of learning in Switzerland and Alsace (*Epistolario*, I, 149–150, 178). There Volta met several savants—particularly H. B. de Saussure and Jean Senebier of Geneva—both of whom would advertise and encourage his work, and help keep him informed about transalpine physics (*Epistolario*, I, 192–193).

Volta's travel grant came from the Austrian government, which then controlled the duchy of Milan, including Como, and which, through its minister Count Carlo di Firmian, was modernizing the educational institutions of the region. Chief among these was the University of Pavia, where the Austrians had been encouraging science, particularly since 1769, when Spallanzani came to the chair of natural history.

In 1777 Pavia had two professorships of physics, both occupied by clerics: a "general" held by Francesco Luini (Jesuit) and an "experimental" held by Carlo Barletti (Scolopian). In 1778 Firmian, the "immortal Maecenas, benefactor and greatest protector of the university" (*Epistolario*, II, 285), sent Luini to Mantua, translated Barletti to general physics, and gave Volta the post he would hold for almost forty years, the professorship of experimental physics at Pavia (*Epistolario*, I, 298). Volta proved a very popular professor (*Epistolario*, II, 41, 283–284). A new lecture hall was built to house his auditors and the university's ever-increasing collection of instruments, many of which Volta bought at state expense on state-financed trips to France and England in 1781–1782 (*Epistolario*, II, 51–141) and to Germany in 1784 (*Epistolario*, II, 225–273).

As Volta's professional opportunity and acquaintance increased, his style of physics altered, at least in its public form. The change was manifest in 1778 in a published letter to his new friend Saussure on electrical capacity. While *Di vi attractiva* developed a microscopic model of electrical action, which explained but did not guide, and while the account of the electrophore was primarily a description of laboratory manipulations, the letter to Saussure applied new theoretical concepts to the design and explanation of new experiments. These powerful concepts, the macroscopic quantities capacity and tension, also appear in Cavendish's now famous memoir of 1771. There is reason to believe that Volta read this memoir, which most contemporary electricians ignored or misunderstood, and that he derived from it—and perhaps also from the works of Aepinus and even of Barletti, who first acquainted him with Aepinus[3]—the clue for the transformation of his otiose notion, "natural saturation," into a serviceable substitute for the concept of potential.

Volta's thought is that the capacity C of a conductor and the tension T of its charge Q alter with its distance from other conductors (*Opere*, III, 201–229; *Epistolario*, I, 275, 280). For example, as the charged shield is raised from the electrophore cake, electrometer threads attached to it spread, owing to an increase in the tension of its charge; since the quantity of charge does not change, the tension grows because the shield becomes less capacious, less able to hold its naturally expansive charge as it moves farther from the opposite electricity of the cake. The reverse effect occurs with a pair of similarly charged conductors: the capacity of each is enlarged, and its tension lessened, as the distance between them increases. Volta deduced that the "atmospheres" of the various surface elements of the same conductor might inhibit one another, and that, for a given surface area, the longer the conductor the greater the capacity. Perhaps, as Cavendish had suggested, the capacity of a single conductor could be increased to that of a Leyden jar. It was just this expectation that Volta confirmed in his letter to Saussure, who had earlier doubted its possibility (*Opere*, III, 213–215).

In describing the experiment Volta used the old term "electrical atmosphere," by which, however, he now no longer meant an envelope of electrical fluid but, as was becoming commonplace, merely a "sphere of activity" (*Opere*, III, 155, 160, 166–167, 182, 206, 236–240). The point is important, as many commentators, perhaps misled by Biot, have ascribed to Volta a belief in the retrograde literal atmospheres that his work helped to destroy. It is plain from Volta's manuscripts—for example, the beautiful and exact theory of the slow-motion charging of a Leyden jar (*Opere*, III, 248–258), or the *Lezioni compendiose sull'elettricità* (*Opere*, IV, 419)—that soon after, if not before, the letter to Saussure, Volta had freed himself of the ideas that "anything real" passed between bodies interacting electrically beyond sparking range, and that the surplus electrical fluid of a positively charged body resided in the air about it (*Opere*, III, 236, 273; IV, 65–68, 71–74; *Epistolario*, II, 213). Occasionally he represented this sphere of activity as a state of the space or air surrounding charged bodies (*Epistolario*, I, 296, 326–327, 376, 411); a representation not of literal atmospheres but of a crude field theory, which may be traced from Canton and Beccaria through Avogadro and on to Faraday.

Volta embodied the quantities capacity and tension, and the implicit relation that he had established between them ($Q = CT$), in a new instrument, a "condensator" for rendering sensible atmospheric electricity otherwise too weak for detection (*Opere*, III, 271–300). This famous device is nothing but an electrophore with a poor conductor like polished marble or oiled wood as its cake. One runs a wire from an apparently unelectrified atmospheric probe to the shield, waits, removes the wire and raises the shield, which can then affect an electroscope. Volta explained that owing to its great capacity the electrophore soaks up the electricity of the probe as often as it becomes charged, while the separated shield, being of small capacity, can reveal the weak collected electricity. He emphasized that the quantity of charge on a conductor increases as the product of its tension and its capacity, the former being the quantity measured by electrometers (*Opere*, IV, 71–74). Others soon incorporated this insight into ingenious multipliers of weak charges, such as the well-known "doubler" invented by William Nicholson.

Meteorology. Volta's interest in meteorology centered on atmospheric electricity, the study of which began in 1752 with the apparent confirmation of Franklin's hypothesis about the electrical character of lightning. It was quickly discovered, by Beccaria and Canton among others, that the atmosphere exhibited electricity even in fair weather, and that, contrary to Franklin's expectation, it was more often negative than positive. This information was at first deduced from the electrical state of the lower end of an insulated pointed pole or wire, which was thought to exchange electrical fluid with the surrounding air. In fact such probes charge partly by conduction but mainly by induction, and their electricity does not give an unambiguous index of the electrical state of the atmosphere. Among the few to understand and to evade this ambiguity was Saussure, whose work directly inspired Volta's.[4]

Saussure employed not a long pole but a form of the bottle electrometer invented by Cavallo, with silver wires ending in pith balls as the indicator. Saussure would touch the stem and case of the electrometer to the ground and suddenly raise the instrument above his head; the consequent spread of the wires indicated, as he said, the electrical tension of the atmosphere at the site of the electrometer. Saussure carried this device on his famous attempt at Mont Blanc in 1787, which Volta, who was then visiting Geneva, commemorated in no fewer than sixty-six *terzini* (*Aggiunte*, 146–152). When he returned to Pavia, Volta undertook to make Saussure's instrument "more obedient" (*Opere*, V, 88–90). In 1787 he began to announce his results in letters to G. L. Lichtenberg, professor of physics at the University of Göttingen, whom he had met on his trip to Germany in 1784. The nine Lichtenberg letters constitute Volta's chief writings on meteorology.

Alerted to the problematic operation of the pointed pole, Volta hit on a solution quite different from Saussure's: bathing the point in flame, which promoted the exchange of electric fluid and brought the point quickly to the potential of the atmosphere just outside it (*Opere*, V, 88–92, 152–156). Volta found that electrometers armed with flames registered four times the electricity recorded by Saussure's detector under identical circumstances (*Opere*, IV, 71–74). The device was widely used, although probably not fully understood, until William Thomson gave its theory in the 1850's and replaced it by his ingenious water-dripper.[5]

The next business was to make of Saussure's electrometer a sensitive, uniformly calibrated, international standard. Volta improved the sensitivity by replacing Saussure's wires with light straws with large effective repelling surfaces (*Opere*, V, 35–42, 68, 71); the result was an inexpensive form of the exactly contemporaneous gold-leaf electrometer (1786) invented by Abraham Bennet. Uniform calibration, which Volta deemed essential, was obtained by giving the electrometer successive sparks from a capacious Leyden jar kept at a constant potential by a small electrophore (*Opere*, V, 39–42). Taking the intercomparability of thermometers and of hygrometers as his model, Volta proposed the adoption of a fundamental unit of tension, namely that of a standard metal disk hung from one arm of a balance a distance d above a conducting surface, and counterbalanced by a certain weight W. The unit, equivalent to a spread of 350 degrees of Volta's straw electroscope, is about 13,350 volt in modern measure.[6]

In experiments with the unit, Volta found that

the "force of attraction" measured by the weight W was proportional to $(T/d)^2$, T being the tension of the disk according to his straw electrometer. It is most interesting that he took this result, which is correct, as evidence against the universality of Coulomb's law, which gives the same dependence on distance, but for a different geometry (*Opere*, V, 78–79, 81–83). Moreover, other geometries yielded "diverse other laws, as curious as they are novel." Like many of his colleagues, Volta did not have mathematics enough to work from a hypothetical law of interaction of electrical elements to the observed electrical forces between macroscopic bodies.

Volta accepted the Franklinist presumption that the instruments of atmospheric electricity measured the surplus (or deficiency) of electrical fluid in the lower atmosphere; and he had suggested in *De vi attractiva* that the fluid enters (or leaves) the air during evaporation. One of the first tasks he assigned his condensator was the detection of the supposititious electrification during change of state. He was then (1782) in France, and undertook the experiments in collaboration with Lavoisier and Laplace. At first they failed, as they should have, there being no such effect; but shortly before Volta left Paris for London they succeeded, or believed they had, and made much of their success. According to Volta, everything depended on a change in electrical capacity suffered by water droplets in going from the liquid to the vapor state (*Opere*, III, 33–34, 301–305, 364; V, 173–187, 196–197; *Epistolario*, II, 104–105; *Aggiunte*, 21–24). They had probably detected electricity generated by the friction of bubbles against the evaporating pan. The subject was to remain confused for over a century.

Volta's explanation, which differed from that in *De vi attractiva*, doubtless owed something to his adherence to the doctrine of latent heat (*Opere*, VI, 313–316), which became widely known in the early 1780's. It remained the basis of his speculations about meteorological phenomena. For example, according to his much admired theory of hail, evaporation abstracts both heat and electricity from vaporizing droplets, creating charged microscopic ice seeds, which dance about under electrical forces in their parent cloud, growing at the expense of surrounding droplets until they become too heavy for the ballet, and fall to the ground (*Opere*, V, 201–206, 283–307, 421–462).

Pneumatics. Volta's work on gases shows the same genius for instrumentation and measurement,

and the same failure or reluctance to establish general principles, that characterize his work on electrostatics. His first pneumatic studies concerned "inflammable air from marshes" (chiefly methane), which he discovered in November 1776 in Lago Maggiore. It was not a chance find. Inflammable air from metals (hydrogen released from acids) had been known since its isolation by Cavendish in 1766, and Franklin's description of a natural source of inflammable air had just been published by Priestley in a book quickly known in Italy.[7] In the autumn of 1776 Volta's friend P. Carlo Giuseppe Campi had found a natural source near Pavia; and Volta himself, intrigued by the "ever more remarkable and interesting subject of the different kinds of air" (*Opere*, VI, 19), had scoured the countryside for telltale bubbles. The testing of his new gas—new in source, flame color, and combustibility (*Opere*, VI, 30)—led him into the faddish field of eudiometry.

In 1772 Priestley had isolated a "nitrous air," which, when combined with common air over water, left a volume of gas less than the sum of the volumes of the ingredients. He found the reduction to be less the more the common air had been vitiated by respiration or combustion; and he proposed to take the degree of reduction as a measure of the "goodness" of the common air. Priestley's procedure was improved, and his interpretation adopted by two of Volta's friends: Marsilio Landriani, who introduced the term "eudiometry," and Felice Fontana, whose nitrous-air eudiometer won wide acceptance in northern Europe (*Epistolario*, I, 218–219; 258–260; III, 4–8). Both hoped that the instrument would help to identify malarial and other insalubrious regions; and for almost thirty years physicists visited swamps, cesspools, dung heaps, prisons, and hospitals hoping to correlate the reading of their eudiometers with the evident foulness of the air. No consistent correlations emerged. In 1805 Humboldt and Gay-Lussac put an end to the search by showing that the percentage of oxygen in unvitiated air was independent of its source. They succeeded by employing a device of Volta's, who had never believed that the eudiometer could measure the salubrity—as opposed to the respirability (oxygen content)—of the air (*Opere*, VI, 9).

Ever interested in large, reproducible effects, Volta had shifted his attention to hydrogen upon discovering that, when mixed with common air and sparked, Cavendish's inflammable air ignited more readily and burnt more fiercely than his own (*Opere*, VI, 50); whence Volta's famous "inflammable air pistol," filled with hydrogen and air or oxygen, and fired by a portable electrophore (*Opere*, VI, 134–135). To perfect this artillery (which could fire a lead ball with force enough to dent wood at fifteen feet [*Opere*, VI, 155]), he looked for the mixture that destroyed the greatest quantity of gas (*Opere*, VI, 146). He thereby came to the problem of the eudiometer, but from a new side, and with a new eudiometric fluid, hydrogen, which could be obtained purer than the standard nitric oxide (*Opere*, VI, 180–181), and acted much more vigorously (*Opere*, VI, 159–160). Volta's first eudiometric technique was to find the minimum volume of the air under test in which a standard amount of inflammable air could be ignited by a spark; the larger the volume, the poorer the air. As for the optimum explosive mixture, it turned out to be four parts inflammable to eleven parts unvitiated common air (*Opere*, VI, 179), or two parts inflammable to one part dephlogisticated air (oxygen) (*Opere*, VI, 190n). In the definitive form of his eudiometer (*Opere*, VII, 173–213), Volta mixed equal volumes of hydrogen and common air, exploded them, and determined the diminution; the maximum contraction, for the best air, fell out just under 3/5 volume, confirming that, as other of his measurements suggested, the maximum possible reduction in unit volume of common air was about 1/5 (*Opere*, VII, 197).[8] Volta's numerical results were fully confirmed by Humboldt and Gay-Lussac, who found oxygen to occupy about 21 percent of the volume of common air. This should be compared to the results obtained by Humboldt, Lavoisier, and Scheele, using the Fontana nitrous-oxide eudiometer, namely 26 to 28 percent.

Volta's eudiometer set up one of the most important discoveries of the eighteenth century, the composition of water, detected by Lavoisier, among others, by sparking oxygen and hydrogen over mercury (1783). As early as the spring of 1777 Volta had been looking for the residue of the reaction. In his version of phlogistic chemistry, inflammable air (H_2) was phlogiston (ϕ) combined with an unknown "base," which he supposed to be of an "acid" or "saline" character (*Opere*, VI, 150, 342, 400–401). He recognized that, since the base might be soluble in water, the sparking should be done over mercury, but he had not enough for the task (*Opere*, VI, 196–197, 303, 410–411; *Epistolario*, I, 267–270). While working to obtain more, he sparked inflammable and common air over water, and noticed (in 1778) that the walls of the test vessel fogged (*Opere*, VI, 382). While in

Paris in 1782 he told Lavoisier about the fogging; and later in the year Lavoisier, Laplace, and Monge obtained water over mercury by Volta's method (*Opere*, VI, 410–411).

The French, following Lavoisier's ideas, thought they had synthesized water; Volta, remaining faithful to phlogiston, believed that they had analyzed the gases (*Opere*, VI, 342, 411; VII, 87–88, 101, 103):

Inflammable air (water + ϕ) + dephlogisticated air (water + caloric) = water + heat.

Volta did not adopt the new chemistry for many years, perhaps not definitively until after 1800, although he began to speak of it more favorably in the 1790's (*Opere*, VII, 246, 269–270, 284; *Epistolario*, III, 61–62). He later said that the decisive proof was his own calcification of metals in closed vessels by burning mirrors. Calcification proceeded until the volume of the air fell by 1/5, precisely the amount of dephlogisticated air that, according to Volta's earlier measurements, would be available to support the combustion (*Opere*, VII, 285).

Volta's later pneumatic studies centered on the action of heat on gases and vapors. His general conception of heat followed the fluid theories of Crawford and Kirwan (*Opere*, VI, 315; VII, 45–47), with one characteristic exception: whereas his sources ascribed the phenomenon of latent heat to a chemical combination responsible for change of state, Volta made the change primary, and the latent heat the result of a consequent jump in specific heat capacity. This concept, developed in notes to the Italian edition of Macquer's *Dictionnaire de chymie* (1783–1784), derived from Volta's mature conception of electrical capacitance and from an assimilation of the properties of the two fluids: since nothing analogous to latency—the supposed inability of accumulated caloric to affect a thermometer—occurred in electricity, it was difficult for Volta to credit it in the case of heat (*Opere*, VII, 19–20). Consequently he once again opposed Lavoisier, now regarding his claim that evaporation arose from the chemical combination of heat and water (*Opere*, VII, 87–93). Volta also opposed the older theory, already under attack, that evaporation consisted of the "solution" of water in the bases of the atmosphere.

Against this last proposition Volta could adduce his own experiments on what we would now call partial pressures. Already in 1784, in a letter to Lichtenberg, Volta sketched the law usually attributed to Dalton for the case of water vapor. Volta also stated clearly in letters obscurely published in 1795 and 1796 that "the quantity of elastic vapor is the same in a space either void of air or filled with air at any density, and depends only upon the degree of heat" (*Opere*, I, 301; VII, 441). Hence he easily derived an argument fatal to the theory of evaporation by solution. Moreover, Volta anticipated and even went beyond Dalton in measuring the dependence of the density and pressure of water vapor on temperature. The laborious and difficult measurements, made in a heated Torricelli space, gave results in very rough agreement with modern determinations.[9]

Volta was more successful in measuring the dilation of air as a function of heat, or rather of temperature indicated on a mercury thermometer. The proportionality of heat and temperature so measured had been established, to Volta's satisfaction, by Deluc, "a most knowledgeable and accurate experimenter" (*Opere*, VII, 414), whose thermometric example had probably encouraged Volta's comparative electrometry. Between 1772 and 1790, when Volta took up the subject, many physicists had tried to measure α, the percentage increase in volume of a gas per degree of temperature. In his masterful memoir published in 1793 (*Opere*, VII, 347–375), Volta pointed to values of α ranging from 1/85 (Priestley) to 1/235 (Saussure) per degree Réaumur, and to uncertainty whether α varied with temperature between the freezing and boiling points. Volta cut through the uncertainty by observing that the dilation produced when heating a gas over water derives from two causes: (1) the true expansion of the gas and the water vapor it contains, and (2) the generation of additional vapor from the walls of the experimental vessel and from the water used to measure the dilation. Dry the vessel carefully, conduct the experiment over mercury or oil, and, according to Volta, you should get an α for air independent of temperature and equal to about 1/216. This value, which agreed perfectly with those obtained by Deluc and by Lambert from less systematic measurements, differs very little from that now accepted.

The journal to which Volta confided these results had little circulation outside Italy. Once again his priority was ignored, this time in favor of Gay-Lussac, who in 1802 deduced a value of α (1/213) poorer than his predecessor's and based on flimsier data, albeit for more gases. (Gay-Lussac obtained α from the total dilation between freezing and boiling points; Volta had measured it for each degree.) It is possible that Gay-Lussac did not obtain his

number in total ignorance of Volta's.[10] In any case, the proposition, "the coefficient of expansion of air is constant," was restored to Volta by unanimous vote of the international congress of physicists meeting at Como in 1927 in observance of the centennial of his death (*Opere*, VII, 346).

Animal Electricity and Galvanism. In 1791 Galvani, professor of anatomy at the University of Bologna, published his now famous study of the electrical excitation of disembodied frog legs. He explained the jerking of a leg upon completing a circuit through the crural nerve and the leg muscle as the direct result of the discharge of a "nerveo-electrical fluid" previously accumulated in the muscle, which he supposed to act like a Leyden jar. The analogy between muscle and jar did not rest only on the need for a complete discharge circuit. Consider also the following phenomenon: the internal electrode of a charged grounded Leyden jar is pointed and brought near a large electrified insulated conductor; when a spark is drawn from the conductor, a "penicillum" of light flashes from the pointed electrode. According to Galvani, precisely the same sort of discharge occurred during the chance observation that had led him into his odd studies: a freshly prepared frog's leg jumped (that is, its muscle discharged, in analogy to the penicillum) if the circuit were completed at the instant that a spark was drawn from a nearby electrical machine. In Galvani's opinion the structure of the muscle, like the peculiar anatomy of the torpedo or electric eel, effected and retained the accumulation of the nerveo-electrical fluid. As for the fluid, it was similar to but distinct from frictional electricity, an "animal" electricity sui generis.

When Volta learned of Galvani's experiments he dismissed them as "unbelievable" and "miraculous." He had a low opinion of physicians, whom he found to be generally "ignorant of the known laws of electricity"; and he recognized "animal electricity" only in electrical fish, to which, however, he ascribed only the power of manipulating common electrical fluid (*Opere*, I, 10–11, 21–23, 26; *Epistolario*, III, 143–145). Moreover, even as late as 28 March 1792, just after he had first tried the experiments, "with little hope of success," under the urging of his colleagues in pathology and anatomy, his immediate research plans included only meteorology and the dilation of gases. But by 1 April the experiments had succeeded, and Volta had begun the brilliantly planned and executed experiments that step by step brought him to the invention of the pile.

Volta's first instinct was to measure the minimum tension of "artificial" or "frictional" electricity that would cause the frog to jerk: "How can causes be found if one does not determine the quantity as well as the quality of the effects?" (*Opere*, I, 27). Frog legs prepared as directed by Galvani proved to be by far the most sensitive electroscope yet discovered. When placed in a discharge train of a Leyden jar, they responded to a tension of as little as 5/100 degree of Volta's straw electrometer, an amount he could only detect after manipulation by the condensator. He also succeeded in inducing convulsions in a live frog by joining its leg and back externally by a circuit made of dissimilar metals. (Galvani had discovered by chance that prepared frog legs kicked violently and reliably when nerve and muscle were joined by a circuit composed of two kinds of metals.) Volta's discovery, probably made in April 1792, required modification of Galvani's theory. While agreeing that the electrical imbalance detected by the spasms arose from action of the animal, Volta doubted the appropriateness of the analogy to the Leyden jar; rather, it seemed to him that a weak animal electricity constantly circulated through the body of a normal frog, and that artificial circuits brought about convulsions by disturbing the natural flow (*Opere*, I, 15, 30–33).[11]

Volta's use of the whole frog—a move unnatural for an anatomist like Galvani—proved consequential. When the animal was intact it could be made to tremble only when struck by a discharge from a Leyden jar or when part of a bimetallic circuit. Volta inferred that the electricity put in action in the second case arose from the mere contact of dissimilar bodies (*Opere*, I, 55, 64–66, 73–74), a property he had already identified in "electrics" (insulators) but was surprised to meet in metals (*Opere*, I, 136). The fact, however, was plain, as well as the conclusion that animal electricity played no part in spasms inspired by bimetallic arcs. The only true galvanic effect, according to Volta, was the convulsion of a freshly prepared specimen in a circuit completed by a single metal (*Opere*, I, 116–118, 156–157, 180). And even this "beautiful and great discovery" (*Opere*, I, 175), this "truly astonishing experiment" (*Opere*, I, 178), could not occur as Galvani thought; for, as Volta showed, the muscle need not be included in the circuit. Electricity excited the nerve, and the nerve the muscle; there was no room for a Leyden jar fabricated of muscle tissue. To illustrate the office of the nerve Volta thought to excite the sense of taste by a bimetallic arc. With great satisfaction he experienced an unpleasant taste by joining a bit of tin on the tip of his tongue to a silver spoon resting further back (*Opere*, I, 56–57, 62–63, 73–74). It happened that, unknown to Volta, this experiment had been described many years earlier by J. G. Sulzer, who, however, did not associate it with electricity and doubtless—again in contrast to Volta—did not design it as a test of theory (*Opere*, I, 152–154, 196).

The tendency of Volta's results was to restrict more and more the domain of animal electricity. By November 1792, after countless trials on diverse unlucky creatures from insects ("it is very amusing to make a [headless] grasshopper sing" [*Opere*, I, 190–191]) to mammals, Volta had concluded that all galvanic excitations arose from external electrical stimulation. As for the classic case (a freshly killed and stripped frog, highly excitable, joined crural nerve to leg muscle by a single metal), Volta supposed that the electricity came not from animal power but from the contact between the metal and unobserved impurities in it (*Opere*, I, 147, 156–157). Nothing remained of the theory of animal electricity, or so Volta told Galvani's nephew and defender, Giovanni Aldini, professor of physics at the University of Bologna, in an open letter published early in 1793 (*Opere*, I, 149–159).

While the Galvanists pondered their response, Volta ranked the metals according to their electromotive power (*Opere*, I, 214, 234, 304) and tried to determine the seat of the electromotive force. He recognized that an effective circuit contained, besides a bimetallic joint, at least one "moist conductor," namely, the nerve to be excited, and he thought it more probable that the electrical imbalance occurred in the contact between the metals and the moist conductor than in the joint between the metals (*Opere*, I, 205, 212–213, 231–232). This proposition gained plausibility by his discovery in 1793 that the electromotive power of a chain of dissimilar metals depends only upon the nature of the two extreme links, precisely those touching the moist conductor, and that nothing happens if each metal is in contact only with moist conductors (*Opere*, I, 226–227).

Volta was accordingly prepared to answer the counterattack launched in 1794 by Galvani, Aldini, and a resourceful physician, Eusebio Valli, who had always thought the contact theory "ridiculous"; for "how [he said] is it possible for a single shilling to contain electricity sufficient to move the leg of a horse?"[12] Their strongest and most worrisome new evidence was Valli's excitation of spasms in freshly prepared frogs using himself as arc. It appeared that convulsions could be induced without the metallic contact which, in his reply to Galvani and Aldini, Volta had just asserted to be necessary (*Opere*, I, 274, 279, 295n, 308). Although many people conceived that Valli's stroke had saved animal electricity, Volta had no trouble turning it to his advantage. He observed that, as Valli had reported, the experiment worked best when the nerves and muscles were moistened with blood or saliva. As he explained to Sir Joseph Banks in March 1795, and then to A. M. Vassalli, professor of physics at the University of Turin, in a letter printed in 1796, a sequence of dissimilar moist conductors could generate an electrical current by contact forces without the intervention of metals (*Opere*, I, 255–256; 295–297).

Volta's next, and characteristic, step was to determine the "electromotive force" (his words) of various combinations of conductors. He tried to rank moist conductors ("conductors of the second kind") as he had the metals ("conductors of the first kind") (*Opere*, I, 371, 405–406). He confirmed that an electromotive force occurred only via the contact of dissimilar conductors (*Opere*, I, 372, 397, 411–413), and he sought the most powerful combination of "electromotors." The results, in order of decreasing power, expressed in Volta's notation (where capital and small letters signify conductors of the first and second kind, respectively [*Opere*, I, 230, 379–382]); *rABr* (where *r*, the frog, is both a conductor of the second kind and the electroscope); *raAr*; *rabr*; *rAr* and *rar*, both zero (*Opere*, 396–397, 401–402). What about *ABCA*? Volta thought that analogy favored the possibility of a weak finite current in such a circuit. But how to detect it when the only electroscope sensitive enough to register galvanic electricity was itself a conductor of the second kind (*Opere*, I, 377–378)? The difficulty instanced a much more serious one, which had long bothered Volta: that his claim of the identity of galvanic and common electricity rested on experiments in which pieces of animals played an indispensable part (*Opere*, I, 490, 540–555).

The contact of zinc and silver develops about 0.78 volt. Volta's most sensitive straw electrometer marked about 40 volt/degree.[13] By the summer of 1796 he had managed to multiply the charges developed by touching dissimilar metals together enough to stimulate his electrometer (*Opere*, I, 525; *Epistolario*, III, 349, 359). He first succeeded with a Nicholson doubler (*Opere*, I, 420–424) and then with an unaided gold-leaf electroscope (*Opere*, I, 435–436); and he later rendered contact electricity easily sensible by a "condensing electroscope," a straightforward combination of the condensator and the straw electrometer (*Epistolario*, III, 438). All these devices, including the doubler, came directly or indirectly from Volta's earlier work. Note that to obtain contact charges that he could multiply Volta had to change his mind about the principal seat of the electromotive force, which he now located in the junction of metals and not in their union with moist conductors (*Opere*, I, 419, 472). In 1797 he published a full account of his detection of galvanic electricity by electrostatic means (*Opere*, I, 393–447).

It remained to find a way to multiply galvanic electricity directly. Volta discovered soon enough that piling metal disks on one another (say *aABAB* ··· *a*) did not help, and that a circuit made only of metals gave no electromotive force. These results led to the useful rule, a precise version of his result of 1793, that the electromotive force of a pile of disks is equal to what its extreme disks would generate if put into immediate contact (*Opere*, II, 61). How or when Volta hit on the far from obvious artifice of repeating the apparently unimportant secondary conductors in his generator is not known; an anticipation appears in one of the combinations published by Gren in 1797 (*Opere*, I, 398, fig. 13, 400). The definitive pile, *AZaAZaAZa* ··· *AZ*, consisting of pairs of silver and zinc disks separated by pieces of moist cardboard, was first made public in 1800, in a letter addressed to Banks, president of the Royal Society of London, and published in its *Philosophical Transactions* (*Opere*, I, 563–582). The letter also describes an alternative arrangement, a "crown of cups" consisting of a circle of glasses filled with salty or alkaline water and connected by bimetallic arcs dipping into the liquid (*Opere*, I, 568, 571; see also *Opere*, I, 399, 403–404).

Volta represented his discovery as an "artificial electric organ," an apparatus "fundamentally the same" as the natural electrical equipment of the torpedo (*Opere*, I, 556, 582). A medium-size pile, with forty or fifty pairs, gave anyone who touched its extremities about the same sensation he could enjoy grasping an electric fish. In both cases, Volta said, a constant current running externally from top to toe of the electromotor passed through the arms and breast, and agitated the sense of touch. Were it directed at the senses of vision, taste, or hearing, the current would cause light, taste, or sound instead (*Opere*, I, 578–580). Neither the pile nor the torpedo give electrostatic signs because, as Cavendish had argued long before, they operate at too low a tension; their effects derive rather from the quantity of electrical matter they move. The analogy to the torpedo played little part in Volta's discovery; the emphasis upon it in the letter to Banks was intended to silence the Galvanists. As for the cause and continuance of the electricity generated by the contact of dissimilar conductors, Volta feigned no hypothesis: "This perpetual motion may appear paradoxical, perhaps inexplicable; but it is nonetheless true and real, and can be touched, as it were, with the hands" (*Opere*, I, 576; see also *Opere*, I, 489).

It appears that Volta possessed most of the ingredients of the pile by 1796, including even an anticipation of the outstanding key discovery, the constructive combination of the generating pairs. The delay in completing the invention may be explained by external circumstances. First, Volta's marriage, in 1794, to Teresa Peregrini, daughter of a government official in Como, quickly brought him a sizable family (three sons between 1795 and 1798) and many new demands upon his time. Second, during just these years, 1796 to 1800, Volta, like many of his colleagues, was distracted by the French invasion of Italy. In May 1796 he was chosen by the city of Como as one of a delegation to honor Napoleon, then fresh from driving the Austrians from the Milanese. Shortly thereafter he became an official of Como's new government (*Epistolario*, III, 291). But he was not comfortable in the position, which he resigned as soon as possible (*Epistolario*, III, 309–310); for although he did not, like Galvani, refuse to take an oath to the new Cisalpine Republic, he had a lingering loyalty, or rather gratitude, toward the Austrian regime, whose favor he had enjoyed. Moreover, the French authorities had not recommended themselves by allowing their soldiers to sack Pavia and to damage Volta's laboratory (*Epistolario*, III, 294). His coldness toward the French and open opposition to Jacobin colleagues, and also the accusation that he favored a proposal to move the university from Pavia to Milan, led to harassment that drove him from Pavia for some months. These opinions did him no harm when the Austrians returned in 1799 and shut up the university; for the victors only took away his job, and not—as in the case of Barletti, who had welcomed the French—his liberty. Thirteen months later the French were back. Napoleon immediately opened the university, and Volta, having recovered his professorship (*Epistolario*, IV, 8), resigned himself to citizenship in the revived Cisalpine Republic. Indeed, he proposed that he and a colleague,

L. Brugnatelli, professor of chemistry, go to Paris to express the gratitude of the university directly to the First Consul (*Epistolario*, IV, 16–17).

The trip, proposed in September 1800, was put off for a year because of war (*Epistolario*, IV, 24–25). It then turned into more than a mission to "cement an alliance of talent and science for the immortality of the two republics" (*Epistolario*, IV, 52–53); for in the interim Volta's letter to Banks had been published, and the chemical power of the pile discovered. The political mission became a triumphal march. Volta showed his experiments in Geneva, at the home of his friend Senebier; in Arcueil; in Paris, in the laboratories of Fourcroy, Seguin, Lamétherie, and above all, of Charles, where a special commission on galvanism of the Paris Academy met four times to see Volta's electricity; and at the Academy itself, where he performed at three sessions, each attended by Napoleon.[14] These demonstrations brought nothing new. They emphasized the electrostatic detection of the contact tension via a condensator and straw electroscope; used the old value of 1/60 degree of the latter (0.67 volt) as the tension of a single silver-zinc pair (*Opere*, II, 39–40, 50–61); and insisted on distinguishing between high tension/low current devices, like the standard electrical machine, and low tension/high current ones, like the pile, the crown of cups, and the torpedo (*Opere*, II, 72–83). When Volta concluded, Napoleon proposed the award of a gold medal; that, providentially, was also the recommendation of the commission on galvanism, which endorsed Volta's identification of galvanic and common electricity, and showed how to compute the tensions of various arrangements of disks and condensators (*Opere*, II, 113–115).

Napoleon continued to patronize Volta, giving him a pension and raising him to count and senator of the kingdom of Italy. In this there was more than politics. Volta's discoveries captured the imagination of Napoleon, who, to ensure continuance of similar inventions, authorized the Academy of Sciences to award a medal "for the best experiment made each year on the galvanic fluid" and a prize of 60,000 francs "to whoever by his experiments and discoveries makes a contribution to electricity and galvanism comparable to Franklin's and Volta's" (*Opere*, II, 122). But there was politics too. Just before leaving Paris in November 1801, Volta received what amounted to orders (*Epistolario*, IV, 88–89) to go to Lyons, to grace, and so endorse, a meeting at which selected Italian delegates were to be inspired to elect Napoleon president of the Cisalpine Republic. The republic soon disappeared into the kingdom of Italy, of which Napoleon became king. Volta played a small part in the kingdom as president of the Consiglio del Dipartimento del Lario (from 1803) and of the Comense Collegio Elettorale (1812). He retained sufficient confidence in French administration to cast his senatorial vote in 1814 in favor of offering the crown of Italy to Napoleon's stepson, Eugène de Beauharnais.

Napoleon was quite right in predicting that the pile presaged a new era in science. Its chemical power, employed in electrolyzing alkali salts, soon revealed the existence of sodium and potassium, a discovery for which Davy won the medal established by Napoleon. Studies of the properties of the current led to the laws of Oersted, Ohm, and Faraday, and to the beginnings of electrotechnology. In all of this Volta played no part. He was not much interested in the chemical effects of the pile, which he considered to be secondary phenomena (*Opere*, II, 37, 91). What effort he devoted to galvanism after his triumph in Paris went toward refuting the old doctrine of animal electricity, still very much alive. His last memoir on the subject, a lengthy review of his reasons for identifying galvanic and common electricity, was submitted under the name of a student in a prize competition announced in 1805 by the Società Italiana delle Scienze as follows: "Explain with clarity and dignity, and without offending anyone, the question of galvanism disputed by our worthy members Giovanni Aldini and Alessandro Volta" (*Opere*, II, 206). None of the papers submitted won the prize. Volta's memoir, which indeed contained little that was new, was printed in 1814 by his student and successor Pietro Configliachi (*Opere*, II, 205–307). After this competition Volta cut down his academic work. He sought and was refused retirement by the French ("a soldier," Napoleon told him, "should die on the field of honor" [*Epistolario*, IV, 455]); the Austrians, who returned in 1814, let him go in 1819. He spent his retirement chiefly in Como, where he died in 1827.

Volta received many honors besides those bestowed by Napoleon. The Royal Society of London elected him a member in 1791 and three years later gave him its highest prize, the Copley Medal, for setting right the Galvanists. He became a correspondent of the Berlin Academy of Sciences in 1786, and a foreign member of the Paris Academy in 1803. His fame also brought tangible rewards. In 1795 his university salary, 5,000 lire, was only double what he had during his last year at the Como Gymnasium. In 1805 he received an addi-

tional annuity of 4,000 lire from Napoleon, which survived the emperor's fall; and in 1809 he began to enjoy a senatorial salary of 24,000 lire. During the last twenty years of his life he had the income of a wealthy man.

As a scientist, Volta was conservative, yet alert to novelties; a strong theoretician, a "raisonneur sans pareil" (*Epistolario*, II, 268), as Lichtenberg said, yet an exceedingly careful and painstaking experimentalist, who constantly improved and varied his apparatus to exclude adventitious special cases. His uncommon imagination for effective instrumentation extended to anticipations of important practical devices such as the electrical telegraph (*Opere*, III, 194) and the incandescent gas lamp (*Opere*, VI, 150; VII, 155). He was no mathematician. His published work contains little mathematics beyond the rule of three and no evidence (according to Biot) that its author had a "mind fit for establishing rigorous theories"; while his lectures customarily skipped the mathematical parts of physics and omitted optics altogether. For these omissions Volta was bitterly attacked by Barletti in the early 1790's, no doubt partly for political reasons, and perhaps out of jealousy as well. The episode cost Volta much time and annoyance, and ended in an elaborate letter to the ministry in defense of his practice.[15] But despite his preference for the nonmathematical branches of physics, Volta fully understood the need for measurement: "Nothing good can be done in physics [he said] unless things are reduced to degrees and numbers" (*Opere*, I, 27). His mixture of precision in experiment and of indifference to—or ineptness at—general mathematical formulations also characterized several of his close colleagues, notably Saussure and Deluc. For the rest Volta went his own way, an autodidact seldom influenced by the work of others except at the beginning of an investigation.

NOTES

1. Quoted by Volpati, *Alessandro Volta*, 119; see *Epistolario*, V, 387–389.
2. For example, F. K. Achard, *Vorlesungen über die Experimentalphysik*, III (Berlin, 1791), 60.
3. *Epistolario*, I, 121; cf. Barletti, *Dubbi e pensieri* (Milan, 1776), 61–63, 103–119.
4. H. B. de Saussure, *Voyages dans les alpes*, II (Geneva–Neuchâtel, 1786), 212–219; see *Opere*, V, 154–155.
5. W. Thomson, *Reprint of Papers on Electricity and Magnetism*, 2nd ed. (London, 1884), 206–208, 227–229.
6. *Opere*, V, 55–56, 75–79; see Polvani, *Alessandro Volta*, 145.
7. *Opere*, VII, 228; Gliozzi, ed., *Opere scelte*, 248n.

8. The 2 vols. contain about 0.2 vol. O_2 and 1 vol. H_2: the total, therefore, falls by 0.6 vols., or 30 percent.
9. Grassi, "I lavori . . .," 562–563; Polvani, *Alessandro Volta*, 221–231; *Opere*, VII, 423–425.
10. Grassi, "I lavori . . .," 528–533.
11. The spasm occurring during discharge of a neighboring electrical machine brought nothing new; as Volta observed (*Opere*, I, 46–48, 175), it arose from the discharge of electricity induced in the specimen analogous to the return stroke in the case of lightning.
12. According to T. Cavallo in a letter to J. Lind, 23 Nov. 1792, British Museum Add. MS 22898, f. 25–26.
13. One degree of Volta's most sensitive straw electrometer equaled 1/10 of a degree of the Henley quadrant electrometer (*Opere*, V, 37, 52, 81; I, 486), 35 degrees of which marked about 13,350 volt (see note 6 above). Hence, one degree of the straw electrometer indicated about 40 volt. Volta later estimated the tension between zinc and silver at 1/60 degree straw (*Opere*, II, 39), or about 0.7 volt.
14. Z. Volta, *Alessandro Volta a Parigi*, 18–19, 41–47, 53–57, 96–97.
15. L. Magrini, "Notizie biografiche e scientifiche su A. Volta dai suoi autografi recentemente rinvenuti e inediti," in *Atti dell'Istituto lombardo di scienze e lettere*, 2 (1861), 254–283, on pp. 260–262, 272; C. Volpati; "Momenti d'amarezza sul camino della gloria," in *Voltiana*, 1 (1926), 437–447.

BIBLIOGRAPHY

I. Collected Works. The best bibliography is F. Scolari, *Alessandro Volta* (Rome, 1927), which incorporates F. Fossatti, "Bibliografia degli scritti editi di Alessandro Volta," in *Memorie dell'Istituto lombardo di scienze e lettere*, cl. sci. mat. nat., 18 (1900), 181–217. See also "Scritti del Volta o che lo riguardano stampati negli *Atti* del R. Istituto lombardo di scienze e lettere," in *Rendiconti dell'Istituto lombardo di scienze e lettere*, 60 (1927), 580–583. There are two collected works: *Collezione dell'opere*, V. Antinori, ed., 3 vols. in 5 (Florence, 1816), which is incomplete; and the magnificent national edition, *Le opere*, 7 vols. (Milan, 1918–1929), which is referred to in this article. There are also several anthologies: *Briefe über thierische Elektricität (1792)* and *Untersuchungen über den Galvanismus, 1796 bis 1800*, A. J. von Oettingen, ed., which are Ostwalds Klassiker der exakten Wissenschaften, nos. 114 and 118 (Leipzig, 1900); *L'opera di Alessandro Volta*, F. Massardi, ed. (Milan, 1927); and the excellent *Opere scelte*, M. Gliozzi, ed. (Turin, 1967).

Volta's correspondence is available in a national edition, *Epistolario*, 5 vols. (Bologna, 1949–1955), which, with the *Opere* and *Aggiunte alle opere e all'epistolario* (Bologna, 1966), supersedes all earlier editions. For the location of Volta's MSS, see *Opere*, I, x–xxi; Scolari, *Alessandro Volta*, pp. 171–462; and the unsigned "La nuova sede del Cartellario voltiano e la annessa biblioteca," in *Rendiconti dell'Istituto lombardo di scienze e lettere*, 60 (1927), 567–579. For Volta's instruments, most of which perished in a fire in 1899, see Società Storica Comense, *Raccolta voltiana* (Como, 1899) and *Il tempio voltiano in Como* (Como, 1939, 1973). Volta's library of printed books, unfortunately unannotated, is now at the Burndy Library, Norwalk, Connecticut.

II. Important Individual Works. For electrostatics, see *De vi attractiva ignis electrici* (Como, 1769), in *Opere*, III, 21–52; Italian trans. in *Opere scelte*, M. Gliozzi, ed., 49–90; *Novus ac simplicissimus electricorum tentaminum apparatus* (Como, 1771), in *Opere*, III, 53–76; "Lettera . . . al dott. Giuseppe Priestley [sull'elettroforo perpetuo]," in *Scelta d'opuscoli interessanti*, 9 (1775), 91–107; 10 (1775), 87–113, in *Opere*, III, 93–108; "Lettera . . . a Giuseppe Klinkosch . . . sulla teoria dell'elettricità vindice e sull'elettroforo perpetuo," *ibid.*, 20 (1776), 32–67, in *Opere*, III, 131–151; "[Lettera al Saussure] Osservazioni sulla capacità de' conduttori elettrici," in *Opuscoli scelti sulle scienze e sulle arti . . .*, 1 (1778), 273–280, in *Opere*, III, 201–229; "Del modo di render sensibilissima la più debole elettricità," in *Philosophical Transactions of the Royal Society*, 72 (1782), 237–280 (English trans. in Appendix, vii–xxxiii), also in *Opere*, III, 271–300; and "Mémoire sur les grands avantages d'une espèce d'isolement très imparfait," in *Journal de physique*, 22 (1783), 325–350; 23 (1783), 3–16, 81–99, and in *Opere*, III, 313–377.

Volta's works on meteorology include "[Lettere a G. C. Lichtenberg] Sulla meteorologia elettrica," in *Biblioteca fisica d'Europa*, 1 (1788), 73–137; 2 (1788), 103–142; 3 (1788), 79–122; 5 (1788), 79–134; 6 (1788), 137–147; 7 (1789), 81–111; 9 (1789), 129–148; 10 (1789), 39–69; 11 (1789), 33–53; 14 (1790), 61–112; in *Opere*, V, 29–228, 239–307; "Sopra la grandine," in *Istituto nazionale italiano, Memorie*, Cl. fis. mat., 1, pt. 2 (1806), 125–190, also in *Opere*, V, 421–462, and in *Journal de physique*, 69 (1809), 333–360.

On pneumatics, see *Lettere [al p. Campi] sull'aria infiammabile nativa delle paludi* (Milan, 1777), in *Opere*, VI, 17–102; "Lettere . . . al marchese Castelli sulla costruzione di un moschetto e di una pistola ad aria infiammabile," in *Scelta d'opuscoli interessanti*, 30 (1777), 86–109; 31 (1777), 3–24; and in *Opere*, VI, 123–150; "[Lettres à Priestley] Sur l'inflammation de l'air inflammable," in *Journal de physique*, 12 (1778), 365–373; 13 (1779), 278–303, and in *Opere*, VI, 173–215; "[Contributions to the Italian edition of P. G. Macquer, *Dizionario di chimica* (Pavia, 1783–1784)]," in *Opere*, VI, 349–436; VII, 3–105; "Descrizione dell'eudiometro al aria infiammabile," in *Annali di chimica e storia naturale*, 1 (1790), 171–213; 2 (1791), 161–186; 3 (1791), 36–45; and in *Opere*, VII, 173–213; and "Della uniforme dilazione dell'aria per ogni grado di calore," *ibid.*, 4 (1793), 227–294, in *Opere*, VII, 345–375.

Volta's works on animal electricity and galvanism include "[Memorie] sull'elettricità animale," in *Giornale fisico-medico*, 2 (1792), 146–187, 241–270; 3 (1792), 35–73; 4 (1793), 63–81; in *Opere*, I, 13–35, 41–74, 149–159; "Account of Some Discoveries Made by Mr. Galvani of Bologna . . . in Two Letters . . . to Mr. Tiberius Cavallo [in French]," in *Philosophical Transactions of the Royal Society*, **83**, pt. 1 (1793), 10–44, also in *Opere*, I, 171–197; "Nuove osservazioni sull'elettricità animale," in *Giornale fisico-medico*, **5** (1792), 192–196, and also in *Opere*, I, 143–147; "Nuova memoria sull'elettricità animale . . . in alcune lettere al sig. ab. Anton Maria Vassalli," in *Annali di chimica e storia naturale*, **5** (1794), 132–144; **6** (1794), 142–166; **11** (1796), 84–128, also in *Opere*, I, 261–281, 287–301; "[Lettere al prof. Gren] Sul galvanismo ossia sull'elettricità eccitata del contatto de' conduttori dissimili," *ibid.*, **13** (1797), 226–274; **14** (1797), 3–74, in *Opere*, I, 393–447; excerpted in German in *Neues Journal der Physik*, **3** (1797), 479–481; **4** (1797), 107–135; and in French in *Annales de chimie*, **23** (1797), 276–315; **29** (1798), 91–93; "[Lettere al cittadino Aldino] Intorno alla pretesa elettricità animale," in *Annali di chimica e storia naturale*, **16** (1798), 3–88, in *Opere*, I, 519–555; "On the Electricity Excited by the Mere Contact of Conducting Substances of Different Kinds [in French]," in *Philosophical Transactions of the Royal Society*, **90**, pt. 2 (1800), 403–431, in *Opere*, I, 563–582, and *Journal de physique*, 51 (1800), 344–354; English trans. in *Philosophical Magazine*, 7 (1800), 288–311; "Sur les phénomènes galvaniques," in *Journal de physique*, 53 (1801), 309–316, and in *Opere*, II, 35–43; "Memoria sull' identità del fluido elettrico col fluido galvanico," in *Ann. chim. stor. natur.*, 19 (1802), 38–88; 21 (1802), 163–211; in *Opere*, II, 45–84 (in part in *Annales de chemie*, 40 [1802], 225–256); *L'identità del fluido elettrico col cosi detto fluido galvanico* [1805] (Pavia, 1814), in *Opere*, II, 205–307.

III. BIOGRAPHY. The best general assessment of Volta is C. Volpati, *Alessandro Volta nella storia e nell'intimità* (Milan, 1927). Giovanni Cau, *Alessandro Volta*: *L'uomo, la scienza, il suo tempo* (Milan, 1927), is also useful. Older influential notices are J. B. Biot, in *Biographie universelle*, XLIV, 2nd ed., 78–81; F. Arago, *Oeuvres*, J. A. Barral, ed., 2nd ed., I (Paris, 1865), 187–240; M. Monti, *Biografia degli italiani illustri*, F. de Tipalso, ed., IX (Venice, 1844), 258–288; and C. Cantù, *Italiani illustri*, 3rd ed., III (Milan, 1879), 567–602.

On Volta's family and early years, see G. Gemelli, "Geneologia ed arma gentilizia della famiglia Volta," *Raccolta voltiana*, no. 6 (Como, 1899); and Z. Volta, *Alessandro Volta. Parte prima . . . Della giovinezza* (Milan, 1875), and the review *Voltiana* (1926–1927). For Volta's travels, see A. Verrechia, "Lichtenberg und Volta," in *Sudhoffs Archiv für Geschichte der Medizin und der Naturwissenschaften*, **51** (1967), 349–360; C. Volpati, "Amici e ammiratori di Alessandro Volta in Germania," in *Nuova rivista storica*, **11** (1927), 535–570; G. Bilancioni, "Alessandro Volta e Antonio Scarpa," in *Archeion*, **8** (1927), 351–363; M. Cermenati, *Alessandro Volta alpinista* (Turin, 1899); and Z. Volta, *Alessandro Volta a Parigi* (Milan, 1879). For politics, see G. Gallavresi, "Alessandro Volta e l'epopea napoleonica," in *Nuova antologia*, **334** (1927) 201–208; and

Z. Volta, "Alessandro Volta e l'università di Pavia dal 1778 al 1799," in *Archivio storico lombardo*, 24 (1899), 393–447. For Volta's honors, see S. Ambrosoli, "Le medaglie di Alessandro Volta," in *Raccolta voltiana*, no. 7 (Como, 1899); and F. Frigerio, "Saggio di iconografia voltiana," in *Como ad Alessandro Volta* (Como, [1945]), 143–156. There is additional bibliography in Scolari, *Alessandro Volta*, 30–44.

IV. WORK. By far the best general account of Volta's work is G. Polvani, *Alessandro Volta* (Pisa, 1942). On electricity, see also M. Gliozzi, *L'elettrologia fino al Volta*, 2 vols. (Naples, 1937), and "Consonanze e dissonanze tra l'elettrostatica di Cavendish e quella di Volta," in *Physis*, 11 (1969), 231–248; F. Massardi, "Sull'importanza dei concetti fondamentali . . . [in Volta's] *De vi attractiva*," in *Rendiconti dell'Istituto lombardo di scienze e lettere*, 59 (1926), 373–381, and "Concordanza di risultati e formule emergenti da manoscritti inediti del Volta con quelli ricavati della fisicomatematica," *ibid.*, 56 (1923), 293–308; and W. C. Walker, "The Detection and Estimation of Electric Charges in the Eighteenth Century," in *Annals of Science*, 1 (1936), 66–99.

On meteorology, see L. Volta, "Alessandro Volta e la meteorologia, specialmente elettrica," in *Rendiconti dell'Istituto lombardo di scienze e lettere*, 60 (1927), 471–482; and C. Volpati, "L'ultimo episodio della vita scientifica di Alessandro Volta (La questione della difesa contra la grandine)," in *Como ad Alessandro Volta* (Como, 1945), 113–142.

On pneumatics, see R. Watermann, "Eudiometrie," in *Technikgeschichte*, 35 (1968), 293–319; V. Broglia, "Alessandro Volta und die Chemie," in *Chemikerzeitung*, 90 (1966), 628–640; W. A. Osman, "Alessandro Volta and the Inflammable Air Eudiometer," in *Annals of Science*, 14 (1958), 215–242; C. Pirotti, "La pistola di Volta e il motore a scoppio," in *Nuovo cimento*, *4:10* (1927), cxxviii–cxxxv; and F. Grassi, "I lavori del Volta e del Gay-Lussac per l'azione del calore," in *Rendiconti dell' Istituto lombardo di scienze e lettere*, 60 (1927), 505–534, and "I lavori del Volta e del Dalton su le tensione dei vapori," *ibid.*, 535–566.

On animal electricity and galvanism, see P. Sue, *Histoire du galvanisme*, 2 vols. (Paris, 1802); G. Carradori, *Istoria del galvanismo in Italia* (Florence, 1817); A. Mauro, "The Role of the Voltaic Pile in the Galvani-Volta Controversy," in *Journal of the History of Medicine*, 24 (1969), 140–150; and T. M. Brown, "The Electrical Current in Early 19th Century French Physics," in *Historical Studies in the Physical Sciences*, 1 (1969), 61–103.

J. L. HEILBRON

VOLTAIRE, FRANÇOIS MARIE AROUET DE (*b.* Paris, France, 21 November 1694; *d.* Paris, 30 May 1778), *literature*.

Voltaire's importance for the history of science lies particularly in his having composed a famous popularization of Newton, *Éléments de la philosophie de Newton* (1738), while also collaborating with his companion and mistress, Émilie, marquise du Châtelet, on her translation of the *Principia* into French, and more generally in his having referred, with the lightness of touch that made him a serious critic of the human condition, his moral philosophy to what he took to be the Newtonian, and hence the correct, account of physical reality.

Born François Arouet, his father having been a lawyer of the middling bourgeoisie and a notary, he took the pen name Voltaire when setting up as a young poet and playwright prior to 1725, one who soon had a certain success in the world of letters and fashion with his *Oedipe* and *Henriade*. The footing there proved slippery when Voltaire exchanged man-about-town insults in January 1726 with a young nobleman whom he had unrealistically thought to be a friend, the chevalier de Rohan. Instead of being accorded the satisfaction of a duel, Voltaire was beaten in the street by lackeys and was then incarcerated as a nuisance in the Bastille. He was released on condition that he exile himself until the embarrassment that his temerity had caused a great and noble family should be forgotten.

It was thus in the wake of shocking injustice and humiliation that Voltaire was in London between 1726 and 1729. He was present for Newton's funeral in Westminster Abbey in 1727. The first mention of scientific matters in his published work occurs in the *Lettres philosophiques*, or *Lettres sur les Anglais*, which he drafted during his English period, although it did not appear until 1734. Among the many merits of life in England that it celebrates, to the disadvantage by comparison of life in France, was the dignity that Voltaire there found accorded in society to men of science and letters. A well-known passage contrasts the physical picture of Paris, where the world is full of Cartesian vortices, to that of London, where it is empty of all but Newtonian attraction. Voltaire discussed Bacon, Locke, Newton, and inoculation against smallpox approvingly in letters 11 through 17, after praising religious pluralism, commercial enterprise, and representative government, and before turning to the theater and literature.

Voltaire had thus already adopted Newtonianism in principle, and had read in and about science, before his association with Mme du Châtelet, which began in 1733. Both were also friendly with Maupertuis, who had verified what he said about

Newton in the *Lettres philosophiques*, and with Clairaut. Fearful of arrest again, Voltaire took up residence at Mme du Châtelet's château at Cirey near the border of Champagne and Lorraine in 1734. They lived there until her death in 1749, and it was there that he undertook intensive study and correspondence with experts preparatory to writing the *Éléments*, in which (he wrote to a friend) he proposed "to reduce this giant to the measure of the dwarfs who are my contemporaries." A frequent visitor to Cirey was Francesco Algarotti, the success of whose *Il Newtonianismo per le dame* (1737) is often said to have inspired Voltaire to write a more serious work.

Voltaire's title is accurate, whether designedly or no, in that the book is about the philosophy that he read out of (or into) Newton and is not a technical guide to the science, whether mathematics, mechanics, or optics. Part I handles the metaphysical and theological issues of the Leibniz-Clarke correspondence, part II the theory of light and colors, and part III gravity and cosmology. Even in the optical part, only four out of fourteen chapters discuss Newton's actual work. The rest of it consists of an overview of seventeenth-century optics in general, so presented as to make it appear that color perception supports the associationist psychology. It is not perfectly clear from the chapters on the *Opticks* itself that Voltaire had grasped the distinction in Newton's mind between the phenomena of refraction, which established the composite character of white light, and the production of colors in thin transparent media, which exhibited the interaction of light and matter and which were later called interference effects. In any case, it was the latter aspect that Voltaire emphasized, probably for the reason that it could more easily be discussed in connection with his favorite among Newton's principles, the principle of attraction.

The transition that Voltaire made from color to gravity would lead the reader to suppose that Newton had extended this cardinal principle from optics to cosmology and had thus come to explain the system of the world. Discussing the *Principia*, Voltaire did give a qualitative sketch of that last topic, which occupies its third book. Newton had himself advised readers that, in addition to Book III, the minimum requisites for comprehending the *Principia* consisted in a command of the definitions and laws of motion and the first three sections of Book I (motion in conic sections under the influence of central forces). Of that Voltaire gave his readers only a verbal summary of proposition 1.

There is no discussion of physical quantities and no statement of the laws of motion. In general, the technical level is indicated by a remark apropos of the *Opticks* which informs the reader that there is a constant proportion between the sines of the angles of incidence and refraction but dispenses him from an explanation of what a sine is, since that would surpass the mathematical demands to be placed on him.

Evidently Voltaire's book may be taken as an index to what a clever writer thought could be expected scientifically of the literate public. It cannot be supposed to have told technically proficient readers anything substantive about Newton's work. The point needs emphasis since something more positive is often attributed to Voltaire's transmission of Newtonian physics in works of general history, which usually credit him with having converted French opinion—whatever that may mean—from Cartesianism to Newtonianism, and also since Voltaire himself does seem to have entertained briefly the desire to make some small contribution to science. He and Mme du Châtelet installed a laboratory at Cirey and made experiments. In 1736 the Académie des Sciences set the problem of the nature of fire for the prize it proposed to award in 1738. Both partners entered memoirs in the competition, which was won by Euler. Voltaire also wrote a piece defending the Newtonian measure of force in the vis viva controversy (on which issue he disagreed with Mme du Châtelet), composed several essays of natural history, and published clarifications and corrections of the cosmological discussion in the *Éléments*. Errors had found their way into the first edition, he explained, because the Dutch printer had made changes in the text without his knowledge.

None of that made any significant difference, however, either to science or probably to Voltaire, who did not persist in these researches. What really mattered to him about science was the vantage point he thought it offered to intelligence in the battle that did count, that of fact against dogma and illusion, which he waged throughout his life. Scholarship has established that it was almost certainly in 1739 that he composed *Micromégas* (not published until 1752 in London). It was his first fully successful venture in the genre of the *conte philosophique*, the form that he brought to its highest state of perfection. The observations of the extraplanetary visitor from Sirius light-heartedly reduce man to his true proportions in the scheme of things. Voltaire wrote it when his head was full of

the information he had assembled for the *Éléments*. As for his masterpiece, *Candide*, there is nothing of science in that famous tale. But we need to appreciate the reason for Voltaire's admiration of Newton in order to take the full thrust of his scorn for Leibniz in the caricature of Dr. Pangloss. For Pangloss is the personification of mealy-mouthed dogma, denaturing every fact and justifying every illusion, however absurd, in the name of principles—"All is for the best in the best of possible worlds"—that will leave untroubled the beneficiaries of the systematic deceptions that rule in society.

Throughout part I of the *Éléments*, Leibniz is the obstacle to enlightenment in metaphysics, as Descartes had been in physics, and for similar reasons: both had presumed to project their doctrines upon God or nature in the guise of necessities. Not so Newton, who had generalized his laws from phenomena, confirmed them by experience, and restricted them within the scope of mathematical formulation. Nothing pleased Voltaire more than repeating how Newton had made no pretense of stating the cause of attraction and had confined himself to demonstrating its quantity. The modern reader who expects to encounter eighteenth-century skepticism in Voltaire may be surprised to find that the *Éléments* opens with the argument that Newton gave in the General Scholium of the *Principia* for the existence and dominion of God. In further chapters Voltaire developed Newton's view that space and time are attributes of God, who, all unconstrained by Leibniz' principle of sufficient reason, had been perfectly free to constitute things as he saw fit. Thus the Newtonian philosophy, in consequence of which it followed that God had accorded a portion of his infinite liberty to man in the form of free will. Now then, all this about God may very well have been tongue-in-cheek on Voltaire's part, but not the part about liberty. For what he really cared about was improving the possibility that an informed man may have to make reasoned choices in a world of events that are largely indifferent to his wishes. The enemy of such a liberty was dogma and never fact. "Droit au fait" was a favorite among his sayings, fortified by what he understood of science; and as for dogma, reinforced by prejudice and tradition and armed by authority, that was the infamy to be scotched in the injunction "Écrasez l'infâme!" yet more regularly repeated in his later, more political, more moral, and (in the highest sense) more journalistic years.

BIBLIOGRAPHY

I. ORIGINAL WORKS. *Éléments de la philosophie de Newton mis à la portée de tout le monde* (Amsterdam, 1738) was published in a revision in 1741 and in a 2nd ed. in 1745. The latter is the version included in vol. XXXI of the Kehl ed. of his works (1784–1789) and in most later collections. Other writings on "Physique" include (1) the letters and a "Défense" of Newtonianism (1739); (2) the "Essai sur la nature du feu et sur sa propagation" and "Doutes sur la mesure des forces mortices et sur leur nature" (1741); (3) an abstract of Mme du Châtelet's memoir on fire, "Mémoire sur un ouvrage de physique de Madame la Marquise du Châtelet" (1739), and a lengthy commentary on her book about Leibniz, "Exposition du livre des Institutions Physiques" (1740); and (4) writings on natural history: "Relation touchant un Maure blanc amené d'Afrique à Paris en 1744," "Dissertation . . . sur les changements arrivés dans notre globe, et sur les pétrifactions qu'on prétend en être les témoignages" (it was in this essay, sent to the Academy in Bologna in 1746, that Voltaire advanced the opinion that it was more probable that fossils found in the Alps had been dropped by travelers than that revolutionary changes have occurred in the order of nature), "Des singularités de la nature" (1768), and "Les colimaçons du Révérend Père l'Escarbotier . . ." (1768). Voltaire reprinted much of ch. 9, pt. III, of the *Éléments* in the article "Figure de la terre," in his *Questions sur l'Encyclopédie* (1770), taking the occasion to make several corrections. Three further fragments appear in the *Mélanges littéraires* of the Kehl ed., "A.M.***" (1739) and "Courte réponse aux longs discours d'un docteur allemand" (1740), both about Newtonianism, and finally, "Lettre sur la prétendue comète" (1773), the appearance of which was vulgarly supposed to herald the end of the world. The 1827 ed. of *Oeuvres complètes* includes these fugitive pieces in its second *Physique* volume (XLII). Theodore Besterman has edited *Voltaire's Correspondence*, 107 vols., Institut et Musée Voltaire (Geneva, 1953–1965). A convenient modern ed. of the *Leibniz-Clarke Correspondence* is that by H. G. Alexander (Manchester, 1956).

II. SECONDARY LITERATURE. The important work on Newtonianism in France, Pierre Brunet, *L'introduction des théories de Newton en France au XVIIIᵉ siècle avant 1738* (Paris, 1931), was never completed. Ira O. Wade, *The Intellectual Development of Voltaire* (Princeton, 1969), draws on the author's earlier, more specialized studies dealing with scientific themes in Voltaire's work. See, especially, Wade's ed. of *Micromégas* (Princeton, 1950), where the 1739 date of composition is convincingly argued. There is a valuable discussion of *Candide* in Peter Gay, *The Enlightenment: An Interpretation*, I (New York, 1966), 197–203. Robert Walters, "Voltaire and the Newtonian Universe," an unpublished dissertation (1954) in the Princeton Univ. library, is a study of the *Éléments*. See also Martin S. Staum, "Newton and Voltaire: Constructive Skeptics," in *Studies on Voltaire and the Eighteenth Century*, **62** (1968), 29–56; and two articles by Henry Guerlac, "Three 18th-Century Social Philosophers: Scientific Influences on Their Thought," in *Daedalus*, **88** (1958), 12–18; and "Where the Statue Stood: Divergent Loyalties to Newton in the 18th Century," in Earl Wasserman, ed., *Aspects of the 18th Century* (Baltimore, 1965), 317–334. The interpretation of the present article is developed more fully in an essay by the undersigned, "Science and the Literary Imagination: Voltaire and Goethe," in David Daiches and A. K. Thorlby, eds., *Literature of the Western World*, IV (London, 1975), 167–194.

CHARLES C. GILLISPIE

VOLTERRA, VITO (*b.* Ancona, Italy, 3 May 1860; *d.* Rome, Italy, 11 October 1940), *mathematics, natural philosophy.*

Volterra was the only child of Abramo Volterra, a cloth merchant, and his wife Angelica Almagià. His ancestors had lived in Bologna, whence at the beginning of the fifteenth century one of them had moved to Volterra, a small city in Tuscany—the origin of the family's present name. In 1459 this ancestor's descendants opened a bank in Florence. Volterras are remembered as fifteenth-century writers and travelers and as collectors of books and ancient codices. In the following centuries branches of the family lived in various Italian cities, including Ancona in the 1700's.

Volterra was two years old when his father died. He and his mother, left amost penniless, were taken into the home of her brother, Alfonso Almagià, an employee of the Banca Nazionale. Later they lived in Turin and in Florence. Volterra spent the greater part of his youth in Florence and considered himself almost a native of that city. He attended the Scuola Tecnica Dante Alighieri and the Istituto Tecnico Galileo Galilei, both of which had excellent teachers, including the physicist Antonio Roiti, who played an important part in Volterra's career.

Volterra was a very precocious child. At the age of eleven he began to study Bertrand's *Traité d'arithmétique* and Legendre's *Éléments de géométrie*. He formulated original problems and tried to solve them. At thirteen he worked on ballistic problems and, after reading Jules Verne's novel *From the Earth to the Moon*, tried to determine the trajectory of a gun's projectile in the combined gravitational field of the earth and the moon—a restricted version of the three-body problem. In his solution the time is partitioned into small intervals, for each of which the force is considered as a constant and the trajectory is given as a succession of small parabolic arcs. Almost forty years later, at the age of fifty-two, Volterra demonstrated this solution in a course of lectures given at the Sorbonne. The idea of studying a natural phenomenon by dividing into small intervals the time in which it occurs, and investigating the phenomenon in each such interval by considering the causes that produce it as invariable, was later applied by Volterra to many other kinds of problems, such as differential linear equations, theory of functionals, and linear substitutions.

Although Volterra was greatly interested in science, his family, which had little money, urged him to follow a commercial career. There followed a struggle between his natural inclination and practical necessity. The family appealed to a distant cousin, Edoardo Almagià, a civil engineer with a doctorate in mathematics, hoping that he would persuade the boy to interrupt his studies and devote himself to business. The cousin, however, who later became Volterra's father-in-law, was so impressed by his mathematical ability that he tried to persuade the family to let the boy pursue his scientific studies. Roiti, having learned that his most able student was being urged to become a bank clerk, immediately nominated him as assistant in the physics laboratory at the University of Florence, an unusual occurrence since Volterra had not enrolled at the university.

Volterra completed high school in 1878 and enrolled in the department of natural sciences at the University of Florence. Two years later he won the competition to become a resident student at the Scuola Normale Superiore in Pisa. At the University of Pisa he enrolled in the mathematics and physics courses given by Betti, Dini, and Riccardo Felici. At first he was very interested in Dini's work in analysis. In one of Volterra's early papers, published while he was still a student, he was the first to present examples of derivable functions the derivatives of which are not reconcilable with Riemann's point of view. This observation was used much later as a starting point for Lebesgue's research on this subject. Volterra was fascinated most by Betti's lectures, and under his influence he devoted his research to mechanics and mathematical physics.

In 1882 Volterra graduated with a doctorate in physics and was immediately appointed Betti's assistant. The following year, at the age of twenty-three, he won the competition for a professorship of mechanics at the University of Pisa. After Betti's death Volterra succeeded him in the chair of mathematical physics. In 1892 Volterra was appointed professor of mechanics at the University of Turin, and in 1900 he succeeded Eugenio Beltrami in the chair of mathematical physics at the University of Rome. In the same year he married Virginia Almagià, who for over forty years was his devoted companion.

In recognition of his scientific achievements, Volterra was made a senator of the kingdom of Italy in 1905. Although he was never attracted by politics, he spoke frequently in the Senate on important issues concerning university organization and problems. He was active in Italian political life during World War I and, later, in the struggle against Fascist oppression.

When World War I broke out, Volterra felt that Italy should join the Allies; and when Italy entered the war, Volterra, although he was fifty-five, enlisted as an officer in the army corps of engineers, joining its air branch. He perfected a new type of airship, studied the possibility of mounting guns in it, and was the first to fire a gun from an airship. He also experimented with airplanes. For these accomplishments he was mentioned in dispatches and decorated with the War Cross.

At the beginning of 1917 Volterra established the Italian Office of War Inventions and became its chairman. He made frequent trips to France and to Great Britain in the process of wartime scientific and technical collaboration among the Allies. He was the first to propose the use of helium as a substitute for hydrogen in airships.

In October 1922 Fascism came to power in Italy. Volterra was one of the few to understand, from the beginning, its threat to the country's democratic institutions. He was one of the principal signatories of the "Intellectuals' Declaration" against Fascism, an action he took while president of the Accademia dei Lincei. When the proposed "laws of national security" were discussed by the Italian Senate, a small group of opposition senators, headed by Volterra and Benedetto Croce, appeared—at great personal risk—at all the Senate's meetings and always voted against Mussolini. By 1930 the parliamentary government created by Cavour in the nineteenth century was abolished, and Volterra never again attended sessions of the Italian Senate.

In 1931, having refused to sign the oath of allegiance imposed upon professors by the Fascist government, Volterra was dismissed from the University of Rome; and in 1932, for the same reason, he was deprived of all his memberships in Italian scientific academies. In 1936, however, on the nomination of Pope Pius XI he was elected to the Pontifical Academy of Sciences.

After 1931 Volterra lectured in Paris at the Sorbonne, in Rumania, in Spain, in Belgium, in Czechoslovakia, and in Switzerland. He spent only short periods in Italy, mainly at his country house at Ariccia, in the Alban Hills south of Rome. From December 1938 he was afflicted by phlebitis, but his mind remained clear and he continued his passionate pursuit of science until his death.

Volterra's scientific work covers the period from 1881, when he published his first papers, to 1940 when his last paper was published in the *Acta* of the Pontifical Academy of Sciences. His most important contributions were in higher analysis, mathematical physics, celestial mechanics, the mathematical theory of elasticity, and mathematical biometrics. His major works in these fields included the foundation of the theory of functionals and the solution of the type of integral equations with variable limits that now bear his name, methods of integrating hyperbolic partial differential equations, the study of hereditary phenomena, optics of birefringent media, the motion of the earth's poles and elastic dislocations of multiconnected bodies, and, in his last years, placing the laws of biological fluctuations on mathematical bases and establishing principles of a demographic dynamics that present analogies to the dynamics of material systems.

Volterra received numerous honors, was a member of almost every major scientific academy and was awarded honorary doctorates by many universities. In 1921 he received an honorary knighthood from George V of England.

Scientific research did not, however, occupy all of Volterra's activity. He was an intimate friend of many well-known scientific, political, literary, and artistic men of his time. He has been compared to a typical man of the Italian Renaissance for the variety of his interests and knowledge, his great scientific curiosity, and his sensitivity to art, literature, and music.

BIBLIOGRAPHY

I. ORIGINAL WORKS. Volterra's works were collected as *Opere matematiche. Memorie e note*, 5 vols. (Rome, 1954–1962).

His writings include *Trois leçons sur quelques progrès récents de la physique mathématique* (Worcester, Mass., 1912), also in *Lectures Delivered . . . by V. Volterra, E. Rutherford, R. W. Wood, C. Barus* (Worcester, Mass., 1912), and translated into German (Leipzig, 1914); *Leçons sur les équations intégrales et les équations intégro-différentielles*, M. Tomassetti and F. S. Zarlatti, eds. (Paris, 1913); *Leçons sur les fonctions de lignes*, collected and edited by Joseph Pérès (Paris, 1913); "Henri Poincaré: L'oeuvre mathématique," in *Revue du mois*, **15** (1913), 129–154; *Saggi scientifici* (Bologna, 1920); and *Leçons sur la composition et les fonctions permutables* (Paris, 1924), written with J. Pérès.

Additional works are *Theory of Functionals and of Integral and Integro-Differential Equations*, Luigi Fantapié, ed., M. Long, trans. (London–Glasgow, 1930), repr. with a preface by Griffith C. Evans and an almost complete bibliography of Volterra's works and a biography by Sir Edmund Whittaker (New York, 1959); *Leçons sur la théorie mathématique de la lutte pour la vie*, Marcel Brelot, ed. (Paris, 1931); *Les associations biologiques au point de vue mathématique* (Paris, 1935), written with U. D'Ancona; *Théorie générale des fonctionnelles* (Paris, 1936), written with J. Pérès; and *Sur les distorsions des corps élastiques (théorie et applications)* (Paris, 1960), written with E. Volterra, preface by J. Pérès.

II. SECONDARY LITERATURE. Biographies of Volterra and descriptions of his scientific work were published immediately after his death in 1940. A year later Sir Edmund Whittaker published a biography in *Obituary Notices of Fellows of the Royal Society of London*, **3** (1941), 691–729, with a bibliography; an abridged version appeared in *Journal of the London Mathematical Society*, **16** (1941), 131–139.

Other biographies and commemorations of Volterra, listed chronologically, include *Enciclopedia italiana di scienze, lettere ed arti*, XXXV (Rome, 1938), 582–583; Émile Picard, in *Comptes rendus . . . de l'Académie des sciences*, **211** (1940), 309–312; S. Mandelbrojt, in *Yearbook. American Philosophical Society* (1940), 448–451; D'Arcy W. Thompson and Sir Sydney Chapman, in *Nature*, **147** (22 Mar. 1941), 349–350; C. Somigliana, in *Acta Pontificiae Accademiae scientiarum*, **6** (1942), 57–86; C. Somigliana, in *Rendiconti del Seminario matematico e fisico di Milano*, **17** (1946), 3–61, with bibliobraphy; Guido Castelnuovo and Carlo Somigliana, "Vito Volterra e la sua opera scientifica," in *Atti dell' Accademia nazionale dei Lincei* (1947), session of 17 Oct.; and J. Pérès, in *Ricerca scientifica*, **18** (1948), 1–9.

See also *Enciclopedia italiana di scienze, lettere ed arti, seconda appendice 1938–1948* (Rome, 1949); and G. Armellini, *Discorso pronunciato . . . per le onoranze a V. Volterra . . .* (Ancona, 1951); Guido Corbellini, *Vito Volterra nel centenario della sua nascita* (Rome, 1960); Accademia Nazionale dei Lincei, *Vito Volterra nel I centenario della nascita* (Rome, 1961); and Francesco G. Tricomi, "Matematici italiani del primo secolo dello stato unitario," *Memorie dell'Accademia delle scienze di Torino*, Cl. di scienze fisiche, matematiche e naturali, 4th ser., no. 1 (1962), 118.

E. VOLTERRA

VOLTZ, PHILIPPE LOUIS (*b.* Strasbourg, France, 15 August 1785; *d.* Paris, France, 30 March 1840), *geology*.

Voltz came from a poor family, and his parents had to make great sacrifices for his education. He entered the École Polytechnique in 1803 and the École des Mines in 1806. After serving as a mining engineer in the Belgian provinces, he held the post of chief engineer of the Strasbourg mineralogical district from August 1814 until 1836. In this capacity he advised industrialists in eastern France, made an inventory of the mineral resources of Alsace, and began the surveys needed to establish a geological map of the province. Only the map of the southern region (the Haut-Rhin department) was completed, however; it was published in 1833. Greatly interested in minerals and fossils, Voltz devoted much time to the development of Strasbourg's museum of natural history, which, as a result, soon possessed one of France's largest collections concerning stratigraphic paleontology.

Voltz's publications on the stratigraphy of eastern France, particularly on the Triassic, display his remarkable gifts as an observer. "Aperçu de la topographie minéralogique de l'Alsace" (1828), which appeared simultaneously in German, treats of the stratigraphy and paleontology of the province, as well as of its mineralogy. Paleontology increasingly attracted Voltz, who published several studies of fossil mollusks, notably belemnites and Nerinea.

Because of his fame as a paleontologist, new fossil forms were frequently named after Voltz: for example, Adolphe Brongniart's genus *Voltzia*, a gymnosperm abundant in the Triassic.

Voltz was fluent in German and encouraged contact between scientists on both sides of the Rhine. In December 1828 he and some of his friends founded the Société d'Histoire Naturelle de Strasbourg. He was also a member of the Geological Society of London and a corresponding member of the Société Industrielle de Mulhouse.

In 1830 Voltz began to give a free course of lectures in geognosy at the Strasbourg Faculty of Sciences. Among his students were Jules Thurmann and Amanz Gressly, the latter of whom apparently took up the notion of facies that Voltz had introduced into geology in 1828. Voltz organized and presided at the special meeting of the French Geological Society held at Strasbourg and in the Vosges 6–14 September 1834, which was attended by many French and foreign geologists.

Voltz's activities went far beyond geology. He was a municipal councillor of Strasbourg and *counseiller général* of the Bas-Rhin department. While holding these offices he became concerned about the conditions of the poor, and he seems to have been an enthusiastic supporter of the July Revolution of 1830.

Named inspector-general of mines in December 1836, Voltz moved to Paris, where, besides handling his administrative duties, he enriched the paleontological collection of the École des Mines. His health began to deteriorate, however, and he died four years later.

BIBLIOGRAPHY

I. ORIGINAL WORKS. Voltz's writings include "Aperçu de la topographie minéralogique de l'Alsace," in J. F. Aufschlager, ed., *Nouvelle description historique et topographique de l'Alsace* (Strasbourg, 1828), 1–66; "Observations sur les bélemnites," in *Mémoires de la Société du Muséum d'histoire naturelle de Strasbourg*, **1**, no. 1 (1830), 1–70; "Carte géologique du département du Haut-Rhin," *Statistiques générales du département du Haut-Rhin*, no. 46 (1833); and "Notice sur le grès bigarré de la grande carrière de Soultz-les-Bains," in *Mémoires de la Société du Muséum d'histoire naturelle de Strasbourg*, **2**, no. 3 (1836), 1–9.

II. SECONDARY LITERATURE. See G. Dubois, "L'enseignement de la géologie à l'Université de Strasbourg avant 1870," in *Revue d'Alsace*, **85**, no. 552 (1938), 1–60; and W. Fischer, *Gesteins- und Lagerstättenbildung im Wandel der wissenschaftlichen Anschauung* (Stuttgart, 1961), 102, 217.

JEAN-CLAUDE GALL

VON NEUMANN, JOHANN (or **JOHN**) (*b.* Budapest, Hungary, 28 December 1903; *d.* Washington, D.C., 8 February 1957), *mathematics, mathematical physics*.

Von Neumann, the eldest of three sons of Max von Neumann, a well-to-do Jewish banker, was privately educated until he entered the Gymnasium in 1914. His unusual mathematical abilities soon came to the attention of his teachers, who pointed out to his father that teaching him conventional school mathematics would be a waste of time; he was therefore tutored in mathematics under the guidance of university professors, and by the age of nineteen he was already recognized as a professional mathematician and had published his first paper. Von Neumann was *Privatdozent* at Berlin from 1927 to 1929 and at Hamburg in 1929–1930, then went to Princeton University for three years; in 1933 he was invited to join the newly opened Institute for Advanced Study, of which he was the youngest permanent member at that time. At the outbreak of World War II, von Neumann was called upon to participate in various scientific projects related to the war effort; in particular, from 1943 he was a consultant on the construction of the atomic bomb at Los Alamos. After the war he retained his membership in numerous government boards and committees, and in 1954 he became a member of the Atomic Energy Commission. His health began to fail in 1955, and he died of cancer two years later.

Von Neumann may have been the last representative of a once-flourishing and numerous group, the great mathematicians who were equally at home in pure and applied mathematics and who throughout their careers maintained a steady production in both directions. Pure and applied mathematics have now become so vast and complex that mastering both seems beyond human capabilities. In von Neumann's generation his ability to absorb and digest an enormous amount of extremely diverse material in a short time was exceptional; and in a profession where quick minds are somewhat commonplace, his amazing rapidity was proverbial. There is hardly a single important part of the mathematics of the 1930's with which he had not at least a passing acquaintance, and the same is probably true of theoretical physics.

Despite his encyclopedic background, von Neumann's work in pure mathematics had a definitely smaller range than that of Poincaré or Hilbert, or even of H. Weyl. His genius lay in analysis and combinatorics, the latter being understood in a very wide sense, including an uncommon ability to organize and axiomatize complex situations that a priori do not seem amenable to mathematical treatment, as in quantum mechanics and the theory of games. As an analyst von Neumann does not belong to the classical school represented by the French and English mathematicians of the early 1900's but, rather, to the tradition of Hilbert, Weyl, and F. Riesz, in which analysis, while being as "hard" as any classical theory, is based on extensive foundations of linear algebra and general topology; however, he never did significant work in number theory, algebraic topology, algebraic geometry, or differential geometry. It is only in

comparison with the greatest mathematical geniuses of history that von Neumann's scope in pure mathematics may appear somewhat restricted; it was far beyond the range of most of his contemporaries, and his extraordinary work in applied mathematics, in which he certainly equals Gauss, Cauchy, or Poincaré, more than compensates for its limitations.

Pure Mathematics. Von Neumann's work in pure mathematics was accomplished between 1925 and 1940, which might be called his *Sturm und Drang* period, when he seemed to be advancing at a breathless speed on all fronts of logic and analysis at once, not to speak of mathematical physics. This work, omitting a few minor papers, can be classified under five main topics.

Logic and Set Theory. Von Neumann's interest in set theory arose very early: in his second paper (1923) he gave an elegant new definition of ordinal numbers, and in the third (1925) he introduced an axiomatic system for set theory quite different from the one proposed by Zermelo and Fraenkel (it was later adopted by Gödel in his research on the continuum hypothesis). In the late 1920's von Neumann also participated in the Hilbert program of metamathematics and published a few papers on proofs of noncontradiction for parts of arithmetic, before Gödel shattered the hopes for a better result.

Measure Theory. Although it was not in the center of von Neumann's preoccupation, he made several valuable contributions to measure theory. His knowledge of group theory enabled him to "explain" the Hausdorff-Banach-Tarski "paradox," in which two balls of different radii in $\mathbf{R}^n$ ($n \geq 3$) are decomposed into a finite number of (nonmeasurable) subsets that are pairwise congruent (such decompositions cannot exist for $n = 1$ or $n = 2$); he showed that $n = 1$ or $n = 2$ is impossible because the orthogonal group in three or more variables contains free non-Abelian groups, whereas it does not for $n \leq 2$.

Another highly ingenious paper established the existence of an algebra of bounded measurable functions on the real line that forms a complete system of representatives of the classes of almost-everywhere-equal measurable bounded functions (each class contains one, and only one, function of the algebra). This theorem, later generalized to arbitrary measure spaces by Dorothy Maharam, holds the key to the "disintegration" process of measures (corresponding to the classical notion of "conditional probability"). It is a curious coincidence that in "Operator Methods in Classical

Mechanics" von Neumann was the first to prove, by a completely different method, the existence of such disintegrations for fairly general types of measures.

On the borderline between this group of papers and the next lies von Neumann's basic work on Haar's measure, which he proved to be unique up to a constant factor; the first proof was valid only for compact groups and used his direct definition of the "mean" of a continuous function over such a group. The extension of that idea to more general groups was the starting point of his subsequent papers, some written in collaboration with Solomon Bochner, on almost-periodic functions on groups.

Lie Groups. One of the highlights of von Neumann's career was his 1933 paper solving Hilbert's "fifth problem" for compact groups, proving that such a group admits a Lie group structure once it is locally homeomorphic with Euclidean space. He had discovered the basic idea behind that paper six years earlier: the fact that closed subgroups of the general linear group are in fact Lie groups. The method of proof of that result was shown a little later by E. Cartan to apply as well to closed subgroups of arbitrary Lie groups.

Spectral Theory of Operators in Hilbert Space. This topic is by far the dominant theme in von Neumann's work. For twenty years he was the undisputed master in this area, which contains what is now considered his most profound and most original creation, the theory of rings of operators. The first papers (1927) in which Hilbert space theory appears are those on the foundations of quantum mechanics (see below). These investigations later led von Neumann to a systematic study of unbounded hermitian operators, which previously had been considered only in a few special cases by Weyl and T. Carleman. His papers on unbounded hermitian operators have not been improved upon since their publication, yet within a few years he realized that the traditional idea of representing an operator by an infinite matrix was totally inadequate, and discovered the topological devices that were to replace it: the use of the graph of an unbounded operator and the extension to such an operator of the classical "Cayley transform," which reduced the structure of a self-adjoint operator to that of a unitary operator (known since Hilbert). At the same time this work led him to discover the defects of a general, densely defined hermitian operator, which later were seen to correspond to the "boundary conditions" for opera-

tors stemming from differential and partial differential equations.

The same group of papers includes another famous result from von Neumann's early years, his proof in 1932 of the ergodic theorem in its "L^2 formulation" given by B. O. Koopman a few months earlier. With G. D. Birkhoff's almost simultaneous proof of the sharper "almost everywhere" formulation of the theorem, von Neumann's results were to form the starting point of all subsequent developments in ergodic theory.

Rings of Operators. Most of von Neumann's results on unbounded operators in Hilbert space were independently discovered a little later by M. H. Stone. But von Neumann's ideas on rings of operators broke entirely new ground. He was well acquainted with the noncommutative algebra beautifully developed by Emmy Noether and E. Artin in the 1920's, and he realized how these concepts simplified and illuminated the theory of matrices. This probably provided the motivation for extending such concepts to algebras consisting of (bounded) operators in a given separable Hilbert space, to which he gave the vague name "rings of operators" and which are now known as "von Neumann algebras." He introduced their theory in the same year as his first paper on unbounded operators, and from the beginning he had the insight to select the two essential features that would allow him further progress: the algebra must be self-adjoint (that is, for any operator in the algebra, its adjoint must also belong to the algebra) and closed under the strong topology of operators and not merely in the finer topology of the norm.

Von Neumann's first result was the "double commutant theorem," which states that the von Neumann algebra generated by a self-adjoint family $\mathscr{F}$ of operators is the commutant of the commutant of $\mathscr{F}$, a generalization of a similar result obtained by I. Schur for semisimple algebras of finite dimension that was to become one of the main tools in his later work. After elucidating the relatively easy study of commutative algebras, von Neumann embarked in 1936, with the partial collaboration of F. J. Murray, on the general study of the noncommutative case. The six major papers in which they developed that theory between 1936 and 1940 certainly rank among the masterpieces of analysis in the twentieth century. They immediately realized that among the von Neumann algebras, the "factors" (those with the center reduced to the scalars) held the key to the structure of the general von Neumann algebras; indeed, in his last major paper on the subject (published in 1949 but dating from around 1940), von Neumann showed how a process of "direct integration" (the analogue of the "direct sum" of the finite dimensional theory) explicitly gave all von Neumann algebras from factors as "building blocks."

The evidence from classical study of noncommutative algebras seemed to lead to the conjecture that all factors would be isomorphic to the algebra $\mathscr{B}(H)$ of all bounded operators in a Hilbert space H (of finite or separable dimension). Murray and von Neumann therefore startled the mathematical world when they showed that the situation was far more complicated. As in the classical theory, their main tool consisted of the self-adjoint idempotents in the algebra, which are simply orthogonal projections on closed subspaces of the Hilbert space; the novelty was that, in contrast with the classical case (or the case of $\mathscr{B}[H]$), minimal idempotents may fail to exist in the algebra, which implies that all idempotents are orthogonal projections on infinite-dimensional subspaces. Nevertheless, they may be *compared*, the projection on a subspace E being considered as "smaller" than one on a subspace F when the algebra contains a partial isometry V sending E onto a subspace of F. This is only a "preorder"; but when one considers the corresponding order relation (between equivalence classes), it turns out that in a factor this is a total order relation that may be described by a "dimension function" that attaches to each equivalence class of projections a real number ≥ 0 or $+\infty$. Murray and von Neumann showed that after proper normalization the range of the dimension could be one of five possibilities: $\{1, 2, \cdots, n\}$ (type I_n, the classical algebras of matrices), $\{1, 2, \cdots, +\infty\}$ (type I_∞, corresponding to the algebras $\mathscr{B}[H]$), the whole interval $[0, 1]$ in the real line (type II_1), the whole interval $[0, +\infty]$ in the extended real line (type II_∞), and the two-element set $\{0, +\infty\}$ (type III).

It may be said that the algebraic structure of a factor imposes on the set of corresponding subspaces of H (images of H by the projections belonging to the factor) an order structure similar to that of the subspaces of a usual projective space, but with completely new possibilities regarding the "dimension" attached to these subspaces. Intrigued by this geometric interpretation of his results, von Neumann developed it in a series of papers on "continuous geometries" and their algebraic satellites, the "regular rings" (which are to continuous geometries as rings of matrices are to vector spaces). This classification, which required

great technical skill in the handling of the spectral theory of operators, immediately led to the question of existence for the new "factors." Murray and von Neumann devoted many of their papers to this question; and they were able to exhibit factors of types II_1, II_∞, and III by using ingenious constructions from ergodic theory (at a time when the subject of actions of groups on measure spaces was still in its infancy) and algebras generated by convolution operators. They went even further and initiated the study of isomorphisms between factors, succeeding, in particular, in obtaining two nonisomorphic factors of type II_1; only very recently has it been proved that there are uncountably many isomorphism classes for factors of types II_1 and III.

Applied Mathematics. *Mathematical Physics.* Von Neumann's most famous work in theoretical physics is his axiomatization of quantum mechanics. When he began work in that field in 1927, the methods used by its founders were hard to formulate in precise mathematical terms: "operators" on "functions" were handled without much consideration of their domain of definition or their topological properties; and it was blithely assumed that such "operators," when self-adjoint, could always be "diagonalized" (as in the finite dimensional case), at the expense of introducing "Dirac functions" as "eigenvectors." Von Neumann showed that mathematical rigor could be restored by taking as basic axioms the assumptions that the states of a physical system were points of a Hilbert space and that the measurable quantities were Hermitian (generally unbounded) operators densely defined in that space. This formalism, the practical use of which became available after von Neumann had developed the spectral theory of unbounded Hermitian operators (1929), has survived subsequent developments of quantum mechanics and is still the basis of nonrelativistic quantum theory; with the introduction of the theory of distributions, it has even become possible to interpret its results in a way similar to Dirac's original intuition.

After 1927 von Neumann also devoted much effort to more specific problems of quantum mechanics, such as the problem of measurement and the foundation of quantum statistics and quantum thermodynamics, proving in particular an ergodic theorem for quantum systems. All this work was developed and expanded in *Mathematische Grundlagen der Quantenmechanik* (1932), in which he also discussed the much-debated question of "causality" versus "indeterminacy" and concluded that no introduction of "hidden parameters" could keep the basic structure of quantum theory and restore "causality."

Quantum mechanics was not the only area of theoretical physics in which von Neumann was active. With Subrahmanyan Chandrasekhar he published two papers on the statistics of the fluctuating gravitational field generated by randomly distributed stars. After he started work on the Manhattan project, leading to atomic weapons, he became interested in the theory of shock waves and wrote many reports on their theoretical and computational aspects.

Numerical Analysis and Computers. Von Neumann's uncommon grasp of applied mathematics, treated as a whole without divorcing theory from experimental realization, was nowhere more apparent than in his work on computers. He became interested in numerical computations in connection with the need for quick estimates and approximate results that developed with the technology used for the war effort—particularly the complex problems of hydrodynamics—and the completely new problems presented by the harnessing of nuclear energy, for which no ready-made theoretical solutions were available. Dissatisfied with the computing machines available immediately after the war, he was led to examine from its foundations the optimal method that such machines should follow, and he introduced new procedures in their logical organization, the "codes" by which a fixed system of wiring could solve a great variety of problems. Von Neumann devised various methods of programming a computer, particularly for finding eigenvalues and inverses of matrices, extrema of functions of several variables, and production of random numbers. Although he never lost sight of the theoretical questions involved (as can be seen in his remarkably original papers with Herman Goldstine, on the limitation of the errors in the numerical inversion of a matrix of large order), he also wanted to have a direct acquaintance with the engineering problems that had to be faced, and supervised the construction of a computer at the Institute for Advanced Study; many fundamental devices in the present machines bear the imprint of his ideas.

In the last years of his life, von Neumann broadened his views to the general theory of automata, in a kind of synthesis of his early interest in logic and his later work on computers. With his characteristic boldness and scope of vision, he did not hesitate to attack two of the most complex questions in the field: how to design reliable machines using unreliable components, and the construction of self-reproducing machines. As usual he brought remarkably new ideas in the approach to solutions of these problems and must be considered one of the founders of a flourishing new mathematical discipline.

Theory of Games. The role as founder is even more obvious for the theory of games, which von Neumann, in a 1926 paper, conjured—so to speak—out of nowhere. To give a quantitative mathematical model for games of chance such as poker or bridge might have seemed a priori impossible, since such games involve free choices by the players at each move, constantly reacting on each other. Yet von Neumann did precisely that, by introducing the general concept of "strategy" (qualitatively considered a few years earlier by E. Borel) and by constructing a model that made this concept amenable to mathematical analysis. That this model was well adapted to the problem was shown conclusively by von Neumann in the same paper, with the proof of the famous minimax theorem: for a game with two players in a normalized form, it asserts the existence of a unique numerical value, representing a gain for one player and a loss for the other, such that each can achieve at least this favorable expectation from his own point of view by using a "strategy" of his own choosing; such strategies for the two players are termed optimal strategies, and the unique numerical value, the minimax value of the game.

This was the starting point for far-reaching generalizations, including applications to economics, a topic in which von Neumann became interested as early as 1937 and that he developed in his major treatise written with O. Morgenstern, *Theory of Games and Economic Behavior* (1944). These theories have developed into a full-fledged mathematical discipline, attracting many researchers and branching into several types of applications to the social sciences.

BIBLIOGRAPHY

Von Neumann's works were brought together as *Collected Works of John Von Neumann*, A. H. Taub, ed., 6 vols. (New York, 1961). His books include *Mathematische Grundlagen der Quantenmechanik* (Berlin, 1932); and *Theory of Games and Economic Behavior* (Princeton, 1944), written with O. Morgenstern. A memorial volume is "John von Neumann, 1903–1957," which is *Bulletin of the American Mathematical Society*, **64**, no. 654 (May 1958).

J. Dieudonné

VORONIN, MIKHAIL STEPANOVICH (*b.* St. Petersburg, Russia [now Leningrad, U.S.S.R.], 2 August 1838; *d.* St. Petersburg, 4 March 1903), *mycology, phytopathology.*

Voronin entered the natural sciences section of the Faculty of Physics and Mathematics of St. Petersburg University in 1854 and specialized in botany, which he studied under L. S. Tsenkovsky, who influenced him to investigate the lower plants. After graduating in 1858 with a candidate's degree, Voronin worked in Holle's laboratory in Heidelberg and in de Bary's laboratory at Freiburg. His acquaintance with de Bary developed into a close friendship and collaboration, and they later published the journal *Beiträge zur Morphologie und Physiologie der Pilze.* It was in de Bary's laboratory that Voronin did his first botanical work, on the anatomy of the stalk of *Calycanthus.* In 1860 he moved to the Antibes laboratories of the French algologists G. A. Thuret and Édouard Bornet. The results of his research there on the marine plants *Acetabularia* and *Espera* were presented in his master's thesis, which he defended at St. Petersburg University in 1861.

Voronin continued his studies of marine plants after returning to Russia but devoted his later work primarily to fungi. Because he was independently wealthy, he was not obliged to seek paid employment. He was *Privatdozent* in mycology at St. Petersburg University in 1869–1870 and taught general cytology and mycology at the St. Petersburg University for Women from 1873 to 1875 without fee. He used his wealth to organize and support scientific institutions and to publish scientific works.

Novorossysk University in Odessa awarded Voronin the doctorate in 1874 without his having defended a dissertation; and in 1898 he was elected a member of the Russian Academy of Sciences, of which he subsequently headed the section of cryptogamous plants. Although he conducted his research in a modest home laboratory, using only the simplest equipment (a Hartnack microscope, a razor, and needles), his results were included in standard Russian and foreign textbooks of botany and mycology. His work was concerned mainly with the lower plants: he discovered and studied their cycle of development and described many that are biologically, as well as botanically, important. His most important research in algology dealt with *Botrydium* and *Chromophyton.*

It was for his mycological research, however, which was of great practical and theoretical importance, that Voronin acquired an international reputation. He discovered and studied the causal organisms of clubroot, sunflower rust, the mold on apples, and ergotism. The cause of clubroot, which ravaged huge areas and destroyed the harvest, had previously been ascribed to insects. Voronin's study (1874–1877) led him to determine that the causal organism was a slime mold, which he named *Plasmodiophora brassicae*; and he proposed concrete measures for combating the disease. His earlier study of the life history of sunflower rust (1869–1875) resulted in the discovery that it was caused by the fungus *Puccinia helianthi.* While investigating "drunken" rye bread from the southern Ussuri region (1890), which induced headache, vomiting, and vertigo, Voronin isolated fifteen fungi, identifying four as those that cause ergotism. It was for these and related discoveries that many consider Voronin the founder of phytopathology in Russia.

Voronin's most important theoretical works were in mycology, to which he contributed knowledge of a new form of basidiomycete, *Exobasidium vaccinii*, which has no fruiting body; and he provided the basis for classifying the smut fungi (Ustilagineae) according to the germination of the chlamydospores. The outstanding results of his research on *Sclerotinia*, which attacks bilberry plants, was his discovery of the parasite of the ascomycete *Sclerotinia heteroica*, which grows successively on two plants; the parasite had previously been considered only a rust fungus.

Voronin also studied the slime mold *Ceratium* (similar in form to higher fungi) and the ascomycetes in the development of ascous fruiting bodies. While examining the latter, he discovered a peculiar structure of the female sex organ (an archicarp) in the form of a thick curved hypha, now known as Voronin's hypha. He also investigated *Chytridium* (Archimycetes) and *Mucorales* (Mucoraceae). With de Bary he established the genus *Synchytrium*, and in 1867 he described the cycle of development of *Synchytrium mercurialis.* In the last year of his life Voronin investigated the development and the sexual and asexual reproduction of three species of the aquatic fungus *Monoblepharis*, which he discovered in Finland.

BIBLIOGRAPHY

I. ORIGINAL WORKS. Many of Voronin's writings were brought together in *Izbrannye proizvedenia* ("Selected Works"; Moscow, 1961), which includes a bibliography of his works (271–274). They include *Mikologicheskie issledovania* ("Mycological Research";

St. Petersburg, 1869); *Issledovania nad razvitiem rzhav-chinnogo gribka Puccinia helianthi, prichinyayushchego bolezn podsolnechnika* ("Research on the Development of the Rust Fungus *Puccinia helianthi*, Which Causes the Sunflower Disease"; St. Petersburg, 1871); *Plasmodiophora brassicae. Organizm, prichinyayushchy kapustnym rasteniam bolezn, izvestnuyu pod nazvaniem kily* (". . . Organism Causing the Disease of Cabbage Known as 'Kila' [Clubroot]"; St. Petersburg, 1877); "O 'pyanom khlebe' v Yuzhno-Ussuryskom krae" ("On 'Drunken Bread' in the Southern Ussuri Region"), in *VIII sezd russkikh estestvoispytatelei v Peterburge* ("VIII Congress of Russian Natural Scientists in St. Petersburg"; St. Petersburg, 1890), 13–21; and "Sclerotinia heteroica," in *Trudy Imperatorskago S.-Peterburgskago obshchestva estestvoispytatelei*, Botany sec., **25** (1895), 84–91.

II. SECONDARY LITERATURE. See I. P. Borodin, "Pamyati nezabvennogo M. S. Voronina" ("Recollections of the Unforgettable M. S. Voronin"), in *Trudy Botanicheskogo sada. Yurevskogo universiteta*, **4**, no. 4 (1903), 286–292; M. S. Dunin, "M. S. Voronin–klassik mikologii i fitopatologii" ("M. S. Voronin–a Classic of Mycology and Phytopathology"), in Voronin's *Izbrannye proizvedenia* (see above), 3–16; which also includes a list of secondary literature (275); A. S. Famintsyn, "Nekrolog M. S. Voronina" ("Obituary of M. S. Voronin"), in *Trudy Imperatorskago S.-Petersburgskago obshchestva estestvoispytatelei*, **34** (1903), 210–222; N. A. Komarnitsky, "Voronin, M. S.," in S. Y. Lipshits, ed., *Russkie botaniki. Biografo-bibliografichesky slovar* ("Russian Botanists. Biographical-Bibliographical Dictionary"), II (Moscow, 1947), 163–168, with lists of Voronin's works (106 titles) and secondary literature (16 titles); N. I. Kuznetsov, "Sorokaletie nauchnoy deyatelnosti M. S. Voronina" ("Fortieth Anniversary of the Scientific Career of Voronin"), in *Trudy Botanicheskogo sada. Yurevskogo universiteta*, **1**, no. 1 (1900), 47–51; and S. Navaschin, "Michael Woronin," in *Berichte der deutschen botanischen Gesellschaft*, **21**, no. 1 (1903), 36–47.

E. M. SENCHENKOVA

VORONOY, GEORGY FEDOSEEVICH (*b.* Zhuravka, Poltava guberniya, Russia, 28 April 1868; *d.* Warsaw, Poland, 20 November 1908), *mathematics.*

Voronoy's father was superintendent of Gymnasiums in Kishinev and in other towns in the southern Ukraine. After graduating from the Gymnasium in Priluki in 1885, Voronoy enrolled in the mathematics section of the Faculty of Physics and Mathematics of the University of St. Petersburg. He graduated in 1889 and was retained to prepare for a teaching career. In 1894 he defended his master's dissertation, on algebraic integers associated with the roots of an irreducible third-degree equation. He then became professor in the Department of Pure Mathematics at the University of Warsaw. He defended his doctoral dissertation, on a generalization of the algorithm of continued fractions, at St. Petersburg in 1897; both dissertations were awarded the Bunyakovsky Prize of the St. Petersburg Academy of Sciences.

Voronoy subsequently elaborated his own ideas on the geometry of numbers and conducted investigations on the analytic theory of numbers. In 1904 he participated in the Third International Congress of Mathematicians in Heidelberg, where he met Minkowski, who was then working on topics closely related to those in which Voronoy was interested.

Voronoy's work, all of which concerns the theory of numbers, can be divided into three groups: algebraic theory of numbers, geometry of numbers, and analytic theory of numbers.

In his doctoral dissertation Voronoy gave the best algorithm known at the time for calculating fundamental units of a general cubic field, for both a positive and negative discriminant.

Voronoy completed two of a planned series of memoirs in which he intended to apply the principle of continuous Hermite parameters to problems of the arithmetical theory of definite and indefinite quadratic forms. In the first of these works, which dealt with certain characteristics of complete quadratic forms, he solved the question posed by Hermite concerning the precise upper limit of the minima of the positive quadratic forms of a given discriminant of n variables. E. I. Zolotarev and A. N. Korkin had given solutions for $n = 4$ and $n = 5$; with the aid of the methods of the geometrical theory of numbers Voronoy gave a full algorithmic solution for any n. In the second paper, which concerned simple parallelepipeds, Voronoy dealt with the determination of all possible methods of filling an n-dimensional Euclidean space with identical convex nonintersecting polyhedra having completely contiguous boundaries (parallelepipeds). A solution of this problem for three-dimensional space had been given by the crystallographer E. S. Fedorov, but his proofs were incomplete. In 1896 Minkowski demonstrated that the parallelepipeds must have centers of symmetry and that the number of their boundaries did not exceed $2(2^n - 1)$. Voronoy imposed the further requirement that $n + 1$ parallelepipeds converge at each summit and completely solved the problem for these conditions.

In a memoir concerning a problem from the theories of asymptotic functions Voronoy solved Dirichlet's problem concerning the determination of the number of whole points under the hyperbola $xy = n$. Dirichlet had found that the number of such points lying in the area $x > 0$, $y > 0$, $xy \leqslant n$ was expressed by the formula $F(n) = n(\log n + 2C - 1) + R(n)$, where $R(n) = O(\sqrt{n})$. By introducing series similar to a Farey series and by dividing the area of summation into the subsets associated with these series, Voronoy substantially improved the evaluation, obtaining $R(n) = O(\sqrt[3]{n} \cdot \log n)$. His paper served as the starting point for the work of I. M. Vinogradov, and the Farey series that he introduced was employed in the investigation of problems in the additive theory of numbers by Vinogradov, G. H. Hardy, and J. E. Littlewood.

BIBLIOGRAPHY

Voronoy's collected works were published as *Sobranie sochineny*, 3 vols. (Kiev, 1952–1953). Papers mentioned in the article are "Sur un problème du calcul des fonctions asymptotiques," in *Journal für die reine und angewandte Mathematik*, **126** (1903), 241–282; "Sur quelques propriétés des formes quadratiques positives parfaites," *ibid.*, **133** (1908), 97–178; and "Recherches sur les parallelloèdres primitifs," *ibid.*, **136** (1909), 67–179. These papers are reprinted in *Sobranie sochineny*, II, 5–50, 171–238, 239–368. On his work, see B. N. Delone, *Peterburgskaya shkola teorii chisel* ("The St. Petersburg School of the Theory of Numbers"; Moscow–Leningrad, 1947).

I. G. BASHMAKOVA

VRIES, HUGO DE (*b*. Haarlem, Netherlands, 16 February 1848; *d*. Lunteren, Netherlands, 21 May 1935), *plant physiology, genetics, evolution*.

The ancestors of Hugo de Vries[1] had been Baptists since the Reformation. As dissenters they were not eligible for public office, but they found an outlet for their talents and energy in trade; during the seventeenth and eighteenth centuries they were prosperous merchants. When the drastic political changes at the end of the eighteenth century brought more liberal views, the activities of the family also changed; they became professors, lawyers, and statesmen.

Hugo de Vries's paternal grandfather, Abraham de Vries, was a Baptist minister and librarian for the city of Haarlem; he was a noted expert on the history of printing. His maternal grandfather, Caspar Jacob Christiaan Reuvens, was the first professor of archaeology at the University of Leiden and founder of its archaeological museum. An uncle, Matthias de Vries, was professor of Dutch literature at Leiden and a pioneer in Dutch philology. The dictionary he started with Lambert A. te Winkel (1863), completed in 1888, was the authoritative source for Dutch spelling for half a century.

Hugo de Vries's father, Gerrit de Vries, studied law and literature at the University of Leiden. He served as a representative in the Provincial States of North Holland for many years and became the leading expert on legislation concerning water management. In 1862 he was appointed to the Council of State, a position he held until his death. Ten years later he was asked by William III to form a cabinet, and he took the post of minister of justice. De Vries's mother, Maria Everardina Reuvens, came from a family of scholars and statesmen.

De Vries was educated in Haarlem at a private Baptist grammar school and subsequently at the municipal Gymnasium. The area around Haarlem was a botanist's paradise, and it awakened in him a deep love for plants at an early age. During his vacations he roamed the entire country on foot, in search of plants for his herbarium. When he entered the university, he felt that his collection of dried phanerogams of the Netherlands was complete.

In 1862 the family moved to The Hague, where de Vries attended the Gymnasium for four years. Since there was no Baptist community in The Hague, he was sent to Leiden on weekends to receive religious instruction. Here he was soon invited by Willem Suringar, a professor of botany, to help classify the plants in the herbarium of the Netherlands Botanical Society.

Consequently, when de Vries matriculated at the University of Leiden in 1866, he was already an expert on the flora of the Netherlands. He therefore turned to other fields of interest. These he found after reading Sachs's *Lehrbuch der Botanik* (1868), to which he owed his interest in plant physiology, and Charles Darwin's *Origin of Species*, which aroused his interest in evolution. The University of Leiden was ill-equipped for the pursuit of either of these studies; plant physiology was not taught there, and there was no laboratory for experimental work. The experimental work for his doctoral dissertation on plant physiology was done in his attic. Suringar was hostile to the theory of evolution; and this hostility, combined with de Vries's youthful enthusiasm, caused a permanent estrangement between them.

De Vries was not happy with the education he

had received at Leiden, and he decided to continue his studies in Germany. In the autumn of 1870 he went to Heidelberg, where he studied with Hofmeister. In the spring and summer of 1871 he was at Sachs's laboratory in Würzburg, where he finally found what he had been seeking. Sachs took a keen interest in his progress and considered him his best pupil.

Although he intended to work in Würzburg for several years, in September 1871 de Vries accepted an appointment as teacher of natural history at the First High School in Amsterdam.[2] He was still able to spend most of the long summer vacations in Sachs's laboratory at Würzburg; the reports of his experimental work there are found in *Arbeiten des botanischen Instituts in Würzburg*, Sachs's journal.

De Vries's teaching duties became more and more demanding, and began to interfere with his studies. Sachs then recommended him for a position at the Prussian Ministry of Agriculture. In January 1875 de Vries was given the task of writing monographs on agricultural plants that were published in the *Landwirtschaftliche Jahrbücher*. The necessary experimental work was done at Würzburg, in space provided by Sachs in his laboratory. Here de Vries wrote monographs on red clover, the potato, and the sugar beet. In addition, he carried out extensive studies on osmosis in plant cells during this period. He frequently traveled to other university towns to meet with the leading professors of botany.

Sachs showed his continued interest in de Vries's future by recommending him for the post of *Privatdozent* in the physiology of cultivated plants at the University of Halle. To be eligible for the appointment, de Vries had to pass a doctoral examination. He defended a dissertation based on his work on the stretching of cells and received the appointment on 12 February 1877.

The lectures at Halle were not a success. Attendance was poor, and there was no real interest in the subject. Thus de Vries was much relieved when he was appointed lecturer in plant physiology at the newly constituted University of Amsterdam. The Amsterdam Athenaeum was founded in 1632 but did not have the authority to grant degrees; it was necessary for students to pass examinations at an accredited university, usually Leiden or Utrecht. In 1877 the Athenaeum was given university status, and many new teachers were needed. De Vries was the first instructor in plant physiology in the Netherlands. In the summer of 1877, he traveled to England to meet the botanists of that country. The highlight of the trip was a visit with Darwin.

In the autumn of 1878 de Vries was appointed extraordinary professor and, on his birthday in 1881, ordinary professor. Until about 1890 he conducted research on osmosis in plant cells—the famous experiments on plasmolysis. In addition to his teaching and research, he sponsored the research of his pupil J. H. Wakker on the diseases of bulb plants; he investigated the causes of the contamination of the water mains of Rotterdam; and he served on the committee to study the future water supply of the city of Amsterdam. During this period Wakker, J. M. Janse, F. A. Went, H. P. Wijsman, and H. W. Heinsius earned their doctorates under de Vries's guidance.

In addition to his experimental work in plant physiology, de Vries made extensive studies of the theories and literature on heredity and variation in plants. About 1890 he abruptly abandoned the study of plant physiology and devoted himself exclusively to heredity and variation. This period in his career began with his *Intracellular Pangenesis* (1889), in which he reviewed critically the work of Spencer, Darwin, Nägeli, and Weismann, and proposed his own theory that "pangenes" were the carriers of hereditary traits. One of the most important books in the history of genetics, it attracted little attention at the time.

De Vries's experimental work in the 1890's led to the rediscovery of Mendel's laws and the discovery of the phenomenon of mutation. The rediscovery of Mendel's laws was announced almost simultaneously by de Vries, Correns, and Tschermak-Seysenegg—in that order. De Vries certainly knew the segregation laws in 1896, and he deduced these laws from his own experimental work and not from reading Mendel's paper or any reference to Mendel's work in the literature.

The results of his more than ten years of experimentation and study were laid down in de Vries's *Die Mutationstheorie* . . . (1901–1903), in which he described in detail his work on the segregation laws, on phenomena of variation, and on plant mutations. The book made him famous, and he was recognized as one of the foremost botanists of his time.

During the 1890's no doctorates were earned under de Vries's guidance. These were years of hard personal work, but apparently they were not happy ones. In 1896 he succeeded C. A. J. A. Oudemans as senior professor of botany at Amsterdam. He was charged with teaching systematic botany and genetics; instruction in plant physiology and pharmacology was turned over to Eduard Verschaffelt.

De Vries's physiological work was well known on the continent of Europe, less so in England, and hardly at all in the United States. His rediscovery of Mendel's laws and the formulation of the mutation theory, however, became widely known, especially in the United States. During the summers of 1904 and 1906, de Vries was invited to lecture at the University of California at Berkeley; in 1912 he was invited to participate at the opening of the Rice Institute in Houston, Texas. He wrote books about each of his American journeys.

After 1900 a number of students earned their doctorates under de Vries: C. J. J. van Hall, T. Weevers, P. J. S. Cramer, J. A. Lodewijks, A. R. Schouten, J. M. Geerts, J. A. Honing, T. J. Stomps, and H. H. Zeylstra. In that period de Vries received many honors. Eleven honorary doctorates were conferred upon him; he was awarded seven gold medals, and was made a regular or honorary member of most of the major academies and societies.

In 1918 de Vries reached mandatory retirement age. He had already bought a house at Lunteren, a remote village, where the soil was suitable for an experimental garden. He also built a laboratory, and he remained professionally active until his death. He also produced a large number of scientific papers during this time. His rather lonely life in Lunteren was relieved by visits from former pupils, friends, and admirers from all over the world, and several students from the the universities of Amsterdam and Utrecht came to Lunteren to do the experimental work for their dissertations.

Scientific Work. In his doctoral dissertation de Vries reviewed the literature concerned with the influence of temperature on the vital processes of plants. Based on an essay that received a gold medal, the dissertation was supported by original experimental data that affirmed or refuted the statements of various authors.

At Würzburg, Sachs studied plant physiology from a mechanical point of view. Initially he assigned de Vries subjects for study, but later he gave him a free hand in the choice of subjects. In 1871 de Vries discovered that stalks and isolated ribs of leaves usually have a greater growth capability on the upper side than on the underside. He called this phenomenon "epinasty" and the reverse phenomenon, which is sometimes found in young organs, "hyponasty." He claimed that these two phenomena, together with the already recognized phenomena of geotropism and heliotropism, are sufficient to explain all growth patterns of plants. In 1872 he studied the mechanism of tendril curving and found it to be almost exclusively the result of increased growth in the outer region of the tendril. In the same year he studied the mechanism of the movements of climbing plants and established that the nutating shoots of such plants are not irritable and that the nutation is caused by the shoots' having a zone of increased growth parallel to the axis, with the zone slowly rotating around the axis of the organ. Darwin greatly admired this work and praised it in his *Climbing Plants*,[3] which started the correspondence between Darwin and de Vries. The next year de Vries investigated the rate of cell growth at various points on the growing shoot and found that the zone of fastest growth is not located at the tip but farther back on the organ.

As a student in Leiden and in Hofmeister's laboratory, de Vries had shown that the contraction of the protoplast of a plant cell, caused by its introduction into a salt solution of appropriate concentration, did not kill the cell, as was generally believed. In addition, he established that the protoplast is permeable only by water. At Würzburg, while writing his monographs on agricultural plants, de Vries continued this research. He wanted to decide how much of the increase of the cell wall of a growing plant organ was the result of the growth of the cell and how much was the result of the stretching of the cell wall caused by the pressure of the cell fluid—turgor. Annulling the turgor by submerging the plant organ in a suitable salt solution, de Vries found that in young, growing cells the expansion caused by turgor amounts to some 10 percent of the total length. In mature cells there was no turgor expansion. He described this work in the *Habilitationsschrift* submitted at Halle.

As a professor at Amsterdam, de Vries continued his research on the function of the cell contents. He theorized that calcium is a waste product in plants, absorbed for the sake of the needed elements with which it is combined and stored in cells as an organic salt (often calcium oxalate). He formulated a growth theory, stating that growth in plants is caused primarily by extension of the cell walls by turgor, with the extension fixed later. He conjectured that organic acids are the chemical compounds that contribute most to the turgor, a conjecture that he qualified later when he had analyzed the cell fluid of some plants. He applied his growth theory to explain many forms of plant movement, including the movement of tendrils, the erection of lodged grain, and the contraction of roots of biennial plants in autumn.

These and other investigations posed questions. How great is the pressure caused by the turgor in the cell? How much does each of the components of the cell fluid contribute to this pressure? What is the cause of an increase of this pressure in cells? In order to answer these questions, de Vries returned to his observations of the effect of salt solutions on plant cells. In previous experiments he had found that if a plant cell is immersed in successively stronger salt solutions, the cell initially contracts; subsequently the protoplast starts contracting and frees itself from the cell wall until it becomes a globular body within the cell. De Vries called this process "plasmolysis." In the new research he used the plant cell as an indicator, immersing the cell in solutions of increasing strength until he found the concentration at which the protoplast just starts to free itself from the cell wall. At this concentration the osmotic pressures of the solution and of the cell contents are equal or—in de Vries's terminology—"isotonic." He determined the isotonic concentration for the solution to be tested and for a reference solution; three times the concentration of the reference solution, divided by the concentration of the solution to be tested, was called the "isotonic coefficient." Saltpeter (KNO_3) was always used as a reference solution.

After determining the isotonic coefficients of a great many chemicals, de Vries found that isotonic coefficients always have a near integer value—ranging from 2 to 5. Generalizing this, he stated the following rules: for neutral organic compounds and organic acids, the isotonic coefficient is 2; for salts with one alkali atom, 3; with two alkali atoms, 4; with three alkali atoms, 5; with one alkaline earth atom, 2; and with two alkaline earth atoms, 4. This is known as the law of isotonic coefficients. Using this law, de Vries was able to determine the proportional contribution to the total osmotic pressure in the cell for each component of the cell fluid. It appears that for different species, different chemicals in the cell fluid account for the largest part of the osmotic pressure: in *Rheum* it is oxalic acid, in *Rosa* it is glucose, and in *Gunnera* it is calcium chloride.

De Vries's work on the isotonic coefficients of solutions led van't Hoff to his formula for the osmotic pressure of solutions, one of the first results in physical chemistry. Van't Hoff's law in turn enabled de Vries to determine the total osmotic pressure in plant cells. At about the same time, Arrhenius discovered the dissociation of molecules in solution. This explained why de Vries had to use a factor of 3 in his computation of the isotonic coefficient. Even under laboratory conditions the reference solution was only about 50 percent ionized, and different salt solutions dissociate to different degrees. The phenomenon of ionization indicates that de Vries's law of isotonic coefficients cannot be exactly true.

The law of isotonic coefficients enabled de Vries to determine the molecular weight of raffinose during a discussion of that weight at a meeting of the Royal Netherlands Academy of Sciences and, in a few minutes, to settle this long-standing question.

During the late 1880's de Vries studied protoplasm. He found that the inner lining of the cell wall, the protoplast, consists of three layers, not two, as was currently believed. He discovered the innermost of these, the tonoplast. He also established that the vacuoles in the cell have a lining of their own, investigated the aggregation of the protoplasm of insectivorous plants, and studied the ribbon-shaped parietal chloroplasts of *Spirogyra*.

In addition to his physiological research on plants, de Vries conducted an extensive study of the literature on variability and heredity. Based on this research, he wrote nineteen articles for a Dutch agricultural journal. This series, "Thoughts on the Improvement of the Races of Our Cultivated Plants" (1885–1887), resembles Darwin's *Origin of Species* and *Variation of Animals and Plants Under Domestication* in its organization and approach. The study probably was of no great use to the farmers for whom it was written, but it was of great importance to de Vries as a means of formulating a program for future research.

In his first work in this new field of interest, *Intracellulare Pangenesis* (1889), de Vries presented his own theory. He considered the hereditary characteristics of living organisms as units that manifest themselves independently of each other and that can, therefore, be studied separately. Each independent characteristic is associated with a material bearer, which de Vries called a "pangene." The pangene is a morphological structure, made up of numerous molecules, that can take nourishment, grow, and divide to yield two new pangenes. After cell division, each daughter cell receives one set of pangenes from the mother cell. A pangene can be either active or latent. Some characteristics may be represented by more than one pangene. Where conflicting characteristics are possible—for example, red or white flowers—the characteristic represented by the largest number of pangenes is dominant. In each reproductive cell at least one of the representative pangenes, either active or latent, is present.

Using these concepts, de Vries explained all the vital phenomena of an organism: how a cell develops into an organ, how metamorphosis is brought about, and how an offspring becomes and remains uniform with the parents. The characteristics of the genus are caused by large aggregates of pangenes, which remain unchanged in the offspring. It is possible that one (or more than one) pangene starts to multiply in an extraordinary way or is changed during cell division. In such a case the different pangene that is created results in a new characteristic of the organism. This is, according to de Vries, the principal mechanism of evolution.

De Vries's pangene theory is remarkably close to the theory formulated later by geneticists, including T. H. Morgan. The concept that a characteristic is represented by two pangenes, each of which may be active or latent, and the concept that the pangenes are linked in groups (later called chromosomes) were not part of de Vries's theory.

De Vries called his material units pangenes to honor Charles Darwin, whose gemmule theory he rejected, however. The name "gene," given to the hereditary unit by Johannsen, was derived from de Vries's pangene.[4]

The research undertaken by de Vries to follow up his theoretical considerations covered several fields. He studied the causes and hereditary properties of many kinds of monstrosities, including forced tensions (*Zwangsdrehungen*—on which he wrote a monograph), fasciations, symphysis, and virescence. Jules MacLeod, professor at the University of Ghent, may have introduced him to the statistical methods of Quetelet and Galton. They had shown that in the animal kingdom the magnitude of variations (for example, the body length of soldiers) was distributed according to a probability curve (Gauss curve; de Vries used the term Galton curve). De Vries demonstrated that this is often true for the plant kingdom as well. This distribution manifested what he called the normal fluctuation of the considered characteristics. There were many cases where a symmetrical curve was not obtained. In some cases only a half Galton curve was obtained; de Vries expressed the opinion that such a curve shows the emergence of a discontinuous variation. In other cases the distribution curve showed two peaks; de Vries conjectured that such a curve indicates that a mixture of two races is present, and he succeeded in isolating these races by selection.

Because of his pangenesis theory and his work on variability, de Vries decided that experimental work in heredity should be performed with closely allied races or varieties, differing in only one characteristic or, at most, a few characteristics. In 1896 he demonstrated to his advanced students the segregation laws, now known as Mendel's laws, in *Papaver somniferum* var. *Mephisto* and var. *Danebrog*. He examined many species belonging to several families, and found the segregation laws confirmed in each case. He did not publish these results, however, reserving them, with his work on mutations, for a single large book. When he accidentally came across a reprint of Mendel's paper early in 1900, de Vries felt obliged to publish in order to protect his priority. This publication triggered the publications of Correns and Tschermak-Seysenegg. The work of de Vries did not quite parallel the work of Mendel, who had studied only two species, *Pisum* and *Hieracium*, and whose work with the latter had been unsuccessful. De Vries demonstrated the segregation laws in some twenty species. On the other hand, Mendel examined not only monohybrids but also dihybrids and trihybrids, and followed the offspring through a great many generations. In his rediscovery papers, de Vries reported on only two dihybrid experiments, and he followed the offspring of a cross through two generations at most. L. C. Dunn correctly states; "It is clear that de Vries was not a 'rediscoverer' but a creator of broad general principles."[5]

After the rediscovery of Mendel's laws, many investigators took up the subject. De Vries was not among them, however. He believed that hybridization only causes redistribution of existing characters and for that reason cannot explain the appearance of new species. Therefore, he concentrated on the phenomenon of mutation, which he believed explained the origin of new species and therefore gave necessary support to the theory of evolution.

One difficulty in studying the origin of new species was that the concept of "species" was ill-defined. Plants recognized as belonging to the same species often showed marked differences. The French botanist Alexis Jordan had found that, among plants recognized as belonging to the same species, there are subgroups of which the members are exactly alike and breed true under self-fertilization. These subgroups were later called "jordanons," while the traditional species, "which a good naturalist intuitively recognizes,"[6] were later called "linneons." De Vries claimed that the jordanon is the true species. Among specimens of the same jordanon, individual differences, including size of leaves and weight of seeds, are still possible; these "individual variations" follow Galton's law.

In 1886, near Hilversum, de Vries noticed on a plot of formerly cultivated land, overgrown with *Oenothera lamarckiana* (evening primrose), a number of specimens that differed markedly from the others. He took seeds of the normal form and of two differing ones for planting in his experimental garden. He went through considerable trouble to discover the origin of these *Oenotheras* (which had escaped from a nearby garden) and to ascertain the history of the introduction of the species into Europe. Although it was said that the plant had originally been introduced from Texas and was known to Lamarck in 1796 under the name *O. grandiflora*, the plant was unknown in the United States. De Vries was convinced that *O. lamarckiana* was a pure species.

New forms that appeared suddenly and unexpectedly were called "single variations" by de Vries; he later called them "mutations." The *Oenotheras* that he had collected near Hilversum soon started to produce new forms, which he judged to differ sufficiently from the parent species for him to consider them a new species, and hence to give them a binomial name. He obtained a giant form, which he named *O. gigas*; a form with pale-green, delicate, narrow leaves, which he named *O. albida*; one with red veins in the leaves, *O. rubinervis*; one with narrow leaves on long stalks, *O. oblonga*; and a dwarf form, *O. nanella*. These mutants appeared to be constant or almost constant under self-fertilization. Another mutant showed only female flowers and still another yielded, after self-fertilization, the original *lamarckiana* plus some mutants. The two mutants found in Hilversum also produced *lamarckianas* as well as mutants, including some new ones.

In order to explain why only the *Oenothera lamarckiana* produced so rich a harvest of mutants, while only a very few other species were known to product mutants (and then only a few mutants at a time), de Vries postulated that in its evolutionary life a species produces mutants over discrete, comparatively short periods of time only — their mutation periods. In addition, he conjectured that these periods are preceded by premutation periods, during which the latent characters are formed.

On the basis of his *Oenothera* research, de Vries distinguished mutations that supply a useful characteristic, which he called "progressive," and those that supply a useless or even harmful characteristic, which he called "retrogressive." Only the progressive characteristics contribute to the evolution of the species.

De Vries carried out extensive crossings between his *Oenothera* mutants. On the basis of this work and additional work on variability, he distinguished two kinds of crosses: bisexual and unisexual. In bisexual crosses the parents differ in at least one characteristic. These characteristics are all active in one parent and latent in the other. In unisexual crosses only one parent possesses a certain characteristic. De Vries associated these concepts with earlier terminology as follows: variety crossings are bisexual, exhibit a Mendel split, and produce fertile offspring; species crossings are unisexual, do not exhibit a Mendel split, and produce less fertile or even infertile offspring. It must be remembered that these concepts date from before the discovery that a characteristic is represented in the somatic cell by two genes, each of which can be either dominant or recessive.

De Vries's work on variability and mutation, necessarily only briefly sketched above, was reported in *Die Mutationstheorie . . .* (1901–1903), a heroic effort to correlate and explain the existing knowledge in this field and his own discoveries. De Vries's 1904 lectures at Berkeley were published as *Species and Varieties* (1905), a book that is much easier to read than *The Mutation Theory*. In his 1906 lectures in Berkeley, his topic was the application of his doctrines to agricultural and horticultural practice. These lectures were published in *Plant Breeding* (1907).

In 1906 de Vries considered his mutation research finished, and he prepared to study, as his next research project, the adaptation of plants to an adverse environment, such as a desert. That year, however, his *Oenothera* cultures showed "twin hybrids" for the first time; to gather information to explain this phenomenon, he decided to continue his *Oenothera* research for a few years. Circumstances forced him to continue the *Oenothera* study for the rest of his life.

Although the mutation theory was generally enthusiastically received, there were critics. The first of these was William Bateson, who suggested as early as 1902 that the *O. lamarckiana* might well be a hybrid. This idea was vigorously advocated by B. M. Davis, who questioned de Vries's arguments for the provenance and the purity of the *lamarckiana*. Davis tried, without notable success, to synthesize a *lamarckiana* by crossing *O. biennis* with *O. grandiflora* and *O. franciscana*. Zeylstra, a student of de Vries's, declared that *O. nanella* was nothing but a diseased *lamarckiana*; another of his students, J. A. Honing, gave a critique that anticipated the later work of O. Renner.

From about 1908 Morgan, Sturtevant, Hermann J. Muller, and Calvin B. Bridges had been studying the genetics of the fruit fly *Drosophila*. This work, which was first summarized in *The Mechanism of Mendelian Heredity* (1915), provided the essentials of the chromosome theory of heredity as it is known today. In this theory de Vries's pangenes, which he had described as single material units existing in a free state in the cell nucleus, became the genes, grouped on the chromosomes in the cell nucleus.

The first mutant to be explained was *O. gigas*. In 1907 Anne M. Lutz found that this mutant is a tetraploid; it has twenty-eight chromosomes in the somatic cells instead of fourteen, as is common with the *Oenotheras*. In 1912 Lutz discovered that *O. lata* is a triploid and that it has fifteen chromosomes.

In the course of time, new anomalies in *Oenothera* were added to those described by de Vries in his *Mutation Theory*. He himself discovered the "twin hybrids": two true-breeding parents yield two different types in the first-generation offspring. Another phenomenon, the significance of which was not realized until 1914, was the fact that often a large percentage of the seeds obtained in *Oenothera* cultures were infertile. These and other phenomena were studied by Renner, who discovered that *O. lamarckiana* is a permanent heterozygote (hence a hybrid) containing two chromosome complexes, which are transmitted as a whole and which he called "gaudens" and "velans." They are balanced lethals. Hence, of the four combinations formed in equal numbers during the first generation of the offspring of self-fertilized *lamarckianas*— that is, *gv*, *vg*, *vv*, *gg*—the latter two (half the total number of seeds) were not viable, the others having the same phenotype as the parent and hence creating the illusion that the plant breeds true. When a *lamarckiana* was crossed with another *Oenothera* species—for example, *O. muricata*—half of the offspring contained the *mg* combination and the other half the *mv* combination, hence the twin hybrids.

Studies by Renner and others showed that the genetic makeup of *Oenothera* is very unusual and complicated; few genera show such phenomena, and those to a much lesser extent. Because of the

work of a large number of investigators, the genetic properties of *Oenothera* are now quite well known. Among these investigators were de Vries himself and his students T. J. Stomps, D. J. Broekens, K. Boedijn, H. Dulfer, and J. A. Leliveld. Much of the *Oenothera* work was done at the Station for Experimental Evolution at Cold Spring Harbor, New York. De Vries gave the keynote speech at the opening of the station in 1904. A. F. Blakeslee and R. E. Cleland, who showed that the chromosome complexes are ring formations, were leaders in this work. The *Oenothera* problem was solved finally by Sturtevant and Sterling Emerson.

The fact that de Vries's mutants were superseded does not mean that his work on the phenomenon of mutation was valueless. Many true mutations have been discovered in the animal and plant kingdoms and mutation is still the cornerstone of the theory of evolution. Next to the *Drosophila* experiments, the work with *Oenothera* has contributed most to the chromosome theory of heredity.

NOTES

1. Sometimes—for example, in *Isis Cumulative Bibliography*, II, 603—the name is given as Hugo Marie de Vries. The addition of Marie is not justified, for no member of the de Vries family ever had more than one Christian name.
2. A secondary school that emphasized modern languages and science.
3. Charles R. Darwin, *The Movements and Habits of Climbing Plants* (London, 1876), see 9, 22, 160, 165, 181.
4. Wilhelm L. Johannsen, *Elemente der exakten Erblichkeitslehre* (Jena, 1909; 2nd ed., 1913). In the 1st ed. (p. 124) Johannsen ignored the use of "pangene" by de Vries; in the 2nd ed. (p. 143) he corrected this omission. Johannsen made the change from pangene to gene to express his opinion that the hereditary unit to be named, formerly designated by the German *Anlage*, is nonmaterial. When the materiality of the hereditary unit was confirmed, the name gene was retained.
5. L. C. Dunn, *A Short History of Genetics* (New York, 1965), 43.
6. In determining whether a form should be ranked as a species or a variety, the opinion of naturalists having sound judgment and wide experience seems the only guide to follow. Darwin, *The Origin of Species* (London, 1859), 47.

BIBLIOGRAPHY

I. ORIGINAL WORKS. A complete bibliography of de Vries's works contains more than 700 entries. About half of these were contributions to popular literature. The most important scientific papers, selected by de Vries himself, are collected in *Opera e periodicis collata*, 7 vols. (Utrecht, 1918–1927).

His most important scientific books and papers are *De invloed der temperatuur op de levensverschijnselen der planten* (The Hague, 1870); "Sur la perméabilité du protoplasme des betteraves rouges," in *Archives néer-*

landaises des sciences exactes et naturelles, **6** (1871), 117–126; "Sur la mort des cellules végétales par l'effet d'une température élevée," *ibid.*, 245–295; "Ueber einige Ursachen der Richtung bilateral symmetrischer Pflanzentheile," in *Arbeiten des botanischen Institutes in Würzburg*, **1** (1872), 223–277; "Längenwachsthum der Ober- und Unterseite sich krümmender Ranken," *ibid.*, **1** (1873), 302–316; "Zur Mechanik der Bewegung von Schlingpflanzen," *ibid.*, 317–342; "Ueber die Dehnbarkeit wachsender Sprosse," *ibid.*, **1** (1874), 519–545; "Ueber Wundholz," in *Flora, oder allgemeine botanische Zeitung*, **59** (1876), 2–6, 17–25, 38–42, 49–55, 81–88, 97–108, 113–121, 129–139; "Ueber longitudinale Epinastie," in *Flora, oder allgemeine botanische Zeitung*, **60** (1877), 385–391; and *Untersuchungen über die mechanischen Ursachen der Zellstreckung, ausgehend von der Wirkung von Salzlösungen auf den Turgor wachsender Pflanzenzellen* (Leipzig, 1877): "Ueber die Ausdehnung wachsender Pflanzenzellen durch ihren Turgor," in *Botanische Zeitung*, **35** (1877), 2–10; "Beiträge zur speziellen Physiologie landwirtschaftlicher Kulturpflanzen, I. Rother Klee," in *Landwirtschaftliche Jahrbücher*, **6** (1877), 465–514, 893–956; "II. Kartoffeln," *ibid.*, **7** (1878), 19–39, 217–249, 591–682; "III. Zuckerrüben," *ibid.*, **8** (1879), 13–35, 417–498.

Other works include *De ademhaling der planten* (Haarlem, 1878); "Ueber die Verkürtzung pflanzlicher Zellen durch Aufnahme von Wasser," in *Botanische Zeitung*, **37** (1879), 649–654; "Ueber die inneren Vorgänge bei den Wachsthumskrümmungen mehrzelliger Organe," *ibid.*, 830–838; "Ueber die Bedeutung der Pflanzensäuren für den Turgor de Zellen," *ibid.*, 847–853; "Over de bewegung der ranken van Sicyos," in *Verslagen en mededeelingen der Koninklijke Akademie van Wetenschappen, Afdeeling Natuurkunde*, 2nd ser., **15** (1880), 51–174; "Ueber den Antheil der Pflanzensäuren an der Turgorkraft wachsender Organe," in *Botanische Zeitung*, **41** (1883), 849–854; "Ueber die periodische Säurebildung der Fettpflanzen," *ibid.*, **42** (1884), 337–343, 353–358; "Eine Methode zur Analyse der Turgorkraft," in *Jahrbüchern für wissenschaftliche Botanik*, **14** (1884), 427–601; "Beschouwingen over het verbeteren van de rassen onzer cultuurplanten," in *Maandblad van de Hollandsche Maatschappij van Landbouw*, 19 articles (May 1885–July 1887).

Also see "Ueber die Bedeutung der Circulation und der Rotation des Protoplasma für den Stofftransport in den Pflanzen," in *Botanische Zeitung*, **43** (1885), 1–6, 17–24; "Plasmolytische Studien über die Wand der Vacuolen," in *Jahrbücher für wissenschaftliche Botanik*, **16** (1885); 464–598; "Ueber die Periodicität im Säuregehalt der Fettpflanzen," in *Verslagen en mededeelingen der Koninklijke Academie van Wetenschappen, Afdeeling Natuurkunde*, 3rd ser., **1** (1885), 58–123; "Ueber die Aggregation im Protoplasma von *Drosera rotundifolia*," in *Botanische Zeitung*, **44** (1886), 1–11, 17–26, 33–43, 57–64; "Ueber den isotonischen Coefficient des Glycerin," *ibid.*, **46** (1888), 229–235, 245–253;

"Osmotische Versuche mit lebenden Membrane," in *Zeitschrift für physikalische Chemie*, **2** (1888), 415–432; "Détermination du poids moléculaire de la raffinose par la méthode plasmolytique," in *Comptes rendus . . . de l'Académie des sciences*, **106** (1888), 751–753; "Isotonische Koeffizienten einiger Salze," in *Zeitschrift für physikalische Chemie*, **3** (1889), 103–109; and *Intracellulare Pangenesis* (Jena, 1889), English trans. as *Intracellular Pangenesis* (Chicago, 1910), Dutch trans. as *Intracellulaire pangenesis* (Amsterdam, 1918).

Further works are *Die Pflanzen und Thiere in den dunklen Räumen der Rotterdamer Wasserleitung* (Jena, 1890); *Monographie der Zwangsdrehungen* (Berlin, 1891), also in *Jahrbücher für wissenschaftliche Botanik*, **23** (1892), 13–206; "Ueber halbe Galtonkurven als Zeichen diskontinuierlicher Variation," in *Berichte der Deutschen botanischen Gesellschaft*, **12** (1894), 197–207; "Sur l'introduction de l'*Oenothera lamarckiana* dans les Pays Bas," in *Nederlandsch kruidkundig archief*, 2nd ser., **6** (1895), 579–583; "Eine zweigipfliche Variationskurve," in *Archiv für Entwicklungsmechanik der Organismen*, **2** (1895), 52–64; "Eenheid in veranderlijkheid," in *Album der Natuur*, **47** (1898), 65–80; "Over het omkeeren van halve Galtonkurven," in *Botanisch Jaarboek*, **10** (1898), 27–61; "Alimentation et sélection," in *Volume jubilaire de la Société de biologie de Paris* (1899), 17–38; "Ueber Curvenselektion bei *Chrysanthemum segetum*," in *Berichte der Deutschen botanischen Gesellschaft*, **17** (1899), 84–98; and "On Biastrepsis in Its Relation to Cultivation," in *Annals of Botany*, **13** (1899), 395–420.

Subsequent writings include "Sur la loi de disjonction des hybrides," in *Comptes rendus . . . de l'Académie des sciences*, **130** (1900), 845–847; "Das Spaltungsgesetz der Bastarde," in *Berichte der deutschen botanischen Gesellschaft*, **18** (1900), 83–90; "Sur les unités des charactères spécifiques et leur application à l'étude des hybrides," in *Revue générale de botanique*, **12** (1900), 257–271; "Sur l'origine expérimentale d'une nouvelle espèce végétale," in *Comptes rendus . . . de l'Académie des sciences*, **131** (1900), 124–126; "Ueber erbungleiche Kreutzungen (vorläufige Mittheilung)," in *Berichte der Deutschen botanischen Gesellschaft*, **18** (1900), 435–443; "Hybridizing of Monstrosities," in *Journal of the Royal Horticultural Society*, **24** (1900), 69–75; "Over het ontstaan van nieuwe soorten in planten," in *Verslagen van de zittingen der wis- en natuurkundige afdeeling van de Koninklijke Academie van wetenschappen*, **9** (1900), 246–248; *Die Mutationstheorie, Versuche und Beobachtungen über die Entstehung von Arten im Pflanzenreich*, 2 vols. (Leipzig, 1901–1903), English trans. as *The Mutation Theory, Experiments and Observations on the Origin of Species in the Vegetable Kingdom*, 2 vols. (Chicago, 1909–1910), from which translation all discussions of Mendel's segregation law have been omitted, and *Die Mutationen und die Mutationsperioden bei der Entstehung der Arten* (Leipzig, 1901).

Also see "Ueber tricotyle Rassen," in *Berichte der*

Deutschen botanischen Gesellschaft, **20** (1902), 45–54; "La loi de Mendel et les charactères constants des hybrides," in *Comptes rendus . . . de l'Académie des sciences*, **136** (1903), 321–323; "On Atavistic Variation in *Oenothera cruciata*," in *Bulletin of the Torrey Botanical Club*, **30** (1903), 75–82; "Anwendung der Mutationslehre auf die Bastardierungsgesetze," in *Berichte der Deutschen botanischen Gesellschaft*, **21** (1903), 45–82; "Sur la relation entre les charactères des hybrides et leurs parents," in *Revue générale de botanique*, **15** (1903), 241–252; "Bastaardeering en bevruchting," in *De Gids*, 4th ser., **21** (1903), 403–450; "Experimenteele evolutie," in *Onze Eeuw*, **4** (1904), 282–309, 362–393; "The Evidence of Evolution," in *University Record of the University of Chicago*, **9** (1904), 202–209; *Naar Californië, Reisherinneringen* (Haarlem, 1905; 2nd ed., 1906); *Het Yellowstone Park; Experimenteele evolutie* (Amsterdam, 1905); *Species and Varieties* (Chicago, 1905); "Aeltere und neuere Selektionsmethoden," in *Botanisches Zentralblatt*, **26** (1906), 385–395; "Die Neuzuchtigungen Luther Burbanks," *ibid.*, 609–621; and "Burbank's Production of Horticultural Novelties," in *Open Court*, **20** (1906), 641–653.

Additional works are "Evolution and Mutation," in *Monist*, **17** (1907), 6–22; "New Principles in Agricultural Plantbreeding," *ibid.*, 209–219; *Naar Californië II* (Haarlem, 1907); *Plant Breeding, Comments on the Experiments of Nilsson and Burbank* (Chicago, 1907; 2nd ed., 1919), Dutch trans. as *Het veredelen van kultuurplanten* (Haarlem, 1908); "On Twin Hybrids," in *Botanical Gazette*, **44** (1907), 401–407; "Bastarde von *Oenothera gigas*," in *Berichte der Deutschen botanischen Gesellschaft*, **26a** (1908), 754–762; "On Triple Hybrids," in *Botanical Gazette*, **47** (1909), 1–8; "Ueber doppelt reziproke Bastarde von *Oenothera biennis* L. und *Oenothera muricata* L.," in *Botanisches Zentralblatt*, **31** (1911), 97–104; "The Evening Primroses of Dixie Landing," in *Science*, n.s. **35** (1912), 599–601, written with H. H. Bartlett; *Die Mutationen in der Erblichkeitslehre* (Berlin, 1912); "*Oenothera Nanella*, Healthy and Diseased," in *Science*, n.s. **35** (1912), 753–754; *Van Texas naar Florida* (Haarlem, 1913); *Gruppenweise Artbildung* (Berlin, 1913); "L'*Oenothera grandiflora* de l'herbier de Lamarck," in *Revue générale de botanique*, **25b** (1914), 151–166; and "The Probable Origin of *Oenothera lamarckiana*," in *Botanical Gazette*, **57** (1914), 345–361.

Further, see "Ueber künstliche Beschleunigung der Wasseraufnahme in Samen durch Druck," in *Botanisches Zentralblatt*, **35** (1915), 161–176; "The Coefficient of Mutation in *Oenothera biennis* L.," in *Botanical Gazette*, **59** (1915), 169–196; "*Oenothera nanella*, a Mendelian Mutant," *ibid.*, **60** (1915), 337–345; "Die Grundlagen der Mutationstheorie," in *Naturwissenschaften*, **4** (1916), 593–598; "Ueber die Abhängigkeit der Mutations Koeffizienten von äusseren Einflüssen," in *Berichte der Deutschen botanischen Gesellschaft*, **34** (1916), 1–7; "New Dimorphic Mutants of the *Oenotheras*," in *Botanical Gazette*, **62** (1916), 249–280; "Die

endemischen Pflanzen von Ceylon und die mutierenden *Oenotheren*," in *Botanisches Zentralblatt*, **36** (1916), 1–11; "Gute, harte und leere Samen von *Oenothera*," in *Zeitschrift für induktive Abstammungs- und Vererbungslehre*, **16** (1916), 239–292; "The Origin of the Mutation Theory," in *Monist*, **27** (1917), 403–410; "*Oenothera lamarckiana mut. velutina*," in *Botanical Gazette*, **63** (1917), 1–25; "Halbmutanten und Zwillingsbastarde," in *Berichte der Deutschen botanischen Gesellschaft*, **35** (1917), 128–135; and "Ueber monohybride Mutationen," in *Botanisches Zentralblatt*, **37** (1917), 139–148.

Additional works by de Vries are "Kreutzungen von *Oenothera lamarckiana mut. velutina*," in *Zeitschrift für induktive Abstammungs- und Vererbungslehre*, **19** (1918), 1–13; "Mass Mutations and Twin Hybrids of *Oenothera grandiflora* Ait.," in *Botanical Gazette*, **65** (1918), 377–422; "Twin Hybrids of *Oenothera hookeri*, T. and G.," in *Genetics*, **3** (1918), 397–421; "Mutations of *Oenothera suaveolens*, Desf.," *ibid.*, 1–26; "Mass Mutations in *Zea mais*," in *Science*, **47** (1918), 465–467; *Van amoebe tot mensch* (Utrecht, 1918); "*Oenothera lamarckiana mut. simplex*," in *Berichte der Deutschen botanischen Gesellschaft*, **37** (1919), 65–73; "*Oenothera Rubinervis*, a Half Mutant," in *Botanical Gazette*, **67** (1919), 1–26; "*Oenothera lamarckiana erythrina*, eine neue Halmutante," in *Zeitschrift für Induktive Abstammungs- und Vererbungslehre*, **21** (1919), 91–118; "Ueber die Mutabilität von *Oenothera lamarckiana mut. simplex*," *ibid.*, **31** (1923), 313–357; "Ueber sesquiplex Mutanten von *Oenothera lamarckiana*," in *Zeitschrift für Botanik*, **15** (1923), 369–408; "*Oenothera lamarckiana mut. perennis*," in *Flora, oder allgemeine botanische Zeitung*, **116** (1923), 336–345; "Ueber die Entstehung von *Oenothera lamarckiana mut. velutina*," in *Botanisches Zentralblatt*, **43** (1923), 213–224; and "On the Distribution of Mutant Characters Among the Chromosomes of *Oenothera lamarckiana*," in *Genetics*, **8** (1923), 233–238, written with K. Boedijn.

Also see "Die Gruppierung der Mutanten von *Oenothera lamarckiana*," in *Berichte der Deutschen botanischen gesellschaft*, **42** (1924), 174–178, written with K. Boedijn; "Doubled Chromosomes of *Oenothera semigigas*," in *Botanical Gazette*, **78** (1924), 249–270, written with K. Boedijn; "Die Mutabilität von *Oenothera lamarckiana gigas*," in *Zeitschrift für induktive Abstammungs- und Vererbungslehre*, **35** (1924), 197–237; "Sekundäre Mutationen von *Oenothera lamarckiana*," in *Zeitschrift für Botanik*, **17** (1925), 193–211; "Mutant Races, Derived From *Oenothera lamarckiana*," in *Genetics*, **10** (1925), 211–222; "Brittle Races of *Oenothera lamarckiana*," in *Botanical Gazette*, **80** (1925), 262–275; "Die latente Mutabilität von *Oenothera biennis*," in *Zeitschrift für induktive Abstammungs- und Vererbungslehre*, **38** (1927), 141–197; "A Survey of the Cultures of *Oenothera lamarckiana* at Lunteren," *ibid.*, **47** (1928), 275–286, written with R. R. Gates; "Ueber das Auftreten von Mutanten aus *Oenothera lamarck-*

iana," *ibid.*, **52** (1929), 121–190; and "Ueber semi-rezessive Anlagen in *Oenothera lamarckiana,*" *ibid.*, **70** (1935), 222–256.

II. SECONDARY LITERATURE. Discussions of de Vries and his work include G. E. Allen, "Hugo de Vries and the Reception of the 'Mutation Theory,' " in *Journal of the History of Biology*, **2** (1969), 55–87; F. M. Andrews, "Hugo de Vries," in *Plant Physiology*, **5** (1930), 175–180; Annelén [pseud.], "Professor Hugo de Vries en de Amsterdamsche Universiteit," in *Algemeen Handelsblad* (15 Oct. 1927); A. F. Blakeslee, "The Work of Professor Hugo de Vries," in *Scientific Monthly*, **36** (1933), 378–380; and "Hugo de Vries, 1848–1935," in *Science*, **81** (1935), 581–582; J. H. van Burkom, "In Memoriam Prof. Hugo de Vries," in *Natura*, **34** (1935), 161; F. Chodat, "Hugo de Vries, 1848–1935," in *Comptes rendus des séances de la Société de physique et d'histoire naturelle de Genève*, **54** (1937), 7–10; R. Cleland, "Hugo de Vries, 1848–1935," in *Journal of Heredity*, **26** (1935), 289–297; and "Hugo de Vries," in *Proceedings of the American Philosophical Society*, **76** (1936), 248–250; J. C. Costerus, "Professor Hugo de Vries," in *Eigen Haard*, **21** (1895), 261–264; C. F. Cox, "Hugo de Vries on the Origin of Species and Varieties by Mutation," in *Journal of the New York Botanical Garden*, **6** (1905), 66–70; E. O. Dodson, "Mendel and the Rediscovery of His Work," in *Scientific Monthly*, **58** (1955), 187–195; and P. Fröschel, "Einige Briefe von Hugo de Vries," in *Acta botanica neerlandica*, **10** (1961), 202–208.

Also see S. S. Gager, "De Vries and His Critics," in *Science*, n.s. **24** (1906), 81–89; R. R. Gates, "Prof. Hugo de Vries, For. Mem. R. S.," in *Nature*, **136** (1935), 133–134; G. C. Gerrits, *Grote Nederlanders bij de opbouw der natuurwetenschappen* (Leiden, 1947); A. D. Hall, "Hugo de Vries," in *Obituary Notices of Fellows of the Royal Society of London*, **4** (1935), 371–373; J. Heimans, "Hugo de Vries," in *Hugo de Vries, Voordrachten ter herdenking van zijn honderdste geboortedag op 16 Februari 1948* (Amsterdam, 1948), 1–9; *Zeventig jaar pangenenleer* (Amsterdam, 1959); "De herontdekking," in *Honderd jaar Mendel* (Wageningen, 1965), 62–80; and "Gregor Mendel and Hugo de Vries on the Species Concept," in *Acta botanica neerlandica*, **18** (1969), 95–98; H. W. Heinsius, "Hugo de Vries, 16 Februari 1848–1918," in *De Amsterdammer* (16 Feb. 1918); J. van der Hoeven, "In Memoriam Hugo de Vries," in *Verslagen en mededeelingen der Koninklijke Akademie van Wetenschappen, Afdeeling Natuurkunde*, **44** (1935), 59–62; A. A. W. Hubrecht, "Hugo de Vries' mutatietheorie," in *De Gids*, 4th ser., **19** (1901), 492–519; H. T. A. Hus, "The Work of Hugo de Vries," in *Sunset Magazine*, **13** (1904), 39–42; and "Hugo de Vries," in *Open Court*, **20** (1906), 713–725; and W. van Itallie-van Embden, "Sprekende portretten," in *Haagsche post* (19 Dec. 1925), an interview with Hugo de Vries.

Other works on de Vries are Ilse Jahn, "Zur Geschichte der Wiederentdeckung der Mendelschen Ge-setze," in *Wissenschaftliche Zeitschrift der Friedrich Schiller-Universität Jena*, **7** (1957–1958), 215–227; E. Lehmann, *Die Theorien der Oenotheraforschung* (Jena, 1922); and "Die Entwicklung der Oenotheraforschung," in *Hugo de Vries, sechs Vorträge zur Feier seines 80 Geburtstages, gehalten im botanischen Institut, Tübingen* (Stuttgart, 1929), 36–42; (D. Manassen), "Prof. Hugo de Vries," in *Algemeen handelsblad* (18 Nov. 1910); M. Moebius, "Hugo de Vries und sein Lebenswerk," in *Revista sudamericana de botánica*, **2** (1935), 162–168; J. W. Moll, "Hugo de Vries, 16 Februari 1848–1918," in *De nieuwe Amsterdammer* (16 Feb. 1918); and D. Müller, "Drei Briefe über reine Linien, von Galton, de Vries und Yule and Wilhelm Johannsen in 1903 geschrieben," in *Centaurus*, **16** (1972), 316–319.

Further works are H. R. Oppenheimer, "Hugo de Vries als Pflanzenphysiologe," in *Palestine Journal of Botany and Horticultural Science*, **1** (1935–1936), 51–69; P. van Oye, "Julius MacLeod en Edward Verschaffelt," in *Mededelingen van de Koninklijke Vlaamsche academie voor wetenschappen, letteren en schoone kunsten van België*, **23** (1961), 3–20; P. W. van der Pas, "Hugo de Vries als taxonoom," in *Scientiarum historia*, **11** (1969), 148–166; "The Correspondence of Hugo de Vries and Charles Darwin," in *Janus*, **57** (1970), 173–213; "Hugo de Vries visits San Diego," in *Journal of San Diego History*, **17** (1971), 12–23; and "Hugo de Vries in the Imperial Valley," *ibid.*; O. Renner, "Hugo de Vries," in *Erbarzt*, **3** (1935), 177–184; and "Hugo de Vries, 1848–1935," in *Naturwissenschaften*, **24** (1936), 321–324; H. F. Roberts, *Plant Hybridization Before Mendel* (New Haven, 1929); and Elisabeth Schiemann, "Hugo de Vries," in *Züchter*, **7** (1935), 159–161; and "Hugo de Vries zum hundertsten Geburtstage," in *Berichte der Deutschen botanischen Gesellschaft*, **62** (1948), 1–15.

In addition, see A. Schierbeek, "De pangenesis theorie van Hugo de Vries," in *Bijdragen tot de geschiedenis der geneeskunde*, **24** (1943), 64–67; G. H. Schull, "Hugo de Vries at Eighty-five," in *Journal of Heredity*, **24** (1933), 1–6; Sinotō Yositō, "Tabi ni ahishi hitobito," in *Kagaku zassan*, **3** (8), (1933), 295–297, a pilgrimage to famous men; T. J. Stomps, "Aus dem Leben und Wirken von Hugo de Vries," in *Hugo de Vries, Sechs Vorträge zur Feier seines 80 Geburtstages, gehalten im botanischen Institut, Tübingen* (Stuttgart, 1929), 7–16; *Vijf en twintig jaren Mutatietheorie* (The Hague, 1930); "Hugo de Vries," in *Berichte der Deutschen botanischen Gesellschaft*, **53** (1936), 85–96; "Hugo de Vries et la cytologie," in *Revue de cytologie et de cytophysiologie végétales*, **2** (3) (1937), 281–285; and "On the Rediscovery of Mendel's Work by Hugo de Vries," in *Journal of Heredity*, **45** (6) (1954), 293–294; E. Von Tschermak-Seysenegg, "Hugo de Vries, der Begründer der Mutationstheorie," in *Reichspost* (2 Feb. 1936); and "Historischer Rückblick auf die Wiederentdeckung der Gregor Mendelschen Arbeit," in *Verhandlungen der Zoologisch-botanischen Gesellschaft in Wien*, **92** (1951), 25–35.

Also see F. J. van Uildriks, "Professor Hugo de Vries zeventig jaar," in *Aarde en haar volken*, **54** (1918), 45–46; T. W. Vaughan, "The Work of Hugo de Vries and Its Importance in the Study of Problems of Evolution," in *Science*, n.s. **23** (1906), 681–691; J. H. Verduyn de Boer, "Hugo de Vries, de groote Nederlandsche geleerde drie en tachtig jaar," in *Huisgenoot* (6 Feb. 1931); J. H. de Vries, *De Amsterdamsche doopsgezinde familie de Vries* (Zutphen, 1911); T. Weevers, "Hugo de Vries als plantenphysioloog," in *Hugo de Vries, Voordrachten ter herdenking van zijn honderdste geboortedag op 16 Februari 1948* (Amsterdam, 1948), 11–15; and F. A. F. C. Went, "Hugo de Vries," in *Mannen en vrouwen van beteekenis in onze dagen*, **31** (7) (1900), 263–320; "Hugo de Vries en de mutatietheorie," in *Elsevier's maandschrift*, **39** (1905), 35–42; and "Herinneringen aan Hugo de Vries," in *Natura*, **27** (1928), 19–21.

PETER W. VAN DER PAS

VULF, YURI VIKTOROVICH. See **Wulff, Georg.**

VVEDENSKY, NIKOLAY EVGENIEVICH (*b.* Kochkovo, Vologodskaya gubernia, Russia, 28 April 1852; *d.* Kochkovo, 16 September 1922), *physiology.*

After graduating from the Vologod religious seminary, Vvedensky entered St. Petersburg University in 1872. Two years later he was arrested for participation in student revolutionary activities and spent more than three years in prison. He graduated from the university in 1879, having studied physiology with I. M. Sechenov, then worked in physiology laboratories in Germany (1881–1882), Austria (1884), and Switzerland (1887).

After defending his master's thesis in 1884, Vvedensky became *Privatdozent* in the department of physiology; in 1889, after Sechenov moved to Moscow, he became extraordinary professor, and in 1895, professor at St. Petersburg University.

Vvedensky's research was devoted to clarifying the regularities in the reaction of living tissue to various irritants. Having applied the method of telephonic auscultation of the excited nerve, he showed that a living system changes not only under the influence of irritation but also during its normal activity; he thus introduced the time factor into physiology. In his master's thesis, "Telefonicheskie issledovania nad elektricheskimi yavleniami v myshechnykh i nervnykh apparatakh" ("Telephonic Research on Electrical Phenomena in Muscle and Nerve Apparatus"), Vvedensky provided a thorough analysis of the literature on muscle con-

traction and nerve fatigue. In his doctoral dissertation, "O sootnosheniakh mezhdu razdrazheniem i vozbuzhdeniem pri tetanuse" ("On the Relationship Between Stimulus and Excitation in Tetanus"; 1886), he formulated the theory of the optimum and pessimum irritation, on the basis of which he established the law of relative functional movement (lability) of tissue. Vvedensky examined nerve-muscle preparations as heterogeneous formations (consisting of nerve tissue, nerve ends, and muscles), the parts of which possess different lability.

Vvedensky's outstanding achievement was his theory of parabiosis, developed in *Vozbuzhdenie, tormozhenie i narkoz* ("Excitation, Inhibition, and Narcosis"; 1901), in which he generalized his ideas on the nature of the processes of excitation and inhibition, showing their identity.

BIBLIOGRAPHY

Vvedensky's writings were brought together in *Polnoe sobranie sochineny* ("Complete Collected Works"), 7 vols. (Leningrad, 1951–1963).

Secondary literature includes I. A. Arshavsky, *N. E. Vvedensky, 1852–1922* (Moscow, 1950); Y. M. Ufland, *Osnovnye etapy razvitia uchenia N. E. Vvedenskogo* ("Basic States in the Development of the Theory of N. E. Vvedensky"; Moscow, 1952); and E. K. Zhukov, "Evolyutsionny metod v shkole Vvedenskogo-Ukhtomskogo" ("The Evolutionary Method in the School of Vvedensky and Ukhtomsky"), in *Uchenye zapiski Leningradskogo . . . universiteta . . .*, Ser. biolog. nauk, **12**, no. 77 (1944), 437–468.

N. A. GRIGORIAN

VYSHNEGRADSKY, IVAN ALEKSEEVICH (*b.* Vyshni Volochek, Tver gubernia [now Kalinin oblast], Russia, 20 December 1831; *d.* St. Petersburg, Russia, 6 April 1895), *mechanics, engineering.*

The son of a priest, Vyshnegradsky enrolled at the Tver Ecclesiastical Seminary in 1843; but after three years he moved to St. Petersburg and enrolled at the Physics and Mathematics Faculty of the Central Pedagogical Institute, where Ostrogradsky's lectures aroused his interest in mathematics and physics. He graduated in 1851 and began teaching mathematics at the St. Petersburg Military School. Vyshnegradsky received his master's degree at St. Petersburg University in 1854 and became instructor in mathematics at the Mikhaylovsky Artillery Academy, where he taught special technical courses.

In 1860 the Artillery Academy sent Vyshnegradsky abroad to study mechanical engineering and to prepare for a professorship in applied mechanics. He spent about two years in Germany, France, Belgium, and England. In 1862, upon his return to Russia, he was appointed professor of applied mechanics at the Mikhaylovsky Artillery Academy and, shortly thereafter, professor of mechanics at the St. Petersburg Technological Institute as well. Vyshnegradsky was both an outstanding theoretician and a gifted design engineer; he was responsible for the reconstruction of many Artillery Department factories and for the construction of railroads. From 1867 to 1878 he was the mechanics and engineering specialist of the Central Artillery Administration. In 1875 he was appointed director of the St. Petersburg Technological Institute. Vyshnegradsky was elected an honorary member of the St. Petersburg Academy of Sciences in 1888.

Vyshnegradsky taught a generation of Russian mechanical engineers and was the head of the first Russian school of mechanical engineering. His students included N. P. Petrov, the founder of the theory of hydraulic friction; V. L. Kirpichev, engineer and scholar who organized higher technical education in Russia; and A. P. Borodin, who introduced a number of improvements in the steam locomotive. He was named a deputy minister of the Russian Ministry of Finance in 1887 and was minister of finance from 1888 until 1892.

Vyshnegradsky's most significant scientific contributions were in the theory of automatic regulation. Before him, many scholars had studied the regulation of industrial processes, but the regulators that they developed were created experimentally and were not explained on a theoretical basis. Through his research Vyshnegradsky established the mathematical bases for the general scientific principles of automatic regulation. Prior to his work the machine and the regulator had been examined individually, and only the statics of the regulator had been studied.

In 1877 Vyshnegradsky published "O regulyatorakh pryamogo deystvia" ("On Direct-Action Regulators"), in which the conditions of stability of a steam engine equipped with a direct-action centrifugal regulator were explained. The stability condition for a regulating system, as established by Vyshnegradsky, is known in the technical literature as the Vyshnegradsky criterion. His article greatly influenced subsequent development of the theory of regulation. Published in several languages, it received great attention in Germany, France, and the United States.

BIBLIOGRAPHY

I. ORIGINAL WORKS. Vyshnegradsky's writings include *O dvizhenii sistemy materialnykh tochek, opredelyaemoy polnymi differentsialnymi unravneniami* ("On the Motion of a System of Material Points, Which [System] Is Defined by Complete Differential Equations"; St. Petersburg, 1854), his master's thesis; *Publichnye populyarnye lektsii o mashinakh* ("Public Popular Lectures on Machines"; St. Petersburg, 1859); "Neskolko zamechany o parovykh pressakh" ("Some Remarks on Steam [Powered] Presses"), in *Artillerysky zhurnal*, **4** (1860), 237–259; *Mekhanicheskaya teoria teploty* ("The Mechanical Theory of Heat"; St. Petersburg, 1873); *Lektsii o parovykh mashinakh, chitannye v Tekhnologicheskom institute* ("Lectures on Steam Engines Read at the Technological Institute"; St. Petersburg, 1874); "Sur la théorie générale des régulateurs," in *Comptes rendus . . . de l'Académie des sciences*, **83** (1876), 318–321; "Über direktwirkende Regulatoren," in *Civilingenieur*, n.s. **22** (1877), 95–131; "O regulyatorakh pryamogo deystvia" ("On Direct-Action Regulators"), in *Izvestiya Peterburgskogo prakticheskogo tekhnologicheskogo instituta*, **1** (1877), 21–62; "Mémoire sur la théorie générale des régulateurs," in *Revue universelle des mines . . .*, 2nd ser., **4** (1878), 1–38; and **5** (1879), 192–227; and *O regulyatorakh nepryamogo deystvia* ("On Indirect-Action Regulators"; St. Petersburg, 1878).

II. SECONDARY LITERATURE. See A. A. Andronov, *I. A. Vyshnegradsky i ego rol v sozdanii teorii avtomaticheskogo regulirovania* ("Vyshnegradsky and His Role in the Creation of the Theory of Automatic Regulation"; Moscow–Leningrad, 1949); A. T. Grigorian, *Ocherki istorii mekhaniki v Rossii* ("Essays on the History of Mechanics in Russia"; Moscow, 1961), 119–131; and V. L. Kirpichev, "I. A. Vyshnegradsky, kak professor i ucheny" ("Vyshnegradsky as Professor and Scholar"), in *Vestnik Obshchestva tekhnologov*, **6** (1895), 307–322.

A. T. GRIGORIAN

VYSOTSKY, GEORGY NIKOLAEVICH (*b.* Nikitovka, Chernigov gubernia, Russia, 19 February 1865; *d.* Kharkov, U.S.S.R., 6 April 1940), *soil science, forestry*.

Vysotsky entered the Petrovsky Agricultural Academy in 1886 and studied forestry, botany, and soil science. He began his scientific career in 1890 at the Berdyansk forest reserve, studying steppe forestry. In May 1892 he joined Dokuchaev's expedition to Poltava and was manager of the Great Anadolian forest reserve in the Ukraine. The ideas of Dokuchaev and Morozov decisively influenced the formation of Vysotsky's scientific views.

From 1904 to 1913 Vysotsky headed the reorganization in St. Petersburg of experimental forestry and also conducted field research in the region near Samara. In 1913 he directed projects for forestation of an artificial forest reserve on the steppe at Kiev. From 1918 Vysotsky taught at and worked for scientific organizations in Kiev, Simferopol, Minsk, and Kharkov; organized and headed experimental forestry research in Byelorussia and the Ukraine; and created and headed departments of forestry and forest management.

Vysotsky's main achievement was the establishment of the scientific foundations of steppe forestry and forest improvement; in particular, he determined that the vegetation of steppes is primarily a combination of oak and shrubbery and showed that what were then considered "normal" types of steppe forests, with a predominance of elms, involved intensive transpiration of soil moisture. Vysotsky demonstrated the influence of a forest on the microclimate of planted areas and adjacent localities, and classified steppe conditions for local growth according to the degree of suitability for forests, and in relation to topography, snow accumulation, depth of groundwater, and presence of subsoil salts.

Vysotsky revealed why steppes are without forests, and in "Ergenya" he showed the evolution of steppe vegetation under human influence and demonstrated experimentally the influence of steppe forest cultivation on the agricultural harvest. He wrote general works on the moisture cycle in nature and on the main questions of forest hydrology, also elucidating the role of forest in the water cycle of the plains in European Russia. Vysotsky pointed out the regularities of the moisture cycle and the movement of salts in the soil in the steppe and underforest, and developed the theory of gley—the formation of sticky clay layers—as a biochemical oxidation-reduction process.

BIBLIOGRAPHY

I. Original Works. Vysotsky's more than 200 writings include "Rastitelnost Veliko-Anadolskogo uchastka" ("Vegetation of the Great Anadolian District"), in *Trudy ekspeditsii V. V. Dokuchaeya* ("Works of the Dokuchaev Expedition"), II, pt. 2 (St. Petersburg, 1898); "Biologicheskie, pochvennye i fenologicheskie nablyudenia i issledovania v Veliko-Anadole 1892–1893 gg." ("Biological, Soil, and Phenological Observations and Research in Great Anadolia 1892–1893"), in *Trudy opytnykh lesnichestv* ("Works in Experimental Forestry"; St. Petersburg, 1901); *Lesnye kultury v Mariupolskom opytnom lesnichestve, 1886–1900* ("Forest Culture in the Mariupol Forest Reserve . . ."; St. Petersburg, 1901); "O nauchnykh issledovaniakh, kasayushchikhsya stepnogo lesorazvedenia" ("On Scientific Investigations Concerning Steppe Forest Culture"), in *Lesnoy zhurnal* (1901), no. 2; "Stepnoy illyuvy i struktura stepnykh pochv" ("Steppe Illuvium and the Structure of Steppe Soils"), in *Pochvovedenie* (1901), nos. 2–4, and (1902), no. 2; "Mikorizy dubovykh i sosnovykh seyantsev" ("Mycorrhizae of Oak and Pine Seedlings"), in *Lesopromyshlennyi vestnik* (1902), no. 29; and "O stimulakh, prepyatstviakh i problemakh razvedenia lesa v stepyakh Rossii" ("On the Incentives, Obstacles, and Problems of Forest Culture on the Steppes of Russia"), in *Trudy II Sezda deyateley selskokhozyaystvennogo opytnogo dela* ("Works of the II Congress of Workers in Experimental Agriculture"), I (St. Petersburg, 1902).

See also "O karte tipov mestoproizrastania" ("On a Map of Types of Local Growth"), in *Sovremennye voprosy russkogo selskogo khozyaystva* ("Contemporary Questions in Russian Agriculture"; St. Petersburg, 1904); "O vzaimnykh sootnosheniakh mezhdu lesnoy rastitelnostyu i vlagoyu, preimushchestvenno v yuzhnorusskikh stepyakh" ("On the Mutual Relations Between Forest Vegetation and Moisture, Primarily on the Southern Russian Steppes"), in *Trudy opytnykh lesnichestv* ("Works in Experimental Forestry," II; St. Petersburg, 1904); "K voprosu o vlianii lesa na nadzemnuyu vlazhnost v Rossii" ("On the Influence of the Forest on Underground Moisture in Russia"), in *Trudy III Sezda deyateley selskokhozyaystvennogo opytnogo dela* ("Works of the III Congress of Workers in Experimental Agriculture"; St. Petersburg, 1905); "Gley," in *Pochvovedeṅie* (1905), no. 7, also in *Lesnoy zhurnal* (1906), no. 3; "Ob oroklimaticheskikh osnovakh klassifikatsii pochv" ("On the Oroclimatic Bases of Soil Classification"), in *Pochvovedenie* (1906), **8**, no. 1; "Pochvenno-botanicheskie issledovania v yuzhnykh Tulskikh zasekakh" ("Soil-Botanical Research in the Southern Tula Abatis"), in *Trudy opytnykh lesnichestv* ("Works in Experimental Forestry"; St. Petersburg, 1906); and "Ob usloviakh lesoproizrastania i lesorazvedenia v stepyakh Evropeyskoy Rossii" ("On the Conditions of Forest Growth and Forest Culture on the Steppes of European Russia"), in *Lesnoy zhurnal* (1907), nos. 1–2.

Additional works are *O lesorastitelnykh usloviakh rayona Samarskogo udelnogo okruga* ("On Forest Growth Conditions of the Samara Region in a Specific District"), 2 vols. (St. Petersburg, 1908–1909); "Buzuluksky bor i ego okrestnosti" ("The Buzuluk Pine Forest and Its Surroundings"), in *Lesnoy zhurnal* (1909), no. 8; "Pochvoobrazovatelnye protsessy v peskakh" ("Soil-Forming Processes in Sands"), in *Izvestiya Russkogo geograficheskogo obshchestva*, **47**, pt. 6 (1911); "Lesnye kultury stepnykh opytnykh lesnichestv s 1893 po 1907 gg." ("Forest Cultures of the Steppe Experimental Forest Reserves 1893–1907"), in *Trudy po lesnomu opytnomu delu v Rossii* ("Works on Experimental Forestry in Russia"), no. 41 (St. Petersburg, 1912); "O dubravakh v Evropeyskoy Rossii" ("Oak Groves in European Russia"), in *Lesnoy zhurnal* (1913), nos. 1–2; "Ergenya, kulturno-fitologichesky ocherk" ("Ergenya, a Cultural-Phytological Sketch"), in *Trudy byuro po prikladnoi botanike* (1915), **10–11**; "Izokarbonaty" ("Isocarbonates"), in *Russki pochvoved* (1915); and *Lesa Ukrainy i uslovia ikh proizrastania i vozobnovlenia* ("The Forests of the Ukraine and the Conditions of Their Growth and Renewal"; Kiev, 1916).

Also see "Lesovodnye ocherki" ("Forestry Culture Notes"), in *Zapiski Belorusskogo gosudarstvennogo instituta selskogo i lesnogo khozyaistva*, **3** (1924); "Ocherki o pochve i rezhime gruntovykh vod" ("Sketches on Soil and Groundwater Conditions"), in *Byulleten pochvoveda* (1927), nos. 1–8; "O roli lesa v povyshenii urozhaynosti" ("On the Role of the Forest in Increasing the Harvest"), in *Lesnoe khozyaistvo* (1929), 10–11; "Uchenie o lesnoy pertinentsii" ("Theory of Forest"), in *Lesovedenia i lesovodstvo* ("Forestry and Forest Culture"; Leningrad, 1930); *Materialy po izucheniyu vodookhrannoy i vodoreguliruyushchey roli lesov i bolot* ("Materials for the Study of the Water-Retaining and Water-Regulating Roles of Forests and Swamps"; Moscow, 1937); *O gidrogeologicheskom i meteorologicheskom vlianii lesov* ("On the Hydrological and Meteorological Influence of Forests"; Moscow, 1938); and his autobiography, in *Pochvovedenie* (1941), no. 3.

II. Secondary Literature. See E. A. Danilov, "G. N. Vysotsky i stepnoe lesorazvedenie" ("Vysotsky and Steppe Forest Culture"), in *Pochvovedenie* (1935), no. 4; A. G. Isachenko, *G. N. Vysotsky – vydayushchysya otechestvenny geograf* ("Vysotsky – Outstanding Native Geographer"; Moscow, 1953); E. M. Lavrenko, "G. N. Vysotsky," in *Russkie botaniki, biografo-bibliografichesky slovar* ("Russian Botanists, Biographical-Bibliographical Dictionary"), II (Moscow, 1947); P. S. Pogrebnyak, "G. N. Vysotsky," in *Vydayushchiesya deyateli otechestvennogo lesovodstva* ("Outstanding Native Workers in Forest Culture"; Moscow, 1950); and "Georgy Nikolaevich Vysotsky, 1865–1940," in *Lyudi russkoy nauki* ("People of Russian Science"; Moscow, 1963); S. S. Sobolev, "G. N. Vysotsky i ego nauchnaya deyatelnost" ("Vysotsky and His Scientific Career"), in *Pochvovedenie* (1935), no. 4; M. E. Tkachenko, "Veliky agrolesomeliorator, pamyati akademika G. N. Vysotskogo" ("Great Improver of Forest Agriculture, Recollections of Academician G. N. Vysotsky"), in *Lesnoe khozyaistvo* (1940), no. 9; and A. A. Yarilov, "G. N.

Vysotsky – sledopyt-geograf" ("Vysotsky – Pathfinder-Geographer"), in *Pochvovedenie* (1941), no. 3.

Vera N. Fedchina

WAAGE, PETER (*b.* island of Hitterø [now Hidra], near Flekkefjord, Norway, 29 June 1833; *d.* Christiania [now Oslo], Norway, 13 January 1900), *chemistry, mineralogy.*

The son of Peter Pedersen Waage, a shipmaster and shipowner, and Regine Lovise Wattne, Waage was raised on the island of Hitterø, where his forebears had lived as seamen for hundreds of years. Since his father was usually at sea, he grew up mainly under the supervision of his mother, who was his first teacher. When his precocity became known (he was able to read at the age of about four), it was decided that rather than follow the traditional family occupation, he was to have further education.

Waage's first regular schooling, at Flekkefjord, began when he was eleven. The school principal persuaded him to go to the University of Christiania, and to prepare for this he entered the fourth year of the Bergen Grammar School in 1849. He passed his matriculation examination *cum laudabilis* in 1854 and studied medicine during his first three years at the university. In 1857 he turned to mineralogy and chemistry. (As a boy, he had an extensive collection of minerals, plants, and insects, and some of his first publications dealt with mineralogy and crystallography.) In 1858 Waage was awarded the Crown Prince's Gold Medal for "Udvikling af de surstofholdige syreradikalers theori," which was published in 1859, the same year as his book *Outline of Crystallography*, written with H. Mohn.

In 1859, after graduation, Waage received a scholarship in chemistry that enabled him to make a year's study tour in France and Germany, beginning in the following spring. Most of his time was spent with Bunsen at Heidelberg. In 1861 Waage was appointed lecturer in chemistry, and in 1866 he was promoted to the only chair of chemistry then existing at the University of Christiania.

C. M. Guldberg and Waage, whose names are linked for their joint discovery of the law of mass action, were related through marriages to daughters of cabinet minister Hans Riddervold; Waage married Johanne Christiane Tandberg Riddervold, by whom he had five children. His wife died in 1869, and in 1870 he married one of Guldberg's sisters, Mathilde Sofie Guldberg; they had six children.

Their collaboration on the studies of chemical affinity that led to the law of mass action began immediately after Guldberg's return from abroad in 1862. The first report of their results, published in 1864 in *Forhandlinger i Videnskabs-selskabet i Christiania*, remained almost completely unknown to scientists, a fate also suffered by a more detailed description of their theory published in French in 1867. The theory did not become generally known until Ostwald, in a paper published in 1877, adopted the law of mass action and proved its validity by new experiments. In 1878 van't Hoff, apparently without any knowledge of Guldberg and Waage's work, derived the law from reaction kinetics. Although the law had several forerunners, the combined efforts of the theorist Guldberg and the empiricist Waage led to the first general mathematical and exact formulation of the role of the amounts of reactants in chemical equilibrium systems.

After completing his studies with Guldberg on the law of mass action, Waage increasingly concentrated on practical problems and on social and religious work, much of which dealt with nutrition and public health. For example, he discovered a method for preserving milk and developed a process for producing unsweetened condensed milk and sterilized canned milk. Waage also devoted considerable time to the industrial exploitation of the large quantities of fish caught along the Norwegian coast, developing a highly concentrated and excellent fish meal that was used on Norwegian ships and expeditions and was exported to Sweden, Finland, Denmark, and Germany. In Waage's time beer was taxed according to the amount of malt used in the brewing; he proposed, however, that taxation be based on alcoholic content and developed a method for determining the concentration of alcohol in beer by measurement of the boiling point.

Religious work with young people was a major interest of Waage's. He was active in the founding and management of the Christiania Ynglingeforening (later the Oslo YMCA) and the Norwegian Christian Youth Association. He was co-editor of the *Polyteknisk tidsskrift*, an active member and officer of scientific societies, and the recipient of many honors.

BIBLIOGRAPHY

I. ORIGINAL WORKS. Guldberg and Waage's papers on the law of mass action were abridged and translated into German by Richard Abegg as *Untersuchungen über die chemischen Affinitäten*, Ostwald's Klassiker der Exakten Wissenschaften no. 104 (Leipzig, 1899). Their first paper, "Studier over Affiniteten," in *Forhandlinger i Videnskabs-selskabet i Christiania*, **7** (1864), 35–45, appears in facs., along with a number of articles on the law, in Haakon Haraldsen, ed., *The Law of Mass Action: A Centenary Volume 1864–1964* (Oslo, 1964), 7–17.

II. SECONDARY LITERATURE. A biography of Waage and a discussion of his work by Haakon Haraldsen are in *The Law of Mass Action*, 26–32, 32–34; and in *Untersuchungen . . . Affinitäten*, 174–178. For a brief obituary see W. Ramsay, in *Journal of the Chemical Society*, **77** (1900), 591–592.

GEORGE B. KAUFFMAN

WAALS, JOHANNES DIDERIK VAN DER (*b.* Leiden, Netherlands, 23 November 1837; *d.* Amsterdam, Netherlands, 8 March 1923), *physics.*

The son of a carpenter, van der Waals became a primary-school teacher. After training for secondary-school teaching (1866), while a headmaster in The Hague, he studied physics at the University of Leiden. On the basis of his knowledge of the work of Clausius and other molecular theorists, he wrote his dissertation, *Over de continuiteit van den gasen vloeistoftoestand* (1873). As Maxwell said, "this at once put his name among the foremost in science." Using rather simple mathematics, the dissertation gave a satisfactory molecular explanation for the phenomena observed in vapors and liquids by Thomas Andrews and other experimenters, especially the existence of a critical temperature, below which a gas can be condensed to a two-phase system of vapor and liquid; while above it there can be only a homogeneous vapor phase. This was one of the first descriptions of a collective molecular effect, although the kinetic theory of gases was already well known.

The law of corresponding states, which van der Waals developed some years later, allows a somewhat better fit with experimental data and in succeeding years was a useful guide in the work on liquefaction of the "permanent" gases. In 1875 he was elected to the Royal Netherlands Academy of Sciences and Letters; and two years later, after the Amsterdam Athenaeum had become the University of Amsterdam, he occupied the chair of physics. As a teacher van der Waals was much admired, and he inspired his pupils to do both experimental and theoretical work. His scientific publications were mostly on molecular physics and thermody-

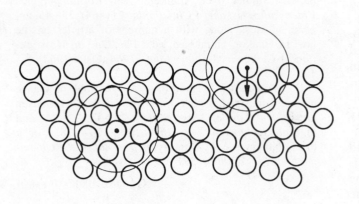

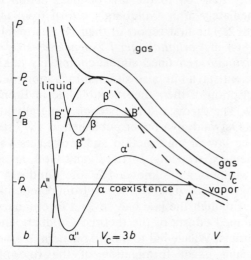

FIGURE 1. In the surface layer the van der Waals attraction sphere is only filled below and so causes a pressure, proportional to $1/V^2$ because both the number of surface molecules and the resultant force on each of them are proportional to the density $\rho = 1/V$.

FIGURE 2. Schematic van der Waals isotherms, one above and two below the critical isotherm T_c. In the coexistence region (inside broken curve) are two horizontal isothermal coexistence line segments, $A'A''$ and $B'B''$, at the two sides of which the surfaces (such as $A''\alpha''\alpha$ and $\alpha\,\alpha'A'$) included by the van der Waals isotherm are equal. (This is Maxwell's rule.)

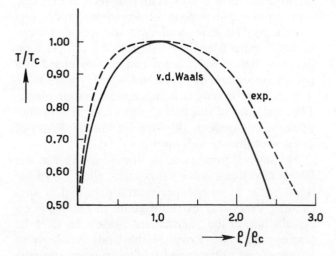

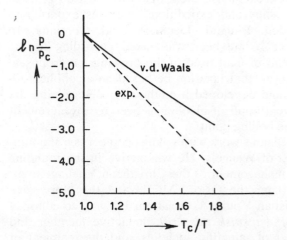

FIGURE 3. Coexistence region in reduced variables according to the van der Waals equation and to experiment for noble gases, and similar substances.

FIGURE 4. Natural logarithm of saturated vapor pressure plotted versus inverse temperature, according to van der Waals equation and to experiment, in reduced variables.

namics. He retired in 1907 and was succeeded by his son, who was named for him. In 1910 he was awarded the Nobel Prize for physics.

The van der Waals equation of state links the pressure P, absolute temperature T, and volume V, using three constants a, b, and R (the last four quantities being proportional to the amount of substance or its square):

$$\left(P+\frac{a}{V^2}\right)(V-b) = RT.$$

The term a/V^2 accounts for the molecular attraction, determined by integrating over the "attraction sphere" (Figure 1). It is supposed (and at medium densities it is sufficiently true) that in the average over time the attraction sphere, outside its central part, is filled rather homogeneously with molecules with a local density equal to the overall density. With very close packing a would no longer remain constant, but gradually increase (by less than a factor of 2). Likewise b, accounting for the non-overlapping of molecules, is equal at low density to four times the "proper" volume of all molecules together, but gradually decreases (by a factor not smaller than 0.5) for very close packing. Moreover, the temperature also affects a and b, since it influences the radial distribution of the molecules around an arbitrary one.

These detailed and, even for modern methods, rather difficult complications were rightly disregarded by van der Waals, although he was aware of them. With constant a and b, the isotherms may be calculated (see Figure 2), giving for the critical point

$$V_c = 3\,b,\ P_c = \frac{a}{27b^2},\ RT_c = \frac{8a}{27b}.$$

Below the critical point the assumption of one homogeneous phase no longer holds, except perhaps for very short moments. Energy relations are such that it is more favorable here for some of the molecules to move closer together (liquid), leaving the other molecules to fill the rest of the volume in a much sparser distribution (vapor). This two-phase system is represented by horizontal line tracks in Figure 2. Most thermodynamic quantities can now be calculated: saturated vapor pressure curve, Joule-Kelvin effect, supercooling, and so on. The comparison with experiment is given in Figures 3 and 4 for noble and pseudo-noble gases (hydrogen, carbon monoxide, nitrogen). The falling of all experimental points for these different substances, in reduced variables, on the same curve is an expression of the law of corresponding states. For other kinds of molecules the divergences from van der Waals's findings may be smaller or larger.

Van der Waals scarcely could have had adequate ideas about the nature of the attractive forces between molecules, so it is historically rather inexact that the London forces (energy proportional to r^{-6}) should often be called "van der Waals forces." It would be somewhat more reasonable to give this name to all forces not of ionic origin, but so loose a terminology would not be useful.

BIBLIOGRAPHY

I. Original Works. Van der Waals's writings include *Over de continuiteit van den gas- en vloeistoftoestand* (Leiden, 1873), his dissertation; articles in *Versl. Kon. Akademie van Wetenschappen*; and *Lehrbuch der Thermodynamik* (Leipzig, 1912), written with P. A. Kohnstamm. See also "The Equation of State for Gases and Liquids," in *Nobel Lectures in Physics, 1901–1921* (Amsterdam, 1967), which also contains a biography.

II. Secondary Literature. See an article in *Physica* (Amsterdam), **4** (1937); S. G. Brush, *Nobel Prizes in Physics* (Milan, 1970); and W. Leendertz, "J. D. van der Waals," in *Gids*, **87** (1923), 151.

J. A. Prins

WACKENRODER, HEINRICH WILHELM FERDINAND (*b.* Burgdorf, near Hannover, Germany, 8 March 1798; *d.* Jena, Germany, 4 September 1854), *pharmacy.*

Wackenroder was the son of Heinrich Wackenroder, a physician and apothecary in Burgdorf, and Charlotte Rougemont. He completed his apothecary's training in nearby Celle and worked for a time in his father's shop. In 1819 he went to Göttingen, where, in addition to pharmacy and natural science, he studied mathematics and medicine. After two and a half years he returned to Burgdorf and in 1824 passed the pharmacy examination. In 1825 Wackenroder became assistant to Friedrich Stromeyer in the pharmacy institute at Göttingen, where he gained experience in teaching and, by accompanying Stromeyer, in the inspection of apothecary shops. In 1827 he received a doctor of philosophy degree at Erlangen. In 1828 Wackenroder became a *Privatdozent* at Göttingen and in the same year accepted an offer to succeed Carl Göbel, as extraordinary professor, at the University of Jena. He was promoted to professor in 1836 and for the rest of his life was director of the pharmacy institute and inspector of apothecary shops for the grand duchy of Saxe-Weimar-Eisenach.

Wackenroder was exceptionally successful as a teacher, researcher, and scientific writer, and, most important, he made pharmacy an independent science. Besides the principal subjects of the curriculum he instituted special courses in forensic chemistry, phytochemistry, zoochemistry, toxicology, pharmaceutical technology, and merchandizing. He was the author of several textbooks, and his *Chemische Tabellen zur Analyse der unorganischen Verbindungen* (1829) went through five editions by 1843. From 1838 to 1854 he was coeditor of *Archiv der Pharmazie.* In this and in other specialized journals he reported on his many experimental investigations, chief among which was his work on phytochemistry. He discovered corydaline in the bulbs of *Corydalis tuberosa,* carotene in carrots, and solanine in potato sprouts. His name is commemorated by "Wackenroder's solution," a solution of polythionic acids formed when diluted sulfurous acid is treated with hydrogen sulfide.

BIBLIOGRAPHY

I. ORIGINAL WORKS. Wackenroder's books include *Chemische Tabellen zur Analyse der unorganischen Verbindungen* (Jena, 1829; 5th ed., 1843); *Anleitung zur qualitativen chemischen Analyse* (Jena, 1836); *Ausführliche Charakteristik der stickstofffreien organischen Säuren nebst Anleitung zur qualitativen chemischen Analyse* (Jena, 1841); and *Chemische Classification der einfachen und zusammengesetzten Körper* (Jena, 1851). For his journal articles, see Royal Society *Catalogue of Scientific Papers,* VI, 219–221; VIII, 1177; and XII, 763–765.

II. SECONDARY LITERATURE. On Wackenroder and his work, see *Allgemeine Deutsche Biographie,* XL, 443–444; Kurt Brauer, "Goethe und die Chemie," in *Zeitschrift für angewandte Chemie,* **37** (1924), 185–189; Fritz Chemnitius, *Die Chemie in Jena von Rolfinck bis Knorr* (Jena, 1929) 32–33; Fritz Ferchl, ed., *Chemisch-Pharmazeutisches Bio- und Bibliographikon* (Mittenwald, 1938), 561–562; H. Ludwig and E. Reichardt, "Biographisches Denkmal für Heinrich Wilhelm Ferdinand Wackenroder," in *Archiv der Pharmazie,* **135** (1856), 101–111; Poggendorff, II, 1237; Eduard Reichardt, "Bericht über die Jubelfeier des Geheimen Hofraths und Professors Dr. H. Wackenroder. . .," in *Archiv der Pharmazie,* **126** (1853), 321–341; Wolfgang Schneider, "Wackenroders Jubiläum," in *Die pharmazeutische Industrie,* **15** (1953), 403–405; and Otto Zekert, *Berühmte Apotheker* (Stuttgart, 1955), 151.

GRETE RONGE

IBN WĀFID, ABŪ AL-MUṬARRIF ʿABD AL-RAHMAN, also known as **Abenguefit, Abenguéfith, Albenguéfith, Abel Nufit** (*fl.* Toledo, Spain, *ca.* 1008–1075), *pharmacology.*

Ibn Wāfid studied the works of Aristotle, Dioscorides, and Galen. At the demand of the king of Toledo, Al-Ma'mūn, he planted a botanical garden in the king's orchard, which extended between the Galiana and Tajo palaces in front of the bridge of Alcántara. Ibn Luengo, a disciple of Ibn Wāfid, and possibly Ibn Baṣṣal studied in the king's garden.

For twenty years Ibn Wāfid worked on the *Kitāb al-adwiya al-mufrada* ("Book of the Simple Medicines"), a synthesis, with some new data, of Dioscorides and Galen. The structure of the book confirms what Ibn Ṣāʿid (Ibn Wāfid's friend and bibliographer) had stated, that is, that Ibn Wāfid did not like to prescribe compound medicines, but simple ones; and, if possible, he abstained from prescribing the latter and tried to cure his patients by following a dietary treatment.

Ibn Wāfid's *Kitāb al-rashshād fī al-ṭibb* ("Guide to Medicine") is a pharmacopoeia and manual of therapeutics. On account of an incorrect reading of the title by Casiri, who confused the letter *rāʾ* for *wāw* thereby reading *wisād,* the title was translated as "Book of the Pillow."

Ibn Wāfid's other works are the following: *Mudjarrabāt fī al-ṭibb* ("Medical Experiences"); *Tadqīq al-naẓar fī ʿilal ḥāssat al-baṣar* ("Observations on the Treatment of Eye Illnesses"), which might be the one preserved in the anonymous manuscript 876 at El Escorial; *Kitāb al-mugīth* ("Book of Assistance"), the title of which alludes to the drug *mugīth,* valuable for the treatment of many diseases; and *Madjmūʿal-filāḥa* ("A Compendium of Agriculture"), which is in a medieval Castilian translation and fragment.

J. M. Millás Vallicrosa found various Arabic manuscripts, in which Ibn Wāfid avoids discussing the pharmacologic properties of plants, and insists on his proper method of tillage. This book was made good use of by Gabriel Alonso de Herrera in his *Agricultura General* (Madrid, 1513; repr., 1819). Ibn Wāfid also wrote a treatise on balneology preserved in a Latin version as *De balneis* (Venice, 1553).

BIBLIOGRAPHY

On Ibn Wāfid and his work, see J. M. Millás Vallicrosa, "La traducción castellana del 'Tratado de Agricultura' de ibn Wāfid," in *Al-Andalus*, **8** (1943), 281–332; and G. Sarton, *Introduction to the History of Science*, I (Baltimore, 1927), 728.

J. VERNET

WAGNER, RUDOLPH (*b.* Bayreuth, Germany, 30 July 1805; *d.* Göttingen, Germany, 13 May 1864), *comparative anatomy, physiology, anthropology.*

Wagner was the son of Lorenz Heinrich Wagner, a Bavarian court councillor and Gymnasium director. He studied at the Gymnasiums in Bayreuth and Augsburg before beginning his medical education at the University of Erlangen in 1822. Two years later he transferred to the University of Würzburg, from which he graduated M.D. in 1826. His interest in the natural sciences then led him to spend eight months studying with Cuvier in Paris, where he received an excellent grounding in comparative anatomy. Returning to Germany, Wagner became a prosector in anatomy at Erlangen; he was made *Privatdozent* in 1829, and professor of comparative anatomy and zoology in 1832. In 1840 he accepted an appointment at Göttingen, where he succeeded J. F. Blumenbach as professor of physiology, comparative anatomy, and general natural history; he also served as curator of Blumenbach's craniological collection and lectured on anthropology.

Wagner conducted research in a number of areas. His most important work concerned mammalian ova and sperm. Purkyně had already, in 1825, discovered the nucleus in the avian egg, while K. E. von Baer had discovered the mammalian ovum (1827), and J. V. Coste had identified its nucleus (1833). It remained for Wagner to discover (1835) an important formation in the ovum of several species of mammals, which he called the *macula germinativa*—later known as the nucleolus. With Dujardin, Wagner was one of the first to use the achromatic microscope to examine sperm; in 1837 he published highly accurate illustrations of spermatozoa, showing the structures that he had actually seen, which he called "seminal threads." His accomplishment was the more noteworthy in that at the same time a number of other biologists believed that spermatozoa were parasitic animals, and even attempted to identify a visceral system in them.

In a series of other microscopical researches, Wagner demonstrated, in 1833, that red blood corpuscles have no nuclei. He made significant contributions to the study of the retina and the choroidea of the eye (1835) and of the electric organ of the torpedo fish (1847), as well as sharing in George Meissner's discovery of the tactile corpuscles of the skin (1852). In 1853 and 1854, Wagner also conducted investigations of the nervous system, by which he was able to show the relation between the peripheral nerve fibers and the ganglion cells of the brain.

Wagner's most important physiological work was his edition of the five-volume *Handwörterbuch der Physiologie mit Rücksicht auf physiologische Pathologie*, published between 1842 and 1853. He intended this work to be a compendium of all the physiological knowledge of the day; it consists of sixty-three extensive review articles by thirty authors, including Lotze, Berzelius, E. H. Weber, Purkyně, Carl Ludwig, Valentin, A. W. Volkmann, and Bidder. Wagner himself contributed a single article on the microscopic structure of the nervous system and an addendum to R. Leuckart's article on reproduction.

In his physiological work, Wagner emphasized the value of microscopic observation. As a leading representative of the histophysiological trend, he considered the microscope to be an essential means of elucidating physiological function, and tended to be somewhat critical of experimentation and pure mensuration. "What the scales are for the chemist, the telescope for the astronomer, so the microscope is for the physiologist," he wrote. His own work exemplifies his theory; his study of the structure of the electric organ of the torpedo fish was designed to explain the production of electric potential, while the discovery of the tactile corpuscles in the skin contributed to the knowledge of the mechanics of the stimulation of sensory nerve endings. A number of Wagner's students worked under similar principles; an investigation of the structure of nerve endings in muscle performed by his pupil W. F. Kühne, for example, led to clarification of the functional transmission of impulses in the motor nerve endings.

In his teaching, Wagner was a captivating lecturer, who emphasized practical instruction. He had an ability to stimulate and help young scientists, and a number of his collaborators, including R. Leuckart, Billroth, Meissner, and Julius Vogel, became prominent in a variety of specialties. He was also able to further his views through the foundation of the Göttingen Physiological Institute.

In addition to his work in anatomy and physiology, Wagner was strongly interested in philosophical problems concerning mind and body, science and society, and morality and materialism. These interests deepened after he suffered a severe pulmonary hemorrhage in 1845, and began to confine his work to the study of the nervous system and to anthropology. His philosophical views were first published in 1851; highly conservative, they proved a source of annoyance to younger scientists. In 1854 Wagner addressed a meeting of German scientists and physicians at Göttingen, and his speech initiated an unpleasant controversy about materialism, in which his chief opponent, Carl Vogt, published a witty and sarcastic critique of Wagner's views on the creation of man and the nature of his soul. The discussion soon thereafter degenerated into personal insult from both sides, and Wagner thereafter confined his attention to more strictly scientific matters.

BIBLIOGRAPHY

I. ORIGINAL WORKS. A chronological list of Wagner's writings, including their translations, is in E. Ehler's paper (see below), 484–488. They include *Zur vergleichenden Physiologie des Blutes. Untersuchungen über Blutkörperchen, Blutbildung und Blutbahn, nebst Bemerkungen über Blutbewegung, Ernährung und Absonderung* (Leipzig, 1832–1833) with *Nachträge* (Leipzig, 1838); *Lehrbuch der vergleichenden Anatomie* (Leipzig, 1834–1835); "Einige Bemerkungen und Fragen über das Keimbläschen," in Müller's *Archiv für Anatomie . . .*, **2** (1835), 373–384; *Prodromus historiae generationis hominis atque animalium* (Leipzig, 1836); "Fragmente zur Physiologie der Zeugung, vorzüglich zur mikroskopischen Analyse des Spermas," in *Abhandlungen der k. Bayerischen Akademie der Wissenschaften,* **2** (1837), 381–416; *Icones physiologicae. Tabulae physiologiam et geneseos historiam illustrantes*, 3 fascs. (Leipzig, 1839), also supp., *Bau und Endigungen der Nerven* (Leipzig, 1847); *Icones zootomicae* (Leipzig, 1841); *Lehrbuch der Physiologie für Vorlesungen und Selbstunterricht,* I. *Specielle Geschichte der Lebensprocesse* (Leipzig, 1839, 1843, 1845); *Elements of Physiology for the Use of Students, and With Special Reference to the Wants of Practitioners*, translated by Robert Willis (London, 1841); *Samuel Thomas von Soemmerings, Leben und Verkehr mit seinen Zeitgenossen*, 2 vols. (Leipzig, 1844); "Ueber den feineren Bau des elektrischen Organs im Zitterrochen," in *Abhandlungen der k. Gesellschaft der Wissenschaften zu Göttingen,* **3** (1848), 141–166; "Semen," in *Todd's Cyclopedia of Anatomy and Physiology*, 1V (London, 1848), written with R. Leuckart; *Neurologische Untersuchungen* (Göttingen, 1854); *Menschenschöpfung und Seelensubstanz* (Göttingen, 1854); *Über Wissen und Glauben mit beson-*

derer Beziehung zur Zukunft der Seelen (Göttingen, 1854); *Der Kampf um die Seele vom Standpunkt der Wissenschaft* (Göttingen, 1857); and *Bericht über die Zusammenkunft einiger Anthropologen im September 1861 in Göttingen . . .* (Leipzig, 1861), written with K. E. von Baer.

II. SECONDARY LITERATURE. "Nekrolog von Rudolph Wagner," by his eldest son, Adolph Wagner, was published in *Nachrichten von der k. Gesellschaft der Wissenschaften zu Göttingen* (1864), 375–399. An interesting critical biography is in E. Ehler, "Göttinger Zoologen," in *Festschrift zur Feier des hunderfünfzigjährigen Bestehens der Königlichen Gesellschaft der Wissenschaften zu Göttingen* (Berlin, 1901), 431–447. Wagner's conception of physiology is discussed in K. E. Rothschuh, *Physiologie. Der Wandel ihrer Konzepte, Probleme und Methoden vom 16. bis 19. Jahrhundert* (Freiburg–Munich, 1968), 260–261, 296–297. On the controversy over materialism, see E. Nordenskiöld, *The History of Biology* (New York, 1928), 450; and H. Degen, "Vor hundert Jahren. Die Naturforscher Versammlung in Göttingen und der Materialismusstreit," in *Naturwissenschaftliche Rundschau,* **7** (1954), 271–277. On anthropology, see Benno Ottow, "K. E. von Baer als Kraniologe und die Anthropologen-Versammlung 1861 in Göttingen," in *Sudhoffs Archiv . . .,* **50** (1966), 43–68.

VLADISLAV KRUTA

WAGNER VON JAUREGG (or WAGNER-JAUREGG), JULIUS (*b.* Wels, Austria, 7 March 1857; *d.* Vienna, Austria, 27 September 1940), *psychiatry.*

Wagner von Jauregg was the son of a civil servant. He entered the University of Vienna medical school in 1874 and, while still a student, worked under Salomon Stricker at the Institut für Allgemeine und Experimentelle Pathologie. After receiving the doctorate in 1880, he became an assistant in Stricker's laboratory, where he met Sigmund Freud, one year his junior. The two young men established a lifelong friendship strong enough to withstand not only the great differences in their personalities and temperaments, but also their later profound disagreements on a number of scientific questions.

Wagner von Jauregg came to psychiatry by chance when, after failing to obtain an assistantship at either of Vienna's teaching hospitals, he seized upon an opportunity to work under Max von Leidesdorf at the university psychiatric clinic in 1883. Although his decision had been a hasty one, he later observed that it had "harmed neither myself nor psychiatry." He quickly found his way in the new subject and by 1885 qualified as a teacher of neurology and, two years later, as a teacher of psychiatry.

During his years as Leidesdorf's assistant, Wagner von Jauregg established the two basic areas of his later research: the pathology of the thyroid gland, and the treatment of general paresis. In 1884 he had observed the remarkable behavior — including aggressive lunging, convulsions, and spasms — of cats that had been subjected to thyroidectomy. In 1889, upon succeeding Krafft-Ebing as professor of psychiatry at the University of Graz, Wagner von Jauregg undertook further studies of the function of the thyroid gland and stated his view that the phenomena of cretinism were caused by the impairment or failure of thyroid function. (These findings were later published in *Jahrbuch für Psychiatrie*, **12** [1894], 102–107, and **13** [1895], 17–36.) He also traveled throughout Styria studying goiter and assessing the effects of treatment by iodine tablets. By 1898 he had become convinced that the regular intake of small amounts of iodine was prophylactic against the disease, and proposed that iodized salt be sold in areas in which goiter was endemic — a measure that the Austrian government put into force in 1923, some years after Switzerland had taken similar measures.

Wagner von Jauregg's work in the treatment of general paresis began in 1887. He then noted that when psychiatric patients contracted infectious, febrile diseases, such as erysipelas or typhus, their mental state was substantially improved after the fever abated. He was consequently led to wonder if psychoses might be treated by inducing fever, and undertook a series of methodical sickbed observations. At this same early date he also speculated whether malarial infection might be used to treat general paresis, or creeping paralysis, as it was then known. In a series of experiments, he began infecting patients with tuberculosis, since Koch had, in 1890, made public his findings on tuberculin, and Wagner von Jauregg thought it too dangerous to induce malaria itself. He had some success, but achieved permanent remission in only a minority of patients.

In 1893 Wagner von Jauregg was called back to Vienna as full professor of psychiatry and director of the Psychiatric and Neurological Clinic. In this post he served as a member of the Austrian board of health, advising on all legislation concerning the mentally ill. It was during his tenure that modern laws, providing exemplary protection to the men-

tally incompetent, were formulated. At the same time, he carried on his own research on the febrile treatment of general paresis, utilizing staphylococci, streptococci, and typhus vaccines, but was unable to improve upon his results.

Despite the advances in the treatment of syphilis made in the early years of the twentieth century, the treatment of general paresis remained uncertain. Ehrlich's Salvarsan was not effective against the disease in its most advanced form, in which it attacked the central nervous system, and paretics still constituted, in the 1910's, some 15 percent of all patients in mental institutions. The life expectancy of such patients was, moreover, only three to four years. Wagner von Jauregg therefore decided to resume his experiments with the malarial treatment of paretics, especially since a number of studies had shown that malaria could be cured by the use of quinine. On 14 June 1917 he for the first time injected a paralytic patient with blood taken from a patient with tertian malaria; this and subsequent trials led to significant improvement of paretic patients and, in some instances, to complete remissions.

The method of fever-therapy was developed systematically from Wagner von Jauregg's findings, and applied from 1919 on. A number of hypotheses were put forward to explain its effectiveness, but Wagner von Jauregg himself believed that the injected malaria acted primarily by strengthening the defense of the organism against the *Spirochaeta pallida* that causes syphilis, thereby increasing its resistance to the poisonous substances produced. The malaria fever-therapy was widely used throughout the 1920's and 1930's, and in 1927 Wagner von Jauregg received the Nobel Prize for his part in its development. It was superseded in the mid-1940's with the introduction of antibiotics, particularly penicillin.

In addition to providing a remedy for a previously incurable disease, Wagner von Jauregg's work served as the basis for the acquisition of new knowledge about the biology of the malarial parasites. Through his experiments it was, for example, learned that *Plasmodium vivax*, the agent of benign tertian malaria, which normally produces the first attack of fever about fourteen days after infection, can have a latency period of as long as forty weeks should the infection occur in autumn, so that the first onset of fever then occurs the following spring or summer.

Wagner von Jauregg retired as director of the Vienna Psychiatric and Neurological Clinic in

1928. He left behind him a great school of psychiatry and neurology; it is characteristic of his generosity of mind that during his tenure the most widely divergent trends in modern psychiatry developed within it. He was tolerant of approaches to which he was not personally sympathetic (as, for example, Freudian psychoanalysis), and the richness and diversity of the psychiatric thinking that flourished in his school is reflected in the names of his students, including Konstantin von Economo, Hans Hoff, Johann Paul Karplus, Otto Kauders, Otto Pötzl, Emil Raimann, Paul Ferdinand Schilder, and Erwin Stransky.

BIBLIOGRAPHY

I. ORIGINAL WORKS. Wagner von Jauregg's works include "Über die Einwirkung fieberhafter Erkrankungen auf Psychosen," in *Jahrbuch für Psychiatrie und Neurologie*, **7** (1887), 94–134; "Zur Reform des Irrenwesens," in *Wiener klinische Wochenschrift*, **14** (1901), 293–296, *passim; Beiträge zur Ätiologie und Pathologie des endemischen Kretinismus* (Vienna, 1910), with Friedrich Schlagenhaufer; *Myxödem und Kretinismus* (Vienna–Leipzig, 1912); "Über die Einwirkung der Malaria auf die progressive Paralyse," in *Psychiatrisch-neurologische Wochenschrift*, **20** (1918–1919), 132–134; *Fieber und Infektionstherapie* (Vienna–Leipzig, 1936); and *Lebenserinnerungen*, L. Schönbauer and M. Jantsch, eds. (Vienna, 1950).

II. SECONDARY LITERATURE. On Wagner von Jauregg and his work, see J. Gerstmann, *Die Malariabehandlung der progressiven Paralyse* (Vienna, 1925); H. Hoff, in *Wiener klinische Wochenschrift*, **62** (1950), 888–889; J. P. Karplus, "Experiment und Klinik," in *Wiener medizinische Wochenschrift*, **82** (1932), 373–375; O. Pötzl, in *Wiener klinische Wochenschrift*, **50** (1937), 277; L. Schönbauer and M. Jantsch, "Julius Wagner Ritter von Jauregg," in K. Kolle, ed., *Grosse Nervenärzte* (Stuttgart, 1956), 254–266, with partial bibliography and secondary literature; E. Stransky, in *Wiener medizinische Wochenschrift*, **77** (1927), 1515–1516; and W. Weygandt, in *Münchener medizinische Wochenschrift*, **74** (1927), 547–548.

ERNA LESKY

WAHLENBERG, GÖRAN (Georg) (*b.* Skarphyttan, Sweden, 1 October 1780; *d.* Uppsala, Sweden, 22 March 1851), *botany.*

Wahlenberg began his studies at Uppsala University at the age of twelve under the guidance of a tutor. He soon decided to study medicine and natural history; botany became his major interest, and under Thunberg he acquired a thorough knowledge of plants. In 1806 Wahlenberg received his medical degree, having defended a dissertation on the sites of medically active substances within the plant body. He advanced very slowly in his career, and his financial situation was correspondingly poor for many years. From 1801 he had several positions, all at very low salaries, at the natural history collections of the university; and in 1814 he was promoted to the poorly paid post of botanical demonstrator. By 1828, when Wahlenberg succeeded Thunberg as professor of botany, he had acquired an international reputation and had completed almost all of his botanical work; his interest now centered on homeopathic cures.

Of the greatest importance for Wahlenberg's scientific development were the many voyages he made during his younger years. In 1799 he traveled to Gotland, in the Baltic Sea, which has remarkable calcareous flora; and he visited Lapland in 1800, 1802, 1807, and 1810. Through these journeys he gained an extensive knowledge of Scandinavian plants and their geographical distribution. In 1811–1814 Wahlenberg traveled in Germany, Switzerland, Austria, and Hungary, comparing the flora of the Alps and the Carpathians with that of the mountains of northern Europe. While visiting Berlin in 1811 he became friends with the botanist C. L. Willdenow and the geologist Leopold von Buch.

Wahlenberg seems to have considered his steadfast defense of the Linnaean tradition in its most limited sense as his most important contribution to botany. His declared ambition was never to alter the limit of a Linnaean species or name and never to abandon the Linnaean sexual system in any detail—however trifling; and his judgment of those whom he considered heretics was severe. His main scientific works were floras, based on his travels, in which he expounded these principles: *Gotlands flora* (1805–1806), *Flora lapponica* (1812), *Flora Carpathorum principalium* (1814), *Flora upsaliensis* (1820), and *Flora suecica* (1824–1826; second edition, 1831–1834). He is better known today, however, for the introductions of these floras than for the works themselves. The introductions place him among the pioneers of plant geography, worthy of comparison with Humboldt.

Wahlenberg's interest in geography dates from early in his career, and his love of cartography is reflected in the beautiful maps included in his works, which he drew. His geographical writings include *Geografisk och ekonomisk beskrifning om Kemi lappmark* (1804), *Berättelse om lappska fjällens höjd och temperatur* (1808; also in German,

1812), *Rön om springkällors temperatur* (1811), and *De vegetatione et climate in Helvetia septentrionali* (1813). Together with his botanical works they constitute his considerable contribution to phytogeography.

A keen observer of the distribution of plants, Wahlenberg made a definitive analysis of the stratification of vegetation on mountains with summits above the snow line and with treeless strata above regions dominated successively by different species of trees. His knowledge of the Alps, Carpathians, and mountains of Lapland enabled him to make comparisons that emphasized both the similarities and the differences in stratification. In Scandinavia, especially in Sweden, Wahlenberg also distinguished the main regions of the lower parts of the country, both those in the north-south succession and those oriented to the east or west according to their distance from or nearness to the coast.

In discussing the causes and circumstances of the differentiation of vegetation, Wahlenberg considered the most important factors to be the influence of climate (temperature and precipitation) and its dependence on latitude, altitude, and nearness to the sea; thus he was aware of oceanic as well as continental floreal elements. According to Wahlenberg the distribution of temperature during the seasons—not the mean annual temperature—was what determined the vegetation of a region. He also indicated the importance of soil temperature, which he measured in springs that did not freeze in winter, together with its difference from the temperature of the air—a difference varying, for instance, in relation to altitude.

Wahlenberg's awareness of the influence of climate on vegetation did not preclude his ascribing great importance to the soil. He also considered the complicated relationship between climate and soil, noting, for instance, that certain calcareous plants of the Carpathians could be found in noncalcareous soil of the Alps and Lapland.

Wahlenberg was also conscious of the time factor in phytogeography and maintained that the present distribution of plants is due to migrations occurring at various times and originating from different sites. Many areas were once submerged and consequently were invaded by their present vegetation only after emerging from the water. The distribution in other countries of Swedish plants indicated to Wahlenberg their migration either from the south, over the Danish islands, or from the vast forests of Finland and Siberia to the northeast.

It is difficult to assess Wahlenberg's true impor-

tance. His views on plant history were obviously influenced by Willdenow, who in his *Grundriss der Kräuterkunde* had stressed mountain regions (considered as former islands on an otherwise submerged earth) as centers of distribution of the plants found in lower regions. Priority for the idea of the influence of climate and geology on vegetation must be shared by Wahlenberg, Humboldt, and Buch. Their ideas, soon disseminated among younger phytogeographers, came to be considered so self-evident that few cared who had introduced them.

Wahlenberg had great difficulty in collaborating with other botanists. Considered odd and egocentric, he went his solitary way, isolated during his last twenty years from botany as well as from others. He never married. No student was allowed to enter the botanical garden, and in winter he forbade skating on the pond. Iron from the skates, he believed, would remain in the water when the pond thawed. His grazing cows would then ingest the iron with their water and thus, according to his homeopathic convictions, render the milk dangerous to drink.

BIBLIOGRAPHY

I. ORIGINAL WORKS. There is a complete list of Wahlenberg's botanical works in T. O. B. N. Krok, *Bibliotheca botanica suecana* (Uppsala–Stockholm, 1925), 741–745. An MS on phytogeography is in S. Borgman, "Göran Wahlenbergs handskrift 'Svensk växtgeografi,'" in *Svensk botanisk tidskrift*, **49** (1955), 337–347. Many letters and MSS are at the library of the University of Uppsala.

II. SECONDARY LITERATURE. There is a biography by E. Wikström in *Kungliga Svenska vetenskapsakademiens handlingar* (1851), 431–505, with bibliography. See also H. Krook, "Den siste linneanen," in his *Angår oss Linné?* (Stockholm, 1971), 104–115; and "Den unge Göran Wahlenberg," in *Nationen och hembygden*, **8** (1960), 188–211. Further works are A. Engler, "Die Entwickelung der Pflanzengeographie in den letzten hundert Jahren und weitere Aufgaben derselben," in *Wissenschaftliche Beiträge zum Gedächtniss der hundertjährigen Wiederkehr des Antritts von Alexander von Humboldt's Reise nach Amerika* (Berlin, 1899), 9–10, 164; and S. Lindroth, in *Kungliga Svenska Vetenskapsakademiens historia*, II (Stockholm, 1967), 424–429.

GUNNAR ERIKSSON

IBN WAHSHIYYA, ABŪ BAKR AHMAD IBN ᶜSALĪ IBN ĀL-MUKHTĀR (*b.* Qussīn, near Janbalā, Iraq, *ca.* 860; *d.* Baghdad, *ca.* 935), *agronomy,* *botany, alchemy, astrology, mysticism, medicine, toxicology, sorcery.*

Little is known about Ibn Waḥshiyya's life except that he was descended from the Nabataeans, the ancient inhabitants of Iraq (known in Arabic as the Nabaṭī). He was skilled and eloquent in their language, one of the West Aramaic group, and very proud of their culture and intellectual contributions. In view of their accomplishments in agriculture, commerce, arts, and applied sciences, Ibn Waḥshiyya said, the Nabataeans for centuries enjoyed a high degree of prestige.

He also practiced astrology in Baghdad during the period when it was a great cosmopolitan city that was a center of both intellectual and economic activity. He used talismans, charms, and incantations to tell fortunes and to heal the sick, and he wrote several books in this field.

Ibn Waḥshiyya was a contemporary of al-Rāzī, who, like him, upheld the art of the alchemist. They make no mention of each other in their writings, however. This is understandable because al-Rāzī was of a different class and a much more skillful physician, alchemist, and philosopher. As is evident from the titles of his books, Ibn Waḥshiyya's alchemical writings were full of sorcery, magic, symbolism, and talismans; while al-Rāzī's books, such as *Sirr al-asrār* and *al-Asrār*, were objective and free from magic and jugglery. Some doubt on Ibn Waḥshiyya's veracity and integrity is cast by his practice of legerdemain, expulsion of devils, and humbug—in addition to his exaggerated statements about the accomplishments of his forebears, the nicknames he included in his family tree (ibn Galatia, ibn Britania, and so on), and his contempt for other civilizations, even the Islamic. Ibn al-Nadīm listed his biobibliography under the section devoted to sorcerers who practiced "bad methods."

Ibn Waḥshiyya's best-known works are *al-Filāḥa al-nabaṭiyya*, on agriculture (allegedly claimed to be a translation from ancient Nabataean writings), and *al-Sumūm wa'l-tiryāqāt*, on poisons and their antidotes. The *Filāḥa* supposedly was completed in 904; but it was not dictated or copied until 930 by a student and associate, Aḥmad ibn al-Ḥusayn ibn ʿAlī ibn Aḥmad ibn Muḥammad ibn ʿAbd al-Malik al-Zayyāt (*d. ca.* 978), who also disseminated it. It is not clear whether al-Zayyāt contributed to the final copy of the *Filāḥa* as well as to *al-Sumūm*.

Both works supposedly were translations from ancient Aramaic texts, the author of which seems to have known similar, earlier writings in Sanskrit, Greek, and Persian. These two works contained significant ideas on agricultural practice and toxicology, and influenced later works on these topics in medieval Islam. The reference by Thomas Aquinas to the works of Ibn Waḥshiyya suggests that one or more of his writings, translated into Latin in the late twelfth or early thirteenth century, were influential in the West as well.

BIBLIOGRAPHY

I. Original Works. Ibn Waḥshiyya's two best-known works, *al-Filāḥa* and *al-Sumūm*, exist in several MSS, some incomplete; at the national libraries in Algiers, Berlin, Cairo, and Paris, the Süleymaniye Library at Istanbul, the Bodleian Library at Oxford, and the British Museum library. Ernest Renan, in *An Essay on the Age and Antiquity of the Book of Nabathaean Agriculture* (London, 1862), explained that Thomas Aquinas referred to Ibn Waḥshiyya's work in a Latin trans. (possibly *al-Filāḥa*), of which abstracted copies were known in Arabic.

His other important work, *al-Sumūm wa'l-tiryāqāt*, was translated into English with useful introduction and indexes by Martin Levey as "Medieval Arabic Toxicology, the Book on Poisons of Ibn Waḥshiyya and Its Relation to Early Indian and Greek Texts," which is *Transactions of the American Philosophical Society*, **56**, pt. 7 (Nov. 1966). I personally examined MSS of the *Sumūm* in Br. Museum, Add. 23, 604—see Charles Rieu, *Catalogus codicum manuscriptorum Orientalium qui in Museo Britannico asservantur*, II, *codices arabicos amplectens*, II (London, 1871), 461–462, 630–631—and in Zahiriyah Library, Damascus, gen. no. 9575, containing chs. 2–17.

The first to list Ibn Waḥshiyya's works was Ibn al-Nadīm of Baghdad, in his *Fihrist*, completed in 987 (Cairo, 1929), 447, 518–519. He mentioned some nine books on sorcery, talismans, and idol worship, most of which are lost (one is preserved in the Bodleian Library). Ibn Waḥshiyya's treatise *al-Asmā'*, or *al-Shawq al-mustahām fī ma'rifat rumūz al-aqlām*, was edited by J. Hammer as *Ancient Alphabets and Hieroglyphic Characters Explained* (London, 1806) and reprinted by Sylvestre de Sacy in Millins *et al.*, eds., *Magasin Encyclopédique*, XVI (1810), 145–175. Most of Ibn Waḥshiyya's approximately five books on alchemy and symbolism are extant in rare MSS and await evaluation.

His book on mysteries of planets and the firmament, attributed to Tankalūshā the Chaldean (Babylonian), is described in D. Chwolson, "Über die Überreste der altbabylonischen Literatur," in *Mémoires de l'Académie impériale des sciences de St.-Pétersbourg*, 6th ser., **8** (1859), 329–524; and Carlo Nallino, *Arabian Astronomy, Its History During the Medieval Times* (Rome,

1911), 198–210. The "Nabataean Agriculture" is discussed in A. von Gutschmid, "Die nabatäische Landwirtschaft und ihre Geschwister," in *Zeitschrift der Deutschen morgenländischen Gesellschaft*, **15** (1861), 82–89, and his *Kleine Schriften*, II (Leipzig, 1890), 677–678, 686–688. Three other works by Ibn Waḥshiyya on medical therapy, natural history, and theology are mentioned; but no extant MSS are known.

II. Secondary Literature. Ibn Abī Uṣaybiʿa alludes to Ibn Waḥshiyya's works on sorcery, specifically the *al-Adwār*, and on alchemy, in his *ʿUyūn al-Anbāʾ*, Būlāq ed., II (Cairo, 1882), 181, 203–204. Much later Ḥājjī Khalīfa, in his *Kashf al-ẓunūn*, II (Cairo, 1893), 101, 203, referred to Ibn Waḥshiyya's work on agriculture and on sorcery. Interest was renewed in Ibn Waḥshiyya's writings in the nineteenth century. See Lucien Leclerc, *Histoire de la médecine arabe*, I (Paris, 1876), 307–315; Ernst Meyer, *Geschichte der Botanik*, III (Königsberg, 1856), 43–88; and T. Nöldeke, "Noch Einiges über die nabatäische Landwirthschaft," in *Zeitschrift der Deutschen morgenländischen Gesellschaft*," **30** (1875), 445–455.

Later biobibliographies besides Levey's "Medieval Arabic Toxicology" (see above) are Carl Brockelmann, *Geschichte der arabischen Litteratur*, I (Leiden, 1943), 279–281, and *Supplement*, I, 430–431; and George Sarton, *Introduction to the History of Science*, I (Baltimore, 1927), 634–635.

On mathematics see L. C. Karpinski, "Hindu Numerals Among the Arabs," in *Bibliotheca mathematica*, n.s. **13** (1913), 97–98; and on his alleged translation of the "Nabataean Agriculture," see E. Wiedemann, "Zur nabatäischen Landwirtschaft," in *Zeitschrift für Semitistik*, **1** (1922), 201–202; and *Encyclopaedia of Islam*, II (Leiden, 1931), 427. Regarding his influence on medical botany and pharmacy see Sami Hamarneh, *Catalogue of Arabic Manuscripts on Medicine and Pharmacy at the British Library* (Cairo, 1975), 60–64.

Sami K. Hamarneh

WALCH, JOHANN ERNST IMMANUEL (*b.* Jena, Germany, 29 August 1725; *d.* Jena, 1 December 1778), *theology, philology, paleontology.*

Walch was the son of Johann Gottlob Walch, professor of theology at Jena, and Charlotte Katharina Buddeus. He received private instruction at home until the age of seventeen, when he entered the University of Jena to study theology and philology. He received his master's degree in 1745 and began lecturing on theology the following year. During 1747 and 1748 he traveled throughout Europe and met scholars from many universities. In 1750 he became assistant professor and in 1755 full professor of logic and metaphysics. Appointed professor of oratory and poetry in 1759, he was twice rector and eight times dean of the university, becoming privy councillor in Weimar in 1770. He declined appointments offered by several other universities.

Walch has been described as a friend of everything beautiful and good, a lecturer who aroused the enthusiasm of his large audiences, and a scholar who spent his life in the service of noble endeavor and scientific work. He was a member of many German and foreign scientific societies. Most of his scholarly publications concern the classical languages; and his work in this field has frequently been valuable in New Testament exegesis, his *Introductio in linguam graecam* being of especial importance. As crucial as these studies were for the advance of philology, however, they are rarely consulted nowadays.

The situation is different with regard to Walch's paleontological works, which are of continuing value. How little progress the subject had made toward specialization at that time is confirmed by the disparate subjects that he was able to examine simultaneously. In his travels Walch had visited natural history collections and had begun to assemble one himself, composed primarily of rocks and fossils, for which he outlined an exact system that he published in 1761–1764. During this period a copperplate engraver in Nuremberg, Georg Wolfgang Knorr (1705–1761), had become famous for his magnificent illustrations, including some in the natural sciences, and had contributed to science even though he was not a scientist. Before his death he had published, with an unscientific text, the first part of a work on fossils. On behalf of Knorr's heirs and using the plates Knorr had prepared, Walch continued its publication under its original title and later with the title *Die Naturgeschichte der Versteinerungen* (1762–1773; the index [1774] is by Johann Samuel Schröter).

In this work Walch presented the first comprehensive paleontology ordered according to the zoological system; it is still occasionally consulted. For him the fossils were not—as had previously been assumed—evidence of the Flood but, rather, of the displacements of the seas. Moreover, in addition to a basic systematics of all the forms known to him, Walch provided a general paleontology (deposition, sedimentary facies, facies distribution), and a history of paleontology that is still worth reading. His recognition that fossils are members of a sequence linked by historical descent was a fundamental perception. Walch made the previously muddled study of fossils into a science.

In the journal *Naturforscher*, which he founded and which continued publication until 1804, Walch rounded out his chief work with many studies of particular problems and findings.

BIBLIOGRAPHY

I. ORIGINAL WORKS. Walch's major publications are *Commentationes quibus antiquorum christianorum doctorum de jureiurando sententiae percensentur et deiudicantur* (Jena, 1744); *De vinculis Paulli apostoli* (1746); *Einleitung in die Harmonie der Evangelisten* (1749); *Marmor Hispaniae antiquum* (Jena, 1750); *Antiquitates Herculanenses litterariae* (Jena, 1750); *Dissertationes in Acta Apostolorum* (1756; 2nd ed., 1759; 3rd ed., 1761); *De arte critica veterum Romanorum* (Jena, 1757); *Die Naturgeschichte der Versteinerungen zur Erläuterung der Knorrischen Sammlung von Merkwürdigkeiten der Natur*, 3 vols. (Nuremberg, 1762–1774); *Das Steinreich systematisch entworfen*, 3 vols. (Halle, 1762–1764); *Introductio in linguam graecam* (1763; 2nd ed., 1772); "Neue lithologische Entdeckungen," in Schröters *Journal für die Liebhaber des Steinreichs . . .*, **1**, no. 2 (1773), 310–320; and "Von dem Schwerdt- oder Sägefisch des Herrn Bürgermeister Bauder in Altdorf," *ibid.*, **2** (1775), 376–378.

II. SECONDARY LITERATURE. Biographical material is in von Dobschütz, "Johann Ernst Immanuel Walch," in *Allgemeine deutsche Biographie*, XL (Leipzig, 1896), 652–655. His importance as a paleontologist is treated in B. von Freyberg, *Die geologische Erforschung Thüringens in älterer Zeit* (Berlin, 1932); and "Johann Friedrich Bauder (1713–1791) und seine Bedeutung für die Versteinerungskunde in Franken," in *Geologische Blätter für Nordost-Bayern . . .*, **8** (1958), 76–106; and K. A. von Zittel, *Geschichte der Geologie und Paläontologie* (Munich–Leipzig, 1899).

B. VON FREYBERG

WALCOTT, CHARLES DOOLITTLE (*b.* New York Mills, New York, 31 March 1850; *d.* Washington, D.C., 9 February 1927), *paleontology.*

Walcott contributed significantly to knowledge of Cambrian faunas and rocks, and was an exceptionally able administrator of science for the federal government. The leader during his time in studies of Cambrian rocks and fossils, he began his scientific career, without benefit of college training, when he moved near Trenton Falls, New York. There he found one of the first occurrences of trilobites with appendages preserved. His work (1875–1881) on trilobites, amplified forty years later, contributed substantially to establishing the zoological importance and position of this group.

After a year's work under James Hall in Albany, New York, Walcott joined the newly formed U.S. Geological Survey in 1879. Although it was here that he made his reputation with studies of the Cambrian, he engaged in other stratigraphic studies as well. His *Paleontology of the Eureka District* (Nevada) (1884) was a standard reference for western fossils.

During the mid-1880's Walcott became involved in the "Taconic" question, concerning the age of rocks at the eastern boundary of New York state. By finding new fossil localities and reinterpreting earlier data, he resolved some problems controversial for half a century. Shortly afterward, he was able to establish that Cambrian rocks at St. John, New Brunswick, had been affected by structural complications. This interpretation brought Cambrian fossil zones of North America into harmony with those established earlier in Europe.

During the 1890's Walcott's fieldwork took him throughout the country, but this period is best marked by a major work on fossil jellyfish (1898). During this and the subsequent decade he completed a number of papers on Cambrian Brachiopoda. This work culminated in the two-volume *Cambrian Brachiopoda* (1912), a worldwide study that considered their biology as well as their stratigraphic position.

Starting in 1907, Walcott extended his fieldwork to the high mountains of Alberta and British Columbia, a rugged area he visited almost annually for the next two decades. His contributions to Cambrian geology and paleontology from this area fill five volumes of *Smithsonian Miscellaneous Collections.* The most significant event was the discovery in 1909 of the Middle Cambrian Burgess shale deposit. Three years of hard work quarrying these rocks resulted in a spectacular collection of numerous soft-bodied organisms of the sort that are almost never preserved as fossils. This has been described by some authorities as the single most important find of fossils, and its discovery and study would certainly have brought him worldwide fame even if it had been the only work in Walcott's career. His work was so voluminous, however, that at the time of his death he had described about one-third of all Cambrian fossils then known; he was probably the second or third most prolific student of American paleontology.

Besides his scientific career, Walcott had a remarkable record as an administrator. He rose through the ranks of the U.S. Geological Survey and in 1894 succeeded John Wesley Powell, becoming the third director of the Survey. From 1902 to 1907 he headed both the Survey and the U.S. Reclamation Service, and for eighteen months (1897–1898) he served as acting assistant secretary of the Smithsonian Institution.

Following the death of S. P. Langley, Walcott was appointed secretary of the Smithsonian Institution in 1907 and resigned as director of the Geological Survey. During the early years of his secretaryship the Museum of Natural History opened; and in the later years he was able to convince C. L. Freer to allow construction of the gallery to house his art collection. This change of plans – Freer had wished to wait until after his death – was in large measure an indication of his confidence in Walcott.

In 1915 Walcott founded the National Advisory Committee for Aeronautics. He was also one of a small group who approached Andrew Carnegie to request support for basic research. This effort eventually led to the founding of the Carnegie Institution of Washington, with Walcott as one of the original incorporators.

F. G. Cottrell had offered his patents to the Smithsonian in 1911. Although they could not legally be accepted, Walcott and Cottrell conceived the idea of a foundation to supply the Smithsonian and other organizations with funds for scientific research, an idea that was developed into the Research Corporation. He also was active in the organization of the National Park Service.

Walcott was president of the National Academy of Sciences from 1917 to 1923 and was a founder of the National Research Council. His scientific accomplishments brought him the presidencies of, and medals from, several societies and a dozen honorary degrees from universities.

BIBLIOGRAPHY

A complete bibliography of Walcott's works may be found in Ellis L. Yochelson, "Charles Doolittle Walcott, 1850–1927," in *Biographical Memoirs. National Academy of Sciences,* **39** (1967), 516–540.

ELLIS L. YOCHELSON

WALD, ABRAHAM (*b.* Cluj, Rumania, 31 October 1902; *d.* India, 13 December 1950), *mathematical statistics, mathematical economics, geometry.*

Wald was born into a family which had considerable intellectual interests but had to earn its livelihood in petty trade because of anti-Jewish restrictions. These restrictions made his education at the University of Cluj, and later at the University of Vienna, very difficult. After Hitler occupied Austria in 1938, Wald moved to the United States, which saved him from death in a German concentration camp – the fate of all but one other member of his numerous family. He later married Lucille Lang, an American, who died with him in an airplane crash.

At Vienna, Wald was a student and a protégé, and later a friend, of Karl Menger. His work in pure mathematics was largely, although not wholly, in geometry. Menger later directed Wald toward mathematical statistics and mathematical economics, so that he was able to find employment with the distinguished economist Oskar Morgenstern.

It seems reasonable to say that Wald's most important work was in statistics, both because of his relative importance in the field and because of the current assessment of the field's importance. One of his great contributions to statistics was to bring to it mathematical precision in the formulation of problems and mathematical rigor in argument. These qualities, which were often lacking when he began his statistical career in 1938, have transformed the subject – although not necessarily to the satisfaction of everyone. It should be emphasized, however, that these accomplishments were a by-product and consequence of his extraordinary ability and the breadth of his statistical interests. Wald wrote lucidly and unambiguously on many statistical subjects, and there is scarcely a branch of modern statistics to which he did not contribute. His writings and lectures were so lucid and so unambiguous because of this precision, and he achieved so much in the way of results, that the superiority of mathematical precision became apparent to all. It is impossible to discuss Wald's statistical results in detail; rather, we shall single out the two most important fields of his work, which he founded and in which his results still dominate: sequential analysis and the theory of decision functions.

In sequential analysis one takes observations seriatim until the evidence is sufficiently strong, bearing in mind certain previously imposed bounds on the probabilities of error. When there are only two possible hypotheses, the "Wald sequential probability ratio test" has the property that it requires the smallest average number of observations under either hypothesis. This famous "optimum property of the sequential probability ratio test" was brilliantly conjectured by Wald in 1943 and proved jointly by him and a colleague in 1948. Wald proved many theorems on the distribution of the required number of observations and obtained many approximations on probabilities of error and average required numbers of observations, that are still used in applications. Most, although not all, of his results were summed up in *Sequential Analysis* (1947). With minor exceptions, the entire contents of this book were obtained by him. Such a phenomenon is rare in mathematical books and indicates the extent to which he founded and dominated the field of sequential analysis.

When Wald began his work in statistics, a large part of the field was concerned with the theory of testing hypotheses. He regarded this theory as, at best (when properly interpreted), one of deciding between exactly two courses of action. Consequently very many statistical problems actually fall outside the scope of this theory. There was no consistent theory for deciding among more than two courses of action, and attempts to force such problems into the framework of the theory of testing hypotheses had yielded very unsatisfactory results. It is interesting that these objections were clearly realized by a theoretician like Wald and not by the practical statisticians in industrial and agricultural laboratories who applied the theory. (For a recent criticism of the theory see Wolfowitz, "Remarks on the Theory of Testing Hypotheses.") Wald's theory of statistical decision functions considers the problem of deciding among any number of (possibly infinitely many) courses of action, both sequentially and nonsequentially. The statistician introduces a loss function that measures the consequences of various actions under different situations. With each statistical procedure (decision function) there is associated a vector, or function, of average loss under the various possible situations (the risk function). The statistical procedures of which the risk functions are not inferior to those of any other form a "complete" class, and the statistician can properly ignore the procedures not in the complete class.

At Wald's death the theory of statistical decision functions was far from the point of application to everyday, practical statistical problems; and little progress has been made in this direction since then. The theory is still of great conceptual and theoretical importance, and provides a logical basis for the formulation of many research problems. Recent research in the theory itself has, however, been chiefly in the direction of very technical mathematical refinements and has not achieved any essential breakthroughs.

Some of Wald's work in statistics originated in economic problems and properly belongs to both subjects. One such example is his work on the identification of economic relations—roughly speaking, the problem whether the distributions, which result from a model of the observed chance variables, uniquely determine all or certain specified parameters of the model. Also included in this category is his work on stochastic difference equations—models involving sequences of chance variables connected by difference equations with "error" chance variables. Wald also proved theorems on the existence of unique solutions for systems of equations for several types of economic systems and studied cost-of-living index numbers, the empirical determination of indifference surfaces, and the elimination of seasonal variation in time series. In all these his methods were ingenious and his contributions very important.

In pure mathematics, Wald's first three published papers and "Zur Axiomatik des Zwischenbegriffes" dealt with the characterization of "betweenness" in metric spaces. He also extended Steinitz's theorem to vectors with infinitely many elements; the theorem states that a divergent series, the elements of which are finite vectors, can, by a permutation of its terms, be made to converge to any element of a linear manifold. Perhaps his best result was the development of a differential geometry that starts from the assumption of a convex, compact metric space that at every point admits what should be called a Wald curvature. From this he was able to derive properties of differential geometry that are postulated in other systems.

Relatively uninterested in mathematical elegance, Wald spent little time in polishing a paper after a problem was solved to his satisfaction. In his masterly hands simple methods sometimes yielded the most amazing results. Although he was readily accessible, he had very few students. With one of these, J. Wolfowitz, who became his friend and colleague, he wrote fifteen joint papers. His American, and largely statistical, period was relatively brief (1938–1950) and extraordinarily productive. During this time he learned mathematical statistics, contributed deeply to it, changed it essentially, and dominated the subject. It has borne his impress since, and the paths he opened are still being pursued.

BIBLIOGRAPHY

I. ORIGINAL WORKS. A comprehensive bibliography of Wald's writings follows Tintner's memoir (see below). His works include "Zur Axiomatik des Zwischenbegriffes," in *Ergebnisse eines mathematischen Kolloquiums*, **4** (1933), 23–24; *Sequential Analysis* (New York–London, 1947); and "Optimum Character of the Sequential Probability Ratio Test," in *Annals of Mathematical Statistics*, **19** (1948), 326–339, written with J. Wolfowitz.

II. SECONDARY LITERATURE. On Wald and his work, see J. Wolfowitz, "Abraham Wald, 1902–1950," in *Annals of Mathematical Statistics*, **23** (1952), 1–13; Karl Menger, "The Formative Years of Abraham Wald and His Work in Geometry," *ibid.*, 14–20; G. Tintner, "Abraham Wald's Contributions to Econometrics," *ibid.*, 21–28; and "The Publications of Abraham Wald," *ibid.*, 29–33, which lists 103 works (1931–1952).

See also J. Wolfowitz, "Remarks on the Theory of Testing Hypotheses," in *New York Statistician*, **18**, no. 7 (Mar. 1967), 1–3.

J. WOLFOWITZ

WALD, FRANTIŠEK (*b*. Brandýsek, near Slaný, Czechoslovakia, 9 January 1861; *d*. Vítkovice [now part of Ostrava], Czechoslovakia, 19 October 1930), *chemistry*.

Wald's father came from Chemnitz (now Karl-Marx-Stadt, D.D.R.) and became foreman in the workshops of the Austrian State Railways at Slaný. His mother was a native German from the Karlovy Vary (Karlsbad) district. Although of German origin, Wald adopted Czech nationality. He attended a Czech municipal school at Kladno, a center of the iron industry, to which his parents moved after the Austro-Prussian War (1866). The thorough grounding in elementary mathematics that he received prepared him well for his essentially nonclassical secondary education at a Czech school in Prague. When he obtained a grant from the Austrian State Railways, Wald was obliged to leave the Czech school and attend its German counterpart. After finishing school he studied chemistry at the German technical university in Prague until 1882 without formally taking a degree, because it was not required for appointment as a technician in industry.

Wald joined the laboratory of Pražská železářská společnost, a leading ironworks at Kladno, and in 1886 he was appointed chief chemist. Although a gifted analytical chemist—he devised ingenious gasometric and other methods of making mining and metallurgical practice more scientific—he gradually came to devote himself to an examination of the theoretical basis of chemistry. Several efforts to procure Wald an appointment to a professorship by, among others, Ostwald and Mach,

were unsuccessful. Finally the chair of theoretical and physical chemistry and metallurgy at the Czech Technical University in Prague was offered to him (1908), and he held it until his retirement twenty years later. During his tenure Wald was twice elected dean of the Faculty of Chemical Technology and was rector of the university (1920–1921).

Dissatisfied with the atomic-molecular interpretation of chemical phenomena, Wald initially turned his attention to the first two laws of thermodynamics. Summarizing his views in *Die Energie und ihre Entwertung* (Leipzig, 1888), he argued that it was an error to elaborate the second law of thermodynamics on the assumption that processes encountered in nature were reversible when actually they were not. For Wald the second law of thermodynamics was based on experience (*Erfahrungsgesetz*) and could be deduced logically, without the aid of mathematics. As for the first law, he believed in the quantitative principle of the conservation of energy but questioned the qualitative equivalence between work and heat. The amount of heat into which a certain amount of work was transformed was not really equivalent to the original amount of work, since it did not possess the same quality or effectiveness (*Wirkungsfähigkeit*). Energy did not disappear, but degenerated; and Wald accepted the inference that heat that could not be usefully transformed into work or other forms of energy was accumulating in the universe. He supposed, however, that the state of uniform temperature or "heat death" to which the universe was tending would be reached only in infinite time.

Wald disapproved of the accepted theoretical basis of chemistry, on the ground that it was too hypothetical. He believed that natural compounds had rather a varying composition. Apparently never conceding the general validity of atomic considerations for chemical theory, Wald attempted to work out a system in which the atom was replaced by the more tangible "phase" as the fundamental concept. He treated the subject extensively in his second book, *Chemie fází* (Prague, 1918). Wald's efforts to establish a general chemical theory on the basis of the phase concept clearly had some relation to his familiarity with problems of phase equilibrium in metallurgical practice. The Russian chemist N. S. Kurnakov, who was thinking along similar lines, valued Wald's work very highly. Wald was deeply impressed by the predictive and controlling faculty of scientific chemistry, which made the chemist as powerful as nature.

Unlike many contemporary scientists, Wald thought that philosophy could not be kept out of science and criticized theoretical chemistry because of its philosophical shallowness. As a convinced idealist he rejected the mechanist interpretation of natural phenomena but, interestingly, admitted both free will and necessity. He recognized the existence of both and did not feel compelled to choose either one or the other. In an article on the theory of chemical operations, written only a year before his death, he proclaimed that this choice, imposed upon man, was unnecessary because each—natural law and human will—had its domain of influence, with conscious practical activity as the mediator. Although it was an insoluble puzzle to many, Wald had no difficulty in reconciling the two seemingly exclusive conceptions and in perceiving the link between them.

BIBLIOGRAPHY

I. Original Works. A list of Wald's publications, compiled by A. Šimek, in *Collection of Czechoslovak Chemical Communications*, **3** (1931), 3–8, does not include Wald's notices and popular articles on scientific and technical topics. "The Foundations of a Theory of Chemical Operations" appeared in Czech in *Přírodovědecký sborník*, **6** (1929); an English trans. appeared in *Collection* (see above), 32–48, and a condensed version in G. Druce, *Two Czech Chemists*, 57–61 (see below).

II. Secondary Literature. Originally planned as a *Festschrift* in honor of Wald's seventieth birthday, *Collection* (see above) contains articles by J. Baborovský on Wald's life (in English), by A. Kříž on Wald's theory of phases and of chemical stoichiometry (in English), and by Q. Quadrát on Wald's contribution to analytical chemistry (in French). These articles served as the main source of information for G. Druce, *Two Czech Chemists: Bohuslav Brauner (1855–1935) František Wald (1861–1930)* (London, 1944). M. Teich discusses Wald's place in the history of chemical thought and practice in Bohemia in *Dějiny exaktních věd v českých zemích*, L. Nový ed. (Prague, 1961), 334–335, 347–351 (in Czech, with Russian and English summaries). See also M. Teich, "Der Energetismus bei Wilhelm Ostwald und František Wald," in *Naturwissenschaft, Tradition, Fortschritt-Beiheft zur Zeitschrift für Geschichte der Naturwissenschaften, Technik und Medizin*, supp. (1963), 147–153. Although neglected it is not quite correct that Wald's work has been almost completely ignored, as Joachim Thiele maintains in "Franz Walds Kritik der theoretischen Chemie (nach Arbeiten aus den Jahren 1902–1906 und unveröffentlichten Briefen)," in *Annals of Science*, **30** (1973), 417–433.

M. Teich

WALDEN, PAUL (also known as PAVEL IVANOVICH VALDEN) (*b.* Rosenbeck parish, Wenden district [now Latvian S.S.R.], Russia, 26 July 1863; *d.* Gammertingen, Germany, 22 January 1957), *chemistry.*

The son of Latvian farmers, Walden was orphaned as a child and was obliged to earn his living as a private tutor. In 1882 he entered the chemistry department of the Riga Polytechnical School, where he began his scientific work under F. W. Ostwald. His first scientific work led to the discovery of the Ostwald-Walden empirical rule, which makes it possible to determine the basicity of multiatomic acids and bases according to molar (gram molecular) electroconductivity (1887). After Ostwald moved to Leipzig, Walden became Carl A. Bischoff's assistant and turned his interest to organic stereochemistry. However, he did not abandon his initial work in electrochemistry; his first doctoral dissertation (1891) was devoted to the determination of the affinities of organic acids by conductometric methods. After graduating in 1889, Walden remained at the Polytechnicum as an assistant (since 1888), becoming professor in 1894. His work with Bischoff, his visit to Adolf von Baeyer, and his frequent visits to Ostwald's laboratory in Leipzig enabled him to combine the viewpoints of organic and physical chemistry, and Walden set as his lifework the synthesis of these two disciplines. He became one of the founders of physical organic chemistry.

Walden's stereochemical research led him to the discovery of "Walden's inversion" (1896), so named by Emil Fischer, in which one optical isomer is converted into its optic antipode by the action of specific reagents so that a change in absolute configuration occurs. Because it did not coincide with existing representations of substitution reactions, the Walden inversion elicited an extended discussion. The mechanism of Walden inversion was clarified in 1934–1937 by E. D. Hughes and C. K. Ingold. They demonstrated that inversion is always involved in nucleophilic substitution reactions involving two steps (S_N2-mechanism).

Walden also conducted detailed studies of autoracemization, sought to relate the degree of specific rotatory power to the chemical structure of an organic molecule, and attempted to substantiate the presence of optically active compounds in crude oil (an argument for biogenesis of petroleum, first mentioned by Walden). These data formed the basis for his second doctoral dissertation, on optical isomerism. After defending it in St. Petersburg in 1899 and becoming professor of inorganic and analytical chemistry at the Riga Polytechnical Institute, Walden embarked upon a study of the electrochemistry of nonaqueous solutions. Between 1900 and 1934 he determined the degree of ionization of about fifty polar nonaqueous solvents, including liquified SO_2, SO_2Cl_2, $SOCl_2$, chlorosulfonic acid, anhydrous sulfuric acid, formamide, nitromethane, esters, and acid anhydrides. He introduced the concepts of solvation and solvolysis, and pioneered in the representation of the ionic mechanism of several organic reactions. In 1906 he introduced a formula, relating the viscosity of the solvent η and the equivalent electroconductivity of a given electrolyte λ, $\lambda_\infty \cdot \eta_\infty = $ constant, where λ_∞ is the equivalent electroconductivity at infinite dilution of a certain ion (electrolyte), and where η_∞ is the viscosity of the solvent at infinite dilution of a certain ion (electrolyte). Walden's work facilitated the rapprochement of the physical and chemical theories of solutions, and the rules that he discovered furthered the construction of the modern theory of acids and bases, the theory of electrolytes, and the study of the mechanism of organic reactions.

Walden earned international recognition for having established empirical laws relating surface tension, critical parameters, and hidden heat of fusion to the molecular weight (degree of association) of liquids, thereby contributing to the study of intermolecular forces in liquids.

Elected a member of the St. Petersburg Academy of Sciences in 1910, Walden combined his professorial duties in Riga with his directorship of the Academy's chemistry laboratory in St. Petersburg (since 1911). He devised a project for establishing an academic institute of chemistry in St. Petersburg, and he was active in scientific education in Russia, furthering the rational use of the nation's natural resources. He wrote the first extensive work on the history of chemistry in Russia (1914, printed in 1917). A founder of Latvia University in Riga, he became its first rector in 1919.

Following the establishment of the postwar nationalist regime in Latvia, Walden emigrated to Germany in August 1919 and became professor of chemistry at the University of Rostock. He continued his research on the electrochemistry of nonaqueous solutions, on which he wrote several monographs. After retiring in 1934 he devoted himself almost exclusively to the history of chemistry. During World War II he moved to Frankfurt-am-Main, and then to Tübingen, where until the age of ninety he lectured on the history of chemistry at the university. His later writings were dedicated to the history, psychology, and logic of chemistry; several were of an autobiographical character. He received honorary doctorates from several universities, was elected a member of scientific academies, and was an honorary member of the Chemical Society of London.

BIBLIOGRAPHY

I. ORIGINAL WORKS. Walden published more than 300 scientific papers. His writings include "Ocherk istorii khimii v Rossii" ("Essay on the History of Chemistry in Russia"), intro. to A. Ladenburg, *Lektsii po istorii razvitia khimii ot Lavuazie do nashikh dney* ("Lectures on the History of the Development of Chemistry From Lavoisier to the Present"; Odessa, 1917, pp. 361–654); *Nauka i zhizn* ("Science and Life"), 3 pts. (Petrograd, 1919–1921); *Optische Umkehrerscheinungen (Waldensche Umkehrung)* (Brunswick, 1919); "Elektrochemie nichtwässriger Lösungen," in *Handbuch der angewandten physikalischen Chemie*, XIII (Leipzig, 1924); "Leitvermögen der Lösungen" (3 pts.), in F. W. Ostwald et al., *Handbuch der allgemeinen Chemie*, IV a, IV b (Leipzig, 1924); *Chemie der freien Radikale* (Leipzig, 1924); and *Geschichte der organischen Chemie seit 1880* (Berlin, 1941).

II. SECONDARY LITERATURE. See W. Hückel, "Paul Walden, 1863–1957," in *Chemische Berichte*, **91** (1958), xix–lxv, with complete bibliography of Walden's writings; J. Stradiņš, *Cilvēki, eksperimenti, idejas* (Riga, 1965), 217–258; *Materialy dlya biograficheskogo slovarya deystvitelnykh chlenov imperatorskoy Akademii nauk* ("Materials for a Biographical Dictionary of Members of the Imperial Academy of Sciences"), I (Petrograd, 1915); J. Stradiņš, "K biografii Paula Valdena" ("On the Biography of Paul Walden"), in *Iz istorii estestvoznania i tekhniki Pribaltiki*, 1 (1968), 157–167; Y. Soloviev and J. Stradiņš, "Paul Valden kak istorik khimii" ("Paul Walden as a Historian of Chemistry"), *ibid*, 5 (1976); and I. Walden-Hollo, "Vospominania ob ottse" ("Recollections of My Father"), in *Nauka i tekhnika* (Riga, 1975), no. 3, 33–35; no. 4, 33–35.

JANIS STRADIŅŠ

WALDEYER-HARTZ, WILHELM VON (*b.* Hehlen, Germany, 6 October 1836; *d.* Berlin, Germany, 23 January 1921), *anatomy.*

Waldeyer was the son of Johann Gottfried Waldeyer, an estate manager, and Wilhelmine von Hartz, the daughter of a schoolteacher. He received his early education at Paderborn, then in 1856 entered the University of Göttingen to study natural sciences. Having attended the lectures of the great anatomist Friedrich Gustav Jacob Henle, however, he changed his course to medicine; indeed, Johannes Sobotta, Waldeyer's best-known student, stated his belief that Waldeyer not only studied medicine but also became an anatomist under Henle's influence. Waldeyer was unable to complete his studies at Göttingen, since it was the university of the kingdom of Hannover and did not, at that time, grant examination certificates to Prussians; he therefore transferred to Greifswald, where he remained until he went to Berlin to qualify. He was drawn there by the reputation of Karl Reichert, the anatomist and embryologist. Believing that a sound knowledge of embryology was essential to an anatomist, Waldeyer finished his studies with Reichert, under whose direction he prepared a doctoral dissertation of the structure and function of the clavicle, published in 1862.

Waldeyer then went to the University of Königsberg as an assistant in the department of physiology. He also taught histology, and became acquainted with the anatomist Friedrich Leopold Goltz. In 1864 he moved on to the University of Breslau, where he had been appointed lecturer in physiology and histology and was also responsible for a service department in pathology. This marked the beginning of his interest in that subject, and he published a number of papers on pathology, including one on the histological changes in muscles following typhoid fever, that led to his appointment, at the age of twenty-nine, as professor of pathology and director of a department of postmortem investigations. In 1866, soon after he received this post, Waldeyer married; in 1868, when he was thirty-two, he was appointed to the chair of pathology. His work at this time was chiefly concentrated on the diagnosis of early cancer, and won him considerable renown; in 1887 he was one of the German doctors called upon to diagnose the Emperor Frederick III's tumor of the vocal cords.

In 1872 Waldeyer went to the University of Strasbourg. The conquest of Alsace by Prussia in the preceding year had resulted in the forced resignation of French professors from that university, and both Waldeyer, who was appointed to the chair of anatomy, and Goltz, who became professor of physiology, were among the Germans who were installed in their stead. Waldeyer remained at Strasbourg for eleven years, then returned to Berlin to succeed Reichert. At Berlin he found an outdated laboratory and a large number of students, but he proved to be a highly successful administra-tor and teacher, and remained there as director of the anatomy department for over thirty-three years. His academic duties required his full time and energies, and after his relocation in Berlin Waldeyer performed little original research; he was an excellent teacher of both anatomy and histology, however, and he and his student Sobotta offered courses that must remain unsurpassed in their careful and varied presentation.

Indeed, Waldeyer's fame as an anatomist derives largely from his brilliant, lucid, and systematic (as they were styled by his contemporaries) lectures. He nevertheless published a significant number of papers on a wide variety of morphological subjects, including studies of the urogenital system, anthropology, the spinal cord of the gorilla, and topographical observations of the pelvis. He was receptive to new ideas, and quickly grasped the importance of, for example, the neurohistological studies of Ramón y Cajal; he himself coined the word "neuron," and helped to lay the foundation upon which the neuron doctrine was established. (He also coined the term "chromosome" to describe the bodies in the nucleus of cells and invented a number of embryological terms, including those that describe the structure of developing teeth, that are still in use.) Waldeyer also published the first description—both embryological and functional—of the naso-oro-pharyngeal lymphatic tissue; Waldeyer's tonsillar ring, the ring of lymphoid tissue formed by the lingual, pharyngeal, and faucial tonsils that encircles the throat or pharynx, is named for him.

Waldeyer remained at the University of Berlin until he was eighty years old, carrying out all the duties that his position imposed. He remained physically and mentally fit until his death, following a stroke, five years later. Of the four children who survived him, none entered medicine or science.

BIBLIOGRAPHY

I. ORIGINAL WORKS. Waldeyer's works include his inaugural dissertation, *De claviculae articulis et functione* (Berlin, 1862); "Untersuchungen über den Ursprung und den Verlauf des Achsenzylinders bei Wirbellosen und Wirbeltieren, sowie über dessen Endverhalten in der quergestreiften Muskelfaser," in *Zeitschrift für rationelle Medicin*, **20** (1863), 193–256; "Über die Endigung der motorischen Nerven in den quergestreiften Muskeln," in *Zentralblatt für die medizinischen Wissenschaften*, **24** (1863), 369–372; *Untersuchungen über die*

Entwicklung der Zähne (Danzig, 1864); "Anatomische und physiologische Untersuchungen über die Lymphherzen der Frösche," in *Zeitschrift für rationelle Medicin*, **21** (1864), 103; "Die Veränderungen der quergestreiften Muskelfasern beim Abdominaltyphus," in *Zentralblatt für die medizinischen Wissenschaften*, **7** (1865), 97–100; *Eierstock und Ei. Ein Beitrag zur Anatomie und Entwicklungsgeschichte der Sexualorgane* (Leipzig, 1870), written with W. Engelmann; "Diffuse Hyperplasie des Knochenmarkes: Leukaemie," in *Virchows Archiv für pathologische Anatomie und Physiologie und für klinische Medizin*, **52** (1871), 305–317; "Die Entwicklung der Carcinome," *ibid.*, **55** (1872), 67–159; "Über die Beziehungen der Hernia diaphragmatica congenita zur Entwicklungsweise des Zwerchfells," in *Deutsche medizinische Wochenschrift*, **14** (1884), 211–212; and "J. Henle. Nachruf," in *Archiv für mikroskopische Anatomie und Entwicklungsmechanik*, **26** (1885), 1–32.

See also "Beiträge zur normalen und vergleichenden Anatomie des Pharynx mit besonderer Beziehung auf dem Schlingweg," in *Sitzungsberichte der Königlich Preussischen Akademie der Wissenschaften zu Berlin*, **12** (1886), 233–250; "Über Karyokinese und ihre Beziehungen zu den Befruchtungsvorgängen," in *Archiv für mikroskopische Anatomie und Entwicklungsmechanik*, **32** (1888), 1–122; "Das Rückenmark des Gorilla, vergleichen mit dem des Menschen," in *Korrespondenzblatt der Deutschen Gesellschaft für Anthropologie, Ethnologie und Urgeschichte*, **19**, no. 10 (1888), 112–113; "Das Gorillarückenmark," in *Abhandlungen der Preussischen Akademie der Wissenschaften* (1889); "Bemerkungen über den Bau der Menschen- und Affenplacenta," in *Archiv für mikroskopische Anatomie und Entwicklungsmechanik*, **35** (1890), 1–51; "Das Gibbongehirn," in *Internationale Beiträge zur wissenschaftliche Medizin*, **1** (1891), 1–40; "Ueber einige neuere Forschungen im gebiete der Anatomie des Centralnervensystems," in *Deutsche medizinische Wochenschrift* (1891), 1213–1219, 1244–1246, 1267–1269, 1287–1289, 1331–1332, 1350–1356; *Beiträge zur Kenntnis der Lage der weiblichen Beckenorgane nebst Beschreibung eines frontalen Gefrierschnittes des Uterus gravidus in situ* (Bonn, 1892), written with F. Cohen; *Das Becken, topographisch-anatomisch mit besonderer Berücksichtigung der Chirurgie und Gynäkologie dargestellt*, II (Bonn, 1899); and "Hirnfurchen und Hirnwindungen, Hirnkommissuren, Hirngewicht," in *Ergebnisse der Anatomie und Entwicklungsgeschichte*, **8** (1899), 362–401.

II. Secondary Literature. On Waldeyer and his work, see J. Sobotta, "Zum Andenken an Wilhelm v. Waldeyer-Hartz. Anatomischer Anzeiger," in *Zentralblatt für die gesamte wissenschaftliche Anatomie*, **56**, nos. 1 and 2 (Vienna, 1922); and W. von Waldeyer-Hartz, *Lebenserinnerungen*, 2nd ed. (Bonn, 1921).

Paul Glees

WALDSEEMÜLLER, MARTIN (*b*. Radolfzell, Germany, 1470; *d*. St.-Dié, France, 1518 [?]), *geography.*

Waldseemüller (he also used the Greek form, Ialocomylus) studied theology at Freiburg im Breisgau, was ordained, and later became canon of St.-Dié, when he settled at the court of Duke René II of Lorraine. The duke's secretary, Gauthier Lud (or Ludd), gathered a small circle of humanists at the court during the opening years of the sixteenth century; calling themselves "Gymnasium Vosagense," they collected and published information on the new world then becoming known through the voyages of discovery. A writer, cartographer, and printer, Waldseemüller appears to have been the most versatile member of the group.

The first, and most important, publication of Lud's group was a slender volume dated 25 April 1507: *Cosmographiae introductio cum quibusdam geometriae ac astronomiae principiis ad eam rem necessariis*. It consisted of a general introduction to cosmography and a Latin translation of the report on Amerigo Vespucci's four voyages. The volume also contained a map and globe gores, representing knowledge gained through the latest discoveries. In a passage appearing on the verso of leaf 103, Waldseemüller suggested that the fourth part of the world should be called the land of Amerigo, or America, since it was discovered by Amerigo Vespucci. To reinforce his suggestion, his maps printed at the time had the name "America" on the southern part of the New World. The maps were soon sold out and lost from view, the sole surviving copy not being discovered until 1901; but the little book made a lasting impression, and Waldseemüller did indeed christen the New World "America."

During the rest of his life, Waldseemüller continued to produce maps, including one of Europe; a world map entitled "Carta marina navigatoria" (1516); and, most important, a set of maps for a new edition of Ptolemy's *Geographia*. Printed at Strasbourg in 1513, this edition is justly called the first modern atlas, for in addition to the traditional Ptolemaic maps, it included twenty *novae tabulae* that brought the *Geographia* up to date. Of these Waldseemüller designed eleven, including an important world map showing the New World but naming it "Land of the Holy Cross," as if the cartographer recognized that Vespucci was not the discoverer of America after all.

BIBLIOGRAPHY

A facs. ed. of the *Cosmographiae introductio*, with English trans. by J. Fischer, F. von Wieser, and C. G. Herbermann, was published as monograph 4 of the U.S. Catholic Historical Society (New York, 1907). An excellent facs. of the 1507 world map was published by J. Fischer and F. R. von Wieser: *Die älteste Karte mit dem Namen Amerika . . . 1507 und die Carta marina . . .1516 des M. Waldseemüller* (Innsbruck, 1905).

Waldseemüller's "Carta marina," repr. after his death, is the subject of a monograph by Hildegard Binder Johnson, *Carta marina—World Geography in Strassburg, 1525* (Minneapolis, 1963).

GEORGE KISH

WALKER, ALEXANDER (*b.* Leith, Scotland, 20 May 1779; *d.* Leith, 7 December 1852), *physiology.*

Walker probably matriculated at the medical school of the University of Edinburgh in 1797 and certainly studied anatomy with John Barclay.[1] He seems not to have completed his studies, since no record of his receiving a medical degree exists.[2] At the age of twenty he went to London, where he was associated with the well-known John Abernethy at St. Bartholomew's Hospital. Following some difficulties there Walker returned in 1808 to Edinburgh,[3] where he lectured in the Assembly Rooms to mixed audiences on general and particular science; his lectures in the Lyceum and elsewhere were well attended by students and medical practitioners. He attracted considerable notice by instructing the students on the mode of cutting down arteries, for which he gave exact mathematical directions. Walker is listed as a lecturer in the "Extra-academical School of Medicine and Surgery" only for the year 1808.[4]

Walker was the founder and editor of *Archives of Universal Science* (1809). Perhaps his earliest writings on neuroanatomy are the two articles published in the *Archives* in April and July 1809. The final issue (July 1809), divided into major sections—sciences, arts, and reviews—contained twelve articles by Walker.[5]

Walker's principal training was in anatomy and his publications in this area are the most controversial; they also are most frequently associated with his name, at least in the medical community. According to P. F. Cranefield, in 1809 Walker suggested that the roots of the spinal nerves differ in function, the anterior root being sensory and the dorsal root being motor—the reverse of the actual state of affairs.[6] Throughout his life Walker maintained his original views, which were proclaimed in *The Nervous System* (1834) and *Documents and Dates of Modern Discoveries in the Nervous System* (1839).[7] Cranefield concludes that Walker was the first to conjecture that one root is sensory and the other motor, while François Magendie was the first to assign the functions correctly and to provide the experimental evidence.

Historian H. T. Buckle cites Walker and others in the use of the deductive method in physiology, although in the case noted it led to erroneous conclusions. Thus Walker was in no sense an experimentalist, as his refusal to learn from the experiments of Magendie well illustrates.[8]

Some years later Walker returned to London, where his major efforts were literary. J. Struthers records that he was connected with several newspapers and was an active founder of the *Literary Gazette*.[9] Walker as a person has remained obscure; but his many books were widely read and reviewed, and many editions appeared both in Britain and in the United States. He pioneered as an authoritative and popular writer on subjects that in the Victorian era received little if any careful attention in print. *Beauty in Woman* (1836) still remains a striking and scholarly work, illustrated with plates from drawings by Henry Howard of the Royal Academy of Arts.[10] This book, together with *Intermarriage* (1838)[11] and *Women Physiologically Considered* (1839), was issued in 1843 as a three-volume collection under the general title *Anthropological Works*. His *Physiognomy Founded on Physiology* (1834) is an excursion into a curious but then popular subject.[12]

In surviving letters Walker refers to his wife and family, but children are not specifically mentioned.[13] Two poignant letters of 1850 have been preserved from the correspondence of Richard Owen, both relating to Owen's support of Walker's request for a Civil List pension from Lord John Russell, the British prime minister.[14]

In 1842 Walker returned to Leith in weakened health and was cared for by James Struthers until his death ten years later.[15] The few surviving records indicate that his final years were extremely difficult.

Walker's dogged determination is evident in much of his writing, particularly in that on the Bell-Magendie controversy. Contributions in the medical area have been well documented, but efforts in other area are little known. His interests ranged widely, and his literary efforts to popularize science (as then understood) were considerable. Walker's creative nature and sensibility to the arts are revealed in a group of works and articles. His lifelong fighting spirit sustained him during many protracted struggles with creditors and other adversaries.

NOTES

1. "Alexander Walker" appears on the matriculation roll of the University of Edinburgh medical school for 1797–1798, 1798–1799, and 1799–1800. Photocopies of the 1797–1798 signature, those from the nine letters of 1809–1812, and two letters of 1850 have been carefully studied (letters, Jean R. Guild, University of Edinburgh library). There are differences as well as similarities; since the latter predominate, the current judgment favors the possibility that it was the subject of this article who signed the matriculation roll. In 1797 Walker was eighteen, certainly an acceptable age for a beginning medical student. The name Alexander Walker is very common in Scotland, however, so caution is essential in the attribution of publications and other materials bearing this name (see P. F. Cranefield, *Alexander Walker*, vii). A. C. P. Callisen, *Medicinisches Schriftsteller–Lexicon*, XX, 329, notes to Alexander Walker 2, "Med. Dr. Edinb. 1832; geb. in Schottland," and his dissertation title: "De calore animali," dated 12 July 1832 at Edinburgh.

   Biographical data are extremely meager for Walker; the only account, by John Struthers (1867), provides no information on lineage or on Walker's immediate family. John Walker (1731–1803) was a professor of natural science at Edinburgh who married late and died without issue (*Fasti ecclesiae scotiacanae*, as noted by Jean R. Guild, University of Edinburgh library, letter, 15 Feb. 1974)—thus any connection with Alexander Walker seems unlikely. Extensive searches for obituary accounts have yielded only the brief *Lancet* note at the time of his death. Searches of Edinburgh newspapers of that date (*The Scotsman*, and others) revealed no obituary (letter, Margaret Deas, National Library of Scotland, 22 Jan. 1974), nor was one found in *Art Journal* for 1852 (letter, A. P. Burton, Victoria and Albert Museum library, 28 Jan. 1974).

   The illustrious Alexander Walker LL.D. (1825–1903) of Aberdeen apparently was not related to the physiologist (obituary, *Aberdeen Daily Journal*, 11 Feb. 1903; letter, Ian M. Smith, (Aberdeen Art Gallery and Museum, 8 Jan. 1975); also see *In Memoriam* (Aberdeen, 1903), 152–161 (letter, C. A. McLaren, Aberdeen University library, 16 Apr. 1975).

2. Walker recognized that the lack of a medical degree reduced his ability to attract students to his lectures. He desired one from "St. Andrews," but no record of receiving it exists (Walker letters, University of Edinburgh, 1809; letter, R. N. Smart, University Library, St. Andrews, 8 Jan. 1974).

3. "In London he had had to leave the school in consequence of showing the students, after lecture, that Abernethy, instead of tying the subclavian artery, had put the ligature round the neighboring nerve-trunk. What position he had occupied at St. Bartholomew's, or in Abernethy's class, I am unaware, but the incident of the nerve being tied instead of the artery (on the dead subject), and Mr. Walker's giving offence and having to leave there, in consequence of pointing it out, I have on good authority" (J. Struthers, *Historical Sketch* . . . 77; also Cranefield, *op. cit.*, v).

4. Struthers, *loc. cit.*; also Cranefield, *loc. cit.*; J. D. Comrie, *History* . . ., 628, 629.

5. Only three issues of *Archives of Universal Science* appeared: Jan., Apr., July 1809. The title pages all state: "By Alexander Walker, Esq., Lecturer on Physiology, Etc." The Journal was printed by Charles Stewart, university printer and a good friend and benefactor to Walker. Most of Walker's surviving letters are to Stewart, and they detail his financial woes.

6. The complicated story of this controversy has been ably analyzed by P. F. Cranefield (1973) in the introductory pages to the reprint of Walker's principal work on this topic, *Documents and Dates* . . . (1839). Cranefield has also published *The Way in and the Way out, François Magendie, Charles Bell and the Roots of the Spinal Nerves* (Mt. Kisco, N.Y., 1974), which reprints all of Walker's writings on the subject.

7. In reference to his controversy with Bell, one reviewer in 1833 stated: "You are a bold man Mr. Walker, and it is to be feared that you think too favorably of yourself. It may be true what you say, but modesty and genius are very usually twins." (*London Monthly Review*, in reference to Walker's priority claims against Bell and Magendie; cited in Allibone, *Critical Dictionary* . . ., 160).

8. H. T. Buckle, *On Scotland* . . . (1970), 23.

9. An examination of early issues of *Literary Gazette* does not record Walker's name among the editorial or other staff listings. He did contribute a letter (signed "W") in 1, no. 9 (22 Mar. 1817); also in 1, no. 10 (29 Mar. 1817).

10. *Beauty in Woman* is an admirable work and, following the 1st ed. (1836), was issued in at least six other printings or eds., the last being the 5th ed. by T. D. Morison (Glasgow, 1892). An exquisitely bound and deluxe copy with extra plates, prepared by the Paris bookbinder Petrus Ruban, is now in the New York Public Library. Ruban may have been active until about 1910 (see *Catalogue de beaux livres* . . . *de Mr. Petrus Ruban* [Paris, 1910]). Such a specially bound work with extra illustrations was a custom of certain nineteenth-century collectors, and the process is also referred to as grangerizing (letter, P. Needham, Pierpont Morgan Library, New York, 26 Feb. 1975).

11. Alexander Walker was a very early writer on heritable variation in man and in domestic animals. *Intermarriage* . . . *and* . . . *and Account of Corresponding Effects in the Breeding of Animals* . . . (1838) includes both a dedicatory letter by Walker to "Thomas Andrew Knight, Esq., F.R.S., F.L.S., President of the Horticultural Society, etc. . . ." and a letter to Walker ". . . respecting his work from George Birkbeck, Esq., M.D., F.G.S."

12. The following has been attributed—perhaps erroneously—to Walker: *Natural System of the History, Anatomy, and Physiology and Pathology of Man, Adapted to the Use of Professional Students, Amatorie, and Artists* (London, 1813), in 4 vols. with atlas of copperplates. No copies of this work are known (listed only in Callisen, *op. cit.*, xxxiii, 328–329). An inquiry failed to locate a copy of the work or to verify Walker's authorship (letter, P. A. Christiansen, University Library, Copenhagen, 3 Sept. 1974).

13. From this period one may note "Mrs. A. Walker" as the author of *Female Beauty as Preserved and Improved by Regimen, Cleanliness and Dress* . . . (London, 1836), also in an American ed. (New York, 1840). The title page of the 1837 ed. reads "by Mrs. A. Walker," yet most book catalogs and library catalog cards list this under "Mrs. Alexander Walker." Among the advertising pages at the end of the 1837 ed. of the work (p. 355) a book is listed by Donald Walker, *Exercises for Ladies*. The 1837 title page also states, "All that regards hygiene and health being furnished by medical friends, and revised by Sir Anthony Carlisle, F.R.S., Vice President of the College of Surgeons. . . ."

   It is possible that this was written by Alexander Walker and issued under his wife's name, although no direct evidence exists for this supposition. The publisher, Hurst, was never utilized by Walker for his many works. One association is interesting, however; in Walker's *Intermarrige* (1839 ed.) a paragraph is quoted from "Sir A. Carlisle in a letter to the author." Anthony Carlisle was professor of anatomy at the Royal Academy of Arts, London, from 1808 to 1824.

One puzzling work is the listing *Walker's Observations on the Constitution of Women* in *The London Catalogue of Books* (London, 1811). The author's full name is not given, but the title reads like a work by Alexander Walker. The title is suggestive of both *Beauty . . . in Woman* and *Female Beauty. . . .*

14. The first is a small handwritten note sent from Seafield, Leith, Scotland, on 16 Jan. 1850 to "Professor Owen": "I kindly & generously supporting my memorial with Lord John Russell it is not to be forgotten that philosophers & medical men have been enlightened by my discoveries during the last forty years; that I am consequently overwhelmed with debts contracted during so long a period; and that a suitable allowance can alone support me in satisfying these, and in closing my life in peace and honour. See the three volumes of Archives of Universal Science, 1809, in the Library of the British Museum, as well as many subsequent publications from which I never deviated. I am, My Dear Sir, Your Most Respectful, Obedient and Obliged Servant, Alexander Walker."

"Professor Owen" almost certainly was Sir Richard Owen, distinguished anatomist and former conservator of the Hunterian Museum at the Royal College of Surgeons. E. H. Cornelius (letter, Royal College of Surgeons, 11 Mar. 1975) notes that this would have been a Civil List pension, that is, a government pension awarded by Queen Victoria on the recommendation of Lord John Russell, prime minister, for services to the nation. It would have been very natural for Owen, as a leading scientist of the day, to support an application on Walker's behalf. The letter, in the collections of the British Library (British Museum) bears a stamp mark "Ex Litt. Ricardi Owen."

The second letter, also to Owen and dated 1 Mar. 1850, was sent from Seafield and is in the Owen Collection, British Museum (Natural History). It further reveals Walker's dire straits and again refers to the pension request.

The pension apparently was not granted, for no record exists in *The Register of Warrants for Civil List Pensions,* T.38/824, or *The Civil List Ledger* (1850) T.38/252. *The Register of Treasury Papers* (1849), T.2/208, records two such requests by Walker. Other funds, however, were made available to him, care of Rev. T. Laing of Leith (*The Minute Book of the Treasury Board* [1849], T.29/539; letter, D. Crook, Public Record Office, London, 24 Apr. 1975).

15. James Struthers M.D. was the brother of John Struthers M.D., author of *Historical Sketch . . .* (1867), which included some biographical notes on A. Walker. John Struthers' information on this point is at variance with a published record: *Edinburgh and Leith Post Office Directory* for 1850–1851 and 1851–1852 listed Alexander Walker as living in Seafield, but there is no mention of his living there in earlier issues of the *Directory* (letter, Margaret Deas, National Library of Scotland, Edinburgh, 14 Apr. 1975).

BIBLIOGRAPHY

I. ORIGINAL WORKS. A bibliography of Walker's publications is provided in the intro. to P. F. Cranefield's facs. repr. (1973) of Walker's *Documents and Dates . . .*, vii–xi; additional items and notes are provided in Cranefield's *The Way in and the Way Out . . .* (Mt. Kisco, N.Y., 1974), 1–3, 24–25. The former lists works of questionable Walker authorship and provides a

record of the many eds. and printings. Only principal works are listed below; those marked with an asterisk are newly revealed (and hence are not listed by Cranefield).

The earliest include *Prospectus of Lectures on the Natural System of Universal Science* (Edinburgh, 1808[?]), a pamphlet; *Prospectus of Two Courses of Lectures*; *One on Anatomy and Physiology*; *the Other on Pathology and the Practice of Medicine* (Edinburgh, 1808), a pamphlet; *Result of the Operation, Publicly Performed, in Order to Refute or to Confirm the Principle of Surgical Operation Proposed by . . .* (Edinburgh, 1808), a pamphlet; "Theory of Phonics, Hearing, etc. Physiological Dissertation on the Functions of the *Ossicula auditus*, and on the Tympanic Muscles in Particular, and on those of the Ear in General," in *London Medical and Physical Journal*, **19** (1808), 385–414; "General Physiology of the Intellectual Organs," in *Archives of Universal Science*, no. 2 (1809), 167–205; "New Anatomy and Physiology of the Brain," *ibid.*, no. 3, 172–179; *"A Critique on the Antique Statues and Those of Michael Angelo, in Which Not Only the Defects in Their Attitudes, but Also the Errors Which They Present With Regard to the Particular Muscles Brought Into Action, Are Pointed out," *ibid.*, no. 3 (July 1809), 224–234; "Sketch of a General Theory of the Intellectual Functions of Man and Animals, Given in Reply to Drs. Cross and Leach," in Thomson's *Annals of Philosophy*, **6** (1815), 26–34, 118–124; and *"An Attempt to Systematise Anatomy, Physiology and Pathology," *ibid.*, 283–292.

Further works are "A Simple Theory of Electricity and Galvanism; Being an Attempt to Prove That the Subjects of the Former Are the Mere Oxygen and Azote of Air, and the Subjects of the Latter the Mere Oxygen and Hydrogen of Water," in Thomson's *Annals of Philosophy*, **8** (1816), 182–189; *a probable Walker letter in *Literary Gazette* (London), **1**, no. 10 (29 Mar. 1817), 146, which refers to the last number of *Thomson's Annals of Philosophy* concerning an article by "Mr. Magendie" on physiological experiments on animals and criticizing Thomson for having published this article, signed "W."; *"Character of the French," in *Blackwood's*, **26** (1829), 309–314; *"Comparison of the Modern With the Ancient Romans," *ibid.*, 314–317; *"Character of the English, Scots, and Irish," *ibid.*, 818–824; *"The Picturesque," in *Arnold's Magazine of the Fine Arts*, n.s. **1**, no. 2 (June 1833), 105–106; *"Cause of the Fine Arts in Greece," *ibid.*, **1**, no. 6 (Oct. 1833), 491–493; and "On the Cause of the Direction of Continents and Islands, Peninsulas, Mountain Chains, Strata, Currents, Winds, Migrations, and Civilization," in *Philosophical Magazine*, 3rd ser., **3** (1833), 426–431.

Also see *Physiognomy Founded on Physiology, and Applied to Various Countries, Professions, and Individuals . . .* (London, 1834), reviewed in *Arnold's Magazine of the Fine Arts*, **4**, no. 9 (July 1834), 255; *The Nervous System, Anatomical and Physiological . . .*

(London, 1834); *Beauty; Illustrated Chiefly by an Analysis and Classification of Beauty in Woman . . .* (London, 1836); *Influence of Natural Beauty, and of Its Defects on Offspring, and Law Regulating the Resemblance of Progeny of Parents* (London, 1837), a pamphlet; *Intermarriage: Or, the Mode in Which, and the Causes Why, Beauty, Health, and Intellect Result From Certain Unions; and Deformity, Disease, and Insanity From Others . . .* (London, 1838); *The New Lavater, or an Improved System of Physiognomy Founded Upon Strictly Scientific Principles . . .* (London, 1839); *Women Physiologically Considered as to Mind, Morals, Marriage, Matrimonial Slavery, Infidelity, Divorce* (London, 1839); the unsigned *Documents and Dates of Modern Discoveries in the Nervous System* (London, 1839), about which Cranefield (1973), in his reprint of this work, notes ". . . it has always been taken for granted that it is by Walker, and there is not the slightest reason to doubt that assumption" (p. iii); *Pathology Founded on the Natural System of Anatomy and Physiology . . .,* 2nd ed. (London, 1841); and "Purification of Edinburgh, etc.," in *Letters on the Sanitary Condition of Edinburgh* (Edinburgh, 1842), cited by Cranefield (1973, p. x) from a copy in the British Museum.

Archival items include a printer's proof of an Edinburgh lecture schedule with markings, and eleven MS letters. Nine MS letters from Walker to Charles Stewart, university printer, are held by the University of Edinburgh library; photocopies have been made available for study (letters, Jean R. Guild, reference librarian, 7 Nov. 1973, 18 Dec. 1973, 15 Feb. 1974). They consist largely of pleas for funds, but a few other matters may be gleaned from them. Only a few letters are dated, but the time span appears to be about 1809–1812. They generally are not individually cited here but are referred to as the "Walker letters."

A possible early signature (1797) occurs on the matriculation rolls of Edinburgh University medical school. No authenticated portrait is known. The National Portrait Gallery (London) notes that a portrait of an Alexander Walker by C. Ambrose was exhibited at the Royal Academy of Arts (London) in 1829 (Walker would have been fifty) and that there is a drawing by Alphonse Legros of a sitter of the same name in the Aberdeen Art Gallery and Museum (letter, M. Rogers, assistant keeper, 2 Dec. 1974). A. Graves, *The Royal Academy of Arts (Exhibitors, 1769–1904)* (1905), 30, lists the works by C. Ambrose exhibited in 1829: no. 24, Alderman Walker; no. 274, David Walker, Esq.; no. 344, Alexander Walker, Esq.; no. 425, Portrait of a Lady. The Scottish National Portrait Gallery (Edinburgh) has no Walker portrait (letter R. E. Hutchinson, keeper, 25 Nov. 1974).

II. SECONDARY LITERATURE. See S. A. Allibone, *A Critical Dictionary of English Literature and British and American Authors . . .,* III (Philadelphia, 1898), 1539, which cites a review of Walker's *Intermarriage* (Philadelphia, 1851) in *Medical Examiner and Record of Medical Science,* 14 (n.s. 7) (1851), 371–372; an anony-

mous death notice of Walker in *Lancet* (1852), 2, 583, dated Leith, Scotland, 7 Dec. 1852; British Museum, *Catalogue of Additions to the Manuscripts, 1931–1935* (London, 1967)—see 843, listing of letter from Alexander Walker, physiologist, to R. Owen (1850), item 42577, F. 273; and *Catalogue of Printed Books,* XXVI, microprint ed. (New York, 1967), 553–554; H. T. Buckle, *On Scotland and the Scotch Intellect,* H. J. Hanham, ed. (Chicago, 1970), 23, taken from *History of Civilization in England,* I (London, 1857); and A. C. P. Callisen, *Medicinisches Schriftsteller-Lexicon,* XX (Copenhagen, 1834), 205–206, and *Nachtrag,* XXXIII (Altona, 1845), 328–329, valuable for the citations to reviews of Walker's many works in the medical journals.

Further works are J. D. Comrie, *History of Scottish Medicine,* 2nd ed., II (London, 1932), 628, 629; P. F. Cranefield, *Alexander Walker, Documents and Dates of Modern Discoveries in the Nervous System,* facs. of the London ed. (1839), with intro. by Cranefield (Metuchen, N.J., 1973)—in notes, p. vi: "Nor should one overlook, in all the polemic and eccentricity, the fact that Walker had a brilliant mind and a wry sense of humor and appears to have begun life as a first rate anatomist, while the many editions of his popular books testify to his skill as a writer"; A. Durel, *Catalogue de beaux livres modernes en éditions de luxe, recouverts de riches reliures composant la bibliothèque de Mr. Petrus Ruban exrelieur* (Paris, 1910), in which Walker's works are not listed; A. Graves, *The Royal Academy of Arts (Exhibitors, 1769–1904),* I (London, 1905), 30; and Sir N. Moore, *The History of St. Bartholomew's Hospital,* II (London, 1918), which makes no mention of A. Walker.

Also see *The New Statistical Account of Scotland,* I, *List of Parishes—Edinburgh* (Edinburgh, 1845), 760–782; J. M. D. Olmsted, *François Magendie—Pioneer in Experimental Physiology and Scientific Medicine in XIX Century France* (New York, 1944), 119, 265; J. Russell, *The Story of Leith* (London, 1922), 334–335, passim; J. Struthers, *Historical Sketch of the Edinburgh Anatomical School* (Edinburgh, 1867), 76–78—these pages quoted verbatim in Cranefield's ed. of Walker's *Documents and Dates . . .,* v–vi; J. Thornton, *John Abernethy. A Biography* (London, 1953), in which no reference to A. Walker is included; University of Edinburgh, *Catalogue of the Library,* III (Edinburgh, 1923), 1106; and W. Wright, "Alexander Walker—Who Was He?" in *Notes and Queries,* 8th ser., 3 (1893), 329; no reply or comment was ever published.

LAWRENCE J. KING

WALKER, JOHN (*b.* Edinburgh, Scotland, 1731; *d.* Edinburgh, 31 December 1803), *geology, botany, religion.*

Walker was born into a family firmly convinced of the value of education. His father was rector of the Canongate Grammar School; and as a youth

Walker was trained in Latin and Greek, reading in those languages at an early age. In addition to his ability in the classics, he developed an interest in minerals before he was fifteen. After finishing studies at his father's school, Walker studied at the University of Edinburgh, where he prepared for the ministry in the (Presbyterian) Church of Scotland.

In the course of his work, Walker became especially interested, through the influence of William Cullen, in chemistry and mineralogy. He soon realized that the classification of minerals had been sadly neglected and therefore traveled throughout the British Isles, sometimes with Cullen, collecting minerals from mines and rock outcrops. Using his own collection as well as that at the University Museum, he had established an "elementa mineralogiae" by the 1750's; this classification was later modified to include 323 genera. Among the most interesting of the minerals that he collected in the 1760's was strontianite, from the mines of Leadhills.

Licensed to preach in 1754, Walker was assigned to his first post at Glencorse in September 1758. He soon met Henry Home, Lord Kames, who became his enthusiastic sponsor. Through this relationship Walker was commissioned to make an extensive study of the Hebrides in 1764. This was neither his first nor his last study of the Hebrides or the Highlands; and it was directly responsible for the preparation of a two-volume book printed posthumously in 1808 by his friend Charles Stewart.

From his first ministry in 1758 until his appointment in 1779 to the chair of natural history at the University of Edinburgh, Walker spent all of his spare time in the study of botany and geology; and his knowledge of Latin and Greek permitted him to read the significant literature in those fields. Walker was greatly influenced by the works of Cronstedt and especially by the contributions of Linnaeus, with whom he corresponded in the 1760's.

Most of Walker's botanical and geological papers were published or prepared between 1758 and 1779. A new phase in his life commenced with his appointment as regius professor at the University of Edinburgh, where his first class was enrolled at least as early as 1781.

Walker was an organizer of the Royal Society of Edinburgh and was appointed first secretary of the Physical Section in 1783; he also was active in the organization of the Natural History Society of Edinburgh in 1782. He was a long-time member of the Highland Society of Scotland and, as a result of his great interest in agriculture, formed the Agriculture Society of Edinburgh in 1792. These groups gave him an opportunity to participate in scientific discussions and provided an outlet for the publication of some of his articles.

In his initial lecture to his first class, Walker said: "I am to teach a science I never was taught." He then proceeded to organize a set of lectures on geology "or the Natural History of the Earth." From then until the end of his life he gave regular lectures on the various aspects of geology and had a great influence in establishing the science as a discipline in higher education. Three of his most famous students were James Hall, John Playfair, and Robert Jameson. In addition, he influenced the early development of geology in America through his student Samuel Latham Mitchill, who became one of the leading men of American science upon his return to the New York area. Walker was a contemporary of James Hutton, and both were members of the Royal Society.

One of Walker's chief contributions was the establishment of geology as an organized classroom subject in an institution of higher learning, and he therefore has a legitimate claim to the title of "Father of Geological Education." Walker's classroom methods were essentially those used today: he lectured; distributed syllabuses; established a laboratory; and brought in rocks, minerals, and fossils, which were studied with a microscope. The laboratory work included the study of polished surfaces and a large suite of minerals and fossils. In the study of minerals he used a hardness scale similar to that devised by Mohs thirty years later.

In his lectures Walker discussed the origin of carbonates and differentiated between limestone and marble. In addition, he considered the origin of igneous and sedimentary rocks. Among the common rocks, he paid particular attention to the origin of basalt and granite and to the environment of deposition of such sediments as sand.

One of his classic works was his essay on peat, in which he made an exhaustive analysis of the organic content and origin of this substance. In other papers he affirmed that petroleum occurred in rocks as a natural substance. His discussion of rock structures included accurate definitions of strike and dip as well as recognition of horizontal strata overlying tilted beds, a condition to which he referred as offlap. Walker described the work of both surface and subsurface water and recognized density stratification of lake water. He wrote about till but did not know of its glacial origin.

Walker collected fossils and used them for demonstration purposes in the laboratory. He classified the methods of fossilization and strongly supported the Linnaean system of binomial nomenclature. He believed that fossils could be used to determine rock chronology and maintained that animals and plants were linked in a common evolutionary chain from the "lowest subject up to the human species . . . all being linked . . . by the most beautiful and regular gradation."

BIBLIOGRAPHY

Walker's writings include "An Essay on Peat, Containing an Account of Its Origin, of Its Chymical Principles and General Properties," in *Prize Essays and Transactions of the Highland Society of Scotland*, **2** (1803), 1–137; *An Economical History of the Hebrides and Highlands of Scotland*, 2 vols. (Edinburgh, 1808); *Essays on Natural History and Rural Economy* (Edinburgh, 1808; repr. London, 1812), published by Charles Stewart; and Harold W. Scott, ed., *Lectures on Geology by John Walker* (Chicago, 1965), with a complete bibliography, including MSS.

Two short biographies of Walker are W. Jardine, "Memoir of John Walker, D.D." in *The Birds of Great Britain and Ireland*, III (London, 1842), 3–50; and George Taylor, "John Walker, D.D., F.R.S.E., 1731–1803, A Notable Scottish Naturalist," in *Transactions of the Botanical Society of Edinburgh*, **38** (1959), 180–203. A portrait is in John Kay, *A Series of Original Portraits and Caricature Etchings*, III (Edinburgh, 1840), 178.

HAROLD W. SCOTT

WALLACE, ALFRED RUSSEL (*b.* Usk, Monmouthshire, Wales, 8 January 1823; *d.* Broadstone, Dorset, England, 7 November 1913), *natural history.*

Wallace was the eighth of nine children born to Thomas Vere Wallace and Mary Anne Greenell. Suffering from constant economic setbacks and having six children to support at the time, his parents had moved across the Severn River to the inexpensive rural environs of Wales less than a mile from the small village of Usk. In the beautiful surroundings beside the Usk River Wallace spent the first five carefree years of his life before moving to Hertford, where he first attended school. In the one-room Hertford Grammar School he studied Latin, French, geography, mathematics, and history. Wallace later had a low opinion of his only formal education, which seems to have been quite pedestrian and dull. More important for his intellectual development, apparently, was the extensive reading of travel works, biographies, novels, classics, and anything else he could find at home.

Early in 1837 Wallace went to London to live temporarily with his brother John. While there he attended lectures at the "Hall of Science" and became acquainted with Robert Owen's socialistic ideas as well as the ideas of religious skeptics. (His agnosticism began at this point and prevented him at a later time from seriously considering orthodox views, largely with religious overtones, on the formation of new species.) The following summer he was apprenticed to his brother William, a surveyor, with whom he worked for the most part until mid-December 1843.

During his survey work, Wallace first began to experience the lure of nature, but not until 1841 did he timidly pursue interests which had barely been awakened in him. The purchase of a cheap book on botany (to assist in beginning a herbarium) marks the beginning of his scientific career, and his interests in botanical explorations and reading continued to grow from that point onward.

About December 1843 his brother's surveying business diminished severely, and Wallace was forced to go to work in 1844 as a master at the Collegiate School in Leicester, where he taught English, arithmetic, surveying, and elementary drawing. At Leicester during 1844–1845, Wallace read widely in the natural sciences; indeed, during the period from 1842 to 1846, he consumed various works by Alexander von Humboldt, T. R. Malthus, Charles Darwin, Robert Chambers, Charles Lyell, and William Swainson. These books profoundly influenced his subsequent intellectual development as did his amateurish explorations in Charnwood Forest in Leicester with his new friend, Henry Walter Bates.

In 1845 his brother William's death forced Wallace to return briefly to surveying and construction work, but he continued reading, collecting, and corresponding with Bates. In 1847 he audaciously suggested to Bates that they transfer their collecting efforts to the forbidding continent of South America and support themselves by collecting objects of natural history. W. H. Edward's *A Voyage up the River Amazon* (1847) prompted them to journey to the Amazon basin, where Wallace explored from 1848 until 1852.

Although he established a scientific reputation for his excellent work in the Amazon, Wallace lost most of his materials, and almost his life, when his ship caught fire and sank in the Atlantic during his return voyage. After his rescue and arrival in

England, Wallace decided to embark on another lengthy expedition, this time to the Malay Archipelago (now Indonesia and Malaysia). During that period of extensive exploration (1854–1862), Wallace formulated the principle of natural selection and made many other fundamental discoveries in biology, geology, geography, ethnography, and other natural sciences.

Upon returning to England in 1862, Wallace enjoyed an enviable reputation as a naturalist. He spent the rest of that decade publishing more articles, culminating with his classic *The Malay Archipelago* (1869), which went through countless editions and was translated into many foreign languages. His interests also began to extend into other nonbiological areas.

During the 1860's Wallace was converted to Spiritualism, which affected his views about natural selection and man. In the late 1870's he became involved in the land nationalization movement and was the first president of the Land Nationalization Society in 1881. In 1890 he publicly announced his acceptance of socialism, which he had thought about since the late 1830's. Early in the twentieth century he supported women's liberation movements through various articles. All these diverse interests were concurrent with his scientific activities, which may be divided into the following categories: natural history exploration, evolutionary biology, ethnology, zoogeography, mimicry and other means of protective colorations, geology, vaccination, and astronomy.

After arriving at Pará (Belém), Brazil, late in May 1848, Wallace and Bates immediately commenced exploring and eventually covered a sizeable portion of the Amazon basin. During the first two years they concentrated their work around Pará, the Tocantins River, and the banks of the Amazon itself as far as Barra (Manaus), where the Amazon and Rio Negro converge. To increase coverage, the young naturalists had already explored a great deal by themselves; and at Barra in March 1850 they separated permanently, with Wallace going to the Rio Negro and Uaupés rivers, and Bates eventually going to the upper Amazon region, where he assembled a spectacular natural history collection before returning to England in 1859.

Wallace penetrated as far north as Javíta in the Orinoco River basin and as far west as Micúru (Mitú) on the Uaupés River. He was deeply impressed by the grandeur of the virgin forest, by the variety and beauty of the butterflies and birds, and by his first encounter with primitive Indians on the Uaupés River area, an experience he never forgot.

The explorations of Wallace were recounted in his fascinating book *A Narrative of Travels on the Amazon and Rio Negro* (1853). Although most of his splendid collections and much else had been lost at sea when his ship sank on 6 August 1852, the book nevertheless displays the keen eyes and mind of a naturalist, by then a mature professional. His notes and drawings on the palm trees of the Amazon fortunately were rescued and appeared in a small but charming book, often quoted in botanical literature on palms.

A major reason for the expedition to South America was to collect information on the variation and evolution of species. In 1845 after reading Chambers' *Vestiges of the Natural History of Creation* (1844), a controversial but extremely stimulating and valuable Victorian work on evolution, Wallace was converted to the belief that species arise through natural laws, rather than by divine fiat. From this point onward he never seriously entertained commonly held views on species, and he apparently convinced Bates that Chambers was right in principle. Their task was to supply scientific details and perhaps uncover a satisfactory evolutionary mechanism.

Since his collections had been lost, Wallace hesitated to declare his views publicly, although glimpses of his ideas may be observed in his early comments on the geographical distribution of monkeys, birds, and insects, as published in articles and in his travel narrative. Also in his narrative are references to the "marvellous adaptation of animals to their food, their habits, and the localities in which they are found" (*A Narrative of Travels on the Amazon and Rio Negro* ([1853], p. 83). He explicitly rejected the orthodox explanations for these phenomena, saying that naturalists were seeking new explanations for species variations. He definitely thought evolution was the explanation but wisely refrained from further comment until 1855.

Since he had not solved the perplexing question of how species evolve while in the Amazon region, Wallace decided to venture once more to the tropics, this time to Southeast Asia and the Malay Archipelago. Securing passage on a government vessel, Wallace departed in 1854 for explorations that lasted eight years and covered between 14,000 and 15,000 miles. The boundaries of the range of his explorations were the Aru Islands to the east; Malacca, Malaya, to the west; the northern tip of Celebes to the north; and as far south as southern Timor.

The enormous quantity of materials gathered

there—about 127,000 specimens of natural history—enabled him to publish scores of fundamental scientific papers on a broad range of topics. These works alone would have established him as one of the greatest English naturalists of his age, but his classic natural history travel book, *The Malay Archipelago* (1869), earned him an international reputation that has endured to this day. On the basis of artistic format, literary style, and scientific merit, it is clearly one of the finest scientific travel books ever written.

From his first arrival in the Malayan region Wallace had decided to gather precise scientific data on groups of animals in order to work out their geographical distribution and consequently to throw light on their origins through evolutionary processes. He kept a notebook on evolution, here designated as his "Species Notebook." His first explicit, published evolutionary statements drew on those materials.

An article by the English naturalist Edward Forbes, Jr., in which he emphatically denied "organic progression" (1854), provoked Wallace to publish a concise synthesis of his ideas on the subject. Like many brilliant works, his "On the Law Which Has Regulated the Introduction of New Species" (September 1855) was based on well-known, acceptable scientific information combined with many personal observations, although he had transformed the mass of facts into an unusually persuasive argument. The evidence was drawn from geology and geography—the distribution of species in time and space—and following nine acceptable generalizations (axioms), Wallace concluded: "*Every species has come into existence coincident both in space and time with a pre-existing closely allied species*" (McKinney, *Lamarck to Darwin: Contributions to Evolutionary Biology, 1809–1859* [1971], pp. 71–72). He claimed that he had explained "the natural system of arrangement of organic beings, their geographical distribution, their geological sequence," as well as the reason for peculiar anatomical structures of organisms. Although the article was carefully worded, Wallace had definitely announced that he was an evolutionist.

Despite this excellent presentation, there were no public replies, although the private comments were quite another matter. Indeed, Edward Blyth, Charles Lyell, and Charles Darwin all read Wallace's article and were greatly impressed by his arguments, but in particular Lyell, who began a complete reexamination of his long-held ideas on species. On 16 April 1856 Lyell discussed Wallace's paper with Darwin, urging him to publish his own views on species as soon as possible. Darwin then began what we now call the long version of the *Origin*, and that version was used as a basis for the *Origin* as published in 1859.

The immediate stimulus for Darwin, however, was a paper written by Wallace entitled "On the Tendency of Varieties to Depart Indefinitely From the Original Type." After publishing his evolutionary paper in 1855, Wallace had continued to search for an evolutionary mechanism. Very ill with malaria while on the large island of Gilolo (Halmahera), some ten miles from Ternate, he formulated the principle of natural selection, the now famous mechanism of evolution; and upon returning to Ternate, he mailed a paper (and covering letter) to Darwin expounding his long-sought discovery (9 March 1858). Evidence now suggests that the "bombshell" arrived at Down House on the third (or fourth) day of June 1858.

Determined that their friend Darwin should receive recognition of priority, Lyell and Hooker decided that Wallace's paper should be presented before the Linnean Society of London along with an excerpt from an essay by Darwin on natural selection and a letter from Darwin to Asa Gray discussing divergence (1 July 1858). Prefatory remarks by Hooker and Lyell emphasized Darwin's priority of discovery, and Wallace's paper was presented last. Wallace was never consulted on these matters and did not learn about the presentation until after the papers were published (20 August 1858). These items focused on natural selection and divergence, not the general arguments for evolution. Darwin's *Origin* appeared late in November 1859.

Wallace subsequently published numerous articles and books supporting evolution with many original and forceful arguments. In 1864 after his return to England he read a paper before the Linnean Society on the variation and distribution of the Papilionidae butterflies of the Malayan region, which demonstrated evolution occurring in nature. In 1870 he published a collection of his evolutionary essays entitled *Contributions to the Theory of Natural Selection*, and in 1876 his monumental *Geographical Distribution of Animals* appeared. He summarized current knowledge of zoogeography and explained the data "by means of established laws of physical and organic change." His *Island Life* (1880) applied evolutionary concepts to insular flora and fauna.

During the 1880's Wallace had given a number of lectures on evolution by means of natural selection, including many while touring the United States in 1886–1887. These mature reflections finally appeared in elaborated form in his important *Darwinism* (1889), which carefully reviewed thirty years of evolutionary biology. While pointing out differences between himself and Darwin, the book actually elaborates a pure form of Darwinian evolution, devoid of Lamarckian elements, and therefore represents (except for the last chapter on man) perhaps the authoritative statement on the subject in the late nineteenth century. The work went through many printings.

From the 1860's onward one of the forceful arguments used by Wallace to support evolution was mimicry. Initially discovered by his traveling companion in South America, Henry Walter Bates, who first published on the subject in 1862, the theory of mimicry was immediately accepted by Wallace since it had been explained in terms of natural selection. Batesian mimicry stated that relatively scarce, unprotected specific forms may resemble other species that are protected by strong smell and bad taste. Resemblance affords protection, and the closer the resemblance, the greater the survival value. In his 1864 paper on the Papilionidae (swallowtailed butterflies) Wallace described mimetic complexes in the Indo-Malayan region, thus forcefully supporting what Bates had observed in South America. Wallace added to those arguments the following two points: a species may have two or more very different forms, and each one may mimic a different model. In the female only, these tend to be polymorphic (Remington [1963], p. 146).

To these views on mimicry Wallace soon added his ideas about numerous other protective resemblances among animals (1867). He observed that resemblances depend upon utility of characters, need for protective concealment, extreme variability of color, and the fact that concealment can most easily be obtained through color modification.

Problems remained, however, for no one had explained why two or more seemingly protected species resemble one another. From a German expatriate living in Brazil came the explanation. In 1878–1879 Fritz Müller observed that if a number of different species are protected chemically or physically in some way, then it is to their advantage in the struggle for existence to resemble one another. Their mutual color patterns warn predators to stay away, and losses during the predator education period are absorbed by a larger group. Wallace had observed but not understood these resemblances, which he thought were due to "unknown local causes." After a second article by

Müller (1881), Wallace heard about these explanations, which had not been enthusiastically supported, and published an article in *Nature* (1882), supporting and accepting Müller's arguments, which afterward gained broad acceptance.

In 1889 Wallace summarized the various ideas on mimicry and proposed an extension of Müller's explanation: that in the same locality several members of the same protected genus may resemble each other; and a scarce edible species can obtain some protection from predators by resembling and intermingling with an abundant edible species (Remington, p. 148). Others, including E. B. Poulton, *The Colours of Animals* (1890), continued the work of Bates, Müller, and Wallace.

While Wallace was one of the founders of modern evolutionary biology, his views on man underwent significant alteration during the 1860's after his return from the Malay Archipelago. Before 1862 he was concerned with man as an animal who had a close kinship with other primates, and his ethnological interests had deep roots, extending back into the 1840's. In a letter to Bates, dated 28 December 1845, Wallace discussed man in an evolutionary context and continued to think of him in the same way for almost twenty years. Immediately before discovering natural selection in 1858, he recalled the work of Malthus, while reflecting on the origins and variations of the indigenous tribes of the Indo-Malayan region.

After returning to England, Wallace's ideas on man underwent modification. The reasons involve his new views about deity and spiritual beings and his decision that natural selection could not adequately explain all aspects of man's development. Although certain points are still unclear, we do know that in 1864 Wallace announced his new view of natural selection, namely that at some point during man's history, his body ceased to change, while his head and brain alone continued to undergo modifications. Man had therefore partially escaped the power of natural selection and could himself influence organic change by selection. Eventually, on land, human selection would supplant natural selection: man's superior intellect had unchained him from an inexorable law of nature, natural selection. As man's social, moral, and intellectual faculties developed, he became a being apart from the ordinary (see Wallace, in *Journal of the Anthropological Society of London*, 2 [1864], clvii–clxxxvii; and Smith [1972], 178–199).

This theme was further developed in 1869 in his review of Lyell's *Principles of Geology* (10th ed., 1867–1868) and *Elements of Geology* (6th ed.,

1865), in which he observed that man's "intellectual capacities and his moral nature were not wholly developed by the same process [natural selection]." Neither natural selection nor evolution can explain the origin of man's intellect. The "moral and higher intellectual nature of man is as unique" as the origin of life on earth. Furthermore, man's brain, speech apparatus, hand, and external form demonstrate that a "Higher Intelligence" had a part in the development of the human race (Wallace, *Contributions to the Theory of Natural Selection*, chapter 10).

This conclusion referring to the necessity of a higher intelligent being agreed at least in principle with the final two pages of Lyell's *Geological Evidences of the Antiquity of Man* (1863) and was a conscious attempt by Wallace to reconcile science with theology, which shocked Darwin, who marked and annotated his copy from Wallace. At one point there are four exclamation marks in the margin (p. 392, lines 6–8). Before reading the review, Darwin had commented to Wallace, "I hope you have not murdered too completely your own and my child." In exactly the reverse direction of a common trend since the eighteenth century, Wallace had *added* deity to his mechanistic, self-regulating universe ("Creation by Law" [October 1867], repr. in *Contributions to the Theory of Natural Selection* [1870]). A watchmaker was necessary after all. These views received full expression in his article "The Limits of Natural Selection as Applied to Man" (*ibid.*, 332–371).

During the early 1860's—clearly no later than 1865—Wallace's statements on man's development (and other topics as well) must be examined within a religious context. What, then, were his religious views and what led him to those views? Wallace's parents were orthodox believers, as was he until 1837 when he lived with his brother John in London and then with his brother William, who was "of advanced liberal and philosophical opinions." While at Leicester in 1844, experiments with phrenology and mesmerism impressed him deeply, leading him to believe in extrasensory phenomena. During his twelve years of natural history exploration, he had heard about Spiritualism and decided to investigate the subject upon his return to England in 1862. Until then, by his own admission, he was an agnostic, a materialist, a philosophical skeptic. That was the situation, he has told us, "up to the time when I first became acquainted with the facts of Spiritualism," that is, July 1865. In two other specific places in this same book, *On Miracles and Modern Spiritualism* (1875), how-

ever, Wallace claimed that he had been an agnostic for "twenty-five years" (thus 1837–1862), and there is additional evidence that appears to support his explicit references to twenty-five years.

In a highly praised paper, "On the Physical Geography of the Malay Archipelago" (read 8 June 1863), Wallace urged that governments and scientific institutions immediately set about assembling the best possible collections of natural history for the purposes of study and interpretation. He then concluded in a manner quite foreign to him in the past:

> If this is not done, future ages will certainly look back upon us as a people so immersed in the pursuit of wealth as to be blind to higher considerations. They will charge us with having culpably allowed the destruction of some of those records of *Creation* which we had in our power to preserve; and *while professing to regard every living thing as the direct handiwork and best evidence of a Creator*, yet, with a strange inconsistency seeing many of them perish irrecoverably from the face of the earth, uncared for and unknown [*Journal of the Royal Geographic Society*, 33 (1863), 234, italics added].

It is therefore possible that Wallace's religious views began to alter as early as 1862/1863, although he claimed emphatically that various facts, "not any preconceived or theoretical opinions," led him in 1865 to accept Spiritualism, but not orthodox Christianity, which he frequently criticized. In 1866 he published a fifty-seven-page booklet, *The Scientific Aspect of the Supernatural*, thereby publicly announcing his support of Spiritualism.

At the moment it must remain a moot point whether he questioned the all-sufficiency of natural selection on purely empirical, scientific grounds or because embryonic religious views had caused him to doubt. Seeds of doubt are evident in Wallace's "The Origin of Human Races and the Antiquity of Man Deduced From the Theory of Natural Selection," but we may be involved in the chicken-egg syndrome. In any event, doubt once raised in Wallace's mind for whatever reason was like an incurable itch until a satisfactory solution was found. Later it was necessary to convince scientists of the weaknesses of natural selection (using the argument of utility) in explaining the evolution of natural phenomena in order to persuade them to consider, as an alternative, psychic phenomena (Smith [1972], pp. 178–199).

Considering his previous history, it is curious that Wallace found satisfaction in Spiritualism, which shocked Darwin and Huxley, but which may have had a profound effect on those scientists with religious views who were unable to resolve their own doubts. Wallace, a discoverer of natural selection, had rescued man from the degradation of evolution and had returned to him his God-given "soul" (intellect). Later elaborations on this theme appeared in *Darwinism* (1889) and particularly *The World of Life* (1910). The full impact of Wallace's conversion has never been assessed.

It has been incorrectly asserted that the facts of geographical distribution led both Darwin and Wallace to accept evolution. In the case of Wallace, the reverse is true. He went to the Amazon to collect facts establishing the case for evolution, but as an amateur naturalist, what facts was he capable of collecting? The answer is clear, especially when we understand that Lyell in his *Principles of Geology* and Chambers in his *Vestiges of the Natural History of Creation*, two major influences on Wallace, stressed the importance of distributional phenomena.

While in South America, he observed variation and the struggle for existence, which he had read about in Lyell, but Wallace's foremost contributions from his Amazon experience originated from observations on the distribution of the palm trees, insects, birds, and monkeys. This evidence substantiated his belief in evolution, and he used it in his important paper "On the Law Which Has Introduced the Introduction of New Species" (1855). Generally speaking, he observed that certain species on opposite sides of river barriers are closely related, but not identical. Since the physical conditions were almost the same, biologists believing in special creation were hard pressed to explain why an omniscient God had created different species on opposite sides of rivers within sight of each other. Wallace thought that after the barriers had been established, evolution had led to the formation of new, but similar, species from the divided parent stock.

Once evolution had been accepted, the facts of geographical distribution could be used to good effect. For example, arguments for the creationist's belief in centers of creation were quickly enlisted as evidence for the evolutionists. Wallace used this technique to good advantage as did Darwin in chapters 11 and 12 on geographical distribution in the *Origin* (1859). Both Wallace and Darwin were able to cite much of their own evidence on this topic in their evolutionary works.

Indeed, it is significant that their works up to 1859 interpreted and applied these facts in a distinctly modern way. Before 1859 most works on natural history had been extremely vague and imprecise in citing the geographical distribution of a specimen. It was not unusual to find merely South America or Brazil as the only locality given. Only after the case for evolution by means of natural selection was presented, particularly in Darwin's *Origin*, did other naturalists begin to give this matter the attention it deserved.

In 1858, however, Wallace was already preparing an announcement of an important zoogeographical discovery, which proposed a boundary line dividing the archipelago into Indo-Malayan (Oriental) and Australian zoological regions. In a paper on the geographical distribution of birds ("Letter From *Mr. Wallace* on the Geographical Distribution of Birds," *Ibis* [1859], 449–454), he first suggested this line and accepted the zoogeographical provinces recommended by P. L. Sclater (1858) on the basis of the zoogeography of bird populations. In 1860 Wallace published a much more elaborate discussion of the zoological geography of the archipelago, before announcing explicitly in 1863 what became known as Wallace's Line, the zoogeographical line that extended between Bali and Lombok in the south and farther north between Borneo and Celebes, and continuing eastward around the Philippines. This line has been shifted many times since 1863 as more zoogeographical information has been accumulated.

Wallace's investigations made it quite clear that zoogeography should be based on a wide range of geographical and geological facts interpreted by evolutionary doctrines. He was also one of the few early zoogeographers to rely on a statistical approach.

The culmination of Wallace's approach was achieved in his monumental two-volume *The Geographical Distribution of Animals* (1876). Relying on data he had collected on families and genera of terrestrial vertebrates, Wallace established evolutionary zoogeography on its modern foundation. While an enormous amount of subsequent data has improved our knowledge and determination of the zoogeographical provinces, few, if any, subsequent works have been more important to the subject.

As was previously the case, Wallace's evolutionary approach to zoogeography provided a rock-solid factual basis for evolutionary biology. While he hoped that his two volumes would elaborate Darwin's two chapters in the *Origin* in much the same way as Darwin's own two-volume work

Animals and Plants Under Domestication extended the first chapter, Wallace's work actually transformed the subject and became the standard authority for many years.

In 1880 Wallace further applied his approach in evolutionary zoogeography to island flora and fauna. Whereas his great work in 1876 dealt with large groups of animals, in his *Island Life* (1880) he focused on species to examine variation, distribution, and dispersal. His discussion of the relevance of the ice ages was extremely important, as was his reemphasis on the interaction and "complete interdependence of organic and inorganic nature." His later works on zoogeography represent summaries of these two great books.

Before going to South America, Wallace knew little about geology except what he had learned from books and what he casually observed as a surveyor. After his two expeditions he published many works utilizing geological information. In his *Geographical Distribution of Animals* Wallace interpreted animal distributions on the basis of geological principles, especially paleontological data. In *Island Life* Wallace presented advanced views on the causes of ice ages, showing the cumulative effects of snow and ice in lowering temperature. He also discussed the general permanence of oceanic and continental areas with a wide range of data. In 1893 Wallace argued vigorously for the action of glaciers in the formation of lake basins.

Early in the 1870's Wallace became acquainted with a group that opposed vaccination, but he did not join the movement until William Tebb introduced him to statistical studies attacking vaccination. Upon personal investigation, Wallace found apparently cogent evidence to renounce his former belief in the efficacy of vaccination, whereupon he published a thirty-eight-page pamphlet "Forty-five Years of Registration Statistics, Proving Vaccination to Be Both Useless and Dangerous" (1885).

Wallace and the anti-vaccination movement forced the appointment of the Royal Commission on Vaccination, and he spent three days presenting evidence (1890). Disregarding his statistical evidence, the commission published a report in 1896 supporting vaccination, which led him in turn to publish a longer work in 1898 (96 pages and 12 diagrams) denouncing the "ignorance and incompetence" of the commission and reiterating his previous opposition with extensive data. In his *The Wonderful Century* (1898), Wallace reprinted his arguments, which he thought would eventually be judged as "one of the most important and truly

scientific of my works." It is perhaps ironic that today we have discontinued smallpox vaccination because more patients die from the vaccination than die from the disease itself.

In *The Wonderful Century* (1898) Wallace had written the chapter "Astronomy and Cosmic Theories," and after the turn of the century he expanded the subject into his *Man's Place in the Universe* (1903). The primary purpose was to establish with extensive scientific data that life as we know it cannot exist elsewhere in the universe. In 1907 he reiterated this theme in his *Is Mars Habitable?*, which was written to refute Lowell's *Mars and Its Canals* (1906). Lowell had presented arguments for advanced life on Mars. Wallace was of course only an intelligent layman, but he corresponded with professional astronomers and presented a strong argument for his views.

Wallace summed up his work on biology in 1910 with his *The World of Life*, in which he accepted the chromosome theory of inheritance and Galton's numerical law of inheritance. Numerical phytogeography and zoogeography also are stressed, and the continuing influence of Spiritualism is evident. This was his last extensive work on scientific matters; his last two books were rehashings of his social ideas.

In April 1866 at the age of forty-three, Wallace married Annie Mitten, the teenage daughter of the English botanist William Mitten. They had three children: Herbert (died age four), Violet, and William.

During his distinguished career Wallace received numerous recognitions of merit, including the Royal Medal of the Royal Society (1868); the Gold Medal, Société de Géographie (1870); LL.D., Dublin, (1882); D.C.L., Oxford (1889); the Darwin Medal, Royal Society (1890); the Founder's Medal of the Royal Geographical Society (1892); election to the Royal Society (1893); the Gold Medal of the Linnean Society of London (1892); the Copley Medal of the Royal Society (1908); Order of Merit (1908); and the first Darwin-Wallace Medal, Linnean Society of London (1908).

Few naturalists have made more important contributions to so many subjects, and yet his views on Spiritualism, vaccination, land nationalization, women's rights, and socialism have combined to diminish his reputation in science. Those who dismiss him as a "crank" forget that cranks often make the machinery go, and in whatever he did, Wallace was one who made things happen. He rarely avoided controversy; indeed, he was at his very best while marshaling evidence for an argument. His brilliant imaginative mind, however, frequently offended lesser spirits, for he did not easily tolerate ignorant, pompous arguments; and while others refrained from the lists, Wallace charged into battle. That he did so greatly enriched science. Ironically, many of his social views, which have long detracted from his scientific contributions, are now widely accepted.

BIBLIOGRAPHY

I. ORIGINAL WORKS. Wallace published more than twenty books. James Marchant, ed., *Alfred Russel Wallace. Letters and Reminiscences* (New York, 1916), 477, provides an incomplete bibliography. Some editions of Wallace's book-length scientific publications (excluding translations) are *Palm Trees of the Amazon and Their Uses* (London, 1853; repr., Lawrence, Kans., 1971); *A Narrative of Travels on the Amazon and Rio Negro* (London, 1853; 2nd ed., 1889; repr. 1971); *The Scientific Aspect of the Supernatural* (London, 1866); *The Malay Archipelago: The Land of the Orang-Utan, and the Bird of Paradise. A Narrative of Travel with Studies of Man and Nature* (London–New York, 1869); *Contributions to the Theory of Natural Selection. A Series of Essays* (London, 1870; 2nd ed., 1871); *The Geographical Distribution of Animals*, 2 vols. (London–New York, 1876; repr., 1962); *Tropical Nature, and Other Essays* (London, 1878); *Australasia. Stanford's Compendium of Geography and Travel*, edited and extended by Wallace (London, 1879; rev., 1893); *Island Life or the Phenomena and Causes of Insular Faunas and Floras Including a Revision and Attempted Solution of the Problem of Geological Climates* (London–New York, 1880; 2nd. ed., 1892); *Darwinism. An Exposition of the Theory of Natural Selection With Some of Its Applications* (London–New York, 1889; 3rd ed., 1912); *Natural Selection and Tropical Nature. Essays on Descriptive and Theoretical Biology* (London–New York, 1891); *Vaccination a Delusion, Its Penal Enforcement a Crime: Proved by the Official Evidence in the Reports of the Royal Commission* (London, 1898); *The Wonderful Century. Its Successes and Its Failures* (London–New York, 1898, 1925); *Studies Scientific & Social*, 2 vols. (London–New York, 1900); *The Wonderful Century Reader* (London, 1901); *Man's Place in the Universe. A Study of the Results of Scientific Research in Relation to the Unity or Plurality of Worlds* (London–New York, 1903; 4th ed., 1904); *My Life, A Record of Events and Opinions*, 2 vols. (London–New York, 1905; 2nd ed., 1908); *Is Mars Habitable?* (London, 1907); Richard Spruce, *Notes of a Botanist on the Amazon and Andes*, 2 vols. (London, 1908), ed. by Wallace; and *The World of Life. A Manifestation of Creative Power, Directive Mind and Ultimate Purpose* (London, 1910; New York, 1911).

Wallace published about 400 articles and reviews, many of which are listed in Marchant (1916), 478–486. Errors and important omissions abound, and no citation should ever be based on Marchant. The Royal Society *Catalogue of Scientific Papers* adds a few missing articles but is still very incomplete. H. Lewis McKinney is doing a bibliography of all of Wallace's works. Wallace's two important evolutionary articles of 1855 and 1858 are reprinted in H. Lewis McKinney, ed., *Lamarck to Darwin: Contributions to Evolutionary Biology, 1809–1859* (Lawrence, Kans., 1971; repr. 1975), 69–82, 89–98.

Many of Wallace's letters appeared in *My Life* (1905). James Marchant, *Alfred Russel Wallace. Letters and Reminiscences*, 2 vols. (London, 1916; American ed., in one vol., New York–London, 1916), is the only other collection of published correspondence, although one by H. Lewis McKinney is in progress. Other letters appear in the lives and letters of Charles Lyell, J. D. Hooker, and Charles Darwin. The most complete list of MSS is in McKinney (1972), 177–179, cited below.

II. SECONDARY LITERATURE. For Wallace's early work up to 1858, concentrating on evolution and natural selection, see H. Lewis McKinney, *Wallace and Natural Selection* (New Haven–London, 1972), which cites most secondary literature up to 1972. For a comparison of Wallace's scientific ideas with currently held ideas, especially in zoogeography, see Wilma George, *Biologist Philosopher. A Study of the Life and Writings of Alfred Russel Wallace* (London–Toronto–New York, 1964). See also Roger Smith, "Alfred Russel Wallace: Philosophy of Nature and Man," in *British Journal for the History of Science*, 6 (1972), 178–199; and M. J. Kottler, "Alfred Russel Wallace, the Origins of Man, and Spiritualism," in *Isis*, 65 (1974), 145–192. For references to mimicry, see Charles Remington, "Historical Backgrounds on Mimicry," in *Proceedings of the XVI International Congress of Zoology*, 4 (1963), 145–149.

H. LEWIS MCKINNEY

WALLACE, WILLIAM (*b.* Dysart, Scotland, 23 September 1768; *d.* Edinburgh, Scotland, 28 April 1843), *mathematics*.

Wallace had no schooling after the age of eleven, when he was apprenticed to a bookbinder; he subsequently taught himself mathematics and became a teacher at Perth. In 1803 he was appointed to the Royal Military College at Great Marlow and in 1819 became professor of mathematics at the University of Edinburgh, where he remained until his retirement in 1838. Wallace wrote many articles for encyclopedias and numerous papers in *Proceedings of the Royal Society of Edinburgh*, including some on mechanical devices. He also played a large part in the establishment of the observatory on Calton Hill, Edinburgh.

The feet of the perpendiculars to the sides of a triangle from a point P on its circumcircle are collinear. This line is sometimes called the pedal line but more often, incorrectly, the Simson line of the triangle relative to P. It was stated by J. S. Mackay that no such theorem is in Simson's published works. The result appears in an article by Wallace in Thomas Leybourn's *Mathematical Repository* (**2** [1799–1800], 111), and Mackay could find no earlier publication. In the preceding volume Wallace had proved that if the sides of a triangle touch a parabola, the circumcircle of the triangle passes through the focus of the parabola, a result already obtained by Lambert. To demonstrate this, Wallace showed that the feet of the perpendiculars from the focus to the sides of the triangle lie on the tangent at the vertex of the parabola, which is equivalent to saying that the pedal line of the triangle is the tangent at the vertex. The close connection of this theorem with the pedal line suggests that Wallace was led to the property of the pedal line from the parabolic property.

In 1804 the following result was proposed for proof in *Mathematical Repository* (n.s. **1**, 22): If four straight lines intersect each other to form four triangles by omitting one line in turn, the circumcircles of these triangles have a point in common. The proposer was "Scoticus," which Leybourn later said was a pseudonym for Wallace. Two solutions were given in the same volume (170). Miquel later proved that five lines determine five sets of four lines, by omitting each in turn; and the five points, one arising from each such set, lie on a circle. Clifford proved that the theorems of Wallace and Miquel are parts of an endless chain of theorems: $2n$ lines determine a point as the intersection of $2n$ circles; taking one more line, $2n + 1$ lines determine $2n + 1$ sets of $2n$ lines, each such set determines a point, and these $2n + 1$ points lie on a circle.

BIBLIOGRAPHY

Two articles by J. S. Mackay in *Proceedings of the Edinburgh Mathematical Society*—**9** (1891), 83–91, and **23** (1905), 80–85—give the bibliography of Wallace's two theorems and later extensions and generalizations with scholarly thoroughness.

For a full account of Wallace's life, see the unsigned but evidently authoritative obituary in *Monthly Notices of the Royal Astronomical Society*, **6** (1845), 31–36.

T. A. A. BROADBENT

WALLACH, OTTO (*b.* Königsberg, Prussia [now Kaliningrad, R.S.F.S.R.], 27 March 1847; *d.* Göttingen, Germany, 26 February 1931), *chemistry.*

Wallach came from a family of lawyers, but his father was a Prussian state official, who was transferred from Königsberg to Stettin and, in 1855, to Potsdam. The boy was educated at the Potsdam Gymnasium, where he acquired his two major interests: chemistry and the history of art. Although he became a professional chemist, he remained an art collector throughout his life and spent many vacations visiting the art galleries of Europe.

In the spring of 1867 Wallach entered the University of Göttingen, where he attended Wöhler's lectures. A short stay in Berlin with A. W. von Hofmann convinced him that Göttingen was a better place for him to work, and he returned there to study for his doctorate with Hans Hübner. He received the degree in 1869 with a dissertation on position isomerism in the toluene series. Wallach worked for a short time in Berlin; but in the spring of 1870 he was called to Bonn as assistant to Kekulé, who was then gradually withdrawing from active laboratory work. He remained at Bonn for nineteen years, with a short interlude of industrial experience at the Aktien Gesellschaft für Anilin-Fabrikation (Agfa) plant in Berlin. His health was never good, and the noxious fumes at the plant soon drove him back to Bonn. In 1889 Wallach succeeded Victor Meyer at Göttingen, where he served as director of the Chemical Institute until his retirement in 1915. He continued experimental work until he was eighty.

Wallach's early work included a number of studies in general organic chemistry. In 1879 he was assigned to teach pharmacy, with which he had had little experience. In teaching this course his attention was drawn to the chemistry of natural compounds. He found in a cupboard at Bonn a number of bottles of plant essential oils that Kekulé had collected but never studied. Wallach decided to investigate these substances, and the rest of his life was devoted to the study of such compounds, on which he published 126 papers.

At the time Wallach began this work, the field was in a state of extreme confusion. No one had obtained truly pure compounds from the natural mixtures; and various names had been proposed for many of the substances thought to be pure. A skilled and patient experimentalist, Wallach set himself the task of characterizing individual compounds beyond doubt and then of determining their relationships. For this purpose he separated the pure substances by careful distillations and studied their reactions with a series of relatively simple reagents.

After three years Wallach had distingished eight pure terpenes, as he called this class of compounds. He suggested that they were composed of five carbon atom fragments known as isoprene units and showed that in many cases it was possible to rearrange one terpene into another by the action of strong acids and high temperatures. He was particularly interested in the relations among the various compounds and was less concerned with preparing new substances, the chief occupation of most organic chemists of the time. Wallach often left the task of synthesizing and determining structures of his compounds to others, for he realized that the field of terpene chemistry was so large that one man could not cover it completely. His methods were so successful and had progressed so far by 1895 that when, in that year, he and others determined the structure of α-terpineol, the structures of an entire series of terpenes were at once established. This accomplishment was an outstanding example of the value of Wallach's experimental methods. After he had made the fundamental discoveries, a number of chemists continued his work, and terpene chemistry became an important branch of organic chemistry. It was soon extended to the biologically important carotenoids and steroids.

Wallach received wide recognition for his work, becoming an honorary member of many universities and scientific societies. These honors culminated in the award of the Nobel Prize in chemistry in 1910.

BIBLIOGRAPHY

I. Original Works. Among Wallach's important papers were his suggestion of isoprene units in terpenes, "Zur Kenntniss der Terpene und ätherischen Oele. IV," in *Justus Liebigs Annalen der Chemie*, **238** (1887), 78–89; and his work on terpineol, "Zur Constitutionsbestimmung des Terpineols," in *Berichte der Deutschen chemischen Gesellschaft*, **28** (1895), 1773–1777. He summarized his work in *Die Terpene und Campher* (Leipzig, 1909; 2nd ed., 1914).

II. Secondary Literature. The longest account of Wallach is the Pedlar lecture by L. Ruzicka, "The Life and Work of Otto Wallach," in *Journal of the Chemical Society* (1932), 1582–1597, repr. in Eduard Farber, ed., *Great Chemists* (New York, 1961), 833–851. A shorter account is William S. Partridge and Ernest R. Schierz, "Otto Wallach: The First Organizer of the Terpenes," in *Journal of Chemical Education*, **24** (1947), 106–108.

An account of the industrial significance of Wallach's work is Albert Eller, "Otto Wallach und seine Bedeutung für die Industrie der ätherischen Öle," in *Zeitschrift für angewandte Chemie . . .*, **44** (1931), 929–932.

Henry M. Leicester

WALLER, AUGUSTUS VOLNEY (*b.* Faversham, England, 21 December 1816; *d.* Geneva, Switzerland, 18 September 1870), *neurophysiology, neurohistology.*

Waller, the son of William Waller, was raised at Nice until his father's death in 1830. He then returned to England, living with Dr. Lacon Lambe and then with William Lambe, a noted vegetarian. Following early training in the physical sciences, Waller studied medicine at Paris, receiving the M.D. degree in 1840. The following year he became a licentiate of the Society of Apothecaries in London and began a successful general practice in Kensington. In 1842 Waller married Matilda Walls; they had two daughters and one son, the physiologist Augustus D. Waller.

The publication of two papers in the *Philosophical Transactions* (1849, 1850) led to Waller's election as a fellow of the Royal Society in 1851. That same year he abandoned his general practice and moved to Bonn in order to devote full time to research; he spent five years there, working principally with the ophthalmologist Julius Budge. The investigations begun in England and continued in Bonn brought Waller the Monthyon Prize of the French Academy of Sciences in 1852 and 1856, and a Royal Society medal in 1860.

After leaving Bonn in 1856 Waller worked in Pierre Flourens's laboratory at the Jardin des Plantes but developed a chronic fever that invalided him for two years. He returned to England and in 1858 was appointed professor of physiology at Queen's College, Birmingham, and physician to the college hospital. A heart condition soon forced him to relinquish these posts, and in the same year he retired to Bruges. Ten years later he moved to Geneva, where he hoped to resume general practice. In the spring of 1870 Waller went briefly to London to deliver the Royal Society's Croonian lecture; he returned to Geneva, where he died suddenly on 18 September.

Waller is best remembered for pioneering a major technique for unraveling the complex structure of the nervous system, the method of secondary or Wallerian degeneration. His interest in the functional anatomy of the nervous system began while he was a medical student in Paris, when he began to study the histology of the frog's tongue. Like other neurohistologists and physiologists in the first half of the nineteenth century, Waller must have found the processes of nerve degeneration and regeneration to be one of the most difficult problems he confronted in trying to elucidate the structure of the nerve fiber. In retrospect, we know today that the large body of erroneous belief about the fine structure of nerve fibers and nerve cells that arose in the eighteenth and nineteenth centuries resulted in part from mistaking the products of nerve degeneration—the structures seen after a nerve was sectioned or otherwise injured—as normal structures. Our knowledge that the nerve cell body is a trophic center, and that a detached nerve cell process consequently will degenerate and die, began with Waller's studies of the frog's tongue.

Using the simple technique of cutting the nerves in the tongue, Waller found in 1849 that degeneration occurred throughout the axon's distal segment and concluded that the nerve cell body is the axon's source of nutriment. His belief that the proximal part of the nerve process and the cell body itself did not degenerate following sectioning of the fiber was subsequently modified through the development of improved staining methods. Waller's study, first reported in the *Philosophical Transactions* (1840), added to the growing evidence that nerve cell bodies and processes were somehow interconnected. By the 1880's, when the origin of nerve fibers from nerve cells had been firmly established with the development of better methods for fixing and staining tissues, the Wallerian method became a major means of tracing the origin and course of nerve fibers and tracts; and in the hands of such investigators as Forel, it helped to establish the neuron theory.

Fundamental studies of the autonomic nervous system, conducted during his five years in Bonn, formed the second major area of Waller's researches. In 1851 and 1852, Budge and Waller published three memoirs in the *Comptes rendus* of the Paris Academy, examining the role of the nervous system on the motion of the eye's iris. In a series of well-designed and carefully executed experiments, they showed the influence of the cervical portion of the sympathetic nerve in dilation of the pupil. They then used the Wallerian method to trace the pathway of the pupillary dilator fibers in the dog, following them in the sectioned sympathetic nerve trunk to the first and second thoracic segments of of the spinal cord. When this region was then stimulated in the intact animal, the pupils of the

eye dilated. When in turn the cervical part of the sympathetic nerve was sectioned unilaterally, electrical stimulation in the thoracic area no longer caused pupillary dilation on the side that had been sectioned. Budge and Waller termed the area of the sympathetic nerve trunk controlling dilation of the pupils the "ciliospinal center."

After receiving the Monthyon Prize for the work on the ciliospinal center, Waller went on to demonstrate the action of the cervical sympathetic nerves on the constriction of blood vessels in the head. His experiments on the vasoconstrictor properties of the nerves from the ciliospinal region confirmed and extended the discoveries of Claude Bernard in 1851 and of Brown-Séquard in 1852.

Waller's work was not confined to the definition of neural structure and function. The readers of *Philosophical Magazine* in the 1840's, for example, found that Waller had turned his microscope upon a variety of objects: "The Microscopic Observations on the Perforation of the Capillaries by the Corpuscles of the Blood" (November 1846), "Origin of Mucus and Pus" (November 1846), and "Microscopic Investigations on Hail" (July and August 1846, March 1847).

BIBLIOGRAPHY

I. ORIGINAL WORKS. Waller's writings include "Experiments on the Section of the Glossopharyngeal and Hypoglossal Nerves of the Frog, and Observations of the Alterations Produced Thereby in the Structure of Their Primitive Fibres," in *Philosophical Transactions of the Royal Society*, **140** (1850), 423–429; "Nouvelle méthode pour l'étude du système nerveux, applicable à l'investigation de la distribution anatomique des cordons nerveux, et au diagnostique des maladies du système nerveux, pendant la vie et après la mort," in *Comptes rendus . . . de l'Académie des sciences*, **33** (1851), 606–611; "Recherches sur le système nerveux. Première partie. Action de la partie cervicale du nerf grand sympathétique et d'une portion de la moelle épinière sur la dilatation de la pupille," *ibid.*, 370–374, written with J. L. Budge; "Recherches expérimentales sur les structures et les fonctions des ganglions," *ibid.*, **34** (1852), 524–527—also papers by Waller on 582–587, 675–679, 842–847; "Septième mémoire sur le système nerveux," *ibid.*, **35** (1852), 301–306; "Huitième mémoire . . .," *ibid.*, 561–564; and "On the Results of the Method Introduced by the Author of Investigating the Nervous System, More Especially as Applied to the Elucidation of the Functions of the Pneumogastric and Sympathetic Nerves," in *Proceedings of the Royal Society*, **18** (1869–1870), 339–343, the Croonian lecture.

II. SECONDARY LITERATURE. See D. Denny Brown, "Augustus Volney Waller," in W. Haymaker, ed., *Foun-*

ders of Neurology (Springfield, Ill., 1953), 95–98; R. Gertler-Samuel, Augustus Volney Waller (1816–1870) als Experimentalforscher (Zurich, 1965); and D'Arcy Power, "Waller, Augustus Volney," in Dictionary of National Biography, XX, 579–580.

JUDITH P. SWAZEY

WALLERIUS, JOHAN GOTTSCHALK (b. Stora Mellösa, Nerke, Sweden, 11 July 1709; d. Uppsala, Sweden, 16 November 1785), chemistry, mineralogy.

Wallerius was the son of Erik Wallerius, a Lutheran minister, and Elisabet Tranaea. At the age of five he studied Latin, Greek, and Hebrew at home with his older brothers. Wallerius entered the Gymnasium in Strängnäs in 1722, enrolled at the University of Uppsala in 1725, and received a master's degree in philosophy in 1731. He next studied medicine and in the same year defended a thesis in anatomy, receiving an assistant professorship in medicine at Lund in 1732. He defended his doctoral dissertation there in 1735; but toward the end of the year he returned to Uppsala, where the Medical Faculty granted him the degree venia docendi. Wallerius practiced medicine and in 1737 was appointed superintendent at Danemarks, a spa near Uppsala, where he analyzed the spring water. As early as 1732 he had shown an interest in mining science and mineraology: en route to Lund through the Swedish mining district he studied mines and blast furnaces and collected mineral specimens. These latter formed the base of his private mineral collection, which ultimately amounted to over 4,000 specimens.

While at Lund, Wallerius studied the extensive mineral collection of Kilian Stobaeus and the famous royal collections in Copenhagen. This experience proved valuable when, at Uppsala, he had to teach chemistry to students of mining science as well as of medicine. For this purpose he installed a small private laboratory when he demonstrated chemical and pharmaceutical reactions for the medical students and where the future mining chemists could practice assaying. All participants in the course were free to ask questions and discuss the experiments. This attitude was new, and the number of students taking the course increased. The interest in chemistry at Uppsala during this period owed much to Wallerius' method of teaching.

In November 1741, Wallerius was appointed assistant professor of medicine at Uppsala, with responsibility for lectures on materia medica and later on physiology and anatomy as well. He also continued his lectures and experiments in chemistry and mineralogy. His research led to Mineralogia eller Mineralriket (1747), his first great work, which was received as an outstanding handbook of contemporary knowledge; never before had such a wealth of minerals been presented so systematically. Wallerius' clear and precise descriptions, which gave more weight to essential chemical properties than to exterior appearance, opened a new epoch in mineralogy. The book became widely known in Europe through translations into German, French, Russian, and (later) Latin, and served as a model for later works. The following year Wallerius published Hydrologia eller Wattu-Riket (1748), in which he tried to classify different kinds of water.

Wallerius was appointed the first professor of chemistry in Sweden, at Uppsala, in 1750; he continued to be responsible for metallurgy and pharmacy. Although chemistry had been taught at Uppsala for more than a century, it remained a minor subject within the Medical Faculty and lacked a spokesman of its own. The authorities used this situation as an excuse not to build a laboratory. When the chair of chemistry was placed within the Philosophical Faculty and the curriculum became part of the examination for the candidate's degree in philosophy, however, the professor also was obliged to examine the medical students and to give courses for future mining chemists. Rooms suitable for lectures had therefore to be furnished, and this need led to the construction of the university's first chemical institute.

Wallerius' courses included laboratory periods, experiments, and lectures. These were later published as Chemia physicae, in three volumes, which represents a summary of contemporary knowledge about chemical substances: acids, alkalies, salts, sulfur, bitumen and other combustible materials, semimetals, and metals. The work begins with a brief history of chemistry and presents the system of chemical symbols then in use, as well as a detailed description of the available chemical apparatus and the procedures. He also published his lectures on metallurgy as Elementa metallurgiae, speciatim chemicae (Stockholm, 1768), "which . . . has cost me innumerable experiments and much trouble."

In accord with the utilitarian tendencies of the time, Wallerius, as professor of chemistry, was called upon to show what his knowledge could contribute to economic life. Mining chemists had already demonstrated the advantages of chemistry applied to mining; and Wallerius' interest in agriculture naturally led him to pursue agricultural chemistry, especially since agriculture was of great importance for the national economy. His research proved so basic and of such scope that he was called the father of agricultural chemistry. His principal work in this field was *Agriculturae fundamenta chemica, Åkerbrukets kemiska grunder* (1761), which was published in Latin and Swedish, in parallel columns, and was later translated into German, French, Spanish, and English. Wallerius established as a fundamental, necessary principle that agricultural chemistry should be based on comparative study of the chemical composition not only of plants but also of the earth in which they grow.

Unsatisfactory working conditions in the laboratory undermined Wallerius' health; and deafness that had begun when he was young increasingly worsened. Finally, serious symptoms of illness forced him to request early retirement, which was granted in 1767. He was allowed to retain his salary in recognition of his thirty-four years devoted to the teaching of chemistry.

After his retirement Wallerius bought a farm outside Uppsala and, by actively applying his chemical theories concerning agriculture, established a model farm. These experiences are collected in *Rön, rörande landtbruket. Om svenska åkerjordarternas egenskaper och skiljemerken samt deras förbättring genom tienlig jordblanning* ("Observations of Agriculture"; 1779), which contains an essay on the qualities and differences of Swedish soils and their improvement by suitable mixing that was awarded a prize by the Royal Swedish Academy of Sciences in Stockholm.

Wallerius was by no means infatuated with innovation in chemistry, and he stubbornly held outdated beliefs. In *Tankar om verldenes, i synnerhet jordenes danande och ändring* ("Thoughts on the Creation and Change of the World, Particularly of the Earth"; 1776), which was translated into many languages, he assigned the highest authority to the biblical account of the history of creation. In chemistry he never entirely freed himself from the Becher-Stahl philosophical-chemical doctrine. Thus he became known not through new discoveries but, rather, for the new ways in which he applied chemistry to agriculture.

BIBLIOGRAPHY

I. ORIGINAL WORKS. Wallerius' numerous publications have appeared in various eds., and some were translated into several languages. J. R. Partington has compiled a good and extensive bibliography in *A History of Chemistry*, III (1962), 169–170, which also includes the titles of the many dissertations over which Wallerius presided and that have been printed in *Disputationum academicarum fasciculus primus cum annotationibus* (Uppsala, 1780) and . . . *fasciculus secundus* (Uppsala, 1781). It can be completed from Wallerius' autobiography, *Curriculum vitae Johan. Gotschalk Wallerii*, which ends with a bibliography that he compiled. This autobiography has been published in its entirety, with intro. and English summary, by Nils Zenzén, in *Lychnos* (1953), 235–259.

II. SECONDARY LITERATURE. See C. E. Bergstrand, *Johan Gottschalk Wallerius som landtbrukskemist och praktisk jordbrukare* (". . . Wallerius as Agricultural Chemist and Practical Farmer"; Stockholm, 1885); T. Frängsmyr, *Geologi och skapelsetro* ("Geology and the Belief in Creation"; Uppsala, 1969); Hugo Olsson, *Kemiens historia i Sverige intill år 1800* (Uppsala, 1971), 108–115, 179–182, 319–321; C. W. Oseen, "En episod i den svenska kemiens historia," in *Lychnos* (1940), 73–85; J. R. Partington, *A History of Chemistry*, III (London, 1962), 169–172; E. Svedmark, "Några anteckningar om Johan Gottschalk Wallerius" ("Some Notes on . . . Wallerius"), in *Geologiska föreningens i Stockholm förhandlingar*, 7 (1885); and Nils Zenzén, "Johan Gottschalk Wallerius and Axel Fredrik Cronstedt," in Sten Lindroth, ed., *Swedish Men of Science* (Stockholm, 1952), 92–104.

UNO BOKLUND

WALLICH, GEORGE CHARLES (*b.* Calcutta, India, November 1815; *d.* Marylebone, London, England, 31 March 1899), *medicine, zoology.*

Wallich was the eldest son of Nathaniel Wallich, superintendent of the botanical gardens at Calcutta from 1815 to 1850. He was sent to Beverley (Yorkshire) and, later, to Reading Grammar School. After attending the arts classes at King's College, Aberdeen, he received an M.D. at Edinburgh University in 1836. Ironically, a classmate was Edward Forbes, proponent of the azoic theory, which held that the absolute depth to which life extended in the seas was 300 fathoms.

Wallich entered the Indian army in 1838, served in the Sutlej (1842) and Punjab (1847) campaigns with distinction, and was field surgeon during the Sonthal rebellion (1855–1856). He was invalided to England in 1857 and spent two years recuperating on Guernsey before settling in Kensington.

In 1860 Wallich shipped as naturalist on H.M.S. *Bulldog*, under the command of Sir Francis Leopold McClintock, to survey the proposed north Atlantic telegraph route between Great Britain and America (2 July–11 November). A single sounding during October in 1,260 fathoms (lat. 59° 27'N., long. 26°41'W.) brought up thirteen living starfishes (*Ophiocomae*) from the bottom—incontrovertible, although at the time generally disregarded, evidence that "The conditions prevailing at great depths . . . are not incompatible with the maintenance of animal life" as well as ". . . the inference that the deep sea has its own special fauna." Although he wrote extensively on this discovery, the majority of Wallich's scientific publications dealt with the Protozoa. He entered the *Bathybius* controversy with trenchancy, correctly opposing Huxley, who, in the course of examining a ten-year-old collection of sea-bottom samples, found a gelatinous substance that he took to be a primitive form of life and named after Ernst Haeckel. The substance was later determined by J. Y. Buchanan to be a precipitate of calcium sulfate caused by the alcohol in which the samples were preserved.

The year before his death, Wallich was awarded the gold medal of the Linnean Society of London "in recognition of his researches into the problems connected with bathybial and pelagic life."

BIBLIOGRAPHY

I. ORIGINAL WORKS. Wallich's writings include *Notes on the Presence of Animal Life at Vast Depths in the Sea* (London, 1860); "Results of Soundings in the North Atlantic," in *Annals and Magazine of Natural History*, 3rd ser., **6** (1860), 457–458; "On the Existence of Animal Life at Great Depths in the Sea," *ibid.*, **7** (1861), 396–399; *The North Atlantic Sea-Bed* (London, 1862); "On the Value of the Distinctive Characters in Amoeba," in *Annals and Magazine of Natural History*, 3rd ser., **12** (1863), 111–151; "On the Vital Functions of the Deep-Sea Protozoa," in *Monthly Microscopical Journal*, **1** (1869), 32–41; "On the Radiolaria as an Order of the Protozoa," in *Popular Science Review*, **17** (1878), 267–281, 368–382; "The Threshold of Evolution," *ibid.*, **19** (1880), 143–155; and "Critical Observations on Prof. Leidy's 'Freshwater Rhizopods of North America,' and Classification of the Rhizopods in General," in *Annals and Magazine of Natural History*, 5th ser., **16** (1885), 317–334, 453–473.

II. SECONDARY LITERATURE. See "Surgeon-Major G. C. Wallich, M.D.," in *Nature*, **60** (4 May 1899), 13; and obituaries in *Indian Medical Gazette*, **34** (1899), 227–228; *Lancet* (8 Apr. 1899), 997; *Journal of the Royal Microscopical Society* (1899), 263–264; and *Transactions and Proceedings of the Botanical Society of Edinburgh*, **21** (1900), 222–224.

DANIEL MERRIMAN

WALLINGFORD. See **Richard of Wallingford.**

WALLIS, JOHN (*b.* Ashford, Kent, England, 3 December 1616; *d.* Oxford, England, 8 November 1703), *mathematics.*

Wallis was the third child of John Wallis and his second wife, Joanna Chapman. His father studied at Trinity College, Cambridge, and after having taken holy orders became minister at Ashford, about 1603. Standing in great esteem and reputation in his town and parish, he died when John was barely six.

Young John grew up, together with his two older sisters and two younger brothers, in the care of his mother. After he had received his first education, he was sent in 1625 to a grammar school at Tenterden, Kent, where, according to his autobiography,[1] he enjoyed a thorough training in Latin. In 1631–1632 Wallis attended the famous school of Martin Holbeach at Felsted, Essex. Besides more Latin and Greek he also learned some Hebrew and was introduced to the elements of logic. As mathematics was not part of the grammar school curriculum, he obtained his first insight into this field during a vacation; he studied what a brother of his had learned in approximately three months as preparation for a trade.

Wallis entered Emanuel College, Cambridge, the "Puritan College," about Christmas 1632 as a pensioner. He not only took the traditional undergraduate courses (obtaining his bachelor of arts degree early in 1637), followed by studies in theology, but he also studied physic, anatomy, astronomy, geography, and other parts of natural philosophy and what was then called mathematics—although the latter "were scarce looked upon, with us, as Academical Studies then in fashion." He was the first student of Francis Glisson to defend the doctrine of the circulation of the blood in a public disputation.

In 1640 Wallis received the degree of master of arts and was ordained by the bishop of Winchester. For some years he earned his living as private chaplain and as minister in London. From 1644, after the outbreak of the Civil War, he also acted as secretary to the Assembly of Divines at Westminster, which was charged with proposing a new form of church government. For about a year he also held a fellowship at Queens' College, Cambridge, in consequence of a Parliamentary ordinance. He gave up this position when he married Susanna Glyde of Northiam, Sussex, on 14 March 1645.

Wallis' appointment as Savilian professor of geometry at Oxford on 14 June 1649 must have come as a surprise to many; his accomplishments thus far, with one exception, had had little to do with mathematics. His predecessor, Peter Turner, was a Royalist who had been dismissed by an order of Parliament; Wallis had rendered valuable services not only as a secretary to the Assembly of Divines but also by his skill in deciphering captured coded letters for the Parliamentarians. Few people in 1649 could have foreseen that within a few years the thirty-two-year-old theologian would become one of the leading mathematicians of his time.

This appointment determined Wallis' career; he held the chair until his death more than half a century later. In addition, in 1657–1658 he was elected–by a somewhat doubtful procedure–custos archivarum (keeper of the archives) to the university, an office he also held for life. In 1654 he had been admitted doctor of divinity. At the Restoration Wallis was confirmed in his offices for having possessed the courage to sign the remonstrance against the execution of King Charles I; he also received the title of royal chaplain to Charles II. When in 1692 Queen Mary II offered Wallis the deanery of Hereford, he declined, hinting that favors for his son and his son-in-law Blencowe would be more welcome signs of recognition of his services to his country.

These achievements include the mathematical works, helping found the Royal Society; his work in the decipherment of code letters for the government; logic; teaching deaf mutes to speak and the related grammatical and phonetical writings; archival studies and his assistance to the university in legal affairs; theological activities as a preacher and author of treatises and books; and the editions (many of them first editions) of mathematical and musical manuscripts of ancient Greek authors.

The first two decades of the Savilian professorship were the most creative period in Wallis' life. He later increasingly turned to editing works of other scientists (J. Horrox, W. Oughtred, and Greek authors) and his own earlier works, and to the preparation of historical and theological discourses. His *Opera mathematica* appeared between 1693 and 1699, financed by and printed at the university.

Wallis enjoyed vigorous health throughout his life. His powers of intellect were remarkable, and he was renowned for his skill in public disputations. But he also possessed a highly contentious disposition and became involved in many violent controversies—the more so since modesty does not seem to have been one of his virtues. Nevertheless he had many devoted friends. It was for Thomas Smith, vice-president of Magdalen College, Oxford, and librarian at the Cottonian Library, London, that Wallis wrote his autobiography in 1697; and Samuel Pepys commissioned Sir Godfrey Kneller to paint a full-length portrait of "that great man and my most honoured friend, Dr. Wallis, to be lodged as an humble present of mine (though a Cambridge-man) to my dear Aunt the University of Oxford."[2] Wallis was interred in St. Mary's, the university church, and an epitaph by his son was placed in the wall near his burial place: "Joannes Wallis, S.T.P., Geometriae Professor Savilianus, et Custos Archivarum Oxon. Hic dormit. Opera reliquit immortalia . . ." ("Here sleeps John Wallis, Doctor of Theology, Savilian Professor of Geometry, and Keeper of the Oxford Archives. He left immortal works. . . ."[3])

Mathematics. Wallis reports in his *Algebra*[4] that his interest in mathematics (beyond the little that he may have learned at Cambridge) was first aroused in 1647 or 1648, when he chanced upon a copy of William Oughtred's *Clavis mathematicae*. After having mastered it in a few weeks, he rediscovered Cardano's solution of the cubic equation (not given by Oughtred) and, continuing where Oughtred had left off, composed in 1648 a *Treatise of Angular Sections*, which remained unpublished until 1685. In the same year, at the request of Cambridge professor of mathematics John Smith, the Platonist (1618–1652), he gave an explanation of Descartes's treatment of the fourth-degree equation. The basic idea, to write the equation as a product of two quadratic factors, could be derived from Harriot's *Artis analyticae praxis* (published posthumously in 1631); yet Wallis repeatedly claimed not to have known this book in 1648. Such was the total evidence of his mathematical talents that Wallis presented when he was made Savilian professor of geometry in 1649.

With a rare energy and perseverance, he now took up the systematic study of all the major mathematical literature available to him in the Savilian and the Bodleian libraries in Oxford. According to the statutes of his chair, Wallis had to give public lectures on the thirteen books of Euclid, on the *Conics* of Apollonius, and on all of Archimedes' work. He was also to offer introductory courses in practical and theoretical arithmetic—with a free choice of textbooks therein. Lectures on other subjects such as cosmography, plane and spherical trigonometry, applied geometry, mechanics, and the theory of music were suggested but not obligatory according to the statutes.

An outcome of his elementary lectures was the *Mathesis universalis, seu opus arithmeticum* (1657). Its treatment of notation, including a historical survey, stressed the great advantages of a suggestive and unified symbolism; yet the influence of Oughtred (who had developed a rather special notation) sometimes makes itself felt—to no great advantage. On the whole, this work reflects the rather weak state of mathematical learning in the universities at the time.

In the treatise *De sectionibus conicis* (1655) Wallis dealt with a classical subject in a new way.[5] He considered the conic sections merely as plane curves, once he had obtained them by sections of a cone, and subjected them to the analytical treatment introduced by Descartes rather than to the traditional synthetic approach. In addition, he employed infinitesimals in the sense of Cavalieri and Torricelli. Here he also first introduced the sign for infinity and used $1/\infty$ to represent, for example, the height of an infinitely small triangle. Although Mydorge in adherence to the ancient methods had obtained a certain simplification of the treatment in 1631, Wallis was rather proud of his achievement; he may not have known Mydorge's *De sectionibus conicis* at the time of writing. Shortly afterward, in 1659, Jan de Witt's valuable treatise *Elementa curvarum linearum*, also employing the analytic symbolism, appeared in Amsterdam. Yet, on the whole, the new viewpoint was accepted only slowly by mathematicians.

Together with his conic sections Wallis published the book on which his fame as a mathematician is grounded, *Arithmetica infinitorum*; the title page is dated 1656, but printing had been completed in the summer of 1655. It resulted mainly from his study of Torricelli's *Opera geometrica* (1644), for Cavalieri's basic work on the methods of indivisibles was unavailable. At first Wallis' attempts to apply these methods to the quadrature of the circle met with failure; and not even a study of the voluminous *Opus geometricum* (1647) of Gregory of St. Vincent, which was devoted to this subject, would help. But then, by an ingenious and daring sequence of interpolations, he produced his famous result[6]

$$\frac{4}{\pi} = \frac{3}{2} \cdot \frac{3}{4} \cdot \frac{5}{4} \cdot \frac{5}{6} \cdot \frac{7}{6} \cdots.$$

Although the method was mistrusted by such eminent mathematicians as Fermat and Huygens, the result was ascertained by numerical computation.

Wallis' main interest lay not with the demonstration, but with the investigation. Actually searching for the value of

$$\int_0^1 (1-x^2)^{\frac{1}{2}} \, dx = \frac{\pi}{4},$$

he considered the generalized integral

$$I(k,n) = \int_0^1 (1-x^{1/k})^n \, dx.$$

Its reciprocal $1 : I(k, n)$ he tabulated first for integral values of k and n (receiving the symmetric array of the binomical coefficients or figurated numbers), then for the fractions $k = \frac{1}{2}, \frac{3}{2}, \frac{5}{2}, \cdots$; for, with $k = n = 1/2$, this should yield $1 : I\left(\frac{1}{2}, \frac{1}{2}\right) = \frac{4}{\pi}$, for which he wrote the symbol $\square$. Then each second value of the row and column which met at $\square$ was a certain (fractional) multiple of $\square$. Assuming that all rows and columns in his table would continually increase, Wallis was able to derive two sequences of upper and lower bounds for $\square$, respectively. When these sequences are continued indefinitely, they yield his famous infinite product. William Brouncker soon transformed it into a regular continued fraction, which Wallis included in his book.

Wallis' method of interpolation—he himself gave it this name, which has become a *terminus technicus*—is based on the assumption of continuity, and, incidentally, seems closely related to the procedure he had to apply when he deciphered coded letters. To preserve this continuity and thereby the underlying mathematical law in his table, Wallis went to the utmost limit. He admitted fractional multiples of the type $A \cdot \frac{0}{1} \cdot \frac{2}{3} \cdot \frac{4}{5} \cdot \frac{6}{7} \cdots$, claiming that A here should be infinite so that the value of the product was a finite number. One must emphasize the kind of "functional thinking" revealed here—not on the basis of geometric curves but of sequences of numerical expressions, that is, tabulated functions.

There are many more remarkable results of a related nature in the *Arithmetica infinitorum*, in the tracts on the cycloid and the cissoid, and in the *Mechanica*.[7] The integral $I(k, n)$ may in fact, by the substitution $x \to y^k$, be transformed into the normal form of the beta integral. He soon derived analytically the integral for the arc length of an ellipse and reduced other integrals to the elliptic one. But more important than the individual problems that Wallis mastered was the novelty of his approach—his analytic viewpoint, in contrast to the traditional geometric one—at a time when the symbolism of analysis had not yet been properly developed. The best documentation of his new "functional thinking" is provided in the *Arithmetica infinitorum*; he finally plots the graphs of the family of functions the values of which he had so far evaluated only for a sequence of distinct points. There he considers not so much the single curves as the sequence of them, since the parameter changes from one integral value to the next. The answer to his question of what the equations of these curves would be for fractional values of the parameter—another type of interpolation and example of "continuous thinking"—was given by Euler by means of the gamma function, the generalized factorial.

The *Arithmetica infinitorum* exerted a singularly important effect on Newton when he studied it in the winter of 1664–1665.[8] Newton generalized even more than Wallis by keeping the upper limit of the integrals $I(k, n)$ variable. He thus arrived at the binomial theorem by way of Wallisian interpolation procedures. In a few cases the binomial expansion could be checked by algebraic division and root extraction; but, just as in the case of Wallis' product, a rigorous justification had to wait until mathematical techniques had been much refined.

The publication of the *Arithmetica infinitorum* immediately provoked a mathematical challenge from Fermat. He directed "to Wallis and the other English mathematicians," some numerical questions: To find a cube, which added to all its aliquot parts will make a square (such as $7^3 + 7^2 + 7 + 1 = 20^2$), and to find a square number, which added to all its aliquot parts, will make a cube.[9] Fermat, lawyer and councillor of the *parlement* in Toulouse, had added: "We await these solutions, which, if England or Belgic or Celtic Gaul do not produce, Narbonese Gaul will." Besides Wallis, Brouncker, later the first president of the Royal Society, participated in the contest on the side of the English. On the Continent, Frenicle de Bessy applied his great skill in handling large numbers. Wallis at first highly underestimated the difficulty as well as the theoretical foundation of Fermat's questions; and Fermat added further problems in 1657–1658. Wallis maintained the number 1 to be a valid solution, and in return drew up some superficially similar questions. His method of solution was more or less that of trial and error, based on intelligent guessing, and in some ways was not unrelated to the procedures employed in his *Arithmetica infinitorum*. Until the end of his life Wallis

had no idea of the number-theoretical insights that Fermat had obtained—nor could he, since his challenger did not reveal them. Afraid that the French mathematicians might reap all the glory from this contest, Wallis obtained permission to publish the letters: the *Commercium epistolicum* appeared in 1658. The last chapter of his *Discourse of Combinations, Alternations, and Aliquot Parts* (1685) deals with "Monsieur Fermat's Problems Concerning Divisors and Aliquot Parts." Finally, among his manuscripts there are also a number of attempts to solve some of Fermat's problems, including the "Theorema Fermatianum Negativum" that $a^3 + b^3 = c^3$ is not possible in integral or rational numbers and another negative theorem that there does not exist a right triangle with square area.[10]

But number theory had no special appeal to Wallis—nor to any other mathematician of the time, Frenicle excepted. This was so partly because it was hardly applicable, as Wallis himself emphasized and partly because it did not suit the taste of seventeenth- and eighteenth-century mathematicians, Euler being a notable exception. Fermat, who had glimpsed the treasures of number theory and had recognized its intrinsic mathematical value, did little to introduce his fellow mathematicians to the subject. Thus the general judgment about the contest had to be based on Wallis' *Commercium epistolicum*, and the editor did not hesitate to underline the achievements he and Brouncker had made. No wonder that his fame was now firmly established throughout Europe.

Wallis also participated in the competition in which Pascal in the summer of 1658 asked for quadratures, cubatures, and centers of gravity of certain figures limited by cycloidal arcs.[11] Neither Wallis nor Lalouvère, who also competed for the prize, satisfied Pascal, and no prize was awarded. This was not quite fair, and in 1659 Wallis replied with *Tractatus duo . . . de cycloide et . . . de cissoide*. Here, as well as in the second part of his voluminous *Mechanica, sive de motu tractatus geometricus* (1669–1671), he again relied on his analytic methods. This second part, on the calculation of centers of gravity, is the major part of the *Mechanica*, and in it Wallis carried on the analytical investigations of the 1650's.

The first part deals with various forms of motion in a strictly "geometrical," that is, Euclidean, manner, starting with definitions followed by propositions. The motion of bodies under the action of gravity is covered in particular. The final chapter of the first part is devoted to a treatment of the balance and introduces the idea of moment, which is essential for the inquiries into the centers of gravity. In the third part, Wallis returns not only to the elementary machines, according to ancient tradition, but above all to a thorough treatment of the problems on percussion. In 1668 percussion and impact were a major topic of discussion at the Royal Society, and Wallis, Wren, and Huygens submitted papers.[12] In the *Mechanica*, Wallis extended his investigations, studying the behavior of both elastic and inelastic bodies. Although in style and subject matter it is not a uniform book, at the time it certainly was one of the most important and comprehensive in its field. It represents a major advance in the mathematization of mechanics, but it was superseded in 1687 by a much greater one—Newton's *Principia.*

Wallis' last great mathematical book was *Treatise of Algebra, Both Historical and Practical* (1685), the fruit of many years' labor.[13] As its title suggests, it was to combine a full exposition of algebra with its history, a feat never previously attempted by any author. The book was Wallis' only major mathematical work to be published in the vernacular. (In 1693 an augmented Latin translation was issued as vol. II of his *Opera mathematica.*)

Of the 100 chapters, the first fourteen trace the history of the subject up to the time of Viète, with emphasis on the development of mathematical notation. The subsequent practical introduction to algebra (chapters 15–63) was based almost entirely on Oughtred's *Clavis mathematicae*, Harriot's *Artis analyticae praxis*, and *An Introduction to Algebra* (1668), Thomas Brancker's translation of J. H. Rahn's *Teutsche Algebra* (1659), with numerous additions by John Pell, Rahn's former teacher. This fact alone signals the great bias Wallis had developed in favor of his countrymen. It becomes even more obvious in the passages where the author claimed that Descartes had obtained his algebraical knowledge from Harriot. Criticisms of Wallis' one-sided account were raised immediately and have continued since. After an insertion concerning the application of algebra to geometry and geometrical interpretations of algebraic facts (chapters 64–72, including an attempt to give a representation of imaginary numbers),[14] Wallis devoted the final twenty-eight chapters to a subject that one would hardly look for in a book on algebra today: a discussion of the methods of exhaustion and of indivisibles, again with reference to the *Arithmetica infinitorum.* Thus the new methods were

still considered as an extension of an old subject rather than as a wholly new field of mathematics.

The *Algebra* also includes an exposition of the method of infinite series and the first printed account, much augmented in the second edition, of some of Newton's pioneering results. Wallis had long been afraid that foreigners might claim the glory of Newton's achievements by publishing some of his ideas as their own before Newton himself had done so. He therefore repeatedly warned his younger colleague at Cambridge not to delay but to leave perfection of his methods to later editions.[15] (Volume III of the *Opera* [1699] contains an *Epistolarum collectio*, of which the most important part is the correspondence between Newton and Leibniz, in particular Newton's famous "Epistola prior" and "Epistola posterior" of 1676.)

Apart from some editions of Greek mathematical classics, the *Algebra* with its several supplementary treatises—*Cono-Cuneus* (a study in analytic three-dimensional geometry), *Angular Sections*,[16] *Angle of Contact*, and *Combinations, Alternations, and Aliquot Parts*—marked the end of the stream of mathematical works. Even without the polemics against Hobbes and some minor pieces, they fill three large volumes.

Wallis helped shape over half a century of mathematics in England. He bore the greatest share of all the efforts made during this time to raise mathematics to the eminence it enjoyed on the Continent. The center of mathematical research and of the "new science" in Galileo's time lay in Italy. It then shifted northward, especially to France and the Netherlands. Because of Wallis' preparative work and Newton's genius, it rested in Britain for a while, until through the influence of Leibniz, the Bernoullis, and Euler it moved back to the Continent.

Nonmathematical Work. Wallis first exhibited his mental powers early in the Civil War (1642 or 1643), when by chance he was shown a letter written in cipher and succeeded in decoding it within a few hours.[17] Because more letters were given to him by the Parliamentarians, rumors were later spread that he had deciphered important royal letters that had fallen into their hands. Wallis strenuously denied the accusation, and it is very unlikely that he revealed anything harmful to the royal family or the public safety—if indeed he came across such information. On the contrary, the confirmation of his offices at the Restoration may well have been a sign of gratitude to him by Charles II. For many years Wallis continued to decipher intercepted letters for the government, especially after the Revolution. In old age he taught the art to his grandson William Blencowe but refused to disclose it when Leibniz on behalf of his government requested information on it.

In his autobiography, written in January 1697, when he was over eighty, Wallis referred to one of his first successes more than half a century earlier:

> Being encouraged by this success, beyond expectation; I afterwards ventured on many others (some of more, some of less difficulty) and scarce missed of any that I undertook, for many years, during our civil Wars, and afterwards. But of late years, *the French Methods of Cipher* are grown so intricate beyond what it was wont to be, that I have failed of many; tho' I did have master'd divers of them.[18]

Of great importance for much of his later scientific work was his introduction, while living in London, to a group interested in the "new" natural and experimental sciences—the circle from which the Royal Society emerged soon after the Restoration.[19] To Wallis we owe one of the few reports on those early meetings that give direct evidence.

> About the year 1645, while I lived in *London* (at a time, when, by our Civil Wars, Academical Studies were much interrupted in both our Universities:) beside the Conversation of divers eminent Divines, as to matters Theological; I had the opportunity of being acquainted with divers worthy Persons, inquisitive into Natural Philosophy, and other parts of Humane Learning; and Particularly of what hath been called the *New Philosophy* or *Experimental Philosophy*.
>
> We did by agreement, divers of us, meet weekly in *London* on a certain day, to treat and discours of such affairs. . . .
>
> These meetings we held sometimes at *Dr. Goddards* lodgings in *Woodstreet* (or some convenient place near) on occasion of his keeping an Operator in his house, for grinding Glasses for Telescopes and Microscopes; and sometime at a convenient place in *Cheap-side*; sometime at *Gresham College* or some place near adjoyning.
>
> Our business was (precluding matters of Theology and State Affairs) to discours and consider of *Philosophical Enquiries*, and such as related thereunto; as *Physick, Anatomy, Geometry, Astronomy, Navigation, Staticks, Magneticks, Chymicks, Mechanicks,* and *Natural Experiments*; with the State of these Studies, as then cultivated, at home and abroad. We there discoursed of the *Circulation of the Blood, the Valves in the Veins, the Venae Lacteae, the Lymphatick vessels, the Copernican Hypothesis, the Nature of Comets, and New Stars, the Satellites of Jupiter, the Oval Shape* (as it then appeared) *of Sat-*

urn, the spots in the Sun, and its Turning on its own Axis, the Inequalities and Selenography of the Moon, the several Phases of Venus and Mercury, the Improvement of Telescopes, and grinding of Glasses for that purpose, the Weight of Air, the Possibility or Impossibility of Vacuities, and Natures Abhorrence thereof; the Torricellian Experiment in Quicksilver, the Descent of heavy Bodies, and the degrees of Acceleration therein; and divers other things of like nature. Some of which were then but New Discoveries, and others not so generally known and imbraced, as now they are; With other things appertaining to what hath been called *The New Philosophy*; which from the times of *Galileo* at *Florence*, and S[r] *Francis Bacon (Lord Verulam)* in *England*, hath been much cultivated in *Italy, France, Germany*, and other Parts abroad, as well as with us in *England.*

> About the year 1648, 1649, some of our company being removed to *Oxford* (first D[r] *Wilkins*, then I, and soon after D[r] *Goddard*) our company divided. Those in *London* continued to meet there as before (and we with them, when we had occasion to be there;) and those of us at *Oxford* . . . continued such meetings in *Oxford*; and brought those Studies into fashion there. . . .
>
> Those meetings in *London* continued, and (after the Kings Return in 1660) were increased with the accession of divers worthy and Honorable Persons; and were afterwards incorporated by the name of *the Royal Society*, etc. and so continue to this day.

While the Royal Society of London did indeed grow and continue, the Oxford offspring suffered a less happy fate. After a period of decline and interruption it seems to have flourished again in the 1680's when Wallis was elected its president and tried to establish closer contacts with the mother society and similar groups in Scotland. But Oldenburg, secretary of the London society, initiated publication of the *Philosophical Transactions* and thereby provided a more permanent means of scientific exchange than personal intercourse and weekly discussions.[20] Wallis made ample use of the *Transactions*; and between 1666 and 1702 he published more than sixty papers and book reviews. The reviews concerned mathematical books, but the papers were more wide-ranging.[21] One of the leading scientists among the early fellows of the Royal Society, he was also one of the most energetic in promoting it and helping it to achieve its goals, at a time when not a few of these virtuosi were men without a real understanding of the scientific experiments conducted and of the complex theories behind them.

Wallis' most successful work was his *Grammatica linguae anglicanae*, with a *Praxis grammatica* and a treatise, *De loquela*, on the production of the sounds of speech. First published in 1652, the sixth, and last, edition in England appeared in 1765; it was also published on the Continent.

In his *History of Modern Colloquial English*, H. C. Wyld emphasized that Wallis "has considerable merits as an observer of sounds, he has good powers of discrimination, nor is he led astray by the spelling like all the sixteenth-century grammarians, and Bullokar, Gill, and Butler in the seventeenth."[22] He then continued to discuss some of Wallis' more noteworthy observations. A much more detailed account is given in M. Lehnert's monograph.[23]

Wallis' *Treatise of Speech* formed a useful theoretical foundation for his pioneering attempts to teach deaf-mutes how to speak. In 1661 and 1662 Wallis instructed two young men, Daniel Whaley and Alexander Popham; the latter had previously been taught by Dr. William Holder. Wallis presented Whaley to the Royal Society on 21 May 1662 and in 1670 reported on his instruction of Popham in the *Philosophical Transactions*—failing to mention Holder's teaching.[24] This unfair act eventually (1678) led to a bitter attack by Holder, to which Wallis replied in no less hostile words.[25]

This was one of the many violent quarrels in which Wallis became involved. Although readily inclined to boast of his achievements and to appropriate the ideas of others for further development, he did not always acknowledge his debt to his predecessors. Furthermore he was often carried away by his temper and would reply without restraint to criticism. He thus quarreled with Holder, Henry Stubbe, Lewis Maydwell, and Fermat; and his longest and most bittered dispute, with Thomas Hobbes, dragged on for over a quarter of a century.[26] Despite, or rather because of, his limited mathematical knowledge, Hobbes claimed in 1655 to possess an absolute quadrature of the circle. Somewhat later he also purported to have solved another of the great mathematical problems—the duplication of the cube. Hobbes's chief transgression, however, was in having dared to criticize Wallis' *Arithmetica infinitorum*. The controversy soon degenerated into the most virulent hostility, which gave rise to wild accusations and abusive language. The quarrel ended only with Hobbes's death in 1679. J. F. Scott has suggested that Wallis' relentless attacks may have been partly motivated by Hobbes's increasing influence, especially as author of the *Leviathan*, and by Wallis' fear that Hobbes's teachings would undermine respect for the Christian religion.

As keeper of the archives, Wallis rendered considerable services to his university. In his brief account of Wallis' life, David Gregory said, "He put the records, and other papers belonging to the University that were under his care into such exact order, and managed its lawsuits with such dexterity and success that he quickly convinced all, even those who made the greatest noise against this election, how fitt he was for the post."[27] A successor as keeper, Reginald L. Poole, also praised Wallis' work: "He left his mark on the Archives in numerous transcripts, but above all by the Repertory of the entire collection which he made on the basis of Mr. Twyne's list in 1664 and which continues to this day the standard catalogue."[28] Wallis' catalogue was not replaced until even later in the twentieth century. Although not a practicing musician, Wallis composed some papers on musical theory that were published in the *Philosophical Transactions*,[29] and he edited works on harmony by Ptolemy, Porphyrius, and Bryennius. One of his papers reports his observation of the "trembling" of consonant strings, while others contain a mathematical discussion of the intervals of the musical scale and the resulting need for temperament in tuning an organ or other keyboard instrument. In an appendix to Ptolemy's *Harmonics*, Wallis attempted to explain the surprising effects attributed to ancient music (which he rendered in modern notation); and he also dealt with these effects in a separate paper. Finally he contributed extended remarks on Thomas Salmon's *Proposal to Perform Musick, in Perfect and Mathematical Proportions* (London, 1688), the forerunner of which, *An Essay to the Advancement of Musick* (London, 1672), had aroused great interest as well as conflicting views.

Theology. From 1690 to 1692 Wallis published a series of eight letters and three sermons on the doctrine of the Holy Trinity, directed against the Unitarians. In order to explain this doctrine he introduced an analogous example from mathematics: a cubical body with three dimensions, length, breadth, and height; and compared the mystery of the Trinity with the cube:

> This longum, latum, profundum, (Long, Broad, and Tall), is but *One* Cube; of *Three Dimensions,* and yet but *One Body*. And this *Father, Son,* and *Holy-Ghost*: Three Persons, and yet but One God.[30]

Wallis' discourses on the Trinity met with marked approval from various theological quarters. It was even used in Pierre Bayle's famous *Diction-*

naire historique et critique in a note to the article on Abailard. Bayle wished to vindicate Abailard of the charge of Tritheism,[31] which had been raised against him for having used an analogy between the Trinity and the syllogism that consists of proposition, assumption, and conclusion. Just as nobody doubts the orthodoxy of Wallis on the basis of his geometrical example, Bayle argued, there was no reason to attack Abailard for his analogy of the syllogism.

Wallis' sermons and other theological works, often praised for their simple and straightforward language, testify that his religious principles were Calvinist, according to the literal sense of the Church of England. He never denied the Puritanism in which he had grown up, although he remained a loyal member of the official church.

From his student days, Wallis sided with the Parliamentarians, and Cromwell is said to have had a great respect for him. As secretary to the Assembly of Divines at Westminster during the Civil War, Wallis became thoroughly familiar with the controversial issues within the Episcopal Church and between the Church and Parliament. Included in his autobiography is a rather long intercalation about this assembly, which was convened to suggest a new form of church government in place of the episcopacy.[32] His interpretation of proceedings carried on half a century earlier might have been somewhat colored by the actual events that followed. The episcopacy was, after all, not abolished; and Wallis had tried to stay on good terms with the bishops and archbishops. Toward the end of the century he strongly opposed the introduction of the Gregorian calendar in England, considering it a kind of submission to Rome. The new calendar was not in fact adopted in Britain until 1752. Some of Wallis' friends and colleagues in the Royal Society exchanged their university posts for careers in the church, but Wallis himself was never given the opportunity. Obviously his trimming politics had made him not totally acceptable to the monarchy, although he did enjoy signs of royal favor. As he himself expressed it, he was "willing whatever side was upmost, to promote (as I was able) any good design for the true Interest of Religion, of Learning, and the publick good."[33]

NOTES

1. C. J. Scriba, "The Autobiography . . .," 24.
2. J. R. Tanner, ed., *Private Correspondence and Miscellaneous Papers of Samuel Pepys, 1679–1703*, II (London, 1926), 257.
3. "S.T.P." is the usual abbreviation for Doctors of Divinity in inscriptions; Wallis was never created professor of theology.
4. J. Wallis, *Algebra*, ch. 46.
5. See H. Wieleitner, "Die Verdienste."
6. For a more detailed description, see Sir T. P. Nunn, "The Arithmetic"; J. F. Scott, *The Mathematical Work*, ch. 4; and D. T. Whiteside, "Patterns of Mathematical Thought in the Later Seventeenth Century," in *Archives for History of Exact Sciences*, 1 (1961), 179–388, esp. 236–243.
7. See W. Kutta, "Elliptische," and A. Prag, "John Wallis," esp. 391–395.
8. D. T. Whiteside, "Newton's Discovery of the General Binomial Theorem," in *Mathematical Gazette*, 45 (1961), 175–180; and *The Mathematical Papers of Isaac Newton*, D. T. Whiteside, ed., I (Cambridge, 1967), 96–111.
9. See G. Wertheim, "P. Fermats Streit . . .," and J. E. Hofmann, "Neues über Fermats. . . ."
10. See C. J. Scriba, *Studien . . .*, chs. 2–3.
11. See K. Hara, "Pascal et Wallis . . .," and J. Hofmann and J. E. Hofmann, "Erste Quadratur der Kissoide," in *Deutsche Mathematik*, 5 (1941), 571–584.
12. See A. R. Hall, "Mechanics and the Royal Society, 1668–1670," in *British Journal for the History of Science*, 3 (1966–1967), 24–38.
13. See J. F. Scott, "John Wallis."
14. See G. Eneström, "Die geometrische Darstellung."
15. See C. J. Scriba, "Neue Dokumente zur Entstehungsgeschichte des Prioritätsstreites zwischen Leibniz und Newton um die Erfindung der Infinitesimalrechnung," in *Akten des Internationalen Leibniz-Kongresses Hannover, 14.–19. November 1966*, II. *Mathematik-Naturwissenschaften* (Wiesbaden, 1969), 69–78.
16. See C. J. Scriba, *Studien*, ch. 1.
17. See D. E. Smith, "John Wallis," and D. Kahn, *The Codebreakers* (New York, 1967), 166–169.
18. See C. J. Scriba, "The Autobiography," 38.
19. Different opinions have been expressed as to whether the Royal Society emerged from the London group described by Wallis or from an independent Oxford group in existence before Wallis came to Oxford in 1649. For a champion of the latter view, see M. Purver, *The Royal Society: Concept and Creation* (London, 1967). A brief review of this is C. J. Scriba, "Zur Entstehung der Royal Society," in *Sudhoffs Archiv für Geschichte der Medizin und der Naturwissenschaften*, 52 (1968), 269–271. There is an extended debate, in three articles by P. M. Rattansi, C. Hill, and A. R. Hall and M. B. Hall, in *Notes and Records. Royal Society of London*, 23 (1968), 129–168, where further references are given. It seems to be without doubt that the London group cannot be ignored. Wallis' report is taken from "The Autobiography," 39–40.
20. Wallis' correspondence with Oldenburg is printed in *The Correspondence of Henry Oldenburg*, A. R. Hall and M. Boas Hall, eds. (Madison, Wis., 1965–   ).
21 For a not quite complete list of Wallis' publications in the *Philosophical Transactions of the Royal Society*, see J. F. Scott, *The Mathematical Work*, 231–233; paper no. 62 is not by Wallis.
22. H. C. Wyld, *A History of Modern Colloquial English*, 3rd ed. (Oxford, 1936; repr. 1953), 170.
23. M. Lehnert, in *Die Grammatik*, criticizes the older work by L. Morel, *De Johannis Wallisii*, as insufficient. See also A. B. Melchior, "Sir Thomas Smith and John Wallis," in *English Studies*, 53 (1972), and his review of John Wallis, "Grammar of the English Language," in *English Studies*, 55 (1974), 83–85.
24. *Philosophical Transactions of the Royal Society*, 5, no. 61 (18 July 1670), 1087–1097 (pagination repeated).
25. W. Holder, *A Supplement*; J. Wallis, *A Defense*.
26. See J. F. Scott, *The Mathematical Work*, ch. 10.
27. Bodleian Library Oxford, MS Smith 31, p. 58; J. Collier, *A Supplement*.

28. R. L. Poole, *A Lecture on the History of the University Archives* (Oxford, 1912), 25.
29. *Philosophical Transactions of the Royal Society*, **12**, no. 134 (23 Apr. 1677), 839–842; **20**, no. 238 (Mar. 1698), 80–84; **20**, no. 242 (July 1698), 249–256; **20**, no. 243 (Aug. 1698), 297–303. See L. S. Lloyd, "Musical Theory in the Early *Philosophical Transactions*," in *Notes and Records. Royal Society of London*, **3** (1940–1941), 149–157.
30. Quoted from R. C. Archibald, "Wallis on the Trinity," 36.
31. See the query by E. H. Neville, "Wallis on the Trinity," 197, who quotes the 5th ed., I (Amsterdam, 1734), 30. In the new ed. (Paris, 1820), it is I, 59–60, note M.
32. See Scriba, "The Autobiography," 31–37.
33. *Ibid.*, 43.

BIBLIOGRAPHY

I. ORIGINAL WORKS. Most of Wallis' publications (including pamphlets and sermons) are listed in the British Museum catalog, but a complete bibliography is still a desideratum. Wallis collected his more important books and some articles in his *Opera mathematica*, 3 vols. (Oxford, 1693–1699), repr. with intro. by C. J. Scriba (Hildesheim–New York, 1972). The table of contents in vol. I contains a list of books that were originally not planned for inclusion in the *Opera mathematica*, which was to consist of two volumes only. A selection of mathematical and nonmathematical works taken from this list and augmented by additional material is included in vol. III.

The *Opera mathematica* should not be confused with the *Operum mathematicorum pars prima* and *pars secunda*, published in 1657 and 1656 [sic], respectively. Vol. I contains *Oratio inauguralis*; *Mathesis universalis, sive arithmeticum opus integrum*; *Adversus Meibomii De proportionibus dialogum, tractatus elenctibus*; and *M. Mersenni locus notatur*. Vol. II contains *De angulo contactus et semicirculo disquisitio geometrica*; *De sectionibus conicis, nova methodo expositis, tractatus*; *Arithmetica infinitorum* (already printed and in some copies distributed in 1655), and the brief *Eclipsis solaris observatio*.

Works cited in the text and in the notes include the reply to W. Holder, *A Supplement to the Philosophical Transactions of July 1670, With Some Reflexions on Dr. John Wallis, His Letter There Inserted* (London, 1678), which Wallis issued under the title *A Defence of the Royal Society, and the Philosophical Transactions, Particularly Those of July 1670, in Answer to the Cavils of Dr. William Holder* (London, 1678); and the voluminous *Treatise of Algebra, Both Historical and Practical, Showing the Original, Progress, and Advancement Thereof, From Time to Time; and by What Steps It Hath Attained to the Height at Which Now It Is* (London, 1685; enl. Latin version in vol. II of the *Opera mathematica*). There is a facs. ed. of the *Grammatica linguae anglicanae* of 1653 (Menston, 1969), and a new ed. with translation and commentary by J. A. Kemp, *Grammar of the English Language* (London, 1972).

Wallis' autobiography was reprinted in C. J. Scriba, "The Autobiography of John Wallis, F.R.S.," in *Notes and Records. Royal Society of London*, **25** (1970), 17–46; this includes a survey of other early biographies of Wallis, including that by David Gregory, which is printed in J. Collier, *A Supplement to the Great Historical, Geographical, Genealogical and Poetical Dictionary . . . Together With a Continuation From the Year 1688, to 1705, by Another Hand* (London, 1705; 2nd ed., 1727).

II. SECONDARY LITERATURE. The book-length monographs on Wallis the mathematician are J. F. Scott, *The Mathematical Work of John Wallis, D.D., F.R.S. (1616–1703)* (London, 1938); and C. J. Scriba, *Studien zur Mathematik des John Wallis (1616–1703). Winkelteilungen, Kombinationslehre und zahlentheoretische Probleme* (Wiesbaden, 1966). Scott surveys Wallis' life and his main published mathematical works; Scriba concentrates on the topics stated in his title, making use also of unpublished MSS, and includes a list of books owned by Wallis, which are now in the Bodleian, as well as a brief survey of the MS material. An index to the correspondence is C. J. Scriba, "A Tentative Index of the Correspondence of John Wallis, F.R.S.," in *Notes and Records. Royal Society of London*, **22** (1967), 58–93.

Wallis the grammarian and phonetician is the subject of L. Morel, *De Johannis Wallisii grammatica linguae anglicanae et tractatu de loquela thesis* (Paris, 1895), which is superseded by M. Lehnert, *Die Grammatik des englischen Sprachmeisters John Wallis (1616–1703)* (Wrocław, 1936). The following articles deal with Wallis or his work: R. C. Archibald, "Wallis on the Trinity," in *American Mathematical Monthly*, **43** (1936), 35–37, and in *Scripta mathematica*, **4** (1936), 202; L. I. Cherkalova, "Sostavnye otnoshenia u Vallisa," in *Doklady na nauchnykh konferentsiakh*, **2**, no. 3 (1964), 153–160; G. Eneström, "Die geometrische Darstellung imaginärer Grössen bei Wallis," in *Bibliotheca mathematica*, 3rd ser., **7** (1906–1907), 263–269; K. Hara, "Pascal et Wallis au sujet de la cycloïde," in *Annals of the Japanese Association of the Philosophy of Science*, **3** (1969), 166–187; J. E. Hofmann, "Neues über Fermats zahlentheoretische Herausforderungen von 1657," in *Abhandlungen der Preussischen Akademie der Wissenschaften*, Math. naturwiss. Kl. (1943), no. 9 (Berlin, 1944); M. Koppe, "Die Bestimmung sämtlicher Näherungsbrüche einer Zahlengrösse bei John Wallis (1672)," in *Sitzungsberichte der Berliner mathematischen Gesellschaft*, **2** (1903), 56–60; F. D. Kramar, "Integratsionnye metody Dzhona Vallisa," in *Istoriko-matematicheskie issledovaniya*, **14** (1961), 11–100; W. Kutta, "Elliptische und andere Integrale bei Wallis," in *Bibliotheca mathematica*, 3rd ser., **2** (1901), 230–234; E. H. Neville, "Wallis on the Trinity," in *Scripta mathematica*, **2** (1934), 197; T. F. Nikonova, "Pervy opyt postroeni istorii algebry anglyskim matematikom Dzhonom Vallisom," in *Uchenye zapiski. Moskovskoi oblastnoi pedagogicheskii institut*, **202** (1968), 379–392; T. P. Nunn, "The Arithmetic of Infinites," in *Mathe-*

matical Gazette, **5** (1910–1911), 345–357, 378–386; H. C. Plummer, "Jeremiah Horrocks and his *Opera posthuma*," in *Notes and Records. Royal Society of London*, **3** (1940–1941), 39–52. Wallis was instrumental in selecting the material for the posthumous ed. of Horrocks' astronomical work.

See also A. Prag, "John Wallis. 1616–1703. Zur Ideengeschichte der Mathematik im 17. Jahrhundert," in *Quellen und Studien zur Geschichte der Mathematik, Astronomie und Physik*, Abt. B, **1** (1931), 381–412, mainly devoted to the *Arithmetica infinitorum* and the *Algebra*, but with many astute remarks on the general state of seventeenth-century mathematics; J. F. Scott, "John Wallis as a Historian of Mathematics," in *Annals of Science*, **1** (1936), 335–357; and "The Reverend John Wallis, F.R.S. (1616–1703)," in *Notes and Records. Royal Society of London*, **15** (1960), 57–67, with selected bibliography by D. T. Whiteside, 66–67; C. J. Scriba, "Wallis and Harriot," in *Centaurus*, **10** (1964), 248–257; "John Wallis' *Treatise of Angular Sections* and Thâbit ibn Qurra's Generalization of the Pythagorean Theorem," in *Isis*, **57** (1966), 56–66; "Das Problem des Prinzen Ruprecht von der Pfalz," in *Praxis der Mathematik*, **10** (1968), 241–246; "Wie läuft Wasser aus einem Gefäss? Eine mathematisch-physikalische Aufzeichnung von John Wallis aus dem Jahr 1667," in *Sudhoffs Archiv*, **52** (1968), 193–210; and "Eine mathematische Festvorlesung vor 300 Jahren," in *Janus*, **56** (1969), 182–190; D. E. Smith, "John Wallis as a Cryptographer," in *Bulletin of the American Mathematical Society*, **24** (1917), 82–96; L. Tenca, "Giovanni Wallis e gli italiani," in *Bollettino dell' Unione matematica italiana*, 3rd ser., **10** (1955), 412–418; G. Wertheim, "P. Fermats Streit mit J. Wallis," in *Abhandlungen zur Geschichte der Mathematik*, **9** (1899), 555–576; H. Wieleitner, "Die Verdienste von John Wallis um die analytische Geometrie," in *Weltall*, **29** (1929–1930), 56–60; and G. U. Yule, "John Wallis, D.D., F.R.S. 1616–1703," in *Notes and Records. Royal Society of London*, **2** (1939), 74–82.

CHRISTOPH J. SCRIBA

WALTER BURLEY. See **Burley, Walter.**

WALTER OF EVESHAM. See **Walter of Odington.**

WALTER OF ODINGTON (*fl.* Evesham and Oxford[?], England, 1280[?]–1330[?]), *alchemy, music.*

As is the case with many medieval alchemists, the details of Walter's life are obscure. There is confusion as to where and when he lived. Astronomical observations in his treatise "Motion of the Eighth Sphere" indicate that he probably was active in the last half of the thirteenth century (*ca.* 1280), but manuscript sources refer to him as having lived in the early fourteenth century. In his manuscripts he is referred to as either Walter of Odington or Walter of Evesham. Odington probably refers to his birthplace, which may be the Oddington in northern Oxfordshire, while Evesham is clearly a reference to the Benedictine abbey at Evesham. Several sources testify to his being a Benedictine monk. Since he was a member of a regular order, it is difficult to explain the sources that place him at Merton College, Oxford, for an extensive period in the early fourteenth century.

Odington wrote treatises on alchemy, optics, arithmetic, and geometry, and a famous work on medieval music, *De speculatione musice*, in which he treated acoustics, the division of the monochord, musical notation, mensurable music, and rules for composition. His most important scientific study was his alchemical investigation *Icocedron* (from "icosahedron").

As the title indicates, the *Icocedron* is divided into twenty chapters; and it follows the general Islamic alchemical-medical-pharmaceutical tradition of the period. The first fourteen chapters present standard alchemical information outlining the basic principles of the art, the methods of preparing the materials, techniques for perfecting the "medicine," and steps to be followed in mixing the elements. In the concluding chapters Walter became more alchemically sophisticated, discussing in some detail metals, the intension and remission of qualities, and the four basic elements of alchemical composition—earth, air, fire, and water. Seeking to present a way of accurately describing alchemical change quantitatively, he assigned quantitative distinctions to each element. Fire is thus hot in the fourth degree and dry near the end of the third degree. Since each degree has sixty minutes, fire therefore has 240 minutes of calidity and 180 minutes of aridity. Through an elaborate procedure of combining the qualities, the secondary quality in a given element could be destroyed and the element reduced to its simplest form.

The most important feature of Walter's alchemical work was his attempt to quantify qualitative intensities. Distinguishing between temperature and (quantity of) heat, he sought to interpret the relationships between qualitative intensities and quantitative amounts. He conceived of qualitative intensity as a magnitude. In the *Icocedron* he presented six rules for these relationships—one stated implicitly, four in tabular form, and one verbally. His verbal rule utilizes functions similar to those of

Bradwardine in his famous "law of motion," while its mathematical expression is similar to that used in the pharmaceutical tradition of al-Kindī by the scholars of Montpellier.

BIBLIOGRAPHY

I. ORIGINAL WORKS. There is no collective ed. of Walter's works. The *Icocedron* is extant in the following MSS: British Museum, Add. MS 15549, fols. $4^r - 20^v$; Cambridge University, Trinity College MS 1122, fols. $177^v - 183^r$; Bodleian Library, Digby MS 119, fols. $142^r - 147^v$; and Murhardsche Bibliothek der Stadt Kassel und Landesbibliothek, Handschrift 2° MS Chem. 8, fols. $240^r - 253^v$. British Museum, Add. MS 15549 has been edited by Phillip Drennon Thomas as "David Ragor's Transcription of Walter of Odington's *Icocedron*," in *Wichita State University Studies*, no. 76 (Aug. 1968), 3 – 24. The incipits and locations for Walter's other scientific works are in Lynn Thorndike and Pearl Kibre, *A Catalogue of Incipits of Mediaeval Scientific Writings in Latin* (Cambridge, Mass., 1963). *De speculatione musice* has been edited in E. Coussemaker, *Scriptorium de musica medii aevi*, I (Paris, 1864).

II. SECONDARY LITERATURE. For a brief biographical sketch, see Henry Davey, "Walter of Evesham or Walter of Odington," in *Dictionary of National Biography*, XX, 702 – 703. Lynn Thorndike presents a cursory examination of Odington's career and works in *A History of Magic and Experimental Science*, III (New York, 1960), 127 – 135. For an examination of his alchemical rules, see Donald Skabelund and Phillip Thomas, "Walter of Odington's Mathematical Treatment of the Primary Qualities," in *Isis*, **60** (1969), 331 – 350; and Phillip D. Thomas, "The Alchemical Thought of Walter of Odington," in *Actes du XIIᵉ Congrès international d'histoire des sciences* (Paris, 1968), 141 – 144.

PHILLIP DRENNON THOMAS

WALTER, PHILIPPE (*b*. Cracow, Poland, 31 May 1810; *d*. Paris, France, 9 April 1847), *chemistry.*

After completing his doctoral studies at Cracow, Walter took part in the unsuccessful popular uprising of 1830 – 1831 against the Russian rulers of Poland. He subsequently found refuge in Paris, where he joined the group associated with Dumas. In 1829 Dumas had been a founder of the École Centrale des Arts et Manufactures, a school for advanced study in the applied sciences; and Walter became a teacher of analytical chemistry there.

Most of Walter's investigations concerned natural plant products. In 1838 he and Pierre Joseph Pelletier isolated toluene as a product of the destructive distillation of pine resin. They called it retinaphtha and correctly determined its composition, noting that it contained the benzoic radical and represented a hydrocarbon sought by chemists, since it was the hypothetical hydrocarbon formed through the replacement of the oxygen atoms of benzoic acid by hydrogen.

In 1840 Pelletier and Walter discovered another important hydrocarbon. By fractional distillation of naphtha they isolated an analogue of ethylene. Determination of its vapor density gave them a formula of C_8H_{16}. Originally named naphthene, it subsequently came to be known as caprylene, after caprylic alcohol, from which it also could be prepared (the modern name is octene).

In 1839 Walter had distinguished between menthol and camphor, finding the molecular formula of menthol by determining its vapor density. This information, together with Dumas's determination of its composition, enabled him to show that menthol was a compound distinct from camphor.

In 1842 Walter observed that the reaction of camphoric anhydride and sulfuric acid yielded sulfocamphoric acid and carbon monoxide. This reaction was the first indication that carbon could be replaced in organic compounds by other elements. Walter noted the importance of this fact, and in an 1843 paper he quoted from a work in which Dumas had suggested that substitution in organic compounds might not be restricted to hydrogen; perhaps oxygen, nitrogen, and even carbon might be replaced. The composition of sulfocamphoric acid clearly showed that an SO_2 group had replaced a carbon atom in the camphoric anhydride. Walter claimed that this work supported Dumas's substitution (1834) and type (1839) theories. In the latter theory organic molecules were considered to be unitary types with properties depending less on the nature of the elements than on their position and arrangement. Thus, the replacement of carbon in camphoric anhydride bolstered Dumas's attempt to establish a novel and controversial theory of organic compounds.

BIBLIOGRAPHY

I. ORIGINAL WORKS. Walter wrote a treatise on Polish chemical nomenclature, *Wyklad nomenklatury chemicznej poskiej i porównanie jej z nomenklaturami lacińską, francuską, angielską i niemiecką* (Cracow, 1842; 2nd ed., 1844). His papers are listed in the Royal Society *Catalogue of Scientific Papers*, VI, 256 – 257. The more important include "Examen des produits provenant du traitement de la résine dans la fabrication du gaz pour l'éclairage," in *Annales de chimie et de physique*, 2nd ser., **67** (1838), 269–303, written with P. J. Pelletier; "Mémoire sur l'essence de menthe poivrée cristallisée," *ibid.*, **72** (1839), 83–109; "Mémoire sur l'action qu'exerce l'acide sulfurique anhydre sur l'acide camphorique anhydre," *ibid.*, **74** (1840), 38–52; "Recherches chimiques sur les bitumes," in *Journal de pharmacie*, **26** (1840), 549–568, written with P. J. Pelletier; and "Mémoire sur l'acide camphorique anhydre," in *Annales de chimie et de physique*, 3rd. ser., **9** (1843), 177–200.

II. SECONDARY LITERATURE. There is a biography of Walter by S. Sekowski and S. Szostkiewicz, *Serce i retorta, czyli zywot chemiiposwiecony* (Warsaw, 1957). See also *Wielka Encyklopedia Powszechna PWN*, XII (Warsaw, 1969), p. 96, and J. R. Partington, *A History of Chemistry*, IV (London, 1964), 340, 367, 558, and 868.

ALBERT B. COSTA

WALTON, IZAAK (*b.* Stafford, England, 9 August 1593; *d.* Winchester, England, 15 December 1683), *zoology.*

Although Walton made no great claims to scientific originality, no author better illustrates the attitude toward natural philosophy in mid-seventeenth-century England. Religion and natural history were then so closely connected and their interactions so numerous that separation must involve severe historical distortion.

Walton's early life is obscure. Son of Gervase Walton of Stafford, a yeoman, Walton probably attended a local grammar school before settling with his sister in London. There he served an apprenticeship and in 1618 gained admission to the Ironmongers' Company. As a prosperous tradesman he was married twice, to Rachel Floud (*d.* 1640) and to Anne Ken (d. 1662), both of whom came from well-connected Anglican families, thereby enhancing Walton's social status and literary connections. His life was long, comfortable, and uneventful, even during the civil wars, which left him untouched in spite of his strong Royalist sympathies.

Walton is primarily remembered as author of *The Compleat Angler*, but his more long-term and extensive literary activity was as a biographer. His valuable contribution to literature and Anglican apologetics developed as a result of a friendship with John Donne, Walton's parish priest. His biographies of Donne (1640) and of another friend, Sir Henry Wotton (1651)—like the subsequent biographies of Thomas Hooker (1665), George Herbert (1670), and Robert Sanderson (1678)—provide a valuable source about a group of figures who considerably influenced the outlook of natural philosophers in the later seventeenth century. The biographies also contain significant information about Bacon, Savile, and Boyle.

Walton's Anglican associations are important for appreciating the origin and bias of *The Compleat Angler*. Angling was a favorite pastime for a whole stream of Anglican divines, from the Elizabethan Dean Alexander Nowell to Walton's friend Archbishop Gilbert Sheldon. Nowell's catechism may even have prompted Walton to adopt the didactic dialogue form for his book. Direct instigation to write on this subject came from Wotton, who himself had intended to compose a treatise in praise of fishing. Composed at a time when the established church greatly needed support, *The Compleat Angler* was a vehicle for many favorite themes in Anglican theology.

Walton's enthusiasm for fishing gave him the keen eye of a naturalist, his knowledge being particularly sound on a wide range of freshwater fish. Contrasting with the stereotyped compendiums on natural history, Walton introduced a wide range of "observations of the nature and breeding, and seasons, and catching of Fish." His own observations were compared with scrupulously acknowledged information drawn from such authors as Gesner, Scala Johann Dubravius, and Bacon. He was much less familiar with authors unavailable in translation, perhaps a reflection on his limited classical education and a notable contrast with the similarly disposed Sir Thomas Browne. Walton also drew material from the small vernacular literature on angling; his debt to the dialogue *The Arte of Angling* (1577) is obvious but extremely difficult to establish precisely. Walton far exceeded his precursors in attaining a balance between natural history and practical advice. A final component of religious, moral, and philosophical digression gave *The Compleat Angler* an extremely wide appeal. Among the topical scientific references was a consideration of Helmont's willow tree experiment, which was cited to support the contention that water was the prime element in nature. After the first modest edition of 1653, the book was revised four times by Walton. Since then almost three hundred editions and translations have appeared, making Walton one of the best-known authors in the English language.

BIBLIOGRAPHY

I. ORIGINAL WORKS. Walton's *Lives* appeared separately between 1640 and 1678; subsequently they appeared in collected editions. The *Lives* and their revisions are discussed in great detail in D. Novarr, *The Making of Walton's Lives* (Ithaca, N.Y., 1958). *The Compleat Angler, or the Contemplative Man's Recreation* was published anonymously in 1653; it was greatly revised and augmented in 1655, and minor revisions were made in 1661, 1668, and 1678. The most detailed edition of *The Compleat Angler* is that of George Washington Bethune (New York–London, 1847). B. S. Horne, *The Compleat Angler 1653–1967* (New York, 1970), lists the editions. For a collected edition of Walton's works, see G. Keynes, *The Compleat Walton* (London, 1929).

II. SECONDARY LITERATURE. No biography is adequate, but see Andrew Lang, "Izaak Walton," in *Dictionary of National Biography*; and Stapleton Morton, *Izaak Walton and His Friends*, 2nd ed. (London, 1904). A more critical evaluation is A. M. Coon, "The Life of Izaak Walton," unpublished Ph.D. diss., Cornell University (1937). On natural history, see R. B. Marston, *Walton and Some Writers on Fish and Fishing* (London, 1894). *The Arte of Angling (1577)*, which was discovered recently, has been edited by G. E. Bentley (Princeton, 1956); its relevance to Walton is examined critically by M. S. Goldman, in *Studies in Honor of T. W. Baldwin*, D. C. Allen, ed. (Urbana, Ill., 1958), 185–204. J. R. Cooper, *The Art of the Compleat Angler* (Durham, N.C., 1968), examines the stylistic sources for *The Compleat Angler*.

CHARLES WEBSTER

WANGERIN, ALBERT (*b.* Greiffenberg, Pomerania, Germany, 18 November 1844; *d.* Halle, Germany, 25 October 1933), *mathematics*.

Wangerin studied mathematics and physics from 1862 to 1866 at the universities of Halle and Königsberg, receiving the Ph.D. from the latter in 1866. Until 1876 he taught in high schools in Posen (now Poznan, Poland) and Berlin. He began to teach on the university level at Easter 1876, when he assumed the post of extraordinary professor at the University of Berlin. In 1882 he was named full professor at the University of Halle, where he remained until his retirement in 1919.

At Königsberg, Wangerin studied under Richelot, a supporter of the Jacobian tradition, and under Franz Neumann. It was Neumann who suggested the subject of his dissertation, and Wangerin later wrote a book (1907) and a highly appreciative article on his former teacher. Wangerin's admiration for Neumann remained an important influence on his choice of research problems. He became an expert on potential theory, spherical functions, and the fields of mathematical physics related to these subjects. For example, in one of his papers he calculated the potential of certain ovaloids and surfaces of revolution. Wangerin also worked, although less intensely, in differential geometry. In 1894 he wrote an article showing how to determine many bending surfaces of a given surface of revolution of constant curvature without knowing its geodetic lines.

Wangerin's importance, however, does not lie in the authorship of enduring scientific works but, rather, in his astonishingly varied activities as university teacher, textbook author, contributor to encyclopedias and journals, editor of historical writings, and president of a scientific academy. While at Berlin he directed his lectures to a fairly broad audience, and even at Halle he continued to be greatly interested in the training of high school teachers. He also wrote a two-volume work on potential theory and spherical functions for the series Sammlung Schubert.

Wangerin wrote two articles for *Encyklopädie der mathematischen Wissenschaften*. The first (1904) deals with the theory of spherical and related functions, especially Lamé and Bessel functions. The second, "Optik; ältere Theorien" (1907), appeared in the physics volume of the *Encyklopädie*. In it Wangerin displays a familiarity with the history of physical theory that is unusual for a mathematician. His sensitivity to historical questions evokes his study, four decades earlier, under Neumann. Wangerin's historical interests are also evident in his editing of works by Gauss, Euler, Lambert, and Lagrange for *Ostwalds Klassiker der exakten Wissenschaften*.

From 1869 to 1924 Wangerin was a coeditor of *Fortschritte der Mathematik*, then the only periodical devoted to reviewing mathematical literature. In this capacity he reviewed almost all the works in his special field published during this period. For 1906 to 1921 Wangerin was president of the Deutsche Akademie der Naturforscher Leopoldina in Halle.

BIBLIOGRAPHY

Wangerin's writings include "Über die Abwicklung von Flächen konstanten Krümmungsmasses sowie einiger anderer Flächen aufeinander," in *Festschrift zur 200-jährigen Jubelfeier der Universität Halle* (Halle, 1894), 1–21; "Theorie der Kugelfunktionen und der verwandten Funktionen, insbesondere der Laméschen und Besselschen (Theorie spezieller, durch lineare Differentialgleichungen definierter, Funktionen)," in *Encyklopädie der mathematischen Wissenschaften*, II, pt. 1 (Leipzig, 1904), 699–759; *Franz Neumann und sein Wirken als Forscher und Lehrer* (Brunswick, 1907);

"Optik, ältere Theorien," in *Encyklopädie der mathematischen Wissenschaften*, V, pt. 3 (Leipzig, 1907), 1–93; *Theorie des Potentials und der Kugelfunktionen*, 2 vols., nos. 58 and 59 in Sammlung Schubert (Leipzig, 1908–1921); "Franz Neumann als Mathematiker," in *Physikalische Zeitschrift*, **11** (1910), 1066–1072; and "Über das Potential gewisser Ovaloide," in *Nova acta Leopoldina*, **6**, no. 1 (1915), 1–80.

Secondary literature includes W. Lorey, "Zum 70. Geburtstag des Mathematikers A. Wangerin," in *Zeitschrift für mathematischen und naturwissenschaftlichen Unterricht*, **46** (1915), 53–57; and "Bericht über die Feier der 80. Wiederkehr des Geburtstages des Herrn Geh. Rats Prof. Dr. Wangerin," in *Jahresberichte der Deutschen Mathematiker-vereinigung*, **34** (1926), 108–111.

WERNER BURAU

WANG HSI-SHAN[1] (*b.* 23 July 1628, registered at Wu-chiang,[2] Soochow prefecture, China; *d.* 18 October 1682), *astronomy.*

Wang was the son of Wang P'ei-chen[3] and his wife, whose maiden name was Chuang.[4] Wang Hsi-shan's epitapher, who would have been expected to cite distinguished ancestors in the preceding few generations, was unable to do so. Wang was designated to continue the family line of a childless uncle, which suggests that he was not the eldest child. Nothing is known of his education except that he was self-taught in mathematics and astronomy. He had no son to arrange for posthumous publication of his writings. One of his few disciples sketched, in twenty-five Chinese characters, the impression Wang made: "emaciated face, protruding teeth, tattered clothes, and shoes burst through the heels. His character made him aloof, as though no one could suit him; but when someone inquired about a scholarly topic, he was forthcoming as a river in flood."[a]

The conventional road to social advancement for the son of an obscure gentry family was the civil service examinations, which required many years of special preparation. This path was clouded by the Manchu conquest. The invaders from the north overran Wang's district in 1645. Whether to collaborate with the alien government was an issue for all Chinese. Wang was only sixteen, but he made it clear that he did not wish to live with the new order: "In a burst of passion, wanting to die, he jumped repeatedly into the river. It always happened that someone was there to save his life. He then refused to take food for seven days, but still did not die. His parents were persistent; he had no choice but to resume eating. Renouncing worldly ambition, he dedicated all of his powers to learning." Hopes for a Ming restoration soon faded, but in letters and manuscripts he never acknowledged the new Ch'ing dynasty. His friends, who included the great transitional figure of neo-Confucianism, Ku Yen-wu[5] (1613–1682), were Ming loyalists.

Wang was not widely traveled; he never met Mei Wen-ting[6] (1633–1721) of Anhwei or the northerner Hsueh Feng-tso[7] (*d.* 1680), now considered the other two great astronomical scholars of the time. Mei acknowledged Wang's preeminence, however, and wrote commentaries (never published) on several of his books.[b]

Wang apparently made an indifferent living by teaching a few disciples. He was supported for a time by a wealthy scholar of his district who was compiling a history of the Ming dynasty (another act of devotion to the lost cause). Wang's career was short, impeded by isolation, and hampered by illness, including partial paralysis of the extremities in his later years. The year before he died, he wrote, in connection with the prediction of a solar eclipse: "Whenever there is a conjunction I have always checked the precision of my computations against that of my observations, despite sickness, cold, or heat, for thirty years and more."[c]

Wang's technical writings circulated in manuscript among astronomers after his death. Their preservation was not guaranteed until a major composition was included in the enormous imperial manuscript collection of rare texts, the *Ssu k'u ch'üan shu*[8] ("Complete Library in Four Repositories," compiled 1773–1785). The descriptive and critical catalog of this collection was printed and brought Wang's contribution to general attention.

The Setting of Wang Hsi-shan's Career. From about 1600 until the papal suppression of the Society of Jesus in 1773, missionaries of that order were practically the sole source of Chinese knowledge about Western astronomy. The Astronomical Bureau was the one part of the Chinese court where groups of foreigners had been employed in positions of trust for some centuries. Indians since the eighth century and Muslims (mostly Central Asians) since the thirteenth century had applied geometrical and trigonometric methods, which the Chinese lacked, to critical computational tasks, especially the prediction of eclipses. Appointments in the Bureau provided the Jesuits with access to the ruling elite, whose conversion was their main object. Mathematical and astronomical treatises demonstrated high learning and proved that the missionaries were civilized and socially acceptable, although religion was not part of the conventional discourse of gentlemen and attempts to convert one's friends were considered bad form. Science was of no direct concern to the missionaries of other orders, who gradually began to proselytize in China, since their clientele was predominantly the poor and forsaken.

The missionaries began publishing on astronomy in the first decade of the seventeenth century. Their earliest writings did not provide what was needed to predict the celestial motions, but merely demonstrated the usefulness of Aristotelian-Ptolemaic cosmology and astronomy as then practiced in Europe. It was mainly the series of treatises presented to the throne in 1631–1635, as part of the campaign to gain operational control of the Astronomical Bureau and institute a calendar reform based on Western techniques, that set out the mathematical rudiments of the calendrical art. The principles of calendar reform were accepted by the government only on the eve of defeat by the Manchus, but the alien dynasty promptly accepted the Jesuits' offer of services. Compiling a new system of calendrical computation was one of many steps usually taken to assert ritually the legitimacy of a new regime. The astronomical treatises, earlier printed individually by the missionaries, were published together by imperial order as *Hsi-yang hsin fa li shu*[9] ("Astronomical Treatises According to the New Methods of the West," presented to the throne in 1646). The prestige of official sponsorship assured them fairly wide distribution, although, unlike other imperial publications, they were never privately reprinted as a set.

Once the Jesuits were established in a secure position to protect their religious activities, there was no need to continue reporting on European developments in astronomy and cosmology—until a series of attacks by their enemies temporarily deposed them from the Astronomical Bureau (1665–1669) and even closed their churches in the provinces. During that time of crisis, the Flemish Jesuit Ferdinand Verbiest wrote several important new books, comprising tables and accounts of instruments and of predictions. The Tychonic cosmology of most of the "New Methods" treatises was not modified until the Keplerian ellipse was quietly introduced—for the solar orbit only—in 1742. The heliostatic world system was not introduced until 1760, after Copernicus' *De revolutionibus* had been removed from the Index (1757), and even then it was only described, without a new computational scheme.

In sum, Chinese astronomers had to form their impression of European astronomy and cosmology from writings that, about 1630, were not untypical of textbooks and handbooks current in the church's educational institutions in Europe but that, as time passed, failed increasingly to reflect the emergence of modern astronomy. It was with this in mind that the distinguished historian of science, Hsi Tse-tsung, remarked, "We can imagine, if Wang Hsi-shan had only come upon [Copernicus' *De revolutionibus*, Galileo's *Dialogo*, and Kepler's *Epitome astronomiae Copernicanae*, all of which the missionaries kept for their private use in Peking], how much greater his contribution to astronomy would have been."[d]

Wang, Mei Wen-ting, and Hsueh Feng-tso were the first scholars in China to respond to the new exact sciences and to shape their influence on their successors. They were, in short, responsible for a scientific revolution. They radically reoriented the sense of how one goes about comprehending the celestial motions. They shifted from using numerical procedures for generating successive angular orientations to using geometric models of successive locations in space. They changed the sense of which concepts, tools, and methods are centrally important, so that geometry and trigonometry largely replaced numerical algebra, and such issues as the absolute sense of rotation of a planet and its relative distance from the earth became important for the first time. They convinced Chinese astronomers that mathematical models can have the power to explain the phenomena as well as to predict them.

This revolution did not reach the same pitch of tension as the one going on in Europe at the same time. It did not burst forth in as fundamental a reorientation of thought about nature. The new ideas and techniques did not arouse in Wang and others a need to cast doubt on all the traditional ideas of what constitutes an astronomical problem and what significance astronomical prediction can have for the ultimate understanding of nature. The traditional idea persisted that explaining the astronomical phenomena could not by itself lead to a synthetic comprehension of the inherent pattern of the cosmos, the Tao.

The limited character of the seventeenth-century breakthrough is perhaps not surprising. The decree of the Congregation of the Index in 1616 denied Chinese access to the fruits of the Copernican revolution and its aftermath in Europe, at least as long as Catholic missionaries were the only foreigners who had reason to write on astronomy in Chinese. Still, the reorientation determined a new style for Chinese astronomy until finally, between 1870 and 1920, that science ceased to exist apart from modern astronomy as an international enterprise.

This train of events has drawn little sympathetic attention, largely because it confounds the widespread assumption that in the encounter between cultures, Western science must assert its dominance by a process so automatic that one need not trouble oneself with a critical examination of instances. On the contrary, in China the new tools were used to rediscover and recast the lost mathematical astronomy of the past and thus to perpetuate traditional values rather than to replace them.[e] The "imperatives of modernization" appear universal to the uncritical merely because the encounter between traditional and modern science in one society after another has been resolved by social change and political fiat, in view of which the comparative appropriateness of each system of science to the cultural environment is beside the point. The same may even be said for early modern technology, which was clearly superior to that of traditional societies, but was superior mainly in applications that did not exist until it generated them.

In a word, there has seldom been a direct encounter between traditional and modern science in East or West. Seventeenth-century China was an exception of great interest because European civilization had no appreciable political or social impact, and astronomy had to make its way on its own merits. Not on its abstract intellectual merits alone, to be sure, for what constituted merit was largely defined by the use the court traditionally made of astronomy.

At the time there were no socially marginal students of astronomy alienated from traditional values and protected by association with privileged foreigners, as would be the case in late nineteenth-century China and elsewhere in the heyday of imperialism. The only astronomers who could respond to the Jesuits' writings were members of the old intellectual elite, who were bound to evaluate innovations in the light of established ideals that they felt an individual responsibility to strengthen and perpetuate.

In order to assure acceptance of the Western methods among people of his own kind, Mei Wenting created the myth that European mathematics evolved out of certain techniques that had originated in China and had been transplanted to the extreme fringes of civilization (Europe and Islam) before losing vitality in their original home.[f] This myth, an appeal to the Chinese tendency to see perfection in high antiquity, connected Western science with certain ambiguous references in ancient historical writing; since it was not intrinsically foreign, it could be taken as more than a curiosity. Mei launched what anthropologists call a "foundation myth" for the institutions (small private groups of masters and disciples) that taught the new astronomy and reorganized themselves around it. There is nothing inherently Chinese about the use of such myths. An analogy that comes to mind is the remarkable European myth that non-European societies could be "discovered," a change of status that authorized their economic despoliation and the systematic destruction of their religious and other customs during the Age of Discovery.

Mei, Wang, and others were aware that, whatever the lost grandeur of archaic times may have been, from about 100 B.C. through the thirteenth century computational astronomy in China had actually continued to grow in power and range within its stylistic limits. Very gradually from the Yüan period (1279–1368), it came to be little practiced outside the Astronomical Bureau, which was dominated by foreign technicians. By 1600 no one was able fully to comprehend the old numerical equations of higher-order, prototrigonometric approximations, applications of the method of finite differences, and other sophisticated techniques. The promise of a renascence seemed to the greatest astronomical figures of the seventeenth century to define the proper field for application of Western knowledge.

Astronomical Work. Wang and his contemporaries were motivated by two central problems. The first was how Western knowledge might be used to revive the lost Chinese exact sciences. Traditional knowledge was recorded, and the perennial problems were set, in the easily accessible standard histories of the various dynasties. Each incorporated a variety of technical treatises, which, among other matters, recorded many of the complete systems of ephemerides computation that had been proposed or accepted for official use since about 100 B.C. Wang was familiar with the chief writings of this sort and, although there is no reason to believe that he met any foreign missionary, with the Jesuit treatises of the early 1630's.

The second problem was how to resolve the internal contradictions of European astronomy. Since circumstances had ruled out a closely unified set of treatises, some discrepancies were due to divergences of approach and varying choices of constants, and some to the missionaries' limitations of skill. The most important source of inconsistency was the different cosmological viewpoints through which European writers tried to convey the best knowledge of their time, before and after the limits of contention about the system of the world were drawn by the decree against the teaching of heliocentricism. Matteo Ricci, writing before 1616, was conventionally "Ptolemaic" (that is, he reflected the doctrines of Aristotle and Ptolemy as understood by the Scholastics of his youth). The writers of about 1630 were Tychonists, except for Johann Schreck (1576–1630). Later Jesuit writers never accounted for the shift to Tychonism—nor for the introduction of Copernicanism in the mid-eighteenth century. Chinese could perceive only the lack of conviction and of unanimity. Misleading statements about Copernicus' contributions, contextless references to alternative systems (for instance, a confusing and unexplained allusion by Schreck to the cosmology of Heraclides Ponticus), and additional isolated innovations in Verbiest's later series of writings muddied even more the question of what should be considered the state of the foreign art. The spotty character of publication made it impossible to take the latest as best; in the missionary writings the latest usually presupposed the earlier without criticizing it.

Wang's response belies the occasional assertions of historians that Chinese were incapable of responding creatively to geometrical models, or that a bias against abstraction would have prevented them from taking up Copernican cosmology had it been available. In adapting the missionaries' version of Tycho Brahe's scheme of the cosmos to his own uses, Wang made considerable adaptations and criticized contradictions in its presentation. He noted, for instance, that a secular diminution in the length of the tropical year had been mentioned, but was ignored in a discussion of the precessional constant (which, by implication, should increase). This was not the modern variation in the length of the year, but one of much greater magnitude that was obsolete in Europe by the time Wang wrote.

Wang's *Hsiao-an hsin fa*[10] ("New Method," completed in 1663) was cast in traditional form, with tables that made only the elementary logistic operations necessary for calculating the ephemerides. It provided, for the first time, methods for predicting planetary occultations and solar transits. Some of Wang's techniques were included along with post-Newtonian data in the *Li hsiang k'ao ch'eng*[11] ("Compendium of Observational and Computational Astronomy," printed 1724), part of a great survey of the mathematical arts sponsored by the K'ang-hsi emperor.

Wang's *Wu hsing hsing tu chieh*[12] ("On the Angular Motions of the Five Planets," completed by the autumn of 1673) was a critically overhauled Tychonic model of the planetary motions, substituting eccentrics for major epicycles and opposing the rotational senses of the superior and inferior planets. This work displays a general familiarity with modern trigonometry. Wang's arguments, unlike those of his Chinese predecessors, were clearly concerned with bodies in motion.

The most original idea in "On the Angular Motions" was Wang's proposal that the planetary anomalies be explained by a force radiating from the outermost moving sphere (*tsung tung t'ien*,[13] or *primum mobile*) and attracting each planet to an extent maximal at apogee. Explanation of celestial motions by forces, instead of the assumption in the old kinematics of compounded circles that the celestial motions were eternal, entered the European debate only through Kepler. There had been no such discussion in China, except for a vague statement by Giacomo Rho (1592–1638) that "The motions of [the sun, Mercury, and Venus] are all due to one potential moving force . . . located in the body of the sun." This assertion was not connected with the remainder of Rho's planetary theory, and no extension to the superior planets was hinted at. Wang's force, although it was not universal like Newtonian gravitation, applied to all the planets known to him and was exerted from the periphery rather than from inside the planetary orbits.[g]

Wang's notion of synthesis went deeper than reconciling ancient schemes of calculation with foreign techniques. The power of Western models not only to predict phenomena but also to exhibit their inherent patterns was what attracted many Chinese. Wang sought to establish metaphysical links for further exploration into celestial reality. This motive lay behind his suggestion that the circle be divided into 384 degrees. The traditional division made each degree (*tu*)[14] equal to one tropical day's mean solar travel (so that in Wang's system there would be 365.2422 degrees). He was aware of the convenience offered by the European system of angular division, especially in manufacturing graduated instruments. He chose the number 384 (3×2^7) in addition to 360 ($2^3 \times 3^2 \times 5$) because, as the number of lines (6×64) in the sixty-four hexagrams of the "Book of Changes," it related astronomical quantities to the fountainhead of conventional speculation about cosmic change and thus uncovered another layer of significance.[h]

Despite his dedication and critical intelligence, it cannot be said that Wang, any more than his contemporaries, succeeded in a mature synthesis of traditional and modern science. They did provide tools and methods as well as a goal. Information from the West was inadequate in many respects, and several generations more were needed to reclaim the traditional corpus of Chinese mathematics and astronomy as part of the astronomer's repertoire.

For many decades students began with the Western writing and went on to study the Chinese technical classics. The latter, as they were successively mastered, increasingly defined the style of research in the exact sciences, even when this research began to be concerned with new problems. By the early nineteenth century, Western mathematics and astronomy were no longer novelties; they had been studied continuously for two hundred years. The basic training in the decades before the Opium Wars (*ca.* 1840) was in the native writings. They served as excellent preparation for up-to-date Western treatises that gradually began to appear as part of a new confrontation — this time a total confrontation — between China and the West.

General Significance for Chinese Thought. Wang Hsi-shan's lifetime was a critical epoch in the evolution of Chinese philosophy. What Western historians call neo-Confucianism was, like the earlier trends it built upon, a search for doctrines of education, self-cultivation, and moral life in society. Its successive new departures depended upon reexamination of antiquity to identify and interpret (differently for each age) the authentic core of Confucian teachings. Well before the late Ming period, expanded scope for self-consciousness, increased blurring of social barriers, and the more penetrating influence of Buddhism and Taoism had deepened religious and moral awareness. This trend affected both the Chu Hsi[15] tradition, which explored the phenomenal world (including the mind and experience recorded in books) to grasp the single coherent pattern inherent in all change, and that of Wang Yang-ming,[16] which strongly emphasized enlightenment through self-awareness, particularly of the mind engaged in conscientious social activity.

The great intellectuals of the dynastic transition were, on the whole, Ming loyalists. That is, they were among the minority who did resist, in the main passively and after the transfer of power had taken place. They were convinced that to plumb the failure of their intellectual predecessors would be to uncover the conditions for philosophical and spiritual reinvigoration, and for responsible engagement in the world of affairs. Among the most influential was Wang Hsi-shan's friend Ku Yen-wu, who saw his late Ming predecessors as distracted from moral commitment and public responsibility by sectarianism, by pedantry and triviality in the Chu tradition, and, in the tradition of Wang, by a subjectivity and individualism ignorant of the authoritarian and hierarchic requirements of social order. Above all, in the view of Ku and other Ch'ing survivors, it was the rivalry of schools — a manifestation of blinding selfishness and pride — that corrupted the neo-Confucian teaching, leaving it unable to rise above the political futility that preceded the debacle at the end of the Ming.[i]

The prescription for the ills of thought was to purge postclassical influences that hid the original principles of Confucius and his orthodox followers. A critical method for the examination of texts was the crucial safeguard; the broad study that printing had made feasible revealed to people of Wang's generation how easily the understanding of their predecessors had been led astray. Some now endeavored to recover the earliest — the uncorrupted — versions and interpretations of the classics, and others studied the working out of canonical moral patterns in the events of history. This work was not intended to replace the quest for a living philosophy, to which the fortunes of the empire had given a new poignancy.

By the mid-eighteenth century, narrowly defined scholarly methodology had become an end in itself, narrow in interpretation and intolerant of the urge to generalize. The call for "social utility, concrete practicality, and tangible evidence," which had promised philosophic regeneration a century earlier, outlived the openness to the unexpected that was implied in its original motivation.[j] Classicism flourished, despite the atrophy of metaphysics, because of the cumulative accomplishments it yielded and because it posed no threat to a state which insisted that collective intellectual activity be apolitical.

This final evolution of disciplinary specialization out of a philosophic renaissance is not of further concern here, but how the Ch'ing style of critical neo-Confucian thought began to take shape in Wang Hsi-shan's lifetime bears examination.

It has often been noticed that certain important neo-Confucians of the early Ch'ing era, especially among those close to the Chu tradition, wrote on mathematics and astronomy. That the extent of this scientific interest has been seriously underestimated becomes clear in an unpublished survey of thirty-six people generally considered major neo-

Confucian figures, from the beginning of the Ch'ing period (1644) to Wang Yin-chih[17] (1766–1834). Of the thirty-six, eighteen left a total of seventy-two books on mathematics and astronomy. A large group of these treatises reconstituted early computational techniques, and about the same number were chronological or other mathematical studies of canonical writings before about 200 B.C.[k]

That overlap of intellectual activity is part of a more general pattern that also connects those known only for philosophy with those known only for science. In the mid-seventeenth century the leaders in both fields were people who eschewed politics and public service. This is perhaps not remarkable, since one expects reevaluation and syncretism to begin with talented and ambitious people who, for one reason or another, remain on the margins of the elite. But in addition to obvious consequences of this social overlap, philosophers and scientists shared important convictions.

To sum up the argument so far, certain critical motifs recurred in neo-Confucianism just after the Manchu conquest, pointing the way toward new departures: emphatic rejection of what were seen as decadent and destructive tendencies at the end of the Ming (Ku Yen-wu located them mainly in the Wang Yang-ming school, but other thinkers were more evenhanded in their apportionment of blame); the belief that those tendencies arose partly because of inadequate study and partly because of Buddhist and other heterodox ideas—as well as a variety of misunderstandings and corruptions—that had insinuated themselves into texts and undisciplined scholarly writings; and the conviction that a sound approach to understanding the inherent patterns of cosmic and human activity (li)[18] and the moral imperatives they imply required critical reexamination of classical literature and history.

All of these ideas also motivated Wang Hsi-shan. We are told by his biographer that after he renounced worldly ambition, "He excoriated heterodoxy [this usually refers to Buddhism, sometimes to Christianity as well], attacked 'innate moral consciousness'[19] [the characteristic doctrine of the Wang Yang-ming school], and accepted the orthodox Confucian tradition of the Chu Hsi line[20] as his personal mission." The preface to his "New Method," instead of conventionally affirming the high antiquity of astronomy, began by taking up questions that had been raised about the authenticity of seven calendars that the historians dated prior to the Han period (206 B.C.), and stated flatly, "There is no doubt that they were forgeries of the

Han." Wang emphasized that astronomers of the recent past comprehended less than their predecessors and that the lost meaning of the technical classics had to be rediscovered.

Earlier scientists had argued, with some consistency, that although mathematical astronomy could provide useful knowledge and advance understanding, the subtle texture of the natural order could ultimately be penetrated only by illumination.[l] Wang did not reject this view, but he saw number as a means toward that penetration: "One who seeks rigor must reach it through computation. Numbers are not themselves the inherent pattern [li]; but because the pattern gives rise to number, through number one may reach enlightenment as to the pattern."[m] This conviction was almost certainly influenced by the argument of Matteo Ricci, in his preface to the Chinese translation of Euclid's Elements (Chi-ho yuan pen, 1607),[21] that geometry is a unique means to knowledge of the inherent pattern, knowledge that does not depend upon individual belief and thus can overcome individual doubts.

These parallels, and others in the writings of Wang's scientific contemporaries, suggest a closer connection than has hitherto been imagined between the scientific revolution of seventeenth-century China and the evolution from philosophy to exact scholarship that took much longer to run its course. In particular, they suggest that Western influence on main currents of early Ch'ing philosophy—on the frontiers of Chinese self-awareness—should not casually be ruled out. Historians have usually ruled it out because they have not studied the scientific literature and because they rely on crude and narrowly defined tests for intellectual influence that ignore the mathematical dimension of human thought.

It is a matter of paramount importance that the first substantial encounter between Chinese traditions and European culture was in mathematical astronomy. Medicine and religion provoked no such response, and confrontations of political ideas were negligible. The style of the response to Western astronomy—how it was assimilated so that it could be understood, how the primacy of traditional values was asserted and the study of the new methods justified in terms of them, how teaching and practice were adapted to existing institutions—set the style of less abstract encounters later, as may be seen in the response to Western military technology in the second half of the nineteenth century. (The broad pattern of response—misconception, ambiguity, interpretation, and piecemeal adaptation—is not utterly unlike that of Americans to Chinese acupuncture in the 1970's.)

This mathematical challenge to values coincided with an even more traumatic challenge, the Manchu invasion. Some shade of ambiguity toward Western science must have come from the Jesuits' prompt tender of services to the Manchus and the immediate official adoption of their astronomical system (although some missionaries accompanied the refugee Ming court south to cover the eventuality of a restoration). It was not, however, characteristic of Chinese to reject what was foreign merely because it was foreign; nationalism had barely been conceived. A willingness to adopt the forms of Chinese culture gave Manchus and Jesuits a right to be where they were—although nothing compelled loyalists to collaborate with either. But Wang Hsi-shan's type of loyalism was only a memory for the following generation, and served as no deterrent to the study of new ideas.

It is merely reasonable to suggest that philosophers were influenced by the early success of astronomers in applying the foreign tools (in eclectic combination with old ones) to the reexamination of classical learning. Reexamination of ancient observations and predictions was an established part of the astronomer's work. Not long after Wang Hsi-shan's lifetime, technical examination of the philosophic classics to fix dates and test authenticity became the explicit end of most astronomical exploration, with the revival of traditional science as an intermediate means that happened to fully occupy many scholars.

Even more important in assessing this channel of European influence is the fact that as time passed, leading neo-Confucian scholars also became mathematicians used to working with Western techniques and concepts. Because these scholars grouped in schools and maintained close relations, even those who never applied European science in their writings were aware of it through discussions with their associates and through reading their monographs.

None of this suggests a simple causal relation between European astronomy as described in the Chinese language in the early seventeenth century and the forms of Chinese thought that became dominant by the middle of the eighteenth. Certainly the relation between astronomy and philosophy was reciprocal as long as philosophy remained vital, despite its moral and social cast and its focus on self-realization. At the beginning the new sci-

ence offered what were seen as powerful tools toward a reformation of thought. Wang Hsi-shan's belief that number could bring ultimate insight into the universal pattern was pregnant in precisely this way; but the consensus formed in more conventional quarters, and was ultimately barren for a new understanding of nature, society, and man. Still, the new techniques could never be mere tools. To use them, as so many thinkers did, was to form habits that reinforced long-held convictions about the usefulness of scholarship in exploring reality. Seventeenth-century European science was not, after all, modern science, least of all as it was artificially perpetuated in China for two centuries by the lack of sources alternative to the missionaries' writings. Chinese philosophers, whose sense of man and the cosmos was in part formed by study of canonic books, responded to the universal explanatory character that this foreign science derived from its Scholastic framework much more than to the grip on direct experience of nature that Wang valued.[n] Astronomy, as it was understood and used by neo-Confucian thinkers, converged with philology, gave it added weight, and obviously played a part in tipping the scale.

It would be premature to suggest any particular line of development between the recourse to Western astronomy among philosophers at the beginning of the Ch'ing period and the eventual swamping of earlier philosophical concerns by exact scholarship—exact scholarship of a kind in which mathematical astronomy finally could be perfectly integrated as one specialty among many. The career of Wang Hsi-shan suggests a range of possible patterns that, tested against many other careers, can throw light on the central enigmas which shroud the failure of imperial China.

NOTES

a. This and the following quotation are from the funerary inscription by Wang Chi in *Sung-ling wen lu*[22] ("Literary Records of Wu-chiang," 1874), 16: 1a–1b.

b. Described in the *Wu-an li suan shu-mu*[23] ("Bibliography of Mei's Writings on Astronomy and Mathematics"; Pai-pu ts'ung-shu chi-ch'eng ed.), 34b–35a. On the relations of Mei and Wang, see the biographical study by Hsi Tse-tsung, "Shih lun Wang Hsi-shan te t'ien-wen kung-tso"[33] (in bibliography). Reliable short biographies of Mei and many other figures mentioned in this article are in Hummel, *Eminent Chinese*.

c. Cited by Hsi Tse-tsung (p. 63) from the MS "Wang Hsiao-an hsien-sheng i-shu pu-pien"[24] ("Supplement to the Posthumous Works of Wang Hsi-shan") in the Peking University Library.

d. Hsi Tse-tsung, *loc. cit.*

e. This point was made by Mikami Yoshio in "'Chūjin den' ron"[25] ("A Study of the *Ch'ou jen chuan*"), in *Tōyō gakuhō*, 16 (1927), 185–222, 287–333, and was repeated in "Chinese

Mathematics," in *Isis*, **11** (1928), 125. It has been developed considerably by Wang P'ing.

f. Wang Hsi-shan accepted this notion. See his *Tsa chu*[26] ("Miscellaneous Essays"), in *Hsiao-an i shu*,[31] XXXV, 1a–2a, 10b–11a. The best discussion of the Chinese origin theory is in Wang P'ing, *Hsi-fang li-suan-hsueh chih shu-ju*,[36] 77–79, 97–103. I see no reason to doubt that Mei, Wang Hsi-shan, and others sincerely believed it.

g. *Shou shan ko ts'ung-shu*[27] ed., 7b, discussed in Sivin, "Copernicus in China," 74–75.

h. *Hsiao-an hsin fa*[10] (in *Hsiao-an i shu*),[31] 2a.

i. I am grateful for this formulation, and for a number of helpful criticisms, to Lynn Struve. I am also thankful for suggestions by Judy Berman, Dianna Gregory, and Yü Ying-shih.

j. William T. de Bary, "Neo-Confucian Cultivation and the Seventeenth-Century 'Enlightenment,'" in *The Unfolding of Neo-Confucianism*, 193.

k. N. Sivin, "What Can the Study of Chinese Science Contribute to Our Understanding of Neo-Confucianism, and How?" working paper for Planning Conference on Early Ch'ing Thought, Berkeley, Calif., 28–31 Aug. 1975.

l. See *DSB* article on Shen Kua.

m. *Tsa chu*, 4a.

n. Willard J. Peterson, in his perceptive "Fang I-chih: Western Learning and the 'Investigation of Things,'" has shown how Fang[51] (1611–1671) used his knowledge of Western sciences to argue for greater emphasis in philosophy upon accumulating knowledge of "physical objects, technology, and natural phenomena." Because Fang's understanding of the exact sciences was mediocre, he responded more enthusiastically than most of his contemporaries to the Scholastic sciences of the body, the earth, weather, and so on that were then becoming obsolete in the West. His influence on scientific thought was negligible, but Peterson suggests (correctly, I believe) an indirect formative influence on early Ch'ing humanists' taste for "building knowledge item by item." He asserts that the tendencies Fang encouraged were "parallel to the secularization of natural philosophy in seventeenth-century Europe" (p. 401); but they were more closely parallel to the antiquated approach of Fang's sources, products of the Counter-Reformation attempt to overcome secularization. Although Fang was no more reluctant than the European Schoolmen to provide an occasional "experiment" to demonstrate a point, he depended as heavily as they upon hearsay and literature, and as little upon personal experience; in his dream of what would now be called a research institute, the only source of knowledge mentioned was "ancient and modern books" (p. 383). In short, the scientific revolution in seventeenth-century China was in the main a response to outmoded knowledge that gave little attention to, and consistently misrepresented, the significance of developments in the direction of modern science. This thesis is fully documented in Sivin, "Copernicus in China."

1. 王錫闡	20. 溓洛溎泗	39. 薮内清，中国の天文暦法
2. 吳江	21. 幾何原本	40. 吉田光邦
3. 培真	22. 王濟，松陵文録	41. 明清時代の科学技術史
4. 莊	23. 勿菴歷算書目	42. 徐宗澤，明清間耶穌会士譯著提要
5. 顧炎武	24. 王曉菴先生遺書補編	43. 丁福保，周雲青，四部総録算法編
6. 梅文鼎	25. 三上義夫，疇人傳論	44. 天文編
7. 薛鳳祚	26. 雜箸	45. 近代中算箸述記
8. 四庫全書	27. 守山閣叢書	46. 清代文集算學類論文
9. 西洋新法歷書	28. 中西算学叢書	47. 年譜
10. 曉菴新法	29. 叢書集成	48. 橋本敬造
11. 歷象考成	30. 大統秝法啓蒙	49. 梅文鼎の暦算学－康熙年間の天文暦算学
12. 五星行度解	31. 遺書	50. 梅文鼎の数学研究
13. 宗動天	32. 木犀軒叢書	51. 方以智
14. 度	33. 席澤宗，試論王錫闡的天文工作	52. 坂出祥伸，方以智の思想
15. 朱熹	34. 科学史集刊	53. 嚴敦杰，伽利略的工作早期在中国的传布
16. 王陽明	35. 疇人傳	54. 全漢昇，清末的西學源出中國説
17. 王引之	36. 王萍，西方歷算学之輸入	55. 嶺南學報
18. 理	37. 李儼，明清之際西算輸入中國年表	56. 殼成
19. 良知	38. 中算史論叢	

BIBLIOGRAPHY

I. ORIGINAL WORKS. Wang Hsi-shan's extant writings are listed in an article by N. Sivin in L. Carrington Goodrich, ed., *Ming Biographical Dictionary* (New York, 1976), 1379–1382. The two most important treatises were *Hsiao-an hsin fa*[10] ("New Methods of Wang Hsi-shan"; completed 1663) and *Wu hsing hsing tu chieh*[12] ("On the Angular Motions of the Five Planets"; completed by the autumn of 1673). Both were first printed in the *Shou shan ko ts'ung-shu*[22] collection (1838) and reprinted in the *Chung-hsi suan-hsueh ts'ung-shu*,[28] 1st ser. (1896) and the *Ts'ung-shu chi ch'eng*[29] collection, 1st ser. (1926). About 1890 the two treatises were combined with *Ta-t'ung li fa ch'i-meng*,[30] an elementary introduction to the Great Concordance system (*Ta-t'ung li*), which had been used throughout the Ming period (1368–1644) for computing the ephemerides, and an assortment of short essays, to form the *Hsiao-an i shu*[31] ("Posthumous Works"), vols. XXXI–XXXV in the *Mu hsi hsuan ts'ung-shu*.[32]

II. SECONDARY LITERATURE. The most thorough study of Wang's life and astronomical work, based on unpublished as well as published sources, is Hsi Tse-tsung, "Shih lun Wang Hsi-shan te t'ien-wen kung-tso"[33] ("An Essay on the Astronomical Work of Wang Hsi-shan"), in *K'o-hsueh-shih chi-k'an*,[34] 6 (1963), 53–65. Its references provide an excellent starting point for further study. The article by Sivin cited above, in a reference book invaluable for the study of Wang's immediate predecessors, is more concerned with biographical and bibliographical matters than is the present essay. The first detailed account of Wang's work, based mainly on excerpts from his writings before they had been printed separately, was in *Ch'ou jen chuan*[35] ("Biographies of Mathematical Astronomers," 1799; Shanghai: Commercial Press, 1935), II, 421–446. This programmatic compendium, which included European as well as Chinese figures, was a major influence on the style of eighteenth-century investigations in the exact sciences.

Although no more than isolated sentences from Wang have been published in translation, N. Sivin has drafted a translation of *Wu hsing hsing tu chieh*[12] for circulation and eventual publication in a source book of Chinese science.

III. EUROPEAN SCIENCE IN SEVENTEENTH-CENTURY CHINA. Little attention has been paid by Western sinologists to the early mathematical encounter of East and West. For instance, Ssu-yu Teng and John K. Fairbank, *China's Response to the West. A Documentary Survey 1839–1923* (Cambridge, Mass., 1954), notes the immediate influence only of "items of practical interest," among which the authors include the calendar. Useful and well-known studies, such as Wolfgang Franke, *China and the West. The Cultural Encounter, 13th to 20th Centuries*, R. A. Wilson, trans. (Oxford, 1967); and Joseph R. Levenson, "The Abortiveness of Empiricism in Early Ch'ing Thought," in *Confucian China and Its Modern Fate* (London–Berkeley, 1958), 3–14, do not reflect knowledge of or curiosity about the technical literature.

The only general history of the Chinese response to European exact sciences is in Wang P'ing, *Hsi-fang li-suan-hsueh chih shu-ju*[36] ("The Introduction of Western Astronomy and Mathematics"), Monographs of the Institute of Modern History, Academia Sinica, 17 (Nankang, Taiwan, 1966), summarized in *Journal of Asian Studies*, 29 (1970), 914–917. This book draws heavily on the biographical articles in *Ch'ou jen chuan* for the seventeenth and eighteenth centuries. A very useful tool for further study of both Jesuit and Chinese

mathematical activities is Li Yen, "Ming-ch'ing chih chi Hsi suan shu-ju Chung-kuo nien-piao"[37] ("A Chronology of the Introduction of Western Mathematics Into China in the Transition Between the Ming and Ch'ing Dynasties"), in *Chung suan shih lun-ts'ung*,[38] vol. III of *Gesammelte Abhandlungen über die Geschichte der chinesischen Mathematik*, rev. ed. (Peking, 1955); 10–68. Jesuit activity has been ably surveyed in Yabuuchi Kiyoshi, *Chūgoku no temmon rekihō*[39] ("Chinese Astronomy"; Tokyo, 1969), 148–174.

There is an important group of studies in Yabuuchi Kiyoshi and Yoshida Mitsukuni,[40] eds., *Min Shin jidai no kaguku gijutsu shi*[41] ("History of Science and Technology in the Ming and Ch'ing Periods"). Research Report. Research Institute of Humanistic Studies, Kyoto University (1970), 1–146. Joseph Needham, in *Science and Civilisation in China*, III (Cambridge, 1959), 437–458, was the first to suggest that the limitations as well as the strengths of the Jesuit missionaries greatly affected the character of the Chinese response. His short and incidental discussion of Wang Hsi-shan (p. 454) includes several errors of fact.

The lives of the Jesuit missionaries and their publications in Chinese have been well-documented by historians of that order. See Henri Bernard, "Les adaptations chinoises d'ouvrages européens. Bibliographie chronologique depuis la venue des Portugais à Canton jusqu'à la Mission française de Pékin, 1514–1688," in *Monumenta serica*, 10 (1945), 1–57, 309–388; Joseph Dehergne, *Répertoire des Jésuites de Chine de 1552 à 1800* (Rome–Paris, 1973); and Louis Pfister, *Notices biographiques et bibliographiques sur les Jésuites de l'ancienne mission de Chine, 1552–1773*, 2 vols. (Shanghai, 1932–1934, completed before Pfister's death in 1891). Dehergne is a comprehensive guide to the extensive literature on missionaries, including archival sources; the last part includes aids to research. See also Henri Cordier, *Essai d'une bibliographie des ouvrages publiés en Chine par les Européens au XVIIe et XVIIIe siècles* (Paris, 1883), based on the collection of the Bibliothèque Nationale. For writings in Chinese, see Hsu Tsung-tse, *Ming Ch'ing chien Ye-su-hui-shih i chu t'i yao*[42] ("Annotated Bibliography of Jesuit Translations and Writings in the Ming and Ch'ing Periods"; Taipei, 1958), with indexes of authors, titles, and subjects.

On European scientific works available to the Jesuits in Peking—and still extant as one of the world's greatest collections of scientific writings of the sixteenth through eighteenth centuries—see H. Verhaeren, *Catalogue of the Pei-t'ang Library*, 3 vols. (Peking, 1944–1948). A list of 251 astronomical books has been excerpted in Henri Bernard-Maitre, "La science européene au tribunal astronomique de Pékin (XVIIe–XIXe siècles)," in *Conférences du Palais de la découverte*, ser. D, 9 (Paris, 1951). See also Boleslaw Szczesniak, "Note on Kepler's *Tabulae Rudolphinae* in the Library of Pei-t'ang in Pekin," in *Isis*, 40 (1949), 344–347.

For studies of Chinese responses, the book of Wang P'ing is especially helpful because of its index, still unusual in Chinese scholarly books. For systematic annotated bibliographies, see Ting Fu-pao and Chou Yun-ch'ing, *Ssu pu tsung lu suan-fa pien*[43] ("General Register of the Quadripartite Library, Section on Mathematics"; Shanghai, 1957) and *Ssu pu tsung lu t'ien-wen pien*[44] ("General Register . . . Section on Astronomy"; Shanghai, 1956), supplemented by Li Yen, "Chin-tai Chung suan chu-shu chi"[45] ("Notes on Books About Chinese Mathematics in Modern Times"), in *Chung suan shih lun-ts'ung*,[38] II (1954), 103–308; and "Ch'ing-tai wen-chi suan-hsueh lei lun-wen"[46] ("Articles That Can Be Classified as Mathematical in Collected Literary Works of Individuals in the Ch'ing Period"), *ibid.*, V (1955), 76–92.

The first resort for biographies of the most prominent Chinese scientific figures is Arthur W. Hummel, *Eminent Chinese of the Ch'ing Period*, 2 vols. (Washington, D.C., 1943–1944); *Ming Biographical History* (see above) will provide similar information about those who reached maturity before the mid-seventeenth century and about a few later people (such as Wang Hsi-shan) not accorded biographies by Hummel. *Ch'ou jen chuan*[35] is composed mostly of long excerpts from technical writings and does not provide a great deal in the way of biography or overview.

A few topical studies throw light on fundamental issues. The genesis, content, and distribution of major Jesuit scientific writings are described in Pasquale d'Elia, "Presentazione della prima traduzione chinese di Euclide," in *Monumenta serica*, 15 (1956), 161–202, with English summary; and Henri Bernard-Maître, "L'encyclopédie astronomique du Père Schall," *ibid.*, 3 (1938), 35–77, 441–527. Early European writings in Chinese on the qualitative sciences are described in Willard J. Peterson, "Western Natural Philosophy Published in Late Ming China," in *Proceedings of the American Philosophical Society*, 117 (1973), 295–322.

The life, associations, and work of Wang Hsi-shan's contemporary Mei Wen-ting have been treated at length in Li Yen, "Mei Wen-ting nien-p'u"[47] ("A Chronological Biography of Mei Wen-ting"), in *Chung suan shih lun-ts'ung*, III, 544–576; and in Hashimoto Keizō,[48] "Bai Buntei no rekisangaku—Kōki nenkan no temmon reki-sangaku"[49] ("The Mathematical Astronomy of Mei Wen-ting—Mathematical Astronomy in the K'ang-hsi Period"), in *Tōhō gakuhō* (Kyoto), 41 (1970), 491–518; and "Bai Buntei no sugaku kenkyū"[50] ("The Mathematical Researches of Mei Wen-ting"), *ibid.*, 44 (1973), 233–279. The thought of Fang I-chih,[51] probably the first Chinese to acquaint himself with the full spectrum of European sciences, has been examined by Sakade Yoshinobu, "Hō Ichi no shisō"[52] ("The Thought of Fang I-chih"), in Yabuuchi Kiyoshi and Yoshida Mitsukuni (see above), 93–134; and by W. J. Peterson, "Fang I-chih: Western Learning and the 'Investigation of Things,'" in W. T. de Bary and the Conference on Seventeenth-Century Chinese Thought, *The Unfolding of Neo-Confucianism*, Studies in Oriental Culture, 10 (New York, 1975), 369–411, an important volume for seventeenth-century thought. Sakade and Peterson should be read together, since Sakade pays comparatively little attention to Fang's treatment of European ideas and techniques; and Peterson, although more concerned with this aspect, is not familiar with the Chinese scientific tradition or with the development of European science.

The introduction of cosmology into China is narrated by Pasquale d'Elia in *Galileo in China, Relations Through the Roman College Between Galileo and the Jesuit Scientist-Missionaries (1610–1640)*, Rufus Suter and Matthew Sciascia, trans. (Cambridge, Mass., 1960), but the emphasis on demonstrating Jesuit accomplishments obscures a number of basic issues. The ambiguities and historic ironies of the Jesuit effort, and the Chinese response after the church's injunction of 1616 limited discussion of the earth's motion, have been examined in detail by N. Sivin in "Copernicus in China," in *Studia Copernicana*, 6 (1973), 63–122, with bibliographical essay, 113–114. A more black-and-white analysis of the same topic, with some important additional information, is Hsi Tse-tsung *et al.*, "Heliocentric Theory in China," in *Scientia sinica*, 16 (1973), 364–376. More limited in scope is Yen Tun-chieh, "Ch'ieh-li-lueh ti kung-tso tsao-ch'i tsai Chung-kuo ti ch'uan-pu"[53] ("The Early Dissemination of Galileo's Work in China"), in *K'o-hsueh-shih chi-k'an*,[34] 7 (1964), 8–27. On the notion that Western mathematics originated in China, see Ch'üan Han-sheng, "Ch'ing-mo ti Hsi-hsueh yuan ch'u Chung-kuo shuo"[54] ("On the Late Ch'ing Theory That Western Science Originated in China"), in *Ling-nan hsüeh pao*,[55] 4 (1935), 57–102; and N. Sivin, "On 'China's Opposition to Western Science During Late Ming and Early Ch'ing,'" in *Isis*, 56 (1965), 201–205. Ch'üan, overlooking the early literature, attributes the Chinese origin theory to Mei Wen-ting's grandson Mei Ku-ch'eng[56] (*ca.* 1681–1763); but his account of the theory's vogue around the turn of the twentieth century deserves attention.

N. SIVIN

WANKLYN, JAMES ALFRED (*b.* Ashton-under-Lyne, Lancashire, England, 18 February 1834; *d.* New Malden, Surrey [now part of London], England, 19 July 1906), *organic and analytical chemistry, public health.*

Wanklyn's career followed that of Edward Frankland, who until 1867 did everything possible to further it—but thereafter much to block it. Wanklyn was the son of Thomas Wanklyn and Ann Dakeyne. After education at the Moravian school in Fairfield, Lancashire, from 1843 until 1849, he was apprenticed for seven years to a Manchester doctor. During the last year of his apprenticeship he was allowed to study chemistry at Owens College, Manchester, with Frankland, whose personal assistant he became in 1856. From 1857 to 1859 Wanklyn studied at Heidelberg with Frankland's former teacher Robert Bunsen; and through Frankland's influence he became Lyon Playfair's demonstrator at Edinburgh University in 1859. Wanklyn settled in London in 1863 and until 1870 was professor of chemistry at the financially impoverished London Institution—the Royal Institution's rival in the City of London. In 1886, after various often stormy engagements as public analyst to Buckingham and its county, Peterborough, Shrewsbury, and High Wycombe, and a lectureship in chemistry and physics at St. George's Hospital, London, from 1877 to 1880, he established a private analytical laboratory and consultancy at New Malden, where he died.

Although Wanklyn, according to Liebig, gained a European reputation for his research on organic synthesis, vapor densities, and qualitative analysis, like J. W. L. Thudichum, whom he assisted in 1869, he was ignored and despised by British academic chemists. Blackballed by the Royal Society, he ostentatiously resigned from the Chemical Society in 1871 and, in 1876, from the Society of Public Analysts, of which he had been a founder in 1874. His only honor (engineered by Thudichum and Liebig) was corresponding membership in the Bavarian Academy of Sciences (1869). (Liebig diplomatically awarded Frankland the honor simultaneously.) Wanklyn's faults were excessive haste to publish and a pugnacious nature (he was involved in several lawsuits); but foremost was his tactless and indomitable controversy with Frankland over water analysis—one of the great Victorian scientific debates that had national implications for public health and that led, in Wanklyn's view, to persecution "for the sake of truth."

As a protégé of Frankland's, Wanklyn was until 1867 concerned principally with synthetic organic chemistry. In 1857 he prepared the organometallic compounds sodium ethyl and potassium ethyl, from which, with carbon dioxide, he synthesized propionic acid (1858), thus apparently confirming the structural views of Kolbe and Frankland that carboxylic acids were alkyl-conjugated oxalic acids. For instance,

$$C_4H_5Na + C_2O_4 = C_4H_5C_2O_3 + ONa.$$

sodium carbon sodium propionate (C=6)
ethyl dioxide

At Edinburgh, Wanklyn improved the Will-Varrentrapp method for estimating organic nitrogen as ammonia by adding alkaline potassium permanganate to increase the oxidative effect of soda lime. In 1866 he began collaborating with the ebullient Ernest Theophron Chapman, who, in a short but brilliant career, shared this interest in organic oxidation. At the London Institution, with Miles H. Smith, they devised a new method for detecting the organic impurity, or sewage, content of water (1867). After free ammonia had been boiled off from a water sample, it was oxidized by alkaline potassium permanganate, and the ammonia evolved (which was estimated colorimetrically with Nessler's reagent) was asserted to be a measure of the organic nitrogen content of the water. This "albuminoid ammonia process" was much simpler and faster than the extremely laborious, albeit more accurate, method promulgated by Frankland and H. E. Armstrong (1867), which analyzed evaporated water residues in vacuo.

The debates over these two methods had serious consequences for both parties, especially for Wanklyn, who supposed, with some evidence, that Frankland used his government position as an analyst of London's water supplies to promote his own, more complex technique. Wanklyn's method tended to underestimate nitrogen content and therefore to underemphasize possible sewage contamination. Hence samples of water from the same supplies often were reported as more salubrious by Wanklyn than by Frankland, who wished to use his results to ensure government action on the purification of water supplies. If Frankland's attitude was politically and socially profitable, his methods were certainly too complex analytically for ordinary public health analysts, who adopted the Wanklyn method. On the other hand, both the analysts and Frankland were prepared to accept bacteriological evidence for insalubrity; whereas Wanklyn, blinded by prejudice, saw this as another of Frankland's "plots." Nevertheless, despite his jaundiced views, Wanklyn's practical manuals on various analytical subjects proved invaluable in training and setting standards for the professional British public health analyst and medical officer of health.

BIBLIOGRAPHY

I. ORIGINAL WORKS. Nearly 150 papers are recorded in the Royal Society Catalogue of Scientific Papers, VI, 262–263; VIII, 1192–1195; XI, 746–747; XIX, 465–466; which, however, ignores a large number of interesting letters to Chemical News and Journal of Physical Science. In addition there are "On the Physical Peculiarities of Solutions of Gases in Liquids," in *Philosophical Magazine*, 6th ser., **3** (1902), 346–348, 498–500; and his contributions to H. Watts, *A Dictionary of Chemistry*, 7 vols. (London, 1863–1875; 2nd ed., 8 vols., London, 1872–1881). Wanklyn's association with John Gamgee's abortive *Milk Journal and Farmers' Gazette. A Monthly Review of the Dairy* . . . (Jan. 1871–Aug. 1872) should also be noted. Patent literature should also be consulted.

Wanklyn's books, which carried the subtitle . . . *a Practical Treatise on* . . ., were *Water Analysis*, written with E. T. Chapman (London, 1868, 1870, 1874, 1876, 1879, 1884, 1889, 1891, 1896 [10th ed.]; a 9th ed. could not be traced)—the 11th, posthumous ed. lacks Chapman's name and was edited by Wanklyn's assistant, William John Cooper (London, 1907)—and a German trans. (Charlottenburg, 1893), with the 4th (1876) to 8th (1891) eds. containing a polemical historical appendix; *A Manual of Public Health*, Ernest Hart, ed. (London, 1874), written with W. H. Mitchell and W. H. Corfield—Wanklyn claimed responsibility for 303–374 only (see *Chemical News*, **29** [1874], 9); *Milk Analysis* (London, 1874; 2nd ed., 1886); *Tea, Coffee and Cocoa* (London, 1874; reissued 1886); *Bread Analysis* (London, 1881; 2nd ed., 1886), written with W. J. Cooper; *The Gas Engineer's Chemical Manual* (London, 1886; 2nd ed., 1888); *Air Analysis* (London, 1890), written with W. J. Cooper; *Sewage Analysis* (London, 1899; 2nd ed., 1905), written with W. J. Cooper; and *Arsenic* (London, 1901).

The Royal Society has a few of Wanklyn's letters. The archives of the London Institution (1805–1912) are housed at Guildhall Library, London.

II. SECONDARY LITERATURE. The 11th ed. of *Water Analysis* (1907) contains a memoir with photograph by W. J. Cooper, an interesting selection of Wanklyn's testimonials, and an appalling bibliography. See also T. E. James, "J. A. Wanklyn," in *Dictionary of National Biography*, supp. I, vol. III, 587–588. His death was conspicuously ignored by the Chemical Society. The Liebig-Thudichum correspondence on Wanklyn's election to the Bavarian Academy of Sciences is reproduced in David L. Drabkin, *Thudichum, Chemist of the Brain* (Philadelphia, 1958), 244–247. The context and significance of Wanklyn's analytical work can be understood from C. A. Mitchell, *Fifty Years of the Society of Public Analysts* (Cambridge, 1932), esp. 1–13.

W. H. BROCK

WARBURG, EMIL GABRIEL (*b*. Altona, near Hamburg, Germany, 9 March 1846; *d*. Grunau, near Bayreuth, Germany, 28 July 1931), *physics*.

Warburg, who came from a wealthy family, grew up in Altona and attended the city's humanistic gymnasium, the Christianeum, where he was almost as interested in languages as in mathematics. His musical education was not neglected, and he became a good pianist.

In 1863, aged seventeen, Warburg began to study science at the University of Heidelberg, which—through the presence of Kirchhoff and Bunsen on its faculty—offered outstanding instruction in physics and chemistry. Warburg was so impressed by Kirchhoff's "magnificent" (*vollendet schöner*) lecture on experimental physics that he decided to change his major from chemistry to physics.

After four semesters at Heidelberg, Warburg transferred to the University of Berlin. He earned his doctorate and qualified as lecturer there, remaining until he received an offer of a professorship. During this period Gustav Magnus attracted many young physicists to his laboratory in Berlin, the only one in Germany besides Franz Neumann's at Königsberg. Warburg soon became friendly with Magnus' assistant, August Kundt; and they remained friends even after Kundt left in 1868 to take up a professorship at the Zurich Polytechnikum.

While at Berlin, Warburg wrote a number of works, most of them on oscillatory problems, including his Latin dissertation, "De systematis corporum vibrantium" (1867). In his *Habilitationsschrift*, "Über den Ausfluss des Quecksilbers aus gläsernen Capillarröhren" (1870), Warburg reported his discovery that no slipping occurs between glass and mercury. He often returned to problems of slipping. Warburg remained a *Privatdozent* for only two years; in 1872 he and Kundt were invited to the newly founded Kaiser Wilhelm University at Strasbourg. Kundt, who brought his assistant Wilhelm Roentgen with him from Würzburg, was named full professor and Warburg was made extraordinary professor.

At Strasbourg, Warburg and Kundt collaborated on two famous studies on the kinetic theory of gases. In 1875 they furnished conclusive experimental confirmation of a consequence that Maxwell had derived from the theory: that the inner friction and the heat conduction of a gas are independent of the pressure, so long as the mean paths of the molecules are negligible with respect to the dimensions of the container. They extended their investigation to very rarefied gases and deduced from the theory the existence of a measurable slipping and of a jump in temperature at the container wall. They also demonstrated the existence of measurable slipping experimentally. Their second prediction, however, was not verified until around the turn of the century, in an experiment carried out at

Warburg's suggestion at Berlin by Marian Smoluchowski and Ernst Gehrcke. In their second joint study (1876) Kundt and Warburg showed that at constant pressure and volume, the specific heats of monatomic gases possess the value 5/3 predicted by the theory.

The explanation of the theoretical relations in these two papers was the work of Warburg, as is evident from a letter mentioned by James Franck. In it Kundt asks Warburg for information on a theoretical point and writes that, since Warburg has developed all the ideas about slipping and has calculated the heat conduction, he ought to help the "thoroughly ordinary experimental physicist" (*ganz gemeinen Experimentalphysiker*) out of a theoretical difficulty. Einstein considered Warburg and Kundt's joint papers of very great significance for the kinetic gas theory. He wrote in 1922:

> This was the first time that a new phenomenon was predicted on the basis of the molecular theory of heat—a phenomenon, moreover, the representation of which on the basis of the theory of continuity of matter was virtually excluded. If the energeticists at the end of the nineteenth century had sufficiently appreciated these arguments, they would have had great difficulty in calling into question the profound validity of the molecular theory.

The collaboration with Kundt ended in 1876, when Warburg obtained a professorship at the University of Freiburg im Breisgau, where he was the sole physicist on the faculty until 1895. At Freiburg he continued his investigation of the kinetic gas theory. It followed from the theory that the friction coefficient is independent of the pressure. He tested this prediction with carbonic acid at high densities and found that the basic notions of the theory were valid. Never losing his interest in this topic, he encouraged his students to work on it and published two comprehensive accounts of it himself. The first, *Über die kinetische Theorie der Gase* (1901), shows that Warburg had mastered the art of good scientific popularization. The second, *Über Wärmeleitung und andere ausgleichende Vorgänge* (1924), is essentially a summary of half a century of research on the subject.

Warburg also undertook research at Freiburg on many other topics. His investigation of elastic aftereffects led him in 1881 to one of his most beautiful results: the experimental discovery and theoretical interpretation of hysteresis in the cyclical magnetization of ferromagnetic materials. Warburg also devoted years of study to electrical conduction in solids, liquids, and gases; and his efforts yielded many discoveries. For example, he ascertained that conductivity in quartz is 100 times greater in the direction of the axis than in the direction perpendicular to it. Another interesting discovery was the electrolytic migration of magnesium and lithium ions through glass. The drifting of electrolytic impurities toward the electrodes acquired significance for electric purification. Warburg's discovery of the cathode fall enabled him to gain important insights in his study of gas discharges. He recognized the significance of the cathode fall for breakdown voltage and measured this characteristic quantity for many gases.

Warburg's works on gas discharges quickly attracted the attention of other scientists. In his unpublished autobiography Philipp Lenard recounts that Heinrich Hertz considered Warburg a leading expert on electric discharges in rarefied gases. Hertz's opinion, which reached the influential Friedrich Theodor Althoff in the Ministry of Education, through Lenard, undoubtedly contributed to Warburg's being invited to Berlin.

Two events remain to be mentioned from Warburg's period at Freiburg: the dedication of the new physics institute in 1891 and the publication of his *Lehrbuch der Experimentalphysik* (1893). This textbook, precise and tersely written, was not easy to read but nevertheless had great success. At age eighty-three, Warburg prepared the twenty-first and twenty-second editions.

In 1895 Warburg succeeded Kundt as professor of experimental physics at the University of Berlin. He thereby obtained the "most eminent chair of physics in Germany" and became a very close associate of Max Planck. He continued to work on his research projects, enlisting the aid of many of his students. While pursuing studies on gas discharges, from 1897 he undertook others on spark discharges and point discharges and on the resulting ozone formation. According to James Franck, Warburg's research on point discharges constituted the basis for the experiments that J. Franck and Gustav Ludwig Hertz conducted on electron collisions.

Warburg's ten years as director of the Berlin physics institute were the most brilliant of his teaching career. The many students who came there constituted what was called the "Warburg school" of experimental physics. The intensive program of research that he and his students conducted is reflected in the 220 publications that originated in the institute during his tenure. Moreover, Franck calculated that around 1930 approximately one-fifth of the professors of experimental physics at German universities and colleges had studied under Warburg. Among the latter was his son Otto Heinrich (1883–1970), who became director of the Kaiser Wilhelm (now Max Planck) Institute for Cell Physiology in 1930 and received the Nobel Prize for physiology or medicine in 1931.

Warburg's teaching activities included a weekly colloquium for professors and students held in the institute's library. Friedrich Kohlrausch, then president of the Physikalisch-Technische Reichsanstalt, was an active participant. Besides his teaching, Warburg rendered important service to physics through his efforts within professional scientific organizations. He was elected chairman of the Berlin Physical Society in 1897; and in 1899 he led this body into the German Physical Society, heading the latter as well until 1905.

In 1905 Warburg left the University of Berlin to succeed Kohlrausch at the Reichsanstalt. Under Warburg's direction the organization of the institute was streamlined and duplication of effort was eliminated. At the same time, however, several new institutes were created within it, including the radioactivity laboratory (1912), in which Hans Geiger developed his *Spitzenzähler* (or point counter). In addition funds were allotted for visiting researchers, who included Einstein and de Haas when they discovered the gyromagnetic effect named for them (1914–1915).

The pace of Warburg's research did not diminish with his move to the Reichsanstalt. He pursued his investigation of point discharges and concurrently (from 1906) undertook photochemical studies that occupied him until shortly before his death. He was one of the founders of quantitative photochemistry and confirmed the fundamental law of the quantum nature of light absorption formulated by Einstein. Further, assisted by several co-workers, Warburg devoted himself to a task especially suited to the facilities at the Reichsanstalt: making precise measurements designed to test Planck's radiation law.

Following his retirement in 1922, Warburg continued his photochemical studies as an independent researcher. Although more than eighty at this time, he wrote three articles: on silent discharge in gases, spark discharge, and photochemistry. He died a few months after his eighty-fifth birthday.

Until well into old age, Warburg followed advances in physics with great attention and impartiality. In 1913, shortly after the discovery of the Stark effect, he was the first to examine its relationship to the equally new Bohr theory. The ex-

periment he devised, although premature and therefore a failure, reflects his openness and quickness of mind.

Of Warburg's approximately 150 publications, only one is of a polemical nature; even then he did not begin the dispute, and it did not concern any scientific matter. All his other writings display a sober objectivity and critical detachment from his own results.

Unlike his contemporary Wilhelm Roentgen, for example, Warburg never achieved the brilliant success that makes a scientist known far beyond the circle of his colleagues. Nevertheless, he produced a wealth of important results that are now part of basic physical knowledge; and he was able to teach many students the procedures of intensive scientific research.

BIBLIOGRAPHY

I. ORIGINAL WORKS. Warburg's writings include *Lehrbuch der Experimentalphysik* (Tübingen, 1893; 22nd ed., 1929); *Über die kinetische Theorie der Gase* (Berlin, 1901); *Helmholtz als Physiker* (Karlsruhe, 1922); "Funkenentladung," in H. Geiger and K. Scheel, eds., *Handbuch der Physik*, XIV (Berlin, 1927), 354–390; "Über die stille Entladung bei Gasen," *ibid.*, 149–170; and "Photochemie," *ibid.*, XVIII (1928), 619–657. A list of his works can be compiled from Poggendorff, III, 1415–1416; IV, 1598; V, 1334–1335; and VI, 2806.

II. SECONDARY LITERATURE. See Albert Einstein, "Emil Warburg als Forscher," in *Naturwissenschaften*, **10** (1922), 823–828, with a list of publications to 1921; James Franck, "Emil Warburg zum Gedächtnis," *ibid.*, **19** (1931), 993–997; Philipp Lenard, "Autobiographie" (unpublished); H. Moser, ed., *Forschung und Prüfung. 75 Jahre Physikalisch-technische Bundesanstalt/Reichsanstalt* (Brunswick, 1962), esp. 8–18; C. Müller, "Emil Warburg 80 Jahre," in *Elektrotechnische Zeitschrift*, **47** (1926), 317; J. Stark, ed., *Forschung und Prüfung. 50 Jahre Physikalisch-technische Bundesanstalt/Reichsanstalt* (Leipzig, 1937), esp. 16–19, 60–63; and Eduard Zentgraf, ed., *Aus der Geschichte der Naturwissenschaften an der Universität Freiburg im Breisgau* (Freiburg im Breisgau, 1957), esp. 18–20.

HANS RAMSER

WARBURG, OTTO HEINRICH (*b.* Freiburg im Breisgau, Baden, Germany, 8 October 1883; *d.* Berlin-Dahlem, Germany, 1 August 1970), *biochemistry.*

Warburg was the son of Emil Gabriel Warburg and Elizabeth Gaertner. He came with his parents in 1896 to Berlin, where his father had been called to the chair of physics at the University of Berlin. The elder Warburg later became president of the Physikalische Reichsanstalt. The family originated in the beautiful little town of Warburg, about thirty miles west of Göttingen. They first appear in the mid-sixteenth century.

Otto Warburg's mother stemmed from a family of public officials and soldiers; her brother, a general in the army, was killed in World War I. Warburg himself served as an officer in the Prussian Horse Guards on the Russian front and was wounded in action. In the early years of this fighting he carried not only a pistol but a medieval lance. Before the war ended, Albert Einstein wrote a remarkable letter to Warburg persuading him to return to the Kaiser Wilhelm Institute for Biology in Dahlem; he had been made a member in 1913 at the instigation of Emil Fischer, his first mentor ten years earlier.

In Berlin, Warburg grew up in two large official residences, both designed by the wife of Helmholtz. Most of the leading scientists of Germany were frequent guests of his parents during this final period of imperial splendor under Wilhelm II.

Later, at the university, Warburg learned chemistry from Fischer, with whom he worked for three years to obtain his doctorate; medicine in the clinic of Ludolf von Krehl at Heidelberg, to whom he was assistant for three years; thermodynamics from Walther Nernst in Berlin, with whom he worked on oxidation-reduction potentials in living systems; and physics and photochemistry from his father, with whom he worked on the quantum requirement of photosynthesis in 1920 in the Physikalische Reichsanstalt.

Warburg's first scientific work (1903–1906), with Fischer, involved splitting of racemic leucine ethyl ester by pancreatin and resolution of the optically active components. Fischer was a severe master, who instructed Warburg, after he had recrystallized the parent compound first three times, and then five more, "Now go ahead twenty-five times more."

This seemingly harsh training stood Warburg in good stead all his life, during which he invariably distinguished between experimentation made "*Für die Wahrheit*" (for the truth) and that made "*Für das Volk*" (to convince others). Having once satisfied himself as to the truth of a discovery, he always proceeded to repeat his experiments twenty to a hundred times before publishing, which explains why he, like Fischer, produced such a mass of virtually error-free and reproducible results.

Warburg learned early that "convincing others" involved much more than steamroller repetition of experimentation. Because of the great number and magnitude of his discoveries, which rank him as the most accomplished biochemist of all time, no biochemist—or scientist—has met with so much controversy, resistance, and delayed acceptance of his work, often lasting (in his own words) ten, twenty, or even fifty years. The reason for this is given by one of his favorite quotations from Hans Fischer (1881–1945), "All science is all too human"; and from Max Planck in his ninetieth year, "A new scientific truth is often accepted, not as a result of opponents becoming convinced and declaring themselves won over, but rather by the opponents dying off, and the oncoming generation of scientists becoming familiar with the new truth right from the start." He was also fond of Darwin's statement in the "Conclusion" of the *Origin of Species*, ". . . I by no means expect to convince experienced naturalists whose minds are stocked with a multitude of facts all viewed, during a long course of years, from a point of view directly opposite to mine . . . but I look with confidence to the future,—to young and rising naturalists, who will be able to view both sides of the question with impartiality." The significance of these quotations increases, of course, with the magnitude of the discovery, since then there is greater upset of previous conceptions.

Warburg endeavored to advance science mainly through his own experimental work, carried out both personally and by technical assistants whom he trained. He believed that many important discoveries were to be made in the laboratory by very simple but heretofore untried variations in experimental conditions. Thus, he discovered the fermentation of tumor cells when he increased by twentyfold the concentration of bicarbonate in the medium. He discovered iron oxygenase (*Atmungsferment*) by raising the pressure of carbon monoxide from 5 to 95 percent or more. He discovered acyl phosphate when in the oxidation-reduction reaction of fermentation the phosphate concentration was increased twenty times; and the energy cycle and one-quantum reaction of photosynthesis was discovered when the light-dark time intervals measured in manometry were shortened from five minutes to one minute.

Among the forty rooms in Warburg's institute, there was no office, no conference room, and no writing room apart from the general library. He never gave lecture courses to students, never served on committees, and never did administrative work. He selected his staff on the basis of technical ability and talent. He preferred to be regarded as an artisan and, as he frequently asserted, a technician. Nevertheless, he was an artist in everything he did, a commanding speaker in English as well as in German, and a uniquely clear writer in both English and German. Warburg's philosophical outlook is summarized by a statement he made in 1964, ". . . a scientist must have the courage to attack the great unsolved problems of his time, and solutions usually have to be forced by carrying out innumerable experiments without much critical hesitation."

Warburg was first and foremost a pioneer in biochemical methodology and in the creation of new tools of investigation—for example, spectrophotometric methods of identification and analysis of cell constituents and enzymes, manometric methods for the study of cell metabolism, numerous microanalytical methods, and methods for the isolation of cell constituents and crystallization of enzymes.

Following is a chronological listing of his major discoveries and fields of interest during more than sixty-five years of research; each item generally involved five to ten publications. Splitting of racemic leucine ethyl ester by pancreatin (first publication 1904); splitting of racemic leucine into its optically active components by means of formyl derivatives (1905), with Emil Fischer; respiration of sea urchin eggs, red blood cells, and grana (1910–1914); development of biochemical manometry (1918–1920–1968); iron catalyses on surfaces, narcotic action—displacement of substrates from surfaces, cyanide action—chemical reactions with iron (1921–1924); quantum requirements of photosynthesis (1920–1924); tissue slice technique (1923); metabolism of tumors (1923–1925); iron, the oxygen-transferring constituent of the respiration enzyme, "iron oxygenase" or *Atmungsferment* (1924); inhibition of cell respiration by carbon monoxide (1925–1926); action spectrum of iron oxygenase (1927–1932); discovery of the yellow enzymes (1932–1933); first crystallization of a flavin, "luminoflavin" (1932); discovery of nicotinamide as the active group of hydrogen-transferring enzymes (1935); nature of coenzyme action and varying degrees of binding with enzymes (1935); development of the optical methods based upon the ultraviolet absorption band of dihydronicotinamide (1935–1937); mechanism of alcohol formation in nature, dihydronicotinamide + acetaldehyde = nicotinamide + ethyl alcohol (1936); stepwise degradation of phosphorylated hexoses to trioses (1936–1937); discovery of the copper of phenol oxidases and its action through valence change (1937); isolation and crystallization of flavin adenine dinucleotide (1938); crystallization of the oxidizing fermentation enzyme and mechanism of the oxidation reaction of fermentation, glyceric aldehyde diphosphate + nicotinamide = phosphoglyceric-*acyl phosphate* + dihydro-nicotinamide (1938); crystallization of enolase and chemistry of fluoride inhibition of fermentation (1941); crystallization of muscle zymohexase (1942); *in vitro* Pasteur reaction with hexosediphosphate and yeast zymohexase (1942); crystallization of the reducing fermentation enzyme from tumors and comparison with the homologous crystallized fermentation enzyme from muscle (1943); fermentation enzymes in the blood of tumor-bearing animals (1943); quinone and green grana (1944); heavy metals as active groups of enzymes and hydrogen-transferring enzymes (1946–1947); manometric actinometer (1948); maximum efficiency of photosynthesis (1949); one-quantum reaction and energy cycle in photosynthesis (1950), with Dean Burk; crystallization of the hemin of iron oxygenase (1951); zymohexase and ascites tumor cells (1952); chemical constitution of the hemin of iron oxygenase (1953); oxidation reaction and enzymes in fermentation (1954–1957); measurement of light absorption in *Chlorella* with the Ulbricht integrating sphere (1954); catalytic action of blue-green light in photosynthesis (1954–1956); oxygen capacity of *Chlorella* (1954–1956); carbon dioxide capacity of *Chlorella* (1956); photochemical water decomposition by living *Chlorella* (1955); origin of cancer cells (1956); functional carbon dioxide in *Chlorella* (1956); role of glutamic acid in photosynthesis (1957–1964); D-lactic acid and glycolic acid in *Chlorella* (1957–1964); Hill reactions in photosynthesis (1958–1968); photosynthesis in green leaves (1958–1963); manometric X-ray actinometer and actions of X rays on various cells (1958–1966); phosphorylation in light (1962); effects of low oxygen pressure on cell respiration, growth, and transformation (1960–1965); healing of mouse ascites cancer with glyceric aldehyde (1963); production of cancer metabolism in normal cells grown in tissue culture (1957–1968); red respiratory enzyme in *Chlorella* (1962–1965); photolyte of photosynthesis, a carbon dioxide-chlorophyll complex (1959–1969); facultative anaerobiosis of cancer cells (1962–1965); prime cause and prevention of cancer (1966–1969); chlorophyll catalysis and Einstein's photochemical law in photosynthesis (1966–1969); action of riboflavin and luminoflavin on growing cancer cells

(1967–1968); role of Vitamin B_1 (thiamine) on changes of normal to cancer cells and vice versa (1970); changes in chlorophyll spectrum in living *Chlorella* upon splitting and resynthesis of the carbon dioxide photolyte by light (1970).

Had Warburg ceased scientific work after the first four decades of his career, his name would now probably be forgotten. When he left his regiment near the end of World War I, his fellow officers said that he would now "return to feeding sea urchins." Few great scientists have ever matured so late.

Which of Warburg's discoveries involved the greatest originality of conception, execution, and proof? According to Warburg himself it was his discovery in 1924 of *Atmungsferment* (iron oxygenase), for which, after several more years of study and controversy, he was awarded the 1931 Nobel Prize in physiology or medicine.

Warburg became convinced in 1926 that iron oxygenase was a hemin compound, as a result of inhibition studies with carbon monoxide, which Claude Bernard had long before shown to be an inhibitor of hemoglobin. In 1926, while experimenting with yeast cells suspended in phosphate solutions containing glucose, Warburg found that carbon monoxide also inhibits cell respiration. By measuring the inhibitions obtained at different oxygen pressures, he found that the action is dependent upon the ratio of CO/O_2 pressures, which indicates that ferrous iron in the enzyme is the point of attack, in contrast to cyanide inhibition of ferric iron long known to occur. But, as Warburg said in his Nobel lecture, "It would never have been possible to reach any certainty of enzyme constitution here, were it not that the carbon monoxide compounds of iron possess in instances the remarkable property of being dissociated by light, as discovered by Mond and Langer in 1891, and, as shown by J. S. Haldane a few years later to alter the equilibrium between hemoglobin, CO, and O_2 in favor of O_2." Thus, by alternating periods of light and darkness for cells respiring in mixtures of CO and O_2, Warburg was able to cause respiration to appear and disappear; in light, carbon monoxide is split from the iron, leaving it free for oxygen activation. In a quantitative examination, Warburg found that Einstein's law of photochemical equivalency was followed, that is, the number of photochemically split Fe—CO groups is equal to the number of light quanta absorbed, independently of wavelength of light. By irradiating with monochromatic light of various wavelengths but of the same intensity, he was able to determine the absorption

spectrum of the iron oxygenase–carbon monoxide complex, as judged by the magnitude of respiration increase. This absorption spectrum showed a remarkable similarity to that of CO—hemoglobin, but with some displacement toward the longer wavelengths, yet clearly identifying the iron oxygenase as a hemin compound, in which the iron is bound to nitrogen by two electron pairs and the porphyrins are cyclic compounds formed, as shown by Hans Fischer, by the linkage of four pyrrole rings through methylene bridges. Finally, the absolute absorption spectrum of the enzyme was determined from time-rate measurements of the light action on respiration, in relation to absolute light absorptions at different wavelengths. It became clear that iron oxygenase has an exceptionally strong light absorption, corresponding to an exceedingly minute concentration in the cell, where it is indeed found in the particulate grana (mitochondria) as adumbrated by some of Warburg's studies prior to World War I. In later decades Warburg succeeded in isolating and analyzing further structural details of the iron oxygenase, the prime cellular respiratory enzyme, and showed that it contains somewhat less nitrogen and iron but more carbon than does blood hemin, and also an unusual hydrocarbon chain whose structure was elucidated by Lynen in 1963. Warburg's subsequent work indicated that the heme pigments of both blood and plants (as in chlorophyll) arise in evolution from the iron oxygenase heme.

Warburg worked more or less continuously for the last fifty years of his life on various aspects of photosynthesis. His first major contribution was to demonstrate that photosynthesis can be made to take place, under appropriate conditions, with almost perfect thermodynamic efficiency. In the equation $CO_2 + H_2O$ = sugar equivalent + O_2, some 110,000 calories of energy are thermodynamically required per mole of CO_2 reduced and O_2 produced; and he found that this could be supplied by no more than four mole quanta of red light of 43,000 calories per mole, corresponding to an efficiency of 112,000 $(4 \times 43,000)$, or 65 percent. In later years, under even better conditions, three mole quanta were found to suffice, corresponding to an efficiency approaching 100 percent conversion of light energy into chemical energy. Again, this quantum requirement was found to be independent of wavelength of light in the visible spectrum, just as was the action of light on the iron oxygenase–carbon monoxide splitting already described.

In 1950 it was found that the mechanism of light

energy conversion proceeded in steps of one quantum each, as required or predicted by the Einstein law of photochemical equivalence. Indeed, when Warburg told Einstein in 1923 about his "four quantum" requirement measurements, Einstein said, "When you get down to one quantum, come back and tell me about it." Between 1923 and 24 October 1950, Warburg worked on the "quantum riddle"; how can four quanta (or three) seemingly act together simultaneously to reduce one molecule each of CO_2 and water to one molecule of O_2 (and sugar equivalent)? On the latter date it was found that $XO_2 + 1$ quantum $= X + 1 O_2$. XO_2 was the substance from which the O_2 developed, and at the same time approximately 1 molecule of CO_2 disappeared. This occurred over a period of about a minute of illumination and was accompanied and followed by a dark reaction in which the substance XO_2 was restored at the expense of two-thirds of the O_2 produced in the light reaction. This dark reaction showed up experimentally as a greatly increased rate of respiration (O_2 consumption and CO_2 production), yielding after three such cycles a net and stable requirement of three quanta for the overall photosynthetic reaction as first written above, and persisting for long periods of time.

It is interesting that the above solution of the quantum riddle was arrived at purely experimentally; no one—not even Einstein—had hypothesized the finally observed quantum mechanism.

Although the experimental solution of the quantum riddle has never been effectively challenged, with respect to the observed one-quantum requirement, nevertheless, the three- or four-quanta requirement for overall photosynthesis, especially during the 1940's and 1950's, was objected to by a host of (but not all) workers who would not or could not adequately reproduce Warburg's experimental conditions. In the late 1950's Warburg proposed to the National Academy of Sciences that a team of selected workers be sent to Dahlem to "see for themselves," but this proposal was not accepted.

Warburg's third great area of endeavor, also on the cancer research, also covered the last fifty years of his life. There were far more scientists and laymen interested in this problem than in cellular respiration and photosynthesis and opposition to many of his findings was much more intense. Beginning in 1922, Warburg discovered the remarkably high production of lactic acid from glucose by cancer cells, both *in vitro* and *in vivo*, as well as aerobically and anaerobically, and, of course, varying in degree over a wide spectrum from cancer to cancer. In contrast, no growing normal tissue in the animal body produced lactic acid from glucose under aerobic conditions. A few non-growing tissues might, but, as in the case of muscle, usually from glycogen and at rates ordinarily far below that of well-developed malignant tumors.

Accompanying the greatly increased glucose fermentation (glycolysis) by cancer cells was an injured respiration, manifested in a variety of ways—decreased rate, uncoupled rate, low succinate oxidative response, loss of Pasteur effect, etc.

These two major findings led Warburg in the 1950's to the following view of cancer causation:

> Cancer cells originate from normal body cells in two phases. The first phase is the irreversible injuring of respiration . . . followed . . . by a long struggle for existence by the injured cells to maintain their structure, in which part of the cells perish from lack of energy, while another part succeed in replacing the irretrievably lost respiration energy by fermentation energy. Because of the morphological inferiority of fermentation energy, the highly differentiated body cells are converted into undifferentiated cells that grow wildly—the cancer cells. . . . Oxygen gas, the donor of energy in plants and animals, is dethroned in the cancer cells and replaced by an energy yielding reaction of the lowest living forms, namely, a fermentation of glucose.[1]

According to Warburg this is the prime cause of cancer, prime cause being defined as "one that is found in every case of the disease." Thus,

> . . . the prime cause of the plague is the plague bacillus, but secondary causes of the plague are filth, rats, and the fleas that transfer the plague bacillus from rats to man. . . . Cancer, above all other diseases, has countless secondary causes. But, even for cancer, there is only one prime cause. . . . There is no disease whose prime cause is better known.[2]

In his famous 1966 Lindau lecture Warburg recommended, for both prevention and treatment of cancer, dietary additions of large amounts of the active groups of the various respiratory enzymes, these active groups constituting first and foremost iron and certain of the B vitamins. This provoked overnight the most widespread controversy, not only throughout Germany but also the Western world.

No account of Warburg's life and work should fail to mention his close association with Jacob Heiss of Kirn, in southern Germany. Heiss was a person of remarkable character, ability, and shrewdness, who from 1918 until Warburg's death entered into virtually all of his activities. He served as administrator, monitor of all scientific papers, financial adviser, and consultant on all affairs, however small, on a daily—even hourly—basis. They were a unique combination, yet the personal character of each was entirely different. Warburg never married; Heiss was his sole heir.

Throughout his life, Warburg was extremely fond of walking, sailing, dogs, and horses, and kept himself in remarkable physical trim. He rode every morning before going to work, for the better part of an hour, during which time he did much of his sustained thinking. He frequently stated that he was a "slow thinker." To many a question put to him he would reply, *"Man muss es überlegen"* (I must think it over), and the answer would be delivered the next day, after a ride. In his eighty-second year, the day after he received his honorary degree from Oxford, he was standing at the Park Lane end of Rotten Row in London, when he saw a riderless horse charging down the Row. He immediately stepped over the low guardrail in front of the oncoming horse and, with both arms upraised, caught it—to cheers of bystanders. One of them, noting Warburg's bowler hat, remarked, "Those boys from the City have something on us West Enders."

Warburg for decades vacationed in England, for whose inhabitants he had an unbounded admiration, and he was an inveterate reader of the London *Times* and *Manchester Guardian*, as well as of innumerable English authors of all sorts, including Churchill, the Mitfords, and various "aristocrats," of whom he considered himself, with amused emphasis, an example par excellence. The American he most admired was Charles Huggins, winner of the 1966 Nobel Prize for physiology or medicine. The person whom he most enjoyed telling playful stories about was himself.

NOTES

1. "On the Origin of Cancer Cells," in *Science*, **123** (1956), 312; *The Prime Cause and Prevention of Cancer*, D. Burk, ed., 2nd ed., rev. (Würzburg, 1969), 6.
2. *Ibid.*, 6; 16.

BIBLIOGRAPHY

Most, although not quite all, of Warburg's original experimental papers to 1961 are, fortunately and conveniently, found in the following books published under his sole authorship: *Ueber den Stoffwechsel der Tumoren* (Berlin, 1926), English trans. by Frank Dickens, *The Metabolism of Tumours* (London–New York, 1930); *Ueber die katalytischen Wirkungen der lebendigen Substanz* (Berlin, 1928); *Schwermetalle als Wirkungsgruppen von Fermenten* (Berlin, 1946); *Wasserstoffuebertragende Fermente* (Berlin, 1948); *Weiterentwicklung der zellphysiologischen Methoden* (Stuttgart–New York, 1962); and *The Prime Cause and Prevention of Cancer*, Dean Burk, trans. (Würzburg, 1969). These collected works contain 200 articles in over 2,000 pages—less than has been written about them by way of reviews, recapitulations, objections, and capitulations.

DEAN BURK

WARD, SETH (*b.* Aspenden, Hertfordshire, England, 5 April 1617; *d.* Knightsbridge [now in London], England, 6 January 1689), *astronomy.*

Ward was the second son of John Ward, attorney, and Martha Dalton Ward. He entered Cambridge in 1632, graduated B.A. in 1637, received the M.A. in 1640, and was elected a fellow of his college. He subsequently became mathematical lecturer (1643) before the ascendancy of the Puritans induced him to leave the university. Only in 1649 did he master his scruples sufficiently to subscribe to the Solemn League and Covenant and return to academic life, this time at Oxford, as a replacement for the ousted Savilian professor of astronomy, John Greaves. Although the Puritan "visitors" succeeded, by such means, in securing the allegiance of the universities, there was a strident group of pamphleteers who insisted that reform should go much deeper, to the very heart of the curriculum—the Aristotelian corpus. In 1654 (the year he received the D.D.), Ward published with John Wilkins *Vindiciae academiarum* in defense of the extent to which the universities had responded to the new learning.

In 1660, on the second occasion of being disappointed in his bid for administrative advancement, Ward abandoned his academic career. In two years he accumulated several church livings and rose to bishop, in which post he proved a zealous administrator of church law and property.

Ward is remembered in the history of astronomy for his formulation of an alternative to Kepler's law of areas. Kepler's law of elliptical motion began to find general acceptance with the publication of Boulliau's *Astronomia philolaica* in 1645. In place of the area law, however, Boulliau postulated a complicated motion described by reference to a cone. Ward, in 1653, showed that Boulliau's scheme amounted to assuming uniform angular motion with respect to the empty focus of the ellipse. An idea with a distinguished pedigree (essentially Ptolemy's bisection of the eccentricity), it presented a very attractive alternative to the intractable Kepler equation. During the following generation, it and various modifications of it were widely used in planetary computations.

BIBLIOGRAPHY

I. ORIGINAL WORKS. Ward's more important writings are *In Ismaelis Bullialdi astronomiae philolaicae fundamenta inquisitio brevis* (Oxford, 1653); *Vindiciae academiarum* (Oxford, 1654), written with John Wilkins; *Astronomia geometrica; ubi methodus proponitur qua primariorum planetarum astronomia sive elliptica sive circularis possit geometrice absolvi* (Oxford, 1656); and *In Thomae Hobbii philosophiam exercitatio epistolica* (Oxford, 1656). He also published a few lesser scientific works and many theological writings.

II. SECONDARY LITERATURE. See Phyllis Allen, "Scientific Studies in the English Universities of the Seventeenth Century," in *Journal of the History of Ideas,* **10** (1949), 219–253; J. L. Russell, "Kepler's Laws of Planetary Motion: 1609–1666," in *British Journal for the History of Science,* **2**, no. 5 (1964), 1–24; and Curtis A. Wilson, "From Kepler's Laws, So-Called, to Universal Gravitation: Empirical Factors," in *Archive for History of Exact Sciences,* **6**, no. 2 (Apr. 1970), 89–170.

VICTOR E. THOREN

WARGENTIN, PEHR WILHELM (*b.* Sunne, Jämtland, Sweden, 11 September 1717; *d.* Stockholm, Sweden, 13 December 1783), *astronomy, demography.*

Wargentin seems to have been destined for science from his early years. His father, Wilhelm Wargentin, had devoted much time to scientific studies and had tried to obtain an appointment as professor of physics at the University of Dorpat before he had been called to a parish in northern Sweden. He taught his eager son the wonders of the skies at an early age, and in 1729 they observed a lunar eclipse. Wargentin continued his astronomical observations as a student at the Gymnasium in Härnösand; and after he entered the University of Uppsala in 1735, astronomy soon became his main interest. Initially, the competent observer Olof Hiorter was his teacher; but in 1737 Anders Celsius, professor of astronomy, returned from his travels and took Wargentin's scientific development in hand. At Celsius' suggestion Wargentin began to concentrate upon calculating the orbits of the moons of Jupiter; in 1741 he

completed a work on this subject (*De satellitibus Jovis*), but the important tables were not published until 1746. Wargentin obtained his master's degree in 1743 and remained at the University of Uppsala, where he was appointed assistant professor on the Philosophical Faculty in 1748. In the fall of 1749 he was offered the position of secretary of the Royal Swedish Academy of Sciences and moved to Stockholm to assume his new duties.

Wargentin was active in three fields: astronomy, where Jupiter's moons remained his specialty; population statistics, of which he is considered one of the modern founders; and the Academy of Sciences, which he served until his death.

Founded in 1739, the Academy of Sciences in Stockholm had already acquired stability and respect; but it remained for Wargentin, as its secretary and moving force for a generation, to extend its activities and to bring it into close contact with the international scientific community. Through it Wargentin became a central figure in the scientific flowering of Sweden in the mid-eighteenth century. He edited the Academy's *Transactions,* published the Swedish almanac, for which the Academy had the license, and actively supported Sweden's introduction of the Gregorian calendar in 1753. As the Academy's astronomer he supervised the construction of its astronomical observatory in Stockholm, which was completed in 1753; the necessary instruments were obtained in London. During the international astronomical years, especially at the times of the transits of Venus of 1761 and 1769, Wargentin organized the Swedish effort and saw to it that the results obtained were immediately communicated to foreign astronomers for publication. His immense correspondence with foreign academicians and scholars—over 4,000 letters that he received have been preserved—constitutes an invaluable source for Swedish as well as European history of science.

Wargentin kept a careful journal of his observations from 1749 on. His main interest remained the moons of Jupiter, which he had begun studying in his youth. The first ephemerides of the satellites of Jupiter had been published by Gian Domenico Cassini in 1666. Although better ones were produced by James Pound and James Bradley in England and by Jacques Cassini in France in 1740, the irregularities of the satellites' movements caused great problems and more exact calculations of the orbits were needed. Here Wargentin made a basic contribution. Working in a purely empirical and statistical manner, he collected a great number of trustworthy observations that he interpreted with intuitive certainty. They were first published as "Tabulae pro calculandis eclipsibus satellitum Jovis" in *Acta Regiae societatis scientiarum Upsaliensis pro 1741* (1746).

Wargentin's values were far more accurate than those of his predecessors, but he nevertheless continued his observations and calculations, which he communicated to Lalande, with whom he corresponded regularly. Wargentin's revised tables of the satellites of Jupiter were published by Lalande in his enlarged edition of *Tables astronomiques de M. Halley* (1759); through new equations he had obtained improved calculations for the movements of the third and the fourth satellites. Wargentin continued to publish new contributions to this subject in Swedish and foreign scientific journals and until his last years he was engaged in improving the theory for the third satellite.

Among his contemporaries Wargentin was considered the outstanding expert in his field, and his tables of Jupiter's moons remained authoritative until the improvement of mathematical analysis made possible exact theoretical solutions of the problems. And his empiricism, even when compared with modern theory, must be considered surprisingly reliable.

Outward circumstances led Wargentin to his other lifelong scientific occupation, population statistics. In 1736 it was decreed that the pastors of Sweden should collect yearly reports on births and deaths, and within the Academy of Sciences the idea grew that the collected material should be submitted to statistical analysis. In 1754 the authorities ordered Wargentin to assume this task. The Royal Table Commission, established two years later with Wargentin as the guiding power, was officially assigned to work with the deposited population tables.

In 1754 Wargentin began publishing his results in a series of demographic articles in the *Transactions* of the Academy. He used both the older and the contemporary pioneers in the field of population statistics (Graunt, Petty, J. P. Süssmilch, Deparcieux); but in his later works he surpassed them and showed a sure, methodical touch. In his most important article, "Mortaliteten i Sverige" (1766), he calculated the mortality rate for different groups in the community: men, women, all inhabitants of Stockholm. He also dealt with birth and mortality rates in different months, the population increase of Stockholm, and the total population of the country. Wargentin may well have been the first to compile mortality tables based on exact figures. His results were of practical importance,

especially for life insurance. Richard Price contacted Wargentin and then published the latter's mortality tables in his *Observations on Reversionary Payments* (1783).

Wargentin received many scientific distinctions and in 1783, shortly before his death, became one of the eight foreign members of the Paris Academy. Although not noted for brilliance or innovation, he had a clear and penetrating mind, even when dealing with mundane matters, almost unlimited energy, and a strong moral integrity.

BIBLIOGRAPHY

I. ORIGINAL WORKS. Wargentin's extensive writings in astronomy, population statistics, and other fields are scattered in many short articles, most of them published in *Kungliga Svenska vetenskapsakademiens handlingar*. His works on the moons of Jupiter are listed by Nordenmark (see below), 224–231. The Royal Swedish Academy of Sciences in Stockholm has his papers, including letters that he received (catalogued by Nordenmark, 425–449).

II. SECONDARY LITERATURE. The basic biography is N. V. E. Nordenmark, *Pehr Wilhelm Wargentin* (Uppsala, 1939), in Swedish. See also Sten Lindroth, in *Kungliga Svenska vetenskapsakademiens historia 1739–1818*, I; pt. 1 (Stockholm, 1967), 48–59, 411–416; and his "Pehr Wilhelm Wargentin," in *Swedish Men of Science* (Stockholm, 1952), 105–112. On the moons of Jupiter, see Bertil Lindblad, "P. W. Wargentins arbeten över Jupitermånarna och modern teori," in *Populär astronomisk tidskrift*, **15** (1934), 9–19. Wargentin as a population statistician has been treated (apart from Nordenmark) by A. R. Cederberg, *Pehr Wargentin als Statistiker* (Helsinki, 1919); and O. Grönlund, *Pehr Wargentin och den svenska befolkningsstatistiken under 1700-talet* (Stockholm, 1946).

STEN LINDROTH

WARING, EDWARD (*b*. Shrewsbury, England, *ca.* 1736; *d*. Plealey, near Shrewsbury, 15 August 1798), *mathematics*.

Little is known of Waring's early life. In 1753 he was admitted to Magdalene College, Cambridge, as a sizar, and his mathematical talent immediately attracted attention. He graduated B.A. as senior wrangler in 1757, was elected a fellow of the college, and in 1760 received the M.A. and resigned his fellowship to accept appointment, on the death of John Colson, as sixth Lucasian professor of mathematics. Although his Lucasian professorship was opposed in some quarters because of his age— he was still in his twenties—Waring soon effectively silenced his critics by publishing, in 1762, his *Miscellanea analytica de aequationibus algebraicis et curvarum proprietatibus*, which gave indisputable proof of his ability and at once established him as a mathematician of the first rank. He was elected a fellow of the Royal Society the following year.

The *Miscellanea* was described by Charles Hutton (in *Mathematical and Philosophical Dictionary*, II [1795], 584) as "one of the most abstruse books written in the abstrusest parts of Algebra." It deals largely with the theory of numbers (some of its chapters are "De fluxionibus fluentium inveniendis," "De methodo incrementorum," and "De infinitis seriebus"), a branch of mathematics for which Waring had a special gift. It contains, without proof, the theorem that every integer is the sum of four squares, nine cubes, nineteen biquadrates, "and so on." In 1770 Waring published *Meditationes algebraicae*, a work that was highly praised by Lagrange; in 1772 he brought out *Proprietates algebraicarum curvarum*; and 1776 saw the publication of *Meditationes analyticae*. In addition to these important treatises, he also, during this period, published a number of learned papers in the *Philosophical Transactions of the Royal Society*. His last major work, *Essay on the Principles of Human Knowledge*, published in 1794, is notable for his application of abstract science to philosophy.

As a mathematician, Waring was unfortunate in working at a time in which English mathematics were in a state of decline. This was in part due to the clumsy notation in which Newton had expounded his calculus and to the geometrical exposition that gave the *Principia* a somewhat archaic appearance and persuaded English readers that the great new mathematical tool forged by Newton and Leibniz (which was then being employed with great vigor and skill on the Continent, particularly by the Bernoullis) was, in fact, not really necessary. This melancholy state of affairs persisted for more than a century, despite the efforts of such distinguished mathematicians as Brook Taylor, Colin Maclaurin, and John Wallis, and led Lalande to observe in a "Notice sur la vie de Condorcet" (*Mercure de France*, 20 Jan. 1796, p. 143) that there was not a single first-rate analyst in all England. (Waring, however, stoutly maintained that his *Miscellanea Analytica* disproved Lalande's charge, and cited its commendation by d'Alembert, Lagrange, and Euler.)

Despite the spectacular improvements in notation by which fundamental mathematical operations were expressed on the Continent, Waring, in his own works, used both the *de*ism of Leibniz and the *do*tage of Newton—the two great rival systems—indifferently, and made no notable contribution to the establishment of a permanent notation in any branch of mathematics. His method of writing exponents (as, for example, on page 8 of the 1785 edition of his *Meditationes analyticae*) was clumsy in the extreme, and in general his presentation is unattractive and his books difficult to follow. He suffered from an apparent lack of intellectual order that rendered his mathematical compositions so confused that they are almost impossible to follow in manuscript, while his published works, perhaps because of his extreme myopia, are riddled with typographical errors. His language, at best, was obscure.

Waring received the Copley Medal of the Royal Society in 1784. He was also elected a member of a number of European scientific societies, notably those of Göttingen and Bologna. He served as Lucasian professor until his death; he was also a commissioner of the important Board of Longitude. Nor were his activities exclusively mathematical; simultaneously with his composition of his books he turned to medicine, and received the M.D. from Cambridge in 1770. He does not appear ever to have practiced medicine, but it is believed that he carried out dissections in the privacy of his Cambridge rooms.

BIBLIOGRAPHY

I. ORIGINAL WORKS. Waring's books include *Miscellanea analytica de aequationibus algebraicis et curvarum proprietatibus* . . . (Cambridge, 1762), his best-known work; *Meditationes algebraicae* (Cambridge, 1770; 3rd ed., 1782); *Proprietates algebraicarum curvarum* (Cambridge, 1772); *Meditationes analyticae* (Cambridge, 1776; 2nd ed., enl., 1785); *On the Principles of Translating Algebraic Quantities Into Probable Relations and Annuities* (Cambridge, 1792); and *Essay on the Principles of Human Knowledge* (Cambridge, 1794).

His papers in the *Philosophical Transactions of the Royal Society* are "Problems," **53** (1763), 294–298; "Some New Properties in Conic Sections," **54** (1764), 193–197; "Two Theorems," **55** (1765), 143–145; "Problems Concerning Interpolations," **69** (1779), 59–67; and "On the General Resolution of Algebraical Equations," *ibid.*, 86–104.

II. SECONDARY LITERATURE. British historians of mathematics have hardly done justice to Waring. *Gentleman's Magazine*, **68**, pt. 2 (1798), 730, 807, contains a brief biography and a list of his principal contributions to mathematics; as does J. A. Venn, *Alumni Cantabrigienses*, pt. 2, IV (Cambridge, 1954), 352. The most exhaustive account of his work is Moritz Cantor, *Vorlesungen über Geschichte der Mathematik*, IV (Leipzig, 1908), 92–95. See also Florian Cajori, *History of Mathematical Notations*, 2 vols. (Chicago, 1928–1929), see indexes and I, 244, which reproduces p. 8 of the 1785 ed. of Waring's *Meditationes analyticae*; and R. T. Gunther, *Early Science in Cambridge* (Oxford, 1937), 60.

J. F. SCOTT

WARMING, JOHANNES EUGENIUS BÜLOW (*b.* Mandø, Denmark, 3 November 1841; *d.* Copenhagen, Denmark, 2 April 1924), *botany.*

Warming, professor of botany at the University of Copenhagen from 1886–1911, laid the foundation of a new branch of botany, ecological plant geography, with the publication of his *Plantesamfund* (1895). During the preceding years he had published many papers on various botanical subjects, several of which rank high in the literature of that time. He also had published two excellent textbooks: *Haandbog i den systematiske botanik* (1879) and *Den almindelige botanik* (1880), both of which have since been enlarged, revised, and translated into several languages.

Warming's father was a Lutheran minister on Mandø, one of the north Frisian Islands. From his childhood he loved the west coast of Jutland, with its marshland and dunes, on which he wrote two volumes of *Dansk plantevaekst: Strandvegetationen* (1906) and *Klitterne* (1909); the third volume was *Skovene* (1919). The work is still important for research on the phytoecology of northwestern Europe.

While still a student, Warming became secretary to the Danish zoologist P. W. Lund, who was excavating fossil Bradypodidae at Lagoa Santa, Minas Gerais, Brazil. He spent 1863–1866 in the tropical savannah, carrying out the most detailed and thorough study of a tropical area undertaken at the time. It took twenty-five years for complete presentation of his large collections in "Symbolae ad floram Brasiliae centralis cognoscendandae," printed in *Videnskabelige Meddelelser fra Dansk naturhistorisk Forening i Kjøbenhavn* (1867–1893). Using the "Symbolae" as a basis, Warming published *Lagoa Santa, et bidrag til den biologiske plantegeografi* (1892) with a lengthy summary in French—perhaps his most outstanding work.

After returning from Brazil, Warming studied for a year under Martius, Naegeli, and Ludwig Radlkofer at Munich and, in 1871, under J. L. von Han-

stein at Bonn. The morphological-organogenetic point of view was then the leading principle in botany, and within a few years Warming became one of the most prominent workers in this branch of botany. His main works during this period were *Er koppen hos vortemaelken (Euphorbia) en blomst eller en blomsterstand?* (1871); *De l'ovule* (1878); and his monograph on purple bacteria; *Om nogle ved Danmarks kyster levende bakterier* (1876).

In the 1870's, however, Warming adopted the theory of evolution. From then on, he became an ardent adherent of the Lamarckian view of the causes of evolution, and his research turned from ontogeny to phylogeny. In 1876 he published his first "Smaa biologiske og morphologiske bidrag" in *Botanisk Tidsskrift*, a series of papers that continued into 1878. They give a masterly account of the morphology and flower biology of numerous species, mostly Danish, pointing out their adaptation to the edaphic factors. Having assumed the difficult task of classifying the plants in a morphological-biological system, Warming published the first results in the monograph *Om skudbygning, overvintring og foryngelse* (1884), based on his examination of Scandinavian species. One of his main works, it illustrates both his comprehensive knowledge and his power to present a subject in an easily understood manner.

Warming was the founder of plant ecology. The term "ecology," first used by Haeckel in 1866, was introduced into botany by H. Reiter in 1885; but it was Warming who made ecology a preferred field of activity for many botanists. In *Plantesamfund* (1895) he formulated the program of his research: "To answer the question: Why each species has its own habit and habitat, why the species congregate to form definite communities and why these have a characteristic physiognomy."

The book created an enormous sensation as a new attempt at grouping and characterizing the plant communities—a new phytogeographical term by which Warming meant a group of species forming a physiognomically well-defined unity, such as a meadow. In all essentials the species of a community are subject to the same external conditions arising from the ecological factors. These factors are of fundamental importance to the ecology of the individual plant and the plant community. Considering water to be the most important factor, Warming divided plant communities into four types: hydrophytic, xerophytic, halophytic, and mesophytic.

Warming's ingenious way of elucidating the relation between the living plant and its surroundings opened an entirely new field of problems, and an immense ecological literature appeared during the following years.

BIBLIOGRAPHY

Warming's *Plantesamfund* (Copenhagen, 1895) was trans. into German as *Lehrbuch der ökologischen Pflanzengeographie*, with additions by E. Knoblauch (Berlin, 1896); later eds. (Berlin, 1902, 1918, 1938) had additions by P. Graebner; into Russian (Moscow, 1901); and into English as *Oecology of Plants. An Introduction to the Study of Plant-Communities* (Oxford, 1909). A complete bibliography of 283 titles is in Christensen (see below).

C. F. A. Christensen, *Den danske botanisk historie*, 3 pts. (Copenhagen, 1924–1926), pt. 1(2), 617–665, 776–806; pt. 2, 367–399, written by a pupil of Warming's, gives a detailed account of his life. *Botanisk Tidsskrift*, **39** (1927), 1–56, contains articles on Warming by L. Rosenvinge, C. Christensen, C. Ostenfeld, A. Mentz, C. Flahault, O. Juel, C. Schröter, and A. Tansley.

D. MÜLLER

WASHBURN, EDWARD WIGHT (*b.* Beatrice, Nebraska, 10 May 1881; *d.* Washington, D.C., 6 February 1934), *physical chemistry*.

Washburn was the son of William Gilmor Washburn, a lumber and brick merchant, and Flora Ella Wight, both of whom had moved to Nebraska from New England. Having taken all the chemistry courses available at the University of Nebraska (1899–1900) while teaching high school (1899–1901), he entered the Massachusetts Institute of Technology in 1901, obtaining the B.S. in chemistry in 1905 and the Ph.D. in 1908 under Arthur A. Noyes. Later that year he became head of the division of physical chemistry at the University of Illinois. In 1910 he married Sophie de Veer of Boston; they had four children. In 1916 Washburn became chairman of the university's department of ceramic engineering.

In 1920 the International Union of Pure and Applied Chemistry was founded. One of its first projects was to compile the *International Critical Tables of Numerical Data, Physics, Chemistry and Technology*. Washburn was named editor-in-chief in 1922 and moved to Washington. In 1926 he became head of the Division of Chemistry of the National Bureau of Standards.

Washburn was chairman of the Division of Chemistry and Chemical Technology of the Na-

tional Research Council in 1922–1923, chairman of the International Commission on Physico-Chemical Standards, and a member of the National Academy of Sciences. From 1920 to 1922 he was editor of the *Journal of the American Ceramic Society*.

Washburn's application, as a graduate student, of physicochemical principles to analytical chemistry had led him to the first thermodynamic treatment of buffer solutions and then to the study of indicators. He was the first to make accurate measurements to determine the value of transference numbers—the fraction of an electric current carried by each ion in an electrolyte solution—and he pioneered the study of the hydration of ions.

At the University of Illinois, Washburn developed thermodynamic treatments of a number of colligative properties and apparatus for the precise measurement of electrical conductance and viscosity. Moving to the university's ceramic engineering department, he applied physicochemical principles to the study of ceramics, to glasses at high temperatures, and to the manufacture of optical glass.

At the National Bureau of Standards, Washburn devised greatly improved techniques for the fractionation and isolation of the chemical constituents of petroleum, and he succeeded in obtaining rubber in crystal form. After Harold C. Urey had separated deuterium, the heavy isotope of hydrogen, from ordinary hydrogen,[1] Washburn suggested that the electrolysis of water should yield gaseous hydrogen and oxygen richer in the lighter isotopes, the residual water thereby becoming richer in the heavier isotopes. The first method for producing deuterium oxide in quantity was thus developed.[2] Washburn found evidence of natural isotope fractionation in water from oceans, the Dead Sea, and Salt Lake, in crystalline hydrate deposits, and in willow sap.[3]

NOTES

1. H. C. Urey, F. G. Brickwedde, and G. M. Murphy, "A Hydrogen Isotope of Mass 2," in *Physical Review*, **39** (1932), 164–165.
2. E. W. Washburn and H. C. Urey, "Concentration of the H^2 Isotope of Hydrogen by the Fractional Electrolysis of Water," in *Proceedings of the National Academy of Sciences of the United States of America*, **18** (1932), 496–498.
3. E. W. Washburn and E. R. Smith, "An Examination of Water From Various Natural Sources for Variations in Isotopic Composition," in *Bureau of Standards Journal of Research*, **12** (1934), 305–311.

BIBLIOGRAPHY

I. ORIGINAL WORKS. Washburn's books are *An Introduction to the Principles of Physical Chemistry From the Standpoint of Modern Atomistics and Thermodynamics* (New York, 1915; rev. ed., 1921), French trans. by H. Weiss and W. Albert Noyes, Jr. (Paris, 1922); and *International Critical Tables of Numerical Data, Physics, Chemistry and Technology*, 7 vols. (New York, 1926–1930), of which he was editor-in-chief.

An almost complete bibliography is in the obituary by W. A. Noyes. A detailed bibliography through 1921 appeared in *Bulletin of the American Ceramic Society*, 1, no. 3 (July 1922), 57–63.

II. SECONDARY LITERATURE. For a detailed biographical memoir, including extensive bibliography, see William Albert Noyes, in *Biographical Memoirs, National Academy of Sciences*, 17 (1937), 67–81. Brief obituaries are T. M. Lowry, in *Nature*, 133 (12 May 1934), 712–713; Lyman J. Briggs, in *Science*, 79 (9 Mar. 1934), 221–222; and an unsigned article in *Bulletin of the American Ceramic Society*, 13, no. 3 (Mar. 1934), 78.

OTTO THEODOR BENFEY

WASHINGTON, HENRY STEPHENS (*b.* Newark, New Jersey, 15 January 1867; *d.* Washington, D.C., 7 January 1934), *geology.*

A distinguished and colorful geologist during the early decades of the twentieth century, Washington pioneered in chemical studies of igneous rocks. He demanded high standards of accuracy in his own analyses and as a result produced the textbook *Manual of Chemical Analysis of Rocks* (1904; 4th ed., 1930), which remained standard for his generation.

A descendant of George Washington, he received the A.B. in 1886, with special honors in natural sciences, and the A.M. in 1888 from Yale College. For the next six years he was involved with the American School of Classical Studies at Athens, participating in archaeological excavations in Attica, Plataea, Argos, and Phillius. This work was influenced by his knowledge of geology in, for example, his determination of the sources of marbles used in Greek sculpture.

In 1891–1892 and 1892–1893 Washington spent the winter semesters at the University of Leipzig, where he received the Ph.D. in 1893 after studying under Zirkel and K. H. Credner. His dissertation was on the volcanoes of the Kula basin in Lydia.

He returned briefly to Yale in 1895 as an instructor in mineralogy. Financially independent,

Washington established a private laboratory at Locust, New Jersey, where he initiated the extensive chemical and mineralogical investigations of igneous rocks that he was to pursue for the rest of his life. Economic reverses forced him to undertake consulting work as a mining geologist from 1906 to 1912. In 1912 he became associated with the geophysical laboratory of the Carnegie Institution of Washington, where he remained, except for 1918–1919, when he was the scientific attaché to the American embassy in Rome.

Washington's chemical analyses of igneous rocks transformed into mineral compositions led to the first serious attempt to classify such substances in collaborative efforts with Whitman Cross, J. P. Iddings, and L. V. Pirsson (the CIPW classification). Although their scheme, published as *Professional Papers. United States Geological Survey*, 14, 28, and 99, achieved neither widespread nor lasting acceptance, it did stimulate an interest among earth scientists in the chemical and mineral compositions of rocks and attempts to produce alternative methods of classification.

In 1917 Washington published an enlarged edition of *Chemical Analyses of Igneous Rocks* (the first edition had appeared in 1903), a monumental assemblage of rock assays drawn from the world literature. He sorted them into superior and inferior classes and pointed out the inadequacies of the latter group. The work was of fundamental importance in establishing standards of analysis and became known throughout the world. Washington's other research spanned a wide spectrum of interests in geology, encompassing volcanism, petrography, isostasy, and geochemistry. He was a member of the committee on nomenclature of the Mineralogical Society of America. His linguistic abilities were used to establish the correct etymologies and pronunciations of mineral names.

BIBLIOGRAPHY

A bibliography of Washington's works is in *Zeitschrift für Vulkanologie*, 16 (1935), 3–6.

Obituaries include C. N. Fenner, in *Science*, 79 (1934), 47–48; and J. Volney Lewis, in *American Mineralogist*, 20 (1935), 179–184.

EDWARD D. GOLDBERG

WASSERMANN, AUGUST VON (*b.* Bamberg, Bavaria, Germany, 21 February 1866; *d.* Berlin, Germany, 15 March 1925), *bacteriology.*

For a detailed account of his life and work, see Supplement.

WATERSTON, JOHN JAMES (*b.* Edinburgh, Scotland, 1811; *d.* near Edinburgh, 18 June 1883), *physics, physical chemistry, astronomy.*

During his lifetime Waterston was considered a minor, somewhat eccentric scientist, known chiefly for his investigations of solar radiation; his other publications on astronomy, physical chemistry, and molecular physics attracted little notice. After his death a manuscript on the kinetic theory of gases that he had submitted in 1845 to the Royal Society of London was discovered in the Society's archives. Had this paper been published when it was first presented, an important branch of physics would have been advanced by ten or fifteen years (in the judgment of Lord Rayleigh and other modern commentators) and Waterston would have been generally recognized as one of its leaders. Instead, Waterston's case has become a classic example of the suppression of originality by an established scientific institution.

Waterston's father, George Waterston, was an Edinburgh manufacturer of sealing wax and other stationery requisites. The family was related to Robert Sandeman, the leader in extending the Sandemanian (or Glasite) religious sect to England and America, and to George Sandeman, founder of the London firm of port wine merchants.

George Waterston was greatly interested in literature, science, and music; his family thus grew up in an atmosphere of culture and came into contact with young literary men. John James was the sixth of nine children, all of whom were educated at the Edinburgh High School, then the leading school in Scotland. Following graduation Waterston became a pupil of Messrs. Grainger and Miller, civil engineers, but also attended lectures at the university, where he took an active part in the student literary society. He studied mathematics and physics under Sir John Leslie and was medalist of his year in Leslie's class. He also attended lectures on anatomy, chemistry, and surgery.

Like John Herapath, another early kinetic theorist, Waterston was interested in the problem of explaining gravity without invoking action at a distance. At the age of nineteen, he published a paper in which he discussed the properties of a system of colliding cylindrical particles, arguing that the latter could generate a gravitational force. Some of the ideas developed in this paper were later utilized in his kinetic theory, particularly the idea that collisions could result in a transfer of energy from the rectilinear to the rotatory mode of motion.

At the age of twenty-one Waterston went to London, where he did drawing and surveying in connection with the rapidly developing British railway system. He became an associate of the Institution of Civil Engineers and contributed a paper to the *Transactions* of that group on a graphical method of estimating the earthwork in embankments and cuttings. In order to have more time free to pursue his scientific interests, he obtained a post in the hydrographers' department of the Admiralty. The head of the department was Captain (afterwards Admiral) Francis Beaufort, who subsequently communicated Waterston s paper on kinetic theory to the Royal Society. On Beaufort's suggestion, and with his backing, Waterston applied in 1839 for the post of naval instructor to the East India Company's cadets at Bombay. He was successful, and found the position satisfactory in that he had sufficient leisure and access to scientific books and journals at the Grant College, Bombay. He taught the theoretical aspects of such subjects as navigation and gunnery.

During his stay in India, Waterston sent home the manuscript of a short book and several scientific papers. The book, an essay on the physiology of the central nervous system, was published anonymously at Edinburgh in 1843. It contains the first expression of Waterston's views on molecules and on the possible application of molecular theory to biology. Some basic principles of the kinetic theory of gases are included, such as "A medium constituted of elastic spherical atoms that are continually impinging against each other with the same velocity, will exert against a vacuum an elastic force that is proportional to the square of this velocity and to this density. . . . The proportion of the whole rectilinear to the whole rotatory momentum is probably constant, and might be found perhaps by calculation." Increase in temperature might correspond to increase of molecular *vis viva*. The distance traveled by a molecule, after hitting one and before encountering another, is inversely related to the density of the medium and to the square of the diameter of the molecules.

These propositions, along with some more fanciful notions, attracted little attention at the time. In December 1845, Waterston presented a more systematic exposition of his theory of gases in a paper entitled "On the Physics of Media That Are Composed of Free and Elastic Molecules in a State of Motion." As a physical justification for his theory he mentioned the wave theory of heat, adopted by analogy with the wave theory of light as a result of the recent experiments by J. D. Forbes and Melloni on radiant heat. He found that "in mixed media the mean square molecular velocity is inversely proportional to the specific weight of the molecules"; this was the first statement of the "equipartition theorem" of statistical mechanics (for translational motion only). Since this conclusion was printed in an abstract of the British Association meeting in 1851, Waterston seems to have established his priority in announcing the theorem even though the rest of his paper was not published until much later (see below). Another original (but quantitatively incorrect) result was that the ratio of the specific heats, at constant pressure and constant volume, for monatomic gases should theoretically be equal to 4/3. (Because of a numerical slip, Waterston failed to obtain the correct value, 5/3.)

Waterston submitted his paper for publication in the *Philosophical Transactions of the Royal Society of London*. At that time the custom of the Society was that a paper submitted by someone not a fellow of the Society could be "read" (officially presented) if it were communicated by a fellow, but it then became the property of the Society and could not be returned to the author even if it was not published. The two referees who examined Waterston's paper recommended that it should not be published. One of them, Baden Powell (professor of geometry at Oxford), said that Waterston's basic principle—that the pressure of a gas is due to impacts of molecules against the sides of the container—was ". . . very difficult to admit, and by no means a satisfactory basis for a mathematical theory." The other referee was the astronomer Sir John William Lubbock, who said, "The paper is nothing but nonsense, unfit even for reading before the Society." These judgments seem rather harsh, not because Waterston's theory was essentially the same as the one successfully proposed in the 1850's by Clausius and Maxwell, but because even by 1845 the physical basis for such a theory—the relation between heat and mechanical energy—was accepted by a substantial portion of the scientific community. Nevertheless, only a brief abstract of Waterston's paper appeared in the *Abstracts of the Papers Printed in the Philosophical Transactions of the Royal Society* in 1846.

Waterston was not able to get his manuscript back, and had failed to keep a copy for himself, so he was unable to publish it elsewhere. He did attempt to draw attention to the paper by privately printing and circulating another abstract of it,

about twelve pages long, and by raising the subject in later papers presented at British Association meetings and in *Philosophical Magazine*. The only immediate response was a critical discussion by W. J. M. Rankine (who preferred a theory of rotating vortices) and an abstract by Helmholtz in *Fortschritte der Physik* that may have had some influence on A. Krönig's revival of the kinetic theory in 1856.

In the paper "On Dynamical Sequences in Kosmos," read at the British Association meeting in 1853, Waterston pointed out that substantial amounts of heat could be generated by the fall of matter into the sun. He thought that the earth might have grown in size over long periods of time by the accretion of such meteoric material, and mentioned other possible astrophysical applications of the theory that heat is equivalent to mechanical energy and may be simply the motion of the elementary parts of bodies. William Thomson adopted a meteoric theory of the sun's heat from this paper, though he later learned that a similar theory had been presented earlier by J. R. Mayer.

In 1857 Waterston resigned his appointment at Bombay and returned to Scotland, apparently having saved enough money to be able to devote his time to scientific work. About this time he published some papers on the experimental measurement of solar radiation, yielding an estimate of about 13 million degrees for the sun's temperature; this figure was frequently quoted in the debate on the sun's temperature during the 1870's. Waterston began experimental work on liquids; and during the next few years he published a series of papers on physical chemistry, mainly in *Philosophical Magazine*. Apparently, he never met any of the scientists who might have recognized the value of his work on the kinetic theory, with the possible exception of Rankine (who spoke at the same session of the British Association meeting at which Waterston presented a paper on gases in 1851).

Among Waterston's chemical papers is one on capillarity and latent heat that reports a calculation of the diameter of a water molecule. The result was 1/214,778,500 inch (approximately 10^{-8} cm.). This estimate was published in 1858, seven years before Joseph Loschmidt's determination of molecular sizes from kinetic theory but forty-two years after Thomas Young's estimate, which was somewhat similar to Waterston's.

In 1878 the Royal Astronomical Society rejected two papers by Waterston. A few months later he resigned, having been a member since 1852. This event reinforced his isolation from the scientific world. According to a memoir by his nephew, Waterston "would not attend the meetings of the Royal Society of Edinburgh though some friends sent him billets, and rather avoided the society of scientific men. . . . We could never understand the way in which he talked of the learned societies, but any mention of them generally brought out considerable abuse without any definite reason assigned."[1]

Waterston's paper on the theory of sound, published in *Philosophical Magazine* in 1858, was the ultimate reason for his posthumous recognition by the scientific community. In 1876, S. Tolver Preston wrote to Maxwell about this paper, noting that Waterston had investigated the kinetic theory of gases as early as 1845, although his work had not yet been published.[2] But Maxwell apparently took no interest in this matter; and it was not until 1891, eight years after Waterston's death, that Lord Rayleigh rediscovered the 1858 paper on sound because of his interest in another of Waterston's papers that cited it. The mention of a manuscript lying in the archives of the Royal Society finally reached the right reader, for Rayleigh was secretary of the Royal Society in 1891, and had no difficulty in retrieving this manuscript. The paper was published in the *Philosophical Transactions* for 1892, with an introduction by Rayleigh, according to whom:

> The history of this paper suggests that highly speculative investigations, especially by an unknown author, are best brought before the scientific world through some other channel than a scientific society, which naturally hesitates to admit into its printed records matter of uncertain value. Perhaps one may go further and say that a young author who believes himself capable of great things would usually do well to secure the favourable recognition of the scientific world by work whose scope is limited, and whose value is easily judged, before embarking on greater flights.

NOTES

1. R. J. Strutt, *Life of John William Strutt, Third Baron Rayleigh* (London, 1924; augmented ed., Madison, Wis., 1968), 171.
2. I am indebted to Dr. C. W. F. Everitt for informing me of this letter.

BIBLIOGRAPHY

I. ORIGINAL WORKS. Most of Waterston's published works are reprinted in *The Collected Scientific Papers of John James Waterston*, edited with a biography by J. S. Haldane (Edinburgh, 1928). This volume omits the following papers: "An Account of an Experiment on the Sun's Actinic Power," in *Monthly Notices of the Royal Astronomical Society*, **17** (1856–1857), 205–206; "On Certain Inductions With Respect to the Heat Engendered by the Possible Fall of a Meteor Into the Sun; and on a Mode of Deducing the Absolute Temperature of the Solar Surface From Thermometric Observation," in *Philosophical Magazine*, 4th ser., **19** (1860), 338–343, and *Monthly Notices of the Royal Astronomical Society*, **20** (1860), 196–202; "Note of an Experiment on Voltaic Conduction," in *Philosophical Magazine*, 4th ser., **31** (1866), 83–84. In addition there are brief reports of his papers presented at British Association meetings in *Athenaeum* (1851), 776; (1852), 980; (1853), 1099–1100. Unpublished materials may be found in the archives of the Royal Society of London.

II. SECONDARY LITERATURE. J. S. Haldane's "Memoir of J. J. Waterston" is printed in the *Papers* (see above); it includes biographical information, portraits, and extensive discussion of Waterston's scientific work and opinions. See also S. G. Brush, "The Development of the Kinetic Theory of Gases. II. Waterston," in *Annals of Science*, **13** (1957), 275–282, and "John James Waterston and the Kinetic Theory of Gases," in *American Scientist*, **49** (1961), 202–214; and E. E. Daub, "Waterston, Rankine, and Clausius on the Kinetic Theory of Gases," in *Isis*, **61** (1970), 105–106.

STEPHEN G. BRUSH

WATERTON, CHARLES (*b.* Walton Hall, Yorkshire, England, 3 June 1782; *d.* Walton Hall, 27 May 1865), *natural history*.

Waterton was the twenty-sixth lord of Walton Hall, being the eldest son of Thomas and Ann Bedingfield Waterton. His family staunchly upheld the Roman Catholic faith and consequently had suffered persecution since the Reformation. Waterton was educated at Stonyhurst College, a Jesuit school, and his detestation of all that was Protestant was a formative and decisive factor in his career. As a boy his energy and high spirits, as well as his pursuit of nature, caused the "holy and benevolent" fathers of Stonyhurst to give up the rod and attempt to tame this incorrigible imp by making him, as he wrote, "rat-catcher to the establishment and also fox-taker, foumart-killer, and crossbow-charger at a time when the young rooks were fledged. Moreover I fulfilled the duties of organ-blower and football maker with entire satisfaction to the public" (*Essays*, xxvii).

Thus was formed the eccentric of later years,

whose indomitable courage and love of nature and travel led him to undertake a famous journey in 1812. Starting from Stabroek (now Georgetown, Guyana), where his family owned plantations, Waterton traveled alone up the Demerara and Essequibo rivers and over the Kanuku Mountains as far as the Rio Branco, a tributary of the Rio Negro; this was an incredibly difficult journey, not made easier by Waterton's proneness to accidents and illness caused chiefly by his own eagerness and temerity. His description of this and other journeys in *Wanderings in South America* gave many almost unbelievable stories which upset the orthodox scientists of the time, whom Waterton described as "closet-naturalists"; he included Audubon, James Rennie, and William Swainson in this category. Swainson, who also collected South American birds, described Waterton's "tendency to clothe fact in the garb of fiction," although Waterton was, above all, sincere and truthful, being carried away only by his impetuosity and enthusiasm. Among his many eccentric adventures were his capture of a live caiman (crocodile) by riding on its back (see *Illustrated London News*, 24 August 1844), and his taking of a live ten-foot boa constrictor by the Watertonian expedient of punching it on the nose and hustling it into a bag before it recovered.

In each instance Waterton's object was the scientific one of dissecting the animal and preserving its skin. This led to the formation of his collection of superbly mounted and preserved specimens at Walton Hall. He invented a new and advanced taxidermic technique of removing the whole interior and preserving only the skin and exterior parts with an alcohol solution of mercuric perchloride. The technique is described by Waterton in *Wanderings* (1825 ed., 307). Sir Joseph Banks wrote of his unrivaled skill in preserving birds. His collection, a large part of which may still be seen at Stonyhurst, was greatly enriched on his third visit to Demerara in 1820, when he took 232 birds, two land tortoises, a sloth, five armadillos, an ant bear, and the caiman, but he was justly annoyed when a customs delay caused the loss of his live *Tinamus* eggs, from which he had hoped to breed this little-known quail.

Waterton's original object in his first journey had been to collect "a quantity of the strongest wourali poison" (*ibid.*,1) and he succeeded, giving an early and accurate account of the preparation of the South American arrow poison curare (p. 54). in which he was preceded only by Alexander von Humboldt in 1800. Waterton's descriptions of the blowpipe and darts also were original (p. 58). He tested his curare on animals and correctly deduced that "the quantity of poison must be proportioned to the animal" (p. 69). Back in London, in 1814 Waterton conducted experiments on donkeys with the aid of the veterinarian William Sewell and the surgeon Benjamin Brodie. The most famous related to Wouralia, the ass whose life was preserved by energetic artificial respiration through a tracheostomy after she had received a large dose of curare (p. 81). This is a technique now revived in modern surgery. These experiments served to draw attention to curare, which later was investigated by Claude Bernard and is now in common medical use. Waterton pronounced himself ready at any time to treat cases of hydrophobia with curare but seems usually to have arrived too late. Later the drug was used successfully by others in the treatment of tetanus.

Waterton's fame as an explorer induced Lord Henry Bathurst, then secretary for the colonies, to offer him an important task, the exploration of the then almost unknown island of Madagascar. True to his eccentric nature, Waterton refused, pleading sickness, but probably influenced by religious prejudice, and so was officially ignored for the rest of his life.

Returning to Walton Hall, Waterton began the project of bird protection for which he deserves to be best remembered. It was Waterton—and not Audubon, as is often thought—who set up the first bird sanctuary; he enclosed Walton Park with a three-mile, seven-foot wall at a cost of £10,000. He banned guns and encouraged the birds to return to a natural state. Among his scientific writings, probably the most interesting is his argument with Audubon and Swainson on the manner in which vultures seek out their food (*Essays*, p. 17).

Waterton achieved fame and notoriety in his lifetime both for his journeys and for his literary skill in describing them. Although his style was akin to that of his hero Tristram Shandy, his biographer Norman Moore states that the *Essays* belong to the literary class of Gilbert White and are not inferior in the quality of their observations. The comparison is not without justice; White was among Waterton's favorite reading.

Waterton's contributions to science, ultimately, were small: some increase in knowledge of curare, an original method of taxidermy, and above all the idea of animal and bird conservation. His nature was eccentric, forthright, outspoken, and lovable;

but since he refused to conform, he remained outside the science establishment.

BIBLIOGRAPHY

Waterton's books are *Wanderings in South America, the North-West of the United States, and the Antilles, in the years 1812, 1816, 1820 and 1824* (London, 1825), which describes his four journeys and has been republished many times; and *Essays on Natural History Chiefly Ornithology* (London, 1838), which includes an autobiographical note. He wrote many articles for Loudon's *Magazine of Natural History*, some of which appear in the *Essays*.

His admiring but verbose biographer, Richard Hobson, wrote *Charles Waterton, His Home, Habits and Handiwork* (London, 1866), which is the chief source for Waterton's eccentricities. Sir Norman Moore, who had known him intimately, wrote a personal tribute in his article on Waterton for *Dictionary of National Biography*, XX, 906–908. An obituary notice is in *Illustrated London News* (17 June 1865). See also Richard Aldington, *The Strange Life of Charles Waterton* (London, 1949); and K. B. Thomas, *Curare, Its History and Usage* (London, 1964), 34–40.

K. Bryn Thomas

WATSON, GEORGE NEVILLE (*b.* Westward Ho!, Devon, England, 31 January 1886; *d.* Leamington Spa, England, 2 February 1965), *mathematics*.

Watson went up to Cambridge University in 1904 as a major scholar of Trinity College, to which he was intensely devoted throughout his life, and held a fellowship there from 1910 to 1916. After a brief period at University College, London, he went to Birmingham in 1918 as professor of mathematics and remained in this post until his retirement in 1951.

Almost all Watson's work was done in complex variable theory. Within this field he was no narrow specialist, his interests ranging widely over problems arising in the theories of difference and differential equations, number theory, special functions, and asymptotic expansions. As a classical analyst Watson showed great power and an outstanding ability to find rigorous and manageable approximations to complicated mathematical expressions; unlike many pure mathematicians, he was not averse to numerical computation, which he performed on his own Brunsviga machine and in which he found relaxation.

Watson wrote over 150 mathematical papers and three books. The first of these books, a Cambridge tract on complex integration, is now rarely consulted; but the remaining two had, and still have, a wide influence, particularly among applied mathematicians and theoretical physicists. The second, *A Course of Modern Analysis*, was written in collaboration with E. T. Whittaker, who had been one of the younger fellows of Trinity when Watson was an undergraduate. The first edition had appeared in 1902 under Whittaker's sole authorship and Watson offered to share the work of preparing the second, which appeared in 1915 and was a considerably expanded version of the original work. The first part of the book develops the basic principles and techniques of analysis and these are applied in the second part to obtain the properties of the many special functions that occur in applications. "Whittaker and Watson" has appeared in several editions and numerous reprints; Watson never lost his interest in it and, in his retirement, embarked upon a much enlarged version, which was never published.

The first fifty of Watson's mathematical papers are concerned mainly with properties and expansions of special mathematical functions. These investigations culminated in the publication of his monumental and definitive *Treatise on the Theory of Bessel Functions* (1922). A second edition, containing only minimal alterations, appeared in 1944; for by then Watson had lost interest in the subject and, unfortunately for the mathematical public, was not prepared to undertake the continuous revision and expansion that would have kept the book up to date. By 1929, also, he had already embarked on his "Ramanujan period"; and during the next ten years a succession of papers appeared in which he proved and extended numerous results that had been stated in the notebooks of the Indian mathematical genius Srinivasa Ramanujan, who had died in 1920. Watson and B. M. Wilson of Liverpool were invited by the University of Madras to become joint editors of a projected work, of an estimated 600 pages, which would contain proofs of Ramanujan's results.

Both editors made considerable progress, and much of their work was published as original papers. The mass of Ramanujan material was so extensive, however, that the fruit of their combined labors never reached the stage of publication in book form. Wilson died in 1935; and by 1939 Watson's impetus had diminished, possibly because of his increased administrative and teaching commitments following the outbreak of World War II. His work not only had provided proofs of formulas and congruences stated by Ramanujan, but also had considerably extended Ramanujan's work on singular moduli and set his work on mock theta functions on a proper foundation. These investigations were admirably suited to Watson's analytical abilities, since they demanded not only great ingenuity but also enormous industry. Much of this work would now be regarded as being outside the main stream of mathematics; but fashions change! His efforts during this period were not devoted solely to problems arising from Ramanujan's notebooks; his important work on what are now called Watson transforms also dates from this time.

With the exception of his investigations on periodic sigma functions, Watson's papers during the last twenty years of his life are of lesser interest.

BIBLIOGRAPHY

A complete list of Watson's mathematical writings is in the obituary notice by R. A. Rankin that appeared in *Journal of the London Mathematical Society*, **41** (1966), 551–565, where a more detailed discussion of some of his work is given. See also the obituary notice by J. M. Whittaker in *Biographical Memoirs of Fellows of the Royal Society*, **12** (1966), 521–530, which supplements the latter and includes a photograph.

Watson's unpublished work on the Ramanujan notebooks is in a collection of MSS deposited in the library of Trinity College, Cambridge.

R. A. RANKIN

WATSON, HEWETT COTTRELL (*b*. Park Hill, Firbeck, Yorkshire, England, 9 May 1804; *d*. Thames Ditton, Surrey, England, 27 July 1881), *phytogeography*, *evolution*, *phrenology*.

Watson was one of ten children born to Holland Watson and Harriett Powell Watson. His father, a magistrate for Cheshire County, planned a military career for his son. While still a youngster, however, Watson crushed his right knee during a game of cricket; he then became destined for the law. It is probable that his limp had a strong influence upon the development of his personality, which was conspicuously hostile.

A small inheritance at the age of twenty-one freed Watson from the necessity of earning a living, and he promptly abandoned his apprenticeship in law in order to collect plants and to study phrenology. The latter interest led him to Edinburgh, where he became intimate with the brothers George and Andrew Combe, and where he studied medicine from 1828 to 1832. Although he won honors as an outstanding student, Watson never took the examinations for a medical degree. There is some evidence that he had a breakdown in either his physical or his mental health.

In 1833 Watson purchased a house at Thames Ditton, near London, where he lived for the rest of his life with a housekeeper. He never married and never held a job, except for one term as a botany instructor in 1837 at the Liverpool School of Medicine. In 1842 William Hooker persuaded him to collect plants for five months in the Azores, which was the only time he ever left Britain.

In his early twenties Watson already had a good knowledge of the geography of British plants, and in 1832 he began publishing both articles on the influence of environmental factors upon the distribution of species and a series of guidebooks to their distribution. This was a promising, though not unusual, start to a career in botany.

At the same time, however, Watson was active in the phrenology movement, and by 1836 phrenology had become his major interest. In that year he published *Statistics of Phrenology, Being a Sketch of the Progress and Present State of That Science in the British Isles*. This book does not contain an application of statistics to phrenological questions; rather, it contains a report on the extent of phrenological activity in Britain. It has been more valuable to historians of phrenology than it was for advancing the movement.

In 1837 Watson purchased the *Phrenological Journal* from the Combes and edited it for three years, hoping to raise the standards of phrenological investigation to the level of a critical science. In this he failed, for two reasons. First, phrenology had developed neither an adequate methodology nor an adequate standard of verification. Merely improving individual phrenological papers was not coming to grips with the fundamental problem. Second, Watson was so blunt and critical in his editorial comments that he succeeded only in arousing the anger of the majority of the journal's readers and contributors. He recognized his failure, concluded that phrenology would never rise to the level of critical science, and in 1840 returned to the study of botany, to which he devoted the remainder of his life.

Watson came to believe in the transformation of species by 1834; and he first defended the idea in a polemical tract, *An Examination of Mr. Scott's Attack Upon Mr. Combe's "Constitution of Man"* (1836). His ideas on the subject were Lamarckian; and although he had aggressively helped to advance phrenological theory, his contributions to evolution were almost devoid of theoretical innovation. Watson agreed with Charles Lyell's judgment that Lamarck had not proved his case, but remained convinced that it could be proved. His subsequent phytogeographical research was motivated in part by a desire to collect evidence that would demonstrate the transformation of species.

Watson published some of his own evidence (1845) in a series of four articles written as a review and a reaction to Robert Chambers' anonymous *Vestiges of the Natural History of Creation* (1844). After revealing the inadequacy of Chambers' knowledge of botany, he went on to present evidence that could stand up to criticism. First, he felt that the paleontological evidence was important; but since this was beyond the scope of his own studies, he did not discuss it. Confining himself to botanical evidence, Watson documented the difficulty of distinguishing the separate species of *Salix, Mentha, Rosa, Rubus,* and *Saxifraga* by citing the divergent estimates of the number of British species of these genera from six manuals on British plants. Other evidence he considered important was the ease with which new varieties could be grown from cultivated species of *Pelargonium, Erica, Rosa, Fuchsia, Calceolaria, Dahlia,* and the pansy. He then pointed to the prevailing confusion among botanists concerning whether domestic fruits and grains were separate species or only varieties of the same species. Watson's next type of evidence was the existence in nature of "species being tied together (so to speak) by a series of intermediate forms." From his own observations he described clusters from a dozen such genera.

Watson's examples of evolution in action—both from his articles of 1845 and from his *Cybele Britannica*—impressed Charles Darwin, who drew upon them to good advantage in chapter 2 of *The Origin of Species.*

Although he was innovative in many minor ways, Watson's isolation from and contempt for his colleagues may have caused him to fall into a rut in his research. Both his phrenological and phytogeographical investigations were correlational in methodology. He never conducted any extensive experiments. In phytogeography he followed Humboldt's example of constructing numerous correlations between environmental factors and distributional patterns. The results, he hoped, would lead to a new understanding of both phytogeography and evolution.

When he read *The Origin of Species,* Watson realized that Darwin had found what he had unsuccessfully sought—a convincing causal explanation of evolution. He immediately wrote to Darwin, "You are the greatest revolutionist in natural history of this century, if not of all centuries." In his later years, however, he became disturbed by Darwin's inability to explain the origin of variations; and he became less certain of the magnitude of Darwin's achievement.

Watson's own work after 1859 was devoted to increasing the accuracy of knowledge of the distribution of British plants, and he helped make the British flora the best-known in the world. The Botanical Society of the British Isles has acknowledged the importance of his contributions by naming its journal *Watsonia.*

BIBLIOGRAPHY

I. ORIGINAL WORKS. Watson's articles on phrenology appeared in *Phrenological Journal* (1829–1840) and can be located in the index of each volume. His articles on botany are listed (but neither completely nor entirely accurately) in the Royal Society *Catalogue of Scientific Papers,* VI, 280–281; VII, 1202. There is a list of his botany books in the account by Boulger (see below).

II. SECONDARY LITERATURE. Besides George S. Boulger, in *Dictionary of National Biography,* XX, 918–920, there are two useful biographical sketches of Watson. The first, anonymous but obviously autobiographical, contains the only known portrait of him as a young man; see *Naturalist,* 4 (1839), 264–269. The other is John G. Baker, "In Memory of Hewett Cottrell Watson," in *Journal of Botany, British and Foreign,* 19 (1881), 257–265, with portrait of Watson in old age; reprinted without portrait in Watson's *Topographical Botany,* 2nd ed. (London, 1883). The details of Watson's life and work can, however, be obtained from the more than 300 letters he wrote that are preserved in British libraries. He burned the letters written to him.

There are no detailed discussions of Watson's contributions to science. For indications of Darwin's use of Watson's knowledge, see the dozen letters from Watson in the Darwin MSS, Cambridge University Library, and R. C. Stauffer, ed., *Charles Darwin's Natural Selection, Being the Second Part of His Big Species Book Written From 1856 to 1858* (Cambridge, 1975), see index. For the background of his work in phrenology, see his own *Statistics of Phrenology*; David Armand De Giustino, "Phrenology in Britain, 1815–1855: A Study of George Combe and His Circle" (Ph. D. diss. Univ. of Wisconsin, 1969); and Charles Gibbon, *The Life of George Combe, Author of "The Constitution of Man,"* 2 vols. (London, 1878). There are ample testimonies in the British botanical literature to the importance of Watson's work. One recent discussion is J. E. Dandy, *Watsonian Vice-Counties of Great Britain* (London, 1969).

FRANK N. EGERTON III

WATSON, RICHARD (*b.* Heversham, Westmorland, England, August 1737; *d.* Windermere, Westmorland, England, 9 July 1816), *chemistry*.

A son of Thomas Watson, headmaster of the grammar school at Heversham, Watson in 1754 entered Trinity College, Cambridge, where he distinguished himself in mathematics. In 1760 he became a fellow of the college, a moderator in 1762, and in November 1764 professor of chemistry, although he was completely ignorant of the subject. After fourteen months of intensive study he began to lecture in 1766. The chair was unendowed, but with the help of influential friends Watson obtained an annual royal grant of £100. In 1769 he was admitted to the Royal Society. Watson's main ambition, however, was to become regius professor of divinity; and when the chair became vacant in October 1771, although unqualified, Watson was able "by hard travelling and some adroitness" to obtain the king's mandate for a doctorate of divinity and was thus elected. In 1782 he became bishop of Llandaff. His theological and political writings were extensive; he defended Christianity against the attacks of Edward Gibbon and Thomas Paine. His unorthodox views made him enemies, however: he was attacked as self-seeking—a charge to which his apparent willingness to accept patronage laid him open.

Watson's most original work in chemistry was an investigation of the phenomena of solution. In 1770 he disproved J. T. Eller's assertion, made about 1750, that the volume of water is not increased when a salt is dissolved in it; and in the severe winter of 1771 he found that the times taken by solutions of a given salt to freeze, starting from the instant that pure water began to freeze when exposed to the air, were proportional to the concentrations. Thus he anticipated Blagden's "law."

Watson is best known for his *Chemical Essays*, some of which are still valued for their lucidity—in particular his account of the phlogiston theory, "Of Fire, Sulphur and Phlogiston" (*Essays*, I, 149–180), which has frequently been cited. He also described the now-classic experiment in which hot water in a tightly stoppered flask can be made to boil by pouring cold water over the air space

("Of Degrees of Heat in Which Water Begins to Part With Its Air and in Which It boils," *Essays*, III, 143–169). His "Of the Saltness and Temperature of the Sea" (*Essays*, II, 93–139) is a perceptive contribution to early marine science.

BIBLIOGRAPHY

I. ORIGINAL WORKS. Watson's chemical works are *Institutionum chemicarum in praelectionibus academicis explicatarum, pars metallurgica* (Cambridge, 1768); *An Essay on the Subjects of Chemistry, and Their General Division* (Cambridge, 1771); and *A Plan of a Course of Chemical Lectures* (Cambridge, 1771). These, together with the papers listed below, are reprinted in *Chemical Essays*, 5 vols. (Cambridge, 1781–1787). There were a number of subsequent eds., published mainly at London, with the same pagination, but the numbering of the eds. is confusing; the last English ed. (London, 1800) is styled "7th edn." Two eds. were published in Dublin—a 1-vol. ("3rd") ed. (1783), containing the essays in vols. I–III, and the complete work in 2 vols. (1791).

Watson's papers are "Experiments and Observations on Various Phoenomena Attending the Solution of Salts," in *Philosophical Transactions of the Royal Society*, **60** (1770), 325–354; "Some Remarks on the Late Cold in February Last," *ibid.*, **61** (1771), 213–220; "Account of an Experiment Made With a Thermometer, Whose Bulb Was Painted Black, and Exposed to the Direct Rays of the Sun," *ibid.*, **63** (1773), 40–41; "Chemical Experiments and Observations on Lead Ore," *ibid.*, **68** (1778), 863–883; "Observations on the Sulphur Wells at Harrogate, Made in July and August 1785," *ibid.*, **76** (1786), 171–188; and "On Orichalcum," in *Memoirs and Proceedings of the Manchester Literary and Philosophical Society*, **2** (1785), 47–67.

II. SECONDARY LITERATURE. The main biographical source is *Anecdotes of the Life of Richard Watson, Bishop of Landaff; Written by Himself at Different Intervals, & Revised in 1814*, published by his son Richard Watson (London, 1817; 2nd ed., 2 vols., 1818), which contains a number of letters. Watson's character and views were scathingly though anonymously attacked in *A Critical Examination of the Bishop of Landaff's Posthumous Volume* . . . (London, 1818). Accounts of Watson are V. Bartow, "Richard Watson, Eighteenth Century Chemist and Clergyman," in *Journal of Chemical Education*, **15** (1938), 103–111; L. J. M. Coleby, "Richard Watson, Professor of Chemistry in the University of Cambridge, 1764–71," in *Annals of Science*, **9** (1953), 101–123; and J. R. Partington, "Richard Watson (1737–1816)," in *Chemistry and Industry*, **56** (1937), 819–821—see also Partington's *History of Chemistry*, II (London, 1961), 765–767, and his *Text-Book of Inorganic Chemistry* (London, 1921), 103, where he first drew attention to Watson's anticipation of Blagden. A letter from Watson to Lord Rockingham and the reply are reproduced and commented on by W. H. G. Armytage in "Richard Watson and the Marquess of Rockingham; an Unpublished Exchange in 1771," in *Annals of Science*, **14** (1961), 155–156. Some account of Watson's theological writings is given by A. Gordon in *Dictionary of National Biography*, XL (1899), 24–27.

E. L. SCOTT

WATSON, SERENO (*b*. East Windsor Hill, Connecticut, 1 December 1826; *d*. Cambridge, Massachusetts, 9 March 1892), *botany*.

Watson was the ninth of the thirteen children of Henry Watson and Julia Reed Watson, both descendants of early Connecticut settlers. Henry Watson was a merchant in the village of East Windsor but moved to the nearby ancestral farm at East Windsor Hill when his father died, shortly after Sereno's birth. The pleasant rural childhood contributed to the boy's love of nature and to his lifelong diffidence.

After preparatory work at East Windsor Hill Academy, Watson attended Yale College, graduating in 1847. He intended to enter medicine and for some years studied under several physicians in New England and New York, and under his brother Louis in Illinois, alternating his studies with various teaching posts. Neither medicine nor teaching appealed to him, so in 1856 he joined another brother in Greensboro, Alabama, where he was secretary of the Planters' Insurance Company until the Civil War began. He left the South to work on the *Journal of Education* in Hartford, Connecticut, from 1861 to 1866 and then attended the Sheffield Scientific School at Yale for a year.

Without definite purpose Watson sailed for California via Panama in 1867. From the Sacramento valley he walked across the Sierra Nevada to volunteer on Clarence King's geological exploration of the fortieth parallel. Among other duties he assisted William Whitman Bailey in botanical collections, since he had enjoyed plant collecting earlier as a minor hobby. Bailey's health was poor; and when he left in March 1868, Watson became the survey's official botanist. He collected extensively in Nevada and Utah. A year later he returned east to study his collections in the herbarium of Daniel C. Eaton at Yale, and in 1870 he removed to Asa Gray's herbarium at Harvard to continue his survey report and botanical studies. In 1874 he became the curator of the Gray Herbarium, and from 1881 to 1884 he was also instructor in phytography at Harvard.

Watson's contributions to botany began with the fifth volume of the fortieth parallel survey (1871), which lists 1,325 plant species and describes and illustrates many of them. This classic was the first account of the distinctive xerophytic and mesophytic vegetation of the Great Basin region. It was also the first example of Watson's painstaking meticulousness in defining the systematics of plants.

In 1873 Watson began compiling the systematic botany of California, started earlier by William H. Brewer, who turned over his material to Watson. This significant work was presented in two volumes in 1876 and 1880. Simultaneously Watson undertook the almost impossible task of indexing all plant species west of the Mississippi River, on which scattered accounts and descriptions had already been published in the accounts of many western explorations. The only completed volume from this undertaking was the very useful one on Polypetalae (1878).

Asa Gray had been revising his classic *Manual of the Botany of Northern United States* before his death in 1888, after which Watson and John M. Coulter completed the work. Most of Watson's other botanical works were published under the title "Contributions to American Botany," in various journals. These, and his separately published works, constitute a fine contribution to plant systematics and relationships.

Watson received an honorary Ph.D. from Iowa College in 1878, and he was elected to the National Academy of Sciences in 1889. He never married and, from lifelong shyness, he never presented his papers in person.

BIBLIOGRAPHY

I. ORIGINAL WORKS. Watson's most significant works, cited in the text, are *Botany*, vol. V of *United States Geological Exploration of the Fortieth Parallel* (Washington, D.C., 1871); *Botany of California*, 2 vols. (Cambridge, Mass., 1876, 1880) (I, *Polypetalae*, was written with W. H. Brewer); *Bibliographical Index to North American Botany . . . Part I, Polypetalae* (Washington, D.C., 1878); and *Gray's Manual of the Botany of the Northern United States . . .*, 6th ed. (New York–Cincinnati–Chicago, 1889; reissued with corrections, 1890). These and Watson's other publications are listed in the biography by Brewer (see below).

II. SECONDARY LITERATURE. William H. Brewer wrote the only extensive account of Watson's life: *Biographical Memoirs. National Academy of Sciences*, **5** (1903), 267–290 (the year of birth is given incorrectly as

1820). A short account by "M.B." appeared in *Scientific American*, **65** (1892), 233–234.

ELIZABETH NOBLE SHOR

WATSON, WILLIAM (*b.* London, England, 3 April 1715; *d.* London, 10 May 1787), *physics, botany, medicine.*

Watson, the son of a cornchandler, obtained a sound basic education at the Merchant Taylors' school before apprenticing himself to an apothecary on 6 April 1731 for a term of eight years.[1] Apothecaries were then enjoying new opportunities and obligations in consequence of having won in 1704, at the expense of their old enemies the physicians, the right to prescribe as well as to compound medicine. An apprentice accordingly had much to learn: botany, chemistry, drug making, and the diagnosis and treatment of common complaints. Watson, "never indolent in the slightest degree" and an "exact economist of his time" (Pulteney, 334), mastered these subjects, especially chemistry and botany, in which he won the annual prize of the Apothecaries Society for skill in identifying plants *in situ* (Pulteney, 297). In June 1738, ten months short of the term of his apprenticeship, he purchased his freedom for two guineas and was sworn into the Apothecaries Society. The same year he married and set up in business for himself.

Through his botanical contacts Watson came to the Royal Society, of which he was to be one of the most productive and influential members. He first attended a meeting on 16 March 1738, as guest of John Martyn, Cambridge professor of botany in absentia, abridger of the *Philosophical Transactions*, and translator of Boerhaave.[2] He was then patronized by Martyn's friend Sir Hans Sloane, who with Martyn, Thomas Birch, and others signed Watson's certificate for admission to the Society, which took place on 9 April 1741. His first communications to the Society dealt either with botany—for instance, an account of the previously undescribed star puffball (*Geaster*), which made him known to Continental authorities (Pulteney, 299)—or with medical oddities encountered in his practice. He continued to write on these matters and, what was more important, to keep his colleagues current with Continental advances in natural history. Among his later botanical papers are studies of the sex of plants, inventories of gardens, and descriptions of useful or poisonous plants. Among his reports those on platinum [7],

on Linnaeus' system, and on Peyssonel's important but neglected demonstration of the animal origin of coral [9], were the most important (Pulteney, 298–309).

Although Watson's natural history papers "do him credit, they would not of themselves have been sufficient to give him celebrity."[3] His reputation came primarily from his studies of electricity, which he began, characteristically, by reproducing and transmitting a discovery made abroad. The discovery, the ignition of warmed spirits by an electric spark, apparently interested Watson as a chemical matter. He successfully extended the operation to all the inflammable liquids he had in stock and showed that they might be fired "repulsively," the spark being drawn across the liquor in an electrified spoon rather than (as was customary) toward the contents of a grounded one [2, 481–487]. These and other small triumphs [3] brought Watson the Copley Medal for 1745 and a public that consumed four editions of his electrical papers [4] before any could be printed in the *Philosophical Transactions*.[4]

Watson's first important discovery, announced in October 1746 [5], was the failure of an expectation authorized by the commonly held effluvial theory of electricity. He anticipated—as did J. N. Allamand, G. M. Bose, and Franklin, all of whom tried the effect independently—that by insulating himself while rubbing a glass tube, he could generate more electricity than if he stood upon the floor and let the effluvia run to ground. Failure suggested to him, as it did to Franklin, that rubbing did not collect electrical matter from the glass, but raised it from the ground; in a word, the tubes and globes acted as "pumps" circulating the electrical "fire." That far Franklin and Watson went independently and together. In extending his system, however, Watson remained within the effluvial frame and did not attain the conception of contrary electrifications. For example, he observed that the usual arrangement, in which the electricity generated by a grounded operator rubbing the globe of an electrical machine runs on to an insulated iron bar, could be inverted by grounding the bar and insulating the operator. But he did not perceive that the bar and the man electrify oppositely, the first positively and second negatively (in Franklin's terminology). Having missed this qualitative difference, Watson developed a crude electrical mechanics that assimilated the effluvia to a subtle, universal, springy "aether." The "pumps" disturb the equilibrium of this ether, which, in straining to regain its balance, brings about electrical attraction and repulsion, much as in the system of the abbé Nollet.

The Leyden jar ruined Watson's theory as it did Nollet's: Neither could account for the opposite electrifications, the condensing action, or the paradoxical role of glass.[5] After an effete try [5, 64], Watson gave up explaining the jar [7, 102] and turned to examining its properties. He found, independently of Le Monnier and Daniel Gralath, that the shock could be increased by arming the bottle with lead[6] or by thinning the glass [5, 31]; that the discharge passed in the most direct way through the conductors forming the external "circuit" [5, 31]; that the discharge occurred in a time too short to measure [6]; and that it might travel great distances through the ground or across water.[7]

Watson's second important electrical discovery was the work of Benjamin Franklin, whose first communication he reviewed before the Royal Society in January 1748 [7, 97–100]. This communication, which introduced the concepts of electrification plus and minus and likened the tube to a pump, may have owed something to Watson's pamphlets of 1745 and 1746 [4, 5]; in any case Watson recommended it as paralleling ideas he himself had recently developed. John Bevis had found that if the operator A and another man B are insulated, A charges whenever B takes a spark from the revolving globe, but never (as Watson had discovered) without B's intervention. B also charges, and A and B can exchange a bigger spark than either can with C, who stands on the ground. A similar experiment had taught Franklin to ascribe to B a greater, and to A a less, than ordinary quantity of electrical matter; Watson concluded that A and B had ether at low and high density, respectively, C in all cases being the standard. Although neither quantity of electrical matter nor its density (a rough forerunner of potential) alone sufficed for an exact electrostatics, Franklin's conception, which retained fewer effluvial trappings, was the more progressive. As it developed further from the common view in Franklin's theory of the Leyden jar, Watson regarded it less favorably [8].

With the successful trial of Franklin's theory of lightning at Marly in 1752, Watson ceased to be the innovating leader of English electricians and assumed instead the same roles in electrical studies that he held in botanical ones: promoter, umpire, reviewer, consultant, and minor contributor. He advertised the English confirmation of Marly, joined with Nollet in exposing the strange delusion of the medicated electrical tubes,[8] abstracted foreign literature, advised on the construction of lightning rods,[9] and reported his own competent experiments on discharge *in vacuo* and medical electricity. He became an adherent of Franklin's theory and a friend of its inventor, with whom he shared political views and an interest in technological innovation (Pulteney, 319–323).

Franklin's success was not the only, or perhaps even the chief, cause of Watson's withdrawal from independent electrical studies in the 1750's. As his reputation rose, so did his sights, and the level and intensity of his medical practice. In September 1757 he received or bought an M.D. from the University of Halle;[10] in December he asked to leave the Society of Apothecaries, which generously disfranchised him, for the large fee of £50, early in 1758. No doubt he sought his release to make himself more acceptable to the Royal College of Physicians, which licensed practitioners with foreign degrees but disdained local apothecaries, however qualified. In 1759 he became a licentiate, and three years later he was chosen physician to the Foundling Hospital.[11] The next step, fellowship in the College, came slowly. The College elevated men without an Oxford or Cambridge M.D. only with great reluctance and after long probation. Watson allied himself with "rebel licentiates" led by John Fothergill, who tried to reform the statutes; but only minor concessions were made, and Watson did not attain the fellowship until 1784, and then *speciali gratia*.[12] He subsequently held the office of censor in the College, and received a knighthood in 1786 in connection with his service. As a physician Watson was compassionate, generous, careful, and well-informed (Pulteney, 338). He interested himself chiefly in epidemic children's diseases, which he saw at the Foundling Hospital, and wrote a useful pamphlet comparing methods of inoculating against smallpox [11].[13]

Watson was a leading member of several institutions. His steady service to the Royal Society, of which he became vice-president under the regime of his friend Sir John Pringle, has been mentioned. He acted as a trustee of the British Museum and arranged its botanical garden (Pulteney, 310). He helped support Priestley's experiments. He was a charter member of the Royal Society (dinner) Club, founded in 1743; a regular at the learned social gatherings at the home of one Watson, a grocer in the Strand;[14] and a member of the Society of Collegiate (Licentiate) Physicians and of the Club of Honest Whigs.[15] He was an able, conscientious, clubbable man, "and his exact observance of the duties of social politeness must ever be remembered with pleasure by all those who enjoyed the happiness of his acquaintance" (Pulteney, 338).

NOTES

1. Society of Apothecaries, Court Minute Books, 8200/6, fol. 61v; Hartog, "Watson," 956; and Pulteney, 296, wrongly begin the apprenticeship in 1730, probably on the strength of Watson's attendance at school (1726–1730), for which see *Register of Merchant Taylors' School*, E. P. Hart, ed., II (London, 1936), 68. Presumably Watson delayed his apprenticeship until just after his sixteenth birthday and then bound himself for eight rather than for the more usual seven years to satisfy the apothecaries' rule that the apprenticeship should not expire before the candidate reached twenty-four.
2. Royal Society, Journal Book, XVI, 212; G. C. Gorham, *Memoir of John Martyn . . .* (London, 1830).
3. Thomas Thomson, *History of the Royal Society* (London, 1812), 434; see tributes from Musschenbroek and Volta in Pulteney, 313 *n*.
4. Cf. H. Baker to P. Doddridge, 24 Nov. 1747, in Doddridge, *Correspondence and Diary*, J. D. Humphreys, ed., V (London, 1831), 28.
5. J. L. Heilbron, "A propos de l'invention de la bouteille de Leyde," in *Revue d'histoire des sciences*, **19** (1966), 133–142.
6. [5, §28]. Watson had this suggestion from John Bevis (see *Dictionary of National Biography*, II, 451–452).
7. [5, §28]. This was a misapprehension, for in such cases each coating of the jar discharges separately to ground.
8. L. Trenngrove, "Chemistry at the Royal Society . . . II," in *Annals of Science*, **20** (1964), 1–57.
9. [10]. Watson was a member of the committee that advised the government on protecting the powder magazine at Purfleet. See D. W. Singer, "Sir John Pringle and His Circle. I," in *Annals of Science*, **6** (1949), 127–180.
10. Pulteney, 326, which also credits Watson with a degree from Wittenberg; Clark, *History . . .*, 565 makes him a graduate of the medical school of the University of Edinburgh.
11. W. Munk, *The Roll of the Royal College of Physicians of London*, 2nd ed., II (London, 1878), 348–349.
12. Clark, *op. cit.*, 571–572; Fox, *Dr. John Fothergill . . .*, 143–151. See Watson to Fothergill, 16 Sept. 1771, in John Thomson, *Account of the Life, Lectures and Writings of William Cullen*, 2nd ed., I (Edinburgh–London, 1859), 657–660.
13. Charles Creighton, *Epidemics in Britain*, II (Cambridge, 1894), 500–503, 514, 705–706. Watson's papers published in Fothergill's compilation, *Medical Observations and Inquiries*, 6 vols. (London, 1757–1784), are noticed in Pulteney, 331–332. Also see Fox, *op cit.*, 141–142.
14. A. Geikie, *Annals of the Royal Society Club* (London, 1917), 11, 141, 160; Singer, "Sir John Pringle," 160.
15. Clark, *op cit.*, 565; Fox, *op. cit.*, 317. See V. W. Crane, "The Club of Honest Whigs," in *William and Mary Quarterly*, **23** (1966), 210–233.

BIBLIOGRAPHY

I. Original Works. An adequate bibliography of Watson's published work may be pieced together from Pulteney (see below) and from P. H. Maty, *A General Index to the Philosophical Transactions* [vols. 1–70] (London, 1787), 788–791. Neither mentions "An Account of Some of the More Rare English Plants Ob-

served in Leicestershire," in *Philosophical Transactions of the Royal Society*, **49** (1755–1756), 803–866; or Watson's reviews of William Brownrigg's *The Art of Making Common Salt* (1748), *ibid.*, **45** (1748), 351–372, and of J. A. Braun's *De admirando frigore artificiale* (1760), *ibid.*, **52** (1761–1762), 156–172.

Among Watson's important writings are [1] "De planta minus cognita, & hactenus non descripta," *ibid.*, **43** (1744–1745), 234–238; [2] "Experiments and Observations Tending to Illustrate the Nature and Properties of Electricity," *ibid.*, 481–501; [3] "Further Experiments and Observations," *ibid.*, **44** (1746), 41–50; [4] *Experiments and Observations Tending to Illustrate . . .*, a reprinting of the two preceding works, which exists in two versions, one by J. Ilive (London, 1745), and the other, differing by the addition of a preface, by C. Davis (London, 1746); [5] "A Sequel to the Experiments and Observations," in *Philosophical Transactions*, **44** (1747), 704–749, preprinted as a pamphlet (London, 1746); [6] "A Collection of Electrical Experiments," *ibid.*, **45** (1748), 49–92; [7] "Some Further Enquiries . . .," *ibid.*, 93–120; [8] "An Account of Mr. Benjamin Franklin's Treatise," *ibid.*, **47** (1751–1752), 202–210; [9] "An Account of a Manuscript Treatise . . . Intituled *Traité du corail*," *ibid.*, 445–469; [10] "Observations Upon the Effect of Lightning," *ibid.*, **52** (1764), 201–227; and [11] *An Account of a Series of Experiments Instituted With a View of Ascertaining the Most Successful Method of Inoculating the Smallpox* (London, 1768).

MSS of Watson's papers and reviews are preserved at the Royal Society and in the Sloane MSS at the British Museum: see A. H. Church, *The Royal Society: Some Account of the "Letters and Papers" of the Period 1741–1806* (Oxford, 1908), 69; and *A Catalogue of the Manuscripts Preserved in the British Museum*, S. Ayscough, ed. (London, 1782). There are some fifty of Watson's private letters in the British Museum, most among the Hardwicke Papers, for which see British Museum, *Catalogue of Additions to Manuscripts . . . in the Years 1894–1899* (London, 1901). Watson's large scientific correspondence, referred to by Pulteney and Wilson (see below), contained many formal reports intended for publication, the following being those that reached the *Phisophical Transactions*: **46** (1749–1750), 470; **47** (1751–1752), 553, 559; **48** (1753–1754), 153, 579, 786; **49** (1755–1756), 16, 371, 558, 579, 668; **50** (1757–1758), 240, 506; **52** (1761–1762), 40, 302; **58** (1768), 58, 136; **59** (1769), 23, 81, 241; **60** (1770), 233; **61** (1771), 136; **62** (1772), 54, 265, 469; **63** (1773), 1, 79.

There is a portrait of Watson at the Royal Society, reproduced in A. Geikie, *Annals of the Royal Society Club* (London, 1917), opp. 24.

II. Secondary Literature. For biographical data see R. Pulteney, *Sketches of the Progress of Botany in England*, II (London, 1790), 295–340, abstracted in H. B. Wilson, *The History of Merchant Taylors' School*, II (London, 1814), *passim*; P. J. Hartog, "Watson, Sir William," in *Dictionary of National Biography*, XX, 956–958; Society of Apothecaries, Court Minute

Books, Guildhall Library MSS, vol. 8200/6, f. 61v, 138v; vol. 8200/7, f. 123v, 125v (information supplied by A. E. J. Hollaender, Keeper of MSS); George Clark, *A History of the Royal College of Physicians*, II (Oxford, 1966), 471, 476–479, 552–573, 586; H. C. Cameron *et al.*, *A History of the Worshipful Society of Apothecaries of London*, I (London, 1963), 78–82, 132–135; and R. H. Fox, *Dr. John Fothergill and His Friends* (London, 1919), 141–151, 215, 317. For Watson's botany see Pulteney. For his electricity see J. Priestley, *The History and Present State of Electricity*, 3rd ed., I (London, 1775), 97–101, 111–118, 130–145, 347–352; I. B. Cohen, *Franklin and Newton* (Philadelphia, 1956), 390–413, 441–452, 501–505; and M. Gliozzi, "Studio comparativo delle teorie elettriche del Nollet, del Watson e del Franklin," in *Archeion*, 15 (1933), 202–215.

J. L. HEILBRON

WATT, JAMES (*b*. Greenock, Scotland, 19 January 1736; *d*. Heathfield, England, 19 August 1819), *engineering, chemistry*.

Although Watt's achievements as an inventor and an engineer have been fully recognized and universally honored, the dependence of his technical work on contemporary science and his own scientific research have long provoked sharp differences of opinion.

Watt's grandfather and father had both followed technical pursuits: the former, Thomas, as a teacher of surveying and navigation ("professor of the mathematicks") and the latter, James, as a shipwright and maker and supplier of nautical instruments. His mother, Agnes Muirhead (or Muireheid), was descended from a family that had at one time been prominent in Scottish life. Owing to his fragile health Watt's attendance at elementary school was somewhat irregular, but he nonetheless attained some proficiency in geometry (in which he showed great interest), Latin, and Greek. Schooling, however, composed only the lesser part of his education; the more consequential portion he received in his father's shop, where he first gained the knowledge and skills of contemporary craftsmanship—woodworking, metalworking, smithing, instrument making, and model making.

At the age of eighteen, having decided to follow the career of scientific instrument maker, Watt left Greenock and took up residence in nearby Glasgow, which was then becoming a center of commerce and industry. In 1775 he went to London, where he spent a year as an apprentice, rapidly mastering the arts and crafts that entered into the making of navigational and scientific instruments.

He found London both disagreeable and a strain on his health, however, and a year later he returned to Scotland. Watt hoped to establish himself in Glasgow as an instrument maker, but he was prevented from doing so by guild restrictions. It was only through the influence of friends on the faculty of the University of Glasgow that he was able in 1757 to evade the jurisdiction of the corporations of tradesmen through an appointment as "mathematical instrument maker to the university." Watt thus found the setting that fostered much of his technical and scientific work. He soon became acquainted with John Robison (who first directed his attention to the steam engine) and Joseph Black; and it was in 1765, during his association with the university, that he made his first and most important invention, the separate condenser for the Newcomen engine. He patented it in 1769 and developed it commercially, first in partnership with John Roebuck and later with Matthew Boulton.

This initial success was followed over the next quarter-century by a remarkable sequence of additional inventions related to the steam engine—the sun-and-planet gearing system to translate the engine's reciprocating motion into rotary motion without employing the common crank (which was entangled in patent claims); the application to the steam engine of the double-acting principle that was then commonly used in pumps; the "expansive principle" whereby Watt recognized that because of its expansive power, steam need not be admitted into the cylinder during the entire stroke; the "parallel motion" with which he connected a rigid piston rod to the overhead beam without causing the rod to wobble; and the "indicator" for determining the pressure in the cylinder during the cycle. Besides these signal contributions to the technology of the atmospheric steam engine, Watt also originated a perspective drawing machine, a letter-copying process, an indicator liquid for testing acidity, and a steam wheel (which he was unable to perfect) for producing rotary motion directly from steam pressure.

In 1766 Watt closed his shop at the university and opened a land surveying and civil engineering office in Glasgow, where he practiced as a civil engineer until 1774. In the latter year he moved to Birmingham and formed the partnership with Boulton whereby he successfully commercialized his improved steam engine design. During the 1790's he was heavily preoccupied with the litigation through which he preserved his separate condenser patent against a series of challenges. And in 1800 both Watt and Boulton retired, turning their business enterprises over to their sons.

Watt became a fellow of the Royal Society of Edinburgh and of London, and was a member of the Lunar Society of Birmingham. He married Margaret Miller, a cousin, and, after her death in 1773, Ann MacGregor, the daughter of a Glasgow merchant. Of the children born from these marriages only a son, James, outlived the father.

Watt's career as a scientist centered on his interest in chemistry. He performed numerous experiments, was in contact with several of the foremost chemists of the day (including Black, Priestley, and Berthollet), and occasionally ventured into the realm of theory. In 1783 he formed the opinion that water is a compound; but his designation of its components was ambiguous, inasmuch as he described them as "dephlogisticated and inflammable air, or phlogiston," where "phlogiston," as he often used the term, signified various gases. During the nineteenth century a spirited debate arose among the partisans of Watt, Cavendish, and Lavoisier over credit for priority in the discovery of the "composition of water." J. R. Partington, the historian of chemistry, after closely evaluating the conflicting claims has lent his authority to the view that while Watt is entitled to credit for first stating that water is not elementary, it was Lavoisier who clearly specified what its components are.

Watt also did experiments during the 1780's that contributed to the commercial application in Britain of the process, which Berthollet had discovered, of bleaching textiles with chlorine. In this case Watt's role as a chemist must be heavily qualified. Unlike Berthollet, whose chemical research was part of a program of theoretical inquiry and who promptly published his discoveries even when they had commercial possibilities, Watt was more akin to what would presently be described as a chemical engineer. His experiments were designed to render the process effective and economical on a commercial scale. Moreover, Watt's father-in-law, James MacGregor, was in the bleaching business; and Watt hoped that by keeping their improvements secret, they would realize substantial profits. He was openly disappointed that Berthollet was conducting his research "earnestly" and was making "his discoveries on it publick." When Watt proposed to Berthollet that, with MacGregor, they acquire a British patent on the process, Berthollet brushed aside the proposal with the remark, "Quand on aime les sciences on a peu besoin de fortune. . . ." These distinctions between the mo-

tivations and purposes of the engineer and the scientist are of great interest in attempting to reach an understanding of the development of modern science.

Twenty years earlier, during the 1760's, Watt had played a similar role in an attempt to commercialize a process for producing alkali using common salt and lime as ingredients. The "theory," according to Watt's own testimony, was formulated by Black; Watt's contribution consisted of experiments designed to find a commercially feasible procedure. Watt unquestionably displayed considerable knowledge of the chemistry of bleaching, dyeing, and alkali production; but in these fields his contributions were to industrial chemistry, not to chemical theory. They were the chemical equivalents of his mechanical inventions (which likewise followed systematic experiments).

In one additional area of his involvement with chemistry, a misunderstanding continues to confound our appreciation of Watt's career as a scientist. Both Robison and Black advanced the claim that the invention of the separate condenser rested upon Watt's understanding of Black's principle of latent heat. Although Watt denied these assertions and presented a convincing description of the events that led to his invention,[1] some writers have not only repeated the claim but have gone further and asserted that Watt discovered or "rediscovered" the principle itself.[2] In fact, however, Watt only noticed the phenomenon (the apparent loss of heat when water is boiled) that is accounted for by the principle of latent heat. Upon describing his observations to Black, he was told of the principle, which Black had been teaching at the University of Glasgow for several years. Watt's own claim was only that he had "stumbled upon one of the material facts by which that beautiful theory is supported."

If we confine our meaning of science to its theoretical dimensions, we must conclude that Watt's inventions were made for the most part independently of science. But there can be no question that, conversely, theoretical science owes much to his inventions. The steam revolution that Watt's work as an inventor promoted, focused the attention of mathematicians and natural philosophers on problems that prompted important research in the theory of heat and in kinematics. Indeed, his "expansive principle" was embodied in the adiabatic expansion phase of Sadi Carnot's heat cycle.[3] And the parallel motion that Watt substituted for the chain and arch head connection stimulated considerable research in pure kinematics.[4]

If, however, we take a wider view of science, we can find still more meaning in Watt's career. For despite the contrast between his modest achievements as a scientist and his extraordinary originality and inventive power as an engineer, his career displays one of the key developments in the history of science—the entrance by engineers into the world of research. During the eighteenth century the traditional affiliation between engineering and craftsmanship was being revised in favor of a merger of engineering with experimental and theoretical science; and in Watt's work in chemistry, in his associations with chemists and natural philosophers, in his employment at the University of Glasgow, and in his membership in the foremost British scientific societies we have one of the earliest and clearest traces of that emerging pattern.

NOTES

1. For a defense of Watt's position, see Donald Fleming, "Latent Heat and the Invention of the Watt Engine," in *Isis*, **43** (1952), 3–5.
2. A. E. Musson and Eric Robinson, *Science and Technology in the Industrial Revolution* (Manchester, 1969), 80. These authors generally claim more for the theoretical content of Watt's work than the present article allows.
3. See Robert Fox, "Watt's Expansive Principle in the Work of Sadi Carnot and Nicolas Clément," in *Notes and Records. Royal Society of London*, **24** (1969–1970), 233–253.
4. See Eugene S. Ferguson, "Kinematics of Mechanisms From the Time of Watt," in *Bulletin of the United States National Museum*, **228**, paper 27 (1962), 185–230.

BIBLIOGRAPHY

I. ORIGINAL WORKS. Watt wrote much but published little. His only publication on his inventions is his ed. of John Robison's *Encyclopaedia Britannica* articles on steam and steam engines: *The Articles Steam and Steam-Engines, Written for the Encyclopaedia Britannica, by the Late John Robison, LL.D., F.R.S.L & E.* (Edinburgh, 1818); this material is reproduced in vol. II of the posthumous collection of Robison's articles, *A System of Mechanical Philosophy*, David Brewster, ed., 4 vols. (Edinburgh, 1822). Two letters by Watt setting forth his views on the composition of water were published by the Royal Society: "Thoughts on the Constituent Parts of Water and of Dephlogisticated Air; With an Account of Some Experiments on That Subject. In a Letter From Mr. James Watt, Engineer, to Mr. De Luc, F.R.S.," in *Philosophical Transactions of the Royal Society,* **74** (1784), 329–353; and "Sequel to the Thoughts on the Constituent Parts of Water and Dephlogisticated Air: In a Subsequent Letter From Mr. James Watt, Engineer, to Mr. De Luc, F.R.S.," *ibid.,* 354–357. Watt's biographer, James Patrick Muirhead,

later reprinted these letters with additional material relevant to the composition-of-water controversy: *Correspondence of the Late James Watt on His Discovery of the Theory of the Composition of Water*, James Patrick Muirhead, ed. (London, 1846).

Watt's interest in the application of pneumatic chemistry to medicine resulted in his collaboration with Thomas Beddoes on the following works: *Considerations on the Medicinal Use of Factitious Airs, and on the Manner of Obtaining Them in Large Quantities* (Bristol, 1794; 2nd ed., 1795; 3rd ed., 1796); and *Medical Cases and Speculations; Including Parts IV and V of Considerations on the Medicinal Powers, and the Production of Factitious Airs* (Bristol, 1796)—Watt's contribution to the first of these was also printed separately as *Description of a Pneumatic Apparatus, With Directions for Procuring the Factitious Airs* (Birmingham, 1795). He also published a note on his test for acidity: "On a New Method of Preparing a Test Liquor to Shew the Presence of Acids and Alkalies in Chemical Mixtures," in *Philosophical Transactions of the Royal Society*, **74** (1784), 419–422.

Some of Watt's multitudinous letters and unpublished papers have been reprinted; vol. II of James Patrick Muirhead, *The Origin and Progress of the Mechanical Inventions of James Watt*, 3 vols. (London, 1854), contains a selection of Watt's correspondence; and recently two systematic collections that include much previously unpublished material have appeared: Eric Robinson and A. E. Musson, *James Watt and the Steam Revolution. A Documentary History* (London, 1969); and Eric Robinson and Douglas McKie, eds., *Partners in Science. Letters of James Watt and Joseph Black* (London, 1970). Many of Watt's letters and notes are preserved among the family papers at Doldowlod, Radnorshire.

II. SECONDARY LITERATURE. Writings on Watt's life and work are voluminous, almost all of them on his engineering rather than his science. For his personal life and especially his family background, see George Williamson, *Memorials of the Lineage, Early Life, Education, and Development of the Genius of James Watt* (Edinburgh, 1856). James Patrick Muirhead's 3-vol. work (see above) is the standard nineteenth-century biography; besides the volume of correspondence (II), vol. I contains a narrative of Watt's life and vol. III patent specifications and information. The narrative is recapitulated in Muirhead's *The Life of James Watt* (London, 1858). Among the more recent biographical works the most valuable is H. W. Dickinson and Rhys Jenkins, *James Watt and the Steam Engine. The Memorial Volume Prepared for the Committee of the Watt Centenary Commemoration at Birmingham 1919* (Oxford, 1927); this work contains a narrative biography, descriptions of many of Watt's technical achievements, reproductions of some of his drawings, and an extensive annotated bibliography. The composition-of-water controversy is summarized and the various claims evaluated in J. R. Partington, *A History of Chemistry*, III (London, 1962), 344–362. Partington's *History* is also useful in connection with Watt's other chemical endeavors. An important study of science in the industrial revolution that bears heavily on Watt's career is A. E. Musson and Eric Robinson, *Science and Technology in the Industrial Revolution* (Manchester, 1969).

The following publications are among those that have recently contributed to a fuller understanding of Watt's place in science: Robert E. Schofield, *The Lunar Society of Birmingham* (Oxford, 1963), 60–82, *passim*; D. S. L. Cardwell, *From Watt to Clausius* (Ithaca, N.Y., 1971), 40–55, *passim*; W. A. Smeaton, "Some Comments on James Watt's Published Account of His Work on Steam and Steam Engines," in *Notes and Records. Royal Society of London*, **26** (1971), 35–42; David F. Larder, "An Unpublished Chemical Essay of James Watt," *ibid.*, **25** (1970), 193–210; and Eric Robinson, "James Watt, Engineer and Man of Science," *ibid.*, **24** (1969–1970), 221–232.

Harold Dorn

WAYJAN IBN RUSTAM. See **Al-Qūhī** (or **Al-Kūhī**), **Abū Sahl Wayjan ibn Rustam.**

AL-WAZZĀN AL-ZAYYĀTĪ AL-GHARNĀTĪ, AL-ḤASAN IBN MUḤAMMAD. See **Leo the African.**

WEBER, ERNST HEINRICH (*b*. Wittenberg, Germany, 24 June 1795; *d*. Leipzig, Germany, 26 January 1878), *anatomy, physiology, psychophysics.*

Weber was the oldest of the three Weber brothers who throughout their lives were closely linked in their scientific activity. Their greatest achievement lay in applying the modern exact methods of mathematical physics to the study of the functioning of various systems of higher animals and man. The leader in this endeavor, Ernst very early drew the attention of the physicist Wilhelm Eduard to the problems of the mechanics of circulation and later influenced the orientation of Eduard Friedrich toward theoretical medicine, helping him to obtain a post at the Leipzig medical school and to remain there as his close collaborator. Eduard was subsequently stimulated and helped by Wilhelm in the study of muscle mechanics.

Their father, Michael Weber, was professor of theology at Wittenberg from 1789 and later—after the fall of the city, a Napoleonic stronghold, in 1814 and the evacuation of the university—at Halle. Ernst, the third of his thirteen children, had been greatly influenced by Ernst Chladni, who of-

ten visited the family and excited the boys' interest in physics as a basis of all natural sciences. Weber attended secondary school in Meissen, where he acquired an excellent knowledge of Latin. In 1811 he began his medical studies at Wittenberg, but the war soon forced him to leave for Leipzig. He received the M.D. in 1815 from the University of Wittenberg, then temporarily evacuated to Schmiedeberg, with a dissertation on comparative anatomy. He could not, however, remain there because the university had no facilities for his anatomical work and its status was uncertain. At Leipzig, Weber became assistant at the medical clinic run by J. C. Clarus, qualified as docent in 1817 with a work on the comparative anatomy of the *nervus sympathicus*, and the following year became extraordinary professor of comparative anatomy. In 1821 he was nominated to the chair of human anatomy, which in 1840 was joined with physiology. In 1865 he gave up physiology and supported the appointment of Carl Ludwig, who established an independent physiological institute that attracted many foreign students. In 1871 Weber retired from the chair of anatomy.

Weber began with research in anatomy and discovered several important structures, some of which still bear his name—for instance, Weber's ossicles, which form a chain of small bones on each side of the air bladder, and the ear atrium of some fishes (the Weberian apparatus). This work marked the beginning of a series of comparative embryological and paleontological studies that led to the discovery of the intermediary stages between the primitive structures of the splanchnocranium and the middle ear auditory ossicles of mammals—a brilliant step in demonstrating the links between isolated facts and continuity in the evolution of structure and function. Weber's injection of the ducts of certain glands showed that their finest branches end blindly in the acini and have no direct communication with the surrounding small blood vessels, as had been supposed despite earlier findings by Malpighi (1686). It proved definitively that the digestive juices are specific products of glands, formed from the material brought by the blood, not just separated from the blood plasma. This finding opened up a new field of physiological and chemical research. Weber's wide experience in both research and teaching enabled him to write a revised edition of G. F. Hildebrandt's *Handbuch der Anatomie*. Its first part, *Allgemeine Anatomie*, entirely rewritten, became a valuable source of information because Weber carefully separated facts from theory and was not satisfied with merely describing structures; rather, he added what was known of their physical properties and chemical composition, as well as an appraisal of their significance. He was convinced that a knowledge of many conditions, not simply anatomical structure, was necessary for understanding the phenomena of life. The disadvantage of Weber's revised edition was that it was completed before the advance brought about by the subsequent development of microscopic research and by the cell theory. He also revised J. C. Rosenmüller's *Handbuch der Anatomie* (1840).

In 1821, assisted by his brother Wilhelm—then only seventeen years old and preparing for his university entrance examination—Weber began a long physical study of the flow and the progress of waves in fluids, particularly in elastic tubes. In their *Wellenlehre* (1825) they formulated the basic laws of hydrodynamics and were the first to apply that branch of physics to the circulation of the blood. Ernst studied—at first with Wilhelm, a precocious genius—the mechanical properties of the arteries, describing them as he would a technical device, the effect of elasticity transforming the pulsatile movement of the blood in the large arteries into a continuous flow into the small ones (1827). He also showed that the pulse is a wave in the arteries caused by the heart action and that its propagation—calculated from the delay of pulsation in a more distant artery—is much faster than the flow of the blood (1834) and that besides dilatation due to the pressure inside an elastic tube, blood vessels also change their diameter under the influence of nerves on the muscle wall (1831). He summarized his findings, the theory of waves in elastic tubes, and the laws of the movement of blood in the vessels in 1850.

Weber also demonstrated the resistance of the capillary bed, the importance of the blood volume, and its influence on the movement and distribution of the blood in the body. His work laid the base for the exact analysis of the movement of fluids in elastic tubes; and although the blood circulation has subsequently been subjected to a thorough research, Weber's work, with some additions but no substantial changes, has remained its foundation.

Another great contribution to the physiology of the blood circulation was the startling discovery by Eduard and Ernst Weber that electrical stimulation of some parts of the brain or of the peripheral end of the vagus nerve slows the action of the heart and can even bring it to a standstill (1845). It was the first instance of nerve action causing inhibition of an autonomic activity, rather than exciting it. It became an important milestone in the evolution of physiology not only for its significance to the circulation but also because its discovery brought to light a hitherto unknown but essential kind of nerve action. The ensuing chain of investigations showed that inhibition is a common phenomenon in the central nervous system and that an adequate balance between excitation and inhibition is indispensable for its normal function.

About 1826 Weber began a long series of remarkable systematic studies of sensory functions, especially of the "lower senses," which had hitherto been one of the most neglected areas of physiology. Physiologists had studied mainly the problems of vision and hearing, which seemed more interesting and promising. In his studies of other physiological problems Weber, a distinguished anatomist, usually followed function in close relation to structure. In this field, however, there was no anatomical basis because the skin, muscle, and visceral receptors were not discovered until later (Meissner, 1852; Krause, 1860). Nonetheless, his physical approach and attempts to determine quantitative relations of the stimulus to its effect, sensation, led to remarkable results despite the very simple methods used in his observations and experiments. An important feature of Weber's examinations and comparisons was the use of the notion of threshold (although this term was not actually used). He was well aware of the significance of its exactly determined values for estimating and comparing the performance of the skin and other sensory organs. A markedly greater ability to distinguish two very slightly different weights when they are lifted from, rather than when placed on, the hand, is explained by a special muscle sense. Examining the sense of touch in great detail, especially the local sense and differential threshold with a compass, Weber determined the characteristics of sensations of pressure and of temperature—positive (warm) and negative (cold)—and stressed the role of adaptation and local differences. Thus he gave sensory physiology a new orientation toward quantitative approach and methods, bringing into prominence both facts (mostly his own findings) and problems. He not only systematically collected facts but also drew rational conclusions about the physiological bases of the observed phenomena. He assumed isolated conduction in nerve fibers and formulated theories of projection and objectification. The division of each nerve fiber into a small circle of nerve endings was the background of local discrimination and of differences in its limen as determined by a compass.

In using his physical considerations as the basis for examining the differential thresholds of skin and muscle sensations, Weber found that two sensations are just noticeably different as long as the ratio between the strengths in each pair of stimuli remains constant. For instance, the smallest appreciable difference between two weights or lengths (usually called "just noticeable difference" or "Weber fraction") is a constant fraction of the weights themselves, approximately 1/30 (a just discriminable increment of intensity).

It was supposed that Weber's law was generally valid, but many discussions and criticisms led to the more moderate view that for most modalities it applies only over a limited range of intensities. Nevertheless, Fechner, assuming that discriminable increments are equal units of sensation, derived the formula

$$S = K \log I + C,$$

where intensity of sensation (S) is a linear function of the logarithm of intensity of the stimulus (I) and K and C are constants. Fechner's derivation has been criticized mainly because the stimulus—a physical factor—can easily be measured, while sensation—a subjective impression—cannot be expressed in physical terms. Quantitative comparisons became possible, however, when modern electrophysiological methods made it possible to follow the response of single sensory fibers—that is, the frequency of the messages from a single receptor. Over a certain range of intensities, it is indeed a linear function of the logarithm of the stimulus, as has been shown for the muscle spindle by B. H. Matthews and for the *Limulus* eye by H. K. Hartline and C. H. Graham. It cannot be stated whether it is fitting for the response of all forms of sense organs, but it seems that Fechner's equation corresponds to a fundamental feature of sense organ behavior.

Weber was the first to draw the attention of physiologists to the skin as the seat of differentiated sense organs directed toward the external world, like other sensory organs, in contrast with the common sensibility (*Gemeingefühl*) directed toward our own body. His research had many philosophical implications and a great impact on further studies of skin senses and some general problems of sensation by physiologists and psychologists.

He began a very fruitful period in the research on senses and is rightly considered as one of the founders of psychophysics. His work on tactile sensations has become classic.

BIBLIOGRAPHY

I. ORIGINAL WORKS. A partial list of Weber's writings was published in *Almanach der K. Akademie der Wissenschaften in Wien*, **2** (1852), 203–211; and, more recently, by P. M. Dawson (see below), 110–113. They include *Anatomia comparata nervi sympathici* (Leipzig, 1817); *De aure et auditu hominis et animalium* (Leipzig, 1820); *Wellenlehre, auf Experimenten begründet* (Leipzig, 1825), written with Eduard Weber; *Zusätze zur Lehre von Bau und Verrichtungen der Geschlechtsorgane* (Leipzig, 1846); "Tastsinn und Gemeingefühl," in R. Wagner, *Handwörterbuch der Physiologie*, III, pt. 2 (Brunswick, 1846, repr. separately 1851), also Ostwalds Klassiker der Exakten Wissenschaften no. 149 (Leipzig, 1905); *Ueber die Anwendung der Wellenlehre auf die Lehre vom Kreislauf des Blutes und insbesondere auf die Pulslehre* (Leipzig, 1850); and "Ueber den Raumsinn und die Empfindungskreise in der Haut und im Auge," in *Berichte über die Verhandlungen der K. Sächsischen Gesellschaft der Wissenschaften*, Math.-phys. Kl. (1852), 85–164, his chief paper on projection and the theory of circles.

Weber's papers were published mainly in *Deutsches Archiv für die Physiologie* and Meckel's *Archiv für Anatomie und Physiologie* (1820–1828), Müller's *Archiv für Anatomie, Physiologie und wissenschaftliche medizin* (1835–1846), and *Berichte über die Verhandlungen der K. Sächsichen Gesellschaft der Wissenschaften zu Leipzig* (1846–1850). Dissertations written under his guidance were collected in *Annotationes anatomicae et physiologicae (Programata collecta)*, 2 fascs. (Leipzig, 1827–1834, 1836–1848), both reed. (1851) with fasc. 3, containing several of his own important papers.

II. SECONDARY LITERATURE. An appreciation of Weber's scientific achievement is C. Ludwig, *Rede zum Gedächtniss an Ernst Heinrich Weber* (Leipzig, 1878). A fairly detailed account in English is P. M. Dawson, "The Life and Work of Ernest Heinrich Weber," in *Phi Beta Pi Quarterly*, **25** (1928), 86–116. With regard to the importance and impact of his work, papers on Weber are rather scarce. See also Ursula Bueck-Rich, *Ernst Heinrich Weber (1795–1878) und der Anfang einer Physiologie der Hautsinne* (inaug. diss., Zurich, 1970); H. E. Hoff, "The History of Vagal Inhibition," in *Bulletin of the History of Medicine*, **8** (1940), 461–496; P. Hoffmann, "Ernst Heinrich Weber's *Annotationes anatomicae et physiologicae*," in *Medizinische Klinik*, **30** (1934), 1250. There are many references to Weber's works on sensory physiology in E. G. Boring, *A History of Experimental Psychology*, 2nd ed. (New York, 1950);

and *Sensation and Perception in the History of Experimental Psychology* (New York, 1942).

VLADISLAV KRUTA

WEBER, HEINRICH (*b.* Heidelberg, Germany, 5 May 1842; *d.* Strasbourg, Germany [now France], 17 May 1913), *mathematics.*

Weber, son of the the historian G. Weber, began the study of mathematics and physics in 1860 at the University of Heidelberg. He then went to Leipzig for a year but subsequently returned to Heidelberg, where he obtained the Ph.D. in 1863. After working at Königsberg under Franz Neumann and F. J. Richelot, he qualified as *Privatdozent* in 1866 at Heidelberg and obtained a post as extraordinary professor there in 1869. He subsequently taught at the Eidgenössische Polytechnikum in Zurich, the University of Königsberg, the Technische Hochschule in Charlottenburg, the universities of Marburg and Göttingen, and, from 1895, at Strasbourg.

Weber was rector of the universities of Königsberg, Marburg, and Strasbourg; member of many German and foreign academies; and recipient of an honorary doctorate from the University of Christiania (now Oslo). He was a cofounder of the Deutsche Mathematiker-Vereinigung and member of the editorial board of *Mathematische Annalen.*

In 1870 Weber married Emilie Dittenberger, daughter of a Weimar court chaplain. Their daughter translated the philosophical writings of Henri Poincaré into German, and their son Rudolf Heinrich became professor of theoretical physics at Rostock. Weber's closest friend was Richard Dedekind, with whom he often collaborated and with whom he edited Riemann's works (1876). Weber's students included Hermann Minkowski and David Hilbert.

An immensely versatile mathematician, Weber focused his research mainly on analysis and its application to mathematical physics and number theory. The direction of his work was decisively influenced by his stay at Königsberg, where Jacobian mathematics still flourished. There he was encouraged by Neumann to investigate physical problems and by Richelot to study algebraic functions. Weber began his research with an examination of the theory of differential equations, which he conducted in Jacobi's manner. Then, building on Carl Neumann's book on Riemann's theory of algebraic functions and on the work of Alfred Clebsch and Paul Gordan on Abelian functions, Weber demonstrated Abel's theorem in its most general form. He also worked on the mathematical treatment of physical problems concerning heat, static and current electricity, the motion of rigid bodies in liquids, and electrolytic displacement. He brought together a portion of this research in *Die partiellen Differentialgleichungen der mathematischen Physik* (1900–1901), a complete reworking and development of a similarly titled book prepared by Karl Hattendorff from Riemann's lectures that had gone through three editions.

Weber investigated important contemporary problems in algebra and number theory, the fields in which he did his most penetrating work. With Dedekind he wrote a fundamental work on algebraic functions that contained a purely arithmetical theory of these functions. One of Weber's outstanding accomplishments was the proof of Kronecker's theorem, which states that the absolute Abelian fields are cyclotomic—that is, they are obtained from the rational numbers through adjunction of roots of unity. In 1891 Weber gave a complete account of the problems of complex multiplication, a topic in which analysis and number theory are inseparably linked. His studies culminated in the two-volume *Lehrbuch der Algebra* (1895–1896), which for decades was indispensable in teaching and research.

Weber was an enthusiastic and inspiring teacher who took great interest in educational questions. In collaboration with Joseph Wellstein and with the assistance of other mathematicians, he edited the *Enzyklopädie der Elementar-Mathematik*, a three-volume work designed for both teachers and students.

BIBLIOGRAPHY

Weber's works include "Ueber singuläre Auflösungen partieller Differentialgleichungen erster Ordnung," in *Journal für die reine und angewandte Mathematik*, **66** (1866), 193–236; "Neuer Beweis des Abelschen Theorems," in *Mathematische Annalen*, **8** (1874), 49–53; "Theorie der algebraischen Funktionen einer Veränderlichen," in *Journal für die reine und angewandte Mathematik*, **92** (1882), 181–290, written with R. Dedekind; *Elliptische Funktionen und algebraische Zahlen* (Brunswick, 1891; 2nd ed., 1908, as vol. III of *Lehrbuch der Algebra*); *Lehrbuch der Algebra*, 2 vols. (Brunswick, 1895–1896; 2nd ed., 1898–1899); *Die partiellen Differentialgleichungen der mathematischen Physik*, 2 vols. (Brunswick, 4th ed., 1900–1901; 5th ed., 1910–1912); and *Enzyklopädie der Elementar-Mathematik*, 3 vols. (Leipzig, 1903–1907), written with Joseph Wellstein *et al.*

There is an obituary by A. Voss in *Jahresberichte der*

Deutschen Mathematiker-Vereinigung, **23** (1914), 431–444, with portrait.

BRUNO SCHOENEBERG

WEBER, MAX WILHELM CARL (*b*. Bonn, Germany, 6 December 1852; *d*. Eerbeek, Netherlands, 7 February 1937), *zoology*.

After attending schools in Oberstein an der Nahe, Neuwied, and Bonn, Weber began the study of medicine at the University of Bonn. His teachers included the zoologists Franz Hermann Troschel and Franz von Leydig (for whom he worked as an assistant) and the anatomist Adolph La Valette St. George. In the winter semester of 1875–1876 Weber studied at Berlin, mainly under the zoologist Eduard von Martens. In 1877 he received the Ph.D. from Bonn for the dissertation "Die Nebenorgane des Auges von einheimischen Lacertiden." Soon afterward he was invited by the anatomist Max Fürbringer to serve as prosector at the anatomy institute of the University of Amsterdam. In 1879 he went to the University of Utrecht as lecturer in anatomy. He was recalled to Amsterdam in 1883 to teach zoology and comparative anatomy.

Weber went on an expedition in the North Atlantic, primarily to study the anatomy of whales. He was aided in this research by his wife, Anna van Bosse, who had studied under Hugo de Vries. Their findings were published in 1886 as the first part of his *Studien über Säugethiere*. In 1888 the couple traveled to the Dutch East Indies, where they visited Sumatra, Java, Celebes, and Flores. On Flores especially, Weber gathered extensive zoological material, as well as ethnographic data. A number of scientists collaborated in publishing descriptions of this material in *Ergebnisse einer Reise nach Niederländisch Ost-Indien* (Leiden, 1890–1907). Weber himself described the freshwater sponges, the trematode genus *Temnocephalus*, and several fish, reptiles, and mammals. With his wife he also investigated the symbiotic algae of the freshwater sponge *Spongilla*. The Webers soon embarked on another voyage, this time to South Africa. Their findings were published in 1897.

The second part of *Studien über Säugethiere* appeared the following year. In 1899 Weber and his wife participated in the Dutch Sibolga expedition, which was sent to examine the marine fauna and flora of Indonesia. Weber himself recorded part of the findings in *The Fishes of the Indo-Australian Archipelago* (Leiden, 1911–1936); the seven-volume work contained descriptions of 131 new species. Weber had acquired an interest in biogeography during his first trip to the Dutch East Indies, and the studies of the Sibolga expedition on freshwater fauna led him to reexamine the well-known differences between the freshwater fauna of Java, Borneo, and Sumatra, on the one hand, and of Celebes, Timor, and Flores, on the other. Weber drew attention to the zoogeographical differences between the northern and southern halves of Celebes and noted the existence of the deepwater zone around the island, the temperature of which differs markedly from that of the rest of the ocean.

During the period of his expeditions and of assessing the material they yielded, Weber also worked on his greatest scientific publication, *Die Säugetiere* (1904), an exposition of both the anatomy and the systematics of the mammals; the second edition is still a standard work.

Weber confined his scientific labors to descriptive zoology. He explicitly defended the old methods of comparative anatomy, although he also recognized the importance of the new experimental areas of biological research.

BIBLIOGRAPHY

A list of Weber's more important works follows D'Arcy Wentworth Thompson's biography in *Obituary Notices of Fellows of the Royal Society of London*, **2** (1938), 347–352, with portrait. They include *Studien über Säugethiere*, 2 pts. (Jena, 1886–1898), comprising "Ein Beitrag zur Frage nach dem Ursprung der Cetaceen" and "Über Descensus testiculorum, Anatomische Bemerkungen über Elephas"; the results of Weber's voyage to South Africa, published as "Zur Kenntnis der Süsswasser-Fauna von Südafrika," in *Zoologische Jahrbücher*, **10** (1897), 135–200; *Der indo-australische Archipel und die Geschichte siener Tierwelt* (Jena, 1902); and *Die Säugetiere* (Jena, 1904), which appeared in a 2nd, enl. ed., 2 vols. (Jena, 1927–1928), completed by a section on paleontology by Othenio Abel.

HANS QUERNER

WEBER, WILHELM EDUARD (*b*. Wittenberg, Germany, 24 October 1804; *d*. Göttingen, Germany, 23 June 1891), *physics*.

Weber was one of twelve children of Michael Weber, professor of theology at the University of Wittenberg. Of four brothers and a sister who lived to an advanced age, the eldest brother became a minister, while the other brothers turned to science and medicine. Ernst Heinrich, who was almost ten years older than Wilhelm, became a leading anatomist and physiologist, and a professor at Leipzig. Eduard, a year and a half younger than Wilhelm, also became professor of anatomy at Leipzig. The interest of the three brothers in science was undoubtedly awakened by the family friends Christian August Langguth, professor of medicine and natural history, in whose house the Webers lived, and the acoustician E. F. Chladni, a fellow lodger.

Langguth's house was burned during the bombardment of Wittenberg by the Prussians in 1813, and in the following year the Webers settled in Halle. Michael Weber became professor of theology at the University of Halle, with which the University of Wittenberg officially merged in 1817. Here Wilhelm began his first scientific work, in collaboration with Ernst Heinrich. The resulting publication, *Wellenlehre, auf Experimente gegründet* (1825), which contains experimental investigations of water and sound waves, made Wilhelm's name known in scientific circles. Some of the experimental work on waves was done before Wilhelm entered the University of Halle in 1822. There he was most influenced by the physicist J. S. C. Schweigger and perhaps by the mathematician J. F. Pfaff. Weber wrote his doctoral dissertation on the theory of reed organ pipes in 1826 under Schweigger. His *Habilitationsschrift* (1827) treated such systems as coupled oscillators, as did four papers in Poggendorff's *Annalen der Physik und Chemie* (1828, 1829). Weber became lecturer and then assistant professor (1828) at Halle. He traveled to Berlin in September 1828 with Ernst Heinrich to attend the seventh meeting of the Gesellschaft Deutscher Naturforscher und Ärzte, organized by Alexander von Humboldt. He delivered a talk on his work on organ pipes that attracted the notice of Humboldt and Gauss. At this time Humboldt interested Gauss in his work on geomagnetism, and Gauss saw in Weber a worthy co-worker if a position became available for him at Göttingen.

In April 1831 the professorship of physics at Göttingen, vacated upon the death of Tobias Mayer, Jr., was offered to Weber; and six years of collaboration and close friendship with Gauss followed. At the end of 1832 Gauss read his paper "Intensitas vis magneticae terrestris ad mensuram absolutam revocata," written with Weber's assistance. In this paper he introduced absolute units of measurement into magnetism; that is, the measurement of the strength of a magnetic property was reduced to measurements of length, time, and mass, and thus became reproducible anywhere without the need of a particular precalibrated magnetic instrument. One of the major themes of Weber's later work was to extend this idea to electrical measurements.

Gauss and Weber founded the Göttingen Magnetische Verein to initiate a network of magnetic observatories and to correlate the resulting measurements. This was to be a more sophisticated version of Humboldt's project. In 1833 they set up a battery-operated telegraph line some 9,000 feet long between the physics laboratory and the astronomical observatory, in order to facilitate simultaneous magnetic observations. This was one of the first practical long-range galvanic telegraphs. A year later, induced currents were used in place of the battery. The *Resultate aus den Beobachtungen des Magnetischen Vereins* for the years of its existence (1836–1841), published from 1837 to 1843, contain mostly articles by Gauss and Weber, although in the later volumes observations were published from many stations throughout the world. Weber's major contribution during this period was the development of sensitive magnetometers and other magnetic instruments.

Busy as he was with magnetism at Göttingen, Weber found time to collaborate with his younger brother Eduard on *Mechanik der menschlichen Gehwerkzeuge* (1836). This work on the physiology and physics of human locomotion represented a continuation of the close bond of the three brothers in scientific research, which had begun with the *Wellenlehre*.

With the death of William IV in 1837, Victoria became queen of England and her uncle, Ernst August, acceded to the rule of Hannover and at once revoked the liberal constitution of 1833. Weber was one of seven Göttingen professors who signed a statement of protest. (The others of the "Göttingen Seven" were F. E. Dahlmann, W. E. Albrecht, Jakob and Wilhelm Grimm, G. Gervinus, and G. H. von Ewald.) At the king's order all seven lost their positions; and Dahlmann, Gervinus, and Jakob Grimm were exiled from Hannover. The Seven received much sympathy from all over Germany; in particular, a committee was formed at Leipzig to raise funds to support them. Despite the loss of his position, Weber continued to work for the Magnetische Verein in Göttingen. Gauss and Humboldt attempted to obtain Weber's reinstatement; but the king insisted on a public retraction, which was unacceptable to Weber. Between March and August 1838 he traveled to Berlin; to London, where he spoke with John Herschel about extending the network of magnetic observing stations; and to Paris, becoming acquainted with many of the leading scientists of his time.

After several years in Göttingen without a university position, Weber became professor of physics at the University of Leipzig in 1843, joining his brothers Ernst Heinrich and Eduard. This position had been held by G. T. Fechner, a close friend of the Webers. Because of severe eyestrain induced by his psychophysical experimentation, which led to temporary blindness, Fechner had to relinquish the post, turning afterward to philosophy and psychology. At Leipzig, Weber formulated his law of electrical force, published in the first of his *Elektrodynamische Maassbestimmungen* (1846). Weber and Fechner, who was a staunch atomist, often discussed scientific matters; and the law is adumbrated in a semiquantitative treatment by Fechner in 1846, which refers to Weber's forthcoming work.

The upheavals of 1848 forced a greater liberality upon Ernst August, and in the following year Weber was able to return to his old position at Göttingen. At his request his replacement, J. B. Listing, was retained, thus creating a double professorship in physics. Weber became director of the astronomical observatory and was closely associated with Rudolph Kohlrausch, a friend for some years who had proposed to test Weber's force law directly, using mechanically accelerated charges. As this was not feasible, Weber in 1856 collaborated with Kohlrausch, then at Marburg, to determine the ratio between the electrodynamic and electrostatic units of charge (the former being greater than the presently used electromagnetic unit by the factor $\sqrt{2}$). This measurement was later used by Maxwell as a crucial support for his electromagnetic theory of light. In 1857 Kohlrausch moved from Marburg to Erlangen and began research with Weber on electrical oscillations, but died in the following year. His son, Friedrich, received the doctorate in 1863 at Göttingen with a thesis on elastic relaxation in metal wires, written under Weber's direction, thus extending an investigation Weber had made on nonmetallic fibers. Friedrich later became lecturer at Göttingen and organized the physical laboratory course at Weber's request.

Weber's later years at Göttingen were devoted to work in electrodynamics and the electrical structure of matter. He retired in the 1870's, relinquishing his duties in physics to his assistant, Eduard Riecke. Toward the end of the century, the latter began the development of the electron theory

of metals from Weber's ideas, a development soon carried to its completion in classical physics by Paul Drude and H. A. Lorentz.

Weber's closest collaborator in his last years was the Leipzig astrophysicist J. K. F. Zöllner, with whom he worked on electrical conductivity. Zöllner envisaged a physics based solely on the interaction of atoms of the two kinds of electricity, a conception taken up by Weber after Zöllner's death in 1882 and left in manuscript as the last of the *Elektrodynamische Maassbestimmungen.*

The career of Hermann von Helmholtz touched that of Weber at several points, and relations between the men were strained. Results in Helmholtz' memoir on the conservation of energy (1847) were at first taken to imply that Weber's force law violated that principle. By introducing a velocity-dependent potential energy, Weber was able to demonstrate that the criticism was unfounded. But in 1870, while investigating the rival electrical theories then extant, Helmholtz found that Weber's law could lead in certain circumstances to states of motion that appeared to be disallowed physically. Weber and his supporters attempted to refute Helmholtz' arguments; and a rather sterile but sometimes bitter dispute lasted for several years, with Zöllner championing Weber's cause with more ardor than tact. At an international congress on the electrical units held in Paris (1881) Helmholtz, the leader of the German delegation, proposed the name "ampere" for the unit of current, although "weber" enjoyed some use at that time. The term "weber" was officially introduced for the practical unit of magnetic flux in 1935.

Weber died peacefully in his garden at the age of eighty-six. He had received many honors from Germany, France, and England, including the title of *Geheimrat* and the Royal Society's Copley Medal. He was described as being friendly, modest, and unsophisticated. His reputation had suffered in the 1870's when Zöllner introduced the American medium, Henry Slade, into the Leipzig circle of which Fechner and the Webers were the leading lights. Weber enjoyed hiking and did much traveling on foot. He never married, his household being sometimes managed by his sister and, in his later years, by his niece.

Wellenlehre, auf Experimente gegründet, which marks the beginning of Wilhelm's scientific career, describes experiments on surface waves in liquids, and on sound and light waves. It is dedicated to Chladni, the family friend who was famed for his experiments on standing waves in plates. The immediate inspiration was the chance observation of standing waves in mercury by Ernst Heinrich. Traveling and standing water waves are described and illustrated in engravings made by the brothers. Using a narrow channel with glass walls, they investigated the dependence of wave velocity on the depth of the water, noted the dispersion of wave packets and the distinction between capillary and gravity waves, and investigated the effect of oil on water and wave interference. The elliptical motion of particles in the water as a wave passes was described. In a historical section they compared their results with contemporary theory, particularly Poisson's. Vortices were treated briefly, without proper comprehension of their peculiarities, an understanding of which did not come until later in the century. The section on sound waves treated the problems connected with resonance in pipes, a field that formed the subject of Wilhelm's next works. Ernst Heinrich also utilized the fruits of this investigation in a later treatise on the circulation of the blood.

Weber's early scientific work—before he was called to Göttingen—centered on acoustics. Several of the earlier papers involve the repetition of experiments of previous investigators. Others develop the discovery in the *Wellenlehre* of the distribution of sound around a tuning fork. His doctoral dissertation (1826) and *Habilitationsschrift* (1827) deal with the acoustic coupling of tongue and air cavity in reed organ pipes. This experimental and theoretical investigation was pursued in papers in *Annalen der Physik und Chemie* (1828–1830). One of the subjects treated was the use of this coupling to maintain constancy of pitch of a pipe under different intensities of blowing, and the possibility that this might provide an improved standard of pitch.

The *Mechanik der menschlichen Gehwerkzeuge* (1836) was a collaborative work by Wilhelm Weber and his younger brother, the anatomist Eduard. It contains an anatomical discussion of the joints used in walking and running, measurements made on living subjects, and a mathematical theory relating the length and duration of a step to anatomical parameters. Drawings were made on the basis of the theory and viewed stroboscopically. Among other results, the work corrected misconceptions about posture and recommended its conclusions to the attention of artists. The introduction suggests the development of a walking machine for traversing rough terrain.

The papers contributed by Weber to *Resultate aus den Beobachtungen des Magnetischen Vereins*

(1836–1841) are concerned partly with the construction of galvanomagnetic instruments, including a beautifully designed portable magnetometer, magnetometers working by electromagnetic induction, and a dynamo. This led him to investigate the dependence of magnetization on temperature and inspired other investigations of unipolar induction and elasticity. Working with the silk fibers used in the magnetometer suspensions, Weber found that aside from the immediate elastic response to a change in load, there was a slow but apparently elastic relaxation (*elastische Nachwirkung*) involving a delayed stretching or shrinking as the stress increased or decreased, respectively. Weber sought to provide a molecular explanation for the phenomenon, and his papers on the subject were published in 1835 and 1841. In the later volumes of the *Resultate*, Weber summarized the results obtained from the various geomagnetic observing stations and helped to create the lithographed maps showing the earth's magnetism. They also contained Weber's first work on extending the idea of absolute units, introduced into magnetism by Gauss, to galvanic measurements.

In the *Resultate* for 1840, Weber defined the absolute electromagnetic unit of current in terms of the deflection of the magnetic needle of a tangent galvanometer. He determined the amount of water decomposed by the flow of a unit of current for one second—that is, by a unit of charge. In the first of the *Elektrodynamische Maassbestimmungen* (1846) he introduced his electrodynamometer, in which a coil is hung by its leads in bifilar suspension in the field of another coil and the current is passed through both. This instrument was used to determine the electrodynamic unit of current, defined in terms of the force between two current elements using Ampère's law, and having a magnitude $\sqrt{2}$ greater than the electromagnetic unit. The response of the electrodynamometer depends on the square of the current and thus is suited for alternating currents. With it Weber measured currents alternating with acoustical frequencies.

In 1846 M. H. von Jacobi circulated an especially prepared copper wire to be used as a resistance standard. Weber was dissatisfied, however, with standards depending on the resistance of a particular object or on the resistivity of a particular substance, and in the *Elektrodynamische Maassbestimmungen* of 1852 he defined an absolute measure for electrical resistance. By use of Ohm's law and the absolute measure of current, the problem is reduced to that of voltage measurement. Weber defined this by the voltage induced in a loop rotating in a given magnetic field. Several practical methods of determining resistance were presented.

In 1855 Weber collaborated with R. Kohlrausch on the measurement of the ratio between the electrodynamic and electrostatic units of charge; their results were published in 1857 as one of the *Elektrodynamische Maassbestimmungen*. A definite small fraction of the charge used was drawn off a large capacitor and measured electrostatically by a Coulomb torsion balance, and the remaining charge was discharged through a ballistic galvanometer. Converting to the ratio between the electromagnetic and electrostatic units, the ratio found was 3.1074×10^8 meters/second, close to the speed of light; but the researchers took no special notice of this.

Weber's greatest theoretical contributions appear in the *Elektrodynamische Maassbestimmungen*, seven long works published from 1846 to 1878, besides a manuscript published posthumously. In the first of these, Weber introduced his dynamometer to test Ampère's law of force between electric current elements, to a degree of precision exceeding Ampère's, and also to investigate electromagnetic ("Volta") induction. Convinced of the validity of Ampère's law, Weber proceeded to a theoretical derivation of a general fundamental law of electrical action, expressing the force between moving charges. Essential to the derivation were the assumptions of central forces and of currents as consisting of the equal and oppositely directed flow of the two kinds of charge. The law contains the expression

$$F = \frac{e_1 e_2}{r^2} \left[1 - \frac{1}{c^2} \left(\frac{dr}{dt} \right)^2 + \frac{2r}{c^2} \frac{d^2 r}{dt^2} \right]$$

where dr/dt is the rate at which the separation r between the charges e_1 and e_2 increases (or the relative radial velocity), while $d^2 r/dt^2$ is the relative radial acceleration, and c is a constant expressing the ratio between the electrodynamic and electrostatic units of charge. On the basis of Maxwell's theory, we know today that c is $\sqrt{2}$ times the speed of light.

The dominant term in the expression is the Coulomb force $e_1 e_2/r^2$. The remaining terms modify this attraction or repulsion when the charges are in motion relative to each other. Thus, envisage, as a simple example, two straight, parallel wires carrying identical currents in the same direction, and consider the forces between elements of the two wires that are side by side. Like charges in the two elements have no motion relative to each other and

will repel with the Coulomb force. Unlike charges, at the point where they move past each other, are neither approaching nor receding from each other; but the acceleration $d^2 r/dt^2$ is positive. This serves to augment the Coulomb attraction between unlike charges in the two wires, which otherwise would simply cancel the repulsion between the like charges. A net attraction between the parallel currents, and hence between the wires, results, in agreement with Ampère's findings.

Suppose, instead, that only one of the wires, A, carries a current initially and that the other wire is moved toward it. Consider the forces on the charges contained in an element B of the second wire. Charges in A that are approaching the point C opposite B are associated with higher values of $(dr/dt)^2$ than charges in A that have moved beyond this point, because of the approach of B to A. Since the velocity-dependent term always diminishes the effect of the Coulomb force, the positive charge in A approaching C has a diminished repulsion on the positive charge in B and the negative charge in A approaching C from the opposite direction has a diminished attraction—that is, more diminished than the forces excited by charges moving away from C. The resultant force on the positive charge in B is opposite to the motion of the current (that is, the positive charge) in A. The force on the negative charge in B is of the same magnitude and in the opposite direction. The net effect on the charges in B is to accelerate them in directions opposite to their counterparts in A—that is, to induce a current in B opposed to the direction of that in A. In fact, Weber's law succeeded in encompassing Ampère's force and the facts of induction, as well as Coulomb's law of electrostatics. Note, however, that if the assumption of the equal and opposite flow of unlike charges in a current is dropped, a constant current would generally exert a force on a static charge, contrary to experience. An abridged version of the paper published two years later also gives a potential function from which the force may be derived.

In the meantime, Helmholtz' memoir on the conservation of energy had appeared, which seemed to disallow velocity-dependent forces. Weber's law depended on the radial components of the relative velocity and acceleration of the charges; but Weber was able to show in 1869, and in greater detail in another of his long papers of 1871, the consistency of his law with energy conservation in a somewhat extended sense. Most decisive for the eventual rejection of Weber's law was its gradual replacement by Maxwell's field theory, particularly with the demonstration of electromagnetic radiation by Hertz in 1888.

From 1848 to 1852 Weber reported his careful quantitative experimental work on the diamagnetism of bismuth. Diamagnetism had been investigated in 1845 by Faraday, who initially interpreted the phenomena in terms of diamagnetic polarity, that is, a reversed magnetic polarity created in the substance when it is introduced into a magnetic field. In 1848 Weber claimed that he had observed induction in a coil caused by the diamagnetism of a piece of bismuth moving in a magnetic field, but at this time he did not distinguish this effect from that of the bulk currents induced in the body of the bismuth. Faraday was unable to observe such an effect attributable to the diamagnetic, rather than to the conducting, property of his samples; and partly as a result, he relinquished his conception of diamagnetic polarity. In research reported in his *Maassbestimmungen* of 1852, however, Weber was able to isolate and demonstrate the existence of the diamagnetic effect. He utilized the effective uniformity of the magnetic field in a long, straight solenoid. A bismuth cylinder moving well inside such a solenoid will not have bulk currents induced in it, but through its motion its diamagnetism will affect surrounding magnetic detectors. In Weber's beautifully designed experiments, such an effect was demonstrated both by the motion of a suspended magnetic needle and by the induction of a current in a surrounding coil.

Weber extended Ampère's theory of magnetism to cover the phenomenon of diamagnetism. In Ampère's theory, ordinary magnetism is accounted for by assuming the existence of permanent molecular electric currents circulating in the molecules of ferromagnetic substances. In an external field these molecules align themselves to give the resulting magnetization. According to Weber, diamagnetism occurs when resistanceless molecular currents are induced in diamagnetic substances. These substances are characterized by molecules that do not contain permanent currents and that have fixed orientation in the substance.

From 1852 Weber attempted to comprehend electrical resistance as a result of the motion of electric fluids or particles. Resistance was presumed to have its cause in the repeated combination and separation of the particles of the two electric fluids, the opposite motions of which composed a current. The existence of permanent Ampèrian currents led Weber to assume that the electric fluids do not interact directly with the material atoms composing the substance, and that in mag-

netic atoms the two kinds of fluid circulate in different, nonintersecting paths about the atoms. At this time he discussed, as a model, a lattice of fixed positive charges about which negative particles rotate in Keplerian ellipses. On application of an electric potential, the negative particles move in widening spirals until they pass over to the region of influence of the neighboring atom, thus migrating along the conductor.

In his article of 1875 and his final, unpublished paper, Weber developed these ideas in an attempt to derive an expression for electrical conductivity in terms of molecular parameters. Success in this was not achieved until the work of Eduard Riecke, Paul Drude, and H. A. Lorentz around the turn of the century; their work introduced the idea of treating the conduction electrons as a gas. Interest in the particulate theory of electricity quickened in the last decades of the century, especially after J. J. Thomson's investigation of cathode rays and Lorentz' interpretation of the Zeeman effect. Nevertheless, Weber's attempts to understand electrical conductivity, as well as thermal conductivity in metals, and thermoelectricity by means of the motion of the electric particles were a very important influence on the later investigations.

In metals, heat presumably was conducted by the jumping of electric particles between the ponderable molecules, those jumping from hotter molecules possessing greater speeds. In insulators, where such a mechanism was ruled out, Weber believed that heat was distributed by radiation through the ether permeating the material (articles of 1862 and 1875). In his doctoral dissertation (1858) Carl Neumann had attempted to explain the magnetic rotation of the plane of polarization of light by assuming an interaction between the Weberian molecular currents and the neighboring ether. Extending these ideas, Weber suggested that the frequency of light emitted by molecules would be the same as the frequency of motion of the electrical particles in the molecular currents. In this connection he developed his planetary model in 1862 and 1871, with charge of one sign fixed to the massive molecule and the oppositely charged electrical particles orbiting around it in accordance with his law of force.

An interesting consequence of Weber's law is that stable, bound orbits exist for two particles of the same sign of charge. Weber speculated that the ether might be composed of particles of like charge bound together, and that the development of the theory of their motion in accord with his law might lead to an understanding of the laws of light and heat radiation. In implementing this view, however, he succeeded no better than Ampère, who had indulged in similar speculations.

Toward the end of his life, Weber developed ideas appearing in Zöllner's *Die Principien einer elektrodynamischen Theorie der Materie* (1876). These included the concept of all matter being compounded of electrically charged particles, held together in various stable configurations by the action of the Weberian law of force. Even gravitation could be subsumed in this unitary picture by adopting in essence the earlier hypotheses of Aepinus and of Mossotti, that the attractive electrostatic forces between unlike charges slightly outbalance the repulsive forces between like charges.

Although he was perhaps most widely known during his life for his law of force, which was discarded with the triumph of Maxwell's field theory, Weber left his more lasting impression on physical theory with his atomistic conception of electric charge and his vision of the role of such charges in determining the electrical, magnetic, and thermal properties of matter.

BIBLIOGRAPHY

I. ORIGINAL WORKS. Weber's writings are collected in *Wilhelm Webers Werke*, 6 vols. (Berlin, 1892–1894). Many articles are in *Resultate aus den Beobachtungen des Magnetischen Vereins*, Carl Friedrich Gauss and Wilhelm Weber, eds., 6 vols. (Göttingen–Leipzig, 1837–1843).

II. SECONDARY LITERATURE. See Eduard Riecke, *Wilhelm Weber*, Rede (Göttingen, 1892); Heinrich Weber, *Wilhelm Weber, eine Lebensskizze* (Breslau, 1893); and K. H. Wiederkehr, *Wilhelm Eduard Weber, Erforscher der Wellenbewegung und der Elektrizität 1804–1891* (Stuttgart, 1967), vol. xxxii of Grosse Naturforscher, H. Degen, ed. The last contains an extensive bibliography.

A. E. WOODRUFF

WEBSTER, JOHN (*b.* Thornton, Craven, England, 3 February 1610; *d.* Clitheroe, England, 18 June 1682), *chemistry, medicine, education.*

Although Webster implied that he studied at Cambridge, there is no record that he was ever a regular student. He also referred to his study of chemistry (*ca.* 1632) under the Hungarian alchemist John Hunyades, who arrived in London sometime after 1623. As with other Renaissance chemists, Webster's interest in chemistry was easily coupled with his concern for religion, and he was ordained a minister sometime after July 1632. Two years later he appears in the records as the curate of Kildwick, in Craven.

Paracelsian chemistry had a special appeal for surgeons, and there is a large iatrochemical literature specifically aimed at the military surgeon. As a Puritan, Webster served both as a surgeon and as a chaplain in the Parliamentary army during the Civil War. By 1648 his opposition to the established church had pushed him into the ranks of the nonconformists; and after the Restoration he was forced to support himself as a "practitioner in Physick and Chirurgery."

It was his concern for those who were preparing for the ministry that led Webster to write the *Academiarum examen*, in which he attacked the English universities. The traditional emphasis on books and disputations, as well as on the "heathen" authors Aristotle and Galen, seemed to him improper for Christians, who should study the glories of the universe (and thus, the Creator) through observation and personal experience. Webster argued against the use of mathematical abstraction in the study of nature, because for him this seemed to emphasize deductive logic. In contrast, the laboratory observations of the chemists offered the proper inductive approach exemplified in the writings of Helmont and Francis Bacon. The *Academiarum examen* is deeply indebted to Robert Fludd's Rosicrucian apology, *Tractatus apologeticus* (1617); and Webster points to Fludd and Bacon as the two authors most to be relied upon in formulating a new philosophy of nature. The most notable reply to Webster's call for educational reform was *Vindiciae academiarum* (1654) of Seth Ward and John Wilkins, in which Webster was taken to task for not having kept abreast of recent changes at the universities that did reflect the new science. He was accused of not having properly understood Bacon and Descartes and also was criticized for his reliance on the chemists. His espousal of Fludd's texts was especially condemned. The conflict between Webster, Ward, and Wilkins clearly points to the sharp division then existing between the chemical philosophers and the early mechanists.

Webster's belief that the aim of true natural magic was to uncover the "secret effects" of nature led him to extend warm support to the foundation of the Royal Society of London; and there is no indication that he was ever disappointed with the course taken by its members. He referred with approval to the society's work in his *Metallogra-*

phia (1671), an interesting compendium of current views on the growth and properties of metals. Here again he indicated his debt to Paracelsus, who made chemistry available to all, and to Helmont, whose work seemed to excel that of all his predecessors. The *Metallographia* was reviewed in the *Philosophical Transactions of the Royal Society*, and Daniel George Morhof later praised it as one of the major published works on minerals.

Webster was no less laudatory to the Royal Society in *The Displaying of Supposed Witchcraft*, completed in 1673 but not published until 1677, in which he attacked the views of Meric Casaubon and Joseph Glanvill. The latter, a member of the Royal Society, had written at length on witchcraft, especially in *Philosophical Considerations Touching Witches and Witchcraft* (1666). In his reply Webster countered that "supernatural" effects supposedly caused by witchcraft would eventually be found to have natural causes.

Although not a scientist of major stature, Webster is significant, for his work reflects important themes germane to the period of the scientific revolution. His conflict with Ward and Wilkins underscored the dispute between the chemical philosophers and the mechanists, his treatise on witchcraft did much to shed light on the meaning of magic and the supernatural in this period, and his work on metals and minerals clearly was considered important by his contemporaries on the Continent as well as in England.

BIBLIOGRAPHY

I. ORIGINAL WORKS. There is a bibliography of Webster's works, including his many sermons, in Bertha Porter's article in the *Dictionary of National Biography*. The most important for the historian of science are *Academiarum examen, or the Examination of Academies* (London, 1654); *Metallographia: or An History of Metals* (London, 1671); and *The Displaying of Supposed Witchcraft* (London, 1677).

Academiarum examen has been reprinted, along with the replies of Ward and Wilkins (*Vindiciae academiarum* [Oxford, 1654]) and Thomas Hall (*Histrio-Mastix. A Whip for Webster* [London, 1654]), by Allen G. Debus, in *Science and Education in the Seventeenth Century* (London–New York, 1970).

Metallographia is reviewed in *Philosophical Transactions of the Royal Society*, **5**, no. 66 (12 Dec. 1670), 2034–2036. Daniel George Morhof, *Polyhistor, literarius, philosophicus et practicus*, 4th ed., II, pt. 2 (Lübeck, 1747), sec. 4, ch. 29, is devoted mainly to *Metallographia*.

II. SECONDARY LITERATURE. There is a discussion of Webster's work, with special reference to the educational problems raised in *Academiarum examen*, in the introductory essay in Allen G. Debus, *Science and Education in the Seventeenth Century*, 1–65. See also Debus' "The Webster-Ward Debate of 1654: The New Philosophy and the Problem of Educational Reform," in *L'univers à la Renaissance: Microcosme et macrocosme*, Travaux de l'Institut pour l'étude de la renaissance et de l'humanisme, IV (Brussels, 1970), 33–51; and "John Webster and the Educational Dilemma of the Seventeenth Century," in *Actes du XIIe Congrès international d'histoire des sciences*, IIIB (Paris, 1970), 15–23. An early, but still useful, discussion of the debate is in Richard F. Jones, *Ancients and Moderns. A Study of the Rise of the Scientific Movement in Seventeenth-Century England* (St. Louis, 1936, 2nd ed., 1961), 101–114.

For references pertinent to education in England and on the Continent, see Debus, *Science and Education in the Seventeenth Century*; but for a further understanding of Webster, see also P. M. Rattansi, "Paracelsus and the Puritan Revolution," in *Ambix*, **11** (1963), 23–32; C. Webster, ed., *Samuel Hartlib and the Advancement of Learning* (Cambridge, 1970), 1–72; and C. Webster, "Science and the Challenge to the Scholastic Curriculum 1640–1660," in *The Changing Curriculum* (London, 1971), 21–35.

ALLEN G. DEBUS

WEBSTER, THOMAS (*b.* Orkney Islands, Scotland, 1773; *d.* London, England, 26 December 1844), *geology.*

Webster came to London early in life and trained as an architect. In 1799 he was appointed clerk of the works at the newly founded Royal Institution and built its famous lecture theater. Later, taking up geology as a profession, he became curator, draughtsman, librarian, secretary, and editor to the Geological Society (founded in 1807). Leaving this employment in 1827, he took up public lecturing and in 1841 became the first professor of geology at University College.

Webster's original geological work was entirely concerned with the stratigraphy of the uppermost Jurassic, the Cretaceous, and the Tertiary rocks of southern England. In his chief paper (1814) Webster recorded a highly important piece of research, perhaps the first of its kind (being detailed and thoroughly scientific) on the geology of a British region. He described the characters of the Tertiary strata of southeast England, and particularly those of the Oligocene (later so called) of the Isle of Wight, in which he recognized an alternation of marine and freshwater formations. Webster compared all of these Tertiary strata with those that had been recently described in the Paris Basin by Cuvier and Brongniart. Of even greater importance was his survey of the Isle of Wight (Cretaceous and Tertiary) and "Isle" of Portland (mostly Jurassic and Cretaceous), with the production of one of the first geological maps of any part of Britain and one that was on a larger scale and more accurate than anything previously attempted. The geology of this classic region was almost completely unknown before Webster sketched, in words and picture, the features of its structure (essentially a monocline with vertical limb) in a series of lucid, forceful, and entertaining "letters" (1816). Unfortunately for the progress of geological knowledge, his duties at the Geological Society pressed so heavily upon him that he was prevented from undertaking any further large-scale investigations.

BIBLIOGRAPHY

I. ORIGINAL WORKS. Webster's autobiography, which was probably written about 1837, is in MS in the library of the Royal Institution, London. Of Webster's ten publications, which are enumerated in Challinor (1964), the following publications are most important: "On the Freshwater Formations of the Isle of Wight, With Some Observations on the Strata Over the Chalk in the Southeast Part of England," in *Transactions of the Geological Society of London*, **2** (1814), 161–254; and "Geological Observations on the Isle of Wight and Adjacent Coast of Dorsetshire in a Series of Letters," in Sir Henry Englefield, *The Picturesque Beauties of the Isle of Wight* (London, 1816), 117–238.

II. SECONDARY LITERATURE. On Webster and his work, see G. S. Boulger, *Dictionary of National Biography*, LX (1899), 126; J. Challinor, "Thomas Webster's Letters on the Geology of the Isle of Wight, 1811–1813," in *Proceedings of the Isle of Wight Natural History and Archaeological Society*, **4** (1949), 108–122; "Some Correspondence of Thomas Webster, Geologist," in *Annals of Science*, **17** (1961), 175–195; **18** (1962), 147–175; **19** (1963), 49–79, 285–297; **20** (1964), 59–80, 143–164; N. Edwards, "Thomas Webster (*circa* 1772–1844)," in *Journal of the Society of Bibliography of Natural History*, **5** (1971), 468–473; E. Forbes, *The Tertiary Fluvio-Marine Formation of the Isle of Wight*, one of the *Memoirs of the Geological Survey of Great Britain* (1856); K. D. C. Vernon, "The Foundation and Early Years of the Royal Institution," in *Proceedings of the Royal Institution of Great Britain*, **39** (1963), 364–402; and H. B. Woodward, *History of the Geological Society of London* (London, 1908), *passim*.

JOHN CHALLINOR

WEDDERBURN, JOSEPH HENRY MACLAGAN (*b*. Forfar, Scotland, 26 February 1882; *d*. Princeton, New Jersey, 9 October 1948), *mathematics*.

Wedderburn was the tenth of fourteen children. His father, Alexander Wedderburn, was a physician in a family of ministers (on his father's side) and lawyers (on his mother's side). In 1898 Wedderburn matriculated at the University of Edinburgh; in 1903 he received an M.A. degree with first-class honors in mathematics. No doubt influenced by the work of Frobenius and Schur, he went to Leipzig and Berlin in 1904. During the same year he proceeded to the United States as a Carnegie fellow at the University of Chicago (E. H. Moore and L. E. Dickson were there). From 1905 to 1909 he was lecturer at the University of Edinburgh and assistant to Chrystal. During this time Wedderburn edited the *Proceedings of the Edinburgh Mathematical Society*, and in 1908 was awarded the doctorate of science.

In 1909 Wedderburn became one of the "preceptors" appointed under Woodrow Wilson at Princeton University. At the outbreak of World War I he enlisted in the British Army and fought in France. After the war he returned to Princeton, where he continued to teach until his retirement in 1945. When the mathematics department at Princeton assumed responsibility for publishing the *Annals of Mathematics*, Wedderburn was its editor from 1912 to 1928. Toward the close of the 1920's, he suffered what appears to have been a nervous breakdown. He led an increasingly solitary life and retired from his university post some years before the normal time. Wedderburn published thirty-eight papers and a textbook, *Lectures on Matrices* (1934), which in the last chapter contains an excellent account of his theorems and their background as well as some original contributions to the subject.

Wedderburn's mathematical work includes two famous theorems, which bear his name; both were established in the years 1905–1908. Before Wedderburn began his investigations, the classification of the semisimple algebras was done only if the ground field was the field of real or complex numbers. This did not lead to deeper insight into hypercomplex numbers (linear associative algebra). Wedderburn attacked the problem in a completely general way and introduced new methods and arrived at a complete understanding of the structure of semisimple algebras over any field. He showed that they are a direct sum of simple algebras and finally—in a celebrated paper ("On Hyper-complex Numbers") that was to be the beginning of a new era in the theory—proved that a simple algebra consists of all matrices of a given degree with elements taken from a division algebra.

Wedderburn's second important contribution concerns the investigation of skew fields with a finite number of elements. The commutative case had been investigated before Moore in 1903, and had led to a complete classification of all commutative fields with a given number of fields. Moore showed that for a given number p^r of elements there exists (apart from isomorphisms) only one field, namely the Galois field of degree r and characteristic k. Since a noncommutative finite field had never been found, one could suspect that it did not exist. In 1905 Wedderburn showed that every field with a finite number of elements is indeed commutative (under multiplication) and therefore a Galois field. This second theorem ("A Theorem on Finite Algebras") gives at once the complete classification of all semisimple algebras with a finite number of elements. But the theorem also had many other applications in number theory and projective geometry. It gave at once the complete structure of all projective geometries with a finite number of points, and it showed that in all these geometries Pascal's theorem was a consequence of Desargues's theorem. The structure of semisimple groups was now reduced to that of noncommutative fields. Wedderburn's theorem had been the special case of a more general Diophantine property of fields and thus opened an entirely new line of research.

BIBLIOGRAPHY

I. ORIGINAL WORKS. Wedderburn's works are "A Theorem on Finite Algebras," in *Transactions of the American Mathematical Society*, **6** (1905), 349–352; "Non-Desarguesian and Non-Pascalian Geometries," *ibid.*, **8** (1907), 379–388, written with O. Veblen; "On Hypercomplex Numbers," in *Proceedings of the London Mathematical Society*, **6**, 2nd ser. (1907–1908), 77–118; "The Automorphic Transformation of a Bilinear Form," in *Annals of Mathematics*, **23**, 2nd ser. (1921–1923), 122–134; "Algebraic Fields," *ibid.*, **24**, 2nd ser. (1922–1923), 237–264; "Algebras Which Do Not Possess a Finite Basis," in *Transactions of the American Mathematical Society*, **26** (1924), 395–426; "A Theorem on Simple Algebras," in *Bulletin of the American Mathematical Society*, **31** (1925), 11–13; "Non-commutative Domains of Integrity," in *Journal für die reine und angewandte Mathematik*, **167** (1931), 129–141; *Lectures on Matrices* (New York, 1934); "Boolean Linear Associative Algebra," in *Annals of*

Mathematics, **35**, 2nd ser. (1934), 185–194; and "The Canonical Form of a Matrix," *ibid.*, **39**, 2nd ser. (1938), 178–180.

II. SECONDARY LITERATURE. See E. Artin, "The Influence of J. H. M. Wedderburn on the Development of Modern Algebra," in *Bulletin of the American Mathematical Society*, **56** (1950), 65–72.

HENRY NATHAN

WEDEL, GEORG WOLFGANG (*b.* Golssen, Germany, 12 November 1645; *d.* Jena, Germany, 6/7 September 1721), *medicine, chemistry.*

After receiving elementary instruction from his father, Pastor Johann Georg Wedel, Wedel entered the famous school in Schulpforta with a scholarship from the Saxon Elector in 1656. He spent five successful years at Schulpforta and then proceeded to the University of Jena, where he studied philosophy and especially medicine. He also participated in disputations, witnessed dissections, acquired an iatrochemical manuscript, all the while maintaining close relations with Guerner Rolfinck. In 1667 Wedel practiced briefly in Landsberg, toured Silesia, visited Wittenberg and Leipzig, and then returned to Jena, where he qualified for his medical license and started giving lectures. Before the year was out, however, he was called to Gotha as a district physician. While practicing medicine there, he took his M.D. at Jena in 1669.

Three years later, expecting to be appointed to the medical faculty at Jena, Wedel took a brief study tour of Holland. In early 1673, shortly after his return, he assumed the chair of anatomy, surgery, and botany. Then, upon the death of his mentor Rolfinck in the spring, Wedel assumed the chair of theoretical medicine. He held this chair until 1719, when the death of another colleague made it possible for Wedel to rise to the chair of practical medicine and chemistry. Meanwhile, he had received many state honors, including personal ennoblement as count palatine in 1694. He even purchased a country estate. After nearly five decades of teaching and writing, Wedel died as the senior member of Jena's entire faculty.

Wedel stood midway between medieval and modern world views, defending astrology and alchemy and championing iatrochemistry. He was a remarkably prolific author, but it was primarily by teaching at one of Germany's largest universities that he influenced a whole generation of physicians, including Hoffmann and Stahl. (Between 1673 and 1721 the average attendance of the

University of Jena was around 940 students. Roughly five percent of the student body was in medicine. See Franz Eulenburg, "Die Frequenz der deutschen Universitäten von ihrer Gründung bis zur Gegenwart," in *Abhandlungen der Sächsischen Akademie der Wissenschaften*, Philologisch-Historische Klasse, **24**, no. 2 [1904].)

BIBLIOGRAPHY

The most complete bibliography of Wedel's publications appears in Johann Heinrich Zedler, ed., *Grosses vollständiges Universal-Lexicon . . .*, LIII (Graz, 1964), 1804–1820. An autobiographical statement written in 1672 is in the Archives of the Deutsche Akademie der Naturforscher, Leopoldina (Halle, German Democratic Republic).

For assessments of his life and work, see Fritz Chemnitius, *Die Chemie in Jena von Rolfinck bis Knorr* (Jena, 1929), 13–53; Ernst Giese and Benno von Hagen, *Geschichte der medizinischen Fakultät der Friedrich-Schiller-Universität Jena* (Jena, 1958), 167–294; and Lynn Thorndike, *A History of Magic and Experimental Science* (New York, 1958), VII, 196, 202; VIII, 146–443.

KARL HUFBAUER

WEDGWOOD, JOSIAH (*b.* Burslem, England, 12 July 1730; *d.* Etruria, England, 3 January 1795), *ceramic technology, chemistry.*

Wedgwood was one of the progressive British industrialists of the eighteenth century whose careers touched the world of science. His father, Thomas Wedgwood, was in the pottery business and his mother, Mary Stringer, was the daughter of a dissenting minister. When Wedgwood was nine years old his father died and, as a result, his schooling ended and his employment in the pottery of his brother Thomas began.

Over the next eleven years Wedgwood mastered the skills of the potter; and after several partnerships he founded his own pottery in 1758. His business affairs soon prospered as his tireless experimental efforts resulted in novel and improved products. In the 1770's these efforts culminated in his greatest success, the jasper ware, which achieved exceptionally pleasant chromatic and textural effects, and which is the product generally brought to mind when "Wedgwood" is used as a description of ceramic products. During the 1760's as Wedgwood enlarged his pottery works, he also built a worker's village, which he named Etruria and where he made his home (Etruria Hall). In 1764 he married a distant cousin, Sarah Wedg-

wood, and the line of descendants (which includes the mother and wife of Charles Darwin) has to this day retained an interest in the Wedgwood potteries.

Wedgwood's position in eighteenth-century science rests on a few minor contributions to experimental chemistry, on his active associations with scientists and scientific societies, and on his general interest in experimental research. During the 1780's he contributed three papers on the measurement of high temperature to the *Philosophical Transactions of the Royal Society of London*. Depending on a property of clay that causes it to shrink as it is heated, the pyrometer ("thermometer for strong fire") that Wedgwood described was seen by him as complementary to the mercurial thermometers that were used to measure low temperature. The device enjoyed some use and caught the interest of both Priestley and Lavoisier. Moreover, as an appendix to one of these papers (1783) Wedgwood described the series of experiments he conducted to evaluate Lavoisier's proposal that heat could be measured by determining the quantity of ice that a warm body could melt.

On several occasions Wedgwood supplied experimental apparatus (pyrometers, retorts, crucibles, and tubing) to various scientists (again including Priestley and Lavoisier) and corresponded with them on experimental procedures. In 1783 Wedgwood was elected a fellow of the Royal Society, but his most significant membership was in the Lunar Society of Birmingham, where he was associated with the foremost British chemists of the period. It was indeed the Lunar Society that reflected the changing structure of industry and technology and anticipated the transformation that saw the affiliation between technology and the crafts loosened in favor of a new affiliation between technology and science.

Wedgwood's interest in experimental chemistry also showed itself in his business and personal affairs. Around 1775 he promoted (unsuccessfully) the formation of an experimental "company" to conduct research on the improvement of porcelain, and he both employed a chemist in his pottery and provided instruction in chemistry for his sons.

Differences of opinion have recently sprung up over the significance of Wedgwood's considerable interest in and knowledge of chemistry. That he read deeply and widely in the field and that he corresponded frequently with chemists on matters of research has been amply demonstrated. There are even indications that in a few instances his technical work on ceramics benefited from his knowledge of the experimental results of chemists. But, despite his habitual use of the terminology of contemporary (phlogiston) chemistry, Wedgwood had only the slightest interest in the cognitive structure of the science, and it seems to have contributed nothing to his industrial exploits. The meaning of his career is rather to be found in its clear statement that during the Industrial Revolution technical men were entering the world of research with the Baconian confidence that technology could learn from science.

BIBLIOGRAPHY

I. ORIGINAL WORKS. The articles that Wedgwood contributed to *Philosophical Transactions of the Royal Society of London* are "An Attempt to Make a Thermometer for Measuring the Higher Degrees of Heat, From a Red Heat Up to the Strongest that Vessels Made of Clay Can Support," **72** (1782), 305–326; "Some Exps. Upon the *Ochra friabilis nigro fusca* of Da Costa Hist. Foss. p. 102; and Called by the Miners of Derbyshire, Black Wadd," **73** (1783), 284–287; "An Attempt to Compare and Connect the Thermometer for Strong Fire, Described in Vol. LXXII of the Philosophical Transactions, With the Common Mercurial Ones," **74** (1784), 358–384; "Additional Observations on Making a Thermometer for Measuring the Higher Degrees of Heat," **76** (1786), 390–408; and "On the Analysis of a Mineral Substance From *New South Wales*," **80** (1790), 306–320.

Many Wedgwood documents are collected in the Wedgwood Museum maintained by Josiah Wedgwood & Sons, Ltd., Barlaston, England. For published collections of Wedgwood's letters, see Ann Finer and George Savage, eds., *The Selected Letters of Josiah Wedgwood* (London, 1965); and Katherine Eufemia Farrer, ed., *Letters of Josiah Wedgwood*, 2 vols. (London, 1903).

II. SECONDARY LITERATURE. The standard biography of Wedgwood is Eliza Meteyard, *The Life of Josiah Wedgwood*, 2 vols. (London, 1865–1866); this work cannot, however, be depended on for an evaluation of Wedgwood's role in science. For an extensive survey of Wedgwood bibliography, see Gisela Heilpern, *Josiah Wedgwood, Eighteenth-Century English Potter: A Bibliography* (Carbondale, Ill., 1967). Wedgwood's association with the Lunar Society is fully discussed in Robert E. Schofield, *The Lunar Society of Birmingham* (London, 1963); see esp. ch. 3 and, for bibliography, p. 455. For a minor dissent from the opinion that Wedgwood was a full member of the Lunar Society, see Eric Robinson, "The Lunar Society: Its Membership and Organization," in *Transactions. Newcomen Society for the Study of the History of Engineering and Technology*, **35** (1964), 153–177.

Many details of Wedgwood's technical work are presented in Robert E. Schofield, "Josiah Wedgwood and the Technology of Glass Manufacturing," in *Technology and Culture*, **3** (1962), 285–297, and in the program of the *Ninth Wedgwood International Seminar: April 23–25, 1964* (New York, 1971), 125–135. Accounts of Wedgwood's role as a chemist are in Eric Robinson, "The Lunar Society and the Improvement of Scientific Instruments: II," in *Annals of Science*, **13** (1957), 1–8; J. A. Chaldecott, "Scientific Activities in Paris in 1791," *ibid.*, **24** (1968), 21–52; and Robert E. Schofield, "Josiah Wedgwood and a Proposed Eighteenth-Century Industrial Research Organization," in *Isis*, **47** (1956), 16–19.

For differing appraisals of the relationship between Wedgwood's chemistry and his technical work, see Robert E. Schofield, "Josiah Wedgwood, Industrial Chemist," in *Chymia*, **5** (1959), 180–192; A. Rupert Hall, "What Did the Industrial Revolution in Britain Owe to Science?", in Neil McKendrick, ed., *Historical Perspectives: Studies in English Thought and Society* (London, 1974), 129–151, esp. 141; Neil McKendrick, "The Role of Science in the Industrial Revolution: A Study of Josiah Wedgwood as a Scientist and Industrial Chemist," in Mikuláš Teich and Robert Young, eds., *Changing Perspectives in the History of Science: Essays in Honour of Joseph Needham* (London, 1973), 274–319; and J. A. Chaldecott, "Josiah Wedgwood (1730–95)—Scientist," in *British Journal for the History of Science*, **8**, no. 28 (1975), 1–16.

HAROLD DORN

WEGENER, ALFRED LOTHAR (*b.* Berlin, Germany, 1 November 1880; *d.* Greenland, November 1930), *meteorology, geophysics.*

Wegener was the son of Richard and Anna Wegener; his father, a doctor of theology, was director of an orphanage. Wegener started his schooling at the Kollnisches Gymnasium in Berlin, studied at the universities of Heidelberg and Innsbruck, and presented a thesis on astronomy at Berlin in 1905. He had meanwhile become interested in meteorology and geology, and a desire to learn at first hand about polar air masses led him to join a Danish expedition to northeastern Greenland in 1906–1908. It was the first of four Greenland expeditions in which he participated, and the exploration of that territory remained one of his dominant interests.

From 1908 to 1912 Wegener was a lecturer in meteorology at the Physical Institute in Marburg. His lectures, noted for their vividness, frankness, and open-mindedness, showed an ability and taste for seizing on broad issues in complicated topics as well as considerable distaste for mathematical detail. In 1912–1913, with Captain J. P. Koch of

Denmark Wegener led his second expedition to Greenland; its emphasis was on glaciology and climatology. From 1914 to 1919 he was mainly a junior military officer. After the war he worked at the meteorological experimental station of the German Marine Observatory at Gross Borstel, near Hamburg. A special professorship in meteorology and geophysics was created for Wegener at the University of Graz in 1924. He went to Greenland as a leader of expeditions in 1929–1930 and 1930–1931. In 1930 he was a member of a party that became lost and suffered severe privations. On 1 November of that year, his fiftieth birthday, he left a base in central Greenland for the west coast and was not seen again.

Wegener's fame today rests on his work as an originator of the idea of continental drift. He stated that he first toyed with the idea in 1910, on noting the degree of apparent correspondence between the shapes of the coasts of the Atlantic on its west and east sides, particularly those of South America and Africa. At first Wegener regarded the idea of drifting continents as improbable, but his interest was rekindled in 1911, when he accidentally learned that evidence of paleontological similarities on both sides of the Atlantic was being used to support the theory that a "land bridge" had once connected Brazil with Africa.

At this time many geologists supported the view that various portions of ocean floors had intermittently risen and fallen in the process of progressive solidification and contraction of the earth from a molten state. This view included the notion that land bridges connecting continents would appear and disappear. Moreover, it was found difficult to reconcile Darwin's theory of evolution with the widely acknowledged similarities in former organic life on different continents except through some connection such as land bridges.

After examining in some detail the paleontological and geological evidence of correspondences on the two sides of the Atlantic, Wegener concluded that the similarities were indeed sufficiently close to demand an explanation, and he linked them to his earlier thinking on continental drift. In a lecture at Frankfurt in January 1912, Wegener announced his theory that continents had actually moved thousands of miles apart during geological time, offering it as an alternative to the land bridge theory. Provisional accounts of the theory appeared in two short papers written in the same year.

During a long sick leave from war service (he was wounded twice), Wegener wrote an extended account of his continental drift theory, which appeared as *Die Entstehung der Kontinente und Ozeane* in 1915. This book was his main work on the subject and incorporated the investigations for which he is now noted. (In the book, Wegener referred to "die Verschiebung der Kontinente," which was accurately translated in the English edition of 1924 as "continental displacement." The term "continental drift" was coined later.)

In his detailed elaboration, Wegener postulated that near the end of the Permian period (about 200 million years ago), there existed a single supercontinent, which he called Pangaea. It subsequently split into several pieces that began to move, generally westward and in some cases toward the equator. In broad terms, America moved westward from Eurasia and Africa to form the Atlantic Ocean, with India later drifting away from Africa and Australia moving toward the equator from Antarctica. At the beginning of the Quaternary period (50 million years ago) Greenland started to separate from Norway. Island arcs, for example, the Antilles, Japan, and the Philippines, were envisaged as detached fragments left in the wakes of drifting continents. Mountain formation was associated with compression of the advancing front of a moving continent against the resistance of the ocean floor.

In his effort to supply a mechanical explanation, Wegener set down an argument, based on a meteorological analogy, to the effect that the continents would drift steadily under an "Eötvös force." He held that such a force could cause a floating body on a rotating planet to drift westward. He interpreted evidence on isostasy as indicating that continents are granitelike bodies that can be regarded as floating on a pliable medium. (Isostasy is connected, inter alia, with the notion of mountain "roots" that are somewhat less dense than the surrounding substrata.) Details of Wegener's interpretation were incompatible with the theory of land bridges, leaving, he thought, continental displacement as the plausible explanation for evidence of paleontological similarities between separated continents. Wegener also suggested *Pohlflucht* (flight from the poles) as a possible mechanism.

At first Wegener looked upon his version of continental drift as a working hypothesis that would undergo some modification as new evidence emerged, and he spent much effort in seeking such evidence. His confidence in this theory grew strongly, and the later editions of his book show the evidence he accumulated. His most important new evidence was from his studies of paleoclimatology; as a meteorologist Wegener had become interested in ancient climates and drew inferences bearing on continental drift from his investigations of their varying patterns over geological time.

Wegener also particularly sought to strengthen his case by precise geodetic observations that involved repeated astronomical position-finding methods and measurements of radio time transmissions across oceans over a number of years. His search for geodetic support was one of the main motives for his third and fourth Greenland expeditions: he hoped to establish that Greenland is now drifting westward from Europe. But no significant results emerged.

Wegener was not the first to propose a form of continental drift. The apparent congruence of the western and eastern Atlantic coasts had attracted notice as early as about 1600, and the notion that the Atlantic continents had drifted apart in earlier times had been put forward specifically by A. Snider-Pellegrini in 1858. In 1908 the geologist H. B. Baker had suggested that all the continents had been grouped around Antarctica 200 million years ago, and in 1910 F. B. Taylor had independently proposed a general drift of continents toward the equator. But Wegener went to greater lengths in expounding the theory and in his sustained efforts to establish it, and it is his name that is principally remembered.

From 1919 to 1928 a great international controversy raged over continental drift. Wegener's arguments on mechanism were shown to be untenable, and the paleontological evidence was held to be inconclusive. In 1928, at a historic gathering of fourteen eminent geologists, five supported the notion of continental drift without reservation and two with reservations, while seven opposed it. From then until after World War II, the theory received comparatively little attention. The most noted early variant of the theory was probably du Toit's proposal in 1937 that instead of a single primordial continent Pangaea, there were two, Laurasia in the northern hemisphere and Gondwanaland in the southern.

Wegener's other scientific contributions included work on the dynamics and thermodynamics of the atmosphere, atmospheric refraction and mirages, optical phenomena in clouds, acoustical waves, and the design of geophysical instruments. In his endeavors to test his continental drift theory through geodetic measúrements, he designed an efficient balloon theodolite for tracking balloons sent up from ships to great heights.

A principal early objection to Wegener's theory was the failure to find an acceptable mechanism. Some time after his death, the idea of convection currents inside the earth's solid upper mantle was developed and a mechanism based on this idea was suggested. The new proposal envisaged an upper mantle that, although behaving like a solid in response to ordinary stresses, flows like a fluid under stresses that have periods comparable with geological time. According to this theory the flow took the form of convection currents and continents were carried, instead of sliding, on top of a convection cell of the moving mantle material.

The early 1950's saw the development of the new science of paleomagnetism, which indicated that remnant rock magnetism has preferred directions persisting over large areas of individual continents but varying from continent to continent. The directions are, moreover, generally different from the lines of magnetic force existing at the earth's surface today. When continental drift, coupled with polar wandering, was invoked to fit the paleomagnetic evidence, interest in Wegener's work, with the modification that convection currents provide the driving mechanism for continental drift, was strongly revived.

A great quantity of evidence on characteristics of midocean ridges was later brought to bear. These ridges, now known to exist on the floors of all the main oceans, are characterized by some unusual properties, including: abnormally high heat flow from the earth's interior, sequences of magnetic and gravity anomalies, thinning of the sediment cover as a ridge is approached, and interesting sequences in ages of ocean-floor material near ridges, giving rise to the notion of "sea-floor spreading." Many geologists have associated the ridges with the top of rising convection currents, and certain ocean trenches with the locations of descending currents. Other recent evidence has indicated that a layer immediately below continents is much weaker than had previously been supposed.

At the same time, older arguments continue to be maintained and newer arguments to be raised against continental drift. Some distinguished geophysicists assert that the currently proposed mechanism does not stand up to fine analysis. Further questions include the following: Why should the east coastline of South America remain largely undistorted although its postulated westerly drift relative to Africa has involved a sideways global distortion of thousands of miles? How reliable are the assumptions made about the earth's magnetic field? What is the reason for the comparative recency, as compared with the age of the earth, of the envisaged drift? Ad hoc answers to all these questions have been put forward by protagonists of the theory, yet again their answers have been questioned.

The enthusiasms of a considerable number of earth scientists lead them to assert, sometimes with a religious fervor, that continental drift is now established. It can at least be said that, whether large-scale continental drift in the envisaged sense has occurred or not, and whatever the finer detail may turn out to be, it is now widely recognized that movements and distortions of the earth's outer layers over geological time must have been substantial. The theory of continental drift also is stimulating many scientists to gather new observations of much value to ideas on the evolution of the earth.

BIBLIOGRAPHY

I. ORIGINAL WORKS. The principal publication for which Wegener is now noted is *Die Entstehung der Kontinente und Ozeane* (Brunswick, 1915; rev. eds., 1920, 1922, and 1929), the 3rd ed. (1922) translated into English, French, Spanish, Swedish, and Russian. A 5th ed., rev. by Kurt Wegener, appeared in 1936. His other publications include *Thermodynamik der Atmosphäre* (Leipzig, 1911); "Die Enstehung der Kontinente und Ozeane," in *Mitteilungen aus Justus Perthes geographischer Anstalt,* **58** (1912), 185–195, 253–256, 305–309; *Die Klimate der geologischen Vorzeit* (Berlin, 1924), written with W. Köppen; "Denkschrift über Inlandeis-Expedition nach Grönland," in *Deutsche Forschung* (Berlin), **42** (1928), 181; and *Mit Motorboot und Schlitten in Grönland* (Leipzig, 1930).

II. SECONDARY LITERATURE. For a full list of Wegener's publications, see H. Benndorf, "Alfred Wegener," in *Beiträge zur Geophysik,* **31** (1931), 337–377. Further details of his life are in Else Wegener, *Alfred Wegener* (Wiesbaden, 1960); J. Georgi, "Memories of Alfred Wegener," in *Continental Drift,* S. K. Runcorn, ed. (London, 1962), 309–324; and K. Wolcken, "Alfred Wegener," in *Meteoros* (1955), 379–382. An important contemporaneous review of Wegener's work on continental drift was published by the American Association of Petrologists in *Theory of Continental Drift; a Symposium on the Origin and Movement of Land Masses, Both Inter-continental and Intra-continental, as Proposed by Alfred Wegener,* E. De Golyer, ed. (London, 1928).

K. E. BULLEN

WEHNELT, ARTHUR RUDOLPH BERTHOLD (*b.* Rio de Janeiro, Brazil, 4 April 1871; *d.* Berlin, Germany, 15 February 1944), *physics.*

Wehnelt was the son of Berthold Wehnelt, an engineer and factory owner, who died at an early age, and Louise Muckelberg. He attended the Louisenstädter Gymnasium in Berlin and the Gymnasium in Landsberg an der Warthe (now Gorzów Wielkopolski, Poland), and graduated from secondary school in the spring of 1892. For a year he studied natural science at the Technische Hochschule in Berlin-Charlottenburg, and then from 1893 to 1897 at the University of Berlin. Next he went to the University of Erlangen, where he received the doctorate in the spring of 1898 under Eilhard G. H. Wiedemann. At Erlangen he became successively assistant at the Physics Institute (1900); *Privatdozent* (after obtaining the *venia docendi* in 1901); and extraordinary professor of physics (1904). In 1906 he was called to the University of Berlin as full professor, and in 1926 he was appointed director of its Physics Institute. He remained at the University of Berlin until 1939, becoming professor emeritus about 1938.

In his dissertation Wehnelt investigated the dark space near the cathode in gas-discharge tubes and established that the high resistance of the dark space of the cathode corresponds to that of a dielectric. Pursuing a remark made by Wiedemann and Ebert in 1891, Wehnelt demonstrated that the cross section of the cathode-ray bundle decreases with pressure, or with a decrease in the diameter of the tube placed around the bundle. These are essentially the characteristics of the epoch-making device (now known as the Wehnelt cylinder) that he later developed. In his *Habilitationsschrift* Wehnelt described the processes occurring in the discharge tube, taking into account the entire discharge from the cathode to the anode, and measured the current and voltage to analyze the discharge processes at various points of the tube. In the preface to his essay Wehnelt defended twelve theses, including the propositions that "electrical lighting systems of 220 volts should be avoided wherever possible," and that "the refinement of modern electrical measuring methods has gone too far."

Wehnelt became well known through his discoveries concerning discharge in rarefied gases. In the course of his research he studied cathode rays, canal rays, and Röntgen rays. He was involved in the technical development of the valve tube (the radio tube), Röntgen tubes, and the Braun tube (the oscilloscope); in this manner he made a fundamental contribution to modern electronics.

Three of Wehnelt's discoveries deserve special mention. While investigating certain light and heat phenomena that had earlier been observed at very small electrodes by Davy, Wehnelt recognized in 1899 that current interruptions (up to 2200 sec^{-1}) originate in rapid gas explosions at the electrode (for example, the platinum point). One such phenomenon, known as disruptive discharge, was demonstrated with the telephone by F. Richarz in 1892 and was applied to an electrical sweating process by Eugène Lagrange and P. Hoho in 1894. Wehnelt first applied an electrolytic interrupter — based on this phenomenon — to the induction coil, and then to short-exposure X-ray photographs.

In 1903 Wehnelt pointed out that a large potential drop exists between the cathode ray and the wall of the tube, and he used this knowledge to alter the hardness of the X rays, which he regulated with the help of an auxiliary tube placed over the cathode. To measure the hardness itself, he utilized the measurements published by Bénoist in 1902, concerning the comparative absorption of silver and aluminum. Wehnelt improved the measurements by replacing Bénoist's aluminum disk with a continuous aluminum wedge (Wehnelt scale).

In 1903–1904 Wehnelt made his most important discovery, the "oxide cathode." He observed a significant decrease in the cathode fall of glow discharge, occurring in the presence of platinum cathodes that had not been carefully enough cleaned; and then he noticed the same thing in the case of metal compounds, but especially in that of oxides. He found that the phenomenon was caused by an increase (which he estimated as 100 times) in the number of negative ions made available by the cathode metal when treated. A practical use for this insight was found in the production of very high current strengths (at 110 volts and 0.01 Hg pressure it is 3 amperes). More important, the exceptionally low exit potential at the cathode enabled Wehnelt to produce slow electrons, as well as soft canal rays. As a result, he could carry out measurements of velocity and of the ratio of charge to mass of very soft cathode rays. Since he knew how to obtain a good focus for the electron beam and how to make it visible at reduced pressure with the aid of the "Wehnelt cylinder," he was especially successful in this undertaking. Wehnelt saw the theoretical interest of his discovery as possible confirmation of the hypothesis put forth by Kaufmann in 1902 and by Abraham in 1903, that is, that the mass of the electron is a purely electromagnetic quantity.

In subsequent research Wehnelt dealt with the photoelectric effect, secondary emission, mass spectra, thin metal layers, and the thermal conductivity of metals.

BIBLIOGRAPHY

I. ORIGINAL WORKS. A fairly complete bibliography of Wehnelt's writings is in Poggendorff, IV, 1608–1609; V, 1345; VI, 2829; and VIIa, 887. See especially *Strom- und Spannungsmessungen an Kathoden in Entladungsröhren. Habilitationsschrift zur Erlangung der Venia docendi der hohen philosophischen Fakultät der Friedrich-Alexander-Universität zu Erlangen* (Leipzig, 1901), which contains the 12 interesting theses that Wehnelt defended.

II. SECONDARY LITERATURE. In addition to the titles cited by Poggendorff, the separately printed edition of Wehnelt's Ph.D. (Leipzig, 1908) is reprinted (without the short biographical sketch originally included) in *Annalen der Physik*, **301** (1898), 511–542. Wehnelt's MSS are held at the Staatsbibliothek Preussischer Kulturbesitz in Berlin; a few MSS also can be found at the Universitätsbibliothek in Erlangen.

HANS KANGRO

WEICHSELBAUM, ANTON (*b*. Schiltern, Austria, 8 February 1845; *d*. Vienna, Austria, 23 October 1920), *pathology.*

The son of a barrelmaker, Weichselbaum attended the gymnasium in Krems, Austria, from 1855 to 1863 and studied medicine at the Imperial Medical Surgical Military Hospital in Vienna, receiving the M.D. in 1869. He subsequently served as assistant to Rokitansky's student Josef Engel in the medical corps and in 1875 became anatomical demonstrator at the First Imperial and Royal Military Hospital in Vienna; in 1877 he received the *venia legendi* for pathological anatomy at the University of Vienna. In 1882 he was named chief demonstrator at the Rudolf Hospital in Vienna, and in 1885 he became associate professor. From 1893 to 1916 he was director of the Pathological-anatomical Institute of the University of Vienna, and in 1912 he became rector of the university.

Weichselbaum was among the first to recognize the importance of bacteriology for pathological anatomy. This fact is reflected in his discovery of the meningococcus and of the diplococcus *lanceolatus pneumoniae*, which bears his name, as well as in his studies on miliary tuberculosis. Weichselbaum was also extremely receptive to the newly developing science of serology. It was, in fact, while serving as assistant in Weichselbaum's laboratory that Karl Landsteiner discovered interagglutination between serum and blood cells.

Moreover, Weichselbaum was one of the first to stress the importance of "constitutional pathology." In his investigations of the pancreas of patients with diabetes mellitus he very early drew attention to the crucial role of the islets of Langerhans, the organs in which insulin was later discovered.

BIBLIOGRAPHY

Weichselbaum's major publications include *Grundriss der pathologischen Histologie* (Leipzig–Vienna, 1892), translated by W. R. Dawson as *The Elements of Pathological Histology* (London, 1895); *Parasitologie* (Jena, 1898); and *Epidemiologie* (Jena, 1899).

On his life and work, see the notice by Siegmund Exner, in *Almanach der Akademie der Wissenschaften in Wien*, **71** (1921), 152–155.

H. CHIARI

WEIERSTRASS, KARL THEODOR WILHELM (*b.* Ostenfelde, Westphalia, Germany, 31 October 1815; *d.* Berlin, Germany, 19 February 1897), *mathematics.*

Weierstrass was the first child of Wilhelm Weierstrass, secretary to the mayor of Ostenfelde, and Theodora Vonderforst, who were married five months before his birth. The family name first appeared in Mettmann, a small town between Düsseldorf and Elberfeld; since the sixteenth century they had been artisans and small merchants. Weierstrass' father, an intelligent, educated man with knowledge of the arts and sciences, could have held higher posts than he actually did; little is known about his mother's family. Weierstrass had a brother and two sisters, none of whom ever married. When Weierstrass was eight his father entered the Prussian taxation service; and as a result of his frequent transfers, young Karl attended several primary schools. In 1829, at the age of fourteen, he was accepted at the Catholic Gymnasium in Paderborn, where his father was assistant and subsequently treasurer at the main customs office.

A distinguished student at the Gymnasium, Weierstrass received several prizes before graduating. Unlike many mathematicians, he had no musical talent; nor did he ever acquire an interest in the theater, painting, or sculpture. He did, however, value lyric poetry and occasionally wrote verses

himself. In 1828, a year after his mother's death, Weierstrass' father remarried. At the age of fifteen Weierstrass reportedly worked as a bookkeeper for a merchant's wife—both to utilize his abilities and to ease the strain of his family's financial situation. A reader of Crelle's *Journal für die reine und angewandte Mathematik* while in his teens, he also gave his brother Peter mathematical coaching that does not seem to have proved helpful: Weierstrass' proofs were generally "knocking," his brother later admitted.

After leaving the Gymnasium in 1834, Weierstrass complied with his father's wish that he study public finance and administration, and entered the University of Bonn. The course of studies that he pursued was planned to permit him to obtain a background in law, administration, and economics—the requisites for those seeking higher administrative posts in Prussia. The study of mathematics or related areas was his first choice, however; and the conflict between duty and inclination led to physical and mental strain. He tried, in vain, to overcome his problems by participating in carefree student life, but he soon came to shun lectures and to restrict himself to studying mathematics on his own, beginning with the *Mécanique céleste* of Laplace. Weierstrass was fortunate in having an understanding adviser in astronomy, mathematics, and physics, Dietrich von Münchow. However, Münchow was of the old school and, because he gave only elementary lectures, was remote from the advances of modern mathematics.

Around this time Weierstrass read Jacobi's *Fundamenta nova theoriae functionum ellipticarum* (1829); the work proved difficult for him, based, as it was, on prior knowledge of Legendre's *Traité des fonctions elliptiques*, published shortly beforehand. A transcript of Christof Gudermann's lecture on modular functions rendered the theory of elliptic variables understandable to him and inspired him to initiate his own research. In a letter to Sophus Lie of 10 April 1882, Weierstrass explained his definitive decision to study mathematics:

For me this letter [from Abel to Legendre], when I became aware of it in Crelle's *Journal* [**6** (1830), 73–80] during my student years, was of the utmost importance. The immediate derivation of the form of representation of the function given by Abel and designated by him by $\lambda(x)$, from the differential equation defining this function, was the first mathematical task I set myself; and its fortunate solution made me determined to devote myself wholly to mathematics; I made this decision in my seventh semester [winter semester 1837–1838], although originally I undertook the study of public finance and administration [*N. H. Abel, Mémorial* (1902), 108].

After eight semesters Weierstrass left the university without taking the examination. Although his father was greatly disappointed, a family friend, president of the court of justice of Paderborn, persuaded him to send Weierstrass to the Theological and Philosophical Academy at Münster, where he would be able to take the teacher's examination after a short time. Weierstrass enrolled on 22 May 1839. Helped and encouraged by Gudermann—Weierstrass was the only university student at his lectures on elliptic functions, he left Münster that autumn to prepare for the state examination.

In January 1840 Weierstrass' father assumed the more remunerative post of director of the saltworks and the family moved to Westernkotten, near Lippstadt. Two months later Weierstrass registered for the examination, and at the beginning of May he received the philological, pedagogical, and mathematical problems for the written examination. The first mathematical problem was one that Weierstrass himself had requested, the representation of elliptic functions. Following Abel, for whose work he had always had the highest regard, Weierstrass presented in his examination an important advance in the new theory of elliptic functions, and this work contains important starting points for his subsequent investigations. Gudermann recognized the significance of his accomplishment and wrote in his evaluation that Weierstrass was "of equal rank with the discoverers who were crowned with glory." The school superintendent was somewhat more restrained. When Weierstrass later read Gudermann's complete critique, he admitted that had he learned of it earlier, he would have published the work immediately and most certainly would have obtained a university chair sooner. He considered it especially fine of Gudermann to have praised him so highly, even though his work contained sharp criticism of Gudermann's method. It is of interest that one of the other mathematical problems that Weierstrass was assigned, on elementary geometry, gave him much difficulty, at least according to his brother's account.

After having passed the oral examinations in April 1841, Weierstrass taught for a one-year probationary period at the Gymnasium in Münster, before transferring to the Catholic secondary school in Deutsch-Krone, West Prussia (1842–1848), and then to the Catholic Gymnasium in Braunsberg, East Prussia (1848–1855). In addition to mathematics and physics, he taught German, botany, geography, history, gymnastics, and even calligraphy. (Reminiscent of this is his peculiar $\mathscr{P}$ of the Weierstrassian p-function.) In recalling the misery of these years, Weierstrass remarked that he had neither a colleague for mathematical discussions nor access to a mathematical library, and that the exchange of scientific letters was a luxury that he could not afford. The "unending dreariness and boredom" would have been unbearable for him without hard work, and his every free minute was devoted to mathematics.

Fortunately he found an understanding senior colleague in Ferdinand Schultz, director of the Braunsberg Institute. Weierstrass was in Deutsch-Krone during the Revolution of 1848. Dirichlet had stated that a mathematician could only be a democrat, and Weierstrass' beliefs were not contrary to this belief. Commissioned to oversee the belletristic section of the local newspaper, he approved reprinting the freedom songs of Georg Herwegh—under the eyes of the censor.

Although not involved in nationalistic struggles, Weierstrass was no stranger to national feelings. He aspired to neither title nor decorations and was reluctant to exchange the simple title of professor for the more pretentious one of privy councillor. His religious views were moderate and tolerant, and he eschewed political as well as religious bigots. Reared a Catholic, he paid homage in a public speech as rector to the cultural significance of the Reformation. In his philosophical outlook he was a frank adherent of Kant and an opponent of Fichte and Schelling.

Weierstrass' first publications on Abelian functions, which appeared in the Braunsberg school prospectus (1848–1849), went unnoticed; but the following work, "Zur Theorie der Abelschen Functionen" (Crelle's *Journal*, **47** [1854], 289–306), elicited enormous interest and marked a decisive turning point in his life. In this memoir he demonstrated the solution to the problem of inversion of the hyperelliptic integrals, which he accomplished by representing Abelian functions as the quotients of constantly converging power series. Many of his results were only hinted at in this work, for since 1850 he had suffered painful attacks of vertigo, which lasted up to an hour and subsided only after a tormenting attack of vomiting. These attacks, which contemporaries called brain spasms, recurred for about twelve years and made it impossible for him to work. Although the 1854 paper was merely a preliminary statement,

Liouville called it "one of those works that marks an epoch in science." On 31 March 1854 the University of Königsberg awarded Weierstrass an honorary doctorate. He was promoted to senior lecturer at Braunsberg and in the fall of 1855 was granted a year's leave to continue his studies. Firmly determined not to return to the school, he applied in August 1855 for the post of Kummer's successor at the University of Breslau—an unusual mode of procedure. (Kummer had been called to Berlin to succeed Dirichlet, who had assumed Gauss's chair at Göttingen.) Weierstrass did not receive the appointment at Breslau.

In the famous "Theorie der Abelschen Functionen" (Crelle's *Journal*, **52** [1856], 285–380), which contains an excerpt from the previously mentioned examination work, Weierstrass proved what previously he had only hinted. According to Hilbert, he had realized one of the greatest achievements of analysis, the solution of the Jacobian inversion problem for hyperelliptic integrals. There was talk of appointment to a post in Austria, but before formal discussions could take place Weierstrass accepted on 14 June 1856 an appointment as professor at the Industry Institute in Berlin, a forerunner of the Technische Hochschule. While he did not have to return to the Gymnasium in Braunsberg, his hopes for appointment to the University of Berlin had not been realized. In September 1856, while attending a conference of natural scientists in Vienna, Weierstrass was offered a special professorship at any Austrian university of his choice. He was still undecided a month later, when he was invited to the University of Berlin as associate professor. He accepted. On 19 November 1856 he became a member of the Berlin Academy. It was not until July 1864 that he was able to leave the Industry Institute and assume a chair at the university.

Having spent the most productive years of his life teaching elementary classes, far from the centers of scientific activity, Weierstrass had found time for his own research only at the expense of his health. Heavy demands were again made on him at Berlin, and on 16 December 1861 he suffered a complete collapse; he did not return to scientific work until the winter semester of 1862–1863. Henceforth he always lectured while seated, consigning the related work at the blackboard to an advanced student. The "brain spasms" were replaced by recurrent attacks of bronchitis and phlebitis, which afflicted him until his death at the age of eighty-one. Nevertheless, he became a recognized master, primarily through his lectures.

He delayed publication of his results not—as has often been charged—because of a "basic aversion to printer's ink" but, rather, because his critical sense invariably compelled him to base any analysis on a firm foundation, starting from a fresh approach and continually revising and expanding.

It was only gradually that Weierstrass acquired the masterly skill in lecturing extolled by his later students. Initially his lectures were seldom clear, orderly, or understandable. His ideas simply streamed forth. Yet his reputation for lecturing on new theories attracted students from around the world, and eventually some 250 students attended his classes. Since no one else offered the same subject matter, graduate students as well as university lecturers were attracted to Berlin. Moreover, he was generous in suggesting topics for dissertations and continuing investigations.

One of Weierstrass' first lectures at Berlin was on the application of Fourier series and integrals to problems of mathematical physics. But the lack of rigor that he detected in all available works on the subject, as well as the fruitlessness of his own efforts to surmount this deficiency, frustrated him to the degree that he decided not to present this course again. It was not until 1885 that he took up the representation of single-valued functions of a real variable by means of trigonometric series, stressing that "he had considered the needs of mathematical physics." Here again are manifest his proverbial striving for the characteristic "Weierstrassian rigor" that virtually compelled him to carry his investigations to an ever higher degree of maturity and completion. His position concerning the applications of his research was clarified in his inaugural speech at the Berlin Academy on 9 July 1857, in which he stated that mathematics occupies an especially high place because only through its aid can a truly satisfying understanding of natural phenomena be obtained. To some degree his outlook approached that of Gauss, who believed that mathematics should be the friend of practice, but never its slave.

Over the years Weierstrass developed a great lecture cycle: "Introduction to the Theory of Analytic Functions"; "Theory of Elliptic Functions," sometimes beginning with the differential calculus, at other times starting with the theory of functions, the point of departure being the algebraic addition theorem; "Application of Elliptic Functions to Problems in Geometry and Mechanics"; "Theory of Abelian Functions"; "Application of Abelian Functions to the Solution of Selected Geometric Problems"; and "Calculus of Variations." Within this cycle Weierstrass erected the entire structure of his mathematics, using as building blocks only that which he himself had proven.

During seven semesters (1864–1873) Weierstrass also lectured on synthetic geometry, thereby honoring his promise to Jakob Steiner before the latter's death in 1863. Steiner's discussions, which Weierstrass had read in Crelle's *Journal* as a student, had especially stimulated his interest; and he was one of the few people in Berlin with whom the old crank had remained on good terms. These lectures were given only out of a sense of obligation, however—not from any interest in the subject; for Weierstrass considered geometric demonstrations to be in very poor taste. If, as has been alleged, he sometimes permitted himself to clarify a point by using a diagram, it was carefully erased.

In addition to lecturing, Weierstrass introduced the first seminar devoted exclusively to mathematics in Germany, a joint undertaking with Kummer at the University of Berlin in 1861. Here again he developed many fruitful concepts that were frequently used by his students as subjects for papers. In his inaugural lecture as rector of the University of Berlin, Weierstrass called for lecturers to "designate the boundaries that had not yet been crossed by science . . . from which positions further advances would then be made possible." The lecturer should neither deny his students "a deeper insight into the progress of his own investigations, nor should he remain silent about his own errors and disappointments."

Weierstrass' students included Heinrich Bruns, Georg Frobenius, Georg Hettner, Ludwig Kiepert, Wilhelm Killing, Johannes Knoblauch, Ernst Kötter, Reinhold von Lilienthal, Hans von Mangoldt, Felix Müller, Eugen Netto, Friedrich Schottky, Ludwig Stickelberger, and Wilhelm Ludwig Thomé. Auditors or participants in the seminar included Paul Bachmann, Oskar Bolza, Friedrich Engel, Leopold Gegenbauer, August Gutzmer, Lothar Heffter, Kurt Hensel, Otto Hölder, Adolf Hurwitz, Felix Klein, Adolf Kneser, Leo Koenigsberger, Fritz Kötter, Mathias Lerch, Sophus Lie, Jacob Lüroth, Franz Mertens, Hermann Minkowski, Gösta Mittag-Leffler, Hermann Amandus Schwarz, and Otto Stolz. The philosopher Edmund Husserl—insofar as he was a mathematician—was also a student of Weierstrass.

Weierstrass was not without his detractors: Felix Klein, for instance, remarked that he and Lie had merely fought for their own points of view in the seminars. Most of Weierstrass' students, however, accepted his theories as an unassailable standard. Doubts were not permitted to arise, and checking was hardly possible since Weierstrass cited very few other sources and arranged his methodical structure so that he was obliged to refer only to himself. Independent opinions, such as Klein's, were the exception.

Weierstrass' criticism of Riemann's basic concept of the theory of functions, namely the application and use of the principle of Dirichlet, resulted in the fact that until the twentieth century his approach to the theory of functions, starting with the power series, was preferred to Riemann's, which originated with complex differentiation. Weierstrass formulated his credo in a letter to his student H. A. Schwarz (3 October 1875):

The more I ponder the principles of function theory—and I do so incessantly—the more I am convinced that it must be founded on simple algebraic truths and that one is therefore on the wrong path if, instead of building on simple and fundamental algebraic propositions, one has recourse to the "transcendental" (to put it briefly), no matter how impressive at first glance, for example, seem the reflections by means of which Riemann discovered so many of the most important properties of algebraic functions. It is self-evident that any and all paths must be open to a researcher during the actual course of his investigations; what is at issue here is merely the question of a systematic theoretical foundation.

Although Weierstrass enjoyed considerable authority at Berlin, he occasionally encountered substantial resistance from his colleagues; and such criticism hurt him deeply. In the late 1870's his relations with his close friend Leopold Kronecker cooled considerably when Kronecker imparted to Weierstrass his antipathy for the work of Georg Cantor. Weierstrass had been one of the first to recognize the value of Cantor's accomplishments and had in fact stimulated his work on the concept of countability. Kronecker, by contrast, proclaimed that he had set himself the task "of investigating the error of every conclusion used in the so-called present method of analysis." Weierstrass' reaction to Kronecker's attack may well have been excessive; but in 1885 he decided to leave Germany and go to Switzerland, believing that everything for which he had worked was near collapse. Determined to prevent such a catastrophe, he resolved to remain in Berlin after all. The choice of his successor and publication of his works were problems still to be resolved—and his successor would have to be endorsed by Kronecker. Kronecker's death in 1891 cleared the path for the appointment of Hermann Amandus Schwarz.

But the publication of Weierstrass' writings was another matter. He was satisfied with neither the circulating transcripts of his lectures nor with the textbooks that followed his concepts and that he had, to some degree, authorized; and his major ideas and methodology remained unpublished. In 1887, having already edited the works of Steiner and Jacobi, Weierstrass decided to publish his own mathematical lifework, assured of the help of the younger mathematicians of his school. He lived to see only the first two volumes appear in print (1894, 1895). According to his wishes, volume IV was given preferential treatment, and it appeared in 1902. The altered title, "Lectures on the Theory of Abelian Transcendentals"—they had always been called the theory of Abelian functions—more accurately reflects the scope of his lectures. Volume III was published the following year. Twelve years elapsed before the appearance of volume V ("Lectures on Elliptic Functions") and volume VI ("Selected Problems of Geometry and Mechanics to be Solved With the Aid of the Theory of Elliptic Functions"), and another dozen years before volume VII ("Lectures on the Calculus of Variations"). All of Weierstrass' efforts to ensure the publication soon after his death of a complete edition of his works were fruitless: volumes VIII–X, intended to contain works on hyperelliptic functions, a second edition of his lectures on elliptic functions, and the theory of functions, remain unpublished.

In 1870, at the age of fifty-five, Weierstrass met the twenty-year-old Russian Sonya Kovalevsky, who had come to Berlin from Heidelberg, where she had taken her first semester under Leo Koenigsberger. Unable to secure her admission to the university, Weierstrass taught her privately; and his role in both her scientific and personal affairs far transcended the usual teacher-student relationship. In her he found a "refreshingly enthusiastic participant" in all his thoughts, and much that he had suspected or fumbled for became clear in his conversations with her. In a letter to her of 20 August 1873, Weierstrass wrote of their having "dreamed and been enraptured of so many riddles that remain for us to solve, on finite and infinite spaces, on the stability of the world system, and on all the other major problems of the mathematics and the physics of the future." It seemed to him as though she had "been close . . . throughout [his] entire life . . . and never have I found anyone who could bring me such understanding of the highest aims of science and such joyful accord with

my intentions and basic principles as you!" Through his intercession she received the doctorate *in absentia* at Göttingen in 1874.

Yet their friendship did not remain untroubled. Her links with socialist circles, her literary career as author of novels, and her advocacy of the emancipation of women strongly biased the judgment of her contemporaries, and resulted in defamation of the friendship. On the other hand, many of his letters to her were unanswered. At one juncture she remained silent for three years. He was instrumental in her obtaining an appointment as lecturer in mathematics at Stockholm in 1883 and a life professorship in mathematics in 1889. The misinterpretation of their relationship and her early death in 1891 brought him additional physical suffering. During his last three years he was confined to a wheelchair, immobile and dependent. He died of pneumonia.

In his inaugural speech to the Berlin Academy, Weierstrass characterized his scientific activity as having centered on the search for "those values of a wholly new type of which analysis had not yet had an example, their actual representation, and the elucidation of their properties." One of his earliest attempts at solving this problem was a treatise (1841) on the representation of an analytic function exhibiting an absolute value that lies between two given boundaries. It contained the Cauchy integral proposition and the Laurent proposition. It was published only fifty-three years later, however, when it became clear that Weierstrass at the age of twenty-six had already had at his disposal the principles of his theory of functions, to the development of which he subsequently devoted his lifework. Yet his contribution to reestablishing the theory of analytic functions ultimately served only to achieve his final aim: the erection of a general theory of Abelian integrals (all integrals over algebraic functions) and the consideration of their converse functions, the Abelian functions.

What Weierstrass considered to be his main scientific task is now held to be less important than his accomplishments in the foundation of his theory. The special functions which he investigated, and the theory of which he lucidly elaborated or transformed, now elicit less interest than his criticism, rigor, generally valid concepts, and the procedures and propositions of the theory of functions. Weierstrass' name remains linked to his preliminary proposition, approximation propositions, double series proposition, proposition of products, and fundamental proposition—as well as the Casorati-Weierstrass proposition. Hundreds of mathematicians were influenced by his uncompromising development of a systematic foundation and his pursuit of a fixed plan after appropriate preparation of detail; and they in turn instilled in their students Weierstrass' concepts of the necessity of clarity and truth, and his belief that the highest aim of science is to achieve general results. Admired by Poincaré for his "unity of thought," Weierstrass was the most important nineteenth-century German mathematician after Gauss and Riemann.

BIBLIOGRAPHY

I. ORIGINAL WORKS. Weierstrass' writings were published as *Mathematische Werke*, 7 vols. (Berlin, 1894–1927). His papers are listed in Poggendorff, II, 1282; III, 1424; IV, 1610; V, 1345; and VI, 2831.

The following letters have been published: to Paul du Bois-Reymond, G. Mittag-Leffler, ed., in *Acta mathematica*, **39** (1923), 199–225; to Leo Koenigsberger, G. Mittag-Leffler, ed., *ibid.*, 226–239; to Lazarus Fuchs, M. Wentscher and L. Schlesinger, ed., *ibid.*, 246–256. Excerpts of Weierstrass' correspondence with Sonya Kovalevsky were included by Mittag-Leffler in his discussion, "Une page de la vie de Weierstrass," in *Compte rendu du deuxième Congrès international des mathématiciens* (Paris, 1902), 131–153; and in "Zur Biographie von Weierstrass," in *Acta mathematica*, **35** (1912), 29–65, which also includes Weierstrass' letters to Mittag-Leffler. His letters to Kovalevsky were also used by Mittag-Leffler in "Weierstrass et Sonja Kowalewsky," *ibid.*, **39** (1923), 133–198; by P. Y. Polubarinova-Kochina, in "K biografii S. V. Kovalevskoy" ("Toward a Biography . . ."), in *Istoriko-matematicheskie issledovaniya*, **7** (1954), 666–712; and by K.-R. Biermann, "Karl Weierstrass," in *Journal für die reine und angewandte Mathematik*, **223** (1966), 191–220, which presents a survey of the known archives and their places of deposition. The Weierstrass letters to Kovalevsky have been published completely in P. Y. Polubarinova-Kochina, *Pisma Karla Weierstrassa K Sof'e Kovalevskoj 1871–1891* (Moscow, 1973).

Propositions for academic elections of Weierstrass are in K.-R. Biermann, "Vorschläge zur Wahl von Mathematikern in die Berliner Akademie," in *Abhandlungen der Deutschen Akademie der Wissenschaften zu Berlin*, Kl. für Mathematik, Physik und Technik (1960), no. 3, 25–34, *passim*. Weierstrass' analysis is discussed in Pierre Duac, "Éléments d'analyse de Karl Weierstrass," in *Archive for History of Exact Sciences*, nos. 1–2, **10** (1973), 41–176.

II. SECONDARY LITERATURE. See Henri Poincaré, "L'oeuvre mathématique de Weierstrass," in *Acta mathematica*, **22** (1899), 1–18; and G. Mittag-Leffler, "Die ersten 40 Jahre des Lebens von Weierstrass," *ibid.*, **39** (1923), 1–57. A number of Weierstrass' students and auditors have published reminiscences of him; see the bibliography to Biermann's memoir (cited above), in *Journal für die reine und angewandte Mathematik*, **223** (1966), 219–220. See also *Festschrift zur Gedächtnisfeier für Karl Weierstrass*, H. Behnke and K. Kopfermann, eds. (Cologne–Opladen, 1966); and especially Heinrich Behnke, "Karl Weierstrass und seine Schule," 13–40; as well as K.-R. Biermann, "Die Berufung von Weierstrass nach Berlin," 41–52. See also the following articles by Biermann: "Dirichlet über Weierstrass," in *Praxis der Mathematik*, **7** (1965), 309–312; "K. Weierstrass und A. v. Humboldt," in *Monatsberichte der Deutschen Akademie der Wissenschaften zu Berlin*, **8** (1966), 33–37; "Karl Weierstrass in seinen wissenschaftlichen Grundsätzen," in *Sudhoffs Archiv*, **50** (1966), 305–309; and the book by Biermann, *Die mathematik und ihre Dozenten an der Berliner Universität 1810–1920* (Berlin, 1973).

KURT-R. BIERMANN

WEIGEL, CHRISTIAN EHRENFRIED (*b*. Stralsund, Germany, 24 May 1748; *d*. Greifswald, Germany, 8 August 1831), *chemistry*.

After receiving instruction from his father, Dr. Bernhard Nicolaus Weigel (inventor of Weigel's medicinal drops) and attending a private school in Stralsund, young Weigel entered the University of Greifswald in 1764. He studied medicine and the natural sciences there for five years, then proceeded to the University of Göttingen, where for the next two years he worked closely with the botanist J. A. Murray, the chemist R. A. Vogel, and the technologist J. Beckmann. Weigel also visited the Harz mining district to collect minerals and to observe metallurgical techniques. In 1771 he took his M.D. at Göttingen with a chemical-mineralogical dissertation.

Weigel then returned to Stralsund, where he practiced medicine and continued his chemical research in his father's laboratory. In 1772 he became an adjunct lecturer and supervisor of the botanical garden at the University of Greifswald; and two years later he was appointed to a new chair of chemistry and pharmacy in the medical faculty of the university. Besides holding this post for the rest of his life, he continued to run the botanical garden until 1781, served on the medical board for Pomerania and Rügen from 1780 to 1806, and directed the chemical institute of the university from 1796 until his death. In 1806 he was ennobled by Emperor Francis II.

Weigel did almost all of his important work during the 1770's and 1780's. In his dissertation he argued at length for J. F. Meyer's pinguic acid theory and published the first diagram of a counter-current condenser. Weigel's two-volume *Grundriss der reinen und angewandten Chemie* (1777) was one of the first German chemistry texts to be directed beyond a medical audience to readers of all classes. In it he first dealt with pure chemistry and then with applications of chemistry in natural philosophy, natural history, medicine, and especially in agriculture, mining, and manufacturing. Weigel also did much to keep German chemists abreast of foreign developments with his translations of works by Wallerius (1776–1780); Guyton de Morveau, Maret, and Durande (1777–1778); H. T. Scheffer (1779); G. v. Engeström (1782); and Lavoisier (1783–1785).

BIBLIOGRAPHY

The most complete list of Weigel's publications is in Fritz Ferchl, ed., *Chemisch-Pharmazeutisches Bio- und Bibliographikon* (Mittenwald, 1937), 571–572.

On Weigel's life and work, see *Neuer Nekrolog der Deutschen*, 9 (1831), 699–705; O. Anselmino, "Nachrichten von früherer Lehrern der Chemie an der Universität Greifswald," in *Mitteilungen des Naturwissenschaftlichen Vereins für Neu-Vorpommern und Rugen in Greifswald*, 38 (1906), 117–130; G. A. Fester, "Zur Geschichte des Gegenstromkühlers," in *Sudhoffs Archiv für Geschichte der Medizin und der Naturwissenschaften*, 45 (1961), 341–350; J. R. Partington, *A History of Chemistry*, III (London, 1962), 148, 175, 372, 521, 594–595, 609; and Johannes Valentin, "Die Entwicklung der pharmazeutischen Chemie an der Ernst Moritz Arndt-Universität in Greifswald," in *Festschrift zur 500-Jahrfeier der Universität Greifswald*, II (Greifswald, 1956), 472–475.

KARL HUFBAUER

WEIGEL, VALENTIN (*b.* Naundorf [near Dresden], Saxony, Germany, 1533; *d.* Zschopau, Saxony, Germany, 10 June 1588), *mysticism, philosophy of nature.*

Weigel occupies an important position in the history of religion and philosophy between Luther and the seventeenth century. He belongs with the mystics who reacted against Lutheran orthodoxy and the rigid institutionalization of the church, advocating instead that all men are created with innate resources that make faith and knowledge independent of the priesthood and even of the spoken or written word. These are merely aids to the awakening of preexisting inward knowledge. Sin being the consequence of contrary will, rebirth and

salvation can occur only when the abandonment of will creates a state of total passivity and submission in the act of union with God. This is the state of *Gelassenheit*, which played an important role in the thought of such early German mystics as Master Eckhart and Johann Tauler, who, with the so-called *Theologia Germanica*, are the chief spiritual fathers of sixteenth-century German mysticism. The early Luther was close to this tradition, which also deeply influenced Thomas Müntzer, Sebastian Franck, and Caspar Schwenckfeld. A generation younger than these figures, Weigel knew and used their work as well as the older sources, in addition to the German mystics including Plotinus, Proclus, Hermes Trismegistus, Dionysius the Areopagite, Johannes Scottus Eriugena, Boethius, and the Neoplatonists of his own century. The special mark of Weigel's thought is the combination of this mystical spiritualism with Paracelsian naturalism, chiefly in his later writings. Weigel founds his theory of knowledge on the doctrine of harmony between the macrocosm and the microcosm; it is by virtue of this harmony that man is capable of knowledge. To know oneself is therefore the beginning of wisdom and faith. Cosmology and anthropology are complementary sources of insight.

The son of poor Catholic parents, Weigel owed his early education to a local patron who secured free schooling for him at the recently established ducal school of St. Afra in Meissen, which Weigel attended from 1549 to 1554. He spent the next ten years at the University of Leipzig, where he gained the B.A. in 1558 and the M.A. the following year. During this time he was among a small number of students who were not only supported by the elector Augustus I (1526–1586), but also given special attention at the university. In addition to theology, Weigel's studies included philosophy, mathematics, natural science, and medicine. He took part in the usual academic disputations and also taught at the university, but in 1564 he transferred to Wittenberg, where he also taught. It is not known why he made the transfer. This long academic career gave Weigel philosophic training and dialectical skill beyond the normal accomplishments of Lutheran pastors. In 1565 he married Katherina Poch, the daughter of a local pastor. In addition to a daughter, they had two sons who grew up to practice medicine. In 1567 the elector appointed him pastor of Zschopau, where he remained until his death, loved and respected by his parishioners for his dedicated work among them.

The years of Weigel's pastorate coincided with a period of growing Lutheran orthodoxy and tightening church discipline. This movement found expression in the Formula of Concord (1577), which enjoined commitment to articles that no spiritual reformer could accept, such as the doctrine of justification and the scriptural principle of the necessity of the outward word. Weigel subscribed to the formula. The ecclesiastical visitors found no grounds to suspect him of unorthodoxy, except on one occasion, in 1572, when he quickly and successfully cleared himself in a written defense that was dedicated to his bishop. In this skillful piece, Weigel relied on both the ideas of the *Theologia Germanica* (first put into print by the young Luther in 1518) and of Luther, thus retaining as much as possible for his own theological position without giving cause for further suspicion.

Underneath this calm official surface, however, Weigel was leading a double life. From 1570 until his last years he wrote a number of works advancing a religious philosophy that constituted a total rejection of the ruling orthodoxy. The large number of still surviving manuscripts indicate that these writings were copied and circulated, perhaps chiefly on the initiative of Weigel's deacon and successor at Zschopau, Benedikt Biedermann, who a dozen years after Weigel's death was disciplined by the church for spreading heretical doctrines. Weigel's writings finally found their way into print between 1609 and 1618, when at least a dozen of his works appeared at Halle and Magdeburg. As a consequence, the authorities took stronger action against Weigelian doctrine, but it was now too late. It seems clear that Boehme during these very years came under the influence of Weigel, whom he resembles in a number of ways, although most clearly in the heavy dependence on Paracelsus. Weigel's thought also had affinity with Rosicrucianism, which during these same years was taking shape in people's imaginations owing to Andreae's mystifications. Weigel's name meant so much that at least a dozen items were falsely published under his name in 1618 alone, a few of them written by other well-known mystics. In consequence Weigel's thought was diffused into a debased form of philosophy, which for the rest of the century was mentioned along with other forms of enthusiasm. Weigel was well known and respected among English Quakers, and like the mystical Protestant tradition in general, his thought was taken up by the Pietists.

It is Weigel's basic doctrine that all men are born with the means of knowing all that pertains to their spiritual welfare. The divine or inward light is infused by God and the Holy Ghost. In conformity with the macrocosm-microcosm doctrine, the light of nature reveals all things pertaining to God's creation, while the light of grace ensures supernatural knowledge. The inward word is the term he commonly uses to refer both to the innate capacity for knowledge and to the innate knowledge itself. The objects of cognition seen by the eye only awaken what is already in the mind; but these objects are necessary just as the kernel of wheat will not grow and bear fruit without being planted in the soil. It is clear that this radical subjectivist doctrine entails a thorough rejection of all outward authority and forms: church, priesthood, sacraments, confession, ceremonies, outward speech or scripture, books, and universities. Even the historical Christ has no meaning; faith and religion do not depend on the revelation of religious events in time and space. With the mystics of his own and the next century, Weigel shared the doctrine that the only true church is invisible as well as the commitment to toleration in all matters. The first three chapters of Genesis constitute an epitome of the scriptural story of creation, fall, and redemption written for this secular life; but in the future world of perfection, there will again be no need of languages, art, and knowledge, just as Adam in his angelic state had no need of them. In that future state, nature as we know it will cease to exist. For the time being, however, we must continue to make the effort to learn by seeking to let the outward world awaken our innate knowledge; the chief obstacle is the false teaching of the pseudotheologians. Weigel's epistemology and cosmology are chiefly set forth in the following works: *Gnothi seauton, nosce teipsum, erkenne dich selbst; Vom Ort der Welt;* and *Der güldene Griff, alle Dinge ohne Irrthum zu erkennen, vielen hochgelehrten unbekannt, und doch allen Menschen nothwendig.*

BIBLIOGRAPHY

I. ORIGINAL WORKS. The canon of Weigel's writings is the subject of Winfried Zeller's *Die Schriften Valentin Weigels: Eine literarkritische Untersuchung* (Berlin, 1940), which is *Historische Studien*, no. 370. In addition to the printed items attributed to Weigel, Zeller lists no less than 130 MSS, of which more than a dozen are judged genuine; a few of these have been printed only recently. The early printings are very rare, but since 1962 Will-Erich Peuckert and W. Zeller have edited the *Sämtliche Werke*, eventually to contain all the genuine works. The following have appeared: *Vom Ort der Welt*, W.-E. Peuckert, ed. (1962); *Von der Vergebung der Sünden oder vom Schlüssel der Kirchen*, W. Zeller, ed.

(1964); *Zwei nützliche Tractate* and *Kurzer Bericht und Anleitung zur Deutschen Theologie*, W. Zeller, ed. (1966); *Dialogus de Christianismo*, Alfred Ehrentreich, ed. (1967); *Ein Büchlein vom wahren seligmachen Glauben*, W. Zeller, ed. (1969)—this last item is Weigel's defense of 1572.

Two of Weigel's works appeared in English trans. during the 17th century: *Of the Life of Christ* (London, 1648) and *Astrologie Theologized* (London, 1649); the latter is presumably a translation of *Gnothi seauton*, pt. II. John Locke's friend, the Quaker Benjamin Furly, with whom Locke stayed while in exile in Rotterdam, left a translation of "A Brief Instruction of the Way and Manner to Know all Things . . . Written by Valentyn Weigelius, in Hyhdutch and New Englished by a Lover of Truth, Benj. Furly 1664" (see W. Zeller, *Schriften*, p. 77).

II. SECONDARY LITERATURE. Three early works have rich bibliographies and important information about the ways in which Weigel and his thought have been judged. There is a long chapter "Vom Weigelianismo," in Ehregott Daniel Colberg's critique of all forms of enthusiasm, *Das Platonisch-Hermetisches Christenthum, begreiffend die historische Erzehlung vom Ursprung und vielerley Secten der heutigen fanatischen Theologie, unterm Namen der Paracelsisten, Weigelianer, Rosencreutzer, Quäker, Böhemisten, Wiedertäuffer, Bourignisten, Labadisten, und Quietisten* (Frankfurt–Leipzig, 1690), 205–264. A very well-informed and sympathetic treatment will be found in Gottfried Arnold, *Unpartheyische Kirchen- und Ketzer-Historie*, 2 vols. in 4 pts. (Frankfurt, 1699–1700), pt. 2, 615–640. *Zedler's Grosses Universal-Lexicon aller Wissenschaften und Künste*, **54** (1747), has two articles (with full bibliographies) that are critical of Weigel's heretical doctrines: "Weigel," cols. 293–304; and "Weigelianer," cols. 304–326.

For a brief bibliography of the secondary literature, see W. Zeller, *Schriften*, pp. 86–87. There is a useful survey of the primary literature, also indicating the subject and argument of each work, in Ludolf Pertz, "Beiträge zur Geschichte der mystischen und ascetischen Literatur. I. Weigels Leben und Schriften," in *Zeitschrift für die historische Theologie* (1857), pp. 3–94; and "II. Weigels Theologie," *ibid.* (1859), pp. 49–123. There is no recent work that compares in scope to that of these two items: Julius Otto Opel, *Valentin Weigel: Ein Beitrag zur Literatur- und Culturgeschichte im 17. Jahrhundert* (Leipzig, 1864); and August Israel, *Valentin Weigels Leben und Schriften* (Zschopau, 1888). For a penetrating and critical review of Israel, see G. Kawerau, in *Theologische Literaturzeitung*, **13** (1888), cols. 594–598.

The author's Lutheran orientation deepens the interest of Hans Maier, "Der mystische Spiritualismus Valentin Weigels," in *Beiträge zur Förderung christlicher Theologie*, **29**, no. 4 (1926), 389–495. Since several of Weigel's works are not readily available, Heinz Längin, "Grundlinien der Erkenntnislehre Valentin Weigels," in

Archiv für Geschichte der Philosophie, **41** (1932), 434–478, is useful not merely for its cogent analyses but also owing to its extensive quotations. See also Winfried Zeller, "Meister Eckhart bei Valentin Weigel. Eine Untersuchung zur Frage der Bedeutung Meister Eckharts für die mystische Renaissance des sechzehnten Jahrhunderts," in *Zeitschrift für Kirchengeschichte*, **57** (1938), 309–355; and Alexandre Koyré, "Un mystique protestant: Valentin Weigel," in *Mystiques, spirituels, alchimistes du XVIᵉ siècle allemand* (Paris, 1955), which is in *Cahiers des Annales*, no. 10, pp. 81–116; originally published in 1930 soon after the same author's monograph on Boehme, this essay has all the virtues of Koyré's insight and clarity, but it should be noted that Koyré accepts *Studium universale* in the Weigel canon, contrary to later and, it would seem, well-founded judgment. Fritz Lieb, *Valentin Weigels Kommentar zur Schöpfungsgeschichte und das Schriftum seines Schülers Benedikt Biedermann* (Zurich, 1962), argues that Biedermann is the author of the pseudo-Weigelian works and that they were written while he was Weigel's deacon, that is, concurrently with Weigel's own works. Lieb also argues that Weigel was the author of the Genesis commentary, "Viererley Auslegung über das erste Capittel Mosis, von der Schöpfung aller Dinge" (see W. Zeller, *Schriften*, pp. 66–67); he shows that it was published early in the 18th century, and that it was translated into Russian during the reign of Catherine II owing to the popularity of Weigel and other mystics in masonic circles. For an important aspect of the relationship between Paracelsus and Weigel, see Kurt Goldammer, "Friedensidee und Toleranzgedanke bei Paracelsus und den Spiritualisten. Franck und Weigel," in *Archiv für Reformationsgeschichte*, **47** (1956), 180–211. Ernst Wilhelm Kämmerer, *Das Leib-Seele-Geist-Problem bei Paracelsus und einigen Autoren des 17. Jahrhunderts* (Wiesbaden, 1971), deals with a Paracelsian doctrine that Weigel took over (esp. pp. 70–76). Two useful recent books place Weigel in the larger context of his century: Siegfried Wollgast, *Der Deutsche Pantheismus im 16. Jahrhundert. Sebastian Franck und seine Wirkungen auf die Entwicklung der pantheistischen Philosophie in Deutschland* (Berlin, 1972), esp. pp. 267–286, and Steven E. Ozment, *Mysticism and Dissent. Religious Ideology and Social Protest in the Sixteenth Century* (New Haven, 1973), pp. 203–245. Weigel's role in Quaker thought is treated in Rufus M. Jones, *Spiritual Reformers in the Sixteenth and Seventeenth Centuries* (New York, 1914), chiefly in the chapter devoted to "Valentine Weigel and Nature Mysticism," pp. 133–150.

HANS AARSLEFF

WEIGERT, CARL (*b.* Munsterberg, Silesia [now Poland], 19 March 1845; *d.* Frankfurt am Main, Germany, 4 August 1904), *pathology, histology, neurology.*

Weigert was born in the same district in Silesia as his cousin Paul Ehrlich, his junior by nine years. The problem of the selective action of dyes on biological materials (microchemical reactions), which led Ehrlich to develop chemotherapy, led Weigert to make revolutionary advances in histological techniques. These advances made it possible for researchers to gain fundamental insights into the fine structure of the central nervous system. Weigert is thus closely associated with brain and spinal cord research and with neurology and psychiatry.

After attending the Gymnasium in Breslau, Weigert studied medicine at the University of Breslau. His teachers included Ferdinand Cohn and Rudolf Heidenhain. Weigert continued his studies in Berlin, where he worked as Virchow's amanuensis. In 1866 he received his medical degree from the University of Berlin for a dissertation, "De nervorum laesionibus telorum ictu effectis." Two years later Weigert became an assistant of Waldeyer-Hartz, professor of pathology at Breslau, who undoubtedly strengthened Weigert's interest in morphology. In 1871 Weigert became clinical assistant to Hermann Lebert, and in 1874 assistant to Julius Cohnheim, under whom he qualified for teaching pathology in 1875. Three years later he followed Cohnheim to the University of Leipzig, where in 1879 he was named extraordinary professor of pathology. For a long time Weigert lectured in place of Cohnheim, who had fallen ill. Cohnheim died in 1884, and when the faculty did not nominate him even as a possible successor, Weigert resigned from his post the following year. He decided to take up medical practice, but he was dissuaded by an offer to become director of the pathological-anatomical institute of the Senckenberg Foundation in Frankfurt. He held this post until his death at age fifty-nine from a coronary embolism. Curiously, he himself had made a major contribution to knowledge of this disease. Weigert never married.

Weigert's most notable personal characteristic was his excessive modesty. He was plagued by doubts about the value of his work and was never satisfied with what he had accomplished. Yet, he was indisputably successful in teaching advanced science students, both in the classroom and in the laboratory. His friend Ludwig Edinger wrote:

> What attracted the many students from all over the world and what persuaded them to persevere in the small, poorly equipped rooms of the Frankfurt institute, which were in no way comparable to the proud university institutes, was the intimate relationship that existed between teacher and students. The door between his work room and the laboratory always stood open. . . . A basic characteristic of his manner of working was never to stick too closely to details. Rather, he always sought to grasp pathological processes as biological processes. . . . Weigert possessed an excellent philosophical training, and philosophical thinking governed his entire way of working. He disciplined himself to renounce any attempt to penetrate what could not be known and was made uneasy by metaphysical speculation. . . . He always viewed the many facts he discovered as mere building stones. In his leisure he amused himself with mathematics.

According to the pathologist O. Lubarsch, Weigert was

> . . . inwardly happy, a truly distinguished and good man, who viewed the weaknesses of those around him with the deep sense of humor of the philosopher and who reacted only mildly against those who wished to harm him. Nothing human was foreign to him, and after a day of hard work he sought relaxation in literature and society, amusing everyone with his warm-hearted humor and his witty conversation. His contact with Scandinavian students prompted him to learn their languages.

With his first major work on the eruption of smallpox on the skin (1874), Weigert opened a new area of research in pathological anatomy—the demonstration of the primary damage of cells and tissues by external influences. First he had to develop the histological techniques necessary to detect this process. With the aid of the improved microtome Weigert dissected pathological tissue into complete serial sections and devised a technique that enabled him to differentiate tissues sharply under the microscope. By 1871 he was able, by staining, to demonstrate the presence of bacteria in tissue sections. This advance was of the greatest importance for the subsequent work of Robert Koch. According to Ehrlich, Weigert's monograph of 1874–1875 already contained "the points of view that guided his work for the rest of his life." In this monograph he began to develop the theory of the "coagulation necrosis" (the term is due to Cohnheim) and to illustrate the reparative (bioplastic) processes of the supporting tissue. Before Roux he had developed also the concept that the cells of the organism are in equilibrium among themselves. Cells cannot disappear without the neighboring cells attempting to take over their place. If elements of the parenchyma disappear, then they are generally replaced by elements of the connective tissue group. Such substitute growth is at first excessive, but gradually a new equilibrium is established. Weigert later applied his theory of the secondary, reparative processes to the substitute growth of neuroglia following atrophy of the nervous parenchyma.

Weigert's experiments on the staining of fibrin (1887) were also important for general pathology, since they exercised a lasting influence on the study of inflammation and on the theory of thrombosis. Weigert's method for staining elastic fibers was also important (although published in 1898, the preliminary work on it began in 1884). In addition, Weigert conducted studies on the genesis of acute general miliarial tuberculosis (affecting the veins and thoracic duct). The studies are remarkable for pointing out its spread from a tubercular source, before the agent of the disease was even known.

Weigert's most important work was in the field of neurohistology. After long preliminary investigation, he presented in 1884 the definitive method of staining medullary sheaths (myelin sheaths). This method enabled scientists to establish a reliable anatomy of the central nervous system. The number of works in this field multiplied within a few years of Weigert's announcement. The research on comparative anatomy of the brain owes its existence to myelin staining; previously, the path of scarcely a single pathway in a lower vertebrate brain had really been established with certainty. Starting about 1886 the new results, which eliminated much speculation, found their way into the textbooks. Recalling this pioneering time, Edinger wrote (1906):

In the first years [of research] fetal brains, which contain only a few medullary sheaths and therefore provide particularly clear views, were a rich source of important facts. Myelin staining finally allowed us, in the middle of the 1880's, to determine, for example, how the afferent nerves reach the brain. Their termination in the nuclei of the dorsal columns was known. From there, as has now been established, a crossed pathway leads to the thalamus. At that time, [however,] people proposed all kinds of uncertain paths leading from those nuclei through the olive into the cerebellum. One of the first works, begun in Weigert's presence, was Lissauer's study of the dorsal roots and the spinal cord, which is still absolutely valid. He and Weigert discovered in the first months that in the disease tabes dorsalis the dorsal (Clarke) columns degenerate.

Weigert sought to find evidence of the selective behavior of the supporting tissue of the nervous system in order to establish the system's pathological histology. This effort led to his successful demonstration in 1887 of the existence of the neuroglia. (His findings were not published until 1890.) He found, again, that the atrophy of the parenchyma is the primary process and results in the growth of the neuroglia as a secondary process. In this area, too, Weigert's work has influenced neurohistological research up to the present, although his notion of the existence of neuroglia fibers "emancipated" from the cell and differentiable by staining did not remain unchallenged.

BIBLIOGRAPHY

I. ORIGINAL WORKS. A chronological list of Weigert's works is in R. Rieder (see below). See also J. Springer, ed., *Gesammelte Abhandlungen von Carl Weigert*, II (Berlin, 1906), which has biographical contributions by R. Rieder, K. Edinger, and P. Ehrlich.

Weigert's works include "Über Bacterien in der Pockenhaut," in *Zentralblatt für die medizinischen Wissenschaften*, 8 (1871), 609–611; *Anatomische Beiträge zur Lehre von den Pocken*, written with Max Cohn, 2 vols.: *I. Die Pockeneffloreszenz der äusseren Haut* (Breslau, 1874), *II. Über pockenähnliche Gebilde in parenchymatösen Organen und deren Beziehung zu Bakterienkolonien* (Breslau, 1875); "Bismarckbraun als Färbemittel," in *Archiv für mikroskopische Anatomie und Entwicklungsmechanik*, 15 (1878), 258–260; *Die Brightsche Nierenerkrankung vom pathologisch-anatomischen Standpunkte*, which is Sammlung klinischer Vorträge No. 162/63 (Leipzig, 1879); "Über Entzündung (Inflammatio, Phlogosis)," in *Eulenburgs Real-Encyclopädie der gesamten Heilkunde* (2nd ed., rev. and enl.), VI (1886), 325–358 (Weigert's contribution was replaced by a corresponding article by Ernst Ziegler in the 3rd ed. [1895]); "Zur Technik der mikroskopischen Bakterienuntersuchungen," in *Virchows Archiv für pathologische Anatomie und Physiologie und für klinische Medizin*, 84 (1881), 275–315; "Über eine neue Untersuchungsmethode des Centralnervensystems," in *Zentralblatt für die medizinischen Wissenschaften*, 20 (1882), 753–757, 772–774; "Thrombose," in *Eulenburgs Real-Encyclopädie der gesamten Heilkunde*, 19 (1889), 638–648, also replaced by Ziegler in 1900; "Über Schnittserien von Celloidinpräparaten des Centralnervensystems," in *Zentralblatt wissenschaftliche Mikroskopie*, 2 (1885), 490–495; "Über eine neue Methode zur Färbung von Fibrin und von Microorganismen," in *Fortschritte der Medizin*, 5 (1887), 228–232; "Zur pathologischen Histologie des Neurogliafasergerüstes," in *Zentralblatt für allgemeine Pathologie und pathologische Anatomie*, 1 (1890), 729–737; "Beiträge zur Kenntnis der normalen menschlichen Neuroglia," in *Abhandlungen der Senckenbergischen Naturforschenden Gesellschaft*, 19 (1895), fasc. 11;

"Die histologische Technik des Centralnervensystems II.2. Die Markscheidenfärbung," in *Ergebnisse der Anatomie und Entwicklungsgeschichte*, **6** (1897), 1–25; "Über eine Methode zur Färbung elastischer Fasern," in *Zentralblatt für allgemeine Pathologie und pathologische Anatomie*, **9** (1898), 289–292; and "Fibrinfärbung," "Markscheiden der Nervenfasern," "Neurogliafärbung," in Paul Ehrlich, Rudolf Krause, Max Mosse, Heinrich Rosin, Carl Weigert, eds., *Enzyklopädie der mikroskopischen Technik*, 2nd ed. (Berlin–Vienna, 1910), I, 457–460; II, 231–238; 298–311.

II. SECONDARY LITERATURE. On Weigert and his work, see L. Edinger, "Carl Weigerts Verdienste um die Neurologie," in R. Rieder (see below), pp. 133–137; P. Ehrlich, "Weigerts Verdienste um die histologische Wissenschaft," *ibid.*, pp. 138–141; G. Herxheimer, "Carl Weigert," in *Zentralblatt für allgemeine Pathologie und pathologische Anatomie*, **15** (1904), 657–662; W. Krücke, "Carl Weigert (1845–1904)," in W. Scholtz, *50 Jahre Neuropathologie in Deutschland* (Stuttgart, 1961), 5–19; W. Krücke and H. Spatz, "Aus den Erinnerungen von Ludwig Edinger," in Ludwig-Edinger-Gedenkschrift (*Schrifter der Wissenschaftlichen Gesellschaft an der Johann Wolfgang Goethe-Universität Frankfurt am Main*, 1st ser. (Wiesbaden, 1959), 19–23; L. Lichtheim, "Karl Weigert," in *Deutsche Zeitschrift für Nervenheilkunde*, **27** (1904), 340–350.

See also O. Lubarsch, "Karl Weigert," in *Deutsche medizinische Wochenschrift*, **30** (1904), 1318–1319; H. Morrison, "Carl Weigert," in *Annals of Medical History*, **6** (1924), 163–177; R. Rieder, *Carl Weigert und seine Bedeutung für die medizinische Wissenschaft unserer Zeit, eine biographische Skizze* (Berlin, 1906); A. Strümpell, "Zur Erinnerung an Carl Weigert," in *Deutsche medizinische Wochenschrift*, **31** (1905), 230–232; and I. H. Talbott, "Carl Weigert (1845–1904)," in *A Biographical History of Medicine* (New York, 1970), 837–840.

G. RUDOLPH

WEINBERG, WILHELM (*b.* Stuttgart, Germany, 25 December 1862; *d.* Tübingen, Germany, 27 November 1937), *human genetics, medical statistics.*

Weinberg was the son of Julius Weinberg, a Stuttgart merchant, and Maria Magdalena Humbert. Weinberg's father was Jewish, and his mother was Protestant; both of his parents died early. Weinberg belonged to the Protestant faith as did his wife, Bertha Wachenbrönner, whom he married in 1896. They had four sons and one daughter. Weinberg studied medicine at the universities of Tübingen and Munich and obtained his M.D. in 1886. After clinical experience in Berlin, Vienna, and Frankfurt, he established himself in Stuttgart

as a general practitioner and obstetrician (1889). For forty-two years he had a large private practice and also acted in public capacities as physician to the poor and to the socially insured. He attended more than 3,500 births, including more than 120 twin births.

Weinberg's discoveries center around four areas: multiple births, population genetics, and ascertainment and medical statistics. His first important paper was the eighty-five-page "Beiträge zur Physiologie und Pathologie der Mehrlingsgeburten beim Menschen" (1901). In the article he established the difference method, which enabled him to derive the proportion of monozygotic and dizygotic twin births from statistical data on the sex combinations of otherwise undifferentiated twin births. He proceeded to discover differences between mono- and dizygotic twins in a variety of traits, including an inheritance of a twinning tendency for dizygotic but not for monozygotic twins.

When Weinberg became aware of Mendelism he asked himself "how different laws of inheritance would influence the composition of the relatives of given individuals." Answers were provided in four extraordinary papers: "Über den Nachweis der Vererbung beim Menschen" (1908), "Über Vererbungsgesetze beim Menschen I and II" (1908–1909), and "Weitere Beiträge zur Theorie der Vererbung" (1910). Weinberg discovered the equilibrium law of monohybrid populations and the varied processes of attainment of equilibria in polyhybrid populations. He had become a founder of population genetics. The equilibrium law was also discovered slightly later by G. H. Hardy. It is now known as the Hardy-Weinberg law. In his studies of population genetics Weinberg's derivations of the correlations between relatives expected under Mendelian heredity took into account both genetic and environmental factors. Indeed, he was the first to partition the total variance of phenotypes into genetic and environmental portions (1909, 1910).

Weinberg recognized early that different types of ascertainment may bias greatly the results of statistical inquiries. In "Über Methode und Fehlerquellen der Untersuchung auf Mendelsche Zahlen beim Menschen" (1912) he furnished methods of correction for various types of ascertainment. One such device is the sib method, in which correct proportions for traits are obtained by finding the ratio of affected to nonaffected traits among the sibs of the affected. Other methods also described by Weinberg are the proband and a priori methods.

Weinberg made detailed studies of mortality statistics and the statistics and genetics of specific diseases. He was the first to construct morbidity tables modeled after the long-known mortality tables.

Weinberg had no personal collaborators or students, although a few contemporary investigators were strongly influenced by him. Only in Weinberg's later years did a new generation again begin to explore the areas in which he had achieved so much.

BIBLIOGRAPHY

I. ORIGINAL WORKS. A bibliography of Weinberg's numerous publications, most of which are journal articles, has been compiled by Eva R. Sherwood; it is on deposit in the biology library of the University of California at Berkeley and has been reprinted in Jh. Ver. Naturkde. Württemberg 118/119 (1964), 61–67.

II. SECONDARY LITERATURE. An obituary (probably written by E. Rüdin) is in *Archiv fur Rassen- und Gesellschaftsbiologie einschliessend Rassen- und Gesellschaftshygiene*, **31** (1937), 54.

See also K. Freudenberg, "Wilhelm Weinberg zum 70. Geburtstage," in *Klinische Wochenschrift*, **12** (1933), 46–47; E. Hübler, "Zum 100. Geburtstag von Wilhelm Weinberg," Jh. Ver. vaterl. Naturkde. Württemberg 118/119 Jahrgang (1964), 57–67; F. J. Kallmann, "Wilhelm Weinberg, M.D.," in *Journal of Nervous and Mental Diseases*, **87** (1938), 263–264; H. Luxenburger, "Wilhelm Weinberg," in *Allgemeine Zeitschrift für Psychiatrie*, **107–109** (1938), 378–381; Curt Stern, "Wilhelm Weinberg, 1862–1937," in *Genetics*, **47** (1962), 1–5; and "Wilhelm Weinberg. Zur hundertjährigen Wiederkehr seines Geburtsjahres," in *Zeitschrift menschliche Vererbungs und Konstitutionslehre*, **36** (1962), 374–382.

CURT STERN

WEINGARTEN, JULIUS (*b*. Berlin, Germany, 2 March 1836; *d*. Freiburg im Breisgau, Germany, 16 June 1910), *mathematics*.

The son of a weaver who had emigrated from Poland, Weingarten graduated from the Berlin municipal trade school in 1852 and then studied mathematics and physics at the University of Berlin and chemistry at the Berlin Gewerbeinstitut. Between 1858 and 1864 he was an assistant teacher at various schools in Berlin. After receiving the Ph.D. from the University of Halle in 1864, Weingarten taught at the Bauakademie in Berlin, where he was promoted to the rank of professor in 1871. His next position was at the newly founded Technische Hochschule in Berlin. In 1902, for reasons

of health, he moved to Freiburg im Breisgau, where he taught as honorary professor until 1908.

Weingarten was inspired by Dirichlet's lectures to study potential theory and later in his career he occasionally published papers on theoretical physics. It was, however, in pure mathematics, particularly in differential geometry, that he made his greatest contribution. Lack of money obliged Weingarten to accept unsatisfactory teaching positions for many years. It was not until he came to Freiburg that, at an advanced age, he found a suitable academic post.

In 1857 the University of Berlin awarded Weingarten a prize for a work on the lines of curvature of a surface, and in 1864 he received the doctorate for the same work. In the meantime he had written other major papers on the theory of surfaces (1861, 1863). This was the most important subject in differential geometry in the nineteenth century, and one of its main problems was that of stating all the surfaces isometric to a given surface. The only class of such surfaces known before Weingarten consisted of the developable surfaces isometric to the plane. These included the cones.

Weingarten was the first to go beyond this stage. For example, he gave the class of surfaces isometric to a given surface of revolution. He had the important insight of introducing those surfaces for which there exists a definite functional relationship between their principal curvatures (1863). These are now called W-surfaces in his honor. Weingarten showed that the one nappe of the central surface of a W-surface is isometric to a surface of revolution and, conversely, that all surfaces isometric to surfaces of revolution can also be obtained in this manner. The W-surfaces are best conceived by considering their spherical image and, operating from this point of view, Weingarten also described various classes of surfaces that are isomorphic to each other. Later he cited classes of this kind in which there are no surfaces of revolution (1884).

In 1886 and 1887 Weingarten studied the infinitesimal deformation of surfaces. Jean-Gaston Darboux, the leading differential geometer of the nineteenth century and author of the four-volume *Leçons sur la théorie générale des surfaces . . .*, stated that Weingarten's achievements were worthy of Gauss. Darboux's work inspired Weingarten to undertake further research, which appeared in a long paper that was awarded a prize by the Paris Academy of Sciences in 1894 and was published in *Acta mathematica* in 1897. In this paper Weingarten reduced the problem of determining all the surfaces isometric to a given surface F to that of determining all solutions of a certain partial differential equation of the Monge-Ampère type.

BIBLIOGRAPHY

Weingarten's writings include "Über die Oberflächen, für welche einer der beiden Hauptkrümmungsmesser eine Funktion des anderen ist," in *Journal für die reine und angewandte Mathematik*, **62** (1863), 160–173; "Über die Theorie der aufeinander abwickelbaren Oberflächen," in *Festschrift der Königlichen Technischen Hochschule zu Berlin* (Berlin, 1884), 1–43; "Über die Deformation einer biegsamen unausdehnbaren Fläche," in *Journal für die reine und angewandte Mathematik,* **100** (1887), 296–310; "Sur la déformation des surfaces," in *Acta mathematica*, **20** (1897), 159–200; and "Mémoire sur la déformation des surfaces," in *Mémoires présentés par divers savants*, 2nd ser., **32** (1902), 1–46.

An obituary is Stanislaus Jolles, "Julius Weingarten," in *Sitzungsberichte der Berliner mathematischen Gesellschaft*, **10** (1911), 8–11.

WERNER BURAU

WEISBACH, JULIUS LUDWIG (*b.* Mittelschmiedeberg, near Annaberg, Germany, 10 August 1806; *d.* Freiberg, Germany, 24 February 1871), *hydraulics.*

The eighth of nine children born to Christian Gottlieb Weisbach, a mine foreman, and Christina Rebekka Stephan, Weisbach received his early education at the lyceum in Annaberg and the Bergschule in Freiberg. In 1822 borrowed funds enabled him to enter the Bergakademie, where Mohs advised him to go on to Göttingen. After two years at the latter university, he followed Mohs in 1829 to the Technical University and University of Vienna, where he studied mathematics, physics, and mechanics. Weisbach spent six months of the following year traveling on foot through Hungary, the Tirol, Bavaria, and Bohemia. From 1831 to 1835 he gradually assumed responsibility for all instruction in mathematics at the Freiberg Gymnasium and, from 1832, that at the Bergakademie as well, despite a low salary and little recognition. In 1832 he married Marie Winkler; their son, Albin, later became professor of mineralogy at the Bergakademie.

The first of Weisbach's numerous publications, *Bergmaschinenmechanik*, appeared in 1835, and the following year he was promoted to full professor of mathematics, mine machinery, and surveying. A trip to the Paris Industrial Exposition in 1839 increased Weisbach's interest in hydraulics and led to his first papers in this field. At the same time he contributed greatly to the development of mine surveying methods, introducing the theodolite in place of compass and protractor. Apparently an indefatigable worker, he assumed responsibility for courses in descriptive geometry, crystallography, and optics, as well as general mechanics.

In 1854 Weisbach was offered a position at the Zurich Polytechnikum (to open in 1855); he chose to remain at Freiberg and the following year assumed the further task of teaching machine design. Also in 1855 he attended the Paris World Exposition, receiving and correcting proof for a new edition of his *Mechanik* en route. During his professional career Weisbach published fourteen books and fifty-nine papers on mathematics, mechanics, and surveying, but primarily on hydraulics. An able experimenter, he presented most of his results in *Experimental-Hydraulik* (Freiberg, 1855); they are also summarized in the hydraulics section of his *Lehrbuch der Ingenieur- und Maschinenmechanik* (Brunswick). The two- (and eventually three-) volume work went through five editions between 1845 and 1901 and was translated into English and other languages. Some of his hydraulic data and formulas are still in use.

From 1850 Weisbach received a series of professional honors, including an honorary doctorate from the University of Leipzig in 1859 and the first honorary membership granted by the Verein Deutscher Ingenieure in 1860. He was a corresponding member of the St. Petersburg Academy of Sciences, the Royal Swedish Academy of Sciences, and the Accademia dei Lincei.

BIBLIOGRAPHY

See H. Undeutsch, *Zum Gedächtnis an Oberbergrat Professor Dr. h. c. Julius Ludwig Weisbach anlässlich seiner hundertjährigen Geburtstagsfeier* (Freiberg, 1906); *Julius Weisbach, Gedenkschrift zu seinem 150. Geburtstag*, Freiberger Forschungsheft Kultur und Technik, 16 (Berlin, 1956), with a complete list of published books and papers; and H. Rouse and S. Ince, *History of Hydraulics* (New York, 1963).

HUNTER ROUSE

WEISMANN, AUGUST FRIEDRICH LEOPOLD (*b.* Frankfurt am Main, Germany, 17 January 1834; *d.* Freiburg im Breisgau, Germany, 5 November 1914), *zoology.*

Weismann's most influential contribution to biological thought was his theory of the continuity of the germ plasm, an explanation of heredity and development. He maintained that the germ plasm, the substance of heredity, was transmitted from generation to generation, distinguishing it from the somatoplasm; he also was foremost in his day in denying that acquired characters were inherited. Keeping abreast of current researches on the cell and the growing understanding of the role of the nucleus and the chromosomes in inheritance, Weismann modified and developed the theory of germinal continuity. Cytology repeatedly confirmed phenomena the existence of which he had presumed from theoretical considerations. He was a strong defender of Darwin's theory of evolution and a leader among Neo-Darwinists arguing for the sufficiency of natural selection.

Weismann was the son of Johann Konrad August Weismann, a classics professor at the Gymnasium in Frankfurt, and Elise Eleanore Lübbren, a talented musician and painter, who understood her son's love of nature. At an early age Weismann gathered butterflies and beetles, bred caterpillars, and assembled an impressive herbarium of the plants in the vicinity of Frankfurt; his pleasure in his butterfly collections and in music was lifelong. He attended the Gymnasium at Frankfurt, and then—despite his preference for physics, chemistry, and botany, and his hopes perhaps to become a chemist—he began the study of medicine at the University of Göttingen; Friedrich Wöhler, a family friend, advised that Weismann take up medicine first, and his father thought he ought to have a practical means of earning a living. Among his professors at Göttingen were Wöhler, Siebold, Lotze, and Henle, who was an especially stimulating teacher to Weismann and one who was wary of the speculation that characterized *Naturphilosophie*. In 1856 Weismann received his medical degree with a dissertation on the formation of hippuric acid in the organism. When he looked back at his education, he felt that detailed research had been emphasized, with little attempt to interrelate the facts and subject matter of the various disciplines, or to deal with broader problems. He was therefore particularly sensitive to the impact of Darwin's *Origin of Species*, which he first read at one sitting in 1861. Thereafter he was a proponent of Darwin's theory of evolution by means of natural selection, and like Fritz Müller and Ernst Haeckel, through his addresses and writings, as well as his lectures to his classes, he was able to draw the theory of descent to the attention of the public.

After graduation Weismann continued his researches, first while he served as an assistant in a clinical hospital in Rostock, then as private assistant to Schulze. A paper on hippuric acid won him a prize at Rostock, as did a chemical investigation of the salt content of the water of the Baltic Sea. He visited Vienna, and late in 1858 he took his examinations and entered the practice of medicine in his native Frankfurt. Medicine did not take up all of Weismann's time, for he now turned to histological investigations stemming from his association with Henle at Göttingen and studied the minute structure of the muscle fibers of the heart. During the summer of 1859 he was a field doctor in Italy but soon returned to zoology, visiting the Jardin des Plantes in Paris, and hearing Geoffroy Saint-Hilaire lecture. Early in 1861 he spent two months at the University of Giessen under Leuckart, who, even during this brief period, exerted a strong influence upon Weismann. It was Leuckart who interested the young physician in developmental studies of insects, the Diptera in particular.

Weismann again practiced medicine for a short while in Frankfurt. Then he served from 1861 to 1863 as the private physician of the Archduke Stephan of Austria at Schaumburg Castle, and meanwhile studied insect development and metamorphosis and pursued researches in his spare time. Then Weismann gave his full attention to zoology. Strongly attracted to the beauty of Freiburg, he habilitated at its university in 1863 and taught comparative anatomy and zoology as a *Privatdozent*. Weismann was appointed extraordinary professor in 1866 and full professor in 1874. The first to hold the chair in zoology, he spent the rest of his career at the University of Freiburg. In 1864 his eyesight became seriously affected, but he continued his work. From 1862 to 1866 six memoirs on insects and another memoir on heart muscle appeared, even when he himself could not use the microscope. Weismann was helped immeasurably by his wife, Marie Dorothea Gruber of Genoa, whom he had met during his 1859 stay in Italy. She read to him and aided in countless ways over the years. They had six children and the family shared many interests; music was a part of their life, and their son became a musician and composer.

It was ten years before Weismann could again use his eyes to make his own observations with the microscope, although a sojourn in Italy from 1869 to 1871, during which his family accompanied him, proved beneficial to his eyesight. Throughout this period and afterward, he was well informed, however, and his lectures continued, although perhaps he was forced more to theorize and to delegate certain researches to others. From insects and their embryology he turned to the small crustaceans, daphnids and ostracods; and with the improvement of his eyesight a series of publications appeared. He then took up the examination of the Hydrozoa and followed the origin of the sexual cells through generations of Hydromedusae. The culmination of this work was *Die Entstehung der Sexualzellen bei den Hydromedusen* (Jena, 1883), and his conclusions, as he traced the fate of the primitive germ cells and those after them, were strong evidence to Weismann that there was a continuity of the germ plasm. When in 1884 his eyesight was again severely limited, his students and other assistants shared his work, his wife helped him, and the research continued under his guidance. Weismann approached his theoretical determinations from an evolutionary standpoint, but he correlated his theories with cytological investigations, and was always alert to the interpretations and arguments of his colleagues. He was not rigid, but, heeding the growing knowledge of the cell, was willing to develop and change his ideas.

In the intervening years Weismann had given the Darwinian theory of evolution his full support, and he called his inaugural lecture, in "justification" of Darwin's ideas, *Über die Berechtigung der Darwin'-schen Theorie* (Leipzig, 1868), a "kind of confession of faith." When the trouble with his eyesight restricted his work, Weismann gave Darwin's theory more of his attention, and in his *Studien zur Descendenztheorie* (Leipzig, 1875–1876)—with a preface by Darwin himself in the English translation (London, 1882)—Weismann treated the seasonal dimorphism of butterflies and questions of evolution and heredity. Although he remained one of the foremost defenders of the Darwinian theory of evolution through natural selection, Weismann—a strict selectionist, more so indeed than Darwin—proceeded to construct his own theory of heredity rather than accept Darwin's hypothesis of pangenesis.

Weismann had come to an early conclusion that the "direction of development," the same as that of the parent, was transferred by means of the protoplasm of the sperm and of the egg cell (1868). Studying the Diptera, he found that the sexual glands had originally been derived from pole cells that were formed by segmentation of the egg. But most crucial to him was the evidence of the Hydromedusae for a continuity of the germ plasm.

Through evolutionary considerations and in probing the question as to whether acquired characters could be inherited, Weismann came to develop his theory of the germ plasm. He decided that acquired characters could not be inherited, for to become inheritable, changes would have to affect the germ plasm itself. Weismann was not the first to conceive of a continuity of the substance of heredity. Galton had outlined it (1872); and Gustav Jäger had actually written of a "continuity of the germ-protoplasm" in 1875, although this was at the time unknown to Weismann. Moritz Nussbaum and August A. Rauber later claimed to have originated the theory; still, it is Weismann with whom the theory of the continuity of the germ plasm is associated, for he first developed it into a coherent explanation of inheritance and brought it into agreement with the new understanding of the cell. While the phenomena of the cell, and specifically of the chromosomes, were being followed in the laboratory, he modified his theory, and in later years it was a presupposition of his views on the sources of variation in evolution.

In 1881 Weismann gave a lecture ("The Duration of Life") before the Deutsche Naturforscher Gesammlung. In the lecture he contrasted the "immortality" of one-celled organisms, which reproduce by division to form two organisms of the same age—this potential immortality was, of course, abrogated by accidents and other vicissitudes—with the division of labor that natural selection had brought about in the more complex forms of life. In the latter forms there was an early separation of the elements that were to form the "immortal" reproductive cells from the elements that were to form the body cells that perished in each succeeding generation. In "On Heredity" (1883) Weismann still conceived of the germ cells as containing configurations of molecules that led to the reproductive cells, as well as other configurations for the somatic cells. The germ cells contained the *Anlagen*—the concept of *Anlagen* was current then, and to Weismann, these were hereditary tendencies or predispositions for certain characters to develop. The *Anlagen* were not affected by the outer conditions that affected the organism, but they were subject to natural selection. Weismann was still dealing with heredity in terms of the entire cell.

In the following years Weismann's concept of the germ plasm changed, and he developed his theory further. He borrowed the term "idioplasm" from the botanist Naegeli, who in 1884 described a continuity of the idioplasm, the protoplasm concerned with inheritance (as distinguished from the rest of the protoplasm) and thereby accounted also for variation. The idioplasm that Naegeli described had a structure of parallel rows of "micelles," which sometimes branched, and formed a network that coursed through the body; cell structure bore little importance in his theory. Since he did not subscribe to the structure Naegeli claimed for the idioplasm, Weismann continued to develop his own concept, but kept in mind the cytological researches then under way.

Following the independent observations (1873) of cell division—and Schneider's illustrations of the stages of division—the previous decade had brought many new discoveries dealing with the cell in division. In 1875 Oscar Hertwig had seen the apparent fusion of the nuclei during fertilization; and in 1879 Walther Flemming had clearly shown that the threads seen in the nucleus split longitudinally, and had pointed to the possible significance of this segmentation. Wilhelm Roux's conclusions influenced Weismann especially; Roux remarked how complicated the process of mitosis was: It seemed to be the means of accomplishing more than a division of the quantity of the substance of the nucleus between the two resulting daughter cells; the mechanism could divide the qualities equally between the new cells that were formed in division. Roux went on to propose that after the first equal segmentation there might possibly be unequal divisions in which the quantity might be the same in each cell, but the distribution of the qualities unequal.

In 1884 several scientists including Weismann independently came to attribute the main role in heredity to the nucleus (as Haeckel had proposed for theoretical reasons in 1866). After studying plant cells and following the reproduction of angiosperms, Strasburger arrived at this conclusion. Shortly thereafter Hertwig published a paper on his researches (Weismann had reached similar views before he read it). Koelliker decided also that the nucleus was involved in inheritance, although he later criticized other tenets of Weismann's theory.

In the essay "The Continuity of the Germ-Plasm" (1885) Weismann reflected upon his new views, for he now located the germ plasm more precisely within the nucleus, and he took up other questions as well. He still had some reservations as to whether the idioplasm was definitely to be identified with the portions of the nuclear filaments that took up chromatin stain, although he had decided that the nucleus indeed contained the idio-plasm of Naegeli. Weismann made his stand on the germ plasm clear: Germ cells did not necessarily lead directly to other germ cells, for germ plasm might be transmitted through a series of cells—its particles remaining discrete—before reproductive cells were again formed.

Weismann was already attempting to find the significance of the process of meiosis that his colleague Edouard Van Beneden had been investigating. He was dissatisfied with the interpretations of Van Beneden, Minot, and Balfour of the apparent casting-out from the egg of some of the hereditary material, for he did not agree that the egg cell was hermaphroditic, shedding the male element preparatory to fertilization. Weismann struggled to solve the problem. At first he maintained that in the maturation of the egg the formation of the polar bodies separated the germ plasm from a "histogenic" plasm in the nucleus, and that this resulted in a necessary preponderance of germ plasm in the reproductive cell. But by 1887 he had decided instead that this was a means of preventing the indefinite increase of ancestral plasms, which each fertilization would otherwise double.

Weismann believed that sexual reproduction led to variation, through the ever-new combinations of *Anlagen*. In the essay "On the Number of Polar Bodies" (1887) he reported on the researches he had undertaken, aided by his pupil and collaborator C. Ishikawa (who later became professor of zoology at Tokyo), when his sight was again severely affected. Although Weismann had for some time assumed that the hereditary qualities were linearly arranged along the "nuclear loops," as he described the chromosomes, and that differentiation occurred by means of qualitative divisions, such as Roux had proposed, he inferred—on theoretical grounds—that there must be a reduction division. He described the mounting complexity of the ancestral plasms, and it was his theory that there were two kinds of division: an equal division or *Aequations-theilung*, whereby the nuclear threads in splitting longitudinally divided the germ plasms equally; and a "reduction division," *Reduktions-theilung*, with half of the loops simply distributed to each of the daughter nuclei. Not only did Weismann theorize that reduction division takes place, but he suggested also that a similar process to that in the maturation of the ovum would occur during spermatogenesis; this was confirmed by cytological investigation.

Meanwhile, Weismann's interest in the problem of the possibility of the inheritance of acquired characters persisted, and he experimented to determine whether environmentally produced changes were inherited. He followed the results of mutilations, cutting off the tails of hundreds of mice, but without any apparent hereditary effects.

Weismann was aware of the attention the thread-like structures in the nucleus—first called chromosomes in 1888 by Waldeyer—were receiving. They were frequently described as rods or loops and seemed to be present in characteristic numbers in certain organisms and tissues. Rabl remarked that the chromosomes must persist throughout the various changes during cell division (1885), and in 1887 he advanced the concept of chromosomal individuality, which was basic to chromosome theory, and which Boveri brought to clearer development and a more definitive statement in the course of his cytological studies. In 1889 de Vries outlined his theory of "intracellular pangenesis" and his concept of the "pangen," an invisibly small vital unit concerned in heredity; this affected Weismann's thinking on the germ plasm, although they disagreed on important points (such as the necessity of a reduction division).

Comparable to de Vries's pangenes were Weismann's "biophors," as he developed his theory of the germ plasm. The biophors, however, did not mix freely as did the hereditary factors that de Vries described. Weismann then postulated the existence of larger, although still submicroscopic units, the "determinants." Weismann proceeded to describe the progression of units that were the "bearers of heredity." In accordance with the appearance of the visible chromosome, there were next the "ids," made up of determinants (the ancestral plasms of his germ plasm theory and linked together in definite groupings). The ids were the disks or microsomes that had been observed beneath the microscope and were linearly arranged on the chromosomes, in Weismann's terms, the "idants." Weismann thought that each id carried the entirety of the ancestral plasms that the individual inherited. Weismann gave his theory with his views of the significance of the laboratory investigations of the cell and of contemporary hypotheses, in *Das Keimplasma. Eine Theorie der Vererbung* (Jena, 1892; English translation, *The Germ-Plasm* [London, 1893]). The theory of the continuity of the germ plasm was now more completely developed and formulated in terms of the cell theory, although necessarily in his own interpretation, for phenomena that were being observed remained to be more clearly understood.

Through these units Weismann also explained the process of differentiation during embryogeny. He continued to maintain the existence of the eternal germ plasm and the mortal somatoplasm, but, in addition, described a sort of breakdown of the complexes of the hereditary units by which specific determinants eventually direct development in the cells concerned. This was predicated on Weismann's belief, based on Roux's opinions years before, that both equal and unequal divisions were possible. But Weismann later made some minor compromises in his theory when eventually he allowed that some changes might occur in the germ plasm itself, although he continued to think in terms of variation, natural selection, and a struggle for existence that took place on various levels.

Weismann's theories elicited praise from some of his contemporaries, criticism from others. Whatever his interpretation of the actual mechanism of the processes he described (the reduction division, for example, which drew much further research and dispute in the 1890's), Weismann brought important questions to active discussion and was especially influential through his stress on chromosome theory and his repudiation of the inheritance of acquired characters. No doubt, too, many of the disagreements in his day were due to the different materials and life histories that were investigated by botanists and zoologists. There seemed to be too many difficulties and exceptions for broad theories to explain them in a way that was completely acceptable. To Oscar Hertwig, for example, Weismann's theory smacked of preformation, and questions of the importance of the relationship between the individual and the environment in determining the direction of development arose.

Early in his career, in advocating the Darwinian theory in Germany, Weismann had joined its most famous popularizers, Müller and Haeckel; but each of the three overstepped the bounds of Darwin's own propositions. According to Darwin's theory, evolution results from the natural selection of heritable, favorable variations that are always occurring by chance. New varieties grow up, and, in time, species are formed by this means. But even Darwin did not rule out some effect of external conditions and use and disuse; in his later years he granted a greater role to these factors than he had in the *Origin of Species* in 1859.

Even as natural selection drew wide attention, naturalists who applied it to specific instances in their experiences and investigations found difficulties and came to various conclusions. Opinions divided on the way in which evolution actually takes place, and differed as to the sources of variation and their relative importance: on whether external changes and acquired characters play a part in inherited variation; and on the role of slow, cumulative change as opposed to saltatory changes. Even if natural selection were accepted, it was difficult to explain the degeneration and loss of organs, and to account for continuing changes below the level at which natural selection could be expected to act at all.

Weismann, who was the most notable of the Neo-Darwinists, claimed that natural selection alone could provide for the formation of varieties and, in time, species, although the Neo-Lamarckians found support in Darwin's own admitted doubts as to the degree to which natural selection alone acted. In addition, there were other variations of the Lamarckian or Darwinian positions, and some naturalists departed from both.

In 1893 and 1894 the *Contemporary Review* carried Weismann's controversy with Herbert Spencer. Weismann contended that natural selection acting on innate variation was the sole factor in establishing varieties and species, and that acquired characters were not transmitted, and therefore were not a source of evolutionary change; Spencer took a Lamarckian view. Weismann's own experiments supported his arguments, and he could point also to the general lack of proof for such inheritance. He became widely known for the stand he took; indeed, it was cited in his day as one of his great contributions to biology.

In his 1894 Romanes lecture, "The Effect of External Influences Upon Development," Weismann dealt with the difficulties of explaining by means of the "all sufficient principle" of natural selection not only the progressive development of some variations, but degenerative changes and the disappearance of useless organs. He maintained that heritable peculiarities occurred among the biophors and cells, and that the peculiarities were actually variations in the primary cell constituents and might be acted upon by natural selection.

Weismann believed that natural selection was adequate even beyond Darwin's hopes—in the light of further researches—but he found it necessary to fortify the theory with some additional, although subsidiary, explanations. Indeed, Weismann was forced to support his arguments for natural selection, first with his theory of "panmixia," then with his further theory of "germinal selection." He thereby extended what he had come to refer to as the "Darwin-Wallace principle of natural selection"; but both panmixia and germinal selection were also based upon Weismann's own theory of

the germ plasm, for he took for granted the theory of idants, ids, determinants, and the smallest units, biophors.

In defending the Darwinian theory, dealing with apparent difficulties or omissions, Weismann "extended" it and expressed his concepts in new terms. He had maintained that sexual reproduction, or to use his term, "amphimixis," was the source of variation; but he also found that he had to explain the disappearance of organs. Arguing that external influences could act only in a selective way on changes that had already occurred in the germ plasm, he ascribed regressive changes to variations that occurred in the primary constituents. But the problem remained. To account for the degeneration and ultimate disappearance of organs or parts, Weismann modified the Darwinian concept of natural selection with his own theory of "panmixia." Resisting Lamarckism, and trying to broaden the usefulness of the theory of natural selection, Weismann now inferred that natural selection not only brings about the development of the part but must actively cause it to be retained. Unless selection continued, the organ that lost its usefulness or specialized importance and became disused would tend to diminish and, in time, to disappear.

But how to explain its complete disappearance? To Weismann, panmixia was still not satisfactory, and his views changed as he looked next to what he called "germinal selection" as a further auxiliary to the selection theory. He went quite beyond Darwin's original account of the way in which natral selection acts and sought the selection process even in the germ plasm itself. Weismann came to see selection as taking place on many levels and in terms of his own later description of the hereditary units. There were fluctuations in nourishment and in conditions of life even among the biophors and determinants, and natural selection determined their chances to reach development and expression in the individual. Thus Weismann had come to conceive of a struggle of parts within the germ plasm. This enabled him to continue to deny that there was a direct effect upon heredity through external conditions, and to insist that, although external causes might eventually lead to variations, this would be through selection acting on the variations that occurred in the internal and innate hereditary tendencies and within the germ plasm itself. Nevertheless, it was quite a change of view for Weismann to admit that internal causes could intensify variational tendencies; and that these were "simply accumulated by natural selection in an ever-growing majority of ids in a germ-plasm through the selection of individuals" (*The Evolution Theory*, II, 332). It was almost impossible at the time for Weismann to separate his considerations of evolution and heredity. De Vries's mutation theory made it necessary for Weismann to define variation more completely in terms of natural selection, and he concluded that no lines could be drawn between variation and mutation, and that ultimately the difference was a matter of the number of similarly varying ids.

For decades Weismann drew a constant stream of students to his laboratory, and they returned to foreign countries with descriptions of his laboratory, reporting him as a teacher of impressive stature. Many students had been drawn to him through hearing of his views; one of his prominent pupils recalled his desire to go to study under Weismann and to hear his theories after having been well-schooled by the opposition. At the University of Freiburg, Weismann saw the Zoological Institute, of which he was the director, installed in a new building with the expansion of the facilities.

Weismann's wife died in 1886, and his second marriage at the age of sixty to Willemina Tesse of Holland lasted but six years. His five surviving children, often with him during summers at his home at Lake Constance, became his greatest comfort. One of his daughters, Hedwig, married the English zoologist W. Newton Parker, who translated *Das Keimplasma* (with Harriet Rönnfeldt). After retiring in 1912 as professor emeritus, Weismann gave his attention to the third edition of his *Vorträge über Descendenztheorie* (1913), which provides an overview of his ideas in an era in which cytology had developed as a science, Mendel's work had become known, and questions of evolution and heredity had been the subject of research and controversy. Following the outbreak of World War I, Weismann was ill and deeply unhappy, with members of his family on both sides during the conflict.

Weismann belonged to the Bavarian Academy of Sciences and was a corresponding member of the Academy at Vienna. He was a foreign member of the Linnean Society and the Royal Society of London, and of the American Philosophical Society.

Weismann traveled throughout Europe, and once to Constantinople. He was often in Italy and on the Riviera, and his work on the Hydromedusae was largely the result of a stay at the Naples zoological station during the winter of 1881–1882.

Weismann took three trips to England but he never met Darwin, although they corresponded.

In Germany the government designated Weismann *Wirklicher Geheimer Rat*, and the king of Bavaria bestowed the Grand Cross of the *Zähringer Löwensorden*. The University of Freiburg gave him the Ph.D. *honoris causa* in 1879. The University of Oxford made him doctor of common law, and the University of Utrecht made him doctor of botany. He received wide recognition when biologists from many countries honored him at the celebration of his seventieth birthday and a bust was given to the Zoological Institute at Freiburg, while the *Zoologische Jahrbücher* marked the anniversary with a *Festschrift*.

Weismann's influence was well summarized in the citations he received in London in 1908. Accompanying the medal awarded him by the Linnean Society at the Darwin-Wallace celebration (which his duties at Freiburg unfortunately prevented his attending) was the following:

> Professor Weismann has played a brilliant part in the development of Darwinian theory, and is indeed the protagonist of that theory in its purest form, retaining all that was the peculiar property of Darwin and Wallace and eliminating the traces of Lamarckism which still survived. . . . his profound knowledge of cytology enabled him to base his theory of heredity on a firmer foundation of fact than had been possible in the case of previous speculations.

The following citation was given him with the award of the Darwin Medal at the anniversary meeting of the Royal Society: "The fact remains that he has done more than any other man to focus attention on the mechanism of inheritance."

BIBLIOGRAPHY

I. ORIGINAL WORKS. Weismann's writings are comprehensively listed in the biography by Gaupp (see below), pp. 290–297. His collected essays are in Edward B. Poulton, Selmar Schönland, and Arthur E. Shipley, trans., *Essays Upon Heredity and Kindred Biological Problems* (Oxford, 1889); and in a two-volume 2nd ed. (Oxford, 1891–1892). For the continued development of his theory of the germ plasm, see *Das Keimplasma* or its translation, noted above. See also *Vorträge über Descendenztheorie* (Jena, 1902, 1904, 1913), trans. by J. Arthur Thomson and Margaret R. Thomson as *The Evolution Theory*, 2 vols. (London, 1904), which provides some account of Weismann's changes in thinking over the years. A short autobiography is Herbert Ernest

Cushman, trans., "Autobiography of Professor Weismann," in *Lamp* (new series of *Book Buyer*), n.s. **26** (1903), 21–26. For a further personal glimpse of Weismann, see his correspondence with Haeckel in Georg Uschmann and Bernhard Hassenstein, "Der Briefwechsel zwischen Ernst Haeckel und August Weismann," in Manfred Gersch, ed., *Kleine Festgabe aus Anlass der hundertjährigen Wiederkehr der Gründung des Zoologischen Institutes der Friedrich-Schiller-Universität Jena im Jahre 1865 durch Ernst Haeckel* (Jena, 1965), 7–68.

II. SECONDARY LITERATURE. The major biography of Weismann is Ernst Gaupp, *August Weismann, sein Leben und sein Werk* (Jena, 1917). Among the articles on his life and work are Edward G. Conklin, "August Weismann," in *Proceedings of the American Philosophical Society*, **54** (1915), iii–xii, and in *Science*, n.s. **41** (1915), 917–923; F. Doflein, "August Weismann," in *Münchener medizinische Wochenschrift*, **61** (1914), 2308–2310; and V. Haecker, "August Weismann," in *Deutsches Biographisches Jahrbuch*, 1914–1916, I (Berlin, 1925), 97–103. See also R. v. Hanstein, "August Weismann," in *Naturwissenschaftliche Wochenschrift*, n.s. **14** (1915), 113–120, 129–136; and the obituary by R. Hertwig, in the *Jahrbuch der bayerischen Akademie der Wissenschaften* (1915), 118–127; Prof. and Mrs. W. N. Parker, "August Friedrich Leopold Weismann," in *Proceedings of the Linnean Society of London*, 129th session (1914–1915), 33–37; and for the recollections of one of Weismann's students, Alexander Petrunkevitch, "August Weismann, Personal Reminiscences," in *Journal of the History of Medicine and Allied Sciences*, **18** (1963), 20–35. Further accounts of his life include E. B. Poulton, "Prof. August Weismann," in *Nature*, **94** (1914), 342–343, and "August Friedrich Leopold Weismann," in *Proceedings of the Royal Society*, ser. B, **89** (1916), xxvii–xxxiv; and H. E. Ziegler, "August Weismann," in *Neue Rundschau*, **26** (1915), 117–124.

Because Weismann's work drew so much attention and controversy, a list of discussions of his views and contributions is necessarily incomplete; contemporary assessments include George John Romanes, *An Examination of Weismannism* (Chicago, 1893); Vernon L. Kellogg, *Darwinism To-Day* (New York–London, 1908), 45–46, 77–78, and *passim,* for Weismann's position as a Neo-Darwinist and his colleagues' stands, and his theories of panmixia and germinal selection; and Yves Delage and Marie Goldsmith, *The Theories of Evolution*, André Tridon, trans. (New York, 1913), 134–162. A later résumé of the course of Weismann's work and of his theories is W. Schleip, "August Weismann's Bedeutung für die Entwicklung der Zoologie und allgemeinen Biologie," in *Naturwissenschaften*, **22** (1934), 33–41. A more recent evaluation of Weismann's work in the context of the growing understanding of cytological phenomena, and his views and those of his contemporaries, is William Coleman, "Cell, Nucleus, and Inheritance: An

Historical Study," in *Proceedings of the American Philosophical Society*, **109** (1965), 126, 149–154. The origins and the development by Weismann of the germ plasm theory, and its implications for him, are discussed in Frederick B. Churchill, "August Weismann and a Break From Tradition," in *Journal of the History of Biology*, **1** (1968), 91–112. For Weismann's theory of the germ plasm against the background of theories of heredity from Darwin's pangenesis, see Gloria Robinson, *A Prelude to Genetics* (Lawrence, Kans., 1976). For an analysis and discussion of Weismann's views on the problem of the reduction division, see Frederick B. Churchill, "Hertwig, Weismann, and the Meaning of Reduction Division Circa 1890," in *Isis*, **61** (1970), 429–457.

GLORIA ROBINSON

WEISS, CHRISTIAN SAMUEL (*b.* Leipzig, Germany, 26 February 1780; *d.* Eger, Hungary, 1 October 1856), *crystallography, mineralogy.*

Weiss's grandfather and father were archdeacons of Nicolai Church. At the age of twelve he began a classical education at the liberal Evangelische Gnadenschule at Hirschberg (now Jelenia Gora, Poland), under the philologist C. L. Bauer. In 1796 he returned to Leipzig to study medicine at the university; but after receiving his baccalaureate degree he switched to chemistry and physics, in which he was awarded the doctorate in 1800, and was then admitted to the faculty. Before teaching, Weiss spent two years in the chemical laboratory of Martin Klaproth (and Valentin Rose the younger) at Berlin, then a center for quantitative mineral analysis. Here he also became acquainted with Dietrich Karsten, curator of the royal mineral collection, and the eminent geologist Leopold von Buch. At their urging he went to the Freiberg Bergakademie for a year with A. G. Werner.

In 1805–1806 Weiss toured areas of geological interest in Austria, Switzerland, and France, spending some months in Paris with René-Just Haüy, André Brochant de Villiers, and Claude Berthollet. From 1803 he taught chemistry, physics, and mineralogy at Leipzig, and in 1808 he was appointed professor of physics. At about that time the University of Berlin was being organized under Wilhelm von Humboldt's leadership with the intention to make it the major center of philosophy and science in Germany by assembling the most eminent faculty. At Buch's instigation Weiss was appointed to the professorship of mineralogy and Klaproth to one in chemistry. Classes at the new university began in 1810, and Weiss occupied the chair of mineralogy until his death. After Dietrich Karsten's death in 1810, Weiss also became curator of the mineralogical museum and was instrumental in getting the government to purchase Buch's priceless collections for the museum in 1853. He served as rector of the university in 1832–1833.

Weiss's ability as a teacher was attested to by generations of students, several of whom made major contributions in science: Gustav Rose, Karl Rammelsberg, Friedrich Quenstedt, Adolph de Kupffer, and particularly Franz Neumann.

In addition to his major contributions to crystallography, Weiss published a number of papers in geology; and with Alexander von Humboldt and Buch he helped lay Werner's neptunist theories to rest.

While he was still a young man, Weiss's many contacts put him at the very center of the quickly developing science of crystallography. At Karsten's suggestion, he early embarked on a translation of Haüy's *Traité de minéralogie*, to which he added lengthy supplements on the process of crystallization. His intimate acquaintance with Werner's very practical view on mineralogy was an effective antidote to Haüy's imaginative but often unsubstantiated speculations. He had high regard for both men, but he did not show either of them the uncritical devotion that each asked of his followers.

Weiss's interpretations of the geometry of crystals were first indicated in his inaugural dissertation for the professorship at Leipzig (1809). They were developed in a long series of papers published in the *Abhandlungen der Königlichen Akademie der Wissenschaften in Berlin* (he was elected a member of the academy in 1815) and in publications of the Gesellschaft Naturforschender Freunde in Berlin. He never published his own textbook of crystallography, although those of Quenstedt (Tübingen, 1840; 1855; 1873) are perhaps derived from his work. His contributions to crystallography were early shaped around the directional aspect of crystals, which he regarded in an abstract, theoretical way as the expression of processes of growth. Haüy's theories of crystallography, then preeminent, interpreted crystals in terms of cleavage-shaped "molecules," which had to be combined in various steps to explain the varieties of crystal forms. By 1815 Weiss had developed the idea of crystallographic axes, which were at once a direction of growth and a basis of classification. In this most important contribution Weiss distinguished crystal systems by the way in

TABLE I
CLASSIFICATION OF CRYSTALS

Weiss[1]	Modern
Three dimensions perpendicular	
All dimensions equal:	
Sphäroëdrisches	Cubic
Homosphäroëdrisches	Holohedral (hexoctahedral class)
Hemisphäroëdrisches	Hemihedral
Tetraëdrisches	Hextetrahedral class
Pentagon-dodekaëdrisches	Didodecahedral class
Two dimensions equal and one different:	
Viergliedrige	Tetragonal
Three dimensions different:	
Zwei-und-zweigliedrige	Orthorhombic
Zwei-und-eingliedrige	Monoclinic
Ein-und-eingliedrige	Triclinic
Three equal dimensions perpendicular to one other dimension	
Sechsgliedrige	Hexagonal
Drei-und-drei gliedrige	Trigonal

which faces were related to such axes: first by whether they resulted in axial intercepts of equal length, and second by whether all or only a fraction of a set of related faces (modern "form") were developed by crystals of a given mineral species.

In the second of these criteria Weiss incidentally provided the first recognition of hemihedrism[2] — that is, a crystal class or point group that displays lower symmetry with a fraction of the faces, while retaining the same basic symmetry or crystal system. It should be clear, however, that while Weiss and others before him implicitly recognized the main rotational axes of symmetry, as well as mirror planes of symmetry, the classification was essentially metrical; and the complete symmetrical classification of crystals into both the seven crystal systems and Hessel's thirty-two crystal classes or point groups (1830) was not completed until 1849 by Auguste Bravais. As shown in the table, we can read into Weiss's listing the crystal systems that are now the primary classification of crystals; but two fundamental faults of Weiss's crystallography continued to be a source of confusion and debate for the next decade or so. First, Weiss insisted on choosing crystallographic axes at right angles, describing as hemihedrisms of the orthorhombic system the crystals now recognized as separate monoclinic and triclinic systems. Second, the axial lengths, while recognized as being unequal in each system except the cubic, were nevertheless thought to have ratios related by square roots of integers.

These two assertions were made plausible by the marked pseudosymmetries of many minerals of low symmetry, such as feldspar. But they were soon disproved by the accurate measurements of interfacial angles made by Kupffer and others using William Wollaston's optical goniometer, by Neumann's demonstration of the variation of angles with temperature, and by Eilhard Mitscherlich's demonstration of the variation of angles in a series of solid solutions. Much of this work had been done at the University of Berlin in Weiss's laboratory and that of his colleague Gustav Rose. But throughout his life Weiss insisted on perpendicular "rational" axes, while increasingly precise measurements pushed him to justify his position with calculations such as an axial ratio for gypsum[3]

$$\text{of } a:b:c = \frac{1}{\sqrt{3^2+1^2}} : 1 : \frac{1}{\sqrt{3^2+2^2+1^2}}.$$

Weiss's philosophical emphasis on important directions in crystals resulted not only in the crystallographic axes and crystal systems but also, even earlier, in the concept of the zone. Although originally conceived as a direction of prominent crystal growth, the term soon was formally defined as the collection of crystal faces parallel to a single line, the zonal direction. The zone concept enabled Weiss to propose replacing Haüy's symbolism for crystal faces with parameters that described the face direction in terms of its intercepts in units of his crystallographic axes. The Weiss symbols were widely used, and eventually they were replaced by the reciprocal symbols known today as Miller (although earlier used by Carl Naumann) indices. Weiss's last important contribution (1820) was the development of algebraic relations among the parameters (e.g., Weiss or Miller indices) of the faces that constitute a zone—Weiss's zone law—which remains a powerful tool in crystallographic calculations. Its application was greatly simplified by Franz Neumann, using projections, in his *Beiträge zur Kristallonomie* (Berlin–Posen, 1823).

From his earliest years Weiss's development as a scientist was strongly influenced by contemporary philosophers, and notably by the theories of nature of Immanuel Kant, Friedrich Schelling, and Johann Fichte—much to the dismay of his clerical father. Weiss was a keen observer but constitutionally disinclined to any experimental work, preferring to base his developments on abstract concepts of mathematical order—to the extent that where experiment was in disagreement, he chose to believe the more orderly theory. It is perhaps ironic that his student Franz Neumann went on to form at Königsberg the most important school of experimental physics (including crystal physics) of the nineteenth century. But Weiss's contributions—crystallographic axes, crystal systems, the zone law, and the concept of hemihedrism—constructed a formal edifice in which much of nineteenth-century crystallography found a compatible home and a place to grow.

NOTES

1. "Uebersichtliche Darstellung . . . der Kristallisations-systeme," table, 336 f. Friedrich Mohs independently worked out an analogous arrangement of crystal systems (but correctly, with inclined axes for the monoclinic and triclinic systems), published in his *Grundriss der Mineralogie*, 2 vols. (Dresden, 1822–1824). David Brewster had already seen the MS of William Haidinger's English trans. of this work—*Treatise on Mineralogy*, 3 vols. (Edinburgh, 1825)—and immediately applied the crystal systems to the interpretation of his observations on double refraction of crystals. A polemical debate on priority ensued in the *Edinburgh Philosophical Journal*, **8** (1823), 103–110 (Weiss) and 275–290 (Mohs). It is an interesting commentary on scientific communications of the day that although Weiss's paper was read at the Berlin Academy of Sciences on 14 Dec. 1815, the *Abhandlungen* for that year was not printed until 1818; and at the end of 1822 a copy was still not in the library of the Bergakademie at Freiberg.
2. Weiss soon recognized other hemihedrisms in the tetragonal and hexagonal systems—see his paper in *Edinburgh Philosophical Journal*, **8** (1823), 103–110.
3. "Über das Gypssystem," in *Abhandlungen der Königlichen Akademie der Wissenschaften in Berlin* (1834), 623–647.

BIBLIOGRAPHY

I. Original Works. Most of Weiss's scientific writings are listed in the Royal Society *Catalogue of Scientific Publications*, IV, 308–310. The most important books are *De indagando formarum crystallinarum charactere geometrico principali* (Leipzig, 1809), trans. into French with commentary by André Brochant de Villiers in *Journal des mines*, **29** (1811), 349–391, 401–444; and *De charactere geometrico principali formarum crys-*

tallinarum octaedricarum pyramidibus rectis basirectangula oblonga commentatio (Leipzig, 1809).

Articles include "Uebersichtliche Darstellung der verschiedenen natürlichen Abteilung der Krystallisations-systeme," in *Abhandlungen der Königlichen Akademie der Wissenschaften in Berlin* (1814–1815), 289–344; "Ueber ein verbesserte Methode für die Bezeichnung der verschiedenen Flächen eines Krystallisations-systeme, nebst Bemerkungen über den Zustand von Polarisirung der Seiten in der Linien der krystallinischer Struktur," *ibid.* (1816–1817), 286–314; and "Ueber mehrere neobeobachtete Krystallflachen des Feldspathes und die Theorie seines Krystallsystems im Allgemeinen," *ibid.* (1820–1821), 145–184.

His translation of Haüy is of primary importance for the original commentaries by Weiss: R. J. Haüy, *Lehrbuch der Mineralogie*, 4 vols., trans. by K. J. B. Karsten and C. S. Weiss (Paris–Leipzig, 1804–1810).

Some of Weiss's teaching materials are preserved at the Deutsche Staatsbibliothek, Berlin, D.D.R.; see Peter Schmidt, "Zur Geschichte der Geologie, Mineralogie und Paläontologie," in *Veroffentlichungen der Bibliothek der Bergakademie Freiberg*, no. 40 (1970), items 675, 807. A large collection of Weiss's letters to his oldest brother, Benjamin, are at the University of Marburg; they were used by both Groth and Fischer in preparing their discussions of Weiss.

II. SECONDARY LITERATURE. Biographical notices of Weiss are by K. F. P. von Martius, in *Akademische Denkreden . . . von Martius* (Leipzig, 1866), 327–344; and in Martin Websky *et al.*, *Gedenkworte am Tage der Feier des hundertjahrigen Geburtstages von Christian Samuel Weiss den 3 März 1880* (n.p., n.d.). A contemporary evaluation of Weiss's crystallography is in Franz von Kobell, *Geschichte der Mineralogie von 1650–1860* (Munich, 1864), 202–214. The most useful and modern critical discussions of Weiss's scientific contributions are Paul Groth, *Entwicklungsgeschichte der mineralo̔gischen Wissenschaften* (Berlin, 1926; repr. Wiesbaden, 1970), 59–76; and Emil Fischer, "Christian Samuel Weiss und seine Bedeutung für die Entwicklung der Krystallographie," in *Wissenschaftliche Zeitschrift der Humboldt-Universität zu Berlin*, Math.-naturwiss. ser., **11** (1962), 249–255. See also Fischer's "Christian Samuel Weiss und die zeitgenössische Philosophie (Fichte, Schelling)," in *Forschungen und Fortschritte*, **37** (1963), 141–143. A more general evaluation of Weiss's influence in the development of crystallography is J. G. Burke, *Origins of the Science of Crystals* (Berkeley, 1966), 147–164.

The founding of the University of Berlin, in which Weiss had a part, is described by Rudolf Virchow in "The Founding of the Berlin University and the Transition From the Philosophic to the Scientific Age," in *Annual Report of the Board of Regents of the Smithsonian Institution* (1894), pt. 1, 681–695.

WILLIAM T. HOLSER

WEISS, EDMUND (*b.* Freiwaldau, Austrian Silesia [now Jesenik, Czechoslovakia], 26 August 1837; *d.* Vienna, Austria, 21 June 1917), *astronomy.*

Weiss, son of Joseph Weiss, a physician, and twin of the botanist Gustav Adolph Weiss, received his earliest education in England (1843–1847). After his father's death, he attended the secondary school at Troppau, Austrian Silesia. In 1855 he entered the University of Vienna, where he studied astronomy, mathematics, and physics, receiving the doctorate in 1860. He was hired as assistant astronomer at the university observatory in 1858 and became an associate astronomer in 1862. In addition to participating in both astronomical and geodetical observations, Weiss became a lecturer in mathematics in 1861 and associate professor in 1869.

When, in 1872, Littrow's idea of building a great modern observatory on the hills of Währing, a suburb of Vienna, approached realization, Weiss was sent to visit new observatories and optical factories in England and the United States. This experience enabled him to contribute substantially to the definitive plans of the new institute. He became full professor of astronomy in 1875 and, eight months after Littrow's death, succeeded him as director of the observatory (1878). He was also president of the Austrian Commission for Geodesy and a member of the Imperial Academy of Sciences.

Weiss organized and participated in expeditions to observe several solar eclipses, the 1874 transit of Venus, and the Leonid meteor showers in 1899. His main contributions to astronomy concern the determination of the orbits of comets, of minor planets, and of meteor showers. In opposition to Schiaparelli's opinion that comets might have been formed by accretion in meteor showers, Weiss proposed in 1868 the now generally accepted view that the latter are products of gradual destruction of comets by tidal forces.

BIBLIOGRAPHY

I. ORIGINAL WORKS. Weiss's books include *Bilder-Atlas der Sternenwelt* (Esslingen, 1892); *Über Kometen mit besonderer Beziehung auf den Halley'schen* (Vienna, 1909); and *Katalogisierung von Argelanders Zonen vom 45. bis 80. Grade nördlicher Deklination* (Vienna, 1919).

Weiss's most important papers were published by the Kaiserliche Akademie der Wissenschaften zu Wien, math. naturwiss. Kl., in its *Denkschriften* and *Sitzungsberichte*, 2nd ser. (here abbreviated, respectively, as *Ak.D.* and *Ak.SB.*): "Bahnbestimmung von (66) Maja," in *Ak.SB.*, **51** (1865), 77–96; "Berechnung der Sonnenfinsternisse der Jahre 1867–1870," *ibid.*, **54** (1866), 796–810, and **56** (1867), 429–454; "Bericht über die Beobachtungen während der ringförmigen Sonnenfinsternis 1867 in Dalmatien," *ibid.*, **55** (1867), 905–944; "Beiträge zur Kenntnis der Sternschnuppen," *ibid.*, **57** (1868), 281–342, and **62** (1870), 277–344; "Beobachtungen während der totalen Sonnenfinsternis 1868 in Aden, etc.," *ibid.*, **58** (1868), 697–720, 882–894; **60** (1869), 326–340; **62** (1870), 873–1016; "Sprungweise Änderungen in einzelnen Reductionselementen eines Instruments," *ibid.*, **64** (1871), 77–104; "Bestimmung der Längendifferenz Wien–Wiener Neustadt durch Chronometer-Übertragung," *ibid.*, **65** (1872), 97–119; "Die praktische Astronomie in Amerika," in *Vierteljahrschrift der Astronomischen Gesellschaft*, **8** (1873), 296–321; "Beobachtung des Venusdurchganges 1874 in Jassy," in *Ak.SB.*, **71** (1875), 185–203; "Bahn der Cometen 1843 I und 1880 a," *ibid.*, **82** (1880), 95–114; "Differentialquotient der wahren Anomalie und des Radiusvectors nach der Excentricität in stark excentrischen Bahnen," *ibid.*, **83** (1881), 466–478; and "Entwicklungen zum Lagrange'schen Reversions-Theorem und Anwendung auf die Lösung der Kepler'schen Gleichung," in *Ak.D.*, **49** (1885), 133–170.

Further articles by Weiss are "Der Binomialreihe verwandte Reihengruppen," in *Ak.SB.*, **91** (1885), 587–596; "Bestimmung von *M* bei Olbers' Methode der Berechnung einer Cometenbahn mit besonderer Berücksichtigung auf den Ausnahmefall," *ibid.*, **92** (1885), 1456–1477; "Berechnung der Präzession etc.," in *Ak.D.*, **53** (1887), 53–80; "Berechnung einer Cometenbahn mit Berücksichtigung von Gliedern höherer Ordnung," in *Ak.SB.*, **100** (1891), 1132–1150; "Systematische Differenzen südlicher Sternkataloge," *ibid.*, **101** (1892), 1269–1406; "Bestimmung der Bahn eines Himmelskörpers aus drei Beobachtungen," in *Ak.D.*, **60** (1893), 345–394; "Höhenberechnung der Sternschnuppen," *ibid.*, **77** (1905), 255–356; "Beiträge zur Kenntnis der atmosphärischen Elektrizität, in *Ak.SB.*, **115** (1906), 1285–1320; "Sichtbarkeitsverhältnisse des Kometen 1905 IV," *ibid.*, **116** (1907), 3–16; "Berechnung einer Ellipse aus zwei Radien und dem eingeschlossenen Winkel," *ibid.*, 345–366; and "Untersuchungen über die Bahn der Kometen 1907 II und 1742," in *Ak.D.*, **84** (1909), 1–14.

Collections of observations of shooting stars were presented by Weiss in *Annalen der Universitäts sternwarte in Wien*, 3rd ser., **20** (1870), 1–114; **23** (1873), 1–113; and **27** (1877), 1–133; Many records of observations, and other minor notes, are in *Astronomische Nachrichten*, **48** (1848)–**176** (1907).

His works are listed in Poggendorff, II, 1290; III, 1429; IV, 1615–1616, 1716; V, 1350–1351.

II. SECONDARY LITERATURE. Joseph von Hepperger wrote three obituaries: in *Astronomische Nachrichten*, **204** (1917), 431; in *Almanach der Akademie der Wissenschaften in Wien*, **68** (1918), 243–248; and *Vierteljahrschrift der Astronomischen Gesellschaft*, **53** (1918), 6–14.

KONRADIN FERRARI D'OCCHIEPPO

WEISS, PIERRE (*b.* Mulhouse, France, 25 March 1865; *d.* Lyons, France, 24 October 1940), *magnetism.*

Weiss's fame derives from the success of his phenomenological theory of ferromagnetism, which he conceived and developed on the basis of a large body of experimental results, many of them obtained by himself or his students. The theory is founded on the hypothesis of a molecular field proportional to the magnetization and acting on the orientation of each atomic moment like a magnetic field of very high intensity. With his theory he was able to account for the known characteristic properties of ferromagnetic bodies (notably the abrupt disappearance of ferromagnetism above a temperature known as the Curie point) and to discover the properties of spontaneous magnetization and magnetocaloric phenomena. Modern quantum theories of ferromagnetism have substantiated Weiss's molecular field hypothesis as a first approximation. According to these theories, the molecular field results from exchange forces of electric origin between the electrons, which bear the atomic magnetic moments.

Weiss came from a petit bourgeois Alsatian family. His father, who owned a haberdashery in Mulhouse, remained there when Alsace was annexed by the German Empire following the Franco-Prussian War. After attending secondary school in Mulhouse, Weiss went to the Zurich Polytechnikum. At his majority he chose French citizenship. In 1887 he graduated, first in his class, from the Polytechnikum with a degree in mechanical engineering. Wishing to undertake basic research, he attended the Lycée St. Louis in Paris to prepare for the competitive entrance examination for the École Normale Supérieure. Admitted in 1888, he was *agrégé* in physical science in 1893 and remained at the school as an assistant (*préparateur*) until 1895. During this period he became friendly with a number of fellow students who later became famous: the mathematicians Élie Cartan, Émile Borel, and Henri Lebesgue and the physicists Aimé Cotton, Jean Perrin, and Paul Langevin.

In 1895 Weiss was named *maître de conférences* at the University of Rennes and, in 1899, at the University of Lyons. In the meantime he had defended his doctoral dissertation, "Recherches sur l'aimantation de la magnétite cristallisée et de quelques alliages de fer et d'antimoine" (1896). In 1902 he returned as professor to the Zurich Polytechnikum, where, in addition to teaching, he directed the physics laboratory until 1918. His stay there was interrupted for two years, at the beginning of World War I, when he worked in Paris for the Office of Inventions, helping to create an acoustical method for locating enemy gun emplacements (Cotton-Weiss method). At Zurich, where his colleagues included Einstein, also a professor at the Polytechnikum, and Peter Debye, professor at the University of Zurich, Weiss gradually developed a great laboratory for magnetic research. He endowed it with a remarkable array of equipment and, above all, trained or attracted many distinguished physicists.

In 1919, following the return of Alsace to France, Weiss went to his native province to create and direct a major physics institute at the University of Strasbourg. Under his guidance, and with the aid of several associates drawn from among the best of his former staff, the laboratory soon surpassed even that of Zurich as a center of magnetic research. Among his numerous students during this period, the most outstanding was Louis Néel. While supervising the many projects undertaken at the institute, Weiss continued to do personal research, even after his retirement in 1936. He was elected to the Paris Academy in 1926.

With the evacuation of Strasbourg at the beginning of World War II, Weiss fled to Lyons, where his best friend, Jean Perrin, also had taken refuge. He died of cancer in October 1940.

In 1898 Weiss married Jane Rancès, whose mother was of English origin. Before her death in 1919, they had had one child, Nicole, who in 1936 married Henri Cartan, son of Élie Cartan and one of the leading mathematicians of his generation. In 1922 Weiss married Marthe Klein, who taught physics in a Paris *lycée*.

Weiss was thin and rather tall. Distinguished-looking and extremely courteous, he wore a pince-nez and wing collar that gave him an air of elegance. His hair and large moustache became completely white when he was still quite young.

Weiss's scientific works, which deal almost exclusively with magnetism, are characterized by great unity. From the time he was an engineering student, Weiss was interested in the complex phenomena of ferromagnetism, and it was to them that he devoted his initial research. He was influenced in this choice by the theoretical studies of Alfred

Ewing and Pierre Curie's "Les propriétés magnétiques des corps à diverses températures" (1895). At first he investigated magnetite and pyrrhotite, hoping that their large natural ferromagnetic monocrystals would enable him to discover the fundamental laws of magnetization. In the case of magnetite (1894–1896), he discovered only that it does not behave as an isotropic medium, even though it is crystallized in the cubic system.

The difficult study of pyrrhotite, the crystals of which are hexagonal prisms, proved much more rewarding (1896–1905). First he discovered that whatever the strength and direction of the magnetic field, the resulting magnetization remains, to a very good approximation, directed in the plane perpendicular to the axis of the crystalline prism. He then found that in this plane there is a direction of easy magnetization, in which saturation is reached in fields of twenty or thirty oersteds, and, perpendicularly, a direction of difficult magnetization, in which saturation has the same value but is reached only in fields exceeding 10,000 oersteds. Finally, he showed that the magnetization produced by an arbitrary field can be determined by vectorially subtracting from this field a "structural field" directed along the axis of difficult magnetization and proportional to the component of the magnetization along that axis. The resulting field assumes the direction of the magnetization, and its strength is linked to that of the magnetization by a relation that is independent of that direction.

In 1905 Paul Langevin published a theory of the paramagnetism of dilute substances. In Langevin's view, one may neglect the interactions between the magnetic moments μ that are assumed to be borne by each molecule of such substances. In the case of weak fields, the theory led to the Curie law, which states that the magnetization is proportional to the magnetic field H and to the reciprocal of the absolute temperature T. For very strong fields, or at very low temperatures, however, the law predicts that the magnetization I will tend toward a limit I_0. According to the theory, this saturation corresponds to the situation in which all the molecular moments are oriented in the direction of the field, despite the thermal agitation tending to vary their directions. Using classical statistical mechanics, Langevin obtained a formula giving the magnetization as a function of the ratio H/T:

(1) $\quad I = I_0 f\left(\dfrac{\mu H}{kT}\right)$, where $f(a) = \coth a - 1/a$.

In order to develop Langevin's ideas, Weiss broadened the concept of structural fields proportional to the magnetization, which he had previously introduced to account for the magnetic anisotropism of pyrrhotite. Langevin's theory led Weiss to conclude that the characteristic properties of the ferromagnetic metals, of which the microcrystalline structure is macroscopically isotropic, result from a global action of the magnetically polarized milieu on each elementary magnetic moment. This orienting action was to be considered equal to that of a magnetic field H_m, called molecular field, proportional to the magnetization ($H_m = NI$). According to Weiss, in the presence of an external field H producing a magnetization I, each atomic moment would be subjected to the total field $H + H_m$; and the mean orientation of these moments, which creates the magnetization I, should be given by Langevin's formula, if H is replaced by $H + H_m = H + NI$. The result is

$$(2) \qquad I = I_0 f\left(\frac{\mu H}{kT} + \frac{\mu N}{kT} I\right).$$

This is the fundamental formula of ferromagnetism in the theory based on the hypothesis of the molecular field. It remains valid when the Langevin function f is replaced by functions arising from the application of quantum statistical mechanics to those moments M that can assume only quantized orientations with respect to this field.

The first consequence of Weiss's formula is that when the temperature T is lower than a certain temperature Θ, a zero magnetization is unstable in the presence of a zero field. In such a case the result will be the appearance of a spontaneous magnetization I_s determined by the implicit relation

$$(3) \qquad I_s = I_0 f\left(\frac{\mu N}{kT} I_s\right).$$

This spontaneous magnetization, equal to I_0 at very low temperatures, at first decreases slowly as the temperature rises, then very rapidly, and finally disappears altogether at the instant that the temperature reaches the critical level of $\Theta = \alpha\,\mu N I_0 / k$ (α being the slope, at the origin, of the curve representing the function f; accordingly, it will be $1/3$ for the Langevin function, or 1 for the quantum function $f(a) = \mathrm{th}\,a$, relative to the magnetic moment associated with the electron spin $1/2$).

The predicted spontaneous magnetization, however, generally is not apparent: most ferromagnetic metals have a zero magnetization in a zero field. The explanation must be that the variously oriented spontaneous magnetizations cancel each other in very small domains (known as Weiss domains, the existence of which was demonstrated much

later). A very strong exterior field is required to render parallel the spontaneous magnetizations of such domains. Further, the saturated macroscopic magnetization at a given temperature differs very little from the spontaneous magnetization, so that the latter quantity can be determined from the former. When the temperature is higher than the temperature Θ, which is that of the Curie point, the spontaneous magnetization is zero. In this case the exterior field H induces a very small magnetization, which according to the general formula (2) assumes the value

$$(4) \qquad I = I_0\,\alpha\frac{\mu H}{k(T - \Theta)} = C\frac{H}{T - \Theta}.$$

The material under examination should then behave like a paramagnetic substance, with a magnetic susceptibility proportional to the reciprocal of the excess of the temperature over that of the Curie point. This relation, known as the Curie-Weiss law, is very well established by experiment. It can even be applied to many more or less concentrated paramagnetic substances with very low Curie temperatures, which are often negative (negative molecular field).

Applying the principles of thermodynamics to ferromagnetic substances, Weiss showed in 1908 that the existence of the spontaneous magnetization should add to the ordinary specific heat a magnetic specific heat proportional to the derivative with respect to the temperature of the square of this spontaneous magnetization. This quantity, therefore, should be zero at absolute zero and increase with the temperature, at first slowly and then more and more quickly up to the Curie point, where it should suddenly vanish. Measurements made on nickel gave quantitative confirmation of this prediction and highlighted, in particular, the discontinuity of the specific heat at the Curie point. In 1918 Weiss also discovered the magnetocaloric effects and showed how thermodynamics can be used to calculate the temperature variation of a magnetic substance placed in a field the intensity of which is altered adiabatically.

The absolute saturation I_0 of the magnetization, deduced from the limit toward which the experimental saturation tends at very low temperatures, yields in a very direct manner the value of the atomic moments μ. Measurements on iron and nickel gave values for their atomic moments the ratio of which was almost exactly that of the whole numbers five and three. This finding led Weiss to postulate in 1911 that the moments of the various magnetic atoms are whole multiples of an elementary moment that he called the magneton. Many other measurements of other ferromagnetic substances—and, through the intermediary of Langevin's theory, of paramagnetic substances—seemed to verify this hypothesis. Several of his contemporaries immediately suggested that his result might well be explained as a quantum effect due to a restriction on the orbital energy of electrons. But Weiss's experimental magneton was found to be approximately equal to one-fifth of the Bohr magneton, which was deduced several years later from the quantification of the electron orbits. Subsequent research in quantum mechanics, however, has provided no grounds for thinking that all atomic moments are whole multiples of the Bohr magneton or of one-fifth of this quantity. It appears that the integral values found by Weiss arose, in general, from an insufficiently founded interpretation of indirect and difficult measurements.

BIBLIOGRAPHY

Weiss's major book is *Le magnétisme* (Paris, 1926), written with G. Foëx.

His earlier articles include "Recherches sur l'aimantation de la magnétite cristallisée et de quelques alliages de fer et d'antimoine," in *Éclairage électrique*, 7 (1896), 487–508, and 8 (1896), 56–68, 105–110, 248–254, his dissertation; "Aimantation de la magnétite cristallisée," in *Journal de physique*, 3rd ser., 5 (1896), 435–453; "Un nouvel électro-aimant de laboratoire donnant un champ de 30.000 unités," in *Éclairage électrique*, 15 (1898), 481–487; "Sur l'aimantation plane de la pyrrhotine," in *Journal de Physique*, 3rd ser., 8 (1899), 542–544; "Un nouveau système d'ampèremètres et de voltmètres indépendants de leur aimant permanent," in *Comptes rendus . . . de l'Académie des sciences*, 132 (1901), 957; "Un nouveau fréquence-mètre," in *Archives des sciences physiques et naturelles*, 18 (1904), 241; "Le travail d'aimantation des cristaux," in *Journal de physique*, 4th ser., 3 (1904), 194–202; "Propriétés magnétiques de la pyrrhotine," *ibid.*, 4 (1905), 469–508, 829–846; "Variation thermique de l'aimantation de la pyrrhotine," *ibid.*, 847–873, written with J. Kunz; "La variation du ferromagnétisme avec la température," in *Comptes rendus . . . de l'Académie des sciences*, 143 (1906), 1136; and "Sur la théorie des propriétés magnétiques du fer au delà du point de transformation," *ibid.*, 144 (1906), 25.

Additional articles are "L'hypothèse du champ moléculaire et la propriété ferromagnétique," in *Journal de physique*, 4th ser., 6 (1907), 661–690; "Sur la biréfringence des liquides organiques," in *Comptes rendus . . . de l'Académie des sciences*, 145 (1907), 870, written with A. Cotton and H. Mouton; "L'intensité d'aimantation à saturation du fer et du nickel," *ibid.*, 1155; "Électro-aimant de grande puissance," in *Journal de physique*, 4th ser., 6 (1907), 353–368; "Mesure du phénomène de Zeeman pour les trois raies bleues du zinc," *ibid.*, 429–445, written with A. Cotton; "Hystérèse dans les champs tournants," *ibid.*, 4th ser., 7 (1908), 5–27, written with V. Planer; "Chaleur spécifique et champ moleculaire des substances ferromagnétiques," *ibid.*, 249–264, written with P. N. Beck; "Sur le rapport de la charge à la masse des électrons," in *Comptes rendus . . . de l'Académie des sciences*, 147 (1908), 968, written with A. Cotton; "Mesure de l'intensité d'aimantation à saturation en valeur absolue," in *Journal de physique*, 4th ser., 9 (1910), 373–397; and "Recherches sur l'aimantation aux très basses températures," *ibid.*, 555–584, written with H. Kamerlingh Onnes.

Also see "Sur l'aimantation du nickel, du cobalt et des alliages nickel-cobalt," in *Comptes rendus . . . de l'Académie des sciences*, 153 (1911), 941, written with O. Bloch; "Étude de l'aimantation des corps ferromagnétiques au-dessus du point de Curie," in *Journal de physique*, 5th ser., 1 (1911), 274–287, 744–753, 805–814, written with G. Foëx; "Sur la rationalité des rapports des moments magnétiques moléculaires et le magnéton," *ibid.*, 900–912, 965–988; "Sur l'aimantation de l'eau et de l'oxygène," in *Comptes rendus . . . de l'Académie des sciences*, 155 (1912), 1234, written with A. Piccard; "Magnetic Properties of Alloys," in *Transactions of the Faraday Society*, 8 (1912), 149–156; "L'aimantation des cristaux et le champ moléculaire," in *Comptes rendus . . . de l'Académie des sciences*, 156 (1913), 1836–1837; "Sur les champs magnétiques obtenus avec un électro-aimant muni de pièces polaires en ferrocobalt," *ibid.*, 1970–1972; "Le spectrographe à prismes de l'École polytechnique de Zurich," in *Archives des sciences physiques et naturelles*, 35 (1913), 5, written with R. Fortrat; and "Sur la nature du champ moléculaire," in *Annales de physique*, 9th ser., 1 (1914), 134–162.

Weiss's later papers include "Ferromagnétisme et équation des fluides," in *Journal de physique*, 5th ser., 7 (1917), 129–144; "Calorimétrie des substances ferromagnétiques," in *Archives des sciences physiques et naturelles*, 42 (1917), 378, and 43 (1917), 22, 113, 199, written with A. Piccard and A. Carrard; "Le phénomène magnétocalorique," in *Journal de physique*, 5th ser., 7 (1917), 103–109, written with A. Piccard; "Sur un nouveau phénomène magnétocalorique," in *Comptes rendus . . . de l'Académie des sciences*, 166 (1918), 352, written with A. Piccard; "Sur les coefficients d'aimantation de l'oxygène, de l'oxyde azotique et la théorie du magnéton," in *Comptes rendus . . . de l'Académie des sciences*, 167 (1918), 484–487, written with E. Bauer and A. Piccard; "Sur le moment atomique de l'oxygène," in *Journal de physique*, 6th ser., 4 (1923), 153–157; "Les moments atomiques," *ibid.*, 5 (1924), 129–152; "Aimantation et phénomène magnétocalorique du nickel," in *Annales de physique*, 10th ser., 5 (1926), 153–213; "Sur les moments atomiques," in *Comptes rendus . . . de l'Académie des sciences*, 187 (1928), 744, written with G. Foëx; "La saturation absolue des ferromagnétiques et les lois d'approche en fonction du champ et de la température," in *Annales de physique*, 10th ser., **12** (1929), 279–374, written with R. Forrer; and "La constante du champ moléculaire. Équation d'état magnétique et calorimétrique," in *Journal de physique*, 7th ser., **1** (1930), 163–175.

Some of his papers appeared in works issued by the Solvay Council: "Les actions mutuelles des molécules aimantées," in *Atomes et électrons* (Paris, 1923), 158–163; "Équation d'état des ferromagnétiques," in *Le magnétisme* (Paris, 1932), 281–323; "L'anomalie de volume des ferromagnétiques," *ibid.*, 325–345; and "Les phénomènes gyromagnétiques," *ibid.*, 347–379.

On his life and work, see Albert Perrier, "In memoriam (Pierre Weiss)," in *Actes de la Société helvétique des sciences naturelles*, **121** (1941), 422–433, with bibliography; and G. Foëx, "L'oeuvre scientifique de Pierre Weiss," in *Annales de physique*, 11th ser., **20** (1945), 111–130.

FRANCIS PERRIN

WEIZMANN, CHAIM (*b.* Motol, White Russia, 27 November 1874; *d.* Rehovot, Israel, 9 November 1952), *organic chemistry, biochemistry.*

"Chemistry is my private occupation. It is this activity in which I rest from my social tasks." Thus did Chaim Weizmann describe the contrapuntal relationship between his lifelong career as a scientist and his leadership of the Zionist movement. In his disciplined mind these two vocations, representing reason and faith, were made harmonious. The intellectual and physical power that science conveys was to help free the Jews for their return to Palestine and was to form a vital, integral part of a revived, modern Jewish culture.

After early religious and secular schooling within the segregated Jewish community of rural Russia, Weizmann, at the age of twelve, entered the Gymnasium at Pinsk and was there exposed to gentile ways and Western European thought. Although his grades were uniformly high, chemistry was his favorite subject. With parental encouragement he set out to gain advanced knowledge in that field at the technical institutes of Darmstadt (1893–1894) and Berlin (1895–1898), where Liebermann and his students (among them Bistrzycki) were investigating polycyclical aromatic compounds of particular interest to dye manufacturers. When Bistrzycki went to the University of Fribourg, Switzerland, Weizmann followed. Soon afterward (1899), Weizmann wrote the dissertation "I. Elektrolytische Reduktion von 1-Nitroanthrachinon. II. Ueber die Kondensation von Phenan-

threnchinon U. 1-Nitroanthrachinon mit einigen Phenolen" and was awarded the Ph.D. *summa cum laude*. He subsequently joined Karl Graebe at the University of Geneva as *Privatdozent*. Extensive research on the naphthacene quinones led to patents that Weizmann sold profitably to French and German dye companies. Meanwhile he was rising to leadership in the world Zionist movement.

Weizmann's decision to move to Manchester in 1904 was prompted by numerous considerations, including greater professional opportunity and a premonition that England could do the most for establishing a Jewish national homeland in Palestine. At Manchester, Weizmann enrolled as a student at the university. The following year, the head of the chemistry department, William H. Perkin, Jr., appointed him research fellow, and in 1907 senior lecturer, in biochemistry. Weizmann secured additional income by serving as consultant for local industry and selling new patents. At this time he also married Vera Chatzmann, a physician.

The university's exceptional scientific faculty stimulated Weizmann in this, the most scientifically productive period of his life. Effective teaching attracted students who did research under his direction. The quest for alizarin-type dyes continued along previous lines: the polyhydroxylation of naphthacene quinone, for example, yielded colors of moderate utility. About 1909 Weizmann added biochemical investigations to his research, seeking to synthesize various naturally occurring peptides; he later studied the photochemical behavior of amino acids, proteins, and ketones. He also began investigating fermentation reactions, searching for a strain of bacteria that would convert carbohydrates into isoamyl alcohol—a precursor, via isoprene, of synthetic rubber. Instead, in 1912, he found the strain *Clostridium acetobutylicum*, which broke starches down into one part ethanol, three parts acetone, and six parts butanol. During World War I, when great quantities of acetone were needed to plasticize the propellant cordite, Weizmann successfully engineered its massive production in Great Britain for the Admiralty and Ministry of Munitions. Plants were also built in India, Canada, and the United States; their production continued after the war, butanol then being the preferred product for use in auto lacquers. Weizmann, in effect, opened the microbiological road to the production of industrial chemicals.

Meanwhile, in 1917, Weizmann secured from Lord Balfour a declaration of British help in establishing a national homeland for the Jews, the subsequent realization of which took so much of

Weizmann's time that he stopped all scientific activity except for promoting the growth of Hebrew University in Jerusalem and his founding of the Daniel Sieff (later Weizmann) Institute of Science in Rehovot (1934), both of which soon became notable centers of scientific learning.

In 1934 Weizmann simultaneously resumed his research at Rehovot and London, adding significantly to lines of investigation begun before 1918, particularly those that were relevant to Palestine's economy: the commercial synthesis of organic compounds from agricultural products or petroleum. Of considerable technical significance was his discovery of several reaction mechanisms by which petroleum fractions could be reduced, by cracking to ethylene and diene fragments, and then recombined into polynuclear aromatics of the type he had used earlier in his dye researches. Previously such dye intermediates could be obtained only from coal tar.

Practical, rather than fundamental, scientific considerations motivated Weizmann's research. Apart from some interest late in his career in reaction mechanisms, his work is generally devoid of theoretical content. With his elevation in 1948 to the presidency of Israel, his career as a Zionist came to a climax and his career as a creative scientist came to an end.

BIBLIOGRAPHY

I. ORIGINAL WORKS. A nearly complete list of Weizmann's 100 or so patents and of his 102 scientific publications can be obtained from the Weizmann Archives, Rehovot, Israel. Abstracts of most of his articles are in *Chemisches Zentralblatt* or *Chemical Abstracts*. The Weizmann Archives hold the great bulk of his papers and continue to expand their collection. The documents are written in English, French, German, Hebrew, Russian, and Yiddish. The nonscientific *Letters and Papers of Chaim Weizmann* are currently being edited in a proposed 25 vols.; vols. I–VII, covering correspondence, 1885–1917, appeared in Hebrew and English eds., the latter published by Oxford University Press (London, 1968–1975). Weizmann's scientific correspondence is being edited for publication under the supervision of Ernst D. Bergmann, of Hebrew University, Jerusalem, and David Lavie of the Weizmann Institute. Weizmann's autobiography, *Trial and Error* (New York, 1949), was written between 1940 and 1948. Less than 5 percent of its 482 pages are devoted to his scientific career.

II. SECONDARY LITERATURE. *Chaim Weizmann, a Biography by Several Hands*, Meyer W. Weisgal and Joel Carmichael, eds. (London, 1962), contains Selman A. Waksman's, "Weizmann as a Bacteriologist." By far the best review of Weizmann's contributions to science is E. D. Bergmann's obituary notice in *Journal of the Chemical Society* (1953), 2840–2844.

JOHN J. BEER

WELCH, WILLIAM HENRY (*b.* Norfolk, Connecticut, 8 April 1850; *d.* Baltimore, Maryland, 30 April 1934), *pathology, bacteriology, public health, medical education.*

Welch was born into a family of physicians who for two generations had practiced medicine in Connecticut. His mother died when he was six months old, and he and his slightly older sister, Emma, were raised with the help of their paternal grandmother. The elder William Welch, a busy family practitioner, was a kind but somewhat distant father. The son prepared for college at a boarding school in Winchester Center and entered Yale in 1866, at the age of sixteen. By the time he graduated in 1870, third in his class, Welch had become interested in the classics and hoped for a position as tutor in Greek.

Not successful in obtaining a post at Yale, Welch accepted a teaching job at an academy in Norwich, New York, for 1870–1871. He then returned home to apprentice himself to his father, thereby beginning medical studies. In the fall of 1871, a very brief exposure to the medical lectures at the College of Physicians and Surgeons in New York City convinced him of the need for science courses. Welch therefore returned to New Haven to spend the academic year 1871–1872 at the Sheffield Scientific School of Yale University, where he concentrated on chemistry. Resuming medical studies at the College of Physicians and Surgeons in the fall of 1872, Welch found little more than a series of didactic lectures. He therefore eagerly accepted a prosectorship in anatomy in 1873, so that at least he could learn by firsthand investigation. Welch's dissertation on goiter was awarded a prize; and several months prior to receiving the M.D. in 1875, he began duty as an intern at Bellevue Hospital, where he had an excellent opportunity to observe and study a large variety of patients. Under the direction of Francis Delafield, Welch soon developed a keen interest in pathology. He also was greatly stimulated by two teachers who had emigrated from Europe, E. C. Seguin and Abraham Jacobi. Welch followed their advice to make a European study tour beginning in the spring of 1876.

During this sojourn of two years Welch visited

and studied at the major medical centers of Strasbourg, Leipzig, Breslau, and Vienna. Two of his research endeavors had special influence in shaping his career. With the physiologist Carl Ludwig at Leipzig, Welch learned to handle living tissue. He investigated the nerve distribution in the auricular septum of the frog heart and visualized the nerve network that later was fully described by Louis Ranvier. At Breslau, under Julius Cohnheim's direction, Welch studied the pathogenesis of pulmonary edema, showing, contrary to Cohnheim's presuppositions. that the condition is primarily mechanical in origin. With the publication of his findings in *Virchows Archiv* in 1878, Welch was launched in the scientific. laboratory-based study of pathology that Virchow, Cohnheim, and other German physicians were developing so fruitfully. For Welch there was no turning back, and he rejected a career in country practice with his father.

Returning to New York in 1878. Welch set out to bring the laboratory tradition in pathology to American medical students. His own school, the College of Physicians and Surgeons, was willing to let him teach a summer course but offered inadequate facilities and no salary. Bellevue Hospital and its medical school, on the other hand, offered to renovate three small rooms and to supply very modest equipment. Here Welch inaugurated the first teaching pathology laboratory in the United States. The student response was heartening but the financial returns meager. To support himself Welch performed autopsies and examined specimens for his medical colleagues. held a popular private class for medical students, wrote a section of the sixth edition (1886) of Austin Flint's *Principles and Practice of Medicine*. and saw a few private patients. It is therefore not surprising that in six years he failed to complete a single piece of pathological research. It was this frustration, as well as the dream of bringing a real science of pathology to America, that had intrigued Welch about the new Johns Hopkins University. From 1876, if not before, Welch had hoped that he might be offered a chair in the proposed medical school faculty. When the offer was made in 1884, Welch accepted the professorship of pathology.

Prior to moving to Baltimore, Welch took an additional year of study in Europe, concentrating on the rapidly emerging field of bacteriology, a subject he had entirely bypassed six years earlier. Study with Carl Flügge, Robert Koch, and Max von Pettenkofer gave Welch the groundwork he needed to bring another aspect of the study of disease to America. When Welch arrived in Baltimore

at the end of 1885 to assume his duties at Johns Hopkins, no hospital or medical school existed. It fell to Welch, with John Shaw Billings; Daniel Coit Gilman, president of the university; and H. Newell Martin, the professor of biology, to recruit the rest of the medical faculty. Welch moved into Martin's laboratory and, with the assistance of William T. Councilman, began a series of pathological and bacteriological studies on thrombosis, embolism, hog cholera. diphtheria, and a number of other projects. In contrast with New York. Welch now had both able assistants and adequate laboratory facilities. which enabled him to bring a number of projects to a successful conclusion. The most renowned discovery coming from Welch's laboratory (opened in 1886) was the correct identification of the gas gangrene bacillus. In 1892, with the help of G. H. F. Nuttall, Welch published his findings regarding the isolation of *Clostridium perfringens*.

With the opening of the Johns Hopkins Hospital in 1889, William Osler, Howard A. Kelly, and William Halsted began residency programs patterned on the German system of postgraduate medical training. This was doubtless one of the most important contributions of the hospital to American medicine. Although not involved directly in clinical training, Welch played a key role in setting the climate for these developments. When the medical school, after overcoming financial difficulties, finally opened in 1893, Welch became its first dean, a post he held until 1898. The students were taken to the wards and were given clinical responsibility. The outstanding basic science chairmen, Franklin P. Mall in anatomy and John J. Abel in pharmacology, had been recruited by Welch. With him, they and the clinical chairmen were instrumental in making basic science and laboratory work, as well as study of patients on the wards, the norm for medical education in America. As much as any individual, Welch made the Johns Hopkins Hospital a new kind of hospital in America. one devoted to science as much as to charity.

It fell to Welch and a handful of colleagues who also had been in the European bacteriological laboratories to alert the American medical profession to the practical applications of the germ theory of disease in relation to medicine and public health. His address to the Medical and Chirurgical Faculty of Maryland in 1887, "Modes of Infection," stressed what he had learned in such cities as Munich. Cholera and typhoid, he pointed out, were caused by specific microorganisms, not a vague miasma. The bacteria could be found in the open sewers still prevalent in Baltimore; and Welch subsequently worked to improve the city's health conditions, serving for many years on the Maryland State Department of Health.

In the spring of 1888 Welch was asked to deliver the Cartwright lectures in New York City. His speech, "On the General Pathology of Fever," presented one of several general reviews for which he became well known. In later years he wrote similar essays on the immune mechanism, thrombosis, and wound infections. Although this work was not derived primarily from his own research, Welch here evidenced his great gift for expression and for cogent summary.

His general summations of current scientific work, his editorship of the *Journal of Experimental Medicine* (1896–1906), and the numerous scientific and civic organizations to which he belonged, more than his own scientific investigations, increasingly made Welch one of the most influential spokesmen for American medicine. Many sought his counsel and he served in some key policy-making positions. He headed the Board of Scientific Directors of the Rockefeller Institute and served as a trustee from 1910 until 1933. At the same time Welch was a member of the board of the Carnegie Institution of Washington. He also served, for shorter periods of time, on the boards or councils of the Milbank Memorial Fund, the Rockefeller Sanitary Commission, and the International Health Board of the Rockefeller Foundation. In 1910 he was president of the American Medical Association and, from 1913 to 1916, president of the National Academy of Sciences.

Long active in the public health movement in America, Welch agreed in 1916 to leave his chair at the school of medicine to become dean of the new School of Hygiene and Public Health at Johns Hopkins. This was the first full-scale school of its kind, although Harvard and M.I.T. had begun a joint effort a few years earlier to train health officers. Welch continued as dean until 1925, when the trustees persuaded him to take the newly established chair in medical history.

Welch's eightieth birthday was celebrated by friends, students, and colleagues. In Washington, President Herbert Hoover called him "our greatest statesman in the field of public health." Welch himself, with characteristic modesty, agreed to accept the accolades only insofar as "I stand here to represent an army of teachers, investigators, pupils, associates, and colleagues, whose work and contributions during this period have advanced the science and art of medicine and public health to the eminent position which they now hold in this country."

BIBLIOGRAPHY

I. ORIGINAL WORKS. A bibliography of Welch's publications was prepared by Walter C. Burket, *Bibliography of William Henry Welch* (Baltimore, 1917); Burket also edited the collected *Papers and Addresses by William Henry Welch*, 3 vols. (Baltimore, 1920). Those few items that appeared after 1920 are listed in the bibliography appended to Simon Flexner's memoir (see below).

The Welch MSS, consisting of many file boxes of letters, diaries, and clippings, indexed and arranged by Simon Flexner and James T. Flexner, are deposited in the William H. Welch Medical Library of Johns Hopkins University, Baltimore.

II. SECONDARY LITERATURE. The most thorough biography is Simon Flexner and James T. Flexner, *William Henry Welch and the Heroic Age of American Medicine* (New York, 1941; repr. 1966). A shorter and more interpretive study is Donald Fleming, *William Henry Welch and the Rise of American Medicine* (Boston, 1954). Two collections of articles are *The Eightieth Birthday of William Henry Welch* (New York, 1930) and a series describing Welch's influence on pathology, public health, medical history, and medical education in a special supp. to *Bulletin of the Johns Hopkins Hospital*, **87**, no. 2, pt. 2 (1950), 1–54.

See also the following listed chronologically: Fielding H. Garrison, "In Memoriam: William Henry Welch (1850–1934)," in *Scientific Monthly*, **38** (1934), 579–582; Harvey Cushing, "The Doctors Welch of Norfolk," in *Connecticut State Medical Journal*, **5** (1941), 557–560; Simon Flexner, "Biographical Memoir of William Henry Welch 1850–1934," in *Biographical Memoirs. National Academy of Sciences*, **22** (1943), 215–231; Owsei Temkin, "The European Background of the Young Dr. Welch," in *Bulletin of the History of Medicine*, **24** (1950), 308–318; Barnett Cohen, "Comments on the Relation of Dr. Welch to the Rise of Microbiology in America," *ibid.*, 319–324; and Carl J. Salomonsen, "Reminiscences of the Summer Semester, 1877, at Breslau," C. L. Temkin, trans., *ibid.*, 333–351.

On Welch's role in the development of the medical institutions of Johns Hopkins University, see Alan M. Chesney, *The Johns Hopkins Hospital and the Johns Hopkins University School of Medicine; a Chronicle*, 3 vols. (Baltimore, 1943–1963); Richard H. Shryock, *The Unique Influence of the Johns Hopkins University on American Medicine* (Copenhagen, 1953); and Thomas B. Turner, *Heritage of Excellence, the Johns Hopkins Medical Institutions, 1914–1917* (Baltimore, 1974).

GERT H. BRIEGER

WELDON, WALTER FRANK RAPHAEL (*b*. London, England, 15 March 1860; *d*. Oxford, England, 13 April 1906), *biometrics*.

One of the founders of biometrics, Weldon was born into a wealthy London family (his father was an industrialist and a Swedenborgian) and was initially educated by private tutors and at fashionable boarding schools. In 1876 he entered University College, London, and began to study zoology under E. Ray Lankester. Two years later he moved to St. John's College, Cambridge, where he continued his study of zoology under Francis Balfour; he graduated in 1881 with a first-class degree in the natural sciences tripos.

At this point Weldon seemed on his way to becoming an orthodox zoologist. He spent some time at the Naples zoological station (1881), was demonstrator for W. T. Sedgwick at Cambridge (1882), completed a dissertation on invertebrate morphology and embryology (1883), and became a fellow of St. John's and a university lecturer in invertebrate morphology (1884). Weldon's style of life was perfectly suited to his profession; he spent two terms of each year in Cambridge, and from June to January he and his wife (whom he had married in 1883) traveled and did research at various marine laboratories. In 1883 he became associated with the laboratory of the Marine Biological Association at Plymouth, and in 1890 he succeeded Lankester as professor at University College.

The move to University College signaled a profound transformation in Weldon's interests. After reading Francis Galton's *Natural Inheritance* (1889), he became convinced that statistical studies of variation would contribute more toward solving the problems of Darwinism than the embryological work in which he had been engaged.

Between 1890 and 1892 Weldon published two papers that were classics of their kind, on variation in the shrimp *Crangon vulgaris*. In the first he demonstrated that body measurements for large populations of shrimp (carapace length, for example) are normally distributed; this was the first normal distribution observed in a wild population subject to the influence of natural selection. The second paper presented the first correlation coefficients derived for a wild population; Weldon demonstrated that in the shrimp pairs of organ lengths are highly correlated in individuals of the same species. Weldon hoped that these correlations would yield quantitative definitions of species and races, replacing the older descriptive definitions, which were based upon a single type specimen. In

1893, in studies of the crab *Carcina moenas*, Weldon found an asymmetrical distribution for frontal breadth. He thus concluded that he was actually measuring two different races of crab that inhabited the same environment but were physically distinguishable.

In 1891 Weldon began to learn probability theory and sought the help of his colleague Karl Pearson. Pearson soon became enthusiastic about the prospect of solving the problems of evolution statistically; he and Weldon began a collaboration that lasted until Weldon's death. One of the first fruits of their combined effort was the formation in 1893 of the Royal Society Evolution Committee (with Galton as chairman), dedicated to large-scale studies of variation.

Weldon's most significant contribution to biometrics was his study of differential death rates in crabs (1894), which was sponsored by the Evolution Committee. He reasoned that if natural selection works by killing "unfit" individuals before they can breed, it should be possible to correlate death rates in youthful populations with physical characteristics. To test this assumption he raised 7,000 young female crabs in jars filled with polluted water from their natural environment, assiduously measured several characteristics of each crab at different times during its growth, and discovered that individuals with greater than normal frontal breadths were more likely to die before reaching reproductive age. From these results Weldon concluded that natural selection can operate on small, apparently insignificant variations and that there is no need to postulate large jumps or discontinuous variations (as had been suggested by Galton in 1889 and by Bateson in 1894) in order to understand how evolution progresses. Weldon knew that his results were tentative, since the experimental procedure was quite faulty; but he was not prepared for the storm of protest that broke about his head. Naturalists were not ready to admit that regression lines and correlation coefficients were relevant in what had been, until then, a purely descriptive science.

The controversy between the advocates of continuous variation and the proponents of discontinuity grew increasingly acrimonious, and eventually the biometricians resigned from the Evolution Committee and founded a journal, *Biometrika* (1901), in which to publish and to propagandize for their new science. After the rediscovery of Mendel's work and the founding of *Biometrika*, two separate schools of genetics developed in England:

the Mendelians, who believed in discontinuous variation and devoted themselves to breeding studies; and the biometricians (including Weldon and Pearson), who believed in continuous variation and devoted themselves to statistical study of variation. In 1900 Weldon moved to Oxford, where he became Linacre professor; but the distance between Weldon and Pearson did not dampen their collaboration. Weldon undertook studies of moths (unpublished), snails, thoroughbred horses, poppies, mice, and men in an effort to find clear-cut cases of evolutionarily significant continuous variation; but none of these studies was as fruitful, either methodologically or substantively, as his earlier work on shrimp and crabs. Perhaps because of his frustration, Weldon worked at a pace and with an intensity that worried his friends. His debate with the Mendelians became even more acrimonious; and a stream of critical articles flowed from his pen, to be published in *Biometrika*. In the midst of an Easter holiday devoted to biometric research Weldon collapsed and died. Many of his colleagues considered his death particularly tragic for having come when he seemed to be entering a very promising phase of his career.

BIBLIOGRAPHY

Weldon's most significant biometric papers are "The Variations Occurring in Certain Decapod Crustacea. I. *Crangon vulgaris*," in *Proceedings of the Royal Society*, **47** (1890), 445–453; "On Certain Correlated Variations in *Crangon vulgaris*," ibid., **51** (1892), 2–21; "On Certain Correlated Variations in *Carcina moenas*," ibid., **54** (1893), 318–329; and "Attempt to Measure the Death-Rate due to the Selective Destruction of *Carcina moenas* With Respect to a Particular Dimension," ibid., **57** (1895), 360–379.

For an understanding of his theoretical dispute with the Mendelians see "Remarks on Variation in Animals and Plants," in *Proceedings of the Royal Society*, **57** (1895), 379–382; "Mendel's Laws of Alternative Inheritance in Peas," in *Biometrika*, **1** (1901–1902), 228–254; "Professor de Vries on the Origin of Species," ibid., 365–374; and "On the Ambiguity of Mendel's Categories," ibid., **2** (1902), 44–55.

The best source for information about Weldon is Karl Pearson, "Walter Frank Raphael Weldon," in *Biometrika*, **5** (1906), 1–50.

Ruth Schwartz Cowan

WELLS, HARRY GIDEON (*b.* Fair Haven [now New Haven], Connecticut, 21 July 1875; *d.* Chicago, Illinois, 26 April 1943), *pathology.*

A distinguished teacher and investigator in chemical and general pathology, Wells was the son of Romanta Wells, a pharmacist. He graduated in 1895, from the Sheffield Scientific School of Yale University, where he was particularly influenced by the biochemist Lafayette B. Mendel. He received the M.D. from Rush Medical College in Chicago in 1898, became assistant there to the pathologist Ludvig Hektoen, and in 1901 entered the department of pathology of the University of Chicago, of which Hektoen was titular chief. Wells was given a free hand in developing the department and was promoted through the ranks to full professor in 1913 and head of the department in 1932. In 1904–1905 he spent a fruitful year with Emil Fischer in Berlin.

Wells was a gifted teacher and a productive investigator, with a genius for succinct compilation of significant literature in his fields of interest. His widely diversified research led to his general acceptance as the country's chief authority on chemical aspects of pathology and immunology, and *Chemical Pathology* (1907) went through five editions.

Like other pathologists Wells found a wealth of important subjects for research in his frequent postmortem examinations. His practice of engaging in personal research on problems about which he felt his knowledge was inadequate led him to studies on fat necrosis, resulting in a clear understanding of this condition; of tissue autolysis, and especially its relation to histological change in the cell nucleus in disease processes; of enzyme changes involved in cell autolysis; of pathological calcification (still recognized as among the country's best studies); of fatty degeneration of the liver, in which his findings are basic to modern knowledge; and unusually productive investigations, in cooperation with Thomas B. Osborne, of the chemical composition of proteins as determined by immunological methods. With Maude Slye and other associates he carried out a long series of investigations of cancer, including its hereditary aspects, that now forms part of the background of this extensive field of medical research.

After 1911, concomitantly with his professorship, Wells was director of medical research at the Sprague Memorial Institute, a medical organization affiliated with the University of Chicago. During and after World War I he was Red Cross commissioner to Rumania, with heavy responsibilities for relief work in the Balkans.

In 1902 Wells married Bertha Robbins; their only son, Gideon R. Wells, became a practicing physician.

BIBLIOGRAPHY

I. ORIGINAL WORKS. Wells's writings include *Chemical Pathology* (Philadelphia, 1907; 5th ed., 1925); *The Chemistry of Tuberculosis* (Baltimore, 1923; 2nd ed., 1932), written with L. M. DeWitt and E. R. Long; and *The Chemical Aspects of Immunity* (New York, 1925; 2nd ed., 1929).

II. SECONDARY LITERATURE. See P. R. Cannon, "H. Gideon Wells, M.D., Ph.D., 1875–1943," in *Archives of Pathology*, **36** (1943), 331–334; and E. R. Long, "Biographical Memoir of Harry Gideon Wells, 1875–1943," in *Biographical Memoirs. National Academy of Sciences*, **26** (1950), 233–263.

ESMOND R. LONG

WELLS, WILLIAM CHARLES (*b.* Charleston, South Carolina, 24 May 1757; *d.* London, England, 18 September 1817), *meteorology, physiology, medicine, natural philosophy.*

Wells was the son of Robert Wells, a printer, and Mary Wells, Scots recently settled in America. At the age of eleven he was sent to Dumfries, Scotland, for schooling; and in 1770 he entered the University of Edinburgh. From 1771 to 1774 he was apprenticed to Alexander Garden, a Charleston physician with an international reputation in botany. He subsequently studied medicine at Edinburgh (1775–1778) and then went to St. Bartholomew's Hospital, London. Wells wrote his thesis, "De frigore," at Leiden and received the M.D. from the University of Edinburgh on 24 July 1780. After practicing at Charleston and at St. Augustine, Florida (1781–1784), he returned to London. He was licensed by the Royal College of Physicians in 1788 and was physician at St. Thomas' Hospital from 1795 until his death. Wells's practice was small, his life austere, and his circle of friends small but distinguished. He suffered from heart failure after 1812 and wrote a memoir on his life in what he correctly thought was his last year. It was published in 1818, together with a collection of his most important works and a violent criticism of the Royal College of Physicians.

Wells's essay "Single Vision With Two Eyes" (1792) led to his becoming a fellow of the Royal Society (1793). In 1795 he published a confirmation of Galvani's report (1791) that muscular contraction could be evoked by weak electrical currents. During the next two decades he wrote on the color of blood, conducted further studies on vision and optics, and provided accurate descriptions of rheumatic heart disease, of proteinuria, hematuria, and edema due to scarlet fever, and

similar cases not due to scarlatina. The studies of albuminuria, promptly translated into French and published at Geneva in 1814, prepared the way for the definitive observations of Richard Bright (1827).

Wells's most important contribution was his meticulous study of the formation of dew and the correct interpretation of his data. He proved that dew is neither invisible rain, falling from heaven, nor "sweat" from plants, but is due to condensation from air in contact with objects that have been cooled by radiating their heat into the cloudless night sky. He showed that a dark substance, charcoal, accumulated more dew than pale material, such as chalk, and that poor conductors of heat, such as plants, were covered with more dew than good conductors, such as metal objects. He also noted that windless nights favored dew formation, because they allowed the air to remain in contact with the cooled objects long enough to deposit its moisture. Although criticized by such eminent men as Thomas Young, the "Essay on Dew" (1814) led to Wells's being awarded the Royal Society's Rumford Medal. This complete and original theory was not generally accepted until its confirmation and extension by John Aitken in 1885.

Charles Darwin considered Wells to have been the first to state the theory of evolution by natural selection of those best fitted to survive in a given environment. His "Observations on the Causes of the Differences in Colour and Form Between the White and Negro Races of Men" was appended to a case report of a white woman with patchy brown discoloration of the skin. He noted how man improves domestic beasts by selection and drew an analogy to the way in which nature effects a similar development of varieties of men best suited to various climates.

Because he pioneered in the study of disease and of the physiology of vision, as well as in natural science, Wells has been claimed as an early American scientist. He was born and remained a loyal British subject, and chose to study, work, and practice at Edinburgh and London. Nevertheless, most articles dealing with his life and work have been written by American physicians, and the New York Academy of Medicine files his publications in its collection of rare Americana.

BIBLIOGRAPHY

I. ORIGINAL WORKS. At Wells's request the works he considered most important were republished in one volume with an autobiographical memoir as *Two Essays:*

Upon Single Vision With Two Eyes. On Dew . . . (London, 1818) (the other titles follow). Most of his papers on medical topics, heart and kidney disease, and so on are in *Transactions of the Society for Improvement of Medical and Chirurgical Knowledge,* **2** (1808) and **3** (1812).

II. SECONDARY LITERATURE. Wells's life is reviewed in *Dictionary of National Biography* and by E. Bartlett, in *Western Journal of Medicine and Surgery,* 3rd ser., **5** (1850), 22–44; W. Dock, in *California and Western Medicine,* **31** (1929), 340–341; and F. S. Pleadwell, in *Annals of Medical History,* n.s. **6** (1934), 128–142.

WILLIAM DOCK

WENDELIN (VENDELINUS), GOTTFRIED (*b.* Herck-la-Ville (or Herk), Belgium, 6 June 1580; *d.* Ghent, Belgium, 1667), *astronomy, meteorology, natural science, humanism, law.*

In the laudatory style of the period, Wendelin was called the Ptolemy of his age. He studied first in his native place and then at Tournai and at Louvain, where at the age of seventeen he observed a lunar eclipse. He subsequently spent time in Nuremberg, Marseilles, Rome, and Digne, before returning to Liège and Herck. He was ordained priest at Brussels and became a curate and a canon of Condé and Tournai, where he was an official of the cathedral. Like many deeply Catholic scientists, he was more attracted by the physical sciences and mathematics than by the biological sciences.

A convinced Copernican, Wendelin upheld his views with a courage that is the more impressive when it is recalled that both Descartes and Galileo were obliged to have their works (respectively, *Discours de la méthode* [1637] and *Discorsi* [1638]) printed in Protestant Holland. The Protestant presses of Leiden had become vital organs in the dissemination of new ideas. Wendelin's audacity appears all the greater in the light of the misfortunes experienced even much later, in 1691, by Martin-Étienne van Velden, a professor at the University of Louvain. A century and a half after Copernicus (1543) and four years after Newton (1687), the rector of the university formally ordered the arrest of van Velden for having attempted to make a student say that one cannot doubt the Copernican system regarding the movement of the planets around the sun.[1]

Wendelin was apparently the first to propose a law of the variation of the obliquity of the ecliptic. According to Bigourdan, Wendelin also observed the influence of temperature on the period of the oscillations of a pendulum, noting that the oscillations are more numerous in winter than in summer. He also recognized—as Galileo had not—that an increase in amplitude increases the period of the oscillations.

Wendelin corresponded with Mersenne, Gassendi, whom he taught astronomy, and Constantijn Huygens.[2] They were all younger than Wendelin. In a letter to Plempius, Descartes solicited Wendelin's opinion of his *Géométrie.*[3] In a letter to Colvius, Descartes wrote: "I'auois aussi desia vu la lampe de Vendelinus [*G. W. Luminarcani . . . Lampas*]; mais elle ne m'a point esclairé."[4] Finally, in a letter to Constantijn Huygens, Descartes praised Wendelin for his *Pluvia purpurea,* calling him "homme sçauant aux Mathematiques, et de tres-bon esprit."[5]

While Wendelin does not appear in the first four volumes of *The Correspondence of Isaac Newton,* he does figure among the seventy-one authors cited in the *Principia* (1687). There he has the honor of being mentioned in the company of Ptolemy, Huygens, Copernicus, Street, Tycho Brahe, and Kepler (book III, proposition 4, theorem 4).

NOTES

1. See J. Pelseneer, in *Biographie nationale publiée par l'-Académie royale de Belgique,* XXVI (1936–1938), cols. 562–567.
2. *Correspondance du P. Marin Mersenne,* C. de Waard, R. Pintard, and B. Rochot, eds. (Paris, 1932–   ); see the indexes to vols. 2–6, 8–10, and 12.
3. See letter of 3 October 1637, in *Oeuvres de Descartes,* C. Adam and P. Tannery, eds., I (Paris, 1897), 411.
4. Letter of 5 September 1643, in *Oeuvres de Descartes, Supplément, Index général* (1913), 16.
5. Letter of 5 October 1646, in *Oeuvres de Descartes,* IV, 516.

BIBLIOGRAPHY

I. ORIGINAL WORKS. Wendelin's works are *Loxias seu de obliquitate solis diatriba . . .* (Antwerp, 1626); *De diluvio liber primus* (Antwerp, 1629); *Id . . . secundus* (incomplete); *Aries seu aurei velleris encomium (ca.* 1632); *De tetracty Pythagorae dissertatio epistolica* (1637); *G. W. Luminarcani . . . Lampas* (Brussels, 1644); *Eclipses lunares ab anno 1573 ad 1643 observatae* (Antwerp, 1644); *Pluvia purpurea Bruxellensis* (Paris, 1647); *Leges salicae illustratae* (Antwerp, 1649); *Luminarcani, Teratologia cometica . . .* (1652); *De causis naturalibus pluviae purpureae Bruxellensis . . .* (London, 1655); *Epistola didactica de calcedonio lapide . . . (ca.* 1655); and *Arcanorum caelestium Sphinx et Oedipus . . .* (Tournai, 1658).

II. SECONDARY LITERATURE. On Wendelin and his

work, see the notice by Lucien Godeaux, in *Biographie nationale publiée par l'Académie royale de Belgique*, XXVII (1938), cols. 180–184, with a bibliography.

JEAN PELSENEER

WENT, FRIEDRICH AUGUST FERDINAND CHRISTIAN (*b*. Amsterdam, Netherlands, 18 June 1863; *d*. Wassenaar, near The Hague, Netherlands, 24 July 1935), *botany*.

Went studied at the University of Amsterdam under Hugo de Vries. As director of a sugarcane experimental station in Kagok, Java (1891–1896), he worked on cane diseases and on the physiology of sugarcane. He established that the first product of photosynthesis is sucrose and determined the sugar concentrations in leaves and stalk during the lifetime of the cane, thereby providing the basis for a method of determining the maturity of cane in the field that is still used. This early experience resulted in a lifelong interest in and promotion of research in tropical agriculture.

Twice during his tenure as professor of botany and director of the botanical laboratory and gardens at the University of Utrecht (1896–1934), the laboratory was rebuilt and enlarged, making it one of the most modern of botanical institutions and a model for many other laboratories.

Went's personal research became increasingly limited because of very heavy teaching duties, but his work (1901) on enzyme formation in the fungus *Monilia* (now named *Neurospora*) was the forerunner of very fruitful work on the biochemistry of *Neurospora*. In later years he became interested in the anatomy and embryology of Podostemonaceae, a family of flowering plants found only in rapids and waterfalls, on which he published extensively (1908–1929).

Went exerted his greatest influence on the development of botany in the first half of the twentieth century through the research of his graduate students. The "Utrecht school" became known for work in many areas of plant physiology, especially temperature responses, tropisms, and auxins. When F. F. Blackman published his important paper on physiological processes and limiting factors (1905), the experimental basis for his theory was meager. Some of the most significant support for Blackman's theory of limiting factors was supplied by Went's students during the next twenty-five years.

The second major contribution of the Utrecht school was work on tropisms, spearheaded by Blaauw's thesis on phototropism. This work for the first time placed tropisms—the responses of plants to environmental factors such as light and gravity—on a quantitative basis, and it became clear that responses to light were explainable strictly as photochemical reactions. An extension of this work explained phototropic responses as differential growth reactions to differential light intensities.

This reduction of phototropism to differential growth initiated the third major research contribution of the Utrecht school, the work on auxins. Went and his colleague Fritz Kögl were the most effective advocates of introducing the concept of plant growth hormones into European biological circles, and the eight theses on auxin published at Utrecht between 1927 and 1934 formed the basis for modern ideas about plant hormones.

Went, especially as president of the Royal Netherlands Academy of Sciences (Amsterdam), contributed immeasurably to improved international understanding among scientists.

BIBLIOGRAPHY

Went's works, published mainly in the *Verhandelingen* of the Royal Netherlands Academy of Sciences, include *De jongste toestanden der vacuolen* (Amsterdam, 1886), his doctoral diss.; and *Untersuchungen über Podostemaceen*, 3 pts. (Amsterdam 1910–1926); he also collaborated on vol. II of the German ed. of S. P. Kostychev, *Lehrbuch der Pflanzenphysiologie* (Berlin, 1931). For a list of his writings, see Royal Netherlands Academy of Sciences (Afd. natuurkunde), *Naamregister van de Verhandelingen en Bijdragen*, I (Amsterdam, 1943), 65; and II (1944), 148–149.

There is an obituary by J. van der Hoeve, in *Verslagen van de gewone vergadering der Afdeeling natuurkunde, K. Akademie van Wetenschappen*, **44** (1935), 90–95.

F. W. WENT

WEPFER, JOHANN-JAKOB (*b*. Schaffhausen, Switzerland, 23 December 1620; *d*. Schaffhausen, 26 January 1695), *medicine, physiology, toxicology*.

Wepfer graduated from the secondary school in Schaffhausen. Among his teachers was Johannes Fabritius of the Palatinate, who taught him natural history and instilled in him a passion for observing the living world. In 1637 Wepfer left Schaffhausen for Strasbourg and then went to Padua, where he studied at the Faculty of Medicine and Pharmacy. In 1647 he received the doctorate in medicine at

Basel and became municipal physician of Schaffhausen, where he remained as physician and scientist. Although Wepfer never occupied a faculty chair—Schaffhausen had no university—he had numerous students, J. C. Payer and J. C. Brunner among them, from throughout Europe. He also became the private physician of several German princes, as well as a famous consultant.

In 1647 Wepfer presented two dissertations, *Disputatio medica inauguralis de palpitatione cordis* and *Oratio de thermarum potu*. In 1648, when he became municipal physician of Schaffhausen, he was given the right to perform autopsies and made extremely complex observations, using a novel method that was not taken up again until the nineteenth century. He first followed the evolution of an illness, carefully noting all its symptoms. He completed his investigations upon cadavers. Wepfer later sought to confirm his hypotheses by performing experiments on animals, which he described in reports published mainly in the *Miscellanea curiosa* issued by the Leopoldina.

Wepfer's major research centered on the brain and, being a skilled experimentalist, he devised new techniques. For instance, he was the first to color cervical vessels through injecting dye. The essentials of his anatomical observations concerning the nervous system are presented in *Historia anatomica de puella sine cerebro nata* (1665). In his classic work, reprinted many times, *Observationes anatomicae ex cadaveribus eorum, quos sustulit apoplexia, cum exercitatione de eius loco affecto* (Schaffhausen, 1658), he collected a large number of original observations based on comparative anatomy of human cadavers. In it he was the first to report that apoplexy involved hemorrhage from blood vessels.

It was in toxicological analysis, however, that Wepfer made his greatest contributions. He systematically studied poisons, with particular attention to the toxic substances synthesized by certain umbellifers, especially the poison and water hemlocks. He was the first to analyze the pharmacological effects of coniine, an alkaloid of hemlock that was not isolated until much later; and his classic description of hemlock poisoning was often cited as the standard. He also experimented upon animals and found an efficacious remedy: the administration of strong emetics. At the same time he noted that coniine, in minute doses, could be useful as an antineuralgic and antispasmodic. He also discovered its remarkable analgesic effect and was the first to use it in minor surgery. One of his publications was *Cicutae aquaticae historia et noxae*

commentario illustrata (Basel, 1679). His numerous discoveries about poisons and their uses made Wepfer an undoubted pioneer in toxicology. He also studied the characteristics of mercury poisoning and was the first to indicate the dangers for workers with this metal who fail to take the proper precautions. This study led him to publish articles on occupational diseases.

After Wepfer's death his heirs, B. and G. M. Wepfer, published some of his writings as *Observationes medico-practicae de affectibus capitis internis et externis* (Schaffhausen, 1727). As a scholarly physician Wepfer made a tremendous contribution to medical treatment and research through his resolute opposition to the influence of dogmatic and traditionalist scientists who stressed ancient texts rather than actual facts.

BIBLIOGRAPHY

Wepfer's works are cited in the text.

Secondary literature includes H. Buess, *Recherches, découvertes et inventions de médecins suisses*, R. Kaech, trans. (Basel, 1946), 25–26; and H. Fischer, *J. Jakob Wepfer* (Zurich, 1931); and *Briefe J. J. Wepfer an seinem Sohn Johann Conrad* (Leipzig, 1943).

P. E. PILET

WERNER, ABRAHAM GOTTLOB (*b.* Wehrau, Upper Lusatia [now Osiecznica, Poland], 25 September 1749; *d.* Dresden, Germany, 30 June 1817), *geology, mineralogy.*

Werner was the only son of Abraham David Werner and Regina Holstein Werner. He had one older sister, Sophia. His family had a long history of association with various ironworks, and his father was inspector of the Duke of Solm's ironworks in Wehrau and Lorenzdorf. The family was well off financially. According to his own "Biographical Notes," Werner received his first formal education from his father, who encouraged his early interest in mineralogy. He also studied with a private tutor before entering the Waisenschule at Bunzlau (now Boleslawiec, Poland) at the age of nine. He remained at the Waisenschule until 1764, when his mother died and his father took him out of the school and made him a *Hüttenschreiber*[1] in the ironworks. After five years of this work, in 1769, Werner was enrolled in the recently founded Bergakademie Freiberg and began studies intended to prepare him for the administration of the Duke of Solm's ironworks.

However, in Freiberg, he was induced to enter

the Saxon mining service; and since no one could expect to achieve an advanced position in the service without a degree in jurisprudence, Werner left the Bergakademie after two years to enter the University of Leipzig, where he studied for three years. During his first two years at the university, he devoted himself mainly to the necessary courses in law, but he became increasingly interested in the study of languages and what is now called historical linguistics, and his interest in mineralogy persisted, until he abandoned the study of law altogether, leaving the university in 1774 without a degree.

In 1773, however, he had written his first book, *Von den äusserlichen Kennzeichen der Fossilien,* which was published in 1774. On the strength of its immediate success, his friend and former teacher K. E. Pabst von Ohain suggested to the Board of the Bergakademie that he be offered a position as teacher of mining and curator of the mineral collection there. Werner accepted the offer, joining the faculty in 1775, and remained at the school for the rest of his life. During his forty-two years there, largely because of his fame as a mineralogist and his skill as a teacher, the little mining academy became one of the most famous schools in the world. And in turn Werner came to be acknowledged as the foremost geologist of his day. He moved in a brilliant circle of friends and was received at the Saxon court. Among his many illustrious students, he counted not only such noted geologists and mineralogists as Leopold von Buch, Alexander von Humboldt, Jean d'Aubuisson de Voisins, Robert Jameson, and Friedrich Mohs, but also such romantic philosophers and writers as Gotthilf Heinrich von Schubert, Henrik Steffens, and Friedrich von Hardenberg (Novalis).

During his lifetime Werner was elected to twenty-two scientific societies, including the Geological Society of London, the Institut National and the Institut Impérial of France, the Imperial Society of Physics and Medicine of Moscow, the Royal Prussian Academy of Sciences, the Royal Stockholm Academy of Sciences, and the Wernerian Society of Edinburgh. During his last years he suffered increasingly from ill health, going frequently to take the waters at various health resorts. He never married, and the will that he dictated shortly before his death bequeathed most of his estate to the Bergakademie Freiberg, which had been such an important part of his life and work.

Although Werner is best known for his contribution to the founding of geology as a science, he first achieved recognition as a mineralogist. He considered mineralogy to be the basis for all study of the earth, dividing it into five branches, of which geognosy (historical geology) was one and oryctognosy (descriptive mineralogy) another. And during all the years in which his theories on geognosy were arousing so much interest and controversy, he continued to work on his mineral system, the final version of which appeared after his death in 1817. His first important mineralogical work, however, *Von den äusserlichen Kennzeichen der Fossilien,* was not a mineral system but a classification of external characteristics of minerals, designed to aid the worker or the student in the field. In it Werner gave an unprecedented number of external characteristics with definitions, usually accompanied by homely examples which could be understood by both the layman and the natural philosopher. He also attempted to establish some standards of quantification and thus to clear away the vagueness in the terminology then in use. As chemistry and crystallography developed, mineralogists came to rely more on chemical analysis and less on external characteristics, but *Von den äusserlichen Kennzeichen der Fossilien,* published when Werner was twenty-five years old, continued to be an important work into the nineteenth century. Thomas Weaver's translation into English was published in Dublin in 1805, and a revised translation by Charles Moxon appeared in 1849.

Werner remained convinced of the importance of external characteristics, not only in the identification of minerals but also in the study of their composition. He reasoned that since the appearance of a mineral changes when its chemical composition is changed, there should be a correlation between chemical composition and external characteristics. On the other hand, he recognized that external characteristics cannot form the basis of a mineral system. He wrote: "One can indeed recognize in the external character of minerals the differences of their composition, provided both are previously determined, but the correlation between these two features cannot be discovered in them."[2] He was convinced that ultimately mineral systems must be based on chemical composition, and to that end he kept abreast of developments in chemistry and helped to bring about the building of a chemical laboratory at the Bergakademie and the engagement of W. A. Lampadius as teacher of chemistry. He himself analyzed minerals in his laboratory and stayed in close contact with M. H. Klaproth, who has been called the founder of quantitative mineral analysis.

In his later years, however, Werner took the position that chemistry was still not sufficiently developed for mineralogy to rely upon it completely. In his own system he retained the four traditional classes—earths, salts, combustibles, and metals—and he began to give priority to geological rather than chemical considerations. A good example is his classification of the diamond among the earths rather than the combustibles, even though he was well aware that the diamond is a carbon. He wrote:

> The diamond, . . ., is by nature, according to its exterior, characterized wholly as an earthy mineral, as a stone. Its geognostic occurrence also speaks for its place among the earths, because the diamond, as far as is known, occurs only with and among stones, and not among combustible minerals, among which it has recently been classed. All uses which are made of it are as a stone. And finally, its identification is not aided in any way by placing it, in lectures and mineral collections, among earth pitch, the three coal species, graphite, and so forth; but it is helpful to place it with the far more similar zircon and the other gems. Let the mineralogical chemist regard this stone as one of the coals and place it among them; but he should permit the oryctognost to act according to the purpose he has in mind in placing the diamond in an oryctognostic system.[3]

Werner considered crystallography to be only a branch of mineralogy, which, although important to the study of mineralogy, is unsuitable as a basis for a mineral system and of limited practical value. However, he did study crystallography himself. He emphasized its importance in his lectures, and urged his students to study it. He was well acquainted with the work of Romé de l'Isle and Haüy, being especially interested in the study of primary crystal forms, especially with what Haüy came to call laws of decrement (*décroissement*). In his own system he incorporated crystal form as an external characteristic.

In spite of his abiding interest in the practical aspects of mineralogy, Werner was not merely a practical mineralogist. Throughout his scientific life, he was concerned with the philosophical aspects of classification in general and of mineral classification in particular. His fullest exposition of his ideas on this subject was published without his permission in 1816 under the title "Werner's oryctognostische Classifikationslehre." In a manuscript of this treatise, Werner wrote that the work represented more than forty years of reflection and research.

The article is especially concerned with the various classificatory categories which Werner understood and the three fundamental tasks of the classification of minerals: *Gattierung, Gradierung,* and *Reihung*—that is, the determination of species, which he considered to be the cornerstone of a mineral system; the establishment of categories more general and less general than species; and the establishment, wherever possible, of kinships among the members of a particular classificatory category. Werner believed that there are only two possible kinds of kinship among minerals. One leads to a complete transition, or *Übergang*, in which "the crystallization suite of one species is so closely related to that of another that both are able to cross over completely into the other."[4] In the other, which he called *Aneinanderstossen* ("coming in touch with one another"), the minerals are related to one another interruptedly, making a complete transition impossible.

Werner's mineral system, complete as it stood at the time, was published three times, once in 1789, once in 1816, and again in 1817. In addition, parts of it appeared incorporated in other works. In 1780 Werner's partial translation of Axel Cronstedt's *Försök til Mineralogie* was published. Werner believed that, at the time, Cronstedt's work was the best available on the subject. He translated only the portion dealing with earths and stones, however, correcting errors, adding information on the constitution of minerals, and making extensive additions concerning external characteristics. His comments and additions so enlarged Cronstedt's work that his translation became a textbook of mineralogy in its own right and was widely used as a teaching aid and reference work.

In 1791–1793 Werner's two-volume catalog of Pabst von Ohain's mineral collection appeared. In this work Werner not only incorporated his mineral system but also put into practice his ideas of what should constitute a complete mineral cabinet, a subject which he had discussed in a 1778 article, "Von den verschiedenerley Mineraliensammlungen, aus denen ein vollständiges Mineralienkabinet bestehen soll." In the article Werner emphasized that a mineral collection should be more than a systematic arrangement of minerals: it should further the understanding of the entire mineral kingdom. He therefore cataloged Pabst von Ohain's collection in five separate collections according to external characteristics, the natural order of minerals in a mineral system, the historical development of the earth's crust, the places of origin of minerals, and the uses of minerals. Von Ohain's collection was ultimately sold to the government of Portugal and shipped to Brazil, where it was used in the teaching of geology and mineralogy in Rio de Janeiro. And the catalog, which was widely used in Europe, was one of the important avenues through which Werner's influence on mineralogy and geology was spread.

None of the complete editions of Werner's system was prepared by Werner himself. The 1789 version was prepared under his supervision by his student C. A. S. Hoffman and was published with Werner's permission. It was also Hoffman who, along with another student, A. W. Köhler, revised the system in 1812 (this is the version which was published in 1816). The 1817 edition, which was published posthumously by order of the Saxon government, was prepared from Werner's notes by his students J. C. Freiesleben, August Breithaupt, and Köhler.

It is interesting to look at the 1789 and the 1817 versions together, for they show not only the changes in Werner's system but also the progress that mineralogy had made in the intervening years. The most striking difference between them is that the earlier work covers only 183 species, whereas the later one covers 317. Of these 317, Werner had independently discovered eight and had given names to numerous others. The names of the eight minerals that he discovered, as well as twenty-six other names which he employed, are still used today to designate the same minerals to which Werner applied them.[5]

Werner's scientific life spanned a time of unusual interest in mineralogy, an interest not confined to scientists but fostered to a large extent by romantic conceptions on the one hand and utilitarian considerations on the other. The store of mineralogical knowledge was rapidly increasing; and advances in chemistry, crystallography, and geology were opening new paths for the study of mineralogy. Through his own research Werner added to the knowledge of mineralogy and helped to systematize that knowledge. Through his teaching and writing he contributed greatly to the dissemination of knowledge. Conservative and cautious, he was always hesitant to add a "new" mineral to his system until he was certain that it was really a mineral and really new; and, although he was willing to employ new methods, he was always reluctant to abandon old ones which he felt had proved their worth. Thus, Werner was a steadying influence at a time of great and varied activity. His work represents the culmination of a long development in mineralogy and the beginning of a new mineralogy, of which he was fully conscious.

Although many earlier writers had speculated on the origin of the earth's crust, Werner is rightly called the father of historical geology, for he was the first to work out a complete, universally applicable geological system. It was he who, more than any other, made geology into a science and an academic discipline. According to William Brande, ". . . to him belongs the principal merit of pointing up the order of succession which the various natural families of rocks are generally found to present, and of having himself developed that order to a considerable extent, with a degree of accuracy which before his time was unobtainable. . . ."[6]

Werner's interest in historical geology stemmed partly from his interest in mineralogy and partly from his interest in mining. But it also reflected his interest in history, for he believed that natural history is an important branch of the history of man, and he felt that the earth's crust is a more reliable source of history than written histories. His system was based on the principle of geological succession, as Brande indicates, and all his geological work was consistent with this historical principle as he saw it.

Werner was well versed in the mineralogical and geological literature of his day, being familiar with the writings of the leading exponents of both fire and water as the major agent in the creation of the earth's crust. A list of writers on geology that he prepared includes among others Steno, Lehmann, Ferber, Hamilton, Füchsel, Saussure, Buffon, and Moro. His own ideas were undoubtedly influenced in one way or another by what he had read and, in fact, his theories bear a rather striking resemblance to those of Steno, with whose work he was apparently familiar.[7] But whatever the background of his theories, Werner thought, on the basis of the geological knowledge of his day, that they were firmly supported by the evidence—a fact which goes far to explain the popularity of his system. Unlike Steno, Lehmann, Moro, and many other earlier and even contemporary writers on geology, he felt no need to fit his theories into the biblical story of creation. There is no indication in his writings, published or unpublished, that any of the floods which are an important part of his theory was the biblical flood. His early religious background was Pietist, at the university he was accused of being an atheist, and in general his attitudes reflected the deism of the eighteenth century. Thus, although his theories, being basically neptunistic, were more acceptable to the defenders of the biblical account of creation than those of the vulcanists, he himself was in no way engaged in the religious aspects of the controversy.

The two basic postulates of the Wernerian theory were that the earth was once enveloped by a universal ocean and that all the important rocks that make up the crust of the earth were either precipitates or sediments from that ocean. Werner placed the rocks in four (later five) classes according to the period in which they were formed, believing that characteristics of the rocks were the result of the depth, content, and conditions of the universal ocean at the time when they were formed. His classification was basically historical. As he himself put it,

> I had to be guided completely in the classificatory presentation or tabulation of these masses by the discoverable time sequence of the particular formations if I wanted to remain true to my plan to sketch through this classification a foundation for a complete canvass of the universal formation of these masses.[8]

Although he did not conceive of the immensity of geological time on the same scale as present-day scientists do, he did write of a time "when the waters, perhaps 1,000,000 years ago, completely covered our earth . . ."; and in his lectures he spoke of the history of the earth "in contrast to which written history is only a point in time."[9] In order to discover the time sequence of rock formation, he used various means, such as compositional and textural features and, especially, the structure of rocks and stratigraphic relations, which he considered the most important clues to the understanding of the history of the earth's crust. His theory included two unexplained general risings of the universal ocean as well as some local floods; but he believed that, in general, the waters had receded very slowly but steadily. The four periods of formation and their corresponding classes of rocks were the primitive, the floetz, the volcanic, and the alluvial.

At the beginning of the primitive period, according to Werner's theory, the universal ocean was very deep and calm; and the first rocks were chemical precipitates which adhered to an originally uneven surface, granite being the first rock formed. Gradually the waters became less calm, so that later rocks of the primitive period are not as crystalline as the older ones; and toward the end of the period there was a general rising of the waters, followed by a comparatively rapid recession, which explains the position of some of the later primitive rocks relative to the older ones. No life existed during the primitive period, and thus the primitive rocks are entirely free of fossils.

The floetz period was characterized by storms in the then low-standing ocean and by the development of life in great abundance, the storms destroying much of the newly developed life as well as some of the older rocks. Once again there was a general inundation, with the waters this time reaching a greater height than ever before. It is with these variations from stormy to calm to stormy and the general inundation that Werner explains the relative position of the floetz rocks and their often broken stratification.

The volcanic and alluvial periods are almost contemporaneous and both extend into the present (as does the floetz period). The volcanic and alluvial rocks, however, are not deposits from the universal ocean but the result of local conditions.

Werner added the transition period and the class of transition rocks to his system after the discovery that some rocks which he had previously classified as primitive contain fossils. According to his explanation, the relatively low-standing waters toward the end of the primitive period were calm at first, but gradually they became increasingly stormy, destroying some of the previously formed rocks as well as some living organisms, which had just begun to develop. Some of the rocks formed from these stormy waters are chemical precipitates and some mechanical depositions.

Thus the Wernerian system in its final form included five periods of formation: primitive, transition, floetz, volcanic, and alluvial. The rocks of the first three periods, which constitute most of the earth's crust, were precipitates or deposits from the universal ocean, those of the two later periods the result of local conditions.

Since Werner believed that the contents of the universal ocean had varied from time to time and from place to place, his theory could account for variations from the general principle that the rocks had been laid down by the universal ocean in layers one above the other. For instance, if the essential contents had at some time been missing from some part of the ocean, an entire formation might be missing from the corresponding area of the earth's crust. Also, there is nothing in the theory to preclude the formation of similar rocks at different times. Thus, it is only in an idealization of the system that the rocks of the earth's crust can be envisioned as enveloping the earth in layers much like the layers of an onion. In fact, the theory was flexible enough that, with the addition of factors such as differential settling and the subsequent effects of erosion, cave-ins, etc., it could explain virtually all the phenomena which were observable in Werner's time. Through his personal magnetism and skill as a teacher, Werner was able to inspire a host of eager students and admirers to go out to attempt that complete canvass of the earth's crust which he had hoped that his system would make possible.

Werner's geological theories were first included in his teaching in the introduction to the course on mining, which he had taught since his arrival at the Bergakademie. By the academic year 1778–1779 he had recognized that he could not cover this theoretical part of the course as thoroughly as he wished in one year and at the same time give sufficient practical instruction. He therefore announced that he would offer the theoretical introduction as a separate course entitled "Lehre von den Gebirgen." Although this was the course which eventually attracted students from all over Europe and from the Americas, bringing fame to the Bergakademie and spreading Werner's theories, it did not attract many students at first; and it was not until the academic year 1786–1787 that it began to be offered yearly. In the meantime, Werner had written his "Kurze Klassifikation und Beschreibung der verschiedenen Gebirgsarten," which was published in the 1786 volume of *Abhandlungen der Böhmischen Gesellschaft der Wissenschaften* and subsequently in at least two pamphlet editions.

Short though it is, the "Kurze Klassifikation" is important to the history of geology for a number of reasons. Although it contains no discussion of Werner's theories, it exemplifies them; and it is the only printed presentation of those theories to come from Werner's own hand. The principle of geologic succession is implicit in it. The rocks are classified according to the period of formation, and virtually all are assumed to be of aqueous origin. The "Kurze Klassifikation" was the first work to separate rock classification from the classification of minerals, and thus it did much to establish petrography as an independent branch of the geological sciences. It gave clear definitions of rocks, many of which had not previously been generally agreed upon. And it inspired the research of many geologists, including many who did not accept Werner's theories, well into the nineteenth century. But the "Kurze Klassifikation" is also important in another way. In a note on the section dealing with volcanic rocks, Werner asserted for the first time that all basalt is of aqueous origin, thus precipitating the great basalt controversy.

Until the end of the eighteenth century, it was generally agreed that granite is of aqueous origin; and many other rocks now considered to be mag-

matic played only a minor role in discussions of the origin of volcanoes. Therefore, for a long time, the whole question of the relative importance of fire and water as agents in the creation of the earth's crust revolved about the origin of basalt, since basalt is so abundant and so widely distributed. There had long been differences of opinion on the matter; but at the time when Werner entered the debate, the weight of opinion seemed to favor igneous origin. Werner's assertion received a great deal of publicity, however; and the debate was resumed with a fervor not shown before, as a host of geologists rushed into the field to seek evidence for one theory or the other.

As early as 1776 Werner had maintained that not all basalt is volcanic in origin. He had previously felt that the theory of the volcanic origin of basalt was "paradoxical," and an examination of the basalt mountain at Stolpen in Saxony had convinced him that that formation at least was of aqueous origin. Further investigations had strengthened his convictions, so that by the time of the publication of the "Kurze Klassifikation" he apparently felt prepared to defend his position. In the spring of 1787, he examined the basalt deposit at Scheibenberg, in the Erzgebirge, where he found layers of sand, clay, and wacke below basalt. He took this as indisputable evidence of the correctness of his assumption and subsequently wrote an article explaining his discovery. This article appeared in the *Intelligenzblatt* of the *Allgemeine Litteraturzeitung* of Jena in the autumn of 1788. In the meantime, the *Magazin für die Naturkunde Helvetiens* had offered a prize for the best essay in answer to the question "What is basalt?" Two of Werner's students entered the competition: J. C. W. Voigt, who advocated the volcanic origin of basalt, and J. F. W. Widenmann, who defended aqueous origin. With the appearance of Werner's article, Voigt wrote a letter to the *Intelligenzblatt* "correcting" Werner; Werner replied with some heat. Subsequently Werner wrote seven more short articles on the subject, all but one of which appeared in *Bergmännisches Journal* in the spring of 1789.[10] In September of the same year the *Magazin für die Naturkunde Helvetiens* published both Voigt's and Widenmann's essays in the same issue that carried Werner's article on the origin of volcanoes.[11] Widenmann won the prize; but neither essay settled the controversy, and the matter continued to be debated and investigated with keen interest for many years. Werner, however, took no further part in it except to explain in his *Neue Theorie von der Entstehung der Gänge* (1791) that basalt veins, like others, are the result of settling from above.

At the time, neither side had any means of proving conclusively that it was right. Petrography alone could not provide sufficient proof; microscopic methods were not then available; and chemical analysis, which showed great constitutional uniformity among basalts but great diversity among lavas, was hardly convincing. Werner's theory was better substantiated by evidence and reasoning than those of his opponents, at least until the 1790's. And it is to his credit as a teacher and investigator that his own students, trained in his methods of research, who had originally gone out to prove him right, were in many instances in the forefront of the investigations that ultimately proved him wrong.

Werner could never bring himself to place basalt among the volcanic rocks. He shifted it from the primitive to the floetz period; but to go further would have been to remove one of the cornerstones of his system, something that few scientists have ever been willing to do.

As a result of his interest in mineralogy and his long association with mining and smelting, Werner was of course interested in ores and ore deposits. But in order to remain faithful to his idea of a universally applicable geological system into which all observable phenomena would fit, he had to work out a theory of the origin of ore deposits which would be consistent with his general theory of the origin of the earth's crust. The result of his work in this area was the *Neue Theorie von der Entstehung der Gänge*, published in 1791.

In this work Werner defined veins as "particular mineral depositories of tabular shape, which in general traverse the strata of rocks . . . and are filled with mineral masses differing more or less from the rocks in which they occur."[12] He distinguished between veins and ore beds, giving an explanation which is in principle historical: minerals which occur in veins are very diverse and give every indication that they were formed during different periods than the surrounding rocks, whereas those in ore beds have the same direction as the strata among which they are found, indicating that they are of contemporaneous origin. He also gave a historical definition of the concept of a vein formation: "I designate all veins of one and the same origin as a vein formation . . . whether they are close together in one region or widely separated from one another in distant countries. . . ."[13]

Werner built his theory of the formation of veins on two major premises. The first of these is that "all true veins are really rents which (of necessity) were originally open and were only later filled from above."[14] He supported this premise on the basis of the structure of veins, comparisons of the structure of veins with that of the country rock, the structure of druses in veins, fragments of country rock in veins, analogy with existing rents, and the laws of mechanics. He explained the formation of rents as a result of diagenetic settling—compaction of the originally wet rock masses and the simultaneous loss of the support of the high-standing waters as these receded—shrinkage, and earthquakes. He thought that on the basis of his first premise the relative age and order of succession of veins, metals, ores, and vein minerals could be determined; and he formulated three criteria for determining the relative age of veins and vein stuff: (1) a vein is always newer than another which it traverses; (2) the materials in the center of a vein are newer than those near its walls, and those that are in its upper parts are newer than those in its lower parts; (3) a mineral which occurs above others in a specimen is newer than the others, and one that appears to be grown into others is older than they.

The second premise is that "the same depositions from water that formed the beds and strata of rock masses, and among these produced many ore-bearing ones, also formed the vein stuff; this took place during the time when the solution which contained such substances was standing above the already existing rents, which were wholly or partly open."[15] Werner pointed out that the materials in veins are structurally different from the same materials in beds and strata. In veins they are usually coarser and better crystallized because in veins they were not so much affected by the activities of the waters, and thus the deposition of materials in veins could proceed much more calmly than the depositions in beds and strata. On the basis of the structure of the vein stuff and the pattern of association of certain minerals, metals, and ores in veins, he tried to establish their relative ages and the sequence of their formation. Thus, he considered tin to be one of the oldest metal formations, since it occurs in granite, and bog iron-ore to be the newest, since it occurs in the alluvial lowland formations.

Neither of Werner's major premises was new, and his theory met with opposition even in his own day and was later discarded. However, many of its elements were of lasting value. Werner formulated basic questions about the origin and history of veins and their contents, established criteria for determining the relative age of veins and vein materials, and presented a comparative study of the structure of veins and rock masses. His student Breithaupt was probably the first to stimulate widespread research on the paragenesis of minerals, but it was Werner who set up the problem and gave impetus to a search for a solution. Perhaps the most important contribution of *Von der Entstehung der Gänge*, however, was that it made the study of vein formation an integral part of historical geology.

After the appearance of the second volume of his catalog of Pabst von Ohain's mineral cabinet in 1793, Werner published little on geology. In 1794, he published a fifty-page article, "Über den Trapp der Schweden," and a lecture which he had given before the Gesellschaft für Mineralogie zu Dresden was published after his death under the title "Allgemeine Betrachtungen über den festen Erdkörper." This work, however, was nothing but the introduction to his course on geognosy. A collection of works on mining and ferrous metallurgy, for which Werner had written three articles and coauthored another, appeared in 1811; however, all the articles had been written much earlier, before 1785.

In his later years, Werner devoted himself to his teaching and his duties as Councillor of Mines. He was always surrounded by students and received numerous visitors. The manuscripts that he left to the Bergakademie are extensive; but during the last twenty years of his life, his contributions to geology were made known largely by word of mouth. Yet he remained a towering figure in his field. Probably no other geologist has ever been so extensively eulogized by followers and opponents alike as he was during the two decades following his death.

NOTES

1. A *Hüttenschreiber* was something of a combination bookkeeper, secretary, assayer, and payroll clerk.
2. *Äusserliche Kennzeichen*, 26–27.
3. S. G. Frisch, *Lebensbeschreibung . . . Werners*, 62–63.
4. Werner MSS.
5. Guntan and Rösler. *Werner Gedenkschrift*, 56–57.
6. W. T. Brande, *Outlines of Geology*, 20.
7. Two copies of Steno's *Prodromus*, one the 1763 ed., are among the books that Werner left to the Bergakademie library.
8. Unpublished reply to a review of the "Kurze Klassifikation," Werner MSS.
9. Geognosy, Werner MSS.
10. Of the six articles published in the *Bergmännische Journal* (1789), four were letters annotated by Werner, in which other geologists gave examples supporting his position. The seventh article, "Von den Butzen-Wacken zu Joachimsthal," was published in Crell's *Chemische Annalen*, **1** (1789).

11. Werner admitted that the concept of volcanoes resulting from the inflammation of coal beds was old, but he maintained that his elaboration of this concept and the proofs that he offered to support it were new.
12. Werner, *Neue Theorie von der Entstehung der Gänge*, 2–3.
13. *Ibid.*, 5–6.
14. *Ibid.*, 51–52.
15. *Ibid.*, 52.

BIBLIOGRAPHY

I. ORIGINAL WORKS. Werner's writings include *Von den äusserlichen Kennzeichen der Fossilien* (Leipzig, 1774), English trans. by Thomas Weaver (Dublin, 1805), rev. trans. by Charles Moxon (London, 1849); "Von den verschiedernerley Mineraliensammlungen, aus denen ein vollständiges Mineralienkabinet bestehen soll . . .," in *Sammlungen zur Physik und Naturgeschichte von einigen Liebhabern dieser Wissenschaften*, **1** (1778), 387–420; *Axel Kronstedts Versuch einer Mineralogie. Aufs neue aus dem schwedischen übersetzt und nächst verschiedenen Anmerkungen vorzüglich mit aeussern Beschreibungen der Fossilein vermehrt von Abraham Gottlob Werner*, **2**, pt. 1 (Leipzig, 1780); "Kurze Klassifikation und Beschreibung der verschiedenen Gebirgsarten," in *Abhandlungen der Böhmischen Gesellschaft der Wissenschaften*, **2** (1786), 272–297; "Bekanntmachung einer am Scheibenberger Hügel über die Entstehung des Basaltes gemachten Entdeckung," in *Allgemeine Litteraturzeitung, Intelligenzblatt* (1788), no. 57, 484–485; "Antwort auf Herrn Bergsekretär Voigts im *Intelligenzblatte* der allgemeinen Litteraturzeitung . . . eingerückte sogennante Berichtigung meiner . . . neuen Entdeckung," *ibid.* (1789), no. 23, 179–184; "Mineralsystem des Herrn Inspektor Werners mit dessen Erlaubnis herausgegeben von C. A. S. Hoffman," in *Bergmännisches Journal*, **1** (1789), 369–398; and "Versuch einer Erklärung der Entstehung der Vulkanen durch die Entzündung mächtiger Steinkohlenschichten, als ein Beytrag zu der Naturgeschichte des Basalts," in *Magazin für die Naturkunde Helvetiens*, **4** (1789), 239–254.

Further works are *Neue Theorie von der Entstehung der Gänge, mit Anwendung auf den Bergbau besonders den freibergischen* (Freiberg, 1791); *Ausführliches und systematisches Verzeichnis des Mineralienkabinets des Herrn Karl Eugen Pabst von Ohain*, 2 vols. (Freiberg, 1791–1793); *Kleine Sammlung Berg- und Hüttenmännischer Schriften* (Leipzig, 1811); "Mineral-System des Herrn Bergrath Werner vom Jahre 1812," in *Neues Bergmännisches Journal*, **4** (1816), 204–231; "Werners oryctognostische Classifikationslehre," in *Hesperus* (1816), 345–349, 377–381, 414–416, 428–430; *Abraham Gottlob Werners letztes Mineral-System. Aus dessen Nachlass auf oberbergamtliche Anordnung herausgegeben und mit Erläuterungen versehen* (Freyberg, 1817); and "Allgemeine Betrachtungen über den festen Erdkörper," in *Auswahl aus den Schriften der unter Werners Mitwirkung gestifteten Gesellschaft für Mineralogie zu Dresden*, **1** (1818), 39–57.

A complete collection of Werner MSS is in the archives and library of the Bergakademie Freiberg. These have now been cataloged. Photostatic copies of some of these MSS are in the History of Science Collections of the University of Oklahoma, in the library of Oklahoma State University, and in the author's private collection. The catalog of the Bergakademie Freiberg collection has been published: Karl-Fritz Zillman, *Bestandsübersicht des handschriftlichen wissenschaftlichen Werner-Nachlasses*, publication no. 24 of the Library of the Bergakademie Freiberg (Freiberg, 1967).

II. Secondary Literature. See *Abraham Gottlob Werner Gedenkschrift aus Anlasz der Wiederkehr seines Todestages nach 150 Jahren am 30. Juni 1967* (Leipzig, 1967), which contains the most extensive published bibliography on Werner; Richard Beck, "Abraham Gottlob Werner. Eine kritische Würdigung des Begründers der modernen Geologie. Zu seinem hundertjährigen Todestage," in *Jahrbuch für das Berg- und Hüttenwesen im Königreich Sachsen* (1917), A3–A50; J. P. van Berghem-Berthout and J. H. Struve, *Principes de minéralogie, ou exposition succincte des caractères extérieurs des fosiles, d'après les leçons du Professeur Werner, augmentés d'additions manuscrites fournies par cet auteur* (Paris, 1795); Heinrich Bingel, *Abraham Gottlob Werner und seine Theorie der Gebirgsbildung* (Marburg, 1934); Karl August Blöde, "Kurzer Nekrolog Abraham Gottlob Werners," in *Auswahl aus den Schriften der unter Werners Mitwirkung gestifteten Gesellschaft für Mineralogie zu Dresden*, 2 (1819), 252–304; William Thomas Brande, *Outlines of Geology* (London, 1829); Leopold von Buch, *Leopold von Buch's Gesammelte Schriften*, J. Ewald, J. Roth, and H. Eck, eds., 4 vols. (Berlin, 1867–1885); Albert Carozzi, trans., *On the External Characters of Minerals*, by A. G. Werner (Urbana, Ill., 1962); and Jean François d'Aubuisson de Voisins, *Traité de géognosie, ou exposé des connaissances actuelles sur la constitution physique et minérale du globe terrestre*, 2 vols. (Strasbourg, 1819; 2nd ed., Paris, 1828–1835).

See also Walther Fischer, "Abraham Gottlob Werner," in *Mitteilungen des Roland*, nos. 4–5 (July–Oct. 1936), 54–60; and *Mineralogie in Sachsen von Agricola bis Werner. Die ältere Geschichte des Staatlichen Museums für Mineralogie und Geologie zu Dresden (1560–1820)*, (Dresden, 1939); Samuel Gottlob Frisch, *Lebensbeschreibung Abraham Gottlob Werners* (Leipzig, 1825); C. A. S. Hoffman and August Breithaupt, *Handbuch der Mineralogie*, 8 vols. (Freiberg, 1811–1817); and Traugott L. Hasse, *Denkschrift zur Erinnerung an die Verdienste des in Dresden am 30. Juni 1817 verstorbenen K. S. Bergrath's Werner und an die Fortschritte bei der Bergakademie zu Freiberg nebst einer übersichtlichen Nebeneinanderstellung der Mineralsysteme Werners und seiner Nachfolger bei dieser Akademie . . .* (Dresden, 1848).

Further works are Robert Jameson, *System of Mineralogy, Comprehending Oryctognosy, Geognosy, Mineralogical Chemistry, Mineralogical Geography, and Oeconomical Mineralogy*, 3 vols. (Edinburgh, 1804–1808); John Murray, *A Comparative View of the Huttonian and Neptunian Systems of Geology: In Answer to the Illustrations of the Huttonian Theory of the Earth, by Professor Playfair* (Edinburgh, 1802); Alexander M. Ospovat, "Abraham Gottlob Werner's Influence on American Geology," in *Proceedings of the Oklahoma Academy of Science*, 40 (1960), 98–103; "Abraham Gottlob Werner and His Influence on Mineralogy and Geology" (doctoral diss., Univ. of Oklahoma, 1960), available from University Microfilms, Inc., Ann Arbor, Mich.; "Abraham Gottlob Werners Gedanken über Wissenschaft und Bildung," in *Neue Hütte*, 12 (1967), 308–313; and trans. of Abraham Gottlob Werner, *Short Classification and Description of the Various Rocks*, with intro. and notes (New York, 1971); Franz Reichetzer, *Anleitung zur Geognosie insbesondere zur Gebirgskunde. Nach Werner für die K. K. Berg-Akademie* (Vienna, 1812); Franz Ambrosius Reusz, *Lehrbuch der Mineralogie nach des Herrn O. B. R. Karsten mineralogischen Tabellen*, 4 pts. in 8 vols. (Leipzig, 1801–1806); and Otfried Wagenbreth, "Abraham Gottlob Werner und der Höhepunkt des Neptunistenstreits um 1790," in *Freiberger Forschungsheft*, ser. D, 11 (1955), 183–241.

Alexander Ospovat

WERNER, ALFRED (*b*. Mulhouse, France, 12 December 1866; *d*. Zurich, Switzerland, 15 November 1919), *chemistry.*

Life and Work. The founder of coordination chemistry, Werner was the fourth and last child of Jean-Adam Werner, an ironworker, and his second wife, Salomé Jeanette Tesché. Although the family decided to remain in Mulhouse after Alsace was annexed to the German Empire in 1871, they continued to speak French at home and their sympathies remained entirely with France. The spirit of rebellion and resistance to authority, so much a part of Werner's childhood, may well have contributed to the revolutionary and iconoclastic character of the theory with which his name is associated. Despite his great reverence for German science—most of his articles appeared in German journals—Werner's political and cultural ties were to France.

His mother had been converted from Protestantism to Catholicism, and at the age of six Werner was enrolled at the Catholic École Libre des Frères (Bruderschule), where the dominant traits of his personality—a remarkable self-confidence and a stubborn independence that made it impossible for him to submit blindly to authority—became evident. The religious teachings of the brothers apparently had little effect on him, for in later life his interest in religion was minimal. From 1878 to 1885 Werner attended the École Professionelle (Höhere Gewerbeschule), a technical school where he studied chemistry. During this time he built his own laboratory in the barn behind his house.

Even at this early stage Werner was preoccupied with classification, systematization, and isomeric relationships. His earliest known scientific work, "Contribution de l'acide urique, des séries de la théobromine, caféine, et leurs derivés," a holograph manuscript that he submitted in September 1885 to Emilio Noelting, director of the Mulhouse Chemie-Schule, was banal in style and unsound in its chemical thinking; but its broad scope and daring attempts at systematization foreshadowed the intellectual heights that Werner attained only a few years later. During 1885–1886 Werner served his year of compulsory military duty in the German army at Karlsruhe, where he audited courses in organic chemistry at the Technische Hochschule. He then entered the Polytechnikum in Zurich, where he studied under Arthur Hantzsch, Georg Lunge, Heinrich Goldschmidt, and Emil Constam.

Werner was a typical nonquantitative genius. At the Polytechnikum he failed his courses in mathematics, and throughout his career his contributions were essentially of a qualitative nature; even his celebrated conductivity studies with Arturo Miolati were only semiquantitative. His failure in descriptive geometry, however, is surprising inasmuch as his coordination theory represents an inspired and ingenious application of geometry to chemistry.

On 3 August 1889 Werner was awarded a degree in technical chemistry. During 1889 and 1890 he served as an unsalaried assistant in Lunge's chemical-technical laboratory while carrying out research under Hantzsch for which he received the doctorate on 13 October 1890.

In three short but eventful years (1890–1893) Werner produced his three most important theoretical papers. His doctoral dissertation, "Über räumliche Anordnung der Atome in stickstoffhaltigen Molekülen," was his first publication and remains his most popular and important work in organic chemistry. By extending the Le Bel and van't Hoff concept of the tetrahedral carbon atom (1874) to the nitrogen atom, Werner and Hantzsch simultaneously explained a great number of puzzling cases of geometrically isomeric trivalent nitrogen derivatives (oximes, azo compounds, hydroxamic acids) and for the first time placed the stereochemistry of nitrogen on a firm theoretical basis. Despite attacks by Victor Meyer, Karl von

Auwers, Eugen Bamberger, and others that extended several decades into the twentieth century, the Werner-Hantzsch theory has withstood the test of time. Today, with only slight modification, it takes its rightful place beside the Le Bel–van't Hoff concept of the tetrahedral carbon atom as one of the cornerstones of stereochemistry.

Werner spent the next two years working on his *Habilitationsschrift*, "Beiträge zur Theorie der Affinität und Valenz," in which he chose to attack the supreme patriarch of structural organic chemistry, August Kekulé. In this work Werner attempted to replace Kekulé's concept of rigidly directed valences with his own more flexible approach, in which he viewed affinity as a variously divisible, attractive force emanating from the center of an atom and acting equally in all directions. By the use of this new concept and without assuming directed valences, Werner was able to derive the accepted van't Hoff configurational formulas. Although this important paper contains the seeds that later flowered in the primary valence (*Hauptvalenz*) and secondary valence (*Nebenvalenz*) of the coordination theory, it deals exclusively with organic compounds. Unfortunately, it was published in a rather obscure journal of limited circulation, where it elicited little notice until brought to the attention of the scientific world in 1904 by a discussion of its concepts in Werner's first textbook.

During the winter semester of 1891–1892 Werner worked on thermochemical problems with Marcellin Berthelot at the Collège de France. Except for the publication of an admittedly minor work on a basic nitrate of calcium and the incorporation of thermochemical data into Werner's later lecture notes, this *Wanderjahr* had little effect on him. The acceptance of Werner's *Habilitationsschrift* by the Swiss authorities early in 1892 permitted him to return to Zurich as a *Privatdozent* at the Polytechnikum. He did not remain there long, for in the fall of 1893 he became associate professor as successor to Viktor Merz at the University of Zurich, where he remained for a quarter-century. In 1894 Werner married Emma Wilhelmine Giesker, a resident of Zurich, and became a Swiss citizen. The following year he was promoted to full professor. His appointment at the University of Zurich originally came about largely because of the almost overnight fame that he had received as a result of the publication of his most important theoretical paper, "Beitrag zur Konstitution anorganischer Verbindungen" (1893), in which he had proposed the basic postulates of his epochal and controversial coordination theory.

The circumstances surrounding the creation of the coordination theory provide a classic example of the "flash of genius" that ranks with Kekulé's dreams of the self-linking of carbon atoms (1858) and of the benzene ring (1865). At the time (late 1892 or early 1893) Werner was a comparatively unknown twenty-six-year-old *Privatdozent* whose primary interest was organic chemistry and whose knowledge of inorganic chemistry was extremely limited. Yet one morning he awoke at two with the solution to the riddle of "molecular compounds," which had come to him like a flash of lightning. He arose from his bed and wrote so quickly and steadily that by five that afternoon he had finished his most important paper.

For the next decade Werner's attention was divided between organic and inorganic chemistry. He had originally been called to the University of Zurich to teach organic chemistry, and it was not until the winter semester of 1902–1903 that he was finally assigned the main lecture course in inorganic chemistry, which he continued to teach along with organic chemistry throughout his career. Although he became increasingly preoccupied with coordination chemistry, more than one-quarter of his publications deal with such organic topics as oximes; hydroxamic and hydroximic acids; phenanthrenes; carboxonium and carbothionium salts; hydroxylamines; azo, azoxy, hydrazo, and nitro compounds; dyestuffs; and the Walden inversion.

Nevertheless, Werner's fame is securely grounded in inorganic chemistry. He began with a study of metal-ammines, hydrates, and double salts; but his ideas soon encompassed almost the whole of systematic inorganic chemistry and even found application in organic chemistry. He was the first to show that stereochemistry is a general phenomenon and is not limited to carbon compounds, and his views of valence and chemical bonding stimulated subsequent research on these fundamental topics.

The coordination theory, with its concepts of coordination number, primary and secondary valence, addition and intercalation compounds, and octahedral, square planar, and tetrahedral configurations, not only provided a logical explanation for known "molecular compounds" but also predicted series of unknown compounds, the eventual discovery of which lent further weight to Werner's controversial ideas. Werner recognized and named many types of inorganic isomerism: coordination, polymerization, ionization, hydrate, salt, coordination position, and valence isomerism. He also postulated explanations for polynuclear complexes, hydrated metal ions, hydrolysis, and acids and bases.

The average chemist probably has become familiar with Werner's views more through his books than through his journal articles. His first, *Lehrbuch der Stereochemie* (1904), never achieved the popularity of his second, *Neuere Anschauungen auf dem Gebiete der anorganischen Chemie* (1905), which went through five editions. As Werner's fame grew and the value of his views became recognized, he received a number of offers from Continental universities, all of which he declined. Honorary memberships and degrees were extended by many European and American universities and scientific societies. In 1913 he became the first Swiss to be awarded the Nobel Prize in chemistry, "in recognition of his work on the linkage of atoms in molecules, by which he has thrown fresh light on old problems and opened new fields of research, particularly in inorganic chemistry." Soon afterward he began to show the signs of a chronic degenerative disease (arteriosclerosis of the brain, aggravated by excessive drinking) that progressively destroyed his physical health and mental faculties. On 15 October 1919 he was forced to resign from his laboratory and teaching duties. Exactly one month later he died after prolonged suffering.

Today, when the practical and theoretical significance of coordination compounds is unquestioned, it is clear that the foundations of modern structural inorganic chemistry were erected by Werner, who has justly been called the inorganic Kekulé.

Coordination Theory. Although Werner was awarded the Nobel Prize in 1913 specifically for his monumental work on coordination compounds, the implications and applications of his research extend far beyond the confines of inorganic chemistry. In fact, they have been of inestimable value in biochemistry and in analytical, organic, and physical chemistry, as well as in such related sciences as mineralogy and crystallography. Even before Werner began his extensive series of experimental researches on "molecular compounds," an almost unprecedented tour de force requiring a quarter of a century, he was vitally concerned with one of the most basic problems of chemistry—the nature of affinity and valence. "Molecular compounds" provided him with a challenging and exciting means to explore this question.

It may come as a surprise that the Kekulé valence theory, so flexible and fruitful in organic chemistry, proved to be a virtual straitjacket when applied to inorganic chemistry. Yet, by his own admission, Kekulé's concept of constant valence proved "embarrassing to the chemist." Instead of abandoning this obviously untenable belief, however, he compounded the error by invoking a still more unsatisfactory concept, that of "molecular compounds," in order to maintain it.

An example or two will suffice to illustrate Kekulé's concept of "molecular compounds." Since he regarded the valences of nitrogen and phosphorus as invariably three, Kekulé was forced to consider ammonium chloride and phosphorus pentachloride as "molecular compounds" with the formulas $NH_3 \cdot HCl$ and $PCl_3 \cdot Cl_2$, respectively. At most, Kekulé's artificial division of compounds into "molecular" and "valence" compounds on the basis of their amenability or nonamenability to the doctrine of constant valence had some limited value as a formal classification, but it in no way explained the nature or operation of the forces involved in the formation of "molecular compounds" by the combination of "valence compounds."

Whereas Kekulé disposed of metal-ammines by banishing them to the limbo of "molecular compounds," other chemists developed highly elaborate theories in order to explain the constitution and properties of these intriguing substances. Probably the most successful and widely accepted of such theories was the one proposed in 1869 by Christian Wilhelm Blomstrand, professor of chemistry at the University of Lund. This "chain theory" was subsequently modified during the 1880's

and 1890's by the chemist destined to become Werner's principal scientific adversary, Sophus Mads Jørgensen, professor of chemistry at the University of Copenhagen.

Under the predominant influence of organic chemistry during the latter half of the nineteenth century, Blomstrand suggested that ammonia molecules could link together as $—NH_3—$ chains in a manner analogous to $—CH_2—$ chains in hydrocarbons. Provision was also made for the observed differences in reactivities of various atoms and groups in metal-ammines. For example, halogen atoms that could not be precipitated immediately by silver nitrate were regarded as bonded directly to the metal atom, while those that could be precipitated were considered to be bonded through the ammonia chains. Despite the theory's admitted limitations, a considerable amount of empirical data could be correlated by its use.

In his revolutionary theory, which marked an abrupt break with the classical theories of valence and structure, Werner postulated two types of valence—primary or ionizable (*Hauptvalenz*) and secondary or nonionizable (*Nebenvalenz*). According to the theory, every metal in a particular oxidation state (primary valence) has a definite coordination number—that is, a fixed number of secondary valences that must be satisfied. Whereas primary valences can be satisfied only by anions, secondary valences can be satisfied not only by

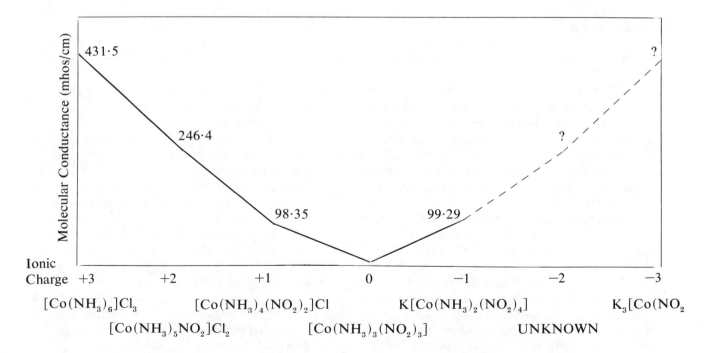

FIGURE 1.   Conductivities of cobalt(III) coordination compounds (Werner and Miolati, 1894).

TABLE I

Class of Compound	BLOMSTRAND-JØRGENSEN		WERNER	
	Formula	No. of Ions	Formula	No. of Ions
Hexammines MA$_6$	Co$\diagdown$NH$_3$—NO$_2$ / NH$_3$—NO$_2$ / NH$_3$—NH$_3$—NH$_3$—NH$_3$—NO$_2$	4	[Co(NH$_3$)$_6$](NO$_2$)$_3$	4
	↓ —NH$_3$		↓ —NH$_3$	
Pentammines MA$_5$B	Co$\diagdown$NO$_2$ / NH$_3$—NO$_2$ / NH$_3$—NH$_3$—NH$_3$—NH$_3$—NO$_2$	3	[Co(NH$_3$)$_5$NO$_2$](NO$_2$)$_2$	3
	↓ —NH$_3$		↓ —NH$_3$	
Tetrammines MA$_4$B$_2$	Co$\diagdown$NO$_2$ / NO$_2$ / NH$_3$—NH$_3$—NH$_3$—NH$_3$—NO$_2$	2	[Co(NH$_3$)$_4$(NO$_2$)$_2$]NO$_2$	2
	↓ —NH$_3$		↓ —NH$_3$	
Triammines MA$_3$B$_3$	Co$\diagdown$NO$_2$ / NO$_2$ / NH$_3$—NH$_3$—NH$_3$—NO$_2$	2	[Co(NH$_3$)$_3$(NO$_2$)$_3$]	0
			↓ —NH$_3$	
Diammines MA$_2$B$_4$	Unaccountable	—	K[Co(NH$_3$)$_2$(NO$_2$)$_4$]	2
			↓ —NH$_3$	
Monoammines MAB$_5$	Unaccountable	—	Unknown for Cobalt	(3)
			↓ —NH$_3$	
Double Salts MB$_6$	Unaccountable	—	K$_3$[Co(NO$_2$)$_6$]	4

anions but also by neutral molecules such as ammonia, water, organic amines, sulfides, and phosphines. These secondary valences are directed in space around the central metal ion (octahedral for coordination number 6, square planar or tetrahedral for coordination number 4); and the aggregate forms a "complex," which should exist as a discrete unit in solution.

The acknowledged test of a scientific theory is its ability to explain known facts and to predict new ones. In examining the success of Werner's coordination theory in meeting these criteria, we shall consider two aspects of the metal-ammines: constitution (how the constituent atoms and groups are bonded) and configuration (the spatial arrangement of these atoms and groups). Although we shall confine ourselves primarily to compounds of coordination number 6 [cobalt (III)], we should bear in mind that Werner used similar arguments to prove the constitution and configuration for compounds of coordination number 4.

Werner's first published experimental work in support of his coordination theory was a study of conductivities, carried out during 1893–1896 in collaboration with Arturo Miolati. According to the new theory, the charge of a complex ion should be equal to the algebraic sum of the charges of the central metal ion and of the coordinated groups. Consequently, as neutral molecules of ammonia (A) in a metal-ammine (MA$_6$) are successively replaced by anions (B), the number of ions in the resulting compounds should progressively decrease until a nonelectrolyte is formed and then should increase as the complex becomes anionic.

Friedrich Kohlrausch's principle of the additivity of equivalent conductivities of salts (1879) pro-

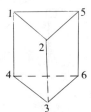

Trigonal Prismatic

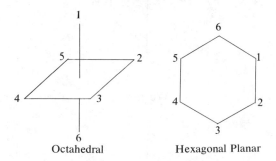

Octahedral   Hexagonal Planar

FIGURE 2. Configurational possibilities for coordination number 6.

vided Werner and Miolati with a convenient method for determining the number of ions in various complexes. After having established the ranges of conductivities to be expected for salts of various types, they were able to demonstrate the complete agreement in magnitude, variation, and pattern between their experimentally measured conductivities (Figure 1) and those predicted according to the coordination theory. Their results were also concordant with the number of precipitable halogen atoms. The constitutions and predicted numbers of ions according to the two theories are contrasted in Table I.

For compounds of the first three classes, the electrolytic character predicted by the two theories is in complete agreement, and conductivity data do not permit a choice between the two. For triammines, however, the ionic character differs radically according to the two theories; and the conductivities of these compounds became an important

and bitterly contested issue. For some nonelectrolytes, unfortunately, Werner and Miolati's conductivity values were not always zero because of aquation reactions:

$$[Co(NH_3)_3Cl_3]^0 + H_2O \rightleftarrows$$
$$[Co(NH_3)_3(H_2O)Cl_2]^+ + Cl^-.$$

Jørgensen immediately seized upon such "discrepancies" in an attempt to discredit their results. But in its explanation of anionic complexes and its demonstration of the existence of a continuous transition series (*Übergangsreihe*) between metal-ammines (MA_6) and double salts (MB_6), the Werner theory succeeded in an area in which the Blomstrand-Jørgensen theory could not pretend to compete.

The technique of "isomer counting" as a means of proving configuration admittedly did not originate with Werner. The idea of an octahedral configuration and its geometric consequences with respect to the number of isomers expected had been considered as early as 1875 by van't Hoff, and the general method probably is most familiar through Wilhelm Körner's work of 1874 on disubstituted and trisubstituted benzene derivatives. Yet the technique of comparing the number and type of isomers actually prepared with the number and type theoretically predicted for various configurations probably reached the height of its development with Werner's work. By this method he was able not only to discredit completely the rival Blomstrand-Jørgensen chain theory but also to demonstrate unequivocally that trivalent cobalt possesses an octahedral configuration rather than another possible symmetrical arrangement, such as hexagonal planar or trigonal prismatic. The method is summarized in Figure 2 and Table II.

In most cases the number and type of isomers prepared corresponded to the expectations for the octahedral arrangement, but there were a few exceptions; and Werner required more than twenty

TABLE II. Predicted Number of Isomers

Compound Type	Octahedral	Hexagonal Planar	Trigonal Prismatic*
MA_6	One	One	One
MA_5B	One	One	One
MA_4B_2	Two (1,2; 1,6)	Three (1,2; 1,3; 1,4)	Three (1,2; 1,3; 1,4)
MA_3B_3	Two (1,2,3; 1,2,6)	Three (1,2,3; 1,2,4; 1,3,5)	Three (1,2,3; 1,2,5; 1,2,6)
$M(\overline{AA})_3$	Two *Optical* Isomers	One	Two *Geometrical* Isomers

*Coordination compounds with this configuration have recently been synthesized.

Praseo

Co—NH₂—NH₂—NH₂—NH₂—Cl

(Jørgensen structure with Cl, CH₂—CH₂ groups)

Jørgensen

Violeo

Co—NH₂—NH₂—NH₂—NH₂—Cl

(Jørgensen violeo structure with Cl, CH₂—CH₂ groups)

Werner

(Werner octahedral structure with Co, Cl, NH₂, CH₂ groups, charge +, Cl⁻)

(Werner octahedral structure with Co, Cl, NH₂, CH₂ groups, charge +, Cl⁻)

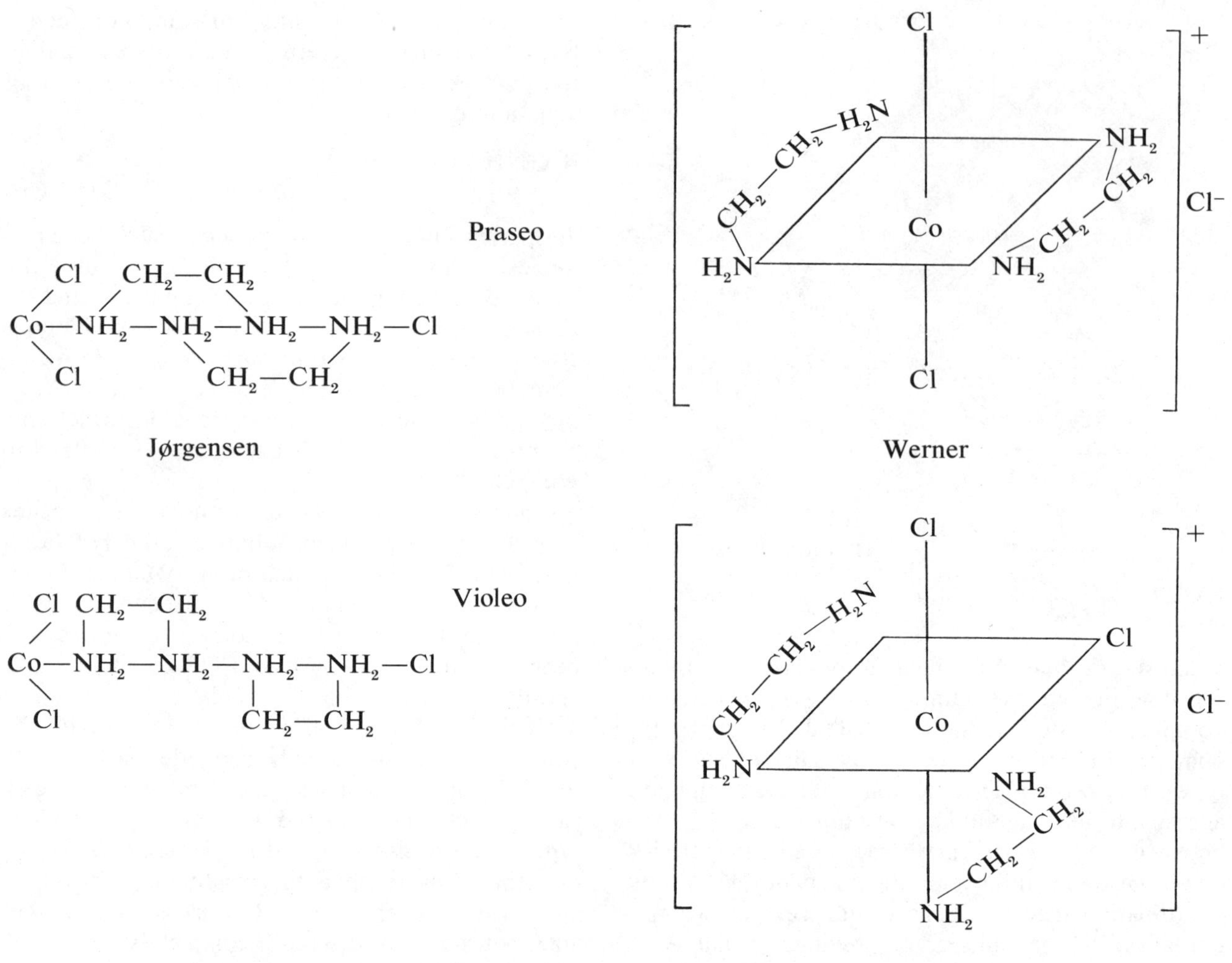

FIGURE 3. Jørgensen and Werner's formulas for praseo and violeo ethylenediamine isomers.

years to accumulate a definitive proof for his structural ideas. For example, the best known case of geometrical (cis-trans) isomerism was observed (by Jørgensen) not among simple tetrammines MA_4B_2 but among salts $M(\overline{AA})_2B_2$, in which the four ammonia molecules have been replaced by two molecules of the bidentate (chelate) organic base, ethylenediamine (en); that is, among the so-called praseo (green) and violeo (violet) series of formula $CoCl_3 \cdot 2en$. Jørgensen regarded the difference in color as due to structural isomerism connected with the linking of the two ethylenediamine molecules, whereas Werner regarded the compounds as stereoisomers, compounds composed of the same atoms and bonds but differing in the orientation of these atoms and bonds in space (Figure 3).

If this type of isomerism were merely a geometrical consequence of the octahedral structure, as Werner maintained, it should also be observed among simple tetrammines MA_4B_2, which do not contain ethylenediamine. Yet for compounds $[Co(NH_3)_4Cl_2]$ X, only one series (praseo) was known. Jørgensen, a confirmed empiricist, quite correctly criticized Werner's theory on the ground that it implied the existence of unknown compounds. It was not until 1907 that Werner succeeded in synthesizing the unstable, highly crucial violeo tetrammines, cis-$[Co(NH_3)_4Cl_2]$ X, which were a necessary consequence of his theory but not of Jørgensen's (Figure 4). His Danish opponent immediately conceded defeat.

Even though the discovery of the long-sought violeo salts convinced Jørgensen that his own views could not be correct, Werner's success in preparing two—and only two—isomers for compounds of types MA_4B_2 and MA_3B_3 was not sufficient for conclusive proof of his octahedral configuration. Despite such "negative" evidence, it could still be argued logically that failure to isolate a third isomer did not necessarily prove its non-

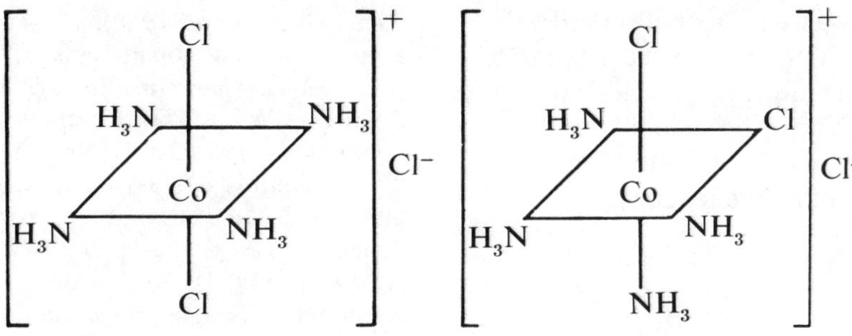

Praseo (trans; 1, 6)
Gibbs and Genth (1857)

Violeo (cis; 1, 2)
Werner (1907)

FIGURE 4.   Ammonia praseo and violeo isomers.

existence. A more "positive" type of proof was necessary.

As early as 1899, Werner recognized that the resolution into optical isomers of certain types of coordination compounds containing chelate groups, which can span only cis positions, could provide the "positive" proof that he needed. After many unsuccessful attempts, in 1911 he succeeded. His resolution, with his American student Victor King (1886–1958), of cis-chloroamminebis(ethylenediamine)cobalt(III) salts by means of the resolving agent silver d-α-bromocamphor-π-sulfonate was sufficient to prove conclusively the octahedral configuration for cobalt(III) (Figure 5). Yet because of the prevalent view that optical activity was almost always connected with carbon atoms, a number of Werner's contemporaries argued that the optical activity of these and the many other mononuclear and polynuclear coordination compounds subsequently resolved by him was somehow due to the organic chelate groups present, even though these symmetrical ligands were all optically inactive. Any vestige of doubt was finally dispelled by Werner's resolution in 1914 of completely carbon-free coordination compounds—the tris[tetrammine-μ-dihydroxocobalt(III)]cobalt(III) salts,

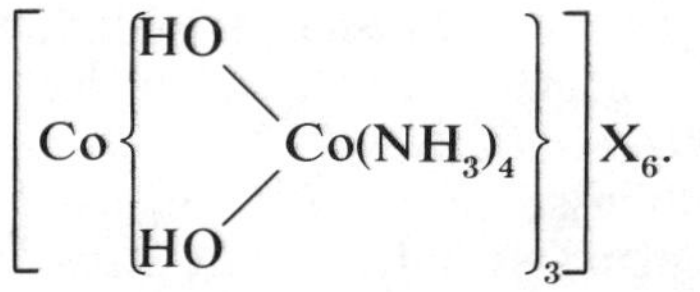

These salts are compounds of the $M(\overline{AA})_3$ type, in which $\overline{AA}$ is the inorganic bidentate ligand

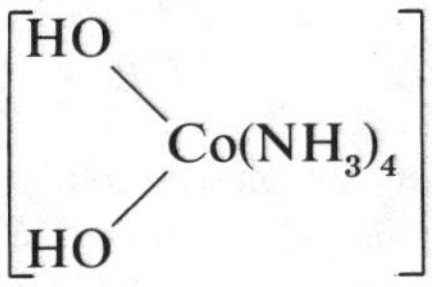

At the beginning of his career Werner had destroyed the monopoly of the carbon atom on geometrical isomerism. In his doctoral dissertation he had explained the isomerism of oximes as due to the tetrahedral configuration of the nitrogen atom. Now, at the peak of his career, he had likewise forced the tetrahedron to relinquish its claim to a monopoly on optical isomerism. One of the major goals of his lifework, the demonstration that stereochemistry is a general phenomenon not limited to carbon compounds and that no fundamental difference exists between organic and inorganic compounds, had been attained.

Finally, we must note that the validity of Wer-

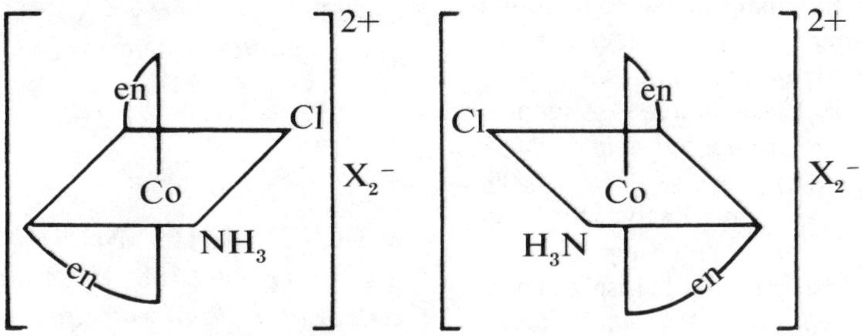

FIGURE 5.   Optical antipodes of cis-[Coen$_2$NH$_3$Cl]X$_2$.

271

ner's structural views was amply confirmed by X-ray diffraction studies. Yet, despite the advent of more direct modern techniques, his classical configurational determinations by simple indirect methods remain a monument to his intuitive vision, experimental skill, and inflexible tenacity.

BIBLIOGRAPHY

I. ORIGINAL WORKS. Werner's writings include "Über räumliche Anordnung der Atome in stickstoffhaltigen Molekülen," in *Berichte der Deutschen chemischen Gesellschaft,* **23** (1890), 11–30, English trans. in G. B. Kauffman, "Foundation of Nitrogen Stereochemistry: Alfred Werner's Inaugural Dissertation," in *Journal of Chemical Education,* **43** (1966), 155–165; "Beiträge zur Theorie der Affinität und Valenz," in *Vierteljahrsschrift der Naturforschenden Gesellschaft in Zürich,* **36** (1891), 129–169, discussed in G. B. Kauffman, "Alfred Werner's Habilitationsschrift," in *Chymia,* **12** (1967), 183–187, English trans. in G. B. Kauffman, "Contributions to the Theory of Affinity and Valence," *ibid.,* 189–216; "Sur un nitrate basique de calcium," in *Annales de chimie et de physique,* **27** (1892), 6th ser., 570–574, also in *Comptes rendus . . . de l'Académie des sciences,* **115** (1892), 169–171; "Beitrag zur Konstitution anorganischer Verbindungen," in *Zeitschrift für anorganische Chemie,* **3** (1893), 267–330, repr. as Ostwald's Klassiker der Exakten Wissenschaften no. 212 (Leipzig, 1924), English trans. in G. B. Kauffman, *Classics in Coordination Chemistry,* Part I, *The Selected Papers of Alfred Werner* (New York, 1968), 5–88; "Beiträge zur Konstitution anorganischer Verbindungen. I," in *Zeitschrift für physikalische Chemie,* **12** (1893), 35–55, "Beiträge . . . II," *ibid.,* **14** (1894), 506–521, and "Beiträge . . . III," *ibid.,* **21** (1896), 225–238—Italian trans. in *Gazzetta chimica italiana,* 2nd ser., **23** (1893), 140–165, **24** (1894), 408–427, and **27** (1896), 299–316, and English trans. of the first two papers in G. B. Kauffman, *Classics in Coordination Chemistry,* Part I (New York, 1968), 89–139; "Beitrag zur Konstitution anorganischer Verbindungen. XVII, Über Oxalatodiäthylendiaminkobaltisalze ($Co_{en_2}^{C_2O_4}$)x," in *Zeitschrift für anorganische Chemie,* **21** (1899), 145–158; *Lehrbuch der Stereochemie* (Jena, 1904)—Werner's views on structural organic chemistry may also be found in an extremely rare monograph by E. Bloch, *Alfred Werners Theorie des Kohlenstoffatoms und die Stereochemie der karbocyklischen Verbindungen* (Vienna–Leipzig, 1903); *Neuere Anschauungen auf dem Gebiete der anorganischen Chemie* (Brunswick, 1905, 1909, 1913, 1920, 1923), 2nd ed. trans. into English by E. P. Hedley as *New Ideas on Inorganic Chemistry* (London, 1911); "Über 1.2-Dichloro-tetrammin-kobaltisalze (Ammoniakvioleosalze)," in *Berichte der Deutschen chemischen Gesellschaft,* **40** (1907), 4817–4825, English trans. in G. B. Kauffman, *Classics in Coordination Chemistry,* Part I (New York, 1968), 141–154; "Zur Kenntnis des asymmetrischen Kobaltatoms. I," in *Berichte der Deutschen chemischen Gesellschaft,* **44** (1911), 1887–1898, English trans. in G. B. Kauffman, *Classics in Coordination Chemistry,* Part I (New York, 1968), 155–173; "Zur Kenntnis des asymmetrischen Kobaltatoms. XII. Über optische Aktivität bei kohlenstofffreien Verbindungen," in *Berichte der Deutschen chemischen Gesellschaft,* **47** (1914), 3087–3094, English trans. in G. B. Kauffman, *Classics in Coordination Chemistry,* Part I (New York, 1968), 175–184; and "Über die Konstitution und Konfiguration von Verbindungen höherer Ordnung," in *Les prix Nobel en 1913* (Stockholm, 1914), trans. into English as "On the Constitution and Configuration of Compounds of Higher Order," in *Nobel Lectures in Chemistry, 1901–1921* (Amsterdam, 1966), 256–269.

II. SECONDARY LITERATURE. A full-length biography by G. B. Kauffman, *Alfred Werner—Founder of Coordination Chemistry* (Berlin–Heidelberg–New York, 1966), deals primarily with Werner's life and career but also includes brief discussions of his work. G. B. Kauffman, *Classics in Coordination Chemistry,* Part I, *The Selected Papers of Alfred Werner* (New York, 1968), presents English translations of Werner's six most important papers together with critical commentary and biographical details. For other papers on various aspects of Werner and his work by G. B. Kauffman, see *Journal of Chemical Education,* **36** (1959), 521–527, and **43** (1966), 155–165, 677–679; *Chemistry,* **39** (1966), no. 12, 14–18; *Education in Chemistry,* **4** (1967), 11–18; *Chymia,* **12** (1967), 183–187, 189–216, 217–219, 221–232; *Naturwissenschaften,* **54** (1967), 573–576; and *Werner Centennial,* Advances in Chemistry series, no. 62 (Washington, D.C., 1967), 41–69. Articles, mostly obituaries, by others include P. Karrer, in *Helvetica chimica acta,* **3** (1920), 196–224, with bibliography of Werner's publications; G. T. Morgan, in *Journal of the Chemical Society* (London), **117** (1920), 1639–1648; P. Pfeiffer, in *Zeitschrift für angewandte Chemie und Zentralblatt für technische Chemie,* **33** (1920), 37–39; and in *Journal of Chemical Education,* **5** (1928), 1090–1098; and J. Lifschitz, in *Zeitschrift für Elektrochemie und angewandte physikalische Chemie,* **26** (1920), 514–529.

The figures and tables in this article are reprinted, by permission, from G. B. Kauffman, "Alfred Werner's Coordination Theory—A Brief Historical Introduction," in *Education in Chemistry,* **4** (1967), 11–18.

GEORGE B. KAUFFMAN

WERNER, JOHANN(ES) (*b.* Nuremberg, Germany, 14 February 1468; *d.* Nuremberg, May [?] 1522), *astronomy, mathematics, geography.*

While still a student in Nuremberg, Werner was drawn to the exact sciences and later said that he was intended for the study of mathematics from his early childhood. He enrolled at the University of Ingolstadt on 21 September 1484; and in 1490 he was appointed chaplain in Herzogenaurach. While studying in Rome (1493–1497) Werner was ordained a priest and met Italian scholars. By then his knowledge of mathematics, astronomy, and geography had increased; and he was allowed to inspect scientific manuscripts. He owned a Menelaus manuscript and was acquainted with unpublished works by Jābir ibn Aflaḥ (Geber) and Theodosius. Werner probably acquired his excellent knowledge of Greek in Italy. After his return to Nuremberg he celebrated his first mass in the church of St. Sebald on 29 April 1498. Probably in response to the requests of Empress Bianca Maria, in 1503 he was appointed priest at Wöhrd, just outside Nuremberg. In 1508 he was serving at St. Johannis Church in Nuremberg, where he remained until his death (between 12 March and 11 June 1522).

Werner was reputed to have been "very diligent" in carrying out all official responsibilities. Since his pastoral duties were rather limited, he devoted much of his time to scientific study. His works brought him recognition from such Nuremberg scholars as Willibald Pirkheimer (1470–1530), Sebald Schreyer (1446–1520), and Cardinal Matthäus Lang (1468–1540). He was friendly with Bernhard Walther (*ca.* 1430–1504) and the choirmaster Lorenz Beheim (1457[?]–1521) from Bamberg, as well as Albrecht Dürer, who occasionally asked his advice on mathematical problems. Werner enjoyed an excellent reputation even among scholars from Vienna: in 1514 the mathematician and imperial historiographer Johannes Stabius arranged the publication of a collection of writings on geography that included works by Werner. The humanist Konrad Celtis, whom Werner regarded as his "most beloved teacher," tried in 1503 to have Werner transferred to Vienna. Emperor Maximilian I appointed him chaplain at his court.

Not all of Werner's numerous works were published during his lifetime. Some remain unprinted, and others have been lost. Besides the 1514 collection containing Werner's and other authors' writings on geography, a collection of mathematical and astronomical works was published at Nuremberg in 1522. A handwritten remark in the Munich copy of the latter work leads us to believe that Werner died while the work was being printed. His meteorological treatise appeared after his death, and his works on spherical trigonometry and me-

teoroscopes were not published until 1907 and 1913, respectively.

Astronomy. In a sense Werner can be regarded as a student of Regiomontanus, for he had access to the latter's writings. Although a skilled maker of astronomical instruments, he showed less talent in theoretical work. The Germanisches Nationalmuseum in Nuremberg possesses a gold-plated brass astrolabe from 1516, probably made by Werner (see Zinner, pl. 25, 4). The clock on the south side of the parish church in Herzogenaurach and the two sundials in the choir of the church at Rosstal may be by Werner. He improved the Jacob's staff that had been used by Regiomontanus to measure interstellar distances, and he described it in the 1514 collection ("In eundem primum librum . . . Ptholomaei . . .," ch. 4, annotations 3–5).

Werner also invented an instrument that he called a "meteoroscope" to solve problems in spherical astronomy. It consists of a metal disk divided into quadrants with a pointer attached. Like the saphea, the first and fourth quadrants contain a stereographic projection of the circles of latitude and longitude, while the second and third quadrants have two different types of sine divisions. The device, which is only known from the description in "De meteoroscopiis," was built not for observational purposes but to replace as many mathematical tables as possible. Even here Werner proved to be a student of Regiomontanus, although his meteoroscope had nothing in common with the device of the same name built by his famous predecessor: Regiomontanus' instrument, the directions for use of which were published by Werner, is an armillary sphere. Werner's treatises on sundials and on astronomical and geographical problems that can be solved by methods of spherical trigonometry have been lost.

A manuscript dated 1521 concerning the making of a device designed to determine the latitudes of planets and one containing tables for the five planets are among Werner's unpublished works, as is a letter to Sebald Schreyer about the comet of 1500. Several horoscopes give evidence of Werner's work in astrology. They were cast for Ursula Gundelfinger, Erasmus Topler, Willibald Pirkheimer, Christoph Scheurl, and Sebald Schreyer.

Werner had less success with his treatises on the movement of the eighth sphere, which constitute the last section of the collection of works published in 1522. He maintains that the so-called precession of the stars would be an irregular movement, thus showing that he was a disciple of the Arab trepidation theory. In a letter to the canon of

Cracow cathedral, Bernhard Wapowski, Copernicus attacked the treatise vigorously; and Tycho Brahe criticized it by accusing Werner of having failed to observe accurately enough the three stars that Werner took as the basis of the movement of the eighth sphere.

Mathematics. Werner's mathematical works are in spherical trigonometry and the theory of conic sections. His principal work on spherical triangles was printed in 1907 from the copy in Codex Vaticanus Reginensis Latinus 1259 (fols. 1r–184r). A second copy of the autograph that contains figures was later discovered by Ernst Zinner (Landesbibliothek Weimar, no. f 324, fols. 1–103). Rheticus intended to publish the two writings, but only the letter of dedication to Ferdinand I of Bohemia and Hungary appeared (Cracow, 1557). The work, in four parts, which was written between 1505 and 1513, was not revised for publication by Werner. A fifth part for which he collected material has been lost. Although the treatise is incomplete, Werner's work was the best of its kind at the time, and its presentation surpassed that in Regiomontanus' books on triangles. In comparison with Regiomontanus' treatise, Werner's work is notable for its methodical presentation and practical applicability.

The various types of triangles are systematically described in part I. The following parts, which probably were written earlier, contain a theory of triangular calculation suitable for practical purposes. The basic formulas of spherical trigonometry and instructions for the solution of right spherical triangles are given in part II, and parts III and IV concern the calculation of oblique-angled triangles. In part IV Werner uses formulas that correspond to the cosine formula. Thus in proposition 5 he does not use the cosine formula as it is known today,

$$\cos b = \cos a \cos c + \sin a \sin c \cos B$$

(angle B is sought; a, b, c are the sides), but writes

$$\frac{\frac{1}{2}(\sin[90° - a + c] - \sin[90° - c - a])}{\sin(90° - b) - \sin(90° - c - a)} = \frac{r}{\sin \mathrm{vers}(180° - B)}.$$

This means that Werner implicitly used the formula

$$2 \sin a \sin c = \cos(a - c) - \cos(a + c),$$

whereby he could replace multiplication and division with addition and subtraction. This method, which later became known as prosthaphaeresis,

was first used by Werner, but mathematicians soon realized that it simplified calculation. Perhaps through Rheticus, Tycho Brahe learned of this procedure, which was used until the introduction of logarithms.

The treatise containing twenty-two theorems on conic sections was intended as an introduction to his work on duplication of the cube. For that reason Werner dealt only with the parabola and hyperbola but not with the ellipse. In a manner similar to the methods of Apollonius, Werner produced a cone by passing through the points of the circumference straight lines that also pass through a point not in the plane of the circle. In contrast to the ancients he did not consider the parabola and the hyperbola to be defined as plane curves but regarded them in connection with the cone by which they were formed. He proved the theorems on conic sections through geometrical observations on the cone.

Werner's report on duplication of the cube contained nothing new, being only a revision of the eleven solutions to this problem found in classical antiquity; they were known to Werner from the translation of the commentary by Eutocius on Archimedes prepared by Giorgio Valla. Werner added twelve supplementary notes to his treatise. The first ten dealt with the transformation of parallelepipeds and cylinders. In the eleventh note Werner proved that the sun's rays fall on the earth in parallel, and in the twelfth he showed that the rays are gathered in one point on a parabolic mirror.

The third writing in the collection of works dated 1522 also contained an Archimedean problem already treated by Eutocius, in which a sphere is to be cut by a plane so that the volumes of the two spherical sections are in a given proportion to each other (*De sphaera et cylindro* II, 4). Werner added his own solution, in which a parabola and hyperbola intersect each other, to those of Dionysodorus and Diocles.

Some mathematical works by Werner have been lost, including one on arithmetic, a work that apparently was influenced by Euclid's *Data*, and a translation of Euclid's *Elements* into German that Werner completed at the request of Pirkheimer and Sebald Beheim for the sum of 100 taler.

Geography and Meteorology. The collection dated 1514 contains Werner's works on mathematical geography. In the commentary on the first book of Ptolemy's *Geography*, Werner explains the basic concepts of spherical geography and then turns to the measurement of degrees on the sphere. When determining the declination of the sun, he refers to the tables compiled by Georg von Peuerbach and Domenico Maria. Werner's method is interesting in that it determines simultaneously the longitude and the latitude of a place (ch. 3, annotation 8): For the first time it was possible for two sites the locations of which are being sought to be found by a combined series of observations. Since for the determination of latitude it is necessary merely to observe the upper and lower culmination of a circumpolar star, but not the position of the sun, quite a few sources of errors were removed. The fourth chapter deals with the determination of the difference in longitude of two places, which can be obtained by simultaneous observation of a lunar eclipse. Another method is based on the determination of the distance of a zodiac star from the moon as seen from two places (ch. 4, annotation 8). This method of calculating the distances to the moon requires only the determination of the angular distances, which can be carried out by means of the Jacob's staff, and the precise knowledge of the true and mean motions of the moon. This method soon replaced the older ones and was then used as the principal method for determining longitude in nautical astronomy.

The methods used by Werner enabled him to improve or to explain certain details of the ancient geographers, especially those of Marinus. Werner's remarks in chapters 7–10 refer to Marinus' determination of places, which he proves to be often incorrect, or to the sea voyages mentioned and explained by Marinus. Werner demonstrated a knowledge of the existence and direction of the trade winds and explained their origin. In addition, he tried to present a theoretical proof of approximate formulas for the determination of distances that were used in navigation.

Werner's contributions to cartography are based on his criticism of Marinus; they can be found at the end of the commentary on Ptolemy and in the "Libellus quatuor terrarum orbis . . . ." The remarks on chapter 24 of the *Geography* lead us to believe that Werner understood the two projections used by Ptolemy (simple conic projection and modified spherical projection) and developed them. The treatise on four other projections of the terrestrial globe, which is dedicated to Pirkheimer, contains more new ideas. In it Werner outlines the principles of stereographic projection and emphasizes that any point on the surface of the sphere can be chosen as the center of projection. In addition, Werner develops three cordiform map projections that resemble one another; the second gives

an equal-area projection of the sphere. The idea of an equivalent projection occurred earlier in the works of Bernard Sylvanus, but Werner and Johannes Stabius were the first to work it out mathematically. Later, Oronce Fine, Peter Apian, and Gerardus Mercator adopted the cordiform projection. It is not known whether Werner designed a map of the world.

Werner's work in geography gained widespread recognition. Peter Apian, in particular, was a student of Werner's in theoretical cartography. The treatises contained in the collection dated 1514 were included almost unchanged in Apian's *Introductio geographica* (1533); Apian even used the proof sheets from the beginning of "In eundem primum librum . . . argumenta" to the end of "Joannis de Regiomonte epistola . . . de compositione et usu cuiusdam meteoroscopii," and admits in several places in his writings how much he had learned from Werner.

In meteorology Werner paved the way for a scientific interpretation. Meteorology and astrology were connected, but he nevertheless attempted to explain this science rationally. A short text on weather forecasting is still available in the manuscript "Regula aurea . . . ." The "guidelines that explain the principles and observations of the changes in the atmosphere," published in 1546 by Johann Schöner, contain meteorological notes for 1513–1520. The weather observations are based mainly on stellar constellations, and hence the course of the moon is of less importance. Although Werner did not collect the data systematically, as Tycho Brahe did, he attempted to incorporate meteorology into physics and to take into consideration the geographical situation of the observational site. Thus he can be regarded as a pioneer of modern meteorology and weather forecasting.

Other Works. The manuscript Codex Guelf. 17.6 Aug. 4° (Herzog-August Bibliothek, Wolfenbüttel) is an autograph in which words occasionally are crossed out and numerous addenda appear in the margin. The treatise gives an annal-like presentation of important events that occurred in Nuremberg between 1506 and 1521, most of which were political. On folios 41v and 70v two astronomical drawings that refer to Nuremberg are incorporated into the text.

BIBLIOGRAPHY

I. ORIGINAL WORKS. Writings published during Werner's lifetime were a collection on geography, *In hoc opere haec continentur . . . .* (Nuremberg, 1514), which contained as well as "Nova translatio primi libri geographiae Cl. Ptolomaei," "In eundem primum librum geographiae Cl. Ptolomaei argumenta, paraphrases, quibus idem liber per sententias ac summatim explicatur, et annotationes," "Libellus de quatuor terrarum orbis in plano figurationibus ab eodem I.V. novissime compertis et enarratis," "Ex fine septimi libri eiusdem geographiae Cl. Ptolomaei super plana terrarum orbis descriptione a priscis instituta geographis," "De his quae geographiae debent adesse Georgii Amirucii Constantinopolitani opusculum," "In idem Georgii Amirucii opusculum appendices," and "Ioannis de Regiomonte epistola ad reverendissimum patrem et dominum Bessarionem de compositione et usu cuiusdam meteoroscopii"—all re-edited by Peter Apian in his *Introductio geographica in doctissimas in Verneri annotationes* (Ingolstadt, 1533); and works on mathematics and astronomy, *In hoc opere haec continentur . . .* (Nuremberg, 1522), which contains "Libellus super vigintiduobus elementis conicis," "Commentarius seu paraphrastica enarratio in undecim modos conficiendi eius problematis quod cubi duplicatio dicitur," "Commentatio in Dionysodori problema, quo data sphaera plano sub data secatur ratione," "Alius modus idem problema conficiendi ab eodem I.V. novissime compertus demonstratusque," "De motu octavae sphaerae tractatus duo," and "Summaria enarratio theoricae motus octavae sphaerae."

Published after Werner's death were *Canones sicut brevissimi, ita etiam doctissimi, complectentes praecepta et observationes de mutatione aurae clarissimi mathematici Ioannis Verneri,* J. Schöner, ed. (Nuremberg, 1546); *Ioannis Verneri de triangulis sphaericis libri quatuor,* A. A. Björnbo, ed., XXIV. pt. 1, of the series Abhandlungen zur Geschichte der Mathematischen Wissenschaften mit Einschluss Ihrer Anwendungen (Leipzig, 1907); and *Ioannis Verneri de meteoroscopiis libri sex,* Joseph Würschmidt, ed., XXIV, pt. 2, of the same series (Leipzig, 1913). Some of Werner's letters are published in H. Rupprich, ed., *Der Briefwechsel des Konrad Celtis* (Munich, 1934).

The following writings remain in MS: "Tabulae latitudinum Saturni, Jovis, Martis, Veneris et Mercurii" (1521), Codex Oxoniensis Digby 132, fols. 1r–28v; "Compositiones et usus organorum latitudinum lunae et quinque planetarum" (1521), *ibid.,* fols. 29r–64r; "Judicium de cometa anni 1500 ad Sebaldum Clamosum alias Schreyer civem Nurembergensem" (1500), Vienna, Codex Vind. lat. 4756, fols. 143r–146v; two horoscopes, *ibid.,* 5002, fols. 104r–111v; "Regula aurea de aeris dispositione diiudicanda singulis diebus," *ibid.,* 5212, fols. 5v–6v; astrological writings, *ibid.,* 10534, fols. 248v, 260r; horoscopes (1498), *ibid.,* 10650, fols. 42r–87r; horoscope (1513), Munich, Clm 27083, fols. 1r–8v; horoscope, Paris, BN Reg. 7417; and "Historicus diarius inde ab anno 1506–1521 Johannis Verneri presbiteri Bambergensis diocesis et vicarii seu rectoris cappelle beatorum Johannis baptiste et Johannis evangeliste Norinbergensis," Wolfenbüttel, Codex Guelf. 17.6 Aug. 4°.

Lost works written before 1514 are "Liber de multimodis tam in astronomia quam in geographia problematis, quae ope arteque horum quinque librorum (Libri quinque de triangulis) absolvuntur"; "Opusculum de nonnullis scioteris, quibus linea meridiana, sublimitas axis mundani et hora diei sub omni climate per umbram solis simul examinantur"; "Tractatus resolutorius, qui prope pedissequus existit libris *Datorum* Euclidis"; and "Libellus arithmeticus, qui complectitur quaedam commenta numeralia." Also lost is a translation of Euclid's *Elements* into German.

II. SECONDARY LITERATURE. On Werner's life, see Johann Gabriel Doppelmayr, *Historische Nachricht von den Nürnbergischen Mathematicis und Künstlern* (Nuremberg, 1730), 31–35; Siegmund Günther, in *Allgemeine deutsche Biographie,* XLII (Leipzig, 1897), 56–58; Abraham Gotthilf Kästner, *Geschichte der Mathematik seit der Wiederherstellung der Wissenschaften bis ans Ende des 18. Jahrhunderts,* II (Göttingen, 1797; repr. Hildesheim–New York, 1970), 52–64; Hans Kressel, "Hans Werner. Der gelehrte Pfarrherr von St. Johannis. Der Freund und wissenschaftliche Lehrmeister Albrecht Dürers," in *Mitteilungen des Vereins für Geschichte der Stadt Nürnberg,* **52** (1963–1964), 287–304; Karl Schottenloher, "Der Mathematiker und Astronom Johann Werner aus Nürnberg. 1466 [*sic*]–1522," in *Festgabe an Hermann Grauert* (Freiburg im Breisgau, 1910), 147–155; and Ernst Zinner, "Die fränkische Sternkunde im 11, bis 16. Jahrhundert," in *Bericht der Naturforschenden Gesellschaft in Bamberg,* **27** (1934), 111–113. A biography of Werner is being prepared by Kurt Pilz as part of his work *600 Jahre Astronomie in Nürnberg.*

Werner's achievements in astronomy are discussed in Ernst Zinner, *Verzeichnis der astronomischen Handschriften des deutschen Kulturgebietes* (Munich, 1925), nos. 11641–11650; and *Deutsche und niederländische astronomische Instrumente des 11.–18. Jahrhunderts* (2nd ed., Munich, 1967), 148, 151 f., 208 f., 584.

Copernicus' criticism of Werner is presented in Maximilian Curtze, "Der Brief des Coppernicus an den Domherrn Wapowski zu Krakau über das Buch des Johannes Werner *De motu octavae sphaerae,*" in *Mittheilungen des Coppernicus-Vereins für Wissenschaft und Kunst, Thorn,* no. 1 (1878), 18–33; Siegmund Günther, "Der Wapowski-Brief des Coppernicus und Werners Tractat über die Praecession," *ibid.,* no. 2 (1880), 1–11; and *Three Copernican Treatises,* translated with intro. and notes by Edward Rosen (2nd ed., New York, 1959), 7–9, 91–106.

Brahe's criticism of Werner appears in *Tychonis Brahe opera omnia,* J. L. E. Dreyer, ed., VII (Copenhagen, 1924), 295, ll. 23–42.

Werner's achievements in mathematics are treated in Moritz Cantor, *Vorlesungen über Geschichte der Mathematik,* II, *Vom Jahre 1200 bis zum Jahre 1668* (2nd ed., Leipzig, 1900; repr. New York–Stuttgart, 1965), 452–459; C. J. Gerhardt, *Geschichte der Mathematik in Deutschland,* which is vol. XVII of Geschichte der

Wissenschaften in Deutschland (Munich, 1877; repr. New York – London, 1965), 23 – 25; and Johannes Tropfke, *Geschichte der Elementar-Mathematik in systematischer Darstellung*, 2nd ed., V (Berlin – Leipzig, 1923), 62, 107 – 110.

On the history of the text of Werner's trigonometry, see A. A. Björnbo, *Ioannis Verneri*, 140 – 175; E. Zinner, *Deutsche . . . Instrumente . . .*, 358; and Karl Heinz Burmeister, *Georg Joachim Rheticus 1514 – 1574. Eine Bio-Bibliographie* (Wiesbaden, 1967), I, 78, 134, 159 f., 181 f., and II, 78.

Werner's achievements in geography and meteorology are discussed in Siegmund Günther, *Studien zur Geschichte der mathematischen und physikalischen Geographie* (Halle, 1878), 273 – 332.

Menso Folkerts

WERNICKE, CARL (*b*. Tarnowitz [now Tarnowskie Gory], Upper Silesia, Germany [now Poland], 15 May 1848; *d*. Dörrberg im Geratal, Germany, 15 June 1905), *neuropsychiatry, neuroanatomy.*

After receiving a medical degree from the University of Breslau in 1870, Wernicke worked there under Heinrich Neumann, qualifying in psychiatry in 1875. During this period he spent six months in Vienna with Theodor Meynert and from 1876 to 1878 was Karl Westphal's assistant in the psychiatric and neurologic clinic at the Charité Hospital in Berlin. He then established a private neuropsychiatric practice in Berlin until 1885, when he became associate professor of neurology and psychiatry at Breslau. He obtained the chair at Breslau in 1890 and in 1904 went to Halle as full professor. He died a year later from injuries received in a bicycling accident.

Wernicke and his teachers Meynert and Westphal were part of the prominent tradition of nineteenth-century German neuropsychiatry stemming from Wilhelm Griesinger. These neuropsychiatrists made no distinction between diseases of the "mind" and diseases of the "brain." As Griesinger had announced, "Mental diseases are affections of the brain." Accordingly, Wernicke called his first major work, *Der aphasische Symptomencomplex* (1874), a "psychological study on an anatomical basis." In that monograph he developed his concept of cerebral localization, fully described sensory aphasia (Wernicke's aphasia) for the first time, and located the lesion causing this condition in the posterior portion of the first left temporal convolution. Wernicke's work also helped to establish firmly the notion of right and left cerebral dominance.

Between 1881 and 1883 Wernicke published his three-volume *Lehrbuch der Gehirnkrankheiten.* This comprehensive survey included a number of original anatomical, pathological, and clinical observations, such as the subsequently confirmed postulation of the symptoms resulting from occlusion of the posterior inferior cerebellar artery. Wernicke also first described (II, 229 – 242) a condition he called acute hemorrhagic superior polioencephalitis. Since renamed Wernicke's encephalopathy, the syndrome is manifested by particular disturbances of consciousness and gait and paralysis of the eye muscles. Wernicke's encephalopathy is frequently seen in chronic alcoholics, but it is also observed in other pathologic states. His original description was in a case of sulfuric acid ingestion.

In his later work Wernicke remained concerned both with brain anatomy and pathology and with clinical neuropsychiatry. With four collaborators he published the three-part *Atlas des Gehirns* between 1897 and 1903. His clinical studies are contained principally in *Grundriss der Psychiatrie in klinischen Vorlesungen* (1894; 2nd ed., 1906) and in *Krankenvorstellungen aus der psychiatrischen Klinik in Breslau* (1899 – 1900). Wernicke also continued to write on aphasia; his last summary of this "symptom complex" appeared in 1903 in the multivolume system of medicine edited by Ernst von Leyden and Felix Klemperer, *Die deutsche Klinik am Eingange des 20. Jahrhunderts.*

Wernicke excelled in careful neuropsychiatric description of patients, but he contended that psychiatry was not sufficiently developed to demarcate specific psychiatric syndromes. This attitude brought him into conflict with Emil Kraepelin, the great psychiatric nosologist of the period.

BIBLIOGRAPHY

I. Original Works. Most of Wernicke's major publications are mentioned in the text. Others include "Das Urwindungssystem des menschlichen Gehirns," in *Archiv für Psychiatrie*, **6** (1876), 298 – 326; "Ueber das Bewusstsein," in *Allgemeine Zeitschrift für Psychiatrie*, **35** (1879), 420 – 431; and "Die neueren Arbeiten über Aphasie," in *Fortschritte der Medizin*, **3** (1885), 824 – 830, and **4** (1886), 371 – 377, 463 – 482. Many of Wernicke's collected papers on neuropathology were published as *Gesammelte Aufsätze und kritische Referate zur Pathologie des Nervensystems* (Berlin, 1893).

Wernicke's 1903 summary of aphasia was translated into English in A. Church, ed., *Diseases of the Nervous System* (New York – London, 1908), 265 – 324.

II. Secondary Literature. The best general treatment of Wernicke is the article by Karl Kleist in Kurt

Kolle, ed., *Grosse Nervenärzte*, II (Stuttgart, 1959), 106–128. There are also short biographical essays by Kurt Goldstein in Webb Haymaker, ed., *Founders of Neurology* (Springfield, Ill., 1953), 406–409; and by H. Liepmann in Theodor Kirchhoff, ed., *Deutsche Irrenärzte*, II (Berlin, 1924). Other articles on Wernicke include H. Liepmann, "Über Wernickes Einfluss auf die klinische Psychiatrie," in *Monatschrifte für Psychiatrie und Neurologie*, **30** (1911), 1–37; P. Schröder, "Die Lehren Wernickes in ihrer Bedeutung für die heutige Psychiatrie," in *Zeitschrift für die gesamte Neurologie und Psychiatrie*, **165** (1939), 38–47; and O. M. Marx, "Nineteenth-Century Medical Psychology. Theoretical Problems in the Work of Griesinger, Meynert, and Wernicke," in *Isis*, **61** (1970), 355–370.

There is a good concise treatment of nineteenth-century German neuropsychiatry in Erwin H. Ackerknecht, *A Short History of Psychiatry* (2nd ed., New York, 1968), 60–71; and several references to Wernicke's work in Lawrence C. McHenry, Jr., *Garrison's History of Neurology* (Springfield, Ill., 1969), 358–361.

Wⁱˡˡⁱᵃᵐ F. Bʸⁿᵘᵐ

WILLIAM F. BYNUM

WERTHEIM, ERNST (*b*. Graz, Austria, 21 February 1864; *d*. Vienna, Austria, 15 February 1920), *gynecology*.

Son of the professor of chemistry at the University of Graz, Wertheim studied medicine there and received his medical degree on 29 February 1888, before becoming an assistant in the department of general and experimental pathology. Under the supervision of Rudolf Klemensiewicz he learned the bacteriological and histological techniques that later enabled him to conduct his fundamental research on gonorrhea in the female genital tract. The first result of his work at Graz, however, was a paper on fowl cholera.

Wertheim left Graz on 30 April 1889 and studied until mid-November under Otto Kahler at Vienna's Second University Clinic. In the same year he became interested in gynecology, the field to which he was to devote a lifetime of research. At the time both of Vienna's women's clinics possessed excellent facilities for specialized postgraduate training in obstetrics and gynecology, where pupils studied for two years at state expense. Upon leaving Kahler, Wertheim entered the institute attached to the Second Vienna Women's Clinic, which was headed by Rudolf Chrobak, and remained there as a student until 30 September 1890. He then moved to Prague as assistant to Friedrich Schauta at the university's women's clinic, returning to Vienna in 1891 when Schauta was appointed head of the First Vienna Women's Clinic.

In his years as an assistant, Wertheim focused his research on gonorrhea in the female genital tract. His training in experimental and bacteriological methods allowed him to give a unique, definitive explanation of the then much-disputed path of the gonorrheal infection. In 1890 he demonstrated the existence of gonococci in tissue of the Fallopian tubes; previously they had been detected only in smear samples. More important, in a series of papers containing the results of bacteriological and histological studies and of experiments on animals, Wertheim showed that the gonococcus affects not only the cylindrical epithelium—as Ernst Bumm maintained in his widely accepted theory (1895)—but also the squamous epithelium (the peritoneum). Through these papers, especially in "Die aszendierende Gonorrhoe beim Weibe. Bakteriologische und klinische Studien zur Biologie des *Gonococcus neisser*" (*Archiv für Gynäkologie*, **42** [1892], 1–86), Wertheim placed his theory of ascending gonorrhea in the female on a firm foundation.

Wertheim gained world fame for systematically developing—to the point where it became standard practice—a radical abdominal operation for cervical cancer. In 1897 he was named chief surgeon of the gynecological service (Bettina Pavilion) of the Elisabeth Hospital and thereby obtained his own operating facilities. By then he had long realized that the customary method of vaginal extirpation was highly unsatisfactory because of its bad aftereffects. (The abdominal method had been abandoned because of its mortality rate of 72 percent.) Consequently, elaborating the work of his predecessors Emil Ries, Theodor Rumpf, and J. G. Clark, Wertheim decided to place the operation for cervical cancer on a modern surgical basis. No longer satisfied with the extirpation of the diseased organ, he sought to remove as much as possible of the organ's surroundings, the perimetrium, along with the neighboring lymph glands. He recognized, as had Julius Massari in 1878, that the greatest possible attention must be directed to protecting the ureters during their exposure and preparation.

On 16 November 1898 Wertheim performed his first radical abdominal operation based on these principles. For his first twenty-nine operations he reported a mortality rate of 38 percent (*Archiv für Gynäkologie*, **61** [1900], 627–668). Tirelessly working to improve his surgical technique, he succeeded in reducing the mortality rate to 10 percent. This assured universal acclaim for his radical abdominal surgery. He gave a full account of his success in *Die erweiterte abdominale Operation bei Carcinoma colli uteri (auf Grund von 500 Fällen)* (Berlin–Vienna, 1911).

In 1910 Wertheim was appointed director of the First University Women's Clinic in Vienna. Firmly determined to replace unsatisfactory surgical procedures with improved ones, he returned his attention to the treatment of uterine prolapse. He had already made a contribution in this area in 1899 with his interposition method, which involved covering the uterus with vaginal lobes. He then developed a version of the operation that is known as suspension and superposition; his richly illustrated book on it appeared a year before his death as *Die operative Behandlung des Prolapses mittelst Interposition und Suspension des Uterus* (Berlin, 1919). Wertheim was as difficult in his personal relations as he was brilliant in his research. Held in great esteem by his colleagues, he was a corresponding or honorary member of many foreign learned societies. At Vienna he created a distinguished school of gynecological surgeons; and his work was carried on there, as well as in Prague and Berlin, by his most outstanding students, Wilhelm Weibel and Georg August Wagner.

BIBLIOGRAPHY

Many of Wertheim's works are mentioned in the text. See also the bibliography in the obituary by Weibel (below).

Secondary literature includes J. Artner and A. Schaller, *Die Wertheimsche Radikaloperation. Anfänge, Fortschritte, Ergebnisse. 1898–1968* (Vienna, 1968); R. Elert, "Zum Gedenkjahr des doppelten Jubiläums der abdominellen Radikaloperation des Gebärmutterkrebses. W. A. Freund (1878)–Wertheim (1898)," in *Klinische Medizin*, **4** (1949), 249–262, very detailed; F. Kermauner, in *Wiener klinische Wochenschrift*, **33** (1920), 183–185; W. Latzko, in *Wiener medizinische Wochenschrift*, **70** (1920), 545–549; E. Lesky, "Die Wiener geburtshilflich-gynäkologische Schule," in *Deutsche medizinische Wochenschrift*, **87** (1962), 2096–2102; and *Die Wiener medizinische Schule im 19. Jahrhundert* (Graz–Cologne, 1965); E. Navratil, "Die Entwicklung der Operationsmethoden zur Entfernung der karzinomatösen Gebärmutter," in *Wiener klinische Wochenschrift*, **60** (1948), 233–238; G. Reiffenstuhl, *ibid.*, **82** (1970), 554–559; W. Weibel, in *Monatsschrift für Geburtshülfe und Gynäkologie*, **61** (1920), 271–279, with bibliography, in *Archiv für Gynäkologie* (Berlin), **113** (1920), v–xvi; in *Zentralblatt für Gynäkologie*, **44** (1920), 281–285; and "25 Jahre 'Wertheimscher' Carcinomoperation," in *Archiv für Gynäkologie* (Berlin), **135** (1929), 1–57; P. Werner and J. Sederl, *Die Wertheimsche Radikaloperation bei Carcinoma colli uteri* (Vienna–Innsbruck, 1952); and T. Antoine, "Letter From Austria," in *Obstetrical and Gynecological Survey*, **24** (1969), 1129–1137.

ERNA LESKY

WESSEL, CASPAR (*b.* Vestby, near Dröbak, Norway, 8 June 1745; *d.* Copenhagen, Denmark, 25 March 1818), *surveying, mathematics.*

Although he was born when Norway was part of Denmark and spent most of his life in Denmark, Wessel is regarded as a Norwegian. (Niels Nielsen, in his *Matematiken i Danmark, 1528–1800* [Copenhagen–Christiania, 1912], gives his birthplace as Jonsrud in Akershus.) His father, Jonas Wessel, was a vicar in the parish of which his grandfather was the pastor; his mother was Maria Schumacher.

After attending the Christiania Cathedral School in Oslo from 1757 to 1763, Wessel spent a year at the University of Copenhagen. In 1764 he began work on the cartography of Denmark, as an assistant with the Danish Survey Commission operating under the Royal Danish Academy of Sciences. He passed the university examination in Roman law in 1778 and became survey superintendent in 1798. He continued as a surveyor and cartographer even after his retirement in 1805, working on special projects until rheumatism forced him to stop in 1812. He was frequently in financial difficulty; but since he would accept no remuneration for special maps of Schleswig and Holstein requested by the French government, the Royal Danish Academy awarded him a silver medal and a set of its *Mémoires* and maps. He was made a knight of Danebrog in 1815.

Wessel's fame as a mathematician is based entirely on one paper, written in Danish and published in the *Mémoires* of the Royal Danish Academy, that established his priority in publication of the geometric representation of complex numbers. John Wallis had given a geometric representation of the complex roots of quadratic equations in 1685; Gauss had had the idea as early as 1799 but did not explicitly publish it until 1831. Robert Argand's independent publication in 1806 must be credited as the source of this concept in modern mathematics because Wessel's work remained essentially unknown until 1895, when its significance was pointed out by Christian Juel. Despite its lack of influence upon the development of mathematics, Wessel's publication was remarkable in many ways: he was not a professional mathematician; Norway and Denmark were not mathematically productive or stimulating at that time; he was not a member of the Royal Danish Academy (he had been helped and encouraged by J. N. Tetens, councillor of state and president of the science section of the Academy); and yet his exposition was, in some respects, superior to and more modern in spirit than Argand's.

The title of Wessel's treatise calls it an "attempt" to give an analytic representation of both distance and direction that could be used to solve plane and spherical polygons. The connection of this goal with Wessel's work as a surveyor and cartographer is obvious. The statement of the problem also suggests that Wessel should be credited with an early formulation of vector addition. In fact, Michael J. Crowe, in *A History of Vector Analysis* (University of Notre Dame Press, 1967), defines the first period in that history as that of a search for hypercomplex numbers to be used in space analysis and dates it from the time of Wessel, whom he calls the first to add vectors in three-dimensional space. Wessel's first step was to note that two segments of the same line, whether of the same or opposite sense, are added by placing the initial point of one at the terminus of the other and defining the sum to be the segment extending from the initial point of the first to the terminal point of the second. He immediately defined the sum of two nonparallel segments in the same way and extended this definition to apply to any number of segments.

For multiplication of line segments, Wessel drew his motivation from the fact that, in arithmetic, a product of two factors has the same ratio to one factor as the other factor has to 1. Assuming that the product and the two factors are in the same plane and have the same initial point as the unit segment, he reasoned that the product vector should differ in direction from one factor by the same amount by which the other factor differed from unity. Wessel then designated two oppositely directed unit segments having the same origin by $+1$ and -1, and assigned to them the direction angles $0°$ and $180°$. To unit segments perpendicular to these he assigned the symbols $+\epsilon$ and $-\epsilon$ and the angles $90°$ and either $270°$ or $-90°$. Wessel immediately pointed out that multiplication of these numbers corresponded to addition of their angles and gave a table in which $(+\epsilon) \cdot (+\epsilon) = -1$. He then noted that this means that $\epsilon = \sqrt{-1}$, and that these operations do not contradict the ordinary rules of algebra. From this and his definition of addition of vectors, Wessel next wrote $\cos v + \epsilon \sin v$ as the algebraic formula for a unit segment and then derived the algebraic formula $(a + \epsilon b) \cdot (c + \epsilon d) = ac - bd + \epsilon(ad + bc)$ for the product of any two segments from the formula for the product of two unit segments derived from the equation $(\cos v + \epsilon \sin v)(\cos u + \epsilon \sin u) = \cos(u + v) + \epsilon \sin (u + v)$, using trigonometric identities.

Thus Wessel's development proceeded rather directly from a geometric problem, through geometric-intuitive reasoning, to an algebraic formula. Argand began with algebraic quantities and sought a geometric representation for them.

Wessel was more modern than Argand in his recognition of the arbitrary nature of the definitions of operations that Argand initially attempted to justify by intuitive arguments. Wessel also sought to extend his definition of multiplication to lines in space. T. N. Thiele's view that Wessel should be credited with anticipating Hamilton's formulation of quaternion multiplication, however, seems to exaggerate the extent of his work, which Wessel himself recognized as incomplete.

Wessel used in his development trigonometric identities that Argand derived by means of his definitions of operations on complex numbers. Argand presented a greater variety of applications of his work, including a proof of the fundamental theorem of algebra and of Ptolemy's theorem. Wessel worked at his original problem of applying algebra to the solution of plane and spherical polygons, after expanding his discussion to include formulas for division, powers, and roots of complex numbers.

Wessel's initial formulation was remarkably clear, direct, concise, and modern. It is regrettable that it was not appreciated for nearly a century and hence did not have the influence it merited.

BIBLIOGRAPHY

I. ORIGINAL WORKS. Wessel's paper "Om directionens analytiske betegning, et forsøg, anvendt fornemmelig til plane og sphaeriske polygoners opløsning," read to the Royal Danish Academy of Sciences on 10 Mar. 1797, was printed in 1798 by J. R. Thiele and was incorporated in *Nye samling af det Kongelige Danske Videnskabernes Selskabs Skrifter*, 2nd ser., **5** (1799), 496–518. Almost a century later the Academy published a French trans., *Essai sur la représentation analytique de la direction, par Caspar Wessel* (Copenhagen, 1897), with prefaces by H. Valentiner and T. N. Thiele.

In the meantime, Sophus Lie published a reproduction in *Archiv for mathematik og naturvidenskab*, **18** (1896). The English trans. by Martin A. Nordgaard of portions of the original paper first appeared in D. E. Smith, ed., *A Source Book in Mathematics* (New York, 1929), 55–66, repr. in H. O. Midonick, ed., *The Treasury of Mathematics* (New York, 1965), 805–814.

II. SECONDARY LITERATURE. The most complete discussions are Viggo Brun, *Regnekunsten i det gamle Norge* (Oslo, 1962), 92–111, with English summary on 120–122; and "Caspar Wessel et l'introduction géométrique des nombres complexes," in *Revue d'histoire des sciences et de leurs applications*, **12** (Jan.-Mar. 1959), 19–24; and Webster Woodruff Beman, "A Chapter in the History of Mathematics," in *Proceedings of the American Association for the Advancement of Science*, **46** (1897), 33–50, which includes a survey of the development of the concept of a complex number with especial emphasis on graphical representation, particularly the work of Wessel—French trans. in *Enseignement mathématique*, **1** (1899), 162–184.

Much information on Wessel's cartographic and surveying activity is in Otto Harms, "Die amtliche Topographie in Oldenburg und ihre kartographischen Ergebnisse," in *Oldenburger Jahrbuch*, **60**, pt. 1 (1961), 1–38.

PHILLIP S. JONES

WEYL, HERMANN (*b*. Elmshorn, near Hamburg, Germany, 9 November 1885; *d*. Zurich, Switzerland, 8 December 1955), *mathematics, mathematical physics.*

Weyl attended the Gymnasium at Altona and, on the recommendation of the headmaster of his Gymnasium, who was a cousin of Hilbert, decided at the age of eighteen to enter the University of Göttingen. Except for one year at Munich he remained at Göttingen, as a student and later as *Privatdozent*, until 1913, when he became professor at the University of Zurich. After Klein's retirement in 1913, Weyl declined an offer to be his successor at Göttingen but accepted a second offer in 1930, after Hilbert had retired. In 1933 he decided he could no longer remain in Nazi Germany and accepted a position at the Institute for Advanced Study at Princeton, where he worked until his retirement in 1951. In the last years of his life he divided his time between Zurich and Princeton.

Weyl undoubtedly was the most gifted of Hilbert's students. Hilbert's thought dominated the first part of his mathematical career; and although later he sharply diverged from his master, particularly on questions related to foundations of mathematics, Weyl always shared his convictions that the value of abstract theories lies in their success

in solving classical problems and that the proper way to approach a question is through a deep analysis of the concepts it involves rather than by blind computations.

Weyl arrived at Göttingen during the period when Hilbert was creating the spectral theory of self-adjoint operators, and spectral theory and harmonic analysis were central in his mathematical research throughout his life. Very soon, however, he considerably broadened the range of his interests, including areas of mathematics into which Hilbert had never penetrated, such as the theory of Lie groups and the analytic theory of numbers, thereby becoming one of the most universal mathematicians of his generation. He also had an important role in the development of mathematical physics, the field to which his most famous books, *Raum, Zeit und Materie* (1918), on the theory of relativity, and *Gruppentheorie und Quantenmechanik* (1928), are devoted.

Weyl's first important work in spectral theory was his *Habilitationsschrift* (1910), on singular boundary conditions for second-order linear differential equations. The classical Sturm-Liouville problem consists in determining solutions of a self-adjoint differential equation

$$(1) \qquad (py')' - (q - \lambda)y = 0$$

in a compact interval $0 \leq x \leq l$, with $p(x) > 0$ and q real in that interval, the solutions being subject to boundary conditions

$$(2) \qquad y'(0) - wy(0) = 0$$

$$(3) \qquad y'(l) - hy(l) = 0$$

with real numbers w, h; it is known that nontrivial solutions exist only when λ takes one of an increasing sequence (λ_n) of real numbers ≥ 0 and tending to $+\infty$ (the spectrum of the equation). Weyl investigated the case in which $l = +\infty$; his idea was to give arbitrary complex values to λ. Then, for given real h, there is a unique solution satisfying (2) and (3), provided w is taken as a complex number $w(\lambda, h)$. When h takes all real values, the points $w(\lambda, h)$ are on a circle $C_l(\lambda)$ in the complex plane. Weyl also showed that when l tends to $+\infty$, the circles $C_l(\lambda)$ (for fixed λ) form a nested family that has a circle or a point as a limit. The distinction between the two cases is independent of the choice of λ, for in the "limit circle" case all solutions of (1) are square-integrable on $[0, +\infty]$, whereas in the "limit point" case only one solution (up to a constant factor) has that property. This was actually the first example of the general theory of defects of an unbounded Hermitian operator, which was created later by von Neumann. Weyl also showed how the classical Fourier series development of a function in a series of multiples of the eigenfunctions of the Sturm-Liouville problem was replaced, when $l - +\infty$, by an expression similar to the Fourier integral (the spectrum then being generally a nondiscrete subset of **R**); he thus anticipated the later developments of the Carleman integral operators and their applications to differential linear equations of arbitrary order and to elliptic linear partial differential equations.

In 1911 Weyl inaugurated another important chapter of spectral theory, the asymptotic study of the eigenvalues of a self-adjoint compact operator U in Hilbert space H, with special attention to applications to the theory of elasticity. For this purpose he introduced the "maximinimal" method for the direct computation of the nth eigenvalue λ_n of U (former methods gave the value of λ_n only after those of $\lambda_1, \lambda_2, \cdots, \lambda_{n-1}$ had been determined). One considers an arbitrary linear subspace F of codimension $n - 1$ in H and the largest value of the scalar product $(U \cdot x \mid x)$ when x takes all values on the intersection of F and the unit sphere $\|x\| = 1$ of H; λ_n is the smallest of these largest values when F is allowed to run through all subspaces of codimension 1. This method has a very intuitive geometric interpretation in the theory of quadrics when H is finite-dimensional; it was used with great efficiency in many problems of functional analysis by Weyl himself and later by Richard Courant, who did much to popularize it and greatly extend its range of applications.

Weyl published the famous paper on equidistribution modulo 1, one of the highlights of his career, in 1916. A sequence (x_n) of real numbers is equidistributed modulo 1 if for any interval $[\alpha, \beta]$ contained in $[0,1]$, the number $\nu(\alpha, \beta; n)$ of elements x_k such that $k \leq n$ and $x_k = N_k + y_k$, with $\alpha \leq y_k \leq \beta$ and N_k a (positive or negative) integer, is such that $\nu(\alpha, \beta; n)/n$ tends to the length $\beta - \alpha$ of the interval when n tends to $+\infty$. Led to such questions by his previous work on the Gibbs phenomenon for series of spherical harmonics, Weyl approached the problem by a completely new — and amazingly simple — method. For any sequence (y_n) of real numbers, write $M((y_n))$ the limit (when it exists) of the arithmetic mean $(y_1 + \cdots + y_n)/n$ when n tends to $+\infty$; then to say that (x_n) is equidistributed means that $M((f(x_n))) = \int_0^1 f(t)dt$ for any function of period 1 coinciding on $[0,1]$ with the characteristic function of any interval $[\alpha, \beta]$. Weyl's familiarity with harmonic analysis enabled him to conclude (1) that this property was equivalent to the existence of $M((f(x_n)))$ for all Riemann integrable functions of period 1 and (2) that it was enough to check the existence of that limit for the particular functions $\exp(2\pi ikx)$ for any integer $k \in Z$. This simple criterion immediately yields the equidistribution of the sequence $(n\alpha)$ for irrational α (proved independently a little earlier by Weyl, Bohl, and Wacław Sierpiński by purely arithmetic methods), as well as a quantitative form of the Kronecker theorems on simultaneous Diophantine approximations.

Weyl's most profound result was the proof of the equidistribution of the sequence $(P(n))$, where P is a polynomial of arbitrary degree, the leading coefficient of which is irrational; this amounts to showing that

$$s_N = \sum_{n=0}^{N} \exp(2\pi i P(n)) = o(N)$$

for N tending to $+\infty$. To give an idea of Weyl's ingenious proof, consider the case when $P(n) = \alpha n^2 + \beta n$ with irrational α. One writes

$$|s_N|^2 = s_N \bar{s}_N = \sum_{m,n=0}^{N} \exp(2\pi i (\alpha(m^2 - n^2) + \beta(m - n)))$$

$$= \sum_{r=-N}^{N} \sigma_r \cdot \exp(2\pi i (\alpha r^2 + \beta r)),$$

where $\sigma_r = \sum_{n \in I_r} \exp(2\pi i \alpha rn)$, I_r being the interval intersection of $[0,N]$ and $[-r, N - r]$ in **Z**. This yields $|s_N|^2 \leq \sum_{r=-N}^{N} |\sigma_r|$. One has two majorations, $|\sigma_r| \leq N + 1$ and $|\sigma_r| \leq 1/\sin(2\pi \alpha r)$. For a given $\epsilon \in [0, \frac{1}{2}]$, the number of integers $r \in [-N,N]$ such that $2\alpha r$ is congruent to a number in the interval $[-\epsilon, \epsilon]$ has the form $4\pi \epsilon N + o(N)$ by equidistribution; hence it is $\leq 5\epsilon N$ for large N. Applying to these integers r the first majoration, and the second to the others, one obtains

$$|s_N|^2 \leq 5\epsilon N(N + 1) + (2N + 1) / \sin(\pi \epsilon) \leq 6\epsilon N^2$$

for large N, thus proving the theorem. The extension of that idea to polynomials of higher degree d is not done by induction on d, but by a more elaborate device using the equidistribution of a multilinear function of d variables. Weyl's results, through the improvements made later by I. M. Vinogradov and his school, have remained fundamental tools in the application of the Hardy-Littlewood method in the additive theory of numbers.

Weyl's versatility is illustrated in a particularly

striking way by the fact that immediately after these original advances in number theory (which he obtained in 1914), he spent more than ten years as a geometer—a geometer in the most modern sense of the word, uniting in his methods topology, algebra, analysis, and geometry in a display of dazzling virtuosity and uncommon depth reminiscent of Riemann. His familiarity with geometry and topology had been acquired a few years earlier when, as a young *Privatdozent* at Göttingen, he had given a course on Riemann's theory of functions; but instead of following his predecessors in their constant appeal to "intuition" for the definition and properties of Riemann surfaces, he set out to give to their theory the same kind of axiomatic and rigorous treatment that Hilbert had given to Euclidean geometry. Using Hilbert's idea of defining neighborhoods by a system of axioms, and influenced by Brouwer's clever application of Poincaré's simplicial methods (which had just been published), he gave the first rigorous definition of a complex manifold of dimension 1 and a thorough treatment (without any appeal to intuition) of all questions regarding orientation, homology, and fundamental groups of these manifolds. *Die Idee der Riemannschen Fläche* (1913) immediately became a classic and inspired all later developments of the theory of differential and complex manifolds.

The first geometric problem that Weyl attempted to solve (1915) was directly inspired by Hilbert's previous work on the rigidity of convex surfaces. Hilbert had shown how the "mixed volumes" considered by Minkowski could be expressed in terms of a second-order elliptic differential linear operator L_H attached to the "Stützfunktion" H of a given convex body; Blaschke had observed that this operator was the one that intervened in the theory of infinitesimal deformation of surfaces, and this knowledge had enabled Hilbert to deduce from his results that such infinitesimal deformations for a convex body could only be Euclidean isometries. Weyl attempted to prove that not only infinitesimal deformations, but finite deformations of a convex surface as well, were necessarily Euclidean isometries. His very original idea, directly inspired by his work on two-dimensional "abstract" Riemannian manifolds, was to prove simultaneously this uniqueness property and an existence statement, namely that any two-dimensional Riemannian compact manifold with everywhere positive curvature was uniquely (up to isometries) imbeddable in Euclidean three-dimensional space. The bold method he proposed for the proof was to proceed by

continuity, starting from the fact that (by another result of Hilbert's) the problem of existence and uniqueness was already solved for the ds^2 of the sphere, and using a family of ds^2 depending continuously on a real parameter linking the given ds^2 to that of the sphere and having all positive curvature. This led him to a "functional differential equation" that he did not completely solve, but later work by L. Nirenberg showed that a complete proof of the theorem could be obtained along these lines.

Interrupted in this work by mobilization into the German army, Weyl did not resume it when he was allowed to return to civilian life in 1916. At Zurich he had worked with Einstein for one year, and he became keenly interested in the general theory of relativity, which had just been published; with his characteristic enthusiasm he devoted most of the next five years to exploring the mathematical framework of the theory. In these investigations Weyl introduced the concept of what is now called a linear connection, linked not to the Lorentz group of orthogonal transformations of a quadratic form of signature (1, 3) but to the enlarged group of similitudes (reproducing the quadratic form only up to a factor); he even thought for a time that this would give him a "unitary theory" encompassing both gravitation and electromagnetism. Although these hopes did not materialize, Weyl's ideas undoubtedly were the source from which E. Cartan, a few years later, developed his general theory of connections (under the name of "generalized spaces").

Weyl's use of tensor calculus in his work on relativity led him to reexamine the basic methods of that calculus and, more generally, of classical invariant theory that had been its forerunner but had fallen into near oblivion after Hilbert's work of 1890. On the other hand, his semiphilosophical, semimathematical ideas on the general concept of "space" in connection with Einstein's theory had directed his investigations to generalizations of Helmholtz's problem of characterizing Euclidean geometry by properties of "free mobility." From these two directions Weyl was brought into contact with the theory of linear representations of Lie groups; his papers on the subject (1925–1927) certainly represent his masterpiece and must be counted among the most influential ones in twentieth-century mathematics.

In the early 1900's Frobenius, I. Schur, and A. Young had completely determined the irreducible rational linear representations of the general linear group $\mathbf{GL}(n,\mathbf{C})$ of complex matrices of order n; it was easy to deduce from Schur's results that all rational linear representations of the special linear group $SL(n,C)$ (matrices of determinant 1) were completely reducible—that is, direct sums of irreducible representations. Independently, E. Cartan in 1913 had described all irreducible linear representations of the simple complex Lie algebras without paying much attention to the exact relation between these representations and the corresponding ones for the simple groups, beyond exhibiting examples of group representations for each type of Lie algebra representations. Furthermore, Cartan apparently had assumed without proof that any (finite-dimensional) linear representation of a semisimple Lie algebra is completely reducible.

Weyl inaugurated a new approach by deliberately focusing his attention on global groups, the Lie algebras being reduced to the status of technical devices. In 1897 Hurwitz had shown how one may form invariants for the orthogonal or unitary group by substituting, for the usual averaging process on finite groups, integration on the (compact) group with respect to an invariant measure. He also had observed that this yields invariants not only of the special unitary group $SU(n)$ but also of the special linear group $SL(n,C)$ (the first example of what Weyl later called the "unitarian trick"). Using Hurwitz's method, I. Schur in 1924 had proved the complete reducibility of all continuous linear representations of $SU(n)$ by showing the existence, on any representation space of that group, of a Hermitian scalar product invariant under the action of $SU(n)$; by using the "unitarian trick" he also was able to prove the complete reducibility of the continuous linear representations of $SL(n,C)$ and to obtain orthogonality relations for the characters of $SU(n)$, generalizing the well-known Frobenius relations for the characters of a finite group. These relations led to the explicit determination of the characters of $SL(n,C)$, which Schur had obtained in 1905 by purely algebraic methods.

Starting from these results, Weyl first made the connection between the methods of Schur and those of E. Cartan for the representations of the Lie algebra of $SL(n,C)$ by pointing out for the first time that the one-to-one correspondence between both types of representations was due to the fact that $SU(n)$ is simply connected. He next extended the same method to the orthogonal and symplectic complex groups, observing, apparently for the first time, the existence of the two-sheeted covering group of the orthogonal group (the "spin" group, for which Cartan had only obtained the linear rep-

resentations by spinors). Finally, Weyl turned to the global theory of all semisimple complex groups. First he showed that the "unitarian trick" had a validity that was not limited to the classical groups by proving that every semisimple complex Lie algebra g could be considered as obtained by complexification from a well-determined real Lie algebra g_u, which was the Lie algebra of a compact group G_u; E. Cartan had obtained that result through a case-by-case examination of all simple complex Lie groups, whereas Weyl obtained a general proof by using the properties of the roots of the semisimple algebra. This established a one-to-one correspondence between linear representations of g and linear representations of g_u; but to apply Hurwitz's method, one had to have a compact Lie group having g_u as Lie algebra and being simply connected. This is not necessarily the case for the group G_u, and to surmount that difficulty, one had to prove that the universal covering group G_u^* of G_u is also compact; the a priori proof that such is the case is one of the deepest and most original parts of Weyl's paper. It is linked to a remarkable geometric interpretation of the roots of the Lie algebra g_u relative to a maximal commutative subalgebra t, which is the Lie algebra of a maximal torus T of G_u. Each root vanishes on a hyperplane of t, and the connected components of the complement of the union of these hyperplanes in the vector space t are polyhedrons that are now called Weyl chambers; each of these chambers has as boundary a number of "walls" equal to the dimension of t.

Using this description (and some intuitive considerations of topological dimension that he did not bother to make rigorous), Weyl showed simultaneously that the fundamental group of G_u was finite (hence G_u^* was compact) and that for G_u the maximal torus T played a role similar to that of the group of diagonal matrices in $SU(n)$: every element of G_u is a conjugate of an element of T. Furthermore, he proved that the Weyl chambers are permuted in a simply transitive way by the finite group generated by the reflections with respect to their walls (now called the Weyl group of g or of G_u); this proof gave him not only a new method of recovering Cartan's "dominant weights" but also the explicit determination of the character of a representation as a function of its dominant weight.

In this determination Weyl had to use the orthogonality relations of the characters of G_u (obtained through an extension of Schur's method) and a property that would replace Frobenius' fundamental theorem in the theory of linear representations of finite groups: that all irreducible representations are obtained by "decomposing" the regular representation. Weyl conceived the extraordinarily bold idea (for the time) of obtaining all irreducible representations of a semisimple group by "decomposing" an infinite-dimensional linear representation of G_u. To replace the group algebra introduced by Frobenius, he considered the continuous complex-valued functions on G_u and took as "product" of two such functions f, g what we now call the convolution $f*g$, defined by $(f*g)(t) = \int f(st^{-1})g(t)dt$, integration being relative to an invariant measure. To each continuous function f the operator $R(f)$: $g \to f*g$ is then associated; the "decomposition" is obtained by considering the space of continuous functions on G_u as a pre-Hilbert space and by showing that for suitable f (those of the form $h*h$, where $\bar{h}(t) = \overline{h(t^{-1})}$), $R(f)$ is Hermitian and compact, so that the classical Schmidt-Riesz theory of compact operators can be applied. It should be noted that in this substitute for the group algebra formed by the continuous functions on G_u, there is no unit element if G_u is not trivial (in contrast with what happens for finite groups); again it was Weyl who saw the way out of this difficulty by using the "regularizing" property of the convolution to introduce "approximate units"—that is, sequences (u_n) of functions that are such that the convolutions u_n*f tend to f for every continuous function f.

Very few of Weyl's 150 published books and papers—even those chiefly of an expository character—lack an original idea or a fresh viewpoint. The influence of his works and of his teaching was considerable: he proved by his example that an "abstract" approach to mathematics is perfectly compatible with "hard" analysis and, in fact, can be one of the most powerful tools when properly applied.

Weyl had a lifelong interest in philosophy and metaphysics, and his mathematical activity was seldom free from philosophical undertones or afterthoughts. At the height of the controversy over the foundations of mathematics, between the formalist school of Hilbert and the intuitionist school of Brouwer, he actively fought on Brouwer's side; and if he never observed too scrupulously the taboos of the intuitionists, he was careful in his papers never to use the axiom of choice. Fortunately, he dealt with theories in which he could do so with impunity.

BIBLIOGRAPHY

Weyl's writings were brought together in his *Gesammelte Abhandlungen*, K. Chandrasekharan, ed., 4 vols. (Berlin–Heidelberg–New York, 1968). See also *Selecta Hermann Weyl* (Basel–Stuttgart, 1956).

J. DIEUDONNÉ

WHARTON, GEORGE (*b.* Strickland, near Kendal, Westmorland, England, 4 April 1617; *d.* London, England, 12 August 1681), *astronomy.*

To his contemporaries George Wharton was renowned as a mathematician and almanac calculator whose main interests lay in the astrological possibilities of astronomy. Although he was not a systematic student of natural science, his beliefs and approach to nature probably placed him closer to the mainstream ideas of the educated gentleman of his time than to many of the followers of the "New Philosophy."

Wharton has been alternatively described as the son of a gentleman of good estate and as the son of a blacksmith who was brought up by a genteel relative. In any case his childhood was comfortable, and in 1633 he went to Oxford. How long he remained there is not known; but he became acquainted with the Durham astronomer William Milbourne, who "addicted" him to astronomy.

With the outbreak of the Civil War in 1642, Wharton raised a troop of horse for the king; but it was wiped out at Stow-on-the Wold, Gloucestershire, in 1645. He then joined the Royalist artillery at Oxford, where he met Elias Ashmole, and remained in the city until it fell to the Parliamentarian army in 1646.

Almanac calculation required little scientific originality but, rather, routine astronomical computation to determine information regarding the calendar, eclipses, and astrology. In 1641, when Wharton began to issue his almanacs, books were beyond the means of most of the population; and the threepenny almanacs were among the few pieces of secular literature to pass through the hands of most people. The influence of these almanacs came not from their astronomical content but from the astrological interpretations drawn therefrom. Wharton's fiercely Royalist interpretation of celestial phenomena was directed to embarrass the parliamentary side. This purpose soon brought him into a vitriolic pamphlet battle with its two leading astrologers, William Lilly and John Booker. The conflict illustrates the widely differing interpretations that astrologers could place on the same phenomenon, depending upon their viewpoint. During the Civil War astrology was a serious business that influenced the morale of armies.

Wharton's world view was that of the astrologer. He showed great deference to Ptolemy and Aristotle; and although respectful toward Copernicus, he apparently considered the earth to be immovable in space. Although not explicit on the subject of the earth's motion, he probably subscribed to the Tychonic world scheme. Wharton saw the cosmos as possessing both physical and metaphysical dimensions that operated through the macrocosm and microcosm. This balancing of qualities runs through all his ideas: just as a planet had two qualities, natural (physical) and astrological, so man's personality had natural and supernatural aspects that operated within a hierarchically conceived universe in which the moral and physical orders ran parallel. They also help explain Wharton's political ideas, for he saw the king as appointed by God within this hierarchy; and rebellion against him would lead not merely to social, but also to philosophical and cosmic, anarchy.

After 1648 Wharton's writings resulted in his being arrested several times by the now triumphant Parliamentarian party. Each time he escaped from prison but was sentenced to death upon his capture in 1650. Fortunately, however, the charitable William Lilly, who had no wish to see his rival hanged, succeeded in securing his pardon.

During the Interregnum, Wharton lived in obscurity although still pursuing his studies and collaborating with his friend, the Oxford antiquarian, Elias Ashmole. Problems of the calendar were an abiding interest; and some of his surviving manuscript material, in the Bodleian Library, discusses the errors in the Julian calendar and their suggested corrections. Elsewhere, Wharton relates his formula for determining the astrological influence of an eclipse. At the Restoration his loyalty was rewarded; and in 1677 he was created a baronet.

Wharton seems to have shown little interest in the scientific movement that developed in the 1660's. He never joined the Royal Society, nor did he have any concern for the new experimental philosophy, in which emphasis tended to be upon other criteria than the astrological. He did, however, on one occasion visit the Royal Greenwich Observatory.

BIBLIOGRAPHY

I. ORIGINAL WORKS. Wharton's principal works were collected by John Gadbury and published as *The Works*

of That Late Most Excellent Philosopher and Astronomer Sir George Wharton (London, 1683). The Bodleian Library, Oxford, has a few of Wharton's surviving astrological and other MSS in Ashmole 242. Some of Wharton's correspondence with Ashmole is reproduced in C. H. Josten, *Elias Ashmole*, 5 vols. (Oxford, 1966). This work also includes Ashmole's diaries, which record his acquaintance with Wharton from the 1640's on.

II. SECONDARY LITERATURE. Edward Shereburne includes a short life of Wharton in his "Catalogue of Astronomers," in the *Appendix* to his *The Sphere of Marcus Manilus* (London, 1675). A full bibliography of Wharton's works is reproduced in the life of Wharton in Anthony Wood's *Athenae Oxoniensis* II (London, 1691), 509–510.

ALLAN CHAPMAN

WHARTON, THOMAS (*b*. Winston-on-Tees, Durham, England, 31 August 1614; *d*. London, England, 15 November 1673), *anatomy, endocrinology*.

The son of John Wharton and Elizabeth Hodson, Wharton studied at Pembroke College, Cambridge, from 1637 to 1642. He subsequently moved to Trinity College, Oxford; spent three years in further study at Bolton, Lancashire; and graduated M.D. at Oxford on 7 May 1647. Thereafter he had a medical practice in London, where he was elected a fellow of the Royal College of Physicians on 23 December 1650. Wharton served as one of its censors six times between 1658 and 1673 and gave the Goulstonian lectures in January 1654. In 1656 he published, at his own expense, his Latin treatise *Adenographia*, "a description of the glands of the entire body," which he dedicated to the College of Physicians. Wharton was appointed physician to St. Thomas's Hospital on 20 November 1657 and practiced in the City through the epidemic of bubonic plague in 1665.

Adenographia gave the first thorough account of the glands of the human body, which Wharton classified as excretory, reductive, and nutrient. He differentiated the viscera from the glands and explained their relationship, describing the spleen and pancreas. He discussed in turn the abdominal and thoracic glands, those of the head, and the reproductive glands, concluding with a section on pathology. His approach was physiological, like Harvey's; but his explanations were often teleological—he suggested, for instance, that one function of the thyroid was "to fill the neck and make it shapely."

Wharton discovered the duct of the submaxillary salivary gland and the jelly of the umbilical cord, both of which are named for him; he also provided the first adequate account of the thyroid and gave it that name. He explained the role of saliva in mastication and digestion but considered that the function of certain glands, such as the adrenals and the thyroid, was to restore to the veins certain humors that were not useful to the nerves. In discussing the reproductive glands Wharton corroborated Harvey's account of the placenta and gave a clear description of the mucoid jelly of the umbilical cord, which keeps it supple and conveys and cushions the fetal vessels. He noted that this jelly does not extend beyond the umbilicus.

Much of Wharton's research was performed on animals: he mentions dissection of calves, and Izaak Walton published his description of an anglerfish (*Lophius*). He also acknowledged the opportunities for human dissection afforded him by the physicians in charge of the hospitals. Most of the glands had been mentioned in general treatises on anatomy, and the lacteals were much discussed at the time when Wharton wrote; but *Adenographia* was the first comprehensive survey. The widely acclaimed book was reprinted six times in Europe. Boerhaave wrote that Wharton "was not a great thinker, but uniquely trustworthy with his scalpel." Although special aspects of glandular anatomy were further explored by some of Wharton's younger contemporaries, the glands were little studied for nearly 200 years.

Wharton's sympathies in the Civil War were republican. In the preface to *Adenographia* he thanked Cromwell's physician, John French, and his surgeon, Thomas Trapham, for their help in his research. He also named Francis Glisson "for shared experiments," George Ent "for advice and help," and his senior hospital colleagues Francis Prujean and Edward Emily. Wharton was also friendly with the mathematician William Oughtred, and he helped Elias Ashmole to compile the catalog of John Tradescant's museum. He was not connected with Gresham College, disliking its "new brood of virtuosi" who founded the Royal Society.

On 25 June 1653 Wharton married Jane Asbridge, who died in 1669. Their son, Thomas II, became a clergyman; but their grandson George and great-grandson Thomas III, George's nephew, were prominent London physicians.

BIBLIOGRAPHY

I. ORIGINAL WORKS. *Adenographia: Sive glandularum totius corporis descriptio* (London, 1656; repr. Amsterdam, 1659; Nijmegen, 1664; Wesel, 1671; Leiden, 1679; Geneva, 1685; Düsseldorf, 1730) contains descriptions of "Wharton's duct" on 128–137 and of "Wharton's jelly" on 243–244. The Royal College of Physicians, London, owns Wharton's "letter book," which contains autograph copies of letters written in English in the last months of his life (Mar.–Oct. 1673). It also has almanacs for 1663–1666, in which Wharton had written miscellaneous notes, including a few prescriptions, case histories, and copies of letters.

II. SECONDARY LITERATURE. Izaak Walton, *The Compleat Angler* (1653), ch. 19, describes Wharton as "a man of great learning and experience and of equal freedom to communicate it." Thomas Bartholin, *Spicilegia bina ex vasis lymphaticis* (Amsterdam, 1661), pt. 2, ch. 5, praises Wharton's discoveries and "incomparable accuracy." Girolamo Barbato, discoverer of blood serum, mentions Wharton's work several times in his *Dissertatio . . . de sanguine et eius sero* (Paris, 1667). Hermann Boerhaave praises *Adenographia* in his *Method of Studying Physick* (London, 1719), 228. Elias Ashmole, *Autobiographical and Historical Notes*, C. H. Josten, ed., 5 vols. (Oxford, 1966), contains numerous personal references from Ashmole's MSS in the Bodleian Library.

H. D. Rolleston, *The Endocrine Glands With an Historical Review* (Oxford, 1936), discusses Wharton's accounts of the thyroid (p. 142) and the adrenals (p. 317); H. Speert, "The Jelly of the Umbilical Cord," in *Obstetrics and Gynecology*, **8**, no. 3 (1956), 380–382, translates and comments on Wharton's description; K. F. Russell, *British Anatomy* (Melbourne, 1963), nos. 854–859, records the editions of *Adenographia*. Biographies are J. F. Payne, "On Some Old Physicians of St. Thomas's Hospital," in *St. Thomas's Hospital Reports*, n.s. **26** (1897), 8–15, with portrait; and Bertha Porter, in *Dictionary of National Biography*, with references to sources and earlier studies.

WILLIAM LEFANU

WHATELY, RICHARD (*b.* London, England, 1 February 1787; *d.* Dublin, Ireland, 1 October 1863), *logic*.

Whately's father, Joseph Whately, was a minister and a lecturer at Gresham College. Shortly before his death in 1797, he placed his son in a private school at Bristol. Whately then went to Oriel College, Oxford, where he studied under Edward Copleston. He received the B.A. in 1808 and was elected fellow of Oriel in 1811. In his first well-known work, the pamphlet *Historic Doubts Relative to Napoleon Buonaparte* (1819), Whately offered a reductio ad absurdum disproof of Hume's challenge to the belief in miracles, arguing that if Hume was right in claiming that one should never believe in a miracle, then, for the same reasons, one also should not believe that Napoleon ever existed.

After marrying in 1821, Whately left Oxford to serve as a minister in Suffolk but returned to Oxford in 1825 to serve as principal of St. Alban Hall. He contributed two famous articles to the *Encyclopedia Metropolitana*, one on logic and the other on rhetoric. Both were reprinted as books: *The Elements of Logic* (1826) and *The Elements of Rhetoric* (1828). He was appointed Drummond professor of political economy in 1829 but resigned in 1831 in order to accept an appointment as archbishop of Dublin, after which he was primarily involved in local politics. His major academic efforts, until the time of his death, consisted in editing the writings of Francis Bacon, Copleston, and Paley.

Whately made no significant technical contributions to logic. His importance is due, instead, to his having been the first English logician to correct a mistaken conception of the nature and function of logic that had dominated English thought since the time of Locke and had led to the sterility of that discipline in England for over 150 years. Whately's work laid the philosophical foundations for the revolutionary developments in logic (notably Boole's algebra of logic) that took place in England during the nineteenth century.

The sterility of eighteenth-century English logic is reflected in two of the most popular texts of that period, Isaac Watts's *Logick, or the Right Use of Reason* (1725) and William Duncan's *Elements of Logick* (1748), neither of which emphasized the formal analysis of the conditions for the validity of reasoning. Duncan's book barely presented any of the traditional formal analysis; and Watts, while including some of it, always prefaced it with the apology that it had little significance. In place of the traditional formal analysis Duncan substituted a description of Locke's views on psychology and epistemology; Watts, who had little interest in theoretical issues, discussed the ways in which men abuse their intellectual faculties and offered practical advice.

Watts and Duncan, following Locke, rejected the traditional formal analysis of reasoning apparently because they felt that it was not helpful in guiding man in the proper use of his intellectual faculties. Since they believed that this sort of practical guidance was the proper role of logic, they concluded that logic must be radically reformed by deemphasizing the traditional formal analysis and by replacing it with the material found in their books.

Maintaining that this mistaken conception of logic had been responsible for its decline, Whately devoted most of *The Elements of Logic* to refuting objections to the traditional formal analysis. According to Whately, logic is concerned with an analysis of the forms of all valid reasoning, that is, with providing forms to which all valid arguments can be reduced. If a formal analysis provides these forms, then it has succeeded in fulfilling the proper role of logic. It would then be totally irrelevant to object to it on the grounds that it does not provide practical rules for the process of reasoning.

This Lockean tradition of logic as a guide for reason was not the only obstacle to traditional formal analysis in eighteenth-century Britain; Thomas Reid and other Scottish philosophers of common sense had also raised a set of objections. Taking their point of departure from Francis Bacon, they argued that deductive reasoning, of the type formally analyzed by the traditional logicians, was of little importance in the acquisition of knowledge, which increases only by observations and experimentation. This point was reinforced by Dugald Stewart's acute observation, long before John Stuart Mill, that deductive arguments are in some sense circular and cannot yield new knowledge because the premises of a valid deductive argument presuppose the truth of its conclusion. If the deductive mode of reasoning that had been formally analyzed by the traditional analysis was of such little value, these philosophers argued, then it would seem to follow that its formal analysis also had little worth.

Whately realized that any defense of logic as the formal analysis of the conditions for the validity of deductive reasoning would have to contain a reply to this set of criticisms, and he began his reply by granting Stewart's point that there is a perfectly good sense in which we learn nothing new in a deductive argument. He claimed, however, that it is nevertheless true that deductive reasoning plays an important role in our cognitive activities. Its purpose is to enable us to discover previously unnoticed consequences of propositions the truth of which we have already established. Since deductive reasoning does play this important role, a formal analysis of the conditions for its validity obviously is of great significance.

For Whately, the much more serious objection to the traditional formal analysis was that it did not present a formal analysis of the conditions for the legitimacy of inductive reasoning. His response to this was twofold. He began by claiming that any inductive inference is really a deductive one: its first premise is a summary of the evidence that certain objects of a given type have a given property, and the second premise is a claim of the form "The property had by the examined individuals is had by all members of that type." Therefore, the conditions for the legitimacy of inductive inferences are given by any adequate account of the conditions for the validity of deductive inferences. Whately admitted, however, that there is a special mode of inference by which we establish the second premise, a mode of which the conditions for legitimate use are not analyzed by an analysis of deductive reasoning. He felt, however, that there could be no formal analysis of these conditions and that the question of the legitimacy of such an inference would have to be decided on independent grounds in each case.

Not all of Whately's responses to the eighteenth-century critiques of logic as a formal analysis of deductive reasoning are valid, and his final remarks about induction are particularly dubious. Most of his writings, however, are quite sound and—more important from the historical point of view—seemed quite valid to his contemporaries. The re-evaluation of formal deductive logic stimulated by Whately's works resulted in tremendous progress in deductive logic in England during the nineteenth century, culminating in the formulation of the algebra of logic.

BIBLIOGRAPHY

I. ORIGINAL WORKS. The only writings by Whately that are still of interest are *Historic Doubts Relative to Napoleon Buonaparte* (London, 1819); *The Elements of Logic* (London, 1826); and *The Elements of Rhetoric* (London, 1828).

II. SECONDARY LITERATURE. The main works are B. A. Brody, *The Rise of the Algebra of Logic* (Ann Arbor, Mich., 1967); M. Prior, "Richard Whately," in P. Edwards, ed., *Encyclopedia of Philosophy*, VIII (New York, 1967), 287–288; and E. J. Whately, *Life and Correspondence of Richard Whately, D.D.*, 2 vols. (London, 1866).

B. A. BRODY

WHEATSTONE, CHARLES (*b.* Gloucester, England, 6 February 1802; *d.* Paris, France, 19 October 1875), *physics*.

Wheatstone was an experimenter and pioneering inventor in acoustics, optics, electricity, and telegraphy. He came from a family of musical instrument makers and dealers, a background that was relevant for much of his early acoustical research. He did not have any formal scientific education. In 1816 he was apprenticed to an uncle in the music business in London. Although Wheatstone became directly involved in the music business in 1823, he worked mainly on practical musical inventions (he patented the concertina in 1829) and on experimental studies of acoustic vibrations. His early work in acoustics became known (some of his papers were translated into French and German), and he was appointed professor of experimental physics at King's College, London, in 1834. He gave some lectures on sound, but most of his work consisted of research in electricity and in optics. Many of his results were communicated by Faraday. In 1836 he became a fellow of the Royal Society, he was knighted in 1868, and in 1873 he was made a foreign associate of the Paris Academy of Sciences. He received many other scientific honors.

Wheatstone's interest in acoustics was basically inspired by his desire to understand properties of a tone, such as timbre, in terms of vibration.[1] During the fifteen years that he worked on acoustics, he investigated the mechanical transmission of sound, visible demonstrations of vibrations, and properties of the vibrating air column (the Wheatstone family was involved in the making of flutes). When he was nineteen, he publicly exhibited an "enchanted lyre" that was activated by the vibrations from a remote piano transmitted to it along a wire. In Wheatstone's kaleidophone (1827) the free end of a vibrating rod was illuminated to provide a visual display of vibration. Because of persistence of vision, one saw intricate curves characteristic of the vibrational modes. Wheatstone was interested in Chladni's sand pattern technique for displaying the vibrational modes of a plate, and in 1833 he tried to use the sand patterns of a square plate to demonstrate the superposition of vibrational modes. In 1832 Wheatstone demonstrated that in the case of a standing wave in an open pipe, the motions at the ends are in opposite directions.[2] He used a pipe bent into a circle so that its ends were on either side of a vibrating square plate; when the ends faced the same region of the plate there was no resonance, but when they faced different regions (which moved in opposite directions), the resonance was strong.

In electricity one of Wheatstone's earliest and most important works (1834) was the measurement of the velocity of an electrical discharge through a wire. Wheatstone had the idea of studying very rapid motions by reflection from a rotating mirror. After trying unsuccessfully to "draw out" the spark produced by an electric discharge, he used the rotating mirror technique to observe the intervals between sparks produced by a single discharge across three spark gaps, located side by side and connected to each other by quarter-mile lengths of copper wire. From the displacement of the middle spark relative to the other two, he estimated the velocity of electricity to be over 250,000 mi./sec. (1.3 times the velocity of light[3]). In 1838 Arago suggested that Wheatstone's rotating mirror technique be used to compare the velocities of light in air and in water—an experiment performed by Foucault and Fizeau in 1850. The technique was also used by Lissajous from 1855, to study acoustic vibrations.

In 1843 Wheatstone published an experimental verification of Ohm's law, helping to make the law (already well known in Germany) more familiar in England. In connection with the verification he developed new ways of measuring resistances and currents. In particular, he invented the rheostat and popularized the Wheatstone bridge, originally invented by Samuel Christie.

In the early years of his acoustic experiments Wheatstone had speculated on the possibilities of conducting sound (including speech) over long distances,[4] and his experiments with the electric telegraph date from the early 1830's. In 1837 he and W. F. Cooke obtained their first patent for a telegraph, the "five needle" instrument, in which each letter was indicated directly by the deflections of two (magnetic) needles. Wheatstone subsequently did considerable work to develop the telegraph into a practical device. His letter-showing dial telegraph and his automatic transmitting and receiving system were particularly important. Wheatstone also worked on submarine telegraphy, performing the first experiments in 1844.

Beginning with his work on the kaleidophone, Wheatstone maintained an interest in vision and optics throughout his career. As the inventor of the stereoscope, later developed by Brewster, Wheatstone found himself—to his own surprise—the first since Leonardo da Vinci to discuss depth perception in terms of the different image received by each eye. He associated the spectrum of the light of a discharge with the metals that constituted the electrodes,[5] and in 1848 he invented the polar clock, which determines the position of the sun from the angle of polarization of sunlight.

Wheatstone also did work on audition, vowel sounds, electrical recording devices, and the dynamo. He was interested in cryptography, deciphering certain historic manuscripts and inventing a cryptograph instrument.[6]

NOTES

1. Wheatstone indicated this in his first published paper, "New Experiments on Sound" (1823). Wheatstone, *Scientific Papers*, 6.
2. *Report of the British Association for the Advancement of Science*, **2** (1832), 558.
3. The estimate was high, probably because the wire was looped back and forth, not straight. Edmund Whittaker, *A History of the Theories of Aether and Electricity*, I (New York, 1960), 228.
4. "On the Transmission of Musical Sounds." Wheatstone. *Scientific Papers*, 62–63.
5. *Report of the British Association for the Advancement of Science*, **5** (1835), pt. 2, 11–12.
6. David Kahn, *The Codebreakers* (New York, 1967), 196–198.

BIBLIOGRAPHY

I. ORIGINAL WORKS. Wheatstone's papers include "Description of the Kaleidophone, or Phonic Kaleidoscope," in *Quarterly Journal of Science, Literature and the Arts*, n.s. **1** (1827), 344–351; "On the Transmission of Musical Sounds Through Solid Linear Conductors, and on Their Subsequent Reciprocation," in *Journal of the Royal Institution*, **2** (1831), 223–238; "On the Figures Obtained by Strewing Sand on Vibrating Surfaces, Commonly Called 'Acoustic Figures,'" in *Philosophical Transactions of the Royal Society*, **123** (1833), 593–634; "An Account of Some Experiments to Measure the Velocity of Electricity and the Duration of Electric Light," *ibid.*, **124** (1834), 583–591; "Contribution to the Physiology of Vision: Part 1. On Some Remarkable and Hitherto Unobserved Phenomena of Binocular Vision," *ibid.*, **128** (1838), 371–394; and "An Account of Several New Instruments and Processes for Determining the Constants of a Voltaic Circuit," *ibid.*, **133** (1843), 303–328.

Most of Wheatstone's published papers are listed in Royal Society *Catalogue of Scientific Papers*, VI, 343–344, and VIII, 1227–1228. Almost all of them are in *The Scientific Papers of Sir Charles Wheatstone* (London, 1879), published by the Physical Society of London.

II. SECONDARY LITERATURE. There are essays on Wheatstone in *Minutes of Proceedings of the Institution of Civil Engineers*, **47** (1876–1877), pt. 1, 283–291; *Proceedings of the Royal Society*, **24** (1875–1876), xvi–xxvii; *Nature*, **13** (1875–1876), 501–503; and *Proceedings of the American Academy of Arts and Sciences*, **81** (1951–1952), 92–96, which contains references to other works on Wheatstone.

Some of Wheatstone's work in musical acoustics is discussed in W. G. Adams, "On the Musical Inventions and Discoveries of the Late Sir Charles Wheatstone," in *Proceedings of the Musical Association*, **2** (1875–1876), 85–93. An account of Wheatstone's verification of Ohm's law is H. J. J. Winter, "The Significance of the Bakerian Lecture of 1843," in *London, Edinburgh, and Dublin Philosophical Magazine and Journal of Science*, 7th ser., **34** (1943), 700–711.

SIGALIA DOSTROVSKY

WHEELER, WILLIAM MORTON (*b.* Milwaukee, Wisconsin, 19 March 1865; *d.* Cambridge, Massachusetts, 19 April 1937), *entomology*.

Because of his misbehavior in public school, Wheeler's parents, Julius Morton and Caroline Anderson Wheeler, sent him to the Engelmann German Academy in Milwaukee, which was noted for its severe discipline. He continued at the attached German-American Normal College, where he haunted the small museum until he knew every specimen. He graduated in 1884. A chance visit by H. A. Ward, of Ward's Natural Science Establishment in Rochester, New York, with specimens for the museum led to an offer to Wheeler to work for Ward in 1884. In the year Wheeler worked at the establishment he began his lifelong friendship with Carl Akeley.

In 1885 Wheeler returned to Milwaukee to teach German and physiology in the Milwaukee High School, of which George W. Peckham was principal. The biology courses of the school were unusually advanced for the day, and Peckham's own principal interest was arachnids and social insects. When E. P. Allis founded the Lake Laboratory not far from the Milwaukee High School, Wheeler met the inspiring C. O. Whitman. From William Patten, also at the Lake Laboratory, Wheeler acquired an interest in the embryology of insects and began work on a detailed study of the development of the cockroach. From 1887 to 1890 Wheeler was custodian of the Milwaukee Public Museum and absorbed embryology on his own. In 1890 he accepted a fellowship at Clark University in Worcester, Massachusetts, to study under Whitman. After receiving his Ph.D. in 1892, Wheeler studied for a year in Europe and then followed Whitman to the University of Chicago as instructor in embryology; he advanced to assistant professor in 1897.

Until this time Wheeler had emphasized embryology and morphology but also had shown considerable interest in insects; he had published on all of these fields. When he became professor of zoology at the University of Texas in 1899, he concentrated on entomology, although he always retained a strong interest in broad aspects of biology. In 1903 he became curator of invertebrate zoology at the American Museum of Natural History in New York City, where in addition to his research he designed the spectacular Hall of the Biology of Invertebrates. The appeal of an academic life drew him to his final position: professor of economic entomology at the Bussey Institution of Harvard University, at Forest Hills, Massachusetts. He preferred the title professor of entomology, to which he had it changed in 1926, and he retired in 1934. As dean of the faculty from 1915 to 1929, Wheeler led the Bussey Institution to a position of excellence in biological research. In 1931 he moved with the institution to its new quarters in the biological laboratories in Cambridge. Wheeler was president of the Entomological Society of America in 1908, and in addition was a member of many academies and the recipient of several honorary degrees.

In his early career Wheeler joined the trend of biology to morphology and embryonic development, beginning with his classic report on the cockroach (1889) and continuing with his brief entry in 1893 into marine biology at Naples, where he studied *Myzostoma*. About the turn of the twentieth century he began to concentrate on ants, and went on to become the foremost expert on ants in particular and social insects in general. "Ants interest me" was his simple explanation. As ecology became specialized, Wheeler contributed to it in his analyses of the structure within ant colonies and the relationships of the colonies to their environment; but he did not consider himself an ecologist. His own change of title at Bussey showed his preference for research for its own sake. He was not a participant in the many projects to control insects in agriculture. With single-minded attention to his research, Wheeler devoted himself to the collection, classification, structure, distribution, habits, and social life of ants. His collected specimens are chiefly at the American Museum of Natural History and the Museum of Comparative Zoology (Harvard).

Outside his immediate research Wheeler wrote also on broader aspects of biology. He considered the animal society essentially an organism from which development could evolve. A firm believer in organic evolution, he nevertheless insisted that Lamarckism had never been disproved, thus incurring the wrath of Neo-Darwinists. He extended his knowledge of social insects to discussions of man as a social animal. Fascinating in conversation and facile in writing, Wheeler attracted good students and co-workers and inspired many of the next generation of entomologists.

BIBLIOGRAPHY

I. ORIGINAL WORKS. Wheeler's complete list of almost 500 publications is in Parker's memoir (see below). Although some of the publications are reviews and other descriptive articles are short, the total list is an impressive record. A summary of his studies appeared as *Ants: Their Structure, Development and Behavior*, Columbia University Biological Series, IX (New York, 1910). In addition to many taxonomic papers, Wheeler's most valuable books are: *Social Life Among the Insects* (New York, 1923); *The Social Insects, Their Origin and Evolution* (New York, 1928); *Emergent Evolution and the Development of Societies* (New York, 1928); and *Foibles of Insects and Men* (New York, 1928).

II. SECONDARY LITERATURE. A fine memorial to Wheeler was written by George Howard Parker, in *Biographical Memoirs. National Academy of Sciences*, **19** (1938), 201–241, which includes his youth, accomplishments, and a complete bibliography. Some autobiographical material is in Wheeler's paper, "Carl Akeley's Early Work and Environment," in *Natural History*, **27** (1927), 133–141, part of which is included in Parker. Wheeler's place in entomology is summarized in E. O. Essig, *A History of Entomology* (New York, 1931; repr., 1965); and in Herbert Osborn, *Fragments of Entomological History*, pt. 1 (Columbus, Ohio, 1937), pt. 2 (1946).

ELIZABETH NOBLE SHOR

WHEWELL, WILLIAM (*b*. Lancaster, England, 24 May 1794; *d*. Cambridge, England, 6 March 1866), *history and philosophy of science, physical astronomy, science education.*

Whewell was the eldest son of a master carpenter, who hoped his son would follow him in his trade. Early displays of intellectual ability convinced the father, however, to send him to the grammar school at Heversham, in Westmorland. In 1812 Whewell began a lifelong career at Trinity College, Cambridge, where he received a classical literary education and what he soon recognized as outdated training in mathematics and science. After election as fellow of the college in 1817, Whewell took his M.A. degree in 1819 and his D.D. degree in 1844. He was ordained deacon around 1825 and then priest (1826) in the Church of England. In 1841 Whewell was appointed master of Trinity, a post he held until his death. He was

named vice-chancellor of the University of Cambridge in 1842 and again in 1855. In 1841 he married Cordelia Marshall, who died in 1855; three years later he married Lady Evering Affleck.

Tall and massively built, Whewell enjoyed good health throughout his life. Friends and foes alike admired his intelligence and breadth of scholarship, his capacity for profound affection, and his generosity. Whewell was self-consciously awkward in dealing with others, however, and he cannot be said to have been a generally popular figure at Cambridge. There can be no doubt about his quick-tempered resentment of criticism, his autocratic and often arbitrary exercise of academic power, and his jealous defense of his own position. Nevertheless, he was widely recognized as one of the central figures in Victorian science. He was a member or honorary member of at least twenty-five British and foreign scientific societies, including the Royal Society, the Royal Astronomical Society, the Royal Irish Academy, and the Royal Society of Edinburgh.

The range of Whewell's scholarly and scientific interests was immense. He composed sermons, English hexameter verses, translations of German literary works, and essays on architecture, theology, philosophy, political economy, and university education. He translated Plato's dialogues into English.

In response to the need for more accurate instruments for use in meteorology Whewell invented a self-registering anemometer that measured the direction, velocity, and temporal duration of the velocity of the wind. Anemometers in use at the time measured direction and pressure of the wind, but did not permit charting of the total movement of the air as did Whewell's device. Whewell's instrument failed to show accurate results in measuring slow movements of the air. The technical problem was solved by the Reverend T. R. Robinson in 1846 by modifying Whewell's instrument by the introduction of the now-familiar windmill with hemispherical cups.

Whewell was especially adept in coining new scientific terms. In correspondence with Michael Faraday he contributed "ion," "anode," and "cathode," among others. To geology he contributed "Eocene," "Miocene," and "Pliocene," and he introduced the terms "physicist" and "scientist."

Apart from his teaching, Whewell's major work in the decade beginning with 1819 was in science education, architecture, experimental physics, and mineralogy. As a member of a group of reformers in which John Herschel, Charles Babbage, and

George Peacock were prominent, Whewell contributed to the attempt to bring the mathematical methods of the French analysts into Cambridge scientific education. In textbooks on mechanics and dynamics, he introduced the calculus for solving problems, while insisting that analysis is no substitute for experimental physics. The reformers were successful; as a college tutor in the 1820's, and later as a major figure in guiding educational changes at Cambridge, Whewell contributed to the development of British physics and to the centrality of Cambridge in that development. In architecture Whewell attempted to refute the contention that the pointed arch is the defining property of Gothic style, arguing that in the history of German architecture the flying buttress, not the pointed arch, completed the transition from Romanesque to high Gothic.

In 1826 and 1828 Whewell and Airy made unsuccessful attempts to measure the density of the earth at a copper mine in Cornwall. Of greater importance was Whewell's work in mineralogy. In a paper read before the Royal Society in 1824, Whewell, according to Herbert Deas, "laid the foundations of mathematical crystallography." His system for calculating the angles of planes of crystals assumed that crystals are aggregates of small rhomboids that can be thought to shrink below the level of possible measurement, thus suggesting that crystals are latticelike. In 1825 Whewell visited Mohs in Germany. In 1828, the year in which Whewell became professor of mineralogy, he published a revision of Mohs's system of mineralogical classification.

Between 1833 and 1850 Whewell published fourteen memoirs on the tides. Before the 1830's little reliable observation of the tides had been undertaken; and the two leading theories, the Newton-Bernoulli equilibrium theory and the Laplace dynamical theory, were largely untested. The British Admiralty and the British Association initiated work on the tides that soon became international in scale. Whewell, with the help of John Lubbock, received and interpreted observations from all over the world; their work earned them the Royal Medal. Whewell's investigation of the tides began with an attempt to apply Thomas Young's idea of cotidal lines to the world's oceans. Had the idea applied, it would have allowed plotting the movements of tidal waves through all the oceans on the basis of initial observations of simultaneous high tides at different places. Whewell abandoned the idea in its general application, however, although the application of the idea did obtain some results for small, confined bodies of water and for shorelines.

Whewell stressed the "diurnal inequality" of the tides ("that which makes the tide of the morning and evening of the same day at the same place, differ both in height and time of high water, according to a law depending on the time of the year"). He showed large variations in this effect in accordance with local circumstances and thought the inequality to be more basic than other features of the tides. From the failure of the idea of cotidal lines and the empirical prominence of the daily inequality, Whewell concluded that no theory of physical astronomy could account for tidal phenomena in a general way. Instead, the variety and multiplicity of the data suggested that detailed study of each individual shoreline was required. Given this conclusion, it is not surprising that Whewell's work did not contribute directly to theory of the tides. Consistent with the principles of his own philosophy of science, however, he could regard himself as having fathered the science of the tides. He thought that the beginning of a science involved the laborious collection and organization of data; full theoretical generality, if any, would come later.

From the late 1830's until his death, Whewell worked mainly in the history and philosophy of science. His three-volume *History of the Inductive Sciences* appeared in 1837; in 1838 he was appointed professor of moral philosophy; and the first edition of his two-volume *The Philosophy of the Inductive Sciences, Founded Upon Their History* was published in 1840. Both the *History* and the *Philosophy* were ambitious works, and together they constitute Whewell's major scholarly achievement. The *History* had no rivals in its day and remains, despite unevenness, one of the important surveys of science from the Greeks to the nineteenth century. Whewell appreciated the importance of Greek science, especially astronomy, but showed typical disregard for the contributions of medieval scientists. His assessment of the importance of contributions of such major figures as Galileo and Descartes suffers from a heavy intrusion of religious and philosophical biases. But his treatment of Newton and other modern mathematical scientists is fair and sometimes brilliant, and is based throughout upon detailed considerations of texts. Whewell's *Philosophy* stimulated major philosophical exchanges between its author and Sir John Herschel, Augustus De Morgan, Henry L. Mansel, and John Stuart Mill. Alongside Mill's *System of Logic* and Herschel's *Preliminary Dis-*

course on the Study of Natural Philosophy, the work ranks as one of the masterpieces of Victorian philosophy of science.

Whewell's effort in these works was unique in his attempt to derive a philosophy of science from the general features of the historical development of empirical science. The importance of this attempt has not been fully appreciated. Whewell thought that the history of science displayed a progressive movement from less to more general theories, from imperfectly understood facts to basic sciences built upon a priori foundations that he called "Fundamental Ideas." All science was theoretical in that no body of data comes to us self-organized; even collection of data involves the imposition of a guiding interpretive idea. Major advances in science occur in what Whewell called an "Inductive Epoch," a period in which the basic ideas of a science are well understood by one or more scientists, and in which the generality and explanatory power of a science are seen to be much more illuminating than those of rival theories. Each such "Epoch" had a "Prelude," a period in which older theories experienced difficulties and new ideas were seen to be required, and a "Sequel," a period in which the new theory was applied and refined.

Largely ignoring the British tradition of empiricist philosophy and methodology, Whewell erected a philosophy of science upon his understanding of history that derived partly from Kant and Plato, and partly from an anachronistic theological position. Like his British predecessors, he thought that induction was the basic method of science. He understood induction not as a form of inference from particulars to generalizations, but as a conceptual act of coming to see that a group of data can best be understood and organized (his term was "colligated") under a certain idea. Furthermore, induction was demonstrative in that it yields necessary truths, propositions the logical opposites of which cannot be clearly conceived. The zenith of the inductive process was reached when a "consilience of inductions" took place—when sets of data previously considered disjoint came to be seen as derivable from the same, much richer theory. Although Whewell thought that the paradigm form of a scientific theory was deductive, he departed from the orthodox hypothetico-deductivist view of science by claiming that tests of the acceptability of given theories are extraevidential, based on considerations of simplicity and consilience. He made some attempt to justify the necessity of the conclusions that induction yields by arguing for the identity of facts and theories, and for the theological view that we know the world the way it is because that is the way God made it.

In physical astronomy Whewell's work on the tides ranks second only to that of Newton. Also of great importance was his lifelong effort to modernize and improve science education at Cambridge. The achievement in history and philosophy of science probably is less significant, although recent revival of interest in Whewell has centered mainly upon his insights in philosophy of science and methodology. Interest is growing in the interrelations of history and philosophy of science; and so long as this interest continues to be fruitful, it will be well worthwhile considering what Whewell had to say on the nature of scientific discovery, inductive methodology, and the characteristics of scientific progress.

BIBLIOGRAPHY

I. Original Works. Whewell's papers are in the Wren Library, Trinity College, Cambridge. A catalog of these papers is available (to libraries) from the Royal Commission on Historical Manuscripts, Quality House, Quality Court, Chancery Lane, London WC2A 1 HP. Unfortunately no complete bibliography of his works exists, although fairly complete ones in history and philosophy of science are available. The projected 10-vol. collection of facsimiles, *The Historical and Philosophical Works of William Whewell*, G. Buchdahl and L. L. Laudan, eds. (London, 1967–   ), will help make his works in those areas more easily available. A selection of Whewell's central works in methodology appears in *William Whewell's Theory of Scientific Method*, Robert E. Butts, ed. (Pittsburgh, 1968).

Whewell's major writings in science include *An Elementary Treatise on Mechanics* (Cambridge, 1819); *A Treatise on Dynamics* (Cambridge, 1823); "A General Method of Calculating the Angles Made by Any Planes of Crystals, and the Laws According to Which They Are Formed," in *Philosophical Transactions of the Royal Society*, **115** (1825), 87–130; *An Essay on Mineralogical Classification and Nomenclature; With Tables of the Orders and Species of Minerals* (Cambridge, 1828); *Architectural Notes on German Churches, With Remarks on the Origin of Gothic Architecture* (Cambridge, 1830); *Analytical Statics* (Cambridge, 1833); "Essay Towards a First Approximation to a Map of Cotidal Lines," in *Philosophical Transactions of the Royal Society*, **123** (1833), 147–236; "On the Results of an Extensive System of Tide Observations Made on the Coasts of Europe and America in June 1835," *ibid.*, **126** (1836), 289–341; "On the Diurnal Inequality Wave Along the Coasts of Europe," *ibid.*, **127** (1837), 227–244; *The Mechanical Euclid* (Cambridge, 1837);

and "On the Results of Continued Tide Observations at Several Places on the British Coasts," in *Philosophical Transactions of the Royal Society*, **140** (1850), 227–233.

Major writings in history and philosophy of science include "On the Nature of the Truth of the Laws of Motion," in *Transactions of the Cambridge Philosophical Society*, **5** (1834), 149–172; *History of the Inductive Sciences*, 3 vols. (London, 1837); *The Philosophy of the Inductive Sciences, Founded Upon Their History*, 2 vols. (London, 1840), the 3rd ed. of which appeared as 3 separate vols.: *The History of Scientific Ideas*, 2 pts. (London, 1858), *Novum organon renovatum* (London, 1858), and *On the Philosophy of Discovery* (London, 1860).

II. SECONDARY LITERATURE. There is very little informed and up-to-date commentary on Whewell's scientific achievements; in recent years his philosophy of science has begun to receive the attention it deserves. There are two biographies: Mrs. Stair Douglas, *Life and Selections From the Correspondence of William Whewell* (London, 1881), on his personal, including university, life; and Isaac Todhunter, *William Whewell* (London, 1876), which surveys his scientific and scholarly work. Both works contain large collections of letters; Todhunter is the best source of bibliography. Of considerable interest are Robert Robson, "William Whewell, F.R.S. (1794–1866), I. Academic Life," and Walter F. Cannon, "II. Contributions to Science and Learning," in *Notes and Records. Royal Society of London*, **19**, no. 2 (Dec. 1964), 168–191. Cannon's paper is the first attempt at a general assessment of Whewell's scientific achievements. Robert Willis, *Remarks on the Architecture of the Middle Ages, Especially of Italy* (Cambridge, 1835), extends and improves upon Whewell's work in architecture. George Airy, "Tides and Waves," in *Encyclopaedia metropolitana*, V (London, 1845), secs. VII and VIII, esp. arts. 496 and 571, praises Whewell's work on the tides, especially his methods of graphical representation of results of observations. Airy preferred the Laplace theory, however, and argued against Whewell's continuing reliance upon the Bernoulli equilibrium theory. Herbert Deas, "Crystallography and Crystallographers in Early 19th-Century England," in *Centaurus*, **6** (1959), 129–148, presents a sympathetic evaluation of Whewell's work in that area.

Whewell's philosophy attracted no disciples; and except for various references to his work in the writings of C. S. Peirce, his system received no serious study until the early 1930's. There are two book-length studies: Robert Blanché, *Le rationalisme de Whewell* (Paris, 1935); and Silvestro Marcucci, *L' "idealismo" scientifico di William Whewell* (Pisa, 1963). British and American studies of Whewell's philosophy in the context of contemporary problems in philosophy of science have taken the form of monographs and papers on specific problems. A potentially quite productive exchange of views on Whewell's concept of consilience of inductions exemplifies the richness and novelty of his insights in

methodology. Among the relevant papers are Robert E. Butts, "Whewell's Logic of Induction," in Ronald Giere and Richard Westfall, eds., *Foundations of Scientific Method: The Nineteenth Century* (Bloomington, Ind., 1973), 53–85; Mary Hesse, "Consilience of Inductions," in Imre Lakatos, ed., *The Problem of Inductive Logic* (Amsterdam, 1968), 232–247; and Larry Laudan, "William Whewell on the Consilience of Inductions," in *Monist*, **55**, no. 3 (1971), 368–391.

ROBERT E. BUTTS

WHISTON, WILLIAM (*b.* Norton, Leicester, England, 9 December 1667; *d.* Lyndon, Rutland, England, 22 August 1752), *mathematics, cosmogony, theology.*

Whiston's father, Josiah Whiston, who was also his first teacher, was a pastor. Whiston studied mathematics at Cambridge, where he earned the master's degree in 1693. He was then, successively, tutor to the nephew of John Tillotson; chaplain of the bishop of Norwich; and rector of Lowestoft and Kessingland, Suffolk. Isaac Newton, who liked and admired Whiston, engaged him as his assistant lecturer in mathematics at Cambridge and in 1701 arranged for Whiston to succeed him as Lucasian professor. But the two men became estranged because of a difference of opinion concerning the interpretation of Biblical chronology. Whiston published several theological works in which he defended heterodox opinions and supported Arianism against the dogma of the Trinity. In 1710 he was deprived of his chair and driven from the university. Newton did nothing at all to help him, even though he himself was secretly anti-Trinitarian.

Whiston moved to London, where he led a bohemian life, while continuing to occupy himself with literature and theology. He often had no money, but he nevertheless frequented the court and high society. He wrote many theological and scientific works in this period, and fell into mystic and prophetic trances. At the age of eighty he became an Anabaptist. He retired to the home of his daughter in Lyndon, where he died in 1752.

Whiston's scientific writings include several mathematical treatises, notably a Latin edition of Euclid (Cambridge, 1703), and *Praelectiones astronomicae* (1707). His most important work is *A New Theory of the Earth, From Its Original to the Consummation of All Things. Wherein the Creation of the World in Six Days, the Universal Deluge, and the General Conflagration, as Laid Down in the Holy Scriptures, Are Shewn to be Perfectly Agreeable to Reason and Philosophy* (London,

1696), which went through six editions, an indication of considerable success. It was dedicated to Newton, and its goal was to redo, with the aid of Newtonian cosmology, what Burnet had done with the aid of Descartes in *Telluris theoria sacra* (London, 1681). Whiston prefaced his book by a long dissertation entitled "A Discourse Concerning the Nature, Stile, and Extent of the Mosaick History of the Creation." In its ninety-four separately numbered pages he set forth the principles of a very free interpretation of *Genesis*. In particular, like Burnet, he contended that the Mosaic account (except for the general introduction consisting of the first verse: "In the beginning God created the heaven and the earth") concerns only the earth, and not even the entire solar system. Again like Burnet, Whiston thought that Moses, whose audience consisted of illiterate Jews, was not able to give a scientific account of the formation of the earth.

Seeking to give his arguments a geometric rigor, Whiston presented the theory itself in four books entitled *Lemmata*, *Hypotheses*, *Phaenomena*, and *Solutions*. According to the theory, the earth was originally a comet, revolving around the sun in a very eccentric orbit. This is the situation commonly described by the term "chaos." Then one day God decided to make the earth a planet, and the chaos vanished; this is the transformation recounted in Genesis. From this time and until the Flood, the earth revolved around the sun in a perfectly circular orbit; the axis of its poles was perpendicular to the plane of the ecliptic, and there were no seasons and no daily rotation. The Flood, which put an end to this state of affairs, was produced by a comet guided by God. The head of the comet, by its attraction or by its impact, broke the surface layer of the earth, causing the waters of the "great abyss" to overflow; the vapors of the tail of the comet condensed to form torrential rains. The oblique impact of the comet displaced the axis of the poles, transformed the circular orbit into an ellipse, and imparted to the earth its rotational movement. Like Woodward, Whiston thought that the layers of sedimentary rocks and the marine fossils discovered on the continents resulted from this flood. Whiston's exposition of his system lacks clarity, and he sometimes contradicted himself.

Like Newton, but less cautiously, Whiston pictured God as intervening in nature, not only to create matter and endow it with gravitation, but also to direct the course of the history of the earth. Whiston's view was that God intervenes both directly (for example, in the creation of man) and

through the intermediary of physical agents (such as a comet). Whiston explicitly stated that these two modes ultimately amount to the same thing. The ideas expressed on this point in *A New Theory of the Earth* were taken up again and made more precise in *Astronomical Principles of Religion Natural and Reveal'd*, which Whiston published in London in 1717. His thinking was similar to that of Richard Bentley and Samuel Clarke but displayed less precision and clarity. Whiston also attempted to justify his hypotheses by an interesting theory of scientific knowledge, derived from Newton but also showing the deep influence of Burnet and Cartesianism.

Whiston was more than simply a representative of an age and of a group of scientists who sought to reconcile science and Revelation. As in the case of Burnet, from whom he took a great deal, his writings were much disputed but also widely read, throughout the entire eighteenth century, and not just in England. For example, Buffon, who summarized Whiston's theory in order to ridicule it, borrowed more from him than he was willing to admit and thus unconsciously promoted the spread of his ideas. It may be said that all the cosmogonies based on the impact of celestial bodies, including that of Jeans, owed something, directly or indirectly, to Whiston's inventions.

BIBLIOGRAPHY

On Whiston and his work, see Paolo Casini, *L'universo-macchina. Origini della filosofia newtoniana* (Bari, 1969); Hélène Metzger, *Attraction universelle et religion naturelle chez quelques commentateurs anglais de Newton* (Paris, 1938); and Victor Monod, *Dieu dans l'univers* (Paris, 1933).

JACQUES ROGER

WHITE, CHARLES (*b*. Manchester, England, 4 October 1728; *d*. Sale, Cheshire, England, 20 February 1813), *obstetrics, surgery*.

White received his early education in Manchester and was apprenticed in medicine to his father, Thomas White. He subsequently studied in London, where he was greatly influenced by John and William Hunter, and in Edinburgh. He then joined his father in practice and soon achieved a reputation in surgery and obstetrics. White helped to found the Manchester Infirmary (1752) and served as its chief surgeon until 1790. He took a leading part in establishing the Lying-in Charity Hospital

(1790), now known as St. Mary's Hospital, and served on its staff. He was also a founder of the Manchester Literary and Philosophical Society (1781). On 18 February 1762 White was elected a fellow of the Royal Society and a member of the Corporation (now the Royal College) of Surgeons of London. The first to lecture on anatomy in Manchester, he eventually became the most eminent surgeon in the north of England. In 1803 an eye infection affected his vision; and in 1811 he retired to Sale, where he died completely blind.

White possessed great stamina, an acute and agile mind, and a forceful character tinged with arrogance. His contributions to surgery were extensive, and he introduced conservative techniques. Stimulated by John Hunter, he studied gradation in animals and plants and in 1799 published a suggestive treatise on evolution, unknown to Darwin, in which he rejected the idea that acquired characteristics could become hereditary. For the study of skulls upon which this work is based, White has been called the founder of anthropometry. His main fame, however, derives from his work in obstetrics. Alexander Gordon (1795), Oliver Wendell Holmes (1843), and Ignaz Semmelweis (1847) are correctly given credit for discovering the infectious nature of puerperal fever. Nevertheless, White, although unaware of its causative agent, recognized some of the associated etiological factors and instituted prophylaxis and therapy accordingly. He was the first to insist on absolute cleanliness during delivery and was thus a pioneer in aseptic midwifery. Together with his astute account of puerperal fever (1773), his recognition of "white leg" (1784), and his enlightened approach to obstetrics in general, it brought him widespread recognition.

BIBLIOGRAPHY

I. ORIGINAL WORKS. White's main works were *A Treatise on the Management of Pregnant and Lying-in Women . . .* (London, 1773, 1777, 1784, 1791), repr. in J. George Adami, *Charles White of Manchester (1728–1813), and the Arrest of Puerperal Fever* (London, 1922), also in French (Paris, 1774), American (Worcester, Mass., 1793), and German eds. (Leipzig, 1775); and *An Inquiry Into the Nature and Cause of That Swelling in One or Both of the Lower Extremities Which Sometimes Happens to Lying-in Women*, 2 vols. (Warrington, 1784–London, 1801).

A list of his surgical writings is in J. E. Dezeimeris, *Dictionnaire historique de la médecine*, IV (Paris, 1839), 402–403. He also published *An Account of the Regular Gradation in Man, and in Different Animals and Vegetables and From the Former to the Latter* (London, 1799); and three papers of less importance in *Philosophical Transactions of the Royal Society*, **51** (1760) and **59** (1769).

II. SECONDARY LITERATURE. Adami (see above), with portrait, makes unwarranted claims for White, as do Charles J. Cullingworth, *Charles White, F.R.S., a Great Provincial Surgeon and Obstetrician of the 18th Century* (London, 1904), and *Lancet* (1903), **2**, 1071–1076. Edward A. Schumann, "Charles White and His Contribution to the Knowledge of Puerperal Sepsis," in *Medical Life*, **36** (1929), 257–270, gives a more balanced judgment of his work. See also E. M. Brockbank, *Sketches of the Lives and Work of the Honorary Staff of the Manchester Infirmary* (Manchester, 1904), 28–65, with a list of local biographical sources; *Dictionary of National Biography*, xxi, 33–34; and H. Thoms, *Classical Contributions to Obstetrics and Gynecology* (Springfield, Ill., 1935), 170–178, with extracts from his book of 1773.

EDWIN CLARKE

WHITE, CHARLES DAVID (*b.* near Palmyra, New York, 1 July 1862; *d.* Washington, D.C., 7 February 1935), *geology.*

A childhood spent on the farm of his parents, Asa Kendrick and Elvira Foster White, instilled in White a love of botany and the outdoors. He prepared for college at nearby Marion Collegiate Institute, where its principal, Daniel Van Cruyningham, a former hired hand on the White farm, strengthened his interest in botany. After teaching in rural schools for two years, David—as he came to be known—won a scholarship to Cornell University (B.S., 1886); and attracted by the courses of Henry Shaler Williams, he turned to paleobotany. Williams recommended him for his ability in drawing fossil plants to Lester F. Ward of the U.S. Geological Survey, and in 1886 White began his forty-nine-year research and administrative career with that organization.

From 1910 to 1912 White was head of the Section of Eastern Coalfields and then served as chief geologist for ten years. In 1922 he returned to his own research but gave considerable time to advisory committee appointments for the National Academy of Sciences, the National Research Council, and professional societies. He was also curator of paleobotany at the U.S. National Museum from 1903 to 1935.

In addition to his many memberships (some honorary) in professional societies, White received numerous awards: D.Sc. from the University of Rochester (1924), the University of Cincinnati

(1924), and Williams College (1925); the Walcott Medal and the Mary Clark Thompson Medal of the National Academy of Sciences, of which he became a member in 1912; the Penrose Medal of the Society of Economic Geologists; and the Boverton Redwood Medal of the Institute of Petroleum Technologists of London. He served as president of the Paleontological Society, the Washington Academy of Sciences, the Geological Society of Washington, and the Geological Society of America.

White's work on fossil plants began with a restudy of the Lacoe collection of Carboniferous specimens, followed by fieldwork on Paleozoic plants in the Appalachian trough and elsewhere. By 1896 he was convincingly able to correlate stratigraphy on the basis of paleobotany and to persuade doubting geologists of the vast extent of the Pennsylvanian Pottsville formation. Turning to the origin of peats and coals and then to the new field of petroleum, he soon established himself as the foremost authority on carbonaceous deposits. Coal and oil, he said in 1908, were the products of sedimentary deposits of organic materials, changed into carbon by chemical and physical action over a long period of time and varied by the original composition, the length of time, and fluctuations in pressure and temperature. The plants forming them grew under uniform humid tropical conditions, and coals formed directly from these in place. One of the first strong advocates of the theory that carbonaceous sediments are progressively of higher grade with increasing metamorphism, White presented contoured maps of Appalachian regions that showed the increases in coal hardness as the intensity of deformation increased (1909). This theory proved to be of considerable economic value in the search for coal.

In 1915 White presented his most valuable contribution, the carbon-ratio theory: As the percentage of fixed carbon in coals increases with higher temperature and pressure, the accompanying oils became increasingly lighter (that is, of higher grade) until, above 70 percent fixed carbon, only gas or neither oil nor gas is found. Ignored by the profession until Myron L. Fuller revived it ten years later, this theory, despite known exceptions, has saved considerable useless drilling. White enlarged upon it in 1935 and revised the "dead line" to 60 percent fixed carbon.

Deeply involved in petroleum research through his Geological Survey position, White led a drive to estimate the nation's oil reserves, especially because of World War I commitments. He pioneered in the investigation of oil shales for future use, instigated studies of temperature records in deep bore holes and mines, and was the first to apply gravimetry to finding anticlinal structure (Damon Mound, Texas). Through the National Research Council he founded a program of basic research on petroleum.

In addition to his many publications on Paleozoic flora, White also dealt with climates of that era and concluded that the extensive Pennsylvanian Pottsville represented its most luxuriant vegetation. He postulated that diastrophism, in creating geographic alterations, is the major cause of climatic change. In his final years he began studies of the Precambrian lime-secreting algae of the Grand Canyon and Glacier National Park.

BIBLIOGRAPHY

I. ORIGINAL WORKS. White's 200 publications deal most significantly with coal and petroleum geology. The comprehensive "The Origin of Coal," *Bulletin of the United States Bureau of Mines,* **38** (1913), was written with Reinhardt Thiessen. The carbon-ratio theory was first expounded in "Some Relations in Origin Between Coal and Petroleum," in *Journal of the Washington Academy of Sciences,* **5** (1915), 189–212; and was amplified, just prior to White's death, in "Metamorphism of Organic Sediments and Derived Oils," in *Bulletin of the American Association of Petroleum Geologists,* **19,** no. 5 (1935), 589–617. His mature views on coals and oils were effectively stated in "The Carbonaceous Sediments," in W. H. Twenhofel, ed., *Treatise on Sedimentation* (Baltimore, 1926; 1932), 351–430.

White produced a number of early papers on paleobotany, of which the monumental "Fossil Flora of the Lower Coal Measures of Missouri," *U.S. Geological Survey Monograph,* no. 37 (1899), is an outstanding example. A long-intended monograph on the Pottsville flora was never completed, but the MS of "Fossil Flora of the Wedington Sandstone Member of the Fayetteville Shale" was published posthumously as *U.S. Geological Survey Professional Paper* no. 186-B (1937).

His studies of plants led White to theories of climate changes, mainly of the Paleozoic, as in "Permo-Carboniferous Climatic Changes in South America," in *Journal of Geology,* **15** (1907), 615–633; and in his significant "Permian of Western America From the Paleobotanical Standpoint," in *Proceedings of the Pan-Pacific Science Congress, 1923,* II (Melbourne, 1924), 1050–1077. His discussion of diastrophism as the major cause of climatic change appeared in "Upper Paleozoic Climate as Indicated by Fossil Plants," in *Scientific Monthly,* **20,** no. 5 (1925), 465–473; and is elaborated in his paper on more recent changes, "Geologic Factors Affecting and Possibly Controlling Pleistocene Ice Sheet Development in North America," in *Journal of the Washington Academy of Sciences,* **16,** no. 3 (1926), 69–72, an abstract.

Further papers on these and his other varied interests, including the estimates of petroleum reserves, are listed in the bibliographies cited below.

II. SECONDARY LITERATURE. White dropped the name of Charles in 1886 and thus is commonly found in indexes as David White. Charles Schuchert summarized his early life and contributions to geology in *Biographical Memoirs. National Academy of Sciences*, **17** (1935–1937), 187–221, with bibliography; see also W. C. Mendenhall, "Memorial of David White," in *Proceedings. Geological Society of America* for 1936 (1937), 271–292, with bibliography. Hugh D. Miser, "David White," in *Bulletin of the American Association of Petroleum Geologists*, **19**, no. 6 (1935), 925–932, covers especially White's work in petroleum geology.

ELIZABETH NOBLE SHOR

WHITE, GILBERT (*b.* Selborne, Hampshire, England, 18 July 1720; *d.* Selborne, 26 June 1793), *zoology, botany, horticulture.*

Edmund Gosse, the distinguished poet and critic, wrote, "The literature of the 18th century has left us no model of innocence, delicacy and alert natural piety more perfect than was the spirit of Gilbert White of Selborne . . . a man who has done more than any other to reconcile science with literature."

White was the eldest son of John White and Anne Holt, daughter of a wealthy clergyman from Streatham, south London. After preliminary schooling in Farnham and then at Basingstoke, he entered Oriel College, Oxford, where he received the B.A. in 1743. White then spent another year at Oxford prior to his election as fellow, finally taking the master's degree in October 1746. He received deacon's orders the following year and became curate at Swarraton, the parish of his uncle, the Reverend Charles White. Three years later White was ordained a priest by the bishop of Hereford, but, not being ambitious for a clerical career nor anxious to move far from Selborne, where his parents had made the family home, he refused all preferments offered him except that at Moreton Pinkney, Northamptonshire, a living belonging to and offered by his own college at Oxford. It provided him with a steady income but did not require him to reside within its boundaries, as it was administered by the curate from the next parish. This situation left White free to accept the curacy at Selborne or of a nearby parish whenever one was vacant, thus permitting his continued stay at the family home, "The Wakes"—a large house in the center of Selborne village (later bequeathed to him

through his uncle)—and to indulge his pursuit of the study of natural history.

It was in his study at "The Wakes" that White entered in diary form daily notes on natural phenomena observed in the garden and during walks that took him into the countryside near Selborne or into adjacent parishes. Much of this material was shared in letters to his many correspondents, in particular with the zoologist Thomas Pennant and with the Honorable Daines Barrington, a well-to-do barrister and keen amateur naturalist. Despite repeated requests by Pennant and others, diffidence prevented White from being persuaded to edit his letters and have them issued in book form until four years before his death, when a selection of a lifetime of observations was published as *The Natural History and Antiquities of Selborne*. The work not only has become a classic of the English literature of the eighteenth century but also has inspired many a tyro in the study of natural history. It has been published in about 200 English editions and reprints and been translated into several foreign languages.

The *Natural History* makes abundantly clear how intense was White's interest in bird life. Unlike many contemporary ornithologists, who confined themselves to studying avian anatomy or to describing plumage, White made detailed notes on bird habits and habitats. He was the first to recognize the difference between the three British *Phylloscopi* (leaf warblers), previously considered as a single species first described by Linnaeus. The plumage is almost identical, but White pointed out that their songs are quite different. He also devoted much time to studying the habits of the nightjar; the parasitical egg-laying of the cuckoo in another species' nest; and whether swallows migrated or hibernated in Britain. Satisfactory explanations of the life styles of both cuckoo and swallow were not put forward until long after White's day. He conjectured that the domestic pigeon stemmed from the blue rock pigeon and not, as had been thought, from either the wood pigeon or the stock dove. This hypothesis was later elaborated by Charles Darwin as part of his marshaling of evidence bearing on his theory of evolution.

White also was the first to recognize Britain's smallest mammal—the harvest mouse—and he added the noctule bat to the British list, although it had earlier been described by Daubenton on the Continent. In the reptilian kingdom White's observations on a pet tortoise named Timothy are best known, although he drew attention to the fact that the blindworm could not be a snake because it produced viviparous young. He also noted many aspects of insect biology, his observations on the habits of the field, house, and mole crickets being his most significant contribution to entomology.

White recorded many species of wild flowers found around Selborne and kept phenological notes each year on their times of flowering and seeding. He was also a keen gardener, as is evident from his manuscript "Garden Kalender" and his "Naturalist's Journal," some entries from which were finally incorporated in the *Natural History.* He also noted various aspects of local agriculture, folk life, weather lore, and even archaeology and astronomy. In fact, few facets of natural phenomena observable around Selborne escaped White's attention, although wider issues affecting the country as a whole drew no comment from him. White was not unaware of the suffering brought about by wars and the industrial revolution, especially among the poor, for after his death many local inhabitants vouched for his manifold kindness and understanding. It was perhaps, rather, that he felt that in the study of natural history he could give of his best; and thus his environment at Selborne became increasingly the center of his world.

Why, then, has a single published volume written by a country cleric developed in generations of its readers so much happiness from studying natural history or engendered within them an undeniable love for the countryside? The answer has best been crystallized in the words of another zoologist, L. C. Miall, written in 1901 but nevertheless still true. Of White he wrote:

> . . . his personal knowledge of nature was great not in relation to knowledge accumulated in books but in comparison with the direct experience of most other naturalists of any age. Here is the one great difference between him and the imitators who have hoped to succeed by mere picturesque writing. White is interesting because nature is interesting; his descriptions are founded upon natural fact, exactly observed and sagaciously interpreted. Very few of his observations . . . need correction more than a hundred years after his death.

White died a bachelor at the age of seventy-two. He willed that his grave in St. Mary's churchyard, Selborne, should not be elaborate. In keeping with his wishes and his unpretentious nature, its plain headstone bears the simple inscription "G. W. June 26, 1793."

BIBLIOGRAPHY

I. ORIGINAL WORKS. White's fame stems from his one book, *The Natural History and Antiquities of Selborne* (London, 1789; facs. repr., Menston, 1970). Many eds. of this work have appeared, some with footnotes that have enhanced the original text, or with copious illustrations. Others have included further material extracted from White's MSS or have given details of his life. The three best are those by Thomas Bell, 2 vols. (London, 1877); R. Bowdler Sharpe, 2 vols. (London–Philadelphia, 1900); and Grant Allen (London–New York, 1900). All eds. published up to 1930 have been collated by E. A. Martin in his *Bibliography of Gilbert White* (London, 1934; repr. Folkestone–London, 1970).

Of White's other MS material, his "Garden Kalender" was published in vol. II of R. Bowdler Sharpe's ed. (see above); the "Calender of Flora, 1766" was published in facs. by the Selborne Society, edited with notes by W. M. Webb (London, 1911); and a further selection from White's "Naturalist's Journal" was edited by W. Johnson and published as *Gilbert White's Journals* (London, 1931; repr. Newton Abbot, Devon, 1970).

II. SECONDARY LITERATURE. Rashleigh Holt White, *The Life and Letters of Gilbert White*, 2 vols. (London, 1901), should be consulted by those who wish to know more of his correspondence and MSS. Two other texts of value are Rev. Walter Sidney Scott, *White of Selborne* (London, 1950); and W. Johnson, *Gilbert White: Pioneer, Poet and Stylist* (London, 1928). C. S. Emden, *Gilbert White in His Village* (London, 1956), and Anthony Rye, *Gilbert White and His Selborne* (London, 1970), both accounts of his life and times, may also be useful.

ERIC W. GROVES

WHITE, ISRAEL CHARLES (*b.* Monongalia County, Virginia [now West Virginia], 1 November 1848; *d.* Baltimore, Maryland, 25 November 1927), *geology.*

The son of Michael White, a progressive farmer, Israel received his early education in private schools. In 1867 he entered what is now West Virginia University, where he studied under John J. Stevenson, who later became an eminent geologist. White graduated in 1872 with highest honors, received the A.M. three years later, and in 1919 was awarded an honorary LL.D. Shortly after his twenty-ninth birthday he became head of the department of geology at West Virginia University. In 1880 Arkansas Industrial University (now the University of Arkansas) awarded him the Ph.D., and in 1921 he received the D.Sc. from the University of Pittsburgh.

White worked on the Second Geological Survey of Pennsylvania and on the U.S. Geological Survey, making a general survey of the coalfields of Pennsylvania, West Virginia, and Ohio that formed the basis for all subsequent detailed study of bituminous coals of the Appalachian region. As an authority on coal he was selected by the government of Brazil to survey and prepare a report on the coals of that country. He advocated and secured the establishment in 1897 of the West Virginia Geological and Economic Survey and headed it for thirty years as state geologist.

White's most notable accomplishment was the practical application of the anticlinal theory of oil and natural gas accumulation. In this theory, which he promulgated in 1885, he demonstrated the important part played by specific gravities of the fluids in the separation of oil and gas into commercial pools in conjunction with anticlinal and domal types of geologic structures. While others had arrived at the same conclusion independently, credit must be given to White for the successful application of the theory and for convincing the industry of its importance.

BIBLIOGRAPHY

I. ORIGINAL WORKS. White wrote over 170 publications. His most important books and papers are "The Geology of Natural Gas," in *Science*, **5** (1885), 521–552, in which he promulgated his anticlinal theory; "Stratigraphy of the Bituminous Coal Field of Pennsylvania, Ohio, and West Virginia," *Bulletin of the United States Geological Survey*, no. **65** (1891); "The Mannington Oil Field, West Virginia, and the History of Its Development," in *Bulletin of the Geological Society of America*, **3** (1892), 187–216; *Report on the Coal Measures and Associated Rocks of South Brazil*, I (Rio de Janeiro, 1908), in English and Portuguese; "Petroleum Fields of Northeastern Mexico Between the Tamesi and Tuxpan Rivers," in *Bulletin of the Geological Society of America*, **24** (1913), 253–274, 706; and "The Anticlinal Theory," in *Report of Proceedings. American Mining Congress*, **19** (1917), 550–556.

II. SECONDARY LITERATURE. See Lloyd L. Brown, "The Life of Dr. Israel Charles White" (M.A. thesis, Univ. of West Virginia, 1936); Herman L. Fairchild, "Memoirs of Israel C. White," in *Bulletin of the Geological Society of America*, **39** (1928), 126–145; and Ray V. Hennen, "Israel C. White, Memorial," in *Bulletin of the American Association of Petroleum Geologists*, **12** (1928), 339–351.

PAUL H. PRICE

WHITE, THOMAS (*b.* Runwell, Essex, England, 1593; *d.* London, England, 6 July 1676), *natural philosophy, theology.*

Very little is known of White's early life. His father, Richard White, married Mary Plowden, daughter of the Catholic lawyer Edmund Plowden. White was sent to the Continent for a carefully supervised Catholic education. He studied first at the English College at St.-Omer; but by the fall of 1609 he had become a member of St. Albans College at Valladolid, where he spent three years before transferring to the English college at Seville. White went to Louvain in 1614 and completed his last years of study for the priesthood at Douai; he was ordained at Arras on 25 March 1617 with the name of Blacklow. In later years he wrote under the names Blacklow, Blacloe, Vitus, Albius, and Anglus.

White became a teacher of philosophy at Douai in 1617; and the following year, after receiving his baccalaureate degree, he began to teach theology. He studied canon law at Paris in 1624–1625 and in the spring of 1626 was sent to Rome as a representative of the secular clergy of England, a duty he fulfilled until 1630. From 1631 to 1633 White was president and professor of theology at the English College of Lisbon. In May 1633 he returned to England; in the following year he was involved in the internal controversies of the English Catholics and became a candidate for the English bishopric. While in England he became a close friend of Sir Kenelm Digby. Through the late 1640's, White lived in Paris. The last two decades of his life were those of White's greatest scholarly productivity and his most sustained involvement in intellectual controversy. In 1662 he returned to England from Douai, where he had taught since about 1650, and remained there until his death. The movement embodying his theological positions, "Blackloism," maintained his ecclesiastical opinions for several decades after his death.

Although he was remarkably productive in philosophy and science, White's ideas were not acceptable to the papacy. On 17 November 1661 the Holy Office condemned eight of his books explicitly (and implicitly all of his other writings, both past and future). Theologically his thought was similar to Jansenism, for in his writings he continually condemns the Jesuits and skeptics. A devoted follower of Aristotle, White viewed the skepticism of the late sixteenth century as the principal hindrance to scientific advancement. His scientific treatises contain modifications and revisions of

Aristotle's thought; his *De mundo* of 1642, for instance, was an analysis and amplification of Aristotelian cosmology. In his *Institutionum peripateticarum* (1646) White presented the most detailed description of his philosophical and scientific approach to the study of nature. His view of nature was qualitative, and he sought spiritual demonstrability in the physical world. In 1657 and 1658 he published *Euclides physicus* and *Euclides metaphysicus*, in which he examined and amplified Aristotle's theory of causation.

White used science as a weapon with which to confront skeptics and as a tool for compounding the certitude of faith. His scientific thought was subordinate to his desire to render theology scientifically verifiable.

BIBLIOGRAPHY

I. ORIGINAL WORKS. White's principal scientific works are *De mundo dialogi tres . . .* (Paris, 1642); *Institutionum peripateticarum* (Lyons, 1646), also in English trans. (London, 1656); *Sonus buccinae sive tres tractatus . . .* (Paris, 1654); *Euclides physicus sive de principiis naturae* (London, 1657); *Euclides metaphysicus sive De principiis sapientiae* (London, 1658); *Exercitatio geometrica* (London, 1658); *Scirri sive scepticis & scepticorum a jure disputationis* (London, 1663); and *An Exclusion of Scepticks From All Title to Dispute* (London, 1665).

II. SECONDARY LITERATURE. The most detailed study of White is Robert I. Bradley, "Blacklo: An Essay in Counter-Reform" (Ph.D. diss., Columbia Univ., 1963). For an examination of White's activities as an English recusant, see Robert I. Bradley, "Blacklo and the Counter-Reformation: An Inquiry Into the Strange Death of Catholic England," in Charles H. Carter, ed., *From the Renaissance to the Counter Reformation* (New York, 1965), 348–370. There are biographical notices in *Dictionary of National Biography* and in various Catholic encyclopedias.

PHILLIP DRENNON THOMAS

WHITEHEAD, ALFRED NORTH (*b.* Ramsgate, Kent, England, 15 February 1861; *d.* Cambridge, Massachusetts, 30 December 1947), *mathematics, mathematical logic, theoretical physics, philosophy.*

Education, religion, and local government were the traditional interests of the family into which Whitehead was born, the son of a southern English schoolteacher turned Anglican clergyman. As a child Whitehead developed a strong sense of the enduring presence of the past, surrounded as he was by relics of England's history. The school to which he was sent in 1875, Sherborne in Dorset, traced its origin to the eighth century. At Sherborne, Whitehead excelled in mathematics, grew to love the poetry of Wordsworth and Shelley, and in his last year acted as head of the school and captain of games. In the autumn of 1880 he entered Trinity College, Cambridge. Although during his whole undergraduate study all his courses were on pure or applied mathematics, he nevertheless developed a considerable knowledge of history, literature, and philosophy. His residence at Cambridge, first as scholar, then as fellow, and finally as senior lecturer in mathematics, lasted from 1880 to 1910. During the latter part of this period he used to give political speeches in the locality; these favored the Liberal party and often entailed his being struck by rotten eggs and oranges. In 1890 he married Evelyn Willoughby Wade, whose sense of beauty and adventure fundamentally influenced Whitehead's philosophical thought. Three children were born to them between 1891 and 1898: Thomas North, Jessie Marie, and Eric Alfred, who was killed in action with the Royal Flying Corps in 1918.

In 1910 Whitehead moved to London, where he held a variety of posts at University College and was professor at the Imperial College of Science and Technology. During this period, while active in assisting to frame new educational programs, he turned his reflective efforts toward formulating a philosophy of science to replace the prevailing materialistic mechanism, which in his view was unable to account for the revolutionary developments taking place in science.

In 1924, at the age of sixty-three, Whitehead became a professor of philosophy at Harvard University. There his previous years of reflection issued in a rapid succession of philosophical works of first importance, principally *Process and Reality: An Essay in Cosmology* (1929). He retired from active teaching only in June 1937, at the age of seventy-six. Whitehead died in his second Cambridge ten years later, still a British subject, but with a great affection for America. He had enjoyed the rare distinction of election to fellowships both in the Royal Society and in the British Academy. In 1945 he was also awarded the British Order of Merit.

Whitehead's life and work thus fall naturally into three periods which, although distinct, manifest a unity of development in his thought. At Cambridge University his writings dealt with mathematics and logic, although his thought already displayed those more general interests that would lead him to philosophy. In his second, or London, period, Whitehead devoted himself to rethinking the conceptual and experiential foundations of the physical sciences. He was stimulated in this work by participating in the discussions of the London Aristotelian Society. The writings of his third, or Harvard, period were distinctly philosophical, commencing with *Science and the Modern World* (1925), and culminating in *Process and Reality* (1929) and *Adventures of Ideas* (1933). These three works contain the essentials of his metaphysical thinking. Noteworthy among his several other books are *The Aims of Education* (1929) and *Religion in the Making* (1926), in which he combines a sensitivity to religious experience with a criticism of traditional religious concepts.

Although Whitehead's intellectual importance lies mainly in philosophy itself, he did significant work in mathematics, mathematical logic, theoretical physics, and philosophy of science.

Mathematics and Mathematical Logic. Whitehead's mathematical work falls into three general areas, the first two of which belong to his residence at Cambridge University, the third to his London period. The first area, algebra and geometry, contains his writings in pure mathematics, chief among which is his first book, *A Treatise on Universal Algebra* (1898). Other examples are papers on "The Geodesic Geometry of Surfaces in Non-Euclidean Space" (1898) and "Sets of Operations in Relation to Groups of Finite Order" (1899). The second area consists in work that would today be termed logic and foundations. It includes work on axiomatics (projective and descriptive geometry), cardinal numbers, and algebra of symbolic logic; it culminates in the three-volume *Principia Mathematica*, written with Bertrand Russell. The third area—less relevant from a mathematical point of view—contains the mathematical work that overlaps other fields of Whitehead's scientific activity, mainly his physics and his philosophy of mathematics. His paper "On Mathematical Concepts of the Material World" (1906) is typical of the former; his *Introduction to Mathematics* (1911) lies in the border area between mathematics and the philosophy of mathematics.

Algebra and Geometry. Whitehead's first book, *A Treatise on Universal Algebra*, seems at first glance entirely mathematical. Only in view of his subsequent development are several of his introductory remarks seen to have a philosophical import. This lengthy book, begun in 1891 and published in 1898, formed part of that nineteenth-century pioneering development sometimes referred to

as the "liberation of algebra" (from restriction to quantities). Although the movement was not exclusively British, there was more than half a century of British tradition on the subject (George Peacock, Augustus De Morgan, and William Rowan Hamilton), to which Whitehead's mathematical work belonged.

Whitehead acknowledged that the ideas in the *Universal Algebra* were largely based on the work of Hermann Grassmann, Hamilton, and Boole. He even stated that his whole subsequent work on mathematical logic was derived from these sources, all of which are classical examples of structures that do not involve quantities.

After an initial discussion of general principles and of Boolean algebra, the *Universal Algebra* is devoted to applications of Grassmann's calculus of extension, which can be regarded as a generalization of Hamilton's quaternions and an extension of arithmetic. Major parts of the modern theory of matrices and determinants, of vector and tensor calculus, and of geometrical algebra are implied in the calculus of extension. Whitehead's elaboration of Grassmann's work consists mainly in applications to Euclidean and non-Euclidean geometry.

Although the *Universal Algebra* displayed great mathematical skill and erudition, it does not seem to have challenged mathematicians or to have contributed in a direct way to further development of the topics involved. It was never reprinted during Whitehead's lifetime. It is plausible to think that, by the time the mathematical world became aware of the many valuable items of the work, these had been incorporated elsewhere in more accessible contexts and more modern frameworks.

Logic and Foundations. Confining itself to the algebras of Boole and Grassmann, the *Universal Algebra* never became what it was intended to be, a comparative study of algebras as symbolic structures. Whitehead planned to make such a comparison in a second volume along with studies of quaternions, matrices, and the general theory of linear algebras. Between 1898 and 1903 he worked on this second volume. It never appeared, and neither did the second volume of Bertrand Russell's *Principles of Mathematics* (1903). The two authors discovered that their projected second volumes "were practically on identical topics," and decided to cooperate in a joint work. In doing so their vision expanded, and it was eight or nine years before their monumental *Principia Mathematica* appeared.

The *Principia Mathematica* consists of three volumes which appeared successively in 1910,

1912, and 1913. A fourth volume, on the logical foundations of geometry, was to have been written by Whitehead alone but was never completed. The *Principia* was mainly inspired by the writings of Gottlob Frege, Georg Cantor, and Giuseppe Peano. At the heart of the treatment of mathematical logic in the *Principia* lies an exposition of sentential logic so well done that it has hardly been improved upon since. Only one axiom (Axiom 5, the "associative principle") was later (1926) proved redundant by Paul Bernays. The development of predicate logic uses Russell's theory of types, as expounded in an introductory chapter in the first volume. The link with set theory is made by considering as a set all the objects satisfying some propositional function. Different types, or levels, of propositional functions yield different types, or levels, of sets, so that the paradoxes in the construction of a set theory are avoided. Subsequently several parts of classical mathematics are reconstructed within the system.

Although the thesis about the reduction of mathematics to logic is Russell's, as is the theory of types, Russell himself stressed that the book was truly a collaboration and that neither he nor Whitehead could have written it alone.[1] The second edition (1925), however, was entirely under Russell's supervision, and the new introduction and appendices were his, albeit with Whitehead's tacit approval.

Taken as a whole, the *Principia* fills a double role. First, it constitutes a formidable effort to prove, or at least make plausible, the philosophical thesis best described by Russell in his preface to *The Principles of Mathematics*: "That all pure mathematics deals exclusively with concepts definable in terms of a very small number of fundamental logical concepts, and that all its propositions are deducible from a very small number of fundamental logical principles." This thesis is commonly expressed by the assertion that logic furnishes a basis for all mathematics. Some time later this assertion induced the so-called logicist thesis, or logicism, developed by Wittgenstein—the belief that both logic and mathematics consist entirely of tautologies. There is no evidence that Whitehead ever agreed with this; on the contrary, his later philosophical work indicates a belief in ontological referents for mathematical expressions. The thesis that logic furnishes a basis for all mathematics was first maintained by Frege but later (1931) refuted by Kurt Gödel, who showed that any system containing arithmetic, including that of the *Principia*, is essentially incomplete.

The second role of the *Principia* is the enrichment of mathematics with an impressive system, based on a thoroughly developed mathematical logic and a set theory free of paradoxes, by which a substantial part of the body of mathematical knowledge becomes organized. The *Principia* is considered to be not only a historical masterpiece of mathematical architecture, but also of contemporary value insofar as it contains subtheories that are still very useful.

Other Mathematical Work. At about the time Whitehead was occupied with the axiomatization of geometric systems, he turned his attention to the mathematical investigation of various possible ways of conceiving the nature of the material world. His paper "On Mathematical Concepts of the Material World" (1906) is just such an effort to create a mathematical although qualitative model of the material world. This effort differs from applied mathematics insofar as it does not apply known mathematics to situations and processes outside mathematics but creates the mathematics *ad hoc* to suit the purpose; yet it resembles applied mathematics insofar as it applies logical-mathematical tools already available. The paper conceives the material world in terms of a set of relations, and of entities that form the "fields" of these relations. The axiomatic mathematical system is not meant to serve as a cosmology but solely to exhibit concepts not inconsistent with some, if not all, of the limited number of propositions believed to be true concerning sense perceptions. Yet the system does have a cosmological character insofar as it tries to comprehend the entire material world. Unlike theoretical physics the paper is entirely devoid of quantitative references. It is thus an interesting attempt to apply logical-mathematical concepts to ontological ones, and is an early indication of Whitehead's dissatisfaction with the Newtonian conception of space and time. In a qualitative way the paper deals with field theory and can be regarded as a forerunner of later work in physics.

The delightful little book *An Introduction to Mathematics* (1911) is another early example of Whitehead's drifting away from the fields of pure mathematics and logic, this time more in the direction of philosophy of mathematics. The book contains a fair amount of solid although mainly fundamental and elementary mathematics, lucidly set out and explained. The object of the book, however, "is not to teach mathematics, but to enable students from the very beginning of their course to know what the science is about, and why it is necessarily the foundation of exact thought as applied to natural phenomena" (p. 2). In it Whitehead stresses the three notions of variable, form, and generality.

Theoretical Physics. Whitehead's contributions to relativity, gravitation, and "unified field" theory grew out of his preoccupations with the principles underlying our knowledge of nature. These philosophical considerations are presented chiefly in *An Enquiry Concerning the Principles of Natural Knowledge* (1919), *The Concept of Nature* (1920), and *The Principle of Relativity* (1922). A. S. Eddington, in his own book *The Nature of the Physical World* (Cambridge, 1929), comments: "Although this book may in most respects seem diametrically opposed to Dr. Whitehead's widely read philosophy of Nature, I think it would be truer to regard him as an ally who from the opposite side of the mountain is tunnelling to meet his less philosophically minded colleagues" (pp. 249–250).

In a chapter on motion in the *Principles of Natural Knowledge*, Whitehead derives the Lorentz transformation equations, now so familiar in Einstein's special theory of relativity. Whitehead's derivation, however, was based on his principle of kinematic symmetry,[2] and was carried through without reference to the concept of light signals. Consequently the velocity c in the equations is not necessarily that of light, although it so happens that in our "cosmic epoch," c is most clearly realized in nature as the velocity of light. There are three types of kinematics, which Whitehead termed "hyperbolic," "elliptic," or "parabolic," according to whether c^2 is positive, negative, or infinite. Whitehead pointed out that the hyperbolic type of kinematics corresponds to the Larmor-Lorentz-Einstein theory of electromagnetic relativity and that the parabolic type reduces to the ordinary Newtonian relativity (Galilean transformation). He rejected the elliptic type as inapplicable to nature.

In *The Principle of Relativity* Whitehead challenged the conceptual foundations of both the special and general theories of Einstein by offering "an alternative rendering of the theory of relativity" (page v). One of Whitehead's fundamental hypotheses was that space-time must possess a uniform structure everywhere and at all times—a conclusion that Whitehead drew from a consideration of the character of our knowledge in general and of our knowledge of nature in particular. He argued that Einstein's view that space-time may exhibit a local curvature fails to provide an adequate theory of measurement:

> Einstein, in my opinion, leaves the whole antecedent theory of measurement in confusion when it is confronted with the actual conditions of our perceptual knowledge. . . . Measurement on his theory lacks systematic uniformity and requires a knowledge of the actual contingent field before it is possible.[3]

Whitehead proposed an action-at-a-distance theory rather than a field theory. He relieved the physicist of the task of having to solve a set of nonlinear partial differential equations. J. L. Synge, who ignored any consideration of the philosophical foundations of the theory, has clearly presented the mathematical formulas of Whitehead's gravitational theory in modern notation.[4]

Using Synge's notation, the world lines of test particles and light rays in Whitehead's theory may be conveniently discussed by the Euler-Lagrange equations:

$$\frac{d}{d\lambda}\frac{\partial L}{\partial \dot{x}_p} - \frac{\partial L}{\partial x_p} = 0, \tag{1}$$

where $2L = -1$ for test particles; $2L = 0$ for light rays; and $\dot{x}_p = \frac{dx_p}{d\lambda}$. The Lagrangian L is defined:

$$L = \frac{1}{2}g_{mn}\dot{x}_m\dot{x}_n, \tag{2}$$

where g_{mn} is a symmetrical tensor defined by

$$g_{mn} = \delta_{mn} + \frac{2mG}{c^2w^3}y_my_n. \tag{3}$$

In equation (3) δ_{mn} is the Kronecker delta; G is the gravitational constant; c is a fundamental velocity; and m is the mass of a particle with a world line given by $x'_n = x'_n(s')$, where s' is the Minkowskian arc length such that $ds'^2 = -dx'_n dx'_n$; $y_m = x_m - x'_m$; and $w = -y_n\frac{dx'_n}{ds'}$. The parameter λ in equation (1) is such that $d\lambda = (-g_{mn}dx_m dx_n)^{\frac{1}{2}}$. Latin suffixes have the range 1, 2, 3, 4. Thus Whitehead's theory of gravitation is described in terms of Minkowskian space-time with $x_4 = ict$ (where $i = \sqrt{-1}$ and c is the speed of light in a vacuum).[5] The basic physical laws of the Whitehead theory are invariant with respect to Lorentz transformations but not necessarily with respect to general coordinate transformations. Whitehead invoked neither the principle of equivalence nor the principle of covariance.

Clifford M. Will has challenged the viability of Whitehead's theory by arguing that it predicts "an anisotropy in the Newtonian gravitational constant G, as measured locally by means of Cavendish experiments."[6] Using Synge's notation, Will calculated Whitehead's prediction of twelve-hour sidereal-time earth tides, which are produced by the galaxy, and found Whitehead's prediction in disagreement with the experimentally measured value of these geotidal effects. In Whitehead's theory the anisotropy in G is a result of the uniform structure of space-time demanded by the theory.

In order to understand the relation of the anisotropy to uniformity we must recognize that in Whitehead's theory gravitational forces are propagated along the geodesics of the uniform structure of space-time, while electromagnetic waves are deflected by the contingencies of the universe.[7] This restriction in the propagation of gravity produces the variation in the gravitational constant. While Whitehead's mathematical formulas imply this restriction, it is not demanded by his philosophy of nature. For Whitehead, gravitational forces share in the contingency of nature, and may therefore be affected, as electromagnetic waves are, by the contingencies of the universe.

In addition to the consideration of gravitation, in chapter 5 of *The Principle of Relativity* Whitehead extends his equations of motion to describe the motion of a particle in a combined gravitational and electromagnetic field. As Rayner points out,[8] this is not a "true" unified field theory since it does not interpret gravitational and electromagnetic phenomena in terms of a single primitive origin.

It is possible to demonstrate, as did Eddington[9] and Synge,[10] that the predictions of Whitehead's theory and those of Einstein's general theory of relativity are equivalent with respect to the four tests of relativity: the deflection of a light ray, the red shift, the advance in the perihelion of a satellite, and radar time delay. The equivalence of the two theories with respect to these tests rests in the remarkable fact that both theories, when solved for a static, spherically symmetrical gravitational field, produce the Schwarzschild solution of the field equations.

In accordance with his usual practice, Whitehead assembled *Relativity* from lectures that he delivered at the Imperial College, the Royal Society of Edinburgh, and Bryn Mawr College. He did not publish in the journals of physical science nor enter into active discourse with members of the scientific community. His gravitational theory is not referred to in the formal treatments of relativity given by such authors as Bergmann, Einstein, and Pauli. The mathematical physicists who studied and extended Whitehead's physical theories in the 1950's had difficulty understanding his esoteric language and his philosophical ideas.

While the two ends of Eddington's tunnel have not yet been joined under the mountain, considerable progress has been made by the careful exposition of Whitehead's philosophy of science by Robert M. Palter.[11]

In 1961 C. Brans and R. H. Dicke developed a modified relativistic theory of gravitation apparently compatible with Mach's principle.[12] It is significant that the Einstein, Whitehead, and Brans-Dicke theories represent distinct conceptual formulations, the predictions of which with regard to observational tests are all so close that it is not yet possible on this basis to make a choice among them. New experiments of high precision on the possible Machian time variation of G and on the precession of the spin axis of a gyroscope,[13] as well as theoretical considerations such as the "parametrized post-Newtonian" (PPN) formalism,[14] may be decisive. At present the Einstein theory is regarded as the most influential and elegant; the Brans-Dicke theory has perhaps the most attractive cosmological consequences;[15] and the Whitehead theory, although clearly the simplest, suffers from its obscurity.

Philosophy of Science. Whitehead once remarked that what worried him was "the muddle geometry had got into" in relation to the physical world.[16] Particularly in view of Einstein's theory of relativity, it was unclear what relation geometrical space had to experience. It was therefore necessary to find a basis in physical experience for the scientific concepts of space and time. These are, Whitehead thought, "the first outcome of the simplest generalisations from experience, and . . . not to be looked for at the tail end of a welter of differential equations."[17] The supposed divorce of abstract scientific concepts from actual experience had resulted in a "bifurcation of nature," a splitting into two disparate natures, of which one was a merely apparent world of sense experience, the other a conjectured, causal world perpetually behind a veil. Aside from extrinsic quantitative relations, the elements of this latter world were presumed to be intrinsically self-contained and unrelated to one another. Somehow this conjectured, monadically disjunctive nature, although itself beyond experience, was supposed to account causally for the unified nature of experience. Whitehead rejected this view as incoherent and as an unsatisfactory foundation for the sciences. According to Whitehead, "we must reject the distinction between nature as it really is and experiences of it which are purely psychological. Our experiences of the apparent world are nature itself."[18]

In his middle writings Whitehead examined how space and time are rooted in experience, and in general laid the foundations of a natural philosophy that would be the necessary presupposition of a reorganized speculative physics. He investigated the coherence of "Nature," understood as the object of perceptual knowledge; and he deliberately, although perhaps unsuccessfully, distinguished nature as thus known from the synthesis of knower and known, which falls within the ambit of metaphysical analysis.

Two special characteristics of Whitehead's analysis are of particular importance: his identification of noninstantaneous events as the basic elements of perceived nature, and the intrinsically relational constitution of these events (as displayed in his doctrine of "significance"). Space and time (or space-time) are then shown to be derivative from the fundamental process by which events are interrelated, rather than a matrix within which events are independently situated. This view contrasts sharply with the prevalent notion that nature consists in an instantaneous collection of independent bodies situated in space-time. Such a view, Whitehead thought, cannot account for the perception of the continuity of existence, nor can it represent the ultimate scientific fact, since change inevitably imports the past and the future into the immediate fact falsely supposed to be embodied in a durationless present instant.

Whitehead's philosophy of nature attempts to balance the view of nature-in-process with a theory of elements ingredient within nature ("objects"), which do not themselves share in nature's passage. Whitehead's boyhood sense of permanences in nature thus emerged both in his mathematical realism and in his philosophic recognition of unchanging characters perpetually being interwoven within the process of nature.

Method of Extensive Abstraction. "Extensive abstraction" is the term Whitehead gave to his method for tracing the roots within experience of the abstract notions of space and time, and of their elements.

In this theory it is experienced events, not physical bodies, that are related; their fundamental relation lies in their overlapping, or "extending over," one another. Later Whitehead recognized that this relation is itself derivative from something more fundamental.[19] The notions of "part," "whole," and "continuity" arise naturally from this relation of extending-over. These properties lead to defining an "abstractive set" as "any set of events that possesses the two properties, (i) of any two

members of the set one contains the other as a part, and (ii) there is no event which is a common part of every member of the set."[20] Such a set of events must be infinite toward the small end, so that there is no least event in the set. Corresponding to the abstractive set of events there is an abstractive set of the intrinsic characters of the events. The latter set converges to an exactly defined locational character. For instance, the locational character of an abstractive set of concentric circles or squares converges to a nondimensional but located point. In analogous fashion, an abstractive set of rectangles, all of which have a common length but variable widths, defines a line segment. With the full development of this technique Whitehead was able to define serial times, and, in terms of them, space. He concluded that all order in space is merely the expression of order in time. "Position in space is merely the expression of diversity of relations to alternative time-systems."[21]

In general Whitehead held that there are two basic aspects in nature. One is its passage or creative advance; the other its character as extended — that is, that its events extend over one another, thus giving nature its continuity. These two facts are the qualities from which time and space originate as abstractions.

The purpose of the method of extensive abstraction is to show the connection of the abstract with the concrete. Whitehead showed, for instance, how the abstract notion of a point of instantaneous space is naturally related to the experience of events in nature, which have the immediately given property of extension. Whitehead's procedure, however, is easily subject to misunderstanding. Most Whitehead scholars agree that Whitehead was trying neither to deduce a geometry from sense experience, nor to give a psychological description of the genesis of geometric concepts. Rather, he was using a mathematical model to clarify relations appearing in perception. Another misinterpretation would be to assume that Whitehead took as the immediate data for sense awareness some kind of Humean sensa instead of events themselves.

In his notes to the second edition of the *Principles of Natural Knowledge* Whitehead suggested certain improvements in his procedure. The final outcome of extensive abstraction is found in part 4 of *Process and Reality*, "The Theory of Extension," in which Whitehead defines points, lines, volumes, and surfaces without presupposing any particular theory of parallelism, and defines a straight line without any reference to measurement.

Uniformity of Spatiotemporal Relations. In the Preface to *The Principle of Relativity* Whitehead states:

> As the result of a consideration of the character of our knowledge in general, and of our knowledge of nature in particular, . . . I deduce that our experience requires and exhibits a basis of uniformity, and that in the case of nature this basis exhibits itself as the uniformity of spatio-temporal relations. This conclusion entirely cuts away the casual heterogeneity of these relations which is the essential of Einstein's later theory.

The mathematical consequences of this conclusion for Whitehead's theory of relativity have already been noted. It remains to indicate summarily the reasons that persuaded Whitehead to adopt this view.

Consonance with the general character of direct experience was one of the gauges by which Whitehead judged any physical theory, for he was intent on discovering the underlying structures of nature as observed. Further, he maintained the traditional division between geometry and physics: it is the role of geometry to reflect the relatedness of events; that of physics to describe the contingency of appearance. He also claimed that it is events, not material bodies, that are the terms of the concrete relations of nature. But since for Whitehead these relations were essentially constitutive of events, it might seem that no event can be known apart from knowledge of all those other events to which it is related. Thus, nothing can be known until everything is known—an impossible requirement for knowledge.

Whitehead met this objection by distinguishing between essential and contingent relations of events. One can know that an event or factor is related to others without knowing their precise character. But since in our knowledge no event discloses the particular individuals constituting the aggregate of events to which it is related, even contingently, this relatedness must embody an intrinsic uniformity apart from particular relationships to particular individuals. This intrinsic and necessary uniformity of the relatedness of events is precisely the uniformity of their spatiotemporal structure.

Whitehead provided an illustration of this in a discussion of equality.[22] Equality presupposes measurement, and measurement presupposes matching (not vice versa). It must follow that "measurement presupposes a structure yielding definite stretches which, in some sense inherent in the structure, match each other."[23] This inherent matching is spatiotemporal uniformity.

It is well known that in his later philosophy Whitehead came to hold—contrary to his earlier belief—that nature is not continuous in fact, but "incurably atomic." Continuity was recognized to belong to potentiality, not to actuality.[24] It has even been claimed that this later revision removes the basic difference between Einstein and Whitehead, so that the Whitehead of *Process and Reality* offers only an alternative interpretation of Einstein's theory of relativity, not an alternative theory.[25] This claim, however, has not found wide support.

Despite some recent interest in it, Whitehead's theory of relativity has been mainly ignored and otherwise not well understood. *The Principle of Relativity* has long been out of print, and it is impossible now to say whether it has a scientific future.

NOTES

1. Bertrand Russell, "Whitehead and Principia Mathematica," *Mind*, n.s. **57** (1948), 137–138.
2. For a discussion and derivation, see C. B. Rayner, "Foundations and Applications of Whitehead's Theory of Relativity," University of London thesis, 1953; "The Application of the Whitehead Theory of Relativity to Non-static, Spherically Symmetrical Systems," in *Proceedings of the Royal Society of London*, **222A** (1954), 509–526.
3. *The Principle of Relativity*, p. 83.
4. J. L. Synge, in *Proceedings of the Royal Society of London*, **211A** (1952), 303–319.
5. Whitehead's requirement that space-time be homogeneous is not violated by a space-time of constant curvature. This extension of Whitehead's theory has been carried out by G. Temple, "A Generalisation of Professor Whitehead's Theory of Relativity," in *Proceedings of the Physical Society of London*, **36** (1923), 176–193; and by C. B. Rayner, "Whitehead's Law of Gravitation in a Space-Time of Constant Curvature," in *Proceedings of the Physical Society of London*, **68B** (1955), 944–950.
6. Clifford M. Will, "Relativistic Gravity in the Solar System . . .," p. 141.
7. Misner, Thorne, and Wheeler, *Gravitation*, p. 430. Whitehead's theory is termed a "two metric" theory of gravitation. The first metric defines the uniform structure of space-time; the second, the physically contingent universe.
8. Rayner, "Foundations and Applications . . .," p. 23.
9. Sir A. S. Eddington, "A Comparison of Whitehead's and Einstein's Formulae," p. 192.
10. J. L. Synge, *The Relativity Theory of A. N. Whitehead* (1951). In ch. 13 of *The Principle of Relativity* Whitehead obtains a red shift that disagrees with Einstein's by a factor of 7/6. This is in disagreement with the terrestrial Mössbauer experiments (see R. V. Pound and G. A. Rebka, Jr., "Apparent Weight of Photons," *Physical Review Letters*, 4 [1960], 337–341). Synge observes, however, that the discrepancy lies in Whitehead's use of a classical rather than a quantum mechanical model of an atom and is not due to Whitehead's gravitational theory. See also C. B.

Rayner, "The Effects of Rotation of the Central Body on Its Planetary Orbits, After the Whitehead Theory of Gravitation," in *Proceedings of the Royal Society of London*, **232A** (1955), 135–148.

11. Robert M. Palter, *Whitehead's Philosophy of Science*.
12. C. Brans and R. H. Dicke, in *Physical Review*, **124** (1961), 925–935.
13. L. I. Schiff, "Experimental Tests of Theories of Relativity," in *Physics Today*, **14**, no. 11 (November 1961), 42–48.
14. C. M. Will, *op. cit.*
15. R. H. Dicke, "Implications for Cosmology of Stellar and Galactic Evolution Rates," in *Review of Modern Physics*, **34** (1962), 110–122.
16. Lowe, *Understanding Whitehead*, p. 193.
17. *Principles of Natural Knowledge*, p. vi.
18. *The Principle of Relativity*, p. 62.
19. *Principles of Natural Knowledge*, p. 202.
20. *The Concept of Nature*, p. 79; *Principles of Natural Knowledge*, p. 104.
21. *The Principle of Relativity*, p. 8.
22. *Ibid.*, ch. 3.
23. *Ibid.*, p. 59.
24. Leclerc, "Whitehead and the Problem of Extension."
25. Seaman, "Whitehead and Relativity."

BIBLIOGRAPHY

I. ORIGINAL WORKS. A chronological list of all Whitehead's writings may be found in P. A. Schilpp (see below). The following works are of most scientific importance: *A Treatise on Universal Algebra, With Applications* (Cambridge, 1898); "On Mathematical Concepts of the Material World," in *Philosophical Transactions of the Royal Society of London*, **205A** (1906), 465–525, also available in the Northrop and Gross anthology (see below): *Principia Mathematica*, 3 vols. (Cambridge, 1910–1913), written with Bertrand Russell; *An Introduction to Mathematics* (London, 1911); "Space, Time, and Relativity," in *Proceedings of the Aristotelian Society*, n.s. **16** (1915–1916), 104–129, also available in the Johnson anthology (see below); *An Enquiry Concerning the Principles of Natural Knowledge* (Cambridge, 1919); *The Concept of Nature* (Cambridge, 1920); *The Principle of Relativity, With Applications to Physical Science* (Cambridge, 1922), which is out-of-print but may be obtained from University Microfilms, Ann Arbor, Mich.; also, pt. 1, "General Principles," is reprinted in the Northrop and Gross anthology; *Science and the Modern World* (New York, 1925); *Process and Reality: An Essay in Cosmology* (New York; 1929), which is of scientific interest chiefly insofar as it gives Whitehead's final version of his theory of extensive abstraction; and *Essays in Science and Philosophy* (New York, 1947), a collection of earlier essays.

Two useful anthologies of Whitehead's writings are F. S. C. Northrop and Mason W. Gross, eds., *Alfred North Whitehead: An Anthology* (New York, 1961); and A. H. Johnson, ed., *Alfred North Whitehead: The Interpretation of Science, Selected Essays* (Indianapolis, 1961).

II. SECONDARY LITERATURE. Paul Arthur Schilpp, ed., *The Philosophy of Alfred North Whitehead*, Library of Living Philosophers Series (New York, 1951), contains Whitehead's "Autobiographical Notes," a complete chronological list of Whitehead's writings, and essays pertinent to Whitehead's science by Lowe, Quine, and Northrop. Victor Lowe, *Understanding Whitehead* (Baltimore, 1962), is a valuable tool, especially pt. 2, "The Development of Whitehead's Philosophy," which is an enlargement of Lowe's essay in the Schilpp volume. Robert M. Palter, *Whitehead's Philosophy of Science* (Chicago, 1960), is a perceptive mathematical exposition of Whitehead's views on extension and on relativity. In 1971 appeared *Process Studies* (published at the School of Theology at Claremont, California), a journal devoting itself to exploring the thought of Whitehead and his intellectual associates. The fourth issue of vol. 1 (Winter 1971) contains a bibliography of secondary literature on Whitehead, to be periodically updated.

The following are cited as examples of the influence of Whitehead's thought on scientists or philosophers of science. In *Experience and Conceptual Activity* (Cambridge, Mass., 1965), J. M. Burgers, a physicist of some distinction, presents for scientists a case for a Whiteheadian rather than a physicalistic world view. Also, a strong Whiteheadian perspective dominates Milič Čapek, *The Philosophical Impact of Contemporary Physics* (New York, 1961).

Whitehead's later metaphysics, although consistent with and developed out of his reflections on science, forms another story altogether. For a more general introduction to his thought and to the literature, see the article on Whitehead in Paul Edwards, ed., *The Encyclopedia of Philosophy*, VIII (New York–London, 1967), 290–296.

On Whitehead's mathematics and logic, see Granville C. Henry, Jr., "Whitehead's Philosophical Response to the New Mathematics," in *Southern Journal of Philosophy*, 7 (1969–1970), 341–349; George L. Kline, ed., *Alfred North Whitehead: Essays on His Philosophy*, pt. 2 (Englewood Cliffs, N.J., 1963); J. J. C. Smart, "Whitehead and Russell's Theory of Types," in *Analysis*, **10** (1949–1950), 93–96, which is critical of the theory of types; Martin Shearn, "Whitehead and Russell's Theory of Types: A Reply," *ibid.*, **11** (1950–1951), 45–48.

On Whitehead's theoretical physics, see Sir A. S. Eddington, "A Comparison of Whitehead's and Einstein's Formulae," in *Nature*, **113** (1924), 192; Charles W. Misner, Kip S. Thorne, John Archibald Wheeler, *Gravitation* (San Francisco, 1973); C. B. Rayner, "Foundations and Applications of Whitehead's Theory of Relativity" (Ph.D. thesis, University of London, 1953); A. Schild, "Gravitational Theories of the Whitehead Type and the Principle of Equivalence," in *Proceedings of the International School of Physics*, "Enrico Fermi," course 20 (Italian Physical Society and Academic Press, 1963), 69–115; Francis Seaman, "Discussion: In Defense of

Duhem," in *Philosophy of Science*, **32** (1965), 287–294, which argues that Whitehead's physical theory in *Process and Reality* illustrates the assumption of geometric, without physical, continuity; J. L. Synge, *The Relativity Theory of Alfred North Whitehead* (College Park, Md., 1951); Clifford M. Will, "Relativistic Gravity in the Solar System, II: Anisotropy in the Newtonian Gravitational Constant," in *Astrophysical Journal*, **169** (1971), 141–155; and "Gravitation Theory," in *Scientific American*, **231**, no. 5 (1974), 24–33, which compares competing theories.

On Whitehead's philosophy of science, see Ann P. Lowry, "Whitehead and the Nature of Mathematical Truth," in *Process Studies*, **1** (1971), 114–123; Thomas N. Hart, S. J., "Whitehead's Critique of Scientific Materialism," in *New Scholasticism*, **43** (1969), 229–251; Nathaniel Lawrence, "Whitehead's Method of Extensive Abstraction," in *Philosophy of Science*, **17** (1950), 142–163; Adolf Grünbaum, "Whitehead's Method of Extensive Abstraction," in *British Journal for the Philosophy of Science*, **4** (1953), 215–226, which attacks the validity of Whitehead's method (see Lowe's reply in *Understanding Whitehead*, pp. 79–80); Caroline Whitbeck, "Simultaneity and Distance," in *Journal of Philosophy*, **66** (1969), 329–340; Wolfe Mays, "Whitehead and the Philosophy of Time," in *Studium generale*, **23** (1970), 509–524; Robert R. Llewellyn, "Whitehead and Newton on Space and Time Structure," in *Process Studies*, **3** (1973), 239–258; Ivor Leclerc, "Whitehead and the Problem of Extension," in *Journal of Philosophy*, **58** (1961), 559–565; Robert M. Palter, "Philosophic Principles and Scientific Theory," in *Philosophy of Science*, **23** (1956), 111–135, compares the theories of Einstein and Whitehead.

See also Francis Seaman, "Whitehead and Relativity," in *Philosophy of Science*, **22** (1955), 222–226; A. P. Ushenco, "A Note on Whitehead and Relativity," in *Journal of Philosophy*, **47** (1950), 100–102; Dean R. Fowler, "Whitehead's Theory of Relativity," in *Process Studies*, **5** (1975), which treats the philosophical foundations of Whitehead's theory of relativity; and Richard J. Blackwell, "Whitehead and the Problem of Simultaneity," in *Modern Schoolman*, **41** (1963–1964), 62–72. The extent to which applications of Whitehead's philosophical scheme agree with modern quantum theory has been discussed by Abner Shimony, "Quantum Physics and the Philosophy of Whitehead," in *Boston Studies in the Philosophy of Science*, II (New York, 1965), 307–330; and by J. M. Burgers, "Comments on Shimony's Paper," *ibid.*, pp. 331–342. Henry J. Folse, Jr., "The Copenhagen Interpretation of Quantum Theory and Whitehead's Philosophy of Organism," in *Tulane Studies in Philosophy*, **23** (1974), 32–47, challenges Shimony's conclusions.

WILLIAM A. BARKER
KAREL L. DE BOUVÈRE, S.C.J.
JAMES W. FELT, S.J.
DEAN R. FOWLER

WHITEHEAD, JOHN HENRY CONSTANTINE (*b.* Madras, India, 11 November 1904; *d.* Princeton, New Jersey, 8 May 1960), *mathematics.*

Whitehead is perhaps best remembered for his idea of developing the theory of homotopy equivalence by the strictly combinatorial method of allowed transformations. He built up an important school of topology at Oxford.

Whitehead was the son of the Right Reverend Henry Whitehead, from 1899 to 1922 bishop of Madras, and of Isobel Duncan of Calne, Wiltshire. She had been one of the first undergraduates to study mathematics at Lady Margaret Hall, Oxford. Bishop Whitehead was the brother of the mathematician Alfred North Whitehead.

Sent to England before he was two, Whitehead saw little of his parents until his father's retirement to England in 1922. He was educated at Eton and Balliol College, Oxford. His Balliol tutor was J. W. Nicholson, who had studied under A. N. Whitehead. Whitehead boxed for the university, was a good cricketer, and an even better poker player. After graduating in mathematics he joined a firm of stockbrokers, but in 1928 he returned to Oxford to do further mathematical work. There he met Oswald Veblen, on leave from Princeton University, and it was arranged that Whitehead should visit Princeton on a Commonwealth fellowship. He was there from 1929 to 1932, when, having taken a Ph.D., he returned to Oxford and a fellowship at Balliol. In 1934 Whitehead married Barbara Shiela Carew Smyth, a concert pianist. They had two sons.

From 1941 to 1945 Whitehead worked at the Admiralty and Foreign Office. He was elected a fellow of the Royal Society in 1944, and Waynflete professor of pure mathematics and fellow of Magdalen College, Oxford, in 1947. He was president of the London Mathematical Society from 1953 to 1955. He died of a heart attack during a visit to Princeton.

On Whitehead's first visit to Princeton he took up the studies that were to occupy the remainder of his life. There he collaborated with S. Lefschetz on a proof that all analytic manifolds can be triangulated (*Mathematical Works*, II, no. 15 [1933]; see bibliography for details of the edition). He offered a proof (*ibid.*, no. 16 [1934]; corrected in no. 18 [1935]) of the Poincaré hypothesis that a simply connected 3-manifold, compact and without boundary, is a topological 3-sphere. Although Whitehead soon found his proof to have been erroneous, work on it committed him to topology. One memorable early discovery was of a counterexample for open 3-manifolds (*ibid.*, no. 20 [1935]). Before turning to topology, Whitehead had made an important study of the geometry of paths. A monograph on the foundations of differential geometry, written with Veblen, contains the first precise definition, through axioms, of a differential manifold (*Mathematical Works*, I, no. 7 [1932]). This definition was much more precise than the concept of a global differential manifold offered, for example, by Robert König (1919) and E. Cartan (1928). In another work written with Veblen (*ibid.*, no. 6 [1931]) the independence of the axioms is proved.

Under the influence of Marston Morse, Whitehead studied differential geometry in the large, and his paper "On the Covering of a Complete Space by the Geodesics Through a Point" (*Mathematical Works*, I, no. 17 [1935]) marks a turning point in this subject. Assuming an analytic manifold with a Finsler metric, he discussed the relationship between different concepts of completeness in the manifold. He also made a detailed investigation of the properties of the locus of characteristic points of a given point. Other notable work in differential geometry includes his new and elegant proof of a theorem first stated by E. E. Levi and of an important analogue (*ibid.*, no. 22 [1936], and no. 36 [1941]).

After 1941 Whitehead was mainly concerned with topology. He had never lost his early interest in the subject, and J. W. Milnor describes his "Simplicial Spaces, Nuclei and *M*-Groups" (*Mathematical Works*, II, no. 28 [1939]) as the paper that will probably be remembered as his most significant work. Milnor discusses this and related work at length (*Mathematical Works*, I, xxv–xxxiii). The 1939 paper was a brilliant extension of the strictly combinatorial type of topology developed by J. W. Alexander and M. H. A. Newman between 1925 and 1932. (Whitehead had met Newman on his first visit to Princeton.) The contents of the paper were characterized by Whitehead's idea of using the strictly combinatorial method of allowed transformations to solve problems in the theory of homotopy equivalence.

Whitehead's interests gradually shifted toward algebraic topology as a result of his search for invariants to characterize the homotopy type of complexes, and for methods of computing their homotopy groups. Newman explains how Whitehead's discovery of certain mistakes he had made in a paper written in 1941 persuaded him to avoid a free "geometrical" style of composition. Whitehead therefore undertook a complete restatement of his earlier work on homotopy, in a way expertly

outlined by Newman. In the last three years of Whitehead's life there was a revival of geometrical topology, which led him to offer, jointly with A. Shapiro, a proof of Dehn's lemma much simpler than the one given in 1957 by C. D. Papakyriakopoulos (*Mathematical Works*, IV, no. 84 [1958]). Here, and in his elaboration of methods laid down by B. Mazur (1958) and Morton Brown (1960), there is ample evidence that Whitehead died at the height of his mathematical powers.

BIBLIOGRAPHY

I. ORIGINAL WORKS. Ninety papers, some of them lengthy memoirs, are collected in *Mathematical Works of J. H. C. Whitehead*, I. M. James, ed., 4 vols. (Oxford, 1962). References in text are to this collection, the numbers corresponding to the list of Whitehead's publications in I, ix–xiii, and the original year of publication being added in brackets. The papers are classified as follows: vol. I, differential geometry; vol. II, complexes and manifolds; vol. III, homotopy theory; vol. IV, algebraic and classical topology. Whitehead collaborated with Oswald Veblen on *The Foundations of Differential Geometry* (Cambridge, 1932), included in *Mathematical Works*, I. His Oxford lectures on Riemannian geometry and linear algebras, which were separately duplicated and circulated by the Mathematical Institute, Oxford, in 1959, are not included in the collected edition.

II. SECONDARY LITERATURE. Vol. I of the *Mathematical Works* is prefaced by a biographical note by M. H. A. Newman and Barbara Whitehead, and by a mathematical appreciation by John W. Milnor. Two other valuable surveys of Whitehead's work are M. H. A. Newman's obituary notice in *Biographical Memoirs of Fellows of the Royal Society*, 7 (1961), 349–363; and P. J. Hilton, "Memorial Tribute to J. H. C. Whitehead," in *L'enseignement mathématique*, 2nd ser., 7 (1961), 107–124.

J. D. NORTH

WHITEHURST, JOHN (*b.* Congleton, Cheshire, England, 10 April 1713; *d.* London, England, 18 February 1788), *geology.*

Whitehurst was a practical natural philosopher in many fields but was particularly celebrated as a clockmaker. In pure science he is important chiefly as a geological pioneer who did work in Derbyshire that was published in a well-known book (of which the first part is entirely speculative) in 1778. He established for the first time the succession of the Carboniferous strata: limestone, Millstone grit (named by him), and coal measures.

Whitehurst formulated the general proposition of

a worldwide orderly superposition of strata, each with its characteristic lithology and fossils. Although the proposition was somewhat vaguely imagined, he here hit on the most significant of all geological generalizations. He investigated the origin of the Derbyshire "toadstones," associated with the limestones, and in so doing examined by implication the origin of all rocks of a like kind. Whitehurst found the rock to be so similar to specimens of recent lavas he had seen that he had no hesitation in assigning to it a volcanic origin, although it was situated in a region—indeed, in a country—that showed not the slightest sign of any recent volcanic activity. He was among the first to recognize the true nature and origin of this great class of rocks, the basalts, and thus to establish the fact of volcanism in past geological times.

Whitehurst went further, however, and realized the possibility of igneous intrusion and recorded an instance of contact thermal metamorphism. In the second edition (1786) of his book he described the basaltic rocks of the Giant's Causeway, on the north coast of Ireland, recognizing their volcanic origin and making a reasonable suggestion as to how they might have been erupted in that region.

BIBLIOGRAPHY

I. ORIGINAL WORKS. Whitehurst's chief writings are *An Inquiry Into the Original State and Formation of the Earth, to Which Is Added an Appendix Containing Some General Observations on the Strata in Derbyshire* (London, 1778; 2nd ed., enl., 1786; repr. as 3rd ed., 1792) and *An Attempt Towards Obtaining Invariable Measures of Length, Capacity, and Weight, From the Mensuration of Time* (London, 1787). His writings are collected in *The Works of John Whitehurst, F.R.S., With Memoirs of His Life and Writings*, C. Hutton, ed. (London, 1792).

II. SECONDARY LITERATURE. See E. I. Carlyle, in *Dictionary of National Biography*; J. Challinor, "From Whitehurst's Inquiry to Farey's Derbyshire: A Chapter in the History of British Geology," in *Transactions and Annual Report. North Staffordshire Field Club*, **81** (1947), 52–88, esp. 53–65; and "The Early Progress of British Geology—II," in *Annals of Science*, **10** (1954), 1–19, see 13–16; T. D. Ford, "Biographical Notes on Derbyshire Authors: John Whitehurst, F.R.S. 1713–1788," in *Bulletin of the Peak District Mines Historical Society*, **5** (1974), 362–369; and W. D. White, "Derbyshire Clockmakers Before 1850; The Whitehurst Family," supp. to *Derbyshire Miscellany*, **1** (1958).

JOHN CHALLINOR

WHITFIELD, ROBERT PARR (*b.* Willowvale, near New Hartford, New York, 27 May 1828; *d.* Troy, New York, 6 April 1910), *invertebrate paleontology, stratigraphy*.

Whitfield was a second-generation American, son of William Fenton Whitfield, an immigrant English maker of mill spindles, and Margaret Parr. His only formal instruction, which he received principally at a Stockport Sunday school and its library during six years spent with his parents in England, ended when the family returned to New York in 1841. He then worked for seven years in his father's trade. In 1847 he married Mary Henry of Utica; of their four children, three survived him. At twenty Whitfield was employed by a Utica "philosophical" instruments firm, where he developed his considerable mechanical skills and drafting ability, serving as manager after 1849.

Participation in the Utica Society of Naturalists and collecting local fossils brought Whitfield to the attention of James Hall, whose assistant he became in 1856, succeeding Fielding Meek in 1858. In 1870 he was appointed principal assistant curator of the New York State Museum. At Albany, Whitfield undertook official field studies in New York and western states. More significantly, he drew thousands of superb illustrations and made preliminary analyses of fossil brachiopods, crinoids, and graptolites for the *Palaeontology of New York* and for the various state surveys of which Hall was either head or contract paleontologist. Much of Whitfield's best work was produced during his twenty-year association with Hall. It included studies of Paleozoic bivalve mollusks, internal structures of fossil brachiopods, the Paleozoic-Mesozoic paleontology of Nevada and Utah for Clarence King's Fortieth Parallel Survey, and descriptions of the Black Hills fossils from the Newton-Jenney survey of 1875.

Although he wrote nine papers with Hall during this interval, like his sometime fellow assistants Meek, Charles White, and William Gabb, Whitfield received less credit in authorship than was his due.[1] From 1872 to 1875 he also lectured informally at Rensselaer Polytechnic Institute, fulfilling Hall's nominal commitment;[2] he served as professor of geology there in 1876–1878, after Hall's retirement.

Early in 1877 Whitfield became the first curator of the newly organized American Museum of Natural History, with initial charge of geology. Its holdings had just been expanded by the purchase of an immense collection of invertebrate fossils from Hall. Whitfield's long-continued task of curation led to a protracted, lively correspondence with Hall as to exactly what percentage of the collection had been merely loaned or sold outright.[3] Whitfield remained at the Museum, with varying curatorial titles and additional responsibilities, until his retirement in December 1909. He was chiefly responsible for the establishment of the Museum's *Bulletin* in 1881, to which he contributed its first five articles. He subsequently wrote numerous descriptions and comparisons of an array of fossil invertebrates, faunas, and their stratigraphic relations. Some were brief, as were many works of late nineteenth-century American paleontologists who often dealt with biotas new to science.

Whitfield's systematic paleontology included investigations of sponges, brachiopods, mollusks, trilobites, scorpions, crustaceans, and crinoids from diverse locales and periods. Some studies involved new or continued contracts from the state surveys of Minnesota, Wisconsin, Indiana, Ohio, and New Jersey. Of his three quarto investigations of Cretaceous and Miocene invertebrates in New Jersey, which were significant contributions to the paleontology of the Atlantic coast, two were published by both the New Jersey and the U.S. Geological Surveys. Whitfield's published discussions of species variability and transmutation reflect the American neo-Lamarckian emphasis on environmental modifying influences and inheritance of acquired characters.

NOTES

1. A controversy over coauthorship of an unsigned preliminary paper on New York Devonian bivalves, inevitably inferred to be by Hall, embittered Whitfield during his last years at Albany. See G. Arthur Cooper, "Concerning the Authorship of the Preliminary Notice of the Lamellibranch Shells of the Upper Helderberg, Hamilton and Chemung Groups, etc., Part 2," in *Journal of the Washington Academy of Sciences*, 21, no. 18 (1931), 459–467.

2. Complaints in the student yearbook, *Transit*, and Rensselaer archival records suggest the position was essentially a sinecure for Hall. "WH-TF---D *alias* TRILOBITE. First differential coefficient of James H-ll, a function of the Ecozoic fossils . . ." appears in the April 1874 *Transit*. Whitfield is credited with formal service in 1877–1878 by Palmer C. Ricketts, *History of the Rensselaer Polytechnic Institute, 1824–1914* (New York–London, 1914), 228. Ricketts, then president of the Institute, as an undergraduate had coedited "the boisterously critical *Transit* of 1874." Samuel Reznick, *Education for a Technological Society. A Sesquicentennial History of Rensselaer Polytechnic Institute* (Troy, N.Y., 1968), 176, 181.

3. Whitfield sold his own personal collection, including more than 100 type specimens, to the University of California at Berkeley in 1886; see Joseph H. Peck, Jr., and Herdis B.

McFarland, "Whitfield Collection Types at the University of California," in *Journal of Paleontology*, 28, no. 3 (1954), 297–309, pl. 29.

BIBLIOGRAPHY

I. ORIGINAL WORKS. The James Hall Papers (KW 13835) in the Manuscripts and History Department, New York State Library, Albany, contain Whitfield's letters to Hall written between 1856 and 1893. Record Unit 33 and the uncatalogued collection of the Geology Department in the American Museum of Natural History Archives form the principal depository of Whitfield's correspondence during the 1896–1910 portion of his curatorship. John James Stevenson's letter to Fielding Bradford Meek of 10 October 1875 discusses Whitfield's search for a new position, as do Whitfield's letters to Meek, Fielding B. Meek Papers, Record Unit 7062, Smithsonian Institution Archives.

In addition to the bibliographies in the memorials by Gratacap and Clarke, 97 of Whitfield's 110 listed publications are cited in John M. Nickles, "Geologic Literature on North America 1785–1918," in *U.S. Geological Survey Bulletin* no. 746, pt. I (1923), 1103–1106. Three additional papers published before 1890 are among those noted in Nelson H. Darton, "Catalogue and Index of Contributions to North American Geology," *ibid.*, no. 127 (1896), 1010–1011.

II. SECONDARY LITERATURE. Articles on Whitfield are John M. Clarke, "Biographical Memoir of Robert Parr Whitfield," in *Bulletin of the Geological Society of America*, 22 (1911), 22–32, with bibliography by Louis Hussakof; Louis P. Gratacap, "Professor Robert Parr Whitfield," in *American Journal of Science*, 4th ser., 29 (1910), 565–566; and "Biographical Memoir of Robert Parr Whitfield," in *Annals of the New York Academy of Sciences*, 20 (1911), 385–398, with bibliography by Louis Hussakof; Edmund O. Hovey, "Robert Parr Whitfield," in *American Museum Journal*, 10 (1910), 119–121; Henry B. Nason, ed., *Biographical Record of the Officers and Graduates of the Rensselaer Polytechnic Institute, 1824–1886* (Troy, N.Y., 1887), 158–161; Chester A. Reeds, "Robert Parr Whitfield," in *Dictionary of American Biography*, XX (1936), 134–135; and the unsigned "Robert Parr Whitfield," in *National Cyclopaedia of American Biography*, V (1907), 92–93.

CLIFFORD M. NELSON

WHITMAN, CHARLES OTIS (*b.* North Woodstock, Maine, 14 December 1842; *d.* Chicago, Illinois, 6 December 1910), *zoology*.

Whitman grew up on a farm, where at an early age he became interested in natural history, particularly pigeons. He was exceptionally skilled as a self-taught taxidermist and built up quite a museum in his father's house. By tutoring and teaching in private schools he earned enough to enter Bowdoin College in 1865, receiving the B.A. in 1868. From 1868 to 1872 he was principal of Westford Academy in Massachusetts, and in 1872–1874 he taught at English High School in Boston. The latter post was of crucial importance for his career because it was in Boston that Whitman came under the influence of Louis Agassiz. As a result he was a participant in the first course in marine biology on Penikese Island, conducted by Agassiz in June and July 1873.

In 1875 Whitman went to Europe, first to Dohrn in Naples and then to Leuckart in Leipzig, where he learned the modern methods of microscopy and embryology. His Ph.D. dissertation (he received the degree in 1878) was on the embryology of *Clepsine* (*Glossiphonia*). In 1879 he went for two years to the Imperial University of Tokyo, as professor of zoology. Since eight of his students later became well-known zoologists, four of them holding major chairs, he has rightly been called the father of zoology in Japan. From November 1881 to May 1882, Whitman was again at the Zoological Station in Naples, working on the embryology, life history, and classification of the dicyemids, publishing a standard reference work on these mesozoans in 1883. One of the most productive periods in his life was the period 1882–1886, when he was assistant in zoology at the Museum of Comparative Zoology at Harvard, under Alexander Agassiz. From 1886 to 1889 Whitman served as director of the Allis Lake Laboratory at Milwaukee, Wisconsin, where he founded the *Journal of Morphology*, the first periodical in America devoted to zoology and anatomy. It served as a model for other publications founded in later years.

In 1889 Whitman accepted the chair of zoology at the newly founded Clark University in Worcester, Massachusetts, where he stayed until 1892. In that year he and most of his colleagues in the science departments moved to the newly established University of Chicago.

Whitman played a leading role in the founding of the Marine Biological Laboratory at Woods Hole in 1888 and served as its first director (1893–1908), developing the policies that have made this institution such a signal success. He resigned because his research on the heredity and behavior of pigeons was seriously impeded by his spending every summer at Woods Hole.

Being extremely unselfish, a person of complete integrity, and dedicated to science, Whitman played an important role as an organizer and first director of new institutions, and as the founder and first editor of new journals. Many of them are still flourishing, such as the Woods Hole Marine Biological Laboratory, the American Society of Zoologists (founded in 1890 as the American Morphological Society), *Journal of Morphology*, and *Biological Bulletin*.

Whitman's research was of unusual breadth. For instance, he became interested in leeches as material for embryological research; but he soon turned to the study of their anatomy and taxonomy, and finally of behavior, all of these interests resulting in publications. He found abundant evidence that the development of the leech egg was completely predetermined but—of course—not preformed in a homunculus-type way. He fought the extreme cell-lineage ("mosaic") interpretation, stressing the contribution to development made by the interaction of cells. Whitman's exceptionally careful work and perceptive interpretation had a profound impact on the embryology of his period. His discovery of the sensilla, segmental sense organs in the leeches, greatly facilitated the study of leech morphology and taxonomy.

When, after 1900, the great split occurred between evolutionists who ascribed evolutionary change to the pressure of a few major mutations and those who ascribed it to selection, Whitman was emphatically on the side of selection. He found numerous (now known to be polygenic) characters in his pigeon crosses that did not mendelize in the simple manner claimed by de Vries and Bateson. For more than ten years he bred pigeons, hybridizing 200 domestic varieties and 40 wild species. The results of his crosses are recorded in two posthumously published volumes (1919). Seeing the similarity of variation in related species and the unmistakable trends of evolutionary change from the most primitive to the most advanced species, Whitman developed a theory of orthogenetic evolution. At first held to be totally erroneous, it is now considered far less of a failure since it has been realized that the potential for variation in a phyletic line is very narrowly prescribed by the existing genotype. Whitman's emphasis was ahead of its time. Some of his studies dealt with the analysis of sexual dimorphism and led to Oscar Riddle's endocrinological research.

Whitman was one of the pioneers of ethology. His paper "Animal Behavior" (1898) contains many well-chosen examples of innate (nonlearned) behavior. In his later work he analyzed particularly the relation between innate and learned behavior and the ability of animals to adjust their behavior to new experiences. His posthumous "The Behavior of Pigeons" (1919) is an extraordinarily detailed analysis. In particular, courtship and breeding behaviors of some forty species are compared. With Oskar Heinroth's pioneering work on ducks (1911), it was the first extensive study in comparative ethology.

BIBLIOGRAPHY

I. ORIGINAL WORKS. A full bibliography of Whitman's contributions (67 titles) is given by Lillie (see below). The more important ones are "The Embryology of *Clepsine*," in *Quarterly Journal of Microscopical Science*, **18** (1878), 215–315; "A Contribution to the Embryology, Life History and Classification of the Dicyemids," in *Mitteilungen aus der Zoologischen Station zu Neapel*, **4** (1883), 1–89; "A Contribution to the History of the Germ Layers in *Clepsine*," in *Journal of Morphology*, **1** (1887), 105–182; "A Series of Lectures on Bonnet and the History of Epigenesis and Preformation," in *Biological Lectures. Marine Biological Laboratory, Woods Hole, Mass.* (1894), 205–272; "Animal Behavior," *ibid.* (1898), 285–338; and the posthumous works edited by Oscar Riddle (vols. I and II) and Harvey A. Carr (vol. III): "Orthogenetic Evolution in Pigeons," *Publications. Carnegie Institution of Washington*, no. 257 (1919), vol. I; "Inheritance, Fertility, and the Dominance of Sex and Color in Hybrids of Wild Species of Pigeons," *ibid.*, vol. II; and "The Behavior of Pigeons," *ibid.*, vol. III.

II. SECONDARY LITERATURE. Information on Whitman's life and work is in C. B. Davenport, "The Personality, Heredity and Work of Charles Otis Whitman," in *American Naturalist*, **51** (1917), 5–30; F. R. Lillie, "Charles Otis Whitman," in C. O. Whitman memorial volume, *Journal of Morphology*, **22** (1911), xv–lxxvii; and E. S. Morse, in *Biographical Memoirs. National Academy of Sciences*, **7** (1912), 269–288.

ERNST MAYR

WHITNEY, JOSIAH DWIGHT (*b.* Northampton, Massachusetts, 23 November 1819; *d.* Lake Sunapee, New Hampshire, 19 August 1896), *geology*.

Whitney was the oldest of the eight children of Josiah Dwight Whitney and Sarah Williston. His father, whose ancestors had come to Massachusetts in 1635, was a prosperous banker; his mother, the daughter of a minister, was a teacher. His parents placed a strong emphasis on education, and Whitney attended the Round Hill School founded at Northampton by George Bancroft and Joseph Green Cogswell, then Phillips Academy at Andover, Massachusetts, before entering Yale College in 1836. While at Yale he attended Benjamin Silliman's lectures on chemistry and Denison Olmsted's course on astronomy, which awakened his interest in science. Upon graduation in 1839, Whitney worked for a time on Charles T. Jackson's geological survey of New Hampshire, but it was a lecture on geology given in Boston by Sir Charles Lyell that determined him to be a scientist, and in May 1842 he left for Europe for advanced training. During the next five years he studied with Élie de Beaumont at the Paris École des Mines, with Karl F. Rammelsberg and Heinrich Rose in Berlin, and with Justus von Liebig in Giessen.

In May 1847 Whitney returned to the United States as a fully trained professional geologist, and was immediately employed by Jackson as an assistant in the latter's geological survey of Michigan. He remained with the survey for two years, then established himself as a mining consultant with an office first in Brookline, then in Cambridge, Massachusetts. He incorporated the experience that he gained into his *The Metallic Wealth of the United States*, published in 1854. This work was a milestone in the literature of ore deposits, and remained a standard text for two decades; one of the first systematic texts in the field, it stimulated serious research on mineral ores and helped to establish mining geology as a scientific discipline. The book enhanced Whitney's national reputation, and he was soon appointed professor of chemistry at the University of Iowa; during the same years, 1855 to 1858, he also served as a member of the Iowa state geological survey (under James Hall), the Illinois survey (under Amos H. Worthen), and (again with Hall) the Wisconsin survey. His work in Illinois dealt largely with zinc and lead deposits, while in Wisconsin he was primarily concerned with lead deposits alone.

Whitney was thus well suited to assume, in 1860, the directorship of the new California geological survey, which he served, intermittently, over the next fourteen years. He also participated in founding the California Academy of Sciences, the University of California, and Yosemite National Park. A number of the young scientists that he trained—including William H. Brewer, James G. Cooper, William M. Gabb, and Clarence King—later became famous, and one of the methods developed by Whitney's group for topographical mapping by triangulation was widely adopted.

In 1865 Whitney was granted a leave of absence from the California survey in order to assume an appointment as Sturgis-Hooper professor at Harvard College, where he was also to be responsible for establishing a school of mines. The school (which was later merged with the Lawrence Scientific School) opened in 1868, the same year in which the California geological survey was suspended after the state legislature refused to pass an appropriation for its continuation. Although Whitney remained as director of the nominal survey until 1874, only three volumes of its findings were published by the state; the Harvard Museum of Comparative Zoology aided him in publishing, in 1880, the important *The Auriferous Gravels of the Sierra Nevada of California*, and he himself brought out a volume of general geological observations in 1882.

Whitney returned to Cambridge permanently in 1875, and was reappointed to the Sturgis-Hooper professorship. He continued to teach at Harvard for the rest of his life, although his gruffness in the classroom and his cool, even unfriendly, personality limited his effectiveness as a teacher. His last major work, *Climatic Changes in Later Geological Times*, published in 1882, drew largely upon the researches he had conducted in the West.

Whitney was, perhaps, not so prominent a leader of American science as some of his contemporaries, and received fewer honors. He became a member of the American Philosophical Society and the National Academy of Sciences in 1865, and was later one of the few American foreign members of the Geological Society of London. Mt. Whitney, the highest point in the contiguous United States, is named in his honor. Whitney died at his summer retreat in New Hampshire; his wife, Louisa Goddard Howe, whom he had married in 1854, and his only child, a daughter, had both predeceased him in 1882.

BIBLIOGRAPHY

I. ORIGINAL WORKS. Whitney's scientific publications include *Report of a Geological Survey of the Upper Missouri Lead Region* (Albany, 1862); *Earthquakes, Volcanoes, and Mountain Building* (Cambridge, Mass., 1871); *Geology and Geological Surveys* (Cambridge, Mass., 1875); *The Auriferous Gravels of the Sierra Nevada of California* (Cambridge, Mass., 1880); *The Climatic Changes of Later Geological Times; a Discussion Based on Observations Made in the Cordilleras of North America* (Cambridge, Mass., 1882); and *The Azoic System and its Proposed Subdivisions* (Cambridge, Mass., 1884), a contribution to classification.

The Whitney Family Manuscripts Collection at Yale University has nearly one thousand letters between Jo-

siah Dwight Whitney and his brother William Dwight Whitney. The Josiah Dwight Whitney MSS at the Bancroft Library of the University of California contain more than six hundred letters to his close associate William H. Brewer, dealing with both scientific and personal matters. The William H. Brewer manuscripts at Yale University relate to Whitney's California years, as does a published account, *Up and Down California in 1860-1864: the Journal of William H. Brewer*, Francis P. Farquhar, ed. (New Haven, 1930).

II. SECONDARY LITERATURE. The fullest account is the somewhat uncritical Edwin T. Brewster, *Life and Letters of Josiah Dwight Whitney* (Boston, 1909). See also William H. Goetzman, *Exploration and Empire: the Explorer and the Scientist in the Winning of the American West* (New York, 1966), a fine survey of Whitney's activities in the West; George P. Merrill, *The First One Hundred Years of American Geology* (Washington, 1924); and Gerald T. White, *Scientists in Conflict: the Beginnings of the Oil Industry in California* (San Marino, 1968), which deals with the controversy between Whitney and Benjamin Silliman, Jr., over oil in California, and is unduly critical of Whitney. Gerald D. Nash, "The Conflict Between Pure and Applied Science in Nineteenth Century Public Policy: the California State Geological Survey, 1860-1874," in *Isis*, **64** (1963), 217-228, summarizes Whitney's career as director of the California Geological Survey.

GERALD D. NASH

WHITTAKER, EDMUND TAYLOR (*b*. Birkdale, Lancashire, England, 24 October 1873; *d*. Edinburgh, Scotland, 24 March 1956), *mathematics, physics, philosophy.*

Whittaker was educated at Manchester Grammar School and Trinity College, Cambridge. He was bracketed second wrangler in the mathematical tripos of 1895, was elected a fellow of Trinity College the following year, and was first Smith's prizeman in 1897. In 1905 he was elected a fellow of the Royal Society, and was awarded the Sylvester and Copley medals of the society in 1931 and 1954 respectively. In 1906 he became astronomer royal for Ireland and from 1912 until his retirement in 1946 was professor of mathematics at the University of Edinburgh. From 1939 to 1944 Whittaker was president of the Royal Society of Edinburgh, and was an honorary member of several learned societies. In 1935 Pope Pius XI conferred on him the cross *pro ecclesia et pontifice* and a year later appointed him to the Pontifical Academy of Sciences. In 1945 Whittaker was knighted and in 1949 became an honorary fellow of Trinity College, Cambridge.

In 1901 Whittaker married Mary Boyd, daughter of the Reverend Thomas Boyd of Cambridge; they had three sons and two daughters. The second son, J. M. Whittaker, became a mathematician and was vice-chancellor of the University of Sheffield. Whittaker's elder daughter married the mathematician E. T. Copson.

Whittaker's deepest interest was in fundamental mathematical physics, and consequently much of his earlier work was concerned with the theory of differential equations. Perhaps his most significant paper in this field was the one published in 1902 in which he obtained the most general solution of Laplace's equation in three dimensions, which is analytic about the origin, in the form

$$\int_0^{2\pi} f(x\cos\alpha + y\sin\alpha + iz, \alpha)\,d\alpha$$

and the corresponding solution of the wave equation in the form

$$\int_0^{\pi}\int_0^{2\pi} f(x\sin\alpha\cos\beta + y\sin\alpha\sin\beta + z\cos\alpha + ct,$$

$$\alpha, \beta)\,d\alpha\,d\beta.$$

The discovery of the general integral representation of any harmonic function brought a new unity into potential theory; the integral representations of Legendre and Bessel functions, for example, were immediate consequences. Moreover, entirely new fields of research in the theory of Mathieu and Lamé functions were opened up. Whittaker also made a detailed study of the differential equation obtained from the hypergeometric equation by a confluence of two singularities, and he introduced the functions $W_{k,m}(z)$, which now bear his name. Another lifelong interest of Whittaker's was the theory of automorphic functions and the standard English book on the subject by L. R. Ford owes much to Whittaker. He also wrote a few papers on special problems in algebra and on numerical analysis.

Whittaker had an intense interest in the theory of relativity and from 1921 onward wrote ten papers on the subject. In one of the papers he gave a definition of spatial distance in curved space-time, which is both mathematically elegant and practical. In other papers he extended well-known formulas in electromagnetism to general relativity, gave a relativistic formulation of Gauss's theorem, and dealt with the relation between tensor calculus and spinor calculus.

Whittaker will long be read, since his textbooks on several diverse branches of mathematics have become classics. *Modern Analysis* (1902) was the first book in English to present the theory of functions of a complex variable at a level suitable for undergraduate and beginning graduate students. Forsyth's *Theory of Functions* had appeared in 1893, but its contents had not penetrated to the general body of mathematicians. *Modern Analysis* was extensively revised and enlarged in 1915 in collaboration with G. N. Watson, whose name was then added to the title page. Whittaker's *Analytical Dynamics*, which was published in 1904, was the first book to give a systematic account in English of the superbly beautiful theory that springs from Hamilton's equations; and it was of fundamental importance in the development of the quantum theory. Then, in 1910 there appeared *The History of the Theories of Aether and Electricity*. In 1951 a revised version of the book was published and constituted the first volume of a new treatise with the same title; it deals with the history up to the end of the nineteenth century. The second volume, which appeared in 1953, describes the developments made between 1900 and 1926 and is concerned mainly with relativity and quantum theory. The two volumes together form Whittaker's *magnum opus*. A contemplated third volume dealing with later theories was never completed.

Notwithstanding the excellence of *Aether and Electricity*, the chapter in the second volume dealing with the special theory of relativity has been criticized for the emphasis it places on the work of Lorentz and Poincaré, and for the consequent impression it gives that the work of Einstein was of minor importance. The consensus is that Whittaker made an error of judgment. As early as 1899 Poincaré had thought it possible that there might not be such a thing as absolute space, and in 1904 he had discussed without mathematics the possibility of a new mechanics in which mass would depend on velocity and in which the velocity of light would be an upper limit to all physically possible velocities. Also, Lorentz had derived the transformation that now bears his name before Einstein published his paper in 1905, but Lorentz interpreted it in terms of absolute space and time, concepts that, according to Born, he was still clinging to a few years before his death in 1928. Likewise, Poincaré seemed to regard the Lorentz transformation (which he discussed in a mathematically impressive paper in 1906) as physically important only because Maxwell's equations are invariant under it. It was Einstein (who had doubts about the ultimate validity of Maxwell's equations) who derived the transformation law from more fundamental physical principles.

Soon after his arrival at the University of Edinburgh, Whittaker instituted a mathematical laboratory and lectured on numerical analysis. His book *The Calculus of Observations*, written with G. Robinson, grew out of these lectures and was published in 1924. At that time very little of its content was to be found in any other book in English.

Although Whittaker expended a tremendous effort on advanced study and research, he regarded his undergraduate teaching as of paramount importance and, in addition to lecturing to the honors classes, he lectured once a week to the first-year class on the history and development of mathematics. He was an outstanding lecturer and by his dignified bearing, his great command of language, his eloquent delivery, and his obvious mastery of his subject, he made a tremendous impression upon young students. They knew at once that they were in the presence of a scholar and teacher of the first rank and in all his prelections they saw at work a mind of astonishing accuracy and force, ranging at will over the whole field of ancient and modern mathematics and presenting with insight and great persuasive power the profundities there disclosed.

Whittaker was a deeply religious man all through his life and, after having belonged to several branches of the Protestant faith—including the Church of Scotland, of which he was an elder—he was received into the Roman Catholic Church in 1930. After retiring from his chair at Edinburgh, Whittaker spent much of his time studying the philosophical aspects of modern physics and the repercussions that recent developments might have on theology. He expounded his views in *The Beginning and End of the World* (1942), *Space and Spirit* (1947), *From Euclid to Eddington* (1949), and in a large number of papers. He wrote from an orthodox Roman Catholic point of view with great emphasis on natural theology and the work of Thomas Aquinas. He deplored that in modern life "the sense of creatureliness and dependence has passed away, and God is left out of account." He was undoubtedly one of the few men of his time who could speak with authority on both physics and theology.

BIBLIOGRAPHY

An extensive account of Whittaker's life and work is in the Whittaker Memorial Number of *Proceedings of the Edinburgh Mathematical Society*, **11**, pt. 1 (1958), 1–70, which includes a general biographical notice and articles by five contributors on different aspects of Whittaker's work. See also biographical notices by G. F. J. Temple, in *Biographical Memoirs of Fellows of the Royal Society*, **2** (1956), 299–325; and by W. H. McCrea, in *Journal of the London Mathematical Society*, **32** (1957), 234–256.

The question concerning the origin of the special theory of relativity is discussed by G. Holton, in *American Journal of Physics*, **28** (1960), 627–636; and M. Born, *The Born-Einstein Letters* (New York, 1971), 197–199.

DANIEL MARTIN

WHYTLAW-GRAY, ROBERT (*b.* London, England, 14 June 1877; *d.* Welwyn Garden City, Hertfordshire, England, 21 January 1958), *physical chemistry.*

Whytlaw-Gray designed and utilized precision techniques for weighing both aerosols and gases, notably radon. The second son of a prosperous Australian businessman, he was educated at Glasgow. From 1896 to 1903 he studied and conducted research under Ramsay and Morris W. Travers at University College, London, beginning a lifelong career in the exact manipulation of gases. In 1903 Gray went to Bonn to continue his redetermination of Stas's atomic weight of nitrogen; he obtained a value of 14.01 (O = 16), compared with the then current standard of 14.04 and with the more recent standard of 14.008. After obtaining the Ph.D. in 1906, he returned to University College, becoming assistant professor in 1908. In 1911 he incorporated the matronymic Whytlaw to his name, possibly to distinguish himself from a colleague, J. A. Gray.

From 1906 to 1914 Gray measured the physical constants of gases in order to determine their atomic weights. Collaborating with Ramsay in the 1910–1911 classic determinations of the density of niton (now called radon), Gray used a modified Steele-Grant microbalance to weigh the minute quantity available—less than 0.10 cubic mm. The new gravity balance, announced in 1909, was constructed of fused quartz having a counterpoised sealed quartz bulb containing a known quantity of air. Balance was effected by varying the external pressure within the case. Absolute weight as a function of buoyancy could be determined by the original instrument with an accuracy of 10^{-7} gram, the instrument being about 100 times more sensitive than the Nernst microbalance. In 1910 their experiments yielded an average atomic weight of 220 for radon; they modified this value to 223 on the basis of their 1911 results. By this means they had well estimated the correct order of magnitude.

On the theoretical side, following the then accepted atomic weight for radium of 226.5, Ramsay and Gray in 1910 suggested on genetic considerations the atomic weight of 222.5 for radon. Using the microbalance again the following year, they redetermined the atomic weight of radium to be 226.36, confirming the results of Curie, and derived thereby an independent check on their proposed atomic weight of 222.4 for radon. It was the value they expected (given their value for radium) if radon and helium were the only products of the disintegration of radium. However, after the official adoption in 1916 of 226.0 for radium, based, rather, upon the 1912 results of Hönigschmid, the value 222.0 was eventually assigned to radon.

From 1915 to 1922 Gray was science master at Eton. In 1917 he began a twenty-year series of confidential investigations concerning toxic and other smokes for the War Office. Gray further improved the design of the microbalance and used an ultramicroscope to count the number of smoke particles, determine their sizes, and study their lifetimes. He modified Smoluchowski's theory concerning the rate of coagulation of homogeneous sols so that it would apply rigorously to gaseous systems. He was also interested in the structure of the coagulating particles and examined the effects of electrification and photophoresis.

In October 1923 Gray succeeded Arthur Smithells at Leeds. While continuing government-supported research on aerosols, he resumed some of his investigations on the compressibilities and densities of simple gases. From 1939 to 1945 Gray led a government inquiry into defense against possible chemical warfare, in addition to serving on the International Committee on Atomic Weights. He retired pro forma in 1942, remaining in office until 1945 and at Leeds until 1950. He continued to improve the microbalance and extended his research to include complex organic gases and vapors. After a brief rest at Coventry, Gray, at seventy-five, resumed his research and consultation at Imperial Chemical Industries in Welwyn Garden City.

BIBLIOGRAPHY

I. ORIGINAL WORKS. An almost complete list of Gray's nearly 100 papers and reports is part of the notice by E. G. Cox and J. Hume, in *Biographical Memoirs of Fellows of the Royal Society*, **4** (1958), 327–339. The best-known papers are the joint communications with Ramsay concerning the dramatic series of density determinations of radon: "La densité de l'émanation du radium," in *Comptes rendus . . . de l'Académie des sciences*, **151** (1910), 126–128; and "The Density of Ni-

ton ('Radium Emanation') and the Disintegration Theory," in *Proceedings of the Royal Society*, **84A** (1911), 536–550. His work on aerosols is summarized in *Smoke: A Study of Aerial Disperse Systems* (London, 1932), written with H. S. Patterson.

II. SECONDARY LITERATURE. F. Challenger, who presented Gray for a degree *honoris causa* at Leeds in 1950, wrote a note in *Nature*, **181** (1958), 527; and R. S. Bradley wrote a detailed obituary notice in *Proceedings of the Chemical Society* (Jan. 1959), 18–20. There is also a brief, unsigned notice in *Chemistry and Industry* (1958), 134.

The original microbalance is described in B. D. Steele and K. Grant, "Sensitive Micro-Balances and a New Method of Weighing Minute Quantities," in *Proceedings of the Royal Society*, **82A** (1909), 580–594; and W. A. Tilden and S. Glasstone, *Chemical Discovery and Invention* (London, 1936), 58–61. The designation "radon" was first suggested by C. Schmidt, in "Periodisches System und Genesis der Elemente," in *Zeitschrift für anorganische und allgemeine Chemie*, **103** (1918), 79–118. The value 235 for the atomic weight of radon was suggested by P. B. Perkins, in "A Determination of the Molecular Weight of Radium Emanation. . . ," in *American Journal of Science*, 4th ser., **25** (1908), 461–473, on the basis of diffusion evidence.

For a discussion of the nitrogen problem as it appeared at the time, see I. Freund, *The Study of Chemical Composition* (Cambridge, 1904; repr. New York, 1968), 313–316. A contemporary consideration of Gray's work on radon is in A. T. Cameron, *Radiochemistry* (London, 1910), *passim*; and W. Ramsay, *The Gases of the Atmosphere*, 4th ed. (London, 1915), 283–291. S. C. Lind, in "The Atomic Weight of Radium Emanation (Niton)," *Science*, **43** (1916), 464–465, argued on genetic considerations that radon should be assigned the atomic weight of 222.0 once 226.0 had been established for radium.

A general treatment of Gray's early work is included in the account by his brother-in-law, M. W. Travers, *A Life of Sir William Ramsay, K.C.B., F.R.S.* (London, 1956), *passim*.

THADDEUS J. TRENN

WHYTT, ROBERT (*b.* Edinburgh, Scotland, 6 September 1714; *d.* Edinburgh, 15 April 1766), *medicine, neurophysiology.*

Remarkably rational in an age in which reaction of a muscle to artificial nerve stimulus was considered magical, Robert Whytt (pronounced "White") was a practitioner of physic, teacher of medicine, and the foremost neurologist of his time. The first to demonstrate reflex action in the spinal cord, he also localized the site of a single reflex (Whytt's reflex), wrote the first important treatise on neurology after Thomas Willis, and gave the first clear description of tuberculous meningitis in children.

The second son of Robert Whytt, an advocate of Bennochy, and Jean Murray, of Woodend, Perthshire, Whytt was born six months after his father died; and his mother died when he was six years old. When he was fourteen, he succeeded to the family estate when his older brother died. After education in the public school at Kirkcaldy, in Fife, he went to St. Andrews University, where he received a degree in arts in 1730. For the next four years he studied medicine at the newly organized medical faculty of Edinburgh, which included Alexander Monro (Primus), Andrew St. Clair (Sinclair), John Rutherford, John Innes, and Andrew Plummer.

In 1734 Whytt went to London, where he studied under William Cheselden and walked the wards of the hospitals; from there he traveled to Paris and the wards of the Charité and Hôtel Dieu, also attending lectures and private dissections by Jacques Benigne Winslow, who condensed and simplified the study of anatomy. Finally, following Edinburgh tradition, he studied under the aged Hermann Boerhaave at Leiden, where he also had the advantages of anatomical instruction under Bernhard Siegfried Albinus. On his way home in 1736, Whytt tarried three days at Rheims to acquire the M.D. from the university after separate Latin examinations "for a considerable space," in anatomy, physiology, and the diagnosis and treatment of various diseases. After his return to Scotland, St. Andrews also awarded him an M.D. on 3 June 1737. Not quite three weeks later (21 June 1737) he became licentiate of the Royal College of Physicians of Edinburgh and set up practice as a physician. He was admitted a fellow of the College on 27 November 1737.

About this time Mrs. Joanna Stephens was stirring up public excitement with her well-publicized sovereign remedy for urinary bladder stones; and after enriching herself for several years, she offered to sell the secret of her nostrum for £5,000. Since it was enthusiastically endorsed by *Gentleman's Magazine*, a popular journal of the period, such highly regarded persons as Horace Walpole, dukes, earls, bishops, and even doctors of medicine offered to contribute to a fund to buy the secret; but not enough money was forthcoming. Parliament then appointed a commission that included such experts as Stephen Hales, William Cheselden, Caesar Hawkins, and other highly respected scientists; and upon their recommendation the government paid the sum asked in order that this medicine might be sold cheaply to the poor.

It was then revealed that the secret consisted of calcined egg and snail shells, with "alicant" soap. This finding induced Whytt to carry out elaborate experiments with limewater and soap, from which he concluded that this mixture had considerable power of disintegrating calculi *in vitro*; he thereupon tested courses of injections into the bladders of patients at the Royal Infirmary of Edinburgh suffering from the stone. His results were first published as "Essay on the Virtues of Lime-Water and Soap in the Cure of the Stone" (1743). This was followed by subsequent reports for nearly a score of years. An important result of this work on alkalies in the treatment of urinary calculi was that it led Joseph Black, then at Glasgow, to do his historic series of experiments on the chemistry of magnesia alba, quicklime, and other alkaline substances in search of a better solvent for stones. In 1754, in the course of that work, Black discovered the first known gas, "fixed air" (carbon dioxide). Soap was recommended as a lithontriptic, especially when dissolved in limewater, well into the latter half of the nineteenth century.

The rebellion of 1745 produced great confusion throughout Scotland, but by the winter of 1746–1747 affairs settled down and the medical faculty at Edinburgh was reorganized. Innes had died, and Whytt was elected to succeed him as professor of the institutes of medicine; and on 26 August 1747 he was elected professor of the practice of medicine, taking over the duties of John Rutherford, who thereafter devoted himself entirely to clinical lectures at the Royal Infirmary. Andrew Sinclair, in failing health, seems to have ceased lecturing entirely, Whytt having officiated for him at the university for some time before this; and Andrew Plummer, another member of the original medical faculty, devoted himself to teaching only chemistry. Whytt was associated with Alexander Monro and William Cullen at the Royal Infirmary, where he gave clinical lectures in 1760 and where he treated many of the patients upon whose clinical records he based much of his speculations and publications.

During his teaching career from 1747 to 1766, Whytt attracted throngs of students to his lecture theater. A practicing physician, he taught physiology in the modern spirit, lecturing in English instead of the customary Latin. One of the first doctors in Scotland to do research in the modern sense, he demonstrated his experiments to his classes and spent considerable time in experimental work on animals. Like his predecessors, Whytt at first used Boerhaave's *Institutiones medicae*, which dealt with physiology, as his textbook; but in 1762 he switched to *Institutiones pathologiae medicinalis* by Boerhaave's disciple Hieronymus David Gaubius (originally published in 1750), which was less limited in its approach.

In 1751 Whytt published *The Vital and Other Involuntary Motions of Animals*, a classic in neurophysiology, which attracted wide attention. After numerous vivisections he concluded that the capacity for muscle movement is preserved for some time after death. He referred to decapitated frogs and other animals moving in a coordinated manner—with some degree of intelligence, as it were—and concluded that the brain cannot be the only center of neurological activity. In 1649 Descartes had explained reflex action as exemplified in blinking of the eyelids on the sudden approach of an object; Robert Boyle had shown that a viper wriggles when pricked even several days after decapitation; and Stephen Hales had shown that destruction of the spinal cord prevents reflex action. The animistic system of Georg Ernst Stahl, based upon the premise that the soul is the *principium vitae*, was not abandoned until Albrecht von Haller convinced the scientific world with his arguments, backed by experimental demonstrations, concerning sensibility and irritability of parts of the human body in 1752. Whytt, however, was not yet able to divert his thinking of the soul as the basis of life and vitality so that he ascribed involuntary movement to the effect of a stimulus acting upon an unconscious "sentient principle." "By the sentient principle," he explained, "I understand the mind or soul in man, and that principle in brutes which resembles it."

Haller called Whytt a "semi-animist," but Whytt was actually opposed to the views of Stahl, Paracelsus, and others that there was a conscious soul in each living thing to direct its vitality and movement. Stahl and his followers looked upon the spinal cord as a simple conductor of nerve impulses; Whytt, however, demonstrated conclusively that centers for involuntary action could be located in "the brain of the spinal marrow." For the first time in the history of physiology, he presented a clear description of what Marshall Hall a century later called "reflex action." Whytt gave admirable accounts of various kinds of reflexes, even of what is now known as the "stretch reflex." He hypothesized the continuity of nerve fibers from the brain and the identity of separate nerve fibrils. Johann August Unzer, who first differentiated between voluntary and involuntary movements and who had experimented upon beheaded people in 1746, denied the intervention of any soul in reflex actions; and Haller, busily engaged in similar physiological pursuits, praised Whytt's book while criticizing it severely in a review published by the Royal Society of Sciences of Göttingen.

In 1753 in the *Commentaries* of this society Haller published his "De partibus corporis humani sensibilibus et irritabilibus," in which he laid down the principle that only certain parts of the body possess sensibility and contended that "irritability," or power of muscle contraction, was an innate property (*vis insita*) of muscle fibers independent of nervous influence and having no connection with sensation or stimulus. Haller had much greater practical experience than Whytt; but Whytt was a more brilliant philosopher, given to freer speculation and gifted with shrewd logic and extraordinary insight, so that his ideas served as the starting points for later physiologists.

Whytt subsequently published *Physiological Essays* (1755), consisting of "An Inquiry Into the Causes Which Promote the Circulation of the Fluids in the Very Small Vessels of Animals" and "Observations on the Sensibility and Irritability of the Parts of Men and Other Animals: Occasioned by M. de Haller's Late Treatise on These Subjects." The first essay concerned the peristaltic action of peripheral blood vessels, which assists the pumping action of the heart, a theory opposed by Haller and not generally accepted until capillary contractility was clearly demonstrated more than a century later. In the second essay Whytt contended that all muscle action was governed by nervous control. Admittedly uncertain about the minute structure of nerves, he nevertheless asserted that sensation, motion, and other functions are brought about through nervous connections between all parts of the body. This was long before Charles Bell (1811) and François Magendie (1822) proved the separate existence of sensory and motor nerve paths.

The dispute between Whytt and Haller attracted much attention on the Continent. By showing that lasting dilatation of the pupil could be due to compression of the optic thalamus, he was the first to localize a reflex (Whytt's reflex). He likewise showed that only a portion of the spinal cord suffices for reflex action. In addition he made the first attempts since Galen to localize the seat of reflex action.

Observations on . . . Nervous, Hypochondriac or Hysteric Disorders (1764) reveals great clinical acumen and provides vivid accounts of a wide range of neurological and psychiatric patients whom Whytt attended at the Royal Infirmary. He declared that disorders variously called flatulent, spasmodic, hypochondriac, hysteric, and, more recently, nervous, had become the wastebasket diagnosis for those conditions about which physicians were ignorant; and therefore he set out "to wipe off this reproach" and to throw some light on these ailments. He resorted to his previous work to explain the nature of these diseases, emphasizing the "sentient and sympathetic power of the nerves," and described instances of referred pain—anticipating, by his explanations of the causes, modern demonstrations of the reasons for them. Whytt clarified Thomas Willis' term "nervous," already in use for over 100 years, and explained such physical phenomena as blushing, lacrimation, and sweating, brought on by emotion or passion, as owing to some change made in the brain or nerves by the mind or sentient principle. This work added significant contributions to scientific medicine.

Through his *Observations on the Dropsy in the Brain*, published posthumously in 1768 by his son, Whytt achieved lasting remembrance in the history of pediatrics for the first clear description of tuberculous meningitis. It is a masterpiece of clinical observation, the finest first description of a disease to appear until then. Brief and lucid, the monograph is based upon about a dozen cases in which everything of clinical value that could be detected without modern laboratory apparatus is recorded. Monro's foramen, connecting the lateral and third ventricles of the brain, was first observed greatly dilated in one of the cases here described, which Monro (Secundus) and Whytt saw in 1764 during a consultation. Allusions to tuberculous meningitis before Whytt, usually included under a general heading of "phrenitis," generally went unnoticed. Whytt's description is a pediatric milestone that gave great impetus to the study of meningitis.

The Works of Dr. Whytt (1768) included papers on a range of subjects which indicate Whytt's versatility. These included "The Difference Between Respiration and Motion of the Heart, in Sleeping and Waking Persons," "Cure of Fractured *tendo achilles*," "Use of Bark in Dysenteries," "Hoarseness After Measles," and "Anomalous and True Gout."

In 1752, upon the recommendation of his friend and former classmate Sir John Pringle, Whytt was elected a fellow of the Royal Society of London. He was appointed physician to the king in Scotland in 1761 and, two years later, was elected president of the Royal College of Physicians of Edinburgh. His first wife was Helen Robertson, the sister of James Robertson, governor of New York. The two children of his first marriage died in infancy. Whytt's interest in colonial America was also enhanced by correspondence with Alexander Garden, Lionel Chalmers, John Moultrie, and John Lining, all living in Charleston, South Carolina, and all of Scottish extraction. His second wife, Louisa Balfour of Pilrig, Midlothian, whom he married in 1743, bore him fourteen children, of whom six survived.

Whytt was not of a robust constitution. He became ill in 1765 and died the next year of symptoms suggestive of diabetes. His grave, in Old Greyfriars Church, is marked by a handsome monument. When Whytt died, William Cullen vacated the chair of chemistry at Edinburgh in favor of Joseph Black, in order to succeed Whytt as professor of the theory of medicine.

BIBLIOGRAPHY

I. ORIGINAL WORKS. "Essay on the Virtues of Lime-Water and Soap in the Cure of the Stone," in *Observations and Essays in Medicine by a Society in Edinburgh*, **2**, pt. 2 (1743), followed by "An Essay Towards the Discovery of a Safe Method for Dissolving the Stone," in *Medical Essays and Observations Revised and Published by a Society in Edinburgh*, **5** (1744), 667–750, and **5**, pt. 2 (1747), 156–242, was a long tract attempting to develop a rational form of treatment for urinary calculus. *Essay on the Virtues of Lime-Water in the Cure of the Stone; With an Appendix, Containing the Case of The Right Honourable Horatio Walpole, Written by Himself* (Edinburgh, 1752; 2nd ed., 1755; 3rd ed., Edinburgh, 1761; Dublin, 1762), contains in the 3rd ed. a further account of the Walpole case, as well as that of the bishop of Llandaff; there was also a French trans. of the 2nd ed. by M. A. Roux (Paris, 1757).

Essays and Observations, Physical and Literary, Read Before the Philosophical Society in Edinburgh, **1** (1754), contains "Of the Difference Between Respiration and the Motion of the Heart, in Sleeping and Waking Persons" (436–446); "On the Various Strengths of Different Lime-Waters" (372–385); and "Of the Anthelmintic Virtues of the Root of the Indian Pink, Being Part of a Letter From Dr. John Lining, Physician at Charlestown, South Carolina, to Dr. Robert Whytt" (386–389). The same journal, **2** (1756), includes "Description of the Matrix or Ovary of the *Buccinum ampullatum*" (8–10), concerning the hermit crab in its shell; and "Some Experiments Made With Opium on Living and Dying Animals" (280–316).

These *Essays and Observations Physical and Literary* were republished in 3 vols. (Edinburgh, 1770–1771). Vol. I contains Whytt's essay on the strengh of different kinds of limewater, read to the Philosophical Society in Edinburgh in 1751 (240); the article by John Lining on Indian pink, addressed to Whytt (436); and the difference in respiration and pulse in people asleep and awake (492). Vol. II includes "The Description of a New Plant by Alexander Garden, Physician at Charleston in South Carolina," which described to Whytt the gardenia found in 1753 and in 1754 "about a mile from the Town of New York, in New England" (1–7); the opium experiments on animals (307–346); and "A Description of the American Yellow Fever in a Letter From Dr. John Lining, Physician at Charlestown in South Carolina, to Dr. Robert Whytt," dated 14 Dec. 1753 and read to the Society on 7 Mar. 1754 (404–432). In vol. III are a critique read to the Philosophical Society by Alexander Monro in 1761 concerning Whytt's opinions regarding the effects of opium on the nervous system of animals (299); "Use of the Bark in Dysenteries, and a Hoarseness After the Measles," read in Jan. 1761 (366–379); and observations (on *arthritis anomala*, or imperfect gout), in reference to a similar article on this subject by another author (466–470).

Other works are *An Essay on the Vital and Other Involuntary Motions of Animals* (Edinburgh, 1751; 2nd ed., 1763); and *Physiological Essays* (Edinburgh, 1755; repr. 1757; 2nd ed., 1759; repr. 1761; 1763; 3rd ed. 1766). The National Library of Medicine, Bethesda, Md., has a MS copy of Whytt's clinical lectures, transcribed by one of his students at Edinburgh in 1761.

Observations on the Nature, Causes and Cure of Those Diseases Which are Commonly Called Nervous, Hypochondriac or Hysteric (Edinburgh, 1764) is the first important English work on neurology. The 2nd ed., corrected, with the author's name spelled "Whyte" (Edinburgh, 1765), was entitled *Observations on the Nature, Causes and Cure of Those Disorders Which Have Been Commonly Called Nervous, Hypochondriac or Hysteric* and includes "Remarks on the Sympathy of the Nerves." The 3rd ed. is dated 1767, in which year a French trans. also appeared at Paris (2nd French ed., 1777); there also are a German ed. (Leipzig, 1766) and a Swedish ed. (Stockholm, 1786). Whytt's last separately and posthumously printed work was *Observations on the Dropsy in the Brain; to Which Are Added His Other Treatises Never Hitherto Published by Themselves* (Edinburgh, 1768).

A collected ed., *Works of Robert Whytt* (Edinburgh, 1768), also was issued by his son, who was assisted in collecting and editing the material by Sir John Pringle; a letter to Pringle dated 10 Nov. 1758, entitled "Account of an Epidemic Distemper at Edinburgh and Several Other Parts in the South of Scotland in the Autumn of 1758," is on 747–752. The collected *Works* was also translated into German as *Saemmtliche zur Practischen Arzneykunst Gehoerige Schriften* (Leipzig, 1771). It in-

cludes the research on limewater and soap in the cure of the stone, 1–238; "Nervous, Hypochondriac or Hysteric Disorders," 239–616; "Cure of a Paralysis by Electricity," 619–623; vesicatories in pulmonary congestion, 623–637 (addressed to the Royal Society in Feb. 1758 and published in the *Philosophical Transactions*, **50**, pt. 2 [1758], 569–578); the epidemic distemper of 1758, 637–646, read by Pringle to the Medical Society of London on 12 Feb. 1759; a work on the uses of the corrosive sublimate of quicksilver in the cure of phagadenic ulcers (in a letter to Pringle, 10 Jan. 1757, read 16 Feb. 1759 and published in *Medical Inquiries and Observations*, **2** [1762]) and additional case reports, 646–662; and "Observations on the Dropsy in the Brain," 662–696. The *Collected Works Relating to Theoretical Medicine* were translated by Johann Ephraim Lietzau as *Sämmtliche zur theoretischen Arzneikunst gehoerige Schriften* (Berlin–Stralsund, 1790) and included "Vital and Involuntary Motions in Animals," "Circulation of Fluids in the Small Vessels," "Sensibility and Irritability," and "Experiments With Opium on Living and Dying Animals."

II. SECONDARY LITERATURE. See Rachel Mary Barclay, *The Life and Work of Robert Whytt, M.D.* (Edinburgh, 1922), M.D. diss.; Alexander Bower, *History of the University of Edinburgh*, II (Edinburgh, 1817), 345, 355; Charles W. Burr, "Robert Whytt," in *Medical Life*, **36** (1929), 109; John D. Comrie, "An Eighteenth Century Neurologist," in *Edinburgh Medical Journal*, n.s. **32** (1925), 755; and *History of Scottish Medicine*, 2nd ed., I (London, 1932), 306; *Documents and Dates of Modern Discoveries in the Nervous System* (London, 1839), 112, 152, 162, 203, which reprints excerpts from Whytt's writings; Roger K. French, *Robert Whytt, the Soul, and Medicine*, Publications of the Wellcome Institute, Historical Monograph ser., no. 17 (London, 1969); John F. Fulton, *Muscular Contraction and the Reflex Control of Movement* (Baltimore, 1926), 32; and *History of Physiology* (London, 1931), 43, 90; Fielding H. Garrison, *History of Medicine*, 4th ed. (Philadelphia, 1929), 326; and Alexander Grant, *Story of the University of Edinburgh*, II (London–Edinburgh, 1884), 401.

See also Heinz Hürzeler, *Robert Whytt (1716–1766) und seine physiologischen Schriften* (Zurich, 1973); Max Neuburger, *Die historische Entwicklung der experimentellen Gehirn- und Rückenmark-Physiologie* (Stuttgart, 1897), 122, 174–192; John Ruräh, *Pediatrics of the Past* (New York, 1925), 401; William Seller, "Memoir of the Life and Writings of Robert Whytt," in *Transactions of the Royal Society of Edinburgh*, **23** (1861–1862), 99–131; George Frederic Still, *History of Paediatrics*, repr. ed. (New York, 1965), 443; G. Stronach, in *Dictionary of National Biography*; John Thomson, *Life of William Cullen*, I (Edinburgh, 1859), 241–258, which presents an extensive evaluation of Whytt; Ilza Veith, *Hysteria: The History of a Disease* (Chicago–London, 1965), 159; and Joseph I. Waring, *History of Medicine in South Carolina 1670–1825*

(Charleston, 1964), 57 and *passim*, which discusses Whytt's relationships with physicians in Charleston and includes excerpts from John Lining's account of yellow fever.

SAMUEL X. RADBILL

WICKERSHEIMER, ERNEST (*b.* Bar-le-Duc, France, 12 July 1880; *d.* Strasbourg, France, 6 August 1965), *history of medicine.*

Wickersheimer's father, of Alsatian origin, was a military physician who had studied at the Strasbourg Faculty of Medicine and at the École du Service de Santé Militaire. Wickersheimer himself studied medicine at Paris, becoming *externe des hôpitaux* in 1899. Quickly drawn to the history of medicine, he defended a doctoral dissertation in this field, *La médecine et les médecins en France à l'époque de la Renaissance* (1905), which established his reputation. (It was cited as early as 1915 in the second edition of J. L. Pagel's *Einführung in die Geschichte der Medizin.*) This early work contained the outline of the research that occupied more than sixty years of a highly productive life in science.

Wickersheimer's dissertation oriented him toward both aspects of his career, for he became a librarian as well as a historian. In 1906 he began his training at the library of the Paris Faculty of Medicine, and the following year he spent six months at the University of Jena. At about the same time he studied under Karl Sudhoff at Leipzig. Wickersheimer furthered his training by working as a librarian at the Sorbonne in 1909, and in 1910 he was named librarian of the Académie de Médecine. Also in 1910 he made his first trip to the United States, where he established contacts with American libraries and formed close relationships with American colleagues. During World War I he served as a military physician.

From 1920 to 1950 Wickersheimer was administrator of the Bibliothèque Universitaire et Régionale de Strasbourg, which became the Bibliothèque Nationale et Universitaire in 1926. He reorganized it in 1919, upon the return of Alsace to France, and in 1945, after World War II. In the latter year the medical division of the library was in a particularly poor state, having been almost totally destroyed in a fire.

Despite his very heavy responsibilities, in 1920 Wickersheimer again took up his historical research. In 1936 he published *Dictionnaire biographique des médecins en France au Moyen-Âge,* an authoritative book on which he continued to work throughout his life. An impressive bibliography of 230 items (which is incomplete, since it stops in 1961), published on the occasion of his scientific jubilee, shows the abundance and variety of the subjects Wickersheimer treated: medicine in the Middle Ages and in the Renaissance, notably at the school of Salerno; medicine in society; medical schools; the teaching of surgery; medical doctrines; hospitals; and therapy—in brief, almost all aspects of the profession. He also made an inventory of the Latin books of the High Middle Ages in French libraries. A distinguished local historian as well, he studied various details relating to the hospitals of Strasbourg.

Wickersheimer received many honors. Elected an officer of the Legion of Honor in 1948, he had earlier received the Croix de Guerre 1914–1918 and was also an officer of public instruction and commander of the Order of Carlos Finlay (Republic of Cuba, 1928). He was awarded the silver medal of the Paris Faculty of Medicine for his dissertation in 1905 and was granted an honorary M.D. by the Johann Wolfgang Goethe University at Frankfurt am Main on 12 July 1960. Posthumously he was awarded a prize by the Académie de Médecine for the body of his work on the history of medicine (see *Bulletin de l'Académie de médecine,* **149** [1965], 784).

A secretary or member of numerous French and foreign learned societies, Wickersheimer served as perpetual secretary of the Académie Internationale d'Histoire des Sciences and honorary president of the Académie Internationale d'Histoire de la Médecine and of the Société Française d'Histoire de la Médecine.

BIBLIOGRAPHY

I. ORIGINAL WORKS. Very extensive, but not complete, bibliographies of Wickersheimer's writings are in "Jubilé scientifique du Dr. Wickersheimer," in *Histoire de la médecine,* spec. no. 1960 (1961), 102–109; and in M. T. d'Alverny, "Travaux du Dr. E. Wickersheimer," in E. Wickersheimer, *Les manuscrits latins de médecine du haut Moyen-Âge dans les bibliothèques de France* (Paris, 1966), 236–248.

II. SECONDARY LITERATURE. See M. T. d'Alverny, "L'oeuvre scientifique du Dr. E. Wickersheimer," in *Humanisme actif, mélanges d'art et de littérature offerts à Julien Cain,* II (Paris, 1968), 299–307; "Jubilé scientifique du Dr. Wickersheimer" (see above), 89–109; and M. Klein, "Le Dr. E. Wickersheimer," in *Clio medica,* **1** (1966), 351–356.

Many French and foreign journals published obituaries of Wickersheimer. The Bibliothèque Nationale et

Universitaire of Strasbourg has, in its Alsatian division, numerous documents written by or pertaining to Wickersheimer.

MARC KLEIN

WIDMAN (or **WEIDEMAN** or **WIDEMAN**), **JOHANNES** (*b.* Eger, Bohemia [now Czechoslovakia], *ca.* 1462; *d.* Leipzig, Germany, after 1498), *mathematics.*

The little known about Widman's life is based on the records of the University of Leipzig. He was entered in the matriculation register in 1480 as "Iohannes Weideman de Egra."[1] He received the bachelor's degree in 1482 and the master's degree in 1485.[2] He then lectured on the fundamentals of arithmetic, on computation on lines, and on algebra, as can be seen from the announcements for and invitations to his courses.[3] Widman's algebra lecture of 1486, the first given in Germany, is preserved in a student's notebook.[4] In this lecture he discussed the twenty-four types of equations generally treated by the Cossists and illustrated them with many problems. He employed the Cossist signs for the powers of the unknowns, as well as symbols for plus, minus, and the root.[5] As Widman explicitly stated elsewhere,[6] he considered computation with irrational numbers and polynomials (*De additis et diminutis*) to be part of the subject matter of algebra. He also treated fractions and proportions in order to prepare his students for the study of algebra.

The work for which Widman is best known, *Behend und hüpsch Rechnung uff allen Kauffmanschafften*, appeared in 1489. After the Trent *Algorismus* (1475) and the Bamberg arithmetic books (1482, 1483), it was the first printed arithmetic book in German; and it far surpassed its predecessors in the scope and number of its examples.[7] It also was notable for containing the first appearance in print of the plus and minus signs. Widman dedicated the book to Sigismund Altmann of Schmidtmühlen, who also enrolled at Leipzig in 1480.[8] There are no direct reports of Widman's activities after 1489; his brief mathematical works that were printed later appeared anonymously and without date of publication. Yet, according to Conrad Wimpina, Widman was still working on mathematical topics in 1498.

Widman's knowledge of arithmetic was based on the *Algorismus Ratisbonensis* and the Bamberg arithmetic book of 1483, as can be seen by comparing the problems treated in these works with those in his own. His arithmetic book of 1489 went through several editions until 1526,[9] when it was superseded by those of Köbel, Adam Ries, and others.

Widman learned algebra primarily from a volume of manuscripts he owned (now known as Codex Dresdensis C 80)[10] that later came into the possession of Georg Sturtz of Erfurt, who about 1523 placed it at the disposition of Ries. A compilation of all that was then known about arithmetic and algebra, the volume contained, in particular, a German algebra of 1471 and one in Latin.[11] The Latin algebra, in the margins of which Widman entered further examples of the twenty-four types of equations, was the basis of his algebra lecture of 1486.[12] The lecture also is partially preserved in another Dresden manuscript (C 80^m) and in manuscripts from Munich and Vienna.[13] This manuscript (C 80) was also the source of Widman's writings that were printed at Leipzig about 1495.[14] Ries borrowed problems for his *Coss* from Widman's algebra, but he was not aware of the author's identity.[15] Following the appearance of printed works on algebra by Grammateus, Rudolff, and Stifel at the beginning of the sixteenth century, Widman's writings fell into neglect.

NOTES

1. See G. Erler, *Die Matrikel der Universität Leipzig*, I, 323. Widman was a member of the Natio Bavarorum.
2. *Ibid.*, II, 228, 289. Master Widman was allowed to live outside the dormitory (*petivit dimissionem burse et obtinuit*).
3. See W. Kaunzner, "Über Johannes Widmann von Eger," 1 f., 45; and E. Wappler, "Zur Geschichte der deutschen Algebra im 15. Jahrhundert," 9 f.; and "Beitrag zur Geschichte der Mathematik," 149, 167.
4. The fee for the lecture was 42 groschen (2 florins). See Kaunzner, *op. cit.*, 45.
5. On the root symbol see Wappler, "Zur Geschichte der deutschen Algebra im 15. Jahrhundert," 13. On the earliest use of the minus sign see Kaunzner, "Deutsche Mathematiker des 15. und 16. Jahrhunderts und ihre Symbolik," 22 f.
6. In Codex Lipsiensis 1470, fol. 432. On this point see Kaunzner, "Über Johannes Widmann von Eger," 41, 92 f.
7. Widman did not use line reckoning; he did, however, present a thorough treatment of proportions using the traditional terminology.
8. He was *Dr. utriusque juris* and rector in 1504.
9. See D. E. Smith, *Rara arithmetica*, 36 f.
10. Widman knew the Regensburg algebra of 1461, and he took certain problems from it (*Regula dele cose super quartum capitulum*). On this point see M. Curtze, in *Abhandlungen zur Geschichte der Mathematik*, 7 (1895), 72; and Wappler, "Zur Geschichte der deutschen Algebra," 540.
11. A description of Codex Dresdensis C 80 is in Kaunzner, "Über Johannes Widmann von Eger," 27–39.
12. See Codex Lipsiensis 1470, fols. 479–493. On this point see Kaunzner, "Über . . . Widmann . . .," 45.

13. See Kaunzner, "Deutsche Mathematiker des 15. und 16. Jahrhunderts und ihre Symbolik," 21.
14. Wappler, "Beitrag zur Geschichte der Mathematik," 167, proposes 1490 as the year of publication. All six treatises have the same format, the same type, and the same size pages and length of lines. They all appeared anonymously, without date and, with one exception (Leipzig), without city. Wimpina (*b*. 1460; enrolled at Leipzig in 1479) enumerated all these works except *Regula falsi*; at the time he made his list the works were commercially available.
15. They are problems that Widman had entered in the margins of Codex Dresdensis C 80. See Wappler, "Zur Geschichte der deutschen Algebra," 541 ff.

BIBLIOGRAPHY

I. Original Works. Widman's *Behend und hüpsch Rechnung uff allen Kauffmanschafften* appeared at Leipzig in 1489. The rest of his works, published anonymously and without city or date, are *Algorithmus integrorum cum probis annexis*; *Algorithmus linealis*; *Algorithmus minutiarum phisicarum*; *Algorithmus minutiarum vulgarium*; *Regula falsi apud philozophantes augmenti et decrementi appellata;* and *Tractatus proportionum plusquam aureus*. On these latter works, see Klebs (below), 35, 36, 281, 324; Wimpina (below), 50; and Wappler, "Beitrag zur Geschichte der Mathematik" (below).

II. Secondary Literature. See M. Cantor, *Vorlesungen zur Geschichte der Mathematik*, 2nd ed., II (Leipzig; 1913), 228 ff.; M. W. Drobisch, *De Ioanni Widmanni Egeriani compendio arithmeticae mercatorum* (Leipzig; 1840); G. Erler, *Die Matrikel der Universität Leipzig*, 3 vols. (Leipzig, 1895–1902), I, 323; II, 228, 289; W. Kaunzner, "Über Johannes Widmann von Eger. Ein Beitrag zur Geschichte der Rechenkunst im ausgehenden Mittelalter," *Veröffentlichungen des Forschungsinstituts des Deutschen Museums für die Geschichte der Naturwissenschaften und der Technik*, ser. C., no. 4 (1968); and "Deutsche Mathematiker des 15. und 16. Jahrhunderts und ihre Symbolik," *ibid.*, ser. A, no. 90 (1971); A. C. Klebs, "Incunabula scientifica et medica," in *Osiris*, **4** (1938), 1–359; D. E. Smith, *Rara arithmetica* (Boston–London, 1908), 36, 40, 44; E. Wappler, "Zur Geschichte der deutschen Algebra im 15. Jahrhundert," in *Programm Gymnasium Zwickau* (1887), 1–32; "Beitrag zur Geschichte der Mathematik," in *Abhandlungen zur Geschichte der Mathematik*, **5** (1890), 147–169; and "Zur Geschichte der deutschen Algebra," *ibid.*, **9** (1899), 537–554; and C. Wimpina, *Scriptorum insignium, qui in celeberrimis praesertim Lipsiensi, Wittenbergensi, Francofurdiana ad Viadrum academiis, a fundatione ipsarum usque ad annum Christi MDXV floruerunt centuria, quondam ab J. J. Madero Hannoverano edita, ex mspto autographo emendata, completa, annotationibusque brevibus ornata*, J. F. T. Merzdorf, ed. (Leipzig, 1839), 50 f.

Kurt Vogel

WIDMANNSTÄTTEN (or WIDMANSTETTER), ALOYS JOSEPH BECK EDLER VON (*b*. Graz, Austria, 13 July 1754; *d*. Vienna, Austria, 10 June 1849), *mineralogy*.

Widmannstätten succeeded his father, Johann Andreas, in the printing trade in 1764. From 1650 the family had enjoyed the exclusive privilege of printing for the province of Steiermark. In 1784, however, they were deprived of this monopoly as a consequence of the introduction of the freedom of the press. Widmannstätten then lost interest in his concern, leased it, and sold it in 1807 to his chief competitor in Graz. He nevertheless maintained an active role in the technical arts, and was often consulted because of his experience. In 1807 Widmannstätten became a director of Emperor Francis I's private technology collection in Vienna. He made several journeys (some on government order) to Germany, France (1815), England (1816), and Italy. For these activities Widmannstätten became a member of the Société d'Encouragement pour l'Industrie Nationale, and in 1817 he was granted a pension.

The "Widmannstätten figures" named after him were discovered by Widmannstätten in 1808 in an iron meteorite from Zagreb. By etching polished sawing planes with diluted nitric acid, he showed a regular pattern of slightly affected intersecting bands, between narrow frames, and angular fields, which were deepened by the acid. The pattern of bands, which is repeated on a small scale by the interstices, corresponds to the traces of octahedral planes in the etched section. Chemically the frames are enriched in nickel. The rugged interstices, having a higher iron content than the bands, constitute fine lamellar aggregates of nickel-poor and nickel-rich phases. Widmannstätten later distinguished these patterns in meteorites from Mexico (1810), Elbogen (now Loket, Czechoslovakia; 1812), and Lénarto (1815).

In 1813 Widmannstätten decided to make direct imprints of such etched surfaces with printer's ink and to publish them. But it was not until 1820 that Carl von Schreibers, at whose institute Widmannstätten made his investigations, published the prints.

BIBLIOGRAPHY

Widmannstätten published no writings.

Biographies are J. K. Hofrichter, "Alois Beck von Widmannstätten," in *Mittheilungen des historischen Ve-*

reins für Steiermark, **2** (1851), 144–150; and C. Wurzbach, in Biografisches Lexicon für Oesterreich, LV (Vienna, 1887), 258–261.

Descriptions of Widmannstätten figures are in W. Haidinger, "Bemerkungen über die zuweilen im geschmeidigen Eisen entstandene krystallinische Structur, verglichen mit jener des Meteoreisens," in Sitzungsberichte der Wiener Akademie der Wissenschaften, math.-naturwiss. Cl., **15** (1855), 354–361; C. von Schreibers, Beyträge zur Geschichte und Kenntniss meteorischer Stein- und Metall-Massen (Vienna, 1820), 70–73, with one print of Widmannstätten figures; R. Vogel, "Physikalisch-Chemisches über Meteoreisen," in C. A. Doelter, Handbuch der Mineralchemie, III, pt. 2, C. Doelter and H. Leitmeier, eds. (Dresden–Leipzig, 1926), 566–567.

JOYCE WEVERS

WIECHERT, EMIL (b. Tilsit, Germany, 26 December 1861; d. Göttingen, Germany, 19 March 1928), physics, geophysics.

Wiechert was the only son of Johann Christian Wiechert, a merchant who died when his son was very young. His mother devoted herself to providing for Wiechert's education and moved with him to Königsberg. Wiechert never left her until she died in 1927, although he married the daughter of a Göttingen lawyer named Ziebart in 1908.

Wiechert graduated in physics from Königsberg University in 1889 and became a lecturer in physics there in 1890. During the following seven years he carried out research on the atomic structure of electricity and matter. In 1897 he was appointed to the University of Göttingen and became the founder of one of the world's most famous schools of geophysics. From 1897 to 1914 some of the most far-reaching results concerning the internal structure of the earth emerged from the work of Wiechert and his pupils, who included B. Gutenberg, K. Zöppritz, L. Geiger, and G. Angenheister.

When Wiechert took up his appointment at Göttingen, seismology was beginning to develop as a quantitative science. In 1892 John Milne had produced the first seismograph capable of providing global coverage of ground motions resulting from large earthquakes. Following a study trip to Italy in 1899 Wiechert decided that he could greatly improve the seismographs then in use, and by 1900 he had produced his famous inverted-pendulum seismograph and had worked out its theory. This instrument was radically different in design from all previous seismographs and consisted essentially of a heavy mass (up to several tons in some versions) that can oscillate about a pivot below it, the mass being held near the equilibrium position by the pressure of thin springs at the top. The records which were made on smoked paper were outstandingly clear and gave a far closer representation of the ground motion than any previous seismographs had done. Wiechert's first seismograph recorded horizontal components of the motion, but he later designed instruments to measure vertical components as well. Wiechert seismographs are still used in some of the world's observatories and continue to supply valuable information.

The Göttingen school of geophysics produced much important work under Wiechert's leadership including the early tables prepared by Zöppritz that gave the travel times of earthquake waves through the interior of the earth, and Gutenberg's calculation giving the value of 2,900 kilometers for the depth of the earth's core. Wiechert himself had contributed to the mathematical theory underlying these calculations and, with Gustav Herglotz, he evolved the basic mathematical process whereby the velocities of seismic waves deep in the interior of the earth can be derived from the travel-time tables. Wiechert had, moreover, been one of the first to suggest the presence of a dense core in the earth and had produced the first theoretical model of the planet that allowed for it.

Wiechert played an important part in the world organization of seismology as one of a small group responsible for founding the International Association of Seismology in 1905, an organization that still flourishes. He also set up geophysical observational centers in the German colonies before World War I, a network that contributed vitally to the early development of seismology.

In addition to his work on seismology, Wiechert contributed to other branches of geophysics, particularly atmospheric electricity. Under his direction various methods were developed for measuring potential gradients and conductivity in the atmosphere.

When World War I came, Wiechert turned his attention to the transmission of sound waves through the atmosphere, applying his knowledge of earthquake theory to observations of sound waves with a view to determining features of the stratification of the atmosphere. He also began investigating the fine structure of the crust of the earth, using specially designed portable seismographs to record waves from artificial explosions. In this research he was one of the pioneers of geophysical prospecting by seismic methods.

Wiechert carried out his work with indefatigable energy and tenacity despite extreme deafness in his later years and the shadow of a serious illness, to which he succumbed at the age of sixty-six.

BIBLIOGRAPHY

Wiechert's most important publications are "Theorie der automatischen Seismographen," in *Abhandlungen der K. Gesellschaft der Wissenschaften zu Göttingen*, Math.-phys. Kl., n.s. **2**, no. 1 (1903), 1; "Über Erdbebenwellen I, II," in *Nachrichten von der Gesellschaft der Wissenschaften zu Göttingen*, Math.-phys. Kl. (1907), 415–549, written with K. Zöppritz; "Our Present Knowledge of the Earth," in *Report of the Board of Regents of the Smithsonian Institution* (1908), 431–449; and "Bestimmung des Weges der Erdbebenwellen im Erdinneren," in *Physikalische Zeitschrift*, **11** (1910), 294–311, written with L. Geiger.

An account of Wiechert's life and scientific work is in "Zum Gedenken Emil Wiecherts anlässlich der 100. Wiederkehr seines Geburtstages," in *Veröffentlichungen des Institutes für Bodendynamik und Erdbebenforschung in Jena*, no. 72 (1962), 5–21, which includes a full list of Wiechert's publications.

K. E. BULLEN

WIED, MAXIMILIAN ZU (*b*. Neuwied, Germany, 23 September 1782; *d*. Neuwied, 3 February 1867), *natural history, ethnology.*

Alexander Philip Maximilian, prince of Wied-Neuwied, was the eighth child and second son of Prince Friedrich Karl, the ruler of a small principality near Koblenz in Rhenish Prussia. His mother encouraged his interest in natural history when he was a youth, and in later years he became a student of Johann Friedrich Blumenbach, from whom derived Wied's interest in studying man as an object of natural history.

Wied had planned to visit America in 1803, but the political and military confusion of the Napoleonic era prevented the trip. He served with the Prussian army at the battle of Jena, during which he was captured by the French. He was exchanged and returned to Neuwied. Resuming military service, he rose to the rank of major general and was with the allied army when it entered Paris in 1814.

As soon as possible after the cessation of hostilities, Wied retired from active military service and began to pursue the study of natural history. With the United States in the process of recovering from the War of 1812, he postponed his trip there and sailed instead for South America. His two-year journey along the coast of Brazil enabled him to observe the primitive Indian tribes of the Brazilian forests and to record the manners and life styles of the Purí, Botocudos, Patachos, and Camacans. In his journals Wied noted the distinctive native flora and fauna of the region between Rio de Janiero and Bahía. After returning to Germany, he arranged his collections and prepared his notes for publication: *Reise nach Brasilien in den Jahren 1815 bis 1817* (1820–1821). This work, soon translated into Dutch, French, and English, established his reputation as a naturalist.

In 1832 Wied sailed for the United States, intending to compare the Indians of South America with those of North America and to journey as far west as the Rocky Mountains in order to examine the flora, fauna, and aboriginal peoples of the trans-Mississippi area. He was accompanied on this trip by the young Swiss artist Karl Bodmer, who had been hired by Wied to illustrate his scientific journals. After spending some time in Boston, New York, and Philadelphia, Wied crossed the Appalachian Mountains and wintered at New Harmony, Indiana, where he met Thomas Say, Charles Alexandre Lesueur, and William and Robert Dale Owen.

In the spring of 1833, Wied received the support of the American Fur Company and was allowed to travel up the Missouri River on their steamers. At Fort McKenzie, below the great falls of the Missouri, he began his investigation of Indian life, collecting data about Indian languages; he later published vocabularies for the Arikaras, Assiniboine, Blackfoot, Cheyenne, Crow, Mandan, and Sioux tribes. Besides studying Indian life, Wied collected plant and animal specimens. With the approach of winter, he returned to Fort Clark on the Missouri, where he remained until the spring of 1834, making an intensive study of the Mandans. Wied's studies of the Indians of the upper Missouri are characterized by a sincere attempt to portray them not as savages but as civilized individuals with acquired skills and mores ideally suited for life in a wilderness. His detailed ethnographical description of these tribes assumes additional importance when it is noted that the smallpox epidemic of 1837–1838 destroyed several of these tribes and substantially reduced the populations of others.

Wied returned to Europe in the summer of 1834. Although many of the animal and plant specimens that he had laboriously collected were lost in a river accident, he published a meticulously edited and handsomely illustrated account of his travels in the American West in his *Reise in das innere Nord-America in den Jahren 1832 bis 1834.*

The rest of Wied's life was devoted to cataloging and studying the specimens he had collected on his trips to North and South America. His primary contribution to science was his detailed ethnographical descriptions of the native tribes of Brazil and the upper Missouri, for he sensitively recorded a way of life that was soon to disappear. His zoological collections were purchased by the American Museum of Natural History. In the field Wied was a skilled observer who carefully recorded the dimensions and habitat of the flora and fauna he collected. Nevertheless, his ethnographical observations are the more valuable aspect of his work.

BIBLIOGRAPHY

I. ORIGINAL WORKS. Wied's account of his trip to Brazil is in *Reise nach Brasilien in den Jahren 1815 bis 1817,* 2 vols. (Frankfurt, 1820–1821), and *Beiträge zur Naturgeschichte von Brasiliens,* 4 vols. in 6 (Weimar, 1825–1833). His account of his American tour, with beautiful plates by Karl Bodmer, is in *Reise in das innere Nord-America in den Jahren 1832 bis 1834,* 2 vols. and atlas (Koblenz, 1839–1841). An English trans. of this work was published in Reuben Gold Thwaites' *Early Western Travels 1748–1846,* XXII–XXV (Cleveland, 1906). Wied's scientific papers are listed in Royal Society *Catalogue of Scientific Papers,* VI, 357–358, and VIII, 1235. The Joslyn Art Museum in Omaha, Nebraska, has a collection of Wied papers that includes MS diaries, correspondence, a scientific journal, and the originals of many of Bodmer's paintings.

II. SECONDARY LITERATURE. There is no definitive study of Wied's life and work. For general surveys of his life, see Vernon Bailey, "Maximilian's Travels in the Interior of North America, 1832 to 1834," in *Natural History,* **23,** no. 4 (July–Aug. 1923), 337–343; the very popular account in Bernard DeVoto, *Across the Wide Missouri* (New York, 1947), 133–146; and Joseph Röder, "The Prince and the Painter," in *Natural History,* **64,** no. 6 (June 1955), 326–329. For a more detailed account consult Philipp Wirtgen, *Zum Andenken an Prinz Maximilian zu Wied, sein Leben und wissenschaftliche Thätigkeit* (Neuwied–Leipzig, 1867); and *Allgemeine deutsche Biographie,* XXIII (1886), 559–564.

PHILLIP DRENNON THOMAS

WIEDEMANN, GUSTAV HEINRICH (*b.* Berlin, Germany, 2 October 1826; *d.* Leipzig, Germany, 23 March 1899), *physics, physical chemistry.*

Wiedemann was one of the outstanding members of the group of physicists trained in Berlin by H. G. Magnus. He held the first professorship of physical chemistry created in Germany (at Leipzig); and he became famous through his works on electromagnetism, especially his textbook on the theory of electricity, and through his long editorship of *Annalen der Physik und Chemie,* known as *Wiedemanns Annalen.*

Wiedemann came from a merchant family. His father died when the boy was two and his mother when he was fifteen, and subsequently he lived with his grandparents. From 1838 to 1844 he studied at the Köllnische Realgymnasium in Berlin. At that time the school's director was the physicist E. F. August, and its teachers included Louis Seebeck. Wiedemann also received much encouragement and help in his studies from an uncle in Berlin, the mechanic C. A. Gruel.

From 1844 to 1847 Wiedemann studied natural sciences at the University of Berlin. His professors included the mathematicians Dirichlet and Joachimsthal; the chemists Heinrich Rose, F. L. Sonnenschein, and E. Mitscherlich; and the physicists Magnus and Dove. Magnus allowed students to use his private laboratory; and it was there that Wiedemann met Helmholtz, who became a lifelong friend. Since Magnus, in conscious rejection of the nineteenth-century *Naturphilosophie,* concerned himself exclusively with experimental physics and neglected all theoretical questions, Helmholtz and Wiedemann decided to work through Poisson's work on the theory of elasticity. In 1847 Wiedemann received the doctorate from Berlin for a dissertation on the urea derivative biuret. In 1850 he qualified as a lecturer with a work on the turning of the polarization plane of light discovered by Faraday. He lectured at the university on selected topics in theoretical physics.

In 1851 Wiedemann married Mitscherlich's daughter Clara; they had a daughter and two sons: Eilhard, who became a physicist and historian of science, and Alfred, who became an Egyptologist. In 1854 Wiedemann was appointed professor of physics at the University of Basel, where he worked closely with the chemist C. F. Schönbein. He returned to Germany in 1863 to accept a post at the Polytechnische Schule in Brunswick. In 1866 he moved to the Polytechnische Schule in Karlsruhe, where, in addition to his duties as professor of physics, he organized meteorological observations for the state of Baden. His works on physical chemistry brought him an offer from Leipzig in 1871 to occupy Germany's first chair of physical chemistry. Upon the retirement of Wilhelm Hankel in 1887, Wiedemann succeeded him as professor of physics at Leipzig, and Wilhelm Ostwald was given the chair of physical chemistry.

In 1877, following Poggendorff's death, Wiedemann became editor of the *Annalen der Physik und Chemie*, a position he held for the rest of his life. During his editorship the *Annalen* became one of the most distinguished German physics periodicals. He was able to increase its usefulness by publishing the *Beiblätter zu den Annalen der Physik*, of which his son Eilhard became editor.

At Berlin, Wiedemann devoted his first studies to electrical conductivity on the surfaces of various metals, to the rotation of the plane of polarization of light under the influence of electric current, and to the thermal conductivity of metals. In 1853, collaborating with Rudolph Franz, he discovered the physical law named for them. It states that at a constant, not very low temperature T, the electrical conductivity κ of metals is approximately proportional to their thermal conductivity λ; that is,

$$\frac{\lambda}{\kappa} = \text{constant } T.$$

At Basel, Wiedemann continued the studies he had begun at Berlin on endosmosis, in which he established the dependence of the osmotic pressure on the current intensity and composition of the solutions. He also performed experiments on torsion and on the magnetization of steel and iron, and examined the influence of temperature on both phenomena. In later years he returned to the problem of the relations between magnetic and mechanical phenomena. This research also constituted the preliminary steps toward Wiedemann's greatest work, *Die Lehre vom Galvanismus* (1861–1863), a systematic presentation of everything known about the subject. This textbook was quickly recognized as the standard work on galvanism and was widely read.

At Brunswick, Wiedemann discovered the additive law for the magnetism of chemical compounds. He also investigated the vapor pressures of salts containing water of crystallization and demonstrated that they depend solely on temperature. At Karlsruhe he worked with R. Rühlmann on an exhaustive study of the processes involved in gas discharge and thereby became a pioneer in that field as well.

Pursuing his early research on magnetism while at Leipzig, Wiedemann posited the existence of "magnetic molecules." His new, reliable determinations of the unit of electrical resistance, the ohm, won the admiration of his colleagues. In making these measurements Wiedemann used the large induction coil designed by Wilhelm Weber. In addition he improved the existing techniques for measuring temperature and devised a new galvanometer that bears his name.

Through both his teaching and his research, Wiedemann made important contributions to physics and chemistry, especially in electromagnetism. In addition, as an editor and as an organizer, he possessed a virtually universal knowledge of physical science and performed many valuable services for his fellow physicists.

BIBLIOGRAPHY

I. ORIGINAL WORKS. A bibliography of Wiedemann's numerous scientific papers is in Poggendorff, II, 1319; III, 1441; IV, 1631; and A. von Harnack, *Geschichte der Königlichen Preussischen Akademie der Wissenschaften zu Berlin*, III (Berlin, 1900), 288 (Wiedemann's academic writings only).

His papers include his dissertation, "De nova quodam corpore ex urea producto" (Berlin, 1847); "Ueber die Wärmeleitungsfähigkeit der Metalle," in *Annalen der Physik und Chemie*, **89** (1853), 457–531, written with R. Franz; "Ueber den Einfluss der Temperaturänderungen auf den Magnetismus des Eisens und Stahls," *ibid.*, **122** (1864), 346–358; "Magnetische Untersuchungen über den Magnetismus der Salze der magnetischen Metalle," *ibid.*, **126** (1865), 1–38; "Ueber den Magnetismus der chemischen Verbindungen," *ibid.*, **135** (1868), 177–237; "Ueber den Durchgang der Elektricität durch Gase," *ibid.*, **145** (1872), 235–259, 364–399; and **158** (1876), 71–87, 252–287, written with R. Rühlmann; "Ueber die Dissociation der gelösten Eisenoxydsalze," *ibid.*, n.s. **5** (1878), 45–83; "Ueber die Torsion," *ibid.*, **6** (1876), 71–87, 252–287, written with Rühlmann; "Ueber die Dissociation der gelösten Eisenoxydsalze," 452–461; and **37** (1889), 610–628; and "Ueber die Bestimmung des Ohm," *ibid.*, **42** (1891), 227–256, 425–449, first published as *Abhandlungen der Preussischen Akademie der Wissenschaften*, phys. Kl. (1884), no. 3.

His book is *Die Lehre vom Galvanismus und Elektromagnetismus nebst technischen Anwendungen*, 2 vols. (Brunswick, 1861–1863; 2nd ed., 1872–1873; 3rd ed., *Die Lehre von der Elektricität*, 4 vols., 1882–1885; 4th ed., 1893–1898).

II. SECONDARY LITERATURE. See H. von Helmholtz, "Gustav Wiedemann beim Beginn des 50. Bandes seiner *Annalen der Physik und Chemie* gewidmet," in *Annalen der Physik*, n.s. **50** (1893), iii–xi; F. Kohlrausch, "Gustav Wiedemann," in *Verhandlungen der Deutschen physikalischen Gesellschaft*, **1** (1899), 155–167; W. Ostwald, "Zur Erinnerung an Gustav Wiedemann," in *Berichte der Kgl. Sächsischen Gesellschaft der Wissenschaften*, math.-phys. Kl., **51** (1899), lxxvii–lxxxiii; M. Planck, "Gustav Wiedemann, dem Herausgeber der Annalen zum Fünfzigjährigen Doctorjubiläum gewidmet," in *Annalen der Physik und Chemie*, n.s. **63** (1897), vii–xi; and H. Reiger, "Wiedemann, Gustav Heinrich," in *Allgemeine deutsche Biographie*, LV (1910), 67–70.

HANS-GÜNTHER KÖRBER

WIEDERSHEIM, ROBERT (*b.* Nürtingen, Baden-Württemberg, Germany, 21 April 1848; *d.* Lindau im Bodensee, Germany, 23 July 1923), *comparative anatomy, embryology.*

Wiedersheim was greatly influenced by his father, who practiced general medicine and was also a naturalist and collector of zoological specimens; his mother, Berta Otto Wiedersheim, died after his birth. At the age of fifteen Wiedersheim devoted most of his time to the microscopic study of the freshwater hydra *fusca* and *viridis.* Consequently his final school report in classical languages was a minor disaster. Although he wanted to study zoology, at the insistence of his father he began medical studies in October 1868 at the University of Tübingen, where he was a pupil of Leydig. In 1871 Wiedersheim transferred to the University of Würzburg to study under A. Koelliker and C. Hasse; the latter proposed to Wiedersheim a thesis on the structure of the stomach of birds, which was completed in 1872.

Wiedersheim finished his medical studies at the University of Freiburg, and became a university demonstrator and lecturer at Würzburg, where he worked under Koelliker until 1876. At Würzburg he established himself as an excellent teacher of systematic and comparative anatomy. At the same time he also published papers on comparative vertebrate anatomy. At the end of 1876 Wiedersheim was appointed associate professor of anatomy at Freiburg, and, in 1883, full professor and director of the Institute of Anatomy and Comparative Anatomy. During his early years at Freiburg, Wiedersheim began the work that led to his world-famous textbook on comparative anatomy *Vergleichende Anatomie der Wirbeltiere*—based on the comparison of vertebrates, and their embryologic and phylogenetic development. English editions of *Comparative Vertebrate Anatomy* by E. N. Parker appeared in 1886 and 1897. In 1907 a rewritten and revised third edition was published. Wiedersheim's fame as a morphologist rests on his book, and numerous students from Europe and the United States were thus attracted to the Institute.

When Wiedersheim retired at the age of seventy, he had firmly established the teaching of comparative anatomy. The respect and affection that his pupils and colleagues had for him is evident in the special issue of the *Zeitschrift für Morphologie und Anthropologie* (1924) published in honor of his seventy-fifth birthday. Wiedersheim was survived by his wife, Tilla Gruber, and one son.

BIBLIOGRAPHY

I. ORIGINAL WORKS. Wiedersheim's works include "Die feineren Strukturverhältnisse der Drüsen im Muskelmagen der Vögel," in *Archiv für mikroskopische Anatomie und Entwicklungsmechanik*, **8** (1872), 435–452; "Beiträge zur Kenntnis der württemberg. Höhlenfauna," in *Verhandlungen der Würzburger physikalische medizinische Gesellschaft*, **4** (1873), 207–222; "Über den Mädelhofener Schädelfund in Unterfranken," in *Archiv für Anthropologie*, **8** (1874), 225–238; "Salamandrina perspicillata und Geotriton fuscus. Versuch einer vergleichenden Anatomie der Salamandrinen," in *Annali del Museo civico di storia naturale Giacomo Doria*, **7** (1875), 5–206; "Bemerkungen zur Anatomie des Euproctus Rusconii," *ibid.*, 545–568; "Zur Anatomie und Physiologie des Phyllodactylus europaeus mit besonderer Berücksichtigung des Aquaeductus vestibuli der Ascalaboten im allgemeinen," in *Morphologisches Jahrbuch*, **1** (1876), 495–534; "Die Kopfdrüsen der geschwänzten Amphibien und die Glandula intermaxillaris der Anuren," in *Zeitschrift für wissenschaftliche Zoologie*, **27** (1876), 1–50; "Die ältesten Formen des Carpus mit Tarsus der heutigen Amphibien," in *Morphologisches Jahrbuch*, **2** (1876), 421–434; "Über Neubildung von Kiemen bei Siren Lacertina," *ibid.*, **3** (1877), 630–631; "Zur Fortpflanzungsgeschichte des Proteus anguineus," *ibid.*, 632; "Das Kopfskelett der Urodelen," *ibid.*, 352–448; "Labyrinthodon Rütimeyeri," in *Abhandlungen der Schweizerischen paläontologischen Gesellschaft*, **5** (1878), 1–56; "Ein neuer Saurus aus der Trias," *ibid.*, **6** (1879), 75–124; "Die spinalartigen Hirnnerven von Ammocoetes und Petromyzon Planeri," in *Zoologischer Anzeiger*, **3** (1880), 446–449; "Über den sogenannten Tentakel der Gymnophionen," *ibid.*, 493–496; "Über den Tarsus der Saurier," *ibid.* (1880), 496; *Morphologische Studien* (1880); "Über die Vermehrung des Os centrale im Carpus und Tarsus des Axolotl," in *Morphologisches Jahrbuch*, **6** (1880), 581–583; "Zur Histologie der Dipnöer-Schuppen," in *Archiv für mikroskopische Anatomie und Entwicklungsmechanik*, **18** (1880), 122–129; "Zur Anatomie des Amblystoma Weismanni," in *Zeitschrift für wissenschaftliche Zoologie*, **32** (1880), 214–236; "Vomero-Nasal (Jacobson's) Organ," in *Comparative Anatomy of Vertebrates*, **3** (1907), 271–273; and "Über das Becken der Fische," in *Morphologisches Jahrbuch*, **1** (1881), 326–327.

II. SECONDARY LITERATURE. See "Robert Wiedersheim, Festschrift zu seinem 75. Geburtstag von seinen Schülern und Freunden," in *Zeitschrift für Morphologie und Anthropologie*, **24** (1924).

PAUL GLEES

WIEGLEB, JOHANN CHRISTIAN (*b.* Langensalza, Germany, 21 December 1732; *d.* Langensalza, 16 January 1800), *pharmacy, chemistry.*

From the end of the seventeenth century, apothecaries played an increasingly active role in scientific research, especially in chemistry. The expansion of the number of medicines through the addition of chemical preparations, begun by Paracelsus, was at first supported by physicians. Later this development stimulated apothecaries to undertake investigations of their own, however, many of which went far beyond the confines of their professional interests. Wiegleb was one of the most important figures among this steadily growing group.

His father, a lawyer, died while Wiegleb was still young; and his mother remarried—again a lawyer—in order to provide a good upbringing for her children. While still at school Wiegleb was an assistant to his stepfather, and thus acquired considerable skill in the use of language and knowledge of legal matters.

Wiegleb was related to several local apothecaries and, by observing them at work, he became interested in entering the profession. Although his family would have preferred him to become a theologian, they allowed him to make his own decision. In 1748, at age sixteen, Wiegleb began his apprenticeship at the Marienapotheke, in Dresden. Unfortunately, the proprietor took virtually no interest in his training, leaving this matter to his assistants, who were far from equal to the task. As a result Wiegleb learned only manual skills. He therefore had to study on his own; and in order to master the relevant technical writings, most of which were in Latin, Wiegleb had to better his scanty knowledge of the language. This he did with great enthusiasm, practicing on the scholarly books he found in the shop, although many of them were outdated. In the beginning he was especially drawn to alchemical writings, which for a time he considered to be the summit of human knowledge.

Wiegleb's apprenticeship lasted six years; but since no successor could be found for him, he remained in Dresden for another six months. The next year he was an assistant to an apothecary in Quedlinburg but, receiving no more instruction there than in his previous position, he returned to Langensalza. The owner of the apothecary shop there had just died; and Wiegleb, now aged twenty-six, was offered the opportunity to manage it for the man's widow. But, wishing to be fully independent, he declined the offer and, with money inherited from his parents, built his own pharmacy. The new building, which was finished in 1759, contained a splendid shop and a model laboratory.

During this period Wiegleb became friendly with Ernst Baldinger, a physician practicing in Langensalza who was very interested in scientific research and who later held professorships at Jena and Marburg. The two men conducted chemistry experiments together, and in the course of their work Baldinger acquainted Wiegleb with the latest developments in chemistry. In 1767 Wiegleb published his first book, which contained a number of brief writings. This was followed by a work on fermentation. He then investigated the alkaline salts found in plants and confirmed Marggraf's discovery of this type of substance. In a number of papers Wiegleb displayed an excellent knowledge of analytic procedures. For instance, he repeated with remarkable exactness the analyses of Torbern Bergman and corrected a considerable number of the latter's errors.

In 1775 Wiegleb published a German translation of R. A. Vogel's *Institutiones chemiae,* a work he greatly admired. It probably was while preparing the book for publication that Wiegleb decided to create an institute for training pharmacists, and he founded such a school in 1779. There the students, after attending lectures, were expected to participate in laboratory experiments and to do independent work. They had the use of an extensive library and could obtain instruction in languages. Later critics believed that the teaching was more suited to training chemists than to providing a general knowledge of all branches of pharmacy, but this view seems exaggerated. In the approximately twenty years of its existence, the school trained about fifty students, including J. F. A. Göttling (later a professor of chemistry at Jena), Klaproth, Hermbstädt, and the botanist Willdenow—which suggests that the curriculum was, in fact, broadly based.

Wiegleb's reputation extended far beyond northern Germany: scientists from throughout Europe visited him. Besides purely scientific questions, he was much concerned with improving technology. Whenever he could, he supported such crafts as dyeing and brewing. Wiegleb even studied the economic problems of Langensalza, and he was elected both a member of the city council and municipal treasurer.

Very little is known about Wiegleb's family. He had nine children, four of whom survived him. Two of the children were deaf and dumb. The last years of his life were difficult, for in 1789 an accident with fulminate of mercury almost blinded him. In addition, he sold his apothecary shop but lost most of the purchase price.

Wiegleb's critical attitude in assessing scientific questions earned him high esteem in learned circles. After several years of work he published *Historisch-kritische Untersuchung der Alchemie* (1777), which went through a second edition. In this work he stated:

> The best accounts from the period when the name alchemy is encountered, . . . are examined, and it is thereby demonstrated that they are, taken together, incapable of confirming the reality of alchemy. Then, the strongest proof is adduced to show that the entire imaginary art of alchemy is impossible according to all known, certain natural laws of human art; thus [it is shown] that it has never truly been practiced by anyone.

Wiegleb carefully examined famous reports of the transformation of metals and pointed out their deficiencies; in a short time his work became widely known. His motto was "To doubt is the beginning of knowledge," so it is all the more astonishing that Wiegleb was a convinced proponent of Stahl's phlogiston theory throughout his life. He believed that phlogiston was a "subtle but destructible" substance that never can be produced in the pure state. Somewhat later he conjectured that phlogiston is a kind of hydrogen. Even after the triumph of Lavoisier's ideas, Wiegleb did not abandon the phlogiston theory. For a long time he contended that phlogiston possessed a negative weight. In the ensuing debates he argued objectively with his opponents and even adopted several of their experimental findings—for instance, he considered the two hypothetical "substances" light and phlogiston to be identical. On the whole, however, he clung to Stahl's theory.

After his book on alchemy, Wiegleb published a great many works. In 1779 he reported that oxalic acid is a separate compound. He also devised an improved method of preparing Glauber's salt and took special pains to work out new and improved mineral analyses. Wiegleb was interested in the history of chemistry and wrote a two-volume work on the subject, as well as a history of gunpowder.

Except for an educational trip to almost all the countries of Europe, Wiegleb never left his native city. Nevertheless, at his death he was one of Germany's best-known men of learning.

BIBLIOGRAPHY

I. ORIGINAL WORKS. Wiegleb's writings include *Kleine chymische Abhandlungen von dem grossen Nutzen der Erkenntniss des acidi pinguis bey der Erklärung vieler chymische Erscheinungen* (Langensalza, 1770), which contains "Entstehung des Glases," "Grüne Flamme borhalt. Alkohols," and "Über die rote Farbe des Zinnobers"; *Vertheidigung der Mayerischen Lehre vom acido pingui gegen verschieden dawider gemachte Einwendungen* (Altenburg, 1770); *Chymische Versuche über die alkalischen Salze* (Berlin, 1774); *Fortgesetzte kleine chymische Abhandlungen* (Langensalza, 1770), which contains "Farbe des Quecksilberoxydes" and "Zerlegung des Salmiaks durch Eisen"; *Rudolf Aug. Vogels Institutiones chemiae als Lehrsätze der Chemie* (Weimar, 1775; 2nd ed., 1785); *Neuer Begriff von der Gährung und den ihr unterwürfigen Körpern* (Weimar, 1776); *Geschichte der Alchemie: Historisch-kritische Untersuchung der Alchemie oder der eingebildeten Goldmacherkunst, von ihrem Ursprunge als Fortgang* . . . (Weimar, 1777; 2nd ed., 1793); *Revision der Grundlehren von der chemischen Verwandtschaft der Körper* (Erfurt, 1777); "Untersuchungen der Waffen der Bronzezeit," in *Acta Academiae Electonum Moguntiaca scientiarum utilis* (Erfurt, 1777); and *Die natürliche Magie* (Berlin, 1779; 1782–1786), completed by G. E. Rosenthal (1805).

Further works are "Über Oxalsäure," in *Chemisches Journal*, **2** (1779); *Handbuch der Allgemeine Chemie* (Berlin, 1781; 2nd ed., 1787; 3rd ed., 1796), also translated into English by C. R. Hopson as *General System of Chemistry Theoretical and Practical, Digested and Arranged With a Particular View to Its Applications. . .* (London, 1789); *P. v. Musschenbroeks Elementa Chemiae als Anfangsgründe der Chemie* (Berlin, 1782); *Onomatologia curiosa artificiosa et magica oder natürliches Zauberlexikon* (Nuremberg, 1784); *Unterhaltende Naturwunder* (Erfurt, 1788), written with F. Kroll; "Über Phlogiston," in *Chemische Annalen*, **2** (1791); *Geschichte des Wachstums und der Erfindungen in der Chemie in der ältesten und mittleren Zeit, aus dem Lateinischen Werk Bergmans mit Zusätzen* (Berlin, 1792); *Deutsches Apothekerbuch*, 2 vols. (Gotha, 1793; 4th ed., 1804), written with J. C. T. Schlegel; "Herstellung der Soda aus Kochsalz mit Hilfe von Vitriol," in *Chemische Annalen* (1793); "Chemische Nomenclatur," *ibid.*, **2** (1796); and "Verkalken des Bleies," *ibid.*, **1** (1797).

Wiegleb also published many essays in booklet form for analytic chemists. In addition, he translated works by Boerhaave and Demachy, and edited writings by G. A. Hoffmann, Dorothea Erxleben, and others.

II. SECONDARY LITERATURE. See H. Gutbier, *Beiträge zur Geschichte der Apotheken in Langensalza* (Langensalza, 1929); and R. Möller, "Ein Apotheker und Chemiker der Aufklärung," in *Pharmazie* (Berlin), **20** (1965), 230–239.

GUNTHER KERSTEIN

WIELAND, HEINRICH OTTO (*b*. Pforzheim, Germany, 4 June 1877; *d*. Starnberg, Germany, 5 August 1957), *organic chemistry*.

Wieland was the son of Theodor Wieland, a pharmaceutical chemist. He studied chemistry at the University of Munich in 1896, the University of Berlin in 1897, and at the Technische Hochschule at Stuttgart in 1898. The following year he returned to Munich and in 1901 received the Ph.D. for his research in organic chemistry under the direction of Johannes Thiele. In 1904 Wieland was appointed *Privatdozent* at the University of Munich and in 1913 received a senior lectureship in organic chemistry. He remained at Munich until 1917, during which time he devoted his research chiefly to the chemistry of organic nitrogen compounds.

Wieland's earliest work concerned the mechanism of addition of the oxides of nitrogen to olefins and the mechanism of the nitration of aromatic hydrocarbons. He was able to isolate the intermediate nitro compounds and show the similarity between the two classes of reactions. At this time Wieland also undertook a study of fulminic acid, and in a review published in 1909 he summarized his investigations of its polymerization and its step-by-step synthesis from ethanol and nitric acid.

Wieland's most significant work during his early career was the chemistry of the hydrazines, a project that led him to the discovery of the first known nitrogen free radicals. In 1911 Wieland prepared tetraphenylhydrazine from the oxidation of diphenylamine. He showed that when heated in toluene, tetraphenylhydrazine dissociates into two diphenylnitrogen free radicals, characterized by the green color that they impart to the solution. Wieland then undertook an extensive study of the effect of ring substituents on the production of radicals from substituted tetraphenylhydrazines.

In 1917 Wieland accepted a position at the Technische Hochschule in Munich, but during 1917–1918 was given a leave of absence to take part in chemical warfare research under the direction of Fritz Haber at the Kaiser-Wilhelm Institute in Berlin-Dahlem. After the war Wieland returned to the Technische Hochschule in Munich, where he remained until 1921, when he accepted a position at the University of Freiburg. In 1924 Willstätter resigned his post as director of the famous Baeyer laboratory at the University of Munich and recommended Wieland to be his successor. Wieland returned to Munich in 1925 and directed the Baeyer laboratory for twenty-five years until his

appointment as emeritus professor in 1950. During this time he became more interested in the structural determination of natural products. He had already shown considerable interest in biochemistry, for in 1912 he first proposed his theory of biological oxidation. In his subsequent studies on the mechanisms of oxidation reactions published in over fifty papers from 1912 to 1943, Wieland was able to demonstrate that many biological oxidation reactions proceed through dehydrogenation. While director of the Baeyer laboratory Wieland and his students worked on the isolation and structural determination of many natural products, including morphine alkaloids, lobeline alkaloids, strychnine alkaloids, pterins that he first isolated from butterfly pigments, mushroom poisons, and cardioactive toad poisons.

Wieland's best-known work, however, concerned the structure of bile acids, for which he received the 1927 Nobel Prize in chemistry. Wieland's research on this subject began in 1912, and for twenty years he and his collaborators sought to gain insight into the complicated structure of cholic acid and other bile acids related to cholesterol, through oxidation of specific portions of the molecule. Gradually information was assembled from such studies performed by Wieland's research group and also those carried out by other chemists, especially Windaus, who received the 1928 Nobel Prize for his work on the constructions of sterols. The following structures were generally accepted for cholic acid and cholesterol when Wieland and Windaus presented their Nobel lectures on 12 December 1928. Only two carbon atoms (numbered 15 and 16) remained to be assigned with certainty, and they were provisionally placed on carbon atom number 10, as shown.

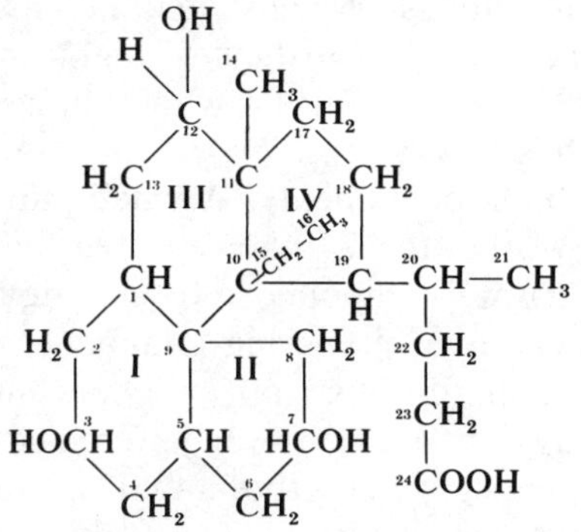

FIGURE 1.   Cholic acid (Wieland, 1928)

During the next four years Wieland and his co-workers at Munich tried to establish the location of

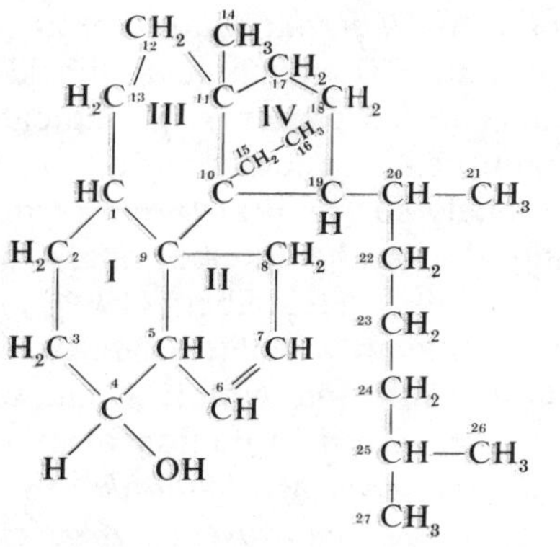

FIGURE 2.   Cholesterol (Windaus, 1928)

these two carbon atoms, yet they met with little success. In 1932 new evidence of the molecular size of steroids, gained from X-ray crystallographic analysis, cast doubt on the basic structure. By reconsidering the data collected over the previous twenty years, Wieland and the British chemists O. Rosenheim and H. King independently arrived at the presently accepted structure of cholic acid.

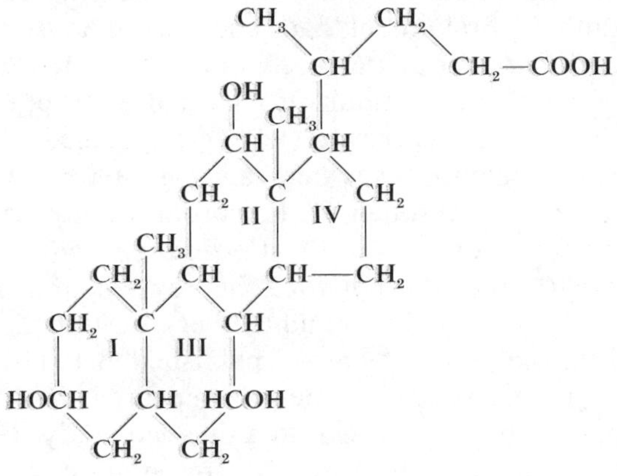

FIGURE 3.   Cholic acid (Wieland, 1932)

Wieland's research continued until his retirement in 1950. He served as editor of the *Annalen der Chemie* for over twenty years and received the Otto Hahn Prize in 1955.

BIBLIOGRAPHY

I. ORIGINAL WORKS. Wieland published over 350 papers and several books, which are listed in Poggendorff, V, 1366–1367; VI, 2876–2877; VII, 985–987. Some of his most important publications are "Die Knallsäure," in *Sammlung chemischer und chemisch-technischer Vorträge*, 14 (1909), 385–461; *Die Hydrazine* (Stuttgart, 1913); "Die Chemie der Gallensäuren," in *Zeitschrift für angewandte Chemie und Zentralblatt für technische Chemie*, 42 (1929), 421–424; "Recent Researches on Biological Oxidation," in *Journal of the Chemical Society* (1931), 1055–1064; *On the Mechanism of Oxidation* (New Haven, Conn., 1932); and "Die Konstitution der Gallensäuren," in *Berichte der Deutschen chemischen Gesellschaft*, 67 (1934), 27–39.

II. SECONDARY LITERATURE. An autobiographical sketch appeared in *Nachrichten aus Chemie und Technik* (1955), 222–223. A useful summary of his work as a tribute on his 65th birthday was given by Elisabeth Dane, "Die Arbeiten H. Wieland auf dem Gebiet der Steroide," in *Die Naturwissenschaften*, 30 (1942), 333–342; Wilhelm Franke, "H. Wielands Arbeiten zum Mechanismus der biologischen Oxydation," *ibid.*, 342–351; Friedrich Klages, "Die Stickstoffarbeiten von H. Wieland," *ibid.*, 351–359; and Clemens Schöpf, "Die Arbeiten Heinrich Wielands über stickstoffhaltige Naturstoffe (Alkaloide und Pterine)," *ibid.*, 359–373.

Other accounts include Rolf Huisigen, "The Wieland Memorial Lecture," in *Proceedings of the Chemical Society* (1958), 210–219; Gulbrand Lunde, "The 1927 and 1928 Nobel Chemistry Prize Winners, Wieland and Windaus," in *Journal of Chemical Education*, 7 (1930), 1763–1771; and Adolph Windaus, "The Chemistry of the Sterols, Bile Acids, and Other Cyclic Constituents of Natural Fats and Oils," in *Annual Review of Biochemistry*, 1 (1932), 109–134.

DANIEL P. JONES

WIELAND (or GUILANDINUS), MELCHIOR (*b.* Königsberg, Germany [now Kaliningrad, R.S.F.S.R.], *ca.* 1520; *d.* Padua, Italy, 8 January 1589), *botany.*

Very little is known about this remarkable scholar, polemicist, able botanist, and traveler. His name probably was Wieland—it certainly was latinized as Guilandinus. He began his studies at the University of Königsberg and continued them in Rome. He traveled as far as Sicily, supporting himself by selling medicinal herbs. He journeyed through many parts of Asia, Palestine, and Egypt with financial support and letters of recommendation from Senator Marino Cavalli, one of the reformers of the Padua Studium. Wieland's return to Italy was not without danger: after being captured by Algerian pirates and being shipwrecked he landed at Genoa, then went to Venice. On 20 September 1561, because of his fame as a scholar, he was asked to succeed Anguillara as director of the botanical garden at Padua. He was equally successful in his university career and was reappointed several times to the chair of "lecture and demonstration of medicinal herbs," which combined botany and pharmacognosy. Wieland, who is buried in the cloister of the Basilica of Saint Anthony in Padua, left his library to the Venetian Republic (it is now in the library of San Marco) and most of his possessions to Benedetto Zorzi.

A scholar of vast knowledge, Wieland was an outstanding director of the Padua botanical garden, into which he introduced many rare plants and a machine for irrigation (1575). He left no writings of particular value. His scientific observations are contained in *epistolae*, letters on botany that include descriptions of now-forgotten and little-known plants.

In keeping with his reputation as a polemicist, Wieland aroused violent enmities, such as that of Mattioli, and formed friendships equally strong, such as that with Falloppio.

BIBLIOGRAPHY

A. von Haller, *Bibliotheca botanica*, I (Zurich, 1771), 320–321; J. J. Mangetus, *Bibliotheca scriptorum medicorum, veterum et recentiorum*, I, pt. 2 (Geneva, 1781), 539; R. de Visiani, *L'Orto botanico di Padova* (Padua, 1842), 9–12; and G. B. de Toni, "Melchiorre Guilandino," in A. Mieli, ed., *Gli scienziati italiani*, I (Rome, 1933), 73–76.

LORIS PREMUDA

WIELEITNER, HEINRICH (*b*. Wasserburg am Inn, Germany, 31 October 1874; *d*. Munich, Germany, 27 December 1931), *mathematics, history of mathematics.*

Wieleitner received his higher education at the Catholic seminaries at Scheyern and Freising but subsequently decided to study mathematics (rather than classical languages and theology) at the University of Munich. Since his parents lived in simple circumstances, C. L. F. Lindemann proposed that Wieleitner be allotted the Lamont stipend for Catholic students of mathematics in 1895. This enabled the gifted young man to complete his studies in 1897 with excellent marks. Three years later he obtained the doctorate with a dissertation on third-order surfaces with oval points, a subject suggested to him by Lindemann.

Meanwhile, Wieleitner had become a high school teacher, his first appointment being at the Gymnasium at Speyer. In 1909 he was made *Gymnasialprofessor* at Pirmasens; in 1915 he returned to Speyer as headmaster of the *Realschule*; in 1920 he moved to Augsburg, and in 1926 he

was promoted to *Oberstudiendirektor* at the Neue Realgymnasium in Munich, a post he held until his death. Parallel to his career as an educator, Wieleitner established a reputation as a geometer and—increasingly so—as a historian of mathematics. Probably during the International Congresses of Mathematicians at Heidelberg (1904) and Rome (1908), he met Italian geometers. He translated an article by Gino Loria and, with E. Ciani, contributed to the revised German edition of *Pascals Repertorium der höheren Mathematik*.[1] In 1905 his *Theorie der ebenen algebraischen Kurven höherer Ordnung* had been published, and in 1908 it was supplemented by *Spezielle ebene Kurven*. In 1914 and 1918 the two volumes of Wieleitner's *Algebraische Kurven* followed. Wieleitner's books were noted for their simple, straightforward presentation and the author's great didactic skill, which made ample use of geometric intuition and insight.

Although always interested in the history of mathematics, Wieleitner would most probably not have become involved in the field had Anton von Braunmühl not died in 1908. With Siegmund Günther, Braunmühl had undertaken to write a *Geschichte der Mathematik* in two volumes. Günther's volume (antiquity to Descartes)[2] appeared in 1908, but his partner left an unfinished manuscript. Wieleitner was persuaded to step in. Thoroughly going through G. Eneström's many critical remarks about Cantor's *Vorlesungen über Geschichte der Mathematik*,[3] he revised and completed part I of Braunmühl's work (arithmetic, algebra, analysis), which was published in 1911; part II (geometry, trigonometry) appeared in 1921. Apart from being based on a detailed study of primary sources, Wieleitner's presentation always stressed the notion of development and progress of mathematics. Giving only minor attention to individual biographies, the author brought the leading ideas to the fore, and wrote a history of mathematical ideas. He followed the same general concept in his *Geschichte der Mathematik*, published in two small volumes in the Sammlung Göschen in 1922–1923.

Shortly after moving to Munich in 1928, Wieleitner, at Sommerfeld's suggestion, was made *Privatdozent*, and in 1930 honorary professor, at the university. Since 1919 he had been corresponding member of the Deutsche Akademie der Naturforscher Leopoldina, and in 1929 he was elected member of the Académie Internationale d'Histoire des Sciences.

Wieleitner published about 150 books and articles and more than 2,500 book reviews. Many of his papers and books—in geometry and the history of mathematics—were addressed to teachers and students of mathematics. In inexpensive source booklets he presented carefully chosen excerpts from mathematical classics for classroom use. His work in the history of mathematics was continued in the same spirit and with the same close connection to mathematical education by Kurt Vogel and J. E. Hofmann.

NOTES

1. *Pascals Repertorium der höheren Mathematik*, 2nd, completely rev. German ed., E. Salkowski and H. E. Timerding, eds., II, pt. 1, *Grundlagen und ebene Geometrie* (Leipzig – Berlin, 1910).
2. Siegmund Günther, *Geschichte der Mathematik*, I, *Von den ältesten Zeiten bis Cartesius* (Leipzig, 1908).
3. Moritz Cantor, *Vorlesungen über Geschichte der Mathematik*, 4 vols. (I: Leipzig, 1880; 2nd ed., 1894; 3rd ed., 1907; II: 1892; 2nd ed., 1899–1900; III: 1894–1898; 2nd ed., 1900–1901; IV: 1908).

BIBLIOGRAPHY

I. ORIGINAL WORKS. Wieleitner's most important books are *Theorie der ebenen algebraischen Kurven höherer Ordnung* (Leipzig, 1905), Sammlung Schubert no. 43; *Spezielle ebene Kurven* (Leipzig, 1908), Sammlung Schubert no. 56; *Geschichte der Mathematik*, II, *Von Cartesius bis zur Wende des 18. Jahrhunderts*, 2 vols. (Leipzig, 1911–1921), Sammlung Schubert nos. 63, 64; *Algebraische Kurven*, 2 vols. (Berlin–Leipzig, 1914–1918; I: 2nd ed., 1919; 3rd ed., 1930; II: 2nd ed., 1919), Sammlung Göschen nos. 435, 436; *Geschichte der Mathematik*, 2 vols. (Berlin–Leipzig, 1922–1923), Sammlung Göschen nos. 226, 875; *Die Geburt der modernen Mathematik*, 2 vols. (Karlsruhe, 1924–1925); and *Mathematische Quellenbücher*, 4 vols. (Berlin, 1927–1929). A combined Russian trans. of the 2 vols. of *Geschichte der Mathematik*, II (Sammlung Schubert nos. 63, 64) and of *Geschichte der Mathematik*, II (Sammlung Göschen no. 875), 53–147, was edited under the title *Istoria matematiki ot Dekarta do serednii XIX stoletia* by A. P. Youschkevitch (Moscow, 1966).

II. SECONDARY LITERATURE. The most extensive obituaries (including bibliographies and a portrait) are J. E. Hofmann, in *Jahresbericht der Deutschen Mathematiker-vereinigung*, **42** (1933), 199–223, with portrait; and J. Ruska, in *Isis*, **18** (1932), 150–165.

CHRISTOPH J. SCRIBA

WIEN, WILHELM CARL WERNER OTTO FRITZ FRANZ (*b*. Gaffken, near Fischhausen, East Prussia [now Primorsk, R.S.F.S.R.], 13 January 1864; *d*. Munich, Germany, 30 August 1928), *theoretical and experimental physics, philosophy of science.*

Life. Wien was the only child of Carl Wien, a farmer with land in Gaffken, and Caroline Gertz,[1] both of whose families were descended from ancestors in Mecklenburg. When Wilhelm was two, they left Gaffken, which was no longer capable of supporting them, and moved to a smaller farm, Drachenstein, in the district of Rastenburg, East Prussia, where Wien spent his youth.[2] He frequently rode through the fields with his father, who was confined to a wagon because of a spinal ailment, and thus Wien early learned about agriculture—in which his mother assumed the bulk of the family's responsibilities.

Wien was especially close to his mother, whose excellent knowledge of history and literature stimulated his interest in those subjects. An introvert, like his father, he made no friends during his early childhood. He learned to ride, swim, and skate; and, as was then customary, a woman was engaged to give him private lessons in French, which he spoke before he was able to write his native language. In 1875 Wien's parents sent him to the Gymnasium in Rastenburg; but he had little inclination for study, preferring to wander through the fields. Furthermore, his preparation, especially in mathematics, was deficient; and in 1880 he was taken out of this municipal school, which was considered by Wien to be democratic, and sent home to learn agriculture. He compensated for his lack of academic instruction through private tutoring (in mathematics he had an outstanding teacher in Switalski) and then entered the Altstädtisches Gymnasium in Königsberg, from which he graduated in 1882, in less time than was usual.

Encouraged mainly by his mother, Wien enrolled at the University of Göttingen in the summer of 1882 to study mathematics and natural sciences. He was not, however, very much taken with what he learned in his mathematics course. Also, being of an independent spirit, he found the lavish life in the student societies distasteful and left the university after only one semester to travel through the Rhineland and Thüringen. Wien returned home with the intention of becoming a farmer but was soon discontented with the training required. He therefore resumed his studies, this time in mathematics and physics at the University of Berlin. In the winter of 1883–1884, after two semesters, he entered Hermann von Helmholtz' laboratory, where he "really came into contact with physics for the first time." During the summer semester of 1884 Wien learned "a great deal" studying under

G. H. Quincke at Heidelberg, then resumed his training under Helmholtz in Berlin. In his second semester as a physics student Wien was given the subject of his doctoral dissertation, the diffraction of light when it strikes a grating. After two more semesters, in 1886, he was awarded the doctorate, although he did not receive a good grade on his final examination.

In the summer of 1886 Wien went to Drachenstein to help his parents reconstruct some buildings that had been destroyed in a fire. Once again he began to consider becoming a farmer. August Kundt, who in 1888 became Helmholtz' successor at the University of Berlin, and Helmholtz himself, who at the same time had been appointed the first president of the newly founded Physikalisch-Technische Reichsanstalt (PTR), reinforced Wien's doubts about physics, maintaining that as an only son he should take over his parents' property; if he wished, he could always pursue scientific research as a hobby. Fate soon decided the issue for Wien, who did not feel capable either of buying a horse or of communicating with farm workers: in 1890 drought forced his parents to sell the farm. Wien thereupon became an assistant to Helmholtz at the PTR in Charlottenburg and his parents moved to Berlin-Westend. Wien, moreover, felt a period of his life come to an end, because his mother fell seriously ill, his father died suddenly the following year, and Bismarck was dismissed by the Emperor.

Even during the time he spent on the farm, Wien continued to study theoretical physics, arbitrarily selecting the problems he would investigate. At the Reichsanstalt, which in 1890 became the center of his professional activities, he conducted an unsuccessful series of experiments, employing platinum foil, that sought to establish a new unit of light. In this dual concern with theory and experiment lay the seed of Wien's development into the rare physicist who possesses equally good knowledge of both areas. Wien in 1892 received the *venia legendi* at the University of Berlin with a work on the localization of energy. From 1894 to 1897, at the suggestion of Helmholtz, he also considered problems of hydrodynamics: specifically, the theory of sea waves and of cyclones.

Wien's independence enabled him to make his own choice of problems for study and soon bore fruit. In 1893 he demonstrated, in a highly original manner, the constancy of the products $\lambda \cdot \theta$, given a shift of the wavelength λ and the corresponding change in temperature θ. In fact, his findings refuted Helmholtz' initial view that the radiation could no longer be treated in exclusively thermodynamic terms. Wien also published, in 1896, the theoretical derivation of a law of the energy distribution of the radiation, which differs only slightly from the currently accepted Planck law.[3]

In collaboration with his friend Ludwig Holborn, Wien executed a series of high- and low-temperature measurements from which he derived considerable satisfaction. Nevertheless, he was happy to receive an offer in 1896 from the Technische Hochschule of Aachen because Friedrich Kohlrausch, who had succeeded Helmholtz as head of the PTR, had drawn up a rigid plan of research that ran counter to Wien's need for freedom. Soon after reaching Aachen, in its gay society, Wien met Luise Mehler; they were married in 1898 and had four children: Gerda, Waltraut, Karl, and Hildegard. In Aachen he continued research begun in Berlin, on Röntgen and cathode radiation, using the apparatus left by his predecessor, Philipp Lenard. The investigation of vacuum radiation of this kind was to constitute the principal area of Wien's research.

In 1899 Wien accepted a post as full professor at the University of Giessen but left after only six months to take up a similar position at the University of Würzburg, where he spent the next twenty years—the most eventful of his scientific career. His experiments, conducted with the aid of the rapidly improving methods of high-vacuum technology, encouraged him to turn his attention to a study of the decay periods of excited atoms (and ions), a topic that occupied the final years of his career.

Wien visited Norway, Spain, Italy, England (1904), Greece (1912), and the Baltic region (1918), in the latter of which he gave several lectures. In 1911 he was awarded the Nobel Prize in physics "pour des découvertes concernant les lois de la radiation de la chaleur"; in his acceptance speech he voiced serious doubts about Planck's radiation theory.[4] In the spring of 1913 he went to Columbia University to deliver six lectures on recent problems of theoretical physics and also visited Harvard and Yale universities. In addition Wien went to Washington to see Arthur Day (Kohlrausch's son-in-law), whose high-temperature measurements—with those of Wien and Holborn—had furnished the data used to confirm the energy distribution law of Wien and later Planck. While at Würzburg, Wien did not confine his attention exclusively to physics. He also studied the subjects to which his mother had introduced him—history and especially foreign literature—as well as the fine arts.

World War I affected Wien deeply, but the struggle against the "Bolsheviks, incited by literary figures (*Literaten*)" in Germany during 1918 and 1919 had an even greater impact. In 1920 he assumed his last post, at the University of Munich, where he had a new physics institute built and served as rector from 1925 to 1926. Whereas in the 1890's Wien had chosen problems arbitrarily according to their appeal to him, he took pains at Munich to select the topics first before trying them together with his students. Although satisfied with his scientific achievements, he took a gloomy view of Germany's situation in the 1920's, marked by "war tribute and socialism."[5]

Scientific Work. In his first scientific publication, when he was twenty-one, Wien demonstrated that when very bright light (whether white or colored) strikes a single metallic edge (grids failed), it is bent far into the geometric shadow of the intercepting screen. He also found that the diffracted light is polarized parallel to the edge. Further, comparison of the color formation in this type of diffraction with the absorption of the diffracting material yielded the complementary color (dissertation, Berlin, 1886). Wien perceived that the difficulties of explaining the phenomenon on the basis of the previous theories lay in their failure to take into account the oscillation of the molecules of the diffracting edge. Turning to optics, in 1888 Wien demonstrated experimentally, using the bolometer, that the dependence of the transmission of a metal on the conductance does not follow—at least in the case of silver—from Maxwell's theory of light and that the latter required further work.

Following his move to the Reichsanstalt in 1890, Wien made temperature measurements and with Holborn developed a thermoelectric temperature scale, in the form of a function of the third power of the temperature, from the "electromotive force" of the thermocouples (1892). He also extended his earlier exploration of the "flux of light" to a study of the energy of thermal radiation.[6] In 1890 and in his habilitation essay of 1892 Wien linked, very generally, J. H. Poynting's "energy flux" of electric currents with the concept of the entropy of radiation. Further, by analogy with the continuous change in position of matter in motion, he also established the motion of the energy of electrodynamic radiation, pursuing the question raised by Hertz of whether this energy can be localized at all during movement. A year later, using theoretical considerations, Wien found a characteristic of electrodynamic radiation, the displacement $\theta \cdot \lambda = \theta_0 \cdot \lambda_0$—for any wavelength λ at the same position on the x-axis—of any two temperature curves characterized by the different temperatures θ and θ_0;[7] he also formulated in words the constancy of the products. In this derivation Wien started from Boltzmann's finding (1884) that the expression for the electrodynamic radiation pressure could be equated with an expression for the thermodynamic radiation pressure. From this equality Boltzmann had formulated a proof of Stefan's law, which Wien completed by considering the wavelength with the aid of Doppler's principle. At the same time, as a subsidiary result, Wien determined that the individual values ϕ of the energy of two temperature curves are related as the ratio of the fifth power of their temperatures:

$$\phi = \phi_0 \frac{\theta^5}{\theta_0^5}.$$

On the basis of Wien's findings, Paschen in 1896 derived the temperature independence of the expression

$$(I_{\lambda\max}, T) : (I_\lambda, T) = f\left(\frac{\lambda}{\lambda_{\max}}\right).$$

Wien then discovered other characteristics of the still unknown Kirchhoff energy distribution function $F(\lambda, T)$. The first was that the radiation energy disappears as wavelengths increase even within the region of finite lengths (1893); the second, that two curves for different temperatures do not intersect and that, beyond the energy maximum, they decrease no more rapidly than in proportion to λ^{-5} (1894). Both phenomena occur because of the inviolability of the second law of thermodynamics.

Starting from these regularities and others (on the whole six),[8] Wien used theoretical considerations to achieve his energy distribution law, which he published in June 1896:

$$\phi_\lambda = C\lambda^{-5} \exp\left(\frac{c}{\lambda\theta}\right),$$

where ϕ_λ is the energy at a given small interval of the abscissa and θ is the temperature. Paschen had previously communicated to Wien his experimentally derived formulation

$$\phi_\lambda = C\lambda^{-\alpha} \exp\left(\frac{c}{\lambda\theta}\right), \text{ with } \alpha = 5.67.$$

Wien replied that he had already derived his law but that, on theoretical grounds, α must equal 5. To support his derivation, Wien referred to W. A. Michelson's "ansatz" (1887) that the radiation function should be handled in accordance with Maxwell's statistical treatment of the velocities v. Starting from Maxwell's distribution law for the number u_v of atoms with velocity v,

$$u_v = \frac{4N}{\sqrt{\pi}} \alpha^3 \exp\left(\frac{-v^2}{\alpha^2}\right) v^2 \, dv,$$

Wien replaced the velocity independent factor $\frac{4N}{\sqrt{\pi}} \alpha^3$ with an initially unknown function $F(\lambda)$ for v^2 on the assumption that the wavelength of radiation from molecules with velocity v is a function only of v^2. By integration and application of Stefan's law, Wien found $F(\lambda)$ to be λ^{-5}, a result that depended also on setting Maxwell's mean energy α^2 proportional to temperature. The exponent was found by the displacement law in such a manner that $v^2 = f(\lambda)$ and $\alpha^2 \propto \theta$. Thus, in Wien's treatment, the fragment $v^2 \exp\left(\frac{-v^2}{\alpha^2}\right)$ of Maxwell's distribution becomes

$$F(\lambda) \exp\left(\frac{-f[\lambda]}{\theta}\right).$$

In this derivation Wien relinquished the assumption he had held since 1890 of the existence of a pure vacuum radiation, replacing it with the "hypothesis" that such radiation enters an empty space from a gas outside that space. This device, which he adopted in order to use Maxwell's gas statistics, was as much a concrete illustration and guide as was Planck's hypothesis of concrete Hertzian resonators. In 1900, and again later, physicists attacked Wien for this new interpretation of the quantities of the Maxwell distribution and, in general, for the reintroduction of gas—instead of his earlier mere cavity radiation. Nevertheless, Wien's treatment of the theory proved to be a masterstroke in the application of his well-founded suppositions of 1893–1894.

In 1897 Wien found confirmation that cathode rays were particles and of their very high velocity (about one-third the velocity of light); and in the following year he determined that they are negatively charged. Around the same time, with the help of combined electric and magnetic deflection, he also discovered the corpuscular nature, the positive charge, and the velocity (about 3.6×10^7 centimeters per second) of the positive rays. This new aspect of Wien's research, inspired by Lorentz' views on the electrostatic origin of gravitation, reached a logical conclusion in Wien's "Über die Möglichkeit einer elektromagnetischen Begründung der Mechanik" (1900). This publication constitutes the high point in the discussion of the change in size arising from the high velocity of the electrons and records the strong doubts concerning the constancy of mass.

While at Würzburg, Wien also continued his experiments with vacuum tubes. In 1905 he determined the lower boundary of the mass of the "positive electron" (called "Kanalstrahlen") as being that of the hydrogen ion. As early as 1908, following Stark's discovery of the Doppler displacement of radiation from these "electrons," Wien examined the mechanism by which they emit light; he pursued these experiments, involving the measurement of the decay time, for the rest of his life. Believing this decay to be much smaller than was supposed, Wien in 1919 devised a way of observing it in the vacuum tube that did not involve collisions: separating the space in which the positive rays are produced from the space in which they are observed. (Unlike cathode rays, they do not penetrate metal foils.) Wien allowed the positive rays to enter through a narrow slit in the vacuum tube, which he had emptied by means of the diffusion air pump recently invented by W. Gaede. (Wien used ten of these pumps.) The production space was maintained at an arbitrary, constant pressure through the gas flow method (Durchströmungsmethode) developed in vacuum processing. This separation technique later gained importance in the construction of elementary particle accelerators. Wien calculated the decay constant $2a$ for the decrease $\exp(-2at)$ of the light intensity of the luminous particle as being approximately 5×10^7 s^{-1}, which was in good agreement with the known value of the beam velocity. In 1922 Wien successfully applied his technique to the separation of arc lines (light from uncharged atoms) from spark lines (light from ions the charge of which could be shown by electrostatic deflection).

In this connection Wien in 1916 demonstrated the existence—which accords with the relativity principle—of a phenomenon that is the inverse of the Stark effect (that is, of the line splitting of a stationary light source in an electric field), experimentally showing the corresponding splitting in the case of a moving light source in a magnetic field. Also in the realm of radiation physics, Wien in 1907 sought to ascertain the lengths of Röntgen waves by measuring the impulse width. His assumption was that these waves arise through the slowing of electrons in the electromagnetic field. In the same year, working with quantum theoretical assumptions, on which the photoelectric effect was based recently, he obtained the good value $\lambda = 6.75 \times 10^{-9}$ cm, five years before Max von Laue, and suggested that Röntgen wavelengths could be measured by means of crystal lattice.

In theoretical physics Wien won recognition for his conceptual experiments.[9] In his view, "such imagined processes . . . ought to correspond to realization with an unlimited degree of approximation" (1893). In 1911 Wien wrote: "Thought experiments" are devised "because for practical reasons it is often impossible to carry them out, and yet they lead to reliable results." He maintained, however, that to posit these imagined "processes . . . the lawlike manner in which they take place . . . must be fully known," although he granted that it is permissible to "idealize."

In *Ziele und Methoden der theoretischen Physik* (1914), Wien distinguished mathematical from theoretical physics.[10] The former, he held, should furnish the mathematical tools—just as mathematics establishes exact relationships between numerical quantities. The latter should seek to determine quantitative laws, for which it must develop hypotheses; it can, however, attain only approximate exactness. Wien held that the laws of nature are simpler than is generally supposed by scientists, who see the infinite variety of their intricate effects. Only quantitative verification through comparison with observed data can protect the theoretician—who generally does not feel bound by experiments—from the many unsuitable ideas he generates. This quantitatively controlled interaction between theory and experiment excludes a possible carelessness in the use of mathematical expressions.

NOTES

1. For these biographical data the author is indebted to Wien's daughter, Waltraut Wien, of Munich.
2. Wien was stimulated to record this review of his career by a letter of 17 Aug. 1927 from an unidentified American living in Denver, who solicited Wien's response to use for pedagogical purposes. See *Aus dem Leben und Wirken eines Physikers*, 1–50.
3. Wien's laws were assimilated so quickly into physics that when Wien went to England in 1904, people expected to see an older man instead of a "young" man of forty years of age.
4. In 1915 Wien again expressed reservations concerning Planck's quanta; see "Theorie der Wärmestrahlung," 217–220.
5. In 1927 Wien believed he could observe "the encroaching

Americanization of all of life that is now taking place in Europe"; see *Aus dem Leben und Wirken eines Physikers*, 74: letter to E. Schrödinger, 1 May 1927.
6. For details of this and of the following contributions by Wien to the study of thermal radiation, see H. Kangro, *Vorgeschichte des Planckschen Strahlungsgesetzes, passim.*
7. The term "displacement law" (*Verschiebungsgesetz*) was coined by Otto Lummer and Ernst Pringsheim in 1899.
8. See H. Kangro, *Vorgeschichte des Planckschen Strahlungsgesetzes*, ch. 3; 4.2; and 5.3.
9. For more on this topic, see H. Kangro, *Vorgeschichte des Planckschen Strahlungsgesetzes, passim.*
10. Wien equated theoretical physics with the English "natural philosophy": see *Aus der Welt der Wissenschaft*, 170. Similarly, Helmholtz and G. T. Wertheim gave the title *Handbuch der theoretischen Physik* to the German translation of William Thomson and P. G. Tait's *Natural Philosophy.*

BIBLIOGRAPHY

I. ORIGINAL WORKS. Karl Wien, ed., *Aus dem Leben und Wirken eines Physikers* (Leipzig, 1930), includes a nearly complete bibliography of his writings; "Ein Rückblick," pp. 1–50, was written about 1927 and contains a mine of information given by Wien himself; a selection of Wien's correspondence with his mother, wife, Professor Oseen (possibly Carl Wilhelm Oseen), G. Mie, A. Sperl, W. Ostwald, Beggerow, and E. Schrödinger is on 51–76. See also Poggendorff, IV, 1633; V, 1368–1369; and VI, 2879.

The first Memoir of the student W. Wien is "Über den Einfluss der ponderablen Theile auf das gebeugte Licht," in *Sitzungsberichte der Königlich Preussischen Akademie der Wissenschaften zu Berlin*, II (1885), 817–819; and "Über die Energievertheilung im Emissionsspectrum eines schwarzen Körpers," in *Annalen der Physik*, **294** (June 1896), 662–669, also in English trans. as "On the Division of Energy in the Emission-Spectrum of a Black Body," in *Philosophical Magazine*, 5th ser., **43** (1897), 214–220. Wien's views on heat radiation are summarized in "Theorie der Strahlung," in *Encyklopädie der Mathematischen Wissenschaften mit Einschluss ihrer Anwendungen*, V: *Physik* (Leipzig, 1909), pt. 3, 182–357, and in "Theorie der Wärmestrahlung," in *Kultur der Gegenwart*, pt. 3, sec. 3, vol. 1 (Berlin, 1915), 217–220.

Wien's early memoirs include "Über die Messung hoher Temperaturen," in *Zeitschrift für Instrumentenkunde*, **12** (1892), 257–266, 296–307; also in *Annalen der Physik*, **283** (1892), 107–134; and **292** (1895), 360–396, both written with Ludwig Holborn; and "Über die Messung tiefer Temperaturen," *ibid.*, **295** (1896), 213–228, also in *Sitzungsberichte der Königlich Preussischen Akademie der Wissenschaften zu Berlin* (1896), 673–677, both written with Holborn.

Subsequent writings are "Untersuchungen über die electrische Entladung in verdünnten Gasen," in *Annalen der Physik*, **301** (1898), 440–452; "Zur Theorie der Strahlung schwarzer Körper. Kritisches," *ibid.*, **308** (1900), 530–539, dated 12 Oct. 1900 – cf. Planck's response, *ibid.*, 764–766, and Wien's answer, "Zur Theorie der Strahlung; Bemerkungen zur Kritik des Herrn Planck," *ibid.*, **309** (1901), 422–424, dated 29 Dec. 1900; "Über die Möglichkeit einer elektromagnetischen Begründung der Mechanik," in *Archives néerlandaises des sciences exactes et naturelles*, 2nd ser., **5** (1900), 96–107; also in *Annalen der Physik*, **310** (1901), 501–513; and "Die Temperatur und Entropie der Strahlung," in *Physikalische Zeitschrift*, **2** (1901), 111 (as *Vorträge und Diskussionen von der 72. Naturforscherversammlung zu Aachen*); which includes remarks on the discussions thereon held at the 72nd Conference of the Gesellschaft Deutscher Naturforscher at Aachen, 20 Oct. 1900.

Later works are "Über die Natur der positiven Elektronen," in *Annalen der Physik*, **314** (1902), 660–664; "Über eine Berechnung der Wellenlänge der Röntgenstrahlen aus dem Planckschen Energie-Element," in *Nachrichten von der Königlichen Gesellschaft der Wissenschaften zu Göttingen*, Math.-naturwiss. Klasse, (1907), 598–601; and "Über die Berechnung der Impulsbreite der Röntgenstrahlen aus ihrer Energie," in *Annalen der Physik*, **327** (1907), 793–797.

Like Planck, Wien gave six lectures in the United States on theoretical physics; they were published as *Vorlesungen über neuere Probleme der Theoretischen Physik, gehalten an der Columbia-Universität in New York im April 1913* (Leipzig–Berlin, 1913); they were followed by *Ziele und Methoden der theoretischen Physik* (Würzburg, 1914), (=Festrede zur Feier des 332, jährigen Bestehens der Julius-Maximilian-Universität in Würzburg am 11. März 1914), also in *Jahrbuch der Radioaktivität und Elektronik*, **12** (1915), 241–259.

On Wien's last main field of research, see "Über Messungen der Leuchtdauer der Atome und die Dämpfung der Spektrallinien," in *Annalen der Physik*, **365** (1919), 597–637; and II, **371** (1921), 229–236; and "Über eine Methode zur Trennung der Bogen- und Funkenlinien der Emissionsspektra," *ibid.*, **374** (1922), 325–334.

Wien's general lectures and papers are included in *Aus der Welt der Wissenschaft* (Leipzig, 1921).

Wien's curricula vitae are the MS "Vita," composed probably in the early 1890's, at the Staatsbibliothek Preussischer Kulturbesitz, Berlin-Dahlem; and "Vita," only printed in his original dissertation on *Untersuchungen über die bei der Beugung des Lichtes auftretenden Absorptionserscheinungen. Inaugural-Dissertation zur Erlangung der Doctorwürde . . . nebst beigefügten Thesen öffentlich zu verteidigen am 3. Februar 1886* (Berlin, n.d. [1886]).

On Wien's MSS, see T. S. Kuhn *et al.*, eds., *Sources for the History of Quantum Physics* (Philadelphia, 1967), 97a. The American Philosophical Society, Philadelphia, has some of Wien's letters to Planck; and nearly 150 letters from Planck to Wien are at the Staatsbibliothek Preussischer Kulturbesitz. The Deutsches Museum, Munich, recently acquired various other papers of Wien.

II. SECONDARY LITERATURE. There is a nearly complete bibliography in Poggendorff, VI, 2879; and VIIa, 991; Max von Laue, "Wilhelm Wien," in *Deutsches Biographisches Jahrbuch, X, das Jahr 1928* (Stuttgart–Berlin, 1931), 302–310, is a useful sketch; see also the obituary notices contributed by various authors in Wien's *Aus dem Leben und Wirken eines Physikers*, 139–189. Other sources include K. Reger, "Wilhelm Wien," in *Nobelpreisträger auf dem Wege ins Atomzeitalter* (Munich–Vienna, 1958), 233–246; and Max Steenbeck, *Wilhelm Wien und sein Einfluss auf die Physik seiner Zeit* (Berlin, 1964) (=Deutsche Akademie der Wissenschaften zu Berlin, Vorträge und Schriften, Heft 94, 1–21). On Wien's early scientific studies, see H. Kangro, *Vorgeschichte des Planckschen Strahlungsgesetzes . . .* (Wiesbaden, 1970), *passim*, esp. ch. 5.

HANS KANGRO

WIENER, LUDWIG CHRISTIAN (*b.* Darmstadt, Germany, 7 December 1826; *d.* Karlsruhe, Germany, 31 July 1896), *mathematics, physics, philosophy.*

In mathematics Christian Wiener did important work in descriptive geometry and the construction of mathematical models. As a physicist he studied chiefly molecular phenomena and atmospheric radiation. In his philosophical writings he advocated a point of view based on the methodology of natural science.

The son of a judge, Wiener attended the gymnasium in Darmstadt and from 1843 to 1847 studied engineering and architecture at the University of Giessen, where he passed the state architecture examination. In 1848 he obtained a post as teacher of physics, mechanics, hydraulics, and descriptive geometry at the Höhere Gewerbeschule (later the Technische Hochschule) of Darmstadt. Two years later he earned the Ph.D. and qualified as a *Privatdozent* in mathematics at the University of Giessen. To further his education he attended the Technical University in Karlsruhe, working for about a year under Ferdinand Redtenbacher, the professor of mechanical engineering. He returned to Giessen in the autumn of 1851; but the following year he accepted a professorship of descriptive geometry at the Technische Hochschule in Karlsruhe, retaining the position until 1896.

An able and respected teacher, Wiener trained a great number of students while conducting important research. Elected rector of the Technische Hochschule three times, he was also a member of the Gewerbeschulrat and the Oberschulrat of the state of Baden. Wiener was liked and esteemed for his upright character, his sense of justice, and his kindliness.

In his mathematical works Wiener frequently used direct intuition as an aid in carrying out proofs. This led him into the realm of aesthetics, as can be seen from his philosophical essay "Über die Schönheit der Linien" (1896), which contains an appendix on the relationship between mathematical continuity and the regularity of forms.

Wiener's chief work was the two-volume *Lehrbuch der darstellenden Geometrie* (1884–1887), based on his teaching experience and numerous publications on descriptive geometry. In the introduction to the *Lehrbuch* he presented a valuable historical survey, based on a firsthand study of the sources, that constituted an important supplement to Chasles's *Aperçu historique sur l'origine et le développement des méthodes en géométrie* (1837). Wiener treated the basic problems of descriptive geometry by a single method: a varied use of the principal lines of a plane. He also sought to simplify individual problems as much as possible and to find the easiest graphical solutions for them. He was not, however, concerned merely with graphical methods, of which he was a master. He was also interested in the problems and their solutions (such as shadow construction and brightness distribution), as well as in the development of the necessary geometric aids. For example, he used imaginary projection and developed a grid method that can be derived from the theory of cyclically projected point series.

Wiener also became known for his mathematical models. In 1869, at the suggestion of R. F. A. Clebsch, he constructed a plaster-of-Paris model of the third-order surface. He displayed his models at expositions of mathematical teaching aids in London (1876), Munich (1893), and Chicago (1893). In analysis he discussed and drew the Weierstrass function, which is everywhere continuous and yet at no point has a derivative.

Extending his works on descriptive geometry into physics, Wiener investigated the illumination conditions for various bodies. Thus, he calculated the amounts of solar radiation received at different latitudes and during the varying lengths of days in the course of the year. His numerical values are still fundamental for the study of atmospheric optics and of the effect of radiation on the earth's climate. In a posthumously published article Wiener examined the total radiation received by the atmosphere and considered problems related to color theory and strengths of perceptions.

In his studies on molecular physics, Wiener demonstrated by extremely careful observations that Brownian movement is an "internal motion peculiar to the liquid state." He developed an atomistic cosmology, which he set forth in *Atomlehre* (1869), the first volume of his chief philosophical work, *Die Grundzüge der Weltordnung*. He presupposed the causality of all natural phenomena and the existence of a real external world but, in accordance with a view widely held at the time, he still accepted the existence of an ether. In his treatment of crystalline forms Wiener developed the concept of the regular point system, which became important in crystallography.

Among the topics Wiener discussed in his writings on moral philosophy were will and morality. He defined free will as independence from external, determining circumstances only, thus precluding full independence—that is, absolute freedom. He opposed the view of some of his contemporaries that scientific research, with its analytic methods, could become a danger to man's sense of morality and beauty. Unlike his other publications, Wiener's philosophical works found only a limited audience.

BIBLIOGRAPHY

I. ORIGINAL WORKS. Bibliographies of Wiener's approximately 100 scientific books and papers are in Poggendorff, II, 1322, and III, 1442; and in the unsigned *Zur Erinnerung an Dr. Christian Wiener* (see below).

Mathematical works include *Über Vielecke und Vielflache* (Leipzig, 1864); *Stereoskopische Photographie des Modells einer Fläche dritter Ordnung mit 27 reellen Geraden* (Leipzig, 1869); "Direkte Lösung der Aufgabe: Einen durch fünf Punkte oder durch fünf Tangenten gegebenen Kegelschnitt auf einen Umdrehungskegel zu legen," in *Zeitschrift für Mathematik und Physik*, **20** (1875), 317–325; "Geometrische und analytische Untersuchung der Weierstrass'schen Funktion," in *Journal für die reine und angewandte Mathematik*, **90** (1880), 221–252; and *Lehrbuch der darstellenden Geometrie*, 2 vols. (Leipzig, 1884–1887).

Writings in physics are "Erklärung des atomistischen Wesens des tropfbar-flüssigen Körperzustandes und Bestätigung desselben durch die sogenannten Molekularbewegungen," in *Annalen der Physik und Chemie*, **118** (1863), 79–94; "Über die Stärke der Bestrahlung der Erde durch die Sonne in ihren verschiedenen Breiten und Jahreszeiten," in *Zeitschrift für Mathematik und Physik*, **22** (1877), 341–368, also abridged in *Österreichische Zeitschrift für Meteorologie*, **14** (1879), 113–129; and "Die Helligkeit des klaren Himmels und die Beleuchtung durch Sonne, Himmel und Rückstrahlung,"

H. Wiener, O. Wiener, and W. Möbius, eds., 2 pts., *Nova acta Leopoldina*, **73**, no. 1 (1900), and **91**, no. 2 (1909).

On philosophy, see *Die Grundzüge der Weltordnung* (Leipzig–Heidelberg, 1863), 2nd ed., 2 vols.: I, *Atomlehre* (1869), II, *Die geistige Welt und der Ursprung der Dinge* (1869); and "Über die Schönheit der Linien," in *Abhandlungen des Naturwissenschaftlichen Vereins in Karlsruhe*, **11** (1896), 47–73.

II. SECONDARY LITERATURE. See the unsigned *Zur Erinnerung an Dr. Christian Wiener* (Karlsruhe, 1896); A. Brill and L. Sohnke, "Christian Wiener," in *Jahresberichte der Deutschen Mathematiker-Vereinigung*, **6** (1897), 46–69; and H. Wiener, "Wiener, Christian," in *Allgemeine deutsche Biographie*, XLII (1897), 790–792.

HANS-GÜNTHER KÖRBER

WIENER, NORBERT (*b.* Columbia, Missouri, 26 November 1894; *d.* Stockholm, Sweden, 18 March 1964), *mathematics.*

Wiener was the son of Leo Wiener, who was born in Byelostok, Russia, and Bertha Kahn. Although a child prodigy, he matured into a renowned mathematician rather slowly. At first he was taught by his father. He entered high school at the age of nine and graduated two years later. After four years in college, he enrolled at the Harvard Graduate School at the age of fifteen in order to study zoology. That soon turned out to be a wrong choice. He next tried philosophy at Cornell. "A philosopher in spite of himself," Wiener took a Ph.D. at Harvard in 1913 with a dissertation on the boundary between philosophy and mathematics. A Harvard traveling fellowship paid his way to Europe. Bertrand Russell was his chief mentor at Cambridge and advised him to learn more mathematics. Neither the examples of Hardy and Littlewood at Cambridge, however, nor those of Hilbert and Landau at Göttingen, converted him to mathematics. Back in the United States in 1915, Wiener tried various jobs teaching philosophy, mathematics, and engineering. In the spring of 1919 he got a position in the mathematics department of the Massachusetts Institute of Technology, not then particularly distinguished in that discipline. An assistant professor in 1924, associate in 1929, and full professor in 1932, he remained at MIT until his retirement. Although his genius contributed to establishing the institute's present reputation, he could never comfort himself over the failure of other American universities, and particularly of Harvard, to show much interest in him. He traveled a great deal, to Europe and to Asia, and his visits to Germany in the interwar years left their traces in many anecdotes told in Continental circles. His *Cybernetics* made him a public figure. President Lyndon Johnson awarded him the National Medal of Science two months before his death. He died during a trip to Sweden and left two daughters. His wife was the former Margaret Engemann.

In appearance and behavior, Norbert Wiener was a baroque figure, short, rotund, and myopic, combining these and many qualities in extreme degree. His conversation was a curious mixture of pomposity and wantonness. He was a poor listener. His self-praise was playful, convincing, and never offensive. He spoke many languages but was not easy to understand in any of them. He was a famously bad lecturer.

Wiener was a great mathematician who opened new perspectives onto fields in which the activity became intense, as it still is. Although most of his ideas have become standard knowledge, his original papers, and especially his books, remain difficult to read. His style was often chaotic. After proving at length a fact that would be too easy if set as an exercise for an intelligent sophomore, he would assume without proof a profound theorem that was seemingly unrelated to the preceding text, then continue with a proof containing puzzling but irrelevant terms, next interrupt it with a totally unrelated historical exposition, meanwhile quote something from the "last chapter" of the book that had actually been in the first, and so on. He would often treat unrelated questions consecutively, and although the discussion of any one of them might be lucid, rigorous, and beautiful, the reader is left puzzled by the lack of continuity. All too often Wiener could not resist the temptation to tell everything that cropped up in his comprehensive mind, and he often had difficulty in separating the relevant mathematics neatly from its scientific and social implications and even from his personal experiences. The reader to whom he appears to be addressing himself seems to alternate in a random order between the layman, the undergraduate student of mathematics, the average mathematician, and Wiener himself.

Wiener wrote a most unusual autobiography. Although it conveys an extremely egocentric view of the world, I find it an agreeable story and not offensive, because it is naturally frank and there is no pose, least of all that of false modesty. All in all it is abundantly clear that he never had the slightest idea of how he appeared in the eyes of others. His account of the ill-starred trip to Europe in 1926–1927 is a particularly good example. Although he says almost nothing about the work of the mathematicians whom he met, he recalled after twenty-five years meeting J. B. S. Haldane and setting him straight over an error in his book *The Gold-Makers*: Haldane had used a Danish name for a character supposed to be an Icelander (*I Am a Mathematician*, 160). In his autobiography Wiener comes through as a fundamentally good-natured person, realistic about his human responsibilities and serious enough to be a good friend, a good citizen, and a good cosmopolite. Despite his broad erudition, the philosophical interludes are no more than common sense, if not downright flat. Unlike many autobiographers, he never usurps the role of a prophet who long ago predicted the course that things have taken. A good biography ought to be written of him, one that would counterbalance his autobiography and do him more justice than anyone can do in a book about himself.

According to his own account, Wiener's understanding of modern mathematics began in 1918, when he came across works on integration, functionals, and differential equations among the books of a young Harvard student who had died. At that time he met I. A. Barnett, who by suggesting that he work on integration in function spaces, put Wiener on the track that would lead him to his greatest achievements, the first of which was differential space. It was already characteristic of Wiener's openness of mind that, rather than being satisfied with a general integration theory, he looked for physical embodiments to test the theory. The first he tried, turbulence, was a failure; but the next, Brownian motion (1921), studied earlier by Einstein, was a success. Wiener conceived a measure in the space of one-dimensional paths that leads to the application of probability concepts in that space (see *Selected Papers*, no. 2). The construction is surprisingly simple. Take the set of continuous functions $x(t)$ of $t \geqq 0$ with $x(0) = 0$ and require that the probability of x passing for t_i between α_i and β_i ($i = 1, \ldots, k$) is provided by the Einstein-Smoluchowski formula that gives for the probability density of a point at x staying at y after a lapse of time t the expression

$$(2 \pi t)^{\frac{1}{2}} \exp (-[y-x]^2/2t).$$

In later work Wiener made this measure more explicit by a measure-preserving mapping of the real number line on function space. He also proved that almost all paths are nondifferentiable and that almost all of them satisfy a Lipschitz condition of any degree $< 1/2$, although almost none does so with such a condition of degree $> 1/2$. "Differential space" is a strange term for this function space

with a measure, promising a measure defined not by finite but by differential methods. Although vaguely operative on the background, this idea was never made explicit by Wiener when he resumed use of the term "differential space" in later work.

In 1923–1925 Wiener published papers that greatly influenced potential theory: Dirichlet's problem, in its full generality (see *Selected Papers*, no. 3). The exterior problem of a compact set K in 3-space led him to the capacitory potential of a measure with support K as a basic tool.

From Brownian motion Wiener turned to the study of more general stochastic processes, and the mathematical needs of MIT's engineering department set him on the new track of harmonic analysis. His work during the next five years culminated in a long paper (1930) on generalized harmonic analysis (see *Selected Papers*, no. 4), which as a result of J. D. Tamarkin's collaboration is very well written. Rather than on the class L^2, Wiener focused on that of measurable functions f with

$$\Phi(x) = \lim_{T \to \infty} \frac{1}{2T} \int_{-T}^{T} f(x+t)\bar{f}(t)\,dt$$

existing for all x, which is even broader than that of almost periodic functions. He borrowed the function Φ from physics as a key to harmonic analysis and connected it later to communication theory. Writing Φ as a Fourier transform,

$$\Phi(t) = (2\pi)^{-\frac{1}{2}} \int_{-\infty}^{\infty} e^{-itu}\,dS(u),$$

he obtained what is now called the spectral distribution S. The most difficult step was to connect S to the integrated Fourier transform g of f by an analogue of the classical formula $S(t) = \int_{-\infty}^{t} |g(\lambda)|^2\,d\lambda$. A brilliant example is: If $f(x) = \pm 1$ for $x_n \leqq x < x_{n+1}$, where the signs are fixed by spinning a coin, then the spectral distribution of f is almost certainly continuous.

A key formula in this field was placed by Wiener on the cover of the second part of his autobiography:

$$\lim_{\epsilon \to 0} \frac{1}{2\epsilon} \int_{-\infty}^{\infty} (g(\omega+\epsilon) - g(\omega-\epsilon))^2\,d\omega$$

$$= \frac{2}{\pi} \lim_{A \to \infty} \frac{1}{2A} \int_{-A}^{A} |f(t)|^2\,dt.$$

When Wiener attempted to prove this, A. E. Ingham led him to what Hardy and Littlewood had called Tauberian theorems; but Wiener did more than adapt their results to his own needs. He gave a marvelous example of the unifying force of mathematical abstraction by recasting the Tauberian question as follows (see *Selected Papers*, no. 5): To prove the validity of

$$\lim_{x \to \infty} \int_{-\infty}^{\infty} K(x-y)f(y)dy = A \int_{-\infty}^{\infty} K(x)dx,$$

by which kind of more tractable kernel K_1, K may be replaced. The answer is (for K and $K_1 \in L_1$): If the Fourier transform of K_1 vanishes nowhere, the validity with K_1 implies that with K. Tauberian theorems have lost much of their interest today, but the argument by which Wiener proved his theorem is still vigorous. Wiener showed that in L_1 the linear span of the translates of a function is dense if its Fourier transform vanishes nowhere. This, again, rests on the remark that the Fourier transform class L_1 is closed with respect to division (as far as possible). Wiener's work in this area became the historical source of the theory of Banach algebras. The "Wiener problem," that is, the problem of deciding whether it is true that in L_1 a function f_1 belongs to the closure of the span of the translates of f_2 if and only if the Fourier transform of f_1 always vanishes together with that of f_2, greatly influenced modern harmonic analysis; it was proved to be wrong by Paul Malliavin in 1959.

Fourier transforms and Tauberian theorems were also the subject of Wiener and R. E. A. C. Paley's collaboration, which led to *Fourier Transforms* (1934). Another cooperative achievement was the study of the Wiener-Hopf equation (see *Selected Papers*, no. 6),

$$f(x) = \int_0^{\infty} K(x-y)f(y)dy,$$

generalizing Eberhard Hopf's investigation on radiation equilibrium. In *I Am a Mathematician* (p. 177), Wiener remarked that although originally accounting for the discontinuity of two physical media at $x = 0$, it can even better serve to embody the discontinuity of knowledge at the boundary of future and past. The previous work on the Wiener-Hopf equation became influential in Wiener's prediction theory.

Until the late 1930's stochastic processes, as exemplified by Brownian motion, and harmonic analysis were loose ends in the fabric of Wiener's thought. To be sure, they were not isolated from each other: the spectrally analyzed function f was thought of as a single stochastic happening, and the earlier cited example shows that such a happening

could even be conceived as embedded in a stochastic process. Work of others in the 1930's shows the dawning of the idea of spectral treatment of stationary stochastic processes; at the end of the decade it became clear that the "Hilbert space trick" of ergodic theory could serve this aim also. Initially Wiener had neglected ergodic theory; in 1938–1939 he fully caught up (see *Selected Papers*, nos. 7–8), although in later work he did not avail himself of these methods as much as he might have done.

Communication theory, which for a long time had been Wiener's background thought, became more prominent in his achievements after 1940. From antiaircraft fire control and noise filtration in radar to control and communication in biological settings, it was technical problems that stimulated his research. Although linear prediction was investigated independently by A. N. Kolmogorov, Wiener's approach had the merit of dealing with prediction and filtering under one heading. If on the strength of ergodicity of the stationary stochastic process $f(f_t \in L_2)$, the covariances $\varphi(t) = (f_t, f_0)$ are supposed to be provided by the data of the past, linear predicting means estimating the future of f by its projection on the linear span of the past f_t. On the other hand, linear filtering means separating the summands "message" and "noise" in $f_t = f_t^1 + f_t^2$, where again the autocovariances and cross covariances $\Phi(t) = (f_t, f_0)$ and $\Phi_1(t) = (f_t^1, f_0)$ are supposed to be known and the message is estimated by its projection on the linear span of the past signals f_t. Both tasks lead to Wiener-Hopf equations for a weighting distribution w,

$$\varphi(t+h) = \int_0^{\infty} \varphi(t-\tau)dw(\tau) \qquad (t \geqq 0),$$

$$\varphi_1(t+h) = \int_0^{\infty} \varphi(t-\tau)dw(\tau) \qquad (t \geqq 0),$$

respectively.

The implications of these fundamental concepts were elaborated in a wartime report that was belatedly published in 1949; it is still difficult to read, although its contents have become basic knowledge in communication theory. Nonlinear filtering was the subject of Wiener's unpublished memorandum (1949) that led to combined research at MIT, as reported by his close collaborator Y. W. Lee (see *Selected Papers*, pp. 17–33). A series of lectures on this subject was published in 1958. One of its main subjects is the use of an orthogonal development of nonlinear (polynomial) Volterra functionals by R. H. Cameron and W. T. Martin (1947) in a spectral theory and in the analysis and synthesis of nonlinear filters, which, rather than with trigonometric inputs, are probed with white Gaussian inputs.

After this brief exposition of Wiener's mathematics of communication, it remains to inspect the broad field that Wiener himself vaguely indicated as cybernetics; he tells how he coined this term, although it had not been unusual in the nineteenth century to indicate government theory. While studying antiaircraft fire control, Wiener may have conceived the idea of considering the operator as part of the steering mechanism and of applying to him such notions as feedback and stability, which had been devised for mechanical systems and electrical circuits. No doubt this kind of analogy had been operative in Wiener's mathematical work from the beginning and sometimes had even been productive. As time passed, such flashes of insight were more consciously put to use in a sort of biological research for which Wiener consulted all kinds of people, except mathematicians, whether or not they had anything to do with it. *Cybernetics, or the Control and Communication in the Animal and the Machine* (1948) is a rather eloquent report of these abortive attempts, in the sense that it shows there is not much to be reported. The value and influence of *Cybernetics*, and other publications of this kind, should not, however, be belittled. It has contributed to popularizing a way of thinking in communication theory terms, such as feedback, information, control, input, output, stability, homeostasis, prediction, and filtering. On the other hand, it also has contributed to spreading mistaken ideas of what mathematics really means. *Cybernetics* suggests that it means embellishing a nonmathematical text with terms and formulas from highbrow mathematics. This is a style that is too often imitated by those who have no idea of the meaning of the mathematical words they use. Almost all so-called applications of information theory are of this kind.

Even measured by Wiener's standards, *Cybernetics* is a badly organized work—a collection of misprints, wrong mathematical statements, mistaken formulas, splendid but unrelated ideas, and logical absurdities. It is sad that this work earned Wiener the greater part of his public renown, but this is an afterthought. At that time mathematical readers were more fascinated by the richness of its ideas than by its shortcomings. Few, if any, reviewers voiced serious criticism.

Wiener published more writings of this kind. The last was a booklet entitled *God and Golem, Inc.* It would have been more appropriate as the swan song of a lesser mathematician than Wiener.

BIBLIOGRAPHY

I. ORIGINAL WORKS. Many of Wiener's writings were brought together in his *Selected Papers* (Cambridge, Mass., 1964), which includes contributions by Y. W. Lee, N. Levinson, and W. T. Martin. Among his works are *Fourier Transforms in the Complex Domain* (New York, 1934), written with Raymond E. A. C. Paley; *Cybernetics, or the Control and Communication in the Animal and the Machine* (Paris–Cambridge, Mass., 1948); *Extrapolation, Interpolation and Smoothing of Stationary Time Series, With Engineering Applications* (Cambridge, Mass.–New York–London, 1949); *Ex-Prodigy—My Childhood and Youth* (New York, 1953; Cambridge, Mass., 1955); *I Am a Mathematician—the Later Life of a Prodigy* (Garden City, N.Y., 1956; repr. Cambridge, Mass., 1964); *Nonlinear Problems in Random Theory* (Cambridge, Mass.–New York–London, 1958); *God and Golem, Inc.* (Cambridge, Mass., 1964); and *Differential Space, Quantum Systems, and Prediction* (Cambridge, Mass., 1966), written with Armand Siegel, Bayard Rankin, and William Ted Martin.

II. SECONDARY LITERATURE. See "Norbert Wiener," *Bulletin of the American Mathematical Society*, spec. iss., **72**, no. 1, pt. 2 (1966), with contributions by N. Levinson, W. Rosenblith and J. Wiesner, M. Brelot, J. P. Kahane, S. Mandelbrojt, M. Kac, J. L. Doob, P. Masani, and W. L. Root, with bibliography of 214 items (not including posthumous works). See also Constance Reid, *Hilbert* (Berlin, 1970), esp. 169–170.

HANS FREUDENTHAL

WIENER, OTTO (*b.* Karlsruhe, Germany, 15 June 1862; *d.* Leipzig, Germany, 18 January 1927), *physics.*

Wiener, whose ancestors included clergymen and jurists, was the son of Christian Wiener, a professor of descriptive geometry at the Technische Hochschule in Karlsruhe. His mother, the former Pauline Hausrath, was the sister of a Protestant theologian and died of typhus when Wiener was three.

Wiener studied physics first at Karlsruhe, then at Berlin, and finally at the University of Strasbourg, where he earned the Ph.D. in 1887 under August Kundt, whose private assistant he was. In 1890 he qualified as a lecturer with "Stehende Lichtwellen." The following year he was named *Dozent* for physics at the Technische Hochschule in Aachen, and in 1894 he was promoted to extraordinary professor. In the same year he married Lina Fenner, daughter of *Geheimrat* Georg Fenner of Hesse-Homburg. In 1895 Wiener accepted an offer of a full professorship at the University of Giessen, where all his efforts were absorbed in the construction and organization of a new physics institute. His experience in this undertaking subsequently proved very useful, when he became involved in a similar project at Leipzig, upon succeeding Gustav Wiedemann as professor of physics.

Wiener reached the summit of his scientific career at its beginning and spent the second half pursuing what proved to be a mirage: "a fundamental law of nature," as he put it, according to which all physical events could be derived from a universal ether and from the velocities and differences in velocities of its parts. In this view, even electrons and protons were considered to be only definite forms of motion: rotating ether rings. Wiener's publications on this subject, more sketches than reports of results, brought more opposition and ridicule than recognition and made him seem an "anti-Einstein" to many of his colleagues.

Wiener's name is linked with the experimental demonstration of standing light waves. In 1888 Heinrich Hertz, working in the physics institute of the Technische Hochschule of Karlsruhe, proved the existence of electromagnetic waves. Those he detected had lengths of about eight meters. A year later Wiener performed a similar experiment with light waves—electromagnetic waves—approximately ten million times shorter than those used by Hertz. In the simplest case these waves arise in front of a plane metal mirror from the interference of incident monochromatic waves with the reflected ones.

Wiener's research in this area was a result of the work he did for his doctoral dissertation, "Über die Phasenänderung des Lichtes bei der Reflexion und Methoden zur Dickenbestimmung dünner Blättchen" (1887). In the latter he had, at Kundt's suggestion, measured light absorption in transparently thin metal plates, obtained by cathode-ray evaporation. In order to evaluate the measurements, however, it was necessary to know the thickness of the plates and the change in the vibration phase resulting from reflection. This change could be determined only when the light was obliquely incident. Through this research Wiener became a pioneer in the physics and techniques of thin plates, a field of great importance today.

A major question remained unanswered: how the vibration phase changes when light is incident perpendicularly. Inspired by Hertz's work, Wiener hoped to find an answer and, if possible, to demonstrate standing light waves. He did, in fact, succeed in making visible nodes and antinodes separated by intervals of about $2 \cdot 10^{-5}$ centimeters in front of a plane silver plate on which monochromatic light shone perpendicularly. He achieved this with a suitably mounted photosensitive plate, like those used in photography, the thickness of which was about 1/30 the wavelengths. Wiener demonstrated conclusively that it was the nodes of the resulting light vibrations, and not antinodes, that lay in the mirror surface. Accordingly, the reflection of light must take place with phase inversion: this was the answer to his question. In addition the experiment revealed that only the electric portion of the electromagnetic light waves blackens the silver chloride in the photosensitive layer. Wiener's amazing success was acknowledged as a masterpiece of experimentation.

The standing waves soon found an application in Gabriel Lippmann's color photography. In this process the silver contained in suitably prepared photosensitive plates is separated into parallel planes by the standing light waves. When viewed in daylight, these planes transmit to the eye only those colors having wavelengths that match the distances between the planes, while the other wavelengths are eliminated through interference. Wiener worked on color photography and proved that the color effects observed in the plates produced by Daguerre (silver plates with a silver iodide layer) arose in the same way they do in Lippmann's plates. Wiener also had a predilection for technical problems, especially of bird flight, and was very interested in the developing subject of aeronautics.

BIBLIOGRAPHY

I. ORIGINAL WORKS. A bibliography of Wiener's writings is in *Berichte. Sächsische Akademie der Wissenschaften*, **79** (1927), 119–121. Among his works are "Über die Phasenänderung des Lichts bei der Reflexion und Methoden zur Dickenbestimmung dünner Blättchen," in *Annalen der Physik und Chemie*, n.s. **31** (1887), 629–672, his doctoral dissertation; "Stehende Lichtwellen und die Schwingungsrichtung polarisierten Lichts," *ibid.*, n.s. **40** (1890), 203–243, 744; and "Farbenphotographie durch Körperfarben und mechanische Farbenanpassung in der Natur," *ibid.*, n.s. **55** (1895), 225–281. The Saxon Academy of Sciences, Leipzig, possesses more than 1,000 pages of a MS "Grundgesetz der Natur."

II. SECONDARY LITERATURE. See K. Lichtenecker, "Otto Wiener," in *Physikalische Zeitschrift*, **29** (1928), 73–78, with portrait; W. Möbius, "Otto Wiener gestorben," in *Zeitschrift für technische Physik*, **8** (1927), 129–131; and L. Weikmann, "Nachruf auf Otto Wiener," in *Berichte. Sächsische Akademie der Wissenschaften*, **79** (1927), 107–120, with portrait and bibliography.

FRITZ FRAUNBERGER

WIESNER, JULIUS VON (*b*. Tschechen, Moravia [now Czechoslovakia], 20 January 1838; *d*. Vienna, Austria, 9 October 1916), *plant anatomy, plant physiology*.

Wiesner was the youngest of eight children of Karl Wiesner, a shipping agent in Tschechen, and of Rosa Deutsch. Shortly after his birth the family moved to Brno, where Wiesner spent his youth and attended secondary school. He began his higher education at the Technical University of Brno and continued it at the University of Vienna, where he studied botany under E. Fenzl and Franz Unger, chemistry under Anton von Schrötter, physiology under Ernst Brücke, and physics under Andreas von Ettingshausen. He received the Ph.D. from Jena in 1860 and a year later qualified as academic lecturer at the Imperial-Royal Polytechnic Institute (now the Technical University) for Physiological Botany in Vienna, where in 1868 he was appointed associate professor. He became professor of plant physiology at the Forestry Institute of Mariabrunn in 1870 and, three years later, professor of plant anatomy and physiology at the University of Vienna, where he remained active until 1909.

Wiesner was rector of the University of Vienna during the academic year 1898–1899. Many scientific academies and learned societies elected him to membership: Vienna, Berlin, Paris, Munich, Rome, Turin, Göttingen, Uppsala, and Christiania (Oslo). He was awarded honorary doctorates by the technical universities of Brno and Vienna and by the universities of Uppsala and Glasgow. For many years Wiesner was a member of the Upper House, and when he retired from teaching in 1909, he was elevated to the hereditary nobility.

The Institute of Plant Physiology of the University of Vienna, of which Wiesner became director in 1873, was still located in a private house. Not until the fall of 1885 was it transferred to the newly built university, where it is still located. There Wiesner could bring to bear his talents as organizer and administrator, and the institution soon became known as one of the world's finest and best equipped plant physiology laboratories.

Led by his preference for applied research, Wiesner at first did work in technical microscopy and plant raw materials. His investigations on the microscopic characteristics of various fibers, on wood types and on how to demonstrate that a substance is wood, on tanning agents and on dyes, as well as on latexes, rubbers, resins, and balsams, were crowned by his integrated treatment of economically valuable plant materials in *Die Rohstoffe des Pflanzenreichs* (1873).

Wiesner also contributed work of enduring value in plant physiology with his studies of transpiration, the movements of plant organs, growth, and other phenomena of plant life. For decades he investigated the relationship between plants and light. In *Der Lichtgenuss der Pflanzen* he summarized his findings on the influence of the intensity and duration of sunlight in natural habitats on the distribution of plants and the development of their organs. Wiesner modified Bunsen and Roscoe's photographic method of light measurement to make it applicable to the requirements of plant physiology. He then undertook light measurements in Java, Egypt, Norway, and Spitsbergen, as well as in various parts of North America, thus providing the first survey of the light climate of the earth.

Wiesner also made valuable contributions to other fields. By examination of old Arabic and Central Asian papers, he demonstrated that six hundred years before the Arabs, the Chinese had known how to make paper from rags.

Wiesner had a lifelong interest in problems of natural philosophy. His book *Erschaffung, Entstehung, Entwicklung* was published a few weeks before his death.

BIBLIOGRAPHY

Wiesner published more than 200 articles in various journals. His books are *Einleitung in die technische Mikroskopie* (Vienna, 1867); *Gummiarten, Harze und Balsame* (Erlangen, 1869); *Die Entstehung des Chlorophylls in der Pflanze* (Vienna, 1877); *Elemente der wissenschaftlichen Botanik*, 3 vols. (I, Vienna, 1881; 2nd ed., 1885; 3rd ed., 1890; II, 1884; 2nd ed., 1891; III, 1889; 2nd ed., 1902; 3rd ed., 1913); *Die Elementarstruktur und das Wachstum der lebenden Substanz* (Vienna, 1892); *Die Rohstoffe des Pflanzenreichs*, 2 vols. (Leipzig, 1873; I, 2nd ed., 1900; II, 2nd ed., 1903; 3rd ed. enl., 3 vols. by J. Möller; 4th ed., by P. Krais and W. von Brehmer, 1927; 5th ed., by C. von Regel, Weinheim, 1962); *Der Lichtgenuss der Pflanzen* (Leipzig,

1907); and *Erschaffung, Entstehung, Entwicklung* (Leipzig, 1916).

A complete bibliography follows Hans Molisch's obituary of Wiesner in *Berichte der Deutschen botanischen Gesellschaft*, **34** (1916), 71–99.

RICHARD BIEBL

WIGAND, ALBERT JULIUS WILHELM (*b.* Treysa, Electoral Hesse, Germany, 21 April 1821; *d.* Marburg, Germany, 22 October 1886), *botany.*

The son of Johann Heinrich Friedrich Wigand, an apothecary, Wigand began the study of mathematics, science, and German philology at the University of Marburg in 1840. After a short period at the University of Berlin, where he studied botany under Karsten, he moved to Jena and became a pupil of Schleiden's. In fact, Wigand can be considered the last and most important member of Schleiden's school of botany. In 1846 he returned to Marburg and published his inaugural dissertation. In the same year he was appointed external university lecturer; in 1851, extraordinary professor; and in 1861, full professor of botany and director of the Botanical Garden and the pharmacognostic institute.

Wigand was active in various areas of botany; and all of his publications are characterized by a philosophical outlook that originated in his strong religious beliefs, although he always attempted to proceed inductively. In his inaugural dissertation he discussed the teratology of plants in the light of a general theory of metamorphosis, a subject to which he subsequently returned.

Strongly opposed to the view that ferns might have generative organs, Wigand mistakenly believed that both the antheridia and the archegonia must be functionless. With greater success he defended his views concerning the cuticula, the intercellular substance, and the structure of the cell wall. He stated that the wall between two cells is the result of chemical processes that lead to deposition of new material (apposition), an interpretation that appeared to be correct. His use of chemical substances for these microscopical investigations renders him a pioneer of microchemical staining techniques.

Wigand also was active in plant physiology, particularly in the study of tannin and the pigments of flowers, plant morphology, and plant systematics. In microbiology Wigand developed a theory of fermentation in which bacteria were morphologically and physiologically independent units, originating from the protoplasm of animal and plant cells in a state of decomposition (anamophosis of protoplasm). This theory was proposed because—on religious grounds—he could not accept the idea of spontaneous generation.

For similar reasons Wigand was one of Darwin's most ardent opponents in Germany, although he always tried to oppose Darwinian theory exclusively on scientific grounds. His own ideas on evolution were developed in *Genealogie der Urzellen* (1872).

Wigand's *Lehrbuch der Pharmakognosie* (1863), a manual for apothecaries, was written primarily from the practical point of view; another important pharmacognostic publication dealt with the origin of gums and resins (1863).

BIBLIOGRAPHY

I. ORIGINAL WORKS. Many of Wigand's papers appeared in *Botanische Zeitung*, **7–29** (1849–1871), and in *Botanische Hefte*, **1–3** (1885–1888), the latter published by the Botanical Garden at Marburg. See also Royal Society *Catalogue of Scientific Papers*, VI, 363; VIII, 1238; XI, 806; XII, 783; and XIX, 608.

His earlier books include *Kritik und Geschichte der Lehre von der Metamorphose der Pflanze* (Leipzig, 1846), his inaugural dissertation; *Grundlegung der Pflanzenteratologie* (Marburg, 1850); *Intercellularsubstanz und Cuticula* (Brunswick, 1850); *Botanische Untersuchungen* (Brunswick, 1854); *Der Baum* (Brunswick, 1854); *Flora von Kurhessen und Nassau* (Marburg, 1859; 3rd ed., Kassel, 1879); *Lehrbuch der Pharmakognosie* (Berlin, 1863; 4th ed., 1887); and *Der botanische Garten von Marburg* (Marburg, 1867; 2nd ed., 1880).

Later publications include *Die Genealogie der Urzellen* (Brunswick, 1872); *Ueber die Auflösung der Arten durch natürliche Zuchtwahl* (Hannover, 1872); *Mikroskopische Untersuchungen* (Stuttgart, 1872); *Der Darwinismus und die Naturforschung Newtons und Cuviers*, 3 vols. (Brunswick, 1874–1877); *Die Alternative: Teleologie oder Zufall?* (Kassel, 1877); *Der Darwinismus, ein Zeichen der Zeit* (Heilbronn, 1878); *Entstehung und Fermentwirkung der Bakterien* (Marburg, 1884); and *Grundsätze aller Naturwissenschaft* (Marburg, 1886).

II. SECONDARY LITERATURE. See E. Dennert, "Julius Wilhelm Albert Wigand," in *Flora*, n.s. **44** (1886), 531–539; F. G. Kohl, "Albert Wigand," in *Botanisches Zentralblatt,* **28** (1886), 350–352, 381–384; A. Tschirch, "Julius Wilhelm Albert Wigand," in *Berichte der Deutschen botanischen Gesellschaft*, **5** (1887), xli–li; and B. Lehmann, *Julius Wilhelm Albert Wigand (1821–1886). Professor der Botanik und Pharmakognosie zu Marburg* (Marburg, 1973).

P. SMIT

WILBRAND, JOHANN BERNHARD (*b*. Clarholz, Germany, 8 March 1779; *d*. Giessen, Germany, 9 May 1846), *physiology*.

Wilbrand was one of the best-known adherents of Schelling's *Naturphilosophie*. Doggedly determined to accept only those facts compatible with his philosophical principles, he went so far as to deny Harvey's blood circulation and the gaseous exchange that occurs in the lungs. Despite his highly speculative ideas, however, he called for the comprehensive and factual observation of nature, because he deemed such empirically derived information essential in fleshing out the philosophical framework of his physiology.

Wilbrand was the only son of farmers who were serfs of a nearby cloister. After education by a local priest, he was sent to a Jesuit gymnasium in Münster. In 1800 he began to study theology and philosophy at the University of Münster, in order to secure a living as a clergyman. A year later Wilbrand transferred to the medical school to pursue his growing interest in the natural sciences. In the ensuing years he studied closely with the chemist Johann Bernhard Bodde (1760–1833), an ardent supporter of *Naturphilosophie*.

After being released from his serfdom in 1803, Wilbrand went to the University of Würzburg, ostensibly for clinical training at the Julius Hospital, since his previous medical education had been purely theoretical. Recommended by Bodde, he met the physician Ignaz Döllinger and the circle of students who attended the philosophical lectures given by Schelling, then a newly appointed member of the faculty. Wilbrand graduated in 1806 with a dissertation on respiration in which he rejected the existence of oxygen and carbon dioxide as independent substances. After receiving his medical degree, he spent several weeks at the Bamberg Hospital, studying the therapeutic methods based on John Brown's system of medicine (the Brunonian system).

Under the auspices of Count Spiegel Zum Desenberg, later archbishop of Cologne, and others, Wilbrand traveled to Paris, where he studied with Cuvier and Lamarck. Upon his return to Münster in 1807 he became an instructor at the medical school, where he lectured on medico-philosophical subjects.

In 1808 Wilbrand was appointed titular professor of comparative anatomy, physiology, and natural history at the University of Giessen. He became a prolific writer and a busy teacher, lecturing on botany, zoology, anatomy, physiology, and *Naturphilosophie*. His preeminence led to honors and contacts with leading German intellectuals, including Goethe, who was interested in Wilbrand's use of the concept of metamorphosis.

In 1817 Wilbrand became director of the botanical gardens and zoo at Giessen. A few years later he received a medal from Friedrich Wilhelm III of Prussia for his schematic depiction of nature in atlas form: *Gemälde der organischen Natur in ihrer Verbreitung auf der Erde* (Giessen, 1821).

Wilbrand's work on human physiology admirably reflects the ideas expressed by the followers of Schelling's *Naturphilosophie* and the methods used to seek their verification. He based his highly speculative physiology on the belief that the organism was a complete psychophysiological entity endowed with opposite or contrasted principles responsible for its vital motions (polarity) and constantly undergoing structural transformations. The latter, in a sense, represented a metabolic process of solidification and liquefaction of organic matter. Wilbrand's methodology stressed the supremacy of the investigator's "mental eye," which could discern the fundamental pattern within nature from the confusing and abundant facts of observation, largely through the use of analogies.

During the 1820's and 1830's Wilbrand regularly attended the yearly meetings of the Gesellschaft Deutscher Naturforscher und Ärzte and often presented his papers at the sessions. As the empirical and inductive method gradually gained ascendancy in German medical circles, audiences became less sympathetic to his faltering philosophical efforts to maintain Schelling's grandiose conception of nature. Risking hostility and even ridicule, Wilbrand adamantly remained opposed to the new scientific research in biochemistry and biophysics, which he considered to be a distorted and piecemeal analysis. Paradoxically, his call for the observation of nature both in breadth and in depth was being heeded by the new generation of physicians, who refused to fetter their conclusions to *Naturphilosophie*.

BIBLIOGRAPHY

I. Original Works. Wilbrand's best-known work is *Physiologie des Menschen* (Giessen, 1815). Among his philosophically oriented writings are *Darstellung der gesammten Organisation*, 2 vols. (Giessen, 1809), and *Ueber den Zusammenhang der Natur mit dem Uebersinnlichen* (Mainz, 1843). He also wrote two widely read textbooks: *Handbuch der Naturgeschichte des Thierreiches* (Giessen, 1829) and *Handbuch der vergleichenden Anatomie in ihrer nächsten Beziehung auf die Phy-*

siologie (Darmstadt, 1838). Wilbrand's *Selbstbiographie* (Giessen, 1831) provides glimpses of his career.

A number of Wilbrand's original documents exist in the family's archives in Darmstadt and have been catalogued by Dr. Axel Murken.

II. Secondary Literature. Brief biographical sketches of Wilbrand are A. Murken, "Johann Bernhard Wilbrand (1779–1846), ein Mediziner und Philosoph aus Clarholz," in *Gütersloher Beiträge,* **8** (1967), 171–175; and K. E. Rothschuh, "Johann Bernhard Wilbrand, ein Münsterländer, Naturforscher und Arzt im Zeitalter der Romantik," in *Westfälische Nachrichten Beilage* (20 Nov. 1954), 93–95, 104; (24 Oct. 1957), 31–32.

Wilbrand's ideas regarding the circulation of the blood are mentioned in E. Hirschfeld, "Romantische Medizin," in *Kyklos,* **3** (1930), 29–31; and in Werner Leibbrand, *Die spekulative Medizin der Romantik* (Hamburg, 1956), 130–132. A more detailed analysis of his basic physiological ideas is C. Probst, "Johann Bernhard Wilbrand (1779–1846) und die Physiologie der Romantik," in *Sudhoffs Archiv,* **50** (1966), 157–178. Wilbrand's relationship with Goethe has been examined by Axel Murken, "Johann Bernhard Wilbrand (1779–1846), ein Naturwissenschaftler der Romantik und seine Beziehung zu J. W. von Goethe," in *Medizinische Monatsschrift,* **24** (1970), 165–170.

Guenter B. Risse

WILCKE, JOHAN CARL (*b.* Wismar, Germany, 6 September 1732; *d.* Stockholm, Sweden, 18 April 1796), *physics.*

Like many of the Swedish savants of the eighteenth century, including Samuel Klingenstierna and Mårten Strömer, his physics professors at the University of Uppsala, Wilcke came from a clerical family. His father, Samuel Wilcke, the son of a Pomeranian shoemaker, had educated himself for the ministry with the aid of generous patrons, especially F. A. Aepinus, professor of theology at the University of Rostock, whose children he tutored. In 1739 Samuel was called to minister to the German-speaking community in Stockholm, where he spent the remainder of his life.

Wilcke received his secondary education at the German school associated with his father's church. In 1750 he entered the University of Uppsala to prepare for the ministry. It was not theology, however, but mathematics and physics that aroused his interest; and for three terms he followed lectures on algebra, spherical trigonometry, mechanics, and experimental physics.

Hoping, perhaps, to save his son from science, Samuel Wilcke agreed to Johan Carl's wish to study at Rostock. The elder Aepinus having died,

Samuel counted on his former pupil, A. I. D. Aepinus, now holder of the Rostock chair of oratory, to urge the merits of the ministerial life. The scheme backfired. At Aepinus' home, where Wilcke boarded, lived the rhetorician's younger brother Franz, who, having rejected the family's plan to make him a physician,[1] taught mathematics at the university. The bond of common interest and sympathy that soon developed between Franz Aepinus and Wilcke accelerated Wilcke's drift from theology; and in 1753, when he matriculated at Göttingen, he no longer inscribed himself "theologus," as he had at Rostock, but as "mathematicus." Two years later A. I. D. Aepinus brought Samuel Wilcke to acquiesce in the *fait accompli*, perhaps made more palatable by the success of Franz Aepinus, who in the spring of 1755 became the astronomer of the Berlin Academy of Sciences.

Electricity. Wilcke joined Franz Aepinus in Berlin, where he devoted his newfound freedom to the study of physics, particularly to the agitated question of the contrary electricities. Were Dufay's vitreous and resinous electrifications differences only in degree, as Nollet insisted, or in kind, as the Franklinists claimed? Aepinus initially inclined toward Nollet, while Wilcke remained uncommitted, predisposed toward Franklin by certain experiments with the Leyden jar but arrested by Nollet's "apparently unanswerable" demonstration of the permeability of glass.[2] To resolve these uncertainties, Wilcke repeated all the experiments urged on either side; he prepared an annotated translation of Franklin's letters; and he found that, in most cases, Nollet's objections rested on misinterpretations of obscure, imprecise, or abbreviated passages in Franklin's work. As for Aepinus, he became an enthusiastic Franklinist when an experiment he designed to confirm Nollet failed.[3]

Not that Franklin's theory was unexceptionable. In his doctoral dissertation, defended at Rostock in 1757, and again in notes to his edition of Franklin, Wilcke showed that absolute insulation did not exist, that any electric per se could act the part of glass in the Leyden experiment, that the charges on the two coatings of the jar are not quite equal, and that substances are not innately vitreous or resinous.[4] This last point Wilcke owed to Canton, who had found that rough glass might be made minus or plus by rubbing with flannel or oiled silk, respectively. Recognizing that friction set up a competition for electrical matter, Wilcke hit on the idea of drawing up a winner's list, the entries being so placed that a given one became positive (or negative) when rubbed by those placed beneath (or above) it. His sequence, the first triboelectric series, consisted of smooth glass, wool, quills, wood, paper, sealing wax, white wax, rough glass, lead, sulfur, and metals other than lead.[5]

The most important result of Wilcke's Berlin period was the invention of the air condenser. Wilcke had consulted Aepinus—who had been studying the electricity of the tourmaline—about Franklin's version of Canton's induction experiments. Aepinus saw that the experimental arrangement amounted to an imperfect Leyden jar with air as dielectric; to check his insight he and Wilcke built a large air condenser (fifty-six square feet) that gave a shock comparable to that from a well-charged bottle. This demonstration threatened the already moribund theory of electrical atmospheres, which Franklin himself had not entirely discarded. While the repulsive force of the upper plate certainly reached the lower, its redundant electrical matter as certainly did not: for in that case the condenser, being shorted internally, could not have charged. Aepinus concluded for an instrumentalist theory of electricity, freely admitting action at a distance without specifying its cause. Indeed, he said, Franklinism must end in agnosticism: to save the simplest electrical phenomena, the particles of common matter must be supposed to repel one another at the same time that, according to the gravitational theory, they are mutually attractive.[6]

Wilcke did not embrace his friend's teachings altogether. He tentatively accepted the most bizarre of the new postulates, the mutual repulsion of matter particles, in order to conquer the enigma of the repulsion between negatively charged bodies; but he continued to ascribe the reciprocal recession of positive bodies to the pressure of their atmospheres. In Wilcke's asymmetric concept, positive atmospheres are material bodies, while negative ones are mere spheres of activity, spaces distorted by the presence of a deficient object.[7] Several years later (1763) Wilcke resolved this asymmetry by accepting and even championing the dualistic theory of Robert Symmer, which replaced the Franklinist negative state, or absence of electrical matter, with the presence of a second electrical fluid [6].

The productive collaboration ended in 1757 when Aepinus left Berlin for St. Petersburg. Wilcke had also received a Russian offer but declined it when Klingenstierna contrived to find him a position in Sweden, a lectureship (which in 1770 became a professorship) in experimental physics at the Royal Swedish Academy of Sciences. The position paid so poorly that Wilcke had to tutor for

room and board. Not until 1777, when his salary had tripled, did he feel he could marry; he chose his housekeeper, Maria Christina Setterberg, who bore him no children to increase his expenses. Not until 1784, when he became secretary of the Academy, did his financial difficulties end. Some of Samuel Wilcke's misgivings about his son's career had been well taken.

Wilcke continued to work on electricity during his first year in Sweden [3]. His most characteristic efforts [4], on the location of charge in a dissectible plate condenser, anticipated the invention of the electrophorus; he observed that, having electrified the condenser, a plate could be removed, discharged, returned, grounded, and again removed, discharged, and so on, "without further electrification by the machine . . . as often as the trial is made." Wilcke explained these effects as inductive in 1762, some thirteen years before Volta described similar experiments without explanation; when he learned of the electrophorus, Wilcke immediately supplied its theory [13]. He acknowledged Volta's invention of a useful machine but rightly asserted priority in discovering its principle, a claim supported by most German-speaking electricians.[8]

Among Wilcke's other electrical researches his lengthy studies of the tourmaline [7] and of cyclones and waterspouts [15] deserve mention. The former are distinguished by careful examination of a multitude of delicate cases that established the validity of Aepinus' concept of electrical poles and corrected many previous errors of detail. The latter, although they do not in fact concern electricity, stemmed from Wilcke's conjecture that cyclonic winds might be driven by atmospheric electricity. With his usual care he gathered all available data on waterspouts and compared them with the behavior of vortices and whirlpools generated in the laboratory. His knowledge of the phenomena was not superseded during the nineteenth century.

Heat. Wilcke's best-known work was his independent discovery of latent heat [11], which, he said, followed from a chance observation made early in 1772. Wishing to remove snow from a small courtyard and expecting that, in obedience to Richmann's law (which states that the temperature R of a mixture of two measures of water, m_1 and m_2, initially at temperatures T_1 and T_2, is $R = [m_1T_1 + m_2T_2]/[m_1 + m_2]$), hot water would melt more than its weight of snow, he was surprised to find the water of very little efficacy and concluded that the law did not hold for mixtures of ice and water. He therefore looked for a new rule. In a typical experiment Wilcke mixed hot water at temperature T and melting snow, measured the resultant temperature θ, and computed the difference between θ and R, the final temperature to be expected from Richmann's law if water at zero degrees had been used in place of the snow.

In the simplest case, when all masses were equal, the mean loss of heat $R - \theta$ was 36 3/28 degrees; hence, as Wilcke concluded, it requires somewhat more than seventy-two degrees of heat to melt unit mass of snow at zero degrees.[9] He observed that these seventy-two degrees disappear or, as we would say, become latent, in liquefying the ice, and that liquefaction occurs without change of temperature. Physically (according to Wilcke) the matter of heat, which, like Franklin's electrical fluid, is made up of mutually repellent particles attracted by common matter, insinuates itself between contiguous ice particles, transforming them into water; further heating causes the water to expand and raises its temperature.[10]

These experiments probably owed less to chance than Wilcke represented. In 1769, while pursuing an old hobby, the study of the shapes of snowflakes and ice crystals, he had made the "paradoxical" observation that water cooled below zero degrees warms on freezing [10]. As Oseen observes, the melting of the courtyard snow with warm water was probably an attempt to study the paradox: it was an experiment, not an accident.[11]

Wilcke returned to the problems of heat in the winter of 1780–1781, interrupting his study of waterspouts in order to follow up Joseph Black's concept of specific heat as reported in J. H. Magellan's *Nouvelle théorie du feu élémentaire* (1780). Wilcke had probably hit upon the same idea (although not the term) a few years earlier, perhaps in pursuing a note in Klingenstierna's *Inledning til naturkunnigheten* (1747), a translation of Musschenbroek's *Elementa physicae*. The note criticized Boerhaave's opinion, approved by Musschenbroek, that at equal temperatures all bodies contain equal amounts of heat by volume. In experiments apparently done in the 1770's, Wilcke showed that the sensible heats in bodies in thermal equilibrium were proportional neither to volume nor to mass; and, after a few false starts, he found how to measure relative heat capacities.[12] Immerse a mass of metal at temperature T in an equal mass of ice-cold water and record the resultant equilibrium temperature θ. Next calculate by Richmann's formula the amount of water w at temperature T that, when mixed with the same quantity of ice-

cold water (taken as unity for convenience), would yield the same resultant θ:

$$\theta = \frac{wT + 1 \cdot 0}{w + 1}, \quad w = \frac{\theta}{T - \theta}.$$

Wilcke probably obtained w for gold and lead before 1780. After seeing Magellan's book, he measured it for ten other substances [16]. Although his numerical results were not good (as in the experiments on latent heat he ignored the heat capacity of the calorimeter[13]), he understood that the w's were the specific heat capacities that he and Black had sought. He also saw in them a further analogy between the properties of the matters of heat and electricity: for not only were they all subtle, elastic, and apparently weightless fluids, but each was retained in ponderable bodies by specific forces dependent upon the nature of the bodies [17].

Miscellaneous Researches. Between 1763, when he finished the electrical studies begun in Berlin, and 1772 Wilcke worked at terrestrial magnetism. He began by inventing a new declination compass [5] and immediately became its slave [14]; a few years later, about 1766, he designed a dipping needle that proved itself on a voyage to China [12]. Encouraged by its performance, he undertook the difficult task of selecting reliable data from conflicting published measurements of dip made with other instruments. The result, an important contribution [9], was a systematic isoclinal chart that showed a magnetic equator and indicated positions for the poles that approximated those obtained from mapping declination. In another important work [8], Wilcke showed that a soft iron needle may be magnetized naturally by placing it in the magnetic meridian, or artificially by setting it near a powerful lodestone; in either case the needle magnetized more readily if the discharge from a Leyden jar passed through it first. Wilcke explained that the discharge helped to rearrange the internal parts of the needle.

Like many leaders of Sweden's eighteenth-century scientific renaissance, Wilcke had a taste and talent for applied science. He improved many standard instruments: the magnetic needles, the air pump, the micrometer, the barometer, the eudiometer. He also made suggestions for ventilating ships, for cooking under pressure ("Papin's digester"), for life preservers, and—at the request of the government—for fortifying the harbor of Landskrona.

Wilcke was a dry, unsociable man, happiest when at work or when reading in the several nonscientific subjects that interested him: theology, travel, belles lettres, music. These qualities made his tenure as secretary of the Academy (1784–1796) a mixed success. A responsible and diligent bureaucrat, he kept up the Academy's correspondence, publications, and records, and tried to maintain its high standards, as exemplified in his own scientific work. But he lacked the spark and influence of his predecessor, the astronomer Pehr Wargentin. Wilcke was not the man to win the Academy public support or to change its direction when, in the 1780's, it grew increasingly anachronistic and isolated. It fell into a decline that, however, did not approach bottom until after his time.[14]

As a physicist Wilcke is distinguished, apart from his substantive contributions, by his emphasis on measurement, exactness, and reproducibility of results. Although not a mathematical physicist in the modern sense, he insisted upon the utility of mathematics and mathematical formulations in experimental philosophy. In these emphases he was by no means unique or original, but he was one of the first physicists to demonstrate their fruitfulness in his own work.[15]

NOTES

1. J. C. Koppe, *Jetzlebendes gelehrtes Mecklenburg,* I (Rostock–Leipzig, 1783), 9–15.
2. F. U. T. Aepinus, *Recueil de différents mémoires sur la tourmaline* (St. Petersburg, 1762), 134; Wilcke, [2], intro., 388–389. The boldface numerals refer to items in the bibliography.
3. Aepinus, *Recueil,* 134–137; Wilcke, [2], 280–286, 348.
4. Wilcke, [2], 219–221, 271–272, 290, 308–309; and [1], 59–60, 81–83.
5. Wilcke, [1], 44–64.
6. Aepinus, *Tentamen theoriae electricitatis et magnetismi* (St. Petersburg, 1759), 5–7, 35–40, 75–83, 257–259; Wilcke, [2], 306–309.
7. Wilcke, [2], 221–224, 233–236, 262–263, 270–271, 307, 340–341.
8. G. C. Lichtenberg, *Briefe,* A. Leitzmann and C. Schüddekopf, eds., III (Leipzig, 1904), 203.
9. Let L be latent heat, M the mean of the experiments; then (since $R = T/2$), $T - 2\theta = 2M$, and $T - \theta = L + \theta$, whence $L = 2M = 72$ and 3/14 degrees. The value should be near 80.
10. Wilcke, [11], 105, 111; [16], 52.
11. Oseen, *Wilcke,* 156, 174–177.
12. This is an undated MS analyzed by Oseen, *Wilcke,* 232–234, 247–248.
13. McKie and Heathcote, *Specific and Latent Heats,* 86–87.
14. Lindroth, *Historia,* II, 20–26.
15. See Wilcke to W. C. G. Karsten, 1 July 1785, in Karsten, *Physisch-chemische Abhandlungen,* I (Halle, 1786), 118–119.

BIBLIOGRAPHY

I. Original Works. A bibliography of Wilcke's printed work and a catalog of his scientific MSS held at the Royal Swedish Academy of Sciences are given in C. W. Oseen, *Johan Carl Wilcke Experimental-fysiker* (Uppsala, 1939), 369–391. The Academy also has much administrative and scientific correspondence from Wilcke's secretaryship. Some scientific correspondence, notably that with C. W. Scheele, is printed in Oseen's biography. All Wilcke's scientific papers were published in Swedish in *Kungliga Svenska vetenskapsakademiens handlingar* (abbreviated below as *Handl.*) and translated into German in A. G. Kaestner, *Der königl. schwedischen Akademie der Wissenschaften, Abhandlungen aus der Naturlehre* (abbreviated below as *Abh.*); in the following bibliography the pages in Kaestner are given after the citation to the *Handlingar.*

Wilcke's most important works are [1] his thesis, *Disputatio physica experimentalis de electricitatibus contrariis* (Rostock, 1757); [2] his ed. of Franklin's letters, *Des Herrn Benjamin Franklins Esq. Briefe von der Electricität . . . nebst Anmerkungen* (Leipzig, 1758); [3] "Electriska rön och försök om den electriska laddningens och stötens åstadkommande vid flera kroppar än glas och porcellain," *Handl.,* **19** (1758), 250–282 (*Abh.,* **20,** 241–268); [4] "Ytterligare rön och försök om contraira electriciteterne vid laddningen och därtil hörande delar," *Handl.,* **23** (1762), 206–229, 245–266 (*Abh.,* **24,** 213–235, 253–274); [5] "Beskrifning på en ny declinations-compass," *Handl.,* **24** (1763), 143–153 (*Abh.,* **25,** 154–164); [6] "Electriska försök med phosphorus," *Handl.,* **24** (1763), 195–214 (*Abh.,* **25,** 207–226); [7] "Historien om tourmalin," *Handl.,* **27** (1766), 89–108, and **29** (1768), 3–25, 97–119 (*Abh.,* **28,** 95–113, and **30,** 3–26, 105–128); [8] "Afhandling om magnetiska kraftens upväckande genom electricitet," *Handl.,* **27** (1766), 294–315 (*Abh.,* **28,** 306–327); [9] "Försök til en magnetisk inclinations-charta," *Handl.,* **29** (1768), 193–225 (*Abh.,* **30,** 209–237); and [10] "Nya rön om vattnets frysning til snö-like is-figurer," *Handl.,* **30** (1769), 89–111 (*Abh.,* **31,** 87–108).

Also see [11] "Om snöns kyla vid smältningen," *Handl.,* **33** (1772), 97–120 (*Abh.,* **34,** 93–116); [12] "Om magnetiska inclinationen, med beskrifning på tvänne inclinations-compasser," *Handl.,* **33** (1772), 287–306 (*Abh.,* **34,** 285–302); [13] "Undersökning om de vid Herr Volta's nya elettrophoro-perpetuo förekommande electriska phenomener," *Handl.,* **38** (1777), 56–83, 128–144, 216–234 (*Abh.,* **39,** 54–78, 116–130, 200–216); [14] "Rön om magnet-nålens årliga och dagelige ändringar i Stockholm," *Handl.,* **38** (1777), 273–300 (*Abh.,* **39,** 259–284); [15] "Försök til uplysning om luft-hvirflar och sky-drag," *Handl.,* 2nd ser., **1** (1780), 1–18, 83–102, **3** (1782), 3–35, **6** (1785), 290–307, and **7** (1786), 3–20 (*Neue Abh.,* **1,** 3–18, 81–97, **3,** 3–31, **6,** 271–286, and **7,** 3–27); [16] "Rön om eldens specifiska myckenhet uti fasta kroppar, och des afmätande," *Handl.,* **2** (1781), 49–78 (*Neue Abh.,* **2,** 48–79; also *Journal de physique,* **26** [1785], 256–268, 381–389); and [17] "Rön om varmens spänstighet och fördeling, i anledning af ångors upstigande och kyla, uti förtunnad luft," *Handl.,* **2** (1781), 143–163 (*Neue Abh.,* **2,** 146–164).

II. SECONDARY LITERATURE. The standard biography is Oseen's; for Wilcke's activities at the Academy, see also N. V. E. Nordenmark, *Pehr Wilhelm Wargentin* (Uppsala, 1939), and S. Lindroth, *Kungliga svenska vetenskapsakademiens historia 1739–1818* (Stockholm, 1967). A brief notice by Anna Beckman appears in S. Lindroth, ed., *Swedish Men of Science* (Stockholm, 1952), 122–130. For Wilcke's work in general, see Oseen; on heat, see also D. McKie and N. H. de V. Heathcote, *The Discovery of Specific and Latent Heats* (London, 1935), 78–108; and on electricity, J. Priestley, *The History and Present State of Electricity*, 3rd ed. (London, 1775), I, 272–276, 358–362, and II, 35–37; and E. Hoppe, *Geschichte der Elektrizität* (Leipzig, 1884).

J. L. HEILBRON

WILCZYNSKI, ERNEST JULIUS (*b.* Hamburg, Germany, 13 November 1876; *d.* Denver, Colorado, 14 September 1932), *mathematics.*

Wilczynski was a son of Max Wilczynski and Friederike Hurwitz, who settled in Chicago when he was young. He returned to Germany for advanced study, receiving the Ph.D. from the University of Berlin in 1897 with a dissertation entitled "Hydrodynamische Untersuchungen mit Anwendung auf die Theorie der Sonnenrotation." Upon returning to the United States, he spent a year as a computer in the Office of the Nautical Almanac in Washington. In 1898 Wilczynski went to the University of California as an instructor; he rose to the rank of associate professor and served there until 1907. From 1903 to 1905 he was in Europe as assistant and associate of the Carnegie Institution of Washington, which provided the financial support that enabled him to write *Projective Differential Geometry of Curves and Ruled Surfaces* (1906). In 1906 he married Countess Ines Masola, of Verona. He was associate professor at the University of Illinois from 1907 to 1910 and at the University of Chicago from 1910 to 1914, achieving full professorship in the latter year.

Wilczynski's main work was in projective differential geometry, a subject of which he is generally considered the creator. A prolific worker, he published seventy-seven books and papers. He was also active in scientific organizations, serving as vice-president of the American Mathematical Society, as a member of the council of the Mathematical Association of America, and as an associate editor of the *Transactions of the American Mathematical Society.* Wilczynski won a prize (and was named laureate) of the Royal Belgian Academy in 1909, and in 1919 he was elected a member of the National Academy of Sciences.

What is now called classical differential geometry studied the local metric properties of geometrical configurations; projective differential geometry proposed similarly to study the local properties invariant under projective transformations. When Wilczynski started his work, about 1900, Halphen's projective differential geometry of curves already existed; but Wilczynski devised new methods, deepened the theory for curves, extended it to surfaces, and brought it to its present form.

In 1900, classical differential geometry was already a century old. Although it could still provide much interesting detail, it had lost its vitality; and by 1920 it had been declared dead. E. T. Bell has suggested that classical differential geometry lacked method and aim. Projective differential geometry, although it contained new points of view, was only a part of that larger subject and, therefore, shared its fate, although classical metric differential geometry is still a staple university course.

BIBLIOGRAPHY

See E. T. Bell, *Development of Mathematics* (New York, 1940), 332; and E. P. Lane, "Ernest Julius Wilczynski," in *American Mathematical Monthly*, **39** (1932), 567–569, see also 500; and "Ernest Julius Wilczynski—In Memoriam," in *Bulletin of the American Mathematical Society*, **39** (1933), 7–14, with bibliography of 77 works published by Wilczynski from 1895 to 1923.

A. SEIDENBERG

WILD, HEINRICH (*b.* Uster, Zurich canton, Switzerland, 17 December 1833; *d.* Zurich, Switzerland, 5 September 1902), *meteorology.*

Wild studied at the universities of Königsberg and Heidelberg, and in 1857 received the Ph.D. at Zurich. In November 1858 he was appointed professor of physics and director of the observatory at the University of Bern. Ten years later he was invited to join the Academy of Sciences in St. Petersburg and to become director of its Central Astrophysical Observatory. In 1876 he founded the magnetometeorological observatory at Pavlovsk and remained its director until 1895.

Wild was an active meteorologist who played a major part in the development of the science in the latter half of the nineteenth century. He improved several instruments, including the anemograph, anemometer, atmometers, barometer, rain gauges, thermograph, several forms of theodolite, instruments for measurement of terrestrial magnetism, polarization photometer, and polaristrobometer. His modifications of these instruments significantly improved techniques for weather observation. He also was directly responsible for significantly extending the network of meteorological observation stations—almost as many in Switzerland as in Russia.

Wild was a member of several international meteorological commissions, including the International Polar Commission (1879–1891), and was largely responsible for the preparation of their reports. He also served on international commissions on the meter and on the reform of chronometric methods.

Wild wrote extensively, in both German and Russian, primarily on meteorological instruments and improved techniques for meteorological observations.

BIBLIOGRAPHY

Wild's writings include "Études métrologiques," in *Mémoires de l'Académie impériale des sciences de St.-Pétersbourg*, 7th ser., **18** (1872), no. 8; *Bestimmung des Werthes der Siemen'schen Widerstands Einheit in absolutem electromagnetischen Masse* (St. Petersburg, 1884); and *Das Konstantinow'sche und magnetische Observatorium in Pawlowski* (St. Petersburg, 1895).

ASIT K. BISWAS
MARGARET R. BISWAS

WILEY, HARVEY WASHINGTON (*b.* Kent, Indiana, 18 October 1844; *d.* Washington, D.C., 30 June 1930), *chemistry.*

The son of Preston Prichard Wiley and Lucinda Maxwell, Wiley received his early education primarily from his father, who ran a subscription school during the seasons when farm work was not pressing. He received the B.A. from Hanover College in 1867, the M.D. from Indiana Medical College in 1871, and the B.S. from Lawrence Scientific School, Harvard, in 1873. In 1874 Wiley became professor of chemistry at Purdue University and served as Indiana state chemist, except for a short interlude of European study, until 1883, when he was appointed chief of the Division (later Bureau) of Chemistry of the U.S. Department of Agriculture.

By then Wiley had become well-known as an analytical chemist with expertise in sugar chemistry and technology. Although he continued the sugar studies in Washington, he was becoming concerned about the widespread adulteration of syrups available in the marketplace. This concern quickly spread to other foods. His agency undertook extensive analysis of commercial foods and reported widespread adulteration. Wiley became active in a campaign to bring about passage of pure food legislation by Congress, but his efforts were repeatedly frustrated.

In 1902 the Bureau of Chemistry undertook studies on the physiological effects of various chemical additives in human foods. These studies, made on human volunteers, raised doubts regarding the safety of salicylates, borates, formaldehyde, benzoates, saccharin, and copper salts in foods. Termed the "Poison Squad" experiments by newsmen, the studies attracted widespread interest.

The Food and Drug Act was finally passed in 1906, after scandals in the drug trade and in the meat-packing industry brought heavy public pressure on Congress for remedial action. The Bureau of Chemistry was charged with enforcement of the new law; but Wiley's efforts were frustrated as a consequence of industrial pressures on Secretary of Agriculture James Wilson, who steadily handicapped Wiley's work with bureaucratic obstructions. Of particular significance was the creation of the Referee Board of Consulting Scientific Experts, headed by Ira Remsen. This board repeated the studies on benzoates and other food additives, arriving at the conclusion that benzoates and saccharin were safe for use in foods, at least in limited amounts.

In 1912, shortly after being exonerated by a congressional committee of charges of alleged misuse of funds, Wiley decided that his enforcement powers had been undermined to such a degree that he could no longer be effective within the government. He resigned to become director of the Bureau of Foods, Sanitation, and Health for the magazine *Good Housekeeping*. Although he had hoped to use this position to educate the public, his efforts were largely ineffective.

Although he had a great deal of personal charm and was an effective public speaker, Wiley was also a forceful, determined, and uncompromising fighter for what he considered the best interests of the public. His firmness brought him many enemies, and his effectiveness declined steadily in the last three decades of his life. Nevertheless, his overall accomplishments were impressive. Besides his work for pure foods and drugs, Wiley was very active in the development of agricultural analysis and was a founder of the Association of Official

Agricultural Chemists in 1884, serving as its president in 1886. He served two terms as president of the American Chemical Society.

BIBLIOGRAPHY

I. ORIGINAL WORKS. Wiley meticulously saved letters, diaries, notebooks, lecture MSS, newspaper clippings, and other papers. His personal papers are now preserved in the Manuscript Division of the Library of Congress. There also are extensive holdings of Wiley letters and related official material in the Bureau of Chemistry records in the National Archives. The National Archives also holds relevant material in the files of the secretary of agriculture, the office of the solicitor general, and the Food and Drug Administration; and there is some related material in the papers of presidents Theodore Roosevelt and William Howard Taft. For information on these holdings, see Oscar E. Anderson, Jr., *The Health of a Nation, Harvey Wiley and the Fight for Pure Food* (Chicago, 1958), *passim*, esp. 280–282.

There is no complete bibliography of Wiley's extensive published works. As chief of the Bureau of Chemistry he had responsibility for all of its publications and wrote many of them. The early work on sugar production is treated in the annual reports of Purdue University and in early bulletins and reports of the Bureau of Chemistry, U.S. Department of Agriculture (*Bulletin* nos. 2, 3, 5, 6, 8, 14, 17, 21). The extensive studies of food adulteration were published as "Foods and Food Adulterants," *Bulletin. Bureau of Chemistry. United States Department of Agriculture*, no. 13, 10 pts. (1887–1899). The "Poison Squad" experiments were published as "Influence of Food Preservatives and Food Adulterants," *ibid.*, no. 84, 6 pts. (1904–1908).

Wiley's work on analytical procedures was published in standard scientific journals and government bulletins until 1884, when he was active in organizing the Association of Official Agricultural Chemists. The Association's *Official Methods* were published by the Bureau of Chemistry as *Bulletin* no. 7 (1885) and were reprinted as *Bulletin* no. 107, with revisions, for many years thereafter. Through Wiley's influence, the Bureau of Chemistry provided extensive support to the Association for publication of proceedings, as well as manpower for checking proposed analytical methods. When Wiley was honorary president of the Association between 1912 and 1930, his annual addresses were published in *Journal of the Association of Official Agricultural Chemists*. His *Principles and Practice of Agricultural Analysis*, 3 vols. (Washington, D.C., 1894–1897), went through rev. eds. in 1906–1911 and 1926.

Other books by Wiley are *Foods and Their Adulteration* (Philadelphia, 1907; 2nd ed., 1911; 3rd ed., 1917); *1001 Tests of Foods, Beverages and Toilet Accessories* (New York, 1914); *The Lure of the Land* (New York, 1915); *Not by Bread Alone, The Principles of Human Nutrition* (New York, 1915); *Beverages and Their Adulteration* (Philadelphia, 1919); *History of a Crime Against the Food Law* (Washington, 1929); and *Harvey W. Wiley, an Autobiography* (Indianapolis, 1930). *History of a Crime*, which is strongly autobiographical, was written late in life, when Wiley was very ill, and reflects a personal bitterness that might have been more tempered had it been written earlier. The *Autobiography*, which was finished with the aid of O. K. Armstrong, gives less attention to the enforcement period, is less belligerent, and is perhaps more representative of the real Wiley, who combined firmness with charm.

II. SECONDARY LITERATURE. The best biography of Wiley is Oscar E. Anderson, Jr., *The Health of a Nation, Harvey W. Wiley and the Fight for Pure Food* (Chicago, 1958). M. Natenberg, *The Legacy of Dr. Wiley and the Administration of His Food and Drug Act* (Chicago, 1957), was written as a propaganda piece and has only minor value. The 1931 meeting of the Association of Official Agricultural Chemists commemorated Wiley—the memorials read by W. W. Skinner, C. A. Browne, W. G. Campbell, *et al.*, are published in *Journal of the Association of Official Agricultural Chemists*, **14** (1931), iii–xxii. Useful short biographies are W. D. Bigelow, "Harvey Washington Wiley," in *Industrial Engineering Chemistry*, **15** (1923), 88; C. A. Browne, "Harvey Washington Wiley," in *Dictionary of American Biography*, XX, 215–216; E. J. Dies, *Titans of the Soil* (Chapel Hill, N.C., 1949), 151–158; and A. J. Ihde, in E. Farber, ed., *Great Chemists* (New York, 1961), 813–819. There is extensive background material on the passage of the Pure Food and Drug Act in Mark Sullivan, *Our Times, The United States, 1900–1925* (New York, 1927), II, 471–551; and James Harvey Young, *The Toadstool Millionaires* (Princeton, 1961), 226–246. On the role of Wiley in the early enforcement of the Pure Food and Drug Act, see O. E. Anderson, Jr., "The Pure-Food Issue: A Republican Dilemma, 1906–1912," in *American Historical Review*, **61** (1956), 550–573.

AARON J. IHDE

WILHELM IV, LANDGRAVE OF HESSE (*b.* Kassel, Germany, 24 June 1532; *d.* Kassel, 25 August 1592), *astronomy.*

A contemporary of Peter Apian, Copernicus, Tycho Brahe, and Kepler, Wilhelm lived in the age of the greatest astronomical revolution since antiquity. He played scarcely any role in the resulting debates, however, for he was concerned primarily with the refining of the techniques of astronomical observation. At the court in Kassel he was tutored, chiefly in mathematics, by Rumold Mercator, the son of the geographer Gerardus Mercator. He became interested in astronomy after reading Apian's *Astronomicum Caesareum*, a splendid book although it was conceived within the conceptual framework of the old geocentric view. Wilhelm's manuscript copy of this work, which is still extant, contains handwritten planetary tables that obviously were drawn up at his request by Andreas Schöner. On the model of Apian's system of movable cardboard disks, Wilhelm devised an arrangement of metal plates that made possible the construction of the *Wilhelmsuhr* (1560–1561). This mechanical astronomical clock was so precise that the ephemerides could be read directly from it. Similar clocks were later produced in great quantities.

Satisfying the love of display then flourishing in most princely courts was not, however, Wilhelm's goal; his main goal was to further the study of astronomy. In the course of his own astronomical observations, he noted the great differences between the true positions of the stars and those calculated on the basis of Ptolemaic theory. He therefore decided to establish a new star catalog derived from actual observations, a project not realized since the time of Hipparchus. Wilhelm began making observations at his private observatory in Kassel and continued until 1567, when he became landgrave. Tycho Brahe, who was in Kassel for a few days in 1575, urged him to hire assistants to carry on the work. Accordingly, Wilhelm invited Christoph Rothmann to come to his court as mathematician and observer, and Joost Bürgi as mechanic.

Both these men did work of considerable scientific distinction while at Kassel. Bürgi constructed globe clocks, pendulum clocks, and mechanical computing devices. Later he became known as Kepler's friend at the court of Rudolf II in Prague. Rothmann was an industrious observer and computer and a resolute supporter of the Copernican world view, as can be seen from his correspondence with Tycho Brahe. The accuracy of the observations made by Wilhelm and Rothmann is astonishing. Their determination of the latitude of Kassel (51°19′) required a correction of only 10″ in the heyday of astronomical geography at the beginning of the nineteenth century.

Only a small part of Wilhelm's project of the Hessian star catalog was realized. Ultimately it included 179 of the 1,032 stars originally planned. (Wilhelm's observations furnished the data for 58 of these and Rothmann's for 121.) Wilhelm nevertheless deserves great credit for undertaking a program that became one of the major tasks of observational astronomy in the following decades, and without which the later development of celestial mechanics would have been inconceivable. He also introduced a new method for determining stel-

lar positions: with his azimuthal quadrant he could determine the moment when a given star reaches a certain altitude. This process, in a slightly altered form, later developed into the basic method for determining stellar positions. Its superiority became evident, however, only with the development of much more accurate clocks than were available to Wilhelm. It was for this reason that Tycho Brahe criticized it.

Many of the instruments and clocks made for Wilhelm's Kassel observatory are preserved at the Astronomisch-Physikalische Kabinett in Kassel.

BIBLIOGRAPHY

I. ORIGINAL WORKS. Tycho Brahe published selections of his scientific correspondence with Wilhelm IV and Rothmann in *Tychonis Brahe Dani epistolarum astronomicarum . . .* (Uraniborg, 1596). Portions of Wilhelm's political correspondence was printed in various other publications. A list of this material was prepared by W. Ribbeck for the article in *Allgemeine deutsche Biographie*, XLIII (Leipzig, 1898), 39. MS material can also be found under the heading "Landgräfliche Personalia Wilhelm IV Astronomica" in the state archive at Marburg.

II. SECONDARY LITERATURE. See P. A. Kirchvogel, "Landgraf Wilhelm IV von Hessen und sein astronomisches Automatenwerk," in *Index zur Geschichte der Medizin, Naturwissenschaften und Technik*, **1** (1953), 12–18; "Wilhelm IV, Tycho Brahe and Eberhard Baldewein—the Missing Instruments of the Kassel Observatory," in *Vistas in Astronomy*, **9** (1968), 109–121, which contains an extensive bibliography; F. Krafft, "Tycho Brahe," in *Die Grossen der Weltgeschichte*, V (Zurich, 1974), 297–345; B. Sticker, "Landgraf Wilhelm IV und die Anfänge der modernen astronomischen Messkunst," in *Sudhoffs Archiv*, **40** (1956), 15–25; and "Die wissenschaftlichen Bestrebungen des Landgrafen Wilhelm IV," in *Zeitschrift des Vereins für Hessische Geschichte und Landeskunde*, **67** (1956), 130–137; R. Wolf, *Geschichte der Astronomie* (Munich, 1877), 266 ff.; and F. X. von Zach, "Landgraf Wilhelm IV," in *Monatliche Correspondenz zur Beförderung der Erd- und Himmelskunde*, **12** (1805), 267–302.

DIETER B. HERRMANN

WILHELMY, LUDWIG FERDINAND (*b.* Stargard, Pomerania [now Poland], 25 December 1812; *d.* Berlin, Germany, 18 February 1864), *physics, chemistry.*

After completing his early schooling, Wilhelmy left Pomerania to study pharmacy in Berlin. He subsequently purchased an apothecary shop in his native state and joined his father in business. His desire for pure scientific research led him to sell the shop in 1843, however, and to study chemistry and physics at Berlin, Giessen, and Heidelberg. In 1846 Wilhelmy received the doctorate from Heidelberg on the basis of a dissertation on heat as a measure of cohesion. After traveling chiefly in Italy and Paris, where he studied with Regnault, he returned to Heidelberg and became a *Privatdozent* in 1849. He remained at the university for only five years; he then returned to private life in Berlin after a six-month stay in Munich, keeping busy with philosophical, mathematical, and physical studies. Wilhelmy never married, preferring to devote his entire attention to expanding his knowledge in all areas of learning. A skilled businessman, he had a warm heart and was always willing to assist a friend. Although shy with strangers, in his small circle of colleagues he was cheerful and witty, defending his peculiar views on many subjects with surprising liveliness.

As a student in Berlin, Wilhelmy joined Magnus in forming a physics colloquium that became the Physical Society in 1845. Among the members of this small circle of young investigators were Paul du Bois-Reymond, Clausius, Helmholtz, and Werner Siemens. Upon his return to Berlin ten years later, Wilhelmy found few of the original group remaining; and he consequently took the lead in directing the younger members. As leader of the Physical Society, he converted part of his Berlin home, as well as his summer villa in Heidelberg, into physics laboratories in 1860. His studies on capillary action, unfinished at his death, were carried out at his home laboratory.

Wilhelmy is best known as the first person to measure the velocity of a homogeneous chemical reaction. In 1850 he published a paper on the law of the action of acids on cane sugar, which went virtually unnoticed until Wilhelm Ostwald called attention to it thirty-four years later. Wilhelmy's procedure involved following the reaction with a polarimeter (widely used at the time of his investigations), which did not disturb the conditions of the reacting system. In the presence of a large amount (considered constant) of water, he found that the amount of sugar changed in any instant was proportional to the amount present, the acid being unchanged. In mathematical terms Wilhelmy presented the familiar law

$$- dZ/dt = MZS,$$

where Z is the concentration of sugar, t is time, S is the acid concentration (presumed unchanging throughout the reaction), and M is a constant today called the reaction velocity constant. He also investigated the temperature dependence of the reaction and assumed that it followed the same exponential law as concentration.

Wilhelmy's earlier physical studies, dealing with heat and utilizing differential equations, prepared the way for his key 1850 paper. In his dissertation (1846) he used Regnault's coefficients of expansion to calculate the force of cohesion and concluded that molecules are acted upon by two forces, heat and cohesion, the former tending to annihilate the latter. His conclusions are reminiscent of Lavoisier's ideas about *calorique*. Wilhelmy added to these conclusions the concept of the numerical equivalence of heat and energy, following J. R. Mayer's 1842 study of this equivalence. In a book published in 1851 Wilhelmy attempted to derive several general relationships among physical properties of compounds. He suggested, for example, that isomeric compounds of equal specific gravity and equal boiling points have equal coefficients of expansion. If the boiling points are different, the coefficients of expansion are inversely proportional to them. The book was meant to be an introduction to a work that would provide a complete understanding of the essence of natural forces, but this ambitious project was never completed.

BIBLIOGRAPHY

I. Original Works. Wilhelmy's important paper on the rate of inversion of sugar, "Ueber das Gesetz, nach welchem die Einwirkung der Säuren auf den Rohrzucker stattfindet," in *Annalen der Physik und Chemie*, **81** (1850), 413–433, 499–526, was reprinted as Ostwald's Klassiker der exacten Wissenschaften, no. 29 (Leipzig, 1891), which contains a complete list of Wilhelmy's ten publications. Portions of the paper were translated into English in Henry M. Leicester and H. S. Klickstein, *A Source Book in Chemistry 1400–1900* (New York, 1952), 396–400. In addition to his book on heat, *Versuch einer mathematisch-physikalischen Wärmetheorie* (Heidelberg, 1851), he published one other book, *Zur physikalischen Begründung der Physiologie und Psychologie* (Heidelberg, 1852).

II. Secondary Literature. A short biographical sketch of Wilhelmy by Georg Quincke is in Ostwald's Klassiker (see above), 45–47. His work is discussed in Eduard Farber, "Early Studies Concerning Time in Chemical Reactions," in *Chymia*, **7** (1961), 135–148.

Sheldon J. Kopperl

WILKINS, JOHN (*b*. Northamptonshire, England, 1614; *d*. London, England, 19 November 1672), *theology, science, scientific and academic administration and organization.*

Wilkins' career coincides with the most eventful period in modern English history—the years just before the Long Parliament to the decade after the Restoration and the formation of the Royal Society. It was not an easy time for an active man to retain influence and office, but Wilkins managed owing to his habit of prudence and a spirit of moderation and tolerance. In 1643 he subscribed to the Solemn League and Covenant and in 1649 he took the engagement of loyalty to the English Commonwealth. He was trusted by Cromwell, whom he advised on the need for a national church and episcopacy against presbytery. After the return of Charles II in 1660, he submitted to the Act of Uniformity and soon enjoyed the favor of the restored monarchy. Still, only the most unforgiving royalists ever questioned his integrity. Throughout his life, he gained and retained the friendship and respect of men of the most diverse political and religious persuasions. No doubt such personal qualities as charm, ready conversation, and energy played their part in his success, but the deeper reason would seem to lie in his commitment to beliefs that transcended the exclusive interests of any particular faction. From the first to the last, all his writings advocate scientific and religious views that by the time of his death had proved that they represented the temper of the times. The new science had triumphed, and the liberal Anglican theology known as latitudinarianism was, thanks to him, on the rise under such men as John Tillotson, Edward Stillingfleet, and Simon Patrick.

Both in print and action, Wilkins was committed to a set of principles and beliefs—generally known as natural theology—which he was the first fully to formulate and advocate in England. He never questioned the importance of the Bible and revelation as sources of faith, and in this respect his thought differs from what later became known as deism. But his writings are devoted to the argument that moral and religious philosophy can be grounded on natural religion, by which he understood what "men might know, and should be obliged unto, by the mere principles of reason, improved by consideration and experience, without the help of revelation."[1]

Owing to the omnipotence, benevolence, and wisdom of God, both the universe and man are so admirably contrived that man can ensure the welfare of his soul by the mature exercise of the faculty of reason, which is the defining quality of his nature. This faculty reveals to man the natural principles that govern creation, thus providing him with knowledge that "may conduce to the proving of a God, and making men religious," by making him understand that "such a great order and constancy amongst" the heavenly bodies "could not at first be made but by a wise providence, nor since preserved without a powerful inhabitant, nor so perpetually governed without a skillful guide."[2]

Similarly, man is endowed with a natural principle that makes him seek moral good "as a rational voluntary free agent,"[3] owing to his steady inclination "to seek his own well-being and happiness," so that "nothing properly is his duty, but what is really his interest," which is another argument "that the author of his being must be infinitely wise and powerful."[4] Man's natural desire for happiness is as certain as the descent of heavy bodies,[5] an example that Wilkins also used to illustrate the fixed laws that rule nature. Both man and nature are governed by laws that ensure the harmony of religion and science.

Consistent with these arguments, Wilkins stated the deistic principle that the salvation of the heathen is not a problem for man to decide; since "God has not thought fit to tell *us* how he will be pleased to deal with such persons, it is not fit for us to tell *Him* how he ought to deal with them."[6] In his writings, Wilkins often used the wise testimony of the ancients to support the knowledge and arguments advanced by the new science. Whether we call some of his writings scientific and others religious is a matter of emphasis; they all have the same aim: to guide man's conduct toward moral virtue, religious devotion, and ultimately the hope of salvation. The pursuit of happiness, even comfort, in this world is man's legitimate interest.

But reason alone is not sufficient. Man is also naturally "a sociable creature . . . having only these two advantages to his protection, Society and Reason . . . Adam in the state of innocence could not be happy, though in Paradise, without a companion."[7] This is a theme Wilkins stresses again and again; it is the foundation of his constant advocacy of conciliation, moderation, and tolerance, often in contexts that refer to "all that confusion and disorder, which seem to be in the affairs of these times."[8] The instrument that ensures the benefits of social intercourse is language: "Every rational creature, being of an imperfect and dependent happiness, is therefore naturally endowed with an ability to communicate his own thoughts and intentions; that so by mutual services, it might the better promote itself in the prosecution of its own well-being."[9] As useful knowledge, both natural and moral, is a function of cooperation, so successful cooperation is a function of communication; the improvement of natural knowledge and language is the response to the "two general curses inflicted on mankind," after the fall of Adam, "the one upon their labors, the other upon their language."[10] After the anniversary meeting of the Royal Society on 30 November 1667 (in which the annual election of officers also took place), Pepys recorded that some members went out for dinner, he himself choosing to sit next to Wilkins "and others whom I value." With his last work, *An Essay Towards a Real Character and a Philosophical Language*, then in the press, Wilkins stated that "man was certainly made for society, he being of all creatures the least armed for defence, and of all creatures in the world the young ones are not able to do anything to help themselves . . . and were it not for speech man would be a very mean creature." Wilkins is the chief source of the Royal Society doctrines about language and style; knowledge based on mere words and phrases has "in it this intrinsical imperfection, that 'tis only so far to be esteemed, as it conduces to the knowledge of things," words themselves being merely "the images of matter." To treat them otherwise is to fall into "Pygmalion's phrenzy."[11]

Wilkins' view of useful knowledge determined his attitude toward the three chief sources of authority: the Bible, antiquity, and books. Using arguments that today are perhaps best known from Galileo's *Letter to the Grand Duchess Christina*, Wilkins repeatedly rejected scriptural authority in natural philosophy, a principle to which all the new scientists were committed; if theology is allowed interference with philosophy, then the status of the latter is endangered as an independent source of the wisdom of the creator. In his first publication, Wilkins stated the principle in these terms: "It is not the endeavor of Moses or the prophets to discover any mathematical or philosophical subtleties; but rather to accommodate themselves to vulgar capacities, and ordinary speech, as nurses are wont to use their infants."[12] On scientific matters, he was also fond of citing contradictory scriptural passages, just as he criticized those among his contemporaries "who upon the invention of any new secret, will presently find out some obscure text or other to father it upon, as if the Holy Ghost must

needs take notice of every particular which their partial fancies did over-value."[13]

He treated classical authors in much the same way as the Bible, using citations to suit his purposes both for and against his own principles, those in the latter category being dismissed as contrary to reason and experience. But he rejected outright the superior authority of antiquity: "In such learning as may be increased by fresh experiments and new discoveries, it is we are the fathers, and of more authority than former ages, because we have the advantage of more time than they had."[14] He was aware that the vast public structures of the Egyptians, Hebrews, Greeks, and Romans might be used to argue against the inferiority of their mechanical knowledge; he answered that if we have nothing of the sort nowadays, the reason does not lie in our knowledge, for "mechanical discoveries are much more exact now," but rather in the fact that "we have not either the same motives to attempt such works, or the same means to effect them as the ancients had." By this he meant that great wealth and power, then concentrated in the hands of a few, were now more widely diffused. "There is now a greater equality amongst mankind and the flourishing of arts and sciences has so stirred up the sparks of men's natural nobility, and made them of such active and industrious spirits, as to free themselves in a great measure from that slavery, which those former and wilder nations were subjected unto."[15]

The belief in the leveling and ennobling effect of the new knowledge found expression in Wilkins' attitude toward "bookish" men and mere bookish learning. Antiquity having slighted the mere manual and practical arts as "base and common," such studies had come to be neglected for hundreds of years, with grave consequences for the well-being of man. But the mechanical arts are just as worthy as the old and honored liberal arts such as logic and rhetoric, indeed "that discipline which discovers the general causes, effects, and proprieties of things, may truly be esteemed as a species of philosophy." Since all studies ought "to conduce to practice as their proper end," book learning is often rightly considered mere "pedantry." Wilkins was eager to overcome the prejudice that studies pertaining to the mind deserve greater respect than those that deal with material things. It was in this spirit that he devoted his *Mathematical Magick* to practical mechanical devices and labor-saving inventions "whereby nature is in any way quickened or advanced in her defects," for these are in fact "so many essays, whereby men do naturally attempt to restore themselves from the first general curse inflicted upon their labors." Wilkins' scientific writings are all of a popular nature, written not for the learned, but for "such common artificers, as are well skilled in the practice of these arts, who may be much advantaged by the right understanding of their grounds and theory." For this reason he wrote in English, referring on the authority of Ramus to the German practice of public lectures given in the vernacular, "for the capacity of every unlettered ingenious artificer."[16] Though he defended the universities on several occasions, Wilkins was aware that they must justify their teaching in terms of real use and benefit to mankind, a view that made him one of the principal advocates of university reform at Oxford and Cambridge.[17]

Wilkins' scientific writings constitute a single, well-conceived educational program to reach a larger audience outside the confines of traditional learning, both to promote natural philosophy and to lend dignity to the practical arts. He announced this program in the opening of his first publication, saying that it was his desire to "raise up some more active spirit to a search after other hidden and unknown truths: since it must needs be a great impediment unto the growth of sciences, for men still to plod on upon beaten principles, as to be afraid of entertaining anything that may seem to contradict them."[18] In this task of popular education, Wilkins' importance can hardly be overestimated. He laid the foundation for the wide participation and interest that the Royal Society enjoyed during its formative years.

The means of this success was pedagogical flair, shown both in his capacity for clear and interesting exposition, always without any suggestion of condescension, and in the choice of subjects, which in the context of the times were sensational. Was the moon inhabited? Could man find a means of flying to it? Was it much like the earth with mountains and oceans? Was the earth a planet? Could man navigate under water, lift heavy weights with little effort, or communicate effectively by other means than ordinary speech? The very titles were catchy—he did not shun the title *Mathematical Magick*, although it was certainly against his principles to suggest that there was any magic in the study of natural philosophy.[19] His more serious purpose was to gain acceptance for the new science, to bring the work of Copernicus, Kepler, Galileo, Gilbert, Mersenne, and others to the attention of his countrymen. Against the authority of the Bible, antiquity, and book learning, he answered that "we must labor to find out what things are in themselves, by our own experience, and a thorough examination of their natures, not what another says of them." Natural religion will prevail; disorder, strife, and sectarianism will vanish when disputes are resolved by giving "soft words but hard arguments."[20] There is no important principle in Thomas Sprat's *History of the Royal Society* that had not earlier been argued by Wilkins. "The universal disposition of this age," wrote Sprat, "is bent upon a rational religion." In his first work Wilkins said that the opponents of new views too often submitted to authority, a point he enforced by saying that "our opposites . . . too often do *jurare in verba magistri*," thus citing the well-known line in Horace from which the Royal Society drew its motto *Nullius in verba*.[21]

There is, finally, another aspect of Wilkins' character that bears some relation to his career and influence: unlike most of his scientific and ecclesiastical associates, he was a man of the world. After their first meeting, Robert Boyle remarked that Wilkins' "entertainment did as well speak him a courtier as his discourse." Anthony à Wood observed that Wilkins was "bred in the court, and was a piece of a traveller, having twice seen the prince of Orange's court at the Hague, in his journey to, and return from, Heydelburg, whither he went to wait upon the prince elector palatine, whose chaplain he was in England."[22] Without such social attainments, Wilkins' sphere of activity would hardly have reached so far beyond his humble origins.

Early Career. Wilkins was born at the Northamptonshire house of his maternal grandfather, the puritan divine John Dod, who was known for an exposition of the Ten Commandments. His mother, Jane Dod, had four children in her marriage to Walter Wilkins, an Oxford goldsmith who died in 1625. John Aubrey reports that the father was "a very ingenious man with a very mechanical head. He was much for trying experiments, and his head ran much upon the perpetual motion." In a second marriage, to Francis Pope, Jane Dod had a son, Walter, who remained close to Wilkins.[23]

After schooling at home, Wilkins began grammar school at the age of nine under the noted Greek and Latin scholar Edward Sylvester, and in May 1627 he matriculated at New Inn Hall, Oxford (later united with Balliol College). He soon transferred to Magdalen Hall, where his tutor was the Baptist divine John Tombes. He graduated B.A.

20 October 1631, and gained the M.A. degree on 11 June 1634; at this time Wilkins was tutor in his college, one of his students being Walter Charleton, who thereby "profited much beyond his years in logic and philosophy."[24] A few years later he was ordained and became vicar of Fawsley. At this time he is reported to have become chaplain to William Fiennes, first viscount Saye and Seale, who was then a supporter of the Puritans and later sat in the Westminster Assembly. But in 1641 Wilkins dedicated his *Mercury* to George Lord Berkeley (1601–1658), signing himself "your lordship's servant and chaplain." His desire to move in high places was further gratified when he became chaplain to Charles Louis, the prince elector Palatine, the king's nephew. The elector lived in England during a good part of the 1640's, befriending the parliamentary party in the hope of securing the restitution of his lost possessions. During the early months of 1646, Wilkins was officially engaged as preacher at Gray's Inn; during these years he also preached at the Savoy.[25]

On 13 April 1648, the Parliamentary Visitors made Wilkins warden of Wadham College. The holder of this office was required to take the degree of doctor of divinity, but on 5 March 1649, the Visitors gave him a year's dispensation, since Wilkins was "at this time in attendance on the prince elector, and cannot in regard of that service have time to do his exercise, and all other things necessary unto that degree."[26] He took the degree on 19 December the same year. Since this occurred at the time when Charles Louis was returning to Heidelberg to take possession of the lands that had been restored to him as a consequence of the Peace of Westphalia, we may surmise that it was at this time that Wilkins made his visits to the Continent and to The Hague.[27]

Beyond these sparse facts, we have little information about Wilkins' life during his formative years. No doubt he spent most of them in Oxford and London. It was in London that he participated in the meetings that were devoted, as John Wallis recorded, to "what has been called the New Philosophy or Experimental Philosophy," these meetings having been convened at the suggestion of Theodore Haak. It is an interesting conjunction that they began during the Westminster Assembly, of which Wallis was then secretary. For a better view of Wilkins' early career, we have his writings and some reasonable conjectures about his associations.

Although published two years apart, the *Discovery* (1638) and the *Discourse* (1640) can be considered a single work. Addressed to the common reader, the primary aim was to make known and to defend the new world picture of Copernicus, Kepler, and Galileo by showing its agreement with reason and experience against subservience to Aristotelian doctrines and literal biblical interpretation. Kepler and especially Galileo's *Siderius nuncius* (1610) and Matthias Bernegger's Latin translation (1635) of the *Dialogue Concerning the Two Chief World Systems* are frequently cited, along with a wealth of other references from the literature that had appeared within the last generation. The work is polemical, but unlike Campanella's *Apologia pro Galileo* (1622), which is cited with approval, it constantly turns the reader's attention to the positive arguments that may be drawn from rational interpretation of observable phenomena. The central argument was borrowed from Galileo: the moon is not a shining disk or whatever else men have imagined, but a world with natural features much like the earth. And if so, then the moon might also be inhabited, although Wilkins does not find sufficient grounds to say what sort of beings the inhabitants are, thus neatly avoiding the touchy question of whether they are descendants of Adam. Further, if the moon shares natural features with the earth, then the argument could be extended to form a uniformitarian view of the constitution of the entire universe, thus breaking down the Aristotelian doctrine of fixed, hierarchical spheres that obey laws other than those of the sublunar world. In both the first and the second work, Wilkins is careful to warn the reader at the outset that he is not pretending to write a precise treatise expounding unquestionable truths; but though much might still be doubtful, he is confident that the hypotheses he defends will, against all prejudice, be granted conformity with observable phenomena and with simplicity of explanation. In the 1640 edition of the *Discovery*, Wilkins added the sensational idea that it might be possible to contrive a way of flying to the moon, thus taking up a suggestion already known in England from Francis Godwin's *Man in the Moone* (1638). In the latter part of the second work, Wilkins supports his argument for the movement of the earth by reference to William Gilbert's suggestion that the earth is a lodestone. Bacon had argued against Gilbert on that point. Both works make few and only general references to Bacon, quite insufficient to attribute any important inspiration to him.

The *Discovery* and the *Discourse* have a wealth of references to recent literature—at least some thirty in each, of which nearly a dozen are new in the second work. They suggest that Wilkins found his occasion in the controversy that grew up in the wake of Philip van Lansberge's *Commentationes in motum terrae, diurnum et annuum* (1630). This work was opposed by Libertus Fromondus both in *Anti-Aristarchus, sive orbis-terrae immobilis* (1631) and in *Vesta, sive Ant-Aristarchi vindex adversus Jac. Lansbergium* (1634), in which he defended the proscription of Copernican doctrine first issued by the congregation of cardinals in 1616 and reiterated in 1633. Fromondus was Wilkins' chief anti-Copernican opponent in both works; only the second work contains Alexander Ross's *Commentum de terrae motu circulari* (1634), which opposes both Lansberge and Nathaniel Carpenter. With a wide and mature command of the literature, Wilkins was engaged in international controversy. There can be no doubt that he succeeded in his aim of gaining acceptance for Copernicus, Kepler, and Galileo in England.[28]

We may wonder why Wilkins, still only in his middle twenties, took up the controversy with so much energy and conviction. In the *Discovery*, the "Epistle to the Reader" states that the work is "but the fruit of some lighter studies," finished in a few weeks; but the extensive reading adduced in both works could hardly have been so quickly mastered. The subject must have required longer preparation, perhaps during his student days and while he was tutor in his college. Henry Briggs, who died in 1630, was the first Savilian professor of geometry; in London he had been close to William Gilbert and Edward Wright, and in Oxford he became acquainted with John Pell and Theodore Haak, who was in Oxford during the later 1620's. Briggs was a strong Copernican and scorned astrology as "a system of groundless conceits," a view that was shared by his Savilian colleague in the astronomy chair, 1621–1643, John Bainbridge, who in London had belonged to the circle of Briggs and Nathaniel Carpenter. Both had been professors at Gresham College before coming to Oxford. It seems reasonable to assume that Wilkins had learned something from either or both of these men, who most closely illustrate the interest and orientation that characterized his career from the beginning.

Only a year later, in 1641, Wilkins published another book on a popular subject, entitled *Mercury, or the Secret and Swift Messenger, Showing How a Man May With Privacy and Speed Communicate His Thoughts to a Friend at Any Distance*. It mentions such old tricks as baking secret messages into loaves of bread, but Wilkins' chief interest was cryptography, of which he gives a wealth of examples, all ready for use. But he also deals with cryptology or secret communication by speaking, either by involving the sense in metaphors and allegories or by changing old words or inventing new ones as is done by thieves, gypsies, and lovers; and with "semeology," that is communication by signs and gestures, as used for instance by deaf-mutes. Thus *Mercury* is not merely a practical guide in the use and decoding of ciphers, but a broadly based discussion of the means of communication, or what today would be called semiotics. The opening chapter states the basic principle that men are born with a natural ability to communicate, capable of learning any language in the same manner as they can master "other arts and sciences"; but men are not born with a single language that is natural to all mankind, for if this were so men would retain it so that all men would have a "double language, which is evidently false." In other words, like Mersenne, Wilkins rejected the natural-language doctrine then advocated by Robert Fludd. Wilkins ridiculed cabalistic interpretations of the sort that was again to occupy him in controversy with John Webster, who attacked the universities for neglecting Jacob Boehme's mystical linguistic doctrines. At the same time, Wilkins saw that the Babelistic multiplicity of languages was a great hindrance to the promotion of arts and sciences, men now wasting much time merely learning words instead of addressing themselves directly to the study of things. Citing such well-known instances as Arabic numerals, astronomical and chemical signs, and musical notes, he devoted a chapter to the possibility of creating a universal character as a remedy for the confusion. It outlines the principles he was later to follow in his final work. At the end of *Mercury*, Wilkins notes that though his work can be used to serve unlawful purposes, it can also be used to uncover them. If the abuse of useful inventions is a reason for suppressing them, he observes, "there is not any art or science which might be lawfully professsed."[29]

After dealing with communication and the second curse on mankind in *Mercury*, Wilkins next turned to the remedies for the first curse, inflicted upon man's labors. This pattern shows how closely Wilkins, with most of his contemporaries, related his concerns to the biblical story of man's terrestrial life. His *Mathematical Magick* (1648) is divided in two parts: "Archimedes or Mechanical Powers" and "Daedalus or Mechanical Motions." These titles might suggest an emphasis on the theoretical problems that had occupied much of the literature on mechanics during the previous generation, but the work is almost wholly devoted to the practical uses of mechanical devices with only enough theory to give the reader a sense of scientific initiation and understanding. The address "To the Reader" explains that the present work forms part of the same educational efforts as Wilkins' previous publications by showing how "a divine power and wisdom might be discerned, even in those common arts which are so much despised." The book's aim was "real benefit," both for gentlemen in the improvement of their estates, as in the draining of mines and coalpits, and for "common artificers" in gaining a "right understanding of the grounds and theory" of the arts they practice. It is therefore a short book, a compendium of knowledge otherwise only available in large, expensive volumes in Latin rather than the vernacular, "for which these mechanical arts of all other are most proper."

The first part deals with the balance, lever, wheel, pulley, wedge, and screw in that order, all illustrated with line drawings and pictures. Then follow chapters that show how the combination of these devices may produce "infinite strength" so as to "pull up any oak by the roots with a hair, lift it up with a straw, or blow it up with one's breath," all illustrated with rather sensational pictures. The second part treats a miscellaneous collection of strange devices and possibilities, such as flying machines, moving and speaking statues, artificial spiders, the imitation of sounds made by birds and man, a land vehicle driven by sails, a submarine, Archimedes' screw, and perpetual motion. This is a strange, almost baroque assembly, but all of these subjects had already been discussed in the extensive literature on which Wilkins drew and a few years later a speaking statue was among the wonders shown to visitors at Wadham College. Automata were a legitimate scientific interest. There is little theory here, even scant hope of practical success, but much excitement. Learned fancies were being shared with a lay audience. It would be a mistake, however, to think that Wilkins was being frivolous. Even in the 1660's the Royal Society was not averse to the pursuit of such projects. There was as yet no clear distinction between what we consider good science and technology as opposed to fruitless speculation. The same scientific success that brought about the disenchantment of the universe also raised technological hopes that entered the realm of magic. Wilkins knew that wonder is the chief impulse to serious study and experiment.

A closer look at the sources of *Mathematical Magick* yields interesting information both about Wilkins' orientation and about the dating. It can easily be seen that many of the line drawings and illustrations are taken from other works along with the principles and devices they illustrate. The most recent work cited is John Greaves's description of the Egyptian pyramids, *Pyradomographia* (1646). But the works on which he chiefly relied were Guidobaldo del Monte's *Liber mechanicorum* (1577) and Marin Mersenne's *Cogitata physicomathematica* (1644).[30] The use of Mersenne is much too extensive to have been introduced in a late revision; if therefore we take seriously Wilkins' statement in the dedication to Charles Louis that "this discourse was composed some years since, at my spare hours in the university," we must conclude that he devoted a good part of his time to university affairs during the mid- and late 1640's, a fact that may explain his sudden appointment to the wardenship of Wadham in 1648. Yet those affairs left him time to write the book, perform his official preaching duties in London, attend the early scientific meetings there, and serve as chaplain to the elector. Wilkins clearly managed his diverse functions with considerable energy.

Wilkins' explanation and illustration of the six traditional mechanical devices relied chiefly on Guidobaldo; a mere visual comparison of the handsome pages of the *Liber mechanicorum* with Wilkins' modest book makes this dependence obvious. Following Pappus, Guidobaldo had reduced all these devices to the same working principle as the lever—with the exception of the wedge, which he also discussed in terms of the inclined plane without making a clear choice between the two. Wilkins altogether omitted the inclined plane, but did not reduce the wedge to the lever principle as he did for the balance, wheel, pulley, and screw, presumably because he did not wish to burden his lay readers with the finer points of theory in a work which in any event limited to the barest minimum the mathematical principles offered by his sources.[31] In the order of the six devices, however, Wilkins followed Mersenne by treating the wheel before the pulley, but he did not use Mersenne's somewhat more complicated analyses. Thus the reader of *Mathematical Magick* would not have gained a sense of the long controversy over the proper understanding of these devices, revived in 1634 by Mersenne's *Les méchaniques de Galilée*.[32] From Mersenne, Wilkins also borrowed his account of the "glossocomus" or "engine of many wheels," with the analysis and illustration that show how it works like a series of interlocking levers.[33] In addition he cited works other than the *Tractatus mechanicus* from the *Cogitata*: on the bending and power of bows,[34] on the flattening of a bullet fired against a wall,[35] and on the submarine.[36]

Wilkins' debt to Mersenne is so heavy that it deserves closer attention. Mersenne is cited in the *Discovery*, the *Discourse*, and in *Mathematical Magick*. He is not mentioned in *Mercury*, but the general subject of this work forms the very core of Mersenne's own enquiries: the phenomena of communication, language, and the possibility of creating a philosophical language. It would be correct to say that Wilkins' scientific writings together present a popular version of Mersenne. The affinity of interests and orientation was too close to stem from common reliance on the same literature. The plurality of worlds was the only subject that separated them, but for Wilkins this was only a tentative suggestion of no systematic importance, confined to the *Discovery* and not repeated. Mersenne's position on the Copernican doctrine was sufficiently ambiguous not to create any problem.[37]

Mersenne and Wilkins shared the conviction that religion and morality have a rational basis, that the grounds of religious belief are not tied to the retention and defense of Aristotelian doctrines, that a rational explanation of nature is possible when firmly based on sense experience and experiment, that this explanation would be mechanical and quantitative, that man is essentially different from the animals by virtue of possessing reason, that man alone is capable of language and communication, and that the growth of knowledge is a function of communication. Both were opposed to magic and the irrational, and for this reason they opposed the belief in the magical and occult powers of words, a doctrine then chiefly associated with Jacob Boehme and Robert Fludd. Language is not part of nature, it can tell us nothing about the essences of things, and thus cannot give "real knowledge" about the things of creation. It is conventional and man-made—"a man is born without any of them, but yet capable of all," Wilkins said. If this were not so, then it would not be possible to maintain that reason and experience together form the exclusive source of scientific knowledge. Thus the nature of language is the crucial problem in the epistemology of the new science. This fact explains some evident similarities between Mersenne, Wilkins, and Locke; as Mersenne felt bound to engage in a sustained critique of Fludd, so Locke argued against Boehme and his English disciple John Webster with his doctrine of "innate

notions."[38] On these grounds Mersenne repeatedly argued that only God can know the essences of things and their true causes. Like Locke, he was convinced that certainty cannot be achieved in physics, "for we do not know the true reason of the effects we clearly see, and which we submit to our uses."[39]

Wilkins stated the same principle in 1649: "In our natural enquiries after the *efficient* causes of things, when our reason is at a stand, we are fain sometimes to sit down and satisfy ourselves in the notion of occult *qualities*, and therefore much more should be content to be ignorant of the *final cause* of things, which lie more deep and obscure than the other."[40] On this central doctrine, Mersenne and Wilkins disagreed with Bacon's goal of penetrating into "the nature of things." This principle severely limits the extent to which Bacon can be said to have guided and informed the new science in England. Bacon in fact played a small role in Wilkins' thought, in no way comparable to Mersenne's role. Mersenne and Wilkins also admired Gilbert on points that Bacon did not accept. As *Mathematical Magick* shows, Wilkins also followed Mersenne in taking an interest in automata; they focused attention on interesting problems. In all their conduct and affairs, both Mersenne and Wilkins showed admirable openness and tolerance, of men as well as of opinions. In spite of the dramatic outward differences of their lives, they offer a beautiful example of the unifying, even irenic effect of the new science, in accordance with their mutual aim.[41]

If with Wilkins' contemporaries we grant that he was the chief promoter of the new science in England—not only by virtue of his writings, but also owing to his personal encouragement of individuals and his success in the shaping of scientific organization before and after the official formation of the Royal Society—then his alliance with Mersenne has far-reaching consequences for the belief that the Rosicrucian enlightenment was the seed-bed of the sort of natural philosophy that it was the aim of the Royal Society to promote. No attempt to assess Wilkins' importance can ignore these problems. Fludd and Mersenne do not go together. The groups they represent are not separated by their interest in a philosophy of nature, but they are set apart by their basic methods and principles, and it is this latter criterion that is crucial. Neither does one owe anything to the other regarding the need for formal cooperation and exchange of knowledge in a college (whether invisible or not) or an academy, for this need had been advocated by Mersenne as early as 1623; it was met by Théophraste Renaudot's conferences as early as 1629 and by Mersenne's own Academia Parisiensis at least by 1635. The ubiquitous presence of Hartlib and others shows nothing except a shared interest in natural philosophy and its results, although this presence has been the chief prop of the Rosicrucian argument. The wide tolerance of men like Mersenne and Wilkins should not be construed to mean positive approval. It has been argued that Continental influences reached England through The Hague, owing to the presence there of the exiled Queen Elizabeth of Bohemia, who for well-known reasons made some political use of such men as Hartlib and John Dury (Durie) as well as their contacts with circles that may, at least in part, be called Rosicrucian. In these matters the queen relied heavily on the services of the roving ambassador Sir Thomas Roe. On these grounds it has been argued that John Wallis' account of the first London scientific meetings in 1645 "seems to give a curiously 'Palatinate' coloring to the origins of the Royal Society."[42]

The weakness of this argument is obvious: it ignores the fact that The Hague was the home of a very different intellectual group that had lively contacts with London. It was through these contacts that Mersenne became more widely known in England. During these years, from 1633 until his death in 1649, the English ambassador at The Hague was Sir William Boswell, whose chief business of course was not with the exiled Palatinate queen, but with the court of the House of Orange. A strong royalist and a Laudian, he was successful in preventing Dutch intervention in the Civil War during the 1640's. At the center of this group in The Hague was Constantijn Huygens, whose political, cultural, and intellectual importance is well known. Huygens' correspondence shows that he was on intimate terms with Boswell,[43] and they shared many scholarly interests, including musicology. As secretary to Prince Frederic Henry of Orange, Huygens was Boswell's main contact with the court. He corresponded with both Descartes and Mersenne, as did Boswell although those letters are lost. Huygens regularly transmitted mail from Mersenne in Paris to recipients in Holland, including Descartes; Boswell occasionally did likewise. Between mid-summer of 1639 and August 1640, Boswell lived in London, and it was during this period that Haak initiated his lively correspondence with Mersenne at the encouragement of Boswell, "with whom Haak seems to have enjoyed a long-lasting and close acquaintance," beginning in 1638.[44] As was to be expected, it is evident that the contents of Mersenne's letters became widely known in London, just as these contacts were in part responsible for Mersenne's close English ties during the early 1640's.[45]

Having already cited Mersenne in his first two publications, Wilkins may have written *Mercury* on a hint from Mersenne transmitted through Haak. At the beginning of this book, Wilkins tells the reader that it was occasioned by a reading of Francis Godwin's *Nuncius inanimatus, or The Mysterious Messenger* (1629), which he had mentioned in the *Discovery*. It is tempting to think that his renewed interest in speedy and secret communication was related to the fact that Haak had sent Mersenne a copy of Godwin's little book, soon receiving the well-founded judgment that it "was indeed very animated because it teaches us nothing, saying not a word about its secret of communication. What is the use of writing, 'I know such and such things,' but not tell; that is to make fun of the readers."[46] In line with this critique, Wilkins' purpose in *Mercury* was precisely to remove linguistic mystification and the secrecy of ciphers by bringing the technique out in the open. It is no wonder that Wilkins kept informed about Mersenne, so that soon after its publication in 1644 he made the *Cogitata physico-mathematica* the main source of his *Mathematical Magick*. It was at this time, in 1645, that Haak called the first London meetings, which not only discussed scientific subjects but also performed experiments. Wallis' list of the topics shows no Rosicrucian inclination, and the meetings themselves were most likely suggested by the success of Mersenne's Academia Parisiensis.[47] It was the group around Huygens and Boswell at The Hague that exerted a decisive influence in England. The chief foreign vehicle of this influence was Mersenne, its chief beneficiary was Wilkins. The Royal Society is in large measure the record of the nature and success of this influence.[48]

The Oxford Years. In 1648 Wilkins entered upon the second stage of his career. Oxford had come under increasingly severe strains during the 1640's. College finances were in disarray, new admissions dropped precipitously, teaching duties were only fitfully performed, and the academic community was torn into factions aligning royalists and men of the old stamp against Parliamentarians, feuding over religious observances, the inviolability of college statutes, the curriculum, the proper conduct and morals of students and teachers, and even proper modes of personal appearance and attire. This situation was intensified by the frothy presence of extreme Anabaptist agitators who acknowledged no authority but their own private revelations. The crisis came to a head after the victorious Parliamentary forces under Fairfax entered the town. On 1 May 1647, Parliament passed an ordinance which empowered a committee to look after "the better regulating and reformation of the University of Oxford, and the several colleges and halls in the same, and for the due correction of offences, abuses, and disorders, especially of late times committed there."

Within the next year the Parliamentary Visitors came to Oxford, ejected the old warden of Wadham College, and appointed Wilkins, who took charge on 13 April 1648. It proved a wise choice. At the young age of thirty-four, he must have impressed the authorities by his accomplishments in the university and in his varied public offices as well as by his forceful advocacy of new learning, his moderation in religious affairs, his energy, and his extensive connections. Under the guidance of a man who was not considered a bigot, the college admissions soon rose steeply, including a large number of country gentlemen and "cavaliers," a fact that may also have helped improve the finances. It is universally acknowledged that Wadham was a distinguished college during Wilkins' wardenship. Among the new fellows of Wadham who came to Oxford from Cambridge were Seth Ward and Lawrence Rooke, "who was much addicted to experimental philosophy." They were joined by other men migrating from London and the scientific meetings there to continue their work in Oxford. They met at various places, including Wadham, where Wilkins created a laboratory. They included the nucleus of the future Royal Society: John Wallis, Jonathan Goddard, William Petty, Ralph Bathurst, Thomas Willis, and Robert Boyle, to whom Wilkins wrote on 6 September 1653: "I should exceedingly rejoice in your being stayed in England this winter, and the advantage of your conversation at Oxford, where you will be a means to quicken and direct our enquiries." Not long after, Boyle took up residence in Oxford.[49] The meetings were also attended by some of the able students who came to Wadham. The most brilliant was Christopher Wren, Wilkins' special protégé in his early career. Among the others were Wilkins' half-brother Walter Pope, Thomas Sprat, William Lloyd, William Neile, and Samuel Parker.

These men and their activities created an air of modernity and intellectual excitement in the university which suited Wilkins' desire to introduce the new philosophy in a manner that at the same time demanded discipline and significant achievement. He would hardly have been disturbed that his circle was in low repute among the Aristotelians, Galenists, and "those of the old stamp, that had been eminent for school and polemical divinity, and disputations and other polite parts of learning, [who] look upon them very inconsiderably, and their experiments as much below their profound learning and the professors of them."[50] This was precisely what reform was about and why so many sought Wilkins' advice and encouragement. When Oldenburg in the spring of 1656 settled in Oxford for a while, he was glad to find lodgings near Wilkins and Wadham, waxing poetic in his description of the new garden's "design and cultivation, where pleasure rivals utility and ingenuity industry."[51] Created at no small expense, the expansion and layout of this formal garden was one of Wilkins' first innovations. It was exquisitely executed with various mechanical wonders, a Doric temple, and, on a mound, a statue of Atlas carrying the world on his shoulders. The garden shows a characteristic aspect of Wilkins' knowledge and orientation, as does his fondness for music.[52] When the warden's friend, the royalist John Evelyn, visited Wadham in July 1654, he was fascinated by the curiosities he was shown. There were not only scientific instruments, but also a "hollow statue which gave a voice and uttered words" and transparent, elaborately adorned apiaries built in the shape of castles and palaces, but constructed so as to make it possible to take out the honey without destroying the bees.[53] In those days science and ingenuity were visual. While still at the Westminster School, Robert Hooke received a copy of *Mathematical Magick* as a gift from the author; and when a few years later he became a student at Oxford, he attended the scientific meetings and sought Wilkins' advice on his experiments on the art of flying and the making of artificial muscles.[54] Ten years later Hooke concluded the preface to *Micrographia* with an eloquent tribute to Wilkins, describing him as many must have seen him during those years:

There is scarce any one invention, which this nation has produced in our age, but it has some way or other been set forward by his assistance. . . . He is indeed a man born for the good of mankind, and for the honor of his country. In the sweetness of whose behavior, in the calmness of his mind, in the unbounded goodness of his heart, we have an evident instance, what the true and the primitive unpassionate religion was, before it was soured by particular factions. . . . So I may thank God, that Dr. Wilkins was an Englishman, for wherever he had lived, there had been the chief seat of generous knowledge and true philosophy.

In the midst of this busy life, Wilkins was also a member of several influential university committees, including the delegacy to which the governance of the university was entrusted by its chancellor, Oliver Cromwell, on 16 October 1652. In this work, Wilkins successfully sought to regain for the university and the colleges their lost autonomy, to mediate between contending factions, and to maintain order and discipline. He especially defended the university against the attacks of radical religious factions, both on the governance of the university and its curriculum. One such attack was Webster's *Academiarum examen* (1654), which Wilkins and Ward answered the same year in *Vindicae academiarum*. It opened with a letter by Wilkins, outlining and rejecting the three main charges. Contrary to Webster's accusations, the university was not a slavish follower of Aristotle but freely opposed him "as any contrary evidence does engage them, being ready to follow the banner of truth by whomsoever it shall be lifted up." Further, the university did not presume to teach what can proceed only from the spirit of God as Webster had charged. And it did not intend to direct its teachings according to the mystical linguistic doctrines of Boehme and "the highly illuminated fraternity of the Rosicrucians." Webster's trust in these authorities, said Wilkins, "may sufficiently convince what a kind of credulous fanatick reformer he is like to prove." Wilkins remained committed to the principles he shared with Mersenne.[55]

There appears to be good reason to accept Tillotson's assessment of Wilkins' achievement in the life of the university: "It is so well known to many worthy persons yet living, and has been so often acknowledged even by his enemies, that in the late times of confusion, almost all that was preserved and kept up of ingenuity and learning, of good order and government in the University of Oxford, was chiefly owing to his prudent conduct and encouragement."[56]

In the spring of 1656, Wilkins married Cromwell's sister, Robina French, which is said to have strengthened his hand with the Lord Protector in the interests of the university.[57]

Cambridge. In 1659 Wilkins made a sudden change of the sort that energetic men, confident of their powers, are prone to make when they, after success in one place, see an opportunity to apply their talents in new territory. After Cromwell's death, Wilkins had become a close adviser to Richard Cromwell, who appointed him master of Trinity College, Cambridge, "thinking he would be as serviceable in that, as he had been in the other university."[58] He took possession in late summer, resigning from the wardenship of Wadham on 3 September 1659. His tenure lasted barely a year. After the king's return to England in May 1660, Henry Ferne was made master, having successfully pressed a claim on the basis of a promise made by Charles I. The reason given was that the statutes did not allow a married master, but without Ferne's intervention this circumstance would hardly have prevented continuation. In a letter of July 1660, "numerously signed," the fellows of Trinity both offered their congratulations on the restoration and requested the reconfirmation of Wilkins, "appointed at their earnest petition, on the death of Dr. Arrowsmith, in 1658."[59]

During his brief association with Cambridge, Wilkins entered the circle of a group of men with whom he, in spite of some differences, had so much in common that he came to be considered one of them. With the Cambridge Platonists, he shared the outlook that was just then coming to be known as latitudinarianism: a commitment to tolerance and comprehension in church affairs, respect for learning, and the principle that the right understanding of religion, both revealed and natural, is essentially governed by reason. At the time of the Act of Uniformity a few years later, Richard Baxter wrote a succinct description of these men. He divided the conformists into three groups: the zealots, those who submitted for a variety of personal and other reasons, and

> those called latitudinarians, who were mostly Cambridge men, Platonists or Cartesians, and many of them Arminians with some additions, having more charitable thoughts than others of the salvation of the heathens and infidels. . . . These were ingenious men and scholars, and of universal principles, and free; abhorring at first the imposition of these little things, but thinking them not great enough to stick at when imposed.[60]

Wilkins' departure from Cambridge was felt as a loss by many, one of them being Isaac Barrow, whom Wilkins helped to the geometry professorship at Gresham College in 1662, the year before Barrow assumed the Lucasian chair at Cambridge. With an uncertain future behind him, Wilkins now gravitated to London and the culmination of his career as the energetic center of the Royal Society.

The Royal Society and the Last Years. In 1660 began the third and last stage of Wilkins' career.

He did not have to wait long for ecclesiastical preferment. On 28 January 1661, he was again elected preacher at Gray's Inn,[61] and at the end of the year George Lord Berkeley (1628–1698) presented him with the living of Cranford, Middlesex.[62] On 11 April 1662 he became vicar of St. Lawrence Jewry in London, a living that was in the king's gift; thus he soon gained royal favor.[63] During the 1660's, he held a plurality of other ecclesiastical offices until in 1668 he became bishop of Chester.[64] Wilkins preached regularly at St. Lawrence Jewry, but his main sphere of activity was elsewhere.

During the late 1650's scientific meetings were held at Gresham College. After attending a lecture by Wren on 28 November 1660, the group gathered to discuss a plan for the founding of "a college for the promoting of physico-mathematical experimental learning." It is an unmistakable sign of Wilkins' importance that he was on this occasion appointed to the chair; within the next two weeks, Oldenburg wrote that Wilkins had been elected "president of the new English Academy very recently founded here under the patronage of the king for the advancement of the sciences."[65] Wilkins was still styled president in the first months of the new year, but on 6 March 1661 Sir Robert Moray was chosen president, no doubt owing to his close associations with the king, whose favor was eagerly and successfully sought during the first years. The rest is a familiar story. The society gained its first official charter under royal patronage a few years later, many new members joined, and an astonishing and ceaseless round of activities got under way, lasting with undiminished energy until about the time of Wilkins' death in November 1672, when attendance at meetings began to drop off and a state of seeming exhaustion set in, no doubt in part owing to a financial crisis. It is hard to say whether this decline was related to the loss of Wilkins, but the coincidence is striking.[66]

The records of these years show that Wilkins was busier than any other member in the affairs of the society. From the beginning until his death, he was each year reelected to the council, being also one of the two secretaries, another elective office, until he became bishop of Chester. He was occasionally called vice-president, although the statutes made no provision for such an office. While secretary, he attended practically every meeting and at most of them he was busy doing something: providing recent information, proposing experiments, being put in charge of this and that, appointed to special committees, asked for advice, engaged in fund-raising, and preparing suitably interesting doings for the king's visits. He proposed a very large number of candidates for membership, suggested that Robert Hooke be made curator of the collections, and proposed Nehemiah Grew as curator for the anatomy of plants.[67] At the same time he also supervised the writing of Sprat's *History of the Royal Society* (1667).[68] During the plague in the summer of 1665, Wilkins, Hooke, and William Petty removed to Durdans near Epsom in Surrey to carry out experiments on "improved chariots" and other mechanical devices; their results were reported to the society the following year. This was one of the several subjects of *Mathematical Magick* that occupied the society during the 1660's.[69]

At the beginning of 1668, Wilkins once more became involved in church affairs. After the fall of Clarendon, during the closing months of the previous year, the way was open for an attempt to bring at least some groups of nonconformists into communion with the church, a policy Wilkins had long supported in accordance with the promise made by the king in the Declaration of Breda shortly before his return to England. It was also advocated by the duke of Buckingham, now the king's first minister. Richard Baxter was approached, but he found himself unable to accept the initial terms of negotiation and requested instead that "two learned peaceable divines" be nominated "to treat with us, till we agreed on the fittest terms." One of them was Wilkins, who drew up a proposal that was revised during further deliberations. Baxter's detailed account shows that Wilkins was a skillful negotiator who tried his best to find a compromise that would satisfy all parties. This proved impossible, and when it became known that a bill for comprehension was ready, Parliament refused to accept it.[70] But Wilkins had Buckingham's patronage, and when the see of Chester fell vacant in August, he was soon appointed and duly consecrated on 14 November 1668.[71] In a diocese known for its large number of Dissenters, he was as lenient to nonconformists as his predecessor had been severe, many being brought into communion with the church owing to his "soft interpretation of the terms of conformity," while others who did not conform were still allowed to preach.[72] Early in 1669, Pepys heard that Wilkins, "my friend . . . shall be removed to Winchester and be Lord Treasurer." Although he discounted this rumor, he added that Wilkins was "a mighty rising man, as being a Latitudinarian, and the Duke of Buckingham's great friend."[73] In the midst of all his activities during the 1660's, Wilkins had also found time to prepare his greatest work, *An Essay Towards a Real Character and a Philosophical Language*, which with the official imprimatur of the Royal Society was presented to it on 7 May 1668.[74]

The *Essay* is the largest and most complete work in a long tradition of speculation and effort to create an artificial language that would, in a contemporary phrase, "repair the ruins of Babel." On one level a mere universal language would accomplish this aim by removing the obstacle that ordinary languages place in the way of common communication, whether in religion, commerce, or science. The universal use of a single language, for example, Latin, would meet this problem, but as Latin lost ground during the early half of the seventeenth century, especially in scientific writings, the need for other solutions was felt with greater urgency. As knowledge grew, in large measure aided by the introduction of common, conceptual, nonverbal symbols (much like Arabic numerals), there seemed to be new hope for the idea of a different sort of language, generally traced back to Ramón Lull, which would refer directly to what knowledge and thought are about, rather than using the imperfect medium of ordinary languages. There was wide agreement with Bacon that in these languages words were a perpetual source of philosophical error, being "framed and applied according to the conceit and capacities of the vulgar sort."[75]

The traditional model for such a language, often cited in the seventeenth century, was the language Adam spoke when he named the animals in his perfect state of knowledge before the fall. In the cabalist tradition, in Boehme and Fludd, it was believed that this language could somehow be recaptured. It was, for instance, seriously believed by some that it could be found by a sort of etymological distillation from all existing languages of the hitherto hidden but original elements of the Adamic language, on the assumption that this language was Hebrew, that Hebrew was the source of all other languages, and that these elements expressed the natures or essences of things. This was the mystical way, repeatedly rejected by Mersenne as nonsense; only God can know the essences of things.

But granting that man can grasp the order of creation by sense experience and reason, it would seem possible for man to comprehend and codify this knowledge in an artificial language based on the study of things. Within the more limited range of fallen man, this language would be a substitute for the lost Adamic language; if complete, it would express all man's knowledge in a methodical, rationally ordered fashion that mirrored the fabric of nature. It would be philosophical and scientific without error. On the practical level, it could be expressed in written or spoken symbols or both. Unlike a universal language, in which knowledge was still tied to the "cheat or words," to use another contemporary phrase, it would deal directly with things. This, it was hoped, would not only make knowledge easier and quicker to attain; it would cause a vast increase in knowledge.

These hopes were sustained by an optimism for which nothing seemed unattainable, similar to other expectations that strike us as equally chimerical, for instance the perpetuum mobile and the squaring of the circle. During the first half of the seventeenth century, a wealth of texts toyed with the possibility of a philosophical language, most of them on the level of groping speculation which never reached articulate statement of basic principles. In addition to these texts, there were many rumors about men who were working on such projects. They were typically surrounded by great secrecy, and there were several instances of offers to reveal the secret for great sums of money. The philosophical language was the exact equivalent of the philosopher's stone. Leibniz brought more conviction, energy, and intelligence to this problem; yet even he never spelled out its full meaning.[76]

Wilkins based his plan on a few basic principles. He assumed that "as men do generally agree in the same principle of reason, so do they likewise agree in the same internal notion or apprehension of things." Now, if the common notions of men could be tied to common marks, written or spoken, then mankind would be "freed from that curse in the confusion of tongues, with all the unhappy consequences of it." These marks would "signify things, and not words," conjoined "with certain invariable rules for all such grammatical derivations and inflexions, and such only, as are natural and necessary," all contrived so "as to have such a dependence upon, and relation to, one another, as might be suitable to the nature of the things and notions which they represented." Thus the various marks, with their modifications, would follow an ordered and rational analysis of knowledge. The advantage would be immense, for "besides [being] the best way to helping the memory by a natural method, the understanding likewise would be highly improved; and we should, by learning the character and the names of things, be instructed likewise in their natures."[77]

Wilkins decided, somewhat arbitrarily he admitted, on forty basic genera, which with "differences" and "species" would produce the marks that would give an inventory of the world, so to speak. Thus "world" is a genus (in the "effable" language represented by *da*), which by addition of the second difference, denoting "celestial" (with the effable sign *d*) produces the notion "heaven" (*dad*). "Earth" has the same elements, but to it must be added the mark for the seventh species, denoting this "globe of sea and land." This mark is *y*, so that the effable sign for earth is *dady*. As was soon observed by several critics, this entire system was after all closely tied to English words. Yet, postulating that it followed a natural method, Wilkins believed that it could be mastered in one month.[78] This belief reveals something about the *Essay*'s ancestry, for this was precisely the claim being made by mystical projectors, who, however, had the good reason for their claim that they assumed a strict interpretation of the macrocosm-microcosm harmony. For them, once the Babelistic confusion of ordinary words and false concepts was stripped away, man would regain the Adamic nakedness of pure and complete knowledge. With pure intellect thus restored, the need for memory would vanish; the small traces of it still required would be caused by the last imperfections in the system, much as friction cannot be entirely overcome.

The *Essay* was tainted by its ancestry. In *Mercury*, Wilkins had outlined some of its principles, although only for the creation of a universal language. In the *Vindicae academiarum*, having ridiculed Webster's mystical advocacy of a genuinely natural, Adamic language, Seth Ward suddenly, as if unrelated to the subject, had said: "It did presently occur to me, that by the help of logic and mathematics this might soon receive a mighty advantage." He then briefly outlined the plan Wilkins executed. "Such a language as this," Ward said, "where every word were a definition and contained the nature of the thing, might not unjustly be termed a natural language, and would afford that which the cabalists and Rosicrucians have vainly sought for in the Hebrew, and in the names assigned by Adam."[79] The evidence shows that it was soon after and with the help of Ward that Wilkins began work on his philosophical language, as he openly admits in the "Epistle to the Reader" in the *Essay*. In rather awkward fashion Wilkins straddled two traditions that in the minds of most

observers could not be brought together. Mersenne had clearly outlined the plan of such a language, but stayed clear of the mystical implications; and, in the event, he seems not to have had faith in its practicality, although he took an interest in its theoretical aspects, much as he did in automata.[80] In the *Essay* Wilkins also modified his optimistic statements with great diffidence about the entire plan and avowals of its tentative, incomplete execution, inviting the Royal Society to appoint a committee to examine it and make suggestions for its improvement. It was fortunate for his reputation that the *Essay* came at the end of Wilkins' career.[81]

The publication of the *Essay* put the Royal Society in a difficult situation. Written by one of its best-known members, encouraged and published under its auspices, it caused a crisis of prestige. It had been much talked about before publication, and it was soon distributed both in England and on the Continent. Yet none of the scientific members of the society had much, if any, faith in it, with the exception of Hooke, who mastered it and continued to take great interest in it.[82] Following Wilkins' wishes, the society immediately set up a committee to report on the *Essay*, but within the society this committee was never heard from again.[83] It was, however, decided that the society's "repository" under Hooke would be organized according to the *Essay*.[84] In its outward relations, the society talked up the *Essay* with much exaggeration. Thus after Christiaan Huygens had voiced his doubts to Moray, the latter quickly wrote back that the character was easy to master; the king had already done so and everyone was now following his example.[85]

Outside the Royal Society, a group of men (some of whom were fellows) continued to seek to improve and perfect the philosophical language, but with the exception of Hooke, these were men without scientific prestige in the society.[86] Having himself already written on similar plans, Leibniz soon learned about the *Essay*; he admired it greatly, although he still found it short of his own requirements. In 1680 he wrote of this admiration to Haak, but added that something "much greater and more useful could be made of it, insofar as algebraic characters are superior to chemical signs."[87] But so far as the Royal Society was concerned, the *Essay* was quietly forgotten.

The *Essay* did have one important effect; it set John Ray to work on botanical classification. Wilkins had lost all his belongings in the Great Fire of London, including part of the as yet unpublished manuscript of the *Essay*.[88] But eager to finish it, he enlisted the help of Francis Willoughby and John Ray in October 1666. They prepared the zoological and botanical tables. Ray was at the time perhaps Wilkins' most intimate and devoted friend; he immediately went to work, spending much of the next year helping Wilkins, on several occasions spending extended periods with him at Chester. But he admitted at the same time that the project did not suit him.

> I was constrained in arranging the tables not to follow the lead of nature, but to accommodate the plants to the author's prescribed system. . . . What possible hope was there that a method of that sort would be satisfactory, and not manifestly imperfect and ridiculous? I frankly and openly admit that it was, for I care for truth more than for my own reputation.[89]

It is a good question whether Wilkins knew of this criticism, which went to the heart of the matter; the *Essay* did not, as he had intended, follow the "method of nature." After publication, Ray helped Wilkins in amending the tables of natural history, just as he also at Wilkins' request made a Latin translation.[90] Later Ray brought his classifications to a perfection that he had not found it possible to achieve within the system of the *Essay*.[91]

Wilkins was now spending most of his time at Chester, with frequent journeys to London. Suffering from "fits of the stone," he unsuccessfully sought a cure at Scarborough Spa during the summer of 1672. On 10 August 1672, Lord Berkeley, recently arrived from Dublin, was nobly entertained by Wilkins at dinner in the bishop's palace at Chester.[92] On 30 October Wilkins was in London, where he attended, for the last time, a meeting of the Royal Society.[93] But the attacks persisted. Hooke and others administered medication, but to no avail. On 19 November 1672, Wilkins died at the house of John Tillotson, who had married his stepdaughter. At his death he is reported to have said that he was "prepared for the great experiment." The funeral sermon was preached by William Lloyd at the Guildhall Chapel on 12 December; the funeral was attended by a very large crowd, "though it proved a wet day, yet his corpse was very honorably attended . . . there were above forty coaches with six horses, besides a great number of others." He was buried in the church of St. Lawrence Jewry.[94]

In his own time Wilkins' stature and influence were very considerable. He was committed to a policy of tolerance that allowed compromise both in political and ecclesiastical affairs, based on the conviction that natural and revealed religion together with the new science proved a benevolent, providential order which, if rightly understood, ensured that mankind could live happily and peacefully, even prosperously, in this world. For this reason, his influence was divided between such men as Hooke, Boyle, and Ray on the one hand, Tillotson, Stillingfleet, and Patrick on the other. In this sense he shaped the temper of England in the latter half of the seventeenth century and left a significant impression on the eighteenth. His influence was acknowledged by John Ray both in the *Wisdom of God Manifested in the Works of the Creation* (1691) and *A Persuasive to a Holy Life* (1700), with the telling subtitle, "From the Happiness Which Attends It Both in This World and in the World to Come." In science, Hooke's tribute in the *Micrographia* leaves no doubt of Wilkins' importance, although he did not make any direct contribution to science. Even those, like Anthony à Wood, whose party loyalties made them caustic critics of men with similar careers, were sparing in their criticism of Wilkins. The age is full of testimonies that are echoed in Gilbert Burnet's summary of Wilkins' character: "He was naturally ambitious, but was the wisest clergyman I ever knew. He was a lover of mankind, and had a delight in doing good."

NOTES

1. *Of the Principles and Duties of Natural Religion*, 8th ed. (London, 1722), p. 34; 1st ed. (London 1675). It was published from Wilkins' papers by his literary executor, John Tillotson, who in the preface explains that the first 12 chapters (pp. 1–165) were left ready for the press by Wilkins. They constitute the greater part of bk. I, entitled *Of the Reasonableness of the Principles and Duties of Natural Religion*. The rest was put together by Tillotson from "the materials left for that purpose," including all of bk. II, *Of the Wisdom of Practicing the Duties of Natural Religion*. There are two references (pp. 48, 55) to Tillotson's sermon *Of the Wisdom of Being Religious* (1664), but these may be insertions and thus do not necessarily determine the time of composition. William Lloyd's *Sermon Preach'd at the Funeral of the Right Reverend Father in God, John Wilkins, D. D., Late Bishop of Chester* is included.
2. *A Discourse Concerning a New Planet, Tending to Prove, That 'Tis Probable Our Earth Is One of the Planets* (London, 1640), in *The Mathematical and Philosophical Works*, 2 vols. (London, 1802), I, 257. The *Discourse* comprises I, 131–261; it was published anonymously.
3. *Principles and Duties*, p. 17.
4. *Ibid.*, p. 73.
5. *Ibid.*, p. 17. Marin Mersenne had used the same metaphor: "Les Méchaniques peuvent enseigner à bien vivre, soit en imitant les corps pesans qui cherchent tousjours leur centre dans celuy de la terre comme l'esprit de l'homme doit chercher le sien dans l'essence divine qui est la source de tous les esprits." Dedication in *Les méchaniques de Galilée*, Bernard Rochot, ed. (Paris, 1966), p. 14.
6. *Ibid.*, p. 346.

7. *Sermons Preach'd Upon Several Occasions*, 2nd ed. (London, 1701), p. 236, 1st ed. (London, 1677, repr. 1680, 1682). There is a preface by the editor, John Tillotson. The axiom that man is a sociable creature is credited to Aristotle and, as often in Wilkins, supported by reference to the Stoics, especially Seneca.

8. *A Discourse Concerning the Beauty of Providence in all the Rugged Passages of It* (London, 1649), p. 65. Similar references occur in *Sermons*. The text of the 9th sermon (pp. 263–287) is Ecclesiastes 4:9—"Two are better than one." Its opening words call Ecclesiastes "a discourse from the most profound principles of reason and philosophy." Like Isaac Barrow, Wilkins had a marked preference for the Wisdom Books (see H. R. McAdoo, *The Spirit of Anglicanism* [London, 1965], p. 239). The 11th sermon (pp. 327–357) and 12th sermon (pp. 359–390) inculcate public spiritedness and cooperation; the theme of the 13th sermon (pp. 391–427) is moderation, followed by a sermon on the evils of vengeance and wrath.

9. *Mercury, or the Secret and Swift Messenger* (London, 1641; 2nd ed., 1694), in *Mathematical and Philosophical Works*, II, 1. *Mercury* comprises II, 1–87; it was published anonymously.

10. *Mercury*, II, 53; these are the opening words of ch. 13, "Concerning an Universal Character, That May Be Legible to All Nations and Languages."

11. *Sermons*, p. 184. The nature of language and the sociability of man were discussed in one of Théophraste Renaudot's conferences, 21 May 1635, with views that agree with Mersenne and Wilkins. *Recueil général des questions traictées ès Conférences du Bureau d'Adresse*, II (Paris, 1660), 458–463; 1st ed. (Paris, 1636). Wilkins had great influence on prose style, both in scientific discourse and in sermons. This is succinctly pointed out by Gilbert Burnet, *History of his Own Time*, 6 vols., 2nd ed. enlarged (Oxford, 1833), I, 347–348. See also Francis Christensen, "John Wilkins and the Royal Society Reform of Prose Style," in *Modern Language Quarterly*, 7 (1946), 179–187, 279–290, and esp. W. S. Howell, *Eighteenth-Century British Logic and Rhetoric* (Princeton, 1971), pp. 448–502. Wilkins' basic stylistic doctrine is already stated in the last section of *Ecclesiastes, or a Discourse Concerning the Gift of Preaching as It Falls Under the Rules of Art*. This section, "Concerning Expression," says that "obscurities in the discourse is an argument of ignorance in the mind. The greatest learning is to be seen in the greatest plainness. The more clearly we understand anything ourselves, the more easily we can expound it to others. When the notion itself is good, the best way to set it off, is in the most obvious plain expression," 3rd ed. (1651), p. 128; 1st ed. (1646). This was Wilkins' most popular work, often reprinted and steadily expanded, also after his death, having reached at least ten printings and its 7th ed. by 1693.

12. *The Discovery of a World in the Moon, Or, a Discourse Tending to Prove, That 'tis Probable There May Be Another Habitable World in That Planet* (London, 1638), in *Mathematical and Philosophical Works*, I, 19. The *Discovery* comprises I, 1–130; it was published anonymously. The 1640 printing contains chapter 14 on the possibility of flying to the moon. Since 1640, the *Discovery* and the *Discourse* have been published together; there was a 5th ed. in 1684. As Wilkins indicates, the words quoted here are taken from Edward Wright's preface to William Gilbert's *De Magnete* (1600). On the same point, Wilkins also refers to John Calvin's *Commentaries on the First Book of Moses, Called Genesis* (see the translation by John King [Edinburgh, 1847], pp. 84–87, 141, 177, 256). I see no evidence that Wilkins knew Galileo's *Letter* with its closely similar arguments, first published in Italian with Latin translation in 1636. In 1640 Wilkins devoted chs. 3–6 of the *Discourse* (I, 149–203) to the same issue, again citing Calvin (now

including the *Commentary on the Psalms*), many passages from the Bible and the Church Fathers, and also such modern writers as Girolamo Zanchi, Franciscus Valesius, Christoph Clavius, Gaspar Sanctius, and Mersenne. Their religious and scientific allegiances were diverse: Sanctius and Clavius were Jesuits, the latter a friend of Galileo but opponent of Copernican astronomy; Zanchi studied at Padua and died at Heidelberg where he served the Palatine rulers; Valesius was a Spanish physician; Mersenne, often cited by Wilkins, took an ambiguous attitude toward Copernicus, but found no scriptural evidence for a charge of heresy, as Wilkins pointed out in the *Discourse* (I.160). Cf. William S. Hine, "Mersenne and Copernicanism," *Isis*, 64 (1973), 18–32. Zanchi (1516–1590) was a Reformed theologian of pronounced irenic tendencies. His use by Wilkins at this time is noteworthy because he was also, along with especially Hugo Grotius, an authority with William Chillingworth in the *Religion of Protestants* (1638). See the excellent study by Robert R. Orr, *Reason and Authority, the Thought of William Chillingworth* (Oxford, 1967). There are other suggestive similarities between Chillingworth and Wilkins. Thus *Principles and Duties*, p. 27, cites the last section in bk. II of Grotius' *De veritate religionis Christianae* for the very same purpose as Chillingworth in *Religion*, ch. 6, sect. 51.

13. *Discourse*, I, 172.

14. *Ibid.*, I, 138. Cf. *Discovery*, "To the Reader": "It is a false conceit for us to think that amongst the ancient variety and search of opinions, the best has still prevailed." (In the *Mathematical and Philosophical Works* [1802] this "To the Reader" is placed at the front of vol. I, before "The Life of the Author.") Mersenne makes the same point in *Questions inouyes* (Paris, 1634), pp. 144–148.

15. *Mathematical Magick, or the Wonders That May Be Performed by Mechanical Geometry* (London, 1648), in *Mathematical and Philosophical Works*, II, 127,131. *Mathematical Magick* comprises II, 89–260, but the dedication to the prince elector Palatine and "To the Reader" are placed at the very front of vol. I. There was a 4th ed. in 1691.

16. *Mathematical Magick*, "To the Reader." For the other points, often repeated in his writings, see the opening chapters, *ibid.* (II, 91–97); cf. *Sermons*, p. 254.

17. *Sermons*, p. 254.

18. *Discovery*, "To the Reader."

19. *Mathematical Magick*, "To the Reader," points out that the title was suggested by Cornelius Agrippa, *De vanitate scientiarum*, ch. 42.

20. *Discourse*, I, 136–137, 134.

21. *Discovery*, I, 14. The full line in Epistle I, 14, reads *Nullius addictus jurare in verba magistri* ("Not pledged to echo the opinions of any master") but the entire context of lines 10–18 is relevant. It was John Evelyn who suggested the motto. In "Praefatio ad lectorem" of the *Quaestiones in Genesim*, Mersenne had recalled the same Horatian passage for precisely the same purpose, against Aristotelian authority and in favor of our own experience of phenomena; Wilkins cited this work in the *Discovery* and in the *Discourse*. See Robert Lenoble, *Mersenne ou la naissance du mécanisme* (Paris, 1943), p. 224; cf. p. 82.

22. R. E. W. Maddison, *The Life of the Honourable Robert Boyle* (London, 1969), p. 85 (Boyle to Hartlib, 14 September 1655); Anthony à Wood, *Athenae Oxonienses*, Philip Bliss, ed., III (London, 1817), col. 971. Wood's information is also in Walter Pope, *Life of Seth Ward* (London, 1697), p. 29.

23. The information often given that Wilkins was born at Fawsley, Northamptonshire, is not certain; see Barbara J. Shapiro, *John Wilkins 1614–1672. An Intellectual Biography* (Berkeley, 1969), pp. 12–13, 254–255.

24. Wood, *op. cit.*, IV (1820), col. 752. Edward Sylvester also taught Chillingworth.

25. Reginald J. Fletcher, *The Pension Book of Gray's Inn, 1569–1669* (London, 1901), pp. 355–357. There is good reason to accept the explanation that it was Wilkins' "skill in the mathematics that chiefly recommended him" to Charles Louis, "his Electoral highness being a great lover and favourer of those sciences, in which he must needs have been very agreeable to his Chaplain, who was entirely of the same turn and temper." See vol. VI (1756), 4266, in *Biographia Britannica*, 7 vols. (London, 1747–1766); this very full and well-informed article is the best biographical account of Wilkins (it covers pp. 4266–4275 and was most likely the work of Thomas Birch).

26. Montague Burrows, ed., *The Register of the Visitors of the University of Oxford from AD 1647 to AD 1658* (London, 1881), p. 22, Camden Society, n.s. 29.

27. It is not clear whether Wilkins made two journeys during 1648–1649, or whether one of them occurred earlier or, less likely, later. Charles Louis spent most of the years between 1644 and his return (May 1649) in England. In 1644 he was invited to attend the sessions of the Westminster Assembly (Bulstrode Whitelocke, *Memorials* [London, 1732], p. 108). Wilkins was formally accepted by the Assembly on 25 September 1643.

28. It is hard to accept Grant McColley's argument that Campanella's *Apologia* is the main source of both the *Discovery* and the *Discourse*. The reason is not merely that the two writers had little in common except their defense of Galileo, but especially that Wilkins used the important literature published since the *Apologia* (1622), including Galileo's own *Dialogue* in the Latin translation (1635). See "The Debt of Bishop Wilkins to the *Apologia pro Galileo* of Tomaso Campanella," in *Annals of Science*, **4** (1939), 150–168. Campanella, *The Defence of Galileo*, tr. by Grant McColley, in *Smith College Studies in History*, **22,** nos. 3–4 (April–July 1937), intro. McColley, "The Ross-Wilkins Controversy," in *Annals of Science*, **3** (1938), 153–189. All these items have much useful information, although they are committed to a view of conflict between science and religion that is now outmoded. Hartlib's "Ephemerides" indicate that Campanella was in London during 1635. Ross answered Wilkins in *The New Planet no Planet* (London, 1646). The entry on Wilkins in *Biographia Britannica* plausibly suggests that the *Discourse* was not merely a treatise on the new astronomy but written as a defense of Galileo: "It was the first just treatise of its kind, and more effectually exposed the folly and absurdity as well as cruelty of the proceedings in the Inquisition by taking no direct notice of them" (*op. cit.*, p. 4268). It is remarkable that Wilkins' defense on the question of biblical authority uses the same arguments as Galileo in the *Letter to the Grand Duchess*, which was presumably not known to Wilkins.

29. Like his two previous books, *Mercury* cites a wealth of sources, both ancient and modern, with some fifty in the latter category. Among the most important are Johannes Trithemius, *De polygraphia* and *De stenographia*, Hermannus Hugo, *De origine scribendi* (1617), and Gustaphus Selenus, *De cryptographia* (1624), the name is a pseudonym for the learned Duke August of Braunschweig-Lüneburg. In 1630 John Pell had written "'A Key to Unlock the Meaning of Johannes Trithemius' in His Steganography; Which Key Mr. Pell the Same Year Imparted to Mr. Samuel Hartlib." (See Wood, *Fasti Oxonienses*, Philip Bliss, ed. [London, 1815], I, 463.) Like the *Discovery* and the *Discourse*, *Mercury* was published anonymously, but the dedication is signed "J. W." It has five commendatory poems at the front, two of them addressing the author as their friend: Richard Hatton, who entered Magdalen Hall, Oxford, on 7 July 1637; and Richart West, who matriculated at Christ Church, Oxford, on 15 February 1633; both presumably knew Wilkins at Oxford, which adds a little to the sparse information we have of Wilkins' life during those

years. Another poem is by Sir Francis Kynaston, the center of a literary coterie at court, who in 1635 founded Musaeum Minerva, an academy for young noblemen. Wilkins was clearly getting known in wider circles.

30. Among other recent works are Pierre Gassendi, *Vita Peireskii* (1641), A. Kircher, *De magnete* (1643), and Mario Bettini, *Apiaria universae philosophia mathematicae, quibus paradoxa et nova pleraque machinamenta ad usus eximios traducta et facillimis demonstrationibus confirmata exhibentur*, 2 vols. (Bologna, 1641–1642).

31. Wilkins mentions Guidobaldo among his chief sources. An abbreviated version of the *Mechanicorum Liber* is in *Mechanics in Sixteenth-Century Italy*, tr. and annotated by Stillman Drake and I. E. Drabkin (University of Wisconsin Press, 1960), pp. 239–328. It includes, on a reduced scale, the line drawings and illustrations of the original. In the final pages of *Mathematical Magick*, Wilkins discussed Archimedes' screw with reference to Guidobaldo's *De cochlea* (1615). This device also interested Mersenne.

32. The only point on which Wilkins may be indebted to Galileo is the subject "concerning the proportion of slowness and swiftness in mechanical motions" (*Mathematical Magick*, II, 146–148), which shows similarity with chapters 1 and 5 of *Les méchaniques* (see Rochot, ed., pp. 23–25, 32–34), but it is possible that Wilkins could have found this in some other source. In that work Galileo did not deal with the wedge, but explained the rest on the principle of the lever. The Mersenne work in question is *Tractatus mechanicus theoricus et practicus* (96 pp.) contained in the *Cogitata physico-mathematica*, which was ready from the press on 1 April 1644. This collective volume also contains other pieces to which Wilkins refers. Mersenne explained the screw in terms of the inclined plane, the balance and the wheel in terms of the lever, and the pulley and the wedge in terms that combined the lever and the inclined plane. During the 1630's, Descartes also treated these devices in a number of letters to Mersenne (about August-October 1630 and again at greater length on 13 July 1638) (see C. de Waard *et al.*, eds., Mersenne, *Correspondance*, II [1937], 602–620, and VII [1962], 347–375); and in the letter to Constantijn Huygens 5 October 1637 (Descartes, *Correspondance*, Ch. Adam and G. Milhaud, eds. II [Paris, 1939], 31–41). These letters do not all offer the same explanations, but Descartes had a low opinion of Guidobaldo's reduction of the pulley to the lever principle, while Galileo found Guidobaldo the best of all writers on these subjects (see Rochot, ed., p. 77).

33. *Mathematical Magick*, II, 137, 135, 138, 148; cf. *Tractatus*, pp. 39–43. Mersenne's term is *glossocomus*. With the same name, this device was also discussed and explained on the principle of interlocking levers, with illustration, in Bettini, *Apiaria*, I, pt. 4, 31–34, with reference to the source in bk. VIII of Pappus, *Mathematicae collectiones* (1588). This book gave an account of the mechanics of Hero of Alexandria, of which the full text was not known until the late nineteenth century. Pappus attributed the term *glossocomus* to Hero, who is also the source of other terms in the technical vocabulary of mechanics. First published in the late sixteenth century, both his *Automata* and *Pneumatics* were very influential, clearly seen, for instance, in Salomon de Caus, *Les raisons des forces mouvantes avec diverses machines tant utilles que plaisantes. Aus quelles sont adjoints plusieurs desseings de grottes et fontaines* (Frankfurt, 1615). Book I, theorem XVI, on the lifting of heavy burdens by the multiplication of forces, has an illustration that bears a striking resemblance to Wilkins' illustration in *Mathematical Magick*, II, 143. De Caus' garden designs found expression in the garden at Wadham College, for instance the mound with a statue (cf. de Caus, bk. II, problem X; bk. I, problem XII, deals with perpetual motion). In John Bate, *The Mysteries of Nature and Art* (London, 1634), bk. I, "Of Water Works," is a popular ex-

position of Hero's *Pneumatics*, with illustrations from the Italian edition, showing how to make mechanical chirping birds and the like, all subjects that also fascinated Mersenne and Wilkins, who was clearly much indebted to this tradition stemming from Hero. De Caus was active in England and Heidelberg in the early seventeenth century. On de Caus, see C. S. Maks, *Salomon de Caus 1576–1626* (Paris, 1935).

34. *Ibid.*, II, 162; cf. Mersenne, *Ballistica et acontismologia* in *Cogitata*.

35. *Ibid.*, II, 174; cf. *De hydraulicus et pneumaticus phaenomenis*, pp. 149–153.

36. *Ibid.*, II, 188–194, "Concerning the Possibility of Framing an Ark for Submarine Navigations"; cf. *De hydraulicus*, pp. 207–208, and *Tractatus de magnetis proprietatibus*, pp. 251–259. In the former, Mersenne, like Wilkins, referred to the submarine constructed by Cornelis Drebbel, who was also known for his work on other devices, including the perpetuum mobile; the name recurs elsewhere in Mersenne. Already in 1634, Mersenne had asked in question 21 of the *Questions inouyes*, pp. 84–89, "Peut-on faire des navires, et des bateaux qui nagent entre deux eaux." The same work opened with one of Wilkins' favorite topics, "A sçavoir si l'art de voller est possible," a problem that recurs in the *Cogitata* (e.g., *Tractatus mechanicus*, p. 41). It is curious that Wilkins already in the *Discovery* (I, 118) had discussed why a man under water does not feel the weight of the water above him, a subject Mersenne treated in the *De hydraulicus*, pp. 204–206. *Mathematical Magick* (II, 192) credits information about an especially accomplished French diver to a note to *Tractatus de magnetis*, placed in the pagination of *Harmoniae liber*, p. 368 (also part of the *Cogitata*). For the greater part of his career, Drebbel was active in England, where he died in 1633. During his stay in London in the early 1620's, Constantijn Huygens was intimately acquainted with Drebbel's projects and inventions, which were also widely discussed later in the century by Boyle, Wren, and Hooke, in addition to Wilkins. See Gerrit Tierie, *Cornelis Drebbel (1572–1633)* (Paris–Amsterdam, 1932); and L. E. Harris, *The Two Netherlanders Humphrey Bradley and Cornelis Drebbel* (Cambridge, 1961), pp. 119–227.

37. See Hine, cited at end of note 12. Some time around 1660, Isaac Newton took extensive notes from the *Mathematical Magick*; see Frank E. Manuel, *A Portrait of Isaac Newton* (Cambridge, Mass., 1968), pp. 11, 49. The same notebook also has long excerpts from Bate, *Mysteries*, bk. III, "Of Drawing, Washing, Limming, Painting, and Engraving." See E. N. da C. Andrade, "Newton's Early Notebook," in *Nature*, **135** (1935), 360.

38. The term is in Webster's *Academiarum examen* (1654); see Aarsleff, "Leibniz on Locke on Language," in *American Philosophical Quarterly*, **1** (1964), 180.

39. *Questions inouyes*, pp. 69–74, where Mersenne also argues that certainty is possible in mathematics since it deals with quantities, it is "une science de l'imagination, ou de pure intelligence, comme la métaphysique, qui ne se soucie pas d'autre objet que du possible absolut."

40. *Discourse Concerning the Beauty of Providence*, p. 71. Belonging to the year of the king's execution, this sermon argued that, "we may infer, how all that confusion and disorder, which seems to be in the affairs of these times, is not so much in the things themselves, as in our mistake of them" (p. 65); it is characteristic of Mersenne and Wilkins that moral and religious arguments jostle statements of scientific principle. In this text Wilkins often cites the Stoics, especially Seneca.

41. For an excellent introduction to Mersenne, see A. C. Crombie's article in *Dictionary of Scientific Biography*, IX (1974), 316–322.

42. Frances Yates, *The Rosicrucian Enlightenment* (London, 1972), p. 182; cf. p. 183: "We have thus here a chain of tradition leading from the Rosicrucian movement to the antecedents of the Royal Society." See also p. 175 and the reference to H. R. Trevor-Roper there.

43. Boswell has a brief entry in the *Dictionary of National Biography*; there is a much fuller life in *Autobiography of Thomas Raymond and Memoirs of the Family of Guise of Elmore* (London, 1917), G. Davis, ed., pp. 69–80 (Camden Society, third series, vol. 18). Boswell was one of the literary executors of Bacon's estate, possessing among other things the important writings edited by Isaac Gruter, *Francisci Baconi de Verulamio scripta in naturali et universali philosophiâ* (Amsterdam, 1653). In 1651, Gruter published another manuscript in Boswell's possession, William Gilbert, *De mundo nostro sublunari philosophia nova*, often known as "Physiologia nova." Bacon used this work in some of his writings, though without citation. Mersenne knew of this work, writing to John Pell, on 20 January 1640, that Gilbert had written on "Selenography or the geography of the moon, which however has not been published" (*Correspondance*, IX [1965], 52). The most likely source of this information is surely Boswell. Boswell also had a collection of John Dee's papers, some of which he intended to publish himself (C. H. Josten, ed., *Elias Ashmole 1617–1692*, 5 vols. [Oxford, 1966], II, 1242; IV, 1372). This was known to Hartlib, who recorded it in the "Ephemerides" in 1639; he said there and later repeated (see Davies, p. 77) that Boswell attributed "all his proficiency in learning whatever it be, to the goodness" of Dee's Preface to Euclid. There is no compelling reason to believe that respect for that Preface means commitment to cabalistic doctrines; it is perhaps wiser to accept Leibniz' opinion that Edward Kelley was an impostor who abused Dee. Boswell was secretary to Lord Herbert of Cherbury in 1620 while the latter was ambassador at Paris. There are references to Boswell in *De Briefwisseling van Constantijn Huygens* (1608–1687), J. A. Worp, ed., 6 vols. (The Hague, 1911–1917). (These are vols. XV, XIX, XXI, XXIV, XXVIII, XXXII in the series *Rijks geschiedkundige Publicatiën*.) The Mersenne *Correspondance* is of primary importance.

44. Pamela Barnett, *Theodore Haak* (The Hague, 1962), p. 32. Wood, *Athenae Oxonienses*, IV, 280, has an instructive list of Haak's "many great and learned acquaintance," including John Williams, John Selden, Henry Briggs, John Pell, Wilkins, and Boswell, "who encouraged him to keep and continue his correspondence with the learned Mersennus, and others of later time." Wood says of Boswell: "He was a learned man, a great encourager of learning, zealous for the Church of England, faithful in the execution of his embassy, and highly valued by eminent persons" (*Fasti*, I, 332). In the 1640's, Haak and Boswell helped Pell to academic appointments in Holland.

45. In 1639 and 1640, Hartlib's "Ephemerides" show knowledge of the Mersenne-Haak correspondence; during Boswell's stay in London at this time, there is also information about him. For information about Haak, see the entry in *Dictionary of Scientific Biography*, IV (1972), 606–608.

46. *Correspondance*, XI (1970), 412 (to Haak, 4 September 1640). Mersenne also wrote to Haak on other subjects that occur in Wilkins, e.g., universal language, underwater navigation, and flying (XI, 417, 408, 435). On 16 November 1640, he wrote to Haak: "Vous avez raison de dire, que ni Dieu, ni les sciences ne sont point liées aux langues, et en effet, chacune est capable d'expliquer toute chose." This statement expresses both his own and Wilkins' rejection of mystical linguistic doctrines (XI, 420).

47. Christoph J. Scriba, "The Autobiography of John Wallis," in *Notes and Records of the Royal Society*, **25** (1970), 40.

48. During the mid-1640's both Wilkins and Haak, himself a native Palatine, were associated with Charles Louis. Both his mother's and his own letters have been extensively published; the letters give no indication that Rosicrucian influ-

ence could have come from that source, or even that the writers had any interest in it. An informative recent article is G. A. Benrath, "Die konfessionellen Unionsbestrebungen des Kurfürsten Karl Ludwigs von der Pfalz (*d.* 1680)," in *Zeitschrift für die Geschichte des Oberrheins*, **116** (1968), 187–252.

49. Boyle, *Works*, Thomas Birch, ed., 6 vols. (London, 1772), VI, 633; this is one of the few Wilkins letters on record. At this time, Wilkins found a place at Wadham for the instrument maker Christopher Brooke (or Brookes), "purposely to encourage his ingenuity" (see Wood, *Fasti*, I, 403; also E. G. R. Taylor, *The Mathematical Practitioners of Tudor and Stuart England* [Cambridge, 1954], p. 234; this book has a valuable alphabetical collection of brief biographies [pp. 165–307], followed by a list of works in chronological order [pp. 311–441]).

50. Anthony à Wood, *The History of the Antiquities of the Colleges and Halls in the University of Oxford*, John Gutch, ed., 2 vols (Oxford, 1792–1796), II, pt. 1, 633–634. Though seen with a somewhat prejudiced eye, this is one of the chief sources for the history of Oxford in this period, with the relevant material on pp. 501–708. Another important source is Montague Burrows, *Register of the Visitors*. The handiest narrative source is Charles Edward Mallet, *A History of the University of Oxford*, 3 vols. (London, 1924–1926). See esp. vol. II, *The Sixteenth and Seventeenth Centuries* (1924).

51. *Correspondence*, A. R. and M. B. Hall, eds., I, 94 (letter to Edward Lawrence, April 1656).

52. See T. G. Jackson, *Wadham College* (Oxford, 1843), on the gardens (with illustration), pp. 211–212, on music, p. 117; there is an account of a famous musical evening at Wadham in "The Life of Anthony à Wood," in *Athenae Oxonienses*, I (1813), xxxii.

53. *Diary*, E. S. de Beer, ed., III, 105–110 (1–13 July 1654).

54. R. T. Gunther, *Early Science at Oxford*, VI (Oxford, 1930), *The Life and Work of Robert Hooke*, pp. 5–9.

55. The two pieces have been reprinted in Allen G. Debus, *Science and Education in the Seventeenth Century: the Webster-Ward Debate* (London, 1970). In his letter to Ward, Wilkins nearly verbatim repeats some passages from the opening chapter of his *Discourse Concerning the Gift of Prayer* (1651; 9th ed., 1718) on the three gifts requisite in a minister.

56. "To the Reader," in *Sermons*. Tillotson was specifically rejecting some critical remarks in Wood's *Historia et Antiquitates Universitatis Oxoniensis* (1674), which was a Latin version done by John Fell from Wood's English manuscript. Wood was much displeased with this version, both because of its bad Latin and because Fell had taken the liberty of inserting his own comments, of which the depreciation of Wilkins was one. In the late summer of 1654, some of the Wadham fellows made official complaint about Wilkins' conduct of college affairs, but after due consideration the charges were rejected by the Visitors; it is not clear what the issue was. See Burrows, *Register*, pp. 394–397.

57. She had previously been married to Peter French of Christ Church, also a man of some importance in the university. Tillotson married a daughter of that marriage. It is an often repeated error that Wilkins on this occasion gained permission to marry from Cromwell, then chancellor of the university; the Wadham statutes had already been altered in 1651 so as to permit the warden to marry—one wonders whether Wilkins contemplated marriage at that time or whether he was acting on principle. See Jackson, *Wadham College*, p. 116. In June 1670, Wilkins was the only bishop to favor a divorce act, then pending (see Edmund Ludlow, *Memoirs*, C. H. Firth, ed., 2 vols. [Oxford, 1894], II, 503). Robina Wilkins died in 1689; she and Wilkins had no children.

58. Mark Noble, *Memoirs of the Protectoral-House of Cromwell*, 2 vols. (London, 1787), I, 314.

59. *Calendar of State Papers, Domestic*, 1660. In preparation for the appointment, Wilkins had been incorporated doctor of divinity at Cambridge on 18 March 1659.

60. *Reliquiae Baxterianae* (London, 1696), pt. I, p. 386. Baxter especially sought the churches where he "heard a learned minister that had not obtruded himself upon the people, but was chosen by them, and preached well (as Dr. Wilkins, Dr. Tillotson . . .)" (*ibid.*, p. 537). Gilbert Burnet made the same point, counting Benjamin Whichcote, Ralph Cudworth, Henry More, and John Worthington along with Wilkins among "the divines called Latitudinarians." "At Cambridge," he wrote, Wilkins "joined with those who studied to propagate better thoughts, to take men off from being in parties, or from narrow notions, from superstitious conceits, and a fierceness about opinions" (*History of His Own Time*, I, 340). I see no reason at all for the opinion, heard in the eighteenth century and repeated by John Tulloch, that Wilkins "was a Calvinist . . . of a somewhat strict type" (*Rational Theology and Christian Philosophy in England in the Seventeenth Century*, 2 vols. [Edinburgh, 1872], II, 442). The terms "latitude-men," "latitudinarian," and "latitudinarianism" first occurred in the 1660's in a pejorative sense, but were soon adopted as the common term. In 1662, the term was used to refer to the men we call the Cambridge Platonists with stress on the connection between them and the mechanical philosophy. (The generic term Cambridge Platonists did not occur until after the middle of the nineteenth century.) See the pamphlet by S. P., *A Brief Account of the New Sect of Latitude-Men Together With Some Reflections on the New Philosophy*. (S. P. is traditionally identified as Simon Patrick, who was also the first English translator (1680) of Grotius' *De veritate*.) There is an illuminating contemporary account in Edward Fowler, *The Principles and Practices of Certain Moderate Divines of the Church of England (Greatly Misunderstood) Truly Represented and Defended* (London, 1670). Thus this book was published soon after the failure of the bill for comprehension. Fowler calls the latitudinarians "persons of great moderation" and says they are also called "rational preachers" and "moral preachers." He names More, Cudworth, John Worthington, Joseph Mede, and Chillingworth. John Beardmore said that Wilkins "was looked upon as the head of the *Latitudinarians*, as they were then stiled." See "Some Memorials of the Most Reverend Dr. John Tillotson . . . Written Upon the News of His Death [1694] by J. B.," in Thomas Birch, *The Life of the Most Reverend Dr. John Tillotson*, 2nd ed. (London, 1753), p. 390. The term "latitude" is given prominence by Chillingworth: "This Deifying of our own Interpretations, and tyrannous inforcing them upon others; this Restraining of the World of God from that latitude and generality, and the Understandings of Men from that liberty, wherein Christ and the Apostles left them, is, and hath been the only Fountain of all the Schisms of the Church . . . the common Incendiary of Christendom" (*Religion of Protestants*, ch. 4, sect. 16; in this passage Chillingworth cites the agreement of Zanchi). The two chief influences on Tillotson were Chillingworth and Wilkins. Ernst Cassirer's *Die Platonische Renaissance in England und die Schule von Cambridge* (Leipzig–Berlin, 1932), opens with the surprising opinion that the Cambridge Platonists were hostile to the new mechanical philosophy and had little understanding of it. With characteristic misjudgment, R. F. Jones believed that Samuel Parker's *A Free and Impartial Censure of the Platonick Philosophy* (Oxford, 1666) was "a vigorous attack" on the Cambridge Platonists; in agreement with the common use of the term "Platonic" at that time, it was a critique of the chief opponents of the new philosophy, *i.e.*, enthusiasts and Rosicrucians of the sort illustrated by John Webster, whose *Academiarum Examen* Jones, astonishingly, calls "the most important expression of the new scientific outlook between Bacon and the Restoration" (*Ancients and Moderns* [paper-

back ed., 1965], pp. 188, 108). For reliable information and interpretation, see Marjorie Nicolson, "Christ's College and the Latitude-Men," in *Modern Philology*, 27 (1929), 35–53, and McAdoo, *The Spirit of Anglicanism*.

61. Fletcher, *Pension Book*, 435–436.

62. White Kennett, *Historical Register* (London, 1706), p. 576. Wilkins succeeded Thomas Fuller on 10 December 1661. The appointment shows Wilkins' life-long association with the Berkeley family, this George Berkeley being the son of the man to whom *Mercury* was dedicated.

63. Kennett, p. 658; Wilkins succeeded Seth Ward, who became bishop of Exeter.

64. Some of these offices are listed in R. B. Gardiner, *The Registers of Wadham College*, pt. I, 1613–1719 (London, 1889), p. 171.

65. *Correspondence*, I, 406 (Oldenburg to Boreel, 13 December 1660). The term "physico-mathematical" may have been heard before, but it brings to mind the title of Mersenne's *Cogitata physico-mathematica*; was it perhaps Wilkins who had brought in the proposal?

66. Wilkins willed £400 to the Society.

67. Among the candidates Wilkins proposed were Haak, John Hoskins, Francis Willoughby, Edward Bysshe, George Smyth, Thomas Sprat, Henry Power, Henry More, John Ray, and Anthony Lowther; they were all elected; he also proposed Ralph Cudworth, who for some reason never joined. *The Record of the Royal Society of London*, 4th ed. (London, 1940), does not list Cudworth among the members, contrary to statements in the recent literature, e.g., J. A. Passmore, *Ralph Cudworth* (Cambridge, 1951), p. 2; and McAdoo, p. 121.

68. I have dealt with that important function in the entry on Thomas Sprat in *Dictionary of Scientific Biography*, XII (1975), 580–587.

69. Thomas Birch, *History of the Royal Society*, 4 vols. (London, 1756–1757), II, 30, 41, 60, 63, 66, 74, 89. Durdans was the property of Lord Berkeley. On his return from Oxford on 7 September 1665, Evelyn stopped at "Durdans by the way, where I found Dr. Wilkins, Sir William Petty and Mr. Hooke contriving chariots, new rigs for ships, a wheel for one to run races in, and other mechanical inventions, and perhaps three such persons together were not to be found elsewhere in Europe, for parts and ingenuity." Samuel Pepys was interested in the same matter; see entries in his *Diary* under 11 and 22 January 1666. See also letter from Hooke to Boyle, 8 July 1665, in Gunther, *Early Science*, VI, 248.

70. Baxter, *Reliquiae*, pt. III, pp. 23 ff; Burnet, *History*, I, 477. On the Comprehension scheme, see Norman Sykes, *From Sheldon to Secker. Aspects of English Church History 1660–1678* (Cambridge, 1959), pp. 71–75. At this time Sir Matthew Hale and Wilkins "came to contract a firm and familiar friendship," so close that "there was an intimacy and freedom in [Hale's] converse with Bishop Wilkins that was singular to him alone." See Gilbert Burnet, *The Life and Death of Sir Matthew Hale, Kt. Sometime Lord Chief Justice of His Majesty's Court of King's Bench* (London, 1700). Hale was also close to James Ussher and Baxter.

71. Burnet, *History*, I, 464; Evelyn's description in *Diary* under that date. Benjamin Whichcote succeeded Wilkins as vicar of St. Lawrence Jewry.

72. Kennett, *Register*, pp. 815, 817, 921.

73. Pepys, *Diary*, 16 March 1669.

74. In a meeting of the Royal Society on 29 October 1662, "Dr. Wilkins was put in mind to prosecute his design of an *universal language*" (Birch, *History*, I, 119).

75. Aarsleff, "Leibniz on Locke on Language," p. 178.

76. Albert Heinekamp, "Ars characteristica und natürliche Sprache bei Leibniz," in *Tijdschrift voor Filosofie*, 34 (1972), 452; this article is an excellent treatment of the subject. The classic treatment is Louis Couturat, *La logique de Leibniz* (Paris, 1901), esp. chs. 2–5. A briefer discussion is

found in L. Couturat and L. Leau, *Histoire de la langue universelle* (Paris, 1907), with a section on Wilkins, pp. 19–22. Paolo Rossi, *Clavis universalis, arti mnemoniche e logica combinatoria da Lullo a Leibniz* (Milan, 1960), is the best history of the subject. In the literature, both primary and secondary, the *locus classicus* for the philosophical language is Descartes's letter to Mersenne, 20 November 1629 (Mersenne, *Correspondance*, II, 323–339), written in response to a project of which Mersenne had sent him a copy. The subject is often mentioned in the Mersenne correspondence, but unfortunately the notes, usually so informative, attached both to the Descartes letter and others on the same subject are very confused; this confusion has gradually been cleared up in recent volumes. It is an index of the low conceptual level of much recent secondary writing on this popular topic that it fails to make the distinction between a merely universal and a philosophical language; this has made it possible for some to argue that the philosophical language came about by a sort of evolutionary growth of stenography.

77. Wilkins, *Essay*, pp. 20–21.

78. *Ibid.*, pp. 51–52, 398, 454.

79. Debus, pp. 214–216 (original pagination, also given there, pp. 20–22). Ward's basic outline does not state anything that had not been said earlier.

80. For the relevant passages in Mersenne, see references given in notes 50–51 to the entry on Mersenne in *Dictionary of Scientific Biography*, IX, 322. Cf. Lenoble, *Mersenne*, pp. 514–518; Eberhard Knobloch, "Marin Mersenne's Beitrag zur Kombinatorik," in *Südhoffs Archiv*, 58 (1974), 356–379.

81. There are useful illustrations in E. N. da C. Andrade, "The Real Character of Bishop Wilkins," in *Annals of Science*, 1 (1936), 1–12. An informative account is Jonathan Cohen, "On the Project of a Universal Character," in *Mind*, 63 (1954), 49–63. Although weak on the intellectual and philosophical context there is much useful detail on contemporary projects in Vivian Salmon, *The Works of Francis Lodwick. A Study of his Writings in the intellectual Context of the Seventeenth Century* (London, 1972). In a monograph entitled *Zum Weltsprachenproblem in England im 17. Jahrhundert. G. Dalgarno's 'Ars signorum' und J. Wilkins' 'Essay' (1668)* (Heidelberg, 1929; *Anglistische Forschungen*, Heft 69), Otto Funke argued that Bacon was the inspiration for such projects and that Wilkins was in large measure indebted to Dalgarno. Funke does not consider Mersenne. In a useful article on "The Evolution of Dalgarno's *Ars Signorum*," in *Studies in Language and Literature in Honour of Margaret Schlauch* (Warsaw, 1966), pp. 353–371, Vivian Salmon, along somewhat similar lines, argued that "without Dalgarno, Wilkins would never have begun the task which led to the *Essay*" (p. 370); the evidence, including troublesome questions of dating and personal relationships, does not warrant that conclusion. Neither Funke nor Salmon considered the cogent contemporary discussion in Robert Plot's *Natural History of Oxfordshire* (London, 1677), pp. 282–285, which concludes that the question must be left open. Anthony à Wood is the source of the persistent belief that Wilkins cribbed from Dalgarno's *Ars signorum*, saying that the author showed it, before it went to press, to Wilkins, "who from thence taking a hint of a greater matter, carried it on, and brought it up to that which you see extant" (*Athenae Oxonienses*, III, 970; this opinion is repeated in the entry on Wilkins in the *Dictionary of National Biography*). Benjamin DeMott has unconvincingly argued for strong Comenian influence on Wilkins ("Comenius and the Real Character in England," in *Publications of the Modern Language Association*, 70 [1955], 1068–1081; "The Sources and Development of John Wilkins' Philosophical Language," in *Journal of English and Germanic Philology*, 57 [1958], 1–13). He rests his argument chiefly on the claim for the irenic religious effect of

the philosophical language, but this is a common claim that cannot be used for such identification, but the fundamental difficulty is that what Comenius had to say on this subject was not original. DeMott ignores Mersenne. Salmon argues against DeMott in "Language-Planning in Seventeenth-century England," *In Memory of J. R. Firth* (London, 1966), pp. 370–397. R. F. Jones, "Science and Language in England of the Mid-Seventeenth Century," in Jones, *The Seventeenth Century* (Stanford, 1951; original publ. 1932), was always a poor guide and is now thoroughly outmoded. Jorge Luis Borges' quaint essay "The Analytical Language of John Wilkins" has brought Wilkins and the *Essay* to the attention of the literati (in *Other Inquisitions 1937–1952*, Ruth L. C. Sims, trans. [New York, 1966], pp. 106–110). See also the entries on Boehme, in *Dictionary of Scientific Biography*, II (1970), 222–224; and on Comenius, *ibid.*, III (1971), 359–363.

82. R. T. Gunther, *Early Science in Oxford*, vol. VIII, *The Cutler Lectures of Robert Hooke* (Oxford, 1931), pp. 150–152, with illustration (reproduced in Andrade). Hooke found it "so truly philosophical, and so perfectly and thoroughly methodical, that there seems to be nothing wanting to make it have the utmost perfection." Hooke's faith in the philosophical language is closely related to his belief in demonstrability in natural science, a belief not shared by his scientific colleagues in the Royal Society.

83. Since this is true also of other committees appointed by the Society during these years, the failure to report cannot be taken as evidence one way or the other.

84. This had been suggested by Wilkins in the "Epistle dedicatory" of the *Essay*. See also Sprat, *History of the Royal Society*, p. 251. Hooke called memory a "repository." His conception intimates a link with the mnemonic tradition; in John Willis' *The Art of Memory* (London, 1621; later reissued), *repository* is the word for the memory device of "an imaginary house or building."

85. Huygens, *Oeuvres*, VI, 397 (Huygens to Moray, 30 March 1669); *ibid.*, p. 425 (Moray to Huygens, 16 April 1669).

86. This correspondence is in the Aubrey MSS in the Bodleian Library, Oxford. They have recently been examined by Vivian Salmon in "John Wilkins' *Essay* (1668): Critics and Continuators," in *Historiographia Linguistica*, 1 (1974), 147–163. Great efforts were made to elicit a plan from Seth Ward, but when it finally came it was found disappointing, inclining "too much to Lullius" (MS Aubrey 13, fol. 113v, Thomas Pigott to Aubrey at Hooke's, Oxford, 14 April 1678).

87. Leibniz, *Philosophische Schriften*, C. I. Gerhardt, ed., 7 vols. (Berlin, 1875–1890), VII, 16.

88. On this occasion Seth Ward helped Wilkins to a precentorship at Exeter; see Pope, *Life of Seth Ward*, p. 56.

89. Ray to Martin Lister (7 May 1669), quoted in Charles E. Raven, *John Ray* (London, 1950), p. 182. Ray repeats this judgment in several other letters of the same years.

90. Several Continental scholars, including Leibniz, had called for a translation. Ray's translation is known to have been in the archives of the Royal Society for more than a century, but has since been lost. As late as May 1678, Aubrey wrote to Ray: "I have at length gotten my desire, viz. an able Frenchman to translate the real Character into French. It is Dr. Lewis du Moulin." W. Derham, *Philosophical Letters of Ray* (London, 1718), p. 144.

91. There is an illuminating discussion of these problems in Phillip R. Sloan, "John Locke, John Ray, and the Problem of Natural System," in *Journal of the History of Biology*, 5 (1972), 1–53. Locke said in the *Essay*: "I am not so vain to think that anyone can pretend to attempt the perfect reforming the languages of the world, no, not so much as of his own country, without making himself ridiculous" (Book III, ch. II, paragraph 2). This represents the general view of the Royal Society. There is cogent criticism of Wilkins'

Essay in *Reflections Upon Learning* (1699) by the antiquary and critic of the new science, Thomas Baker; see *Reflections*, 4th ed. (1708), pp. 21–22.

92. *Calendar of State Papers, Domestic*, 1672.

93. *Diary of Robert Hooke*, 1672–1680, Henry W. Robinson and Walter Adams, eds. (London, 1935), p. 11.

94. The death is reported in Hooke's *Diary* under 19 November: "Lord Bishop of Chester died about 9 in the morning of a suppression of the urine." On the next day, he had more details: "Dr. Needham brought in account of Lord Chester's having no stoppage in his uriters nor defect in the kidneys. There was only found 2 small stones in one kidney and some little gravel in one uriter but neither big enough to stop the water. 'Twas believed his opiates and some other medicines killed him, there being no visible cause of his death, he died very quickly and with little pain, lament of all." The cause of Wilkins' death continued to be a matter of debate. In 1695, the physician Edward Baynard published "An Account of the Probable Causes of the Pain in Rheumatisms; as also of the Cure of a Total Suppression of Urine, not caused by a Stone, by the Use of Acids," in *Philosophical Transactions of the Royal Society*, 19 (Jan.–Feb. 1695), 19–20. Baynard suggests that Wilkins' case was falsely diagnosed.

BIBLIOGRAPHY

I. ORIGINAL WORKS. In addition to the works in the notes, see the following: *A Sermon Preached Before the King on March 7, 1669* (London, 1669). *A Sermon Preached Before the King on March 19, 1671* (London, 1671). These two sermons are not reprinted in Tillotson's collection of fifteen sermons.

When the *Discovery* and the *Discourse* were first published together, in 1640, they appeared under the title *A Discourse Concerning a New World and Another Planet in Two Books*. Several of Wilkins' works have been issued in reprints in recent years. *The Mathematical and Philosophical Works* (London, 1708) is the first collection of the works covered by that title. They are here placed in chronological order of publication with separate paginations and title pages. This edition opens with a "Life of the Author and an Account of His Writings," and closes with "An Abstract of Dr. Wilkins' *Essay Towards a Real Character and a Philosophical Language*." The contents are the same as in the 1802 edition.

II. SECONDARY LITERATURE. This literature is given in the notes. Our knowledge of Wilkins' life derives chiefly from the early biographical writings: William Lloyd's funeral sermon; Walter Pope, *The Life of Seth Ward . . . With a Brief Account of Bishop Wilkins, Mr. Lawrence Rooke, Dr. Isaac Barrow, Dr. Turberville, and Others* (London, 1697); John Aubrey, "John Wilkins," in *Aubrey's Brief Lives*, Oliver Lawson Dick, ed., (Ann Arbor, 1957), pp. 319–320; A. à Wood, *Athenae Oxonienses*, Bliss, ed., III (1817), cols. 967–971, but this rich source has much relevant information scattered throughout the four volumes. This is also true of Wood, *Fasti Oxonienses*, and *History of the Antiquities of the Colleges and Halls in the University of Oxford*, John Gutch, ed. See also Pierre Bayle, *A General Dictionary*,

Historical and Critical. . . , John Peter Bernard, Thomas Birch, John Lockman, eds., 10 vols. (London, 1734–1741), X, 160–164. The best biographical entry on Wilkins is the one in *Biographia Britannica*; see n. 25. It is much better than the entry in the *Dictionary of National Biography*. (In the article on Thomas Sprat in the *Dictionary of Scientific Biography* I attributed independent value to the biographical notice of Wilkins entered at the time of his death in Birch, *History of the Royal Society*, III, 67–68; I now believe that this notice was inserted by Birch.) These sources have formed the bases of entries in biographical reference works since the eighteenth century, with the accretion of more or less reliable anecdotal matter from other sources.

Since Wilkins was so widely known in his own time, he is mentioned in most contemporary records, some published long ago and some only recently, such as the diaries of John Evelyn and Samuel Pepys, Birch's *History of the Royal Society*, the correspondence of Henry Oldenburg, and *The Diary and Correspondence of Dr. John Worthington*, James Crossley, ed., 2 vols in three parts (Manchester, 1847, 1855, 1886, with vol. II, part II edited by R. C. Christie). (These are vols. 13, 36, and 114 in the publications of the Chetham Society.)

On Wilkins and Wadham College, the most important treatment is Jackson's *Wadham College*, but see also J. Wells, *Wadham College* (London, 1898), pp. 69–87. Patrick A. W. Henderson, *The Life and Times of John Wilkins* (London, 1910), is chiefly about Wadham College. The best modern biography is Dorothy Stimson, "Dr. Wilkins and the Royal Society," in *Journal of Modern History*, **3** (1931), 539–563. Neither J. G. Crowther, *Founders of British Science* (London, 1960), nor E. J. Bowen and Sir Harold Hartley, "John Wilkins," in *The Royal Society, Its Origins and Founders*, Sir Harold Hartley, ed. (London, 1960), pp. 47–56, offer anything new, and they are not reliable. For some reason, the subject of Wilkins at large has proved an open field for guesswork, partisan interpretation, and free anecdotal accretion. The intellectual history of England in the mid-seventeenth century has been treated in a number of recent books that show great diversity of interpretation, e.g., Christopher Hill, *Intellectual Origins of the English Revolution* (Oxford, 1965), and Frances A. Yates, *The Rosicrucian Enlightenment* (London, 1972); this literature tends to be occupied with polemics rather than substance. Barbara J. Shapiro, *John Wilkins 1614–1672. An Intellectual Biography* (Berkeley, 1969), has some new biographical information, but does not meet its claim to being an intellectual biography. Based on a small part of the relevant literature, Henry G. van Leeuwen presents an illuminating discussion of his subject in *The Problem of Certainty in English Thought 1630–1690* (The Hague, 1963). He argues that Chillingworth's discussion was followed by Tillotson, which is correct, but then postulates that Wilkins and Glanvill, learning from Tillotson, "secularized" the argument for the benefit of science

and the Royal Society. Simple chronology is enough to refute that interpretation. The deeper problem, however, is that van Leeuwen ignores Wilkins' early writings except *Mercury*, and that he makes a distinction between religion and science (as is also clearly shown in the notion of secularization) that is not warranted by the texts and the intellectual framework of the time; the term "natural religion" should be a sufficient reminder of that fact. (Van Leeuwen also states that *Mathematical Magick* was composed, "like most of [Wilkins'] earlier works, during his school days" [p. 56].) Shapiro rightly argues against van Leeuwen (pp. 232–316); see also Shapiro, "Latitudinarianism and Science in Seventeenth-Century England," in *Past and Present*, no. 40 (July 1968), 16–41. Marjorie Hope Nicolson's *Voyages to the Moon* (New York, 1948) is the classic treatment of a subject that has come to be associated with Wilkins. There is an excellent account of Wilkins in H. R. McAdoo, *The Spirit of Anglicanism. A Survey of Anglican Theological Method in the Seventeenth Century* (London, 1965), esp. pp. 203–231; it is the most important recent treatment of Wilkins. See also the bibliographies under the entries for Theodore Haak and Thomas Sprat in the *Dictionary of Scientific Biography*.

I have also used material contained in five lectures given under the auspices of the Program in the History and Philosophy of Science at Princeton University in the spring of 1964, entitled "Language, Man, and Knowledge in the 16th and 17th Centuries."

HANS AARSLEFF

WILKS, SAMUEL STANLEY (*b.* Little Elm, Texas, 17 June 1906; *d.* Princeton, New Jersey, 7 March 1964), *mathematical statistics.*

Wilks was the eldest of the three children of Chance C. and Bertha May Gammon Wilks. His father trained for a career in banking but after a few years chose to operate a 250-acre farm near Little Elm. His mother had a talent for music and art and instilled her own lively curiosity in her three sons.[1] Wilks obtained his grade-school education in a one-room schoolhouse and attended high school in Denton, where during his final year he skipped study hall regularly in order to take a mathematics course at North Texas State Teachers College, where he received an A.B. in architecture in 1926.

Believing his eyesight inadequate for architecture, Wilks embarked on a career in mathematics. During the school year 1926–1927, he taught mathematics and manual training in a public school in Austin, Texas, and began graduate study of mathematics at the University of Texas. He continued his studies as a part-time instructor in 1927–1928, received an M.A. in mathematics in

1928, and remained as an instructor during the academic year 1928–1929.

Granted a two-year fellowship by the University of Iowa, in the summer of 1929 Wilks began a program of study and research leading to receipt, in June 1931, of a Ph.D. in mathematics. National research fellowships enabled him to continue research and training in mathematical statistics at Columbia University (1931–1932), University College, London (1932), and Cambridge University (1933). Wilks's scientific career was subsequently centered at Princeton, where he rose from instructor in mathematics (1933) to professor of mathematical statistics (1944).

Wilks married Gena Orr, of Denton, in September 1931; they had one son, Stanley Neal Wilks. He was a member of the American Philosophical Society, the International Statistical Institute, and the American Academy of Arts and Sciences, and a fellow of the American Association for the Advancement of Science. He also belonged to most major societies in his field.

Wilks's education was extraordinary for the number of prominent people involved in it. At the University of Texas, his first course in advanced mathematics was set theory, taught by R. L. Moore, noted for his researches in topology, his unusual methods of teaching, and his contempt for applied mathematics. Having a strong practical bent, however, Wilks was more interested in probability and statistics, taught by Edward L. Dodd; known for his researches on mathematical and statistical properties of various types of means, Dodd encouraged Wilks to pursue further study of these subjects at the University of Iowa (now the State University of Iowa).

At Iowa, Wilks was introduced by Henry L. Rietz to "the theory of small samples" pioneered by "Student" (W. S. Gossett) and fully developed by R. A. Fisher, and to statistical methods employed in experimental psychology and educational testing by E. F. Lindquist.

Wilks chose Columbia University for his first year of postdoctoral study and research because Harold Hotelling, a pioneer in multivariate analysis and the person in the United States most versed in the "Student"-Fisher theory of small samples, had just been appointed professor there in the economics department. At Columbia, Wilks attended the lectures at Teachers College of Charles E. Spearman, considered the father of factor analysis, and became acquainted with the work at Bell Telephone Laboratories of Walter A. She-

what, originator of statistical quality control of manufacturing processes.

Wilks spent the first part of his second year writing a joint paper with Egon S. Pearson in the department of Karl Pearson at University College, London. At Cambridge University he worked with John Wishart who had been a research assistant to both Karl Pearson and Fisher, and whose work in multivariate analysis was close to Wilks's main interest.

Wilks's first ten published papers were contributions to the branch of statistical theory and methodology known as multivariate analysis, and it was to this area that he made his greatest contributions. His doctoral dissertation, written under Henry L. Rietz, provided the small-sample theory for answering a number of questions arising in use of the technique of "matched" groups in experimental work in educational psychology. It was preceded by a short note, "The Standard Error of the Means of 'Matched' Samples" (1931). This note and dissertation are the first in a long series of papers on topics in multivariate analysis suggested to Wilks by problems in experimental psychology and educational testing.

It was, however, his paper, "Certain Generalizations in the Analysis of Variance," that immediately established Wilks's stature. In this paper he defined the "generalized variance" of a sample of n individuals from a multivariate population, constructed multivariate generalizations of the correlation ratio and coefficient of multiple correlation, deduced the moments of the sampling distributions of these and other related functions in random samples from a normal multivariate population from Wishart's generalized product moment distribution (1928), constructed the likelihood ratio criterion for testing the null hypothesis that k multivariate samples of sizes $n_1, n_2, \cdots, n_k$ are random samples from a common multivariate normal population (now called Wilks's Λ criterion) and derived its sampling distribution under the null hypothesis, and similarly explored various other multivariate likelihood ratio criteria.

Three other papers written in 1931–1932 concerned derivation of the sampling distributions of estimates of the parameters of a bivariate normal distribution from "fragmentary samples"—that is, when some of the individuals in a sample yield observations on both variables, x and y, and some only on x, or on y, alone; derivation of the distribution of the multiple correlation coefficient in samples from a normal population with a nonzero multiple correlation coefficient directly from Wishart's generalized product moment distribution (1928) without using the geometrical notions and an invariance property utilized by Fisher in his derivation (1928); and derivation of an exact expression for the standard error of an observed "tetrad difference," an outgrowth of attending Spearman's lectures.

"Methods of Statistical Analysis . . . for k Samples of Two Variables" (1933), written with E. S. Pearson, and "Moment-Generating Operators for Determinants of Product Moments . . ." (1934) are the products of Wilks's year in England. The first consists of elaboration in greater detail for the bivariate normal case of the techniques developed for the multivariate normal in his "Certain Generalizations . . .," and reflects his and Pearson's growing interest in industrial applications by including a worked example based on data from W. A. Shewhart (1931). The second may be regarded as an extension of the work of J. Wishart and M. S. Bartlett, who had just completed an "independent" derivation of Wishart's product moment distribution "by purely algebraic methods" when Wilks arrived in Cambridge. His next important contribution to multivariate analysis, "On the Independence of k Sets of Normally Distributed . . . Variables" (1935), appears to have been written to meet a need encountered in his work with the College Entrance Examination Board, as do many of his later contributions to multivariate analysis.

In addition to his extensive and penetrating studies of likelihood ratio tests for various hypotheses relating to multivariate normal distributions, Wilks made similar investigations (1935) relating to multinomial distributions and to independence in two-, three-, and higher-dimensional contingency tables. He also provided (1938) a compact proof of the basic theorem on the large-sample distribution of the likelihood ratio criterion for testing "composite" statistical hypotheses—that is, when the "null hypothesis" tested specifies the values of, say, only m out of the h parameters of the probability distribution concerned. Jerzy Neyman's basic paper on the theory confidence-interval estimation appeared in 1937. The following year Wilks showed that under fairly general conditions confidence intervals for a parameter of a probability distribution based upon its maximum-likelihood estimator are, on the average, the shortest obtainable in large samples.

In response to a need expressed by Shewhart,

Wilks in 1941 laid the foundations of the theory of statistical "tolerance limits," which actually are confidence limits, in the sense of Neyman's theory—not, however, for the value of some parameter of the distribution sampled, as in Neyman's development but, rather, for the location of a specified fraction of the distribution sampled. He showed that a suitably selected pair of ordered observations ("order statistics") in a sample of sufficient size from an arbitrary continuous distribution provides a pair of limits (statistical "tolerance limits") to which there corresponds a stated chance that at least a specified fraction of the underlying distribution is contained between these limits, thus providing the "distribution-free" solution needed when the assumption of an underlying normal distribution of industrial production is unwarranted. Wilks also derived the corresponding parametric solution of maximum efficiency in the case of sampling from a normal distribution (based on the sample mean and standard deviation) and an expression for the relative efficiency of the distribution-free solution in this case.

In 1942 Wilks developed formulas for the probabilities that at least a fraction N_0/N of a second random sample of N observations from an arbitrary continuous distribution (a) would lie above the rth "order statistic" (rth observation in increasing order of size), $1 \le r \le n$, in a first random sample of size n from the same distribution, or (b) would be included between the rth and sth order statistics, $1 \le r < s \le n$, of the first sample; and illustrated the application of these results to the setting of one- and two-sided statistical tolerance limits. This work was Wilks's earliest contribution to "nonparametric" or "distribution-free" methods of statistical inference, an area of research of which he provided an extensive review in depth in "Order Statistics" (1948).

Wilks was a founder of the Institute of Mathematical Statistics (1935) and remained an active member. The Institute took full responsibility for the *Annals of Mathematical Statistics*, and Wilks became editor, with the June 1938 issue.[2] He served through the December 1949 issue, guiding the development of the *Annals* from a marginal journal, with a small subscription list, to the foremost publication in its field.

Although Wilks became an instructor in the department of mathematics at Princeton University at the beginning of the academic year 1933–1934, he did not give a formal course in statistics at Princeton until 1936, owing to a prior commitment that the university had made with an instructor in the department of economics and social institutions who had been sent off at university expense to develop a course on "modern statistical theory" two years before; and owing to the need for resolution by the university's administration of an equitable division of responsibility for the teaching of statistics between that department (which theretofore had been solely responsible for all teaching of statistics) and the department of mathematics.[3] Wilks was promoted to assistant professor in 1936. In the fall term he taught a graduate course, the substance of which he published as his *Lectures . . . on . . . Statistical Inference, 1936–37 . . .*; and in the spring of 1937 he gave an undergraduate course, quite possibly the first carefully formulated college undergraduate course in mathematical statistics based on one term of calculus.

Wilks's service to the federal government began with his appointment in 1936 as a collaborator in the Soil Conservation Program of the U.S. Department of Agriculture. He continued to serve the government as a member of the Applied Mathematics Panel, National Defense Research Committee, Office of Scientific Research and Development; chairman of the mathematics panel, Research and Development Board, Defense Department; adviser to the Selective Service System and the Bureau of the Budget; a member of various committees of the National Science Foundation, the National Academy of Sciences, and NASA; and an academic member of the Army Mathematics Advisory Panel. In 1947 he was awarded the Presidential Certificate of Merit for his contributions to antisubmarine warfare and the solution of convoy problems.

Wilks was deeply interested in the whole spectrum of mathematical education. In "Personnel and Training Problems in Statistics" (1947) he outlined the growing use of statistical methods, the demand for personnel, and problems of training, and made recommendations that served as a guide in the rapid growth of university centers of training in statistics after World War II. Drawing on his experience at Princeton, he urged, in "Teaching Statistical Inference in Elementary Mathematics Courses" (1958), teaching the principles of statistical inference to freshmen and sophomores, and further proposed revamping high school curricula in mathematics and the sciences to provide instruction in probability, statistics, logic, and other modern mathematical subjects. During his last few years he worked with an experimental program in a

school at Princeton that introduced mathematics at the elementary level, down to kindergarten.

NOTES

1. An unfortunate consequence of the father's predilection for alliteration in naming his sons is that publications of Samuel Stanley and Syrrel Singleton Wilks (a physiologist and expert in aerospace medicine) are sometimes lumped together under "S. S. Wilks" in bibliographic works, such as *Science Citation Index*.
2. For a fuller account of the founding and early years of the *Annals of Mathematical Statistics*, see the letter from Harry C. Carver, dated 14 Apr. 1972, to professor [W. J.] Hall, reproduced in *Bulletin of the Institute of Mathematical Statistics*, **2**, no. 1 (Jan. 1973), 11–14; and Allen T. Craig, "Our Silver Anniversary," in *Annals of Mathematical Statistics*, **31**, no. 4 (Dec. 1960), 835–837.
3. The background of this delay and its ultimate resolution are discussed in detail by Churchill Eisenhart, in "Samuel S. Wilks and the Army Experiment Design Conference Series," an address at the twentieth Conference on the Design of Experiments in Army Research, Development and Testing, held at Fort Belvoir, Va., 23–25 Oct. 1974, published in the *Proceedings* of this conference (U.S. Army Research Office Report 75–2 June 1975), 1–47. This account also contains material unavailable elsewhere on Wilks's family and early career, together with extensive notes on the American institutions and personages that played important roles in it.

BIBLIOGRAPHY

I. Original Works. "The Publications of S. S. Wilks," prepared by T. W. Anderson, in *Annals of Mathematical Statistics*, **36**, no. 1 (Feb. 1965), 24–27, which gives bibliographic details for five books, forty-eight articles, and twelve "other writings," appears to be complete with respect to the first two categories but not to the last. All forty-eight articles are repr. in T. W. Anderson, *S. S. Wilks: Collected Papers – Contributions to Mathematical Statistics* (New York, 1967), as are Anderson's lists of Wilks's publications, in rearranged form (xxvii–xxxiii). Particulars on thirty-one additional "other writings" are given by Churchill Eisenhart, "A Supplementary List of Publications of S. S. Wilks," in *American Statistician*, **29**, no. 1 (Feb. 1975), 25–27.

Among the more important of Wilks's publications are three holograph books: *Lectures by S. S. Wilks on the Theory of Statistical Inference 1936–1937, Princeton University* (Ann Arbor, Mich., 1937); *Elementary Statistical Analysis* (Princeton, 1948), quite conceivably the first carefully developed undergraduate course in mathematical statistics based on one term of calculus; and *Mathematical Statistics* (New York, 1962), a far more advanced, comprehensive treatment – *Mathematical Statistics* (Princeton, 1943) was an early version of some of the same material, prepared partly with the help of his students. He also wrote *Introductory Probability and Statistical Inference: An Experimental Course*

(New York, 1957; rev. ed., Princeton, 1959; Spanish trans., Rosario, Argentina, 1961), with E. C. Douglas, F. Mosteller, R. S. Pieters, D. E. Richmond, R. E. K. Rourke, and G. B. Thomas; and *Introductory Engineering Statistics* (New York, 1965; 2nd ed., 1971), with Irwin Guttman (2nd ed. with Guttman and J. S. Hunter).

Of his research papers, the most notable are "The Standard Error of the Means of 'Matched' Samples," in *Journal of Educational Psychology*, **22,** no. 3 (Mar. 1931), 205–208, repr. as paper 1 in *Collected Papers*; "On the Distributions of Statistics in Samples From a Normal Population of Two Variables With Matched Sampling of One Variable," in *Metron*, **9,** nos. 3–4 (Mar. 1932), 87–126, repr. as paper 2 in *Collected Papers*, his doctoral dissertation; "Certain Generalizations in the Analysis of Variance," in *Biometrika*, **24,** pts. 3–4 (Nov. 1932), 471–494, repr. as paper 6 in *Collected Papers*; "Methods of Statistical Analysis Appropriate for *k* Samples of Two Variables," *ibid.*, **25,** pts. 3–4 (Dec. 1933), 353–378, repr. as paper 7 in *Collected Papers*, written with E. S. Pearson; "Moment-Generating Operators for Determinants of Product Moments in Samples From a Normal System," in *Annals of Mathematics*, 2nd ser., **35,** no. 2 (Apr. 1934), 312–340, repr. as paper 8 in *Collected Papers*; "On the Independence of *k* Sets of Normally Distributed Statistical Variables," in *Econometrica*, **3,** no. 3 (July 1935), 309–326, repr. as paper 9 in *Collected Papers*; "The Likelihood Test of Independence in Contingency Tables," in *Annals of Mathematical Statistics*, **6,** no. 4 (Dec. 1935), 190–196, repr. as paper 11 in *Collected Papers*; "The Large-Sample Distribution of the Likelihood Ratio for Testing Composite Hypotheses," *ibid.*, **9,** no. 1 (Mar. 1938), 60–62, repr. as paper 14 in *Collected Papers*; and "Weighting Systems for Linear Functions of Correlated Variables When There Is No Dependent Variable," in *Psychometrika*, **3,** no. 1 (Mar. 1938), 23–40, repr. as paper 16 in *Collected Papers*.

See also "Shortest Average Confidence Intervals From Large Samples," in *Annals of Mathematical Statistics*, **9,** no. 3 (Sept. 1938), 166–175, repr. as paper 17 in *Collected Papers*; "An Optimum Property of Confidence Regions Associated With the Likelihood Function," *ibid.*, **10,** no. 4 (Dec. 1939), 225–235, repr. as paper 20 in *Collected Papers*, written with J. F. Daly; "Determination of Sample Sizes for Setting Tolerance Limits," *ibid.*, **12,** no. 1 (Mar. 1941), 91–96, repr. as paper 23 in *Collected Papers*; "Statistical Prediction With Special Reference to the Problem of Tolerance Limits," *ibid.*, **13,** no. 4 (Dec. 1942), 400–409, repr. as paper 26 in *Collected Papers*; "Sample Criteria for Testing Equality of Means, Equality of Variances, and Equality of Covariances in a Normal Multivariate Population," *ibid.*, **17,** no. 3 (Sept. 1946), 257–281, repr. as paper 28 in *Collected Papers*; "Order Statistics," in *Bulletin of the American Mathematical Society*, **54,** no. 1 (Jan. 1948), 6–50, repr. as paper 32 in *Collected Papers*; and "Multivariate Statistical Outliers," in *Sankhya*, **25A,** pt. 4 (Dec. 1963), 407–426, repr. as paper 48 in *Collected Papers*.

Two important papers on teaching and training in statistics are "Personnel and Training Problems in Statistics," in *American Mathematical Monthly*, **54,** no. 9 (Nov. 1947), 525–528; and "Teaching Statistical Inference in Elementary Mathematics Courses," *ibid.*, **65,** no. 3 (Mar. 1958), 143–152.

Following Wilks's death, his "working papers on subjects requiring statistical analysis; letters, reports and papers relating to professional organizations," were donated by his widow and Princeton University to the American Philosophical Society; for further details see *Guide to the Archives and Manuscript Collections of the American Philosophical Society* (Philadelphia, 1966), 146. Another dozen items of correspondence (1946, 1961–1962) are preserved in the Leonard J. Savage Papers (MS group 695), Sterling Memorial Library, Yale University. Wilks's professional books and journals have been placed in the S. S. Wilks Room in New Fine Hall, Princeton University.

II. SECONDARY LITERATURE. The biography of Wilks by Frederick Mosteller in *International Encyclopedia of the Social Sciences*, XVI (New York, 1968), 550–553, provides an informative summary of the highlights of Wilks's life, work, and impact in diverse professional roles. Wilks's research contributions and other writings are reviewed in the comprehensive obituary by T. W. Anderson in *Annals of Mathematical Statistics*, **36,** no. 1 (Feb. 1965), 1–23 (repr. in *S. S. Wilks: Collected Papers*), which is preceded by a photograph—not in *Collected Papers*—of Wilks at his desk. A less technical but equally full account of Wilks's life and work is Frederick Mosteller, "Samuel S. Wilks: Statesman of Statistics," in *American Statistician*, **18,** no. 2 (Apr. 1964), 11–17; there is some additional illuminating information in the obituaries by W. G. Cochran, in *Review of the International Statistical Institute*, **32,** nos. 1–2 (June 1964), 189–191; and John W. Tukey, in *Yearbook, American Philosophical Society* for 1964 (1965), 147–154. The obituary in *Estadística* (Washington, D.C.), **22,** no. 83 (June 1964), 338–340, tells of his activities in connection with the Inter-American Statistical Institute.

The eight articles that constitute "Memorial to Samuel S. Wilks" in *Journal of the American Statistical Association*, **60,** no. 312 (Dec. 1965), 938–966, are rich sources of further information, insight, and perspective: Frederick F. Stephan and John W. Tukey, "Sam Wilks in Princeton," 939–944; Frederick Mosteller, "His Writings in Applied Statistics," 944–953; Alex M. Mood, "His Philosophy About His Work," 953–955; Morris H. Hansen, "His Contributions to Government," 955–957; Leslie E. Simon, "His Stimulus to Army Statistics," 957–962; Morris H. Hansen, "His Contributions to the American Statistical Association," 962–964; W. J. Dixon, "His Editorship of the *Annals of Mathematical Statistics*," 964–965; and the unsigned "The Wilks Award," 965–966.

Other publications cited or mentioned in the text are: R. A. Fisher, "On the Mathematical Foundations of Theoretical Statistics," in *Philosophical Transactions of the Royal Society*, **222A**, no. 602 (19 Apr. 1922), 309–368; and "The General Sampling Distribution of the Multiple Correlation Coefficient," in *Proceedings of the Royal Society*, **121A**, no. A788 (1 Dec. 1928), 654–673; E. F. Lindquist, "The Significance of a Difference Between 'Matched' Groups," in *Journal of Educational Psychology*, **22** (Mar. 1931), 197–204; J. Neyman, "Outline of a Theory of Statistical Estimation Based on the Classical Theory of Probability," in *Philosophical Transactions of the Royal Society*, **236A**, no. 767 (30 Aug. 1937), 333–380, repr. as paper no. 20 in *A Selection of Early Statistical Papers of J. Neyman* (Cambridge–Berkeley–Los Angeles, 1967); J. Neyman and E. S. Pearson, "On the Use and Interpretation of Certain Test Criteria. Part I," in *Biometrika*, **20A**, pts. 1–2 (July 1928), 175–240, repr. as paper no. 1 in *Joint Statistical Papers of J. Neyman and E. S. Pearson* (Cambridge–Berkeley–Los Angeles, 1967); "On the Use and Interpretation of Certain Test Criteria. Part II," *ibid.*, pts. 3–4 (Dec. 1928), 263–294, repr. as paper no. 2 in *Joint . . . Papers*; and "On the Problem of *k* Samples," in *Bulletin international de l'Académie polonaise des sciences et des lettres*, no. 6A (June 1931), 460–481, repr. as paper no. 4 in *Joint . . . Papers*; Walter A. Shewhart, *Economic Control of Quality of Manufactured Product* (New York, 1931), 42; J. Wishart, "The Generalized Product Moment Distribution in Samples From a Normal Multivariate Population," in *Biometrika*, **20A**, pts. 1–2 (July 1928), 32–52; and J. Wishart and M. S. Bartlett, "The Generalized Product Moment Distribution in a Normal System," in *Proceedings of the Cambridge Philosophical Society. Mathematical and Physical Sciences*, **29**, pt. 2 (10 May 1933), 260–270.

Churchill Eisenhart

WILLDENOW, KARL LUDWIG (*b.* Berlin, Germany, 22 August 1765; *d.* Berlin, 10 July 1812), *botany, historical phytogeography.*

As a botanist in the Linnaean tradition, Willdenow was concerned chiefly with description and classification, although he also helped lay the foundations for the new field of phytogeography that emerged in the early nineteenth century. He was introduced to the study of plants by his father, Karl Johann Willdenow, a Berlin apothecary, and by the botanist Johann Gottlieb Gleditsch. He also was taught chemistry by his father's friend and colleague Martin Heinrich Klaproth. After receiving his early formal education at a Berlin Gymnasium,

Willdenow studied at the pharmacy school conducted by Johann Christian Wiegleb at Langensalza. He completed the course in 1785 and then went on to study medicine at Halle, receiving the M.D. in 1789. In 1790 he married Henriette Louise Habermass and took over his father's establishment in Berlin.

By this time Willdenow had already published a flora of Berlin (1787), and for several years he had been conducting informal botanical lessons and field trips in and around the city. One of those whom he introduced to the subject, in 1788, was Alexander von Humboldt, who became a lifelong friend and occasional scientific collaborator. Willdenow greatly extended his educational influence in 1792 with the publication of *Grundriss der Kräuterkunde*, a basic textbook intended to replace Linnaeus' obsolete *Philosophia botanica* (1751). The *Grundriss* was a great success and long remained a standard text, going through numerous editions in several languages.

Willdenow's growing reputation brought him membership in the Berlin Academy of Sciences in 1794, and in 1801 he became its principal botanist. He was named professor of natural history at the Berlin Medical-Surgical College in 1798, at which time he gave up his apothecary shop. In 1801 Willdenow became curator of the Berlin Botanical Garden, which he developed from modest proportions into one of the most comprehensive in Europe, introducing numerous exotic species into its collection. He also continued to add to his own herbarium, which numbered more than 20,000 dried specimens at the time of his death and is still preserved at the Berlin Botanical Garden. In 1810 he was named professor of botany at the new University of Berlin, but he died before formally taking up the duties.

Willdenow lived in an age of worldwide geographical exploration; and although he rarely traveled outside Germany, and never outside Europe, he corresponded with many explorers, who provided him with thousands of botanical specimens for his collections and research. Humboldt, for example, gave him the nearly 400 plants that he collected in Spain in 1799; and it was to Willdenow that he turned for assistance in describing and classifying the thousands of new species that he and Bonpland brought back from their long South American expedition. In 1810 Willdenow traveled to Paris to help Humboldt; but the work was interrupted by an illness that forced him to return to Berlin, where he died in 1812. Willdenow also had undertaken a thorough revision of Linnaeus' *Species plantarum*, a massive project that was incomplete in five volumes at the time of his death.

Although Willdenow's own researches were primarily taxonomic, he recognized that plants also fall into distinct geographical groups and advocated the systematic investigation of various regularities of plant distribution. Some aspects of the subject had been treated by earlier botanists, especially Linnaeus and his students; but Willdenow was one of the first to conceive of a separate and clearly historical botanical discipline dealing with plant distribution in relation to climatic, geographical, geological, migrational, and other factors. He discussed this discipline under the heading "History of Plants" in the first edition of *Grundriss der Kräuterkunde*, and substantially revised and expanded the section in subsequent editions. (The following account is based primarily on the second edition.)

Willdenow understood the scope of the history of plants "to include the influence of climate on vegetation, the changes that plants have probably endured during the revolutions of our globe, the distribution of plants over the earth, their migrations, and, finally, the means by which Nature has provided for their preservation." Under the first heading he included the role of climate in defining the great floral regions of the earth, the climatic adaptations of plant species as limiting factors to their distribution, and the ability of climate to influence the relative number of plant species found in a given region as well as their general characteristics, such as size and shape. In discussing the changes of plants, Willdenow refuted Linnaeus' theory that new plant species can arise through the hybridization of old ones, and he defended the idea that many species have become extinct over the course of time as a result of geological and climatic change.

Regarding the distribution of species, Willdenow proposed a theory that attempted to relate the present division of the earth into floristic regions to its geological history, as then understood. Accepting the view that the earth was originally covered by a vast sea from which only the highest mountain ranges emerged, he suggested that God had populated these mountain archipelagoes with all the plant species that would ever exist, creating each species in only one place. Then, as the seas receded, the plants descended from the mountains and spread to the surrounding lowlands until they encountered some barrier to further migration. It is because each mountain range had a unique set of species that the earth is still divided into more or less distinct floral regions that cannot be explained through climate alone. As Willdenow demonstrated at length, however, there are numerous means by which plants and their seeds can be transported over even the most formidable barriers, so that considerable mixing of the original floras has occurred and continues to occur. He thus recognized that the present distribution of plants is the result of historical processes and suggested that by studying the present pattern it should be possible to reconstruct both the primeval distribution and the processes that have led to the present one. He himself did not pursue this and other aspects of the history of plants in great detail; but it was partly due to his influence that others, especially Humboldt, went on to establish the field of phytogeography on a comprehensive basis.

BIBLIOGRAPHY

I. ORIGINAL WORKS. Willdenow's principal books include *Florae Berolinensis prodromus* (Berlin, 1787); *Tractatus botanico-medicus de Achilleis* (Halle, 1789); *Historia amaranthorum* (Zurich, 1790); *Grundriss der Kräuterkunde* (Berlin, 1792; 2nd. ed., 1798), also in English as *Principles of Botany* (Edinburgh, 1805); *Berlinische Baumzucht* (Berlin, 1796); *Species plantarum*, 5 vols. (Berlin, 1797–1810); *Anleitung zum Selbststudium der Botanik* (Berlin, 1804); *Hortus Berolinensis*, 2 vols. (Berlin, 1806–1809); and *Enumeratio horti regii botanici Berolinensis* (Berlin, 1809; supp., 1813). Most of Willdenow's numerous articles are listed in the Royal Society *Catalogue of Scientific Papers*, VI, 372–374.

II. SECONDARY LITERATURE. Willdenow's friend and colleague D. F. L. von Schlechtendal gave an account of his activities as head of the Berlin Botanical Garden in his intro. to the supp. of the *Enumeratio* (1813), iii–x; he published a fuller *éloge* in *Magazin für der neuesten Entdeckungen in der gesammten Naturkunde*, 6 (1814), v–xvi. Also useful are Clemens König, "Karl Ludwig Willdenow," in *Allgemeine deutsche Biographie*, XLIII (Leipzig, 1898), 252–254; and Max Lenz, *Geschichte der königlichen Friedrich-Wilhelms-Universität zu Berlin*, I (Halle, 1910), 247–249. John Ise and Fritz G. Lange, eds., *Die Jugendbriefe Alexander von Humboldts 1787–1799* (Berlin, 1973); and E. T. Hamy, ed., *Lettres américaines d'Alexandre de Humboldt (1798–1807)* (Paris, 1905), contain a number of important letters from Humboldt to Willdenow as well as many references to him. See also Hanno Beck, *Alexander von Humboldt*, 2 vols. (Wiesbaden, 1959–1961), *passim*, esp. I, 16–17, and II, 65–68; Wolfgang-Hagen Hein, "Alexander von Humboldt und Karl Ludwig Willdenow," in *Pharmazeutische Zeitung*, **104** (1959), 467–471; and Clemens König, "Die historische Entwicklung der pflanzengeographische Ideen Humboldts," in *Naturwissenschaftliche Wochenschrift*, **10** (1895), 77–81, 95–98, 117–124 (not examined).

JEROME J. BYLEBYL

WILLIAM HEYTESBURY. See Heytesbury, William.

WILLIAM OF AUVERGNE, also known as **Guilielmus Arvernus** or **Alvernus** (*b*. Aurillac, Auvergne [now Cantal], France, between 1180 and 1190; *d*. Paris, France, 30 March 1249), *philosophy, theology.*

After a brief teaching career at Paris, where he was made a master of theology in 1223, William was named bishop of Paris in 1228, a post he held until his death; thus he is sometimes called William of Paris. Like a famous successor, Étienne Tempier, he meddled in the affairs of the university, but generally with a more positive attitude toward the pagan learning that was then being introduced there. He did not think highly of the Jews, however, and was among those responsible for the public burning of the Talmud at Paris in June 1242.

As a Christian philosopher William may be characterized as the last eminent French theologian, completing the tradition of Abailard and Bernard of Clairvaux; and as the first great scholastic, setting the stage for Alexander of Hales, Albertus Magnus, and their disciples. Alexander and Albertus were at Paris with him, as was Roger Bacon, who mentions having heard him lecture on the active intellect.[1] William was also friendly with Robert Grosseteste, bishop of Lincoln, and is mentioned as a possible source of the latter's "metaphysics of light." Because of his insistence on the primacy of being and his use of an Avicennian teaching on essence and existence taken over later by Thomas Aquinas, William is sometimes seen as a forerunner of Aquinas. Actually William was more in the tradition of Augustine, Boethius, and the School of Chartres; and many of his teachings were combated energetically by Aquinas.

While setting himself to destroy the errors of the pagans, William insisted on a careful study of their writings. He was particularly interested in Ibn Sīnā, although he also cited al-Karajī, al-Ghazzali, al-Fārābī, Ibn Rushd, al-Battānī, Abū Ma'shar, Altāf Husain Hālī, al-Biṭrūjī, and al-Farghānī; and he knew of the writings ascribed to Hermes Trismegistus. References to these authors occur mainly in William's *De universo*, the part of his monu-

mental *Magisterium divinale* devoted to the world of nature, composed around 1231–1236. *De universo* serves as an intermediary between the early medieval writings on cosmology of Isidore of Seville and Bede and the great encyclopedias of Vincent of Beauvais and Albertus Magnus that appeared later in the century. Apart from the insight it provides into William's sparse knowledge of astronomy and cosmology, it is of considerable value for its accounts and critiques of medieval magic and so-called experimental science; William, in fact, speaks often of "experiments," but without any of the modern connotations of this expression.

The *De universo* is divided into two parts, each containing three subdivisions. The first part considers creation as a whole and treats in some detail of the material universe, while the second part is devoted almost entirely to the spiritual universe, that is, the world of intelligences, angels, and demons. William upholds the Christian doctrine of creation against the Manichaeans, insisting on the unity of the universe and arguing against a plurality of worlds and any void existing between them.[2] He teaches that the heavens and the elements were created at the same time,[3] and that the heavenly bodies have no proper movements independent of the celestial spheres in which they are imbedded,[4] the latter consideration leading him to discourse on the relative thickness of the various spheres.[5] In general he endorses the astronomical system of al-Biṭrūjī, against that of Ptolemy, without manifesting an adequate grasp of either.[6] There are also Platonic overtones in his exposition, as when he explains the motion of the heavens in terms of a power emanating from al-Biṭrūjī's ninth sphere, somewhat like the World Soul of the *Timaeus*,[7] and which he also likens to the phenomenon of magnetic induction.[8] He is explicit that the Holy Spirit is in no way to be identified with the soul of the world.[9]

The second part of *De universo* opens with an account of Aristotle's intelligences and a repudiation of them as movers of the heavens.[10] William teaches also that angels are not necessary to account for the differences in velocities of the heavenly bodies.[11] While discussing the motion of separated substances, he treats incidentally of the speed of illumination, holding that the operation of light is not instantaneous.[12] He inveighs against the excesses of judicial astrology, although he admits the existence of remarkable phenomena, "which some physicians and certain natural philosophers refer to as empirical."[13] William distinguishes between natural magic and black magic, allowing that the former is based on the hidden properties of natural substances and is wrong only if used for evil purposes, whereas the latter involves the intervention of demons and is itself evil.[14] Like most medievals William is exceedingly credulous in accepting and recounting the many marvelous and occult workings of nature of which he has heard or read in earlier authors. He lists detailed prescriptions against the practice of magic in another part of his *Magisterium divinale* entitled *De legibus*; this prohibits the cult of the stars and heavenly bodies and the idolatry of elements, statues, and similar objects.[15]

NOTES

1. *Opus tertium*, cap. 23; see *The 'Opus Majus' of Roger Bacon*, J. H. Bridges, ed., I (Oxford, 1897), p. xxvii.
2. *De universo*, Primae partis prima pars, cc. 13–16, in *Opera omnia* (Paris, 1674), I, pp. 607–611.
3. *Ibid.*, cc. 28–29, pp. 624–625.
4. *Ibid.*, c. 44, pp. 648–653.
5. *Ibid.*, c. 45, p. 654.
6. *Ibid.*, c. 44, pp. 651–653.
7. *Ibid.*, Primae partis tertia pars, c. 28, pp. 798–801.
8. *Ibid.*, c. 29, pp. 801–803; see Duhem, *Le système du monde*, III, pp. 258–260.
9. *De universo*, Primae partis tertia pars, c. 33, p. 806.
10. *Ibid.*, Secundae partis prima pars, c. 45, p. 843.
11. *Ibid.*, Secundae partis secunda pars, c. 97, pp. 951–952.
12. *Ibid.*, c. 101, pp. 953–954.
13. ". . . exemplis occultarum operationum et mirabilium, quaeque nonnulli medicorum et etiam quidam philosophorum naturalium empirica vocant," *ibid.*, c. 76, p. 929.
14. *Ibid.*, Secundae partis tertia pars, cc. 7–8, pp. 1029–1035; c. 18, pp. 1049–1050.
15. *Opera omnia* (Paris, 1674), vol. I, pp. 18–102, esp. 44, 77, 81, and 86.

BIBLIOGRAPHY

I. ORIGINAL WORKS. William's *Opera omnia* has appeared in various Latin editions, all incomplete: (Nuremberg, 1496); 2 vols. (Paris, 1516, 1574); 2 vols. (Venice, 1591); 2 vols. (Orleans, 1674); and 2 vols. (Paris, 1674; repr., Frankfurt am Main, 1963). The sermons contained in vol. II of the latter edition and attributed to William of Auvergne are actually those of the Dominican William Peraldus (Perrauld or Perault). An important treatise has been edited by J. R. O'Donnell, as "Tractatus magistri Guillelmi Alvernensis *De bono et malo*," in *Mediaeval Studies*, **8** (1946), 245–299, and **16** (1954), 219–271; other writings and sermons still remain unedited.

II. SECONDARY LITERATURE. For a brief biography of William and a bibliography, see J. R. O'Donnell, in *New Catholic Encyclopedia*, XIV (New York, 1967), 921. On William's philosophy, see Étienne Gilson, *History of Christian Philosophy in the Middle Ages* (New York–London, 1955), 250–258, 658–660, with bibliogra-

phy. William's contributions viewed in relation to the history of science are detailed in George Sarton, *Introduction to the History of Science*, II, pt. 2 (Baltimore, 1931), 588; E. J. Dijksterhuis, *The Mechanization of the World Picture*, C. Dikshoorn, trans. (Oxford, 1961), 139, passim; L. Thorndike, *A History of Magic and Experimental Science*, II (New York, 1923), 338–371; Pierre Duhem, *Le système du monde*, III (Paris, 1915), 249–260, and V (Paris, 1917), 260–285; and *Études sur Léonard de Vinci*, II (Paris, 1909), 408–410. A valuable study of William's teaching on the soul and on psychology is E. A. Moody, "William of Auvergne and His Treatise *De Anima*," in his *Studies in Medieval Philosophy, Science, and Logic, Collected Papers 1933–1969* (Berkeley, Calif., 1975), 1–109.

WILLIAM A. WALLACE, O.P.

WILLIAM OF MOERBEKE. See Moerbeke, William of.

WILLIAM OF OCKHAM. See Ockham, William of.

WILLIAM OF SAINT-CLOUD (*fl.* France, end of the thirteenth century), *astronomy*.

All of the very little that is known about William of Saint-Cloud comes from his own writings. The earliest recorded date of his activity is 1285, when he observed a conjunction of Saturn and Jupiter (28 December), an event to which he alludes in his *Almanach*. He was undoubtedly well received in French court circles, for his calendar is dedicated to Queen Marie of Brabant, widow of Philip III (The Bold); and he translated it into French at the request of Jeanne of Navarre, wife of Philip IV (The Fair).[1] There is no substantial evidence to support the hypothesis that William of Saint-Cloud is identical with a certain Simon of Saint-Cloud, canon of Meaux and a steward of the queen. Simon is mentioned several times in archival documents. Nor are there grounds for calling William of Saint-Cloud by the name of Lefebvre.[2]

William of Saint-Cloud's known works are devoted entirely to astronomy. The treatise on the *Directorium*, or "adrescoir," is the oldest of his preserved writings; it is referred to in the *Calendrier de la reine*. The instrument described in it is a magnetic compass with a graduation in unequal hours; it is provided with a table for computing the duration of diurnal arcs.

The text accompanying Queen Marie's calendar deals with problems relating to the daily movement of the sun and to the astronomy of the *primum mobile* (inequality of days and nights according to the season and the geographic latitude; division of the inhabited world into "climats"), as well as to the nineteen-year lunar cycle. Beyond the information usually found in such works (the number of the decemnovennial cycle coordinated with the day of the month in which the new moon occurs during the year designated by this number; ferial letters; saints' days; entry of the sun into the signs of the zodiac), the calendar also furnishes more technical data: the height of the sun at noon, duration of diurnal and nocturnal arcs, and hours of the new moon. The most notable aspect of this work is William's firm resolve to establish his calendar on a purely astronomical basis. As a result, he contradicted the ecclesiastical calendrical computation, emphasizing its inadequacy and errors. For example, he presented—along with the traditional decemnovennial cycle (which he was still obliged to give since the rules of the ecclesiastical computation had not been abrogated)—another cycle that conformed to the scientific data of the astronomical tables and that he designated by the letters from *a* to *t*. Toward the same end, he appended a table that permits the user to make corrections—beyond the first solar cycle of four years—in the dates of the entry of the sun into the signs of the zodiac. The base year of the calendar is 1292 (not, as P. Duhem stated, 1296).

The starting point of William's *Almanach* is also the year 1292. The purpose of this work was to furnish the effective positions of the planets, in contradistinction to the astronomical tables which gave only the elements for computing these positions. The introductory text is neither a theory of the planets, nor, properly speaking, canons (which, moreover, are scarcely necessary in an almanac), but rather an account of the observations and considerations on which the book is based. William takes this opportunity to point out the errors he has detected in the astronomical tables he used and to show how he has corrected them. The tables that he subjects to this criticism are those of Toledo (used in the Arab calendar) and of Toulouse (in which the preceding tables are applied to the Christian calendar). Nevertheless, in the end he adopted the tables of Toulouse, although not without making slightly arbitrary corrections in some of the mean movements listed in them.

In discussing the movement of the eighth sphere, William verified by calculation and observation the value he assigned to it for the year 1290. Noting that this value differs by nearly a degree from the one that would result from applying Thābit ibn Qurra's theory of accession and recession, he rejected the latter and opted for the Ptolemaic theory of simple precession. He recorded, however, the different values that astronomers assigned to the amplitude of that precession.

The *Almanach* gives the daily planetary positions for a period of twenty years beginning with 1292. For the sun and the moon, all the positions are accurate to within a minute. This same degree of accuracy is achieved for the positions of the outer planets every ten days and for the positions of the inner planets every five days. The latitude of the moon is also given.

William's *Almanach* makes no reference to the Alphonsine tables. Yet Duhem, on the basis of a text printed with the works of Nicolas of Cusa, which he attributed to William, concluded that the latter was among the first, if not the first, in Paris, to know of and use the Alphonsine tables. (William supposedly learned about them a few years after composing his *Almanach*.) Nothing is less certain. Duhem's analysis of this short text is insufficiently critical. Moreover, it seems certain that the Alphonsine tables were not introduced into Parisian astronomy before 1320,[3] after which date it is not certain that William of Saint-Cloud was still alive.

The works of William had only a limited diffusion, undoubtedly because they were based on astronomical data (the tables of Toledo and Toulouse) that was soon superseded by the Alphonsine tables. Nonetheless, William holds a very significant place in the history of medieval astronomy. The many observations he made and the conclusions he drew from them, together with his criticism of the available tabular material, make him a genuine precursor, perhaps even the chief inspiration, of the Parisian astronomers of the first half of the fourteenth century.

NOTES

1. E. Zinner, *Verzeichnis der astronomischen Handschriften des deutschen Kulturgebietes* (Munich, 1925), p. 413, mistakenly speaks of a Queen Elisabeth. Further, Zinner (*ibid.*, nos. 2578–2594) confused William of Saint-Cloud with William the Englishman, the Marseilles author of *De urina non visa* and of several other astronomical texts, who lived in the first half of the thirteenth century.

2. A. d'Avezac, "Note sur Guillaume de Saint-Cloud," in *Mémoires de l'Institut national de France, Académie des inscriptions et belles-lettres*, **29**, pt. 1 (1877), 8–10, and in *Académie des inscriptions et belles-lettres, comptes-rendus des séances 1869*, 29–31.

3. See the article on John of Murs, in *Dictionary of Scientific Biography*, VII.

BIBLIOGRAPHY

I. ORIGINAL WORKS. The treatises of William of Saint-Cloud are all unpublished. The MS of the directorium or "adrescoir" ("Presens ingenium directivum (*sic*) vocitavi . . .") is Paris Arsenal 1037, fol. 7v–8v, together with the French trans. ("Très haute dame ci sont les proufis que l'en puet avoir . . ."). The Latin text is in verse. Another very summary French version, without the table of hours, can also be found in Paris Arsenal 2872, fol. 21v.

The calendar in Latin ("Testante Vegecio in libro suo de re militari . . .") is in: Florence Laur. XXX.24, fol. 99–109v and 110–123; Paris Arsenal 534, fol. 91–106; Paris lat.7281, fol. 145–148 (text only); Paris lat. 15171, fol. 88–101. Calendar in French ("Si come Vegeces tesmoigne en son livre . . .") in: Paris Arsenal 2872, fol. 1–21. Another French version, with a reduced commentary, is in Rennes 593, fol. 1–7.

The MSS of the almanac ("Cum intentio mea sit componere almanac . . .") contain either the text only or the tables only; the text is in: Cues 215, fol. 24–31v; Paris lat.7281, fol. 141–144v; Paris nouv. acq. lat. 1242, fol. 41–44; summary note in Cues 212, fol. 405–406; the tables are in: Vatican lat. 4572; Paris lat. 16210.

The MS Erfurt 4° 355, fol. 44v preserves a star table "verified at Saint-Cloud near Paris in 1294" that may well be by William.

II. SECONDARY LITERATURE. On William and his work, see E. Littré, in *Histoire littéraire de la France*, XXV (Paris, 1869), 63–74; P. Duhem, *Le système du monde*, IV (Paris, 1916), 10–24.

EMMANUEL POULLE

WILLIAM OF SHERWOOD, also **Shyreswood, Shirewode** (*fl.* Oxford, thirteenth century), *logic.*

William was active as a master of logic at a time when that subject was making remarkable progress. He probably was born in Nottinghamshire between 1200 and 1210 and is most likely to have studied at Oxford or Paris, or both. His only extant works are treatises on logic explicitly attributed to him in the manuscripts: the *Introductio in logicam*, a compendium af Aristotelian-Boethian logic together with the main topics of terminist logic; and *Syncategoremata*, a more advanced treatise on the semantic and logical properties of syncategorematic words (words that have special logical-semantic effects on subjects, predicates, or their combinations). The Paris manuscript B.N. Lat. 16.617 contains, besides the works mentioned above, four logical treatises, three of which may in all probability be ascribed to William: the first of two treatises entitled *De insolubilibus* (works dealing with the paradoxes of self-reference, as in the proposition "What I am saying now is false"); *De obligationibus*, a work on the rules to be observed in formal disputations (its authenticity is certain); and *Petitiones contrariorum*, on the solution of logical puzzles, called *sophismata* in medieval usage, that arise from hidden ambiguity in the premises of an inference. The last three works have not been printed so far.

William's teaching at the University of Paris is uncertain and cannot be deduced, as is usually done, from any influence on Paris logicians and metaphysicians, since such influence cannot be shown. It is certain, however, that he was an active master at Oxford sometime before 27 January 1249, when his name is found in a deed. He certainly was there during the great disturbance between the northern and the Irish scholars. William was treasurer of the cathedral church of Lincoln at least from about 1256 to 1265. He is mentioned as rector of Aylesbury, in Buckinghamshire, in October 1266 and of Attleborough, Norfolk. He must have died sometime between 1266 and 1271.

As a writer of a compendium on logic, William can be compared with two other authors of such works, Peter of Spain (*ca.* 1205–1277) and Lambert of Auxerre, who wrote his compendium between 1253 and 1257. Accurate investigations have shown that they worked quite independently of each other; the dissimilarities are numerous, and the resemblances can be explained by the authors' sharing a common tradition of logical teaching.

William's compendium, unlike that by Peter of Spain, was not very influential in later times. His impact on his contemporaries can be deduced only from a passage in Roger Bacon's *Opus tertium* (1267), in which William is described as "one of the famous wise men in Christendom" who was "much wiser than Albert the Great in what is called *philosophia communis*" (logic); Bacon even calls him the greatest logician. Elsewhere (*Compendium studii philosophie*, written in 1271 or 1272) Bacon mentions William as one of the wise and solid philosophers and theologians of the older generation. His evaluation of William as a first-rate logician is the more remarkable since his opinion of contemporary philosophers tended to be disdainful. The importance of William's work in the development of logic has not yet been fully investigated.

BIBLIOGRAPHY

I. ORIGINAL WORKS. The *Introductiones in logicam* was edited by M. Grabmann as *Sitzungsberichte der Bayerischen Akademie der Wissenschaften zu München*,

Phil.-hist. Abt. (1937), no. 10 (unreliable), and was translated by Norman Kretzmann as *William of Sherwood's Introduction to Logic* (Minneapolis, 1968), with introd. and notes. The *Syncategoremata* was edited by J. R. O'Donnell, in *Mediaeval Studies*, **3** (1941), 46–93.

II. SECONDARY LITERATURE. Kretzmann's introd. to his trans. is fundamental for evidence on William's career. On William as a logician, see the work of Kretzmann and also L. M. de Rijk, *Logica modernorum*, II, *The Origin and Early Development of the Theory of Supposition* (Assen, Netherlands, 1967), 567–591; and "The Development of *Suppositio naturalis* in Mediaeval Logic, I: Natural Supposition as Non-Contextual Supposition," in *Vivarium*, **9** (1971), 71–107.

L. M. DE RIJK

WILLIAMS, HENRY SHALER (*b*. Ithaca, New York, 6 March 1847; *d*. Havana, Cuba, 31 July 1918), *paleontology, stratigraphy.*

Williams was a member of a family long prominent in Ithaca. Interested in natural science from an early age, he received the Ph.B from the Sheffield Scientific School in 1868 and the Ph.D. from Yale in 1871. After a few years in his father's business he joined the faculty of geology at Cornell University. Of an independent mind regarding geology, he vigorously commenced paleontological and stratigraphical studies on the Devonian rocks of central New York; and during the next twelve years he not only described these rocks and their fossils in detail but also demonstrated that the true relationships to time (correlation) could be shown only by careful paleontological analysis. Williams quickly became aware of the connection between facies changes and the evolution and shifting of fossil faunas, a theme that occupied much of his work and one on which he published many papers, culminating in his monograph on recurrent *Tropidoleptus* zones (1913). He extended his work on the Middle Paleozoic strata over much of the eastern United States as an associate of the U.S. Geological Survey.

Williams' major contribution to the philosophical aspects of paleontology and stratigraphy was his book *Geological Biology* (1895). Ahead of its time, its ideas had little impact; but it is now recognized as a minor classic. From analyses of the fossil record Williams made a perceptive and striking estimate of the relative lengths of time involved in the great geological periods, estimates that correspond closely to those based on the radiometric methods of the twentieth century.

In 1892 Williams returned to Yale as Silliman professor of geology, succeeding James Dwight Dana. In 1904 he returned to Cornell, where he continued his research until he retired in 1912. In 1888 he was one of the leading founders of the Geological Society of America, two years after he had organized the Society of the Sigma Xi.

BIBLIOGRAPHY

I. ORIGINAL WORKS. Williams wrote more than ninety books and papers, among which the most important are "The *Cuboides* Zone and Its Fauna; a Discussion of the Methods of Correlation," in *Bulletin of the Geological Society of America*, **1** (1890), 481–500; "Correlation Papers: Devonian and Carboniferous," in *Bulletin of the United States Geological Survey*, **80** (1891); "Dual Nomenclature in Geological Classification," in *Journal of Geology*, **2** (1894), 145–160; *Geological Biology, an Introduction to the Geological History of Organisms* (New York, 1895); "Fossil Faunas and Their Use in Correlating Geological Formations," in *American Journal of Science*, **163** (1902), 417–432; "Shifting of Faunas as a Problem of Stratigraphic Geology," in *Bulletin of the Geological Society of America*, **14** (1903); and "Recurrent *Tropidoleptus* Zones of the Upper Devonian in New York," in *Professional Papers. United States Geological Survey*, **79** (1913).

II. SECONDARY LITERATURE. See H. F. Cleland, "Memorial of Henry Shaler Williams," in *Bulletin of the Geological Society of America*, **30** (1919), 47–65, with portrait and bibliography of published works; H. E. Gregory, "Professor Williams at Yale," in *Science*, **49** (1919), 63–65; Charles Schuchert, "Henry Shaler Williams, an Appreciation of His Work in Stratigraphy," in *American Journal of Science*, **196** (1918), 682–687; and Stuart Weller, "Henry Shaler Williams," in *Journal of Geology*, **26** (1918), 698–700.

JOHN W. WELLS

WILLIAMS, ROBERT RUNNELS (*b*. Nellore, India, 16 February 1886; *d*. Summit, New Jersey, 2 October 1965), *chemistry, nutrition.*

Williams was the son of Robert Runnels Williams, a Baptist missionary, and Alice Evelyn Mills. He was educated at Ottawa University in Kansas and the University of Chicago, receiving the B.S. from Chicago in 1907 and the M.S. in 1908. An honorary D.Sc. was conferred on him by Ottawa University in 1935; six other universities so honored him later. In 1909 he became a chemist with the Bureau of Science in Manila, where he undertook a search for the substance in rice polishings that was a curative for beriberi. The search, identification, and medical use of the substance became the major objective of his entire career.

By the time he returned to the United States in 1914, to take a position with the Bureau of Chemistry of the U.S. Department of Agriculture, Williams had established that the active substance was a nitrogenous base. When the United States entered World War I, he was briefly involved with chemical warfare investigations and was later to work for the Bureau of Aircraft Production. After the war he left government service, because the low salary made it difficult to support his wife and four children.

Williams worked briefly in 1919 for the Melcho Chemical Company of Bayonne, New Jersey, developing a process for recovering chemicals from petroleum refinery gases. Between 1919 and 1925 he did research on submarine insulating materials for the Western Electric Company, and in 1925 he became director of the chemistry laboratories at the Bell Telephone Laboratories, a position he held until 1946.

During the 1920's and 1930's Williams pursued his research on the anti-beriberi factor (ultimately named vitamin B_1) in his spare time. Working at night in a laboratory in his garage, he slowly improved the concentration of his products. In the meantime, B. C. P. Jansen and W. F. Donath obtained the pure substance in very low yield (1926). In 1927 Williams secured research support from the Carnegie Corporation and was given laboratory space at Columbia University. By 1933 he and his co-workers had obtained pure crystals of vitamin B_1. By using quinine for the elution of the vitamin from fuller's earth, he developed a procedure that gave better yields than those obtained by other workers in the field.

Once the pure compound was available, Williams found that sulfurous acid cleavage gave a pyrimidine and a thiazole fraction. These were identified, and in 1935 a tentative formula for the structure was published; soon thereafter the compound, named thiamine, was synthesized. Commercial synthesis soon followed.

In 1933 Williams had severed his connections with the Carnegie Corporation as a result of a disagreement regarding patent policy. He subsequently assigned his patents to Research Corporation, a foundation set up by Frederick Cottrell in 1912 to develop scientific patents and use the proceeds for further research. In 1946 Williams became director of grants for Research Corporation, and in 1951 he was made director. Personal profits from Williams'

patents were set aside in the Williams-Waterman Fund for the Combat of Dietary Disease, the income being used to support nutritional research and field programs to combat malnutrition.

In 1940 Williams became chairman of the Cereal Committee of the Food and Nutrition Board of the National Research Council. As a wartime measure he promoted the enrichment of flour and bread with thiamine, other vitamins, iron, and calcium.

Williams' younger brother, Roger John (b. 1893) discovered pantothenic acid and did extensive nutritional research, especially on the vitamin B complex. The brothers were jointly honored by receipt of the Charles Frederick Chandler Medal of Columbia University in 1942.

BIBLIOGRAPHY

I. ORIGINAL WORKS. *Toward the Conquest of Beriberi* (Cambridge, Mass., 1961), although a historical account of the study of the disease and the vitamin preventing it, is somewhat autobiographical because Williams was deeply involved in the isolation and synthesis of thiamine. As autobiography, however, it is limited almost entirely to his vitamin research and the use of the synthetic product in controlling the disease, especially in the Philippines; there is very little other personal history. The story of the patents and the use of the income therefrom is treated in *The Williams-Waterman Fund for the Combat of Dietary Diseases: A History of the Period 1935 Through 1955* (New York, 1956). Williams was coauthor with Tom D. Spies of *Vitamin B_1 (Thiamine) and Its Use in Medicine* (New York, 1938). The program on flour enrichment is described in "Enrichment of Flour and Bread. A History of the Movement," *Bulletin of the National Research Council*, no. 110 (1944), written with R. M. Wilder. The research on the isolation, proof of structure, and synthesis of thiamine was published mostly in *Journal of the American Chemical Society*. Particularly significant are "Larger Yields of Crystalline Antineuritic Vitamin," in *Journal of the American Chemical Society*, **56** (1934), 1187–1191, written with R. E. Waterman and J. C. Keresztesy; "Structure of Vitamin B_1," *ibid.*, **57** (1935), 229–230, and **58** (1936), 1063–1064; "Studies of Crystalline Vitamin B_1. III. Cleavage of the Vitamin With Sulfite," *ibid.*, 536–537, written with R. E. Waterman, J. C. Keresztesy, and E. R. Buchman; "Studies . . . B_1. VIII. Sulfite Cleavage. Chemistry of the Acidic Product," *ibid.*, 1093–1095, written with E. R. Buchman and A. E. Ruehle; and "Studies . . . B_1. XI. Sulfite Cleavage. Chemistry of the Basic Part," *ibid.*, 1849–1851, written with E. R. Buchman and J. C. Keresztesy.

II. SECONDARY LITERATURE. The only significant short biography of Williams appears in *National Cyclo-*

paedia of American Biography, current vol. F (1942), 204–205. Also see *McGraw-Hill Modern Men of Science*, I (New York, 1966), 537.

AARON J. IHDE

WILLIAMSON, ALEXANDER WILLIAM (*b.* Wandsworth, London, England, 1 May 1824; *d.* Hindhead, Surrey, England, 6 May 1904), *organic chemistry.*

Although he published little, Williamson was the most influential chemist in Great Britain during the period 1850–1870, two critical decades in which chemists released themselves from the stranglehold of Berzelius' electrochemical dualism, forged a unitary system of inorganic and organic chemistry, created a rational system of atomic weights, developed concepts of valence and structure, and organized themselves professionally. In all these changes and developments Williamson was a leader, as researcher, teacher, critic, and elder statesman.

Williamson was the second of three children of Alexander Williamson, a clerk at East India House who was a friend of the economist James Mill, and Antonia McAndrew, a merchant's daughter. Throughout his life he was racked by severe physical disabilities: a semiparalyzed left arm, a blind right eye, and a myopic left one. These deficiencies undoubtedly promoted his later disenchantment with detailed laboratory work and encouraged his theoretical and speculative powers, which had been stimulated by his philosophical education. In 1840, following schooling and private instruction at Kensington, Paris, and Dijon, Williamson began medical training at the University of Heidelberg, where he was encouraged to study chemistry by Leopold Gmelin. From 1844 to 1846 he worked with Liebig at Giessen, where he published his first papers on bleaching salts, ozone, and Prussian blue. Apparently he had independent means, and from 1846 to 1849 he established a private research laboratory in Paris, where he fraternized with Laurent, Gerhardt, Wurtz, and Dumas and, on the recommendation of his older childhood friend John Stuart Mill, took private lessons in mathematics from Auguste Comte. The latter regarded Williamson as one of his most promising converts to positivism, but in England Williamson proved a disappointing disciple and advocate.

In 1849, encouraged by Graham and supported by Liebig, Dumas, Laurent, and Hofmann, Williamson applied for the vacant chair of practical chemistry at University College, London. On Graham's retirement from the chair of general chemistry in 1855 Williamson, rather unfortunately, took both titles. He remained at University College until his retirement to farming in the countryside at Hindhead in 1887. As far as active research was concerned, however, Williamson's retirement dated from the completion of his etherification studies in 1854. The reasons for this are complex. Certainly the falling off of his research was not due to any loss of mental power, nor entirely to his absorption in academic politics (where his agitation for university science degrees was successful in 1870), nor yet to his involvement in the affairs of the Chemical and Royal Societies or the British Association for the Advancement of Science. The apparent decline of his work did, however, accompany the development of other practical and engineering interests that ultimately proved fruitless technically and financially, and for which little documentary evidence survives. The only positive results of these enthusiasms were pedagogic: Williamson insisted that his chemistry classes take conducted tours of industrial plants, and he was instrumental in creating a chair of applied chemistry (chemical engineering) at University College.

Many private letters refer to Williamson's superior and acute intellectual powers. Kekulé, who was in London from 1854 to 1855, found his friendship and ideas "excellent schooling for making the mind independent," while Odling was always proud to have followed in his footsteps. Acquaintances were sometimes repelled by his tendency to make cutting remarks; and in arguments he constantly interrupted, so that his opponent's meaning could not be fully expressed. He was a forceful and dogmatic critic of papers at the Chemical Society, for which he established the valuable system of monthly abstracts of British and foreign papers in 1871. Basically, however, he was a kindly man whose sociability, and that of his wife Emma Catherine Key, made him the natural choice as British host to the first Japanese noblemen who came to England to learn Western ways in 1863.

Williamson announced his elegant theory of etherification to the British Association at Edinburgh in August 1850. At this time there were various rival theories concerning the structures of alcohol and ethyl ether, but in all cases ether was supposed to be formed by the loss of water from alcohol. Williamson's initial intention, however, was not to clarify a muddled theoretical situation but to develop practical methods for preparing the homologous higher alcohols. To his "astonishment," when he used Hofmann's alkyl radical substitution technique and reacted ethyl iodide with a solution of potassium in alcohol (potassium ethylate [ethoxide]), he obtained ordinary ethyl ether instead of an ethylated alcohol (Williamson's synthesis).

Influenced by his familiarity with the work of Laurent and Gerhardt, Williamson saw that the relationship between alcohol and ether could not be one of the loss or addition of water but, rather, of substitution, since ether contains two ethyl radicals but the same quantity of oxygen as alcohol. Since equal molecular magnitudes were involved, the formulas of these substances had to be expressed in terms of the French chemists' formula for water, H_2O (instead of HO or H_4O_2):

$$C^2H^5I + \frac{C^2H^5}{K}O = \frac{C^2H^5}{C^2H^5}O + KI \ (C = 12).$$

Williamson saw, however, that this result might still be explained according to a four-volume formulation (H_4O_2) if it were supposed that both potassium ethylate and ethyl iodide contained ether:

$$C^4H^{10}O + C^4H^{10}I^2 = 2(C^4H^{10}O) + 2KI.$$
$$K^2O$$

This possibility was disproved by using methyl (instead of ethyl) iodide, in which case a mixed ether was formed, not a mixture of ethyl ether and methyl ether:

$$CH^3I + \frac{C^2H^5}{K}O = \frac{C^2H^5}{CH^3}O + KI,$$

$$\text{not} \quad C^2H^6I^2 + C^4H^{10}O = C^4H^{10}O + C^2H^6O + 2KI.$$
$$K^2O \qquad \text{ethyl} \quad \text{methyl}$$
$$\text{ether} \quad \text{ether}$$

These views were confirmed independently by G. C. Chancel in 1850.

Williamson also explained the process of continuous etherification by the action of sulfuric acid on alcohol. According to the contact theory of Mitscherlich and Berzelius, the sulfuric acid acted merely as a catalyst in this reaction; but according to Liebig's chemical theory, ether was produced only after the intermediate formation of ethyl hydrogen sulfate. The latter, argued Williamson, played a role in the double decomposition analogous to that of potassium ethylate in his synthesis of ether. He represented the exchanges in two stages:

$$\text{I.} \quad \frac{H}{H}SO^4 \quad \frac{H}{C^2H^5}O \qquad \frac{C^2H^5}{H}SO^4 \quad \frac{H}{H}O \qquad \text{II.} \quad \frac{C^2H^5}{H}SO^4 \quad \frac{H}{C^2H^5}O \qquad \frac{H}{C^2H^5}SO^4 \quad \frac{C^2H^5}{C^2H^5}O$$

The sulfuric acid produced in II was recycled for further decompositions.

Williamson's impressive study has been rightly seen as laying the foundation for twentieth-century mechanistic studies. Historically it had a number of important consequences. First, Williamson completely rejected the notion of a catalytic force and opted for chemical intermediates in catalyzed reactions. In this stand he echoed Comte, for whom catalysis was a metaphysical fancy improper to the positive stage that chemistry was achieving. Second, Williamson was led to visualize atoms and molecules in motion, and not as the static particles of traditional Daltonism. The mechanism of etherification was inconceivable unless it was viewed as a process of continuous atomic exchange. Such a viewpoint proved to be a major step toward the reunification of chemistry with physics via the kinetic theory of gases, the ionic theory of electrolytes, and the revival of Berthollet's law of mass action. Williamson developed his views on dynamic atomism only in a series of lectures; confirmatory experiments with slow double decompositions were abandoned after a severe explosion. Finally, and most important, the study suggested that analogies for the structures of both organic and inorganic substances should be based on the inorganic type, water.

Echoing Laurent's use in 1846 of a water analogy, Williamson argued in 1851 that "water may be assumed as a very general type and standard of comparison, by viewing other bodies as formed from it by the replacement of one or more atoms of hydrogen in water by the equivalent of various simple or compound radicals" (*Papers on Etherification*, 40). For instance,

$$\frac{H}{H}O \qquad \frac{C_2H_5}{H}O \qquad \frac{C_2H_5}{C_2H_5}O \qquad \frac{C_2H_3O}{H}O \qquad \frac{C_2H_3O}{C_2H_5}O.$$
$$\text{water} \quad \text{ethyl} \quad \text{ethyl} \quad \text{acetic} \quad \text{ethyl}$$
$$\text{alcohol} \quad \text{ether} \quad \text{acid} \quad \text{acetate}$$

Through the work of Odling and Gerhardt this formal analogy with water completed the unification of organic and inorganic chemistry (the "new type theory"); and through the admission of valence by Frankland, Kekulé, and Odling it permitted the emergence of the real structural formulas that Williamson saw as the ultimate goal of positive chemistry. The use of multiples of water, suggested by Williamson in 1851, gave a ready explanation for differences of basicity and was confirmed by him in 1856, when the chlorination of sulfuric acid produced chlorosulfonic acid:

$$\begin{bmatrix} H \\ H \\ H \\ H \end{bmatrix} O \quad SO_2 \begin{matrix} H \\ O \\ O \\ H \end{matrix} O \xrightarrow{PCl_5} SO_2 \begin{matrix} H \\ O \\ Cl \end{matrix} O .$$

Similarly, his "prediction" of a class of anhydrous organic acids formed by the replacement of the H of acetic acid was dramatically confirmed by Gerhardt in 1852, when he reacted acetyl chloride with potassium acetate:

$$C_2H_3O \atop K \!\! O + C_2H_3OCl = C_2H_3O \atop C_2H_3O \!\! O + \quad KCl.$$
$$\text{acetic anhydride}$$

During the 1860's Williamson, who was a devout atomist, did much to eradicate the predominant skepticism of his fellow chemists. His proselytism culminated in a famous clash with his friend B. C. Brodie, Jr., at the Chemical Society in 1869. Brodie's interest in notation and nomenclature was shared by Williamson, who faced the problems involved when writing his chemistry textbook (1865). In 1864 he introduced parentheses into formulas to enclose reaction-invariant groups, such as $Ca(CO_3)_3$; abolished the Berzelian "plus" sign in compounds; and proposed the suffix "-ic" for the base of all salts (including hydrogen) to avoid circumlocution—for instance, "sodic chloride" for "chloride of sodium" and "hydric sulphate" for "sulphuric acid." Although the latter convention was adopted by several British chemists, it did not survive into the twentieth century.

BIBLIOGRAPHY

I. ORIGINAL WORKS. Thirty-five papers by Williamson are recorded in Royal Society, *Catalogue of Scientific Papers*, VI, 379–380; VIII, 1244–1245; XI, 817; XIX, 637. The major ones are conveniently collected as *Papers on Etherification and on the Constitution of Salts*, Alembic Club Reprint no. 16 (Edinburgh, 1902; reiss., 1949). This reprint unfortunately does not include "On Dr. Kolbe's Additive Formulae," in *Journal of the Chemical Society*, 7 (1855), 122–129, published in reply to A. W. H. Kolbe, "Critical Remarks on Williamson's Water, Ether and Salt Theories," *ibid.*, 111–121—German original in *Justus Liebigs Annalen der Chemie*, **90** (1854), 44–61. These papers exemplify the clash between the type and radical theories. Williamson published two books: a political tract written with his father-in-law,

T. Hewitt Key, *Invasion Invited by the Defenceless State of England* (London, 1858); and *Chemistry for Students* (Oxford, 1865; 2nd ed., 1868; 3rd ed., 1873), for which *Problems From Williamson's Chemistry With Solutions* (Oxford, 1866) was also issued.

Williamson's correspondence with Brodie is printed in W. H. Brock, ed., *The Atomic Debates* (Leicester, 1967), 95–96, 119–120. The bulk of Williamson's papers, which were in the possession of his son, Dr. Oliver Key Williamson, were stolen during a native rebellion in Africa (private information from J. Harris, 1963). For surviving letters and referee's reports, consult the Royal Society, Royal Institution, and Imperial College Archives, London.

II. SECONDARY LITERATURE. There are two detailed and very fine obituaries by Williamson's pupils: Edmund Divers, in *Proceedings of the Royal Society*, **78A** (1907), xxiv–xliv, with portrait; and G. Carey Foster, in *Journal of the Chemical Society*, **87** (1905), 605–618; the German trans. in *Berichte der Deutschen Chemischen Gesellschaft*, **44** (1911), 2253–2269, has an unusual photograph. Williamson's relationship with Comte is discussed by W. M. Simon, "Comte's English Disciples," in *Victorian Studies*, **8** (1964–1965), 161–162; and by Brock (see above), 145–152, who also treats Williamson's atomism extensively. For Williamson's friendship with Laurent and Gerhardt, see the nonindexed E. Grimaux and C. Gerhardt, Jr., *Charles Gerhardt: sa vie, son oeuvre, sa correspondance* (Paris, 1900), 218, 220–221 (letter of 1851), 240–243, 249–250, 263–264, 412–413, and 558; and with Kekulé, see R. Anschütz, *August Kekulé*, I (Berlin, 1929), *passim*. The context of Williamson's work is fully discussed in J. R. Partington, *A History of Chemistry*, IV (London, 1964), ch. 14; J. S. Rowe, "Chemical Studies at University College, London" (Ph.D. diss., London, 1955), 211–328e; and C. A. Russell, *A History of Valence* (Leicester, 1971), ch. 3. See also J. Harris and W. H. Brock, "From Giessen to Gower Street: Towards a Biography of Williamson," in *Annals of Science*, **31** (1974), 95–130.

W. H. BROCK

WILLIAMSON, WILLIAM CRAWFORD (*b.* Scarborough, Yorkshire, England, 24 November 1816; *d.* Clapham Common, England, 23 June 1895), *botany, geology, zoology, paleontology.*

Williamson's father, John Williamson, was an accomplished gardener and naturalist who for many years served as curator of the Scarborough Museum, Yorkshire. His close friendships with the geologists John Phillips and William Smith greatly aided and influenced young Williamson. His mother was Elizabeth Crawford, the daughter of a jeweler and lapidary of Scarborough. As a boy Williamson was devoted to his maternal grandfather, from whom he learned the art of cutting and polishing stones.

Williamson was married in June 1842 to Sophia Wood, who died in 1871. Three years later he married Annie C. Heaton; their son Herbert became a painter. A member and officer of several leading scientific societies, he was elected to fellowship of the Royal Society in 1854 and received its gold medal in 1874.

Williamson's activity ranged widely over the field of natural history in his research, teaching, and popular lecturing. With considerable success he bridged two ages in science—from the early 1800's, when a natural scientist was expected to deal with nearly all natural phenomena, to the close of the century, with its rapidly developing specialization. He is best remembered today by students of paleobotany for his series of studies of the fossil plants of the British coalfields, in which he laid a large part of the foundations of our knowledge of the earliest pteridophytes as well as of the early seed plants.

Most of Williamson's scientific publications present data that he discovered and recorded with brilliance and efficiency. In addition, his *Reminiscences* constitutes a delightful and informative record of a vanished way of life that had a real bearing on the development of biological and geological science in the nineteenth century.

Because of poor health, Williamson's education was inadequate. At the age of six he was sent to William Potter's school and studied Latin and English. He received his only real instruction from the Reverend Thomas Irving at the Thornton grammar school, which he attended for only six months. When he was fifteen Williamson's parents decided that he would benefit from education in France. He arrived at the school, which was located at Bourbourg, near Calais, but found that there were no vacancies; and it was only with difficulty that he obtained admission as a special student. The school was attended mostly by English boys and English was the spoken language. He learned little and returned home in less than a year, disillusioned.

In the meantime Williamson's parents had determined that he should prepare for a medical career, a decision that had somewhat happier results than his French venture. His medical studies began in 1832 with a three-year apprenticeship to Thomas Weddell, a general practitioner in Scarborough. Such doctors were rarely paid for visits to their patients; rather, their income was derived from the sale of the drugs that were prescribed. The preparation of these medications was one of the chief responsibilities of the apprentice. His other duties consisted of delivering the drugs and preparing the annual bills. Although he claimed that he learned little with Weddell that could not have been mastered in a few weeks in an apothecary's shop, Williamson did have considerable free time to extend his field studies of rocks, fossils, and plants of the surrounding area and to read scientific literature.

Through his father, Williamson had met, at a very early age, such great naturalists as Murchison, Sedgwick, Lyell, and Buckland. At the age of sixteen or seventeen, while apprenticed to Weddell, he was invited to prepare many of the illustrations for Lindley and Hutton's *Fossil Flora of Great Britain*, a renowned compilation of the day that is still used as a reference work.

In 1835, through the aid of a Dr. Phillips of Manchester, Williamson was appointed curator of the Manchester Natural History Society, with an annual stipend of £110. He resigned this position in June 1838 in order to enter medical school in Manchester. Faced with the necessity of obtaining funds to pay his school fees and living expenses, Williamson gave a series of lectures in nearby towns. Teaching, particularly popular lecturing, was clearly a great joy to him; and he was adept at presenting his vast scientific knowledge in an understandable fashion to a variety of audiences.

Medical instruction in Manchester was mediocre at best and, determined to avail himself of what seemed to be the best that Britain could offer, Williamson entered University College, London, in September 1840. He was not disappointed and quickly became aware that with such teachers as Quain in anatomy, Sharpey in physiology, Robert Liston and Astley Cooper in surgery, C. B. Williams in medicine, Graham in chemistry, and Lindley in botany, "no man with brains could fail to learn" (*Reminiscences*, p. 86). Although Williamson had prepared drawings for Lindley some years before, the two had never met; and Lindley was quite surprised to find that he had had so youthful a collaborator. Williamson's medical studies progressed well, but his other interests were not abandoned; he found time to attend meetings of the Geological Society of London, where he furthered his acquaintances with Sedgwick, Murchison, Greenough, James Yates, and Basil Hall.

On 1 January 1842 Williamson returned to Manchester, mounted a brass plate on the door of a house at the corner of Wilton Street and Oxford Road, and started his medical practice. While attending an afternoon tea with a friend of his wife, he met a young man who asked whether he had read Mantell's recently published *Medals of Creation*. Later, in reading this work, Williamson was attracted by the author's report that native chalk consisted largely of microscopic fossil shells; this observation led him to initiate one of his most important scientific studies of diatoms, desmids, and Foraminifera, and opened up other lines of research involving microscopy.

In 1851 Williamson was elected first professor of natural history and geology at the newly founded Owens College in Manchester; and for about nineteen years he taught the courses in botany, comparative anatomy, geology, and paleontology. As the amount of knowledge and the college enrollment increased, he was relieved of teaching geology by the appointment of William Boyd Dawkins in 1872; and in 1880 Arthur Milnes Marshall joined the staff as zoologist, leaving Williamson free to devote more time to botany.

He continued to conduct his busy medical practice along with his academic duties. Having a special interest in ear ailments, Williamson spent some time in Ménière's consulting rooms at Paris, learning his techniques and the use of new instruments; and he also studied in London. After returning to Manchester, he helped raise funds for the establishment of an institution for aural ailments.

One of Williamson's earliest investigations was concerned with a tumulus—a Bronze Age grave—found at Gristhorpe Cliff, a few miles south of Scarborough. Although others were also involved, Williamson played a leading part in the excavation and laboratory treatment of the remains; and his report was published in Scarborough in 1834—before he was eighteen. The article attracted the attention of Buckland, who reprinted part of it in a weekly journal. Interest in the paper led to the publication of a second edition in 1836, and in 1872 Williamson prepared a third, revised edition.

In 1842 Williamson was given a small sample of sediment from the Levant in which he found Foraminifera. Similar material was received from other sources, including samples from Charles Darwin, who had recently returned from the *Beagle* voyage. These studies produced an article published in the *Memoirs of the Manchester Literary and Philosophical Society* that was a pioneering contribution to the understanding of the part that Foraminifera played in the formation of geological deposits. In 1851–1852 Williamson conducted a very meticulous and time-consuming study of *Volvox globator*, a motile, spherical colony of green cells, barely visible with the naked eye, that is commonly found in fresh waters. He observed asexual reproduction, apparently for the first time, and discovered basic facts about the mode of connection of the several hundred cells that make up the plant.

In 1849 and 1851 Williamson published two papers in the *Philosophical Transactions of the Royal Society* on the cellular structure and development of the teeth and bones of fish. As a developmental study this pioneering research was highly regarded and was instrumental in his election to the Royal Society.

Williamson's greatest contribution dealt with the petrified plants found in the Upper Carboniferous coal seams of Lancashire and Yorkshire. Sometime in the 1840's collectors began to bring him specimens of "coal balls," aggregations of petrified plants found in the coal itself that included fragments of stems, leaves, seeds, and other reproductive parts sometimes preserved in excellent cellular detail. As representative samples of the vegetation of the Carboniferous, "coal balls" presented an unparalleled source of information, and Williamson took full advantage of it. The challenge was to prepare the plant materials for study and to fit the parts together so as to reconstruct the trees that forested so much of the northern hemisphere some 200 million years ago.

Williamson's contribution in this area is the more remarkable because he prepared his own thin sections of the fossil material and drew most of the numerous illustrations for his text. The main body of this research was published in the *Philosophical Transactions of the Royal Society* in nineteen parts (1871–1893). At the outset Williamson was probably unaware of the magnitude of the task; and he encountered some difficulty when he submitted the first part under the title "On the Organization of the Fossil Plants of the Coal-Measures. Part I," because an editor complained that it would obligate the publication of more than one part. His "coal ball" investigations supplied much basic information on the early evolution of the pteridophytes and the more primitive seed plant groups, and were continued by many investigators in Europe and the United States. The nineteen papers may not contain outstanding writing, but they are a sound record on which later workers have been able to reconstruct, to a considerable degree, the vegetation of the Carboniferous landscapes.

Williamson served as president of the Manchester Scientific Students' Association, the Manchester Literary and Philosophical Society, and the

Union of Yorkshire Naturalists. He received the gold medal of the Royal Society in 1874 and the Wollaston Medal of the Geological Society in 1890. The University of Edinburgh awarded him the LL.D. in 1883, and he was elected an honorary member of the Göttingen Academy of Sciences and the Royal Swedish Academy of Sciences.

It is sad that in 1892, at the age of seventy-six and after forty-one years of service to Owens College, Williamson's application for a pension was refused by the College Council—on the grounds that it might establish a precedent. He retired to the London suburb of Clapham Common. As his strength declined, and with much fossil plant research left to do, Williamson invited D. H. Scott, a young botanist from Kew Gardens, to collaborate with him. Scott carried on the work for many years and became a major figure in paleontology.

BIBLIOGRAPHY

I. ORIGINAL WORKS. Williamson's autobiography is *Reminiscences of a Yorkshire Naturalist* (London, 1896), edited by his wife. In addition to several popular lectures on scientific subjects, he also published *On the Recent Foraminifera of Great Britain* (London, 1858); and *A Monograph on the Morphology and Histology of Stigmaria ficoides* (London, 1887). The Royal Society *Catalogue of Scientific Papers*, VI, 380–381; VIII, 1245–1246; XI, 817–818; and XIX, 638–639; lists 106 articles, including two papers written with D. H. Scott.

Works cited in the text are *Description of the Tumulus Lately Opened at Gristhorpe* (Scarborough, 1834; 2nd ed., 1836; 3rd ed., rev., 1872); "On Some of the Microscopical Objects Found in the Mud of the Levant and Other Deposits, With Remarks on the Mode of Formation of Calcareous and Infusorial Siliceous Rocks," in *Memoirs of the Manchester Literary and Philosophical Society*, 2nd ser., **8** (1848), 1–128; "On the Microscopic Structure of the Scales and Dermal Teeth of Some Ganoid and Placoid Fishes," in *Philosophical Transactions of the Royal Society*, **139** (1849), 435–476; "The Structure and Development of the Scales and Bones of Fishes," *ibid.*, **141** (1851), 643–702: "On the *Volvox globator*," in *Memoirs of the Manchester Literary and Philosophical Society*, 2nd ser., **9** (1851), 321–339; and in *Transactions of the Royal Microscopical Society of London*, n.s. **1** (1853), 45–56; and "On the Organization of the Fossil Plants of the Coal-Measures," 19 pts., in *Philosophical Transactions of the Royal Society*, **161–184B** (1871–1893), his most important work.

II. SECONDARY LITERATURE. On Williamson and his work, see Charles Bailey, "Memoir of Professor Williamson," in *Report and Proceedings of the Manchester*

Scientific Students' Association for 1886, 1–8; the unsigned "In Memoriam. William Crawford Williamson," in *Proceedings of the Yorkshire Geological and Polytechnic Society*, n.s. **8** (1899), 95–111; H. D. Scott, "Williamson's Researches on the Carboniferous Flora," in *Science Progress*, **4** (1895), 253–272; and L. F. Ward, "Saporta and Williamson and Their Work in Paleobotany," in *Science*, n.s. **2** (1895), 141–150.

Further biographical details may be found in *Mémoires de la Société de physique et d'histoire naturelle de Genève*, **32** (1894–1897), pt. 2, xi–xii; *Geological Magazine*, **2** (1895), 383–384; *Journal of Botany, British and Foreign*, **33** (1895), 298–300; *Journal of the Royal Microscopical Society* (1895), 478; *Leopoldina*, **31** (1895), 169; *Nature*, **52** (1895), 441–443; *Canadian Record of Science*, **6** (1896), 443–447; *Memoirs and Proceedings of the Manchester Literary and Philosophical Society*, 4th ser., **10** (1896), 112–125; and *Proceedings of the Royal Society*, **60** (1897), xxvii–xxxii.

HENRY N. ANDREWS

WILLIAM THE ENGLISHMAN (*fl.* France, thirteenth century), *astronomy, astrology.*

All that is known of William's life is that he was a physician who lived in Marseilles during the first half of the thirteenth century. Of the works that have been variously attributed to him, four, *De urina non visa*, an *Astrologia*, a *Summa super quarto libro metheorum*, and a treatise on the *astrolabium Arzachelis*, or saphea, may be considered to be his with some certainty. All four may be dated between 1220 and 1231. A number of other works may be less clearly ascribed to William.

Of the tracts known to be by William (Guillelmus Anglicus civis Marsiliensis), the first, *De urina non visa* ("Ne vel ignorancie vel potius invidie redarguar mi germane. . ."), is a brief argument that an astrologer's prognostication of an illness, based on astral conjunctions, may have the same validity as that of a physician who has been able to observe the symptomatology of the patient. William here cites the case of a patient whom he himself had examined, and for whom he had made the astrological prediction that he would live for another two months and eight days; his prediction proved correct. The planetary positions upon which he based it indicate that the diagnosis could have been made only in the last days of the year 1219; the *De urina non visa* cannot have been written before 1220, and the date 1219 that occurs in one manuscript must have meant *anno completo*.[1]

The text of William's *Astrologia* is undated, but a manuscript in the Biblioteca Capitular Colombina in Seville ascribes it to 1220. This treatise, which begins "Quoniam astrologie speculatio prima figuram ipsius. . .," is a theory of the planets that follows, like a commentary, the canons of the Toledan Tables, although they are never referred to specifically. The work opens with a description of the astrolabe and some of its uses, then reviews the principles of stereographic projection, together with a note on the construction of an *instrument azimutal* for taking meridian altitudes. William does not give any exact information about the eccentricities of the planets, the length of the radii of the epicycles, or the duration of their revolutions; his materials on the conjunctions of the sun and moon and on eclipses is, however, developed at some length. The examination of the movement of the eighth sphere – that is, the motion of "accession and recession" – was discussed at the end of the treatise, as in the canons of al-Zarqālī.

The only known manuscript of the *Summa super quarto libro metheorum*, which begins "Rerum corruptibilium effectus ut ad nutum et voluntatem . . .," gives 1230 as the date of the text. This rather brief text is organized into three parts, of which the first deals with terrestrial phenomena, both those originating within the earth (water, gems, and minerals) and those found upon its surface (as, for example, alums and salts). The second section treats of aerial, or meteorological matters, including evaporation, dew, rain and snow, ice, winds, earthquakes, and rainbows, while the concluding portion discusses the ethereal region of shooting stars, thunder, lightning, and comets. William drew the hypothesis that the Flood had resulted from the joining together of the waters that fall from the sky and the waters that well up from the depths of the earth. He further attributed the *passiones etheris* to vapors rising from the earth, the nature of the phenomenon being dependent upon the nature and form of the ascendant vapor – thus, dry vapors moving directly upward create shooting stars.

William's treatise on the saphea, or *astrolabium Arzachelis*, beginning "Siderei motus et effectus motuum speculator . . .," represents the introduction of this instrument into the Latin West. Invented in the eleventh century by 'Alī ibn Khalaf or by al-Zarqālī,[2] the saphea is, like the astrolabe, conceived on the principle of the stereographic projection of both the movable sphere of the stars and the zodiac and a fixed sphere of celestial reference; it differs from the astrolabe in that the pole of projection is one of the points of intersection of the ecliptic and the zodiac, while the plane of projection is that of the colure of the solstices.

Contrary to what is generally believed, William's treatise on the saphea is not an abridgment of al-Zarqālī's, which became known in the Latin West only through Ibn Tibbon's translation of 1263. It is likely that William knew of this kind of projection only by hearsay and had never seen it put into practice, since the method he himself proposes – and upon which he says he worked for six years – is at best rather clumsy. William attempted to construct the almucantars of the horizon by piercing small holes on the limb of his instrument and holding strings parallel to the diameter by which this horizon was projected. Thus the instrument he describes is characterized by both the stereographic projection proper to the saphea and by the orthographic projection invented by J. de Rojas in the middle of the sixteenth century.[3] The saphea devised by al-Zarqālī, which became known to the West in 1263, used the almucantars of the equator as a provisional coordinate system: a reference point could then be selected on this system and moved to that of the horizon by means of a ruler. He remained true to the tradition of the astrolabe, inscribing twenty stars on his saphea although they have no purpose on it. Their positions were taken directly from al-Zarqālī's table, which had been established according to Ptolemy's positions, with the addition to the longitudes of a constant of 14°7′.[4]

In addition to these works, a number of others attributed to William remain problematical. Confusions and improbable identifications with other Williams, notably Guillaume Grisaunt and William of Aragon[5] have further wrongly extended the list. A *De virtute aquile* attributed to a "Guillelmus Anglicus" is known in only one manuscript.[6] While the sixteenth-century bibliographer John Bale composed a list of nine titles, of which six are given with their incipits,[7] only one corresponds to a genuine work, the *Astrologia*. The other works ascribed by Bale to William are all to some degree suspect. A *De quadratura circuli* ("Aristoteles in eo qui de categoriis . . .") is found in other manuscripts only under the name of Campanus;[8] a *De motu capitis* ("Motum accessionis et recessionis. . .") and a *De magnitudine solis* ("Dico quod sol apparet mag. . .") have not been located; a *De qualitatibus signorum* ("Cum humana corpora sint omnia. . .") has been erroneously cited by E. Zinner in a manuscript at Bamberg which does not contain the incipit given;[9] a *De significatione signorum* cannot be identified for lack of an incipit; a *De urina non visa* has an aberrant incipit ("De corpore quidem humano Satur. . ."); and a *De judicio patientis* and a *De causa ignorancie* are probably doublets of the *De urina*.[10]

The attribution to a "W. A." by Wellcome manuscript 175 of a pseudotranslation of Abū-l-Qāsim 'Ammor's (Conamusoli) *De infirmitate oculorum* likewise cannot be maintained,[11] while the *De stellis fixis* that P. Duhem has put forward as William's translation of a treatise by al-Zarqālī is in fact only the star table appended to the treatise on the saphea. The *Scripta Marsiliensis* or *Scripta super canones Azarchelis* (beginning "Cum," or "Quoniam, cujuslibet actionis; liber iste scilicet canones tabularum. . ."),[12] a commentary on the canons of the Toledan Tables, cannot be by William, although sometimes ascribed to him, since it alludes to the turquet, and therefore cannot have been written earlier than the end of the thirteenth century. (The first treatise on this instrument was in fact written by either Francon of Poland in 1284 or by Bernard of Verdun at a slightly earlier date.)[13] It is further worth noting that the *Scripta Marsiliensis* suggests, especially in certain aspects of the development of its treatment of the movement of the eighth sphere, the commentary on these same canons written by John of Sicily in 1291.[14]

NOTES

1. L. Thorndike, *A History of Magic and Experimental Science*, II, 485–486, n. 5. Simon de Phares, *Recueil des plus célèbres astrologues* . . ., E. Wickersheimer, ed. (Paris, 1929), 180, 191, devoted two accounts to William the Englishman: one for the beginning of the twelfth century, in which he also attributed to the author of the *De urina non visa* a prediction of the destruction of Liège, and one for 1219, in which he alludes to a "book of astrology" the first words of which ("De ignorantie . . .") recall those of *De urina*.

2. J.-M. Millás Vallicrosa, *Estudios sobre Azarquiel* (Madrid–Granada, 1943–1950), 438–447. On the theory of the saphea, see S. García Franco, *Catalogo crítico de astrolabios existentes en España* (Madrid, 1945), 64–65; H. Michel, *Traité de l'astrolabe* (Paris, 1947), 95–97.

3. Michel, *op. cit.*, 20, 105–107; and F. Maddison, *Hugo Helt and the Rojas Astrolabe Projection* (Coimbra, 1966; *Agrupamento de Estudos de Cartografia Antiga, no. 12*).

4. The MS tradition of William's star table is so mediocre that its origin is difficult to uncover; on this table see P. Kunitzsch, *Typen von Sternverzeichnissen in astronomischen Handschriften des 10. bis 14. Jahrhunderts* (Wiesbaden, 1966), 77.

5. Thorndike, *op. cit.*, 301.

6. *Ibid.*, 487.

7. J. Bale, *Scriptorum illustrium Majoris Brytannie*, I (Basel, 1557), 446; the same information can be found in Bale's *Index Britanniae scriptorum*, R. L. Poole and M. Bateson, eds. (Oxford, 1902), 114–115.

8. This attribution is examined and the text is edited by M.

Clagett, *Archimedes in the Middle Ages*, I (Madison, Wis., 1964), 581–609; to the twelve MSS cited add Columbia University Smith add. 1, fols. 138v–139.

9. E. Zinner, *Verzeichnis des astronomischen Handschriften des deutschen Kulturgebietes* (Munich, 1925), 4034.

10. J. Bale, *Scriptorum*, II (1559), 46, also notes a Guillelmus Anglicus who is the author of a *De incarnatione verbi* and of a *Commentarium de anima* that are hardly likely to be by the citizen of Marseilles. C. H. Lohr, "Medieval Latin Aristotle Commentaries," in *Traditio*, **24** (1968), 194, who has not identified this William, does not indicate the author of the commentary on the fourth book of the *Meteorology*.

11. See G. Sarton, *Introduction to the History of Science*, I (Baltimore, 1927), 729.

12. A short extract was published by M. Curtze, "Urkunden zur Geschichte des Trigonometrie im christlichen Mittelalter," in *Bibliotheca mathematica*, 3rd ser., **1** (1900), 321–416: no. 3 (347–353), "Aus den Scripta Marsiliensis super canones Azarchelis." MSS are Berlin F.246, fols. 144–154v; Erfurt F.394, fols. 111v–119.

13. E. Poulle, "Bernard de Verdun et le turquet," in *Isis*, **55** (1964), 200–208.

14. The following text (188–190) that Sédillot published does not belong to the treatise of William the Englishman; it is the first chapter of Ibn Tibbon's translation (in 1263) of the treatise by al-Zarqālī.

BIBLIOGRAPHY

I. ORIGINAL WORKS. The *De urina non visa* is preserved in numerous MSS; to those cited in L. Thorndike, *A History of Magic and Experimental Science*, II (New York, 1923), 485–486, add Berlin F.246, fols. 252v–254v; Cracow 551, fols. 122–124; Oxford, Hertford Coll. 4, fols. 44–46v. A French trans. of this text is cited by J. Camus, "Un manuscrit namurois du XVᵉ siècle," in *Revue des langues romanes*, **38** (1895), 31–32. The *Astrologia* is preserved in the following MSS: Erfurt F.394, fols. 136–140v; Erfurt 4° 357, fols. 1–21; Paris lat. 7298, fols. 111v–124v; Seville 5-I-25, fols. I–33; Vienna 5311, fols. 42–51v. Of the *Summa super quarto libro metheorum*, only one MS is known: Paris lat. 6552, fols. 39v–41v.

As for the treatise on the saphea, the section on its construction was published by L. A. Sédillot, "Mémoire sur les instruments astronomiques des Arabes," in *Mémoires présentés par divers savants à l'Académie des inscriptions et belles-lettres*, 1st ser., **1** (1844), 185–188, text repr. in R. T. Gunther, *The Astrolabes of the World*, I (Oxford, 1932), 259–262; and that on its uses by P. Tannery, "Le traité du quadrant de maître Robert Anglès," in *Notices et extraits des manuscrits de la Bibliothèque nationale*, **35**, pt. 2 (1897), 75–80, repr. in his *Mémoires scientifiques*, V, 190–197.

II. SECONDARY LITERATURE. Brief accounts of William are in P. Duhem, *Le système du monde*, III (Paris, 1915), 287–291; L. Thorndike, *A History of Magic and Experimental Science*, II (New York, 1923), 485–487; E. Wickersheimer, *Dictionnaire biographique des médecins en France au moyen âge* (Paris, 1936), 224–225;

and C. H. Talbot and E. A. Hammond, *The Medical Practitioners in Medieval England* (London, 1965), 381–382.

EMMANUEL POULLE

WILLIS, BAILEY (*b.* Idlewild-on-Hudson, New York, 31 March 1857; *d.* Palo Alto, California, 19 February 1949), *geology*.

The early schooling of Bailey Willis, son of poet Nathaniel Parker Willis, was a haphazard mixture of private tutoring, some classroom instruction, and listening to the literary repartee engendered by his father's friends and work. After Nathaniel Willis' death in 1868, the boy's maternal grandfather, Arctic explorer Joseph Grinnell, decided that his bright grandson should have the advantage of more rigorous schooling. So at the age of thirteen Bailey was enrolled in a German boarding school under stern Prussian professors who encouraged concentration on studies by liberal application of the rattan switch. Fortified by four years of this preparation, young Willis returned to New York in 1874, and enrolled at Columbia University, where he received degrees in mining engineering (1878) and civil engineering (1879).

In 1880 the Northern Pacific Railway hired Willis as geologist, and sent him to explore for coal in the forest wilderness of Washington Territory. East of his area towered Mount Rainier, then known by its Indian name, Tacoma. This glacier-clad volcano claimed Willis' interest for the rest of his life; his first two scientific publications are about it, and he pioneered a new climbing route to its summit via glaciers on its north side. In 1894 Willis submitted documents to Congress that led, in 1899, to establishment of Mount Rainier National Park.

In 1882 Willis married Altena Holstein Grinnell; she died four years later, leaving an infant daughter. In 1898 Willis married Margaret Delight Baker; three children were born of this union. Willis was frequently away from home, engaged in geologic field work or foreign exploration. Margaret Baker Willis saved all his letters written during these travels. From these letters, after his retirement, came three partly autobiographical books: *Living Africa* (1930), *A Yanqui in Patagonia* (1947), and *Friendly China* (1949).

Willis was employed by the U.S. Geological Survey from 1882 to 1915, except for one year (1903–1904) spent in geologic exploration in China, and four years (1910–1914) helping the Argentine government start a geological survey and plan irrigation projects. During his career with the U.S. Geological Survey Willis published more than sixty scientific papers including several long monographs. One of many papers that received wide attention came from Willis' deep interest in the complexly folded rocks of the Appalachian Mountains. Trained as an engineer, Willis longed to simulate the great Appalachian folds in the laboratory. He built a "pressure box" and began experiments. After initial failures Willis recognized that materials used in his small pressure box must be scaled down in strength if they were to behave like the huge masses of rock deformed by compression in the crust of the earth. Using plastic materials, and also by loading his artificial strata under a cover of loose lead shot, he made artificial "mountains" with folds that closely resembled those of the Appalachians. His report "The Mechanics of Appalachian Structure" (1893) brought Willis international fame, especially in Europe, where experimental geology was gaining in interest.

In 1915 Willis joined the Stanford University faculty as professor and chairman of the geology department. During his years as professor, and those after retiring at sixty-five, Willis' geologic interests became more firmly entrenched in structural geology and seismology. His textbook *Geologic Structures* (1923) and more than seventy papers dealing with continental genesis, rift valleys, faults, and earthquakes were published during this period. So familiar did Willis become in many California towns on his search for knowledge about earthquakes, that he was known throughout the state as the "earthquake professor." He tried to get California politicians to pass an enforceable building code outlawing the shoddy construction that had compounded damage during earthquakes. His rueful assessment of this effort was "I didn't get the building code, but I certainly did increase earthquake insurance rates."

Willis received his first honorary doctorate in 1910 from the University of Berlin for the Appalachian experiments and for his research in China. Later he was elected to the National Academy of Sciences and the American Philosophical Society. Belgium awarded him the Legion of Honor in 1936, and he won the Penrose Medal of the Geological Society of America in 1944.

BIBLIOGRAPHY

I. ORIGINAL WORKS. A complete bibliography of Willis' publications is in *Bulletin of the Geological Society* of America, **73** (1962), 68–72. Some of his better-known works are "Canyons and Glaciers. A Journey to the Ice Fields of Mount Tacoma," in *Northwest*, **1**, no. 2 (1883), repr. with slight modification, in E. S. Meany, *Mount Rainier. A Record of Exploration* (New York, 1916), 142–149; "Mount Tacoma in Washington Territory," in *Proceedings of the Newport Natural History Society*, **2** (1884), 13–21; "Conditions of Sedimentary Depositions," in *Journal of Geology*, **1** (1893), 476–520; "Conditions of Appalachian Faulting," in *American Journal of Science*, **46** (1893), 257–268, written with C. W. Hayes; "The Mechanics of Appalachian Structure," in *Report of the United States Geological Survey*, **13**, pt. 2 (1893), 211–281; "Some Coal Fields of Puget Sound," *ibid.*, **18**, pt. 3 (1898), 393–436; "A Symposium on the Classification and Nomenclature of Geologic Time-divisions," in *Journal of Geology*, **6** (1898), 345–347; "The Mount Rainier National Park," in *Forester*, **5** (1899), 97–102; "Individuals of Stratigraphic Classification," in *Journal of Geology*, **9** (1901), 557–569; "Stratigraphy and Structure, Lewis and Livingston Ranges, Montana," in *Bulletin of the Geological Society of America*, **13** (1902), 305–352; "Physiography and Deformation of the Wenatchee-Chelan District, Cascade Range," in *Professional Papers. United States Geological Survey*, no. 19 (1903), 41–97; *Carte géologique de l'Amérique du Nord*, prepared for Congrès Géologique Internationale, 10th session (Mexico, 1906); and "Research in China," in *Publications. Carnegie Institution of Washington*, no. 54 (1907), written with E. Blackwelder, R. H. Sargent, and F. Hirth.

Other works are *Outline of Geologic History With Special Reference to North America* (Chicago, 1910), written with R. D. Salisbury; "Index to the Stratigraphy of North America," in *Professional Papers. United States Geological Survey*, no. 71 (1912); *Northern Patagonia* (New York, 1914); "Discoidal Structure of the Lithosphere," in *Bulletin of the Geological Society of America*, **31** (1920), 247–302; Fault Map of the State of California (1922), compiled with H. O. Wood, from data assembled by the Seismological Society of America; *Geologic Structures* (New York, 1923); "Dead Sea Problem: Rift Valley or Ramp Valley?" in *Bulletin of the Geological Society of America*, **39** (1928), 490–542; "Continental Genesis," *ibid.*, **40** (1929), 281–336; "African Rift Valleys—a Geological Study," in *Carnegie Institution Washington News Service Bulletin*, no. 2 (1930), 27–34; *Living Africa. A Geologist's Wandering Through the Rift Valleys* (New York, 1930); "Isthmian Links," in *Bulletin of the Geological Society of America*, **43** (1932), 917–952; "African Plateaus and Rift Valleys," in *Publications. Carnegie Institution of Washington*, no. 470 (1936); "Asthenolith (Melting Spot) Theory," in *Bulletin of the Geological Society of America*, **49** (1938), 603–614; "San Andreas Rift, California," in *Journal of Geology*, **46** (1938), 1017–1057; "Eruptivity and Mountain Building," in *Bulletin of the Geological Society of America*, **52** (1941), 1643–1683, written with Robin Willis; *A Yanqui in Patagonia* (Stanford, 1947);

and *Friendly China: Two Thousand Miles Afoot Among the Chinese* (Stanford, 1949).

II. SECONDARY LITERATURE. On Willis and his work, see E. Blackwelder, "Bailey Willis, 1857–1949," in *Biographical Memoirs. National Academy of Sciences*, **35** (1961), 333–350; Philip B. King, "Bailey Willis," in *Dictionary of American Biography*, supp. 4 (1974), 896–897; and Aaron C. Waters, "Memorial to Bailey Willis, 1857–1949," in *Bulletin of the Geological Society of America*, **73** (1962).

AARON C. WATERS

WILLIS, ROBERT (*b.* London, England, 27 February 1800; *d.* Cambridge, England, 28 February 1875), *engineering, medieval archaeology.*

The son and grandson of distinguished physicians, Willis was educated privately until he entered Gonville and Caius College, Cambridge, where he obtained the B.A. and was elected fellow in 1826. The next year he was ordained deacon and priest. In 1837 he succeeded William Farish as Jacksonian professor of natural and experimental philosophy at Cambridge, a position he held until his death. His major contributions were to the kinematics of mechanisms and the study of the architecture of the Middle Ages.

As Jacksonian professor, Willis delivered lectures on "mechanical philosophy" and applied his considerable inventive skills to the improvement of a set of machine parts for the construction of demonstration models. By modifying Farish's original concepts, Willis reportedly was able to popularize what may well have been the first mechanical model-building kit. His lectures formed the basis for his book, *Principles of Mechanism* (1841).

Willis was one of the first clearly to enunciate the importance of excluding causal forces in the study of the motion of machinery. He introduced the term "kinematics," as an anglicized version of the French *cinématique*, thereby originating the English name of the branch of mechanics that deals with the geometry of motion without regard to the forces causing it. He felt it was important to change the study of machinery from a descriptive to an analytic science. Although his analytical treatments were modest by modern standards, Willis' advocacy of a systematic approach to the design of machine mechanisms anticipated and encouraged the subsequent development of kinematic analysis and synthesis of mechanisms. He organized mechanisms according to an original classification scheme, little of which is in use today although it was quite popular until almost 1900.

By 1870, when an enlarged second edition of his book was published, Willis was able to point out that all (thirteen) books on mechanisms published (in England and France) in the previous twenty-two years utilized his classifications and nomenclature. The present ideas on classification, however, are derived mainly from Franz Reuleaux, who published his landmark book in the year of Willis' death. Of lasting value was Willis' chapter on gear teeth, taken from an earlier paper (1838). In this outstanding work he introduced the idea of manufacturing sets of interchangeable gears, pointed out the convenience of 14.5° involute gear teeth (in common use today), and described his invention of a device to facilitate the approximate layout of gear teeth. (He coined the name "odontograph," which has become the generic term for devices used to assist in that process.) His book remains the earliest attempt to develop a complete treatment of the science of machine kinematics.

Willis studied architecture and archaeology with great enthusiasm and expertise, becoming an authority on medieval Latin and construction techniques. In 1835 he published an essay that is reputed to have been the first work to call serious attention to the Gothic style of architecture. He then invented a device (the "cymograph") for copying architectural moldings. Subsequently he published studies detailing the original construction and the later modifications to the cathedrals of Hereford, Canterbury, Winchester, York, Chichester, and Worcester and the abbey churches of Glastonbury and Sherborne. To all these works he brought such great skills in antiquarian research and keen insight that most of these works are still regarded as the definitive studies of these churches. He continued his archaeological writings until his health became seriously impaired, several years before his death.

Willis' earliest works include an analysis of a well-known mechanical chess player, which he correctly proved was a hoax (1821), and three papers applying mathematical analysis to the flow of air. The last of these (1828–1829) describes the action of the larynx. As a result of this work he was elected to the Royal Society (1830). Willis served on government commissions in such diverse fields as structures and astronomy, and participated in two international expositions. He held honorary membership in and medals from various learned societies, and in 1862 was president of the British Association for the Advancement of Science.

BIBLIOGRAPHY

I. ORIGINAL WORKS. A complete list of works published during Willis' lifetime is given in his book, *Principles of Mechanism*, 2nd ed., enl. (London, 1870), the title page of which contains a list of his honors and more prominent affiliations. Eleven years after Willis' death his nephew and biographer, John Willis Clark, completed and published Willis' *The Architectural History of the University and Colleges of Cambridge*, 4 vols. (Cambridge, 1886), written with T. W. Clark.

II. SECONDARY LITERATURE. *Dictionary of National Biography*, XXI, 492–494, contains an excellent biography by John Willis Clark. Abbreviated biographical data are in J. A. Venn, *Alumni Cantabrigienses*, pt. II, VI (Cambridge, 1954). Venn includes some items that Clark omits, and differs with Clark on several minor dates. For a historical perspective on Willis' mechanism work see R. S. Hartenberg and J. Denavit, *Kinematics of Linkages* (New York, 1964), 16, 70–75; and E. S. Ferguson, "Kinematics of Mechanisms From the Time of Watt," *Contributions. Museum of History and Technology, United States National Museum*, paper 27 (Washington, D.C., 1962). For historical perspective on Willis' architectural studies one apparently must consult books on the various cathedrals. See, for example, the intro. to H. R. Williamson, *Canterbury Cathedral* (London, 1953).

BERNARD ROTH

WILLIS, THOMAS (*b*. Great Bedwyn, Wiltshire, England, 27 January 1621; *d*. London, England, 11 November 1675), *anatomy, medicine.*

Thomas was the eldest of three sons of Rachel Howell and Thomas Willis, the steward of the manor at Great Bedwyn. Before his mother's death in 1631, the family moved to North Hinksey, Berkshire, where the mother had property. The proximity to Oxford (a mile and a half) enabled young Willis to be schooled there with Edward Sylvester, who numbered John Wilkins among his former pupils. He matriculated in the university from Christ Church on 3 March 1637, and worked as a servitor to one of the cathedral canons while proceeding B.A. (19 June 1639) and M.A. (18 June 1642). His father's death in 1643 left Thomas head of the family, and his own partisan military service in the losing royalist cause during the siege of Oxford forestalled his career in the church. He turned to medicine, taking his B.Med. and license to practice on 8 December 1646.

Almost from the beginning of his medical career, Willis evinced an interest in science, first in mathematics but increasingly in chemistry and anatomy.

He and two Trinity fellows, Ralph Bathurst and John Lydall, carried out chemical experiments in Willis' Christ Church rooms in the late 1640's. When other eminent scientists—Wilkins, William Petty, John Wallis, Seth Ward, and Jonathan Goddard—were appointed to Oxford positions during the years 1648–1651, Willis and his friends joined them and others in forming a philosophical "Clubb" to meet weekly and perform experiments *in rota*. By the mid-1650's the club had grown to include Robert Boyle, Christopher Wren, Thomas Millington, and Robert Hooke, the last of whom Willis hired out of Christ Church as a chemical assistant, later recommending him for the same position with Boyle.

Willis and his friends also did anatomical dissections. A clinical notebook kept by Willis (*ca.* 1651) recorded not only his careful case notes, but also the results of occasional postmortems. In the famous case of Anne Green, convicted and hanged at Oxford for infanticide in late 1650, Willis, Petty, Bathurst, and others assembled for a dissection, only to find the cadaver still very much alive.

Working within this milieu, Willis completed by 1656 his first scientific work, *De fermentatione*. During the late 1650's he supplemented this with a major work on fevers, *De febribus*, and a shorter piece on urine, *Dissertatio epistolica de urinis*, written to Bathurst; they were published in 1659 as *Diatribae duae medico-philosophicae*.

In *De fermentatione* Willis argued that all bodies are composed of five kinds of particles: those of spirit, sulfur, salt, water, and earth, in order of decreasing activity. Any body containing a mixture of these particles is capable of fermentation, which Willis defined as an intestine (internal) motion of a body's chemical particles leading to the perfection or transformation of that body. According to this process must becomes wine and wort becomes beer; liquids coagulate, or solids precipitate from them; food is converted into chyme, and thence into blood. Animals and plants grow by the process of fermentation, just as they are corrupted by it after their deaths.

But for the physician, the most important kinds of fermentation take place in the fluids of the human body, most especially the blood. Accepting Harvey's theory, he proposed a reason for the circulation: as the blood and its dissolved food pass through the heart, a ferment implanted there excites a fermentation, or "accension," by which heat is generated and the food converted into nutrient blood.

Willis' theories were new versions of ideas widely discussed on the Continent during the 1640's and 1650's, but with the important difference that Willis cast his explanations into the atomistic and chemical terms that Boyle had made so popular among his Oxford friends. Helmont had proposed that numerous physiological processes are carried out by fermentation, but had used that term to denote an animistic function of the soul. Descartes and the Dutch Cartesian Hogelande had written of a ferment in the heart, but not in such chemical and corpuscular detail.

But Willis was a practitioner as well as a theorist, and the second tract of 1659, *De febribus*, exemplifies what was to be the enduring characteristic of his published works: the concern to use anatomy, physiology, and chemistry to explain clinical findings. Harvey's discovery of the circulation, Willis said, necessitated a new theory of fevers based upon knowledge of fermentation. Fever was nothing but the natural exothermic fermentation of the heart, excited to a preternatural degree by foreign materials introduced into the blood. After speculations on the mechanisms that might be involved in this derangement, Willis filled the remainder of the tract with detailed descriptions of fevers drawn from his own casebooks. His characterizations of epidemics were particularly acute, reporting in detail the first English outbreak of war-typhus, among the Oxford troops in 1643, cases of plague in 1645, measles and smallpox in 1649 and 1654, and influenza in 1657 and 1658. He also recorded what seems to be the first reliable clinical description of typhoid fever. To a great degree he, rather than Sydenham or Morton, began the tradition of English epidemiology.

Willis' life changed radically after the restoration of Charles II in 1660. Wilkins and other friends moved to London and founded the Royal Society. Willis remained in Oxford, and both his growing scientific reputation, and his unswerving loyalty to king and church, were rewarded by his appointment at the university as Sedleian professor of natural philosophy. He graduated D.Med. on 30 October 1660, and began regular lectures on natural philosophy and medicine, which attracted a large audience.

The notes for some of these lectures were copied out by Willis' assistant from Christ Church, Richard Lower, and later extracted by Lower's friend John Locke. They show the degree to which Willis ignored the statutory injunction to teach only from Aristotle. Rather, he lectured on neurological topics: sense and motion, the cerebellum, sleeping and waking, pleasure and pain, as well as the clinical effects of neurological changes in diseases such as convulsions, epilepsy, hysteria, vertigo, lethargy, and paralysis.

But, as he remarked later in the preface to *Cerebri anatome*, he was dissatisfied with the excessively speculative nature of these lectures, and in late 1661 he and Lower began a series of dissections of the brain with a view toward clarifying such questions. Lower wrote to Boyle in early 1662 that Willis found "most parts of the brain imperfectly described," and intended "to make a whole new draught thereof, with the several uses of the distinct parts." Thomas Millington contributed to the discussions, and just before the completion of the book in July 1663, Wren executed a magnificent series of drawings to illustrate the text.

The *Cerebri anatome*, published early in 1664, is the foundation document of the anatomy of the central and autonomic nervous systems. It greatly surpassed, in the detail and precision of its descriptions, the fragmentary treatments of the brain that had preceded it. As a text it continued to be used until the late eighteenth century, and was mandatory background reading for neuroanatomists until the mid-nineteenth century.

Willis' description and classification of the pairs of cranial nerves superseded those of Falloppio (1561), and remained in widespread use until those of Sömmering (1778) replaced them in the late eighteenth century. Willis recognized ten such nerves. His first six are those used today: olfactory, optic, oculomotor, trochlear, trigeminal, and abducens. His seventh cranial nerve included both the facial and auditory (VII and VIII), while his eighth combined the glossopharyngeal and vagus (IX and X) with the cranial root of the spinal accessory (XI). Willis' ninth cranial nerve is the hypoglossal (XII), and his tenth is the modern first cervical. He described and delineated the spinal root of the accessory (XI) nerve, not numbering it separately, but pointing out how it accompanies and then diverges from the vagus. The distribution of all cranial nerves is described in great detail.

Willis' aim in tracing out the cranial nerves was rather more physiological than anatomical: they fitted closely into his ideas of cerebral and cerebellar localization. Most classical and Renaissance anatomists had believed that the three commonly recognized mental functions, sense, imagination, and memory, were carried out by animal spirits inhabiting the cerebrospinal fluid that filled the system of cerebral ventricles. While he accepted the action of animal spirits, Willis rejected ventricular localization on the grounds that distinctions of function could better be maintained by animal spirits acting within the solid portions of the brain.

Willis believed that voluntary functions are localized in the cerebrum. Animal spirits intended for these functions are generated in the grey cerebral cortex from the arterial blood, which continually bathed the cerebrum. These spirits pass inward into the white medullary matter, where they are differentiated and distributed into tracts for separate kinds of voluntary action. Those concerned with sense are localized in the corpora striata, those with imagination (intelligence) in the corpus callosum, and those with memory distribute back outward into the cerebral cortex.

According to Willis, sense impressions are carried inward to the corpora striata, where an inward perception arises. If the impression is carried farther on to the corpus callosum, then imagination results. If the fluctuation of spirits are struck back out to the cortex, memory of the event or idea is created. If impressions are reflected back out to the voluntary muscles directly from the corpora striata, then a "reflex" action could occur without conscious volition—a concept of the reflex considerably more sophisticated than that which Descartes had propounded a few years earlier.

In performing these functions the spirits transmit their information as successive wave fronts along predetermined medullary tracts. And just as ripples from several sources in a pond can cross unchanged, so also can the same tracts carry both sensory (afferent) and motor (efferent) impulses.

Involuntary functions are carried out in an analogous way by the cerebellum and its attendant structures, the corpora quadrigemina and the medulla oblongata. Spirits for these functions are generated from the blood in the cerebellar cortex, then flow into the underlying medullary structures, where they interact to regulate heartbeat, respiration, and digestion. Therefore it is proper that four pairs of cranial nerves (modern V–X) which have involuntary functions should have their origins near the cerebellum.

These involuntary actions are performed especially by what Willis called the "intercostal" and "vagal" nerves—respectively the modern sympathetic and parasympathetic divisions of the autonomic nervous system. The former Willis believed to have an intracranial origin, arising indirectly from the V and VI nerves, while the latter come directly from the vagus (X). He traced in great detail how these two systems, and their attendant spinal ganglia, innervate all the major organs of the thorax and abdomen. By this pathway the cerebellum and medulla could effect the regulation of involuntary functions and, in turn, the states of the viscera could affect the higher conscious functions of the central cerebrum and the higher brain stem.

Good follower of Harvey that he was, Willis did not neglect to trace how all these structures are bathed and nourished by the circulating blood. In preparing the book, Lower wrote in a letter to Boyle (4 June 1663), Willis had been especially struck by a postmortem in which a man dying of an unrelated disease exhibited a completely occluded right carotid artery. Yet blood continued to flow to both cerebral hemispheres, and the patient had complained only of a headache on the left side, where the carotid artery had been enlarged by the increased blood flow. To account for this, Willis traced out the circle of anastomosed arteries at the base of the brain by which, if any carotid or vertebral arteries were blocked, the remaining ones could maintain full blood flow to all parts of the brain. Willis and Lower confirmed this by tying both carotids in a spaniel, to no ill effect. They further demonstrated it by using a technique of injection developed in Oxford a few years earlier by Wren and Boyle; they syringed ink into one artery, and observed it flowing out from the others. The "circle of Willis," as it has since been known, is clearly delineated in the Wren drawings in the *Cerebri anatome*. Although an anatomical description of the circle had been published by Wepfer in 1658, Willis was the first to grasp and demonstrate its physiological and pathological significance.

Many of Willis' deepest insights derived from an unparalleled knowledge of the comparative anatomy of the nervous system. In writing *Cerebri anatome* he drew conclusions from dissections of fish, birds, and more than a dozen different mammals. From these he suggested that the convolutional complexity of the human cerebral cortex is correlated with man's greater intelligence. He observed that the cerebellum has a uniformity of appearance in mammals that accords well with its function as a source of animal spirits for involuntary actions. He rejected the Cartesian suggestion that the pineal gland is the seat of the soul because he saw its presence not just in man, but also in other quadrupeds, birds, and fish.

The *Cerebri anatome*, and to a lesser degree the *Diatribae duae*, established the lines of research with which Willis occupied the remainder of his scientific life. The action of the nervous system and the composition and function of the blood were his two primary foci. He explored anatomical structures (usually with the assistance of a junior collaborator such as Lower, or his successors Edmund King and John Masters), postulated a series of conclusions about their functions, deduced from them his explanations of malfunction in the course of disease, and illustrated his conclusions with case histories and postmortems.

Before leaving Oxford in late 1667 to set up a large and lucrative practice in London, Willis brought out his *Pathologiae cerebri et nervosi generis specimen*, the clinical study that he had promised as a companion volume to the *Cerebri anatome*. He based his analysis of convulsive diseases upon the belief that muscle contraction results from the explosive mixing of two types of particles: saline-spirituous particles from the nerves, and nitrosulfurous particles from arterial blood. When spirits of too heterogeneous a nature are supplied by the nervous system, the muscle contraction is too powerful and uncontrolled, thus causing convulsive symptoms. Thus, he argued, epilepsy originates in the central cerebrum, not in the meninges. Or too few particles could be supplied by the nerves, resulting in weak contractions, a condition he illustrated with the first clinical description of myasthenia gravis. Convulsive coughs and asthma, hysterical and hypochondriacal disorders, even scurvy, he saw as nervous afflictions.

In 1670 he published another set of tracts elaborating his earlier ideas on metabolic heat and muscular contraction. He accepted Lower's contention, advanced in the *Tractatus de corde* (1669), that the heart is merely a muscle and does not have an innate ferment. Therefore, the body's heat must come from an "accension" or fermentative process lodged in the blood itself. This process demands both a sulfurous fuel, derived from food, and nitrous particles derived from the air. It is exactly analogous to inorganic combustion. Here Willis was both elaborating his and Bathurst's earlier ideas on the metabolic function of air, and adding to them ideas about aerial "nitrous particles" published by his Oxford confreres Boyle, Hooke, Lower, and John Mayow, during the 1660's.

These same themes, joined to others also adumbrated in the *Cerebri anatome*, were the core of his anatomical and clinical study *De anima brutorum*, published in 1672. Man, he said, has two souls: a corporeal, mortal soul which he shares with animals, and a rational, immortal one which is uniquely human. The corporeal, or "brutish," soul consists of two parts: one lodged in the blood is responsible for vital functions, the other located in the nervous system is responsible for functions of action and sensation. The vital soul in the blood performs nourishment by taking up and distributing food particles, and produces heat and vitality by "burning" some of these with the nitrous particles derived from the air. In explaining the functions of the sensitive soul, he recapitulated many of the neurophysiological concepts first introduced in *Cerebri anatome*, especially those of localization, and extended these to invertebrates with some of the first detailed dissections of the earthworm, oyster, and lobster. He traced in detail how vibratory impressions from the five senses are transmitted through the plenum of animal spirits which inhabit the nervous system, and how these impulses are interpreted, processed, and stored in specialized parts of the cerebrum and medulla oblongata.

As in previous books, Willis was not satisfied with anatomical investigation and speculative interpretation. He goes on to argue, with the aid of extensive case histories and numerous postmortems, how a broad range of disorders are due to derangement of the neural portion of the corporeal soul. Sleeping and waking, headache, lethargy, narcolepsy, coma, nightmare, vertigo, apoplexy, delirium, frenzy, and paralysis—all are of neurological, rather than supernatural or humoral, origins.

Willis' last work, the *Pharmaceutice rationalis*, was cast from the same mold. In its two parts, brought out in 1674 and 1675, he summarized the anatomy and physiology of the thoracic and abdominal organs, hypothesized mechanisms of their pathology, and filled pages with case histories, therapies, and postmortems. Many observations testify to his acute clinical judgment. He discovered the superficial lymphatics of the lungs, distinguished acute tuberculosis from the chronic fibroid type, and gave the first clinical and pathological account of emphysema. He described extrasystoles of the heart, aortic stenosis, heart failure in chronic bronchitis, and emboli lodging in the pulmonary artery. He was the first European to note the sweet taste of the urine in diabetes mellitus, and described the pains and weakness of diabetic polyneuritis. He made original observations on the muscle layers of the stomach wall, and devised the use of a whalebone probang to treat achalasia of the cardia.

Unfortunately, Willis was scarcely able to enjoy the acclaim that greeted his later works. Tending a busy London practice, he was unable—and perhaps disinclined—to participate in the activities of the Royal Society and the Royal College of Physicians. His personal life was touched with tragedy: six of his eight children died before adolescence; his first wife, the sister of John Fell, dean of Christ Church, died in 1670; and both of his brothers predeceased him. Willis died of pneumonia on 11 November 1675, and was buried in the north transept of Westminster Abbey. He was survived by his second wife.

Willis has often been castigated for the unremittingly speculative nature of much of his writings, but that is to judge him by the scientific taste of the present century. He saw himself as a physician whose lasting contribution would be to formulate a series of corpuscular explanations that would link anatomical fact with clinical practice. A number of these hypotheses were, even if incorrect, extremely fruitful. His notion of animal heat arising from a fermentation in the blood, fed by a nitrous aerial agent, was elaborated by Mayow into a concept of respiration and metabolism that foreshadowed Lavoisier's discovery of oxygen a century later. Willis' ideas of cerebral localization were the impetus for a line of experimental work traceable into the early nineteenth century. His notion of the corporeal soul in the nervous system, and the disorders to which it was prone, was both a contribution to comparative psychology and the beginning of modern concepts of neurology. His speculations on the involuntary functions of the "intercostal" and "vagal" nerves provided the foundation of our knowledge of the autonomic nervous system. Yet if these ideas were more subject to correction than his easily verifiable conclusions on cerebral and cerebellar structure, cerebral circulation, and the cranial nerves, they were no less important a part of his *oeuvre*. Willis accomplished much, not in spite of his penchant for speculation, but because of it. He attempted, with an energy and insight unsurpassed in the seventeenth century, to construct a medical system that encompassed not only his own anatomical discoveries and acute clinical observations, but set them within the emerging Harveian physiology and the new corpuscular natural philosophy.

BIBLIOGRAPHY

I. ORIGINAL WORKS. Willis' clinical notebook (*ca.* 1651) is MS 799ᴬ in the Wellcome Medical Historical Library, London. *Diatribae duae medico-philosophicae, quarum prior agit de fermentatione . . . altera de febribus* (London, 1659) had a 2nd ed., with additions, in 1662. Locke's extracts from Lower's transcript of Willis' lectures are in the Bodleian Library, Oxford, MS

Locke f. 19, pp. 1–82, *passim. Cerebri anatome: cui accessit nervorum descriptio et usus* (London, 1664) was published that year first in a quarto and then in an octavo ed.; a trans. is available in modern facs. (Montreal, 1964). *Pathologiae cerebri, et nervosi generis specimen* (Oxford, 1667) contains a frontispiece portrait of Willis, aetatis suae 45, drawn by Loggan. The two tracts *De sanguinis accensione* and *De motu musculari* were published with *Affectionum quae dicuntur hystericae & hypochondriacae pathologia spasmodica vindicata* (London, 1670). *De anima brutorum* (Oxford, 1672) is available in a facs. of a 17th-century trans. *Pharmaceutice rationalis* was published in two pts.: I (Oxford, 1674), II (Oxford, 1675), published posthumously. Willis' *Opera omnia* was published immediately after his death (Geneva, 1676) and reprinted several times in the succeeding decades. These often contain a tract *De ratione motus muscularum* which was misattributed to Willis; it was published anonymously by William Croone (London, 1664). All of Willis' works, with the exception of *Affectionum*, were translated by Samuel Pordage and published in one volume as *Practice of Physick* (London, 1684).

II. Secondary Literature. Hansruedi Isler, *Thomas Willis, 1621–1675: Doctor and Scientist* (New York, 1968), is the only full-length historical treatment of Willis. Audrey B. Davis, *Circulation Physiology and Medical Chemistry in England 1650–1680* (Lawrence, Kans., 1973), sheds much light on Willis' chemical ideas. The fundamental study by Alfred Meyer and Raymond Hierons, "On Thomas Willis's Concepts of Neurophysiology," in *Medical History*, 9 (1965), 1–15, 142–155, has an extensive bibliography which provides the best entree into the literature on specialized topics of Willis' life and work.

Robert G. Frank, Jr.

WILLISTON, SAMUEL WENDELL (*b.* Roxbury, Massachusetts, 10 July 1851; *d.* Chicago, Illinois, 30 August 1918), *vertebrate paleontology, entomology, medicine.*

The fourth child of Samuel Williston—an unschooled blacksmith—and Jane A. Turner, Williston developed an early passion for books and education. Infected by the concern of New Englanders over the fate of Kansas in the turmoil leading to the Civil War, the family moved in 1857 to Manhattan, Kansas, where the young man entered Kansas State Agricultural College in 1866. The inspiring professor of natural sciences Benjamin Franklin Mudge introduced him to science. At the college "there were no laboratories of any kind, no microscopes and but few instruments. The college catalogue of about that time . . . gravely mentions

an electrical machine, three Leyden jars and six test-tubes!" (Williston, "Recollections" [1916]). Williston discovered Darwin's writings on evolution and lectured on them, for which he was denounced by the local church.

In 1874, two years after Williston received his B.S., Mudge invited him to join a fossil-collecting trip to western Kansas for Othniel Charles Marsh of Yale. Two summers of collecting led to Williston's direct employment by Marsh until 1885, as a leader of fossil-collecting expeditions in the western United States and as a laboratory assistant in New Haven, Connecticut. He worked at each of the first three American quarries of giant dinosaurs in Colorado and Wyoming soon after their almost simultaneous discovery in 1877.

Torn by uncertainties over which career to pursue, Williston, while working for Marsh, also earned an M.D. (1880) and a Ph.D. in entomology (1885), both from Yale. He married Annie Isabel Hathaway in 1881; they had five children.

Williston's contributions to medicine were chiefly in public health and education. From 1886 to 1890 he taught anatomy at Yale. In 1887 and 1888 he was health officer of New Haven; and from 1888 to 1890 he studied the pollution of rivers for the state of Connecticut. In 1891 he served on the Kansas State Board of Health and helped to establish the licensing and medical registration of doctors there. He was instrumental in establishing the medical school at the University of Kansas and became its first dean (1898–1902).

In 1890 Williston became a professor of geology and paleontology at the University of Kansas. He immediately began his own expeditions for vertebrate fossils, concentrating on the productive Cretaceous deposits of western Kansas and later on those of eastern Wyoming.

Williston entered paleontology when it was advancing beyond the stage of descriptive classification and was ready for synthesis. Georges Cuvier had described fossil remains, and Richard Owen had proposed the term *Dinosauria* before 1850. But the fossil remains in America were more extensive than those of Europe, and Darwin's theory of evolution provided impetus to the exploration of the past. Joseph Leidy had described American fossils, mainly from the east coast. When advancing railroads opened the American West to travel in the 1850's and 1860's, Marsh and Edward Drinker Cope competed for collections of fossil vertebrates from many different geologic horizons. They described many species and began the classification of the early reptiles and mammals.

In Kansas, Williston concentrated on mosasaurs, plesiosaurs, and pterodactyls, assembling the known information and describing new species. His detailed observations of the structure and anatomy of these animals enabled him to discuss their probable habits. He also found and described many other single fossils, publishing extensively in the *Kansas University Quarterly* throughout the 1890's. He contributed summary articles on the fossil birds, dinosaurs, crocodiles, mosasaurs, and turtles to the geological survey of Kansas (1898).

His acceptance of a professorship at the University of Chicago in 1902 soon changed the direction of his research, perhaps because of the Permian fossils already in the university collection. Williston collected extensively in the Permian red beds of Texas and New Mexico, ably aided by his preparator, Paul C. Miller. Rich pockets of fossils produced a wealth of labyrinthodont amphibians and reptiles from the early beginnings of the reptiles; indeed, some of the genera discussed by Williston (e.g., *Seymouria*) cannot be positively assigned to either one of these classes. His descriptions and synthesis of the classification of these primitive vertebrates stand as his major contribution to paleontology. Always fascinated by anatomy, he presented a number of carefully drawn restorations of the animals he described, many of his papers appearing in the *Journal of Geology* between 1902 and 1918.

While at the University of Chicago, Williston did not ignore the fossil groups that he had previously studied; in addition to single papers, he published his highly readable *Water Reptiles of the Past and Present* (1914). His knowledge of the broad spectrum of the reptiles led him to surveys of the evolution of the class and provided a firm foundation for later paleontologists. *The Osteology of the Reptiles* (1925), in preparation at the time of his death and completed at his request by William K. Gregory, summarized his classification.

Williston was president of the Kansas Academy of Science in 1897, president of Sigma Xi from 1901 to 1904, and delegate to the Ninth International Congress of Zoology at Monaco in 1913. He also received an Sc.D. from Yale in 1913, and in 1915 was elected to the National Academy of Sciences.

Williston's early indecision on his own career, combined with what he considered the tyranny of Marsh as an employer (Marsh did not want his employees to publish articles on paleontology), had led him into entomology and medicine. He began collecting beetles first as a diversion on field trips

in 1876, but he soon concluded that the order was already too widely known for a beginner to make a name for himself. Quite arbitrarily he selected flies as a hobby for study and he found his diversion highly rewarding. The Diptera of North America had been scarcely touched at this time (the 1870's), and Williston, through persistence and research in European publications, was able to classify the multitudinous members of the order. He also found his subjects readily available: on picnics, field trips, vacations, and even in the classroom while he lectured on paleontology. Although he never taught courses in entomology, he published a great many papers on Diptera and presented his classification of the order in three successive editions of *The Manual of North American Diptera* (1888; 2nd ed., 1896; 3rd. ed., 1908), the last edition of which was liberally illustrated by himself. He was consulted by entomologists throughout the world and received many specimens for identification.

Williston was an outstanding teacher and beloved by his students. His broad interests and keen enthusiasm influenced many students who later became prominent in a variety of fields, such as Clarence E. McClung, Ermine Cowles Case, and Barnum Brown.

BIBLIOGRAPHY

I. ORIGINAL WORKS. Williston published more than 300 papers and books; about half on paleontology and geology; about 100 on entomology; and the remainder on education, zoology, and public health.

Many of the short papers in paleontology are of value for descriptions of species and for the discussions of phylogenetic relationships. His individual papers on fossil birds, dinosaurs, crocodiles, mosasaurs, and turtles of Kansas in 1898 (*University Geological Survey of Kansas*, 4 [1898]) are of special value.

Significant review papers on specific Cretaceous vertebrate fossil groups include "Kansas Pterodactyls, Part I," in *Kansas University Quarterly*, 1 (1892), 1–13; "Kansas Pterodactyls, Part II," *ibid.*, 2 (1893), 79–81; "Range and Distribution of the Mosasaurs, With Remarks on Synonymy," *ibid.*, A ser., 6 (1897), 177–185; "Mosasaurs" (1898), cited above; "North American Plesiosaurs, Part I," in *Publications. Field Museum of Natural History*, Geological ser., 73 (1903), 1–77; and "North American Plesiosaurs: Trinacromerum," in *Journal of Geology*, 16 (1908), 715–736.

On Permian amphibians and reptiles, see *American Permian Vertebrates* (Chicago, 1911), a summary of the then-known information on this group morphologically and taxonomically; "Primitive Reptiles: A Review," in *Journal of Morphology*, 23 (1912), 637–663, enlarges the relationships of these animals to a worldwide scale and presents the characters necessary to the classification; "Permocarboniferous Vertebrates From New Mexico," in *Publications. Carnegie Institution of Washington*, **181** (1913), offers the results of Williston's collections and description of species in the New Mexico red beds; "Synopsis of the American Permocarboniferous Tetrapoda," in *Contributions From the Walker Museum*, **1** (1916), 193–236, is a final summary of the group that Williston knew better than anyone else at the time of his death.

On reptile classification, see *Water Reptiles of the Past and Present* (Chicago, 1914), with detailed information on the classification, habits, and special adaptations of all aquatic reptiles in their respective geologic periods; "The Phylogeny and Classification of Reptiles," in *Journal of Geology*, **25** (1917), 411–421, which contains Williston's final published graphical classification of the terrestrial vertebrates; and *The Osteology of the Reptiles*, W. K. Gregory, ed. (Cambridge, 1925), a valuable posthumous work, illustrated by extensive line drawings done by him in his final years.

Williston's major contributions in entomology were the three successive editions of *The Manual of North American Diptera* (published in New Haven, with slightly different titles in 1888 and 1896; 3rd ed., 1908). His earlier "Synopsis of North American Syrphidae," in *Bulletin. United States National Museum*, no. 31 (1886), is an exhaustive treatise of that family. South American, Central American, and West Indian Diptera were described in several monographs, reference to which can be found in Shor and in Lull (see below).

II. SECONDARY LITERATURE. Williston's childhood, youth, and the circumstances leading to his entry into paleontology are contained in his MS "Recollections," written in 1916; the MS is included in Elizabeth N. Shor's *Fossils and Flies* (Norman, Okla., 1971), as is an account of the remainder of Williston's life, scientific accomplishments, and a complete bibliography. Brief summaries of Williston's life are in "A Tribute to the Life and Work of Samuel Wendell Williston," *Sigma Xi Quarterly*, **7**, no. 1 (1919); and in Richard S. Lull, "Bibliographical Memoir, Samuel Wendell Williston," in *Memoirs of the National Academy of Sciences*, **17** (1924), 115–141. The latter includes a detailed summary of Williston's paleontological contributions and an almost complete bibliography.

All memorials to Williston prior to Shor give the date of his birth as 1852 instead of 1851, because of an indecipherable date in "Recollections." A birth certificate from Boston (which now includes Roxbury), Mass., confirms the birth date of 1851.

ELIZABETH NOBLE SHOR

WILLSTÄTTER, RICHARD (*b.* Karlsruhe, Germany, 13 August 1872; *d.* Locarno, Switzerland, 3 August 1942), *organic chemistry*.

Willstätter was the son of Max and Sophie Ulmann Willstätter. Raised in a well-to-do German-Jewish mercantile family, Willstätter attended school in Karlsruhe and then at the Realgymnasium in Nuremberg. Upon graduation in 1890, he entered the University of Munich. It was there that he came under the influence of Adolf von Baeyer, who took great interest in Willstätter's career, and recommended him to his colleague A. Einhorn, with whom Willstätter did his research for the Ph.D. (1894). Two years later Willstätter became *Privatdozent*, and in 1902 he was appointed extraordinary professor in Baeyer's institute. The following year he married Sophie Leser; they had two children before her untimely death in 1908. In 1905 he became professor of chemistry at the Eidgenössische Technische Hochschule in Zurich, but left in 1912 to become director of the new Kaiser-Wilhelm Institute of Chemistry in Berlin-Dahlem.

In 1916 Willstätter succeeded Baeyer as professor of chemistry at the University of Munich, but resigned in 1924 in protest of the vote of the faculty to deny V. M. Goldschmidt appointment as the successor of Groth. After 1925 most of Willstätter's research was conducted by Margarete Rohdewald in a laboratory provided by Wieland, Willstätter's successor at Munich. With the rise of Hitler, and the threat of arrest during the anti-Semitic campaign of 1938, Willstätter had to leave Germany; after considerable difficulty, he succeeded in emigrating to Switzerland in March 1939.

Willstätter's doctoral work with Einhorn dealt with the structure of cocaine, for which the latter had proposed a formula in 1893. After the completion of his Ph.D., Willstätter continued to work on the tropine alkaloids related to cocaine; and through a series of brilliant chemical degradative and synthetic researches during 1894–1898, he demonstrated that earlier formulas were incorrect and that these alkaloids belonged to a family of bicyclic compounds having a seven-member ring. During succeeding years he returned occasionally to the tropine alkaloids; thus, in 1918, he described an elegant synthesis of ecgonine. His studies in this field also led him to prepare cyclooctatetraene in 1913.

Another area of organic chemistry that Willstätter illuminated early in his career was the chemistry of quinones and quinone imines. In 1905 he prepared the hitherto unknown *o*-benzoquinone, and the succeeding seven years saw a series of important studies on the Wurster dyes, formed by the oxidation of *p*-phenylenediamines, and on "aniline black," a product of the oxidation of aniline.

The high point of Willstätter's chemical work was attained during 1905–1914, in his studies on chlorophyll and the anthocyanins. Although there had been much earlier work on chlorophyll—and its chemical relation to the porphyrin of hemoglobin was appreciated—it was Willstätter who laid the groundwork for the later complete elucidation of the structure of chlorophyll by Hans Fischer. Willstätter separated chlorophyll from green plants into two components (chlorophyll *a* and *b*), and showed them to be magnesium complexes of a porphyrin derivative (pheophytin) in which one of the two carboxyl groups is esterified by a long-chain alcohol (phytol). Through a series of chemical degradations with acid and alkali, Willstätter and his associates produced a series of well-characterized chemical intermediates on the way to the simpler porphyrins. During the course of his work on chlorophyll, he found that in the presence of ethanol, an enzyme present in plant tissue (chlorophyllase) catalyzes the transesterification of phytol by ethanol. This observation awakened his interest in enzymes, the principal subject of his last researches. The work on the chemistry of chlorophyll was followed, during 1915–1916, by studies on its role in photosynthesis, and on the assimilation by green plants of carbon dioxide; but these researches did not lead to clear-cut conclusions.

Willstätter's studies on the anthocyanin pigments of flowers, in particular on cyanin and pelargonin, led to his establishment of their structures as oxonium salts derived from hydroxychromanes, whose hydroxyl groups are linked to sugar units. These brilliant studies were terminated by the advent of World War I but served as the basis for the further development of the subject by R. Robinson. In 1915 Willstätter was awarded the Nobel Prize for chemistry in recognition of his "pioneer researches on plant pigments, especially chlorophyll."

With the resumption of his work after the war, Willstätter turned his attention to the nature of enzymes. From 1918 to 1925 he and his associates attempted to develop methods for the purification of a variety of enzymes: peroxidase, lipase, trypsin, and amylase. An important consequence of these efforts was the introduction, into enzyme chemistry, of reproducible adsorption methods (for example, the use of aluminum hydroxide) and of improved assay procedures for the determination of enzymic activity. Willstätter advocated the idea that enzymes are substances of low molecular weight that are merely adsorbed on colloidal carriers; he also questioned J. B. Sumner's claim (1926) to have isolated the enzyme urease in the form of a crystalline protein the integrity of which was essential for catalytic activity. Although some of Willstätter's students (notably E. Waldschmidt-Leitz) continued to defend his view into the 1930's, the isolation of pepsin and other enzymes as crystalline proteins led to its rejection. In his final years of research, with M. Rohdewald, Willstätter studied the biochemical transformation of glycogen and the role of polysaccharides in alcoholic fermentation; this work did not have a significant impact.

BIBLIOGRAPHY

I. ORIGINAL WORKS. Willstätter's works include *Untersuchungen über Chlorophyll* (Berlin, 1913), written with Arthur Stoll, trans. by F. M. Schertz and A. R. Merz, as *Investigations on Chlorophyll* (Lancaster, Pa., 1928). Willstätter and Stoll also collected their papers on photosynthesis, in *Untersuchungen über die Assimilation der Kohlensäure* (Berlin, 1918). The papers on enzymes are collected in *Untersuchungen über Enzyme*, 2 vols. (Berlin, 1928); see also *Problems and Methods in Enzyme Research* (Ithaca, N.Y., 1927). Willstätter wrote an autobiography, *Aus meinem Leben*, A. Stoll, ed. (Weinheim, 1948), trans. by L. S. Hornig as *From My Life* (New York, 1965).

II. SECONDARY LITERATURE. Appreciations of Willstätter's work are R. Huisgen, in *Journal of Chemical Education*, **38** (1961), 10–15; R. Robinson, in *Journal of the Chemical Society* (1953), 999–1026; and in *Obituary Notices of Fellows of the Royal Society*, **8** (1953), 609–634; and J. Renz, *Helvetica Chimica Acta*, **56** (1973), 1–14.

JOSEPH S. FRUTON

WILLUGHBY, FRANCIS (*b*. Middleton, Warwickshire, England, 22 November 1635; *d*. Middleton, 3 July 1672), *natural history*.

Willughby was the third child but only son of Sir Francis Willughby and his wife Cassandra, daughter of Thomas Ridgeway, earl of Londonderry. The Willughbys[1] of Wollaton, Nottinghamshire, and Middleton, Warwickshire, were country gentry with estates in many English counties. Francis was educated at Sutton Coldfield School, and in September 1652 entered Trinity College, Cambridge, where John Ray was a lecturer and where their lifelong association began. Upon graduation in 1656, Willughby continued his scientific studies, possibly at Cambridge, in the late 1650's[2]; and for a time in 1660 he was reading books on natural history at the Bodleian Library in Oxford. In 1663 he became one of the original fellows of the Royal Society of London.

During the summer of 1660, Willughby and Ray probably carried out their first tour together through northern England and the Isle of Man collecting botanical specimens. Accompanied by Philip Skippon, they made another tour through Wales and the west country in 1662. These tours were the prelude to a more ambitious journey. Their intentions had now widened into a comprehensive consideration of plants, birds, fish, animals, and insects. Sailing from Dover to Calais on 18 April 1663, Willughby, with Ray, Skippon, and Nathaniel Bacon, traveled through the Low Countries and Germany into Italy, where they spent the winter of 1663–1664 at Padua. It was there that they visited the botanical gardens and studied anatomy at the university, where Willughby matriculated in January 1664. At Naples, Willughby left the party and traveled through Spain before returning to England late in 1664. Besides collecting specimens, he purchased paintings or engravings of flowers, birds, fish, small mammals, and reptiles.

While Ray and Skippon were still abroad in March 1665, Willughby tapped birch trees and noted the behavior of the rising sap.[3] In 1669 and succeeding years he expanded and continued these experiments with Ray. Willughby reported his observations on leaf-cutting bees and on ichneumon wasps to the Royal Society in 1670 and 1671. When his father died in December 1665, Willughby continued to live with his mother at Middleton, where in the winter of 1666–1667 Ray assisted him in the arrangement and labeling of the collection of specimens. Resuming their field explorations in the summer, they toured southwest England.

In January 1668 Willughby married Emma Barnard, the younger daughter of Sir Henry Barnard; they had two sons and a daughter. Willughby had suffered periods of ill health and was naturally inclined to a studious life, but this did not deter him from undertaking strenuous journeys. At the time of his death he was planning to visit North America to study the animals there.

Willughby's work in natural history is inseparable from that of Ray, Skippon, and Francis Jessop. His name is associated particularly with birds, fish, animals, and insects, but his surviving collection shows that his botanical work was not insignificant. When Willughby died Ray was at Middleton,

and he remained there until the winter of 1675–1676, ostensibly as a trustee under Willughby's will and as tutor to his children, but primarily working on his collections. When compiling Willughby's *Ornithologia* (1676) and *Historia piscium* (1686), Ray supplemented Willughby's material with that of himself and other naturalists. Similarly, Ray's *Historia plantarum* and more especially his writings on animals and insects incorporated Willughby's observations. They were following in the footsteps of Jean and Gaspard Bauhin, Aldrovandi, Gesner, and Rondelet, to whose nomenclature they largely adhered. They consciously adopted a systematic approach, compiling detailed descriptions based on personal observation. Willughby's contribution was twofold: he gave encouragement and financial support to Ray; and he contributed his own fieldwork, experiments and observations, and his collection of plants, birds, and fishes. Willughby's work was incomplete partly on account of his early death, and partly, perhaps, on account of the wide-ranging interests of the man who could compile notes on the history of his family or gather information on contemporary games with the same assiduity as he collected his specimens.[4] It was left to Ray, with the help of Skippon and Jessop, his cotrustees, to assemble the material into a methodical presentable form.

NOTES

1. There were many variants of the name. The modern spelling "Willoughby" was adopted by the naturalist's son Thomas, first Baron Middleton.
2. Willughby's commonplace book on a wide variety of religious, classical, and scientific subjects includes chemical experiments, some headed "Mr. Wrays," dated 1658 and 1659. Middleton MS MiLM 15.
3. *Ibid.*, botanical notes.
4. Middleton MSS MiLM 13 (notes based on the family archives) and MiLM 14 (a book on games); compare his lists of vocabularies in many languages compiled on the Welsh and continental tours, Mi 4/149a/3.1–16.

BIBLIOGRAPHY

I. Original Works. Original MSS of and relating to Willughby, and his botanical specimens, are in the Middleton collection deposited in the Library Manuscripts Department of the University of Nottingham. Reports on the flow of sap in trees, leaf-cutting bees, and ichneumon wasps are in *Philosophical Transactions of the Royal Society of London*, **4**, no. 48 (1669), 963; no. 57 (1670), 1165; **5**, no. 58 (1670), 1199; no. 65 (1670), 2100; **6**, no. 70 (1671), 2125; no. 74 (1671), 2221; no. 76 (1671), 2279.

Other works are *Observations, Topographical, Moral and Physical Made in a Journey Through the Low-Countries, Germany, Italy and France . . . by John Ray, Fellow of the Royal Society. Whereunto Is Added a Brief Account of Francis Willughby Esq.; His Voyage Through a Great Part of Spain* (London, 1673); *Francisci Willughbeii de Middleton in agro Waricensi, armigeri, e Regia Societate, Ornithologiae, libri tres . . . totum opus recognovit, digessit, supplevit Joannes Raius* (London, 1676), trans. into English as *The Ornithology of Francis Willughby of Middleton in the County of Warwick, Esq., Fellow of the Royal Society in Three Books . . . by John Ray, Fellow of the Royal Society* (London, 1678); and *Francisci Willughbeii armig. de Historia piscium libri quatuor . . . totum opus recognovit, coaptavit, supplevit, librum etiam primum et secundum integros adjecit Joannes Raius et Societate Regia* (Oxford, 1686).

II. SECONDARY LITERATURE. On Willughby and his work, see W. Blunt, *The Art of Botanical Illustration* (London, 1950), 68, which describes the book of flower paintings (Middleton MS MiLM22); G. S. Boulger in *Dictionary of National Biography,* XXI, 525–528; C. Brown, *Lives of Nottinghamshire Worthies* (London, 1882), 207–211; L. C. Miall, *The Early Naturalists, Their Lives and Work (1530–1789)* (London, 1912), 99–130; C. E. Raven, *John Ray Naturalist, His Life and Works* (Cambridge, 1942; 2nd ed., 1950); M. A. Welch, "Francis Willoughby, F. R. S. (1635–1672)," in *Journal of the Society for the Bibliography of Natural History,* 6, pt. 2 (1972), 71–85, which includes a descriptive archival list of the material in the Middleton collection; and A. C. Wood, ed., *The Continuation of the History of the Willoughby Family by Cassandra Duchess of Chandos* (Windsor, 1958), which is a printed edition of the contemporary account by the naturalist's daughter (Middleton MS MiLM 37).

MARY A. WELCH

WILSING, JOHANNES (*b.* Berlin, Germany, 8 September 1856; *d.* Potsdam, Germany, 23 December 1943), *astronomy.*

Wilsing received his doctorate from the University of Berlin in 1880 and the following year became assistant at the Potsdam Astrophysical Observatory. In 1893 he received the post of observer and, in 1898, chief observer. He retired in 1921.

Wilsing conducted many observations concerning problems in astrophysics, a relatively new branch of astronomy at the time. He observed the velocity of rotation of the sun and offered a hydrodynamic explanation of its variation with latitude. Several of his studies deal with the influence on astrophysical measurements of systematic errors, such as atmospheric dispersion or optical and mechanical deficiencies of telescopes. He also observed novae, nebulae, and double stars.

Several of Wilsing's publications deal with methods, such as the derivation of the surface temperature of a star from photometric measurements of its spectrum. In calculating the diameters of stars, he used the laws of radiation and the measured values of the surface temperature, from which the radiating area can be computed. Although this method was not new in principle, Wilsing was the first to apply it systematically. His results were confirmed some years later, when the first interferometric measurements of stellar diameters were made.

BIBLIOGRAPHY

I. ORIGINAL WORKS. Wilsing's writings include *Determination of the Mean Density of the Earth by Means of a Pendulum Principle,* J. H. Gore, trans. (Washington, 1890); "Über die Helligkeitsverteilung im Sonnenspektrum nach Messungen an Spektrogrammen," which is *Publikationen des Astrophysikalischen Observatoriums zu Potsdam,* 22, no. 66 (1913); and "Messungen der Farben, der Helligkeiten und der Durchmesser der Sterne mit Anwendung der Planckschen Strahlungsgleichung," *ibid.,* 24, no. 76 (1920).

See also the Royal Society *Catalogue of Scientific Papers,* XIX, 644–645, which lists 38 memoirs published to 1900; and Poggendorff, IV, 1645–1646; V, 1376–1377; VI, 2896; and VIIa, 1015.

II. SECONDARY LITERATURE. See E. von der Pahlen, *Lehrbuch der Stellarstatistik* (Leipzig, 1937); and M. Waldmeier, *Ergebnisse und Probleme der Sonnenforschung* (Leipzig, 1941).

F. SCHMEIDLER

WILSON, ALEXANDER (*b.* St. Andrews, Scotland, 1714; *d.* Edinburgh, Scotland, 18 October 1786), *astronomy.*

Wilson was the son of Patrick Wilson, the town clerk of St. Andrews, and of Clara Fairfoul. He was very young when his father died, and he was brought up under the care of his mother. He studied at the College of St. Andrews, receiving an M.A. in 1733. He then was apprenticed to a surgeon-apothecary, first in St. Andrews, later in London. A chance visit to a typefoundry brought about a change in his career. Struck by an idea for an improved method of making type, he returned to St. Andrews in 1739 and set up a typefoundry there in 1742 with the assistance of a friend. The foundry was enlarged and moved to Camlachie, near Glasgow, in 1744. Since his student days Wilson had maintained an active interest in astronomy, and in 1760 was appointed—mainly through the influence of the duke of Argyll—first professor of practical astronomy at the University of Glasgow. He retained this post until 1784.

In 1774 Wilson published some observations, which showed that sunspots were depressions in the luminous matter surrounding the sun. This was not an entirely original hypothesis, for it had been suggested earlier by Christoph Scheiner, Philippe de La Hire, and Jacques Cassini. Nevertheless, Wilson's use of strict geometrical reasoning in his demonstration made his argument very forceful, and led to a renewed burst of enthusiasm for sunspot observations.

By carefully studying the apparent change in appearance of a spot as it crossed the solar disk, Wilson observed that the penumbra appeared narrowest on the side of it that was nearest the center of the sun, and widest on the side nearest the edge. He noted that this could be explained as an effect of perspective if the spot were a funnel-shaped depression, with the umbra corresponding to the bottom of the funnel and the penumbra to the sloping sides. Going beyond his observational data, Wilson conjectured that the sun was an immense dark globe surrounded by a thin shell of luminous matter. According to this view, sunspots were excavations in the luminous matter caused "by the working of some sort of elastic vapour, which is generated within the dark globe."

Wilson's interpretation of sunspots was challenged by Lalande in France, but supported by Sir William Herschel in England. Herschel then developed the interpretation into a general description of the solar constitution, which remained standard until the advent of spectroscopic investigations.

Wilson also speculated on a question posed by Newton in his *Opticks* (4th ed. [London, 1730], query 28): "What hinders the fixed stars from falling upon one another?" His answer, published in a short anonymous tract entitled *Thoughts on General Gravitation*, was that the entire universe partook in a periodic motion around some "grand centre of general gravitation."

Wilson was awarded an honorary M.D. from St. Andrews in 1763, and was one of the original members of the Royal Society of Edinburgh. In 1752 he married Jean Sharp. His portrait, a medallion by James Tassie, hangs in the National Portrait Gallery, Edinburgh.

BIBLIOGRAPHY

I. ORIGINAL WORKS. Wilson's works are "Observations of the Transit of Venus Over the Sun," in *Philo-sophical Transactions of the Royal Society*, **59** (1769), 333–338; "An Account of the Remarkable Cold Observed at Glasgow, in the Month of January, 1768," *ibid.*, **61** (1771), 326–331; *A Specimen of Some of the Printing Types Cast in the Foundry of Alexander Wilson and Sons* (Glasgow[?], 1772); "Observations on the Solar Spots," in *Philosophical Transactions of the Royal Society*, **64** (1774), 1–30; "An Improvement Proposed in the Cross Wires of Telescopes," *ibid.*, **64** (1774), 105–107; *Thoughts on General Gravitation, and Views Thence Arising as to the State of the Universe* (n.p., 1777[?]); and "An Answer to the Objectives Stated by M. De la Lande, in the Memoirs of the French Academy for the Year 1776, Against the Solar Spots Being Excavations in the Luminous Matter of the Sun, Together With a Short Examination of the Views Entertained by Him Upon that Subject," in *Philosophical Transactions of the Royal Society*, **73** (1783), 144–168.

II. SECONDARY LITERATURE. For a brief biographical sketch of Wilson's life, see the article by George Stronach in the *Dictionary of National Biography*, XXI, 545–546.

Good but brief accounts of Wilson's theories can be found in Agnes M. Clerke, *A Popular History of Astronomy During the Nineteenth Century* (Edinburgh–New York, 1886), and Robert Grant, *History of Physical Astronomy From the Earliest Ages to the Middle of the Nineteenth Century* (London, 1852).

HOWARD PLOTKIN

WILSON, ALEXANDER (*b*. Paisley, Scotland, 6 July 1766; *d*. Philadelphia, Pennsylvania, 23 August 1813), *ornithology*.

Wilson's background was so remote from scientific interests that his emergence in the last five years of his life as an ornithologist, and as founder of the science in America, is one of the most remarkable aspects of his extraordinary career. His father was a smuggler who, after his marriage to Mary McNab—comely and pious "and in every way (in a good sense) a superior person"—reformed and become a prosperous silk gauze weaver and loom operator in Paisley. Alexander, their third child and only son, was baptized by Paisley's most famous citizen, the Reverend John Witherspoon, later president of Princeton and a leading figure in the American Revolution. The boy studied at the Paisley Grammar School, and since he was precocious was intended for the ministry and placed in charge of a divinity student—the only formal education he received. From an early age he read widely, since the relatively high cultural standard of Paisley made books readily available.

Wilson's mother died when he was ten years old, and his father, marrying again almost immediately, returned to smuggling. Wilson was placed on a farm as a herd boy, and at thirteen was apprenticed to a weaver, William Duncan, the husband of his sister Mary. When he ended his apprenticeship at the age of sixteen, he became a peddler, tramping country roads across Scotland with a pack of cloth that he had woven with his brother-in-law. He wrote poetry, especially after the publication of Burns's first book in 1786 awakened Scottish intellectuals to the poetry of the common life and language around them; and he sketched and made his own designs for the cloth he wove. Both efforts were evidence of a powerful creative impulse seeking an outlet, but could scarcely be said to foreshadow the assured prose and the superb bird paintings that distinguished *The American Ornithology*. Nor was Wilson's home environment one that fostered scrupulously exact observation. His father took over an ancient, half-ruined castle, the Tower of Auchinbathie, near Lochwinnoch, where he operated illegal stills and hired weavers to work smuggled silk. Smuggling was not so sternly condemned in the west of Scotland as to make the household disreputable, but secrecy and the rural underworld prevented the development of disciplined habits such as the *Ornithology* was to require of Wilson.

After painful struggles with finances and his self-distrust, Wilson published a volume of poetry in 1790. It gained him favorable notice without improving his station in life. Thomas Crichton, Paisley's most eminent man of letters, characterized Wilson's poems accurately: "For original ideas, a masculine superiority of language, high graphic and descriptive character—especially his Scottish poems—they will stand a fair comparison with any of our Scottish poets, Burns not excepted. But Wilson is far short of that poet in fine poetic imagination." Wilson's masterpiece, *Watty and Meg*, a popular favorite for generations, was attributed to Burns, a fair indication of its hold on the public.

Watty and Meg was published anonymously, Wilson being in prison at the time. He was jailed during an obscure dispute, in a period of great social stress, with William Sharp, a wealthy Paisley mill owner. Wilson published a poem "The Shark" accusing Sharp of stealthily lengthening the measuring devices by which his employees were paid, all weaving then being piecework. Shortly before the poem appeared Sharp received an anonymous letter containing an offer to suppress the poem for a payment of five guineas, which made the charge against Wilson not libel, but blackmail. Wilson was arrested, roughly handled, convicted, ordered to beg the pardon of God and Mr. Sharp, to burn the poem in the public square, and to pay fines and damages amounting to £60 sterling—more than a weaver's annual earnings. These were reduced on appeal, but Wilson was in and out of court and jail for two years; and his friends, who signed peace bonds, were threatened with ruin if he became involved in conflict with any of His Majesty's servants, something that became increasingly likely in that time of riot and disorder. His love affair with Martha McLean, a Paisley girl of a well-to-do family, was broken off. Sir William Jardine in his otherwise laudatory biography of Wilson held the Sharp episode to be the only disreputable act of Wilson's career. Alexander Grosart, a Paisley historian, concluded that the charge in the original poem had been true, and would have been aired in a trial for libel; the charge of blackmail prevented any such disclosure. Wilson was utterly disheartened when he sailed for America in May 1794, telling Crichton, "I must get out of my mind."

Against the waste and disorder of his years in Scotland, Wilson's achievement in *The American Ornithology* became phenomenal. The development of the work in his mind can be traced in personal notes scattered throughout the *Ornithology* and in Wilson's poems and letters after he settled near Philadelphia. He had an initial interest in birds as game and the folklore and hunting skills they involved, inspired by the immense flights of ducks and geese over the school in which he taught (1796–1801) at Milestown near the Delaware. There was also a growing awareness of the wealth and variety of the bird life in the wilderness, marked especially during his long journeys on foot to a farm in western New York, which he had purchased for the family of his sister Mary, whose husband had abandoned her. As he sketched from life the birds his students brought him, Wilson developed his skill in drawing. His observations became more exact as he studied the hummingbirds, orioles, owls, grosbeaks, finches, and hawks that he kept as pets and watched in the woods and fields. As early as 1803 he wrote to Crichton that he was beginning to draw all of America's finest birds. During the four years (1802–1806) he taught at Gray's Ferry, his vague plans came into focus through his association with the venerable naturalist William Bartram, whose home in Bartram's garden, Kingsessing, Pennsylvania, was near his school; but he was still so little prepared that he asked Bartram to identify some of the birds he sketched. "I am miserably deficient in many requirements," he wrote Bartram, in reference to the *Ornithology*. "Botany, Mineralogy and Drawing I most ardently wish to be instructed in. . . . Can I make any progress in Botany, sufficient to enable me to be useful, and what would be the most proper way to proceed?"

Wilson was forty years old when he left teaching to try to classify scientifically and to describe accurately and picture in faithful color all the species of birds in America. His plan called for a ten-volume work to be sold by subscription for $120 a set. Samuel Bradford, a Philadelphia publisher, agreed to bring out one volume, and to continue the series if Wilson could secure 200 subscribers on the strength of that sample. Volume I, which included such familiar birds as the bluejay, the Baltimore oriole, and the robin, with two to six birds pictured on each of ten colored plates, appeared in the fall of 1808. With this in hand Wilson set out through the northeastern states, signing up subscribers, the first of the great journeys that carried him more than ten thousand miles in the next five years. The southern trip that followed was more successful. With the encouragement of President Thomas Jefferson, who subscribed and urged others to do so, Wilson made his way through the South to Savannah, signing up 250 subscribers and collecting specimens to be pictured in later volumes. He also formed lasting relations with naturalists, including Stephen Elliott and John Abbot, who provided him with specimens and accurate information he could not otherwise have obtained. Forty-two birds were included in Volume II, and when that book was in the hands of the printer in January 1810, Wilson set out on an amazing journey, six wilderness months down the Ohio by rowboat and over the Natchez Trace by horseback, to New Orleans, some three thousand miles, at a cost of $455, but returning a treasury of heretofore unknown species, and enough new subscribers to bring the total to more than 450.

At Louisville, according to Audubon's later recollections, Wilson tried to induce him to become a subscriber, and became depressed when Audubon showed him his own portfolio of bird drawings that were superior to those in Wilson's book. Wilson himself made no mention of the encounter in his letters or in the long account of his journey he published after his return to Philadelphia in *The Port Folio*. Doubt was cast on Audubon's account during his long conflict with George Ord, one of Wilson's literary executors. In a pioneering study of Wilson, Elsa Guerdrum Allen concluded that

Audubon's ambitions were awakened by his first view of Wilson's book. In any case it is unlikely that Wilson would have reacted so strongly at the sight of a more accomplished artist's work in his own field; John Abbot was also a technically trained artist whose bird paintings are of very high quality, and Wilson remained on close terms with Abbot. The distinction of Wilson's work is in the unity of his paintings and his text, and when individual works are inferior to Audubon's splendid and spectacular plates (although in many cases, such as the snowy owl and the Mississippi kite, Wilson's work is plainly superior), Wilson's birds are always birds rather than decorations, supplementing and adding to the text, in an unparalleled catalog of nature: his birds are really wild.

Wilson's third volume appeared in February 1811, and the fourth only seven months later; both were editions of 500 copies. "I have sacrificed everything to print my *Ornithology*," Wilson wrote to the botanist André Michaux. Except for the engraver Alexander Lawson, who cut most of the plates, Wilson worked almost without assistance. He oversaw all details; hired and supervised the colorists; secured virtually all of the subscribers; familiarized himself with the scientific literature of each species; pictured each bird; and composed his brief, engaging, and exact descriptions that are often masterpieces of nature writing.

Before his death of dysentery at the age of forty-seven, Wilson had completed eight volumes of the *Ornithology*, and the drawings for the ninth volume; he had painted and described 264 species. He added forty-eight new species to those previously known to exist in the United States, prepared good life histories for ninety-four species, and maintained a standard so exacting that in a century and a half only a score of minor errors have been found in the *Ornithology*. Francis Herrick, the biographer of Audubon, wrote, "When we consider that Wilson's entire working period on the *Ornithology* was not over ten years . . . the achievement of this man is little short of marvelous"—an accurate appraisal, except that the period was nearer five years than ten.

During the economic stress of the War of 1812, Wilson's colorists left him, Bradford's interest in the *Ornithology* ebbed, and Wilson, forced to color many of the plates himself, also acted as a collector of the money due from subscribers. His social life had long since ceased to exist (although he was elected to membership in the American Philosophical Society in 1813) apart from his friendship with the family of Jacob Miller, a wealthy landowner he had known in his schoolteaching days at Miles-town. The only intimate relation during days of unceasing but often inspired work was with Sarah Miller, the daughter of the family, fifteen years younger than Wilson, to whom he was reportedly engaged and to whom he left everything he owned, including the rights to the *Ornithology*.

BIBLIOGRAPHY

I. ORIGINAL WORKS. Virtually all of Wilson's nature writing is included in *The American Ornithology*, 9 vols. (Philadelphia, 1808–1813), and most of his journals, travel accounts, and many letters are in *Poems and Literary Prose of Alexander Wilson*, the Reverend Alexander B. Grosart, ed., 2 vols. (Paisley, 1876).

II. SECONDARY LITERATURE. George Ord prepared a biographical introduction to the 9th vol. of the *Ornithology*, 2nd ed. (1824–1825), which was supplanted by a much fuller work by Sir William Jardine as the introduction to the 3-vol. edition of the *Ornithology*, prepared "with additions" by Charles Lucien Bonaparte (London, 1832).

Other biographies include Elsa Guerdrum Allen, *The History of American Ornithology Before Audubon* (New York, 1969); Robert Cantwell, *Alexander Wilson, Naturalist and Pioneer* (Philadelphia, 1961); Thomas Crichton, *Biographical Sketch of the Late Alexander Wilson* (Paisley, 1819); W. M. Hetherington, *Memoir of Alexander Wilson* (Edinburgh, 1831); William B. O. Peabody, *Life of Alexander Wilson* (Boston, 1839); Emerson Stringham, *Alexander Wilson, a Founder of Scientific Ornithology* (Kerrville, Tex. 1958); and James Southall Wilson, *Alexander Wilson, Poet-Naturalist* (New York, 1906).

ROBERT CANTWELL

WILSON, BENJAMIN (*b.* Leeds, England, 1721; *d.* Bloomsbury, London, England, 6 June 1788), *electricity*.

Wilson was born in the latter part of 1721, the youngest of the fourteen children of Major Wilson, "the most considerable merchant in Leeds," and Elizabeth Yates. His father's house at Mill Hill, near Leeds, was decorated by Jacques Parmentier, a French artist, and it was to his influence that Wilson attributed the origin of his own interest in art. Later in his youth Wilson studied for a year with another French artist, Longueville, who was working on commissioned historical paintings in the neighborhood.

The Wilson family became impoverished while Benjamin was still under twenty, and the boy went to London, on foot, to find employment. He worked as a clerk, in poor circumstances, continuing his artistic studies whenever occasion offered. During this period, Wilson, by his own account, read widely in the field of experimental philosophy. This interest was channeled toward the novel science of electricity through his friendship with the apothecary William Watson, who was awarded the Copley Medal of the Royal Society in 1745 for his electrical experiments. Wilson also corresponded from 1745 with John Smeaton, the civil engineer. As his own ideas on electricity developed, Wilson came to know Martin Folkes, president of the Royal Society from 1741 to 1753, who advised the young painter to begin his career in Ireland, so that he could return to London a master of his craft. Consequently, Wilson went to Ireland for a short period in 1746, and again from 1748 to 1750.

While in Dublin in 1746, Wilson was allowed to use the experimental room in Trinity College, which resulted in his first publication, *An Essay Towards an Explication of the Phaenomena of Electricity Deduced From the Aether of Sir Isaac Newton*. His second, longer stay in Ireland permitted the writing of *A Treatise on Electricity* (1750), published in London after his return. It was no doubt as a result of this work that Wilson was elected a fellow of the Royal Society on 5 December 1751.

Upon his return from Ireland, Wilson took a seven-year lease of the house in Great Queen Street, Lincoln's Inn Fields, previously occupied by Sir Godfrey Kneller. Wilson was mainly employed in portrait painting, having as sitters many of the men of science whom he came to know through his interest in electricity, including at least eight fellows of the Royal Society. He also painted the actor David Garrick and the poet Thomas Gray, whom he had met at Cambridge in 1747. Wilson also showed some skill at engraving, and produced a famous caricature in February 1766 (at the time of the repeal of the Stamp Act), which, selling for one shilling a copy, brought him £100 in four days. Wilson won the patronage of the duke of York, and upon the death, in 1764, of William Hogarth, who was a friend of Wilson's, the duke gave him Hogarth's appointment of sergeant-painter. Wilson's career as a painter was both successful and remunerative, but he was, unfortunately, fond of speculation, and was declared a defaulter on the Stock Exchange in 1766.

Wilson's scientific interests were almost exclusively concerned with electricity. Following his two publications in 1746 and 1750, he invented and exhibited a large electrical apparatus. With the physician Benjamin Hoadly, he carried out electrical research, the results of which were published in *Observations on a Series of Electrical Experiments* (1756). The purpose of these three books was to assert the identity of electricity, in particular the Franklinian single electric fluid, with the Newtonian aether, as postulated in the English edition of the *Opticks*. In 1757 Wilson visited France and repeated many of his experiments at St. Germain-en-Laye. The culmination of this period of research was the award to Wilson of the Royal Society's Copley Medal in 1760.

The most remarkable of Wilson's scientific activities was his public controversy with Franklin on the question of whether lightning conductors should be round or pointed at the top. Wilson held that "thunder rods" should be round-headed, for, recognizing quite correctly that a pointed metal rod attracts lightning, he believed that if these rods were erected on buildings, they would actually cause the lightning to strike. Wilson was nominated by the Royal Society to serve on a committee to regulate the erection of lightning conductors on St. Paul's Cathedral, and was later asked by the Board of Ordnance to inspect the gunpowder magazines at Purfleet. In 1773, a Royal Society committee, on which he also sat, considered the problem of the magazine, and finally advised the erection of a pointed rod on the summit of the building, Wilson being the sole dissenter. Wilson continued the dispute, publicly disagreeing with the opinion of the Royal Society, and the arguments of such noted scientists as Franklin, Cavendish, and Nairne. Finally, in July 1777, Wilson arranged a huge demonstration before King George III in the Pantheon in Oxford Street. He certainly convinced the king, who declared that Wilson's arguments were sufficient to persuade the apple-women in the street. The scientific world took a different view, for Wilson had continued the dispute beyond the bounds of reason, and the editors of the abridgment of the *Philosophical Transactions* were strongly critical: "But he has been chiefly distinguished as the ostensible person whose perverse conduct in the affair of the conductors of lightning produced such shameful discord and dissensions in the Royal Society, as continued for many years after, to the great detriment of science" (*The Philosophical Transactions of the Royal Society, Abridged, 1755 to 1763*, XI [London, 1809], p. 15).

One commission that united Wilson's interests in science and in painting was the task, entrusted to him by James Short, of producing a map of the moon. Wilson would have received 100 guineas

had he completed the task, but, because of the strain on his eyesight and because working at night gave him constant colds, he was unable to finish the map. The contact between the two men was not, however, without practical outcome, for Wilson painted a portrait of Short, and also one of his fellow telescope-maker, John Dollond.

Wilson's electrical studies brought him into correspondence with foreign scientists throughout Europe. He was a member of four European academies, including the Istituto delle Scienze ed Arti Liberali at Bologna, where he was the first Englishman to be so honored.

In 1771 Wilson married a Miss Hetherington, and the couple had seven children. Wilson's third son was General Sir Robert Thomas Wilson, whose *Life*, published in 1862 by Herbert Randolph, contains an abridgment of Benjamin Wilson's manuscript autobiography, which he had most strictly directed should not be published.

BIBLIOGRAPHY

I. ORIGINAL WORKS. Wilson initially published most of his work in *Philosophical Transactions of the Royal Society*; some of these papers have been reprinted, and are included here. See *An Essay Towards an Explication of the Phaenomena of Electricity, Deduced From the Aether of Sir Isaac Newton* (London, 1746); *A Treatise on Electricity* (London, 1750); *Observations on a Series of Electrical Experiments* (London, 1756), written with Benjamin Hoadly; *A Letter to Mr. Apinus on the Electricity of Tourmaline, With Observations on Mr. Aepinus's Work on the Same Subject* (London, 1764); *A Letter to the Marquess of Rockingham, With Some Observations on the Effects of Lightning* (London, 1765); *Observations Upon Lightning, and the Method of Securing Buildings From Its Effects, in a Letter*, by B. Wilson and Others (London, 1773); *Further Observations Upon Lightning, Together With Some Experiments* (London, 1774); *A Series of Experiments on the Subject of Phosphori, and Their Prismatic Colours: in Which Are Discovered, Some New Properties of Light* (London, 1775; 2nd ed., 1776); *A Letter, From F. Beccaria to Mr. Wilson, Concerning the Light Exhibited in the Dark by the Bologna Phosphorus, Made According to Mr. Canton's Method, and Illuminated Through Coloured Glasses*, printed with *To the Reverend Father Beccaria, Professor of Natural Philosophy at Turin. B. Wilson London 23 September 1776* (n.p., n.d.); *An Account of Experiments Made at the Pantheon, On the Nature and Use of Conductors: to Which Are Added, Some New Experiments With the Leyden Phial* (London, 1778; 2nd ed., 1788); *A Letter To Mr. Euler, Professor of Philosophy, and Member of the Imperial Academy of Sciences at Petersbourg, . . .* (London, 1779); and *A Short View of Electricity* (London, 1780).

There is a typescript copy of Wilson's *Memoirs* on deposit at the National Portrait Gallery, London. These *Memoirs* Wilson never intended for publication, but they were drawn on by H. Randolph (see below). A volume of letters to and papers of Wilson is in the British Museum, Add. MS. 30094.

II. SECONDARY LITERATURE. See Herbert Randolph, *The Life of Sir Robert Wilson* (London, 1862); G. J. Symons, ed., *Lightning Rod Conference. Report of the Delegates* . . . (London, 1882); G. L'E. Turner, "A Portrait of James Short, F.R.S., Attributable to Benjamin Wilson, F.R.S.," in *Notes and Records of the Royal Society of London*, **22** (1967), 105–112. For an account of Wilson's theory of electricity, and for a contemporary criticism, see R. W. Home, "Some Manuscripts on Electrical and Other Subjects Attributed to Thomas Bayes, F.R.S.," *Notes and Records of the Royal Society*, **29**, no. 1 (October 1974), 81–90 (especially pp. 84–87).

G. L'E. TURNER

WILSON, CHARLES THOMSON REES (*b.* near Glencorse, Midlothian, Scotland, 14 February 1869; *d.* Carlops, Peeblesshire, Scotland, 15 November 1959), *atomic physics, meteorological physics*.

Wilson's father, John Wilson, was well-known in Scotland for his experiments in sheep farming; his mother, the former Annie Clark Harper, came from a Glasgow family of thread manufacturers. John Wilson died when Charles, the youngest of his eight children by two marriages, was four years old; and Mrs. Wilson then moved to Manchester. The family was not well off, and Wilson owed his university education to the kindness and financial support of his half brother William, a successful businessman in Calcutta.

Wilson attended Greenheyes Collegiate School in Manchester; and even then he showed an interest in natural science, preparing specimens for observation under the microscope. At the age of fifteen he entered Owens College, Manchester, registering as a medical student but taking a B.Sc. degree when he was eighteen. A further year was spent studying philosophy and the classics, after which Wilson won an entrance scholarship to Sidney Sussex College, Cambridge. By this time all thought of medicine had been abandoned, and he had determined to become a physicist. The years following receipt of his degree and the death of his half brother William (1892) were difficult ones, for he had to help support his mother, yet longed to devote himself to research. Wilson taught for a short time at Bradford Grammar School in York-

shire but was drawn again to continue his experimental work at Cambridge, making just enough to live on by serving as demonstrator for medical students. It was at this time that Rutherford, Townsend, and McClelland became research students at the Cavendish Laboratory; and Wilson joined them in the famous discussions over tea.

In 1896 Wilson was awarded the Clerk Maxwell studentship for three years. After a year of work on atmospheric electricity problems for the Meteorological Council, he was elected in 1900 a fellow of Sidney Sussex College, and was appointed a university lecturer and demonstrator. For the two university posts his annual salaries were, respectively, £100 and £50; his duties were to take charge of the teaching of advanced practical physics and to lecture to the part II physics class on light, which was a new course. Wilson's influence on the teaching of experimental physics at Cambridge was considerable, his chief innovation being to give his students minor research problems to solve in the laboratory, rather than carry out textbook experiments. From 1925 to 1934 Wilson was Jacksonian professor of natural philosophy at Cambridge.

The Royal Society elected Wilson a fellow in 1900, and awarded him the Hughes Medal in 1911, a Royal Mcdal in 1922, and the Copley Medal in 1935. A Nobel Prize for physics was awarded jointly to Wilson and A. H. Compton in 1927 for their work on the scattering of high-energy photons. Wilson was appointed Companion of Honour by the king in 1937, and held honorary degrees from Aberdeen, Glasgow, Manchester, Liverpool, London, and Cambridge.

At the age of thirty-nine Wilson married Jessie Fraser Dick; the couple had a son and two daughters. Soon after his retirement from the Jacksonian chair, Wilson left Cambridge and returned to Scotland. The twenty-three years of his retirement were extremely active. He continued climbing well into his eighties, and at eighty-six traveled in an airplane for the first time. He died after a short illness at his cottage a few miles from his birthplace.

Wilson attributed the shaping of his research career to his experiences on holiday in the Highlands:

In September 1894 I spent a few weeks in the Observatory . . . on the summit of Ben Nevis . . . . The wonderful optical phenomena shown when the sun shone on the clouds surrounding the hill-top, and especially the coloured rings surrounding the sun (coronas) or surrounding the shadow cast by the hilltop or observer on mist or cloud (glories), greatly excited my interest and made me wish to imitate them in the laboratory.[1]

Elsewhere he wrote: ". . . It is hardly necessary for me to say that these experiments might have had little result had it not been that they were made in the Cavendish Laboratory at the beginning of the wonderful years of the discovery of the electron, X-rays and radioactivity."[2] Such was the initial impetus behind Wilson's work. J. J. Thomson assessed his achievement thus:

> This work of C. T. R. Wilson, proceeding . . . since 1895, has rarely been equalled as an example of ingenuity, insight, skill in manipulation, unfailing patience and dogged determination. . . . The beautiful photographs that he published [of the tracks of atomic particles] required years of unremitting work before they were brought to the standard he obtained. . . . It is to him that we owe the creation of a method which has been of inestimable value to the progress of science.[3]

Wilson well exemplifies the British experimental scientist whose inspiration was found not in mathematical concepts but in the observation of natural phenomena. Early in 1895 he posed himself a set of questions on cloud formation, and in March of that year he began to build the first apparatus to condense water vapor in dust-free air (see Fig. 1). By August he had established that the critical volume-ratio limit for drop formation in clean conditions was $V_2/V_1 = 1.25$. In February 1896, shortly after the discovery of X rays by Röntgen, Wilson used a primitive X-ray tube, made by J. J. Thomson's assistant, to irradiate the expansion chamber. At the same volume ratio as before, a dense fog was produced by the X rays, which led Wilson to suppose that the condensation nuclei were ions, to which the conductivity of a gas exposed to X rays was attributed by Thomson and Rutherford. By the spring of 1899 he wrote a summary of his work: "General results of all the experiments. *Negative* ions *begin* to be caught about $V_2/V_1 = 1.25$ and *all* appear to be caught when $V_2/V_1 = 1.28$. Density of negative fog shows no increase from this point onwards. *Positive* ions begin to be visible about $V_2/V_1 = 1.31$. Fogs are constant and identical with the negative from 1.35 upwards."[4]

The phenomena discovered empirically by Wilson may, briefly, be explained as follows. When air saturated with water vapor is suddenly cooled by

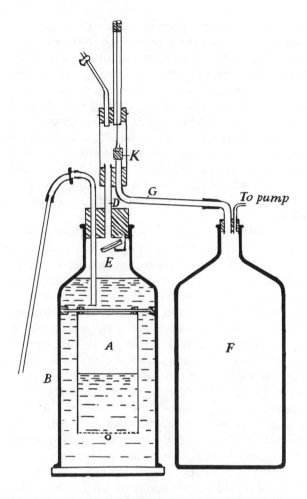

FIGURE 1.   Wilson's 1895 apparatus. The gas to be expanded is in the glass vessel *A*, which itself is placed inside a glass bottle *B*, which is partially filled with water so as to trap the gas in the inner vessel. The air above the water in the bottle is connected with an evacuated vessel *F* by tubes *D* and *G*, to which are fitted valves *E* and *K*, the latter of which is normally closed. When this valve is quickly opened, the air at the top of the bottle *B* rushes into the evacuated vessel *F* and the water in *B* rises until it fills the top of the bottle, and by doing so, closes the valve *E*, so stopping further expansion of the gas in *A*. By suitably adjusting the initial volume of the gas in *A* and the amount of water in *B*, the relative expansion of the gas in *A* can be precisely controlled.

an adiabatic expansion, it becomes supersaturated. In this condition, condensation into droplets will occur, provided there are nuclei present. Dust particles allow drops to form immediately, and so Wilson carefully eliminated all gross matter from his apparatus. Negative ions act as nuclei at an expansion ratio of 1.25 (fourfold supersaturation), and positive ions become nuclei at 1.31 (sixfold supersaturation). At about 1.38 (eightfold supersaturation) air molecules themselves will act as drop nuclei in the absence of all others. The vapor pressure of a spherical drop is greater than that of a plane surface in inverse proportion to the radius of the drop, so that if a very small drop forms, it will reevaporate immediately. A nucleus gives the necessary larger radius to assist the persistence of a drop. The surface tension of the liquid of the drop also is important, because it acts to contract the drop and thus reduce the radius. If a drop carries an electric charge, this acts contrary to the surface tension, tending to enlarge the drop. Because of a characteristic of the skin of a water drop, negative ions are more effective than positive ions in nucleation.

Wilson continued to experiment with ultraviolet radiation and other techniques for producing condensation effects, but soon concentrated on atmospheric electricity, not returning to the cloud chamber until December 1910. He designed an improved chamber with new methods of illumination and the possibility of photographing the results. At this time Wilson realized that it might be possible to reveal the track of an α ray by condensing water drops onto the ions produced by its passage. During March 1911 he saw this effect produced in his apparatus. Thus, the elucidation of phenomena seen in the Scottish hills led to the possibility of studying the processes of radioactivity, and the Wilson cloud chamber became an important piece of laboratory equipment. But it was in the study of cosmic rays that it achieved its full power, particularly in the refined form developed by Patrick Blackett, in which it was possible to study particles of very high energy and the production of electron-positron pairs with the chamber situated in a strong magnetic field.

The study of atmospheric electricity was dramatically thrust upon Wilson by the experience of his hair standing on end while at the summit of Ben Nevis in June 1895. The subsequent lightning flash impressed on him the magnitude of the electric field of a thundercloud. In his experiments he used captive balloons and kites to measure the strength of the electric field at various heights. He also developed a sensitive electrometer and voltameter, as well as a capillary electrometer for the measurement of the earth's electric field and air-earth currents. In fine weather there is always an electric field directed toward the earth that has a potential gradient of 100–200 volts per meter. The total negative charge on the whole earth is about 500,000 coulombs. The current from the upper atmosphere to the earth is sufficient to discharge the earth in a matter of minutes, so the problem is to account for the maintenance of the earth's charge.

Equilibrium probably is kept by the thunderstorm, the global incidence of which is about two thousand at any given time.

The theory put forward by Wilson to explain the electric structure of a thundercloud implied that the top would be positively charged and the bottom negatively charged. He thought that larger drops would be found on negatively charged nuclei, causing them to fall faster than those positively charged. Wilson's last scientific paper, "A Theory of Thundercloud Electricity," was communicated to the Royal Society in 1956, when he was the oldest fellow. Although his theory is not complete, it certainly was a crucial contribution to a problem that has yet to be fully solved.

Work on the conductivity of air was done by Wilson in 1900, using very well insulated electroscopes. They always showed a residual leakage that was the same in daylight and in darkness, and for positive or negative charge. Having described his results, Wilson made the significant statement: "Experiments were now carried out to test whether the production of ions in dust-free air could be explained as being due to radiation from sources . . . outside our atmosphere, possibly radiation like Röntgen rays or like cathode rays, but of enormously greater penetrating power."[5] This ingenious hypothesis was tested in 1911 by Victor Hess, who took an electroscope up in a balloon, thereby discovering that after an initial fall, the conductivity of air increased with altitude. To explain this effect, Hess postulated the existence of "cosmic radiation."

NOTES

1. Nobel lecture, Stockholm, 12 Dec. 1927.
2. "Ben Nevis Sixty Years Ago," in *Weather*, **9** (1954), 310.
3. J. J. Thomson, *Recollections and Reflections* (London, 1936), 419–420.
4. From laboratory notebook A3 in the library of the Royal Society, cited by Dee and Wormell, "Index . . .," 57.
5. "On the Leakage of Electricity Through Dust-Free Air," in *Proceedings of the Cambridge Philosophical Society*, **11** (1900–1902), 32.

BIBLIOGRAPHY

I. ORIGINAL WORKS. Wilson published 45 papers between 1895 and 1956, a third of them in *Proceedings of the Cambridge Philosophical Society*, and the majority of the remainder in *Proceedings of the Royal Society*, ser. A. A complete list of the papers is printed in Blackett (below), 294 f., and in Dee and Wormell (below), 65 f. An autobiographical account is "Reminiscences of My

Early Years," in *Notes and Records. Royal Society of London*, **14** (1960), 163–173. MS laboratory records for 1895–1940 are in the library of the Royal Society.

II. SECONDARY LITERATURE. See P. M. S. Blackett, "Charles Thomson Rees Wilson 1869–1959," in *Biographical Memoirs of Fellows of the Royal Society*, **6** (1960), 269–295; and P. I. Dee and T. W. Wormell, "An Index to C. T. R. Wilson's Laboratory Records and Notebooks in the Library of the Royal Society," in *Notes and Records. Royal Society of London*, **18** (1963), 54–66.

G. L'E. TURNER

WILSON, EDMUND BEECHER (*b.* Geneva, Illinois, 19 October 1856; *d.* New York, N.Y., 3 March 1939), *cytology, embryology, heredity.*

Wilson was among the most important and prolific biologists in the last part of the nineteenth and first part of the twentieth centuries. As an investigator of remarkable observational and analytical skill, he contributed significantly to an understanding of the structure and function of the cell. As a meticulous and exhaustive encyclopedist, he brought together and organized vast quantities of research related to the cell—its structural, hereditary, and developmental aspects. Wilson's *The Cell in Development and Inheritance* (1896) is a monument to his comprehensive and profound understanding of major biological problems of the time, many of which are still unsolved. Born three years before the publication of Darwin's *Origin of Species*, Wilson grew up in an era during which biology was transformed from a science dominated by natural history into one that was more and more concerned with rigorous and quantitative experimental analysis. His own work played a significant part in this transition.

Wilson was the son of Caroline Clarke and Isaac G. Wilson. His father, a graduate of Brown University and Harvard Law School, had gone west in the 1840's to open a law practice. In later years he served as county judge, circuit court judge, and finally as chief justice of the Appellate Court of Chicago. Wilson's maternal ancestors were descended from Thomas Clarke, reputed mate of the Mayflower who had settled at Plymouth, Massachusetts, in 1623. After the financial crash of 1837, Scotto Clarke (Wilson's grandfather) moved from Boston to Geneva, Illinois, with his four children, of whom Caroline was the second youngest. She and Isaac Wilson were married in 1843 and had five children: Frank, Ellen, Charles, Edmund, and Harriet.

Wilson grew up in a cultured atmosphere that encouraged his two lifelong interests: the study of living things and music. When his father was appointed a circuit court judge in 1859, Wilson's parents moved to Chicago and he was "adopted" by his mother's sister, Mrs. Charles Patten of Geneva. Thus, from an early age Wilson had two homes, both of which encouraged his varied interests and provided him with, as he wrote, "four parents between whom I hardly distinguished in point of love and loyalty." During his childhood he spent considerable time in the countryside around Geneva collecting specimens, which he stored in a special room provided for him in the Patten house. In the fall of 1872 his uncle suggested that young Wilson (he was not quite sixteen) take over the small country school that his brother Charles had taught the year before. Living with his aunt and uncle in Oswego, Illinois, he spent a year teaching everything from arithmetic to reading. It was a rewarding experience and strongly supported his desire for further education. Inspired by a cousin, Samuel Clarke, who was then attending Antioch College in Ohio, Wilson decided in the summer of 1873 to apply for admission to that institution. At Antioch he received his first formal instruction in zoology, botany, Latin, geometry, trigonometry, and chemistry, the latter with laboratory work that he found especially appealing. He paid his way partly by odd jobs, one of which was manufacturing the gas by which the college was lit. Instead of returning to Antioch the following fall, however, Wilson decided to begin a career in science by attending the Sheffield Scientific School at Yale, about which he had heard so much from Sam Clarke, who was enrolled there. Realizing that he lacked the proper background to enter Yale, he decided to live with his parents in Chicago for the next year (1874–1875) and attend the (old) University of Chicago for additional preparation.

Wilson entered the Sheffield Scientific School in 1875, and during his first year he took courses in zoology with A. E. Verrill and in embryology with Sidney I. Smith. At Yale he also had his first real exposure to the study of heredity and evolution, through a series of lectures given by William Henry Brewer, who made an indelible impression on Wilson and his classmates by lecturing "with the utmost fire and vehemence."[1] After receiving the bachelor's degree (Ph.B.) in 1878, Wilson was invited to remain at Sheffield as a graduate student and assistant, which he did for one year; but soon Samuel Clarke once again found new and exciting horizons, this time at the newly opened Johns

Hopkins University, where he was then enrolled. Clarke's letters were so enthusiastic that Wilson and his close friend William T. Sedgwick, who also completed his studies at Yale, applied for and received fellowships to Hopkins beginning in the fall of 1878.

Wilson's three years at Hopkins opened up a wholly new world—that of original investigation: he studied with the physiologist H. Newell Martin and the morphologist William Keith Brooks, both of whom emphasized research by continually pointing out the many unsolved problems in contemporary biology. Wilson received his Ph.D. in 1881, then remained at Hopkins for another year as assistant. In the spring of 1882 he went abroad for further study. For several months he was at Cambridge, where he met Michael Foster, William Bateson, and T. H. Huxley. Wilson then proceeded to Leipzig, where he worked in the laboratory of the invertebrate zoologist Rudolf Leuckart and attended lectures by the mechanistic physiologist Carl Ludwig. After leaving Leipzig he headed south to the zoological station at Naples. Through an arrangement with Williams College, where Samuel Clarke was then teaching, Wilson obtained a table in the laboratory for part of the year 1882–1883. The Naples station made a deep and lasting impression on him, for he met and became close friends with Anton Dohrn, the director, and with a number of embryologists and invertebrate zoologists, including Edouard Meyer and Arnold Lang. Like his friend T. H. Morgan, Wilson found that his first experience at the Naples station was one of the most exciting of his life, and set the direction for much of his future thinking about biological research. Many years later he wrote: "It was a rich combination of serious effort, new friendships, incomparable beauty of scenery, a strange and piquant civilization, a new and charming language, new vistas of scientific work opening before me; in short, a realization of my wildest, most unreal dreams."[2]

On his return from Naples, Wilson taught for one year (1883–1884) at Williams College, replacing his cousin Sam Clarke, who was on leave to spend the year at Naples. The following year (1884–1885) he held a lectureship at the Massachusetts Institute of Technology, where he worked closely with his friend Sedgwick on a biology textbook they had begun planning several years earlier. In 1885 Wilson accepted an offer to head the biology department at Bryn Mawr College, where he remained until 1891, when he was appointed adjunct professor of zoology and chairman of the zoology department at Columbia. He remained at Columbia for the rest of his career, retiring as Da Costa professor of zoology in 1928.

Before assuming his official duties at Columbia, Wilson spent a year abroad, the first half with Theodor Boveri at Munich and the second half at Naples with the experimental embryologists Hans Driesch and Curt Herbst. During this and later years Wilson spent considerable time at marine stations and on collecting trips. His long association with the Marine Biological Laboratory, Woods Hole, Massachusetts (both as investigator and as trustee), and the Chesapeake Zoological Laboratory of Johns Hopkins were part and parcel of the importance he attached to studying living specimens, and especially marine forms as material for basic biological investigation.

A well-liked and eminently respected teacher, Wilson was known for his deep personal interest in his students. His lectures were highly polished and meticulously researched, possessing a balanced structure that demonstrated his strong sense of organization and aesthetics. Wilson's success as a teacher stemmed partly from his enormous erudition and from the warm and articulate manner in which he conveyed his enthusiasms. He taught students at Columbia, both graduate and undergraduate, to see biology as a whole, as a series of fields—such as heredity, evolution, and embryology—at a time when many workers saw only separate disciplines. Among Wilson's graduate students (or those who took some courses with him) at Columbia were G. N. Calkins, A. P. Mathews, C. E. McClung, H. J. Muller, Franz Schrader, and W. S. Sutton.

Music was of intense interest to Wilson throughout his life. He was a cellist of outstanding accomplishment, being, in the words of one contemporary musician, "the foremost non-professional player in New York."[3] To Wilson music was a solace, no less nor more beautiful than a living organism or a cell—something to which he owed, in his own words, some of the greatest pleasures of his life. In addition he loved sailing and skippered numerous collecting and pleasure expeditions out of Woods Hole, Bermuda, and other ports. He was also a linguist of considerable ability, with a knowledge of German, French, Italian, Spanish, and Arabic.

In 1904 Wilson married Anne Maynard Kidder, daughter of Jerome Henry Kidder, a friend of Spencer Fullerton Baird. Wilson had first met her at Woods Hole, where her family spent nearly every summer. The Wilsons had one daughter, Nancy, who became a professional cellist.

Although Wilson always worked concurrently on a variety of problems, his career can be divided roughly into three periods, each of which was dominated by a particular set of interests: 1879–1891, descriptive embryology and morphology (including studies of cell lineage); 1891–1903, experimental embryology (including the organization of the egg, the effects of various substances on differentiation, and artificial parthenogenesis); and 1903–1938, heredity (including the relation of Mendelism to cytology, sex determination, and evolution). To Wilson these various topics converged in a single problem: How does the individual organism lie implicit in the fertilized (or even unfertilized) egg? This problem could be broken down into a number of subsidiary and more specific questions: What is the mechanism by which the likeness of the parents is transmitted to the offspring? How is hereditary information transformed into a complete adult during embryonic development? How do the cell nucleus and its hereditary components direct the day-to-day activity of cells? How does the interaction between parts—nucleus and cytoplasm, egg and sperm, one embryonic tissue layer and another, the whole embryo and its environment—influence the final form of the adult organism? Early in his career Wilson saw that answers to all of these questions bring the investigator down to the level of the cell. He felt it was impossible fully to understand larger problems, such as those occurring on the tissue, organ, organismic, or population level, without a thorough knowledge of the cell—its structure, organization, and functions.

As a student of William Keith Brooks, Wilson was schooled in the aims and methods of morphology. Morphology, a discipline prominent in the late nineteenth century, utilized a variety of areas— embryology, systematics, comparative anatomy, cytology, heredity, and physiology—to determine phylogenetic relationships. Problems of embryology, for example, were not considered so much for their own value but, rather, for whatever light they might throw on the evolutionary history of various species. Although Wilson's later work, particularly after 1891, gradually moved away from such overriding concern with phylogeny, his early papers strongly showed the influence of Brooks, whom he found an inspiring teacher. Brooks let his students alone, and Wilson was able to pursue whatever leads he wanted in the laboratory. Brooks also taught his students to think of biology in terms of problems still to be solved, rather than as a static and accumulated body of facts. He had a distinct philosophical bent that led him to think of problems—biological or nonbiological—in a large framework. He seldom accepted any conclusion on its own, always examining not only the evidence on which it was based but also the underlying philosophical assumptions and points of view. Wilson wrote of his experience with Brooks: "It was through informal talks and discussions in the laboratory, at his house, and later at the summer laboratories by the sea that I absorbed new ideas, new problems, points of view, etc. . . . From him I learned how closely biological problems are bound up with philosophical considerations. He taught me to read Aristotle, Bacon, Hume, Berkeley, Huxley; to think about the phenomena of life instead of merely trying to record and classify them."[4]

Descriptive Embryology and Morphology: 1879– 1891. Although Wilson published two papers on the systematics of Pycnogonida (sea spiders) in 1879 and 1881, the result of work he had carried out for the Ph.B. at Yale, his earliest work of importance involved studies on the embryology and morphology of the coelenterate *Renilla*. This work was undertaken for his doctoral dissertation and consisted of comparing serial sections of embryos to determine the cellular changes occurring during development. Among other things, he observed that despite the regular division of the nuclei, the cytoplasmic cleavage of the egg was variable, either definitely segmenting the egg surface from the beginning or being relatively unexpressed until as late as the fourth division, when simultaneous formation of all cell boundaries might occur. By observing the development of various members of the *Renilla* colony (not all polyps were the same morphologically), Wilson drew some interesting physiological, ontogenetic, and phylogenetic conclusions. His presentation of the *Renilla* work won the commendation of Huxley when Wilson was in England in 1882, and Huxley had the young man read the paper before the Royal Society (it was published in the *Philosophical Transactions* in 1883).[5]

During the two years following Wilson's return to the United States, his teaching duties at Williams College and M.I.T. provided little opportunity for continuing his research. He did, however, complete the writing of a textbook, *General Biology*, with Sedgwick, based upon ideas they both developed from observing H. Newell Martin's approach to introductory biology at Johns Hopkins. *General Biology*, published in 1886, was an attempt to treat the study of living organisms from a more analytical and integrated viewpoint than had been customary. To this end Wilson and Sedgwick treated life as a manifestation of chemical and physical laws; the properties of life were a result of the properties of its constituent atoms and molecules. They also included both plant and animal material in their discussion, and tried to show how all organic processes involved an interaction of the living system with its environment. *General Biology* provides one of the earliest examples of Wilson's broad perspective on biological problems, and as a textbook it was influential in bringing a new approach to the taxonomically and phylogenetically oriented introductory courses offered in most universities around the turn of the century.

After taking up his duties at Bryn Mawr in 1885, Wilson continued his studies on the cellular and morphological basis of early development with work on two annelids: *Lumbricus*, the earthworm, and *Nereis*, a marine polychaete. In his reports on *Lumbricus* (1887, 1889, 1890), he focused particularly on the origin of the mesoderm. He traced early development (cleavage through gastrula) in detail and demonstrated that the mesoderm is formed in a spiral or, as it was called, "mosaic" manner—that is, certain cells were set aside quite early to form the mesodermal tissues. These cells began to proliferate at the gastrula stage and all mesodermal tissues originated from them. Wilson's work on *Lumbricus* settled an existing controversy on the nature of mesoderm origin and showed, in conjunction with the subsequent work on *Nereis*, that spiral cleavage probably was characteristic of all annelids.

The earthworm proved to be a less than satisfactory organism for such studies, however, because it was difficult to follow the cells during successive cleavages. Following the lead of E. A. Andrews of Johns Hopkins, who had first pointed out the favorable nature of *Nereis* larvae (obtained at Woods Hole) for early embryological studies (these organisms show highly precise, regular, and easily observable cleavage patterns), Wilson carried out an exhaustive study of these forms. The *Nereis* work, published in 1890 and 1892 (although carried out mostly between 1885 and 1890), was a landmark both in the history of modern biology and in Wilson's career.

To study early cleavage Wilson developed to a high degree a method known as "cell lineage." It involved following the cell-by-cell development of young embryos from fertilization to blastula, cataloging the exact position of every daughter cell. From such studies it was possible to determine the exact ancestry of every cell in a blastula, and thus to determine the pattern by which cell division had occurred. Cell lineage studies are enormously intricate and detailed, and require considerable patience and observational skill (see Figure 1). The purpose of these studies was to apply the methods of comparative embryology to very early stages of development in different species. By accurately determining which cells in the early embryo came from which "lineage," for example, Wilson was able to show that triploblastic animals (those having three germ layers) fall into two large groups in terms of the mode of mesodermal formation. One group, including the annelids, arthropods, and mollusks, showed the spiral or mosaic pattern he had observed in the earthworm. The other group, including the echinoderms, primitive chordates, and vertebrates, showed a pattern called "radial," in which the mesoderm originates from pouches in the archenteron of the gastrula. Thus, cell lineage provided a means of establishing homologies in very early embryonic development that often were obscured in later stages. The work on *Lumbricus* and *Nereis* confirmed the study of cell lineage as an important embryological and morphological tool. It also established Wilson's reputation as a biologist of considerable observational skill and interpretive ability.

Wilson's detailed work also showed, however, that the problem of homologies, as many biologists were beinning to suspect, was more complex than had originally been thought when Ernst Haeckel proposed the biogenetic law in 1866. Although there might be many similarities among large animal phyla in cleavage patterns, there were some very important differences: structures obviously homologous in later embryonic stages sometimes derived from noncorresponding cells of earlier stages. Cell lineage patterns, like any other embryonic patterns, were not absolute criteria, and could suggest phylogenetic relationships only in the broadest outlines. Wilson recognized that embryonic processes (and structures) undergo evolution just as adults do, and that the present pattern of an organism's development is not a fossilized repetition of its ancestral history.

The choice of problems in Wilson's early work was largely influenced by the aims and methods of the morphologists, such as Brooks, under whom he was trained. Yet whatever problem he studied

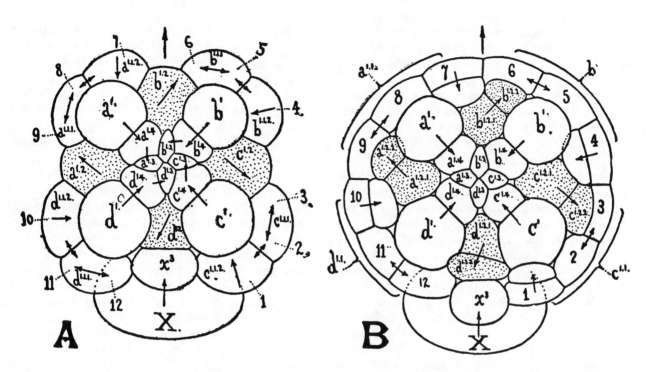

FIGURE 1.   Cell lineage study of young embryo of the marine worm *Nereis*. Letters and numbers indicate generational relationships; arrows indicate cleavage patterns. SOURCE: E. B. Wilson, "Cell Lineage of *Nereis*," in *Journal of Morphology*, 6 (1892), 396.

always showed the distinct mark of his personality: a meticulous attention to detail and an eye for larger issues. At the same time, interested as he was in the grander problems of evolution, or the "nature" of life, he rigorously avoided flights of fancy or ungrounded speculation. He could work on cell lineage as a means of understanding evolutionary relationships, without committing himself unalterably to a strict and rigid interpretation of embryological homologies. To Wilson, the primary process in biological investigation was the accurate determination of what happened in any phenomenon. Once the process or structure was described and observed to be repeatable, then it could be related to larger issues and theories—the understandings—of how life is organized.

Experimental Embryology and Cytology: 1891–1903. Wilson's first year as adjunct professor of zoology at Columbia was spent on leave in Europe (1891–1892). During the first half he worked with Theodor Boveri at Munich, and during the second half at the zoological station in Naples. From Boveri he learned much about the chromosomes and their relation to normal or abnormal development. He imbibed Boveri's concern for the importance of the chromosomes in cell division and in determining the course of development. By the time Wilson reached Munich, Boveri had al-

ready concluded that the nucleus—specifically the chromosomes—was the most important element in determining an organism's heredity. Further, Boveri argued that the chromosomes should be expected to influence development specifically—a thesis he was finally able to demonstrate in 1901 with experiments on doubly and triply fertilized sea urchin eggs. Particularly important during his stay in Munich were what Wilson learned about cytology and the strong personal relationship he developed with Boveri. To Wilson, Boveri was "far more than a brilliant scientific discoverer and teacher. He was a many-sided man, gifted in many directions, an excellent musician, a good amateur painter, and we found many points of contact far outside of the realm of science."[6] Wilson dedicated his major work, *The Cell in Development and Inheritance* (1896), to Boveri; and they remained close friends until Boveri's death in 1915.

At Naples, Wilson was exposed to the new experimental embryology through the work of Hans Driesch and Curt Herbst, both of whom were testing Wilhelm Roux's mosaic theory of development, put forth in 1888. Roux claimed that during cleavage, hereditary material is qualitatively divided among the daughter cells so that by the time the organism is fully differentiated, each cell has only one type of determinant (muscle cells have only muscle determinants, liver cells only liver determinants, and so on). Driesch, on the other hand, isolated cells from very young embryos—two-, four-, or eight-cell stage—and observed that each could develop into a normal larva, contradicting Roux's premises. To Driesch, these findings emphasized the plasticity of the embryo and of the embryonic process. The embryo remained able to reconstruct itself, which suggested that all cells retained the full complement of hereditary information, which was not qualitatively restricted, as Roux supposed. The Roux-Driesch controversy focused on a new and important question: the mechanism of cell differentiation. Although this problem had been recognized for many generations (in fact, it had been implicit in preformation and epigenesis arguments from the seventeenth century on), it had been eclipsed as a prominent biological question by the increased interest in evolutionary questions (by morphology) in the latter half of the nineteenth century.

The questions of cell differentiation raised by the Roux-Driesch controversy greatly stimulated Wilson's imagination, and turned his attention away from the more morphologically oriented studies and toward more critical questions of experimental embryology. Although he never abandoned the older methods of observation or the study of cell lineage, he began to see that embryology had important questions in its own right. Characteristically, Wilson kept his balance in the cross fire between the Roux and Driesch camps. He realized from his previous work on *Nereis* that although developmental processes are indeed determined, they also are plastic—they constantly reveal an interplay between the hereditary characteristics of the organism and the total environmental conditions to which it is exposed. The manner of development of a part, Wilson wrote in 1899, is "a manifestation of the general formative energy acting at a particular point under given conditions— the formative processes in special parts [being] definitely correlated with the organization of the entire mass."[7] He was acutely aware that embryonic parts continually interact, to differing degrees at various stages and varying from species to species.[8] Although he reserved his final judgment until more critical evidence was available (unlike many other biologists, who quickly took sides), from the early 1890's Wilson tended to agree with Driesch in opposing the seemingly very simplistic nature of the mosaic theory.[9] The Roux theory was mechanically possible and logically consistent, but

to Wilson it could not account for all the facts—for example, Driesch's results or the phenomena of regeneration. Like Driesch, he saw that embryos have enormous abilities to restore themselves to normal function even if profoundly disturbed by experimental conditions. Unlike Driesch, however, Wilson did not take ultimate refuge in mystical forces and entelechies to explain these restorative capacities.

Exposure to the problems and methods of the new school of experimental embryology (called by Roux's term, *Entwicklungsmechanik*, roughly translated as "developmental mechanics," or simply as "experimental embryology") raised in Wilson's mind the question of how differentiation *does* take place if it is not the result of a simple segregation of hereditary material among daughter cells. He reasoned that if differentiation were not a mosaic process, the key to both its regularity and its amazing flexibility somehow must reside in the organization of the egg cell, particularly the cytoplasm. Assuming, as Boveri and others had shown, that every daughter cell receives the same number and kind of chromosomes as the parent cell, he concluded that differentiation must be triggered by variations in the cytoplasm in which each nucleus lies. Thus, the egg cell's cytoplasm must be "preorganized" in such a way that regional localization of substances exists before cleavage begins. How the cytoplasm became structured in the first place was utter speculation, yet on the assumption of preorganization in the egg, Wilson could explain why some species seemed to show mosaic, and others nonmosaic, patterns of development. Those species appearing to have mosaic development simply showed cytoplasmic regionalization at a much earlier time in the embryo's life than those that seemed to be nonmosaic.

The concept of "prelocalization," or "formative substances," was strongly criticized by numerous workers, including T. H. Morgan, who objected that the idea simply pushed the problem of differentiation back another step by postulating that it had already taken place within the cytoplasm of the egg. Since there was no satisfactory explanation of how this localization was attained, the concept of cytoplasmic differentiation in the ovum seemed to be without substance. Yet Wilson retained a conviction that prelocalization was to some extent a reality, for his studies suggested that the cytoplasm was not a homogeneous mix but was, in fact, quite diverse in its local composition. Although it is recognized today that eggs (indeed, all cells) have regional localization, such as polarity or gradients of distribution of certain substances, it has become clear that in many species the organization of the cytoplasm has little to do with the process of differentiation.

As a result of his work with Boveri, Wilson developed a strong interest in the cytological events surrounding cell division, particularly those involved in the maturation of the egg. On returning to the United States, he took up the study of chromosome movements, particularly spindle formation and the origin of the centrosomes (today called cell centrioles). In a lengthy study done with his pupil A. P. Mathews (1895), Wilson produced solid evidence against Hermann Fol's widely held theory of the "quadrille of the centers." (Fol maintained that the sperm and egg centrosomes fuse after fertilization, then divide, moving through the cytoplasm, like dancers changing partners in the eighteenth-century square dance quadrille, to form the two poles of the spindle apparatus.) Wilson showed that in echinoderms (especially the sea urchin) the poles were formed only by division of the sperm's centrosome. He went on to demonstrate in later papers that the centrosomes were formed within the cytoplasm, not within the nucleus, as had previously been thought. Close observation of the movements and doubling of centrosomes convinced Wilson that the replication of these bodies did not cause, and was not caused by, the replication of the chromosomes. The doubling of both sets of structures probably responded, he maintained, to some underlying rhythm in the cell's activity. He demonstrated this clearly by comparing rhythmic changes in protoplasmic activity in fragments of fertilized eggs of the marine mollusk *Dentalium*. Parallel rhythmic changes could be observed in the fragments and in nucleated portions of the same egg, even though the two parts were no longer physically associated.

During his initial decade at Columbia, Wilson prepared the first edition of *The Cell in Development and Inheritance* (1896), the basis of which was a series of lectures he gave in 1892–1893. *The Cell* was much more than a compilation of all the relevant information on various parts of the cell and various cell processes. It was not only a synthesis of a great deal of information (the bibliography in itself represents a prodigious effort) but also reflected Wilson's wide-ranging and balanced views of contemporary problems, and his special emphasis on the function of cytology in elucidating such topics as embryology, heredity, evolution,

and general physiology. The primary aim of the work, according to Wilson, was no less grandiose than "to bring the cell-theory and the evolution-theory into organic connection."[10] He believed that a fundamental understanding of the cell in all its aspects (structure, development, and physiological functions) would provide a better understanding of those fundamental processes—heredity, variation, and differentiation—on which evolution was ultimately based.

The book is organized to lead the reader to appreciate the role of the cell—the nucleus and chromosomes—in heredity. The opening chapter deals with cell structure in broad overview, the second with cell division. The following three chapters deal with the germ cells: their structure and mode of origin, the phenomenon of fertilization, and the maturation divisions by which the gametes are prepared for fertilization. The sixth chapter deals with cell organization—the structure of chromosomes and the evidence for their "individuality"—and with centrosome origin and astral formation. The seventh chapter reviews the physiological properties of the cell, and the eighth treats the maturation of the ovum and the general laws of cell division of which it is an expression. In the final chapter Wilson considers the basic phenomena of development (as elucidated by Roux, Driesch, Herbst, Chabry, and others) in terms of cell structure and function.

The central conclusion of the book is in the eighth and ninth chapters, where Wilson focused on an in-depth study of the cleavage of the fertilized egg and on the various experimental results that shed light on the underlying processes by which cleavage could produce cell differentiation. Thus, at the end of the book Wilson was able to muster many lines of evidence to demonstrate the key point: "that the nucleus contains the physical basis of inheritance; and that chromatin, its essential constituent, is the idioplasm postulated on Nägeli's theory."[11] Several lines of evidence led Wilson to place the seat of heredity within the cell nucleus and particularly in the chromosomes. First, the persistent accuracy with which the chromosomes replicate and are distributed, in contrast with the often random division of the cytoplasm by region, indicates the importance of ensuring that each daughter cell receives a full complement of chromosomes. Although Wilson had not abandoned the idea of cell prelocalization, he recognized that compartmentalization of the cytoplasm through cleavage was a much less precise process than distribution of the chromosomes. The greater precision of the chromosome distribution mechanism suggested that it was intimately related to the hereditary process, which by definition must be a regular and highly accurate phenomenon.

Second, the work of Boveri in particular (1887) had suggested that chromosomes maintain their individuality and continuity from one cell generation to the next. Contrary to an idea prevalent in the 1870's and 1880's, he and others had shown that the chromosomes do not disintegrate between divisions, but have the same spatial arrangement after interphase as before. Although Wilson was not willing to conclude that the physical structure of the chromosomes was necessarily maintained unbroken from interphase to interphase, he did argue that all evidence pointed to the maintenance of hereditary integrity.

Third, abundant cytological evidence showed that while sperm and egg had enormously different cytoplasmic components (the sperm has virtually no cytoplasm), they seemed, on the whole, to affect the heredity of the offspring equally. Thus it would appear, Wilson pointed out, that the cytoplasm had relatively little hereditary function.

Fourth, experiments by M. Nussbaum, Gruber, Verworn, and others on many different types of cells (including Protozoa) indicated that enucleated cells did not function normally. Whatever the exact function of the nucleus, it was necessary to the normal maintenance of cell activity. It seemed evident that the control that the nucleus appeared to exert over the entire cell must be an expression of the cell's heredity.

Even in 1896 Wilson recognized that the control of the nucleus over the cytoplasm was ultimately a matter of chemical interactions. Too little was known about the chemistry of chromatin for him to formulate a specific idea about how this might work, but he did maintain that the nucleus was the seat of constructive (anabolic), and the cytoplasm of destructive (catabolic), processes. This view came directly from Claude Bernard, who some twenty years earlier (1878) had postulated a similar division of chemical labor between nucleus and cytoplasm.[12] To Wilson, inheritance (associated with the nucleus) "is the recurrence through the transmission from generation to generation of a specific substance or idioplasm which we have seen reason to identify with chromatin. If the nucleus be the formative center of the cell, if nutritive substances be elaborated by or under the influence of the nucleus while they are built into the living fabric, then the specific character of the cytoplasm is determined by that of the nucleus."[13]

Yet Wilson was aware enough of biological phenomena in general to recognize that the cytoplasm also must profoundly influence the nucleus. The nucleus could not function, after all, if it did not have a cytoplasm upon which to act. But more than that, he saw that as the nucleus altered the cytoplasm (by building up certain substances), of necessity it also altered its own environment. Chemical change—interaction and modification—was always occurring between the nucleus and cytoplasm in a living cell. Development was nothing more than a highly ordered example of this interaction, in which the expression of hereditary information in each cell nucleus was successively altered as cleavage and morphogenetic changes occurred. To Wilson, the nucleus and the cytoplasm were intimately involved in the cell's chemistry, heredity, and development. They had different but complementary functions, and had to be understood in relation to each other. Neither could, or should, be viewed in isolation.

The Cell went through three editions and numerous reprintings. It is estimated that this book has been the single most influential treatise on cytology during the twentieth century. Many of the problems that Wilson clearly outlined (such as the relationship between the nucleus and cytoplasm in cell differentiation) are still being investigated. And no one has succeeded in posing those problems more clearly than he did in his many writings, particularly in *The Cell*. In reading *The Cell*, one is impressed not only by Wilson's skill as a summarizer (an encyclopedist in the best sense of the word) but also with the enormous patience and effort on the part of hundreds of other workers as well, who over the past century have contributed to the growing knowledge of cell structure and function.

Studies on Chromosomes and Heredity: 1903–1912. Historically, one of the most important functions of *The Cell* was to pave the way for a more rapid acceptance of Mendelian theory, once it was reintroduced to the scientific community in 1900. By focusing attention on the cell nucleus, and particularly on the chromosomes as the seat of heredity, Wilson prepared many biologists—especially cytologists—to see the relationship between Mendel's laws and the events of maturation of the sperm and egg. Recognition of the possible parallel between the segregation and random assortment of Mendelian "factors" (later called "genes") and the chromosome reduction division (meiosis) that occurs during gametogenesis ultimately provided a material basis for the science of genetics.

Although two of the rediscoverers of Mendel's work—Carl Correns and Hugo de Vries—had offered a chromosomal interpretation of their own and Mendel's findings shortly after 1900, it was not based on much observational evidence. In 1900, however, H. von Winiwarter reported the occurrence of synapsis in early reduction division; and in 1901 T. H. Montgomery discovered that the chromosomes are present in germ cell nuclei (prior to reduction division) as pairs of homologues. With these observations Wilson and his group were ready by 1901 to seek possible connections between Mendelism and cytology. In fact, it was one of Wilson's graduate students, Walter S. Sutton, who made the connection first and most cogently (1902). In studying synapsis (the intertwining of the two chromosomes in a homologous pair of chromosomes), Sutton showed that the visible behavior of the chromosomes afforded an explanation of the first and second Mendelian laws. His careful studies of chromosomal pairing provided cytological evidence that the chromosomes segregating in reduction division are the two members of a homologous pair, not any two random chromosomes. Thus each chromosome could be considered as the counterpart of a Mendelian factor, or at least a bearer of one factor. Wilson quickly came to see the importance of Sutton's work, and supported his conclusions.

In 1902 another former student of Wilson's, Clarence E. McClung, pointed out that the unpaired "accessory" chromosome (later called the X by Wilson), long known to exist in the males of some arthropods, might offer a clue to how sex was inherited. Wilson was intrigued by McClung's work and set out to study the occurrence and distribution of the accessory chromosome in a number of species, mostly insects. In 1905 Wilson and, independently, Nettie M. Stevens of Bryn Mawr published extensive cytological evidence suggesting a chromosomal basis for sex determination.[14] These works provided the missing link between cytology and heredity. Wilson and Stevens concluded that females normally have a chromosome complement of XX and males have one of XY. In oögenesis and spermatogenesis, the X and X (for oögenesis) and the X and Y (for spermatogenesis) separate, and end up, by meiotic division, in separate gametes. All eggs thus have a single X chromosome, while sperm can have either an X or a Y. When a Y-bearing sperm fertilizes an egg, the off-spring is a male (XY); when an X-bearing sperm fertilizes an egg, the offspring is a female (XX).

Wilson and Stevens recognized that a few groups of organisms have variations (or reversals) of this scheme—for instance, species that normally lack a Y or in which the females are XY and the males XX (the latter case is true for moths, butterflies, and birds). The 1905 papers by Wilson and Stevens not only cleared up a long-standing controversy on the nature of sex determination (for example, whether it was hereditarily or environmentally induced) but also were the first reports that any specific hereditary trait (or set of characteristics, such as those associated with sex) could be identified with one specific pair of chromosomes.

Wilson pursued studies on the chromosomes, particularly in relation to sex inheritance, over the next seven years (1905–1912), producing a series of eight papers entitled "Studies on Chromosomes." In general these papers worked out the chromosomal theory of sex determination (essentially as it is understood today) in great detail, and supported its Mendelian nature. Among other things, Wilson showed that the Y chromosomes in different insect species are of widely different sizes in comparison with the X; in some species the X and Y are of virtually equal size, whereas in others the Y is very small, and in still others it is nonexistent. He also observed that in a species where the female is normally XX, some females have the combination XXY and some males have only a single X and no Y. Wilson interpreted these cases as having resulted from the failure of the X and the Y to separate during spermatogenesis in the organism's male parent. When the same phenomenon was observed in *Drosophila* by C. B. Bridges in 1913, he and Wilson jointly coined the term "nondisjunction" for the failure of two homologues to segregate during meiosis. These and other results led Wilson to postulate that the Y chromosome had degenerated over the course of evolutionary history. He felt it represented either inactive chromatin material or an excess that was duplicated elsewhere in the chromosome group.[15] Wilson considered the X to be the active member of the sex chromosome pair, and therefore the causal agent of sex determination. Although we know today that the matter is not so simple, he was essentially correct in judging that the Y has little actual hereditary function, in relation to sex or anything else.

Wilson speculated that the difference between

the X and the Y might be due to the presence on the X (or associated with it) of a specific chemical substance (perhaps an enzyme) that produces a definite reaction on the part of the developing individual.[16] Surprisingly, his idea that a chromosome might carry out its hereditary function by producing enzymes is close to the modern conception that genes (located on chromosomes) code for enzymes that catalyze specific biochemical reactions. Although Wilson was the first to point out, repeatedly, that too little was known about the chemistry of living cells to formulate any meaningful theory of how characteristics were determined, he saw the importance of phrasing hereditary or developmental problems in chemical terms.

Wilson's studies on chromosomes provided the important cytological foundation upon which T. H. Morgan's later chromosome theory of inheritance was based. In 1910 Morgan, Wilson's close friend and colleague in the zoology department at Columbia, discovered a white-eyed male *Drosophila* in his laboratory culture (*Drosophila* normally have red eyes). Although initially skeptical of the Mendelian theory, Morgan found that Mendel's assumptions provided the best means of accounting for the hereditary pattern observed in the white-eye condition. Moreover, he saw that white eyes seemed to occur mostly but not exclusively in males, a fact that could be explained only by assuming that the "factor" for eye color was located on the X chromosome. Wilson quickly saw the implications of this work, and in 1911 he used Morgan's findings as further support for a chromosomal interpretation of sex. He also saw immediately what later came to be called sex-linked inheritance. Thus, the keystone to the chromosome theory of inheritance was laid in the Columbia laboratory, where Morgan from the animal-breeding side, and Wilson from the cytological side, provided evidence that hereditary units exist as material entities located on chromosomes in the nucleus.

Aside from his intellectual contribution to the chromosome theory of heredity, Wilson was influential as a strong supporter of Morgan and his group as they expanded the *Drosophila* work. As head of the zoology department, he encouraged all the "fly room" workers, especially graduate students, by his persistent interest in the work and by the obvious connections it bore to his earlier work on chromosomes. Most of the students who worked with Morgan (Muller, A. H. Sturtevant, Bridges, Edgar Altenberg) had been Columbia undergraduates and had taken Wilson's courses or used his textbook in the introductory course taught

by G. N. Calkins and James H. McGregor. In his teaching from 1906 on, Wilson particularly emphasized the relations between Mendelian heredity, chromosomes, and evolutionary theory, using as the text for his second-level one-semester course on heredity R. H. Lock's provocative book *Variation, Heredity and Evolution* (1906). Far ahead of its time, Lock's book treated Mendelian heredity, cytology, and Darwinian evolution in a completely integrated fashion, something few biologists did until after 1915.

Thus many of the students who came to work with Morgan after 1910 had been prepared by Wilson to see clearly the relationships between chromosomes, Mendelian theory, and the concept of natural selection in a way that not even Morgan himself was able to accomplish at the time. Wilson also supported the *Drosophila* group by incorporating its findings into his own work, and by championing the new ideas long before many other biologists began to follow Morgan's lead. Although neither Wilson, Morgan, nor any other biologists at the time could "see" that Mendelian genes were parts of chromosomes, or that crossing-over and exchange of chromosome parts actually took place as the *Drosophila* group postulated, Wilson felt that the conclusions were sound because they were consistent and fitted all the data. In a lecture in 1913, he pointed out that although the hypothesis of crossing-over and chromosome mapping techniques based on it were bold ventures, they were justified because, pragmatically, they worked; that is, they accounted for all the data better than any other explanations. To Wilson it was by just such venturesome ideas that new possibilities of discovery were opened.[17]

Recognizing the importance to Morgan's work of testing the assumption of crossing-over (in meiotic divisions), Wilson set out in 1912 (study VIII) to examine the cytological evidence for such a process. Obtaining preparations of germ tissue from F. A. Janssens (who had originated the hypothesis of crossing-over in 1909), A. and K. E. Schreiner, and McClung, Wilson showed that synapsis seemed to be a real phenomenon—that is, homologous chromosomes do appear to come together and wrap around each other prior to the first meiotic division. The evidence was not clear enough, however, to show any signs of the actual exchange of chromosome parts. It was not until 1931 that techniques were developed sufficiently for such workers with animals as Curt Stern and with plants as Barbara McClintock and Harriet Creighton to observe actual exchanges between strands and thus provide final proof that crossing-over was a real phenomenon.

The years between 1902 and 1912 marked the zenith of Wilson's creative period. The eight studies on chromosomes were brilliant examples of his observational and analytical skill. In this work his broad-reaching mind incisively drew the connections between Mendelian theory and cytology, long before many other workers (including Morgan) were prepared to make the same bold leaps. The chromosomal concept of Mendelian heredity was a logical view for Wilson to maintain because it provided the link he had intuitively held for many years between the cell, heredity, and development. The main theme enunciated in *The Cell* (1896) was being realized in actuality by the parallel studies on chromosomes in Wilson's laboratory and on the process of heredity in Morgan's.

Later Work: 1912–1938. Wilson's studies after 1912 were variations of a single basic question: What cell constituents other than the chromosomes affect the hereditary process? There were really two aspects to this question. One was that of extrachromosomal inheritance: the replicative function of such organelles as chloroplasts or mitochondria, which by 1920 were known to be able to reproduce themselves without nuclear control. The other question was the effect of the cytoplasm on the expression of genetic potential in the nucleus. In investigating the former, Wilson studied various cytoplasmic bodies in scorpions and insects: the Golgi bodies, chondriosomes (today called mitochondria), and vacuomes (today, vacuoles). From these observations he concluded that at least the Golgi bodies and chondriosomes increase in size and fragment, so that daughter cells get equal numbers with each division. Nevertheless, he maintained that these cytoplasmic bodies do not have genetic individuality as chromosomes do—they are all very much alike, and must be derived from the same genetic ancestry, with little or no divergence.[18] In investigating the effect of the cytoplasm on expression of genetic potential, Wilson steadfastly stated his earlier view that it is impossible to make any rigid distinction between the nucleus and cytoplasm. What was important was to attempt to determine the precise ways in which the nucleus influenced the cytoplasm, and vice versa; in that way not only would it be possible to achieve a clearer understanding of how the cell functions to maintain itself day by day, but also it would be possible to see more clearly how such phenomena as cell differentiation might be brought about.

In his later years Wilson came more and more to view the cell as a plastic, ever-changing structure, continually building itself up and tearing itself down, a constant dialectic between stability and change, heredity and variation, maintenance and differentiation. In the flux of materials and changing structures, the constancy in cell life was an underlying organization of molecules and intramolecular associations that produced life.

The culminating work of Wilson's later years was the complete revision and expansion of *The Cell* into a third edition (1925). So much information had accumulated regarding cytological phenomena in the quarter-century since the second edition that only Wilson, with his encyclopedic mind and incisive judgment, could have undertaken, let alone completed, such a task. Although Wilson's health showed signs of decline after 1920, he nonetheless worked tirelessly on what was a monumental undertaking for a man in his mid-sixties. The completed volume of over 1,200 pages was published when Wilson was sixty-nine. "In it," wrote H. J. Muller, "virtually the whole of cytology from the time of its birth more than half a century before, stood integrated."[19] The revised edition, under the title *The Cell in Development and Heredity*, was awarded the Daniel Giraud Elliot Medal of the National Academy of Sciences (1928) and the gold medal of the Linnean Society of London (1928). Perhaps the most significant testimony to its value is that it is still found on the bookshelves of cytologists, not only as a useful reference source for older literature but also because many of the problems it posed are still being investigated. The electron microscope, with all the resolution of fine detail and astounding discoveries it has made possible, has not yet solved many of the questions and problems that Wilson raised. Developments in cytology since 1925 have expanded upon, but not contradicted, the underlying concepts that he wove so skillfully into the fabric of *The Cell*.

In all three editions of *The Cell* Wilson related the phenomena of heredity, cell structure and function, and development to organic evolution and adaptation. To him the central problem of evolution as posed by Darwin was how hereditary variations come about. In 1896 he recognized the importance of August Weismann's conception of the continuity of the germ plasm, and much of his future cytological work on chromosomes served to support the basic idea of a separation of germ and somatoplasm. Wilson's insight into this problem lay in his recognition that heredity was a cellular phenomenon—something that Darwin and his followers also had recognized. The Darwinian theory of pangenesis was, after all, only a cellular mechanism for how variations could occur. From his early studies of cells and his growing awareness that the nucleus was the locale of a cell's heredity, Wilson rejected the pangenesis theory. Because cell heredity was localized in the nucleus, specifically in the chromosomes, and because each set of chromosomes had continuity—that is, it transmitted its effects only vertically from one generation to the next—somatic variations could not be transmitted to the germ cells of the same organism.

By his further work on chromosome structure and variation, approached cytologically and framed in Mendelian terms, Wilson tried to show how new heritable variations arose and could be acted upon by selection. The most significant support that he gave to the Darwinian theory was his conviction that heredity (and its correlate, variation) was ultimately a cellular (chromosomal) phenomenon. By emphasizing this relationship and by providing, through cytological studies, a material basis for Mendelian heredity, Wilson paved the way for a comprehensive theory of evolution, which emerged after 1930.[20]

Despite his strong interest in evolution, Wilson was somewhat skeptical of certain aspects of the Darwinian theory of natural selection. Like many of his contemporaries, he greatly admired Darwin's work as a naturalist and his synthetic powers as expressed in *The Origin of Species*. He felt, however, that Darwin had placed too much emphasis on evolution by the accumulation of small variations (what Darwin called "individual differences") that could not be shown to be inherited. By failing to distinguish adequately between inherited and acquired variations, Darwin had not provided a mechanism for the origin of adaptations. Wilson also was bothered by the emphasis that Darwin's theory placed on "chance" in the origin of species. Strongly influenced as he was by his old teacher, he stated in 1907 that Brooks's epigram was true, that "the essence of life is not protoplasm but purpose."[21] Wilson did not believe in teleological principles or in vitalistic driving forces in evolution. Nevertheless, like many biologists raised in the era of Haeckel, Weismann, and the other German Darwinians, it was difficult for him to believe that purpose was altogether lacking in evolutionary processes. The evolution of adaptations as intricate and functional as the vertebrate eye, simply by the accumulation of numerous *chance* variations, seemed to defy reason. As late as 1930 Wilson wrote: "[I am] not yet quite ready

to admit that higgledy-piggledy can provide an adequate explanation of organic adaptations."[22] Yet when confronted with alternative explanations of adaptation, Wilson always found himself forced to return to Darwin. In 1915 he wrote, "We have made it the mode to minimize Darwin's theory . . . but . . . we should take heed how we underestimate the one really simple and intelligible explanation of organic adaptation, inadequate though it may now seem, that has thus far been placed in our hands."[23]

Wilson's Scientific Methodology. Although Wilson was trained as a morphologist, he embraced the quantitative and experimental side of biology early in his career, following his 1891–1893 stay in Europe, particularly at the Naples station. Along with a number of younger biologists around the end of the nineteenth century, he felt that for biology to make any progress, it had to avoid vapid speculation and the construction of all-embracing theories that had no basis in empirical evidence. To him careful observation, hypothesis formulation, and experimentation were the only true means of reaching valid conclusions. Experiment alone was never enough—experiments had to be designed to test something, and that something was a particular hypothesis. In Wilson's view, however, hypotheses had to be testable. If they were not, then oversimplified and misleading ideas could gain a vast following, as had happened with the speculative theories of Haeckel and Weismann. The strong advocacy that Wilson and many of his contemporaries (including T. H. Morgan, Ross G. Harrison, Jacques Loeb) made in behalf of experimental biology was in some part a reaction to the nonexperimental, speculative methodology characteristic of a previous generation.

Behind Wilson's experimentalism lay a firm belief that "the scientific method is the mechanistic method."[24] By mechanistic he meant, as did most of the younger workers at the time, that phenomena should be subject to experimental analysis and that biological processes should be investigated in physicochemical terms. Wilson did not argue that the only meaningful explanation of a biological process was one couched in terms of chemical equations. He knew full well that such explanations were not then possible for most biological phenomena, yet he firmly believed that they should be sought as much as chemical theory and technology would allow. He was not a crude mechanist and could not share the extreme mechanical bias of Jacques Loeb. Nevertheless, he felt that until biologists could understand the complex events characterizing cell life in chemical terms, they could not gain much understanding of the nature of life or its immense complexities.

Philosophically, Wilson believed that the scientific method was the only way to understand the world—inside or outside the laboratory; but he did not think that scientific truths are final truths, for truth itself, he claimed, is relative. To Wilson the fundamental concepts of science had no finality. "The profound significance of what we call natural laws lies in the fact that they tersely sum up our experience of the world at any given moment . . . ."[25] Science was for him a creative process, the ideas running in advance, to some extent, of the hard facts. Wilson's own artistic and aesthetic sense allowed him to see science as no different, in its creative aspects, from music, art, or literature: "At every point the material world overflows with half-revealed meanings about which science is forever weaving her imaginative fabrics; and at their best these have all the freedom, boldness and beauty of true works of art."[26] Wilson the musician and Wilson the cytologist were one and the same person, applying different skills at different times but with the same aesthetic delight and by the same intellectual methods. For Wilson internal beauty had to be matched with external reality. Music to him was not simply a theory—it was the reality of notes played on an instrument, filling a room and affecting human beings. Science also was not theories—fossilized answers—it was the reality of what could be observed, predicted, and repeated by living, imaginative people. Both music and biology had their theoretical sides, but the theories had meaning only as they were applied in practice on a day-to-day basis.

In his teaching, as in his investigation and his writing, Wilson emphasized that science was not accumulated knowledge, but a process of reasoning, understanding, and testing that understanding against natural phenomena. He saw life as a whole, in all its manifest complexities, harmonies, and apparent inconsistencies. He was willing to let an issue rest unresolved rather than propose a solution that was untested or untestable. Yet in his emphasis on process, he saw the human side of science—that it was ultimately the activity of human beings, not monuments of static information. Externally, he lived the life of the classic reserved scholar; but beneath this formality was a fire that, as Muller put it, was rigorously channeled into self-discipline. Nevertheless, it was this fire that shone through even the most meticulous pages of cell lineage studies, or the most detailed analysis of mitotic patterns. It transformed details into the comprehensive and exciting fabric of biological ideas that Wilson wove throughout his life.

NOTES

1. T. H. Morgan, "Edmund Beecher Wilson," 318.
2. *Ibid.*, 320.
3. H. J. Muller, "Edmund B. Wilson—an Appreciation," 166.
4. Morgan, *op. cit.*, 319.
5. E. B. Wilson, "The Development of *Renilla.*"
6. Morgan, *op. cit.*, 321.
7. Muller, *op. cit.*, 17.
8. E. B. Wilson, "Amphioxus and the Mosaic Theory of Development," in *Journal of Morphology*, **8** (1893), 579–638, esp. 636–638.
9. The "mosaic theory" of Roux and Weismann is not to be confused with the "mosaic" or spiral pattern of cleavage Wilson had discussed in his earlier (1887, 1889, 1890) work on annelid embryology. The Roux-Weismann "mosaic" theory held that differentiation during ontogeny was caused by the qualitative nuclear division of hereditary material during cleavage; as development progressed, cells gradually lost more and more hereditary potential in terms of the kinds of adult tissues they could form.
10. E. B. Wilson, *The Cell in Development and Inheritance*, intro., 11.
11. *Ibid.*, 302.
12. *Ibid.*, 247; Bernard's work is in his *Leçons sur les phénomènes de la vie . . .*, I (Paris, 1878), 523.
13. Muller, *op. cit.*, 35.
14. E. B. Wilson, "The Chromosomes in Relation to the Determination of Sex in Insects"; N. M. Stevens, "Studies in Spermatogenesis With Especial Reference to the 'Accessory Chromosome,'" *Publications. Carnegie Institution of Washington*, no. 36 (1905).
15. E. B. Wilson, "Studies on Chromosomes. V. The Chromosomes of *Metapodius*, a Contribution to the Hypothesis of the Genetic Continuity of Chromosomes."
16. *Ibid.*
17. Muller, *op. cit.*, 156.
18. *Ibid.*, 160.
19. *Ibid.*, 161.
20. For Wilson's more explicit attempts to discuss cytology and Darwinism, see his "The Cell in Relation to Heredity and Evolution"; and his "Biology."
21. Muller, *op. cit.*, 153.
22. *Ibid.*, 153–154.
23. *Ibid.*, 153.
24. *Ibid.*, 24.
25. E. B. Wilson, "Science and Liberal Education," in *Science*, **42** (1915), 625–630.
26. *Ibid.*

BIBLIOGRAPHY

I. ORIGINAL WORKS. There is no complete, or even selected, collection of Wilson's writings, although a bound set of his reprints, with only minor omissions, is in the library of the Marine Biological Laboratory, Woods Hole, Mass. A complete bibliography is in T. H. Morgan's obituary (see below).

Wilson's books are *General Biology* (New York, 1886), written with William T. Sedgwick; *An Atlas of Fertilization and Karyokinesis of the Ovum* (New York,

1895); *The Cell in Development and Inheritance* (New York, 1896; 2nd ed., 1900; 3rd ed., rev. and enl., *The Cell in Development and Heredity*, 1925)—1st ed. repr., with intro. by H. J. Muller, as Sources of Science, no. 30 (New York, 1966).

Among his most important journal articles are "A Problem of Morphology as Illustrated by the Development of the Earthworm" (abstract), in *Johns Hopkins University Circulars* (May 1880), 66; "The Development of *Renilla*," in *Philosophical Transactions of the Royal Society*, **174** (1883), 723–815; "The Embryology of the Earthworm," in *Journal of Morphology*, **3** (1889), 387–462; "The Cell-Lineage of *Nereis*. A Contribution to the Cytogeny of the Annelid Body," *ibid.*, **6** (1892), 361–480; "The Mosaic Theory of Development," in *Biological Lectures. Marine Biological Laboratory, Woods Hole, 1893* (1894), 1–14; "Maturation, Fertilization, and Polarity in the Echinoderm Egg. New Light on the 'Quadrille of the Centers,'" in *Journal of Morphology*, **10** (1895), 319–342, written with A. P. Mathews; "On Cleavage and Mosaic-Work," in *Archiv für Entwicklungsmechanik der Organismen*, **3** (1896), 19–26; "The Structure of Protoplasm," in *Biological Lectures, Woods Hole Marine Biological Laboratories for 1898* (1899), 1–20; "Cell Lineage and Ancestral Reminiscence," *ibid.*, 21–42; "Some Aspects of Recent Biological Research," in *International Monthly* (June 1900), 1–22; "Mendel's Principles of Heredity and the Maturation of the Germ-Cells," in *Science*, **16** (1902), 991–993; "Mr. Cook on Evolution, Cytology and Mendel's Laws," in *Popular Science Monthly* (Nov. 1903), 188–189; "The Problem of Development," in *Science*, **21** (1905), 281–294, presidential address, New York Academy of Sciences, 19 Dec. 1904; and "The Chromosomes in Relation to the Determination of Sex in Insects," *ibid.*, **22** (1905), 500–502.

Wilson's eight "Studies on Chromosomes" are in *Journal of Experimental Zoology*, **2** (1905), 371–405; II. "The Paired Microchromosomes, Idiochromosomes and Heterotropic Chromosomes in *Hemiptera*," *ibid.*, 507–545; III. "The Sexual Differences of the Chromosome-Groups in *Hemiptera*, With Some Considerations on the Determination and Inheritance of Sex," **3** (1906), 1–40; IV. "The 'Accessory' Chromosome in *Syromastes* and *Pyrrochoris* With a Comparative Review of the Types of Sexual Differences of the Chromosome Groups," **6** (1909), 69–99; V. "The Chromosomes of *Metapodius*, a Contribution to the Hypothesis of the Genetic Continuity of Chromosomes," *ibid.*, 147–205; VI. "A New Type of Chromosome Combination in *Metapodius*," **9** (1910), 53–78; VII. "A Review of the Chromosomes of *Nezara*; With Some More General Considerations," **12**, (1911), 71–110; VIII. "Observations on the Maturation-Phenomena in Certain *Hemiptera* and Other Forms, With Considerations on Synapsis and Reduction," **13** (1912), 345–448.

Other papers are "Mendelian Inheritance and the Purity of the Gametes," in *Science*, **23** (1906), 112–113; "Recent Studies of Heredity," in *Harvey Lectures*

(1906–1907), 200; "Notes on the Chromosome Groups of *Metapodius* and *Banasa*," in *Biological Bulletin*, **12** (1907), 303–313; "Differences in the Chromosome-Groups of Closely Related Species and Varieties, and Their Possible Bearing on the 'Physiological Species,'" in *Proceedings, Seventh International Congress of Zoology* (1909), 1–2; "The Cell in Relation to Heredity and Evolution," *Fifty Years of Darwinism* (New York, 1909), 92–113; "The Chromosomes in Relation to the Determination of Sex," in *Science Progress*, **16** (1910), 570–592; "Some Aspects of Cytology in Relation to the Study of Genetics," in *American Naturalist*, **46** (1912), 57–67; "Observations on Synapsis and Reduction," in *Science*, **35** (1912), 470–471; "The Bearing of Cytological Research on Heredity," in *Proceedings of the Royal Society*, **88** (1914), 333–352, the Croonian lecture for 1914; "Chiasmatype and Crossing Over," in *American Naturalist*, **54** (1920), 193–219, written with T. H. Morgan; "The Physical Basis of Life," in *Science*, **57** (1923), 277–286; and "Biology," in *A Quarter Century of Learning: 1904–1929* (New York, 1931), 241–260.

II. Secondary Literature. The standard obituary is T. H. Morgan, "Edmund Beecher Wilson, 1856–1939," in *Biographical Memoirs. National Academy of Sciences*, **21** (1941), 315–342, condensed in *Science*, **89** (1939), 258–259; also *New York Times* (4 Mar. 1939), 15, col. 1.

Perhaps the best single evaluation of Wilson's life and especially of his career, and the source that has been particularly helpful in preparing this article, is H. J. Muller, "Edmund B. Wilson—an Appreciation," in *American Naturalist*, **77** (1943), 5–37, 142–172. Muller's familiarity with Wilson's work is thorough, and his assessment of Wilson's place in the history of twentieth-century biology is authoritative. A shorter version of this "appreciation" is in Muller's intro. to the 1966 repr. of *The Cell in Development and Inheritance*.

Despite his importance to twentieth-century biology (and American science especially), there is a surprising lack of biographical or critical material on Wilson's life and work.

Garland E. Allen

WILSON, EDWIN BIDWELL (*b.* Hartford, Connecticut, 25 April 1879; *d.* Brookline, Massachusetts, 28 December 1964), *mathematics, physics, statistics, public health.*

The son of a schoolteacher, Wilson graduated B.A. from Harvard in 1899 and Ph.D. from Yale two years later. He studied for a while in Paris, taught mathematics at Yale, and then moved to the Massachusetts Institute of Technology, becoming head of the department of physics there in 1917. Five years later he was appointed professor of vital statistics at the Harvard School of Public Health. Wilson's work in that capacity earned him two presidencies in 1929: of the American Statistical Association and the Social Sciences Research Council, New York. Following his retirement in 1945 he acted as consultant to the Office of Naval Research. Throughout his long and varied career (among other things he was managing editor of the *Proceedings of the National Academy of Sciences*, Washington, for half a century) Wilson combined a quiet if somewhat crotchety Yankee charm with a firm sense of high standards in research and exposition.

In each of his fields Wilson made characteristic contributions. As a student of Willard Gibbs at Yale, he codified the great physicist's lectures on vector analysis into a textbook. This beautiful work, published when Wilson was only twenty-two years old, had a profound and lasting influence on the notation for and use of vector analysis. Meantime, Wilson's mind and pen began to range over many other areas of mathematics, including the foundations of projective and differential geometry; and in 1903 he criticized, with bold sharpness, Hilbert's "so-called foundations" of geometry. In 1912 Wilson published a comprehensive text on advanced calculus that was the first really modern book of its kind in the United States. Immediately successful, it had no rival for many years. Wilson's interest in theoretical physics, inspired by Gibbs, resulted in papers on mechanics and relativity. World War I led him to study aerodynamics, in which he gave a course; and he did research on the theory of the effects of gusts on airplane flight. Outcomes of this work were the publication of a book on aeronautics in 1920 and the stimulation of a group of students who were to make a mark in that field.

Early in the 1920's Wilson began to think carefully about probability and statistics. Because of his Harvard professorship he naturally focused on vital statistics, but he also pondered the theory of errors and its relation to quantitative biology and astronomy. In this field he was both innovative and evangelical—constantly drawing attention to the role of statistics in biology and urging the recruitment of full-time statisticians.

A major contribution to inferential statistics was Wilson's restructuring of interval estimation. For long before his time it had been vaguely implicit that the attachment of a standard error to a point estimate was a crude interval estimate. Thus, noting, say, that a series of observations yielded 129 ± 22 mm. as the mean length of a sample of Armadillidiidae, the researcher could add that the true (parametric) value lay, with a probability of about 2/3, in the interval 107–151. In an admirably concise note published in 1927, Wilson pointed out that, logically, a true value cannot have a probable location. He also showed how a rigorous and unelliptic statement could be made about the probability that an estimated interval will embrace the (fixed) parameter. This interval was essentially what became known as a confidence interval, as rediscovered and developed by Jerzy Neyman and his school. The priority must, however, be given to Wilson.

In studying cumulative population growth, and in handling quantal-response bioassay (which involves "all-or-none" reactions of members of a biological population to an agent), Wilson was an early and effective advocate of the logistic function, $P = (1 + \exp[-(\alpha + \beta X)])^{-1}$, where P is the probability of response to the amount X of the agent, and α and β are parameters. He published methods of handling data that fitted this function, and thus of estimating the potency of the agent.

Wilson exhibited a constructively critical mind, quick to expose flaws and errors. Each of his books was an effective and timely exposition of a major subject, and his best papers made lasting impressions. He contributed to many disciplines other than his specialties, including epidemiology, sociology, and economics. His greatest originality may have been reached in his papers on statistics—which, interestingly, was a subject he did not explore deeply until middle age.

BIBLIOGRAPHY

Wilson's three important books are *Vector Analysis* (New York, 1901); *Advanced Calculus* (Boston, 1912); and *Aeronautics* (New York, 1920). Some noteworthy papers are "The So-Called Foundations of Geometry," in *Archiv der Mathematik und Physik*, **6** (1903), 104–122; "The Space-Time Manifold of Relativity; the non-Euclidean Geometry of Mechanics and Electromagnetics," in *Proceedings of the American Academy of Arts and Sciences*, **48** (1912), 389–507, written with G. N. Lewis; "Differential Geometry of Two-Dimensional Surfaces in Hyperspace," *ibid.*, **52** (1916), 270–386, written with C. L. E. Moore; "Probable Inference, the Law of Succession, and Statistical Inference," in *Journal of the American Statistical Association*, **22** (1927), 209–212; "Periodogram of American Business Activity," in *Quarterly Journal of Economics*, **48** (1934), 375–417; and "The Determination of LD-50 and Its Sampling Error in Bioassay," in *Proceedings of the National Academy of Sciences of the United States of America*, **29** (1943), 79–85, 114–120, 257–262, written with Jane Worcester.

A full account of Wilson's life and work, by Jerome Hunsaker and Saunders Mac Lane, is in *Biographical Memoirs. National Academy of Sciences*, **43** (1973), 285–320, with bibliography.

NORMAN T. GRIDGEMAN
SAUNDERS MAC LANE

WILSON, JOHN (*b*. Applethwaite, Westmorland, England, 6 August 1741; *d*. Kendal, Westmorland, 18 October 1793), *mathematics*.

Wilson was educated at Kendal and at Peterhouse, Cambridge, where in the mathematical tripos of 1761 he was senior wrangler. He was elected a fellow of Peterhouse in 1764 and a fellow of the Royal Society in 1782. As an undergraduate he attracted notice in the university by his defense of Waring, then Lucasian professor of mathematics, against adverse criticism of the latter's *Miscellanea analytica* (1762).

As a private tutor at Cambridge, Wilson had a high reputation; but after a short period of teaching, he was called to the bar in 1766 and acquired a considerable practice on the northern circuit. In 1786 he was raised to the bench of the Court of Common Pleas; later he served for a short time as one of the commissioners for the great seal, between the retirement of Lord Edward Thurlow from the office of lord chancellor and the appointment of Lord Loughborough.

Wilson's name is given to the theorem that if p is a prime number, then $1 + (p-1)!$ is divisible by p. The first published statement of the theorem was by Waring in his *Meditationes algebraicae* (1770), although manuscripts in the Hannover Library show that the result had been found by Leibniz. Waring ascribed the theorem to Wilson but did not prove it; the first published proof was given by Lagrange (1773), who provided a direct proof from which Fermat's theorem (1640), first proved by Euler in 1736, can be deduced: If p is a prime and a is not divisible by p, then $a^{p-1} - 1$ is divisible by p. Lagrange also showed that Wilson's theorem can be deduced from Fermat's theorem, and that the converse of Wilson's theorem is true: if n divides $1 + (n - 1)!$, then n is a prime.

In a series of letters exchanged between Sir Frederick Pollock and Augustus De Morgan, published by W. W. Rouse Ball, Pollock described the mathematical work done at Cambridge in the first decade of the nineteenth century, and asserted that Wilson's theorem was a guess that neither he nor Waring could prove.

Wilson's result has been generalized to provide a series of theorems relating to the symmetric functions of the integers $1, 2, \cdots, p - 1$, and in other ways. The history of the theorem and its generalizations is given in detail by L. E. Dickson.

BIBLIOGRAPHY

For Wilson's life, see *Dictionary of National Biography*, XXI, p. 578; and Atkinson, *Worthies of Westmorland*, II (London, 1850); for personal details, Augustus De Morgan, *Budget of Paradoxes*, 2nd ed. (Chicago–London, 1915); W. W. Rouse Ball, *A History of the Study of Mathematics at Cambridge* (Cambridge, 1889).

For Wilson's theorem, see the following, listed chronologically: E. Waring, *Meditationes algebraicae* (Cambridge, 1770); J. L. Lagrange, in *Nouveaux mémoires de l'Académie de Berlin* (1773); and L. E. Dickson, *History of the Theory of Numbers*, I (repr. New York, 1934), ch. 3.

T. A. A. BROADBENT

WINCHELL FAMILY (founded in America by British immigrant Robert Winchell, who lived in Windsor, Connecticut, from 1635 until his death in 1669). The descendants spread from Connecticut to eastern New York, where the eighth-generation brothers Alexander Winchell (1824–1891) and Newton Horace Winchell (1839–1914) were born in Dutchess County. Alexander married and settled in Michigan. Newton Horace joined his older brother in Michigan, married, and eventually moved to Minnesota. Among the five children of the ninth generation two sons were born: Horace Vaughn Winchell (1865–1923), who married one of Alexander Winchell's daughters, and Alexander Newton Winchell (1874–1958).

The family is unique in its contributions to geology over four generations. They were leaders in the broad organization of geology as a professional science in America and in the establishment of the first American journal devoted solely to geology. Members have been prominent in education, state and national government, and civic activities.

They have published numerous scientific papers, reports, and books, as well as popular communications, in geology, mineralogy, petrology, mining, mining law, archaeology, ethnology, and religion.

BIBLIOGRAPHY

N. H. Winchell and A. N. Winchell, *The Winchell Genealogy, the Ancestry and Children of Those Born to the Winchell Name in America Since 1635*, 2nd ed. (Minneapolis, Minn., 1917).

WINCHELL, ALEXANDER (*b.* Northeast, New York, 31 December 1824; *d.* Ann Arbor, Michigan, 19 February 1891), *geology, education.*

Winchell was the son of Horace and Caroline McAllister Winchell, both of whom were schoolteachers. After graduation from Wesleyan University in 1847, he was appointed teacher of natural science at Pennington Male Seminary in New Jersey, where he studied the local flora and languages, and conducted experiments with electricity. Subsequent teaching positions were Amenia Seminary (New York), Newbern Academy (Alabama), and Mesopotamia Female Seminary (Eutow, Alabama). Winchell was president of Masonic University (Selma, Alabama) and was professor at the University of Michigan, first of physics and engineering (1853–1855), and later, when a chair was established, of geology, zoology, and botany (1855–1873). During his tenure at the University of Michigan he served on the State Geological Survey (1859–1861, 1869–1871). While chancellor and then professor of geology at Syracuse University (1873–1874), Winchell accepted a professorship in geology, zoology, and botany at Vanderbilt University. The latter chair was abolished in 1878, allegedly for economic reasons; it is believed, however, that the action was taken by the university board of trust because of his views on evolution. In 1879 he was recalled to the University of Michigan as professor of geology and paleontology, and remained there until his death.

Winchell's main impact was his role in organizing geology as a science in America. By popularizing its principles as having both economic and cultural value, he influenced legislation establishing geological surveys. In addition to his teaching, lecturing, and voluminous writing, he made important scientific observations in a great many fields, mainly stratigraphy and paleontology. He described seven new genera and 304 new species of organisms, mostly fossil. Winchell established the basin shape of the strata in Michigan and predicted the economic development of salt in the Saginaw Valley. Much effort went into the description of a series of strata called the Marshall group that encompassed many previously described beds and required a strong defense. He was particularly interested in the oil-bearing formation of Michigan and the ancient (Archean) rocks of Minnesota. Other geological work included studies on glacia-

tion, pedology, geochronology, hydrology, and sedimentology.

Winchell's second major contribution was in bridging the alleged gap between science and religion. He defended the Christian Scriptures and sought rational interpretations that harmonized with scientific observations. He revived the seventeenth-century idea of preadamites and presented an anthropological account of the evolution of the human family without, in his view, contravening the Scriptures. Winchell's greatest endeavor resulted in a highly imaginative world history that brought together cosmology and geology. His published bibliography lists 255 titles, but his personal list of compositions numbers 566. Other scientific subjects to which he contributed include astronomy, climatology, meteorology, and zoology.

Winchell's brother Newton Horace described him as a man of strong personality and convictions, physically strong, and deft in mechanical construction. Audiences were inspired and entertained by his popular scientific lectures. He wrote poetry, mostly unpublished, and mastered at least seven languages. Winchell has been called the father of the Geological Society of America, founded in 1888, of which he was president in 1891, and was one of the founders of *American Geologist* (1888). He married Julia F. Lines of Utica, New York, a teacher of instrumental music, on 5 December 1849. Of their six children only two daughters lived to maturity.

BIBLIOGRAPHY

Winchell's writings on geology are *First Biennial Report of the Progress of the Geological Survey of Michigan, Embracing Observations on the Geology, Zoology, and Botany of the Lower Peninsula* (Lansing, 1861); "On the Saliferous Rocks and Salt Springs of Michigan," in *American Journal of Science*, 2nd ser., **34** (1862), 307–311; "Description of Fossils From the Marshall and Huron Groups of Michigan," in *Proceedings of the Academy of Natural Sciences of Philadelphia*, **14** (1862), 405–430; *The Oil Region of Michigan. Description of the Baker Tract, Situated in the Heart of the Oil Region of Michigan* (Detroit, 1864); *The Grand Traverse Region. A Report on the Geological and Industrial Resources of the Counties of Antrim, Grand Traverse, Benzie, and Leelanaw, in the Lower Peninsula of Michigan* (Ann Arbor, 1866); "On the Geological Age and Equivalents of the Marshall Group," in *Proceedings of the American Philosophical Society*, **11** (1871), 245–260; *Geological Studies, or Elements of Geology* (Chicago, 1886); "The Taconic Question," in *American Geologist*, **1** (1888), 347–363; and "American Opinion on the Older Rocks," in *Report of the Minnesota Geological Survey*, **18** (1891), 65–226.

His works on evolution are *Sketches of Creation* (New York, 1870); *The Doctrine of Evolution* (New York, 1874); *Reconciliation of Science and Religion* (New York, 1877); *Preadamites* (Chicago, 1880); *Sparks From a Geologist's Hammer* (Chicago, 1881); and *World Life or Comparative Geology* (Chicago, 1883).

WINCHELL, ALEXANDER NEWTON (*b*. Minneapolis, Minnesota, 2 March 1874; *d*. New Haven, Connecticut, 7 June 1958), *mineralogy, petrology, education.*

Winchell was the youngest son of Newton Horace and Charlotte Sophia Imus Winchell. He received the B.S. degree in 1896 and the M.S. in 1897 from the University of Minnesota, where his mineralogical studies were directed by Charles P. Berkey. Following a year as instructor in physics at Central High School, Minneapolis, he married and went to the University of Paris to pursue advanced studies in mineralogy and petrology under Alfred Lacroix. Winchell received the D.Sc. from the University of Paris in 1900 and then took a post at the Montana School of Mines. In 1907 he moved to the University of Wisconsin, becoming full professor in 1908, and remained there until his retirement in 1944. In 1934 he was given a semester's leave of absence to study X-ray methods under Linus Pauling at the California Institute of Technology and under W. H. Taylor and W. L. Bragg at the University of Manchester. His promotion to chairman of the department of geology at the University of Wisconsin also was effective in 1934.

Winchell was associated with the U.S. Geological Survey from 1901 to 1910. After moving to Connecticut in 1948, he was made an honorary fellow in geology at Yale University. He served as visiting professor of mineralogy at the University of Virginia (1948–1949) and at Columbia University (1949–1950), and was resident mineral consultant at the Stamford laboratory of the American Cyanamid Company for three years. His remaining years were devoted to reviewing geological literature and revising his books.

Winchell's major contribution to geology is his well-known, and still widely used, three-volume *Elements of Optical Mineralogy*. The first volume was an outgrowth of a book that his father and himself published in 1909. That book was the first presentation in English of the principles and methods of optical mineralogy, and its expanded revision quickly became a major textbook. Four additional revised editions followed. The second volume dealt with the description of minerals and underwent three revisions, the last in collaboration with his youngest son, Horace. The third volume, on determinative tables, was revised once. These volumes contain the compilation and correlation of vast amounts of data relating the optical and physical properties of crystals to their composition. Three-fourths of the approximately 120 diagrams in the book giving graphical representation of these relations were developed by Winchell. He made significant contributions to the understanding of such major mineral groups as the feldspars, pyroxenes, melilites, amphiboles, micas, chlorites, zeolites, and scapolites. He also compiled a text on the optical properties of synthetic minerals and one on organic crystals, as well as an elementary textbook on mineralogy. These books made the methods of mineralogy available to chemists, ceramists, and other research scientists.

Winchell's second major contribution to geology was in the field of petrology. He studied Keweenawan igneous rocks, the relationship of igneous rocks to the occurrence of ores, and limestone alteration, and devised a graphical classification of rocks. Petrological field observations from Montana, Oregon, Nevada, and Utah are recorded.

Winchell devoted forty-four years to the teaching of mineralogy. He was an inspiring teacher with boundless patience, always ready and available to answer students' questions. He was friend and counselor to students, earnest in discussion and a source of support.

Winchell served as president of the Mineralogical Society of America in 1932 and as vice-president of the Geological Society of America in the same year. The former society awarded him its highest award, the Roebling Medal, in 1955.

Serious games were Winchell's principal hobbies: chess, bridge, Russian bank, anagrams, and crossword puzzles. He pursued his early university interest in history and was an active member of a poetry club. After retirement he enjoyed traveling in North America.

He married Clare Edith Christello of Minneapolis on 29 May 1898. They had five children, most of whom carried on the family tradition in geology directly or indirectly, through marriage to geologists. Two years after the death of his wife in 1932, he married Florence Mabel Sylvester, granddaughter of Alexander Winchell.

BIBLIOGRAPHY

Winchell's textbooks are *Elements of Optical Mineralogy*, 3 vols.: I, *Principles and Methods* (5th ed., New York, 1937), II, *Descriptions of Minerals* (4th ed., New York, 1951), written with H. Winchell; III, *Determinative Tables* (2nd ed., New York, 1939); *Microscopic Characters of Artificial Inorganic Solid Substances or Artificial Minerals* (New York, 1931); *Optical Properties of Organic Compounds*, 2nd ed. (New York, 1954); and *Elements of Mineralogy* (New York, 1942).

On the optical properties of major mineral groups, see "Studies in the Pyroxene Group," in *American Journal of Science*, 5th ser., **6** (1923), 504–520; "The Composition of Melilite," *ibid.*, **8** (1924), 375–384; "Studies in the Amphibole Group," *ibid.*, **7** (1924), 287–310; "Studies in the Mica Group," *ibid.*, **9** (1925), 309–327, 415–430; "Studies in the Feldspar Group," in *Journal of Geology*, **34** (1925), 714–727; and "Chlorite as a Polycomponent System," in *American Journal of Science*, 5th ser., **11** (1926), 282–300.

Petrological studies are "Mineralogical and Petrographic Study of the Gabbroid Rocks of Minnesota," in *American Geologist*, **19** (1900), 336–339; "Review of Nomenclature of Keweenawan Igneous Rocks," in *Journal of Geology*, **16** (1908), 765–774; "Discussion of Igneous Rocks as Related to Occurrence of Ores," in American Institute of Mining Engineers, *Ore-Deposits* (New York, 1913), 303–304; "Rock Classification on Three Co-ordinates," in *Journal of Geology*, **21** (1913), 208–223; *Petrology and Mineral Resources of Jackson and Josephine Counties, Oregon* (Salem, 1914); and "Petrographic Studies of Limestone Alteration at Bingham," in *Transactions of the American Institute of Mining and Metallurgical Engineers*, **70** (1924), 884–903.

Historical studies are "Minnesota's Northern Boundary," in Minnesota Historical Society Collections, **8**, pt. 2 (1896), 184–212; and "Minnesota's Eastern, Southern, and Western Boundaries," *ibid.*, **10** (1905), 1–11.

WINCHELL, HORACE VAUGHN (*b*. Galesburg, Michigan, 1 November 1865; *d*. Los Angeles, California, 28 July 1923), *geology, mining engineering*.

Winchell was the oldest son of Newton Horace and Charlotte Sophia Imus Winchell. He studied at the University of Minnesota and then at the University of Michigan, from which he graduated in 1889. Apparently he was greatly influenced by his uncle Alexander Winchell, under whom he studied, as well as by his father to follow the natural sciences. Winchell's first geological work was for the Minnesota State Geological Survey; his father, the state geologist, assigned him to study the Mesabi Range, where promising sources of iron ore had just been discovered.

Winchell next worked for the Minnesota Mining Company, until the panic of 1893 terminated exploration. He and F. F. Sharpless then formed a partnership as consulting geologists; however, depression in the mining industry and work in the West led to its early dissolution. In 1898 he became geologist for Anaconda Copper Mining Company and, in 1906, geologist for the Northern Pacific Railroad. In 1908 Winchell broadened his consulting practice and established an independent office in Minneapolis. His assignments took him to Alaska, Mexico, South America, Europe, and Russia, where he witnessed the Kerensky revolution in February 1917. He retired to Los Angeles in 1921.

Winchell's contributions to science differ greatly from those of his father and his uncle even though his early career was devoted to the observational and theoretical aspects of geology, in close association with his father. Field studies on the drift-covered Mesabi Range outlined the importance of that area as a source of iron ore and led to an acceptable theory of the origin of the ores. A *Bulletin of the Minnesota Geological Survey* on the iron ores of the state, published with his father (1891), and another report published in 1893 especially on the Mesabi Range, influenced the development of the most productive iron ore deposits in America. The iron-bearing strata originated, according to Winchell and his father, as a chemical precipitate from Precambrian ocean waters; however, the algal and bacterial structures found later suggest some involvement of organic life forms. Winchell favored the view that the soft ore bodies resulted from the alteration of the strata by surface waters (secondary enrichment), and his early experience on the iron ranges influenced his thinking with regard to other ore deposits around the world.

In 1898 Winchell turned to the practical applications of the then-evolving science of geology to mining and engineering problems, and eventually to mining law. A geological staff was organized at the Anaconda Copper Mining Company, and his success in advising the engineers on the development of the Butte, Montana, mine no doubt encouraged other companies to set up similar departments. Winchell's policies of geological investigation and continuity of geological mapping in connection with mining operations proved highly profitable. In 1903 he published the conclusion that the upper levels of the Butte copper ore were the result of secondary enrichment, basing that view on geological field studies and laboratory experiments carried out with the assistance of C. F. Tolman, Jr. The concept of secondary enrichment had already been well established, but Winchell's role in its development was recognized.

Turning to private consulting practice in 1908, Winchell became more involved in the legal problems accompanying mining. He was particularly interested in the statutes attending extralateral rights, those governing mining claims, and ownership of mineral properties in the public domain. His concern for the public good and appreciation of the need for development of new resources was great. One of his friends and biographers records his occasional bitterness toward the hostile and critical attitude of some of his fellow scientists, who held such service to humanity in low regard in comparison with the contributions of pure science.

Winchell helped found, and for about five years served as editor of, *Economic Geology*, which was in effect a continuation of the *American Geologist* (the latter established primarily through the efforts of his father in 1888). He also was one of three American associate editors of *Zeitschrift für praktische Geologie* from 1896 until the United States entered World War I. He was president of the American Institute of Mining Engineers in 1919, succeeding Herbert Hoover.

In his younger years Winchell gained national recognition as a whist player and represented Minneapolis in championship contests. Golf was his chief outdoor recreation. He acquired a valuable library on mining geology that after his death was given to the Engineering Societies Library in New York by his wife and the Anaconda Company. He married his cousin, Ida Belle Winchell, on 15 January 1890. One child was born but survived only six months.

BIBLIOGRAPHY

Scientific observations are "On a Possible Chemical Origin of the Iron Ores on the Keewatin in Minnesota," in *American Geologist*, **4** (1889), 291–300, written with N. H. Winchell; "The Iron Ores of Minnesota; Their Geology, Discovery, Development, Qualities, Origin, and Comparison With Those of Other Districts," *Bulletin of the Minnesota Geological and Natural History Survey*, no. 6 (1891), written with N. H. Winchell; "The Mesabi Iron Range," in *Transactions of the American Institute of Mining Engineers*, **21** (1893), 644–686; "The Lake Superior Iron Ore Region, United States of America," in *Transactions of the Federal Institute of Mining Engineers* (London), **13** (1897), 493–562; "Synthesis of Chalcocite and Its Genesis at Butte, Montana,"

in *Bulletin of the Geological Society of America*, **14** (1903), 269–276; and "The Genesis of Ores in the Light of Modern Theory," in *Popular Science Monthly*, **72** (1908), 534–542.

Legal contributions are "Mining Laws," in *Transactions of the Canadian Mining Institute*, **15** (1912), 535–551; "Why the Mining Laws Should Be Revised," in *Transactions of the American Institute of Mining Engineers*, **48** (1915), 361–385; "Apex Litigation. Jim Butler Versus West End," in *Mining and Scientific Press*, **110** (1915), 763–765; "Mining Laws," in Robert Peele, ed., *Mining Engineers' Handbook* (New York, 1917), 1465–1514; and "Uniform Mining Law for North America," in *Transactions of the American Institute of Mining Engineers*, **61** (1919), 696–705.

WINCHELL, NEWTON HORACE (*b*. Northeast, New York, 17 December 1839; *d*. Minneapolis, Minnesota, 2 May 1914), *geology, archaeology*.

Winchell was the son of Horace and Caroline McAllister Winchell, both of whom were schoolteachers. At the age of sixteen he began teaching in public schools and continued that career after entering the University of Michigan in 1857. His studies and teaching were interrupted by military service in the Civil War, and he graduated in 1866. Winchell served as principal and superintendent of local schools in Michigan between 1866 and 1870. He accepted a position as assistant to his brother Alexander on the Michigan State Geological Survey and, later, to John Strong Newberry on the Ohio State Geological Survey. In 1872 he was appointed state geologist of Minnesota, a post he held for twenty-eight years. While state geologist Winchell was also professor of geology and curator of the museum at Michigan State University; he relinquished most of those duties, however, after about seven years. The last nine years of his life were devoted to the Minnesota Historical Society.

In contrast with his brother Alexander, Winchell confined his scientific work mainly to original research rather than teaching and lecturing. His principal contributions to geology are recorded in twenty-four *Reports* and ten *Bulletins of the Minnesota Geological and Natural History Survey*. With a group of assistants he surveyed, mapped, and reported on every county in the state. The great iron ore deposits of the Mesabi and Vermilion ranges, as well as the Marquette, Gogebic, and Cuyuna ranges, were studied in detail. In addition, building-stone resources, copper deposits, lignite and anthracite beds, water supply, salt wells, and drift soils were examined in the course of the survey. Winchell's interpretations of the structure of the Lake Superior region and of the origin of the iron ores was in conflict with that of the geologists of the U.S. Geological Survey.

Winchell also contributed to the ornithological, entomological, and botanic studies of Minnesota. Although his geological work was carried out almost exclusively in that state, he prepared the first geologic map of the interior of the Black Hills as a member of Custer's expedition in 1874. With the aid of his younger son, Alexander Newton, Winchell wrote *Elements of Optical Mineralogy*, revised editions of which are still used. Aid on other projects was given by his older son, Horace Vaughn Winchell, and his son-in-law, Ulysses S. Grant, the geologist.

Contributions to archaeology were among Winchell's earliest studies, and these were related to his glaciological work. His detailed and accurate conception of the waning of the great ice sheets and the contemporaneous existence of man produced considerable debate. Most noteworthy was his estimate of the time of the last stage of the Glacial Period by approximating the rate of recession of St. Anthony's Falls, cutting the Mississippi River gorge from Fort Snelling to the present site of the falls in Minneapolis. The duration of postglacial time was estimated as about eight thousand years, assuming that the gorge was the result of postglacial erosion. Winchell assembled data on thousands of Indian mounds and concluded that they were the work of the immediate ancestors of the present Indian tribes and were not constructed by a prior race of distinct culture. His geological knowledge aided him in determining that paleolithic man in America antedated the Kansan stage of glaciation.

Winchell was appointed by President Grover Cleveland as a member of the Federal Assay Commission in 1886. He was one of the founders of the Minnesota Academy of Sciences in 1873 and served three terms as its president. The first steps in the organization of the Geological Society of America were instigated by Winchell in 1881, and he served as its president in 1902. In cooperation with his brother Alexander and other geologists, in 1888 he established the *American Geologist*, which he edited for eighteen years. The initial purpose of the journal was to provide a nonpartisan publication free from the influence of the national geological survey, which was viewed as encroaching on the domain of the state geological surveys. The causes for concern evaporated, and the magazine turned to then-current geological problems. In 1905 the magazine was incorporated

with, enlarged, and published as *Economic Geology*, the title it still bears.

Winchell married Charlotte Sophia Imus of Galesburg, Michigan, on 24 August 1864. Of their five children, their sons Alexander Newton and Horace Vaughn became prominent geologists and one daughter, Avis, married a successful geologist, Ulysses Sherman Grant.

BIBLIOGRAPHY

Winchell's writings include "Report of a Reconnaissance of the Black Hills of Dakota in 1874," in *Geological Report* . . . (1875), 21–65; *The Geology of Minnesota; Final Report of the Geological and Natural History Survey of Minnesota,* 6 vols. (Minneapolis, 1884–1901); "The So-Called Huronian Rocks in the Vicinity of Sudbury, Ontario," in *Bulletin of the Minnesota Academy of Sciences,* **3** (1889), 183–185; "The Iron Ores of Minnesota; Their Geology, Discovery, Development, Qualities, Origin, and Comparison With Those of Other Mining Districts," *Bulletin of the Minnesota Geological and Natural History Survey,* no. 6 (1891), written with H. V. Winchell; "Was Man in America in the Glacial Period?," in *Bulletin of the Geological Society of America,* **14** (1903), 133–152; *Elements of Optical Mineralogy,* 3 vols. (New York, 1909), written with A. N. Winchell; *The Aborigines of Minnesota* (St. Paul, Minn., 1911); and "The Weathering of Aboriginal Stone Artifacts; a Consideration of the Paleoliths of Kansas," *Collections of the Minnesota Historical Society,* **16**, pt. 1 (1913).

H. S. YODER, JR.

WINDAUS, ADOLF OTTO REINHOLD (*b.* Berlin, Germany, 25 December 1876; *d.* Göttingen, Germany, 9 June 1959), *chemistry.*

Windaus came from a family that had a strong technological background. The ancestors of his father, Adolf, had been weavers and clothing manufacturers for two hundred years. The family of his mother, Margarete Elster, consisted mostly of artisans and craftsmen. A few of his ancestors had held academic positions, but none had been physicians or scientists.

The boy received his elementary education at the French Gymnasium in Berlin, where almost no science was taught and the emphasis was on literature. In his final year, however, Windaus learned of the bacteriological work done by Koch and Pasteur, and was much impressed with the benefits of their studies for humanity. He therefore decided upon a career in medicine, thereby disappointing his widowed mother, who had hoped that he would enter the family business.

Windaus entered the University of Berlin in 1895 and almost at once began to attend Emil Fischer's lectures on chemistry. He was particularly struck by the physiological applications of Fischer's work and began to develop what later became the basis for his approach to chemistry, an interest in the general physiological mechanisms of the compounds he studied. Windaus was awarded the bachelor's degree in 1897 and then decided to attend the University of Freiburg im Breisgau, where Heinrich Kiliani taught chemistry. For a time Windaus continued his medical studies, but he began to neglect medicine more and more, and finally gave it up entirely. At Kiliani's suggestion he started to study the chemistry of the glycosides of digitalis, and in 1899 he received the Ph.D. with a dissertation on those substances. The next year was spent in military service at Berlin, but during the summer semester Windaus was able to assist Fischer in a study of the formation of quaternary ammonium compounds from aniline. At the end of 1901 he returned to Freiburg to devote himself entirely to chemistry.

Kiliani suggested that Windaus begin a study of cholesterol, a compound about which almost nothing was known. Windaus felt that a substance so widely distributed in animal cells, and in related forms in plants, must have close connections with other physiologically important compounds. He thus entered on studies that occupied much of the rest of his life. In 1903 he presented his inaugural dissertation, "Cholesterin," and became a *Privatdozent* at Freiburg, with promotion to a professorship three years later.

In 1913 Windaus accepted the chair of medical chemistry at Innsbruck, where he remained for two years. In 1915 he was called to the University of Göttingen as successor to Otto Wallach. Until his retirement he was professor of chemistry and director of the chemical laboratory at Göttingen. Windaus continued active research even during the period of National Socialism. Although he was not in sympathy with the Hitler regime, he was allowed to continue his work because of the reputation he had established. He ceased active investigation in 1938, and after his retirement in 1944 he published no further papers.

The course of Windaus' scientific activity was determined by his early studies on digitalis and cholesterol. His work always had some relation to natural products; and although he was mainly concerned with the structure of cholesterol, he investigated many other sterols and established the membership of these substances in a group that he called the "sterines." In 1908 Windaus found that cholesterol formed an insoluble compound with digitonin. This explained the action of cholesterol in preventing the hemolytic activity shown by the saponins, of which digitonin was one. Thereafter he included studies of the structure of the saponins in his research program.

During this period Windaus' friend Heinrich Wieland, at Munich, was studying the structure of the bile acids; and among the derivatives of these substances he had prepared a compound that he called cholanic acid. In 1919 Windaus prepared the same acid from cholesterol, thus demonstrating the close chemical relationship between the sterines and the bile acids. The results obtained by the workers at Göttingen and Munich could now be combined. Active study at the two institutions finally led to the determination of the correct structure for the sterol ring in 1932.

Even before the structure of cholesterol was completely established, there had been indications that the substance was involved in some way in vitamin activity. By the early 1920's it was known that rickets could be cured by administration of certain fish liver oils. The study of vitamins was widespread at the time, and it was assumed that the liver oils contained a specific substance, called vitamin D, that was responsible for the cure. It also had been found, however, that exposure of the patient to ultraviolet light could bring about a cure. The studies of A. F. Hess had established this fact firmly. A dilemma thus arose: Was a chemical or a physical process responsible for the favorable effect? Most physiologists assumed that two different processes were involved. It was believed that the vitamin cured rickets specifically, just as vitamin C cured scurvy, but that exposure to ultraviolet light raised the general level of resistance to the disease.

In 1924 Harry Steenbock and Alfred Hess independently showed that exposure of certain foods to ultraviolet light made them active in curing rickets. This indicated that some compound was photochemically converted into vitamin D, and thus the concept of a provitamin was developed. Hess found that the provitamin occurred in the sterine fraction of the irradiated foods, and in 1925 he sought Windaus as a collaborator in determining the chemical nature of the vitamin and its precursor. Windaus was eager to accept the invitation because of his approach to the general chemistry of natural products. Although many chemists believed that he was concerned only with cholesterol, he said that he was not interested in the chemical composition of any particular substance, but only in the major relationships between natural products. The collaboration between Hess (in New York City) and Windaus resulted in the development at Göttingen of an active center for vitamin research. Rosenheim (in London) soon joined the project, and the results of studies in the three cities confirmed each other.

It was at first thought that cholesterol was the provitamin, since irradiation of a supposedly pure sample produced an active product. When a more highly purified sample failed to yield the same result, it was recognized that this idea was incorrect. Robert Pohl, working at Göttingen, showed by a study of absorption spectra that a very small amount of an impurity was present in the original cholesterol sample; and in 1927 Hess and Windaus identified the impurity as the fungus sterol ergosterol. Windaus soon demonstrated that the conversion of ergosterol to the vitamin involved an isomerization. Attempts to isolate the pure vitamin apparently were successful when a crystalline compound was obtained, but later it was shown that this was a molecular compound of the vitamin with another sterol. By this time the molecular compound had been so well established that the name vitamin D_1 was applied to it; and when a pure vitamin finally was isolated from irradiated ergosterol, it was called vitamin D_2, or calciferol.

It was assumed that ergosterol was the only provitamin, but Windaus continued to seek other sterols that could serve as precursors of vitamin D. In 1932 he and his co-workers prepared 7-dehydrocholesterol and showed that it also was a provitamin. The name vitamin D_2 was retained for the substance obtained from ergosterol, and the new vitamin was named D_3. It proved to be even more important than vitamin D_2, since it was obtained by activation of a sterol synthesized by the animal body. Hans Brockmann, working in Windaus' laboratory, confirmed this fact when he isolated pure vitamin D_3 from tuna liver oil.

Knowledge of the structure of the various D vitamins soon followed. During the rest of his working life, Windaus was occupied with a study of the structural features necessary for a sterol to qualify as a provitamin, and with determining the course of the photochemical reactions by which activation of the provitamins occurs. He identified and characterized the other compounds formed in these reactions: lumisterol, tachysterol, and the suprasterols.

By 1927 the work on the structure and chemistry of the sterols and of vitamin D, although by no means complete, had proceeded so far that the importance of the results was clear. The brilliance of the contributions made by Wieland and Windaus was equally obvious. In 1927 Wieland was awarded the Nobel Prize in chemistry for his work on the bile acids, and in 1928 the same prize was given to Windaus for his studies on the constitution of the sterols and their connection with other substances occurring in nature.

As soon as the sterol ring structure was determined (1932), it became possible to assign structures to many other biologically important sterols. Adolf Butenandt, an assistant to Windaus, was able almost at once to present the structures of the male and female sex hormones, even though he had only 25 mg. of the male hormone available for study. Adrenal cortical hormones, saponins, glycosides, and even the poisonous substances found in the skin of certain toads (substances that Windaus had done much to characterize) were found to belong to the same group; and the term "sterine" was replaced by the more significant name "steroid."

Another important result of Windaus' studies on steroid structure arose from his discovery that in saturated derivatives of cholesterol there is a type of isomerism due to the cis- or trans-fusion of two saturated rings. This opened the field of the stereochemistry of condensed ring systems, a subject developed in detail by another student of Windaus', Walther Hückel.

At the time of his early work on cholesterol, Windaus also had undertaken a study, in collaboration with the biochemist Franz Knoop, of the reaction of sugars with ammonia. They hoped to convert the sugars into amino acids, and thus to establish the possibility of converting carbohydrates into proteins. To Windaus' surprise, when he treated glucose with ammoniacal zinc hydroxide, he obtained derivatives of imidazole. From a study of these compounds he discovered that the amino acid histidine was an imidazole derivative. In the course of this work he discovered the physiologically very important compound histamine, which became commercially available as a result of his work. These investigations brought Windaus into contact with the chemical industry, and he retained close connections with industrial chemists for the rest of his life. Chemical concerns supplied him with many of the substances he needed in his research and often suggested problems.

After his early work on the imidazoles, Windaus abandoned this line of research until 1929. He was induced to return to the field because two Dutch chemists, B. C. P. Jansen and W. F. Donath, had claimed that the antineuritic vitamin, B_1 or thiamin, contained an imidazole ring. In their analysis, however, they had overlooked the presence of sulfur in the compound. On the basis of a correct analysis, Windaus was able to assign the proper empirical formula and to show that the compound contained not an imidazole, but a thiazole and a pyrimidine ring. He isolated the pure vitamin B_1 from yeast, and his work helped greatly in the final synthesis of the vitamin by Robert R. Williams.

Windaus' other major field of research was the determination of the structure of colchicene, a substance that proved to have strong mutagenic properties for plants.

Although Windaus had been almost alone in his early investigations on cholesterol, the field he had pioneered quickly became a major branch of organic chemistry and biochemistry. He always worked closely with his colleagues in Germany and abroad, and gave his students great freedom in their research, as well as full credit for their contributions.

Besides the Nobel Prize, Windaus received many honorary degrees and memberships in scientific societies. He was awarded the Baeyer, Pasteur, and Goethe medals. Windaus served on the editorial board of *Justus Liebigs Annalen der Chemie*; and volumes **603** and **604** of that journal were dedicated to him in 1957, to celebrate his eightieth birthday.

BIBLIOGRAPHY

Windaus' scientific papers are listed in Poggendorff, V, 1380; VI, 2901–2902; VIIa, 1016–1018.

The most complete biography is the Windaus memorial lecture by Adolf Butenandt, in *Proceedings of the Chemical Society* (1961), 131–138. Further details are in Gulbrand Lunde, "The 1927 and 1928 Chemistry Prize Winners, Wieland and Windaus," in *Journal of Chemical Education*, **7** (1930), 1767–1777. A survey of the significance of the work on steroids and its relation to the work of others is H. H. Inhoffen, "50 Jahre Sterin-Chemie," in *Naturwissenschaften*, **38** (1951), 553–558.

HENRY M. LEICESTER

WING, VINCENT (*b.* North Luffenham, Rutland, England, 19 April 1619; *d.* North Luffenham, 30 September 1668), *astronomy.*

Wing's father, for whom he was named, was a small landowner. Young Wing had little formal education and began earning his living at an early age as a surveyor, almanac compiler, astrologer, and prolific writer of astromical works. His almanacs were the most popular of their time; and, in Flamsteed's judgment, Wing produced "our exactest ephemerides." He was an eager polemicist and frequently was involved in public disputes over astronomical and astrological matters.

Wing's career as an astronomer mirrors the development of astronomical thought during the seventeenth century. His first book, *Urania practica* (1649), asserted the stability of the earth and was Ptolemaic in spirit. A published attack on it by Jeremy Shakerley may have led to Wing's conversion to Copernicanism. By 1651 he had accepted the fundamentals of Keplerian astronomy as modified by Ismael Boulliau.

Like many astronomers in the second half of the seventeenth century, Wing, following Boulliau and Seth Ward, opted for an "empty-focus" variant of Kepler's second law, holding that a planet moving in an elliptical orbit describes equal angles in equal times about the focus not occupied by the sun. In works published in 1651 and 1656 Wing, adopting Boulliau's method, had his elliptical orbits, including that of the moon, generated in purely geometrical fashion by circles and epicycles. In his posthumously published *Astronomia Britannica*, however, he discarded the epicycles in favor of a refined version of the theory proposed by Ward in the latter's *Astronomia geometrica* (1656), in which the elliptical orbits were assumed to be physically generated. Wing's celestial mechanics contained a mixture of Cartesian and Keplerian components, with a rotating sun and celestial vortex pushing the planets around in their orbits.

BIBLIOGRAPHY

I. ORIGINAL WORKS. Wing produced a great many almanacs, ephemerides, and astrological pamphlets. His chief works are *Urania practica* (London, 1649; 2nd ed., 1652); *Ens fictum Shakerley, His In-artificial Anatomy of Urania practica* (London, 1649), written with William Leybourn; *Harmonicon coeleste: Or the Coelestial Harmony of the Visible World* (London, 1651); *Astronomia instaurata: Or a New and Compendious Restauration of Astronomy* (London, 1656); *Geodates practicus: Or the Art of Surveying* (London, 1664); *Examen astronomiae Carolinae* (London, 1665); and *Astronomia Britannica* (London, 1669).

II. SECONDARY LITERATURE. See J. B. J. Delambre, *Histoire de l'astronomie moderne*, II (Paris, 1821), 519–524; and John Gadbury, *A Brief Relation of the Life and Death of the Late Famous Mathematician and Astrologer, Mr. Vincent Wing* (London, 1670).

WILBUR APPLEBAUM

WINKLER, CLEMENS (*b*. Freiberg, Germany, 26 December 1838; *d*. Dresden, Germany, 8 October 1904), *technical chemistry, analytical chemistry.*

Winkler was the son of Kurt Winkler, a pupil of Berzelius, N. G. Sefström, and Gahn, who managed a large cobalt works. Before entering the School of Mines at Freiberg, he spent his school vacations in his father's laboratory, where he learned accuracy and cleanliness in the Berzelius tradition.

Winkler entered the cobalt trade and soon became interested in the use of sulfur gases from the smelting furnaces as a raw material in preparing sulfuric acid. This involved refining the methods of industrial gas analysis, and to this end he developed the Winkler gas burette; he also published the first comprehensive book on the subject (1876). Extending his work on sulfuric acid, Winkler prepared oleum (required by dyestuff makers) from chamber acid by means of a contact process. Chamber acid was decomposed at red heat to yield oxygen and sulfur dioxide, which were converted to sulfur trioxide over platinized asbestos. This was a significant advance, since previous workers had attempted to use finely divided metals or oxides without the asbestos support. The weakness of Winkler's process lay in the gases being formed and used in stoichiometric proportions; his paper, however, was published in 1875, four years before the law of mass action was enunciated in its final form.

In 1885 a rich vein of argyrodite was discovered, and the mineral was submitted to Winkler for complete analysis. In repeated experiments his totals were consistent at 93 percent, and his early training in mineral analysis under his father convinced him that some hitherto unrecognized element must be present. In 1886 he showed that the substance was Mendeleev's predicted ekasilicon, which he named germanium.

Winkler was professor of analytical and technical chemistry at the Freiberg School of Mines from 1873 to 1902. At a time when most German chemical effort was concentrated in the organic field, he made many contributions to inorganic and analytical chemistry through his patient teaching and his prolific writing.

BIBLIOGRAPHY

I. ORIGINAL WORKS. The Royal Society *Catalogue of Scientific Papers*, VI, 396, VIII, 1251–1252, XI, 824–825, and XIX, 662, lists 91 publications. The following are relevant to the present article: "Versuche über die Ueberführung der schwefligen Säure in Schwefelsäureanhydrid durch Contact wirkung behufs Darstellung von rauchender Schwefelsäure," in *Dinglers polytechnisches Journal*, **218** (1875), 128, on the contact process for sulfuric acid; and "Germaniun, Ge, ein neues, nichtmetallisches Element," in *Berichte der Deutschen chemischen Gesellschaft,* **19** (1886), 210, on germanium. The book on gas analysis is *Anleitung zur chemischen Untersuchung der Industrie-gase* (Freiberg, 1876), which is supplemented by "Beiträge zur technischen Gasanalyse," in *Zeitschrift für analytische Chemie,* **28** (1889), 269–289. Those who wish to savor Winkler's literary style should read "Über die Entdeckung neuer Elemente," in *Berichte der Deutschen chemischen Gesellschaft,* **30** (1897), 6–21.

II. SECONDARY LITERATURE. There is no complete biography in English. The definitive life is Otto Brunck, in *Berichte der Deutschen chemischen Gesellschaft,* **39** (1906), pt. 4, 4491–4548. A shorter version by Brunck is in G. Bugge, *Das Buch der grossen Chemiker,* II (Berlin, 1930), 336–350.

The best English account of the sulfuric acid process is in G. Lunge, *Sulphuric Acid and Alkali*, 4th ed., I (London, 1913), in which 46 index references are given. Winkler's work on gas analysis has 25 index references in G. Lunge, *Technical Gas Analysis* (London, 1914). The discovery of germanium is described in M. E. Weekes, *Discovery of the Elements*, 4th ed. (Easton, Pa., 1939), 319–323.

W. A. CAMPBELL

WINKLER, LAJOS WILHELM (*b*. Arad, Hungary [now Rumania], 21 May 1863; *d*. Budapest, Hungary, 14 April 1939), *chemistry.*

One of nine children of Vilmos Winkler, a wholesaler, Winkler worked as an assistant pharmacist in a chemist's shop in Arad and then studied chemistry and pharmacology at the University of Budapest. He received the degree in pharmacology in 1885 and soon afterward was offered an assistant professorship at the university's chemical institute by Károly Than, a former student of Bunsen and a founder of scientific chemistry in Hungary.

Research at the institute was then focused primarily on the chemical composition of Hungarian mineral waters, the main concern being dissolved gases. Winkler's own interest centered on the quantitative determination of dissolved oxygen in water, which provided the subject of his doctoral dissertation at the University of Budapest in 1888. The Winkler method, as it is now known, laid the foundation of his scientific reputation.

Winkler subsequently participated in the preparation of volumes II–IV of the *Pharmacopoeia Hungarica*, of which Than was editor-in-chief. For this work he developed analytical methods that can be used even in simply furnished laboratories, and his practical sense and keen critical ability enabled him to create a number of techniques of enduring value, in particular his colorimetric titration method of determining the contamination of water by metals, such as iron, lead, and copper, and by ammonia, hydrogen sulfide, silica, and other contaminants, and his method of determining iodine concentration. In 1896 he was elected corresponding member of the Hungarian Academy of Sciences and, in 1922, full member. Than's department was divided into two professorships following his death in 1908; Winkler became professor of analytics and pharmacology, as well as director of the first chemistry department at the University of Budapest.

Winkler opened new areas in analytical chemistry, and the methods that he elaborated were summarized in *Die chemische Analyse* (1931–1936). In addition to his work in high-precision gravimetry and the analysis of gases, water, and pharmaceuticals, he also investigated the absorption coefficients of gases in various solvents and devised instruments for their measurement that still provide reliable data. The results of his work on gas, which covered nearly twenty years, became internationally known with their inclusion in Hans Landolt and Richard Börnstein's *Physikalisch-chemische Tabellen* (1883), replacing Bunsen's values. His figures have not been superseded.

To a considerable extent, the success of Winkler's methods for the determination of halogens can be regarded as responsible for the subsequent development of halogen analysis as an almost exclusively Hungarian field of investigation, and many of his pupils conducted work in that area. Winkler's techniques reflect the importance of strictly maintaining the prescribed experimental conditions in all reproducible methods. The application of the fundamental principle of reproducibility—in analytical chemistry as in the natural sciences—constitutes one of his most outstanding accomplishments.

BIBLIOGRAPHY

A list of more than 200 of Winkler's works, compiled by L. Szebellédy, was published in *Winkler Lajos dr. emlékezete* (Budapest, 1940), 17–26. His writings include his dissertation, "Die Bestimmungen des in Wasser gelösten Sauerstoffs," in *Berichte der Deutschen chemischen Gesellschaft*, **21** (1889), 2843–2855; *Die chemische Analyse*, XXIX, XXX (Stuttgart, 1931–1936); and *Ausgewählte Untersuchungsverfahren für das chemische Laboratorium*, 2 vols. (Stuttgart, 1931–1936).

On Winkler's life and work, see E. Schulek, "L. W. Winkler, 1863–1939," in *Talanta*, **10** (1963), 423–428.

I. DE GRAAF BIERBRAUWER-WURTZ

WINLOCK, JOSEPH (*b.* Shelby County, Kentucky, 6 February 1826; *d.* Cambridge, Massachusetts, 11 June 1875), *astronomy, mathematics.*

Immediately upon graduation from Shelby College in 1845, Winlock was appointed professor of mathematics and astronomy in that school. At the meeting in 1851 of the American Association for the Advancement of Science, Winlock met Benjamin Peirce; thenceforth he was esteemed and promoted by the scientific lazzaroni. In 1852 he moved to Cambridge, as a computer for the *American Ephemeris and Nautical Almanac*. Using the refracting telescope from Shelby set up in the Cloverden Observatory in Cambridge, Winlock and B. A. Gould made astronomical observations. In 1857 Winlock was appointed professor of mathematics at the U.S. Naval Observatory. The following year he was promoted to superintendent of the Nautical Almanac office, which position he resigned in 1859 to take charge of the mathematics department of the U.S. Naval Academy. Following the outbreak of the Civil War, Winlock again took superintendence of the Nautical Almanac. In 1863 he was made an original member of the National Academy of Sciences. In 1866, backed by the "Coast Survey Clique," he was appointed Phillips professor of astronomy and director of the Harvard College Observatory. In 1871 the professorship of geodesy in the Lawrence Scientific School was added to his duties.

While at Harvard, Winlock's primary concern was to develop and obtain more accurate and efficient instruments. In this he was quite successful. Had he lived longer he would probably have used them to even better advantage than he did. Troughton & Simms, after extensive collaboration with Winlock, supplied a large and improved meridian circle. With this the Harvard zone of stars for the Astronomische Gesellschaft was determined; after Winlock's death the observations were continued by William A. Rogers, and the computations by Winlock's oldest daughter, Anna.

Among Winlock's several contributions to solar photography was the development of fixed-horizontal long-focus refracting telescopes. One of these instruments, installed at Harvard in 1870, took daily pictures of the sun. Similar telescopes were used by the eight U.S. government-sponsored expeditions to record the 1874 transit of Venus. Also worthy of mention are Winlock's particularly detailed photographs of the solar corona during the eclipse of 1869. Celestial spectroscopy also received Winlock's attention, and he obtained several fine spectroscopes for the observatory. To take the fullest possible advantage offered by the 1870 solar eclipse, Winlock devised a mechanical method of recording the positions of spectral lines. Throughout his tenure at Harvard he collaborated with the Coast Survey on both astronomical and geographical projects.

BIBLIOGRAPHY

The Royal Society of London, *Catalogue of Scientific Papers*, Poggendorff, and the *Bibliographie Générale de l'Astronomie* list about two dozen Winlock papers. Winlock's influence as an astronomer, both his own work and the work he inspired in others, is best seen in the *Annals* of the Harvard College Observatory. Vols. **5**, **6**, **7**, published by Winlock, contain work done by his predecessors, W. C. Bond and G. P. Bond. Published after Winlock's death, vol. **8**, pt. 1, "Historical Account of the Astronomical Observatory of Harvard College," details the instrumental additions and improvements engineered by Winlock; pt. 2 contains "Astronomical Engravings of the Moon, Planets, etc.; Prepared [by L. Trouvelot] at the Astronomical Observatory of Harvard College Under the Direction of the Late Joseph Winlock, A.M." Vol. **9** contains C. S. Peirce's photometrical researches, also supported by Winlock. Vols. **10**, **12**, **14**, **16**, **25**, **35**, and **36** contain astrometric catalogs prepared by William A. Rogers under the direction of Winlock and his successor, E. C. Pickering (in many cases these observations had been begun by Winlock). Vol. **13** contains micrometric observations from 1866 to 1881, made under, and in some cases by, Winlock and Pickering.

The most extensive obituary is in *Proceedings of the American Academy of Arts and Sciences*, **11** (1875–1876), 339–350, republished verbatim in *Biographical Memoirs of the National Academy of Sciences*, **1** (Washington, D.C., 1875). See also *Nature*, **12** (1875), 191–192; *American Journal of Science*, **10** (1875), 159–160 (quoted in *Scientific American*); *Dictionary of American Biography*, XX (New York, 1936); and the various published histories of the Harvard College Observatory.

DEBORAH JEAN WARNER

WINSLØW, JACOB (or **JACQUES-BÉNIGNE**) (*b.* Odense, Denmark, 17 April 1669; *d.* Paris, France, 3 April 1760), *anatomy*.

Winsløw was the eldest of the thirteen children of Peder Jacobsen Winsløw, dean of the Protestant Church of Our Lady in Odense, and Martha Bruun, whose own father had held the same post. He received his early education from his father, who was learned in both linguistics and archaeology, and at the Odense secondary school. He entered the University of Copenhagen to study theology in 1687. Although he delivered several sermons, Winsløw was soon attracted to the natural sciences, inspired by Oliger Jacobaeus and Caspar Bartholin the younger. From 1691 to 1696 he attended Borch's College and worked under the county barber-surgeon Johannes de Buchwald. Although Buchwald was the best surgeon in Copenhagen, Winsløw concentrated on anatomy, since the sight of blood alarmed him; he himself never performed an operation. He soon became Bartholin's prosector, and the latter was so pleased with his public anatomical demonstrations that he promoted him anatomicus regius, a post held by Winsløw's granduncle, Niels Stensen, some twenty years before.

In 1697 Winsløw was awarded a royal grant and accompanied Buchwald to the Netherlands, where he not only studied anatomy, but also received practical training in clinical medicine, surgery, and obstetrics, including private instruction with a midwife. These studies, together with his association with a number of Dutch scientists—including Johann Rau, Pieter Verduyn, and Hendrik van Deventer—convinced him of the value of the practical application of basic anatomical and physiological investigations.

Winsløw stayed in the Netherlands for fourteen months, then moved to Paris, where he began to study anatomy and surgery with J.-G. Duverney. A spiritual crisis intervened, however, inspired by discussions with his friend Ole Worm and by the treatises of Jacques-Bénigne Bossuet. After a series of conversations with the latter, in 1699 Winsløw converted to Roman Catholicism (taking his baptismal name Bénigne from Bossuet), whereupon his subsidy from the Danish government was terminated. With the help of Bossuet and other Catholic patrons he was soon able to resume his work with Duverney, and in 1704 Winsløw became a medical licentiate at the Hôtel Dieu and was authorized to practice as a physician in the city of Paris. Duverney made him his assistant in anatomy and surgery at the Jardin du Roi.

Winsløw became a member of the Académie Royale des Sciences in 1708; he also maintained a busy medical practice, was appointed physician at the Hôpital Général and at Bicêtre in 1709, assumed Duverney's duties at the Jardin du Roi in 1721, and was made *docteur-régent* of the Paris Faculty of Medicine in 1728. In 1743 Winsløw became full professor of anatomy at the Jardin du Roi; he held this post until 1758, when he was obliged to retire because of extreme deafness. On 18 February 1745 Winsløw dedicated the new anatomical theater of the Paris Faculty of Medicine, a building that still stands at 13 rue de la Bûcherie. Although in the address he made upon that occasion he referred to himself as being merely the successor of Riolan, Bartholin, and Stensen, he was in fact regarded as the greatest European anatomist of his day, and attracted a number of able students, including Albrecht von Haller.

Winsløw's own anatomical studies combined a talent for making observations with systematic thoroughness. Between 1711 and 1743 he published nearly thirty treatises, on a variety of subjects, in the *Mémoires de l'Académie royale des sciences*. Among these works was a series of investigations, published between 1715 and 1726, of the course of the various muscles, in which Winsløw showed that a single muscle does not function alone as a flexor or supinator, but rather that muscles work in groups as synergists, and always in relation to antagonists. In another tract of 1715 he described the foramen between the greater and lesser sacs of the peritoneum that is now named for him.

In "Sur les mouvements de la tête, du col et du reste de l'épine du dos" of 1730, Winsløw was the first to describe exactly the function of the small intervertebral joints, while in " . . . certaines mouvements avec les deux mains à la fois . . . plus facilement en sens contraire qu'en même sens" (1739), he noted that this effect is caused by nerve crossings in the brain and spinal cord, and not by any action of the muscles. In 1742 he published an account, based on comparative anatomical studies, of the function of the digastric muscles in opening the mouth through lowering the mandible. He also found occasion, in two articles published between 1740 and 1742, to inveigh against the formidable corsets worn by women at that time, and between 1733 and 1743 published a series of treatises on

monsters, in which he demonstrated that congenital malformations resulted from faulty predispositions and were not lesions of a normal fetus.

Winsløw's best-known work was his *Exposition anatomique de la structure du corps humain*, first published in 1732, then in a large number of subsequent editions and translations. The *Exposition* was the first treatise on descriptive anatomy, and, in its elimination of extraneous physiological details and hypothetical explanations, represented a pioneer work of exact scientific research. It was used by students and surgeons well into the following century. In it, following Stensen, Winsløw introduced a number of exact new terms, still used today, including the *corpora quadrigemina anteriora et posteriora* (formerly called *nates* and *testes*) and, for the ganglion chain, the "grand sympathetic nerve," the smaller branches being the "lesser sympathetic nerves."

Winsløw remained in Paris for the rest of his life, although he was invited on several occasions to return to Denmark. Only one of his treatises, *Mortis incertae signa* (1740), was translated into Danish (1868).

BIBLIOGRAPHY

I. ORIGINAL WORKS. There is a full list of Winsløw's works in H. Ehrencron-Müller, *Forfatterlexikon*, IX (Copenhagen, 1932), 124–128. His major writings are *Exposition anatomique de la structure du corps humain* (Paris, 1732; many later eds., 4 vols., 1732–1776); English trans. by G. Douglas, *An Anatomical Exposition of the Structure of the Human Body*, 2 vols. (London, 1733); German trans. by Georg Matthiae, *Anatomische Abhandlung von dem Bau des menschlichen Leibes*, 4 vols. (Berlin, 1733); enl. ed. by Bernhard Siegfried Albinus, 5 vols. (Basel, 1754); Latin trans. by E. Gallico, *Expositio anatomica structurae corporis humani*, 4 vols. (Frankfurt–Leipzig, 1758), also in 2 vols. (Venice, 1758); Italian trans., *Esposizione anatomica della struttura del corpo humano*, 4 vols. (Bologna, 1743; Naples, 1746, 1763, 1775; Venice, 1747, 1767); and *Quaestio medico-chirurgica . . . an mortis incertae signa minus incerta a chirurgicis, quam ab aliis experimentis* (Paris, 1740), also in French (Paris, 1742), Italian (Naples, 1744, 1775), Swedish (Stockholm, 1751), German (Leipzig, 1754), and Danish (Sorø, 1868). His autobiography is *L'autobiographie de J.-B. Winsløw*, Vilhelm Maar, ed. (Paris–Copenhagen, 1912).

II. SECONDARY LITERATURE. See E. Hintzsche, *Albrecht v. Haller Tagebuch der Studienreise nach London, Paris* (Bern, 1968), 35; V. Maar, "Lidt om J.-B. Winsløw," in *Festskrift til Julius Petersen, V. Meisen, Prominent Danish Scientists* (Copenhagen, 1932), 53–60; R. Schär, *Albrecht von Hallers neue anatomisch-*

physiologische Befunde (Bern, 1958), 50; E. Snorrason, *L'anatomiste J.-B. Winslow, 1669–1760* (Copenhagen, 1969); and T. Vetter, "La vie active de Jacques-Benigne Winslow," in *Nordisk medicin* (1971), 107–129.

E. SNORRASON

WINTHROP, JOHN (*b.* Groton, Suffolk, England, 12 February 1606; *d.* Boston, Massachusetts, 5 April 1676), *natural philosophy, medicine.*

Winthrop, who has frequently been called John Winthrop, Jr., by historians to distinguish him from his father, was the son of John Winthrop and Mary Forth. Born into the Puritan landed gentry, he studied for two years at Trinity College, Dublin, read law at the Inner Temple, and toured Europe. In 1631 he married his cousin Martha Fones and emigrated to the Massachusetts Bay Colony, following his father, who had been chosen governor of the Puritan company.

Thereafter the younger Winthrop held various colonial offices, culminating in a long tenure as governor of Connecticut. His first wife died in 1634, and he married Elizabeth Reade the following year. Winthrop's moderate but fluctuating means gave him a certain degree of independence, and in his spare time he was able to undertake a wide range of scientific activities and to carry on an extensive correspondence with other investigators.

To meet the needs of New England's settlers and in the hope of providing commodities for export, Winthrop frequently searched for mineral resources. He used processes familiar from reading and observation to produce iron, salt, indigo, saltpeter, and other substances, and he promoted the development of a graphite mine. With the exception of his infant iron and graphite industries, which eventually were brought to fruition by others, these efforts did not achieve lasting success. Nevertheless, Winthrop is widely recognized as one of the founders of American industrial chemistry.

Winthrop was a devoted student of the Hermetic philosophy, which helped to form his early attitudes toward science. Little is known about his alchemical experiments, which began during his residence at the Inner Temple and continued, at least peripatetically, for a long while. Although there is no evidence that Winthrop ever claimed the alchemical secret, he had the reputation of an "adept." Circumstantial evidence has involved him in the problem of the authorship of the treatises on the theory and practice of alchemy published under the pseudonym "Eirenaeus Philalethes," even though George Starkey probably used Winthrop only as an inspiration for the American adept from whom he claimed to have obtained some of the manuscripts. Winthrop amassed a large collection of alchemical and other scientific books within a general library of considerable extent. Portions of it are in various repositories.

Winthrop's medical records show that although he dispensed a wide variety of herbal preparations, he depended heavily on chemical medicines, especially antimonials and niter. At first his practice was limited to family and friends, but as word of his skill and willingness spread, he received medical requests from many parts of New England. Frequently his remedies were given free to the poor; and at his death Winthrop was undoubtedly New England's foremost physician.

His astronomical observations were of little consequence, although in 1660 Winthrop was operating what was probably the first large telescope in the American colonies, a ten-foot refractor. Several years later he was using a smaller instrument, and in 1668 he was attempting to perfect a telescope with a focal length of eight or ten feet.

Winthrop's letters reveal his interest in scientific phenomena as diverse as waterspouts and the metamorphosis of insects. During one of his voyages to England and Europe, he was admitted in 1662 to the group soon to be chartered as the Royal Society, and he was the first fellow resident in North America. While in London (1661–1663) he read papers on diverse subjects at the Society's meetings and was a faithful correspondent after returning to New England. Several of his communications were printed in the *Philosophical Transactions.* Although Winthrop contributed little to the history of scientific thought, he was the first scientific investigator of note in British America.

BIBLIOGRAPHY

I. ORIGINAL WORKS. Most of Winthrop's scientific observations were reported in his correspondence, now being printed by the Massachusetts Historical Society as part of *The Winthrop Papers* (Boston, 1929–   ). Earlier selections of his letters are cited below. Cromwell Mortimer, secretary of the Royal Society, asserted that Winthrop wrote "several learned Pieces . . . in Natural Philosophy; which indeed his innate Modesty would not suffer him to publish immediately, and when prevailed on by Friends to impart some of them to the Public, he concealed his Name, not being solicitous of the Reputation they might reflect on their Author" (*Philosophical*

Transactions of the Royal Society, **40** [1737–1738]). The only publications traditionally credited to Winthrop are excerpts from his letters in *Philosophical Transactions of the Royal Society,* **5** (1670), 1151–1153; and **6** (1671), 2221–2224; as well as a paper, "The Description, Culture, and Use of Maiz," *ibid.,* **12** (1678), 1065–1069, presented to the Royal Society in 1662.

II. SECONDARY LITERATURE. The first extensive biography is Robert C. Black, *The Younger John Winthrop* (New York–London, 1966). The only full-length study of Winthrop's scientific activities is Ronald S. Wilkinson, *The Younger John Winthrop and Seventeenth-Century Science* ([London], 1975). E. N. Hartley, *Ironworks on the Saugus* (Norman, Okla., 1957), examines his ironmaking endeavors. Among other specific modern studies are Ronald S. Wilkinson, " 'Hermes Christianus': John Winthrop, Jr. and Chemical Medicine in Seventeenth Century New England," in Allen Debus, ed., *Science, Medicine and Society in the Renaissance: Essays to Honor Walter Pagel* (New York, 1972), I, 221–241; John W. Streeter, "John Winthrop, Junior, and the Fifth Satellite of Jupiter," in *Isis,* **39** (1948), 159–163; supplemented by Ronald S. Wilkinson, "John Winthrop, Jr. and America's First Telescopes," in *New England Quarterly,* **35** (1962), 520–523; and Ronald S. Wilkinson, "The Alchemical Library of John Winthrop, Jr. and His Descendants in Colonial America," in *Ambix,* **11** (1963), 33–51, and **13** (1966), 139–186.

RONALD S. WILKINSON

WINTHROP, JOHN (*b.* Boston, Massachusetts, 19 December 1714; *d.* Cambridge, Massachusetts, 3 May 1779), *astronomy, mathematics.*

One of sixteen children of Adam Winthrop and Anne Wainwright, John Winthrop was born into a New England family that was already famous both politically and scientifically. His great-granduncle and namesake, the son of Winthrop the elder, who immigrated to Massachusetts in 1630, was a founding member of the Royal Society of London, and governor of Connecticut from 1660 until his death in 1676. He was a notable administrator of the new settlements and a practical student of chemistry. It is interesting to note that one of his communications to the Royal Society concerned a fifth satellite of Jupiter. Here he anticipated his descendant's far more extensive astronomical studies.

Winthrop attended the Boston Latin School and Harvard College, from which he graduated in 1732. For the next six years he lived at home and studied privately to such effect that in 1738, at the age of twenty-four, he was appointed the second Hollis professor of mathematics and natural philosophy at Harvard, succeeding Isaac Greenwood.

His duties included giving illustrated public lectures and taking charge of the considerable collection of philosophical instruments in Harvard Hall, used for demonstrations.

During his long tenure of the Hollis chair, which ceased only with his death, Winthrop established the first experimental physics laboratory in America; taught the laws of mechanics, optics, and astronomy according to Newton's principles; and introduced into the mathematics curriculum the study of the calculus. Perhaps his most important work for Harvard followed the disastrous fire that destroyed Harvard Hall on the night of 24 January 1764. The fire gutted the last of Harvard's original buildings and wiped out the valuable collection of scientific instruments. It fell to Winthrop to arrange for the replacement of the collection, which he was well equipped to do, both because of his scientific knowledge and because of his family connections and many friends on both sides of the Atlantic. The most active and influential of these friends was Franklin, who knew many of the finest instrument makers in London. The first orders for new apparatus went to London in June 1764, and over the next few years, instruments bearing such names as John Ellicott, Jeremiah Sisson, James Short, Peter Dollond, Benjamin Martin, Edward Nairne, and George Adams were dispatched to Harvard. The two major shipments were valued together at about £540. Among the instruments were two telescopes produced by Short. Winthrop himself owned a telescope by Short (made *ca.* 1755), which appears in the portrait of him painted by John Singleton Copley about 1773.

After 1739 Winthrop carried out many astronomical observations, the majority of which were reported in the *Philosophical Transactions of the Royal Society.* He observed the transits of Mercury in 1740, 1743, and again in 1769; and he used his observations to help determine the difference in longitude between Cambridge, Massachusetts, and Greenwich, England. In April 1759 he delivered lectures on the return of Halley's comet of 1682. Perhaps his most important astronomical work was concerned with the two transits of Venus in 1761 and 1769, which engaged astronomers all over the world. For the 1761 transit Winthrop organized an expedition from Harvard to St. John's, Newfoundland, which provided the material for one of his most important papers. In 1769 he published the results of further work in *Two Lectures on the Parallax and Distance of the Sun, as Deducible From the Transit of Venus.* Winthrop was also interested in magnetism and meteorology, and carried out systematic observations over a period of twenty years, reporting in 1756 on the effects of the severe earthquake in New England.

A number of honors were awarded to Winthrop in his later years. On 27 June 1765 he was proposed as a fellow of the Royal Society at the instigation of Franklin; Short was another of his supporters. Winthrop must have been closely associated with both men at this time, over replacement instruments for his college, and in work on the transits of Venus. Not only did Short make telescopes for observatories throughout the world, but he was also closely concerned with the Royal Society's plans for observing the phenomena. Winthrop's election was delayed until February 1766, when the ballot finally took place. Franklin signed a bond for his contributions, and the Harvard records show that his fees, not exceeding fifty-two shillings, were paid out of the treasury of the society in return for his placing a volume of the *Philosophical Transactions* annually in the library. In 1769 Winthrop became a member of the American Philosophical Society. He received the honorary degrees of LL.D. from the University of Edinburgh and from Harvard in 1771 and 1773, respectively.

Winthrop's first wife, whom he married in 1746, was Rebecca Townsend; and three years after her death in 1756, he married Hannah Fayerweather, a widow, who survived him. Winthrop was an ardent patriot, and a friend and adviser of George Washington. His career maintained the family tradition of public service allied with learning.

BIBLIOGRAPHY

I. ORIGINAL WORKS. Winthrop's works include "Concerning the Transit of Mercury Over the Sun, April 21, 1740 and of an Eclipse of the Moon, Dec. 21, 1740," in *Philosophical Transactions of the Royal Society*, **42** (1742–1743), 572–578; "An Account of the Earthquake Felt in New England, and the Neighbouring Parts of America, on the 18th of November 1755," *ibid.*, **50** (1757–1758), 1–18; "An Account of a Meteor Seen in New England, and of a Whirlwind Felt in That Country," *ibid.*, **52** (1761–1762), 6–16; "An Account of Several Fiery Meteors Seen in North America," *ibid.*, **54** (1764), 185–188; "Extract of a Letter . . . to James Short," *ibid.*, 277–278, on longitude and the equation of time; "Observations on the Transit of Venus, June 6, 1761, at St. John's, Newfoundland," *ibid.*, 279–283; "Cogitata de Cometis," *ibid.*, **57** (1767), 132–154; "Observations of the Transit of Venus Over the Sun, June 3, 1769," *ibid.*, **59** (1769), 351–358; "Observations of the Transit of Mercury Over the Sun, Oc-

tober 25, 1743," *ibid.*, 505–506; "Extract of a Letter . . . to B. Franklin," *ibid.*, **60** (1770), 358–362, on the transit of Venus and the aberration of light; "Observations of the Transit of Mercury Over the Sun, November 9th, 1769," *ibid.*, **61** (1771), 51–52; and "Remarks Upon a Passage in Castillione's Life of Sir Isaac Newton," *ibid.*, **64** (1774), 153–157.

Some of the material in the above papers was published separately including *Relation of a Voyage From Boston to Newfoundland for the Observation of the Transit of Venus, June 6, 1761* (Boston, 1761); and *Two Lectures on the Parallax and Distance of the Sun, as Deducible From the Transit of Venus. Read in Holden-Chapel at Harvard-College in Cambridge, New England, in March 1769* (Boston, 1769).

II. SECONDARY LITERATURE. On Winthrop and his work, see I. Bernard Cohen, *Some Early Tools of American Science. An Account of the Early Scientific Instruments and Mineralogical and Biological Collections in Harvard University* (Cambridge, Mass., 1950), *passim* (225 ff.); Raymond Phineas Stearns, "Colonial Fellows of the Royal Society of London, 1661–1788," in *Notes and Records of the Royal Society of London*, **8** (1951), 178–246; Raymond Phineas Stearns, *Science in the British Colonies of America* (Urbana, Ill., 1970), esp. 642–670; G. L'E. Turner, "The Apparatus of Science," in *History of Science*, **9** (1970), 129–138, an essay review of David P. Wheatland, *The Apparatus of Science at Harvard 1766–1800. Collection of Historical Scientific Instruments, Harvard University* (Cambridge, Mass., 1968), written with Barbara Carson; and *Dictionary of American Biography*, X, pp. 414–416.

G. L'E. TURNER

WINTNER, AUREL (*b*. Budapest, Hungary, 8 April 1903; *d*. Baltimore, Maryland, 15 January 1958), *mathematics.*

Wintner studied mathematics at the University of Leipzig from 1927 to 1929. During that period he was an editorial assistant for *Mathematische Zeitschrift* and *Jahrbuch über die Fortschritte der Mathematik*, serving under the direction of Leon Lichtenstein, who for many years was editor of those journals. This period of apprenticeship had a profound influence on Wintner, and he often expressed his gratitude for his training under Lichtenstein.

Wintner's mathematical reputation was established by a series of papers on the Hill lunar theory that gave the first mathematically rigorous proof of the convergence of George Hill's method involving infinitely many unknowns. He received the Ph.D. at Leipzig in 1929, then spent a semester in Rome as a Rockefeller fellow and another in Copenhagen, where he worked with Elis Stromgren. As a result of that collaboration, Wintner was able to provide a theoretical basis for Stromgren's "natural termination principle" for orbit periods, which was an empirical analysis of the degeneration of periodic orbits.

In 1929 Wintner published *Spektraltheorie der unendlichen Matrizen*, which contains the first proofs of the basic facts in Hilbert space—the fundamental mathematical construct in the then-developing physical theory of quantum mechanics. Unfortunately, Wintner's fundamental contributions to this subject were (and are) not adequately appreciated because he formulated his results in the language of matrices rather than in the more abstract language of operators, made popular by von Neumann. This lack of recognition embittered Wintner and made him suspicious of the (genuine) merits of the more abstract developments in recent mathematics.

In 1930 Wintner married the daughter of Otto Hölder, one of his teachers at Leipzig. In the same year he joined the faculty of the Johns Hopkins University, where he remained until his death. In 1944 he became an editor of *American Journal of Mathematics*, to which he devoted most of his energy, both through his scientific contributions (a substantial part of his most valuable work after he came to America was published there) and through his editorial work.

Wintner's work in America covered the entire range of classical analysis, from probability and analytic number theory to differential equations and basic questions in local differential geometry. Much of his work from 1936 to 1958 was done in collaboration with his student and colleague Philip Hartman. He published several papers with Norbert Wiener in a branch of probability theory that is now coming back into fashion. He also produced works with several other mathematicians. In 1941 Wintner published *Analytical Foundations of Celestial Mechanics*, which combines great astronomical and mathematical scholarship with deep and meticulous analysis. He is best known for this work.

Wintner, a man of high moral principles, opposed direct government support of scholarly research, for fear of interference. He not only accepted considerable financial hardship by personally refusing such support but also was willing to forgo fruitful scientific collaboration in order to maintain his ideals.

SHLOMO STERNBERG

WISLICENUS, JOHANNES (*b.* Klein-Eichstedt, near Querfurt, Germany, 24 June 1835; *d.* Leipzig, Germany, 5 December 1902), *chemistry.*

Wislicenus was a student at the University of Halle in 1853 when his father, a Lutheran pastor of liberal religious and political views, was ordered arrested for the publication of a biblical study. The family fled to the United States, where Wislicenus became an assistant in the analytical laboratory at Harvard. When the family returned to Europe in 1856, he completed his studies at Halle. Subsequently he was professor of chemistry at the Zurich Oberen Industrieschule (1861), associate (1864) and full professor (1867) at the University of Zurich, and professor (1870) at the Eidgenössische Technische Hochschule in Zurich. He succeeded Adolf Strecker at Würzburg in 1872 and Kolbe at Leipzig in 1885.

Wislicenus' first papers were joint publications with Wilhelm Heintz, professor of chemistry at Halle. They studied the condensation of aldehydes with ammonia and isolated the base oxytetraldin in 1858. At Zurich he and his colleague Adolf Fick investigated the origin of muscle energy. According to Liebig, proteins produced force and their oxidation furnished the energy for muscle power. Carbohydrates and fats produced only heat. In 1865 Wislicenus and Fick tested this theory by climbing the Faulhorn in the Swiss Alps, calculating the work done during the ascent and measuring the amount of nitrogen in the urine excreted. They proved that the oxidation of protein contributed little to muscle energy and concluded that protein was used mainly in the growth and maintenance of tissues, carbohydrate and fat oxidation being the source of muscle energy.

Between 1863 and 1873 Wislicenus studied lactic and paralactic acids. These two acids and a third, hydracrylic acid, were monobasic acids with the formula $C_3H_6O_3$. Wislicenus observed that lactic acid was optically active but paralactic acid was not. In 1863 he represented them according to the type theory, their radicals being in a different order:

$$\begin{cases} CH_3 \\ CH(OH) \\ CO(OH) \end{cases} \quad \begin{cases} CH_2(OH) \\ CH_2 \\ CO(OH) \end{cases}$$

When structural formulas came into use during the 1860's, he proposed that the two acids had identical structures. By 1873 Wislicenus established that lactic acid and paralactic acid were both α-hydroxypropionic acid, while hydracrylic acid was β-hydroxypropionic acid. He was the first to establish the structural identity of two different substances, and argued that ordinary structural formulas were inadequate: the two acids must be represented by three-dimensional formulas that indicate the different arrangement of the atoms in space. Wislicenus called this type of isomerism "geometrical isomerism."

In 1874 van't Hoff presented his theory of the tetrahedral carbon atom and asserted that it occurred to him after reading Wislicenus' paper of the previous year. Wislicenus enthusiastically accepted van't Hoff's theory and wrote to him in 1875 for permission to have *La chimie dans l'espace* translated into German. He later contributed an introduction to the German edition.

Wislicenus was the leader in applying and extending the ideas of van't Hoff and Le Bel, and his successes helped to bring chemists to the new field of stereochemistry. There was, however, no serious attempt to apply the theory of the tetrahedral carbon atom to cases other than optical isomers. Van't Hoff had suggested that doubly linked carbon atoms could be represented by two tetrahedrons with one edge in common and that possibilities for isomerism occurred when two or more of the radicals attached to these carbon atoms differed.

In 1887 Wislicenus published an important paper on the stereoisomerism of unsaturated carbon compounds, extending the hypothesis along the lines suggested by van't Hoff and also considering the attractive-repulsive forces of the atoms in order to determine the most probable geometric configuration of the atoms in the molecule. He showed how the interpretation of maleic and fumaric acids as geometric isomers explained their chemical transformations. In addition he determined the configurations of many unsaturated isomeric carbon compounds and investigated geometric isomerism in cyclic compounds.

Wislicenus contributed to several areas of organic chemistry. He introduced molecular silver as a synthetic agent, preparing adipic acid from β-iodopropionic acid in 1869. Other syntheses include hydantoin (1873), glutaric acid (1878), vinyl ether (1878), cyclic ketones (1893), and vinyl acetic acid (1899). In a long series of researches on acetoacetic ester and its derivatives, he established the conditions for their hydrolysis, showing that acid hydrolysis produced a ketone, alcohol, and carbon dioxide, and alkaline hydrolysis a fatty acid and alcohol (1878). Wislicenus elucidated the structure of acetoacetic ester by replacing hydrogen with sodium. The ethyl sodioacetoacetate combined with an alkyl iodide gave high yields of a substituted acetoacetic ester that could accept another sodium atom and exchange it for a second alkyl group.

BIBLIOGRAPHY

I. ORIGINAL WORKS. Wislicenus' study of the geometric isomerism of unsaturated compounds is "Über die räumliche Anordnung der Atome in organischen Molekülen und ihre Bestimmung in geometrischisomeren ungesättigten Verbindung," in *Abhandlungen der K. Sächsischen Gesellschaft der Wissenschaften*, math.-phys. Kl., **14** (1887), 1–78, translated by George M. Richardson in *Foundations of Stereochemistry: Memoirs by Pasteur, Van't Hoff, Le Bel and Wislicenus* (New York, 1901), 65–132. Wislicenus rewrote Strecker's textbook, originally based on an earlier work of Regnault, *Regnault-Strecker's Kurzes Lehrbuch der Chemie*, 2 vols. (Brunswick, 1874–1881); there is an English version of the organic chemistry part: *A. Strecker's Short Textbook of Organic Chemistry by Dr. J. Wislicenus*, translated and edited by W. R. Hodgkinson and A. J. Greenaway (London, 1881).

Significant papers include "Über ein basisches Zersetzungsproduck des Aldehydammoniaks," in *Annalen der Physik und Chemie*, **105** (1858), 577–597, written with W. Heintz; "Studien zur Geschichte der Milchsäure und ihrer Homologen," in *Justus Liebigs Annalen der Chemie*, **125** (1863), 41–70; **133** (1865), 257–287; and **146** (1868), 145–161; "On the Origin of Muscular Power," in *Philosophical Magazine*, 4th ser., **31** (1866), 485–503, written with A. Fick; "Synthetische Untersuchungen über die Säuren der Reihe $C_nH_{2n}(CO \cdot OH)_2$," in *Annalen der Chemie*, **149** (1869), 215–224; "Über die isomeren Milchsäuren," *ibid.*, **166** (1873), 3–64; "Über die optisch-active Milchsäuren der Fleischflüssigkeit, die Paramilchsäure," *ibid.*, **167** (1873), 302–346; "Über Acetessigestersynthesen," *ibid.*, **186** (1877), 161–228; "Spaltung des Acetessigester und seiner Alkylsubstitutionsproducte durch Basen," *ibid.*, **190** (1878), 257–281; "Über Vinyläthyläther," *ibid.*, **192** (1878), 106–128; "Untersuchungen zur Bestimmung der räumliche Atomlagerung," *ibid.*, **246** (1888), 53–96, **248** (1888), 281–355, and **250** (1889), 224–254; and "Über Ringketone," *ibid.*, **275** (1893), 309–382.

II. SECONDARY LITERATURE. There are two detailed accounts of Wislicenus' work: Ernst Beckmann, "Johannes Wislicenus," in *Berichte der Deutschen chemischen Gesellschaft*, **37** (1904), 4861–4946, which includes a bibliography; and William Henry Perkin, Jr., "The Wislicenus Memorial Lecture," in *Memorial Lectures Delivered Before the Chemical Society*, II (London, 1914), 59–92, which originally appeared in *Journal of the Chemical Society*, **87** (1905), 501–534.

ALBERT B. COSTA

WISTAR, CASPAR (*b*. Philadelphia, Pennsylvania, 13 September 1761; *d*. Philadelphia, 22 January 1818), *anatomy*.

Fifth of eight children of Richard Wistar, proprietor of a glass factory at Salem, New Jersey, and his wife, Sarah Wyatt, both of whom were Quakers, Wistar attended Friends' schools in Philadelphia and, moved, it is said, by the sufferings of the wounded at the battle of Germantown in 1777, determined to study medicine. After a preceptorship with Dr. John Redman, he entered the University of the State of Pennsylvania in 1779, and was graduated Bachelor of Medicine three years later. Although he had some difficulty in obtaining permission of the Quaker monthly meeting to go abroad, because he had fought a duel, Wistar went to England, where he studied anatomy in London under John Hunter. At Edinburgh he so impressed his instructors that, had university regulations not prevented, they would have allowed him to take a degree after only one year's study. Wistar was equally respected by his fellow students, who elected him an annual president of both the Royal Medical Society (a student organization) and the Edinburgh Natural History Society, to the latter of which he read a paper on moisture in the atmosphere. His graduation thesis, *De animo demisso* (1786), was said to have given "unspeakable relief" to several Scottish hypochondriacs; and Dr. Charles Stuart of Edinburgh urged that it be translated and printed in London. After a short visit to the Continent, Wistar returned home in 1787.

As a practitioner Wistar was slow and deliberate, but accurate; he lacked the temperament for surgery. During the Philadelphia yellow fever epidemic of 1793 he remained in the city, giving Benjamin Rush aid and support; but, falling ill himself, he allowed Adam Kuhn to treat him. Thereafter he abandoned Rush's heroic prescriptions of bleeding and purging and so lost his friendship. Wistar was an attending physician of the Philadelphia Dispensary in 1786, a physician to the Pennsylvania Hospital in 1793–1810, and in 1809 founder of a vaccine society that vaccinated 1,102 persons in its first year. Ill health forced him to give up much of his practice several years before his death.

As a medical teacher Wistar was professor of chemistry at the College of Philadelphia[1] in 1789, then adjunct professor of anatomy, surgery, and midwifery at the new University of Pennsylvania in 1792; after 1810 he taught only anatomy. Through hard work he made himself an effective lecturer, and introduced two innovations in instruction: in class he employed large-scale models of the small parts of the human structure; and he divided the students into small groups ("bone classes"), giving each an assortment of bones and preparations to study and identify. As a further aid he prepared *A System of Anatomy for the Use of Students of Medicine* (2 vols., Philadelphia, 1811). Drawing heavily on John Innes, Monro, and Bichat, this manual went into nine editions and was pronounced by Charles Caldwell, who was not given to praising his contemporaries, as "without rival, in any language." It was John Syng Dorsey's judgment of Wistar that "no one could fail to become an Anatomist who diligently attended his lectures." Wistar's only original contribution to anatomy was an account, read in 1814, of the sphenoid sinuses ("Observations on Those Processes of the Ethnoid Bone Which Originally Form the Sphenoidal Sinuses," in *Transactions of the American Philosophical Society*, n.s. **1** [1818], 371–374). In his latter years he wrote several papers on paleontological remains from Big Bone Lick (near Burlington), Kentucky.

Wistar married twice: in 1788 Isabella Marshall, who died in 1790; and in 1798 Elizabeth Mifflin, who died in 1840, by whom he had a daughter and two sons, one of whom became a physician. Concerned since young manhood about slavery, Wistar was president of the Pennsylvania Abolition Society when he died. He was equally sensitive to the plight of the American Indians and was one of the first in the United States to appreciate their eloquence, examples of which he used to read aloud to his family. In the American Philosophical Society, to which he was elected in 1787, he succeeded Thomas Jefferson as president (1815–1818). His portrait was painted by Bass Otis, and William Rush carved a bust; and he is otherwise remembered by the lovely flowering shrub that Thomas Nuttall named for him,[2] as well as for his agreeable informal suppers for friends and distinguished strangers, which, institutionalized after his death as the Wistar Association, continue little changed to this day.

NOTES

1. The College of Philadelphia was chartered in 1755, and flourished until 1779, when it was abolished by legislative act and its endowments and property were given to a new institution called the University of the State of Pennsylvania. In 1789 the confiscated properties were restored to the ousted trustees of the defunct College, and the institution was revived under its old name. Thus in 1789 there were two collegiate institutions in Philadelphia: the College of Philadelphia and the University of the State of Pennsylvania. In 1791

the two institutions were united under a new charter and name as the University of Pennsylvania.

2. Nuttall named the plant for his friend Caspar Wistar, but spelled it Wister. That would not cause any complication but for the fact that one branch of the family in Philadelphia spells its name Wister; and in the mid-nineteenth century, after the principals were dead, Charles J. Wister, Jr., claimed that Nuttall had named the plant for *his* father, who (to make the argument plausible) was an amateur of natural history. The generally accepted version is that the wisteria was named for Caspar Wistar and really should be spelled wistaria.

BIBLIOGRAPHY

I. ORIGINAL WORKS. In addition to those cited in the text, Wistar's writings include *Eulogium on Doctor William Shippen . . .* (Philadelphia, 1818); several articles in *Transactions of the American Philosophical Society*; and one (on scurvy) in *Philadelphia Medical and Physical Journal*, **2**, pt. 2 (1806). His letter "To the Physicians of Philadelphia" on yellow fever is in the Philadelphia *General Advertiser* (26 Sept. 1793); the controversy with Rush continues in letters of 4 Oct. and 1 and 8 Nov. 1793. Miscellaneous MS letters and papers are at the Historical Society of Pennsylvania, the College of Physicians of Philadelphia, and the American Philosophical Society Library.

II. SECONDARY LITERATURE. Contemporary eulogies and memoirs by Charles Caldwell, *An Elogium on Caspar Wistar, M.D.* (Philadelphia, 1818), David Hosack, *Tribute to the Memory of the Late Caspar Wistar, M.D.* (New York, 1818), and William Tilghman, *An Elogium in Commemoration of Dr. Caspar Wistar* (Philadelphia, 1818), are the principal sources. Within a few weeks of Wistar's death Tilghman asked professional colleagues, friends, and family for information and judgments; the replies, preserved at the College of Physicians of Philadelphia, contain more data than the printed eulogy. Sketches of Wistar in Samuel D. Gross, ed., *Lives of Eminent American Physicians and Surgeons of the Nineteenth Century* (Philadelphia, 1861); and James Thacher, *American Medical Biography* (Boston, 1828), add little to the 1818 eulogies. William S. Middleton, "Caspar Wistar, Junior," in *Annals of Medical History*, **4** (1922), 64–76, contains the essential facts. On Rush's relations with Wistar, see *Letters of Benjamin Rush*, L. H. Butterfield, ed., 2 vols. (Princeton, 1950).

WHITFIELD J. BELL, JR.

WITELO (*b.* Poland, *ca.* 1230/1235; *d. after ca.* 1275), *optics, natural philosophy.*

Life. Very little is known of Witelo's life. His homeland and national origins must be inferred from scattered remarks in his *Perspectiva*. There he refers to "our homeland, namely Poland" and mentions the city of Vratizlavia (Wrocław) and the nearby towns of Borek and Liegnitz,[1] thus reveal-

ing an intimate knowledge of the environs of Breslau (Wrocław) in Silesia, which suggests that he probably was born and raised there. In the preface to the *Perspectiva*, Witelo refers to himself as "the son of Thuringians and Poles," from which it may be gathered that on the paternal side he was descended from the Germans of Thuringia who colonized Silesia in the twelfth and thirteenth centuries, while on the maternal side he was of Polish descent.

Witelo's education and adult life likewise must be reconstructed from the most fragmentary evidence. It may be surmised, from a reference to time spent in Paris and a description of a nocturnal brawl that occurred there in 1253, that he received his undergraduate education at the University of Paris in the early 1250's. He must, therefore, have been born in the early or middle 1230's. In the 1260's Witelo was studying canon law at Padua, as revealed by his reference to an event that occurred in Padua in 1262 or 1265.[2] His presence in Padua also is indicated by the explicit of his *Tractatus de primaria causa penitentie et de natura demonum* (written during his stay in Padua), in which he is referred to as "Witilo, student in canon law."[3] It is evident, however, that at Padua he was not totally preoccupied with his legal studies, for he wrote the *Tractatus* during an Easter recess, and, according to Birkenmajer, it reflects the teachings of Plato, Galen, Ibn Sīnā, Aristotle, Ibn Rushd, Euclid, and Ibn al-Haytham (Alhazen).[4]

Late in 1268 or early in 1269, Witelo appeared in Viterbo, where he became acquainted with William of Moerbeke, papal confessor and translator of philosophical and scientific works from Greek to Latin, to whom Witelo later dedicated the *Perspectiva*. We know nothing further of Witelo's movements unless he is to be identified with the person of that name who served as chaplain to King Ottocar II of Bohemia and who was sent on a mission to Pope Gregory X in 1274.[5] The Bern manuscript of Witelo's *Perspectiva* refers to the author as "Magister Witelo de Viconia," which has given rise to speculation that Witelo retired to the Premonstratensian abbey of Vicogne during his declining years.[6]

It is apparent that Witelo's *Perspectiva* was not composed before 1270, since it draws on Hero of Alexandria's *Catoptrica*, the translation of which was completed by Moerbeke on 31 December 1269. Because such an immense work probably was not written in less than several years, it is unlikely that Witelo died before the mid-1270's.

There continues to be a good deal of confusion

regarding Witelo's name. In the printed editions of the *Perspectiva*, the author's name is spelled "Vitellio" or "Vitello"; and a number of historians have adopted this orthography. Maximilien Curtze and Clemens Baeumker have demonstrated, however, that early manuscripts of the *Perspectiva* give overwhelming support to the form "Witelo."[7] They have argued, further, that "Witelo" is a diminutive of "Wito" or "Wido," a given name commonly encountered in Thuringian documents of the thirteenth century. Family names were uncommon in thirteenth-century Poland, and there is no evidence to suggest that Witelo had one.

Works. Witelo's known extant works are *Perspectiva* and *De primaria causa penitentie et de natura demonum*. In addition he refers in the *Perspectiva* to several other treatises of his, none of which can now be identified: *De elementatis conclusionibus, Philosophia naturalis, Scientia motuum celestium, Naturales anime passiones*, and *De ordine entium*.[8] These titles reveal the range of Witelo's interests in natural philosophy. Nevertheless, since only the *Perspectiva* has been the object of detailed study, his reputation rests almost solely on his work in optics.

Optics in the latter half of the thirteenth century was hardly (if at all) an experimental endeavor;[9] the principal task was to master an abundance of literature on the subject. By far the most important optical treatise in Witelo's day was Ibn al-Haytham's *Optics* or *De aspectibus*, rendered into Latin by an unidentified translator late in the twelfth or early in the thirteenth century. Although Witelo never refers to Ibn al-Haytham by name, there can be no doubt that the latter was his chief source: Witelo normally treats the same topics in the same fashion and sometimes even in the same words; occasionally he omits or inserts a topic, and often he seeks to clarify or supplement one of Ibn al-Haytham's points by further elaboration or an improved demonstration, but in very few respects does he escape the general framework inherited through the latter's *Optics*.[10]

Yet other influences are evident. It is beyond dispute that Witelo used the *Optica* of Ptolemy, whose table of refraction he reproduces; the *Catoptrica* of Hero, whose principle of minimum distance he employs to explain reflection at equal angles; and the *De speculis comburentibus* (anonymous in the thirteenth century, but now attributed to Ibn al-Haytham), from which he drew his analysis of paraboloidal mirrors. There can be little doubt that he also was familiar with the widely circulated *Optica (De visu)* of Euclid, *Ca-toptrica (De speculis)* of Pseudo-Euclid, *De aspectibus* of al-Kindī, and the physiological and psychological works of Galen, Ḥunayn ibn Isḥāq, Ibn Sīnā, and Ibn Rushd. As for Latin authors, Alexander Birkenmajer has argued that Witelo was strongly influenced by Robert Grosseteste's *De lineis angulis et figuris* and Roger Bacon's *De multiplicatione specierum*.[11] In addition, it is certain that he knew Bacon's *Opus maius* and possible that he knew John Pecham's *Perspectiva communis*.[12] Witelo also relied on a number of ancient mathematical works, including those of Euclid and Apollonius of Perga, and perhaps of Eutocius, Archimedes, Theon of Alexandria, and Pappus.

Witelo's *Perspectiva* is an immense folio volume of nearly five hundred pages in the three printed editions, and no detailed analysis of its contents has ever been made. It will be possible, in the remainder of this article, only to trace its most significant features. The scope of the *Perspectiva* is revealed by the following outline of its contents: Book I consists of definitions, postulates, and 137 geometrical theorems, which provide the mathematical principles required for the optical demonstrations of the remaining nine books. In this book Witelo skillfully summarizes the aspects of the geometrical achievement of antiquity that are relevant to his own geometrical optics. Book II deals with the nature of radiation, the propagation of light and color in straight or refracted lines, shadows, and the problem of pinhole images. Book III is concerned with the physiology, psychology, and geometry of monocular and binocular vision by means of rectilinear radiation.

Book IV treats the perception of the twenty visible intentions other than light and color, including size, shape, remoteness, corporeity, roughness, darkness, and beauty. It also deals with errors in the perception of these intentions—principally errors in judging distance, shape, and relative size. This book is thus largely psychological in tone, although it includes a number of matters that fall into the realm of traditional geometrical perspective. In book V, Witelo considers vision by reflected rays, beginning with the nature and geometrical laws of reflection and proceeding to a detailed analysis of plane mirrors. Image formation in curved mirrors occupies books VI through IX of the *Perspectiva*—convex spherical mirrors in book VI, convex cylindrical and conical mirrors in book VII, concave spherical mirrors in book VIII, and concave cylindrical, conical, and paraboloidal mirrors in book IX. Book X is concerned with vision by rays refracted at plane or spherical interfaces; it also includes a discussion of the rainbow and other meteorological phenomena.

The most essential feature of any optical system might seem to be its theory of the nature of light. Witelo's concerns were principally geometrical, however, and he formulated no systematic account of the nature of light. From scattered remarks throughout the *Perspectiva* (particularly its preface) one can hope at the very most to classify him within a broad tradition on this question. He writes in the preface: "Sensible light is the intermediary of corporeal influences"; "Light is a corporeal form"; and "Light is the first of all sensible forms." It is apparent from such remarks that light is regarded as the intermediary in certain natural actions—an instance of the multiplication of forms. Light is thus one particular manifestation of a more general phenomenon, the propagation of force or influence from one natural body to another. But although light is only one instance of natural action, it is the instance most accessible to the senses and most amenable to analysis; therefore it serves, for Witelo, as the paradigm for the investigation of all natural actions. Thus he writes, at the conclusion of his quantitative analysis of refraction, "These are the things that occur to lights and colors and universally to all forms in their diffusion through transparent bodies and in the refraction that occurs in all of them."[13] And in the preface he remarks, "The investigation [in general, of the action of one body on another] properly proceeds by means of visible entities." It is evident, then, that Witelo falls very generally into the Neoplatonic tradition traceable from Plotinus through Ibn Gabirol to Grosseteste and Bacon. For Witelo, as for these predecessors, every natural body propagates its power to surrounding bodies, of which propagation light is the principal example. Moreover, Witelo would seem to follow Grosseteste and Bacon in perceiving that optics thus becomes the fundamental science of nature.

A second essential feature of any optical system, about which Witelo says somewhat more, is the propagation of light or visible forms. According to Witelo, light is always propagated rectilinearly unless it encounters a reflecting or refracting surface. This fact, he claims, can be verified experimentally; and he even describes the required apparatus. The same apparatus had already been described by Ibn al-Haytham, however, and there is no reason to believe that Witelo personally verified the rectilinear propagation of light by experimental means. Witelo is uninformative on the physical mechanism of propagation, but one can surmise from his use of

terms like "multiplication" and "diffusion" that his view was not far from that of Roger Bacon. He departs from Ibn al-Haytham and Bacon and most of the ancient optical tradition on the temporal aspects of propagation, arguing that light requires no time for propagation through an extended medium. He proceeds on logical grounds, reducing to absurdity the claim that the propagation of light requires time. Witelo is unable to maintain this position, however, and later admits that "every light passing through a transparent body tranverses it with an exceedingly swift and insensible motion. And yet the motion occurs more swiftly through more transparent bodies than through less transparent bodies."[14]

The applicability of geometry to optical problems follows from the principle of rectilinear propagation: light proceeding along straight lines, subject only to the rules of reflection and refraction, clearly is amenable to geometrical analysis. Witelo draws a careful distinction, however, between the one-dimensional lines employed in a geometrical analysis of optical phenomena and actual rays (or radial lines) of light. The latter are real physical lines traversed by the smallest visible light, and "in the least light that can be supposed, there is width. . . . Therefore in a radial line along which light is diffused, there is some width."[15] Nevertheless, "in the middle of that [radial line] is an imaginary mathematical line, parallel to which are all the other mathematical lines in that natural line."[16] And since the mathematical lines always fall within the natural radial lines, the former adequately represent the actual path of light, and it is proper to employ them in optical demonstrations.[17]

It was still a matter of debate in the thirteenth century whether rays issue only from the visible object or whether, in addition, there is an emission from the observer's eye that assists in the act of sight. Witelo follows Ibn al-Haytham (and departs from Grosseteste, Bacon, and Pecham) in acknowledging no emission of visual rays from the eye; sight is due solely to the forms of light and color issuing in all directions from every point (or small part) of the visible object and entering the observer's eye to produce visual sensations.

Witelo also follows Ibn al-Haytham (and the entire ancient and medieval optical tradition) in declaring that the sensitive organ of the eye is the glacial humor (or crystalline lens), which occupies the central position. Sight occurs, therefore, when the forms of light and color are arranged on the surface of the crystalline lens in the same order as the points of the visible object from which they issued: on the surface of the crystalline lens there is a "union of the visible forms and the soul's organ,"[18] which constitutes the act of sight. But how is it possible for light to be arranged on the surface of the lens exactly as on the surface of the object, since light issues in all directions from every point of the object? Witelo supplies precisely the same answer as Ibn al-Haytham, Bacon, and Pecham: only unrefracted light is strong enough to be efficacious in sight, and there is but one unrefracted ray issuing from each point of the visible object—the ray proceeding toward the center of curvature of the humors and tunics of the eye. The collection of all such unrefracted rays maintains its configuration between the visible object and the glacial humor and consequently forms, on the surface of the latter, an exact image (albeit reduced in size) of the visible object. Yet the act of sight is not completed in the glacial humor; the forms it receives pass through to the optic nerve and thence to the anterior part of the brain, where the nerves from the two eyes intersect to form the "common nerve," the residence of the visual power or *ultimum sentiens*, where a final judgment is made.

The geometrical structure that Witelo builds upon the conception of rays and mathematical lines naturally encompasses problems of geometrical perspective and image formation by reflection and refraction, but it also extends to the anatomy of the eye and the act of sight. He describes the eye in traditional terms, as a composite of three humors—glacial or crystalline, vitreous, and albugineous (aqueous)—and four tunics—uvea, cornea, conjunctiva or consolidativa, and aranea or retina. Geometrical considerations predominate in Witelo's descriptions of these tunics and humors: all are spherical in form; all tunics and humors anterior to the glacial humor must have concentric surfaces, so that a ray perpendicular to one is perpendicular to the rest and passes through all of them without refraction; and the glacial and vitreous humors have precisely the necessary shapes and relative densities to refract the rays converging toward the center of the eye before they actually intersect, and to conduct them through the vitreous humor and optic nerve without alteration or inversion.

Books V–IX of the *Perspectiva* are devoted to the science of catoptrics. The foundation of this science is the law of reflection, which Witelo derives from the principle of minimum distance: since nature does nothing in vain, it "always acts along the shortest lines."[19] Following Hero of Alexandria, Witelo demonstrates that the shortest lines connecting two points and a reflecting surface

are those that make equal angles with the surface.[20] He also argues that the plane formed by the incident and reflected rays is perpendicular to the surface of reflection (or, in the case of curved mirrors, its tangent), and that an object seen by reflection appears to be located where the backward extension of the ray incident on the eye intersects the perpendicular dropped from the visible object to the reflecting surface.

Employing these three rules and the principles of geometry, Witelo proceeds to solve a series of very abstruse problems in reflection, drawn primarily (but not entirely) from Ptolemy and Ibn al-Haytham. This is the most substantial section of the *Perspectiva*, occupying much of books V–IX and some 200 pages in the printed editions. Witelo deals skillfully with such problems as inversion and reversal of images, determination of precise size and location of images formed by concave and convex mirrors of various shapes, and computation of the number of images of a given object visible in a concave spherical mirror. Not until the seventeenth century was his catoptrics excelled in the West.

Book X of the *Perspectiva* deals with the refraction of light. In book II, Witelo had described an instrument for gathering quantitative data on the propagation of light, and in book X he claims to have used the same instrument in the formulation of tables of refraction. In fact, there is ample evidence that this claim is untrue. In the first place, the upper half of the table is taken directly from Ptolemy's *Optica*.[21] Second, the values appearing in the table are not those given by experiment, but sets of numbers conforming to a regular progression—the differences between successive angles of refraction (corresponding to angles of incidence taken at 10° intervals) form an arithmetic progression with a common difference of one-half degree. Finally, the lower half of the table was computed by Witelo from the values in the upper half by erroneous application of the reciprocal law; consequently it includes preposterous results, such as angles of refraction (measured from the perpendicular) greater than 90° and no recognition whatsoever of total internal reflection.

Nevertheless, at the qualitative level Witelo is fully cognizant of the principal phenomena of refraction: light passing obliquely from a less dense to a more dense medium is refracted toward the perpendicular, while light passing in the reverse direction is refracted away from the perpendicular. But Witelo is not content with a quantitative or qualitative description of the geometrical phenomena of refraction; he also presents a mechanical explanation based on the varying resistance offered to the passage of light by different transparent substances, the idea that ease of traversing a medium is associated with proximity to the perpendicular, and the principle that light is so refracted at a transparent interface as to most nearly preserve uniformity of strength or action in the two media. In the course of this analysis, Witelo resolves the oblique motion of light into components perpendicular and parallel to the refracting interface.[22]

Influence. It is difficult to separate Witelo's influence on the history of late medieval and early modern optics from that of Ibn al-Haytham, particularly after their works were published in a single volume in 1572. One can affirm in general that their writings, along with John Pecham's *Perspectiva communis*, served as the standard textbooks on optics until well into the seventeenth century. More specifically, it is possible to establish Witelo's influence on Henry of Hesse, Blasius of Parma, and Nicole Oresme in the fourteenth century; Lorenzo Ghiberti, Johannes Regiomontanus, and Leonardo da Vinci in the fifteenth century; Giambattista della Porta, Francesco Maurolico, Giovanni Battista Benedetti, Tycho Brahe, William Gilbert, Simon Stevin, and Thomas Harriot in the sixteenth century; and Kepler, Galileo, Willebrord Snell, Descartes, and Francesco Grimaldi in the seventeenth century.[23]

NOTES

1. Bk. X, theor. 74; bk. IV, theor. 28.
2. The former date is in Alexander Birkenmajer's "Witelo e lo studio di Padova," 156; the latter is in Lynn Thorndike's *History of Magic and Experimental Science*, V, 86.
3. Birkenmajer, *op. cit.*, 160. Thorndike regards this as a single work, while Birkenmajer treats it as two separate treatises.
4. *Ibid.*, 162.
5. *Ibid.*, 157.
6. Alternatively, it has been suggested that "Viconia" may be a misreading of "Vitovia," a Polish village (which is not, however, near Wrocław); see Clemens Baeumker, "Zur Biographie," 360.
7. Maximilien Curtze, "Sur l'orthographe," 49–66; Baeumker, *Witelo*, 190–200.
8. Some of these titles are less than certain: in several cases one cannot be sure that Witelo is claiming authorship; in others it is not clear whether the title applies to a chapter or to an entire treatise. For more detail, see Baeumker, *Witelo*, 239–244. *De intelligentiis*, formerly ascribed to Witelo, is now generally attributed to Adam Pulchrae Mulieris (*fl.* 1225). On the possibility that certain MSS of Euclid's *De visu* actually contain Witelo's recension of this treatise, see Wilfred R. Theisen, "The Medieval Tradition of Euclid's *Optics*" (Ph.D. diss., Univ. of Wis., 1972), 58–60; David C. Lindberg, intro. to facs. repr. of Risner ed., xi, xxvii.
9. If there is an exception to this generalization, it seems to be in the science of the rainbow, where both Witelo and Dietrich von Freiberg claim to have used spherical containers filled with water to simulate a raindrop.
10. The appellation "Alhazen's ape," later applied to Witelo by Giambattista della Porta, is unfortunate, since it ignores Witelo's frequent attempts to revise or supplement Ibn al-Haytham and the constant use of his own critical powers.
11. Alexander Birkenmajer, "Robert Grosseteste and Richard Fournival," in *Mediaevalia et humanistica*, 5 (1948), 36.
12. In bk. X, theor. 78, Witelo refers to those who maintain that the sum of the altitudes of the sun and rainbow is 42°, the precise value given by Bacon in pt. VI of the *Opus maius*. As I have argued in my ed. of Pecham's *Perspectiva communis* and elsewhere, it is possible that Witelo borrowed from Pecham, but more likely that Pecham borrowed from Witelo.
13. Bk. X, theor. 8.
14. Bk. II, theor. 47.
15. Bk. II, theor. 3.
16. *Ibid.*
17. Yet Witelo recognizes that not only the imaginary mathematical lines, but even radial lines, are fictions and that radiation is actually continuous; see, for example, bk. X, theor. 3. Nevertheless, analysis into rays and imaginary lines is a legitimate technique for the solution of most optical problems.
18. Bk. III, theor. 6.
19. Bk. V, theor. 5.
20. Bk. I, theor. 17–18; bk. V, theor. 18.
21. Witelo's table appears in bk. X, theor. 8, a trans. of which appears in Edward Grant, *A Source Book in Medieval Science* (Cambridge, Mass., 1974), 424–426.
22. Bk. II, theor. 47.
23. Witelo's influence is treated more fully in the intro. to the facs. repr. of the Risner ed., xxi–xxv.

BIBLIOGRAPHY

I. ORIGINAL WORKS. Witelo's most important extant work is the *Perspectiva*, first published under the title *Optica* (Nuremberg, 1535; 1551) and, with Ibn al-Haytham's *Optics*, in a volume edited by Friedrich Risner, entitled *Opticae thesaurus* (Basel, 1572; facs. repr., New York, 1972). There is no modern ed. except for a few sections edited by Clemens Baeumker in his *Witelo* and a critical ed. and English trans. of bk. I (with analysis and commentary) by Sabetai Unguru, "Witelo as a Mathematician: A Study in XIIIth Century Mathematics" (Ph.D. diss., Univ. of Wis., 1970). The extant MSS of the *Perspectiva* are listed in the intro. to the facs. repr. of Risner's *Opticae thesaurus*. A comparison of early MSS with the three printed eds. reveals that the latter are quite accurate.

Witelo's *De primaria causa penitentie et de natura demonum* (regarded as two separate works by Birkenmajer) is extant in British Museum, MS Sloane 2156 (15th century), fols. 148r–154v; and Paris, B.N. MS Lat. 14796 (15th century), fols. 86v–97v (abbreviated version). Birkenmajer has edited the latter MS in his "Études sur Witelo, I" (see below).

II. SECONDARY LITERATURE. The major studies of Witelo are Clemens Baeumker, *Witelo, ein Philosoph und Naturforscher des XIII. Jahrhunderts*, which is *Beiträge zur Geschichte der Philosophie des Mittelalters*,

III, pt. 2 (Münster, 1908); and Aleksander Birkenmajer, "Études sur Witelo," which appear in his *Études d'histoire des sciences en Pologne* (Wrocław, 1972), 97–434. Portions of Baeumker's analysis are vitiated by his erroneous attribution of *De intelligentiis* to Witelo, and it is absolutely essential that close attention be paid to Birkenmajer's cautions, reservations, and corrections. The first and fourth of Birkenmajer's "Études sur Witelo" were previously published as "Studja nad Witelonem, I," in *Archiwum komisji do badania historji filozofji w polsce*, **2**, pt. 1 (1921), 1–149; and "Witelo e lo studio di Padova," in *Omaggio dell'Accademia polacca di scienze e lettere all'Università di Padova nel settimo centenario della sua fondazione* (Cracow, 1922), 147–168.

On Witelo's life, in addition to Baeumker's *Witelo* and Birkenmajer's "Études sur Witelo," see Baeumker, "Zur Biographie des Philosophen und Naturforschers Witelo," in *Historisches Jahrbuch der Görres-Gesellschaft*, **33** (1912), 359–361; Maximilien Curtze, "Sur l'orthographe du nom et sur la patrie de Witelo (Vitellion)," in *Bullettino di bibliografia e di storia delle scienze matematiche e fisiche*, **4** (1871), 49–77; and David C. Lindberg, "Lines of Influence in Thirteenth-Century Optics: Bacon, Witelo, and Pecham," in *Speculum*, **46** (1971), 72–75, 77–83; and intro. to the facs. repr. of the Risner ed. (New York, 1972), vii–xiii.

Studies of particular aspects of Witelo's thought appear in Carl B. Boyer, *The Rainbow: From Myth to Mathematics* (New York, 1959), ch. IV; A. C. Crombie, *Robert Grosseteste and the Origins of Experimental Science 1100–1700* (Oxford, 1953), 213–232; David C. Lindberg, "The Theory of Pinhole Images From Antiquity to the Thirteenth Century," in *Archive for History of Exact Sciences*, **5**, pt. 2 (1968), 154–176; and "The Cause of Refraction in Medieval Optics," in *British Journal for the History of Science*, **4** (1968–1969), 23–38; and Sabetai Unguru, "Witelo and Thirteenth-Century Mathematics: An Assessment of His Contributions," in *Isis*, **63** (1972), 496–508. Witelo's *De natura demonum* has been summarized briefly by Lynn Thorndike in *A History of Magic and Experimental Science*, V (New York, 1941), 86–89.

DAVID C. LINDBERG

WITHAM, HENRY (*b.* Minsteracres, Northumberland, England, 1779; *d.* Lartington Hall, Yorkshire, England, 28 November 1844), *geology, paleobotany.*

Witham was the second son of John Silvertop and Catherine Lawson of Brough, Yorkshire. He married Eliza Witham of Headlam, a niece and coheiress of William Witham of Cliffe, Yorkshire, and took the name and arms of Witham. He became the first Roman Catholic high sheriff of County Durham; and his son Thomas, a Roman Catholic priest, inherited the Lartington estate after the deaths of his three brothers.

Witham's interest in geology and paleobotany was expressed in the founding in 1829 of the Natural History Society of Northumberland, Durham, and Newcastle upon Tyne, of which he was a founder-member and vice-president. He was also a member of the Wernerian Natural History Society and of the Royal Society of Edinburgh. Witham read the paper "On the Vegetation of the First Period of an Ancient World" before the Wernerian Society on 5 December 1829. In Edinburgh he became acquainted with William Nicol, whose method of making thin rock sections led to a revolution in both paleobotany and petrology. This technique was a development of that used by George Sanderson, an Edinburgh lapidary. Witham's Edinburgh connections also led to his "Description of a Fossil Tree Discovered in the Quarry of Craigleith" (1831). Three specimens were found, and the first was removed to the grounds of the Natural History Museum at South Kensington. Witham employed the Aberdeen botanist William MacGillivray to illustrate this paper and his *Observations on Fossil Vegetables* (1831).

In Newcastle, Witham met N. J. Winch and W. Hutton, both interested in botany and geology. Winch wrote on the geology of Northumberland (1816) and the Tweed banks (1831). Witham subscribed to J. Lindley and Hutton's *Fossil Flora* (1831–1837) and dedicated his *Observations on Fossil Vegetables* to Hutton. Their common interest was partly stimulated by the great development of mining at that time.

Witham was the first investigator of *Lepidodendron harcourtii, Pitus withami* (the Craigleith tree), *P. antiqua* and *P. primaeva* (the Lennel Braes trees), *Cordaites brandlingi* (the Wideopen tree), and *Anabathra pulcherrima* (a petrified xylem cylinder of *Stigmaria*).

His interpretation of the *Pitus* trees was that they were fossil gymnosperms and not vascular cryptogams. Hence his greatest achievement was to show that gymnosperms were prevalent in Lower Carboniferous rocks.

BIBLIOGRAPHY

I. ORIGINAL WORKS. Witham's writings include "Vegetation of the First Period of an Ancient World," in *Philosophical Magazine*, 2nd ser., **7** (1830), 28–29; "On the Vegetable Fossils Found at Lennel Braes, Near Coldstream, Upon the Banks of the River Tweed in Berwickshire," *ibid.*, **8** (1830), 16–21; "On the Red

462

Sandstones of Berwickshire, Particularly Those at the Mouth of the River Tweed," in *Transactions of the Natural History Society of Northumberland . . .*, **1** (1831), 172–183; "Description of a Fossil Tree Discovered in the Quarry of Craigleith," *ibid.*, 294–301; *Observations on Fossil Vegetables, Accompanied by Representations of Their Internal Structure as Seen Through the Microscope* (Edinburgh–London, 1831), repr. as *The Internal Structure of Fossil Vegetables Found in the Carboniferous and Oolitic Deposits of Great Britain, Described and Illustrated* (Edinburgh–London, 1833); "On the *Lepidodendron harcourtii*," in *Transactions of the Natural History Society of Northumberland . . .*, **2** (1838), 236–238; and "On the Effects Produced by a Greenstone Dyke Upon the Coal, in Passing Over Cockfield Fell, in the County of Durham," *ibid.*, 343–345.

II. SECONDARY LITERATURE. See J. Lindley and W. Hutton, *Fossil Flora of Great Britain*, 3 vols. (London, 1831–1837); A. G. Long, "The Fossil Plants of Berwickshire: Part 1," in *History of the Berwickshire Naturalists' Club*, **34** (1959), 248–273; A. G. MacGregor and R. J. A. Eckford, "The Upper Old Red and Lower Carboniferous Sediments of Teviotdale and Tweedside, and the Stones of the Abbeys of the Scottish Borderland," in *Transactions of the Edinburgh Geological Society*, **14** (1948), 230–252; F. W. Oliver, ed., D. H. Scott, "William Crawford Williamson (1816–95)," in *Makers of British Botany* (Cambridge, 1913), 243–260; D. H. Scott, *Extinct Plants and Problems of Evolution* (London, 1924), 150; and N. J. Winch, "Observations on the Geology of Northumberland and Durham," in *Transactions of the Geological Society*, **4** (1816), 1–101; and "Remarks on the Geology of the Banks of the Tweed," in *Transactions of the Natural History Society of Northumberland . . .*, **1** (1831), 117–131.

ALBERT G. LONG

WITHERING, WILLIAM (*b*. Wellington, Shropshire, England, March 1741; *d*. Birmingham, England, 6 October 1799), *medicine, botany, natural history*.

Withering was the only son of a prosperous Wellington apothecary. He entered the University of Edinburgh in 1762, graduating M.D. in 1766. In 1767 he settled into a relatively quiet country practice at Stafford. Upon the death of Dr. William Small in 1775, Withering removed to Birmingham and soon had one of the largest provincial practices of his day. He was active in Birmingham's vigorous Lunar Society, other members of which included Joseph Priestley, Erasmus Darwin, Josiah Wedgwood, Matthew Boulton, and James Watt. Withering was elected a fellow of the Royal Society in 1784. He was also a fellow of the Linnean

Society and a foreign corresponding member of the Royal Academy of Sciences of Lisbon. He visited Portugal twice in search of a salubrious climate which would slow the progressive deterioration of the chronic pulmonary condition (probably tuberculosis) from which he suffered the last fifteen years of his life. It ultimately caused his death at the age of fifty-eight.

Withering always remained primarily a practicing physician. Nevertheless, he had broad scientific interests and published significant work in botany, mineralogy, chemistry, and medicine. His botanical investigations began as a systematic collection of the flora indigenous to the Stafford area. He eventually extended his herbarium to include plants from all parts of Great Britain. His first major publication, *A Botanical Arrangement of all the Vegetables Naturally Growing in Great Britain* (1776), was little more than a translation of the portions of Linnaeus' writings relevant to English botany. As Withering acquired more botanical experience, however, his *Botanical Arrangement* became increasingly based on his first-hand observations. In the last edition published during his lifetime (1796), Withering effected a number of important taxonomic changes in the Linnaean system. He also surveyed the British cryptogams, a class of plants imperfectly described by Linnaeus.

As a botanist Withering rarely penetrated beyond the descriptive level; yet his *Botanical Arrangement* was the product of many years' patient, methodical study. It remained a standard British flora long after his death. Between 1805 and 1830 Withering's son added four editions to the three that Withering had published.

Although Withering's botanical interests were confined principally to indigenous British plants, his reputation on the Continent was such that the French botanist L'Héritier de Brutelle named a genus of plants (of the Solanaceae family) *Witheringia*. In 1796 the German mineralogist Abraham Gottlob Werner also commemorated Withering's name when he called barium carbonate "witherite." Withering had in 1782 first demonstrated that naturally occurring barium carbonate is a compound distinct from other barium salts, such as the sulfite and the oxide. Withering also published several other chemical and mineralogical papers, chiefly in the *Philosophical Transactions*, including "Experiments on Different Kinds of Marl" (1773) and "An Analysis of Two Mineral Substances, Viz. the Rowley Rag-Stone and the Toad-Stone" (1784). His chemical investigations were obviously nourished by his close friendship with

Priestley, who communicated some of Withering's early papers to the Royal Society. Withering was opposed to the phlogiston theory, which he satirized in a verse essay "The Life and Death of Phlogiston," read before the Lunar Society in 1796. He never published his experiments on phlogiston, however, and though he mentioned in a letter having "given up my pursuits upon *Phlogiston* to Dr. Priestley," the latter obviously found Withering's strictures on the theory unconvincing.

In 1783 Withering translated Torbern Bergman's *Sciagraphia regni mineralis* as the *Outlines of Mineralogy*, to which he added notes. He also chemically analyzed the waters at various spas in England and Portugal.

Withering maintained a lifelong interest in climate. He kept an extensive meteorological journal, from which his son printed extracts in 1822. Withering was especially interested—personally and professionally—in the effects of different climates on patients suffering from consumption. He rather grimly concluded (after twice wintering in Portugal) that the dry Portuguese weather is ineffectual in such cases.

In addition to his scientific publications, Withering left two significant medical treatises. In his *Account of the Scarlet Fever and Sore Throat* he moved from a brief description of a 1778 Birmingham epidemic of scarlet fever to a more general consideration of the causes, diagnosis, and treatment of the disease. He insisted on its contagiousness and noted the occasional development of generalized edema shortly after the disappearance of the fever.

In 1785 Withering published his *Account of the Foxglove, and Some of Its Medical Uses*, which is a genuine classic of clinical medicine. In it he summarized a decade's careful study of digitalis, the cardiotonic glycoside obtained from the leaves of the foxglove (*Digitalis purpurea*). Withering honestly recorded both successes and failures in his trials with the drug, and the gradual development of his skill in using digitalis may be followed in the chronological series of 163 cases reported in his book. Withering learned to employ digitalis only in selected cases of edema (dropsy). He stressed that care must be taken in adjusting the dose, and he accurately described the signs and symptoms of digitalis toxicity and established clear guidelines for its rational use. Despite Withering's modest but definite claims for the efficacy of the foxglove, the drug became for nineteenth-century clinicians a kind of panacea. It was prescribed (in dangerously large doses) for a variety of conditions. Only in the past few decades has the real merit of Withering's work on the foxglove been recognized. The place of digitalis in the contemporary pharmacopoeia remarkably vindicates Withering's prediction that "TIME will fix the real value upon this discovery."

His published medical writings amply demonstrate that Withering's reputation as a practitioner was justified. The breadth of his extraprofessional interests made him a proper member of the group of savants who constituted the Birmingham Lunar Society.

BIBLIOGRAPHY

I. ORIGINAL WORKS. Withering's principal publications include the following: *A Botanical Arrangement*, 2 vols. (London, 1776); the 3rd ed., 4 vols. (Birmingham, 1796), is the most important. *An Account of the Scarlet Fever and Sore Throat* (London, 1779; 2nd ed. 1793; German trans., 1781). *An Account of the Foxglove, and Some of Its Medical Uses* (Birmingham, 1785). A limited facs. ed. was brought out in 1948. *A Chemical Analysis of the Water at Caldas da Rainha* (Lisbon, 1795).

There are in addition minor medical tracts and letters, Withering's translation of Bergman's *Sciagraphia regni mineralis*, and papers in the *Philosophical Transactions*, the *Transactions of the Linnean Society*, and *Annals of Medicine*. A convenient bibliography is given in *The Miscellaneous Tracts of the Late William Withering, M.D. F.R.S.*, ed. by his son, 2 vols. (London, 1822), 207–209. This work reprints virtually all of Withering's writings, with the exception of the *Botanical Arrangement* and Withering's Edinburgh thesis *De angina gangraenosa* (Edinburgh, 1766). The latter has recently been translated by Charles D. O'Malley, *Journal of the History of Medicine*, **8** (1953), 16–45.

II. SECONDARY LITERATURE. The standard account of Withering's life remains the *Memoir* by his son, included in the *Miscellaneous Tracts* (see above). There is a full-scale modern biography by T. Whitmore Peck and K. Douglas Wilkinson, *William Withering of Birmingham* (Bristol–London, 1950). It includes a number of previously unpublished letters and a short bibliography. A recent paper has summarized Withering's work on digitalis, J. W. Estes and P. D. White, "William Withering and the Purple Foxglove," in *Scientific American*, **212** (1965), 110–119. See also John F. Fulton, "The Place of William Withering in Scientific Medicine," in *Journal of the History of Medicine*, **8** (1953), 1–15. For the later history of digitalis therapy, see E. H. Ackerknecht, "Aspects of the History of Therapeutics," in *Bulletin of the History of Medicine*, **36** (1962), 389–419. Some of Withering's work on mineralogy is discussed by Frederick D. Zeeman, "William Withering as a Mineralogist, the Story of Witherite," *ibid.*, **24** (1950), 530–538.

Withering's relations with various members of the

Lunar Society may be best approached through Robert Schofield's study, *The Lunar Society of Birmingham* (Oxford, 1963), with a splendid bibliography of published and manuscript materials.

WILLIAM F. BYNUM

WITT, JAN DE (*b.* Dordrecht, Netherlands, 24 September 1625; *d.* The Hague, Netherlands, 20 August 1672), *mathematics.*

De Witt was the son of Jacob de Witt, burgomeister of Dordrecht, and Anna van de Corput. Both families were prominent members of the regent class which governed the towns and provinces of the Netherlands. He entered Dordrecht Latin school in 1636, and went to the University of Leiden in 1641. There he studied law, leaving for France in 1645 to take his degree at Angers. At Leiden he studied mathematics privately with Frans van Schooten the Younger, and received from him an excellent training in Cartesian mathematics. De Witt was a talented mathematician who had little time to devote to mathematics. He became pensionary of Dordrecht in 1650, and grand pensionary of Holland in 1653, making him the leader of the States Party, and, in effect, the prime minister of the Netherlands. He was a statesman of unusual ability and strength of character who guided the affairs of the United Provinces during the twenty-year interregnum in the Stadtholdership during the minority of William of Orange. This was one of the most critical periods in Dutch history, with the three Anglo-Dutch wars; the hostility of the Orange faction culminated in the murder of de Witt and his brother Cornelis by a mob in 1672.

De Witt's most important mathematical work was his *Elementa curvarum linearum*, written before 1650 and printed in Van Schooten's second Latin edition of Descartes's *Géométrie* (1659–1661). It is in two books: the first, a synthetic treatment of the geometric theory found in the early books of Apollonius' *Conics*; and the second, one of the first systematic developments of the analytic geometry of the straight line and conic. In the first book the *symptomae* (expressed as proportions) of the parabola, ellipse, and hyperbola are derived as plane loci, rather than as sections of the cone. His locus definitions of the ellipse are familiar to us today: the eccentric angle construction (a point fixed with respect to a rotating segment); the trammel construction (a fixed point on a given segment moving on two intersecting lines); and the "string" construction, based on the two-focus definition. For the hyperbola and parabola the locus is con-

structed as the intersection of corresponding members of two pencils of lines, one parallel and one concurrent. In modern terms these are interesting unintentional examples of the Steiner-Chasles projective definition of the conics, where the vertex of one pencil is at infinity.

De Witt is credited with introducing the term "directrix" for the parabola, but it is clear from his derivation that he does not use the term for the fixed line of our focus-directrix definition. Given fixed lines *DB* and *EF* intersecting at *D*, with *B* the pole and *EF* the directrix: for any point *H* on *EF*, if ∠*HBL* is constructed equal to ∠*FDB*, a line through *H* parallel to *BD* cuts *BL* in *G*, a point on the locus. *AC* is drawn through *B* with ∠*DBC* = ∠*BDF*, cutting *HG* in *I*, and *GK* is drawn parallel to *AC*. Since triangles *BDH* and *GKB* are similar, $(BI)^2 = (BD)(BK)$ or $y^2 = px$, a parabola with vertex at *B*, abscissa *BK* = *x*, and ordinate *KG* = *y*. If *EF* is perpendicular to *DB*, a rectangular coordinate system results, but *EF* is not our directrix.

In the first book of the *Elementa* de Witt not only freed the conics from the cone with his kinematic constructions, but satisfied the Cartesian criteria of constructibility. This book was written, as he reported to van Schooten, to give a background for the new analytic development of the second book. He began the analytic treatment by showing that equations of the first degree represent straight lines. As was usual at the time he did not use negative coordinates, graphing only segments or rays in the first quadrant. He carefully explained the actual construction of the lines for arbitrary coeffi-

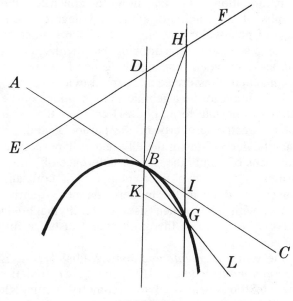

FIGURE 1

cients since they would be needed in his transformations reducing general quadratic equations to type conics. For each conic de Witt began with simplified equations equivalent to his standard forms in book I, and then used translations and rotations to reduce more complicated equations to the canonical forms. For example, in the hyperbola

$$yy + \frac{2bxy}{a} + 2cy = \frac{fxx}{a} + ex + dd$$

he lets

$$z = y + \frac{b}{a}x + c$$

and then

$$v = x + h$$

where h is the coefficient of the linear term in x after the first substitution, giving

$$\frac{aazz}{fa + bb} = vv - hh + \frac{aadd + aacc}{fa + bb},$$

a standard hyperbola which cuts the new v or z axes according as hh is greater than or less than $\frac{aadd + aacc}{fa + bb}$. Although de Witt seems to be aware of the characteristic of the general quadratic equation in choosing his examples, he does not explicitly mention its use to determine the type of conic except in the case of the parabola. There he states that, if the terms of the second degree are a perfect square, the equation represents a parabola.

The last chapter is a summing up of the various transformations showing how to construct the graphs of all equations of second degree. Each case of positive and negative coefficients must be handled separately in a drawing, but the discussion for each curve is completely general, and both original and transformed axes are drawn.

In addition to the algebraic simplifications of the curves to normal form, book II contains the usual focus-directrix property of the parabola, and the analytic derivations of the ellipse and hyperbola as the locus of points the sum or difference of whose distances from two fixed points is a constant. These are done in the modern manner, squaring twice, with the explicit use of the Pythagorean theorem in place of the more recent distance formula.

De Witt's *Elementa* and John Wallis' *Tractatus de sectionibus conicis* (1655) are considered the first textbooks in analytic geometry. Although Wallis raised the question of priority, their approaches were different and completely independent. Wallis first defined the conics as second-degree equations and deduced the properties of the curves from the equations, while de Witt defined them geometrically in the plane, and then showed that quadratic equations could be reduced to his normal forms.

Christiaan Huygens once wrote John Wallis of de Witt: "Could he have spared all his strength for mathematical works, he would have surpassed us all." His geometry was his only contribution to pure mathematics, but he turned his mathematical interests to the financial problems of the province of Holland throughout his long tenure as grand pensionary. The chief means of raising money for the States was by life or fixed annuities. In 1665 de Witt succeeded in reducing the interest rate from 5 to 4 percent and established a sinking fund with the interest saved by the conversion accumulated at compound interest to be applied to the debt of Holland, which could thus be paid in forty-one years. The second Anglo-Dutch War (1665–1667), however, defeated this scheme. The English wars were a perpetual financial drain, and more than half of the expenditure of the province of Holland (which had to defray the costs of the war almost alone) was swallowed up in interest payments.

In April 1671 it was resolved to negotiate funds by life annuities, thereby limiting the debt to one generation. De Witt prepared a treatise for the States of Holland demonstrating mathematically that life annuities were being offered at too high a rate of interest in comparison with fixed annuities. For many years the rule-of-thumb rates for life annuities had been twice the standard rate of interest. Holland had recently reduced the rate of interest to twenty-five years' purchase (4 percent) and was selling life annuities at fourteen years' purchase ($7\frac{1}{7}$ percent). De Witt wanted to raise the price to sixteen years' purchase ($6\frac{1}{4}$ percent). His *Waerdye van Lyf-renten naer proportie van Los-renten* (July, 1671) is certainly among the first attempts to apply the theory of probability to economic problems. It was written as a political paper, and remained buried in the archives for almost two hundred years. Since its discovery and publication by Frederick Hendriks in 1852 there have been many articles (some of which are listed in the bibliography) explaining or criticizing it on the basis of modern actuarial science. It is actually a very simple and ingenious dissertation based only on the use of the principle of mathematical expectation to form equal contracts.

De Witt listed the present values at 4 percent of annuity payments of 10,000,000 stuyvers (to avoid decimals) per half year, and summed the mathematical expectations using hypothetical mortality rates for different ages. He first presupposed that a man is equally likely to die in the first or last half of any year, and then, since annuities were generally purchased on young lives, extended this to any half year of the "years of full vigor" from age three to fifty-three. For simplicity he considered the first hundred half years equally destructive or mortal, although he stated that the likelihood of decease is actually smaller in the first years. So too, he stopped at age eighty, although many live beyond that age. In the next ten years, fifty-three to sixty-three, the chance of dying does not exceed more than in the proportion of 3 to 2 the chance of dying in the first period; from sixty-three to seventy-three, the chance of dying is not more than 2 to 1; and from seventy-three to eighty, not more than 3 to 1.

De Witt gives many examples to explain the use of the concept of mathematical expectation. The following one is basic to his later calculations, and has been overlooked by many commentators. Consider a man of forty and a man of fifty-eight. According to his presuppositions the chances of the older man dying compared with the younger man are as 3 to 2. An equal contract could be devised: if the person of fifty-eight dies in six months, the younger man inherits 2,000 florins, but if the man of forty dies in six months, the elder inherits 3,000 florins. That is, the chance of the man of fifty-eight gaining 3,000 florins. is as 2 to 3, or, in terms of de Witt's annuity calculations, the chance of receiving a particular annuity payment in the second period is two-thirds that in the first period.

From this reasoning de Witt's calculations are straightforward: He sums the present values for the first hundred half years; two-thirds the present values for the next twenty half years; for the next twenty, one-half the present values; and one-third for the last fourteen. All these are summed and the average taken, giving a little more than sixteen florins as the present value of one florin of annuity on a young and healthy life. If the method had been applied to actual mortality tables, the labor would have been formidable. Later in 1671 de Witt and Jan Hudde corresponded on the problem of survivorship annuities on more than one life, and here both used actual mortality figures taken from the annuity records of Holland. Working with several groups of at least a hundred persons of a given age de Witt developed appropriate rates for annuities on two lives. These were extended a posteriori to any number of lives by a Pascal triangle, with a promise to Hudde to establish the results a priori. This was the culmination of de Witt's work with annuities, but for political reasons he suggested to Hudde that the public not be informed of the results of their study, since they were willing to buy annuities on more than one life at the current rate, which was favorable to the government.

BIBLIOGRAPHY

I. ORIGINAL WORKS. *Elementa curvarum linearum*, in Frans van Schooten's Latin ed. of Descartes's *Géométrie, Geometria a Renato Descartes* (Amsterdam, 1659–1661). *Waerdye van Lyf-renten naer proportie van Los-renten* (The Hague, 1671; facs. ed. Haarlem, 1879). Six volumes of letters in *Werken van het Historisch Genootschap te Utrecht*, 3d ser., XVIII, XXV, XXXI, XXXIII, XLII, XLIV (1906–1922). Volume XXXIII contains letters to and from mathematicians including the letters to Jan Hudde on annuities on more than one life.

II. SECONDARY LITERATURE. Of the many biographies of de Witt, Nicolaas Japikse, *Johan de Witt* (Amsterdam, 1915), is indispensable. Still valuable is G. A. Lefèvre-Pontalis, *Jean de Witt, Grand Pensionnaire de Hollande*, 2 vols. (Paris, 1884); English trans., S. F. Stephenson and A. Stephenson (London, 1885). For a reliable discussion of the period, and the relations between de Witt and William III, see Pieter Geyl, *The Netherlands in the Seventeenth Century, Part Two 1648–1715* (London, 1964), and his *Oranje en Stuart* (Utrecht, 1939), English trans., Arnold Pomerans (London, 1969). For the geometry see P. van Geer, "Johan de Witt als Wiskundige," in *Nieuw Archief voor Wiskundige*, 2nd ser., 11 (1915), 98–126; and C. B. Boyer, *History of Analytic Geometry* (New York, 1956).

An English translation of the work on life annuities can be found in Frederick Hendriks, "Contributions to the History of Insurance . . . a Restoration of the Grand Pensionary De Witt's Treatise on Life Annuities," in *The Assurance Magazine* (now *Journal of the Institute of Actuaries*), 2 (1852), 230–258. Vols. 3 (1901), 10 (1908), and 11 (1909) of the *Archief voor Verzekeringe Wetenschap* contain articles offering varying criticisms and explanations of de Witt's writings on annuities.

JOY B. EASTON

WITTGENSTEIN, LUDWIG (JOSEF JOHANN) (*b.* Vienna, Austria, 26 April 1889; *d.* Cambridge, England, 29 April 1951), *philosophy.*

Wittgenstein was one of the most imaginative and original thinkers of the twentieth century, a legend during his lifetime and an enduring influence since. To his numerous admirers and followers, his work marks a decisive turn in the history of philosophy and in all fields of investigation to which philosophical method is pertinent.

Ludwig Wittgenstein, as he always called himself, was the youngest of eight children. His father, an engineer and a successful steel magnate, was a prominent patron of the arts in Vienna. Wittgenstein was never at home in this worldly and sophisticated setting; and his life and work alike show the imprint of a deeply serious temperament, radically at odds with the compromises of bourgeois society.

Educated privately until he was fourteen, Wittgenstein spent only three years at school (in Linz) before entering the Technical Institute at Berlin-Charlottenburg with a view to becoming an engineer. As a research student at Manchester University (1908–1911) he made original contributions to the design of a jet-reaction propeller for airplanes. His interests having turned to the foundations of mathematics and to logic, in 1911, on the advice of Gottlob Frege, he became a student of Bertrand Russell's at Cambridge University. In 1913–1914, while living in solitude in Norway, he was already composing the *Tractatus*, although it was not published until 1921. During the period 1919–1926 Wittgenstein studied for and obtained a diploma qualifying him for elementary school teaching, and eventually taught in a number of small village schools in Austria. For a while he worked as a gardener's assistant at a convent near Vienna. He also designed and built, for one of his sisters, a remarkable house that is still standing (at the time this article was written) in the Kundmanngasse (and declared a national monument by the Austrian government). In 1929, Wittgenstein returned to Cambridge, was made a fellow of Trinity College, and began the famous succession of informal classes through which his philosophical views gradually became known. Ten years later he was appointed professor of philosophy in succession to G. E. Moore. He worked in a medical school and a medical laboratory during World War II and resigned his professorship in 1947. He died of cancer four years later.

The last two decades of Wittgenstein's life were filled with unremitting intellectual work. His many manuscripts include, in addition to his masterpiece, the *Philosophical Investigations* (which he left almost ready for press), several full-length books, and thousands of pages of additional materials. His last finished piece of work, *On Certainty* (composed in 1950–1951), shows him in full possession of penetrating powers of insight and expression.

Wittgenstein's later work stands in sharp contrast with and opposition to the conceptions presented in the *Tractatus*. That book, written in short, epigrammatic paragraphs carefully arranged in quasi-logical form (with a special system of decimal references marking the relative subordination of successive items), remains cryptic on essential points and lends itself to a variety of different interpretations. A central theme is the delineation of the essential characteristics that any language or symbol system must manifest. It would therefore not be unfair to call it a "Critique of Pure Language." Wittgenstein's celebrated "picture theory of language" insists upon the presence in language, as the root of its semantic power, of an isomorphism between sentences and the possible states of affairs to which they ultimately refer. Reality must be composed of "facts"—patterned clusters of ultimate simples or "objects"—each standing in one–one correspondence to the simple names that underlie the superficial complexity of ordinary language. Thus the "logical form" of reality (roughly speaking, the patterns of possible co-occurrence of the simple "objects") must be reflected in the "logic of language" (the corresponding patterns of co-occurrence of the semantic elements).

It was part of the originality of this version of "logical atomism" to reject any possibility of the representation, from some external standpoint, of the "logical form" itself. The "logic" of reality and its linguistic mirror must "*show* itself," through the impossibility of "saying" what *cannot* be said: the limits of language are the "limits of thought." What philosophers have tried to say about metaphysics, transcendental ethics and aesthetics, and theology turns out to consist of pseudo propositions that are "nonsense." The book accordingly ends with the much quoted line, "Whereof one cannot speak, one must be silent." (This article must necessarily omit reference to Wittgenstein's important technical contributions to the foundations of logic, focusing on the notion of "tautology"; to probability theory; and to the philosophy of science.)

Some hostile critics, such as Karl Popper, have regarded the conclusion of the *Tractatus* as a self-refutation, which reduces the book itself to the sort of "nonsense" that cannot be "said"; other readers, notably early members of the Vienna Circle, have sought to purge the *Tractatus* of its allegedly irrelevant "mystical" intrusions and to quarry from it a positivistic critique of metaphysics. But a more sympathetic reading would treat it as a peculiar sort of demonstration ("showing") of how a powerful conception of the necessary relations between symbolism and reality, pushed to its logical consequences, results in an impasse, from which there is no escape except through a revolution in perspective and approach. From this standpoint the *Tractatus* is a prime example of what Wittgenstein later came to call a "metaphysical cramp," an obsession with a single conception of what the metaphysical situation *must be*—and the natural springboard for his subsequent revolution in method.

Although there is considerable continuity between the *Tractatus* and the later masterpiece, the *Philosophical Investigations* (completed some twenty-five years later), the second work reads at first sight like a wholesale rejection of the earlier methodology. In the *Investigations*, the earlier interest in the one and only "logical form," manifested in every adequate linguistic or symbolic system, is rejected as arising from a distorted metaphysical conception. Attention shifts to language as it is used in concrete social practices, constituted partly by rules of syntax and application, but even more importantly by a background "agreement in the form of life" that shows itself in practice but is not reducible to formal principles. The a priori considerations that dominated the *Tractatus* are replaced by meticulous attention to the "natural history" of language, the complex and various ways in which men actually communicate and express their thoughts. The prime philosophical error is to impose upon this motley of speech practices some a priori model of what language *must* be like. Wittgenstein shows, by detailed discussion of questions that have been the staple of philosophical dispute for two thousand years, how such oversimplified impositions generate *insolubilia*. He hoped to have shown how such "philosophical sickness" can yield to rational treatment.

Wittgenstein's later work introduced a number of special notions that continue to be of high value, despite their often cryptic and controversial character. Among them are the notions of a "language game" (a deliberately simplified model of speech practice, introduced for the sake of comparison), of a "criterion" of use, and of "family resemblances" (the overlapping pattern of relations that hold together the items referred to by some general term).

Wittgenstein's later methods of investigation are "dialectical," in the sense of proceeding repeatedly from the real or fancied philosophical difficulties of an imaginary interlocutor. His writings provide tantalizing glimpses of his incomparable style of face-to-face philosophizing with friends and pupils.

Despite a lifelong interest in science and its relations to philosophy, Wittgenstein did comparatively little work on the philosophy of science (although the *Tractatus* contains some important contributions). On the philosophy of mathematics, on the other hand, he left voluminous manuscripts, still in process of publication and critical evaluation.

It is misleading to assign to Wittgenstein, as is too often done, the stock labels "behaviorist" or "positivist." His life was devoted, with exemplary single-mindedness, to discovering a radically new way of leading men out of the darkness of conceptual confusion.

BIBLIOGRAPHY

I. ORIGINAL WORKS. All of Wittgenstein's works, except the first and third, were published posthumously. Since he composed in German, translations are, at his desire, published with the original German text facing. Exceptions to this are indicated below.

1. *Tractatus Logico-Philosophicus*, translated by D. F. Pears and B. F. McGuinness, with intro. by Bertrand Russell (London, 1961). The trans. by C. K. Ogden in the original English ed. (London, 1922), although faulty in places, still deserves attention. There also have been translations into Italian, Russian, French, Finnish, Swedish, Danish, and Chinese.

2. *The Blue and Brown Books*, with a preface by Rush Rhees (Oxford, 1958). Originally dictated in English (1933–1935) for the use of Wittgenstein's pupils. Although superseded by the *Investigations*, still the best introduction to the later work.

3. *Notebooks 1914–1916*, translated by G. E. M. Anscombe, edited by Anscombe and G. H. von Wright (Oxford, 1961). Surviving parts of the notebooks used in preparing the *Tractatus*. An indispensable aid to the study of that work.

4. *Philosophical Investigations*, translated by G. E. M. Anscombe, edited by Anscombe and Rush Rhees (London, 1953). The great masterpiece of Wittgenstein's later thought.

5. *Remarks on the Foundations of Mathematics*, translated by G. E. M. Anscombe (Oxford, 1956). Compiled from MSS by the literary executors, G. H. von Wright, Rush Rhees, and G. E. M. Anscombe.

6. *Philosophische Bemerkungen* (Oxford, 1964). German text only. Composed 1929–1930.

7. *Zettel*, translated by G. E. M. Anscombe (Oxford, 1966). Based on notes arranged as for a book.

8. *Philosophische Grammatik* (Oxford, 1969), translated by Anthony Kenny, edited by Rush Rhees, as *Philosophical Grammar* (Berkeley, Calif., 1974).

9. *On Certainty*, translated by G. E. M. Anscombe (Oxford, 1969). Composed in 1950–1951.

10. *Ludwig Wittgenstein und der Wiener Kreis*, B. F.

McGuinness, ed. (Oxford, 1967). Conversations with Moritz Schlick, based upon verbatim shorthand reports by Friedrich Waismann.

11. *Lectures and Conversations on Aesthetics, Psychology and Religious Belief*, Cyril Barrett, ed. (Oxford, 1966).

Other sets of lecture notes, some of them transcribed verbatim, are in private circulation.

Almost all of Wittgenstein's voluminous MSS are preserved in the library of Trinity College, Cambridge. The entire *Nachlass* has been microfilmed by Cornell University Library, Ithaca, New York, from which microfilm copies and Xeroxes can be purchased. A detailed guide to the Cornell collection is in G. H. von Wright, "The Wittgenstein Papers," in *Philosophical Review*, **78** (1969), 483–503.

A very full bibliography of primary and secondary writings is in K. T. Fann, *Wittgenstein's Conception of Philosophy* (Oxford–Berkeley, 1969), with a supp. by Fann in *Revue internationale de philosophie*, **23** (1969), 363–370.

II. SECONDARY LITERATURE. For Wittgenstein's life and teaching, see especially Norman Malcolm, *Ludwig Wittgenstein: A Memoir*, rev. ed. (London, 1966), which also contains a biographical sketch by G. H. von Wright and a photograph. An authorized biography by B. F. McGuinness is in course of preparation.

Among the many commentaries on the *Tractatus* are G. E. M. Anscombe, *An Introduction to Wittgenstein's Tractatus* (London, 1959), the earliest and in some ways the most useful; Max Black, *A Companion to Wittgenstein's Tractatus* (Cambridge–Ithaca, N.Y., 1964), an elaborate exegesis; J. Griffin, *Wittgenstein's Logical Atomism* (London, 1964), which stresses the influence of Heinrich Hertz; and E. Stenius, *Wittgenstein's Tractatus* (Oxford, 1960), a penetrating but controversial analysis.

For the later work; see especially Norman Malcolm, "Wittgenstein's *Philosophical Investigations*," in *Philosophical Review*, **63** (1954), 530–559; and Peter Winch, ed., *Studies in the Philosophy of Wittgenstein* (London, 1969). A useful comprehensive anthology is K. T. Fann, ed., *Ludwig Wittgenstein: The Man and His Philosophy* (New York, 1967).

Possible applications to science are well illustrated in W. H. Watson, *Understanding Physics Today* (Cambridge, 1963).

MAX BLACK

WITTICH (or **WITTICHIUS**), **PAUL** (*b.* Breslau, Silesia [now Wrocław, Poland], 1555 [?]; *d.* Breslau, 9 January 1587), *mathematics.*

Little is known about Wittich's life. In the summer of 1580, with a letter of introduction from Hagecius, he went for a short time to Uraniborg to work with Tycho Brahe.[1] He soon showed himself to be a skillful mathematician, for with Tycho he discovered—or, more precisely, rediscovered—the method of prostaphaeresis, by which the products and quotients of trigonometric functions appearing in trigonometric formulas can be replaced by simpler sums and differences.[2] The two formulas involved in this method are $\sin a \cdot \sin b = \frac{1}{2}(\cos [a - b] - \cos [a + b])$ and $\cos a \cdot \cos b = \frac{1}{2}(\cos [a - b] + \cos [a + b])$.

The individual contributions of Tycho Brahe and Wittich cannot be established with certainty, but that of Wittich, who was the better mathematician, was probably the greater.[3] A letter is extant in which Tycho reported on this period of collaboration, during which each freely and fully shared his results with the other.[4] It is therefore understandable that he was very angry with Wittich when he learned that the method of prostaphaeresis had become known in Kassel, which Wittich had visited in 1584, and that Nicolai Reymers Bär (Ursus) had published it as his own discovery in 1588.[5]

Actually, Ursus also had been at Uraniborg in 1584 and had secretly noted down the method, although he did not discover the proof, which Brahe kept more carefully concealed. Wittich, who taught mathematics in Breslau from 1582 to 1584, was in Kassel in 1584.[6] There he described to Joost Bürgi, the clockmaker for Landgrave Wilhelm IV, the instruments used by Tycho in his observatory,[7] which Bürgi reproduced and improved. Wittich also showed him the proof of prostaphaeresis; and it was from Bürgi that Ursus learned the *mysteria triangulorum*. Wittich left Kassel before 1586 and died in Breslau at the beginning of 1587. On learning of Wittich's death, Tycho regretted that he had doubted his honesty.[8]

The method of prostaphaeresis originated with Johann Werner, who developed it in conjunction with the law of cosines for sides of a spherical triangle. In Regiomontanus' formulation the law reads: sinvers A: sinvers a — sinvers $[b - c] = r^2$: $(\sin b \cdot \sin c)$.[9] If one eliminates the *sinus versus* and takes $r = sinus\ totus = 1$, the result is

$$\frac{1}{\cos A} = \frac{\sin b \cdot \sin c}{\cos a - \cos b \cdot \cos c}.$$

Here Werner, who preserved the *sinus versus*, used the first formula of prostaphaeresis in handling the term $\sin b \cdot \sin c$; whereas Tycho and Wittich also knew the cosine law with the term $\cos b \cdot \cos c$.[10] They also used prostaphaeresis for problems that Werner solved without this method.[11]

It is unlikely that Tycho and Wittich ever saw Werner's *De triangulis sphaericis libri quatuor*. Its manuscript, which Rheticus wanted to publish at Cracow in 1557, was not printed until 1907.[12] On the other hand, Tycho knew that such a work existed and sought, unsuccessfully, to obtain a copy of it.[13] He and Wittich might, therefore, have been encouraged by this knowledge to work out the details of such a method, which Ursus stole and published in 1588.[14]

In 1580 Tycho and Wittich probably had not seen Viète's *Canon mathematicus* of 1579. Further evidence on this point is provided by Longomontanus, who was Tycho's assistant at Uraniborg from 1589 to 1597.[15] The method of converting products to sums or differences was further developed by Bürgi, Clavius, and Jöstel, among others. Specifically, Bürgi took as his starting point the relationship between arithmetic and geometric series and introduced logarithms.[16] He thereby definitively replaced the older method with an improved one that Pitiscus called *modus Byrgii*.[17] Kepler, who was thoroughly acquainted with Tycho's computations, mentions the *negotium prostaphaereticum Wittichianum* in his book on optics (1604).[18]

NOTES

1. Wittich, who left Uraniborg because of a matter concerning an inheritance, carried with him a letter from Brahe to Hagecius dated 4 Nov. 1580. Since Brahe received no answer, Wittich was suspected of not having delivered it. Hagecius later cleared up the matter (23 Sept. 1582). On this point see Brahe, *Opera*, VII, 72.
2. In the letter of 4 Nov. 1580 Brahe speaks of their efforts to develop this method, "quae per προσταφαίρεσιν procedit absque taediosa multiplicatione et divisione." *Ibid.*, 58.
3. See R. Wolf, *Handbuch der Astronomie*, 227 f.
4. See Brahe's letter to Hagecius of 14 Mar. 1592. Brahe, *Opera*, VII, 323.
5. See Brahe's letter to Hagecius of 1 July 1586, in which he writes that Wittich "agit sane minus sincere mecum." *Ibid.*, 108.
6. See J. L. E. Dreyer, *Tycho Brahe*, 121, n. 4.
7. Among these were the mural quadrant with transverse calibration. On this see C. D. Hellman, "Brahe," 405.
8. See his letter of 14 Jan. 1595: "nec vivum nec defunctum suis privavi honoribus." *Opera*, VI, 327.
9. See A. von Braunmühl, *Vorlesungen über Geschichte der Trigonometrie*, 131; and J. Tropfke, *Geschichte der Elementar-Mathematik*, 139 ff.
10. In the process *cos* is replaced by the *sin* of the complement. See A. A. Björnbo, "Ioannis Verneri . . . ," 169; and M. Cantor, *Vorlesungen über Geschichte der Mathematik*, 642f. Ibn Yūnus also knew prostaphaeresis for the operation involving cos *a* · cos *b*. See G. Sarton, *Introduction to the History of Science*, 716 ff. Traces of the law can be found in a special case in Indian mathematics (see Braunmühl, 41), and in a more complete form in the work of al-Battānī (see Sarton, *op. cit.*, 603; and Braunmühl, *op. cit.*, 53). Peuerbach also knew this law and derived it independently in *Compositio tabule altitudinis solis ad omnes horas* (in Codex Vindobonensis 5203, fols. 54r–55r).

11. See Björnbo, *op. cit.*, 169.
12. It was published by Björnbo with the preface by Rheticus that was printed in 1557.
13. See Tycho's letter to Hagecius of 25 Aug. 1585. *Opera*, VII, 95.
14. See Björnbo, *op. cit.*, 171.
15. See Dreyer, *op. cit.*, 361, n. 3, and 383.
16. Even before Stifel's *Arithmetica integra* (1544) others, including Chuquet and Heinrich Grammateus, compared the two types of series. On this point see L. Nový, "Bürgi," 602.
17. See Bartholomeo Pitiscus, *Trigonometriae sive de dimensione triangulorum libri quinque*, 3rd ed. (Frankfurt, 1612), 177.
18. Kepler, *Gesammelte Werke*, II (Munich, 1939), 336.

BIBLIOGRAPHY

See A. A. Björnbo, "Ioannis Verneri De triangulis sphaericis libri quatuor, de meteoroscopiis libri sex cum prooemio Georgii Ioachimi Rhetici. I. De triangulis sphaericis," in *Abhandlungen zur Geschichte der mathematischen Wissenschaften*, **24** (1907), 150–175; A. von Braunmühl, *Vorlesungen über Geschichte der Trigonometrie*, I (Leipzig, 1900), 256; 260; and "Zur Geschichte der prosthaphaeretischen Methode in der Trigonometrie," in *Abhandlungen zur Geschichte der Mathematik*, **9** (1899), 15–29; M. Cantor, *Vorlesungen über Geschichte der Mathematik*, 2nd ed., II (Leipzig, 1913), 937; J. L. E. Dreyer, *Tycho Brahe, a Picture of Scientific Life and Work in the Sixteenth Century* (Edinburgh, 1890; New York, 1963), 405; and *Tychonis Brahe Opera omnia*, XV (Copenhagen, 1913; 1929), 50; C. D. Hellman, "Brahe," in *Dictionary of Scientific Biography*, II, 401–416; L. Nový, "Bürgi," *ibid.*, 602–603; G. Sarton, *Introduction to the History of Science*, I (Baltimore, 1927); J. Tropfke, *Geschichte der Elementar-Mathematik*, 2nd ed., V (Berlin–Leipzig, 1923), 108 ff; and R. Wolf, *Handbuch der Astronomie, ihrer Geschichte und Litteratur*, I (Zurich, 1890).

KURT VOGEL

WOEPCKE, FRANZ (*b.* Dessau, Germany, 6 May 1826; *d.* Paris, France, 25 March 1864), *mathematics, Oriental studies*.

Woepcke was the son of Ernst Woepcke, the Wittenberg postmaster, and Karolina Chapon. He studied mathematics and physics at Berlin from 1843 to 1847, receiving the Ph.D. *magna cum laude* in the latter year for a work on sundials in antiquity. In addition to pure mathematics, he was particularly interested in its history, a subject that Humboldt encouraged him to pursue.[1] In the mid-nineteenth century very little was known of the Arab contribution to the development of mathematics. Many Latin translations from the Arabic had existed since the twelfth century; but the texts themselves were not accessible, and further research was thus effectively blocked.[2] Woepcke therefore went to Bonn in 1848 to learn Arabic.[3] After qualifying as *Privatdozent* in the spring of 1850, he went to Leiden, where there were many Arabic manuscripts, and in May of the same year, to Paris, then the center of Oriental studies in Europe.[4] In Paris he studied Persian (under Julius von Mohl) and Sanskrit (under P. O. Foucaux), as well as mathematics (under J. Liouville).

Woepcke interrupted his stay in Paris only from 1856 to 1858, when he taught mathematics and physics at the French Gymnasium in Berlin. He resigned his post because it left him no time for research. His tireless work on Arabic and Persian manuscripts enabled Woepcke to publish some thirty texts.[5] His edition of al-Khayyāmī's *Algebra*, which appeared in 1851, was followed in 1853 by a selection from al-Karajī's *Algebra*. In 1861 and 1863 Woepcke worked on manuscripts in Oxford and in London. He was obliged to return to Paris in December 1863 because his health, always weak, was failing rapidly. He died at the age of thirty-seven and was buried in Père Lachaise cemetery.

Woepcke's contemporaries praised him as modest but confident in his own judgment and as an enemy of all superficiality.[6] He valued only facts and left the working out of unproved conclusions to others.

A member of many learned societies, Woepcke made an outstanding contribution to the knowledge of Eastern contributions in the history of mathematics. Although he investigated many specific problems in various fields, his studies centered on the algebra of the Arabs (its symbolism and the determination of its Greek and Indian components) and on the Indian and Arab influence on the West (the spread of Hindu numerals and methods and the sources of the work of Leonardo Fibonacci).[7] He also attempted to reconstruct lost texts of Apollonius and Euclid on the basis of Arab manuscripts.[8]

Woepcke's own mathematical research dealt mainly with curves and surfaces, equations of the *n*th degree, and function theory. He also translated into French works by J. Steiner (central curves) and by Weierstrass (theory of Abelian functions). Because of Woepcke's early death, many of his editorial projects—for which he had already copied or translated the Arabic texts—were left unfinished or were continued by others.[9] Among the material that came into the possession of Boncompagni were 174 letters and a codex with twenty-five unpublished works, including selections from texts, translations, and notes.[10]

NOTES

1. In 1851 Woepcke translated Humboldt's *Über die bei verschiedenen Völkern üblichen Systeme von Zahlzeichen* (1829) into French.
2. Exceptions were the *Algebra* of al-Khwārizmī, edited by Frederic Rosen (London, 1831), and the *Jawāmiʿ* of al-Farghānī edited by Jacob Golius (Amsterdam, 1669).
3. He studied Arabic under G. W. F. Freytag and J. Gildemeister, as well as astronomy under F. W. A. Argelander.
4. Among the scholars who had worked there were Silvestre de Sacy, J. J. Sédillot, and L. A. Sédillot. See G. Sarton, *Introduction to the History of Science*, I, 665, 667, 717; II, 622.
5. See E. Narducci, "Intorno alla vita ed agli scritti di Francesco Woepcke," 123.
6. *Ibid.*, 123 ff.
7. His investigations include Leonardo Fibonacci's solution of the third-degree equation, two Arab approximation methods for determining sine 1°, Indian methods for calculating the sine, ancient methods of multiplication, and astrolabes. On Woepcke's works see Sarton, *op. cit.*, I, 600 (Thābit ibn Qurra), 663 (number symbols in a MS of 970), 667 (Abu'l-Wafāʾ), 718 (Abū Jaʿfar ibn al-Ḥusain), 719 (al-Karkhī); II, 401 (Muḥammad ibn al-Ḥusain); III, 1765 (al-Qalasādī), 1766 (spread of Hindu numerals and algebraic symbolism).
8. *Ibid.*, I, 154, 174; also R. C. Archibald, *Euclid's Book on Divisions of Figures With a Restoration Based on Woepcke's Text and on the Practica Geometriae of Leonardo Pisano* (Cambridge, 1915), 9–13.
9. Baron de Slane (see Sarton, I, 665; II, 401) edited works on al-Qūhī, al-Sijzī, and Muḥammad ibn al-Ḥusain (dealing with universal compasses for all conic sections); and Aristide Marre (see Sarton, II, 1000) edited the *Talkhīṣ* of Ibn al-Bannāʾ—for Woepcke's preliminary work on this text see Narducci, *op. cit.*, 129, 151.
10. See Narducci, *op. cit.*, 151 f.

BIBLIOGRAPHY

I. ORIGINAL WORKS. There are complete bibliographies in Narducci (see below), 133–152; and in Poggendorff, II, 1353–1354, and III, 1458. His writings include *Disquisitiones archaeologico-mathematicae circa solaria veterum* (Berlin, 1847), his doctoral dissertation; *L'Algèbre d'Omar Alkhayyāmī* (Paris, 1851); *Extrait du Fakhrī, Traité d'algèbre par Aboū Bekr Mohammed ben Alhaçan Alkarkhī* (Paris, 1853); *Sur l'introduction de l'arithmétique indienne en Occident*, (Rome, 1859); and "Recherches sur plusieurs ouvrages de Léonard de Pise . . . et sur les rapports qui existent entre ces ouvrages et les travaux mathématiques des arabes," in *Atti dell'Accademia pontificia dei Nuovi Lincei*, **10** (1856–1857), 236–248; **12** (1858–1859), 399–438; and **14** (1860–1861), 211–227, 241–269, 301–356.

II. SECONDARY LITERATURE. See J. Fück, *Die ara-*

bischen Studien in Europa (Leipzig, 1955), 204; E. Narducci, "Intorno alla vita ed agli scritti di Francesco Woepcke," in *Bullettino di bibliografia e di storia delle scienze matematiche e fisiche*, **2** (1869), 119–152; and G. Sarton, *Introduction to the History of Science*, 3 vols. (Baltimore, 1927–1948).

KURT VOGEL

WÖHLER, AUGUST (*b*. Soltau, Germany, 22 June 1819; *d*. Hannover, Germany, 21 March 1914), *material testing*.

Wöhler came from a Protestant family from the small town of Soltau in the Luneburg Heath. His father, Georg Heinrich Wöhler, was a teacher at and rector of the local school, where Wöhler studied for a time. To encourage his mathematical gifts, the boy was sent to the technical college (which later became the Technische Hochschule) in Hannover, headed at the time by the technologist Karl Karmarsch. Having very successfully completed his studies with the aid of a scholarship, Wöhler began a short period of practical training. In 1840 he joined the Borsig engineering works in Berlin, where he gained practical experience in the construction of railway lines. After returning briefly to Hannover in 1843, he went to Belgium to learn how to operate a locomotive and then became an engineer on the first Hannoverian railroad (Hannover to Lehrte). In 1844 Wöhler was promoted to administrative engineer, and in 1847 he became chief superintendent of rolling stock on the Lower Silesia-Brandenburg Railroad (Berlin to Frankfurt-an-der-Oder to Breslau), which was taken over by the state in 1854. His brilliant work in this post was of lasting influence. In 1874 he was named imperial railway director and was placed on the newly created management board of the Imperial Railways in Strasbourg, on which he remained until 1889. His many awards included an honorary doctorate in engineering.

Wöhler's important scientific achievements originated in the problems he encountered while working for the railroads. They consisted in the study and description of the dynamic strength of engineering materials and, more generally, in the creation of modern techniques for testing materials.

In 1852 the Prussian minister of commerce, industry, and public works appointed Wöhler to a commission established to investigate the causes of axle breakage and train derailments. Wöhler's first publications dealt with the theory of elasticity. In 1855 he derived the formulas for calculating the sag of lattice girders, commonly called the equation of three moments (*Zeitschrift für Bauwesen*, **5**, 122–166); these formulas were published two years before the work of Clapeyron, for whom they are often named. At the same time Wöhler recommended that bridge girders be supported at one end on roller bearings to absorb thermal expansion—a precaution that became universal practice.

Wöhler gained broad recognition through his fatigue bending tests, for which he constructed the experimental apparatus. In them the tested material, usually iron or steel, was subjected to a sequence of stresses that bent or turned it back and forth millions of times. Wöhler distinguished between static, increasing, and alternating loads; and he arrived at universally valid results, known as Wöhler's laws. They can be understood from a consideration of the fatigue strength diagram, a curve that shows the dependence of the time strength of a given construction material on the number of load-application cycles borne. The first of Wöhler's four laws states (in his formulation of 1870): "The failure of the material can . . . occur through constantly repeated vibrations, no single one of which reaches the absolute rupture limit. The differences of the tensions, which frame the vibrations, are decisive for failure of cohesion" (*Zeitschrift für Bauwesen*, **20**, 83).

Wöhler's results, illustrated by pieces broken in fatigue tests, were presented at the Paris Exposition in 1867, at which his work first came to the attention of the English (see *Engineering*, **2** [1867], 160). Also in 1867 Wöhler urged the "introduction of a government-approved classification for iron and steel" and the establishment of a government bureau for testing materials. These proposals were not implemented until sometime later, and then only gradually; but they greatly promoted the development of uniform quality in the manufacture of construction materials, as well as progress toward an "honest trade" in them.

BIBLIOGRAPHY

I. ORIGINAL WORKS. A list of Wöhler's writings is in R. Blaum, "August Wöhler" (see below). Between 1851 and 1898 Wöhler published 42 articles in German technical periodicals. The most important include "Theorie rechteckiger eiserner Brückenbalken mit Gitterwänden und mit Blechwänden," in *Zeitschrift für Bauwesen*, **5** (1855), 122–166; "Resultate der in der Central-Werkstatt der Niederschlesisch-Märkischen Eisenbahn zu Frankfurt a. d. O. angestellten Versuche über die relative Festigkeit von Eisen, Stahl und Kupfer," *ibid.*, **16** (1866), 67–84; and "Über die Festigkeitsversuche mit Eisen und Stahl," *ibid.*, **20** (1870), 73–106.

II. Secondary Literature. See R. Blaum, "August Wöhler," in *Beiträge zur Geschichte der Technik und Industrie*, **8** (1918), 35–55, with bibliography and portrait; and "August Wöhler," in *Deutsches biographisches Jahrbuch*, Überleitungsband I (Stuttgart, 1925), 103–107; F. G. Braune, "Zum 150. Geburtstag von August Wöhler," in *Technik*, **24** (1969), 400–402; A. J. Kennedy, "Fatigue Since Wöhler: A Century of Research," in *Engineering* (London), **186** (1958), 781–782; L. Troske, "August Wöhler," in *Zentralblatt der Bauverwaltung*, **34** (1914), 242–244, with portrait and bibliography; and W. Ruske, "August Wöhler (1819–1914) zur 150. Wiederkehr seines Geburtstages," in *Materialprüfung*, **11** (1969), 181–188.

Ludolf von Mackensen

WÖHLER, FRIEDRICH (*b.* Eschersheim, near Frankfurt-am-Main, Germany, 31 July 1800; *d.* Göttingen, Germany, 23 September 1882), *chemistry.*

For three generations the Wöhlers had been equerries to the electors of Hesse. Wöhler's mother, Anna Katharina Schröder, was the daughter of a professor of philosophy; his father, Anton August Wöhler, left Hesse and became a farmer, then a court official and leading citizen at Frankfurt. Wöhler attended the public school but had extra instruction in Latin, French, and music so that he could attend the Gymnasium from 1814 to 1820. After a year at the University of Marburg and two years at Heidelberg he qualified in 1823 for the M.D., specializing in gynecology. In 1828 he married a cousin, Franziska Wöhler, who bore him two children before her death in 1832; in 1834 he married Julie Pfeiffer, the daughter of a banker, by whom he had four children.

Wöhler's teaching career began at an industrial school in Berlin (1825–1831). He next became professor at a similar institution in Kassel (1831–1836), and he finally settled as professor of chemistry at Göttingen (1836–1882). He was elected foreign member of the Royal Society in 1854 and was awarded its Copley Medal in 1872. From 1864 he was a foreign associate of the Institut de France and an officer of the Legion of Honor.

From childhood Wöhler had a passionate interest in practical chemistry and the collection of minerals. At Heidelberg, Leopold Gmelin encouraged him to experiment on cyanates and, since it

was essential for Germans to go abroad for systematic training in chemistry, he recommended Wöhler to Berzelius in Stockholm. Here Wöhler received a year's rigorous training in mineral analysis and formed a firm friendship with Berzelius. The friendship lasted until the latter's death in 1848 and can be followed through a voluminous correspondence. For over twenty years Wöhler translated Berzelius' influential and often controversial annual reports, occasionally modifying their polemical tone but not their content. He also translated three editions of Berzelius' *Lehrbuch der Chemie*, a labor undertaken for both love and money. Berzelius' influence depended on accurate and prompt translation of his work, and Wöhler wrote an elementary textbook of inorganic chemistry that had fifteen editions in his lifetime and was translated into French, Dutch, Danish, and Swedish; a textbook of organic chemistry that reached thirteen editions; and a textbook of analytical methods that was translated into English.

Wöhler became acquainted with his lifelong friend Liebig as a result of what seemed in 1825 to be a minor squabble over the interpretation of analytical results but became a classic example of a new phenomenon that Berzelius in 1830 called isomerism: both silver cyanate and silver fulminate correspond to the empirical formula $AgCNO$. Liebig had studied the explosive fulminate and at first rejected Wöhler's 1824 results for the stable cyanate. In the 1820's most chemists assumed that only one chemical compound corresponded to one set of analytical percentages. Wöhler and Liebig exchanged letters, often visited each other, and sometimes took vacations together from 1829 until Liebig's death in 1873.

Wöhler's interest in cyanates led to a historic preparation of "artificial" urea, the circumstances of which are best described in his letter to Berzelius of 22 February 1828:

> I can no longer, as it were, hold back my chemical urine; and I have to let out that I can make urea without needing a kidney, whether of man or dog: the ammonium salt of cyanic acid is urea.
>
> Perhaps you can remember the experiments that I performed in those happy days when I was still working with you, when I found that whenever one tried to combine cyanic acid with ammonia a white crystalline solid appeared that behaved like neither cyanic acid nor ammonia . . . . I took this up again as a subject that would fit into a short time interval, a small undertaking that would quickly be completed and— thank God—would not require a single weighing.
>
> The supposed ammonium cyanate was easily ob-

tained by reacting lead cyanate with ammonia solution. . . . Four-sided right-angled prisms, beautifully crystalline, were obtained. When these were treated with acids, no cyanic acid was liberated, and with alkali, no trace of ammonia. But with nitric acid lustrous flakes of an easily crystallized compound, strongly acid in character, were formed; I was disposed to accept this as a new acid because when it was heated, neither nitric nor nitrous acid was evolved, but a great deal of ammonia. Then I found that if it were saturated with alkali, the so-called ammonium cyanate reappeared; and this could be extracted with alcohol. Now, quite suddenly I had it! All that was needed was to compare urea from urine with this urea from a cyanate.

The letter goes on to describe how the discovery adds to the pairs of substances of similar composition but of different properties already known. "It is noticeable that in making cyanates (and in making ammonia) we always have to start with an organic substance. . . ."

In his published paper Wöhler referred to his work of 1823, in which he had shown that cyanogen and aqueous ammonia yielded oxalic acid and a white crystalline solid that he now realized was urea.[1] This, and his new method, he considered to be remarkable examples of the preparation "by art" of a substance of animal origin from inorganic materials.

In a widely quoted obituary, A. W. Hofmann grossly exaggerated the impact of Wöhler's discovery on his contemporaries; there are in fact few references to the discovery in papers or letters of the time. J. H. Brooke and others refer to the literature and indicate the place of Wöhler's discovery in the history of the decay of vitalism and the establishment of isomerism.[2]

Wöhler's period in Berlin was remarkable not only for the preparation of urea but also for the extraction of aluminum.[3] In Lavoisier's *Traité* of 1789 there is a list of earths, including alumina: although he was confident that the fixed alkalies (soda and potash) were compound, he was less certain of the earths. In 1807 Davy succeeded in decomposing soda and potash, and in 1808 he attempted electrolysis of the earths. In his footnotes to the printed lecture the history is reviewed.[4] Meanwhile, Berzelius succeeded in using mercury as a cathode for the electrolysis of most lime and barytes; but alumina would not yield aluminum for him or for Davy, even though the latter tried many ingenious variations. With hindsight one can perhaps see that Davy almost certainly obtained aluminum in several very impure forms: as an amal-

gam, as a solution in potassium, fused into molten glass, and in iron alloys. Faraday and James Stodart in 1822 made an aluminum-in-iron alloy by reduction with coal and iron. In 1824 Berzelius succeeded in extracting silicon by heating potassium with potassium fluosilicate, the potassium probably being provided by Wöhler; the method failed for aluminum.

In 1825 H. C. Oersted showed a specimen of metal, which he believed was aluminum, to the Academy of Sciences in Copenhagen (the specimen is not now available). He certainly prepared aluminum chloride by a new method, but it seems unlikely that the metal was pure aluminum; an alloy of aluminum and potassium seems probable. Oersted, a friend of Wöhler's, never published a claim and made no objection when Wöhler tacitly assumed priority in 1827 or when he published further details in 1845. Recently, however, claims have been made for Oersted's priority.[5]

On 10 October 1827 Wöhler wrote to Berzelius: "Oersted has told me that he does not intend to carry on with his experiments with aluminum chloride. I have already made a first repetition of his researches. . . ." Later he said, "Like Oersted, I decomposed it [aluminum chloride] with potassium amalgam and distilled the product. When the mercury had gone, an iron-black lump of metal remained; but on strong heating it distilled as a green vapor." In a paper of 1827, Wöhler emphasized that he was not saying the extraction was impossible by Oersted's method, but that he could not repeat the process and had a better one. Oersted had, he stated, given up the work and had given him permission to go ahead—an important ethical consideration. He praised Oersted's "most ingenious method" for the preparation of the aluminum chloride from which the metal could be extracted.

Wöhler's technique was to cover a small quantity of potassium in a platinum crucible with excess aluminum chloride and to heat the covered crucible gently to start the vigorous reaction. Nothing remained to react with water to produce alkalies when the reaction mixture was put into water. Other workers possibly had failed with similar methods because alkalies would react with the finely divided metallic product. Wöhler proved that his metal contained no potassium and described its chemical properties in considerable detail, especially its reactions with other elements, acids, and alkalies; footnotes contain many references to other metals.

In 1845 Wöhler published a supplementary paper to amplify his descriptions of 1827.

Later in the nineteenth century, when aluminum became a common metal, Wöhler was honored by Napoleon III. Twentieth-century controversies over priority in the extraction of pure aluminum have added little to history except nationalistic claims. Wöhler subsequently used the same technique to extract beryllium and what he thought was yttrium from their chlorides.

From the first letters exchanged between Wöhler and Liebig the possibility of a joint work was discussed; but when it came, it was not, as might have been expected, on cyanates but, rather, on mellitic acid. In 1825 Wöhler became interested in the mineral honeystone sent to him by Heinrich Rose. (Its true structure, as the aluminum salt of mellitic acid—$C_6(COOH)_6$—was established by Adolf von Baeyer in 1865.) He isolated the pure acid, a calcium salt, and other metallic derivatives; and qualitative observations led him to believe that the acid contained little or no hydrogen and might be related to benzoic acid. (By "the acid," workers at that time usually meant the acid anhydride.) Liebig took over the practical work in 1829, and a joint paper eventually was published. Wöhler returned to the subject in 1839; was helped by Berzelius with supplies of the rare material in 1840; and prepared the ammonium salt, an amide, euchronic acid, and some unidentified colored derivatives.

In January 1830, Liebig wrote to Wöhler that he would prefer their joint work to be on cyanates instead of honeystone. While he was still a student Wöhler had already published four papers on cyanates; one on mercuric thiocyanate ("Pharaoh's serpent") and perthionic acid, two more after his year with Berzelius, and a fourth in 1829 on the products of dry distillation of urea and uric acid. Urea yielded an acid that he recognized as cyanic acid, already discovered and named by Serullas. But whereas Serullas gave $C_2N_2O_2$ as an empirical formula, Wöhler decided that the acid of the cyanates must be C_2N_2O. Liebig and Wöhler set to work separately, Wöhler doing the preparation and Liebig the analyses. They struggled for a year and were remarkably successful in sorting out the complex products of dry distillation of urea. In 1845 they returned to the study of the product obtained when cyanic acid reacts with alcohol and decided, with irony, to call it allophanic ether (literally, "unexpected" ether).

The correspondence shows that a joint paper consisting of miscellaneous observations in inorganic chemistry was based solely on Wöhler's work. From the letters that each exchanged with Berzelius it can be seen that at this time both men were considering a study of oil obtained by pressing bitter almonds. Wöhler suggested it to Liebig as a joint study, and they began to collect materials. After Wöhler's first wife died in June 1832, he went for about seven weeks to Giessen and worked feverishly with Liebig: although they published other joint works, this was the only time they actually shared a laboratory. A letter to Berzelius shows that in the first four weeks, they had carried out most of the experimental work and tentatively drawn what were to be their final conclusions.

In 1832 the two workers left open the relationship between the oil and the material from which it was extracted. In their classic paper—which was actually written by Wöhler although Liebig is listed as coauthor—they summarized their achievements: ". . . we make the general assertion that as a result of our experiments, it is established that there is a body, composed of three elements, that remains stable in the presence of reagents and that can be regarded not only as the radical of benzoic acid but, perhaps with slight variations, as the radical of a large number of similar compounds."[6]

Wöhler and Liebig overcame considerable practical and theoretical difficulties. They had, for example, no thermometer that could measure temperatures over 130° C. Misunderstanding over the relative atomic weights of heavy metals and oxygen led to formulas double the modern ones (for instance, $C^{14}H^{10}O^2$ instead of C_7H_5O for the benzoyl radical); their analysis of benzoic acid therefore gave a result different from that established by Berzelius. The last of these difficulties was overcome after correspondence with Berzelius and analysis of the silver salt. They established the relation between the oil (now called benzeldehyde) and its aerial oxidation product (already known as benzoic acid). Then the reaction of the oil with potassium hydroxide was studied and a new "oil" (now called benzyl alcohol) extracted with potassium benzoate.

By passing chlorine through the oil, Wöhler and Liebig obtained benzoyl chloride, which they converted into the bromide and iodide; they also studied its reaction with alkalies, water, alcohol, and dry ammonia gas. Among many derivatives they obtained were benzoyl sulfide, benzoyl cyanide, benzamide, ethyl benzoate, benzonitrile, and benzoin. They did not analyze all of these, and some of their tentative formulas are incorrect; but the paper established beyond doubt what they claimed: the existence of a body that was constant from one compound to another. Incidentally, many of the compounds they first prepared and described (such as benzoyl chloride) were important in the future development of organic chemistry. Berzelius was prepared to accept the benzoyl radical, oxygen included; but he later felt that oxygen could not possibly be present in an electropositive radical and withdrew his support for such a view.

When Wöhler moved to Göttingen in 1836, he wrote to Liebig and proposed a joint study of amygdalin. Within two days of sending this letter he mailed another showing that he had virtually solved the problem. In 1830 P. J. Robiquet and Antoine Boutron-Charlard had found that crushed bitter almonds smell of bitter almonds only when moistened and that, from the crushed nuts, fats, a resin, a liquid sugar, and the substance they called amygdalin could be extracted by addition of boiling alcohol. Wöhler showed in 1836 that the crystalline amygdalin could be decomposed by a vegetable emulsion, providing the emulsion had not been coagulated by boiling. Liebig took over the quantitative work and showed that a sugar was the other product of the decomposition. Amygdalin, the first example of a glycoside, was the subject of a joint paper.

Wöhler and Liebig collaborated on one more major piece of work, a study of uric acid. Wöhler suggested the subject, and the idea seems to have come from his medical interests. Uric acid was not easily obtainable—snake excrement was the only substantial source—and relationships with urea and allantoin were suspected by Wöhler. As a student he had won a prize in 1828 for an essay on the conversion in the human body of chemicals taken orally and excreted in urine. The technique adopted by Liebig and Wöhler was to subject uric acid, and the derivatives they prepared, to oxidation and reduction by reagents of different concentrations and strengths. Wöhler seems to have been the first to heat reagents together in sealed glass tubes, but after an explosion he thought metal ones safer.

Their 100-page paper described fourteen new compounds and their preparation and analysis.[7] An attempt to establish a new radical called "uril" ($C_8N_4O_4$) was less successful. Perhaps even more significant than the sophisticated, practical and theoretical organic chemistry was the new spirit revealed. Writing to Berzelius in 1828, Wöhler was doubtful whether animal substances could be prepared in the laboratory. In 1832 he began the pa-

per on the benzoyl radical with a description of organic chemistry as "the dark region of organic nature." But in 1838 his work with Liebig led him to write (at Liebig's suggestion):

> The philosophy of chemistry will conclude from this work that it must be held not only as probable but [as] certain that all organic substances, insofar as they no longer belong to the organism, will be prepared in the laboratory. Sugar, salicin, morphine will be produced artificially. It is true that the route to these end products is not yet clear to us, because the intermediaries from which these materials develop are still unknown, but we shall learn to know them.

Although the two friends published further joint papers, they did no more major investigations together, Liebig turning to agricultural and physiological chemistry and Wöhler to inorganic studies. Although at the age of forty Wöhler had published only a quarter of the papers he was to present, none of the later ones was as important for the development of chemistry as those before 1840. As professor of chemistry and pharmacy, director of the laboratories, and inspector general of all the apothecaries in the kingdom of Hannover, he had to spend a great deal of time from 1836 to 1848 on inspection tours of apothecary shops. During these years translation of Berzelius' texts also took much time. The school of chemistry at Göttingen grew steadily, and Wöhler estimated that about 1,750 students heard his lectures between 1845 and 1852, 2,950 between 1853 and 1859, and 3,550 between 1860 and 1866. He thus had all the duties of a government official, translator, and teacher, as well as father of a growing family; and later in his life he told Kolbe that he had not kept up with theoretical developments. Nevertheless, during his forty-six years as professor at Göttingen he produced a stream of interesting papers.

Wöhler was always fascinated by geological samples, which were sent to him from all over the world by friends and ex-students. Meteorites were equally absorbing, and he published some fifty papers on minerals, meteorites, and their analyses: he noted the passivity of meteoric iron. Wöhler's lively curiosity and ingenuity shine through these papers, as well as through the fifteen papers on general analytical methods.

In organic chemistry Wöhler studied quinone and hydroquinone, established the relationship between them, and discovered quinhydrone (which he compared for beauty of color with the feathers of the hummingbird). He distrusted theoretical speculation and in 1840 published a paper under the pseudonym S. C. H. Windler satirizing Dumas's substitution theory. Had he allowed the developing theories to guide his research, it is possible that his time spent on organic chemistry might have been more fruitful: many of his papers are on topics already absorbed into the current theories (such as chloroform) or too far beyond the tide of research to affect it (such as alkaloids). Like other workers, including Berzelius, he spent time on ill-characterized substances of biological or medicinal interest that failed to yield clear chemical results even to his superb manipulative technique.

Mid-nineteenth-century inorganic chemistry was relatively static, and many workers simply collected data and prepared new compounds. Their interest was in the materials themselves: rocks, crystals, or chemicals. There is hardly a metal for which Wöhler did not prepare new salts, and he was particularly fond of colored derivatives. Some of his methods, such as the preparation of phosphorus by heating calcined bones with sand, have since been developed industrially. He was the first to make acetylene from calcium carbide (1862), by heating together zinc, calcium, and carbon. Copper-colored cubic crystals from blast-furnace slag, which had been thought to be metallic titanium, were shown by Wöhler in 1850 to be a compound of titanium, carbon, and nitrogen.

Working with Deville, Wöhler used aluminum to extract crystalline boron from boric acid, and crystalline silicon from potassium fluosilicate. Heinrich Buff consulted Wöhler about the gas, which ignited explosively and spontaneously in air, that he obtained during the electrolysis of dilute acids with aluminum electrodes. Wöhler realized that the aluminum contained silicon and went on to discover and describe silicon hydride. Wöhler and Buff were the first to prepare organosilicon compounds, silicon chloroform, iodoform, and bromoform.

Unlike his close friends Liebig and Berzelius—and, indeed, many of the eminent chemists of the time—Wöhler rarely made enemies, kept a cool good humor, avoided rancorous argument, and did his best to bring together even such immovable objects and irresistible forces as Berzelius and Liebig. His efforts to counsel Liebig to lead a quieter life were unsuccessful.

Wöhler was popular with students from throughout the world. An account of his American pupils, many of whom obtained important chemical posts, was published.[8] His most distinguished student probably was Hermann Kolbe, who became professor at Marburg and remained a close friend until

Wöhler's death. Unlike Liebig, Wöhler remained interested in what went on his laboratory even in old age.

NOTES

1. "Über künstliche Bildung des Harnstoffs," in *Annalen der Physik und Chemie*, 2nd ser., **12** (1828), 253–256.
2. John H. Brooke, "Wöhlers Urea and Its Vital Force?—A Verdict From the Chemists," in *Ambix*, **15** (1968), 84–114. The significance of the preparation of urea from inorganic sources has been pressed by (among others) W. H. J. Warren, in "Contemporary Reception of Wöhler's Discovery of the Synthesis of Urea," in *Journal of Chemical Education*, **5** (1928), 1539–1552; and dismissed by the following: D. McKie, "Wöhler's Synthetic Urea and the Rejection of Vitalism, a Chemical Legend," in *Nature*, **153** (1944), 608–610; P. Mendelssohn-Bartholdy, "Wöhler's Work on Urea," *ibid.*, **154** (1944), 150–151; and E. Campaigne, "Wöhler and the Overthrow of Vitalism," in *Journal of Chemical Education*, **32** (1955), 403, an attempt to reestablish the significance of the discovery. A balanced review of the literature is in T. O. Lipman, "Wohler's Preparation of Urea and the Fate of Vitalism," *ibid.*, **41** (1964), 452–458.
3. "Über das Aluminium," in *Annalen der Physik und Chemie*, 2nd ser., **11** (1827), 146–161.
4. Humphry Davy, "Electrochemical Researches on the Decomposition of the Earths . . .," in *Philosophical Transactions of the Royal Society*, **98** (1808), 333–370.
5. N. Bjerrum's claim for Oersted in "Die Entdeckung des Aluminiums," in *Zeitschrift für angewandte Chemie*, **39** (1926), 316–317, was countered by K. Goldschmidt in "Nochmals die Entdeckung des Aluminiums," *ibid.*, 375–376. Oersted's notes were published by Kirstine Meyer, in *H. C. Oersted Naturvidenskabelige skrifter*, II (Copenhagen, 1920), 467–470.
6. "Untersuchungen über das Radikal der Benzoesäure," in *Justus Liebigs Annalen der Chemie*, **1** (1832), 249–282, written with Liebig; also reprinted separately as Ostwalds Klassiker der Exacten Wissenschaften, no. 22 (Leipzig, 1891).
7. "Untersuchungen über die Natur der Harnsäure," in *Justus Liebigs Annalen der Chemie*, **26** (1838), 241–340, written with Liebig.
8. H. S. van Klooster, "Friedrich Wöhler and His American Pupils," in *Journal of Chemical Education*, **21** (1944), 158–170.

BIBLIOGRAPHY

I. ORIGINAL WORKS. A. W. Hofmann's complete bibliography of Wöhler's publications (see below) includes details of the translations of his books. The Royal Society *Catalogue of Scientific Papers*, VI, 410–419; VIII, 1258–1259; XI, 836–837; and XII, 790; lists all the translations of the papers.

Wöhler's books include *Grundriss der unorganischen Chemie* (Berlin, 1831; 15th ed., 1873); *Grundriss der organischen Chemie* (Berlin, 1840; 13th ed., 1882); *Beispiele zur Übung in der analytischen Chemie* (Göttingen, 1849), published anonymously; *Practische Übungen in der chemischen Analyse* (Göttingen, 1853), also published anonymously; and *Die Mineralanalyse in Beispielen* (Göttingen, 1861), published as a 2nd ed. of the last work.

Wöhler's translation of Berzelius' *Lärbok i kemien* appeared as *Lehrbuch der Chemie*, 4 vols. (Dresden, 1825–1831); the 3rd ed., rev., appeared in 10 vols. (Dresden, 1833–1841); and the 4th ed. in 10 vols. (Dresden–Leipzig, 1835–1841). Vols. 4–20 of Berzelius' annual surveys of progress in the sciences, *Årsberättelser öfver Vetenskapernas Framsteg*, 27 vols. (Stockholm, 1822–1848), were also translated by Wöhler.

The voluminous correspondence between Wöhler and Berzelius was published almost in its entirety by O. Wallach, ed., *Briefwechsel zwischen J. Berzelius und F. Wöhler*, 2 vols. (Wiesbaden, 1901), with scholarly footnotes.

Extracts from the Liebig-Wöhler correspondence were published six years after Wöhler's death. Wöhler selected and often polished the extracts, probably intending publication as a tribute to his friend: A. W. Hofmann, ed., *Aus Justus Liebig's und Friedrich Wöhler's Briefwechsel 1829–1873*, 2 vols. (Brunswick, 1888). The extracts represent perhaps a quarter of the total correspondence, the bulk of which is at the Bayerische Staatsbibliothek, Munich. Wöhler's (and Liebig's) publisher, Vieweg, in Brunswick, has about 100 letters exchanged between the two men. The Deutsches Museum, Munich, and the University Library, Göttingen, hold letters exchanged by Wöhler and other chemists, such as Kolbe and Bunsen.

II. SECONDARY LITERATURE. A. W. Hofmann, "Zur Erinnerung an Friedrich Wöhler," in *Berichte der Deutschen chemischen Gesellschaft*, **15** (1882), 3127–3290; and Johannes Valentin, *Friedrich Wöhler* (Stuttgart, 1949), 159–170, both include bibliographies of Wöhler's writings. See also Th. Kunzmann, *Die Bedeutung der wissenschaftlichen Tätigkeit Friedrich Wöhler's für die Entwicklung der Deutschen chemischen Industrie* (Berlin, 1830), which gives a list of papers published by Wöhler's students between 1837 and 1863; and Poggendorff, VII A (1970), 779–783, with a full list of papers published on Wöhler and his work.

ROBIN KEEN

WOLF, CHARLES JOSEPH ÉTIENNE (*b.* Vorges, near Laon, Aisne, France, 9 November 1827; *d.* St.-Servan, Ille-et-Vilaine, France, 4 July 1918), *astronomy, history of science.*

Wolf, whose family included a number of teachers and professors, entered the École Normale Supérieure in 1848. *Agrégé* in science three years later, he taught at the *lycée* in Nîmes and later at the one in Metz. His initial research, a study of capillarity as a function of temperature, earned him a doctorate in physical sciences in 1856. He was then named professor of physics at the Faculty of Sciences of the University of Montpellier, where

in 1862 he demonstrated that the spectra of incandescent bodies, then thought to be rigorously stable, vary when the temperature of the body rises.

Also in 1862, Le Verrier had Wolf named astronomer at the Paris observatory. Assigned at first to the Service Méridien, Wolf studied the personal equation affecting meridian observations and built an apparatus for determining it. He also worked on the electric synchronization of astronomical clocks, perfecting a device that was later adopted for the clocks of Paris.

Wolf was transferred to the Service des Équatoriaux, where, in collaboration with Rayet, he photographed the penumbra of the moon during the eclipse of October 1865. Shortly afterward he observed the spectrum of a nova and noted that it contained bright lines, a new phenomenon that he subsequently sought to detect in the spectra of other stars. For this purpose, Wolf devised a direct-view spectroscope, with neither slit nor lens, that could immediately be substituted for the eyepiece of a telescope when a star was sighted. With this instrument the user could quickly carry out a spectroscopic exploration of the sky. In 1867 Wolf and Rayet discovered three stars exhibiting the phenomenon they were seeking; these were the first examples of what are called Wolf-Rayet stars, also known as stars of spectral type W.

Of Wolf's thirty or so published notes and articles we shall mention only two of the most important. In 1869, exploiting data gathered during the transit of Mercury in the preceding year, he definitively solved the problem of the "black drop," a phenomenon that occurs at the moment when the image of a planet comes into contact with the solar limb, making it difficult to determine the instant of contact. He showed that the phenomenon, which Lalande had attributed to irradiation, is purely instrumental and can be eliminated by the use of a sufficiently large objective that is free from aberration. From 1873 to 1875 Wolf studied the Pleiades, which serve as standards for astrometrical measurements. He established the first general catalog of the cluster (1877), containing the positions and magnitudes of 571 stars.

Appointed *professeur suppléant* of astronomy at the Paris Faculty of Sciences in 1875, Wolf was named titular professor in 1892. He left the observatory in the latter year and retired from teaching in 1901. Toward the end of his career Wolf became especially interested in the history of science. In the *Bulletin astronomique* of 1884 and 1885 he published seven articles on cosmogonic hypotheses and later collected them in a book that also contained his own complete French translation of Kant's *Universal Natural History and Theory of the Heavens*. He also studied ancient standards of weights and lengths and the history of the pendulum. His most remarkable work, *Histoire de l'Observatoire de Paris*, is still the only full-length account of that institution. Drawing on original documents, he related "the history of the buildings and of their successive transformations, of the instruments used there, of the astronomers who lived there, and of the administration under which they lived."

Wolf was deeply religious, austere but kindly. He often spent vacations in the house where he was born. Obliged to leave it in 1914 because of the German invasion, he took refuge in St.-Servan, where he died a few months before his native city was reconquered. Wolf was elected to the Académie des Sciences in 1883 and served as its president in 1898.

BIBLIOGRAPHY

I. ORIGINAL WORKS. Wolf's physical works are "Influence de la température sur les phénomènes qui se passent dans les tubes capillaires," in *Annales de chimie et de physique*, **49** (1857), 230–279; "Sur les spectres des métaux alcalins," in *Comptes rendus . . . de l'Académie des sciences*, **55** (1862), 334–336, written with M. Diacon; and "Sur le pouvoir réflecteur des miroirs en verre argenté," in *Journal de physiùe théorique et appliquée*, **1** (1872), 81–86.

His writings on instruments and their use include "Recherches sur l'équation personnelle . . .," in *Annales de l'Observatoire de Paris, Mémoires*, **8** (1865), 153–208; "Description d'un nouveau spectroscope," in *Comptes rendus . . . de l'Académie des sciences*, **65** (1867), 292–293; "Description du sidérostat de Foucault," in *Annales scientifiques de l'École normale supérieure*, 2nd ser., **1** (1872), 51–84; "Les applications de l'électricité à l'astronomie," in *Bulletin de la Société internationale des électriciens*, **2** (1885), 105–125; and "Comparaison des divers systèmes de synchronisation des horloges astronomiques," in *Comptes rendus . . . de l'Académie des sciences*, **105** (1887), 1155–1159.

Among his astronomical writings are "Sur le passage de Mercure du 4 novembre 1868 . . .," in *Comptes rendus . . . de l'Académie des sciences*, **68** (1869), 181–183, written with C. André; "Description du groupe des Pléïades," in *Annales de l'Observatoire de Paris, Mémoires*, **14**, pt. 1 (1877), A1–A81; *Les hypothèses cosmogoniques* (Paris, 1886); and *Astronomie et géodésie*, H. Le Barbier and P. Bourguignon, eds. (Paris, 1891). The discovery of the Wolf-Rayet stars is discussed in Jacques R. Lévy, "Rayet," in *DSB*, XI, 319–321. Various observations are reported in some twenty notes in *Comptes rendus . . . de l'Académie des sciences*, **62–107** (1866–1888).

The history of science is treated in "Étalons de poids et mesures de l'Observatoire . . .," in *Annales de l'Observatoire de Paris*, **17** (1883), C1–C78; "Rôle de Lavoisier . . . système métrique," in *Comptes rendus . . . de l'Académie des sciences*, **102** (1886), 1279–1284; *Travaux relatifs à la théorie et aux applications du pendule*, 2 vols. (Paris, 1889–1891); and *Histoire de l'Observatoire de Paris de sa fondation à 1793* (Paris, 1902). Wolf also wrote seven notes, mainly on ancient standards of length, that appeared in *Comptes rendus . . . de l'Académie des sciences*, **95–125** (1882–1897).

II. SECONDARY LITERATURE. See G. Bigourdan, "Notice sur la vie et les travaux de M. Ch. Wolf," in *Comptes rendus . . . de l'Académie des sciences*, **167** (1918), 46–48; and an anonymous "Notice nécrologique," in *Astronomie*, **32** (1918), 255–256; and P. Painlevé, "Annonce de la mort de Ch. Wolf," in *Comptes rendus . . . de l'Académie des sciences*, **167** (1918), 45–46.

JACQUES R. LÉVY

WOLF, JOHANN RUDOLF (*b.* Fällanden, near Zurich, Switzerland, 7 July 1816; *d.* Zurich, 6 December 1893), *astronomy, history of science.*

Wolf, the son of Johannes Wolf, a minister, and Regula Gossweiler, came from an old Zurich family, the Windeggen-Wolfs, who had been citizens of that city since the fourteenth century. He was educated at the Zurich Industrieschule and at the newly founded university, where his teachers included K. H. Gräffe and J. L. Raabe. He continued his studies at Vienna (1836–1838) under J. J. von Littrow and J. A. von Ettingshausen, and at Berlin (1838) under Encke, Dirichlet, and Steiner. In 1839 Wolf went to Bern to teach mathematics and physics, and from 1844 to 1855 he also was professor of astronomy at the university. In the latter year he accepted a double appointment at Zurich as professor of astronomy at the Eidgenössische Technische Hochschule and the university. Through his efforts an observatory was constructed at Zurich and opened in 1864.

Wolf's most important achievement was the reliable determination of the lengths of sunspot periods and of their relationship to the variation in terrestrial magnetism. After Schwabe announced that he had detected a period of about ten years, Edward Sabine, Alfred Gautier, and Wolf discovered, simultaneously and independently of each other, that this period paralleled the variations in the earth's magnetic elements. By exploiting older data, Wolf gathered enough solar observations to establish, in 1852, a mean value of 11.1 years for the duration of a period. He also determined the epochs of all the maxima and minima from 1610 on; and until his death he continued to publish regular reports on his determinations of the relative numbers of the sunspots (see his *Geschichte der Astronomie*, secs. 188, 235).

Wolf also made a significant contribution to the study of the history of science. His works in this field are a rich source of historical data, and by virtue of his astonishingly wide-ranging studies they are exceptionally reliable. Moreoever, they are very concise. Among his scholarly achievements was the discovery that the correspondence of Johann I Bernoulli, augmented by that of the younger Bernoullis, had been sold shortly before 1800, in part to the Stockholm Academy and in part to the prince of Gotha. As a result, O. Spiess was able to make the correspondence accessible to scholars. Wolf's own correspondence was stored with his other papers at the Zurich observatory and, unfortunately, his successor, lacking space, disposed of it (see Spiess's preface to *Der Briefwechsel von Johann Bernoulli*, 35, n. 1).

Wolf founded the *Vierteljahrsschrift der Naturforschenden Gesellschaft in Zürich* in 1856 and served as its editor until his death. Upon the establishment of the Zurich Polytechnikum he was named head librarian; during his tenure he assembled a valuable collection of early printed books on astronomy, mathematics, and other branches of science.

BIBLIOGRAPHY

I. ORIGINAL WORKS. Wolf's writings include "Über den gelehrten Briefwechsel der Bernoulli," in *Mitteilungen der Naturforschenden Gesellschaft in Bern*, no. 109 (1848), 1–7; *Handbuch der Mathematik, Physik, Geodäsie und Astronomie*, 2 vols. (Zurich, 1869–1872); "Die Correspondenz von Johannes Bernoulli," in *Vierteljahrsschrift der Naturforschenden Gesellschaft in Zürich*, 21 (1876), 384–386; *Geschichte der Astronomie* (Munich, 1877), vol. XVI of *Geschichte der Wissenschaften in Deutschland*; and *Handbuch der Astronomie, ihre Geschichte und Literatur*, 2 vols. (Zurich, 1890–1893).

The Zurich observatory has Wolf's journal for 1816–1841, which consists of 141 MS pages. Wolf's contributions to *Biographien zur Kulturgeschichte der Schweiz*, 4 vols. (Zurich, 1858–1862), were continued in 475 articles subsequently published in *Vierteljahrsschrift der Naturforschenden Gesellschaft in Zürich*, 6–39 (1861–1894).

II. SECONDARY LITERATURE. See Heinz Balmer, "Rudolf Wolf und seine Briefsammlung," in *Librarium*, 8, no. 2 (1965), 95–105; and Alvin Jaeggli, "Die Berufung des Astronomen Joh. Rudolf Wolf nach Zürich 1855," Eidgenössische Technische Hochschule (Zurich), Bibliothek, Schriftenreihe no. 11 (1968).

Obituaries are collected in *Reden, gehalten bei den Trauerfeierlichkeiten für Herrn Dr. J. R. Wolf* (Zurich, 1894). A list of additional obituaries, some with bibliography and portrait, is in *Vierteljahrsschrift der Naturforschenden Gesellschaft in Zürich*, 46 (1901), 333.

See also *Der Briefwechsel von Johann Bernoulli*, O. Spiess, ed., I (Basel, 1955), foreword, 35, n. 1.

J. J. BURCKHARDT

WOLF, MAXIMILIAN FRANZ JOSEPH CORNELIUS (*b.* Heidelberg, Germany, 21 June 1863; *d.* Heidelberg, 3 October 1932), *astronomy*.

Wolf was the son of Franz Wolf and Elise Helwerth. He became interested in astronomy in his youth and in 1885 his father, a rich physician, constructed a private observatory for him. Wolf received the Ph.D. at Heidelberg in 1888, with a dissertation on celestial mechanics written under the direction of Leo Königsberger; he then went to Stockholm, where for the next two years he continued his studies with Gylden. His dissertation on asteroids, influenced by Gylden, was published in Stockholm, and won him a post as academic lecturer when he returned to the University of Heidelberg. He remained there for the rest of his life, and was professor of astrophysics and astronomy from 1901 until 1932.

Working in his private observatory, Wolf soon became famous for his innovative photographic methods. He used a wide-angle lens to investigate the diffuse nebulae of our galaxy and invented a technique for discovering asteroids through the streaks they made on time-exposure plates. These investigations brought him into contact with a number of American astronomers, especially E. E. Barnard, and in 1893 Wolf visited United States observatories. He returned with plans to build a new observatory at Heidelberg, since his own, located in the center of town, was inadequate; the grand duke of Baden was interested in Wolf's ideas, and under his patronage an observatory was begun on the Königstuhl. Catherine Wolfe Bruce, of New York City, who was also interested in Wolf's work, made him the generous gift of the sixteen-inch double telescope that provided the foundation for his continuing investigations.

At the new Baden Observatory, Wolf and his

collaborators discovered hundreds of new asteroids and determined their positions by means of a visual refractor—a method that produced more exact results than those derivable photographically from plates made with a wide-angle objective. From 1906 on, Wolf also used a reflector to obtain spectrographs of the galactic nebulae. He began by studying the shapes of the gaseous nebulae—as they were then called—but soon became interested in their relationships to surrounding stars. Observing that many of the extended dark nebulae show dark patches, or "cavities," as he called them, Wolf counted the number of stars in such dark areas and demonstrated by the statistics of stellar magnitudes that the cavities were clouds of cosmic dust. Since it was possible for him to make spectrophotographs of only the brighter single stars, he made use of red filters on his reflector to view an extended celestial region.

As a result of these investigations Wolf was early able to recognize the difference between the gaseous and planetary nebulae, on the one hand, and the spiral nebulae, on the other. He reached this distinction through both spectral analysis and the study of the systematic distribution of the spiral nebulae relative to the mean plane of the galaxy. By 1911 he had offered a number of conclusions that were later generally adopted.

Wolf also studied single stars, comparing two plates of the same celestial region, photographed at different times, to find variable stars of substantial proper motion. In these investigations he often made use of the stereocomparator, an instrument that he and Pulfrich had invented. Wolf was of the view that, in general, the instrument creates the science, as could best be seen in the development of astronomy; throughout his career, he therefore paid particular attention to instrumentation. He further wished to establish an observatory in a more favorable climate than that of central Europe, and, with his friend A. F. Lindemann (the father of the British physicist Lord Cherwell), investigated sites around the Mediterranean—without, however, finding any more suitable.

Wolf was an exceptional teacher as well as researcher, and attracted students from all over the world, including Luigi Carnera, R. S. Dugan, August Kopff, and Heinrich Vogt. His lectures were vivid, and he often illustrated them with slides. In addition to his students, Wolf had friends all over the world, particularly in the United States; he was therefore much affected by World War I, and following the war was active in attempting to restore scientific relationships between America and Germany.

Wolf received many honors and awards. He was also highly esteemed by amateur astronomers, from whom he often received requests for celestial photographs. It is characteristic of his generosity that he was often able to oblige them, even though he prepared all copies and slides of his photographs himself.

BIBLIOGRAPHY

I. ORIGINAL WORKS. Wolf's works include "Photographic Observations of Minor Planets," in *Astronomy and Astrophysics*, **12** (1893), 779; "Reflector and Portrait Lens in Celestial Photography," in *Nature*, **55** (1897), 582; *Königstuhlnebellisten*, I–XVI (1902–1928); "Stereoskopische Bestimmung der relativen Eigenbewegung von Fixsternen," in *Astronomische Nachrichten*, **171** (1906), 321–326; "Spektren von Gsnebeln," in *Vierteljahrsschrift der Astronomischen Gesellschaft*, **43** (1908), 208; "Die Spektra zweier planetarischer Nebel, Heidelberg," in *Sitzungsberichte der Akademie der Wissenschaften in Wien*, pt. 2A (1911); "Auffindung und Messung von Eigenbewegungen durch Stereoeffekt," in *Vierteljahrsschrift der Astronomischen Gesellschaft*, **51** (1916), 113; "Die Sternleeren beim Amerikanebel," in *Astronomische Nachrichten*, **223** (1924), 89; and "Die Sternleeren bei S. Monocerotis," in *Seeliger-Festschrift* (1924), 312.

II. SECONDARY LITERATURE. On Wolf and his work, see H.-C. Freiesleben, Max Wolf, *Der Bahnbrecher der Himmelsphotographie*, Grosse Naturforscher, no. 26 (Stuttgart, 1962), which includes a comprehensive bibliography of Wolf's publications, 232–238.

H.-CHRIST. FREIESLEBEN

WOLFF, CASPAR FRIEDRICH (*b.* Berlin, Germany, 18 January 1734; *d.* St. Petersburg, Russia [now Leningrad, U.S.S.R.], 22 February 1794), *biology.*

For a detailed account of his life and work, see Supplement.

WOLFF, CHRISTIAN (*b.* Breslau, Silesia [now Wrocław, Poland], 24 January 1679; *d.* Halle, Germany, 9 April 1754), *philosophy.*

During his school years at Breslau, Wolff became acquainted with Cartesian ideas, although he concentrated at first on the writings of the Scholastics. He then became interested in logic, which ultimately left him dissatisfied because it lacked any sustained account of an "art of discovery." This view of logic, together with a lifelong search for certainty in matters scientific and philosophical, led to his interest in mathematics, not for its own sake but for its methodological implications. After three years at Jena, Wolff received the master's degree from Leipzig in 1702, becoming first a lecturer in mathematics and then, in 1706, professor of mathematics and natural science at the University of Halle. He was recommended for the latter post by Leibniz, with whom he had established a correspondence and whose philosophical ideas, although somewhat modified and vulgarized, subsequently became the cornerstone of his own philosophical writings.

At Halle, Wolff lectured on mathematics and algebra, building and fortification, as well as experimental and theoretical physics; a glimpse of the kind of courses given may be obtained from one of the earliest writings of this period, his popular handbook *Anfangsgründe aller mathematischen Wissenschaften* (1710). Gradually the interest in logic supervened, leading in 1713 to publication of *Vernünftige Gedanken von den Kräften des menschlichen Verstandes* (the so-called "German Logic"); and by 1719 his philosophical lecturing, which had become the focus of his university activities, found its first full expression in *Vernünftige Gedanken von Gott, der Welt und der Seele des Menschen* . . . (the "German Metaphysics"), which testifies to the leading influence of Leibniz. Although the form of these works is characteristically Scholastic, the importance of their publication in German, rather than Latin, cannot be overrated; by creating a German philosophical vocabulary, it led to a great spread of philosophical interest in eighteenth-century Germany that reinforced the general movement toward deism, determinism, and free thought incipient in these writings.

Indeed, Wolff's deterministic tendencies led to his dismissal from Halle in 1723, after which he taught at the University of Marburg, where he published another set of writings, this time in Latin, many of them corresponding to the earlier German versions but more formal and Scholastic in appearance and with an impressive complex of definitions, theorems, and demonstrations, as instanced in the important volumes on ontology and general cosmology. As Wolff's fame spread, he received invitations to return to Prussia and to go to Berlin; but he finally settled again at Halle,

where he continued to write on law, moral philosophy, and related subjects.

Wolff was essentially a popularizer and (to some extent inspired by Leibniz) sought to effect a formal synthesis between Scholasticism, the new mathematical methods, and more recent scientific conceptions. From Leibniz he also inherited the emphasis on certain philosophical ideas, such as the principles of contradiction and of sufficient reason, as well as the central attention given to the notion of possibility in their metaphysical writings. Round these conceptions Wolff organized a vast philosophical system; if it was not original, and was rather eclectic, it nevertheless set the tone and produced the form in which questions were to be debated by contemporaries and successors down to the time of Kant. The tone is that of a seeming rationalism that nevertheless tries to incorporate the empirical and theoretical results of recent scientific and mathematical innovations. Indeed, it was Wolff's respect for the mathematical method, as he understood it, that inspired the form of his writing, with its strict definitions and syllogistic development.

Limiting ourselves here to aspects of Wolff's philosophy of physical science, we find one of his basic models, both in ontology and in methodology, to be analysis and synthesis. Analysis yields the set of irreducible predicates of a thing which provide the ground or reason for its possibility; that there must be a ground is postulated by the principle of sufficient reason. This principle in turn falls under the principle of contradiction, since it would be self-contradictory (Wolff holds) to posit anything without a sufficient reason. He thus fails to distinguish in principle between logical and empirical possibility. Although mere possibility of finite things does not entail their existence, existence is stated to be merely "the complement of possibility," God being the ground of both actuality and possibility. To know that something exists, however, requires recourse to experience, both direct and inferential, through the giving of reasons; the reference to reason again permits a convenient slide from *ratio cognoscendi* to *ratio essendi*, and from logical to real possibility.

These doctrines quite naturally lead to Wolff's deterministic formulations of his cosmological principles, which emphasize the rational connections between things, given as sequences or coexistences; these formal themes were later directly echoed in Kant's writings. The visible world is a machine, operating in accordance with the laws of motion: almost one-third of the *Cosmologia generalis* treats these laws. In his physics Wolff is an outspoken corpuscularian, although the ultimate elements, the *atomi naturae*, are neither extended nor divisible. All that can be said a priori is that the properties of composites derive from their elementary constituents; empirical knowledge is limited to the properties of the composites. Thus the a priori part evidently provides no more than the mechanist-determinist theme, although modified by the Leibnizian idea of a competing teleological explanation of things.

Wolff's doctrine of space as the order of things existing simultaneously, although having some resemblance to Leibniz's theories, is more uncompromisingly kinetic. Space is mere phenomenon, both in the sense that it is secondary and ontologically derivative from coexisting substances, and in that it is perceived only "confusedly." Also, since the notion of substances as coexisting presupposes their mutual interaction, it is the latter conception that is ontologically basic. Wolff's bodily substances, being essentially centers of action, are also more uncompromisingly purely physical than Leibniz's monads; in all this, Wolff's views foreshadow the basic positions taken by Kant in his early writings down to about 1760. Similarly, Wolff's theory of time makes time reducible to the order of successive things in a continuous series; time is not given without the latter, he states expressly.

BIBLIOGRAPHY

I. ORIGINAL WORKS. The latest standard edition of Wolff's works is Christian Wolff, *Gesammelte Werke*, J. École, J. E. Hofmann, M. Thomann, and H. W. Arndt, eds. (Hildesheim, 1962–   ), German writings in 11 vols., Latin writings in 35 vols., containing major bibliographies of and on Wolff's writings.

Wolff's chief writings bearing on mathematics and the methodology and philosophy of science include the following in German: *Vernünftige Gedanken von den Kräften des menschlichen Verstandes* (Halle, 1713); *Auszug aus den Anfangsgründen aller mathematischen Wissenschaften* (Halle, 1717); *Vernünftige Gedanken von Gott, der Welt und der Seele der Menschen . . .* (Frankfurt–Leipzig, 1720); *Vernünftige Gedanken von den Wirkungen der Natur* (Halle, 1723); *Vernünftige Gedanken von den Absichten der natürlichen Dinge* (Frankfurt, 1724); and *Vernünftige Gedanken von dem Gebrauch der Theile in Menschen, Tieren und Pflanzen* (Frankfurt, 1725).

Latin works are *Philosophia rationalis sive logica* (Frankfurt–Leipzig, 1728); *Philosophia prima, sive ontologia* (Frankfurt, 1729); and *Cosmologia generalis* (Frankfurt, 1731).

An English trans. is *Discursus preliminaris de philosophia in genere*, translated by R. J. Blackwell (Indianapolis, 1963).

II. SECONDARY LITERATURE. The following, listed chronologically, concern Wolff's scientific and methodological ideas: J. E. Erdmann, *Grundriss der Geschichte der Philosophie*, II (Berlin, 1866), §290, 187–196; E. Kohlmeyer, *Kosmos und Kosmogonie bei Christian Wolff* (Göttingen, 1911); H. Lüthje, "Christian Wolffs Philosophiebegriff," in *Kant-Studien*, 30 (1923), 39–56; H. J. de Vleeschauwer, "La genèse de la méthode mathématique de Wolff," in *Revue belge de philologie et d'histoire*, 11 (1931), 651–677; M. Campo, *Christian Wolff e il razionalismo precritico*, 2 vols. (Milan, 1939); H. Heimsoeth, "Christian Wolffs Ontologie und die Prinzipienforschung Immanuel Kants," in *Studien zur Philosophie Immanuel Kants*, supp. no. 71 (1956), 1–92; J. École, "Un essai d'explication rationelle du monde ou la *Cosmologia generalis* de Christian Wolff," in *Giornale di metafisica*, 18 (1963), 622–650; and "Cosmologie wolffienne et dynamique leibnitienne," in *Études philosophiques*, n.s. 19 (1964), 3–9; J. V. Burns, *Dynamism in the Cosmology of Christian Wolff* (New York, 1966); L. W. Beck, *Early German Philosophy* (Cambridge, Mass., 1969), ch. 11, 256–272; and Tore Frängsmyr, "Christian Wolff's Mathematical Method," in *Journal of the History of Ideas*, 36 (1975), 653–668.

GERD BUCHDAHL

WOLLASTON, FRANCIS (*b*. London, England, 23 November 1731; *d*. Chislehurst, Kent, England, 31 October 1815), *astronomy*.

Wollaston was the eldest son of Mary Fauquier and Francis Wollaston. With his brother Charlton, he matriculated at Sidney Sussex College, Cambridge, in June 1748. He graduated LL.B. in 1754. With the intention of practicing law, he entered Lincoln's Inn on 24 November 1750, but he soon decided to enter the church, and was ordained deacon in 1754 and priest in 1755. In 1758 Simon Fanshawe presented him to the living of Dengie, in Essex, and in the same year he married Althea Hyde, by whom he had ten daughters and seven sons, one of whom was William Hyde Wollaston. In 1761 he became rector of East Dereham, in Norfolk (where his father had a summer residence). In 1769 he was made rector of Chislehurst in Kent and elected a fellow of the Royal Society. Other ecclesiastical benefices followed, but he continued to live at Chislehurst. His wife died there in 1798.

Wollaston's first book, *Address to the Clergy of the Church of England in Particular and to All Christians in General Proposing an Application for Relief, etc.*, was offered in support of a parliamentary bill of 1772, proposing to remove the obligation placed on members of the universities to subscribe to the Thirty-nine Articles of religion, and to replace it with a simple declaration of faith in the scriptures. The obligation had been established by the Ecclesiastical Commission of 1562, which agreed on the articles. Wollaston's support was of no avail and the university tests were not abolished until 1871. Two other books written during the next two years with a view to mild reform of the church seem likewise to have been little noticed, but not to have stood in the way of preferment.

Wollaston's serious interest in astronomy was, as he explained, calculated to remove him to a "distance from the misrepresentations of narrow-minded bigots."[1] At Chislehurst he had a private observatory built at the top of a square brick house. Here he used a telescope with a triple object-glass, made for him in 1771 by Peter Dollond. The telescope passed to his son, and thence to the Royal Astronomical Society. With the Dollond telescope Wollaston saw and described the great spot and belts of Jupiter, although he recorded no colors (1772). He equipped his observatory with a thermometer and barometer, and presented papers to the Royal Society on the variation in the rate of his astronomical clock with corresponding atmospheric conditions. He was not able to correlate these quantities in any significant way.

Wollaston long entertained the hope that astronomers might collaborate on a general plan for improving star catalogues and drafting them in a way that would facilitate the measurement of small stellar movements. In 1789 he published a very substantial collection of comparative catalogues with a preface announcing his plan and discussing the many previous catalogues on which he based his coordinates, which were reduced to 1 January 1790. As an essay in history, what he wrote was not altogether reliable, as S. P. Rigaud pointed out in connection with Wollaston's remarks on Bradley.[2] His catalogue was much used by William Herschel.[3]

Wollaston produced a number of ideas for new instruments, but he tended to make exaggerated claims for them, and none was of any great moment. He saw the merits of the transit circle, having worked with a small one (fourteen-inch focal length) from before 1772, but tried in vain to persuade first Jesse Ramsden and then Edward Troughton to make a larger one to his design. Finally, in 1781, William Cary began work, according to Wollaston's plan, on an altazimuth instrument (for the method of equal altitudes). Another instrument of his was a "universal meridian dial," for any latitude, on which he wrote a pamphlet; but

in doing so he added little to the art of dialing. Wollaston's most important contribution to astronomy was not made at a fundamental level, but was rather his publication of two or three useful practical aids to the ordinary astronomer and navigator, one of which was the comparative catalogue already mentioned. The second was a catalogue of circumpolar stars (1800), made with his transit instrument, by way of practicing what he had preached in his first book of 1789. The instrument is described at the end of the catalogue, together with explanations of the tables and formulas for calculating from them. His third important contribution, a collection of ten plates depicting the heavens "as they appear to the naked eye," was published by John Cary Sr., whose firm was renowned for its maps, atlases, and globes.

Although the date of birth given above is that which Wollaston himself gives in his autobiography, he makes a curious remark on the second page of his *Fasciculus astronomicus* of 1800, to the effect that the observations it embodies were made with "the eye of an old man, turned threescore before he engaged in the work."

NOTES

1. *The Secret History of a Private Man*, p. 54.
2. *Miscellaneous Works of James Bradley* (Oxford, 1832), p. 59.
3. J. L. E. Dreyer, ed., *Scientific Papers of Sir William Herschel* (London, 1912), p. 40.

BIBLIOGRAPHY

Wollaston's chief works are *A Specimen of a General Astronomical Catalogue, Arranged in Zones of North Polar Distance . . .* (London, 1789); *Directions for Making an Universal Meridian Dial Capable of Being Set to Any Latitude, Which Shall Give the Mean-Solar Time of Noon, by Inspection, Without any Calculation Whatsoever* (London, 1793); *Fasciculus astronomicus, Containing Observations of the Northern Circumpolar Region; Together With Some Account of the Instrument With Which They Were Made . . .* (London, 1800); *A Portraiture of the Heavens as They Appear to the Naked Eye* (London, 1811).

References to papers in the *Philosophical Transactions* of the Royal Society will be found in the printed books. In 1795 Wollaston printed privately for his friends a short autobiography written in the third person, *The Secret History of a Private Man*, which is rare. It contains an interesting account of the way in which his proposed reforms of 1772 were received in different quarters and is largely a justification of his behavior as a minister of the English church. There are letters from him in the British Museum (Add. MSS. 32887, f. 501; 32888, f. 198; 32892, f.155; 32896, f. 360; 32902, f. 330). A combined entry on Francis and his youngest brother, George, written by E. I. Carlyle, is in *Dictionary of National Biography*. There are references to genealogies of the family in the *DNB* entry on William Wollaston.

J. D. NORTH

WOLLASTON, WILLIAM HYDE (*b*. East Dereham, Norfolk, England, 6 August 1766; *d*. London, England, 22 December 1828), *chemistry, optics, physiology*.

Wollaston's family had become well known through their interests in science and theology. His great-grandfather, William Wollaston, was the author of *Religion of Nature Delineated*, a widely read work on natural religion published in 1724. His father, Francis Wollaston, a vicar and fellow of the Royal Society, was interested in astronomy and compiled a catalog of stars, *Fasciculus astronomicus*, which appeared in 1800. The famous physician William Heberden was his uncle. His father's brother, Charlton Wollaston, who died before William's birth, was a physician to the royal household.

William went to school at Charterhouse and in 1782 entered Caius College, Cambridge, as a medical student. There he pursued his favorite field, botany, but also studied some astronomy and, most important for his future work, became interested in chemistry. He attended the lectures of Isaac Milner, Jacksonian professor of chemistry, and performed experiments in the laboratory of his elder brother, Francis, who then held a lectureship in mathematics and who later lectured in chemistry, succeeding Milner in 1792. His interest in chemistry was also stimulated by Smithson Tennant, who was also studying medicine. William graduated in 1787 and then completed his medical studies in London. He first practiced in Huntingdon in 1792, but after a few months he went to Bury St. Edmunds. He became a fellow of the Royal Society in 1793. Four years later he moved to London. In 1800, either because of his failure in a contest for the appointment of physician to St. George's Hospital or through his dislike of the profession, he abandoned medicine and turned his attention to the other sciences. In 1802 he was awarded the Copley Medal of the Royal Society for his published papers. He became secretary of the Royal Society in 1804. As a member of numerous committees he gave advice on matters of sci-

entific interest. He was associated with the attempts to bring uniformity into the system of weights and measures and recommended the introduction of the imperial gallon, which was accepted in 1824. Between 1818 and 1828 he was an active member of the Board of Longitude, and was particularly concerned with nautical instruments. In 1820 he was president of the Royal Society, for the interim period before Humphry Davy's election. In 1823 he was elected foreign associate of the Académie des Sciences. Shortly before his death on 22 December 1828, he made notable donations for scientific research. He gave two thousand pounds to the Royal Society for promoting research, so initiating the Donation Fund. He also invested one thousand pounds in the name of the Geological Society, of which he had been a member since 1812. The proceeds from the first year's income were used to cast a die for a medal bearing Wollaston's head. The "Wollaston Medal," first awarded to William Smith in 1832, has continued to be an annual prize of the Society.

In the same year that he left the medical profession Wollaston formed a partnership with Tennant which was to bring him fame and wealth. Tennant had traveled to Sweden and met J. G. Gahn, an adept of chemical analysis on the small scale. This may well have been the source of Wollaston's practice of working with unusually small quantities, a distinctive feature of his chemical operations. When Berzelius visited England he was astonished at the extent to which Wollaston had developed this art. In a letter to Gahn[1] he remarked that the whole of Wollaston's chemical apparatus consisted of no more than a few bottles standing on a small wooden board with a handle. The bottles contained the common reagents and were so stoppered that their contents could be extracted in drops. Substances were investigated on a small piece of glass. A good example of Wollaston's small-scale chemistry was his introduction of the standard laboratory test for magnesium by the precipitation of magnesium ammonium phosphate, assisted by the scratching of a glass point.[2] But his skill was best demonstrated in his important investigations on the platinum metals.

Wollaston and Tennant were both interested in platinum, which continued to resist the efforts of chemists (particularly intensive since the middle of the eighteenth century) to produce it in a satisfactory malleable state in which it might be worked. Tennant bought a large quantity of crude platinum ore, and the partners began work on the intractable metal. Tennant was soon able to announce his discovery of osmium and iridium, new elements in the crude ore; but Wollaston was the harder worker, and it was through his continuing experiments, conducted in his private laboratory, into which he was reluctant to admit anyone, that the difficult practical problem was solved.

It had become common practice to refine the crude ore by dissolving it in aqua regia and then to precipitate platinum by means of ammonium chloride, with which it forms an insoluble complex salt. To recover any platinum still in solution Wollaston added bars of iron, and treated the precipitate as before with aqua regia and ammonium chloride. Adding iron for the second time, he obtained a precipitate with unexpected properties. When it was treated with nitric acid, a red solution formed. This gave an amalgam when treated with mercury, which in turn was decomposed by heat, leaving a white metal. The new metal, which he had discovered by July 1802, he first called "ceresium" after the recently discovered asteroid. But he soon changed the name to "palladium," after Pallas, another asteroid.

Instead of reporting his discovery openly Wollaston sent out anonymous printed notices in April 1803, describing the properties of the new metal and advertising its sale at a Soho shop. This attracted the attention of Richard Chenevix, a chemist, who suspected fraud from the way in which the discovery was announced. He bought the advertised stock and performed many experiments. In spite of his conviction that palladium was an alloy of known metals, none of his many attempts to analyze it succeeded. He claimed, however, that he had synthesized palladium by mixing a solution containing mercuric oxide and platinum in aqua regia with a solution of ferrous sulfate. When heated, this mixture produced a precipitate that fused into a button, supposedly indistinguishable from palladium, though it was in fact a compound of platinum with silicon and boron contained in the powdered charcoal used for the fusion. Chenevix concluded that palladium was an alloy of platinum and mercury. He felt he had found the key to reducing the number of the elements, whose recent rapid increase had led him to suspect their real simplicity. One critic, congratulating Chenevix, pointed out that the pursuit of alchemical transmutations was not as ridiculous as it had seemed.[3] Wollaston replied, again anonymously, offering a prize of twenty pounds to anyone who could synthesize palladium. The repeated failures to achieve this result soon convinced chemists that palladium was a genuine new metal. In 1804 Wollaston announced his discovery of rhodium in the crude platinum ore. Yet he withheld the identity of the discoverer of palladium until February 1805. He mentioned his fears of competing workers anticipating his discoveries, but he never fully explained his curious behavior, which according to Banks, the president of the Royal Society, had brought him into disfavor with scientists who were "open and communicative."

It was Wollaston's skill in working with small quantities that made possible the isolation and characterization of the new metals rhodium and palladium. For these metals are only present in platinum ore in small amounts. From one thousand grains of crude ore he had extracted five grains of palladium and four grains of rhodium. Vauquelin, who was working with much larger quantities of platinum ore at the same time, thought that Wollaston's achievement "seems at first incredible."[4]

In 1805 Wollaston stated that he had at last found a way to make platinum malleable, but he gave no details of his process until shortly before his death in 1828. His paper earned him the Royal Medal of the Royal Society. No one had yet succeeded in fusing platinum in large quantities. Previous workers had tried the effects of heat and pressure on the platinum sponge, obtained by the ignition of the complex ammonium salt. Through trial and error, and a careful attention to detail in the treatment of his material, Wollaston brought remarkable refinements to this method. His techniques included the slow thermal decomposition of the ammonium salt, the avoidance of burnishing by gently powdering the platinum sponge, sieving, and sedimentation. This process produced a uniform powder, essential to the production of malleable platinum. Impurities were removed by washing and forming a compact mass under water. The cake so formed was powerfully compressed by a toggle press. Finally the compact metal was carefully dried and forged. These details of Wollaston's process constitute the fundamental procedures of modern powder metallurgy. His process was not immediately adopted in industry; but it was followed, at least in part, by Liebig at the Giessen laboratory.[5] Today it is recognized as a standard method for producing compact metals from powder.

Wollaston sold the laboratory apparatus which he made from his malleable platinum. He drew very fine platinum wires by a process that is still used, and superintended the construction of platinum vessels for the concentration of sulfuric acid. These are the earliest platinum boilers known. They were sold to manufacturers. According to one estimate, Wollaston's profit up to 1826 from the sale of articles of platinum and the other platinum metals was £15,000.[6]

In theoretical chemistry Wollaston influenced the way in which the new atomic theory of Dalton was received. His own attitude to atomic chemistry varied remarkably between bold speculation and complete skepticism. In 1808 he described his experiments on carbonates, sulfates, and oxalates, which proved that the composition of these substances was regulated by the law of multiple proportions. These additional instances of the law were easily verifiable and were often mentioned as standard examples. Wollaston accepted that his findings were merely particular instances of Dalton's assertion that the atoms of elements united one to one, or by some simple multiple relation. He speculated on the possible atomic composition of the oxalates of potash. With brilliant intuition he predicted that arithmetical relations between atoms would be insufficient to explain chemical combination, and that spatial considerations would have to be introduced. He stated that a compound of four particles of one type and one of another would be stable if the four surrounding particles were arranged tetrahedrally. This surmise was confirmed much later in the century with the development of the stereochemistry of the carbon atom.

Wollaston therefore appeared to accept Dalton's theory, pointing to its possible extension. Yet already there was a hint of reservation in his statement that the "virtual extent"[7] of the particles was spherical. He discussed this idea in more detail in his paper on the structure of crystals, which was read in 1812. He remarked that the existence of ultimate physical atoms was not established and that virtually spherical particles, consisting of mathematical points surrounded by forces of attraction and repulsion, would explain the structure of crystals equally well.[8] This theory of unextended point centers of force, invented by Bošković in the eighteenth century, had already interested Davy. Later Faraday would accept it in favor of the extended massy atoms of Dalton.

In 1813 Wollaston discussed the atomic theory in a way that was to have a surprisingly wide appeal. His tone was totally different from that of his earlier treatment of the subject. He complained justifiably that there was no known way to establish the numbers of atoms present in particular compounds, but he went on to say that in any case such questions were "purely theoretical" and unnecessary for practical chemistry. He therefore proposed to draw up a scale, based on the most reliable analyses available, which would express the proportions in which the common chemical substances combined. This summary of chemical facts would provide chemists with immediate answers to the routine problems of laboratory work. Referring all combinations to a standard oxygen unit of 10, he calculated the combining proportions of various substances, and distributed their names and values on a sliding rule, along a line logarithmically divided from 10 to 320. He was thus able to compute mechanically chemical proportions that before had been obtained only by lengthy multiplication and division. Chemists were not yet employing tables of logarithms for their calculations. According to Wollaston the numbers that he had given to each substance were reliable and not "warped" by the atomic theory. He called these values "equivalents." This use of the term earlier introduced by Cavendish was unfortunate for it implied that every chemical has a fixed equivalent, an erroneous conception that persisted until Laurent, thirty years later, pointed out how chemical equivalence varies with function.

Abandoning atoms and conjectures Wollaston had attempted to strip chemistry of all but the factual content of experimental results. There appeared to remain a purely descriptive chemistry, a body of recipes for producing desired effects, summarized on an instrument. Yet this was an illusion. Chemists, particularly in England, succumbed to this apparently factual presentation. They did not detect the intrusion of hypothesis, which later prevented Comte from recognizing Wollaston's treatment as fully positivistic, saying that it amounted to no more than a "mere artifice of language."[9] In his calculation of representational numbers Wollaston had in fact made assumptions about composition of exactly the same arbitrary nature that he had objected to in Dalton. For example he assumed that the two oxides of carbon consisted of one equivalent of carbon united to one and two equivalents of oxygen. The same hypotheses crept, apparently unnoticed, into Davy's calculations of "proportional numbers"; they too relied on tacit suppositions on the constitution of oxides. Like Wollaston, he presented his numbers as deduced from experiment and free from theoretical assumptions. The skepticism of Davy and Wollaston was accepted by English chemists as embodying a sound philosophy and for many years dictated their reactions to Dalton's theory. A typical statement came from William Brande, who welcomed Wollaston's treatment of chemistry as "divested of all hypothetical aspect."[10] One fellow of the Royal Society objected that Wollaston and not Dalton should have been given the Royal Medal, for he had done for the atomic theory what Watt had done for the steam engine: he had rendered it useful.[11] Perhaps the clearest indication of Wollaston's influence appeared in the new *Chemical Dictionary* of Andrew Ure, containing the entry "Atomic Theory. See Equivalents (Chemical)."[12] Wollaston's "equivalents" continued to be used in this sense until the middle of the century. Chemists were convinced that equivalents expressed the unalterable facts of chemical proportions. Reluctant to introduce theories of matter into their science, or to accept calculations of atomic weights (Berzelius conceded these were based on unproved suppositions of atomic constitution and were therefore subject to revision), they felt the language of equivalents was safest. This circumstance accounts for the preference later given to Gmelin's equivalents over Berzelius' system of atomic weights. There was even a tendency to use "equivalent" and "atom" synonymously. According to this usage, "atom" was regarded as a convenient alternative to "proportion" or "equivalent," and carried no theoretical implications. It was left for later generations of chemists to distinguish between equivalents, atoms, and molecules, and to show how atomic weights could be unambiguously determined.

Wollaston's chemical slide rule, in some form, was in general use in laboratories for over twenty years. The instrument was reportedly sold in the bookstores of New York and Vienna. Schweigger, the editor of the *Journal für Chemie und Physik*, reproduced two copies of Wollaston's scale in one issue, so that one copy could be cut out and pasted on a slider.[13] Berzelius said that he used the instrument constantly. Faraday, in his practical manual, described it as a commonly used calculating device.[14] But the instrument began to fall into disuse around 1840 on account of the increasing demands for more accurate calculations. In 1842 Thomas Graham said that Wollaston's instrument was "not of much practical value" and gave instead the logarithms of atomic weights.[15]

In 1822, in spite of his earlier firm skepticism, Wollaston returned by a most unusual route to the full acceptance of Dalton's theory. With startling boldness he asserted that conclusive tests on the existence of atoms could be made through the observation of planets. He argued that the particles of the atmosphere of the earth were subject to the opposing forces of their mutual repulsion and gravity. If there were a limit to the divisibility of atmo-

spheric matter, the weight of these ultimate particles would prevent further atmospheric expansion. But if matter were endlessly divisible into lighter and lighter particles, the force of repulsion would overcome gravity. Then the atmosphere of the earth would not terminate at a finite height, but would expand freely into celestial space and collect about the planets through gravitational attraction. Wollaston therefore believed that the classical problem of the divisibility of matter could be decided by a crucial test in astronomy. In May 1821 Venus was passing very close to the sun in superior conjunction. He carefully followed the path of the planet with a small telescope. He was unable to detect any apparent retardation in the motion of Venus that might be attributable to refraction by the solar atmosphere. He added that Jupiter possessed no sensible atmosphere, since the occultation of its satellites was also unretarded. He was in no doubt that the atmosphere of the earth was of finite extent. Therefore he concluded that it was composed of ultimate atoms. He argued that since the laws of definite proportions were true for all kinds of matter, not just the elastic atmosphere but all substances could be regarded as composed of indivisible atoms. He asserted that the equivalents of chemistry really did express the relative weights of atoms, but curiously he made no mention of the problem, which had earlier troubled him, of estimating the numbers of atoms that entered into combination.

There was a surprising delay before the weakness of Wollaston's logic was exposed. Meanwhile the popular expositions of the atomic theory given by Turner and Daubeny accepted the attractive new argument as a clear proof of atomism. Graham pointed out that the atmosphere of the earth could be limited simply through condensation at low temperatures and that Wollaston's explanation in terms of atoms was unnecessary.[16] But it was not realized until much later that Wollaston had put forward a circular argument.[17] He had assumed from the start that if there were atmospheric particles of limited divisibility, these must be the ultimate Daltonian atoms that participated in chemical change; but the particles of oxygen and nitrogen in the atmosphere need not be monatomic. The carbon dioxide and water vapor of the atmosphere were clearly not chemically simple particles. In fact, the height of an atmosphere is controlled by the weight of polyatomic molecular particles, and by the temperature.

In general, chemists did not share Wollaston's concern to test the divisibility of matter. They were content to deal with combining weights as they found them, without speculating on their further divisibility outside of chemistry. Wollaston's paper continued to be referred to in connection with the existence of a universal ether.[18]

Some of Wollaston's best work was in crystallography, another field of study intimately connected with the structure of matter. The fundamental laws of crystallography had been discovered toward the end of the eighteenth century. In part this was the work of Haüy, who had created a system of crystallography in a spirit of mathematical idealism. It was a problem for Haüy's contemporaries to determine how far the details of his thought, inspired by a belief in the simplicity of nature, were representative of reality. Haüy had constructed algebraic formulas that related the various occurring crystalline forms of a given substance to the primitive form, which could be extracted by mechanical cleavage from each of them. For example, the primitive form of calcium carbonate was a rhomboid, which could be extracted from the secondary forms with hexagonal and pentagonal faces by appropriate cleavages. Once the dimensions of the primitive form were known it was possible to deduce the angles of any related secondary form. Where the primitive form was a regular solid, such as the cube, the required dimensions could be inferred from considerations of symmetry; but with less regular forms such as the obtuse rhomboid the dimensional ratios had to be calculated from measurements of angles. This was approximately performed by the contact goniometer, which consisted of a hinged pair of arms attached to a protractor. It was difficult to make the arms coincide with the crystal faces, and Haüy never claimed an accuracy beyond twenty or thirty minutes of arc. Within this wide margin afforded by approximate measurement, Haüy was able to consider various possible dimensional ratios and select those that most satisfied his metaphysical beliefs. He asserted that the dimensions of the regular solids, which were correctly expressed as ratios of square roots of small integers, reflected Nature's simplicity. This, he argued, must be discernible in the irregular forms also. Accordingly, in his discussion of these he chose the simplest possible ratios consonant with measurement, and then adjusted the angles by calculation.

In 1809 Wollaston described his newly invented reflective goniometer, which allowed a far greater accuracy in the measurement of crystals. It consisted of a graduated circle, vertically fixed on a horizontal axle. The crystal was attached by wax to a small leveling device joined to the axle. An object was viewed by reflection in one face of the crystal, and then the crystal was rotated until the same object appeared in the adjacent face. The angle through which the graduated circle had moved was read off. This procedure gave the angle of the crystal to the nearest five minutes. In this way Wollaston showed the angle of rhomboidal calcium carbonate to differ by over thirty minutes from that given by Haüy. He detected even greater discrepancies in other carbonates, but if he had shown the way he was not prepared to carry out the extensive determinations needed to correct Haüy's data. Doing so was largely the work of William Phillips, a printer and bookseller, whom Wollaston had instructed in the use of his instrument. Employing the new goniometer with graduations of half-minutes, Phillips compiled the most accurate body of crystal data that had hitherto existed. The results showed that Haüy's values for the irregular primitive forms, based on conceptions of simplicity, were incorrect. By 1824 several continental authorities, including Mohs and Mitscherlich, had rejected Haüy's data. It was fitting that John Herschel later mentioned Wollaston's goniometer as an illustration of the influence of instrumentation on the progress of science.[19] The modern goniometer is the result of extensive refinements of Wollaston's original design.

Haüy had concluded, from the polyhedral fragments produced in cleavage, that the crystal kingdom was constructed from three molecular forms: the tetrahedron, the triangular prism, and the parallelepiped. His conclusions were criticized by others, including Wollaston, who objected that a stable crystal could not result from such arrangements as the grouping of tetrahedral particles hanging together at their edges. Regarding this as precarious masonry, Wollaston in 1812 proposed alternative spherical crystal units, joined together as closely as possible by mutual attraction. His close-packed formations of triangularly arranged spheres imitated the commonly occurring crystal forms. From a rhomboid of spheres, tetrahedral groups of spheres could be removed, leaving an octahedron. This accounted for the cleavage of rhomboidal fluorspar. Wollaston was surprised to learn that the beginnings of his theory were to be found in the thirteenth observation of Hooke's *Micrographia*. He also constructed other forms from spheroids, earlier considered by Huygens. The most original part of his theory concerned the cubic form. He explained this in terms of two different kinds of sphere, which he referred to as "black and white balls," so arranged that each black ball was equidistant from all surrounding white balls; balls of the same type were also equidistant from each other. This produced a cube from two interpenetrating tetrahedra. In the twentieth century the lattice structure of sodium chloride was shown to be of this type; but in 1812 Wollaston's theory was an unverifiable speculation.

The most enthusiastic supporter of Wollaston's theory was John Daniell, professor of chemistry at King's College, London. He brought forward various arguments, none of which was successful, to show that this theory was the only one that would explain the facts of crystallography. For example, he tried to interpret his observations on crystal etching in this way. Etched forms are indicative of crystal symmetry, but they could not have provided the crucial data on internal structure that Daniell believed he had found. The American mineralogist James Dana also adopted Wollaston's spheres and spheroids but grouped them differently according to supposed discrete polarities. This development represented some steps toward the conception of the space lattice.

Discussions of the type initiated by Haüy and Wollaston were not favorably received in the early nineteenth century. At the time crystallography was largely concerned with the geometrical treatment of external symmetry. The important physical study of internal structure was not revived until the middle of the century. Wollaston's speculations provided an early example of how, in the absence of direct experimental investigation, remarkably close approximations to the actual internal structure of crystals could be derived through the arrangement of spheres.

The mineral wollastonite was named by a French admirer of Wollaston's work in crystallography.[20]

A large part of Wollaston's published work was devoted to optics, notably in the design of instruments. His early papers on atmospheric refraction discussed the phenomena of the mirage. His theoretical treatment was muddled and made no advance on existing theories, which had similarly exaggerated the effects of water vapor. But his imitation of the phenomena by mixing liquids of different density was frequently referred to.[21] Further, his own careful observations across heated surfaces and description of a mirage, which he was surprised to observe while sitting in a boat near Chelsea, provided Biot with data for his mathematical treatment of the phenomenon.[22] Wollaston was particularly interested in such irregular refraction

for the difficulties it created in navigation. Altitudes were taken with reference to the horizon, and the necessary dip correction was difficult to assess in cases of unusual refraction close to the horizon. He therefore designed a dip sector, a modified sextant, which allowed the dip to be measured by simultaneous observation of opposite points of the horizon. The commissioners of the Admiralty directed Ross, and later Parry, to make observations on the dip of the horizon during their arctic voyages, taking with them Wollaston's instrument.[23] But they reported that the dip sector was of limited use, since the atmospheric conditions were not uniform on the opposite sides of the horizon. Wollaston Island, in Baffin Bay, was named for him by Ross; it was the first of several arctic christenings in his honor.[24]

In 1802 Wollaston introduced the important method of determining refractive indices by total internal reflection. The alternative method of minimum deviation, however, continued to be used in the early part of the century. His observations on an impure spectrum led him to conclude that there were only four colors in the solar spectrum. This influenced Thomas Young, who was led by it to alter his own theory of color vision. At the same time Wollaston discovered the dark lines in the solar spectrum, later to be known as Fraunhofer lines. Also in 1802 he presented convincing experimental support for Huygens' wave theory. Using the technique he had invented, he measured the refractive index of Iceland spar in different directions and showed that for different planes of incidence the extraordinary ray was refracted exactly as Huygens' theory predicted. He did not commit himself, however, to a firm statement of belief in the wave theory. This later brought Wollaston charges of timidity and undue caution; but there was no reason why he should have gone further, particularly since Young's impressive evidence had not yet appeared.

In 1803 he described his "periscopic spectacles," designed to allow clear vision in oblique directions. He substituted meniscus lenses for the generally used biconvex and biconcave forms. Contemporary opticians were mistaken in supposing the elimination of spherical aberration to be the prime consideration in the design of spectacles. As Wollaston correctly pointed out, the eye looks through only a small part of the lens at any instant, and the chief requirement is sharp vision in all directions. Meniscus lenses had been recommended since the seventeenth century. It is uncertain to what extent Wollaston's spectacles were used; the general introduction of meniscus spectacles did not occur until after the work of F. Ostwalt at the close of the nineteenth century.

In 1807 Wollaston described his camera lucida, a quadrilateral glass prism, which by two total internal reflections sent horizontal rays from an object vertically upward into the eye viewing above the prism. This device was widely used as an aid in drawing.[25] It was also commonly attached to the eyepiece of microscopes for sketching images. He also improved the camera obscura by introducing a meniscus lens and an aperture, so reducing the curvature of image of a laterally extended object. In this form it was employed as an early camera by Niepce and Daguerre.

His well-known microscopic doublet was described posthumously. Dissatisfied with the performance of the compound microscope, which was soon to receive essential improvements from Lister, Wollaston proposed the use of a combination of two planoconvex lenses to reduce the aberration of the simple microscope. This suggestion had been occasionally adopted since the seventeenth century, and more recently John Herschel had worked out formulas for the aberration of spherical surfaces in combination. But Herschel's suggestion of a biconvex lens combined with a concavo-convex type presented difficulties in grinding. Wollaston's doublets were easier to make, particularly since two surfaces were plane. The improved resolution impressed workers and led to further developments. A diaphragm was placed between the lenses, and triplet combinations were introduced. While there were continuing attempts to improve the compound microscope, the simple microscope, improved through Wollaston's suggestions, continued to be used. His improvements in illumination, involving the reduction of glare by a type of field stop, were also adopted.

Although he left the medical profession Wollaston continued to be interested in physiology. In 1797 he characterized the principal constituents of urinary calculi, and in 1812 identified a new and rare type of stone, which he called "cystic oxide" since it occurred in the bladder. This was later renamed cystine, the first of the amino acids to be discovered. Fourcroy and Vauquelin reported similar investigations, but unaccountably gave no recognition to Wollaston. This led Alexander Marcet, a physician, to set matters right in a popular work dedicated to Wollaston, his friend.[26]

In his Croonian lecture of 1809 Wollaston stated for the first time the vibratory character of muscular action. He had been led to this discovery by considering the sound heard when a finger is put in the ear, which he compared to the sound of distant carriages. He reproduced the sound by rubbing a pencil over notches on a board and thus determined the frequency of the vibration. With one finger against his ear he found the number of notches which the pencil had to pass over in five seconds to produce the same sound. His value of 20–30 vibrations per second was later shown by Helmholtz to be the first overtone of the fundamental frequency of muscular murmur.

In 1811 he announced that in spite of the known presence of sugar in the urine of diabetics he had failed to detect it in the blood taken from victims of this disease. He adopted the eccentric theory, proposed by Charles Darwin (d. 1778), the son of Erasmus Darwin, that there existed an unknown route between the stomach and bladder, allowing the sugar to avoid the blood and pass directly into the urine. The failure to detect the sugar was due to the use of stale specimens. Claude Bernard later pointed out that quick tests on fresh serum were essential, since sugar is unstable in blood.

In 1820 Wollaston read an interesting paper on the physiology of the ear. Hiding in the library of Sir Henry Bunbury he had produced high notes from pipes and looked to see which of the company present had heard them. He discovered that there was a sharply defined upper limit to audibility and that this varied noticeably with the individual. He also speculated that some insects might communicate by high notes inaudible to humans. His correct conclusions were challenged by the French authority Savart, who erroneously stated that the inaudibility of the high notes was due, not to their frequency, but to their low intensity.

In 1824 Wollaston discussed a traditional problem in physiology, that of binocular vision. In the eighteenth century it had been debated whether this faculty of combining two images was inherited or acquired. Newton had argued for the former possibility in the fifteenth query of his *Opticks*, postulating an arrangement of the optic nerves in which corresponding points of the retinas were connected by nerve fibers that joined before entering the brain. The same theory was proposed by Wollaston, who supposed it to be new. The intricate structure of the human chiasma was still not known. Wollaston was led to adopt the correct arrangement of "semi-decussation," also given by Newton, as a result of his experiences of hemianopia, a disease in which there is a loss of sight in symmetrical parts of each eye. This relation of hemianopia to semi-decussation had also been noticed before; but Wollaston's was the fullest description of the disease that had yet appeared. As a theory of binocular vision it was opposed by those who favored alternative nervous arrangements in the chiasma, and by those who continued to insist that an explanation in terms of acquisition was required. The invention of Wheatstone's stereoscope emphasized the psychological character of binocular vision which had been ignored in the physiological explanations of Newton and Wollaston.

Wollaston and Humphry Davy both died within a few months of each other. It became common to contrast the soaring poetic imagination of Davy with the cautious approach of Wollaston. It is clear from his letters to Young that Wollaston warned against speculation and advised staying close to the facts. It is equally clear from his work that he did not practice what he preached. George Peacock, the Victorian biographer of Young, said of Wollaston that "posterity is not likely to maintain the same high estimate of his powers which was made by his contemporaries."[27] There is growing evidence that this prediction will not stand.

NOTES

1. *Jac. Berzelius Bref*, H. D. Söderbaum, ed., IV, pt. 9 (Uppsala, 1912–1941), p. 73.
2. W. Saunders, *A Treatise on the Chemical History and Medical Powers of Some of the Most Celebrated Mineral Waters*, 2nd ed. (London, 1805), 391–392.
3. "Inquiries Concerning the Nature of a Metallic Substance, Lately Sold in London . . .," in *Edinburgh Review*, 4 (1804), 168.
4. N. Vauquelin, "Mémoire sur le palladium et le rhodium," in *Annales de chimie*, 88 (1813), 170.
5. J. Pelouze, "Note sur la fabrication du platine," in *Annales de chimie et de physique*, 62 (1836), 443–444.
6. L. F. Gilbert, "W. H. Wollaston MSS at Cambridge," in *Notes and Records of the Royal Society*, 9 (1952), 326.
7. Wollaston, "On Super-acid and Sub-acid Salts," in *Philosophical Transactions of the Royal Society*, 98 (1808), 101.
8. Wollaston, "The Bakerian Lecture. On the Elementary Particles of Certain Crystals," *ibid.*, 103 (1813), 61.
9. A. Comte, *Cours de philosophie positive*, III (Paris, 1830–1842), 149.
10. "Proceedings of the Royal Institution," in *Quarterly Journal of Literature, Science and the Arts*, 21 (1826), 109–110.
11. "On the Recent Adjudgment of the Royal Medals . . .," *ibid.*, 1 (1827), 15.
12. A. Ure, *A Dictionary of Chemistry* (London, 1821).
13. "Synoptische Scale der chemischen Aequivalente," in *Journal für Chemie und Physik*, 12 (1814), 105.
14. Faraday, *Chemical Manipulation* (London, 1827), p. 551.
15. T. Graham, *Elements of Chemistry* (London, 1842), 117, 1071.
16. T. Graham, "On the Finite Extent of the Atmosphere," in *Philosophical Magazine*, 1 (1827), 107–109.
17. G. Wilson, "On Wollaston's Argument From the Limitation of the Atmosphere, as to the Finite Divisibility of Matter," in *Transactions of the Royal Society of Edinburgh*, 16 (1849), 79–86.
18. A. v. Humboldt, *Kosmos*, III (Stuttgart–Tübingen, 1845–1862), p. 52.
19. J. F. W. Herschel, *Preliminary Discourse on the Study of Natural Philosophy* (London, 1833), p. 354.
20. S. Léman, "Meionite," in *Nouveau dictionnaire d'histoire naturelle*, XX (Paris, 1816–1819), 28–31.
21. A similar treatment is given in the fifty-eighth observation of Hooke's *Micrographia*.
22. J. Biot, "Recherches sur les réfractions extraordinaires," in *Mémoires de l'Institut national des sciences et arts*, 10 (1810), 6–7.
23. J. Ross, *A Voyage of Discovery . . .* (London, 1819), pp. xviii, 10, app.
24. *Ibid.*, p. 206.
25. B. Hall, *Forty Etchings From Sketches Made With the Camera Lucida in North America in 1827 and 1828* (Edinburgh–London, 1829).
26. A. Marcet, *An Essay on the Chemical History and Medical Treatment of Calculous Disorders* (London, 1817).
27. G. Peacock, *Life of Thomas Young* (London, 1855), p. 470.

BIBLIOGRAPHY

I. ORIGINAL WORKS. In 1949 a collection of Wollaston's notebooks was discovered in the Department of Mineralogy and Petrology, Cambridge. A report containing the essential new information is L. F. Gilbert, "W. H. Wollaston MSS. at Cambridge," in *Notes and Records of the Royal Society of London*, 9 (1952), 311–332. There are also some of his notes and letters in the Science Museum, London, concerning his work on the production of rhodium alloys, platinum wires, and boilers. Additional information on the controversy over palladium is contained in the letters between Chenevix and Banks in volumes XIV and XV of the Dawson Turner copies of the Banks correspondence, Natural History Museum, London. Wollaston's activities on the Board of Longitude are recorded in the minutes of the board at the Royal Greenwich Observatory, Herstmonceux Castle, Sussex. The Royal Society possesses some of Wollaston's letters to Thomas Young, in which he stated his reluctance to speculate. One of these has been published in D. Turner, "Thomas Young on the Eye and Vision," in *Science, Medicine and History*, E. A. Underwood, ed., II (Oxford, 1953), 251. Other letters, in which he commented on the wave theory of light, can be found in T. Young, *Miscellaneous Works*, G. Peacock, ed., I (London, 1855), 233, 261.

Wollaston's published scientific work appeared in the journals. The list given in the Royal Society *Catalogue of Scientific Papers* is almost complete. The following are not mentioned there: "On Gouty and Urinary Concretions," in *Philosophical Transactions of the Royal Society*, 87 (1797), 386–400; "Report from the Select Committee on Weights and Measures," in *Parliamentary Papers*, 3 (1813–1814), 140–141; and "Instructions for the Adjustments and Use of the Instruments Intended for the Northern Expeditions," in *Journal of Science and the Arts*, 5 (1818), 223–226. Some interesting information on the background of Wollaston's work was related by a close acquaintance, the Reverend H. Hasted, in his "Reminiscences of Dr. Wollaston," in *Proceedings of the Bury and West Suffolk Archaeological Institute*, **1** (1849), 121–134.

II. SECONDARY LITERATURE. No satisfactory study of Wollaston's work as a whole has yet been published. A discussion of his work on the platinum metals is included in D. Mc. Donald's excellent *A History of Platinum from the Earliest Times to the Eighteen-eighties* (London, 1960). Useful as general introductions to nineteenth-century crystallography are L. Sohncke, *Entwickelung einer Theorie der Krystallstruktur* (Leipzig, 1879), pp. 5–18; and P. H. von Groth, *Entwickelungsgeschichte der Mineralogischen Wissenschaften* (Berlin, 1926). Wollaston's work in crystallography has been discussed in D. C. Goodman, "Problems in Crystallography in the Nineteenth Century," in *Ambix*, **16** (1969), 152–166. For his fluctuating views on atoms, see D. C. Goodman, "Wollaston and the Atomic Theory of Dalton," in *Historical Studies in the Physical Sciences*, **1** (1969), 37–59. Wollaston's meniscus lenses have been discussed by M. v. Rohr, who states that Wollaston's spectacles were sold in Vienna by Voigtländer. See his "Der grosse Streit bei des Einführung des periskopischen Brillenglases," in *Central-Zeitung für Optik und Mechanik*, **43** (1922), 490–491; "Contributions to the History of English Opticians in the First Half of the Nineteenth Century (With Special Reference to Spectacle History)," in *Transactions of the Optical Society*, **28** (1926–1927), 117–144; and "Meniscus Spectacle Lenses," in *British Journal of Physiological Optics*, **6** (1932), 183–187. Another useful article is H. C. King. "The Life and Optical Work of W. H. Wollaston," in *British Journal of Physiological Optics*, **11** (1954), 10–31. A Herschel MS, which tells of the intrigues involving Wollaston in the election of a successor to Banks, has been discussed in L. F. Gilbert, "The Election to the Presidency of the Royal Society in 1820," in *Notes and Records of the Royal Society*, **11** (1955), 256–279.

D. C. GOODMAN

WOLTMAN, REINHARD (*b.* Axstedt, Germany, December 1757; *d.* Hamburg, Germany, 20 April 1837), *hydraulics*.

The son of a farmer, Woltman, while quite young, taught school, presumably at Axstedt. Within a short time, however, he was transferred to the part of the Ritzebüttel district that borders on the North Sea between the Elbe and Weser rivers, an area where the shore-erosion problem constitutes a challenge to the best hydraulic engineers. During the previous century its protection facilities had been destroyed by storms and floods at least seven times. Developing stronger means of protection became an early and primary objective of Woltman's career.

In May 1779, Woltman was appointed an under-inspector and clerical employee in the office responsible for erecting and maintaining the erosion-control structures. The following year he began taking courses in mathematics and architecture at Hamburg, and later at the universities of Kiel and Göttingen. Woltman subsequently made a journey during which he met skilled workers in hydraulics at Frankfurt, Strasbourg, Paris, Cherbourg, Calais, Dover, London, and Holland. Upon his return to Ritzebüttel on 20 November 1784, he began work on the local erosion problem.

Woltman's first book, *Theorie und Gebrauch des hydrometrischen Flügels* (1790), drew attention to the extensive use in England and Holland of wind and water power. That, he contended, should be done in the Hamburg area, and added that if his idea was adopted, frequent measurements of wind and water velocities could provide data of great value. The remainder of his book was devoted largely to the instruments he had designed for that purpose.

The four-volume *Beiträge zur hydraulischen Architektur* (1791–1799) immediately attracted attention in the scientific community with the first volume, which concerned the management of dikes and reinforcing shorelines. In 1792 Woltman became a member of the Hollandsche Maatschappij der Wetenschappen, at Haarlem; and the Bataafsch Genootschap der Proefondervindelijke Wijsbegeerle, at Rotterdam; and was offered membership in the Königliche Gesellschaft der Wissenschaften, at Prague. In the following year he received an offer to join the Königliche Akademie der Wissenschaften, at Göttingen.

On 1 October 1797 Woltman married Johanna Schuback, the daughter of his first patron. They had five children.

Antoine Chézy (1718–1798) and Pierre Du Buat, contemporaries of Woltman's, conceived what may have been the earliest valid equations concerning the velocity of flowing water. Woltman probably never heard of Chézy, whose formula attracted little attention until 1897, when Clemens Herschel wrote an article about it. Du Buat's work (published in 1779 under the title *Principes d'hydraulique*), however, received immediate and highly favorable attention. Its velocity equation was so complicated and so difficult to apply that Woltman recommended a much simpler one. Much of its simplification was made possible by experiments that Woltman had conducted and that justified the use of powers (exponents) for velocity ranging from 1.75 (for pipes) to 2 (for open channels).

Those values, as noted by Hunter Rouse, "are precisely the limits now accepted."

The invention of Woltman's water current meter in 1790 brought him the most lasting fame. One of the original models has been preserved at the Deutsches Museum in Munich, and a replica of it is on display at the National Museum of History and Technology in Washington, D.C.

Just before he constructed the current meter, Woltman had designed two anemometers, one of which was intended to be mounted on a standard and the other to be held in one's hand. These, he admitted, were largely patterned after C. G. Schober's anemometer (described in the 1752 editions of the *Hamburgisches Magazin*). After numerous futile attempts to evaluate, both experimentally and mathematically, the relationship between the speed of their rotors and the velocity of the wind, he decided to calibrate them in still water. That procedure was successful, and it convinced him that with only a minor change (reducing the rotor to about one-third of its original size), the device would be suitable for measuring the velocity of water flowing in rivers. He thereupon ordered his mechanic (a man named Steinmetz, in Cuxhaven) to build such a model. The ancestry of practically all of the propeller-type current meters in use today can be traced to that particular model.

For many years after Woltman announced that meter's invention, most of the improved versions of it were called "Woltman meters" as a courtesy to him (a commendable practice, but one that unfortunately has resulted in his having erroneously been credited in many modern textbooks and magazines with having built a better instrument than he actually designed). A somewhat similar practice is presently being carried on with respect to the type of meters that measure the amount of water flowing through pipes supplying homes and office buildings. In some instances modern engineers have applied the Woltman principle of operation to such meters, and have identified them as "Woltman meters."

BIBLIOGRAPHY

I. ORIGINAL WORKS. Woltman's first published work was *Theorie und Gebrauch des hydrometrischen Flügels oder eine zuverlässige Methode die Geschwindigkeit der Winde und strömenden Gewässer zu beobachten* (Hamburg, 1790). His other major work was *Beiträge zur hydraulischen Architektur*, 4 vols. (Göttingen, 1791–1799).

II. SECONDARY LITERATURE. *Allegemeine deutsche Biographie*, XLIV, 192–199, lists the titles of several additional works by Woltman relating to canals, navigation, and shore protection. Also see Hegell, "Die Urform des Woltmanschen Flügels," in *Zeitschrift des Verbands deutscher Architekten- und Ingenieurvereine* (11 Apr. 1914), 129–130; Steponas Kolupaila, *Bibliography of Hydrometry* (Notre Dame, Ind., 1961), 12, 282, 326, 328–329; and Hunter Rouse and Simon Ince, *History of Hydraulics* (Ann Arbor, Mich., 1957), 134–136, 141.

ARTHUR H. FRAZIER

WOOD, HORATIO C (*b*. Philadelphia, Pennsylvania, 13 January 1841; *d*. Philadelphia, 3 January 1920), *pharmacology, therapeutics.*

Wood's father, Horatio Curtis Wood, was a successful Philadelphia businessman. His mother, Elizabeth Head Bacon Wood, was descended, like her husband, from English immigrants of the seventeenth century. Wood attended Quaker schools, then in 1862 graduated with an M.D. from the medical department of the University of Pennsylvania. After residencies in the Philadelphia Hospital (Blockley) and the Pennsylvania Hospital, followed by medical service with the Union army, he entered private practice in 1865 or 1866. In 1866 he assumed the chair of botany in the auxiliary faculty of medicine at the University of Pennsylvania. In the same year he married Elizabeth Longacre, by whom he had a daughter and three sons, two of whom became physicians.

Wood was a pioneer in experimental pharmacology and therapeutics, supplemented by other contributions to the field of materia medica. About fifty of his nearly three hundred publications dealt with pharmacology, experimental therapeutics, and physiology. A high proportion of this segment of his papers involved laboratory study of the physiological action of drugs in animals, still exceptional at the time. Moreover, Wood understood the import and potential of work by such contemporaries as Oswald Schmiedeberg, A. R. Cushny, Benjamin Ward Richardson, Alexander Crum Brown, Thomas R. Fraser, and Thomas Lauder Brunton. Thus Wood became an early American exponent of the study of the physiological action of drugs under laboratory conditions and of classifying medicines according to such actions, thereby supplanting the primacy of empirical clinical experience as the basic guide to progressive therapeutics.

At the age of twenty Wood published his first paper, "Contributions to the Carboniferous Flora of the United States" (1861). During the 1860's he published at least eleven papers on freshwater algae and fourteen on the myriopods, thereafter devoting himself increasingly to clinical pathology (especially neurology and psychiatry), experimental pharmacology, and therapeutics. In the latter field his first important experimental paper reported "On the Medical Activity of the Hemp Plant [Marijuana], as Grown in North America" (1869). Among his most important subsequent investigations, as Wood himself believed, were studies of the physiology and treatment of sunstroke, the mechanism and treatment of fever, the discovery of the physiological and therapeutic action of hyoscine, and the treatment of accidents of anesthesia. Wood appears more important as an American exponent of animal experimentation and of the applications of new findings and a new outlook to therapeutics than as an originator of new methods.

Wood's laboratory did not serve to train a new school of pharmacologists as did that of John J. Abel. The reasons, although never seriously assessed, may relate to their time, funding of the laboratories, and even the personalities of the two men. Abel told G. B. Roth that during the summer of 1884 he served in Wood's laboratory as "a research assistant, without pay," then left to study in Ludwig's laboratory at Leipzig.

Beyond his scientific papers Wood's strong national influence was mediated by prolific editorial work. His *Treatise on Therapeutics . . .* (1874), which went through thirteen editions, helped foster the transition from case-based therapeutics to what Wood considered an experiment-based "applied science." He published five other medical books. He served as senior editor of the ubiquitous *Dispensatory of the United States of America* for five editions (from the fifteenth edition of 1883), edited at least three medical journals at various times, and from 1866 edited American editions of the British *Manual of Materia Medica and Therapeutics* (Frederick J. Farre, Robert Bentley, and Robert Warington's updated abridgment of Jonathan Pereira's *Elements of Materia Medica and Therapeutics*).

The Pharmacopeia of the United States, which generates legally enforceable standards for drugs, had the benefit of Wood's services from 1890 to 1910 as president of the policy-setting U.S. Pharmacopeial Convention. In 1902 he and Frederick B. Power represented the United States government in Brussels at the International Conference for the Unification of the Formulae of Heroic Medicines.

In 1906 poor health forced Wood to retire from academic work; and he considered that the presidential address he sent in 1910 from his sickbed to the U.S. Pharmacopeial Convention brought his career to a close.

If Wood's professional work was more forward-looking than daring, it accorded with a personality basically conservative, although critically evaluative of the old empirical therapeutics. If the practical dominated the theoretical in his work, he offered in a transitional period a combination of originality, productivity, and literary skill that influenced the shift of American pharmacology and therapeutics toward experimental, quantifiable science.

BIBLIOGRAPHY

I. Original Works. A "Bibliographic Record 1860–1911" of Wood's publications appears in *Transactions and Studies of the College of Physicians of Philadelphia*, 3rd ser., **42** (1920), 242–257, which lists seven books on aspects of clinical medicine and materia medica, and published papers that may be classified as 142 on clinical pathology, neurology, medicine, and therapeutics; fifty-four on experimental pharmacology, physiology, and pathology; forty-four published addresses and lectures; fifteen papers on botany; fourteen on entomology; thirteen on medical jurisprudence and toxicology; and ten articles primarily for the laity. Some MS material is in the library of the College of Physicians of Philadelphia. Wood's useful "Reminiscences . . .," written "toward the close of his life," were published in the *Transactions and Studies of the College of Physicians of Philadelphia*, 3rd ser., **42** (1920), 195–234, with nine topical sections. Most of his publications are likewise in the library of the College of Physicians, and a large majority of them also are in the National Library of Medicine. For orientation to Wood's level of work, thought, and style, see *A Treatise on Therapeutics, Comprising Materia Medica and Toxicology, with Especial Reference to the Application of the Physiological Action of Drugs to Clinical Medicine* (Philadelphia, 1874); "Hyoscine; Its Physiological and Therapeutic Action," in *Therapeutic Gazette*, **9** (1885), 1–10; and "On the Medical Activity of the Hemp Plant [Marijuana], as Grown in North America," in *Transactions of the American Philosophical Society*, **11** (1869), 226–232. Wood's first publication was "Contributions to the Carboniferous Flora of the United States," in *Proceedings of the Philadelphia Academy of Natural Sciences*, no. 1 (1860), 236–240, no. 2 (1860), 519–552.

II. Secondary Literature. Roth explains that Horatio C Wood had no middle name, hence insisted that no period be used after the "C" of his name. The varying result in the literature is further aggravated by his father being named Horatio Curtis Wood and his son, Horatio Charles Wood, Jr.

There is no definitive biography of Wood. A reliable

article, emphasizing his professional life, is George B. Roth, "An Early American Pharmacologist: Horatio C Wood (1841–1920)," in *Isis*, **30** (1939), 37–45. More revealing of his personality are Hobart Amory Hare, "Horatio C. Wood, The Pioneer in American Pharmacology," in *Therapeutic Gazette*, **44** (1920), 322–324; and G. E. de Schweinitz, "Dr. H. C. Wood as a Medical Teacher," in *Transactions and Studies of the College of Physicians of Philadelphia*, 3rd ser., **42** (1920), 235–241. Of supplemental usefulness are G. E. de Schweinitz, "Memoir of Dr. H. C Wood," *ibid.*, pp. 155–165; F. X. Dercum, "Memoir of Dr. H. C Wood," *ibid.*, pp. 166–169; Charles K. Mills, "Reminiscences of Dr. Horatio C Wood," *ibid.*, pp. 175–186; Henry Beates, Jr., "Professor Horatio C. Wood," in *American Journal of Pharmacy*, **77** (1905), 376–379; Henry Beates, Jr., "Horatio C. Wood, M.D.," in *Medical Record*, **98** (1920), 393–396; and *Family Sketches, Compiled and Arranged by Julianna R. Wood* (Philadelphia, 1870).

GLENN SONNEDECKER

WOOD, ROBERT WILLIAMS (*b.* Concord, Massachusetts, 2 May 1868; *d.* Amityville, New York, 11 August 1955), *experimental physics.*

Wood's chief contributions to science lay in physical optics, particularly in spectroscopy, in which he obtained experimental results of great significance for the advance of atomic physics during the first third of the twentieth century. He was an extremely versatile laboratory worker, and his insatiable curiosity took him into many other scientific and technical fields, such as the photography of sound waves, properties of ultrasonic radiation, color photography, molecular physics, the manufacture of high-precision diffraction gratings, fluorescence, and scientific crime detection. His book *Physical Optics* (1905) became the classic treatise on the experimental aspects of the subject in its day, and went through three editions.

Wood did his undergraduate work at Harvard College, where in 1891 he received the B.A. degree with a major in chemistry. His academic record was undistinguished in the required fields of languages and mathematics, but he early showed enthusiasm for all aspects of science. His graduate work took him to Johns Hopkins, Chicago, Berlin, and the Massachusetts Institute of Technology. This graduate study never led to an earned doctorate, and Wood had to content himself with honorary degrees. Early in his career he shifted from chemistry to physics; and by the time he began teaching at the University of Wisconsin in 1897, he had definitely settled down to the life of a physicist. In 1901 Wood succeeded Henry A. Rowland

as professor of experimental physics at Johns Hopkins, where he remained for the rest of his life, retiring officially in 1938 but continuing in an honorary capacity until his death.

Before he assumed his professorial duties at Johns Hopkins, Wood had published more than thirty papers covering a wide range of investigations in physics and chemistry. This early work clearly indicated his zest for experimentation and his great ingenuity in devising relatively simple ways to exhibit spectacular effects. Color always fascinated him and entered vitally into much of his work. Zeal for mathematical symbolism seems to have been missing from his make-up. He preferred to express his experimental results in terms of physical pictures that he felt he (and many others) could understand better than mathematical equations, which he found rather boring. This trait, which persisted throughout his life, in no way interfered with the fundamental logic of his physical reasoning.

At Johns Hopkins Wood's teaching duties were light and he devoted himself mainly to research, which for over three decades was concentrated primarily on the optical properties of gases and vapors, a field in which he soon established himself as an internationally recognized authority. His work on sodium vapor was especially noteworthy. The extensive precision measurements of atomic spectra stimulated much similar work in this area, and all of it proved to be of the utmost importance in connection with the growing interest in atomic models during the early years of the twentieth century. The Bohr theory, in particular, leaned very heavily on spectroscopic data such as those provided by Wood and others whose work was suggested by him. This was particularly true of Wood's fundamental experimental work from 1903 to 1920 on fluorescence and resonance radiation of vapors and also the effect of electric and magnetic fields on spectrum lines. Wood also greatly stimulated research in optical spectroscopy by his improvements in the diffraction grating and his determination to ensure a steady supply of high-quality gratings to investigators throughout the world by keeping Rowland's ruling engines steadily at work.

Throughout his professional career Wood was continually on the lookout for interesting new phenomena to study, particularly those involving striking effects. During World War I and in the late 1920's he became interested in high-frequency sound waves and their physical and biological properties. Wood performed many experiments in collaboration with A. L. Loomis at the latter's Tuxedo Park laboratory. These experiments received considerable publicity through the Colver lectures at Brown University in 1937 and the book based on them, *Supersonics, the Science of Inaudible Sounds* (1939). This work, although popular in presentation, aroused great interest and undoubtedly stimulated research in the important field of acoustics now known as ultrasonics.

A master lecturer with highly developed showmanship, Wood much enjoyed talking to large audiences and demonstrating his ideas with graphic experiments—the more spectacular the better. His name on the program of a meeting of the American Physical Society always guaranteed the attendance of a large crowd to hear his paper. Much more confident in experimental results than in abstract, mathematical reasoning, Wood was never satisfied until an experiment showed with complete clarity the idea he was trying to communicate. He often carried his showmanship into the perpetration of practical jokes; this led to the invention of a host of stories about his exploits that have become legendary among American physicists.

In his teaching of optics, Wood early felt the need for a book that would emphasize, to a greater extent than available standard texts, the experimental aspects of twentieth-century research on light. His *Physical Optics* satisfied an existing need for more information about the kind of optical techniques he himself was introducing. A fourth edition was being prepared at the time of his death.

Inevitably, as a result of his zeal for experimentation, Wood made many inventions. He was usually too restless, however, to follow them through the development stage into practical, salable devices. He was more successful financially as a legal consultant in cases involving scientific and related technical matters, as well as with his delightful little book *How to Tell the Birds From the Flowers, and Other Woodcuts*. First published in 1917, it went through twenty editions. Written to amuse Wood's children, it entertained hosts of others who knew nothing of his scientific achievements. Also worthy of mention is his experimentation with humorous scientific verse, of which his "Contemptuary Science" (relativity and Michelson optics) is a good example.

Wood traveled widely both in the United States and abroad. He received many honors in recognition of his accomplishments. In addition to his membership in the National Academy of Sciences, he was one of the relatively few foreign members of the Royal Society. He also belonged to numerous other foreign academies and societies, and received six honorary degrees and numerous medals.

BIBLIOGRAPHY

I. ORIGINAL WORKS. Wood's complete scientific bibliography includes 2 books and 227 articles. The complete list is in Dieke's memoir (below). The following is a selection intended to illustrate the breadth of his work.

His books are *Physical Optics* (New York–London, 1905; 2nd ed., 1911; 3rd ed., 1934) and *Supersonics, the Science of Inaudible Sounds* (Providence, R.I., 1939).

Wood's earliest articles are "The Kingdom of the Dream. Experience With Hasheesh," in New York *Sunday Herald* (23 Sept. 1888); "Effects of Pressure on Ice," in *American Journal of Science*, 3rd ser., **41** (1891), 30–33; "Eine einfache Methode, die Dauer von Torsions-schwingungen zu bestimmen," in Wiedemann's *Annalen der Physik und Chemie*, n.s. **56** (1895), 171–172; "Ueber eine neue Form der Quecksilber Luftpumpe und die Erhaltung eines guten Vacuums bei Röntgen'schen Versuchen," *ibid.*, **58** (1896), 205–208; "On the Absorption Spectrum of Solutions of Iodine and Bromine Above the Critical Temperature," in *Philosophical Magazine*, 5th ser., **41** (1896), 423–431; *Zeitschrift für physikalische Chemie*, **19** (1896), 689–695; "A New Form of Cathode Discharge and the Production of X-Rays, Together With Some Notes on Diffraction," in *Physical Review*, **5** (1897), 1–10; "Phase-Reversal Zone-Plates and Diffraction Telescope," in *Philosophical Magazine*, 5th ser., **45** (1898), 511–522; "The Anomalous Dispersion of Cyanin," *ibid.*, **46** (1898), 380–386; "An Application of the Diffraction Grating to Colour Photography," *ibid.*, **47** (1899), 368–372; and "Photography of Sound Waves by the 'Schlieren Methode,'" *ibid.*, **48** (1899), 218–227.

Papers written early in the twentieth century are "Zone Plate Photography," in *Photographic Journal*, **24** (1900), 248–250; "The Photography of Sound Waves and the Demonstration of the Evolutions of Reflected Wave Fronts With the Cinematograph," in *Philosophical Magazine*, 5th ser., **50** (1900), 148–157; *Chemical News*, **81** (1900), 103; *Proceedings of the Royal Society*, **A66** (1900), 283–290; *Report of the Board of Regents of the Smithsonian Institution* for 1900 (1901), 359; "An Application of the Method of Striae to the Illumination of Objects Under the Microscope," in *Philosophical Magazine*, 5th ser., **50** (1900), 347–349; "Vortex Rings," in *Nature*, **63** (1901), 418–420; "The Problem of the Daylight Observation of the Corona," in *Astrophysical Journal*, **12** (1901), 281–286; "The Nature of the Solar Corona," *ibid.*, **13** (1901), 68–79; "The Anomalous Dispersion of Carbon," in *Philosophical Magazine*, 6th ser., **1** (1901), 405–410; "On the Production of a Bright-Line Spectrum by Anomalous Dispersion and Its

Application to the 'Flash-Spectrum,'" *ibid.*, 551–555; *Naturwissenschaftliche Rundschau*, **16** (1901), 394; *Astrophysical Journal*, **13** (1901), 63–67; "Anomalous Dispersion of Sodium Vapour," in *Proceedings of the Royal Society*, **69** (1901), 157–171; "The Invisibility of Transparent Objects," in *Physical Review*, **15** (1902), 123–124; "Absorption, Dispersion, and Surface Colour of Selenium," in *Philosophical Magazine*, 6th ser., **3** (1902), 607–622; "On a Remarkable Case of Uneven Distribution of Light in a Diffraction Grating Spectrum," *ibid.*, **4** (1902), 396–402; and "The Kinetic Theory of the Expansion of Compressing Gas Into a Vacuum," in *Science*, **16** (1902), 908–909.

Also see "Screens Transparent Only to Ultra-Violet Light and Their Use in Spectrum Photography," in *Philosophical Magazine*, 6th ser., **5** (1903), 257–263; *Physikalische Zeitschrift*, **4** (1903), 337–338; and *Astrophysical Journal*, **17** (1903), 133–140; "On the Anomalous Dispersion, Absorption and Surface Colour of Nitroso Dimethyl Aniline With a Note on the Dispersion of Toluene," in *Philosophical Magazine*, 6th ser., **6** (1903), 96–112; *Records of the American Academy of Arts and Sciences*, **39** (1903), 51–66; "Electrical Resonance of Metal Particles for Light Waves. Third Communication," in *Philosophical Magazine*, 6th ser., **6** (1903), 259–266; "Fluorescence and Absorption Spectra of Sodium Vapour," *ibid.*, 362–374; and *Astrophysical Journal*, **18** (1903), 94–111; written with J. H. Moore; "The N-Rays (Letter Exposing Delusion)," in *Nature*, **70** (1904), 530–531; "Apparatus to Illustrate the Pressure of Sound Waves," in *Physical Review*, **20** (1905), 113–114; and *Physikalische Zeitschrift*, **6** (1905), 22; "The Magnetic Rotation of Sodium Vapour," in *Physical Review*, **21** (1905), 41–51, written with H. W. Springsteen; "The Magneto-Optics of Sodium Vapour and the Rotatory Dispersion Formula," in *Philosophical Magazine*, 6th ser., **10** (1905), 408–427; "The Fluorescence of Sodium Vapour and the Resonance Radiation of Electrons," *ibid.*, 513–525; "Abnormal Polarization and Colour of Light Scattered by Small Absorbing Particles," *ibid.*, **12** (1906), 147–149; "Die Temperaturstrahlung des Joddampfes," in *Physikalische Zeitschrift*, **8** (1907), 517; "Polarized Fluorescence of Metallic Vapors and the Solar Corona," in *Astrophysical Journal*, **28** (1908), 75–78; "An Extension of the Principal Series of the Sodium Spectrum," in *Philosophical Magazine*, 6th ser., **16** (1908), 945–947; "The Mercury Paraboloid as a Reflecting Telescope," in *Astrophysical Journal*, **29** (1909), 164–176; and "The Ultra-Violet Absorption, Fluorescence, and Magnetic Rotation of Sodium Vapour," in *Philosophical Magazine*, 6th ser., **18** (1909), 530–535.

During the second decade of the century, Wood wrote "Determination of Stellar Velocities With the Objective Prism," in *Astrophysical Journal*, **31** (1910), 376–377; "The Echelette Grating for the Infra-Red," in *Philosophical Magazine*, 6th ser., **20** (1910), 770–778; "The Resonance Spectra of Iodine," *ibid.*, **21** (1911), 261–265;

"Diffraction Gratings With Controlled Groove Form and Abnormal Distribution of Intensity," *ibid.*, **23** (1912), 310–317; and *Physikalische Zeitschrift*, **13** (1912), 261–264; "Selective Absorption of Light on the Moon's Surface and Lunar Petrography," in *Astrophysical Journal*, **36** (1912), 75–84; "Method of Obtaining Very Narrow Absorption Lines for Investigations in Magnetic Fields," in *Physikalische Zeitschrift*, **14** (1913), 405, written with P. Zeeman; "The Satellites of the Mercury Lines," in *Philosophical Magazine*, 6th ser., **25** (1913), 443–449; and *Physikalische Zeitschrift*, **14** (1913), 273–275; "The Effect of Electric and Magnetic Fields on the Emission Lines of Solids," in *Philosophical Magazine*, 6th ser., **30** (1915), 316–320, written with C. E. Mendenhall; "Monochromatic Photographs of Jupiter and Saturn," in *Astrophysical Journal*, **43** (1916), 310–319; and "Condensation and Reflection of Gas Molecules," in *Philosophical Magazine*, 6th ser., **32** (1916), 364–371.

In the 1920's Wood produced "Light Scattering by Air and the Blue Colour of the Sky," in *Philosophical Magazine*, 6th ser., **39** (1920), 423–433; "Extension of the Balmer Series of Hydrogen, and Spectroscopic Phenomena of Very Long Vacuum Tubes," in *Proceedings of the Royal Society*, **A97** (1920), 455–470; "On the Influence of Magnetic Fields on the Polarization of Resonance Radiation," *ibid.*, **A103** (1923), 396–403, written with A. Ellett; "Fine Structure, Absorption and Zeeman Effect of the 2536 Mercury Line," in *Philosophical Magazine*, 6th ser., **50** (1925), 761–774; and *Nature*, **115** (1925), 461; "Improved Grating for Vacuum Spectrographs," in *Philosophical Magazine*, 7th ser., **2** (1926), 310–312, written with T. Lyman; "The Physical and Biological Effects of High Frequency Sound Waves of Great Intensity," *ibid.*, **4** (1927), 417–436, written with A. L. Loomis; "Anti-Stokes Radiation of Fluorescent Liquids," *ibid.*, **6** (1928), 310–312; "Raman Spectra of Scattered Radiation," *ibid.*, 729–743; and "Improved Technique for the Raman Effect," in *Physical Review*, **33** (1929), 294, and **36** (1930), 1421–1430.

His last works were "Absorption Spectra of Salts in Liquid Ammonia," in *Physical Review*, **38** (1931), 1648–1650; "The Purple Gold of Tut-Ankhamun," in *Journal of Egyptian Archaeology*, **20** (1934), 62–65; "Fluorescence of Chlorophyll in Its Relation to Photochemical Processes in Plants and Organic Solutions," in *Journal of Chemical Physics*, **4** (1936), 551–560, written with J. Franck; and "Improved Diffraction Gratings and Replicas," in *Journal of the Optical Society of America*, **34** (1944), 509–516.

II. SECONDARY LITERATURE. See the biographical sketch by G. H. Dieke in *Biographical Memoirs of Fellows of the Royal Society*, **2** (1956), 327–345; and William Seabrook, *Doctor Wood—Modern Wizard of the Laboratory* (New York, 1941).

R. B. LINDSAY

WOODHOUSE, ROBERT (*b*. Norwich, England, 28 April 1773; *d*. Cambridge, England, 28 [23?] December 1827), *mathematics*.

Woodhouse was a critic and reformer. The son of Robert Woodhouse, a linen draper, and of the daughter of J. Alderson, a nonconformist minister, Woodhouse attended the grammar school at North Walsham. In 1790 he was admitted to Caius College, Cambridge, and four years later graduated with the B.A., as senior wrangler and first Smith's prizeman. In 1798 he received the M.A. from the university, and was successively fellow (1798–1823), Lucasian professor of mathematics (1820–1822), and Plumian professor of astronomy and experimental philosophy (1822–1827). Woodhouse also served as the first superintendent of the astronomical observatory at Cambridge. In 1802 he was elected a fellow of the Royal Society. He married Harriet Wilkens in 1823; they had one son, Robert.

Woodhouse was primarily interested in what was then called the metaphysics of mathematics; that is, he was concerned with questions such as the proper theoretical foundations of the calculus, the role of geometric and analytic methods, the importance of notation, and the nature of imaginary numbers. Many of these questions are discussed in his *Principles of Analytical Calculation* (1803), a polemic aimed primarily at the fellows and professors at Cambridge. In this work Woodhouse defended analytic methods, the differential notation, and a theory of calculus based, like that of Lagrange, on series expansions. It does not appear to have had much influence in the introduction of continental methods at Cambridge. His elementary text on trigonometry (1809), however, was widely used. George Peacock, who himself played a decisive role in the reform of mathematical studies at Cambridge, considered this work to be of major importance in achieving this goal. It was not polemical, but used analytic methods and the differential notation throughout.

Woodhouse's other writings include a history of the calculus of variations (1810), a treatise on astronomy (1812), and a work on the theory of gravitation, somewhat misnamed *Physical Astronomy* (1818). In all these works Woodhouse presented the results of continental research from the time of Newton up to his own time.

BIBLIOGRAPHY

I. ORIGINAL WORKS. Woodhouse's papers include "On the Necessary Truth of Certain Conclusions Ob-tained by Means of Imaginary Quantities," in *Philosophical Transactions of the Royal Society*, **91** (1801), 89–119; and "On the Independence of the Analytical and Geometrical Methods of Investigation; and on the Advantages To Be Derived From Their Separation," *ibid.*, **92** (1809), 85–125. His books are *Principles of Analytical Calculation* (Cambridge, 1803); *A Treatise on Plane and Spherical Trigonometry* (Cambridge, 1809; 5th rev. ed., 1827); *A Treatise on Isoperimetrical Problems and the Calculus of Variations* (Cambridge, 1810), reprinted as *A History of the Calculus of Variations in the Eighteenth Century* (New York, n.d.); *Treatise on Astronomy* (Cambridge, 1812); and *Physical Astronomy* (Cambridge, 1818).

II. SECONDARY LITERATURE. The fullest account of Woodhouse's life and work is in Augustus DeMorgan, "Robert Woodhouse," in *Penny Cyclopaedia*, XXVII (London, 1843), 526–527. Woodhouse's influence is considered in Elaine Koppelman, *Calculus of Operations: French Influence in British Mathematics in the First Half of the Nineteenth Century* (Ph.D. diss., Johns Hopkins University, 1969).

ELAINE KOPPELMAN

WOODWARD, JOHN (*b*. Derbyshire, England, 1 May 1665; *d*. London, England, 25 April 1728), *geology*, *mineralogy*, *botany*.

Woodward was said to have been the son of a man of good family from Gloucestershire. He was educated at a country school, where he became proficient in Latin and Greek. About 1680, at the age of sixteen, he was apprenticed to a linen draper in London, but he abandoned this occupation to pursue a further course of study. A few years later he became acquainted with Peter Barwick, physician in ordinary to Charles II. Barwick was impressed by Woodward's ability and about 1684 took him into his household to study medicine; Woodward remained there about four years. During this period, while on a visit to Sherborne, in Gloucestershire, he studied botany in the surrounding country. While on these excursions he learned for the first time that rocks may contain fossil animal remains—they are particularly common in the Jurassic rocks in that neighborhood. These fossils greatly interested Woodward, and he resolved to investigate their occurrence in other parts of the country. This he undoubtedly did, although no detailed account of the course of his investigation has survived. Later in Oxford he made the acquaintance of two well-known naturalists, Robert Plot, keeper of the Ashmolean Museum, and his assistant, Edward Lhwyd, who also were interested in fossils. On this occasion he seems to have told Lhwyd that he had already formed a theory explaining the origin of fossils, then still a matter for debate. In 1690 Lhwyd wrote to a friend stating that Woodward seemed well informed for his age but that he doubted that he was sufficiently experienced to satisfy others on such a debatable matter. This visit to Gloucestershire, however, led Woodward to adopt the study of geology and mineralogy as one of his major interests for the rest of his life. Meanwhile he continued to study medicine and botany.

In 1692, at the age of twenty-seven, Woodward was appointed professor of physic at Gresham College, London. His candidature had been supported by Barwick and Plot, among others. Evidently his abilities had now become more widely known. Barwick stated in his testimonial that Woodward "had made the greatest advance not only in physick, anatomy, botany, . . . but likewise in all other useful learning of any man I ever knew of his age . . .," and that he was "very much respected upon this account by persons of the greatest judgement and learning." Not long after his appointment Woodward took up residence at Gresham College. He was elected fellow of the Royal Society in 1693, and in 1695 was awarded the degree of doctor of medicine by special dispensation of Thomas Tenison, archbishop of Canterbury. In 1696 Woodward was granted an M.D. by the University of Cambridge, and he was elected fellow of the College of Physicians in March 1703. As a lecturer in Gresham College his obligations were relatively light, and he established himself as a practicing physician at least as early as 1709. In 1718 he published his only medical work, *The State of Physick and of Diseases . . . More Particularly of the Smallpox*. The treatment he recommended contradicted the views of some eminent contemporary physicians, notably Richard Mead, and led to a duel between the latter and Woodward.

Woodward continued to reside at Gresham College until his death. He spent much time cataloging his geological and mineralogical specimens, and his museum was often visited by other naturalists, who have left accounts of it. Contemporary records show that, at least in later life, Woodward, although not without friends, was a man of unattractive character, conceited, quarrelsome, and dogmatic. He quarreled with the council of the Royal Society and made enemies of other naturalists. His contributions to science were nevertheless of considerable importance.

Woodward made a valuable contribution to bo-

tanical science as a result of a series of systematic experiments on plant nutrition carried out in 1691 and 1692, a detailed account of which was published by the Royal Society in 1699. The most important result of this investigation was a clear demonstration that the greater part of the water absorbed by a growing plant is exhaled through its pores into the atmosphere. This was the first demonstration of transpiration. Woodward also claimed that the food of plants is not water, but the mineral substances dissolved in the water.

Woodward is chiefly remembered for his contributions to the earth sciences. In his first contribution, *Essay Toward a Natural History of the Earth* (1695), which he claimed was based on his own observations, Woodward assumed that the earth formerly had been submerged beneath a universal deluge, the waters of which had originated in a central abyss within the earth. These waters had dissolved, or disintegrated and held in suspension, all the stony and mineral matter forming the outer crust of the earth. At the same time all the animals and plants then living were submerged in the waters but were not destroyed beyond recognition. From the confused mass that had formed, the matter in suspension, both organic and inorganic, subsided in an order determined, so far as was possible, by the specific gravity of the individual components. Thus a stratigraphic succession was formed in which the specific gravity of both the organic remains and the rock matrix in which they occurred decreased gradually in passing upward in the succession.

Woodward asserted unequivocally that the fossil organic remains that had been found in rocks were definitely the remains of living animals or plants, a view not universally held at that time. The term "fossil" was then widely used to denote anything that was dug out of the earth, whether mineral substances or organic remains. In Woodward's terminology stones and minerals were "native" fossils, and organic remains were "extraneous" fossils.

Woodward had observed that particular rock formations might contain a different assemblage of extraneous fossil forms to those occurring in beds above or below the formation. While he realized that this observation required explanation, it is perhaps not surprising that he failed to recognize the true explanation; and the one he offered was soon criticized.

Woodward's *Essay* was widely read both in Great Britain and, in translation, in other European countries. The great Swiss naturalist J. J. Scheuchzer was converted by the *Essay* to belief in the organic origin of animal remains in rocks; and he translated the *Essay* into Latin. The information contained in the *Essay* and in Woodward's later works also contributed toward establishing that strata throughout the world are, generally speaking, similar in character, a conclusion necessary before any acceptable theory of the origin of the rocks of the crust of the earth could be formulated.

In 1696 Woodward published an anonymous twenty-page pamphlet entitled *Brief Instructions for making Observations in All Parts of the World: as Also for Collecting, Preserving, and Sending Over Natural Things*. This work is of considerable interest, for even today it might, with little emendation, serve the purpose for which it was intended. Woodward circulated the work, and by this means and through correspondents in many countries ultimately formed a large collection of fossils and minerals. The value of this collection lay in the fact that he studied his specimens carefully, and later published pioneer attempts to classify both minerals and fossils systematically.

Woodward first discussed minerals in Section 4 of his *Essay*, where he admitted the difficulties in determining the nature of individual minerals, inevitable at that time, and made greater because he included rocks with his minerals. His first attempt at a classification appeared in 1704, under the head "Fossils," in John Harris' *Lexicon chemicum*. Here he divided minerals into six classes: "earths, stones, salts, bitumina, metallick minerals and metals." This was reprinted with little change, in his *Naturalis historia telluris* (1714). Later Woodward greatly enlarged this classification. His *Fossils of All Kinds Digested Into a Method* (1728) contains a large folding table setting out his mineral classification systematically. It retains the six major classes, but adds many more subdivisions. The text includes fifty-six pages in which some two hundred minerals are described, together with their mode of occurrence in some cases. Some specimens were examined under the microscope, and Woodward mentions the characteristics used in making his determinations.

Woodward's last work, published posthumously in 1729, was *An Attempt Towards a Natural History of the Fossils of England*. This was a detailed catalogue of his collection of both British and foreign minerals and fossils, some 4,000–5,000 in number, occupying about 600 pages of small print. Localities of specimens are given with, frequently, the names of correspondents who supplied them, and the use of some minerals is discussed.

In this catalogue Woodward enlarged his systematic classification of minerals to include eleven classes. He also classified animal and vegetable fossils, the former into fifteen classes, and the latter into five groups. His classification of fossils was more elaborate and rational than that used by Lhwyd in his *Lithophylacii Britannici ichnographia* (1699). Woodward's catalogue was used by geologists for almost a century after its publication. The specimens he collected are now preserved in the Sedgwick Museum, Cambridge, in their original cases.

Woodward's last contribution to the advancement of geological science was to leave in his will funds to establish at the University of Cambridge a lectureship bearing his name; this eventually evolved into the Woodwardian chair in geology, the earliest such post in the subject in any British university.

Woodward was the first British author to publish a systematic classification and description of minerals based on his own observations; and he emphasized the importance to his country of its mineral wealth. His classification of fossil organic remains was one of the earliest of its kind. While Woodward does not rank high as an original thinker, his contributions to the advancement of geology were important in their time, in relation to contemporary knowledge. The true value of his work can only be assessed by examining all of his geological publications. His works were widely read and must have done much to stimulate interest in the earth sciences.

BIBLIOGRAPHY

I. ORIGINAL WORKS. For a complete and annotated bibliography of Woodward's works, including translations, see V. A. Eyles, "John Woodward, F.R.S., F.R.C.P., M.D. (1665–1728): A Bio-bibliographical Account of His Life and Work," in *Journal of the Society for the Bibliography of Natural History*, **5** (1971), 399–427.

Woodward's books of geological and mineralogical interest are *Essay Toward a Natural History of the Earth* (London, 1695; 2nd ed.; 1702; 3rd ed., 1723); *Brief Instructions for Making Observations in All Parts of the World* (London, 1696; repr. with an introduction by V. A. Eyles, by the Society for the Bibliography of Natural History, London, 1973); *Naturalis historia telluris illustrata & aucta* (London, 1714); *Natural History of the Earth, Illustrated, Enlarged and Defended* (London, 1726), a translation by B. Holloway, of *Naturalis historia*, with some other papers by Woodward; *A Supplement and Continuation of the Essay Towards a*

Natural History of the Earth (London, 1726), which is identical with the previous work, except for the title; *Fossils of All Kinds Digested Into a Method* (London, 1728); and *An Attempt Towards a Natural History of the Fossils of England . . .* (London, 1729).

Woodward's botanical paper, "Some Thoughts and Experiments Concerning Vegetation," is in *Philosophical Transactions of the Royal Society,* **21** (1699), 193–227.

II. SECONDARY LITERATURE. The principal biographical source is J. Ward, *Lives of the Professors of Gresham College* (London, 1740), 283–301. See also J. W. Clark and T. McKenny Hughes, *Life and Letters of Adam Sedgwick . . . Woodwardian Professor of Geology,* I (Cambridge, 1890), 166–189, for the establishment of the Woodwardian chair in geology.

For contemporary commentaries on Woodward's *Essay,* see J. A[rbuthnot], *An Examination of Dr. Woodward's Account of the Deluge, &c.* (London, 1697), and J. Harris, *Remarks on Some Late Papers, Relating to the Universal Deluge: and to the Natural History of the Earth* (London, 1697). A critical commentary on Woodward's geological views in relation to those of his contemporaries and eighteenth-century successors is in K. B. Collier, *Cosmogonies of Our Fathers* (New York, 1934; repr. 1968).

Woodward's quarrel with the Royal Society is described by G. R. De Beer in *Sir Hans Sloane and the British Museum* (London, 1953), 90–91. His botanical work is discussed in H. H. Thomas, "Experimental Plant Biology in Pre-Linnean Times," in *Bulletin of the British Society for the History of Science,* **2** (1955), 20–21.

Unpublished correspondence and MSS of Woodward are preserved in the archives of the Royal Society of London, and in the British Museum and Bodleian libraries. His correspondence with Cotton Mather and other naturalists in North America is discussed by R. P. Stearns, in *Science in the British Colonies of America* (Urbana, 1970).

V. A. EYLES

WOODWARD, ROBERT SIMPSON (*b.* Rochester, Michigan, 21 July 1849; *d.* Washington, D.C., 29 June 1924), *applied mathematics, geophysics.*

Woodward was part of the tradition of mathematical physics that saw the earth as the great object of study. In 1904 he asserted, "The earth is thus at once the grandest of laboratories and the grandest of museums available to man." To this laboratory and museum Woodward brought a great skill in mathematics and an insistence on obtaining data of the highest precision in a form suitable for computation. The last point greatly influenced the young John Hayford, who worked with Woodward at the Coast and Geodetic Survey.

After receiving a degree in civil engineering in 1872 from the University of Michigan, Woodward worked for ten years with the Lake Survey of the U.S. Corps of Engineers. From 1882 to 1884 he was an astronomer with the U.S. Transit of Venus Commission. Woodward next served for six years with the U.S. Geological Survey, successively occupying the posts of astronomer, geographer, and chief geographer.

His most notable scientific contributions occurred during this period. For G. K. Gilbert he calculated the effects on shore lines of the removal of superficial masses by means of potential theory. In this work and his passing consideration of isostasy, Woodward considered thermal effects, clearly related to the concern with how heat influenced base bars and other instruments of precision. In a series of papers in 1887–1888, Woodward explored the cooling of homogeneous spheres and the diffusion of heat in rectangular masses. The findings were applied to Kelvin's work on the age of the earth. By 1889 Woodward criticized Kelvin for the "unverified assumption of an initial uniform temperature and a constant diffusivity." As the data for Kelvin's calculations were derived from observations of continental areas, Woodward felt the probabilities were against obtaining satisfactory numerical results for the entire earth. His position strengthened the opposition of many geologists, at least in America, to Kelvin's constriction of geological time.

From 1890 to 1893 Woodward was with the U.S. Coast and Geodetic Survey. In 1893 he became professor of mechanics and mathematical physics at Columbia University; and in 1895 he was named dean of the College of Pure Science. From 1904 through 1920 Woodward was president of the Carnegie Institution of Washington, succeeding D. C. Gilman. As chairman of two advisory committees, he had previously played a role in the development of the policies of the Institution. He was a strong administrator and largely responsible for the direction taken by the Carnegie Institution.

BIBLIOGRAPHY

F. E. Wright's memoir in *Biographical Memoirs. National Academy of Sciences,* **19** (1938), 1–24, has a good bibliography. The archives of the Carnegie Institution of Washington and the papers of many of his contemporaries contain manuscripts by or about Woodward. The correspondence of J. McK. Cattell in the Library of Congress and T. W. Richards in the Harvard

University Archives are valuable for his views on the policies of the Carnegie Institution. The correspondence of Presidents Low and Butler in the Office of the Secretary, Columbia University, contains a small number of interesting items.

NATHAN REINGOLD

WOOLLEY, CHARLES LEONARD (*b.* Upper Clapton, London, England, 17 April 1880; *d.* London, 20 February 1960), *archaeology.*

Woolley was the son of the Reverend George Herbert Woolley and Sarah Cathcart. He received a degree from New College, Oxford, and intended to become a schoolmaster. W. A. Spooner, warden of New College, decided, however, that Woolley was to be an archaeologist. Woolley therefore became assistant to Sir Arthur Evans at the Ashmolean Museum in 1905, and did his first fieldwork and digging at Corbridge, Northumberland. He wrote in his autobiography, *Spadework* (1953): "I know only too well that the work there would have scandalized, and rightly . . . any British archaeologist of today. It was however typical of what was done forty-five years ago, when field-archaeology was, comparatively speaking, in its infancy and few diggers in this country thought it necessary to follow the example of that great pioneer, Pitt Rivers."

Woolley's work in the Near East began in 1907, when he dug with D. Randall-MacIver in Nubia. In 1912 he was appointed to succeed Reginald Campbell Thompson as director of the British Museum expedition to Carchemish. He was accompanied by T. E. Lawrence, with whom, after a subsequent six-week archaeological reconnaissance, he wrote *The Wilderness of Zin* (1915). During World War I, Woolley served as an intelligence officer on the British General Staff in Egypt.

The war had ended excavation in the Middle East, but work was recommenced as soon as hostilities ceased. Even before the 1918 armistice Thompson began digging at Ur and Eridu, under the auspices of the British Museum. On the strength of his finds, the museum sent an expedition under H. R. Hall to dig at both sites. Hall also found al'Ubaid, a new site four miles west of Ur. In 1922 a joint expedition of the British Museum and the Museum of the University of Pennsylvania, directed by Woolley, continued Thompson and Hall's work. Woolley began at Ur, then transferred his attention to al'Ubaid, returning to Ur in 1926 and the following years. It was in 1926 that the great prehistoric cemetery at Ur, with its "royal

tombs," was discovered and excavated. The discovery of these tombs, with their splendid treasures of gold and lapis lazuli and their remarkable evidence of funerary ritual, caused a sensation comparable with Schliemann's discoveries at Mycenae and those of Lord Carnarvon and Howard Carter of Tutankhamen's tomb. Woolley dug at Ur, Eridu, and al'Ubaid for thirteen years, publishing preliminary reports in *Antiquaries Journal* and the full report in a series of volumes between 1928 and 1938.

At Ur, Woolley demonstrated his remarkable insight into the methods used by early craftsmen and builders. The joint expedition under Woolley not only inaugurated a brilliant revival of excavation in Mesopotamia in the 1920's and 1930's; it also was responsible for widespread popular interest in Mesopotamian archaeology and the origins of civilization there. In 1900 very few people had heard of the Sumerians, but by 1930 the Sumerians had been added to the collection of prehistoric peoples of whom almost everyone knew something. This was due in part to the sensational nature of the Ur excavations, but also to the clear and helpful popular accounts published by Woolley, notably *Ur of the Chaldees* (1929), *The Sumerians* (1930), *Abraham, Recent Discoveries and Hebrew Origins* (1936), and *Excavations at Ur: A Record of Twelve Years Work* (1954).

Like W. M. Flinders Petrie and A. H. Pitt-Rivers, Woolley believed passionately that the results of archaeology must be communicated to the general public in readable form. It has been said, in criticism of his work, that he ran ahead of what could reasonably be inferred; but he was always certain of what he was doing in the scholarly popularization of archaeology. As Sir Max Mallowan said, "If his imagination sometimes outran the facts, that to him was preferable to allowing knowledge to lie dormant and inconclusive." The freshness and relevance of his popularization can be seen not only in the books on Ur and the Sumerians but also in *Spadework*, written when he was in his seventies, and *Dead Towns and Living Men*, written in 1920.

His great work in Mesopotamia over, Woolley dug at al-Mina, near Antioch, and at Atchana during 1937–1939 and 1946–1949. Atchana-Alalakh was revealed as the ancient Hittite capital of the province of Hatay in Turkey (now the sanjak of Alexandretta): it was destroyed in 1200 B.C. but had nine periods of occupation dating back from the latter year to the twentieth century B.C. The results were published in a popular account, *A

Forgotten Kingdom (1953), and in *The Alakh Excavations and Tell Atchana* (1958). These works again showed Woolley's strength and purpose: he always published his results in exemplary and scholarly form, and also communicated them to the general public. As Sir Max Mallowan said: "To have dug so much and left nothing unwritten was indeed a phenomenal record."

In 1938 Woolley was invited by the government of India to advise it on the development and organization of archaeology there; his plans, not implemented until the 1940's, were carried out by Sir Mortimer Wheeler, director general of archaeology for India. During World War II, Woolley was specially employed in the safeguarding of museums, libraries, archives, and art galleries. Knighted in 1935, he continued working in the last quarter-century of his life, producing with Jacquetta Hawkes the first volume of the UNESCO *History of Mankind* (1963).

Woolley married Katharine Elizabeth Keeling in 1927.

BIBLIOGRAPHY

Woolley's principal works are mentioned in the text. In addition see article in *Dictionary of National Biography* and obituary in *The Times* (22 Feb. 1960).

GLYN DANIEL

WORM, OLE (or **OLAUS WORMIUS**) (*b.* Aarhus, Jutland, Denmark, 13 May 1588; *d.* Copenhagen, Denmark, 7 September 1654), *natural history.*

Worm was the son of the mayor of Aarhus and a descendant of refugees from religious persecution in Holland. He received his basic education in Aarhus, then attended schools and universities including Marburg, Montpellier, Strasbourg, and Padua and received the doctorate in medicine at Basel in 1611. He then practiced medicine in London until 1611, when he was appointed professor of humanities at the University of Copenhagen. In 1615 he became professor of Greek, and in 1624 professor of medicine, which chair he retained until his death thirty years later. He also was several times elected rector of the university. Worm continued to practice medicine throughout his life and was personal physician to King Christian V. He discovered and described the small bones that occasionally occur along the lambdoid suture of the human skull; they are still called Wormian bones.

A conscientious physician, Worm remained in Copenhagen to tend his patients during epidemics when many had fled the city; in the plague of 1654 he caught the disease and died of it.

A gifted polymath, Worm collected many types of objects, especially those of natural history and man-made artifacts, which he carefully arranged and classified, following a rigorous method; he also prepared a detailed catalog, published in 1655 by his son William as *Museum Wormianum.* His museum, which became one of the great attractions of Copenhagen, is illustrated in *Museum Wormianum*: an assortment of bizarre and exotic objects, antiques, and stuffed animals. It included the skull of a narwhal properly described; previously narwhal tusks had been supposed to be the horns of unicorns. There were many prehistoric stone implements, but Worm did not conclude that they belonged to a stone age and were artifacts; he labeled them "*Cerauniae,* so called because they are thought to fall to earth in flashes of lightning"—a belief widely held at that time. This is curious, because Worm recognized the tip of a stone harpoon point embedded in a marine animal found in Greenland, and also knew of stone tools and weapons from America. On his death, Worm's museum passed to King Frederik III and was installed in the old castle at Copenhagen. The king planned a new building for Worm's collections and library opposite Christiansborg Palace; but the second story, housing the museum, was not finished until after the king's death in 1680. It was open to the public on payment of an admission charge, and was one of the first such museums.

Worm was interested in Danish antiquities and published accounts of them in his *Monumenta Danica* and his *Fasti Danici* (1643). Particularly interested in runic inscriptions, he traveled extensively to visit runic sites, and collected information through correspondence. In 1626 he arranged for a royal circular to instruct all clergy to submit a report on any runic inscriptions, burial sites, or other historical remains known in their parishes. In 1639 a gold horn was discovered in Jutland; and Worm, with his great knowledge of runes and antiquities, was asked to describe and study it; he did so, publishing *De aureo cornu* in 1641. This gold horn, and another discovered a hundred years later, were stolen from the royal collections in 1802 and destroyed. Worm's account of the horn, and the runes and designs on it, are therefore of great importance.

GLYN DANIEL

WORSAAE, JENS JACOB (*b*. Vejle, Jutland, Denmark, 14 March 1821; *d*. Copenhagen, Denmark, 15 August 1885), *archaeology.*

Worsaae, the son of the county sheriff at Vejle, became an avid collector of antiquities while quite young. As a schoolboy he and a porter employed by a local merchant traveled the countryside, collecting pottery and bronzes and driving shafts into prehistoric barrows. By the time he entered school in Copenhagen, he was said to have the best and largest collection of antiquities from Jutland. In 1836, when he was only fifteen, Worsaae contacted Christian Jurgensen Thomsen, director of the Museum of Northern Antiquities, now the National Museum. Thomsen used him as an unpaid assistant; and soon Worsaae was cataloging artifacts, conducting groups through the museum, and helping Thomsen with his correspondence.

Worsaae was concerned with aspects of prehistoric archaeology that did not interest Thomsen, who was essentially a museum man. Worsaae was keenly interested in excavation. During the years 1836–1840 Worsaae spent his Sundays and half-holidays digging barrows north of Copenhagen; and in the vacations he did fieldwork and excavation in Jutland, where his parents paid two laborers to assist him. He drove broad trenches through the barrows and observed changes of structure and secondary burials; thus he became one of the first archaeologists to study and understand the stratigraphy of barrow construction. In 1840 Worsaae published an article on the grave mounds of Denmark, dividing them into those of the Stone Age, the Bronze Age, and the Iron Age. It was a revolutionary achievement, for he had taken Thomsen's three-age system, originally devised as a museum classification; had applied it to antiquities found in the field; and had shown, in barrows and in peat bogs, that Thomsen's theory of the three successive ages of the prehistoric past was supported by stratigraphy. This sequence was later demonstrated in the Swiss lake-dwellings. The Danish technological model of the past thus became a proven fact of prehistory.

In 1841, Christian VIII of Denmark gave Worsaae a travel grant to study antiquities in southern Sweden. The result was his book *Danmarks oldtid oplyst ved Oldsager og Gravhøje* (1843), which was also published in German and was translated into English as *The Primaeval Antiquities of England and Denmark* (1849). It is undoubtedly one of the half-dozen most important and influential books on archaeology published in the nineteenth century. Worsaae then traveled in England, Ire-land, and Scotland, studying the national collections and lecturing to archaeologists on the Danish three-age system.

Upon returning to Denmark from his British tour, Worsaae was appointed to the specially created post of inspector of ancient monuments. In 1854 he was made professor of archaeology at the University of Copenhagen, and in 1865 he succeeded Thomsen as director of the National Museum, a post he held until his death. Worsaae traveled throughout Europe, attending international conferences and making and continuing friendships. He was both a representative of Scandinavian archaeology and one of the first pan-European archaeological travelers and scholars.

Worsaae recognized the truth and importance of the Danish three-age system stated by Thomsen, calling it "the first clear ray . . . shed across the Universal prehistoric gloom of the North and the World in general." His contribution was to confirm the truth of the system through his excavations and his careful stratigraphical observations. Worsaae was interested not only in the archaeological record of the past but also in the interpretation of that record in terms of human history. For instance, he wondered whether the succession of the Stone, Bronze, and Iron ages meant cultural evolution, or whether they were caused by the arrival of new people. In all his writings he was keenly aware of the issue of independent invention versus diffusion, which became highly controversial during the nineteenth and early twentieth centuries. Worsaae himself believed that the transition from Stone Age to Bronze Age must have meant the arrival in northern Europe of new people from southeastern Europe or the Near East; he argued, however, that the change from Bronze Age to Iron Age could have been effected by trade and the movement of small groups without an invasion of new people.

Worsaae has often been described as the first full-time professional archaeologist. Johannes Brøndsted, a mid-twentieth-century director of the National Museum at Copenhagen, described him as "the actual founder of antiquarian research as an independent science." A pioneer in excavation, interpretation, synthesis, and exposition, he was undoubtedly one of the great archaeologists of the nineteenth century.

BIBLIOGRAPHY

O. Klindt-Jensen, *A History of Scandinavian Archaeology* (London, 1975), has a complete bibliography of Worsaae's writings. See also G. Bibby, *The Testimo-*

ny of the Spade (London, 1957); G. E. Daniel, *A Hundred Years of Archaeology* (London, 1950); G. E. Daniel, *The Idea of Prehistory* (London, 1962); and G. E. Daniel, *The Origins and Growth of Archaeology* (London, 1967).

GLYN DANIEL

WOTTON, EDWARD (*b.* Oxford, England, 1492; *d.* London, England, 5 October 1555), *medicine, natural history.*

Wotton was the son of Richard Wotton, beadle of the University of Oxford. He was educated at "the Grammar School joining to Magdalen College" and, from 1506, at Magdalen College, graduating 9 February 1514 and becoming a fellow in 1516. In January 1521 he followed John Claymond to Corpus Christi College, becoming *socio compar* then (or, more probably, in 1523). He was later permitted to travel in Italy for from three to five years to "improve his learning, and chiefly to learn Greek." He also studied at the medical school of Padua, graduating M.D. in 1526. He was awarded the same degree at Oxford on his return later that year. Wotton subsequently moved to London. He was admitted fellow of the Royal College of Physicians on 8 February 1528 and was active in its administration, becoming president (1541–1543) and censor (1552–1553, 1555). As befitted his standing at the College of Physicians, he numbered among his patients the duke of Norfolk and Margaret Pole, countess of Salisbury; and he is said to have also been physician to Henry VIII.

Wotton's scientific reputation rests on *De differentiis animalium libri decem* (Paris, 1552). Each of the ten books in this work deals with a major topic. Books 1–3 are devoted to generalities—the parts of, functions of, and differences between animals; book 4 deals with man; book 5, with quadrupeds that bear live young; book 6, with quadrupeds that lay eggs; book 7, with birds; book 8, with fishes; book 9, with insects; and book 10, with crustaceans, squids, and mollusks. The entries in the index refer to page numbers and a letter. Each text page is lettered every eleven lines (A–D on recto; E–H on verso), a practical and advanced method of indexing, considering the date at which it was printed.

Wotton dedicated his book to King Edward VI, then in the last year of his six-year reign. In this dedication he explains his purpose in compiling the book and briefly describes its preparation, mentioning his practice of studying books of writers whose work could contribute to his knowledge of medicine and telling how he compiled commentaries on them. His discovery of the writings of the French botanist Jean Ruel (*De natura stirpium libri tres*, 1536) and of Agricola (*De natura fossilium*, 1546) came during his study of earlier authors; but feeling that he could not surpass their work in these fields, Wotton concentrated on zoology.

It is as a compilation that Wotton's *De differentiis animalium* must be viewed. C. E. Raven called it an "astonishing mosaic of extracts from every sort of Graeco-Roman writer . . .," but this should not necessarily be seen as criticism. The book is the production of a man whose education had been that of a classical scholar and whose professional ethos was firmly rooted in ancient writers. Nothing was more natural than for Wotton to produce virtually a pandect of the classical writers on zoology. In this work, as in his sources, he was reflecting the development of Renaissance science. The invention of printing had made well-produced, indexed, and often illustrated editions of the writings of the classical authorities widely available. This rendered the knowledge of classicists of Wotton's generation more precise and paved the way for the great Renaissance encyclopedists in the natural sciences. Wotton was the first of these writers (although Adam Lonicerus' *Naturalis historiae opus novum . . .* was published a year earlier, Wotton's work had been long in preparation), who include Konrad Gesner and Ulisse Aldrovandi. Modern naturalists may view Wotton's book, like Aldrovandi's and much of Gesner's, as overweighted with information from the literature and short on original observation. But the encyclopedists' purpose was to bring together the literature; and it was on their foundation that the earliest field naturalists, such as William Turner (*ca.* 1508–1568) and Pierre Belon, built.

De differentiis animalium had a considerable influence on later naturalists. Gesner praised it as a complete and clearly written digest of zoological knowledge, but its reputation was greatest in entomological circles. In book 9 Wotton discussed at some length the complexities of insect metamorphoses but was unable to reconcile them with his implicit belief in spontaneous generation (lice formed from human sweat, and caterpillars from plants). This book, in which he gave evidence for some original observation, was quoted—and in places relied upon—in Thomas Moffett's *Theatrum insectorum*, an elaborate and creditable work based partly on the manuscript of his friend and fellow physician Thomas Penny. Moffett's book, published posthumously in 1634 by Theodore Mayerne, was the first entomological book to appear in England. It was later translated into English by John Rowland and appeared in 1658 as the third volume of Edward Topsell's *History of Four-Footed Beasts and Serpents . . .*, a book that, because of its use of English and despite its inclusion of fabulous beasts, strongly influenced popular interest in natural history in England.

BIBLIOGRAPHY

See A. F. P[ollard], "Wotton, Edward," in *Dictionary of National Biography*, LXIII, 48–49; C. E. Raven, *English Naturalists From Neckam to Ray* (London, 1947); and G. Sarton, *The Appreciation of Ancient and Medieval Science During the Renaissance (1450–1600)* (Philadelphia, 1955).

ALWYNE WHEELER

WOULFE, PETER (*b.* Ireland [?], 1727 [?]; *d.* London, England, 1803), *chemistry*.

The familiar two-necked bottle generally known as a Woulfe's bottle has long been a standard item of equipment in most chemical laboratories. The apparatus has been traced back to J. R. Glauber, and its attribution to Woulfe seems to stem from his use of a vessel with two outlets in a series of distillation experiments described in 1767. His "new method" was designed to prevent the escape of fumes "very hurtful to the lungs" by passing them through a tube into water.

Woulfe's origins are obscure and little seems to be known of him before his election to the Royal Society in 1767, when he was described as a "Gentleman well skilled in Natural Philosophy and particularly Chymistry." He was then living at Clerkenwell, London, and apparently spent most summers in Paris. What little is known of Woulfe, apart from his published work, relates mainly to his eccentricities, which were described by W. T. Brande, who cited him as a belated believer in transmutation: "He had long vainly searched for the elixir, and attributed his repeated failures to the want of due preparation by pious and charitable acts." Humphry Davy, in his *Collected Works* (IX, 367), said that Woulfe attached prayers to his apparatus. A. N. von Scherer, writing in 1801 (*Allgemeines Journal der Chemie*, 5, 128–129), hints at mental derangement "that must concern all friends of philosophy and natural knowledge"; and it seems probable that these aberrations were char-

acteristic of Woulfe's later years. His writings are lucid and show him to have been a competent chemist. Priestley frequently referred to him as a valued friend who lent him apparatus, offered guidance, and suggested experiments.

In the paper of 1767, Woulfe described the preparation of "marine ether" (ethyl chloride) by mixing alcohol vapor with that from the reaction of sulfuric acid with common salt (hydrogen chloride). He also made "nitrous ether" (ethyl nitrate) by distilling alcohol with nitric acid, and in a paper of 1784 he reported a better yield by what he claimed to be a hitherto untried method, using niter and sulfuric acid instead of nitric acid. In 1771 he investigated "mosaic gold" (stannic sulfide; see J. R. Partington, in *Isis*, **21** [1934], 203–206), and in the same paper he described the use of indigo. After mentioning the known method of making a blue solution of the dye in sulfuric acid, Woulfe described its solution in nitric acid. He claimed to have obtained a yellow substance that would dye wool and silk, the first recorded preparation of picric acid.

In December 1775 Woulfe was appointed to give the first lecture under the terms of the will of Henry Baker, who had bequeathed £100 to the Royal Society, the interest from which was to finance an annual discourse, provided the first was given within a year of payment. The Council accepted the bequest on 13 July 1775, and Woulfe delivered the first Bakerian lecture on 20 June 1776.

BIBLIOGRAPHY

I. ORIGINAL WORKS. Woulfe's main papers are "Experiments on the Distillation of Acids, Volatile Alkalies, &c. Shewing How They May Be Condensed Without Loss, & How Thereby We May Avoid Disagreeable and Noxious Fumes," in *Philosophical Transactions of the Royal Society*, **57** (1767), 517–536; "Experiments to Shew the Nature of Aurum Mosaicum," *ibid.*, **61** (1771), 114–130; "Experiments on a New Colouring Substance From the Island of Amsterdam in the South Sea," *ibid.*, **65** (1775), 91–93; "Experiments Made in Order to Ascertain the Nature of Some Mineral Substances; and, in Particular, to See How Far the Acids of Sea-Salt and of Vitriol Contribute to Mineralize Metallic and Other Substances," *ibid.*, **66** (1776), 605–623, the first Bakerian lecture; "Experiments on Some Mineral Substances," *ibid.*, **69** (1779), 11–34; and "A New Method of Making Nitrous Aether, by Which Means a Greater Quantity of the Aether Is Obtained With Less Expense and Trouble Than by Any Other Process Heretofore Described," read 5 Feb. 1784—the paper was not published in *Philosophical Transactions*, but the MS is in the Royal Society archives, *Letters & Papers*, decade VIII (vol. **73**), no. 57; a French trans. was published in *Observations sur la physique, sur l'histoire naturelle et sur les arts*, **25** (1784), 352–354. A few of Woulfe's later papers (of little importance) were published in the *Observations*.

II. Secondary Literature. No adequate biography exists; what little material is available seems to depend on W. T. Brande's short account, based on hearsay, in the historical introduction to his *Manual of Chemistry*—for instance, 6th ed., I (London, 1848), xvii–xviii. P. J. Hartog, in *Dictionary of National Biography*, LXIII (1900), 63–64, gives some useful references; see also J. R. Partington, *History of Chemistry*, III (London, 1962), 300–301. The origin and history of "Woulfe's bottle" is discussed in W. A. Campbell, "Peter Woulfe and His Bottle," in *Chemistry and Industry* (1957), 1182–1183.

E. L. Scott

WREN, CHRISTOPHER (*b*. East Knoyle, Wiltshire, England, 20 October 1632; *d*. London, England, 25 February 1723), *mathematics, architecture*.

Wren came from a family with strong ecclesiastical traditions. His father, for whom he was named, was rector of East Knoyle, chaplain to Charles I, and later (1634) dean of Windsor. His uncle, Matthew Wren, was successively bishop of Hereford, Norwich, and Ely. Wren was frail as a child, yet even in his earliest years he manifested an interest in the construction of mechanical instruments that included a rain gauge and a "pneumatic engine." He was educated at Westminster School, whence he proceeded in 1649 to Wadham College, Oxford. There he became closely associated with John Wilkins, who was later bishop of Chester and a member of that distinguished group whose activities led to the formation of the Royal Society. At Wadham College, Wren's talent for mathematical and scientific pursuits soon attracted attention. He graduated B.A. in 1651, and three years later received the M.A. He was elected a fellow of All Souls College, Oxford, in 1653 and remained in residence there until 1657.

Wren's interest in astronomy appears to have manifested itself about that time, and it led to his appointment, as professor of astronomy at Gresham College in 1657. In his inaugural lecture, after mentioning the relation of astronomy to mathematics, to theology in the interpretation of the Scriptures, to medicine, and above all to navigation, he praised the new liberty in the study and observation of nature, and the rejection of the tyranny of ancient opinions. He retained this professorship until 1661, when he was appointed Savilian professor of astronomy at Oxford, a post he occupied until 1673.

Wren is best remembered as an architect. His fame as the most distinguished architect England has produced probably has obscured his accomplishments in other branches of science. He was perhaps the most accomplished man of his day. While at Oxford he ranked high in his knowledge of anatomy; and his abilities as a demonstrator in that subject were acknowledged with praise by Thomas Willis in his *Cerebri anatome*, for which Wren made all the drawings. Wren also is said to have been the pioneer in the physiological experiments of injecting various liquids into the veins of living animals (Weld, *History*, I, 273).

Wren made important contributions to mathematics; and Newton, in the second edition of his *Principia* (1713), classed him with John Wallis and Christiaan Huygens as the leading geometers of the day ("Christopherus Wrennus, eques auratus, geometrarum facile principes," p. 19). Chief among his contributions was his rectification of the cycloid. This curve, because of its singularly beautiful properties, had long been a favorite of geometers since its discovery early in the sixteenth century. Many of its properties had been discovered by Pascal; its rectification, the finding of a straight line equal to an arc of the curve, was effected by Wren in 1658 and also by Fermat.

In 1668 Oldenburg asked Wren, along with Wallis and Huygens, to inform the Royal Society of his research into the laws of impact. In a terse paper read on 17 December 1668 and published on 11 January 1669 in the *Philosophical Transactions*, Wren offered a theoretical solution based on the model of a balance beam on which the impacting bodies are suspended at distances from the point of impact proportional to their initial speeds. Equilibrium in the model corresponds to an impact situation in which bodies approach one another at speeds inversely proportional to their sizes and, Wren postulated as a "Law of Nature," rebound at their initial speeds, which Wren termed their "proper speeds." In cases in which the center of motion does not coincide with the center of gravity of the system, Wren postulated that impact shifts the center of motion to a point equidistant from the center of gravity on the opposite side. Employing the further postulate that the speed of approach equals the speed of separation, Wren set forth rules of calculation that yield the center of gravity from the known sizes and initial speeds of the bodies, and then use the speeds and the center of gravity to compute the final speeds. The close fit of these results with experiment seems to have been the basic source of Wren's confidence in his solution. Wren also made a number of pendulum experiments, and Wilkins declared that he was the first to suggest the determination of standard measure of length by means of the oscillation of the pendulum (Weld, *History*, I, 196).

Even as a boy Wren had shown that he had the capacity to become a draftsman of exceptional ability. He probably applied himself to the serious study of the subject when he was commissioned to submit plans for the building of the chapel of Pembroke College, Cambridge, which was completed in 1663. His next major achievement was the building of the Sheldonian Theatre, Oxford, a model of which was exhibited before the newly formed Royal Society in April 1663. It was completed in 1669, and in that year Charles II appointed Wren surveyor of the royal works, a post he retained for half a century.

Meanwhile, the Great Fire had given Wren a unique opportunity to display his skill as an architect. Much of the City of London had been destroyed in the conflagration, including the old St. Paul's. This building, ancient and ruinous, had long been in urgent need of repairs; and just before the fire Wren had been invited by the dean to prepare plans for the building of a new cathedral. Wren's original plans were not approved, so he prepared a second scheme, having meanwhile obtained the concession that he might make such alterations as he deemed advisable. This second scheme was accepted, and a warrant for the building of the cathedral was issued in 1675. The first stone was laid on 21 June 1675, and after many delays the cathedral was finished in 1710.

Much of the City having been destroyed, Wren was invited to submit plans for the rebuilding of some fifty churches consumed in the flames. (These are described in *Parentalia*, 309–318.) At Oxford he built, in addition to the Sheldonian Theatre, the Tom Tower of Christ Church and Queen's College Chapel. At Cambridge, besides the chapel at Pembroke College, he built the library of Trinity College.

Wren received many honors. The University of Oxford conferred upon him the degree of doctor of Civil Laws; Cambridge awarded him the LL.D. In 1673 he was knighted. Wren also represented many constituencies in Parliament at different periods. In 1669 he married Faith Coghill, of Blechingdon, Oxford, by whom he had two sons, one of whom survived him. On the death of Lady Wren he married Jane Fitzwilliam, by whom he had a son and a daughter.

Wren played a prominent part in the formation of the Royal Society of London, which arose out of the informal gatherings of the votaries of experimental science that took place about the middle of the seventeenth century. These gatherings doubtless were inspired by the growing desire for learning that had been stimulated by the writings of Francis Bacon, notably the *Novum Organum*; but they also owed much to the institution founded under the will of Sir Thomas Gresham, according to which seven professors were employed to lecture on successive days of the week on divinity, astronomy, geometry, physic, law, rhetoric, and music. Of those whose enthusiasm prompted them to associate themselves with the new venture, the best-known, besides Wren, were Robert Boyle, John Wilkins, John Wallis, John Evelyn, Robert Hooke, and William Petty. The meetings, held at Gresham College, were suspended during the troubled times that followed the Civil War. On the return of Charles II in May 1660, they were revived and the need for a more formal organization was at once recognized. Accordingly, on 28 November 1660 the following memorandum was drawn up: "These persons following . . . mett together at Gresham Colledge to heare Mr. Wren's lecture." At the end of Wren's lecture it was proposed that the meetings should continue weekly. A list was drawn up of those interested; and at a meeting held on 19 December 1660, it was ordered that subsequent meetings should be held at Gresham College.

The charter of incorporation passed the great seal on 15 July 1662 (which thus is the date of the formation of the Royal Society); Wren is said to have prepared its preamble. A Council was formed, with Wren as one of the members. He was the Society's third president, serving from 30 November 1680 to 30 November 1682. *The Record Book of the Royal Society of London* (1940, 18) pays tribute to Wren's zeal and encouragement despite the difficulties facing the young organization: "To him the Royal Society owes a deep debt of gratitude for the constant and loyal service which he rendered to it in its early days."

Wren also studied meteorology long before it had become an exact science through the work of Mariotte, Boyle, and Hooke. He was one of the earliest naturalists to investigate, by means of the microscope, the structure of insects; and his remarkable skill as a draftsman enabled him to make accurate drawings of what he saw.

Wren was largely instrumental in arranging for the (unauthorized) publication of Flamsteed's *Historia coelestis Britannica* (1712), which had been financed by Prince George, Queen Anne's consort, but had ceased with his death in 1708. When at length printing was resumed, many obstacles were placed in Flamsteed's way. Wren had been appointed a member of the committee to oversee the printing of the work; and despite much opposition, he gave Flamsteed great encouragement. Nevertheless, Flamsteed's wishes met with little response; and after the work eventually appeared under Halley's editorship, Flamsteed managed to secure three hundred of the four hundred copies printed and at once consigned them to the flames.

In 1718 Wren was superseded as surveyor of the royal works, after more than fifty years of active and laborious service to the crown and the public. He then retired to Hampton Court, where he spent the last five years of his life. He is buried in St. Paul's Cathedral, where a tablet to his memory has been erected.

BIBLIOGRAPHY

I. ORIGINAL WORKS. Among Wren's papers that appeared in *Philosophical Transactions of the Royal Society* are "Lex collisionis corporum" (Mar. 1669); "Description of an Instrument Invented Divers Years Ago for Drawing the Out-line of Any Objective in Perspective" (1669); "The Generation of an Hyperbolical Cylindroid Demonstrated, and the Application Thereof to the Grinding of Hyperbolical Glasses" (June 1669); "A Description of Dr. Christopher Wren's Engin Designed for Grinding Hyperbolical Glasses" (Nov. 1669); and "On Finding a Straight Line Equal to That of a Cycloid and to the Parts Thereof" (Nov. 1673).

II. SECONDARY LITERATURE. See Sir Harold Hartley, ed., *The Royal Society: Its Origins and Founders* (London, 1960); *Parentalia, or Memoirs of the Family of Wrens* (London, 1750), compiled by his son Christopher and published by his grandson Stephen; *Record Book of the Royal Society of London* (1940); and C. R. Weld, *History of the Royal Society*, 2 vols. (London, 1848).

For Wren's work on impact, see A. R. Hall and M. B. Hall, eds., *The Correspondence of Henry Oldenburg*, V (Madison, Wis., 1968), 117–118, 125, 134–135, 193, 263, 265, and in particular 319–320 (Wren's paper in the original Latin) and 320–321 (an English translation).

J. F. SCOTT

WRIGHT, ALMROTH EDWARD (*b.* Middleton Tyas, near Richmond, Yorkshire, England, 10 August 1861; *d.* Farnham Common, Buckingham-shire, England, 30 April 1947), *pathology, bacteriology, immunology.*

Wright was the second son of an Irish Presbyterian minister, Charles Henry Hamilton Wright, and Ebba Johanna Dorothea Almroth, daughter of a Swedish chemistry professor. While his father was a minister at Dresden, Germany (1863–1868); at Boulogne, France (1868–1874); and at Belfast, Ireland (1874–1885), Almroth was educated by his parents and tutors, and at Belfast Academical Institution. At age seventeen he entered Trinity College, Dublin, and earned a B.A. in modern literature in 1882 and a B.M. in medicine in 1883. On a traveling scholarship he studied medicine at the University of Leipzig.

Upon his return to London, Wright read law briefly, then became an Admiralty clerk with free time for medical research at the Brown Institution (University of London) in Wandsworth. Subsequently he was demonstrator of pathology at Cambridge University in 1887; and on a scholarship he studied pathological anatomy at Marburg and physiological chemistry at Strasbourg. Wright was demonstrator of physiology at the University of Sydney, Australia, in 1889–1891; and, back in England, he worked briefly in the laboratory of the College of Physicians and Surgeons, London, in the latter year.

In 1892 Wright was appointed professor of pathology at Army Medical School, Netley, where for ten years he did original work in blood coagulation and in bacteriology, achieving notable success in developing a vaccine against typhoid fever. The vaccine was tested on soldiers in India and Wright promoted it as a member of the India Plague Commission (1898–1900), for which he wrote the report. The War Office used it on a voluntary and poorly supervised basis during the Boer War, but the vaccine seemed effective; and Britain alone entered World War I with troops largely immunized against typhoid fever. Differences with the army caused Wright to resign and enabled him to undertake his major work as professor of pathology at St. Mary's Hospital, London (1902–1946).

At St. Mary's Hospital, Wright continued his pioneer work in immunization, aided by a remarkable team of research workers that included Alexander Fleming, who later discovered penicillin. Wright initially had a large private practice and told his research workers to do the same, partly because the laboratory pay was small but mainly because he believed it made one sensitive to the human aspects of the laboratory's work. Wright's enthusiasm inspired his staff, who willingly worked long hours on difficult problems. At frequent midnight teas in the laboratory, he dominated the robust discussion. Fond of quoting long passages of poetry, he was admired by colleagues and disciples, although others found him difficult.

By 1908 the laboratory had expanded into a considerable research institute that received important visitors, including Robert Koch, Paul Ehrlich, and Élie Metchnikoff. In 1911 Wright was invited to South Africa to help prevent pneumonia among Africans in the Rand gold mines. He introduced prophylactic inoculations that greatly reduced the pneumonia death rate. In 1913 Wright was offered, and accepted, the directorship of the department of bacteriology of the Medical Research Committee (later Council) laboratory at Hampstead. World War I intervened, however; and from 1914 to 1919 Wright served in France, heading a research laboratory concerned with wound infections. He locally applied a hypertonic salt solution as an osmotic agent to draw lymph into wounds and also provided scientific justification for the early closure of such wounds. Honors for his war work included the Le Conte Prize of the Académie des Sciences (1915), the Buchanan Medal of the Royal Society (1917), and a special medal of the Royal Society of Medicine (1920) "for the best medical work in connection with the war." For his previous achievements in typhoid vaccine, he had been knighted in 1906.

Rather than resume the government position interrupted by the war, Wright in 1919 returned to St. Mary's Hospital for more than a quarter-century of work in immunology, retiring as principal of the research institute in 1946. Although a prodigious worker in the years between the wars, his influence declined in the medical world, partly because his bluntly stated and unconventional views caused some antagonism. These views were not all on medical questions. He was vigorously antifeminist and wrote letters to the press and a book to show that women were biologically and psychologically inferior. George Bernard Shaw wrote a devastating reply to Wright's views on the subject; but despite this difference of views, Shaw admired Wright, praised the quality of his writings, and got the idea for his play *The Doctor's Dilemma* from discussions with Wright and his colleagues at St. Mary's Hospital. The play's leading character, Sir Colenso Ridgeon, is modeled on Wright. His biographer, Leonard Colebrook, attributes some of Wright's unorthodox views to his long seclusion in the laboratory and to the austeri-

ty of his scientific life. Nevertheless, he had a fairly regular home life, having married Jane Georgina Wilson in 1889. They had two sons and a daughter.

Wright's place in science as one of the founders of modern immunology is probably not far behind that of Pasteur, Ehrlich, and Metchnikoff. Working independently, Wright in England, and Richard F. J. Pfeiffer and Wilhelm Kolle in Germany, shared in the discovery of a successful typhoid vaccine (1896). Wright also originated vaccines against enteric tuberculosis and pneumonia; proved the worth of inoculating with dead microbes; and made valuable contributions to the study of opsonins, blood substances that help to overcome bacteria. Although typhoid vaccine was his best-known discovery, members of the research team he trained made important contributions, and his lifetime work in immunology was considerable.

BIBLIOGRAPHY

I. ORIGINAL WORKS. Prominent among Wright's more than 150 scientific writings are *A Short Treatise on Anti-Typhoid Inoculation* (London, 1904); *Principles of Microscopy* (London, 1906); *Studies in Immunization* (London, 1909; 2nd ser., 1944); *Technique of the Teat and Capillary Glass Tube* (London, 1912, 2nd ed., 1921); *The Unexpurgated Case Against Woman Suffrage* (London, 1913); *Prolegomena to Logic Which Searches for the Truth* (1941), the first part of a larger philosophical work never completed but continued posthumously in *Alethetropic Logic* (London, 1953); *Pathology and Treatment of War Wounds* (London, 1942); and *On Induction* (London, 1943). Seventeen of his works are listed in an obituary, "Sir Almroth Wright," in *British Medical Journal*, no. 4506 (17 May 1947), 699–700.

II. SECONDARY LITERATURE. An evaluative sketch appears in M. Marquardt, "Pioneer in Vaccine Therapy," in *Life and Letters*, **50** (Sept. 1946), 127–130. A book-length biography is Leonard Colebrook, *Almroth Wright; Provocative Doctor and Thinker* (London, 1954); also see his "Almroth Edward Wright," in *Lancet*, **252**, no. 6454 (10 May 1947), 654–656; and "Wright, Sir Almroth Edward," in *Dictionary of National Biography 1941–1950*, 976–978. Other obituary accounts are in *Who Was Who 1941–50* (London, 1952), 1265; New York *Times* (1 May 1947), 25; *Times* of London (1, 3, 6, 9, 13 May 1947), all on 7; *Obituary Notices of Fellows of the Royal Society of London*, **17** (1948), 297–314; and R. T. Mummery, "Sir Almroth Edward Wright, 1861–1947," in *Nature*, **159** (31 May 1947), 731–732. Wright also is mentioned in L. J. Ludovici, *Fleming, Discoverer of Penicillin* (Bloomington, Ind., 1952); and André Maurois, *The Life of Sir Alexan-*

der Fleming, Discoverer of Penicillin (New York, 1959), esp. chs. 3, 4.

FRANKLIN PARKER

WRIGHT, EDWARD (*b.* Garveston, Norfolk, England, October 1561; *d.* London, England, November 1615), *mathematics, cartography.*

Details of Wright's life are unusually scanty and must be supplemented from facts about his relatives and friends. His father, Henry, was described as "mediocris fortunae, deceased" when an elder brother, Thomas, entered Gonville and Caius College, Cambridge, as a pensioner in 1574. Probably both boys were taught by John Hayward at a neighboring school in Hardingham. Edward joined his brother at Caius College, Cambridge, in December 1576; but Thomas' support for him was short-lived, since he died early in 1579. Wright's academic career closely paralleled that of John Fletcher: both graduated B.A. in 1581 and M.A. in 1584, and obtained their fellowships on Lady Day 1587. Fletcher had returned for his fellowship after teaching for a few years at Dronfield Grammar School, Derbyshire, so it is possible that Wright was also away from Cambridge in 1581–1584. Fletcher had a reputation as a medical writer, collaborator with Sir Christopher Heydon on his *Defence of Judiciall Astrologie* (1603), and as mathematics teacher to Henry Briggs. Contemporary with Briggs at St. John's College was Thomas Bernhere, later Wright's brother-in-law; both graduated M.A. in 1585, became fellows of their college, and were closely associated with Henry Alvey, one of the leaders of Cambridge Puritanism.[1]

In 1589 Wright received royal permission to absent himself from Cambridge in order to accompany George, earl of Cumberland, on an expedition to the Azores that was intended to acquire booty from Spanish ships. In 1599 Wright wrote that "the time of my first employment at sea" was "more than tenne yeares since." This suggests that he may have been to sea previously, possibly in 1581–1584. There seems little doubt that he had already acquired a reputation in mathematical navigation, and none at all that his 1589 voyage contributed greatly to his main achievements (described below). Wright returned to Cambridge at the end of 1589 and prepared a draft of his most important book, *Certaine Errors in Navigation*, in the next year or so. The 1599 printed version incorporates results obtained from observations made at London in the period May 1594–November 1597. It would seem that Wright had moved from Cambridge before the expiration of his fellowship in 1596, and that he had married the sister of Thomas Bernhere in 1595. Their only son, Samuel, entered his father's college after schooling in London but died before graduation, "a youth of much promise."

The succession of London mathematical lecturers is confused, but it is probable that Wright had some such employment after leaving Cambridge. These lecturers had been supported by Sir Thomas Smith and Sir John Wolstenholme, two rich city merchants closely connected with several trading companies. Thomas Hood was an early lecturer, and it has been suggested that Wright succeeded him. The position was complicated by the starting of Gresham College, where Henry Briggs was first professor of geometry.

There was, however, still a need for lecturers in navigation; Wright was serving in this capacity in 1614 when the East India Company took over the patronage and paid him an annual salary of £50. He may have held this post from 1612, the year of the death of Prince Henry, whom he had tutored in mathematics and whose librarianship he had been expecting. About the same time Wright was surveyor for the New River project, under Sir Hugh Myddleton, for bringing water to London.[2] During this London period he also wrote and published a number of mathematical tracts.

The publishing history of Wright's main work, *Certaine Errors in Navigation*, is complex. Wright himself outlined the impetus for the chief feature of this work, the justification of the so-called Mercator map projection, described as "the greatest advance ever made in marine cartography."[3] He criticized the usual sea charts as "like an inextricable labyrinth of error," offering as an instance his own experience in 1589: land was sighted "when by account of the ordinary chart we should have beene 50 leagues short of it." He admitted that his development had been prompted by Mercator's 1569 map of the world, but stated that neither Mercator nor anybody else had shown him how to do it. Wright's principle was very simple: to increase the distance apart of the parallels of latitude to match the exaggeration arising from the assumption that they were equally long. Since the lengths of the parallels varied according to a factor $\cos \lambda$, the correction factor was $\sec \lambda$ at any point. In order to plot the parallels on the new charts, Wright had effectively to perform the integration $\int_0^\lambda \sec \lambda\, d\lambda$. This was done numerically—in his

513

own words, "by perpetual addition of the Secantes answerable to the latitudes of each point or parallel into the summe compounded of all the former secantes. . . ."

Wright's development of the Mercator projection was first published by others. *Thomas Blundevile His Exercises Containing Six Treatises* (1594) was an important navigation compilation, the first to describe the use of the sine, tangent, and secant trigonometric functions. The author was at a loss to explain the new (Mercator) arrangement, which had been constructed "by what rule I knowe not, unless it be by such a table, as my friende M. Wright of Caius College in Cambridge at my request sent me (I thanke him) not long since for that purpose, which table with his consent, I have here plainlie set down together with the use thereof as followeth." The table of meridional parts was given at degree intervals.[4]

Two years later, following his publication of a Dutch version of Emery Molyneux's globe, Jodocus Hondius published at Amsterdam the well-known "Christian-Knight" maps of the world and of the four continents. These were based on Wright's theory of Mercator's projection, but were issued without acknowledgment. It seems that when he was in England, Hondius had been allowed to see the manuscript of Wright's *Certaine Errors*. In 1598–1600 Richard Hakluyt published his *Principal Navigations*, which contains two world charts on the new projection, that of 1600 a revision of the first. Although there is no attribution, it is clear that Wright was a major collaborator; further revisions in Hakluyt's work were made for versions in the 1610 and 1657 editions of his *Certaine Errors*.

Before the Hakluyt maps, William Barlow had included in his *The Navigator's Supply* (1597) a demonstration of Wright's projection "obtained of a friend of mine of like profession unto myself." This evidence of interest in his work was brought home to Wright when the earl of Cumberland showed him a manuscript that had been found among the possessions of Abraham Kendall and was being prepared for the press. Wright was surprised to find it was a copy of his own *Certaine Errors*, an experience that convinced him it was time to publish the work himself.

Ultimately Wright included his "The Voyage of the Earl of Cumberland to the Azores," which had been printed in Hakluyt's second volume. With it was a chart of the Azores on the new projection, showing Cumberland's route; this has been judged to be more significant than the world charts, since it was large enough to be used. *Certaine Errors* discussed other navigation problems, and was considerably extended in the second edition (1610), dedicated to Prince Henry. Wright also contributed to two seminal works. In 1600 he helped, particularly with a preface, to produce William Gilbert's *De magnete*. His translation of John Napier's *Mirifici logarithmorum canonis descriptio*, *A Description of the Admirable Table of Logarithmes*, appeared posthumously. It was approved by Napier and brought out by his friend Henry Briggs after the death of Wright's son Samuel, who contributed the dedication to the East Indies Company. The book marks the lifelong collaboration between Briggs and Wright, and the latter's efforts to spread a better understanding of navigation.[5] Nobody had done more to "set the seal on the supremacy of the English in the theory and practice of the art of navigation at this time."

NOTES

1. The baptism at Garveston took place on 8 October 1561; the father's will, dated 17 January 1573, left his house to his wife, Margaret, and then to Edward. (This information was provided by the Norfolk and Norwich Record Office.) J. Venn, *Biographical History of Gonville and Caius College*, I (Cambridge, 1897), 88–89; "From the Library," in *Midland Medical Review*, 1 (1961), 185–187; H. C. Porter, *Reformation and Reaction in Tudor Cambridge* (Cambridge, 1958).

2. See J. E. C. Hill, *Intellectual Origins of the English Revolution*, 39–40; D. W. Waters, *The Art of Navigation . . .*, 239, 278. It is possible that Wright lectured at Trinity House, Deptford, since he dedicated his 1599 translation to its master, Richard Polter.

3. Waters, *op. cit.*, 121.

4. R. C. Archibald, in *Mathematical Tables and Other Aids to Computation*, 3 (1948), 223–225, ignores these earlier eds. of the table, as well as the (independent) MS calculations by Thomas Harriot, discussed by Waters.

5. The final quotation is from Waters (p. 219), who also mentions several mathematical instruments that Wright helped to develop.

BIBLIOGRAPHY

I. Original Works. Wright's main work, *Certaine Errors in Navigation, Arising Either of the Ordinarie Erroneous Making or Using of the Sea Chart, Compasse, Crosse Staffe, and Tables of Declinations of the Sunne, and Fixed Starres Detected and Corrected* (London, 1599; 2nd ed., enl., 1610; 3rd ed., Joseph Moxon, ed., 1657), includes, at the end, "The Voyage of the . . . Earle of Cumberland to the Azores," also printed by Hakluyt (1599) and at Lisbon (1911). Other writings are *The Haven-Finding Art*, translated from the Dutch of Simon Stevin (London, 1599), repr. in *Certaine Errors* (1657) and, in part, by H. D. Harradon in *Territorial Magazine*, **50** (Mar. 1945); *Description and*

Use of the Sphaere (London, 1614; 1627); *A Short Treatise of Dialling: Shewing the Making of All Sorts of Sun-Dials* (London, 1614); and *A Description of the Admirable Tables of Logarithmes*, translated from the Latin of John Napier (London, 1616; 1618).

The MSS at Dublin are briefly listed by T. K. Abbott, *Catalogue of the Manuscripts in the Library of Trinity College* (Dublin, 1900).

II. SECONDARY LITERATURE. See W. W. R. Ball, *A History of the Study of Mathematics at Cambridge* (Cambridge, 1889), 25–27; F. Cajori, "On an Integration Ante-dating the Integral Calculus," in *Bibliotheca mathematica*, 3rd ser., **14** (1914), 312–319; and "Algebra in Napier's Day and Alleged Prior Inventions," in C. G. Knott, ed., *Napier Tercentenary Memorial Volume* (Edinburgh, 1915), 93–109; J. E. C. Hill, *Intellectual Origins of the English Revolution* (Oxford, 1965); C. Hutton, *A Philosophical and Mathematical Dictionary*, new ed., II (London, 1815), 619–620; J. K. Laughton, *Dictionary of National Biography*, LXIII; E. J. S. Parsons and W. F. Morris, "Edward Wright and His Work," in *Imago mundi*, **3** (1939), 61–71; Helen M. Wallis, "The First English Globe: A Recent Discovery," in *Geographical Journal*, **108** (1951), 275–290; "Further Light on the Molyneux Globes," *ibid.*, **121** (1955), 304–311; and "World Map in Principal Navigations, 1599: Evidence to Suggest That Edward Wright was the Main Author," an unpublished note (1972); and D. W. Waters, *The Art of Navigation in England in Elizabethan and Early Stuart Times* (London, 1958).

P. J. WALLIS

WRIGHT, FREDERICK EUGENE (*b.* Marquette, Michigan, 16 October 1877; *d.* Sagastaweka Island, near Gananoque, Ontario, Canada, 25 August 1953), *petrology.*

Wright was the son of Charles Eugene Wright, pioneer geologist of Michigan, and Carolyn Alice Dox. He was awarded the Ph.D. by the University of Heidelberg in 1900 after work in petrology under Harry Rosenbusch and in crystallography under Victor Goldschmidt. On returning to America, he spent three years as instructor at the Michigan College of Mines, Houghton, Michigan. About 1905 Wright became associated with the United States Geological Survey and he retained a connection for a dozen years. He was appointed petrologist at the Geophysical Laboratory of the Carnegie Institution of Washington in 1906, working there until his retirement in 1944 and retaining an office at the Institution until his death.

Wright's primary scientific contributions were in the development of the petrographic microscope and its applications, improvement of the techniques for the manufacture of high-grade optical glass during World War I, design and construction of a torsion gravity meter of high precision for field use (with J. L. England), and the remote determination of the nature of materials on the surface of the moon by optical methods. His book *The Methods of Petrographic-Microscope Research* (1911) greatly influenced the promotion of quantitative measurement of the optical properties of crystals. Similarly, Wright's reports on the systematic measurement of the amount of plane polarization and the relative spectral intensities of the reflected rays from various regions of the moon's surface, from which the nature of the materials on the surface were deduced, were pioneer efforts.

Wright's petrological field studies were made in the Upper Peninsula of Michigan, Alaska, South Africa, and the Columbia River in Washington and Oregon. Gravity surveys were undertaken in collaboration with Vening Meinesz in the Caribbean by means of a submarine and in Guatemala, in an effort to determine the relationship between gravity anomalies and volcanism. Data for his lunar field studies were collected at the United States Naval Observatory in Washington, D. C., and the Mt. Wilson Observatory in California.

Service to the sciences was performed through Wright's administrative work in the National Academy of Sciences (elected 1923) as a vice-president (1927–1931) and as home secretary (1931–1951). He also held high offices in the American Mathematical Society, Geological Society of America, Mineralogical Society of America, and Optical Society of America.

BIBLIOGRAPHY

Optical studies by Wright are "On the Measurement of Extinction Angles in the Thin Section," in *American Journal of Science*, 4th ser., **26** (1908), 349–390; "The Methods of Petrographic-Microscopic Research: Their Relative Accuracy and Range of Application," *Publications. Carnegie Institution of Washington*, no. 158 (1911); "The Formation of Interference Figures: A Study of the Phenomena Exhibited by Transparent Inactive Crystal Plates in Convergent Polarized Light," in *Journal of the Optical Society of America*, **7** (1923), 779–817; and "Computation of the Optic Axial Angle From the Three Principal Refractive Indices," in *American Mineralogist*, **36** (1951), 543–556.

A work on the glass industry is *The Manufacture of Optical Glass and of Optical Systems: A War-time Problem*, Paper on Optical Glass no. 40, Ordnance Dept. Document no. 2037 (Washington, D.C., 1921).

Field studies include *Notes on the Rocks and Minerals of Michigan* (Houghton, Mich., 1905); "The Intrusive Rocks of Mount Bohemia, Michigan," in *Annual Report of the Michigan Geological Survey* (1908), 361–402; "The Ketchikan and Wrangel Mining Districts," *Bulletin of the United States Geological Survey* no. 347 (1908), written with C. W. Wright; and "The Hot Springs of Iceland," in *Journal of Geology*, **32** (1924), 462–464.

Works on mineralogy are "Quartz as a Geologic Thermometer," in *American Journal of Science*, 4th ser., **27** (1909), 421–447, written with E. S. Larsen; optical studies in G. A. Rankin, "The Ternary System CaO-Al_2O_3-SiO_2," *ibid.*, **39** (1915), 1–79; and "Afwillite, a New Hydrous Calcium Silicate, From Dutoitspan Mine, Kimberley, South Africa," in *Mineralogical Magazine*, **20** (1925), 277–286, written with J. Parry.

Lunar studies include "Polarization of Light Reflected From Rough Surfaces With Special Reference to Light Reflected by the Moon," in *Proceedings of the National Academy of Sciences of the United States of America*, **13** (1927), 535–540; and "The Surface of the Moon," *Publications. Carnegie Institution of Washington* no. 501 (1938), 59–74.

Writings on gravity are "The Gravity Measuring Cruise of the U. S. Submarine S-21," in *Publications of the United States Naval Observatory*, 2nd ser., no. 13 (1930), app. 1, written with F. A. Vening Meinesz; "An Improved Torsion Gravity Meter," in *American Journal of Science*, **35A** (1938), 373–383, written with J. L. England; and "Gravity Measurements in Guatemala," in *Transactions. American Geophysical Union*, **22** (1941), 512–515.

H. S. YODER, JR.

WRIGHT, GEORGE FREDERICK (*b*. Whitehall, New York, 22 January 1838; *d*. Oberlin, Ohio, 20 April 1921), *geology*.

Wright was the fifth of the six children of Walter Wright, a farmer, and Mary Peabody Colburn. The attraction of the Wright family to the New School Calvinism, with its emphasis on personal regeneration and humanitarian reform, led Walter Wright to send five of his children to Oberlin College. The election of Charles Grandison Finney, one of the leaders of the New School movement, to the presidency of the college in 1851 was a sign to the Wrights that Oberlin was sound. After preparation at Castleton Academy in Vermont, Wright entered Oberlin in 1855. After receiving the B.A. degree in 1859, he entered the Oberlin Theological Seminary. Although his studies were interrupted by Civil War service, Wright received the M.A. in 1862. Just after graduation he married Huldah Maria Day and took her to Bakersfield, Vermont, where he became pastor of the Congregational church. In 1872 he moved to Andover, Massachu-

setts, where he began a nine-year ministry in the Free (Congregational) Church. In 1904, five years after the death of his first wife, who had borne him four children, Wright married Florence Eleanor Bedford.

Wright's interest in geology began during his childhood and gradually transformed from avocation to profession. He was an avid reader; and before he was twelve, he had gone through John C. Frémont's *Report* of his Rocky Mountain expedition. Oberlin's classical education furnished Wright with more than a smattering of science courses. In each winter of his college years he taught at a district school, a task that allowed him to travel through various parts of Ohio searching for fossils and collecting rock specimens. At Bakersfield, Wright read Lyell's *Antiquity of Man* and Darwin's *Origin of Species*. He explored portions of the Green Mountains and became an authority on the effects of glaciation in the locality. His friend Charles H. Hitchcock, Dartmouth geologist and administrator of the New Hampshire Geological Survey, brought Wright's work to the attention of a number of naturalists.

While he was at Andover, Wright worked closely with several Harvard scientists. It was his insistence that persuaded Asa Gray to publish *Darwiniana* (1876). Receptive to Darwin's views, Wright was the foremost early champion of a Christian Darwinist theology. His initial religious and scientific interests in the problem of the antiquity of man continued to stimulate his geological research for the rest of his life. His theory was that Andover's "Indian Ridge" was of glacial rather than marine origin, and in 1875 he demonstrated that the formation was one of a series of prominent eskers in New England. Clarence King, who soon became first director of the United States Geological Survey, gave Wright's conclusions an endorsement that brought his published work to the attention of James Dwight Dana, William M. Davis, and other geologists, with whom Wright continued to be associated.

In 1880 Wright and Clarence King noted the existence of a glacial boundary south of the Massachusetts shoreline; and with Henry C. Lewis and Peter Lesley, Wright followed the line of morainic deposits through New Jersey. In 1881 he worked with Lewis as an assistant on the Second Geological Survey of Pennsylvania, under Lesley's direction. Late in that year Wright accepted a professorship at Oberlin, where he taught courses in theology and glacial geology for the next twenty-seven years, in 1892 becoming professor of the harmony of science and revelation. Soon he had identified the drift margin in Ohio, and as an assistant on the United States Geological Survey (1884–1892) he completed the work through Indiana and Illinois. Thus he had personally traced the drift margin from the Atlantic Ocean to the Mississippi River.

In the summer of 1886 Wright conducted a series of pioneer investigations at Glacier Bay, Alaska. A result of his extensive glacial investigations in North America was an invitation to lecture before the Lowell Institute in Boston in the winter of 1887–1888. The revised lectures were published in 1889 as *The Ice Age in North America and Its Bearings Upon the Antiquity of Man.* He was twice again a Lowell lecturer: in 1891–1892 and in 1896–1897.

In the last three decades of his life, Wright's scientific work consisted mostly of explanations and refinements of his earlier conclusions. He investigated Greenland glaciation in 1894, and between 1892 and 1908 he visited Europe four times to observe effects of the Pleistocene there. Although he retired from teaching in 1907, he remained a tireless worker. From 1883 until his death he was editor of *Bibliotheca sacra,* a major theological quarterly. He also aided his son, Frederick Bennett Wright, in the publication of the thirteen volumes of *Records of the Past,* a journal of archaeology. Wright was instrumental in the crusade to preserve the nation's prehistoric earthworks.

Wright brought amazing energy to his work. Dedicated to the necessity for firsthand observation, he realized a long-held ambition in 1900, when he went to China to begin an arduous journey through Manchuria and Siberia. Traveling by mule, river steamer, train, and even hundreds of miles by horse-drawn cart, he recorded observations of elevated shorelines, loess deposits, and other consequences of Pleistocene action from Vladivostok to the Black Sea.

Wright was a founder of the Geological Society of America and was active in many other organizations, such as the American Association for the Advancement of Science, the Boston Society of Natural History, the Essex Institute of Salem, the American Anthropological Association, and the Arctic Club. He was president of the Ohio State Archaeological and Historical Society from 1907 to 1918. In 1887 Wright received two honorary degrees: the D.D. from Brown University and the LL.D. from Drury College, Springfield, Missouri.

Wright belonged to a generation of American geologists whose endeavors laid the basis for modern glacial theory. He was the most vigorous propo-

nent of several hypotheses that glaciologists debated, sometimes heatedly. On the basis of his work at the Niagara gorge and elsewhere, Wright advocated the relatively late end of the Ice Age, approximately ten thousand years ago. Further, he contended that there had been only one Ice Age. According to Wright, the Ice Age did not consist of alternating periods during which much of the northern hemisphere was covered by glaciers, and intermittent periods in which there was no extraordinary glaciation—the multiple glaciation theory, which set the duration of the Ice Age at far over one hundred thousand years. Rather, he espoused a unitary theory: the Ice Age, he believed, consisted of the alternate ebbing and flowing of glaciers over a period of time not to exceed ninety thousand years. Wright conceded the existence of certain interglacial deposits, but argued that these represented merely local recessions, not distinct interglacial periods.

The most controversial of Wright's positions was his unqualified affirmation of the existence of man in North America during the Pleistocene. Thomas C. Chamberlin, W. J. McGee, and John W. Powell opposed him, thus setting off a heated debate that involved some of the major scientific societies and periodicals, the departments of geology of the University of Chicago and other schools, and the United States Geological Survey for over two years. The spark was the publication of Wright's *Man and the Glacial Period* (1892), a careful description of alleged evidences of glacial man. Wright's side was supported in whole or in part by Warren Upham, Newton H. and Alexander Winchell, Nathaniel S. Shaler, Frederic W. Putnam, James D. Dana, and other naturalists; but it was a later generation that achieved a fuller appreciation of his efforts to establish the relatively short period since the end of the Ice Age, and to substantiate the existence of Pleistocene man in North America.

BIBLIOGRAPHY

I. ORIGINAL WORKS. Among Wright's 16 books and almost 600 articles not cited in the text are "Some Remarkable Gravel Ridges in the Merrimack Valley," in *Proceedings of the Boston Society of Natural History,* **19** (1877), 47–63; "The Glacial Phenomena of North America and Their Relation to the Question of Man's Antiquity in the Valley of the Delaware," in *Bulletin of the Essex Institute,* **13** (1881), 65–72; "Recent Investigations Concerning the Southern Boundary of the Glaciated Area of Ohio," in *American Journal of Science,*

3rd ser., **26** (1883), 44–56; "The Niagara River and the Glacial Period," *ibid.*, **28** (1884), 32–35; "The Terminal Moraine in Ohio, Kentucky, and Indiana," in H. C. Lewis et al., *Report on the Terminal Moraine in Pennsylvania and Western New York*, Second Geological Survey of Pennsylvania, vol. Z (Harrisburg, Pa., 1884), 203–243; "The Muir Glacier," in *American Journal of Science*, 3rd ser., **33** (1887), 1–18; "The Glacial Boundary in Western Pennsylvania, Ohio, Kentucky, Indiana, and Illinois," *Bulletin of the United States Geological Survey* no. 58 (1890); "Prehistoric Man on the Pacific Coast," in *Atlantic Monthly*, **67** (1891), 501–513; "Excitement Over Glacial Theories," in *Science*, **20** (1892), 360–361; "Unity of the Glacial Epoch," in *American Journal of Science*, 3rd ser., **44** (1892), 351–373; "Continuity of the Glacial Period," *ibid.*, **47** (1894), 161–187; and *Greenland Icefields and Life in the North Atlantic: With a New Discussion of the Causes of the Ice Age* (New York, 1896), written with Warren Upham.

Other works are "Recent Geological Changes in Northern and Central Asia," in *Quarterly Journal of the Geological Society of London*, **57** (1901), 244–250; "Evidence of the Agency of Water in the Distribution of the Loess in the Missouri Valley," in *American Geologist*, **33** (1904), 205–222; "Postglacial Erosion and Oxidation," in *Bulletin of the Geological Society of America*, **23** (1912), 277–296; *Origin and Antiquity of Man* (Oberlin, Ohio, 1912); and *Story of My Life and Work* (Oberlin, Ohio, 1916). The 5th ed. (1911) of *Ice Age in North America* was a complete revision.

II. SECONDARY LITERATURE. See "George Frederick Wright: In Memoriam," in *Ohio Archaeological and Historical Society Quarterly*, **30** (1921), 162–175; and Warren Upham, "Memorial of George Frederick Wright," in *Bulletin of the Geological Society of America*, **33** (1922), 14–30. For an investigation of Wright's work with Asa Gray, see Michael McGiffert, "Christian Darwinism: The Partnership of Asa Gray and George Frederick Wright, 1874–1881" (Ph.D. diss., Yale University, 1958). The most extensive study of Wright is William J. Morison, "George Frederick Wright: In Defense of Darwinism and Fundamentalism, 1838–1921" (Ph.D. diss., Vanderbilt University, 1971), which is based on Wright's correspondence and other papers (some 15,000 items) in the Oberlin College Archives.

WILLIAM J. MORISON

WRIGHT, THOMAS (*b.* Byers Green, near Durham, England, 22 September 1711; *d.* Byers Green, 25 February 1786), *astronomy.*

Thomas Wright "of Durham" was the third son of John Wright, a yeoman and carpenter who had a small holding near Durham. His early schooling was cut short by an impediment of speech, and at the age of thirteen he was apprenticed to a clockmaker. He used his leisure time to study astronomy with such alarming dedication that his father burned his books in an effort to frustrate this enthusiasm. In 1729 Wright was involved in a scandal that forced him to flee from his master, and after some adventures he reached home and was released from his apprenticeship. Failing to find employment, he studied navigation and noted: "Reflecting almost upon every object, conseive may find Ideas of ye Deaty and Creation."

After sampling various occupations, Wright began to teach navigation in the seaport of Sunderland; but in the spring of 1731 he was unemployed, owing to the departure of the seamen. He was evidently a natural teacher; and he used his leisure to prepare mathematical and astronomical publications, often in the form of wall sheets or "schemes." One such sheet, covering twenty-four square feet and accompanied by a "key" in the form of a substantial quarto volume, *Clavis coelestis*, appeared in 1742. Wright also spent much of the 1730's surveying the estates of the aristocracy and giving private and public classes in the physical sciences. Evidently he achieved something of an international reputation, for in 1742 he was offered £300 a year to become professor of navigation at the Imperial Academy of Sciences at St. Petersburg. His journal notes that "His Proposals of £500 were sent to Russia," but nothing came of these negotiations.

In 1746 Wright went to Ireland for several months to assemble drawings of antiquities for his most successful work, *Louthiana*. (A sequel to *Louthiana* and a volume on the antiquities of England remain in manuscript.) In 1750 Wright published his most significant work, *An Original Theory or New Hypothesis of the Universe*. By 1755, when his *Universal Architecture* appeared, his thoughts were turning again to his birthplace, perhaps because his noble patrons were beginning to die off. The following year he laid the foundation of a house at Byers Green and in 1762 retired there, to "Prosicute my Studies." He died there in 1786, unmarried but survived by a daughter.

Wright's early reflections on "Ideas of ye Deaty and Creation" mark the beginning of his lifelong preoccupation with the reconciliation of his religious and scientific views of the universe. The telescope revealed to the observer the structure of our locality in the universe, but religion alone could provide the cosmological overview. In its simplest form this overview comprised a unique divine center, the abode of God and the angels; a "Region of Mortality," consisting of the sun and the other stars forming a spherical assemblage surrounding the divine center; and an outer darkness or other spatial realization of the punishment of the wicked.

Wright's first attempt to effect this reconciliation is found in a manuscript dated 1734 that appears to be the text of a lecture-sermon accompanying a vast visual aid, now lost. In this lecture the divine center (the center in the moral order) was also the gravitational center (the center in the physical order), and thus Wright required the sun and the other stars to be moving in orbit about this center in order to avoid gravitational collapse. He found evidence of this circulation among the stars in Halley's 1718 paper on proper motions (*Philosophical Transactions*, **30**, 736–738), the significance of which had escaped more powerful minds than Wright's.

In his efforts to bring home to his audience their precarious moral position, Wright in his visual aid portrayed a cross section of creation, one that passed through the divine center and the solar system. With artistic license he represented the visible portion of the universe as it actually appears to us, beginning with the sun and moon and extending to the Milky Way, which he considered to be the effect of innumerable distant stars in the plane of the cross section. It was only sometime after 1742 that Wright realized that his model of the universe would produce such a milky way in *any* of the possible cross sections, whereas the visible Milky Way is unique. In 1750 *An Original Theory* met this difficulty by making the shell of stars thin, so that the Milky Way is the plane tangent to the shell at the position occupied by the solar system; because the shell is thin, no milky effect is produced when the observer looks either inward or outward. Alternatively (but with the loss of spherical symmetry), Wright would permit the stars to form a flattened ring, like a large-scale Saturn's ring; the Milky Way was then in the plane of the ring. This latter version, in which the stars lie in a plane and orbit their center as the planets orbit the sun, appealed to Immanuel Kant, who, not realizing that the center of Wright's system was supernatural, credited Wright with originating a disk-shaped model of the galaxy.

In fact, *An Original Theory* proposes a multiplicity of star systems, each with its own supernatural center; and the punishment of the wicked is not provided for. These defects may have prompted Wright to compose *Second Thoughts*, which never reached publication in his lifetime. In this late manuscript the universe consists of an infinite sequence of concentric shells surrounding the divine center; our sky is one of these solid shells and, viewed from without, appears as a large version of the sun. Viewed from within, it is studded with volcanoes, which we see as the stars and the Milky Way. A good life is rewarded by promotion to a more spacious sphere for our next existence; an evil life is punished by demotion to a more cramped sphere that, although nearer the divine center in terms of miles, is still infinitely many spheres away from it.

With *Second Thoughts*, Wright achieved his reconciliation of science and religion: the observations had at last been fitted into a universe that had symmetry about the center and provision for rewards and punishments. *Second Thoughts*, with its solid sky in which the stars are volcanoes, appears retrograde to modern readers, but only because we have been accustomed to judge Wright on our terms rather than his.

BIBLIOGRAPHY

I. ORIGINAL WORKS. Wright's autobiographical notes (for the period ending 1746) are in British Museum, Add. 15627, most recently edited by Edward Hughes as "The Early Journal of Thomas Wright of Durham," in *Annals of Science*, **7** (1951), 1–24. His principal books are *Clavis coelestis* (London, 1742), facs. ed. by M. A. Hoskin (London, 1967); *Louthiana* (London, 1748, 1758); *An Original Theory or New Hypothesis of the Universe* (London, 1750), facs. ed. by M. A. Hoskin, with transcription of Wright's MS "A Theory of the Universe" (1734) (London, 1971); *Universal Architecture* (London, 1755); and *Second or Singular Thoughts Upon the Theory of the Universe*, edited from the MS by M. A. Hoskin (London, 1968). Eight vols. of Wright MSS are in the Central Library, Newcastle-upon-Tyne, and others are in the Royal Society, the Royal Astronomical Society, and Durham University; the Durham collection includes the unique copy of his *The Universal Vicissitude of Seasons* (1737) and other rarities. A broadsheet, "The Universal Vicissitude of Seasons," is in the possession of Harrison Horblit.

II. SECONDARY LITERATURE. A list of books and MSS by Wright and a bibliography of secondary literature is appended to "Thomas Wright of Durham and Immanuel Kant," in Herbert Dingle and G. R. Martin, eds., *Chemistry and Beyond. A Selection From the Writings of the Late Professor F. A. Paneth* (New York, 1964), 93–119; note that *Pannauticon* is not a book but an instrument, and that Paneth fundamentally misunderstands Wright's cosmology. For an account of the development of Wright's cosmology, see M. A. Hoskin, "The Cosmology of Thomas Wright of Durham," in *Journal for the History of Astronomy*, **1** (1970), 44–52; and the intros. to the modern eds. of his works.

MICHAEL A. HOSKIN

WRIGHT, WILBUR (*b.* Millville, Indiana, 16 April 1867; *d.* Dayton, Ohio, 30 May 1912), and **WRIGHT, ORVILLE** (*b.* Dayton, Ohio, 19 August 1871; *d.* Dayton, Ohio, 30 January 1948), *aeronautics.*

Wilbur and Orville Wright, the sons of Milton Wright, a bishop of the United Brethren Church, and Susan Catherine Koerner, had two older brothers, Reuchlin and Lorin, and a younger sister, Katharine. Their upbringing in a family where liberality of thought and individual initiative and expression were encouraged, contributed markedly to their later achievements. Although their formal education did not go beyond high school, they were widely read, especially in the technical literature of the day, and taught themselves mathematics and smatterings of French and German. Both were of medium stature, trim, and athletic, and from boyhood showed powers of physical endurance and mechanical skill and ingenuity. After youthful ventures in editing and printing small neighborhood newspapers, they set up the Wright Cycle Company in 1892 and for the next decade made their living by the design, manufacture, and sale of bicycles.

The death, on 10 August 1896, of German aviation pioneer Otto Lilienthal, from injuries suffered in a gliding accident, led the Wrights to the serious study of flight. By 1899 they had carried their theory of lateral balance (aileron control) to the point of a practical demonstration made by Wilbur, in August, using a five-foot-span biplane kite. Equilibrium was maintained and maneuver made possible by varying the air pressures at the wing tips through adjustment of the angles of attack on the two sides. With this action and an adjustable horizontal surface (elevator), later (1902) combined with the compensating action of a movable vertical rudder, they achieved control about the three axes of the airplane. The system was patented in 1906 and has been used on all airplanes ever since.

Discovering—from field experiments and tentative gliding trials at Kitty Hawk, North Carolina, in the summers of 1900 and 1901—that almost all existing aerodynamic data were erroneous, the Wrights designed a small wind tunnel in which, in the fall of 1901, they tested several hundred model airfoils and obtained reliable lift and drag measurements as well as many other essential aerodynamic data.

With this knowledge, in October 1902 they began the construction of a powered airplane. The all-up weight, including pilot, was 750 pounds. The engine and propellers were of their own design and manufacture, and the propellers were based entirely on theories they originated. With this machine four successful flights were made from the level sand near the Kill Devil Hills, North Carolina, on 17 December 1903. The final, longest flight lasted for fifty-nine seconds and covered a distance of 852 feet; this represented about half a mile through the air.

The Wrights devoted the next five years to improving both their invention and their skill as pilots. In 1905, with the airplane nearing the state of practical utility, they offered their patent and their scientific data to the United States War Department, which rejected the overture. Convinced that the first use of the airplane would be in war, the Wrights sought markets abroad. In 1908, after many rebuffs, they received purchase offers from a French syndicate and from the United States government.

Demonstration trials in the two countries took place concurrently, with Orville flying in the United States and Wilbur in France. All doubt of the Wrights' mastery of the air evaporated, and the honors and adulation of two continents were heaped upon them. In 1909 Wilbur flew at Rome and Orville at Berlin.

The culmination of the Wrights' achievements came with Wilbur's two flights at New York in 1909. On 29 September, taking off from and landing at Governors Island, he made a circuit of the Statue of Liberty; on 4 October he flew a twenty-one-mile course to Grant's Tomb and back.

After their triumph the brothers quietly turned to teaching others to fly and to directing the Wright Company. They now had many imitators and rivals, and were forced to defend their pioneer patent in the courts. Under the strain, Wilbur contracted typhoid fever and died suddenly on 30 May 1912. Having divested himself of his interest in the Wright Company in 1915, Orville, after World War I, confined his aviation activities mainly to research, including membership in the National Advisory Committee for Aeronautics. He survived his brother by nearly thirty-six years. On the twenty-fifth anniversary of the first flight he witnessed the laying of the cornerstone of the Wright Brothers National Memorial at Kill Devil Hills—the only United States national monument erected during the lifetime of a man so honored.

BIBLIOGRAPHY

The letters of the Wrights have been collected in *Miracle at Kitty Hawk; the Letters of Wilbur and Orville Wright*, Fred C. Kelly, ed. (New York, 1951); their papers in *The Papers of Wilbur and Orville Wright, Including the Chanute-Wright Letters and Other Papers of Octave Chanute*, M. W. McFarland, ed., 2 vols. (New York–Toronto–London, 1953). Orville Wright wrote *How We Invented the Aeroplane* (New York, 1953), edited with a commentary by F. C. Kelly. See also Wilbur Wright's first rebuttal deposition contained in the complainant's record in the case of Wright Company v. Herring Curtiss Company and Glenn H. Curtiss, U.S. District Court, Western District of New York, 1912, vol. 1.

A biography of the Wrights is F. C. Kelly, *The Wright Brothers: A Biography Authorized by Orville Wright* (New York, 1943). See also C. H. Gibbs-Smith, *The Invention of the Aeroplane (1799–1909)* (New York, 1966); W. Langewiesche, "What the Wrights Really Invented," in *Harper's Magazine*, **200** (June 1950), 102–105; and M. W. McFarland, "When the Airplane Was a Military Secret: A Study of National Attitudes Before 1914," in *U.S. Air Services*, **39** (Sept. 1954), 11, 14, 16; (Oct. 1954), 18, 20–22; and "The Fame of Wilbur Wright," *ibid.*, **40** (Dec. 1955), 4–6.

MARVIN W. McFARLAND

WRIGHT, WILLIAM HAMMOND (*b.* San Francisco, California, 4 November 1871; *d.* San Jose, California, 16 May 1959), *astronomy*.

Wright was a skillful designer of astronomical equipment, which he used to photograph the spectra of stars and nebulae. He was also the first to use six-color photography in studying the planet Mars.

Wright's parents were Joanna Maynard Shaw and Selden Stuart Wright. After attending public schools in San Francisco, Wright went to the University of California, where he received a B.S. in civil engineering in 1893. He remained in Berkeley for two years of graduate study, specializing in astronomy, and then transferred for a year to the University of Chicago, where George Ellery Hale taught him the latest techniques for photographing spectra.

In 1897 Wright returned to California, as an assistant astronomer at the Lick observatory, to help W. W. Campbell in his studies of solar motion. For this purpose Wright photographed the spectra of many stars, to get line-of-sight velocities. In 1903 he was sent to Santiago, Chile, with a 36½-inch telescope, which he installed on Cerro San Cristobal and used to acquire similar data for stars in the Southern hemisphere. He was accompanied by his wife, Elna Warren Leib, whom he had married in 1901. Wright remained in charge of this Southern station of Lick until 1906.

Back in California, Wright was promoted to astronomer in 1908, a post he held until 1944. In 1935 he was also appointed director of the Lick observatory, succeeding R. G. Aitken.

Wright investigated nebulae, particularly those called planetaries, and the high temperature stars that they surround. His data later helped I. S. Bowen to unravel the "nebulium" mystery.

Wright also continued work on novae, begun earlier with Campbell. His photographs of the spectrum of Nova Geminorum 1912, taken on every possible night during more than nineteen months, provided a wealth of details basic to the understanding of such complicated explosive events.

Beginning in 1924, while Mars was well situated for observation, Wright used the Crossley 36-inch reflector at Lick to photograph this planet (and also Jupiter, Saturn, and Venus) on suitably sensitized emulsions exposed through a series of filters, in colors ranging from ultraviolet through the visible to the infrared. The enhanced contrast, particularly for Mars in the infrared, revealed many unsuspected details.

Wright's final project was a set of about 1,300 large-scale celestial photographs, to be repeated for comparison after several decades and thus provide information on motions within our galaxy. Interrupted by World War II, it was finally begun in 1947 by C. D. Shane, who succeeded Wright as director of Lick.

The National Academy of Sciences elected Wright to membership in 1922, and awarded him its Draper Medal in 1928, the same year that he received the Janssen Medal of the Paris Academy of Sciences. He was elected a fellow of the Royal Astronomical Society (London) in 1927, and was awarded its Gold Medal in 1938. Wright's two advanced degrees were both honorary: a D.Sc. from Northwestern University (1929) and an LL.D. from the University of California (1944).

BIBLIOGRAPHY

I. ORIGINAL WORKS. An early paper by Wright on instrumentation, "The Auxiliary Apparatus of the Mills Spectrograph for Photographing the Comparison Spectrum," is in *Astrophysical Journal*, **12** (1900), 274–278. For a brief account of Wright's work in Chile, see "On Some Results Obtained by the D. O. Mills Expedition to the Southern Hemisphere," *ibid.*, **20** (1904), 140–145. His ideas concerning the central stars of planetary nebulae are described in "The Relation Between the Wolf-Rayet Stars and the Planetary Nebulae," *ibid.*, **40** (1914), 466–472. Wright's photographs of Mars and other planets are described and illustrated in his George Darwin Lecture, "On Photographs of the Brighter Planets by Light of Different Colours," in *Monthly Notices of the Royal Astronomical Society*, **88** (1928), 709–718, with 34 figs. on 3 plates.

"The Spectrum of Nova Geminorum (1912)" appears in *Publications of the Lick Observatory*, **14** (1940), 27–91.

II. SECONDARY LITERATURE. Contemporary appraisals of Wright's work include the address delivered by H. Spencer Jones when awarding the Gold Medal to Wright, in *Monthly Notices of the Royal Astronomical Society*, **98** (1938), 358–374. An obituary notice by Paul W. Merrill is in *Publications of the Astronomical Society of the Pacific*, **71** (1959), 305–306, with a photograph of Wright; another by C. D. Shane can be found in *American Philosophical Society Yearbook for 1959*, 150–153.

SALLY H. DIEKE

WRÓBLEWSKI, ZYGMUNT FLORENTY VON (*b.* Grodno, Lithuania, Russia, 28 October 1845; *d.* Cracow, Poland, 19 April 1888), *physics.*

The son of a lawyer, Wróblewski achieved his fame as an experimental physicist mainly through work on the liquefaction of gases that he did with K. S. Olszewski. After attending the Gymnasium in Grodno, he entered the University of Kiev in 1862 but was banished to Siberia in 1863 for participating in the January Revolution in Poland. He was amnestied in 1869 and allowed to travel to Berlin for treatment at the eye clinic run by Albrecht von Graefe. Wróblewski began to study natural science in Berlin and continued his studies at Heidelberg and then at Munich, where in 1874 he earned the Ph.D. and worked briefly as an assistant. In 1875–1876 he was an assistant at the University of Strasbourg, where he qualified as a lecturer. A stipend from the Cracow Academy of Sciences enabled Wróblewski to continue his training in Paris under Henri Sainte-Claire Deville and to visit London, Oxford, and Cambridge. He was named professor of physics at Jagiellonian University, Cracow, in 1882. In 1883 he became corresponding member of the Cracow Academy and, in 1887, of the Vienna Academy of Sciences, which awarded him the Baumgartner Prize for his work on the liquefaction of air.

At this time the problem of liquefying the "permanent" gases was attracting much attention.[1] Wróblewski did research in this field with Olszewski. Improving upon a method devised by L. P. Cailletet and using ethylene in a vacuum as the cooling agent, they reached the critical temperature of air, thereby becoming the first to obtain oxygen, nitrogen, and carbon monoxide as waterlike fluids. Wróblewski reported their results to the Vienna Academy on 12 April 1883. During their brief collaboration Wróblewski and Olszewski, stimulated by mutual competition, also liquefied hydrogen, at least in the dynamic state.

Wróblewski, who had first investigated electrical phenomena, found during his research on low temperatures that the electrical conductivity of copper displays "extremely remarkable properties" at such temperatures. He drew attention to the great importance of this phenomenon and specifically to the greatly increased conductivity.[2] This extremely promising research came to an abrupt end when Wróblewski suffered fatal burns after overturning his kerosine lamp.

NOTES

1. At this time the term "permanent" was applied to gases that could not be transformed into other states of matter—that is, gases the critical temperature of which had not yet been determined. Among these were air and its components. Gases that could be liquefied or produced by sublimation were called—in contradistinction to the "permanent" gases—"coercible."
2. The existence of this superconductivity was demonstrated in 1911 by Kamerlingh-Onnes.

BIBLIOGRAPHY

I. ORIGINAL WORKS. Wróblewski's numerous scientific papers include his doctoral dissertation, *Untersuchungen über die Erregung der Electricität durch mechanische Mittel* (Munich, 1874); *Die Diffusion der Gase durch absorbierende Substanzen* (Strasbourg, 1876), his *Habilitationsschrift*; "Ueber die Gesetze, nach welchen die Gase sich in flüssigen, festflüssigen und festen Körpern verbreiten," in *Annalen der Physik und Chemie*, n.s. **2** (1877), 481–513; "Ueber die Natur der Absorption der Gase," *ibid.*, **8** (1879), 29–52; "Untersuchungen über die Absorption der Gase durch

Flüssigkeiten unter hohen Drucken," *ibid.*, **17** (1882), 103–128; and **18** (1883), 290–308, on carbon monoxide; "Ueber die Verflüssigung des Sauerstoffs, Stickstoffs und Kohlenoxyds," *ibid.*, **20** (1883), 243–257, written with K. S. Olszewski; "Ueber das specifische Gewicht des Sauerstoffs," *ibid.*, 860–870; "Ueber den Gebrauch des siedenden Sauerstoffs, Stickstoffs, Kohlenoxyds, sowie der atmosphärischen Luft als Kältemittel," *ibid.*, **25** (1885), 371–407; "Ueber den electrischen Widerstand des Kupfers bei den höchsten Kältegraden," *ibid.*, **26** (1885), 27–31; "Ueber das Verhalten der flüssigen atmosphärischen Luft," *ibid.*, 134–144; and "Ueber die Darstellung des Zusammenhangs zwischen dem gasförmigen und flüssigen Zustand der Materie durch Isopyknen," *ibid.*, **29** (1886), 428–451. See also "Über den Gebrauch des siedenden Sauerstoffs etc.," in *Sitzungsberichte der Akademie der Wissenschaften in Wien*, Math.-phys. Kl., **91** (1885), pt. 2, 667–711; and "Über die Condensation der schwer coërciblen Gase," *ibid.*, **92** (1886), pt. 2, 639–651. Wroblewski also published many of his scientific papers in *Anzeiger der Akademie der Wissenschaften Krakau*.

Bibliographies are in Academy of Sciences, Cracow, *Katalog der akademischen Publikationen seit 1873 bis 1909* (Cracow, 1910); and Poggendorff, III, 1468; and IV, 1672.

II. SECONDARY LITERATURE. Writings in German include M. von Smoluchowski, "Karl Olszewski," in *Naturwissenschaften*, **5** (1917), 738–740, an obituary that includes a biographical note on Wróblewski; and E. Suess, "Siegesmund von Wroblewski," in *Almanach der Akademie der Wissenschaften in Wien*, **38** (1888), 190–192, an obituary.

Articles in Polish, details of which were supplied by Dr. I. Stroński, of Cracow, include T. Estreicher, "Zygmunt Wróblewski," in *Wszechświat* (1948), 215–219, which commemorates the 60th anniversary of Wróblewski's death; *Kronika Uniwersytetu Jagiellońskiego 1864–1887* (Cracow, 1887), 179–182; A. Pasternak, "Karol Olszewski (1846–1915) i Zygmunt Wróblewski (1845–1888)," in *Polscy badacze przyrody* ("Polish Investigators of Nature"; Warsaw, 1959), 174–203; and T. Piech, "Zarys historii katedr fizyki Uniwersytetu Jagiellońskiego" ("Compendium of the History of the Chairs of Physics at the Jagiellonian University"), in *Studia ad universitatis Iagellonicae Cracoviensis facultatis mathematicae, physicae, chemicae cathedrarum historiam pertinentia* (Cracow, 1964), 223–270.

Further biographical literature is listed in the *DSB* article on Olszewski, X, 206–207.

HANS-GÜNTHER KÖRBER

WROŃSKI, JÓZEF MARIA. For a detailed account of his life and work, see **Hoëné-Wroński (or Hoehne), Józef Maria,** in the Supplement.

WU, HSIEN (*b.* Foochow, Fukien, China, 24 November 1893; *d.* Boston, Massachusetts, 1 August 1959), *biochemistry, nutrition.*

Wu, the son of Hsiao-chien Wu and Liang Shih Wu, achieved worldwide recognition for his early studies in the United States and became China's foremost biochemist and nutrition scientist. His name is particularly associated with analytical procedures known as the Folin and Wu methods. Born to a scholarly family, he received tutorial training in the Chinese classics, starting at age six, advanced rapidly through high school, and received a scholarship in naval architecture at the Massachusetts Institute of Technology in 1911.

During his first summer vacation, on a farm in New England, Wu became fascinated by the new horizons in biology opened by T. H. Huxley's "On the Physical Basis of Life." He changed his major field to chemistry and biology while at M.I.T. and then transferred to Harvard in 1917 for graduate studies (Ph.D., 1919) in biochemistry with Otto Folin. His interest in architecture was transformed into a lifelong hobby in the field of art, and his background in mathematics, physics, and organic analysis furnished a sound basis for his new career.

In medical laboratories it had been customary to require large samples of blood for diagnostic testing and metabolic research—a practice with disadvantages for both the physician and the patient. In his doctoral dissertation, "A System of Blood Analysis," Wu developed techniques that permitted quantitative measurements of the major constituents of blood with only 10 ml. samples. The methods included a particularly good procedure for measuring the sugar content in blood or urine in a sample as small as one drop. Wu remained with Folin for a year of postdoctoral work and subsequently accepted an appointment at the Peking Union Medical College. By 1923 Wu had organized an outstanding teaching and research program in Peking, and was promoted to associate professor and head of the department of biochemistry in 1924. He retained a full professorship from 1928 until January 1942, when the college was taken over by the Japanese. At Shanghai on 20 December 1924, Wu married Daisy Yen, a graduate student in biochemistry. They traveled to New York, where he worked with Donald van Slyke at the Rockefeller Institute and Mrs. Wu resumed her studies with Henry C. Sherman at Columbia University.

Wu's more than 150 research papers included many contributions on the functions of electrolytes, immunochemistry, biochemical analysis, food composition, and the behavior of proteins in solution and the changes involved in protein denaturing, including the first suggestion that the change was characterized from a globular to an open structure.

After his return to Peking, Wu urged that greater attention be paid to the effects of food habits on health; and his nutritional research on experimental animals was accompanied with studies of eating habits of human beings. Wu served as a member of the National Committee on Standardization of Scientific Terminology of China (1921–1927). In 1926 he assisted in organizing the Chinese Physiological Society and later served as president and on the editorial board (1926–1941). He was elected adviser to the Institute of Physiology of the Academia Sinica in 1930, a member of the administrative committee that directed Peking Union Medical College in 1935–1937, and a fellow of the Academia Sinica.

Wu's growing international recognition brought him membership in the American Society of Biological Chemists, honorary membership in the Deutsche Akademie Naturforscher Leopoldina, the advisory board of *Biochemica et biophysica acta*, and the Standing Advisory Committee on Nutrition of the Food and Agriculture Organization of the United Nations (1948–1950).

During the first two years of the Japanese occupation of Peking, Wu lived in retirement at his home. In March 1944 the Chinese government, then at Chungking, invited him to organize the National Nutrition Institute there; he developed plans for the Institute and was appointed director. Some three months later the government sent him to the United States as nutrition expert for a commission to study postwar problems of rehabilitation and reconstruction. After a year of negotiating for equipment and for dried milk in food shipments, and economic study at the Brookings Institution, Wu returned to Chungking to report on his mission and to submit further plans for the Nutrition Institute.

In 1946 the government invited Wu to direct a branch of the National Institute of Health in Peking as well as to continue as director of the Nutrition Institute, which had been reestablished in Nanking. But an invitation from UNESCO to be one of the six Chinese delegates to the International Physiological Congress at Oxford in July 1947 permitted Wu to revisit the United States. His discussions with his friend T. P. Hou encouraged him to plan for an Institute of Human Biology in China. He gathered equipment for research, including a mass spectrometer and instruments for isotope research; and to familiarize himself with its use he served as visiting scholar for more than a year at the biochemistry department of the College of Physicians and Surgeons of Columbia University. He also assisted in the purchase of equipment and supplies for the National Institute of Health in Peking and shipped library materials for the projected Research Institute of Human Biology.

Meanwhile, in China the Communists had surrounded Peking. With extreme difficulty Mrs. Wu escaped with their five children in January 1949 and reached San Francisco six months later. In September of that year Wu was appointed visiting professor of biochemistry at the Medical College of the University of Alabama, where he continued his research, assisted by his wife.

In 1952 Wu suffered a heart attack that led to his retirement from Alabama in August 1953. He recovered almost completely and moved to Boston, where he continued to write. A second coronary thrombosis in April 1958 was followed by his death the following year.

BIBLIOGRAPHY

Wu's books are *Principles of Nutrition* (Shanghai, 1929), in Chinese; and *Principles of Physical Biochemistry* (Peking, 1934), in English. Of his 157 scientific memoirs, the most representative are "A System of Blood Analysis," in *Journal of Biological Chemistry*, **38** (1919), 81–110, written with Otto Folin; "Studies of Gas and Electrolyte Equilibria in the Blood. V. Factors Controlling the Electrolyte and Water Distribution in the Blood," *ibid.*, **56** (1923), 765–849, written with D. D. van Slyke and F. C. McLean; "Composition of Antigen-precipitin Precipitate," in *Proceedings of the Society for Experimental Biology and Medicine*, **25** (1928), 853–855, written with L. H. Cheng and C. P. Li; and **26** (1929), 737–738, written with P. P. T. Sah and C. P. Li; "Studies on Denaturation of Proteins. XIII. A Theory of Denaturation," in *Chinese Journal of Physiology*, **5** (1931), 321–344; "Nutritional Deficiencies in China and Southeast Asia," in *Proceedings of the Fourth International Congress on Tropical Medicine and Malaria*, II (Washington, 1948), 1217–1223; and "Interpretation of Urinary N^{15}-Excretion Data Following Administration of an N^{15}-labeled Amino Acid," in *Journal of Applied Physiology*, **14** (1959), 11–21, written with Julius Sendroy, Jr., and Charles W. Bishop.

On his life and work, see Daisy Yen Wu, *Hsien Wu, 1893–1959, In Loving Memory* (Boston, 1959), which includes a complete bibliography of his writings and tributes from his colleagues.

C. G. KING

WULFF, GEORG (Yuri Viktorovich) (*b*. Nezhin, Russia, [now Ukrainian S.S.R.], 10 June 1863; *d*. Moscow, U.S.S.R., 25 December 1925), *crystallography*.

Wulff's father, Viktor Konstantinovich, was director of the boy's Gymnasium in Nezhin. Wulff received his secondary education in Warsaw and entered the natural sciences section of the Faculty of Physics and Mathematics at Warsaw in 1880, specializing in mineralogy and crystallography with A. E. Lagorio and N. G. Egorov. His interest in crystallography was reflected in two student publications, on the morphology (1883) and physical properties of crystals (1884); and he was awarded a gold medal by the university for his research on the physical properties of quartz. After graduating in 1885, he was retained to prepare for a professorship in the department of mineralogy.

Wulff's growing interest in physical crystallography and the structure of crystals led him to investigate the optical anomalies of crystals and the cause of rotation of the plane of polarization in crystals, as well as the piezoelectrical properties of quartz. From 1889 to 1892 he visited E. S. Fyodorov and V. I. Vernadsky at St. Petersburg, P. Groth at Munich, and M. A. Cornu at Paris, becoming acquainted with the most recent developments in crystallography. He was the first to win recognition outside of Russia for Fyodorov's classical research on the theory of the structure of crystals.

After defending his master's dissertation in 1892 at Warsaw, Wulff became lecturer in the department of mineralogy there. Four years later he defended his doctoral dissertation at the University of Odessa and was appointed professor of mineralogy at the University of Kazan, where he remained for three semesters. In 1898 he became head of the department of mineralogy at the University of Warsaw; and in 1908, at Vernadsky's invitation, he began teaching crystallography, crystal chemistry, and crystal optics at the University of Moscow. He transferred to the Shanyavsky University in 1911, heading the laboratory of crystallography, and in 1916 became director of the department of mineralogy and crystallography at the Moscow University for Women. From 1918 until his death he was professor at the University of Moscow. He started teaching and research in crystallography at the Faculty of Physics. His teaching chiefly concerned questions of geometrical crystallography, the growth of crystals and crystallophysics. He became a corresponding member of the Academy of Sciences of the U.S.S.R. in 1921, was president of the P. N. Lebedev Physical Society, and was a member of the All-Union Mineralogical Society.

Pierre Curie's discovery of piezoelectricity in quartz (1880) and his work on the equilibrium morphology of crystals (1885) stimulated Wulff to investigate false pyroelectrical quartz (1886) and the velocity of growth and dissolution of crystals (1896, 1901). The results of this work led to the formulation of the Curie-Wulff principle (1916). Curie had demonstrated an inverse proportion between the sizes of crystal faces and their specific surface energies (K_i). In his 1895 thesis Wulff showed that for a constant volume, total surface energy would be minimized when the specific surface energies for each face (K_i) were proportional to the perpendicular distances (n_i or Wulff vectors) from a central point (the Wulff point) to each face$-K_1 : K_2 : K_3 : \cdots : = n_1 : n_2 : n_3 : \cdots$. In modern studies of crystal growth the equilibrium form derived from the theorem is known as the Wulff construction.

Wulff also conducted detailed studies of the influence of concentration currents on the morphology of growing crystals (1895), investigated liquid crystals (1909), and invented a rotating crystallizer that, by removing the influence of concentration currents, made possible the formation of perfectly formed crystals. Fyodorov's and Arthur Schönflies' theoretical investigations of crystal symmetry and structure were incorporated in Wulff's works on the theory of crystal habit (1908) and on the structure of quartz, and in optical research on pseudosymmetric crystals (1887–1890). One of the first to recognize the superiority of the Fyodorov-Goldschmidt theodolitic goniometry (the two-circle goniometer), Wulff developed methods of measuring and computing for it. In 1909 he proposed the Wulff net, the stereographic projection of a sphere with its meridians and parallels, oriented with polar axis horizontal; it is still widely used in optical, X-ray, and morphological crystallography. Wulff's goniometric research showed an essential deviation between ideal theory and the real crystal.

In 1896 Wulff presented crystal symmetry in an original manner, by using consecutive reflection only in planes, which was incorporated in his subsequent textbooks (1923, 1926). Laue's discovery of X-ray diffraction in 1912 drew Wulff's attention to X-ray structural research on crystals. Concurrently with W. H. and W. L. Bragg (1913), he independently developed the diffraction relationship $n\lambda = 2d \sin \theta$ (the Bragg-Wulff equation), the basis for the structural analysis of crystals. Wulff founded the first X-ray laboratory in Russia and conducted

X-ray diffraction research there on crystal structure.

Wulff's lectures on crystallography made extensive use of microprojection, which made possible the immediate reproduction and visual demonstration of phenomena arising in the growth and development of crystals. His students included Sigmund Weiberg, A. V. Shubnikov, E. E. Flint, A. B. Mlodseevsky, and S. T. Konobeevsky.

BIBLIOGRAPHY

I. ORIGINAL WORKS. Some of Wulff's writings on crystallophysics and crystallography were collected as *Izbrannye raboty po kristallofizike i kristallografii* ("Selected Works in Crystallophysics and Crystallography"; Moscow, 1952). Separately published early works include "Opytnoe issledovanie elektricheskikh svoystv kvartsa" ("Experimental Investigation of the Electrical Properties of Quartz"), in *Varshavskia universitetskia izvestia*, no. 3 (1886), 1–17; "O stroenii kristallov kvartsa" ("On the Structure of Quartz Crystals"), in *Zapiski Imperatorskago mineralogicheskago obshchestva*, 2nd ser., 25 (1889), 341–342; "Ob uproshchenii kristallograficheskikh vychisleny" ("On the Simplification of Crystallographic Computation"), *ibid.*, 29 (1892), 58–64; "Svoystva nekotorykh psevdosimmetricheskikh kristallov v svyazi s teoriey kristallicheskogo stroenia veshchestva" ("Properties of Certain Pseudosymmetrical Crystals in Relation to the Theory of the Crystal Structure of Matter"), *ibid.*, 29 (1892), 65–130, his master's thesis; "K voprosu o skorostyakh rosta i rastvorenia kristallicheskikh graney" ("On the Question of the Velocity of Growth and Dissolution of Crystal Faces"), in *Varshavskia universitetskia izvestia*, 7 (1895), 1–40; 8 (1895), 41–56; n.s., 1 (1896), 57–88; 2 (1896), 89–122, his doctoral diss.; also in German, in *Zeitschrift für Kristallographie und Mineralogie*, 34 (1901), 449–530; "Die Symmetrieebene als Grundelement der Symmetrie," *ibid.*, 27 (1897), 556–558; "Untersuchungen im Gebiete der optischen Eigenschaften isomorpher Krystalle," *ibid.*, 36 (1902), 1–28; "Ein Beitrag zur Theodolithmethode," *ibid.*, 37 (1903), 50–56; and "Untersuchungen über die Genauigkeitsgrenzen der Gesetze der geometrischen Krystallographie," *ibid.*, 38 (1904), 1–57; and *Rukovodstvo po kristallografii* ("Guide to Crystallography"; Warsaw, 1904).

Subsequent works are *Simmetria i ee proyavlenia v prirode* ("Symmetry and Its Appearance in Nature"; Moscow, 1908); "Über die Krystallisation des Kaliumjodids auf dem Glimmer," in *Zeitschrift für Kristallographie und Mineralogie*, 45 (1908), 335–345; "Zur Theorie des Krystallhabitus," *ibid.*, 433–472; "Über die Natur 'flüssiger' und 'fliessender' Krystalle," *ibid.*, 46 (1909), 261–265; "Über die Kristallröntgenogramme," in *Physikalische Zeitschrift*, 14 (1913), 217–222; "O kapillyarnoy teorii formy kristallov" ("On the Capillary Theory of Crystal Forms"), in *Zhurnal Russkago fiziko-khimicheskago obshchestva pri Imperatorskom St-Peterburgskom universitete*, 48 (1916), 337–349; *Kristally, ikh obrazovanie, vid i stroenie* ("Crystals, Their Formation, Appearance, and Structure"; Moscow, 1917; 2nd ed., ed. and annotated by M. M. and S. Sabshnikov, 1926); *Zhizn crystallov* ("The Life of Crystals"; Moscow, 1918); *Osnovy kristallografii* ("Principles of Crystallography"; Moscow, 1923, 2nd ed., 1926); *Praktichesky kurs geometricheskoy kristallografii so stereograficheskoy setkoy* ("Practical Course in Geometrical Crystallography With a Stereographic Net"; Moscow, 1924); and "O molekulyarnoy strukture muskovita" ("On the Molecular Structure of Muscovite"), in *Trudy Instituta prikladnoi mineralogii i metallurgii*, 25 (1926), 22–29.

II. SECONDARY LITERATURE. On Wulff and his work, see G. G. Leleyn and G. A. Kirsanov, "Khronologichesky ukazatel trudov Y. V. Vulfa" ("Chronological Guide to the Works of Y. V. Vulf"), in *Trudy Instituta kristallografii. Akademiya nauk SSSR* (1951), no. 6, 15–24; and M. von Laue, "Der Wulffsche Satz für die Gleichgewichtsform von Kristallen," in *Zeitschrift für Kristallographie und Mineralogie*, 105 (1943), 124–133.

V. A. FRANK-KAMENETSKY

WUNDT, WILHELM (*b.* Neckarau, Baden, Germany, 16 August 1832; *d.* Gross Bothen, Germany, 31 August 1920), *psychology.*

Wundt described his father, a Lutheran pastor, as cheery and impractical, and his mother as more aggressive. His earliest memories were painful: falling down a flight of cellar stairs, and being roused from fantasy by a paternal box on the ear. Play with other children was rare; visits to sympathetic oldsters frequent; daydreaming a passion. Father, mother, and maternal grandfather (who resided at Heidelberg) all had a hand, not infrequently heavy, in his early education. At the age of eight he acquired as tutor a kindly young vicar with whom he shared a room, and to whom he was soon more attached than to his parents. When the vicar obtained his own parish, heartbroken Wundt, now twelve years of age, was permitted to go with him. At thirteen Wundt endured a traumatic year at a Catholic Gymnasium, where he was advised to seek some honorable calling—such as the postal service—which did not require an education. Transferred to Heidelberg, sharing a room with a much older brother and a cousin, he at last acquired friends and the effective work habits that were to distinguish his adult life.

Because his scholastic record remained too poor to qualify for a scholarship, Wundt's first university year was at Tübingen, where an uncle was brain anatomist. Thus accidentally he was pointed toward medicine. Completing his studies at the University of Heidelberg, he passed the state examinations with distinction. Apprehensive about medical practice, he spent six months as assistant at a hospital, then a semester at Berlin under Johannes Müller and du Bois-Reymond. In 1857 Wundt finally attained academic shelter as *Privatdozent* at Heidelberg, and the following year he published *Die Lehre von der Muskelbewegung*, begun under du Bois-Reymond. Ominously, the mentor to whom the work was dedicated did not even acknowledge it.

Also in 1858 Wundt published the first of six experimental reports on sensory perception, which became the *Beiträge* of 1862, and was appointed assistant to Helmholtz, just called to Heidelberg. Helmholtz largely ignored and evidently disdained his assistant, assigning him to supervise a routine laboratory course. In 1863 Wundt resigned the unrewarding post, started lecturing on psychology, and published the *Vorlesungen*. The following year he became *ausserordentlicher Professor*.

In the introduction to the *Beiträge* Wundt calls for an inductive psychology, but E. B. Titchener points out that in this he leans on Mill's *Logic*. Wundt mainly emphasizes social data—reflecting the influence of Steinthal and Lazarus, and current interest in moral statistics—but sees experiment as essential because unconscious determinants of thinking are not accessible to introspection (see Helmholtz on "unconscious inference"). Insistence that philosophy must reflect the findings of science foreshadows his later *Logik* and *Ethik*. Finally, there is the immodest claim to a great discovery revealed more fully in the *Vorlesungen*: experimental determination of the "natural unit of time" as the duration of the "swiftest thought," along with experimental demonstration of the unity of consciousness. Wundt found that an observer cannot precisely note the position of a moving pointer at the instant when a click is heard—a problem arising out of the concern of astronomers with errors in fixing the moment of transit for a star. (Later, without confession of error, Wundt wrote disdainfully of the outdated "needle's eye theory of consciousness.") It is problematic whether, by his bold interpretation, Wundt perhaps sought to place himself alongside Helmholtz, who had measured the speed of the nerve impulse, and Fechner, whose *Elemente der Psychophysik* (1860) opened the era of quantitative psychology.

Helmholtz left Heidelberg in 1871. Although Wundt had published prolifically (despite four years in the Baden legislature) and since 1867 had been lecturing on "physiological psychology," he was passed over in the selection of a successor. Soon, however, he was developing those lectures into his *magnum opus*, the *Grundzüge der physiologischen Psychologie*, which attempted "to define the limits of a new science." The preface of the work is dated March 1874, and in that month he was surprised by a call to Zurich as professor of inductive philosophy, offering an escape from his humiliating post at Heidelberg. Wundt had been recommended by Friedrich A. Lange, who had held the post previously, and whom he had met once long before at a conference on workers' education. (Lange must have read the first half of the *Grundzüge* when it appeared in 1873. He refers approvingly to it in his *Geschichte des Materialismus* [1874]; Wundt was not mentioned in the 1866 edition, which already contained the famous phrase "eine Psychologie ohne Seele," which is sometimes falsely attributed to Wundt.)

This election to a post in philosophy facilitated the more important call to Leipzig in 1875, which Wundt owed to the enthusiasm of Zöllner (later an adherent of the medium Henry Slade) and, as Wundt relates, to the readiness of an indifferent faculty to hire two obscure candidates for the price of one man of distinction. Thus did Wundt arrive at the post in which he would attain international renown.

The flow of books continued, and Wundt became the most popular lecturer of the University of Zurich. At forty-three, the ugly-duckling physiologist had become a resplendent swan philosopher. Wundt's fame, however, is rooted in what has been called the first psychology laboratory—the first, indeed, in which the instrumentation familiar to physiologists was domiciled in halls of philosophy, and called upon to monitor controlled introspections. It started as a demonstration laboratory, but in the winter semester of 1879–1880 a student, Max Friedrich, performed an experiment on "apperception time," which was reported in the first issue of Wundt's new journal, *Philosophische Studien*. It was the start of an avalanche. The immense influence of Wundt's laboratory can be traced in three ways: first, in successive revisions of the *Grundzüge* (1880, 1887, 1893, 1902, 1908–1911) which, as it fattened, provided more and more information, including detailed illustrations of experimental apparatus, constituting virtual manuals for those aspiring to found new centers of "brass instrument psychology"; second, in the increasingly cosmopolitan authorship of reports in the *Studien*, as Wundt attracted foreign, and most conspicuously American, students; and third, in the dozens of new laboratories, again especially in the United States, directed by his former students. In 1903 Cattell found eighteen Wundt students among fifty leading psychologists in the United States. Most would have subscribed to H. C. Warren's sentiment: "Coming to him as I did from an atmosphere of philosophical speculation, the spirit of his laboratory was a God-send. I owe much to Wundt for the change he wrought in my life-ideals" (*Psychological Review*, **28**, p. 169).

Not everyone admired Wundt. Carl Stumpf, who suffered some of Wundt's intolerant diatribes, deplored his influence in a letter to William James, who replied consolingly in 1887:

> He aims at being a sort of Napoleon of the intellectual world. Unfortunately he will never have a Waterloo, for he is a Napoleon without genius and no central idea. . . . Cut him up like a worm, and each fragment crawls; there is no *noeud vital* in his mental medulla oblongata, so that you can't kill him all at once. . . . He has utilized to the uttermost fibre every gift that Heaven endowed him with at his birth, and made of it all that mortal pertinacity could make. He is the finished example of how much mere *education* can do for a man [R. B. Perry, *The Thought and Character of William James*, II (Boston, 1935), 68f.].

The diffuseness of Wundt's thought, and his practice of changing his views without specific acknowledgement other than to warn against reliance on earlier editions, make it impossible to give a brief synopsis of his system. For example, in 1874 he wrote that physiological psychology "cannot sidestep" the question "how internal and external existence are ultimately related." In 1902, however, he dismissed this view as one which "has been mistakenly asserted," and declared that physiological psychology does not seek to "derive or explain the phenomena of mental life from those of physical life." Wundt also had little tolerance for results not conforming to his own theories. G. S. Hall wrote that Wundt "seems to wish to be the last in fields where he was the first, instead of taking pleasure in seeing successors arise who advance his lines still further" (*Founders of Modern Psychology*, p. 419). When L. Lange discovered in 1888 the important distinction between sensorial and motor attitudes in reaction time experiments, Wundt, still aspiring to measure by the "subtractive method" the duration of the supposed psychic component of a voluntary act, dictated an absurd interpretation in terms of "complete" and "incomplete" reactions, thus rejecting as invalid all research in which reactions were too fast. When Oswald Külpe and his Würzburg group made advances in experimental analysis of the thought process, Wundt rejected their work, and his own early commitment, to insist that these processes could be studied only in the history of social institutions. (The ten-volume "folk psychology," which Wundt devoted to this problem, is cited only as a monument to his industry.) Wundt's later views tended to alienate his followers. Ever faithful to the Wundtian method, Titchener could find no empirical basis for the tridimensional theory of feeling, and many rejected Wundt's theory of apperception as not merely mistaken, but as a betrayal of scientific principles.

At an opportune moment of history, Wundt proclaimed himself the commander of a crusade, and enthusiasts flocked to his banner. Socially shy and intellectually arrogant, he often confused the defense of his command with advancement of the cause. His name is linked to no significant finding, no theory that did not prove to be flagrant error, no problem freshly defined. Yet Wundt constituted an important rallying point for the generation of young men who saw experimental psychology as a new avenue to man's self-understanding.

BIBLIOGRAPHY

I. ORIGINAL WORKS. For a complete bibliography of Wundt's writings, see Eleonore Wundt, ed., *Wilhelm Wundts Werk, ein Verzeichnis seiner sämtlichen Schriften* (Munich, 1927).

His early works include *Die Lehre von der Muskelbewegung* (Brunswick, 1858); *Beiträge zur Theorie der Sinneswahrnehmung* (Leipzig–Heidelberg, 1862); *Vorlesungen über die Menschen- und Thierseele*, 2 vols. (Leipzig, 1863; 6th ed., Leipzig, 1919); *Lehrbuch der Physiologie des Menschen* (Erlangen, 1865; 4th ed., Stuttgart, 1878); *Handbuch der medicinischen Physik* (Erlangen, 1867); *Untersuchungen zur Mechanik der Nerven und Nervencentren*, 2 vols. (Erlangen, 1871–1876); *Grundzüge der physiologischen Psychologie* (Leipzig, 1873–1874; 6th ed., 3 vols., Leipzig, 1908–1911); *Logik*, 2 vols. (Stuttgart, 1880–1883; 5th ed., 3 vols., 1923–1924); *Philosophische Studien*, 20 vols. (1881–1903); *Essays* (Leipzig, 1885; 2nd ed., Leipzig, 1906).

Subsequent writings include *Ethik* (Stuttgart, 1886; 4th ed., 3 vols., Stuttgart, 1912); *System der Philosophie* (Leipzig, 1889; 4th ed., 2 vols., 1919); *Grundriss der Psychologie* (Leipzig, 1896; 11th ed., Leipzig, 1913); *Völkerpsychologie* (Leipzig, 1900; 2 vols., 1904; 10 vols., 1911–1920); *Einleitung in die Philosophie* (Leipzig, 1901; 9th ed., Leipzig, 1922); *Einführung in die Psychologie* (Leipzig, 1911); *Elemente der Völker-psychologie* (Leipzig, 1912); *Reden und Aufsätze* (Leipzig, 1913); *Sinnliche und übersinnliche Welt* (Leipzig, 1914); *Die Nationen und ihre Philosophie* (Leipzig, 1915); and *Erlebtes und Erkanntes* (Stuttgart, 1920), his autobiography.

II. SECONDARY LITERATURE. On Wundt and his work, see E. G. Boring, *A History of Experimental Psychology*, 2nd ed. (New York, 1950), 316–347; R. Eisler, *Wundt's Philosophie und Psychologie* (Leipzig, 1902); G. S. Hall, *Founders of Modern Psychology* (New York–London, 1912), 311–458; A. Heussner, *Einführungen in Wilhelm Wundts Philosophie und Psychologie* (Göttingen, 1920); Arthur Hoffmann, ed., "Wilhelm Wundt, eine Würdigung," which is *Beiträge zur Philosophie des deutschen Idealismus*, **2**, nos. 3–4 (1922); E. König, *W. Wundt, seine Philosophie und Psychologie* (Stuttgart, 1901); Willi Nef, *Die Philosophie Wilhelm Wundts* (Leipzig, 1923); Peter Petersen, *Wilhelm Wundt und seine Zeit* (Stuttgart, 1925); E. B. Titchener, "Wilhelm Wundt," in *American Journal of Psychology*, **32** (1921), 161–178; W. Wirth, "Unserem grossen Lehrer Wilhelm Wundt in unauslöschlicher Dankbarkeit zum Gedächtnis," in *Archiv für die gesamte Psychologie*, **40** (1921), i–xvi; and "In Memory of Wilhelm Wundt," in *Psychological Review*, **28** (1921), 153–188, which includes reminiscences by seventeen of his American students.

SOLOMON DIAMOND

WURTZ, CHARLES-ADOLPHE (*b.* Wolfisheim, near Strasbourg, France, 26 November 1817; *d.* Paris, France, 12 May 1884), *chemistry.*

Wurtz spent the earliest years of his life in Wolfisheim, a small village near Strasbourg, where his father, Jean-Jacques Wurtz, was Lutheran pastor. He grew up in a rather modest but cultured home that was intellectually stimulating and part of a healthy agricultural community. His mother, the former Sophie Kreiss, came from a well-educated family and she appears to have enjoyed a more good-humored disposition than her husband. Their son was intelligent and gifted with an artistic nature; but he attended the Protestant school in Strasbourg, from July 1826, without showing exceptional promise. By the age of seventeen, however, he had developed a sufficiently strong interest in chemistry to offend his father with experiments in the house. The idea of a career in the ministry appealed less and less to young Wurtz, who eventually was allowed to embark on a medical course at the University of Strasbourg. This promised both security and an opportunity to cultivate his interest in chemistry.

Such was his preoccupation with chemistry that for his doctorate in medicine (1843) Wurtz wrote a thesis on fibrin and albumin, in which he described a method for the purification of soluble albumin and argued for a difference between albumin of blood and the albumin of an egg. After graduating from Strasbourg it was a natural step to move to Giessen, where he could study with Liebig and where he soon met A. W. Hofmann, who later complemented his work on the amines and became his biographer. At Giessen, Wurtz began his research with a study of hypophosphorous acid in an attempt to decide between the conflicting formulas of Dulong and Rose. He found time to translate Liebig's papers into French for the *Annales de chimie*, work that brought him into contact with the leading chemists in Paris. In 1844 he moved to that city, where he soon joined Dumas at the laboratory associated with the Faculty of Medicine.

It was there that Wurtz succeeded Dumas as lecturer in organic chemistry (1849), as professor (1853), and as dean (1866). In the last, more administrative, role he did much to improve the scientific education of medical students and to ensure that clinical professors in the hospitals had better laboratory facilities. From the Faculty of Medicine, Wurtz transferred to a chair of organic chemistry which had been specially created for him at the Sorbonne in 1874. Henceforth he could leave behind his heavy administrative responsibilities and enjoy his real vocation as a teacher.

Throughout his life Wurtz remained true to his Lutheran heritage. He displayed no trace of anti-clericalism and found little difficulty in harmonizing his science with his faith. A staunch defender of the atomic theory against the skeptical positivism of Berthelot, he gave the theory a teleological interpretation in perfect accord with his natural theology. He was greatly respected for the diplomacy he demonstrated while dean of the Faculty of Medicine and for the liberalism that enabled him to defend his socialist colleagues, Alfred Naquet and Robin, with whom he had few political sympathies. His liberalism also manifested itself in his campaign for the admission of women students to the Faculty of Medicine. Intensely patriotic, he was deeply affected by the capture of his native Alsace during the Franco-Prussian War. He helped to found a society for the protection of the refugees who crowded into Paris from Alsace, and he was active when the capital fell under siege.

In 1852 Wurtz married a well-to-do childhood friend; they had four children, only one of whom pursued a scientific career. Wurtz became one of the most enthusiastic and outstanding teachers of his generation. There never was a school of chemists in France to compare with Liebig's school at Giessen, but Wurtz probably came closest to realizing one. Since he spoke French, German, and English, he was able to surround himself with the most distinguished chemists of the day. Both Couper and Butlerov—pioneers of structural organic chemistry—studied in his laboratory; around him, during the 1870's, there gathered such giants as Le Bel, van't Hoff, and Charles Friedel; and it was in his laboratory that Boisbaudran discovered gallium in 1875.

For twenty years, from 1852, Wurtz was responsible for the section devoted to foreign literature in the *Annales de chimie*. He became a member of the Academy of Medicine in 1856, vice-president in 1869, and president in 1871. He was elected a foreign member of the Royal Society in 1864 and subsequently won the Copley Medal. Twice he was awarded the Jecker Prize by the Institut de France; and after having been admitted to the chemical section of the Académie des Sciences in 1867, he was its president in 1881. Although the idea for the Société Chimique de France may not have originated with Wurtz, he became its secretary, its guiding spirit, and several times its president (1864, 1874, 1878). It was during his presidency (1864) that a number of separate publications, including Wurtz's own *Répertoire de chimie pure*, were amalgamated to produce the official bulletin of the society. Wurtz was equally active in establishing the Association Française pour l'Avancement des Sciences, a new organization designed to foster science in the provinces. Modeled on the British Association, with which Wurtz was impressed, the French Association held its first meeting at Bordeaux in 1872. When it met at Lille in 1874, Wurtz delivered the presidential address. He was greatly honored in public as well as in academic life; he was elected mayor of the seventh *arrondissement* of Paris, and a member of the Senate.

Wurtz began his career at a time of great crisis in organic chemistry. The dualistic approach of Berzelius appeared to be collapsing under the attacks of Dumas, Laurent, and Gerhardt, who favored a unitary conception of organic molecules based on their novel concepts of substitution and double decomposition. Wurtz had no hesitation in aligning himself with Laurent and Gerhardt; he was the first teacher in France to champion their ideas, and became the architect of a new chemical system that embraced their antidualist concepts as well as the emerging concepts of atomicity (valence) and chemical types.

Remarkable among Wurtz's earliest discoveries was his fulfillment of Liebig's prediction that there might be organic compounds analogous to ammonia and derivable from it by the replacement of hydrogen. Having prepared substituted ureas from derivatives of cyanogen, Wurtz investigated the action of potassium hydroxide on his products and thus obtained the primary amines (1849). This outstanding discovery was soon extended by his friend Hofmann, who demonstrated that the remaining two equivalents of hydrogen in ammonia could be replaced by alkyl groups. This provided support for the thesis that Gerhardt was developing: that it might be possible to regard all organic compounds as derivable from a small number of inorganic types, such as water or hydrogen chloride. The value of ammonia as a primitive type was now revealed:

$$
\begin{array}{cccc}
H & R^1 & R^1 & R^1 \\
\diagdown & \diagdown & \diagdown & \diagdown \\
H{-}N, & H{-}N, & R^2{-}N, & R^2{-}N. \\
\diagup & \diagup & \diagup & \diagup \\
H & H & H & R^3
\end{array}
$$

It was thus clear to Wurtz that organic radicals could replace hydrogen without destroying the basic structure or type. There remained, however, at least one serious bone of contention between the dualist school, now represented by Kolbe in Germany, and the new French school. Could the hydrocarbon radicals, postulated by the dualists as constituents of organic acids and alcohols, actually be isolated in the free state? Kolbe and the English chemist Frankland thought they had succeeded in isolating the methyl and ethyl radicals, thereby corroborating the conservative view that acetic acid, for example, contained the methyl radical, then written as (C_4H_6), in precisely the same way that sulfuric acid contained sulfur:

$$(C_4H_6) \cdot O_3 + H_2O \text{ and } S \cdot O_3 + H_2O.$$

Laurent and Gerhardt, however, argued that the supposed verification was circular, and that what Kolbe and Frankland had isolated were dimers of the hypothetical radicals. The stability, vapor density, and boiling point of these controversial hydrocarbons favored the dimerization thesis, but a strictly chemical proof was required. It was Wurtz who provided this proof when he applied sodium to a mixture of alkyl iodides (1854). When the iodides were those of ethyl and butyl, he obtained some ethyl butyl; when they were those of butyl and amyl, he obtained some butyl amyl, thereby confirming that what was isolated contained two equivalents of the radicals, and was not a free radical itself.

Wurtz ingeniously had not only found a new method for synthesizing alkanes but also had embarrassed the dualism of Kolbe, with whom, as with Berthelot, he was frequently at loggerheads. Earlier in his career (1844), Wurtz had observed the curious fact that a hydride of copper, when treated with hydrochloric acid, generated a quantity of hydrogen double that contained by the hydride alone. This clearly suggested the possibility that each molecule of hydrogen might comprise two equivalents or atoms of hydrogen—one from the hydride and one from the acid. His work on the controversial hydrocarbons now endorsed this interpretation, which in turn ratified Avogadro's molecular hypothesis, so long neglected. Consequently, Wurtz became a leading advocate of the presuppositions underlying the atom-molecule distinction and, by examining the dissociation of anomalous vapors such as phosphorus pentachloride, he was able to rehabilitate the use of vapor density measurements for the determination of molecular weights.

One of Wurtz's most popular works was *La théorie atomique* (1879). Its title denoted more than the atomic-molecular theory of Avogadro or Ampère; it designated a theory that incorporated the idea of combining power or atomicity of the atoms—a new concept for which Wurtz had helped to clear the ground. He had done so by contributing to the notion of polyatomic organic radicals and by clarifying the distinctions between affinity, basicity, and atomicity. It was one of his outstanding contributions to chemistry that he succeeded in the preparation of the first dihydroxy alcohol, ethylene glycol, by the hydrolysis of ethylene diiodide (1856). It followed that the radicals of ethyl alcohol, glycol, and glycerin had ascending capacities of combination, and could be called monatomic, diatomic, and triatomic, respectively. This concept of characteristic combining power, when applied to the elements, precipitated the notion of valence, credit for which belongs to several of Wurtz's contemporaries as well as to him. The carbon—carbon bond eluded Wurtz—perhaps because of his conviction that the elements (including carbon) might exhibit more than one valence. Although he was justified in this belief, he differed from Kekulé, whose commitment to an exclusive tetravalence for carbon inexorably led to the vital carbon—carbon bond. The evidence suggests, moreover, that Wurtz simply found the structural formulas of Kekulé and Couper too arbitrary and unnecessarily pretentious.

By dehydrating glycol, Wurtz procured ethylene oxide—a missing link that permitted him to construct a comprehensive series of analogies between organic and inorganic oxides that was based on the twin concepts of atomicity and type. Ethylene oxide was hailed as an analogue of the oxides of diatomic calcium and barium, just as glyceryl oxide could be represented as an analogue of the oxides of triatomic antimony and bismuth. On the basis of such analogical argument—and not, it should be noted, on the basis of organic synthesis—Wurtz proclaimed the unification of chemistry (1862).

Wurtz excelled as a practical chemist, and almost all his contributions were of lasting value. Among his many miscellaneous methods of synthesis were those for the production of phosphorous oxychloride, of neurine from ethylene oxide, of aldol from acetaldehyde, of phenol from benzene, and of esters from alkyl halides and the silver salts of acids.

BIBLIOGRAPHY

I. ORIGINAL WORKS. Several of Wurtz's major works were translated into English and German. The following references are to the original French eds.: *Répertoire de chimie pure en France et à l'étranger*, 4 vols. (Paris, 1858–1862); *Chimie médicale* (Paris, 1864); *Leçons de chimie professées en 1863 par MM. A. Wurtz, A. Lamy, L. Grandeau* (Paris, 1864); *Leçons de philosophie chimique* (Paris, 1864); *Leçons élémentaires de chimie moderne* (Paris, 1867–1868); *Histoire des doctrines chimiques depuis Lavoisier* (Paris, 1868), the intro. to *Dictionnaire de chimie pure et appliquée*, 14 vols. (Paris, 1868–1878); English trans. by H. Watts, *A History of Chemical Theory* (London, 1869); *Les hautes études pratiques dans les universités allemandes* (Paris, 1870); *La théorie atomique* (Paris, 1879); and *Traité de chimie biologique* (Paris, 1880).

Besides these volumes, Wurtz published prolifically. His work included at least 140 papers, of which the following contain his most important contributions to the advance of chemistry:"Recherches sur la constitution de l'acide hypophosphoreux," in *Annales de chimie et de physique*, 3rd ser., 7 (1843), 35–50; "Sur l'hydrure de cuivre," *ibid.*, 11 (1844), 250–252; "Recherches sur les éthers cyaniques et leurs dérivés," in *Comptes rendus . . . de l'Académie des sciences*, 27 (1848), 241–243; "Sur une série d'alcalis organiques homologues avec l'ammoniaque," *ibid.*, 28 (1849), 223–226; "Recherches sur les urées composées," *ibid.*, 32 (1851), 414–419; "Sur l'alcool butylique," *ibid.*, 35 (1852), 310–312; "Sur la théorie des amides," *ibid.*, 37 (1853), 246–250 and

357–361; "Sur une nouvelle classe de radicaux organiques," in *Annales de chimie et de physique*, 3rd ser., **44** (1855), 275–313; "Sur le glycol ou alcool diatomique," in *Comptes rendus . . . de l'Académie des sciences*, **43** (1856), 199–204; "Sur l'acétal et sur les glycols," *ibid.*, 478–481; "Sur la formation artificielle de la glycérine," in *Annales de chimie et de physique*, 3rd ser., **51** (1857), 94–101; "Recherches sur l'acide lactique," in *Comptes rendus . . . de l'Académie des sciences*, **46** (1858), 1228–1232; and "Sur l'oxyde d'éthylène," *ibid.*, **48** (1859), 101–105. Also see "Observations sur la théorie des types," in *Répertoire de chimie pure*, **2** (1860), 354–359; "On Oxide of Ethylene, Considered as a Link Between Organic and Mineral Chemistry," in *Journal of the Chemical Society*, **15** (1862), 387–406; "Sur l'oxyde d'éthylène et les alcools polyéthyléniques," in *Annales de chimie et de physique*, 3rd ser., **69** (1863), 317–355; "Nouveau mode de formation de quelques hydrogènes carbonés," in *Comptes rendus . . . de l'Académie des sciences*, **54** (1862), 387–390; "Sur l'atomicité des éléments," in *Bulletin de la Société chimique de Paris*, **2** (1864), 247–253; "Sur les densités de vapeur anomales," in *Comptes rendus . . . de l'Académie des sciences*, **60** (1865), 728–732, and **62** (1866), 1182–1186; "Transformation des carbures aromatiques en phénols," *ibid.*, **64** (1867), 749–751; "Synthèse de la névrine," *ibid.*, **65** (1867), 1015–1018; "Sur la densité de vapeur du perchlorure de phosphore," in *Comptes rendus de l'Association française pour l'avancement des sciences*, **1** (1872), 426–445; "Nouvelles recherches sur l'aldol," in *Comptes rendus . . . de l'Académie des sciences*, **76** (1873), 1165–1171; "Recherches sur la loi d'Avogadro et d'Ampère," *ibid.*, **84** (1877), 977–983; "Sur la notation atomique," *ibid.*, 1264–1268; "Sur les densités de vapeur," *ibid.*, 1347–1349; "Sur le ferment digestif du *Carica papaya*," *ibid.*, **89** (1879), 425–429; "Sur la papaïne; contribution à l'histoire des ferments solubles," *ibid.*, **90** (1880), 1379–1385; and "Sur la préparation de l'aldol," *ibid.*, **92** (1881), 1438–1439.

The above selection is taken and corrected from the bibliography appended to C. Friedel, "Notice sur la vie et les travaux de Charles-Adolphe Wurtz," in *Bulletin de la Société chimique*, **43** (1885), i–lxxx, which also was published in the introduction to the 1886 ed. of Wurtz's *La théorie atomique*.

II. SECONDARY LITERATURE. Besides Friedel's, biographies include A. W. Hofmann, in *Berichte der Deutschen chemischen Gesellschaft*, **17**, no. 1 (1884), 1207–1211, and **20**, no. 3 (1887), 815–996; A. Gautier, in *Revue scientifique*, **55** (1917), 769–770; M. Tiffeneau *et al.*, *ibid.*, **59** (1921), 573–602; and A. Williamson, in *Proceedings of the Royal Society*, **38** (1885), xxiii–xxxiv.

The following works contain invaluable information for an accurate appraisal of Wurtz's contributions to chemistry: R. Anschütz, *August Kekulé* (Berlin, 1929); J. H. Brooke, "Organic Synthesis and the Unification of Chemistry— a Reappraisal," in *British Journal for the*

History of Science, **5** (1971), 363–392; G. V. Bykov and J. Jacques, "Deux pionniers de la chimie moderne, Adolphe Wurtz et Alexandre M. Boutlerov, d'après une correspondance inédite," in *Revue d'histoire des sciences*, **13** (1960), 115–134; E. Farber, "The Glycol Centenary," in *Journal of Chemical Education*, **33** (1956), 117; H. Hartley, *Studies in the History of Chemistry* (Oxford, 1971), ch. 8; D. Larder, "A Dialectical Consideration of Butlerov's Theory of Chemical Structure," in *Ambix*, **18** (1971), 26–48; A. Metz, "La notation atomique et la théorie atomique en France à la fin du XIXᵉ siècle," in *Revue d'histoire des sciences*, **16** (1963), 233–239; J. R. Partington, "The Chemical Society of France, 1857–1957," in *Nature*, **180** (1957), 1165; and *A History of Chemistry*, IV (London, 1964), 477–488; C. A. Russell, *The History of Valency* (Leicester, 1971); and G. Urbain, "J. B. Dumas and C. A. Wurtz—leur rôle dans l'histoire des théories atomiques et moléculaires," in *Bulletin de la Société chimique*, 5th ser., **1** (1934), 1425–1447.

JOHN HEDLEY BROOKE

WYMAN, JEFFRIES (*b.* Chelmsford, Massachusetts, 11 August 1814; *d.* Bethlehem, New Hampshire, 4 September 1874), *anatomy, physiology*.

Wyman was almost a model of the nineteenth-century scientist who had no life independent of his calling. His father, Rufus Wyman, was a physician who named his son after John Jeffries of Boston, his medical teacher; Wyman's mother was Ann Morrill. An elder brother, Morrill Wyman, was known as the leading physician in nineteenth-century Cambridge. In 1818 Dr. Rufus Wyman became physician for the McLean Asylum for the Insane, then in Charlestown, Massachusetts, where Jeffries had his early schooling. After preparation at an academy in Chelmsford and at Phillips Academy at Andover, Wyman entered Harvard in 1829, graduating in 1833. Not a distinguished scholar in general, he continued while an undergraduate a very early interest in natural history and in sketching, major tools for a future anatomist. In his senior year Wyman contracted pneumonia, which continued as what Oliver Wendell Holmes called "the pulmonary affection that kept him an invalid, and ended by causing his death." This condition, from which he suffered for over forty years, set important limits and opened important opportunities for the entire span of his adult career, especially its requirement that he spend each winter in the South.

After graduating from Harvard College, Wyman apprenticed himself to a John C. Dalton in Chelmsford and attended the Harvard Medical

School in Boston. On receiving the M.D. in 1837, he failed to find a lucrative post in a country town and was thus forced to open a Boston office and to accept the poorly paid post of demonstrator of anatomy to John C. Warren, the Hersey professor at Harvard Medical School.

Wyman's turn from medicine to science was largely financed by the few wealthy men in Boston who figured in most scientific careers of the community at that time. John Amory Lowell made Wyman curator of the Lowell Institute, and in the winter of 1840–1841 he delivered a course of lectures on comparative anatomy and physiology. With the proceeds Wyman made a somewhat truncated tour of Europe, the standard preparation for an American career in science. He spent the summer of 1841 in Paris, attending lectures in human anatomy, comparative anatomy, physiology, and zoology. In London he worked on the Hunterian collections at the Royal College of Surgeons and became acquainted with Richard Owen before being called back by the death of his father.

On returning to Boston, Wyman took his place in the scientific community as a fellow of the American Academy of Arts and Sciences and curator of reptiles and fishes at the Boston Society of Natural History. For a living and as an excuse to go south, he was professor of anatomy and physiology at Hampden-Sydney Medical College in Richmond, Virginia, from 1843 to 1848. During that period he published some of the most significant papers of his career.

In 1847 Wyman's friends at Harvard made an arrangement that sufficed to give him a secure institutional base for the rest of his life and gave the university a markedly increased capability in zoology and anatomy. Wyman took the Hersey professorship, which hitherto had been at the medical school, to Cambridge, while Oliver Wendell Holmes occupied a new chair, the Parkman professorship, in Boston. Thus anatomy and physiology became subjects for undergraduates, not just for medical students, and Wyman had scope to work on species other than *Homo sapiens*. Since Asa Gray, the Fisher professor of natural history, wished to concentrate on botany, he expected Wyman to hold recitations in elementary zoology, especially when Gray was on leave—an arrangement that significantly advanced the differentiation of botany and zoology at Harvard. With the appointment, at essentially the same time, of Louis Agassiz to the Lawrence Scientific School, Wyman was able to confine himself largely to anatomy and physiology, and to undertake the development of a

museum of anatomical specimens that he continued to the end of his life.

Specialization was still in transition in the mid-nineteenth century, however, and Wyman was associated with his brother Morrill Wyman in a private medical school at Cambridge from 1857 to 1866. From 1856 to the end of his life he was the beneficiary of gifts from two wealthy Bostonians that made possible his winter research in warm climates. He often went to Europe or Florida. He traveled to Surinam in 1856; and despite the fever he contracted there, was ready by 1858 to go on the La Plata expedition of J. M. Forbes, which crossed the Andes and returned from Chile by way of Peru and the Isthmus of Panama. This type of invalid's rest played a part in Wyman's becoming professor of American archaeology and ethnology in 1866, and effectively the first director of the museum endowed by George Peabody. Thus in the last years of his life he built an archaeological museum alongside his anatomical museum in Boylston Hall at Harvard.

Wyman gained national recognition as president of the American Association for the Advancement of Science and an original member of the National Academy of Sciences, but he did not serve as the one and soon resigned from the other. He was married twice, in 1850 and 1861. His first wife died in 1855, leaving two daughters; the second died in 1864, leaving a son. The combination of delicate health and steady exertion remained constant until Wyman's last days in September 1874. He was on an annual visit to the White Mountains, to escape the autumnal catarrh, when he died suddenly of a hemorrhage from the lungs.

Of Wyman's 175 papers, mostly on anatomy, the ones that attracted most notice were a series on the gorilla. With the memoir (written with Thomas S. Savage) "Notice of the Characters, Habits, and Osteology of *Troglodytes gorilla*, a New Species of Ourang From the Gaboon River" (*Boston Journal of Natural History*, 5 [1845–1847], 417–442), Wyman established himself as the peer of Richard Owen in elaborating the anatomical features of the higher primates. In 1859 he worked on a large collection of gorilla skins and skeletons sent him by the explorer Paul Belloni du Chaillu.

Despite his pose of being above the fray in both scientific politics and intellectual controversy, Wyman followed with informed care the great issues of his day that swirled about the work of Charles Darwin and Louis Pasteur. In his study of the gorilla, he generalized only to the extent of saying that the

. . . difference between the cranium, the pelvis, and the conformation of the upper extremities in the negro and Caucasian sinks into comparative insignificance when compared with the vast difference that exists between the conformation of the same parts in the negro and the orang. Yet it cannot be denied, however wide the separation, that the negro and orang do afford the points where man and brute, when the totality of their organization is considered, most nearly approach each other [p. 441].

From 1860 to 1866 Wyman corresponded with Darwin, providing a number of examples of possible natural selection. One was the action of light favoring black pigs over white ones after they had eaten a Florida plant called paint-root (*Lachnanthes tinctoria*). Another concerned the perfection of cells formed by bees, on which Wyman made and published careful measurements. These cases appear in *Origin of Species*. Other examples provided by Wyman include blind fishes of Mammoth Cave, Natá cattle of South America, malformed codfish, and rattlesnakes. Wyman never expressed himself publicly on the *Origin of Species*; but clearly he seriously considered the problems set by Darwin, and the testimony of his friends indicates that he was privately pro-Darwin and not a vitalist (and equally privately a theist).

In 1862 Wyman began a series of experiments on spontaneous generation, paralleling those conducted by Pasteur and Pouchet. He reported the presence of infusoria even in flasks that had been boiled, under carefully controlled conditions, for four hours. When the flasks were boiled beyond five hours, no infusoria appeared. Yet Wyman did not press to any conclusion, either for or against spontaneous generation, being content to let his experiments stand as examples of logically conceived and elegantly executed laboratory exercises.

Wyman's influence lived on in a number of students. B. G. Wilder carried on his theories of symmetry and homology in limbs. S. Weir Mitchell paid tribute to his inspiration. Perhaps the most effective projector of Wyman's influence, even into the twentieth century, was the young man who as a student had come to respect him over Agassiz and who succeeded him in his course in anatomy and physiology at the time of his death—William James.

BIBLIOGRAPHY

A collection of MSS long held by the Wyman family is at the Countway Library, Harvard Medical School.

Biographical notices include Asa Gray, "Jeffries Wyman Memorial Meeting . . . October 7, 1874," in *Proceedings of the Boston Society of Natural History*, **17** (1875), 96–124; Oliver Wendell Holmes, "Memoir of Professor Jeffries Wyman," in *Proceedings of the Massachusetts Historical Society*, **14** (Apr. 1875), 4–24; and A. S. Packard, "Memoir of Jeffries Wyman, 1814–1874," in *Biographical Memoirs. National Academy of Sciences*, **2** (1886), 75–126, which includes a list of Wyman's works.

Also see R. N. Doesch, "Early American Experiments on 'Spontaneous Generation' by Jeffries Wyman (1814–1874)," in *Journal of the History of Medicine and Allied Sciences*, **17** (1962), 326–332; A. H. Dupree, ed., "Some Letters From Charles Darwin to Jeffries Wyman," in *Isis*, **42** (1951), 104–110; and "Jeffries Wyman's Views on Evolution," *ibid.*, **44** (1953), 243–246; and G. E. Gifford, ed., "Twelve Letters From Jeffries Wyman, M.D.: Hampden-Sydney Medical College, Richmond, Virginia, 1843–1848," in *Journal of the History of Medicine and Allied Sciences*, **20** (1965), 309–333; and "An American in Paris, 1841–1842: Four Letters From Jeffries Wyman," *ibid.*, **22** (1967), 274–285.

A. HUNTER DUPREE

XENOCRATES OF CHALCEDON (*b.* Chalcedon [now Kadiköy], Bithynia [now Turkey], 396/395 B.C.; *d.* Athens, 314/313 B.C.), *philosophy, mathematics.*

Xenocrates was a student of Plato and, as head of the Academy from 339 B.C. to 314/313 B.C., was one of the founders of the ancient Academic tradition. He entered the Academy (in 378 B.C. at the earliest or 373 B.C. at the latest) and about ten years later accompanied Plato to Syracuse, the latter's second or third voyage to that city.

After Plato's death, Xenocrates and Aristotle were invited to Assos, where they remained until the overthrow of Hermias of Atarneus, in 342 B.C. Plato's successor, his nephew Speusippus, headed the Academy until his death in 340/339 B.C. He sent for Xenocrates, who was not in Athens at this time, and designated him his successor. Nevertheless, an election was held, which Xenocrates won by only a few votes. The opposing candidates, Heraclitus Ponticus and Menedemus of Pyrrha, thereupon left the Academy. (Aristotle was not a candidate.)

In 322 B.C. Xenocrates was appointed to an Athenian legation sent to negotiate with Antipatrus of Macedonia, but it was compelled, instead, to acknowledge Athens' submission. Since Xenocrates was not a citizen of Athens, Antipatrus did not recognize his status as a legitimate ambassador. Both before and after this incident, Xenocrates refused to seek Athenian citizenship, since he disapproved of the city's close relations with Macedonia. On this point he was at odds with the political views of his predecessor Speusippus, who had supported Athens' pro-Macedonian policies. After Xenocrates' death, the leadership of the Academy passed to Polemo.

According to a number of less substantiated anecdotes about Xenocrates, he is depicted as good-natured, gentle, and considerate—but it is also quite clear that he lacked the *charis* ("graciousness") of his teacher Plato. Along with these traits, Xenocrates is reported to have displayed singular diligence. The list of his writings, which is entirely preserved in Diogenes Laërtius (IV, 11–14), contains about seventy titles. The biographical anecdotes state that he never left the Academy (where all his work was done) more than once a year.

None of Xenocrates' writings has survived. They presumably were never published—that is, copied; rather, the single copies in his own hand were deposited at the Academy. When Athens was stormed by Sulla's troops in March 86 B.C., the Academy, located in front of the western gate, was destroyed together with its priceless library.[1]

Unlike Aristotle, Xenocrates did not wish to develop philosophy in new ways but considered it his task to maintain Plato's theory as he had received and understood it. This motivation underlay his extensive literary activity, in which enterprise Xenocrates relied primarily upon his memory—recalling, in some instances, what he had learned as much as twenty-five years earlier. Curiously, Xenocrates did not respect an important request made by Plato, who considered language unreliable and conducive to misunderstanding. Plato apparently insisted that the contents of philosophical doctrines be communicated only to those who would strictly observe definite precautions, and consequently he often veiled his meaning through the use of metaphors and myths. Xenocrates, seeking to render Plato's theories teachable without recourse to such means, established a system of doctrinal propositions. Xenocrates' lifework consisted in producing a kind of codification—and thus, of necessity, a transformation—of Plato's philosophy. But it immediately became apparent that others, especially Aristotle, understood Plato in a wholly different way with respect to certain key questions. Crantor of Soli, for example—himself the pupil of Xenocrates—was to some extent justified in reproaching him for having taught things that differed from elements of Plato's surviving written works. Although Xenocrates did not attribute to Plato anything that he had not actually taught, he did choose single aspects—often overly narrow ones—from among the many possible ways of presenting Plato's conceptions and raised them to the status of official doctrine.

Xenocrates probably found himself confronted with a difficult situation. After it had been decided to preserve Plato's teachings, everything that could not be encompassed in words—that is, everything that Plato had not wished to set forth in words—had to be abandoned in the interest of a clear and systematic presentation. In the process of achieving this sort of systematization, Xenocrates sacrificed Plato's dialectical approach.

Xenocrates was no more a dualist than Plato. Rather, he belongs to the long line of those who have attempted to overcome a type of dualism within the philosophical tradition. In particular, in his comprehensive outline, Xenocrates provided for a gradation of all the elements that make up ontology, physics, ethics, and epistemology. This scheme assumed major importance in the development of Neoplatonism. Xenocrates' real legacy lies in the conception he sketched of a hierarchy of all existing things that culminates in a single, highest point, the One.

NOTES

1. Reports of Xenocrates' teaching have survived only in the works of Aristotle and in Cicero and other Roman authors. The so-called fragments have been collected by Richard Heinze (see below). All knowledge of Xenocrates was probably based on an indirect tradition. Reports of his teaching are only rarely supported by mention of the title of one of his works (cf. Heinze, 157, 158). The only quotation that seems to bear the stamp of authenticity is in Themistius' (A.D. 317[?]–*ca*. 388) commentary on Aristotle's *De anima*; see *Commentaria in Aristotelem Graeca*, V, pt. 3, R. Heinze, ed. (Berlin, 1899); and Heinze, *Xenocrates*, frag. 61. In Themistius' reference to the fifth book of περὶ φύσεως, it is possible to glean an indication that Xenocrates may have introduced the definition of soul as number. In this instance it cannot be ruled out that περὶ φύσεως had been published either in whole or in part. But such a quotation may also have originated in Crantor's interpretation of the *Timaeus*, where he suddenly attacks Xenocrates' theory of the soul as number. On this point, see H. F. Cherniss, *Aristotle's Criticism of Plato and the Academy*, I (Baltimore, 1944; repr. New York, 1962), p. 399, n. 325.

BIBLIOGRAPHY

See H. F. Cherniss, *The Riddle of the Early Academy* (Berkeley, Calif., 1945; repr. New York, 1962), esp. 31–59; H. Dörrie, "Xenokrates," in Pauly-Wissowa, *Real Encyclopädie der classischen Altertumswiss-*

enschaft, 2nd ser., IX, 1512–1528; R. Heinze, *Xenocrates. Darstellung der Lehre und Sammlung der Fragmente* (Leipzig, 1892; repr., Hildesheim, 1965); S. Mekler, ed., *Academicorum philosophorum index Herculanensis* (Berlin, 1902), esp. 38–39; Diogenes Laërtius, *De vitis philosophorum*, IV, ch. 2—also in English, R. D. Hicks, trans., *Lives of Eminent Philosophers*, I (Cambridge, Mass.–London, 1966), 380–393; and U. von Wilamowitz-Moellendorff, *Platon, sein Leben und seine Werke*, 4th ed., I (Berlin, 1948), 579–581.

H. DÖRRIE

XENOPHANES (*b.* Colophon, Ionia, *ca.* 580–570 B.C.; *d. ca.* 478 B.C.), *theology, epistemology.*

It is generally believed that Xenophanes was born about 570 B.C. in Colophon, a Greek city in Asia Minor. He left Ionia after 545, the time of the Persian conquest, in order to live in the western part of the Greek world, in southern Italy and Sicily. He died after 478. If it is true, as has recently been suggested,[1] that Xenophanes did not leave Colophon at the time he left Asia but, rather, ten years earlier, then the date of his birth can be pushed back in accordance with the ancient chronology (580–577). On this hypothesis, he was banished from Colophon in 555, when the city came under a tyrannical regime (at that time it was still under the control of the kingdom of Lydia).

Xenophanes seems to have opposed this regime openly and to have been known after 555 as a poet fighting for the restoration of his native city's ancient civil liberties. It was probably toward this end that he devoted an epic to the origins of Colophon and wrote a poem in honor of Elea. He may have joined the Phocaeans who founded the latter city on the coast of southern Italy (540–535).

Xenophanes' deep personal involvement in political matters is inseparable from his intellectual activities. He profoundly influenced Greek thought in at least two respects, through his criticism of the anthropomorphic beliefs upheld by traditional religion and through his "monist" definition of God. The principal surviving fragments of his elegies clearly show that the intellectual and moral reform to which he dedicated himself had a political objective. He believed that the thinker, through his statements, should clear the way for a strengthening of communal life within the framework of the city-state.

To further this goal, Xenophanes extended his critique of anthropomorphism to all the attitudes and activities attributed to the gods, judging these incompatible with a just conception of divine reality. He rejected the picture of the divine world and its organization propagated by Homer and Hesiod; he objected to certain ritual practices; and he denied that the gods intervene physically either in divination or in meteorological phenomena. To this refutation of accepted views which he elaborated in his *Satires* (Σίλλοι) Xenophanes joined a description of the attributes of God. These are such, he asserted, as reason conceives them when it has cast off the hold of mythology and popular beliefs. Thus, starting from the notion of omnipotence, Xenophanes derived the concepts of God's unity (that is, unicity or wholeness) and eternity. God, he stated, is present everywhere and acts without intermediary and without displacement or movement, solely by means of His mind's will.

Did Xenophanes apply his ideas concerning the attributes of divine reality to the universe? Did he identify God with the cosmos, as has often been supposed?[2] It does not seem that he did. This pantheistic interpretation (given by Theophrastus and already proposed by Aristotle) appears incompatible—despite the opposing views of certain authors[3]—with the wording of the existing fragments. Still, these attributes endow Xenophanes' God with an ontological status remarkably similar to that later enunciated in certain propositions of Eleatic logic. According to tradition, Xenophanes was the teacher of Parmenides; and the latter could indeed have found in the conception of a unique, eternal, and omnipotent God the starting point for his deduction of the properties of being.

Xenophanes' monotheism did not entail a denigration of man. On the contrary, he affirmed man's autonomy in material progress and civilization. But he did make a distinction of great epistemological significance: God alone possesses complete knowledge, whereas man can attain genuine knowledge only within the limits assigned to the combined activity of his senses. That is, he can really know only particular objects or partial aspects of the world. With regard to the totality of things, the universe (and God himself), man must be satisfied with a probable knowledge, which is incapable of verifying the truth of what it grasps. This restriction has given rise to much discussion. Some authors, including a few modern ones, have claimed that Xenophanes meant to apply it to empirical knowledge itself, thus portraying him as an advocate of radical skepticism.[4] This view is incorrect. He thought that human knowledge was limited, not with respect to things but relative to God's omniscience.

Xenophanes did not conceive or set forth a

complete doctrine of the physical world,[5] although he occasionally touched on physical questions in his polemical writings—alluding, for example, to Thales, Anaximander, and Pythagoras. He was neither a philosopher of nature nor a "sage" in the primary sense of the word. Highly independent and curious about everything (something for which he was reproached by Heraclitus), Xenophanes was a poet and a thinker who played a major role in the intellectual adventure of his age. He stimulated the emancipation of reason in Greek ethical and religious discourse and thus contributed, although indirectly, to the triumph of systematic thinking in science and philosophical reflection.

NOTES

1. See P. Steinmetz, "Xenophanesstudien," sec. entitled "Zur Datierung."
2. Most recently, by M. Untersteiner, *Senofane*, clxxxix–cciii; and W. K. C. Guthrie, *A History of Greek Philosophy*, 381–383. Those who disagree or reserve opinion include W. Jaeger, *The Theology of the Early Greek Philosophers*, 43 and n. 23; H. Fränkel, *Dichtung und Philosophie des frühen Griechentums*, 378; A. Lumpe, *Die Philosophie des Xenophanes von Kolophon*, 22–26; and G. S. Kirk, in Kirk and Raven, *The Presocratic Philosophers*, 171–172.
3. Especially Guthrie, *loc. cit.*,
4. Particularly E. Heitsch, who opposes Fränkel. Compare K. von Fritz, "Xenophanes," cols. 1557–1559.
5. Summaries of the controversy over the *De natura* attributed to him are in Untersteiner, *op. cit.*, ccxlii–ccl; and in Reale's note in E. Zeller and R. Mondolfo, *La filosofia dei Greci* . . ., 69–71. There are some perceptive remarks in Steinmetz, *op. cit.*, 54–68 ("Ein Lehrgedicht des Xenophanes?").

BIBLIOGRAPHY

Fragments and testimonia are in H. Diels and W. Kranz, *Die Fragmente der Vorsokratiker*, 6th ed., I (Berlin, 1951), 113–139; the fragments alone are in E. Diehl, *Anthologia lyrica Graeca*, fasc. 1, *Poetae elegiaci*, 3rd ed. (Leipzig, 1949), 64–76; and M. L. West, *Iambi et Elegi Graeci ante Alexandrum cantati*, II (Oxford, 1972), 163–170. There is abundant information in M. Untersteiner, *Senofane, testimonianze e frammenti* (Florence, 1955); and E. Zeller and R. Mondolfo, *La filosofia dei Greci nel suo sviluppo storico*, pt. 1, I *Presocratici*, III, *Eleati*, G. Reale, ed. (Florence, 1967), 1–164.

Recent writings include H. Fränkel, *Dichtung und Philosophie des frühen Griechentums* (New York, 1951; 2nd ed., Munich, 1962), 371–384; K. von Fritz, "Xenophanes," in Pauly-Wissowa, *Real-Encyclopädie der classischen Altertumswissenschaft*, 2nd ser., IX (1967), cols. 1541–1562; W. K. C. Guthrie, *A History of Greek Philosophy*, I (Cambridge, 1962), 360–402; E. Heitsch, "Das Wissen des Xenophanes," in *Rheinisches Museum für Philologie*, **109** (1966), 193–235; H. Herter, "Das

Symposion des Xenophanes," in *Wiener Studien*, **69** (1956), 33–48; W. Jaeger, *The Theology of the Early Greek Philosophers* (Oxford, 1947), 38–54; G. S. Kirk, in G. S. Kirk and J. E. Raven, *The Presocratic Philosophers* (Cambridge, 1957), 163–181; A. Lumpe, *Die Philosophie des Xenophanes von Kolophon* (Munich, 1952); A. Rivier, "Remarques sur les fragments 34 et 35 de Xénophane," in *Revue de philologie*, 3rd ser., **30** (1956), 37–61; P. Steinmetz, "Xenophanesstudien," in *Rheinisches Museum für Philologie*, **109** (1966), 13–73; and M. Untersteiner, intro. and commentary to his *Senofane* (see above).

ANDRÉ RIVIER

YAḤYĀ IBN ABĪ MANṢŪR (*d.* near Aleppo, Syria, 832), *astronomy*.

Yaḥyā was a member of an important family of Persian scientists. His father, Abū Manṣūr Abān, was an astrologer; his son ʿAlī bin Yaḥyā (*d.* 888) was eminent in Baghdad and had a great library in which Abū Maʿshar studied; his grandson, Hārūn ibn ʿAlī (*d.* 900), also was an astronomer.

Yaḥyā spent his life casting horoscopes (one is given by Ibn al-Qifṭī, *Taʾrīkh al-ḥukamāʾ*, pp. 358–359) and seeking methods to determine the positions of the stars with maximum precision. His first work as an astrologer was in the service of al-Faḍl ibn Sahl, vizier of Caliph al-Maʾmūn. After al-Faḍl's assassination in February 818, he entered the service of al-Maʾmūn and converted to Islam. We know that Yaḥyā was an official at *bayt al-ḥikma*, who may have controlled funds for astronomy. He taught the Banū Mūsā and died while accompanying the caliph on an expedition against Tarsus.

Yaḥyā was appointed director of the group of scholars who by order of al-Maʾmūn (828) established an observatory in the Shamāsiya quarter of Baghdad and the observatory at the monastery of Dayr Murrān in Damascus. These centers were intended to make observations that would improve and correct existing astronomical tables. The Damascus observatory was headed by Ḥabash al-Ḥāsib (*d.* 864/874), who sent the results obtained there to Baghdad for further elaboration. This would explain why the tables attributed to Ḥabash are closely connected with *Zīj al-mumtaḥan*.

Yaḥyā's team of scientists included al-Marwarrūdhī; al-Khwārizmī, who collaborated with Yaḥyā in 828; and Sanad ibn ʿAlī (*d. ca.* 864), who was in charge of improving the observational instruments, some of which were unusually large. Most of the instruments were graduated by the leading expert of the time, ʿAlī ibn ʿIsā al-Asṭurlābī, but were not very dependable. The group also included the Banū Mūsā and al-Jawharī; the latter corrected the data concerning the positions of the planets and of the sun and moon, incorporating the results into his own tables, on the margin of the *Zīj al-mumtaḥan*.

This same group measured one degree of the meridian by using two different processes: the measurement *in situ* of one degree on the earth's surface, and the corroboration of this value by means of a trigonometric process based on measuring the dip of the horizon, with an astrolabe, from a mountaintop. The latter method seems to have been used for the first time by Sanad ibn ʿAlī.

The numerical results of the observations were recorded in the *Zīj al-mumtaḥan* (*Tabulae probatae* in Latin). It should be noted that the words *mumtaḥan* and *probatae* are generic and indicate any table based on observation; thus the tables of Yaḥyā's group are not the only ones to be so designated. The observations terminated abruptly at the almost simultaneous deaths of the caliph and of Yaḥyā. A written copy of the completed work was deposited at the library of the palace at Baghdad. Only one manuscript (Escorial 927) is known to contain these tables, but it is badly bound and comprises many folios that are not from Yaḥyā's work and are explicitly attributed to astronomers of the tenth and eleventh centuries. Internal criticism seems to indicate that the first folios are by Yaḥyā's group and the later ones are intermingled with works written long afterward.

Because of this disorder the manuscript can be analyzed only by means of an arbitrary arrangement. It contains a full explanation of calendars (Coptic, Greek, Jewish, Muslim) and chronological eras, the majority of which are primitive. Many of the tables were compiled before those of Yaḥyā, the date being indicated at the top of each. It is difficult to state precisely to what extent they used all the trigonometric functions, as Ḥabash did in his works. The elements for the calculation of ephemerides were generally primitive, as were two of the tables of star positions. The inferior planets are treated numerically (without theoretical explanations) as satellites of the sun—an approach equivalent to the system of Heraclides and Tycho Brahe. This model may have been suggested by an ancient text, perhaps one by Theon of Smyrna. The margins contain tables and rules that are very difficult to group and date because of their lack of unity.

Yaḥyā's tables exerted a great influence on astronomy: Thābit ibn Qurra (*d.* 901) wrote an introduction to them that drew on data supplied by Ḥabash, and took these tables into consideration in his works on eclipses. Ibn Yūnus adapted them for use in Egypt; and al-Zarqālī derived from them the value of the inclination of the ecliptic and certain other values used in the calculation of ephemerides.

BIBLIOGRAPHY

See E. S. Kennedy, "A Survey of Islamic Astronomical Tables," in *Transactions of the American Philosophical Society*, n.s. **46**, no. 2 (1956), nos. 15, 51. MS Escorial 927 is analyzed by J. Vernet, in "Las ʿTabulae probatae,'" in *Homenaje a Millás-Vallicrosa*, II (Barcelona, 1956), 501–522. Vernet has also published partial studies of this MS: "Los símbolos planetarios rumíes," in *al-Andalus*, **16** (1951), 493; and "Un antiguo tratado sobre el calendario judío en las *Tabulae probatae*," in *Sefarad*, **14** (1954), 59–78.

The Arabic sources are listed in W. Hartner, "Ḥabash al-Ḥāsib al-Marwazī," in *Encyclopedia of Islam*, new ed., III (Leiden–London, 1971), 8–9; Aydin Sayili, *The Observatory in Islam* (Ankara, 1960), 50–87 and index; and H. Suter, *Die Mathematiker und Astronomen der Araber und ihre Werke* (Leipzig, 1900), nos. 14, 22.

J. VERNET

YAʿĪSH IBN IBRĀHĪM. See **Al-Umawī, Abū ʿAbdallāh Yaʿīsh ibn Ibrāhīm ibn Yūsuf ibn Simāk al-Andalusī.**

YANG HUI (*fl.* China, *ca.* 1261–1275), *mathematics.*

The thirteenth century was perhaps the most significant period in the history of Chinese mathematics. It began with the appearance of Ch'in Chiu-shao's *Shu-shu chiu-chang* in 1247, and the following year Li Chih issued an equally important work, the *Ts'e-yüan hai-ching*. These two great algebraists were later joined by Yang Hui (literary name, Yang Ch'ien-kuang), whose publications far surpassed those of his predecessors, and of whom we have absolutely no knowledge. The golden age of Chinese mathematics came to an end after the appearance of Chu Shih-chieh's *Ssu-yüan yü-chien* in 1303. Of the works of these four great Chinese mathematicians, those by Yang Hui have, until very recently, been the least studied and analyzed.

Nothing is known about the life of Yang Hui,

except that he produced mathematical writings. From the prefaces to his works we learn that he was a native of Ch'ien-t'ang (now Hangchow). He seems to have been a civil servant, having served in T'ai-chou, and he had probably visited Su-chou (modern Soochow). His friends and acquaintances included Ch'en Chi-hsien, Liu Pi-chien, Ch'iu Hsü-chü, and Shih Chung-yung, the last having collaborated with him on one of his works; but we know nothing else about their personal history. Yang Hui also names as his teacher another mathematician, Liu I, a native of Chung-shan, of whom nothing is known.

In 1261 Yang Hui wrote *Hsiang-chieh chiu-chang suan-fa* ("Detailed Analysis of the Mathematical Rules in the Nine Chapters"), a commentary on the old Chinese mathematical classic *Chiu-chang suan-shu*, by Liu Hui. The present version of the *Hsiang-chieh chiu-chang suan-fa*, which is based on the edition in the *I-chia-t'ang ts'ung-shu* (1842) collection, is incomplete and consists of only five chapters; two additional chapters have been restored from the *Yung-lo ta-tien* encyclopedia. Besides the original nine chapters of the *Chiu-chang suan-shu*, Yang Hui's *Hsiang-chieh chiu-chang suan-fa* included three additional chapters, making a total of twelve. According to the preface written by Yang Hui, he had selected 80 of the 246 problems in the *Chiu-chang suan-shu* for detailed discussion. The now-lost introductory chapter of the *Hsiang-chieh chiu-chang suan-fa*, as we learn from the *shao-kuang* ("Diminishing Breadth") chapter of the text and from a quotation in another of Yang Hui's works, *Suan-fa t'ung-pien pen-mo*, contained diagrams and illustrations.

Chapter 1, according to what Yang Hui describes in *Suan-fa t'ung-pien pen-mo* and *Ch'eng-ch'u t'ung-pien suan-pao*, dealt with the ordinary methods of multiplication and division. This chapter is also lost, but two of its problems have been restored from the *Yung-lo ta-tien* encyclopedia by Li Yen; it was entitled *Chu-chia suan-fa*. Chapter 2, *Fang-t'ien* ("Surveying of Land"), is now lost. Chapter 3, *Su-mi* ("Millet and Rice") is also lost, but three problems have been restored from the *Yung-lo ta-tien* encyclopedia. Chapter 4, *Ts'ui-fen* ("Distribution by Progression"), is no longer extant; but eleven of its problems have been restored from the *Yung-lo ta-tien* encyclopedia. Chapter 5, *Shao-kuang* ("Diminishing Breadth"), also has been partially restored from the *Yung-lo ta-tien* encyclopedia. As for Chapter 6, *Shang-kung* ("Consultations on Engineering Works"), the *I-chia-t'ang ts'ung-shu* collection contains thirteen

problems (fifteen problems are missing). Chapter 7, *Chün-shu* ("Impartial Taxation"); chapter 8, *Ying-pu-tsu* ("Excess and Deficiency"); chapter 9, *Fang-ch'eng* ("Calculation by Tabulation"); chapter 10, *Kou-ku* ("Right Angles"); and chapter 11, *Tsuan-lei* ("Reclassifications") remain more or less intact in the *I-chia-t'ang ts'ung-shu* collection, except for one missing problem in chapter 7 and four in chapter 9.

In 1450 the Ming mathematician Wu Ching wrote the *Chiu-chang hsiang-chu pi-lei suan-fa* ("Comparative Detailed Analysis of the Mathematical Rules in the Nine Chapters"), in which the "old questions" (*ku-wen*) are referred to. Yen Tun-chieh has shown that Wu Ching's "old questions" were based on Yang Hui's *Hsiang-chieh chiu-chang suan-fa*, and he has been engaged in restoring this text. A substantial part of the *I-chia-t'ang ts'ung-shu* edition of the text has been rendered into English by Lam Lay Yong of the University of Singapore.

Yang Hui published his second mathematical work, the two-volume *Jih-yung suan-fa* ("Mathematical Rules in Common Use"), in 1262. This book is no longer extant. Some sections have, however, been restored by Li Yen from the *Chu-chia suan-fa* in the *Yung-lo ta-tien* encyclopedia. The book seems to be quite elementary.

In 1274 Yang Hui produced the *Ch'eng-ch'u t'ung-pien pen-mo* ("Fundamental Mutual Changes in Multiplications and Divisions") in three volumes. The first volume was originally known as the *Suan-fa t'ung-pien pen-mo* ("Fundamental Mutual Changes in Calculations"); the second as *Ch'eng-ch'u t'ung-pien suan-pao* ("Treasure of Mathematical Arts on the Mutual Changes in Multiplications and Divisions"); and the last volume, written in collaboration with Shih Chung-yung, was originally called *Fa-suan ch'ü-yung pen-mo* ("Fundamentals of the Applications of Mathematics"). The next year Yang Hui wrote the *T'ien-mou pi-lei ch'eng-ch'u chieh-fa* ("Practical Rules of Mathematics for Surveying") in two volumes. This was followed in the same year by the *Hsü-ku chai-ch'i suan-fa* ("Continuation of Ancient Mathematical Methods for Elucidating the Strange Properties of Numbers"), written after Yang Hui had been shown old mathematical documents by his friends Liu Pi-chien and Ch'iu Hsü-chü. Subsequently all seven volumes that Yang Hui wrote in 1274–1275 came to be known under a single title, *Yang Hui suan-fa* ("The Mathematical Arts of Yang Hui"). The work was first printed in 1378 by the Ch'in-te shu-t'ang Press and was reprinted in

Korea in 1433. A handwritten copy of the Korean reprint was made by the seventeenth-century Japanese mathematician Seki Takakazu. (A copy of the Korean reprint is in the Peking National Library.) Li Yen had Seki Takakazu's handwritten copy of the *Yang Hui suan-fa* dated 1661; it became the property of the Academia Sinica after his death in 1963. At the beginning of the seventeenth century Mao Chin (1598–1652) made a handwritten copy of the fourteenth-century edition of the *Yang Hui suan-fa*.

All of Yang Hui's writings, including the *Yang Hui suan-fa*, seem to have been forgotten during the eighteenth century. Efforts were made in 1810 to reconstruct the text from the *Yung-lo ta-tien* encyclopedia by Juan Yuan (1764–1849), but they were confined to a portion of the *Hsü-ku chai-ch'i suan-fa*. In 1776 Pao T'ing-po included a portion of the *Hsü-ku chai-ch'i suan-fa* in his *Chih-pu-tsu ts'ung-shu* collection that may have come from the restoration by Juan Yuan. In 1814 Huang P'ei-lieh discovered an incomplete and disarranged copy of the *Yang Hui suan-fa*. He and the mathematician Li Jui put the text in order. Lo Shih-lin had a handwritten copy made of this corrected text. This version of the *Yang Hui suan-fa*, consisting of only six volumes, was incorporated into the *I-chia-t'ang ts'ung-shu* collection by Yu Sung-nien (1842). Later reproduction of the text in the *Ts'ung-shu chi-ch'eng* collection (1936) is based on the version in the *I-chia-t'ang ts'ung-shu* collection. Some of the textual errors in the book have been corrected by Sung Ching-ch'ang in his *Yang Hui suan-fa cha-chi*. A full English translation and commentary of the *Yang Hui suan-fa* was made by Lam Lay Yong in 1966 for her doctoral dissertation at the University of Singapore.

The *Hsiang-chieh chiu-chang suan-fa* is perhaps the best-known, but certainly not the most interesting, of Yang Hui's writings. In it he explains the questions and problems in the *Chiu-chang suan-shu*, sometimes illustrating them with diagrams, and gives the detailed solutions. Problems of the same nature also are compared with each other. In the last chapter, the *Tsuan lei*, Yang Hui reclassifies all the 246 problems in the *Chiu-chang suan-shu* in order of progressive difficulty, for the benefit of students of mathematics. Some examples of algebraic series given by Yang Hui in this book are

$$m^2 + (m+1)^2 + \cdots + (m+n)^2 = \frac{m+1}{3}$$

$$\left\{ m^2 + (m+n)^2 + m(m+n) + \frac{1}{2} \left[(m+n) - m \right] \right\}$$

$$1 + 3 + 6 + \cdots + \frac{n(n+1)}{2} = \frac{1}{6}n \, (n+1) \, (n+2)$$

$$1^2 + 2^2 + 3^2 + \cdots + n^2 = \frac{1}{3}n \, (n + \frac{1}{2}) \, (n+1)$$

$$1^2 + (a+1)^2 + \cdots + (c-1)^2 + c^2$$
$$= \frac{1}{3}(c-a)\left(c^2 + a^2 + ca + \frac{c-a}{2} \right).$$

The portions restored from the *Yung-lo ta-tien* encyclopedia contain the earliest illustration of the "Pascal triangle." Yang Hui states that this diagram was derived from an earlier mathematical text, the *Shih-so suan-shu* of Chia Hsien (*fl. ca.* 1050). This diagram shows the coefficients of the expansion of $(x + a)^n$ up to the sixth power. Another diagram showing coefficients up to the eighth power was later found in the early fourteenth-century work *Ssu-yüan yü-chien* of Chu Shih-chieh. Other Chinese mathematicians using the Pascal triangle before Blaise Pascal were Wu Ching (1450), Chou Shu-hsüeh (1588), and Ch'eng Ta-wei (1592).

The *Tsuan-lei* also quotes a method of solving numerical equations higher than the second degree taken from Chia Hsien's *Shih-so suan-shu*. The method is similar to that rediscovered independently in the early nineteenth century, by Ruffini and Horner, for solving numerical equations of all orders by continuous approximation. A method called the *tseng-ch'eng k'ai-li-fang fa* for solving a cubic equation $x^3 - 1860867 = 0$ is given in detail below.

The number 1860867 is set up in the second row of a counting board in which five rows are used—the top row (*shang*) is for the root to be obtained, the second row (*shih*) is for the constant, the third row (*fang*) is for the coefficient of x, the fourth row (*lien*) is for the coefficient of x^2, and the last row (*hsia-fa*) is for the coefficient of x^3. Thus 1 is placed in the last row, and this coefficient is shifted to the left, moving two place values at a time until it comes in line with the number 1860867, at the extreme left in this case, as shown in Figure 1a.

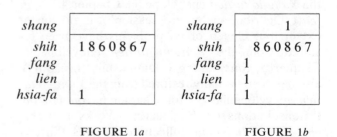

shang		*shang*	1
shih	1 8 6 0 8 6 7	*shih*	8 6 0 8 6 7
fang		*fang*	1
lien		*lien*	1
hsia-fa	1	*hsia-fa*	1

FIGURE 1a   FIGURE 1b

540

Since x lies between 100 and 200, 1 is placed at the hundreds' place of the first row. Multiplying this number by the number in the last row yields 1, which is entered on the fourth row as the *lien*. Again, multiplying the number 1 on the top row by the *lien* gives 1, which is entered in the third row as the *fang*. The number in the *fang* row is subtracted from the number in the same column in the *shih* row. The result is shown in Figure 1*b*. The number on the last row is multiplied a second time by the number in the top row and the product is added to the *lien*. The number in the *lien* is multiplied by the number on the top row and added to the *fang*. The number on the last row is multiplied a third time by the number on the top row, and the product is added to the number in the fourth row. The result is shown in Figure 1*c*.

shang	1
shih	860867
fang	3
lien	3
hsia-fa	1

FIGURE 1*c*

shang	1
shih	860867
fang	3
lien	3
hsia-fa	1

FIGURE 1*d*

The number in the third row (*fang*) is moved to the right by one place, that in the fourth row (*lien*) is moved to the right by two places, and that in the fifth row (*hsia-fa*) is moved to the right by three places, as in Figure 1*d*.

For the next approximation, x is found to lie between 120 and 130. Hence 2 is placed in the upper row in the tens' place. The same process is repeated, using 2 as the multiplier. We have $2 \times 1 = 2$, which is then added to the *lien*, giving a sum 32; and $2 \times 32 = 64$, which, when added to the *fang*, gives 364. Then 364 is multiplied by 2, giving 728, which is subtracted from 860867 to give 132867, as shown in Figure 1*e*.

shang	12
shih	132867
fang	364
lien	32
hsia-fa	1

FIGURE 1*e*

shang	12
shih	132867
fang	364
lien	36
hsia-fa	1

FIGURE 1*f*

Then the *hsia-fa* is multiplied a second time by 2 and added to the *lien*, giving a sum of 34, which again is multiplied by 2 and added to the *fang*, giv-

ing 432. The *hsia-fa* is multiplied a third time by 2 and added to the number 34 in the *lien* row to give 36, as shown in Figure 1*f*. The number in the *fang* row is now shifted one place, that in the *lien* row two places, and that in the *hsia-fa* row three places to the right, as shown in Figure 1*g*.

shang	12
shih	132867
fang	432
lien	36
hsia-fa	1

FIGURE 1*g*

shang	123
shih	0
fang	44289
lien	363
hsia-fa	1

FIGURE 1*h*

The last digit of x is found to be 3. This is placed on the top row in the units' column. Three times the *hsia-fa* gives 3, which, when added to the *lien*, gives 363. $3 \times 363 = 1089$, which, when added to the *fang*, gives 44289. Then $3 \times 44289 = 132867$, which, when subtracted from the *shih* row, leaves no remainder, as shown in Figure 1*h*. The root is therefore 123.

The *Hsiang-chieh chiu-chang suan-fa* also contains a method for solving quartic equations called the *tseng san-ch'eng k'ai-fang fa*. This involves the equation $x^4 - 1,336,336 = 0$. The method used is similar to that employed above for cubic equations. The solution is presented below in a slightly modified form, in order to show the resemblance to Horner's method.

$$x^4 - 1336336 = 0, \quad x = 34$$

$$
\begin{array}{l}
1(10)^4 + \quad 0 \times (10)^3 + \quad\quad 0 \times (10)^2 + \quad\quad 0 \times (10) - 1336336 \quad \lfloor 30 \\
\quad\quad\quad + \; 30 \times (10)^3 + \; 900 \times (10)^2 + \; 27000 \times (10) + \; 810000 \\
\hline
1(10)^4 + \; 30 \times (10)^3 + \; 900 \times (10)^2 + \; 27000 \times (10) - \; 526336 \\
\quad\quad\quad + \; 30 \times (10)^3 + 1800 \times (10)^2 + \; 81000 \times (10) \\
\hline
1(10)^4 + \; 60 \times (10)^3 + 2700 \times (10)^2 + 108000 \times (10) \\
\quad\quad\quad + \; 30 \times (10)^3 + 2700 \times (10)^2 \\
\hline
1(10)^4 + \; 90 \times (10)^3 + 5400 \times (10)^2 \\
\quad\quad\quad + \; 30 \times (10)^3 \\
\hline
1(10)^4 + 120 \times (10)^3 \\
1(10)^4 + 120 \times (10)^3 + 5400 \times (10)^2 + 108000 \times (10) - 526336 \quad \lfloor 4 \\
\quad\quad\quad\quad\; 4 \times (10)^3 + \; 496 \times (10)^2 + \; 23584 \times (10) + 526336 \\
\hline
1(10)^4 + 124 \times (10)^3 + 5896 \times (10)^2 + 131584 \times (10) + \quad\quad 0
\end{array}
$$

Both the few remaining problems of the *Jih-yung suan-fa* restored from the *Yung-lo ta-tien* encyclopedia and its title *Jih-yung*, meaning "daily or common use," suggest that the book must be of an elementary and practical nature, although we no longer have access to its entire text. Two examples follow:

a. A certain article actually weighs 112 pounds.

How much does it weigh on the provincial steelyard? Answer: 140 pounds. (Note that on the provincial steelyard 100 pounds would read 125 pounds.)

b. The weight of a certain article reads 391 pounds, 4 ounces, on a provincial steelyard. What is its actual weight? Answer: 313 pounds.

The first volume of the *Ch'eng-ch'u t'ung-pien pen-mo* (the *Suan-fa t'ung-pien pen-mo*) gives a syllabus or program of study for the beginner that is followed by a detailed explanation of variations in the methods of multiplication. In it Yang Hui shows how division can be conveniently replaced by multiplication by using the reciprocal of the divisor as the multiplier. For example, $2746 \div 25 = 27.46 \times 4$; $2746 \div 14.285 = 27.46 \times 7$; and $2746 \div 12.5 = 27.46 \times 8$. Sometimes he multiplies successively by the factors of the multiplier—for example, $274 \times 48 = 274 \times 6 \times 8$—and at other times he shows that the multiplier can be multiplied by the multiplicand—for example, $247 \times 7360 = 7360 \times 247$. Of special interest are the "additive" and "subtractive" methods that are applied to multiplication. These methods are quite conveniently used on the counting board, or even on an abacus, where the numbers are set up rather than written on a piece of paper. If the multiplier is 21, 31, 41, 51, 61, 71, 81, or 91, multiplication can be performed by multiplying only with the tens' digit and the result, shifted one decimal place to the left, is added to the multiplicand. Yang Hui gives a number of examples to illustrate this method, such as $232 \times 31 = 232 \times 30 + 232$; $234 \times 410 = 234 \times 400 + 2340$. In the "subtractive" method of multiplication, if the multiplier x is a number of n digits, the multiplicand p is first multiplied by 10^n, and from the result the product of the multiplicand and the difference between 10^n and the multiplier x is subtracted, that is, $xp = 10^n p - (10^n - x)p$. Yang Hui gives the example $26410 \times 7 = 264100 - (2641 \times 3) = 1848700$. The book ends with an account of how division can be performed.

In the second volume of the *Ch'eng-ch'u t'ung-pien pen-mo* (the *Ch'eng-ch'u t'ung-pien suan-pao*), Yang Hui proceeds further in showing how division can be avoided by multiplying with the reciprocal of the divisor. He also elaborates on the "additive" and "subtractive" methods for multiplication. He states the rule that in division, the result remains unchanged if both the dividend and the divisor are multiplied or divided by the same quantity. The examples he gives include the following:

$$(a)\quad 274 \div 6.25 = (274 \times 16) \div (6.25 \times 16)$$
$$= 2.74 \times 16$$
$$= 27.4 + (2.74 \times 6)$$

$$(b)\quad 342 \times 56 = \frac{342}{2} \times 112$$
$$= \tfrac{1}{2}[34200 + 3420 + (342 \times 2)]$$

$$(c)\quad 247 \times 1.95 = 247 \times 1.3 \times 1.5$$
$$= [247 + (247 \times 0.3)] \times 1.5$$
$$= [247 + (247 \times 0.3)] + [247 + (247 \times 0.3)] \times 0.5$$

$$(d)\quad 107 \times 10600 = 1070000 + (107 \times 600)$$

$$(e)\quad 19152 \div 56 = \frac{19152}{4} \div \frac{56}{4}$$
$$= 4788 \div 14$$

$$(f)\quad 9731 \div 37 = (9731 \times 3) \div (37 \times 3)$$
$$= 29193 \div 111$$

The "subtractive" method for division is applied to cases (*e*) and (*f*). The steps for solving (*e*) are shown below:

```
            3 4 2
     14 4 7 8 8
        3 0
        1 7 8 8
        1 2
          5 8 8
          4 0
          1 8 8
          1 6
            2 8
            2 0
              8
              8
```

Yang Hui also states the rule that in multiplication, the result remains unchanged if the multiplicand is multiplied by a number and the multiplier by the reciprocal of the same number. Then he shows how to apply the rule to make the methods of "additive" and "subtractive" multiplication applicable. For example, $237 \times 56 = \frac{237}{2} \times (56 \times 2) = 118.5 \times 112 = 11850 + 1185 + (118.5 \times 2)$. The last part of the volume contains the division tables, the first instance of such tables in Chinese mathematical texts. They were later used by the Chinese in division operations involving the abacus.

The last volume of the *Ch'eng-ch'u t'ung-pien pen-mo* (the *Fa-suan ch'ü-yung pen-mo*) gives various rapid methods for multiplication and division for multipliers and divisors from 2 to 300 that are based on the rules described in the first two volumes. For example, when the multiplier is 228, Yang Hui and his collaborator Shih Chung-yung recommend the use of the factors 12 and 19 and the successive application of the "additive" method of multiplication; and when the multiplier is 125, they recommend shifting the multiplicand three places to the left and then halving it three times successively.

The *T'ien-mou pi-lei ch'eng-ch'u chieh-fa*, interesting mainly for its theory of equations, consists of two chapters. The first begins with a method for finding the area of a rectangular farm that is extended to problems involving other measures—weights, lengths, volumes, and money. These problems indicate that the length measurements for the sides of a rectangle can be employed as "dummy variables." Yang Hui was hence on the path leading to algebra, although neither he nor his Chinese contemporaries made extensive use of symbols. The text shows that Yang Hui had a highly developed conception of decimal places, simplified certain divisions by multiplication with reciprocals, and avoided the use of common fractions and showed his preference for decimal fractions. The words *chieh-fa* (literally, "shorter method") in the title must have referred to these and other simplified methods that he introduced. Three different values for the ratio of the circumference to the diameter of a circle are used: 3, 22/7, and 3.14. The rest of the first chapter deals with the area of the annulus, the isosceles triangle, and the trapezium; series; and arithmetic progressions exemplified by problems involving bundles of arrows with either square or circular cross sections.

The second chapter of the *T'ien-mou pi-lei ch'eng-ch'u chieh-fa* contains the earliest explanations of the Chinese methods for solving quadratic equations. For equations of the type $x^2 + 12x = 864$, Yang Hui recommends the *tai tsung k'ai fang* method, literally the method of extracting the root by attaching a side rectangle (*tsung*). The constant 864 is called *chi* ("total area"). If $x = 10x_1 + x_2$, $10x_1$ is called the *ch'u shang* ("first deliberation") and x_2 the *tz'u shang* ("second deliberation"), then $(10x_1)^2$ is the *fang fa*, x_2^2 the *yü*, and $10x_1x_2$ the *lien*. Also, $12x = 12(10x_1 + x_2)$, with $12(10x_1)$ being called *tsung fang* and $12x_2$ the *tsung*. Five rows on the counting board are used: *shang, chih, fang-fa, tsung-fang,* and *yü* or *yü suan,* in descending order. The constant 864 is first placed in the second row (*chih*), then the coefficient of x on the fourth row (*tsung-fang*), and the coefficient for x^2

on the last row (*yü suan*). The coefficient of x is moved one place to the left and that of x^2 two places to the left, as shown in Figure 2a. The value of x lies between 20 and 30. The number 20, called the *ch'u shang*, is placed on the top row. Taking the number 2 of the *ch'u shang* as the multiplier, the product with the *yü suan* is 20. This is entered in the third row (*fang-fa*), as shown in Figure 2b. The number 2 of the *ch'u shang* is again used as the multiplier to find the products of the *fang-fa* and the *tsung-fang*. The sum of these two products (640) is subtracted from the *chih*, giving a remainder 224. The third row, now known as *lien*, and the fourth row, now known as *tsung*, are moved one place to the right, while the *yü suan* is moved by two places, as shown in Figure 2c.

shang			
chih	8	6	4
tsung-fang	1	2	
yü suan	1	0	

FIGURE 2a

shang	2		
chih	8	6	4
fang-fa	2	0	
tsung-fang	1	2	
yü suan	1	0	

FIGURE 2b

shang		2	
chih	2	2	4
lien		4	0
tsung		1	2
yü suan			1

FIGURE 2c

shang		2	4
chih	2	2	4
lien yü		4	4
tsung		1	2
yü suan			1

FIGURE 2d

The "second deliberation" (*tz'u shang*) is found to be 4. This is placed in the first row after the number 2. The product of the "second deliberation" and the *yü suan*, called *yü*, is added to the third row, which then becomes known as the *lien yü* (44 in this case). See Figure 2d. The sum of the products of the "second deliberation" and each of the *lien yü* and the *tsung* (224), when subtracted from the *chih*, leaves no remainder. Hence $x = 24$.

The solution is also illustrated by Yang Hui in a diagram as shown in Figure 3. If $x = (10x_1 + x_2)$, where $x_1 = 2$ and $x_2 = 4$, then from the equation $(10x_1 + x_2)^2 + 12(10x_1 + x_2) = 864$ we obtain $100x_1^2 + 20x_1x_2 + x_2^2 + 120x_1 + 12x_2 = 864$. Here the *fang-fa* is given by $100x_1^2$, the two *lien* by $20x_1x_2$, *yü* by x_2^2, *tsung-fang* by $120x_1$, *tsung* by $12x_2$, and the total *chih* by 864.

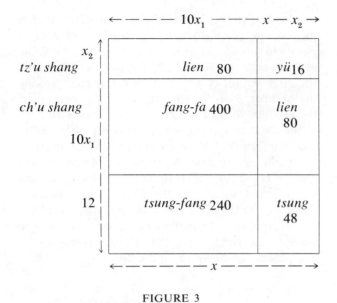

←———— $10x_1$ ———— $x - x_2$ →

x_2

tz'u shang | lien 80 | yü 16

ch'u shang | fang-fa 400 | lien 80

$10x_1$

12 | tsung-fang 240 | tsung 48

←—————— x ——————→

FIGURE 3

For equations of the type $x^2 - 12x = 864$, Yang Hui gives two different methods: the *i chi k'ai fang* (extracting the root by increasing the area) and the *chien ts'ung k'ai fang* (extracting the root by detaching a side rectangle, or *ts'ung*). For finding the smaller root of an equation of the type $-x^2 + 60x = 84$, Yang Hui recommends either of the two methods of *i yu* ("adding a corner square") and *chien ts'ung* ("detaching a side rectangle"). Finally, for finding the larger root of the same equation, he gives the *fan chi* ("inverted area") method. The geometrical illustrations of these methods suggest that means for solving quadratic equations may have been first derived geometrically by Yang Hui. Negative roots are also discussed, and the general solutions given are similar to Horner's method.

In the second chapter of his *T'ien-mou pi-lei ch'eng-ch'u chieh-fa*, Yang Hui also describes a method of solving equations of the type $x^4 = c$ as given by Liu I and Chia Hsien. There are also quadratic equations in which either the product and the difference of the two roots, or the product and the sum of the two roots, are given, such as an equation in the form

$$(x+y)^2 = (x-y)^2 + 4xy.$$

In addition he gives a general formula for the positive roots in the form

$$x = \frac{-b \pm \sqrt{(b^2 - 4ac)}}{2a}.$$

The methods used by Yang Hui for solving quadratic equations appear to be more flexible than those used in the West. He also demonstrated how to solve a biquadratic equation of the form

$$-5x^4 + 52x^3 + 128x^2 = 4096$$

by means of the *san-ch'eng-fang* ("quartic root") method, which is very similar to the method rediscovered in the early nineteenth century by Horner and Ruffini. The rest of the chapter deals with dissection of areas: a rectangle cut off from a larger isosceles triangle, a trapezium cut away from a larger trapezium, and an annulus cut off from a larger annulus.

The *Hsü-ku chai-ch'i suan-fa* consists of two chapters. The first has recently aroused considerable interest because of its magic squares. It is found only in the rare Sung edition and is missing from the *I-chia-t'ang ts'ung-shu* version, which is more commonly available. The *Hsü-ku chai-ch'i suan-fa* is the earliest Chinese text extant that gives magic squares higher than the third order and magic circles as well. In the preface Yang Hui disclaims originality in regard to its contents, saying that the material was among the old manuscripts and mathematical texts brought to him by his friends Liu Pi-chien and Ch'iu Hsü-chü. After Yang Hui, magic squares were discussed by Ch'eng Ta-wei in his *Suan-fa t'ung-tsung* (1593), by Fang Chung-t'ung in his *Shu-tu yen* (1661), by Chang Ch'ao in his *Hsin-chai tsa tsu* (*ca.* 1670), and by Pao Ch'i-shou in his *Pi-nai-shan-fang chi* (*ca.* 1880). In 1935 Li Yen published a paper on Chinese magic squares and reproduced the entire section on magic squares in Yang Hui's book, but with some misprints. In 1959 a subsection on magic squares was included in volume III of Joseph Needham's *Science and Civilisation in China*. Through the courtesy of Li Yen, a microfilm of his personal copy of the Sung edition of *Yang Hui suan-fa*, containing the full chapter on magic squares, was obtained for the preparation of Lam Lay Yong's doctoral dissertation. Some of Yang Hui's magic squares are shown in Figure 4a–4h.

④ ⑨ ②

③ ⑤ ⑦

⑧ ① ⑥

2	16	13	3
11	5	8	10
7	9	12	6
14	4	1	15

FIGURE 4a. The *lo-shu*, magic square of order 3.

FIGURE 4b. One of the two magic squares of order 4.

12	27	33	23	10
28	18	13	26	20
11	25	21	17	31
22	16	29	24	14
32	19	9	15	30

FIGURE 4c. One of the two magic squares of order 5.

13	22	18	27	11	20
31	4	36	9	29	2
12	21	14	23	16	25
30	3	5	32	34	7
17	26	10	19	15	24
8	35	28	1	6	33

FIGURE 4d. One of the two magic squares of order 6.

46	8	16	20	29	7	49
3	40	35	36	18	41	2
44	12	33	23	19	38	6
28	26	11	25	39	24	22
5	7	31	27	17	13	45
48	9	15	14	32	10	47
1	43	34	30	21	42	4

FIGURE 4e. One of the two magic squares of order 7.

61	3	2	64	57	7	6	60
12	54	55	9	16	50	51	13
20	46	47	17	24	42	43	21
37	27	26	40	33	31	30	36
29	35	34	32	25	39	38	28
44	22	23	41	48	18	19	45
52	14	15	49	56	10	11	53
5	59	58	8	1	63	62	4

FIGURE 4f. One of the two magic squares of order 8.

31	76	13	36	81	18	29	74	11
22	40	58	27	45	63	20	38	56
67	4	49	72	9	54	65	2	47
30	75	12	32	77	14	34	79	16
21	39	57	23	41	59	25	43	61
66	3	48	68	5	50	70	7	52
35	80	17	28	73	10	33	78	15
26	44	62	19	37	55	24	42	60
71	8	53	64	1	46	69	6	51

FIGURE 4g. Magic square of order 9 (the number 43 is incorrectly written in Seki Takakazu's copy as 42).

1	20	21	40	41	60	61	80	81	100
99	82	79	62	59	42	39	22	19	2
3	18	23	38	43	58	63	78	83	98
97	84	77	64	57	44	37	24	17	4
5	16	25	36	45	56	65	76	85	96
95	86	75	66	55	46	35	26	15	6
14	7	34	27	54	47	74	67	94	87
88	93	68	73	48	53	28	33	8	13
12	9	32	29	52	49	72	69	92	89
91	90	71	70	51	50	31	30	11	10

FIGURE 4h. Magic square of order 10.

Besides magic squares and magic circles, the first chapter deals with problems on indeterminate analysis, calendar computation, geometrical progressions, and volumes and areas of objects of various regular shapes. For indeterminate analysis Yang Hui also gives the common names used in his time, such as *Ch'in Wang an tien ping* ("the Prince of Ch'in's secret method of counting soldiers"), *chien kuan shu* ("method of cutting lengths of tube"), and *fu shê chih shu* ("method of repeating trials"). (For further details on the Chinese method of indeterminate analysis, see "Ch'in Chiushao.")

The first problem of the second chapter reads: "A number of pheasants and rabbits are placed together in the same cage. Thirty-five heads and ninety-four feet are counted. Find the number of each." Besides simultaneous linear equations of two unknowns, this chapter deals with three unknowns. The chapter then considers miscellaneous examples taken from several mathematical texts, including Liu Hui's *Chiu-chang suan-shu* and *Hai-tao suan-ching*, the *Sun-tzu suan-ching*, the *Chang Ch'iu-chien suan-ching*, the *Ying-yung suan-fa*, the *Chih-nan suan-fa*, and the *Pien-ku t'ung-yuan*. The last three mathematical texts were printed during the eleventh and twelfth centuries in China, but are now lost. It is only through the works of Yang Hui that fragments of them and some of the other texts printed in the same era are extant. In the last problem of the chapter, Yang Hui gives a detailed analysis of the method employed by Liu Hui in his *Hai-tao suan-ching*. It is interesting that Yang Hui's first publication, the *Hsiang-chieh chiu-chang suan-fa*, is a study of Liu Hui's *Chiu-chang suan-shu*, which had been authoritative in China for about 1,000 years. Thus, with the last problem in his last publication, Yang Hui had completed a total analysis of Liu Hui's writings.

BIBLIOGRAPHY

See Schuyler Cammann, "The Evolution of Magic Squares in China," in *Journal of the American Oriental Society*, **80**, no. 2 (1960), 116; and "Old Chinese Magic Squares," in *Sinologica*, 7, no. 1 (1962), 14; Ch'ien Pao-tsung, *Chung-kuo Shu-hsüeh-shih* (Peking, 1964); Ch'ien Pao-tsung *et al.*, *Sung Yuan shu-hsüeh-shih lun-wen-chi* (Peking, 1966); Hsü Shun-fang, *Chung-suan-chia ti tai-shu-hsüeh yen chiu* (Peking, 1952); Lam Lay Yong, *The Yang Hui Suan Fa, a Thirteenth-Century Chinese Mathematical Treatise* (Singapore); Li Yen, "Chung-suan-shih lun-ts'ung," in *Gesammelte Abhandlungen über die Geschichte der Chinesischen Mathematik*, II and III (Shanghai, 1935); and *Chung-kuo suan-hsüeh-shih* (Shanghai, 1937; rev. ed., 1955); Li Yen and Tu Shih-jan, *Chung-kuo ku-tai shu-hsüeh chien-shih*, II (Peking, 1964); Yoshio Mikami, *Mathematics in China and Japan* (1913; repr. New York, 1961); and Joseph Needham, *Science and Civilisation in China*, III (Cambridge, 1959).

HO PENG-YOKE

IBN YA'QŪB. See **Ibrāhīm ibn Ya'qūb al-Isrā'īlī al-Turṭushi.**

YA'QŪB IBN ISḤĀQ. See **Al-Kindī, Abū Yūsuf Ya'qūb ibn Isḥāq al-Ṣabbāḥ.**

YA'QŪB IBN ṬĀRIQ (*fl.* Baghdad, second half of eighth century), *astronomy.*

Ya'qūb ibn Ṭāriq was the astronomer most closely connected with al-Fazārī in introducing the *Zīj al-Sindhind* to Islamic scientists; he seems, in fact, to have collaborated personally with the Indian astronomer who came to Baghdad with an embassy from Sind in 771 or 773. The most important of his works in this connection were *Zīj maḥlūl fī al-Sindhind li daraja daraja* ("Astronomical Tables in the Sindhind Resolved for Every Degree"), *Tarkīb al-aflāk* ("Composition of the Spheres"), and *Kitāb al-'ilal* ("Book of Causes").

Evidently the most prominent feature of the *Zīj* was that the interval between the entries in the columns of arguments for the tables was one degree. Its basic parameters were very similar to those of the *Zīj al-Sindhind al-kabīr* of al-Fazārī, except that Ya'qūb completely accepted the equa-

tions of the center from the *Zīj al-Shāh* while mixing in some equations of the anomaly from the *Zīj al-Arkand* (the *ārdharātrika* system; see essay in Supplement).

In the *Tarkīb al-aflāk*, Ya'qūb also drew upon the *Zīj al-Sindhind* and the *Zīj al-Arkand*, as well as on his conversations with the Indian astronomer. The subjects covered in this work, insofar as they can be determined, were the geocentric distances of the planetary orbits, geography, the computation of the *ahargaṇa*, and perhaps the geometric models of planetary motion. The *Kitāb al-'ilal* is known only from citations by al-Bīrūnī in his work *On Shadows*; the extant fragments deal exclusively with rules for employing the gnomon. Like al-Fazārī, Ya'qūb is inconsistent, adopting whatever formula comes to hand without regard for its relation to the other formulas in his books. Also like al-Fazārī, he is significant primarily for his role in the transmission of Indian science to Islam.

BIBLIOGRAPHY

The fragments by Ya'qūb ibn Ṭāriq are collected and discussed in D. Pingree, "The Fragments of the Works of Ya'qūb ibn Ṭāriq," in *Journal of Near Eastern Studies*, **27** (1968), 97–125; and in E. S. Kennedy, "The Lunar Visibility Theory of Ya'qūb ibn Ṭāriq," *ibid.*, 126–132.

DAVID PINGREE

YĀQŪT AL-ḤAMAWĪ AL-RŪMĪ, SHIHĀB AL-DĪN ABŪ 'ABDALLĀH YĀQŪT IBN 'ABD AL-LĀH (*b.* Rūm, Byzantine empire [now Turkey], 1179; *d.* Aleppo, Syria, 20 August 1229), *transmission of knowledge, geography.*

Probably of Greek parentage, Yāqūt was taken prisoner as a young boy and was brought to Baghdad, where he was sold as a slave to a merchant named 'Askar ibn Ibrāhīm al-Ḥamawī (after whom Yāqūt was also called al-Ḥamawī; he assumed his other names later). When Yāqūt grew up, 'Askar gave him a little education; for being uneducated himself, he needed someone to assume the secretarial work of his trade. He subsequently engaged Yāqūt and sent him on commercial tours to what is now Qeys Island in the Persian Gulf and to Syria. In 1199, after a disagreement, Yāqūt was freed by his master; he then copied and sold manuscripts and studied Arabic language and grammar under al-'Ukbarī (*d.* 1219) and Ibn Ya'īsh (*d.* 1245). After a reconciliation, Yāqūt rejoined his former master in the latter's trade activities. After ʿAskar's death, Yāqūt settled in Baghdad as a bookseller. He held strong Kharijite views that he expressed in Damascus (1215) during a public argument with a supporter of ʿAlī ibn Abī Ṭālib. The crowd could not tolerate his attack on ʿAlī and assaulted him. He later escaped from Aleppo; went to Mosul; and then, by way of Irbil, reached Marw, where he remained for two years, consulting the rich libraries and collecting material for his books. Toward the end of 1218, Yāqūt went to Khorezm; he encountered the invading Mongol armies in 1219–1220 and escaped to Khurāsān, leaving behind all his belongings. In 1220 he reached Mosul and in 1222 finally arrived in Aleppo, where he remained under the patronage of Abuʾl-Ḥasan ʿAlī ibn Yūsuf al-Qifṭī (d. 1248) until his death. Yāqūt spent most of his life traveling in the Islamic world: Syria, Palestine, Egypt, Iran, Iraq, Khurāsān, and Khorezm.

Besides earning his livelihood as a bookseller, Yāqūt seems to have spent much time as an author. Only four of his several known works have been discovered: *Muʿjam al-buldān* ("Dictionary of the Lands"); *Kitāb irshād al-arīb ilā maʿrifat al-adīb* (or *Irshād al-alibbāʾ ilā maʿrifat al-udabāʾ*) known as *Muʿjam al-udabā* (or *Ṭabaqāt al-udabāʾ*) ("Dictionary of the Learned Men"); *Kitāb al-mushtarik waḍʿan waʾl-mukhtalif ṣaqʿan*, containing selections from *Muʿjam al-buldān* that list only the titles that applied to several places that had the same name; and *Al-Muqtaḍab min kitāb jamharat al-nasab* (on the genealogy of the Arabs). Among his other works are *Kitāb al-mabdaʾ waʾl-maʾāl* and *Kitāb al-duʾal* (both on history); *Akhbār al-shuʿarāʾ al-mutaʾakhkhirīn waʾl-qudamāʾ*, probably identical with Yāqūt's *Muʿjam al-shuʿarāʾ*, a biographical dictionary of poets in forty-two volumes; *Kitāb Akhbār al-Mutanabbī*, on the life of the poet al-Mutanabbī; *Majmūʿ kalām Abī ʿAlī al-Fārisī*, a collection of the sayings of al-Fārisī (d. 987); and *ʿUnwān kitāb al-aghānī*, probably an introduction to the famous *Book of Songs* by Abuʾl-Faraj al-Iṣfahānī (d. 967).

In the absence of his works on history, it is difficult to judge Yāqūt's ability as a historian; but as a biographer he was one of the outstanding scholars of medieval Islam, possessing encyclopedic knowledge. Distinguishing a man of letters (*adīb*) from a scholar (*ʿālim*), he quotes a saying: ". . . the man of letters selects the choicest from everything and then composes it, while a scholar selects a particular branch of knowledge and then improves upon it."[1] For Yāqūt the science of *akhbār* ("usages of the Prophet") was the fountainhead of all knowledge and wisdom, and was superior to all other sciences. Quoting Abuʾl-Ḥasan ʿAlī ibn al-Ḥasan, who quoted earlier writers on the subject, Yāqūt says that if scholars had not concerned themselves with *akhbār* and with the works of *ʾāthār* (usages of the companions of the Prophet), the beginnings of knowledge would have become corrupt and its ends would have perished. It had been said since ancient times, he points out, that *nasab* (genealogy) and *akhbār* were the sciences of the kings and the nobility.[2] With these precepts in mind, Yāqūt compiled his monumental *Dictionary of the Learned Men,* which covers the lives of men of letters, grammarians, linguists, genealogists, famous readers (of the Koran), historians, and secretaries, with preference given to poets; authors of prose whose poetry was of a secondary order came later. This work and *Akhbār al-shuʿarāʾ* covered, in Yāqūt's words, most of the information relating to men of letters, including scholars and poets.[3] The work, in more or less alphabetical order, reveals the author's deep interest in Arabic literature and his vast knowledge of the subject.

Equally concerned with geography, Yāqūt believed in its intrinsic relationship with history and emphasized the importance of orthography of place names. Arranged in alphabetical order, the *Muʿjam al-buldān* attempts to fix the spellings of place names and gives their geographical positions, boundaries, and coordinates. It covers cities and towns, rivers and valleys, mountains and deserts, and seas and islands. For each place Yāqūt gives information on eminent natives, including anecdotes and interesting facts. He was aware that earlier writers had not paid sufficient attention to the two important aspects of a place name: orthography and geographical position of the place. This earlier inattention often misled scholars and men of letters.[4] The inspiration to produce a geographical dictionary that would serve as a travel guide to Muslims, however, came from the teachings of the Koran and other religious works.[5] Yāqūt believed that such a work was essential not only for the traveler but also for the jurist, the theologian, the historian, the physician, the astrologer, and the savant.[6]

Besides utilizing a variety of sources for this work, including biographers, geographers, and historians, Yāqūt enhanced its value by adding his own experiences and observations gathered during his travels, as well as information acquired from those he met. An important aspect of the work is that Yāqūt preserved a number of passages from works that have only recently become available. That he was fully conversant with the various concepts of Muslim geographers relating to mathematical, physical, and regional geography is amply evident in the long introduction, which includes discussions of the geographical and legal terms used in the book.[7]

Ever since its publication more than seven centuries ago, the *Muʿjam* has been an important historical and geographical reference work for scholars in the Islamic world as well as for Orientalists in the West. Because of its size, an abridgment was prepared in the fourteenth century by ʿAbd al-Muʾmin ibn ʿAbd al-Ḥaqq, under the title *Marāṣid al-iṭṭilāʿ ʿalā asmāʾ al-amkina waʾl-biqāʿ*, that included only the geographical material.

Yāqūt represented an age when knowledge in medieval Islam had almost reached its zenith. It was a period of consolidation of the knowledge acquired during the preceding centuries by the Muslim scientists, and scholars like Yāqūt had begun producing comprehensive dictionaries, biographies, and general surveys in specific aspects of the arts and sciences. It was about this time that the center of Muslim academic and intellectual activity had shifted from Baghdad, which played its role for over four centuries, to places like Aleppo, Damascus, and Cairo. It is for this reason that we find in Yāqūt's works a variety of information, including ethnology, folklore, literature, and other features of medieval Islamic society. In this respect, Yāqūt may be considered one of the most outstanding transmitters of knowledge of the medieval period.

NOTES

1. See D. S. Margoliouth, *Yāqūt's Dictionary of the Learned Men*, I, 17.
2. *Ibid.*, 27–28.
3. *Ibid.*, 5–6.
4. See Wadie Jwaideh, *The Introductory Chapters of Yāqūt's Muʿjam al-buldān*, 3–4.
5. *Ibid.*, 1–3.
6. *Ibid.*, 4–9.
7. *Ibid.*; 19 f.

BIBLIOGRAPHY

I. ORIGINAL WORKS. The *Kitāb al-mushtarik waḍʿan waʾl mukhtalif ṣaqʿan* was edited by F. Wüstenfeld as *Jacut's Moschtarik, das ist: Lexicon geographischer Homonyme* (Göttingen, 1846).

Muʿjam al-buldān was edited by F. Wüstenfeld as *Jacut's geographisches Wörterbuch, Kitāb Muʿjam al-buldān*, 6 vols. (Leipzig, 1866–1873); see also *Sachin-dex zu Wüstenfeld's Ausgabe von Jaqut's Muʿgam al-buldān* (Stuttgart, 1928). Other eds. of the *Muʿjam* are F. J. Heer, *Die historischen und geographischen Quellen in Jāqūt's geographischem Wörterbuch* (Strasbourg, 1898); a version by Aḥmad Amīn al-Shanqīṭī, 8 vols. (Cairo, 1906), with a 2-vol. supp. by Muḥammad Amīn al-Khānjī (Cairo, 1907); and Wadie Jwaideh, *The Introductory Chapters of Yāqūt's Muʿjam al-buldān* (Leiden, 1959), an excellent analysis of the intro. The abridgment of this work, *Marāṣid al-iṭṭilāʿ ʿalā asmāʾ al-amkina waʾl-biqāʿ*, was edited by T. W. Juynboll, 6 vols. (Leiden, 1851–1864); other abridgments are listed in Jwaideh, ix–x.

Kitāb irshād al-arīb ilā maʿrifat al-adīb (Muʿjam al-udabā) was edited by D. S. Margoliouth as *Yāqūt's Dictionary of the Learned Men*, no. 6 in the E. J. W. Gibb Memorial Series, 6 vols. (Leiden, 1907–1931).

Al-Muqtaḍab min kitāb jamharat al-nasab is discussed in Fuat Sezgin, *Geschichte des arabischen Schrifttums*, I (Leiden, 1967), 269.

II. SECONDARY LITERATURE. See R. Blachère, "Yāqūt al-Rūmī," in *Encyclopaedia of Islam*, 1st ed., IV, pt. 2, 1153–1154; C. Brockelmann, *Geschichte der arabischen Literatur*, I (Leiden, 1943), 480, and *Supp.*, I (Leiden, 1937), 880; Ḥājjī Khalīfa, *Kashf al-ẓunūn*, I (Istanbul, 1941), 64, 363, and II (Istanbul, 1943), 1580, 1691–1692, 1733–1734, 1734–1735, 1793; *Ibn Khallikan's Biographical Dictionary*, translated by Baron MacGuckin de Slane, IV (Paris, 1871), 9–22; I. Y. Krachkovsky, *Arabskaya geograficheskaya literatura*, IV (Moscow–Leningrad, 1957), 330–342, Arabic trans. by Ṣalāḥ al-Dīn ʿUthmān Hāshim as *Taʾrīkh al-abab al-jughrāfī al-ʿArabī*, I (Cairo, 1963), 335–344; and Ibn al-Qifṭī, *Inbāh al-ruwāh*, facs. Arabic text concerning Yāqūt, published by Rudolf Sellheim in his "Neue Materialien zur Biographie des Yāqūt," in Wolfgang Voigt, ed., *Forschungen und Fortschritte der Katalogisierung der orientalischen Handschriften in Deutschland* (Wiesbaden, 1966), tables xvi–xxxiv.

S. MAQBUL AHMAD

YATIVṚṢABHA (*fl.* India, sixth century), *cosmography, mathematics.*

Yativṛṣabha (in Prākrit, Jadivasaha) was a Jain author who studied under Ārya Mañkṣu and Nāgahastin, and compiled several works in Prākrit expounding Jain traditions. One of these, the *Tiloyapaṇṇattī*, a description of the universe and its parts, is of some importance to historians of Indian science because it incorporates formulas representative of developments in Jain mathematics between the older canonical works and the later texts of the ninth and following centuries. Unfortunately, almost nothing is known of Yativṛṣabha himself; his lifetime is fixed by the general character of his work, as well as by his references to the *Loyav-*

ibhāga (probably that written by Sarvanandin in 458) and by the reference to him by Jinabhadra Kṣamāśramaṇa (*fl.* 609). Yativṛṣabha's statement that the end of the Gupta dynasty occurred after 231 years of rule must refer to a contemporary event in Gupta Era 231 (A.D. 551).

BIBLIOGRAPHY

The *Tiloyapaṇṇattī* was edited by Ādinātha Upādhyāya and Hīrālāla Jaina with a Hindi paraphrase by Balchandra, 2 vols. (Śolāpura, 1943–1951; I, 2nd ed., 1956). There is a study in Hindi of the mathematics of Yativṛṣabha by Lakṣmīcandra Jaina, *Tiloyapaṇṇattī kā gaṇita* (Śolāpura, 1958).

DAVID PINGREE

YAVANEŚVARA (*fl.* western India, 149/150), *astrology, astronomy.*

The word Yavaneśvara is a title (meaning "Lord of the Greeks"), not a proper name; it or its equivalent, Yavanarāja, was borne by several officials in western India during the rule of the Western Kṣatrapas (*ca.* 78–390), and perhaps earlier under Aśoka, in the middle of the third century B.C. Their function was evidently to act as leader of the Greek merchants settled in the area. The one in whom we are interested was responsible for the translation of a Greek text on astrology into Sanskrit prose in 149/150, during the reign of Rudradāman, the most powerful of the Western Kṣatrapas. The Greek original was composed in Egypt (probably at Alexandria) in the first half of the second century B.C., and therefore is one of the earliest Greek astrological texts known to us in a substantially complete form, although not in the original. Unfortunately, even Yavaneśvara's prose translation is no longer available, and we must be content with the versification of it made by Sphujidhvaja in 269/270.

In translating this text Yavaneśvara did his best to make it appeal to an Indian audience. He interpreted illustrations of the deities of the Decans and Horās that appeared in the Greek manuscript, for instance, in terms of Śaivite iconography; and he introduced the caste system and some elements of the older Indian astral-omen texts and of *āyurveda* into his work. Yavaneśvara also included a version of a Greek adaptation of Babylonian planetary theory, in an attempt to make it possible for Indians to become astrologers (see essay in Supplement). He was extremely successful, and the basic methodology of all of Indian horoscopy can be traced back to his translation and another, lost translation from the Greek that was known to Satya (in the third century?).

BIBLIOGRAPHY

An ed. of the *Yavanajātaka* of Sphujidhvaja by D. Pingree, in the Harvard Oriental Series, will contain all available information about Yavaneśvara.

DAVID PINGREE

YERKES, ROBERT MEARNS (*b.* Breadysville, Pennsylvania, 26 May 1876; *d.* New Haven, Connecticut, 3 February 1956), *comparative psychology.*

Yerkes was the oldest child of Silas Marshall Yerkes and Susanna Addis Carrell Yerkes, members of well-established farm families residing north of Philadelphia. As a boy he lived close to nature, and his familiarity with domestic and wild animals eventually contributed to his vocational choice. He attended an ungraded rural school and at the age of fifteen entered the State Normal School at West Chester, transferring in 1892 to Ursinus College, first as a preparatory and then as a collegiate student (A.B., 1897). In 1897 Yerkes entered Harvard, where he was awarded an A.B. after one year and proceeded into graduate work without pause, working with Hugo Münsterberg. He received the A.M. in 1899 and the Ph.D., in psychology, in 1902. By the latter year he had already launched his scientific career with publications and was appointed instructor in psychology at Harvard.

Yerkes was one of the first of a new breed of comparative psychologists who worked with their animal subjects in a laboratory setting. He was preceded by only a few years by E. L. Thorndike (whom he assisted one summer at Woods Hole) and was a collaborator (mostly by mail) with a slightly younger man, John B. Watson. Thorndike's work led him into learning theory and educational psychology. Watson's researches led him into methodological reformism and antimentalism in general psychology. Yerkes, in contrast with these eminent peers, recognized the kinship of the animal psyche to that of man; but whatever his speculations on that kinship, in his research he studied animal behavior for its own sake, assuming that he was dealing with animal mentality as such. Relatively early in his career he was convinced that observing the animals with mental processes most like those of man, the apes, would be of great importance to psychology. He envisaged the founding of an institute in which these animals might be studied, but decades passed before he realized this project.

Yerkes became the most versatile psychologist of his generation, responding in large part to the stimulation of the Harvard environment in the years that he spent there as a faculty member, until 1917. Under the aegis of Harvard psychiatrist E. E. Southard, Yerkes spent half his time from 1913 to 1917 as psychologist to the Boston Psychopathic Hospital. Although he published relatively little based on that experience, he became deeply involved in mental testing as it was then practiced. As a result, from 1917 to 1919 Yerkes was in effect largely in charge of the psychological testing of U.S. Army personnel in World War I. The large-scale use of psychological devices at that time substantially established the profession of psychology in a way quite different from its tendency toward academic isolation before the war. Yerkes, through his promotional and organizing abilities, was one of the key figures responsible for the conception and execution of the program.

Although appointed chairman of the psychology department at the University of Minnesota in 1917, Yerkes never took up residence in Minneapolis. From 1919 to 1924 he stayed in Washington with the National Research Council and continued to exploit his ability to facilitate the work of others. He headed N.R.C. programs to explore the characteristics of various types of humans who migrate (with immigration-exclusion legislation in mind) and to encourage and support scientific investigation of sexuality. He also was elected president of the American Psychological Association for 1916–1917, while still an assistant professor at Harvard. Yerkes thus spent the period of the war and afterward in the heart of the American scientific establishment, helping to make policy and channel funds. Although he did not feel himself to be one of the scientist-politicians, he served on committees and advisory groups and exerted great influence on the institutions and programs of American science in the 1920's and 1930's.

Not until 1924 did Yerkes return to his career as comparative psychologist, when Yale offered him an opportunity to join its new Institute of Psychology and devote himself primarily to advanced work with nonhuman primates. By 1929 he had founded an experimental station near Orange Park, Florida, later named the Yerkes Laboratories of Primate Biology, which was the nucleus of the present Yerkes Regional Primate Research Center. As preparation for this work he and his wife, Ada Watterson Yerkes, published a monumental book, *The Great Apes: A Study of Anthropoid Life* (1929), which for decades remained the standard work on the biology and psychology of these mammals.

Yerkes had originally wanted to become a physician, but he was turned away from medicine by his interest in laboratory research. Nevertheless, throughout his life his basic approach to comparative psychology remained close to biology. Indeed, at Yale his appointment ultimately was in the department of physiology. In psychology, Yerkes' formal stance was very traditional—influenced, he said, by the structuralism of E. B. Titchener of Cornell. Yerkes' early research work, dealing with invertebrates, involved testing organic reactions to sensory input. He soon began working with vertebrates, giving attention not only to sensations but also to instinct, imitation, and learning. His classic study of the behavior of the dancing mouse (1907) did much to establish the rat and mouse as standard laboratory animals in psychology. Many of his most important publications in science involved new techniques for psychological investigation. With his wide-ranging interests, Yerkes also was responsible for contributions as diverse as (with S. Morgulis) publishing the first important American notice on Pavlov's work (1909) and a discussion of the application of psychological findings to illumination engineering (1911). In 1915 Yerkes spent half a year with G. V. Hamilton's primate colony in Santa Barbara, California; and after that period, except for the war and N.R.C. service, devoted himself largely to the monkeys and apes. The staff at the Orange Park laboratories employed both experimental and observational methods, mostly utilizing chimpanzees for subjects, and produced a prodigious number of publications. The topics covered involved not only thinking processes and adaptation but also physiology and even medicine.

Yerkes was a determined, persistent leader. A large part of his contribution to science consisted of organizing or directing it. He was, for example, a prime mover in the reform of the structure of the American Psychological Association in 1943. He was founding editor of the *Journal of Animal Behavior* in 1911 and held other important editorial positions. Yerkes felt that he had a limited amount of energy that was always overtaxed, so what resources he had were used efficiently. He retired as director of the Orange Park laboratories in 1941, and in 1944 retired from Yale. He continued to publish, although at a diminishing rate, and was still active in organizing psychology in the World War II defense effort.

BIBLIOGRAPHY

I. ORIGINAL WORKS. The number of Yerkes' published works is very large. The most complete bibliography is in Ernest R. Hilgard, "Robert Mearns Yerkes, May 26, 1876–February 3, 1956," in *Biographical Memoirs. National Academy of Sciences*, **38** (1965), 412–425. Probably no bibliography will ever be complete, including all of the editorial material and book reviews contributed to various journals, especially those he edited. Yerkes published a short autobiography, "Robert Mearns Yerkes, Psychobiologist," in Carl Murchison, ed., *A History of Psychology in Autobiography*, II (Worcester, Mass., 1932), 381–407; and, posthumously, some reminiscences, "Creating a Chimpanzee Community," Roberta W. Yerkes, ed., in *Yale Journal of Biology and Medicine*, **36** (1963), 205–223.

The Yerkes Papers, in the Yale University Medical School Library, constitute one of the chief sources of the history of American psychology in the first half of the twentieth century. They include an unpublished autobiography, a bibliography, and a large number of letters and personal papers covering not only Yerkes and his career but also all of the activities in which he was interested.

II. SECONDARY LITERATURE. The most authoritative account of Yerkes' life is Hilgard's (see above), 385–425. Also see E. G. Boring, "Robert Mearns Yerkes (1876–1956)," in *Yearbook. American Philosophical Society* (1956), 133–140; Leonard Carmichael, "Robert Mearns Yerkes, 1876–1956," in *Psychological Review*, **64** (1957), 1–7; and Richard M. Elliott, "Robert Mearns Yerkes (1876–1956)," in *American Journal of Psychology*, **69** (1956), 487–494.

JOHN C. BURNHAM

YERSIN, ALEXANDRE (*b.* Aubonne, near Lausanne, Switzerland, 23 September 1863; *d.* Nha Trang, Annam, Vietnam, 1 March 1943), *medicine, bacteriology.*

Yersin received his secondary education in Lausanne before entering the Academy there; he subsequently attended the University of Marburg and the Paris Faculty of Medicine. Having cut himself while performing an autopsy on a patient who had died of rabies, Yersin immediately contacted Émile Roux, one of Pasteur's most brilliant pupils, who gave him an injection of a new therapeutic serum that saved his life. This incident brought Yersin into close contact with Roux, who hired him as his assistant in 1888 and with whom he conducted research on rabies. He then worked with Robert Koch, in Berlin, collaborating with the noted microbiologist in his research on the tubercle bacillus. Upon his return to Paris, Yersin began his own research with Roux, at the Institut Pasteur, on the toxic properties of the diphtheria bacillus. In 1889, however, he suddenly embarked as ship's doctor on a steamer bound for Saigon and Manila. He returned to Paris and left again for Indochina; and during three dangerous expeditions into the interior, he discovered the high plateau of Langbiang, where he founded a small colonial village. The area soon became a vacation center for Europeans, and the city of Dalat was developed there. In 1935 the municipal authorities established the Lycée Yersin at Dalat.

In 1894 Yersin became a medical officer in the French colonial service and conducted research on the bubonic plague epidemic that was sweeping through China, in order to determine the measures that should be taken to prevent its spread into Indochina. At a small bacteriological research laboratory, set up for him in Hong Kong, he discovered the plague bacterium, practically at the same time that Kitasato did so independently; and after much work he isolated an effective serum. In 1904 he was recalled to Paris and continued his research at the Institut Pasteur, of which Roux had become director. With Albert Calmette and Amédée Borrel he made the important observation that certain animals can be immunized against the plague through the injection of dead plague bacteria. He then returned to Nha Trang, where a branch of the Institut Pasteur had been established under his direction. There, in modest laboratories, Yersin perfected an antiplague serum that made it possible to reduce the death rate from 90 percent to about 7 percent.

With the assistance of Paul Doumer, then governor-general of Indochina, a medical school was founded at Hanoi; Yersin directed this center of study and research for many years. Through Yersin's work Indochina was able to control the epidemics that beset the country, especially malaria. In recognition of his medical achievements, the French government appointed Yersin honorary director of the Institut Pasteur.

Besides his activity in science and medicine in Indochina, Yersin conducted research in agronomy. His interest in the cultivation of grains and in soil conditions led him to initiate a series of ecological studies. He also reflected on the natural history of Indochina, having become fascinated by the flora and fauna of his adopted country. Yersin became deeply concerned over the needs of the sick and the poor and fought hard against the exploitation of the lower classes.

BIBLIOGRAPHY

I. ORIGINAL WORKS. Yersin's dissertation for the M.D. at Paris is *Sur le développement du tubercule expérimental* (Paris, 1888). Most of his papers were published in *Annales de l'Institut Pasteur*; they include "Contributions à l'étude de la diphtérie," **2** (1888), 629–661; **3** (1889), 273–288; and **4** (1890), 385–426, written with Émile Roux; "La peste bubonique à Hong-Kong," *ibid.*, **8** (1894), 662–667; and "Sur la peste bubonique: Sérothérapie," *ibid.*, **11** (1897), 81–93.

II. SECONDARY LITERATURE. See N. Bernard, "A. Yersin (1863–1943)," in *Annales de l'Institut Pasteur*, **69** (1943), 129–134, with portrait; N. Bernard, P. Hauduroy, and G. Olivier, *Yersin et la peste* (Lausanne, 1944); H. Buess, *Recherches, découvertes et inventions de médecins suisses* (Basel, 1946); and P. Hauduroy, "Les découvertes de Yersin et les méthodes pastoriennes," in *Schweizerische medizinische Wochenschrift*, **24** (1943), 750–751.

P. E. PILET

YOUDEN, WILLIAM JOHN (*b.* Townsville, Australia, 12 April 1900; *d.* Washington, D.C., 31 March 1971), *mathematical statistics*.

Youden was the eldest child of William John Youden, an English engineer, and Margaret Hamilton of Carluke, Scotland. In 1902 the family returned to the father's birthplace—Dover, England—and resided there until 1907, when they left for America. They lived for a while in Connecticut and at Niagara Falls, New York, where Youden attended public school, before moving to Rochester, New York, in 1916. Youden attended the University of Rochester from 1917 to 1921, except for three months in the U.S. Army in 1918, receiving a B.S. in chemical engineering in 1921. He began graduate work in September 1922 at Columbia University, earning an M.A. in chemistry in 1923 and a Ph.D. the following year.

Immediately after receiving his doctorate, Youden joined the Boyce Thompson Institute for Plant Research in Yonkers, New York, as a physical chemist. He held this post until May 1948, when he joined the National Bureau of Standards as assistant chief of the Statistical Engineering Laboratory, Applied Mathematics Division. Three years later he became a consultant on statistical design and analysis of experiments to the chief of this division, a position he retained until his retirement in 1965.

He was an honorary fellow of the Royal Statistical Society (1965), and was awarded the Medal of Freedom in 1964, the 1969 Samuel S. Wilks Memorial Medal of the American Statistical Association, and the 1969 Shewhart Medal of the American Society for Quality Control.

In 1922 Youden married Gladys Baxter of Rochester, New York; they had two sons, William Wallace (1925–1968) and Robert Hamilton, both of whom chose careers in the computer field. In 1938 he married Grethe Hartmann of Copenhagen, Denmark; they had one son, Julius Hartmann, now a teacher in Copenhagen. In 1957 Youden married Didi Stockfleth of the Norwegian Embassy staff in Washington, D.C. Survivors of his immediate family include his widow, Didi; two sons; and eight grandchildren. Youden is buried in the National Cemetery, Gettysburg National Military Park, Gettysburg, Pennsylvania, in deference to his expressed wishes to remain in his adopted country.

Youden began his professional career as a physical chemist. His first paper to exhibit any knowledge of statistical methods, "A Nomogram for Use in Connection With Gutzeit Arsenic Determinations on Apples" (September 1931), was expository in nature: he noted that the differences between repeated determinations of a physicochemical property of a particular biological material reflect not only the "probable error" of the method of chemical analysis but also the "probable error" of the technique of sampling the material under investigation, and then, with guidance from a 1926 paper of W. A. Shewhart, outlined the requisite theory, illustrated its application, furnished a nomogram to facilitate correct evaluation of the precision of a particular procedure, and pointed out statistical errors that marred a number of earlier publications on sampling of apples for determination of arsenical sprayed residue. This paper marks the beginning of Youden's "missionary" efforts to acquaint research workers with statistical methods of value in their work.

About 1928 Youden "obtained one of the 1050 copies . . . of the first edition" of R. A. Fisher's *Statistical Methods . . .* (1925). At that time he was "so discouraged by what was called 'measurement' in biology that [he] was on the point of resigning" his post at Boyce Thompson. "Fisher's book opened a ray of hope," however, and

Youden soon realized that at Boyce Thompson "he had the opportunity to perform agricultural experiments, both in the field and in the greenhouse, and to try out the early experiment designs" and Fisher's new small-sample methods of data analysis. The publicity for the visit of Fisher to Iowa State College in 1931 came to Youden's attention and aroused his curiosity, but he was unable to attend. When Fisher visited Cornell on his way home from Iowa, Youden "drove there . . . to show him an experimental arrangement."[1] During the academic year 1931–1932, he commuted to Columbia University to attend Harold Hotelling's lectures on statistical inference.[2] During the next few years he published a number of mathematical-statistical papers describing the application of statistical techniques to problems arising in studies of seeds, soils, apples, leaves, and trees.

During this period Youden devoted increasing attention to statistical design of experiments, in order to cope with the enormous variability of biological material; he found many of the standard experiment designs developed for agricultural field trials to be directly applicable in greenhouse work, a situation that led to vastly improved precision. Recognizing that the limited number of leaves per plant thwarted the use of Latin square designs to effect precise within-plant comparisons of a large number of proposed treatments for plant ailments, Youden devised new symmetrically balanced, incomplete block designs that had the characteristic "double control" of Latin square designs but not the restriction that each "treatment" (or "variety") must occur once, and only once, in each row and column. He brought these new designs to R. A. Fisher's attention in 1936 and subsequently completed "Use of Incomplete Block Replications in Estimating Tobacco Mosaic Virus" (1937), in which he presented and illustrated the application of four of the new rectangular experimental arrangements. In a subsequent paper (1940) he gave eight additional designs and, for six of these, the complementary designs. Youden's new rectangular experiment designs, called "Youden squares" by R. A. Fisher and F. Yates in the introduction to their *Statistical Tables* . . . (1938, p. 18), were immediately found to be of broad utility in biological and medical research generally; to be applicable but of less value in agricultural field trials; and, with the advent of World War II, to be of great value in the scientific and engineering experimentation connected with research and development. To gain further knowledge of statistical theory and methodology, Youden audited the courses on

mathematical statistics and design of experiments, by Harold Hotelling and Gertrude M. Cox respectively, and took part in a number of statistical seminars at North Carolina State College in 1941.

Youden served as an operations analyst with the U.S. Army Air Force (1942–1945), first in Britain, where he directed a group of civilian scientists seeking to determine the controlling factor in bombing accuracy; in the latter part of the war, he conducted similar studies in India, China, and the Marianas, preparatory to the assault on Japan. He displayed exceptional skill in the invention of novel, and adaptation of standard, statistical tools of experiment design and analysis to cope with problems arising in the study of bombing accuracy.

In a lecture delivered at the National Bureau of Standards in 1947 Youden exposed the fallaciousness of the all-too-common belief that statistical methods—and a statistician—can be of no help unless one has a vast amount of data; demonstrated that one can obtain valuable information on the precision—and even on the accuracy—of a measurement process without making a large number of measurements on any single quantity; and showed how in routine work one can often obtain useful auxiliary information on personnel or equipment at little or no additional cost through skillful preliminary planning of the measurements to be taken.

On joining the National Bureau of Standards in 1948, Youden revealed the advantages of applying carefully selected statistical principles and techniques in the planning stages of the Bureau's experiments or tests. In the course of these demonstrations he noticed, and was one of the first to capitalize on, important differences between experimentation in the biological and agricultural sciences and in the physical and chemical sciences.[3]

Youden began to devise new forms of experimental arrangements, the first of which were his "linked blocks" (1951) and "chain blocks" (1953), developed to take advantage of special circumstances of spectrographic determinations of chemical elements carried out by comparing spectrum lines recorded on photographic plates.[4] In 1952 Youden and W. S. Connor began to exploit a special class of experiment designs having block size 2 in the thermometer, meter-bar and radium-standards calibration programs of the Bureau. The thermometer calibration application involved observation of the reading of each of the two thermometers forming a "block"; the meter-bar application involved observation of only the difference of the lengths of the two meter bars making up a block, and marked the start of the development of a system of "calibration designs" that Youden did not develop further and exploit as a special subclass until 1962; and the radium-standards application, observation of only ratios of pairs of standards. Meanwhile, Youden and J. S. Hunter had originated a class of designs that they termed "partially replicated Latin squares" (1955), to provide a means of checking whether the assumption of additivity of row, column, and treatment effects is valid. In an epochal 1956 address, "Randomization and Experimentation," Youden introduced a technique for constrained randomization that obviates "the difficult position of the statistician who rules against . . . a systematic sequence when advanced on the grounds of convenience and insists on it when it pops out of the hat" (p. 16).

In the early 1960's Youden exploited a class of selected experiment designs with the specific purpose of identifying and estimating the effects of sources of systematic error. "Systematic Errors in Physical Constants" (1961) contains his first "ruggedness test" based on the observation that some of the fractional factorial designs, developed a decade earlier for optimum multifactor experiments, were ready-made for testing the "ruggedness" (insensitivity) of a method of measurement with respect to recognized sources of systematic error and changes of conditions likely to be encountered in practice.

Youden also originated at least three new statistical techniques: an index for rating diagnostic tests (1950), the two-sample chart for graphical diagnosis of interlaboratory test results (1959), and an extreme rank sum test for outliers (1963), devised to test the statistical significance of outlier laboratories in interlaboratory collaborative tests. The two-sample chart has the same advantage as Shewhart's control chart: simplicity of construction, visual pinpointing of trouble spots, and comparative ease of more refined analysis.[5] Although developed in the setting of test methods for properties of materials, it has become a standard tool of the National Conference of Standard Laboratories in its nationwide program of searching for and rectifying systematic differences in the most accurate programs for instrument calibration.[6] This technique is now used in all measurement fields where interlaboratory agreement is important, and the term "Youden plot" specifies not only the plotting technique but also the experimental procedure for sampling the performance of each laboratory through the results obtained on paired test items.[7] In the case of the extreme rank sum test, Youden's

ideas had been anticipated by R. Doornbos and H. J. Prins (1958), but it was characteristic of Youden that he had independently conceived his test primarily as a device to dramatize and clarify the messages contained in experimental results, rather than as a contribution to distribution-free statistical methods.

By his publications and by his example, Youden contributed substantially to the achievement of objectivity in experimentation and to the establishment of more exact standards for drawing scientific conclusions. "Enduring Values," his address as retiring president of the Philosophical Society of Washington, is an exposition of schemes for incorporating investigations of systematic errors into experimental determinations of fundamental physical constants and a plea for efforts by scientists to accumulate objective evidence for the description of the precision and accuracy of their work. Shortly before his death Youden completed the manuscript of another "missionary" effort, *Risk, Choice and Prediction* (1974), formally intended to familiarize students in the seventh grade and above with basic statistical concepts but actually meant "for anyone . . . who wants to learn in a relatively painless way how the concept and techniques of statistics can help us better understand today's complex world" (p. vii).

NOTES

1. The first and last quotations are from W. J. Youden, "Memorial to Sir Ronald Aylmer Fisher," in *Journal of the American Statistical Association*, **57**, no. 300 (Dec. 1962), 727; the others, from Youden's "The Evolution of Designed Experiments," 59.

2. At this time Hotelling was the person in the United States most versed in the Student-Fisher theory of small samples.

3. Of paramount importance, he noted, is the difference in the magnitude of the errors of measurement: in agricultural and biological experimentation unavoidable variation is likely to be large, so the early experiment designs developed for application in these fields compensate by incorporating many determinations of the quantities of principal interest; physical measurements, in contrast, can often be made with high precision and the experimental material usually is comparatively homogeneous, so that the quantities of interest often can be determined with acceptably small standard errors from as few as two or three, or even from a single, indirect determination. Also, in many experimental situations in the physical sciences and engineering, a "block" and the "plots" within a block are sharply and naturally defined, and often are quite distinct; this is in marked contrast with the arbitrary division of a given land area into "blocks" in agricultural field trials, and the subdivision of a block into contiguous "plots." Consequently, various "interactions" commonly present in agricultural field trials are often absent or negligible in physical-science experimentation.

4. "Linked block" designs are incomplete block designs for which every pair of blocks has the same number of treat-

ments in common; they were subsequently shown to be special cases of partially balanced incomplete block designs with two associate classes of triangular type. "Chain block" designs were developed for situations in which the number of treatments considerably exceeds the block size while, within blocks, comparisons are of such high precision that at most two replications are needed, some treatments occurring only once. Chain block designs with two-way elimination of heterogeneity were subsequently devised by Mandel (1954).

5. See, for example, Acheson J. Duncan, *Quality Control and Industrial Statistics*, 3rd ed. (Homewood, Ill., 1965), pt. 4.

6. See *Proceedings of the 1966 Standards Laboratory Conference*, National Bureau of Standards Miscellaneous Publication 291 (Washington, D.C., 1967), 19, 20, 27–29, 42, 45, 48, 51, 61, 62.

7. "Graphical Diagnosis of Interlaboratory Test Results" (May 1959) is the basic reference on the "Youden plot." A condensed version appeared in *Technical News Bulletin. National Bureau of Standards*, **43**, no. 1 (Jan. 1959), 16–18; and its evolution can be followed in four columns in *Industrial and Engineering Chemistry*: "Presentation for Action," **50**, no. 8 (Aug. 1958), 83A–84A; "Product Specifications and Test Procedures," *ibid.*, no. 10 (Oct. 1958), 91A–92A; "Circumstances Alter Cases," *ibid.*, no. 12 (Dec. 1958), 77A–78A; and "What Is Measurement?" **51**, no. 2 (Feb. 1959), 81A–82A.

BIBLIOGRAPHY

I. ORIGINAL WORKS. Brian L. Joiner and Roy H. Wampler, "Bibliography of W. J. Youden," in *Journal of Quality Technology (JQT)*, **4**, no. 1 (Jan. 1972), 62–66, lists 5 books and 110 papers (including book chapters, encyclopedia articles, editorials, and military documents, but excluding the 36 columns "Statistical Design" [see below], 9 book reviews, and a foreword to a book by another author). It appears to be complete and correct, except for omission of the 1969 and 1970 papers noted below, a premature date for his posthumous book, and the unfortunate substitution of "Quality" for the third word in the title "Simplified Statistical Quantity Control" (1963). Dedicated to Youden, this issue of *JQT* also contains a portrait; a biographical essay by Churchill Eisenhart; "Summary and Index for 'Statistical Design,'" prepared by Mary G. Natrella, which covers his column "Statistical Design," in *Industrial and Engineering Chemistry* (1954–1959); and reproductions of several papers and other materials.

Youden's doctoral dissertation, *A New Method for the Gravimetric Determination of Zirconium*, was privately printed (New York, 1924). As a member of the staff of the Boyce Thompson Institute, he published 15 research papers on chemical and biological studies and instrumentation pertinent to the work of the Institute in *Contributions From Boyce Thompson Institute*. The items below that are marked "repr. in *JQT*" were reprinted in *Journal of Quality Technology*, **4**, no. 1; and those marked "repr. in *SP 300-1*," in Harry H. Ku, ed., *Precision Measurement and Calibration: Selected Papers on Statistical Concepts and Procedures*, National Bureau of Standards Special Publication 300, vol. 1 (Washington, D.C., 1969).

Youden's earlier works include "A Nomogram for Use in Connection With Gutzeit Arsenic Determinations on Apples," in *Contributions From Boyce Thompson Institute*, **3**, no. 3 (Sept. 1931), 363–373; "Statistical Analysis of Seed Germination Data Through the Use of the Chi-Square Test," *ibid.*, **4**, no. 2 (June 1932), 219–232; "A Statistical Study of the Local Lesion Method for Estimating Tobacco Mosaic Virus," *ibid.*, **6**, no. 3 (July–Sept. 1934), 437–454, written with Helen P. Beale; "Relation of Virus Concentration to the Number of Lesions Produced," *ibid.*, **7**, no. 1 (Jan.–Mar. 1935), 37–53, written with H. P. Beale and J. D. Guthrie; "Field Trials With Fibre Pots," *ibid.*, **8**, no. 4 (Oct.–Dec. 1936), 317–331, written with P. W. Zimmerman; "Use of Incomplete Block Replications in Estimating Tobacco Mosaic Virus," *ibid.*, **9**, no. 1 (Nov. 1937), 41–48 (repr. in *JQT*), the paper in which the first four members of a new class of rectangular experimental arrangements, now called "Youden squares," appeared—the paper that catapulted him to fame; "Selection of Efficient Methods for Soil Sampling," *ibid.*, 59–70; "Experimental Designs to Increase the Accuracy of Greenhouse Studies," *ibid.*, **11**, no. 3 (Apr.–June 1940), 219–228; "Burette Experiment," part of the lecture "A Statistical Technique for Analytical Data," delivered at the National Bureau of Standards, 29 Apr. 1947, repr. in *JQT* from pp. 344–346, 350, of Churchill Eisenhart, "Some Canons of Sound Experimentation," in *Bulletin de l'Institut international de statistique*, **37**, no. 3 (1960), 339–350; and "Technique for Testing the Accuracy of Analytical Data," in *Analytical Chemistry*, **19**, no. 12 (Dec. 1947), 946–950.

Later works are "Index for Rating Diagnostic Tests," in *Cancer*, **3**, no. 1 (Jan. 1950), 32–35; "Linked Blocks: A New Class of Incomplete Block Designs" (abstract only), in *Biometrics*, **7**, no. 1 (Mar. 1951), 124; *Statistical Methods for Chemists* (New York, 1951); also in Italian (Genoa, 1964), his first book; "Statistical Aspects of Analytical Determinations," in *Analyst* (London), **77**, no. 921 (Dec. 1952), 874–878 (repr. in *JQT*); "The Chain Block Design," in *Biometrics*, **9**, no. 2 (June 1953), 127–140, written with W. S. Connor; "Sets of Three Measurements," in *Scientific Monthly*, **77**, no. 3 (Sept. 1953), 143–147 (repr. in *JQT*); "Making One Measurement Do the Work of Two," in *Chemical Engineering Progress*, **49**, no. 10 (Oct. 1953), 549–552 (repr. in *JQT*), written with W. S. Connor; "New Experimental Designs for Paired Observations," in *Journal of Research of the National Bureau of Standards*, **53**, no. 3 (Sept. 1954), 191–196 (repr. in *SP 300-1*), written with W. S. Connor; "Instrumental Drift," in *Science*, **120**, no. 3121 (22 Oct. 1954), 627–631 (repr. in *SP 300-1*); "Comparison of Four National Radium Standards: Part 2. Statistical Procedures and Survey," in *Journal of Research of the National Bureau of Standards*, **53**, no. 5 (Nov. 1954), 273–275 (repr. in *SP 300-1*), written with W. S. Connor; "Partially Replicated Latin Squares," in *Biometrics*, **11**, no. 4 (Dec. 1955), 399–405, written with J. S. Hunter; "Graphical Diagnosis of Interlaboratory Test Results," in *Industrial Quality Control*, **15**, no. 11 (May 1959), 24–28 (repr. in *JQT* and *SP 300-1*), the basic reference on Youden's two-sample procedure and diagram, collectively known as the "Youden Plot"; and "Measurements Made by Matching With Known Standards," in *Technometrics*, **1**, no. 2 (May 1959), 101–109, written with W. S. Connor and N. C. Severo.

See also *Statistical Design* (Washington, D.C., 1960), a collection of 36 articles published in a column with this title in *Industrial and Engineering Chemistry*, **46**, no. 2 (Feb. 1954)–**51**, no. 12 (Dec. 1959); "Physical Measurements and Experiment Design," in *Colloques internationaux du Centre national de la recherche scientifique* (Paris, 1961), no. 110, le Plan d'Expériences, 115–128 (repr. in *SP-300-1*); "Systematic Errors in Physical Constants," in *Physics Today*, **14**, no. 9 (Sept. 1961), 32–42, repr. in *Technometrics*, **4**, no. 1 [Feb. 1962], 111–123, and in *SP-300-1*; "Experimental Design and ASTM Committees," in *Materials Research and Standards*, **1**, no. 11 (Nov. 1961), 862–867 (repr. in *SP-300-1*); *Experimentation and Measurement* (New York, 1962); "Uncertainties in Calibration," in *I.R.E. Transactions on Instrumentation*, I-11, nos. 3–4 (Dec. 1962), 133–138 (repr. in *JQT* and *SP-300-1*); "Ranking Laboratories by Round-Robin Tests," in *Materials Research and Standards*, **3**, no. 1 (Jan. 1963), 9–13 (repr. in *SP-300-1*); "Measurement Agreement Comparisons," in *Proceedings of the 1962 Standards Laboratory Conference*, National Bureau of Standards Miscellaneous Publication 248 (Washington, D.C., 1963), 147–151 (repr. in *SP-300-1*), the paper that inspired the work of Bose and Cameron (1965, 1967) and Eicke and Cameron (1967) on "calibration designs." "The Collaborative Test," in *Journal of the Association of Official Agricultural Chemists*, **46**, no. 1 (Feb. 1963), 55–62 (repr. in *SP-300-1*); and "The Evolution of Designed Experiments," in *Proceedings of the [1963] IBM Scientific Computing Symposium on Statistics* (White Plains, N.Y., 1965), 59–67 (repr. in *JQT*).

Additional works are *Statistical Techniques for Collaborative Tests* (Washington, D.C., 1967); "How Mathematics Appraises Risks and Gambles," in T. L. Saaty and F. J. Weyl, eds., *The Spirit and Uses of Mathematics* (New York, 1969), 167–187; "A Revised Scheme for the Comparison of Quantitative Methods," in *American Journal of Clinical Pathology*, **54**, no. 3 (Sept. 1970), 454–462, written with R. N. Barnett; "Enduring Values," in *Technometrics*, **14**, no. 1 (Feb. 1972), 1–11; "Randomization and Experimentation," *ibid.*, 13–22; and *Risk, Choice and Prediction: An Introduction to Experimentation* (North Scituate, Mass., 1974).

Copies of all of Youden's journal articles and book reviews, and many of his book chapters, published in 1924–1965, are bound together, generally in chronological order, along with copies of his Air Force manuals, his patents, and his paperbound books *Statistical Design* and *Experimentation and Measurement*, in "W. J. Youden Publications," 2 vols., in the Historical Collection of the Library of the National Bureau of Standards at Gaithersburg, Md. His publications of 1966–1972 have been assembled for a 3rd vol. There is also a vol. containing abstracts of his talks given during 1949–1965. In addition, among the records of the Bureau's Applied Mathematics Division are eight boxes of "Youdeniana" covering 1920–1971: personal records given by his widow in 1973, professional correspondence, reports, commendations, certificates of awards, photographs, handwritten notes and computations, and a considerable number of Youden's papers, lectures, and speeches.

II. SECONDARY LITERATURE. Youden's career and contributions to statistical theory and practice are summarized briefly in an unsigned obituary in *American Statistician*, **25**, no. 3 (June 1971), 51; and more fully by Churchill Eisenhart, in *Journal of Quality Technology*, **4**, no. 1 (Jan. 1972), 1–6. His contributions to the theory of statistical design and analysis of experiments are given special attention by Churchill Eisenhart and Joan R. Rosenblatt, in *Annals of Mathematical Statistics*, **43**, no. 4 (Aug. 1972), 1035–1040. A biography of Youden, with portrait, is scheduled for publication in *National Cyclopedia of American Biography*, LVI (1975), 99–100. A biographical essay on Youden's contributions to statistical theory, methodology, by Harry H. Ku is to appear in a volume tentatively titled *Statistics: Articles From the International Encyclopedia of the Social Sciences* (New York, 1976 or 1977).

Other publications are R. N. Barnett, "A Scheme for the Comparison of Quantitative Methods," in *American Journal of Clinical Pathology*, **43**, no. 6 (June 1965), 562–569; R. C. Bose and J. M. Cameron, "The Bridge Tournament Problem and Calibration Designs for Comparing Pairs of Objects," in *Journal of Research of the National Bureau of Standards*, **69B**, no. 4 (Oct.–Dec. 1965), 323–332; "Calibration Designs Based on Solutions to the Tournament Problem," *ibid.*, **71B**, no. 4 (Oct.–Dec. 1967), 149–160; Willard H. Clatworthy, *Tables of Two-Associate-Class Partially Balanced Designs*, National Bureau of Standards Applied Mathematics Series, no. 63 (Washington, D.C., 1973); W. G. Eicke and J. M. Cameron, *Designs for Surveillance of the Volt Maintained by a Small Group of Saturated Standard Cells*, National Bureau of Standards Technical Note 430 (Washington, D.C., 1967); R. A. Fisher, *Statistical Methods for Research Workers* (Edinburgh–London, 1925; 14th ed., Edinburgh–London–Darien, Conn., 1970); R. A. Fisher and F. Yates, *Statistical Tables for Biological, Agricultural, and Medical Research* (Edinburgh–London, 1938; 6th ed., London–New York, 1963); H. O. Halvorson and N. R. Ziegler, "Application of Statistics to Problems in Bacteriology. I. A Means of Determining Bacterial Population by the Dilution Method," in *Journal of Bacteriology*, **25**, no. 2 (Feb. 1933), 101–121; J. Mandel, "Chain Block Designs With Two-Way Elimination of Homogeneity," in *Biometrics* **10**, no. 2 (June 1954), 251–272; Benjamin L. Page, "Calibration of Meter Line Standards of Length at the National Bureau of Standards," in *Journal of Research of the National Bureau of Standards*, **54**, no. 1 (Jan. 1955), 1–14; Gary J. Sutter, George Zyskind, and Oscar Kempthorne, "Some Aspects of Constrained Randomization," *Aeronautical Research Laboratories Report 63-18* (Wright-Patterson Air Force Base, Ohio, 1963); and Walter A. Shewhart, "Correction of Data for Errors of Measurement," in *Bell System Technical Journal*, **5**, no. 1 (Jan. 1926), 11–26.

CHURCHILL EISENHART

[Contribution of the National Bureau of Standards, not subject to copyright.]

YOUNG, CHARLES AUGUSTUS (*b*. Hanover, New Hampshire, 15 December 1834; *d*. Hanover, 3 January 1908), *astronomy.*

The family into which Young was born had strong ties to Hanover, New Hampshire, the site of Dartmouth College. His mother was Eliza M. Adams, whose father, Ebenezer Adams, occupied the chair of mathematics and philosophy (later natural philosophy and astronomy) at Dartmouth from 1810 to 1833 and was succeeded by Ira Young, Charles's father, who held the post until 1858. Charles began his higher education at Dartmouth at the age of fourteen; four years later he graduated with distinction, first in a class of fifty. Upon graduation he took a post teaching classics at Phillips Academy, Andover, Massachusetts. He taught there full-time for two years and part-time for another year while he was enrolled in the Andover Theological Seminary.

Fortunately for astronomy, in 1856 Young changed his plans to become a missionary and instead took the post of professor of mathematics and natural philosophy at Western Reserve College in Hudson, Ohio. He began his duties there in January 1857 and the following August married Augusta S. Mixer, by whom he had three children.

In 1862 Young took four months from his teaching at Hudson in order to captain Company B, 85th Regiment, Ohio Volunteers. The duties of the command involved only the guarding of prisoners, yet he returned to civilian life with his health impaired.

Although offered a professorship of mathematics at Dartmouth, Young remained in Hudson until 1866, when he accepted the Appleton professorship at Dartmouth—the chair that had been held by his grandfather and father.

The promise of modern equipment and less classroom time induced Young to move to the College of New Jersey (now Princeton University) in 1877. He held that post until retiring to Hanover in 1905.

Young started serious research soon after taking his post at Dartmouth. The spectroscope was just beginning to become a powerful tool in astronomy, and the Appleton fund was sufficient to provide him with the equipment he requested. In 1869 he began publishing a series of "spectroscopic notes" in the *Journal of the Franklin Institute*. The nine papers dealt with spectra of the solar chromosphere and sunspots, design and use of equipment, and observations of prominences, including possibly the first photograph ever taken of one. During 1871–1872 he compiled a catalog of bright spectral lines in the sun; and in 1876, using the Doppler principle, he measured the rotational velocity of the sun. (Although Vogel had made a similar measurement in 1871, Young's results were more accurate.)

His interest in solar research naturally led Young to make many eclipse expeditions. While on the U.S. Naval Observatory expedition to Burlington, Iowa, 7 August 1869, he devised a method for using the spectroscope to observe first contact; he later suggested the application of this method for the transit of Venus in 1874. And during an expedition to Jerez, Spain, 22 December 1870, he found that the dark lines of the sun's spectrum are momentarily reversed just at totality; hence, he is credited with the important discovery of the "reversing layer." He led other expeditions to Colorado (July 1878), Russia (August 1887), and North Carolina (May 1900).

Although Young had a conspicuous career as a researcher, he was equally talented as a teacher. Besides teaching at Western Reserve, Dartmouth, and Princeton, he lectured at Mount Holyoke Female Seminary (now College) from 1868 to 1903, at Bradford Academy from 1872 to 1898, and at Williams College from 1873 to 1875; in addition, he gave numerous talks to the public.

In addition to teaching, Young wrote some of the most famous and widely read textbooks on astronomy of his era. These books aided the education of several generations of scientists and affected the writing of astronomy texts for the next half-century. *General Astronomy* (1888), which was widely adopted, sold over thirty thousand copies by 1910—an amazing accomplishment at that time. His *Elements of Astronomy*, a more basic text, appeared in 1890. A version for younger students, *Lessons in Astronomy* (1891), had sold sixty thousand copies by 1910. Finally, in 1902, he issued an intermediate-level text that became perhaps his most famous book—his *Manual of Astronomy.* (Incidentally, the renowned text by H. N. Russell, R. S. Dugan, and J. Q. Stewart [1926] was a revision of the *Manual.*) More than any other individual, Young profoundly influenced the nature of American texts in the field.

Young received many honorary degrees and awards; he was also a member and officer of leading astronomical and scientific societies in the United States and abroad.

BIBLIOGRAPHY

I. ORIGINAL WORKS. Young's books include *The Sun,* in International Science Series (New York, 1881; rev. ed., 1895); *A Text-book of General Astronomy for Colleges and Scientific Schools* (Boston, 1888; rev. ed., 1898); *Elements of Astronomy* (Boston, 1890); *Lessons in Astronomy* (Boston, 1891); and *Manual of Astronomy* (Boston, 1902). Some of his most important papers are "American Astronomy—Its History, Present State, Needs and Prospects," in *Proceedings of the American Association for Advancement of Science* (1876), 35–48; "Pending Problems in Astronomy," *ibid.* (1884), 1–27; "Ten Years' Progress in Astronomy, 1876–1886," in *Sidereal Messenger,* **6** (1887), 4–41; "Spectroscopic Notes," in *Journal of the Franklin Institute,* **58** (1869), 141–142, 287–288, 416–424; **60** (1870), 64–65, 232a–232b, 331–340, 349–351; **62** (1871), 348–360, 430; "On the Solar Corona," in *American Journal of Science,* 3rd ser., **1** (1871), 311–320; and "Observations on the Displacement of Lines in the Solar Spectrum Caused by the Sun's Rotation," *ibid.,* **12** (1876), 321–328.

II. SECONDARY LITERATURE. Obituaries and biographical sketches of Young are the following, listed chronologically: New York *Times* (5 Jan. 1908), pt. I, 11; Hector MacPherson, Jr., in *Observatory,* **31** (1908), 122–125; John M. Poor, in *Popular Astronomy,* **16** (1908), 218–230; Sidney D. Townley, in *Publications of the Astronomical Society of the Pacific,* **20** (1908), 46–47; Edwin B. Frost, in *Astrophysical Journal,* **30** (1909), 323–338, with portrait; Henry Norris Russell, in *Monthly Notices of the Royal Astronomical Society,* **69** (1909), 257–260; and Edwin B. Frost, in *Biographical Memoirs. National Academy of Sciences,* **7** (1910), 89–114, with portrait and complete bibliography. Further background information can be found in Agnes M. Clerke, *A Popular History of Astronomy* (London, 1893); A. Pannekoek, *A History of Astronomy* (London, 1961); Otto Struve and Velta Zebergs, *Astronomy of the Twentieth Century* (New York, 1962); and Reginald L. Waterfield, *A Hundred Years of Astronomy* (New York, 1938).

RICHARD BERENDZEN
RICHARD HART

YOUNG, JOHN RICHARDSON (*b.* Hagerstown, Maryland, 1782 [?]; *d.* Hagerstown, 8 June 1804), *physiology.*

Young received his early education at the College of New Jersey (now Princeton), graduating in 1799. Before beginning his formal medical education at the University of Pennsylvania in November 1802, he evidently served as apprentice to his father, Dr. Samuel Young, a native of Ireland who is reported to have received his medical education in Edinburgh. Few other biographical facts are known about John Young or his family beyond the information on their gravestones. His mother, Ann Richardson Young, died at the age of thirty-one and two sisters at ages twenty-one and thirty. John himself died in his twenty-second year, after a two-month illness. There is a family tradition that he and his sisters died of tuberculosis. An obituary, describing Young's fatal illness in detail, suggests this disease. All were survived by the father, who lived to the age of 108, not dying until 1838.

Young's claim to historical notice rests on his inaugural dissertation for the M.D. degree, *An Experimental Inquiry, Into the Principles of Nutrition, and the Digestive Process* (Philadelphia, 1803). Dedicating this student effort to his father and to Dr. Benjamin Smith Barton, his professor of materia medica, Young argued that "acid" is not nutritious, and that digestion is a process whereby foodstuffs are dissolved in the stomach, are mixed with bile and pancreatic juices in the duodenum, and then converted into chyle by a secretory process in the ducts of the lacteals. He rejected the notion that digestion involves fermentation and supported the view that the gastric juice naturally contains phosphoric acid. His opinions were supported by a number of animal experiments.

A variety of claims have been made for Young's contributions to our understanding of gastric physiology and to our methods of experimentation. A study of the knowledge of his time and the work of other scientists and medical students, however, does not support the view that Young's work should be singled out for special acclaim. There is clear evidence, on the other hand, that even his contemporaries recognized in him a young man of "uncommon talents and great industry," and, in his work, "a very ingenious Thesis."

BIBLIOGRAPHY

I. ORIGINAL WORKS. Note reference in the text. This thesis was republished several times, and recently as facsimile reprint no. 1 in the history of science sponsored by the History of Science Society of the University of Illinois, with an introductory essay by William C. Rose (Urbana, 1959). His only other writings were two letters published posthumously: "A Case of Tetanus Cured by Mercury," in *Philadelphia Medical and Physical Journal*, **1** (1804), 47–51, and one on the use of Saccharum Saturni in three cases of uterine hemorrhage, *ibid.*, 145.

II. SECONDARY LITERATURE. See Howard A. Kelly, "John R. Young, Pioneer American Physiologist," in *Bulletin of The Johns Hopkins Hospital*, **29** (1918), 186–191; and *Dictionary of American Medical Biography* (New York–London, 1928), 1352. D. G. Bates, "American Therapeutics in 1804: the Case of John R. Young," in *Bulletin of the History of Medicine*, **38** (1964), 226–240, quotes in full the obituary notice of Young and the details of his last illness. In "The Background to John Young's Thesis on Digestion," *ibid.*, **36** (1962), 341–361, Bates evaluates Young's work in the light of the knowledge and research of his contemporaries.

DONALD G. BATES

YOUNG, JOHN WESLEY (*b.* Columbus, Ohio, 17 November 1879; *d.* Hanover, New Hampshire, 17 February 1932), *mathematics, education.*

Young's father, William Henry Young, a lieutenant colonel during the Civil War and professor of ancient languages at Ohio University, served as United States consul in Karlsruhe, Germany, from 1869 to 1876. His mother was Marie Widdenhorn.

After graduating from the Gymnasium in Baden-Baden, Young entered Ohio State University and earned a Ph.B. there in 1899. He received the A.M. (1901) and the Ph.D. (1904) from Cornell University. In 1907 he married Mary Louise Aston of Columbus. After teaching at Northwestern University (1903–1905), Princeton (preceptor, 1905–1908), the University of Illinois (1908–1910), the University of Kansas (head of the mathematics department, 1910–1911), and the University of Chicago (summer of 1911), he settled for the rest of his life at Dartmouth College, where he modernized and humanized the mathematics curriculum.

Young was influential in many learned societies, both in the United States and in Europe. He served as an editor of *Mathematics Teacher; Bulletin* and *Colloquium Publications, American Mathematical Society*; and *Carus Mathematical Monographs*. His active participation in the American Mathematical Society included membership on its Council (1907–1925) and vice-presidency

(1928–1930). He was also instrumental in the founding of the Mathematical Association of America, of which he was vice-president in 1918 and president in 1929–1931. As chairman of its Committee on Mathematical Requirements (1916–1924), he edited a 652-page report, *The Reorganization of Mathematics in Secondary Education* (1923), which circulated widely and profoundly influenced educational thought and practice.

Throughout his professional career three themes dominated Young's publications: the concept of generalization, the presentation of advanced mathematics from an elementary viewpoint, and, in conjunction with these, the "popularization" of mathematics. His *Lectures on the Fundamental Concepts of Algebra and Geometry* (1911) is excellently written and is still highly regarded. At Dartmouth his synoptic course gave the nonspecialist an understanding of topics in advanced mathematics.

In 1908, Young, with Oswald Veblen, created a set of postulates for projective geometry that embodied the first fully independent set of assumptions for that branch of geometry. This formulation served as the basis for the first volume of the classic *Projective Geometry* (1910), written with Veblen.

BIBLIOGRAPHY

I. ORIGINAL WORKS. Young's A.M. thesis, "On the Homomorphisms of a Group," in *Transactions of the American Mathematical Society*, **3** (1902), 186–191; and "On a Certain Group of Isomorphisms," in *American Journal of Mathematics*, **25** (1903), 206–212, were written under the direction of G. A. Miller. His Ph.D. dissertation, "On the Group of Sign (0, 3; 2,4,∞) and the Functions Belonging to It," in *Transactions of the American Mathematical Society*, **5** (1904), 81–104, used methods developed by Klein in treating the elliptic modular group. Other papers are "The Use of Hypercomplex Numbers in Certain Problems of the Modular Group," in *Bulletin of the American Mathematical Society*, **11** (1905), 363–367; "A Class of Discontinuous ζ-Groups Defined by the Normal Curves of the Fourth Order in a Space of Four Dimensions," in *Rendiconti del Circulo matematico di Palermo*, **23** (1907), 97–106; "A Fundamental Invariant of the Discontinuous ζ-Groups Defined by the Normal Curves of Order n in a Space of n Dimensions," in *Bulletin of the American Mathematical Society*, **14** (1908), 363–367; "A Set of Assumptions for Projective Geometry," in *American Journal of Mathematics*, **30** (1908), 347–380, written with O. Veblen; "The Discontinuous ζ-Groups Defined by Rational Normal Curves in a Space of n Dimensions," in *Bulletin of the American Mathematical Soci-*

ety, **16** (1910), 363–368; "The Geometries Associated With a Certain System of Cremona Groups," in *Transactions of the American Mathematical Society*, **17** (1916), 233–244, written with F. M. Morgan; and "A New Formulation for General Algebra," in *Annals of Mathematics*, 2nd ser., **29** (1921), 47–60.

Lectures on the Fundamental Concepts of Algebra and Geometry (New York, 1911) was translated into Italian by L. Pierro (Naples, 1919). *Projective Geometry* (Boston, 1910) consisted of 2 vols.: vol. I written with O. Veblen and vol. II, published under the names of Veblen and Young, but written by Veblen alone. *Projective Geometry* was Carus Mathematical Monograph no. 4 (Chicago, 1930). Young wrote a number of elementary mathematical textbooks, of which *Elementary Mathematical Analysis* (New York, 1918), written with F. M. Morgan, a pioneer text in the reorganization of freshman college courses, is structured around the unifying concept of function.

Other works are *The Reorganization of Mathematics in Secondary Education* (Oberlin, Ohio, 1923); "The Organization of College Courses in Mathematics for Freshmen," in *American Mathematical Monthly*, **30** (1923), 6–14; "Geometry," (in part) in *Encyclopaedia Britannica*, 14th ed. (1929), X, 174–178; and "The Adjustment Between Secondary School and College Work," in *Journal of Engineering Education*, **22** (1932), 586–595, his last paper, which is typical of a number of papers on collegiate mathematical teaching. Young's retiring presidential address, "Functions of the Mathematical Association of America," in *American Mathematical Monthly*, **39** (1932), 6–15, is an excellent example of the range of his interests and involvement in mathematics.

II. SECONDARY LITERATURE. K. D. Beetle and C. E. Wilder, "John Wesley Young: In Memoriam," in *Bulletin of the American Mathematical Society*, **38** (1932), 603–610, is a good biography complete with bibliography. Another, by E. M. Hopkins, L. L. Silverman, and H. E. Slaught, is in *American Mathematical Monthly*, **39** (1932), 309–314. For additional material on Young's role in American mathematics and the work and impact of the National Committee on Mathematical Requirements, see K. O. May, ed., *The Mathematical Association of America: Its First Fifty Years* (1972), 26–30, 39–40, 44; and the *American Mathematical Monthly*, **23** (1916), 226, 283; **24** (1917), 463–464; **25** (1918), 56–59; **26** (1919), 223–234, 279–280, 439–440, 462–463; **27** (1920), 101–104, 145–146, 194, 341–342, 441–442; **28** (1921), 357–358; **29** (1922), 46; and **32** (1925), 157.

HENRY S. TROPP

YOUNG, SYDNEY (*b.* Farnworth, near Widnes, Lancashire, England, 29 December 1857; *d.* Bristol, England, 8 April 1937), *physical chemistry*.

Young was a pioneer in the separation and specification of pure organic compounds, and clarified crucial thermodynamical relationships for solids and liquids. The third son of a merchant, he spent five years at a private school in Southport and two years at the Royal Institution School in Liverpool. After being in business for over two years, he entered Owens College, Manchester, in 1876, studying chemistry under Henry Roscoe and Carl Schorlemmer. In 1880 Young graduated B.Sc. of London University. He then became elected an associate of Owens College in 1881. Young spent the semester 1881–1882 at Strasbourg, where he was one of the first to assist Rudolph Fittig in an extended investigation "of the relations between lactones and the corresponding acids."[1] In 1883 he obtained the D.Sc. from London University and in 1893 he was elected fellow of the Royal Society.

Young had been appointed lecturer and demonstrator of chemistry at University College, Bristol, under William Ramsay in the autumn of 1882, and he succeeded to that professorship in 1887. Young married Grace M. Kimmins in 1896 and they had twin sons.

Young was stimulated by Thomas Carnelley's observation of "hot" ice in 1881. He proved that the volatilizing point of ice depended upon the pressure and showed "that the volatilising point-pressure curve is identical with the vapour pressure curve constructed from the data calculated on theoretical grounds by James Thomson and by Kirchhoff."[2] This result for ice, whereby the graph representing the vapor pressure at different temperatures also represents the volatilizing point as a function of pressure, was extended to other volatile solids as well as to substances such as benzene. Young and Ramsay in partnership at Bristol published an exhaustive series of researches concerning the vapor pressures of liquids, thereby clarifying the vexed question of whether Kopp's quantitative laws held at all pressures. In 1885 Ramsay and Young derived an empirical equation that relates the ratio of absolute temperature of two liquids at a given vapor pressure, say T_a/T_b, to the same at another vapor pressure, say $T_{a'}/T_{b'}$, so that

$$T_a/T_b = T_{a'}/T_{b'} + c(T_a - T_{a'}),$$

where c is constant.

Showing that c is very small or even negligible for two closely related chemical substances (they used chloro- and homobenzenes), one can say that $T_a/T_b = T_{a'}/T_{b'}$. The significance is that this ratio is constant at all vapor pressures and that the ratio of the boiling points of two similar substances is the same, no matter what the external pressure. Later, Young tested van der Waals' equation of state noting that it was "strictly true" only for closely related substances, such as the four halogen derivatives of benzene. In the case of non-associated substances "the ratio of the actual to the theoretical density at the critical point should be the same for all unassociated substances"[3], and van der Waals had determined the ratio RT_c/p_cV_c = 2.67. The actual values of this ratio tend to range between 3 and 4 being much higher in the case of associated substances. Modifying the equation of state, Conrad Dieterici in 1899 succeeded in obtaining the theoretical ratio 3.695.[4] For this work Young devised a new method, improving upon the law of Cailletet and Mathias, for determining the critical densities of liquid and saturated vapor.

From 1885, the potential for extracting paraffin hydrocarbons from petroleum and a general need for substances in the purest possible state led Young to devise improved techniques of fractional distillation. He designed and constructed his own efficient still heads, having become an expert glassblower while at Strasbourg. He examined the vapor pressures, boiling points, and behavior on distillation of mixed liquids including azeotropic mixtures. He also developed a new method of dehydrating ethyl alcohol using benzene and made fundamental contributions concerning the separation and specification of the various hydrocarbons of petroleum.[5] In 1904 Young proposed that the difference in boiling points between two successive homologues of most unassociated compounds is not constant, as Kopp had maintained in 1842, but is a function of the absolute temperature, approximating the empirical formula

$$\Delta = 144.86/T^{0.0148\sqrt{T}},$$

where T is the absolute boiling point of the lower member. Most of Young's publications appeared during his twenty-one years at Bristol. Succeeding James Emerson Reynolds at Trinity College, Dublin, in 1903, he assumed a post not conducive to continued research, but he did advance the development of the practical applications of his separation techniques.[6] Although Young established no "school," he was the first to prove that pure organic compounds are no less unique than inorganic elements; and through his techniques he exerted a profound, albeit often indirect, influence on all subsequent research concerning their purification.[7]

NOTES

1. E. von Meyer, *A History of Chemistry* (London, 1891), 410–411.
2. "On the Volatilisation of Solids," in A. Keitner, *Menschen und Menschenwerk*, II, 453.
3. *Ibid.*, 453–454.
4. A. Findlay, *Introduction to Physical Chemistry*, 3rd ed. (London, 1953), p. 85.
5. The industrial implications of this patented dehydration process using a third liquid seem first to have been appreciated and adopted by the firm of C. A. F. Kahlbaum, Chemische Fabrik, Berlin. Young received special recognition in 1933 from the Petroleum Division of the American Chemical Society, which expressed high appreciation of his work on distillation, on the composition of petroleum, and on the specification of numerous hydrocarbons.
6. Young elaborated a quantitative analysis of the process in *Fractional Distillation* (1903; 1918) and included industrial applications in *Distillation, Principles and Processes* (1922).
7. J. Timmermans, "Sydney Young," 13–14.

BIBLIOGRAPHY

I. Original Works. Most of Young's more than 100 publications are listed in W. R. G. Atkins, "Sydney Young," in *Obituary Notices of Fellows of the Royal Society of London*, **2** (1938), 370–379, also abbreviated in *Dictionary of National Biography, 1931–1940*, 932–933. His books include *Fractional Distillation* (London, 1903); *Stoichiometry* (London, 1908; 2nd ed., 1918); and *Distillation, Principles and Processes* (London, 1922), written with Ernest Briggs, T. H. Butler, T. H. Durrans, F. R. Henley, James Kewley, and Joseph Reilly; 2nd ed. condensed and translated into German as *Theorie und Praxis der Destillation* (Berlin, 1932). His scientific "autobiography" is "On the Volatilisation of Solids and the Thermic Properties of Mixtures of Liquids," in A. Keitner, ed., *Menschen und Menschenwerke*, II (Vienna, 1925), 453–454, preceded by a German version, 451–453, and followed by a French version, 455–456. This is preceded by a brief autobiographical résumé, 450–451, in the three languages. Two autobiographical letters, dated 8 Nov. 1892 and 13 Feb. 1893, accompanied by his list of publications, are in the Krause *Album*, IV, held in the Sondersammlungen of the library of the Deutsches Museum, Munich.

II. Secondary Literature. Young's successor at Bristol, F. Francis, considered his work in detail in *Journal of the Chemical Society* (1937), 1332–1336; and there is an anonymous obituary in *Nature*, **180** (1957), 1451. His work on critical constants is considered in W. Nernst, *Theoretische Chemie* (Stuttgart, 1921), 241–246. The importance of Young's research and influence on the purification of organic compounds is stressed in J. Timmermans, "Sydney Young," in *Endeavour*, **6** (1947), 11–14. A discussion of subsequent developments pertaining to the distillation of hydrocarbons at high pressures is in the article on M. Benedict in *Modern Men of Science*, I (New York, 1966), 31.

Young's research on boiling points in relation to the work of his contemporaries is considered in S. Smiles, *The Relations Between Chemical Constitution and Some Physical Properties* (London, 1910), ch. 7. The significance of Young's experimental results for theoretical chemistry is considered in H. Davies, "On Some Applications of the Law of the Rectilinear Diameter," in *Philosophical Magazine*, 6th ser., **24** (1912), 418–421; and G. Le Bas, "The Unit-Stere Theory," *ibid.*, **14** (1907), 340–346.

THADDEUS J. TRENN

YOUNG, THOMAS (*b.* Milverton, Somerset, England, 13 June 1773; *d.* London, England, 10 May 1829), *natural philosophy*.

Young made many discoveries in natural philosophy and physiological optics, and he was one of the first persons to translate Egyptian hieroglyphics. He is most famous, however, for his attempt to win acceptance for an undulatory theory of light. His scientific colleagues initially did not share his interest in that theory, nor were they ultimately convinced by his discoveries or arguments. Young's achievements, as well as his failures, must be understood in the context of his education, heterogeneous career, and personality as well as in terms of contemporary scientific evaluations of his theories.

The eldest son of Thomas Young, a mercer and banker, and of Sara Davis, Young was raised as a member of the Society of Friends and was largely self-educated in languages and natural philosophy.[1] He learned to read at age two; and by the time he was six, he had read twice through the Bible and had started the study of Latin. Between 1780 and 1786 he attended two boarding schools, where he learned elementary mathematics and gained a reading knowledge of Latin, Greek, French, and Italian. He also had begun independent study of natural history, natural philosophy, and fluxions, and had learned to make telescopes and microscopes. In 1786 Young began independent study of Hebrew, Chaldean, Syriac, Samaritan, Arabic, Persian, Turkish, and Ethiopic. Shortly thereafter he became tutor to his lifelong friend and biographer Hudson Gurney, who was a member of the Gurney banking family. By 1792 Young had become a proficient Greek and Latin scholar; had mastered the fluxionary calculus; and had read Newton's *Principia* and *Opticks*, Lavoisier's *Elements of Chemistry*, Joseph Black's manuscript lectures on chemistry, and Boerhaave's *Methodus studii medici*, in addition to plays, law, and politics.

Between 1792 and 1799 Young studied medicine at London, Edinburgh, and Göttingen (M.D. 1796). In January 1794 he was one of the assistants in experiments of the Society for Philosophical Experiments and Conversations organized by the chemist Bryan Higgins.[2] At the end of 1794 and in 1795, when he was at Edinburgh, Young began to enjoy music, dancing, and the theater, and in general to abandon the practices of the Society of Friends. Later in his life he formally became a member of the Church of England. In 1797–1803 he was enrolled at Emmanuel College, Cambridge (M.B. 1803; M.D. 1808), partly to fulfill the requirements of the Royal College of Physicians and partly to satisfy the desires of his maternal uncle, Dr. Richard Brocklesby, who sponsored Young's medical education and his election to the Royal Society (19 June 1794). Brocklesby also introduced Young to such influential men as Edmund Burke, the duke of Richmond, and William Herschel. When Brocklesby died in 1797, he left Young his London house, his library and paintings, and £10,000.

Young moved from Cambridge to London in 1800 and attempted to establish a medical practice. He was never very successful as a doctor, however, probably because he did not inspire confidence in his patients. Young's undemanding practice gave him the opportunity for regular attendance at meetings of the Royal Society. As a result he became known to the Society's president, Joseph Banks, and to Benjamin Thompson, Count Rumford, founder of the Royal Institution.

In the summer of 1801 Rumford was looking for a professor of natural philosophy. On Banks's and Rumford's recommendations, Young was employed on 3 August 1801 as "Professor of Natural Philosophy, Editor of the Journals, and Superintendent of the House" at an annual salary of £300. As professor his first task was to prepare popular lectures on natural philosophy and the mechanical arts for the Institution's members. Young delivered these lectures in 1802 and 1803, and published them in revised form in 1807.[3] These lectures were erudite and at times contained the results of his recent researches. They also were obscure, technical, and too detailed for a popular audience. At the same time, in contrast, his colleague Humphry Davy's lectures on chemistry were brilliantly successful. Rumford left England in May 1802, and Young seems to have become unpopular with the Institution's managers. They probably forced his resignation, which was effective 4 July 1803.

Young then resumed his medical practice in London and later, during the summers, in the fashionable resort of Worthing. On 4 June 1804 he married Eliza Maxwell, who was related to the Scottish aristocracy through the family of Sir William Maxwell of Calderwood.

In November 1808 Young gave the endowed Croonian lecture to the Royal Society, and he was elected to the Royal College of Physicians in December 1809. During that winter at the Middlesex Hospital, he delivered a course of lectures on physiology, chemistry, nosology and general practice, and materia medica that were published in 1813. In January 1811 Young at last obtained a lifetime professional position, being elected physician to St. George's Hospital. He was chosen to give the other endowed Croonian lecture, to the Royal College of Physicians, in 1822–1823; and in March 1824 he was appointed inspector of calculations and physician to the Palladium Insurance Company, at an annual salary of £500.[4]

During much of his life Young received a substantial income as an anonymous author of a wide variety of articles. In 1808 and 1809 he wrote for *Retrospect of Philosophical, Mechanical, Chemical, and Agricultural Discoveries*; in 1810, an article for the *London Review*; between 1809 and 1818, at least twenty-one articles for the *Quarterly Review*; and between 1816 and 1824, over sixty articles for the *Supplement* to the *Encyclopaedia Britannica*. These last works included biographies as well as many pieces on optics, mechanics, and the mechanical arts.[5]

Young held several public offices related to science. From 22 March 1804 until his death he was foreign secretary of the Royal Society, and after 1806 he was also a member of its Council.[6] He was at times a consultant to the Admiralty and was secretary to the Royal Commission on Weights and Measures from 1816 to 1821. Young was secretary of the Board of Longitude at an annual salary of £300, from 1818 until its abolition in 1828. He was also superintendent of the *Nautical Almanac*, at an additional salary of £100, from 1818 until his death. The last two posts were among the very few salaried scientific offices in England; and with Young's appointment to them, his combined income became commensurate with his social status. In 1827 Young's work in science received international recognition when he was elected a foreign associate of the French Academy of Sciences.

Young was a scholar with the scholar's love of knowledge and the search for truth, however esoteric. Shortly before his death he persisted, against

advice, in compiling his Egyptian dictionary and expressed great satisfaction that he had not yet spent an idle day in his life.[7] His self-education in many difficult fields shows that Young had the successful autodidact's persistence and self-confidence. He never seems, however, to have developed much sensitivity, in his professional relations, to other people's emotions or differing perspectives. In his professional life he seems to have been formal almost to coldness and self-assured almost to being cocksure. Young's writing was frequently both prolix and obscure; at other times it was concise almost to incomprehensibility. Despite these limitations, as his varied career continued to open new vistas of scholarship, his education and independent income gave him the opportunity to be an intellectual dilettante.

As a result of his uncle's influence and his marriage, Young's social position was high. He was fond of music, dancing, riding, and conversation; and he valued and sought friendship and acquaintance with persons of culture and social status. After abandoning his Quakerism he seems to have been impeccably conservative at a time when the English establishment was reacting conservatively to threats of radicalism at home and abroad. In sum, Young had the attitudes, personality, ability, money, and influence to be a gentleman scholar and to pursue careers as a physician, writer, administrator of science, and Egyptologist as well as a natural philosopher. Consequently his diverse writings are related to each other more by his love of scholarship and the accidents of his life than by any dominant research or theoretical concerns. But, unlike most gentlemen scholars, Young was exceptionally intelligent. As a result he often showed clear insight into problems and made several important discoveries.[8]

Young once remarked that, as a natural philosopher, ". . . acute suggestion was . . . always more in the line of my ambition than experimental illustration."[9] In the course of his very diverse writings Young did make many "acute suggestions," as well as many ingenious "experimental illustrations" in physiological optics, the theory of light, mechanics, and Egyptian linguistics. In none of these fields, however, did he systematically develop his discoveries, hypotheses, or suggestions. Nor did he fully confront their implications. His failure to do this, despite the importance of some of his discoveries, partly accounts for his limited influence in science.

From 1791 to 1801 Young published most of his experiments and theories in physiological optics.

In 1793, when he wrote his first important paper, "Observations on Vision," there was no consensus about the mechanism of the accommodation of the eye. Various earlier authors had argued that the eye adjusted its focus to different distances by changing either its length or the curvature of the cornea or the crystalline lens. Young conjectured that the lens is composed of muscle fibers. Nerve impulses are sent ". . . by the ciliary processes to the muscle of the crystalline [lens], which, by the contraction of its fibers, becomes more convex. . . ."[10] Measurements that he had made on the lens from an ox's eye indicated that the necessary changes in focal length could easily be achieved by the possible changes in the shape of the lens.[11] In 1796, in his doctoral dissertation, Young temporarily abandoned this hypothesis when he learned of investigations by Everard Home, John Hunter, and Jesse Ramsden that appeared to show that accommodation was achieved by changes in the curvature of the cornea and the length of the eyeball.[12] These researchers also had reported that persons who had had their lenses removed retained the ability to change the focus of their eyes.

Late in 1800, however, Young reaffirmed his first hypothesis after performing new experiments that refuted Home and Ramsden. In his paper "On the Mechanism of the Eye," he first derived a formula for the path of a ray refracted through a variable medium such as the crystalline lens.[13] Then he described his improvement of the optometer, analyzed the optics of the eye, made optometric measurements of its dimensions, and computed the change in focal length that would be required to accommodate for near and far vision.[14] Next Young determined the amount of change that would have to occur in the curvature of the cornea or the length of the eyeball, singly or jointly, for accommodation.[15] Then he made a series of observations that were sensitive enough to detect as little as one-fourth of the required change in the cornea. He found no change.[16] His most compelling experimental demonstration, however, involved immersing his eye in water, which eliminated refraction at the cornea. His power of accommodation was unchanged. This result eliminated any role for the cornea in accommodation.[17] Furthermore, Young could perceive no change in his ability to accommodate even when he made the length of his eye almost invariable by rotating it to the side and applying pressure.[18] Finally he reported that he had examined five persons without crystalline lenses. None of them had any power of accommodation.[19]

To show that changes in the shape of the lens do take place, Young made use of his own astigmatism. By viewing an out-of-focus point he first produced a "star" image on his retina; then, when he passed light from the point through horizontal or vertical slits, he observed straight-sided bands with a relaxed eye and curved bands when he accommodated. From this he concluded that the only way such curvature could be produced was by a change in the shape of the lens during accommodation. Young then formally readopted his opinion of 1793 that the lens changed shape because it was a muscle. His final conclusion was ". . . whether we call the lens a muscle or not, it seems demonstrable that such a change of figure takes place [in it] as can be produced by no external cause."[20] In this paper Young clearly demonstrated his ability to formulate a testable hypothesis, design and perform delicate experiments, and draw convincing conclusions that did not go beyond his evidence. Moreover, he was able to refute the opinions of persons as renowned as Home.

In 1801 Young also suggested that the retina responded to all colors in terms of variable amounts of three "principal colours." He believed that ". . . it is probable that the motion of the retina is rather of a vibratory than an undulatory nature. . . ." "Now as it is almost impossible to conceive each sensitive point of the retina to contain an infinite number of particles, each capable of vibrating in consonance with every possible undulation, it becomes necessary to suppose the number limited, for instance, to the three principal colours, red, yellow, and blue. . . ."[21] In 1807 he reaffirmed this hypothesis but eliminated any reference to undulatory theories:

> It is certain that the perfect sensations of yellow and blue are produced, respectively, by mixtures of red and green and of green and violet light, and there is reason to suspect that those sensations are always compounded of the separate sensations combined; at least, this supposition simplifies the theory of colours: it may, therefore, be adopted with advantage, until it is found to be inconsistent with any of the phenomena. . . ."[22]

Part of Young's "reason to suspect" presumably was based on John Dalton's blindness to red. Dalton thought it ". . . probable that [his] vitreous humour is of a deep blue tinge; but this has not been observed by anatomists, and it is much more simple to suppose the absence or paralysis of those fibres of the retina which are calculated to perceive red. . . ."[23] Young published nothing more than these "acute suggestions" on tricolor vision. Maxwell and Helmholtz later modified and extended his speculations into what has come to be called the Young-Helmholtz theory of color sensation.[24]

Young's most sustained interest in natural philosophy was his attempt to gain acceptance for an undulatory theory of light. His failure was partly the result of several hostile reviews of his first papers by Henry Brougham and partly the result of the inherent limitations of his work.[25] Young never worked out a detailed mathematical theory; nor were his suggestions, except possibly for the principle of interference, influential on the man who did, Augustin Fresnel. Young's colleagues, however, quickly acknowledged the principle of interference as a major discovery.

Young's interest in the nature of light probably began with his investigations on the formation of the human voice for his Göttingen dissertation and lecture. During the three years after his return to England from Göttingen he wrote an essay on the human voice. This led him to make theoretical and experimental investigations of vibrating strings, musical pipes, beats, and the motions of fluids.[26] During this work he was "forcibly impressed" with the probability of a very close "analogy" between the vibrations of a series of organ pipes and the colors of thin plates. This analogy was later dismissed by his biographer George Peacock as "fanciful and altogether unfounded," but it seems at least to have impressed Young with the need to reexamine the accepted theory of the nature of light.[27]

For many years before 1800, when Young published his first discussion of the nature of light, most men of science had affirmed that light is particulate. Newton had argued that if light were a vibration in a material medium, it ought to bend around corners; and the vibrations would have to have "sides" to account for polarization. Finally, Newton's arguments that the hypothetical Cartesian interplanetary vortices would cause the planets to spiral into the sun seemed to apply with equal force to an interplanetary ether.[28] In part, Newton was also writing against the opinions of Descartes that light was a "pression" in a fluid and the ideas of Hooke and Huygens that light was a random succession of pulses in a Cartesian plenum. During the middle of the eighteenth century Euler had argued that there was a strong analogy between light and sound, and that light must therefore be a vibratory motion.[29] In 1792 Abraham Bennet suggested to the Royal Society that light must be caused by vibrations in "the universally

diffused caloric or matter of heat or fluid of light."[30] Neither Euler nor Bennet attracted any large following, presumably because the same conclusive objections seemed to apply to their particular hypotheses that had to Descartes's, Hooke's, and Huygens'. In contrast, the emission theory did not seem to be plagued by similar difficulties. In fact, in 1796 and 1797 Brougham had shown how the emission theory could be extended to account for "inflection" and dispersion.[31]

In January 1800, in a paper he submitted to the Royal Society, Young reopened this old debate. He began by stating that ". . . some considerations may be brought forwards, which may tend to diminish the weight of objections to a theory similar to the Huygenian. There are also one or two difficulties in the Newtonian system, which have been little observed."[32] The difficulties were not new: light has a uniform velocity regardless of its origin; light is partially reflected from every refracting surface. In turn, one of the prime "objections" to the vibratory system was that it was founded on the supposed existence of some kind of "luminous ether." Young answered this objection by arguing that ". . . a medium resembling, in many properties, that which has been denominated ether, does really exist, is undeniably proved by the phaenomena of electricity. . . ."[33] Because the existence of the "electric medium" was manifest, Young argued, it was legitimate to assume the existence of an analogous medium for the transmission of light. He then went on to assume that, except for its medium and frequency, the undulations of light are the same as those of sound. Making these assumptions, Young was then able to "solve" the problems of the uniform motion of light that seemed to plague the emission system because ". . . all impressions are known to be transmitted through an elastic fluid with the same velocity."[34]

Earlier in this paper Young had shown that sound waves had very little tendency to diverge. Hence, he argued, ". . . in a medium so highly elastic as the luminous ether must be supposed to be, the tendency [for light] to diverge must be infinitely small, and the grand objection to the system of vibration will be removed."[35] By assuming that "all refracting media" contain the same mechanical ether composed of elastic particles of matter, but with different densities of ether of the same "elasticity," he then gave qualitative explanations of ". . . partial and total reflection, refraction, and inflection [diffraction]."[36] He concluded this discussion by asserting that the "colours of light consist in the different frequencies of vibration of the luminous ether. . . ." He based this statement on his supposed analogy between the sounds of organ pipes and the colors of thin plates.[37] There were no published scientific reactions to this article, perhaps because Young's arguments seemed familiar, speculative, and unsupported by any new experiments.

In May 1801 Young discovered his principle of interference ". . . by reflecting on the beautiful experiments of Newton. . . ."[38] By November he had completed what became his second Bakerian lecture, "On the Theory of Light and Colours," in which he partially announced this principle.[39] His statement, however, was entirely hypothetical and not at all experimental: "When two Undulations, from different Origins, coincide either perfectly or very nearly in Direction, their joint effect is a Combination of the Motions belonging to each."[40] He then went on to report an experiment that exhibited four orders of fringes produced by sunlight reflected from a set of parallel grooves in glass.[41] Using Newton's measurements of the spacing of ring colors and the principle of interference, Young calculated the wavelengths and frequencies of the visible spectrum and gave qualitative explanations of the colors of thin and thick plates, of color or blackness associated with total internal reflection, and of the colors produced by inflection.[42] On the basis of these reasonings he concluded: "Radiant Light consists in Undulations of the luminiferous Ether."[43]

In July 1802 Young made the first full announcement of his principle of interference: ". . . whenever two portions of the same light arrive at the eye by different routes, either exactly or very nearly in the same direction, the light becomes most intense when the difference of the routes is any multiple of a certain length, and least intense in the intermediate state of the interfering portions; and this length is different for light of different colours."[44] He then used the principle to explain the fringes produced by thin fibers and the "colours of mixed plates." In the case of the fibers, Young argued that the fringes were caused by the interference of one portion of light reflected from the fiber with another ". . . bending around its opposite side. . . ."[45] He then calculated the difference in path lengths for the two portions of red light that produced the first fringe. His measurements were crude, and his result was only within 11 percent of what he had expected on the basis of Newton's data.

Nonetheless, Young concluded that ". . . this coincidence, with only an error of one-ninth of so minute a quantity [is] sufficiently perfect to warrant completely the explanation of the phenomenon, and even to render a repetition of the experiment unnecessary. . . ."[46] Young was easily persuaded. He used the same "bending" explanation to account for the halos that can be seen when wool is held to the light and also for ". . . coloured atmospherical halos. . . ."[47] Young's suggestions about fringes appear to have been unknown to Fresnel, who, in 1818 or 1819, was the first to work out a physically and mathematically satisfactory explanation for the fringes that did not depend on the implausible (and nonexistent) reflecting and bending of rays.[48]

In this same paper Young's discussion of the colors of mixed plates is more persuasive than his discussion of fibers. First he showed that the diameter of the colored circles was inversely dependent on the "refractive density" of the interposed liquids. Next he asserted that this and other experiments necessarily implied that if light is an undulation, then its velocity increases as the medium becomes denser. Young then attempted to describe an experiment that would decide between the two theories. Although the results he obtained from this experiment were equivocal, he asserted that they confirmed his prediction.[49]

Young presented his first really convincing evidence that fringes are produced by interference in his third Bakerian lecture, "Experiments and Calculations Relative to Physical Optics" (November 1803). First he exposed a small piece of paper to sunlight diverging from a pinhole. The shadow exhibited not only fringes of color, but ". . . the shadow itself was divided by similar parallel fringes. . . ." Then, by inserting a small screen into either edge of the shadow, he was able to make the fringes disappear, ". . . although the light inflected on the other side was allowed to retain its course. . . . [Hence] these fringes were the joint effects of the portions of light passing on each side of the slip of card, and inflected . . . into the shadow."[50] In the next section of the paper Young compared the different "characteristic lengths" (wavelengths) that were implied by Newton's observations of fringes produced by knife edges and by a hair, and in his own similar experiments with the "analogous interval, deduced from the experiments of Newton on thin plates." These results were within about 13 percent of each other. Young optimistically concluded that "this appears to be a coincidence fully sufficient to authorise us to attribute these two classes of phenomena to the same cause."[51]

Young did not stop after presenting his evidence and conclusions. Rather, he applied his "principle" to the explanation of supernumerary rainbows, to the colors of natural bodies, and to an "Argumentative Inference respecting the Nature of Light."[52] Young's argument was that because the lengths and phenomena in his experiments on interference and in the colors of thin plates described by Newton were similar, they were the result of the same phenomenon: interference. Newton's experiments showed that the denser the medium, the smaller the intervals; but, according to the emission theory, they ought to be larger. Therefore light must move more slowly in the denser medium, which was contrary to the assumption made to explain refraction on the emission theory. Therefore light must be an undulation in the "luminous ether."[53]

Young summarized and completed this phase of work on his undulatory theory by 1807 for the published version of his lectures for the Royal Institution. To the published version he added a description of his two-slit experimental demonstration of interference. He also withdrew his speculation that the colors of halos were produced by interference.[54] At this time Young had convincingly demonstrated the fact of the interference of light, but he had by no means demonstrated that light was the longitudinal undulation of a mechanical, luminiferous ether.

By early 1810 Young had learned of Malus's discovery that, in addition to its being polarized by transmission through Iceland spar, light could be polarized by reflection.[55] This implied that light could acquire ". . . properties independent of its direction . . ., but exclusively relative to the sides of the . . . ray. . . ."[56] Young concluded that "the general tenor of these phenomena is such, as obviously to point to some property resembling polarity, which appears to be much more easily reconcileable with the Newtonian ideas than with those of Huygens."[57] Between 1811 and the early 1820's Arago, Biot, and Brewster had made many additional discoveries indicating that light must have some property very much like "sides" or "poles."[58] In January 1817 Young suggested to Arago that polarization might be accounted for by assuming that it was a minute transverse component added to the longitudinal undulation of the ether that he assumed to be light. "But its inconceivable minuteness suggests a doubt as to the possibility of its producing any sensible effects; in a physical sense it is almost an evanescent quantity, although not in a mathematical one."[59]

By September 1817, when he wrote his article "Chromatics" for the *Supplement* to the *Encyclopaedia Britannica*, Young had worked out a qualitative suggestion of how a small transverse component of vibration might explain the facts of partial reflection. He was not confident that it did, however, nor did he express any idea that the vibrations might be entirely transverse. All that he felt willing to suggest was ". . . a mathematical postulate, in the undulatory theory, without attempting to demonstrate its physical foundation, that a transverse motion might be propagated in a direct line . . . ." Rather than urge that this modification was true, Young was careful to emphasize the minuteness of this transverse component. As far as the discovery of the true theory of light was concerned, he continued to believe at this time that ". . . the general phenomena of polarisation . . . cannot be said to have been explained on any hypothesis respecting the nature of light."[60] Young's theory at this time also did not adequately predict the spacing of the exterior fringes of diffraction or the rectilinear propagation of light. Less than a year later Fresnel had completed a mathematical solution to both problems.

In July 1818 Fresnel reported his solution to the problem of explaining the exterior fringes. In it he demonstrated their production by a combination of the principles of interference and Huygens' principle that a wave front may be considered to consist of an infinite number of point sources of new waves. Fresnel calculated their combined effect by means of the integral calculus; and in passing he also derived a solution to the problem of the rectilinear propagation of light, which had been Newton's chief objection to an undulatory theory. Fresnel also designed and performed interference experiments to test his theory. The difference between the observed and calculated positions of the successive minima were never greater than 0.05 millimeter, or 7.4 percent, and they averaged about 0.006 millimeter, or about 0.9 percent. Probably independently of any knowledge of Young's work or any use of transverse undulations, Fresnel had given a far more persuasive argument than had Young that light was an undulation.[61]

As early as 1816, and before Young made his suggestion, Fresnel seems to have become convinced that polarization might be explained by some kind of transverse undulation.[62] From then until his death in 1827, he applied this assumption successfully to a detailed mathematical analysis of most known optical phenomena: reflection, diffraction, partial reflection, single and double refraction, and polarization. Only in the cases of dispersion, elliptical polarization by reflection from metals, and absorption did he fail to derive explanations and formulas the predictions of which were repeatedly confirmed by experiments. What Young had been unable to do, even conceptually, Fresnel did mathematically and experimentally—and probably with little or no assistance from Young's suggestions. Young claimed the priority of suggestion, and Fresnel agreed; but Fresnel disagreed with Young's contention that he had planted the tree and Fresnel had picked the apples: "I am personally convinced that the apple would have appeared without the tree, for the first explanations which occurred to me of the phenomena of the coloured rings, or of the laws of reflection and refraction, I have drawn from my own resources, without having read either [Young's] work or that of Huygens."[63] Nor was Fresnel plagued by doubt that light was indeed the transverse undulations of an elastic solid ether. In contrast, Young realized that if there were to be transverse undulations, the undulating particles had to have lateral adhesion. But, he concluded in 1823, if they did, ". . . it might be inferred that the luminiferous ether, pervading all space, and penetrating almost all substances is not only highly elastic, but absolutely solid!!!"[64]

Young's hesitation to affirm the existence of transverse undulations, which were now essential for the success of any undulatory theory, depended on his sense of the physical requirements they imposed on the ether. At this time there was no alternative to a mechanical ether. Young then abandoned his work on light and returned to some of his older lines of research at least in part because he had been unable to reduce the theory of light to propositions in mechanics.

In these investigations on light, Young based his hypotheses on physical analogies or comprehensible physical entities such as mechanical ethers and imponderable fluids. If such a hypothesis led to a plausible qualitative or semi-quantitative explanation, he was quick to make rather extravagant claims that his explanations were true. He had mathematical ingenuity; but he had taught himself fluxions, and he lacked skill in the use of differential and integral calculus. His discovery of interference convinced him that light must be at least periodic and probably an undulation; but his conviction that there must be some mechanical entity to undulate brought him to an impasse, for he was convinced that a mechanical luminiferous ether that would support transverse undulations had contradictory physical properties. Ironically, Young's faith in the use of mechanical analogies, which had led him to his discoveries, kept him from accepting their consequences. Perhaps this was because he lacked Fresnel's skill and faith in mathematics. Fresnel either solved or ignored the problems that checkmated Young. His skillful solutions were immensely persuasive to the scientific community that Young never persuaded. Young's work in mechanics had limited influence for some of these same reasons.

Young described many of his mechanical discoveries and "suggestions" in his first book, *A Course of Lectures on Natural Philosophy and the Mechanical Arts*. Its organization is typical of numerous popular "courses," "systems," and "syllabi" that were published frequently in eighteenth-century Britain. In it Young juxtaposed lectures on motion, forces, and "passive strength" with ones on drawing, writing, modeling, and engraving; lectures on hydrostatics, hydraulics, and the friction of fluids with ones on hydraulic and pneumatic machines such as pumps, steam engines, and firearms; and lectures on astronomy, the physics of matter, electricity, and magnetism with ones on the study of meteors, vegetation, and animals.

Unlike those of the typical "course," however, Young's arguments and illustrations were much more complicated and detailed, his prose much less easy to understand, and his scholarship intimidating. Moreover, the size of the work—two thick quarto volumes totaling more than 1,500 closely printed pages—made it a work of reference for a few persons rather than a popular text for many. Indeed, much of the second volume is devoted to a long index and to an annotated bibliography of about twenty thousand references on all the subjects in his lectures, as well as reprints of several of his papers. Anyone who had the patience and time to explore Young's *Course* might have found several "acute suggestions." At this time, however, several relatively new scientific journals had begun to supplant books as the means to announce discoveries that were not presented to the philosophical societies.[65]

In his lecture "On Collision," Young was probably the first person to suggest substituting the term "energy" for "living force" or *vis viva*. His use of the term, however, indicates that he did not generalize this concept. Rather, he used it only twice for what is now known as kinetic energy. As he stated it: "The term energy may be applied, with great propriety, to the product of the mass or weight of a body, into the square of the number expressing its velocity."[66] The word "energy" did not become widely used until after its revival in the early 1850's by Rankine and William Thomson in their writings on the conservation of energy.[67]

Young defined a "modulus of elasticity" in his lecture "Passive Strength and Friction." It is almost impossibly obscure: ". . . we may express the elasticity of any substance by the weight of a certain column of the same substance, which may be denominated the modulus of its elasticity, and of which the weight is such that any addition to it would increase it in the same proportion, as the weight added would shorten, by its pressure, a portion of the substance of equal diameter."[68] In 1867 Thomson and Tait reexpressed this modulus in its present form: the ratio between the stressing force and the resultant strain. They also pointed out that this is the modulus of elasticity of stretch and that there is an additional modulus of rigidity or shear modulus that is not the same as Young's.[69]

In his *Lectures*, Young also began developing his theory of the tides. His first discussion outlined how an analogy to a pendulum might be used to develop such a theory.[70] By 1811 he had completed its mathematical development; and he finally published it, anonymously, in 1813. His analysis used fluxions and was based on another analogy to a pendulum. The vibrating pendulum also had a vibrating suspension point, and it experienced resistance proportional to the first and second powers of the velocity. In this way Young attempted to account for the effects of friction on the times of the tides.[71] In 1824 he published his complete theory in the *Supplement* to the *Encyclopaedia Britannica*.[72] His theory was probably the most complete when it was published, but it was soon superseded by the work of Airy, which was published in 1842.[73] Airy appears not to have read any of Young's writings on the tides.[74]

As a result of his work on the tides and on the Commission on Weights and Measures, Young had a continuing interest in the pendulum. As secretary to the Commission, he wrote its report, which recommended that the standard of length be a "pendulum vibrating seconds of mean solar time in London, on the level of the sea, and in a vacuum."[75] Later he published articles on errors affecting this standard, on measurements of the resistance of the air, and reduction of pendulum lengths to sea level.[76]

After he ended his work on the theory of light, Young returned to his long-standing interest in languages.[77] In 1813 he had started his attempts to decipher the Egyptian hieroglyphics, and by the following year he had translated the "enchorial" or demotic running script of the Rosetta Stone and had concluded that the enchorial was derived from the hieroglyphic. He published very little at this time, and most of what did appear was published anonymously. During 1817 and 1818 Young returned to the subject in preparing his unsigned article "Egypt" for the *Supplement* to the *Britannica*. This article was published in December 1819. His other commitments for the *Britannica*, as well as the demands of his new duties with the Board of Longitude and the *Nautical Almanac*, seem to have prevented Young from doing much more on hieroglyphics than corresponding with other translators. In March 1823 he published, under his own name, a comparison of his and J. F. Champollion's alphabets, intending to assert his own priority. His final work was *Enchorial Egyptian Dictionary*, published in 1830.

Young's need to defend his priority as a translator of hieroglyphics, as well as his limited influence in natural philosophy (quite apart from the technical limitations of his work), were the results of his personality, his methods of communication, and his career. He left his "acute suggestions" for others to develop and complete. He published sporadically; often anonymously; frequently in obscure, unwieldy, or unlikely publications; and almost always using weak analogies, awkward prose, or inadequate mathematics. Young's frequent changes of occupation prevented sustained concentration and favored dilettantism. Late in his life, when he added the demands of two salaried scientific posts to his other activities, his principal avocation had become hieroglyphics. Young was a sporadically brilliant, gentleman natural philosopher who lived to see other men receive the credit and fame for completing what he had begun.

NOTES

1. Biographical details of Young's life are drawn from the accounts of François Arago, Hudson Gurney, George Peacock, Thomas Pettigrew, and Alexander Wood. Also see H. S. Rovell, "Thomas Young and Gottingen."
2. J. R. Partington, *History of Chemistry*, III, 728–729.
3. *The Archives of the Royal Institution of Great Britain in Facsimile*, II and III, *passim*. My account of Young's career at the Royal Institution is based on G. N. Cantor, "Thomas Young's Lectures at the Royal Institution." Also see Morris Berman, "The Early Years of the Royal Institution, 1799–1810: A Re-evaluation"; and Henry Bence Jones, *The Royal Institution*, 188–257.

4. William Munk, *The Roll of the Royal College of Physicians of London*, III, 80–88; and *Works of Sir Benjamin Collins Brodie*, I, 90–93.

5. Young's anonymous publications are listed in Hudson Gurney, *Memoir of the Life of Thomas Young*, 51–62. Hill Shine and Helen Shine, in *The Quarterly Review Under Gifford*, list a few other articles that may be by Young, *passim*.

6. *Record of the Royal Society of London* (1897), 208, and Sir Henry Lyons, *The Royal Society 1660–1940*, 220, 223, 227, 243, 245.

7. Gurney, *op. cit.*, 42.

8. G. N. Cantor's articles on Young have been very influential on my thinking, but I alone am responsible for this interpretation of Young's personality.

9. George Peacock, *Life of Thomas Young*, p. 397.

10. Young, "Observations on Vision," *Works*, I, 5.

11. *Ibid.*, 6.

12. Everard Home, "On the Mechanism Employed in Producing Muscular Motion," 1.

13. Young, *Works*, I, 20–21.

14. *Ibid.*, 21–36.

15. *Ibid.*, 37.

16. *Ibid.*, 37–40.

17. *Ibid.*, 41.

18. *Ibid.*, 41–43.

19. *Ibid.*, 46–48.

20. *Ibid.*, 51.

21. "On the Theory of Light and Colours," *ibid.*, 146, 147. "I use the word undulation, in preference to vibration, because vibration is generally understood as implying a motion which is continued alternately backwards and forwards, by a combination of the momentum of the body with an accelerating force, and which is naturally more or less permanent; but an undulation is supposed to consist in a vibratory motion, transmitted successively through different parts of a medium, without any tendency in each particle to continue its motion, except in consequence of the transmission of succeeding undulations, from a distinct vibrating body; as, in the air, the vibrations of a chord produce the undulations constituting sound" [p. 143].

22. *A Course of Lectures on Natural Philosophy and the Mechanical Arts*, I, 439.

23. *Ibid.*, II, 315.

24. Alexander Wood, *Thomas Young, Natural Philosopher 1773–1829*, 113. Also see James P. C. Southall, ed., *Helmholtz's Treatise on Physiological Optics*, II, 143–146.

25. For a discussion of Brougham's reviews, see the traditional accounts by Peacock and Wood in their biographies of Young. Their interpretation has recently been challenged by G. N. Cantor, "Henry Brougham and the Scottish Methodological Tradition"; and Edgar W. Morse, "Natural Philosophy, Hypotheses, and Impiety," ch. 2.

26. Peacock, *op. cit.*, 90–91; Andrew Dalzel, *History of the University of Edinburgh*, I, 144, 161; Young, *Works*, I, 199.

27. Young, *Works*, I, 199, 81, 81n.

28. Isaac Newton, *Opticks*, 362–370 (query 28). Also see A. W. Badcock, "Physical Optics at the Royal Society, 1660–1800." Badcock's article should be compared with the interpretation in Henry Steffens, "The Development of Newtonian Optics in England, 1738–1831." Steffens' thesis is one of the first systematic reinterpretations of the history of eighteenth-century optics. His conclusions are similar to mine.

29. Sir Edmund Whittaker, *A History of the Theories of Aether and Electricity*. I, *The Classical Theories*, 97–98.

30. Quoted in Badcock, *op. cit.*, 102.

31. Henry Brougham, "Experiments and Observations on the Inflection, Reflection, and Colours of Light"; and "Further Experiments and Observations on the Affections and Properties of Light."

32. "Outlines of Experiments and Inquiries Respecting Sound and Light," *Works*, I, 79.

33. *Ibid.*

34. *Ibid.*

35. *Ibid.*, 78–79.

36. *Ibid.*, 80.

37. *Ibid.*, 81.

38. "A Reply to the Animadversions of the Edinburgh Reviewers . . .," in *Works*, I, 202.

39. *Works*, I, 140–169.

40. *Ibid.*, p. 157. Original in italics.

41. *Ibid.*, 159.

42. *Ibid.*, 160–166.

43. *Ibid.*, 166.

44. "An Account of Some Cases of the Production of Colours not Hitherto Described," *ibid.*, 170.

45. *Ibid.*, 171.

46. *Ibid.*, 171–172.

47. *Ibid.*, 172–173. Young later designed an instrument, which he called an eriometer, that used these interference halos as a measure of the average size of the fibers in wool or particles suspended in a fluid. See Young, *Works*, I, 343, 172, 305.

48. Augustin Fresnel, "Mémoire sur la diffraction de la lumière," 282–364. Also see A. Rubinowica, "Thomas Young," which gives a different interpretation of Young's explanation. On the colors of the halos see Carl Boyer, *The Rainbow*, esp. 245–246. These halos are caused by refraction in ice crystals. The smaller halo was first explained by Mariotte in 1679 and the larger by Henry Cavendish in a conversation with Young. See Young, *Course of Lectures*, I, 443–444.

49. Young, *Works*, I, 174–175.

50. *Ibid.*, 180.

51. *Ibid.*, 181–184.

52. Supernumerary rainbows are the result of interference effects; the colors of natural bodies are not. See Boyer, *op. cit.*, *passim*, for a discussion of supernumerary rainbows.

53. Young, *Works*, I, 187–188.

54. Young, *Course of Lectures*, I, 457–471, 443–444.

55. Young, *Works*, I, 247–254.

56. "Popular Statement of the Beautiful Experiments of Malus . . .," 345.

57. Young, *Works*, I, 251.

58. Alexander Wood, *Thomas Young*, 181–183. Also see Morse, *loc cit.*

59. Young, *Works*, I, 383.

60. *Ibid.*, 332–334.

61. Augustin Fresnel, "Mémoire sur la diffraction de la lumière," 262–335.

62. Fresnel, "Mémoire sur l'influence de la polarisation," 394n.

63. Young, *Works*, I, 401–402. Translation in Alexander Wood, *Thomas Young*, 199.

64. Young, *Works*, I, 415.

65. S. Lilley, " 'Nicholson's Journal' (1792–1813)." Also see David M. Knight, *Natural Science Books in English 1600–1900*, esp. ch. 11.

66. Young, *Course of Lectures*, I, 78.

67. The term used by Helmholtz and the other "discoverers" of the conservation of energy had been "conservation of force." Young's work seems to have been rooted in the *vis viva* controversies of the eighteenth century. Also see D. S. L. Cardwell, "Early Development of the Concepts of Power, Work, and Energy."

68. Young, *Course of Lectures*, I, 137. Also see Young, *Works*, II, 129.

69. W. Thomson and P. G. Tait, *Natural Philosophy*, secs. 686, 687; W. Thomson, "Elasticity," secs, 41, 42. Also see the discussion by Isaac Todhunter in *History of the Theory of Elasticity*, I, 82–83.

70. Young, *Course of Lectures*, I, 583–588. His principal innovation in the *Lectures* was to devise a map of the times

of simultaneous high water around the British Isles (I, pl. 38, fig. 21). He did not, however, discuss this concept or its potential for generalization. The term "cotidal map" now used was coined in 1833 by Whewell: "Essay Towards . . . a Map of Co-tidal Lines."

71. Young, *Works*, II, 262–290.
72. *Ibid.*, 291–335.
73. "Tides and Waves," in *Encyclopaedia metropolitana* (1842).
74. Young, *Works*, II, 262n.
75. *Supplement* to the *Encyclopaedia Britannica*, VI, 788.
76. Young, *Works*, II, 8–28, 93–98; 99–101; *Quarterly Journal of Science, Literature and the Arts*, **22** (1827), 365–367.
77. Wood, *Thomas Young*, chs. 9, 10. Young, *Works*, III, is devoted almost entirely to his writings on languages. It also includes correspondence related to his controversies about priorities.

BIBLIOGRAPHY

I. ORIGINAL WORKS. Many of Young's writings are reprinted in *Miscellaneous Works of the Late Thomas Young, M.D., F.R.S., . . .*, George Peacock and John Leitch, eds., 3 vols. (London, 1855). An appendix in Alexander Wood and Frank Oldham, *Thomas Young, Natural Philosopher 1773–1829* (Cambridge, 1954), reproduces the bibliography of his articles that is given in the Royal Society *Catalogue of Scientific Papers*; it is limited, however, to articles published after 1799. An indispensable supplementary bibliography, probably taken from Young's own records, is given in Hudson Gurney, *Memoir of the Life of Thomas Young* (London, 1831). Young's articles for the *Supplement* to the *Encyclopaedia Britannica* are listed in Wood, *Young*, 258. (Wood's discussion implies that Young also wrote the article "Craniology." According to the signature on the article, however, it was written by Peter Mark Roget.) Some of Young's biographical articles are mentioned in Wood and reprinted in Young's *Works*. The full list is in Gurney's biography. The most complete list of Young's anonymous articles in the *Quarterly Review* is in Hill and Helen Shine, *The Quarterly Review Under Gifford. Identification of Contributors 1809–1824* (Chapel Hill, N.C., 1949).

With few exceptions Young's MSS are not in public ownership. Some of his correspondence on hieroglyphics is in the British Museum. A few letters, and his official papers as foreign secretary, are in the library of the Royal Society. Microfilms of the papers of the Board of Longitude (1714–1829) are in the National Maritime Museum, Greenwich.

Young's books include *A Course of Lectures on Natural Philosophy and the Mechanical Arts*, 2 vols. (London, 1807; facs. ed., New York, 1971); and *An Introduction to Medical Literature, Including a System of Practical Nosology* (London, 1813).

Among his articles are "Observations on Vision," in his *Works*, I, 1–11; "Bakerian Lecture on the Mechanism of the Eye," *ibid.*, 12–63; "Outlines of Experiments and Inquiries Respecting Sound and Light," *ibid.*, 64–98; "Bakerian Lecture on the Theory of Light and Colours," *ibid.*, 140–169; "An Account of Some Cases of Colours not Hitherto Described," *ibid.*, 170–178; "Experiments and Calculations Relative to Physical Optics," *ibid.*, 179–191; "A Reply to the Animadversions of the Edinburgh Reviewers," *ibid.*, 192–215; "Review of the Memoirs of Arcueil," *ibid.*, 234–259; "Chromatics," *ibid.*, 279–342; letter from Young to Arago, *ibid.*, 380–384; "Polarisation. Addendum," *ibid.*, 412–417; "Remarks on the Probabilities of Error in Physical Observations, and on the Density of the Earth, Considered With Regard to the Reduction of Experiments on the Pendulum," *ibid.*, II, 8–28; "On the Resistance of the Air. Determined From Captain Kater's Experiments on the Pendulum," *ibid.*, 93–98; "A Theory of Tides, Including the Consideration of Resistance," *ibid.*, 262–290; "Tides," *ibid.*, 291–335; "Weights and Measures," in *Supplement* to the *Encyclopaedia Britannica*, VI (1824), 785–796; and "Note on Professor Svanberg's Reduction of the Length of the Pendulum," in *Quarterly Journal of Science, Literature and the Arts*, **22** (1827), 365–367.

II. SECONDARY LITERATURE. Besides the biographies cited above, see François Arago, "Thomas Young," in *Biographies of Distinguished Scientific Men*, translated by W. H. Smith, Baden Powell, and Robert Grant (London, 1857), 472–518; George Peacock, *Life of Thomas Young, M.D., F.R.S., . . .* (London, 1855); and Thomas Pettigrew, "Thomas Young," in *Medical Portrait Gallery* (London, 1840).

Also see George Biddell Airy, "Tides and Waves," in *Encyclopaedia metropolitana* (1842); *The Works of Sir Benjamin Collins Brodie With an Autobiography*, collected and arranged by Charles Hawkins, 3 vols. (London, 1865); Henry Brougham, "Experiments and Observations on the Inflection, Reflection, and Colours of Light," in *Philosophical Transactions of the Royal Society*, **86** (1796), 227–277; and "Further Experiments and Observations on the Affections and Properties of Light," *ibid.*, **87** (1797), 352–385; Andrew Dalzel, *History of the University of Edinburgh From Its Foundation. With a Memoir of the Author*, 2 vols. (Edinburgh, 1862), I, 68, 118, 120, 137–140, 142–144, 148–150, 159–162, 191–193, 205–208, 212–214, 223–225; Augustin Fresnel, "Mémoire sur la diffraction de la lumière," in *Oeuvres complètes d'Augustin Fresnel*, I (Paris, 1866), 247–382; and "Mémoire sur l'influence de la polarisation dans l'action que les rayons lumineux exercent les uns sur les autres," *ibid.*, 385–409; Everard Home, "The Croonian Lecture. On the Mechanism Employed in Producing Muscular Motion," in *Philosophical Transactions of the Royal Society*, **85** (1795), 1; Sir Isaac Newton, *Opticks* (New York, 1952); "Popular Statement of the Beautiful Experiments of Malus, in Which He Has Developed a New Property of Light," in *Nicholson's Journal of Natural Philosophy*, **33** (1812), 344–348; *The Archives of the Royal Institution of Great Britain in Facsimile. Minutes of Managers' Meetings, 1799–1900*, 4 vols. in 3 (Menton, England, 1971–1973), II, 203, 205; III, 129, 148, 154; and William Whewell, "Essay Towards a First Approximation to a Map of Co-tidal Lines," in *Philosophical Transactions of the Royal Society*, **123** (1833), 147–236.

The standard histories of optics are Ernst Mach, *The Principles of Optics, an Historical and Philosophical Treatment* (New York, 1953); Vasco Ronchi, *Histoire de la lumière*, translated by Juliette Taton (Paris, 1956); and Sir Edmund Whittaker, *A History of the Theories of Aether and Electricity*, I, *The Classical Theories* (New York, 1960).

Other works that deal with aspects of Young's career include A. W. Badcock, "Physical Optics at the Royal Society, 1600–1800," in *British Journal for the History of Science*, **1** (1962), 99–116; Morris Berman, "The Early Years of the the Royal Institution, 1799–1810; A Re-evaluation," in *Science Studies*, **2** (1972), 205–240; Carl Boyer, *The Rainbow From Myth to Mathematics* (New York, 1959); Geoffrey N. Cantor, "The Changing Role of Young's Ether," in *British Journal for the History of Science*, **5** (1970), 44–62; "Thomas Young's Lectures at the Royal Institution," in *Notes and Records. Royal Society of London*, **25** (1970), 87–112; and "Henry Brougham and the Scottish Methodological Tradition," in *Studies in History and Philosophy of Science*, **2** (1971), 69–89; D. S. L. Cardwell, "Early Development of the Concepts of Power, Work, and Energy," in *British Journal for the History of Science*, **3** (1967), 209–224; *Helmholtz's Treatise on Physiological Optics*, James P. C. Southall, ed., 3 vols. in 2 (New York, 1962); Henry Bence Jones, *The Royal Institution: Its Founder and Its First Professors* (London, 1871), 188–257; David M. Knight, *Natural Science Books in English 1600–1900* (London, 1972); S. Lilley, " 'Nicholson's Journal' (1787–1813)," in *Annals of Science*, **6** (1948), 78–101; and Sir Henry Lyons, *The Royal Society 1660–1940; a History of Its Administration Under Its Charters* (New York, 1968), 220, 243.

Also see Edgar W. Morse, "Natural Philosophy, Hypotheses, and Impiety: Sir David Brewster Confronts the Undulatory Theory of Light" (Ph. D. diss., Univ. of California, Berkeley, 1972); William Munk, *Roll of the Royal College of Physicians of London*, III (London, 1878), 80–88; James R. Partington, *A History of Chemistry*, III (London, 1962), 728–729; *Record of the Royal Society of London* (London, 1897), 208, 251, 369; H. S. Rovell, "Thomas Young and Gottingen," in *Nature*, **88** (1912), 516; A. Rubinowica, "Thomas Young and the Theory of Diffraction," *ibid.*, **180** (1958), 160–162; Henry Steffens, "The Development of Newtonian Optics in England, 1738–1831" (M. A. thesis, Cornell Univ., 1965); Sir William Thomson, "Elasticity," in *Encyclopaedia Britannica*, 9th ed.; Sir William Thomson and Peter Guthrie Tait, *Treatise on Natural Philosophy* (Oxford, 1867); and Isaac Todhunter, *A History of the Theory of Elasticity and of the Strength of Materials From Galilei to the Present Time*, 2 vols. (Cambridge, 1886–1893), I, 82–83.

EDGAR W. MORSE

YOUNG, WILLIAM HENRY (*b.* London, England, 20 October 1863; *d.* Lausanne, Switzerland, 7 July 1942), *mathematics.*

Young was the eldest son of Henry Young and Hephzibah Jeal. The Young family had been bankers in the City for some generations. Young went to the City of London School, of which the headmaster, Edwin A. Abbott, author of the mathematical fantasy *Flatland*, recognized his flair for mathematics. Young entered Peterhouse, Cambridge, in 1881. In the mathematical tripos of 1884 he was expected to be senior wrangler but was placed fourth. In later years he related that he refused to restrict his interests (intellectual and athletic) to the intensive training in mathematics necessary for the highest place in the order of merit. The first books he borrowed from the College library were the works of Molière. Instead of writing a mathematical essay for a Smith's prize, he competed for and won a prize in theology. He was of Baptist stock and, at Cambridge, was baptized into the Church of England.

Young was a fellow of Peterhouse from 1886 to 1892, but he held no official position in the college or the university. It is surprising that between the ages of twenty-five and thirty-five he did not turn to research, but deliberately set himself to earn a large income and accumulate savings by private teaching of undergraduates from early morning until late at night.

In 1896 Young married Grace Emily, daughter of Henry W. Chisholm. She had taken the mathematical tripos and was ranked equal to a wrangler; as Grace Chisholm Young, she became a mathematician of international reputation. At the end of their first year together, she said, " . . . he proposed, and I eagerly agreed, to throw up lucre, go abroad, and devote ourselves to research." They lived mainly in Göttingen until 1908 and then in Switzerland, first in Geneva and later in Lausanne.

In striking contrast with most mathematicians, Young did hardly any research until he was over thirty-five, but between 1900 and 1924 he wrote more than two hundred papers. At the turn of the century, the theory of real functions was subject to artificial and unaesthetic restrictions. For instance, the standard process (Riemann's) of reconstructing an integral from its derivative required the continuity of the derivative. In the late 1890's the Paris school, led by Baire and Borel, laid the foundations of an essentially more powerful theory, based on the concept of the measure of a set of points. Lebesgue's famous thesis, "Intégrale, longueur, aire," appeared in 1902. Young, working independently, arrived at a definition of integration, different in form from, but essentially equivalent to, Lebesgue's. He was anticipated by about two years, and it must have been a heavy blow to one who had become conscious of his power to make fundamental discoveries; but he bore the disappointment magnanimously, and himself called the integral that of Lebesgue. Many aspects of the later development are Young's own, notably his method of monotone sequences as used in the Stieltjes integral.

Young showed supreme power in two other fields of analysis. The first is the theory of Fourier series. In 1912 he established the connection between the sum of the qth powers of the Fourier constants of a function f and the integral of f^p, where p and q are conjugate indices and q is an even integer. The completion for unrestricted q was achieved after eleven years by Hausdorff. Young proved many other theorems, some of striking simplicity and beauty, about Fourier series and more general orthogonal series. The second field — in which lay what was probably Young's most far-reaching work — was the basic differential calculus of functions of more than one variable. The best tribute to it is that, since 1910, every author of an advanced calculus textbook has adopted Young's approach.

Every word and every movement of Young gave evidence of restless vitality. His appearance was striking; after his marriage he grew a beard — red in contrast with his dark hair — and wore it very long in later years. Of his three sons and three daughters, Professor Laurence Chisholm Young and Dr. Rosalind Cecily Tanner continued their parents' work in pure mathematics. The eldest son was killed flying in France in 1917.

Young held part-time chairs at Calcutta (1913–1916) and Liverpool (1913–1919), and he was professor at Aberystwyth from 1919 to 1923. More than once electors to a chair passed him over in favor of men less powerful as mathematicians but less exacting as colleagues. He was an honorary doctor of the universities of Calcutta, Geneva, and Strasbourg; and his honors included the Sylvester Medal of the Royal Society (1928). He was president of the International Union of Mathematicians in 1929–1936.

When France fell in 1940 he was at Lausanne, cut off from his family, and he had to remain there, unhappy and restive, for the last two years of his life.

BIBLIOGRAPHY

I. ORIGINAL WORKS. Young wrote more than 200 papers; for a list of the most important of them see the obituary notices below. His books are *The First Book of Geometry* (London, 1905), written with Grace Chisholm Young, an excellent and original book doubtless composed for the education of their children; *The Theory of Sets of Points* (Cambridge, 1906), written with Grace Chisholm Young; and *The Fundamental Theorems of the Differential Calculus* (Cambridge, 1910).

II. SECONDARY LITERATURE. See *Obituary Notices of Fellows of the Royal Society of London*, **3** (1943), 307–323, with portrait; *Journal of the London Mathematical Society*, **17** (1942), 218–237; and *Dictionary of National Biography, 1941–1950* (1959), 984–985.

J. C. BURKILL

YULE, GEORGE UDNY (*b.* Morham, near Haddington, Scotland, 18 February 1871; *d.* Cambridge, England, 26 June 1951), *statistics*.

Yule was the son of Sir George Udny Yule and his wife, Henrietta Peach. Sir George was an administrator in the Indian Civil Service and a member of an old Scottish farming family with a history of some government, military, and literary distinction. In 1875 Sir George moved his family to London; and Yule was sent first to a day school there, and then to a preparatory school near Rugby. He was subsequently educated at Winchester College and University College, London, where, between 1887 and 1890, he read civil engineering but did not take a degree, there being none in the subject at the time. After two years' training in a small engineering works, however, Yule decided against engineering as a career and spent a year under Heinrich Hertz at Bonn, investigating the passage of electric waves through dielectrics. This was the subject of his first published paper.

Yule returned to London in 1893, at the invitation of Karl Pearson, to become demonstrator (lecturer) at University College. In 1896 he was promoted to assistant professor of applied mathematics. Three years later he married May Winifred Cummings, but the marriage was annulled in 1912. One consequence of his marriage was that Yule felt obliged to earn a higher salary, and he accepted a dreary administrative post at the City and Guilds of London Institute. Between 1902 and 1909 he held concurrently a lectureship in statistics at University College, delivering evening lectures that were the basis of his first book, *An Introduction to the Theory of Statistics* (1911). This was for long the only comprehensive textbook on the subject (the fourteenth edition, revised by M. G. Kendall, appeared in 1958). The *Introduction*, and his reputation as a lecturer, led to his being offered the newly created lectureship in statistics at Cambridge in 1912.

Yule soon had a range of practical statistical experience. He was statistician to the School of Agriculture at Cambridge while he was university lecturer; and during World War I he was statistician first to the director of army contracts, and later to the Ministry of Food. In 1922 Yule became a fellow of both St. John's College, Cambridge, and the Royal Society. He resigned his university position in 1931. The following year he obtained a pilot's license; but a serious heart ailment obliged him to spend most of his retirement in a quieter pursuit, a study of the statistical aspects of literary style.

Yule's principal achievements in statistical theory concern regression and correlation, association, time series, Mendelian inheritance, and epidemiology. His early memoirs on correlation (1897, 1907) and association (1900) have proved to be fundamental. In the first of these he introduced the concept of partial ("net") correlation, and in 1907 he demonstrated that the sampling distributions for partial correlation coefficients are of the same form as those for total correlation coefficients. In the paper of 1900, Yule presented the coefficient of association for the measurement of the degree of association in 2×2 contingency tables. His introduction of this coefficient led to a long controversy with his former mentor and friend Karl Pearson, who joined forces with David Heron in a protest that M. G. Kendall said was "remarkable for having missed the point over more pages (173) than perhaps any other memoir in statistical history" (*The Advanced Theory of Statistics*, I [1943], 322).

From 1912 Yule and Major M. Greenwood laid the foundations of the theory of accident distributions. In 1921 he wrote on time correlation and began work leading to a well-known paper on sunspots (1927) that marked the beginning of the modern theory of oscillatory time series.

Yule introduced many new ideas into statistical theory and corrected many errors, especially in biometrics. He made important studies of the mathematics of biological evolution and of the statistics of agricultural field trials. He was perhaps the first to consider (1902) whether the observed correlations between parents and offspring could be accounted for by multifactorial Mendelian inheritance, a problem taken up later by R. A. Fisher.

BIBLIOGRAPHY

Yule wrote two books: *An Introduction to the Theory of Statistics* (London, 1911) and *The Statistical Study of Literary Vocabulary* (Cambridge, 1944). A bibliography of 71 scientific papers, plus 12 on other subjects, is in the notice by F. Yates, in *Obituary Notices of Fellows of the Royal Society of London*, **8** (1952), 309–323, with portrait.

J. D. NORTH

IBN YŪNUS, ABUʾL-ḤASAN ʿALĪ IBN ʿABD AL-RAḤMĀN IBN AḤMAD IBN YŪNUS AL-ṢADAFĪ (*d.* Fusṭāṭ, Egypt, 1009), *astronomy, mathematics.*

Ibn Yūnus was one of the greatest astronomers of medieval Islam. He came from a respected family, his great-grandfather Yūnus having been a companion of the famous legal scholar al-Shāfiʿī and his father, ʿAbd al-Raḥmān, being a distinguished historian and scholar of *ḥadīth* (the sayings of Muḥammad). Besides being famous as an astronomer and astrologer, Ibn Yūnus was widely acclaimed as a poet, and some of his poems have been preserved. Unfortunately nothing of consequence is known about his early life or education.

We know that as a young man Ibn Yūnus witnessed the Fatimid conquest of Egypt and the foundation of Cairo in 969. In the period from 977 to 996, which corresponds roughly to the reign of Caliph al-ʿAzīz, he made astronomical observations that were renewed by order of Caliph al-Ḥakim, who succeeded al-ʿAzīz in 996 at the age of eleven and was much interested in astrology. Ibn Yūnus' recorded observations continued until 1003.

Ibn Yūnus' major work was *al-Zīj al-Ḥākimī al-kabīr*, *zīj* meaning an astronomical handbook with tables. It is a particularly fine representative of a class of astronomical handbooks, numbering perhaps 200, compiled in medieval Islam. The *Zīj* of Ibn Yūnus was dedicated to Caliph al-Ḥakim and was aptly named *al-kabīr* ("large"). The text of the first forty-four of the eighty-one chapters of the original work is twice the length of the text of the *Zīj* by al-Battānī and contains more than twice as many tables as the earlier work. The only extant chapters of the *Ḥakimī zīj* are in two unpublished manuscripts at Leiden and Oxford, comprising about three hundred folios. A manuscript in Paris contains an anonymous abridgment of part of the *Zīj* and is a source for some additional chapters up to chapter 57, and chapters 77–81.

The importance of Ibn Yūnus was realized in the West when the Leiden manuscript was first seriously studied. In 1804, Armand-Pierre Caussin de Perceval published the text of Ibn Yūnus' observational reports with a French translation. He also included the introduction to the *Zīj*, which contains the titles of the eighty-one chapters. J.-J. Sédillot's translation (now lost) of the Leiden and Paris manuscripts was summarized by Delambre in 1819. The German scholar Carl Schoy published several articles containing translations and analyses of individual chapters of the *Zīj* relating to spherical astronomy and sundial theory.

The *Ḥākimī zīj* deals with the standard topics of Islamic astronomy but is distinguished from all other extant *zījes* by beginning with a list of observations made by Ibn Yūnus and of observations made by some of his predecessors, quoted from their works. Despite the critical attitude of Ibn Yūnus toward these earlier scholars and his careful recording of their observations and some of his own, he completely neglects to describe the observations that he used in establishing his own planetary parameters—nor does he indicate whether he used any instruments for these observations. Indeed, the *Ḥākimī zīj* is a poor source of information about the instruments used by Ibn Yūnus. In his account of measurements of the latitude of Fusṭāṭ and of the obliquity of the ecliptic from solar meridian altitudes at the solstices, Ibn Yūnus states that he used an instrument provided by Caliph al-ʿAzīz and Caliph al-Ḥākim. Although he describes it only by mentioning that the divisions for each minute of arc were clearly visible on its scale, the instrument was probably a large meridian ring. His only other references to instruments used for simple observations are to an astrolabe and a gnomon.

In view of the paucity of this information, it is remarkable that the statement that Ibn Yūnus worked in a "well-equipped observatory" is often found in popular accounts of Islamic astronomy. A. Sayili, *The Observatory in Islam*, has shown how this notion gained acceptance in Western literature.

There are two sources, however, that might cast a little more light on the situation if their reliability could be established. First, the historian Ibn Ḥammād (*fl. ca.* 1200) mentions a copper instrument, resembling an astrolabe three cubits in diameter, that a contemporary of his had seen and associated with the Ḥākimī observations. Likewise the Yemenite Sultan al-Ashraf (*fl. ca.* 1290), who was an astronomer, records that al-Ḥākim had an armillary sphere consisting of nine rings, each of which weighed 2,000 pounds and was large enough

for a man to ride through on horseback. The possibility that this large instrument was that known to have been constructed in Cairo about 1125—over a century after the death of Ibn Yūnus—cannot yet be discounted.

There is evidence that al-Ḥākim had a house on the Muqaṭṭam hills overlooking Cairo, which may have contained astronomical instruments: Ibn Yūnus is known to have visited this house on one occasion to make observations of Venus. Nevertheless, al-Ḥākim's unsuccessful attempt to build an observatory in Cairo took place after Ibn Yūnus' death; and the only locations mentioned by Ibn Yūnus in his own accounts of his observations are the Mosque of Ibn Naṣr al-Maghribī at al-Qarāfa, and the house of his great-grandfather Yūnus, in nearby Fusṭāṭ. A note written in the fifteenth century on the title folio of the Leiden manuscript of the *Ḥākimī zīj* states that Ibn Yūnus' observations were made in the area of Birkat al-Ḥabash in Fusṭāṭ.

Ibn Yūnus explains in the introduction to his *Zīj* that the work is intended to replace the *Mumtaḥan zīj* of Yaḥyā ibn Abī Manṣūr, prepared for the Abbasid Caliph al-Maʾmūn in Baghdad almost 200 years earlier. Ibn Yūnus reports the observations of some astronomers before his own time, in which what was observed was at variance with what was calculated with the tables of the *Mumtaḥan zīj*. When reporting his own observations, Ibn Yūnus often compares what he observed with what he had computed with the *Mumtaḥan* tables.

From the introduction and chapters 4, 5, and 6 of the *Ḥākimī zīj*, which contain the observation accounts, it is clear that Ibn Yūnus was familiar with the *zījes* of Ḥabash al-Ḥāsib, al-Battānī, and al-Nayrīzī, as well as the *Mumtaḥan zīj*. The observations made by Ḥabash that Ibn Yūnus quotes are not in the two extant versions of Ḥabash's *Zīj*. Ibn Yūnus also records observations made by al-Māhānī, whose works are not extant. He lists the planetary parameters of the *Mumtaḥan zīj*, and this has enabled the positive identification of at least the planetary tables in the only extant manuscript of this early work, which contains considerable spurious material. Ibn Yūnus also quotes observations made by the Banū Amājūr family in Baghdad; their five *zījes* are not extant. Other works quoted by Ibn Yūnus, although not necessarily directly, are the *zījes* of al-Nihāwandī, Ibn al-Adamī, the Banū Mūsā, Abū Maʿshar, Ibn al-Aʿlam, al-Ṣūfī, and Muḥammad al-Samarqandī; none of these works is extant, and Ibn Yūnus' references provide valuable information about them.

The observations described by Ibn Yūnus are of conjunctions of planets with each other and with Regulus, solar and lunar eclipses, and equinoxes; he also records measurements of the obliquity of the ecliptic (chapter 11) and of the maximum lunar latitude (chapter 38). All of these accounts are notable for their lack of information on observational procedures. The following passage is a translation of one of Ibn Yūnus' accounts of a planetary conjunction that he had observed:

A conjunction of Venus and Mercury in Gemini, observed in the western sky: The two planets were in conjunction after sunset on the night whose morning was Monday, the thirteenth day of Jumādā II 390 Hegira era. The time was approximately eight equinoctial hours after midday on Sunday, which was the fifth day of Khardādh, 369 Yazdigird era. Mercury was north of Venus and their latitude difference was a third of a degree. According to the *Mumtahan Zīj* their longitude difference was four and a half degrees [A. P. Caussin de Perceval, "Le livre de la grande table Hakémite," in *Notices et extraits des manuscrits de la Bibliothèque nationale*, **7** (1804), p. 217].

The Sunday mentioned was 19 May 1000, and computation with modern tables confirms that there was a conjunction in longitude that evening and that Mercury was indeed one-third degree north of Venus. About forty such planetary conjunctions observed by Ibn Yūnus are described in the *Zīj*.

The following passage is a translation of Ibn Yūnus' account of the lunar eclipse that occurred on 22 April 981 (Oppolzer no. 3379):

This lunar eclipse was in the month of Shawwāl, 370 Hegira era, on the night whose morning was Friday, the third day of Urdibihisht, 350 Yazdigird era. We gathered to observe this eclipse at al-Qarāfa, in the Mosque of Ibn Naṣr al-Maghribī. We perceived first contact when the altitude of the moon was approximately 21°. About a quarter of the lunar diameter was eclipsed, and reemergence occurred about a quarter of an hour before sunrise [A. P. Caussin de Perceval, p. 187].

Some of the thirty eclipses reported by Ibn Yūnus were used by Simon Newcomb in his determination of the secular acceleration of the moon. More recently, other observations recorded in the *Ḥākimī zīj* have been used by R. Newton.

The first chapter of the *Zīj* is the longest of the extant chapters and deals with the Muslim, Coptic, Syrian, and Persian calendars. There are detailed instructions for converting a date in one calendar

to any of the other calendars, and extensive tables for that purpose. There are also tables for determining the dates of Lent and Easter in both the Syrian and the Coptic calendars. Such tables are found in several Islamic *zījes*.

Chapters 7 and 9, on planetary longitudes, contain instructions for determining true longitudes from the tables of mean motion and equations. No theory is described, but the theory underlying the instructions and tables is entirely Ptolemaic. The mean motions differ from those used by Ibn Yūnus' predecessors, and his values for the sun and moon were deemed sufficiently reliable by al-Ṭūsī to be used in the *Īlkhānī zīj* 250 years later. Ibn Yūnus' planetary tables are computed for both the Muslim and Persian calendars, and define the mean positions of the sun, moon, and planets, as well as the astrologically significant "comet" *al-kayd*, for over 2,700 Muslim and 1,800 Persian years from the respective epochs 622 and 632.

For the year 1003, Ibn Yūnus gives the solar apogee as Gemini 26;10° and the maximum solar equation as 2;0,30°, corresponding to a double eccentricity of 2;6,10° (where the solar deferent radius is 60). No solar observations made by Ibn Yūnus are recorded in the *Zīj*. He changes the values of the lunar epicyclic radius and the eccentricity from Ptolemy's 5;15° and 10;19°, also used in the *Mumtahan zīj*, to 5;1,14° and 11;7°, respectively (the latter is not used consistently), again without explanation. His planetary equation tables are identical with those of Ptolemy's *Handy Tables* and the *Mumtahan zīj* for Saturn, Jupiter, and Mars. For Venus, Ibn Yūnus assumes an eccentricity exactly half that of the sun and uses an epicyclic radius of 43;42 rather than Ptolemy's 43;10. For Mercury he adopts a maximum equation of 4;2°, an Indian parameter previously used in the *Zīj* of al-Khwārizmī, rather than Ptolemy's 3;2°. Ibn Yūnus' tables of equations for the moon, Venus, and Mercury contain the same inconsistencies as al-Battānī's tables for Venus, in that some of the columns are not adjusted for the new parameters; this is a fairly common feature in Islamic *zījes*. There is evidence that Ibn Yūnus was not altogether satisfied with his determination of the planetary apogees: the *Ḥākimī zīj* contains three different sets of values (chapters 6, 8, and 9).

In his discussion of solar and lunar distances (chapters 55, 56), Ibn Yūnus assumes a maximum solar parallax of 0;1,57°, instead of Ptolemy's value of 0;2,51°. Chapters 59–75, on parallax and eclipse theory and the associated tables, are not in the known manuscripts; and their recovery in other sources would be extremely valuable for the study of Islamic astronomy.

In chapter 38, on lunar and planetary latitudes, Ibn Yūnus states that he found the maximum lunar latitude to be 5;3°. Although he says that he measured it many times and repeatedly found this value, he does not say how the measurements were made. He did not pursue the suggestion of the Banū Amājūr that he quotes: that the maximum lunar latitude was not constant. His planetary latitude tables are derived from those in the *Almagest*, except in the case of Venus, for which he used values originally taken from the *Handy Tables*.

Ibn Yūnus measured the position of Regulus as Leo 15;55° in 1003. His value for the motion of the fixed stars is 1° in 70 1/4 Persian years (of 365 days) and apparently was computed by using his own observation of Regulus and that made by Hipparchus; it is the most accurate of all known Islamic values. He had information at his disposal from which he might have deduced that the motion of the planetary apogees was different from the motion of the fixed stars, but he chose to conclude that the apogees moved at the same rate as the stars (chapter 8).

The trigonometric functions used by Ibn Yūnus are functions of arcs rather than angles, and are computed for radius 60, as was standard in Islamic works. Chapter 10 of the *Zīj* contains a table of sines for each 0;10° of arc, computed to four significant sexagesimal digits. The values are seldom in error by more than ±2 in the fourth digit. Ibn Yūnus determined the sine of 1° to be 1;2,49,43,28 (to base 60), using a method equivalent to interpolating linearly between the values of $\sin x/x$ for $x = 15/16°$ and $9/8°$. He then improved this value by a rather dubious technique to obtain 1;2,49,43,4. The accurate value to this degree of precision is 1;2,49,43,11. Ibn Yūnus' younger contemporary al-Bīrūnī was able to calculate the chord of a unit circle subtended by an angle of 1° correctly to five significant sexagesimal digits. Although in chapter 11 of the *Zīj* Ibn Yūnus tabulates the cotangent function to three sexagesimal digits for each ten minutes of arc, he does not take full advantage of it. Many of the methods he suggests throughout the *Zīj* require divisions of sines by cosines; and he uses the cotangent function, which he calls the shadow, only when the argument is an altitude arc.

In spherical astronomy (chapters 12–54) Ibn Yūnus reached a very high level of sophistication. Although none of his formulas is explained, it seems probable that most of them were derived by means of orthogonal projections and analemma

constructions, rather than by the application of the rules of spherical trigonometry that were being developed by Muslim scholars in Iraq and Persia. Altogether, there are several hundred formulas outlined in the *Zīj*, many of which are trivially equivalent. These are stated in words and without recourse to any symbols. For each method outlined, Ibn Yūnus generally gives at least one numerical example. The problems of spherical astronomy discussed in the *Ḥākimī zīj* are more varied than those in most major Islamic *zījes*, and the following examples are intended to illustrate the scope of the treatment.

Ibn Yūnus describes several methods for computing right and oblique ascensions (chapters 13, 14). He also computes both, the latter for each degree of the ecliptic and for each degree of terrestrial latitude from 1° to 48° (beyond which limit, according to Manṣūr ibn 'Irāq, "there is no one who studies this sort of thing or even thinks about it"). Ibn Yūnus discusses in great detail the determination of time and solar azimuth from solar altitude, and it will be clear from the tables mentioned below that he devoted much effort to these problems. Certain functions that he discusses in the text are also tabulated, such as the solar altitude in the prime vertical and the rising amplitude of the sun (that is, the distance of the rising sun from the east point). The problem of finding solar altitude from solar azimuth (chapter 24) is not so simple as the inverse problem; but Ibn Yūnus solves it in several ways, including the use of an algebraic method. He also tabulates the solar altitude for certain azimuths, such as that of the *qibla*, the direction of Mecca (chapter 28), and ten different azimuths (chapter 24), to be used for finding the meridian. Several geometric solutions to the problem of determining the *qibla*, a favorite of the Islamic astronomers, are also outlined. One of Ibn Yūnus' solutions is equivalent to successive applications of the cosine rule and sine rule for spherical triangles, but is derived by a projection method that was also used by the contemporary Egyptian scholar Ibn al-Haytham.

Particularly elegant solutions are presented for finding the meridian from three solar observations on the same day (chapter 23) and for finding the time between two solar observations on the same day (chapter 33). The latter problem is solved by a direct application of the cosine rule for plane triangles, the earliest attested use of this rule. Ibn Yūnus transforms ecliptic to equatorial coordinates (chapter 39) by a method equivalent to the cosine rule for spherical triangles but probably derived by means of an analemma construction. His sundial theory (chapters 26, 27, 35) is also of considerable sophistication. It deals with horizontal and vertical sundials, the latter oriented in the meridian, the prime vertical, or a general direction inclined to both. He proves geometrically that for a horizontal sundial the gnomon shadow measures the altitude of the upper rim of the solar disk, and stresses the precautions to be taken when setting the gnomon on a marble slab to ensure that it is aligned correctly.

The chapters of the *Zīj* dealing with astrological calculations (77–81), although partially extant in the anonymous abridgment of the work, have never been studied. Ibn Yūnus was famous as an astrologer and, according to his biographers, devoted much time to making astrological predictions. His *Kitāb bulūgh al-umniyya* ("On the Attainment of Desire") consists of twelve chapters devoted to the significance of the heliacal risings of Sirius when the moon is in any of the twelve zodiacal signs, and to predictions based on the day of the week on which the first day of the Coptic year falls.

In chapter 10 of the *Ḥākimī zīj* Ibn Yūnus states that he had prepared a shorter version of his major work; this, unfortunately, is no longer extant. There are, however, numerous later *zījes* compiled in Egypt, Persia, and Yemen that are extant and contain material ultimately due to Ibn Yūnus. For example, the thirteenth-century Egyptian *Muṣṭalaḥ zīj*, as well the *Īlkhānī zīj* of al-Ṭūsī and the *Zīj* of Muḥyi'l-Dīn al-Maghribī, both compiled at the observatory in Maragha, Persia, in the thirteenth century, relied on the *Ḥākimī zīj*. Likewise, the *Mukhtār zīj* by the thirteenth-century Yemenite astronomer Abu'l-ʿUqūl is based mainly on a *zīj* by Ibn Yūnus other than the *Ḥākimī*; and an anonymous fourteenth-century Yemenite *zīj* is adapted from the *Ḥākimī zīj*.

There are other sets of tables preserved in the manuscript sources and attributed to Ibn Yūnus that are distinct from those in the *Ḥākimī zīj* but based on them. First, Ibn Yūnus appears to be the author of tables of the sine and tangent functions for each minute of arc, as well as tables of solar declination for each minute of solar longitude. These sine tables display values of the sine function to five sexagesimal digits, which is roughly equivalent to nine decimal digits. The values are often in error in the fourth sexagesimal digit, however, so that it was a premature undertaking. Indeed, over four centuries passed before the compilation of the trigonometric tables in the *Zīj* of Ulugh Beg in Samarkand, in which values are also given to five sexagesimal digits for each minute of arc—but are generally correct. Second, it appears that Ibn Yūnus was the author of an extensive set of tables, called *al-Taʿdīl al-muḥkam*, that display the equations of the sun and moon; the latter are of particular interest. They are based on those in the *Ḥākimī zīj* but are arranged so as to facilitate computation of the lunar position: the equation is tabulated as a function of the double elongation and mean anomaly, both of which can be taken from the mean motion tables; thus there is no need to find the true anomaly. The table, which accurately defines the Ptolemaic lunar equation for Ibn Yūnus' parameters, contains over 34,000 entries.

Ibn Yūnus' second major work was part of the corpus of spherical astronomical tables for timekeeping used in Cairo until the nineteenth century. It is difficult to ascertain precisely how many tables in this corpus, which later became known as the *Kitāb ghayat al-intifaʿ* ("Very Useful Tables"), were actually computed by Ibn Yūnus. Some appear to have been compiled by the late thirteenth-century astronomer al-Maqsī. The corpus exists in numerous manuscript sources, each containing different arrangements of the tables or only selected sets of tables; and in its entirety the corpus consists of about 200 pages of tables, most of which contain 180 entries. The tables are generally rather accurately computed and are all based on Ibn Yūnus' values of 30;0° for the latitude of Cairo and 23;35° for the obliquity of the ecliptic.

The main tables in the corpus display the time since sunrise, the time remaining to midday, and the solar azimuth as functions of the solar altitude and solar longitude. Entries are tabulated for each degree of solar altitude and longitude, and each of the three sets contains over ten thousand entries. The remaining tables in the corpus are of spherical astronomical functions, some of which relate to the determination of the five daily prayers of Islam.

The times of Muslim prayer are defined with reference to the apparent daily motion of the sun across the sky and vary throughout the year. The prayers must be performed within certain intervals of time, which are variously defined. The following general definitions underlie the tables in the corpus. The day is considered to begin at sunset, and the evening prayer is performed between sunset and nightfall. The permitted interval for the night prayer begins at nightfall. The interval for the morning prayer begins at daybreak and the prayer must be completed by sunrise. The period for the noon prayer begins when the sun is on the meridian, and that for the afternoon prayer begins when the shadow of any object is equal to its midday shadow plus the length of the object.

Examples of functions relating to the prayer times, which are tabulated in the corpus for each degree of solar longitude, include the following:

1. The length of morning and evening twilights, defining the permitted times for the morning and evening prayers, based on the assumption that twilight appears or disappears when the sun reaches a particular angle of depression below the horizon. (The angles suggested by Ibn Yūnus in the *Ḥākimī zīj* are 18° for both phenomena, but in a later work he suggests 20° and 16° for morning and evening, respectively. The main twilight tables in the corpus are based on 19° and 17°.)

2. The time from nightfall to daybreak, defining the permitted interval for the night prayer.

3. The time from sunrise to midday.

4. The time from midday to the beginning of the time for the afternoon prayer, defining the interval for the noon prayer; and the time from the beginning of the afternoon prayer to sunset, defining the interval for the afternoon prayer.

5. Corrections to the semidiurnal arc for the effect of refraction at the horizon, apparently based on the assumption that the true horizon is about 2/3° below the visible horizon. (These corrections, which are specifically attributed to Ibn Yūnus, represent the earliest attested quantitative estimate of the effect of refraction on horizon phenomena.)

6. The solar altitude in the azimuth of Mecca, and the time when the sun has this azimuth. (Such tables were used to establish the direction of prayer and the orientation of *miḥrābs* in mosques.)

Virtually all later Egyptian prayer tables until the nineteenth century were based on those in this main corpus. In certain cases the original tables were well-disguised, the entries being written out in words for each day of the Coptic year or a given Muslim year. The impressive developments in astronomical timekeeping in thirteenth-century Yemen and fourteenth-century Syria, particularly the tables of Abu'l-ʿUqūl for Taʿizz and of al-Khalīlī for Damascus, also owe their inspiration to the main Cairo corpus.

It is clear from the biography of Ibn Yūnus by his contemporary al-Musabbiḥī, preserved in the writings of later authors, that Ibn Yūnus was an eccentric. Al-Musabbiḥī describes him as a careless and absent-minded man who dressed shabbily and had a comic appearance. One day, when he was in good health, he predicted his own death in seven days. He attended to his personal business, locked himself in his house, and washed the ink off his manuscripts. He then recited the Koran until he died — on the day he had predicted. According to his biographer, Ibn Yūnus' son was so stupid that he sold his father's papers by the pound in the soap market.

BIBLIOGRAPHY

I. ORIGINAL WORKS. Ibn Yūnus' works are the following:

1. *al-Zīj al-Ḥākimī al-kabīr*: MS Leiden Cod. Or. 143 contains chs. 1–22; MS Oxford Hunt. 331 contains chs. 21–44; MS Paris B.N. ar. 2496 is an anonymous abridgment containing some additional chs. up to 57 and chs. 77–81; MS Leiden Cod. Or. 2813 contains part of ch. 1. Extracts from Ibn Yūnus' mean motion tables are in numerous later sources, such as MS Princeton Yahuda 3475, fols. 16r–21r; and MS Cairo Dār al-Kutub, *mīqāt* 116M.

2. Other *zījes* are not extant. A treatise on the compilation of solar, lunar, and planetary ephemerides, which appears to be taken from a *zīj* by Ibn Yūnus other than the *Ḥākimī*, survives in MS Cairo Dār al-Kutub, *mīqāt* 116M, fols. 8v–9r; and probably in MS Berlin Ahlwardt no. 5742, pt. 2. A fragment of an Egyptian *zīj* containing tables due to Ibn Yūnus is MS Berlin Ahlwardt 5733. The late thirteenth-century Yemenite *Mukhtār zīj*, extant in MS British Museum 768 (Or. 3624), appears to be based on a *zīj* by Ibn Yūnus compiled prior to the *Ḥākimī zīj*.

The following *zījes* are incorrectly attributed to Ibn Yūnus on their title folios: MS Aleppo Awqāf 947; MS Cairo Talʿat, *mīqāt* 138; MSS Paris B.N. ar. 2520 and 2513. The first two are quite unrelated to the Egyptian astronomer. The two Paris MSS are copies of the thirteenth-century Egyptian *Muṣṭalaḥ zīj* and a later recension: they both contain material due to Ibn Yūnus. Two treatises purporting to be commentaries on a *zīj* by Ibn Yūnus—MS Gotha Forschungsbibliothek A1401 and MS Cairo Dār al-Kutub, *mīqāt* 1106—are based on the *Muṣṭalaḥ zīj*.

Short notices on topics in spherical astronomy attributed to Ibn Yūnus are in MS Milan Ambrosiana 281e (C49) and MS Paris B.N. ar. 2506.

3. *Kitāb ghāyat al-intifāʿ* ("Very Useful Tables," a later title given to the corpus). The following sources contain most of the tables: MS Dublin Chester Beatty 3673 and MS Cairo Dār al-Kutub, *mīqāt* 108.

Ibn Yūnus' original solar azimuth tables, entitled *Kitāb al-samt*, are extant in MS Dublin Chester Beatty no. 3673, pt. 1; MS Gotha Forschungsbibliothek no. A1410, pt. 1; MS Cairo Dār al-Kutub, *mīqāt* 137M; and MS Cairo Azhar, *falak* no. 4382, pt. 2.

The tables of time since sunrise, entitled *Kitāb al-dāʾir* and associated with al-Maqsī (*fl.* 1275), are in

MS Gotha Forschungsbibliothek A1402. The hour-angle tables and numerous prayer tables in the version by Ibn al-Kattānī (fl. 1360) are preserved in MS Istanbul Kiliç Ali Paša 684. The hour-angle tables, entitled *Kitāb fadl al-dā'ir*, are copied separately in MS Cairo Taymūriyya, *riyādiyyāt* 191; and MS Cairo Azhar, *falak* no. 4382, pt. 1; they are copied together with the tables of time since sunrise in MS Dublin Chester Beatty no. 3673, pt. 2; and MS Dublin Chester Beatty 4078. The edition of the corpus by al-Bakhāniqī (fl. 1350) is extant in MS Cairo Dār al-Kutub, *mīqāt* 53 and 108.

There are literally dozens of MSS that contain extracts from the corpus in varying degrees of confusion.

MSS Cairo Taymūriyya, *riyādiyyāt* 354; and Dār al-Kutub, *mīqāt* 1207, together constitute a corpus of tables for timekeeping computed for the latitude of Alexandria. In the first the tables are falsely attributed to Ibn Yūnus.

4. *Kitāb al-jayb* (sine tables) are extant in MS Berlin Ahlwardt no. 5752, pt. 1; and MS Damascus Zāhiriyya 3109.

5. *Kitāb al-zill* (cotangent tables) apparently are not extant. The tangent tables in MS Berlin Ahlwardt no. 5767, pt. 3, attributed to Ibn Yūnus are not based on the cotangent tables in the *Hākimī zīj*.

6. *Kitāb al-mayl* (solar declination tables) are MS Berlin Ahlwardt 5752,2.

7. *Kitāb al-ta'dīl al-muhkam* (solar and lunar equation tables) are extant in MS Cairo Dār al-Kutub, *mīqāt* 29, which contains the complete set of lunar tables; MS Gotha Forschungsbibliothek no. A1410, pt. 2, which contains an incomplete set; and MS British Museum Or. 3624, fols. 111v–129r, 113v–151r, which contains the solar tables and a related set of lunar tables.

8. A short treatise on a candle clock is in MS Beirut St. Joseph, Arabe 223/12. This is attributed to Ibn Yūnus al Misrī (the Egyptian) in the introduction but is attributed by the Syrian engineer al-Jazarī (fl. ca. 1200) to Yūnus al-Asturlābī (the astrolabe maker), who may not be identical with the celebrated Egyptian astronomer. On the clock itself, see E. S. Kennedy and W. Ukashah, "The Chandelier Clock of Ibn Yūnis," in *Isis*, **60** (1969), 543–545.

9. *Kitāb bulūgh al-umniyya fī mā yata'allaq bitulū' al-Shi'rā l-yamāniyya* (astrological treatise) is in MS Manchester Mingana 927 (916); MS Gotha Forschungsbibliothek A1459; and MS Cairo Dār al-Kutub, *majāmī'* 289.

10. The poem on the times of prayer attributed to Ibn Yūnus in MS Cairo Dār al-Kutub, *mīqāt* 181M, fols. 46v–48r, also occurs in two corrupt versions attributed to the Imām al-Shāfi'ī in MSS Berlin Ahlwardt 5700, fol. 11r, and 5820, fol. 65r. Poems by Ibn Yūnus are found in several medieval Arabic anthologies.

II. SECONDARY LITERATURE. Early studies of the *Hākimī zīj* are A. P. Caussin de Perceval, "Le livre de la grande table Hakémite," in *Notices et extraits des manuscrits de la Bibliothèque nationale*, **7** (1804),

16–240, on the observation accounts; and J.-B. Delambre, *Histoire de l'astronomie du moyen âge* (Paris, 1819, repr. New York–London, 1965), containing a summary of the contents of the *Zīj*. The major studies by Carl Schoy on Ibn Yūnus are his "Beiträge zur arabischen Trigonometrie," in *Isis*, **5** (1923), 364–399; *Gnomonik der Araber*, which is I, pt. 6, of E. von Bassermann-Jordan, ed., *Die Geschichte der Zeitmessung und der Uhren* (Berlin–Leipzig, 1923); and *Über den Gnomonschatten und die Schattentafeln der arabischen Astronomie* (Hannover, 1923). Articles by Schoy on individual chapters of the *Hākimī zīj* were published in *Annalen der Hydrographie und maritimen Meteorologie*, **48** (1920), 97–111; **49** (1921), 124–133; and **50** (1922), 3–20, 265–271. A more recent study of the spherical astronomy in the *Hākimī zīj* is D. A. King, "The Astronomical Works of Ibn Yūnus" (Ph.D. diss., Yale University, 1972). The tables entitled *Kitāb al-ta'dīl al-muhkam* are discussed in D. A. King, "A Double-Argument Table for the Lunar Equation Attributed to Ibn Yūnus," in *Centaurus*, **18** (1974), 129–146.

On the observatories in medieval Cairo, see A. Sayili, *The Observatory in Islam* (Ankara, 1960), 130–156, 167–175.

For modern studies relying on data from Ibn Yūnus' observational accounts, see S. Newcomb, "Researches on the Motion of the Moon. Part I . . .," in *Washington Observations for 1875* (1878), app. 2; and R. Newton, *Ancient Astronomical Observations and the Acceleration of the Earth and Moon* (Baltimore, 1970).

On the corpus of spherical astronomical tables for Cairo attributed to Ibn Yūnus, see D. A. King, "Ibn Yūnus' *Very Useful Tables* for Reckoning Time by the Sun," in *Archive for History of Exact Sciences*, **10** (1973), 342–394. The problems of their attribution, and all other known medieval tables for regulating the times of prayer, are discussed in D. A. King, *Studies on Astronomical Timekeeping in Medieval Islam*, which is in preparation.

DAVID A. KING

ZABARELLA, JACOPO (*b*. Padua, Italy, 5 September 1533; *d*. Padua, 15 October 1589), *natural philosophy, scientific method.*

Zabarella was born into an old and noble Paduan family, the son of Giulio Zabarella, from whom, as firstborn son, he inherited the title of palatine count. After humanistic education he entered the University of Padua, where he studied logic with Bernardino Tomitano and natural philosophy with Marcantonio de' Passeri, among others, receiving a degree in 1553. In 1564 he succeeded Tomitano in the first chair of logic and four years later moved to the more prestigious and more lucrative chair of natural philosophy, a position he held for the remainder of his life.

Zabarella must be considered one of the major figures of the revival of Aristotelian philosophy in the sixteenth and seventeenth centuries. He is a prime representative of a specifically Italian form of Aristotelianism in which the teaching of philosophy was closely tied to the needs of medical education. Consequently, his writings epitomize a "naturalistic" approach to philosophy rather than the more theological and metaphysical orientation that had developed in the universities of northern Europe during the later Middle Ages. Both Zabarella's published writings and his teaching focused upon an interpretation of Aristotle's works on logic and natural philosophy. Especially in the latter subject he displayed a strongly empirical approach to understanding the physical and biological world. Observation and experience played an important role in his attempt to comprehend nature, although a very rudimentary "experimental method" was developed in his works. Like most philosophers of the period, Zabarella was more concerned to understand the organic, biological world of natural change than the more abstract realm of what later came to be called "physical science." Consequently, he gave little attention to the possible uses of mathematics as a tool for understanding the physical world.

According to most interpreters, Zabarella's lasting contribution lies in his work on logic and scientific method. It was also in this field that he gained an enormous and authoritative reputation during his lifetime and in the first half of the seventeenth century. Although he wrote on many aspects of logic, it was to methodological questions that he devoted his major effort and on which he wrote most penetratingly. Following an interpretation of Aristotle that goes back to the Greek commentators, Zabarella insisted that logic is not, strictly speaking, a part of philosophy itself but, rather, is an instrumental discipline (*instrumentum*) that furnishes a useful tool of inquiry for the arts and sciences. Expanding on and clarifying Aristotle's doctrine of scientific method, as found in the *Posterior Analytics*, he distinguished "demonstrative method" (*methodus demonstrativa*) from "resolutive method" (*methodus resolutiva*). Both of these are syllogistic in form. The former proceeds from causes to effects, and the latter from effects to causes.

These notions are analyzed in great detail in Zabarella's *De methodis*, one of the works in his *Opera logica* (1578). The same collection contains his *De regressu*, which attempts to work out a specific form of demonstration to be applied to the investigation of natural science. In it Zabarella explains his notion of "regress" (*regressus*), a concept that for many years had been discussed in writings on logic and natural philosophy by his Italian Aristotelian predecessors. "Regress" uses both demonstrative and resolutive methods. It is a technique by which one proceeds from a particular effect to its cause and then returns to a consideration of the effect, thereby having gained a fuller understanding of it and its relation to its cause. The procedure involves several distinct steps, including a careful intellectual analysis (*examen mentale*) of the situation and a final attempt to relate cause to effect in a fuller manner than was possible at the beginning of the analysis.

Although it has been suggested that Zabarella and the methodological tradition of Italian Aristotelianism that he represented were a major influence on the development of Galileo's scientific methodology, concrete evidence has not been adduced to establish a direct connection. In fact, the Aristotelian terminology and doctrines that Zabarella and Galileo share (for instance, Zabarella, *methodus resolutiva*; Galileo, *metodo resolutivo*) seem for the most part to have been commonplaces of late medieval and Renaissance thought, in mathematics and medicine as well as in logic and natural philosophy. Moreover, Zabarella's application of these methods was unwaveringly cast in a syllogistic form, whereas Galileo repeatedly rejected the use of the syllogism in scientific investigation. It must also be noted that—quite contrary to Galileo—Zabarella in no way suggested a systematic application of mathematics to the study of the natural world. On the other hand, it seems evident that the clarity of Zabarella's thought, the precision of his distinctions, and his sharp focus on problems of natural science contributed materially to the progressive clarification of the place of the sciences in the cultural complex of the seventeenth century.

In addition to his logical works, Zabarella wrote commentaries on several treatises of Aristotle (including *Posterior Analytics, De anima, De generatione et corruptione, Meteorology*, and several books of the *Physics*). An important and influential collection of short treatises on various topics of natural philosophy is in his *De rebus naturalibus* (1590). In it he treats many specific problems (including the motion of heavy and light bodies, reaction, the regions of the air, mixture, and elementary qualities), often showing the acuteness of the logical works, although he often relies on traditional solutions to the problems. This collection contains a typical late sixteenth-century approach to natural philosophy. Little attention is paid to the peculiarly medieval natural philosophy that had been dominant from the early fourteenth century until the early sixteenth century (such as the Merton and Paris schools), but much is devoted to a study of the Greek text of Aristotle and a consideration of the opinions of his Greek expositors. Zabarella was an excellent Greek scholar and devoted much effort to presenting what he considered to be the true meaning of Aristotle's text. He also drew extensively upon such Greek commentators as Alexander of Aphrodisias, Themistius, Olympiodorus, Philoponus, and Simplicius. Besides relying on earlier authorities, Zabarella occasionally displayed a strongly empirical bent that allowed him to utilize his personal experiences to reject traditional views. Throughout his writings the approach to natural philosophical problems is qualitative and bears little similarity to either that developed at Oxford and Paris during the fourteenth century or to that employed by Galileo and others in the seventeenth century.

The important collections of Zabarella's writings, especially the *Opera logica* and the *De rebus naturalibus*, began to exert influence throughout Europe soon after the initial publication. This lasted at least until the middle of the seventeenth century, when the dominant position of the Aristotelian tradition finally began to wane. Besides Italy, Zabarella was particularly influential in Germany, where his works were frequently reprinted, and in the British Isles, where the Scholastic revival of the early seventeenth century owed much to his writings. The full impact of Zabarella and the extent of his influence on later philosophy and science have yet to be worked out in detail.

BIBLIOGRAPHY

I. ORIGINAL WORKS. The most important of Zabarella's logical writings are in the *Opera logica* (Venice, 1578; repr. Hildesheim, 1966; at least 14 later eds.) and *In duos Aristotelis libros posteriorum analyticorum commentarii* (Venice, 1582; repr. Frankfurt, 1966; several later eds.). *De rebus naturalibus* (Venice, 1590; repr. Frankfurt, 1966; at least 8 later eds.) contains 30 short works on natural philosophy. Among the commentaries on Aristotle are *In libros Aristotelis Physicorum commentarii* (Venice, 1601; Frankfurt, 1602), the latter ed. also containing commentaries on *De generatione et corruptione* and *Meteorology*, and *In tres Aristotelis libros De anima commentarii* (Venice, 1605; Frankfurt, 1606, 1619). For a more exhaustive listing of printed works and of MSS, see Edwards' dissertation (below). See also M. Dal Pra, "Una *oratio* programmatica di G. Zabarella," in *Rivista critica di storia della filosofia*, **21** (1966), 286–290.

II. SECONDARY LITERATURE. The most comprehensive studies of Zabarella are William F. Edwards, "The Logic of Iacopo Zabarella (1533–1589)," Ph.D. diss. Columbia University, 1960; summary in *Dissertation Abstracts*, **21** (1961), 2745–2746 (with an extensive bibliography of Zabarella's works and secondary literature published before 1960), and Antonino Poppi, *La dottrina della scienza in Giacomo Zabarella* (Padua, 1972). Other useful general treatments are Edwards' article in *Enciclopedia filosofica*, 2nd ed., VI (Florence, 1967), 1187–1189; Eugenio Garin, *Storia della filosofia italiana* (Turin, 1966), 548–558; and Giuseppe Saitta, *Il pensiero italiano nell'umanesimo e nel Rinascimento*, 2nd ed., II (Florence, 1960), 400–423.

Works more specifically oriented toward Zabarella's logic or methodology include F. Bottin, "La teoria del *regressus* in Giacomo Zabarella," in *Saggi e ricerche su Aristotele . . . Zabarella . . .* (Padua, 1972), 49–70; E. Cassirer, *Das Erkenntnisproblem*, 2nd ed., I (Berlin, 1911), 136–143; A. Corsano, "Per la storia del pensiero del tardo Rinascimento. X: Lo strumentalismo logico di I. Zabarella," in *Giornale critico della filosofia italiana*, **41** (1962), 507–517; A. Crescini, *Le origini del metodo analitico: Il Cinquecento* (Udine, 1965), 168–188; W. F. Edwards, "The Averroism of Iacopo Zabarella," in *Atti del XII Congresso internazionale di filosofia*, IX (Florence, 1960), 91–107; N. W. Gilbert, *Renaissance Concepts of Method* (New York, 1960), 167–176; and "Galileo and the School of Padua," in *Journal of the History of Philosophy*, **1** (1963), 223–231; P. Ragnisco, "Una polemica di logica nell'Università di Padova nelle scuole di B. Petrella e G. Zabarella," in *Atti del Istituto veneto di scienze, lettere ed arti*, 6th ser., **4** (1886), 463–502; and "La polemica tra Francesco Piccolomini e Giacomo Zabarella nell'Università di Padova," *ibid.*, 1217–1252; J. H. Randall, *The School of Padua and the Emergence of Modern Science* (Padua, 1961), 15–68; W. Risse, *Die Logik der Neuzeit*, I (Stuttgart–Bad Cannstatt, 1964), 278–290; C. B. Schmitt, "Experience and Experiment: A Comparison of Zabarella's View With Galileo's in *De motu*," in *Studies in the Renaissance*, **16** (1969), 80–138; and C. Vasoli, *Studi sulla cultura del Rinascimento* (Manduria, 1968), 308–342.

CHARLES B. SCHMITT

ZACH, FRANZ XAVER VON (*b*. Pest [now part of Budapest], Hungary, 4 June 1754; *d*. Paris, France, 2 September 1832), *surveying, astronomy*.

Zach came of a noble and distinguished family. His father was a well-known physician who practiced at Pressburg (now Bratislava), Czechoslova-kia, and later at Pest. Authorities disagree in which of these cities it was that Zach was born.

Although the transit of Venus in 1769 and the comet of the same year had awakened his interest in astronomy, Zach joined the Austrian army as an engineering officer but soon left to participate in the survey of Austria. He spent the next few years in Berlin and London as tutor to the children of Graf Moritz von Brühl, the ambassador of Saxony, who was an amateur astronomer and possessed a private observatory. In 1786 Zach entered the ser-vice of Duke Ernst II of Saxe-Coburg, with the ti-tle of "Oberst Wachtmeister." The duke erected an observatory for him on the Seeberg near Gotha, and Zach remained in charge of it until 1806. Upon the duke's death Zach became chief steward to the duchess and in this capacity traveled with her to Italy. Greatly troubled by kidney stones in his later years, he went to Paris to seek relief from this ailment and died there of cholera.

While on the Seeberg, Zach published a series of observations, as well as solar tables and star cata-logs. In 1798–1799 he and F. J. Bertuch edited *Geographische Ephemeriden*, in which the fore-most travelers of the period recounted their ex-periences. This led to the founding of *Monatliche Correspondenz zur Beförderung der Erd- und Him-melskunde* (1800), in which the latest astronomical news was published. Zach edited this publication until 1813 and later, while in Genoa, recommenced it in French, as *Correspondance astronomique, géographique, hydrographique et statistique* (1818–1826). Toward the end of the eighteenth century he formed an association of twenty-four astronomers, each of whom was assigned a celes-tial zone to be searched methodically, especially for new comets and planets, a project that culmi-nated in the discovery of the asteroids.

BIBLIOGRAPHY

Among Zach's most important writings are *Explicatio et usus tabellarum solis, explicatio et usus catalogi stel-larum fixarum* (Gotha, 1792); *Novae et correctae tabu-lae motuum solis* (Gotha, 1792); *Nachricht von der preussische trigonometrische und astronomische Auf-nahme von Thüringen, usw.*, pt. I (Gotha, 1806); *Tabu-lae speciales aberrationis et nutationis . . .*, 2 vols. (Gotha, 1806–1807); and *L'attraction des montagnes et ses effets sur le fil à plomb* (Avignon, 1814). There is a complete list of his books and articles in Poggendorff, II, 1387–1389.

For biographical information, see the notice by Günther, in *Allgemeine deutsche Biographie*, XLIV (Leipzig, 1898), 613–615.

LETTIE S. MULTHAUF

ZACUTO (or **ZACUT**), **ABRAHAM BAR SAMUEL BAR ABRAHAM** (*b.* Salamanca, Spain, *ca.* 1450; *d.* Portugal, *ca.* 1522), *astrology, astronomy.*

Zacuto, a Jew, studied medicine and astrology at the University of Salamanca. He became a re-nowned astrologer, and several historians have claimed that he held a chair there; but no docu-ment has yet been found that can adequately sup-port this assertion. After the publication of the law requiring the conversion or expulsion of the Span-ish Jews, Zacuto went to Portugal, where he was welcomed by John II (1492). The king, who was interested in developing the art of navigation, quickly profited from Zacuto's presence to refine a number of rules pertaining to sailing.

Gaspar Correia (*Lendas da India*, I [Lisbon, 1858], 261) stated that Zacuto introduced the as-trolabe into Portugal and that he was the author of the quadrennial tables of solar declination. Al-though the first part of this claim is surely incorrect (the astrolabe had been known in Portugal since at least the twelfth century), Correia was right about the second. Zacuto was certainly the author of the tables prepared for the voyage of Vasco da Gama (for the period 1497–1500), fragments of which are preserved in Andre Pires' *Livro de marinharia* (edited by Luis de Albuquerque [Coimbra, 1963], 34–81). These tables were computed on the basis of elements taken from Zacuto's *Almanach perpe-tuum*. Two editions of this work were printed at Leiria in 1496; and in one of them the preliminary note, containing explanations of the use of the ta-bles, was translated into Spanish. Until 1537 – the year of the publication of Pedro Nuñez Salaciense's tables – all solar tables for navigation prepared in Portugal were computed from the figures fur-nished by the *Almanach perpetuum*.

BIBLIOGRAPHY

Zacuto's most important work, *Almanach perpetuum* (Leiria, 1496), was reprinted several times. His other writings, which deal primarily with astrology, include *Mixtapé ha' isteganin* ("Judgments of Astrology") and *Haajibbun hagadol* ("The Great Compilation"). *Trata-do de las ynfluencias del cielo* and *Dos eclipses del sol y la luna* were first published by Joaquim de Carvalho in *Estudos sobre a cultura portuguesa do século XVI,* I (Coimbra, 1947), 109–177, 177–183.

Luis de Albuquerque

AL-ZAHRĀWĪ, ABUᵓL-QĀSIM KHALAF IBN ᶜABBĀS, also known as **Abulcasis** (*b.* al-Zahrāᵓ, near Córdoba, Spain, *ca.* 936; *d.* al-Zahrāᵓ, *ca.* 1013), *medicine, pharmacology.*

The epithet al-Zahrāwī derives from the fact that he spent most of his life in his native city as a practicing physician-pharmacist-surgeon. Although references have been made to his contributions to theology and the natural sciences, none of his writings in these fields, if any, is known.

We know nothing about al-Zahrāwī's parentage except that his forebears were of the Anṣār, who presumably came from Arabia (al-Anṣār) with the Muslim armies that conquered and inhabited Spain, later forming the aristocracy in the larger Moorish cities and in the capital, Córdoba. Little is known of al-Zahrāwī himself, but his life-span coincided with the golden age of Moorish Spain, when intellectual activities, including the natural and mathematical sciences, reached their first peak. Córdoba and al-Zahrāᵓ then formed a metropolitan area unmatched for excellence in Europe, except for Constantinople.

Al-Zahrāwī was first mentioned very briefly by Futuḥ al-Ḥumaydī, Ibn Ḥazm, and Ibn Abī Uṣaybiᶜa. His only known literary contribution, *al-Taṣrīf li-man ᶜajiza ᶜan al-taᵓlīf,* a medical encyclopedia in thirty treatises, sheds some additional light on his life and personality. He seems to have traveled very infrequently. His *Taṣrīf,* completed about A.D. 1000, was the result of almost fifty years of medical education and experience. In it he discussed not only medicine and surgery, but also midwifery, pharmaceutical and cosmetic preparations, materia medica, cookery and dietetics, weights and measures, technical terminology, medical chemistry, therapeutics, and psychotherapy.

Al-Zahrāwī attempted to separate medical practice from alchemy, theology, and philosophy, advocating specialization in the health professions: "Too much branching and specializing in many fields before perfecting one of them causes frustration and mental fatigue." He also called for upholding the high ethical standards of the healing art, the return to and reliance on nature, and recognition that "time plays an important part in the treatment and cure of diseases."

Al-Zahrāwī was the first to recommend surgical removal of a broken patella and the first to practice lithotomy on women. He introduced what is now known as the Walcher position in obstetrics and devised new obstetrical forceps. He gave original descriptions for manufacturing and using probes, surgical knives, scalpels, and hooks of various shapes and designs. He invented several types of true surgical scissors ending in recurved or ring extremities, as well as grasping forceps. He described lachrymal fistula and other eye operations in which he employed pointed blades, speculums, and hooks. For scaling teeth he used long-handled scrapers fluted for good grip. He was the first to describe accurately aural polyps as well as lithotomy using a special scoop and lancets. His illustrations of surgical instruments are the earliest known to be intended for use in teaching and for demonstration of the method of manufacture. Before Paré, he ligatured arteries and recommended several types of threads and catguts in suturing. Al-Zahrāwī applied plasters and bandages to ordinary fractures; described hydatid cysts, hemophilia, and the extraction of a polyp; and gave a very interesting explanation of a case of hydrocephaly resulting from a congenital defect caused by blocked drainage of cerebral fluid: "I have seen a baby boy whose head was abnormally enlarged with prominence of the forehead and sides to the point that the body became unable to hold it up." Al-Zahrāwī's surgery was the most advanced in the Middle Ages until the thirteenth century. Although its influence in Arab lands has been limited (Ibn al-Quff in the thirteenth century was an exception), his surgical and chemopharmaceutical writings were highly regarded in the West after they were translated into Latin by Gerard of Cremona, Rogerius Frugardi, Rolandus Parmensis, Arnald of Villanova, and others. His emphasis on the importance of human anatomy and physiology generated special interest in their study by later doctors. He observed, for example, that the brain includes the three functions of the intellect: imagination, thought, and memory.

Al-Zahrāwī was not only one of the greatest surgeons of medieval Islam, but a great educator and psychiatrist as well. He devoted a substantial section in the *Taṣrīf* to child education and behavior, table etiquette, school curriculum, and academic specialization. He encouraged the study of medicine by intelligent and gifted students after completion of their primary education in language, religion, grammar, poetry, mathematics, astronomy, logic, and philosophy. Following the Hippocratic tradition, he divided man's life-span into four stages: early age up to twenty years, youth up to forty, adulthood or maturity to sixty, and old age over sixty.

Al-Zahrāwī emphasized hygienic measures, special diets for the sick and the healthy, and effective, high-quality drugs for the benefit of patients. He promoted bedside clinical medicine and strong doctor-patient relationships: "Only by repeated visits to the patient's bedside can the physician follow the progress of his medical treatment."

As a natural scientist and applied chemist, al-Zahrāwī described Spanish fauna and flora and simples of plant, animal, and mineral origins, where they are found, their cultivation, and their preservation. He also discussed technical methods of preparing and of purifying for medicinal uses such chemical substances as litharge, ceruse (white lead), iron pyrite (crystalline marcasite), vitriols, and verdigris. He likewise recommended the use of minerals, elements, and precious stones—individually or compounded with other simples—for medical treatment. In his psychiatric treatment, al-Zahrāwī used drugs to induce hallucinations, thrills, and happiness. For example, he manufactured an opium-based remedy that he called "the bringer of joy and gladness, because it relaxes the soul, dispels bad thoughts and worries, moderates temperaments, and is useful against melancholy."

BIBLIOGRAPHY

I. ORIGINAL WORKS. Al-Zahrāwī's *al-Taṣrīf* exists in fragments in treatises, and in its entirety in numerous MSS. A list of most of these extant copies, with bibliography, can be found in Sami Hamarneh and Glenn Sonnedecker, *A Pharmaceutical View of Abulcasis al-Zahrāwī in Moorish Spain* (Leiden, 1963), 130–133, 137–151. There are partial translations in Spanish, Hebrew, and Latin (see bibliographies below). Another important ed. (although with numerous errors) is the Arabic-Latin copy of the surgical treatise by Johannes Channing, *Albucasis De chirurgia arabice et latine*, 2 vols. (Oxford, 1778). The French trans. by Lucien Leclerc, *La chirurgie d'Abulcasis* (Paris, 1861), with a useful intro., was very influential in making al-Zahrāwī's surgery better known to modern historians of science. The definitive ed. by M. S. Spink and G. L. Lewis, *Abulcasis on Surgery and Instruments* (Berkeley, Calif., 1973), includes the Arabic text with English trans. and commentary. There is also a rare lithographed copy of the surgical treatise (Lucknow, 1878). For details see also George Sarton, *Introduction to the History of Science*, I (Baltimore, 1927), 681–682; Carl Brockelmann, *Geschichte der arabischen Literatur*, I (Leiden, 1943), 276–277, and Supp., I (Leiden, 1937), 425; and Sami

Hamarneh, *Catalogue of Arabic Manuscripts on Medicine and Pharmacy at the British Library* (Cairo, 1975), 90–93.

II. SECONDARY LITERATURE. The first to write a separate biography of al-Zahrāwī was Muhammad ibn Futūḥ al-Ḥumaydī (1029–1095), *Jadhwat al-Muqtabis fī Dhikr Wulāt al-Andalus* (Cairo, 1952), 195. Abū Muḥammad Ibn Ḥazm (994–1064) of Córdoba, in his epistle defending the scholars of his native country, quoted by Aḥmad al-Maqqarī in the first part of his *Nafḥ al-Ṭīb*, mentions him in passing as a great surgeon-physician. Al-Dabbī and Ibn Bashkuwāl quote al-Ḥumaydī with no additional information. Ibn Abī Uṣaybiʿa, ʿAyūn al-anbāʾ, Būlāq ed., II (Cairo, 1882), 52, mentions al-Zahrāwī's interest in medical therapy and his knowledge of the materia medica.

For a thorough exposition of his general contributions see Lucien Leclerc, *Histoire de la médecine arabe*, I (Paris, 1876), 437–457; and Sami Hamarneh, *Ẓāhirīyah Index* (Damascus, 1969), 147–170. For his surgery, see Zaki Aly, "La chirurgie arabe en Espagne," in *Bulletin de la Société française d'histoire de la médecine*, **26** (1932), 236–243; George J. Fisher, "Abul-Casem . . . al-Zahravi," in *Annals of Anatomy and Surgery*, **8** (1883), 21–29, 74–82, 124–131; Ernst F. Gurlt, *Geschichte der Chirurgie und ihrer Ausübung*, I (Berlin, 1898), 620–649; Sami Hamarneh, "Drawings and Pharmacy in al-Zahrāwī's 10th Century Surgical Treatise," in *Contributions. Museum of History and Technology, United States National Museum*, no. 228 (1961), paper 22, 81–94; and Tewfick Makhluf, *L'oeuvre chirurgical d'Abul-Cassim . . . ez-Zahrawi* (Paris, 1930).

For his contribution to medicine and obstetrics, see M. S. Abu Ganimah, *Abul-Kasim ein Forscher der arabischen Medizin* (Berlin, 1929); Henri Paul J. Rénaud, "La prétendue 'hygiène d'Albucasis' et sa véritable origine," in *Petrus nonius* (Lisbon), **3** (1941), 171–179; and Martin S. Spink, "Arabian Gynaecology," in *Proceedings of the Royal Society of Medicine*, **30** (1937), 653–671.

For his contribution to weights and measures, cosmetology, materia medica, and chemotherapy, see Sami Hamarneh, "Climax of Chemical Therapy in 10th-Century Arabic Medicine," in *Islam* (Berlin), **38** (1963), 287–288; "The First Known Independent Treatise on Cosmetology in Spain," in *Bulletin of the History of Medicine*, **39** (1965), 309–325; and *A Pharmaceutical View of . . . al-Zahrāwī . . .* (Leiden, 1963), 37–126, written with Glenn Sonnedecker; and H. Sauvaire, "Traité sur les poids et mesures par ez-Zahrawy," in *Journal of the Royal Asiatic Society of Great Britain and Ireland*, n.s. **16** (1884), 495–524.

SAMI HAMARNEH

ZAKARIYĀ IBN MUḤAMMAD IBN MAḤMŪD. See Al-Qazwīnī, Zakariyā ibn Muḥammad ibn Maḥmūd, Abū Yaḥyā.

ZALUŽANSKÝ

ZALUŽANSKÝ ZE ZALUŽAN, ADAM (*b*. Mnichovo Hradiště, Bohemia [now Czechoslovakia], *ca*. 1558; *d*. Prague, Bohemia, 8 December 1613), *botany, medicine.*

The son of a clerk, Zalužanský studied at Wittenberg and at the Charles University in Prague, where he received the bachelor's degree in 1581 and the M.A. in 1584; he was awarded the M.D. at Helmstedt in 1587. He subsequently became lecturer in the Greek classics at the Charles University; in 1591 he was elected dean of the Faculty of Philosophy and, in 1593, rector of the university. After his marriage the following year he was obliged to leave the university. He established a medical practice in Prague and died during the plague. Active in the medicine, pharmacy, poetry, politics, and religion of his time, Zalužanský was also instrumental in reforming the Charles University.

Zalužanský's three-volume botanical treatise, *Methodi herbariae*, represented an important departure from the customary practice of publishing herbals mainly to illustrate the herbs used in medicine. Zalužanský's work had a purely scientific purpose, like Cesalpino's *De plantis* (1583), and included no illustrations. Botany, he wrote, should be investigated as an independent branch of natural history.

In book I, "De aetiologia plantarum," Zalužanský included a chapter on the sexuality of plants, in which he drew on Aristotle, Theophrastus, and especially on Pliny's principles that all plants have sexuality—although most of them do not display sexual differentiation. Like the ancient writers, he was completely unaware of the functions of stamens and pistils, understanding sexuality as male and female principles inherent in the plant as a whole. His idea of sex connected in a single individual, however, was the first speculation about monoecism in plants.

In book II, "De historia plantarum," Zalužanský attempted to establish a natural system of plants. Beginning with the fungi and mosses, he progressed to grasses, herbs, and finally to species of wood. Since his method was primarily physiognomical, based on habitual features, he was rarely able to discern the natural groupings that he found in the Papilionaceae, Compositae, and Malvaceae, for example; his other groupings are considerably heterogeneous.

The last book, "De exercitio plantarum," deals with botanical methodology. Zalužanský termed the first phase of botanical investigation "analysis," believing that it should lead to the description of the plant under examination and to the determination of its name and properties. The second phase, "genesis," was based on the data thus obtained and would demonstrate the properties of plants that are useful to man.

Despite his emphasis on the study of all stages of plant evolution, Zalužanský frequently used the empirical findings of other authors, such as L'Obel, Dodoens, Mattioli, and even Pliny. His overall conception was still Aristotelian and Scholastic, rather than that of a natural historian seeking new knowledge. However, because he conceived of botany as an independent science and presented his treatise completely along these lines, he deserves to be considered one of the first researchers whose methodological concepts were capable of influencing the evolution of botany.

BIBLIOGRAPHY

I. ORIGINAL WORKS. Zalužanský's writings are *Rzad apathekařský* (Prague, 1592); *Methodi herbariae libri tres* (Prague, 1592; Frankfurt, 1604), facs. repr., K. Pejml, ed. (Prague, 1940); *Oratio pro anatomia et restauratione totius studii medici in inclyto regno Bohemiae* (Prague, 1600); and *Animadversionum medicarum in Galenum et Avicennam libri septem* (Frankfurt, 1604).

II. SECONDARY LITERATURE. On Zalužanský and his work, see L. Čelakovský, "Adam Zalužanský ze Zalužan ve svém poměru k nauce o pohlaví rostlin" ("Zalužanský in His Relation to the Theory of Sexuality in Plants"), in *Osvěta*, **6** (1876), 33–54; and K. Pejml, "Adam Zalužanský de Zalužany, sa personnalité et son oeuvre en tenant compte de son ouvrage *Methodi herbariae libri tres*," in *Summa dissertationum Facultati rerum naturalium Universitati Carolinae anno 1946*, no. 183 (Prague, 1948), 11–13.

VĚRA EISNEROVÁ

ZAMBECCARI

ZAMBECCARI, GIUSEPPE (*b*. Castelfranco di Sotto, Italy, 19 March 1655; *d*. Pisa, Italy, 13 December 1728), *medicine.*

At age eighteen Zambeccari was admitted to the Ducal College of Pisa (which later became the university), where he studied medicine. The college already had won renown as the producer of the previous centuries' most brilliant scientists, and among his professors was the anatomist Lorenzo Bellini.

After graduating in 1679, Zambeccari moved to Florence, where he continued his studies under Francesco Redi, who encouraged him to improve his clinical knowledge by working as an intern in the wards of the Hospital of Santa Maria Novella. According to the custom of the time, Zambeccari lodged in the house of his professor and there in 1680 conducted his most important experiments in physiology, which consisted in removing various internal organs from live animals (mainly dogs) in order to acquire a better understanding of what functions they performed in relation to the whole organism.

One of the first series of experiments dealt with the removal of the spleen; several of the animals operated upon survived, and a few months later they were killed and carefully examined in order to discover what anatomical, pathological, and physiological changes had been caused by the removal of the organ. Obviously, no conclusive results could be obtained from these experiments, because nothing was then known of the function of the spleen and there were no means of carrying the investigation further.

Turning to the study of other organs, Zambeccari performed unilateral nephrectomies and discovered that the animal apparently was not incommoded by the operation. In other experiments he tied the common bile duct and thus demonstrated that bile is not formed in the gall bladder, as was then the common belief. Encouraged by the results of his experiments, he not only removed the bile duct but also fragments of hepatic tissue, and even entire lobes of the liver, always finding that a good percentage of the animals survived the operation. Zambeccari performed a resection of the cecum and finally went so far as to remove the pancreas and to ligate the mesenteric veins. He also studied the eyes and noted that pricking the cornea of various animals rapidly leads to the reconstitution of the aqueous humor.

Despite their importance in the history of experimental physiology, Zambeccari's studies had no immediate impact on biology—in part because too little was known for them to be really useful. Nevertheless, the book in which he described his experiments was for a time rather successful and went through several editions; yet it does not appear to have inspired others to use the same approach. Later the work was forgotten completely, until Murri in the nineteenth century brought it to the attention of scientists.

After his period of experimentation in the house of Francesco Redi, Zambeccari returned to Pisa, where he was offered the chair of practical medicine and, in 1689, that of medicine proper. In 1704 he succeeded Bellini in the chair of anatomy.

Other works—manuscripts and letters by Zam-beccari, some of which are of considerable interest—either are still unpublished or were discovered only long after his death. Among the most interesting is *Del sonno della vigilia e dell'uso dell'oppio*, written in 1685, but published, by C. Fedeli, only in 1914. An essay of deductive rather than experimental character, it deals with the physiology of nerve transmission and of muscle contraction, and is based on Galenic assumptions on anatomy and physiology, and on the iatromechanical concepts of Borelli and Galileo.

BIBLIOGRAPHY

I. Original Works. Zambeccari's writings include *Esperienze del Dottor Giuseppe Zambeccari intorno a diverse viscere tagliate a diversi animali viventi . . .* (Florence, 1680); *Del sonno, della vigilia, e dell'uso dell'oppio* (Pisa, 1914); and *Breve trattato de'bagni di Pisa e di Lucca* (Padua, 1712).

II. Secondary Literature. See U. Calamida, "Di un carteggio inedito di Giuseppe Zambeccari," in *Atti del III Congresso della Società italiana di storia delle scienze mediche e naturali* (Venice, 1925), 120–127; C. Fedeli, *Giuseppe Zambeccari, lettera sulle separazioni* (Pisa, 1927), reviewed by A. Corsini in *Rivista di storia critica delle scienze mediche e naturali*, **18** (1927), 320; P. Ferrari, *Giuseppe Zambeccari* (Pontremoli, 1925), reviewed by A. Corsini, *ibid.*, **17** (1926), 112–113; and S. Jarcho, "A Seventeenth-Century Pioneer in Experimental Physiology and Surgery," in *Bulletin of the History of Medicine*, **9** (1941), 144–176; and "Experiments of Doctor Joseph Zambeccari Concerning the Excision of Various Organs From Different Living Animals," *ibid.*, 311–331.

Carlo Castellani

ZAMBONINI, FERRUCCIO (*b.* Rome, Italy, 17 December 1880; *d.* Naples, Italy, 12 January 1932), *chemistry, mineralogy.*

Zambonini was the son of Ersilia Zuccari and Gustavo Zambonini di Montebugnoli, descended from a noble Bolognese family reduced to near poverty by financial misfortunes. Obliged to work in order to continue his studies after his father's death, Ferruccio gave private lessons and copied documents for lawyers. His first scientific works were published in 1898, while he was still a student. Zambonini graduated in natural sciences from the University of Rome in 1903 and immediately became assistant in chemistry at the Turin Polytechnic, where he acquired a considerable knowledge of general analytical chemistry. In 1906 he worked at the University of Naples with Arcangelo Scacchi, the leading Italian mineralogist of the nineteenth century.

Zambonini became professor at the University of Sassari in 1909; in 1911 he moved to Palermo, and two years later to Turin. In 1923 he obtained the chair of general chemistry at the University of Naples, where he continued his mineralogical and chemical studies, and was twice elected vice-chancellor of the university.

Zambonini was the leading mineralogist in Italy during the first half of the twentieth century, a worthy successor of Scacchi and Quintino Sella. He contributed substantially to the knowledge of many minerals; one of his most important works, fundamental to the study of volcanic products, is "Mineralogia vesuviana" (1910), a collection of data on more than 250 minerals from Vesuvius and Monte Somma, with Zambonini's own minutely detailed descriptions and commentary. This work won him the annual award of the Royal Academy of Sciences of Naples. An appendix appeared in 1912; and in 1935, after his death, Emanuele Quercigh compiled a second edition that included all the new minerals discovered by Zambonini and his students during the last ten years of his life.

Zambonini extended knowledge of the dehydration of minerals by studying the role of water in hydrated silicates, especially zeolites, and interpreting it in the light of the most advanced theories of colloidal chemistry.

In general mineralogy, Zambonini's isomorphism anticipated modern theories of crystal chemistry. In the important "Sulle soluzioni solide dei composti di calcio . . . ," he demonstrated that the rare earths in trivalent rare-earth compounds could be replaced with alkaline earths and lead. He also studied their solubility limits in the solid state and applied the results to simplifying and clarifying the formulas of many minerals.

Zambonini gave a very interesting explanation of the concomitant replacement, in plagioclase, of part of the calcium by sodium, and silicon by aluminum. He attributed this substitution to the closeness of the ionic radii of sodium and calcium, and of aluminum and silicon. In Italy the possibility of isomorphism between ions of similar radii is called Zambonini's rule, thus indicating that the solid-state substitution of elements is a function of ionic radii, rather than of chemical properties.

Zambonini conducted pioneering research on the mixed crystals of the epidote-clinozoisite series. Through chemical analyses and a study of optical properties of the series, he discovered the existence of mixed stereoisomeric crystals, which have the same chemical composition but optical properties that vary according to whether it is the free aluminum or the aluminum bound to hydroxyls that is replaced by ferric iron.

While working on mixed crystals of molybdates and tungstates of calcium, barium, strontium, and lead, along with the analogous compounds of rare earths, Zambonini noticed that their angular values and optical properties did not vary regularly with the variation in composition of the mixed crystals; there are values not included among those of the two components. He observed the same phenomenon in pairs of artificial compounds and concluded that there are important exceptions to the principle that the optical properties of mixed crystals vary regularly with their composition. In 1922 Zambonini discovered the isomorphism between potassium fluoborate and permanganate, two compounds that are similar neither in their chemical composition nor even in the sum of their valences.

In 1924 Zambonini was one of the first to apply Bohr's atomic theory in order to explain the isomorphism of the trivalent rare earths with alkaline earths and with lead. Similarly, in 1911 he had been an innovator when he used the amount of lead and uranium in some minerals to date rocks, thus anticipating modern methods. For the calculations he used Rayleigh's formula.

In the last decade of his active life Zambonini contributed substantially to the chemistry of the rare earths. With collaborators he studied an isotherm of the binary systems of rare-earth sulfates and alkaline sulfates. He had obtained the lanthanides for this research through a long and difficult process of fractionation, carried out with very limited means at the Institute of Mineralogy in Turin.

For his scientific achievements Zambonini was elected a member of the Accademia Nazionale dei Lincei, the Royal Academy of Physical and Mathematical Sciences of Naples, the Academy of Sciences of Turin, and the Accademia dei Quaranta. He was president of the Geological Society of Italy, vice-chancellor of the University of Naples, and editor of *Zeitschrift für Kristallographie, Kristallgeometrie, Kristallphysik, Kristallchemie,* and was awarded the Wilde Prize of the Institut de France.

BIBLIOGRAPHY

A comprehensive bibliography of Zambonini's writings (162 works) is included with F. Giordani, "Commemorazione di Ferruccio Zambonini," in *Rendiconti dell'Accademia delle scienze fisiche e matematiche*, 4th ser., **3** (1933), 8–19, with portrait. "Mineralogia Vesuviana" was published as *Atti dell'Accademia delle scienze fisiche e matematiche*, 2nd ser., **14**, no. 6 (1910); 2nd ed., compiled by E. Quercigh, *ibid.*, supp. **20** (1935). "Sulle soluzioni solide dei composti de calcio, stronzio, bario e piombo con quelli delle 'terre rare' . . . ," appeared in *Rivista italiana di mineralogia*, **45** (1915), 1–185.

GUIDO CAROBBI

ZANOTTI, EUSTACHIO (*b.* Bologna, Italy, 27 November 1709; *d.* Bologna, 15 May 1782), *astronomy, geometry.*

Like the astronomer Eustachio Manfredi, his godfather, Zanotti belonged to a prominent family distinguished in the arts, letters, and sciences. The son of Gian Pietro Zanotti and Costanza Gambari, he was educated by the Jesuits and entered the University of Bologna, becoming Manfredi's assistant at the Institute of Sciences in 1729. He graduated in philosophy in 1730 and obtained his first university post, as reader in mechanics at Bologna, in 1738, after presenting his trial lecture on the Newtonian theory of light. The following year he succeeded Manfredi as director of the Institute observatory, a post to which he dedicated himself almost exclusively for the next forty years, never marrying and declining all offers from other universities. He began teaching hydraulics at the university in 1760, having been requested by the government to supervise works on rivers and waterways. His publications in this field include a work on the characteristics of riverbeds near the sea (1760) that remained in print for almost a century. Zanotti wrote the last part of Manfredi's *Elementi della geometria*, "according to the method of indivisibles"; and his lucid and informative *Trattato teorico-pratico di prospettiva* (1766) was intended for painters as well as mathematicians.

Zanotti established a reputation as an astronomer even before Manfredi's death, through the discovery of two comets, to the second of which (1739) he attributed a parabolic orbit. In 1741, under his direction, the new instruments that Manfredi had ordered from Sisson's were installed at the Bologna observatory: a mural quadrant 1.2 meters in radius and a transit instrument with a focal length of about one meter. In 1780 he added a movable equatorial telescope made by Dollond's. With the acquisition of Sisson's instruments, Zanotti's observatory became one of the finest in Europe. In 1748 and 1749, with his assistants G. Brunelli and Petronio Matteucci, he carried out repeated observations of the sun and planets, and compiled a catalog of 447 stars, all but thirty-three of them within the zodiac. The work was published with additions in 1750 as an appendix to the new edition of Manfredi's introductory volume to his ephemerides. Zanotti continued to publish the ephemerides with scrupulous care: three volumes covered the period 1751–1774, and a fourth was published posthumously by Matteucci in 1786.

Zanotti's principal observations and descriptions, including some on occultations of stars by the moon, concern six comets (1737, 1739, 1742, 1743–1744, Halley's comet of 1758, and 1769), four lunar eclipses (December 1739, January 1740, November 1745, June 1750), three solar eclipses (August 1738, July 1748, January 1750), the aurora borealis (December 1737, March 1739), and transits of Mercury (1743, 1753) and of Venus (1751) on the sun.

In 1750 Zanotti was invited by the Paris Academy of Sciences to participate in a major international research project, the main purpose of which was to measure the lunar parallax. His observations provided the program with some of its most accurate results.

Zanotti's accomplishments also included the restoration in 1776 of Gian Domenico Cassini's sundial in the church of San Petronio. The displaced perforated roofing slab forming the gnomon was raised slightly, restoring the instrument to its original height. The old deformed iron ship representing the meridian was removed and a solid foundation was laid as a base for new level marble slabs with the new brass meridian strip. Accurate geodetic and topographic measurements made in 1904 and 1925 have verified that the instrument has remained as Zanotti left it, that is, in the position that perfectly reproduces Cassini's original conditions of construction.

According to L. Palcani-Caccianemici, his collaborator and principal biographer, Zanotti was also a pioneer in the study of variable stars, a little-understood phenomenon that was then considered to represent an error of vision or an effect caused by the intervening atmosphere. Zanotti, however, maintained that changes of light occur even when the possibility of such causes is entirely ruled out. "If you observe with a telescope two stars extremely near to each other," he said, "you will see that one remains exactly the same and that the other, altered in intensity, no longer appears as before."

Unfortunately, no trace of these observations appears in Zanotti's published writings, possibly because he did not wish to seem to be questioning the incorruptibility and constancy of the heavens — a subject about which the Aristotelians who controlled the University of Bologna were particularly sensitive and uncompromising.

BIBLIOGRAPHY

I. ORIGINAL WORKS. Most of Zanotti's astronomical memoirs appeared in the *Commentarii* of the Istituto e Accademia delle scienze di Bologna, beginning with **2**, pt. 1 (1745); the last one appeared in **7** (1791) — see index. There are also three papers, in English, in *Philosophical Transactions of the Royal Society*: on the aurora borealis, **41** (1741), 593–601; on the comet of 1739, *ibid.*, 809; and on the 1761 transit of Venus, **52** (1761), 399–414.

His most important separately published works are *Ephemerides motuum coelestium ex anno 1751 ad annum 1786*, 3 vols. (Bologna, 1750–1774); *Trattato teorico-pratico di prospettiva* (Bologna, 1766); and *La meridiana del tempio di San Petronio rinnovata l'anno 1776* (Bologna, 1779).

II. SECONDARY LITERATURE. See the following, listed chronologically: L. Palcani-Caccianemici, *De vita Eustachii Zanotti commentarius* (Bologna, 1782), translated into Italian by G. A. Maggi as "Elogio di Eustachio Zanotti" and prefixed, with a bibliography, to the new ed. of Zanotti's *Trattato teorico-pratico di prospettiva* (Milan, 1825); *Vita di Eustachio Zanotti* (n.p., n.d.), an extract from *Giornale dei letterati* (Pisa), **58** (1785), 175–197; G. Fantuzzi, *Notizie degli scrittori bolognesi*, VIII (Bologna, 1790; repr. 1965), 265–270; Noël Poudra, *Histoire de la perspective* (Paris, 1864), 529–533; and P. Riccardi, *Biblioteca matematica italiana*, II (Modena, 1873–1876; repr. Milan, 1952), 651–657.

More recent works include M. Rajna, "L'astronomia in Bologna," in *Memorie della Società degli spettroscopisti italiani*, **32** (1903), 241–250, esp. 245; Federigo Guarducci, *La meridiana di Gian Domenico Cassini nel tempio di San Petronio di Bologna riveduta nel 1904 e nel 1925* (Bologna, 1925); E. Bortolotti, *La storia della matematica nella Università di Bologna* (Bologna, 1947), 177–178; G. Horn d'Arturo, *Piccola enciclopedia astronomica*, II (Bologna, 1960), 364; and Anna Maria Matteucci, *Carlo Francesco Dotti e l'architettura bolognese del settecento*, 2nd ed. (Bologna, 1969), 44–49, *passim*.

GIORGIO TABARRONI

ZARANKIEWICZ, KAZIMIERZ (*b.* Czestochowa, Poland, 2 May 1902; *d.* London, England, 5 September 1959), *mathematics*.

Zarankiewicz's contributions to mathematics were in topology, the theory of graphs, the theory of complex functions, number theory, and mathematical education. In addition, he founded and for several years headed the Polish Astronautical Society.

Zarankiewicz was born and raised in a moderately well-to-do family. After obtaining his baccalaureate degree in 1919 in Bedzin, near Czestochowa, he studied mathematics at the University of Warsaw. He was awarded a Ph.D. there in 1923 for a dissertation (published in 1927) on the cut points in connected sets, and in 1924 he was made assistant to the professor of mathematics at the Warsaw Polytechnic. In 1929, following publication of his habilitation essay on a topological property of the plane (concerning the mutual cuttings of three regions and three continua), Zarankiewicz became *Dozent* in mathematics. He spent the academic year 1930–1931 working with Karl Menger at Vienna and with Richard von Mises at Berlin, where he collaborated with Stefan Bergman and other mathematicians. Upon returning to Warsaw, Zarankiewicz was assigned to teach a course in rational mechanics at the Polytechnic and courses in mathematics and statistics at the High Agricultural College. In 1936 he was invited to the University of Tomsk to lecture for a semester on conformal mappings, particularly on several problems that he had solved. He was named substitute for the professor at the Warsaw Polytechnic in 1937; but his nomination to the professorship (1939) was not confirmed until after the war.

During the German occupation of Poland, Zarankiewicz taught mathematics, clandestinely, to underground groups of high school and college students. In 1944 he was deported to a forced labor camp in Germany. Returning to the ruins of Warsaw in 1945, he resumed his courses at the Polytechnic and continued them for the rest of his life.

Zarankiewicz was appointed full professor in 1948 and spent several months of that year working at Harvard and at several other American universities. From 1949 to 1957 he supervised in Poland the Mathematical Olympics for high school students, remaining a member of its central board thereafter. At the same time he was a member of the editorial committees of *Applied Mechanics Reviews* and *Matematyka*, a Polish journal written primarily for secondary-school teachers. From 1948 to 1951 Zarankiewicz headed the Warsaw section of the Polish Mathematical Society. He maintained a long-standing active interest in astronautics and in the organization in Poland of a society founded in 1956 devoted to this field. Zarankiewicz died in London while presiding over a plenary session of the Tenth Congress of the International Astronautical Federation, of which he was vice-president. His funeral was held in Warsaw, and one of the city's streets is named for him. Zarankiewicz's topological writings deal mainly with cut points, that is, those points which disconnect (and locally disconnect) the continua, and with the continua that disconnect the spaces. In 1926 he showed, among other things, that if C is a locally connected continuum (a continuous image of an interval), then the set $\tau(C)$ of all the cut points of C possesses a special structure, which he characterized. In particular, the closures of the constituents of $\tau(C)$ are dendrites, that is, one-dimensional, acyclic, locally connected continua. In his doctoral dissertation Zarankiewicz introduced and investigated the important notion of the continua of convergence. He characterized the locally connected continua by the equivalence, for their closed subsets F, of the connectedness and of the semicontinuity of the set $C \setminus F$ between all the pairs of its points. He also characterized the dendrites C by the structure of the set $C \setminus \tau(C)$ and—independently of Pavel Urysohn—the hereditarily locally connected continua by the absence, among their subsets, of continua of convergence (1927). In 1932 and in 1951 Zarankiewicz resumed and extended the study of the set $\tau(C)$ by a series of remarkable theorems. In particular, he gave a new definition of the cyclic element in G. T. Whyburn's sense, for the locally connected continua.

Zarankiewicz's studies of the cutting of spaces by continua were concerned especially with local cuttings of the plane or, which is topologically equivalent, of the sphere. In publications of 1927, 1929, and 1932, he established interesting topological characterizations of the circumference, of the straight line, and of several other lines with the aid of the number of their connectedness points, another notion that he originated. These theorems, which are more quantitative than qualitative, reflect Zarankiewicz's tendency to seek numerical solutions in every field of mathematics in which he worked. For example, generalizing R. L. Moore's theorem concerning triods in the plane, he showed that in Euclidean spaces of more than two dimensions, every family of disjoint continua each of which locally cuts the space at a point that cuts locally itself (*doppelt zerlegender Punkt*) is, at most, countable (1934).

Zarankiewicz's last publication in topology (1952), written with C. Kuratowski, deals with a problem that is still unsolved: Given n disjoint regions in the plane (or on the sphere) with connected boundaries $R_1, R_2, \cdots, R_n$ and k continua $C_1, C_2, \cdots, C_k$, each of which meets each of these regions, what is the minimum number $s_{k,n}$ of couples i,j such that C_i cuts R_j (where $i = 1, 2, \cdots, k$ and $j = 1, 2, \cdots, n$)? Zarankiewicz's conjecture is that $s_{k,n} = (k - 2)(n - 2)$ for all integers $k \geq 2$ and $n \geq 2$. In 1928 he showed that $s_{3,3} = 1$ and, in the joint work of 1952, that the presumed formula holds for all $k \geq 4$ and $n \leq 4$.

In theory of graphs Zarankiewicz developed, among other topics, a criterion for the existence of a complete subgraph of highest possible order in every graph of a given order in which the minimum number of edges arising from a vertex is sufficiently high (1947). Later, Pál Turán improved this criterion and devoted an interesting study to it. Zarankiewicz, in publications of 1953 and 1954, solved, independently of K. Urbanik, a problem posed by Turán by showing that if A and B are finite sets of the plane, composed of a and b points, respectively, and if each point of A is joined by a simple arc to all the points of B in such a way that outside of A and B every point of intersection belongs to two arcs, then the number of these points of intersection is at least equal to

$$E(\tfrac{1}{2}a) \cdot E(\tfrac{1}{2}a - 1) \cdot E(\tfrac{1}{2}b) \cdot E(\tfrac{1}{2}b - 1),$$

and that this minimum is attained. In 1951 Zarankiewicz posed the problem of finding the least number $k_j(n)$ such that every set of $k_j(n)$ points of a plane net of n^2 points (where $n > 3$) contains j^2 points on j lines and on j columns. Several other authors have subsequently treated this problem.

Zarankiewicz's works on complex functions (1934, 1938, 1956) deal principally with the kernel (*Kernfunktion*) and its applications. Given a complete system $\{\phi_\nu(z)\}$, where $\nu = 1, 2, \cdots$, of orthonormal analytic functions in a domain D of the complex plane of $z = x + iy$, the function $K_D(z,\bar{\zeta}) = \sum_{\nu=1}^{\infty} \phi_\nu(z) \overline{\phi_\nu(\zeta)}$ is called the kernel of the domain D. It is known that it exists for all z and ζ of D and depends only on D; that if D is simply connected—that is, if the boundary of D is connected—the function $W(z) = \int_0^z K_D(z,\bar{\zeta})dz$ transforms D onto the interior of the circle $|W| < c$, where c is a constant; that the function K_D is a relative invariant—that is,

$$K_D(z,\bar{z}) = K_{D^*}(z^*[z], \overline{z^*[z]}) \cdot |dz^*(z)/dz|^2$$

for every analytic function z^*z mapping the domain D of the complex z-plane onto the domain D^* of the complex z^*-plane—and, consequently, that the formula

$$ds_D^2(z) = K_D(z,\bar{z}) \cdot |dz|^2 =$$
$$K_{D^*}(z^*,\overline{z^*}) \cdot |dz^*|^2 = ds_{D^*}^2(z^*)$$

represents the square of the length of the line or element of an invariant metric; that the curvature of the metric

$$J_D(z,\bar{z}) = -\frac{2}{K_D(z,\bar{z})} \cdot \frac{d^2 \log K_D(z,\bar{z})}{dz d\bar{z}}$$

is an absolute invariant of the conformal mappings; and that if the boundary of D is connected, $J_D(z,\bar{z})$ is a constant.

Zarankiewicz showed that when the boundary of D is doubly connected—that is, has exactly two components—the function $J_D(z,\bar{z})$ is no longer constant; in this case he represented it by doubly periodic functions. He established a criterion for recognizing, with the aid of this representation, when it is that a boundary domain with two components can be transformed conformally into another domain of this type. (P. P. Kufarev determined the minimum domain, that is, into which every domain of which the boundary consists of two components is transformable by the function $W[z]$.) This Zarankiewicz result played an important role in the development of the theory of the kernel and its generalizations to several variables, notably to pseudo-conformal transformations in space of more than three dimensions.

In number theory Zarankiewicz devoted particular attention to what are called triangular numbers—that is, triplets of integers equal to the lengths of the sides of right triangles (1949). His ideas inspired a work by Sierpiński (1954), at the end of which the author reproduces an ingenious example, inspired by Zarankiewicz, of a decomposition of the set of natural numbers into two disjoint classes, neither of which contains a triplet of consecutive numbers or an infinite arithmetic progression.

BIBLIOGRAPHY

I. ORIGINAL WORKS. A list of Zarankiewicz's 45 publications is in *Colloquium mathematicum*, **12** (1964), 285–288. The most important are "Sur les points de division dans les ensembles connexes," in *Fundamenta mathematicae*, **9** (1927), 124–171; "Über eine topologische Eigenschaft der Ebene," *ibid.*, **11** (1928), 19–26; "Uber die lokale Zerschneidung der Ebene," in *Monat-shefte für Mathematik und Physik*, **39** (1932), 43–45; "Sur la représentation conforme d'un domaine doublement connexe sur un anneau circulaire," in *Comptes rendus . . . de l'Académie des sciences*, **198** (1934), 1347–1349; "Über doppeltzerlegende Punkte," in *Fundamenta mathematicae*, **23** (1934), 166–171; "O liczbach trójkątowych" ("On Triangular Numbers"), in *Matematyka*, **2** (1949), nos. 4–5; "Sur un problème concernant les coupures des régions par des continus," in *Fundamenta mathematicae*, **39** (1952), 15–24, written with C. Kuratowski; and "On a Problem of P. Turán Concerning Graphs," *ibid.*, **41** (1954), 137–145.

II. SECONDARY LITERATURE. See S. Bergman, R. Duda, B. Knaster, Jan Mycielski, and A. Schinzel, "Kazimierz Zarankiewicz," in *Wiadomości matematyczne*, 2nd ser., **9** (1966), 175–184 (in Polish), also in *Colloquium mathematicum*, **12** (1964), 277–288 (in French), which contains a list of Zarankiewicz's works.

BRONISLAW KNASTER

AL-ZARQĀLĪ (or **AZARQUIEL**), **ABŪ ISḤĀQ IBRĀHĪM IBN YAḤYĀ AL-NAQQĀSH** (d. Córdoba, Spain, 15 October 1100),[1] *astronomy.*

During his lifetime al-Zarqālī was known in Spain as Azarquiel; the correct form of this word, al-Zarqiyāl, was preserved by Ibn al-Qifṭī.[2] The name is composed of the Arabic article *al*, the adjective *zarqā'* ("blue"; "the blue-eyed one"), and *ellus/el*, the diminutive form in Spain.

The few known facts of al-Zarqālī's life can be established from the autobiographical passages in his works. He must have been born in the first quarter of the eleventh century to a family of artisans.[3] His manual skill led him to enter the service of Cadi Ibn Ṣāʿid of Toledo[4] as a maker of the delicate instruments needed to continue the astronomical observations begun about 1060—possibly to emulate those carried out by Yaḥyā ibn Abī Manṣūr—perhaps by order of al-Maʾmūn of Toledo. Al-Zarqālī's intelligence encouraged his clients to supply him with the books he needed to educate himself; and around 1062 he became a member of the group of which he soon became director. He constructed the water clocks of Toledo, which al-Zuhrī[5] has described; and they must have aroused great admiration, for Moses benʿEzra (d. ca. 1135) dedicated a poem to them that begins: "Marble, work of Zarquiel." The clocks were in use until 1133, when Hamis ibn Zabara, having been given permission by Alfonso VII to try to discover how they worked, took them apart and could not reassemble them. They constituted a very precise lunar calendar and were, to some extent, the predecessors of the clocks or planetary calendar devices that became fashionable in seventeenth-century Europe.

Al-Zarqālī lived in Toledo until the insecurity of the city, repeatedly attacked by Alfonso VI, obliged him to move to Córdoba sometime after 1078. There he determined the longitude of Calbalazada (Regulus) in 1080 and the culmination of the planets a year later; in 1087 he carried out his last observations. This fact has led some writers to establish 1087 as the year of his death. Very little is known about his students, who included Muḥammad ibn Ibrāhīm ibn Yaḥyā al-Sayyid (d. 1144). Considerably more is known about his indirect influence on such later authors as Ibn al-Kammād, al-Biṭrūjī, Abu'l-Ḥasan ʿAlī, Ibn al-Bannā', and Abraham IbnʿEzra.

We shall consider seven works definitely known to be by al-Zarqālī.[6] The first is the Toledan Tables. The original Arabic version has been lost, but two Latin versions have survived: one by Gerard of Cremona and one by an unknown author, perhaps John of Seville, who presents a shorter text than Gerard's. It deals only with a collective work directed by Cadi Ibn Ṣāʿid, in which al-Zarqālī participated and of which he wrote the definitive account. The Latin version, analyzed by Delambre in the nineteenth century and by Millás-Vallicrosa more recently, deals with a combination of processes and methods.[7] It follows the table of al-Khwārizmī in the determination of the right ascensions, and the equations of the sun and moon and of the planets; al-Battānī's table in the oblique ascension, ascendant, parallax, eclipses, and the setting of planets; Hermes' table in the equation of houses; and Thābit ibn Qurra's table in the theory of trepidation or accession and recession. Such Indian processes as the kardaga are used side by side with the sine, cosine, tangent, and cotangent. The table of stellar positions is apparently based on an older one corrected in precession.[8] The Toledan Tables were extraordinarily successful in the Latin world: the Marseilles Tables (ca. 1140) were based on them, and by the twelfth century they were used throughout Europe, ultimately displaced only by the Alfonsine Tables. They also influenced the Islamic West, in the works of Ibn al-Kammād, for example.

Almanac of Ammonius[9] was elaborated by al-Zarqālī in 1089, using material that predated 800, as M. Boutelle[10] has demonstrated. Millás-Vallicrosa identified the Aumenius Humeniz in various texts with Ammonius, son of Hermias, a disciple

of Proclus, who restored the Platonic school of Alexandria in the late fifth and early sixth centuries. Study of the tabular values, unique in the medieval Arabic literature, shows that the Almanac deals with the combination of planetary values and Ptolemaic parameters with the Babylonian doctrine of the limit years, calculated according to the linear system A by Nabu-Rimannu, as van der Waerden[11] has demonstrated. Drawing on the works of Hipparchus and Ptolemy,[12] al-Zarqālī's Almanac was known in Europe as part of al-Biṭrūjī's corpus until the fifteenth century.

The trigonometrical portion of the Almanac presents the same minglings of sources and contains tables of sines, cosines, versed sines, secants, and tangents. The work was translated into Latin (John of Pavia, 1154; William of Saint Cloud, 1296), Hebrew (Jacob ibn Tibbon, 1301), Portuguese, Catalan, and Castilian. Regiomontanus may be considered one of the last to express al-Zarqālī's views.

Suma referente al movimiento del sol has been lost.[13] Its subject is known, since the identically entitled work of Thābit ibn Qurra, written about two centuries earlier, is extant, and al-Zarqālī refers to the latter work in his Tratado . . . de las estrellas fijas. It is based on twenty-five years of observations in which he discovered the proper motion of the solar apogee, which he set as 1° every 299 common years (12.04″ annually) counted in the same direction as the zodiacal signs. This discovery is shown in the Marseilles Tables (ca. 1140), and Abu'l-ḤasanʿAlī (fl. 1260) attempted to explain it by means of an epicycle, which he sought to provide.[14]

Tratado relativo al movimiento de las estrellas fijas is preserved in the Hebrew translation by Samuel ben Yehuda (called Miles of Marseilles).[15] Known by Ibn Rushd,[16] the Tratado sought to demonstrate mathematically the trepidation theory according to which the movement of the sphere of the fixed stars is determined by the movement of a straight line that joins the center of the earth with a movable point on a base circle or epicycle. Comparing his observations with those of earlier authors, al-Zarqālī explained the trepidation theory according to three models that situate the epicycle (1) in a meridian plane, (2) in the plane of the ecliptic, and (3) with two equal epicycles centered in the mean equinoctial points normal to the equator. He always took the beginning of Aries as a movable point on the epicycle and referred its motion to the vernal equinox. He thus justified the accession and retrocession of the fixed stars, studied and calculated their longitudinal movement, and determined the dimensions of the epicycle (radius) and period of trepidation in the three models. Having critically studied the results to which the three models led him, al-Zarqālī concluded that the third conforms most to the observational data and, relying on it, he constructed tables of mean movement at the beginning of Aries in Christian, Arab, and Persian years. In the same work he studied the variation in the obliquity of the ecliptic by the action of two small circles, one concentric to an equator of 23°43' diameter and the other with its center in a point of the first, 10' in diameter.

Tratado de la azafea[17] concerns the *azafea*. Al-Zarqālī constructed one superior to the universal sheet of ʿAlī ibn Khalaf.[18] The latter, which exerted only limited influence in the Muslim world — and none in the Latin world — contained the stereographic projection of the sphere on a plane normal to the ecliptic that cuts it along the solstitial line Cancer–Capricorn. In his *Tratado de la azafea* (*al-ṣafīḥa*) al-Zarqālī presented the stereographic projections of the equatorial circle and of the circle of the ecliptic at the same time.

The construction of the apparatus and the formulation of the corresponding rules occurred in several stages. Before 1078 a draft of the *Tratado* was dedicated to al-Maʾmūn of Toledo (*azafea maʾmūniyya*); it was not transmitted to Alfonso X. The *azafea ʿabbādiyya* (dedicated to al-Muʿtamid ibn ʿAbbād) subsequently appeared in two versions: the major one, comprising 100 chapters, was translated into Castilian at the court of Alfonso X[19] and exerted little influence in the Latin world; the minor one, of sixty-one chapters, was transmitted through Jacob ibn Tibbon, Moshe Galino, and William the Englishman to influence Gemma Frisius, Juan de Rojas (*fl.* 1550), and Philippe de La Hire.[20]

The back of the copy of this instrument, at the Fabra Observatory of Barcelona,[21] presents the orthographical projection of the sphere and, in the fourth quadrant, a representation of sines that Millás-Vallicrosa called the quadrant *vetustísimo* ("very ancient"), which is believed to date from the mid-tenth century in the Iberian Peninsula.[22]

Arabic treatises on the *azafea* frequently contain a description of the *azafea shakāziyya*, which is not now known. It was the forerunner, however, of the *shakāzī*[23] quadrant invented by Ibn Tibūgā (1358–1447), who used the same projection system on its face as in al-Zarqālī's *azafea*, the only difference being the omission of the ecliptic projec-

tion and the major circles of longitude and the minor circles of latitude. Also, there was an alteration of the quadrant of the umbra, which was at the back of the *azafea* and was determined by an arc of extensive or convex umbras parallel to the arc of altitude. There was an ordinary zodiacal calendar on the back, another of right ascension, and the projection of the fixed stars that made it possible to determine, by means of the alidade, the equatorial coordinates of any fixed star through simple reading. The projection system used seems to be the polar stereographic one of the ordinary astrolabe.

Tratado de la lámina de los siete planetas, dedicated to al-Muʿtamid, was written in 1081 and surpasses the book on the sheets of the seven planets by Ibn al-Samh (*d.* 1035)[24]; it is a predecessor of the *Aequatorium planetarium* of the Renaissance. The importance of the Arabic text lies in its clarification of one of the most debated passages in medieval astronomy, for in the graphic representation included in the Castilian translation ordered by Alfonso X (The Wise) the orbit of Mercury is not circular.[25] On this basis it has been alleged that al-Zarqālī anticipated Kepler in stating that orbits — the orbit of Mercury in this case — are elliptical. Although the Arabic text merely states that an orbit is *bayḍī* ("oval"), it shows that al-Zarqālī treated Mercury in the same deductive way that Kepler dealt with Mars in his *Astronomia nova*. Before establishing his first law, Kepler considered the possibility of elliptical orbits; it is not known whether he knew al-Zarqālī's text.[26]

Influencias y figuras de los planetas is an astrological treatise of no particular importance.

NOTES

1. Ibn al-Abbār, *Takmila*, Bel-Ben Cheneb, ed. (Algiers, 1920), no. 358, p. 169.
2. *Tarīkh al-ḥukamāʾ*, J. Lippert, ed. (Leipzig, 1903), 57.
3. Isḥāq Israeli, *Yesod ʿolam* (Berlin, 1848), IV, 7.
4. *Tabaqāt al-umam*, Luis Cheikho, ed. (Beirut, 1912), 74; French trans. by Régis Blachère (Paris, 1935), 138–139.
5. Castilian trans. by J. M. Millás-Vallicrosa, *Estudios sobre Azarquiel* (Madrid–Granada, 1943–1950), 6–9.
6. The order of the works is that of Millás-Vallicrosa, *op. cit.*, the fundamental work on the subject.
7. E. S. Kennedy, "A Survey of Islamic Astronomical Tables," in *Transactions of the American Philosophical Society*, n.s. **46**, no. 2 (1956), no. 24.
8. Analysis in Millás-Vallicrosa, *op. cit.*, 22–71.
9. Edition of the Arabic canons, Castilian trans., and corresponding tables, *ibid.*, 72–237.
10. M. Boutelle, "The Almanac of Azarquiel," in *Centaurus*, **12**, no. 1 (1967), 12–19.
11. "The Date of Invention of Babylonian Planetary Theory,"

in *Archive for History of Exact Sciences*, **5**, no. 1 (1968), 70–78.

12. *Almagest*, IX, 3.

13. See Millás-Vallicrosa, *op. cit.*, 239–247.

14. This discovery must have been made by al-Zarqālī. See Willy Hartner, "Al-Battānī," in *DSB*, I, 507–516.

15. Edition and Castilian trans. by Millás-Vallicrosa, *op. cit.*, 239–343.

16. See the Castilian trans. by Carlos Quirós Rodríguez, *Compendio de metafísica de Averroes* (Madrid, 1919); *Kitāb mā baʿd al-ṭabīʿa* (Hyderabad, 1945), 135–136; and esp. O. Neugebauer, "Thâbit ben Qurra, 'On the Solar Year' and 'On the Motion of the Eighth Sphere,' " in *Proceedings of the American Philosophical Society*, **106**, no. 3 (1962), 264–299; B. R. Goldstein, "On the Theory of Trepidation," in *Centaurus*, **10** (1964), 232–247; and J. D. North, "Medieval Star Catalogues and the Movement of the Eighth Sphere," in *Archives internationales d'histoire des sciences*, **20** (1967), 71–83.

17. See Millás-Vallicrosa, *op. cit.*, 425–455.

18. The instructions for the construction and use of this instrument appear in *Los libros del saber de astronomía*, M. Rico y Sinobas, ed., III (Madrid, 1864), 1–132.

19. *Ibid.*, 135–237.

20. Jacob ibn Tibbon, *Tractat de l'assafea d'Azarquiel*, ed. of the Hebrew and Latin texts and Catalan trans. by J. M. Millás-Vallicrosa (Barcelona, 1933); E. Poulle, "Un instrument astronomique dans l'Occident latin: La 'saphea,' " in *A. Giuseppe Ermini* (Spoleto, 1970), 491–510; and F. Maddison, "Hugo Helt and the Rojas Astrolabe Projection," in *Revista da Faculdade de ciências, Universidade de Coimbra*, **39** (1966).

21. J. M. Millás-Vallicrosa, "Un ejemplar de azafea árabe de Azarquiel," in *al-Andalus*, **9**, no. 1 (1944), 111–119.

22. See J. Vernet, "La ciencia en el islam y occidente," in *XII Settimane di studio del centro italiano di studi sull'alto medioevo*, II (Spoleto, 1965), 555–556.

23. See Julio Samsó Moya, "Un instrumento astronómico de raigambre zarqalí: El cuadrante šakāzī de Ibn Ṭībugā, in *Memorias de la Real Academia de buenas letras de Barcelona*, **13** (1971–1975), 5–31.

24. See Millás-Vallicrosa, *op. cit.*, III, 241–271.

25. *Ibid.*, 272–284.

26. See Willy Hartner, *Oriens, Occidens* (Hildesheim, 1968), 474–478, 486.

J. Vernet

ZAVADOVSKY, MIKHAIL MIKHAYLOVICH (*b.* Pokrovka, Kherson guberniya [now Kirovograd], Russia, 29 July 1891; *d.* Moscow, U.S.S.R., 28 March 1957), *biology.*

Zavadovsky graduated in 1914 from the natural sciences section of the Faculty of Physics and Mathematics of Moscow University, with a dissertation on the lipoid semipermeable covering of the eggs of *Ascaris megalocephala.* From 1915 to 1918 he was assistant to N. K. Koltsov at the Moscow University for Women and, after receiving his master's degree in 1918, became privatdocent of experimental biology at Moscow University. He worked at the Askania Nova Zoo and at the University of the Crimea from 1919 to 1921, when he resumed teaching at Moscow University. From 1922 to 1924 he was professor at the Karl Liebknecht Institute of National Education in Moscow.

He assumed the post of head of the department of general biology at the Second Moscow State University (1924–1928) and from 1925 was director of the Moscow zoo, where he organized a laboratory of experimental biology. In 1929 he became head of the laboratory of the physiology of growth of the Institute of Livestock Breeding of the Lenin All-Union Academy of Agricultural Sciences, and from 1930 to 1948 he headed the department and laboratory of the dynamics of development at the First Moscow State University.

In his studies of the developmental conditions of parasitic worms (Ascaris, Enterobius, Trichostrongylidae) Zavadovsky paid special attention to the importance of external factors, such as oxygen and the chemical constitution of the environment, with the aim of developing measures to combat infection in man and domestic animals. He published about forty articles on experimental parasitology.

Zavadovsky also analyzed the development of sexual characteristics. Detailed studies of the effects of castration and the transplantation of sex glands in chickens led him to conclude that some of the secondary sexual characteristics depend on their development in the hormones of the sex glands and that some are independent of it. These results were confirmed in his studies of ducks, pheasants, antelopes, and horned cattle. The similarity of gelded males and spayed females testified to the equipotentiality of the soma of both sexes.

Zavadovsky established that the monosexuality of female birds and the bisexuality of the male—and the opposite among mammals and amphibians—correspond to the distribution of sex chromosomes: XY in female and XX in male birds, and the reverse among mammals and amphibians. Zavadovsky investigated the interrelationships between secondary sexual characteristics and the sex glands and studied the interaction of the endocrine glands. He concluded that a ± mutual influence exists: an organ that stimulates another also is inhibited by it. This was an important application to biology of the cybernetic principle of positive or negative feedback.

On the basis of his study of the sexual cycle of laboratory and farm animals, Zavadovsky suggested the possibility of hormonal stimulation of multiple pregnancy in sheep, by introducing the blood serum of a mare in foal. The applications of his proposal were especially important in the breeding of Karakul sheep.

BIBLIOGRAPHY

I. ORIGINAL WORKS. Zavadovsky's writings include "O lipoidnoy polupronitsaemoy obolochke yaits *Ascaris megalocephala*" ("On the Lipoid Semipermeable Covering of the Eggs of *Ascaris megalocephala*"), in *Uchenye zapiski universiteta im. Shanyavskogo*, **1–2** (1914–1915); *Pol i razvitie ego priznakov. K analizu formoobrazovania u zhivotnykh* ("Sex and the Development of Its Signs. Toward an Analysis of Formation in Animals"; Moscow, 1922); *Pol zhivotnykh i ego prevrashchenie* ("Sex of Animals and Its Transformation"; Moscow, 1923); *Ravnopotentsialna li soma samtsa i samki u ptits i mlikopitayushchikh?* ("Are the Male and Female Soma Equipotent in Birds and Mammals?"; Moscow, 1923); "Hängt der Alterdimorphismus von der Geschlechtsdrüsen ab?" in *Biologia generalis*, **2** (1926), 631–638; "The Bisexual Nature of the Hen and Experimental Hermaphroditism in Hens," *ibid.*, **3** (1927); and "Priroda skorlupy yaits askarid raznykh vidov" ("The Nature of the Shells of the Eggs of Various Kinds of Ascarids"), in *Trudy Laboratorii eksperimentalnoi biologii Moskovskogo zooparka*, **4** (1928), 201–207; and *Vneshnie i vnutrennie faktory razvitia* ("External and Internal Factors of Development"; Moscow, 1928).

See also *Dinamika razvitia organizma* ("Dynamics of Development in the Organism"; Moscow, 1931); "Printsip ± vzaimodeystvia v razvitii osobi" ("Principle of ± Mutual Interaction in the Development of the Individual"), in *Uspekhi sovremennoi biologii*, **2** (1933), 86–103; "Upravlenie polovym tsiklom krolikov, ovets, i korov" ("Control of the Sexual Cycle of Rabbits, Sheep, and Cows"), in *Trudy po dinamike razvitiya*, **11** (1939), 15–24; "Opyt eksperimentalnogo mnogoplodia ovets" ("An Experimental Attempt at Multiple Pregnancy in Sheep"), *ibid.*, 94–112; *Estestvennoe i eksperimentalnoe mnogoplodie korov* ("Natural and Experimental Multiple Pregnancy in Cows"; Alma Ata, 1947): and *Teoria i praktika gormonalnogo metoda stimulyatsii mnogoplodia selskokhozyaystvennykh zhivotnykh* ("Theory and Practice of the Hormonal Method of Stimulating Multiple Pregnancy in Agricultural Animals"; Moscow, 1963).

II. SECONDARY LITERATURE. See N. I. Vavilov, "Prof. M. M. Zavadovskomu (Po povodu 20-letia ego nauchnoy deyatelnosti)" ("To Prof. M. M. Zavadovsky [On the 20th Anniversary of His Scientific Career]"), in *Trudy po dinamike razvitiya*, **10** (1935), 9–11.

L. J. BLACHER

ZAVARZIN, ALEKSEY ALEKSEEVICH (*b.* St. Petersburg, Russia [now Leningrad, U.S.S.R.], 25 March 1886; *d.* Leningrad, 25 July 1945), *histology, biology, embryology.*

Zavarzin completed his secondary studies at the technical high school in St. Petersburg in 1902 and graduated from the natural sciences section of the Faculty of Physics and Mathematics at St. Petersburg in 1907. Retained in the histology department of A. S. Dogel to prepare for a teaching career, he passed his master's examination in zoology and comparative anatomy in 1910 and defended his dissertation, on the structure of the sensory nervous system and the optical ganglia of insects, three years later. He was subsequently professor of histology and embryology at the University of Perm (1916–1922), at the Military Medical Academy in Leningrad (1922–1936), at the First Leningrad Medical Institute (1936–1945), and at the University of Tomsk (1941–1944). From 1932 to 1945 he headed the department of general morphology of the All-Union Institute of Experimental Medicine and in 1944–1945 was director of the Institute of Cytology, Histology, and Embryology of the U.S.S.R. Academy of Sciences.

In a series of works on the histology of the nervous system of insects (1911–1924) Zavarzin established the morphological similarity of the optical centers and of the trunk brain of systematically distant animals (mammals and insects) and formulated a theory of parallelism of histological structure that he subsequently stated in "Ob evolyutsionnoy dinamike tkaney" ("On the Evolutionary Dynamics of Tissues," 1934). In a two-part monograph devoted to the structure of blood cells and connective tissue (1945–1947) Zavarzin further developed the evolutionary trend in histology, transforming it from a purely descriptive into a dynamic discipline.

Zavarzin also worked on problems of general biology—the origin of multicelled organisms, the biological basis of inflammation, the theory of embryonic layers, the theory of the cellular structure of organisms, the significance of remote sensory organs in the formation of the encephalic section of the central nervous system, the relationship between form, function, and development as three aspects of a unified biological process, and the coordination of changes of histological structures in onto- and phylogenesis.

BIBLIOGRAPHY

I. ORIGINAL WORKS. Zavarzin's textbooks, the fruit of almost thirty years' teaching, are *Kratkoe rukovodstvo po embriologii cheloveka i pozvonochnykh zhivotnykh* ("A Short Guide to the Embryology of Man and the Vertebrates"; Leningrad–Moscow, 1929; 4th ed.,

Leningrad, 1939); *Kurs mikroskopicheskoy anatomii* ("Course in Microscopic Anatomy"; Moscow–Leningrad, 1930); *Kurs obshchey gistologii* ("Course in General Histology"; Leningrad, 1932); *Kurs gistologii* ("Course in Histology"), 2 pts. (Moscow, 1933); *Kurs gistologii i mikroskopicheskoy anatomii* ("Course in Histology and Microscopic Anatomy"; 5th ed., Leningrad, 1939); *Kurs gistologii*, 6th ed. (Moscow, 1946), written with A. V. Rumyantsev; and *Isbrannye trudy* ("Selected Works"), 4 vols. (Moscow–Leningrad, 1950–1953).

Specialized works include "Histologische Studien über Insekten," in *Zeitschrift für wissenschaftliche Zoologie*, **97** (1911), 481–510; **100** (1912), 245–286, 447–458; **108** (1913), 175–257; and **122** (1924), 97–115, 323–424; and "Der Parallelismus der Strukturen als ein Grundprinzip der Morphologie," *ibid.*, **124** (1925), 118–212.

Among his monographs are *Gistologicheskie issledovania chuvstvitelnoy nervnoy sistemy i opticheskikh gangliev nasekomykh* ("Histological Research on the Sensory Nervous System and the Optical Ganglia of Insects"; St. Petersburg, 1913), his master's thesis; *Ocherki po evolyutsionnoy gistologii nervnoy sistemy* ("Sketches in the Evolutionary Histology of the Nervous System"; 1941); and *Ocherki po evolyutsionnoy gistologii krovi i soedinitelnoy tkani* ("Sketches in the Evolutionary Histology of the Blood and Connective Tissue"), 2 vols. (Moscow, 1945–1947).

II. SECONDARY LITERATURE. On Zavarzin and his work, see A. I. Abrikosov, "Aleksey Alekseevich Zavarzin," in *Vestnik Akademii meditsinskikh nauk SSSR* (1946), no. 1, 65–67, an obituary; D. N. Nasonov and A. A. Zavarzin (his son), *Aleksey Alekseevich Zavarzin* (Moscow, 1951), with bibliography; and G. A. Nevmyvaka; *Aleksey Alekseevich Zavarzin* (Leningrad, 1971).

L. J. BLACHER

AL-ZAYYĀTĪ AL-GHARNĀṬĪ, AL-ḤASAN IBN MUḤAMMAD AL-WAZZĀN. See Leo the African.

ZEEMAN, PIETER (*b.* Zonnemaire, Zeeland, Netherlands, 25 May 1865; *d.* Amsterdam, Netherlands, 9 October 1943), *physics*.

Zeeman is best remembered for his observations in 1896 of the magneto-optic phenomenon that almost immediately was named the Zeeman effect. His experimental discovery was not fortuitous, but the fruition of theoretical views that had motivated attempts over a span of thirty-five years to detect some such interaction between magnetism and light. Zeeman's initial observations were beautifully comprehended by H. A. Lorentz' electromagnetic theory, which also served to guide Zeeman in

the very early refinement and extension of his discovery. As a result Zeeman and Lorentz shared the 1902 Nobel Prize for physics in recognition of their accomplishment and of the promise, since overwhelmingly fulfilled, of the Zeeman effect for contributing to the understanding of spectra and the particulate structure of matter.

Following his elementary education in the small village of Zonnemaire, Zeeman was sent by his parents—Catharinus Farandinus Zeeman, a Lutheran minister, and Wilhelmina Worst—to the secondary school at Zierikzee, five miles away. He subsequently studied classical languages for two years at the gymnasium in Delft in order to satisfy requirements for the university. His early scientific education seems to have been adequate, for he published an account of the aurora borealis from Zonnemaire and impressed Kamerlingh Onnes, whom he met at Delft, with his grasp of Maxwell's investigations on heat. Zeeman entered the University of Leiden in 1885 and spent nearly a dozen years there, working with Kamerlingh Onnes and Lorentz and becoming the latter's assistant in 1890. For his careful measurements of the Kerr effect, Zeeman won the gold medal of the Netherlands Scientific Society of Haarlem in 1892 and was awarded the doctorate in 1893. After a semester at the Kohlrausch Institute in Strasbourg, he returned to the University of Leiden as a *Privatdozent*. In 1895 he married Johanna Elisabeth Lebret; they had a son and three daughters.

From January 1897 until his retirement in 1935, Zeeman was associated with the University of Amsterdam. Appointed a lecturer, he was promoted to professor of physics in 1900, succeeded the retiring J. D. van der Waals as director of the Physical Institute in 1908; in 1923 he also became director of the new Laboratorium Physica, later renamed the Zeeman Laboratory. Zeeman the master experimentalist was most effective as a teacher in his regular informal discussions with advanced degree candidates concerning the progress and problems of their laboratory research. He received many awards and honors, including honorary degrees from at least ten universities; was a member of a number of academies; and was *associé étranger* of the Paris Academy of Sciences. He served as secretary of the Mathematics-Physics Section of the Royal Netherlands Academy of Sciences, Amsterdam, and was a knight of the Order of Orange-Nassau and commander of the Order of the Netherlands Lion.

Zeeman's was the third magneto-optic effect to be discovered. In 1845 Faraday had observed the first, which related magnetism and propagated light, and served as the experimental basis for subsequent attempts to discover a magnetic influence upon a source of light. The theoretical basis was provided by William Thomson (Lord Kelvin) and by Maxwell's establishment of the electromagnetic nature of light. Zeeman interrupted his measurements of the Kerr effect (the second magneto-optic effect to be discovered) to improvise an experiment, which proved unsuccessful, seeking some change in the spectrum of a sodium flame that was burning in a magnetic field. About a year later Zeeman learned that Faraday had attempted this experiment without success in 1862. His reaction was that if Faraday, whom he considered the greatest experimental genius of all time, had thought the experiment worth doing, then he could well afford to repeat it again, using the best spectroscopic apparatus and a specially designed electromagnet. Thus equipped, Zeeman observed that the D-lines of the sodium spectrum were decidedly broadened. Within a few weeks he obtained this broadening for other spectral lines and for the related absorption spectra, and convincingly demonstrated that the broadening was a direct effect of the magnetic field.

These results were presented to the Amsterdam Academy of Sciences on 31 October 1896. At the next meeting of the Academy, four weeks later, Zeeman reported that the Lorentz theory not only comprehended his initial findings but also had been used to predict that the light from the edges of the magnetically broadened lines should be polarized—and, further, that he had observed this polarization. The Lorentz theory also provided the equation

$$\frac{T' - T}{T} = \frac{e}{m} \cdot \frac{HT}{4\pi},$$

where T is the natural period of vibration of an ion of charge e and mass m, and T' is its period in a magnetic field of strength H. This equation enabled Zeeman, from his measurements of H, T, and T', to calculate the charge-to-mass ratio of the vibrating "ion" of the sodium atom. His next paper established that the "ion" was negatively charged.

In the spring of 1897, after his move to the University of Amsterdam, Zeeman resolved a magnetically "broadened" spectral line into the triplet of distinct polarized components that the Lorentz theory predicted for a sufficiently intense magnetic field. This in a very real sense was the peak of the Zeeman-Lorentz investigation of the Zeeman effect. For his more exacting measurements at this time, Zeeman had to travel to the University of Groningen to use Hermann Haga's superior spectroscopic apparatus; and by the end of 1897, the limitations of his research facilities at Amsterdam proved decisive for Zeeman. The main deficiency, which persisted until the building of his own laboratory, was in the mountings of his spectroscope. The most interesting and demanding measurements required an isolated and rigid mounting system to ensure sharp definition in the photographs. Zeeman's attempts in this regard were usually spoiled (less than one photograph in thirty was usable) by vibrations due either to human movement on the same floor as his laboratory or to the traffic of Amsterdam—even in the middle of the night. After a promising but qualitative anticipation of a fundamental relationship between Zeeman effect patterns and the laws of spectral series (subsequently called Preston's law), Zeeman felt compelled to abandon this suggestive investigation and turn to less exacting studies of the Zeeman effect and related matters. For these researches, which fully engaged him over the next fifteen years, the magneto-optic theory of Woldemar Voigt performed the same roles that Lorentz' theory had for Zeeman's earlier investigations.

During World War I, Zeeman initiated a systematic redetermination of the velocity of propagation of light in moving transparent media. Early in the nineteenth century Fresnel's optical theory had required that light propagated longitudinally with respect to moving glass would suffer a velocity change of $\left(1 - \frac{1}{\mu^2}\right)v$, where the factor $\left(1 - \frac{1}{\mu^2}\right)$ was the Fresnel coefficient, μ was the index of refraction, and v was the velocity of the glass. In the middle of the century Fizeau had used interference techniques to obtain experimental support for the Fresnel coefficient in the case of light traversing flowing water. Thirty-five years later Michelson and Morley repeated Fizeau's experiment, and with their more precise interferometer they obtained a value in closer agreement with the Fresnel coefficient. Zeeman's interest in this question was generated by the theoretical investigations that Lorentz conducted in 1895 and subsequently reconsidered in terms of relativity theory. By taking into account the dispersion of light in the medium, Lorentz deduced that the coefficient must be $\left(1 - \frac{1}{\mu^2} - \frac{\lambda}{\mu} \cdot \frac{d\mu}{d\lambda}\right)$, where λ is the wavelength of light. In two papers of 1915 and 1916, Zeeman essentially repeated the Michelson-Morley experiment with water and showed that the experimental value of the coefficient *did* vary with the wavelength of the light used, and that within his limits of experimental error it confirmed the Lorentz rather than the Fresnel expression. In 1919 and 1920 Zeeman collaborated with others to communicate three additional papers dealing with the same measurements, but for quartz and glass rather than water. The use of rapidly moving solid substances imposed extraordinary experimental difficulties that Zeeman painstakingly surmounted in order to obtain experimental results further supporting the Lorentz refinement of the Fresnel coefficient.

In 1918 Zeeman published the results of another extremely meticulous set of experiments that also carried profound implications for relativity theory. These measurements, which Zeeman conducted at his country home after determining that they could not be made in the laboratory because of the ever-present vibrations, established an equality of the inertial and gravitational mass for certain crystals and radioactive substances to within one part in twenty or thirty million.

With the facilities of his new laboratory available after 1923, Zeeman, in collaboration with others, finally turned to experiments involving precision measurements of the Zeeman effect. Most notable were a series of investigations of the magnetic resolution of the spectral lines of certain noble gases and, in 1932, a detailed and beautifully presented study of the hyperfine structure and Zeeman effect of the strong spectral lines of rhenium, both of which confirmed the value of the nuclear moment of the two rhenium isotopes.

BIBLIOGRAPHY

I. Original Works. Zeeman's main magneto-optic papers, written in the period 1896–1913, were collected and republished in the commemorative volume *Verhandelingen van Dr. P. Zeeman over magneto-optische Verschijnselen*, H. A. Lorentz, H. Kamerlingh Onnes, I. M. Graftdijk, J. J. Hallo, and H. R. Woltjer, eds. (Leiden, 1921). In *Researches in Magneto-Optics* (London, 1913) Zeeman discussed his own and others' contributions during this same period to Zeeman effect and closely related phenomena, and appended a very valuable bibliography. Nearly all of Zeeman's published papers are cataloged in Poggendorff, IV, 1682; V, 1404–1405; and VI, 2957–2958.

II. Secondary Literature. For information on Zeeman see the "Biography" in *Nobel Lectures. Physics, 1901–1921* (Amsterdam, 1967), 41–44; Lord Rayleigh, "Pieter Zeeman 1865–1943," in *Obituary Notices of Fellows of the Royal Society of London*, **4** (1944), 591–595; and H. Kamerlingh Onnes, "Zeeman's Ont-dekking van het naar hem genoemde Effect," in *Physica*, **1** (1921), 241–250.

James Brookes Spencer

ZEILLER, RENÉ CHARLES (*b.* Nancy, France, 14 January 1847; *d.* Paris, France, 27 November 1915), *paleobotany.*

A number of Zeiller's ancestors were graduates of the École Polytechnique and his family environment was propitious to his intellectual development. He early developed an interest in botany through contact with his maternal grandfather, Charles Guibal, who took him on excursions through the Lorraine countryside. After attending the Lycée Bonaparte in Paris and then the *lycée* in Nancy, Zeiller studied at the École Polytechnique in Paris and at the École des Mines in Nancy, where he obtained his degree in 1870.

At the beginning of 1871 Zeiller was named engineer of the *sous-arrondissement* of Tours and assigned to supervise work on a portion of the Orléans railway. He returned to Paris in 1874, still working for the same administration and holding the same rank for ten years, until his promotion to chief engineer. In 1882 he transferred to the Service de Topographie Souterraine des Bassins Houillers de France. Rising steadily through the hierarchy of the Conseil Général des Mines, he ultimately became its president in 1911. Admired for his scrupulousness in fulfilling his duties, he was appointed, in addition, a member of the Commission des Appareils à Vapeur.

Zeiller's first publications dealt with technical subjects, namely with the application of geology to the detection of metal-bearing deposits. Although he faithfully executed all his administrative tasks, he still found time to pursue his interest in the study of fossil plants. In this research he was able to profit from his knowledge of both botany and geology and to draw conclusions useful in one or the other of these fields. In 1878 he was appointed *chargé de cours* of plant paleontology at the École des Mines in Paris. Named curator of the school's paleontology collections in 1881, he made such important additions to the collections that scientists came from all over the world to study them.

Zeiller studied fossil plants in order to determine their structures and relationships. At the same time he viewed them as constituents of large groups, the relative ages and geographic distribution of which he attempted to establish. His various memoirs in the series *Gîtes minéraux de la France* (published by the Ministry of Public Works) are models of good scientific publications. In them he dealt also with purely botanical questions. For example, he investigated the mode of fructification of fossil ferns, a subject that long had held his attention. In the course of answering questions raised by leading scientists who held opposing views, Zeiller elaborated several brilliant demonstrations. He proved that *Sigillaria*, in spite of their centrifugal secondary wood and the composition of the vascular strands of their leaves, are cryptogams. This conclusion, a result of his discovery of cones, was in opposition to the French school of Brongniart but in accord with English authors. Through his study of carbonaceous impressions of *Sphenophyllum* fructifications, Zeiller arrived at the conclusion that these plants have no true relation with any living type. His description of several species of *Psaronius* constitutes a model of anatomical research.

Zeiller was especially interested in the Cycadofilicales, and his clear and precise account of the "Fougères à graines" is still worth reading. Finally, after his trenchant critique, the existence of a group of so-called proangiosperms (proposed by Gaston de Saporta) could no longer be accepted.

Zeiller did not restrict his numerous publications to material gathered in France. Through his study of the *Glossopteris* floras (he established that they come from the Permian-Triassic period, as the geologists held) and his attribution of different ages—and not always from the Carboniferous—to coals of varied origins (Tonkin, Chile, New Caledonia), he was the author of several revolutionary ideas.

Zeiller had an impressive capacity for work. Beyond his professional activities and personal research, he also wrote a remarkable treatise on paleobotany and regularly published bibliographic analyses containing abundant new critical commentary. In spite of an incurable disease, he retained his kindly manner until the very end.

BIBLIOGRAPHY

On Zeiller and his work, see G. Bonnier, "René Zeiller," in *Revue générale de botanique*, **28** (1916), 354–367, with portrait, and *ibid.*, **29** (1917), 33–55, which lists his works; A. Carpentier, "René Zeiller (1847–1915). Son oeuvre paléobotanique," in *Bulletin. Société botanique de France*, **25**, 5th ser. (1928), 46–67, with portrait and list of Zeiller's works; and D. H. Scott, "Charles René Zeiller," in *Proceedings of the Linnean Society of London* (1916), 74–78.

F. STOCKMANS

ZEISE, WILLIAM CHRISTOPHER (*b.* Slagelse, Denmark, 15 October 1789; *d.* Copenhagen, Denmark, 12 November 1847), *chemistry.*

The son of Friedrich Zeise, a pharmacist, and Johanna Helena Hammond, Zeise developed an interest in the natural sciences while attending secondary school, which he left in 1805 without graduating. He then was admitted to a pharmacy in Copenhagen—the customary way in Denmark of beginning the study of the natural sciences; the pharmacist for whom he worked was extraordinary professor of chemistry at the University of Copenhagen. After a few months poor health obliged Zeise to return to the family pharmacy in Slagelse, where he continued his studies. The influence of Lavoisier's concepts was so great that by 1806 he had rearranged his father's pharmacy according to the antiphlogistic nomenclature, officially introduced in the Danish pharmacopoeia the previous year.

After returning to Copenhagen in 1806, Zeise lived with Oersted and his family. Oersted, who had recently become extraordinary professor of physics and chemistry at Copenhagen, appointed Zeise his lecture assistant. In 1809 he began the study of medicine, physics, and chemistry; and in 1815 he graduated with a degree in pharmacy. The following year he received his master's degree, and in 1817 he defended his doctoral dissertation, on the action of alkalies on organic substances.

Zeise visited chemical laboratories in Göttingen and Paris in 1818 and in 1819 returned to Copenhagen, where, under Oersted's influence, he established one of the first laboratories in Europe for analytical and organic chemistry. In 1822 he became extraordinary professor of chemistry at the University of Copenhagen and from 1829 until his death was professor of organic chemistry at the Polytechnic Institute of Copenhagen, which had been established on Oersted's initiative. In 1824 he was elected a member of the Royal Danish Academy of Sciences.

Zeise's investigation of organic sulfur compounds led to the discovery of a new class of organic compounds that he named xanthogenates (usually called xanthates), because they were isolated as yellow potassium salts in 1823. Other classes of sulfur compounds that he discovered include the thioalcohols (thiols), in 1833, for which he coined the name mercaptan because they form insoluble mercury salts (*corpus mercurium captans*), and the sulfides (thioethers), in 1836.

Zeise's work on organic platinum compounds in the early 1830's involved him in the controversy between Dumas and Liebig. Zeise believed that his own elemental analysis of these compounds supported Dumas's etherin theory and his rules of substitution, but Liebig considered Zeise's analysis to be incorrect. Vehemently objecting to Liebig's insinuations, Zeise repeated the analysis and completely verified the composition as first established. Curiously, his investigations of mercaptans and sulfides decided the dispute between Liebig and the French chemists in Liebig's favor.

Zeise belongs to the group of organic chemists who laid the foundations of scientific organic chemistry in the first half of the nineteenth century. He also studied the composition of the products obtained by the dry distillation of tobacco and tobacco smoke (1843) and undertook one of the earliest investigations of carotene (1846).

BIBLIOGRAPHY

I. ORIGINAL WORKS. A complete bibliography of Zeise's writings is included in Veibel (see below); see also the Royal Society *Catalogue of Scientific Papers*, VI, 494–496. His books include *Udførlig Fremstilling af Chemiens Hovedlærdomme, såavel i theoretisk som i praktisk Henseende* (Copenhagen, 1829); and *Haandbog i de organiske Stoffers almindelige Chemie* (Copenhagen, 1847).

Among his memoirs are "Om Svovelkulstoffets Forbindelser med Æskene," in *Oversigt over det K. Danske Videnskabernes Selskabs Forhandlinger* (1821–1822), 12–13, and (1822–1823), 10–16; "Die Xanthogensäure nebst einigen Producten und Verbindungen derselben," in *Journal für Chemie und Physik*, 36 (1822), 1–67; "Forsøg over Virkningen mellem Chlorplatin og Viinaand," in *Oversigt . . . Forhandlinger* (1830–1831), 24–25; "Wirkung des Platinchlorids auf Alkohol und daraus hervorgehende Produkte," in *Journal für Chemie und Physik*, 62 (1831), 393–441, and 63 (1831), 121–135; also in *Annalen der Physik*, 21 (1831), 497–541; "Nye undersøgelser af Svovelforbindelser," in *Oversigt . . . Forhandlinger* (1833–1834), 9–16; "Über das Mercaptan," in *Annalen der Physik*, 31 (1834), 369–431; "Mercaptanet, med Bemærkninger over nogle andre nye Producter af Svovelvinsyresaltene," in *Kongelige Danske Videnskabernes Selskabs naturvidenskabelige og mathematiske Afhandlinger*, 4th ser., 6 (1837), 1–70; "Nye Undersøgelser over det brændbare Chlorplatin," *ibid.*, 333–356; "Undersøgelser over Producterne af Tobakkens tørre Destillation og om Tobaksrøgens chemiske Beskaffenhed," in *Oversigt . . . Forhandlinger* (1843), 13–17; "Über die Produkte der trockenen Destillation des Tabaks und die Bestandtheile des Tabakrauches," in *Journal für praktische Chemie*, 29 (1843), 383–395; and "Beretning om nogle Forsøg over Carotinet," in *Oversigt . . . Forhandlinger* (1847), 101–103.

II. SECONDARY LITERATURE. See the obituary in *Oversigt over det K. Danske Videnskabernes Selskabs Forhandlinger* (1848), 19–30; and Stig Veibel, *Kemien i Danmark*, 2 vols. (Copenhagen, 1939–1943), I, 155–160, 180–188, and II, 488–494, with complete bibliography.

STIG VEIBEL

ZELINSKY, NIKOLAY DMITRIEVICH (*b.* Tiraspol, Kherson province [now Moldavian S.S.R.], Russia, 6 February 1861; *d.* Moscow, U.S.S.R., 31 July 1953), *chemistry.*

After graduating from the University of Odessa in 1884, Zelinsky was sent in 1885 to Germany, where he worked with Johannes Wislicenus at Leipzig and Victor Meyer at Göttingen. While trying to obtain tetrahydrothiophene in Meyer's laboratory he synthesized di-(β-chloroethyl) sulfide (mustard gas)—and became its first victim, receiving serious burns. In 1889 at Odessa he defended his master's thesis, on isomers in the thiophene series, and, in 1891, his doctoral dissertation, on stereoisomers in the series of saturated carbon compounds. From 1893 to 1953 Zelinsky was professor at Moscow University, except for the period 1911–1917, when he headed the central laboratory of the Ministry of Finances in St. Petersburg and taught at the Polytechnical Institute.

In connection with the use of poison gases in World War I, Zelinsky in 1915 developed a method of obtaining activated charcoal and a universal gas mask that was used by the Russian and Allied armies. In 1918–1919 he devised a process for obtaining gasoline by cracking higher-boiling petroleum fractions in the presence of aluminum chloride. He was elected a corresponding member of the U.S.S.R. Academy of Sciences in 1926 and, in 1929, a full member. In 1934 he became department head at the Institute of Organic Chemistry (named for him in 1953) of the U.S.S.R. Academy of Sciences.

Zelinsky's most important work dealt with the chemistry of hydrocarbons and with organic catalysis. From 1895 to 1905 he was the first to synthesize many hydrocarbons of the cyclopentane and cyclohexane series, used as standards in the study of the composition of petroleum fractions. He subsequently synthesized other hydrocarbons, including cyclopropanes and cyclobutanes, as well as bicyclic spirane and bridged hydrocarbons, and studied their catalytic transformations, many of which are considered classical. In 1911 Zelinsky discovered the smooth dehydrogenation of cyclohexane (and its homologues) into benzene in the presence of platinum and palladium catalysts at 300°C. In the 1920's and 1930's he studied the selective character of this reaction, used it extensively to determine the content in gasoline of cyclohexane hydrocarbons and kerosene fractions, and proposed it as an industrial process for obtaining aromatic hydrocarbons from petroleum.

In the 1930's Zelinsky investigated the reaction (that he discovered in 1911) of the disproportionation of hydrogen in cyclohexene with the simultaneous formation of benzene and cyclohexane. This reaction, which he termed "irreversible catalysis" and which occurred at room temperature in the presence of platinum and palladium, is peculiar to cyclohexane hydrocarbons containing double bonds, including many terpenes. In 1934 Zelinsky discovered the hydrogenolysis of cyclopentane hydrocarbons with their transformation into alkanes in the presence of platinized carbon and hydrogen at 300–310°C.

Zelinsky showed the intermediate formation of methylene radicals in many heterogenous catalytic reactions: in the decomposition of cyclohexane, in the Fischer-Tropsch synthesis on a cobalt catalyst, in the hydrocondensation of olefins with carbon monoxide, and in the hydropolymerization of olefins in the presence of small quantities of carbon monoxide. A pioneer in the study of the reciprocal isomerization of cyclopentane and cyclohexane hydrocarbons in the presence of aluminum chloride and aluminum bromide, Zelinsky showed in 1939 that cyclohexene and its homologues are almost completely isomerized (in the presence of oxides of aluminum, beryllium, or silicon at 400–450°C.) into homologues of cyclopentene. As early as 1915 he used oxide catalysts in the cracking of petroleum, which led to a lowering of the temperature of the process and to an increase in the quantity of aromatic hydrocarbons formed. In 1915 he was the first to show that in the transformation of organic compounds the reason for the poisoning of the catalyst is that a layer of carbon is deposited on its surface. By oxidizing it, the catalyst can be regenerated. This method found wide application in connection with the extensive use twenty years later of oxide catalysts in the petroleum processing industry.

To confirm the organic theory of the origin of petroleum, Zelinsky in the 1920's and 1930's obtained artificial petroleum from plant and animal materials—cholesterol, fatty acids, beeswax, and abietic acid under the action of aluminum chloride. He developed the cyanohydrin method for obtaining alpha-amino acids and was the first to obtain a number of hydroxy amino acids. In 1912 Zelinsky achieved the hydrolysis of proteins using dilute acids and, in addition to amino acids, obtained their cyclic anhydrides, diketopiperazines. In connection with this he proposed the theory of the cyclic structure of protein molecules.

BIBLIOGRAPHY

I. ORIGINAL WORKS. Zelinsky's writings were published as *Izbrannye trudy* ("Selected Works"), 2 vols. (Moscow–Leningrad, 1941); *Sobranie trudov* ("Collection of Works"), 4 vols. (Moscow–Leningrad, 1954–1957); and *Izbrannye trudy* ("Selected Works"; Moscow, 1968); all three eds. include biography, bibliography, and a sketch of his career.

II. SECONDARY LITERATURE. On Zelinsky and his work, see the following: A. A. Balandin, "Akademik Nikolay Dmitrievich Zelinsky," in *Vestnik Akademii nauk SSSR*, **16**, nos. 5–6 (1946), 80–90, published on his eighty-fifth birthday; N. A. Figurovsky, *Ocherk razvitia ugolnogo protivogaza* ("A Sketch of the Development of the Charcoal Gas Mask"; Moscow, 1952); and *Zamechatelnnoe russkoe izobretenie (k 40-letiyu izobretenia ugolnogo protivogaza N. D. Zelinskogo* ("A Remarkable Russian Invention [on the Fortieth Anniversary of the Invention of the Charcoal Gas Mask by Zelinsky"]; Moscow, 1956); B. A. Kazansky. "Raboty N. D. Zelinskogo i ego shkoly v oblasti kataliticheskikh prevrashcheny uglevodorodov" ("Work of Zelinsky and His School in the Field of Catalytic Transformations of Hydrocarbons"), in *Yubileyny sbornik, posvyashchenny tridtsatiletiyu Velikoy Oktyabrskoy sotsialisticheskoy revolyutsii* ("Jubilee Collection Dedicated to the Thirtieth Anniversary of the Great October Socialist Revolution"), pt. 1 (Moscow–Leningrad, 1947); and *Vydayushchysya sovetsky ucheny akademik Nikolay Dmitrievich Zelinsky* ("The Distinguished Soviet Scientist Academician. . ."; Moscow, 1951, published on his ninetieth birthday; B. A. Kazansky, A. N. Nesmeyanov, and A. F. Platé, "Raboty akademika N. D. Zelinskogo i ego shkoly v oblasti khimii uglevodorodov i organicheskogo kataliza" ("The Work of Academician Zelinsky and His School in the Field of the Chemistry of Hydrocarbons and Organic Catalysis"), in *Uchenye zapiski Moskovskogo gosudarstvennogo universiteta*, no. 175 (1956), 5–53; Y. G. Mamadaliev, *Akademik Nikolay Dmitrievich Zelinsky* (Baku, 1951); A. F. Platé, "N. D. Zelinsky i sovremennoe razvitie neftekhimii" ("Zelinsky and the Contemporary Development of Petrol Chemistry"), in *Neftekhimia*, **1** (1961), 7–14, published on the centenary of his birth; V. M. Rodionov, "Nikolay Dmitrievich Zelinsky (vospominania i vstrechi) (". . . [Reminiscences and Meetings"]), in *Soobshchenie o nauchnykh rabotakh chlenov Vsesoyuznogo Khimicheskogo obshchestva im. D. I. Mendeleeva*, no. 12 (1951; N. I. Shuykin, "Pamyati akademika Nikolaya Dmitrievicha Zelinskogo" ("Memories of

Academician . . ."), in *Zhurnal obshchei khimii*, **31** (1961), i–vii, published on the centenary of his birth; V. A. Volkov and A. N. Shamin, "Novye dokumenty o deyatelnosti N. D. Zelinskogo" ("New Documents on the Activities of Zelinsky"), in *Voprosy istorii estestvoznania i tekhniki* (1975), no. 1, 54–56; and Y. K. Yuriev and R. Y. Levina, *Zhizn i deyatelnost akademika Nikolaya Dmitrievicha Zelinskogo* ("Life and Activities of Academician Zelinsky"; Moscow, 1953), also in English (1958).

A. F. PLATÉ

ZEMPLÉN, GÉZA (*b.* Trencsén, Hungary [now Trenčín, Czechoslovakia], 26 October 1883; *d.* Budapest, Hungary, 24 July 1956), *organic chemistry.*

Zemplén was the son of János Zemplén, a postal employee, and of Janka Wittlin. His older brother Viktor was appointed professor of physics at the Technical University of Budapest in 1913 at the age of thirty-three, but his career was cut short three years later by his death in World War I. Zemplén's ability is reflected in his having been named professor at the Technical University in 1913—when he was only twenty-nine. With his appointment the university created the first institute for organic chemistry in Hungary.

After attending secondary school in Fiume, Zemplén studied chemistry, physics, and biology at the University of Budapest, earning the doctorate in physics in 1904. He then became an assistant in chemistry at the Mining and Forestry Academy of Selmec, where he conducted analytic studies of natural substances. In 1909 Zemplén obtained a three-year scholarship; and until 1912 he worked in Emil Fischer's laboratory in Berlin, then one of the most famous centers for the study of organic chemistry. He worked on the synthesis of amino acids and of the synthetic disaccharide cellobiose, publishing several papers on these topics in collaboration with Fischer.

Appointed professor in 1913, Zemplén began a period of intense, uninterrupted research and also trained many students who later achieved distinction as organic chemists. His principal field of research was the carbohydrates. His most important results were the deacetylation of sugar acetates with sodium methoxide (Zemplén's saponification) and the successive (Zemplén) degradation of sugars to derivatives containing increasingly less carbon—which at the time was considered the best method for establishing the structure of the disaccharides. Zemplén also devised the so-called mercury acetate catalytic method for the production of oligosaccharides. His contributions to organic chemistry were printed in more than 200 publications, most of which appeared in German in *Berichte der Deutschen chemischen Gesellschaft.*

Zemplén's successful career earned him membership in many scientific societies. His life was not, however, free from personal tragedy. Upon returning home from a year spent at the University of Washington (1947), he fell ill with cancer and spent his remaining years bedridden.

BIBLIOGRAPHY

A complete list of Zemplén's publications is in an obituary, in German, by R. Bognár, in *Acta chimica Academiae scientiarum hungaricae*, **19** (1959), 121. See also L. Mora, *Géza Zemplén, 1883–1956* (Budapest, 1974), with complete bibliography; "The Degradation of Sugars," in *Berichte der Deutschen chemischen Gesellschaft*, **59** (1926), 125; and "The Mercury Acetate Method," *ibid.*, **62** (1929), 990.

FERENC SZABADVÁRY

ZENODORUS (*b.* Athens [?]; *fl.* early second century B.C.), *mathematics.*

Zenodorus is known to have been the author of a treatise on isoperimetric figures—plane figures of equal perimeter but differing areas, and solid figures of equal surface but differing volumes.[1] This has not survived as such, but it is epitomized in Pappus' *Collection*, in the commentary by Theon of Alexandria on Ptolemy's *Almagest*, and in the anonymous *Introduction to the Almagest.*

Older writers placed the date of Zenodorus in the fifth century B.C., but this was through a mistaken identification with a Zenodorus who is said by Proclus to have belonged "to the succession of Oenopides."[2] From several references by Zenodorus to Archimedes, Nokk rightly concluded that he must have flourished after, say, 200 B.C.[3] Because Quintilian showed awareness of isoperimetry, F. Hultsch and M. Cantor conjectured a lower limit of A.D. 90 for his life; but Zenodorus made no claim to have been the only, or even the first, person to have written on the subject and the deduction is erroneous.[4] Until recently all that could be said with certainty was that he lived after Archimedes and before Pappus, say 200 B.C.–A.D. 300; but it is now established that he must have flourished in the early part of the second century B.C. A fragment from a biography of the Epicurean philosopher Philonides, found in the Herculaneum papyrus roll no. 1044, mentions among his acquaintances a Zenodorus at least once and perhaps twice. In publishing the fragment, W. Crönert identified him with the mathematician.[5] G. J. Toomer, in an elaborate study of occurrences of the name, concluded that unless Zenodorus was a Hellenized Semite (which is not impossible), the comparative rarity of the name confirms Crönert's identification.[6] This is made certain by the fact that in the Arabic translation of Diocles' treatise *On Burning Mirrors*, which has been discovered and edited by Toomer, Zenodorus is mentioned as having posed a problem to Diocles. Toomer's literal translation reads:

> The book of Diocles on burning mirrors. He said: Pythion the geometer, who was of the people of Thasos, wrote a letter to Conon in which he asked him how to find a mirror surface such that when it is placed facing the sun the rays reflected from it meet the circumference of a circle. And when Zenodorus the astronomer came down to Arcadia and was introduced [?] to us, he asked us how to find a mirror surface such that when it is placed facing the sun the rays reflected from it meet a point and thus cause burning. So we want to explain the answer to the problem posed by Pythion and to that posed by Zenodorus; in the course of this we shall make use of the premises established by our predecessors.[7]

It is no bar to the identification of this Zenodorus with the author of the isoperimetric propositions that he is here called an astronomer. There was considerable overlap between mathematics and astronomy—Euclid, Archimedes, and Apollonius are notable examples—and Zenodorus may well have written astronomical works of which we have no knowledge. A Vatican manuscript gives a catalog of astronomers—οἱ περὶ τοῦ πόλου συντάξαντες—which includes the name Zenodorus.[8] In accordance with the principle of not multiplying entities unnecessarily, it would seem that he, too, should be identified with the mathematician.

The Herculaneum fragments mention two visits by Zenodorus to Athens. On the onomastic evidence he could be from Cyrene, or Ptolemaic Egypt, or possibly from Chios or Erythrae. But the name is attested eight or nine times in Athens; and on the assumption that he was an Athenian, Toomer has plausibly identified him with a member of the Lamptrai family mentioned in an inscription that lists contributions for some unknown purpose during the archonship of Hermogenes (183–182 B.C.).[9]

It is only in Theon's commentary that the iso-perimetric propositions are specifically attributed to Zenodorus, but the passages in Pappus' *Collection* and the *Introduction* are so similar that they also must be derived from him. They are not, however, simply lifted from Zenodorus: there are considerable differences in the order and wording of the propositions in the three sources, and the question which is nearest to the original has given rise to some discussion. In all probability Pappus, like Theon, reproduced the propositions of Zenodorus at the relevant point in his commentary on the first book of the *Almagest*, where Ptolemy says, "Among different figures having an equal perimeter, since that which has the more angles is greater, of plane figures the circle is the greatest and of solid figures the sphere."[10] If so, this may have been the most exact reproduction of Zenodorus' text, ascribed to him by name, as in Theon; when he came to compile his *Collection*, Pappus varied the presentation, added the proposition "Of all segments of a circle having equal circumferences, the semicircle is the greatest in area," and proceeded to a disquisition on the semiregular solids of Archimedes.[11] Theon would have drawn upon Pappus, and the anonymous author of the *Introduction* upon both.

It would appear that Zenodorus' treatise contained fourteen propositions. There is agreement in the three versions that the first was "Of regular polygons having the same perimeter, the greater is that which has the more angles." The final proposition, stated but not actually proved, was almost certainly "If a sphere and a regular polyhedron have the same surface [area], the sphere is the greater." In between came such propositions as the following:

"If a circle and a regular polygon have the same perimeter, the circle is the greater."

"If on the base of an isosceles triangle there be set up a non-isosceles triangle having the same perimeter, the isosceles triangle is the greater."

"Given two similar right-angled triangles, the square on the sum of the hypotenuses is equal to the sum of the squares on the corresponding sides taken together."

"If on unequal bases there be set up two similar isosceles triangles, and on the same bases there be set up two dissimilar isosceles triangles having together the same perimeter as the two similar triangles, the sum of the similar triangles is greater than the sum of the dissimilar triangles."

"Among polygons with an equal perimeter and an equal number of sides, the regular polygon is the greatest."

"If a regular polygon [with an even number of sides] revolves about one of the longest diagonals, there is generated a solid bounded by conical surfaces that is less than the sphere having the same surface."

"Each of the five regular solids is less than the sphere with equal surface."

There is no little subtlety in the reasoning; indeed, rigorous proofs of the isoperimetric properties of the circle and sphere had to wait until H. A. Schwarz provided them in 1884.[12]

NOTES

1. The Greek title is given by Theon, *Commentaires de Pappus et de Théon d'Alexandrie sur l'Almageste*, A. Rome, ed. II, 355.4, as Περὶ ἰσοπεριμέτρων σχημάτων. The earlier editors had read ἰσομέτρων, but Rome showed that this was a variant reading in some MSS. "Isoperimetric" makes better sense than "isometric" and is confirmed by the comment of Simplicius in his *Commentarium in Aristotelis de caelo*, J. L. Heiberg, ed., in *Commentaria in Aristotelem Graeca*, VII (Berlin, 1894), 412; δέδεικται . . . παρὰ Ἀρχιμήδους καὶ παρὰ Ζηνοδώρου πλατύτερον, ὅτι τῶν ἰσοπεριμέτρων σχημάτων πολυχωρητότερός ἐστιν ἐν μὲν τοῖς ἐπιπέδοις ὁ κύκλος, ἐν δὲ τοῖς στερεοῖσιν ἡ σφαῖρα.
2. Proclus, *In primum Euclidis*, G. Friedlein, ed. (Leipzig, 1873; repr. Hildesheim, 1967), 80.15–16; English trans. by Glenn R. Morrow, *Proclus: A Commentary on the First Book of Euclid's Elements* (Princeton, 1970), 66. There is one genuine reference to Zenodorus in Proclus, Friedlein ed., 165.24, where it is asserted that there are four-sided triangles, called "barblike" by some but "hollow-angled" (κοιλογώνια) by Zenodorus. The reference is to a quadrilateral with one angle greater than two right angles, called a reentrant angle. It was formerly believed that this word occurred in Theon's version, and Nokk (see note 3) used this as an argument for believing Theon's text to be the nearest the original; but Rome, *op. cit.*, 371, has shown that the word is interpolated and may, indeed, have been interpolated before Proclus read Theon. It remains in the *Introduction to the Almagest: Pappi Alexandrini Collectionis quae supersunt . . .*, F. Hultsch, ed., III, 1194.12, 13, 16.
3. Nokk, *Zenodorus' Abhandlung über die isoperimetrischen Figuren*, 27–29.
4. Quintilian, *De institutione oratoria* (I.10, 39–45), L. Radermacher, ed. (Leipzig, 1907; 6th ed., enl. and corr. by V. Buchheit, 1971), I, 63.12–64.12; also M. Winterbottom, ed. (Oxford, 1970), I, 65.28–66.31. But B. L. van der Waerden, *Science Awakening*, 2nd English ed., 268, is in error in saying that Quintilian "mentions" Zenodorus. Also see F. Hultsch, *op. cit.*, III, 1190; and M. Cantor, *Vorlesungen über Geschichte der Mathematik*, 3rd ed., I (Leipzig, 1907), 549.
5. W. Crönert, "Der Epikureer Philonides." The name occurs in fr. 31, ll. 4–5 (Crönert, 953–954) and probably in fr. 34, l. 1 (Crönert, 954).
6. G. J. Toomer, "The Mathematician Zenodorus," 186.
7. *Ibid.*, 190–191. In both cases where the name occurs, Zenodorus is an emendation, but Toomer regards it as certain.
8. Vaticanus Graecus 381, published by Ernst Maass in *Hermes*, 16 (1881), 388, and more definitively in his *Aratea* (Berlin, 1892), 123. Maass himself identified the Zenodorus of the catalog with the mathematician.
9. Toomer, *op. cit.*, 187–190.
10. Ptolemy, *Syntaxis mathematica (Almagest)*, I.3: *Claudii Ptolemaei Opera quae exstant omnia*, J. L. Heiberg, ed., I (Leipzig, 1898), 13.16–19.
11. If Pappus in his commentary gave credit to Zenodorus, as Theon did, this would help to explain why Zenodorus is not mentioned in the *Collection*; there is no question of Pappus' trying to appropriate another's work as his own.
12. H. A. Schwarz, "Beweis des Satzes, dass die Kugel kleinere Oberfläche besitzt, als jeder andere Körper gleichen Volumens," in *Nachrichten von der Gesellschaft der Wissenschaften zu Göttingen* (1884), 1–13, repr. in Schwarz's *Gesammelte mathematische Abhandlungen*, II (Berlin, 1890), 327–340.

BIBLIOGRAPHY

I. ORIGINAL WORKS. Zenodorus' one known work was entitled Περὶ ἰσοπεριμέτρων. It has not survived as such but is epitomized in three other works: Pappus, *Collection*, V. 3–19: *Pappi Alexandrini Collectionis quae supersunt*, F. Hultsch, ed., I (Berlin, 1876), 308.2–334.21; Theon of Alexandria, *Commentary on the Almagest* I.3: *Commentaires de Pappus et de Théon d'Alexandrie sur l'Almageste*, A. Rome, ed., II, *Théon d'Alexandrie* (Vatican City, 1936), 354.19–379.15, which was translated into Latin and collated with the passages in Pappus in Hultsch, *op. cit.*, III (Berlin, 1878), 1189–1211; and an anonymous work usually known as the *Introduction to the Almagest* and published in F. Hultsch, *op. cit.*, III, as "Anonymi commentarius de figuris planis isoperimetris," 1138–1165.

II. SECONDARY LITERATURE. See the following, listed chronologically: Nokk, *Zenodorus' Abhandlung über die isoperimetrischen Figuren nach den Auszügen welche uns die Alexandriner Theon und Pappus aus derselben überliefert haben* (Freiburg im Breisgau, 1860); James Gow, *A Short History of Greek Mathematics* (Cambridge, 1884), 271–272; W. Crönert, "Der Epikureer Philonides," in *Sitzungsberichte der Preussischen Akademie der Wissenschaften zu Berlin* (1900), 942–959; W. Schmidt, "Zur Geschichte der Isoperimetrie im Altertum," in *Bibliotheca mathematica*, 3rd ser., 2 (1901), 5–8; T. L. Heath, *A History of Greek Mathematics*, II (Oxford, 1921), 207–213; W. Müller, "Das isoperimetrische Problem im Altertum," in *Sudhoffs Archiv*, 37 (1953), 39–71, with a German trans. of Theon's epitome; B. L. van der Waerden, *Science Awakening*, English trans. of *Ontwakende Wetenschap* with author's additions, 2nd ed. (Groningen, n.d.), 268–269; and G. J. Toomer, "The Mathematician Zenodorus," in *Greek, Roman and Byzantine Studies*, 13 (1972), 177–192.

IVOR BULMER-THOMAS

ZENO OF CITIUM (*b.* Citium, Cyprus, *ca.* 335 B.C.; *d.* Athens, 263 B.C.), *philosophy*.

Cyprus was colonized by Greeks, but had many Phoenician inhabitants. Zeno's father was a merchant called Mnaseas, perhaps a Greek version of the Phoenician Manasse or Menahem; Zeno is commonly referred to as "the Phoenician" by ancient writers. His education, however, was Greek; he studied in Athens and eventually set up his own school there. The most important of his teachers were Polemo, then head of Plato's Academy; Stilpo the Megarian; and Crates the Cynic. He was well soaked in the Greek tradition of philosophy; and his own style of philosophizing, although strikingly original, shows clear traces of the influence of Heraclitus, Socrates, Plato, and Aristotle.

Zeno established his own school about 300 B.C., perhaps in deliberate opposition to the school of Epicurus, which had recently been founded. He taught in the Stoa Poikile, or "Painted Colonnade"; and the name "Stoics" supplanted "Zenonians" for his pupils. At the end of his life he was given public honors by the Athenians. The headship of the school passed first to Cleanthes of Assos, and then to Chrysippus of Soli: the individual contributions of the three to the school's doctrine are hard to disentangle, in the absence of any complete writings, and this article does not attempt the task. The school survived until at least A.D. 260.

The main emphasis of Stoic teaching was moral. Zeno preached a morality that could claim to be an interpretation of the message of Socrates. The peculiarity of his doctrine is his refusal to allow that there is anything good but the virtuous state of a man's soul, coupled with an "all or nothing" definition of virtue. Virtue is wisdom: the wise man is wholly good, and everything he does is good, whereas the rest of the world and all its doings are sunk in iniquity. The wisdom that constitutes virtue includes an understanding of nature —indeed, that is its most important component, since the wise man's goal is "to live in accordance with nature." A particular physics, a particular interpretation of the physical universe and man's place in it, is an essential part of Stoicism, just as the atomic world picture was an essential part of the rival Epicurean doctrine.

The Stoic world picture was the lineal descendant of Plato's *Timaeus* and Aristotle's *De caelo*, with some features of the Platonic-Aristotelian cosmology exaggerated or changed, perhaps in conscious opposition to Epicurus. It was a picture of the cosmos as a single material continuum. Outside the cosmos there is void space: this was a modification of Aristotelian doctrine, according to which there is nothing whatever outside the cosmos, not even void space.

The cosmos is regarded as a single whole substance. It is held together by an unexplained natural tendency of matter to contract upon its own center. Thus Aristotle's cosmology is altered in two respects. The center toward which matter tends to move is its own center, the center of the cosmic body, whereas in Aristotle's system it is the center of the universe; second, all matter in the Stoic system, including the two light, centrifugal elements of the Aristotelian cosmology, air and fire, tends toward the center. These two elements, according to the Stoics, are less heavy than earth and water; and for that reason they tend to stay outside the central spheres composed of the heavy elements. They are also characterized as the active elements: together they make up *pneuma*, which permeates the whole cosmos and sets up a tension ($\tau\acute{o}\nu o\varsigma$) in it. The elements are not chemically differentiated but are defined by the tension set up by these motions. In whatever region of the cosmic body, itself nothing but an undifferentiated medium, there is a strong motion toward the center, we can say that that is an earthy region; wherever there is a strong tendency to resist the centripetal motion, we can say there is fire; and so on.

Recent interpreters of Stoic physics (J. Christensen and S. Sambursky) have pointed out its similarity to a field theory. The whole cosmos is a field within which various motions occur, and within it one field can be distinguished from another according to the motions that occur there. The nature of the tension thus set up determines the properties possessed by the region in question. The Stoics lacked a language with which this idea could be expressed coherently, however, and their retention of expressions appropriate to a metaphysics of "substance" led to many paradoxes. In particular, since any region of the universe, in their theory, may contain motions of many different kinds; and since it is these motions that produce identifiable characters or "bodies," they had to say that many bodies could coexist in the same region. They were widely criticized in antiquity for this theory of "total mixture" ($\kappa\rho\hat{a}\sigma\iota\varsigma\ \delta\iota'\ \ddot{o}\lambda o\upsilon$).

Since the cosmos is a single continuous field, motions in one part of it may affect those in any other part: the Stoics used the term "sympathy" for this feature. Certain peculiarities of Stoicism follow from it. Astrology and divination, for example, are thus given a rationale; and they receive more emphasis than in any previous philosophical system. Moreover, the interconnection of all events in the cosmic continuum is referred to as fate. From an ideally complete description of the structure of the cosmos up to a given time, one can theoretically predict all subsequent events: all that happens is in accordance with fate. With regard to human actions, the Stoics held the position that since the individual's state of mind enters into the conditions that determine an action, human "freedom" is preserved despite the doctrine of universal fate (their Peripatetic opponents replied that in that case, a stone that is dropped also falls "freely"; and the distinction between "free" and "forced" is lost).

The active elements in the cosmos are also referred to collectively as "God." By this designation the Stoics attributed many characteristics to the cosmos. It is ever-living, in the sense that it will never cease to exist, although it undergoes the cyclical transformations described below. It is rational: that is, the structure of the cosmos is patterned on principles that the human mind, being endowed with reason (λόγος), can recognize. (The divine λόγος that permeates and directs the Stoic cosmos is referred to in the famous opening words of St. John's gospel.) Moreover, the structure of the cosmos, being determined by divine reason, is good. The Stoics exploited the argument from design to the full, and found evidences of divine craftsmanship everywhere in nature. Stoic ethics and physics are thus in full accord with one another: the good life for man is to be an assenting part of the cosmos, "to live in accordance with nature."

Periodically the cosmos loses all its differentiation and is wholly consumed by and assimilated to the divine fiery *pneuma*. After this ἐκπύρωσις, the cosmos is formed again. Since the structure is determined by the divine *pneuma*, which remains constant throughout the conflagration, the structure of the reborn cosmos is identical in every detail with that of the previous one; and every event is repeated in every cycle.

In logic the Stoics made important innovations (perhaps, however, these should be attributed not to Zeno but to Chrysippus). Aristotle's logic of predication was adapted to his view of science as demonstrating the connections between the properties of substances. The basic entities of Stoic philosophy are not substances but motions or events. Its logic, correspondingly, is a logic of propositions. The Stoics worked out five "indemonstrable" inference patterns, and tested the validity of other schemata by trying to reduce them to these five. They used ordinal numbers to stand for propositions, thus: "If the first, then the second; but the first, therefore, the second." "If the first, then the second; but not the second, therefore, not the first." We may conjecture that there is some

connection between this development of the logic of inference and the growth of experimental science in the Alexandrian schools, beginning in the third century B.C.

As a moral system, Stoicism had a wide acceptance in the Roman republic and the empire, Marcus Aurelius being the most notable follower. The physical system, however, soon became contaminated with elements of Platonism and Aristotelianism; Stoic ideas occur frequently but rather unsystematically in the work of Galen, the Neoplatonists, and the Peripatetic commentators. The unified and consistent world picture worked out by Zeno, which was in fact a remarkable achievement, lost its clear outlines and merged with the general amalgam of Aristotelianism, Neoplatonism, and Christianity that dominated the intellectual centers of late antiquity.

BIBLIOGRAPHY

There is no complete work by Zeno or any other of the old Stoics extant. The standard collection of fragments is J. von Arnim, *Stoicorum veterum fragmenta*, 4 vols. (Leipzig, 1905–1924). There is an ancient life of Zeno and other Stoics in Diogenes Laërtius, *Lives of Eminent Philosophers*, VII, text and trans. by R. D. Hicks in the Loeb Classical Library (London, 1953).

The fullest modern account of Stoicism is Max Pohlenz, *Die Stoa*, 2nd ed., 2 vols. (Göttingen, 1959). Johnny Christensen, *An Essay on the Unity of Stoic Philosophy* (Copenhagen, 1962), gives a short, original, and highly stimulating summary. S. Sambursky, *Physics of the Stoics* (London, 1959), translates selected texts into English and gives a comprehensive account of the cosmology, not altogether free from anachronism. The most important books on the logic are Benson Mates, *Stoic Logic* (Berkeley, 1953), and Michael Frede, *Die stoische Logik* (Göttingen, 1974). For epistemology and ethics, also see A. A. Long, ed., *Essays on Stoic Philosophy* (London, 1970). There is a recent monograph on Zeno, by Andreas Graeser, *Zenon von Kition* (Berlin, 1975).

DAVID J. FURLEY

ZENO OF ELEA (*b.* Elea, Lucania, *ca.* 490 B.C.; *d.* Elea, *ca.* 425 B.C.), *philosophy, mathematics.*

Zeno became a friend and disciple of Parmenides, with whom, according to Plato's dialogue *Parmenides*, he visited Athens in the middle of the fifth century B.C. Some later Greek authors, however, considered this visit an invention of Plato's. According to a widespread legend with many greatly differing versions, Zeno was tortured and

sion] at all and big to the extent of infinite extension [H. Diels and W. Kranz, *Die Fragmente der Vorsokratiker*, 7th ed., I, no. 29, p. 266].

The use of the word ἀπέχειν, which usually means "to be away from" or "to be at a distance from," has induced some commentators to interpret Zeno's argument in the following way: If there are many things, they must be distinguished from one another. If they are distinguished, they must be separate. If they are separate, there must be something between them. Since Zeno (according to his cosmology) denied the existence of empty space, what separates things must itself be a thing, which in its turn must be separated by another thing that separates it from the things that it separates from one another, and so on ad infinitum. But this interpretation is hardly reconcilable with what precedes and what follows in Simplicius' account. The argument as a whole is understandable only on the assumption that ἀπέχειν is used as a synonym of προέχειν, the term that is used in the remainder of the argument. The gist of the argument then appears to have been: What has size is divisible. What is divisible is not a real One, since it has parts. But any part that "lies beyond" or protrudes from a given part of something that has size, has size in its turn; hence it has parts, and so on ad infinitum, so that it becomes both small and big beyond all measure. If this is the meaning of the argument, we are back with the problem of the continuum.

Vlastos has claimed that in the last sentence of the fragment quoted, Zeno committed "a logical gaffe" by assuming that through infinitely continued division, one finally ends with particles "of no size!" But on closer inspection it seems clear that Zeno, at least in the first half of his last argument, committed no such logical error. Far from assuming that by infinitely continued division one finally comes to particles "of no size," his argument is based on the very opposite assumption: however far the division may have proceeded, what remains still always has size, hence is further divisible, hence has parts, hence is not really One. Therefore, in order to be really One (indivisible)—and this conclusion, if one starts from Zeno's assumption, is perfectly sound—it must be without size. But—and here the preliminary argument reported by Simplicius is brought in—what has no size does not make a thing to which it is added bigger, nor a thing from which it is subtracted smaller, and hence appears to be nothing.

This interpretation and analysis of the argument is also in perfect agreement with a statement elsewhere attributed to Zeno, to the effect that if someone could really explain to him what the One is, then he would also be in a position to explain plurality. At the same time it shows that Plato was right when he reported that Zeno did not try to give direct support to Parmenides' doctrine that only the One exists, but merely tried to show that from the assumption of a plurality of things, no less strange conclusions could be drawn than from the assumption that there is nothing but the One.

Granting this, one may still contend that Zeno committed a "logical gaffe" in the second part of his last argument, where he speaks of an infinite number of parts that would make the size of the object composed of them grow beyond all measure; this statement appears to be at variance with one of the most elementary applications of the theory of convergent series: that the sum of the infinite series $1/2 + 1/4 + 1/8 + 1/16 + \cdots = 1$. But this mathematical formula is a convenient symbol for the fact that infinitely continued bisection of a unit cannot exceed the unit, a fact of which Zeno, as other fragments clearly show, was perfectly aware. What he obviously did try to point out is that it is not possible for the human mind to build up the sum of such an infinite series starting, so to speak, from the other end, the end with the "degenerative elements," as Grünbaum calls them. When building up a sum, one has always to start with elements that have size. The difficulty is essentially that which H. Fränkel so lucidly described in regard to motion.

The other paradoxes of Zeno that have been preserved by ancient tradition are not so profound and can be resolved completely. One of them is that of the falling millet: If a falling bushel of millet makes a noise, so must an individual grain; if the latter makes no noise, neither can the bushel of grain, for the size of the grain has a definite ratio to that of the heap. The same must then be true of the noises. The resolution here lies in the limitation of perception, which also plays a role in the modern discussion of the perception of time. Interestingly, Zeno argues on the basis of the mathematical argument that there must be a definite proportion.

Another is the paradox of the moving blocks: If four blocks *BBBB* of equal size move along four blocks *AAAA* of the same size which are at rest, and at the same time four blocks *CCCC*, again of the same size and the first two of which have arrived below the last two of the row *AAAA*, move with the same speed as *BBBB* in the opposite direction from *BBBB*, then *BBBB* will pass two blocks of *AAAA* in the same time in which they pass four blocks of *CCCC*. But since their speed remains the same, and yet time is measured by the distance traveled at the same time, half the time is equal to the double time.

Alexander of Aphrodisias made the following diagram to illustrate Zeno's moving blocks argument:

$$AAAA$$
$$BBBB \rightarrow$$
$$\leftarrow CCCC$$

It is interesting that this argument contains the first glimpse in ancient literature of an awareness of the relativity of motion. It is the only Zenonian paradox preserved that has nothing to do with the problem of the continuum, although there have been some attempts in modern times by Paul Tannery and R. E. Siegel to show that there is a connection.

Concerning Zeno's importance for the development of ancient Greek mathematics, the most various views have been held and are still held by modern historians of science. Tannery was the first to suggest that Zeno's relation to the philosophy of Parmenides may have been less close than ancient tradition affirms and that Zeno was much more deeply influenced by problems arising from the discovery of incommensurability by the Pythagoreans. On the basis of the same assumption, H. Hasse and H. Scholz tried to show that Zeno was the "man of destiny" of ancient mathematics. They attempted to prove that the Pythagoreans, after having discovered the incommensurability of the diameter of a square with its side, had tried to overcome the resultant difficulties by assuming the existence of infinitely small elementary lines (*lineae indivisibiles*). It was against this inaccurate handling of the infinitesimal that Zeno protested, thus forcing the next generation of Pythagorean mathematicians to give the theory a better and more accurate foundation.

Other scholars (W. Burkert, A. Szabó, J. A. Philip) contend that since, according to ancient tradition, the Pythagoreans engaged in a rather abstruse number mysticism such a profound mathematical discovery as that of incommensurability cannot have been made by them, but must have been made by "practical mathematicians" influenced by Zeno's paradoxes. But there is no direct road leading from Zeno's paradoxes to the proof of incommensurability in specific cases, whereas some of the speculations supporting the Pythagoreans (when carried through with the consistency char-

acteristic of the philosophers of the first century) must almost inevitably have led to the discovery, although we do not know exactly how it was first made; and there is no tradition concerning an effect of Zeno's speculations on the development of mathematics in the second half of the fifth century B.C. B. L. van der Waerden has shown that what we know of mathematical theories of the second half of the fifth century B.C. — when the discovery of incommensurability undoubtedly was made — is rather at variance with the assumption that Zeno had any considerable influence on the development of mathematics at that time.

This, however, does not necessarily mean that Zeno's name has to be stricken from the history of ancient Greek mathematics. In all likelihood he received the first impulse toward the invention of his paradoxes not from mathematics but, as attested by Plato, from the speculations of Parmenides, and did not immediately have a strong influence on the development of Greek mathematics. But it is hardly by chance that Plato wrote his dialogue *Parmenides*, in which he refers to Zeno's paradoxes, around the time that Eudoxus of Cnidus, who revised the theory of proportions in such a way as to enable him to handle the infinitesimal with an accuracy that has remained unsurpassed, spent some years at Athens and was a member of Plato's academy. Zeno's paradoxes can hardly have failed to have been thoroughly discussed then, and so Zeno may still have had some influence on Greek mathematics at that decisive point in its development.

BIBLIOGRAPHY

I. ORIGINAL WORKS. An extensive bibliography is in W. Totok, *Handbuch der Geschichte der Philosophie*, I, *Altertum* (Frankfurt, 1964), 123–124. The text ed. most convenient for English-speaking readers is H. D. P. Lee, *Zeno of Elea. A Text With Translation and Commentary* (Cambridge, 1936). In the secondary literature, however, the fragments are usually quoted according to H. Diels and W. Kranz, *Die Fragmente der Vorsokratiker*, 7th ed., I, no. 29, 247–258.

II. SECONDARY LITERATURE. See Guido Calogero, *Studi sull'Eleatismo* (Rome, 1932); and *Storia della logica antica*, I (Bari, 1967), 171–208; H. Fränkel, "Zeno of Elea's Attacks on Plurality," in *American Journal of Philology*, 63 (1942), 1–25, 193–206; Adolf Grünbaum, "A Consistent Conception of the Extended Linear Continuum as an Aggregate of Unextended Elements," in *Philosophy of Science*, 19 (1952), 288–305; "The Nature of Time," in *Frontiers of Science and Philosophy*, 1 (1962), 149–184; and *Modern Science and Zeno's Par-*

adoxes (Middletown, Conn., 1967); H. Hasse and H. Scholz, *Die Grundlagenkrisis der griechischen Mathematik* (Berlin–Charlottenburg, 1928); J. A. Philip, *Pythagoras and Early Pythagoreanism* (Toronto, 1966), 206–207; R. E. Siegel, "The Paradoxes of Zeno," in *Janus*, 48 (1959), 42 ff.; A. Szabó, *Anfänge der griechischen Mathematik* (Munich–Vienna, 1939), 333 ff.; P. Tannery, "Le concept scientifique du continu. Zénon d'Élée et Georg Cantor," in *Revue philosophique de la France et de l'étranger*, 20 (1885), 385–410, esp. 393–394; and *Pour l'histoire de la science hellène*, 2nd ed. (Paris, 1930), 248 ff.; B. L. van der Waerden, "Zenon und die Grundlagenkrise der griechischen Mathematik," in *Mathematische Annalen*, 117, no. 2 (1940), 141–161, esp. 151 ff.; and G. Vlastos, "A Note on Zeno's Arrow," in *Phronesis*, 11 (1966), 3–18; and "Zeno's Race Course," in *Journal of the History of Philosophy*, 4 (1966), 95–108.

Concerning the influence of Zeno and of the methods developed by Eudoxus on the nineteenth-century attempts to give the calculus a more exact foundation, see M. Black, "Achilles and the Tortoise," in *Analysis*, 11 (1951), 91–101; J. M. Hinton and C. B. Martin, "Achilles and the Tortoise," *ibid.*, 14 (1953), 56–68; G. E. L. Owen, "Zeno and the Mathematicians," in *Proceedings of the Aristotelian Society*, n.s. 58 (1957–1958), 199–222; L. E. Thomas, "Achilles and the Tortoise," in *Analysis*, 12 (1952), 92–94; R. Taylor, "Mr. Black on Temporal Paradoxes," *ibid.*, 38–44; and J. O. Wisdom, "Achilles on a Physical Racecourse," *ibid.*, 67–72. There is also an especially instructive earlier paper by M. Dehn, "Raum, Zeit, Zahl bei Aristoteles vom mathematischen Standpunkt aus," in *Scientia*, 40 (1936), 12–21, 69–74, which deals with Aristotle's criticism of Zeno's paradoxes and its importance for modern mathematics.

Concerning the general problems underlying Zeno's paradoxes, see also P. Beisswanger, *Die Anfechtbarkeit der klassischen Mathematik* (Stuttgart, 1965); P. Bennacerraf, "What Numbers Could Not Be," in *Philosophical Review*, 74 (1965), 47–73; and Hermann Weyl, "Über die neue Grundlagenkrise der Mathematik," in *Mathematische Zeitschrift*, 10 (1921), 39–79.

KURT VON FRITZ

ZENO OF SIDON (*b.* Sidon, *ca.* 150 B.C.; *d.* Athens, *ca.* 70 B.C.), *philosophy, mathematics, logic.*

According to ancient tradition, Zeno of Sidon was a very prolific writer who discussed theory of knowledge, logic, various aspects of ancient atomic theory, the fundamental differences of the sexes (from which it follows that they have different diseases), problems of Epicurean ethics, literary criticism, style, oratory, poetry, and mathematics. Very little is known of the contents of these writings except those on mathematics and logic, which are of great interest.

Epicurus had been a very severe critic of mathematics as a science; but what he said about it is very superficial and shows that he did not understand what mathematics is. This is not at all the case with Zeno's criticism of Euclid's axiomatics. In his commentary on Euclid, Proclus says that Zeno attacked the first theorem of the *Elements* (the construction of an equilateral triangle) on the ground that it is valid only if one assumes that two straight lines cannot have more than one point in common, and that Euclid has not set this down as an axiom. On the same ground he attacked Euclid's fourth postulate, which asserts the equality of all right angles, observing that it presupposes the construction of a right angle, which is not given until I, 11. In addition, Proclus and Sextus Empiricus mention several criticisms of Euclid that they attribute to an unnamed Epicurean and that are similar to Zeno's criticisms: for instance, that there is no axiom establishing the infinite divisibility of curves, which is connected with a discussion of various consequences following from the assumption that curves are not infinitely divisible but, rather, are composed of the smallest units of indivisible lines. There is also a criticism anticipating Schopenhauer's of Euclid's method of superimposition, by which he proves the first theorem of congruence and a few other theorems: namely, that only matter can be moved in space.

On the basis of these criticisms of Euclid's axiomatics, E. M. Bruins has claimed that Zeno of Sidon was the first to discover the possibility of non-Euclidean geometry. This claim appears exaggerated, since there is not the slightest tradition indicating that Zeno elaborated his criticism in such a way as to show positively how a non-Euclidean geometrical system could be built up. Zeno's criticisms of Euclid are pertinent, however, and if any of the ancient philosophers and mathematicians who tried to refute them had been able to grasp their full implications, the development of mathematics might have taken a different turn.

Lengthy fragments of a treatise by the Epicurean philosopher Philodemus of Gadara have been found on a papyrus from Herculaneum (no. 1065), and most of those preserved contain a report on a controversy between Zeno and contemporary Stoics over the foundations of knowledge. In this dispute Zeno defended the old Epicurean doctrine that all human knowledge is derived exclusively from experience. What makes it interesting, however, is that he bases his defense on a theory that he calls "transition according to similarity" ($\mu\epsilon\tau\acute{\alpha}\beta\alpha\sigma\iota\varsigma$ $\kappa\alpha\vartheta$' $\acute{o}\mu o\iota\acute{o}\tau\eta\tau\alpha$) or "transition from the apparent to the not apparent" ($\mu\epsilon\tau\acute{\alpha}\beta\alpha\sigma\iota\varsigma$ $\grave{\alpha}\pi\grave{o}$ $\tau\tilde{\omega}\nu$ $\phi\alpha\iota\nu o\mu\acute{\epsilon}\nu\omega\nu$ $\acute{\epsilon}\varsigma$ $\tau\grave{\alpha}$ $\grave{\alpha}\phi\alpha\nu\tilde{\eta}$), but that is essentially an anticipation of John Stuart Mill's theory of induction.

In contrast to Aristotle's theory of induction, according to which the most certain kind of induction is that in which one case is sufficient to make it evident that the same must be true in all similar cases, and in opposition to the Stoic doctrine that no number of cases ever permits the conclusion that the same must be true in all cases, Zeno insisted that all knowledge is fundamentally derived by inference to all cases from a great many cases without observed counter-instance. He carried this principle to the extreme by asserting that the knowledge that the square with a side of length 4 is the only square in which the sum of the length of the sides (16) is equal to the contents ($4 \times 4 = 16$) was derived from measuring innumerable squares, although here it is evident that the result—insofar as it is correct, one-dimensional measures being equated with two-dimensional measures—can be derived from a simple deduction and that nobody will be so foolish as to "verify" it in innumerable squares. The recent proof by computers that the principle is not altogether applicable to mathematics and number theory shows that certain theorems of Pólya's that had been considered universally valid because they had been proved up to very high numbers were not valid beyond higher numbers unreachable by human calculation.

The details of the controversy between Zeno and the Stoics is extremely interesting because sometimes the positions become curiously reversed, and because it provides a kind of phenomenology of induction going beyond most modern works.

BIBLIOGRAPHY

I. ORIGINAL WORKS. Extracts of Zeno's lectures are in T. Gomperz, *Herkulanische Studien*, I (Leipzig, 1865), 24–27; and P. H. and E. A. de Lacy, eds., *Philodemos: On Methods of Inference*, which is Philological Monographs of the American Philological Association, no. 10 (Philadelphia, 1941)—see index. (Pp. 22–66, columns Ia, 1–XIX, 4, are mostly extracts from Zeno's lectures, but it is not certain how far they are literal.)

II. SECONDARY LITERATURE. See Ludger Adam, "Das Wahrheits- und Hypothesenproblem bei Demo-krit, Epikur und Zeno, dem Epikureer" (Ph.D. diss., Göttingen, 1947); E. M. Bruins, *La géométrie non-euclidienne dans l'antiquité,* Publications de l'Université de Paris, D121 (Paris, 1967); and G. Vlastos, "Zeno of Sidon as a Critic of Euclid," in *The Classical Tradition: Literary and Historical Studies in Honor of Harry Caplan* (New York, 1966), 148–159.

On problems arising in connection with Zeno's theory of induction, see C. B. Haselgrove, "A Disproof of a Conjecture of Pólya," in *Mathematica,* 5 (1958), 141; K. von Fritz, "Die $\acute{\epsilon}\pi\alpha\gamma\omega\gamma\acute{\eta}$ bei Aristoteles," in *Sitzungsberichte der Bayerischen Akademie der Wissenschaften zu München,* Phil.-hist. Kl. (1964), no. 3, 40 ff., 62 ff.; and R. Queneau, "Conjectures fausses en théorie des nombres," in *Mélanges Koyré,* I (Paris, 1964), 475 ff.

KURT VON FRITZ

ZERMELO, ERNST FRIEDRICH FERDINAND (*b.* Berlin, Germany, 27 July 1871; *d.* Freiburg im Breisgau, Germany, 21 May 1953), *mathematics.*

The son of Ferdinand Rudolf Theodor Zermelo, a college professor, and Maria Augusta Elisabeth Zieger, Zermelo received his secondary education at the Luisenstädtisches Gymnasium in Berlin, where he passed his final examination in 1889. He subsequently studied mathematics, physics, and philosophy at Berlin, Halle, and Freiburg, taking courses taught by Frobenius, Lazarus Fuchs, Planck, Erhard Schmidt, H. A. Schwarz, and Edmund Husserl. In 1894 he received the doctorate at Berlin with the dissertation *Untersuchungen zur Variationsrechnung.* Zermelo went to Göttingen and in 1899 was appointed *Privatdozent* after having submitted the *Habilitationsschrift* "Hydrodynamische Untersuchungen über die Wirbelbewegungen in einer Kugelfläche." In December 1905, shortly after his sensational proof of the well-ordering theorem (1904), Zermelo was named titular professor at Göttingen. In 1910 he accepted a professorship at Zurich, which poor health forced him to resign in 1916. A year after he had left Göttingen, 5000 marks from the interest of the Wolfskehl Fund was awarded him on the initiative of David Hilbert in recognition of his results in set theory (and to enable him to recover his health). After resigning his post at Zurich, Zermelo lived in the Black Forest until 1926, when he was appointed honorary professor at the University of Freiburg im Breisgau. He renounced connection with the university in 1935 because of his disapproval of the Hitler regime. After the war he requested reinstatement, which was granted him in 1946.

Zermelo had a lively interest in physics and a

keen sense for the application of mathematics to practical problems. He prepared German editions of Glazebrook's *Light* and Gibbs's *Elementary Principles in Statistical Mechanics*; and after having shown in "Ueber einen Satz der Dynamik" how application of Poincaré's recurrence theorem leads to the nonexistence of irreversible processes in the kinetic theory of gases, he had a penetrating discussion with Boltzmann on the explanation of irreversible processes.

In Zermelo's dissertation, which dealt with the calculus of variations, he extended Weierstrass' method for the extrema of integrals over a class of curves to the case of integrands depending on derivatives of arbitrarily high order, at the same time giving a careful definition of the notion of neighborhood in the space of curves. Throughout his life he was faithful to the calculus of variations, on which he often lectured and to which he contributed a report on its progress written with H. Hahn for the *Encyklopädie der mathematischen Wissenschaften* (1904) and the paper "Über die Navigation . . ." (1929).

Further examples of his original contributions to practical questions are his method for estimating the strength of participants in tournaments ("Die Berechnung der Turnier-Ergebnisse," 1929), which has been used in chess tournaments, and his investigation of the fracture of a cube of sugar ("Über die Bruchlinien zentrierter Ovale," 1933).

As an assistant at Göttingen, Zermelo lectured during the winter semester of 1900–1901 on set theory, to the development of which he was to contribute decisively. He had studied Cantor's work thoroughly, and his conversations with Erhard Schmidt led to his ingenious proof of the well-ordering theorem, which states that every set can be well-ordered; that is, in every set a relation $a \prec b$, to be read as "a comes before b," can be introduced, such that (1) for any two elements a and b, either $a = b$ or $a \prec b$ or $b \prec a$; (2) if for three elements a, b, c we have $a \prec b$ and $b \prec c$, then $a \prec c$; (3) any nonvoid subset has a first element. In a commentary to his own proof, Zermelo pointed out the underlying hypothesis that for any infinite system of sets there always are relations under which every set corresponds to one of its elements. The proof stirred the mathematical world and produced a great deal of criticism—most of it unjustified—which Zermelo answered elegantly in "Neuer Beweis" (1908), where he also gave a second proof of the theorem. His answer to Poincaré's accusation of impredicativity is of some historical interest because he points out certain consequences and peculiarities of the predicative position that have played a role in the development of predicative mathematics.

Also in 1908 Zermelo set up an axiom system for Cantor's set theory that has proved of tremendous importance for the development of mathematics. It consists of seven axioms and uses only two technical terms: set and $\in$, the symbol for the "element of" relation. Zermelo emphasized the descriptive nature of the axioms, starting with a domain B of objects and then specifying under what conditions (the axioms) an object is to be called a set. With the exception of the null set introduced in axiom 2, every set a is an object of B for which there is another object b of B such that $a \in b$.

Axiom 1 (extensionality): $m = n$ if and only if $a \in m$ is equivalent to $a \in n$.

Axiom 2 (elementary sets): There is a null set, having no element at all. Every object a of B determines a set $\{a\}$ with a as its only element. Any two objects a, b of B determine a set $\{a,b\}$ with precisely a and b as elements.

Axiom 3 (separation): If a property E is definite for the elements of a set m, then there is a subset m_E of m consisting of exactly those elements of m for which E holds.

Axiom 4 (power set): To any set m there is a set $P(m)$ that has the subsets of m for its elements.

Axiom 5 (union): To any set m there is a set $\cup m$, the union of m, consisting of the elements of the elements of m.

Axiom 6 (axiom of choice): If m is a set of disjoint nonvoid sets, then $\cup m$ contains a subset n that contains exactly one element from every set of m.

Axiom 7 (infinity): There is a set z that has the null set as an element and has the property that if a is an element of z, then $\{a\}$ is also an element of z.

In order to avoid the paradoxes, particularly Russell's paradox, which would render the system useless, Zermelo restricted set formation by the condition of definiteness of the defining property of a subset. A property E definite for set m is explained as one for which the basic relations of B permit one to decide whether or not E holds for any element of m. Although this condition seemed to preclude contradictions in the system, Zermelo explicitly left aside the difficult questions of independence and consistency. This was a wise decision, as one may realize after having seen the solutions of the questions of relative consistency and independence of axiom 6 by Kurt Gödel (1938) and P. J. Cohen (1963), respectively.

Because of its generality the notion of definite property is very elegant. It is rather difficult to apply, however, because it does not yield a general method for proving a proposed property to be definite.

Although nonaxiomatic Cantorian set theory was then flourishing, particularly the branch that developed into point-set topology, there was no progress in axiomatic set theory until 1921, when A. Fraenkel, in his attempts to prove the independence of the axiom of choice, pointed out some defects in Zermelo's system. Fraenkel's objections were threefold. First, the axiom of infinity is too weak; second, the system is by no means categorical; and third, the notion of definite property is too vague to handle in proofs of independence and consistency. These remarks led Fraenkel to add the powerful axiom of replacement, which adds to any set s its image under some function F, while the notion of function is introduced by definition. Another way of obtaining a similar result was achieved by T. Skolem, who specified a definite property as one expressible in first-order logic.

After having realized the importance of Fraenkel's and Skolem's remarks, Zermelo set out in "Über den Begriff der Definitheit in der Axiomatik" (1929) to axiomatize this notion by describing the set of definite properties as the smallest set containing the basic relations of the domain B and satisfying certain closure conditions. He admitted that the reason for doing so was methodological: to keep to the "pure axiomatic" method, in avoidance of the genetic method and the use of the notion of finite number. Since there is no categoricity, an investigation of the structure of the possible domains b—models for axiomatic set theory—makes sense. In "Über Grenzzahlen und Mengenbereiche" (1930) Zermelo investigated the structure of models of an axiom system consisting of his earlier axioms 1, 4, 5, the last part of 2, the unrestricted form of 3, a liberal axiom of replacement, and an axiom of well-foundedness (with respect to $\in$) stating that every subdomain T of domain B contains at least one element t_0 that has no element t in T.

Zermelo's fragmentary attempt, in "Grundlagen einer allgemeinen Theorie der mathematischen Satzsysteme" (1935), to abolish the limitations of proof theory has not been of great consequence because his conception of a proof as a system of theorems, well-founded with respect to the relation of consequence, seems too general to lead to results of sufficient interest.

BIBLIOGRAPHY

I. ORIGINAL WORKS. Zermelo's writings include *Untersuchungen zur Variationsrechnung* (Berlin, 1894), his dissertation; "Ueber einen Satz der Dynamik und die mechanische Wärmetheorie," in *Annalen der Physik und Chemie*, n.s. **57** (1896), 485–494; "Ueber mechanische Erklärungen irreversibler Vorgänge. Eine Antwort auf Hrn. Boltzmann's 'Entgegnung,'" *ibid.*, **59** (1896), 793–801; *Das Licht. Grundriss der Optik für Studierende und Schüler*, his trans. of R. T. Glazebrook's *Light* (Berlin, 1897); "Ueber die Bewegung eines Punktsystemes bei Bedingungsungleichungen," in *Nachrichten von der K. Gesellschaft der Wissenschaften zu Göttingen*, math.-phys. Kl. (1899), 306–310; "Über die Anwendung der Wahrscheinlichkeitsrechnung auf dynamische Systeme," in *Physikalische Zeitschrift*, **1** (1899–1900), 317–320; "Ueber die Addition transfiniter Cardinalzahlen," in *Nachrichten von der K. Gesellschaft der Wissenschaften zu Göttingen*, math.-phys. Kl. (1901), 34–38; "Hydrodynamische Untersuchungen über die Wirbelbewegungen in einer Kugelfläche," in *Zeitschrift für Mathematik und Physik*, **47** (1902), 201–237, his *Habilitationsschrift*; and "Zur Theorie der kürzesten Linien," in *Jahresberichte der Deutschen Mathematikervereinigung*, **11** (1902), 184–187.

Further works are "Über die Herleitung der Differentialgleichung bei Variationsproblemen," in *Mathematische Annalen*, **58** (1904), 558–564; "Beweis, dass jede Menge wohlgeordnet werden kann," *ibid.*, **59** (1904), 514–516; "Weiterentwickelung der Variationsrechnung in den letzten Jahren," in *Encyklopädie der mathematischen Wissenschaften*, II, pt. 1 (Leipzig, 1904), 626–641, written with H. Hahn; *Elementare Grundlagen der statistischen Mechanik*, his trans. of Gibbs's work (Leipzig, 1905); "Neuer Beweis für die Möglichkeit einer Wohlordnung," in *Mathematische Annalen*, **65** (1908), 107–128; "Untersuchungen über die Grundlagen der Mengenlehre I," *ibid.*, 261–281; "Sur les ensembles finis et le principe de l'induction complète," in *Acta mathematica*, **32** (1909), 185–193; "Die Einstellung der Grenzkonzentrationen an der Trennungsfläche zweier Lösungsmittel," in *Physikalische Zeitschrift*, **10** (1909), 958–961, written with E. H. Riesenfeld; and "Ueber die Grundlagen der Arithmetik," in *Atti del IV Congresso internazionale dei matematici*, II (Rome, 1909), 8–11.

See also "Über eine Anwendung der Mengenlehre auf die Theorie des Schachspiels," in *Proceedings of the Fifth International Congress of Mathematicians*, II (Cambridge, 1913), 501–504; "Über ganze transzendente Zahlen," in *Mathematische Annalen*, **75** (1914), 434–442; "Ueber das Masz und die Diskrepanz von Punktmengen," in *Journal für die reine und angewandte Mathematik*, **158** (1927), 154–167; "Über den Begriff der Definitheit in der Axiomatik," in *Fundamenta mathematicae*, **14** (1929), 339–344; "Die Berechnung der Turnier-Ergebnisse als ein Maximumproblem der Wahrscheinlichkeitsrechnung," in *Mathematische Zeit-*

schrift, **29** (1929), 436–460; and "Über die Navigation in der Luft als Problem der Variationsrechnung," in *Jahresberichte der Deutschen Mathematikervereinigung*, **39** (1929), 44–48.

Additional works are "Über Grenzzahlen und Mengenbereiche. Neue Untersuchungen über die Grundlagen der Mengenlehre," in *Fundamenta mathematicae*, **16** (1930), 29–47; "Über die logische Form der mathematischen Theorien," in *Annales de la Société polonaise de mathématique*, **9** (1930), 187; "Über das Navigationsproblem bei ruhender oder veränderlicher Windverteilung," in *Zeitschrift für angewandte Mathematik und Mechanik*, **11** (1931), 114–124; "Über mathematische Systeme und die Logik des Unendlichen," in *Forschungen und Fortschritte*, **8** (1932), 6–7; "Über Stufen der Quantifikation und die Logik des Unendlichen," in *Jahresberichte der Deutschen Mathematikervereinigung*, **41** (1932), 85–88; "Über die Bruchlinien zentrierter Ovale. (Wie zerbricht ein Stück Zucker?)," in *Zeitschrift für angewandte Mathematik und Mechanik*, **13** (1933), 168–170; "Elementare Betrachtungen zur Theorie der Primzahlen," in *Nachrichten von der Gesellschaft der Wissenschaften zu Göttingen*, Fachgruppe 1, **1** (1934), 43–46; and "Grundlagen einer allgemeinen Theorie der mathematischen Satzsysteme. (Erste Mitteilung)," in *Fundamenta mathematicae*, **25** (1935), 136–146.

A collection of papers left by Zermelo is in the library of the University of Freiburg im Breisgau. A short description, furnished by H. Gericke, is as follows: a set of copies of articles by Zermelo and other mathematicians, a collection of letters and MSS and sketches of published papers, lecture notes in shorthand, parts of a translation of Homer in German verse, the second part of his *Habilitationsschrift*, and a sketch of a patent application "Kreisel zur Stabilisierung von Fahr- und Motorrädern" (gyroscope for stabilizing bicycles and motorcycles).

II. SECONDARY LITERATURE. Quite a number of relevant papers on set theory, including three memoirs by Zermelo, are reprinted in J. van Heijenoort, *From Frege to Gödel* (Cambridge, Mass., 1967), which also contains references to the literature up to 1966. See also P. J. Cohen, *Set Theory and the Continuum Hypothesis* (New York, 1966); S. Fefermann, "Systems of Predicative Analysis," in *Journal of Symbolic Logic*, **29** (1964), 1–30; A. Fraenkel, "Über die Zermelosche Begründung der Mengenlehre," in *Jahresberichte der Deutschen Mathematikervereinigung*, **30** (1921), 97–98; "Zu den Grundlagen der Cantor-Zermeloschen Mengenlehre," in *Mathematische Annalen*, **86** (1922), 230–237; "Der Begriff 'definit' und die Unabhängigkeit des Auswahlsaxioms," in *Sitzungsberichte der Preussischen Akademie der Wissenschaften zu Berlin* (1922), 253–257; and *Foundations of Set Theory* (Amsterdam, 1958), written with Y. Bar-Hillel; H. Gericke, *Beiträge zur Freiburger Wissenschafts- und universitätsgeschichte*, VII, *Zur Geschichte der Mathematik an der Universität Freiburg i.Br.*, J. Vincke, ed. (Freiburg im Breisgau, 1955), 72–73; K. Gödel, *The Consistency of the Continuum Hypothesis* (Princeton, N.J., 1940); G. Kreisel and J. L. Krivine, *Elements of Mathematical Logic* (Amsterdam, 1967); M. Pinl, "Kollegen in einer dunklen Zeit," in *Jahresberichte der Deutsche Mathematikervereinigung*, **71** (1969), 167–228, esp. 221–222; C. Reid, *Hilbert* (Berlin, 1970); and J. Barkeley Rosser, *Simplified Independence Proofs* (New York, 1969).

Also see T. Skolem, "Logisch-kombinatorische Untersuchungen über die Erfüllbarkeit oder Beweisbarkeit mathematischer Sätze nebst einem Theoreme über dichte Mengen," in *Skrifter utgitt av Videnskapsselskapet i Kristiania*, I. Mat.-naturvid. kl. (1920), no. 4; "Einige Bemerkungen zur axiomatischen Begründung der Mengenlehre," in *Matematiker kongressen i Helsinfors den 4–7 Juli 1922* (Helsinki, 1923), 217–232; "Über einige Grundlagenfragen der Mathematik," in *Skrifter utgitt av det Norske videnskaps-akademi i Oslo*, I. Mat.-naturvid. kl. (1929), no. 4; and "Einige Bemerkungen zu der Abhandlung von E. Zermelo: 'Über die Definitheit in der Axiomatik,' " in *Fundamenta mathematicae*, **15** (1930), 337–341; and L. Zoretti and A. Rosenthal, "Die Punktmengen," in *Encyklopädie der mathematischen Wissenschaften*, II, pt. 3 (Leipzig, 1923), 855–1030.

B. VAN ROOTSELAAR

ZERNIKE, FRITS (*b*. Amsterdam, Netherlands, 16 July 1888; *d*. Naarden, near Amsterdam, 10 March 1966), *theoretical physics, technical physics*.

Zernike's father, headmaster of an elementary school, was well-known for his textbooks on arithmetic. While a chemistry student at Amsterdam University, Zernike won two gold medals for prize questions in mathematics and physics. In 1913 he became assistant to the astronomer J. C. Kapteyn at the University of Groningen, where he held various academic positions until his retirement at the age of seventy.

Zernike's dissertation, "L'opalescence critique, théorie et expériments" (Amsterdam, 1915), is still worth reading. In 1915 he succeeded L. S. Ornstein as lecturer in theoretical physics at Groningen; in 1920 he became full professor; and in 1941 his chair was extended to include mathematical and technical physics and theoretical mechanics. He became a member of the Royal Netherlands Academy of Sciences at Amsterdam in 1946, and seven years later he won the Nobel Prize in physics.

Widely read and wide-ranging in his work, Zernike paid especial interest to three main areas: statistical physics and fluctuation phenomena, the construction of instruments, and interference of light waves. He was an able speaker and possessed an extraordinary combination of mathematical and instrument-making skill, always using these abilities to bring out, in the simplest way, the essential physical principles of a problem or the characteristics of an instrument. Later his methods often found wider application.

For instance, in the wave theory of light he introduced the set of polynomials orthogonal on a circle that is widely used by mathematicians under the name of Zernike polynomials. In molecular statistics he introduced the concept of a radial distribution function $g(r)$ giving the mean number density of molecular centers around an arbitrary molecular center. Through Fourier inversion it leads to exact expressions for the scattering of light or the diffraction of X rays in liquids, and its use has been extended to other fields.

In constructing instruments Zernike always started from first principles and worked out the significant mathematical consequences. This procedure often led to unexpected results—for instance, the discovery that for a sensitive moving-coil galvanometer, the moment of inertia of the mirror has to be roughly three times that of the moving coil. The usual technique of instrument makers had been just the opposite: making the mirror quite small compared with the moving coil. He also worked on the ultracentrifuge and thermoelectrical devices.

Experimental and mathematical skill also formed the base of Zernike's best-known contribution, the method of phase contrast in wave theory. It is now generally used in microscopy but has a much wider application: it was used, for instance, in his study of errors in telescope mirrors and in the Groningen Rowland grating (in the winter of 1930–1931, the first application of phase contrast). It led Zernike to study the "degree of coherence" in light and to approach what is now called holography. The essential point is that in the "primary" diffraction pattern, already studied by Abbe, a phase difference between the central part and the wings of the pattern exists—and can be manipulated to increase the contrast in the image or to reconstruct the object in three dimensions.

It has rightly been said that the spirit of Zernike's work is reminiscent of Lord Rayleigh's, to whom Zernike often referred in his lectures.

BIBLIOGRAPHY

Lists of Zernike's writings are included in the obituaries by Tolansky and by Prins and Nijboer (see below). An autobiographical article is "How I Discovered Phase Contrast," in Les prix Nobel en 1953 (Stockholm, 1953), 107–114.

See H. Brinkman, in Nederlands tijdschrift voor natuurkunde, 24 (1958), 139; J. A. Prins, in Jaarboek der Koninklijke Nederlandsche akademie van wetenschappen (1965–1966), 370–377; J. A. Prins and B. R. A. Nijboer, in Nederlands tijdschrift voor natuurkunde, 19 (1953), 314–328; S. Tolansky, in Biographical Memoirs of Fellows of the Royal Society (London), 13 (1967), 393–402, with portrait; and N. G. van Kampen, in Nature, 211 (1966), 465—also see 172 (1953), 938.

J. A. Prins

ZEUNER, GUSTAV ANTON (b. Chemnitz [now Karl-Marx-Stadt, German Democratic Republic], 30 November 1828; d. Dresden, Germany, 17 October 1907), *mechanical engineering, thermodynamics.*

The son of a cabinetmaker, Zeuner completed an apprenticeship in his father's trade (1846). At the local trade school and through private study he acquired the background to enter the mining academy at Freiberg, Saxony. Working closely with Julius Weisbach, professor of mechanics and mining machinery and a notable hydraulics engineer, he became interested in applied mathematics and mechanical engineering and decided upon a teaching career. In the first years after his graduation (1851–1855), Zeuner was unable to find permanent employment in his native Saxony because of his participation in the 1849 uprisings in Dresden. In 1851 he visited Paris, where he met Poncelet and Regnault. Impressed by Foucault's pendulum, he wrote a theoretical study of it, for which the University of Leipzig awarded him the Ph.D. in 1853. In 1854, with Weisbach and C. R. Bornemann, he founded the important journal Der Civil-Ingenieur, of which he was the first editor (until 1857) and a major contributor.

In 1855 Zeuner was appointed professor of mechanics and theory of machines, and head of the mechanical department of the Federal Polytechnicum at Zurich. His years in Zurich were his most productive; he served as deputy director (1859–1865, 1868–1871) and director (1865–1868) of the institution, and wrote his most important scientific works. In later years Zeuner was preoccupied with administrative duties, serving as director of the Freiberg Mining Academy from 1871 to 1875, and of the Dresden Polytechnical Institute from 1873 to 1890. He resigned the latter post in 1890, after reorganizing the school into an institute of technology, and retired from teaching in 1897.

Reflecting the influence of Weisbach, Zeuner's earliest publications dealt with problems of hydraulics and water turbines, a subject to which he returned in his later years. Most of his works, however, were devoted to theoretical aspects of the steam engine. In a book on steam-engine valve gears (1858) he proposed a graphical treatment of valve motion that was soon internationally accepted as the Zeuner diagram; other monographs dealt with steam injectors and the dynamic imbalances in the motion of locomotives. His main work, a comprehensive text on thermodynamics (1860), presented the first synthesis into a consistent system of the newly formulated first and second laws of thermodynamics and the improved understanding of the properties of steam (resulting from Regnault's experiments). This book, which had lasting international success, was distinguished by its emphasis upon theoretical principles and by its deductive approach. It was later criticized for its failure to do justice to certain practical problems of the steam engine.

Zeuner also made pioneer contributions to mathematical statistics and insurance mathematics, in which he had become interested through a study of miners' insurance systems done while he was a student.

Zeuner has been praised as a naturally gifted lecturer and teacher. Among his most prominent students were the engineers Carl von Linde and Hans Lorenz and the physicist Wilhelm Röntgen, whose doctoral dissertation (1869) he supervised.

BIBLIOGRAPHY

I. ORIGINAL WORKS. Lists of Zeuner's publications are in Poggendorff, II, 1407–1408; III, 1481–1482; IV, 1689; and V, 1410. His books are *Die Schiebersteuerungen* (Freiberg, 1858; 6th ed., Leipzig, 1904), translated by J. F. Klein as *Treatise on Valve Gears* (London–New York, 1884) and by A. Debize and E. Merijot as *Traité des distributions par tiroir* (Paris, 1869); *Grundzüge der mechanischen Wärmetheorie* (Freiberg, 1860; 2nd ed., 1866), 3rd ed., enl., *Technische Thermodynamik*, 2 vols. (Leipzig, 1887–1890; 5th ed., 1905–1906), translated by Maurice Arnthal as *Théorie mécanique de la chaleur* (Paris, 1869) and by J. F. Klein as *Technical Thermodynamics* (New York, 1907); *Das Locomotiven-Blasrohr* (Stuttgart, 1863); *Abhandlungen aus der mathematischen Statistik* (Leipzig, 1869); and *Vorlesungen über Theorie der Turbinen* (Leipzig, 1899).

II. SECONDARY LITERATURE. Biographical information on Zeuner, listed chronologically, is A. Slaby *et al.*, "Gustav Zeuner," in *Zeitschrift des Vereins deutscher Ingenieure*, 51 (1907), 2049–2050; R. Mollier, "Gustav Zeuner," *ibid.*, 52 (1908), 1221–1224; Verein Deutscher Ingenieure, *Gustav Zeuner, sein Leben und Wirken* (Berlin, 1928); and Gustav Zeuner-Schnorf, "Als junger Professor an die Hochschule berufen," in *Neue Zürcher Zeitung*, no. 2812 (22 Oct. 1955); and "Röntgens Doktorvater in Zürich," in *Technische Rundschau* (1958).

OTTO MAYR

ZEUTHEN, HIERONYMUS GEORG (*b*. Grimstrup, West Jutland, Denmark, 15 February 1839; *d*. Copenhagen, Denmark, 6 January 1920), *mathematics, mechanics, history of mathematics*.

The son of a minister, Zeuthen received his earliest education in Grimstrup and at the age of ten entered the secondary school in Sorø, where his father had been transferred. From 1857 to 1862 he studied pure and applied mathematics at the University of Copenhagen. After passing the examination for a master's degree, he received a stipend in 1863 to travel to Paris for further study with Chasles. Having become familiar with his writings, Zeuthen followed Chasles's lead in his own work on enumerative methods in geometry and also in undertaking research on the history of mathematics.

Zeuthen found in enumerative methods in geometry ("number geometry") a fertile area for research. His first work on this subject was his doctoral dissertation at the University of Copenhagen, *Nyt Bidrag til Laeren on Systemer af Keglesnit* (1865), which was also published in French in *Nouvelles annales de mathématiques* (2nd ser., 5 [1866]) as "Nouvelle méthode pour déterminer les caractéristiques des systèmes de coniques." In this work Zeuthen adhered closely to Chasles's theory of the characteristics of conic systems but also presented new points of view: for the elementary systems under consideration, he first ascertained the numbers for point or line conics in order to employ them to determine the characteristics. Arthur Cayley presented a thorough discussion of the relationships and an exposition of the entire theory in "On the Curves Which Satisfy Given Conditions" (*Philosophical Transactions of the Royal Society*, 158 [1868], 75–143).

The first decade of Zeuthen's scientific activity was devoted entirely to enumerative methods in geometry, and his works were published in *Tidsskrift for Mathematik*, of which he was editor from 1871 to 1889; he was also a contributor to *Mathematische Annalen* and other European scientific journals. A summary of this work was presented in

Lehrbuch der abzählenden Methoden der Geometrie (1914); and as a leading expert in the field, he was chosen to write "Encyklopädiebericht über abzählende Methoden" for the *Encyklopädie der mathematischen Wissenschaften* (III, pt. 2 [1905], 257–312).

In 1871 Zeuthen became assistant professor and in 1886 full professor at the University of Copenhagen, where he remained until his death, serving as rector in 1896. While at the university he also taught at the nearby Polytechnic Academy and for many years was secretary of the Royal Danish Academy of Sciences.

After 1875, in addition to teaching, Zeuthen wrote on mechanics, geometry, and the history of mathematics. In his first major work on this subject, "Kegelsnitlaeren in Oltiden" (1885), he presented an exposition of Apollonius of Perga's theory of conic sections, in which he showed that Apollonius had employed oblique coordinates in deriving the properties of conics. Zeuthen also found in his work the projective production of the conics from two pencils of lines.

In a second, larger work (1896), Zeuthen traced the development of mathematics to the Middle Ages, presenting the influences of the Greek tradition that were transmitted to medieval mathematics through the Arabs and the rediscovery of the original works. He continued his historical studies in *Geschichte der Mathematik im 16. und 17. Jahrhundert* (1903), a large portion of which is devoted to Descartes and Viète, with regard not only to algebra and analytic geometry but also to the history of analysis, the development of which Zeuthen traced from its beginnings to Newton and Leibniz. Zeuthen also emphasized the importance of Barrow's works in the emergence of this discipline.

Although in these works Zeuthen naturally drew on the findings and references of other authors, his results were based essentially on careful study of original texts. Moreover, he strove to attune his thinking to the ancient forms of mathematics, in order to appraise the value of the resources and methods available in earlier periods. Although he was criticized for not providing full details concerning his sources, it is widely conceded that Zeuthen was the foremost historian of mathematics of his time, perhaps superior to Moritz Cantor and Siegmund Günther.

Zeuthen saw things intuitively: he constantly strove to attain an overall conception that would embrace the details of the subject under investigation and afford a way of seizing their significance. This approach characterized his historical research equally with his work on enumerative methods in geometry.

A *Festschrift* was dedicated to Zeuthen on his seventieth birthday, and in honor of his eightieth birthday a medal with his likeness was struck.

BIBLIOGRAPHY

I. ORIGINAL WORKS. A list of Zeuthen's 161 published writings is in M. Noether, "Hieronymus Georg Zeuthen" (see below), 15–23. Among his most important monographs are *Grundriss einer elementargeometrischen Kegelschnittslehre* (Leipzig, 1882); "Kegelsnitlaeren in Oltiden," which is *Kongelige Danske Videnskabernes Selskabs Skrifter*, 6th ser., **3**, no. 1 (1885), 1–319, 2nd ed. by O. Neugebauer (Copenhagen, 1949); German trans. by R. Fischer-Benzon as *Die Lehre von den Kegelschnitten im Altertum* (Copenhagen, 1886), 2nd ed. by J. E. Hofmann (Hildesheim, 1966); *Forelaesning over Mathematikens Historie: Oldtig i Middelalder* (Copenhagen, 1893), German trans. as *Geschichte der Mathematik im Altertum und Mittelalter* (Copenhagen, 1896), French trans. by J. Mascart (Paris, 1902); *Geschichte der Mathematik im 16. und 17. Jahrhundert* (Copenhagen, 1903), also in German (Leipzig, 1903) and Russian (Moscow–Leningrad, 1933); and *Lehrbuch der abzählenden Methoden der Geometrie* (Leipzig–Berlin, 1914).

II. SECONDARY LITERATURE. See Johannes Hjelmslev, "Hieronymus Georg Zeuthen," in *Matematisk Tidsskrift*, ser. A (1939), 1–10; and Max Noether, "Hieronymus Georg Zeuthen," in *Mathematische Annalen*, **83** (1921), 1–23. Luigi Berzolari, "Bericht über die allgemeine Theorie der höheren ebenen algebraischen Kurven," in *Encyklopädie mathematischen Wissenschaften*, III, pt. 2 (Leipzig, 1906), 313–455, contains many references to Zeuthen's work and results.

KARLHEINZ HAAS

ZHUKOVSKY, NIKOLAY EGOROVICH (*b.* Orekhovo, Vladimir province, Russia, 17 January 1847; *d.* Moscow, U.S.S.R., 17 March 1921), *mechanics, mathematics.*

The son of a communications engineer, Zhukovsky completed his secondary education at the Fourth Gymnasium for Men in Moscow in 1864 and graduated in 1868 from the Faculty of Physics and Mathematics of the University of Moscow, having specialized in applied mathematics. In 1870 he began teaching at the Second Gymnasium for Women in Moscow, and at the beginning of 1872 he was invited to teach mathematics at the Moscow Technical School, at which he also lectured on theoretical mechanics from 1874. Two years later he defended a dissertation at the Technical School on the kinematics of a liquid and was awarded the degree of master of applied mathematics; a separate chair of mechanics was subsequently established for him at the school. In 1882 he defended his doctoral dissertation, on the stability of motion, at Moscow University and four years later became head of the department of mechanics there. In 1894 he was elected corresponding member of the St. Petersburg Academy of Sciences, and in 1900 he was promoted to member. Unwilling to leave his teaching posts in Moscow and undertake the requisite move to St. Petersburg, however, Zhukovsky withdrew his candidacy. A member of the Moscow Mathematical Society, he also served as vice-president from 1903 to 1905, and as president from 1905 until his death he proved to be an outstanding administrator.

Zhukovsky's approximately 200 publications in mechanics and its applications to technology reveal the wide range of his interests. Several works are devoted to the motion of a solid around a fixed point, in particular, to the case of Sonya Kovalevsky, for which he gave an elegant geometrical interpretation. He also wrote on the theory of ships, on the resistance of materials, and on practical mechanics. From the beginning of the twentieth century his interest focused primarily on aerodynamics and aviation, to which he devoted himself exclusively in his later years.

In his clear and well-organized lectures Zhukovsky made extensive use of geometric methods, which he valued highly. His lectures on hydrodynamics were standard works for many years, and his course on the theory of regulation of mechanical action (1908–1909) was the first rigorous presentation in Russian of the fundamentals of that subject. His lectures at the Moscow Technical School on the theoretical basis of aeronautics (1911–1912) were the world's first systematic course in aviation theory and were based largely on his own theoretical research and on experiments conducted in laboratories that he had established. During World War I Zhukovsky and his students taught special courses for pilots at the Technical School.

Instrumental in the development of Soviet aviation, Zhukovsky was named head of the Central Aerohydrodynamics Institute, established in 1918. The school of aviation that subsequently developed from it was based on his teaching and be-

came the N. E. Zhukovsky Academy of Military and Aeronautical Engineering in 1922.

Zhukovsky is considered the founder of Russian hydromechanics and aeromechanics. In his master's thesis (1876) he made extensive use of geometric, as well as analytic, methods to establish the kinematic laws of particles in a current. In 1885 he was awarded the N. D. Brashman Prize for a major theoretical work on the motion of a solid containing a homogeneous liquid. The methods that he developed in this memoir made it possible to solve certain problems of astronomy, concerning the laws of planetary rotation, and of ballistics, on the theory of projectiles having liquid cores. In a work dealing with a modification of the Kirchhoff method for determining the motion of a liquid in two dimensions with constant velocity and an unknown line of flow (1890), Zhukovsky used the theory of functions of a complex variable to elaborate a method for determining the resistance of a profile having any number of critical points. In addition to solving the problems studied by Kirchhoff, he resolved others, the solution of which had been extremely complicated with the use of existing methods. A memoir written with S. A. Chaplygin (1906) gave a precise solution to the problem of the motion of a lubricant between pin and bearings, and stimulated a number of other investigations.

In hydraulics, Zhukovsky in 1888 undertook theoretical research on the movement of subsurface water and studied the influence of pressure on water-permeated sand, establishing the relation between changes in the water level and changes in barometric pressure. Showing that the variation in the water level depends on the thickness of the water-bearing layer, he introduced formulas to determine the undergound water supply, using experimental data extensively. This research was summarized in a work on hydraulic shock in water pipes (1898), in which Zhukovsky established that the reason for damage to water mains was the sudden increase in pressure that followed the rapid closing of the valves. Extensive experiments enabled him to present the physical nature of hydraulic shock, to give a formula for determining the time needed for safe closing of the mains, and to elaborate a method for preserving them from damage effected by hydraulic shock. Zhukovsky acquired an international reputation for this theory, which has remained fundamental to problems of hydraulic shock.

Zhukovsky's other works in hydrodynamics

concern the formation of riverbeds (1914) and the selection of a river site for constructing dams and for withdrawing water used to cool machines at large power stations (1915).

Known as "the father of Russian aviation," Zhukovsky became interested in the late 1880's in flight in heavier-than-air machines, a basic problem of which was lift. The experimental data that had been obtained proved useful only in particular cases; attempts to determine lift on the basis of theoretical premises—especially existing theories of jet stream—yielded results that differed considerably from experimental findings.

Considering it necessary to first establish a physical picture of lift, Zhukovsky in 1890 considered the possibility that it can result from certain vortical motions caused by the viscosity of the surrounding medium. His subsequent experiments with disks rotating in an air current (1890–1891) anticipated his concept of bound vortices, the basis of his theory of lift. In 1891 Zhukovsky began studying the dynamics of flight in heavier-than-air machines, theoretically substantiating the possibility of complex motion of an airborne craft, in particular the existence of loops. In 1890–1891 Zhukovsky undertook experiments designed to study the changing position of the center of pressure of a wing with the simplest profile, a flat disk. By that time he had already turned his attention to the question of stability and was conducting tests of gliders and kites. In studying propeller thrust, Zhukovsky considered heavier-than-air craft powered by flapping wings, multipropellered helicopters, and screw propellors. In 1897 he presented a method of computing the most efficient angle of attack of a wing.

Zhukovsky's works on the motion of a substance in a liquid, published in the 1880's and 1890's, included a memoir on the paradox of Du Buat (1734–1809), for which he gave a physical explanation. In 1779 Du Buat had shown experimentally that the resistance of an immobile disk in a moving liquid is greater than the resistance of a disk moving at the same speed in a stationary liquid—a phenomenon that seemed to contradict the general laws of mechanics. Zhukovsky explained the discrepancy by the fact that, in practice, turbulence always occurs on the walls and the free surface of a liquid. To support his explanation he constructed a small device by means of which he showed that when there is no turbulence the pressure remains the same in both cases.

Zhukovsky established that lift results from the flow in an airstream of an immobile bound vortex (or system of vortices) by which the object can be replaced. From this starting point, he derived a formula for lift, equal to the product of the density of air, the circulation velocity of the surrounding airstream, and the velocity of the body. The theorem was confirmed in experiments with rotating oblong disks, conducted in 1905–1906 at the Aerodynamics Institute at Kuchino, near Moscow.

The formulation in 1910 of the Zhukovsky-Chaplygin postulate, concerning the determination of the rate of circulation around a wing, made it possible to solve the problem of lift, to determine its moment, and to develop a profile for airplane wings. Zhukovsky also investigated the profile of resistance of a wing and established the existence of resistance caused by the flow of turbulence from the wing's sharp leading edge.

In high-speed aerodynamics, Zhukovsky in 1919 presented a theory of the distribution of high-velocity plane and spherical waves, and demonstrated its possible application to determine the resistance of projectiles. His work in airplane stability included a major monograph (1918) in which he considered the construction of airplanes on the assumption that the longerons bear uniform loads arising from the weight of the wings and from the air pressure.

Zhukovsky initiated the study in Russia of the theory of bombing from airplanes. In 1915 he offered a method of determining the trajectory and bomb velocity when the air resistance is proportional to the square of the velocity; he provided a method of calculating the change of air density from a given altitude; and he examined various practical methods for using bombing and sighting apparatus.

S. A. Chaplygin was the most distinguished member of Zhukovsky's school, which included A. I. Nekrasov, L. S. Leybenzon, V. P. Vetchinkin, B. N. Yuriev, and A. N. Tupolev.

BIBLIOGRAPHY

I. ORIGINAL WORKS. Zhukovsky's complete collected works were published as *Polnoe sobranie sochineny*, 9 vols. (Moscow–Leningrad, 1935–1937). Other collections are *Izbrannye sochinenia* ("Selected Works"), 2 vols. (Moscow–Leningrad, 1948); and *Sobranie sochineny* ("Collected Works"), 7 vols. (Moscow–Leningrad, 1948–1950).

II. SECONDARY LITERATURE. On Zhukovsky and his work, see V. A. Dombrovskaya, *Nikolay Egorovich*

Zhukovsky (Moscow–Leningrad, 1939), with bibliography of his writings; V. V. Golubev, *Nikolay Egorovich Zhukovsky* (Moscow, 1947); A. T. Grigorian, *Ocherki istorii mekhaniki v Rossii* ("Sketches of the History of Mechanics in Russia"; Moscow, 1961); "Vklad N. E. Zhukovskogo i S. A. Chaplygina v gidro-dinamiky i aerodinamiku" ("The Contribution of Zhukovsky and Chaplygin to Hydrodynamics and Aerodynamics"), in *Evolyutsia mekhaniki v Rossii* ("Evolution of Mechanics in Russia"; Moscow, 1967); and *Mekhanika ot antichnosti do nashikh dney* ("Mechanics From Antiquity to Our Time"; Moscow, 1971); A. A. Kosmodemyansky, "Nikolay Egorovich Zhukovsky," in *Lyudi russkoy nauki* ("Men of Russian Science"; Moscow, 1961), 169–177; and L. S. Leybenzon, *Nikolay Egorovich Zhukovsky* (Moscow–Leningrad, 1947).

A. T. GRIGORIAN

ZININ, NIKOLAY NIKOLAEVICH (*b.* Shusha, Transcaucasia [now Azerbaydzhan S.S.R.], Russia, 25 August 1812; *d.* St. Petersburg, Russia, 18 February 1880), *chemistry.*

After graduating from Kazan University in 1836, Zinin studied abroad from 1837 to 1840 and worked in Liebig's laboratory at Giessen for about a year. From 1841 to 1847 he was professor of technical chemistry at Kazan University and, from 1848 until his retirement in 1874, was professor of chemistry at the St. Petersburg Academy of Medicine and Surgery. In the mid-1860's Zinin worked mainly at the St. Petersburg Academy of Sciences, to which he had been elected in 1855. He was also the first president of the Russian Chemical Society. His students included Borodin and Butlerov.

Zinin's work in chemistry concerned the aromatic compounds. His doctoral dissertation, prepared in Liebig's laboratory and defended at St. Petersburg University in 1841, was devoted to obtaining benzoin by the condensation of benzaldehyde, and benzyl by the oxidation of benzoin; he returned to the study of these compounds in the 1850's. Zinin developed a method for the reduction of organic compounds using hydrogen at the moment of separation (acid + metal). In particular he studied the reduction of benzyl into benzoin and benzaldehyde into hydrobenzoin, obtaining numerous other reaction products by the use of these compounds.

Zinin is known primarily for his research on the reduction of nitro compounds into amino derivatives by the action of ammonium sulfide. In 1842 he described the reduction of nitrobenzene into aniline, and α-nitronaphthalene into the corresponding aminonaphthalene; in 1845, by reducing azobenzene, he obtained benzidine. His discovery of the reaction for obtaining numerous representatives of the amino derivatives was later of great importance for the creation of the aniline dye industry.

Zinin also obtained valuable results in his research on allyl derivatives; he was the first to synthesize allyl alcohol, allyl mustard oil, and the allyl esters of a number of organic acids. During the Crimean War Zinin also studied nitroglycerine as an explosive substance.

BIBLIOGRAPHY

I. ORIGINAL WORKS. Zinin's writings include *O soedineniakh benzoila i ob otkrytykh novykh telakh, otnosyashchikhsya k benzoilovomu rodu* ("On the Compounds of Benzoyl and on the Discovery of New Bodies, Related to the Benzoyl Type"; St. Petersburg, 1840), his doctoral diss.; "Organische Salzbasen aus Nitronaphthalos und Nitrobenzid mittelst Schwefelwasserstoff entstehend," in *Annalen der Chemie und Pharmacie*, **44** (1842), 283–287; "Ueber die Einwirkung des ätherischen Senföhls auf die organischen Basen," *ibid.*, **84** (1852), 346–349; and "Ueber die Einführung von Wasserstoff in organische Verbindungen," *ibid.*, **119** (1861), 179–182. See also the Royal Society *Catalogue of Scientific Papers*, VI, 512; VIII, 1304; and XI, 890; which lists 34 memoirs by Zinin.

II. SECONDARY LITERATURE. On Zinin and his work, see A. M. Butlerov and A. P. Borodin, "Nicolaus Nicolajewitsch Zinin," in *Berichte der Deutschen chemischen Gesellschaft*, **14** (1881), 2887–2908; N. A. Figurovsky and Y. I. Soloviev, *Nikolay Nikolaevich Zinin. Biografichesky ocherk* (Moscow, 1957), with bibliography; A. W. von Hofmann's report (8 Mar. 1880), in *Berichte der Deutschen chemischen Gesellschaft*, **13** (1880), 449–450; H. M. Leicester, "N. N. Zinin, an Early Russian Chemist," in *Journal of Chemical Education*, **17** (1940), 303–306; and B. N. Menshutkin, *Nikolay Nikolaevich Zinin* (Berlin, 1921), in Russian.

G. V. BYKOV

ZINSSER, HANS (*b.* New York, N.Y., 17 November 1878; *d.* New York, 4 September 1940), *bacteriology, immunology.*

Zinsser was the youngest son of August Zinsser, a German immigrant who had founded a prosperous chemical products business. The household retained many Old World features, and German was Zinsser's primary language until he was ten years old. He was also steeped in French literature and culture from his childhood. His early education at the fashionable private school run by Julius Sachs was supplemented by frequent trips to Europe, as well as by a year of study at Wiesbaden, Germany.

In 1895, Zinsser entered Columbia College, where George Edward Woodberry, professor of comparative literature, cultivated his poetic imagination. His enthusiasm for science was the result of courses under Edmund B. Wilson and Bashford Dean. He never lost either enthusiasm, but science became his profession and poetry a seriously practiced avocation. Zinsser graduated from the College of Physicians and Surgeons of Columbia University in 1903, receiving both the M.D. and the M.A. He interned at Roosevelt Hospital from 1903 to 1905, and after three years of desultory medical practice accepted a full-time appointment at Columbia as instructor in bacteriology. From 1907 until 1910 he was also assistant pathologist at St. Luke's Hospital. In 1910 Zinsser went to Stanford University, where in 1911 he was appointed professor of bacteriology and immunology. He was recalled to a similar position at Columbia in 1913. From 1923 until his death Zinsser taught at the Harvard Medical School, becoming Charles Wilder professor of bacteriology and immunology in 1925.

In 1905 Zinsser married Ruby Handforth Kunz; they had a son and a daughter. He received numerous honors and awards, including honorary doctorates from Columbia, Western Reserve, Lehigh, Yale, and Harvard. He was decorated with the French Legion of Honor and the American Distinguished Service Medal, and served as president of both the American Association of Immunologists (1919) and the Society of American Bacteriologists (1926).

Zinsser's professional interests ranged from theoretical questions involving the physicochemical nature of the antigen-antibody reaction to practical problems of military sanitation and the epidemiology of infectious diseases. He visited Serbia in 1915 as a member of the American Red Cross Sanitary Commission and studied at first hand an outbreak of epidemic typhus. His field investigations of typhus later took him to the Soviet Union (1923), Mexico (1931), and China (1938). Zinsser described these experiences in two engaging books, *Rats, Lice and History* (1935) and *As I Remember Him* (1940). The former book, which he called a "biography of the life history of typhus," was a Rabelaisian mixture of history, wit, philosophy, and science that achieved instant popularity. *As I Remember Him*, although written in the third person, is autobiographical. Its subject—"R.S."—is Zinsser himself, those letters forming the pseudo-

nym under which he published poems in the *Atlantic Monthly* and other periodicals. In his autobiography Zinsser exhibited much of the playfulness and penchant for digression that characterized the earlier popular study of typhus. Writing the book while suffering from lymphocytic leukemia, he movingly recorded his subjective reactions to the disease that was causing his death.

Zinsser dedicated *Rats, Lice and History* to his friend Charles Nicolle, the French bacteriologist, novelist, and philosopher who received the 1928 Nobel Prize in medicine or physiology for his studies on the natural history of typhus. Nicolle and his colleagues had proved that epidemic typhus is louse-borne, and had infected monkeys and guinea pigs with the causative organism (*Rickettsia prowazekii*), thus providing convenient laboratory models for studying host response to the disease. The relationship of louse-borne epidemic typhus to the sporadic, endemic variety of the disease prevalent in the southeastern regions of the United States was poorly understood. Endemic typhus was generally grouped with another form of typhus first described in 1898 by N. E. Brill, who found it among immigrants in New York City. In the 1920's and 1930's Hermann Mooser and others proved that the etiologic agent of murine (endemic) typhus is another species of *Rickettsia* (since named *R. mooseri*). The rat flea, rather than the louse, serves as the principal vector in transmitting this variety of typhus. In the early 1930's Zinsser and his associates suggested that Brill's disease is clinically distinct from murine typhus. Zinsser then demonstrated that the causative organism in Brill's disease is *R. prowazekii* rather than *R. mooseri*. He hypothesized that Brill's disease represented recrudescent typhus in patients who have already recovered from a primary attack of epidemic typhus. His hypothesis has since been confirmed epidemiologically and serologically, and Brill's disease has been renamed Brill-Zinsser's disease.

Zinsser and his associates, who included M. Ruiz Castañeda, Harry Plotz, S. H. Zia, and J. F. Enders, also worked on the production of an effective vaccine against typhus. From these studies came important new tissue culture methods for growing *Rickettsia*. Zinsser and his associates also developed improved staining techniques for *Rickettsia* in both smears and tissue cultures.

Zinsser's name is thus intimately connected with the development of modern knowledge of rickettsial diseases, an association cemented by the popularity of *Rats, Lice and History*. Nevertheless, his typhus studies represent only a portion of his scientific output. Zinsser also did important work on the nature of the antigen-antibody reaction, the etiology of rheumatic fever, the phenomena of delayed hypersensitivity and allergy, the measurement of virus size, and the host response to syphilis.

His concern with the fundamental problems of immunology began about the time Zinsser went to Stanford (1910). He was convinced that physical chemistry would ultimately provide the means of understanding the reactions between antigens and antibodies. Accordingly, he attempted to rectify the deficiencies in his own mathematical and chemical competence. This conviction also led to his collaboration with Stewart Young, professor of colloid chemistry at Stanford, with whom he investigated colloidal aspects of the precipitin reaction. Zinsser's first work on the influence of heat on antigens also dates from his time at Stanford. From the solutions of bacteria and their metabolic products used to immunize laboratory animals, he identified various heat-resistant fractions with pronounced antigenic properties. Since proteins are denatured by heat (and acid), he thus showed that other classes of compounds besides proteins can be antigenic. Zinsser's studies of tuberculin hypersensitivity and allergy led him to stress the importance of nonprotein bacterial products in the immune response. He was the first to formulate clearly the distinction between the tuberculin type of allergic reaction and classic anaphylactic shock.

In addition to his work on immunologic aspects of tuberculosis, Zinsser studied various hyperimmune and allergic phenomena associated with streptococcal and pneumococcal infections. He was one of several scientists in the 1910's and 1920's to suggest that diseases such as rheumatic fever and glomerulonephritis result from hypersensitivity to toxins produced by certain strains of *Streptococci*.

Between 1926 and 1930 Zinsser published a number of papers on the herpes virus and on herpes encephalitis. He and Fei-fang Tang undertook to measure virus sizes by passing viruses through graded filters. They obtained good approximations, although more sensitive techniques have since been developed.

Zinsser's research interests thus covered a spectrum of bacteriological and immunological problems. He also produced two systematic treatises. *A Textbook of Bacteriology* was first published in 1910, in collaboration with Philip H. Hiss, Jr., a Columbia bacteriologist. This book passed through eight editions during Zinsser's lifetime and was translated into several foreign languages, including Chinese. After Hiss's death in 1913, Zinsser collaborated with Stanhope Bayne-Jones in the production of the seventh and eighth editions, and through Bayne-Jones and others, the *Textbook* reached the fourteenth edition (1968).

Infection and Resistance (1914) was Zinsser's other major book. His own contributions to immunology may be traced through the successive editions of this work, last published in 1939 in collaboration with John Enders and Henry D. Fothergill as *Immunity: Principles and Application in Medicine and Public Health*.

Literature, history, politics, education, art, music, and philosophy also came within Zinsser's ken. He played an active role in university life at Harvard in the 1920's and 1930's, and taught his pupils and research associates far more than the principles of bacteriology.

BIBLIOGRAPHY

I. ORIGINAL WORKS. The various eds. of Zinsser's *Textbook of Bacteriology* (New York–London, 1910), written with Philip H. Hiss, Jr., and *Infection and Resistance* (New York, 1914) give good accounts of his own work in relation to the developing disciplines of bacteriology and immunology. The following papers deal with particular aspects in greater detail: "On the Possible Importance of Colloidal Protection in Certain Phases of the Precipitin Reaction," in *Journal of Experimental Medicine*, 17 (1913), 396–408, written with Stewart Young; "Studies on the Tuberculin Reaction and on Specific Hypersensitiveness in Bacterial Infection," *ibid.*, 34 (1921), 495–524; "On the Significance of Bacterial Allergy in Infectious Disease," in *Bulletin of the New York Academy of Medicine*, 2nd ser., 4 (1928), 351–383; "Studies in Ultrafiltration," in *Journal of Experimental Medicine*, 47 (1927), 357–378, written with Fei-fang Tang; "The Bacteriology of Rheumatic Fever and the Allergic Hypothesis," in *Archives of Internal Medicine*, 42 (1928), 301–309, written with H. Yu; and "Varieties of Typhus Fever and the Epidemiology of the American Form of European Typhus Fever (Brill's Disease)," in *American Journal of Hygiene*, 20 (1934), 513–532.

Books by Zinsser not mentioned in the text are *A Laboratory Course in Serum Study* (New York, 1916), written with J. G. Hopkins and Reuben Ottenburg; and *Spring, Summer and Autumn* (New York, 1942), a volume of poems.

A full bibliography of Zinsser's writings is appended to Simeon Burt Wolbach's memoir in *Biographical Memoirs. National Academy of Sciences*, 24 (1947), 323–360.

II. SECONDARY LITERATURE. In addition to the Wolbach memoir, other valuable obituaries include those of

S. Bayne-Jones, in *Archives of Pathology*, **31** (1941), 269–280; J. F. Enders, in *Harvard Medical Alumni Bulletin*, **15**, no. 1 (1940), supp., 1–15; J. H. Mueller, in *Journal of Bacteriology*, **40** (1940), 747–753. Zinsser's contributions to the study of typhus were summarized by P. K. Olitsky, "Hans Zinsser and His Studies on Typhus Fever," in *Journal of the American Medical Association*, **116** (1941), 907–912.

WILLIAM F. BYNUM

ZIRKEL, FERDINAND (*b.* Bonn, Germany, 20 May 1838; *d.* Bonn, 11 June 1912), *geology, petrography, mineralogy.*

Zirkel was the son of Joseph Zirkel, professor of mathematics at a Gymnasium in Bonn. He remained single all his life, taking care of his mother and widowed sister. He enrolled at the University of Bonn in 1855, studying geology, mineralogy, and chemistry to prepare for a career as a mining geologist. After several semesters of practical work in mines, he returned to graduate work at Bonn, mainly with C. G. Bischof, J. J. Nöggerath, and Gerhard vom Rath.

In 1860 Zirkel traveled to the Faeroes and Iceland and remained for a time in Scotland and England, visiting the major ore deposits. His collection of volcanic rocks from Iceland was the basis of his doctoral dissertation, for which he obtained his degree on 14 March 1861. With this work Zirkel abandoned his original intention of becoming a mining geologist, although his later position on the Board of the Mansfelder Kupferschiefer Company kept him in touch with economic geology.

Very important to Zirkel was his friendship with Henry Clifton Sorby, whom he met in England and who started him off in the work that made his reputation in a field of which he was a founder, petrographic microscopy. Zirkel spent two years in Vienna with Haidinger, working out a monograph on bournonite (1862) and a paper entitled "Mikroskopische Untersuchungen von Gesteinen und Mineralien" (1862), the latter amplifying Sorby's discovery of fluid inclusions in minerals.

On the basis of his early systematic work on microscopic petrography, Zirkel was appointed associate professor in Lemberg (now Lvov) in 1863 and full professor in 1865. He obtained a professorship in Kiel in 1868 and in Leipzig in 1870, as successor to K. F. Naumann. He remained at Leipzig until his retirement in 1909.

Zirkel's fame was established in 1873 with *Die mikroskopische Beschaffenheit der Mineralien und Gesteine*. His most influential work was *Lehrbuch der Petrographie* (1866).

In 1894–1895 Zirkel visited Ceylon and the United States. A member of many European academies of science, he received an honorary doctorate from Oxford in 1907. He served as rector of the University of Leipzig and was a member of the Sächsische Akademie der Wissenschaften.

BIBLIOGRAPHY

I. ORIGINAL WORKS. Zirkel's books include *De geognostica Islandiae constitutione observationes* (Bonn, 1861), his dissertation; *Reise nach Island* (Leipzig, 1862), written with W. Preyer; *Lehrbuch der Petrographie*, 2 vols. (Bonn, 1866; 2nd ed., 3 vols., Leipzig, 1893–1894); *Untersuchungen über die mikroskopische Zusammensetzung und Struktur der Basaltgesteine* (Bonn, 1870); *Die mikroskopische Beschaffenheit der Mineralien und Gesteine* (Leipzig, 1873); and his eds. of K. F. Naumann's *Elemente der Mineralogie* (10th ed., Leipzig, 1877; through the 15th ed., 1907).

Among his articles are "Die trachytischen Gesteine der Eifel," in *Zeitschrift der Deutschen geologischen Gesellschaft*, **11** (1859), 507–540; "Versuch einer Monographie des Bournonit," in *Sitzungsberichte der K. Akademie der Wissenschaften in Wien*, **45**, sec. 1 (1862), 431–466; "Über die mikroskopische Zusammensetzung der Phonolithe" in Poggendorff's *Annalen der Physik*, **131** (1867), 298–336; "Beiträge zur geologischen Kenntnis der Pyrenäen," in *Zeitschrift der Deutschen geologischen Gesellschaft*, **19** (1867), 68–215; "Mikroskopische Untersuchung über die glasigen und halbglasigen Gesteine," *ibid.*, 737–802; "Microscopical Petrography," in *United States Geological Exploration of the Fortieth Parallel* (Washington, D.C., 1876), also published in German as "Über die Krystallischen Gesteine längs des 40. Breitegrades in Nordwest-Amerika," in *Math. Phys. Ber.*, **29** (1877), 156–243; "Über Urausscheidungen in rheinischen Basalten," in *Abhandlungen der K. Sächsischen Gesellschaft der Wissenschaften*, math.-phys. Kl., **28**, no. 3 (1903), 103–198; and "Über die gegenseitigen Beziehungen zwischen der Petrographie und angrenzenden Wissenschaften," in *Journal of Geology*, **12**, no. 6 (1904), 485–500, in German.

II. SECONDARY LITERATURE. See R. Brauns, "Ferdinand Zirkel," in *Zentralblatt für Mineralogie, Geologie, und Paläontologie* (1912), 513–521; F. Rinne, "Nachruf auf F. Zirkel," in *Berichte über die Verhandlungen der Sächsischen Gesellschaft der Wissenschaften zu Leipzig*, math.-phys. Kl., **64** (1912), 501–508; and Felix Wahnschaffe, "Ferdinand Zirkel," in *Zeitschrift der Deutschen geologischen Gesellschaft*, B, *Monatsberichte*, **64**, no. 7 (1912), 353–363, the most complete and accurate obituary.

G. C. AMSTUTZ

ZITTEL, KARL ALFRED VON (*b.* Bahlingen, Baden, Germany, 25 September 1839; *d.* Munich, Germany, 5 January 1904), *paleontology, geology, history of geology.*

Zittel is recognized as the leading teacher of paleontology in the nineteenth century and as the only encyclopedist of the subject. The youngest son of Karl Zittel, a liberal and politically active Protestant minister, he grew up in an intellectually stimulating atmosphere. At the University of Heidelberg he studied under the geologist Carl Caesar von Leonhard, the mineralogist Johann Reinhard Blum, and the paleontologist Heinrich Georg Bronn. He also worked without pay in a shop that sold natural history specimens. After 1860 Zittel traveled in Scandinavia for three months before completing his training in Paris under Edmond Hébert.

In 1862 Zittel moved to Vienna, then a center of geological studies. At first he worked as a volunteer at the Geologische Reichsanstalt and participated in the mapping of Dalmatia. The following year he qualified as a lecturer at the University of Vienna, at which Eduard Suess had just begun to teach. Zittel declined an offer from the University of Lemberg (Lvov) in order to continue his studies of the extensive Viennese paleontological collections and to produce his first publications. From 1863 to 1866 he was professor of mineralogy, geognosy, and paleontology at the technical college in Karlsruhe.

In 1866 Zittel accepted a post at the University of Munich, attracted by its exceptionally rich paleontological collections. As successor to Albert Oppel, he dealt with a stage that the latter had established: the Tithonian or Portland stage of the Upper Jurassic. Aided by his students, Zittel not only described Tithonian fauna in terms of guide fossils but also presented a thorough biological (paleozoological) discussion of the material. While at Munich, Zittel devoted great energy to expanding the paleontological collections, thereby establishing the basis for his lifework: the creation of a systematics of the organic fossil record. The results of twenty years of almost superhuman effort were presented in *Handbuch der Palaeontologie.* The sole author of the four volumes on paleozoology, Zittel covered all forms of fossils, from the protozoans to the mammals.

In order to further his great project, Zittel undertook intensive research into inadequately examined areas, devoting particular attention to relationships between fossil and recent forms of life. He was the first to investigate fossil sponges by zoological methods—previously they had been described solely according to external characteristics. Zittel exposed their skeletons through cauterization and applied to siliceous sponges the distinction between Lithistida and Hexactinellida that Oscar Schmidt had discovered in 1870. Having initially found lithistid skeletal elements in the hexactinellid *Coeloptychium,* he first assumed on this evidence that both groups were of common origin. He soon learned, however, that these lithistic elements were drifted secondarily by water transportation and, therefore, Hexactinellida and Lithistida were distinctly separated in their fossilized state.

Under the influence of Darwin's work, Zittel became a pioneer of evolutionary paleontology. Nonetheless, his experience with the sponges made him very cautious with regard to all phylogenetic speculation. Although his writings obviously are based on the presupposition of the continuity of the evolutionary process in nature, he constantly stressed the lacunae in the evidence for this continuity. This cautious attitude, however, has endowed Zittel's work with lasting value.

Zittel's exposition of systematics in the *Handbuch* is characterized by unsurpassed clarity at every level, down to that of the individual taxonomic diagnoses. He sustained this lucidity although he knew that it is impossible to determine the relationships between fossil types with certainty and that, therefore, every attempt to establish boundaries between them is necessarily precarious: this constitutes his greatest achievement in the *Handbuch.* In this respect, Zittel profited from the pre-Darwinian heritage of his student years, when he learned to give precise linguistic expression to boundaries that supposedly were found fixed in nature. He respected the concepts of systematics as a historical validity and was aware that their ability to provide continuity and a synoptic view would be threatened if they were made to depend too heavily on changing phylogenetic interpretations. Zittel adhered to these principles in his two-volume *Grundzüge der Paläontologie,* which was translated into several foreign languages and earned him the title "Linnaeus of Paleontology." His primarily systematic treatment might appear one-sided to the modern paleontologist, accustomed to an increasingly differentiated discipline and a marked ecological orientation. Nevertheless, using this approach, Zittel endowed the subject with a firm basis that is still indispensable.

As a paleontologist, Zittel naturally took an interest in historical geology and also contributed to general and regional geology. For example, he was the first to investigate closely the evidence of the existence of diluvial moraines on the Bavarian plateau. Further, while participating in Gerhard Rohlf's expedition to the Libyan desert (1873–1874), he recognized that the desert sand was a product of wind erosion and did not originate in a Quaternary Sahara sea. (Some scientists had held that European Quaternary glaciation was influenced by the existence of such a sea.)

Zittel's mastery of geology and paleontology and of its history is reflected in his *Geschichte der Geologie und Paläontologie*. (As an encyclopedic historical survey it still remains an indispensable reference for the history of geology, being especially strong on Continental European developments of the nineteenth century. The English translation of 1901 is currently in print.) In this book he could proudly state that during his years at the University of Munich, it became "a center of paleontological and geological studies, where a considerable number of researchers from all parts of the world received their training."

Elected president of the Bavarian Academy of Sciences in 1899, Zittel was an honorary member of many scientific societies in Europe and elsewhere.

BIBLIOGRAPHY

I. ORIGINAL WORKS. Zittel's writings include "Analyse des Arendaler Orthits," in *Annalen der Physik*, **108** (1859), his dissertation; "Die Bivalven der Gosaugebilde in den nordöstlichen Alpen," in *Denkschriften der Akademie der Wissenschaften* (Vienna), Math.-phys. Kl., **24** (1864), 105–177; and **25** (1866), 77–198; "Palaeontologische Studien über die Grenzschichten der Jura- und Kreideformation im Gebiete der Karpathen, Alpen und Apenninen. I. Die Cephalopoden der Stramberger Schichten," in *Palaeontologische Mitteilungen aus dem Museum des K. bayerischen Staates*, **2** (1868); II. "Die Fauna der älteren Cephalopoden führenden Tithonbildungen," in *Palaeontographica*, supp. 2, pts. 1 and 2 (1870); III. "Die Gastropoden der Stramberger Schichten," in *Palaeontologische Mitteilungen aus dem Museum des K. bayerischen Staates*, **2**, pt. 3 (1873), 311–491; *Aus der Urzeit* (Munich, 1872; 2nd ed., 1875); "Über Gletschererscheinungen in der bayerischen Hochebene," in *Sitzungsberichte der Bayerischen Akademie der Wissenschaften zu München*, Math.-phys. Kl., **4** (1874), 252–283; "Uber *Coeloptychium*. Ein Beitrag zur Kenntnis der Organisation fossiler Spongien," in *Abhandlungen der Bayerischen Akademie der Wissenschaften*, Math.-phys. Kl., **12**, pt. 3 (1876), 1–80; and "Studien über fossile Spongien," *ibid.*, **13**, pts. 1 and 2 (1877–1878).

Additional works are "Über die Flugsaurier (Pterodactylen) aus dem lithographischen Schiefer Bayerns," in *Palaeontographica*, **29** (1882), 47–80; "Beiträge zur Geologie und Paläontologie der Libyschen Wüste und der angrenzenden Gebiete von Ägypten, I. Geologie," *ibid.*, **30** (1883), 1–153; *Handbuch der Palaeontologie*, pt. 1, 4 vols. (Munich–Leipzig, 1876–1893); pt. 2, *Palaeophytologie*, by W. P. Schimper and A. Schenk (Munich–Leipzig, 1890), French trans. by Charles Barrois as *Traité de paléontologie*, 5 vols. (Paris, 1883–1894), English trans. by Charles Eastman as *Text-book of Paleontology* (London, 1896; 2nd ed., 1913); "The Geological Development, Descent and Distribution of the Mammalia," in *Geological Magazine*, **30** (1893), 401–412, 455–468, 501–514; *Grundzüge der Paläontologie (Paläozoologie)*, 2 vols. (Munich–Leipzig–Berlin, 1895); "Ontogenie, Phylogenie und Systematik," in *Comptes rendus du Congrès international de géologie, 1894* (Lausanne, 1897), 134–136; *Geschichte der Geologie und Paläontologie bis Ende des 19. Jahrhunderts* (Munich–Leipzig, 1899), trans. by Maria M. Ogilvie-Gordon as *History of Geology and Paleontology* (London, 1901; repr. New York, 1962); and "Über wissenschaftliche Wahrheit," in *Festrede der Bayerischen Akademie der Wissenschaften* (1902).

II. SECONDARY LITERATURE. See Charles Barrois, "Notice nécrologique sur K.-A. von Zittel," in *Bulletin de la Société géologique de France*, 4th ser., **4** (1904), 488–493; Wilhelm Branco, "Karl Alfred von Zittel," in *Monatsberichte der Deutschen geologischen Gesellschaft*, **56** (1904), 1–7; Wilhelm Deecke, "Karl Alfred von Zittel," in *Badische Biographien*, VI (Heidelberg, 1935), 380–387; Archibald Geikie, "Anniversary Address," in *Quarterly Journal of the Geological Society of London*, **60** (1904), 1v–1ix; Otto Jaekel, "K. A. von Zittel, Der Altmeister der Paläontologie," in *Naturwissenschaftliche Wochenschrift*, n.s. **3**, no. 23 (1904), 1–7; F. L. Kitchin, "Professor Karl Alfred von Zittel," in *Geological Magazine*, **41** (1904), 90–96, with complete bibliography; J. F. Pompeckj, "Karl Alfred von Zittel . . .," in *Palaeontographica*, **50** (1903–1904), 3–28, with portrait; August Rothpletz, "Gedächtnisrede auf Karl Alfred von Zittel," in *Sitzungsberichte der Bayerischen Akademie der Wissenschaften zu München*, **35** (1905), 3–23; and Charles Schuchert, "Karl Alfred von Zittel," in *Annual Report of the Board of Regents of the Smithsonian Institution* for 1904 (1905), 779–786.

HELMUT HÖLDER

ZÖLLNER, JOHANN KARL FRIEDRICH (*b.* Berlin, Germany, 8 November 1834; *d.* Leipzig, Germany, 25 April 1882), *astrophysics.*

Zöllner's father was a patternmaker and later a cotton printer. Although Zöllner had displayed outstanding talent for constructing instruments and conducting experiments by the age of sixteen, the death of his father (1853) obliged him to take over the direction of his factory. But Zöllner was not temperamentally suited to a business career; and shortly after assuming that post, he gave it up and resumed his education. He failed the final secondary-school examination, however, because of his poor marks in languages. In 1855 he began to study physics and other sciences at the University of Berlin, where his teachers included H. G. Magnus and H. W. Dove. While still a student Zöllner published "Photometrische Untersuchungen" in Poggendorff's *Annalen der Physik und Chemie*. He also worked on developing electric motors; but considering the great success later achieved in this area by Werner von Siemens, Zöllner's efforts proved to be of little significance.

Zöllner erected a small private observatory in the tower of his father's factory in Schöneweide (now part of Berlin), and thus was able to test his ideas concerning the photometry of celestial bodies. In 1857 Zöllner went to Basel, where his teachers included G. H. Wiedemann. He received the Ph.D. in 1859 for a work on photometric problems: "Photometrische Untersuchungen, insbesondere über die Lichtentwickelung galvanisch glühender Plantindrähte."

Exploiting a chance observation that he had made at Basel, Zöllner invented the astrophotometer, which was constructed in the Kern optical and mechanical workshop in Aarau. (The accompanying drawing illustrates the instrument's operating principle.) Using this instrument, he investigated fundamental problems of photometry, made critical comparisons of other photometers, and soon

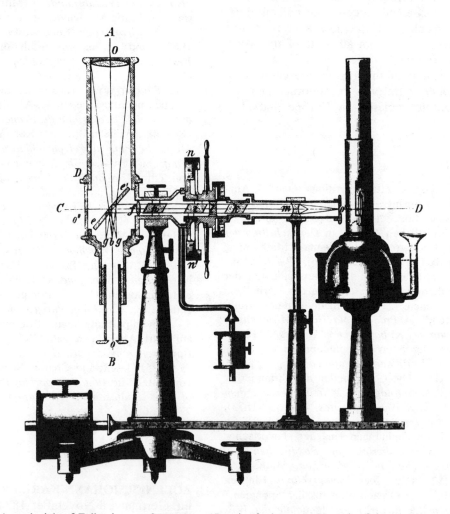

FIGURE 1.   Operating principle of Zollner's astrophotometer. *AB*, axis of telescope; *CD*, axis of the arrangement for production of the artificial star; *m*, biconcave lens; *k*, Nicol prism; *l*, optical crystal; *i,k*, Nicol prisms; *f*, convex lens; *eé*, glass plate; *F*, petroleum lamp; *O*, objective; *o*, eyepiece; *b*, image of the star; *gg*, image of the artificial star. SOURCE: G. Müller, *Die Photometrie der Gestirne* (Leipzig, 1897), p. 247.

amassed a considerable body of material. On the advice of Mitscherlich and Wiedemann, Zöllner decided to compete for a prize offered by the Vienna Academy of Sciences. The jury found that Zöllner had not measured the brightness of enough stars to receive the prize. (The other two papers submitted were also found not to merit the prize, and thus none was awarded.) Zöllner's entry, which he published in 1861, is a classic of astrophysics. The photometer described in it far surpassed all its predecessors and, in a modified form, found wide application. The Potsdam observatory used it to obtain data of unsurpassed precision for its *Photometrische Durchmusterung des nördlichen Himmels*, the publication of which was begun at Zöllner's suggestion.

In 1862 Zöllner moved to Leipzig and published *Photometrische Studien mit besonderer Rücksicht auf die physische Beschaffenheit der Himmelskörper* (1865), which contains his *Habilitationsschrift*, "Theorie der relativen Lichtstärken der Mondphasen." With his appointment as professor at the University of Leipzig in 1866, Zöllner's financial situation was sufficiently improved for him to resume important experimental research.

Among Zöllner's main achievements was the design of the reversion spectroscope (1869), another instrument that demonstrates his experimental ingenuity. It is based on the principle of the heliometer: two beams of rays are conducted through two direct-vision prism systems arranged so that the dispersion within them occurs in mutually opposed directions. With this device Zöllner intended to improve the precision of measurements of Doppler shifts in the spectra of objects with velocities having a high radial component. Hermann Vogel used the instrument to determine the rotational velocity at the solar equator. The reversion spectroscope later lost its importance with the adoption of more exact methods.

Another important device designed by Zöllner is the horizontal pendulum, which in improved form was widely used in geophysical research. Inspired by the attempts of Janssen and Lockyer to observe solar protuberances, Zöllner devised the first method that made these phenomena easily amenable to study. These inventions, which make Zöllner a pioneer in astrophysics, brought him membership in the Saxon Academy of Sciences at Leipzig.

Zöllner also made an intensive study of theoretical questions, including solar theory, sunspots, and solar rotation, and Olber's paradox. One product of these rather speculative inquiries was especially important for the development of spectroscopy, the memoir "Über den Einfluss der Dichtigkeit und Temperatur auf die Spektra glühender Gase" (1870). Also of far-reaching significance was Zöllner's theory of comets, in which he correctly assumed that elements of the nucleus of a comet gradually vaporize as it nears the sun. Beyond this, the book on comets contains a wealth of penetrating remarks on the subject announced in the subtitle, *Beiträge zur Geschichte und Theorie der Erkenntnis*. This portion of the book is notable for a number of at least partly original ideas and critical comments. Attacking abuses in the scientific profession of his time, Zöllner lashed out with great vehemence against the vanity of scientists, ridiculing scientific careerism and contending that these vices are harmful to the progress of science.

Zöllner continued this polemic in such works as *Das deutsche Volk und seine Professoren. Eine Sammlung von Citaten ohne Kommentar* (Leipzig, 1880). The scientific community responded with counterattacks, and there began a long period of controversies that drove Zöllner into ever greater isolation. The excessive irony and somewhat biased approach of which he was guilty were, however, only partially responsible for this isolation. In 1875 Zöllner had met William Crookes in London and begun an intensive study of Spiritualism, to which he ultimately became a convert. In the following years some scientists did not hesitate to attribute Zöllner's views to an increasingly serious mental illness. Even sympathetic friends, such as Otto Struve, were much dismayed by his adherence to this doctrine, and W. Foerster termed this change in his life both remarkable and painful. Zöllner, however, refused to be dissuaded from pursuing his unscientific speculations; and he saw in his supposed proofs of the existence of a "transcendental world" a support for theology. In the last years of his life he produced little work of scientific significance. He died—presumably of a stroke—while preparing the preface to the third edition of his book on comets.

BIBLIOGRAPHY

I. ORIGINAL WORKS. A full list of Zöllner's major works can be found in F. Koerber, *Zöllner*, 106–107 (below). The papers by Zöllner in the *Astronomische Nachrichten* are cited in *Generalregister der Bände 41–80 der Astronomischen Nachrichten*, H. Kobold, ed. (Kiel 1938), col. 118; and in *Generalregister der Bände 81–120 der Astronomischen Nachrichten*, A. Krueger, ed. (Kiel, 1891), col. 132. Zöllner collected his most important shorter works as *Wissenschaftliche Abhandlungen*, 4 vols. (Leipzig, 1878–1881).

II. SECONDARY LITERATURE. See W. Foerster, *Lebenserinnerungen und Lebenshoffnungen* (Berlin, 1911), 95–98; D. B. Herrmann, "Karl Friedrich Zöllner und die 'Potsdamer Durchmusterung,'" in *Sterne*, **50** (1974), 170–180; and "Ein eigenhändiger Lebenslauf von Karl Friedrich Zöllner aus dem Jahre 1864," *Mitteilungen der Archenhold-Sternwarte Berlin-Trepton*, **97** (1974); R. Knott, "Zöllner," in *Allgemeine deutsche Biographie*, XLV (Leipzig, 1900), 426–428; F. Koerber, *Karl Friedrich Zöllner* (Berlin, 1899), no. 53 of Sammlung Populärer Schriften, edited by the Gesellschaft Urania, Berlin; S. (probably W. Scheibner), "Todes-Anzeige," in *Astronomische Nachrichten*, **102** (1882), cols. 175–176; and M. Wirth, *Friedrich Zöllner. Ein Vortrag mit Zöllners Bild und Handschrift* (Leipzig, 1882).

DIETER B. HERRMANN

ZOLOTAREV, EGOR IVANOVICH (*b*. St. Petersburg, Russia [now Leningrad, U.S.S.R.], 12 April 1847; *d*. St. Petersburg, 19 July 1878), *mathematics.*

Zolotarev was the son of a watchmaker. After graduating in 1863 with a silver medal from the Gymnasium, he enrolled at the Faculty of Physics and Mathematics of St. Petersburg, where he attended the lectures of Chebyshev and his student A. N. Korkin. He graduated with the candidate's degree in 1867 and the following year became assistant professor there. In 1869 he defended his master's dissertation, on an indeterminate third-degree equation; his doctoral dissertation (1874) was devoted to the theory of algebraic integers. In 1876 he was appointed professor of mathematics at St. Petersburg and junior assistant of applied mathematics at the St. Petersburg Academy of Sciences.

On two trips abroad Zolotarev attended the lectures of Kummer and Weierstrass, and met with Hermite. He shared his impressions of noted scholars and discussed mathematical problems with Korkin, whose collaborator he subsequently became. Zolotarev died at the age of thirty-one, of blood poisoning, after having fallen under a train.

The most gifted member of the St. Petersburg school of mathematics, Zolotarev produced fundamental works on mathematical analysis and the theory of numbers during his eleven-year career. Independent of Dedekind and Kronecker, he constructed a theory of divisibility for the whole numbers of any field of algebraic numbers, working along the lines developed by Kummer and elaborating the ideas and methods that now comprise the core of local algebra. He operated with the

numbers of the local ring Z_p and its full closure in the field $Q(\theta)$ and, in essence, brought under examination the semilocal ring O_p. In modern terminology Zolotarev's results consisted in proving that (1) the ring O_p is a finite type of Z_p-modulus and (2) O_p is a ring of principal ideals. In his local approach to the concept of a number of the field $Q(\theta)$ Zolotarev demonstrated that the ring O of the whole numbers in $Q(\theta)$ is the intersection of all semilocal rings O_p. Zolotarev defined ideal numbers in O as essentially valuations and found the simple elements of O_p with the aid of a lemma that is the analog of the theory of expansion into Puiseux series.

Zolotarev employed a theory that he had constructed for determining, with a finite number of operations, the possibility of selecting a number, A, such that the second-order elliptical differential $(x + A)\ dx\ /\sqrt{R(x)}$, where $R(x)$ is a fourth-degree polynomial with real coefficients, can be integrated in logarithms. Abel demonstrated that for an affirmative solution it is necessary and sufficient that $\sqrt{R(x)}$ be expandable into a periodic continuous fraction; but because he did not give an evaluation of the length of the period, his solution was ineffective. Zolotarev provided the required evaluation, applying the equation of the division of elliptic functions.

With Korkin, Zolotarev worked on the problem posed by Hermite of determining the minima of positive quadratic forms of n variables having real coefficients; they gave exhaustive solutions for the cases $n = 4$ and $n = 5$.

Among Zolotarev's other works are an original proof of the law of quadratic reciprocity, based on the group-theoretic lemma that Frobenius had called "the most interesting," as well as solutions of difficult individual questions in the theory of the optimal approximation of functions. Thus, Zolotarev found the nth-degree polynomial, the first coefficient of which is equal to unity and the second coefficient of which is fixed, that deviates least from zero.

BIBLIOGRAPHY

I. ORIGINAL WORKS. Zolotarev's complete writings were published by the V. A. Steklov Institute of Physics and Mathematics as *Polnoe sobranie sochineny*, 2 vols. (Leningrad, 1931–1932); see esp. II, 72–129; and "Sur la théorie des nombres complexes," in *Journal de mathématiques pures et appliquées*, 3rd ser., **6** (1880), 51–84, 129–166.

II. SECONDARY LITERATURE. On Zolotarev's life and

work, see I. G. Bashmakova, "Obosnovanie teorii deli-mosti v trudakh E. I. Zolotareva" ("Foundation of the Theory of Divisibility in Zolotarev's Works"), in *Istoriko-matematicheskie issledovaniya*, **2** (1949), 231–351; N. G. Chebotarev, "Ob osnovanii teorii idealov po Zolotarevu" ("On the Foundation of the Theory of Ideals According to Zolotarev"), in *Uspekhi matematicheskikh nauk*, **2**, no. 6 (1947), 52–67; B. N. Delone, *Peterburgskaya shkola teorii chisel* ("The St. Petersburg School of the Theory of Numbers"; Moscow–Leningrad, 1947); R. O. Kuzmin, "Zhizn i nauchnaya deyatelnost Egora Ivanovicha Zolotareva" ("Zolotarev's Life and Scientific Activity"), in *Uspekhi matematicheskikh nauk*, **2**, no. 6 (1947), 21–51; and E. P. Ozhigova, *Egor Ivanovich Zolotarev* (Moscow–Leningrad, 1966).

I. G. BASHMAKOVA

ZOSIMUS OF PANOPOLIS (*b.* Panopolis [now Akhmīm], Egypt; *d.* Alexandria, Egypt; *fl. ca.* A.D. 300), *alchemy.*

Zosimus appears to be the earliest genuine historical figure mentioned as an author in the Greek alchemical texts. Almost nothing is known about his life. That he came from Panopolis, in the Thebaid (Upper Egypt), is known from the extant texts, as well as from two of the three nonalchemical authors who mention him: Photius (ninth century) and Georgius Syncellus (eighth–ninth century). The third source, the *Suda* (formerly called the *Suidas*, about 950), refers to him as Alexandrian, but this undoubtedly means that he later lived in Alexandria. No source gives his dates. He is generally thought to have lived somewhat earlier than the alchemist Synesius, whose dates have been established by his having sent a book to Dioscorus, the high priest of the Serapeum in Alexandria, which was destroyed in 389. Accordingly, Zosimus is presumed to have lived around A.D. 300.

Books in which the author's name is given as Zosimus are preserved in Greek, Syriac, and Arabic. Syncellus mentions the book *Imuth* and cites from it a story (known from the Book of Enoch, chapters 6–8) in which alchemy, along with other arts, is revealed to mortal women by the fallen angels, who seek to win their favor in this way. This story is repeated in one of the Syriac texts by Zosimus, and the title *Imuth* occurs in various places in the Syriac texts (see *La chimie au moyen âge*, II, 238). The citations in the latter are inconsistent, however, and it is therefore impossible to infer anything with certainty about the nature of this book.

The *Suda* mentions a much more famous work, the twenty-eight books addressed to Zosimus' (spiritual) sister Theosebeia, each of which is sup-posedly designated by a different letter of the alphabet. The book *Omega* is found among the surviving Greek texts; and it, along with the books called *Kappa* and *Sigma*, is cited in certain Greek texts. In various books Theosebeia is addressed by name and by the phrase "O woman." Her name also appears in the Syriac texts, as does the second-person feminine form of address. Since the Greek alphabet contains only twenty-four letters, the designation of four of the books remains problematic.

Also unclear is the relationship between these twenty-eight books and the thirty-five chapters addressed to Eusebius, which are cited in the list of alchemical writings in *Codex Marcianus* 299. This list, which obviously records the contents of the codex (the oldest one preserved—it dates from the eleventh century) as it existed at an earlier date, mentions several works of Zosimus that are preserved. Among these are the book *On "Arete"* (outstanding quality or, perhaps, peculiarity), *On the Composition of Waters*, and fifteen chapters addressed to Theodorus, known as *On Tools and Ovens*. Another extant work in the list is the commentary by Olympiodorus on Zosimus' κατ' ἐνέργειαν, but it cannot be determined from the commentary how this title ought to be translated. Berthelot classified the book among the "traités démocritains."

The grouping of texts in Berthelot's *Collection des anciens alchimistes grecs* does not give a clear idea of which texts were really written by Zosimus. Among those under the name "Zosimus," several contain only citations from Zosimus; and for others not even this much is true. On the other hand, many of the texts attributed to Zosimus include citations from later authors. Further, Berthelot did not collect the citations from Zosimus that appeared in other books not published under his name, nor did he carry out a systematic comparison of the Syriac and Greek texts. Finally, the Syriac texts are given only in a French abridgment, not in the original.

As a result, it is not clear which texts can rightly be attributed to Zosimus. And the incomprehensibility of many of the writings makes textual criticism all the more difficult.

Of the approximately twenty Arabic book titles that F. Sezgin has assigned to Zosimus, some sound as if they might be translations from the Greek, while others appear to be Arabic forgeries. Something is known about the contents of only one of these books—thanks to H. E. Stapleton; the assertion found in this book that it was translated in the year A.H. 38 (A.D. 659) does not fit its context and is obviously false. Many citations from Zosimus can be found in published Arabic texts or in their Latin translation, sometimes with his name and sometimes, as in the *Turba philosophorum*, without it. The citations in the latter book are mentioned by J. Ruska in his notes to the translation. Sezgin's remarks on the individual titles require more careful examination.

Given the absence of a critical monographic study of Zosimus, it is scarcely possible to present an accurate account of his teachings. His alchemical statements sometimes take the form of visions, in which chemical apparatus is represented as a temple and the metals are personified (for example, as "lead man" and "copper man"). The alchemical operation itself is viewed as a sacrificial act. Elsewhere, his remarks bear the stamp of a kind of natural philosophy but are confined to general discussions. In any case, because of Zosimus' frequently allegorical style, it is not evident whether his subject is alchemy or religion. On the whole, his works seem to be the expression of a mystic religion that is almost never entirely eliminated even in the seemingly technical sections. He warns Theosebeia about deceitful prophets and against according the art too high a value. He advises her to cool her passion and to resist desire (for gold). Genuine and natural change of color is obtained by public worship. In this connection he cites two treatises from the *Corpus Hermeticum*: the *Poimandres* (I) and the *Crater* (IV).

Despite this attitude, Zosimus fulfills Theosebeia's wish and describes the apparatus and the ovens used in alchemical work. He warns her, however, to beware of persons who have misled her into adopting a contrary method of alchemical practice—people who love gold more than reason. He derides a priest named Nilus and is especially pleased with his lack of success.

Zosimus, who in the later texts is called "the Old One," "the Divine," and "the Most Learned," shows in all his writings a great reverence for the ancient masters of the art: Hermes, Agathodaemon, Zoroaster, Democritus, Ostanes, and Maria. On the whole, he gives the impression of having been a pagan, and E. Riess is certainly correct in ascribing the Christian passages in his books to later writers. Zosimus calls himself unoriginal, simply a compiler and commentator; but this is not necessarily sincere. He may have been seeking to win more confidence in his writings or perhaps was following an established literary practice.

Zosimus' works have attracted the interest of a number of historians of religion, who have investi-

gated his relations to Gnosticism and especially to the Hermetic tradition (for example, R. Reitzenstein, W. Scott, H. Jonas, and A. J. Festugière). Ruska has been particularly concerned with the Hermetic tendencies in Zosimus' thought; he and Festugière have edited and translated some of his works, making many improvements in Berthelot's texts. The extensive use of pseudonyms in these writings makes it impossible to give an account of Zosimus' own chemical theories until further progress is made in the study of the entire corpus of Greek alchemical texts. Only then will scholars be able to judge the originality of Zosimus' ideas—a problem that has been studied especially by I. Hammer Jensen.

BIBLIOGRAPHY

See M. Berthelot, *Les origines de l'alchimie* (Paris, 1885); *Collection des anciens alchimistes grecs* (Paris, 1888; repr. London, 1963); and *La chimie au moyen âge*, II and III (Paris, 1893); *Catalogue des manuscrits alchimiques grecs*, II (Brussels, 1927); A. J. Festugière, "Alchymia," in *Antiquité classique*, **8** (1939), 71 ff.; and *La révélation d'Hermès Trismégiste*, I (Paris, 1944; 2nd ed., 1950); W. Gundel, "Alchemie," in *Real-Lexikon für Antike und Christentum*, I (1950); I. Hammer Jensen, *Die älteste Alchymie* (Copenhagen, 1921); H. Jonas, *Gnosis und spätantiker Geist* (Göttingen, 1934–1964); E. O. von Lippmann, *Entstehung und Ausbreitung der Alchemie* (Berlin, 1919), 75–92; A. D. Nock and A. J. Festugière, eds. and trans., *Corpus Hermeticum*, IV (Paris, 1954), 117–121; R. Reitzenstein, *Poimandres* (Leipzig, 1904); E. Riess, "Alchemie," in *Real-Encyclopädie der classischen Altertumswissenschaft*, I (1894); J. Ruska, *Tabula smaragdina* (Heidelberg, 1926); "Zosimos," in G. Bugge, ed., *Das Buch der grossen Chemiker* (Berlin, 1929); and *Turba philosophorum* (Berlin, 1931); W. Scott, *Hermetica*, IV (Oxford, 1936); F. Sezgin, *Geschichte des arabischen Schrifttums*, IV (Leiden, 1971), 73–77; and H. E. Stapleton and R. F. Azo, "An Alchemical Compilation of the Thirteenth Century A.D.," in *Memoirs of the Asiatic Society of Bengal*, **3**, no. 2 (1910).

M. PLESSNER

ZSIGMONDY, RICHARD ADOLF (*b.* Vienna, Austria, 1 April 1865; *d.* Göttingen, Germany, 24 September 1929), *colloidal chemistry*.

Zsigmondy was a figure of paramount importance on colloid chemistry during the first quarter of the twentieth century. His receipt of the Nobel Prize in 1925, for invention of the ultramicroscope and his work on colloids, was the first time this fledgling science had been so honored. In 1926 work by J. B. Perrin and Theodor Svedberg that followed directly from Zsigmondy's achievement was recognized by the Nobel Prizes in physics and in chemistry, respectively. No Nobel Prize since then has been awarded for work solely in colloid chemistry.

Zsigmondy was the son of Adolf Zsigmondy, a dentist, and Irma von Szakmáry. He spent his childhood, his school years, and his first years as a university student in Vienna. He took his Ph.D. in organic chemistry at the University of Munich in 1890. With this, his activity in organic chemistry ended. Neither was he influenced by the great schools of physical chemistry developing in the Netherlands under van't Hoff and at Leipzig under Ostwald. At Berlin, Zsigmondy worked on inorganic inclusions in glass with the physicist A. A. Kundt (1891–1892) and then, at Graz, as a lecturer on chemical technology at the Technische Hochschule until 1897, when he joined the Schott Glass Manufacturing Company in Jena. There he was concerned with colored and turbid glasses, and invented the famous Jena milk glass. Zsigmondy left industrial work in 1900 to pursue private research that led to the invention of the ultramicroscope and his classic studies on gold sols. His achievements during this period led to his being called to Göttingen as professor of inorganic chemistry in 1907.

Zsigmondy became interested in colloids through work with glasses that owed their color and opacity to colloidal inclusions. He soon recognized that the red fluids first prepared by Faraday through reduction of gold salts are largely analogues of ruby glasses, and he developed techniques for preparing them reproducibly. These gold sols became the model systems used in much of his work.

The presence of colloidal particles is apparent from the cone of scattered light known as the Tyndall beam. Zsigmondy's great contribution was the invention of the ultramicroscope, which rendered the individual particles visible. In the ultramicroscope, ordinary illumination along the microscope's axis is replaced by illumination perpendicular to the axis. With such dark-field illumination the individual particles are rendered luminous by the scattered light that reaches the eye of the observer, in much the same fashion that moving dust particles are illuminated in a sunbeam. This achievement had been rejected for particles much smaller than the resolving power of the microscope. Certainly the exaltation Zsigmondy experi-

enced could hardly have been less than that of any other intrepid explorer who reveals a new universe. "A swarm of dancing gnats in a sunbeam will give one an idea of the motion of the gold particles in the hydrosol of gold. They hop, dance, jump, dash together, and fly away from each other, so that it is difficult in the whirl to get one's bearings."

Although dark-field illumination had long been a recognized procedure in microscopy, many difficult technical problems remained. Zsigmondy was assisted by H. F. W. Siedentopf, a physicist with the Zeiss Company of Jena, in the design and construction of the apparatus. The company's director, Ernst Abbe, placed the facilities of the Zeiss plant at their disposal even though Zsigmondy was not associated with the company. Indeed, this activity came at a time in his career when he had no professional attachment.

Much of Zsigmondy's research was devoted to the investigation of gold sols and particularly purple of Cassius. Although it had been investigated by a number of noted chemists, no decision had been reached whether this peculiar preparation, valued as a glass paint, is a mixture or a compound. In 1898, Zsigmondy was able to show that it is a mixture of very small gold and stannic acid particles, and later directly confirmed the correctness of this finding with his ultramicroscope.

Zsigmondy investigated the color changes that occur in gold sols upon the addition of salts and studied the inhibition of these effects upon the addition of such protective agents as gelatin and gum arabic. With the aid of the ultramicroscope, he demonstrated that the color changes reflected alteration of particle size due to coagulation and that the protective agents acted so as to inhibit the coagulation.

At Göttingen, Zsigmondy was occupied with ultrafiltration, which he developed as another useful tool for the investigation of colloidal systems. He explored a broad range of substances, especially silica gels and soap gels.

A lover of nature and an avid mountain climber, Zsigmondy acquired an estate near Terlano, in the southern Tirol, to which he retreated frequently. He married Laura Luise Müller, the daughter of Wilhelm Müller, in 1903; they had two daughters.

BIBLIOGRAPHY

Zsigmondy's books include *Zur Erkenntnis der Kolloide* (Jena, 1905), translated by Jerome Alexander as *Colloids and the Ultramicroscope* (New York, 1909);

Kolloidchemie; ein Lehrbuch (Leipzig, 1912), 5th ed., 2 vols. (Leipzig, 1925–1927), translated by E. B. Spear as *The Chemistry of Colloids* (New York, 1917); and *Das kolloide Gold* (Leipzig, 1925), written with P. A. Thiessen.

His memoirs and other writings are listed in Poggendorff, IV, 1695–1696; V, 1414; and VI, 2971; for a comprehensive bibliography of secondary literature, see VIIa supp., 796.

MILTON KERKER

ZUBOV, NIKOLAY NIKOLAEVICH (*b*. Izmail, Russia [now Ukrainian S.S.R.], 23 May 1885; *d*. Moscow, U.S.S.R., 11 November 1960), *oceanography*.

The son of an officer, Zubov followed family tradition and enrolled in the Naval Cadet Corps, from which he graduated in 1904. He then attended the Naval Academy, where his interest turned to hydrography. After graduating in 1910, Zubov served as navigator and commander of a torpedo boat. In 1912, on a cruise in the Barents Sea aboard the messenger ship *Bakan*, Zubov carried out his first scientific work, a plane-table and hydrographic survey of Matyushikha Inlet, on Novaya Zemlya, and of the lower course and mouth of the Pesha River, at the Cheshska Gulf. The following year he entered civil service as hydrographer in the office of commercial ports of the Ministry of Trade and Industry. Soon afterward he participated in the scientific expedition to the Baltic Sea of the schooner *Utro*, and in 1914 he was sent to Norway to continue his education at the geophysics institute at Bergen.

World War I interrupted Zubov's studies and he returned to military service as commander of the torpedo boat *Burny*; after the war he served at naval headquarters. Zubov was one of the first members of the Floating Marine Scientific Institute (later the State Institute of Oceanography), founded in 1921 to study the Soviet Union's northern seas; he was also director of the hydrological section. In 1923 the institute acquired its own expedition ship, the *Persia*, which, with Zubov's participation, completed four hydrographic sections of the southwest Barents Sea by 1928, using only modest equipment (two bathometers, each containing a thermometer, and two rotors). Zubov presented his scientific conclusions in a series of articles (1929–1932) that were later generalized in a fundamental memoir (1932); this work deals with the three major problems that retained his interest

throughout his life: the vertical mixing of seawater, ocean currents, and sea ice.

Zubov explained the origin of intermediate layers of warm and cold seawater and gave a theoretical basis and practical method for estimating the intensity of the vertical circulation during the winter. His findings were immediately applied to explain the degree of aeration of deep-sea layers, data needed in estimating fish reserves and their food resources. His subsequently elaborated concept of the index of freezing was important in compiling forecasts of sea ice.

Zubov also reexamined and reinterpreted the theoretical basis of Wilhelm Bjerknes' indirect method of determining the elements of ocean currents. Whereas Bjerknes believed the heterogeneity of density to be the cause of currents, Zubov considered it, to a much greater extent, a consequence of them. Zubov demonstrated that Bjerknes' method failed to reckon fully with density currents (or "convection currents" as they were then known) and, in general, with stationary currents of any origin. Zubov called his method of calculating currents according to the distribution of density the "dynamic method of processing oceanologic observations." His practical handbook on the subject (1935) led to a series of "dynamic maps" of the Soviet Union's northern seas that gave the first sufficiently accurate picture of their currents.

In his 1932 memoir Zubov also included his theory of climatic changes of currents, described their influence on sea ice, and gave the basis for forecasting ice in the polar seas. Internal waves and high tides were also considered; the latter subject was also dealt with in an important monograph (1933) that includes a rigorous exposition of the modern theory of high tides.

As secretary of the Soviet committee of the Second International Polar Year (1932–1933), Zubov was responsible for introducing into the program Soviet studies of the Arctic Ocean and of its northern seas, from the Greenland to the Chukchi seas, as well as the Bering and Okhotsk seas and the Sea of Japan. During the summer of 1932 he participated in a cruise aboard the small sail-and-motor-powered trawler, the *N. M. Knipovich*. The excursion confirmed the accuracy of Zubov's forecast that ice conditions would be extremely light in the Barents Sea, even at the highest latitudes; and the *Knipovich*—although completely unfit for high-latitude navigation—attained the latitude of 82°05′ and was the first ship ever to circumnavigate Franz Joseph Land.

In 1935 Zubov headed the scientific section of the first Soviet high-latitude expedition of the ice-breaker *Sadko*, which explored the upper latitudes of the Greenland, Barents, and Kara seas, and charted the paths of the deep northern currents that penetrate the warm waters of the Atlantic in the Kara Sea.

He also engaged in aerial reconnaissance of Arctic ice, proposing early spring surveys of separate regions and reconnaissance of an entire route "with only the naked eye." He completed his last flight over the Arctic practically on the eve of his seventieth birthday. Like the others it yielded new information used to maintain navigation through the northern sea routes.

Having returned to naval service at the beginning of World War II, Zubov was sent to Arkhangelsk, charged with maintaining an open passage through the ice at the mouth of the Northern Dvina River for the White Sea military flotilla. In an original work on the subject (1942) Zubov started from the assumption that in certain cases ice may be considered not as an elastic, but as a plastic, body. His formula based on this proposition for determining the flexure of ice under a given load was subsequently confirmed in practice. In his important monograph on Arctic ice (1945) Zubov presented the laws of the origin, growth, movement and hummocking, weakening, and melting of the Arctic ice cover. Central to the work is his conception of the unity of the "air-ice-water" system, of the close interdependence of phenomena in these areas, and of the continually changing condition of the ice. Of special importance to navigation were his method for predicting the depth of ice, his rule correlating the drift of ice with isobars, and his method of calculating the winter vertical circulation of water and the index of freezing. Certain conclusions that were not based on direct observation at the time they were formulated have subsequently been fully confirmed. For example, beginning with his general laws of ice drift, Zubov determined the presence, in that sector of the central polar basin bordering the Pacific Ocean, of an anticyclonic system of ice drift along a gigantic and, in some instances, closed curve. The presence of this circular motion was established over a decade later by the drifting of Soviet and American stations at the North Pole.

Zubov harbored a lifelong dislike of the term "oceanography," considering it equivalent only to a descriptive study of the sea. He considered himself an "oceanologist," who attempted to penetrate the very essence of the processes that he studied. In 1931 he created the world's first department of oceanology at the Hydrometeorological Center of the U.S.S.R. From 1949 to 1960 he was professor of oceanology at the University of Moscow.

Zubov's abiding concern with observation was expressed in his more than 200 publications. In September 1959, a year before his death, the results of his last two studies were presented to the First International Oceanographic Congress in New York. They dealt with the relation between barometric relief and sea level, and condensation as a result of the intermingling of seawater (written with K. D. Sabinin).

BIBLIOGRAPHY

I. ORIGINAL WORKS. An edition of Zubov's selected works on oceanology was published on his seventieth birthday as *Izbrannye trudy po okeanologii* (Moscow, 1955). His earlier writings include "Gidrologischeskie raboty Plavuchevo morskogo nauchnogo instituta v yugo-zapadnoy chasti Barentseva morya letom 1928 g. na e/s *Persey*" ("Hydrological Work of the Floating Marine Scientific Institute in the Southwest Part of the Barents Sea in the Summer of 1928 Aboard the Expedition Ship *Persia*"), in *Trudy Gosudarstvennogo okeanograficheskogo instituta*, **2**, no. 4 (1932); *Elementarnoe uchenie o prilivakh v more* ("An Elementary Study of High Tides"; Moscow, 1933); *Okeanograficheskie tablitsy* ("Oceanographic Tables"; Moscow, 1931), based on Knudsen's *Hydrographische Tabellen*; 2nd ed., enl., as *Okeanologicheskie tablitsy* ("Oceanographic Tables"; 1940), with supplementary *Okeanologicheskie grafiky* ("Oceanologic Graphs"; 1941), written with K. M. Sirotov; 3rd ed., rev. and enl. (Moscow–Leningrad, 1957); *Dinamichesky metod obrabotki okeanologicheskikh nablyudeny* ("The Dynamic Method of Processing Oceanologic Observations"; Moscow, 1935); *Morskie vody i ldy* ("Seawater and Sea Ice"; Moscow, 1938); *Morya zemnogo shara* ("Seas of the Terrestrial Globe"), appendix to the index of geographic names in the *Great Soviet World Atlas* (Moscow, 1940), written with A. V. Everling; *Ldy Arktiki* ("Arctic Ice"; Moscow–Leningrad, 1945); also in English trans. (San Diego, Calif., n.d. [1963]); *Dinamicheskaya okeanologia* ("Dynamic Oceanology"; Moscow–Leningrad, 1947); *V tsentre Arktiki* ("In the Middle of the Arctic"; Moscow–Leningrad, 1948); *Otechestvennye moreplavateliissledovateli morey i okeanov* ("Native Navigator-Investigators of the Seas and Oceans"; Moscow, 1954); *Dinamichesky metod vychislenia elementov morskikh techeny* ("Dynamic Method of Calculating the Elements of Ocean Currents"; Leningrad, 1956); *Osnovy uchenia o prolivakh Mirovogo okeana* ("Basic Study of Straits of the World's Oceans";

Moscow, 1956); and *Uplotnenie pri smeshenii morskikh vod raznoy temperatury i solenosti* ("Condensation as a Result of Mixing Seawater of Different Temperatures and Salinity"; Leningrad, 1957).

II. SECONDARY LITERATURE. On Zubov and his work, see A. D. Dobrovolsky, "N. N. Zubov—Okeanolog" ("N N. Zubov—Oceanologist"), in Zubov's *Izbrannye trudy* (see above), 5–11; and "N. N. Zubov—odin iz krupneyshikh sovetskikh okeanologov" (". . . One of the Most Outstanding Soviet Oceanologists"), in *Okeanologia*, **1** no. 2 (1961), 355–359; and B. L. Lagutin, A. M. Muromtsev, and A. A. Yushchak, "Pamyati Nikolaya Nikolaevicha Zubova" ("In Memory of . . ."), in *Meteorologia i gidrologia* (1961), no. 5, 59–60.

A. F. PLAKHOTNIK

ZUBOV, VASILY PAVLOVICH (*b.* Aleksandrov, Ivanovo province [now Ivanovskaya oblast], Russia, 1 August 1899; *d.* Moscow, U.S.S.R., 8 April 1963), *history of science.*

The son of a professor of chemistry at the University of Moscow, Zubov graduated in 1922 from the Faculty of History and Philology at Moscow; he subsequently worked at the Academy of Artistic Sciences and, from 1935 to 1945, at the Academy of Architecture. His main interest, the artistic, technical, and architectural ideas of the Middle Ages and the Renaissance, was focused on Alberti, Barbaro, and Leonardo da Vinci, whose works he helped to publish in Russian. He received the doctorate of art in 1946 for his research on Alberti's theory of architecture.

Zubov's interest had gradually turned to the scientific and philosophical literature of that time, and in 1945 he transferred to the Institute of the History of Sciences and Technology of the U.S.S.R. Academy of Sciences. In December 1946 he presented a critique of Duhem's concept of medieval science and proposed a program of further work in this area. In carrying it out, he conducted many investigations in the history of atomic theory and in the development of mathematics, physics, and mechanics. Related to these studies were his specialized work on the development of philosophical thought in Russia from the eleventh to the seventeenth centuries and a number of general investigations of the physics of antiquity, the Middle Ages, and the Renaissance. His extensive editorial activity included his work as editor of a three-volume history of the natural sciences in Russia (1957–1962).

Zubov was elected corresponding member of the International Academy of the History of Science in 1958 and became an active member two years later. In 1963 he was posthumously awarded the George Sarton Medal of the History of Science Society.

BIBLIOGRAPHY

I. ORIGINAL WORKS. Zubov's more than 200 writings include *Istoriografia estestvennykh nauk v Rossii* ("Historiography of the Natural Sciences in Russia"; Moscow, 1956); *Ocherki razvitia osnovnykh fizicheskikh idey* ("Sketches of the Development of Basic Physical Ideas"; Moscow, 1959); *Leonardo da Vinci* (Leningrad, 1961), English trans. by D. H. Kraus (Cambridge, Mass., 1968); *Ocherki razvitia osnovnykh ponyaty mekhaniki* ("Sketches in the Development of Basic Concepts of Mechanics"; Moscow, 1962), written with A. T. Grigorian; *Aristotle* (Moscow, 1963); and *Razvitie atomisticheskikh predstavleny do nachala XIX veka* ("Development of Atomic Ideas to the Beginning of the Nineteenth Century"; Moscow, 1965), with a detailed bibliography of Zubov's works, 360–370.

II. SECONDARY LITERATURE. On Zubov and his work, see A. T. Grigorian, B. G. Kuznetsov, and A. P. Youschkevitch, "Vassili Pavlovitch Zoubov," in *Archives internationales d'histoire des sciences*, **16** (1963), 305–306; C. Maccagni, "Vasilij Pavlovic' Zubov, 1899–1963," in *Physics*, **5** (1963), 333–339; and A. P. Youschkevitch, "Vassili Zoubov, homme et savant," in *Actes del' XIe Congrès international d'histoire des sciences*, I (Wrocław–Warsaw–Cracow, 1968), 34–40.

A. T. GRIGORIAN
A. P. YOUSCHKEVITCH

ZUCCHI, NICCOLÒ (*b.* Parma, Italy, 6 December 1586; *d.* Rome, Italy, 21 May 1670), *mathematics, theology.*

Zucchi taught rhetoric, and later theology and mathematics, at the Jesuits' Roman College, of which he was also rector and from which he moved to the one in Ravenna. Returning to Rome, he held the office of preacher in the Apostolic Palace for at least seven years and was in charge of his religious order's mother house. Because of the esteem in which he was held, he was a member of the retinue of the papal legate sent to the court of the Emperor Ferdinand II, where he met Kepler; Zucchi considered this event one of the most important in his life.

Zucchi is remembered today for his research, "partly the fruit of experiment and partly of reasoning," in optics. In 1616 (or perhaps 1608) he had constructed an apparatus in which an ocular lens was used to observe the image produced by reflection from a concave metal mirror. This was one of the earliest reflecting telescopes, in which the enlargement is obtained by the interaction of mirrors and lenses.

Later, in *Optica philosophia* . . ., Zucchi described the apparatus, from which, wittingly or not, the most improved models of a slightly later date were derived (those of Gregory and Newton, for instance).

This apparatus enabled Zucchi to make a more thorough examination of the spots on Mars (1640), observed four years earlier by F. Fontana, and thus to supply material for Cassini's discovery of the rotation of that planet (1666).

Zucchi worked in a period of contradictory thought and scientific investigation. Alongside the clarity of Galileo's ideas were beliefs at once highly ingenuous and abstruse, as well as extravagant errors.

Hence, Zucchi accepted strange astronomical theories, which he expounded with the utmost certainty in his sermons. In the cathedral of Pisa (Galileo's native city) he asserted in 1638 that the sun is further from the earth during the summer than in winter and that this is proved by the need to alter the length of the telescope in those seasons in order to be able to observe sunspots. But this was not enough: he stated that Venus is nearer the sun than Mercury is, because the former represents beauty and the latter skill. Such statements elicited laughter in some circles but were simultaneously accepted in others as a sign of profound doctrine.

BIBLIOGRAPHY

Zucchi's main work is *Optica philosophia experimentalis et ratione a fundamentis constituta* . . ., 2 vols. (Leiden, 1652–1656). He was also author of *Nova de machinis philosophia* (Rome, 1649).

LUIGI CAMPEDELLI

IBN ZUHR, ABŪ MARWĀN ᶜABD AL-MALIK IBN ABIᵓL-ᶜALĀᵓ (Latin, **ABHOMERON** or **AVENZOAR**) (*b.* Seville, Spain, *ca.* 1092; *d.* Seville, 1162), *medicine, toxicology, medical botany, theology.*

Ibn Zuhr was the patronymic of a family of famous scholars and physicians from the Arabian tribe of Iyād who had settled in Moorish Spain in the tenth century, if not earlier. Most important and influential among them was the physician Abū Marwān ibn Zuhr. He first studied medicine under his father, Abuᵓl-ᶜAlāᵓ Zuhr, and excelled at an early age. Like his father, Ibn Zuhr served the Murābiṭ dynasty (Almoravids, 1090–1147) in Spain and was well received in their courts. He was then called to serve in the palace of his patron, ᶜAlī ibn Tashfīn (reigned 1106–1143) at Marrakesh, Morocco.

Apparently as the result of a misunderstanding, his patron insulted Ibn Zuhr, removed him from his office about 1141, and threw him in prison. As a result of the many indignities he suffered, Ibn Zuhr retained both physical scars and bad feelings after his eventual pardon and release. Therefore, as he said in his later writings, it was not difficult, after the fall of the Murābiṭs, to establish friendships with their enemies, the Muwaḥḥids (Almohads). The new ruler, Abū Muḥammad ᶜAbd al-Muᵓmin (*d.* 1163), welcomed Ibn Zuhr and appointed him not only as his court physician but also as a counseling vizier. Ibn Zuhr dedicated two works to him: his treatise on theriaca, *al-Tiryāq al-sabᶜīnī*, and one on diet, *al-Aghdhiya.*

During this later period, Ibn Zuhr accumulated much prestige and wealth and became a very close friend of Ibn Rushd, to whom he dedicated his best-known book, *al-Taysīr*. Ibn Rushd had asked Ibn Zuhr to write this book on the treatment of particular diseases of the organs of the body and methods of therapy. He personally compiled and wrote his *al-Kulliyyāt*, on the generalities of medicine, as a supplement to *al-Taysīr*, as he explained in the introduction.

Ibn Zuhr's daughter became one of the better-known midwives in Islam, and a son became a physician, poet, and man of letters. On one occasion, when Ibn Zuhr was away from his office, the son treated patients. In recognition of the son's excellent performance and to encourage him, Ibn Zuhr dedicated to him *al-Tadhkira*, a book on therapeutics, fevers, and the careful use of laxatives—which he considered to be poisons when abused.

After a career in medical teaching, practice, and writing, Ibn Zuhr died of a malignant tumor and was buried outside the victory gate in Seville. He exerted a considerable influence on Western as well as Arabic medicine, after his works were translated and widely circulated in Latin and Hebrew. Although a true follower of Hippocrates and Galen, he developed numerous original ideas through his medical experimentation and observation. Ibn Zuhr wrote on the therapeutic value of good diets and on antidotes against poisons, and cautioned against deliberate uses of purgatives in treating the sick, who needed curing medications, not "poisons." He urged physicians to use mild drugs and to watch the reactions of the patient, especially for the first three days; if the drug was found useful, a larger dose could then be administered. He explained that drugs mixed with honey or sugar were carried to the liver, which reacted to these substances.

Ibn Zuhr described in more detail than his predecessors mediastinal tumors and the appearance of abscesses on the pericardium, paralysis of the pharynx, scabies, inflammation of the middle ear, and intestinal erosions. He also recommended tracheotomy, first described and illustrated by Abulcasis al-Zahrāwī, almost a century and a half earlier, artificial feeding through the gullet or the rectum, and the use of cold water to reduce fevers. Ibn Zuhr realized the noxiousness of air coming from marshes and, like Ḥunayn ibn Isḥāq (d. 877), Galen's competent translator, he emphasized the importance of clean, "good" air for health. As a clinician and medical therapist, he was one of the best Muslim physicians in Moorish Spain; and his influence on medicine in the West continued until the Renaissance.

BIBLIOGRAPHY

I. Original Works. All of Ibn Zuhr's nine known works, which were widely circulated in the twelfth century and after, were medical. A century later Ibn Abī Uṣaybiʿa, in ʿAyūn al-anbāʾ, Būlāq ed., II (Cairo, 1882), 66–67, mentioned only seven:

1. al-Tiryāq al-sabʿīnī. This apparently lost work is on theriaca, which incorporates 70 drugs, and its abstraction into seven and/or ten ingredients, known also as al-Antala theriaca. He prepared this antidote for his patron, ʿAbd al-Muʾmin, as a safeguard against poisoning by his enemies.

2. Fī al-Zīna. Little is known about this lost work, except that the title suggests recipes for beautification, cosmetics, and skin medication. He wrote it during his early life as a medical author and practitioner and was ashamed of some of its contents when he matured in experience and knowledge.

3. al-Aghdhiya. Ibn Zuhr wrote this work on diet at the request of his patron, ʿAbd al-Muʾmin, to provide information on accessible foods and their therapeutic advantages. The writer of this article consulted a 14th-century MS of it in Istanbul (Ahmad III Library at the Suleimaniye Library) in 58 fols. Other copies of this dietetic text are also extant.

4. Fī ʿIlal al-kilā. Lost treatise on kidney diseases that Ibn Zuhr wrote at the request of colleagues in Seville.

5. Fī ʿIllatay al-Baras waʾl-ʾbahaq. A lost treatise on leprosy and vitiligo (known also as piebald skin and leukoderma), how they differ, and their treatment.

6. al-Tadhkira. A thesaurus for his son, then a young doctor, concerning the treatment of diseases. Studied by Gabriel Colin in his Avenzoar, sa vie et ses oeuvres (Paris, 1911).

7. al-Taysīr fiʾl-mudāwāt waʾl-ʾtadbīr (Latin, Alteisirʾ [or Teissir], scilicet regiminis et medelae), Ibn Zuhr's best-known medical text in 30 treatises, written at the instigation of Ibn Rushd, who copied it. A few copies exist in Arabic and several more survive in Latin and Hebrew, an indication of its wide circulation in Europe. See the list in Ludwig Choulant, Handbuch der Bücherkunde für dir ältere Medicin (Leipzig, 1841), 375–376. Copies examined by the author were in the National Library, Rabat (Q159), and the Royal Library, Rabat (no. 1538). In it Ibn Zuhr, as was the practice of the time, mixed astrology with pharmacological and experimental observations, and superstitions with rational and objective reasoning. The book was highly praised by Ibn Rushd as the best available on particulars in medicine and therapeutics. See Kitāb al-Kulliyyāt (Larache, Morocco, 1939), 230. Nonetheless, Ibn Zuhr was definitely influenced by the works of such predecessors as Ḥunayn ibn Isḥāq, al-Rāzī, and al-Zahrāwī.

8. al-Iqtiṣād fī Iṣlāḥ al-Anfus waʾl-ajsād. This work ("On the Ecology of the Treatment and Healing of Body and Soul") is not mentioned by Ibn Uṣaybiʿa. It was written for the Murābiṭ Prince Ibrāhīm ibn Yūsuf ibn Tashfīn after he moved to Morocco in 1121. It discusses therapeutics and hygiene and was written for the lay reader. At least in spirit and in title this work followed a similar one for treatment of body and soul, Kitāb al Irshād li-maṣāliḥ al-Anfus waʾl-ajsād, by the Egyptian Jewish physician Hibat Allāh ibn Jumayʿ (d. 1198). Two extant copies of al-Iqtiṣād are mentioned in Carl Brockelmann, Geschichte der arabischen Literatur, I (Leiden, 1943), 642, and Supp. I (Leiden, 1937), 890. It was studied by H. P. J. Renaud in "Trois études . . .," in Hespéris, 12 (1931), 91–105; and 20 (1935), 87.

9. Jāmiʿ Asrār al-Ṭibb. The other book not mentioned by Ibn Uṣaybiʿa, a copy of which is in the National Library, Rabat (D 532), is a compendium ("The Comprehensive Text on the Mystery or Secrets of the Healing Art"). It discusses human physiology especially in regards to the digestive system, physical therapy, and dietetics. It also describes the functions of other bodily organs including the liver, spleen, bladder, as well as general diseases such as fevers, gout, and hemorrhoids. It further includes a formulary on syrups, electuaries, and other pharmaceutical preparations, which erroneously was thought to be composed as an appendix to al-Taysīr and of which other copies are extant. Upon a thorough examination of the text this author discovered that it was the contribution of his father Abū al-ʿAlāʾ ibn Zuhr, who died in Cordoba, 525/1131. This same Rabat copy includes another text entitled al-Shāfī min al-Amrāḍ wāʾl-ʿIlal ("The Healer of All Diseases")

on the diagnosis and treatment of diseases. It was dedicated, and most probably by the father, Abū al-ʿAlāʾ ibn Zuhr, to the Murābiṭ prince al-Manṣūr Abū al-ʿAbbās Aḥmad, divided into forty discourses with frequent quotations from ancient sages and religious sayings.

II. Secondary Literature. Ibn Abī Uṣaybiʿa's contemporary, Muḥammad Ibn al-Abbār (1199–1260), gave one of the earliest brief biographies of Ibn Zuhr and other members of his family in *Takmilat al-ṣila*, edited by F. Codera in *Bibliotheca arabico-hispana*, II (Madrid, 1889), 616. He was also mentioned in Abuʾl-Falāḥ ibn al-ʿImād (*d.* 1679), *Shadharāt al-dhahab*, IV (Cairo, 1350 A.H.), 179; and his *Taysīr* and *Jāmiʿ* are listed in Ḥājjī Khalīfa's *Kashf al-ẓunūn*, I (Cairo, 1892), 354, and (Istanbul, 1941) 520. The most useful and reliable later references are the following, listed chronologically: F. Wüstenfeld, *Geschichte der arabischen Aerzte und Naturforscher* (Göttingen, 1840), 90–91; L. Leclerc, *Histoire de la médecine arabe*, II (Paris, 1876), 86–93; Gabriel Colin, "Ibn Zuhr," in *Encyclopaedia of Islam*, II (Leiden, 1927), 430–431; George Sarton, *Introduction to the History of Science*, II (Baltimore, 1931), 231–234; Aldo Mieli, *La science arabe et son rôle dans l'évolution scientifique mondiale*, 2nd ed. (Leiden, 1966), 188–215; R. Arnaldes, "Ibn Zuhr," in *Encyclopaedia of Islam*, new ed., III (Leiden, 1969), 976–979; Sami Hamarneh, *Index of Mss. in the Ẓāhiriyyah* (Damascus, 1969), 174–176, in Arabic; and the commemorative volume, *Al-Ṭabib Ibn Zuhr* (Aleppo, 1972), esp. Michael Khoury, "Banu Zuhr," pp. 159–203.

Sami Hamarneh

ZWELFER, JOHANN (*b.* Rhenish Palatinate, 1618; *d.* Vienna, Austria [?], 1668), *pharmacy, chemistry.*

After sixteen years as a pharmacist in his native region, Zwelfer went to Padua to study medicine. Upon receiving the M.D. degree he went to Vienna, where he apparently spent the rest of his life. Statements that he taught chemistry and was physician to the court are undocumented. He is first mentioned as the author of a book of corrections to the standard German pharmacopoeia and of an original pharmacopoeia, both published in 1652.

Zwelfer has been credited with a few minor chemical innovations: the "purification" of calomel (mercurous chloride) by use of a water wash to remove the violently poisonous mercuric chloride, and the preparation of a pure form of iron oxide (*crocus martis*) by igniting ferrous nitrate. His outstanding contribution, however, was his general influence in reform of the pharmacopoeia. He was the first to write a commentary on a pharmacopoeia—the standard official German work of that genre, the *Pharmacopoeia Augustana*, which had appeared under the auspices of the Collegium Medicum of Augsburg and was used in Vienna. Like other works of this kind, it was an uncritical compilation of recipes, many of them ancient. The principal controversy involved in the compilation of these works had been whether the new chemical remedies associated with Paracelsus should be included; but this question had been settled, with their inclusion, by the time Zwelfer became active. The next issue was the improvement of the recipes, and he may have been the first to raise it.

In this commentary Zwelfer reveals a bellicose nature in the tradition of Paracelsus. Objections (*animadversiones*) are made to almost every recipe—and in a tone of sarcasm and invective that aroused not only the Collegium Medicum of Augsburg, but almost everyone mentioned in the book, in cities from Montpellier to Venice. In Venice, Otto Tachenius was inspired by Zwelfer's criticism of his "volatile viperine salt" to write his *Hippocrates chimicus*, a more famous book than any Zwelfer ever wrote. The Augsburg Collegium was still sufficiently aroused sixteen years after Zwelfer's death to issue, with him in mind, a "renovated and augmented" *Pharmacopoeia Augustana*.

Zwelfer was indeed intemperate, as Lucas Schröck charges in the preface to the renovated *Pharmacopoeia* and as is still evident to the reader of these books. "Stupid" was his word for the *Pharmacopoeia*'s recipe for the preparation of "water of lead oxide" (*aqua lithargyri*) by distillation, and it was a typical comment. But it seems to have been Zwelfer's choice of words that was at fault, for this recipe is absent from the renovated *Pharmacopoeia*. As for Tachenius, he was indignant that "Reformer" (as he called Zwelfer) had called him a "cheat" in his commentary on the *Pharmacopoeia*'s recipe for viperine salt. But Tachenius went on to admit that there was some truth in the remark, for although he personally instructed Zwelfer in the preparation of the salt, he did so only "metaphorically," feeling it best not to reveal the recipe.

BIBLIOGRAPHY

I. Original Works. Zwelfer's works are *Animadversiones in Pharmacopoeia Augustana* (Vienna, 1652) and *Pharmacopoeia regia* (Vienna, 1652).

II. Secondary Literature. Zwelfer is said to have died unlamented, a situation apparently reflected in the standard German and Austrian national biographical dictionaries, in which he is not mentioned. The brief account in C. G. Jöcher, *Allgemeines Gelehrten-Lexicon*, IV (Leipzig, 1751), 2141, has served as the source for most later accounts. J. R. Partington, *A History of Chemistry*, II (London, 1961), 292, 296–297, has a little on his chemistry.

R. P. MULTHAUF